From Neuron to Cognition
via Computational Neuroscience

Computational Neuroscience
Terrence J. Sejnowski and Tomaso A. Poggio, editors

For a complete list of books in this series, see the back of the book and https://mitpress.mit.edu/category/series/computational-neuroscience

From Neuron to Cognition
via Computational Neuroscience

EDITED BY

Michael A. Arbib
and
James J. Bonaiuto

The MIT Press
Cambridge, Massachusetts
London, England

This book was set in Baskerville by Toppan Best-set Premedia Limited. Printed and bound in the United States of America.

Library of Congress Cataloging-in-Publication Data

Names: Arbib, Michael A., editor. | Bonaiuto, James J., editor.
Title: From neuron to cognition via computational neuroscience / edited by Michael A. Arbib and James J. Bonaiuto.
Description: Cambridge, MA : The MIT Press, 2016. | Series: Computational neuroscience | Includes bibliographical references and index.
Identifiers: LCCN 2016013116 | ISBN 9780262034968 (hardcover : alk. paper)
Subjects: LCSH: Neurosciences. | Cognitive neuroscience. | Computational neuroscience.
Classification: LCC QP355.2 .F76 2016 | DDC 612.8/2330285—dc23 LC record available at https://lccn.loc.gov/2016013116

10 9 8 7 6 5 4 3 2 1

CONTENTS

SERIES FOREWORD

Computational neuroscience is an approach to understanding the development and function of nervous systems at many different structural scales, including the biophysical, the circuit, and the systems levels. Methods include theoretical analysis and modeling of neurons, networks, and brain systems and are complementary to empirical techniques in neuroscience. Areas and topics of particular interest to this book series include computational mechanisms in neurons, analysis of signal processing in neural circuits, representation of sensory information, systems models of sensorimotor integration, computational approaches to biological motor control, and models of learning and memory. Further topics of interest include the intersection of computational neuroscience with engineering, from representation and dynamics, to observation and control.

Terrence J. Sejnowski

Tomaso Poggio

PREFACE

From the middle of the nineteenth century onward, analysis of the brain has offered a rapidly increasing set of advances in the quest to understand how we come to know the physical and social world in which we are immersed. Neuroscience has become a very broad field indeed: each year around 30,000 researchers and students from around the world attend the annual meeting of the Society for Neuroscience (SfN), with talks and posters covering an immense range of subfields from the fine details of genetics and molecular biology via neurons, circuits, and brain regions to human cognition, neurological disorders, social interaction, and the ethical implications of research. This book seeks to introduce the reader to the middle range: from neurons, circuits, and brain regions to human cognition, with just enough subcellular details to enrich our insights into how neurons can work together to yield at times surprising results.

We trace a path from neuron to cognition *via computational neuroscience*, but what is computational neuroscience? The use of the computer is now ubiquitous in all fields of science, not just in neuroscience, for the gathering, analyzing, storing, archiving, retrieving, and synthesizing of data. However, we reserve the term *computational neuroscience* for the attempt to understand neural systems by developing *models of the causal interactions* which underlie the observed behaviors of these systems. Although some of these models may be purely conceptual, they will in general be formalized enough to be expressed mathematically. In some cases the resultant mathematical system can be understood by pure reasoning alone. More often, though, the formalism will be the basis for a computer program that can be used to run numerous simulations to explore the behavior of the system and thus assess to what extent the model gives us fresh insight into the actual and potential behavior of the neural (and, in some cases, embodied) system that it represents.

Computational neuroscience has another, perhaps more romantic, meaning that overlaps the model-centered perspective. Here the aim is to explore the notion that "the brain is a computer" without reducing the brain to a classic serial electronic computer, but rather in the attempt to open up the field of computer science (as has indeed happened to some extent in the last decade or two) to the notions of (1) cooperative computation: distributed as well as parallel, rather than serial, computation; (2) adaptive computation exploiting diverse forms of learning and memory beyond passive storage; and (3) an action–perception cycle in which computation is not confined to numerical and other formal structures but enmeshed in the demands of interaction with the external world via sophisticated sensors and effectors. The true aim of this book is to explore "how the brain does it."

From Neuron to Cognition via Computational Neuroscience provides an integrated view of neuroscience with a computational flavor but with a real sense of the empirical data, bridging the divide that too often separates those interested in the details of how neurons work and those who wish to understand the brain's contribution to cognition and behavior. It thus introduces the whole range of computational neuroscience from neurons interacting via synaptic processing to the behavior of networks and the modeling of higher order brain functions.

In this age of access to a dazzling diversity of readings, videos, and more via the World Wide Web, it may seem anachronistic to present a textbook whose confines can provide but a small sample of what those 30,000 SfN attendees are working on. However, that very diversity makes it hard indeed to establish a firm foundation on which to build one's own expertise. This book, then, will provide the student with the core knowledge on which to base future exploration, whether it be drilling deeply into some specialized area or thinking broadly to see the linkages between the topics we explore here and a whole range of other topics in neuroscience, artificial intelligence, robotics, cognitive science, and even philosophy of mind.

Who the Book Is For

Some readers will come to the book simply to learn what a chapter or two has to offer. Others will come to the book for an overview of "how the brain works" and plan to read most of the chapters but will skip many of the equations and other more technical details. If so, we think you will find the book valuable. However, this is a textbook, and it is designed to meet the interests and needs of the student taking a course based on the book and the instructor teaching that course. First, a quick word to the instructor. We imagine that you may have already done research on topics covered in a chapter or two in this book (but it's fine if you haven't) and have perhaps taught material covered in some of the chapters—as well as some topics not in this book—but will also be teaching material that will at first be new to you. It is with you in mind that we have developed the Website. "How to Use This Book and Its Website" describes the tools the Website offers—complementing the text with slides for lectures, a wide range of exercises, project suggestions, links to videos, suggestions for further reading, and access to the Brain Operation Database (BODB) and other neuroinformatics tools.

Since you may not have time to cover all the chapters during the time available for your course, the "Opening Perspective" of chapter 1 will help you design your own road map, deciding which chapters to teach in detail, which to present briefly, and which to leave to self-study by students who find their topics of interest. Of course, the text and slides will work equally well if you are an advocate of the "flipped classroom," having students study the lecture (reading slides and related text in tandem) beforehand, with the class devoted to discussion and working through some of the exercises.

But what of you, the student? We assume that you have some basic knowledge of neuroscience: you know the basic facts about neurons and synapses and have some general ideas about the structure and function of the human brain—the book will quickly fill you in on these as seen from a computational point of view, but it will be good if you have some feel for the biology or psychology that enriches the key concepts. And you will need to know—or your instructor will have to teach you—the basics of differential equations, linear algebra, and probability theory, and have experience with or be able to quickly pick up the elements of programming in MATLAB or Python, the two languages used by the authors in designing the exercises. If you are more adventurous, you may work on one of the suggested projects, and these may require you to master additional skills. But beyond that, you may come from diverse backgrounds, whether in neuroscience itself or nearby fields such as biology or psychology or cognitive science, or you may come from physics, mathematics, or computer science looking for a new application area, or you may even be a student of artificial intelligence wanting to learn more about "how brains do it." In each case we think you will find the book well structured to meet your needs. However, we also think you will find chapter 1 very helpful in developing your own perspective on the diverse materials the book offers. Indeed, you may find it best to read this chapter at least twice—not only before studying the other chapters but also later to help you reassess how the various topics tie together, perhaps then more fully appreciating points that made little impact on a first reading.

To end this section, we stress that neuroscience itself is a field of great breadth and depth, and only some of its subfields have yielded to the methods of computational neuroscience—and no one book can do justice to all these subfields. But when you have mastered this book you will be well prepared to set out on your own journeys, whether it is to read popular accounts of brain research with fresh insight, to learn the basics of subfields not covered here, or to boldly go where no one has gone before.

An Overview of the Chapters

Each chapter of *From Neuron to Cognition via Computational Neuroscience* has been written by a leading expert or experts on its topic, with a range of topics that will ensure that the reader gains a good feel for the tools of computational neuroscience and how indeed they can be used to bridge from neuron to cognition. But even with this focus we have had to omit a number of topics in the attempt to keep the volume to a moderate length. Successive drafts of the chapters were resolutely edited to give the book some coherence in style and development and to ensure that a range of complementary materials are available on the Website. We say "some" coherence because even with our focus on just one subset of neuroscience research, the reader will find here a fascinating diversity of systems and behaviors and the computational tools to model them. With this, let's turn to a quick overview of the book's chapters. A more thorough overview, together with some necessary background, is provided in chapter 1, "From Neuron to Cognition: An Opening Perspective."

The next three chapters provide the underpinnings for the rest of the book. Nicolas Brunel introduces "Basic Neuron and Network Models," including some sense of the inner workings of neurons we need if we are to understand some of the potential dynamics of neural circuitry. Jonathan E. Rubin and John Rinzel then introduce the treatment of "Neurons and Neuronal Networks as Dynamical Systems," with the emphasis on the way in which the mathematics of nonlinear systems can use such notions as equilibria, stability, and oscillations to illuminate general properties of neural system dynamics. Nathan Vierling-Claassen and Stephanie R. Jones then assess "Neural Rhythms" which provide an important form of collective behavior that emerges from interactions within a network of neurons. To round out these foundations, we offer a chapter on "Linking Models with Empirical Data: The Brain Operation Database," stressing the way in which empirical data may combine with experience with prior modeling to ground the design of a new model of a neural system while further empirical data serve for the testing of the model. The Brain Operation Database (BODB) then provides the means for the cumulative development of the theory–experiment cycle.

The remaining chapters could be read in almost any order since an attempt has been made to render them as self-contained as possible but with cross-references supplied to other chapters for any necessary background. The actual order was chosen with two aims in mind: (1) to ensure that, where possible, the necessary background for one chapter was made available in an earlier chapter and (2) to cycle through the various themes to encourage the reader to develop a more and more integrated view of the whole. The table of contents shows the actual order; here we regroup the chapters to make clear the key themes covered in this book.

Action: The brain has its main effect on the world through the ability of motor neurons to cause muscles to contract and allow them to relax as the basis for coordinated behavior. The treatment of "Motor Pattern Generation," by Auke J. Ijspeert, Andrej Bicanski, Jeremie Knuesel, and Jean-Marie Cabelguen, gives pride of place to networks in the spinal cord which can yield rhythmic activity such as that involved in locomotion, but which can be tuned to adjust that locomotion to current circumstances. At the end of his chapter, Ijspeert et al. look at the extra lessons that can be learned when a brain model is implemented as the control system of a robot. This theme is fully developed in the later chapter on "Embodied Models and Neurorobotics" by Tony J. Prescott, Joseph Ayers, Frank W. Grasso, and Paul F.M.J. Verschure. By contrast, Nicolas Schweighofer emphasizes general "Computational Models in Motor Control" whereby feedback and feedforward combine to yield fast and effective interaction with objects in the external world. Ziad M. Hafed focuses on motor control for the eye in his exposition of "Saccades and Smooth Pursuit Eye Movements" while James J. Bonaiuto turns to the arm and hand in modeling the explicit integration of vision and motor control involved in "Reach and Grasp: Control, Development, and Recognition."

Neural Plasticity, Learning, and Memory: Wulfram Gerstner introduces the most basic form of changes in synaptic efficacy with experience, exploring several variants of "Hebbian Plasticity and Learning," where the emphasis is on unsupervised learning. A different form of plasticity for neurons is provided by "Neuromodulation," as explored by Jean-Marc Fellous, Carmen Canavier, and Michael E. Hasselmo. Nathaniel Daw explores "Reinforcement Learning," in which synapses across a network change on the basis of when and whether behavior yields positive or negative reinforcement. We see that the neuromodulator dopamine carries reinforcement-related signals. Stefano Fusi and Xiao-Jing Wang build on these mechanisms to offer models of "Short-Term, Long-Term, and Working Memory."

Vision: Although there are many perceptual systems, this book focuses on just one, namely, vision. (Chapter 1 offers some pointers to research on other systems.) The mammalian visual system starts with the retina, whence it follows two main pathways, one via the superior colliculus of the midbrain and the other via the thalamus and on to visual cortex. As noted above, Ziad M. Hafed links brain stem and cortex, as well as vision and motor control, in modeling "Saccades and Smooth Pursuit Eye Movements." Alan L. Yuille and Daniel Kersten chart the early stages of visual processing in the thalamo-cortical pathway in their exposition of "Early Vision," while James A. Bednar and Christopher K.I. Williams tie low-level vision to neural plasticity as they shift the focus to the developmental timescale with their modeling of "Neural Maps: Their Function and Development." Owen Lewis and Tomaso Poggio take up where Yuille and Kersten leave off, with their introduction to "Object and Scene Perception."

Bridging the Levels: Our knowledge of the brain comes both from animal studies where neuroanatomists and neurophysiologists probe the connectivity and activity of single cells in specific neural circuits (others study genetics and molecular biology, but these themes are at best tangential to this book) and from human studies which may include studies of brain imaging and the degradation of behavior and cognition that follows the ravages of brain lesions and disease. Three chapters bridge these levels. Richard P. Cooper, in "Schema Theory and Neuropsychology," shows how a range of data may be better modeled by the competition and cooperation of functional units called schemas rather than by attempting a reduction to the neuron network level. Complementing this, "Synthetic Brain Imaging," by

Jacqueline A. Griego, Carlos R. Cortes, Ransom Winder, and Malle A. Tagamets, shows how neural network models of the human brain inspired by animal data can be simulated in a way which yields predictions which can be tested against brain imaging data. The chapter on "Neural Rhythms" also bridges the levels, in this case from large-scale rhythms to the activity of neurons. Mohamed A. Sherif and William W. Lytton show how modeling can yield insights into various "Brain Diseases."

A Regional Focus: Many chapters in this book look at models of the functions of one or more interconnected brain regions. Here we note chapters that look explicitly at three important brain regions—hippocampus, basal ganglia (striatum), and cerebellum—and sample some of the functions they support in concert with key brain regions with which they are connected. These are "Hippocampus: From Spatial Processing to Episodic Memory," by Edmund T. Rolls, "Cerebellar Adaptation and Supervised Learning in Motor Control," by John Porrill and Paul Dean, and "Integrative Functions of the Corticostriatal System," by Peter Ford Dominey and Pierre Enel.

Language: We conclude with two chapters that discuss language, the core of the distinctively human mode of cognition. Angelo Cangelosi discusses "Language Processing" primarily in terms of adaptive artificial networks but with links to brain and behavior. The closing chapter offers a coda to the whole book by showing how many of its themes come together in the ongoing attempt to develop models that shed light on "Evolving the Language-Ready Brain."

Michael A. Arbib
La Jolla and Los Angeles

James J. Bonaiuto
London

HOW TO USE THIS BOOK AND ITS WEBSITE

As noted in the preface, this book has been designed to be used as a textbook for an introductory course in computational neuroscience, and has an accompanying website (https://mitpress.mit.edu/neuron-cognition), which contains further instructions on how to access the ancillary materials that accompany the book. Here we first survey the necessary background for readers of this book and the ancillary digital materials available.

Background

This book could be read by individuals with no prior study of the brain if they are content to accept formal models without the larger framework that prior study of neuroscience can provide. Indeed, an attempt is made in each chapter to offer an exposition of key empirical findings that provide the framework for the models presented in that chapter, while chapter 1 offers a wealth of background needed for the book as a whole. Nonetheless, it would be helpful for readers to have some prior acquaintance with basic ideas concerning neurons and the synapses that link them, the neurotransmitters which signal between neurons and the receptors on the postsynaptic membrane that make use of those signals, and axonal transmission, as well as the division of the mammalian brain into structures such as the cerebral cortex, hippocampus, cerebellum, and basal ganglia, the further subdivision of cerebral cortex into cytoarchitectonic areas, and the use of columns and layers as orthogonal ways of examining the cellular structure of cerebral cortex. It would also help to be aware of the use of microelectrodes to study cell firing in certain animal brains and of brain imaging methods such as functional magnetic resonance imaging for studying the human brain.

With respect to a mathematical background, the basic notion of setting up and numerically solving ordinary differential equations is most important. From linear algebra we need (but not often) the notion of matrix and eigenvalue and related results. Finally, from probability theory we need the notions of conditional probability, Bayes law, and Markov chains. However, the latter topics are required only rarely.

No background in computer programming is required for reading the book as such. However, those who want to test their mastery of the material through computer simulation will find a range of exercises on the website that will require use of MATLAB or Python for their solution.

Materials on the Web

Each chapter provides a pedagogically sound treatment of its specified topic. The accompanying website (http://mitpress.mit.edu/books/neuron-cognition) provides the following ancillary materials:

1. A restructuring of the chapter as a sequence of lectures, with a set of slides for each lecture. These slides may be used by readers to obtain an alternative perspective on each chapter but are specifically designed for use by instructors in the classroom, whether or not their expertise in computational neuroscience extends to the given topic.
2. A set of exercises that allows students to test their mastery of the material. Those that require computer simulation are keyed to MATLAB and/or Python.
3. Suggestions for projects which allow detailed exploration and extension of topics from the chapter. In some cases, these will require the student to use a simulation environment other than MATLAB or Python—NEURON is a frequent choice when Hodgkin-Huxley style modeling is involved. The assignment of such projects is thus at the discretion of the instructor.

In addition, many chapters have one or more models documented in the Brain Operation Database (BODB; http://bodb.usc.edu/), providing a high-level view of each model together with summaries of empirical data used in their design and testing, as well as brain regions and brain operating principles and notes on related models. See chapter 5 for an introduction to "Linking Models with Empirical Data: The Brain Operation Database."

The book's website allows these resources to be retrieved on a chapter by chapter basis or via resource category. It will allow for the addition of new exercises, links to literature, supplementary material, and so forth, thus allowing updating to proceed on a faster timescale than successive print editions of the textbook.

1 From Neuron to Cognition: An Opening Perspective

MICHAEL A. ARBIB

1.1 Preamble

The title *From Neuron to Cognition via Computational Neuroscience* indicates two overall biases: to move up from neurons toward an understanding of their role in behavior and cognition (with cognition considered in a broad sense which includes action, perception, and diverse forms of memory, as well as planning and decision-making) and to show how computational modeling can complement empirical research in advancing that understanding. We explore how a brain supports behavior through the linkage between perception and action (the *action–perception cycle*), which in turn rests on learning and memory. The aim of this "Opening Perspective" is to provide a framework for reading the chapters which follow. It will show what aspects of cognition we develop in detail, and what we do not—while sketching how what we do cover provides a firm foundation for exploration of the full range of research in cognitive neuroscience, whether computational or not. And in the end we will address that most specifically human form of cognition, namely, language.

A note to the reader: The chapter opens with three sections that offer a shaping perspective for thinking about the brain: "The Maps Are Not the Territory," "Schema Theory and Action-Oriented Perception," and "A Fast Overview of the Human Brain." Thereafter, each section opens with pointers to the chapters that are most relevant to the topic of that section, and a subsection for each chapter will summarize some of its most important data, modeling, and discussions. However, many of these accounts will also include discussion of relevant material from other chapters and observations of my own on the topic at hand. Thus, this "Opening Perspective" can also be used as a reference to enrich the study of particular chapters. The emphasis is not on summarizing what the authors so expertly present but rather on providing a framework which will, I hope, enable the reader to more fully appreciate the contribution each chapter makes to our overall understanding.

As the reader will discover, there are manifold connections between the topics treated in the individual chapters, and so there is no single path through the chapters that best suits all readers. Thus, where some may choose to read the chapters in numerical order, for others this "Opening Perspective" may instead help them define an optimal path tailored to their own interests. In any case, this "Opening Perspective" may well repay being read twice: first before reading the other chapters to get the general ideas involved in the path from neurons (or other aggregates relevant to understanding the brain) to cognition, and then again once one has studied other chapters to more fully appreciate the connections charted here, enriching the second reading with new insights of one's own.

1.2 The Maps Are Not the Territory

It is a truism that "The map is not the territory," but here I want to emphasize that in general a territory can be the basis for multiple maps, each developed to serve different tasks. Consider, for example, traditional paper maps of a region. One would serve motorists and show various roads. To use it to go from A to B you would locate (representations of) A and B on the map and then find a route between them, perhaps choosing between a fast freeway and a slower route that runs alongside a river. Another would serve hikers and would include contour information (missing on the road map) needed to determine where an apparent path might prove too steep. For those who use smartphones instead of paper maps, there may be a loss of overall perspective (getting a sense of the geographical relation between A and B, and the possibilities that lie between them) but a gain in offloading some of the computations to the phone and getting a continual update on what to do next to continue from the present location toward B.

Maps in the brain: A crucial organizing principle for the primary sensory and motor areas of the brain and for those subcortical areas with which they are tightly linked is provided by the existence of topographic maps within them (their development is the focus of chapter 13, "Neural Maps: Their Function and Development"). In the visual system, for example, one might test a neuron to determine what type of visual feature best stimulates it and where in the visual field that features elicits the greatest response. The positions where stimuli cause the neuron to fire is called its *receptive field*. Neighboring neurons in the visual areas usually respond to similar regions of the image. Hence these areas are roughly retinotopic in the sense that their spatial organization is similar to that of the image at the retina. This retinotopic structure is strongest in primary visual cortex and gets weaker, and receptive fields get larger, as we ascend the visual hierarchy. In addition to *retinotopic* maps in the visual system, the brain has *somatotopic* maps (organized as a "homunculus" matching the surface of the body in somatosensory cortex), tonotopic or cochleotopic maps (corresponding to the frequency gradient in the cochlea), and whisker barrel maps in rodents (see chapter 16). Motor cortex also contains several "homunculus" maps, but mapping to the muscles and joints of the body, or perhaps to motor actions, rather than to touch receptors in the skin.

Two things to note. First, the "real estate" on each map reflects the functional importance of the area it represents, not its size. Thus, the retinotopic map in visual cortex devotes more space to the foveal region than the far periphery. Similarly, a somatotopic map devotes more space to, for example, the hand than the back. Second, as we move away from the sensory or motor periphery, the notion of "map" becomes less useful. Still, we may find it useful to talk of reference frames to distinguish what aspects of the world best characterize the maximal firing of cells—for example, in reaching to grasp, we can distinguish regions of the brain that code the visual position of a target on the retina from the head-centered position from the position relative to the shoulder that best situates the generation of motor commands. Of course, when we enter more abstract domains of cognition, even the notion of reference frame may become less useful.

Maps of the brain: The models developed for computational neuroscience are more akin to maps on a smartphone. Underpinning them is a body of data representing a range of relevant information about the "territory" of interest, but atop them lies a range of "computational routines" that can help us move toward answers to the questions that interest us. However—as is clear from distinguishing the interests of the driver and the hiker—different brain models must be developed to address different issues. The term "neuroscience" suggests a shared interest in neurons, but in fact that interest can take very different forms. Some people might want to develop a drug that ameliorates or cures some disease, and they may find that neurons are too big to interest them and focus all their energies on understanding the role of a specific macromolecule in a synapse, and how genetic defects may impair that role. Another person may find the neuron too small, seeking, for example, to use brain imaging to assess the relative "activity" of different brain regions in some cognitive behavior. Indeed, there are hundreds of relatively distinct niches where a neuroscientist might happily spend a decade or more of his or her career. Each of the 24 subsequent chapters introduces the reader to one or a few such niches, but in each case (as marked by the cross-references) expertise in one area demands at least some knowledge in other areas, and so an education in neuroscience requires the ability to both drill deeply and think broadly.

We saw earlier that a useful map (in the everyday sense of the word) is useful only if it leaves a great deal out so that we can make sense of what we need to know for our current activity. For this reason, a detailed model of all known biochemical processes in the brain (even were it feasible) might obscure rather than illuminate the simulated processes. The strategy in this book is instead to isolate fundamental empirical problems in neuroscience and then seek as simple a model as possible to chart them. As we develop such models, we continually compare them in search of fundamental "brain operating principles." We also assess where features of different models are ripe to be combined into more comprehensive models—but where now the increased complexity is mitigated by our understanding of the way in which the prior brain operating principles find expression in the prior models and thence in the new, more integrated approach.

With our concern for cognition, we must take care to distinguish the brain from the mind (the realm of the "mental," including the cognitive). In great part, brain theory seeks to analyze how the brain guides the behaving organism in its interactions with the dynamic world around it, but much of the control of such interactions is not mental, and much of what is mental is subsymbolic and/or unconscious. Without offering a precise definition of "mental," let's just say that many people can agree on examples of mental activity (perceiving a visual scene, reading, thinking, etc.) even if they take the diametrically opposite philosophical positions of dualism (mind and brain are separate) or monism (mind is a function of brain). They would then agree that some mental activity (e.g., contemplation) need not result in overt interactions with the dynamic external world, and that much of the brain's

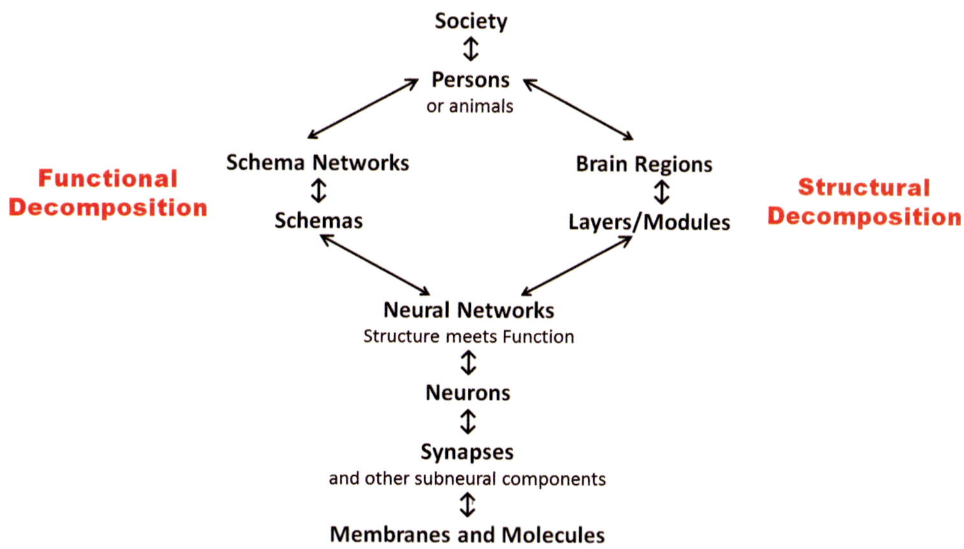

FIGURE 1.1 An overall view of different levels of analysis that may enter into research in neuroscience, proceeding both up and down from neurons, and noting that some overall behavior may be addressed in terms of either constituent *functions* (schemas) or the brain *structures* that implement it, or both. Cautionary note: Behavior in general results from brains within bodies in social interaction. Modeling parts of the brain in isolation can yield important insights but may blind us to key dynamics resulting from brain–body–environment interactions.

activity (e.g., controlling normal breathing) is not mental. This book's hypothesis is that mental processes can all be explained in terms of "brains and bodies in interaction," but many mental phenomena have not yet been explained in this way. This book is a progress report.

Earlier, I placed the term "computational routines" in quotes. This is to emphasize that the way the brain "computes" is very different from the way computation is distributed across the electronic circuits of today's smartphones, computers, and computer networks. Instead, we seek to characterize the way in which the brain's "computations" are distributed across networks of active, interactive, and adaptive entities, whether those be neurons or brain regions or schemas or other structural or functional entities. Nonetheless, just as it is clear that a simulation of an airplane's flight on a digital computer yields invaluable information, so shall we find that our investigations of "how the brain computes" are greatly assisted by the use of simulation on digital computers—without implying that the brain "computes" as such a computer does, any more than we would plan to fly on a computer instead of a plane.

Much of computational neuroscience seeks to provide mathematical abstractions of subneural dynamics—of which the most important is the Hodgkin–Huxley equations—as a basis for creating detailed models of individual neurons which, in turn, underwrite the modeling of relatively specific neural networks. However, subneural modeling will not play the central role in this book (though two chapters will supply the basics of this approach and others will apply them). Instead our primary concern will be with how the brain supports cognition and behavior. Most chapters will focus on how a neural network or a set of interacting networks serves to enable a single human (or animal) to perform just one or two of these functions in interaction with a limited sample of the external world. A quick tour of figure 1.1 will frame our various concerns, and then the rest of the chapter will survey key themes.

The dominant feature of the figure is that it offers two paths to link the "top-down" study of people and animals interacting with their physical and social environment with the "bottom-up" study of how neurons behave when connected into neural networks. The right-hand path of *structural decomposition* is the one more familiar to neuroscientists, dividing the brain into regions such as Brodmann's subdivision of human cerebral cortex (discussed below in relation to figure 1.5), and then dividing the region into subareas, layers, and anatomical modules such as columns. By contrast, the left-hand path of *functional decomposition* starts with some cognitive and behavioral function, such as the visual control of hand movements, and seeks to understand how it can be decomposed into functional units called *schemas*. Such a decomposition may be hierarchical, and the "lowest-level" schemas may be explainable in terms of the activity of neurobiologically defined neural networks. However, in other cases (e.g., in the study of language) this may not yet be possible and matching schemas to the neural level will thus pose challenges for future research.

Working from the bottom up in figure 1.1, our prime concern will be to understand how neurons can be linked together in circuits that do "interesting things." To understand a circuit, we may need to use data on what types of neurons are involved (these may differ in different parts of the brain, and in different animals) and on how they are connected. From the point of view of information processing and learning, the most important synapses are chemical synapses. When an action potential of neuron A arrives at a chemical synapse, a chemical signal (neurotransmitter) is ejected into the synaptic cleft and taken up by receptors embedded into the membrane of the receiving neuron, B. The response of the receiving neuron to presynaptic spike arrival can be measured as an excitatory or inhibitory postsynaptic potential (EPSP or IPSP) at its soma—not a spike so much as a nudge toward (excitation) or away from (inhibition) generation of a spike by neuron B. The strength of a synapse is not fixed but can change (see, e.g., chapter 6, "Hebbian Plasticity and Learning"). In most of the models in this volume, we can simply characterize the synapses by "weights" (a scalar measure of the influence of the presynaptic neuron on the postsynaptic neuron) together with simple learning rules which characterize how those weights may change as the circuit operates. However, there are neurotransmitters that act by changing the dynamic properties of the postsynaptic synapse or neuron. These are called *neuromodulators* (see chapter 7, "Neuromodulation"). One of the most important of these is dopamine, which factors into multiple models because of its role in "Reinforcement Learning" (see chapter 10). A crucial observation is that the effect of a neurotransmitter depends not only on the chemical itself but also on the nature of the receptor in the postsynaptic membrane. Think of a key that opens several locks—each lock may open a door to different pathways. For example, dopamine exerts its post- and presynaptic effects via two families of receptors, the D1 and D2 receptors (and more)—as a result of which a given release of dopamine can yield very different effects in different families of cells, as is documented in chapter 10 and again in chapter 23, "Brain Diseases."

Working from the top down in figure 1.1, we note that animals (and, in particular, humans) have evolved not only to interact with the physical world but also to interact socially with their conspecifics. However, in most of this volume, we will ignore the nascent field of social neuroscience and instead focus on specific functions vital to the animal's interaction with the world and on the brain regions implicated therein. For example, a study of vision might seek to understand how the brain supports schemas for microfunctions like recognizing predator and prey. But this very example makes clear that, while many neuroscientists focus solely on vision, others will seek to learn how the brain can exploit the recognition of predator versus prey to guide the appropriate course of action to capture the prey or escape the predator.

Other neuroscientists may start their investigation with questions inspired by the distinctive anatomy of large-scale structures of the brain. Why, for example, does the cerebellum have its distinctive neural architecture? Such questions become more intriguing when structure meets function. We know that the cerebellum is necessary for us to move gracefully, but we can still move clumsily without it. What other parts of the brain, then, must work in cooperation with the cerebellum in normal behavior? We might address this at the level of human brain imaging, asking how different forms of behavior—some with more emphasis on the planning of action, some with more emphasis on its graceful execution—differentially activate various regions of the brain. Other data, though, will come from animal studies, where single-cell recordings reveal the differential activity of neurons in various parts of the brain during these different conditions.

It is at the confluence of the top-down interest in structure and function and the bottom-up interest in the interactions of neurons that much of our work in this book lies—but, as the rest of this chapter makes clear, by no means all. Our strategy in the next three sections is thus:

- To introduce some key ideas about schemas to establish the mind-set of assessing the processes of competition and cooperation between alternate schemas that may underlie cognition and behavior.
- To provide a quick high-level tour of the human brain to establish the mind-set that the brain involves many different regions with distinct patterns of circuitry, and thus the study of the brain cannot be reduced to the dynamics and adaptive properties of any single formal pattern of neural architecture.
- To look at some classic models of activity in neurons and neural networks to provide basic concepts that will be elaborated upon in later chapters, where complexities will be modeled in light of both bottom-up constraints (e.g., using data from neurophysiology) and top-down constraints (e.g., relative activity of different brain regions, as provided by brain imaging).

In each case, the initial account given here will be enriched by pointers to later chapters. Conversely, some of the linkages suggested here will become more meaningful after the specific viewpoints of several chapters have been understood.

Our cumulative knowledge in neuroscience comes from the study of the brains (or nervous systems) of diverse animals (including humans). We can study the function of human brains through introspection or observation of behavior. This may ground a top-down, schema-style analysis but cannot tell us to what extent the schemas correspond to the function of specific brain regions. The classic data of neuropsychology in the nineteenth and twentieth centuries supported inferences in this direction by studying how function degraded when parts of the brain were damaged, and such observations have been enriched in recent decades by new techniques for imaging the human brain, but such methods look at relative activity in different brain regions, not at the millions of neurons that support that activity. Thus, we turn to other animals for data on neural circuitry. Monkeys, cats, and rats are the source of much neurophysiological data relevant to some aspects of human brain function (though "higher cognitive functions" involve evolutionary variations on general mammalian circuitry) while the classic description of how action potentials propagate in neurons rests on the classic Hodgkin and Huxley experiments on giant axons from squids. It should be added that neuroscientists study the brains (and bodies) of other animals not only to gain lessons for human neuroscience but also to better understand the unique capabilities of other species—whether the flight of birds, the social behavior of ants, or the whisking behavior of rats—and these, in turn, may inform new directions in robotics.

1.2.1 Linking Models with Empirical Data A brain model may be either an explicit *computational* model, together with computer code and simulations, of how a particular brain system operates at, for example, the level of dynamic interactions in biologically realistic brain regions and neural networks; a *mathematical* model to be explored by, for example, the dynamical system methods of chapter 3 (but it may also be implemented to get further results by simulation); or a *conceptual* model, in which case a "simulation" result may simply be an inference from the verbal formulation of the model. The last is simply a sketch of a "computationally possible" model that may not have been implemented but nonetheless guides experiments. It may serve as the basis for later development of computational models but is itself not yet specific enough for implementation. In each case, a question that will be important to ask of a model is this: Where did the assumptions used in building the model come from, and to what extent do results obtained with the model explain extant empirical data or suggest predictions worth testing with new experiments?

In chapter 5, "Linking Models with Empirical Data: The Brain Operation Database," James J. Bonaiuto and I stress the importance of understanding how models relate to empirical data: assessing which empirical data were used to constrain the design of the model and which were used to test the performance of the model, and the results of those tests. In particular, we introduce the Brain Operation Database, BODB (http://bodb.usc.edu/bodb/), which supports the documentation of models of the structure and performance of models in relation to empirical data. In BODB, each model is linked to *Brain Operating Principles* (BOPs), *Summaries of Empirical Data* (SEDs), and *Summaries of Simulation Results* (SSRs).

A BOP is a general principle of brain operation that is not limited to one experimental protocol or simulation but can be found in multiple regions or in multiple tasks. For example, the Winner-Take-All BOP holds when a neural network (whether in an actual brain or simulated) converts a pattern of activation of its inputs to a peak of excitation that corresponds to the locus of maximal excitation on the inputs. This has been applied to modeling diverse brain regions (including frog tectum and monkey superior colliculus). However, the role of a brain region, and thus the relevant BOPs for modeling it, may vary between different tasks. Indeed, modern brain imaging techniques only work because of such differential involvement of brain regions.

Experimental protocols specify the particular conditions under which a given data set (whether brain imaging, or at the neural, subneural or other levels) is gathered. If we want to develop a model of those brain regions to explain the results of diverse experiments, then each experimental protocol must be translated into a *simulation protocol*. The model can then be tested by comparing the SSRs obtained from a particular simulation protocol with the SEDs obtained with the corresponding experimental protocol.

Some brain models build upon earlier works to increase biological plausibility, explain more SEDs (including new experimental results published since the development of earlier models), or both. In other cases, models that were developed independently may provide complementary or even contradictory views of overlapping sets of empirical results. BODB supports model comparison, or *benchmarking*, in a tabular form based on a comparison of the SEDs used by the models to support their development and the SEDs the model can explain. Two models from chapter 20 are chosen for scrutiny in chapter 5—the so-called MNS and MNS2 models of development of mirror neurons—to present an explicit account of the relevant BOPs, and to illustrate how models may be compared by assessing which SEDs each can explain and which SEDs enter into model design. Whether one uses the tools provided by BODB or not (and several chapters in this book include models documented in BODB), such careful comparison of existing models in the context of available data can be a powerful tool in developing specifications for further models in the area of computational neuroscience to which the models under comparison belong. Chapter 5 also sketches the way in which BODB can support collaboration between modelers and experimentalists—the exchange and linking of experiments with the computational models that may support, explain, contradict, or predict empirical evidence. Collaboration should support not only modelers in grounding computational models in empirical results but also experimentalists in allowing them to contribute summaries of their own data that may challenge modelers.

1.3 Schema Theory and Action-Oriented Perception

In figure 1.1, we introduced the idea of a functional analysis in terms of schemas. In this section, we provide a brief introduction to schema theory. We then turn to the notion of action-oriented perception to offer a framework for studying cognition that is very different from one that roots its study in the notion of thought as purely occurring in some abstract symbolic system.

1.3.1 Introducing Schemas Frederic Bartlett (1932) studied how we incorporate new memories into our existing body of knowledge. He asked subjects to read stories and then, later, to recall them. Recalled stories became transformed from the original stories because subjects were using their general knowledge and cultural expectations ("schemas") to reconstruct the forgotten links. However, not all schemas are of this form. More generally, a schema is what is learned (or innately given) about some aspect of the world, combining knowledge with the processes for applying it. For example, a *perceptual schema* might support visually recognizing an apple while also generating predictions about touch and taste. But recognizing an apple may be linked to many different courses of action, and these may be linked to one or more different *motor schemas*—for example, to place the apple in one's shopping basket, to place it in a bowl, to pick it up, to peel it, to cook with it, to eat it, to discard a rotten apple, and so forth. Note that, in the list just given, some items are apple specific whereas others invoke generic schemas for reaching and grasping. There is no one grand "apple schema" which links all "apple perception strategies" to "every action that involves an apple." Moreover, in the schema-theoretic approach, "apple perception" is not mere categorization—"This is an apple"—but may provide access to a range of parameters relevant to interaction with the apple at hand.

Clearly, then, multiple schemas must be linked in supporting any given behavior. An example of such linkage, in what we call a *coordinated control program*, is given in figure 1.2, which provides a specific example of top-down schema analysis, based on the empirical work of Jeannerod and Biguer (1982). They charted the way in which the hand, as it moves to grasp a ball, is preshaped so that it is of the right shape and orientation to enclose the ball prior to gripping it firmly. Moreover, they found that, to a first approximation, the movement can be broken into a fast initial movement and a slow approach movement, with the transition from the fast to the slow phase of transport coming just before closing of the fingers from the preshape so that touch may take over in controlling the final grasp. We will relate these phases to feedforward and feedback in our discussion of chapter 9.

Figure 1.2 models this behavior in terms of interacting schemas. This model anticipates the much later discovery, reported in chapter 20 ("Reach and Grasp: Control, Development, and Recognition") of perceptual schemas for grasping located in a specific area of parietal cortex and motor schemas for grasping situated in a related area of premotor cortex. Chapter 20 also takes this high-level view forward in terms of neurobiologically motivated models of the underlying neural networks.

The schemas of figure 1.2 are akin to the blocks in a conventional block diagram for a control system but have the special property that they can be activated and deactivated. Thus, where control theory usually examines the properties of a fixed control system, schema theory allows the control system to expand and contract, adding and deleting subschemas in a task- and data-dependent manner. Solid lines indicate the transfer of data from one schema to another, and dashed lines indicate the transfer of activation. Crucially, then, schemas can be combined to form such *coordinated control programs* which control the phasing in and out of patterns of schema coactivation and the passing of control parameters from perceptual to motor schemas. Moreover, perceptual and motor schemas may be embedded in assemblages embracing more abstract schemas to yield accounts of cognition and language which link psychology to neuroscience. A corollary to this is that knowledge is distributed across multiple regions in the brain so that a multiplicity of different

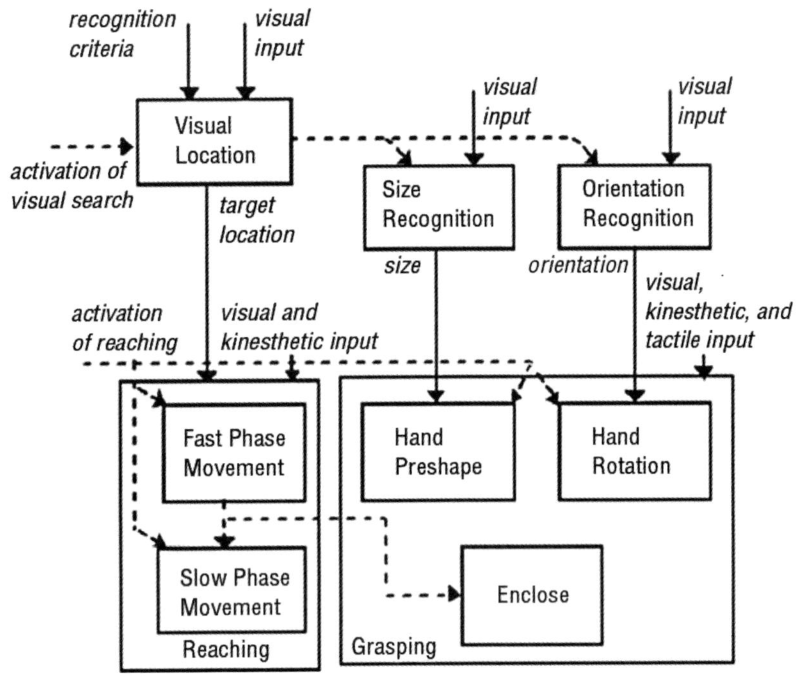

FIGURE 1.2 Hypothetical coordinated control program for reaching and grasping. Different perceptual schemas (top half of figure) provide input for the motor schemas (bottom half of figure) for the control of "reaching" (arm transport) and "grasping" (controlling the hand to conform to the object). Note, too, the timing relations posited here between subschemas within the reaching motor schema and those within the motor schema for grasping. Dashed lines indicate activation signals; solid lines show transfer of data. Reaching was hypothesized to involve a ballistic phase followed by a feedback phase, with the transition between the two also activating the Enclose motor schema. Adapted from Arbib (1981).

representations must be linked into an integrated whole.

Chapter 25 ("Evolving the Language-Ready Brain") introduces a complementary example of schema processing. Here the challenge is to understand a visual scene. This is done by associating schemas with different portions of the scene. However, this leads us to introduce the notion of *schema instance* as an active deployment of a schema. For example, if there are two cars in a scene, we may want to go beyond simply labeling them as cars to associate each one with a different instance of the schema, each specifying values for parameters such as location, size, make, and color. However, in initial processing of a scene, different schema instances may compete to interpret part of a scene (is that a glimpse of sky or part of another building?) or to choose between motor schemas for different actions (will I pick up the coffee mug or keep typing?). Thus, each schema instance has an associated *activity level*. The activity level of a perceptual schema instance signals the credibility of the hypothesis that what the schema represents is indeed present. The activity level of a motor schema instance may signal its "degree of readiness" to control some course of action.

Each schema (instance) may itself involve the integrated activity of multiple brain regions. The method of interaction of schema instances is *cooperative computation*: competition between discordant schema instances may decrease their activity, cooperation based on coherence between instances may increase their activity, and the process continues until an assemblage of schemas wins the competition to direct perception, thought, and/or action. In this way, "computations" that are often seen as the province of traditional symbol-based processing are instead carried out by distributed "neuron-like" methods which do not involve explicit symbolic control. In chapter 14 ("Schema Theory and Neuropsychology") Richard P. Cooper presents a schema-based system—contention scheduling—for routine behavior, the operation of which is modulated in nonroutine situations by a second system—the supervisory attentional system. His model supports the assessment of data on the effects of brain lesions on human patients.

1.3.2 THE ACTION–PERCEPTION CYCLE Figure 1.2 showed how a number of perceptual schemas simultaneously activated to characterize properties of an object could activate motor schemas for the coordinated

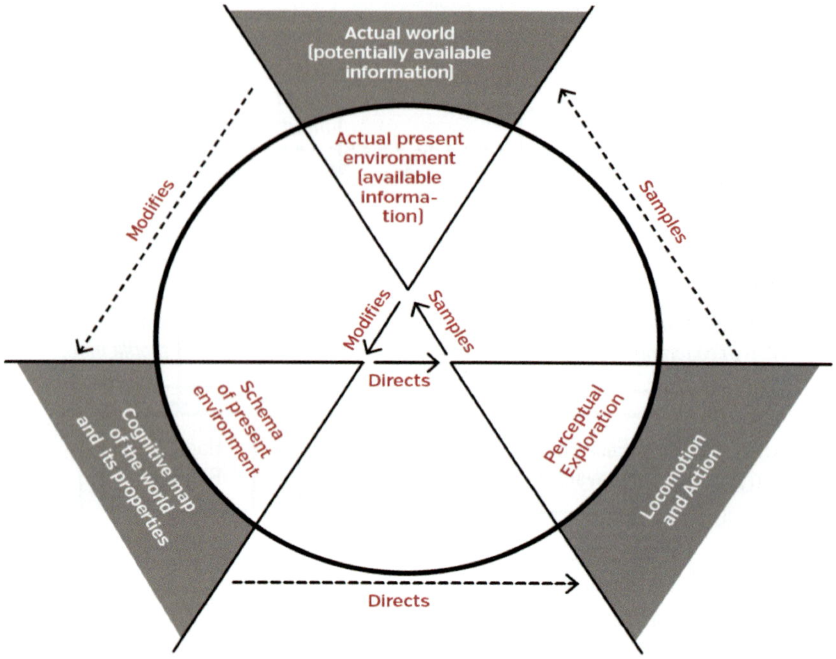

FIGURE 1.3 The action–perception cycle. The point about a cycle is that one can start anywhere. We can, for example, start tracing each cycle in the actions at lower right: in the inner loop, perceptual exploration shifts our attention to different aspects of the world, thus updating the schema assemblage that represents our current environment; the outer loop stresses those interactions (both physical and social) which may modify the world or our relationship with it; these changes may update our cognitive map of the world, and this in turn may update our (not necessarily conscious) plans for action. Adapted from Neisser (1976).

control of reaching for and grasping the object. An important point is that the set of active motor schemas would change as the coordinated actions progress. Our brief pointer to chapter 25 emphasized that visual perception may do more than characterize a single object—visual input can invoke knowledge coded as a network of schemas in long-term memory to create an assemblage of schema instances that represent aspects of the visual environment. Figure 1.3 integrates these observations into a single perspective, the action–perception cycle (Neisser, 1976). Far from being passive responders to the sensory flow, we actively seek information relevant to our current goals (though, of course, some unexpected events may grab our attention and may even modify our goals). Thus, our actions may either serve to focus attention on sensory data of current interest or change our relation with the physical world, including other people and other creatures. Brains mediate social interactions as well as interactions with the physical (including built) environment.

At any time our current perception of the world combines with our current goals, plans, and motivations to activate certain combinations of motor schemas which determine our actions. Each action may change the world and/or our relationship with the world (contrast opening a door vs. turning one's head just to shift the focus of attention). Thus, the input available about the world at any time will change, both because of our actions (both physical and social) and because the world may be changing anyway. As a result, the assemblage of currently active schemas will be updated and will include perceptions of actions and other relationships, as well as objects and their properties or parameters. The rotating arrows in figure 1.3 indicate that memory structures are invoked in updating our perceptions and our plan of action (further variables relate to the motion and internal state of the body) but also indicate that we continually perceive the consequences of our actions or simply observe correlated changes in the external world so that both our working memory and long-term memory may be modified. Thus, *the action–perception cycle is a learning cycle*, too. As we interact with the world, we are constantly learning, not only laying down memories of specific events (*episodic memory*) but also acquiring and refining diverse skills (*procedural memory*) which are coded at least in part by our perceptual schemas, motor schemas, and the larger structures that integrate them.

1.3.3 THREE FLAVORS OF SCHEMA THEORY For brain theory, the top-level schemas must be "large" enough to allow an analysis of behavior at or near the psychological level yet also be subject to successive decomposition down to a level that may, in certain cases, be implemented in specific neural networks. One must distinguish a schema as a *functional* unit from a neural network as a *structural* unit. A given schema may be distributed across several neural networks; a given neural network may be involved in the implementation of several different schemas. A top-down analysis (decomposing a function) may suggest that a certain schema is embedded in a certain part of the brain; we can then marshal the available data from anatomy and neurophysiology to assess whether the circuitry can, indeed, subserve an instance of that schema. It often happens that the empirical data are inadequate. We then make hypotheses for experimental confirmation. Alternatively, bottom-up analysis of a brain region (assembling its constituents) may suggest that it subserves a different schema from that originally hypothesized, and we must then conduct a new top-down analysis in the light of these newfound constraints. This discussion supports the following distinction between two "flavors" of schema theory:

Basic schema theory studies schemas as dynamical, interacting systems which underlie cognition and overt behavior (and not just conscious processes). Basic schema theory is defined at a functional level which associates schemas with specific perceptual, motor, and cognitive abilities and then stresses how our mental life results from the dynamic interaction—the competition and cooperation—of many schema instances. It refines and extends an overly phenomenological account of the "mental level."

Neural schema theory: The "downward" extension of schema theory seeks to understand how schemas and their interactions may indeed be played out over neural circuitry—a move from psychology and cognitive science as classically conceived (viewing the mind "from the outside") to cognitive neuroscience. Neural schema theory analyzes data from neurophysiology, lesion studies, and brain imaging to see how schemas may be restructured to relate to distributed neural mechanisms.

A third flavor comes when we seek to distinguish "schemas in the head" from forms of knowledge and action shared by a community:

Social schema theory: The "upward" extension of schema theory seeks to understand how "social schemas" constituted by collective patterns of behavior in a society may provide an external reality for a person's acquisition of schemas "in the head" in the sense of basic schema theory. Conversely, it is the collective effect of behaviors which express schemas within the heads of many individuals that constitutes, and changes, this social reality. Social schemas represent the collective effect of behavior—whether related to everyday events, language, religion, ideology, myth, or scientific paradigm—governed by related schemas (in the sense of basic schema theory) in the individuals of a community (Arbib and Hesse, 1986).

1.4 A Fast Overview of the Human Brain

Figure 1.4 diagrams the cerebral cortex in relation to some (but not all) of the brain regions of importance to us. Cerebral cortex contains excitatory *pyramidal neurons* and a range of inhibitory *interneurons*. Dissection and staining methods have illuminated how the layered structure of cerebral cortex varies from region to region. This *cytoarchitecture* led Brodmann (1909) to parcellate the cerebral cortex as shown in figure 1.5. A few numbers of these Brodmann areas (BAs) are especially worth remembering:

- Vision: Primary visual cortex (BA17) feeds into secondary visual cortex (BA18, BA19).
- Primary somatosensory cortex (BA1), which mediates touch and body sense, feeds into the secondary areas (BA2, BA3).
- Language areas include BA22, Wernicke's area (BA40), and Broca's area (BA44 and BA45).
- Frontal eye fields (BA8) provide a cortical outpost for the control of eye movements.

Brodmann's map of the cerebral cortex (it does not include other brain regions) does not tell us the connectivity (gross neuroanatomy) between these regions of the cortex and with other regions of the brain. Indeed, each region is part of complex patterns of interaction, including interactions with subcortical regions, and so one must not read too much into the functional labels in the above list. Various chapters in this book will focus on specific functions related to cognition and behavior and explore and model the larger networks needed to support them. In general, such models will require us to go beyond the gross neuroanatomy to explore details of neurons and the circuitry in which they are connected.

Functional properties can be investigated through electrophysiological recording methods that allow us to link the firing of one or more neurons to ongoing action, perception, and/or learning. Such methods are rarely justified in humans but yield a wealth of data about animal brains that lets us form hypotheses about mechanisms operative in the human brain. We can learn about animal brains and about human brains

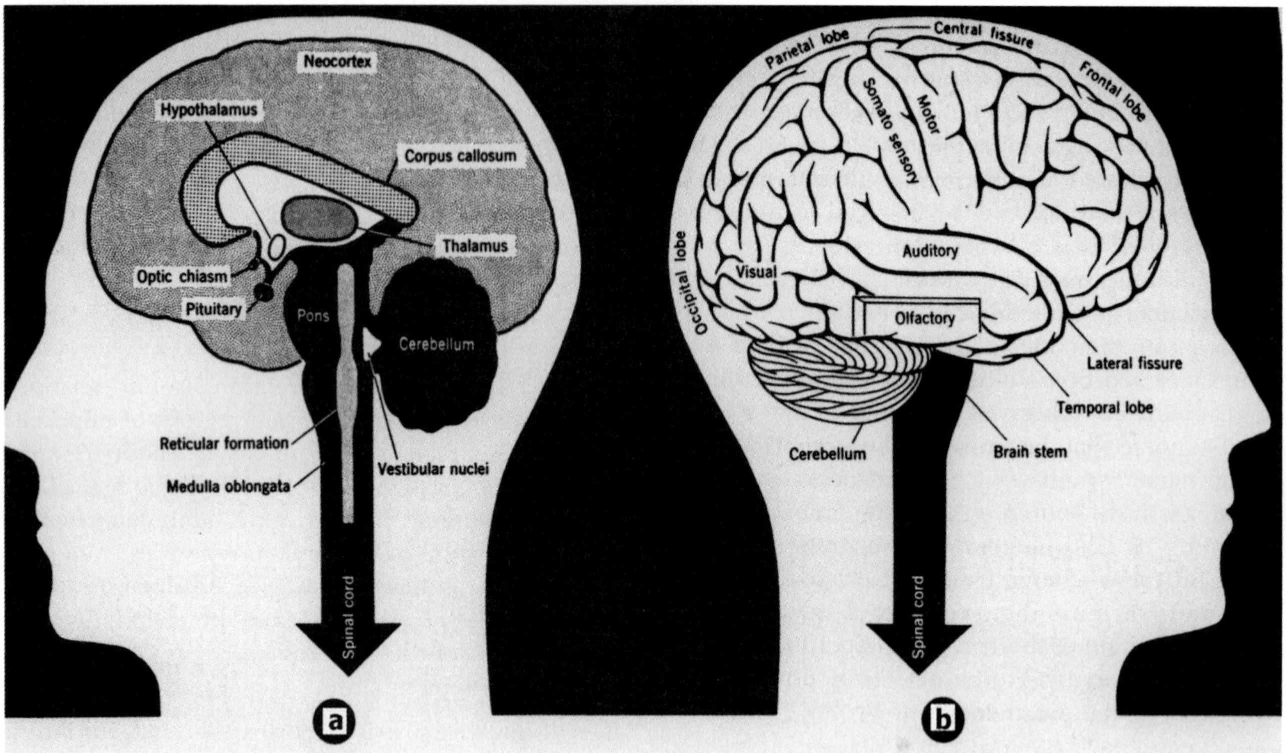

FIGURE 1.4 Medial (a) and lateral (b) views of the human brain. We see the four lobes (occipital, parietal, temporal, and frontal) of the cerebral cortex with the location of primary sensory cortices for vision, audition, and somatosensation (olfactory cortex is a separate, phylogenetically older structure) as well as the primary motor cortex. The medial view reminds us that there is a lot more to the brain than just the cerebral cortex. We will be particularly attentive to the cerebellum and (not shown in the figure) the basal ganglia and the hippocampus.

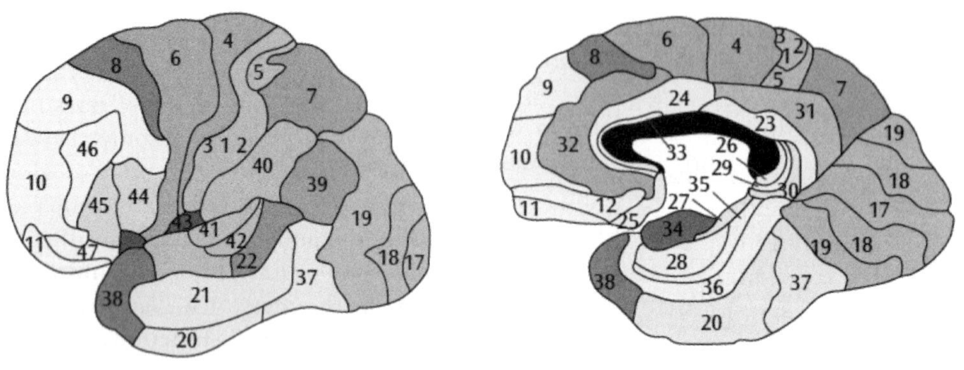

FIGURE 1.5 Brodmann areas in the human brain. A lateral view of cerebral cortex is shown at left, and a medial view at right. The Brodmann areas are all areas of cerebral cortex. There are many other regions of importance to us, including cerebellum, basal ganglia, and hippocampus.

(affected by disease, tumors, neurosurgery) by studying the effects of damage to, or lesion of, a given cluster or structure. In order to demonstrate the relationship between structure and function, damage to a specific part of the brain should produce a measurable functional deficit, and damage to other structures should not produce the same deficit. Note, though, that loss of a function on excision of a region R does not imply that R by itself supports the missing function (though this is true in some cases), but rather that other regions require interaction with R to jointly support it.

Brain imaging methods are the primary methods for identifying function in the living human brain. In *electroencephalography* (EEG), electrodes are attached to the scalp to measure electrophysiological activity of primarily pyramidal neurons and synchronized brain currents, such as *gamma* and *alpha waves* underneath the skull (see chapter 4, "Neural Rhythms"). *Magnetoencephalography* (MEG) also measures brain activity via accompanying magnetic fields. *Diffusion tensor imaging* (DTI) uses magnetic resonance imaging (MRI) technology to assess white matter connectivity and integrity. Of especial importance in cognitive neuroscience, and a main focus of chapter 15, are positron emission tomography (PET) and functional MRI (fMRI). In both methods, the signal obtained is allocated across an image that covers some area of the brain with three-dimensional "cubes," referred to as *voxels*. The voxels are on the order of several millimeters in each dimension. Both PET and fMRI are thought to reflect synaptic activity more than neuron spiking activity. fMRI measurements typically are relative to a baseline, so that two different cognitive states must be compared against each other. Our present knowledge about the neural basis of human behavior and cognition relies largely on the data from human neuroimaging studies and animal brain neurophysiology studies. Chapter 15 ("Synthetic Brain Imaging") provides more detail on these methods and shows how, for example, fMRI data may be related, via modeling, to the much finer scale activity of neurons in the regions under consideration. A plethora of new experimental methods are becoming available—optogenetics, and mapping of gene expression to name just two—and these each challenge us to develop new modeling techniques that can address the new data and seek explanations that integrate them with the insights gained from other experimental methodologies.

We round out our fast overview of the human brain in two subsections, one looking at two pathways of vision and action (with a brief note on the auditory system) and the other introducing basic anatomical substrates of motivation and emotion. In each case, the subsection ends with an evolutionary perspective, suggesting the ways in which the human brain extends capabilities shared with other species, such as monkeys and rats, respectively.

1.4.1 Two Pathways for Visuomotor Integration

Studies of vision have shown that there is really no single "visual system" in the brain but rather "many visual systems" computing such factors as depth, motion, color and so on—even though we are conscious only of a single integrated visual experience. The visual cortex can be decomposed into a number of visual areas based on anatomical and electrophysiological measurements. Like all areas of the cortex, these regions have a standard six-layer structure. Primary visual cortex, V1, receives retinal input via the lateral geniculate nucleus (LGN) of the thalamus. It then projects directly and indirectly to numerous other "visual areas" of cortex such as V2, V4, and medial temporal (MT), medial superior temporal (MST), and the inferior temporal (IT) cortex. There are numerous interactions between these visual areas, but it is common to concentrate on two hierarchical streams: the *ventral stream*, which consists of V1, V2, V4, and IT, and the *dorsal stream*, which goes from V1 to MT to the parietal cortex.

The thalamocortical pathway via LGN and V1 is not the only way that vision can affect action and perception. Another path from the retina leads to the superior colliculus, a region of the midbrain which plays a critical role in the control of eye movements (see chapter 19, "Saccades and Smooth Pursuit Eye Movements"). However, in terms of cortical processing, it is particularly striking that the ability to use the size of an object to preshape the hand before grasping it (recall the schema model of figure 1.2) can be dissociated by brain lesions from the ability to consciously recognize and describe that size.

Ungerleider and Mishkin (1982) conducted experiments with *monkeys*, using lesions to distinguish the role of the *dorsal* and *ventral* pathways. They showed that a ventral lesion would impair memory of the pattern associated with hidden food while a dorsal lesion would impair memory of where the food had been hidden—and thus spoke of the "what" and "where" pathways, respectively. However, in figure 1.6 we refer to a different paradigm—observations of the effect on reaching and grasping of lesions observed in the brains of *human* neurological patients. Here a ventral lesion affects the ability to name the object ("what"), while a dorsal lesion affects the ability to preshape for grasping or otherwise using the object ("how").

How: Goodale et al. (1991) studied a patient (DF) with damage to the ventral path. When asked to indicate the width of a single block either verbally or by means of her index finger and thumb, DF's finger separation bore no relationship to the dimensions of the object and showed considerable trial to trial variability. Yet when she was asked simply to reach out and pick up the block, the peak aperture (well before contact with the object) between index finger and thumb changed systematically with the width of the object. A similar dissociation was seen in her responses to the orientation of stimuli. In other words, DF could preshape accurately, even though she appeared to have no conscious appreciation (expressible either verbally or in pantomime) of the visual parameters that guided the preshape. This aligns the dorsal path with the schema model of figure 1.2.

What: Jeannerod et al. (1994) report a study of impairment of grasping in a patient (AT) with a bilateral

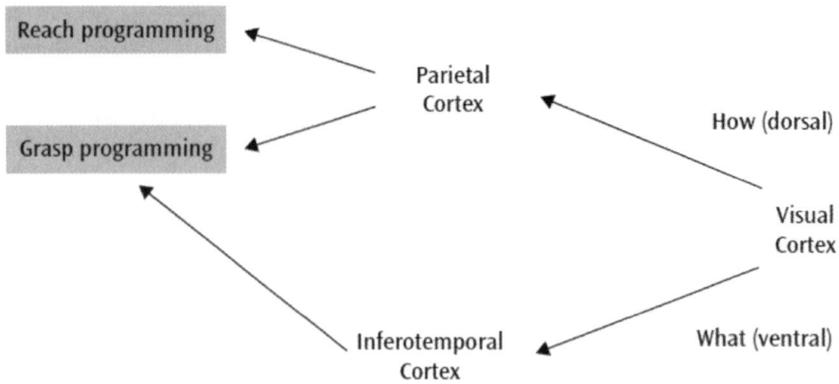

FIGURE 1.6 The "what" and "how" pathways for visual information involved in successful grasping or manipulation of an object. The "how" pathway from visual cortex travels dorsally, via parietal cortex. The "what" (ventral) path travels via inferotemporal cortex.

posterior parietal lesion that grossly impaired the dorsal pathway. AT is the "opposite" of DF—she can use her hand to pantomime the size of a cylinder, and can reach without deficit toward the location of such an object, but cannot preshape appropriately when asked to grasp it. Surprisingly, however, when the stimulus used for the grasp was not a cylinder (for which the "semantics" contains no information about expected size) but rather a familiar object—such as a reel of thread, or a lipstick—for which the "usual" size is part of the subject's knowledge, AT showed a relatively appropriate preshape. This suggests, as indicated by the arrow at bottom left of figure 1.6, that a pathway from IT cortex provides the parietal areas with "default values" of action-related parameters, that is, values which can serve in place of actual sensory data to represent the approximate parameters of a known object to help the parietofrontal system.

We distinguish *praxic* action, in which the hands (or other effectors) are used to interact physically with objects or other creatures, from *communicative* action (both manual and vocal). Waving good-bye and brushing away a cloud of flies might employ the same motion yet they are different actions, the first communicative, the second praxic. We thus see in the AT and DF data a dissociation between parietal and IT pathways, respectively, for the praxic use of size information (which is why we call the dorsal pathway the "how" pathway) and the communicative "declaration" of that information either verbally or through pantomime (which is why we call the ventral pathway the "what" pathway). Chapter 20, "Reach and Grasp: Control, Development, and Recognition," offers a neurobiological model of the pathways in figure 1.6 as they occur in the monkey (where the ventral pathway is implicated in planning, though not communication); chapter 25, "Evolving the Language-Ready Brain," then suggests how capacities for pantomime and language enrich this mix in the human brain.

Although in this book most examples of perception and of the action–perception cycle emphasize vision, other sensory systems (such as audition, somatosensation, touch, taste, and smell) do of course play a crucial role in animal and human behavior, and these systems have been the focus of important modeling in computational neuroscience. Here we simply offer data on the auditory system that complements our study of the dorsal and the ventral pathways emanating from primary visual cortex. The auditory system must also analyze both the identity and the location of the stimuli it detects. Auditory processing of "what" and "where" involves different structures and pathways even before the signals reach the primary auditory cortex. This is situated in the temporal lobe and has a *tonotopic* organization (i.e., it can be mapped in terms of the predominant frequencies of the sounds to which cells respond) while nonprimary auditory cortex responds to complex sounds. Research on the auditory system in human and nonhuman primates (Rauschecker, 1998; Romanski et al., 1999) showed that there are two pathways for processing auditory input emanating from primary auditory cortex. The *dorsal auditory stream* processes spatial information while the *ventral auditory stream* processes auditory pattern and object information. Auditory objects, including speech sounds, are identified in anterior superior temporal cortex which projects directly to inferior frontal regions. Both streams eventually project to the frontal cortex, which integrates auditory spatial and object information with each other and with visual and other modalities.

Rauschecker (1998) suggests that enhancements in analysis of frequency, FM rate, bandwidth, and timing relative to other primate species have occurred during human evolution—augmenting mechanisms that are

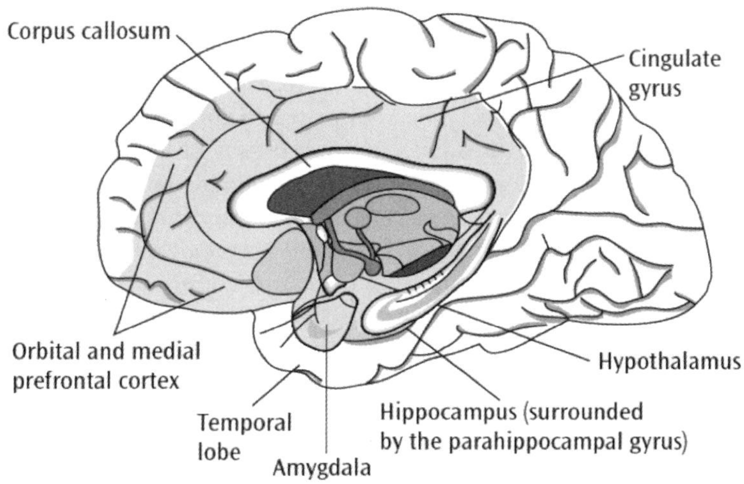

FIGURE 1.7 A diagram of the limbic system (i.e., the brain regions "that throw their limbs" around the thalamus) showing the relative placement of amygdala (named for its almond shape), hippocampus (named for its sea-horse-like appearance in cross-section), cingulate gyrus, medial prefrontal cortex, and temporal lobe in the human brain. Adapted from Williams et al. (2007).

available in, for example, monkey auditory communication systems. On this basis, speech perception is thought to be possible because it combines a high-resolution system for phonological decoding with more efficient memory mechanisms and an ability for abstraction, both residing in a highly developed and expanded frontal cortex.

1.4.2 MOTIVATION AND EMOTION Unfortunately, the present book has no explicit chapter on the neuroscience of emotion (see Fellous and Arbib, 2005, for a collection of articles under the banner *Who Needs Emotions: The Brain Meets the Robot*), but the topic deserves brief mention here. Emotion can be analyzed under two headings:

"External" aspect of emotions: Emotional expression for communication and social coordination. If we see that someone is angry, we will interact with that person more cautiously than we would otherwise, or not at all.

"Internal" aspects of emotions: These frequently contribute to the organization of behavior (prioritization, action selection, attention, social coordination, and learning). For example, the actions one is likely to perform vary greatly depending on whether one is angry or sad.

These two aspects have coevolved. Animals need to survive and perform efficiently within their ecological niche, and in each case the patterns of coordination will greatly influence the suite of relevant emotions (if such are indeed needed) and the means whereby they are communicated. The emotional state sets the framework in which the choice (whether conscious or unconscious) of actions will unfold. But emotions, too, are embedded in the action–perception cycle, so that one's emotions may change as the consequence of one's actions become apparent—and our perception of these consequences may well rest on our perception of the emotional response of others to our behavior.

Figure 1.7 sketches a number of brain regions that support motivation and emotion. These regions constitute "limbs" surrounding the thalamus, forming what is called the *limbic system.* (It has nothing to do, directly, with the control of arms and legs.) Animals (including humans) come with an innate set of basic "drives"—for hunger, thirst, sex, self-preservation, and so forth—and these provide *motivation*, the basic motor for behavior. Motivated behavior not only includes bodily behavior (as in feeding and fleeing, orofacial responses, and defensive and mating activities) but also autonomic output (e.g., heart rate and blood pressure) and visceroendocrine output (e.g., adrenaline, release of sex hormones). These lie at the heart [sic] of our emotional repertoire. However, the emotions that we talk about and perceive in others are both more restricted than this (how many people perceive another's cortisol level?) yet also more subtle, intertwining these basic motivations with our complex cognitions of social role and interactions, as in the cases of jealousy and pride.

The core of the motivation system is provided by the nuclei of a deep-seated region of the brain called the *hypothalamus.* These nuclei are devoted to the elaboration and control of specific survival behaviors such as spontaneous locomotion, exploration, and ingestive, defensive, and reproductive behaviors. Basically, the hypothalamus talks "downward" for basic behavioral control and "upward" to involve the cortex in determining when particular behaviors are appropriate. Indeed, many instances of motivated behavior—eating, drinking, grooming, attack, sleep, maternal behavior, hoarding, copulation—have been evoked by direct electrical or chemical stimulation of the hypothalamus. The lateral hypothalamus plays a critical role in arousal, control of behavioral state, and reward-seeking behavior. It includes what Jim Olds (1969) referred to as the "pleasure center" because rats will press a lever thousands of times per hour to deliver electrical stimulation to this region. Kelley (2005) suggested that neuromodulators (chapter 7) support the "readout" of the phylogenetically old motivational circuitry activity into emotions.

The almond-shaped amygdala, associated with the experience and conditioning of fear (Ledoux, 2000), sits next to the hippocampus and is surrounded by the temporal lobe. Studies of spatial navigation in rats suggests that rat hippocampus provides a "You are here" function. Hippocampus thus plays a role in fear conditioning in rats, letting them use contextual cues to recognize that they are in a place "where bad things happen," and react fearfully as a result.

Another perspective comes from the result of neurosurgery on a patient referred to as H.M. (1926–2008).[1] His hippocampus, the amygdala, and the parahippocampal gyrus were almost completely ablated bilaterally. As a consequence, he became unable to form new *episodic memories* (i.e., memories of episodes or events he experienced) although his intellectual and perceptual functions, his memories of events that occurred before the surgery, and his *working memory* were intact. (See chapter 18, "Hippocampus: From Spatial Processing to Episodic Memory.") However, H.M. was still able to form new *procedural memories* (i.e., he was able to acquire new skills). Such observations led to the following hypotheses:

1. Hippocampus is necessary for the formation of episodic memories. (Note that this goes beyond what we know about H.M., since his surgery removed several adjacent regions as well as the hippocampus.)
2. However, the fact that episodic memory from before the neurosurgery remains available shows that the hippocampus is not necessary for the long-term storage of episodic memories. We thus talk of a process of *consolidation* whereby episodic memories, shaped by the hippocampus, are eventually stored in long-lasting form elsewhere.
3. Hippocampus is *not* necessary for the formation of procedural memories. In fact, we now think that procedural memory involves a partnership between regions of frontal cortex and other regions of the brain called the basal ganglia and the cerebellum. (See chapters 21 and 22.)

Back to emotion. Human emotions are strongly influenced by social milieu—what is embarrassing in one culture (e.g., nudity) may be regarded as perfectly normal in another. The amygdala can influence cortical areas by way of feedback either from proprioceptive or visceral signals or hormones, via projections to various "arousal" networks, and through interaction with the *medial prefrontal cortex*. This area has widespread influences on cognition and behavior and also sends connections to several amygdala regions, allowing cognitive functions organized in prefrontal regions to regulate the amygdala and its fear reactions. Connections of the amygdala with hippocampus and cerebral cortex provide the possible basis for enrichment of "behavioral emotion" by episodic memory and cognitive state.

1.5 Neurons and Neural Networks

Nicolas Brunel introduces "Basic Neuron and Network Models" in chapter 2, which provides some sense of how models capture aspects of the workings of neurons that will help us understand the potential functionality of neural circuitry. Jonathan Rubin and John Rinzel then survey the treatment of "Neurons and Neuronal Networks as Dynamical Systems" (chapter 3) with the emphasis on the way in which the mathematics of nonlinear systems can use such notions as equilibria, stability, and oscillations to illuminate general properties of neural system dynamics. There is something of a chicken-and-egg problem between chapters 2 and 3 in that some of the terminology of dynamical systems theory is needed for the exposition of basic neuron and network models, but it is only with some understanding of such models that one can appreciate the deeper application of dynamical systems theory to neuroscience.

There are radically different types of neurons in the human brain, and endless variations in neuron types of other species. In brain theory, the complexities of real neurons are abstracted in many ways to aid in understanding different aspects of neural network development, learning, or function. In what follows, we assume the reader has had an initial exposure to basic neural components such as dendrites, soma (cell body), axons and synapses, and the propagation of spikes down the branches of an axon to reach synapses on other neurons or on muscles or other effector organs. In chapter 2, the basic neuron and network models include statistical as well as deterministic models of neuronal activity and neural coding. Output of a neuron may be characterized by spike trains (i.e., viewing the output of each neuron as a time series, defined in continuous time, of when the action potential at the axon hillock passes threshold for axonal propagation) or by firing rate (some measure of the rate at which spikes have been generated in a recent time interval). In primary visual cortex, the firing rate of a neuron seems adequate to characterize its relationship to some specific local feature of the visual input (such as an edge of a particular orientation). However, a cell in the inferior olive that sends "training signals" to the cerebellar cortex (chapter 21) fires just a few times a second, and so the timing of each individual spike may be crucial. Moreover, in networks whose activity oscillates in time, information can be provided by the timing of spikes, or variation of firing rate, relative to this overall oscillation (see chapter 4, "Neural Rhythms," and chapter 8, "Motor Pattern Generation"). And what counts as a "neuron" in a model may indeed correspond to one neuron or to a pool of neurons with some shared characteristics. Again, each neuron may be represented as a single point where synaptic activity is gathered to determine spikes or firing rate, or it may be a spatially extended structure with various "compartments."

Neurons are interconnected to form complex networks. Feedforward networks contain no loops; many models of such networks are arranged as a series of layers—input layer, possibly "hidden" layers, output layer—with synaptic connections going only in the direction from input to output. (Note the somewhat related use of the terms feedback and feedforward in chapter 9, "Computational Models in Motor Control": Feedback control involves a loop back to monitor output values; feedforward does not.) Most, if not all, networks in the central nervous system are recurrent networks, that is, they contain loops. A number of models have been based on simple recurrent networks (SRNs). In this case, a simple feedforward network has a layer (the context layer or state layer) whose input comes from one or more of the other layers but whose output is fed back to, for example, the hidden layer. The firing of this extra layer thus offers a state variable which depends on the history of the network and thus influences how the network will respond to its current input.

A key idea in making sense of a neural network is that of a *population code*. In many cases, the variables that we use to characterize behavior or cognition from the outside (and which we might thus be tempted to think would be coded by the firing of a specific neuron in the brain) are in fact distributed over a whole population of neurons. Consider "edge detectors" in primary visual cortex. We might find a cell that fires most for an edge of orientation θ falling on retinal location (x,y), with firing rate dropping off if any of θ, x, or y change. But suppose this cell fires submaximally. Does this mean that there is a θ–edge at (x,y) that is a little bit blurred, or maybe a slightly sharper edge at (x,y) with a somewhat different orientation or a θ–edge located a little off (x,y) or …? Only examining a population of nearby cells will answer the question. Conversely, when we look at motor cortex, each individual cell might be "voting" for or against a specific vector of force for muscles tugging around a given joint, and only the population can yield the time-varying control signal to move the index finger to touch a particular point in the visual field.

We start by listing various simplified models of neurons that will often be used in this book and then look at the detailed type of neural modeling that involves versions of the Hodgkin–Huxley equations, which is usually the basis for compartmental models.

1.5.1 Simplified Models of Neurons and Synapses

Here I just briefly list various models, referring the reader to chapter 2 for equations and further discussion.

The simplest model of the neuron, the *McCulloch–Pitts neuron* (McCulloch and Pitts, 1943), divides time into units comparable to a refractory period (i.e., the interval after one spike when another spike cannot be generated [this notion is somewhat of a simplification]) so that, in each time period, at most one spike can be generated at the axon hillock of a given neuron. We write $y(t) = 1$ if a spike appears at time t, and $y(t) = 0$ if not. Each *synapse* has an attached *weight* w. A synapse is *excitatory* if $w > 0$, and *inhibitory* if $w < 0$. A *threshold* θ is associated with each neuron, and the model posits exactly one unit of delay in the effect of *all* presynaptic inputs on the cell's output, so that a neuron "fires" (i.e., has value 1 on its output line) at time $t + 1$ if the weighted value of its inputs at time t is at least θ. Formally, if at time t the value of the ith input is $x_i(t)$ and the output one time step later is $y(t + 1)$, then

$$y(t+1) = 1 \text{ if and only if } \sum_i w_i x_i(t) \geq \theta.$$

It is a straightforward exercise to show how weights and threshold can be set to yield neurons which realize the logical functions AND, OR, and NOT. As a result, McCulloch–Pitts neurons are sufficient to build networks which can function as the control circuitry for a computer carrying out computations of arbitrary complexity. This discovery played a crucial role in the development of automata theory and in the study of learning machines. While the McCulloch–Pitts neuron no longer plays an active part *in computational neuroscience*, it is still widely used in *neural computing*, the technological application of networks of adaptive artificial neurons, especially when it is generalized so that the input and output values can lie anywhere in the range [0, 1] and the function $f(\sum_i w_i x_i(t))$ which yields $y(t + 1)$ is a continuously varying (usually monotonic) function rather than a step function. However, *it is one thing to define model neurons with sufficient logical power to subserve any discrete computation; it is quite another to understand how the neurons in actual brains perform their tasks.* More generally, the problem is to select just which units to model, and to decide how such units are to be represented. A variant of the McCulloch–Pitts model that has proved attractive to statistical physicists is the *Hopfield model.* It is like the McCulloch–Pitts model save that (1) synaptic weights are symmetrical—the weight w_{ij} from neuron j to neuron i equals w_{ji}—and (2) Updating is asynchronous: at each time step, just one neuron is selected at random to be updated. The key property (see appendix A) is that an "energy" function can be defined for the network and has the property that in the absence of inputs the dynamics of the network will carry it to a state of minimum energy.

When we add "learning rules" (rules for how synaptic strengths change on the basis of neural activity) to such simple neurons, we get "connectionist models," where the interest lies in how the network can learn some appropriate input–output behavior whether or not the artificial "neurons" have any correspondence to the real neurons of an actual brain. A brief introduction to such networks is given in appendix A; then appendix B adds some details on the special case of associator networks. However, in some cases connectionist networks may factor into more realistic brain models when the model's "computational units" are suggested to represent the dynamics of (pools of) identifiable neurons in some actual brain region.

Perhaps the simplest model of a neuron using continuous time and with the temporal distribution of spikes as the axonal output is the *integrate-and-fire neuron*: each synaptic input generates a time course of decaying activity in the postsynaptic neuron; these postsynaptic potentials yield traces which are summed at the axon hillock (initiation zone for axonal spiking); and each time the sum of these traces passes a threshold, a spike is fired down the axon. To capture the notion of refractory period, the threshold in the model immediately rises to a very high value after each spike is initiated, and then it decays back to its resting value.

One of the simplest firing rate models is the *leaky integrator* neuron. The *internal state* of the neuron is described by the membrane potential m(t) at the axon hillock. No description is given of how spikes are generated. Instead, a firing rate $M(t) = \sigma(m(t))$ is specified by a sigmoid function σ of the membrane potential *m*: $\sigma(m)$ increases from 0 to some maximum value as m increases from $-\infty$ to $+\infty$. The time evolution of the cell's membrane potential is given by the differential equation

$$\tau\dot{x} = -m(t) + \sum_i w_i X_i(t) + h$$

where τ is a time constant, $X_i(t)$ is the firing rate at the ith input, the w_i are the synaptic weights, and h is the equilibrium membrane potential in the absence of inputs. The $-m(t)$ term opposes unbounded integration of the inputs over time by a "leakage" of the potential m(t) as it tries to return to its input-free equilibrium h.

It should be noted that, even at this simple level of modeling, there are alternative models. In the foregoing model, we have used subtractive inhibition. However, there are inhibitory synapses that seem better described by *shunting* inhibition which, applied at a given point on a dendrite, serves to divide, rather than subtract from, the potential change passively propagating from more distal synapses. Again, the "lumped frequency" model cannot model the relative timing effects due to differential delays in the effects of different synapses upon the membrane potential at the axon hillock. These might be approximated by introducing appropriate delay terms

$$\tau\dot{x} = -m(t) + \sum_i w_i X_i(t - t_i) + h.$$

Chapter 2 includes discussion of rate models with delays and rate models with slow negative feedback. All this reinforces the observation that there is no modeling approach which is automatically appropriate. Rather, we seek to find the simplest model adequate to address the complexity of a given range of problems. What is the neural code? What computations are performed by a part of the neuron (e.g., dendrites) or a single neuron or require a whole circuit? What do different types of neurons contribute? And how do neurons in different parts of the brain compete and cooperate to yield the organism's overall behavior and the updates to memory that will affect that behavior on future occasions?

One approach to modeling neural structures has been to consider a sheet or array of neurons for each type of neuron in the structure. In this context, firing rate models are used to investigate the dynamics of spatially extended networks, for example, systems which encode continuous stimuli. In such *dynamic field models*, connectivity between neurons depends on the physical distance between their location and/or their selectivity properties. Such a model allows one to study how the dynamics depends on the spatial and/or functional structure of the connectivity. Wilson and Cowan (1973) used such fields to study the emergence of thalamocortical interactions. They employed a *mean field model*, where similar neurons are aggregated into a unit whose activity represents the mean activity (e.g., average firing rate) of the population of cells it represents. Amari and Arbib (1977) used a dynamic field model to explain a *winner-take-all* (WTA) behavior in which the winner (or a few winners) silences other neurons in a population through strong inhibitory connections; they then showed how cooperative linkage of an array of such WTA networks could yield a model of stereo vision. These two efforts formalized models of frog prey selection (Didday, 1976) and stereopsis (Dev, 1975), respectively, but are broadly applicable in diverse brain systems.

1.5.2 Hodgkin–Huxley Modeling The propagation of action potentials is now very well understood. Briefly, the change of membrane potential is mediated by the flow of ions, especially sodium and potassium, across the membrane. In their Nobel Prize–winning work, Hodgkin and Huxley (1952) showed that the *conductance* of the membrane to sodium and potassium ions—the ease with which they flow across the membrane—depends on the transmembrane voltage. They developed elegant equations describing the voltage- and time-dependence of the sodium and potassium conductances. These equations have given us great insight into cellular function. Hodgkin and Huxley used curve fitting from experimental data to determine the terms for conductance change in their model. Subsequently, much research has probed the structure of complex molecules that form *channels* which selectively allow the passage of specific ions through the membrane. This research has demonstrated how channel properties not only account for the terms in the Hodgkin–Huxley equations but also underlie more complex dynamics which may allow even small patches of neural membrane to act like complex computing elements. In such modeling, time is continuous and spikes are "unwrapped" to yield a time course of changes in the membrane potential.

Much mathematical research has gone into studying Hodgkin–Huxley-like equations, showing, for example, that neurons can support rhythmic pulse generation even without input, and explicating triggered long-distance propagation.

Hodgkin and Huxley deduced the dynamical equations of their model by fitting their data on the squid giant axon preparation. In the Hodgkin–Huxley model, the dynamics of the membrane potential $V(t)$ is governed by a first-order differential equation

$$C\frac{dV}{dt} = -I_L(V) - I_{Na}(V) - I_K(V) + I_{ext}(t)$$

where C is the membrane capacitance, $I_L(V)$ is a leak current, $I_L(V) = g_L(V - V_L)$ where g_L is the leak conductance and $V_L \sim -70\ mV$ is the leak potential, $I_{Na}(V)$ is a sodium current, $I_K(V)$ is a potassium current, and $I_{ext}(t)$ is an external current, which can describe a current injected through an electrode, or synaptic currents. Hodgkin and Huxley found that the dynamics of this current could be well described by the following system of equations for the sodium current,

$$I_{Na}(V) = \bar{g}_{Na} m^3 h (V - V_{Na})$$

$$\tau_m(V)\frac{dm}{dt} = -m + m_\infty(V)$$

$$\tau_h(V)\frac{dh}{dt} = -h + h_\infty(V),$$

and these for the potassium current,

$$I_K(V) = \bar{g}_K n^4 (V - V_K)$$

$$\tau_n(V)\frac{dn}{dt} = -n - n_\infty(V).$$

Thus, the conductances (inverse of resistances, $g = 1/R$) are not fixed, but vary with voltage:

$$g_{Na} = \bar{g}_{Na}\ m^3 h,$$

$$g_K = \bar{g}_K\ n^4.$$

What is striking is that the data fitting started with an argument (based on *qualitative* curve fitting) for the m^3h form in the I_{Na} equation and the n^4 form in the I_K equation and was then carried out quantitatively on the m, h, and n equations, not on the primary data on spike propagation—and was done before electronic computers had replaced tedious computations using hand-cranked machines for addition and multiplication. Their formalism has been highly influential for its capacity to quantitatively reproduce generation of the action potential (and its propagation for a spatially extended version), and a similar formalism has been used to describe quantitatively a large number of additional voltage-gated currents. However, its complexity makes its mathematical characterization difficult, and simulations of large numbers of such neurons expensive. As noted in chapter 2 ("Basic Neuron and Network Models"), various strategies exist for simplifying the equations to get useful approximations like the FitzHugh Nagumo and Morris–Lecar models.

In some models, the neuron will be treated as a single compartment modeled by Hodgkin–Huxley-style equations; in other cases, the neuron may be modeled by several linked compartments (e.g., representing major sections of the dendrites, and one each for soma and axon). In the latter case, we may speak of a *compartmental model* of the neuron (Rall, 2003). In either case, much extra work is required to model the synapses which convert the membrane potential in the axon of one neuron into an appropriate change in one or more compartments in other neurons. The activity of ionic channels is explicitly modeled and determines the membrane potential and the generation of action potentials.

A little more on synapses: when spikes along an axon reach presynaptic boutons, they initiate release of neurotransmitters in the synaptic cleft which bind to receptor channels on the postsynaptic membrane. This triggers opening of the channels, which leads to ion entry in the postsynaptic cell, and consequently postsynaptic currents (PSCs). Common types of excitatory receptor channels are the AMPA receptors which lead to a fast excitatory PSC (EPSC) and NMDA receptors, which lead to a slower EPSC. Synaptic currents induced by NMDA receptors have a complex voltage dependence since they are blocked by magnesium ions in a voltage-dependent way. NMDA-receptor mediated EPSPs are critical for the maintenance of information in working memory. They allow a network to maintain its active state without the need for synaptic modification. The most common inhibitory receptor channel is the GABA receptor which leads to an inhibitory PSC (IPSC). Quantitatively, analysis of these currents may include their time course, their voltage dependence, and their history dependence, and whether they are deterministic or stochastic. Moreover, many types of synapses are plastic, changing over time. Furthermore, as noted earlier, there are neurotransmitters that act by changing the dynamic properties of the postsynaptic synapse or neuron (see chapter 7, "Neuromodulation").

In line with "The map is not the territory," chapter 8 ("Motor Pattern Generation") offers models at three different levels for the networks underlying central pattern generators (CPGs): the most detailed model uses Hodgkin–Huxley-style neuron models. Other network models use simpler neuron models such as leaky-integrator neurons or integrate-and-fire neurons to assess how network properties and interneuron connections affect the various circuits (see chapter 2). At the most abstract level, and specifically for studies when rhythms are the focus of attention, CPGs have been investigated using systems of coupled oscillators, with the dynamics of a whole unit pattern generator circuit captured by a single mathematical oscillator model rather than the dynamics of a more realistic neural network. The dynamics of the whole CPG circuit is then studied using several of these oscillators coupled together. In such a model, the key variable is the phase θ (so this model may be called the theta model of oscillator) which advances at a steady frequency in the isolated neuron, with a spike fired each time θ passes 0, but it can be advanced by excitatory input and retarded by inhibitory input.

Chapter 4 ("Neural Rhythms") also uses a variety of neural models. In one network, they simply couple two theta neurons. One is an excitatory cell, representing a pyramidal cell producing fast AMPA type excitation, and the other an inhibitory cell representing an interneuron producing $GABA_A$ type inhibition. They then explore the effects on the rhythm of each cell when they are coupled by synapses with time constants reflecting the fact that the inhibition takes longer to decay than the excitation. (Chapter 11 notes that models of working memory also exploit the observation that the dynamics of NMDA receptors is qualitatively different from the dynamics of inhibitory [GABA] and fast excitatory [AMPA] conductances.) Chapter 4 shows, for example, that if inhibition is strong and long, the network rhythm will be quite slow, but if inhibition is shorter or weaker, the rhythm will be faster—and, of course, with no inhibition, the network spikes as fast as the intrinsic firing rate of the excitatory. The point of this model, then, is to demonstrate as simply as possible the effects of coupling neurons when we take synaptic time constants into account. However, real neurons often contain several types of intrinsic currents that can produce interesting behaviors, like rapid bursts of spikes that depend on slow calcium currents or spike trains that gradually slow because of additional potassium currents. They thus turn to a more realistic model with larger networks of Hodgkin–Huxley-style neurons in which activation of intrinsic currents via synaptic interactions, or modulation with behavioral state, can cause a single network to express rhythmic activity at several frequencies.

To give a sense of the detail that may go into Hodgkin–Huxley-style modeling, here are the five currents used in a chapter 4 ("Neural Rhythms") model of a two-cell excitatory–inhibitory network which can transition from fast rhythms to slower beta and alpha rhythms, with each cell simulated as a single compartment neuron:

Sodium Current: Depolarizing sodium current that initiates spiking.

Potassium Current: Hyperpolarizing potassium current that terminates individual spikes.

T-Current: A low threshold depolarizing calcium current that can only be activated after hyperpolarization, important for generating bursts of spikes.

h-Current: Depolarizing mixed cation current, activated at hyperpolarized levels and important for bursting.

AHP-Current (also referred to as an m-current): Slowly inactivating hyperpolarizing potassium current, shapes the after-hyperpolarization potential and generates adaptation with repeated spiking, yielding slowing of repeated spikes.

The first two are the voltage-dependent sodium and potassium currents that generate spiking, as studied by Hodgkin and Huxley (1952). The parameters in the equations in the computational model are obtained from intracellular experimental data using curve-fitting techniques. They are often similar to those described by Hodgkin and Huxley but fitted to mammalian cells.

In addition to the five currents whose dynamics constitute the behavior of the compartment, the model includes equations for the time course of synaptic currents whereby one cell affects another:

Synaptic Inhibition: Synaptic inhibitory current, activated by spiking in the presynaptic inhibitory cell.

Synaptic Excitation: Synaptic excitatory current, activated by spiking in the presynaptic excitatory cell.

My aim here is not to explain these currents but rather to alert the reader to how some computational neuroscientists have used detailed modeling at this level to explain network properties that cannot be captured by the simple leaky integrator or integrate-and-fire models. However, many subsequent chapters get by very well with the simplified neurons—so part of the art of modeling neural systems underlying behavior and cognition is to find the simplest model of the individual neurons which allows one to explain the properties of interest through mathematical analysis or computer simulation of networks of such neurons.

Chapter 7 ("Neuromodulation") includes a model of dopaminergic (dopamine-releasing) neurons which also have more complex firing patterns than the original Hodgkin–Huxley model and include multiple compartments (soma, proximal dendrite, distal dendrite). The model is of interest because any drug or phenomenon capable of modulating specific intrinsic conductances of these neurons can change their activity from tonic to bursting and can, in turn, have drastic effects on the neural computations of dozens of brain areas to which the neurons project, thus affecting both ongoing activity and neural plasticity. The model explores the role of Ca^{2+} in the spontaneous pacemaking in dopamine neurons. The model involves two additional voltage-dependent currents, the transient outward K^{+} current, and the L-type Ca^{2+} current. The model also tracks the concentration of certain ionic intracellular free calcium ions and sodium ions in order to more precisely model currents that are dependent on these concentrations. This requires specification of the small conductance calcium-activated potassium current, a calcium pump, and a sodium pump. The sodium and calcium pumps maintain the concentrations of these ions at homeostatic levels and may produce a net current. Finally, the model includes two synaptic currents relevant to synaptic activation of dopaminergic neurons: a GABAergic current and an NMDA current. The GABAergic current represents the level of inhibition. Experiments have shown that NMDA currents contribute to burst firing whereas $GABA_A$ currents reduce the tendency to burst.

As we have mentioned, chapter 8 models central pattern generators (CPGs) at three levels of detail. The biophysically detailed model models a simplified segmental network of the spinal CPG in the lamprey, with each neuron analyzed as five-compartment Hodgkin–Huxley neuron model was used to build. The neurons interact through two types of excitatory synapses (targeting AMPA and NMDA receptors, respectively) and inhibitory synapses. In addition to synapses, AMPA and NMDA "bath conductances" are included to mimic the effect of the addition of receptor-specific pharmacological agents to the solution that bathes the spinal tissue in physiological experiments. In the soma, sodium and potassium channels are accompanied by a calcium channel and two calcium-dependent potassium channels. One is driven by calcium entering through the membrane calcium channel, and one by calcium entering at the NMDA synapse. The latter further exhibits a voltage-dependent magnesium block. The dendrites are passive targets of synaptic connections, with five ionic channels, the spike generating sodium and potassium channels, a calcium channel, and two calcium-dependent potassium channels. As noted in chapter 8, simpler models suffice to assess basic rhythmic activity in appropriately structured networks of excitatory and inhibitory neurons. Hodgkin–Huxley models become relevant when, for example, assessing experimental findings where specific membrane channels have been (partially) blocked—as in the case where applying the drug tetrodotoxin (TTX) blocks the sodium channels that generate action potentials.

1.5.3 The Dynamical Systems Approach to Neuroscience Earlier (in section 1.2.1), I stressed the importance of understanding how models relate to empirical data: assessing, in particular, which empirical data were used to constrain the design of the model and which were used to test the performance of the model, and the results of those tests. Yet, ironically, much important work in computational neuroscience operates at a qualitative level. As chapter 3 ("Neurons and Neuronal Networks as Dynamical Systems") demonstrates, our attempts to understand dysfunctions such as epilepsy, oscillations linked with sleep or cognition, temporal sequences that represent the memories of events, or computation for perception and action can often be enriched by stepping back from specific neural systems to understand more general dynamic processes. Chapter 3 samples the way in which the mathematics of dynamic systems theory can address such questions as the following: How do transitions between dynamical states occur, and what are some generic signatures of transition types that provide insights into neural mechanisms? How do we understand or account for the fine temporal structure of a computation yet not be burdened by tracking every action potential in trying to identify longer timescale structure?

The starting point for the dynamical systems approach is to think of solutions to ordinary differential equations (ODEs) as *trajectories* in *phase space.* A phase space diagram, or *phase portrait,* along with a representative set of trajectories, may illuminate the possible behaviors that can result from different choices of initial conditions. The key to establishing the dynamics of a system will often be to establish states which do not change under the ODEs of the system—these are called *equilibrium points, fixed points, critical points,* or *singularities*—and then to assess whether they are *asymptotically stable,* in which case trajectories from nearby initial conditions approach the critical point as $t \rightarrow \infty$; *stable,* in which case nearby trajectories need not approach but must stay nearby, or *unstable,* in which case, no matter how close we zoom in, there are always one or more directions along which trajectories move away from the critical point. Of course, chapter 3 carries the discussion forward by getting into technical details not mentioned here. These include the use of nullclines as an aid in for determining the behavior of trajectories.

To understand a dynamical system, it is always important to know where the critical points of the model lie within phase space and what behaviors trajectories exhibit in local neighborhoods of these points. A *separatrix* separates trajectories with one form of behavior from trajectories with another, qualitatively distinct behavior (e.g., converging to, or escaping from, different critical points). In many cases, a particular model (set of ODEs) might include some general parameter, such as the value of the threshold for neurons in the system. Intuitively, a network might see a lot of firing when the threshold is low, and little or no firing when the threshold is high. In general, there will be a range of parameter values for which behavior (as characterized, e.g., by the distribution of stable and unstable equilibria) is qualitatively the same; it will then be of interest to find values of a key parameter such that the behavior of the system changes qualitatively as the parameter passes through this value. We say that a *bifurcation* occurs when critical points change stability, or are born or annihilated, as a parameter is varied. Chapter 3 also discusses bifurcations which give rise to oscillations.

Here, briefly, is a sample of the dynamical analyses carried forward in chapter 3:

Developing embryonic chick spinal cord neurons exhibit episodic rhythms: If different variables evolve on multiple timescales, then there is hope of considering separate, lower-dimensional fast and slow subsystems and piecing their dynamics together to understand full model dynamics. This approach has been used with great success in the analysis of neuronal bursting, or alternation between active phases of rapid action potential firing and quiescent phases without spikes. In the example of the embryonic chick spinal cord neurons, population bursts of activity (called episodes) with durations on the order of tens of seconds alternate with long interepisode intervals of 2–10 minutes. Synapses that are normally inhibitory are functionally excitatory at this stage of development. The model describes only the average firing rate of neurons within the population, rather than the instantaneous membrane potential and spikes of the individual neurons. In doing so, it is a mean field model. It assumes a purely excitatory recurrent network in which synaptic coupling is susceptible to both short-term and long-term, very slow, activity-dependent depression. The key dynamical systems method exemplified here is *separation of timescales.* Since certain variables vary very slowly, one can first get a sense of the system by assessing the dynamics of the system when the slow variables are fixed in value; one can then explore the effect of different values of the slow variables on the stability properties of the system. In this case, by noting that the synaptic efficacy is not constant but rather slowly evolves, the model can incorporate the effects of slow synaptic depression.

Alternations in perception of ambiguous scenes: Many readers will have experienced phenomena like the

Necker cube—when we view it for an extended time we experience transitions between different perceptions. Both imaging in humans and electrophysiology in monkeys provide evidence for alternating activity in cortical areas and even subcortical areas. Such phenomena share inevitability (perception will switch) with mutual exclusivity. Moreover, the alternations are not periodic but are stochastically distributed. The intuitive conceptual model involves two neuronal subpopulations corresponding to the two percepts. When one is active, the other is not, and such mutual exclusivity can be achieved through mutual inhibition. This inhibitory interaction provides *competition*: The two populations are competing for percept dominance. The alternations in perception can then be explained, for example, by a slow negative feedback process akin to fatigue.

Intriguingly, such an alternation plays an important role in the study of locomotion, though there the alternation is regular rather than stochastic. In models presented in chapter 8 ("Motor Pattern Generation") the basic rhythm generating unit of the spinal cord combines two identical half-center oscillators (Brown, 1914). Two subnetworks which make up the oscillator alternate their active and silent phases during an oscillation by mutually suppressing each other via inhibitory synapses. By a WTA process, one becomes more active and suppresses the other. Eventually it fatigues, and then the other experiences a postinhibitory rebound so that its activity dominates and inhibits the other for a while.

In both these cases, a model with two competing neuronal populations is used to explain bistability, but in fact bistability or multistability can also arise in single-cell models. The virtue of the dynamic systems approach is that it can "step back" from many of the specific details of a neuron or network to understand why this is so—for instance, by characterizing models in terms of the bifurcations they exhibit as model parameters are varied.

In the above WTA scenario, the stable states are equilibrium points where one dependent variable is large, corresponding to an active percept, and the other is small, corresponding to a suppressed percept. In contrast, some network interactions can be cooperative, promoting the activity of all components of the network. A simple form of cooperation arises with two units that excite one another to implement a short-term memory.

1.5.4 Neural Rhythms We have already been introduced to neural rhythms, describing the Hodgkin–Huxley models of chapter 4 ("Neural Rhythms") and chapter 8 ("Motor Pattern Generation"). In chapter 8, the focus is on the generation of regular rhythms in motor control, such as we see in breathing, or the alternation of flexion and extension in a limb, the coordination between the limbs in walking, and the waves of flexion and extension that travel down the body of a swimming lamprey or salamander. Here, however, we focus on the treatment by Nathan Vierling-Claassen and Stephanie R. Jones in chapter 4 of rhythms in cerebral cortex. These rhythms may be generated locally or involve interaction with other brain areas such as the thalamus. They occur at many scales across the brain, and the varied coupling between rhythms of different frequencies has been correlated with healthy brain function and as a marker of disease states. Here are some of the key rhythms and their correlates, arranged in order of increasing frequency:

Delta (<4 Hz) is expressed during slow wave sleep.

Theta (4–7 Hz) is engaged in memory processes and navigation.

Alpha (7–14 Hz) increases during states of idleness and decreases with attention.

Beta (15–29 Hz) is involved in cross-area communication, correlated with detection, and coordinated with motor actions.

Mu rhythms arise from sensorimotor cortex and contain a mix of alpha and beta frequencies; this sensorimotor rhythm is involved in perception, motor initiation, and sensorimotor integration.

Gamma (30–100+ Hz) is increased poststimulus with attention and sensory processing.

Two separate points are worth making here:

1. While the alpha rhythm is correlated with disengagement from attention, some of the other rhythms are correlated with active cognitive processing. We are thus in very different territory from the pattern generators of, for example, locomotion, where all the neurons are tightly phase locked to the locomotory rhythm, as they control and coordinate flexion and extension of each leg within some overall gait. If the cortex is to mediate cognition, as, for example, in visual perception or in decision-making, then different neurons must fire differentially, both across the cortex and across each cycle of the rhythm, to code different patterns of information—even though, collectively, the whole population contributes (somewhat noisily) to the troughs and peaks of the ongoing rhythm.

2. It is one thing to explain, with this caveat in mind, how a neural rhythm emerges as a population response and seek to understand why different networks may in different situations yield rhythms in the above frequency bands; it is another to understand whether such rhythms are an epiphenomenon of cortical processing

or whether they are crucial in organizing neurons to carry out different cognitive tasks. Nonetheless, a measure of synchrony is hypothesized to be critical for the binding of information and for the effective downstream flow of information in the brain.

Chapter 4 advances models of the following phenomena:

How gamma rhythms create synchronous cell assemblies robust to heterogeneity and noise: This model extends the theta neuron model introduced earlier (caution: this theta has nothing to do with the theta rhythm) to a larger network and demonstrates how the same inhibitory mechanism that creates the gamma period also promotes synchrony among cells, even in conditions of sparse random connectivity and random noise. Rhythms in these larger networks can emerge at slower frequencies if specific parameters are changed. However, when the rhythm is slower, it can become more susceptible to disruption.

How gamma rhythms link attention and stimulus selection: Experimental work on the interaction between attention and visual orientation selectivity in the macaque visual cortex (Reynolds et al., 1999) demonstrates that even with exactly the same sensory stimuli, attention can shape the neural response so that the firing rate is high in one case but low in the other. Chapter 4 shows how the gamma rhythm model could provide the context for such an effect. Interestingly, this variation in firing rates must mean that the cells studied cannot be at the rhythm frequency all the while but instead fire faster for a while (and, of course, other cells show little if any response to the given visual stimuli). Thus, even when a network is synchronized overall, certain cells can "break step" to code specific data while other cells are just "chugging along" because their features are not currently significant.

Alpha and beta rhythms in the somatosensory mu complex: Here the focus is on computational modeling of alpha and beta rhythms that appear in awake humans (in contrast to the prominent slow rhythms present in sleep). This model was specifically designed to study the magnetic currents that are recorded with MEG which are presumed to come from the net summed electrical current flow in large spatially aligned pyramidal neuron dendrites in cortex. The alpha and beta components of the rhythm in primary somatosensory cortex (SI) influence perception, are modulated by attention, and change with healthy aging and repeated exposure to the task. Across all spontaneous data, high power epochs of the alpha and beta overlap on short time windows only 50% of the time and otherwise occur independently, suggesting their generation may involve separable processes or distinct mechanisms. To model this signal, the cortical network contains multicompartment pyramidal neurons with spatially extended dendrites, from which current flow within the cells is calculated. These cells are divided into supragranular (cortical layers II/III, near the cortical surface) and infragranular (layer V/VI, deeper in the cortex) layers, and the current dipole signal is calculated as the current flow inside the pyramidal neuron dendrites in a direction parallel to the apical dendrites, summed across the population. Multiple compartments are needed here to estimate the current dipole signal, where the rhythm is generated by subthreshold activity within the pyramidal neuron dendrites. The model generated the prediction that the alpha and beta components of the cortical SI mu rhythm arise from the interaction of two stochastic rhythms at approximately 10 Hz from sources outside of SI. Chapter 4 provides a step-by-step description of how the external drives interact within the pyramidal neuron dendrites to generate the alpha and beta components of the SI mu rhythm. The modeled current dipole signal produces many features in agreement with the MEG data, including the probability of simultaneous occurrences of alpha and beta.

Chapter 4 closes with a discussion of a number of interesting studies of neural rhythms, including use of multicompartment neural modeling to replicate persistent gamma oscillations, sleep spindles, and epileptogenic bursts in a single detailed thalamocortical model. Here we focus on two specific issues:

Binding: Singer and Gray (Engel et al., 1991; Singer and Gray, 1995) offered temporal coding by coherent oscillations as a potential solution to the binding, problem: namely, how is it that in a visual scene the features which belong to each object are linked together? For example, if we are presented with a blue square and a red triangle, what stops us seeing a red square and a blue triangle? The Singer–Gray hypothesis is that gamma rhythms can be coherent across spatially distinct visual areas despite changes in rhythm frequency, and that neurons coding features of a given object tend to fire at the same phase with respect to the overall rhythm. I confess that I remain skeptical since no convincing account has been given as to how this phase coding might be interpreted in the understanding of complex scenes—but, in any case, the binding problem remains an important issue for cognitive neuroscience.

Hippocampus: The hippocampus shows robust rhythmic activity in the gamma and theta bands, with evidence that gamma often emerges at specific phases of theta activity (but see chapter 18 for a rhythm-free account of "Hippocampus: From Spatial Processing to Episodic Memory" based on associator networks). This activity has been proposed to aid coding the memory and retrieval of an animal's place in physical space, via anatomically distributed gamma "cell assemblies" expressing transient synchrony gated by a theta rhythm.

Chapter 7 ("Neuromodulation") offers a model of cholinergic modulation of the theta rhythm, noting that it could involve hippocampal interneurons which are directly depolarized by acetylcholine (Ach).

Finally, computational models of disrupted rhythmicity in the striatum/basal ganglia network in Parkinson's disease may be used to guide deep brain stimulation paradigms that help alleviate symptoms. See chapter 23 for further approaches via computational neuroscience to "Brain Diseases."

1.6 Action

The treatment of "Motor Pattern Generation" (chapter 8) by Auke Ijspeert, Andrej Bicanski, Jeremie Knuesel, and Jean-Marie Cabelguen gives pride of place to networks in the spinal cord which can yield rhythmic activity such as that involved in locomotion but which can be tuned to adjust that locomotion to current circumstances. At the end of their chapter, Ijspeert et al. look at the extra lessons that can be learned when a brain model is implemented as the control system of a robot. This theme is further developed in chapter 16, "Embodied Models and Neurorobotics," by Tony J. Prescott, Joseph Ayers, Frank W. Grasso, and Paul F.M.J. Verschure. By contrast, Nicolas Schweighofer emphasizes "Computational Models in Motor Control" (chapter 9) whereby feedback and feedforward combine with inverse and forward models to yield fast and effective interaction with objects in the external world. Ziad M. Hafed links brainstem and cortex, as well as vision and motor control, in modeling "Saccades and Smooth Pursuit Eye Movements" (chapter 19). James J. Bonaiuto focuses on arm and hand to tackle the explicit integration of vision and motor control involved in "Reach and Grasp: Control, Development, and Recognition" (chapter 20).

Figure 1.3 introduced the action–perception cycle while the attendant discussion added that it is a learning cycle too. Thus, just as we can enter our analysis of the action–perception cycle at any point in the cycle, we can enter the discussion of cognitive function at any one of action, perception, or memory (and there are other topics to be discussed as well). In this "Opening Perspective" we adopt the order action, memory, and perception (with a primary emphasis on vision) for our exposition, but any other order might work as well. Here we choose to emphasize that the brain has its main effect on the world through the ability of neurons to control coordinated behavior. However, even though our emphasis is on action, we will repeatedly see the importance of the perception that guides that action (and action that gathers information for perception), as well as noting the importance of learning for feedforward control.

1.6.1 Central Pattern Generators/Motor Pattern Generators Many people, when thinking about the brain, imagine that all of cognition resides in cerebral cortex. However, just as our consciousness is the "tip of the iceberg" of neural processing, so can the role of cerebral cortex best be understood in relation to diverse "subcortical" regions, from the spinal cord on up to hippocampus (chapter 18), cerebellum (chapter 21), and the basal ganglia (chapter 22).

One of my favorite examples of the surprising power of the spinal cord is provided by the discovery by Fukson et al. (1980) that the spinal frog (i.e., a frog which has received a lesion that disconnects the spinal cord from the rest of the nervous system) takes the body schema (in this case an encoding of the disposition of the various parts of the body) into account during the wiping reflex. The hind limb of the spinal frog produces a wiping reflex evoked by electrically or chemically stimulating distal skin of the forelimb. The frog first places the hind limb in a posture irrespective of forelimb position, but the next movement depends on forelimb position—in other words, the spinal cord can on its own aim the hind limb to remove an irritant from a particular place on its forelimb even though that place changes relative to the hind limb as the forelimb moves and the body schema is adjusted accordingly.

Stimulating a specific point in the frog spinal cord appears to activate a *force field* toward a specific rest posture; stimulating another point activates a force field toward another posture; and stimulating the two points simultaneously leads to a linear superposition of the two force fields, so that a whole range of movements can be generated by changing activation levels of a few force fields. Kargo and Giszter (2000) showed how a finite set of force fields could account for the natural wiping reflex in the frog, with these discrete motor pattern generators serving as *motor primitives* to produce natural behaviors (Bizzi and Cheung, 2013) . However, the issue remains of determining what circuitry can muster the appropriate primitives in the appropriate combinations with the appropriate timing. The result of Fukson et al. shows that for the wiping reflex the circuitry is contained within the spinal cord, but for many behaviors (e.g., those depending on visual or auditory cues) that cannot be the case. (More on this in our discussion of chapter 9, "Computational Models in Motor Control," below.)

Turning to rhythmic behaviors like locomotion, even a completely isolated spinal cord can produce "fictive locomotion." If a cat with a cut across the midbrain is placed on a moving treadmill, repetitive electrical stimulation can suddenly trigger walking, and increasing the strength of the stimulus can evoke a trot or gallop. We thus distinguish a CPG (*central pattern generator*) as the part of the motor circuit that can generate an "open-loop" pattern, and a *motor pattern generator* (MPG) as the complete motor circuit that includes the CPG, sensory feedback loops, and modulation from descending pathways. There can, however, be motor patterns without involving spinal CPGs, for example through direct cortical connections onto motoneurons as for finger movements in primates.

Chapter 8 observes that rhythmic CPGs are distributed networks of multiple oscillatory pools coupled together, with typically one or two oscillatory pools, "unit burst generators," per degree of freedom in the body. They are capable of producing coordinated rhythmic patterns in isolation without phasic (i.e., rhythmic) inputs from descending pathways or from sensory feedback, but they are shaped and entrained by sensory feedback as part of the larger circuitry of MPGs. During rhythmic movements some proprioceptive inputs may shape the pattern of muscle activities during a cycle while others may affect the timing of the phases of the cycle. Moreover, feedback from the limbs plays a crucial role in interlimb coordination during locomotion, for example, in gait transitions. Of course, locomotion is usually toward some target or avoiding some obstacle or predator, and so in general one requires vision and other senses as input to path planning, which will, in turn, affect the MPGs. Indeed, the basic components of the turn motor pattern are centrally generated with steering of locomotion (including swimming) obtained by differentially activating left–right descending pathways.

We have already reviewed the three levels of neural modeling employed in chapter 9, ranging from highly detailed neuronal models with Hodgkin–Huxley neurons to highly abstract models of coupled oscillators. The CPG models based on these neurons all share the property that low-dimensional inputs (e.g., speed of locomotion, how much to turn) are transformed into high-dimensional motoneuron output signals (the spinal cord has many segments). Pattern generators may thus be seen as the inverse of *filters* (like the feature-detecting networks in the visual cortex) which take high-dimensional inputs and produce low-dimensional outputs.

We now turn to the various structures in which the neural networks are embodied. Chapter 8 uses a model by Ekeberg (1993) to demonstrate the interaction between spinal cord and body—the circuitry supporting fictive swimming is of little use unless the intact fish can actually swim! Ekeberg combined a CPG model with a muscle model and a mechanical simulation of the lamprey body in water, showing that the muscle activation signals generated by the CPG could indeed lead to forward swimming. The lamprey uses an anguilliform (eel-like) swimming mode in which a traveling wave of lateral body undulation is propagated from head to tail. The wavelength of the traveling wave corresponds roughly to one body length and is maintained approximately constant at different speeds of swimming (to which different bursting frequencies correspond). Applying extra excitation to the most rostral segments leads to small phase lags between segments and, therefore, to a wave of neural activity that travels from head to tail.

Further research on CPGs, this time in the salamander, was enhanced by linking the neural networks to a robot salamander, as we explain in the next section.

1.6.2 Neurorobotics Various groups have studied diverse locomoting robots including hexapod and octopod robots inspired by insect locomotion, anguilliform swimming of lamprey/eel robots, quadruped robots, and biped robots. CPG models have presented several interesting properties that make them useful for the control of locomotion in robots. Nonetheless, chapter 9 cautions that it is still an art to design a CPG for a particular type of robot and a particular type of locomotion. It is not yet clear how to obtain rich motor skills with CPGs.

However, a complementary issue is to explore what robots can do for neurobiological research. The salamander, an amphibian, was studied as a link between the lamprey and research on tetrapods. Salamanders swim like the lamprey, with fast axial undulations that travel from the head to tail. The limbs are folded to the body. Out of the water, salamanders switch to a stepping gait in which the body makes an S-shaped standing wave with nodes at the girdles so that diagonally opposed limbs move together. Ijspeert et al. (2005, 2007) used a salamander-like robot to show that their CPG model could produce effective gaits in water and on ground with variable speed and heading. Such a study requires a "body" for validation and cannot be studied only at a neuronal level. It also allowed them to compare the generated gaits to those of the real salamander and to demonstrate that the coordination between body undulation and limb movements during walking allows the salamander to increase its walking velocity compared to gaits that do not use body undulations.

This general theme of what robots can do for neurobiology—*neurorobotics* or *embodied computational neuroscience*, in which simulated (software or hardware) brain circuits are embedded in robotic bodies—is the focus of chapter 16, "Embodied Models and Neurorobotics." It is important to stress that, just as Ijspeert et al. used a specially designed salamander robot, off-the-shelf robots would seldom be helpful for such a study—special-purpose sensors and/or effectors have to be developed once their relevance is established to the current investigation. The aim, then, is to explore how special-purpose robots may help researchers go beyond pure simulation in order to understand better what is going on inside the brains of various animals. One of the trade-offs in modeling is to decide when the benefits of having an embodied model, in terms of access to real-world physics, justifies the cost of building and running a bespoke physical system. The strategy is to choose a key set of morphological constraints, rather than mimicking an entire creature. Chapter 16 considers several examples:

Plume-tracking in a robot lobster: American lobsters, as well as several other crustacean species, are able to locate the source of an attractive smell by walking to it, following the odor over distances of many meters. The challenge here was to design a chemical sensor which can be employed underwater to base the robotic calculations on actual dispersal of an odor plume. A complementary effort developed CPGs/MPGs for a lobsterbot whose locomotion can parallel that of the real lobster and then assess the transformation from olfaction to motor control.

The mammalian vibrissal system: Rats and other creatures can map their environment by "whisking," moving their whiskers back and forth many times per second against objects and surfaces of interest. The force pattern from bending each whisker can be sensed separately and sent to a region of somatosensory cortex called *barrel cortex* in which each "barrel" is a distinct neural cluster receiving input from a single whisker. This motivates assessing how robotic models that can capture some of the physics of how whiskers interact with surfaces in an exploring animal might shed light on the nature of the processing occurring in this part of the mammalian sensorimotor system. The challenge here was to build an approximation to the whisker processing apparatus found in the animal. These artificial whiskers have been symmetrically placed on the "heads" of two rodent robots. Whisker deflection patterns and corresponding activity in modeled rapidly adapting primary afferent neurons were tested for the *Whiskerbot* robot in relation to switching of feedback control of sensor movement. The whiskered robot *Shrewbot* has recently been deployed to perform further neurorobotic investigations of vibrissal touch including tactile simultaneous localization and mapping and even modeling of whisker-based predator–prey pursuit.

Scaling up to the human brain? I place a question mark here since this section of chapter 16 "seeks to extend embodied modeling to encapsulate the control architecture of the vertebrate brain" by introducing the *Distributed Adaptive Control* (DAC) proposed by Paul F.M.J. Verschure and colleagues. It is claimed that DAC is an "elaborate robot-based theory," but no connection is made with robots presented earlier in the chapter, though section 16.5.1 does present a DAC approach to the problem of how an animal builds a map of the world and locates itself accordingly—akin to the *simultaneous localization and mapping* (SLAM) problem in robotics. The model of hippocampus offered there offers a counterpoint to that of chapter 18, "Hippocampus: From Spatial Processing to Episodic Memory." Nonetheless, the emphasis in chapter 16 is on DAC as a general framework in which specific models of specific brain systems can be developed, informed by the general view that the brain (and not just in humans) optimizes four fundamental objective functions to generate action:

> *Why*: The motivation for action in terms of needs, drives, and goals.
>
> *What*: The objects in the world that actions pertain to.
>
> *Where*: The location of objects in the world and the self.
>
> *When*: The timing of action relative to the dynamics of the world.

One could add social interaction to DAC's focus on objects.

The centerpiece of the discussion of DAC is a figure which comprises four tightly coupled layers: *Soma, Reactive, Adaptive,* and *Contextual.* Three functional columns of organization are distinguished across these layers: *exosensing,* defined as the sensation, perception, and retaining of states of the world; *endosensing,* which is detecting, signaling, and retaining states of the physically instantiated self; and the *interface to the world* through action. Clearly, such considerations inform diverse models in this book to varying degrees, even though few are linked to neurorobotics.

As we discuss in more detail near the end of this "Opening Perspective," chapter 24, on *Language*

Processing, offers a further take on neurorobotics, approaching developmental acquisition of language by replicating a study of development in children with humanoid robot experiments.

1.6.3 General Issues in Motor Control Consider the task of reaching to grasp a target object (recall figure 1.2). One must (1) determine the position of the target in the environment, often through vision; (2) determine the current position and shape of the hand; (3) compute the spatial relationship between the target and the hand, transforming sensory information into a common frame of reference; (4) form a plan that specifies the trajectory the hand should take to the target location; and, finally, (5) compute the motor commands that cause the hand to reach the target with sufficient precision and speed and limited energetic cost. Chapter 20 ("Reach and Grasp: Control, Development, and Recognition") focuses on cortical structures relevant to this specific task while chapter 19 emphasizes midbrain structures involved in "Saccades and Smooth Pursuit Eye Movements." But first we see how chapter 9, "Computational Models in Motor Control," explores general computational approaches to solving such a problem in motor control rather than the brain structures which implement them.

Chapter 2 views neurons and neural nets as dynamical systems. In considering motor control, the "plant," that is, some body part together with the muscles that control it, is a dynamical system, as is the controller (brain subsystem) that controls the muscles. Thus, successful motor control involves the coupling of a neural dynamical system with a biomechanical dynamical system (recall our earlier discussion of Ekeberg's model of the lamprey). A crucial property of the plant is its *state,* for example, the position and angular velocity about each joint of an arm. The dynamical system must thus be characterized in terms of its inputs, states, and outputs, with equations representing how the state changes in response to different inputs and how the output depends on the current state.

Highly nonlinear dynamics cannot be ignored when generating accurate and fast arm movements. The system is nonlinear and combines kinematic variables from different joints; nonetheless, the nervous system can perform fast and accurate arm movement thanks to a sophisticated combination of feedback and feedforward controllers. In a *feedback controller,* the variable being controlled is measured and "fed back" into the controller to compute the motor command. In *negative feedback,* the observed output value is subtracted from the desired value with control thus depending only on error signals. A problem here is that the controller must operate with a feedback signal that is delayed relative to the time at which the relevant outputs occurred. *Feedforward control* simply precomputes the control for the whole trajectory without waiting for feedback. This strategy does not suffer from the delay problem but requires a high level of accuracy. Feedforward controllers must thus be adaptive to learn about the controlled system from experience. Although such a controller does not use feedback in generating control signals, it does employ "learning feedback"—it uses feedback on any errors in a movement to update itself to increase future accuracy. Because of noise and disturbances, movements resulting from a purely feedforward motor command will differ from what was planned. By combining both feedforward and feedback control, however, the nervous system can overcome these challenges and generate a motor command that brings the arm on target. Feedforward control produces fast movements in the face of long delays in neural transmission and in general brings the system close enough to its target so that feedback control can rapidly correct movements when they deviate from the intended goal (compare the fast and slow phases of figure 1.2).

Computational neuroscience develops models *of* the brain. Here we look at models (as distinct from maps) *in* the brain. Specifically, we look at how the controller exploits forward and inverse models of the plant. The information flow of the plant is from input (the commands provided by the controller) to the output (the behavior of the plant which may be monitored in part by feedback signals). A *forward model* represents this causal flow: it predicts the effect of applying certain control signals to the plant. A forward model is implemented and learned within neural circuitry within the brain, presumably in the cerebellum and parietal cortex, whereas the plant is an actual biomechanical system, such as an arm or an eye together with the attached muscles, or a part of the body together with some object upon which it is acting. Sensory prediction can cancel out the sensory changes induced by self-motion, thereby enhancing more relevant sensory information. In a related fashion, forward models may be important in the sense of self, as they allow the actor to determine if an external force was responsible for a movement.

An *inverse model* operates in the reverse direction: given a desired behavior, it computes a command that will help the controller cause the plant to yield that behavior. Unlike forward models, inverse models are ill-posed: There may be many different commands that yield the same behavior.

The role of the *state estimator* is to correct the predictions of the state derived from the forward model by computing the error between the actual sensory output and the predicted sensory output. Section 9.5 of chapter 9 explains how Kalman filtering is used for this purpose. Sections 9.6 and 9.7 then examine motor control and learning, respectively.

Actual hand paths during point-to-point reaching movements are highly stereotypical within and across individuals, typically close to a straight line and with an approximately bell shaped velocity profile. Why and how would the nervous system generate such stereotypical movements? More generally, how do humans choose one particular pattern of muscle coordination over the infinitely many other possible patterns that achieve the same goal? There is mounting evidence that the nervous system has devised solutions that are near optimal with respect to some cost. For example, the near straight trajectory with bell-shaped velocity can be generated by a controller that gets from A to B while minimizing jerk (the derivative of acceleration). A feedback/feedforward controller can then ensure that the motor commands are appropriate for the arm to implement this trajectory. But is this the right cost function for optimal control? The minimum jerk ignores dynamics, considering only kinematics in its planning. One refinement is to consider a cost that penalizes torque change. Another approach considers only the endpoint variance of the reach while assuming that the noise in the nervous system is signal dependent—the greater the motor command, the greater the noise.

However, low-level motor commands (e.g., muscle activations) for complex motor acts (a tennis serve, for instance) cannot be learned from pure trial and error with learning methods such as reinforcement learning (chapter 10) because the credit assignment problem (e.g., which muscle activation at what moment caused my serve to fail) is hard in high dimensions. A hierarchical optimal control scheme may be able to solve this problem, with a higher level of low dimension that monitors progress and exploits the many different ways of achieving the goal, and a lower level of high dimensionality that provides automatisms and corrections. As an alternative, there might be a "language" of more abstract building blocks in motor control, such as the motor primitives in the frog spinal cord we considered earlier. Using primitives, discrete and rhythmic movement can be modeled with nonlinear differential equations that form stable point and limit cycle attractors, respectively. Such equations can be learned with optimization methods, such that even rather complex movements such as a tennis stroke can be represented by such differential equations.

Because the properties of the sensorimotor system change throughout life or with fatigue, internal models must be continually updated based on experience. Errors resulting from a movement are typically available in extrapersonal space and do not tell the learner what muscle activation is needed to achieve the goal. Furthermore, the motor commands are often in much higher dimension than the sensory measurements. This is known as the distal credit assignment problem: because the errors in motor commands are not directly available for learning, the inverse model is a trickier proposition than learning the forward model. Chapter 9 closes by reviewing feedback error learning (which has been applied to the cerebellum) as well as noting that current human motor control research is exploring how reinforcement learning can lead to (near) optimal control of movements such as reaching. However, it also notes that nervous systems do not follow mathematical formalisms. The n-dimensional vector of the mathematical representations are coded via activities distributed over a population of m neurons, with $m \gg n$. For instance, in the motor cortex a population of neurons that are only broadly tuned to a range of movement vectors can, as a population, uniquely determine the direction of reaching movements in three-dimensional space (Georgopoulos et al., 1986).

1.6.4 Saccades and Smooth Pursuit Eye Movements Chapter 19 presents computational models for some of the brain mechanisms responsible for "Saccades and Smooth Pursuit Eye Movements." Saccades are rapid changes in eye position that quickly and accurately align the fovea with a new region in visual space, whereas smooth pursuit eye movements are used instead to track moving objects. As distinct from chapter 9, chapter 19 characterizes control models in the frequency domain using Laplace transform techniques.

Basic systems analysis shows that a saccade requires more than a simple step-input (relaxing to a new equilibrium is too slow). Instead, Robinson (1964) predicted that saccades are achieved by having an initial powerful pulse accelerate the eye with a subsequent slide in activity occurring until reaching the control value needed to maintain the eye at its new orbital position. This control principle was validated through neurophysiological recordings from oculomotor neurons. The model suggests how the holding value can be obtained through mathematical integration of this velocity command. Thus, the oculomotor drive signal (oculomotor neuron activity) necessary for saccades may be achieved through a superposition of two related commands: a velocity (burst) command, and a position (tonic) command which also serves the purpose of "gaze holding" in general. Models differ in their predictions of how control of such velocity and position commands is achieved upstream of the oculomotor neurons.

Visual correction of motor error is too slow to take effect during these extremely short eye movements. Later experimental observations revealed that saccade generation is in fact achieved in a closed-loop, feedback manner. However, this assertion is somewhat misleading—the loop uses "estimated feedback," an estimate of eye position exploiting a forward model to compensate for the delay in registering actual eye position. As the saccade proceeds, the local feedback pathway updates the estimate of eye displacement, resulting in a reduced dynamic motor error. Further modeling suggests how the retinotopic code in the superior colliculus (the key midbrain visual region) is converted into the time course of firing in the brainstem neurons that constitute the saccade generator. As modeled in chapter 22, "Integrative Functions of the Corticostriatal System," cortical and subcortical areas also play a role in saccade generation.

With this, we turn to smooth pursuit eye movements. If smooth pursuit waited for visual information, it would lag (predicted) changes in target trajectory by too long. Thus, mathematically, smooth pursuit deals with extrapolation of time series, as in classic engineering Wiener-type filters (Wiener, 1949) or the Kalman-type filters (Kalman, 1960) already introduced in chapter 9. If the eye lags its target, catch-up saccades are triggered intermittently to foveate the moving target and eliminate the position error. Thus, smooth pursuit and saccades may be considered two parts of a single system in which coordination between the smooth and saccadic components is crucial.

Smooth pursuit models emphasize the difference between target velocity and eye velocity, often referred to as retinal slip velocity. In one model, slip velocity is passed through a control element that generates an eye acceleration command. This command is then integrated to provide input to a simplified version of the eye plant model. Adding pursuit control signals based on slip acceleration helps make the pursuit response sensitive to either fast or slow changes in target velocity. If a target moves periodically in a predictably sinusoidal manner, smooth pursuit eventually will no longer lag the target and instead show a sinusoidal change in eye position that is almost perfectly synchronized with the target. Moreover, if a target of pursuit is transiently blanked (because of, e.g., occlusion), pursuit can proceed for some time and can show evidence of expectations and predictions depending on the behavioral state of the subject.

Signals from mediotemporal cortex (MT) are involved in estimating motion speed and direction, which are input to the smooth pursuit control system. MT uses a distributed retinotopic map: individual neurons respond to motion signals within a small prescribed region in retinal space, and these neurons also exhibit specific tuning for motion speed and direction. According to the vector averaging model of MT decoding, the motion of the smooth pursuit target may be described as the average of the motions encoded by the active population of neurons in MT, weighted by the activity of each neuron. (This echoes the Georgopoulos code for motor cortex.) Other cortical areas, too, contain signals related to the classic systems-level models of smooth pursuit, while downstream of the cerebral cortex neurons in the brainstem and cerebellum modulate descending motion-related signals to generate the required eye velocity command.

Chapter 19 complements the discussion of saccades and smooth pursuit with comments on fixational eye movements, target selection, decision-making, and attention for eye movements, and the influence of eye movements on visual perception.

1.6.5 Reach and Grasp: Control, Development, and Recognition As in the transition from retinotopic codes in superior colliculus to temporal codes in the brainstem, brain regions may code information in different ways. Moreover, different cortical regions encode the spatial location of objects in different reference frames: visual cortex encodes object location retinotopically while various parietal regions seem to encode object locations relative to the hand and must adjust their representations as the hand moves. Moreover, the brain not only must encode the position of an object but also must encode properties of the object relevant to how, not just where, it is to be grasped—and this depends upon the perception and utilization of *affordances*—opportunities for action that are directly perceivable without recourse to higher-level cognitive functions. Once contact with the object is made, haptic (touch) sensations are used to stabilize the grasp, but the role of vision is emphasized here.

Modeling Reaching and Grasping Most of the *reaching* models reviewed in chapter 20 do not attempt to localize the constituent computations to any particular brain region but instead analyze processes that may underlie these computations. The first issue for reaching is planning the trajectory of the endpoint of the arm (e.g., the wrist or a fingertip). Chapter 9 introduced the notion of the minimum jerk as one possible optimality criterion for point-to-point trajectories. But how does the brain generate minimum-jerk trajectories? A number of (computational, rather than neural) models are offered for the underlying mechanisms. A further model generates arbitrarily curved movements by using dynamic motor primitives, dynamical systems with parameters that can be set to generate any trajectory, akin to those considered in chapter 9.

A complementary topic is that of development of reaching in infants. Models here are based on reinforcement learning (chapter 10), with intrinsically motivated goal-directed actions based on exploratory movements, or motor babbling. In these models, movements are randomly generated in response to a target and the mechanisms generating the movements are modified via reinforcement. This process of variability and "opportunistic selection" introduces new behaviors into the motor repertoire. One model discussed here hypothesizes that infants decompose reaching movements into a series of submovements in order to compensate for errors due to uncertainty in the effects of motor commands. Simulation results accord with data showing that as infants develop, they reach with fewer, but longer, submovements.

In contrast, most of the models of grasping and action recognition described in this chapter link schemas to different brain regions, filling in neural details and extending the analysis. In some sense, then, chapter 20 takes over where figure 1.2 leaves off. Much of the modeling is based on neurophysiology of the macaque. Premotor cortex region F5 plays a key role in driving primary motor cortex, referred to as F1 in the macaque studies. F5 contains so-called *canonical neurons* selective for grasp type and phase which are thus thought to encode a high-level representation of the grasp motor schema while F1 neurons (or, at least, some of them) encode the component movements via a population code for specific muscle activity. Intriguingly, F5 was also found to contain what are called *mirror neurons*. Each fires not only during the performance of a specific type of grasp (even in the dark) but also when the monkey observes another individual performing a similar grasp. While F5 canonical neurons are thought to encode a high-level representation of a planned grasp, F5 mirror neurons are thought to recognize another individual's grasping actions by mapping the visual features of an observed grasp onto the same high-level grasp representation. Chapter 20 presents several models. Of these, the FARS model of grasping and the mirror neuron system (MNS) models of mirror neurons and action recognition will play an important role in chapter 25 ("Evolving the Language-Ready Brain"), which suggests how manual gesture (both action and recognition) might have constituted a crucial step on the path to language.

The FARS Model: The basic planning and control system for grasping is modeled by the FARS model, which embeds F5 canonical neurons in a larger system. Here, the dorsal visual stream (recall figure 1.6) passes through a parietal area called AIP that extracts affordances relevant to the control of hand movements and then on to the canonical neurons of F5. The dorsal stream does not know "what" the object is, it can only see the object as a set of possible affordances; whereas the ventral stream (from primary visual cortex to inferotemporal cortex, IT) is able to recognize what the object is. This information is passed to prefrontal cortex (PFC), which can then, on the basis of the current goals of the organism and the recognition of the nature of the object, bias AIP to "favor" the affordance appropriate to the task at hand.

The TRoPICALS Model: The FARS model was based heavily on data on the macaque brain, but it was hypothesized that it also describes how homologous regions in the human brain support the visually guided reach to grasp, and synthetic brain imaging (chapter 15) has been used to make predictions about PET imaging of the brains of humans reaching to grasp different objects. The *two route, prefrontal instruction, competition of affordances, language simulation* (TRoPICALS) model builds on FARS *considered as a model of the human brain.* While FARS lacks a computational account of reach planning, TRoPICALS includes a dorsal stream reach network. Not only does it show how grasps appropriate to the affordances extracted by different objects can be selected, it also ties this in with a simple model of language in order to study the effects of verbal cuing of grasps. (For further discussion, see chapter 24, "Language Processing.")

Grasp Development: Two developmental models are presented that focus on learning: one demonstrates the use of population codes and WTA dynamics with reinforcement learning to learn the motor parameters needed to plan successful grasps; the other uses competitive supervised learning to extract affordances from low-level object features and map them to the motor parameters learned by the former model.

The discussion of models of grasp control is rounded out by discussion of online control of grasp patterns and approaches to reach and grasp coordination, but we now turn to models of action recognition and the mirror system.

Modeling Mirror Neurons The discovery of the mirror neuron system in the 1990s was exciting in that it potentially provided a neural basis for action understanding—though it might better be said that mirror neurons provide a motoric basis for enriching such understanding as supported by the ventral stream. In any case, this discovery has played a crucial role in the development of social neuroscience, including discussion of "theory of mind" (what one person thinks they know about the mind of another) and such disorders as autism. Chapter 25 looks at mirror neurons in relation to imitation, gesture, and language while stressing that mirror neurons do not support these capabilities by themselves but only in concert with other systems "beyond the mirror." Here we focus on chapter 20's review of modeling the development of mirror neurons—but note that even here diverse regions beyond F5 are necessary for the models.

The MNS family of models: These simulate the shaping of mirror neuron response properties through experience. The MNS model of the monkey mirror system used F5 canonical neuron activity encoding the type of grasp as the training signal to develop F5 mirror neurons that could associate the grasp type with visual input encoding the trajectory of observed hand–object relations. The MNS2 model added audio input to model data on *audiovisual mirror neurons*, that is, mirror neurons that respond to the sight and sound of actions with characteristic sounds such as paper tearing and nut cracking, and added a spatial working memory with *dynamic remapping* of hand position to explain the response of mirror neurons when the target object was recently visible but is currently hidden. This general notion of updating an internal representation on the basis of expected changes brought about by ongoing action will also play an important role in the study of saccadic eye movements in chapter 22.

Imitation: Demiris and Hayes (2002) employed a battery of behaviors paired with forward models (recall chapter 9). Each "behavior" is a control system which takes current state and goal and provides the motor output (it is thus what an "inverse model" models) while its forward model transforms the code for the motor output (and the goal) to generate what the next state would be if the corresponding behavior were executed. The task of the architecture is to observe a particular trajectory and determine which behavior in its own repertoire best matches it. This, then, is imitation of a behavior one already knows how to execute; it is to be distinguished from the use of imitation to master a novel behavior not yet in one's repertoire.

1.7 Neural Plasticity, Learning and Memory

We have already seen that learning plays an essential role in a variety of models which focus on different forms of action. We now turn the spotlight on the learning mechanisms themselves and the forms of memory they support. Wulfram Gerstner introduces the most basic form of changes in synaptic efficacy with experience, exploring several variants of "Hebbian Plasticity and Learning" (chapter 6) with the emphasis on unsupervised learning. A different form of plasticity for neurons is provided by "Neuromodulation" (chapter 7) as explored by Jean-Marc Fellous, Carmen Canavier, and Michael E. Hasselmo. This focuses especially on dopaminergic and cholinergic modulation. Dopamine plays a crucial role in reinforcement learning and so reappears when Nathaniel Daw explores "Reinforcement Learning" (chapter 10). Here synapses across a network change on the basis of when and whether behavior yields positive or negative reinforcement. Finally, Stefano Fusi and Xiao-Jing Wang build on these insights into neural plasticity to offer models of "Short-Term, Long-Term, and Working Memory" (chapter 11).

We distinguished *episodic* learning, being able to recall something of earlier aspects of our lives (recall the sad case of H.M.) from *procedural* learning, the development of a new skill such as riding a bike. And the list can be extended, such as memories for facts and faces. The cerebral cortex is implicated in all of these, but subcortical structures play an essential role as partners in memory formation, with the hippocampus (chapter 18) proving essential to episodic memory while cerebellum (chapter 21) and the basal ganglia (chapter 22) are each crucial for procedural learning. However, we defer attention to these brain regions till later, focusing for now on the general mechanisms of synaptic plasticity that make learning in all its variety possible.

1.7.1 Hebbian Plasticity and Learning The strength of a synapse from a presynaptic neuron to a postsynaptic neuron can be measured as the amplitude of the postsynaptic potential which is generated if a spike emitted by the presynaptic neuron arrives at the synapse. If, because of some stimulation protocol, the neurons are "active together" for some time, the synapse is strengthened. This can be measured as an increase in the amplitude of the postsynaptic potential. The changes of the strength (or other properties) of synapses are called *synaptic plasticity*. A strengthening of synapses induced by the coactivation of two neurons is called *Hebbian plasticity*. Donald Hebb (1949) gave the most cited formulation of correlation-based plasticity: "When an axon of cell A is near enough to excite cell B and repeatedly or persistently takes part in firing it, some growth process or metabolic change takes place in one or both cells such that A's efficiency, as one of the cells firing B, is increased." Since no other neuron C is mentioned here, this learning rule is "local," and since repeated joint activity can be measured by a correlation function, we may say that Hebbian learning is sensitive to correlations between the activity of A and B.

Hebb's idea was that the repeated coactivation of different neurons in a network (e.g., each time we encounter a banana) will ensure that the connections between the simultaneously active neurons will be strengthened. In this case, the memory concept "banana" could be formed and represented by the set of strongly connected cells, called a *cell assembly* (Hebb, 1949).

Short-term plasticity lasts only for one or a few seconds. It is induced rapidly, but it also disappears rapidly. Chapter 6 introduces a protocol called *spike timing dependent plasticity* (STDP) for which the effect persists for minutes and even hours. In the classic STDP protocol applied to cortical pyramidal cells (other neurons in other regions may exhibit different forms of plasticity), if the presynaptic spike arrives at the synapses systematically before the postsynaptic neuron fires, then the synapse is strengthened and *long-term potentiation* (LTP) occurs. If the postsynaptic neuron fires before the presynaptic one, then the synapse is weakened and *long-term depression* (LTD) occurs. STDP protocols show that plasticity can depend on precise spike timing, not just loosely on "Two neurons are active together."

Hebbian plasticity can be incorporated in both rate-based and spike-based neuron models A rate-based Hebbian protocol defines joint activity (or "correlations") on a timescale of a few hundred milliseconds. A spike-timing-based Hebbian protocol studies joint activity on the timescales of a few milliseconds. A problem with the original Hebbian rule is that it only increases synaptic strength. Clearly, then, a reformulation is required to limit synapse growth and to weaken some synapses (LTD) while strengthening others for each neuron to become a useful pattern processor. Oja's rule uses a second-order term to ensure that if a certain subgroup of synapses is strengthened, then other synapses to the same postsynaptic neuron have to be weakened. This seems more elegant than simply normalizing the weights to a neuron in the simulation that follows Hebbian changes by an explicit step returning the sum of all weights converging onto the same postsynaptic neuron to a given value.

The Bienenstock–Cooper–Munro (BCM) rule (originally developed to account for observations in visual cortex during developmental learning) offers one neurobiologically plausible way to combine synaptic potentiation and depression. If the presynaptic neuron is inactive, the synaptic weight does not change. If the presynaptic neuron is active while at the same time the postsynaptic neuron is firing at a high rate, the synaptic weight increases since both neurons are jointly active. Hence it is a Hebbian rule. However, if the presynaptic neuron is firing, there is a threshold point between the induction of LTD and LTP. To avoid having the postsynaptic firing rate blow up or decay to zero, BCM turns the threshold into an adaptive variable in such a way that the learning rule exhibits competition. As a result, the postsynaptic neuron becomes specialized to a subset of its inputs.

From the point of view of a neuroscientist, one particularly intriguing result is that Hebbian learning can lead to the development of receptive fields, whereas from the algorithmic point of view, the most interesting result is that Hebbian learning is akin to principal component analysis. Chapter 6 presents the evidence for each claim. One example is provided by a specific model of the development of visual receptive fields in which the evolution of the synaptic weights leads to asymmetrical receptive fields and thus to orientation selectivity. Moreover, neuronal response properties vary continuously across the cortex. The neurons are said to form a map for orientation. The formation of cortical maps is the topic of chapter 13 and has a long-standing history in computational neuroscience.

Traditional models of Hebbian learning have been formulated at the level of firing rates or abstract variables describing the activity of pre- and postsynaptic neuron. In addition to treating such rules (as we have just described), chapter 6 also formulates Hebbian learning rules at the level of spikes, showing that this necessarily leads to STDP with a window of simultaneity that describes the co-occurrence of spikes. Rather than the detailed biophysics based mostly on calcium, the focus is on phenomenological models in which STDP can be understood as a causal interaction between traces (e.g., neurotransmitter) left at the synapses by spikes of presynaptic neurons and action potentials propagating backward into the dendrite of the postsynaptic neuron. However, the voltage also plays an important role when there is no action potential. An experimentally found STDP window shows two phases: if a presynaptic spike occurs repeatedly a few milliseconds before the postsynaptic one, the synaptic weight increases; for reverse timing the weight decreases. The Hebbian formulation that the presynaptic neuron "takes part in firing" the postsynaptic one suggests a causal relation and corresponds to a temporal order "pre-before-post." Hence, simultaneity in the Hebbian sense should be defined as a temporal window that is shifted slightly to the left. The STDP model complements the asymmetrical learning window by a negative part for post-before-pre timing. Just as for the rate models of Hebbian learning discussed earlier, a rule for depression of synapses can induce competition between synapses so that some synapses grow at the expense of others, making Hebbian learning more selective.

1.7.2 NEUROMODULATION Most of the models in this book view synaptic activity purely in terms of excitatory or inhibitory effects on the postsynaptic neuron, with learning rules modifying the synaptic "weights." In chapter 7, though, the focus shifts to *neuromodulation*: here certain neurotransmitters in some way modulate the functionality of the postsynaptic neuron, rather than just contributing an EPSP or an IPSP. The focus of the chapter is on *extrinsic* neuromodulation where the neuromodulatory signals originate from an area outside the neural substrate whose computation is under study. The computational effects of such extrinsic neuromodulation may be widespread in that they can simultaneously influence many functionally different sites—though different classes of receptors in the postsynaptic membrane will allow different neurons to respond to the signal in somewhat distinctive ways. Although the chapter briefly reviews intriguing effects of neuromodulation in invertebrates and the role of diverse neuromodulators in mammals, the focus of the chapter falls on two specific neuromodulators in the mammalian brain, dopamine (DA) and acetylcholine (ACh). One focus is the dopaminergic output of the ventral tegmental area and substantia nigra on the basal ganglia while a second is the influence of the cholinergic output of the septum on hippocampal computations. Note then the importance demonstrated here of subcortical regions to the overall function of the brain.

Many computational studies that feature neuromodulation implement it as a slow and diffuse process, tonically changing membrane or synaptic properties. Such models may use volume transmission in which a neuromodulator is released in a diffuse manner to cause relatively uniform activation of receptors throughout the tissue. However, fast changes are also important. Acetylcholine levels in the hippocampus can involve slow changes, with high acetylcholine levels during active exploration or sleep, sustained attention, and movement-related theta oscillations. However, fast, phasic changes in acetylcholine have also been seen after stimulus events. Dopamine cells are able to burst (fast timescale) and fire tonically (slower timescale) depending on the behavioral or cognitive demands. The neuromodulatory effects of tonic levels of dopamine are thought to be important in both movement and cognition. Figure 1.8 charts the distinction between the direct and indirect paths through the basal ganglia (we will meet these again in chapters 10, 22, and 23). Tonic levels of dopamine are hypothesized to facilitate actions that are appropriate for the current state of the animal by increasing excitability in the direct pathway that

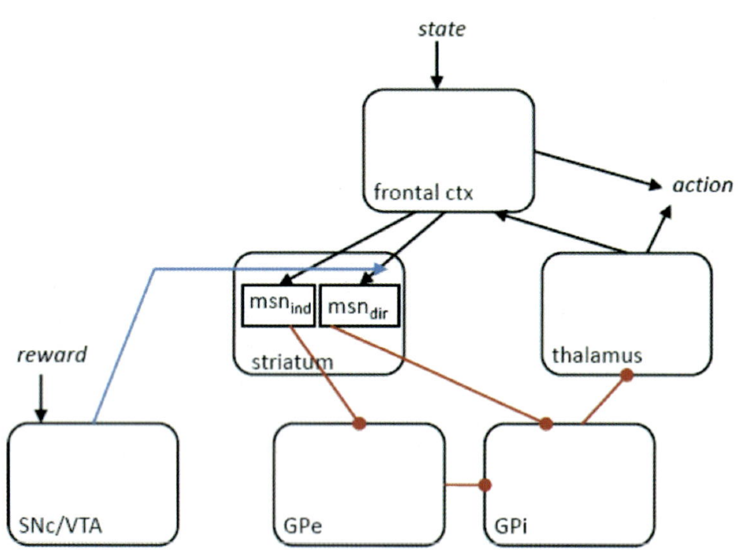

FIGURE 1.8 Hypothesized functional anatomy of dopaminergic influence on action learning (from chapter 10, after Alexander and Crutcher, 1990, and Frank et al., 2004). Medium spiny neurons in striatum project into either the "direct" pathway (msn_{dir}) (D1 receptors) to external segment of the globus pallidus (GPe) or the "indirect" pathway (msn_{ind}) (D2 receptors) to the internal segment of the globus pallidus (GPi), ultimately reciprocating their input from cortex (ctx) via thalamus. Because the direct and indirect pathways from striatum to thalamus contain three and two inhibitory synapses (red), respectively, increased activity of medium spiny neurons in the indirect pathway has a net inhibitory effect on thalamus and cortex whereas activity into the direct pathway is disinhibitory. These effects are thought to inhibit and facilitate, respectively, thalamic and cortical movement commands. By influencing plasticity on either set of medium spiny neurons, dopamine may drive learning to produce or inhibit actions. VTA, ventral tegmental area; SNc, substantia nigra pars compacta.

facilitates action and decreasing excitability in the indirect pathway that inhibits action (it can do this because there are different dopamine receptors in the two pathways) whereas in the PFC they appear to stabilize representations that are important in attention, cognitive control, and working memory.

But computations may also be influenced by fast modulation of dopamine. In a classic series of experiments in monkeys, Schultz and his colleagues showed that during the learning of an association between a neutral stimulus and a reward, dopaminergic neurons of the ventral tegmental area (VTA) or substantia nigra (SNC) exhibited transient increases in activity lasting less than 150 ms in response to behaviorally relevant signals associated with reward delivery. The major theoretical framework used to explain the phasic behavior of DA neurons is that of temporal difference learning (see chapter 10, "Reinforcement Learning"). There is a broad consensus that phasic dopamine signal bursts serve as a training signal for classical and instrumental conditioning, probably via the well-documented neuromodulatory effects of dopamine on the late form of LTP at corticostriatal and hippocampal synapses.

Dopaminergic Modulation The modeling focus here in chapter 7 is not on the modulation but on the modulator, namely, the firing of dopaminergic (dopamine-releasing) neurons. As already mentioned in our discussion of Hodgkin–Huxley-style modeling, chapter 7 includes a model of these DA neurons, which can exhibit complex firing patterns. The model is of interest because any drug or phenomenon capable of modulating specific intrinsic conductances of these neurons can change their activity from tonic to bursting and can, in turn, have dramatic effects on the neural computations of dozens of brain areas to which the neurons project. Some salient hypotheses regarding dopamine neurons are as follows. (1) In the absence of afferent input, dopamine neurons exhibit spontaneous regular pacemaker firing; this background level of tonic firing ensures that both positive and negative signals can be transmitted encoding the difference between expected and actual rewards; (2) dopamine neurons can only fire rapid bursts in the presence of NMDA synaptic activation, a key player in synaptic plasticity; and (3) the complex contributions of the intrinsic currents to the firing dynamics allow for rich modulation of the propensity of DA neurons to transmit tonic or phasic signals.

Modeling the dependence of tonic and bursting firing modes on intrinsic and synaptic properties of DA neurons is followed in chapter 7 by a survey of postsynaptic effects and behavioral functions of dopamine. Behavioral roles range from purely motor to highly cognitive. The loss of midbrain dopamine neurons results in the symptomatology of Parkinson's disease, and abnormal dopaminergic signaling has been implicated in schizophrenia and addiction (see chapter 23, "Brain Diseases"). Dopamine neuromodulation is involved in the neural substrate of persistent firing and working memory (chapter 11, "Short-Term, Long-Term, and Working Memory") and in the manner in which multiple systems interact at various phases of reinforcement learning.

As noted earlier, the effects of dopamine can differ greatly depending on the receptors (D1 vs. D2) in the postsynaptic membrane, and this is true in cortex as well as basal ganglia. Models suggest that D1-mediated effects not only enhance strong, persistent activity associated with ongoing representations of specific information in working memory that are supported by strong recurrent network activity but also diminish the effects of weaker or brief inputs, minimizing the effect of distractors. D2 receptor activation in general has the opposite effect on representations in working memory, destabilizing them and potentially allowing the working memory buffers to incorporate new information.

Although no explicit computational model has been proposed, dopamine has been suggested to act in concert with other neuromodulators to achieve higher level forms of learning (Doya, 2002; Seymour et al., 2012). One suggestion is that dopamine controls the magnitude of the reward prediction error, serotonin controls the reward discounting amount (how much the reward value is decreased if it is expected late), noradrenaline controls the randomness of the policy (i.e., mapping from the state of the animal to the next action to take), and acetylcholine controls the overall learning rate at which synapses are modified.

Cholinergic Modulation Cortical structures receive broadly distributed cholinergic innervation from populations of cholinergic neurons in the basal forebrain, and these basal forebrain structures, in turn, are regulated by brain-stem nuclei containing cholinergic neurons. The cortex contains a tiny number of local cholinergic interneurons, but there are large populations of cholinergic interneurons in the basal ganglia.

High levels of ACh enhance memory encoding and attention by increasing the spiking response to afferent input and reducing excitatory feedback in the cortex. Lower levels of ACh result in less influence of afferent input and stronger influence of excitatory feedback that could enhance consolidation of previously encoded memories. These changes in circuit dynamics result from effects at different subtypes of ACh receptors that include two major categories named after individual examples of drugs that selectively activate them. *Nicotinic receptors* respond to the drug nicotine (as well as other nicotinic drugs) but do not respond to the drug muscarine. *Muscarinic receptors* respond selectively to the drug muscarine (and other muscarinic drugs), but not to the drug nicotine. Activation of nicotinic receptors increases the amplitude of excitatory synaptic potentials induced by thalamic input to neocortex but not at excitatory feedback (i.e., intracortical) synapses, while ACh acting at muscarinic presynaptic receptors inhibits the release of the transmitter glutamate at specific excitatory synapses within cortical circuits, thereby reducing excitatory synaptic potentials.

We have already reviewed a number of different neural rhythms and related modeling (chapter 4). Here we focus on the theta rhythm. This rhythm is most prominent when animals explore the environment or attend to behaviorally relevant stimuli and decreases when an animal performs behaviors such as eating or grooming. The presence of theta rhythm correlates with better learning in conditioning tasks and encoding of memories, and the loss of theta rhythm correlates with impaired encoding. Theta rhythm is reduced by damage to the cholinergic (Lee et al., 1994) and GABAergic input to the hippocampus from the medial septum. Cholinergic modulation of theta rhythm could involve hippocampal interneurons, which are directly depolarized by ACh. The combination of depolarization and presynaptic inhibition appears paradoxical, but computational modeling demonstrates that these combined effects reduce background activity while heightening response to suprathreshold sensory stimuli.

After a brief discussion of cellular models of persistent spiking, chapter 7 presents a network model of cholinergic enhancement of encoding. The modeling helps clarify how the physiological effects of ACh may set appropriate dynamics for encoding new memories and modify cortical dynamics for retrieval. The model presented has the structure of a standard associative memory model. (For more on associative networks, see appendix B, and for models addressing different aspects of hippocampal function, see chapter 18.) A key factor is to include a term $c(t)$ for the selective presynaptic inhibition by ACh of the modifiable synapses (but not afferent input). The cholinergic modulation takes the value of 1 when ACh is present (causing presynaptic inhibition of glutamatergic transmission) or 0 when ACh is not present (causing no presynaptic inhibition). Interference during encoding can appear in any model in which Hebbian synaptic modification of excitatory feedback connections encodes associations between different elements of the afferent input. The solution is to suppress the retrieval of previously encoded associations during the encoding of new associations. The model fits the physiological data, as the same synapses that show muscarinic enhancement of synaptic modification commonly show muscarinic presynaptic inhibition of glutamatergic synaptic transmission.

Behavioral data (again, recall H.M.) demonstrate that recently encoded memories are more sensitive than older memories to the effects of partial lesions of the hippocampus. This supports a two-stage theory of memory formation in which memories are initially encoded by Hebbian synaptic modification in the hippocampus during the first stage (active waking), and then memories are subsequently reactivated during a second stage (quiet waking or slow-wave sleep) to allow gradual strengthening of additional synaptic connections in the cortex (Buzsaki, 1989; McClelland et al., 1995). High levels of ACh during waking could prevent interference from previously stored memories whereas lower levels of ACh during sleep would allow certain memories to reactivate in hippocampus and spread back to the cortex for consolidation. In this framework, the modulatory effects of ACh could switch the network between a state of acquiring new episodic information in the hippocampus during active waking and a state of updating the semantic representations in neocortex during the occurrence of sharp waves in quiet waking and slow-wave sleep.

1.7.3 Reinforcement Learning We may distinguishes three learning styles:

Unsupervised learning detects correlations in the stream of input signals. Hebbian learning, as discussed above, is an unsupervised learning rule: there is no notion of good or bad, successful or unsuccessful, rewarding or painful. Hebbian learning is suitable to detect correlations in the input and can therefore be used to explain developmental processes such as the formation of receptive fields. Unsupervised learning can serve to find the first principal component of the signal, an independent component, or the center of a cluster of data points.

In *supervised learning* the "teacher" has to give a target output for every input signal in a specified training set. Learning is to yield input–output behavior as close as possible to that provided by the teacher, while generalizing to novel inputs as well.

Reinforcement learning (Sutton and Barto, 1998) is a form of trial-and-error learning that serves to increase the chance of gathering rewards and avoiding punishments from a "critic"—all in the absence of a teacher who specifies the "correct" strategy. This is the problem that faces an animal foraging for food or a human practicing to ride a bicycle.

Chapter 6 closes with a section on reward-based Hebbian learning which can thus be seen as a bridge to chapter 10, "Reinforcement Learning." Hebbian learning can be used for reward-based learning if the Hebbian learning rule is modulated by reward signals. In related fashion, stable memory storage and retrieval suggests the need for a gating mechanism which decides which perception is novel or interesting and worth storing. Hebbian learning at a synapse depends, then, not only on the activities of its two neurons but also on the presence of a more global neuromodulator that enables the initiation or stabilization of Hebbian weight changes (as in the cholinergic modulation effects discussed in chapter 7).

Recall the famous classical conditioning experiments of Pavlov in which dogs could be trained, by repeated pairings in which a bell was sounded before food was presented, that the bell predicted food. Here, salivation served as a direct behavioral readout of the learned association—the trained animal is salivating because it expects food (the "unconditioned stimulus"). The Rescorla–Wagner model (Rescorla and Wagner, 1972) of classical conditioning could explain this and a range of other effects. However, it could not explain *second-order conditioning*, a laboratory model of how outcomes (e.g., money) other than primary biological rewards can nonetheless serve as surrogates for them. Chapter 10 emphasizes that this failure points the way toward a refined model which, in turn, paves the way for the theory of reinforcement learning.

Many computational models of this learning process assume that decisions between options are made based on predictions the agent makes about the value of each option (e.g., the amount of food or the chance of winning). On this view, control depends on prediction: that is, learning to predict biologically important events is an essential subcomponent of the full reinforcement learning problem of learning to make good decisions. This leads to the notion that learning is error driven—an idea that receives powerful support from an apparent neural correlate of a prediction error signal. Dopamine-containing neurons in the midbrain, when recorded while animals perform classical conditioning tasks, carry a signal that appears similar to the prediction error formalized in the Rescorla–Wagner model. Moreover, differences between the observed signal and that predicted by the Rescorla–Wagner model suggested elaborated theories that explain second-order conditioning and connect these computations to a more sophisticated form of action choice.

Dopaminergic neurons innervate cortex, particularly frontal cortex, and the basal ganglia. The brain's densest region of dopaminergic innervation is the basal ganglia's input nucleus, the striatum. With this, we return to the way dopamine can have differential effects depending on the presence of D1 or D2 receptors in the postsynaptic neurons. In figure 1.8, two sets of medium spiny neurons in the striatum are distinguished not only by the paths to which they project (direct vs. indirect) but also by their differential response to dopamine from the VTA or SNc (substantia nigra pars compacta). Due to different numbers of inhibitory synapses along each path, the direct and indirect pathways are believed to have a net facilitatory or inhibitory impact, respectively, on their cortical targets, in turn facilitating or inhibiting movements that motor neurons in cortex ultimately control. In a classic view of motor disorders (DeLong, 1990), imbalance between these two pathways can cause either poverty of movement—as in Parkinson's disease (see chapter 23)—or excess, as in tics or other undesirable movements seen in other disorders involving the basal ganglia. This conceptual model neglects the important role of neural oscillations and related physical tremor in Parkinson's disease. Nonetheless, it is a useful starting point.

With this we turn to sequential predictions and temporal difference learning. In many real world decisions, the consequences for reward of any individual choice may be deferred and also depend on subsequent choices and events. In such tasks the decision maker must take into account not just the reward that is expected immediately for a choice but how that choice might set the stage for later reward or punishment. One would like to choose the action that maximizes the reward accumulated over the long run. The expected sum of rewards following some state may be written as:

$$V(s) = E[r_t + \gamma r_{t+1} + \gamma^2 r_{t+2} + \cdots \ |s_t = s],$$

where the operator indicates the expectation is over the array of possible outcomes, r_t is the reinforcement at time t, and γ is a discount factor—the later we receive a reward, the less valuable it seems to us now. Thus,

$$V(s) = E[r_t] + \gamma \ V(s_{t+1}).$$

If at time t we take the actual r_t as an unbiased estimator of what would have earlier been the expectation $E[r_t]$, we can use $r_t + \gamma V(s_{t+1})$ as our updated estimate of V(s). The idea of temporal difference learning is to train up an "adaptive critic" that learns to produce a prediction P(s) of the actual expected reinforcement V(s). In view of the previous analysis, the *temporal difference* (TD)

$$(r_t + \gamma P(s_{t+1})) - P(t)$$

may be viewed as a prediction error—marking how much the prediction at time t differs from the updated estimate available at time t + 1 after receipt of r_t.

If TD is positive, then the current reward expectation is higher than the previous prediction had anticipated. In this case, the previous value was too pessimistic and should be increased. Conversely, if TD is negative, future reward expectancy has dropped unexpectedly, and the previous prediction was too optimistic. Chapter 10 develops this notion of a system that replace the actual intermittent reinforcement signal with a continually updated estimate of *expected reinforcement* (expected sum of rewards) that is available at each step. Neurophysiological data suggest that the dopamine signal reaching the striatum correlates with the TD prediction error. However, there are other views of dopaminergic activity. For example, dopamine projections to PFC are involved in working memory (chapter 11). However, since dopamine neurons may respond to a variety of surprising events that are not strictly reward related, Redgrave et al. (2008) have suggested that dopamine might have a role in discovering actions that have a causal effect on the sensory world, rather than reward per se.

Chapter 10 also treats dopamine and sequential choice, as well as model-based reinforcement learning and goal-directed behavior. It closes with a brief discussion of theories which seek to bring mechanisms for reinforcement learning into contact with higher level functions such as cognitive control. Another crucial function for control is the management of working memory, which O'Reilly and Frank (2006) have attempted to recast as itself a reinforcement learning problem. The idea here is that strategically "gating" pieces of information in and out of working memory can be viewed as a set of actions, not unlike the actions in any other reinforcement learning problem except that they operate internally instead of externally. Then the same reinforcement learning machinery can be used to learn a gating policy that adaptively manages the brain's working memory resources, maintaining information that is relevant for future reward seeking and purging information that is not. Finally, the same reinforcement learning mechanisms used to learn simple problems can also be applied to control problems across a hierarchy comprising multiple levels of abstraction. And this marks a good transition point to the themes of chapter 11.

1.7.4 Long-Term, Short-Term, and Working Memory Atkinson and Shiffrin (1968) introduced the three-stage model of memory that motivates the title of chapter 11: selected components of sensory information are transferred to a short-term store with limited capacity. Only a selection of these reaches a long-term memory store. There is a simple model of memory consolidation, in which memories are first in a labile state and only later become long lasting. *Working memory* complements short- and long-term memories. Short-term memory is assumed to be the result of a relatively passive process. In contrast, working memory is more "active" as there are mental operations performed on the information currently held in memory. Therefore, working memory involves manipulation as well as storage of information for a short while. This concept was introduced by Miller et al. (1960) and developed by Baddeley in the 1970s in cognitive psychology (Baddeley, 2003).

Memories are initially labile and subject to rapid degradation because they are stored in fast dynamical variables. With time, such memories are consolidated because they are transferred to slow dynamical variables, and they become stable and highly resistant to memory interference (i.e., to the storage of correlated information that can disrupt similar previously stored memories). Consolidation occurs both in the form of molecular processes at individual synapses and in the form of systems consolidation, which is a reorganization of the memory trace within different brain systems. These mechanisms protect memories both from the degradation due to the passage of time and from the overwriting caused by the storage of other memories. Chapter 11 links synaptic mechanisms of memory consolidation to alteration of some of the molecules that are locally present at the synapse. These modifications are induced by the activity of the presynaptic neuron, and they depend on the state of the postsynaptic neuron, as in the case of Hebbian plasticity. In particular, the chapter discusses how a complex molecule called CaMKII affects the strength and the number of AMPA receptors in an excitatory synapse and, hence, the strength of the synaptic connection. These molecules can then regulate gene transcription and protein synthesis which then lead to permanent changes in the morphology of the synaptic connections.

Modeling Long-Term Memory Considerable attention is paid to a number of rather abstract memory models inspired by work in statistical physics. For example, in models of magnets, one may look at the cooperative effects mediated by atomic magnets, each characterized as "spin up" and "spin down." This inspires a similar statistical analysis of neural networks operating on a discrete timescale, with each neuron firing or not firing at each time step. Such models sidestep the rich associations that constitute much of human memory and instead assume that the memories to be stored are basically patterns of synaptic modifications which are random and uncorrelated, and that each memory is stored in a single exposure. A key question in this approach is to determine memory capacity, that is, how many random and uncorrelated memories can be stored in a neural circuit of a given size. The approach is motivated by questions such as these: How does the memory capacity scale with the total number of synapses? Can we build a synaptic model that can take advantage of the huge number of synapses that are available in biological brains? Interestingly, there are ways of answering these questions that do not depend on the details of the type of learning that is considered.

Memory capacity in the case of unbounded synapses scales linearly with the total number of synapses. Unfortunately, most of the memory models are plagued by the problem of catastrophic forgetting as soon as we impose boundaries on the synaptic weights. The method is to choose a generic memory and estimate its strength immediately after it is stored and then during the process of storage of other memories. In the case of a Hopfield network (see appendix A), when the critical capacity is surpassed, there is an abrupt transition to a regime in which none of the memories, not even the most recently stored, can be retrieved. This is called the blackout catastrophe and is clearly in contrast with what happens in biological brains in which old memories are gradually forgotten to make room for the new ones. The storage capacity depends also on the representations of the memories. For example, if memories are stored as sparse patterns of activity (i.e., patterns of activity in which only a small fraction f of neurons is active), then the amount of information to be stored per memory decreases and the memory capacity increases.

Memory storage capacity with bounded synapses in online learning is modeled in the extreme case of synapses with two stable states, which is the minimal number of states to have memory. These models reveal a stability–plasticity dilemma (improving one degrades the other). One solution exploits a diversity of timescales. When information about memories is transferred from fast synapses to slow synapses, it is possible to extend lifetimes of memories with no interactions without sacrificing the initial memory trace. The information transfer can occur within a synapse (e.g., in the form of metaplasticity, reflecting the cascade of biochemical processes which underlies the expression of long-term modifications) or across synapses in different areas (memory consolidation at systems level) as in the case in which memories are initially stored in a temporary repository where the synapses are fast (e.g., in the hippocampus) and then are consolidated when transferred to brain areas (e.g., cortex) where the synapses are slow. Recall the related discussion in chapter 7.

Modeling Working Memory Different persistent activity patterns represent distinct items of information, and working memory requires the ability to rapidly switch on and off between any of these memory states and a background state. Such stimulus-selective persistent activity has been observed widely in single neurons of the PFC, posterior parietal cortex, and other brain areas while recording from behaving monkeys performing tasks in which a behavioral response depends on working memory of a stimulus across a time delay of a few seconds. This persistence time is orders of magnitude longer than the biophysical time constants (tens of milliseconds) of fast electrical signals in neurons and synapses. For this reason, persistent activity is often modeled via "dynamical attractors," that is, any self-sustained and stable state of a neural network (e.g., a stable equilibrium or a stable oscillation or limit cycle). We are thus back to the dynamical systems view of neurons and neural nets of chapter 3. Here, the memory states are assumed to correspond to multiple attractors in such a way that a memory state can be switched on and off by transient inputs. In a discrete attractor network, the memory states are separate equilibria each with its own basin of attraction. Input can drive the network to a new state within the appropriate basin of attraction; the system then follows a trajectory toward the corresponding attractor state, the "retrieved memory." As shown in appendix A, a Hopfield network can be associated with an energy function. Each energy minimum is an attractor state, in that the network perturbed slightly away will converge back to it.

Peter Milner (a former colleague of Hebb) expressed doubts about the attractor theory, on the ground that a reverberatory neural assembly is liable to fire out of control. However, working memory activity in the cortex occurs at moderate rates (typically 10 to 50 Hz), significantly above the spontaneous firing rate (a few Hertz), but well below the maximum firing capability of cortical neurons. Milner (1999) suggested that inhibitory damping might prevent reverberation altogether. This observation provides the jumping off point for the introduction of related models in chapter 11. The classic Wilson–Cowan model (Wilson and Cowan, 1973) couples a population of excitatory neurons with a population of inhibitory neurons to support analysis of stability and instability of different overall patterns of excitation and inhibition. More realistic models with integrate-and-fire neurons and synaptic dynamics revealed that cortical circuits with multiple attractor states are significantly more stable if the recurrent excitation is slow compared to inhibition, for example, partially mediated by the NMDA receptors. The dynamics of NMDA receptors is qualitatively and quantitatively different from the dynamics of inhibitory (GABA) and fast excitatory (AMPA) conductances. Inhibition stabilizes the network dynamics, gives rise to stimulus selectivity, and induces resistance against distractors. The prediction of a critical role of NMDA receptors in a working memory system such as the PFC has gained empirical support. For example, blocking NMDA receptors in the recorded neurons of behaving monkeys abolished stimulus-selective persistent activity. As chapter 11 shows, other cellular and synaptic processes may contribute to the generation of persistent activity and have been incorporated in models of working memory. Moral: While many successful models ignore synaptic dynamics (other than cumulative weight changes), some phenomena demand modelers attend to their details

While point attractors make sense for encoding a set of discrete memories, they do not make sense when some continuously varying parameter needs to be stored. Here one turns for models to line attractor networks, where all trajectories on the energy landscape flow down the walls of a trough toward a line of minimum energy at the bottom, each point of which is a stable equilibrium, so that an analog quantity can be stored parametrically by position on the line. For example, in the delayed oculomotor response task, a subject is required to remember a visual cue (such as a directional angle or a visual target location), across a delay period in order to perform a memory-guided saccade (more on this in chapter 22, "Integrative Functions of the Corticostriatal System"). A network model for such spatial working memory employs a "Mexican-hat" architecture with localized recurrent excitation between pyramidal cells with similar preference to spatial cues, and broader inhibition mediated by interneurons. A transient drives a local subpopulation of cells to fire at high rates. Internal excitation is large enough to sustain elevated activity, so that the firing pattern persists after the stimulus is withdrawn. Synaptic inhibition ensures that the activity does not spread to the rest of the network. At the end of the delay period the cue information can be retrieved by reading out the peak location of the persistent activity pattern, and the network is reset back to the resting state.

Mental states must be actively maintained in order to carry out tasks in our daily lives, and the PFC is known to play a major role in rule representation and rule learning. Chapter 11 concludes by exploring how attractor network models of the type presented above can be generalized to represent mental states. Simple, low-dimensional neuronal representations, which are typical of many attractor models, have proved inadequate for describing rule-based behavior. One possible solution is to note that a network can be designed in which randomly connected neurons are naturally activated by a combination of synaptic inputs from external stimuli as well as "rule-coding neurons" and such mixed selectivity is exactly what is needed to solve the task of appropriately scheduling behavior according to the current rule. This is related to the *reservoir computing* model for learning sequences given in chapter 22 ("Integrative Functions of the Corticostriatal System"). Heterogeneity and mixed selectivity are salient yet puzzling characteristics of frontal cortical neurons recorded from behaving animals, but they actually provide a signature of the high dimensionality of the neural representations. This suggests that mixed selectivity is computationally desirable as it allows representation of facts, memories, events, and importantly their combinations. The last are critically important for enabling the PFC to subserve flexible context- and rule-dependent behavior.

1.8 Vision

Although there are many perceptual systems, and we give some consideration to audition, olfaction, proprioception, and "whisking," this book emphasizes just one, namely, vision. The mammalian visual system starts with the retina whence it follows two main pathways, one to the superior colliculus of the midbrain and the other via the thalamus and on to visual cortex—and there, in turn, divides into the dorsal and ventral pathways within cortex. We have already seen how Ziad M. Hafed studies the role of the collicular pathway in "Saccades and Smooth Pursuit Eye Movements" (chapter 19). Alan L. Yuille and Daniel Kersten chart the early stages of visual processing in the thalamocortical pathway in their exposition of "Early Vision" (chapter 12). James A. Bednar and Christopher K. I. Williams then provide models for the emergence of many of the visual processes charted in chapter 12 as they shift the focus to the developmental timescale with their modeling of "Neural Maps: Their Function and Development" (chapter 13). Owen Lewis and Tomaso Poggio take up where Yuille and Kersten leave off, introducing high-level vision with their analysis of "Object and Scene Perception" (chapter 17).

We distinguish between those models of vision which attempt to describe how humans or animals see by accounting for behavioral or neural data and those, called computer vision, whose goals are to perform visual tasks effectively on computers without necessarily attempting to model how humans, or animals, perform them. In either case, vision researchers break vision down into different visual tasks which can be studied separately. These tasks include image segmentation, depth estimation, and object recognition. We often proceed as if these are performed by semi-independent modules which output the appropriate representations. This simplification can be questioned, but in any case we need to explore how the tasks or modules interact with each other. We classify visual tasks into three levels:

1. *Low-level vision*, which "preprocesses" the image to extract salient local features.
2. *Mid-level vision*, which estimates the structure and properties (including distance and motion) of geometric surfaces.

These two levels, early vision, are the focus of chapter 12.

3. *High-level vision.* In chapter 17, the emphasis is on recognizing objects and getting the gist of visual scenes. Chapter 25 ("Evolving the Language-Ready Brain") briefly reviews a schema-theoretic account of scene understanding, in which the disposition of and relation between multiple objects is taken into account.

Each level provides information which is passed on to the next level, but each level may also be open to "top-down" influences which may emphasize processing relevant to the task at hand.

Our theme of action-oriented perception (recall figure 1.3) reminds us that vision is not an end in itself (just seeing what is "out there") but also a basis for action. For example, chapter 20 ("Reach and Grasp: Control, Development, and Recognition") showed how visual perception of the affordances of an object could guide actions and also examined the challenges of recognizing actions of the self and others.

1.8.1 LOW- AND MID-LEVEL VISION Low-level vision exploits statistical properties of images which are true for most images (e.g., that images tend to be piecewise smooth). It also includes estimating the local motion of images. This involves finding the correspondence between points in images taken at different times.

Mid-level vision builds on low-level vision to exploit general properties of geometry, materials, and lighting, but not data about specific objects or scene structures. Mid-level vision includes reasoning about depth ordering of surfaces and, in particular, knowing that surfaces can partially occlude each other. The human visual system assumes that most images contain a ground plane with objects standing on it (e.g., a man standing on a lawn). By the laws of perspective projection, if there are parallel lines in the image (such as the tracks of a railway line) then the projection of these lines in the image will converge at a vanishing point. Humans can use vanishing points to estimate the orientation of the ground plane. Such estimates can be misleading, as evidenced by optical illusions (Gregory, 1967).

Binocular stereo is another vision module which estimates the depth and orientation of surfaces. It requires solving a correspondence problem between features in the two eyes which relate to the same point in space. Humans can also get three-dimensional shape information of surfaces from shading, texture, and even contours. In addition, the ability to detect material properties and textures of objects is very useful both when performing actions (e.g., is this object slippery?) and as a precursor for object recognition and scene understanding. These are often studied as separate modules, but their integration may involve processes of cooperation (strengthen cues from different modules that may be in agreement) and competition (if cues disagree, then one may come to predominate, though in other cases some synthesis may resolve the apparent discrepancy).

Receptive Fields Chapter 12 charts a large range of models. The first set employs linear models of neurons to describe the *center–surround* receptive field properties in the retina and LGN. The retina contains two types of ganglion cells (output cells) with very different temporal characteristics. M-cells (magno cells) have receptive fields which are spatially large but temporally small and project to the ventral stream; P-cells (parvo cells) have receptive fields which are spatially smaller but temporally larger and project to the dorsal stream. (Each stream is spatially segregated in LGN on its way to V1.) Linear models also describe the *simple cells* in visual area V1, that is, cells which respond preferentially to edges of a particular orientation in a small receptive field. The chapter provides background material on *linear filtering* and *Fourier analysis* which helps us understand the properties of these neural models. It also uses functional analysis in which an image, or any signal, can be expressed uniquely as a weighted sum of basis functions. This implies *for a linear system* that if we know the response to each basis function, then we can predict the response to any input. (Recall the discussion in section 1.6.1 of assembling trajectories out of motor primitives.)

More generally, simple cells have been modeled as *Gabor wavelets* (named for Dennis Gabor, the inventor of holography). Rather than a single edge, the receptive field is described by a spatial grating, regularly spaced bars at a specific orientation, spatial frequency, and phase (i.e., offset from the center of the receptive field). Moreover, the receptive field drops off according to a two-dimensional Gaussian. *Complex cells* in V1 are also orientation tuned but have large receptive fields. Chapter 12 describes a model where the complex cell receives input from two Gabor cells which are 90 degrees out of phase but of the same orientation and spatial frequency. Some researchers question whether complex cells receive input from simple cells, arguing that the computations could be done by nonlinear neurons which exploit the complexity of the dendritic tree.

Other perspectives include the use of sparsity to suggest receptive field properties based on the statistics of natural images and also the idea of matched filters. This reverts to an older idea of receptive fields as feature detectors. Barlow (1961) proposed sparsity as a general principle for modeling the brain using ideas from communication theory (Shannon and Weaver, 1949) to interpret the observation that typically only a small number of neurons are active. Sparsity can be used to derive the properties of receptive fields of cells in V1 if we assume that these cells are designed to represent properties of natural images. Here, instead of hypothesizing a particular form of receptive fields (e.g., Gabor filters), we can try to predict these receptive fields from studying images. This requires learning suitable basis functions from a set of natural images. This criterion has been applied to natural images, and the resulting basis functions include filters which look like Gabor functions, but they also include other types of filters which are also observed in experiments.

Where do the receptive fields of simple cells come from? The most plausible explanation is that they are learned in an unsupervised self-organizing manner, but it is debatable whether this happens before birth because of spontaneous activity in the brain and/or in response to natural stimuli observed after birth. Studies based on modifications of the Hebb learning rule have some experimental support (recall chapter 6). The basic findings are that center–surround, orientation-selective, quadrature pairs, and disparity-sensitive cells (precursors to cells which can estimate depth from binocular stereo) could all be obtained by variants of the same learning rule, Oja's rule (already encountered in chapter 6).

Segmentation, Stereo and Motion Linear filter models of receptive fields can be used to perform local estimates of *binocular stereo* and *motion*. These involve having filter banks, populations of filters, which are tuned to different properties of the stimuli so that estimates of depth and motion can be extracted from the population.

The disparity energy model estimates disparity based on local properties of the image; nonlocal context can be used to improve disparity estimation. The model is formulated using Gabor filters and has some claim to biological plausibility. Related models can be applied to estimate motion for sequences of images. Spatiotemporal filters are biologically plausible ways to measure motion which agree with properties of cells in the visual cortex (and recall the discussion of MT in chapter 19, "Saccades and Smooth Pursuit Eye Movements"). The computation must solve the *aperture problem.* If we move an edge of an object behind a window, we can only detect the motion normal to the edge. It is only by integrating across the different receptive fields examining local edges of an object that we can determine the motion of the object itself. A two-stage model solves this, where in the first population of neurons each neuron (i.e., filter) is sensitive to the local spatiotemporal frequency of the input image; a second population of neurons then extracts the motion information from the first population.

Chapter 12 then turns to the use of probabilities and decision theory. The approach also illustrates the importance of knowing whether filter responses, hence visual cues for the task, are dependent or independent. *Decision theory* was developed as a way to make decisions in the presence of uncertainty. The key ideas of decision theory are developed here by first assessing detection of edges based on local evidence in the image; the use of nonlocal or contextual information is considered later in the chapter. *Bayesian decision theory* is deployed to manage the trade-off between two types of errors: false positives, which are nonedge stimuli which the decision rule mistakenly decides to be an edge, and false negatives, which are edge stimuli which are mistakenly classified as not being edges. Bayesian decision theory is a framework for making optimal decisions in the presence of uncertainty; signal-detection theory (which is often used as a framework to model how humans make decisions when performing visual, auditory, and other tasks) can be obtained as a special case. Two types of context are considered: the first uses spatial context and is low- or mid-level since it depends only on generic properties of images and surfaces; the second type is high-level and object specific. For example, if we detect a face in an image, then our knowledge about faces enables us to detect the boundaries that delimit eyes, nose, and mouth (an example of top-down influence from high-level vision on the low and middle levels).

Psychophysicists have discovered many perceptual phenomena demonstrating that human perception of local regions of an image can be strongly influenced by their spatial context. For example, local image regions which differ from their neighbors tend to "pop out" and attract attention while, conversely, similar image features which form spatially smooth structures tend to get "grouped" together to form a coherent percept. Image properties such as color tend to spread out, or fill in regions, until they hit a boundary. In general, there is a tendency for low-level vision to group together similar image features and make breaks at places where the features change significantly. Segmenting an image into different regions is one of the first stages of object recognition (in the ventral stream) and a precursor to estimating the three-dimensional structure of objects, or surfaces, in order to grasp them or avoid them (dorsal stream). Since low-level vision interacts with high-level vision, it is unlikely that tasks like segmentation are performed entirely by low-level processes.

Macaque single-electrode studies show that the activities of neurons on monkey area V1 appear to involve spatial interactions with other neurons. When monkeys are shown structured stimuli, their responses over the first 60 ms are similar to those predicted by classic models, but their later activity spreads in from the boundaries, roughly similar to predictions of computational models. To address such data, chapter 12 introduces probabilistic models of neurons and shows how the previous linear filter models can be derived as approximations. It then introduces neural network models and shows their relationship to probability models. Topics modeled include detection of edges using spatial context. This relates to a range of phenomena known as association fields. For this model, we have a set of neurons at every spatial position, each tuned to a different angle and a default cell (no edge). One interaction term encourages edges to be continuous and their directions to change smoothly; another uses explicit line process variables which "break" images into regions where the intensity is piecewise smooth.

The modeling then turns to stereo vision. The key problem is to solve the *correspondence problem* between the inputs in the two eyes: for example, if two features detected via the left eye are similar, which of them should be matched to a given feature detected via the right eye? Once this problem is solved, the depth of the points in space can be estimated by (the neural equivalent of) trigonometry. Classic models assume that the surface is piecewise smooth. This leads to a Markov field model which includes excitatory connections, imposing the geometric constraints, with inhibitory connections which prevent points from one eye having more than one match in the second eye. Cooperation implements the excitatory constraints; competition deals with the inhibitory constraints. (In an early example of stability analysis of dynamic neural fields, Amari and Arbib, 1977, reformulated the original scheme of this kind, Dev, 1975.) Attention then turns from depth mapping to motion extraction, emphasizing that local estimates of motion must be integrated with contextual cues using, for example, the priors that most motion is slow and smooth to take past history into account and not merely use the change of image from frame to frame.

Surfaces, Objects, and Context At the behavioral level, psychophysicists have studied how humans combine different visual cues (such as shading, texture, binocular stereo, structure from motion) to estimate depth, surface geometry, and other surface properties. How can these cues can be combined? The most straightforward manner is to use a separate module for each cue to compute different estimates of the properties of interest, for example, the surface geometry, and then merge these estimates into a single representation. However, theoretical analysis suggests that we need to distinguish between situations where the cues are statistically independent of each other and cases where they are not. We need also to determine whether cues are using similar, and hence redundant, prior information. These considerations leads to a distinction between weak and strong coupling between the modules. To understand strong coupling, it is helpful to consider two types of situations. Firstly, some image cues only apply to parts of the image, and we need a process to select those parts (e.g., shape from texture cues are only valid within certain parts of an image). Secondly, for some images there may be several alternative ways to generate them, yielding several possible interpretations, and we need to select the most probable. For example, perspective cues give very strong perceptions of shape, but they rely on the presence of straight lines in the image, so they are valid for images of Manhattan but will not work in the jungle. These considerations show that cue combination often requires *model selection*, in order to determine in what parts of the image, if any, the cue is applicable.

Chapter 12 presents a number of models within this overall framework. First, simple models are presented which assume that the cues compute representations independently and then combine their outputs by taking linear weighted combinations. In some specific cases, these weights can be derived as measures of the uncertainly of the cues. Bayesian analysis of probabilities on structured representations is then introduced and used to model neurons and groups of neurons. The general probabilistic framework has many applications, including sensorimotor control and cognition, and provides a common language for describing vision problems, mathematical techniques for modeling them, and algorithms that can be applied to solve them. Two examples of strong coupling are considered. The first concerns the perception of texture while the second deals with multisensory integration with structural uncertainty. Visual and auditory cues can have a common cause (e.g., you see a dog moving and hear it barking) whereas in other situations the auditory and visual cues are due to different causes (e.g., a cat moves and a nearby dog barks). We thus see the importance of top-down processing to complement bottom-up processing from retina through LGN and V1 and on into the dorsal and ventral streams. Before moving on to discuss high-level vision, we take a break to focus on how maps in the visual system may be formed. In the process, we will encounter developmental models for the emergence of some of the visual processes charted in chapter 12.

1.8.2 Neural Maps: Their Function and Development A *neural map* is a sheet of neurons systematically related to another population of neurons, including the special but common case of a *topographic map* as a spatially organized mapping of this type. A *feature map* is then a topographic mapping of some underlying features or patterns in the input data. How do they develop? Systematic maps first emerge in a sketchy form in sensory areas through genetic specification of initial topography, based on molecular guidance cues. The maps are then shaped by activity-dependent and experience-dependent processes which typically result in locally smooth maps. However, the apparent absence of continuous topographic or laminar organization for orientation in rodent V1 suggests that the continuity of the map may not be crucial for function. The *coverage* of the maps, that is, the extent to which they fully tile the multidimensional space of all combinations of features (e.g., orientation and retinotopic position) seems to be of more functional relevance. The models in chapter 13 focus on achieving good coverage via locally continuous, topographically organized feature maps, even though continuity does not apply to all maps in all species, because local continuity does provide important constraints on the possible map formation mechanisms. Relaxing constraints on map formation in these models can account for results in rodents.

In the simplest case, a map model consists of two sets of neurons (input and map [output]), each arranged as a 2D neural *sheet* (array of units) with topographic connections between them. More realistic models include multiple sheets and sets of connections. In nearly all cases a neuron is modeled as a single-compartment (point) unit, with the response of a neuron given by its firing rate. Each unit typically also represents many real units, often collapsing a vertical column of similar neurons into one computational unit for efficiency.

The focus of the chapter is on maps in the visual system though similar principles have been applied to model nonvisual modalities such as somatosensory cortex, whisker-barrel (not whisky barrel—see chapter 16) maps, and sparse-coding models of a wide range of sensory neuron types. There are also corresponding data and modeling for maps and plasticity in the motor cortex. Different aspects of the stimuli will lead to very different map shapes and properties, so different organizations are expected for each modality. For instance, in auditory cortex there is a primary organization by tonotopy (matching the arrangement of receptors on the cochlea), but it is not yet well established which other dimensions of auditory stimuli might be represented. Although modeling has replicated the basic organization of somatosensory cortex, how the various touch, temperature, and pain receptors distributed across the skin surface affect the organizations in somatosensory cortex is not well understood.

Model maps are considered largely in isolation in chapter 13, and primarily in terms of feedforward processing, but (as we have seen in our discussion of low- and mid-level vision) real maps are situated in the context of numerous other maps that are simultaneously involved in neural processing, including widespread feedforward and feedback connections between them. An open challenge, then, is to better understand how neural maps incorporate top-down inputs from higher-level units which may identify longer-range structure, and feed back information about this to lower-level units (see the discussion of context and model selection in chapter 12). With these caveats in mind, we now turn to bottom-up processes in the formation of maps in the visual system.

As a starting point, consider that retinal ganglion cell axons from the eye form an ordered projection onto the optic tectum of a frog or the superior colliculus of a mammal (a *retinotectal map*). Initial ideas of absolute addressing eventually gave way to relative addressing, to account for experiments in which, for example, half of the retina was ablated and the remaining half innervated the entire tectum. We now know there is a chemical matching process driven by orthogonal spatial gradients of families of molecules and receptors: axons originating from neurons in different locations on the retina have different levels of EphA and EphB, forming a two-dimensional retinal coordinate system, and prefer corresponding levels of ephrins-A and B on the targets. These "sketch" the map. Once the axons arrive near their targets, neural activity appears to be important for local refinement of the projection, because disruptions to spontaneous retinal activity change the specific patterns of connectivity at the topographically mapped location. This process also appears to require some form of competition for targets, such as axon–axon interactions.

Chapter 13 uses a model containing only the activity-dependent subset of these mechanisms to highlight similarities and differences between explanations for topographic maps and feature maps. The self-organizing-map (SOM) algorithm uses a set of 2D input patterns presented to a network with initially random local connectivity, some form of lateral interactions in the target region that provide competition and cooperation between neurons, and a local Hebbian-based synaptic modification rule. The SOM algorithm has two alternating phases: (1) find the most active neuron for the current input; then (2) adjust the weights of the winning neuron and, to a lesser extent, its distance-weighted neighbors to have weights that are more similar to the input pattern. This gradually drives neurons to become selective and drives nearby neurons to become more similar in their response preferences.

The SOM model provides nothing to ensure the overall orientation of the map, nor does it provide explanations for how maps can develop under activity blockade, nor for the specific changes when genes related to the chemical gradients are knocked out. Addressing these limitations requires adding mechanisms explicitly incorporating chemical gradients and chemoaffinity processes which establish the overall orientation of the map and the initial topography. Most of the models in chapter 13 pick up where the gradient-based Eph/ephrin explanations leave off, modeling activity-dependent feature map formation once the overall topography has been established.

Models of the development of feature maps exploit the SOM network again, but now focusing on the mapping between the LGN and V1. The underlying manifold is now of a *higher* dimensionality than the SOM map, and modeling helps explain the patterns of emergent feature maps, not just the initial topography. Accounting for eye preference results in stripy or patchy patterns like ocular dominance maps while *cyclic* quantities like orientation generate realistic pinwheel-based orientation maps, in part because of geometric constraints on achieving map continuity for cyclic quantities in a 2D plane. However, it is difficult to relate the SOM model to biological processes at any level of detail.

Several closely related models have been proposed for map formation, each using firing-rate single-compartment neurons, local Hebbian synaptic learning rules, and explicit patterns of connectivity rather than global picking of winners or optimization of abstract quantities. The gain-control/adaptive/lateral (GCAL) model of cortical topographic feature maps replaces implausible mechanisms inherited from SOM with well-established local mechanisms such as homeostatic plasticity. Retinotopy is built in; GCAL simulates only the activity-dependent local reorganization in connectivity after the gradient-based processing has completed. In addition to the topographically mapped afferent connections, the V1 cells also receive lateral excitatory and lateral inhibitory input from their neighbors with a Mexican-hat configuration. The result is to focus an initially diffuse pattern of activation around one or more hotspots that receive more excitation than inhibition, gradually resulting in spatially discrete "bubbles" of activity in response to a slowly changing input. These interactions are a locally implemented version of the global winner-picking and neighborhood kernel from the SOM, and just as for the SOM, they ensure that the maps achieve good coverage and continuity. Unlike the SOM, there can be many spatially distinct "winning" patches simultaneously, each learning properties from the retinotopically corresponding inputs.

Once neurons have developed selectivity, treating them as a set of disconnected spatial filters is at best a useful first step. Cortical neurons also exhibit complex changes in receptive field and response properties when stimuli are present in locations surrounding the "classical" receptive field: perceptual completion or suppression effects, illusory contours, "pop out" of certain stimuli from their background, and a variety of other contextual phenomena (compare chapter 12). Such connectivity emerges automatically from the GCAL model, driven by Hebbian learning that strengthens connections between neurons with correlated responses—patchy lateral connectivity patterns, with neurons connected by similarity of responses to the multidimensional input space, not only by locality in the cortical space. Work with related but more complex models suggests that correlation-based connections can account for many of the nonclassical phenomena listed above.

Chapter 13 then considers a different class of models in which functional criteria are incorporated explicitly into the definition of the model. A series of such models starts from the assumption that the computational goal of a sensory system is to recover information about the underlying structure of the world, presumably in order to guide actions relating to that structure. Given raw sensory input, the objective would be to *infer* a set of external causes that would explain the observed data. We must then ask, as chapter 13 does not, how the dorsal and ventral pathways develop differentially to support different types of inference. Indeed, many features of the world can only be discovered through interaction and thus require loops involving the motor system, placing them outside the scope of chapter 13. Instead, it addresses the task of finding neurons that support *basis vectors* such that each small patch of an image can be obtained as a linear combination of the basis vectors. (Again, recall chapter 12.) In other words the challenge here is to produce a *generative* model, that is, one whose activity pattern can be used directly to generate an input pattern. The generative approach allows explicit expression of an *objective function*, where a representation is iteratively optimized to minimize or maximize a mathematically specified criterion. For instance, the objective function might specify a low reconstruction error. Note that specifying that reconstruction error should be low does not necessarily mean assuming that the visual system actually does reconstruct the input in practice—it simply specifies a criterion that the map's representation of the input should be rich enough to allow that in some form. For one of these models (Olshausen and Field, 1996) the basis functions for a set of patches of whitened natural images resemble the receptive fields shaped like Gabor functions found in V1 in mammals, leading to a hypothesis that primary visual cortex neurons are wired to form a basis set that provides good coverage of the space of natural image patches using sparse coding, that is, with only a small number of active units at any one time. (Again, compare chapter 12.) Models based on independent components analysis also lead to Gabor-shaped weights (Bell and Sejnowski, 1997). However, each of these algorithms results in a relatively narrow range of receptive field shapes, missing many of the receptive field types commonly observed in animals. More recent work has shown how to account for the full range of monkey receptive field shapes by using an alternative definition of sparseness.

1.8.3 OBJECT AND SCENE PERCEPTION In chapter 17, *object recognition* is viewed as the task, given an image and a list of category labels such as "car," "dog," or "house," of determining which label(s) correctly describes the image's content. In general, though, a scene might contain multiple objects, so one can focus the problem by applying some segmentation process (chapter 12) to extract a region which is a candidate for being an object, and then seek a label for that region. As our discussion of the two visual pathways and modeling of the reach to grasp suggests, the assignment of a label to a scene might be only one step toward extracting parameters relevant to the particularities of the object or to interaction with it.

When one turns to scenes, one may distinguish two complementary challenges. *Gist recognition* is akin to a variant of object recognition: Is the scene a Parisian scene, a freeway scene, a view of mountains, and so forth? Chapter 12 addresses this problem in some detail but does not address the more general problem of *scene understanding*—given a scene, what are some of the key objects in the scene and what are the relations between them? For example, recognizing that a bear is in a scene would in real life just be the start of the analysis: Is the bear moving away from me or looking right at me? My behavior would vary greatly, accordingly. The action recognition of chapter 20 is a case in point. Chapter 25 recalls a classic artificial intelligence (AI) system for scene understanding, the VISIONS system, and claims that its competition and cooperation between instances of visual schemas is "in the style of the brain."

Object Recognition Solutions to the object recognition problem have two stages, representation and classification. The type of representation advocated in chapter 17 is one in which two images with the same semantic content (same individual, same object or scene class) are close together; that is, distance in representation space should reflect distance in semantic space. A new image can then be classified based on where its representation falls relative to the representations of images for which the label is known. Classification can be thought of as dividing the representation space into regions where all images in each region have the same label. The chapter focuses almost entirely on the representation problem.

As we have seen, visual cortex forms a hierarchy of interconnected areas, so that useful representations can be built up across the layers, with low-level areas contributing building blocks to the more semantically meaningful representations found further upstream. However, top-down influences may drive attention not only in the form of eye movements (chapter 19, "Saccades and Smooth Pursuit Eye Movements") but also in affecting the activity of feature detectors early in the visual pathway. (Recall the study of the interaction between attention and visual orientation selectivity in the macaque visual cortex which is modeled in chapter 4, "Neural Rhythms," in relation to the gamma rhythm.)

Chapter 17 focuses on the ventral stream: V1, V2, V4, and IT. As information passes along the stream, each area computes a new representation, encoded in the activities of that area's cells. The higher the area, the better is its representation able to support categorization. A number of experimental techniques are available to study the brain's activity in general, and the representations computed along the ventral stream in particular. At the single-cell level, the tool of choice, electrophysiology, involves implanting electrodes in the brain of an animal (e.g., in the macaque) and recording the electrical activity induced by presentation of different stimuli. Once neurophysiology has captured neural responses within an area, it remains to characterize the information that this representation contains. Suppose an experiment showed images $I_1, \ldots, I_n$, from categories $c_1, \ldots, c_n$, and that image I_i elicited neural response R_i. The *classification problem* is then to learn a mapping directly from neural response to category label. As in machine learning, the set of pairs (R_i, c_i) is divided into a training set used to train a classifier and a test set used to validate its performance. If good classification is possible, one may conclude that the representations R_i contain category information. Generally, decoding studies use linear classifiers, extracting category information. Two important caveats: (1) establishing that the area supports patterns rich enough to make a classification does not establish that the area computes the categories or even provides information that other areas use to compute the categories, and (2) even if the area does compute the categories, the computations engaged in by the neural network model may be very different from those employed by the neural circuitry of the brain.

The V1 representation can support edge detection and object segmentation, for instance, but it lacks the high-level semantic content and robustness to directly support effective categorization without further transformations. V2 contains cells responsive to angles and junctions formed by combinations of the bar-like features preferred by V1 cells. Further up, V4 represents still more complex curvature features. IT contains cells responsive specifically to certain high-level object classes. At each stage, though, most neurons are responsive to a variety of stimuli, arguing against a one-to-one correspondence between cells and high-level features. It is not the firing of a single cell that encodes the scene but the activity of a population.

The relevant models fall into two classes. The first studies what happens during a first, entirely feedforward pass through the ventral stream, from V1 to IT. The second explores the role of feedback processing, in which areas communicate bidirectionally to converge on an image interpretation. A classic feedforward model is the Neocognitron, "A self-organizing neural network model for a mechanism of pattern recognition unaffected by shift in position" (Fukushima, 1980). Here a learning rule tunes alternating layers of simple (localized receptive fields) and complex (broad receptive fields) cells to better and better classify input patterns. Chapter 17 gives special emphasis to the HMAX architecture. It consists of four processing layers, corresponding to areas V1 though IT of the ventral stream. Similarly to the Neocognitron, each area except the last has simple and complex sublayers.

Gist Processing Object recognition is supplemented by a distinct set of processes housed in a distinct set of brain areas that quickly extract the global geometric properties of a scene that support fast recognition of its gist. Evidence for *gist processing* comes from neuroimaging studies which have found at least three areas that preferentially respond to scenes or places over objects: the parahippocampal place area (PPA), retrosplenial cortex (RSC), and the occipital place area (OPA). In a particularly striking finding, PPA activation to an image of a room was more or less unchanged by removing all the moveable objects (furniture etc.), evidence of a global rather than object-based encoding.

Oliva and Torralba (2006) modeled the role of global image features in extracting the gist of a scene. The model begins with a V1-like representation given by the output of a bank of Gabor filters spanning a range of orientations and scales and positioned densely across the image. A first reduction step coarsens the spatial resolution of this representation, averaging the output of each filter within each window. The second reduction step assesses this representation against a database of natural images and applies principal component analysis. Each principal component is a linear combination of the averaged filter output of the original representation. These linear combinations are then "fused" into what the authors call global feature templates which span the whole image.

An image's gist category may contain a lot of information about which objects it is likely to contain and where they are likely to appear. A street scene, for instance, is likely to contain multiple cars while a mountain or forest scene probably contains none. Furthermore, cars appear at predictable locations in street scenes; a car is unlikely to appear in the sky. A further model used scene-level information to adjust the outputs of local object detectors, ensuring that the final detections contain a plausible number of objects of each class, and that these objects appear in plausible locations—so here we do see the beginning of a transition from gist extraction to more detailed understanding of the scene.

Indeed, a related model uses scene features to predict human eye movements during visual search. The model combines mechanisms for computing the inherent saliency of image regions with scene-based tools for predicting where objects of the target class are likely to appear. Experiments showed that including the scene-level features lead to significantly improved predictions of subjects' fixations. This provides a top-down complement to the bottom-up processes studied in chapter 19, "Saccades and Smooth Pursuit Eye Movements."

It is generally agreed that feedback connections (i.e., connections proceeding top-down in the visual hierarchy) mediate attentional processing, carrying signals about stimulus salience and task relevance from frontal and parietal areas. Chapter 17 thus notes a class of models which views feedback connections as primarily generative in nature, enabling the brain to synthesize as well as consume images. (Sufferers of Charles Bonnet syndrome report extremely vivid hallucinations which they have trouble distinguishing from reality, and which imaging studies have shown to have very similar neural signatures to normal visual processing.) In 1867 Hermann von Helmholtz advanced the theory of "perception as unconscious inference." The goal of visual cortex is then to infer from an image to key aspects of the world state that caused it. This method is also called *analysis by synthesis*. A generative model here takes the form of a probabilistic graphical model, a methodology that also played a role in chapter 12. In one model, higher visual areas play the role of the receiver with lower areas as sender. Higher areas use feedback connections to transmit their generative predictions of the next lower level's activity, and lower areas send up the errors in these predictions via feedforward connections.

Neural Bayesian inference provides another approach to inference in generative vision. This is a message-passing algorithm based on communication between nodes of a graphical model. For example, we may consider the graph and make use of potential functions which measure the compatibility of pairs of nodes. Belief propagation plays the key role. As the model's beliefs evolve, many become overwhelming unlikely, as manifested in low weights. At each iteration, therefore, a new set is formed by resampling with replacement according to these weights. This will tend to reinforce those that are doing well while killing off ones that are doing poorly.

An alternative approach invokes image parsing models. Just as a sentence is composed of systematically combined constituents, images are made up of substructures. A scene is a collection of objects, which are themselves collections of parts, and so on. Given a grammar describing how natural images might be built up from visual primitives, image parsing seeks to invert this process.

The mainstream view is of a first, purely feedforward pass sufficient for relatively straightforward categorization tasks followed by a more sustained feedback process supporting more difficult or ambiguous determinations. An EEG study found category-selective information arising in the PFC in only 150 ms (but after visual cortex). This timing has been a challenge for feedback models, given their more extensive processing demands. However, if recurrent processing occurs continuously in local feedback loops, rather than requiring multiple complete top-to-bottom passes, then recognition at the 150-ms timescale is not unrealistic.

Interpreting Visual Scenes: The VISIONS System VISIONS, a scene-understanding system (chapter 25), moves from a color photograph to a labeling of the various objects contained in the scene. This implemented AI system demonstrated how a schema network can be used to model some portion of world knowledge (how to recognize objects from low-level perceptual features extracted from an image and how they relate spatially to their parts or frequently associated objects) and how this knowledge can be retrieved on the basis of activation values of already instantiated schemas (see figure 1.9).

VISIONS illustrates *cooperative computation*, the competition and cooperation of a multitude of schema instances in coming together to compose a schema assemblage appropriate for the current scene.

Low- to mid-level vision performs *segmentation*: competition and cooperation at the level of local image features grows edges and regions to yield a first-pass subdivision of the image to ground semantic analysis.

High-level vision brings in *perceptual schemas*. Some schema instances may be activated and linked to a region by bottom-up (data-driven) processes, but spatial and other relations between objects may be encoded in the schema network in long-term memory to allow schemas to compete and cooperate to link their schema instances to different regions of the image via top-down (hypothesis-driven) processes, to yield the interpretation of the scene in terms of the most activated schema instances.

Note that the initial segmentation of the scene (and passing of various parameters for these segments) may not match the division of the scene into objects (some edges may be missed; shadows may yield false segments), and so working memory may need to send requests to low-level processes to update their analysis of the image.

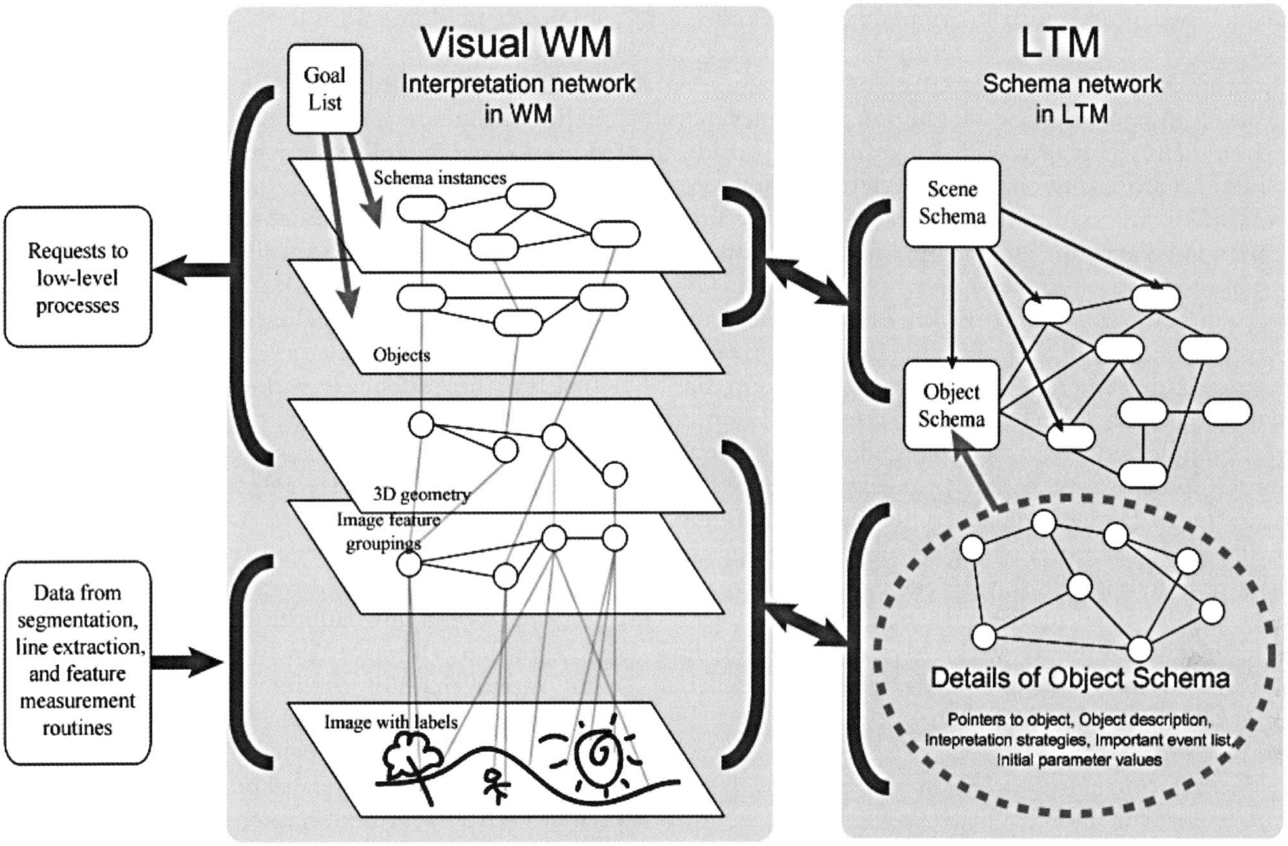

FIGURE 1.9 Segmentation of a scene into candidate regions (provides the bridge between the original image and the interpretation of a scene in VISIONS. The visual working memory (WM) of VISIONS interprets the current scene by a network of parameterized instances of schemas from long-term memory (LTM). These schema instances are linked to the visual world via the intermediate database that offers an updatable analysis of the division of the world into regions that are candidates for interpretation as agents and objects, possibly in relation with each other.

1.9 *Bridging the Levels*

Our knowledge of the brain comes both from animal studies where neuroanatomists and neurophysiologists probe the connectivity and activity of single cells in specific neural circuits (others study genetics and molecular biology, but these themes are at best tangential to this book) and from human studies which may include studies of brain imaging and the degradation of behavior and cognition that follows the ravages of brain lesions and disease. Three chapters bridge these levels. Richard P. Cooper, in "Schema Theory and Neuropsychology" (chapter 14), shows how a range of data may be better modeled by the competition and cooperation of schemas rather than by attempting a reduction to the neuron network level. Complementing this, "Synthetic Brain Imaging" (chapter 15), by Jacqueline A. Griego, Carlos R. Cortes, Ransom Winder, and Malle A. Tagamets, shows how neural network models of the human brain inspired by animal data can be simulated in a way which yields predictions which can be tested against brain imaging data. In chapter 23, Mohamed A. Sherif and William W. Lytton bridge from the level of neural and neurochemical mechanisms to a variety of high-level symptoms to show how modeling can yield insights into various "Brain Diseases."

1.9.1 SCHEMA THEORY AND NEUROPSYCHOLOGY We have already introduced the notion of *perceptual* and *motor schemas* and noted that chapter 25 ("Evolving the Language-Ready Brain") introduces the VISIONS system in which competition and cooperation between instances of perceptual schemas develops an interpretation of a visual scene. Chapter 14 first develops our insight into schema theory by providing historical background and then examining schema-based models of approach and avoidance behavior in frog and toad (Arbib, 1987) to show how lesion data may help us adjudicate between competing models of a given range of behaviors. However, the main thrust of the chapter is to show the relevance of schema theory to making sense of data from neuropsychology and to compare a schema-based model with a competing connectionist model.

Neuropsychology studies the effects of brain injury on aspects of human behavior. In addition to its relevance for treatment, neuropsychological evidence provides additional constraints on theories of the neural processes underlying cognition and behavior because such theories must account not only for normal operation but also for the effects of brain injury on behavior. Localization evidence is at its strongest when data from neuropsychology and brain imaging can be integrated. In some circumstances, hypotheses on localization may relate to specific (sets of) schemas in the brain. In other cases, chapter 14 cautions that it may not be the schemas themselves but processes that operate over schemas that may be localized. In either case, specifying schemas may provide a crucial intermediary in the analysis of the underlying neural networks. Moreover, while there are similarities in the abstract nature of perceptual, event, linguistic, and motor schemas, there are also domain-specific differences. (Chapter 25 discusses their similarities and differences in an evolutionary framework.)

Higher-Level Schemas for the Control of Thought and Action Norman and Shallice (1980) viewed routine behavior as the province of a schema-based system, contention scheduling, which is modulated in nonroutine situations by a supervisory system. In our earlier vocabulary, chapter 14 views schemas as hierarchical, so that contention scheduling may invoke a schema which is structured like the coordinated control program of figure 1.2, whereas unusual situations require planning (perhaps unconscious) by the supervisory system (which may well be distributed across the brain) to assemble the schemas needed to handle the situation. The general setup is illustrated by considering the *Prepare instant coffee* schema, which includes subschemas for boiling water and adding the various ingredients to a mug. A schema's activation may be affected by several factors, including top-down excitation either from the supervisory system or from selected superordinate schemas, bottom-up excitation (or "triggering") from the internal model of the environment, and inhibitory influences from competing schemas. Object representations excite those schemas that can use the objects in proportion to their activation, and vice versa (recall the notion of *affordance*). This means that highly active schemas will tend to excite the representations of relevant objects, while highly salient objects will tend to excite the schemas which might be applied to them. When a schema is selected, the most active object representations are then assigned, or "bound," to its argument roles.

In nonroutine situations the operation of contention scheduling is modulated by the supervisory system. This is held to be a complex of subsystems which work together to create a "temporary schema." This then feeds into the contention scheduling system to produce deliberate or planned behavior. The primary interface between the supervisory system and contention scheduling is a special-purpose working memory, which holds a representation of the temporary schema. Execution of the schema is then monitored in case it fails to yield the intended result.

Much of the evidence in support of the theory comes from two sources: the behavior of neurologically healthy individuals when they are distracted and the behavior of patients with specific deficits following brain injury. A number of errors of action can be understood in terms of behavior being "captured" by an unintended schema. Neuropsychological evidence includes *utilization behavior*, where a patient's behavior appears to be driven primarily by schemas triggered by the environment rather than by an internalized plan (schema assemblage). Utilization behavior is a relatively rare syndrome, but it has been associated with damage to PFC, and particularly with bilateral damage to superior medial prefrontal areas. Chapter 14 introduces further such data from neuropsychology, including action disorganization syndrome, ideational apraxia, and Parkinson's disease. (Chapter 23, "Brain Diseases," looks at cellular and neuromodulatory changes associated with Parkinson's disease.)

The Interactive Activation and Competition (IAC) Model of Contention Scheduling Chapter 14 then presents the interactive activation and competition (IAC) model of contention scheduling (Cooper and Shallice, 2000) which was implemented for the task of beverage preparation. A patient with frontal lobe damage produced a range of errors when attempting to prepare instant coffee within the institutional setting, such as pouring cream into his orange juice, adding breakfast cereal to the coffee cup, and attempting to pour from a sealed container. The initial goal of the simulation work was to provide a qualitative account of these and related errors.

Schemas at the lowest level of the hierarchy were assumed to correspond directly to basic actions such as "pick up" or "unscrew," but if the model were to be interfaced with a model of the motor system, these basic actions could well be implemented in terms of schemas that could be directly related to neuronal processing (see chapter 20).

Each schema has an associated goal (basically a state of the world) that will be achieved if the schema is performed successfully. Internally, schemas consist of a partially ordered set of subgoals. Partial order is specified through preconditions associated with each subgoal. Preconditions and postconditions provide a flexible way to allow for variation in the state of the world when a schema is selected. (Compare the augmented competitive queuing [ACQ] model to be introduced below as a topic in chapter 25.) Each schema within the schema hierarchy has an associated activation value which varies over time between zero and one in response to the net excitatory (positive) or inhibitory (negative) input to the schema. The network is hierarchical, with the top-down flow of activation within the hierarchy of schemas. A schema is selected when its activation exceeds a threshold and remains selected until all postconditions of all of its subgoals are satisfied, or the corresponding action is complete. On deselection, schemas are inhibited until their activation falls to rest. Schemas have argument roles, and an internal model (established through perceptual schemas and possibly updated by motor schemas) makes object and effector representations available for binding to those argument roles. It is moreover assumed that there is only a small set of such argument roles (paralleling the argument roles of agent, patient, theme, etc. from linguistic theory), and that object and effector representations compete for these roles through additional interactive activation and competition networks.

The basic IAC model does not contain any learning mechanism, but other work has demonstrated how standard reinforcement learning techniques (chapter 10) could augment the model.

Evidence from neuropsychology suggests that three brain regions are of particular interest within routine sequential action selection: frontal cortex, left temporoparietal cortex, and the basal ganglia. It is suggested that superior medial PFC corresponds to the source of input from the supervisory system to contention scheduling. Damage to right lateral frontal cortex may be understood in functional terms as an impairment of monitoring. Intuitively, monitoring is involved in the IAC model in checking whether subgoal preconditions and postconditions are met. Left temporoparietal cortex may be involved in the triggering of schemas by object representations. The IAC model is not committed to any specific function of the basal ganglia, but one possibility is that competition between schemas is resolved by the basal ganglia, which participate in corticosubcortical loops (see chapter 22, "Integrative Functions of the Corticostriatal System").

The Simple Recurrent Network Model of Routine Sequential Action Selection In a direct response to the IAC model of contention scheduling, Botvinick and Plaut (2004) developed a simple recurrent network (SRN) model of the beverage preparation task. An SRN is a feedforward network augmented with a layer whose output is fed back into the system, so that its firing provides a state variable which depends on the history of the network. In this case, activation of hidden units is determined by the weighted sum of activations from input units, together with a weighted sum of recurrent activation from the hidden units on the previous time step. This recurrence allows the network to generate sequential outputs where the output on each cycle is a function of current input and the state of the hidden layer on the previous cycle.

Input to the model represents the current state of the environment: what, if anything, is held by the model's hand and what is the model fixating on. Two of these fixation units indicate the task instruction—whether the model is required to make coffee or tea. Output units represent the possible actions within a simulated world. Thus, one output node represents "fixate-mug," while another represents "pick-up." Given this approach to encoding action, if the model is to generate a sequence of actions to stir the contents of a mug, then it must perform a sequence of five actions: fixate on the spoon, pick it up, fixate on the mug, stir it, and then signal that stirring is complete. The network is trained by backpropagation through time (see appendix A). In an elaboration of the SRN model, the hidden/context layer is developed into a series of layers that learn increasingly abstract task representations. This is consistent with some current views of the PFC in which regions closer to the front of the brain operate on more abstract representations than regions closer to the or the output (i.e., motor cortex). This work implies that the SRN model should be broadly conceived of as frontally localized, but beyond this the SRN model makes no commitment to localization.

The SRN model has one great advantage over the IAC model—it acquires action sequence knowledge on the basis of experience with a training set. However, it takes 20,000 trials (more than 18 years at three cups a day!!) to learn to make tea and coffee. Behavioral data suggests that both the IAC and SRN models can account for basic facts relating to pathologies of action selection, but that the IAC model can account for a greater range of effects. Crucially, the IAC representation can be almost trivially modified from, for example, making coffee with a spoonful of instant coffee to making tea with a tea bag without involving an extended learning period. Indeed, one of the virtues of schema theory is that one can rapidly learn a skill that involves combining known schemas in a novel way (even if some practice is required to "automatize" it). Similarly, if one has learned to make tea with teabags in a box, it is trivial—for IAC, not for SRN—to adjust to a new box that opens in a somewhat different way. At the same time, one should note that the way the brain's neural networks support this flexibility remains a mystery—to me, at least (Arbib, 1988). We need to implement something like finite automata in neural nets to represent schemas. This means we need a working memory for the current state of each automaton. A new input both updates this state representation and triggers readout of the next output. The hidden layer of an SRN acts like a state representation, but the point missing in SRNs is the different timescale of perceptual and motor acts (fractions of a second to several seconds—and longer for hierarchical plans) and neural actuations. Chapter 11 suggests that mixed selectivity in neural networks allows representation of facts, memories, events, and the combinations that are critically important for enabling the PFC to subserve context- and rule-dependent flexible behavior. However, this still seems to leave open many questions about the activation, deactivation, and message passing between schemas in hierarchical assemblages.

Modeling Neurological Deficits Within a computational cognitive model a lesion may be simulated in a variety of ways. One approach in models that include numerical processes subject to noise is to increase the variance of that noise. An alternative for models that include multiple interacting subsystems is to impair the level or degree of interaction between subsystems, or just to remove a subsystem. Chapter 14 applies this methodology to a range of deficits.

If the schema-based IAC model is lesioned by reducing the top-down influence within the schema network, then the model exhibits utilization behavior. That is, hierarchically structured schema selection breaks down, and behavior is driven more by the representation of the environment than by intentions. However, the SRN cannot show utilization behavior that is a function of the presence of objects in the environment. The model has no concept of salience, and there is no way in which objects that are differentially salient might have differential effects on behavior.

Within the context of the IAC model, action disorganization associated with closed head injury affecting the frontal lobes is analyzed as following from nonspecific dysfunction of activation propagation between schema nodes within contention scheduling. It is further argued that this may be modeled though the addition of noise to the activations of nodes within the schema network.

Further analyses concern ideational apraxia, disorders of rate, and even effects reminiscent of the slowing in action selection associated with Parkinson's disease.

1.9.2 Synthetic Brain Imaging How might we use insights from detailed animal neurophysiology to "fill in the details" and provide more detailed models that can help us understand the results of human brain imaging? Chapter 15 introduces "Synthetic Brain Imaging" (SBI) as a means to exploit animal data to answer questions regarding the interpretation of human brain imaging data at both local and global levels. Spanning the levels from single-cell data to brain imaging depends on the assumption that a region in the animal brain has a homologue in the human brain. Alternatively, one may combine, for example, macaque neurophysiological insights with evolutionary hypotheses to derive neural network models of human brain function that can be tested against human brain imaging (see chapter 25 for further discussion). Insights may also be obtained from comparative fMRI studies of macaque and human.

As background, the chapter offers a comprehensive review of the relationship of activations obtained with PET and fMRI to physiological measures. The authors sometimes use the generic term *regional cerebral blood flow* (rCBF), whether they are analyzing fMRI or PET; on other occasions they use the term BOLD for the blood-oxygen-level dependent signal that is specific to fMRI. In both methods, the image is composed of three-dimensional *voxels* that reflect the signal in the corresponding cubes of brain on the order of several millimeters in each dimension. One estimate puts 630,000 neurons in a typical 3-mm^3 voxel. Both PET and fMRI are thought to reflect synaptic activity more than neuron spiking activity—both excitatory and inhibitory synaptic activity may increase PET and BOLD activation—and this observation is at the heart of SBI. This background also discusses experimental designs and analysis methods for imaging. *Effective connectivity* assesses "connectivity" between two brain regions by measuring correlations between the noisy signals obtained from each region during a task. Unlike *anatomical* connectivity, effective connectivity is *task dependent*. For example, inhibition from another region may block effective connectivity between two regions in one task yet be absent, thus allowing this effective connectivity, in another task. It is the goal of SBI to add models of the dynamic interactions within each brain region that support detailed information processing that subserves the observed task while explaining the PET or fMRI data. In some sense, then, it introduces "really causal" dynamic modeling to the study of brain imaging.

As the metabolic demands of inhibition require energy, inhibitory synaptic input might also be expected to increase blood flow. Most connections in the cerebral cortex are excitatory, and most processing takes place in local circuits with a high degree of excitatory recurrence. Afferents from other regions account for a relatively small proportion of the synapses. These characteristics mean a small afferent push into a local circuit is amplified within the circuit; in other words, there is a nonlinearly increasing response to input. The first model in chapter 15 suggests that, given a moderate to high recurrent excitatory local connectivity, inhibition is more likely to result in increases of rCBF/metabolism in a context where the region does not receive excitation from other sources. The model integrates absolute synaptic activity over the time course for the study, for example, approximately 1 minute for PET and about 1–2 seconds for fMRI. The simulation results suggest that when a region does not receive extensive excitation from other sources, and given the local circuit properties, synaptic activity from afferent inhibition can exceed the total reduction of local synaptic activity. The result is a rise in rCBF, even though spiking is reduced in local neurons. Further assessment of the model suggests that under conditions that are similar to those found in cortical circuits, it is possible that the net effect of inhibition on imaging results can depend in subtle ways on the nature of the tasks and type of modulation. A complementary study modeled the computational features of basal ganglia processing loops. Simulations of the caudate signal demonstrated that presynaptic inhibition could account for the reduced caudate signal in one contrast whereas postsynaptic inhibition accounted for an increase that was observed in putamen.

Animal Data and Neuronal Dynamics Cognitive tasks require multiple interacting brain regions, so adequate explanations need to include these global aspects. Tagamets and Horwitz (1998) modeled a *delayed match-to-sample* (DMS) task in a large-scale model that includes multiple regions of the brain. The goal of the model was to reproduce the electrical activity as recorded from animals, to generate outputs that indicate whether the stimuli match, and to generate synthetic imaging data similar to that seen in the human experiments. The model includes four regions along the *ventral* ("what") visual pathway: primary visual cortex (V1), visual association cortex (V4), inferotemporal cortex (IT/TEO), and the PFC. In the model, the PFC circuit performs the DMS task and was intended to capture the function of working memory with neural responses that mimic the dynamics of single cells found in monkey PFC during different task phases.

Connections from LGN to V1 and from V1 to V4 were arranged to implement selectivity for either horizontal or vertical lines, respectively. (Other types of feature detector were ignored.) In order for the model to achieve electrical response characteristics similar to those observed in monkey, with TEO/IT tuning and shift invariance for specific shapes, an unsupervised competitive Hebbian learning rule was used to train feedforward connections from area V4 to area TEO/IT. During training the stimuli were presented on the LGN input layer for 25 iterations, with a two-pixel shift of the entire stimulus in a random direction every 5 iterations. This procedure allowed the TEO/IT area to learn a shift invariant representation of the different stimuli. The trained network allowed the model to distinguish among different stimuli during the PET simulation of the match-to-sample task. A memory circuit allowed the model to maintain representations of stimuli during the delay period and to generate a response if the current stimulus matched the previous one. The model showed appropriate behaviors during each stage of a DMS task. The results suggested that the attentional component affects a particular type of neuron that responds both to a stimulus and during the delay which has been identified in delayed-response tasks in the monkey.

In order to simulate an imaging experiment, two tasks are compared: the one-back matching task and passive viewing. As before, the absolute values of all synaptic activities were summed separately for all the neurons in each region and integrated over time, separately for each task. This provides the quantitative data for matching imaging studies. The simulations produced quantitatively similar PET activation to the human data across the different regions.

Another study (Edin et al., 2007) addressed the developmental relationship between working memory maturation and neuronal activity. Differences due to developmental effects were examined by creating several different models of a child brain and a single model of an adult brain and comparing brain activations for each of the developmental effects of interest. Analogous experiments were run with groups of children and adults, and the modeled results were compared to the human data in order to determine which model best fitted the experimental data. Synaptic pruning was modeled by redistributing synapses without changing total strength either within or between regions. In effect, loss of synaptic strength due to removal of some synapses was offset by increasing the strength of the remaining synapses. Changes in myelination were modeled by a parameter that controlled the speed of transmission between the two regions of the model. Comparison of brain imaging data with the synthetic data suggested that a combination of synaptic strength increases and synaptic pruning best distinguished the adult from the child brain activations. Neither of these effects alone was sufficient, nor did changes in myelination reproduce the experimental data.

Behavioral and Cognitive States With this we turn to the relationship between imaging data and tasks, behaviors, cognitive states, and performance. This can include trying to explain empirical human imaging tasks from a cognitive perspective, or linking neurophysiological animal data to behavior, status, or performance in conditions that are comparable across species. When choosing regions to model, there may not be any appropriate physiological data to guide selection, as, for example, in language tasks. The model must nonetheless produce performance measures that can be related to human performance on the imaging task, which may be a challenging design problem.

This framework was exemplified by a model of how attention modulates regional responses as a function of specialization of object recognition in the "what" pathway and location and spatial relationships between objects in the "where" pathway (Corchs and Deco, 2004). The model was used to simulate fMRI data from a study of whether feature-based attention globally affects visual cortical responses to unattended stimuli that have overlapping feature with attended stimuli. It instantiated the biased competition hypothesis that attention modulates visual processing by enhancing features or locations of the attended stimulus and in concert reduces the suppressive interactions from neurons representing nearby distractors. (See chapter 4 for a model of related data.) Comparisons between the simulated and actual fMRI BOLD response were qualitative. This means the rank order of the response strength by conditions was the same but the simulated BOLD and actual BOLD were in arbitrary units and not subject to quantitative evaluation. The simulated BOLD agreed with the fMRI finding that feature-based attention can influence the visual cortical response outside the attended location.

Synthetic Imaging of Disease, Connectivity, and Data-Driven Learning Considerable effort has been invested in imaging of brain-related disorders, such as schizophrenia, autism, bipolar disorder, and dyslexia, just to name a few. These studies typically compare groups of patients with controls in order to identify brain regions or connections that differ between the groups in order to isolate function-specific abnormalities. Each disorder is associated with a particular set of symptoms, cognitive or psychological abnormalities, and patterns of brain activation or connectivity changes. Chapter 15 examines connectivity changes in schizophrenia to exemplify the relevance of SBI to examining disease processes. After introducing data-driven training methods for matching model parameters to imaging data, it shows how manipulating other model parameters can inform interpretations of imaging findings such as reduced interregional connectivity.

The relevant study used a large-scale model to examine factors that might underlie effective connectivity reductions in schizophrenia (Winder et al., 2007). This work had two primary goals: (1) to examine how local circuit properties affect functional and effective connectivity and (2) to examine potential mechanisms that underlie the reduced temporofrontal connectivity that is frequently seen in schizophrenia. The Tagamets and Horwitz (1998) model was extended by adding a second hemisphere and then an additional region, the hippocampus, to each hemisphere. The working memory circuit was also extended to include both "match" and "mismatch" units, which respond when a stimulus either matches or does not match the one held in memory. The task was adapted to address an fMRI imaging study of a one-back matching task using words and other strings of letters. Before using the model, connections from area V4 to area IT were trained on four-letter words using an unsupervised Hebbian learning rule so that these stimuli would be as "familiar" to the model as words are to humans.

One approach to explaining specific differences between tasks or groups in an fMRI study is to begin with a model of one condition that matches the experimental data for that condition and then examine model parameter changes that can explain the other condition. In order to match the modeled data to the experimental data, a supervised learning method was introduced, referred to as *gains learning*. During gains learning, the interregional path weights are rescaled, without changing their topography, until the modeled fMRI signals in all regions match data derived from an experimental study. Since gains learning modifies all individual connections between two regions by the same proportion (not necessarily a correct assumption), it preserves the topography and thus the coding properties among units in a region.

In order to examine possible mechanisms for differences between healthy controls and persons with schizophrenia, gains learning was applied to data from each of the two groups. The result was that the left-sided connection strength modeled between temporal cortex (TEO/IT) and frontal cortex (PFC) in schizophrenia was less than 30% of that in the controls. The other connections were largely similar to those of the controls. This is consistent with imaging studies. Further results with the model suggest that reduced recurrent connectivity in local prefrontal circuits may underlie the working memory deficits frequently found in schizophrenia. This result on seeking to model neural correlates of schizophrenia segues well into the discussion of "Brain Diseases" in chapter 23.

1.9.3 **Brain Diseases** Chapter 23 emphasizes that brain diseases can start at a molecular or cellular level and then produce effects all the way up across different temporal and spatial scales till reaching the systems level of the whole organism. Alternatively, disease may start at a high level (e.g., traumatic brain injury and stroke) and descend down to produce alterations at cellular and molecular levels. The chapter examines four diseases: epilepsy, Parkinson's disease, stroke, and schizophrenia. The emphasis is on multiscale modeling because of its potential clinical applicability through new approaches to developing pharmacological treatments.

Epilepsy The location of a localized (focal) epileptic seizure determines the brain function that is affected. Generalized seizures involve the whole brain and result in loss of consciousness. Usually, the abnormal firing during seizures will stop after a period of time, and so the symptoms will stop too.

Occurrence of seizures is related not only to increased excitability of individual neurons, but also to altered neuronal dynamics within the network. Multiscale modeling helps us understand how these many elements interact to produce the disorder. Because the presenting "behavior" (a convulsion) is extremely simple, the correlation between basic science and clinical phenomenology is closer in epilepsy than in other brain disorders.

The existence of widespread distinct oscillatory frequencies in the EEG (recall chapter 4, "Neural Rhythms") suggests that dynamical subsystems are detectable even at the highest scale. Hence, it has been proposed that the enormous dimensionality of the brain produces trajectories that lie in much lower-dimensional subspaces that can then be modeled by equivalent low-dimensional dynamical systems. Following this hypothesis, low-dimensional mean field (neuron population models or lumped models) have been widely utilized to simulate seizures. Seizures can spread in a series of stages: a focal seizure with secondary generalization.

One relevant model focused on the CA3 region of the hippocampus. It consisted of connected modules, with each module consisting of a lumped excitatory population and a lumped inhibitory population. The excitatory population activation was described by Hodgkin–Huxley-style equations which depended on calcium, potassium, and leak currents. Activity spread in the model took place at different timescales based on different mechanisms. On a fast timescale, the activity of a module projected to other modules. At a slower timescale, activity results in increased extracellular potassium level, which then diffused to nearby neuronal groups. A key parameter determines the duration before the activity is transmitted from one group to another. Exploring the range of its values revealed two points of bifurcation. Below a certain value, all of the modules in the network would fire together, producing a seizure. Above that value, the network entered a periodic mode, with the modules firing out of phase. This mode was spatially dependent, such that closer modules fired at closer phase than those farther away. At the second point of bifurcation, the network entered a mode of random firing. The modeling also explored the effects of inhibition on the firing synchronicity by a variable which governed the ratio of current going into excitatory and inhibitory modules.

Other models relate the spread of seizures to network connectivity. Cortex is suborganized into subnetworks or modules, with more connections within the module (intramodular) than between the modules (intermodular), suggestive of hierarchical connectivity. The connections at various levels are described in terms of *connectomes* at various levels, within the column, between columns (within area), between areas (within system), and between systems.

Another model consisted of a cortical module and a thalamic module. Both cortical and thalamic modules consisted of interconnected excitatory and inhibitory populations. System dynamics could produce either a physiological or a pathological seizure-like activation pattern. The transition from the normal nonseizure state to the seizure-like activity occurred when a threshold for an inhibitory synaptic activation threshold (on the excitatory population of either the cortical or thalamic modules) was reached, generating augmented inhibition and higher amplitude oscillations. Noise played an important role in switching between attractors in this type of model, suggesting why there might be limitations in the predictability of seizures.

Contrasting with the lumped models described above, more detailed neural-level models incorporate individual neurons but still must simplify in many different ways. For example, one simplified network consisting of excitatory, inhibitory, and driver neurons was used to investigate the transition from normal activity to a clonic seizure. The *driver neurons* drove the network through their spontaneous firing. Two patterns of firing were noted: low-amplitude activity, and prolonged continuous firing, suggestive of tonic epileptiform activity.

It was observed that altering the firing timing of the driving neurons, without changing any of the other parameters, resulted in switching between these two patterns. Crucially, it was possible to grossly model the effect of anti-epileptic medications, showing how blocking voltage-dependent sodium channels—simulated by reducing the excitability of the individual cells—removed the tonic seizure without affecting baseline activity.

Parkinson's Disease Parkinson's disease is one of several neurodegenerative disorders which affect a particular population of neurons whose degeneration deprives the rest of the brain of a needed projection or needed neuromodulator (recall chapter 7, "Neuromodulation"). In the case of Parkinson's disease, this population is the neurons of the SNc in the brainstem. These neurons release the neurotransmitter dopamine (DA) and project strongly to the basal ganglia. (Compare the treatment of the basal ganglia in chapter 10, "Reinforcement Learning," and chapter 22, "Integrative Functions of the Corticostriatal System.") Their degeneration results in reduction of the dopamine levels in the basal ganglia, producing a group of symptoms which are initially primarily motoric. The disease then progresses to affect nonmotoric brain function as well. The major symptoms include difficulty initiating movements, slowed movement (bradykinesia) with concomitant slowed thought (bradyphrenia), increased muscle tone around the joints (rigidity), resting tremors, and difficulty coordinating bodily movements (postural instability).

The initial treatment of Parkinson's disease involves exogenous replacement of missing DA. This can be done by either providing a DA precursor (L-dopa) which gets converted into DA in the brain, or providing agonist drugs to activate DA receptors. Initially, this therapy produces improvement of several aspects of the disease, for example, reduced rigidity, faster spontaneous movements, and reduced tremor. Interestingly, there is sometimes an initial dramatic response to replacement, with the patient reporting a seemingly miraculous return to normal function. This is impressive since the original phasic release of the DA has been replaced by a continuous tonic neurotransmitter bath. Unfortunately, later on, the symptoms will start appearing again despite the treatment, and then may be more severe than they otherwise would have been, perhaps because of downregulation of DA receptors.

The basal ganglia consists of different nuclei and is divided anatomically into striatum, the major input portal to the basal ganglia, and globus pallidus, the output route. The main cells in the striatum are GABAergic neurons—medium spiny projection neurons—that project to the globus pallidus interna (GPi) and substantia nigra pars reticulata (SNr) but not to SNc, the site of DA production. As we saw in figure 1.8 on the role of the basal ganglia in action selection, the basal ganglia has been conceptualized as instantiating Go and No-go circuits linked to two classes of striatal cells, with either D1 or D2 DA receptors. D1 cells in striatum (on the direct pathway which inhibits GPi/SNr) results in a Go signal while stimulation of D2 cells in the striatum (on the indirect pathway which disinhibits GPi/SNr through globus pallidus externa; GPe) results in a No-go signal: the *task* is less likely to be executed. Since D2 receptors have higher affinity for dopamine than Dl receptors, DA depletion in Parkinson's disease results in D2 receptors' being more active, which causes the indirect pathway (responsible for the No-go signal) to be more active.

As we saw in chapter 10, the basal ganglia is also involved in reinforcement learning for which variation in phasic release of DA serves as a reward prediction error. This is part of the reason why patients suffering from Parkinson's disease suffer from learning deficits and have impairment in tasks related to trial-by-trial error feedback.

Parkinson's disease starts at the cellular and molecular levels with degeneration of the neurons in the SNc, with the resultant reduction of DA levels at the basal ganglia. However, the symptoms appear at the behavioral level in the form of problems in motoric functions and in learning. Treatment takes place at a molecular level, the level of DA receptors, while the improvement of symptoms takes place at a behavioral level.

Chapter 23 shows how computational neuroscience can bridge these spatial and temporal scales. One model focuses on action selection and shows how reduction of DA (as in Parkinson's disease) would result in dysfunction because of limited ability to select responses to stimuli. High levels of DA (e.g., as a result of overtreatment using L-dopa or DA agonists) would also result in dysfunction because of inability to respond to a more salient stimulus presented after a less salient stimulus. Another computer model investigating the effects of dopamine depletion and repletion on both tonic and phasic dopamine inputs showed that Parkinson's disease networks exhibited more errors in two modeled tasks in comparison to the intact networks. Simulating an overmedicated network yielded more errors in a probabilistic reversal task during the part that required reversal of acquired learning. This happened because the high levels of DA suppressed the No-go units and so impaired the network's ability to learn No-go representations during negative feedback, important to allow for learning reversal. This model was able to replicate the phenomenon of patients suffering from Parkinson's disease being prone to different types of error even when medicated.

Stroke A stroke develops when neurons die following the cessation or reduction of blood supply to a brain region. Note that such a region is determined by the blood supply and has no relation to the extent of Brodmann areas. Nearby neurons show reduced metabolic and neural function and are sensitive to further insult either through further reduction in their metabolic supplies or through additional excessive excitation. This excessive excitation, termed excitotoxicity, can occur either through synaptic activation or through nonspecific stimulation of glutamate receptors by glutamate released into the extracellular space from other damaged or dead cells.

The effect of a stroke varies over time as the brain reorganizes, more or less effectively. The principles are akin to those whereby cortical representations of different body regions change in response to alterations of received stimulus characteristics. (Related phenomena and modeling are discussed in chapter 13, "Neural Maps: Their Function and Development.") Following a stroke, parts of a sensory cortical region may be missing, and it is hypothesized that changes in the receptive fields of surviving cells may play a role in recovery: the function of the cells that are missing may be replaced in part by local cells with similar or overlapping receptive fields. The alternative, but not mutually exclusive, hypothesis is that other remote areas of the brain take over functions once handled by the ischemic area.

A detailed model of the MT (middle temporal cortex, or V5) region, which is involved in representing the motion of different targets in the visual field, provides an interesting example. A one-layer recurrent model of MT received convergent inputs from a receptive layer representing lower visual areas (such as V1) and had lateral connections in the form of near excitatory and far inhibitory connections ("Mexican hat" distribution of connections, but with no self-excitation). Loss of the more distant inhibitory connections resulted in substantial increased activation (unmasking) such that receptive inputs that were previously unable to activate a given MT unit could now do so, having less inhibition to counter. Expansion and contraction in the receptive fields was shown to represent an expanding wave of dynamical consequences resulting from the loss of projections from the lost neurons. However, the contractions seen were much less than those found in the experimental animal model. When the histology of the ischemic regions was looked at more carefully, it was found that there is reduced expression of $GABA_A$ receptors in the vicinity of stroke, as well as increased excitability in these regions. This suggests the presence of a halo of disinhibition around the ablated area. When this halo was included in the model, it resulted in augmentation of both expansion and contraction of the receptive fields. Further modeling is described to account for the observation that overall receptive field changes appear to take place in two phases. The first phase involves the immediate changes in neuronal dynamics described in this model which were the immediate consequence of cell death. During the second phase, synaptic plasticity changes take place over the course of hours to days.

Computer models for the recovery process which takes place with physiotherapy provide clinical predictions. Subjects with a stroke affecting one hand will tend to use it, and therapy may constrain the good hand to encourage the patient's use of the impaired hand. My group (Han et al., 2008) developed a model of the recovery process which predicted a functional threshold, beyond which improved use of the affected limb would lead to greater spontaneous use. This would then provide a virtuous circle which would provide continued improvement without the need for further explicit rehabilitation strategies.

Schizophrenia While the prior diseases are considered neurological, schizophrenia is considered psychiatric. However, schizophrenia is in fact a brain disease, and any treatment will have to address problems in the brain as well as trying to ameliorate learned behaviors. Different patients with schizophrenia can present different symptoms and etiologies can include genetic susceptibility, intrauterine exposure to infections, and stressful life situations. Progress in understanding and treating schizophrenia has been slow, in large part because of this multifactorial nature. We saw one model relating schizophrenia to connectivity in chapter 15. Chapter 23 focuses on a set of theories concerning pathogenesis that builds on evidence from animal studies and neuropsychological testing of cognitive coordination—the view (hardly controversial for readers of this volume) that different regions and networks of the brain are able to perform their function through coordinating their activities within themselves and between each other. The hypothesis, then, is that symptoms of schizophrenia arise because of failure of the coordination between brain networks (cognitive discoordination).

Ketamine is an NMDA receptor antagonist that is used as an anesthetic. When a subanesthetic dose is given to nonpsychotic human subjects, they start experiencing hallucinations, delusions, and thought-process disturbances. When ketamine is given to patients with a history of schizophrenia, it exacerbates and worsens their symptoms. Ketamine also produces changes in brain oscillations. In humans, EEG changes have been recorded in healthy volunteers under subanesthetic doses of ketamine. The changes were increased power of gamma-band oscillations and decreased power of delta-band oscillations. Psychotic symptoms were found to be associated with similar changes in EEG power spectra. Based on this, subanesthetic doses of ketamine have been used to produce an animal model for schizophrenia (or psychosis in general). By an animal model we mean here an animal with schizophrenia-like symptoms, not a computational model of the animal.

The change in the power of different frequency bands could be one of the mechanisms responsible for generation of psychotic symptoms. Hippocampus and frontal cortex are two areas primarily affected in schizophrenia. To investigate how ketamine produces the observed changes in the power spectrum, a neural-level model of hippocampal circuitry was developed, consisting of pyramidal neurons, inhibitory fast-spiking basket cells, and inhibitory slow-spiking oriens lacunosum moleculare (OLM) cells. Theta drive from medial septum (MS) was simulated by providing periodic inhibitory input to OLM and basket cells. (See chapter 4 for other models of "Neural Rhythms.") To simulate the blocking by ketamine at a particular site, conductance of NMDA receptors was set to zero at that site. The simulations suggested a sequence of pathophysiological alterations that led from a reduction in activity at a particular receptor to spectral changes at the network level. Selective current injection into the OLM cell population was shown to recover the control oscillations by direct opposition at the primary pathological focus. Here a tonic activation was used to replace a missing periodic effect, suggestive of the situation in Parkinson's disease where tonic dopamine replaces the phasic release that has been lost. Such a prediction points to the role computational neuroscience can play in pointing out possible mechanisms for new treatments even in a complex illness such as schizophrenia.

1.10 A Regional Focus

Many chapters in this book look at models of the functions of one or more interconnected brain regions. Here we review chapters that look explicitly at three important brain regions—hippocampus, basal ganglia (including striatum), and cerebellum—and sample some of the functions they support in concert with key brain regions with which they are connected. These are "Hippocampus: From Spatial Processing to Episodic Memory," by Edmund T. Rolls (chapter 18), "Cerebellar Adaptation and Supervised Learning in Motor Control," by John Porrill and Paul Dean (chapter 21), and (complementing models of the basal ganglia considered above) "Integrative Functions of the Corticostriatal System," by Peter Ford Dominey and Pierre Enel (chapter 22).

1.10.1 Hippocampus: From Spatial Processing to Episodic Memory We earlier discussed the case of H.M., for whom removal of large sections of his temporal lobes, including the hippocampus, destroyed his ability to form new *episodic memories* but left unimpaired his *working memory* and his ability to form new *procedural memories*. Chapter 18 presents an extended computational theory due to Edmund T. Rolls and his colleagues of how the hippocampus operates in episodic memory. Episodic memory almost always has a spatial component, and there is a spatial representation in the hippocampus that can be updated by self-motion to produce path integration. The theory is based on the remarkable neural architecture of the hippocampus, on the effects of damage to it, and on neuronal activity recorded in it. Once memories have been stored in the hippocampus, they may later need to be recalled to the neocortex, and a theory of the recall mechanism is part of the overall theory. Once recalled to the neocortex, the memories of particular events or episodes can be reported verbally, and hence this is a type of declarative memory. The recalled information may also be combined with other information to be reorganized and stored semantically in the neocortex, that is, stored in a form that reflects meaning and structure in contrast to the episodic memories in the form captured as discrete memories by the hippocampus.

Here are some key facts:

1. Tasks that require objects to be associated with the place in which they are located are sensitive to hippocampal damage as is temporal order memory for a sequence of places or objects, a functionality that seems important in temporally linking a sequence of events within an episodic memory.

2. In rodents, *place cells*, which respond when a rat is near a particular place, are found in the hippocampus. In monkeys, there is a prominent representation of place as the location at which the primate is looking rather than the place where the monkey is located in the room. (Similarly, a human can remember where an object or person has been seen based on looking at the place even without having been at that place.)

3. In the medial entorhinal cortex, there are *grid cells*, which have regularly spaced peaks of firing in an environment, so that a single neuron increases then decreases its firing a number of times as the rat traverses the environment.

The primate hippocampus receives inputs, via the entorhinal cortex (BA 28) and the highly developed parahippocampal gyrus (areas TF and TH of temporal cortex) as well as the perirhinal cortex, from many processing streams of the cerebral association cortex, including the visual and auditory temporal lobe association cortical areas, the PFC, and the parietal cortex. Through its connections, the hippocampus is thus potentially able to associate object and spatial representations. In addition, the entorhinal cortex receives inputs from the amygdala, and the orbitofrontal cortex, which could provide reward/emotion-related information to the hippocampus

Internally, the hippocampus links the dentate gyrus (DG) with CA1, CA2, and CA3 (the so-called cornus ammonis regions). The primary output from the hippocampus to neocortex originates in CA1 and projects to subiculum, entorhinal cortex, and parahippocampal structures (areas TF–TH) as well as PFC. These are the pathways that are likely to be involved in the recall of information from the hippocampus back to the rest of the neocortex.

The theory of the operation of hippocampal circuitry as a memory system rests on the general theory of associator networks (see appendix B for an introduction).

The DG is modeled as performing pattern separation by competitive learning to produce sparse representations, producing, for example, neurons with place-like fields from entorhinal cortex grid cells.

The hippocampal CA3 system is modeled as a single attractor or autoassociation network to enable formation of rapid, one-trial associations between inputs originating from very different parts of the cerebral cortex—such as associations between information originating in the temporal lobe visual cortex about the presence of an object and information about spatial location (place in rodents, or spatial view in primates) of an object and/or reward. Moreover, the network can complete the whole memory during recall from a part of it. The theory is extended to associations between time and object or reward to implement temporal order memory, also important in episodic memory.

Each event might be stored in as little as 1 second on the basis of a vector of neurons firing at the inputs to the hippocampus. An episodic memory might be a single event or a sequence of events. Sparse connectivity in autoassociation networks in the brain, for example those in CA3 and those implemented by recurrent collaterals in the neocortex, may be advantageous by reducing the probability of multiple connections between any pair of neurons, which, if present, distort and considerably reduce the number of memories that can be stored (recall the discussion of storage capacity in chapter 11, "Short-Term, Long-Term, and Working Memory").

Both heterosynaptic LTD (in which a synapse decreases in strength when the presynaptic term is low and the postsynaptic term is high) as well as LTP (an increase in synaptic strength when both the pre- and postsynaptic terms are high) are required for effective learning (see chapter 6, "Hebbian Plasticity and Learning").

Evidence that the CA3 system may not be required during recall in a reference memory (previously learned) spatial task is that CA3 lesioned rats are not impaired during recall of a previously learned *water maze task* (not really a maze—the rat must learn how to use cues around a tank of opaque water to locate a platform beneath the water on which it can stand). However, if completion from an incomplete cue is needed, then CA3 NMDA receptors are necessary even in a reference memory task. Thus, the CA3 system appears to be especially needed in rapid, one-trial object-place recall and when completion from an incomplete cue is required.

A model using a discrete sparse distributed representation with graded firing rates to encode and store representations of objects showed that attractor networks can store both continuous patterns and discrete patterns. Such "hybrid" attractor networks can thus be used to store, for example, the location in (continuous, physical) space (e.g., the place "out there" in a room represented by spatial view cells) where an object (a discrete item) is present. (Recall the line attractors from chapter 11).

Acetylcholine turns down the efficacy of the recurrent collateral synapses between CA3 neurons. Thus, cholinergic modulation (see chapter 7, "Neuromodulation") may help to allow external inputs rather than the internal recurrent collateral inputs to dominate the firing of the CA3 neurons during learning, as the current theory proposes. If cholinergic activation at the same time facilitated LTP in the recurrent collaterals (as it appears to in the neocortex), then cholinergic activation could have a useful double role in facilitating new learning at times of behavioral activation and emotional arousal, when presumably it may be particularly relevant to allocate some of the limited memory capacity to new memories. Acetylcholine may also facilitate memory storage (vs. recall) by enhancing firing in dentate granule cells. Moreover, by acting as a competitive network, the DG would perform pattern separation, enabling the hippocampus to store different memories of even similar events.

The entorhinal cortex contains grid cells, which have high firing in a rat in a two-dimensional spatial grid as the rat traverses an environment, with larger grid spacings in the ventral entorhinal cortex (Hafting et al., 2005). This may be a system optimized for path integration which may self-organize during locomotion with longer time constants producing more widely spaced grids in the ventral entorhinal cortex. How are the grid cell representations, which would not be suitable for association of an object or reward with a place to form an episodic memory, transformed into a place representation that would be appropriate for this type of episodic memory? Computational approaches to this system model it as a set of linked-ring continuous attractors (i.e., line attractors but with the line linked back to form a circle to yield the periodic nature of the grid). The notion is that as the rat locomotes, the peak of the firing in the continuous attractor moves across the grid cells, and after a certain distance has been traversed, the peak returns to the same set of neurons, completing the ring. (I must confess that I do not [yet?] understand why there would be an evolutionary advantage in interposing diverse periodicities before the place cells.)

The CA3 cells connect to the CA1 cells by the Schaeffer collateral synapses. The associative modifiability in this connection helps the full information present in CA3 to be retrieved in the CA1 neurons. Part of the hypothesis is that the different subparts of an episodic memory, which must be represented separately in CA3 to allow for completion, can be combined together by competitive learning in CA1 to produce an efficient retrieval representation for the recall via the backprojection pathways to the neocortex of memories stored in the neocortex. It is suggested that the modifiable connections from the CA3 neurons to the CA1 neurons allow the whole episode in CA3 to be produced in CA1. The CA1 neurons would then activate, via their termination in the deep layers of the entorhinal cortex, at least the pyramidal cells in the deep layers of the entorhinal cortex. These entorhinal cortex layer V neurons would then, by virtue of their projections back to the parts of cerebral cortex that originally provided the inputs to the hippocampus, mediate recall of the episode.

The theory by McClelland et al. (1995) described in chapter 6 is similar to the theory described above, except that it holds that the last set of synapses that are modified rapidly during the learning of each episode are those between the CA3 and the CA1 pyramidal cells. That theory also emphasizes that the hippocampal and neocortical memory systems may be quite different, with the hippocampus specialized for the rapid ("one-shot") learning of unstructured single events or episodes, and the neocortex for the slower learning of semantic representations (structured representations in which the components are linked) which may necessarily benefit from the many exemplars needed to shape the semantic representation, a process that is helped by the recall of episodic memories from the hippocampus.

Chapter 18 also offers some general ideas on how to model the temporal order memory in the hippocampus that may underlie episodic memory. However, since consolidated episodic memories can survive in cortex in the absence of hippocampus, one must further address the issue of how such consolidation preserves temporal relations (not absolute timing but the ability to recall that one event may have occurred some minutes, hours, or even longer before another).

Finally, the chapter provides a brief account of navigation and the hippocampus. It is noted that a model of navigation based on place cells in rodents could not fully model its role in primate navigation in view of the presence of spatial view cells in primates. However, a shared property is that hippocampal place cells in rodents and spatial view cells in primates can be updated by idiothetic (self-motion) inputs. This is akin to the dynamic remapping of hand position in the MNS2 model of chapter 20 ("Reach and Grasp: Control, Development, and Recognition") and of saccade targets in chapter 22 ("Integrative Functions of the Corticostriatal System"). The chapter closes with a model of goal-directed navigation based on forward linear look-ahead probes of trajectories in a network of head direction cells, grid cells, place cells, and PFC cells (Erdem and Hasselmo, 2012). An alternative model (not in the chapter) links an affordance-based model of locomotion to the World Graph model (presumably stored in PFC) of the paths linking significant places in the environment (Arbib and Bonaiuto, 2012; Guazzelli et al., 1998). Different paths can be generated depending on the motivational state of the animal, employing a variant of temporal difference learning (chapter 10) called spatial difference learning.

1.10.2 Integrative Functions of the Corticostriatal System We have already examined the push–pull relation between the direct and indirect paths through the basal ganglia (see figure 1.8) in chapter 10 ("Reinforcement Learning") and chapter 23 ("Brain Diseases"). Chapter 22 introduces a range of models of the loops linking cortex and basal ganglia. There are several different corticostriatal loops in addition to the oculomotor loop. The dorsolateral prefrontal circuit is associated with high cognitive function, working memory, and cognitive control. The lateral orbitofrontal circuit is concerned in aspects of motivation, emotion, and decision-making (Bechara et al., 2000). However, the notion of segregation is a simplification, and information from different input sources can overlap. For example, work on the monkey has demonstrated overlapping projections from cognitive and reward-related cortical areas in the rostral and dorsal striatum which may provide the basis for incentive-based learning.

The principal influence of the basal ganglia on the thalamocortical system is to inhibit the system and then selectively, under the right conditions as indicated by cortex, to release that inhibition. In the domain of motor control, this will allow action selection based on the behavioral context that is defined by cortical activity and its projection to the striatum and will also implement working memory via corticothalamocortical reciprocal interaction. The system is inherently adaptive so that through learning, different behavioral contexts can lead to the selection of appropriate action. The primary mechanism for plasticity in the corticostriatal system is at the level of the corticostriatal synapse, where the influence of dopamine can produce long-term synaptic changes. As we have seen (chapter 7, "Neuromodulation"; chapter 10, "Reinforcement Learning") release of dopamine at the corticostriatal synapse is intricately linked to reward. Thus, corticostriatal plasticity, under the control of reward-related dopamine, should provide a major component in the adaptive behavioral repertoire of the primate.

Working Memory and Dynamic Remapping in the Saccade System Chapter 19 reviewed the basic mechanisms in brainstem and superior colliculus involved in eye movements. Chapter 22 reviews data on the neurophysiology of cerebral cortex and basal ganglia that relate to extending the cognitive control of eye movements. The oculomotor loop has a region of striatum known as the caudate as its input nucleus linking to the SNr as its output nucleus. Crucially, the SNr provides an inhibitory projection to superior colliculus which can block or reduce superior colliculus triggering of saccades. However, activity in striatum can selectively inhibit loci in SNr which as a result will disinhibit the topographically related area of superior colliculus.

The first model explored in chapter 22 explains how all these brain regions together with the frontal eye fields (FEF) of the frontal cortex and lateral intraparietal sulcus (LIP) in parietal cortex can act together to produce three kinds of saccade: *simple saccades*, where the saccade to a target will be inhibited during the presence of a fixation point; *delay saccades*, in which a saccade is generated to a target presented at some time within fixation period, but the saccade is initiated only after the fixation point is removed (this requires a model of *working memory*); and *double saccades*, in which two targets are presented in order during the fixation period, and the monkey saccades to these (no longer visible) targets in the same order once the fixation point is terminated. The simulations were performed using leaky integrator neurons.

Working memory is modeled via a memory loop between FEF and thalamus, supporting sustained memory-related responses in the FEF, caudate, SNr, and thalamus, as observed experimentally. Thus, we see an illustration of corticostriatal function for action selection and working memory. The basal ganglia can disinhibit the memory loop to allow a target to be maintained in working memory and can restore inhibition when a target has been acquired and thus memory of it need no longer be maintained.

When a target is visible, it is its position on the retina (its retinotopic code, serving as a motor command or "motor error") that informs the superior colliculus of the size and direction of the saccade to be made (compare chapter 19). During a saccade to one target, the position of any other target on the retina shifts in the equal and opposite direction. The subtlety of the double saccade is that the saccades occur when the targets are no longer visible. Thus, after the saccade to the first target, A, the initial retinal representation of the second target, B, no longer corresponds to what the retinotopic code for B would have become were B still visible. The model thus provides a *dynamic remapping* process. Neural activity reflecting this dynamic remapping has been observed in the superior colliculus of the cat and macaque and in the FEF and parietal cortex of the macaque. The model updates the retinotopic code for B in LIP during the first saccade and makes this code available to superior colliculus to control the second saccade. Lateral connections in a two-dimensional layer of neurons are modulated by corollary discharge from the superior colliculus command for the saccade, thus compensating for that movement.

Plasticity and Integrative Functions of the Corticostriatal System An extension of the previous model demonstrates the learning of conditional visual discrimination tasks in which the use of a specific color/shape cue as fixation point can be associated with choosing to saccade specifically to the left or right of two targets.

The model codes aspects of the *what* (ventral) and *where* (dorsal) visual pathways. In the *where* pathway, retinal topography that corresponds to motor error for targets is preserved throughout. In the *what* pathway, shape and color are represented as feature vectors in V4 and IT. The key hypothesis is that corticostriatal projections from IT to caudate are modifiable by reward-related dopamine from SNc. The model again uses leaky integrator neurons that model the membrane potential of the neurons. The system employs a simplified form of reinforcement learning in which (as distinct from the temporal difference learning of chapter 10), immediate reward, coded as the dopamine signal, is either offered (positive reinforcement) or denied (negative reinforcement) at each trial. Suppose that a blue square as fixation point is the cue for a saccade to the left of two targets when the fixation point is terminated. Then the learning scheme will (1) strengthen all synapses that link units in IT that are active in coding a blue square to all those units in caudate that are active in disinhibiting a leftward, rewarded saccade (Hebbian learning) but (2) will weaken all such synapses if the action is rightward and thus unrewarded (anti-Hebbian learning). The simulation shows that this cumulatively changes the IT → caudate synapses to yield consistently correct choices for each color/shape cue.

An extension of this approach provides a model of how the corticostriatal system can contribute to the sequential organization of behavior. The starting point is the observation that (as seen in the macaque) PFC has an abundance of local recurrent connectivity. Indeed, such local recurrent connectivity is a general property of cortex. This was exploited to model sequence learning as follows: PFC is modeled as a network of recurrently connected leaky integrator neurons which encode sequence states. In this case, it is shown how learning can associate a cue that is to be associated with a specific sequence A-B-C ... with a distinctive initial state x1 from which PFC will (without any special training) pass through successive states x2, x3, x4, ... Reinforcement of connections from PFC to the caudate can then reward those cases in which x2 stimulates caudate to disinhibit the desired action A, action x3 stimulates caudate to disinhibit B, and action x3 stimulates caudate to disinhibit C, ... An extension of this "temporal recurrent network" model to an "abstract recurrent network" model showed its applicability to abstract sequence processing (i.e., learning what sequences ABCBAC and DEFEDF have in common). This work, in turn, led to the development of a model of corticostriatal function in language processing in which a dorsal cortical system including pars opercularis would contribute to abstract structure processing, and a ventral system including pars triangularis would contribute to the integration of semantic content in language processing (a brief account of this language model is given in chapter 24, "Language Processing").

A powerful sequence-learning capability can thus be provided by simulating PFC as a population of neurons linked by fixed recurrent connections, with corticostriatal connections modifiable by a reward-related dopamine signal. Sequence context is encoded in the recurrent network, and this context is read out in the caudate. There is a deep relation between this view of the corticostriatal system and a domain of machine learning referred to as *reservoir computing*—we saw another use of such a strategy in chapter 11. Reservoir computing is characterized by two central notions. The first is that a network of computing units (rate coding or spiking neurons) that are connected with fixed recurrent connections can generate a form of dynamical system that encodes the history of its inputs. The second notion is that the contents of this representation can be "read out" by neurons connected to the reservoir units by modifiable synapses, in order to learn the appropriate transformation from the input signal to the output signal.

Dopamine Function and System-Level Models of the Corticostriatal System The reinforcement learning paradigm has (as we saw in chapter 10) provided insight into the function of dopamine as "prediction error" in the context of the temporal-difference learning paradigm which addresses the issue of .predicting expected reinforcement when actual (positive or negative) reinforcement is intermittent.

Gurney et al. (2001) contest the traditional view of the direct (striatum–SNr–thalamus) and indirect (striatum–GPe–STN–SNr–thalamus) pathways in the basal ganglia, in which the direct pathway allows cortical inputs to yield a net excitatory effect on thalamus, and the indirect pathway a net inhibitory effect on thalamus (via SNr). They instead propose a new configuration with distinct pathways for action selection and control. Prescott et al. (2006) used an extended version of this model in an embodied robotics context. (For diverse approaches to the linkage of computational neuroscience and robotics, see chapter 8, "Motor Pattern Generation"; chapter 16, "Embodied Models and Neurorobotics"; and chapter 24, "Language Processing.")

Frank et al. (2004) propose a model in which the direct Go pathway is responsible for action execution via disinhibition of the thalamus, and the indirect No-go pathway inhibits action via the inhibition of the thalamus due to the inclusion of an additional inhibitory component (the GPe). They exploit the respective excitatory and inhibitory effects of dopamine on D1 and D2 receptors in the direct and indirect pathway. A subsequent model reexamines this model in the context of new data on the role of the STN in a "hyperdirect" pathway.

Despite the diversity of these and other models, chapter 22 concludes that there is a baseline set of features that underlies most of these approaches. First, dopamine plays a crucial role in adaptive behavior, including its role in plasticity at the corticostriatal synapse and its role as a neuromodulator (see chapter 7). Second, disinhibition is important, with SNr/GPi holding a tonic inhibition over superior colliculus in the oculomotor loop, and over, for example, motor cortex via thalamus in the skeletomotor loop that is selectively released based on specific cortical input to the striatum. Third, the notion of the role of the direct and indirect pathways is less agreed upon across models. Fourth, a number of models exploit the notion of using recurrence for working memory and/or sequence context, either at the cortical level and/or the cortex–basal ganglia–thalamus–cortex loop. An important development is the extension of such models from simple action selection to higher cognitive function including learning in tasks that are considered to require executive function, setting internal contexts for making future choices.

Emphasizing neurons neglected in the above models, Graybiel et al. (1994) documented how corticostriatal projections from a given node of sensorimotor cortex form a distributed set of projection modules in the striatum and suggested that processing in these distributed modules can be coordinated under the control of tonically active neurons (TANs) in the striatum. Graybiel's group has recently demonstrated that TANs in the ventral striatum provide a bidirectional output signal during learning—increasing firing for rewarded trials and decreasing for unrewarded trials during acquisition and retaining only the negative responses after learning (Atallah et al., 2014). Another issue is for the modeler to explore the potential of "loop overlap," the divergence of topographically related areas onto distributed striatal modules and how it might favor some associations (easy to exploit) whereas others will be weak and thus exploitable only through extensive training.

1.10.3 Cerebellar Adaptation and Supervised Learning in Motor Control The evolution of cerebellar capacity in primates seems to have tracked that of the cerebral cortex, indicating that the cerebellum must supply some essential computational features that are not duplicated elsewhere. The cerebellum seems to allow high-level control systems to issue simple task-related commands that initiate smooth, coordinated series of movements. Current computational theories in which the cerebellum implements high-level modules such as *forward and inverse models* (see chapter 9, "Computational Models in Motor Control") are extensions of this basic idea. The cerebellum combines a cerebellar cortex with the deep cerebellar nuclei, but the phylogenetically oldest pathway travels from cerebellar cortex via the vestibular nuclei located in the brainstem. In chapter 21, the term "cerebellum" is in most instances used as a synonym for cerebellar cortex. This cortex consists of repeating microcircuits arranged in a "quasi-crystalline" structure. It is three layered, and consists of the granular layer, the Purkinje cell (PC) layer, and the molecular layer. The mossy fibers excite small neurons in the granular layer called granule cells. These neurons take input from four to five mossy fibers. There is an enormous number of granule cells (more than 10^{11} in humans). Axons from granule cells rise up through the PC layer, first making excitatory synaptic contact with nearby PCs and then bifurcating into parallel fibers (PFs) which extend laterally in the molecular layer up to a few millimeters along the folia making further excitatory contacts with the dendrites of the PCs over which they pass (figure 1.10 suggests the complexity of each PC). Conjoint activity along the PFs can elicit so-called "simple spikes" on the PC axon. Each PC makes approximately 150,000 synaptic contacts with passing parallel fibers, and each fiber will contact about 700 PCs. PCs also receive excitation from one (yes, just one) climbing fiber (CF) emerging from a cell in a region called the inferior olive, whose firing causes obligatory firing (a complex spike) of the PC. Each innervates tens of PCs, but each individual PC only takes input from a single climbing fiber, which fires only a few times per second (clearly needing a spiking model).

PCs are inhibitory, provide the only output of the cerebellar cortex, and project to a cerebellar or vestibular nucleus. PFs also synapse with inhibitory interneurons which then synapse on PCs to provide an alternative (inhibition of inhibition) effect of mossy fibers on the cerebellar nuclei. Thus, the only way the cerebellar cortex can affect the rest of the brain, and thus behavior, is by "inhibitory sculpting" of activity in the nuclei. Cerebellar cortex can be subdivided into smaller scale zonal stripes, which have recently been shown to have distinct molecular identity and specific afferent inputs, and then into *microzones*, which are small areas of cortex with specific climbing fiber connectivity and output targets, believed to be associated with specific functions. These are the units of analysis in chapter 21.

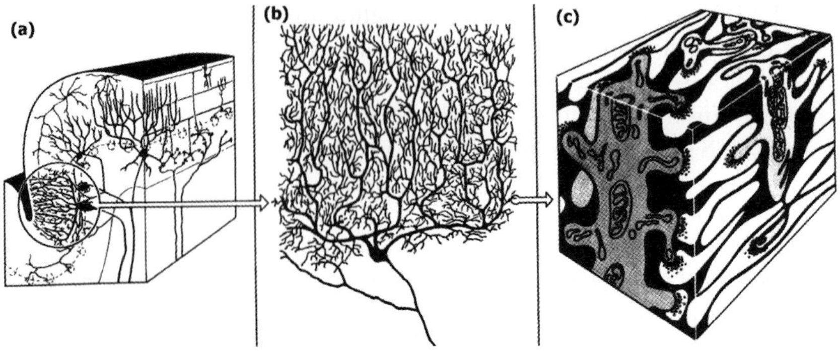

FIGURE 1.10 (a) A small piece of cerebellum. (b) A Purkinje cell. (c) Just a few of the synapses of parallel fibers on a branch of the Purkinje cell dendritic tree.

Chapter 21 notes some of the features that stimulate study of the cerebellum, including the dramatic morphology of the PC, the exceptionally large number of PF inputs, the unusual CF input and 1–1 firing of complex spikes it causes, the exceptionally low rate of CF activity, the organization into microzones, and more. But perhaps the greatest stimulus to cerebellar modeling was the book *The Cerebellum as a Neuronal Machine* (Eccles et al., 1967), which did not itself offer computational models but provided the data in a way that became the touchstone for various initial attempts at modeling at that time. It was shortly followed by two papers that together provided the core idea for considering the cerebellum as a learning machine. David Marr and James Albus independently proposed a similar mechanism for this, usually referred to as the Marr–Albus model (Albus, 1971; Marr, 1969). Basically, Marr proposed that conjoint firing of a PF from a granule cell and the CF from the inferior olive onto the same PC strengthens the synapse between the PF and the PC, while Albus proposed that the synapse is weakened. This inspired experimental work by Masao Ito and his colleagues (Ito et al., 1982) that established the Albus version (LTD) as the predominant effect.

Chapter 21 rings the changes on one powerful reformulation of the basic Marr–Albus framework model, the *adaptive filter model*, which emphasizes temporal processing because it is "a good example, given our present level of understanding, of what a really good computational model of a brain region might look like." The adaptive filter model has four basic assumptions:

1. The granule cell layer provides an efficient recoding of the input signal.
2. The PC acts as a synthesis filter, and its informative output is carried by simple spike activity.
3. The purpose of the climbing fiber input is to supply a teaching signal: simultaneous PF and CF activity supports LTD of synapses.
4. Performance on a broad range of tasks can be improved by plugging in the microzone considered as a "cerebellar chip" when supplied with the relevant inputs on the mossy fibers, an appropriate teaching signal on the CFs, and with its output connected to an appropriate target nucleus.

To cope with environmental variations such as varying loads, and internal changes such as fatigue or damage, motor systems must be able to adapt. The CF teaching signal hypothesis gives the cerebellum a biologically plausible signal to drive supervised learning. Chapter 21 develops several examples of such learning:

Using the PC for Adaptive Synthesis: When the PFs carry a bank of signals which are responses to a trigger signal presented at time t = 0 carried on the mossy fibers, a PC can be trained to produce a given output profile. To allow PFs to convey negative signal values, it is assumed that signals are rate coded by modulating a tonic firing rate, allowing both positive and negative signals to be transmitted. The PF inputs are assumed to convey information about a set of basis functions with the PC learning the appropriate weights to combine them in its firing pattern. The key challenge met by the model is to find a suitable teaching signal for the CF. In this model, the difference between actual and desired PC output is used. With this learning rule the filter quickly learns to synthesize a good approximation to the required profile. Despite its simplicity, this example does demonstrate that the ability to combine a large number of inputs with adaptable weights conveys great computational power.

Using the difference between actual and desired PC output as the error seems hard to reconcile with the low spiking rate of the CF. The authors' suggestion is that an event-triggered average of CF activity over many trials could provide sufficient high-frequency content in the teaching signal.

Covariance Learning: A first covariance rule derived from gradient descent is a batch learning rule. As such it requires the storage of long sequences of data before the expected value can be calculated, making it unsuitable for biological learning systems. A more plausible continuous time learning rule is obtained by dividing the long batch of data into small intervals and then taking the limit to give a continuous covariance learning rule. An appropriate combination of LTD and LTP is summarized as an STDP (recall chapter 6) profile in which a strong LTD dip for nearly coincident spikes is surrounded by weak LTP side lobes. It is noted that although there is a tendency to associate STDP with precise spike timing, its role as a mechanism for the implementation of correlational learning rules for rate coded signals is probably just as important.

Another issue is that the teaching signal is not instantaneously available for learning at the relevant synapses. The transformation from motor error to CF signal can be subject to large transmission delays. Since we cannot advance the CF signal, we must delay the effect of the PF signal in the learning rule. A filter which affects such a delay is called an *eligibility trace,* and it can be incorporated into the STDP profile by situating the LTD dip at an interspike time corresponding to the transmission delay.

Noise Cancellation: The next example addresses noise cancellation to highlight the ability of the adaptive filter to synthesize outputs from more general context signals than simple triggers. It uses the outputs from a randomly connected recurrent inhibitory network or "reservoir"—but this reservoir is proposed to be provided by the granular layer of cerebellum (Yamazaki and Tanaka, 2005) rather than the PFC (as was the case in chapter 11, "Short-term, Long-term, and Working Memory," and chapter 22, "Integrative Functions of the Corticostriatal System"). Since there is plasticity in the granular layer, it is suggested that this may be aimed at producing optimal, uncorrelated, basis functions which can speed learning.

Adaptive noise cancellation can be regarded as learning a forward model of the noise channel. The model exploits the basic assumption that the noise signals and hence also the error, which is a combination of these noise signals, are independent of the signal of interest. The external wiring of the "cerebellar chip" required for this noise cancellation task is thus as follows: (1) PFs carry the noise sources which are the context for this task, (2) PC output goes to a comparator to cancel the noise in the signal, and (3) a negative copy of the comparator output is used as teaching signal on the CF. The context noise signals are supplied on PF inputs to the apical dendrites of the PC, but the noisy signal is also supplied as an input on basal dendrites. The comparator stage can thus occur within the cell, with neuronal output as the teaching signal, eliminating the need for a CF. The required learning rule is homosynaptic and anti-Hebbian. Simulation shows that learning can take place in real time. This allows the filter to follow slow changes in the amplitude and phase of the contaminating signal if the learning rate is well chosen. This characteristic of the adaptive filter lends it much of its computational power.

The Oculomotor System: The function of the vestibulo-ocular reflex (VOR) is to maintain a stable retinal image by counter-rotating the eyes in response to movements of the head signaled by the vestibular system. Since the eye movement produced by the reflex has no effect on the input vestibular signal, this reflex must operate in feedforward or open-loop mode. Long-term calibration is therefore required to ensure accuracy. Two main pathways contribute to the reflex: the first is a three-neuron arc via the brainstem which causes vestibular signals to produce oppositely directed head movements; the second is a side path via the floccular region of the cerebellum. The task of the cerebellar side path is to learn to compensate for the limitations in brainstem processing. An inadequate VOR will result in the eye moving with the head relative to the world, and this produces movement of the image across the retina, called retinal slip, and there is evidence that the CF carries retinal slip information. Chapter 21 shows how the adaptive filter model can be deployed to provide suitable adaptability. Note that the cerebellum does not here learn a complete inverse model, but only a partial inverse model which supplements the processing already available in the brain-stem path.

The final section of chapter 21 is concerned with how current and future research can close the gap between the (possibly beautiful) abstraction of a mathematically secure system-level model, and the (possibly messy) details of actual neuronal chemistry, firing, and connectivity. Two strategies are employed: (1) to determine how far the model is consistent with experimental evidence concerning the cerebellar microcircuit itself and (2) to assess whether the model is consistent with experimental evidence concerning how the external connectivity of the cerebellum underpins behavior. Clearly, these strategies are of general applicability throughout computational neuroscience, and there are examples of each of them (whether or not their role is made so explicit) throughout the book. For the cerebellum, these strategies are applied to microcircuit properties (e.g., why are there so many granule cells?) and grouped into three categories: those that are explained, those that are not explained but may be irrelevant, and those which are not explained by the adaptive filter model and may indeed pose a problem for the future success of the model as a neurobiological explanation of the cerebellum. The last category includes recent data on granular-layer processing and PC firing.

Different parts of the cerebellum have unique connectivity, in contrast to the generally uniform structure of the cerebellar microcircuit itself. Each zone also projects to its own target region in the deep cerebellar or vestibular nuclei, each of which projects both to a specific set of target structures in the rest of the brain and spinal cord and to a specific region of the inferior olive. It has been shown that this olivary region is identical to the region that projects to the original cortical zone, so that a loop is formed suggesting a functional subunit. Such consideration led Ito as an experimentalist and my own group as modelers to place the emphasis on the *microcomplex*—which integrates microzones of cerebellar cortex with patches of cerebellar nuclei and the inferior olive with which they are integrated, linking each to a specific MPG, whether located upstream in cortex or downstream in spinal cord, as in chapter 8. In one study (Arbib et al., 1995), we modeled the role of the cerebellum in prism adaptation. This involves coordination as well as adaptation. The notion was that (1) a microcomplex for a single MPG can (by PC inhibitory sculpting of the output of the cerebellar nucleus) provide the adaptation to tune the parameters of the MPG to provide more effective feedforward for the metrics and that (2) moreover, because PFs are relatively long compared to the width of a microcomplex, PFs may be shared between MPGs and so adaptation of the PF–PC synapses of these shared fibers may respond to CF signaling of errors resulting from poor coordination between the MPGs and reduce those errors accordingly.

1.11 Language

We conclude with two chapters that discuss language, the core of the distinctively human mode of cognition. Angelo Cangelosi discusses "Language Processing" (chapter 24) primarily in terms of adaptive artificial ("connectionist") networks, but with links to brain and behavior. The closing chapter offers a coda to the whole book by showing how many of its themes come together in the ongoing attempt to offer an integrative perspective on "Evolving the Language-Ready Brain" (chapter 25). Both chapters investigate (to varying degrees) the simultaneous consideration of sensorimotor processes, abstract symbol handling, and social knowledge.

1.11.1 Language Processing Much (by no means all) of chapter 24 adopts a connectionist approach (see appendix A) using networks of adaptive units which, though called "neurons," need have no direct relation to specific neurons or groups of neurons in the human brain. The key question for the chapter is whether learning in such a network can more or less adequately yield the input–output behavior of one of the many processes relevant to language. One of the network structures considered here is the SRN (recall the SRN model of a neuropsychological syndrome in chapter 14). This is a standard three-layer feedforward perceptron with an additional set of units called a "memory" or "context" layer connected to form a loop. These memory nodes copy the state of activation of the hidden layer nodes at time step t and then send it back as input to the same hidden layer together with the new input stimuli at time $t + 1$. This is then used to predict the next output at time step $t + 2$. Other connectionist approaches use evolutionary algorithms applied to neural networks, or Kohonen SOMs (considered at some length in chapter 13, "Neural Maps: Their Function and Development"). The chapter is enriched by embodied and neurorobotic models of action and language integration (compare chapter 8, "Motor Pattern Generation," and chapter 16, "Embodied Models and Neurorobotics"). The chapter offers a brief review of the neuroanatomy of language processing, and shows how the integration of connectionist models into hybrid, neuroanatomically constrained modular architectures is making it possible to simulate neural processing data such as ERP and fMRI using something like the "Synthetic Brain Imaging" of chapter 15 but without neurophysiological constraints on the underlying networks.

Connectionist Models of Language Learning The English past tense of verbs can take the regular morphological form "add *–ed* to the stem" (e.g., love → loved; wonder → wondered) or be irregular (e.g., go → went; do → did; wear → wore). A classic feedforward model was trained to learn *all* these forms simply by adaptation of synaptic weights with no explicit appeal to the rule for regular verbs. This model was used by Rumelhart and McClelland (1986) to reject the dual-process theory of language morphology learning (i.e., plug in the general rule unless you know the verb is irregular), and the model had the additional merit that for many nonwords similar in form to irregular verbs, the neural network would indeed form a similar irregular form as would a human asked to treat the nonword as a verb. I would add, though, an extra twist to the story: when learning a second language, we often do learn a rule to "get started," though we may eventually succeed in automatizing production of the correct form. It is as if our brain contains a circuit that quickly stores the rule, and this circuit then acts as the "internal teacher" for the network that carries out the task without invoking the rule. Learning to automatize certain classes of behavior is an important linkage of conscious and unconscious processes.

Elman (1990) used SRNs to simulate both the prediction of the next sound (letter) in words as well as the prediction of the next word in a sentence. (Of course, this needs to be considerably augmented to explain the way context lets us predict with some confidence the first word we will see on turning to the next page of a book.) With only 29 words used in the sentences that provided the time sequence that was the input to the model, hierarchical cluster analysis of the hidden layer activations demonstrated that the SRN has extracted the lexical classes of the 29 words used by simply relying on word order information. The hidden layer representations separated nouns and verbs and separated nouns for inanimate (food) and animate (humans and animals) entities and so forth. In a follow-up study, complex nested sentences were used, involving relative clauses and number agreement, as in "Boys who chase dogs see girls."

Deep dyslexia is an acquired reading disorder characterized by the production of semantic errors, as well as visual and orthographic ones. For example, "NIGHT" might be read as "sleep." Detailed simulation of a variety of symptoms of deep dyslexia patients demonstrated that a connectionist network could explain dissociation of various symptom complexes without having to rely on modular theories of normal and impaired cognitive capabilities. However, such claims need to be subjected to the type of comparative analysis offered in chapter 14 of an SRN versus schema-theoretic model and the fact that lesions in different parts of the brain can yield different symptoms. Nonetheless, the key point is secure: different lesions to even a single brain region may yield different symptoms.

There are models which, while basically connectionist, nonetheless are "neuroanatomical" in that they contain subnetworks which are viewed as representing different regions of the brain. Garagnani, Wennekers, and Pulvermüller (2008) showed how Hebbian learning in connections within a chain of six such "brain regions" can link simple perceptual (hearing a word) and motor (saying the word) functions of language, and do so in a way that yielded a qualitative correlate with some EEG data (the mismatch negativity with latency 100–250 ms) for psycholinguistics experiments with familiar words and unfamiliar pseudoword stimuli. Husain et al. (2004) developed a large-scale, neurobiologically realistic network model of auditory pattern recognition in the primary auditory cortex and PFC. Although it is not cited there, it is an example of the type of modeling described in chapter 15, "Synthetic Brain Imaging." It models the response to acoustic stimuli and the processing in auditory cortical areas. It is relevant to neural models of language in that it can be extended to investigate phonological aspects of language processing.

In chapter 22, "Integrative Functions of the Corticostriatal System," we met a model of how interaction between PFC and basal ganglia could provide a basis for sequence learning. Dominey and collaborators extended this to offer a neuroanatomical model of language learning that has been applied to processing grammatical constructions as well as to nonlinguistic sequence learning. For the linguistic (syntactic comprehension) task, the model focuses on the assignment of thematic roles to noun phrases using (contra the above deep dyslexia model) a modular neural architecture constrained by data on the left-hemisphere brain areas involved in auditory sentence processing.

Language, Brain and Body Chapter 24 next turns to studies supporting the view that at least some key aspects of language are intrinsically linked to embodiment and sensorimotor knowledge—the *grounding* of language in perception and action. (Compare/contrast the view of "Embodied Models and Neurorobotics" in chapter 16.) A strong link between embodiment and language has been hypothesized to be crucial in the early stages of language acquisition. Chapter 25 ("Evolving the Language-Ready Brain") argues for a strong role for the visual control of hand movements in laying the basis for the emergence of the capacity to acquire and use language. Of course, seeing a role for embodiment need not imply that all aspects of language are embodied—indeed a capacity for abstraction may suggest that novel mechanisms (or novel modes of performance for old mechanisms) must be invoked (Arbib et al., 2014). Chapter 24 reviews various forms of action–language compatibility effects from the psycholinguistic literature. It then shows how neurorobotic studies support the modeling of embodiment factors both for higher order cognitive capabilities, such as language, as well as navigation and foraging tasks.

One study directly addressed embodiment factors in the developmental acquisition of language by replicating a child development study with humanoid robot experiments. In this *developmental robotics* approach (Cangelosi and Schlesinger, 2015), a robot equipped with a connectionist neural control system gradually acquires linguistic and sensorimotor skills through interactions with the physical world and with a human tutor. The tutor, in front of the robot, shows two novel objects respectively in the left and right location of a table. Initially the robot moves its posture to look at each object and learns to categorize it according to its visual features such as its color. Subsequently the tutor hides both objects, directs the robot's attention toward the right side where the first object was shown, and says a new word: "Modi." In the test phase, both objects are presented simultaneously in the center of the table, the robot is asked "Find the Modi," and must now look and point at the object that was presented in the right location. The learning process was based on three SOMs: (1) a "color map" used to categorize objects according to their color, (2) an "auditory map" used to represent the words heard by the robot, and (3) the "body-hub map" which is the key component for embodiment which in this case is based on the posture of eye and head position when the robot has to look to the left and to the right of the scene. Another model used SOMs for object–word associations to provide a neurocomputational account of fast mapping in early word learning, that is, the learning of words after one or very few training examples. Moreover, chapter 24 goes beyond SOMs to offer the basic concept of *a map connecting two or more other maps* as the building block of a hierarchical set of interconnected brain areas. This can be adapted to investigate specific neural theories of language and cognition integration.

Another field of language study with a strong focus on embodiment is that of the multi-agent and robotic models of the evolution of language, stressing cultural transmission and thus complementing the attention to biological evolution in chapter 25. In particular, Steels (2011) has developed a model of the cultural evolution of syntax, based on fluid construction grammar, directly inspired by developmental constructivist theories of language. The approach has been tested in numerous experiments in which two robots communicate with each other in a shared environment and develop a shared lexicon or some aspects of syntax.

The chapter also discusses the TRoPICALS model, which, as also noted in chapter 20 ("Reach and Grasp: Control, Development, and Recognition") extends the FARS model of the interaction of dorsal and ventral routes in grasping by linking it to a simple language system. After training, the robot's reaction times (RTs) are shown to reproduce an action–language compatibility effect seen in humans where a symbolic action has a faster RT if its direction is compatible with the location of the indicated part of the stimulus. This model demonstrates a link of sensorimotor phenomena with other processes in language use and visual categorization.

Another embodied neural model was used to study the consequences of architectural and functional constraints in evolutionary neural networks that learn to process the different classes of nouns and verbs. A variant of synthetic brain imaging (chapter 15) was used to examine the internal organization of the neural networks and qualitatively compare the results with human brain imaging data that indicate that nouns more strongly activate the posterior areas of the brain related to sensory and associative processing while verbs more strongly activate the anterior motor areas.

1.11.2 Evolving the Language-Ready Brain The final chapter uses a Big Question—"How did the human brain evolve so that we can use language?"—to motivate treatment of a range of topics that build on models introduced in earlier chapters while also showing how a surprisingly wide range of experimental techniques and observational studies may be implicated in the development and testing of the diverse models that such a broad topic demands. The general question frames a more specific set of questions which we gather under the heading of *comparative primatology:* What can we learn from studies in present-day monkeys (especially macaques), apes (especially chimpanzees), and humans that can guide us in making hypotheses about (1) the last common ancestors of humans and macaques (LCA-m, perhaps 25 million years ago) and chimpanzees (LCA-c, some 5 to 7 million years ago), and (2) what hypotheses can we then assess for the progression from LCA-m to LCA-c and thence to modern humans such that biological evolution gave us a *language-ready brain*—in the sense that early *Homo sapiens* may have had brains like ours, but did not yet possess the social structures that create a language community, something which might have required tens of millennia of cultural evolution? Factor in the insights from computational neuroscience and we have *computational comparative neuroprimatology*, which shares the issues of chapter 15 ("Synthetic Brain Imaging") on using data from macaque to gain insights into circuitry performing similar functions in the human brain, while adding the need to develop and test evolutionary hypotheses when assessing brain mechanisms for distinctively human capabilities like language.

The chapter then summarizes the mirror system hypothesis on the evolution of the language-ready brain (Arbib, 2012; Arbib and Rizzolatti, 1997) which traces a path from a mirror system for manual action (LCA-m) via simple imitation and the ability to add a limited set of manual gestures (LCA-c) thence to complex imitation and, via pantomime, to protosign and protospeech—with the claim that the path to speech is indirect (via manual gesture prior to sufficently flexible vocal control) and that early humans had protolanguage (protosign and protospeech) but that the capacity for language then developed via cultural evolution in *Homo sapiens* with little if any further change required in the brain's genome. Here we distinguish protolanguage from language—the latter has an increasingly large lexicon coupled with a grammar which provides a compositional semantics which allows words to be combined to form novel utterances which convey new meanings. The parity principle gained by evolving a mirror system (coupled to other systems "beyond the mirror") that can recognize words as articulatory actions allows others to understand (much of) these new meanings. Construction grammar is introduced as a general approach to grammar that combines form and meaning, building up utterances by combining various more or less specialized rules for form–meaning combination called *constructions*. Variants of these due to Steels and Dominey were presented in chapter 24 ("Language Processing").

We then recall macaque brain models from chapter 20 ("Reach and Grasp: Control, Development, and Recognition")—the FARS model of affordance and task-guided grasp and the MNS models for the recognition of manual actions—and build upon them to develop the ACQ model of opportunistic scheduling of actions. ACQ applies temporal difference learning (chapter 10) to manage desirability and executability of actions competing to be carried out next but is especially innovative in that it stresses the utility of the mirror system in learning about one's own actions, not just in observing the actions of others, so that one may observe whether or not one's executed action was indeed the intended action. These macaque models are then used to ground a model of ape gestural development. The treatment introduces *dyadic brain modeling*—modeling how the brains of two simulated apes change as they interact with each other—in such a way that a relatively complex pattern of physical interaction can become reduced to a simple communicative gesture that achieves the same distal goal. The hypothesis is made that it is increased ability to base action on proprioceptive cues that makes this process of *ontogenetic ritualization* available to chimpanzees (and presumably to LCA-c) but not to present-day monkeys or to LCA-m.

With the insights gained from these models, we are able to offer a fresh perspective on the mirror system hypothesis and the brain mechanisms underlying language while suggesting some of the challenges for brain modeling posed by reconsideration of several of the posited steps toward the language-ready brain. In particular, we distinguish a dorsal path for "words as articulatory actions" and a ventral path which associates words with their meanings. Some of the steps remain controversial, and resolution of these controversies will require not only further modeling but new studies of macaque neurophysiology, field studies of nonhuman primate communication, and new experiments in human psycholinguistics and neurolinguistics, as well as comparative neuroanatomy of various species.

The final part of the chapter seeks to use schema theory to provide a (partial) characterization of how the human brain processes language. As part of the effort to integrate the language system with other systems (recall the discussion of embodiment in chapter 24) we focus on the linkage between language and vision, using the classic schema-based VISIONS system for understanding visual scenes as the core of our effort. We start by looking at language production, and more specifically the processes involved in describing a visual scene. The model has two basic ingredients: a SemRep which extracts from an assemblage of spatially anchored and parameterized visual schemas a "semantic representation" of those elements to be factored into the utterance; and a new version of construction grammar, Template Construction Grammar (TCG), which provides the means (fully implemented on a computer) to repeatedly apply constructions to a SemRep to produce one or more utterances, which may be well-formed sentences but might also be fragmentary. Finally, we extend TCG to offer a conceptual model of language comprehension which is to be developed into a neurolinguistic model which includes accounts of the relation of brain lesions to agrammatism and other forms of aphasia.

1.12 *An Invitation to Research*

Various chapters have stepped back from their specific concerns to offer insights into brain modeling in general. Here are just two of them:

> As the problems that are addressed increase in complexity, more attention should be focused on the mapping of problems to neural architectures, and the identification of equivalence classes for types of problems that can be addressed by a given architecture. For example, what is the equivalence class of problems that can be addressed by ... the Reservoir system?... Typically, models have been developed to address specific classes of behavior... [Less attention has been given to the] notion of shifting between different problem sets, or learning how to solve multiple problems and how to determine what is the current problem... (from chapter 22)

> This problem [of a model not quite matching the data] highlights the important role of failure (falsifiability) in modeling—the inability to explain some aspect of reality spurs the

experimental search for additional critical facts about the system. This also highlights a contrast between simulation modeling and traditional closed-form physics modeling. Falsification in simulation generally requires that a model be augmented, whereas in closed-form modeling it may require that the model be entirely abandoned. (from chapter 23)

The former emphasizes the importance of seeking brain operating principles (chapter 5) and seeing how they are instantiated in different models; the latter emphasizes the importance of the cumulative development of new models which explain data more accurately or offer new insights (possibly qualitative in the case of dynamical system models such as those of chapter 3).

In chapter 22 ("Integrative Functions of the Corticostriatal System") we met a model of how interaction between PFC and basal ganglia could provide a basis for sequence learning while chapter 24 ("Language Processing") showed how this had been extended to offer a corticostriatal model of processing grammatical constructions.

Chapter 20 ("Reach and Grasp: Control, Development, and Recognition") first introduced the MNS model of how learning could yield mirror neurons for the recognition of different manual actions and then showed how to extend it to form the MNS2 model, which could support, for example, the development of audiovisual mirror neurons which could recognize actions associated with distinctive sounds through visual and/or auditory cues.

In each case we see an example of cumulative model development, a theme exemplified in chapter 5 ("Linking Models with Empirical Data: The Brain Operation Database") by using the MNS → MNS2 transition as a case study of how neuroinformatics tools (BODB) for comparing the relation of different models to available summaries of empirical data could aid the process of model development.

Of course, more generally, the related models being considered will not have one as an extension of the other but will have been developed by different groups focusing on different modeling tools in their quest to understand related (but not necessarily identical) aspects of brain function. For example, chapter 21 ("Cerebellar Adaptation and Supervised Learning in Motor Control") focused on the adaptive filter model, but I closed my summary with a brief mention of an alternative model. There was no claim that the latter strategy was "better" than the adaptive filter account (which clearly has many successes), but rather my aim was to reinforce the general point (often made implicitly) that has become clear again and again in this "Opening Perspective": The brain is very, very complex. It is engaged in many, many different behaviors. Moreover, even when we focus on specific subsystems, the progress to date often provides alternative approaches to modeling. Much work remains to be done both through experiments and through modeling to resolve, integrate, and/or move beyond these varied but partial successes.

A wealth of exciting research challenges remain.

NOTE

1. H.M. (Henry Molaison) died on December 2nd, 2008, at the age of 82. An informative full-page obituary was printed in the December 18, 2008, edition of *The Economist* weekly newspaper.

REFERENCES

Albus, J. S. (1971). A theory of cerebellar function. *Mathematical Biosciences, 10,* 25–61.

Alexander, G. E., & Crutcher, M. D. (1990). Functional architecture of basal ganglia circuits: Neural substrates of parallel processing. *Trends in Neurosciences, 13,* 266–271.

Amari, S.-I., & Arbib, M. A. (1977). Competition and cooperation in neural nets. In J. Metzler (Ed.), *Systems Neuroscience* (pp. 119–166). New York: Academic Press.

Arbib, M. A. (1981). Perceptual structures and distributed motor control. In V. B. Brooks (Ed.), *Handbook of Physiology—The Nervous System II. Motor Control* (pp. 1449–1480). Bethesda, MD: American Physiological Society.

Arbib, M. A. (1987). Levels of modelling of visually guided behavior (with peer commentary and author's response). *Behavioral and Brain Sciences, 10,* 407–465.

Arbib, M. A. (1988). Schemas and neurons: Two levels of neural computing. In R. Eckmiller & C. v. d. Malsburg (Eds.), *Neural Computers* (pp. 311–20). Berlin: Springer-Verlag.

Arbib, M. A. (2012). *How the Brain Got Language: The Mirror System Hypothesis.* New York: Oxford University Press.

Arbib M. A., & Bonaiuto, J. J. (2012). Multiple levels of spatial organization: World Graphs and spatial difference learning. *Adaptive Behavior, 20,* 287–303.

Arbib, M. A., Gasser, B., & Barrès, V. (2014). Language is handy but is it embodied? *Neuropsychologia, 55,* 57–70.

Arbib, M. A., & Hesse, M. B. (1986). *The Construction of Reality.* Cambridge: Cambridge University Press.

Arbib, M. A., & Rizzolatti, G. (1997). Neural expectations: A possible evolutionary path from manual skills to language. *Communication & Cognition, 29,* 393–424.

Arbib, M. A., Schweighofer, N., & Thach, W. T. (1995). Modeling the cerebellum: From adaptation to coordination. In D. J. Glencross & J. P. Piek (Eds.), *Motor Control and Sensory-Motor Integration: Issues and Directions* (pp. 11–36). Amsterdam: North-Holland Elsevier Science.

Atallah, H. E., McCool, A. D., Howe, M. W., & Graybiel, A. M. (2014). Neurons in the ventral striatum exhibit cell-type-specific representations of outcome during learning. *Neuron, 82,* 1145–1156.

Atkinson, R. C., & Shiffrin, R. M. (1968). Human memory: A proposed system and its control processes. In K. W.

SPENCE & J. T. SPENCE (Eds.), *The Psychology of Learning and Motivation* (Vol. 2, pp. 89–195). New York: Academic Press.

BADDELEY, A. D. (2003). Working memory: Looking back and looking forward. *Nature Reviews. Neuroscience, 4,* 829–839.

BARLOW, H. B. (1961). Possible principles underlying the transformation of sensory messages. In W. A. ROSENBLITH (Ed.), *Sensory Communication* (pp. 217–234). Cambridge, MA: MIT Press.

BARTLETT, F. C. (1932). *Remembering.* Cambridge: Cambridge University Press.

BECHARA, A., DAMASIO, H., & DAMASIO, A. R. (2000). Emotion, decision making and the orbitofrontal cortex. *Cerebral Cortex, 10,* 295–307.

BELL, A. J., & SEJNOWSKI, T. J. (1997). The "independent components" of natural scenes are edge filters. *Vision Research, 37,* 3327–3338.

BIZZI, E., & CHEUNG, V. C. (2013). The neural origin of muscle synergies. *Frontiers in Computational Neuroscience, 7,* 51.

BOTVINICK, M. M., & PLAUT, D. C. (2004). Doing without schema hierarchies: A recurrent connectionist approach to normal and impaired routine sequential action. *Psychological Review, 111,* 395–429.

BRODMANN, K. (1909). *Vergleichende Lokalisationslehre der Großhirnrinde: in ihren Prinzipien dargestellt auf Grund des Zellenbaues.* Leipzig: Ja Barth.

BROWN, T. G. (1914). On the nature of the fundamental activity of the nervous centres; together with an analysis of the conditioning of rhythmic activity in progression, and a theory of evolution of function in the nervous system. *Journal of Physiology, 48,* 18–46.

BUZSAKI, G. (1989). Two-stage model of memory trace formation: A role for "noisy" brain states. *Neuroscience, 31,* 551–570.

CANGELOSI, A., & SCHLESINGER, M. (2015). *Developmental Robotics: From Babies to Robots.* Cambridge, MA: MIT Press.

COOPER, R. P., & SHALLICE, T. (2000). Contention scheduling and the control of routine activities. *Cognitive Neuropsychology, 17,* 297–338.

CORCHS, S., & DECO, G. (2004). Feature-based attention in human visual cortex: Simulation of fMRI data. *NeuroImage, 21*(1), 36–45.

DELONG, M. R. (1990). Primate models of movement disorders of basal ganglia origin. *Trends in Neurosciences, 13,* 281–285.

DEMIRIS, Y., & HAYES, G. (2002). Imitation as a dual-route process featuring predictive and learning components: A biologically plausible computational model. In K. DAUTENHAHN & C. L. NEHANIV (Eds.), *Imitation in Animals and Artifacts* (pp. 327–361). Cambridge, MA: MIT Press.

DEV, P. (1975). Perception of depth surfaces in random-dot stereograms: A neural model. *International Journal of Man-Machine Studies, 7,* 511–528.

DIDDAY, R. L. (1976). A model of visuomotor mechanisms in the frog optic tectum. *Mathematical Biosciences, 30,* 169–180.

DOYA, K. (2002). Metalearning and neuromodulation. *Neural Networks, 15,* 495–506.

ECCLES, J. C., ITO, M., & SZENTÁGOTHAI, J. (1967). *The Cerebellum as a Neuronal Machine.* New York: Springer-Verlag.

EDIN, F., MACOVEANU, J., OLESEN, P., TEGNÉR, J., & KLINGBERG, T. (2007). Stronger synaptic connectivity as a mechanism behind development of working memory-related brain activity during childhood. *Journal of Cognitive Neuroscience, 19*(5), 750–760.

EKEBERG, Ö. (1993). A combined neuronal and mechanical model of fish swimming. *Biological Cybernetics, 69,* 363–374.

ELMAN, J. L. (1990). Finding structure in time. *Cognitive Science, 14,* 179–211.

ENGEL, A. K., KÖNIG, P., KREITER, A. K., GRAY, C. M., & SINGER, W. (1991). Temporal coding by coherent oscillations as a potential solution to the binding problem: Physiological evidence. In H. G. SCHUSTER (Ed.), *Nonlinear Dynamics and Neural Networks* (pp. 149–163). Weinheim, Germany: VCH.

ERDEM, U. M., & HASSELMO, M. (2012). A goal-directed spatial navigation model using forward trajectory planning based on grid cells. *European Journal of Neuroscience, 35,* 916–931.

FELLOUS, J.-M., & ARBIB, M. A. (Eds.). (2005). *Who Needs Emotions: The Brain Meets the Robot.* Oxford: Oxford University Press.

FRANK, M. J., SEEBERGER, L. C., & O'REILLY, R. C. (2004). By carrot or by stick: Cognitive reinforcement learning in parkinsonism. *Science, 306,* 1940–1943.

FUKSON, O. I., BERKINBLIT, M., & FELDMAN, A. G. (1980). The spinal frog takes into account the scheme of its body during the wiping reflex. *Science, 209,* 1261–1263.

FUKUSHIMA, K. (1980). Neocognitron: A self-organizing neural network model for a mechanism of pattern recognition unaffected by shift in position. *Biological Cybernetics, 36,* 193–202.

GARAGNANI, M., WENNEKERS, T., & PULVERMÜLLER, F. (2008). A neuroanatomically-grounded Hebbian learning model of attention-language interactions in the human brain. *European Journal of Neuroscience, 27*(2), 492–513.

GEORGOPOULOS, A. P., SCHWARTZ, A. B., & KETTNER, R. E. (1986). Neuronal population coding of movement direction. *Science, 233,* 1416–1419.

GOODALE, M. A., MILNER, A. D., JAKOBSON, L. S., & CAREY, D. P. (1991). A neurological dissociation between perceiving objects and grasping them. *Nature, 349,* 154–156.

GRAYBIEL, A. M., AOSAKI, T., FLAHERTY, A. W., & KIMURA, M. (1994). The basal ganglia and adaptive motor control. *Science, 265,* 1826–1831.

GREGORY, R. L. (1967). Will seeing machines have illusions? In N. L. COLLINS & D. MICHIE (Eds.), *Machine Intelligence 1* (pp. 169–180). Edinburgh: Oliver & Boyd.

GUAZZELLI, A., CORBACHO, F. J., BOTA, M., & ARBIB, M. A. (1998). Affordances, motivation, and the world graph theory. *Adaptive Behavior, 6,* 435–471.

GURNEY, K., PRESCOTT, T. J., & REDGRAVE, P. (2001). A computational model of action selection in the basal ganglia: I. A new functional anatomy. *Biological Cybernetics, 84,* 401–410.

HAFTING, T., FYHN, M., MOLDEN, S., MOSER, M. B., & MOSER, E. I. (2005). Microstructure of a spatial map in the entorhinal cortex. *Nature, 436,* 801–806.

HAN, C. E., ARBIB, M. A., & SCHWEIGHOFER N. (2008). Stroke rehabilitation reaches a threshold. *PLoS Computational Biology* (www.ploscompbiol.org), *4,* e1000133.

HEBB, D. O. (1949). *The Organization of Behavior.* New York: Wiley.

HODGKIN, A. L., & HUXLEY, A. F. (1952). A quantitative description of membrane current and its application to conduction and excitation in nerve. *Journal of Physiology, 117,* 500–544.

Husain, F. T., Tagamets, M. A., Fromm, S. J., Braun, A. R., & Horwitz, B. (2004). Relating neuronal dynamics for auditory object processing to neuroimaging activity: A computational modeling and an fMRI study. *NeuroImage, 21*(4), 1701–1720.

IJspeert, A. J., Crespi, A., & Cabelguen, J. M. (2005). Simulation and robotics studies of salamander locomotion: Applying neurobiological principles to the control of locomotion in robots. *Neuroinformatics, 3,* 171–195.

IJspeert, A. J., Crespi, A., Ryczko, D., & Cabelguen, J. M. (2007). From swimming to walking with a salamander robot driven by a spinal cord model. *Science, 315,* 1416–1420.

Ito, M., Sakurai, M., & Tongroach, P. (1982). Climbing fibre induced depression of both mossy fibre responsiveness and glutamate sensitivity of cerebellar Purkinje cells. *Journal of Physiology, 324,* 113–134.

Jeannerod, M., & Biguer, B. (1982). Visuomotor mechanisms in reaching within extra-personal space. In D. J. Ingle, R. J. W. Mansfield, & M. A. Goodale (Eds.), *Advances in the Analysis of Visual Behavior* (pp. 387–409). Cambridge, MA: MIT Press.

Jeannerod, M., Decety, J., & Michel, F. (1994). Impairment of grasping movements following a bilateral posterior parietal lesion. *Neuropsychologia, 32,* 369–380.

Kalman, R. E. (1960). A new approach to linear filtering and prediction problems. *Journal of Fluids Engineering, 82,* 35–45.

Kargo, W. J., & Giszter, S. F. (2000). Rapid correction of aimed movements by summation of force-field primitives. *Journal of Neuroscience, 20,* 409–426.

Kelley, A. (2005). Neurochemical networks encoding emotion and motivation. In J. Fellous & M. Arbib (Eds.), *Who Needs Emotions? The Brain Meets the Robot* (pp. 29–77). Oxford: Oxford University Press.

LeDoux, J. E. (2000). Emotion circuits in the brain. *Annual Review of Neuroscience, 23,* 155–184.

Lee, M. G., Chrobak, J. J., Sik, A., Wiley, R. G., & Buzsaki, G. (1994). Hippocampal theta activity following selective lesion of the septal cholinergic system. *Neuroscience, 62,* 1033–1047.

Marr, D. (1969). A theory of cerebellar cortex. *Journal of Physiology, 202,* 437–470.

McClelland, J. L., McNaughton, B. L., & O'Reilly, R. C. (1995). Why there are complementary learning systems in the hippocampus and neocortex: Insights from the successes and failures of connectionist models of learning and memory. *Psychological Review, 102,* 419–457.

McCulloch, W. S., & Pitts, W. H. (1943). A logical calculus of the ideas immanent in nervous activity. *Bulletin of Mathematical Biophysics, 5,* 115–133.

Miller, G. A., Galanter, E., & Pribram, K. H. (1960). *Plans and the Structure of Behavior.* New York: Holt, Rinehart & Winston.

Milner, P. M. (1999). *The Autonomous Brain.* Mahwah, NJ: Erlbaum.

Neisser, U. (1976). *Cognition and Reality: Principles and Implications of Cognitive Psychology.* San Francisco, CA: Freeman.

Norman, D. A., & Shallice, T. (1980). *Attention to Action: Willed and Automatic Control of Behavior. Chip Report 99.* San Diego: University of California.

Olds, J. (1969). The central nervous system and the reinforcement of behavior. *American Psychologist, 24,* 114–132.

Oliva, A., & Torralba, A. (2006). Building the gist of a scene: The role of global image features in recognition. *Progress in Brain Research, 155,* 23–36.

Olshausen, B., & Field, D. (1996). Emergence of simple-cell receptive field properties by learning a sparse code for natural images. *Nature, 381,* 607–609.

O'Reilly, R. C., & Frank, M. J. (2006). Making working memory work: A computational model of learning in the prefrontal cortex and basal ganglia. *Neural Computation, 18,* 283–328.

Prescott, T. J., Gonzalez, F. M. M., Gurney, K., Humphries, M. D., & Redgrave, P. (2006). A robot model of the basal ganglia: Behavior and intrinsic processing. *Neural Networks, 19,* 31–61.

Rall, W. (2003). Perspective on neuron model complexity. In M. A. Arbib (Ed.), *The Handbook of Brain Theory and Neural Networks* (2nd ed., pp. 877–881). Cambridge, MA: MIT Press.

Rauschecker, J. P. (1998). Cortical processing of complex sounds. *Current Opinion in Neurobiology, 8,* 516–521.

Redgrave, P., Gurney, K., & Reynolds, J. (2008). What is reinforced by phasic dopamine signals? *Brain Research. Brain Research Reviews, 58,* 322–339.

Rescorla, R. A., & Wagner, A. R. (1972). A theory of Pavlovian conditioning: Variations in the effectiveness of reinforcement and nonreinforcement. In A. A. Black & W. F. Prokasy (Eds.), *Classical Conditioning II: Current Research and Theory* (pp. 64–99). New York: Appleton-Century-Crofts.

Reynolds, J. H., Chelazzi, L., & Desimone, R. (1999). Competitive mechanisms subserve attention in macaque areas V2 and V4. *Journal of Neuroscience, 19,* 1736–1753.

Robinson, D. A. (1964). The mechanics of human saccadic eye movement. *Journal of Physiology, 174,* 245–264.

Romanski, L. M., Tian, B., Fritz, J., Mishkin, M., Goldman-Rakic, P. S., & Rauschecker, J. P. (1999). Dual streams of auditory afferents target multiple domains in the primate prefrontal cortex. *Nature Neuroscience, 2,* 1131–1136.

Rumelhart, D. E., & McClelland, J. L. (1986). On learning the past tenses of English verbs. In D. E. Rumelhart & J. L. McClelland (Eds.), *Parallel Distributed Processing: Explorations in the Microstructure of Cognition* (pp. 216–271). Cambridge, MA: MIT Press,.

Seymour, B., Daw, N. D., Roiser, J. P., Dayan, P., & Dolan, R. (2012). Serotonin selectively modulates reward value in human decision-making. *Journal of Neuroscience, 32,* 5833–5842.

Shannon, C. E., & Weaver, W. (1949). *The Mathematical Theory of Communication.* Urbana: University of Illinois Press.

Singer, W., & Gray, C. M. (1995). Visual feature integration and the temporal correlation hypothesis. *Annual Review of Neuroscience, 18,* 555–586.

Steels, L. (2011). Modeling the cultural evolution of language. *Physics of Life Reviews, 8,* 339–356.

Sutton, R. S., & Barto, A. G. (1998). *Reinforcement Learning: An Introduction.* Cambridge, MA: MIT Press.

Tagamets, M. A., & Horwitz, B. (1998). Integrating electrophysiological and anatomical experimental data to create a large-scale model that simulates a delayed match-to-sample human brain imaging study. *Cerebral Cortex, 8,* 310–320.

Ungerleider, L. G., & Mishkin, M. (1982). Two cortical visual systems. In D. J. Ingle, M. A. Goodale, & R. J. W.

MANSFIELD (Eds.), *Analysis of Visual Behavior* (pp. 549–586). Cambridge, MA: MIT Press.

WIENER, N. (1949). *Extrapolation, Interpolation, and Smoothing of Stationary Time Series with Engineering Applications.* Cambridge, MA: MIT Press.

WILLIAMS, S. M., WHITE, L. E., & MACE, A. C. (2007). *Sylvius 4: An Interactive Atlas and Visual Glossary of Human Neuroanatomy* (Available only as software). Sunderland, MA: Sinauer Associates (in association with Pyramis Studios).

WILSON, H. R., & COWAN, J. D. (1973). A mathematical theory of the functional dynamics of cortical and thalamic nervous tissue. *Kybernetik, 13,* 55–80.

WINDER, R., CORTES, C. R., REGGIA, J. A., & TAGAMETS, M. A. (2007). Functional connectivity in fMRI: A modeling approach for estimation and for relating to local circuits. *NeuroImage, 34*(3), 1093–1107.

YAMAZAKI, T., & TANAKA, S. (2005). Neural modeling of an internal clock. *Neural Computation, 17,* 1032–1058.

2 Basic Neuron and Network Models

NICOLAS BRUNEL

This chapter presents an overview of basic concepts and tools used in theoretical studies at the single-neuron and network levels. Section 2.1 reviews statistical models and methods used to describe neuronal activity and to describe how this activity encodes external stimuli. Section 2.2 presents simple models for populations of neurons, the so-called rate or neural mass models. Sections 2.3 and 2.4 present models of single neurons and single synapses, respectively. Finally, section 2.5 contains an introduction to neural networks. Note that these sections by no means represent an exhaustive review of the relevant literature, but are rather primers which provide useful entry points to the relevant literature. For more in-depth discussions of these topics, the reader should consult textbooks such as Dayan and Abbott (2001), Ermentrout and Terman (2010), and Gerstner, Kistler, Naud and Paninski (2014).

2.1 *Statistical Models of Neuronal Activity and Neural Coding*

Sensory inputs are encoded in the central nervous system (CNS) by the spike trains of populations of neurons that form the output of the corresponding sensory systems (e.g., the ganglion cells of the retina for the visual system). These spike trains influence the internal state of the CNS and, consequently, motor outputs that are conveyed to target muscles in the form of spike trains. One of the fundamental questions in computational neuroscience is to understand how information about external inputs, internal brain states, and motor outputs is encoded in spike trains of single neurons or populations of neurons.

Responses of neurons are inherently variable from one experimental trial to the next. Thus, descriptions of neuronal activity need to be statistical in nature. We first focus on statistical descriptions of the spike trains for a given experimental condition. One can then study how the statistics of firing is affected by a change in parameters describing these experimental conditions. For example, in sensory systems, the conditional dependence of the statistics of spike trains of single neurons on parameters describing a stimulus can in principle reveal the information contained in the spike train of the recorded neuron about the stimulus.

2.1.1 Spike Trains, Variability, Poststimulus Time Histogram, Tuning Curves Experimentalists characterize the response of neurons using multiple identical trials. This is shown schematically in figure 2.1, which shows the response of a simulated single neuron over multiple trials to two stimuli (A and B), presented in between the vertical gray lines. The top panels show a raster plot (spike times for different trials), that is, the *spike trains* of the neuron for several trials. A spike train consists of a list of spike times, $\{t_{ik}, i = 1, \ldots, N_T, k = 1, \ldots, N_i\}$, where t_{ik} is the time of the k-th spike in trial number i, N_T is the number of trials, and N_i is the number of spikes emitted during trial i. Spike trains are often described formally by a sum of delta functions at times t_{ik}. For practical purposes, one often discretizes time in bins of width Δt. Spike trains are then described by a string of T binary numbers (assuming Δt is small enough so that the probability of observing two spikes in a bin is negligible) $S_i(t)$, $t = 1, \ldots, T$, describing the emission of a spike ($S_i(t) = 1$), or not ($S_i(t) = 0$) in bin t in trial i.

Note that in figure 2.1, spike trains show a great deal of variability—both within each trial and across trials. This variability is typical of in vivo recordings. Variability can be quantified in several ways. The temporal variability within trials can be quantified computing the *distribution of interspike intervals* (ISI) $P(t)$. From this distribution, one can compute the mean ISI (the inverse of the firing rate of the neuron) and the standard deviation of the ISI, which quantifies the variability of firing. To get a dimensionless measure of variability, the standard deviation is divided by the mean, yielding the *coefficient of variation* (CV) of the ISIs, which is the most common measure of the temporal irregularity of spike trains. The variability across trials can be quantified computing the distribution of spike counts across trials in a given temporal window of width Δt. From such a

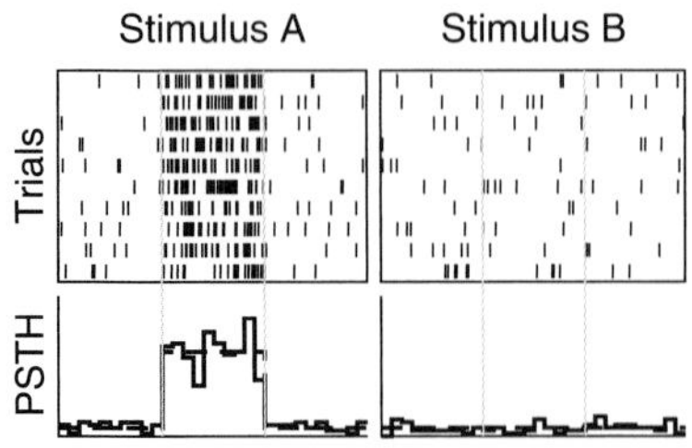

FIGURE 2.1 Response of a simulated neuron to two external stimuli. Top panels show 10 trials. The stimulus is shown in between the gray vertical lines. The neuron emits spikes at 3 spikes per second as a Poisson process, except during presentation of stimulus A, where its firing rate increases to 30 spikes per second. The bottom panel shows a poststimulus time histogram (PSTH; intantaneous firing rate), computed from the 10 trials in 100-ms bins. The true rate is indicated as a dashed line.

distribution, one can compute the *Fano factor*, which is simply the variance of spike counts divided by the mean.

Because of this variability, spike trains are often modeled as *point processes.* A point process is a stochastic process whose realizations consist of a set of isolated points in time. A simple point process that has been extensively used in computational neuroscience is the *Poisson process*, which is characterized by the independence of successive ISIs (the "renewal" property), and an exponential distribution of ISIs, $P(t) = \nu \exp(-\nu t)$, where ν is the *firing rate* (or spike rate), measured in spikes per second. When the firing rate ν is constant in time, a Poisson process is said to be homogeneous, while when the firing rate is time-dependent, it is said to be inhomogeneous. Homogeneous Poisson processes are characterized by a CV equal to 1 (in other words, the standard deviation of the ISI distribution is equal to its mean), and a Fano factor also equal to 1 (the variance of spike counts over a given time window is equal to the mean). Another property of the Poisson process is that the number of spikes n emitted in an interval of duration T is given by the Poisson distribution,

$$P(n) = \frac{(\nu T)^n}{n!} \exp(-\nu T), \qquad (2.1)$$

when ν is constant in time. For an inhomogeneous Poisson process with time-dependent rate $\nu(t)$, the probability of observing n spikes in an interval of duration T is given by equation 2.1, where νT needs to be replaced by the integral of $\nu(t)$ in the corresponding interval. The spike trains shown in figure 2.1 were drawn from a Poisson process with a rate of 3 spikes per second, except during the presentation of stimulus A (in between the two vertical gray bars), where the rate was increased to 30 spikes per second. Note that these firing rates are typical of spontaneous and stimulus-evoked firing rates in mammalian cortex.

The Poisson process has been quite popular in neuroscience because of its simplicity, but also because spike trains recorded in vivo bear similarities with Poisson processes. In particular, CVs of cortical neurons are typically of order one (Softky and Koch, 1993; Compte et al., 2003). However, spike trains also differ in significant ways from Poisson processes—in particular, ISIs cannot be shorter than a minimal interval called the absolute refractory period, due to the biophysics of action potential generation (see section 2.3). For this reason, other types of point processes have been used. Popular examples have been the Poisson process with dead time in which the ISI distribution is zero for intervals shorter than the dead time, and exponential at larger intervals; processes with ISI distributions described by a gamma distribution, or an inverse Gaussian distribution, which are both two-parameter families of distributions that can lead to arbitrary values of the CV (see, e.g., van Vreeswijk, 2010). Spike trains can also be modeled as the output of specific biophysical models (e.g., an integrate-and-fire neuron; see section 2.3) with a noisy input.

A consequence of the variability in spike trains is the need to perform averages over many trials. One particular average is the poststimulus time histogram (PSTH), shown in the bottom panels of figure 2.1. To compute a PSTH, one first needs to divide time in bins (in the example of figure 2.1, the 3-second interval is divided in 30 100-ms bins). One then computes the firing rate in each bin by counting the total number of spikes over all trials in the bin and then dividing this number by the number of trials times the duration of the bin. This gives an estimate of the instantaneous firing rate of the neuron as a function of time. Note that this estimate necessarily deviates from the *true* firing rate because it is computed from a finite number of trials (compare the solid and dashed lines in the bottom panels of figure 2.1).

In our example of figure 2.1, we see that the firing rate increases during the presentation of A but not during the presentation of B. This means that information about which stimulus is shown to the neuron is encoded in the firing rate of the neuron. In particular, it is possible to determine with high probability which stimulus was shown by simply evaluating its firing rate over a sufficiently long period. Stimulus-dependent changes in firing rates of neurons are ubiquitous in sensory systems, and this observation has led to the concept of *rate coding.*

Tuning curves describe how the average firing rate of a neuron depends on a stimulus. Several examples of functions that have been used to fit neuronal data are

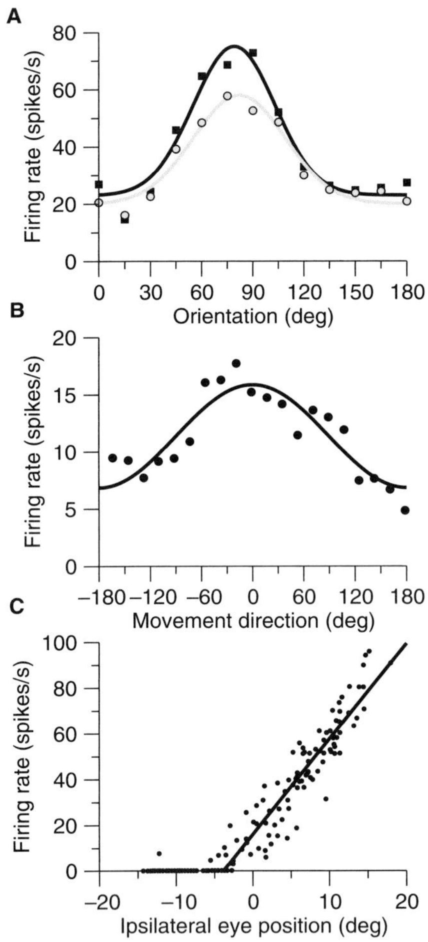

FIGURE 2.2 Tuning curves used to fit neural recordings. (A) Gaussian tuning curve, used to fit the dependence of firing rate on orientation in a V4 neuron (McAdams and Maunsell, 1999). (B) Cosine function, fitting the dependence of firing rate on arm-movement direction in motor cortex (Amirikian et al., 2000). (C) Threshold-linear tuning curve, used in the oculomotor system (firing rate as a function of the position of the eye; Aksay et al., 2000).

shown in figure 2.2. Figure 2.2A shows Gaussian tuning curves that have been used to describe quantitatively the response of neurons in area V4 to orientation (McAdams and Maunsell, 1999). Figure 2.2B shows a cosine tuning curve that has been used to describe quantitatively the response of neurons in motor cortex to the direction of an arm movement (Amirikian et al., 2000). Finally, figure 2.2C shows a threshold-linear tuning curve, used in oculomotor nuclei to describe the relationship of firing rate with the horizontal location of the eye (Aksay et al., 2000).

2.1.2 LINEAR–NONLINEAR POISSON MODELS, SPIKE-TRIGGERED AVERAGE, RECEPTIVE FIELDS For time-dependent sensory inputs, a more dynamic representation of the neuronal response than the one provided by static tuning curves is needed. A simple model to quantify the response of neurons to arbitrary time-dependent inputs is the linear–nonlinear Poisson (LNP) model (see, e.g., Pillow et al., 2005). In this model, the stimulus $s(t)$ is first convolved with a temporal filter $K(t)$ (the linear stage). The output of the filter is then passed through a static nonlinear function Φ (the nonlinear stage), which gives the instantaneous firing rate,

$$r(t) = \Phi\left[\int_{-\infty}^{t} K(t-t')s(t')\right]. \tag{2.2}$$

The temporal filter accounts for the various time delays separating a temporal stimulus from the neuronal response. The nonlinearity is necessary to account for the fact that neuronal firing rates are necessarily bounded—they are inherently non-negative and cannot exceed an upper bound because of the biophysics of action potential generation (see section 2.3).

The spike train of the LNP model is then generated using an inhomogeneous Poisson process (P, for Poisson), with an instantaneous firing rate given by equation 2.2. This model can be used to fit neural data through the following two steps: one first obtains the filter K using *spike-triggered average* of the stimulus (see figure 2.3A, B, and C). One then obtains the static nonlinearity by averaging $r(t)$ for each value of the convolution between the stimulus and the filter K (see figure 2.3D).

In many instances, sensory stimuli depend on space as well as time. In the particular case of the visual system, a stimulus is characterized by the luminance, or light intensity, at each point on the visual field at a given time. Neurons in the early visual system should therefore be described by a filter that depends on space as well as on time, $K(x, y, t)$. The spatial area in which $K(x, y, t)$ is significantly different from zero defines the *receptive field* of the neuron. Sometimes, the function $K(x, y, t)$ is referred to as the "spatiotemporal receptive field" of a neuron. Spatiotemporal receptive fields can sometimes be written as a product of a spatial one and a temporal one,

$$K(x, y, t) = K_s(x, y)K_t(t). \tag{2.3}$$

In the retina and thalamus, spatial receptive fields are approximately circularly symmetrical and are often modeled by differences of Gaussians, with an ON (or OFF) center region, and an OFF (or ON) surround region,

$$K_s(x, y) \propto \pm\left(\frac{1}{2\pi\sigma_c^2}\exp\left(-\frac{x^2+y^2}{2\sigma_c^2}\right) - \frac{B}{2\pi\sigma_s^2}\exp\left(-\frac{x^2+y^2}{2\sigma_s^2}\right)\right) \tag{2.4}$$

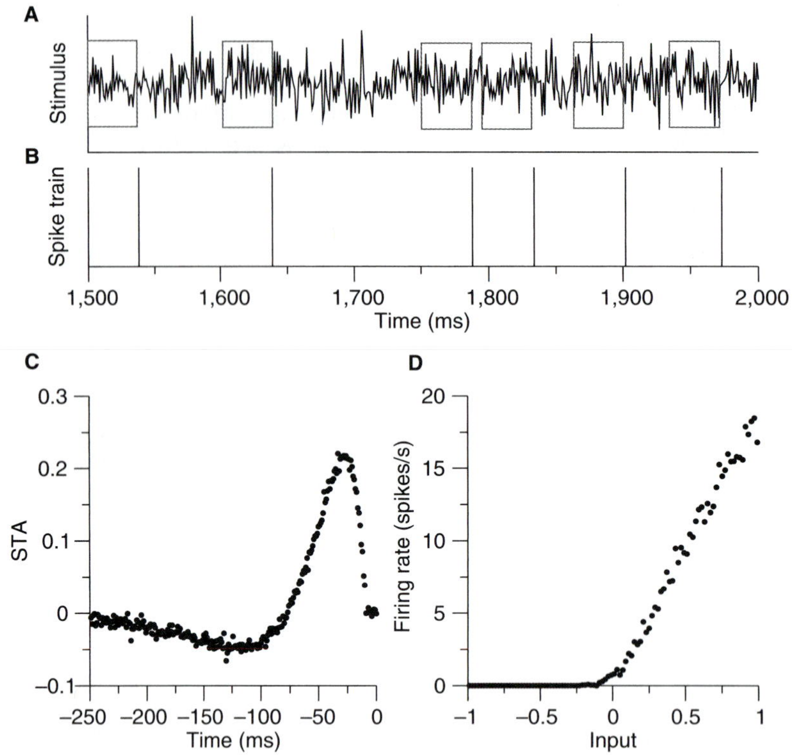

FIGURE 2.3 Characterizing the mapping from a time-dependent stimulus $s(t)$ to an output spike train. (A) White noise stimulus $s(t)$ (discretized in 1-ms bins). (B) Spike train of a simulated neuron in response to the stimulus. The boxes in panel A show time windows preceding each emitted spike. (C) The spike-triggered average (STA) of the stimulus consists in averaging the stimulus at a time t preceding a spike. It corresponds to averaging the stimulus over the boxes preceding the spikes shown in A. The resulting function gives the temporal filter $K(t)$. (D) Static transfer function: firing rate as a function of the filtered input (the stimulus $s(t)$ convolved with $K(t)$).

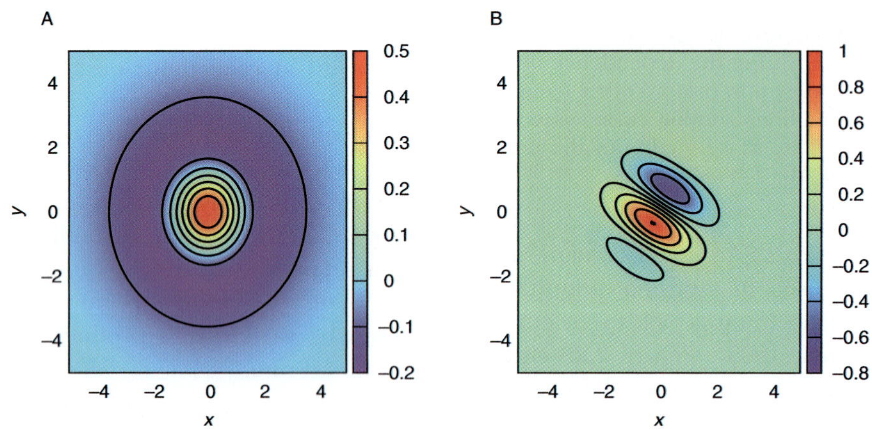

FIGURE 2.4 Examples of two-dimensional functions used to fit receptive fields in the early visual system. (A) Difference of Gaussians, used to fit receptive fields in lateral geniculate nucleus. (B) Gabor function, used to fit receptive fields in V1.

where B is the balance between center and surround, σ_c is the width of the center, and σ_s is the width of the surround. An example of such a receptive field is shown in figure 2.4A. In primary visual cortex, spatial receptive fields are no longer circularly symmetrical and are modeled using Gabor functions,

$$K_s(x,y) \propto \exp\left(-\frac{x'^2+\gamma^2 y'^2}{2\sigma^2}\right)\cos\left(2\pi\frac{x'}{\lambda}+\psi\right) \tag{2.5}$$

where $x' = x\cos(\theta) + y\sin(\theta)$, $y' = -x\sin(\theta) + y\cos(\theta)$, λ is the wavelength, θ the orientation, ψ a phase offset, σ the width of the Gaussian envelope, and γ a spatial

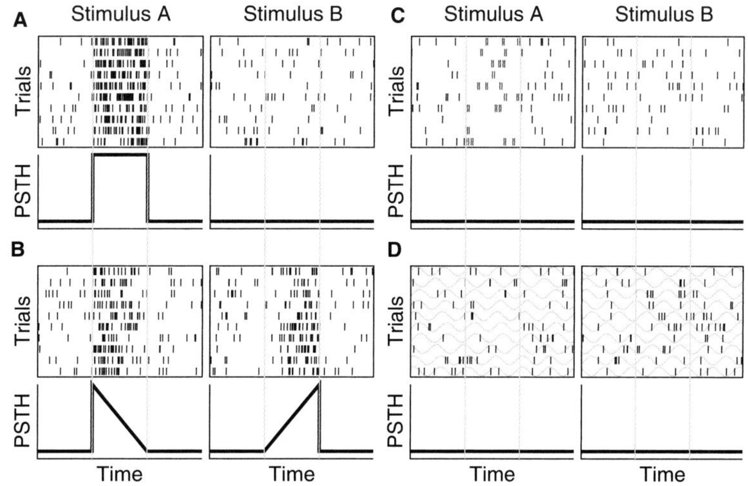

FIGURE 2.5 Single-neuron codes. Each (2 × 2) group of panels shows a particular coding scenario. The response of a single neuron over multiple trials is shown for two stimuli (A and B), presented in between the vertical gray lines (note that in A, B the stimulus induces a change in firing rate while it does not in C, D). For each group, the top panels show a raster plot (spike times for different trials) while the bottom panels show a poststimulus time histogram (PSTH; instantaneous firing rate as a function of time). (A) Neuron responds as a Poisson process with stimulus-dependent stationary firing rate (static firing rate code). (B) Neuron responds as an inhomogeneous Poisson process with time-dependent firing rate (instantaneous firing rate code). (C) Spike pattern code. Firing rates are independent of stimulus and time, and information is contained in higher order statistics of spike trains. Here, the neuron responds with bursts of two spikes (indicated in red) to stimulus A, but not B. (D) Phase-of-firing code. The firing rate of the neuron is again independent of stimulus and time but is modulated by an oscillatory variable, indicated by a gray line (note that the phase of the oscillation with respect to stimulus onset is random). Here, the neuron fires preferentially at the peak/trough of the oscillation when stimulus A/B is presented, respectively.

aspect ratio. An example of such a receptive field is shown in figure 2.4B.

2.1.3 SINGLE-NEURON CODING: FIRING RATES AND BEYOND Statistical characterizations described so far make the implicit assumption that the instantaneous firing rate is the only information-carrying quantity of the neuron. This is illustrated in figure 2.5A (which is identical to figure 2.1, already discussed above) and 2.5B, using the simple case of a single neuron recorded in an experiment during which two static stimuli, A and B, are presented repeatedly during an interval $[0, T]$. In figure 2.5A, the instantaneous firing rate is constant throughout the presentation of the stimulus, and all the information about the stimulus is carried by the firing rate (a "static rate" code). In this example, the neuron emits on average 30 spikes per second in response to A, while it emits on average 3 spikes per second in response to B. This means that a "readout" neuron that counts spikes from this neuron in a sufficiently long interval could infer which stimulus was presented. In figure 2.5B, the temporal dynamics of the instantaneous firing rate provide information about the stimulus. For example, the neuron emits spikes at a higher rate at the beginning of the interval in response to A while it emits spikes at a higher rate at the end of the interval in response to B. This is an example of an "instantaneous firing rate code."

Other quantities beyond the instantaneous firing rate might also carry information about external stimuli. Two scenarios in particular have been proposed. The first is information carried by higher order statistics of the spike train. For example, in figure 2.5C, the neuron emits bursts of two spikes with a specific ISI in response to A while it does not in response to B. In this case, information is provided by a specific spike pattern—hence the term "spike pattern code." The second scenario can be implemented in networks whose activity oscillates in time, in which information can be provided by the timing of spikes relative to this global oscillation. In figure 2.5D, the spikes of our neuron are emitted preferentially at the peak of an ongoing oscillation when stimulus A is presented while they are emitted at the trough of the oscillation when stimulus B is presented. Note that in these two cases, the instantaneous firing rate remains constant in time, independent of the stimulus, and therefore contains no information about the stimulus.

2.1.4 POPULATION CODING We have so far discussed information contained in the spike train of single neurons. To understand how information is encoded in populations of neurons ("population coding"), we again need to obtain an accurate statistical description of the activity of the population. Recording all neurons simultaneously in a given structure was considered until recently a distant dream, but advances in electrophysiological and imaging techniques are now making this feat feasible in some preparations (Pillow et al., 2008; Ahrens & Engert, 2015). In most cases however, one must content oneself with having access to only a small fraction of the neurons in a given structure. Models of population activity that are built from simultaneously recorded neurons can be distinguished by whether they include correlations between neurons, and by the nature of these correlations. The simplest conceivable model of population activity is one in which neurons emit spikes independently with a stimulus-specific firing rate. In many cases, however, an independent model provides a poor description of multi-unit activity. In those cases, models incorporating correlations between spike trains need to be used (see, e.g., Schneidman et al., 2006; Pillow et al., 2008; Ohiorhenuan et al., 2010; Lin et al., 2015).

As in the case of single neurons, different scenarios for population coding have been proposed. Some of those scenarios are illustrated in figure 2.6: in this column, each row in the upper panel shows the spike train for a different neuron during a single trial while the lower panel uses the color code for a neuron to sketch the PSTH that might be obtained over many trials with that cell.

In figure 2.6A, neurons are uncorrelated, and only their (stationary) firing rates carry information about the stimulus, with different cells varying in their responsivity to the two stimuli. In figure 2.6B, the mean firing rates carry no information about the stimulus, but the temporal profile of instantaneous firing rates does. In the particular case shown in figure 2.6B, different stimuli trigger distinct sequences of activity in which neurons fire in different orders, depending on the stimulus that triggered the sequence. In figure 2.6C, correlations between neurons carry information about stimuli. In this particular example, synchronous spiking of the top five neurons signal the appearance of stimulus A while synchronous spiking of the bottom five neurons signal the appearance of stimulus B. Finally, in figure 2.6D, the phase of firing with respect to an ongoing oscillation is the information-carrying quantity. Each neuron fires preferentially at a specific phase of the ongoing oscillation, and the preferred phase depends on the shown stimulus. Note that this scenario

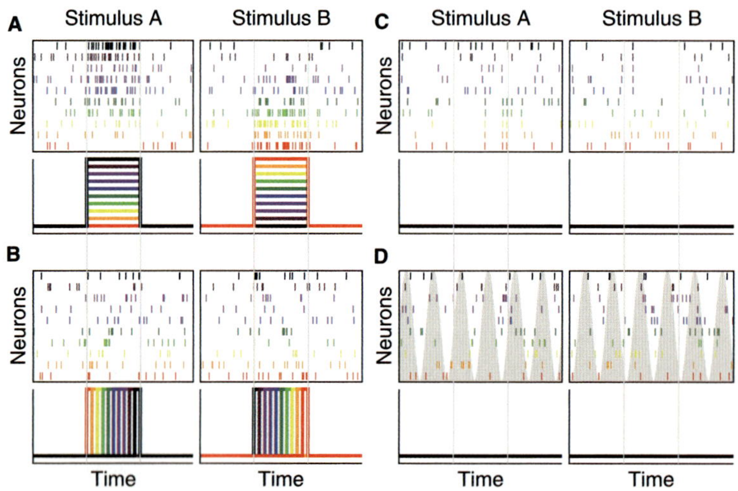

FIGURE 2.6 Population codes. Each (2 × 2) group of panels shows a particular coding scenario. The response of a population of neurons for a single trial (different neurons are shown with different colors) is shown for two stimuli (A and B), presented in between the vertical gray lines (note that in A, B, the stimulus induces a change in firing rates while it does not in C, D). For each group, the top panel shows a raster plot (spike times of the different neurons) while the bottom panel shows a poststimulus time histogram (instantaneous firing rates as a function of time). (A) Neurons respond as a Poisson process with stimulus-dependent stationary firing rates (static firing rate code). (B) Neurons respond as Poisson processes with time-dependent firing rate (instantaneous firing rate code). In the example shown here, each neuron increases its firing rate during a specific time interval, leading to a sequence of activity. The sequence is different for stimuli A and B. (C) Spike pattern code. Firing rates are independent of stimulus and time, and information is contained in higher order statistics of spike trains. Here, selective groups of neurons respond with synchronous spikes. (D) Phase-of-firing codes. Firing rates are again independent of stimulus and time but are modulated by an oscillatory variable, indicated in gray (the phase of the oscillation with respect to stimulus onset is random). Single neurons have specific preferred phases for each stimulus.

bears similarities to the one shown in figure 2.6B, since the population activity takes the form of sequences of activity that repeat at every cycle of the oscillation. Note also that all the scenarios indicated in figures 2.5 and 2.6 represent cases in which a single variable encodes information about spike trains. These scenarios are not mutually exclusive, and information could be encoded by several variables simultaneously, a phenomenon called multiplexing (Friedrich et al. 2004; Panzeri et al., 2010).

In order to quantify information encoding capabilities of neurons and networks of neurons, and compare different coding scenarios, various tools from information theory are available (Cover and Thomas, 1991; Rieke et al., 1999; Dayan and Abbott 2001). Shannon's mutual information quantifies how uncertainty about the input to the neuron is reduced by the observation of its spike train, in bits (one bit meaning the reduction of uncertainty by a factor 2); Fisher information can be used in the case of continuous stimuli and provides a lower bound on the variance of an unbiased estimator of the input.

These methods are increasingly used to characterize coding by single neurons or populations of neurons in various sensory systems. In the vast majority of investigated systems, firing rates have been shown to carry a significant amount of information about external stimuli. In some cases, the timescale at which information is encoded in the instantaneous firing rate has been shown to be significantly shorter than the timescales of the encoded stimuli (Butts et al., 2007; Jacobs et al., 2009; Panzeri et al., 2010), a phenomenon which has sometimes been referred to as "temporal code." In some cases, the phase of spikes with respect to a global oscillatory variable can also be shown to carry information about sensory inputs (Jensen and Lisman, 2000; Kayser et al., 2009).

To conclude this section, it should be noted that the presence of information in specific patterns of activity about specific stimuli does not imply that this information is used by the brain to cause behavior. To prove a causal relationship between specific patterns of activity and behavior, one needs to be able to manipulate selectively these patterns and observe the resulting change in behavior. Though this is in general a daunting task in a behaving animal, recent technological advances (in particular optogenetic tools) might allow such causal relationships to be tested in a not too distant future.

2.2 *Firing Rate Models*

A popular approach in modeling populations of neurons is to describe population activity not in terms of individual neuron variables but rather in terms of the average activity of the different populations of cells that compose the network (Beurle, 1956; Griffith, 1963, 1965; Wilson and Cowan, 1972, 1973; Amari, 1977). This approach in particular makes sense when the goal is to understand phenomena observed using large-scale recording or imaging techniques such as electroencephalogaphy, magnetoencephalography, functional magnetic resonance imaging, or voltage-sensitive-dye imaging, which do not provide access to the single-neuron level but rather to population-averaged activities. These models can also be used on a finer scale when it is hypothesized that neurons can be aggregated as pools of similar neurons, or treated as discrete exemplars of a virtual continuum. Such models are known under various names—"neural mass," "firing rate," "rate," or "Wilson–Cowan" models. "Neural field" models refer to population activity models with a spatial structure.

In the simplest rate model, the dynamics of a population of cells is described by a first-order differential equation which describes the temporal evolution of the average activity r as

$$\tau\frac{dr}{dt} = -r + \Phi(I_{ext} + Jr) \tag{2.6}$$

where τ is the time constant of the firing rate dynamics, I_{ext} is the external input to the population, J is proportional to the strength of the synaptic connections within the population (if $J < 0$, the population is said to be inhibitory, while if $J > 0$, it is said to be excitatory), and Φ is a static transfer function that gives the steady-state activity as a function of the inputs (as can be seen by setting the left-hand side to 0). Since the activity is a non-negative quantity, Φ must be positive or zero for any value of the inputs and is typically a monotonically increasing function. Popular choices for this function have been the threshold-linear transfer function,

$$\Phi(I) = \begin{cases} I - T & I > T \\ 0 & I \leq T \end{cases} \tag{2.7}$$

where T is a threshold; continuous sigmoidal functions such as the logistic function, $\Phi(I) = 1/(1 + \exp(-\beta(I - \theta)))$; transfer functions of specific spiking neuron models, such as the leaky integrate-and-fire neuron (see section 2.3); or transfer functions fitted from electrophysiological data. These transfer functions are illustrated in figure 2.7.

2.2.1 ONE-POPULATION RATE MODELS In a one-population network receiving constant inputs, equation 2.6 always converges to a fixed point starting from any

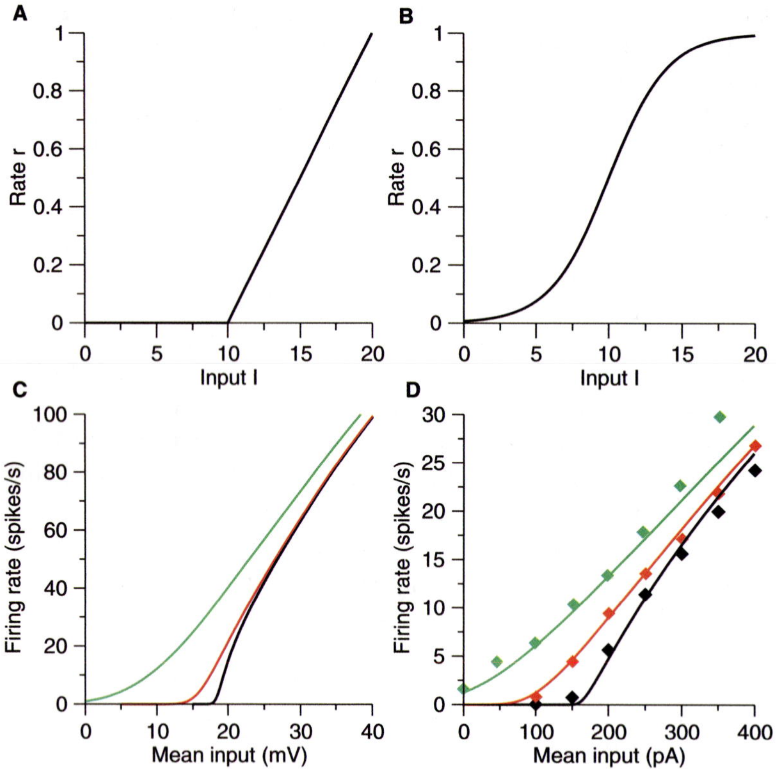

FIGURE 2.7 Transfer functions. (A) Threshold-linear transfer function. (B) Sigmoidal transfer function. (C) Transfer function of a leaky integrate-and-fire neuron in the presence of Gaussian white noise of various intensity levels (black, 1 mV; red, 2 mV; green: 5 mV). (D) Firing rate of a cortical pyramidal cell as a function of the mean injected current I (in pA) for three different values of the noise intensity (symbols). The full lines show a fit with a transfer function of an integrate-and-fire neuron with adaptation. Adapted from Rauch et al. (2003).

initial condition. In an inhibitory network, only a single fixed point exists. In an excitatory network, and with a suitable nonlinear transfer function, the network can exhibit bistability, between a low and a high activity state (see figure 2.8).

The emergence of bistability can be understood by plotting dr/dt as a function of r (see figure 2.8C). Fixed points are the intersections of this curve with the x-axis. In the absence of excitatory feedback, dr/dt is simply a straight line of slope $-1/\tau$ and thus a single fixed point exists, $r = \Phi(I_{ext})$. In the presence of sufficiently large excitatory feedback, the curve dr/dt is no longer monotonically decreasing with r, and beyond a critical value of J (close to $J = 16$ in figure 2.8C) it intersects three times the x-axis. Thus, beyond this critical value, three fixed points exist. The next question is whether these fixed points are stable or unstable. The stability of each fixed point can be readily checked by looking at the sign of dr/dt close to the fixed point. This sign will indicate whether initial conditions that are close to the fixed point tend to converge to it, or on the opposite to drift away from it. Figure 2.8C shows that the fixed points with lowest and highest rates are stable while the fixed point with intermediate rate is unstable. A bifurcation diagram showing how firing rates in stable and unstable fixed points vary with the strength of excitatory feedback J is shown in figure 2.8D.

This model can be thought as a minimal model for bistability due to recurrent excitation. Such a network can keep a short-term memory of a transient input, as shown in figure 2.8A. This short-term memory manifests itself as a persistent activity of a population of excitatory neurons. The phenomenon of persistent activity is believed to be a neuronal correlate of working memory, an active form of short-term memory (Goldman-Rakic, 1995; Wang, 2001). It will be discussed again in section 2.5, and in much greater depth in chapter 11 (Fusi and Wang).

2.2.2 EXCITATORY–INHIBITORY MODEL Neurons in the CNS can be subdivided into many cell types, depending on their morphology, connectivity, electrophysiological properties, and molecular features. The most basic classification separates neurons in two populations, based on their effect on postsynaptic targets: excitatory neurons tend to increase the activity of their

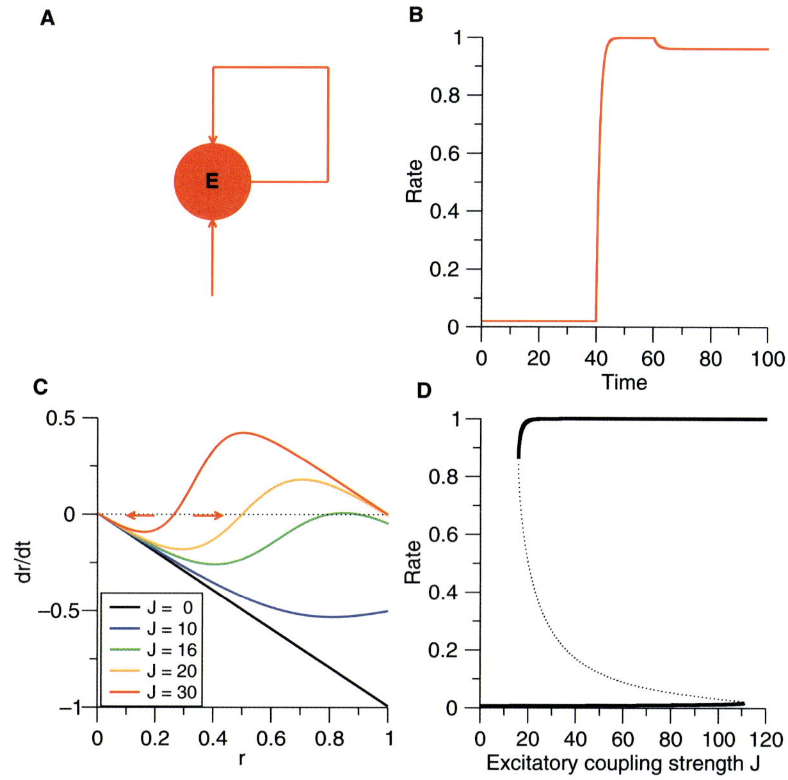

FIGURE 2.8 An excitatory (E) population can be bistable, provided the recurrent coupling is strong enough. (A) Sketch of the excitatory rate model. (B) An external input shown between $t = 40$ and $t = 60$ switches the population from a low activity to a high activity state (here, as well as all other figures of section 2.2, time is measured in units of τ). (C) dr/dt as a function of r for different values of J. For $J = 0$, dr/dt as a function of r is a straight line of negative slope, and therefore there exists a single fixed point $r = \Phi(I)$. For $J > 0$, dr/dt becomes a nonlinear function that may intersect the $dr/dt = 0$ line three times provided J is increased beyond a critical value ~16 mV, leading to three possible fixed points (see the red, orange, and green curves). The stability of these fixed points can be determined by inspecting the sign of dr/dt. When three fixed points exist, the lower and higher fixed points are stable while the intermediate fixed point is unstable. This can be seen by the fact that on the left (resp. right) of the intermediate fixed point, $dr/dt < 0$ (resp. > 0), and therefore the dynamics will tend to decrease (resp. increase) the rate (see corresponding arrows, corresponding to the red curve). (D) Bifurcation diagram, showing the fixed points of the dynamics as a function of J. Stable fixed points are indicated as full curves, while the unstable fixed point is indicated as a dotted curve. Note the large bistable region in which two stable fixed points coexist.

targets, while inhibitory neurons do the opposite. Rate models of two populations of cells, one excitatory, the other inhibitory (Wilson and Cowan, 1973), are described by

$$\tau_E \frac{dr_E}{dt} = -r_E + \Phi_E(I_{EX} + J_{EE}r_E - J_{EI}r_I), \quad (2.8)$$

$$\tau_I \frac{dr_I}{dt} = -r_I + \Phi_I(I_{IX} + J_{IE}r_E - J_{II}r_I) \quad (2.9)$$

where $r_E(t)$, $r_I(t)$ are the rates of excitatory (E) and inhibitory (I) populations, τ_E, τ_I the time constants, and Φ_E, Φ_I the transfer functions of both populations. I_{EX} and I_{IX} are the external inputs to E and I populations while J_{ab} (a, b = E, I) measure the strength of the connectivity from population b to a. In such a model, the interplay between excitation and inhibition can potentially lead to oscillations, provided the feedback between both populations (as measured by the product $J_{EI}J_{IE}$) and the excitatory-to-excitatory connectivity (as measured by J_{EE}) are both strong enough. The mechanism of this oscillation can readily be explained as follows: an increase of excitation leads to an increase in inhibition, followed by a decrease in excitation, followed by a decrease in inhibition, and eventually another increase in excitation which completes the cycle of the oscillation (see figure 2.9). This E–I mechanism has been proposed to account for oscillations in several brain areas (see below for alternate mechanisms, and chapter 4, Vierling-Claassen and Jones for more detailed discussion of oscillations in networks of neurons).

Excitatory–inhibitory networks have additional interesting features. In such networks, the connectivity

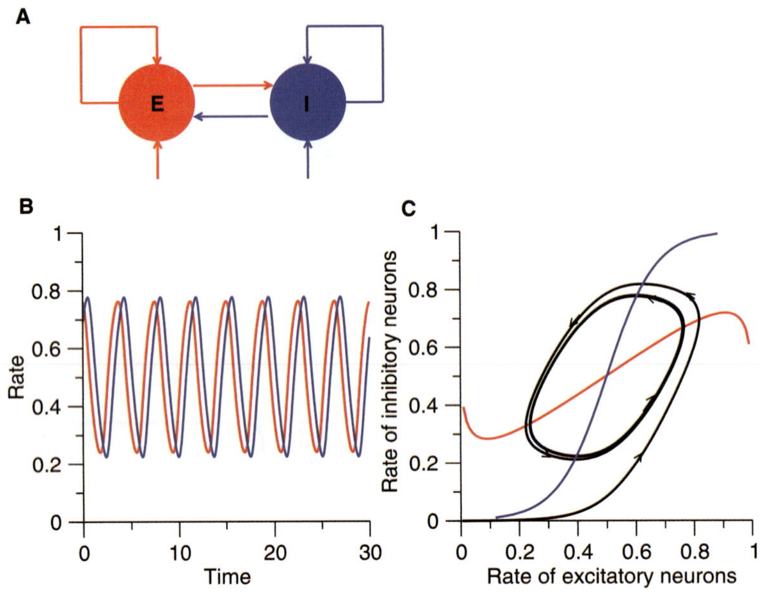

FIGURE 2.9 An excitatory–inhibitory (EI) model can oscillate, provided the feedback between E and I populations, and the E–E coupling, are strong enough. (A) Sketch of the model. (B) Excitatory (red) and inhibitory (blue) rates as a function of time, in the oscillatory regime. (C) r_E–r_I plane, showing the r_E nullcline (red, defined as the set of values of r_E and r_I for which $dr_E/dt = 0$ vanishes) and the r_I nullcline (blue, defined by $dr_I/dt = 0$), together with a trajectory of the model (black), starting from an initial condition $r_E = r_I = 0$. Notice the dynamics converges to a stable limit cycle. Note also that the fixed point (at the intersection of the nullclines) is unstable for the parameters chosen for this plot.

matrix is non-normal, which leads to nonorthogonal eigenvectors. As a consequence, the activity in the network may transiently grow in response to inputs, a phenomenon termed "balanced amplification" by Murphy and Miller (2009). This property has been shown to account for various dynamical phenomena observed in primary visual cortex (Murphy and Miller, 2009; Ozeki et al., 2009).

2.2.3 MULTIPLE INHIBITORY POPULATIONS—WINNER-TAKE-ALL Multistability can also be generated with multiple inhibitory populations, with excitatory external inputs and mutual inhibitory connections between the populations. In this case, a "winner-take-all" behavior emerges, provided the inhibitory feedback is strong enough: the only stable states of the system are states in which a single inhibitory population is active (the winner) and silences the other populations through strong inhibitory connections (see figure 2.10; Amari and Arbib, 1977).

In the simplest case, illustrated in figure 2.10, only two inhibitory populations are present, whose rates evolve according to the following:

$$\tau \frac{dr_1}{dt} = -r_1 + \Phi(I_X - Jr_2), \qquad (2.10)$$

$$\tau \frac{dr_2}{dt} = -r_2 + \Phi(I_X - Jr_1). \qquad (2.11)$$

The dynamics in the (r_1, r_2) plane is illustrated in figure 2.10C, for parameters for which the system has two stable fixed points (in which one of the populations is strongly active while the other is almost silent), and one unstable fixed point. The diagonal dashed line represents the boundary of the basins of attraction of both stable fixed points. Any initial condition above (below) this separatrix converges to the upper left (lower right) fixed point.

2.2.4 RATE MODEL WITH DELAYS In the simplest rate model, there is a single time constant describing the dynamics of the firing rate. This time constant can be thought of as representing the intrinsic properties of the neuron. Alternatively, it could represent in a schematic fashion the synaptic time constants of the network. However, equation 2.6 misses an important property of synaptic currents, that is, delays due to the finite conduction time from the site of action potential initiation to the synaptic site. A delay D can be incorporated in a rate model, by using

$$\tau \frac{dr}{dt} = -r + \Phi[I_{ext} + Jr(t-D)] \qquad (2.12)$$

instead of equation 2.6. Note that the above equation can be generalized to multiple populations that could each have their own time constant, transfer function, and delay. The delay leads to a qualitative change of the

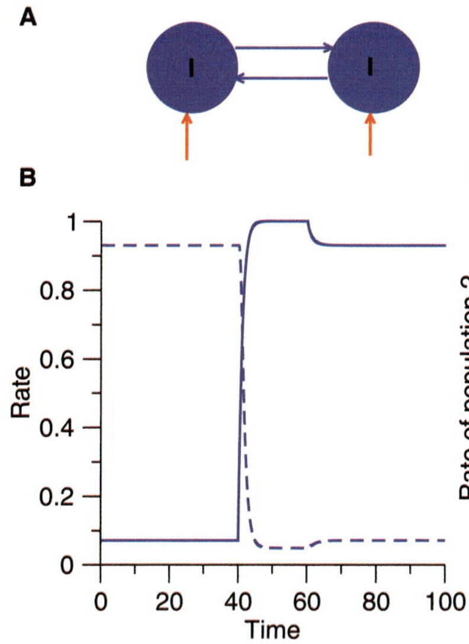

FIGURE 2.10 Two mutually coupled inhibitory populations have a "winner-take-all" behavior, provided the mutual coupling is strong enough. External inputs can switch the system from one state to the other. (A) Sketch of the model. (B) Rates of the two populations (shown as solid and dashed lines) as a function of time, in the winner-take-all regime. The system has two stable states. In both states, the "winner" population suppresses the other. Transient external inputs (here, applied between $t = 40$ and $t = 60$) can switch the system from one state to the other. (C) (r_1,r_2) plane, showing the r_1 and r_2 nullclines (blue). The intersections of the two nullclines are the fixed points of the dynamics. Two fixed points are stable (filled blue circles); one is unstable. Arrows indicate the direction of the flow. The dashed black line indicates the separatrix that forms the boundary of the basins of attraction of the two stable fixed points.

dynamics when interactions are purely inhibitory. In this case, the delay can lead to oscillations, which are not present in the absence of delays. In the presence of delays, an increase in inhibitory firing rate leads to an increase in inhibitory feedback, resulting in a decrease of firing, after a time proportional to the delay. In turn, this decrease in inhibition leads to a disinhibition of the network, and therefore an increase in inhibitory firing rate, after, again, a time proportional to the delay. Such an oscillation is typically faster than an oscillation induced by E–I feedback. In E–I networks, there are therefore two possible mechanisms for generation of oscillations: one due to E–I connectivity, the other due to I–I connectivity (Brunel and Wang, 2003; Ledoux and Brunel, 2011).

2.2.5 RATE MODELS WITH SLOW NEGATIVE FEEDBACK In the models presented above, negative feedback is provided by synaptic inhibition, which typically acts on short time scales. Other forms of negative feedback are present in real neural circuits: at the single-cell level, various types of adaptation currents (see the description of adaptation in section 2.3); at the synaptic level, short-term depression (see section 2.4). These forms of negative feedback typically act on slower timescales than synaptic inhibition. They can be added naturally in firing rate models by adding a new variable which provides a slow negative feedback on firing rates. This slow negative feedback can generate slow oscillations (see Tsodyks et al., 1998; Tabak et al., 2000; Tsodyks et al., 2000; Fuhrmann et al., 2002). This class of models has been used to describe slow oscillations in developing neural networks (Tabak et al., 2000) and is described in detail in chapter 3 (Rubin and Rinzel).

2.2.6 NEURAL FIELD MODELS Rate models are also used to investigate the dynamics of spatially extended networks, or systems which encode continuous stimuli. Such population activity models with a spatial structure are sometimes referred to as "neural field models." In such models, the activity $r(x)$ at a given location x evolves according to

$$\tau \frac{dr(x)}{dt} = -r(x) + \Phi\left(I_{ext}(x,t) + \int dy J(x,y) r(y)\right) \quad (2.13)$$

where the network architecture is characterized by the "synaptic footprint" $J(x, y)$ that specifies the strength of connections from location y to x, while $I_{ext}(x, t)$ describes inputs coming from outside the modeled network (Wilson and Cowan, 1973; Amari, 1977; Ermentrout, 1998). The spatial dependence of J reflects the fact that connectivity between neurons depends on the physical distance between them (Holmgren et al., 2003; Levy and Reyes, 2012) and/or the difference between their selectivity properties (Ko et al., 2011). Such a model

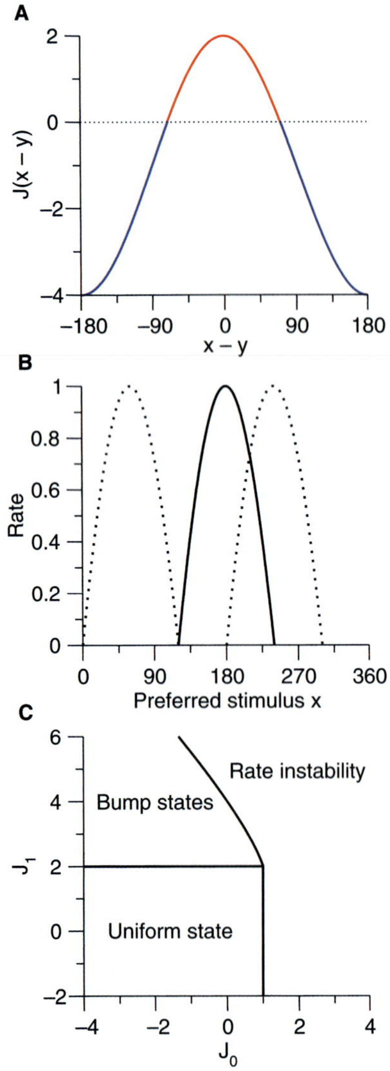

FIGURE 2.11 Ring model. (A) Example of "synaptic footprint" (strength of the connections between neurons at locations x and y, as a function of their "distance" $x - y$) given by equation 2.14, with $J_0 = -1$, $J_1 = 3$, which leads to short-range excitation (red) and long-range inhibition (blue). (B) For the synaptic footprint shown in A, the model converges to spatially localized attractors ("bump attractors"). These attractors can be peaked around any location x. Three examples are shown. (C) "Phase diagram" of the model, showing the three possible regimes described in the text.

allows one to study how the dynamics depends on the spatial and/or functional structure of the connectivity.

The simplest possible spatial structure that has been considered is a ring. This so-called ring model has been studied in many contexts. It has been proposed as a simplified model for a hypercolumn in primary visual cortex, where the variable x represents the preferred orientation of a group of neurons (Ben-Yishai et al., 1995; Ben-Yishai et al., 1997; Hansel and Sompolinsky, 1998; Goldberg et al., 2004). It has also been proposed as a simplified model for prefrontal cortical networks that store the location of a spatial stimulus in working memory (Camperi and Wang, 1998; Itskov et al., 2011). Variants of this model have been proposed as models of head direction cells (Zhang, 1996; Xie et al., 2002), place cells (Tsodyks and Sejnowski, 1995), and grid cells (Burak and Fiete, 2009).

Ben-Yishai et al. (1995) studied a simple scenario in which the transfer function Φ is taken to be threshold-linear (see equation 2.7), the connectivity is taken to be

$$J(x, y) = J_0 + J_1 \cos(x - y) \tag{2.14}$$

where both $x, y \in [0, 2\pi]$ (see figure 2.11A), and the inputs are constant in time and space. In this scenario, one can characterize analytically the "phase diagram" of the model (see figure 2.11C). There exist three qualitatively different regions in the J_0, J_1 plane; when $J_0 < 1$ (for a transfer function of slope 1) and $J_1 < 2$, only a spatially uniform state is stable; from that region, if one increases J_0 beyond 1, the uniform state becomes unstable through a rate instability (rates either diverge or go to zero). If $J_1 > 2$, a Turing instability occurs (Turing, 1952). A "bump state" emerges, in which the activity peaks at a particular location on the ring, and decreases monotonically as a function of distance from the peak. Such bump states, or bump attractors, are shown in figure 2.11B. Bump states are a general feature of "Mexican-hat" connectivity, in which nearby neurons interact in a predominantly excitatory fashion while at large distances connectivity is predominantly inhibitory (Wilson and Cowan, 1973; Amari, 1977; Ermentrout, 1998). The presence of these bump states has been proposed to account to the emergence of "orientation states" during spontaneous activity in V1 (Tsodyks et al., 1999; Goldberg et al., 2004), of "spatial working memory states" in monkey prefrontal cortex (Camperi and Wang, 1998; Compte et al., 2000; Wimmer et al., 2014), and for properties of head direction cells (Zhang, 1996; Xie et al., 2002), place cells (Tsodyks and Sejnowski, 1995), and grid cells (Burak and Fiete, 2009).

2.3 *Neurons*

Brains are composed of two broad classes of cells, neurons and glial cells. Here, we will focus exclusively on neurons. Neurons are traditionally viewed as the basic computational units of the nervous system, while glial cells have been mostly thought to provide support and protection for neurons. Anatomically, neurons are characterized by the long and typically highly branched processes that emerge from the soma. On the input side, the dendrites are typically highly branched structures that can extend up to 1 mm from the soma, on which synaptic contacts from other neurons are made. The number of presynaptic neurons varies greatly from cell to cell, from a single one in unipolar brush cells of the cerebellum to approximately 150,000 in Purkinje cells, with a typical number being in the 1,000–10,000 range in mammalian CNS. On the output side, the axon propagates the spikes emitted by the neuron to postsynaptic neurons. These axons either remain local or can extend up to large distances (as is the case for some pyramidal cells in cortex Purkinje cells in the cerebellum and motor neurons in the spinal cord). Neurons express a large diversity of ionic channels that shape their response to incoming inputs (Johnston and Wu, 1995). Neurons can be modeled at different levels of complexity, and consequently there exist a large number of single-neuron models, from binary neurons to models described by thousands of nonlinearly coupled differential equations. In the following, a few of the more popular models are briefly discussed. (For more details, see, e.g., Dayan and Abbott, 2001; Gerstner and Kistler, 2002; Izhikevich, 2007; Brunel, 2010; Ermentrout and Terman, 2010)

2.3.1 HODGKIN–HUXLEY MODEL Hodgkin–Huxley-type models describe the dynamics of the membrane potential, and a variable number of voltage-gated ionic currents, that are responsible for various dynamical behaviors of a neuron (Ermentrout and Terman, 2010). This class of models dates back to the pioneering work of Hodgkin and Huxley (1952), who deduced the dynamical equations of their model from electrophysiological recordings in the squid giant axon preparation. Their formalism has since been used to describe a large diversity of voltage-gated ionic currents in models of many types of neurons (see, e.g., Huguenard and McCormick, 1994)

In the Hodgkin–Huxley model, the dynamics of the membrane potential $V(t)$ is governed by a first-order differential equation

$$C\frac{dV}{dt} = -I_L(V) - I_{Na}(V) - I_K(V) + I_{ext}(t) \tag{2.15}$$

where C is the membrane capacitance, $I_L(V)$ is a leak current, $I_L(V) = g_L(V - V_L)$ where g_L is the leak conductance and $V_L \sim -70$ mV is the leak potential, $I_{Na}(V)$ is a sodium current, $I_K(V)$ is a potassium current, and $I_{ext}(t)$ is an external current, which can describe a current injected through an electrode, or synaptic currents.

The sodium current $I_{Na}(V)$ is due to influx of sodium ions into the cell, which tends to depolarize the neuron (since sodium ions are positively charged), through sodium channels. Hodgkin and Huxley found that the dynamics of this current could be well described by the following system of equations,

$$\begin{aligned} I_{Na}(V) &= \bar{g}_{Na} m^3 h (V - V_{Na}) \\ \tau_m(V)\frac{dm}{dt} &= -m + m_\infty(V) \\ \tau_h(V)\frac{dh}{dt} &= -h + h_\infty(V), \end{aligned} \tag{2.16}$$

where $\bar{g}_{Na}$ is the maximal conductance of sodium channels, m is an activation variable, whose dynamics is described by a first-order differential equation, where $\tau_m(V)$ is a voltage-dependent activation time constant (shown in figure 2.12A), and $m_\infty(V)$ is a voltage-dependent steady-state activation (shown in figure 2.12A), h is an inactivation variable, whose dynamics is also described by a first-order differential equation, where $\tau_h(V)$ is a voltage-dependent inactivation time constant (shown in figure 2.12A), and $h_\infty(V)$ is a voltage-dependent steady-state inactivation (shown in figure 2.12A). Finally, $V_{Na} \sim +70$ mV is the reversal potential of sodium channels—the Nernst equilibrium potential at which there is an equilibrium between diffusion (that tends to make sodium ions move because of concentration gradients) and the electrical force (that tends to make them move due to the electrical field induced by the membrane potential difference between the intracellular and extracellular sides).

The sodium current is responsible for the initiation of action potentials, thanks to the positive feedback provided by the m variable on the voltage. When the neuron is depolarized by external inputs (either from an electrode or synapses), the membrane potential increases. This increase in membrane potential leads to the opening of sodium channels, as described by the activation variable m. This opening leads to further depolarization of the neuron because the sodium current tends to depolarize the cell when the voltage is lower than the reversal potential $V_{Na} \sim +70$ mV. As a result, the increase in sodium channel activation leads to a rapid increase in depolarization (the upstroke of the action potential; see figure 2.12B). This mechanism explains the upstroke of the action potential.

The next phase of the action potential (the downstroke) is explained by a combination of two factors. The first is inactivation of the sodium current—the h variable decreases to zero as V increases. The second is due to the opening of another voltage-gated ionic

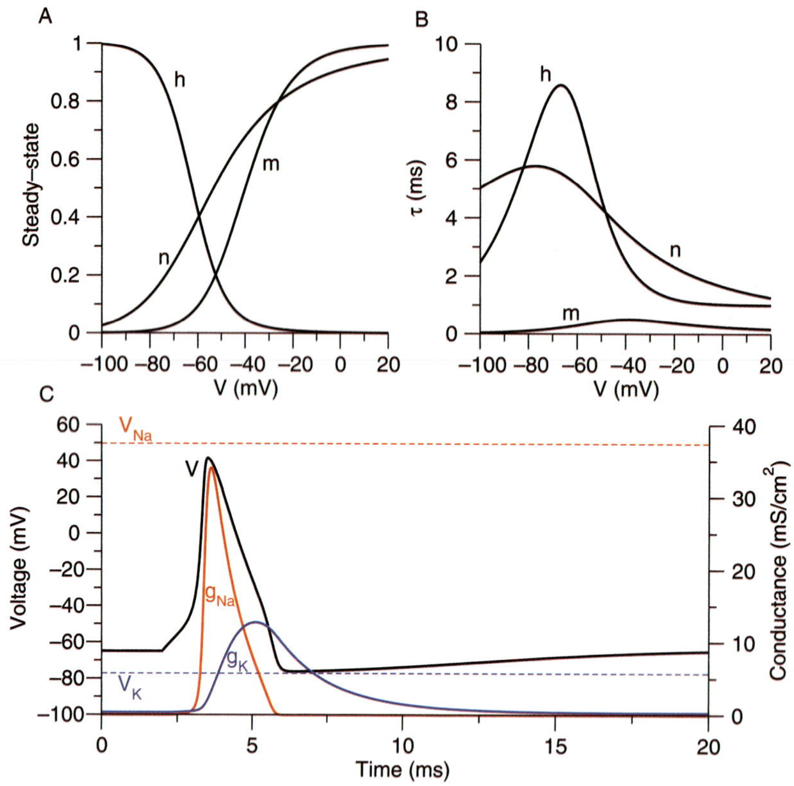

FIGURE 2.12 The Hodgkin–Huxley model. (A) Steady-state activation (m_∞ and n_∞) and inactivation (h_∞) curves as a function of voltage. (B) Time constants of activation and inactivation variables as a function of voltage. (C) Action potential in the Hodgkin–Huxley model. The first stage consists in a sharp increase in the sodium conductance g_{Na} that leads to the upstroke of the action potential (steep rise in the membrane potential V) toward the sodium reversal potential V_{Na}, which is approximately 115 mV above the resting potential. The second stage consists in a decrease in sodium conductance due to inactivation and in an increase in potassium conductance g_K which tends to bring the voltage back toward the potassium reversal potential V_K. Adapted from Hodgkin and Huxley (1952).

current, a potassium current $I_K(V)$ due to potassium ions leaving the cell following the opening of potassium channels. This current was also described quantitatively by Hodgkin and Huxley using the following equations:

$$I_K(V) = \bar{g}_K n^4 (V - V_K),$$
$$\tau_n(V)\frac{dn}{dt} = -n - n_\infty(V)$$

where $\bar{g}_K$ is the maximal conductance of sodium channels, n is an activation variable, whose dynamics is described by a first-order differential equation, where $\tau_n(V)$ is a voltage-dependent activation time constant (shown in figure 2.12A), and $n_\infty(V)$ is a voltage-dependent steady-state activation (shown in figure 2.12A). Note that contrary to the fast sodium current which is *transient*, the potassium current is *persistent*—it does not inactivate (see figure 2.12B). Finally, $V_K \sim -90$ mV is the reversal potential of potassium channels. The activation of the potassium current, with a time constant that is similar to the inactivation of the sodium current, participates in repolarizing the neuron at membrane potentials close to the resting potential (see figure 2.12B).

This model has been highly influential for its capacity to reproduce quantitatively action potential generation (and its propagation in a spatially extended version), and a similar formalism has been used to describe quantitatively a large number of additional voltage-gated currents (see, e.g., Ermentrout and Terman, 2010). However, its complexity makes its mathematical characterization difficult, and simulations of large numbers of such neurons expensive. For this reason, modelers have introduced simplified models. We now turn our attention to such models.

2.3.2 TWO-VARIABLE EXCITABLE MODELS The Hodgkin–Huxley model, and other models with additional ionic currents described with the same formalism, can in some circumstances be approximated by a system of only two differential equations (Krinsky and Kokoz, 1973; Ermentrout and Terman, 2010). In the case of the Hodgkin–Huxley model, the approximation relies on two features of the model: (1) the m variable describing the activation of the fast sodium channels is much faster than all other variables (see figure 2.12A), and therefore one can approximate its dynamics as instantaneous, $m(t) = m_\infty(V(t))$; (2) the h and n variables play a similar role of a negative feedback on the voltage, which leads to a repolarization following action potential generation. Furthermore, $h + n \sim 1$ during action potential generation, and so one of these variables can be dropped.

This leads to a system of two differential equations describing the dynamics of the voltage V, together with a "recovery" variable n:

$$\frac{dV}{dt} = f(V,n), \tag{2.17}$$

$$\frac{dn}{dt} = g(V,n). \tag{2.18}$$

Notable examples of this class of models are the FitzHugh Nagumo model (FitzHugh, 1961; Nagumo et al., 1962), in which a further simplification led to f being cubic in V and linear in n while g is linear in both V and n, and the Morris–Lecar model (Morris and Lecar, 1981; see also Rinzel and Ermentrout, 1998). These models can be analyzed using standard tools from dynamical systems theory. In particular, the dynamics can be visualized on the $V - n$ phase plane; see chapter 3 (Rubin and Rinzel) for examples of phase plane analysis. Due to this simplification, these models have been instrumental in thoroughly understanding the mechanisms of the bifurcations leading to action potential generation in Hodgkin–Huxley models (see, e.g., Ermentrout and Terman, 2010).

2.3.3 Integrate-and-Fire Type Models The Hodgkin–Huxley model can be further simplified by ignoring the dynamics of sodium and potassium currents in the subthreshold range, setting a fixed voltage threshold for spike emission, and a fixed reset for the repolarization following the spike. This leads to the *leaky integrate-and-fire* (LIF) model, characterized by its membrane potential $V(t)$ whose dynamics obey the much simpler equation

$$\tau_m \frac{dV}{dt} = -(V - V_L) + I_{syn}(t) \tag{2.19}$$

where τ_m is the membrane time constant (typical values are in the range 10–20 ms), V_L is the resting potential (typical values are on the order of –70 mV, but it is often set to 0 for simplicity), and $I_{syn}(t)$ are synaptic inputs (described in the next section). Spikes are emitted whenever the membrane potential exceeds a threshold V_t which is typically 10–20 mV above the resting potential. After the spike, the neuron is reset to a reset potential V_r, which is typically in between the resting potential and the threshold (see figure 2.13A). An absolute refractory period τ_{arp} (during which the membrane potential is clamped to V_r) is often added to the model, to capture the fact that action potentials have a finite duration of order 1–2 ms. In this way the firing rate of the neuron cannot grow arbitrarily large but rather saturates for large synaptic inputs to $1/\tau_{arp}$.

This model can be traced back to Lapicque's pioneering work (Lapicque, 1907), but the term "integrate-and-fire" was first introduced by Knight (1972). This model has been taken as the model of choice by theorists interested in characterizing the dynamics of single neurons in stochastic conditions (Stein, 1965; Knight, 1972; Tuckwell, 1988; Burkitt, 2006a, 2006b). It has also been a popular model for large-scale network simulations, as well as analytical studies of network dynamics (Abbott and van Vreeswijk, 1993; Treves, 1993; Brunel and Hakim, 1999; Brunel, 2000).

The simplicity of the LIF model has many advantages, but it comes with the price that the model lacks many of the features of real neurons. The dynamics of the membrane potential close to the action potential threshold leading to spike initiation is ignored. Furthermore, the model shows little of the rich repertoire of dynamics seen in real cells, such as subthreshold resonance, firing rate adaptation, and bursting, to give just a few examples. Several generalizations of this simple model have been proposed that account for such phenomena.

In *nonlinear integrate-and-fire* neurons, spike-generation dynamics is incorporated through a nonlinear function of voltage. Such models are described by the equation

$$\tau_m \frac{dV}{dt} = -V + \psi(V) + I_{syn}(t) \tag{2.20}$$

where $\psi(V)$ is a nonlinear function of voltage (Fourcaud-Trocmé et al., 2003). With a suitable choice of $\psi(V)$, the membrane potential diverges to infinity in finite time if the synaptic inputs exceed some threshold. The time of divergence to infinity defines spike timing (it is also possible to define for practical purposes a large but finite voltage threshold). Then, as in the LIF model, the voltage is reset to some potential V_r after a refractory period τ_{rp}.

Specific choices for the function $\psi(V)$ that have been proposed include a quadratic function (Ermentrout and Kopell, 1986; Ermentrout, 1996; Gutkin and Ermentrout, 1998; Latham et al., 2000), a quartic function (Touboul, 2008), and an exponential function

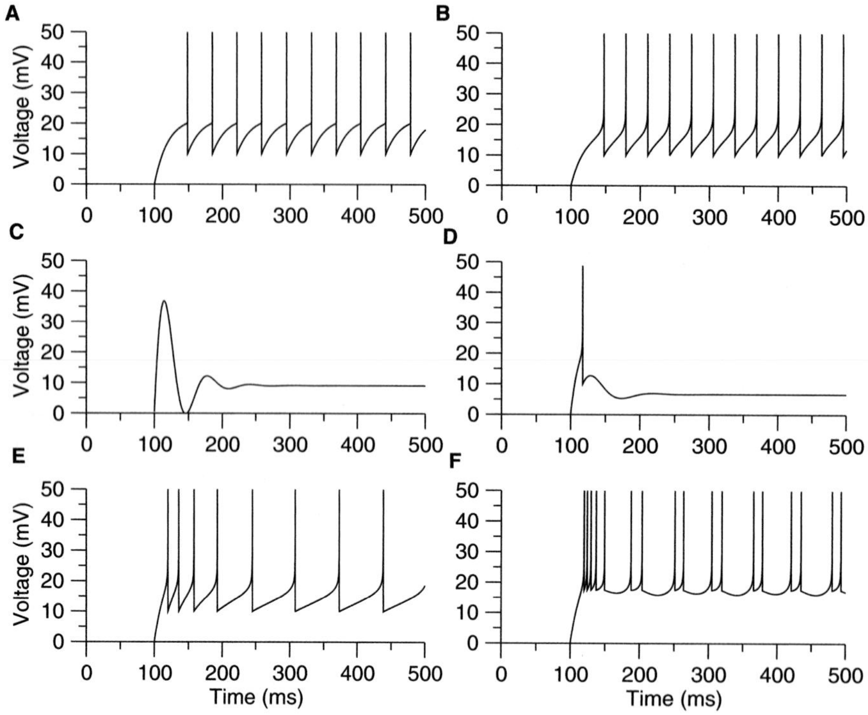

FIGURE 2.13 Leaky integrate-and-fire (LIF) type models. (A) LIF neurons emit spikes whenever the voltage reaches a threshold V_T (here equal to 20 mV) and are then reset to a value V_r (here 10 mV). Spikes have been added to better visualize spike timings. (B) In exponential integrate-and-fire (EIF) neurons the voltage diverges to infinity in finite time whenever the current inputs are strong enough. The voltage is then reset to a value V_r. (C) An adaptive exponential integrate-and-fire (aEIF) with a damped oscillatory response to current step due to the negative feedback provided on the voltage by the "adaptation" variable w. (D) An aEIF can emit a single spike in response to an steady input current. (E) An aEIF with spike-triggered adaptation. (F) An aEIF, that emit bursts of action potential in response to a current step.

(Fourcaud-Trocmé et al., 2003). The *quadratic integrate-and-fire* (QIF) neuron is defined by

$$\psi(V) = \frac{(V - V_T)^2}{2\Delta_T} + V \qquad (2.21)$$

where V_T is the threshold (defined as the highest stable value of the membrane potential in the presence of constant input currents), and Δ_T measures the sharpness of action potential initiation. In this model, the reset can be set to $-\infty$. This model is related through a change of variable to the theta model (Ermentrout and Kopell, 1986). The *exponential integrate-and-fire* (EIF) neuron (Fourcaud-Trocmé et al., 2003) is defined by

$$\psi(V) = \Delta_T \exp\left(\frac{V - V_T}{\Delta_T}\right) \qquad (2.22)$$

where V_T and Δ_T have the same meaning as for the QIF model (see figure 2.13B). Fourcaud-Trocmé et al. (2003) found that an exponential nonlinearity is the choice that best approximates the dynamics of neuron models with a fast sodium current described by the Hodgkin–Huxley formalism.

Badel et al. showed that the exponential nonlinearity fits well the dynamics of both pyramidal cells and fast-spiking interneurons in cortex (Badel et al., 2008a, 2008b). In this study, the authors injected noisy currents $I_{syn}(t)$, and plotted $F(V) = dV/dt - I_{syn}(t)/\tau_m$ as a function of V. The resulting function could be fitted well by a straight line with negative slope (representing the leak current), plus an exponential, exactly like in the EIF model, equation 2.22. This procedure allowed the authors to find parameters of the EIF model that best fit their data, and to show that the EIF model with parameters fitted from data reproduces accurately the spike trains of the recorded neuron when both model and recorded neuron are injected the same realization of a stochastic input current. They further showed that the fit could be improved by adding a second dynamical variable describing the dynamics of the threshold V_T. In this refractory EIF (rEIF) model, the threshold jumps by a fixed amount following a spike and then decays exponentially toward a baseline value, leading to relative refractoriness after a spike.

The rEIF model is a particular case of a model in the class of *adaptive integrate-and-fire* models. In addition to the membrane potential variable, these models include a second variable coupled to the voltage variable. This second variable can be driven by the voltage (Richardson et al., 2003), to capture the dynamics of intrinsic currents that are active in the subthreshold range, or by spikes (Treves, 1993; Liu and Wang, 2001), to capture the effects of firing rate adaptation. Coupled with the nonlinearity in spike generation, these models can generate a wide diversity of subthreshold and suprathreshold behaviors seen in real cells (subthreshold resonance and oscillations, firing rate adaptation, bursting, etc). Notable examples in this class of models are the Izhikevich model (a QIF model with a second variable; Izhikevich, 2003); the adaptive exponential integrate-and-fire (aEIF) model introduced by Brette and Gerstner (2005); and the rEIF model discussed above (Badel et al., 2008a,b).

The aEIF model is defined by the following system of equations,

$$\begin{aligned} \tau_m \frac{dV}{dt} &= -V + \Delta_T \exp\left(\frac{V - V_T}{\Delta_T}\right) - W + I_{syn}(t), \\ \tau_w \frac{dW}{dt} &= -W + aV, \end{aligned} \tag{2.23}$$

where W is the "adaptation" variable that provides a negative feedback on the voltage equation. Equations 2.23 are supplemented by the following reset conditions: when a spike is generated, the voltage is reset to $V = V_r$ while the adaptation variable is increased by an amount b. a is a parameter describing the strength of the coupling between the voltage and the adaptation variable and captures the dynamics of intrinsic currents that are active in the subthreshold range (e.g., the H current), while b describes spike-dependent adaptation, due to calcium-dependent potassium currents, or the M-type potassium current. Figure 2.13C–F shows a few examples of dynamical behaviors of aEIF neurons for different parameter choices. For a large enough a, an aEIF neuron shows damped oscillations of the voltage in response to a current step (figure 2.13C). For large enough current steps, such a neuron can trigger a single spike in response to a prolonged step (figure 2.13D). Figure 2.13E shows firing rate adaptation, due to spike-induced jump b in the adaptation variable. Finally, figure 2.13F shows bursting behavior that occurs for a large enough reset V_r, so that a burst of spikes is emitted before the buildup of the adaptation variable terminates it.

Other generalizations of the LIF model have been proposed. In particular, in *spike-response* models (Gerstner, 1995) the dynamics of the membrane potential is no longer described by a simple differential equation. Rather, the model is described by two kernels: the first kernel describes how the inputs are filtered by the membrane potential; the second describes the effect of the spike of the neuron itself on the membrane potential. In such a way, this model can reproduce arbitrary shapes of post-spike repolarization. This class of models contains the LIF model as a special case.

2.3.4 Binary and Analog Rate Neurons A last drastic simplification consists in reducing a neuron to a binary unit (active or inactive). *Binary neurons* were initially proposed by McCulloch and Pitts in their seminal 1943 paper. In such a model, the neuron output is a binary variable (the state of neuron i, $S_i(t) = 0, 1$, or sometimes $-1, 1$) which is a (deterministic or stochastic) function of the inputs. The inputs $h_i(t)$ are a weighted sum of the activity of all presynaptic neurons,

$$h_i(t) = \sum_j J_{ij} S_j(t) \tag{2.24}$$

where J_{ij} are the synaptic weights from neuron j to neuron i. Neuronal updating can be done either synchronously or asynchronously, and either deterministically or stochastically. For synchronous deterministic updating,

$$S_i(t+1) = \Theta(h_i(t) - T) \tag{2.25}$$

where T is a threshold, and $\Theta(.)$ is the Heaviside function, ($\Theta(x) = 1$ for $x \geq 0$, $\Theta(x) = 0$ for $x < 0$), for all i. For asynchronous updating, a neuron is selected at random at each time step, and equation 2.25 is applied. For stochastic updating, the state of neuron i is drawn randomly, with $\text{Prob}(S_i(t+1) = 1) = \phi(h_i(t))$, where ϕ is a monotonically increasing function from 0 to 1. Binary neurons have been of fundamental importance in many contexts, from the study of supervised learning (see the perceptron model discussed in chapter 1) to the study of dynamics of recurrent networks (see the Hopfield model and balanced network models below).

Analog firing rate neurons (also called leaky integrator neurons) can be seen as a simple generalization of binary neurons, where the output r_i is no longer a binary variable but rather a continuous (firing rate) variable that obeys a simple ordinary differential equation,

$$\tau \frac{dr_i}{dt} = -r_i + \Phi\left(\sum_j J_{ij} r_j(t)\right) \tag{2.26}$$

where Φ is the static transfer function, and J_{ij} are the synaptic weights from neuron j to neuron i. Formally, these neurons are equivalent to the "neural mass/firing rate" models described in section 2.2 above.

The difference is that the basic variable of the model describes the firing rate of individual neurons rather than the average firing rate of a population of functionally similar neurons.

2.3.5 SUMMARY Given the large variety of models which are present in the literature, computational neuroscientists are often faced with the question of which model to use. To understand the biophysical mechanisms of specific firing patterns of neurons, Hodgkin–Huxley-type models have provided invaluable. This formalism has allowed a large number of cell types to be described quantitatively, using specific sets of ionic currents (see, e.g., Huguenard and McCormick, 1994). Such models can then sometimes be reduced, using approximations or mathematical reductions in specific limits, to simpler 2-D excitable models (e.g., Krinsky and Kokoz, 1973) or even simpler integrate-and-fire type models (Abbott and Kepler, 1990; Fourcaud-Trocmé et al., 2003).

Integrate-and-fire type models are the models of choice to understand the impact of noise on the input–output relationship of neurons. They are also the models of choice for network studies, for two reasons: (1) their simplicity makes it possible to use analytical methods to analyze their response to noisy inputs (Tuckwell, 1988; Burkitt, 2006a, 2006b); such methods can be generalized to study the dynamics of networks of such neurons in the presence of noise (Abbott and van Vreeswijk, 1993; Brunel and Hakim, 1999; Brunel, 2000); (2) they make it possible to simulate large scale networks (Izhikevich and Edelman, 2008). Recent studies have shown that these models, even though they do not describe in detail the biophysical properties of real neurons, can nonetheless reproduce quantitatively electrophysiological recordings of real neurons: the frequency versus input current (f–I) curve (Rauch et al., 2003) (see figure 2.7) or even the full spike trains of such neurons upon stimulation by random fluctuating currents (see figure 2.13, Badel et al., 2008a,b; Gerstner and Naud, 2009; Kobayashi et al., 2009).

Finally, binary neurons have been fundamental as a first step to understanding the dynamics of networks with a nontrivial synaptic structure, such as networks whose synaptic structure is shaped by learning (Hopfield, 1982; Tsodyks and Feigel'man, 1988), thanks to methods from statistical physics (see section 2.5).

2.4 *Synapses*

Synapses are the sites of communication between neurons. They can be subdivided into two fundamentally different types, chemical synapses and electrical synapses. Here we focus on chemical synapses, which typically vastly outnumber their electrical counterparts. Communication through chemical synapses is driven by spikes of presynaptic neurons that travel along the axon. When spikes reach presynaptic boutons, the depolarization causes release of neurotransmitters in the synaptic cleft separating the pre- and postsynaptic sides. Neurotransmitters then bind to receptor channels on the postsynaptic membrane, which triggers opening of the channels. This opening leads to ion entry in the postsynaptic cell and, consequently, postsynaptic currents (PSCs). Common types of excitatory receptor channels are AMPA receptors, which lead to a fast excitatory PSC (EPSC), and NMDA receptors, which lead to a slower EPSC. The most common type of inhibitory receptor channel is the GABA receptor, which leads to an inhibitory PSC (IPSC). As for single neuron models, synapses can be modelled at vastly different levels of complexity, from a single number in binary synapses to detailed 3D models that describe synapses at the nanometer scale (see chapters 3, 6, and 8 in de Schutter, 2010). In addition, many types of synapses in the CNS are plastic—their strength is a dynamical variable which maintains a history of its past pre- and postsynaptic spike trains. Here, we focus on simplified models of synapses that are most often used in network studies.

In networks of binary or firing rate neurons, synapses are described by a single number: the synaptic efficacy, which describes the strength with which activity in the presynaptic neuron affects the postsynaptic neuron (J_{ij} in equations 2.24 and 2.26).

In networks of spiking neurons, models for synapses must specify how a presynaptic input spike train (usually described as sums of delta functions; see figure 2.14A) is transformed into a PSC. To describe quantitatively these currents, the following features must be specified: (1) their time course; (2) their voltage dependence; (3) their history dependence; (4) whether they are deterministic or stochastic, and, if stochastic, their statistics. We now consider separately these four aspects.

1. *Time course:* Under the assumption of a linear summation, synaptic currents elicited by spikes of neuron j at times t_j^k can be written as

$$I_j(t) \propto \sum_k S(t - t_j^k) \qquad (2.27)$$

where the function $S(t)$ describes the time course of an individual PSCs triggered by a spike at time $t = 0$. The following choices of the function $S(t)$ have been used:

- A delayed delta function, $S(t) \propto \delta(t - D)$. In this case, the synaptic currents are a sum of delta functions, with a delay D after the presynaptic spikes (see figure 2.14B).

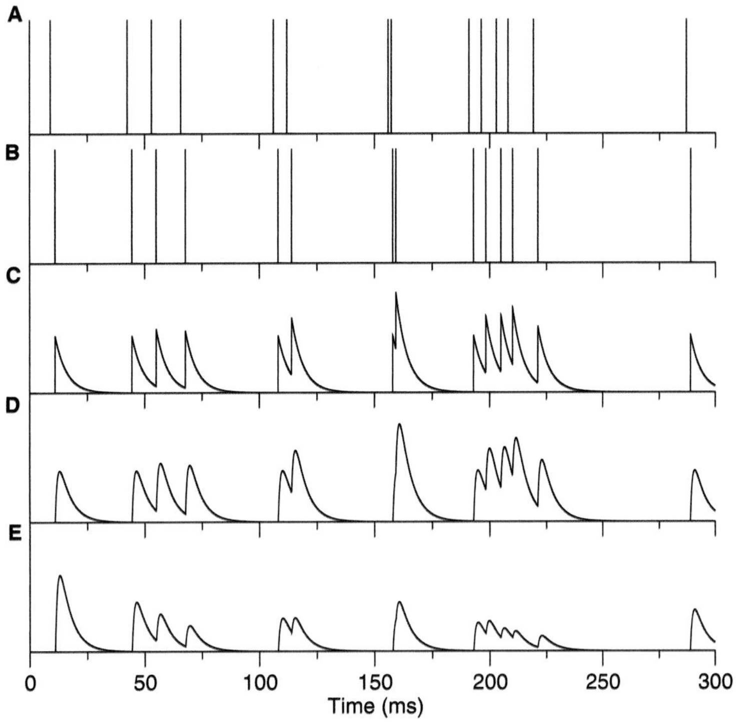

FIGURE 2.14 Synaptic current models. (A) Input spike train (realization of a Poisson process with a 50-Hz rate). (B) Instantaneous synapses with a delay (latency) of 2 ms. (C) Jump after a delay of 2 ms, followed by an exponential decay with time constant 5 ms. (D) Delayed difference of exponentials. (E) Tsodyks–Markram model. Note the pronounced depression of synaptic responses following brief interspike intervals and recovery following long interspike intervals.

- An instantaneous jump followed by an exponential decay with a decay time constant τ_s, $S(t) \propto \exp(-t/\tau_s)\Theta(t)$ where $\Theta(t)$ is the Heaviside function (see figure 2.14C). Such PSCs are obtained when the synaptic currents obey the differential equation

$$\tau_s \frac{dI_j(t)}{dt} = -I_j(t) + J\tau_s \sum_k \delta(t - t_j^k), \qquad (2.28)$$

- A difference of exponentials, $S(t) \propto [\exp(-t/\tau_s) - \exp(-t/\tau_r)]\Theta(t)$ which describes a PSC with a rising phase (with a timescale $\tau_r < \tau_s$) and a decay phase. Such PSCs are obtained when synaptic currents obey a system of two differential equations,

$$\tau_s \frac{dI_j(t)}{dt} = -I_j(t) + x_j(t)$$

$$\tau_r \frac{dx_j(t)}{dt} = -x_j(t) + J\tau_s \sum_k \delta(t - t_j^k)$$

- A delayed exponential function $S(t) \propto \exp(-(t - D)/\tau_s)\Theta(t - D)$, or delayed difference of exponentials, $S(t) \propto (\exp(-(t - D)/\tau_s) - \exp(-(t - D)/\tau_r))\Theta(t - D)$ where Θ is the Heaviside function. This is the simplest function that describes the three typical phases of a PSC (latency, rise, and decay—see figure 2.14D). Such a time course can be obtained using the following differential equations,

$$\tau_s \frac{dI_j(t)}{dt} = -I_j(t) + x_j(t) \qquad (2.28)$$

$$\tau_r \frac{dx_j(t)}{dt} = -x_j(t) + J\tau_s \sum_k \delta(t - t_j^k - D) \qquad (2.29)$$

2. *Voltage dependence:* The simplest choice for synaptic currents is to take them as voltage independent. This leads to a so-called "current-based model" in which

$$I_j(t) = J \sum_k S(t - t_j^k) \qquad (2.30)$$

where J is a constant independent of voltage. Such models are useful for analytical studies of network dynamics because of their simplicity.

Real synaptic currents depend on voltage through the driving force, that is, the difference between the voltage and the reversal potential associated to the receptor channel. Taking into account this driving force leads to a so-called "conductance-based model" in which

$$I_j(t) = g(V(t) - V_{syn}) \sum_k S(t - t_j^k) \qquad (2.31)$$

where g is a constant measuring synaptic strength in conductance units, V_{syn} is the associated reversal potential (typical values for excitatory synapses are 70 mV above the resting potential, while for inhibitory synapses the reversal potential is typically close to or below the resting potential).

Finally, synaptic currents induced by NMDA receptors have a more complex voltage dependence since they are blocked by magnesium ions in a voltage-dependent way. This can be incorporated in the model by adding a sigmoidal dependence of g on the voltage (see, e.g., Jahr and Stevens, 1990; Wang, 1999)

3. *History dependence:* Synaptic currents exhibit various forms of plasticity from short-term plasticity to long-term plasticity. Short-term plasticity has been described by simple phenomenological models. In the simplest such model, introduced by Tsodyks and Markram (1997), the parameter J in equation 2.30 or g in equation 2.31 becomes a dynamical variable $J(t) = \bar{J}ux(t)$, where $x(t)$ is the fraction of available "synaptic resources," which drops by a fixed amount u at each presynaptic spike and then recovers exponentially:

$$\frac{dx}{dt} = \frac{1-x}{\tau_D} - ux\sum_k \delta(t - t_k), \tag{2.32}$$

where τ_D is the time constant of recovery, u measures the fraction of resources that are released at each spike, and the sum of delta functions represent a sum of presynaptic spikes, occurring at times $t = t_k$. Synaptic currents described by the Tsodyks and Markram (1997) model are shown in figure 2.14E. Note that the amplitude of PSCs decreases markedly when presynaptic spikes occur in close succession, and then recover during sufficiently long inter-spike intervals. The opposite phenomenon of short-term facilitation (transient increase in synaptic strength upon repetitive spiking at sufficiently high frequency) can be obtained by transforming the constant u in Eq.(2.32) into a dynamic variable that increases at each presynaptic spike, and relaxes exponentially between successive presynaptic spikes. These models have been used to successfully fit experimental data in a number of preparations (see, e.g., Wang et al., 2006). Models describing long-term plasticity are outside the scope of this chapter and will be described in Wulfram Gerstner's chapter (chapter 6).

4. *Deterministic versus stochastic:* Synaptic currents are highly stochastic, with large fluctuations from one PSC to the next. A standard model to describe these fluctuations is the quantal model initially proposed by Katz and collaborators (del Castillo and Katz, 1954, Johnston and Wu, 1995). In this so-called quantal model, a presynaptic terminal has N independent release sites (with as many releasable vesicles docked to these sites), which can release neurotransmitter with a probability p when an action potential invades the presynaptic terminal. The postsynaptic current is then proportional to q times the number of released vesicles, where q is the response induced by a single vesicle. The probability to observe a current of size nq is therefore given by a simple binomial law, $P(J = nq) = C_N^n p^n (1-p)^{N-n}$, where C_N^ns are binomial coefficients. To fit the distributions of amplitudes of real PSCs, q must itself be a random variable, whose distribution is typically taken to be a Gaussian. These stochastic models can be extended to include short-term depression and facilitation (see, e.g., Mongillo et al., 2012).

2.5 *Networks*

2.5.1 Network Architectures Neurons are interconnected to form complex networks. Mathematically speaking, a network is defined as a set of N neurons, together with a connectivity matrix J_{ij} $(i, j = 1, \ldots, N)$, where J_{ij} is the synaptic strength from neuron j to neuron i (see equations 2.24, 2.26). In other words, a network of neurons is a directed weighted graph, in which the vertices (or nodes) are the neurons, and the directed edges are synapses. In the connectivity matrix the diagonal elements are autapses, which are often set to zero, $J_{ii} = 0$ for all i.

Specific types of networks have been extensively studied by theorists. *Feedforward networks* are networks which contain no feedback loops; special emphasis has been placed on networks which can be arranged as a series of layers, separating an input layer from an output layer, with synaptic connections going only in one direction. In such networks, information propagates in one direction, from the input layer to the output layer. Such networks have been proposed as models for sensory systems, which are traditionally seen as a feedforward series of processing stages.

However, the vast majority of networks in the CNS contain feedback loops. Networks with feedback loops are sometimes called *recurrent networks.* This means that neurons in a given layer are typically interconnected between themselves and may receive inputs from higher layers (i.e., those separated by more synapses from the sensory input). Prototypical networks which contain extensive recurrent connectivity are networks of the neocortex and hippocampus area CA3. To study recurrent networks, theorists have often considered the simplest case of *fully connected networks* (a complete graph where any neuron sends outputs to, and receives inputs from, all other neurons of the network). This is, of course, an oversimplification since connection probability is typically significantly lower than one for nearby neurons (Markram et al., 1997; Sjöström et al., 2001; Holmgren et al., 2003; Levy and Reyes, 2012) and decays with distance beyond a few hundred microns (Holmgren et al., 2003). This fact has led to the study of *randomly connected networks*, in which any synaptic connection is present with a probability which is lower than one. Of particular interest are *sparsely connected networks*, in which the connection probability is vanishingly small in the large N limit.

2.5.2 NETWORKS OF BINARY NEURONS Networks of binary neurons have been studied extensively since they are the most amenable to analysis, using methods from statistical physics. In the particular case of a symmetrical connectivity matrix ($J_{ij} = J_{ji}$ for all $i \neq j$), with no autapses ($J_{ii} = 0$ for all i), such networks can be described by an energy (or Lyapunov) function,

$$E(t) = -\frac{1}{2}\sum_{i,j=1}^{N} J_{ij} S_i(t) S_j(t)\,, \tag{2.33}$$

which is guaranteed to decrease (or remain constant) under asynchronous updating of the network dynamics. To demonstrate this property, suppose at time t the neurons have states $S_i(t)$, with an associated energy $E(t)$. A time $t + dt$ neuron k is updated according to

$$S_k(t+dt) = \mathrm{sign}\left(\sum_{j \neq k} J_{kj} S_j(t)\right). \tag{2.34}$$

Using the symmetry of the connectivity matrix, we can write the new energy as

$$E(t+dt) = E(t) - (S_k(t+dt) - S_k(t))\sum_{j\neq k} J_{kj} S_j(t)\,. \tag{2.35}$$

If the update does not change the state of neuron k, then clearly the energy does not change. If neuron k changes its state, then it is easy to see from equations 2.34, 2.35 that

$$E(t + dt) < E(t). \tag{2.36}$$

Since the energy function is lower bounded, the dynamics must decrease the energy function until it reaches a local minimum, which corresponds to a fixed point of the dynamics. As a consequence, the equilibrium properties of such networks can be described by standard equilibrium statistical physics. These tools have been, in particular, crucial to the understanding of the properties of networks shaped by Hebbian learning, whose prototype is the Hopfield model (Hopfield, 1982).

In the *Hopfield model*, specific patterns of activity $\{\xi_i^\mu\}$ (where ξ_i^μ is the state of neuron $i = 1, \ldots, N$ in pattern $\mu = 1, \ldots, p$) are memorized in the synaptic connectivity matrix through a "Hebbian" learning rule:

$$J_{ij} = \frac{1}{N}\sum_{\mu} \xi_i^\mu \xi_j^\mu. \tag{2.37}$$

When the patterns are uncorrelated and unbiased, they are attractors of the network dynamics when $p << N$. To see why, let us consider the situation where at time t the network is in a state corresponding to one of the patterns, say $\mu = 1$. To check whether this state is an attractor of the dynamics, one needs to compute the total inputs to each neuron in the network,

$$\sum_i J_{ij} S_j(t) = \frac{1}{N}\sum_{i,\mu} \xi_i^\mu \xi_j^\mu \xi_j^1 \tag{2.38}$$

$$= \xi_i^1 + \frac{1}{N}\sum_{i,\mu\neq 1} \xi_i^\mu \xi_j^\mu \xi_j^1 \tag{2.39}$$

where on the second line of the above equation, we have separated in the sum over patterns the terms corresponding to pattern $\mu = 1$ from all the other patterns ($\mu \neq 1$). The first term corresponds to the "signal" due to the imprinting of the pattern in the connectivity matrix while the second term corresponds to the "noise" due to interference with other stored patterns. This second term is a sum of Np random uncorrelated binary variables divided by N and becomes, therefore, in the large N limit a Gaussian random variable with zero mean and variance p/N. This noise term is thus negligible compared to the signal when $p << N$. When the noise can be neglected, equations 2.34, 2.39 show that the pattern is an attractor of the dynamics.

When $p \sim N$, the noise term becomes of the same order as the signal. However, it can be shown that network states close to the stored patterns remain attractors of the network dynamics for $p < 0.14N$ (Amit et al., 1985). Therefore, such networks can store a number of patterns that scale linearly with network size. Sompolinsky (1986) and Derrida et al. (1987) later showed that in randomly diluted networks, in which connections between neurons are present only with some probability, the number of stored patterns scales with the number of modifiable connections per neuron.

The Hopfield model was generalized to a network with 0,1 neurons, storing patterns with low levels of activity, by Tsodyks and Feigel'man (1988). In the *Tsodyks-Feigel'man model*, the connectivity matrix is given by

$$J_{ij} = \frac{1}{f(1-f)N}\sum_{\mu} (\xi_i^\mu - f)(\xi_j^\mu - f) \tag{2.40}$$

where ξ_i^μ is the state of neuron i in pattern μ. ξ_i^μs are drawn randomly and independently for each neuron

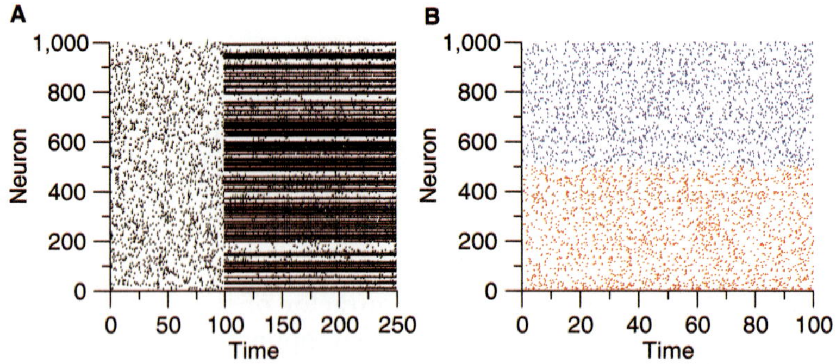

FIGURE 2.15 Dynamics of networks of binary neurons. (A) Raster plot of an associative memory network composed of binary neurons (Tsodyks-Feigel'man model) with stochastic synchronous updating. A dot indicates that the corresponding neuron is active at the corresponding time step. At time $t = 100$, (time is measured in units of the synchronous updatings) one of the stored memories is activated through current injected in 80% of the foreground neurons. The inputs are removed at $t = 150$. A pattern of activity corresponding to the stored memory is a stable attractor of the network dynamics in spite of the noise. (B) Raster plot of a balanced network (two populations of binary neurons, one excitatory, the other inhibitory). Times when neurons flip to the active state are indicated by a red dot (if the neuron is excitatory) or a blue dot (if the neuron is inhibitory). Note the highly irregular nature of neural firing (interspike intervals are highly variable) and the lack of synchronization between neurons.

and each pattern with Prob $(\xi_i^\mu = 1) = f$, where f is the coding level.

The dynamics of such a network is illustrated in figure 2.15A. The network is initialized in a state that is uncorrelated with any stored pattern. Between $t = 100$ and $t = 150$, external inputs are injected to the "foreground" neurons of one of the stored memories. This sets the network in the basin of attraction of the corresponding memory, and the foreground neurons of the corresponding pattern maintain their activity following removal of the stimulus, in spite of the noise.

Thus, this network exhibits selective *persistent activity* following the presentation of one of the stored patterns—neurons selective for the corresponding pattern maintain an elevated activity after the input has been removed. The phenomenon of persistent activity has been observed in various cortical areas of behaving monkeys during delayed response tasks (see, e.g., Wang, 2001). Attractor networks are discussed in more detail in chapter 11 (Fusi and Wang).

Brain networks are not symmetrical. This lack of symmetry has motivated studies of networks of binary neurons with asymmetrical connectivity that can also be analyzed mathematically (see, e.g., Derrida et al., 1987; Ginzburg and Sompolinsky, 1994; van Vreeswijk and Sompolinsky, 1996; van Vreeswijk and Sompolinsky, 1998). Such tools have been used in order to understand the dynamics of networks in which synaptic couplings scale as $1/N$, for which the dynamics of average "firing rates" (probability that a neuron is in an active state) reduce to Wilson–Cowan firing rate-type equations (Ginzburg and Sompolinsky, 1994).

The strength of connections in cortex seems, however, much stronger than a naive $1/N$ scaling (amplitudes of postsynaptic potentials are typically on the order of a few percent of the threshold [Markram et al., 1997, Sjostrom et al., 2001, Holmgren et al., 2003, Lefort et al., 2009], while the typical number of connections of cortical neurons is about 10,000 [Braitenberg and Shütz, 1991]). Furthermore, in vivo recordings show that the membrane potential of cortical neurons exhibit very large fluctuations. This seems inconsistent with the integration of a large number N of very small ($\sim 1/N$) inputs, which by the central limit theorem should lead to very small fluctuations.

This led several groups to study the dynamics of networks of randomly connected excitatory and inhibitory neurons in a strongly coupled regime in which individual synaptic strengths are much stronger than $1/N$ (Tsodyks and Sejnowski, 1995, Amit and Brunel, 1997). In particular, van Vreeswijk and Sompolinsky investigated a "balanced network" composed of two populations of neurons, excitatory and inhibitory in which each neuron receive in average K excitatory and K inhibitory connections, where $K << N$, and in which synapses have strengths that scale as $1/\sqrt{K}$ when K becomes large (van Vreeswijk and Sompolinsky, 1996; van Vreeswijk and Sompolinsky, 1998). Provided conditions on synaptic strengths are satisfied (in particular, inhibitory connections should be strong enough), the network settles in a "balanced state" in which both excitatory and inhibitory inputs are very large but cancel up to terms of order one. In such a network, the temporal fluctuations seen by neurons stay finite in the large K

limit (in contrast to networks with $1/N$ couplings in which such fluctuations vanish) and generate highly irregular firing of neurons of the network (illustrated in figure 2.15B). This model thus naturally explains the high variability of firing seen in cortical neurons in vivo. Another interesting property of this network is that the network timescale becomes infinitely fast in the large N limit, which allows the network to track time-dependent inputs very fast (van Vreeswijk and Sompolinsky, 1996; van Vreeswijk and Sompolinsky, 1998; Renart et al., 2010). The activity of such a network is illustrated in figure 2.15B.

2.5.3 Networks of Spiking Neurons The analysis of networks of spiking neurons is significantly more involved than the analysis of networks of binary neurons. Early studies considered fully connected, homogeneous networks of spiking neurons and studied the conditions for which the network settles into an asynchronous state, or vice versa in a strongly synchronized state (Mirollo and Strogatz, 1990; Abbott and van Vreeswijk, 1993; Treves, 1993; Hansel et al., 1993). Later studies focused on randomly connected networks composed of two populations of spiking neurons, excitatory and inhibitory (Amit and Brunel, 1997; Brunel, 2000; Brunel and Wang, 2003). These studies found that the network can settle in a low rate state in which all neurons fire in a highly irregular fashion (close to a Poisson process), provided the inhibition is strong enough so that it effectively dominates the recurrent inputs (similar to the "balanced network" of binary neurons). However, in networks of spiking neurons low rate states can be of several types. Networks can settle in *asynchronous states* in which the population activity becomes constant in time in the large N limit, and correlations between neurons vanish in this limit (see figure 2.16A for an illustration of such dynamics). Alternatively, they can settle in *synchronous states* in which the population activity oscillates in time, with a frequency that depends on synaptic time constants, and as a result strong correlations exist between neurons. Both states exhibit strong single-neuron irregularity in the inhibition-dominated regime. Interestingly, such networks can switch from an asynchronous state to a synchronous state by increasing the external inputs. This phenomenon could explain the occurrence of transient fast oscillations in many sensory systems in the presence of strong sensory stimulation.

In this chapter, we have focused on some of the most popular models in computational neuroscience. We have seen that at all levels of investigation (from synapse to networks), a broad range of models exist, from the highly simplified to the highly detailed. Highly simplified models (such as the LIF neuron, or networks of binary neurons) have the advantage of allowing for detailed analytical insight that leads to a deep understanding of their behavior. However, such models sometimes fail to capture behaviors shown by real neural systems. Highly detailed models capture a wider diversity of behaviors, but their complexity makes them much harder to study analytically, or through numerical simulations, because of a very large, and consequently hard to explore, parameter space. One of the challenges of research in computational neuroscience lies in choosing the right intermediate level for a given problem—a model that is simple enough as to allow the best possible insights yet does not miss the crucial parameters leading to the behavior one wishes to explain.

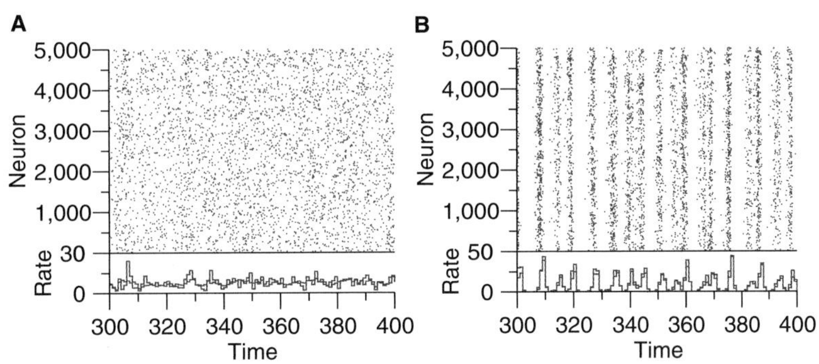

Figure 2.16 Dynamics of randomly connected networks of leaky integrate-and-fire neurons (4,000 excitatory and 1,000 inhibitory neurons). Both A and B show a raster plot (spike times of excitatory neurons in red, inhibitory neurons in blue), and the average spike rate of both populations as a function of time (computed in 1-ms bins). (A) The network is in an asynchronous irregular state. The population averaged spike rate is approximately constant in time, with fluctuations around the mean due to finite size effects. (B) The same network switches to a synchronous irregular state with stronger external inputs. Note the pronounced population oscillation at approximately 150 Hz while single neurons fire irregularly with a rate which is smaller than the population frequency. At each cycle of the oscillation, only a fraction of neurons fire.

REFERENCES

Abbott, L. F., & Kepler, T. (1990). Model neurons: From Hodgkin-Huxley to Hopfield. In L. Garrido (Ed.), *Statistical Mechanics of Neural Networks* (pp. 5–18). Berlin: Springer-Verlag.

Abbott, L. F., & van Vreeswijk, C. (1993). Asynchronous states in a network of pulse-coupled oscillators. *Physical Review E: Statistical Physics, Plasmas, Fluids, and Related Interdisciplinary Topics, 48,* 1483–1490.

Ahrens, M. B., & Engert, F. (2015). Large-scale imaging in small brains. *Current Opinion in Neurobiology, 32,* 78–86.

Aksay, E., Baker, R., Seung, H. S., & Tank, D. W. (2000). Anatomy and discharge properties of pre-motor neurons in the goldfish medulla that have eye-position signals during fixations. *Journal of Neurophysiology, 84,* 1035–1049.

Amari, S. (1977). Dynamics of pattern formation in lateral-inhibition type neural fields. *Biological Cybernetics, 27,* 77–87.

Amari, S.-I., & Arbib, M. A. (1977). *Competition and Cooperation in Neural Nets* (pp. 119–166). New York: Academic Press.

Amirikian, B., Georgopoulos, A. P., & Georgopulos, A. P. (2000). Directional tuning profiles of motor cortical cells. *Neuroscience Research, 36,* 73–79.

Amit, D. J., & Brunel, N. (1997). Model of global spontaneous activity and local structured activity during delay periods in the cerebral cortex. *Cerebral Cortex, 7,* 237–252.

Amit, D. J., Gutfreund, H., & Sompolinsky, H. (1985). Storing infinite numbers of patterns in a spin-glass model of neural networks. *Physical Review Letters, 55,* 1530–1531.

Badel, L., Lefort, S., Berger, T. K., Petersen, C. C., Gerstner, W., & Richardson, M. J. (2008a). Extracting non-linear integrate-and-fire models from experimental data using dynamic I–V curves. *Biological Cybernetics, 99,* 361–370.

Badel, L., Lefort, S., Brette, R., Petersen, C. C., Gerstner, W., & Richardson, M. J. (2008b). Dynamic I–V curves are reliable predictors of naturalistic pyramidal-neuron voltage traces. *Journal of Neurophysiology, 99,* 656–666.

Ben-Yishai, R., Hansel, D., & Sompolinsky, H. (1997). Traveling waves and the processing of weakly tuned inputs in a cortical network module. *Journal of Computational Neuroscience, 4,* 57–77.

Ben-Yishai, R., Lev Bar-Or, R., & Sompolinsky, H. (1995). Theory of orientation tuning in visual cortex. *Proceedings of the National Academy of Sciences of the United States of America, 92,* 3844–3848.

Beurle, R. L. (1956). Properties of a mass of cells capable of regenerating pulses. *Philosophical Transactions of the Royal Society of London. Series B, Biological Sciences, 240,* 5594.

Braitenberg, V., & Shüz, A. (1991). *Cortex: Statistics and Geometry of Neuronal Connectivity.* Springer.

Brette, R., & Gerstner, W. (2005). Adaptive exponential integrate-and-fire model as an effective description of neuronal activity. *Journal of Neurophysiology, 94,* 3637–3642.

Brunel, N. (2000). Dynamics of sparsely connected networks of excitatory and inhibitory spiking neurons. *Journal of Computational Neuroscience, 8,* 183–208.

Brunel, N. (2010). Modeling point neurons: From Hodgkin-Huxley to integrate-and-fire. In E. D. Schutter (Ed.), *Computational Modeling Methods for Neuroscientists* (pp. 161–186). Cambridge, MA: MIT Press.

Brunel, N., & Hakim, V. (1999). Fast global oscillations in networks of integrate-and-fire neurons with low firing rates. *Neural Computation, 11,* 1621–1671.

Brunel, N., & Wang, X.-J. (2003). What determines the frequency of fast network oscillations with irregular neural discharges? *Journal of Neurophysiology, 90,* 415–430.

Burak, Y., & Fiete, I. R. (2009). Accurate path integration in continuous attractor network models of grid cells. *PLoS Computational Biology, 5,* e1000291.

Burkitt, A. N. (2006a). A review of the integrate-and-fire neuron model: I. Homogeneous synaptic input. *Biological Cybernetics, 95,* 1–19.

Burkitt, A. N. (2006b). A review of the integrate-and-fire neuron model: II. Inhomogeneous synaptic input and network properties. *Biological Cybernetics, 95,* 97–112.

Butts, D. A., Weng, C., Jin, J., Yeh, C. I., Lesica, N. A., Alonso, J. M., et al. (2007). Temporal precision in the neural code and the timescales of natural vision. *Nature, 449,* 92–95.

Camperi, M., & Wang, X.-J. (1998). A model of visuospatial short-term memory in prefrontal cortex: Recurrent network and cellular bistability. *Journal of Computational Neuroscience, 5,* 383–405.

Compte, A., Brunel, N., Goldman-Rakic, P. S., & Wang, X.-J. (2000). Synaptic mechanisms and network dynamics underlying spatial working memory in a cortical network model. *Cerebral Cortex, 10,* 910–923.

Compte, A., Constantinidis, C., Tegnér, J., Raghavachari, S., Chafee, M., Goldman-Rakic, P. S., et al. (2003). Temporally irregular mnemonic persistent activity in prefrontal neurons of monkeys during a delayed response task. *Journal of Neurophysiology, 90,* 3441–3454.

Cover, T. M., & Thomas, J. A. (1991). *Information Theory.* New York: Wiley.

Dayan, P., & Abbott, L. (2001). *Theoretical Neuroscience.* Cambridge, MA: MIT Press.

Del Castillo, J., & Katz, B. (1954). Quantal components of the end-plate potential. *Journal of Physiology, 124,* 560–573.

Derrida, B., Gardner, E., & Zippelius, A. (1987). An exactly solvable asymmetric neural network model. *Europhysics Letters, 4,* 167–173.

De Schutter, E. (Ed.). (2010). *Computational Modeling Methods for Neuroscientists.* Cambridge, MA: MIT Press.

Ermentrout, G. B. (1996). Type I membranes, phase resetting curves, and synchrony. *Neural Computation, 8,* 979–1001.

Ermentrout, G. B. (1998). Neural networks as spatio-temporal pattern-forming systems. *Reports on Progress in Physics, 61,* 353–430.

Ermentrout, G. B., & Kopell, N. (1986). Parabolic bursting in an excitable system coupled with a slow oscillation. *SIAM Journal on Applied Mathematics, 46,* 233–253.

Ermentrout, G. B., & Terman, D. H. (2010). *Mathematical Foundations of Neuroscience.* New York: Springer.

FitzHugh, R. (1961). Impulses and physiological states in theoretical models of nerve membrane. *Biophysical Journal, 1,* 445–466.

Fourcaud-Trocmé, N., Hansel, D., van Vreeswijk, C., & Brunel, N. (2003). How spike generation mechanisms determine the neuronal response to fluctuating inputs. *Journal of Neuroscience, 23,* 11628–11640.

Friedrich, R. W., Habermann, C. J., & Laurent, G. (2004). Multiplexing using synchrony in the zebrafish olfactory bulb. *Nature Neuroscience, 7*, 862–871.

Fuhrmann, G., Markram, H., & Tsodyks, M. (2002). Spike frequency adaptation and neocortical rhythms. *Journal of Neurophysiology, 88*, 761–770.

Gerstner, W. (1995). Time structure of the activity in neural network models. *Physical Review E: Statistical Physics, Plasmas, Fluids, and Related Interdisciplinary Topics, 51*, 738–758.

Gerstner, W., & Kistler, W. M. (2002). *Spiking Neuron Models: Single Neurons, Populations, Plasticity*. Cambridge: Cambridge University Press.

Gerstner, W., Kistler, W. M., Naud, R., & Paninski, L. (2014). *Neuronal Dynamics: From Single Neurons to Networks and Models of Cognition*. Cambridge: Cambridge University Press.

Gerstner, W., & Naud, R. (2009). Neuroscience: How good are neuron models? *Science, 326*, 379–380.

Ginzburg, I., & Sompolinsky, H. (1994). Theory of correlations in stochastic neural networks. *Physical Review E: Statistical Physics, Plasmas, Fluids, and Related Interdisciplinary Topics, 50*, 3171–3191.

Goldberg, J. A., Rokni, U., & Sompolinsky, H. (2004). Patterns of ongoing activity and the functional architecture of the primary visual cortex. *Neuron, 42*, 489–500.

Goldman-Rakic, P. S. (1995). Cellular basis of working memory. *Neuron, 14*, 477–485.

Gutkin, B., & Ermentrout, G. B. (1998). Dynamics of membrane excitability determine interspike interval variability: A link between spike generation mechanisms and cortical spike train statistics. *Neural Computation, 10*, 1047–1065.

Hansel, D., Mato, G., & Meunier, C. (1993). Phase dynamics for weakly coupled Hodgkin-Huxley neurons. *Europhysics Letters, 23*, 367–370.

Hansel, D., & Sompolinsky, H. (1998). Modeling feature selectivity in local cortical circuits. In C. Koch & I. Segev (Eds.), *Methods in Neuronal Modeling* (2nd ed., pp. 499–567). Cambridge, MA: MIT Press.

Hodgkin, A. L., & Huxley, A. F. (1952). A quantitative description of membrane current and its application to conductance and excitation in nerve. *Journal of Physiology, 117*, 500–544.

Holmgren, C., Harkany, T., Svennenfors, B., & Zilberter, Y. (2003). Pyramidal cell communication within local networks in layer 2/3 of rat neocortex. *Journal of Physiology, 551*, 139–153.

Hopfield, J. J. (1982). Neural networks and physical systems with emergent collective computational abilities. *Proceedings of the National Academy of Sciences of the United States of America, 79*, 2554–2558.

Huguenard, J., & McCormick, D. (1994). *Electrophysiology of the Neuron: An Interactive Tutorial.* New York: Oxford University Press.

Itskov, V., Hansel, D., & Tsodyks, M. (2011). Short-term facilitation may stabilize parametric working memory trace. *Frontiers in Computational Neuroscience, 5*, 40.

Izhikevich, E. M. (2003). Simple model of spiking neurons. *IEEE Transactions on Neural Networks, 14*, 1569–1572.

Izhikevich, E. M. (2007). *Dynamical Systems in Neuroscience.* Cambridge, MA: MIT Press.

Izhikevich, E. M., & Edelman, G. M. (2008). Large-scale model of mammalian thalamocortical systems. *Proceedings of the National Academy of Sciences of the United States of America, 105*, 3593–3598.

Jacobs, A. L., Fridman, G., Douglas, R. M., Alam, N. M., Latham, P. E., Prusky, G. T., et al. (2009). Ruling out and ruling in neural codes. *Proceedings of the National Academy of Sciences of the United States of America, 106*, 5936–5941.

Jahr, C. E., & Stevens, C. F. (1990). Voltage dependence of NMDA-activated macroscopic conductances predicted by single-channel kinetics. *Journal of Neuroscience, 10*, 3178–3182.

Jensen, O., & Lisman, J. E. (2000). Position reconstruction from an ensemble of hippocampal place cells: Contribution of theta phase coding. *Journal of Neurophysiology, 83*, 2602–2609.

Johnston, D., & Wu, S. M. S. (1995). *Foundations of Cellular Neurophysiology.* Cambridge, MA: MIT Press.

Kayser, C., Montemurro, M. A., Logothetis, N. K., & Panzeri, S. (2009). Spike-phase coding boosts and stabilizes information carried by spatial and temporal spike patterns. *Neuron, 61*, 597–608.

Knight, B. W. (1972). Dynamics of encoding in a population of neurons. *Journal of General Physiology, 59*, 734–766.

Ko, H., Hofer, S. B., Pichler, B., Buchanan, K. A., Sjostrom, P. J., & Mrsic-Flogel, T. D. (2011). Functional specificity of local synaptic connections in neocortical networks. *Nature, 473*, 87–91.

Kobayashi, R., Tsubo, Y., & Shinomoto, S. (2009). Made-to-order spiking neuron model equipped with a multi-timescale adaptive threshold. *Frontiers in Computational Neuroscience, 3*, 9.

Krinsky, V. I., & Kokoz, Y. M. (1973). Analysis of equations of excitable membranes. *Biophysics, 18*, 533–539.

Lapicque, L. (1907). Recherches quantitatives sur l'excitabilité électrique des nerfs traitée comme une polarisation. *Journal de Physiologie et de Pathologie Générale, 9*, 620–635.

Latham, P. E., Richmond, B. J., Nelson, P. G., & Nirenberg, S. (2000). Intrinsic dynamics in neuronal networks: I. Theory. *Journal of Neurophysiology, 83*, 808–827.

Ledoux, E., & Brunel, N. (2011). Dynamics of networks of excitatory and inhibitory neurons in response to time-dependent inputs. *Frontiers in Computational Neuroscience, 5*, 25.

Lefort, S., Tomm, C., Floyd-Sarria, J. C., & Petersen, C. C. (2009). The excitatory neuronal network of the C2 barrel column in mouse primary somatosensory cortex. *Neuron, 61*, 301–316.

Levy, R. B., & Reyes, A. D. (2012). Spatial profile of excitatory and inhibitory synaptic connectivity in mouse primary auditory cortex. *Journal of Neuroscience, 32*, 5609–5619.

Liu, Y. H., & Wang, X.-J. (2001). Spike-frequency adaptation of a generalized integrate-and fire model neuron. *Journal of Computational Neuroscience, 10*, 25–45.

Markram, H., Lubke, J., Frotscher, M., Roth, A., & Sakmann, B. (1997). Physiology and anatomy of synaptic connections between thick tufted pyramidal neurones in the developing rat neocortex. *Journal of Physiology, 500*, 409–440.

McAdams, C. J., & Maunsell, J. H. (1999). Effects of attention on orientation-tuning functions of single neurons in macaque cortical area V4. *Journal of Neuroscience, 19*, 431–441.

McCulloch, W. S., & Pitts, W. A. (1943). A logical calculus of the ideas immanent in nervous activity. *Bulletin of Mathematical Biophysics, 5,* 115–133.

Mirollo, R. E., & Strogatz, S. H. (1990). Synchronization of pulse-coupled biological oscillators. *SIAM Journal on Applied Mathematics, 50,* 1645–1662.

Mongillo, G., Hansel, D., & van Vreeswijk, C. (2012). Bistability and spatiotemporal irregularity in neuronal networks with nonlinear synaptic transmission. *Physical Review Letters, 108,* 158101.

Morris, C., & Lecar, H. (1981). Voltage oscillations in the barnacle giant muscle fiber. *Biophysical Journal, 35,* 193–213.

Murphy, B. K., & Miller, K. D. (2009). Balanced amplification: A new mechanism of selective amplification of neural activity patterns. *Neuron, 61,* 635–648.

Nagumo, J. S., Arimoto, S., & Yochizawa, S. (1962). An active pulse transmission line simulating nerve axon. *Proceedings of the IRE, 50,* 2061–2070.

Ohiorhenuan, I. E., Mechler, F., Purpura, K. P., Schmid, A. M., Hu, Q., & Victor, J. D. (2010). Sparse coding and high-order correlations in fine-scale cortical networks. *Nature, 466,* 617–621.

Ozeki, H., Finn, I. M., Schaffer, E. S., Miller, K. D., & Ferster, D. (2009). Inhibitory stabilization of the cortical network underlies visual surround suppression. *Neuron, 62,* 578–592.

Panzeri, S., Brunel, N., Logothetis, N. K., & Kayser, C. (2010). Sensory neural codes using multiplexed temporal scales. *Trends in Neurosciences, 33,* 111–120.

Pillow, J. W., Paninski, L., Uzzell, V. J., Simoncelli, E. P., & Chichilnisky, E. J. (2005). Prediction and decoding of retinal ganglion cell responses with a probabilistic spiking model. *Journal of Neuroscience, 25,* 11003–11013.

Pillow, J. W., Shlens, J., Paninski, L., Sher, A., Litke, A. M., Chichilnisky, E. J., et al. (2008). Spatio-temporal correlations and visual signalling in a complete neuronal population. *Nature, 454,* 995–999.

Rauch, A., Camera, G. L., Lüscher, H.-R., Senn, W., & Fusi, S. (2003). Neocortical pyramidal cells respond as integrate-and-fire neurons to *in vivo*-like input currents. *Journal of Neurophysiology, 90,* 1598–1612.

Renart, A., de la Rocha, J., Bartho, P., Hollender, L., Parga, N., Reyes, A., et al. (2010). The asynchronous state in cortical circuits. *Science, 327,* 587–590.

Richardson, M., Brunel, N., & Hakim, V. (2003). From subthreshold to firing-rate resonance. *Journal of Neurophysiology, 89,* 2538–2554.

Rieke, F., Warland, D., de Ruyter van Steveninck, R., & Bialek, W. (1999). *Spikes: Exploring the Neural Code.* Cambridge, MA: MIT Press.

Rinzel, J., & Ermentrout, G. B. (1998). Analysis of neural excitability and oscillations. In C. Koch & I. Segev (Eds.), *Methods in Neuronal Modeling* (pp. 251–291). Cambridge, MA: MIT Press.

Schneidman, E., Berry, M. J., Segev, R., & Bialek, W. (2006). Weak pairwise correlations imply strongly correlated network states in a neural population. *Nature, 440,* 1007–1012.

Sjöström, P. J., Turrigiano, G. G., & Nelson, S. (2001). Rate, timing, and cooperativity jointly determine cortical synaptic plasticity. *Neuron, 32,* 1149–1164.

Softky, W. R., & Koch, C. (1993). The highly irregular firing of cortical cells is inconsistent with temporal integration of random EPSPs. *Journal of Neuroscience, 13,* 334–350.

Sompolinsky, H. (1986). Neural networks with nonlinear synapses and a static noise. *Physical Review A., 34,* 2571–2574.

Stein, R. (1965). A theoretical analysis of neuronal variability. *Biophysical Journal, 5,* 173–194.

Tabak, J., Senn, W., O'Donovan, M. J., & Rinzel, J. (2000). Modeling of spontaneous activity in developing spinal cord using activity-dependent depression in an excitatory network. *Journal of Neuroscience, 20,* 3041–3056.

Touboul, J. (2008). Bifurcation analysis of a general class of nonlinear integrate-and-fire neurons. *SIAM Journal on Applied Mathematics, 68,* 1045–1079.

Treves, A. (1993). Mean-field analysis of neuronal spike dynamics. *Network, 4,* 259–284.

Tsodyks, M., & Feigel'man, M. V. (1988). The enhanced storage capacity in neural networks with low activity level. *Europhysics Letters, 6,* 101–105.

Tsodyks, M., Kenet, T., Grinvald, A., & Arieli, A. (1999). Linking spontaneous activity of single cortical neurons and the underlying functional architecture. *Science, 286,* 1943–1946.

Tsodyks, M., Uziel, A., & Markram, H. (2000). Synchrony generation in recurrent networks with frequency-dependent synapses. *Journal of Neuroscience, 20,* RC50.

Tsodyks, M. V., & Markram, H. (1997). The neural code between neocortical pyramidal neurons depends on neurotransmitter release probability. *Proceedings of the National Academy of Sciences of the United States of America, 94,* 719–723.

Tsodyks, M. V., Pawelzik, K., & Markram, H. (1998). Neural networks with dynamic synapses. *Neural Computation, 10,* 821–835.

Tsodyks, M. V., & Sejnowski, T. (1995). Associative memory and hippocampal place cells. *International Journal of Neural Systems, 6,* 81–86.

Tuckwell, H. C. (1988). *Introduction to Theoretical Neurobiology.* Cambridge: Cambridge University Press.

Turing, A. M. (1952). The chemical basis of morphogenesis. *Philosophical Transactions of the Royal Society of London. Series B, Biological Sciences, 237,* 37–72.

van Vreeswijk, C. (2010). Stochastic models of spike trains. In S. Grün & S. Rotter (Eds.), *Analysis of Parallel Spike Trains* (pp. 3–20). New York: Springer.

van Vreeswijk, C., & Sompolinsky, H. (1996). Chaos in neuronal networks with balanced excitatory and inhibitory activity. *Science, 274,* 1724–1726.

van Vreeswijk, C., & Sompolinsky, H. (1998). Chaotic balanced state in a model of cortical circuits. *Neural Computation, 10,* 1321–1371.

Wang, X.-J. (1999). Synaptic basis of cortical persistent activity: The importance of NMDA receptors to working memory. *Journal of Neuroscience, 19,* 9587–9603.

Wang, X.-J. (2001). Synaptic reverberation underlying mnemonic persistent activity. *Trends in Neurosciences, 24,* 455–463.

Wang, Y., Markram, H., Goodman, P. H., Berger, T. K., Ma, J., & Goldman-Rakic, P. S. (2006). Heterogeneity in the pyramidal network of the medial prefrontal cortex. *Nature Neuroscience, 9,* 534–542.

Wilson, H. R., & Cowan, J. D. (1972). Excitatory and inhibitory interactions in localized populations of model neurons. *Biophysical Journal, 12*, 1–24.

Wilson, H. R., & Cowan, J. D. (1973). A mathematical theory of the functional dynamics of cortical and thalamic nervous tissue. *Kybernetik, 13*, 55–80.

Wimmer, K., Nykamp, D. Q., Constantinidis, C., & Compte, A. (2014). Bump attractor dynamics in prefrontal cortex explains behavioral precision in spatial working memory. *Nature Neuroscience, 17*, 431–439.

Xie, X., Hahnloser, R. H. R., & Seung, H. S. (2002). Double-ring network model of the head direction system. *Physical Review E: Statistical, Nonlinear, and Soft Matter Physics, 66*, 041902.

Zhang, K. (1996). Representation of spatial orientation by the intrinsic dynamics of the head-direction cell ensembles: A theory. *Journal of Neuroscience, 16*, 2112–2126.

3 Neurons and Neuronal Networks as Dynamical Systems

JONATHAN E. RUBIN AND JOHN RINZEL

3.1 *Neurons: Dynamical Systems of the Brain*

3.1.1 The Dynamical Systems Approach to Neuroscience Neuronal systems are dynamic, reflecting the interactions of numerous processes that evolve over a broad range of temporal and spatial scales. In some cases, neuronal networks produce dynamics internally. Examples include rhythms that drive motor patterns, activity patterns associated with dysfunctions such as epilepsy or Parkinson's disease, oscillations linked with normal behaviors including sleep and cognition (delta waves, theta and gamma oscillations), and temporal sequences that represent the memories of events. Alternatively, stimulus-driven signal processing involves dynamical computation for feature detection and representation and for action recruitment. Neuromodulators, brain state, context, and other factors affect all of these dynamic processes.

The observation of these rich and diverse phenomena has led to many questions: When and how are different dynamical states recruited, modulated, and coordinated? How do transitions between dynamical states occur and what are some generic signatures of transition types that provide insights on mechanisms? How do we understand or account for the fine temporal structure of a computation without the burden of tracking every action potential in trying to identify longer timescale structure? The concepts of dynamical systems provide us with a foundation for characterizing the underlying structure of dynamical states and transitions between different states in neuronal models. Although different neuronal computations will be implemented with or supported by specific neural mechanisms at the cell and circuit level, the methodology of dynamical systems theory harnesses the idea that there can be generic features that provide a common framework for certain phenomena. By understanding the qualitative commonalities of dynamical features underlying neuronal rhythms, for example, we can classify rhythmic states, predict transition behaviors associated with parameter variations, and develop intuition to guide further exploration. This theory-based analysis contrasts with sweeping computationally across the parameter space of a model to achieve a pseudo-empirical characterization of possible states and the switching between them.

Dynamical systems methods can also impose structure on empirical observations. For example, in 1948, Hodgkin, with an empirical approach, classified three types of cellular excitability (Hodgkin, 1948). He described some dynamical properties of axonal spiking, namely, the onset and threshold behavior of repetitive firing as a steady stimulus was tuned beyond threshold, extracted from numerous observations. Rinzel and Ermentrout some 40 years later (Rinzel and Ermentrout, 1989) provided the mathematical framework based on dynamical systems theory for Hodgkin's classifications of repetitive firing. The theory supported this classification and offered the means to identify and interpret the conditions associated with Hodgkin's activity classes within models and neurons as well as to make predictions about additional properties. While that application of dynamical systems theory related to cellular excitability, here we present case studies dealing with some features of firing rate models for neuronal networks, treated as dynamical systems.

We begin by surveying a few examples of dynamics extracted from neuronal systems, captured in mathematical models, and analyzed mathematically using dynamical systems theory. In section 3.2, we concisely present concepts from dynamical systems that are most useful for understanding solutions to models of neuronal systems, with a focus on attractors and other geometric structures that organize solution behavior. We also apply such concepts to understand the dynamic responses of the model neuronal networks introduced in section 3.1. We link solution behavior to mathematical structure, showing how solution properties emerge from certain features of the model equations that can be identified and understood with tools from dynamical systems. In section 3.3, we extend our case study of neuronal network models to encompass additional classes of network behavior, and we follow up with some concluding remarks in section 3.4.

3.1.2 Phase Space The starting point for the dynamical systems approach is to think of solutions to ordinary differential equations (ODEs) as *trajectories* in *phase space*. We will give the basic ideas relating to these concepts here and then present more details and examples from neuroscience later in the chapter. A brief, more abstract and mathematically precise introduction to these basic ideas is provided in the appendix (section 3.5). To begin, consider an ODE describing the rate of change of a quantity $x(t)$ that has n components, say, $x(t) = (x_1(t), x_2(t), \ldots, x_n(t))$. Such an equation would be written $\dot{x}(t) = f(t, x(t))$, where the function f provides a rule for generating the time derivative $\dot{x}(t) = dx/dt$ at a time t, given the current state $x(t)$. Often, f is not directly dependent on t, meaning that the rate of change of x depends only on the current state of x, not on the time when this state is achieved, and we will assume henceforth that this is the case. In theory, we could solve the ODE and plot each component $x_i(t)$ versus t. It is often useful, however, to plot the x_i against each other, without an explicit time axis. To do so, we define an n-dimensional *phase space* in which each axis corresponds to the value of a particular component of $x(t)$ and we consider the behavior of solutions in this space. For example, if the solution to our ODE is $x(t) = (\cos(t), \sin(t))$, then the phase space is $\mathbb{R}^2$ and the solution is represented as a circle in this space, called the *trajectory* for this solution. If we choose any initial point on this circle, then the solution at future times will be given by other points on the circle; that is, the trajectory for the solution describes the full behavior of this solution for all time.

One nice feature of phase space is that we can use a single phase space diagram, or *phase portrait*, to clearly represent the behavior of many solutions. The types of ODEs that we will consider have the uniqueness property, meaning that, from any particular starting value of x, there is only a single possible future behavior of x (note that we are ignoring effects of noise here). However, different starting values $x(0)$, or *initial conditions*, lead to different trajectories. The uniqueness property implies that these trajectories do not intersect, and thus we can draw multiple trajectories in a single phase portrait without any confusion. Since any point x in a phase space may be selected as an initial condition, we cannot illustrate all trajectories for an ODE in the same diagram. Typically, we draw a representative set of trajectories, showcasing the possible behaviors that can result from different choices of initial conditions.

Certain structures often prove quite useful in generating trajectories in phase space. *Nullclines* are curves of points where $\dot{x}_i = 0$ for some $i \in \{1, \ldots, n\}$. Even though nullclines are not themselves trajectories, sketching them is often useful for determining the behavior of trajectories. For example, if x_1 is on the horizontal axis of a two-dimensional phase portrait, then trajectories must cross the x_1-nullcline in the vertical direction (i.e., with no change in x_1). Points that belong to the nullclines of all components of an ODE, such that $f(x) = 0$, are called *critical points*, also known as *equilbrium points*, *fixed points*, or *singularities*, and will be discussed in more detail in section 3.2. For now, it suffices to mention that a critical point may be *asymptotically stable*, in which case trajectories from nearby initial conditions approach the critical point as $t \to \infty$, *stable*, in which case trajectories from nearby initial conditions stay within a neighborhood of the critical point, or *unstable*, in which case no matter how small a circle one draws about the critical point, there is always an initial condition in the circle for which the trajectory evolves away from the critical point. Critical points are crucial to organizing the behavior of all trajectories for an ODE and thus are an essential part of any phase portrait.

To conclude this introductory dynamical systems material, we now consider how two basic neuronal behaviors that can be captured by two-dimensional ODEs are represented in the corresponding two-dimensional phase spaces, also known as *phase planes*. One natural activity pattern for a neuron is repetitive spiking. The simplest representation of repetitive spiking is a periodic behavior. To get periodic repetitive spiking in the phase plane, some trajectories must avoid becoming attracted to a critical point and must instead approach a trajectory for a periodic solution. Just as in the $(\cos(t), \sin(t))$ example above, the trajectory for a periodic solution, say, with period $T > 0$, is necessarily a closed curve in phase space since all variables must come back to their original values at times $T, 2T, 3T, \ldots$. On a closed trajectory in the phase plane, each variable must increase for some time period (with $\dot{x}_i > 0$) and decrease for some other time period (with $\dot{x}_i < 0$). Thus, a closed trajectory must cross the nullclines for both variables. An example of a periodic orbit with associated nullclines appears in figure 3.1.

A second natural neuronal behavior is *excitability*. A neuron is considered excitable if it generates a subthreshold response to a weak transient input, generates a spike in response to a sufficiently strong transient input, and after responding returns to a nonspiking rest state in either case. To achieve excitability in an ODE model, we must have an asymptotically stable rest critical point, corresponding to a neuron's resting

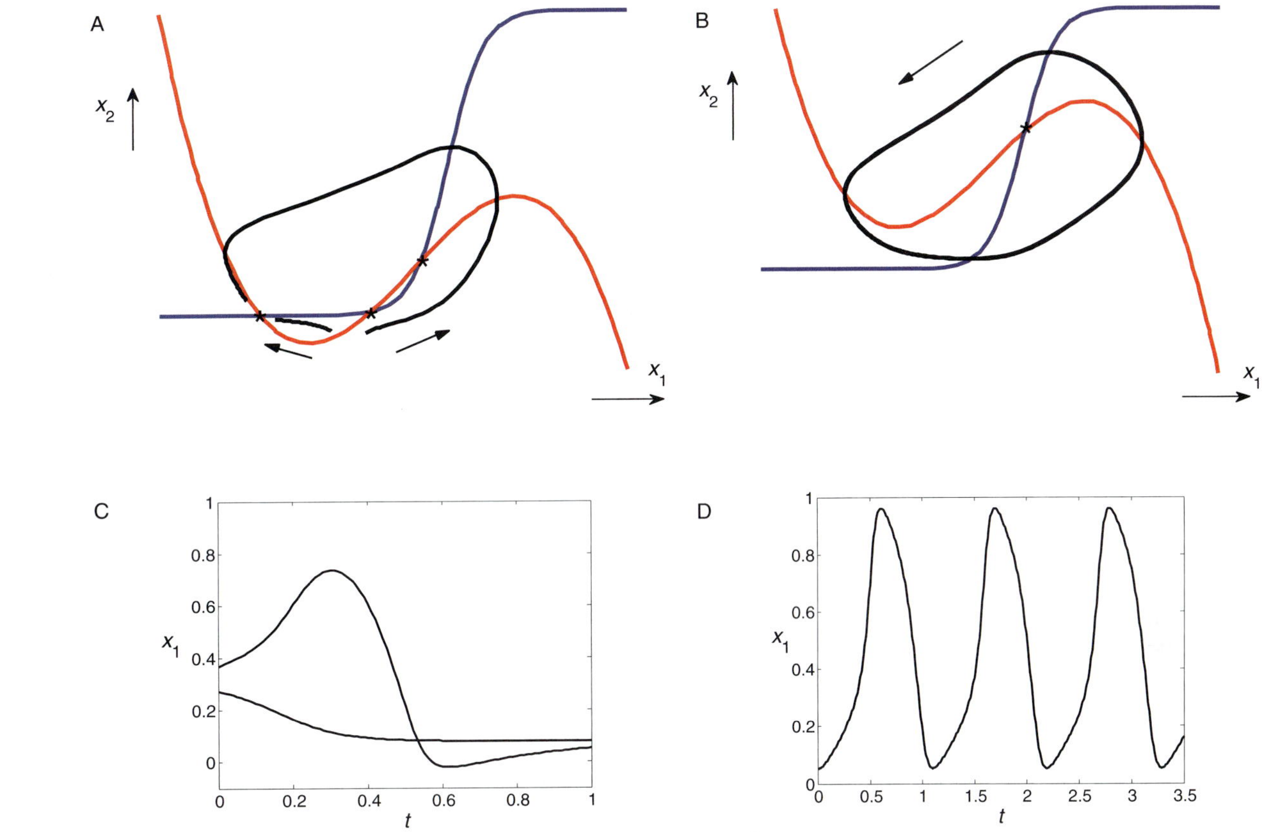

FIGURE 3.1 Dynamics represented in a two-dimensional phase plane and comparison to time courses. If we interpret this system as a reduced neuronal model, then x_1 plays the role of a neuron's membrane potential. x_1-nullclines are red, x_2-nullclines are blue, and trajectories are black, with directions of motion shown by arrows. Critical points are indicated by small black stars at each intersection of the nullclines. (A) Excitable regime, with two different trajectories (from different initial conditions on either side of the middle branch of the x_1-nullcline) shown. One trajectory starts to the left of this middle branch and simply decays to the rest state, corresponding to a subthreshold response. The other trajectory starts to the right of the middle branch and makes a large excursion, corresponding to a spike, before converging back to rest. (B) Periodic orbit in the phase plane. (C) Excitable system responses from (A) shown as functions of time (in arbitrary time units). The subthreshold response is the monotonically decreasing curve while the spike rises and then falls again. (D) Time course of the periodic solution from (B) (arbitrary time units).

potential, in the absence of input, so that trajectories can return to a rest state after a spike. If there is also another critical point of what is known as a *saddle* type, then structures associated with that critical point can set up a threshold surface (of dimension one less than that of the phase space). If a trajectory is started on one side of the threshold, then it returns to rest directly, whereas if it starts on the other side, it undergoes a significant excursion, representing a spike, before returning to rest. These possibilities are illustrated in figure 3.1, where the nullclines have three intersections, including a rest critical point at lowest x_1 and a saddle point (which is approached by trajectories along one special direction but pushes trajectories away in all other directions) at intermediate x_1. Here, the input is viewed as occurring instantaneously, and its strength sets the starting position of x_1. Excitability can also occur with noninstantaneous inputs. Stimulation with a steady depolarizing current for a finite time, for example, would change the model equations in a way that would move the red x_1-nullcline upward, possibly initiating a train of spikes. How much the nullclines move depends on the strength of the input, and this nullcline movement determines whether the resulting trajectory will be subthreshold or will yield spikes, continuing until the input is turned off.

3.1.3 SOME MOTIVATING EXAMPLES OF NEURONAL DYNAMICS In this subsection we present an introductory discussion of two examples of models for neuronal circuit dynamics that embody mathematical structures that are seen in various systems, from motor control to perception. In particular, the introduction of these examples offers the opportunity to consider some

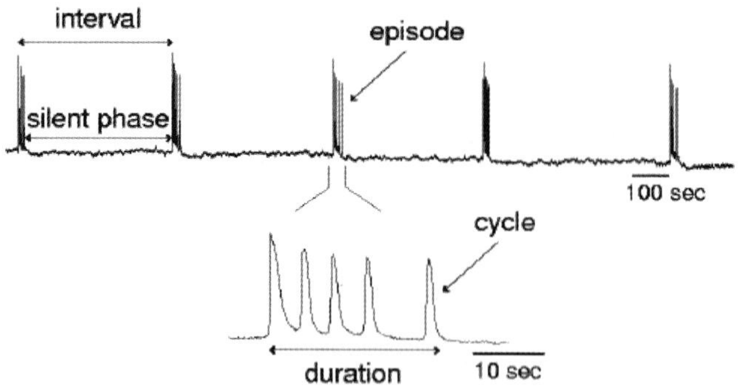

FIGURE 3.2 Population activity of neurons in a segment of an embryonic chick spinal cord (adapted from Tabak et al., 2000). The mechanisms underlying the population dynamics will be discussed later in the section.

dynamics in phase space. Later in the chapter, we will give a more detailed mathematical presentation of some dynamical systems ideas and show how they can be used to study these models more carefully.

3.1.3.1 Episodic Rhythms in Developing Spinal Cord Many developing neuronal systems exhibit spontaneous activity that is crucial for development. For example, in embryonic chick spinal cord the neurons exhibit multimodal oscillatory behavior: population bursts of activity (called episodes) with episode durations on the order of tens of seconds alternate with long interepisode intervals of 2–10 minutes (Tabak and Rinzel, 2005). Figure 3.2 shows an example of such a recording taken from the excised spinal cord. The recording, made from a bundle of nerve fibers that carry the (motoneuron) output of a segment of the spinal cord, represents the combined activity of a large population of cells. The bursting-like pattern is network driven, thought not to be generated intrinsically by individual neurons. Interestingly, the synapses that are normally inhibitory are functionally excitatory at this stage of development. Experiments also show that the size of the evoked synaptic potentials in the network is transiently reduced after an episode of spontaneous activity (Tabak and Rinzel, 2005).

These basic observations have led to an idealized model for the network's rhythmic behavior. As in many neuronal systems, excitability and oscillations arise from the basic mechanisms of a fast regenerative process to create a sudden rise or upstroke in activity and slower negative feedback to quench the activity. In this model the regenerative effect is due to the mutual excitation among the cells. If the network activity level is low and some cells become excited, they can recruit other cells, and so on, leading to a surge in activity. There are multiple timescales of negative feedback: a moderately slow synaptic depression that leads to the up–down cycles within an episode and a very slow synaptic depression that terminates an episode and recovers between episodes, allowing the next episode to begin.

The model describes only the average firing rate of neurons within the population rather than the instantaneous membrane potential and spikes of the individual neurons. In doing so, it is a mean field model. It assumes a purely excitatory recurrent network in which synaptic coupling is susceptible to both short-term and long-term, very slow, activity-dependent depression. The model is described by three coupled nonlinear differential equations for the activity or firing rate of the network, a, and for short- and long-term depression, d and s, respectively.

The separation of timescales allows us to develop the model and to gain insight into the underlying dynamical mechanisms by freezing successively the slow variables, d and s. For example, we can understand the onset of an episode by holding d and s fixed and asking about recruitment in the excitatory recurrent network. A basic model for $a(t)$ is

$$\tau_a \dot{a} = a_\infty(input) - a.$$

Here the overdot denotes differentiation with respect to time, and a_∞ is the input–output relation of a neuron, taken to be an increasing sigmoidal function $a_\infty(x) = 1/\{1 + \exp[(x - \theta)/\sigma]\}$ with $\sigma < 0$. τ_a is an effective time constant, reflecting the recruitment time in the network. Because the network is recurrent, the input is proportional to the activity of the network itself:

$$\tau_a \dot{a} = a_\infty(na) - a. \tag{3.1}$$

The coefficient n in equation 3.1 represents the efficacy of coupling and is here treated as a constant. Steady states of activity occur where $\dot{a} = 0$, given by intersections of the identity line with the sigmoidal curve $a_\infty(na)$, as shown in figure 3.3A for different values of n. For low n (weak efficacy, e.g., $n = 0.3$), there is only one stable steady state and it is at a low activity level; for n high (e.g., $n = 0.9$), there is a sole stable state, at high activity. For intermediate values of n (e.g., $n = 0.5$), there are three steady states, the middle of which is unstable and the lower and upper of which are stable; such a regime, in which there are two different stable states, is referred to as *bistable*. The unstable middle state represents a threshold of activity level that, if surpassed, will lead to recruitment of self-sustained activity, analogous to the upstroke event of an episode (figure 3.3B).

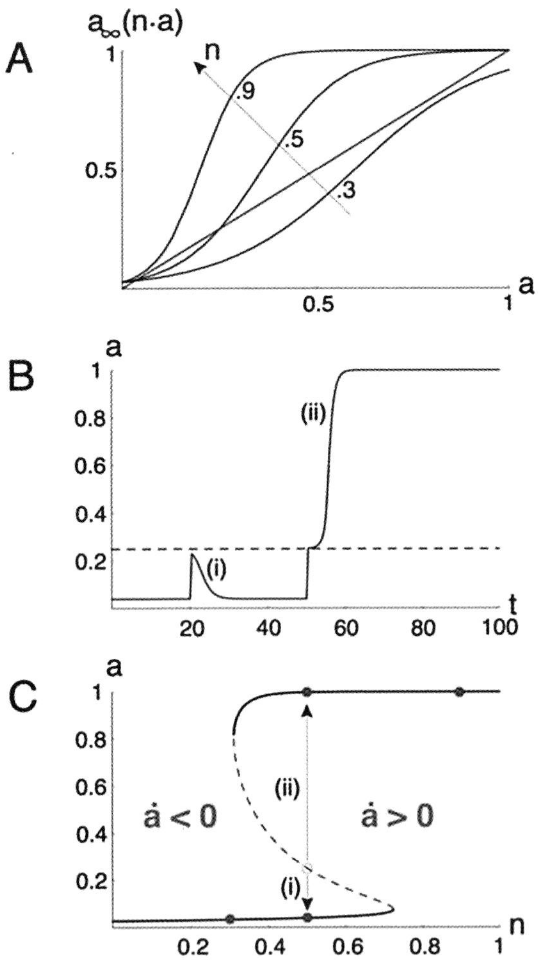

FIGURE 3.3 Understanding the dynamics of a recurrent excitatory network (adapted from Tabak et al., 2000). (A) Population input–output function $a_\infty(na)$ plotted versus a for different values of n. Intersections of each curve with the identity line $y = a$ determine the steady states of firing for that value of n. In the case of three steady states, the low and high activity states are stable and the middle state, which is unstable, corresponds to the network threshold. (B) Time courses for brief activation of a fraction of neurons in the population by subthreshold (i) and superthreshold (ii) stimuli, with $n = 0.5$. The latter leads to self-sustained, high activity. (C) Steady states plotted in the $n - a$ plane, labeled as stable (solid) or unstable (dashed). Dots represent steady states for the particular n values shown in (A). Trajectories labeled in (C) correspond to the time courses shown in (B). If n is allowed to slowly vary, then trajectories follow the stable branches shown, with n increasing as synapses recover on the low-a branch and n decreasing as synapses depress on the high-a branch; at the folds, trajectories make rapid jumps between branches.

In actuality, the synaptic efficacy is not constant but rather slowly evolves, incorporating the effects of slow synaptic depression. Thus, we can include a second differential equation for n (although we do not write it out explicitly here) and represent model dynamics in the (n, a) phase plane. Figure 3.3C gives a hint of how the full model will work. The curve drawn there is the a-nullcline, with $\dot{a} < 0$ to its left and $\dot{a} > 0$ to its right. Initial conditions off the a-nullcline tend to quickly approach the a-nullcline because a evolves faster than n. If we start from an initial condition on the lower branch of the a-nullcline, since a is low, n is increasing and the trajectory moves rightward along the lower branch of the a-nullcline, corresponding to synaptic recovery. When the trajectory reaches the right knee of the a-nullcline, the a equation pulls it to the upper branch, which corresponds to high a. The population activity is now high, which leads to decreasing n, corresponding to depression of synapses, until the solution reaches the left knee and then crashes back to the lower branch. The underlying bistability of a is a crucial feature of the dynamical mechanism.

To incorporate the depression mechanistically, we represent n as the product of the two depression variables: sd. The equations of the full model become

$$\begin{aligned} \tau_a \dot{a} &= a_\infty(sda) - a, \\ \tau_d \dot{d} &= d_\infty(a) - d, \\ \tau_s \dot{s} &= s_\infty(a) - s. \end{aligned} \tag{3.2}$$

The parameters of the model can be found in Tabak et al. (2000). The functions d_∞ and s_∞ are depression (turn-off) functions. They are sigmoidally decreasing with a from 1 to 0. With these terms in the model, elevated activity decreases d and s, on different timescales quantified by τ_d and τ_s, respectively, and thus decreases a_∞ and $\dot{a}$ as well. Biologically, these effects correspond to the weakening of synapses due to synaptic depression in high-activity states.

Since s is very slow, it makes sense to consider the (a, d)-subsystem with s frozen. While a large n value led to a high-a steady state for the one-dimensional equation 3.1, this state is replaced by oscillations in the two-dimensional (a, d)-subsystem, which mimics the cycling within an episode (figure 3.4A). When s is not frozen, the full, three-variable model produces the episodic rhythmicity seen in experiments (figure 3.4B). During a high-activity state of cycling, s slowly decreases until a critical level is reached, the episode ends, and the system switches to a quiescent state. During the quiescent phase, the synapses slowly recover from depression and s slowly increases to a level that allows a new episode to begin. We will later describe more fully the mathematical mechanisms generating the multiphase solution, using dynamical systems theory.

3.1.3.2 Alternations in Perception of Ambiguous Scenes When observers view for an extended time an

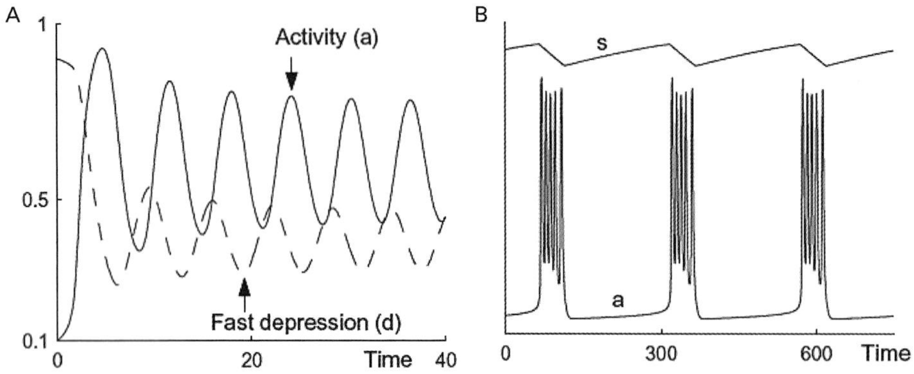

FIGURE 3.4 Dynamics of cycling and episodic rhythmicity (adapted from Borisyuk and Rinzel, 2005) in a spinal cord model. (A) Fast oscillations of cycling in the $a - d$ subsystem with s fixed at $s = 0.9$. (B) Time course of the episodic rhythm of the full system, where s repeatedly drops below the level that supports the pattern of (A) and then slowly recovers.

ambiguous visual scene with two or more different interpretations, they report switching between different perceptions. A familiar example is the vase–faces picture (figure 3.5A). Since the retinal image is constant in time, the alternations reflect neuronal dynamics at higher brain centers. With imaging in humans and electrophysiology in monkeys, one sees evidence for alternating activity in cortical areas and even subcortical areas (Leopold and Logothetis, 1999; Pressnitzer et al., 2008; Sterzer et al., 2009). The various phenomena that have been studied share common features: inevitability (perception will switch), mutual exclusivity, and random variability, presumably linked with stochastic effects (Leopold and Logothetis, 1999). Indeed the alternations are not periodic; the dominance durations have been described as gamma or lognormal distributed (figure 3.5C). For purposes of quantification, the dynamic plaid stimulus is useful (figure 3.5B). The parameters of the two drifting gratings can be varied together (angle, velocity, contrast) or independently (wavelength, contrast). Here, for various ranges of parameters, the behavioral response is bistable in that it is possible to perceive coherence (movement together along the direction of the mean angle) and transparency (sliding over one another) for the same stimulus, alternating in time. (See the book's website for video demos.)

Levelt (1968) described for binocular rivalry some parametric dependencies of the perceptual dynamics for alternations (see Introduction of Seely and Chow, 2011 for a compact summary of Levelt's four propositions). In the paradigm of binocular rivalry each eye is presented a different image, for example, a red drifting grating to the left eye and a green drifting grating to the right eye (with different orientations). The subject reports continuously the percept of a red or green grating, with abrupt switches between the two observations. If the contrast levels, I_1 and I_2, for the red and green gratings, respectively, are identical and increased together, then the mean dominance durations, T_1 and T_2, decrease (figure 3.5D1), which is Levelt's Proposition IV. If the contrast for the left eye's image, I_1, is decreased while holding the other, I_2, fixed, the mean dominance durations change in a curious way: T_1 changes little while T_2 increases significantly (figure 3.5D2), which is Levelt's Proposition II. An analog for Levelt's Proposition II in the case of plaids is seen by decreasing the angle between the gratings, thereby favoring coherence over transparency. We will consider this example in more detail later.

An intuitive conceptual model that embodies general neural mechanisms is the following. Suppose there are two neuronal subpopulations (networks) that correspond to the two percept representations and they are active alternately. When one is active, and the other is not active, the percept corresponding to the active network is dominant and vice versa. Such mutual exclusivity can be achieved through mutual inhibition. This inhibitory interaction is equivalent to competition: the two populations are competing for percept dominance. When there are alternations in perception, we also need a mechanism to allow switching to occur. Typically, we postulate some slow negative feedback process akin to fatigue. It could be spike frequency adaptation (the active neurons' capacity to fire wears down) or synaptic depression (say, the inhibitory synapses deplete their resources). This conceptual model is schematized in figure 3.6A.

Let's put this description into mathematical terms, using a firing rate framework as in the previous section. Denote the mean firing rate of neurons in population j by $r_j(t)$, $j = 1,2$. Suppose $S(input)$ is the sigmoidal input–output relation of a neuron, assumed to be identical for the two populations. Let τ be the effective time

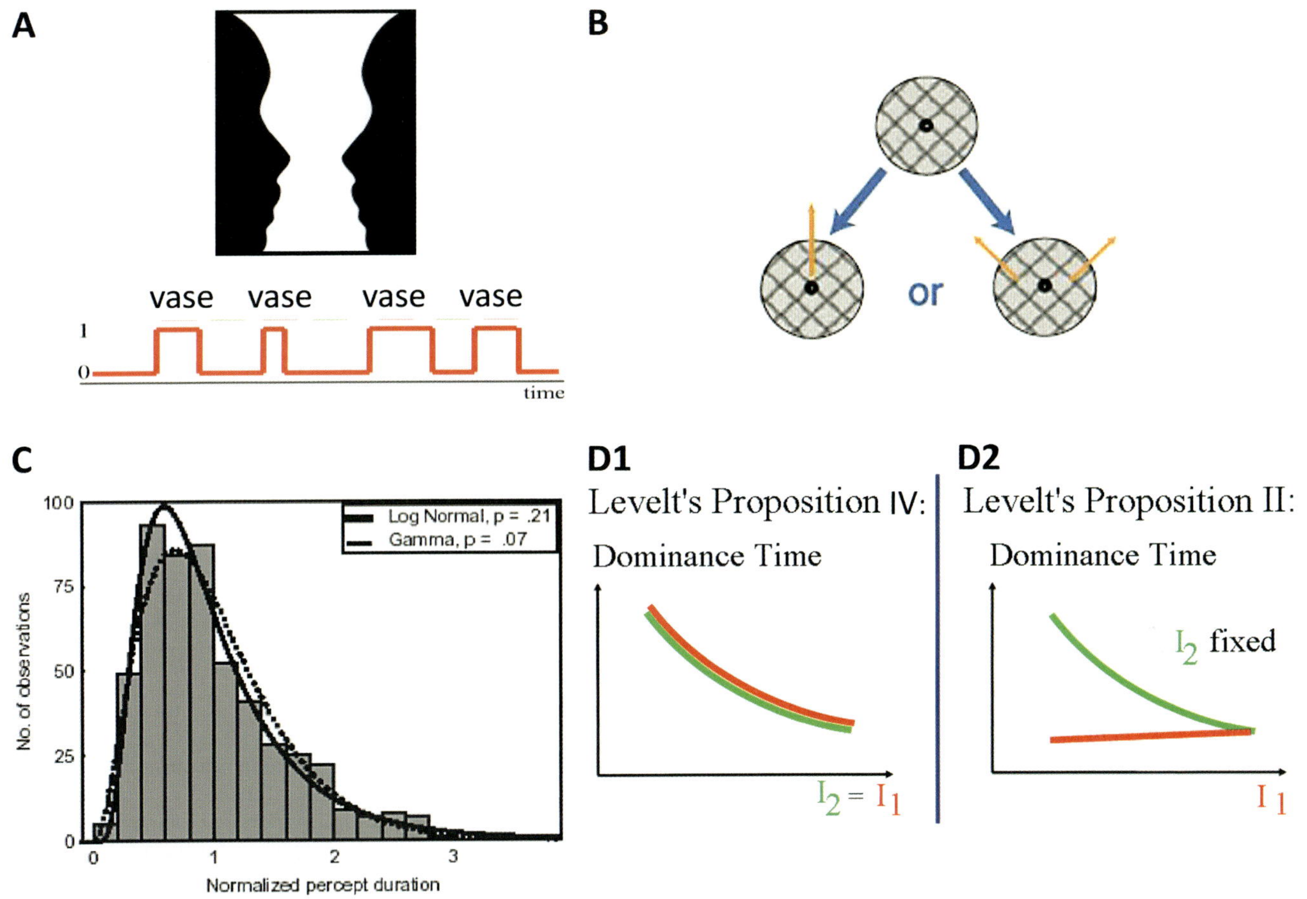

FIGURE 3.5 Perceptual bistability for ambiguous sensory stimuli. (A) The static vase–faces image leads to alternations in perception as reported by a subject, with random durations of a few seconds. (B) The dynamic plaid stimulus consists of two drifting gratings with an angle, α, between the drift directions. The perception may alternate between coherence (moving together along the average direction) and transparency, with the gratings drifting across one another (adapted from Moreno-Bote et al., *J. Vision,* 10: 1–18, 2010. Copyright held by the Association for Research in Vision and Ophthalmology). (C) The dominance durations are described by a gamma or lognormal distribution, here normalized by the mean duration (adapted from Rubin and Hupe, 2004). (D) Dependence on a parameter that reflects the strength of the stimulus, for example, in binocular rivalry (a different image is presented to each eye with same or different strength, say, illumination contrast, denoted by I_1, I_2). (D1) Increasing I_1 and I_2 simultaneously leads to faster alternations. (D2) Decreasing the intensity for, say, eye 1, leads to little change in mean dominance duration for eye 1 but rather induces a significant increase in mean dominance duration for the other eye, eye 2.

constant for buildup/decay of network activity after increases/decreases of input. The mutual inhibition is represented as negative input, proportional to the firing rate of the competing population with coupling strength β. We implement spike frequency adaptation with level $a_j(t)$ as a subtractive term with strength ϕ. We could alternatively implement the slow negative feedback as synaptic depression that could be modeled as multiplicative, using a factor, say $d \leq 1$, on the coupling term.

There are four differential equations for the model, for the firing rate r and slow adaptation variable a in each subpopulation (Laing and Chow, 2002; Shpiro et al., 2007). They take the following form,

$$\begin{aligned}\tau \dot{r}_1 &= -r_1 + S(-\beta r_2 - \phi a_1 + I_1),\\ \tau_a \dot{a}_1 &= f_{a,\infty}(r_1) - a_1,\\ \tau \dot{r}_2 &= -r_2 + S(-\beta r_1 - \phi a_2 + I_2),\\ \tau_a \dot{a}_2 &= f_{a,\infty}(r_2) - a_2,\end{aligned} \qquad (3.3)$$

where $f_{a,\infty}(r)$ represents the steady state level of firing-rate-dependent adaptation. The adaptation is slow, so $\tau << \tau_a$.

This model produces an alternating activity pattern (figure 3.6C). When one population is highly active ("dominant"), its adaptation variable slowly increases. Meanwhile, the suppressed population is recovering

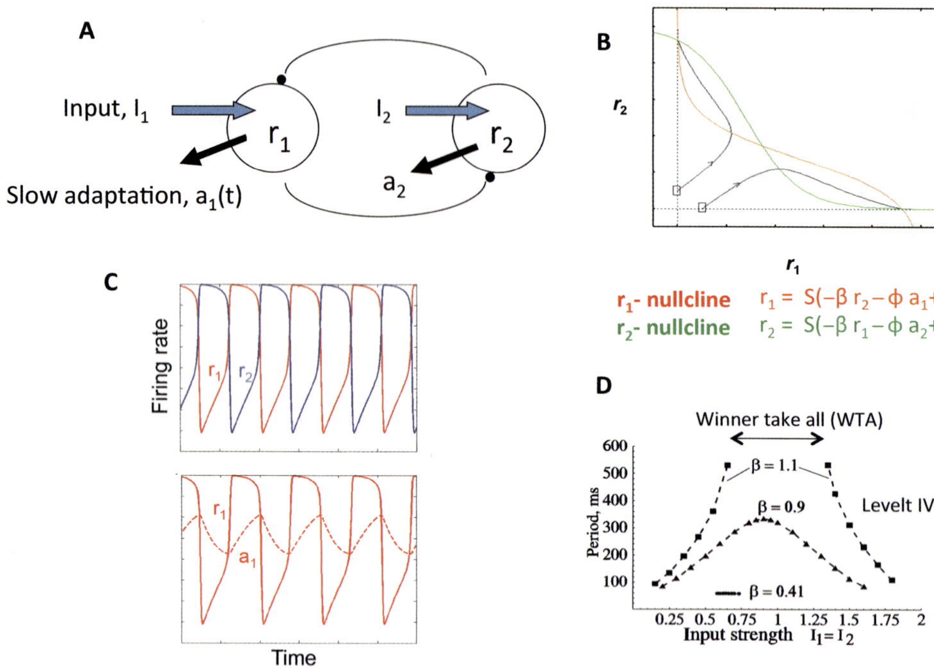

FIGURE 3.6 Behavior of a firing rate model for perceptual competition. (A) Schematic of a two-population model competing through mutual inhibition. (B) With the two slow adaptation variables a_1 and a_2 frozen, the phase plane for $r_1 - r_2$ shows bistability. There are two attractors, each corresponding to one of the two percepts dominant with the other one suppressed, the feature of mutual exclusivity. The black curves are trajectories, one of which converges to each of these attractors. (C) Top: The time courses of r_1 and r_2 with sudden switches in dominance. Bottom: When a percept is dominant (high firing rate, $r(t)$) the adaptation, $a(t)$, slowly accumulates, and it decreases slowly when r is low due to suppression by inhibition. (D) Mean dominance duration increases and then decreases as intensities I_1, I_2 are increased together if strength of inhibition, β, is not too strong (adapted from Shpiro et al., 2007). For strong inhibition the mid-range of stimulus intensity leads to winner-take-all behavior. The right branch of the response curves corresponds to Levelt's Proposition IV (see figure 3.5D1).

from adaptation (its a is decreasing). For switching to result, the active population must adapt enough, the suppressed population must recover sufficiently, or both must occur. The time courses show that switching is abrupt, suggesting an underlying bistability. This bistability can be seen by examining the dynamics of the fast subsystem, consisting of the r_1 and r_2 equations with the slow variables a_1 and a_2 frozen. As can be seen by drawing the r_1- and r_2-nullclines in the (r_1, r_2) phase plane with a_1, a_2 fixed at appropriate values, two stable critical points can coexist, one with population 1 dominant while population 2 is suppressed and another that is the opposite (figure 3.6B). This is the phase plane for competition by mutual inhibition alone, and thus we see already that mutual exclusivity, also known as *winner-take-all* dynamics (which also generalizes to multiple populations), is a direct consequence of mutual inhibition.

This simple model captures some of the behaviors seen in psychophysical experiments. Indeed, the mean dominance duration decreases with stimulus intensity as in Levelt's Proposition IV (figure 3.5D1) but only for some range of adequate stimulus strengths (figure 3.6D; Shpiro et al., 2007). For alternate formulations of such models, still employing competition via inhibition, the discrepant behavior (increasing duration with increasing intensity for low input strengths in the model, as in figure 3.6D, but not in experiments) can be reduced or eliminated (Seely and Chow, 2011).

Recent experiments and model simulations (including stochasticity) have led to generalizations of Levelt's Proposition II (figure 3.5D2; Brascamp et al., 2006; Moreno-Bote et al., 2010; Huguet and Rinzel, 2014). By covering the full range of the varying stimulus parameter (I_1 above I_2 as well as below I_2), it is found that while the mean dominance time T_1 varies little for the range $I_1 < I_2$, it does vary a lot for the range $I_2 < I_1$. There is a similar dichotomy for T_2 (see also Daun et al., 2009). We can identify "being dominant" with the percept that is reported for the greater fraction of time. Most interesting is that when the switching rate from experiments, say, for the dynamic plaids (figure 3.5B), is plotted as a function of the fraction of time coherence is perceived, we find that the switching rate is maximal at 50% (figure

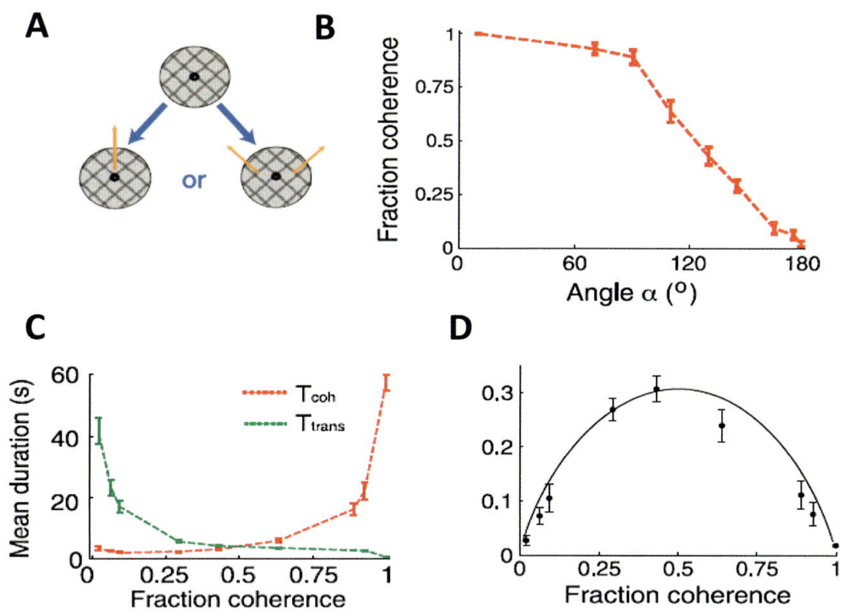

FIGURE 3.7 Results for the drifting grating stimulus (adapted from Moreno-Bote et al., *J. Vision*, 10: 1-18, 2010. Copyright held by the Association for Research in Vision and Ophthalmology). (A) The gratings in the stimulus could be perceived as moving together or in opposite directions. (B) The fraction of time that coherence is perceived (fraction coherence) decreases with the angle, α, between gratings. (C, D) Correspondingly, mean perception durations when plotted against fraction coherence show symmetrical behavior around the 50% point (C), and maximum switching rate occurs at the 50% coherence point (D).

3.7; Moreno-Bote et al., 2010). This has a nice evolutionary interpretation: when the percept is most uncertain, one should frequently explore the alternatives. Finally, it is important to note that the mathematical structure of this model has a long history in models of neuronal systems. Such models were formulated as idealizations to describe the dynamics of neural circuits, called central pattern generators, that generate alternating outputs to drive antagonistic muscle groups involved in repetitive behaviors like breathing and locomotion (e.g., Brown, 1911; Brown, 1914; cf. chapter 8, "Motor Pattern Generation"). Indeed, simplified models such as this one can show surprisingly rich dynamics, with possible relevance to a wide variety of neuronal applications (Yuste et al., 2005).

3.2 Phase Space Analysis

As we have seen in the previous section, illustrated in particular in figures 3.3C and 3.6B, considering a phase plane can be quite useful for understanding how a particular form of dynamics emerges from a model and depends on model parameters. In this section, we build from the starting point given in section 3.1.2 to give a very condensed account of some key ideas that are useful for more precise analysis and understanding of neuroscience models. Many textbooks discuss these and other ideas from dynamical systems theory at a variety of different levels of detail and mathematical sophistication, with examples from a variety of applications (Guckenheimer and Holmes, 1983; Strogatz, 2001; Perko, 2001; Wiggins, 2003), including neuroscience (Izhikevich, 2006; Ermentrout and Terman, 2010). What we are after is the ability to decide, given an ODE model, what are the possible long-term or *asymptotic* behaviors of its solutions, and what are the types of transient responses we might observe on shorter timescales, before the asymptotic state is (approximately) reached? As noted in the previous section, to answer these questions, it is always important to know where the critical points of the model lie within phase space and what behaviors trajectories exhibit in local neighborhoods of these points. In this section, we first treat one-dimensional ODEs, which naturally are the simplest to understand, and then we move on to the analysis of behavior near critical points in higher-dimensional ODEs. More technical details are given in the appendix (section 3.5).

3.2.1 ONE-DIMENSIONAL DYNAMICS To understand how the dynamical systems approach can be used to analyze the flow of an ODE, it is useful to start with one-dimensional systems, described by the scalar equation $\dot{x} = f(x)$. The *critical points* of the system are those values of x for which $f(x) = 0$. Let us assume that f is continuous with a continuous derivative. On an interval $I_{1,2}$ between any successive pair of critical points, say, (x_1, x_2), $f(x)$ has just one sign. If $f(x) > 0$ and we consider a solution $x(t)$ with initial condition $x(0) \in I_{1,2}$, then $x(t)$ increases in time. It is not too hard to show that in this case, $x(t) \in (x_1, x_2)$ for each finite t and that $x(t)$ converges to x_2 from below as $t \to \infty$. Similarly, if $f(x) < 0$ on $I_{1,2}$, then $x(t)$ is a decreasing function that converges to x_1 from above as $t \to \infty$. In summary, in a one-dimensional system, solutions that start away from critical points either increase or decrease monotonically toward critical points or become unbounded as time advances, and the sign of f determines the direction of the change. Note that a particular solution cannot switch between increasing and decreasing over time if the ODE itself is autonomous (i.e., $\dot{x} = f(x)$ rather than $\dot{x} = f(x, t)$). Thus, autonomous one-dimensional ODEs are not adequate models to describe the time course of a neuronal oscillation or of excitable behavior, unless they have a discontinuous right-hand side (e.g., if a threshold jump condition is appended, as in the *integrate-and-fire model*).

Next, return to the case when $f(x) > 0$ on (x_1, x_2). We have $f(x_2) = 0$, and let us assume that $f(x)$ becomes negative on an interval (x_2, x_3) for some $x_3 > x_2$. Thus, if $x(0) \in (x_2, x_3)$, then $x(t)$ converges to x_2 from above as $t \to \infty$. Combined with the above paragraph, this convergence implies that x_2 is asymptotically stable in this case: for any initial condition in the neighborhood (x_1, x_3) of x_2, the corresponding trajectory converges to x_2 as $t \to \infty$. Since f switches from positive to negative as x increases through x_2, we conclude that $f'(x_2) < 0$. In general, the sign of $f'(x)$ determines the stability of the critical point x of a one-dimensional system, with a negative sign yielding asymptotic stability and a positive sign yielding instability, unless $f'(x) = 0$, which is an inconclusive result.

In the situation described above, our phase space is simply the real line. We can easily draw a phase line diagram, which will consist of a line segment with all critical points labeled, with a rightward arrow on the line segment to the left of x_2 showing that trajectories starting below x_2 increase toward it, and with a leftward arrow on the line segment to the right of x_2 showing that trajectories starting above x_2 decrease toward it, and with similar arrows on the other segments of the line (figure 3.8A). We can also draw an analogy between this situation and a system with a single-well potential energy with a minimum at x_2. In such a system, a particle at $x < x_2$ has a greater energy than it would at $x = x_2$, and thus its position increases to x_2, with an analogous decrease of x toward x_2 from $x > x_2$ (figure 3.8A). The function $f(x)$ is like the negation of the derivative of the potential: f is positive below x_2 where the potential is

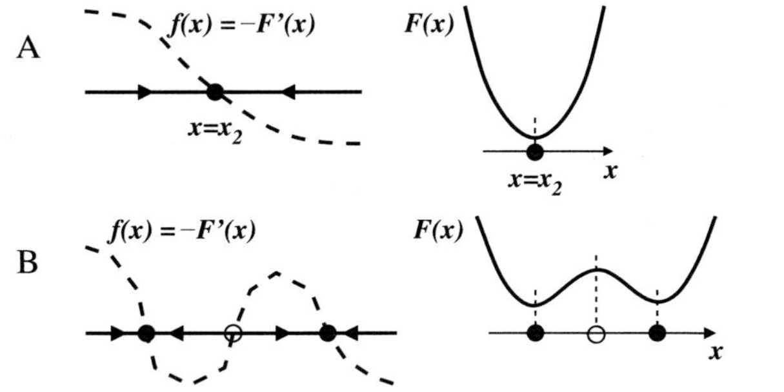

FIGURE 3.8 One-dimensional dynamics is analogous to minimization of a potential. (A) Trajectories of a one-dimensional system $x' = f(x)$ with a single attracting critical point (x_2, filled circle; left) are analogous to paths of particles governed by a single-well potential $F(x)$ (right). (B) Similarly, a one-dimensional system with two attracting critical points separated by an unstable critical point (open circle; left) is analogous to a double-well potential (right).

decreasing, f is negative above x_2 where the potential is increasing, and at $x = x_2$, where the potential has a minimum, $f(x_2) = 0$. The fact that an energy minimum is stable corresponds to the fact that the second derivative of the potential is positive, which is exactly the stability condition that we described above, namely, $f'(x_2) < 0$.

We now discuss a specific example, namely, equation 3.1 from section 3.1.3 (for a similar discussion with a more generic example, see Ermentrout and Rinzel, 2010), which we rewrite here:

$$\tau_a \dot{a} = a_\infty(na) - a.$$

Here, τ_a, $n > 0$ are parameters although τ_a will not play a role in the discussion in this section. As mentioned earlier, the function $a_\infty(x)$ is an increasing sigmoidal function ranging from 0 to 1, of the form

$$1/[1 + \exp((x - \theta)/\sigma)] \tag{3.4}$$

for some parameters $\theta > 0$, $\sigma < 0$. The value of n controls the shape of the sigmoidal function a_∞. As noted in section 3.1.3, for intermediate values of n, say, $n \in (n_1, n_2)$, the sigmoid intersects the identity line at three a values. Correspondingly, if we fix n in this interval, then the equation $a_\infty(na) - a = 0$ has three roots, say, $a_1 < a_2 < a_3$; see figure 3.3A. These roots are the critical points of equation 3.1. When $a \in (a_1, a_2)$, $a_\infty(na) < a$, so $\dot{a} < 0$ and $a(t)$ decreases to $a = a_1$ as $t \to \infty$. Similarly, when $a \in (a_2, a_3)$, $a_\infty(na) > a$, so $\dot{a} > 0$ and $a(t)$ increases to $a = a_3$ as $t \to \infty$.

Note that $da_\infty(na)/da > 0$ and that this derivative is small near $a = a_1$, a_3 and large near $a = a_2$.

Correspondingly, it turns out that the outer two roots a_1, a_3 are asymptotically stable, since $d(a_\infty(na) - a)/da = da_\infty(na)/da - 1 < 0$ at these points, and thus the system represented by equation 3.1 is indeed *bistable*, as noted in the previous section. Similarly, $d(a_\infty(na) - a)/da > 0$ at a_2, such that this critical point is unstable. The unstable critical point $a = a_2$ acts as a *threshold* that separates the 1-D phase space into the *basins of attraction* of $a = a_1$ and $a = a_3$. This bistable state is analogous to a double-well potential (figure 3.8B). In this analogy, each stable critical point lies at the trough of one of the potential wells. Indeed, the negative derivatives $d(a_\infty(na) - a)/da$ at these points correspond to the points a_1, a_3 lying at energy minima while the positive derivative $d(a_\infty(na_2) - a_2)/da$ corresponds to a_2 lying at a local maximum of the energy. At a local energy maximum, any perturbation of the position of a particle will yield a transition away from the maximum to a lower energy state, which captures the instability seen in the dynamics of equation 3.1 at $a = a_2$.

Next, consider what happens as the parameter n is increased. As can be seen in figure 3.3A, since a_∞ is an increasing function, this variation in n lifts $a_\infty(na)$ to attain greater values for each a. Thus, the critical points $a = a_1$ and $a = a_2$ come toward each other. As long as both still exist, the qualitative behavior of the flow of equation 3.1 does not change: the ODE remains a bistable system with a threshold. If n increases enough, however, these two equilibria merge and annihilate, say, at $n = n_2$. For $n > n_2$, the system has qualitatively different dynamics: it is monostable, with only the upper equilibrium point, which is asymptotically stable, remaining. In recognition of this qualitative change, we say that a *bifurcation* takes place at $n = n_2$. Similarly, if n is made sufficiently small, then a bifurcation from bistability to monostability occurs at some $n = n_1$, where the equilibria corresponding to a_2 and a_3 merge and disappear. Figure 3.3C encompasses all of this information, including location of critical points and their stability, in a single plot, called a *bifurcation diagram*. For example, at $n = 0.5$, a cross-section of the diagram illustrates the bistability and threshold behavior that we have discussed.

3.2.2 CLASSIFICATION OF CRITICAL POINTS Now, suppose that

$$\dot{x} = f(x),\ x \in U \subset \mathbb{R}^n \tag{3.5}$$

has a critical point at $x = x_0$. A Taylor expansion for $x \in U$ gives

$$f(x) = f(x_0) + Df(x_0)(x - x_0) + R(x, x_0),$$

where $Df(x_0)$ is the Jacobian matrix of partial derivatives of the components of f with respect to the components of x, evaluated at $x = x_0$, and $R(x, x_0)$ is a remainder term. The remainder term converges to 0 similarly to $||x - x_0||^2$ (where $||x||$ denotes the length of vector x) as x approaches x_0, and $f(x_0) = 0$. Thus, for x near x_0, it is reasonable to approximate equation 3.5 by the *linearized system*

$$\dot{v} = Df(x_0)v,\ v \in \mathbb{R}^n. \tag{3.6}$$

In the case of $n = 1$, for example, we have $Df(x_0) = f'(x_0)$, while in the case of $n = 2$, we write $f = (f_1(x, y), f_2(x, y))$ and $x_0 = (x_{1_0}, x_{2_0})$ to obtain

$$Df(x_0) = \begin{bmatrix} \partial f_1(x_{1_0}, x_{2_0})/\partial x_1 & \partial f_1(x_{1_0}, x_{2_0})/\partial x_2 \\ \partial f_2(x_{1_0}, x_{2_0})/\partial x_1 & \partial f_2(x_{1_0}, x_{2_0})/\partial x_2 \end{bmatrix}.$$

Recall that solutions of constant coefficient linear systems $\dot{v} = Av$ are linear combinations of exponential, sine, and cosine functions and their products, possibly multiplied by polynomial terms if the set of *eigenvalues* of A, which we denote $\sigma(A)$, includes any repeats (Coddington and Levinson, 1955). (The eigenvalues of a square matrix A with n rows and columns are the values λ that satisfy the equation $\det(A - \lambda I) = 0$, where I is the n by n identity matrix; the case $n = 2$ is discussed in more detail below.) In particular, if all eigenvalues of A have nonzero real parts, then the growth and decay of solutions is determined by the signs of these real parts. Based on this idea, the *Hartman–Grobman theorem* tells us that the eigenvalues of $Df(x_0)$ control the behavior of equation 3.5 near $x = x_0$, as long as they all have nonzero real parts (e.g., Perko, 2001).

As noted earlier, we can classify a critical point in general terms as *asymptotically stable*, *stable*, or *unstable*. In fact, we can give a more refined classification of critical points based on the eigenvalues of the matrix $Df(x_0)$. To state precise mathematical definitions associated with this classification, we will now let $\Phi(t, x)$ denote the solution of the initial value problem given by equation 3.5 with initial condition $x(0) = x$.

General stability classification for critical points: A critical point x_0 is

- *stable* if for all $\varepsilon > 0$, there exists a $\delta > 0$ such that if $||x - x_0|| < \delta$, then $||\Phi(t, x) - x_0|| < \varepsilon$ for all $t > 0$,
- *asymptotically stable* if it is stable and there exists $\gamma > 0$ such that if $||x - x_0|| < \gamma$, then $||\Phi(t, x) - x_0|| \to 0$ as $t \to \infty$, and
- *unstable* if it is not stable.

For example, for a harmonic oscillator

$$\begin{aligned} \dot{x} &= y, \\ \dot{y} &= -x, \end{aligned} \tag{3.7}$$

the origin is a critical point that is stable but not asymptotically stable. That is, equation 3.7 yields periodic oscillations from any initial condition not at the origin and correspondingly the origin is surrounded by a nested family of periodic orbits in the phase plane.

Classification of critical points based on eigenvalues of $Df(x_0)$*:* It can be shown (e.g., see Strogatz, 2001; Perko, 2001) that if $Re(\lambda) \neq 0$ for all $\lambda \in \sigma(Df(x_0))$, then the critical point x_0 of equation 3.5 is either asymptotically stable (if all of its eigenvalues have negative real parts) or unstable, and its stability type matches that of the critical point $v=0$ of equation 3.6. This result is the *principle of linearized stability*. In particular,

- if all eigenvalues of $Df(x_0)$ are real and of the same sign, then x_0 is a *node*,
- a node with all eigenvalues negative is a *sink*,
- a node with all eigenvalues positive is a *source*, and
- if all eigenvalues of $Df(x_0)$ are real but both signs are represented, then x_0 is a *saddle*.

Note that when a critical point is a saddle, there are some directions along which solutions contract toward the critical point but other directions along which solutions expand away, as we saw in the excitability phase plane in section 3.1.2.

A more detailed classification of eigenvalues can be done in planar systems, where $n = 2$. There, the characteristic (eigenvalue) equation is quadratic, and eigenvalues are given by

$$\lambda_{\pm} = \frac{\tau \pm \sqrt{\tau^2 - 4\Delta}}{2}$$

where τ is the trace of $Df(x_0)$ and $\Delta = \det Df(x_0)$ (recall that for a matrix

$$A = \begin{bmatrix} a & b \\ c & d \end{bmatrix}$$

the trace of A is given by $a + d$ and the determinant of A equals $ad - bc$). If $\tau^2 - 4\Delta \geq 0$, then both eigenvalues are real. If $\tau^2 - 4\Delta < 0$, then the eigenvalues are complex conjugates, and only then will sines and cosines modulate the trajectories.

Classification of critical points for planar systems: Without loss of generality, suppose that equation 3.5 has a critical point at $x = 0$ and assume $n = 2$. Consider the corresponding linear system given by equation 3.6. This system itself has a critical point at $v = 0$. The critical point $v = 0$ for equation 3.6 is

- a *saddle* if $\Delta < 0$, such that $\lambda_- < 0 < \lambda_+$,
- a *node* if $\Delta > 0$ and $\tau^2 - 4\Delta \geq 0$, such that $\lambda_\pm \in \mathbb{R}$ have the same sign, which is asymptotically stable if $\tau < 0$ and unstable if $\tau > 0$,
- a *spiral* if $\Delta > 0$ and $\tau^2 - 4\Delta < 0$, which is asymptotically stable (and called a *focus*) if $\tau < 0$ and unstable if $\tau > 0$, and
- a *center* if $\Delta > 0$ and $\tau = 0$, such that $\lambda_\pm$ are imaginary.

The terminology and stability properties of saddles, nodes, and spirals carry over to $x = 0$ for equation 3.5, by the principle of linearized stability.

3.2.3 Two-Dimensional Systems, Nullclines, and Separatrices

Here, we start by following the exposition provided by Ermentrout and Rinzel (2010). Surprisingly, much of the behavior of neurons can be captured by looking at two-dimensional dynamics. First, consider the general two-variable system,

$$\begin{aligned} \dot{x} &= f(x, y), \\ \dot{y} &= g(x, y). \end{aligned} \tag{3.8}$$

The right-hand sides of these equations can be considered as rates, such that $(f(x, y), g(x, y))$ specifies a velocity vector at (x, y) for each (x, y) in the two-dimensional phase plane for the system given by equation 3.8. As discussed in section 3.1.2, the points for which $f(x, y) = 0$ define a curve implicitly, called the x-*nullcline*. These points are special because the condition $f(x, y) = 0$ implies that the x-velocity is zero. Thus, when a trajectory crosses the x-nullcline in the phase plane, the flow is vertical; $x(t)$ would typically be at a local maximum or minimum. Similarly, the y-*nullcline*, along which the flow is horizontal, is defined by $g(x, y) = 0$. The two nullclines partition the phase plane into various regions. The locations of the nullclines together with the direction of a representative local velocity vector within each region provide quite a bit of information about the qualitative behavior of the flow for equation 3.8. Moreover, intersections between the two nullclines occur at the points where both $\dot{x}$ and $\dot{y}$ vanish, corresponding exactly to the critical points for the two-dimensional system. Thus, as with the one-dimensional equation 3.1, we can immediately see the critical points of system 3.8, graphically, by plotting its nullclines and observing their intersections.

Unlike the one-dimensional system, critical point stability cannot in general be determined by inspection. Sometimes, in fact, stability does follow from nullcline slopes; more generally, stability can be determined analytically or numerically. That is, suppose that $(x, y) = (x^*, y^*)$ is a critical point, given by $f(x^*, y^*) = g(x^*, y^*) = 0$. From the discussion in section 3.2.2, its stability is determined by the eigenvalues of the matrix

$$A = \begin{pmatrix} \partial f(x^*, y^*)/\partial x & \partial f(x^*, y^*)/\partial y \\ \partial g(x^*, y^*)/\partial x & \partial g(x^*, y^*)/\partial y \end{pmatrix} \equiv \begin{pmatrix} a & b \\ c & d \end{pmatrix}.$$

Indeed, from section 3.2.2, the critical point will be asymptotically stable if and only if $\tau = \text{trace}(A) = a + d < 0$ and $\Delta = \det(A) = ad - bc > 0$.

As in the one-dimensional example in section 3.2.1, bifurcations can occur in which critical points change stability, or are born or annihilated, as a parameter is varied. The trace and determinant conditions tell us what effects parameters must have on a, b, c, d to achieve a change of stability. Functional neural dynamics involve oscillations, and certain bifurcations can give rise to oscillations. Oscillations initially emerging in the neighborhood of a critical point can arise through an *Andronov–Hopf* (AH) *bifurcation*, which occurs when parameter variation causes the eigenvalues of A to be purely imaginary (with $\text{trace}(A) = 0$ and $\det(A) > 0$); recall that these conditions arise in the case of a *center* for a linear system, but they need not yield a center in a nonlinear system since the Hartman–Grobman theorem does not apply when eigenvalues are purely imaginary. An AH bifurcation can switch a system from having a stable critical point to having an unstable critical point surrounded by a stable periodic orbit (*supercritical* AH bifurcation)—that is, a switch from subthreshold oscillations that damp out to oscillations that grow into periodic activity—or from having an unstable critical point to having a stable critical point surrounded by an unstable periodic orbit (*subcritical* AH bifurcation). Although local linear analysis can identify parameter values where an AH bifurcation gives rise to oscillations, the criticality of an AH bifurcation in a particular system cannot be determined simply from the eigenvalues of A (see section 3.2.3.2 below for an example). Interestingly, there are also global mechanisms that give rise to oscillations that cannot be detected by local analysis.

3.2.3.1 Chick Spinal Cord: 2-D Subsystem and Fast–Slow Timescale Separation Now, we revisit the rhythms shown in figure 3.4B but focus on the slow envelope of the oscillations, ignoring the fine structure within each cycle. This slow wave is described by treating d as a parameter in equation 3.2 with s still evolving in time, to obtain

$$\begin{aligned} \tau_a \dot{a} &= a_\infty(sda) - a, \\ \tau_s \dot{s} &= s_\infty(a) - s, \end{aligned} \tag{3.9}$$

with d at a fixed value. Suppose we assume that $\tau_a << \tau_s$, such that a changes much faster than s. In our one-dimensional reduction of the equation, we used $n = ds$ as a parameter. Now that s is slowly evolving, its effect on the system turns out to be equivalent to slowly sweeping the parameter n back and forth in the one-dimensional system.

To understand this idea more precisely, we analyze the dynamics of the two-dimensional system 3.9 using the dynamical systems approach that we have presented (for full details, see Tabak et al., 2000; see also Tabak and Rinzel, 2005). Suppose that we fix d at a value such that the a- and s-nullclines attain the configuration shown in figure 3.9. Their unique intersection point is the only critical point of equation 3.9. After dividing through by the time constants to put the system in the form of equation 3.8, linearization at the critical point yields the matrix

$$\begin{pmatrix} (sda'_\infty(sda) - 1)/\tau_a & ada'_\infty(sda)/\tau_a \\ s'_\infty(a)/\tau_s & -1/\tau_s \end{pmatrix}.$$

We have $a'_\infty(sda) > 0$ and $s'_\infty(a) < 0$. For $\tau_a = \tau_s$, the magnitudes of these derivatives cause both eigenvalues of the matrix to be real and positive (as long as σ_a, the slope factor in equation 3.4 for a, is not too small), while smaller τ_s allows the determinant to grow such that the eigenvalues become complex with positive real part. Since the critical point is unstable, solutions that start away from this point will not approach it. Trajectories that start within the physiologically relevant region $R_{s,a} = \{(s, a): 0 < s, a < 1\}$ must remain bounded, however. To see this fact, note that if $s = 0$, then $s' > 0$, and if $a = 0$, then $a' > 0$; thus, trajectories with initial conditions in $R_{s,a}$ maintain $a, s > 0$ for all time. Similarly, both a_∞ and s_∞ are bounded above by 1, such that $s' < 0$ on $\{s = 1\}$ and $a' < 0$ on $\{a = 1\}$; thus, trajectories in $R_{s,a}$ also maintain $a, s < 1$ for all time.

Therefore, trajectories in $R_{s,a}$ not at the critical point remain bounded but stay away from the critical point for all time. So what becomes of them? The answer is given by a powerful result for 2-D systems, the *Poincaré–Bendixson theorem* (Strogatz, 2001; Perko, 2001): If a trajectory is confined to a compact (i.e., closed and bounded) set in R^2, then it converges to a critical point or to a periodic orbit as $t \to \infty$. In the example we are discussing, the latter possibility applies. The Poincaré–Bendixson theorem provides perhaps the most useful approach for proving the existence of stable periodic orbits for planar systems: show that there is a region in phase space such that trajectories cannot leave once they enter, and show that this region does not contain

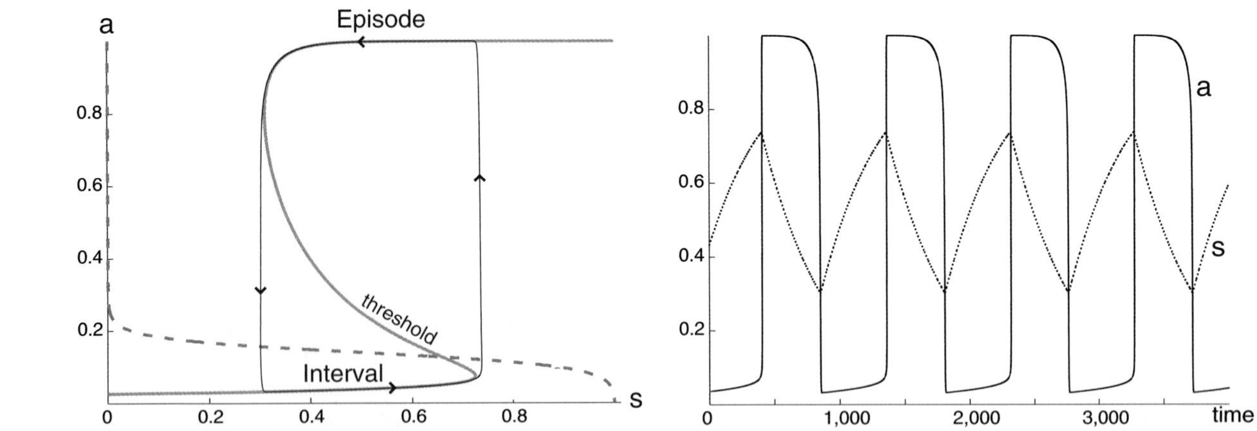

FIGURE 3.9 Model 3.9 in the relaxation oscillation regime (adapted from Tabak et al., 2000). Left: Nullcline configuration. The a-nullcline is solid, the s-nullcline is dashed. The nullclines intersect on the middle branch of the a-nullcline. This middle branch represents a threshold between stable states of high and low a if s is fixed as a parameter. With dynamic s, oscillations occur, with s traversing the range between the left fold where the upper and middle a-nullcline branches come together and the right fold where the middle branch meets the lower branch, such that the extent of the middle branch defines the range of s traversed during these oscillations. Right: Time courses of a and s during relaxation oscillations consisting of slow excursions along the stable (upper and lower) branches of the a-nullcline alternating with very rapid jumps between the branches. The abrupt transitions in a follow from the assumption that $\tau_a << \tau_s$.

a stable critical point, and you can conclude that there must be at least one asymptotically stable periodic orbit inside the region (in fact, there can be several, along with unstable periodic orbits, although the case of a single asymptotically stable periodic orbit arises most commonly in simple models). Note that, unlike the local analysis used to find the onset of oscillations in an AH bifurcation, the Poincaré–Bendixson theorem is a global analysis tool that does not involve linearization and does not require a transition from a regime without oscillations to a regime with oscillations. The periodic solution to system 3.9 corresponds to the slow episodic rhythm in the chick spinal cord, with corresponding time courses of a and s shown in figure 3.9. The oscillations that arise, consisting of prolonged periods of drift or relaxation near the nullcline of the fast variable a alternating with abrupt transitions between nullcline branches (figure 3.9), are called *relaxation oscillations*. Within this regime, restoring the evolution of d would provide a mechanism for the fast spikes within the prolonged active, or high a, part of each episode (see section 3.2.3.2 below).

Next, suppose that d is changed so that the a- and s-nullclines intersect on the lower branch of the a-nullcline, say, at (a^*, s^*). An analysis of the matrix obtained from linearization of system 3.9 about (a^*, s^*) shows that it has real negative eigenvalues. If we consider a trajectory with an initial condition near (a^*, s^*), then the principle of linearized stability implies that the trajectory will converge to (a^*, s^*) as $t \to \infty$. In fact, from similar arguments to those used to apply the Poincaré–Bendixson theorem above, we can see that any initial condition yields a trajectory with this long-term convergence behavior. However, linearization does not tell us what the transient behavior of the trajectory will be before it gets close to (a^*, s^*).

It turns out system 3.9 exhibits *excitable* dynamics when configured to have a stable equilibrium point on the low-a branch of the a-nullcline. As noted in section 3.1.2, in an excitable system, there are initial conditions from which trajectories make prolonged excursions before settling down to equilibrium behavior (e.g., as illustrated in figure 3.1). To understand the excitability in system 3.9 recall that $\tau_a << \tau_s$. This difference in timescales means that $\dot{a} >> \dot{s}$, except near the a-nullcline, where $\dot{a}$ becomes small. In the extreme, we can let $\tau_s \to \infty$, such that s is frozen. This limit takes us back to the 1-D dynamics of a, which we analyzed previously.

Consider the range of s values where the a-nullcline is triple-branched, say, with branches $a_l(s)$, $a_m(s)$, $a_u(s)$ (s between the values where the two vertical excursions occur in figure 3.9). If we fix s in this range and choose $a_l(s) < a < a_m(s)$, then $\dot{a} < 0$ and a approaches $a_l(s)$. If, however, we choose $a_m(s) < a < a_u(s)$, then a increases toward $a_u(s)$. With s fixed, in fact, $a \to a_u(s)$ as $t \to \infty$. If we restore the s dynamics, however, then for an initial condition (a, s) with $a_m(s) < a < a_u(s)$, the corresponding trajectory first quickly approaches $a_u(s)$ and then slowly drifts down along the curve $\{a = a_u(s)\}$, governed by the s dynamics with $\dot{s} < 0$, until $a_u(s)$ meets $a_m(s)$ and both nullcline branches end. This slow excursion is followed by a fast jump down toward $\{a = a_l(s)\}$. In a neighborhood of $a_l(s)$, $\dot{s} > 0$, so finally the trajectory slowly drifts up along $\{a = a_l(s)\}$, and it approaches (a^*, s^*) as $t \to \infty$, as claimed.

The jump up to $a_u(s)$ and drift along $a_u(s)$ constitute the prolonged excursion that characterizes excitability. The description provided here, with switching between fast jumps and slow drifts, can be made more precise by defining *fast* and *slow subsystems*, corresponding to the $\tau_s \to \infty$ limit with different scalings of time. These subsystems can be used to define *singular solutions*, similar to what we described above, and the existence of true solutions $O(1/\tau_s)$-close to these singular solutions can be established rigorously by checking a small number of hypotheses relating to the interactions of the fast and slow dynamics.

To make these ideas more concrete, suppose we have a system

$$\begin{aligned} \dot{x} &= f(x,y), \\ \dot{y} &= \varepsilon g(x,y), \end{aligned} \tag{3.10}$$

where the parameter ε satisfies $0 < \varepsilon << 1$ and differentiation is with respect to time t. In system 3.9, $1/\tau_s$ plays the role of the small parameter that we call ε here. The fast subsystem is obtained by setting $\varepsilon = 0$, such that y becomes a parameter in the ODE for x. To define the slow subsystem, let $\tau = \varepsilon t$ and take derivatives with respect to τ to convert system 3.10 to

$$\begin{aligned} \varepsilon x' &= f(x,y), \\ \varepsilon y' &= \varepsilon g(x,y), \end{aligned}$$

where prime denotes differentiation with respect to τ instead of t. Cancel the ε from both sides of the y equation and let $\varepsilon = 0$ to obtain the differential-algebraic slow subsystem

$$\begin{aligned} 0 &= f(x,y), \\ y' &= g(x,y). \end{aligned}$$

To construct a singular solution for system 3.10 first consider an initial condition (x_0, y_0) such that $f(x_0, y_0) \neq 0$. For ε sufficiently small, $f(x_0, y_0) >> \varepsilon g(x_0, y_0)$. Thus, we use the fast subsystem, with y fixed at y_0, to generate a trajectory, which will be the first piece of our singular solution; call it F_1. Suppose that the equation $f(x, y) = 0$ has solution branches $x = x_i(y)$ for $i = 1, 2, \ldots, n$; note that these are branches of the x-nullcline. Let us assume that F_1 asymptotes to a point $(x_i(y_0), y_0)$ for some $i \in \{1, \ldots, n\}$ as $t \to \infty$. In this limit, the first equation of the slow subsystem is satisfied, and we use the second equation, specifically $y' = g(x_i(y), y)$, to generate the next piece of the singular solution, call it S_1. Suppose that the solution branch $x_i(y)$ of $f(x, y) = 0$ exists for $y \in (y_l, y_u)$, an interval that includes y_0 (similar to the range of s between the values where the two vertical excursions occur in figure 3.9). If the dynamics of the slow subsystem cause y to evolve to reach y_l or y_u, then the condition $f(x, y) = 0$ is lost and we switch back to the fast subsystem. We continue in this manner, alternating between fast subsystem trajectories that provide connections between fast subsystem equilibrium points $(x_i(y), y)$ and $(x_j(y), y)$ for some i, j and slow subsystem trajectories that each drift along a solution branch of $f(x, y) = 0$ (i.e., of the x-nullcline), to obtain a full singular solution, $F_1 \cup S_1 \cup F_2 \cup S_2 \ldots$ which can terminate with a final fast or slow piece. Typical singular solutions of interest either are periodic, returning back to a starting point after a finite number of fast and slow excursions (e.g., the oscillation of system 3.9 discussed above and shown in figure 3.9), or else connect from one equilibrium point of the full system, system 3.10, to another (same or different) equilibrium point, again with finitely many excursions involved. For typical neuronal models away from bifurcations, singular solutions extend to nearby true solutions of system 3.10 over some range of ε values. Additional details and examples relating to the analysis of two-timescale neuronal models can be seen elsewhere (Rinzel, 1987; Rubin and Terman, 2002; Borisyuk and Rinzel, 2005; Ermentrout and Terman, 2010).

3.2.3.2 Chick Spinal Cord: Fast Oscillations For the full model of spinal cord rhythmicity described by equation 3.2, the 2-D simplified version given in system 3.9 captures qualitatively the slow episodic behavior (figure 3.2) but not the fine structure of cycling within each episode. To understand the mechanism for cycle generation, we "look inside" an episode, treating s as a parameter and considering the dynamical interactions of the variables a and d. Phase plane analysis of the a–d subsystem with s frozen reveals three steady states, one stable and two unstable (figure 3.10A). The steady state of high activity corresponds to the attracting high-a branch of the a-nullcline in the left panel of figure 3.9. When d is dynamic, however, this steady state becomes unstable for a range of s values, where it is surrounded by a stable limit cycle. Limit cycle oscillations correspond to the repetitive cycling during an episode. The time course and clockwise flow of the limit cycle's trajectory show that when activity is high, the recurrent connectivity depresses (i.e., $d(t)$ decreases), bringing down the activity, and when the activity is low, there is some recovery from depression (i.e., $d(t)$ increases).

If s were allowed to vary slowly according to the full 3-D model (equation 3.2), then this a–d phase plane would slowly change as s decreased during an episode.

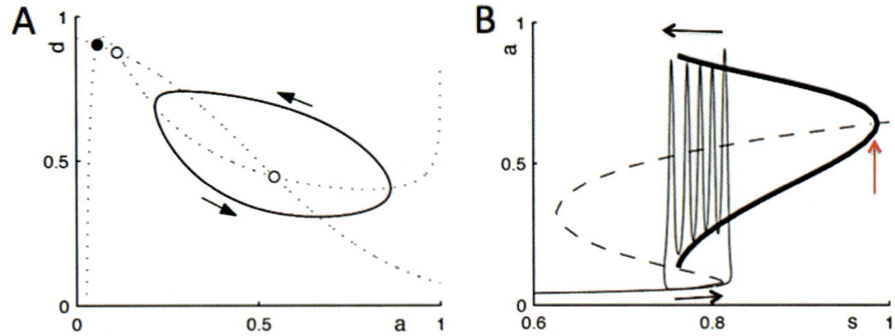

FIGURE 3.10 Cycling in an episodic rhythm. (A) Phase plane for $a - d$ fast subsystem with s frozen at 0.78 and nullclines shown as dashed curves. The nullcline intersections are critical points, two of which are unstable (open circles) and one of which is stable (filled circle). A cycling periodic orbit coexists with the stable low activity steady state (filled circle). (B) Trajectory of episodic rhythm (from the time course in figure 3.4) projected onto the $s - a$ phase plane. The s-shaped curve corresponds to the steady states of the $a - d$ system (as in A), with solid (dashed) corresponding to stable (unstable) states. The heavy solid curve indicates the maximal and minimal a values along the cycling orbits as shown in A. This family of orbits emerges with small amplitude as a supercritical Andronov–Hopf bifurcation (red arrow).

The limit cycle would collide with the "middle" steady state, a saddle point, and disappear. This disappearance coincides with termination of an episode, and it leaves the a–d system with only one attractor, the stable steady state at low a. This steady state corresponds to the quiescent phase of the episodic rhythm. The case of two attractors in the a–d system in figure 3.10A is another type of bistability; it involves coexistence of a stable steady state and a stable limit cycle. The full behavior for the slow episodic rhythm with attracting quiescent phases and episodes of cycling is explained with a fast/slow description in more detail in Tabak et al. (2000).

3.2.3.3 Perceptual Dynamics: 2-D Subsystem and Winner-Take-All The bistability that we described in section 3.2.1 for the 1-D model given by equation 3.1 was based on a regenerative process, the fast recruitment of activity due to recurrent excitation; this was bistability between states of high and low activity. Bistability can also arise without explicit excitation. In the case of neuronal competition between two populations coupled with mutual inhibition (figure 3.6A), an active population suppresses the other. If the competition is symmetrical and populations are identical, either population can be the dominant one, which represents a form of bistability.

Such bistability underlies the slow perceptual alternations of section 3.1.3.2. It is seen in the fast dynamics of r_1 and r_2 in the perceptual dynamics model of equation 3.3, with the slow variables a_1 and a_2 now treated as parameters:

$$\begin{aligned} \dot{r}_1 &= -r_1 + S(-\beta r_2 - \phi a_1 + I_1), \\ \dot{r}_2 &= -r_2 + S(-\beta r_1 - \phi a_2 + I_2), \end{aligned} \tag{3.11}$$

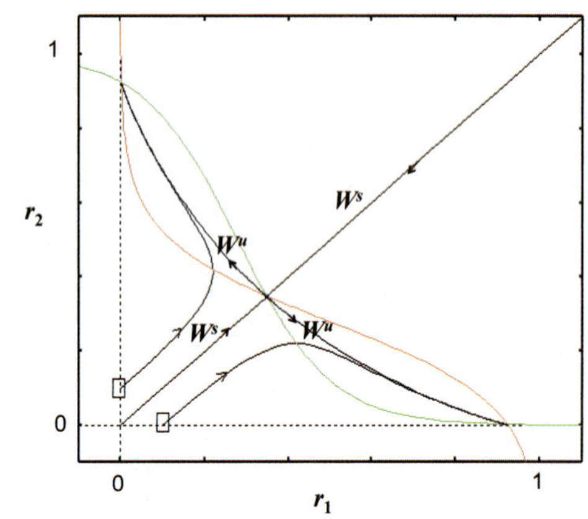

FIGURE 3.11 Phase plane including stable and unstable manifolds of the saddle R_s for system 3.11. The red and green nullclines for r_1 and r_2, respectively, intersect in three critical points, one along each axis and one, R_s, with $r_1 = r_2 > 0$. The stable manifold W^s consists of all solutions that approach R_s exponentially as $t \to \infty$ and comprises two branches that meet at R_s and are aligned along the line $r_1 = r_2$ in this example. The unstable manifold W^u consists of all solutions that approach R_s as $t \to -\infty$ and comprises two branches that emanate from R_s, and each branch projects out toward one of the axes in this example.

where we have set $\tau = 1$ for convenience. From figure 3.6B (see also figure 3.11), we see that for certain values of a_1, a_2, the r_1– and r_2–nullclines intersect in three points, one that we denote R_1 with large r_1 and small r_2, one that we denote R_2 with small r_1 and large r_2, and one that we denote R_s at which both variables take intermediate values. Linearization of system 3.11 shows that R_1 and R_2 are both asymptotically stable nodes, while R_s is a saddle, with eigenvalues $\lambda_- < 0 < \lambda_+$ that we make use of below. Thus, the system exhibits bistability for this choice of (a_1, a_2), corresponding to the observation that either stimulus can be perceived.

Which perception actually occurs depends on the initial state of the system. Interestingly, although it is unstable, it is the saddle R_s that provides the structure for partitioning initial conditions into a set yielding trajectories that converge to R_1 and a set from which R_2 wins. When we solve a linear system of ODEs, $\dot{x} = Ax$, for which the coefficient matrix A has real eigenvalues, the solution $x(t)$ is a linear combination of terms involving the eigenvectors of A. The eigenvectors themselves are invariant: an initial condition on an eigenvector yields a solution that stays on the eigenvector for all time. The eigenvectors for the matrix obtained from linearizing the nonlinear system 3.11 about R_s are not invariant for the nonlinear system itself, however; they provide only a local description of dynamics. The *stable manifold theorem* states that there is exactly one pair of solutions of system 3.11 that converges to R_s as $t \to \infty$, both of which are tangent to the eigenvector v_-, corresponding to λ_-, as $t \to \infty$. These solutions are branches of the *stable manifold* W^s of R_s, as illustrated in figure 3.11. Think of v_- as a line through R_s, partitioned into two rays by R_s; one of branch of W^s approaches R_s along one of the rays (say, from the northeast), and the other approaches along the other ray (from the southwest). There is also an *unstable manifold* W^u of R_s, the two branches of which approach R_s along the eigenvector v_+, corresponding to λ_+, as $t \to -\infty$ (figure 3.11).

As we have stated, the branches of W^s approach R_s as $t \to \infty$. In fact, the combined curve W^s (including R_s itself) forms a *separatrix* for system 3.11. The name separatrix derives from the observation that the curve separates trajectories with one form of behavior from trajectories with another, qualitatively distinct behavior. In this case, the separatrix represents an example of the threshold we referred to in section 3.1.2 and is also the 2-D generalization of the threshold critical point in a 1-D flow. Solutions that start above W^s in the positive quadrant are trapped above it and approach R_2 as $t \to \infty$ while solutions that start below W^s are trapped below it and approach R_1 as $t \to \infty$. This configuration is called a *winner-take-all (WTA)* regime because almost all initial conditions lead to outcomes in which either r_1 or r_2 is elevated and the other component is suppressed. To see the details of how the separatrix is formed, we can look at what happens to the branches of W^s if we follow them *backward* in time. Note that in forward time, the flow along the coordinate axes $\{r_1 = 0\}$ and $\{r_2 = 0\}$ points into the positive quadrant since $\dot{r}_1 = S(-\beta r_2 - \phi a_1 + I_1) > 0$ for $r_1 = 0$ and $\dot{r}_2 = S(-\beta r_1 - \phi a_2 + I_2) > 0$ for $r_2 = 0$, similarly to the case for system 3.9 discussed above. Thus, in backward time, the flow points out of the positive quadrant, and indeed the lower branch of W^s—call it W_l^s—leaves the quadrant through one of the coordinate axes (or through the origin if $-\phi a_1 + I_1 = -\phi a_2 + I_2$ as in figure 3.11). W_l^s is an actual solution of system 3.11, so other solutions cannot cross through it. Similarly, the upper branch W_u^s goes off to infinity as $t \to -\infty$ and forms a barrier for solutions above R_s.

Next, let us return to the unstable manifold W^u of R_s. While its two branches approach R_s as $t \to -\infty$ and thus emanate from R_s in forward time, what becomes of them as t increases? It turns out that each of these curves approaches another critical point, R_1 or R_2, as $t \to \infty$. Trajectories that approach one critical point as $t \to -\infty$ and a different critical point as $t \to \infty$ are called *heteroclinic orbits*. We now sketch the dynamical systems argument for why W^u forms heteroclinic orbits in this example. A calculation shows that the eigenvector v^+ has a slope that is intermediate between the slopes of the r_1– and r_2–nullclines at R_s. Thus, as the branches of W^u emanate from R_s, they lie in the region between these nullclines, one that we call W_2^u in the lobe above and to the left of R_s and one that we call W_1^u in the lobe below and to the right (see figures 3.6B and 3.11). Consider the upper lobe. The lobe is bounded by part of the r_1-nullcline (red) and part of the r_2-nullcline (green). Since the r_1-nullcline segment lies below the r_2-nullcline, $\dot{r}_2 > 0$ on it. Thus, any trajectory from inside the lobe that approaches the r_1-nullcline is pushed back inside the lobe: trajectories cannot leave through the r_1-nullcline. Similarly, the boundary where $\dot{r}_2 = 0$ lies to the right of the r_1-nullcline, where $\dot{r}_1 < 0$. Thus, trajectories cannot leave through this boundary either. In particular, W_2^u is trapped in the lobe for all time. The Poincaré–Bendixson theorem tells us that the bounded trajectory W_2^u must approach a periodic orbit or a critical point as $t \to \infty$, but $\dot{r}_1$, $\dot{r}_2$ do not switch signs inside the lobe, so no periodic orbit can exist there. Thus, $W_2^u \to R_2$ as $t \to \infty$, and an analogous argument implies that $W_1^u \to R_1$ as $t \to \infty$.

Often heteroclinic orbits serve as dynamic attractors within systems of two or more dimensions. For example, most trajectories of system 3.11 that approach R_1 or R_2 do so along W_1^u or W_2^u, respectively. In systems with multiple saddles, we may observe multiphase rhythms in which trajectories travel close to a heteroclinic orbit connecting two saddles, spend a long time near the destination saddle, and eventually emerge along another heteroclinic orbit, which guides them to a new saddle, and so on. Such heteroclinic cycles have been hypothesized to allow for dynamic coding or sequential memory, in which stimulus identification or recall involves a passage through a sequence of activity states (Rabinovich et al., 2001; Rabinovich et al., 2012).

Changing the values at which a_1, a_2 are frozen in system 3.11 alters the positions of the nullclines. For example, as a_1 increases, the critical points R_s and R_1 may merge and annihilate in a saddle-node bifurcation. In such a bifurcation, saddle and node critical points come together and disappear, analogously to what we described earlier for a 1-D system. If this happens, then trajectories in the positive quadrant all tend to the remaining stable critical point R_2. Varying a_2 can induce a similar bifurcation of R_s and R_2. These observations hint at a mechanism for perceptual alternation. That is, when a_1 and a_2 are allowed to vary according to the full system 3.11, the nullclines drift slowly, leading to alternations in saddle-node bifurcations for the disappearance of R_1 and R_2; see section 3.3.2.

Finally, we note that a similar WTA competition mechanism also arises in decision-making models, such as the two-variable reduced model derived by Wong and Wang (2006):

$$\begin{aligned} \dot{S}_1 &= -S_1/\tau_S + (1-S_1)\gamma H(x_1,x_2), \\ \dot{S}_2 &= -S_2/\tau_S + (1-S_2)\gamma H(x_1,x_2), \\ x_1 &= J_{N,11}S_1 - J_{N,12}S_2 + I_0 + I_1 + I_{noise,1}, \\ x_2 &= J_{N,22}S_2 - J_{N,21}S_1 + I_0 + I_2 + I_{noise,2}, \end{aligned} \tag{3.12}$$

where S_j are mean field representations of two neuronal populations that receive baseline input I_0 and individualized inputs that are deterministic, I_j, and stochastic, $I_{noise,j}$, respectively. H represents the input–output relation for cells in each ensemble. The activity variables S_j correspond to the levels of slow synaptic activation that drive recurrent excitation in each population and also activate a common inhibitory pool that provides the interaction mechanism for competition. Such slow synaptic activation typically arises through release and uptake of the neurotransmitter NMDA, for which the postsynaptic receptors exhibit slower kinetics, with more complicated voltage dependence, than are associated with the more straightforward, fast synaptic excitation due to the neurotransmitter AMPA.

An interesting feature noted there is that the two competing states can be bistable, separated by a saddle and the corresponding separatrix, over a range of input values that includes 0. A large enough input to one population, say S_1, allows that population to win the competition by causing a saddle-node bifurcation that annihilates the saddle and the stable state in which the other population is dominant. If this input is removed after a trajectory approaches a neighborhood of the winning equilibrium point, the system remains in the basin of attraction of the equilibrium point with $S_1 >> S_2$. The authors propose that this mechanism can allow for memory of a decision choice to be sustained, even after the stimulus that evokes the decision is gone (note that Amari and Arbib, 1977, also considered such behavior in their WTA model, and how to override it). The case when I_1 and I_2 differ little, in particular, corresponds to a fine discrimination task. When the stimulus is turned on, the system trajectory moves from the origin toward W^u and then, buffeted by noise, migrates leftward or rightward until it surpasses the decision threshold for R_1 or R_2. This essentially 1-D behavior after the initial transient corresponds to evidence accumulation in a drift-diffusion model for decision-making (Bogacz et al., 2006; Eckhoff et al., 2011).

3.2.4 Lessons Learned: Extracting General Features of Neuronal Model Dynamics

We have described a variety of dynamic features that can arise in planar models. These features include oscillations, excitability, fast–slow dynamics, bistability, separatrices, stable and unstable manifolds, and heteroclinic orbits. We used a model with two competing neuronal populations to discuss bistability, but in fact bistability or multistability can also arise in single-cell models, typically between a low state and a high state sustained by a depolarization-activated positive feedback (e.g., Loewenstein et al., 2005; Fransén et al., 2006). Indeed, the dynamics of any model is determined by the types of equilibrium points it supports, by the configuration of its nullclines, and by the direction of the associated flow in the regions of phase space between nullcline crossings. Bistability and excitability in single-cell models are both associated with interactions of a cubic-shaped nullcline and a monotonic nullcline, often together with a separation of timescales. Biologically, these features emerge from the interaction of fast positive feedback and slow negative feedback, which arise commonly from neuronal membrane currents (e.g., Izhikevich, 2006).

Another way to characterize models is in terms of the bifurcations they exhibit as model parameters are varied. For example, the switch from excitable dynamics to sustained oscillations as input current is amplified typically occurs through one of two bifurcation mechanisms, giving rise to what are known as Class I or Class II dynamics, respectively. These bifurcation structures have been extensively elucidated elsewhere (Rinzel and Ermentrout, 1989; Rinzel and Ermentrout, 1998). Model dynamics may become more complex when more equations are included. But as discussed in section 3.2.2, if different variables evolve on multiple timescales, then there is hope of considering separate, lower-dimensional fast and slow subsystems and piecing their dynamics together to understand full model dynamics.

This approach has been used with great success in the analysis of neuronal bursting, or alternation between active phases of rapid action potential firing and quiescent phases without spikes. In this analysis, slow variables are first treated as parameters, and the bifurcation structure of the fast subsystem with respect to these parameters is determined. The slow dynamics is used to generate trajectories that drift along branches of stable states of the fast subsystem, like the drift along a branch of the a-nullcline in the spinal cord model (system 3.9), while the fast dynamics determines the nature of each stable state (e.g., quiescence or repetitive spiking) and provides abrupt transitions between these states (Rinzel, 1987). The full dynamics of the spinal cord model, including the fast spikes within each episode, represents a form of bursting and can be analyzed using this approach. A thorough classification of single neuron bursting dynamics based on possible fast subsystem bifurcation structures has also been provided elsewhere (Izhikevich 2000), and the richness of bursting dynamics continues to provide topics for active research (see, e.g., Coombes and Bressloff, 2005; Rubin et al., 2009a; and many others).

At this point, we have explained the basics of the dynamical systems approach to analyzing small systems of differential equations, we have presented examples of how this approach can be used to link equation features with certain dynamic phenomena observed in neurons, and we have provided references to sources in the literature where particular types of dynamics are treated in detail. There are other forms of dynamics on which we will not comment. For example, chaos, associated with sensitivity to initial conditions and additional properties that ensure mixing of trajectories, arises in neuronal models with at least three variables in the continuous case (e.g., Terman, 1991; Guckenheimer and Oliva, 2002) and can be analyzed with dynamical systems methods (Devaney, 2003; Strogatz, 2001). In the next section, we switch gears and discuss some additional types of dynamics seen in neuronal network models and their interpretations in terms of interacting populations.

3.3 Beyond Winner-Take-All in Network Dynamics

3.3.1 Cooperation In the WTA scenario that we discussed for system 3.11, the stable states are equilibrium points where one dependent variable is large, corresponding to an active percept, and the other is small, corresponding to a suppressed percept. In contrast, some network interactions can be cooperative, promoting the activity of all components of the network. A simple form of cooperation arises with two units that excite one another to implement a short-term memory. One such example, presented in Wilson's book (Wilson, 1999, section 6.4), takes the form

$$\begin{aligned} \dot{E}_1 &= \left(-E_1 + 100(3E_2)^2 / (120^2 + (3E_2)^2)\right) / \tau, \\ \dot{E}_2 &= \left(-E_2 + 100(3E_1)^2 / (120^2 + (3E_1)^2)\right) / \tau, \end{aligned} \tag{3.13}$$

where each E_i represents the activity level of a neuronal population in a memory storage system. The nullclines of system 3.13 intersect in three points along $E_1 = E_2$. Two of these, a rest state at the origin and a state with elevated E_1 and E_2, are asymptotically stable while the intermediate state provides a separatrix between their basins of attraction, analogous to that in the WTA scenario. From the rest state, a brief, pulsatile input that causes a separatrix crossing can induce a switch to the high-activity state, corresponding to detection and storage of a suprathreshold stimulus. Note that the input can be suprathreshold even if it is delivered to just one population, due to the configuration of the separatrix, as long as it is sufficiently large. The stimulated unit recruits the other, and the ensuing positive feedback allows the system to settle at the elevated state. Of course, the rational nonlinearities in system 3.13 can be replaced by sigmoid functions of the form in equation 3.4 as long as the function in the E_1 equation increases with E_2 and vice versa and the parameters θ, σ are chosen to achieve three intersections (e.g., $\theta > 0$ small, $\sigma < 0$ with $|\sigma|$ small).

Interestingly, a model that appears to be competitive can also give rise to a stable cooperative state. For example, Ermentrout (2003) presents the two-population model

$$\begin{aligned} \tau_1 \dot{x}_1 &= -x_1 + f(w_{11}x_1 - w_{12}x_2 + s_1), \\ \tau_1 \dot{x}_2 &= -x_2 + f(-w_{21}x_1 + w_{22}x_2 + s_2), \end{aligned} \tag{3.14}$$

with all $w_{ij} > 0$, which is analogous to the perceptual rivalry model but with the addition of self-excitation (w_{11}, w_{22} terms); see figure 3.12, upper left. A similar structure is in fact present in system 3.12 (Wong and Wang, 2006); see also Eckhoff et al. (2011). (In fact, this type of population-level or mean field description of network activity is due to Wilson and Cowan (1972, 1973), and others—Amari (1977); Amari and Arbib (1977)—who originally considered interacting excitatory and inhibitory populations.)

If f is invertible, then the nullclines of equation 3.14 are given by the equations

$$x_2 = -\left[f^{-1}(x_1) - w_{11}x_1 - s_1\right] / w_{12}, \; (x_1 - \text{nullcline})$$

$$x_1 = -\left[f^{-1}(x_2) - w_{22}x_2 - s_2\right] / w_{21}, \; (x_2 - \text{nullcline})$$

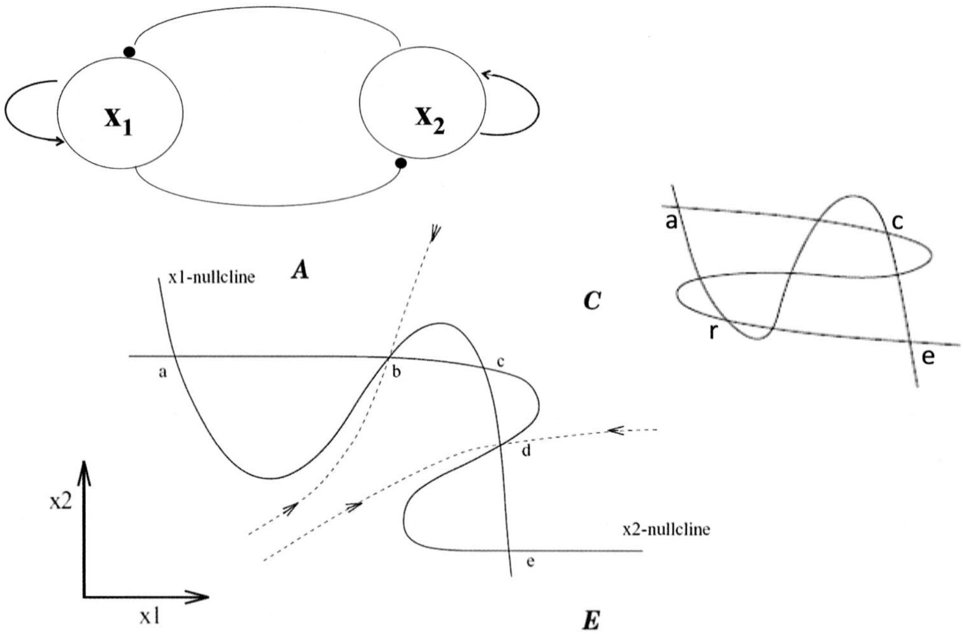

FIGURE 3.12 Competition and cooperation. Architecture of a two-population firing rate model with mutual inhibition and internal recurrent self-excitation (upper left). The mutual inhibition embodies competition to produce the winner-take-all attractors "a" and "e." If the recurrent excitation is adequately strong, there is also an attractor representing a cooperative response with the units in a "high-high" state, "c." A possible resting configuration (upper right) illustrates the coexistence of these three attractors together with the stable resting state "r," making this a simple model for short-term memory. Note that an input can change the configuration of the nullclines (lower left) and annihilate the resting state. Two of the remaining states, "b" and "d," are saddles with associated separatrices (dashed), which form the boundaries between regions where trajectories converge to a, c, or e, denoted by A, C, E, respectively.

As Ermentrout points out, these nullclines become cubic shaped if w_{11} or w_{22}, respectively, becomes sufficiently large. In this case, there can be up to nine nullcline intersections, depending on the values of the other parameters in the model (figure 3.12, upper right). It is not difficult to determine the stability of the equilibria given by these intersection points. If all nine intersections are present, then four are saddles, one is an unstable node, and there are four stable nodes, a rest state with x_1, x_2 low, a cooperative state with both high, and two competitive states with one high and the other low. Introducing a transient stimulus by increasing s_1 and/or s_2 alters the nullcline positions. A sufficiently large stimulus will induce one or more bifurcations that eliminate the rest state (as well as at least one unstable equilibrium). Specifically, if we consider an initial condition near rest, then a transient large increase in s_1 will move the nullclines in a way that puts the resulting trajectory in the basin of attraction of the state with x_1 large and x_2 small (figure 3.12, lower left). A resulting transition to this x_1-dominant state could represent the encoding of the stimulus, and analogous encoding of an x_2 stimulus could be achieved by a transient increase in s_2. Finally, increasing both s_1 and s_2 can push the trajectory to the cooperative state, such that it encodes a combined, coincident-like stimulus.

3.3.2 ALTERNATIVES TO WINNER-TAKE-ALL IN THE RIVALRY MODEL WITH DYNAMIC ADAPTATION We have already encountered a form of competition in the 2-D perceptual rivalry model obtained by freezing the adaptation variables a_1, a_2 in equation 3.3. As is evident in figure 3.6B and was discussed in the previous section, for certain choices of parameters, the r_1- and r_2-nullclines intersect in three equilibrium points. Two of these are stable, corresponding to states in which one or the other percept takes over, and the third is an unstable saddle point. From any starting state $(r_1(0), r_2(0))$, a competition between the stable states ensues, and the position of $(r_1(0), r_2(0))$ relative to the stable manifold of the saddle determines the outcome of this WTA competition.

What happens if we include dynamic adaptation, allowing all four variables in equation 3.3 to evolve? The answer depends on the impact of adaptation on the r_i-nullclines. For example, denote the stable equilibrium at which percept 1 wins by $R_1(a_1, a_2) = (r_1^*(a_1, a_2), r_2^*(a_1, a_2))$, where r_1^* significantly exceeds r_2^* (e.g., $r_1^* \approx 1, r_2^* \approx 0$ in figure 3.6B). If (r_1, r_2) is near (r_1^*, r_2^*), then a_1 will approach the large value of $f_{a,\infty}(r_1)$ and a_2 will approach the small value of $f_{a,\infty}(r_2)$, moving the r_1–and r_2–nullclines in the (r_1, r_2) phase plane. This adaptation of percept 1, and relief from adaptation of percept 2, may or may not cause R_1 to be lost. This outcome depends on whether or not the equations

$$\begin{aligned} r_1 &= S(-\beta r_2 - \phi f_{a,\infty}(r_1) + I_1), \\ r_2 &= S(-\beta r_1 - \phi f_{a,\infty}(r_2) + I_2) \end{aligned}$$

have a solution that is a continuation of the branch (r_1^*, r_2^*). If so, then even with adaptation, state R_1 represents a stable equilibrium corresponding to complete dominance of percept 1. If not, then the trajectory being tracked will jump away from R_1 toward another state, and percept 1 will be lost, at least temporarily.

A very thorough analysis of the dynamics of equation 3.3, with input $I_1 = I_2 = I$ and with $f_{a,\infty}(x) = x$, has been carried out in a sequence of papers (Curtu et al., 2008; Curtu, 2010; Curtu and Rubin, 2011); an earlier work also compared this system to several other similar binocular rivalry models that feature competition (Shpiro et al., 2007). Depending on I and on the values of other parameters, the system can be in one of five main dynamic regimes, only one of which is WTA (figure 3.13); additionally, there are small I intervals in which more exotic dynamics, beyond the five main regimes, can arise. Two of the regimes feature oscillations as described above, corresponding to perceptual rivalry of the type arising in visual illusions.

To summarize part of the analysis done by Curtu et al. (2008), assume that $I_1 = I_2 = I$ and $f_{a,\infty}(x) = x$.

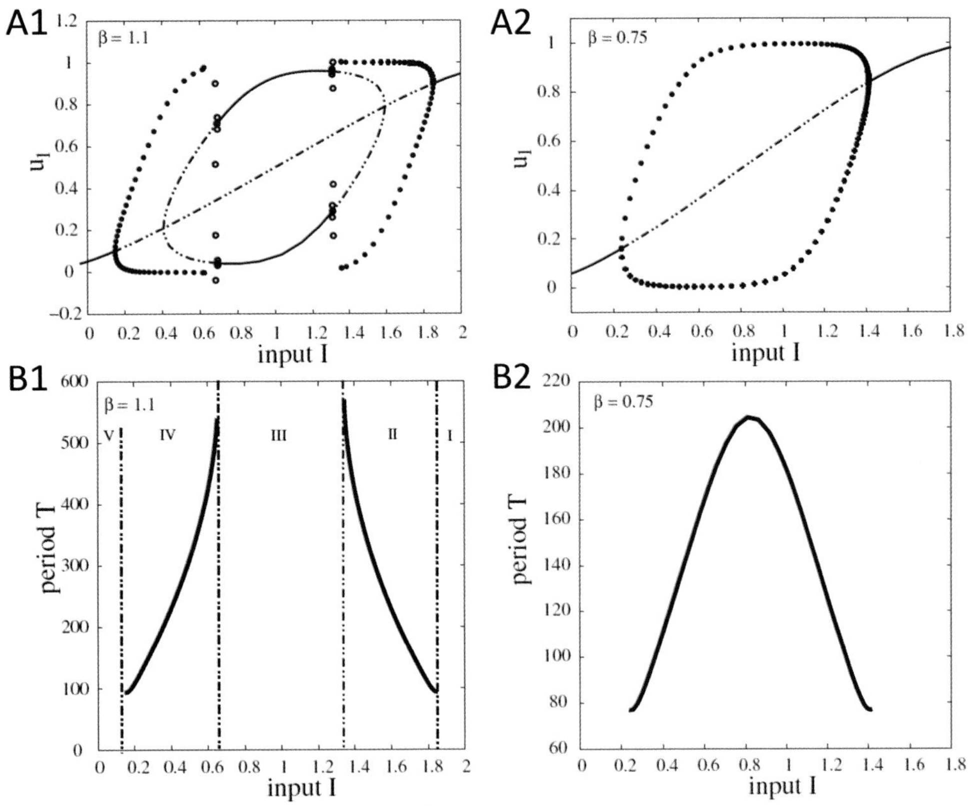

FIGURE 3.13 Bifurcation diagram for equation 3.3 as input parameter $I_1 = I_2 = I$ is varied shows up to five dynamic regimes (u_1 here corresponds to r_1 in the text). (A1) When $\beta = 1.1$, as I grows from 0, a stable critical point corresponding to small values of both u_1 and u_2 (solid) destabilizes in an Andronov–Hopf bifurcation (near $I = 0.2$) that yields stable oscillations (solid circles). As I increases to around 0.4, the now unstable critical point (dash-dotted) undergoes a pitchfork bifurcation, at which two additional unstable critical point branches emerge. These stabilize around $I = 0.7$, giving the winner-take-all (WTA) regime. Similar bifurcations occur with additional increases in I, by symmetry, the first of which restores oscillations and the second of which yields a state in which u_1, u_2 are both elevated. (A2) When $\beta = 0.75$, increases in I do not yield the WTA regime. (B1) In the five-regime case, oscillation period increases monotonically when I is below the WTA regime and decreases monotonically for I above this regime. (B2) In the case without WTA, period varies nonmonotonically with I. Adapted from Curtu et al. (2008). Copyright (c) 2008 Society for Industrial and Applied Mathematics. Reprinted with permission. All rights reserved.

Equation 3.3 has an equilibrium point of the form $R^* = (r^*, r^*, r^*, r^*)$ as long as parameters are chosen such that the equation $r = S(\beta r - \phi r + I)$ has at least one solution; assume that this is true. The stability of the equilibrium point R^* depends on model parameter values. If I is sufficiently small or large, then R^* is asymptotically stable. Stability can be lost through an AH bifurcation or a pitchfork bifurcation as I is varied. The former would give rise to periodic oscillations or perceptual rivalry. In a pitchfork bifurcation, there is a transition from one to three critical points, which in this model represents a symmetry-breaking that leads to bistability of two WTA scenarios, in each of which one percept dominates. A little bit of calculation shows that the AH bifurcation is encountered first as I increases if τ_a and ϕ are relatively large, which is assumed by Curtu et al. (2008) (figure 3.13). In the resulting oscillations, if r_1 is large, then a_1 grows. The system cannot support an r_1-dominant state with a_1 large, and thus a switch eventually occurs to a state with r_1 small and r_2 large. However, this state is later lost when a_2 grows enough. Thus, the dynamics of the full system results in periodic oscillations in which r_1 and r_2 take turns jumping up to a high state, with the corresponding percept dominant, and then back down to a low one.

If I is made even larger, then the WTA scenario emerges: both $R_1(a_1, a_2)$ and $R_2(a_1, a_2)$ persist even in the limit as $a_1 \to f_{a,\infty}(r_1)$, $a_2 \to f_{a,\infty}(r_2)$. Thus, the dynamics leads to the same bistability discussed in the 2-D reduced case. However, since an AH bifurcation had already occurred, the switch from oscillations to WTA is not through a simple pitchfork bifurcation (as a direct transition from the stability of R^* to WTA would have been if it had happened first) but rather involves a more complicated sequence of bifurcations that is beyond the scope of this chapter (Curtu, 2010; Curtu and Rubin, 2011). With additional increases in I beyond the WTA regime, competition is lost again. Recall that for I values below the WTA regime, oscillations resulted because neither r variable was able to settle in the high state; that is, the transiently elevated component eventually *released* the other from suppression. Now, for I values above the WTA regime, oscillations result because neither r can settle in a low state; as its adaptation variable decreases, the nullclines shift such that the low r variable can increase, or *escape* from suppression. This adjustment decreases S in the other r equation because of the $-\beta$ term, causing the other r variable to decrease and eventually leading to a dominance switch. Finally, if I becomes large enough, the symmetrical equilibrium point R^* restabilizes through another AH bifurcation, with a complex conjugate pair of eigenvalues switching from positive to negative real parts. Now, however, this equilibrium occurs with r large: a large input stabilizes the up state for both variables, even in the fully adapted limit.

3.3.3 Transitions by Release and Escape

In the above discussion, we used the phrases *release* and *escape*. These ideas were introduced quite a while ago to label two different modes of transition that can arise in model oscillations (Arbas and Calabrese, 1987; Wang and Rinzel, 1992; Skinner et al., 1993; Skinner et al., 1994; Daun et al., 2009). Which mechanism occurs within a particular model and parameter regime has significant implications for the qualitative behavior of the system in that regime. For a network of two neurons coupled by mutual inhibition, consider an anti-phase or anti-symmetrical initial condition in which one cell is actively spiking and the other is quiet. For oscillations to occur, the two cells much switch roles. One possible way that this switch can occur is that, because of adaptation, the active cell may no longer be able to remain active. If it transitions to a low-voltage quiet state, then it will no longer inhibit the quiet cell, and the removal of inhibition may allow the quiet cell to activate. In this case, we say that a *release* from inhibition has occurred and caused the switch in the cells' roles (figure 3.14A).

Now, suppose that in the release case, a common excitatory input to both cells is a parameter in the model and is increased slightly. To determine the impact of this change, we should consider its effect on cells when they are active, since the active cell controls each switch in the oscillation in this release regime. When a cell is active, the extra input will cause it to stay active longer; that is, a greater level of adaptation will be needed to terminate its active state. Thus, increases in input to systems with oscillations due to release, as long as they are not large enough to destroy the release mechanism, cause increases in the oscillation period (figure 3.13).

An alternative oscillation scenario occurs in similar systems when, again starting from an anti-symmetrical initial condition, the recovery from adaptation eventually allows the quiet cell to activate, despite being fully inhibited by the active cell (figure 3.14B). When the quiet cell becomes active, it strongly inhibits the active cell, which may terminate the cell's activity, especially since the cell has been adapting while active. In this case, we say that an *escape* from inhibition has occurred and is responsible for the transition in which cell is active and which is quiet.

Again, suppose that an excitatory input parameter common to both cells is increased by a small amount, but now consider the escape case. In this case, the quiet cell is in control of the switch in the oscillation phase.

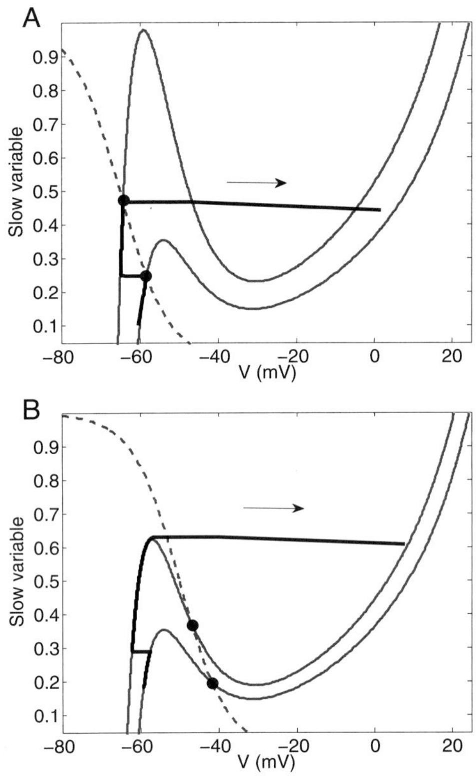

FIGURE 3.14 Release and escape transitions in the phase plane. (A) An example of an inhibited neuron becoming active through release. The position of the V-nullcline (blue solid) depends on the inhibition level—the lower position arises in the absence of inhibition, the elevated position with inhibition present—and in this illustration its intersection with the slow variable nullcline (blue dashed) always yields a stable critical point (black circles). Without inhibition, a trajectory (black solid) approaches the relevant critical point. When inhibition turns on, the trajectory approaches the critical point corresponding to the inhibited V-nullcline. When the cell is subsequently *released* from inhibition, its trajectory is able to jump across to large V (arrow) and an action potential is fired. (B) An example of an inhibited neuron becoming active through escape. All elements are similar to (A) except now the critical points are on the middle branches of the V-nullclines and are unstable. In this case, the inhibited cell is able to *escape* to large V and fire an action potential despite being inhibited.

The increased input will allow the quiet cell to activate after less time in the silent phase, with less recovery from inhibition, than before. Thus, increases in input to systems with oscillations due to escape, which do not destroy the escape mechanism, cause decreases in the oscillation period (figure 3.13).

One can go beyond the basic release/escape classification and distinguish between *synaptic* and *intrinsic* forms of both escape and release (Skinner et al., 1994). This distinction is based on whether a synaptic threshold is crossed during a slow excursion along a fast variable nullcline branch (synaptic mechanism) or a knee of a fast nullcline is reached, resulting in a fast jump through the synaptic threshold θ (intrinsic mechanism). Interestingly, oscillations due to escape, release, and combinations of the two can all be tuned to have similar time courses. Nonetheless, the switching mechanism underlying these oscillations can be revealed by breaking the symmetry in the inputs to the two coupled neurons, say, by increasing g_{app1} while holding g_{app2} fixed. Under this manipulation, if oscillations occur through escape, the duration of the silent phase of neuron 1 will shrink, while the duration of the silent phase of neuron 2 will remain constant. This *independent phase modulation* arises because with oscillations based on escape, the silent neuron is in control of each dominance switch. Strengthening the input to a cell can thus shorten its silent phase, and it can be shown that under some standard conditions, having less time in the active phase does not significantly impact the silent phase duration for the neuron receiving the fixed stimulation level (Daun et al., 2009). In contrast, if oscillations occur through release, then changing g_{app1} while holding g_{app2} fixed has little impact on either cell's silent phase duration. Finally, if oscillations occur through a combination of escape and release, then changing g_{app1} with g_{app2} fixed actually causes silent phase durations for both neurons to change, again under some basic assumptions elaborated elsewhere (Daun et al., 2009). This analysis suggests that escape-based oscillations may play a crucial role in rhythms used by central pattern generators, for example, where it may be desirable to change the duration of one phase of a motor function (e.g., inspiration) without necessarily altering the duration of another phase (e.g., expiration).

3.4 *Concluding Remarks*

In this chapter, we have surveyed some basic ideas of dynamical systems analysis that frequently arise for neuronal models and have illustrated how they can be used to understand a variety of phenomena in several firing rate models of neuronal population activity. As noted in our introduction (section 3.1.1), dynamical systems ideas are useful for providing precise representations of particular phenomena. Concepts like firing an action potential, crossing a decision threshold, and being excited by a synaptic input may sometimes fall into the "I know it when I see it" category, but having a standardized way to represent them allows us to communicate ideas unambiguously, which is critical for advancing our field. We have included some mathematical definitions that perhaps seem abstract, to ensure that readers are aware that mathematical concepts that may sometimes be discussed somewhat loosely really do lie on precise underpinnings. While we have not explicitly treated noise in this chapter, we note that understanding the deterministic dynamics at the core of any stochastic system is an essential first step to analyzing its behavior, so these ideas are useful in stochastic settings as well; for example, the separatrix present in the competition scenario is at the core of decision-making models even when they include stochasticity. Of course, the analysis of the additional effects induced by stochasticity, as well as their relation to the deterministic complexity associated with chaos, are critical, central directions in theoretical neuroscience today (Laing and Lord, 2009; Ermentrout and Terman, 2010; Bressloff, 2012).

In addition to providing a common terminology, dynamical systems analysis really can provide insights that are not evident from examination of time courses alone. Bistability effects that give rise to relaxation oscillations when the relevant "parameter" is equipped with slow dynamics, as we have seen in the chick spinal cord model (system 3.9), are an example of how a complex-seeming phenomenon can be represented very elegantly using bifurcation diagrams and phase planes. The phase plane representations of competition and cooperation that we have discussed show us how varying the strength of an input can influence the dominance of each possible stable equilibrium state, made precise in terms of its basin of attraction. This analysis also explains why there are abrupt transitions in system behavior as input strength is varied since we can clearly see the abrupt loss or gain of nullcline intersections and how these influence system dynamics. The escape and release phenomena that we discussed represent another case where dynamical systems analysis provides insight. In the context of perceptual rivalry, increasing the input to both competing neurons or neuronal populations can cause switching to speed up or slow down, depending on the current input level. This observation is perplexing, and there is no obvious signature in the activity time courses to explain it. However, dynamical systems analysis leads us to think in terms of escape and release transition mechanisms, and this perspective leads to a clear explanation for the disparate effects. Specifically, in the escape regime, the suppressed cell is in control of each transition, such that adding input facilitates dominance switching, while in the release regime, the active cell is in control, such that adding input delays dominance switching (Skinner et al., 1993; Skinner et al., 1994; Curtu et al., 2008). Additional analysis extends these ideas to the case of unequal inputs (Daun et al., 2009) or multiple competing units (Rubin et al., 2009b). Many other works have harnessed dynamical systems analysis to study additional types of synaptic interactions, such as interesting effects of fast threshold modulation and postinhibitory rebound, synchronization and clustering, and a variety of other phenomena. Indeed, the utility of dynamical systems has placed it at the center of the modern theoretical neuroscience tool kit, and as new data on neuronal activation and connectivity patterns become available, dynamical systems theory promises to continue to guide us to new insights and predictions about neuronal network dynamics.

3.5 *Appendix: Basic Mathematics of Dynamical Systems*

Here we provide a mathematically precise introduction to the fundamental ideas of dynamical systems theory. These ideas are presented in more complete detail in various textbooks (Perko, 2001; Teschl, 2012).

ODEs and flows: Consider an autonomous ODE initial value problem for an n-dimensional quantity $x(t)$:

$$\dot{x} = f(x), \; x(t_0) = x_0. \tag{3.15}$$

This is a set of n ODEs for which we specify the phase space as an open subset U of Euclidean n-space, $\mathbb{R}^n$. To ensure that a unique solution exists from any initial condition in U, one must assume that the vector-valued function f has some smoothness properties. Specifically, in mathematical terms, we assume that $x_0 \in U$ and that $f: U \to \mathbb{R}^n$ for $U \subset \mathbb{R}^n$ is a *locally Lipschitz* function (i.e., for all $x \in U$, there are a neighborhood $V \subset U$ containing x and constants $\delta, L > 0$ such that if $x_1, x_2 \in V$ with $|x_1 - x_2| < \delta$, then $|f(x_1) - f(x_2)| < L|x_1 - x_2|$); note that all continuously differentiable (or C^1) functions are locally Lipschitz, but so are some additional functions like $f(x) = |x|$.

From a geometrical viewpoint, we can envision studying a collection of solutions to equation 3.15 by selecting a set of initial points $(t_0^k, x_0^k), k = 1, \ldots, m$, and letting them *flow* forward as specified by equation 3.15, forming curves, or trajectories, in U. Each point $x \in U$ belongs to a trajectory, and the tangent vector to the trajectory (the local velocity vector) at x is given by $f(x)$. Based on this idea, mathematicians define a function called the *flow* associated with the ODE (3.15). Given an initial point x_0 and a time t, the flow Φ prescribes the position in phase space $\Phi(t, x_0)$ at time t of the trajectory that passes through $x = x_0$ at $t = 0$; that is, the flow tells us how points are moved around in phase space. Also, since there is no explicit time-dependence in equation 3.15, we have a time translation property for solutions: $\Phi(s + t, x_0) = \Phi(t, \Phi(s, x_0))$. In the case of a closed orbit, such as a limit cycle, this means we can talk about the phase of the oscillation by choosing a reference point and normalizing the time for one full traverse of the orbit to a fixed value, typically 1 or 2π.

Mathematical properties of flows: For completeness, we include the properties of a flow in more mathematical terms. The same proof that establishes the standard existence and uniqueness theorems for solutions of ODEs (Perko, 2001; Teschl, 2012) implies that there exist a set $\Omega \subset \mathrm{R} \times U$ and a *flow* Φ: $\Omega \to U$, generated by equation 3.15, with the following properties:

1. $(0, x_0) \in \Omega$ for each $x_0 \in U$,
2. for each $x_0 \in U$, there is a maximal open interval of times including $t = 0$ on which $\Phi(t, x_0)$ exists, say $I_{x0} \subset \mathbb{R}$, such that $I_{x0} \times x_0 \subset \Omega$,
3. for each $x_0 \in U$, $\Phi(0, x_0) = x_0$ and $\Phi(t, x_0)$: $I_{x0} \to U$ is a solution to equation 3.15,
4. if $s, t \in \mathrm{R}$ are nonnegative and $(s + t, x_0) \in \Omega$, then $\Phi(s + t, x_0) = \Phi(t, \Phi(s, x_0))$,
5. (uniqueness property) for any solution of equation 3.15, say $\Psi(t, x_0)$ for $x_0 \in U$, and any $T \in \mathrm{R}$ for which $\Psi(T, x_0)$ is defined, there exists $s \in I_{x0}$ such that $\Phi(s, x_0) = \Psi(T, x_0)$ (i.e., Ψ is simply a time translation of Φ), and
6. $\Phi(t, x_0)$ is locally Lipschitz in x_0 (and C^r in x_0 if f is).

ODE models from neuroscience typically have locally Lipschitz right-hand sides except for possible points of jump discontinuity, such as thresholds in integrate-and-fire type models. Such jumps are easily handled; an ODE is solved up to a jump, the jump provides a new initial condition, and a solution is continued from there. Thus, given a neuroscience model, we will have a corresponding flow that represents its solutions.

Organization of flows—critical points and global phenomena: Next we consider the question of what organizes such flows. Recall that a *critical point* of an ODE is any point x such that $f(x) = 0$. Another general theorem, called the *rectification theorem* (Arnold, 1973), says that if we consider any open set in phase space that does not contain a critical point of an ODE, then there is a continuous, invertible mapping that takes trajectories of the flow for the ODE into straight lines that solve the equation

$$(\dot{x}_1, \dot{x}_2, \ldots, \dot{x}_n) = (1, 0, \ldots, 0).$$

In other words, if we zoom in close enough to any noncritical point in the phase space and examine the flow there, we will never find anything interesting! Thus, flows in phase space are organized by local behavior near critical points, and complex solutions representing neuronal behavior arise through a combination of local influences and global phenomena. A periodic solution, as we mentioned earlier, provides a simple example of an important global phenomenon: in a local neighborhood of any point on such a trajectory, there is nothing to indicate that the trajectory is periodic, but if we zoom out far enough, we can see that in fact it is a closed curve. Because they cannot generally be understood through local information, global phenomena are difficult to study, but several examples such as limit cycles and heteroclinic orbits (i.e., orbits converging to two different critical points as $t \to \pm\infty$) arise throughout this chapter, and certain global phenomena are critical to neural dynamics.

Extensions. We conclude with two final remarks about the settings where dynamical systems theory applies. First, note that neuronal models may include time-dependent terms such as time-varying applied current and thus take the form of nonautonomous ODEs, $\dot{x} = f(t, x)$, in contrast to the form $\dot{x} = f(x)$ discussed here. One can always let $s = t$ and convert this nonautonomous ODE to the autonomous one $\dot{y} = g(y)$ where $y = (x, s) \in \mathrm{R}^{n+1}$ for $x \in \mathrm{R}^n$ and $g(y) = (f(s, x), 1)$. Since s is included as a dependent variable, the ODE for y is autonomous, and thus this case is technically covered by our discussion, although it will not have any critical points since $\dot{s} = 1$. Sometimes, time-dependent terms can be understood as perturbations to autonomous systems, and there are methods (e.g., perturbation theory, averaging) for doing this in certain cases (e.g., Keener, 1988; Guckenheimer and Holmes, 1983). Second, the ideas of dynamical systems extend beyond ODEs to certain partial differential equations and other equations. In particular, using a dynamical systems approach to study the existence and stability of a variety of solutions including fronts, waves, and bumps in integrodifferential equation models, such as neural field equations, represents a rich area of mathematical neuroscience research (Ermentrout and Terman, 2010; Bressloff, 2012).

REFERENCES

Amari, S. (1977). Dynamics of pattern formation in lateral-inhibition type neural fields. *Biological Cybernetics*, 27(2), 77–87.

Amari, S.-I., & Arbib, M. (1977). Competition and cooperation in neural nets. In J. Metzler (Ed.), *Systems Neuroscience* (pp. 119–166). New York: Academic Press.

Arbas, E., & Calabrese, R. (1987). Slow oscillations of membrane potential in interneurons that control heartbeat in the medicinal leech. *Journal of Neuroscience*, 7, 3953–3960.

Arnold, V. (1973). *Ordinary Differential Equations*. Cambridge, MA: MIT Press.

Bogacz, R., Brown, E., Moehlis, J., Holmes, P., & Cohen, J. (2006). The physics of optimal decision making: A formal analysis of models of performance in two-alternative forced-choice tasks. *Psychological Review, 113*(4), 700–765.

Borisyuk, A., & Rinzel, J. (2005). Understanding neuronal dynamics by geometrical dissection of minimal models. In C. Chow, B. Gutkin, D. Hansel, C. Meunier, & J. Dalibard (Eds.), *Models and Methods in Neurophysics: Proceedings of les Houches Summer School 2003 (Session LXXX)* (pp. 19–72). Amsterdam: Elsevier.

Brascamp, J., van Ee, R., Noest, A., Jacobs, R., & van den Berg, A. (2006). The time course of binocular rivalry reveals a fundamental role of noise. *Journal of Vision (Charlottesville, Va.), 6*, 1244–1256.

Bressloff, P. (2012). Spatiotemporal dynamics of continuum neural fields. *Journal of Physics. A, Mathematical and Theoretical, 45*, 033001.

Brown, T. (1911). The intrinsic factors in the act of progression in the mammal. *Proceedings of the Royal Society of London. Series B, Containing Papers of a Biological Character, 84*(572), 308–319.

Brown, T. (1914). On the nature of the fundamental activity of the nervous centres; together with an analysis of the conditioning of rhythmic activity in progression, and a theory of the evolution of function in the nervous system. *Journal of Physiology, 48*(1), 18–46.

Coddington, E., & Levinson, N. (1955). *Theory of Ordinary Differential Equations.* New York: McGraw-Hill.

Coombes, S., & Bressloff, P. (Eds.). (2005). *Bursting: The Genesis of Rhythm in the Nervous System.* Singapore: World Scientific.

Curtu, R. (2010). Singular Hopf bifurcations and mixed-mode oscillations in a two-cell inhibitory neural network. *Physica D. Nonlinear Phenomena, 239*(9), 504–514.

Curtu, R., & Rubin, J. (2011). Interaction of canard and singular Hopf mechanisms in a neural model. *SIAM Journal on Applied Dynamical Systems, 10*(4), 1443–1479.

Curtu, R., Shpiro, A., Rubin, N., & Rinzel, J. (2008). Mechanisms for frequency control in neuronal competition models. *SIAM Journal on Applied Dynamical Systems, 7*, 609–649.

Daun, S., Rubin, J., & Rybak, I. (2009). Control of oscillation periods and phase durations in half-center central pattern generators: A comparative mechanistic analysis. *Journal of Computational Neuroscience, 27*, 3–36.

Devaney, R. L. (2003). *An Introduction to Chaotic Dynamical Systems.* Boulder, CO: Westview Press.

Eckhoff, P., Wong-Lin, K., & Holmes, P. (2011). Dimension reduction and dynamics of a spiking neural network model for decision making under neuromodulation. *SIAM Journal on Applied Dynamical Systems, 10*(1), 148–188.

Ermentrout, B. (2003). Phase plane analysis of neural nets. In M. A. Arbib (Ed.), *The Handbook of Brain Theory and Neural Networks* (pp. 881–889). Cambridge, MA: MIT Press.

Ermentrout, G. B., & Rinzel, J. (2010). Differential equations. In E. De Schutter (Ed.), *Computational Modeling Methods for Neuroscientists* (pp. 1–29). Cambridge, MA: MIT Press.

Ermentrout, G. B., & Terman, D. (2010). *Mathematical Foundations of Neuroscience.* New York: Springer.

Fransén, E., Tahvildari, B., Egorov, A., Hasselmo, M., & Alonso, A. (2006). Mechanism of graded persistent cellular activity of entorhinal cortex layer V neurons. *Neuron, 49*(5), 735–746.

Guckenheimer, J., & Holmes, P. (1983). *Nonlinear Oscillations, Dynamical Systems, and Bifurcations of Vector Fields.* New York: Springer.

Guckenheimer, J., & Oliva, R. A. (2002). Chaos in the Hodgkin–Huxley model. *SIAM Journal on Applied Dynamical Systems, 1*(1), 105–114.

Hodgkin, A. (1948). The local electric changes associated with repetitive action in a non-medullated axon. *Journal of Physiology, 107*(2), 165–181.

Huguet, G., & Rinzel, J. (2014). Multistability in perceptual dynamics. In D. Jaeger & R. Jung (Eds.), *Encyclopedia of Computational Neuroscience* (Vol. II, pp. 1–11). New York: Springer.

Izhikevich, E (2000). Neural excitability, spiking and bursting. *International Journal of Bifurcation and Chaos, 10*(6), 1171–1266.

Izhikevich, E. (2006). *Dynamical Systems in Neuroscience: The Geometry of Excitability and Bursting.* Cambridge, MA: MIT Press.

Keener, J. P. (1988). *Principles of Applied Mathematics: Transformation and Approximation.* Redwood City, CA: Addison Wesley, Advanced Book Program.

Laing, C., & Chow, C. (2002). A spiking neuron model for binocular rivalry. *Journal of Computational Neuroscience, 12*, 39–53.

Laing, C., & Lord, G. (Eds.). (2009). *Stochastic Methods in Neuroscience.* Oxford: Oxford University Press.

Leopold, D., & Logothetis, N. (1999). Multistable phenomena: Changing views in perception. *Trends in Cognitive Sciences, 3*(7), 254–264.

Levelt, W. (1968). *On Binocular Rivalry.* The Hague: Mouton.

Loewenstein, Y., Mahon, S., Chadderton, P., Kitamura, K., Sompolinsky, H., Yarom, Y., et al. (2005). Bistability of cerebellar Purkinje cells modulated by sensory stimulation. *Nature Neuroscience, 8*(2), 202–211.

Moreno-Bote, R., Shpiro, A., Rinzel, J., & Rubin, N. (2010). Alternation rate in perceptual bistability is maximal at and symmetric around equi-dominance. *Journal of Vision (Charlottesville, Va.), 10*, 1–18.

Perko, L. (2001). *Differential Equations and Dynamical Systems.* New York: Springer.

Pressnitzer, D., Sayles, M., & Winters, I. (2008). Perceptual organization of sound begins in the auditory periphery. *Current Biology, 18*, 1124–1128.

Rabinovich, M., Friston, K., & Verona, P. (2012). *Principles of Brain Dynamics: Global State Interactions.* Cambridge, MA: MIT Press.

Rabinovich, M., Volkovskii, A., Lecanda, P., Huerta, R., Abarbanel, H., & Laurent, G. (2001). Dynamical encoding by networks of competing neuron groups: Winnerless competition. *Physical Review Letters, 87*(6), 68102.

Rinzel, J. (1987). A formal classification of bursting mechanisms in excitable systems. In A. Gleason (Ed.), *Proceedings of the International Congress of Mathematicians* (pp. 1578–1594). Providence, RI: American Mathematical Society.

Rinzel, J., & Ermentrout, G. (1989). Analysis of neural excitability and oscillations. In C. Koch & I. Segev (Eds.), *Methods in Neuronal Modeling: From Synapses to Networks* (1st ed., pp. 251–291). Cambridge, MA: MIT Press.

Rinzel, J., & Ermentrout, G. (1998). Analysis of neural excitability and oscillations. In C. Koch & I. Segev (Eds.),

Methods in Neuronal Modeling: From Synapses to Networks (2nd ed., pp. 251–291). Cambridge, MA: MIT Press.

Rubin, J., Hayes, J., Mendenhall, J., & Del Negro, C. (2009a). Calcium-activated nonspecific cation current and synaptic depression promote network-dependent burst oscillations. *Proceedings of the National Academy of Sciences of the United States of America, 106*(8), 2939–2944.

Rubin, J., Shevtsova, N. A., Ermentrout, G. B., Smith, J. C., & Rybak, I. A. (2009b). Multiple rhythmic states in a model of the respiratory central pattern generator. *Journal of Neurophysiology, 101,* 2146–2165.

Rubin, J., & Terman, D. (2002). Geometric singular perturbation analysis of neuronal dynamics. In B. Fiedler (Ed.), *Handbook of Dynamical Systems: Vol. 2. Towards Applications.* Amsterdam: Elsevier.

Rubin, N., & Hupe, J.-M. (2005). Dynamics of perceptual bistability: Plaids and binocular rivalry compared. In D. Alais & R. Blake (Eds.), *Binocular Rivalry* (pp. 137–154). Cambridge, MA: MIT Press.

Seely, J., & Chow, C. (2011). Role of mutual inhibition in binocular rivalry. *Journal of Neurophysiology, 106*(5), 2136–2150.

Shpiro, A., Curtu, R., Rinzel, J., & Rubin, N. (2007). Dynamical characteristics common to neuronal competition models. *Journal of Neurophysiology, 97,* 462–473.

Skinner, F., Kopell, N., & Marder, E. (1994). Mechanisms for oscillation and frequency control in reciprocally inhibitory model neural networks. *Journal of Computational Neuroscience, 1*(1), 69–87.

Skinner, F., Turrigiano, G., & Marder, E. (1993). Frequency and burst duration in oscillating neurons and two-cell networks. *Biological Cybernetics, 69*(5), 375–383.

Sterzer, P., Kleinschmidt, A., & Rees, G. (2009). The neural bases of multistable perception. *Trends in Cognitive Sciences, 13,* 310–318.

Strogatz, S. (2001). *Nonlinear Dynamics and Chaos.* Cambridge, MA: Perseus Press.

Tabak, J., & Rinzel, J. (2005). Bursting in excitatory neural networks. In S. Coombes & P. Bressloff (Eds.), *Bursting: The Genesis of Rhythm in the Nervous System* (pp. 273–302). Hackensack, NJ: World Scientific.

Tabak, J., Senn, W., O'Donovan, M., & Rinzel, J. (2000). Modeling of spontaneous activity in developing spinal cord using activity-dependent depression in an excitatory network. *Journal of Neuroscience, 20*(8), 3041–3056.

Terman, D. (1991). Chaotic spikes arising from a model of bursting in excitable membranes. *SIAM Journal on Applied Mathematics, 51*(5), 1418–1450.

Teschl, G. (2012). *Ordinary Differential Equations and Dynamical Systems.* Providence, RI: American Mathematical Society.

Wang, X., & Rinzel, J. (1992). Alternating and synchronous rhythms in reciprocally inhibitory model neurons. *Neural Computation, 4*(1), 84–97.

Wiggins, S. (2003). *Introduction to Applied Nonlinear Dynamical Systems and Chaos.* New York: Springer.

Wilson, H. (1999). *Spikes, Decisions, and Actions: The Dynamical Foundations of Neuroscience.* Oxford: Oxford University Press.

Wilson, H., & Cowan, J. (1972). Excitatory and inhibitory interactions in localized populations of model neurons. *Biophysical Journal, 12*(1), 1–24.

Wilson, H., & Cowan, J. (1973). A mathematical theory of the functional dynamics of cortical and thalamic nervous tissue. *Biological Cybernetics, 13*(2), 55–80.

Wong, K., & Wang, X. (2006). A recurrent network mechanism of time integration in perceptual decisions. *Journal of Neuroscience, 26*(4), 1314–1328.

Yuste, R., MacLean, J., Smith, J., & Lansner, A. (2005). The cortex as a central pattern generator. *Nature Reviews. Neuroscience, 6*(6), 477–483.

4 Neural Rhythms

NATHAN VIERLING-CLAASSEN AND STEPHANIE R. JONES

4.1 *Introduction*

Rhythmic neural activity, produced either by individual neurons or populations of neurons, is a critical component of many basic bodily functions, such as breathing, chewing, digestion and "Motor Pattern Generation" (chapter 8). Although neural rhythms are present and important throughout the body, this chapter focuses on models that describe the mechanisms and functions of neural rhythms in neocortical networks of the mammalian brain. These rhythms may be generated locally or involve interaction with other brain areas such as the thalamus (figure 4.1A).

Hans Berger made the first human electroencephalography (EEG) recording in 1924, detecting neural electrical signals from electrodes placed outside the head. Even in these early human brain recordings, low-frequency "alpha" rhythms (7–14 cycles per second, or Hertz) were detected during resting state signals on primitive electrodes above visual cortex. Since then, neural rhythms of varying frequency and spatial specificity have been recorded throughout the brain. They range from very slow (<1 Hz) to very fast (>300 Hz) depending on the area recorded, recording technique, and behavioral state. The emergence and coordination of rhythmic activity at many scales across the brain, including systematic coupling between rhythms of different frequencies, has been correlated with healthy brain function and as a marker of disease states (see Buzsaki and Draguhn, 2004, and Schnitzler and Gross, 2005, for reviews). Active sensing of the environment by the body's motor and attention systems is often naturally set up to produce rhythmic sampling, which filters sensory inflow in a manner that is coordinated with brain rhythms (Schroeder, Wilson, Radman, Scharfman, and Lakatos, 2010). See table 4.1 for a list of commonly recorded frequency bands and examples of associated brain functions. Substantial research efforts have been exerted to understand the cellular- and network-level neural mechanisms generating these rhythms and evaluate their role in brain function. Computational modeling has proven critical in our developing understanding of the mechanisms and potential functional roles of neural rhythms.

On the finest spatial scale, rhythmic behavior can be measured in individual neurons by intracellular recordings, which use extremely fine-tipped glass electrodes to patch the membrane of a neuron. In this manner, rhythms can be recorded as spiking at regularly timed intervals or by measuring the currents within a single cell and recording subthreshold activity driven by synaptic currents (see recordings in figure 4.2A, bottom traces). On a larger scale, extracellular recordings of local field potential and multi-unit activity can capture rhythms in subthreshold and population spiking activity summed over activity of neurons around the electrode tip, as in figure 4.1B (see also figure 4.2A, top trace). On an even larger scale, measures of rhythmic electrical and magnetic activity can be made from outside the head using electroencephalography (EEG) and magnetoencephalography (MEG), respectively, as shown in figure 4.1C. These methods record field activity produced by large electrical currents, known as current dipoles, generated within the cortex (figure 4.1D). The current dipoles originate from the summed postsynaptic longitudinal currents within large populations of spatially aligned pyramidal neuron dendrites that act like electrical cables (figure 4.1D, right; see also section 4.4) (Hamalainen, Hari, Ilmoniemi, Knuutila, and Lounasmaa, 1993; Ikeda, Wang, and Okada, 2005; Murakami, Zhang, Hirose, and Okada, 2002).

The variation in how neural rhythms are recorded highlights both the difficulty and the power of using computational neural modeling to study brain rhythms. There is some level of "noise" inherent to all recorded data, which adds uncertainty in the interpretation of the precise neural generators. While the model should be tailored to the data at hand, there will inevitably be limitations in how precisely model output can represent the actual data, and dynamic models of the whole brain are still unfeasible. However, there are

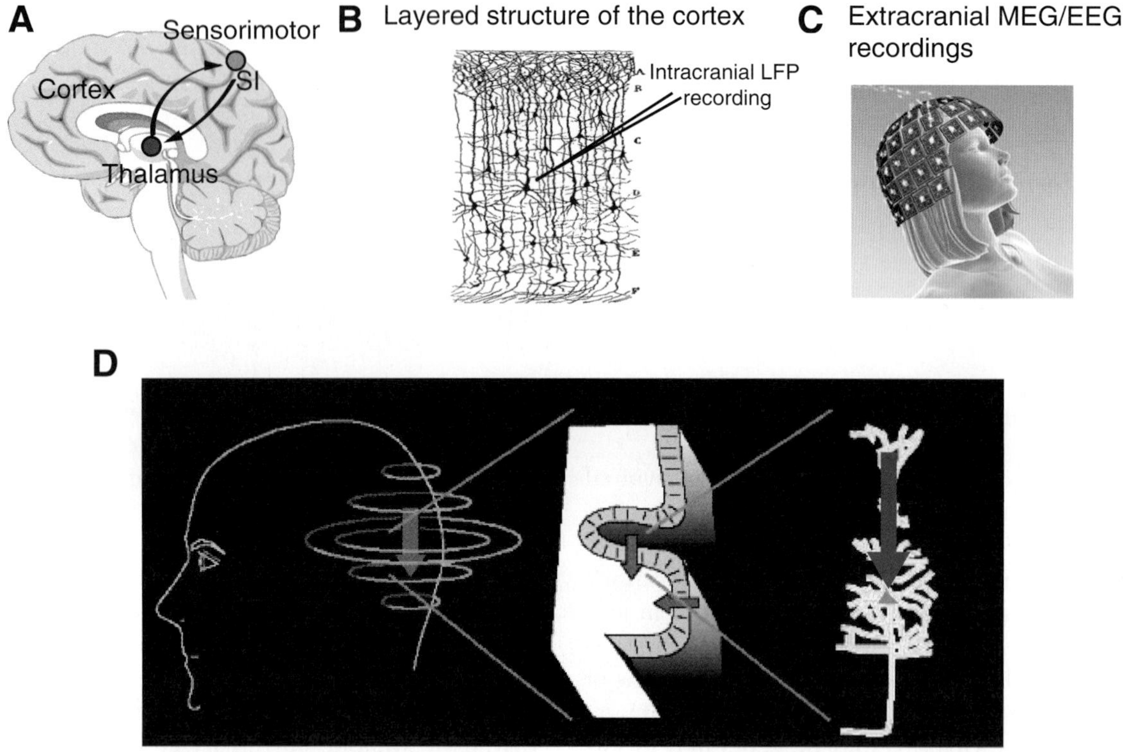

FIGURE 4.1 (A) Human brain showing sensorimotor cortex and thalamic regions. Arrows emphasize reciprocal interactions between thalamus and cortex that may mediate slow-frequency rhythms. Sections 4.2 and 4.3 emphasize neural rhythms local to cortex. Section 4.4 discusses sensorimotor "mu" rhythms that rely on thalamocortical interactions. (B) Depiction of the layered structure of the cortex where cell bodies lie in six layers, labeled A–F in reproduction from Santiago Ramon y Cajal's drawings published in 1904. A schematic sharp electrode depicts how intracranial local field potentials (LFPs) are recorded. (C) Schematic illustration of extracranial magnetoencephalography (MEG) sensors that record macroscopic scale brain signals from sensors outside the scalp, as does EEG. (D) Schematic illustration of the electrical primary current dipole source estimated from MEG and EEG data. Left panel depicts a primary current dipole (arrow) and corresponding generated magnetic field. Middle panel depicts current dipoles oriented tangential (parallel) and radial (perpendicular) to skull. MEG records activity from tangential sources only, EEG records activity from both orientations. Right panel depicts the cellular-level origin of the primary current sources, which are generated by the net summed postsynaptic intracellular electrical current flow within the long spatially aligned apical dendrites of large populations of synchronous cortical pyramidal neurons; see section 4.4.

TABLE 4.1
Commonly recorded brain rhythms and examples of associated brain states

Rhythm Type	Frequency (Hz)	Brain States
Delta	<4	Expressed during slow wave sleep
Theta	4–7	Engaged in memory processes and navigation
Alpha	7–14	Increased during states of idleness; decreased with attention; actively engaged to block distraction
Spindles	Alpha frequency rhythms that wax and wane at delta frequencies	Expressed during REM sleep
Beta	15–29	Involved in cross-area communication; correlated with detection; coordinated with motor actions
Mu	Rhythms from sensorimotor cortex; in magnetoencephalography, contains complex of alpha and beta frequencies (e.g., figure 4.8)	Sensorimotor rhythm involved in perception, motor initiation, and sensorimotor integration
Gamma	30–100+ (e.g., figure 4.2)	Increased poststimulus with attention and sensory processing

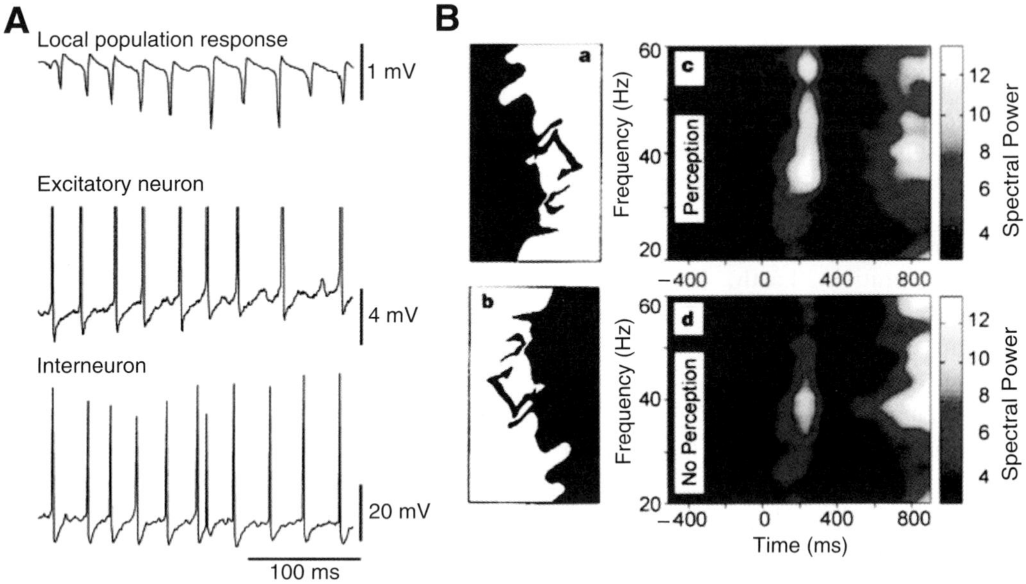

FIGURE 4.2 (A) Example voltage traces showing population activity and cell spiking during gamma rhythms near 40 Hz in an in vitro slice preparation (hippocampal region CA1). Top: extracellular local field potential recording of population activity. Middle and lower: recording from an individual excitatory cell and interneuron. Traces show characteristic activity but were not concurrently recorded. From Whittington et al. (2000). (B) Example of increase in EEG gamma-range (30–100 Hz) power when an ambiguous visual stimulus is perceived as a face. The stimulus is presented either right side up (a), which is perceived by most people as a face, or upside down (b), which is not usually perceived as a face. The right-hand panel shows an average spectrogram of the resulting EEG activity in humans asked to indicate whether or not they saw a face. Lighter gray indicates greater amplitude or strength at a particular frequency (y-axis) over time (x-axis), and the color bar indicates standard deviations from the power during a 500-ms prestimulus baseline. The stimulus is presented at time zero. Note that near 200 ms, gamma power is much stronger when the face is perceived (c) than when it is not (d). The later activity near 800 ms comes from postperceptual processes and closely follows reaction time, and it does not significantly differ with perception. From Rodriguez et al. (1999).

benefits to studying reduced computational models and experimental preparations of various scales. Through modeling, it is possible to make connections between individual cell behavior and the rhythmic behavior of a larger population and to generate testable predictions regarding the role of neural rhythms in brain function.

Computational neural modeling of single cells explores properties of individual neurons that can generate rhythmic spiking behavior. For example, see regular rhythmic firing expressed in integrate-and-fire, Hodgkin–Huxley, and bursting activity in chapters 2 and 3. In the brain, neurons are embedded in a network, synaptically and electrically coupled to hundreds of other neurons, both locally and in other brain areas. In some cases, the rhythmic activity of individual neurons may drive the activity of the ensemble, but in most biologically realistic circumstances, the network structure and interaction among multiple cell types and brain areas influences the rhythm expressed by the population. This chapter explores computational models of rhythms in networks.

Section 4.2 begins by describing how rhythmic activity can emerge in a simplified two-cell network, focusing on the mechanisms and functions of one of the most prominently studied neocortical rhythms, the 30- to 100-Hz gamma rhythm (section 4.2.1). Using a neural model in which a cell's rhythmic activity can be quantified with one equation (known as the "theta" neuron), this section describes how gamma rhythms can emerge from synaptic excitatory and inhibitory interactions between two cells (section 4.2.2). Section 4.2.3 describes how transitions from gamma to lower frequencies can emerge in a two-cell network model including more biophysically detailed neurons that incorporate Hodgkin–Huxley-type dynamics. Section 4.3 introduces how larger more complex network models can be used to begin to study the importance of gamma in brain function, including models that predict a role for gamma rhythms in creation of robust locally synchronous cell assemblies (section 4.3.1) and a possible role in attention-mediated stimulus selection (section 4.3.2). Section 4.4 expands to a study of rhythms that are generated by thalamocortical interactions (see figure 4.1A), highlighting low-frequency alpha and beta rhythms from somatosensory cortex measured with MEG, known as the mu rhythm (section 4.4.1). Section 4.4.2 addresses why simulating this macroscopic-scale human data requires multicompartment neurons and thalamocortical interactions, in contrast to the prior local network models described in the chapter, and section 4.4.3 describes how this strategy led to predictions on low-frequency rhythm generation. Lastly, section 4.4.4 explores how these rhythms may impact the salience of sensory signals in a manner that correlates with perception.

4.2 Network Mechanisms of Rhythm Generation

4.2.1 The Gamma Rhythm: Overview of Recordings and Functional Relevance One of the most explored examples of the computational modeling of neural rhythms is in the study of gamma rhythms (30–100 Hz). Gamma rhythms are present throughout the brain and have been implicated as functionally important for many aspects of cognition including attention, perceptual binding, memory formation, and learning. A key feature of gamma that has been exposed by computational modeling is that neural mechanisms that generate gamma also act to synchronize populations. In so doing, these mechanisms may contribute to the creation of coherent cell assemblies, or subpopulations of cells that are more likely to fire together. These cell assemblies may aid in the binding of information between brain regions that are functionally relevant to a task (Olufsen, Whittington, Camperi, and Kopell, 2003); see also the Discussion section (section 4.5).

As motivation for the study of gamma rhythms, consider two foundational experiments that give insight into the functional role of gamma in human cognition (Rodriguez et al., 1999) and into the cellular mechanism by which gamma is generated (Whittington, Traub, Kopell, Ermentrout, and Buhl, 2000). Rodriguez and colleagues were among the first to show evidence supporting a role for gamma rhythms in human cognition and, in particular, in binding functionally relevant information across the brain. They examined human EEG activity, recorded from 30 electrodes over the scalp during a task in which people were asked to observe ambiguous pictures that are usually perceived as faces when presented right side up but as meaningless when presented upside down (figure 4.2B). Forming the perception of a face in a somewhat ambiguous stimulus is the kind of task that involves "binding"—that is, combining neural information from multiple sources to form a single coherent percept. With this experiment, it is possible to examine neural activity both when a coherent percept is formed and when it is not. The right side of figure 4.2B shows a spectrogram (averaged over all contacts, trials, and 10 subjects) of EEG activity after the visual stimulus is presented. A spectrogram is a visual presentation of how the strengths of rhythm frequencies change over time. Lighter gray indicates more power at a particular frequency (y-axis) over time (x-axis), where the visual stimulus is presented at time zero. Note that near 200 ms, gamma power is stronger when the face is perceived (right, top) than when it is not (right, bottom). Spectrograms are a way to see visually what frequencies are present in a signal (multiple frequencies can be present at once) as well as how they change over time (see also section 4.4). The authors also measured the synchrony of gamma-band activity across different sites and found that even though there was gamma power in both cases, there was substantially more synchrony across cortical areas when the face was perceived. These two findings, that gamma activity was both stronger and more synchronous during this cognitive task, indicate that gamma may play a role in information processing. This is a foundational example of noninvasive experimental work connecting brain rhythms and cognition that has driven some of the more basic biological and computational work aimed at understanding how and why rhythms emerge in the gamma frequency band.

Motivated in part by the association of gamma-range rhythms with cognitive performance, extensive electrophysiological work has been done in animal systems to examine possible cellular-level generators for these rhythms. Many of the initial efforts to study the generators of gamma-frequency activity have used in vitro slice preparations, where slices of brain tissue (usually from rats or mice) including the brain region one wishes to study are kept alive in artificial cerebrospinal fluid. Slice preparations have been particularly useful for dissecting network mechanisms that generate rhythms since individual cells can be recorded, and different pharmacological or electrical manipulations can be used to elicit rhythmic activities and to discern the mechanisms that maintain them. Figure 4.2A shows an example of gamma generated in a hippocampal CA1 slice in a bath of the excitatory neurotransmitter kainite after a brief electrical stimulation. Kainite was used to increase the excitability of the cell and induce spiking activity. Traces from a single excitatory cell, a single inhibitory cell, and the population rhythm are shown, where both the excitatory cell and the inhibitory cell fire on every cycle of a gamma rhythm. This illustrative example presents a case where both pyramidal and interneuron types fire on most cycles of the rhythm, a behavior termed pyramidal–interneuron network gamma (PING) (Whittington et al., 2000), which will be described in detail below. In other conditions, individual cells, particularly excitatory cells, may not participate on all cycles of the rhythm, sometimes called "sparse gamma" (Kilpatrick and Ermentrout, 2011). In other cases, only inhibitory cells may be active, a situation termed interneuron network gamma. A common factor found in each case is that the decay time constant of fast $GABA_A$ inhibition is critical for setting the period of rhythmicity, and $GABA_A$ blockers abolish the rhythm in slice. The modeling described below details how this inhibition interacts with excitation to create PING-type rhythms. Experiments examining gamma rhythms in slice have provided the foundation for discovering mechanisms underlying these rhythms, and in particular the key role of inhibition. Since these initial slice explorations, in vivo experimental work in mice has also provided evidence for the role of $GABA_A$ inhibition in gamma rhythm generation (Cardin et al., 2009), which supports the idea that analogous mechanisms can coordinate gamma in both slice experiments and intact mammalian brains and may subserve the observations in humans.

4.2.2 THE GAMMA RHYTHM: MECHANISMS OF GAMMA GENERATION IN A SIMPLIFIED TWO-CELL THETA NEURON NETWORK MODEL This section starts with a demonstration of a minimal computational model of a network generating rhythmic neural activity, with the aim of applying modeling to understand how two cells can interact to produce gamma-frequency activity like that observed experimentally in figure 4.2A. A one-dimensional oscillator is used to represent each cell, one excitatory and one inhibitory, with dynamic rules for how these oscillators interact (figure 4.3B). Employing a simplified network allows access to all the mathematical details and a thorough description of the exact parameters that control the network rhythm. The oscillator model used here is known as the "Theta" neuron, referring to the phase of the neural oscillation described by the Greek letter theta, not to be confused with the *theta* rhythm of population neural activity (4–7 Hz; table 4.1).

The single cell Theta neuron: In this model, a point traveling along a circle represents the spiking activity of a cell. The dynamic variable that governs cell behavior is the one-dimensional angle θ designating position on the circle, which is also the phase of the oscillator at a given time point. A spike is defined as occurring when $\theta = \pi \ (mod\ 2\pi)$. This is in contrast to the Hodgkin–Huxley model (chapter 2), where the cell travels along a limit cycle in a four- or higher dimensional space. The equation describing the activity of the simulated excitatory and inhibitory cell in the network is as follows:

$$\frac{d\theta_k}{dt} = 1 - \cos\theta_k + (b_k + S_k)(1 + \cos\theta_k) \qquad (4.1)$$

While this model is simplified and the output θ is different from the usual voltage modeled for a neuron, many parameters do have physiological meaning. In particular, b_k is analogous to an external constant applied current that can either drive the cell to spike (if it is positive) or prevent it from spiking (if it is negative). S_k is the synaptic input, that is, the term that describes excitation or inhibition coming from other cells in the network. Overall, time is in milliseconds, k = E is the excitatory cell, representative of a pyramidal cell producing fast AMPA type excitation, and k = I is the inhibitory cell, representative of an interneuron producing $GABA_A$-type inhibition. This canonical "Theta" model can be attained by a change of variables from a quadratic integrate-and-fire model (Borgers and Kopell, 2005) or from a certain class of conductance-based models that spike slowly near threshold, referred to as "Type 1" excitability (Kopell and Ermentrout, 1986).

First consider cell behavior in the absence of synaptic input; that is, assume that $S_k = 0$. For $b_k < 0$, setting the rate of change for θ equal to zero and solving for θ, there are exactly two fixed points at

$$\theta_\pm = \pm\cos^{-1}\frac{1+b_k}{1-b_k} \qquad (4.2)$$

where θ_- is stable and θ_+ is unstable. Thus, for $b_k < 0$ in the absence of input, the cell approaches the fixed point at θ_- regardless of initial conditions (unless it starts at exactly θ_+). It is possible to develop an intuitive understanding of this stability by noting that for initial conditions on a circle $\theta_- < \theta < \theta_+$, equation 4.1 indicates $d\theta/dt$ is negative, that is, solutions will move away from the positive fixed point θ_+ and toward the negative fixed point θ_-. However, for initial conditions that start with an angular value $\theta_+ < \theta < 2\pi + \theta_-$, equation 4.1 indicates that $d\theta/dt$ is positive, and again, solutions move away from θ_+ and toward θ_-, but in the opposite direction around the circle (figure 4.3A, left). As b_k is varied, the system moves through a saddle node bifurcation at b_k = 0 (see chapter 4). That is, as the parameter b_k, increases through zero, there is one fixed point when $b_k = 0$, and there are no fixed points for $b_k > 0$. In this regime, solutions move repeatedly around the circle with period $T = \pi / \sqrt{b_k}$, moving slowly near $\theta = 0$ and very rapidly near $\theta = \pi$, and a spike is defined as occurring when $\theta = \pi \ (mod\ 2\pi)$ (figure 4.3A, right). For this example, set

$b_k = -0.01$ for the interneuron, and $b_k = 0.06$ for the excitatory cell. With these parameters, the interneuron is silent without input, and the excitatory cell fires at 76.6 Hz. Figure 4.3D shows a raster plot of spike times for these disconnected cells (as depicted in figure 4.3B). Each vertical black line represents a spike in the E-cell.

Synaptic coupling of cells generates pyramidal-interneuron-gamma (PING): Next, the cells are connected synaptically via the S_k term (connections shown schematically in figure 4.3B). Reciprocal coupling between the cells changes their behavior, generating a coherent network rhythm.

The synaptic input term is defined as follows:

$$S_k = \alpha_j g_{jk} s_{jk}. \tag{4.3}$$

This represents the input from cell j to cell k, where j and k can be E or I representing the excitatory or inhibitory cell. s_{jk} is a synaptic gating variable, while the synaptic weight g_{jk} represents the strength of the synapse from cell j to cell k and is analogous to the maximal conductance g_{max} in the Hodgkin–Huxley model. For this example, we set $g_{EI} = 0.2$ and $g_{IE} = 0.1$. The α_j term controls excitation and inhibition with

$$\alpha_j = \begin{pmatrix} +1 & \text{if presynaptic cell } j \text{ is excitatory} \\ -1 & \text{if presynaptic cell } j \text{ is inhibitory} \end{pmatrix}. \tag{4.4}$$

The idea of the synaptic gating variable s_{jk} is that it is maximal after a spike is fired and then decays more or less rapidly (figure 4.3C), depending on the parameter τ_j. This simulates the general time course and shape of postsynaptic excitation and inhibition recorded intracellularly in neurons. It is defined by

$$\frac{ds_{jk}}{dt} = -\frac{s_{jk}}{\tau_j} + e^{-\eta(1+\cos\theta_j)}\frac{1-s_{jk}}{\tau_R} \tag{4.5}$$

with $\eta = 5$ and $\tau_R = 0.1$. Provided η is large enough, most of the time, this synaptic variable behaves like exponential decay at a rate controlled by τ_j (on a timescale of milliseconds). However, near values of $\theta_j = \pi$, the second term pushes s_{jk} very rapidly toward 1, at a rate controlled by $\tau_R = 0.1$, analogous to a rate of synaptic rise. Here, the exact value of η does not have a direct physiological meaning; it just needs to be large enough that the second term is negligible when $\theta_j \neq \pi$. Thus, for our purposes, the most important and physiologically relevant parameter in this equation is τ_j, which controls the decay time of excitatory postsynaptic currents (if cell j is excitatory) and inhibitory postsynaptic currents (if cell j is inhibitory). For the E-cell we take $\tau_j = 2$, which is representative of AMPA synaptic kinetics and within the experimentally observed range for AMPA decay in mammalian cortical cells. For the I-cell we take $\tau_j = 10$, which is selected to be representative of $GABA_A$ synaptic decay, which takes longer to decay than AMPA excitation. Figure 4.3C shows example solutions for the synaptic gating variables s_{IE} (inhibition, figure 4.3C, top) and s_{EI} (excitation, figure 4.3C, bottom). Note that the modeled inhibitory synaptic input decays more slowly than the excitatory.

Recall that when these two cells are disconnected, the inhibitory cell is silent and the excitatory cell spikes at 76.6 Hz (figure 4.3D). When we first add the excitatory connection, the inhibitory cell fires at the same rate as the disconnected excitatory cell (figure 4.3E). Figure 4.3F shows what happens when the inhibitory synapse is also added. Now the network slows, because inhibition is suppressing the intrinsic firing of the excitatory cell, locking the network into a new stable rhythm. This is an example of how network interactions can shape a new rhythm from cells that would behave entirely differently if they were disconnected. This behavior requires participation from both cell types.

In summary, the PING rhythm is created by interplay between excitation and inhibition. The rhythm begins when the E-cell fires and excites the I-cell enough to fire. The I-cell then fires and temporarily inhibits the E-cell. The timing of inhibition is critical for controlling the rhythmic behavior of the network as a whole. If inhibition is strong and long, the network rhythm will be quite slow. If, however, inhibition is shorter or weaker, the rhythm will be faster, and in the extreme case of no inhibition, the network spikes as fast as the intrinsic firing rate of the excitatory cell (as in figure 4.3D). Importantly, at biophysically reasonable values of the synaptic time constants, a gamma-frequency PING rhythm where both neurons fire near 40 Hz has been shown to be consistent with experimental results like those shown in figure 4.2A (Whittington et al., 2000). As parameters for the applied currents and synaptic weights are varied, the rhythm can change but remains solidly in the gamma band reported experimentally for a broad range of values. Synaptic weights are also important for pacing the rhythm, but their role relies heavily on how they interact with the decay time constant of $GABA_A$-type inhibition that is critical for regulating the rhythm.

This basic mechanism of interacting excitation and inhibition is employed to some degree in most computational modeling studies of gamma rhythms. However, the interaction between rhythmic spiking in individual neurons and population rhythms (i.e., with many more

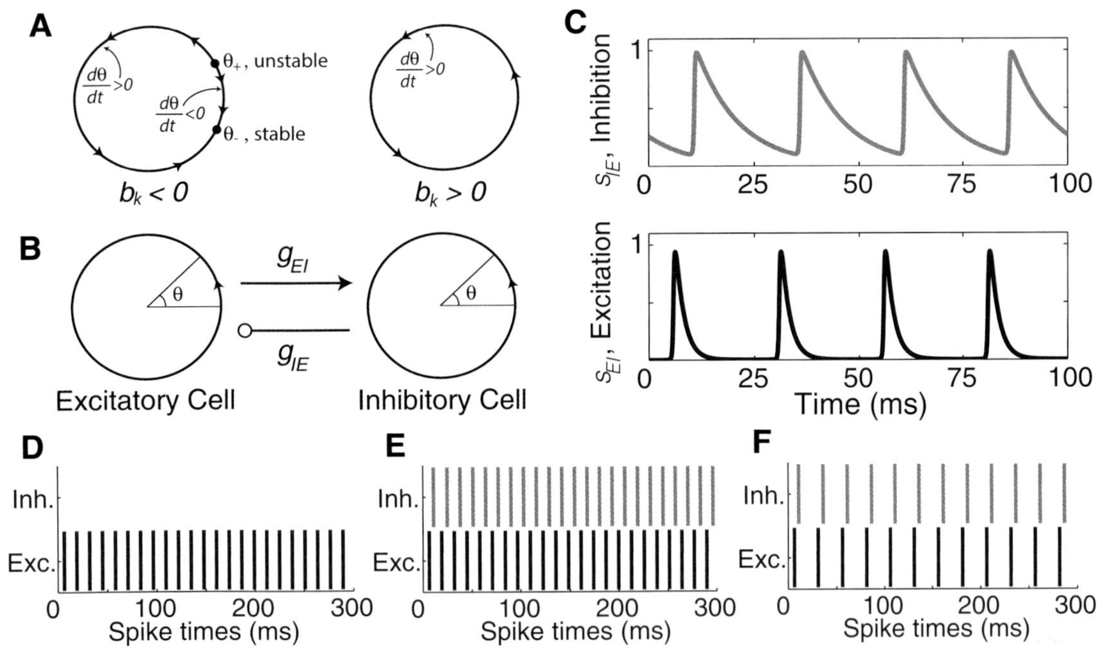

FIGURE 4.3 (A) Solution behavior for single theta model cell (see equation 4.1). When $b_k < 0$, there are two fixed points, one stable and one unstable (left). When $b_k > 0$, there are no fixed points and solutions always move in the positive direction (right). (B) Schematic illustration of a two-cell Theta neuron network. (C) Synaptic inhibition (top) and excitation (bottom). Note that inhibition wears off more slowly than excitation. This is controlled by the parameter τ_j, which is set at 10 (a decay rate of 1/10 ms) for inhibition and 2 (a decay rate of ½ ms) for excitation. (D) Inhibitory (Inh.) cell is silent and excitatory (Exc.) cell spikes when cells are disconnected. (E) Fast firing at 76.6 Hz in both cells with E-cell connected to I-cell. (F) Network slows to 40 Hz when both E-to-I and I-to-E synaptic connections are present.

than two neurons) can be complex, particularly in the presence of noise, and will be explored later in the chapter (section 4.3.1).

4.2.3 TRANSITIONS FROM GAMMA TO LOWER FREQUENCY RHYTHMS IN A TWO-CELL CONDUCTANCE-BASED NEURON NETWORK MODEL For mathematical clarity, the previous example used a simplified description of the activity of individual neurons, describing only how strongly they were driven (or not driven) to fire, without the individual cellular currents that contribute to spiking. This model highlights the importance of synaptic interactions in defining the frequency of rhythms. However, real neurons often contain several types of intrinsic currents that can produce interesting behaviors, like rapid bursts of spikes that depend on slow calcium currents or spike trains that gradually slow because of additional potassium currents. Activation of intrinsic currents via synaptic interactions, or modulation with behavioral state, can cause a single network to express rhythmic activity at several frequencies. Modeling with conductance-based cells containing several different currents allows study of these interactions.

Several mechanisms beyond the synaptic interactions considered for PING are important for generating slower alpha (7–14 Hz) and beta (15–29 Hz) rhythms. Experimental and modeling studies suggest that two currents in cortical pyramidal neurons important for slower frequency alpha rhythms are the h-current, which allows passage of multiple ions, and the low-threshold calcium T-current. The "h-" is short for hyperpolarization-activated cyclic nucleotide (HCN)-gated channel and should not to be confused with the inactivation variable h in the Hodgkin–Huxley sodium current equation. At hyperpolarized levels, h- and T-currents are activated and deinactivated, respectively, and act to depolarize the cell creating rebound spiking or bursting on a timescale near 10 Hz in the alpha band (S. R. Jones, Pinto, Kaper, and Kopell, 2000; Pinto, Jones, Kaper, and Kopell, 2003; Silva, Amitai, and Connors, 1991). These currents may contribute to local alpha generation, as considered in this section, or may interact with alpha frequency drive from thalamus, which can be generated by similar mechanisms (Vijayan and Kopell, 2012). Corticothalamic interactions for alpha and beta rhythms will be considered in section 4.4.

The after-hyperpolarization (AHP)-potassium currents in pyramidal cells are thought to be important for beta-frequency activity. This family of currents pulls the cell voltage below rest after a spike and contributes to spike adaptation. Their importance in beta rhythmicity has been studied with experimental slice recordings (Roopun et al., 2006; Whittington et al., 2000) and neural models (Kopell, Ermentrout, Whittington, and Traub, 2000; Pinto et al., 2003). In these studies, transitions from gamma to beta rhythms can emerge when local inhibitory processes combine with AHP-current activation (see also McCarthy, Brown, and Kopell, 2008; Pervouchine et al., 2006). Descriptions of the dynamic mechanisms and impact on cell behavior of the key ionic currents used in this model are included in table 4.2.

While this next model is not directly compared to specific experimental data, it is illustrative of a single network that can express multiple rhythms and demonstrates how intrinsic currents and synaptic connections can support these transitions. Here, we consider a two-cell excitatory–inhibitory cortical network simulated with Hodgkin–Huxley-type conductance-based neurons that include these additional h-, T-, and AHP-currents, in addition to the voltage-dependent sodium (Na^+) and potassium (K^+) currents that generate spiking. With these more detailed cells, we explore how an isolated cortical network can transition between three rhythmic states. Following the work of Pinto et al. (2003), we show how a single two-cell excitatory–inhibitory network can transition from expressing fast PING rhythms to slower beta and alpha rhythms, as parameters modeling decreases in arousal/attention are altered. Such transitions are consistent with the notion that lower frequency rhythms are more prevalent during lower arousal states. In section 4.4, we describe alternative mechanisms for the generation of alpha and beta activity, specific to human MEG recordings.

TABLE 4.2

Descriptions of the dynamic mechanisms and impact on cell behavior of the key ionic currents in the model for section 4.2.3

Current	Mechanism	Cell Behaviors
Sodium	• Depolarizing current activated at spiking threshold. • Initiates spiking by allowing positively charged sodium ions to quickly enter the cell, increasing voltage toward the sodium reversal potential of E_{Na} (typically 50 mV or higher).	• Initiates spikes.
Potassium	• Hyperpolarizing current that terminates individual spikes. • Terminates spiking by enabling positively charged potassium ions to exit the cell, decreasing voltage toward the potassium reversal potential E_K (typically −100 to −70 mV).	• Terminates individual spikes.
T-current	• Depolarizing calcium activated at a low threshold levels. • When activated, the current increases voltage toward the calcium reversal potential, E_{Ca} (typically 125–150 mV). • Activation requires a two-stage process: (1) the cell has to first be hyperpolarized to allow deinactivation, and then (2) depolarized (e.g., via h-current activity) to allow activation. • This process is described by the dynamics of two gating variables in table 4.3.	• Generates low-threshold calcium spikes. • Important for generating rhythmic bursts of spikes in combination with h-current.
h-current	• Depolarizing mixed cation current, activated at hyperpolarized levels (when the cell is below rest). • When activated, the h-current increases voltage toward a reversal potential of E_h (typically −45 to −10 mV). • Activation of the h-current can generate sufficient depolarization to trigger activation of the T-current.	• In combination with the T-current, the h-current is important for generating rhythmic bursts of spikes.
AHP-current	• Hyperpolarizing potassium current that is rapidly activated by the postspike negative voltage deflection. • The current inactivates slowly, allowing potassium ions to gradually leave the cell. • Stronger AHP currents generate larger postspike hyperpolarizations, which can then activate h-current and deinactivate the T-current.	• Contributes to the shape and depth of the after-hyperpolarization that happens after a sodium spike (AHP). • Causes adaption: slowing of repeated spikes.

Note. Detailed equations for each channel are included in table 4.3. AHP, after-hyperpolarization.

TABLE 4.3

Equations governing intrinsic cellular currents in the model discussed in section 4.2.3, a two-cell network that produces alpha, beta, and gamma rhythmicity

Current	Gate	Rate equations	Kinetics
Sodium Current: Depolarizing sodium current that initiates spiking			
$I_{Na} = g_{Na}m^3h(E_{Na} - V)$	m	$\alpha_m = \dfrac{0.32(V+54)}{1-\exp(-0.25(V+54))}$ $\beta_m = \dfrac{0.28(V+27)}{\exp(0.2(V+27))-1}$	$\dfrac{da}{dt} = \alpha_a(1-a) - \beta_a a$
	h	$\alpha_h = 0.128(\exp((-1/18)*(V+50)))$ $\beta_h = \dfrac{4}{1+\exp(-0.2(V+27))}$	
Potassium Current: Hyperpolarizing potassium current that terminates individual spikes			
$I_K = g_k n^4(E_K - V)$	n	$\alpha_n = \dfrac{0.32(V+52)}{1-\exp(-0.2(V+52))}$ $\beta_n = 0.5(\exp(-0.025*(V+57)))$	$\dfrac{da}{dt} = \alpha_a(1-a) - \beta_a a$
T-Current: A low-threshold depolarizing calcium current that can only be activated after hyperpolarization, important for generating bursts of spikes			
$I_T = g_T m_T^2 h_T(E_{Ca} - V)$	m	$\tau_{mT} = 0.44 + \dfrac{0.15}{\exp(0.1(V+27)) + \exp((-1/15)(V+102))}$ $m_{\infty,T} = 1/(1+\exp((-1/7.4)(V+52)))$	$\dfrac{da}{dt} = (\alpha_\infty - a)/\tau_a$
	h	$\tau_{hT} = 22.7 + \dfrac{0.27}{\exp(0.25(V+18)) + \exp((-1/50)(V+407))}$ $h_{\infty,T} = 1/(1+\exp(0.2(V+80)))$	
h-Current: Depolarizing mixed cation current, activated at hyperpolarized levels and important for bursting			
$I_h = g_h r(E_h - V)$	r	$\tau_r = \dfrac{1}{\exp(-14.59-0.086V) + \exp(-1.87+0.0701V)}$ $r_\infty = 1/(1+\exp((1/5.5)(V+75)))$	$\dfrac{da}{dt} = (\alpha_\infty - a)/\tau_a$
AHP-Current: Slowly inactivating hyperpolarizing potassium current, shapes the after-hyperpolarization potential and generates adaptation with repeated spiking (also referred to as an m-current)			
$I_{AHP} = g_{AHP}w(E_K - V)$		$\tau_w = \dfrac{400}{3.3\exp((V+35)/20) + \exp(-(V+35)/20)}$ $w_\infty = 1/(1+\exp(-0.1(V+35)))$	$\dfrac{da}{dt} = (\alpha_\infty - a)/\tau_a$

Note. All equations as in Pinto et al. (2003). AHP, after-hyperpolarization.

In this model, each cell is simulated as a single-compartment neuron. The cells are coupled with fast excitation and inhibition, representing AMPA and $GABA_A$ synapses, respectively, and contain constant applied currents. Both cells contain intrinsic currents that produce spiking behaviors including sodium, potassium, and leak currents. The excitatory cell contains additional currents that can contribute to slower frequency alpha (I_h, and I_T) and beta (I_{AHP}) rhythms. The current balance equation for the E- and I-cells are as follows:

$$C_m \frac{dV_E}{dt} = I_L + I_{Na} + I_K + I_T + I_h + I_{GABA_A} + I_{appE},$$

$$C_m \frac{dV_I}{dt} = I_L + I_{Na} + I_K + I_{AMPA} + I_{appI},$$

Table 4.4

Equations governing synaptic currents in the model discussed in section 4.2.3, a two-cell network that produces alpha, beta, and gamma rhythmicity

Synaptic Inhibition: **Synaptic inhibitory current, activated by spiking in the presynaptic inhibitory cell**			
$I_{GABA_A} = g_{ie} GABA_A (E_{GABA_A} - V)$	***GABA_A***	$\tau_{GABA_A} = 10(ms)$ $k_{GABA_A} = 2(1 + \tanh(V_I / 4))$	$\frac{da}{dt} = k_a(1-a) - a/\tau_a$
Synaptic Excitation: Synaptic excitatory current, activated by spiking in the presynaptic excitatory cell			
$I_{AMPA} = g_{ei} AMPA(E_{AMPA} - V)$	*AMPA*	$\tau_{AMPA} = 2(ms)$ $k_{AMPA} = 5(1 + \tanh(V_E / 4))$	$\frac{da}{dt} = k_a(1-a) - a/\tau_a$

Note. All equations are as in Pinto et al. (2003). In general, for models like this, the parameters in these equations are fit to intracellular experimental data using curve-fitting techniques. They are often similar to the dynamics described in the original Hodgkin–Huxley model but are adapted to fit the specifics of mammalian cells.

$$I_{ion}(V) = g_{ion} a^p b^q (E_{ion} - V). \tag{4.6}$$

Detailed equations for ionic and synaptic currents are given in tables 4.3 and 4.4 (Pinto et al., 2003). In addition, the passive leak current is given by equations $I_L = g_L(E_L - V)$. The maximal conductance parameters for fixed intrinsic and synaptic currents are $g_L = 0.1$ (ms/cm^2), $g_{Na} = 100$, $g_K = 80$, $g_T = 2.7$, $g_h = 0.25$, $g_{ie} = 1$ (nA) $g_{ei} = 0.2$. Reversal potentials are $E_L = -67$ (mV), $E_{Na} = 100$, $E_K = -100$, $E_{Ca} = 125$, $E_h = -43$, $E_{GABA} = -80$, $E_{AMPA} = 0$. Only three parameters are manipulated to generate the varied behaviors in this network, the maximal conductance for the AHP current, g_{AHP}, and the applied currents for both cells, $I_{app,I}$ and $I_{app,\,E}$. Alterations to the AHP- and applied currents simulate neuromodulatory changes occurring with attention or other arousal mechanisms (see also chapter 7, "Neuromodulation"). Neuromodulation is the physiological process by which a neuron responds to a neurotransmitter that modulates its excitability. Acetylcholine, which is released from the brainstem and basal forebrain during attention and arousal, is one such neurotransmitter that impacts the excitability of the neurons it contacts in a manner similar to changes in the AHP-current conductance. Specifically, a high attentive state can be simulated with (1) decreased AHP-current conductance, thought to occur under conditions of high acetylcholine accompanying attention (Hasselmo, 1995); and (2) increased applied current, for both excitatory and inhibitory cells. Such manipulations control the overall level of depolarization or hyperpolarization of the cells, impacting the I_h, I_T, and I_{AHP} currents that shape the rhythmic behaviors of the cell.

First, consider a simulated state of "high" arousal characterized by a low AHP conductance in the excitatory cell and high values of I_{app} for the excitatory and/or inhibitory cells ($G_{AHP} = 0$, $I_{app,E} = 1.5$, $I_{app,I} = 0.5$). In this state, the I_{AHP}, I_h, and I_T are inactive, allowing the E-cell to act as a cell spiking in a similar fashion to that described with the theta model in section 4.2.2. This network expresses a gamma rhythm via the PING mechanism (figure 4.4A). The E-cell fires, which triggers a spike in the I-cell that, in turn, inhibits the E-cell (the order of spiking is shown more clearly in figure 4.4B). In contrast to the example in section 4.2.2, for these particular parameters the I-cell can fire on its own without excitatory input. However, we still consider this to be a PING mechanism, because the I-cell in isolation fires much more slowly than the E-cell (at about 26 Hz; data not shown), and so, when coupled, an excitatory spike directly triggers each I-cell spike. However, this intrinsic I-cell drive becomes important for the slower rhythms. The time courses for both GABA$_A$ and the AHP conductance during this PING rhythm are shown in figure 4.4C.

When arousal is decreased to a simulated "low" state, by increasing the AHP conductance ($G_{AHP} = 1$, $I_{app,E} = 1.5$, $I_{app,I} = 0.5$), the behavior transitions so that the excitatory cells are less active and express a 20-Hz "beat-skipping" beta rhythm (Whittington et al., 2000) (figures 4.4D and 4.4E). In this case, now that an AHP-current has been added, the E-cells are less active and the applied current to the I-cell is important for setting the pace of the rhythm. The hyperpolarizing AHP potassium current prevents the excitatory cell from firing immediately after GABA$_A$ inhibition wears off, and it takes another full spike cycle of the I-cell before the E-cell fires again (figure 4.4F). This E-cell spike then triggers the I-cell to fire, sooner in this cycle than it would have if driven only by its own applied current. One important thing to note is that even though the

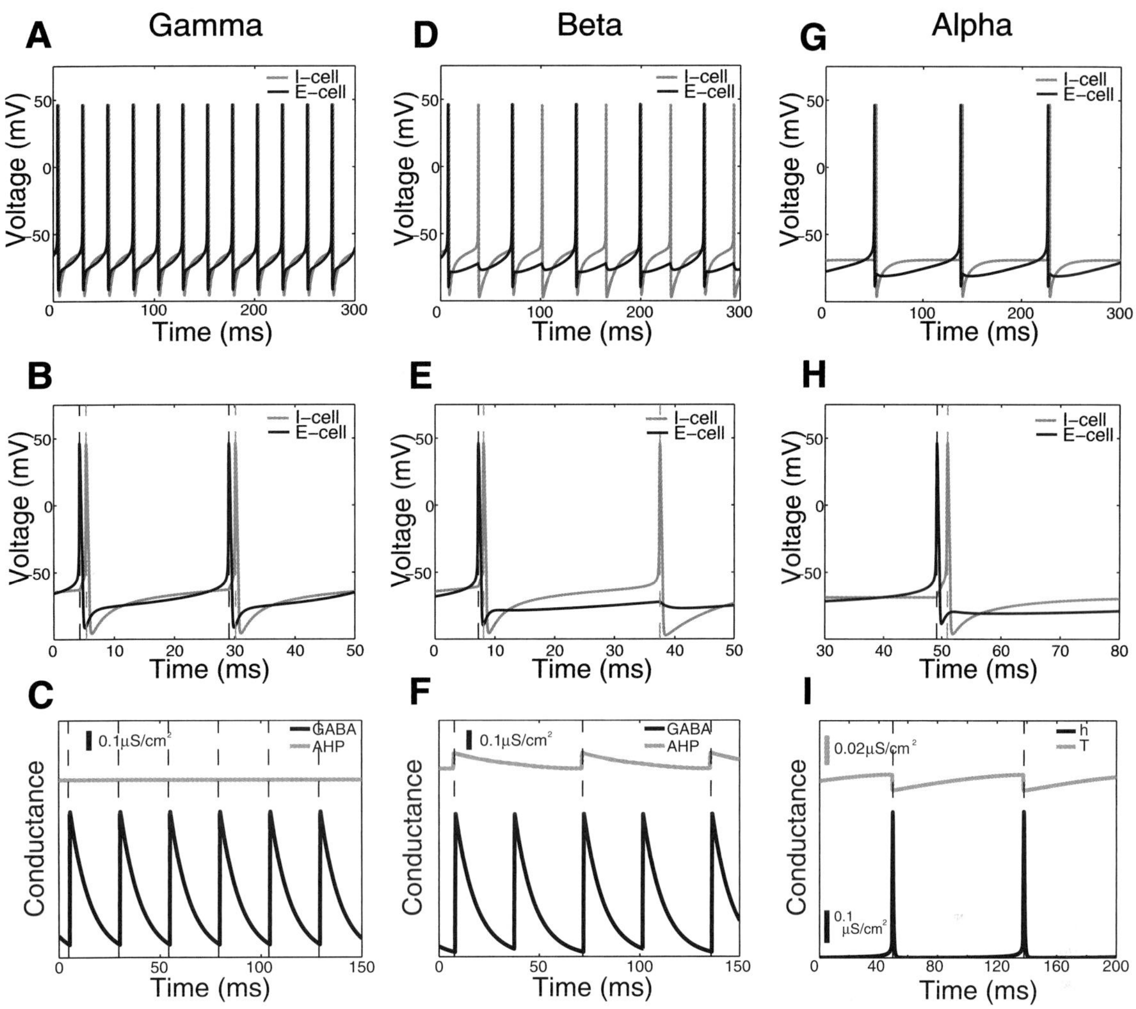

FIGURE 4.4 (A) Synchronous gamma-frequency spiking between the excitatory (black) and inhibitory (gray) cells during a modeled "high" arousal state, with parameters $G_{AHP} = 0$, $I_{app,E} = 1.5$, $I_{app,I} = 0.5$. (B) Behavior during a 50-ms window to show order of spiking (E-cell fires first). (C) $GABA_A$ and after-hyperpolarization (AHP) conductance during gamma rhythm (with $G_{AHP} = 0$). (D) When lowered arousal is modeled as an increased AHP current, the excitatory cell fires at beta (black) and the inhibitory cell at gamma (gray). Parameters are $G_{AHP} = 1$, $I_{app,E} = 1.5$, $I_{app,I} = 0.5$. (E) The order of spiking (E-cell first) over 50 ms in window. (F) $GABA_A$ and AHP conductance during beta rhythm; summed hyperpolarizing effects cause "beat-skipping" in the excitatory cell. (G and H) A synchronous alpha rhythm that emerges with a further decrease in arousal (negative applied currents for both E-cell and I-cell) with parameters $G_{AHP} = 1$, $I_{app,E} = -0.5$, $I_{app,I} = -0.2$. (I) h- and T-conductance during alpha rhythm. These hyperpolarization-activated currents cause rebound spiking at ~10 Hz in the excitatory cell.

intrinsic firing of the I-cell is important to set the pace of the rhythm, it is not the sole driver of the network rhythm. As in the case of gamma, the I-cell fires more often, and the E-cell fires less often than they would in the decoupled network.

If a further decrease in arousal is simulated, by reducing the applied currents to negative hyperpolarizing values, while keeping the increased AHP conductance ($G_{AHP} = 1$, $I_{app,E} = -0.5$, $I_{app,I} = -0.2$.), the model expresses an alpha rhythm with both the E- and I-cells firing synchronously near 10 Hz (figure 4.4G). In this case, the hyperpolarization of the excitatory cell activates the h-current and deinactivates the T-current, respectively, and the period of the rhythm is controlled by their time constants. A spike in the E-cell triggers a spike in the I-cell (spike timing can be seen more clearly in figure 4.4H). The resulting $GABA_A$ inhibition from the I-cell, in combination with the hyperpolarizing applied currents, engages the E-cell h- and T-currents. The h-current slowly depolarizes the cell, ultimately triggering a spike via the T-current about 100 ms after the inhibition. The h- and T-currents over two cycles of the alpha rhythm are shown in figure 4.4I. For this regime, the AHP-current contributes little to the oscillation and, if the h- and T-currents were not present, the E-cell would not spike at all.

Thus, this two-cell network, with the addition of AHP, T-, and h-currents in the pyramidal cell, can express rhythms of different frequencies generated by distinct network mechanisms. This somewhat more detailed modeling of intrinsic currents in individual cells permits expanded insight into transitions between network rhythms and interactions between intrinsic cellular and network properties.

For descriptive purposes, we have considered transitions in rhythms via only a few parameter changes. Pinto and colleagues (2003) map out a wide parameter regime across AHP-current and drive strength over which the rhythms transition. For some transitions, there is a clear bifurcation in the dynamics of the network such that the system sharply changes its rhythm-generating mechanism and frequency expressed (e.g., from gamma to beta as the AHP is increased); for others, there is a more gradual shift in the frequency of the rhythm without a clear change in the dynamics of the network (e.g., from beta to alpha with decreased drive). We refer the reader to the original paper for further detail. In all cases, the exact frequency produced is also dependent on synaptic connectivity, timing, and strength, which have been fixed in this analysis (see section 4.3 for further exploration). The approach of fixing some parameters while examining the impact of others is useful for exploring network dynamics, and a goal is often to consider parameters that are as close to biologically relevant as possible. However, there is evidence that properties of ion channels, synaptic kinetics, and connectivity are not static in nervous systems but can vary to some degree while maintaining optimal rhythmic (and nonrhythmic) behaviors in the systems (Marder, 2011). This has been observed in the crustacean stomatogastric nervous system where preserving specific rhythmic motor behavior is critical to function (Marder, Tobin, and Grashow, 2007). An important and unknown question is whether similar phenomena may exist in regulating neocortical rhythms. Modeling can help reveal the space over which rhythms are preserved and guide us to understanding why.

4.3 The Computational Role of Gamma: Larger Network Models

Section 4.2 considered simplified cases of tight synchrony between just two cells. It is important to note that if such tight synchrony were present across a large network, it would likely impair function, and indeed tightly synchronous activity across a large network is a hallmark of seizure activity (Holt and Netoff, 2013). Nonetheless, some level of synchrony is likely important. Synchrony in local networks, particularly in subpopulations of cells active in a given processing context, is hypothesized to be critical for the binding of information and for the effective downstream flow of information in the brain (Engel, Fries, and Singer, 2001). In the simplified cases of section 4.2, each neuron can be thought of as representative of a subpopulation of cells that participates in the rhythm. We now consider larger populations that exhibit some variation in spiking timing, and generate a more direct link to brain function and cognition. These larger networks are used to examine how synchronous gamma activity can emerge out of noisy and heterogeneous network conditions (section 4.3.1) and how synchronous subpopulations may emerge and interact within a larger network, in a manner that mediates how attention enhances stimulus selection (section 4.3.2).

4.3.1 Gamma Rhythms Create Synchronous Cell Assemblies Robust to Heterogeneity and Noise Using a model similar to that used in the work of Börgers and Kopell (Borgers and Kopell, 2003, 2005), this section extends the previous Theta neuron model from section 4.2.2 to a larger network and demonstrates how the same inhibitory mechanism that creates the gamma period also promotes synchrony among cells, even in conditions of sparse random connectivity and random noise.

Consider a network with 40 excitatory and 10 inhibitory Theta neurons, shown schematically in figure 4.5A, controlled by the same equations as the model in section 4.2.2 (equations 4.1–4.5). Each cell receives a constant level of external drive that is the same for each cell population, I_I, and I_E, respectively. E- and I-cell populations are reciprocally coupled, and connections within a population may or may not be present (E-to-E or I-to-I connections); the probability of connection between populations can vary. The maximal conductance strengths of individual synapses, where i indexes E-cells and j indexes the I-cells, is described by the following additional equations:

$$g_{ij} \frac{g_{EI}}{p_{EI} N_{EI}} \omega_{ij} \qquad (4.7)$$

where N is the number of E-cells, $g_{EI} \geq 0$ is a constant and indicates the total strength of E-to-I connection across the whole network, and $\omega_{ij} = 1$ with probability $0 < p_{EI} < 1$ and is otherwise set to zero. This equation scales the strength of each individual synapse so that the average summed maximal strength of synaptic input

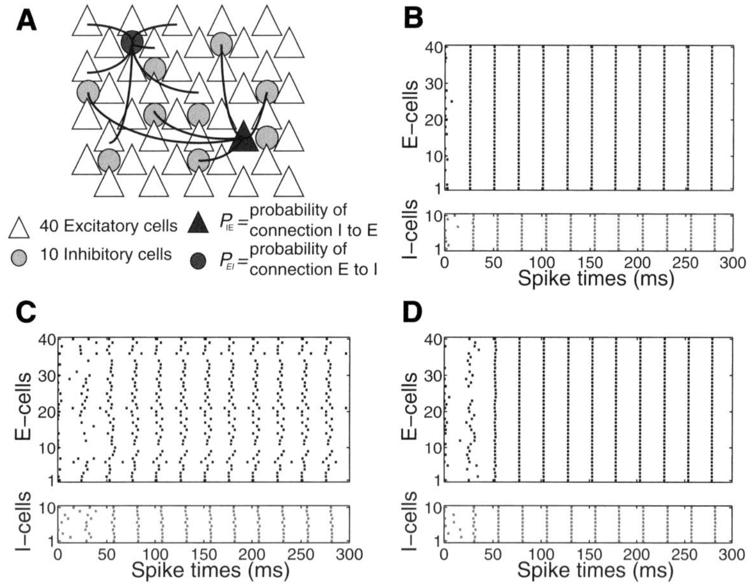

FIGURE 4.5 (A) Schematic of larger network with 40 E-cells and 10 I-cells with example of random input pattern for two cells with random I-to-E and E-to-I connections. (B) Network with all-to-all connectivity E-to-I and I-to-E, but no I-to-I or E-to-E connections, and randomized initial conditions. The network expresses very tight pyramidal–interneuron network gamma, much like the two-cell network shown in figure 4.3F. Note that the vertical lines are composed of aligned dots representing tightly synchronized firing times of individual cells. (C) The same simulation, but with probability of E-to-I and I-to-E connectivity at 0.5 and a variable number of inputs per cell. The network still expresses gamma, though with less synchrony in both the E- and I-cell populations. (D) Simulation still with connectivity of 0.5 (as in panel C), but now with a constant number of inputs for all cells. With this more uniform random connectivity, the network synchronizes tightly.

remains constant, even as the probability of connection is varied. This equation describes E-to-I synapses, but an analogous description holds for I-to-E, E-to-E, and I-to-I synapses.

Gamma synchrony with all-to-all or sparse random connectivity: First consider a network with $p_{EI} = p_{IE} = 1$, that is, connectivity is all-to-all between populations and we set $I_E = 0.1$, $I_I = 0$, $g_{EI} = g_{IE} = 0.25$, $g_{EE} = g_{II} = 0$, $t_E = 2$, $t_I = 10$. This is very much like the two-cell network in section 4.2.2, with E-cells synaptically coupled to I, and vice versa, but no connections from I to I, or from E to E. Even though every cell starts from a different initial condition, the network rapidly synchronizes to a coherent gamma rhythm by the PING mechanism. The raster plot in figure 4.5B shows all cells spiking tightly in unison. Even though this network is larger, it behaves almost exactly like the two-cell PING network in section 4.2.2 because every cell receives uniform input that quickly washes out any of the initial variation in starting position.

Now consider a somewhat more realistic situation, where connections between populations are placed at random, and thus each cell receives different levels of input, setting $p_{EI} = p_{IE} = 0.5$, as depicted schematically in figure 4.5A. The choice of how to place these connections is important. Start by considering all possible connections between the populations, and then randomly select half to have nonzero connection strength, and the other half to be set to a strength of $g_{ij} = 0$. If we consider, for example, the connections from the I- to the E-cells, the mean number of inhibitory nonzero inputs for each cell, taken over all the E-cells, would be 5, or rather, half of the total number of possible inputs (10). However, some E-cells will receive more than 5 inputs, and some less, so the strength of inhibition for each E-cell will vary somewhat across the network. The resulting spiking pattern with this type of connectivity is shown in figure 4.5C. Even with this increased variability in the network structure, synchrony emerges at the same frequency as the all-to-all case but is looser across the population. Both the E-cells and the I-cells spike at roughly the same time but with some variation in the exact timing. While the rhythm is still strong, this is in contrast to the extremely tight millisecond-level synchrony in the case of all-to-all connections.

If, instead, we keep the connection probability at 0.5 but alter the random structure such that each cell receives an exact fixed number of inputs (e.g., an E-cell receives exactly 5 inputs selected randomly from the 10 possible I-cells), then, although there is still randomness in the connectivity, the rhythm tightens (figure 4.5D) to the same level of synchrony as the all-to-all case shown in figure 4.5B. Similar mechanisms of rhythmic stability in the presence of network heterogeneity have been observed in other simple and complex network models validating the robustness of gamma synchrony (Golomb and Hansel, 2000; Vierling-Claassen, Cardin, Moore, and Jones, 2010; Wang and Buzsaki, 1996).

Gamma synchrony in the presence of noise: Next, we examine the robustness of gamma synchrony to noise, considering a network similar to that described in Borgers and Kopell (2005). Starting with the same 50-cell network structure (figure 4.5A), noise is added by driving E-cells to spike at randomly selected times, representing background activation or inputs coming in from outside the modeled region or network. Set connections as all-to-all between the E- and I-cells ($p_{EI} = p_{IE} = 1$), and network parameters to be $g_{EI} = 0.6$, $g_{IE} = 0.2$, $g_{II} = 0$, $I_E = 0.1$, and $I_I = 0$. Figure 4.6A shows the resulting rhythm, in which the network fires in tight gamma synchrony near 40 Hz among E- and I-cells, exactly as before (see figure 4.5B).

Noise is added using one of the most common and simple ways to introduce a source of random spiking to the network, by forcing each E-cell to spike according to a different randomly selected spike train. For each cell, we select a spike train where the interval between these noisy spikes is chosen via a Poisson process. That is, the amount of time between spikes is chosen from an exponential distribution, which has a probability of firing that can be described by the function $\lambda e^{-\lambda x}$ (known as a probability density function) and has the property that the average waiting time between spikes is $1/\lambda$. In this network, Poisson process noise is added such that each E-cell is forced to spike at an average rate of $\nu_{SE} = 7$ Hz (*S* standing for "stochastic"). In this case, the mean interval between spikes is about $1{,}000/7 \approx 142$ ms, which means for this Poisson process, $\lambda \approx 1/142$. With this noise added to the E-cells, the PING rhythm breaks down as shown in figure 4.6B. The breakdown occurs when a large enough fraction of the E-cells spike out of order, driving an additional inhibitory volley, which causes gaps in the E-cell rhythm.

However, this noise-induced asynchrony can be counteracted in the network in two specific ways. First, if the coupling from the E-to-I-cells is decreased, here to $g_{EI} = 0.24$, then the I-cells are less susceptible to spiking in response to out-of-order excitatory input and the rhythm is restored (figure 4.6C). Second, if I-to-I coupling is added, here with $g_{II} = g_{IE} = 0.2$, the rhythm can still emerge, at least in the I-cell population, even in the face of extreme noise (figure 4.6D; here we set $\nu_{SE} = 50$ Hz), though synchrony is certainly weak in the excitatory population. In this case, synchrony persists because I-to-I inhibition prevents inhibitory spikes in response to out-of-order E-cell spikes.

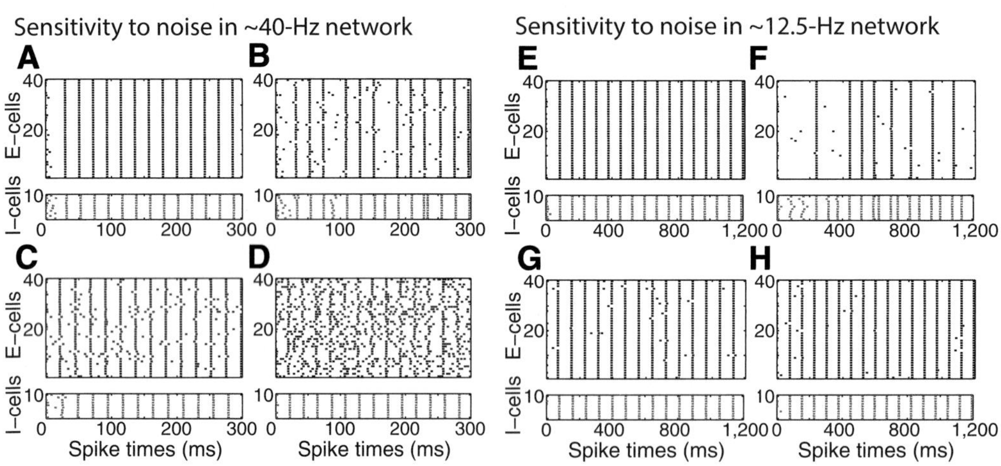

FIGURE 4.6 (A) Synchrony in a 40 E- and 10 I-cell network with all-to-all coupling. (B) Network with random 7-Hz noise in the E-cells that introduces gaps in the E-cell activity and breaks down synchrony. (C) Decreased E-to-I coupling recovers the rhythm in the presence of noise to the E-cells. (D) I-to-I coupling makes rhythm (especially in I-cells) robust to noise even at extremely high rates (50 Hz in this simulation). In comparing panel D to panel B, note that in panel D, the I-cells fire in a very even rhythm, and that even though there are noisy spikes in the E-cells, there is also periodic rhythmic behavior with no extended gaps. (E) 40 E- and 10 I-cell network of Theta cells with all-to-all coupling and low E-cell applied current produces a slow pyramidal–interneuron network gamma (note the simulation time is much longer than in panels A–D). (F) In this slow network, even very slow noise in the E-cells (0.5 Hz) introduces gaps in the E-cell activity and breaks down synchrony. (G) Decreased E-to-I coupling recovers the slow rhythm. (H) I-to-I coupling also recovers the rhythm.

Rhythms controlled by PING-type mechanisms in these larger networks can emerge at slower frequencies if specific parameters are changed. However, when the rhythm is slower, it can become more susceptible to disruption. If we drop the E-cell applied current very low, to $I_E = 0.007$ (from 0.1), but keep the other parameters the same, a very slow PING emerges, at a frequency of about 12.5 Hz (figure 4.6E). This rhythm is very sensitive to noise in the E-cells and breaks down with a very low $\nu_{SE} = 0.5$ in the E-cells (figure 4.6F). The rhythm can be recovered using the same mechanisms that protected the more robust PING considered above (figure 4.6A–D). That is, if E-to-I inhibition is decreased, the I-cells are less sensitive to noisy spikes in the E-cells ($g_{EI} = 0.06$, dropped from 0.6; figure 4.6G). Alternatively, if I-to-I connections are added ($g_{II} = g_{IE} = 0.2$; figure 4.6H), the slower rhythm can be stabilized, but it is robust over a much smaller range of parameters and requires much more careful tuning to emerge.

4.3.2 Gamma Rhythms Link Attention and Stimulus Selection Several models have investigated the link between gamma and attention. This section describes one such example (Borgers, Epstein, and Kopell, 2008). The motivating data for this study was experimental work on the interaction between attention and visual orientation-selectivity in the macaque visual cortex (Reynolds, Chelazzi, and Desimone, 1999). In these experiments, monkeys were shown images of either one or two oriented bars while single neurons were recorded in visual cortex. The monkeys were directed to attend to either one particular location where bars would be presented or to a location away from where the bars were shown. In visual cortex, some neurons are selective for the angle of an oriented bar; for example, a neuron tuned for an angle of 45 degrees will fire very strongly for a bar at this angle but much more weakly for a bar at a different angle (such as 90 degrees).

Figure 4.7 shows an example of the kind of cell behaviors considered in this work. In particular, the cell shown in this figure can respond very strongly to a single vertically oriented bar, which is labeled "Ref," for reference. Even if this vertical bar "Ref" stimulus is presented while the animal is attending to a different area of the screen, the spike rate response is quite high (labeled "Ref Att Away" in figure 4.7A, top dotted line). In contrast, with attention directed away, the cell responds at a low spike rate to a second stimulus, with a different orientation, a horizontal bar (labeled "Probe Att Away" in figure 4.7A, bottom solid line). If both bars are presented at the same time, then the cell fires between these two rates (labeled "Pair Att Away" in figure 4.7A, middle dashed line).

However, if both the "Ref" and "Probe" stimuli are presented, and the animal has been directed to pay particular attention to the location where the vertical "Ref" stimulus is located, then the firing rate for the cell is comparable to what it would be if the extra "Probe" stimulus were not presented at all (labeled "Pair Att Ref" in figure 4.7B, top dotted line). That is, when attention is directed at the "Ref" stimulus location, the cell behaves as if this extra distracting "Probe" stimulus were not present at all. Although figure 4.7A and B shows only one example cell, this pattern played out over many cortical cells.

This example data presents an interesting puzzle. How is it that when presented with exactly the same sensory stimuli, attention alone can shape the neural response so that the firing rate is high in one case but low in the other? This section describes an adaptation of modeling by Börgers and colleagues (Borgers et al., 2008) that predicts a possible link between gamma rhythms in local cortical networks and attention-mediated stimulus selection like that observed in the work of Reynolds et al. First note that the original data do not examine the role of gamma rhythms. Rather, the model examines a possible mechanism by which gamma rhythms may contribute to shifts in firing rate to the same stimulus with attention. In the process, this model generates testable predictions about how the biological underpinnings of gamma rhythmicity may interact with attention, reshaping active neural networks in response to cognitive demand. It is certainly possible that non-rhythm-based methods may be able to generate similar results, but in support of the model predictions, a subsequent study found enhanced gamma-band synchronization in visual cortex with attention in a similar task (Fries, Womelsdorf, Oostenveld, and Desimone, 2008). Additional work also indicated that oscillatory inhibition may be more efficient in producing these firing shifts than asynchronous inhibition (Borgers, Epstein, and Kopell, 2005).

Consider a model network of 160 excitatory and 40 inhibitory cells, using the same Hodgkin–Huxley-type conductance-based cell model as in section 4.2.3. The network contains strong E-to-I, I-to-E, and I-to-I connectivity and starts in a baseline state where the I-cells fail to synchronize and their asynchronous activity suppresses E-cell activity (this regime is explored in exercises looking at noise in I-cells). The network can escape from this suppressed state when drive to the I-cells is lowered or drive to the E-cells is raised. To model presentation of only the primary "Ref" stimuli when attention is directed away, a subset of E-cells, which we will call "Ref Tuned"

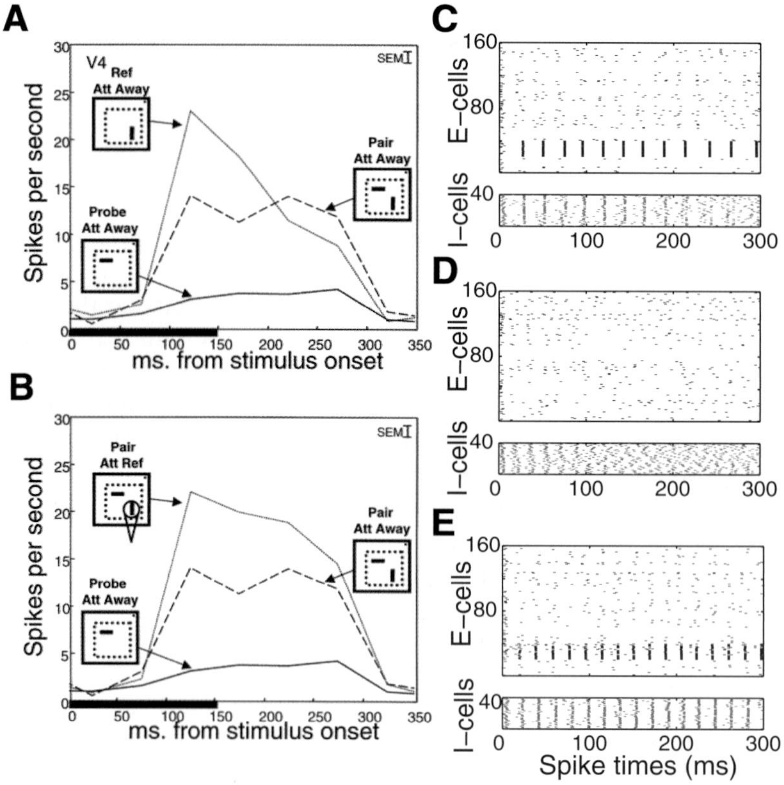

FIGURE 4.7 Here we see the average firing rate responses of a single cell in the V4 area of visual cortex in Macaque for which the reference ("Ref") is a more effective stimulus than "Probe." (A) The firing rate response of the cell when the stimuli are presented, either individually or together, while the monkey was instructed to attend (Att) elsewhere. Note that the response to the stimulus labeled "Ref" is very strong (dotted line), and the response to the stimulus labeled "Probe" is weak but positive (solid black line). When both stimuli are presented, the firing rate of this cell is between these two extremes (dashed line). (B) The same cell's activity when presented with the same stimuli, only now, when both stimuli are presented, the monkey has been instructed to attend to the location where the "Ref" stimulus was presented. Note that with both stimuli presented while attending to the "Ref" stimulus, the firing rate increases compared to when attention was directed elsewhere (compare the dashed line to the dotted line), almost to the same level as when the stimulus was presented alone (as in the dashed line in panel A). (A) and (B) from Reynolds et al. (1999). (C) A raster plot of model activity when only one modeled stimulus is presented. A gamma rhythm including the I-cells and the subset of E-cells strongly driven by the stimulus emerges. However, when a second stimulus is presented (D), the rhythmicity is destroyed, and the firing rate decreases for cells driven by the first stimulus. (E) Depiction of what happens when both stimuli are presented but attention is also added, modeled as stronger drive to the E-cells driven by the first stimulus, and also as increased inhibition over the whole network.

cells, are driven with an applied current, modeling drive to neurons that are strongly tuned to respond to that particular stimulus. In addition, all of the I-cells receive an applied current, and a separate subpopulation of excitatory cells receives noisy background drive, representing cells that are not strongly activated by this stimulus but do have some response. We consider this model to be taking place within the receptive field of the visual stimuli that are presented. That is, the subject's attention is spatial, but for the purposes of our model, we can simply think of attention as directed toward or away from the spatial location.

When this modeled "Ref" stimulus is presented alone, the network quickly synchronizes into a PING rhythm including all the I-cells and the subset of strongly driven E-cells (figure 4.7C). As a result, the excitatory "Ref Tuned" cells most strongly driven by the stimulus fire at high rates. This is analogous to the relatively "high" firing rate of the cell in response to only the "Ref Att Away" stimulus in figure 4.7A (the dotted line). A model of the "Probe" stimulus can be added in exactly the same way, by strongly driving a subpopulation of "Probe Tuned" E-cells (with a slightly different applied current), increasing the applied current to all I-cells and adding noise to another subset of E-cells. When this second stimulus is added, the network fails to produce a rhythm. As a result, the "Probe Tuned" E-cells that initially produced such a robust response have a much lower firing rate (figure 4.7D). This is analogous the "Pair Att Away" middle dashed line in figure 4.7A. Here, consider the firing shown in figure 4.7C–E to represent firing when the rate is highest, that is, during and directly after the visual stimulus. While this model does not directly consider onset and offset of the stimulus, that expansion of the model could be included as a turning "on" and "off" of the modeled stimuli, with some additional consideration of background activity.

However, if we now simulate attention in the network, the E-cells responding to the "Ref" stimulus can once again fire at high rates, as in the experimental data. Attention is modeled with a net reduction in the baseline applied current across the network—mimicking cholinergic modulation—and by selectively strengthening the applied current to the "Probe Tuned" E-cells by 30%, modeling top-down attention. With these changes, PING rhythmicity reemerges for the attended stimulus, and the firing rate of the "Probe Tuned" E-cells increases (figure 4.7E), analogously to the "Pair Att Ref" top dotted line in figure 4.7B.

This mechanism of stimulus selection with competition requires that the inhibitory synapses are strong enough to enable PING rhythms but weak enough to not suppress all noise-driven spiking. Each modeled stimulus must add enough drive to the I-cells so that the two simultaneously presented stimuli desynchronize the I-cells. The increase of inhibition onto to I-cells when attention is modeled, which decreases their activity, and the enhanced drive to the "Probe Tuned" E-cells, must be strong enough to counter the effects of the added excitation of the I-cells from two stimuli. Hence, the attended stimuli must be strengthened by a sufficiently large amount. Within these requirements the mechanism is robust, and changes in other parameters can be varied considerably without qualitatively changing the results.

Although detail has been gradually added to the example models throughout this chapter, we are still at a level where the direct connection between the data and the model is somewhat schematic. We now move on to consider a model with more explicit connections between data and model (e.g., the same units of measure).

4.4 Mechanisms and Functions of Low-Frequency Thalamocortical Rhythms

Section 4.2.3 considered how low-frequency rhythms in the alpha and beta regime can emerge in simplified network models of single-compartment spiking neurons because of synaptic interactions and the intrinsic properties of the individual cells. Notably, these rhythms were self-assembled in a local network, which did not require or interact with rhythmic drive from other brain regions. Many studies suggest that in the in vivo behaving brain, multiple reciprocal interactions between functionally connected brain areas contribute to neural rhythms, particularly for low-frequency rhythms that are slower than gamma. There is evidence that low-frequency rhythms recorded from the cortex often involve reciprocal thalamocortical interactions (see figure 4.1A; see also the discussion of sleep rhythms in section 4.5).

Here, we focus on a computational modeling study of alpha and beta rhythms that appear in awake humans (in contrast to the prominent slow rhythms present in sleep). This model was specifically designed to study the large-scale electrical current signals that are recorded with MEG/EEG (S. R. Jones et al., 2009; S. R. Jones, Pritchett, Stufflebeam, Hamalainen, and Moore, 2007; Lee and Jones, 2013; Ziegler et al., 2010) which are presumed to come from the net summed electrical current flow in large spatially aligned pyramidal neuron dendrites in cortex. These currents are influenced by synaptic drive from the thalamus and other parts of cortex (see figure 4.1D, right, and section 4.4.2). Designing the output of the model to specifically reproduce this signal in close agreement with measured data was essential for generating a different theory of alpha and beta generation than that described for local networks in section 4.2.3.

MEG and EEG are noninvasive recording techniques that measure magnetic and electrical field signals from the surface of the head. Analytic procedures, known as inverse solution techniques (which are beyond the scope of this chapter), work backward from these signals measured outside the head to localize the sources of neurally generated electrical activity inside the brain. An inherent problem with finding inverse solutions is that the solutions are not unique. Various methods have been developed to estimate the actual distribution of the sources from the recorded data, with some methods assuming the activity is generated from a few putative source and others taking into account the entire cortical surface, with various levels of details in how the cortex and head are modeled (see also Barrès, Simons, and Arbib, 2013). The estimated sources, called equivalent current dipoles, are located primarily on the cortex because of the electromagnetic physics of the brain (see figure 4.1D, Hamalainen, et al., 1993). These techniques can be applied to either MEG or EEG data and are often combined with structural magnetic resonance imaging (MRI) images of the brain to more precisely define the location of activity.

Figure 4.1D (middle) shows tangential sources in the sulci (grooved regions of the cortex) and radial sources in the gyri (outer surface regions of the cortex, gray arrows). Because of the physical properties of magnetic fields around the head, MEG detects electrical signals generated in the grooves, or sulci, of the cortex, and in contrast to EEG, the signal is not affected by the skull and scalp. These features aid in the localization and interpretation of the underlying neural sources but mask radial activity (Hamalainen et al., 1993). The complementary information recorded with simultaneous MEG and EEG can increase the accuracy of inverse solution estimates.

Section 4.4.1 gives an overview of the MEG data the model was designed to reproduce. Section 4.4.2 describes the biophysical origin of the electrical current signals measured with MEG/EEG and how the model was specifically designed to reproduce these currents, with equal units of measure. Section 4.4.3 describes how the model produces the combined alpha/beta rhythm, and section 4.4.4 describes the role of alpha and beta rhythms on the salience of sensory evoked responses.

4.4.1 Alpha and Beta in the Somatosensory Mu Complex: Overview of Recordings and Functional Relevance Before describing the model implementation and results, first consider the data that the model was designed to simulate. Figure 4.8A shows an example of a spontaneous rhythm recorded in primary somatosensory cortex in humans with MEG. The middle of the top panel depicts the raw primary current dipole waveform of a source localized to the hand area in primary somatosensory cortex (SI). In practice, we first found the location of the hand area by applying inverse solution techniques to sensor recordings of a strong stimulation to the finger. The inverse techniques estimate the location and temporal profile of the primary electrical current dipole contributing to the recorded response. Inverse solution techniques also estimate the direction of current flow into or out of the cortex (e.g., from supragranular to infragranular or vice versa) at different points in time. We then investigated the spontaneous signal from this location and noted that during spontaneous states it oscillates and the current flow alters up and down the cortical layers. The dipole location is shown in blue in the right top panel, and the blue line depicts the direction of the current during the downward deflections of the oscillations, which corresponded to current flow from supragranular to infragranular layers based on structural MRI imaging. The left top panel displays a time–frequency spectrogram of the unfiltered waveform of the spontaneous signal. This SI-localized rhythm contains a combination of alpha (7–14 Hz) and beta (15–29 Hz) activity, referred to as the "mu" rhythm, which is a term specific to sensorimotor cortex (see figure 4.1A). Jones and colleagues found that the alpha and beta components of the SI rhythm influence perception (S. R. Jones et al., 2009), are modulated by attention (S. R. Jones et al., 2010), and change with healthy aging (Ziegler et al., 2010) and repeated exposure to the task (Kerr et al., 2011; Wan et al., 2011). In the top left panel of figure 4.8A, the spectrogram shows a 1-second example of the temporal structure of this signal where the alpha and beta components do not overlap in time. The red box outlines an example epoch of high beta power that does not occur simultaneously with an epoch of high power alpha activity. Across all spontaneous data, high power epochs of the alpha and beta overlap on short time windows only 50% of the time (see figure 4.9A, right panel inset) and otherwise occur independently, suggesting their generation may involve separable processes or distinct mechanisms. Further, the oscillation waveform was symmetrical around a zero point (figure 4.8A, middle top panel), with the oscillation alternating between positive and negative polarity. The power spectral density of the rhythm has a dominant peak in the alpha band with a smaller peak in the beta band (figure 4.9A, left panel). The remainder of section 4.4 describes a computational model that reproduces these key features of the experimental MEG signal.

4.4.2 Overview of Strategy to Simulate Human Somatosensory Rhythms Measured with MEG/EEG In this section, we focus on current dipoles that have been localized with standard inverse solution techniques to the primary somatosensory cortex, SI (S. R. Jones et al., 2007). The model represents the local circuitry in SI under the influence of external thalamic drive. The data in section 4.4.1 were produced from an estimated current dipole signal located specifically in the hand area of SI, as shown in figure 4.8A, top right panel (S. R. Jones et al., 2009; S. R. Jones et al., 2007). Once the location was determined, the waveform was calculated for this source (figure 4.8A, top middle panel) and frequency analysis was applied to the waveform (figure 4.8A, top left panel; figure 4.9A).

The summed postsynaptic longitudinal current flow inside the dendrites of the large aligned cortical pyramidal neurons induce the current dipole signal (figure 4.1D, right panel; the gray arrow shows current flow in a pyramidal cell) (Hamalainen et al., 1993; Ikeda et al., 2005; Murakami et al., 2002). The more the neurons that participate in synchronous activity, the

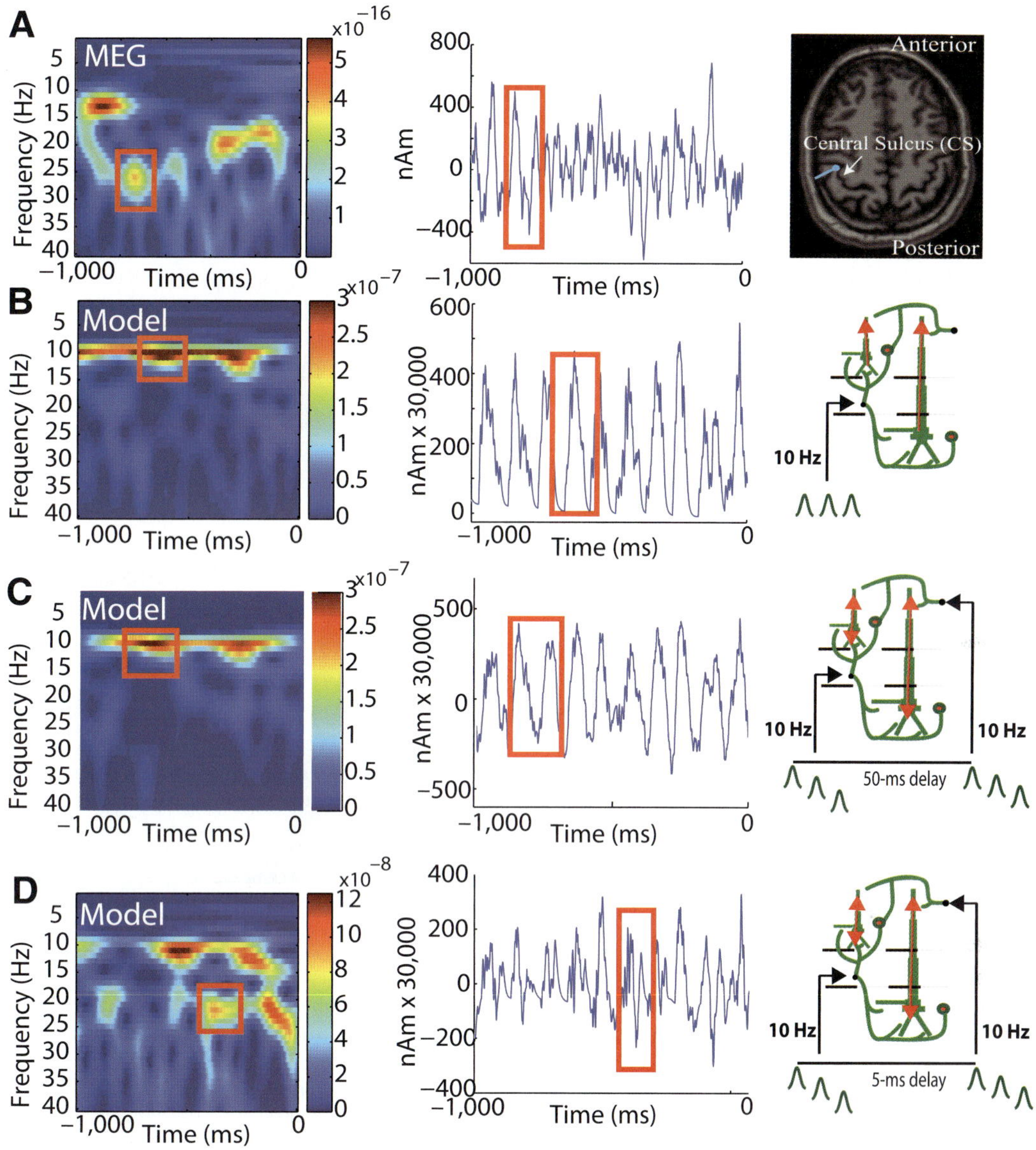

FIGURE 4.8 The "mu" rhythm, a somatosensory cortex (SI)-localized rhythm which contains a combination of alpha and beta activity: (A) Time frequency spectrogram of magnetoencephalograph (MEG) SI mu rhythm (left panel) and corresponding waveform (middle panel) shows short bouts of alpha (7–14 Hz) and beta (15–29 Hz) activity created by an oscillation with positive and negative polarity (above and below zero). This signal is the unfiltered MEG current dipole localized to the hand area of SI (see section 4.4.2). The red boxes highlight a bout of beta-band activity in the MEG data that does not occur simultaneous with high alpha power. Compare this MEG beta oscillation to a beta event reproduced in the model shown in the red box in panel D and notice that the oscillations have the same oscillatory pattern beginning at similar phases. The blue dot and line illustrate the SI current dipole location on a representative subject's structural MRI (right panel). (B) Model SI current dipole signal created by stochastic 10-Hz proximal-only drive (right panel) produces 10-Hz activity with only positive polarity (left and middle panel). The red boxes highlight a cycle of alpha activity in this proximal-only drive case; see figure 4.10B for illustration of drive. (C) Model SI current dipole signal created by antiphase (50-ms mean delay) 10-Hz proximal and distal drive (right panel) produces 10-Hz activity with positive polarity and negative polarity (left and middle panel). The red boxes highlight a cycle of alpha activity with antiphase proximal and distal drive. (D) Model SI current dipole signal created by nearly synchronous (5-ms mean delay) 10-Hz proximal and distal drive (right panel) produces short bouts of alpha and beta activity with positive and negative polarity (left and middle panel) in agreement with the MEG data; compare red boxes in panel D with panel A. Adapted from S. R. Jones et al. (2009).

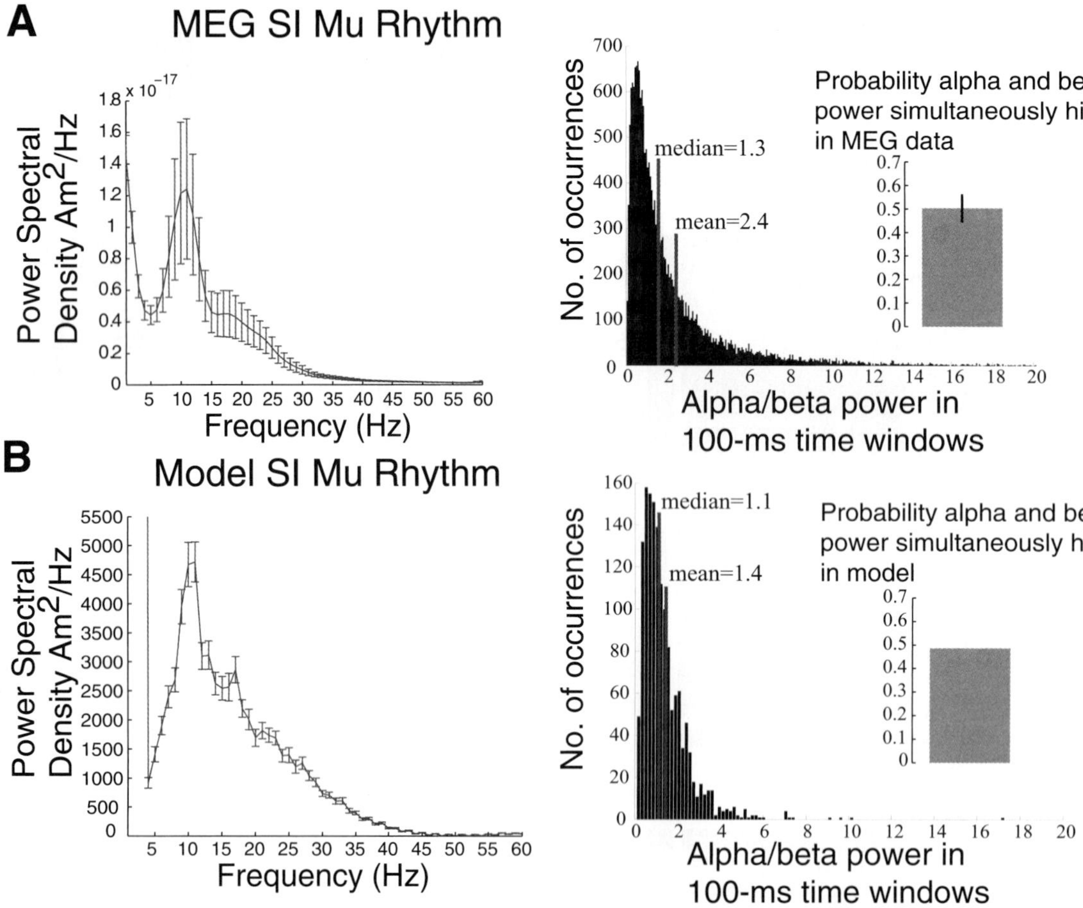

FIGURE 4.9 (A) Power spectral density of magnetoencephalography (MEG)-measured spontaneous somatosensory cortex (SI) signal (mean and standard error across n = 7 subjects), as described in figure 4.8A (left panel). Histogram of the occurrence of alpha and beta simultaneously in high power in short (100-ms) time windows (right panel). Inset shows mean and standard error across all epochs and subjects. Alpha and beta overlap in high power approximately 50% of the time. (B) Corresponding calculations in the model show close agreement with the MEG data. Adapted from S. R. Jones et al. (2009).

bigger the signal will be. The SI finger area studied by Jones and colleagues is oriented such that the long dendrites of the pyramidal neurons lie tangential to the MEG sensors and therefore create a recordable current dipole signal. The orientation of the current flow during the downward deflections of the oscillations (blue line in figure 4.8A, top right panel) suggests that during this part of the oscillation the current flows from the top of the dendrites toward the soma, but during the positive deflections, the current flows in the opposite direction.

To model this signal, the cortical network described here contains multicompartment pyramidal neurons with spatially extended dendrites, from which current flow within the cells is calculated (see chapter 2 for a description of multicompartment cells). These cells are divided into supragranular (layers II/III, near the cortical surface, labeled A and B in figure 4.1B) and infragranular (layer V/VI, deeper in the cortex, labeled D and E in figure 4.1B) layers, and the current dipole signal is calculated as the current flow inside the pyramidal neuron dendrites in a direction parallel to the apical dendrites, summed across the population in nano-ampere-meter (nAm) units (see black arrows in figure 4.10B).

Figures 4.10C and D show the contrast between the simpler single-compartment models we examined in sections 4.2 and 4.3, where rhythms were expressed by spiking in the network, and the multiple compartments needed here to estimate the current dipole signal, where the rhythm is generated by subthreshold activity within the pyramidal neuron dendrites.

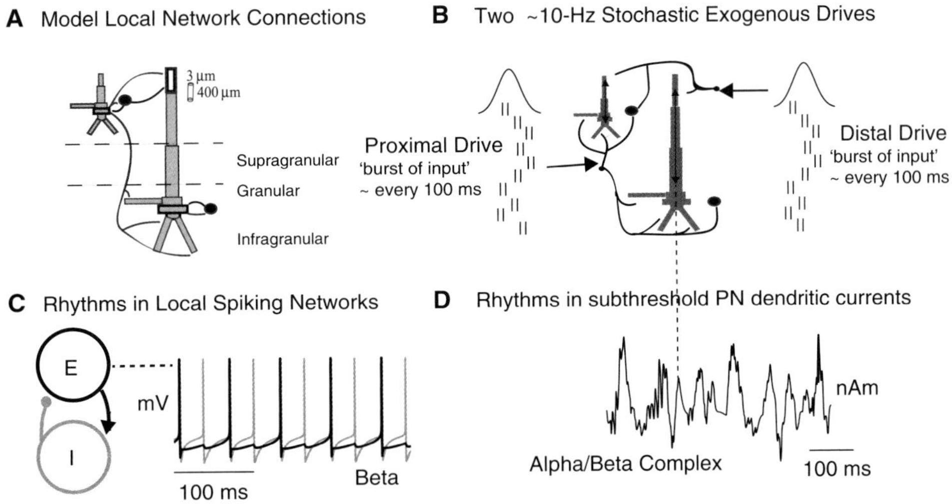

FIGURE 4.10 (A) Schematic of local somatosensory cortex (SI) cortical column model. Multicompartment pyramidal cell (gray) in supragranular (near the cortical surface) and deeper infragranular layers are interconnected with single-compartment interneuron (black) through local AMPA and $GABA_A$ synapses. (B) Schematic of proximal and distal excitatory synaptic drive to the SI network. Proximal drive contacts the portions of the pyramidal dendrites that are near the soma, and distal drive contacts dendrites that are far from the soma. Stochastic spiking patterns of the 10-Hz drive are shown in black and represent bursts of action potentials that drive the network approximately every 100 ms. Black arrows depict intracellular dendritic current flow that generates the model magnetoencephalography (MEG) current dipole. (C and D) Schematic illustration comparing rhythms generated from the membrane potential of spiking neurons (described in sections 4.2 and 4.3) with rhythms generated from subthreshold postsynaptic longitudinal currents in pyramidal dendrites, representing MEG/EEG primary current dipole signals (described in section 4.4). PN, pyramidal neuron.

4.4.3 Mechanisms of Alpha and Beta Generation in the MEG Somatosensory Mu Complex

Summary of Model Results By designing model output to be in close agreement with the recorded MEG data, the model generated the prediction that the alpha and beta components of the cortical SI mu rhythm arise from the interaction of two stochastic rhythms at approximately 10 Hz from sources outside of SI. These external rhythms drive the cortex in distinct layers, namely, infragranular and supragranular layers, referred to as "proximal" and "distal" respectively, because of their postsynaptic location on the pyramidal neurons (figure 4.10B). As detailed below, we assume that these external drives are supplied by the thalamus. When these inputs are both simulated to be approximately 10 Hz and to drive the S1 network nearly simultaneously on each cycle, then alpha and beta components of the SI rhythm emerge with characteristic features that are in close agreement with the MEG data (see figure 4.8D and compare it to figure 4.8A; compare also figure 4.9A and 4.9B). The timing and strengths of the proximal and distal drives control the dominance of alpha or beta expressed at any point in time. Beta oscillations emerge when the rhythms arrive nearly simultaneously (<5-ms mean delay on each cycle) and when the "distal" drive is strong enough; otherwise alpha oscillations emerge (figure 4.8 B–D).

Summary of Cortical SI Model Construction and Calculation of the Net Current Dipoles *SI cortical column model:* The model contains 100 multicompartment pyramidal neurons in the supra- and infragranular layers based on the morphology in Bush and Sejnowski (1993). Each compartment is modeled as a short section of cylindrical cable (see the scale in figure 4.10A), and the current flow between compartments is calculated with Rall's cable equations for electrical current in cables (Carnevale and Hines, 2006; Rall, 1964). Single-compartment inhibitory neurons are evenly interleaved throughout each layer (30 per layer). The pyramidal neurons are synaptically coupled with excitation (AMPA) and inhibition ($GABA_A$), in a manner similar to the other models in this chapter and also contain slow excitation (NMDA) and slow inhibition ($GABA_B$), which decay at a slower rate. The local network connection patterns and postsynaptic contact locations are as shown in figure 4.10A. The electrophysiological properties in each neuron were simulated with Hodgkin–Huxley-type ionic current kinetics, and the cell types contained currents in each compartment tuned to produce spiking behaviors observed in vivo. In addition to the sodium and potassium currents needed to generate spiking, the infragranular pyramidal cells contained h-, T-, and AHP-currents, and supragranular pyramidal cells contained AHP-currents, similar to those described in section 4.2.3. These currents were tuned in the network based on spiking behaviors, but in the spontaneous rhythm generation described below they are activated in a *subthreshold* regime and contribute little to the time constant of current flow within the pyramidal neuron dendrites. Spiking activities in this model are considered in section 4.4.4.

External drive: We at last introduce the thalamus into the circuitry. The proximal drive shown in figure 4.10B is representative thalamic input that specifically targets infragranular layers that relay sensory information to the cortex. The distal drive is representative of input from other cortical areas or input from thalamic areas that specifically contact these superficial layers (E. G. Jones, 1998; Rockland and Pandya, 1979). For the spontaneous rhythms considered here, we hypothesize that the distal drive is also from thalamic nuclei. The exact driving sources were not explicitly modeled since the model results depend only on features of the driving spikes. Instead, their influences are represented as patterns of spiking activity that drive excitatory AMPA synapses about every 100 ms (i.e., 10 two-spike bursts occur each cycle with a mean interval of 100 ms). There is some randomness in the interval, with a variance 400 ms (figure 4.10B). The choice of 10 Hz was motivated by spontaneous thalamic oscillations at 10 Hz that project broadly throughout the cortex (Destexhe and Sejnowski, 2001; Hughes and Crunelli, 2005; Vijayan and Kopell, 2012). Note that while the driving sources may be influenced by cortical feedback (e.g., as part of a thalamocortical loop of activity as depicted in figure 4.1A), here we consider only one direction of influence (thalamus to cortex); see Bonjean et al. (2012) for models incorporating thalamocortical loops in the study of sleep rhythms. A step-by-step description of the cortical model development and all code used can be found in S. Jones (2012); see also the model database website (senselab .med.yale.edu/ModelDB/); and Homework Project 1 is designed to study this model in further detail.

Interacting 10-Hz external drives create alpha and beta components: Here, we give a step-by-step description of how these external drives interact within the pyramidal neuron dendrites to generate the alpha and beta components of the SI mu rhythm. First, consider what happens to the simulated current dipole signal when only the 10-Hz proximal drive is simulated (depicted in figure 4.8B, right panel). This drive initiates current flow up the pyramidal neuron dendrites about every 100 ms, and this is propagated by electrical cable properties and intrinsic ionic currents. The resulting current dipole signal produces a pure alpha frequency (10 Hz) with the polarity of the waveform always positive (figure 4.8B, middle panel). This does not agree with the MEG data, which has both alpha and beta components as well as a waveform that oscillates around positive and negative polarity (compare middle panels in figure 4.8A and B). Now add the second ~10-Hz distal excitatory drive, initially forcing this drive to arrive in antiphase with the proximal drive (i.e., 50-ms mean delay between proximal and distal drive on each cycle; figure 4.8C, right panel). In this case, alternating current flow is generated up and down the pyramidal neuron dendrites creating a signal that oscillates around zero at a pure 10-Hz rhythm (figure 4.8C). This also does not agree with the MEG data. The frequency expressed in this signal depends on the delay between the two drives, and when the distal drive comes in nearly simultaneously with the proximal drive (figure 4.8D shows a 5-ms mean delay), beta-frequency oscillations emerge (red box in figure 4.8D, middle panel, and corresponding spectrogram in left panel). Beta oscillations are created by a break in the proximal induced upward current flow that is replaced by downward current flow from the distal drive. Since in the full model the drives are stochastic on each cycle, beta and alpha emerge on different cycles and do not regularly overlap in time.

In summary, when two equal strength stochastic 10-Hz drives to SI are simulated at proximal and distal locations nearly simultaneously (mean 5-ms delay), the modeled current dipole signal produces many features in agreement with the MEG data, including an oscillation that is symmetrical around zero, similar power spectral density curves, and probability of simultaneous occurrences of alpha and beta (compare figure 4.8A and D, and figure 4.9A and B). A further investigation of how the properties of the drive impact relative alpha and beta expression in the model is part of homework project 1 on the book's website (see also S. R. Jones et al., 2009, for further exploration).

4.4.4 The Computational Role of Alpha and Beta: Impact of Prestimulus Rhythms on the Salience of Sensory Signals Spontaneous rhythms may regulate information processing through a direct impact on the salience of sensory signals. Low-frequency alpha and beta rhythms in particular are thought to act as sensory filters that inhibit the processing of sensory stimuli (Klimesch, Sauseng, and Hanslmayr, 2007). They are actively decreased with selective attention in cortical areas relevant to stimulus processing and simultaneously increased in cortical areas where distracting stimuli are present (S. R. Jones et al., 2010; Worden, Foxe, Wang, and Simpson, 2000).

Impact of the Mu Rhythm on Peaks in Tactile Evoked Responses That Correlate with Detection In MEG recordings taken during a tactile detection task where subjects are asked to report whether they felt a light tap to the finger, detection probabilities are higher when the power of the prestimulus alpha and beta components of the mu rhythm in SI are lower (S.R. Jones et al., 2010). In other words, if strong alpha and beta rhythms are present when the stimulus arrives, then the subject is less likely to feel the stimulus. Figure 4.11A shows the average response from a current dipole in SI elicited from a brief perceptual threshold level tap (stimulus strength maintained at 50% detection level throughout the experiment) to the middle fingertip sorted over trials that were reported to be detected (dark blue) and not detected (light blue) (mean n = 7 subjects). Figure 4.11A shows evoked activity in SI up to 175 ms after the stimulus. The report of detection was made with a button press from the opposite hand well after the initial 175-ms tactile stimulus (cued response at least 500 ms poststimulus). Interestingly, an observable evoked response was elicited even when the tap was not reported as being felt by the subject (light blue). However, there are clear differences in early peaks in this tactile "evoked response" depending on whether the subject detected the finger tap. On detected trials there is a smaller initial peak near 50 ms (labeled M50 in figure 4.11A), followed by a 70-ms peak (M70) that is slightly earlier and larger in magnitude. After 70 ms, the magnitude of the evoked response remains higher than on nondetected trials.

Figure 4.11B shows the average SI evoked responses sorted over prestimulus (1 second) mu power (averaged over both alpha and beta components) rather than over detection. The average of the top half of the sorted trials is labeled high mu (light blue) and the bottom half low mu (dark blue). Differences in the early peaks in the evoked response sorted over mu power are analogous to those sorted over detection (compare the red boxes in figure 4.11A and B), suggesting these correlates of detection are directly impacted by the prestimulus mu rhythm.

The model described in section 4.4.3 can be applied to interpret the cellular-level events that may generate the association between the prestimulus rhythm and the tactile evoked response peaks (figure 4.12B; Jones, Pritchett, et al., 2009). We first describe how the tactile response can be simulated without considering an ongoing rhythm, and then we describe the impact of the prestimulus mu rhythm on this response.

To reproduce the polarity and timing of peaks in the MEG-measured tactile evoked response waveform in the SI model, a sequence of simulated external inputs is generated (spiking inputs that activate postsynaptic receptors). These inputs are given to the SI network through the same "proximal" and "distal" pathways as those described for the spontaneous rhythms (see figure 4.10B), but in this case as a nonrhythmic spiking drive. The timing and location of the simulated "evoked drive" is shown schematically in figure 4.12B and consists of a "proximal" drive at about 25 ms, representative of the initial thalamic input to SI (since it takes about 25 ms for stimulation of the fingertips to reach SI), followed by a "distal" drive at about 70 ms, from cortical area that performs higher level processing, and a late "proximal" drive at about 135 ms, as would arise from a reemergent thalamic input ~110 ms after the initial drive (e.g., one cycle of a stimulus-induced alpha rhythm) Adjusting the magnitude and timing of these drives, which were jittered randomly across model trials (n = 100 trials), this sequence of drive accurately reproduced each of the early peaks (<175 ms) in the SI tactile evoked response for threshold-level stimulus amplitudes (50% detected) (Jones et al., 2007).

Next, we consider how these peaks emerge in the context of studying the impact of the ongoing mu rhythm on the tactile evoked response. High and low prestimulus power states can be simulated in the model

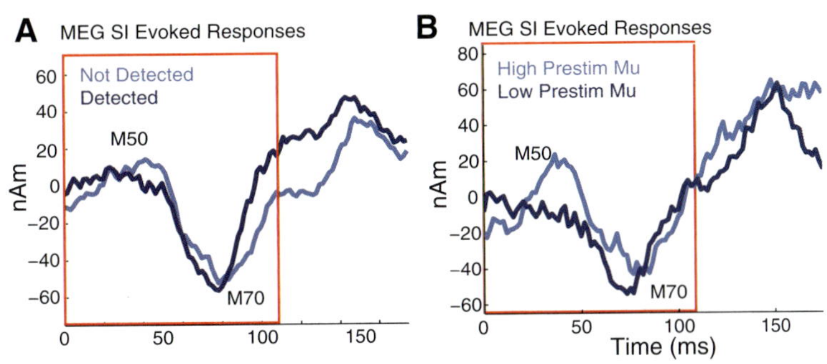

FIGURE 4.11 (A) Magnetoencephalography (MEG) somatosensory cortex (SI) evoked response elicited by a perceptual-threshold-level tap to the finger sorted over trials that were detected (dark blue) and not detected (light blue) (mean n = 7 subjects; 100 trials each per subject). (B) MEG SI evoked response sorted over prestimulus (prestim; 1 second) mu power; dark blue, low power (bottom 50% of sorted trials); light blue, high power (top 50% of sorted trials).

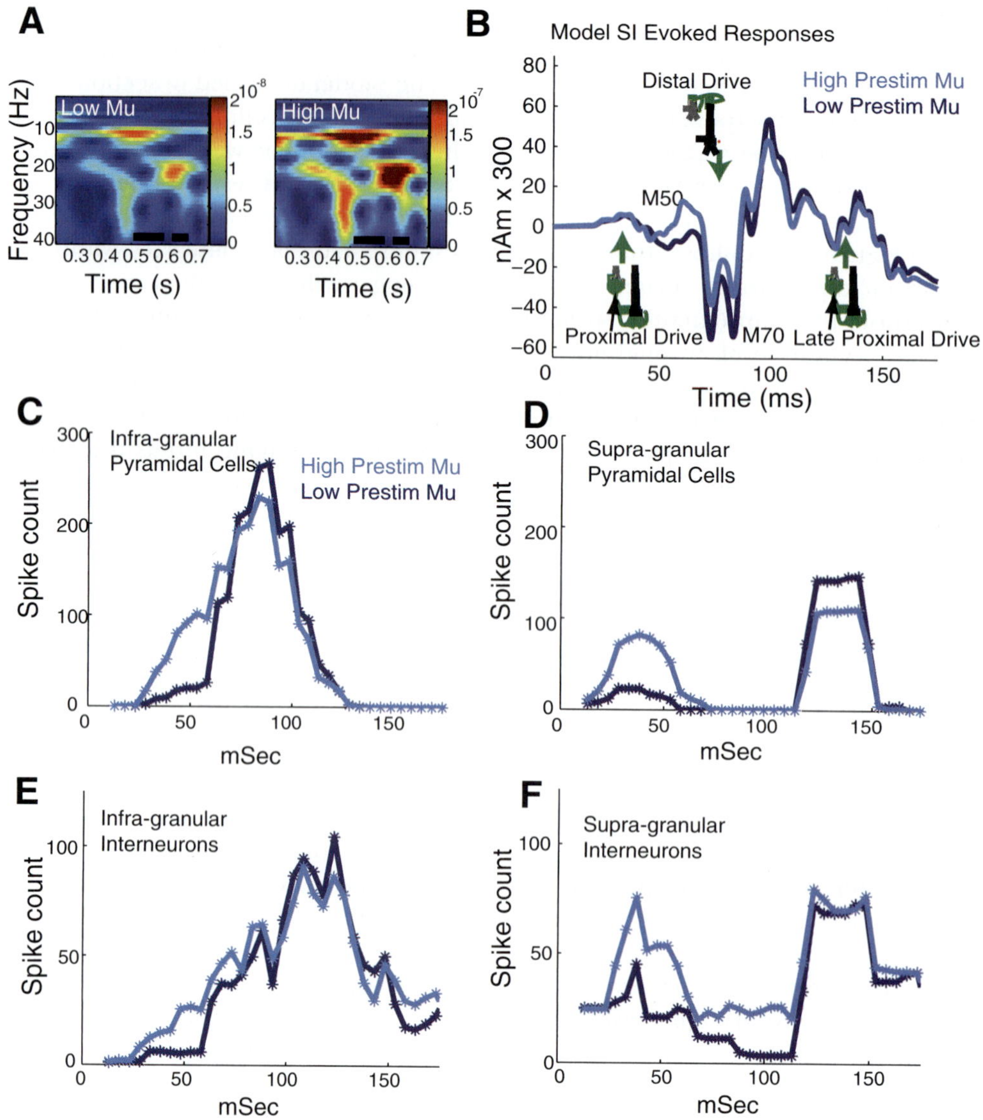

FIGURE 4.12 (A) Spectrograms over which model somatosensory cortex (SI) evoked responses in panel B were calculated. Black bars mark the windows over which the evoked responses were simulated at several phases and averaged (see section 4.4.4). (B) Model SI evoked response simulated during low power (dark blue) and high power (light blue) mu rhythms as depicted in panel A. (C and D) Spiking responses in model pyramidal cells, and interneurons (E and F), in each layer summed over 5-ms bins during simulated evoked responses. Prestim, prestimulus. Adapted from Jones et al. (2009); Jones et al. (2007).

by changing the "efficacy" of the ongoing 10-Hz rhythm external drives described above. Figure 4.12A shows examples where high and low prestimulus alpha and beta components were simulated by adjusting the strength of the maximal postsynaptic excitatory conductances (high mu "proximal" and "distal" conductance were $0.6e^{-4}$ millisiemens (mS); low mu $0.4e^{-4}$ mS), with the mean delay between the ongoing stochastic 10-Hz proximal and distal drives is set to 5 ms, as in figure 4.8D.

Simulating the defined evoked sequence of drive during high or low mu power states allows analysis of the impact of the rhythm on the sensory evoked response. In figure 4.12B, the evoked response sequence was simulated to start at several phases in both alpha and beta components of the ongoing mu rhythm, marked with black lines in figure 4.12A, and the results were averaged. Figure 4.12B shows the model reproduces the net impact of mu power, such that low-power spontaneous rhythms created a smaller initial M50 peak, and a larger subsequent M70 peak, than high-power rhythms. Differences in M70 timing and later activity observed in the MEG data were not reproduced in the model. Thus, the model predicts this later activity may not be impacted directly by the spontaneous rhythm (see Jones, Pritchett, et al., 2007, for further predictions).

What cellular and network mechanisms create the evoked response peaks and their differences with prestimulus power? The initial 25-ms "proximal" drive in the simulated evoked response sequence described above induces spiking in the pyramidal neurons and propagation of current flow back up the apical dendrites to induce the positive M50 peak. The subsequent M70 peak emerges from the "distal" input, which drives current flow down the pyramidal dendrites, creating a negative peak. This drive then causes spiking in the network that again propagates back up the dendrites, creating the M100 peak. The network activity then settles, and a second "proximal" drive at 135 ms creates another wave of spiking, generating the M135 peak (see Jones, Pritchett, et al., 2007). During high prestimulus mu states (figure 4.12C–F, light blue), the network is in a more depolarized ongoing state than in low prestimulus mu states (figure 4.12C–F, dark blue). Thus, when the initial 25-ms "proximal" drive is simulated there is more initial firing in the pyramidal cells, creating a larger M50 peak. Figure 4.12C–F shows the spike rates in each model neuron population. The initial spiking in the pyramidal cells in turn recruits a greater level of poststimulus spiking in the inhibitory cells (compare firing rates of I-cells in figure 4.12E and 4.12F). Hence, when the 70-ms "distal" drive is simulated, the pyramidal neurons are inhibited and the M70 peak is smaller, which decreases subsequent spiking in the infragranular pyramidal neurons.

The deeper infragranular pyramidal cells are known to send signals out from SI to other parts of the brain. A final conjecture based on the model predictions is that decreased infragranular pyramidal spiking beginning at 70 ms during high prestimulus mu corresponds to decreased downstream propagation of information from SI to other areas. This decrease may directly impact detection probability.

4.5 Discussion

This chapter described computational neural modeling used to study the mechanisms and functions of mammalian brain rhythms. The main focus was on rhythms recorded in neocortical networks in awake states, especially mechanisms that generate spiking synchrony at a single frequency (sections 4.2 and 4.3). We have not discussed a large body of computational modeling work examining cross-frequency rhythms and spike-phase relationships, or on neural rhythms in regions including hippocampus, striatum/basal ganglia, thalamus, and cerebellum, several of which are touched upon below. As described throughout this chapter, there is sufficient evidence to suggest rhythms are a signature of many important processes, and their presence can directly impact properties of neural firing (e.g., rate and synchrony). However, much work is yet to be done to precisely define if and what their computational role is in high-level perception and action. Below we briefly discuss and suggest further readings on other computational studies of rhythms.

More on the computational role of gamma: In the study of gamma mechanisms, sections 4.2 and 4.3 focused on generation of synchrony in networks or subpopulations of spiking cells (see also Buzsaki and Wang, 2012). Mechanisms of gamma in macroscopic current source signals as described in section 4.4 have also been studied (Lee and Jones, 2013). Experimental evidence indicates that individual cells may carry information through the relation of their spiking to the dominant rhythm rather than by spiking precisely at a dominant rhythm. Some of the hallmark experimental findings regarding neural rhythms examine phase alignment between spiking of individual cells and the local field, such as the work of Singer and Gray showing visual binding by relative phase of spiking to the local field (Gray, Konig, Engel, and Singer, 1989; Gray and Singer, 1989); see also discussion of theta–gamma coupling in hippocampus below. Recent experimental work has shown that gamma rhythms can be coherent across spatially distinct visual areas despite changes in rhythm frequency, suggesting that gamma coherence enhances communication between areas (Fries, 2009; Jia, Tanabe, and Kohn, 2013; Roberts et al., 2013). Other neural modeling studies have highlighted the possible functional importance of gamma for response gain modulation (Knoblich, Siegle, Pritchett, and Moore, 2010), the flexible routing of information flow (Akam and Kullmann, 2010; Middleton et al., 2008), and as a means of controlling synchrony without large changes in firing rate (Tiesinga and Sejnowski, 2004).

Other models of human thalamocortical rhythms: Section 4.4 described how simplified models can be expanded to include multicompartment cells and input from other brain regions and can be used to study the dynamics of human MEG signals generated by thalamocortical networks. Work by Traub and colleagues makes excellent use of multicompartment modeling techniques for studying human brain rhythms at the microscopic and mesoscopic scale; for example, see Traub et al. (2005). This work replicated many rhythmic neural network behaviors, such as persistent gamma oscillations, sleep spindles, and epileptogenic bursts in a single detailed thalamocortical model. We also refer the reader to an extensive body of work by Destexhe and colleagues, who have studied the thalamocortical interactions regulating sleep rhythms (e.g., Destexhe and Sejnowski, 2001); see also Bonjean et al. (2012) and the book's website.

Some models of macroscopically measured brain rhythms have incorporated a type of modeling known as "neural mass models" that generally assume the MEG/EEG signal is proportional to some superposition of the model neurons' membrane potentials that contributes to the mean state of the network (e.g., David and Friston, 2003; Moran et al., 2011). These models are often embedded in a Bayesian statistical framework that enables detailed studies of complex behavioral paradigms and multibrain area interactions that regulate rhythmicity.

Coupled theta–gamma rhythms in hippocampus: The hippocampus shows robust rhythmic activity in the gamma and theta bands, with evidence that gamma often emerges at specific phases of theta activity. This activity has been proposed to be functionally significant for coding the memory and retrieval of an animal's place in physical space, via anatomically distributed gamma "cell assemblies" expressing transient synchrony gated by a theta rhythm (Harris, Csicsvari, Hirase, Dragoi, and Buzsaki, 2003; see also O'Keefe and Recce, 1993).

In the hippocampus, a specific class of inhibitory cells, namely, the oriens lacunosum-moleculare (OLM) cells, produce longer lasting inhibition and are thought to play an important role in generating the theta rhythmicity in that brain region, while fast-spiking interneurons, like those considered in the neocortical modeling in this chapter, generate gamma rhythmicity via PING mechanisms (see figure 4.2A). Together, these two interneuron populations can generate interlocking gamma and theta patterns observed in hippocampus in spatial navigation and memory tasks (Gloveli et al., 2005; Harris et al., 2002). The coordination of hippocampal gamma cell assembly via these mechanisms has been investigated with computational modeling (Tort, Rotstein, Dugladze, Gloveli, and Kopell, 2007).

Both computational and experimental work indicates this type of cross-frequency interaction may be generally important for organizing information about multiple items, where different phase relationships between the gamma and theta cycles encode separate items, or sequences of events (Lisman and Jensen, 2013). Additional modeling work indicates that interneurons in cortex similar to OLM cells, namely, low-threshold spiking cells, may similarly be important for amplifying or stabilizing slower rhythms in cortex (Vierling-Claassen et al., 2010).

Rhythms in disease: An important application not discussed in this chapter is the use of modeling to understand how rhythms become disrupted in disease states (see chapter 23, "Brain Diseases"). By studying the mechanisms of disrupted rhythmicity, modeling can be a powerful tool to target novel therapeutic strategies. As one example, computational models of disrupted rhythmicity in the striatum/basal ganglia network in Parkinson's disease may be used to guide deep-brain-stimulation paradigms that help alleviate symptoms (McCarthy et al., 2011; Moran et al., 2011). Other studies have targeted biophysically interpretable sources of rhythm disruption that may help guide pharmacological interventions. For example, Vierling-Claassen and colleagues used modeling to study how altered inhibition in schizophrenia may change cortical rhythmicity (Lewis, Hashimoto, and Volk, 2005), inhibiting gamma rhythmicity and possibly enhancing slower beta rhythms (Vierling-Claassen, Siekmeier, Stufflebeam, and Kopell, 2008). Computational modeling of neural rhythms is used to study autism, epilepsy, Alzheimer's disease, Parkinson's disease, and other neurological conditions.

By necessity, this chapter focused somewhat narrowly on gamma, beta, and alpha rhythms centered in cortex in the awake states. However the array of concepts considered, including modeling synaptic interactions, intracellular currents, the robustness and interaction of rhythms in heterogeneous networks, consideration of both single- and multicompartment neural models, and modeling neural rhythmic data at a broad range of scales, from intracellular in vitro recordings to macroscopic human MEG signals, are relevant to applications throughout the brain.

REFERENCES

Akam, T., & Kullmann, D. M. (2010). Oscillations and filtering networks support flexible routing of information. *Neuron, 67*(2), 308–320.

Barrès, V., Simons, A. 3rd, & Arbib, M. (2013). Synthetic event-related potentials: a computational bridge between neurolinguistic models and experiments. *Neural Netw., 37,* 66-92.

Bonjean, M., Baker, T., Bazhenov, M., Cash, S., Halgren, E., & Sejnowski, T. (2012). Interactions between core and matrix thalamocortical projections in human sleep spindle synchronization. *Journal of Neuroscience, 32*(15), 5250–5263.

Borgers, C., Epstein, S., & Kopell, N. J. (2005). Background gamma rhythmicity and attention in cortical local circuits: A computational study. *Proceedings of the National Academy of Sciences of the United States of America, 102*(19), 7002–7007.

Borgers, C., Epstein, S., & Kopell, N. J. (2008). Gamma oscillations mediate stimulus competition and attentional selection in a cortical network model. *Proceedings of the*

National Academy of Sciences of the United States of America, 105(46), 18023–18028.

Borgers, C., & Kopell, N. (2003). Synchronization in networks of excitatory and inhibitory neurons with sparse, random connectivity. *Neural Computation, 15*(3), 509–538.

Borgers, C., & Kopell, N. (2005). Effects of noisy drive on rhythms in networks of excitatory and inhibitory neurons. *Neural Computation, 17*(3), 557–608.

Bush, P. C., & Sejnowski, T. J. (1993). Reduced compartmental models of neocortical pyramidal cells. *Journal of Neuroscience Methods, 46*(2), 159–166.

Buzsaki, G., & Draguhn, A. (2004). Neuronal oscillations in cortical networks. *Science, 304*(5679), 1926–1929.

Buzsaki, G., & Wang, X. J. (2012). Mechanisms of gamma oscillations. *Annual Review of Neuroscience, 35*, 203–225.

Cardin, J. A., Carlen, M., Meletis, K., Knoblich, U., Zhang, F., & Deisseroth, K., et al. (2009). Driving fast-spiking cells induces gamma rhythm and controls sensory responses. *Nature, 459*, 663–667.

Carnevale, N. T., & Hines, M. L. (2006). *The NEURON Book*. Cambridge: Cambridge University Press.

David, O., & Friston, K. J. (2003). A neural mass model for MEG/EEG: Coupling and neuronal dynamics. *NeuroImage, 20*(3), 1743–1755.

Destexhe, A., & Sejnowski, T. J. (2001). *Thalamocortical Assemblies: How Ion Channels, Single Neurons, and Large-Scale Networks Organize Sleep Oscillations*. Oxford: Oxford University Press.

Engel, A. K., Fries, P., & Singer, W. (2001). Dynamic predictions: Oscillations and synchrony in top-down processing. *Nature Reviews. Neuroscience, 2*(10), 704–716.

Fries, P. (2009). Neuronal gamma-band synchronization as a fundamental process in cortical computation. *Annual Review of Neuroscience, 32*, 209–224.

Fries, P., Womelsdorf, T., Oostenveld, R., & Desimone, R. (2008). The effects of visual stimulation and selective visual attention on rhythmic neuronal synchronization in macaque area V4. *Journal of Neuroscience, 28*(18), 4823–4835.

Gloveli, T., Dugladze, T., Rotstein, H. G., Traub, R. D., Monyer, H., Heinemann, U., et al. (2005). Orthogonal arrangement of rhythm-generating microcircuits in the hippocampus. *Proceedings of the National Academy of Sciences of the United States of America, 102*(37), 13295–13300.

Golomb, D., & Hansel, D. (2000). The number of synaptic inputs and the synchrony of large, sparse neuronal networks. *Neural Computation, 12*(5), 1095–1139.

Gray, C. M., Konig, P., Engel, A. K., & Singer, W. (1989). Oscillatory responses in cat visual cortex exhibit intercolumnar synchronization which reflects global stimulus properties. *Nature, 338*(6213), 334–337.

Gray, C. M., & Singer, W. (1989). Stimulus-specific neuronal oscillations in orientation columns of cat visual cortex. *Proceedings of the National Academy of Sciences of the United States of America, 86*(5), 1698–1702.

Hamalainen, M., Hari, R., Ilmoniemi, R. J., Knuutila, J., & Lounasmaa, O. V. (1993). Magnetoencephalography—Theory, instrumentation, and applications to noninvasive studies of the working human brain. *Reviews of Modern Physics, 65*, 413–497.

Harris, K. D., Csicsvari, J., Hirase, H., Dragoi, G., & Buzsaki, G. (2003). Organization of cell assemblies in the hippocampus. *Nature, 424*(6948), 552–556.

Harris, K. D., Henze, D. A., Hirase, H., Leinekugel, X., Dragoi, G., Czurko, A., et al. (2002). Spike train dynamics predicts theta-related phase precession in hippocampal pyramidal cells. *Nature, 417*(6890), 738–741.

Hasselmo, M. E. (1995). Neuromodulation and cortical function: Modeling the physiological basis of behavior. *Behavioural Brain Research, 67*(1), 1–27.

Holt, A. B., & Netoff, T. I. (2013). Computational modeling of epilepsy for an experimental neurologist. *Experimental Neurology, 244*, 75–86.

Hughes, S. W., & Crunelli, V. (2005). Thalamic mechanisms of EEG alpha rhythms and their pathological implications. *Neuroscientist, 11*(4), 357–372.

Ikeda, H., Wang, Y., & Okada, Y. C. (2005). Origins of the somatic N20 and high-frequency oscillations evoked by trigeminal stimulation in the piglets. *Clinical Neurophysiology, 116*, 827–841.

Jia, X., Tanabe, S., & Kohn, A. (2013). Gamma and the coordination of spiking activity in early visual cortex. *Neuron, 77*(4), 762–774.

Jones, E. G. (1998). Viewpoint: The core and matrix of thalamic organization. *Neuroscience, 85*(2), 331–345.

Jones, S. (2012). Biophysically principled computational neural network modeling of magneto-electro-encephalography measured human brain sscillations. In T. Fellin & M. Halassa (Eds.), *Neuronal Network Analysis: Concepts and Experimental Approaches* (pp. 459–485). New York: Humana Press.

Jones, S. R., Kerr, C. E., Wan, Q., Pritchett, D. L., Hämäläinen, M., & Moore, C. (2010). Cued spatial attention drives functionally-relevant modulation of the mu rhythm in primary somatosensory cortex. *Journal of Neuroscience, 30*(41), 13760–13765.

Jones, S. R., Pinto, D. J., Kaper, T. J., & Kopell, N. (2000). Alpha-frequency rhythms desynchronize over long cortical distances: A modeling study. *Journal of Computational Neuroscience, 9*(3), 271–291.

Jones, S. R., Pritchett, D. L., Sikora, M. A., Stufflebeam, S. M., Hamalainen, M., & Moore, C. I. (2009). Quantitative analysis and biophysically realistic neural modeling of the MEG mu rhythm: Rhythmogenesis and modulation of sensory-evoked responses. *Journal of Neurophysiology, 102*(6), 3554–3572.

Jones, S. R., Pritchett, D. L., Stufflebeam, S. M., Hamalainen, M., & Moore, C. I. (2007). Neural correlates of tactile detection: A combined magnetoencephalography and biophysically based computational modeling study. *Journal of Neuroscience, 27*(40), 10751–10764.

Kerr, C. E., Jones, S. R., Wan, Q., Pritchett, D. L., Wasserman, R. H., Wexler, A., et al. (2011). Effects of mindfulness meditation training on anticipatory alpha modulation in primary somatosensory cortex. *Brain Research Bulletin, 85*(3–4), 96–103.

Kilpatrick, Z. P., & Ermentrout, B. (2011). Sparse gamma rhythms arising through clustering in adapting neuronal networks. *PLoS Computational Biology, 7*(11), e1002281.

Klimesch, W., Sauseng, P., & Hanslmayr, S. (2007). EEG alpha oscillations: The inhibition-timing hypothesis. *Brain Research. Brain Research Reviews, 53*(1), 63–88.

Knoblich, U., Siegle, J. H., Pritchett, D. L., & Moore, C. I. (2010). What do we gain from gamma? Local dynamic gain modulation drives enhanced efficacy and efficiency of signal transmission. *Frontiers in Human Neuroscience, 4*, 185.

Kopell, N., & Ermentrout, G. B. (1986). Subcellular oscillations and bursting. *Mathematical Biosciences, 78*(2), 265–291.

Kopell, N., Ermentrout, G. B., Whittington, M. A., & Traub, R. D. (2000). Gamma rhythms and beta rhythms have different synchronization properties. *Proceedings of the National Academy of Sciences of the United States of America, 97*(4), 1867–1872.

Lee, S., & Jones, S. R. (2013). Distinguishing mechanisms of gamma frequency oscillations in human current source signals using a computational model of a laminar neocortical network. *Frontiers in Human Neuroscience, 7*, 869.

Lewis, D. A., Hashimoto, T., & Volk, D. W. (2005). Cortical inhibitory neurons and schizophrenia. *Nature Reviews. Neuroscience, 6*(4), 312–324.

Lisman, J. E., & Jensen, O. (2013). The theta-gamma neural code. *Neuron, 77*(6), 1002–1016.

Marder, E. (2011). Variability, compensation, and modulation in neurons and circuits. *Proceedings of the National Academy of Sciences of the United States of America, 108*(Suppl 3), 15542–15548.

Marder, E., Tobin, A. E., & Grashow, R. (2007). How tightly tuned are network parameters? Insight from computational and experimental studies in small rhythmic motor networks. *Progress in Brain Research, 165*, 193–200.

McCarthy, M. M., Brown, E. N., & Kopell, N. (2008). Potential network mechanisms mediating electroencephalographic beta rhythm changes during propofol-induced paradoxical excitation. *Journal of Neuroscience, 28*(50), 13488–13504.

McCarthy, M. M., Moore-Kochlacs, C., Gu, X., Boyden, E. S., Han, X., & Kopell, N. (2011). Striatal origin of the pathologic beta oscillations in Parkinson's disease. *Proceedings of the National Academy of Sciences of the United States of America, 108*(28), 11620–11625.

Middleton, S., Jalics, J., Kispersky, T., Lebeau, F. E., Roopun, A. K., Kopell, N. J., et al. (2008). NMDA receptor-dependent switching between different gamma rhythm-generating microcircuits in entorhinal cortex. *Proceedings of the National Academy of Sciences of the United States of America, 105*(47), 18572–18577.

Moran, R. J., Mallet, N., Litvak, V., Dolan, R. J., Magill, P. J., Friston, K. J., et al. (2011). Alterations in brain connectivity underlying beta oscillations in parkinsonism. *PLoS Computational Biology*, 7(8), e1002124.

Murakami, S., Zhang, T., Hirose, A., & Okada, Y. C. (2002). Physiological origins of evoked magnetic fields and extracellular field potentials produced by guinea-pig CA3 hippocampal slices. *Journal of Physiology, 544*(Pt 1), 237–251.

O'Keefe, J., & Recce, M. L. (1993). Phase relationship between hippocampal place units and the EEG theta rhythm. *Hippocampus, 3*(3), 317–330.

Olufsen, M. S., Whittington, M. A., Camperi, M., & Kopell, N. (2003). New roles for the gamma rhythm: Population tuning and preprocessing for the Beta rhythm. *Journal of Computational Neuroscience, 14*(1), 33–54.

Pervouchine, D. D., Netoff, T. I., Rotstein, H. G., White, J. A., Cunningham, M. O., Whittington, M. A., et al. (2006). Low-dimensional maps encoding dynamics in entorhinal cortex and hippocampus. *Neural Computation, 18*(11), 2617–2650.

Pinto, D. J., Jones, S. R., Kaper, T. J., & Kopell, N. (2003). Analysis of state-dependent transitions in frequency and long-distance coordination in a model oscillatory cortical circuit. *Journal of Computational Neuroscience, 15*(2), 283–298.

Rall, W. (1964). Theoretical significance of dendritic tree for input-output relation. In R. F. Reiss (Ed.), *Neural Theory and Modeling* (pp. 73–97). Stanford, CA: Stanford University Press.

Reynolds, J. H., Chelazzi, L., & Desimone, R. (1999). Competitive mechanisms subserve attention in macaque areas V2 and V4. *Journal of Neuroscience, 19*(5), 1736–1753.

Roberts, M. J., Lowet, E., Brunet, N. M., Ter Wal, M., Tiesinga, P., Fries, P., et al. (2013). Robust gamma coherence between macaque V1 and V2 by dynamic frequency matching. *Neuron, 78*(3), 523–536.

Rockland, K. S., & Pandya, D. N. (1979). Laminar origins and terminations of cortical connections of the occipital lobe in the rhesus monkey. *Brain Research, 179*(1), 3–20.

Rodriguez, E., George, N., Lachaux, J. P., Martinerie, J., Renault, B., & Varela, F. J. (1999). Perception's shadow: Long-distance synchronization of human brain activity. *Nature, 397*(6718), 430–433.

Roopun, A. K., Middleton, S. J., Cunningham, M. O., LeBeau, F. E., Bibbig, A., Whittington, M. A., et al. (2006). A beta2-frequency (20–30 Hz) oscillation in nonsynaptic networks of somatosensory cortex. *Proceedings of the National Academy of Sciences of the United States of America, 103*(42), 15646–15650.

Schnitzler, A., & Gross, J. (2005). Normal and pathological oscillatory communication in the brain. *Nature Reviews Neuroscience, 6*(4), 285–296.

Schroeder, C. E., Wilson, D. A., Radman, T., Scharfman, H., & Lakatos, P. (2010). Dynamics of active sensing and perceptual selection. *Current Opinion in Neurobiology, 20*(2), 172–176.

Silva, L. R., Amitai, Y., & Connors, B. W. (1991). Intrinsic oscillations of neocortex generated by layer 5 pyramidal neurons. *Science, 251*(4992), 432–435.

Tiesinga, P. H., & Sejnowski, T. J. (2004). Rapid temporal modulation of synchrony by competition in cortical interneuron networks. *Neural Computation, 16*(2), 251–275.

Tort, A. B., Rotstein, H. G., Dugladze, T., Gloveli, T., & Kopell, N. J. (2007). On the formation of gamma-coherent cell assemblies by oriens lacunosum-moleculare interneurons in the hippocampus. *Proceedings of the National Academy of Sciences of the United States of America, 104*(33), 13490–13495.

Traub, R. D., Contreras, D., Cunningham, M. O., Murray, H., LeBeau, F. E., Roopun, A., et al. (2005). Single-column thalamocortical network model exhibiting gamma oscillations, sleep spindles, and epileptogenic bursts. *Journal of Neurophysiology, 93*(4), 2194–2232.

Vierling-Claassen, D., Cardin, J., Moore, C., & Jones, S. (2010). Computational modeling of distinct neocortical oscillations driven by cell-type selective optogenetic drive: Separable resonant circuits controlled by low-threshold spiking and fast-spiking interneurons. *Frontiers in Neuroscience*, (4), 198.

Vierling-Claassen, D., Siekmeier, P., Stufflebeam, S., & Kopell, N. J. (2008). Modeling GABA alterations in schizophrenia: A link between impaired inhibition and altered gamma and beta range auditory entrainment. *Journal of Neurophysiology, 99*, 2656–2671.

Vijayan, S., & Kopell, N. J. (2012). Thalamic model of awake alpha oscillations and implications for stimulus processing. *Proceedings of the National Academy of Sciences of the United States of America, 109*(45), 18553–18558.

Wan, Q., Kerr, C., Pritchett, D., Hamalainen, M., Moore, C., & Jones, S. (2011). Dynamics of dynamics within a single data acquisition session: Variation in neocortical alpha oscillations in human MEG. *PLoS One, 6*(9), e24941.

Wang, X. J., & Buzsaki, G. (1996). Gamma oscillation by synaptic inhibition in a hippocampal interneuronal network model. *Journal of Neuroscience, 16*(20), 6402–6413.

Whittington, M. A., Traub, R. D., Kopell, N., Ermentrout, B., & Buhl, E. H. (2000). Inhibition-based rhythms: Experimental and mathematical observations on network dynamics. *International Journal of Psychophysiology, 38*(3), 315–336.

Worden, M. S., Foxe, J. J., Wang, N., & Simpson, G. V. (2000). Anticipatory biasing of visuospatial attention indexed by retinotopically specific alpha-band electroencephalography increases over occipital cortex. *Journal of Neuroscience, 20*(6), RC63.

Ziegler, D. A., Pritchett, D. L., Hosseini-Varnamkhasti, P., Corkin, S., Hamalainen, M., Moore, C. I., et al. (2010). Transformations in oscillatory activity and evoked responses in primary somatosensory cortex in middle age: A combined computational neural modeling and MEG study. *NeuroImage, 52*(3), 897–912.

5 Linking Models with Empirical Data: The Brain Operation Database

JAMES J. BONAIUTO AND MICHAEL A. ARBIB

In this book, we encounter a number of very general models of brain function (winner-take-all, Hebbian learning, and even more abstract models of, e.g., bifurcation in nonlinear systems) which we may consider to represent general principles exemplified in many different neural systems. However, we are also concerned (in the name of systems neuroscience and cognitive neuroscience) with models of how the neurons in one brain region or multiple brain regions work together to function in particular ways—what we might call neuroanatomically grounded system models. In what follows, we will use the qualification "neuroanatomically grounded system" when we need to stress the distinction from other classes of model, but we will drop it when it is clear from the context. We will introduce a pair of neuroanatomically grounded system models as exemplary of such models, namely, the mirror neuron system (MNS) and MNS2 models of the mirror neuron system, modeling how the brain of one monkey is able to recognize the manual actions of another monkey or a human (see chapter 20, "Reach and Grasp: Control, Development, and Recognition," for more on the relation of these models to other neurobiological modeling). However, this is not an end in itself for this chapter. Rather, we want to help students learn to read journal articles presenting a neuroanatomically grounded system model with close attention to the way in which it does (or does not) register with empirical data: assessing which empirical data were used to constrain the design of the model and which were used to test the performance of the model, and the results of those tests. In particular, we will introduce the Brain Operation Database, BODB (http://bodb.usc.edu/bodb/), which supports the documentation of the structure and performance of models in relation to empirical data. The task of this chapter is to introduce the BODB methodology and use examples (two brain models based on macaque neurophysiology, one human brain imaging study) to illuminate a tour of key features of BODB. As you read the rest of the chapter, it will often prove useful to log in to BODB and try out the various options. Box 5.1 offers a minimal guide to using BODB.

Neuroinformatics, the bringing to bear of informatics on neuroscience, has often focused on the development of databases for empirical data gathered from neurobiological systems (including humans) rather than on the results of modeling. A given neuroinformatics resource in this sense is usually restricted to data in a well-focused domain, such as gene expression or neuroanatomy or brain imaging. Our approach is one of the few to also emphasize the construction of computational models of the dynamic processes studied in systems and cognitive neuroscience.

To further explain what the Brain Operation Database *is*, we first clarify what it *is not* by contrasting it with ModelDB (Migliore et al., 2003; http://senselab.med.yale.edu/modeldb/), a curated database of published models for computational neuroscience, providing access to the code for such models and instructions to replicate published simulation results. By contrast, BODB *is not* a repository of code for models though it does provide a uniform resource locator (URL) field in each model entry for links to working code. Rather (see figure 5.1), BODB is organized around four conceptual entities: *Models, Brain Operating Principles* (*BOPs*), *Summaries of Empirical Data* (*SEDs*), and *Summaries of Simulation Results* (*SSRs*). Many of the SEDs in the current version of BODB are simply entered in natural language. However, the formal structuring of SEDs for various classes of experiments is an ongoing concern as we shall see below. Currently, BODB offers a structured format for entry of brain imaging data using an activation table in Talairach (Talairach and Tournoux, 1988) or Montreal Neurological Institute (MNI; Evans, Collins, and Milner, 1992) coordinates that can then be viewed against a representation of the human brain. We also provide connectivity SEDs derived through federation with CoCoMac, the macaque macroconnectivity database initiated by the late Rolf Kötter (http://cocomac.g-node.org). A full exposition of BODB and the underlying methodology is provided by Arbib et al. (2014) as part

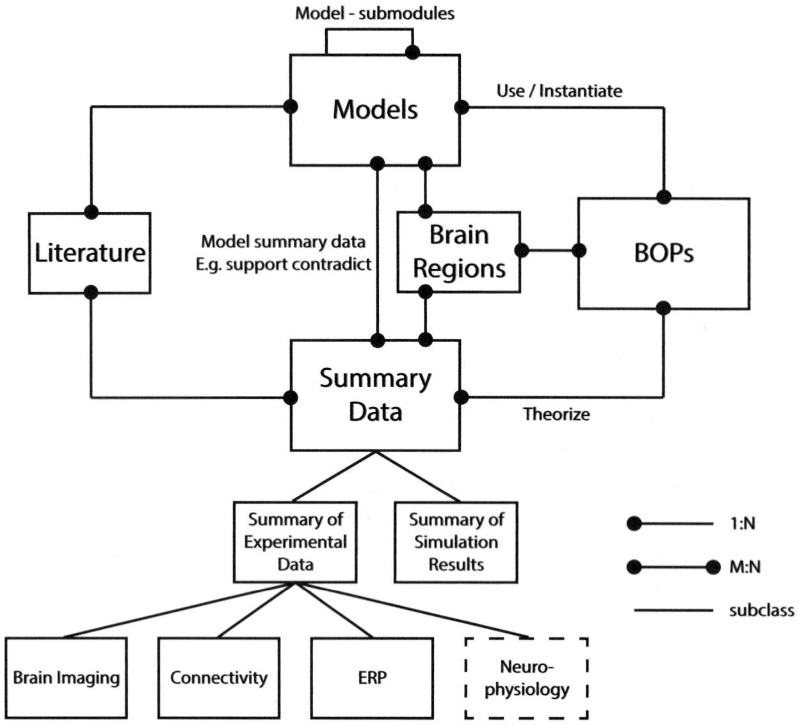

FIGURE 5.1 An overview of the data structures of the Brain Operation Database. The structured format for model descriptions supports sharing and meta-analysis of models. Model entries link to brain regions they simulate, providing a structural ontology. They also link to the Brain Operating Principles (BOPs) that they use or instantiate, providing a functional ontology for model descriptions. The link from model entries to summary data allows simulation results from the model to be compared to summaries of experimental data that they explain or contradict. Specialized data formats have been developed for brain imaging, connectivity, and event-related potential (ERP) data.

of a special issue of the journal *Neuroinformatics* (2014, Vol. 12, Issue 1) on "Action and Language Mechanisms in the Brain: Data, Models and Neuroinformatics."

Whereas other efforts seek to standardize formal descriptions of certain classes of neural network models independently of programming language and simulator (Nordlie, Gewaltig, and Plesser, 2009; e.g., Gleeson et al., 2010), BODB encourages modelers to make explicit the assumptions, explanations, and predictions of their models in a common framework. This allows comparison with other models not only in the usual terms of structures (brain regions, circuits, neurons, etc.) but also in terms of functionality (brain operating principles) and related experimental data. Searches can be conducted on all the above entities. A particularly valuable feature is to use one model to anchor a search for other models which address related empirical data. The individual modeler can take advantage of the BODB framework to review existing models that explain a core data set, identify data sets not addressed by current modeling efforts, search for related models and brain operating principles, and use them to design a model to explain the new data. Having retrieved models related to a current modeling effort, BODB can set tailored benchmarks for developing a new model—one must demonstrate that one's model explains more data than the benchmark models and/or explains them better.

5.1 *The Key Components of BODB*

Our approach to a modeling-centered neuroinformatics involves four key components:

Brain Models: A brain model should be placed clearly within the context of empirical data, including those which support the model's design and those which test the simulation results of the model (based on data-relevant simulation protocols). Ideally, the model should also yield testable predictions (or, at least, stimulate new ideas about mechanisms) that may be developed into new hypotheses for further experiments. A brain model may be either an explicit *computational model*, together with computer code and simulations, of how a particular brain system operates at, for example, the level of dynamic interactions in biologically realistic brain regions and neural networks, or a *conceptual* model, in which case a "simulation" result may simply be an inference from the verbal formulation of the model. The former tries to emulate brain function based on a set of simulation parameters so that simulation results can be compared with a certain collection of experiments or used to predict results of further studies. The latter is simply a sketch of a "computationally possible" model that may not have been implemented but nonetheless guides experiments. It may serve as the basis for later development of computational models but is itself not yet specific enough for implementation.

Brain Operating Principles (BOPs): A BOP is a general principle of brain operation (like the winner-take-all and Hebbian learning principles mentioned above) that is not limited to one experimental protocol or simulation but can be found in multiple regions or in multiple tasks. Here we use *principle* in the sense (Definition II.3.c of the *Oxford English Dictionary*) of "a general or inclusive statement about an aspect of the natural world [in our case, brain and behavior] that has numerous special applications or is applicable in a wide variety of cases." Thus, a BOP may be relevant to the function of multiple brain regions. For example, the "Winner-Take-All" BOP holds when a neural network (whether in an actual brain or simulated) converts a pattern of activation of its inputs to a peak of excitation on its outputs that corresponds to the locus of maximal excitation on the inputs. This has been postulated to apply to diverse brain regions (including frog tectum and monkey superior colliculus). Conversely, the role of a brain region, and thus the relevant BOPs for modeling it, may vary between different tasks.

Summaries of Experimental Data (SEDs): An SED is an encapsulation of empirical findings in a form that can be used to support the design of a model or as a target to be explained by the model. In other words we advocate presentation of empirical data in the form of concise summaries of relevance to the model, rather than unfiltered experimental results. However, since summarization may sometimes obscure crucial details, we also advocate that, where possible, links be provided from SEDs back to resources that store primary data sets so that reanalysis can be supported where necessary.

Summaries of Simulation Result (SSRs): An SSR summarizes key findings from the model. Comparisons of SSRs with SEDs can be used to test the model. Other SSRs may offer predictions to be tested by new experiments.

An approach guided by these ideas allows researchers to place a model of brain function within the proper context of (1) competing models of similar systems, and (2) the data which guide production and testing of these models.

In many cases, a brain model will explicitly represent the interactions of multiple brain regions, with a distinct module for each region of the brain included in the model. Such a brain model can be characterized by a *structural ontology*, namely, characterizing the model by the neuroanatomical structures (brain regions or finer structures) to which it corresponds. However, we also characterize models by a *functional ontology*, further characterizing the model by the Brain Operating Principles, BOPs, which it exemplifies. Thus, one may search for alternative models of particular brain regions, or different neural instantiations of given operating principles.

As noted earlier, we will use the MNS and MNS2 models of mirror neurons to illustrate a number of features of BODB. Although the exposition of the models must wait till section 5.2, we here list the BOPs used in these two models to give the reader a better sense of what is involved in a Brain Operating Principle.

Here are the BOPs related to both the MNS and MNS2 models:

- *Sensorimotor Coupling*: The brain has representations that bridge between perceptual contents and action plans. Such common codes integrate action and perception. Perceptual stimuli are interpreted in part based on the motor representations which are commonly associated with the corresponding sensory experience.
- *Supervised Learning*: Training based on a teaching signal indicating the error on the last performance.
- *Corollary Discharge*: This principle involves the sending of a copy (known as the efferent copy) of a motor output to another brain region or neural circuit.

And here are the BOPs related to the MNS2 model, but not the MNS model.

- *Dynamic Remapping*: Dynamic remapping refers to the updating of a neural map to reflect the result of an executed action.
- *Hebbian Learning*: Hebbian learning is a type of learning for neurons in which a synapse is strengthened if its activity coincides with the firing of the postsynaptic neuron.
- *Sensor Fusion*: Sensor fusion typically refers to the integration of data from different types of sensors, as in combining visual, auditory, and tactile cues, to more accurately extract relevant information from the environment.
- *Temporal Pattern Processing*: Temporal pattern processing is important for various intelligent behaviors,

including hearing, vision, speech, music, and motor control. Because we live in an ever-changing environment, an intelligent system must encode patterns over time, recognizing and generating temporal patterns.

- *Working Memory*: Working memory consists of contextual information that is stored for the duration of the execution span of the current task.

To see how each BOP is related to one of the models, open the model entry in BODB (see box 5.1 for details) and then scroll down to show "Related BOPs." To get a listing of all BOPs in BODB, simply go to the BODB Home Page, click on the "Search" tab, click on the "BOPs" tab, and then (without entering any keywords) click on "Search."

5.2 *The Importance of Protocols*

To proceed further, we may note that many different studies may be conducted on a given brain region or regions, seeing quite different effects as they are active in a range of different tasks. Indeed, modern brain imaging techniques only work because there can be differential involvement of brain regions in different tasks. *Experimental protocols* specify the particular conditions under which a given data set (whether brain imaging, or at the neural, subneural, or other levels) is gathered. The experimentalist observes a system B in the external world, and the protocol defines a class of experiments based on the preparation used, the external stimuli applied, experimental manipulations (Was the animal trained beforehand; were drugs applied? If so, what are the specifics?), and the measurements and observations made (whether through behavior, brain imaging, neurophysiology, chemical assay, etc.).

A brain model, then, might be specifically set up in an attempt to explain how certain brain regions work together to achieve a specific task. But if, more generally, we want to develop a model of those brain regions to explain the results of diverse experiments, then each experimental protocol must be translated into a *simulation protocol* for adding "preparation," "stimulation," and "observation" to a model of the brain regions, as shown in figure 5.2. Since a particular brain system S can be studied with many different protocols, we want to end up with a single model B of S which "works" with a range of different protocols. For each such protocol P, we need an appropriate *stimulus interface* K mapping each P-stimulus to an input to the model and a *response interface* Z decoding each output of B to an output that can be compared with actual measurements on the neurobiological system S.

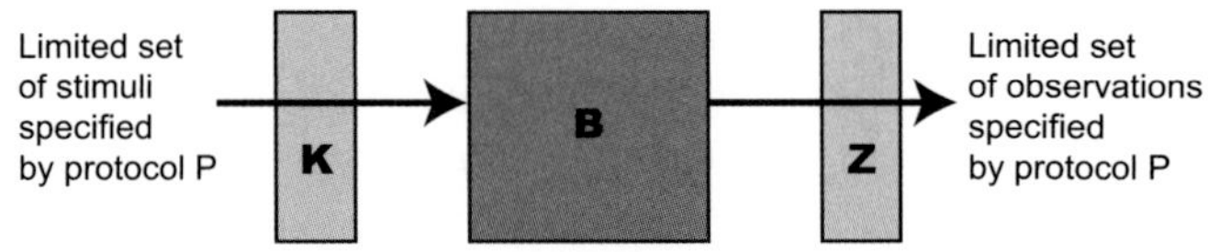

FIGURE 5.2 A model of a brain system as studied using an experimental protocol must be augmented by computational models of the way in which inputs are applied according to that protocol and the way in which the outputs of the model are "instrumented" to yield a representation of the results obtained with that protocol. (See text for notations.)

On this view, a *Summary of Empirical Data* (SED) should encapsulate experimental results along with the associated protocol while a *Summary of Simulation Results* (SSR) summarizes key findings with the model (in each case employing a specific simulation protocol) in a form which supports comparison with SEDs as a means of validating the model. Note the contrast: An SED is independent of any model and may be part of the design and testing of one model, several models, or no model at all. By contrast, an SSR is model specific. In turn, SSRs may link back to SEDs and either *explain* or *contradict* (or be irrelevant to) certain empirical results.

For each model, BODB stores answers to questions like these:

- What is the structure of the model?
- What task does it perform?
- What experimental data were used in its design?
- What are the results of simulations using the model?
- What experimental data does it explain or contradict?
- What does it predict?

This allows researchers to place a model of brain function more easily within the proper context of (1) competing models of similar systems, and (2) the data which guide production of these models. In addition, the BOPs which populate the functional ontology of the database provide a theoretical footing to analyze how these abstract principles are realized in brain tissue. Specifically, BODB is designed to support data entry for brain models and their associated SSRs as well as BOPs and SEDs. Data entry can be accomplished by people who wish to include material concerning published data and models, or it can be performed by researchers as they develop new models. BODB supports searches on these entities, tools for comparison of models, and, as we shall see in due course, much more. In particular, BODB also supports Collaboratory Workspaces where experimentalists and modelers may come together in the testing and designing of new models and the results of testing predictions.

Box 5.1 A Fast Primer for Getting Started on BODB

Readers are encouraged to check out features of the Brain Operation Database, BODB (http://bodb.usc.edu/bodb/), as they read this chapter. The task of this chapter is to introduce the BODB methodology and use examples (two brain models based on macaque neurophysiology, one human brain imaging study) to illuminate a tour of key features of BODB. Much further detail can be found by invoking Help on the various pages of BODB, as well as by consulting the Manual and Tutorial at http://bodb.usc.edu/media/BODB_Manual.pdf.

To get started, go to http://bodb.usc.edu/bodb/ and click "Login" or (if this is the first visit) "Register."

The main page contains a toolbar containing the options "Insert," "Search," "Visualization," "Drafts," "Favorites," "Workspace," and "About." Insert allows users to insert new Literature, Brain Operating Principles (BOP), Model, or Summary of Experimental Data (SED) records. Search allows the user to search all BODB records. Visualization opens Brainsurfer, a three-dimensional visualization tool for human and brain atlases where the user can view coordinates associated with Brain Imaging SEDs. Drafts allows users to view and edit current records that they are working on. Favorites allows users to users to view records that they have bookmarked. Workspace allows users to work on and administer their Collaboratory Workspaces (CoWos). About leads to the page describing the history of BODB and its implementers.

To find the model entries for the MNS and MNS2 models, click on "Search" in the toolbar. On the Search page, click on the "Models" tab to open the options to search for model records. Enter "MNS" in the "Any field" text box to search for any models containing the keywords "MNS." Then Click "Show" next to "Advanced Options" to show the other fields that models can be searched by. To find the MNS model, enter "Oztop" in the Author field or "Bonaiuto" to find the MNS2 model. Click the "Search" button below the search options to perform the search.

To enter a SED, click on "Insert" in the toolbar and under the section "Summary of Experimental Data," click on the type of SED you would like to enter (currently generic, brain imaging, event-related potential [ERP], and connectivity data are available). The generic SED insert page contains the following fields:

Title: The name of the SED entry.

Brief Description: A brief description of the SED. We provide a text input area, but in general the entry should be much shorter (say, 200 words) since a full description can be entered in the "Narrative" field.

Narrative: A full length description of the SED. The narrative must be detailed enough to provide rigorous constraints for the design and/or testing of models.

Tags: A list of keywords associated with this SED entry.

Public: This button will only show if the user has permissions to make an entry public. If so, then the user can click to check this box and the entry will display to the general public and not just logged-in users of BODB.

The page also contains sections to add figures, related BOPs, brain regions, and references. Clicking on the "Show" link next to each section will expand that section, displaying items linked to the current entry and allowing the user to add new items. After expansion, the "Show" link will turn into a "Hide" link that can be used to collapse the section. Click on the "Save Draft" button to save a draft of the present entry to the BODB database. The draft will also be available to edit at a later time and will be listed in the Drafts section for convenience. If the user has permissions to publish the entry, then they can click the "Save" button and publish directly. Without permissions, this button is unavailable and the user must contact the administrator to publish the entry.

5.3 *Case Study: Models of the Mirror System*

To demonstrate these concepts and their implementation in BODB, we interleave discussion of two models of the primate mirror system with an exposition of how relevant material is represented in BODB. The exposition here, by contrast, motivates a range of features of BODB, but to master these features readers should turn to the tutorial, exercises, and projects listed on the book's website while, of course, trying out the various features on the website itself at http://bodb.usc.edu/.

Researchers recording neural activity from single neurons in F5 (an area of premotor cortex) and inferior parietal regions of macaque monkeys found that some neurons fired both when the monkey performed an action, like reaching for a piece of food, and when the monkey watched the experimenter perform the same action (see chapter 20 for references to the primary literature). They called these *mirror neurons.* There are very few studies of activity of individual neurons in humans save in the rare cases appropriate in preparation for neurosurgery. However, brain imaging in the mid-1990s showed that the human brain contains "mirror systems," that is, regions more active both when executing a class of actions and observing them than when simply observing objects which might be acted upon. Since then, hundreds, even thousands, of brain imaging studies—of varying quality—have assessed the properties of various human "mirror systems," though the activations observed in some cases may be due to neurons in the observed regions that are other than mirror neurons.

To date there have been very few neural models of the mirror system. We will discuss what is perhaps the earliest such model, the MNS model (Oztop and Arbib, 2002), and its sequel, the MNS2 model (Bonaiuto,

Rosta, and Arbib, 2007). Both show how the response properties of mirror neurons can be shaped by experience. In these models, F5 mirror neurons acquire their response properties by learning to associate the pattern of activity of F5 *canonical neurons* (those active in executing an action but not responsive to another's execution of the action) with the visual signals generated during self-observation while producing grasps. In this sense, the F5 canonical neurons provide a training signal for F5 mirror neurons, which also receive visual input in the form of highly preprocessed hand–object relations. Once trained, F5 mirror neurons can reproduce this pattern of activity given only the visual signals received during observation of another agent performing a grasp. The present comparison is based on an earlier treatment (Demiris, Aziz-Zadeh, and Bonaiuto, 2014) which also includes assessment of a robot-based application of the mirror system.

5.3.1 The MNS Model The MNS model explains how mirror neurons are developed through learning. The main hypothesis of the model is that the activity of F5 canonical neurons which control the grasp provides a training signal for mirror neurons to associate with the temporal profile of the visual features experienced during self-executed grasps. Thus, grasp learning precedes initial mirror neuron formation developmentally, and the model shows how the mirror system may learn to recognize actions already in the repertoire of the F5 canonical neurons. Because these actions are encoded by the model in an object-centered reference frame, it can recognize another agent's actions since they elicit similar input to the system as self-observation while performing the same actions.

MNS is divided into three submodules (figure 5.3) which are further comprised of networks of modules representing brain regions (figure 5.4). The Reach and Grasp module controls the movement using techniques from robotics (inverse kinematics, gradient descent, and joint angle interpolation). The F5 canonical neurons encode the type of grasp (power, precision, or

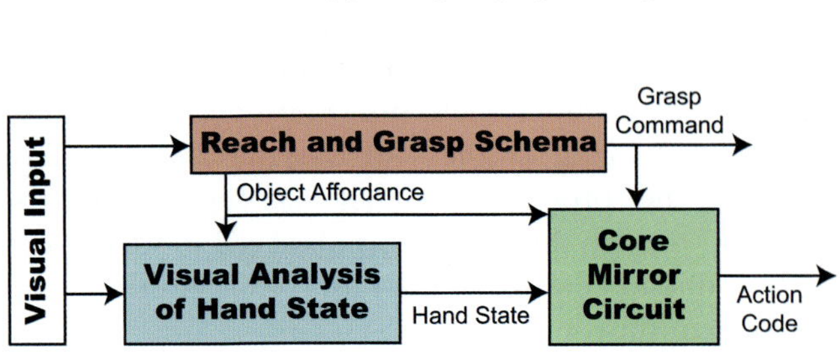

Figure 5.3 A high-level view of the mirror neuron system (MNS) model in terms of three schemas, showing the crucial information that each makes available. Adapted from Oztop and Arbib (2002).

side). This simplifies the modeling effort to more easily achieve the goals of the MNS model—to show how mirror neuron activity can be shaped by canonical neural activity—but leaves open other aspects for future model development. The Visual Analysis of Hand State module recovers the three-dimensional configuration of the hand from the two-dimensional visual input image and outputs a representation of the shape and motion of the hand relative to an object affordance. Note that the emphasis is on the state of the hand *relative to the object affordance* rather than to some other framework such as that formed by the body of the actor or the observer.

Briefly, during a self-guided action, the grasp code (activity of canonical neurons) emitted by the Reach and Grasp schema activates a set of neurons in the Core Mirror Circuit which constitute an Action Code for the current action. Whether observing one's own action or the action of another, Visual Analysis of Hand State can combine visual input concerning the movement and preshaping of the observed hand with an analysis supplied by the Reach and Grasp schema of a specific *affordance* (shape and location of an "opportunity for grasping") to provide the Hand State trajectory (i.e., trajectory of how hand shape and motion relates to the object affordance). The key idea is that, during self-action, an efferent copy of the F5 canonical neuron activity provides a training signal to enable the Core Mirror Circuit to learn to associate Grasp Commands with Hand State trajectories in such a way that, subsequently, the same mirror neurons will be activated, and thus Action Code emitted, when a similar Hand State trajectory is observed during actions by others. The accuracy of the Action Code increases as more and more of the trajectory is observed.

In this model, the vector input of hand–object affordance relations is referred to as the *hand state.* Oztop and Arbib (2002) showed that an artificial neural network corresponding to PF and F5 mirror neurons (PF and F5 are regions of parietal cortex and premotor cortex, respectively) could be trained to recognize the grasp type from the *hand state trajectory,* with correct classification often being achieved well before the hand reached the object. During training, the activity of F5 canonical neurons encodes the grasp currently being executed by the monkey and serves as a training signal for F5 mirror neurons. This allows F5 mirror neurons to respond to hand–object trajectories associated with different types of grasps. After training, the appropriate mirror neurons begin to fire in response to viewing the appropriate trajectories even when unaccompanied by F5 canonical firing. Crucially, this training allows F5 mirror neurons to respond to observation of

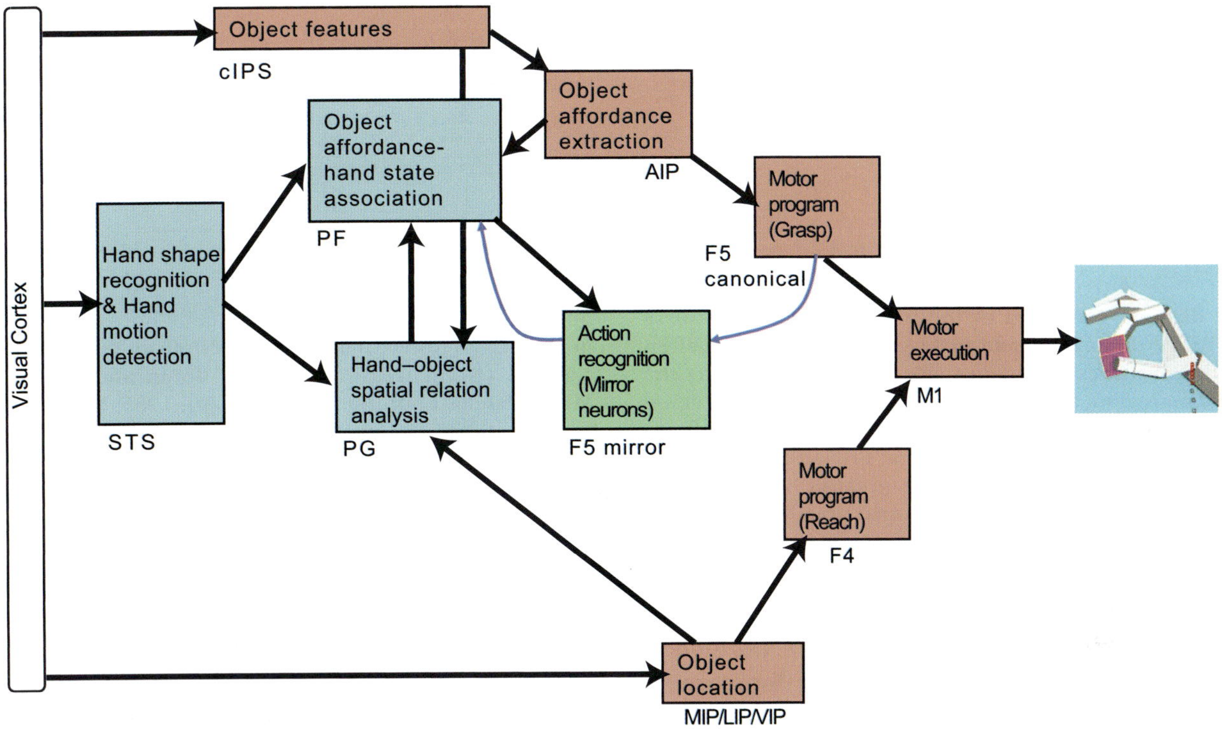

FIGURE 5.4 The mirror neuron system (MNS) model in more detail, linking different subsystems to specific brain regions in the monkey. (1) Top diagonal: Object features are processed by the anterior intraparietal sulcus (AIP) to extract grasp affordances, based on information about the surfaces of an object encoded in the caudal intraparietal sulcus (cIPS). Encoded affordances are sent on to the canonical neurons of F5 that choose a particular grasp. (2) Bottom right: Recognizing the location of the object (this may involve medial, lateral, and ventral areas of the intraparietal sulcus; MIP, LIP, and VIP) provides parameters to the motor programming area F4 (a premotor area adjacent to F4) which computes the reach. The information about the reach and the grasp is taken by primary motor cortex M1 to control the hand and the arm. (3) Grasp observation elements of the MNS model: At the bottom left is a module (putatively in the superior temporal sulcus; STS) combining two schemas, one for recognizing the shape of the hand, and the other for determining the hand's motion. Just to the right of these is the schema for hand–object spatial relation analysis. It takes information about object features, the motion of the hand, and the location of the object to infer the relation between hand and object. (5) The center two regions, PF and F5 mirror, form the core mirror circuit. These regions associate the visual input (hand state) with the motor program input from region F5 canonical neurons during the learning process. Adapted from Oztop and Arbib (2002).

hand–object relational trajectories even during observation of another agent because the hand state is based on the view of a hand moving relative to the object.

F5 canonical neurons were modeled as an array of neurons whose activity determined the type of grasp executed. The learning mechanism used was backpropagation with one hidden layer. The major drawback of the MNS model was its treatment of the hand state trajectory. At each time point, the initial segment of the trajectory up to that time was fitted by a cubic spline and then sampled at 30 times spanning the segment to produce a 210-dimensional input vector to the network. In this way, the temporal representation of hand state was preprocessed such that it could be encoded in a spatial representation for input into the feedforward network. The network was trained on a set of "self-performed" grasps using the activity of F5 canonical neurons as the training signal for the output layer of the network, representing F5 mirror neurons.

MNS makes use of a supervised learning technique, backpropagation, in order to train the mirror neuron response to visual inputs using F5 canonical neuron activity as a training signal. In this way the visual sensory information is coupled with motor representations for grasping. The learning rule is nonbiological, but in more realistic implementations, F5 canonical neuron activity could still provide a training signal or eligibility trace for a more biologically plausible learning rule. The MNS model's use of backpropagation is based on a biologically unrealistic coding of inputs to the mirror neurons, but its structure was designed based on connectivity and functional properties of relevant parts of the primate brain. The network diagram generated by

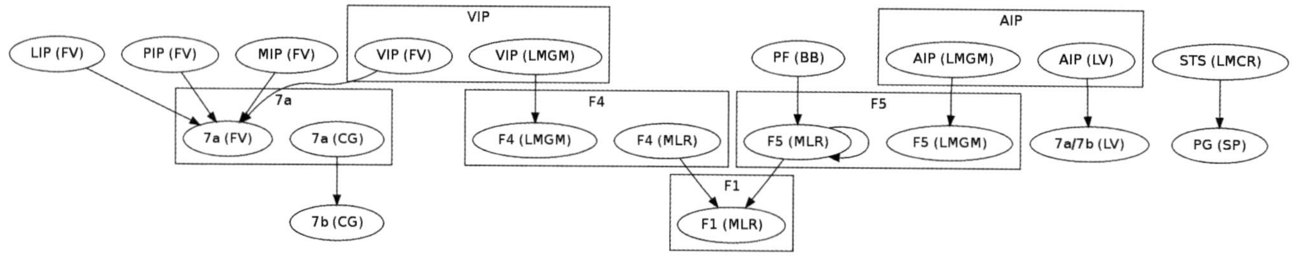

FIGURE 5.5 A connectivity Summary of Empirical Data (SED) diagram generated by the Brain Operation Database (BODB) for the mirror neuron system (MNS) model. Different nomenclatures sometimes give different names to the same brain region, and even when they agree on brain region names, they can draw the boundaries differently. (Brain regions: LIP, lateral intraparietal area; PIP, posterior intraparietal area; MIP, medial intraparietal area; VIP: ventral intraparietal area; AIP, anterior intraparietal area; STS, superior temporal sulcus.) BODB thus groups regions from different nomenclatures that have the same name but currently makes no attempt to resolve naming inconsistencies between nomenclatures. (Nomenclatures: BB, von Bonin and Bailey, 1947; MLR, Matelli, Luppino, and Rizzolatti, 1985; SP, Seltzer and Pandya, 1986; CG, Cavada and Goldman-Rakic, 1989; FV, Felleman and Van Essen, 1991; LMCR, Luppino et al., 1993; LMGM, Luppino et al., 1999; LV, Lewis and Van Essen, 2000.) From Demiris, Aziz-Zadeh, and Bonaiuto (2014).

BODB showing the connectivity SEDs used to build MNS is shown in figure 5.5. (Note that the connectivity tool was not available when the MNS and MNS2 models were first developed. See section 5.4 below for more on connectivity data in BODB and its relation to the CoCoMac database.) This figure demonstrates the difficulty in building a model based on experimental data which refers to different, sometimes conflicting, brain nomenclatures. By this we mean that different neuroanatomists (and thus brain atlases) will agree on what to call certain brain regions but in other cases call them by different names. For example, figure 5.5 shows connectivity for regions 7a and 7b, which are widely agreed to correspond to the parietal areas PG and PF, respectively. To compound the problem, the boundary between two regions (whatever their names) may be drawn somewhat differently in different analyses. In any case, no matter how one resolves this nomenclature dilemma in aggregating a number of different reports, each connectivity SED is represented in this diagram by an arrow connecting two nodes, where the nodes represent brain regions and the direction of the arrow indicates the direction of the projection. Each node contains the abbreviation for the brain region it represents (i.e., MIP stands for medial intraparietal area), followed by the abbreviation for the nomenclature it is defined in (i.e., SP stands for Seltzer and Pandya). Efforts have been made to evaluate the degree of overlap between two regions in different nomenclatures (Bota and Arbib, 2004; Kotter, 2004), but currently BODB groups regions with the same name together.

Model entries function as a repository for structural descriptions of computational models of brain mechanisms, with forms for users to enter a model summary, narrative, architecture diagrams, and overview descriptions of inputs, outputs, and internal states of computational models. Further components document related BOPs, brain regions, SEDs and SSRs. Linkage to BOPs provides a functional perspective on the model; linkage to brain regions provides a structural perspective. Figure 5.6 provides a high-level view of the entry for MNS (with most details hidden; see section 5.3 of the BODB manual, v1.0.0, for details of the fields on the model page). We can see (in more or less detail) the following aspects of the model entry:

Title: Gives the model a distinctive name + list of authors.

Brief Description: Summarizes key points which are expanded upon in the Narrative.

Narrative: This reviews "how the model works" and "what the model does." It will thus refer in summary fashion to the Architecture (various diagrams) as well as the Summaries of Experimental Data (SEDs) and Simulation Results (SSRs) and their relationships.

Tags: Keywords related to the model.

Diagram: Each module should be assigned a name. As exemplified in the transition from figure 5.3 to figure 5.4, a model may be hierarchical, and one may thus choose to use several diagrams. Submodules in figure 5.4 are color-coded to correspond to their parent module in figure 5.3. (In some cases, not illustrated here, a submodule of the overall model may be of such importance that it will be documented as a separate model.)

Caption: This should mention all the modules seen in the diagram(s) and how the interactions between them contribute to the function of the model.

Inputs, Outputs, States, Submodules: The table is to be repeated both for the overall model considered as a single module, and for each submodule contained within it. It is crucial to assess how the internal state

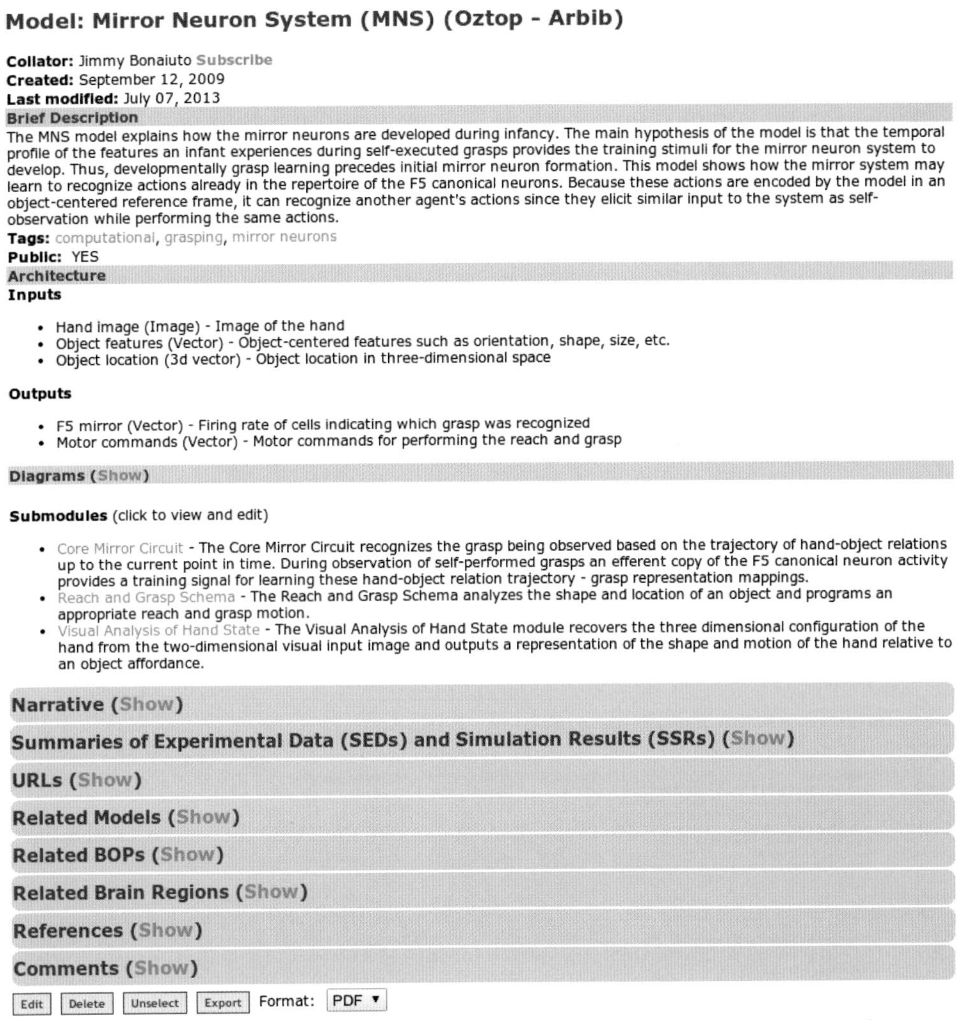

Model: Mirror Neuron System (MNS) (Oztop - Arbib)

Collator: Jimmy Bonaiuto Subscribe
Created: September 12, 2009
Last modified: July 07, 2013
Brief Description
The MNS model explains how the mirror neurons are developed during infancy. The main hypothesis of the model is that the temporal profile of the features an infant experiences during self-executed grasps provides the training stimuli for the mirror neuron system to develop. Thus, developmentally grasp learning precedes initial mirror neuron formation. This model shows how the mirror system may learn to recognize actions already in the repertoire of the F5 canonical neurons. Because these actions are encoded by the model in an object-centered reference frame, it can recognize another agent's actions since they elicit similar input to the system as self-observation while performing the same actions.
Tags: computational, grasping, mirror neurons
Public: YES
Architecture
Inputs

- Hand image (Image) - Image of the hand
- Object features (Vector) - Object-centered features such as orientation, shape, size, etc.
- Object location (3d vector) - Object location in three-dimensional space

Outputs

- F5 mirror (Vector) - Firing rate of cells indicating which grasp was recognized
- Motor commands (Vector) - Motor commands for performing the reach and grasp

Diagrams (Show)

Submodules (click to view and edit)

- Core Mirror Circuit - The Core Mirror Circuit recognizes the grasp being observed based on the trajectory of hand-object relations up to the current point in time. During observation of self-performed grasps an efferent copy of the F5 canonical neuron activity provides a training signal for learning these hand-object relation trajectory - grasp representation mappings.
- Reach and Grasp Schema - The Reach and Grasp Schema analyzes the shape and location of an object and programs an appropriate reach and grasp motion.
- Visual Analysis of Hand State - The Visual Analysis of Hand State module recovers the three dimensional configuration of the hand from the two-dimensional visual input image and outputs a representation of the shape and motion of the hand relative to an object affordance.

Narrative (Show)
Summaries of Experimental Data (SEDs) and Simulation Results (SSRs) (Show)
URLs (Show)
Related Models (Show)
Related BOPs (Show)
Related Brain Regions (Show)
References (Show)
Comments (Show)
Edit Delete Unselect Export Format: PDF ▼

FIGURE 5.6 High-level view of the entry for MNS (the mirror neuron system; with most details hidden; see box 5.1 for how to find this entry in the Brain Operation Database).

captures the history of the module in such a way as to allow the model to capture the way in which current inputs update the state and yield the output. In general, the state of a module comprises the state of all the submodules which comprise it.

Summaries of Experimental Data (SEDs): SEDs are each documented independently of any model—an empirical result holds whether or not it is linked to a model—but the links to SEDs in a model entry must specify how each SED was used in designing or testing the model. The challenge is to decide which SEDs are needed to complete the documentation of the model. Documenting the relevance of an SED forces the collator to sharpen the presentation of the SED to ensure that, for example, it makes the qualitative or quantitative results explicit to ground the SSR comparison. We will shortly provide tables which present information for MNS that has been entered in BODB. The full SED entries are separated from the model entry—the link is provided by entering the Title and Description of each SED in a separate line of the tables provided within a model entry. Before creating a new SED entry, one should conduct a search to see whether that SED has already been entered in BODB (probably with a somewhat different Title or Description from what you envisaged).

Summaries of Simulation Results (SSRs): Unlike the model-independent SEDs, the SSRs report simulations with a specific model and so are components within a particular model entry. An SSR summarizes results obtained by running the model with one or more protocols (though, in some cases—but not for MNS or MNS2—the SSR will summarize results of mathematical investigation of the model, rather than simulations).

Predictions: When simulation results do not link to SEDs based on the extant literature (it may take quite some search in PubMed or elsewhere to feel reasonably confident about this claim), they may be so interesting that it pays to recast the result as an empirical Prediction that may capture the interest of experimentalists. Predictions should outline what experimental protocol would be used, summarize the predictions of what would be found using this protocol, and link to one or more SSRs.

Related Models: This documents which Models available in BODB relate to the way the present Model works. The collator may insert "stubs" for related models that are not yet in BODB. In either case, the entry should explain why the cited model is relevant to the current Model. This might be because it provides an alternative explanation of some key SEDs or contains a module comparable to a module of the current Model.

Related BOPs: Development of a model will be informed by, and/or provide an application of a particular BOP. Being able to search on this entry allows one to find other models using one or more of the BOPs related to the current model, thus allowing one to assess possibly different accounts of how each BOP is realized in neural circuitry.

Related Brain Structures: A brain structure will be relevant because it is in part modeled by a module, or a group of interconnected modules, of the model. While this is not the case for MNS, the description might also list a brain structure which the model does not "really" address and yet which is relevant to extension of the model—it thus acts as a placeholder for future research. For example, there might be a box labeled "cerebellum" about which the model provides no details. While in MNS (and most current entries in BODB) the brain structures are neuroanatomical regions, in other models a brain structure might be a brain region, a named array or column type, a particular type of neuron, or even a specific type of receptor on the postsynaptic membrane.

SEDs used to *build* the MNS model are listed in table 5.1. Figure 5.7 shows how some of these are represented

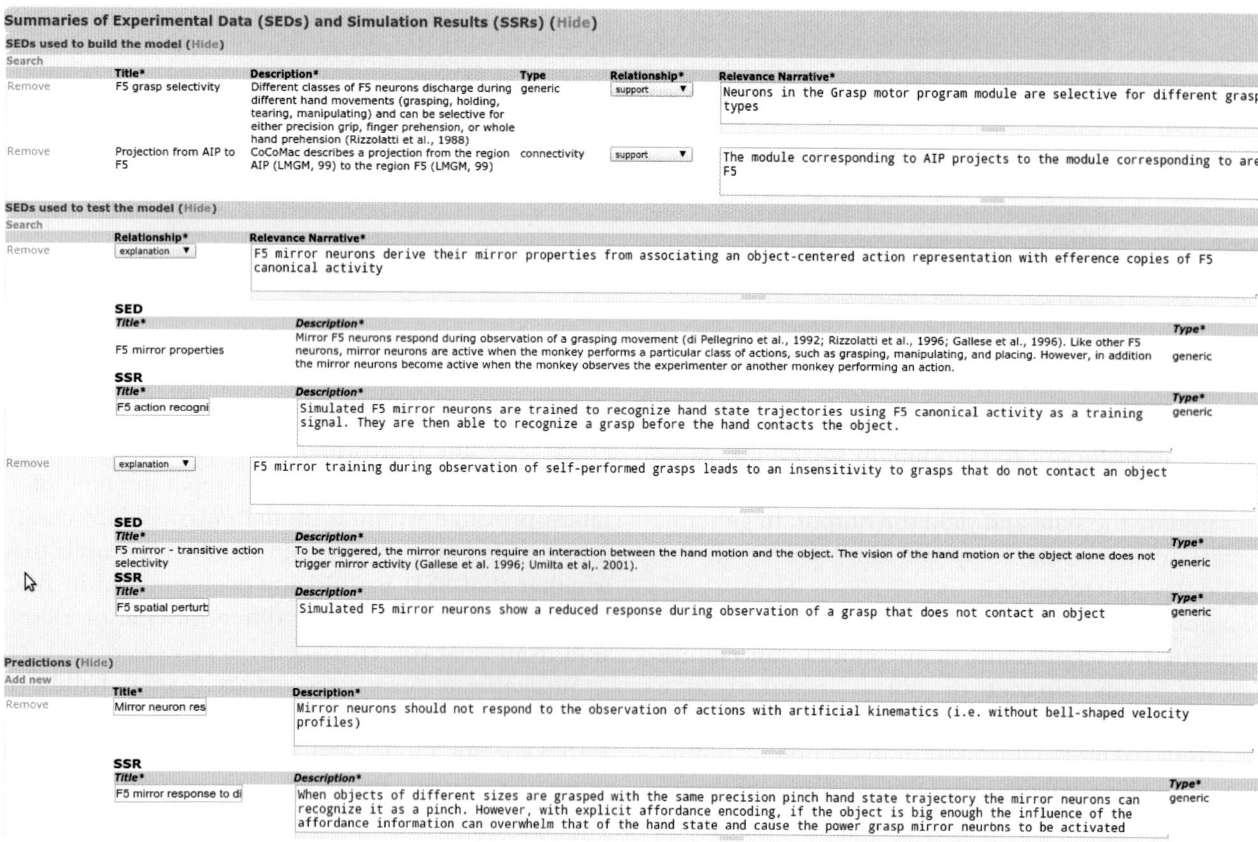
Summaries of Experimental Data (SEDs) and Simulation Results (SSRs) (Hide)

SEDs used to build the model (Hide)

Search

	Title*	Description*	Type	Relationship*	Relevance Narrative*
Remove	F5 grasp selectivity	Different classes of F5 neurons discharge during different hand movements (grasping, holding, tearing, manipulating) and can be selective for either precision grip, finger prehension, or whole hand prehension (Rizzolatti et al., 1988)	generic	support	Neurons in the Grasp motor program module are selective for different grasp types
Remove	Projection from AIP to F5	CoCoMac describes a projection from the region AIP (LMGM, 99) to the region F5 (LMGM, 99)	connectivity	support	The module corresponding to AIP projects to the module corresponding to area F5

SEDs used to test the model (Hide)

Search

Relationship* | Relevance Narrative*

Remove | explanation | F5 mirror neurons derive their mirror properties from associating an object-centered action representation with efference copies of F5 canonical activity

SED

Title*: F5 mirror properties

Description*: Mirror F5 neurons respond during observation of a grasping movement (di Pellegrino et al., 1992; Rizzolatti et al., 1996; Gallese et al., 1996). Like other F5 neurons, mirror neurons are active when the monkey performs a particular class of actions, such as grasping, manipulating, and placing. However, in addition the mirror neurons become active when the monkey observes the experimenter or another monkey performing an action.

Type*: generic

SSR

Title*: F5 action recogni

Description*: Simulated F5 mirror neurons are trained to recognize hand state trajectories using F5 canonical activity as a training signal. They are then able to recognize a grasp before the hand contacts the object.

Type*: generic

Remove | explanation | F5 mirror training during observation of self-performed grasps leads to an insensitivity to grasps that do not contact an object

SED

Title*: F5 mirror - transitive action selectivity

Description*: To be triggered, the mirror neurons require an interaction between the hand motion and the object. The vision of the hand motion or the object alone does not trigger mirror activity (Gallese et al. 1996; Umilta et al,. 2001).

Type*: generic

SSR

Title*: F5 spatial perturb

Description*: Simulated F5 mirror neurons show a reduced response during observation of a grasp that does not contact an object

Type*: generic

Predictions (Hide)

Add new

Title* | Description*

Remove | Mirror neuron res | Mirror neurons should not respond to the observation of actions with artificial kinematics (i.e. without bell-shaped velocity profiles)

SSR

Title*: F5 mirror response to di

Description*: When objects of different sizes are grasped with the same precision pinch hand state trajectory the mirror neurons can recognize it as a pinch. However, with explicit affordance encoding, if the object is big enough the influence of the affordance information can overwhelm that of the hand state and cause the power grasp mirror neurons to be activated

Type*: generic

FIGURE 5.7 Data entry form for entering Summaries of Simulation Results (SSRs) of a model and linking them to Summaries of Empirical Data (SEDs). The SEDs must be entered separately in the Brain Operation Database so they can be linked to different models as needed. The two entries under the section "SEDs used to build the model" describe F5 grasp selectivity and the projection from the anterior intraparietal sulcus area (AIP) to premotor area F5. The entry under "SEDs used to test the model" links the SED describing how mirror neurons are responsive only to hand–object interactions with the SSR describing spatial perturbation simulations with the model that explain this SED. The entry under "Predictions" describes simulations with the model using altered arm kinematics that make predictions for future experiments.

TABLE 5.1

Summaries of Experimental Data (SEDs) used to build the MNS model

Name	Description	Relationship	Relevance Narrative
Separation of dorsal and ventral visual streams	The dorsal visual path extends into the parietal cortex and represents information crucial for object interaction while the ventral stream extends into the inferotemporal cortex and recognizes the identity of an object.	scene setting	The module focuses on the dorsal stream.
AIP Grasp Selectivity	Some neurons in AIP are selective for different types of grasps.	support	The Object affordance extraction module encodes grasp affordance information.
cIPS projection to AIP	The caudal intraparietal sulcus projects to AIP.	support	The module corresponding to cIPS projects to the module corresponding to area AIP.
F4 reach selectivity	F4 represents the arm goal position and sets up the initial reach program.	support	The Reach motor program module programs the reaching movement.
PIP object selectivity	The posterior (or caudal) intraparietal sulcus (PIP or cIPS) codes object-centered information such as object shape and different parameters for different shapes such as cylinder width, surface orientation, and axes orientation.	support	The Object features module encodes object shape information.
VIP location selectivity	Area VIP represents the location and orientation of target objects in a broadly tuned population code in a peripersonal coordinate system.	support	The Object location module encodes the location of the center of the object.
Area 7 is reciprocally connected with STS	Areas 7a, 7ip, and 7b, to a lesser extent, are reciprocally connected with the cortex of the STS.	support	The module corresponding to STS projects to the modules corresponding to areas 7a and 7b.
STS shape-selective cells	Shape-selective cells respond selectively to hand–object interactions and respond similarly to the sight of the monkey's own movements as well as those of others.	support	The Hand shape recognition schema responds to observation of hand–object interactions regardless of who is performing them.
STS visual response	STS contains neurons that respond to the observation of actions such as walking, limb movements, and grasping.	support	The Hand motion detection schema responds to the observation of hand–object actions.
STS visual response	STS contains neurons that respond to the observation of actions such as walking, limb movements, and grasping.	support	The Hand shape recognition schema responds to the observation of hand–object actions.

Table 5.1 (continued)

Name	Description	Relationship	Relevance Narrative
F5 canonical visual properties	Some F5 cells respond to the observation of graspable objects.	support	Neurons in the Grasp motor program module area activated by visual responses in the AIP module.
F5 grasp selectivity	Different classes of F5 neurons discharge during different hand movements (grasping, holding, tearing, manipulating) and can be selective for either precision grip, finger prehension, or whole hand prehension.	support	Neurons in the Grasp motor program module are selective for different grasp types.
F5 mirror—broadly congruent	For some F5 mirror neurons, the congruence between preferred observed and performed actions is quite loose. For example, a mirror neuron may fire during performance of precision pinches and observation of power grasps.	scene setting	The model focuses on strictly congruent mirror neurons.
F5 mirror—strictly congruent	For some mirror neurons, not only must the general action (e.g., grasping) match, but also the way the action is executed (e.g., power grasp) must match as well.	support	After training, neurons in the Action recognition module behave like strictly congruent mirror neurons.

Note. Each SED lists some property of the brain or brain region(s) and can be either scene setting or supporting of the model's design. The relevance narrative describes the relationship between the model and data in further detail. MNS, mirror neuron system; AIP, anterior intraparietal area; cIPS, caudal intraparietal sulcus; PIP, posterior intraparietal area; VIP, ventral intraparietal area; STS, superior temporal sulcus.

Source: From Demiris, Aziz-Zadeh, and Bonaiuto (2014).

in the actual BODB entry for MNS. The first SED, "Separation of dorsal and ventral visual streams," is linked to the MNS model with the *scene setting* relationship, meaning that it was not used for any design decisions but places the focus of the model (the dorsal visual stream) in a larger context. The remaining SEDs describe neurophysiological properties of brain regions included in the model, and the relevance narrative for each describes how these properties influence the functionality of the related MNS modules.

SEDs used to *test* the MNS model are shown in table 5.2. Each test SED is linked to a summary of simulation results (SSR) from the model which either explains the data or is contradicted by it. The relevance narrative for each SED–SSR pair details how the data described in the SED are explained by or contradict the results described in the SSR. In this case, F5 mirror properties and transitive action selectivity are explained by corresponding simulation results from the MNS model. Some SSRs cannot be linked to existing SEDs and therefore make predictions by the model for future experiments. Predictions made by the MNS model and their associated SSRs are listed in table 5.3. Currently, BODB allows predictions to describe what should be found if an experiment analogous to the linked SSRs is performed. In order to facilitate rapid interchange between experimentalists and modelers, future neuroinformatics tools must make explicit the link between experimental protocol and the simulation protocol used to approximate it (see the Discussion).

5.3.2 The MNS2 Model When one has a model M_1 that successfully explains an important set of SEDs, how is one to proceed when new SEDs come to one's attention? The first step is to see if M_1 does explain the new data. If so, BODB should be updated to include the new SEDs and the BODB entry for M_1 should be updated to link these to the new SSRs that explain them. If M_1 does not, then one must change it in some way. One approach is to look to see if there are other models that do

Table 5.2

Summaries of Experimental Data (SEDs) used to test the MNS model, each linked to a Summary of Simulation Results (SSR) generated by the model which either explains the data or is contradicted by it

Name	Relationship	Relevance Narrative
F5 mirror properties from object-centered representation	explanation	F5 mirror neurons derive their mirror properties from associating an object-centered action representation with efference copies of F5 canonical activity.
	SED	
	Name	**Description**
	F5 mirror properties	Mirror F5 neurons respond during observation of a grasping movement (di Pellegrino et al., 1992; Rizzolatti et al., 1996; Gallese et al., 1996). Like other F5 neurons, mirror neurons are active when the monkey performs a particular class of actions, such as grasping, manipulating, and placing. However, in addition the mirror neurons become active when the monkey observes the experimenter or another monkey performing an action.
	SSR	
	Name	**Description**
	F5 action recognition	Simulated F5 mirror neurons are trained to recognize hand state trajectories using F5 canonical activity as a training signal. They are then able to recognize a grasp before the hand contacts the object.
F5 mirror response to intransitive actions	explanation	F5 mirror training during observation of self-performed grasps leads to an insensitivity to grasps that do not contact an object.
	SED	
	Name	**Description**
	F5 mirror—transitive action selectivity	To be triggered, the mirror neurons require an interaction between the hand motion and the object. The vision of the hand motion or the object alone does not trigger mirror activity (Gallese et al., 1996; Umiltà et al., 2001).
	SSR	
	Name	**Description**
	F5 spatial perturbation	Simulated F5 mirror neurons show a reduced response during observation of a grasp that does not contact an object.

Note. The relevance narrative describes the relationship between the simulation results and experimental data in further detail. MNS, mirror neuron system.

Source: From Demiris, Aziz-Zadeh, and Bonaiuto (2014).

explain the available data. If such a model M_2 exists and it explains all the SEDs that M_1 explains as well as the new SEDs, then perhaps the time has come to jettison M_1 in favor of M_2. Otherwise, one must compare the two models and see if there is some way to combine their best features to explain all the SEDs that either one currently explains. However, we now give an example of a different situation. Here, no new model explains the new SEDs (to be described in a moment) that model M_1 (in this case, MNS) explains, and so one may seek to modify M_1 to obtain a new model M_2 (in this case the MNS2 model) which does cover all the SEDs. In this case, the new data demonstrated the existence of audiovisual mirror neurons (figure 5.8A) and mirror neurons that respond to the observation of partially hidden grasps (figure 5.8B).

Various SEDs supported the extension of the MNS model. To address the data of figure 5.8A, this included the SED for an auditory cortex projection to F5 entered in BODB as shown in figure 5.9. MNS2 extended the capacity of MNS by adding an additional network to recognize sounds given by patterns of *auditory nerve*

TABLE 5.3

Predictions made by the MNS model (each prediction can be associated with one or more Summaries of Simulation Results; SSRs)

Name	Description	
Effect of explicit affordance coding on mirror neuron response to hand state trajectory–affordance mismatch	If object affordance information is explicitly available to mirror neurons, it can in some cases overwhelm the influence of the hand state trajectory when it is not appropriate for the object affordance.	
	SSRs	
	Name	**Description**
	F5 mirror response to altered kinematics	Simulated F5 mirror neurons do not respond to observation of reach-and-grasp movements that have constant arm-joint velocities, disrupting the time course (but not the path) of the wrist trajectory.
Mirror neuron response to ambiguous grasps	A grasp with an ambiguous prefix may drive the mirror neurons in such a way that the system will, in certain circumstances, at first give weight to the wrong classification, with only the late stages of the trajectory sufficing for the incorrect mirror neuron to be vanquished.	
	SSRs	
	Name	**Description**
	F5 mirror response to ambiguous grasps	If the beginning of a grasp is ambiguous, mirror neurons selective for other grasps may become active until the grasp can be resolved.
Mirror neuron response to artificial movements	Mirror neurons should not respond to the observation of actions with artificial kinematics (i.e. without bell-shaped velocity profiles).	
	SSRs	
	Name	**Description**
	F5 mirror response to different-sized objects with same hand state trajectory	When objects of different sizes are grasped with the same precision pinch hand state trajectory, the mirror neurons can recognize it as a pinch. However, with explicit affordance encoding, if the object is big enough, the influence of the affordance information can overwhelm that of the hand state and cause the power grasp mirror neurons to be activated.
Mirror neuron response to grasp–object axis mismatches	Mirror neurons should not respond when the orientation of the grasp does not match the axis of the object.	
	SSRs	
	Name	**Description**
	F5 mirror response to grasp–object axis mismatch	Simulated F5 mirror neurons do not respond when the orientation of the observed grasp does not match the axis of the object.
Temporal effect of explicit affordance coding on mirror neuron response	If object affordance information is explicitly available to mirror neurons, they should respond faster when the grasp matches the observed affordance.	
	SSRs	
	Name	**Description**
	F5 mirror response to precision pinch with explicit affordance encoding	With explicit affordance encoding, the mirror neuron selective for precision pinches was activated earlier when the observed pinch was directed toward smaller objects.

Note. MNS, mirror neuron system.
Source: From Demiris, Aziz-Zadeh, and Bonaiuto (2014).

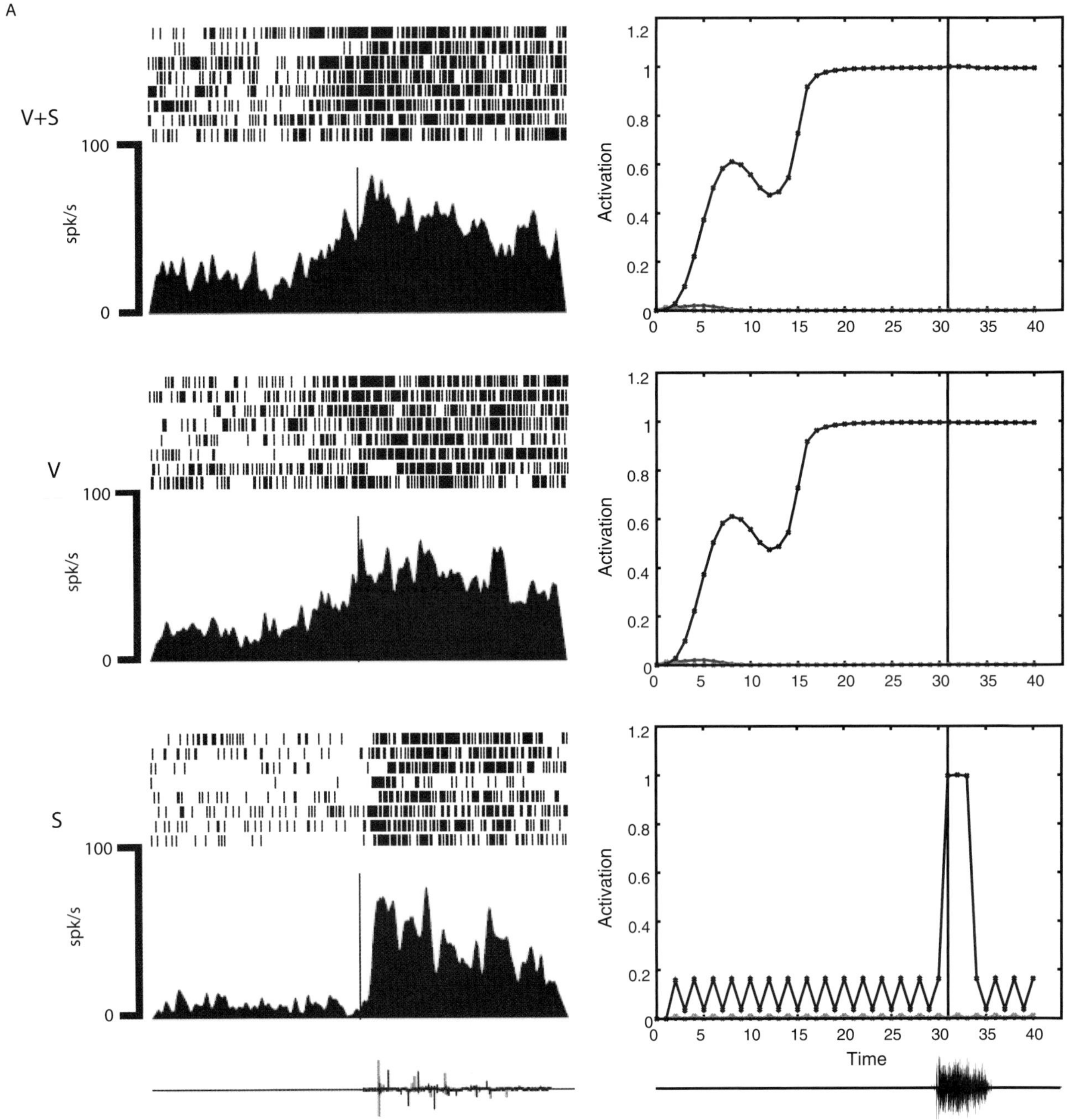

FIGURE 5.8 (A) The left column shows the firing of an *audio-visual mirror neuron,* that is, one that is related to an action with a distinctive sound (like that for breaking a peanut or tearing paper) and is active whether the monkey executes the action or sees and/or hears it being performed (V+S, vision and sound; V, vision only; S, sound only; spk/s, spikes per second; Keysers et al., 2003). The right column shows the simulation results from the mirror neuron system 2 (MNS2) model that explain it.

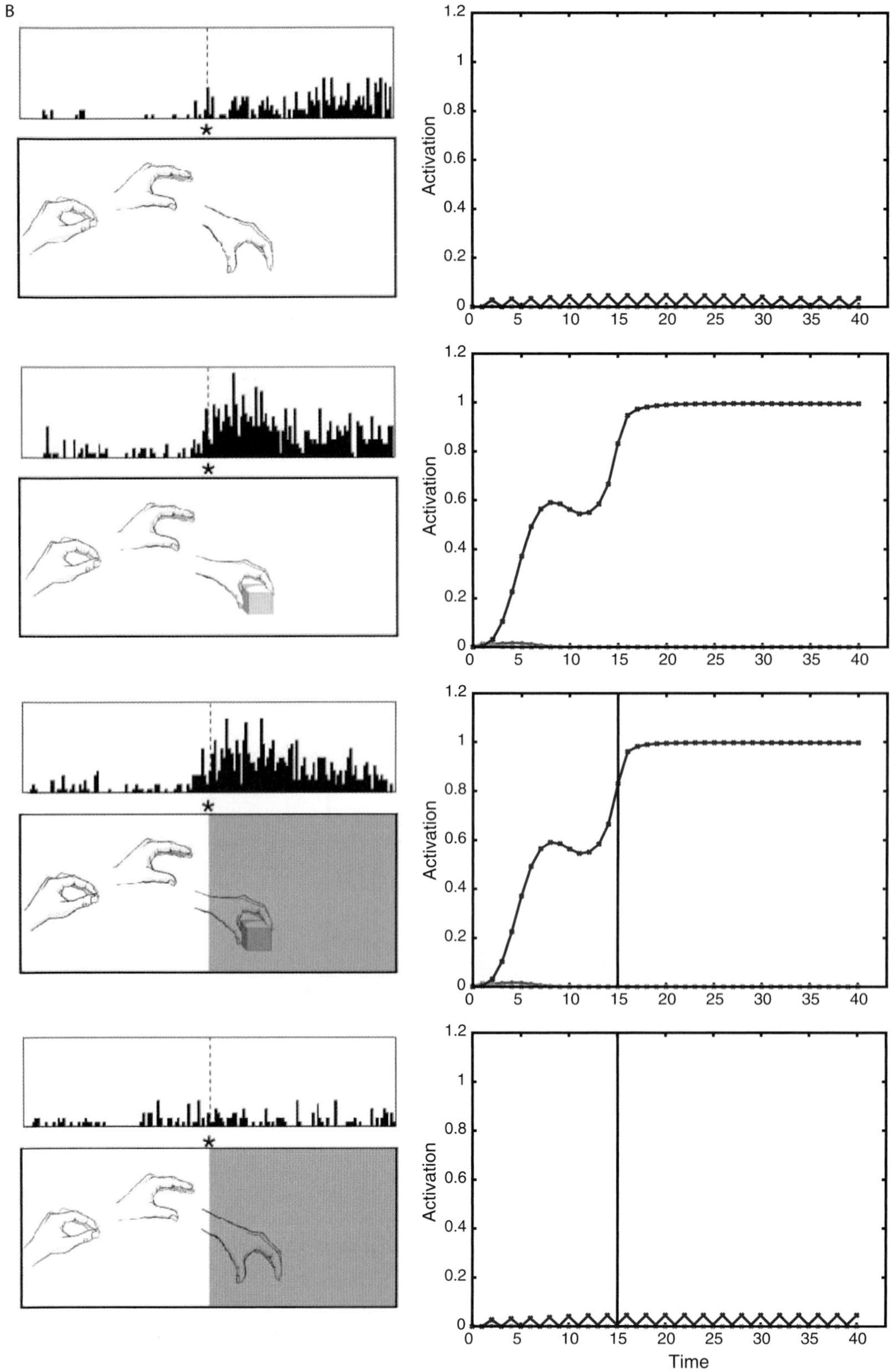

FIGURE 5.8 (continued) (B) The left column shows first that a mirror neuron that is active during observation of the grasping of an object shows little firing if the action is pantomimed in the absence of the object. It further shows that this same relation holds if the monkey has recently seen the object but that object is behind a screen when the actual grasp takes place (Umiltà et al., 2001). The right column shows the simulation results that explain these data. From Bonaiuto, Rosta, and Arbib (2007).

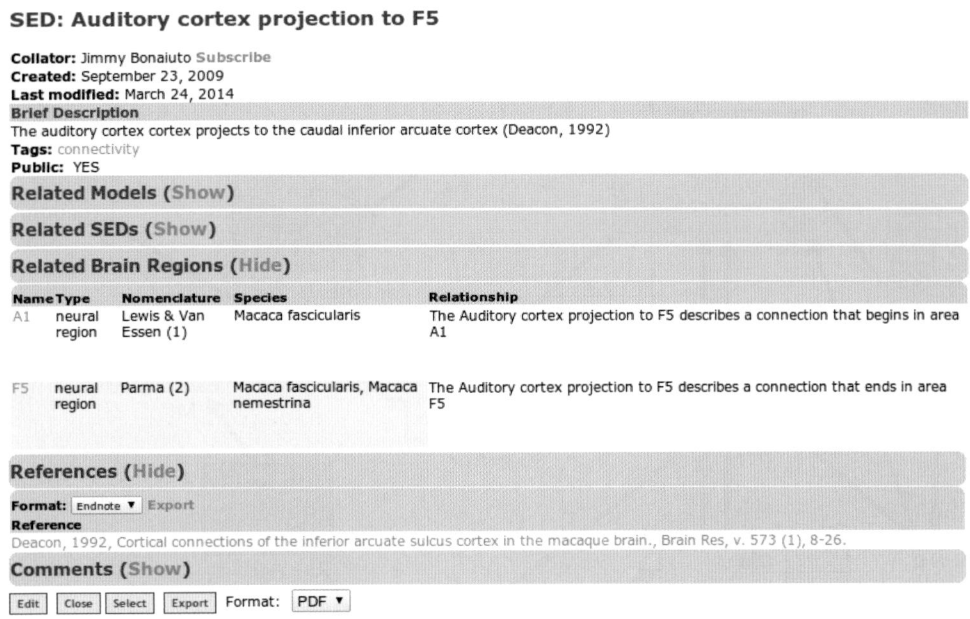
SED: Auditory cortex projection to F5

Collator: Jimmy Bonaiuto Subscribe
Created: September 23, 2009
Last modified: March 24, 2014
Brief Description
The auditory cortex cortex projects to the caudal inferior arcuate cortex (Deacon, 1992)
Tags: connectivity
Public: YES

Related Models (Show)

Related SEDs (Show)

Related Brain Regions (Hide)

Name	Type	Nomenclature	Species	Relationship
A1	neural region	Lewis & Van Essen (1)	Macaca fascicularis	The Auditory cortex projection to F5 describes a connection that begins in area A1
F5	neural region	Parma (2)	Macaca fascicularis, Macaca nemestrina	The Auditory cortex projection to F5 describes a connection that ends in area F5

References (Hide)

Format: Endnote Export
Reference
Deacon, 1992, Cortical connections of the inferior arcuate sulcus cortex in the macaque brain., Brain Res, v. 573 (1), 8-26.

Comments (Show)

Edit Close Select Export Format: PDF

FIGURE 5.9 The Brain Operation Database entry for a Summary of Empirical Data (SED) demonstrating auditory cortex (A1) projections to F5.

firings. The output of this network was associated with the output of the main network (the pattern of mirror neuron activity) so that mirror neurons could respond to the characteristic sound of an action even if the action itself was not currently visible. To address the data of figure 5.8B, MNS2 includes a *working memory* of the position of the object (if it has been observed recently) as well as a *working memory* representation of the arm and hand which is updated by *dynamic remapping* (e.g., updating an internal representation of the position of the no longer visible hand on the basis of observed elbow motion). Appropriately integrated, these features enabled MNS2 to explain the emergence of audiovisual mirror neurons and to recognize partially hidden grasps (again, see chapter 20 for details). These addenda to the MNS model of figure 5.4 are shown in figure 5.10.

In addition to being expanded to address these data, MNS2 was modified from MNS to use a recurrent architecture that avoids the biologically implausible neural encoding required for the learning mechanism of the MNS model. The original model performed a temporal-to-spatial transformation on the hand state trajectory in order to input it to a feedforward neural network. In contrast, the MNS2 model used a recurrent network that allowed the raw hand state to be input at each time step.

The main components of MNS2 are the audio and visual recurrent neural networks (RNNs), cochlear model, hand and object working memory, and the hand/object association module (figure 5.10). The hand/object association schema supplies input to the visual RNN in the form of hand/object relations such as axis disparities, joint angles, distance, and velocity. The hand and object working memory schemas relay the current visual hand and object signals to the hand/object association schema, but if either information is not available because the hand or object is invisible, a stored copy of the last observed values is passed. In the case of the hand working memory, this copy can be remapped using visual information about the movement of the arm. As in the MNS model, F5 canonical neurons encode the type of grasp being performed and their activity is used to train neurons in the output layer of the visual RNN (corresponding to mirror neurons) to classify grasps based on visual information. The training algorithm used was backpropagation through time (Werbos, 1990), allowing the network to learn to associate hand state trajectories with the appropriate motor signal used to generate the observed grasp.

With this we turn to the BODB entries for the MNS2 model, starting with the SEDs used to build the model (table 5.4). As before, the reader interested in the novel terminology of the table is referred to chapter 20 for details. MNS2 extends the original MNS model, and thus it inherits the same SEDs used to build MNS. Table 5.4 therefore only lists additional SEDs which are unique to MNS2.

Because MNS2 extends MNS, it explains some of the same SEDs such as "F5 mirror properties from

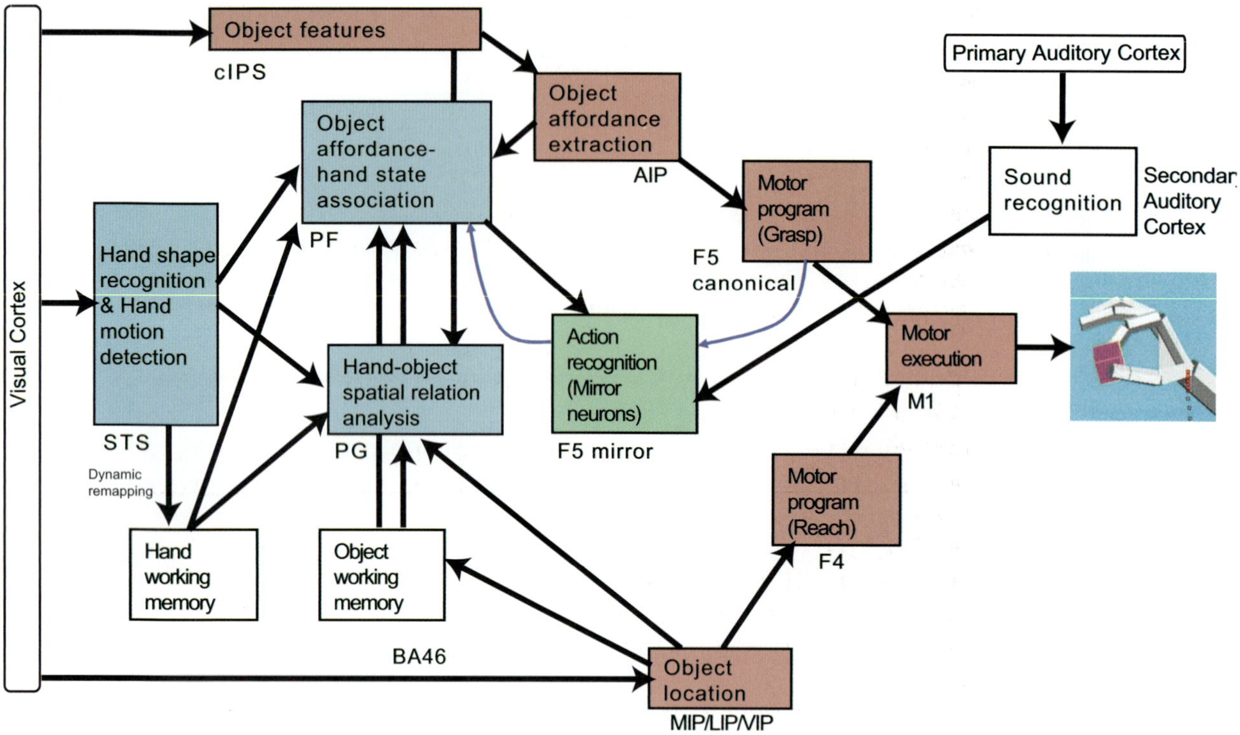

FIGURE 5.10 The mirror neuron system 2 (MNS2) model builds upon the mirror neuron system (MNS) model to address additional experimental data. As in the MNS model, the anterior intraparietal sulcus (AIP) extracts grasp affordances, based on object feature information from the caudal intraparietal sulcus (cIPS), and projects to canonical neurons in premotor area F5. The location of the object is encoded by the medial, lateral, and ventral areas of the intraparietal sulcus (MIP, LIP, and VIP) and provided to premotor area F4 to plan the reach. Visual information about the hand shape and motion is encoded by the superior temporal sulcus (STS), and combined with object information in parietal areas PF and PG before projecting to mirror neurons in premotor area F5. The addition of working memory networks for the hand and object in Brodmann area 46 (BA46) allows mirror neurons to respond to grasps whose final state is occluded from view. A model of the primary and secondary auditory cortices allows the model to recognize sounds and associate them through Hebbian learning with the actions that produce them. Modified from Bonaiuto, Rosta, and Arbib (2007).

object-centered representation" and "F5 mirror response to intransitive actions" but links them to its own SSRs which demonstrate that this new version of the MNS model can also explain these data. Additional SEDs are listed in table 5.5 that test the novel features of the MNS2 model such as audiovisual mirror properties and mirror responses to partially hidden grasps. Predictions made by the MNS2 model are listed in table 5.6. MNS2 predicts that if dynamic remapping is used to update the working memory representation of the hand during observation of partially hidden grasps, mirror neurons should cease firing once the hand over- or undershoots the hidden target object.

5.3.3 BENCHMARKING: MODEL EVALUATION AND COMPARISON Some brain models build upon earlier works to increase biological plausibility, explain more SEDs (including new experimental results published since the development of earlier models), or both. In other cases, models that were developed independently may provide complementary or even contradictory views of overlapping sets of empirical results. BODB supports model comparison, or *benchmarking*, in a tabular form based on a comparison of the SEDs used by the models to support their development and the SEDs the model can explain. Figure 5.11 shows a benchmarking table automatically generated by BODB from the entries for the MNS and MNS2 models (see section 16 of the BODB manual, v1.0.0, for instructions on how to generate model benchmark comparisons). In this case, the comparison clearly reveals how MNS2 extends MNS. In other cases, where the compared models have been developed more independently, comparing what SEDs are needed to support each model and what SEDs are explained by each model can pinpoint areas of disagreement and challenges for further model development.

Table 5.4

Summaries of Experimental Data (SEDs) used to build the MNS2 model

Name	Description	Relationship	Relevance Narrative
Auditory cortex projection to F5	The auditory cortex projects to the caudal inferior arcuate cortex.	support	The module corresponding to the auditory cortex projects to the module corresponding to area F5.
Brodmann area 46 role in working memory	Brodmann area 46 is typically implicated in working memory.	scene setting	The Working Memory schema corresponds to Brodmann area 46.
Indirect auditory cortex projection to premotor cortex	The auditory cortex projects to area 6 via area 8.	support	The module corresponding to the auditory cortex projects to the module corresponding to area F5.
Nonprimary auditory cortex encoding	Macaque nonprimary cortex responds to complex sounds.	support	The output of the Sound recognition schema uniquely identifies different sounds.
Primary auditory cortex organization	The macaque primary cortex has a tonotopic organization.	support	The input to the Sounds recognition schema is a tonotopic representation of the sound.
Working memory activity in area 46 and inferior parietal	There is activity in both the dorsolateral prefrontal cortex and inferior parietal cortex (areas 7A, 7B, 7IP, and 7M) of the macaque monkey during spatial working memory tasks.	support	The Working Memory schema is reciprocally connected with the modules representing areas 7a and 7b.

Note. MNS, mirror neuron system.
Source: From Demiris, Aziz-Zadeh, and Bonaiuto (2014).

Table 5.5

Summaries of Experimental Data (SEDs) used to test the MNS2 model

Name	Relationship	Relevance Narrative
F5 mirror properties from object-centered representation	explanation	F5 mirror neurons derive their mirror properties from associating an object-centered action representation with efference copies of F5 canonical activity.
	SED	
	Name	**Description**
	F5 mirror properties	Mirror F5 neurons respond during observation of a grasping movement (di Pellegrino et al., 1992; Rizzolatti et al., 1996; Gallese et al., 1996). Like other F5 neurons, mirror neurons are active when the monkey performs a particular class of actions, such as grasping, manipulating, and placing. However, in addition the mirror neurons become active when the monkey observes the experimenter or another monkey performing an action.
	SSR	
	Name	**Description**
	F5 action recognition	Simulated F5 mirror neurons are trained to recognize hand state trajectories using F5 canonical activity as a training signal. They are then able to recognize a grasp before the hand contacts the object.

TABLE 5.5 (continued)

Name	Relationship	Relevance Narrative
F5 mirror response to intransitive actions	explanation	F5 mirror training during observation of self-performed grasps leads to an insensitivity to grasps that do not contact an object.
	SED	
	Name	**Description**
	F5 mirror—transitive action selectivity	To be triggered, the mirror neurons require an interaction between the hand motion and the object. The vision of the hand motion or the object alone does not trigger mirror activity (Gallese et al., 1996; Umiltà et al., 2001).
	SSR	
	Name	**Description**
	F5 pantomimed grasp	Simulated F5 mirror neurons do not respond during observation of a pantomimed grasp.
F5 audiovisual mirror properties from Hebbian association	explanation	F5 mirror neurons that are active during observation of an action can come to respond to characteristics sounds of the action through Hebbian association with the output of sound recognition areas.
	SED	
	Name	**Description**
	F5 audiovisual mirror neurons	Kohler et al. (2002) found that some of the mirror neurons in area F5 of the macaque premotor cortex that are responsive for the observation of actions associated with characteristic noises are just as responsive for the sounds of these actions.
	SSR	
	Name	**Description**
	F5 audiovisual action recognition	Hebbian association of object recognition input with F5 mirror activity allows F5 mirror neurons to respond to invisible actions when the sound associated with them is perceived.
F5 mirror response to partially hidden grasps	explanation	A working memory representation of the hand and object as well as dynamic remapping of the hand position using observed arm movement allows mirror neurons to respond to observed grasps that are partially hidden by a screen.
	SED	
	Name	**Description**
	F5 mirror—partially hidden grasps	Mirror neurons selective for grasping do not respond to pantomimed grasps but will respond to grasps that are partially hidden by a screen if an object was recently seen behind it. These neurons will not respond to a grasp directed behind the screen if no object is known to be behind it.
	SSR	
	Name	**Description**
	F5 hidden grasp	Simulated F5 mirror neurons respond to visible and partially hidden grasps but do not respond to visible or partially hidden pantomimed grasps.
IPL visual properties through training	explanation	The model shows how IPL neurons can selectively respond to the observation of different types of grasps through training.
	SED	
	Name	**Description**
	IPL visual properties	The IPL contains neurons that respond to grasp observation, but not production.
	SSR	
	Name	**Description**
	Discriminative grasp hidden units	After training, neurons in the hidden layer corresponding to the IPL selectively respond to the observation of specific grasps.

Note. MNS, mirror neuron system; SSR, Summary of Simulation Results; IPL, inferior parietal lobule.
Source: From Demiris, Aziz-Zadeh, and Bonaiuto (2014).

Table 5.6

Predictions made by the MNS2 model

Name	Description	
F5 mirror response to partially hidden overshot grasps	Mirror neurons should be able to detect when a partially hidden grasp will not actually contact the object and cease responding.	
	SSRs	
	Name	**Description**
	F5 mirror response to partially hidden grasp	With dynamic remapping, a partially hidden grasp that overshoots the object causes mirror neurons to cease responding.

Note. MNS, mirror neuron system.
Source: From Demiris, Aziz-Zadeh, and Bonaiuto (2014).

	Models	
SEDs	Mirror Neuron System (MNS) (Oztop - Arbib)	Mirror Neuron System 2 (MNS2) (Bonaiuto - Rosta - Arbib)
F5 mirror properties	Explanation	Explanation
F5 mirror - strictly congruent	Support	Support
F5 mirror - broadly congruent	Scene Setting	Scene Setting
PF projection to F5	Support	Support
AIP projection to F5	Support	Support
STS visual response	Support	Support
Area 7 is reciprocally connected with STS	Support	Support
F5 mirror - transitive action selectivity	Explanation	Explanation
cIPS projection to AIP	Support	Support
F5 audiovisual mirror neurons		Explanation
Separation of dorsal and ventral visual streams	Scene Setting	Scene Setting
VIP projection to F4	Support	Support
STS shape-selective cells	Support	Support
AIP Grasp Selectivity	Support	Support
Area 7b connection to visual areas	Support	Support
Area 7b connection with AIP	Support	Support
cIPS projection to 7a	Support	Support
F5 grasp selectivity	Support	Support
F5 mirror - partially hidden grasps		Explanation
LIP projection to 7a	Support	Support
MIP projection to 7a	Support	Support
VIP projection to 7a	Support	Support
F4 reach selectivity	Support	Support
VIP location selectivity	Support	Support
Area 46 projection to 7a		Support
Area 46 projection to 7b		Support
F5 canonical visual properties	Support	Support
PIP object selectivity	Support	Support
Auditory cortex projection to F5		Support
Indirect auditory cortex projection to premotor cortex		Support
LIP projection to area 46		Support
STS projection to area 46		Support
IPL visual properties		Explanation
BA 46 role in working memory		Scene Setting
Nonprimary auditory cortex encoding		Support
Primary auditory cortex organization		Support
Working memory activity in area 46 and inferior parietal		Support

Figure 5.11 A benchmarking report automatically generated by the Brain Operation Database from entries for the mirror neuron system (MNS) and mirror neuron system 2 (MNS2) models. Summary data related to these models are listed and colored differently by relation categories. The benchmarking feature allows users to compare and contrast models by their explanatory power. We see that MNS2 uses the Summaries of Empirical Data (SEDs) used by MNS in the same roles but incorporates a range of additional SEDs in designing the extended model as the basis for explaining SEDs left unexplained by MNS. From Demiris, Aziz-Zadeh, and Bonaiuto (2014).

5.4 Brain Imaging Data

Given the diversity of types of data extant in neuroscience, a similar variety of explicit forms is required to structure SED entries in a way which facilitates search for related SEDs. The current release of BODB has a special SED entry form for brain imaging, which we now describe. At present most other new SED entries must be made using a generic form that allows the collator to enter a title, brief description, natural language narrative, uploaded figures, and links to related brain regions. The generic form serves as a means to succinctly describe empirical evidence at just the level of detail required to contribute to the development or validation of a model. It is important that future groups of users can develop new SED entry forms to facilitate meta-analysis of data entries. The formats for SSRs should then be the same as for the SEDs against which they are to be tested.

5.4.1 An Introduction to Brain Imaging

There are many ways of gathering information on activity in the intact human brain, but we are here concerned with functional magnetic resonance imaging (fMRI) and positron emission tomography (PET). Both methods measure neural activity indirectly by measuring the amount of blood flow to various regions of the brain. The cellular processes required to generate postsynaptic potentials require oxygen; therefore, neural activity leads to a series of events resulting in increased blood flow and an oversupply of oxygenated blood, and thus a dramatic change in the ratio of deoxyhemoglobin to oxyhemoglobin. Through the use of radioactive tracers injected in the bloodstream, PET can measure changes in the regional cerebral blood flow. fMRI takes advantage of the magnetic properties of oxygenated and deoxygenated hemoglobin to measure a related signal, the blood oxygenation level dependent (BOLD) response. While the spatial and temporal resolutions of both techniques are quite coarse, they can be used to generate three-dimensional images of global brain activity elicited by various cognitive tasks at a spatial resolution on the order of a few millimeters.

A typical experiment is of the following kind: Scan a subject while performing task A, then scan the subject while performing task B. For each voxel (volume element in the computer reconstruction of the 3-D structure of the subject's brain), compute a number representing the statistical significance of each of the assertions "There is more brain activity in the voxel for task A than for task B" and "There is more brain activity in the voxel for task B than for task A." The result of an analysis (e.g., A > B) can then be visualized in various ways, for example, by mapping each number to a color ranging from blue (low value, but above threshold) to red (high value) and then viewing the color distribution either on slices which contain high values or by projecting the whole brain onto one plane. The main data gathered will be of the kind "When subjects execute task A as compared to task B, then regions R1, R2, et cetera are more significantly activated." The regions R1, R2 clearly depend on the confidence level set for the test of significance. An important caveat is that *a region that is more significantly active in task A than in task B may nonetheless play a critical role in task B.*

The relationship between neural activity and the hemodynamic response has been a topic of intense research, some of which is sampled in chapter 15, "Synthetic Brain Imaging." One hypothesis is that it is integrated synaptic activity that drives increased metabolism and the blood flow response in parallel. A complete account of the path from neural activity to the hemodynamic response will require an understanding of both neurometabolic and neurovascular coupling. In order to completely link neural network models to brain imaging data, a "vampire model" of the neuron (and glia)—how the brain "sucks blood" (i.e., extracts oxygen as part of its metabolism)—is needed. In the absence of such a model, synthetic brain imaging is a technique whereby the blood flow response is simulated by averaging or otherwise transforming the synaptic activity of a neural model to make predictions testable against the results of human brain imaging.

In order to deal with the problem of variability in brain anatomy, standardized coordinate spaces have been defined with which to compare results from different studies—for example, the Talairach (Talairach and Tournoux, 1988) and MNI (Evans, Collins, and Milner, 1992) spaces for humans, and the stereotaxic and MNI spaces for monkeys. A *coordinate space* defines an origin and axes for a reference frame for a chosen "standard brain" for a particular species but does not specify which brain regions encompass which sets of coordinates. A *nomenclature* defines a (typically hierarchical) set of brain region names for a particular species while a *brain atlas* maps the names in a nomenclature onto sets of coordinates in a particular coordinate space. The Talairach atlas maps a list of brain region names onto sets of coordinates in Talairach space. Because the Talairach coordinate system is based on the scan of a single individual while the MNI atlas is based on the averaged scans of many subjects, the size of the brain in each template is slightly different. This makes conversion from one coordinate system to the other less than straightforward, but nonlinear transformations exist to convert coordinates more or less accurately between the two systems.

MRI scans from multiple individuals will vary greatly because of differences in slice orientation and brain features (i.e., brain size and shape vary across individuals). Therefore, to analyze experimental results across subjects, their individual scans are normalized to a standard coordinate space. Normalization is the process of translating, rotating, scaling, and maybe warping a brain to roughly match a standard template image. Both the Talairach and MNI systems use three-dimensional coordinates (x, y, z) to describe distances in millimeters (mm) relative to the origin, which is defined as the anterior commissure—a bundle of white matter connecting the olfactory areas of the two cerebral hemispheres. The x, y, z dimensions refer to left–right, posterior–anterior, and ventral–dorsal respectively. So (38, 64, 58) refers to a point in right posterior dorsal region of the brain. If one has normalized a subject's brain to one matching the brain in the Talairach or MNI atlas, then one can examine the atlas to get a general idea of the region in which a peak of activity occurs. However, because of individual variation, such an association may be probabilistic at best.

5.4.2 A Case Study With this, let's look at a mirror system-related study using human brain imaging. Aziz-Zadeh et al. (2006) studied neural correlates of action observation and reading phrases describing actions. The aim of the study was to determine whether the same brain regions involved in visual recognition of actions are active during linguistic processing of the same actions. There were three visual recognition conditions containing videos of mouth, hand, or foot actions, with five videos in each condition. The mouth videos showed a side view of the mouth of an actor biting five different fruits (e.g., peach or banana). The hand videos showed a hand reaching for and grasping five different objects (e.g., scissors, pen). The foot videos showed the foot of an actor pressing on various objects (e.g., piano pedal, car brake). Activation during these conditions was compared to that during a resting baseline.

Figure 5.12 shows six axial slices through the brain with regions active during each type of condition outlined in red for action observation and blue for reading action phrases. This map of activity is generated by performing a statistical analysis of the signal at each voxel under different experimental conditions and plotting the result of this analysis (T value from a *t* test or Z score, for example), thresholded at a given significance, *p*. The coordinates corresponding to the voxels which have the peak significance in each blob are typically reported in a table along with their significance level (table 5.7).

5.4.3 Brain Imaging Data in BODB BODB's brain imaging SED format summarizes neuroimaging data in

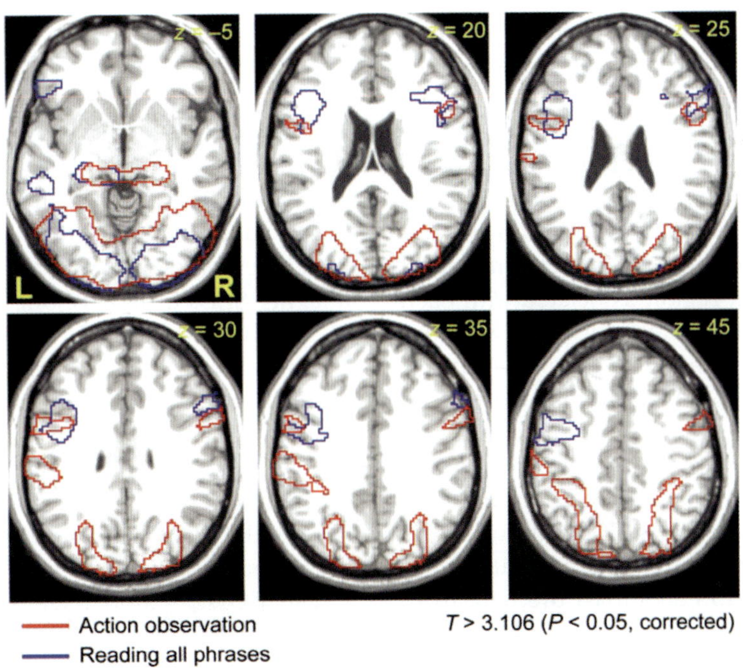

Figure 5.12 Areas activated by observation of actions compared to resting baseline (red) and reading action-related phrases compared to resting baseline (blue). From Aziz-Zadeh et al. (2006).

TABLE 5.7

An example where two separate comparisons from a brain imaging experiment are combined in a single published *table*

Area	MNI coordinates (mm)			Extent (mm^3)	Max T	Cluster p
	x	y	z			
Observation						
Bilateral visual areas, dorsal stream and LGN				244,936		<0.0001
Bilateral V1	6	–92	–2		20.4	
Left IT	–42	–58	–16		16.5	
Right IT	44	–52	–18		13.7	
Left superior parietal	–34	–44	46		7.0	
Right superior parietal	28	–56	48		9.0	
Left supramarginal gyrus	–54	–32	30		8.2	
Right precentral gyrus				5,832		0.0012
Dorsal peak	52	0	48		7.5	
Ventral peak	54	4	28		7.3	
Left precentral gyrus	–52	4	30	3,360	8.2	0.016
Language						
Bilateral visual areas extending to inferior temporal areas				115,648		<0.0001
Left visual	–36	–90	–8		14.3	
Right visual	28	–96	6		16.5	
Left posterior IFG and precentral gyrus				21,192		<0.0001
Precentral gyrus	–40	6	32		8.7	
IFG pars opercularis/pars triangularis	–44	20	18			
IFG pars orbitalis	–54	32	–6			
Right posterior IFG (dorsal pars opercularis)	28	22	18	4,336	5.7	0.0053
Left LGN	–26	–30	–8	3,904	13.4	0.0085

Note. The table shows areas active during action observation compared to resting baseline (the entries following the "Observation" heading), and reading action phrases compared to resting baseline (the entries following the "Language" heading). MNI, Montreal Neurological Institute; LGN, lateral geniculate nucleus; IT, inferotemporal area; IFG, inferior frontal gyrus.

Source: Adapted from Aziz-Zadeh et al., (2006).

the form of coordinates of peaks of activations in different experimental contrasts, allowing collation from published literature as well as derivation from raw data, and easy comparison with model simulation results based on the brain region associated with each coordinate vector. In order to allow users to visualize imaging SEDs, we have developed *BrainSurfer*—a three-dimensional visualization tool for multiple atlases.

Although brain imaging data are often supplied in table format, different articles may differ in the number, labeling, and layout of columns, and differ in the free-form description of the experimental conditions of that table. BODB therefore offers a tool to enter fMRI tables which is flexible enough to encompass the data from most of the tables found in the literature while maintaining an output which is standard for all SEDs, providing a structured input form for representing the table's description and data, and specifying metadata. In general, a single table in an article may require multiple Tables in the BODB format, each one describing a single contrast. For example, the data from the Aziz-Zadeh et al. (2006) study from table 5.7 yields two brain imaging SEDs in BODB, one for the visual observation of actions (figure 5.13) and one for reading action-related phrases (figure 5.14). BODB allows users to define the name of the columns by using both pre-defined column type such as brain region, x, y, z, hemisphere, Z score, T value and Brodmann Area (BA), but one may add and name additional columns (see section 5.4.2 of the BODB manual v1.0.0 for details on entering brain imaging SEDs).

The BODB search page allows users to search for brain imaging SEDs by several fields specific to this entry type: Control Condition, Experimental Condition, Method, Coordinate Brain Region, and Coordinate Range. The search results page will then display a

☆ **Brain Imaging SED: Observation of hand, foot and mouth actions**

Collator: Lisa Aziz-Zadeh Subscribe
Created: January 03, 2014
Last modified by: Lisa Aziz-Zadeh Subscribe
Last modified: January 20, 2014
Tags: action observation, Human, mirror system

Brief Description

Areas activated during action observation (averaged across three effectors) as compared to the resting baseline. Activations extended well into the dorsal visual stream and there were also bilateral activations in ventral premotor cortex.

Public: NO

Method: fMRI

Control condition: Resting baseline

Experimental condition: There were three kinds of blocks containing videos of mouth, hand, or foot actions. There were five videos in each condition. The mouth videos showed a side-on view of the mouth of an actor (including nose and chin) biting five different fruits (e.g., peach, banana). The hand videos showed a hand reaching for and grasping five different objects (e.g., scissors, pen). The foot videos showed the foot (and lower leg) of an actor pressing on various objects (e.g., piano pedal, car brake). Each video lasted for 1.5 s and there were 0.5 s between videos.

Coordinate space: MNI

Data

Brain Region	Hemisphere	x	y	z	T
☑ V1	right	6.0	-92.0	-2.0	20.400
☑ IT	left	-42.0	-58.0	-16.0	16.500
☑ IT	right	44.0	-52.0	-18.0	13.700
☑ superior parietal	left	-34.0	-44.0	46.0	7.000
☑ superior parietal	right	28.0	-56.0	48.0	9.000
☑ supramarginal gyrus	left	-54.0	-32.0	30.0	8.200
☑ precentral gyrus	right	52.0	0.0	48.0	7.500
☑ precentral gyrus	right	54.0	4.0	28.0	7.300
☑ precentral gyrus	left	-52.0	4.0	30.0	8.200

Select All | Unselect All

References (Hide)

Format: Endnote ▾ Export

Reference	Collator
Aziz-Zadeh et al., 2006, Congruent embodied representations for visually presented actions and linguistic phrases describing actions., Curr Biol, v. 16 (18), 1818-23.	Jimmy Bonaiuto

Discussion (Show)

Edit | Delete | Workspace Unselect | Manage Permissions

Export Format: PDF ▾

FIGURE 5.13 The Brain Operation Database brain imaging Summary of Empirical Data (SED) corresponding to the contrast action observation > resting baseline from table 5.7. (Brain regions: IT, inferior temporal cortex.)

list of all imaging SEDs which match the entered criteria (figure 5.15). Imaging SEDs derived through federation (see section 5.5) with the Brede database of brain imaging studies (http://hendrix.imm.dtu.dk/services/jerne/brede/) will have a *View in Brede* link next to them. Users can click this link to view more information on this data set in the Brede database. The checkbox next to each coordinate in the search results allows users to select that coordinate to include in a saved set of coordinates.

Clicking on the "Favorites" link in the main toolbar will bring up a page listing all the entries that you have bookmarked, along with the set of currently selected coordinates (figure 5.16). Clicking on the "Show" link next to the Coordinates section will list the currently selected brain imaging SED coordinates and the SEDs they are linked to under the "Active Coordinate Set" section. Selections can be saved as Saved Coordinate Set by giving the set a name ("Reading vs Observation of hand, foot, and mouth actions" in figure 5.16) and, optionally, a description, and clicking "Save." Previously saved coordinate sets are shown in the Saved Coordinate Sets section to the right and can be loaded as the active coordinate set by clicking on the "Load" link to the left of their name.

5.4.4 BRAINSURFER Sets of selected SED coordinates can be visualized in *BrainSurfer*, the latest in a series of efforts to visualize neuroscientific data from various experiments within BODB. BrainSurfer combines three-dimensional visualization functionality with features such as region search, export functionality, and integration with BODB for data management. Crucially, it adds extensibility for multiple brain atlases: it can be used to search and view any brain for which a coordinate space and nomenclature is provided in an appropriate framework. It currently exploits both the data on the human brain from the Talairach Daemon developed by Peter Fox and Jack Lancaster (Lancaster et al., 1997) and the brain atlas of the macaque brain

☆ **Brain Imaging SED: Reading phrases relating to hand, foot, and mouth actions**

Collator: Lisa Aziz-Zadeh **Subscribe**
Created: January 03, 2014
Last modified by: Jimmy Bonaiuto
Last modified: January 06, 2015
Tags: language, Human, mirror system

Brief Description

Areas activated by reading action phrases (averaged across literal and metaphorical sentences and across the three effectors) as compared to resting baseline.

Public: YES

Method: fMRI

Control condition: Resting baseline

Experimental condition: There were six kinds of language blocks obtained by crossing three effectors (mouth, hand, and foot) by two sentence types (literal and metaphorical). Examples of the phrases included: mouth/literal (e.g., "biting the peach," "biting the banana"); hand/literal (e.g., "grasping the scissors," "grasping the pen"); foot/literal (e.g., "pressing the piano pedal," "pressing the car brake"); mouth/metaphorical (e.g., "biting off more than you can chew," "chewing over the details"); hand/metaphorical (e.g., "grasping the idea," "handling the truth"); and foot/metaphorical (e.g., "kicking off the year," "time is running"). There were five sentences in each condition. Each block consisted of ten sentences (thus the five phrases were repeated twice each) displayed for 1.5 s each, with 0.5 s between phrases. Therefore, the total data for each condition consisted of 40 sentences (five sentences, which were read eight times each: two repetitions per block, by two blocks per run, by two runs).

Coordinate space: MNI

Data

Brain Region	Hemisphere	x	y	z	T
✔ posterior IFG - dorsal pars opercularis	right	28.0	22.0	18.0	5.700
✔ IFG pars orbitalis	left	-54.0	32.0	-6.0	8.700
✔ IFG pars opercularis - pars triangularis	left	-44.0	20.0	18.0	8.700
✔ precentral gyrus	left	-40.0	6.0	32.0	8.700
✔ V1	right	28.0	-96.0	6.0	16.500
✔ V1	left	-36.0	-90.0	-8.0	14.300
✔ LGN	left	-26.0	-30.0	-8.0	13.400

Select All | Unselect All

References (Show)

Discussion (Show)

Edit | Delete | Workspace Unselect | Manage Permissions
Export Format: PDF ▾

FIGURE 5.14 The Brain Operation Database brain imaging Summary of Empirical Data (SED) corresponding to the contrast reading action phrase > resting baseline from table 5.7. (Brain regions: IFG, inferior frontal gyrus, LGN, lateral geniculate nucleus.)

developed by George Paxinos and Charles Watson (Paxinos et al., 2009). Additionally, BrainSurfer adds improved navigation tools, and visualization of brain region isosurfaces. (See box 5.1 for how to access BrainSurfer, and section 8 of the BODB manual, v1.0.0, for further details on its use.)

BrainSurfer's main screen section contains slices through the brain at three orientations—axial, coronal, and sagittal, with images from the selected brain atlas on each slice (figure 5.17). The slice planes can be clicked and dragged, updating the slice image as they move. Users can left-click and drag regions outside of the slice planes to rotate the brain, right-click and drag to translate it, and use the mouse wheel to zoom. The reference axes rotate along with the brain. The View Controls section allows users to select which atlas to view, modify the view angle and axes, tooltips, and slice visibility, and export the visualization in a report or image. The Brain Region Browser allows users to browse the brain regions defined in the currently selected atlas and select which ones appear in the visualization. The SED Coordinate Browser lists the SED coordinates in the user's currently active coordinate set in BODB (see previous section). When an object is clicked on, the Selected Object Controls shows its current attributes and allows users to modify them. When a slice is selected, the Selected Object Controls section allows the user to modify its position using the slider control and to turn on and off the visibility of the coordinate labels along the corresponding axis.

Clicking on the checkbox next to a brain region in the Brain Region Browser makes the region visible as an isosurface mesh (by default) in the main screen and opens its visualization attributes in the Selected Object Controls section. Users can adjust the transparency and color of the region by using the appropriate slider controls. Clicking on the "Show 3D" checkbox toggles between visualizing the brain region in three dimensions as an isosurface mesh (figure 5.18) or as a two-dimensional shaded area on the brain slices (figure

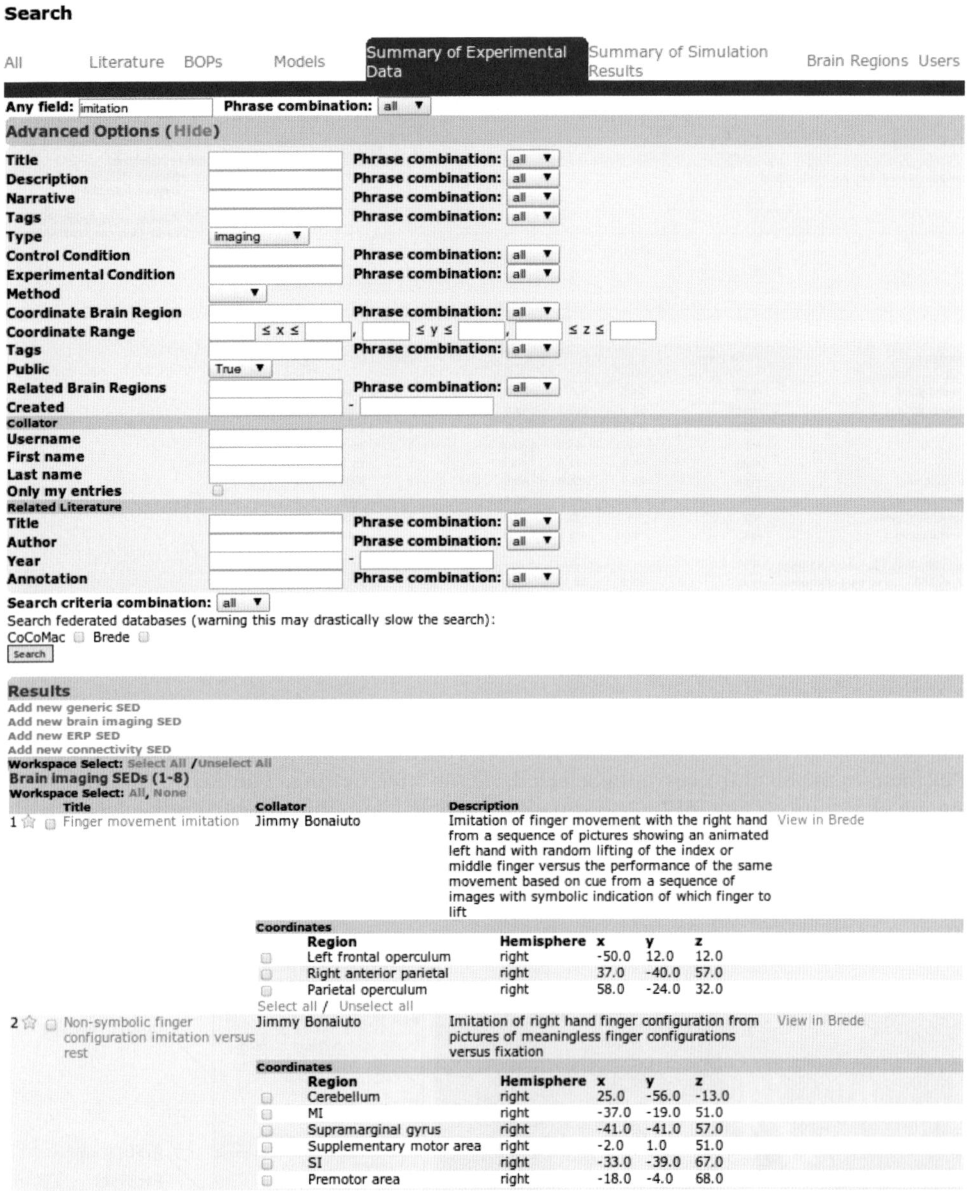

FIGURE 5.15 The search page after searching for brain imaging Summaries of Empirical Data (SEDs) containing the phrase "imitation." Two of the eight returned results are shown. Each contains several coordinates and a link to the more detailed entry in the Brede database.

5.19). Brain region definitions are hierarchical, and clicking on the checkbox next to one will make it and its child regions visible.

When opened, BrainSurfer connects to BODB and loads the user's active coordinate set (see previous section for how this set is defined in BODB before opening BrainSurfer). These coordinates appear in the three-dimensional display as red cubes by default. Users can click on a coordinate in the SED Coordinate Browser to change the transparency, color, and shape in the Selected Object Controls section (figure 5.20). Clicking on the checkbox next to a coordinate will toggle its visibility. Clicking on an SED in the SED Coordinate Browser allows the user to change the visualization attributes of all the coordinates in that SED simultaneously by using the controls in the Selected Object Controls section. This allows the user to distinguish coordinates from different SEDs by assigning them all a distinct color and shape.

BrainSurfer can be used to visualize data in multiple brain atlases and, where appropriate transformations are available, will automatically transform coordinates

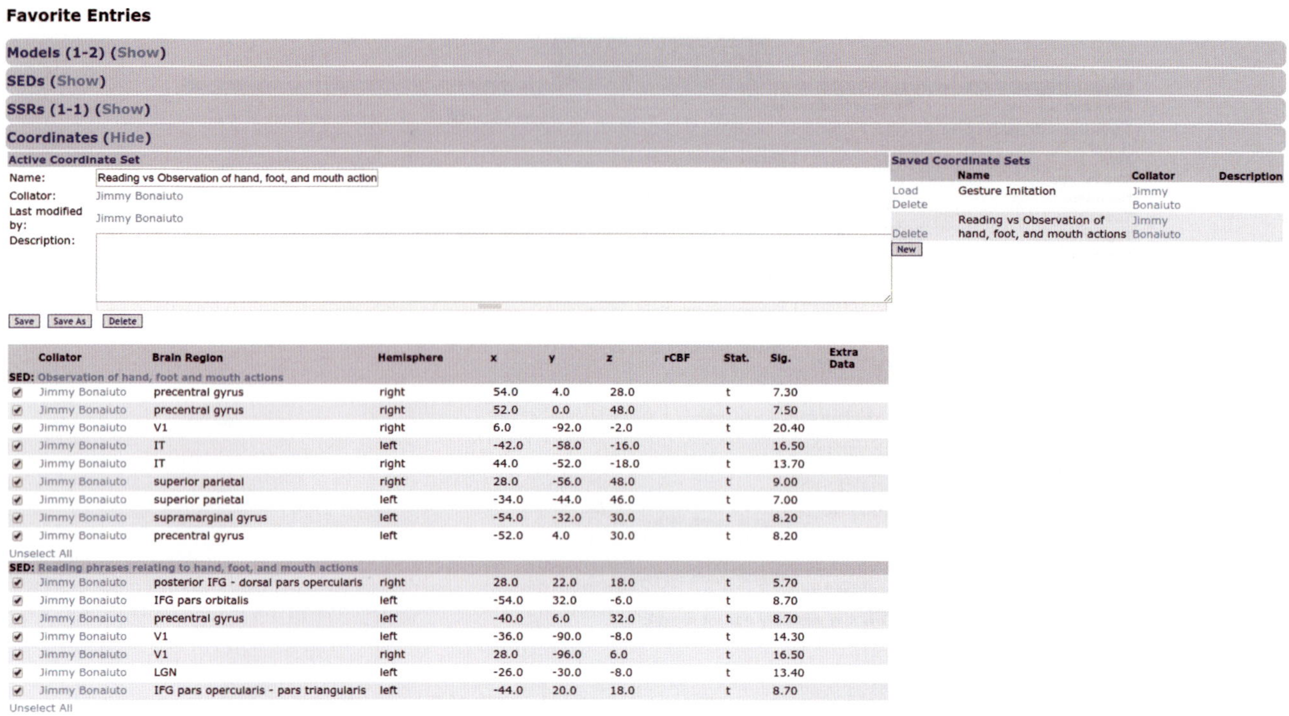

	Collator	Brain Region	Hemisphere	x	y	z	rCBF	Stat.	Sig.	Extra Data
SED: Observation of hand, foot and mouth actions										
✔	Jimmy Bonaiuto	precentral gyrus	right	54.0	4.0	28.0		t	7.30	
✔	Jimmy Bonaiuto	precentral gyrus	right	52.0	0.0	48.0		t	7.50	
✔	Jimmy Bonaiuto	V1	right	6.0	-92.0	-2.0		t	20.40	
✔	Jimmy Bonaiuto	IT	left	-42.0	-58.0	-16.0		t	16.50	
✔	Jimmy Bonaiuto	IT	right	44.0	-52.0	-18.0		t	13.70	
✔	Jimmy Bonaiuto	superior parietal	right	28.0	-56.0	48.0		t	9.00	
✔	Jimmy Bonaiuto	superior parietal	left	-34.0	-44.0	46.0		t	7.00	
✔	Jimmy Bonaiuto	supramarginal gyrus	left	-54.0	-32.0	30.0		t	8.20	
✔	Jimmy Bonaiuto	precentral gyrus	left	-52.0	4.0	30.0		t	8.20	
Unselect All										
SED: Reading phrases relating to hand, foot, and mouth actions										
✔	Jimmy Bonaiuto	posterior IFG - dorsal pars opercularis	right	28.0	22.0	18.0		t	5.70	
✔	Jimmy Bonaiuto	IFG pars orbitalis	left	-54.0	32.0	-6.0		t	8.70	
✔	Jimmy Bonaiuto	precentral gyrus	left	-40.0	6.0	32.0		t	8.70	
✔	Jimmy Bonaiuto	V1	left	-36.0	-90.0	-8.0		t	14.30	
✔	Jimmy Bonaiuto	V1	right	28.0	-96.0	6.0		t	16.50	
✔	Jimmy Bonaiuto	LGN	left	-26.0	-30.0	-8.0		t	13.40	
✔	Jimmy Bonaiuto	IFG pars opercularis - pars triangularis	left	-44.0	20.0	18.0		t	8.70	
Unselect All										

FIGURE 5.16 The favorites page lists the entries that have been bookmarked by the user as well as the currently selected brain imaging Summary of Empirical Data (SED) coordinates (Brain regions: IT, inferior temporal cortex; IFG, inferior frontal gyrus, LGN, lateral geniculate nucleus). Selected coordinates can be grouped into sets and saved with a name for later viewing. The Active Coordinate Set section lists the coordinates that are currently selected and the Saved Coordinate Sets lists previously saved set of coordinates that can be loaded.

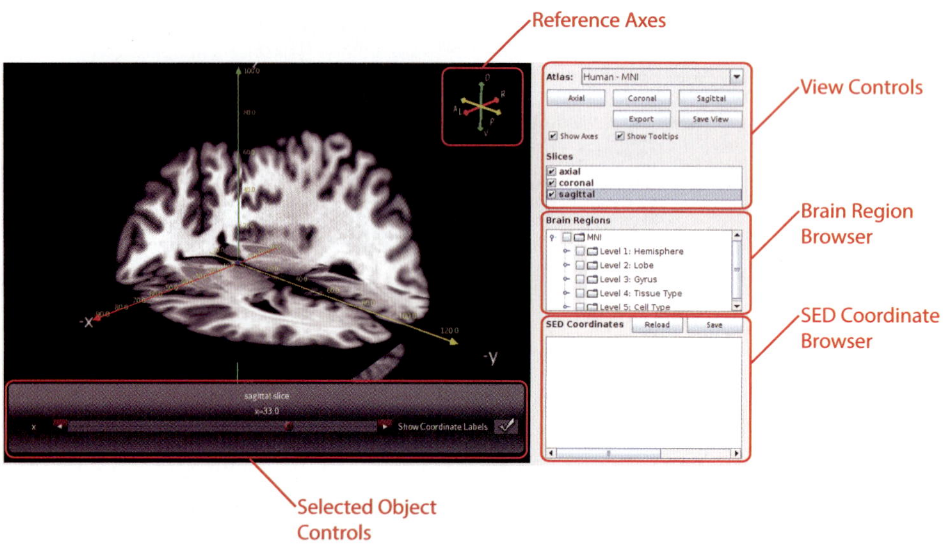

FIGURE 5.17 Main components of the BrainSurfer user interface. The Reference Axes rotate along with the axes in the main section in order to allow the user to keep track of the current orientation when zoomed in. Users can select an object in the main section such as brain slice or region, and the Selected Object Controls section will be populated with options to manipulate the position or appearance of the selected object. The View Controls section allows the user to change the current axes, reset the view to several canonical orientations, export the information shown in BrainSurfer, and set various other viewing options. The Brain Region Browser displays a hierarchical list of brain regions in the current atlas which can be clicked on to display them in the main section as three-dimensional blobs or two-dimensional projections onto each slice. The Summary of Empirical Data (SED) Coordinate Browser lists the SED coordinates that the user has currently selected in the Brain Operation Database.

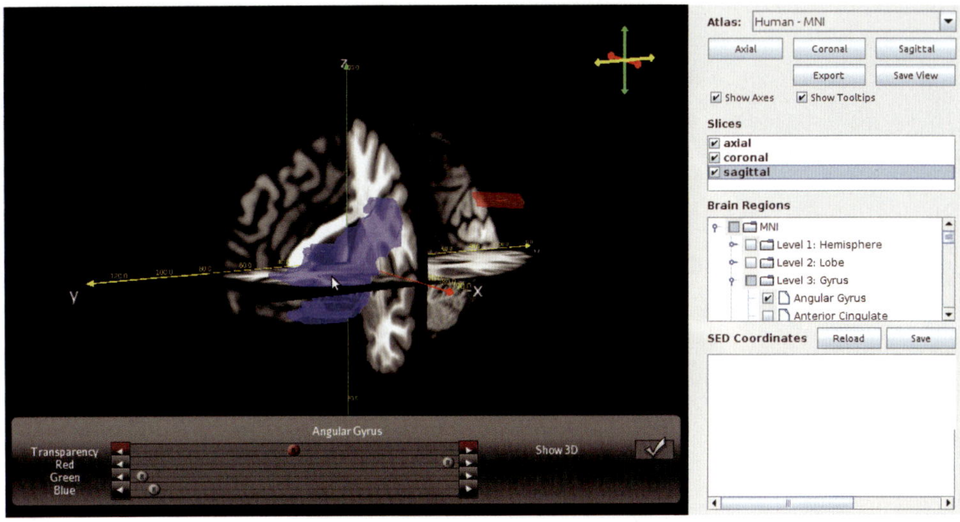

FIGURE 5.18 Three-dimensional visualization of brain regions in BrainSurfer using the Montreal Neurological Institute (MNI) atlas (note that MNI is selected in the Atlas dropdown menu in the View Controls section). Here the inferior frontal gyrus (blue) and angular gyrus (red) have been selected in the Brain Region Browser. The angular gyrus isosurface has been clicked on, which populates controls for modifying its appearance in the Selected Object Controls section. It has been made semi-transparent, with a three-dimensional isosurface representation (the Show 3D checkbox).

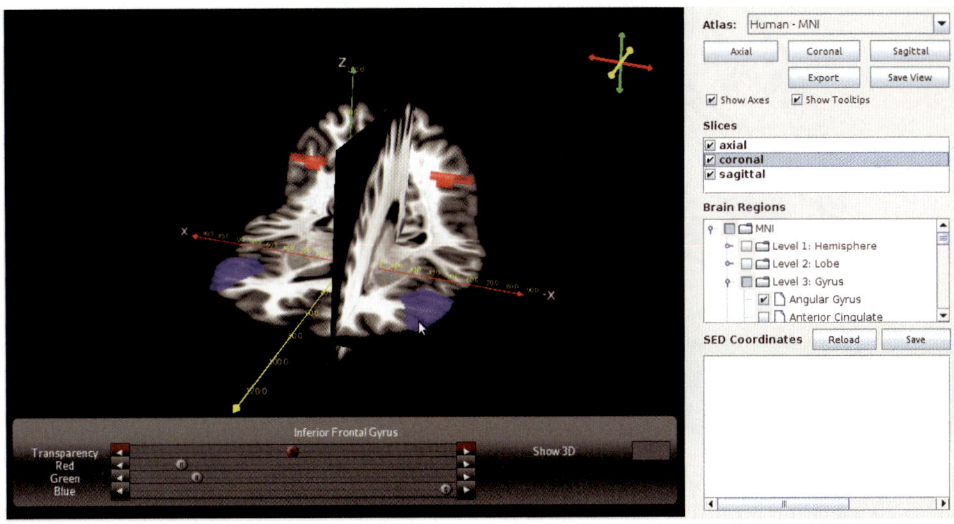

FIGURE 5.19 Two-dimensional visualization of brain regions in BrainSurfer using the Montreal Neurological Institute (MNI) atlas. Here the inferior frontal gyrus (blue) and angular gyrus (red) have been converted to two-dimensional projections on the brain slices by clicking on them and unchecking the "Show 3D" checkbox in the Selected Object Controls section.

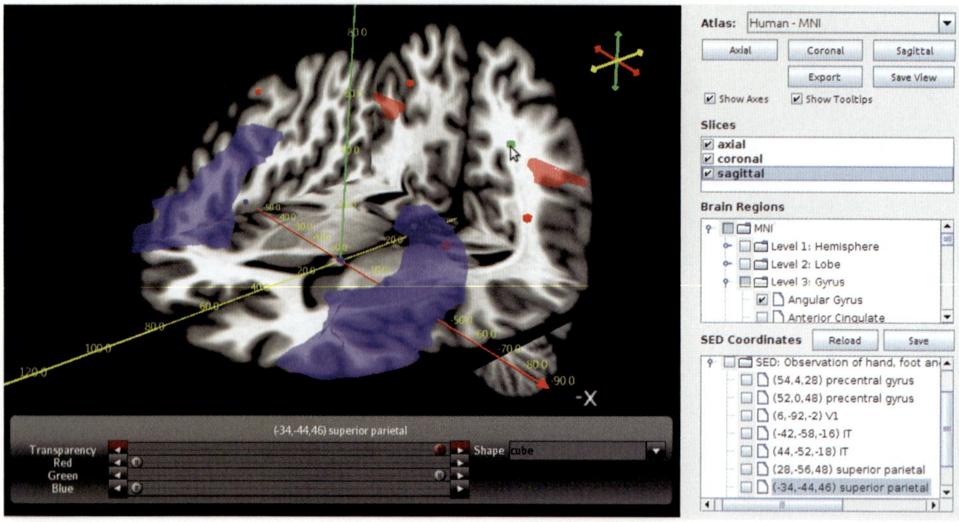

FIGURE 5.20 Visualization of Summary of Empirical Data (SED) coordinates in BrainSurfer using the Montreal Neurological Institute (MNI) atlas. Here we see not only a brain view like those of the previous figures but also a set of colored dots—with dots of any one color showing the 3-D locations of peaks recorded in a single brain imaging table retrieved from the Brain Operation Database. Clicking on a dot will show the coordinates and brain region that it corresponds to in the Selected Object Controls section along with options to change its transparency, color, and shape.

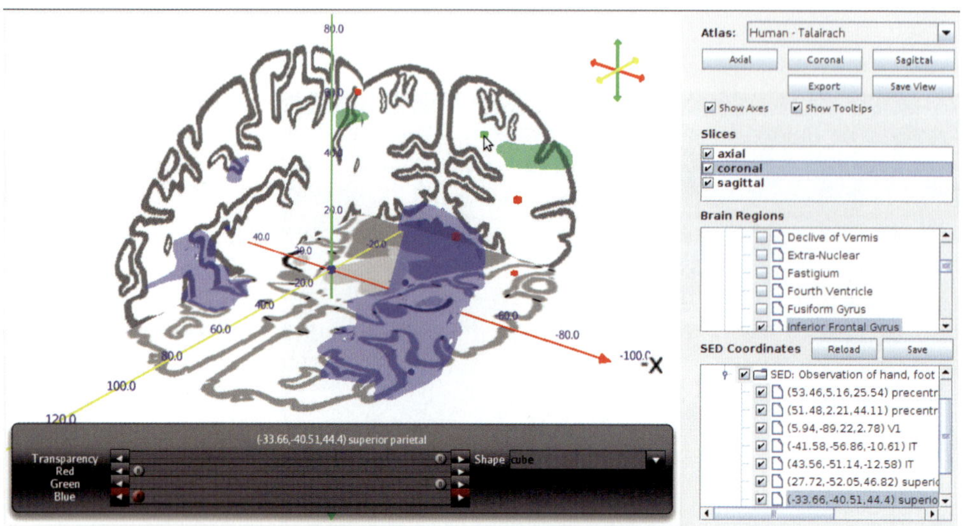

FIGURE 5.21 Visualization in BrainSurfer using the Talairach atlas of the same Summary of Empirical Data (SED) coordinates as displayed in figure 5.20 using the Montreal Neurological Institute (MNI) atlas. Coordinates entered in MNI space are automatically converted to the Talairach space when the Talairach atlas is selected.

between coordinate spaces. Selecting an atlas in the dropdown box of the View Controls section will change the atlas. Figure 5.21 shows the same SED coordinates shown in figure 5.20 using the Talairach atlas. In addition to the MNI and Talairach human brain atlases, BrainSurfer contains the Paxinos–Watson macaque brain atlas (figure 5.22). Future development plans for BODB and BrainSurfer include visualization of macaque neurophysiology. This will require neurophysiologists to report not just the stereotaxic coordinates of their electrode penetrations, which are individual to each animal, but coordinates in a standard atlas space after a structural image (acquired through MRI perhaps) of the animal's brain has been warped into that standard space. We now turn to the current treatment of macaque connectivity data in BODB.

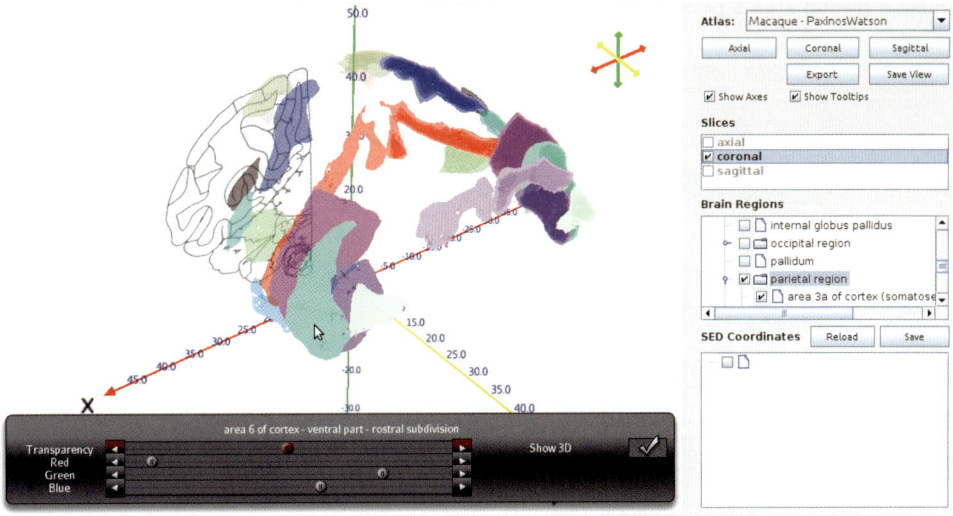

FIGURE 5.22 Visualization of the Paxinos–Watson macaque brain atlas in BrainSurfer. This atlas contains only the coronal brain slice and a subset of brain regions in the complete atlas. Here a set of frontoparietal brain regions have been selected for visualization as 3-D isosurface meshes.

5.5 *Connectivity Data*

The structural connectivity of the macaque brain is typically inferred through tract tracing experiments. These experiments proceed by injecting a tracer substance into a brain region (the injected region) and, after a length of time, examining the other areas of the brain to see where it has spread to (the labeled regions). Anterograde tracers travel from the soma of a neuron along the axon while retrograde tracers travel in reverse toward the soma. Injecting an anterograde tracer into a brain region therefore reveals which other parts of the brain that region projects to while injecting a retrograde tracer reveals which other regions project to it.

CoCoMac (Collations of Connectivity data on the Macaque brain) is a neuroinformatics database dedicated to the storage and description of tract tracing data in the macaque brain (http://cocomac.g-node.org/). CoCoMac stores detailed information about these experiments including the type of tracer used, the density of tracer observed in the labeled regions (which can be used to infer the strength of the connection), and the cortical layers that were labeled with the tracer. Because this data is reported using various brain nomenclatures which can be contradictory and sometimes overlapping, CoCoMac has the ability to translate between regions defined in different nomenclatures where possible.

BODB represents this data in a simplified form of Connectivity SEDs, which define the source region and target region. BODB allows users to search for connectivity SEDs based on the original CoCoMac. Using CoCoMac's web-based API, BODB translates the user's search into a CoCoMac query, parses the returned results, and caches them locally in the form of Connectivity SEDs. Each Connectivity SED derived from CoCoMac retains a link to the original more detailed entry in CoCoMac. The Search Results page displays a list of all connectivity SEDs which match the given criteria. To view detailed information for any projection, users can click on the "View in CoCoMac" link in the appropriate row to open a window linking to the appropriate page in CoCoMac for that projection.

BODB provides tools to dynamically generate directed "network graphs" illustrating a selected set of connections, displayed using Graphviz (http://www.graphviz.org/; Ellson et al., 2002). Nodes in the graph represent brain regions, and the edges represent anatomical projections with arrows indicating the direction of the projection. Users can click on graph nodes to visit pages listing the details of the corresponding brain region, and edges can be clicked on to view the page for the corresponding connectivity SED. Network graphs are generated in BODB in two ways. Users can generate network graphs for a selected set of projections, or they can be generated automatically for each brain region view page showing that region's input and output connectivity (see section 17.3 of the BODB manual, v1.0.0, figure 5.23). In particular, BODB allows users to generate connectivity graphs based on selected connectivity SEDs. These graphs can be used by modelers to design the overall architecture of their model as well as by groups of modelers and experimentalists to define the network of interest for the collaboration.

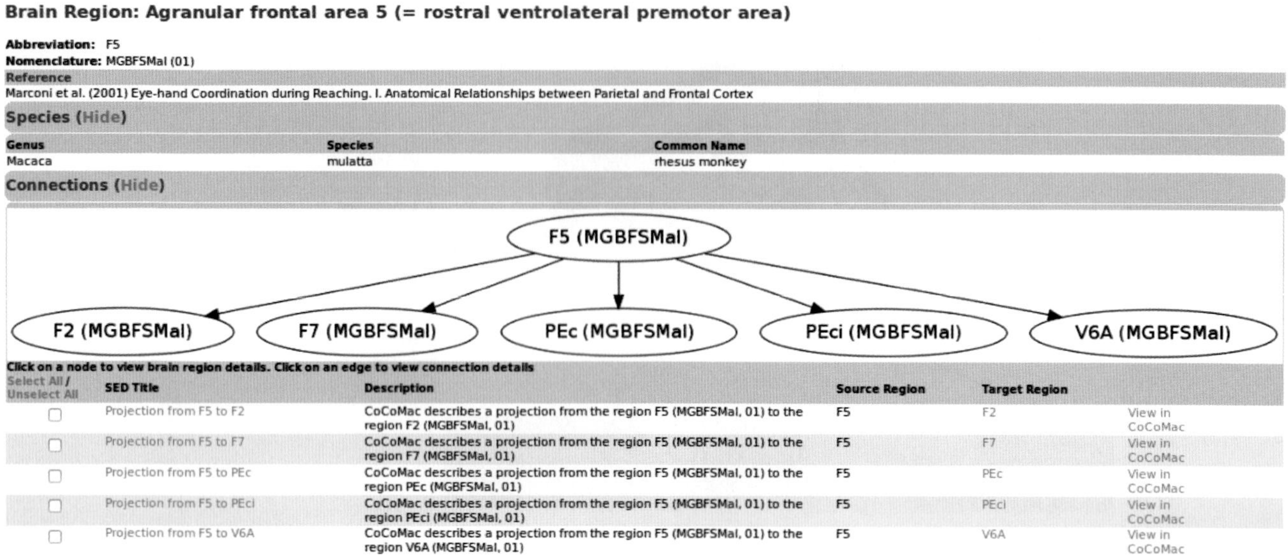

FIGURE 5.23 The brain region view page displays information about the brain region, including the species it is defined in, and all the connectivity Summaries of Empirical Data (SEDs) in which it is included. A network graph is generated from all of its inputs and outputs. F2, F5, PEc, and so forth are labels for regions of the macaque brain; the details are not important here. From Arbib et al. (2014).

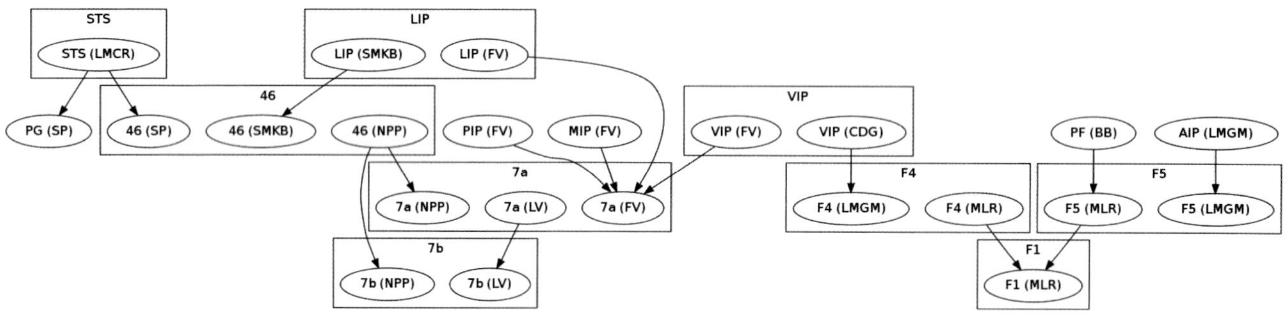

FIGURE 5.24 A connectivity Summary of Empirical Data diagram generated by the Brain Operation Database for the mirror neuron system (MNS2) model (Brain regions: LIP, lateral intraparietal area; PIP, posterior intraparietal area; MIP, medial intraparietal area; VIP: ventral intraparietal area; AIP, anterior intraparietal area; STS, superior temporal sulcus; Nomenclatures: BB, von Bonin and Bailey, 1947; MLR, Matelli, Luppino, and Rizzolatti, 1985; SP, Seltzer and Pandya, 1986; CG, Cavada and Goldman-Rakic, 1989; NPP, Neal, Pearson, and Powell, 1990; FV, Felleman and Van Essen, 1991; CDG, Colby, Duhamel, and Goldberg, 1993; LMCR, Luppino et al., 1993; SMKB, Schall et al., 1995; LMGM, Luppino, Murata, and Matelli, 1999; LV, Lewis and Van Essen, 2000.) From Demiris, Aziz-Zadeh, and Bonaiuto (2014).

The connectivity SED diagram generated by BODB for the MNS2 model is shown in figure 5.24. The modeled network is basically the same as that for MNS (figure 5.5), with the addition of area 46, which is connected with the superior temporal sulcus and inferior parietal areas and implements working memory in the model. The auditory projections used to extend MNS to address audiovisual mirror neurons are largely absent from CoCoMac and are thus listed in table 5.4 along with other generic SEDs used to build MNS2.

5.6 Federation

BODB is one resource among many in neuroinformatics. To supplement its documentation of models and tools for their comparison, we need databases that support comparison and search of empirical data for particular subcommunities of experimentalists and/or field-workers. BODB currently has links to the ModelDB, PubMed, CoCoMac, and Brede databases (figure 5.25), and efforts are currently under way to develop

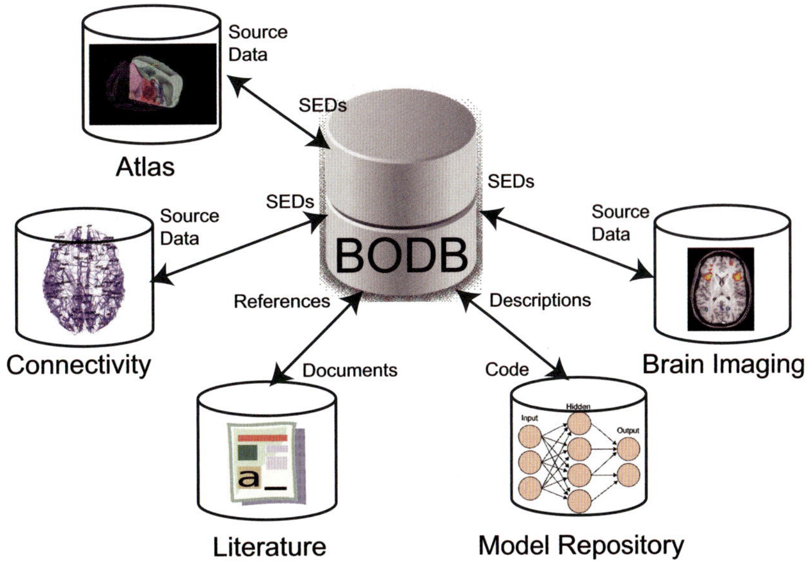

FIGURE 5.25 The Brain Operation Database (BODB) regarded as a node in a federation of databases. Summaries of Experimental Data (SEDs) in BODB link out to more detailed records of source data in external databases. BODB stores model descriptions rather than model code, but the descriptions are linked to code repositories where available.

databases for storing ape gesture and neurophysiology data which will eventually be linked to and summarized in BODB. ModelDB stores model code along with a very brief description of the model using a unique identifier. These entries can be imported into BODB, which will fill in the basic fields of a model entry, allowing the user to complete the description of the model and link to summary data, as well as maintaining a link to the model code in ModelDB. Each literature entry in BODB can be entered by hand, or the user has the option to search and import literature records from PubMed. The PubMed ID is stored along with the literature entry in order to provide a link back to PubMed from BODB. As already noted, connectivity SEDs in BODB are implemented as summaries of more detailed entries in CoCoMac (Kotter, 2004) with links to these entries. We have also developed a system to parse and import summaries of brain imaging experiments from Brede as imaging SEDs. We worked with the author of Brede to develop a mapping from its schema to BODB's brain imaging SED format, which we used to create summarized versions of Brede's experiment descriptions in BODB. We also maintain a mapping from the unique identifiers of the BODB entries to the associated ones in Brede in order to provide bidirectional links between the two sites.

There are currently no dedicated databases for some types of experiments and/or fieldwork. For example, ape gesture researchers typically take hours of video footage of apes in the wild, code the videos for behavioral or gestural events using their own coding schemes, and then publish an analysis of the coded videos. We have been working on an effort to support storing these videos and their coding in a Gesture and Behavior DataBase (GBDB; http://gbdb.usc.edu). When complete, the data will be available in GBDB at a detailed level, and summarized in BODB at a higher level with links to the original entries in GBDB. Similarly, there is currently no widely used database for neurophysiological data. We are working with several mirror neuron experimentalists to develop a database dedicated to storing neural spiking data along with detailed descriptions of the experimental protocols used. Again, this information will be available in summary form in BODB as neurophysiology SEDs with links to the detailed entries in the dedicated database.

The challenge in developing a database dedicated to storing detailed descriptions and data from experiments is coming up with an ontology that is detailed enough to allow interesting analyses to be run across studies but flexible enough to accommodate the widely varying experimental protocols in use. For example, some mirror neuron studies show the monkeys live demonstrations of manual actions to elicit mirror neuron firing while others use videos. The lack of three-dimensional visual information from the video will elicit different activity in visual cortices which may affect mirror neuron activity. Therefore seemingly insignificant differences between experiments could have a large impact on the data and should be captured by the database.

Where databases dedicated to storing experimental data exist, the challenge is summarizing them in BODB at a level useful to modelers. For example, the CoCoMac database stores data from tract tracing experiments at a very fine level of detail including the type of tracer used, how it was injected, how dense the labeling was in effective regions, and so forth. Most modelers are not interested in these details and only want to know if there is indeed a projection from one region to another so that they can construct their model accordingly. BODB therefore represents connectivity data in the form of region *x* projects to region *y*. However, models can be constructed at different levels of granularity, and some modelers may be interested in which layer or cell type is involved in a projection or in estimating the relative strength of the projection. This is where the link from the summary data in BODB to the original data source is useful. BODB users can click on the "View in CoCoMac" link for any connectivity SED derived from CoCoMac in order to see the more detailed information stored there.

Each entry in BODB is associated with a unique ID, generated by the database when the entry is inserted. These unique IDs can be used to reference the entity permanently and retrieve representations of the entity. Representations can be retrieved in HTML format for viewing in a web browser, or as portable document format (PDF) or rich text format (RTF) files. BODB parses incoming requests in order to extract the requested entry type and ID, so representations can be retrieved by supplying the entry type (model, BOP, etc.) and unique ID. For example, the entry for MNS model has the ID 675. The entry can be accessed using the following URL:

http://bodb.usc.edu/bodb/model/675/

Reports in PDF or RTF formats can be accessed using the following URL:

http://bodb.usc.edu/bodb/model/675/report/

This system of unique entry identifiers and dynamic URLs allows us to participate in PubMed's LinkOut program, which provides links from reference information in PubMed to corresponding pages in BODB that list the entities associated with the reference.

5.7 *Collaboratory Workspaces*

At the core of our modeling approach is collaboration—the ontological framework is designed to support the exchange and linking of experiments with the computational models that may support, explain, contradict, or predict empirical evidence. Collaboration should support not only modelers in grounding computational models in empirical results but also experimentalists in allowing them to contribute summaries of their own data that may challenge modelers. We are extending the push for collaboration beyond online content production and knowledge engineering and into the modeling process, which includes the documentation of computational models and the related empirical evidence at an earlier stage of development through the benchmarking of models. Sharing data is an elusive goal for computational neuroscience. However, engaging the experimentalist or the modeler after the fact with little extrinsic motivation has proven difficult—data sharing efforts often struggle to gain a critical mass.

Any individual or group of researchers with a well-focused set of shared interests can define a Collaborative Workspace (CoWo). The Workspace provides an area for users to curate selections of models, SEDs addressed by current models or challenging further model development, imaging coordinates, and SSRs (see section 11 of the BODB manual, v1.0.0). Analysis of models already in the CoWo can reveal challenges for building new models to replace existing models whose relative strengths and weaknesses are revealed by benchmarking. In addition to uses for model development, a shared workspace can offer a point of interaction for experimentalists and modelers researching a common area. Experimentalists can take note of predictions posted by modelers as possible challenges for new experiments and can also post SEDs that they offer as challenges for modelers.

After registration, each user is assigned their own default workspace which is only accessible to them. Each workspace is only visible to users who belong to it, and only users with administrative access to a workspace can add or remove users. Users can add additional workspaces for specific projects and join workspaces to which they have been invited. At the top right corner of the BODB screen, the user's currently active workspace is listed. The BODB search mechanism can be used to search for other workspaces and change the currently active one by clicking on the radio button next to a workspace in the search results. Clicking on the name of a workspace will open the workspace view page, which consists of multiple tabs, one for each section (figure 5.26). Each workspace contains the following tabs: "Activity" lists all of the actions that workspace members have taken within the workspace or on an entry included in the workspace, "Entries" lists all entries included in the workspace, "Coordinates" lists

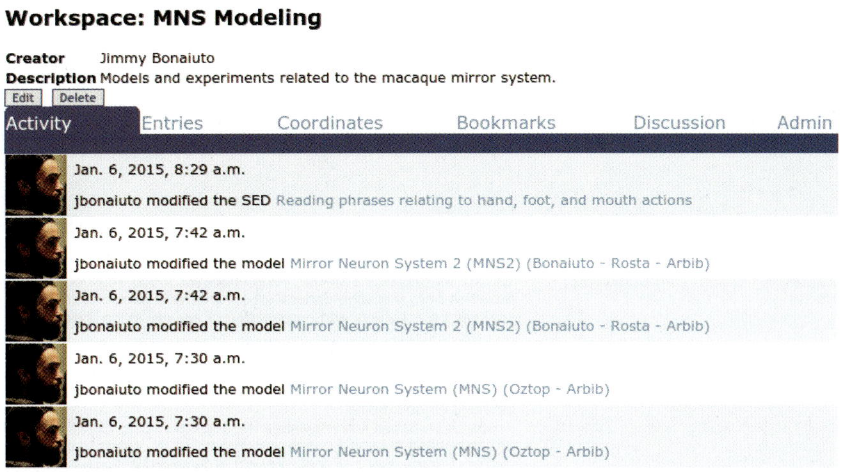

FIGURE 5.26 The main page for an example workspace centered on modeling the mirror system. Administrators of the workspace can edit the description or delete it. The main sections are the Activity trace, Entries list, Coordinates, Bookmarks, Discussion, and Admin sections.

the saved coordinate selections in the workspace (just as users can have their own saved coordinate selections; see section 5.3), "Bookmarks" lists saved bookmarks to external web resources, "Discussion" contains a discussion section, and "Admin" allows workspace administrators to manage access to the workspace.

The core section of a workspace is the "Entries" tab. Figure 5.27 shows the "Entries" tab for the MNS Workspace which lists all of the entries that have been added to that workspace, grouped by type. Each entry has a checkbox next to it which can be used to add or remove it from the workspace (the same checkbox appears in the search results with the same functionality) and star which can be clicked on to add or remove it from the user's personal favorites list. The Models section has been expanded in figure 5.27 to show all the models in the MNS workspace. Clicking on the "Benchmark" button at the bottom of the models list will run the BODB model benchmarking tool on those models (see section 5.2.3 for an example of benchmarking with the MNS and MNS2 models). SEDs in the workspace are also listed in the "Entries" tab, and a reverse benchmark can be run on them showing all the models that use them as building or testing SEDs.

The "Admin" tab is only visible to users with administrative access to the workspace (figure 5.28). The creator of the workspace always has administrative access by default. Here users can turn their private workspace into a Collaboratory Workspace by sending invitations to other users to join. Invited users will receive an email or message (depending on their preferences) with links that they can click on to accept or decline the invitation. Sent invitations are listed at the bottom of the tab along with their status and the option to resend them. Each workspace member is listed in the Members section and can be removed from the workspace or granted administrative access.

5.8 A Sampler of Models in BODB

We have used the MNS and MNS2 models to exemplify the ways in which brain regions, BOPs, SEDs, and SSRs all enter into the documentation of "how a model works" and "what it achieves in relation to empirical data." We close the chapter by sketching two ways forward: (1) how to take advantage of links that exist between models treated in this book and entries in BODB and (2) an invitation to get actively involved in model entry, SED entry, and software development for BODB.

A number of the models in this volume are documented in BODB. Here we give a short high-level view of some of these models and invite you to review these entries as you read the related chapters.

5.8.1 ILGM: HTTP://BODB.USC.EDU/BODB/MODEL/679/ The Infant Learning to Grasp Model (ILGM; Oztop, Bradley, and Arbib, 2004) is a developmental model of grasping described in chapter 20. It simulates how infants learn to plan grasps based on visual object features through a trial-and-error process. The BODB entry for ILGM shows how the model is broken down into Input, Learning to Grasp, Movement Generation, and Movement Evaluation modules. The support building SEDs are "AIP grasp selectivity," "F5 grasp selectivity," and the connectivity SED "AIP

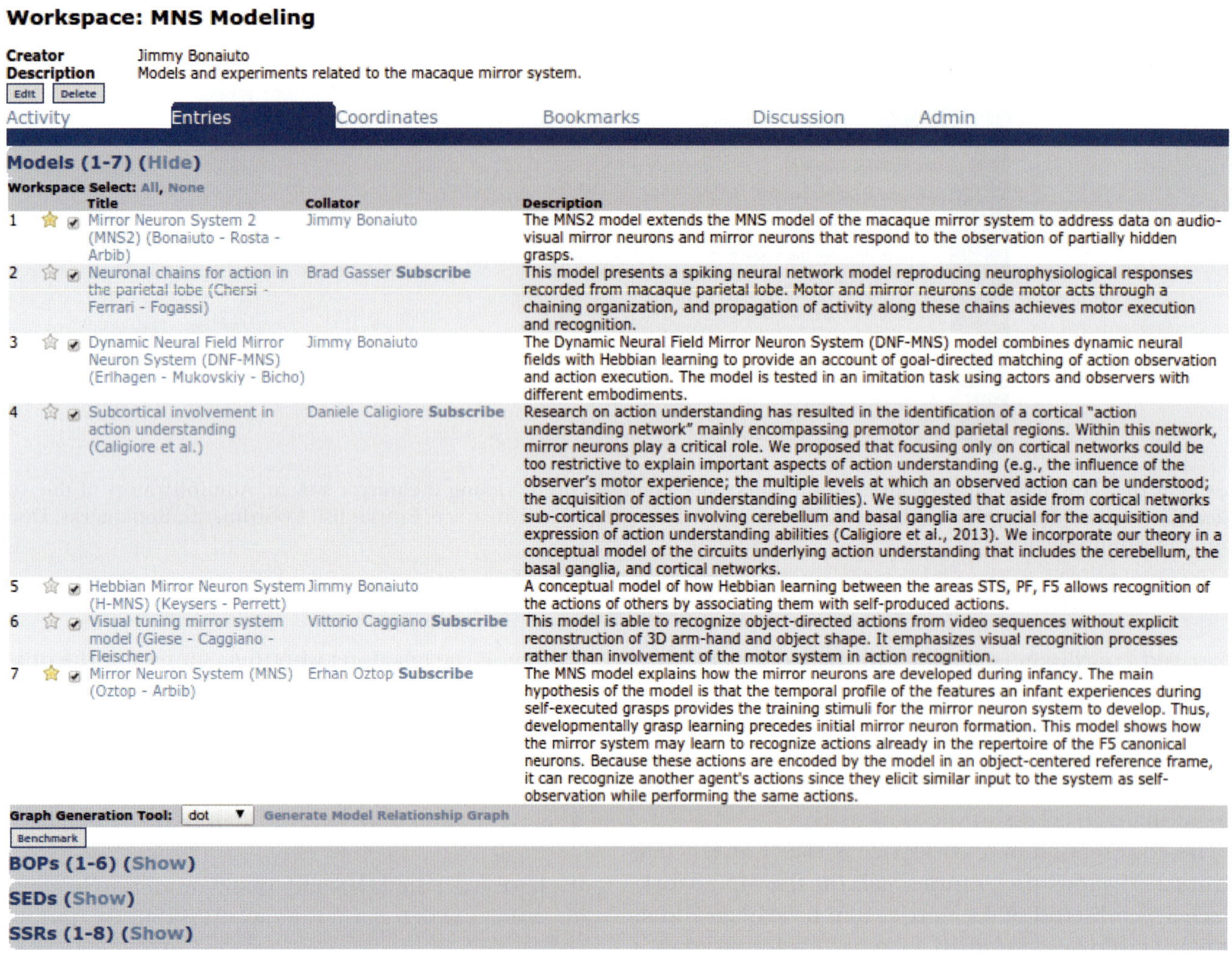

Workspace: MNS Modeling

Creator Jimmy Bonaiuto
Description Models and experiments related to the macaque mirror system.
Edit Delete

Activity | Entries | Coordinates | Bookmarks | Discussion | Admin

Models (1-7) (Hide)
Workspace Select: All, None

	Title	Collator	Description
1	Mirror Neuron System 2 (MNS2) (Bonaiuto - Rosta - Arbib)	Jimmy Bonaiuto	The MNS2 model extends the MNS model of the macaque mirror system to address data on audio-visual mirror neurons and mirror neurons that respond to the observation of partially hidden grasps.
2	Neuronal chains for action in the parietal lobe (Chersi - Ferrari - Fogassi)	Brad Gasser **Subscribe**	This model presents a spiking neural network model reproducing neurophysiological responses recorded from macaque parietal lobe. Motor and mirror neurons code motor acts through a chaining organization, and propagation of activity along these chains achieves motor execution and recognition.
3	Dynamic Neural Field Mirror Neuron System (DNF-MNS) (Erlhagen - Mukovskiy - Bicho)	Jimmy Bonaiuto	The Dynamic Neural Field Mirror Neuron System (DNF-MNS) model combines dynamic neural fields with Hebbian learning to provide an account of goal-directed matching of action observation and action execution. The model is tested in an imitation task using actors and observers with different embodiments.
4	Subcortical involvement in action understanding (Caligiore et al.)	Daniele Caligiore **Subscribe**	Research on action understanding has resulted in the identification of a cortical "action understanding network" mainly encompassing premotor and parietal regions. Within this network, mirror neurons play a critical role. We proposed that focusing only on cortical networks could be too restrictive to explain important aspects of action understanding (e.g., the influence of the observer's motor experience; the multiple levels at which an observed action can be understood; the acquisition of action understanding abilities). We suggested that aside from cortical networks sub-cortical processes involving cerebellum and basal ganglia are crucial for the acquisition and expression of action understanding abilities (Caligiore et al., 2013). We incorporate our theory in a conceptual model of the circuits underlying action understanding that includes the cerebellum, the basal ganglia, and cortical networks.
5	Hebbian Mirror Neuron System (H-MNS) (Keysers - Perrett)	Jimmy Bonaiuto	A conceptual model of how Hebbian learning between the areas STS, PF, F5 allows recognition of the actions of others by associating them with self-produced actions.
6	Visual tuning mirror system model (Giese - Caggiano - Fleischer)	Vittorio Caggiano **Subscribe**	This model is able to recognize object-directed actions from video sequences without explicit reconstruction of 3D arm-hand and object shape. It emphasizes visual recognition processes rather than involvement of the motor system in action recognition.
7	Mirror Neuron System (MNS) (Oztop - Arbib)	Erhan Oztop **Subscribe**	The MNS model explains how the mirror neurons are developed during infancy. The main hypothesis of the model is that the temporal profile of the features an infant experiences during self-executed grasps provides the training stimuli for the mirror neuron system to develop. Thus, developmentally grasp learning precedes initial mirror neuron formation. This model shows how the mirror system may learn to recognize actions already in the repertoire of the F5 canonical neurons. Because these actions are encoded by the model in an object-centered reference frame, it can recognize another agent's actions since they elicit similar input to the system as self-observation while performing the same actions.

Graph Generation Tool: dot Generate Model Relationship Graph
Benchmark
BOPs (1-6) (Show)
SEDs (Show)
SSRs (1-8) (Show)

FIGURE 5.27 The "Entries" tab for the MNS (mirror neuron system) Modeling workspaces. Workspace entries are grouped by type (models; BOPs, Brain Operating Principles; SEDs, Summaries of Experimental Data; SSRs, Summaries of Simulation Results); the models section is expanded here. The checkbox next to each entry allows users to add or remove entries from the workspace. The star can be clicked on to add or remove entries from the user's private favorites list.

projection to F5." These SEDs support the model design decisions to structure the Input and Learning to Grasp modules as well as the projections between them. Several testing SEDs are explained by the SSRs of the module. For example, the SED "Infant precision grasping" is explained by the SSR demonstrating that precision grips can emerge from open-loop grasp learning. The model makes two predictions concerning the posture and locations likely used by infants at different stages of development. This entry shows how predictions can be linked to the SSRs used to make them. The model is linked to several BOPs, such as "Population Coding" and "Reinforcement Learning," for example. This makes the model entry easily searchable by the mechanisms employed by its various modules.

5.8.2 Brainstem Control of Saccades: http://bodb.usc.edu/bodb/model/22712/

The brainstem control of saccades model (described in chapter 19) simulates groups of neurons as a feedback control system for the generation of saccades. The BODB entry breaks this model up into several submodules representing populations of different types of neurons. The model is linked to two building SEDs which describe the kinematics of saccade movements and neurophysiological data from brainstem neurons involved in saccade control. This model is linked to one testing SED that is contradicted by the model's SSRs. The testing SED describes the fact that saccades remain accurate after lesioning a particular type of neuron; however, the model predicts that this lesion should impair saccades. The relevance narrative for this SED

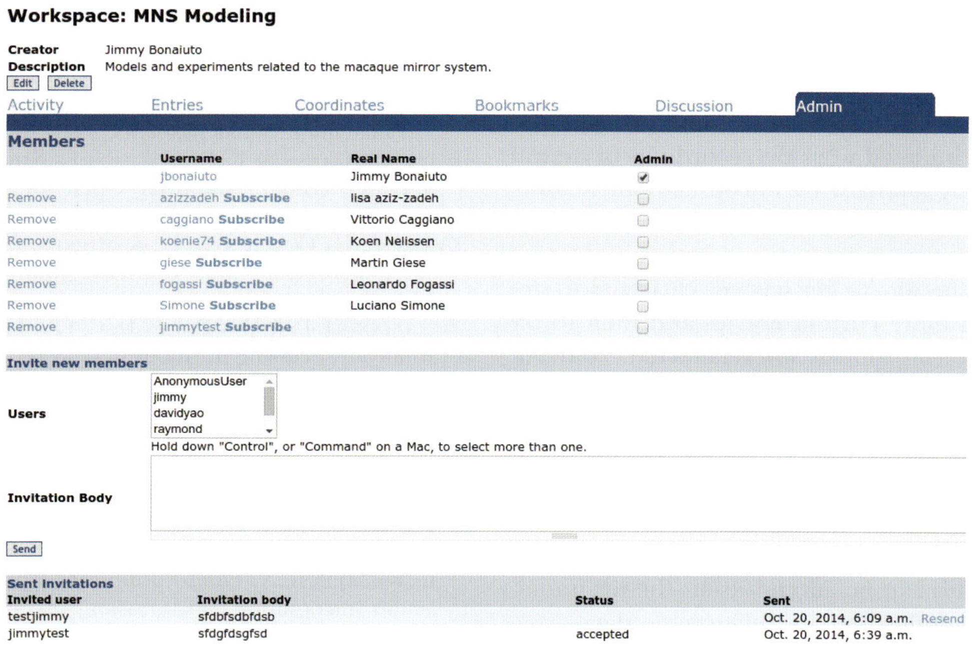

FIGURE 5.28 The workspace "Admin" tab is only visible to users with administrative privileges for that workspace. Administrators can remove users from the workspace, grant them administrative access, or send invitations to other users to join the workspace.

describes how the SED and model could be reconciled through the addition of signals from other structures such as the cerebellum. This suggests a possible avenue for developing a successor model that could explain rather than contradict this SED. In addition to the related BOPs that were previously present in BODB ("Feedforward and Feedback Control" and "Internal Models"), this model entry suggested a new BOP: "Push–Pull Control." This BOP describes the mechanism used by the model to control the eye via excitation and inhibition of antagonist muscles on either side of it.

5.8.3 MEG MU RHYTHM MODEL: HTTP://BODB.USC.EDU/BODB/MODEL/21897/ The magnetoencephalography (MEG) mu rhythm model is a cortical column model that can generate simulated MEG signals measured in human somatosensory cortex (see chapter 4). This BODB entry contains numerous building SEDs which support the design of the model at a very fine level including the maximal conductance of various currents in the model neurons, locations of synapses on pyramidal neurons for different input projections, and laminar specificity of inputs to the somatosensory cortex, as well as the timing of the inputs to the model. The testing SEDs demonstrate how the model explains the spontaneous MEG measured somatosensory rhythm as well as the relationship between the prestimulus mu rhythm and the evoked response to tactile stimuli. The model predicts that inhibition recruited during higher power prestimulus rhythms reduces the subsequent evoked response peaks, and this prediction is supported by simulated tactile evoked responses during high and low prestimulus mu power. This model entry links the model to several BOPs including "Action-Oriented Perception" and "Top-Down/Bottom-Up Hybridization." The "Action-Oriented Perception" BOP states that an organism does not simply react to current stimuli; rather it actively seeks input data relevant to its ongoing actions. The model predicts that decreased ongoing alpha and beta rhythms in somatosensory cortex increase the gain of sensory evoked responses, generating enhanced tactile detection and suggests that these rhythms are shifted with attention to optimize perception. In this model, the ongoing rhythm reflects "top-down" activity (e.g., modulated with attention), which is hybridized in the model with the external sensory evoked drive, reflecting bottom-up activity.

5.9 Getting Involved with BODB

We have shown how brain imaging data and neuroanatomical data may be captured in BODB. At present, many of the SEDs in BODB for other forms of data take the form of purely verbal summaries, ill-suited to formal comparison. As this chapter is being sent to press, we are perfecting SED formats for ERPs and primate gestures, and have initiated work on recordings of neural firing in behaving animals. We encourage our readers to engage in the development of further formats of particular relevance to their own work in computational neuroscience. The way to get started, clearly, is to get experience with more current entries and then to try your hand at entering models and related data yourself. We consider this in two stages.

We encourage you to enter existing models into BODB. Of especial value is to enter models that are either related to models already in BODB or to each other and then carry out benchmarking to assess where each model succeeds and where it fails relative to the other model. This can be an end itself—better understanding each model and its relation to data—but, more importantly, can set the stage for the design of new models that can encompass new data using structures and processes that can be more fully validated by empirical data. Collaborative efforts can then be supported by carrying out these efforts in a Collaboratory Workspace shared by like-minded scholars or students. Further suggestions for how to proceed along these lines can be found at http://bodb.usc.edu/media/BODB_Workspace_Manual.pdf for groups of researchers, both experimentalists and modelers, working together to advance some specific topic in cognitive or systems neuroscience.

REFERENCES

Arbib, M. A., Plangprasopchok, A., Bonaiuto, J., & Schuler, R. E. (2014). A neuroinformatics of brain modeling and its implementation in the Brain Operation Database BODB. *Neuroinformatics, 12*(1), 5–26.

Aziz-Zadeh, L., Wilson, S. M., Rizzolatti, G., & Iacoboni, M. (2006). Congruent embodied representations for visually presented actions and linguistic phrases describing actions. *Current Biology, 16*(18), 1818–1823.

Bonaiuto, J., Rosta, E., & Arbib, M. (2007). Extending the mirror neuron system model: I. Audible actions and invisible grasps. *Biological Cybernetics, 96*(1), 9–38.

Bota, M., & Arbib, M. A. (2004). Integrating databases and expert systems for the analysis of brain structures: Connections, similarities, and homologies. *Neuroinformatics, 2,* 19–58.

Cavada, C., & Goldman-Rakic, P. S. (1989). Posterior parietal cortex in rhesus monkey: II. Evidence for segregated corticocortical networks linking sensory and limbic areas with the frontal lobe. *Journal of Comparative Neurology, 287*(4), 422–445.

Colby, C. L., Duhamel, J. R., & Goldberg, M. E. (1993). Ventral intraparietal area of the macaque: Anatomic location and visual response properties. *Journal of Neurophysiology, 69*(3), 902–914.

Demiris, Y., Aziz-Zadeh, L., & Bonaiuto, J. (2014). Information processing in the mirror neuron system in primates and machines. *Neuroinformatics, 12*(1), 63–91.

Ellson, J., Gansner, E., Koutsofios, L., North, S. C., & Woodhull, G. (2002). Graphviz—Open source graph drawing tools. *Lecture Notes in Computer Science, 2265,* 483–484.

Evans, A. C., Collins, D. L., & Milner, B. (1992). An MRI-based stereotactic atlas from 250 young normal subjects. *Journal of the Society for Neuroscience Abstracts, 18,* 408.

Felleman, D. J., & Van Essen, D. C. (1991). Distributed hierarchical processing in the primate cerebral cortex. *Cerebral Cortex, 1*(1), 1–47.

Gleeson, P., Crook, S., Cannon, R. C., Hines, M. L., Billings, G.O., Farinella, M., et al. (2010). NeuroML: A language for describing data driven models of neurons and networks with a high degree of biological detail. *PLoS Computational Biology, 6*(6), e1000815.

Keysers, C., Kohler, E., Umiltà, M. A., Nanetti, L., Fogassi, L., & Gallese, V. (2003). Audiovisual mirror neurons and action recognition. *Experimental Brain Research, 153*(4), 628–636.

Kotter, R. (2004). Online retrieval, processing, and visualization of primate connectivity data from the CoCoMac database. *Neuroinformatics, 2*(2), 127–144.

Lancaster, J., Summerlin, J., Rainey, L., Freitas, C. S., & Fox, P. T. (1997). The Talairach Daemon: A database server for Talairach Atlas Labels. *NeuroImage, 5*(4 Pt 2), S633.

Lewis, J. W., & Van Essen, D. C. (2000). Corticocortical connections of visual, sensorimotor, and multimodal processing areas in the parietal lobe of the macaque monkey. *Journal of Comparative Neurology, 428*(1), 112–137.

Luppino, G., Matelli, M., Camarda, R., & Rizzolatti, G. (1993). Corticocortical connections of area F3 (SMA-proper) and area F6 (pre-SMA) in the macaque monkey. *Journal of Comparative Neurology, 338*(1), 114–140.

Luppino, G., Murata, A., Govoni, P., & Matelli, M. (1999). Largely segregated parietofrontal connections linking rostral intraparietal cortex (areas AIP and VIP) and the ventral premotor cortex (areas F5 and F4). *Experimental Brain Research, 128*(1–2), 181–187.

Matelli, M., Luppino, G., & Rizzolatti, G. (1985). Patterns of cytochrome oxidase activity in the frontal agranular cortex of the macaque monkey. *Behavioural Brain Research, 18*(2), 125–136.

Migliore, M., Morse, T. M., Davison, A. P., Marenco, L., Shepherd, G. M., & Hines, M. L. (2003). ModelDB. *Neuroinformatics, 1*(1), 135–139.

Neal, J. W., Pearson, R. C., & Powell, T. P. S. (1990). The ipsilateral cortico-cortical connections of area 7b, PF, in the parietal and temporal lobes of the monkey. *Brain Research, 524*(1), 119–132.

Nordlie, E., Gewaltig, M. O., & Plesser, H. E. (2009). Towards reproducible descriptions of neuronal network models. *PLoS Computational Biology, 5*(8), e1000456.

Oztop, E., & Arbib, M. A. (2002). Schema design and implementation of the grasp-related mirror neuron system. *Biological Cybernetics, 87*(2), 116–140.

Oztop, E., Bradley, N. S., & Arbib, M. A. (2004). Infant grasp learning: A computational model. *Experimental Brain Research, 158*(4), 480–503.

Paxinos, G., Huang, X., & Toga, A. W. (2009). *The Rhesus Monkey Brain in Stereotaxic Coordinates.* Amsterdam: Elsevier.

Schall, J. D., Morel, A., King, D. J., & Bullier, J. (1995). Topography of visual cortex connections with frontal eye field in macaque: Convergence and segregation of processing streams. *Journal of Neuroscience, 15*(6), 4464–4487.

Seltzer, B., & Pandya, D. N. (1986). Posterior parietal projections to the intraparietal sulcus of the rhesus monkey. *Experimental Brain Research, 62*(3), 459–469.

Talairach, J., & Tournoux, P. (1988). *Co-planar Stereotaxic Atlas of the Human Brain.* New York: Thieme.

Umiltà, M. A., Kohler, E., Gallese, V., Fogassi, L., Fadiga, L., Keysers, C., & Rizzolatti, G. (2001). I know what you are doing: A neurophysiological study. *Neuron, 31*, 155–165.

von Bonin, G., & Bailey, P. (1947). The neocortex of Macaca mulatta. *Illinois Monographs in the Medical Sciences, 5*(4).

Werbos, P. J. (1990). Backpropagation through time: What it does and how to do it. *Proceedings of the IEEE, 78*(10), 1550–1560.

6 Hebbian Learning and Plasticity

WULFRAM GERSTNER

The elementary processing units in the brain are neurons (see chapter 2) which are connected to each other via cable-like extensions, called axons and dendrites. The contact points between an axon terminal of a neuron A and the dendrite of another neuron B are called synapses. From the point of view of information processing and learning, the most important synapses are chemical synapses. There also exist electrical synapses, which we will not consider further because they are less relevant for learning. When an action potential of neuron A arrives at a chemical synapse, a chemical signal (neurotransmitter) is ejected into the synaptic cleft and taken up by receptors embedded into the membrane of the receiving neuron (figure 6.1A). Upon transmitter binding, an ion channel opens, and ions flow into the cell and cause a change in the membrane potential of neuron B. The response of the receiving neuron to presynaptic spike arrival can be measured as an excitatory or inhibitory postsynaptic potential (EPSP or IPSP) at its soma.

The most important insight for all the rest of chapter is that the strength of a synapse is not fixed but can change. The strength of a synapse from a presynaptic neuron j to a postsynaptic neuron i can be measured as the amplitude of the postsynaptic potential which is generated if the spike emitted by the presynaptic neuron arrives at the synapse (figure 6.1B, top). If, because of some stimulation protocol, neuron i and neuron j are "active together" for some time, the synapse is strengthened, which can be measured as an increase in the amplitude of the postsynaptic potential (figure 6.1B, bottom). The changes of the strength (or other properties) of synapses are called synaptic plasticity. Synaptic plasticity is an important basis of learning (Bliss and Collingridge, 1993; Malenka and Nicoll, 1999; Morris, 2003). A strengthening of synapses induced by the coactivation of two neurons is called "Hebbian" plasticity (Hebb, 1949).

Synaptic plasticity comes in several forms, such as short-term and long-term plasticity; we are mainly interested in long-term plasticity as this is the most relevant one for learning. Section 6.1 reviews these different forms of plasticity. Hebbian plasticity protocols typically induce long-term plasticity. Several experimental paradigms and results can be interpreted in the framework of Hebbian plasticity. The aim of this chapter is to formulate Hebbian plasticity mathematically and study its functional consequences.

6.1 *Introduction to Synaptic Plasticity and Long-Term Potentiation*

Changes of synapses in the network of neurons in the brain are called synaptic plasticity. These changes are thought to be the basis of learning (Bliss and Collingridge, 1993; Malenka and Nicoll, 1999; Morris, 2003). However, before we start with a review of some experiments regarding synaptic plasticity, we need to ask this question: what is its putative functional role? Or, to say it differently, what is learning?

6.1.1 WHAT IS LEARNING? A small child learns to unscrew the lid of a bottle. An older child learns to ride a bicycle, to ski, or to skateboard. Learning activities such as in these examples typically use an internal reward system of the brain: it hurts if we fall from the bicycle, and we are happy if we achieve for the first time a slalom slope with the skis.

Learning is also the basis of declarative and episodic memories: we know the name of the current president of the United States because we have heard it often enough; we know the date of the French Revolution because we have learned it in school. These are examples of declarative memories, or memories of facts. We can remember the first day we went to school; we can recall a beautiful scene encountered during our last vacation. These are examples of episodic memories, which have been generated most often without explicit learning but are still acquired (as opposed to inborn) and have therefore been "learned" in the loose sense of the word.

As a result of learning, a musician is able to distinguish between tones that sound absolutely identical to the ear of a normal untrained human. In experiments

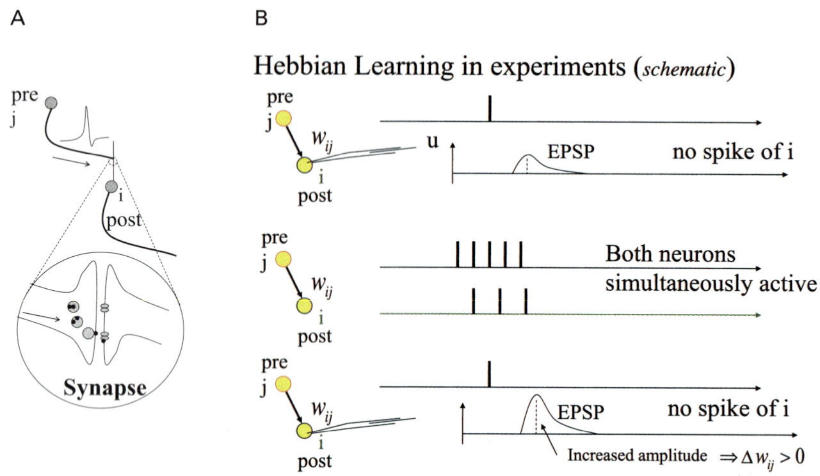

FIGURE 6.1 (A) Chemical synapse. A neuron j sends a spike along the axon to a synapse onto neuron i. The sending neuron is called presynaptic ("pre"), the receiving neuron postsynaptic ("post"). Upon arrival of a spike at the presynaptic terminal (zoomed image, left part), synaptic vesicles release neurotransmitter into the synaptic cleft. If neurotransmitter molecules are captured by receptors of the postsynaptic cell, synaptic ion channels open, causing the influx of electric current into the postsynaptic cell. (B) Hebbian plasticity. Top: The amplitude of the excitatory postsynaptic potential (EPSP) generated at the soma of neuron i in response to the spike arrival at an excitatory synapse is a measure of the synaptic strength w_{ij}. Middle: The neurons are "active together," which triggers Hebbian plasticity. Bottom: The result of Hebbian plasticity is an increase of the synaptic weight w_{ij}, measured again as the amplitude of the postsynaptic potential in response to a single spike arriving at the synapse.

with monkeys measurable differences were found between the auditory areas of animals exposed to specific tones and others living in a normal environment (Recanzone et al., 1993). More generally, it is believed that the cortex adapts itself such that more neurons are devoted to stimuli that appear more frequently or are more important and less neurons to less relevant ones (Buonomano and Merzenich, 1998). This adaptation of cortex (see chapter 13, "Neural Maps: Their Function and Development") is also subsumed under the term of "learning" in the wider sense. For example, developmental learning refers to the fact that during a critical period after birth connectivity between neurons in cortex develops toward its final functional form (Hubel, 1988).

Most likely, all the different forms of learning that we have mentioned (action learning, formation of declarative or episodic memories, adaptation of cortical organization) are in one way or another related to changes in the strength of synapses. These changes, and their relation to "learning," are the topic of this chapter.

6.1.2 SHORT-TERM VERSUS LONG-TERM SYNAPTIC PLASTICITY Changes in synaptic strength can be induced in a controlled way in preparations of brain slices containing subnetworks of connected neurons. Conceptually, the simplest situation is that where the activity of two connected neurons can be controlled by injecting short current pulses.

First, the strength of the connection is measured by generating a single test spike in the presynaptic neuron while recording the postsynaptic potential (or postsynaptic current) in the postsynaptic neuron. Then an appropriate stimulus is given to induce a change of the synapse. Finally, a second test spike is evoked in the presynaptic neuron, and the change in the amplitude of the postsynaptic potential is noted (figure 6.1B).

One important question is whether the induced change will disappear rapidly or persist for a long time (figure 6.2A). For example, a sequence of three, five, or ten spikes at 100 Hz in the presynaptic neuron can increase or decrease the amplitude of the measured EPSP by a factor of two or more. To be specific, let us consider a sequence of three presynaptic spikes. Suppose that the EPSP in response to the first spike arrival has an amplitude of 1.0 mV, the second EPSP an amplitude of 1.2 mV, and the third one an amplitude of 1.4 mV. This increase in EPSP amplitude is an indication that the synapse is plastic, that is, its strength is not fixed. Importantly, if another test pulse is given 5 seconds later, this change has disappeared (Markram et al., 1998; Abbott et al., 1997). Since this type of plasticity lasts only for one or a few seconds, it is called short-term plasticity. Short-term plasticity can lead to a transient potentiation or a transient depression of the synapse (figure 6.2A). It is induced rapidly, but it also disappears rapidly. The effect does not persist.

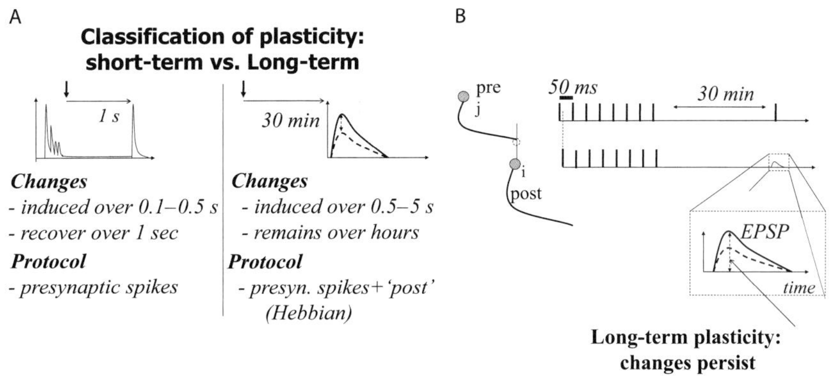

FIGURE 6.2 (A) Left: A sequence of four excitatory postsynaptic potentials (EPSPs) induced by the arrival of four spikes at the synapse. The response to the first spike is strongest. Subsequent EPSPs have smaller amplitude. This effect is called "synaptic depression." Depression has disappeared one second later. Since the synapse recovers rapidly, this effect is called short-term plasticity (or short-term depression). Right: Synaptic potentiation induced by a Hebbian protocol is still present 30 minutes after induction of plasticity. It is therefore called long-term potentiation (LTP). (B) One possible protocol to induce LTP is to experimentally activate both pre- and postsynaptic neurons together, with a small delay pre-before-post. The weight w_{ij} of the synapse from neuron j to i is changed after the LTP protocol, visible as a difference in the amplitude of the postsynaptic potential before and after application of the Hebbian protocol (inset, lower right). pre and post, pre- and postsynaptic neuron.

We now consider a different stimulus. The presynaptic neuron is stimulated (e.g., by injecting 60 short electrical pulses into the neuron) so as to produce 60 spikes at 20 Hz. In parallel the postsynaptic neuron is also stimulated to produce 60 spikes at 20 Hz, but the two stimulation protocols are slightly shifted so that the postsynaptic neuron fires always 5 ms *after* the presynaptic one (figure 6.2B). Note that the stimulus only lasts 3 seconds in total. If we measure the amplitude of the EPSP after the stimulation protocol and compare it with that before, we observe a significant increase in the size of the EPSP. The synapse is said to be "potentiated." Importantly, the effect persists for minutes and even hours. Hence it is an example of persistent potentiation, also called long-term potentiation (LTP). A slight change in the stimulation paradigm can lead to long-term depression (LTD).

We can loosely summarize this subsection by saying that presynaptic activity alone induces short-term plasticity, while the joint activity of pre- and postsynaptic neuron together causes long-term plasticity. Since we are interested in plasticity for learning, we focus in the following on persistent plasticity, that is, LTP and LTD.

6.1.3 EXPERIMENTAL PROTOCOLS FOR LTP, LTD, AND SPIKE TIMING DEPENDENT PLASTICITY The specific protocol discussed above is called spike timing dependent plasticity (STDP) (Markram et al., 1997; Bi and Poo, 1998; Abbott and Nelson, 2000; Sjöström et al., 2001; J. Sjöström and Gerstner, 2010), but it is only one example of a much broader class of stimulation protocols that are all suitable to induce LTP. In the classic STDP protocol both pre- and postsynaptic neuron are driven by precisely timed current pulses to emit spikes at appropriate moments in time. If the presynaptic spike arrives at the synapses systematically *before* the postsynaptic neuron fires, then the synapse is strengthened and LTP occurs. If the sequence is reversed, so that the postsynaptic neuron fires before the presynaptic one, then the synapse is weakened and LTD occurs.

Controlling the spiking of two connected neurons requires advanced experimental techniques that only became available in the 1990s. STDP protocols show that plasticity can depend on precise spike timing, not just loosely on "two neurons are active together." We will come back to STDP protocols further below.

Instead of the postsynaptic neuron being driven to firing by short electrical pulses, it can also be held at a weakly or strongly depolarized level. If the depolarization is combined with presynaptic spike arrival, this causes LTP or LTD depending on the amount of depolarization (Artola et al., 1990; Ngezahayo et al., 2000). These, and similar (Sjöström et al., 2001) experiments, show that the induction of long-term plasticity depends on postsynaptic voltage.

Instead of controlling both presynaptic and postsynaptic neurons precisely, one can also stimulate a bundle of presynaptic fibers by a high-frequency sequence of extracellular current pulses while recording (extracellularly) from postsynaptic neurons. If enough presynaptic fibers are stimulated, the postsynaptic neuron is depolarized or even firing, and LTP can be induced. The first results on LTP and LTD in the 1970s and 1980s used such extracellular stimulation and recording techniques (Bliss and Lomo, 1973; Levy and Stewart, 1983). Such an extracellular stimulation protocol is also particularly useful to confirm that changes induced by LTP or LTD indeed persist for many hours (Frey and Morris, 1998).

We mentioned already a distinction between spike-timing-based protocols on one side and voltage-based or extracellular protocols on the other side. Another important distinction is the presence or absence of neuromodulators during the plasticity-inducing protocol (Reynolds and Wickens, 2002; Pawlak and Kerr, 2008; Seol et al., 2007; Zhang et al., 2009; Pawlak et al., 2010). We will come back to the role of neuromodulators in section 6.5 of this chapter, as well as in chapter 10.

6.2 *Hebbian Learning*

LTP and LTD are thought to be the synaptic basis of learning in the brain. Many experiments on LTP and LTD, and nearly all synaptic theories of learning, have been inspired by Hebbian learning. Hebbian learning under the influence of neuromodulators is linked to reinforcement learning and will be discussed below in section 6.5.

6.2.1 Definition of Hebbian Learning and Hebbian Plasticity Hebb (1949) gave a concise formulation of correlation-based plasticity which has conceptual roots that can in fact be traced back much further into the past (Markram et al., 2011). Hebb postulated that it would be useful to have a rule such that synapses are modified whenever the two neurons that are connected to each other are active together. It is sometimes summarized in the slogan "fire together—wire together," but the exact wording is worth a read:

> When an axon of cell A is near enough to excite cell B and repeatedly or persistently takes part in firing it, some growth process or metabolic change takes place in one or both cells such that A's efficiency, as one of the cells firing B, is increased.

In contrast to the compressed slogan it contains a "causal" tone, because "taking part in firing" the postsynaptic neuron implies that the presynaptic neuron is, at least partly, causing the spike of the postsynaptic one. We will come back to this aspect when we speak more about STDP in section 6.4.

At the moment, two insights are important. First, for the connection from A to B only two neurons are important, namely, A and B, but not any other neuron C that might make a connection onto A or B. We summarize this insight by saying that the learning rule is "*local*": only information that is available at the location of the synapse can be used to change the weight of that synapse. Obviously, information about the activity of A is accessible at the synapse because spikes of neuron A arrive at the synapse. Similarly, the state of neuron B can be sensed at the synapses by a general increase in the membrane potential, by a backpropagating action potential, or by an increase in some chemical substance.

Second, the wording "cell A ... repeatedly ... takes part in firing" cell B implies that both cells have to be active together several times. We summarize this insight by saying that the learning rule must be sensitive to the *joint activity* between the action potentials of the two neurons. Since repeated joint activity can be measured by a *correlation* function, we may say that Hebbian learning is sensitive to correlations between the activity of A and B.

With these two insights in mind, let us now return to the plasticity protocols for LTP that we discussed in the previous subsection. Indeed, in the STDP protocol (with 60 repetitions of pre-before-post, i.e., a presynaptic neuron is made to fire 5 ms before the postsynaptic one) both neurons, the presynaptic neuron A and the postsynaptic neuron B, are repeatedly active together (i.e., the firing of A and B is correlated on a precise temporal scale) while other neurons C do not play a role (i.e., we have locality). Hence, it is a Hebbian protocol.

The traditional extracellular stimulation protocol excites many presynaptic fibers to generated spikes at high frequencies, and these presynaptic spikes are likely to fire the postsynaptic neuron, too. Therefore, we have again a Hebbian protocol. The main difference is that of timescale: a rate-based Hebbian protocol defines joint activity (or "correlations") and a timescale of a few hundred milliseconds (Levy and Stewart, 1983). A spike-timing-based Hebbian protocol studies joint activity on the timescale of a few milliseconds (Markram et al., 1997).

Joint activity seems to require postsynaptic firing. However, one may argue that postsynaptic firing can be (experimentally) replaced by strong postsynaptic depolarization, so that the depolarization protocols mentioned above fall also into the category of Hebbian protocols. We may summarize this by stating that

Hebbian protocols are experimental procedures that induce correlations between the activity of the presynaptic neuron and the state of the postsynaptic one, where the state can be defined by the firing rate, spikes, or membrane potential of the postsynaptic cell. Hebbian plasticity is the result of such a Hebbian protocol.

Before we formulate this insight mathematically, let us ask: why should we care about Hebbian plasticity? What is its functional role?

6.2.2 Why Should Hebbian Learning Be Useful? The short answer is: if connections between neurons reflect past correlated activity, then these stronger connections are useful to represent associations. Let us illustrate this by way of an example.

Suppose that, somewhere in the brain, we have a network of neurons that can mutually excite each other. The network could consist of 10,000 neurons or more, but only a few of these are depicted in figure 6.3. We assume now that, while a human or animal sees a banana for the first time, a subset of these neurons are active. The neurons may represent in an abstract form the different sensations associated with the percept, such as the form, color, or smell of the banana. Maybe somebody standing nearby says in addition the word "banana" or breaks it open and starts to eat it.

If the brain has a Hebbian learning rule, the result of the coactivation of different neurons in our network will be that all the connections between the simultaneously active neurons are strengthened. The claim is that this means that the item "banana" has now been memorized. The memory concept "banana" has been formed and is represented by the set of strongly connected cells, called a cell assembly (Hebb, 1949).

How can we check that the memory concept "banana" works? Let us suppose that, the next day, the subject sees a banana, partly hidden behind another object. The banana as a whole is not visible, but there are some yellow cues, organized into a slightly bent structure. Let us now *suppose* that seeing part of the banana stimulates again a subset of the neurons that have been active the day before when the banana was seen for the first time. Because of the previously established strong connections, the neurons that are part of the "banana concept" but are currently inactive will now be activated by neighboring neurons that send spikes across the strong connections. As a result, after a few iterations, the memory concept "banana" is fully activated and the memory item is retrieved. Moreover, in the presence of spontaneous fluctuations of the activity, different aspects of "bananahood" will flash up. The basic idea of memory retrieval discussed here is at the heart of models of working memory and long-term memory (see chapter 13).

6.2.3 A Family of Hebbian Rules: Toward Mathematical Formulations Hebb was a theoretically inclined psychologist, who developed the essential insights into why the principle, which we now call the Hebb rule, would be useful, but he was not a mathematician and never wrote down a mathematical formulation of his rule. Finding an appropriate mathematical description is our task now. In this subsection we follow closely the treatment in chapter 10.2 of Gerstner and Kistler (2002).

In order to find a mathematically formulated learning rule based on Hebb's postulate we focus on a single synapse with efficacy w_{ij} that transmits signals from a presynaptic neuron j to a postsynaptic neuron i. For the moment we focus on a description in terms of mean firing rates. That means we treat the activity as a continuous variable ν (representing the number of spikes in some time interval T where T is on the order of a few hundred milliseconds). In other words, we work at a

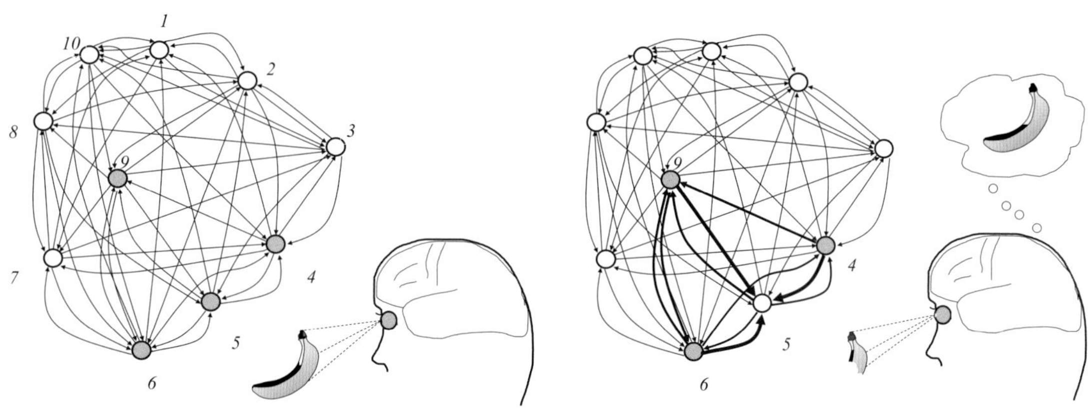

Figure 6.3 A novel percept causes simultaneous activity of a subset of neurons, for example, those numbered as 4, 5, 6, and 9 (left). Under the assumption of Hebbian learning, the connections between these neurons will be strengthened because their activity switches on together, so that their firing is correlated. The set of strongly linked cells representing a common percept is a cell assembly. A partial cue later (right) is sufficient to recall the complete memory item since neurons that are not activated by the sensory cue (e.g., neuron 5) will be excited by the strong lateral connections.

low temporal resolution. In the following, the activity of the presynaptic neuron is denoted by ν_j and that of the postsynaptic neuron by ν_i.

As mentioned before, there are two aspects in Hebb's postulate that are particularly important, namely, *locality* and *correlations*. Locality means that the change of the synaptic efficacy can only depend on local variables, that is, on information that is available at the site of the synapse, such as pre- and postsynaptic firing rate, and the actual value of the synaptic efficacy, but not on the activity of other neurons. Based on the locality of Hebbian plasticity, we can make an educated guess for the change of the synaptic efficacy,

$$\frac{\mathrm{d}}{\mathrm{d}t} w_{ij} = F(w_{ij}; \nu_i, \nu_j). \tag{6.1}$$

Here, $\mathrm{d}w_{ij}/\mathrm{d}t$ is the rate of change of the synaptic coupling strength, and F is a so far undetermined function.

We may wonder whether there are other local variables (e.g., the membrane potential u_i) that should be included as additional arguments of the function F. It turns out that in standard rate models this is not necessary since the membrane potential u_i is uniquely determined by the postsynaptic firing rate, $\nu_i = g(u_i)$, with a monotone gain function g.

The second important aspect of Hebb's postulate, coactivation, or correlation, implies that pre- and postsynaptic neuron have to be active *simultaneously* for a synaptic weight change to occur. "Simultaneous" here can take different meanings depending on whether we consider a rate description or a spike-based description, but let us focus for the moment on the simplest case, that is, a rate model. We can use the notion of simultaneous activity to learn something about the function F. If F is sufficiently well-behaved, we can expand F in a Taylor series about $\nu_i = \nu_j = 0$,

$$\frac{\mathrm{d}}{\mathrm{d}t} w_{ij} = c_0(w_{ij}) + c_1^{\text{post}}(w_{ij})\nu_i + c_1^{\text{pre}}(w_{ij})\nu_j + c_2^{\text{pre}}(w_{ij})\nu_j^2 + c_2^{\text{post}}(w_{ij})\nu_i^2 + c_2^{\text{corr}}(w_{ij})\nu_i\nu_j + \mathcal{O}(\nu^3). \tag{6.2}$$

The term containing c_2^{corr} on the right-hand side of equation 6.2 is bilinear in pre- and postsynaptic activity. This term implements the AND condition necessary to detect correlations between pre- and postsynaptic activity. It is this term which makes Hebbian learning a useful concept.

Naive Hebbian Learning Rule The simplest choice for our function F is to fix c_2^{corr} at a positive constant and to set all other terms in the Taylor expansion to zero. The result is the prototype of Hebbian learning,

$$\frac{\mathrm{d}}{\mathrm{d}t} w_{ij} = c_2^{\text{corr}} \nu_i \nu_j. \tag{6.3}$$

However, this is too simple. Rates ν_i and ν_j are positive. Together with a positive constant c_2^{corr} this would lead to unlimited growth of weights. Synapses with large weight are physically bigger and therefore take more space in the brain than smaller ones. Infinitely large synapses are impossible because the brain as a whole is of finite size.

Limiting Weight Growth The dependence of F on the synaptic efficacy w_{ij} takes care of the fact that w_{ij} has to be bounded. If F were independent of w_{ij}, then the synaptic efficacy would grow without limit if the same potentiating stimulus is applied over and over again. The need for controlling weight growth was recognized as soon as Hebbian rules were put on a computer (Rochester et al., 1956). There are many ways to limit weight growth in simulations. Two simple practical ones are the following.

First, explosion of weights can be avoided by "hard bounds": $c_2^{\text{corr}}(w_{ij}) = \gamma_2 > 0$ for $0 < w_{ij} < 1$ and zero otherwise. In other words, weight growth stops if the weight reaches its maximum value.

Second, a more subtle saturation of synaptic weights can be achieved if the parameter c_2^{corr} in equation 6.2 tends to zero as w_{ij} approaches its maximum value, say, $w^{\max} = 1$, for example,

$$c_2^{\text{corr}}(w_{ij}) = \gamma_2(1 - w_{ij}) \tag{6.4}$$

with a positive constant γ_2. An interpolation between the soft bounds and hard bounds can be implemented by writing $c_2^{\text{corr}}(w_{ij}) = \gamma_2(1 - w_{ij})^{\eta}$ with a parameter $0 \le \eta \le 1$. For $\eta = 0$ we retrieve the hard bounds while for $\eta = 1$ we are back to the linear soft bounds (Gütig et al., 2003).

Correlation-Based Hebbian Rules Obviously, setting all parameters except c_2^{corr} to zero is a very special case of the general framework developed in equation 6.2. Are there other "Hebbian" learning rules in this framework?

First we note that a learning rule with $c_2^{\text{corr}} = 0$ and only first-order terms (such as $c_1^{\text{post}} \neq 0$ or $c_1^{\text{pre}} \neq 0$) would be called non-Hebbian plasticity because pre- or postsynaptic activity alone induces a change of the synaptic efficacy. Hence these learning rules miss the *correlation* aspect of Hebb's principle. Thus, if we stop the expansion after the terms which are quadratic in the rates, a learning rule in the family of equation 6.2 needs a term $c_2^{\mathrm{corr}} > 0$ so as to qualify as Hebbian.

However, complicated Hebbian rules that are sensitive to correlations can be constructed if in addition to the linear terms, and the terms ν_i^2 or ν_j^2, other terms in the expansion of equation 6.2, such as $\nu_i\nu_j^2$, $\nu_i^2\nu_j$, $\nu_i^2\nu_j^2$, and so forth, are included as well. A learning rule with a positive coefficient in front of $\nu_i^2\nu_j$ would also qualify as Hebbian, even if c_2^{corr} vanishes or is negative, because at high postsynaptic firing rates the positive correlations between pre- and postsynaptic activity dominate the dynamics. We will come back to this case below in the context of the Bienenstock–Cooper–Munro rule.

Hebb's original proposal does not contain a rule for a decrease of synaptic weights. In a system where synapses can only be strengthened, all efficacies will finally saturate at their upper maximum value. An option of decreasing the weights (synaptic depression) is therefore a necessary requirement for any useful learning rule (Rochester et al., 1956; Stent, 1973). This can, for example, be achieved by weight decay, which can be implemented in equation 6.2 by setting

$$c_0(w_{ij}) = -\gamma_0 w_{ij}. \tag{6.5}$$

Here, γ_0 is a (small) positive constant that describes the rate by which w_{ij} decays back to zero in the absence of stimulation. Our formulation (equation 6.2) is hence sufficiently general to allow for a combination of synaptic potentiation and depression. If we combine equations 6.4 and 6.5, we obtain the learning rule

$$\frac{\mathrm{d}}{\mathrm{d}t} w_{ij} = c_2^{\mathrm{corr}}(w_{ij})\nu_i\nu_j - \gamma_0 w_{ij}. \tag{6.6}$$

In the absence of stimulation, or if only one of the two neurons is active, the last term leads to an exponential decay of the weight toward $w_{ij} = 0$. In other words, memories in such a model cannot last for long unless each memory is regularly reactivated.

Competition, Normalization, and Oja's Rule Another interesting aspect of learning rules is *competition.* The idea is that synaptic weights can only grow at the expense of others, so that if a certain subgroup of synapses is strengthened, other synapses to the same postsynaptic neuron have to be weakened. Practically, competition can be implemented in simulations by normalizing (in an explicit algebraic step) the sum of all weights converging onto the same postsynaptic neuron (Rochester et al., 1956; von der Malsburg, 1973; Miller and MacKay, 1994). Though this can be motivated by a limitation of common synaptic resources, such a learning rule violates locality of synaptic plasticity.

Much more elegant, however, is a formulation that remains local but makes use of the second-order term ν_i^2 in equation 6.2. Specifically, we take $c_2^{\mathrm{corr}} = \gamma > 0$ and $c_2^{\mathrm{post}} = -\gamma w_{ij}$ and set all other parameters to zero. The learning rule

$$\frac{\mathrm{d}}{\mathrm{d}t} w_{ij} = \gamma[\nu_i\nu_j - w_{ij}\nu_i^2] \tag{6.7}$$

is called Oja's rule (Oja, 1982). If combined with a linear rate model neuron (i.e., the firing rate of the postsynaptic neuron is $\nu_i = \sum_j w_{ij}\nu_j$), Oja's rule converges asymptotically to synaptic weights that are normalized to $\sum_j w_{ij}^2 = 1$ while keeping the essential Hebbian properties of the standard rule of equation 6.3. We note that normalization of $\sum_j w_{ij}^2$ implies competition between the synapses that make connections to the same postsynaptic neuron, that is, if some weights grow, others must decrease. With other learning rules a different normalization $\sum_j w_{ij} = 1$ can be enforced (Miller and MacKay, 1994).

6.2.4 The Bienenstock–Cooper–Munro Rule as an Example of Hebbian Learning

Consider a learning rule in the framework of equation 6.2 with a higher-order term equal to $\eta\nu_i^2\nu_j$ where η is the expansion parameter. We use a parameter c_2^{corr} of the "Hebbian" term equal to $c_2^{\mathrm{corr}} = -\eta\theta$ where we have introduced a new parameter in the form of a reference rate θ and set all other expansion parameters to zero. The expansion then gives

$$\frac{\mathrm{d}}{\mathrm{d}t} w_{ij} = \eta(\nu_i^2 - \theta\nu_i)\nu_j. \tag{6.8}$$

This is an example of the Bienenstock–Cooper–Munro rule (Bienenstock et al., 1982; Cooper et al., 2004), also called the BCM rule, which is, more generally, written as

$$\frac{\mathrm{d}}{\mathrm{d}t} w_{ij} = \eta\phi(\nu_i)\nu_j \tag{6.9}$$

with a nonlinear function $\phi(\nu_i)$ which we take as $\phi(\nu_i) = \nu_i(\nu_i - \theta)$ where θ is the reference rate (Bienenstock et al., 1982).

What does this learning rule do? If the presynaptic neuron is inactive ($\nu_j = 0$), the synaptic weight does not change. Let us now suppose that the presynaptic neuron is active ($\nu_j > 0$) while at the same time the postsynaptic neuron is firing at a high rate $\nu_i > \theta$ (see figure 6.4).

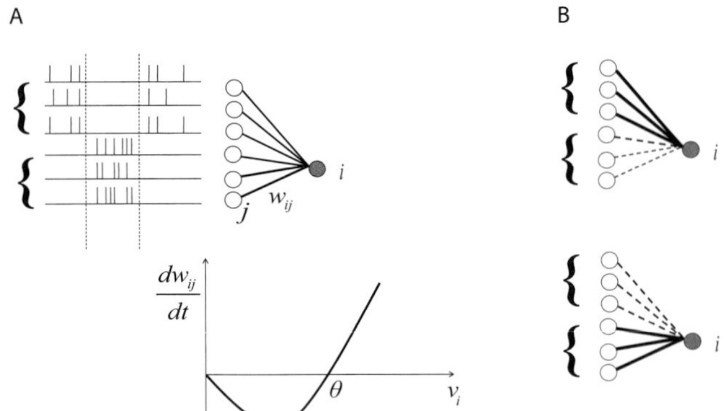

FIGURE 6.4 Learning in the Bienenstock–Cooper–Munro model of synaptic plasticity. (A) A postsynaptic neuron i receives input from several presynaptic neurons; one of them is marked with the index j. The connection w_{ij} from j to i increases or decreases as a function of the postsynaptic firing rate v_i (bottom), provided that the presynaptic neuron j is active. The activity of presynaptic neurons (top left) switches from time to time (vertical lines) between two groups of correlated inputs. (B) During learning one of the input groups develops strong synapses while the synapses of the others are weakened.

The synaptic weight increases since both neurons are jointly active. Hence it is a Hebbian rule. If the presynaptic neuron is firing while the postsynaptic neuron is only weakly active ($v_i < \theta$), then the synaptic weight decreases. Thus, the parameter θ marks the transition point between the induction of LTD and LTP. Bienenstock et al. (1982) proposed this learning rule to account for observations in visual cortex during developmental learning. The nonlinearity of the function ϕ ensures that (1) no plasticity happens if the postsynaptic neuron is inactive, (2) both LTP and LTD are possible, and (3) high postsynaptic activity leads to LTP.

Two important insights can be derived from an analysis of equation 6.9 (see also Exercise 1 on the book's website for an intuitive approach).

First, we can convince ourselves that the weight dynamics have a fixed point at $v_i = \theta$. If all the synaptic weights remain fixed, then the postsynaptic rate (for constant presynaptic input) also has a fixed point. However, if the postsynaptic rate is just slightly above θ, then all active synapses will increase, and so does the postsynaptic rate. Thus, the fixed point of the firing rate is unstable. In order to avoid that the postsynaptic firing rate blows up (or decays to zero), it is therefore necessary to turn θ into an adaptive variable (Bienenstock et al., 1982). More precisely, θ must have a superlinear dependence upon the averaged firing rate $\overline{v}_i$. A good choice is to set $\theta = (\overline{v}_i)^2 / v_{\text{target}}$ where v_{target} is a parameter that plays the role of a target rate (Cooper et al., 2004).

Second, if the input is structured into two groups, for example, we randomly switch between two stimuli, a first one that activates input fibers 1 to 50 and a second one that drives inputs 51 to 100, and if the parameter θ is in an appropriate regime, then the learning rule separates the weights into two groups: some weights increase toward the upper bound at the expense of other weights that decrease to zero (figure 6.4B). Thus, the learning rule exhibits competition.

Competition is the key ingredient to understanding the development of receptive fields: because of synaptic plasticity, the postsynaptic neuron becomes specialized to a subset of the inputs. Functional consequences of Hebbian learning, in particular the development of receptive fields, is the topic of the next section.

6.3 *Functional Consequences of Hebbian Learning*

In the previous section, a family of Hebbian learning rules has been introduced. We will now study functional consequences of Hebbian plasticity. From the point of view of a neuroscientist, one particularly intriguing result is that Hebbian learning can lead to the development of receptive fields. From the algorithmic point of view, the most interesting result is that Hebbian learning is akin to principal component analysis (PCA). Let us discuss these points sequentially.

6.3.1 WHAT ARE RECEPTIVE FIELDS? A given neuron in cortex does not respond to all stimuli, but only to a small subset of potential stimuli. For example, a neuron in primary visual cortex responds to visual stimuli—but not all of them. Suppose a monkey fixates on a red cross in the center of a gray computer screen while an electrophysiologist records the activity of one single neuron in the visual cortex of the monkey (figure 6.5A). On the screen localized light blobs are switched on at various locations, in some random sequence. For most possible locations the neuron remains quiet. However, if the light blob occurs in one specific small region of the screen, the neuron becomes active. The zone in the visual field where the neuron is responsive is called the (visual) receptive field.

Ganglion cells in the retina have circularly symmetrical receptive fields, and so do neurons in the lateral geniculate nucleus (LGN), the subsequent processing step on the way to cortex. Neurons in the primary area of visual cortex have receptive fields that best respond to elongated light bars (Hubel, 1962). Different neurons prefer light bars at different orientations. In other words, each neuron in the primary area exhibits orientation tuning, that is, if the light bar is turned around,

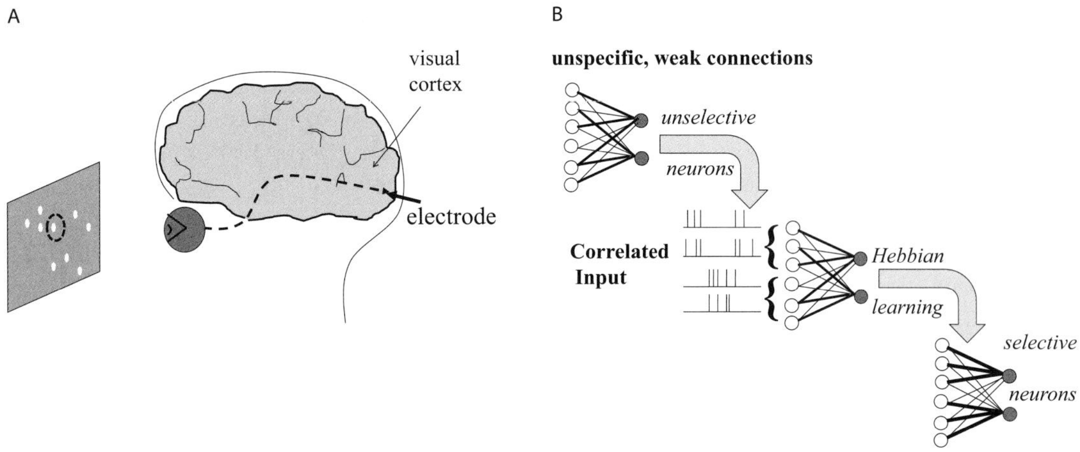

FIGURE 6.5 (A) The receptive field of a visual neuron is the zone on the gray screen (left) in which a small light blob evokes a response. The response is measured with an electrode in proximity to the neuron (right). (B) If model neurons are stimulated with spatially correlated input so that neighboring presynaptic neurons respond similarly to changes in the input, the postsynaptic neurons develop localized receptive fields as a result of Hebbian synaptic plasticity.

a neuron responds vividly to some preferred orientation and only weakly to a stimulus orthogonal to the preferred one. Receptive fields occur not only in vision but also in other modalities such as audition or touch.

6.3.2 APPLICATION TO RECEPTIVE FIELD DEVELOPMENT To understand the concept of receptive fields from the point of view of modeling, let us consider a group of several postsynaptic neurons, each being connected to the same set of presynaptic neurons, say, 1,000 presynaptic neurons. The presynaptic neurons play the role of potential locations of a stimulus.

Suppose that you test the response properties of one specific postsynaptic neuron by giving an input to a single presynaptic neuron. Before plasticity started all presynaptic neurons were connected with approximately the same weak weight to this postsynaptic neuron. Therefore the postsynaptic neuron responds unspecifically to all possible input locations. In other words, the postsynaptic neuron does not exhibit a localized receptive field.

We now switch on Hebbian learning (implemented, e.g., by the BCM rule of equation 6.8 or the Oja rule of equation 6.7) and stimulate the array of presynaptic neurons by "natural stimuli." We assume that natural stimuli consist of objects of finite spatial extension so that neighboring presynaptic neurons receive, most of the time, a similar stimulation. We find that after learning, the postsynaptic neuron responds only to a small subset of the presynaptic neurons. Hence, the neuron has developed a receptive field.

During Hebbian learning, neurons become sensitive to a subset of the inputs, namely, those inputs that show the strongest correlations. We will develop in the next section the analysis tools that will enable us to understand this phenomenon. While, for didactic reasons, we will focus the analysis on a single postsynaptic neuron, both the analysis and simulations can also be done for connected neurons.

Let us assume that model neurons are linear. More precisely, we assume that the gain function is $g(h) = h$. In other words, the firing rate of a cortical neuron i is

$$\nu_i = g(h_i) = h_i = \sum_j w_{ij} x_j + \sum_k b_{ik} \nu_k \qquad (6.10)$$

where w_{ij} denotes the feedforward weights from a presynaptic neuron j (located in the "input space," e.g., the retina or LGN) to a cortical neuron i, and b_{ik} is the lateral connectivity within cortex from neuron k to i. The activity of the input neurons are denoted by the variable x_j.

We assume locally restricted lateral connectivity. Lateral connectivity is fixed, that is, there is no plasticity in the lateral weights b_{ik}. Moreover, we assume that because of genetically encoded chemical gradients, a rough first arrangement of the feedforward weights w_{ij} is prewired at the beginning of cortical development, so that neurons in the upper left corner of the grid have receptive fields located in the upper left region of the input space. However, even the rather rich genetic code does not contain enough information to signal to all the billions of potential synapses between pre- and postsynaptic neurons where to form. Rather synaptic plasticity rules are in place during development that control the formation, strengthening, and weakening of synapses. Therefore we assume in the model that the synaptic weights w_{ij} from the input layer to the grid of postsynaptic neurons are subject to a Hebbian learning rule.

An analysis of such a model is possible (Linsker, 1986b; MacKay and Miller, 1990; Wimbauer et al., 1998). Here, however, we will just summarize some results from a computer simulation consisting of an array of 8 × 8 cortical neurons and two times 20 × 20 LGN neurons. Figure 6.6 shows a typical outcome of such a simulation after stimulation of local clusters of presynaptic neurons at random locations. Each of the small rectangles shows the receptive field of the corresponding cortical neuron. Light gray means that the neuron responds with an increased firing rate to a bright spot at that particular position within its receptive field; dark gray indicates inhibition.

There are two interesting aspects. First, the evolution of the synaptic weights has led to asymmetrical receptive fields, which give rise to orientation selectivity. Second, the structure of the receptive fields of neighboring cortical neurons are similar; neuronal response properties thus vary continuously across the cortex. The neurons are said to form a *map* for, for example,

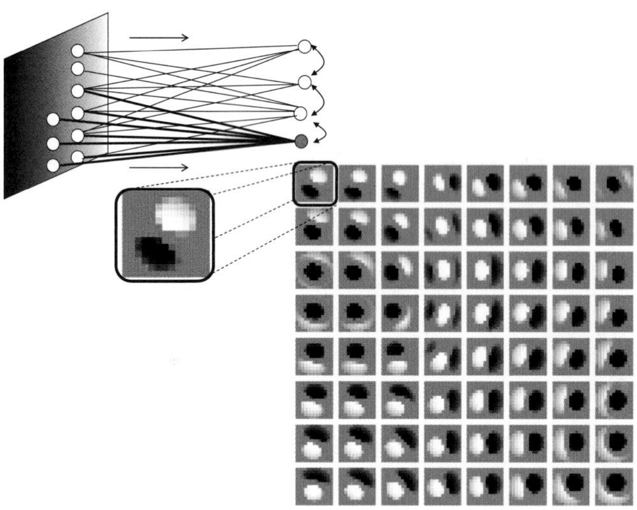

FIGURE 6.6 Top left: An array of sensory neurons projects onto several cortical neurons. Cortical neurons have local lateral connectivity of fixed weight. Main figure: Receptive fields (small squares) of 64 cortical neurons (large grid). Each small square shows the distribution of weights for one of the cortical neurons. The receptive field of the neuron in the upper left corner of the grid is enlarged, and the interpretation of the receptive field in terms of projections from presynaptic "ON" and "OFF" neurons is indicated schematically in the upper left part of the figure. Sensory "ON neurons" respond to bright spots of light while "OFF neurons" respond to dark stimuli. Strong connections from "ON neurons" onto a cortical neuron are represented by white, those from "OFF neurons" as black pixels. Weak or absent connections are shown as gray pixels. Adapted from Wimbauer et al. (1998).

orientation. The formation of cortical maps is the topic of chapter 14 and has a long-standing history as a modeling problem in the field of computational neuroscience (von der Malsburg, 1973; Willshaw and von der Malsburg, 1976; Kohonen, 1982; Linsker, 1986a; Grossberg, 1969).

6.3.3 CORRELATIONS AND PRINCIPAL COMPONENT ANALYSIS We now turn to a formal derivation of the functional properties of Hebbian learning rules. We focus on Hebbian learning in a rate formulation (cf. equation 6.2) and show mathematically that simple models of Hebbian learning are sensitive to correlations in the input (Oja, 1982). Note that this is a nontrivial statement: by definition, Hebbian learning is sensitive to correlations between pre- and postsynaptic neurons; what we are claiming, however, is that it is sensitive to correlations between several presynaptic neurons. More precisely, we will show that a standard Hebb rule combined with a linear rate model for the postsynaptic neuron performs PCA. This section follows closely the text in chapter 11.1 of Gerstner and Kistler (2002).

We analyze the evolution of synaptic weights using the simple Hebbian learning rule of equation 6.3 and then generalize to the Oja rule (equation 6.7). For the analysis, we consider a highly simplified scenario consisting of an analog neuron that receives input from N presynaptic neurons with firing rates ν_j^{pre} via synapses with weights w_j (cf. figure 6.7). We think of the presynaptic neurons as "input neurons" which, however, do not have to be retinal neurons but could be some other sensory neurons. The input layer could, for example, consist of neurons in the LGN that project to neurons in the visual cortex. We will see that the statistical properties of the input control the evolution of synaptic weights.

For the sake of simplicity, we model the presynaptic input as a set of static patterns. Let us suppose that we have a total of p patterns $\{\vec{\xi}^\mu; 1 \le \mu \le p\}$. At each time step one of the patterns $\vec{\xi}^\mu$ is selected at random and presented to the network by fixing the presynaptic rates at $\nu_j^{\text{pre}} = \xi_j^\mu$. We call this the *static-pattern scenario.* The presynaptic activity drives the postsynaptic neuron, and the joint activity of pre- and postsynaptic neurons triggers changes of the synaptic weights. The synaptic weights are modified according to a Hebbian learning rule, that is, according to the correlation of pre- and postsynaptic activity (cf. equation 6.3). Before the next input pattern is chosen, the weights are changed by an amount

$$\Delta w_j = \gamma \nu^{\text{post}} \nu_j^{\text{pre}}. \tag{6.11}$$

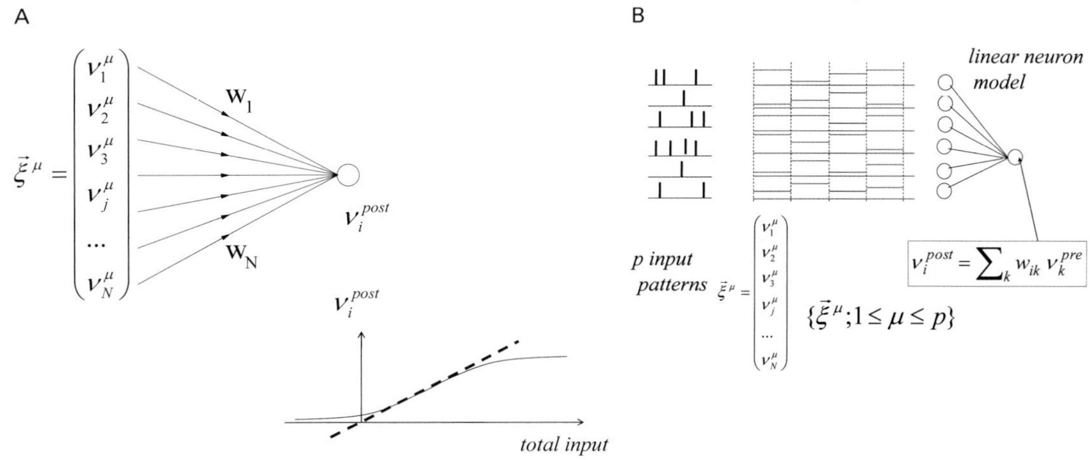

FIGURE 6.7 (A) Patterns $\vec{\xi}^{\mu}$ are applied as a set of presynaptic firing rates ν_j, that is, $\xi_j^{\mu} = \nu_j^{\mu}$ for $1 \le j \le N$. The output rate of the postsynaptic neuron is taken as a linear function of the total input, an approximation to a sigmoidal gain function. Adapted from Gerstner and Kistler (2002). (B) Patterns are applied one after the other in a sequence.

Here, $0 < \gamma \ll 1$ is a small constant called "learning rate." Since there is only one postsynaptic neuron, we have suppressed the index i of the postsynaptic cell. The learning rate in the static-pattern scenario is closely linked to the correlation coefficient c_2^{corr} in the continuous-time Hebb rule introduced in equation 6.3. In order to highlight the relation, let us assume that each pattern $\vec{\xi}^{\mu}$ is applied during an interval Δt. For Δt sufficiently small, we have $\gamma = c_2^{\text{corr}} \Delta t$.

In a general rate model, the firing rate ν^{post} of the postsynaptic neuron is given by a nonlinear function of the total input $\nu^{\text{post}} = g\left(\sum_k w_k \nu_k^{\text{pre}}\right)$, but for the sake of simplicity, we restrict our discussion in the following to a *linear* rate model with

$$\nu^{\text{post}} = \sum_k w_k \nu_k^{\text{pre}}. \tag{6.12}$$

Obviously, this is a highly simplified neuron model, but it will serve our purpose of gaining some insights into the evolution of synaptic weights.

If we combine the learning rule (equation 6.3) with the linear rate model of equation 6.12, we find after the presentation of pattern $\vec{\xi}^{\mu}$ that the synaptic weight w_j connecting the presynaptic neuron j to the postsynaptic cell is changed by an amount

$$\Delta w_j = \gamma \sum_k w_k \nu_k^{\text{pre}} \nu_j^{\text{pre}} = \gamma \sum_k w_k \xi_k^{\mu} \xi_j^{\mu}. \tag{6.13}$$

The evolution of the weight vector $\vec{w} = (w_1, \ldots, w_N)$ is thus determined by the iteration

$$w_j(n+1) = w_j(n) + \gamma \sum_k w_k(n) \xi_k^{\mu_n} \xi_j^{\mu_n}, \tag{6.14}$$

where μ_n denotes the pattern that is presented during the nth time step.

We are interested in the long-term behavior of the synaptic weights. To this end we assume that the weight vector evolves along a more or less deterministic trajectory with only small stochastic deviations that result from the randomness in which new input patterns are chosen. This is, for example, the case if the learning rate is small so that a large number of patterns has to be presented in order to induce a substantial weight change. In such a situation it is sensible to consider the expectation value of the weight vector, that is, the weight vector $\langle \vec{w}(n) \rangle$ averaged over the sequence $(\vec{\xi}^{\mu_1}, \vec{\xi}^{\mu_2}, \ldots, \vec{\xi}^{\mu_n})$ of all patterns that so far have been presented to the network. From equation 6.14 we find

$$\begin{aligned}\langle w_j(n+1) \rangle &= \langle w_j(n) \rangle + \gamma \sum_k \langle w_k(n) \xi_k^{\mu_{n+1}} \xi_j^{\mu_{n+1}} \rangle \\ &= \langle w_j(n) \rangle + \gamma \sum_k \langle w_k(n) \rangle \langle \xi_k^{\mu_{n+1}} \xi_j^{\mu_{n+1}} \rangle \\ &= \langle w_j(n) \rangle + \gamma \sum_k C_{jk} \langle w_k(n) \rangle.\end{aligned} \tag{6.15}$$

The angular brackets denote an ensemble average over the whole sequence of input patterns $(\vec{\xi}^{\mu_1}, \vec{\xi}^{\mu_2}, \ldots)$. The second equality is due to the fact that input patterns are chosen *independently* in each time step, so that the average over $w_k(n)$ and $(\xi_k^{\mu_{n+1}} \xi_j^{\mu_{n+1}})$ can be factorized. In the final expression we have introduced the symmetrical correlation matrix $C_{kj} = C_{jk}$ with,

$$C_{jk} = \frac{1}{p} \sum_{\mu=1}^{p} \xi_j^{\mu} \xi_k^{\mu} = \langle \xi_j^{\mu} \xi_k^{\mu} \rangle_{\mu}. \tag{6.16}$$

The expression of equation 6.15 can be written in a more compact form using matrix notation,

$$\langle \vec{w}(n+1)\rangle = (1+\gamma C)\langle \vec{w}(n)\rangle = (1+\gamma C)^{n+1}\langle \vec{w}(0)\rangle, \qquad (6.17)$$

where $\vec{w}(n) = (w_1(n), \ldots, w_N(n))$ is the weight vector and 1 is the identity matrix.

If we express the weight vector in terms of the eigenvectors $\vec{e}_k$ of C,

$$\langle \vec{w}(n)\rangle = \sum_k a_k(n)\vec{e}_k, \qquad (6.18)$$

we obtain an explicit expression for $\langle \vec{w}(n)\rangle$ for any given initial condition $a_k(0)$, namely,

$$\langle \vec{w}(n)\rangle = \sum_k (1+\gamma\lambda_k)^n a_k(0)\vec{e}_k. \qquad (6.19)$$

Since a symmetrical correlation matrix is positive semi-definite, all eigenvalues λ_k are real and positive. Therefore, the weight vector is growing exponentially, but the growth will soon be dominated by the eigenvector with the largest eigenvalue, that is, the *first principal component*,

$$\langle \vec{w}(n)\rangle \xrightarrow{n\to\infty} (1+\gamma\lambda_1)^n a_1(0)\vec{e}_1\,. \qquad (6.20)$$

Recall that the output of the linear neuron model (equation 6.12) is proportional to the projection of the current input pattern $\vec{\xi}^\mu$ on the direction $\vec{w}$. For $\vec{w} \propto \vec{e}_1$, the output is therefore proportional to the projection on the first principal component of the input distribution. A Hebbian learning rule such as that of equation 6.13 is thus able to extract the first principal component of the input data; see figure 6.8.

From a data-processing point of view, the extraction of the first principle component of the input data set by a biologically inspired learning rule seems to be very compelling. There are, however, a few drawbacks and pitfalls. First, the above statement about the Hebbian learning rule is limited to the *expectation value* of the weight vector. This is not a real problem, however, since if the learning rate is sufficiently low, then the actual weight vector is at every time step very close to the expected one and exhibits a small jitter-like movement around the expected one.

Second, principal components are meaningful only if the input data are normalized to zero mean, that is, distributed around the origin. This requirement is *not* consistent with a rate interpretation because rates are usually positive. This problem, however, can be overcome by learning rules with appropriately chosen linear terms c^{pre} and c^{post} in equation 6.2.

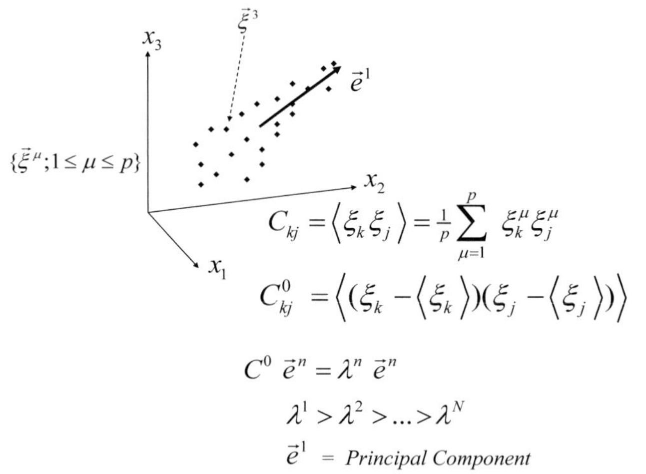

FIGURE 6.8 A cloud of data points $\vec{\xi}^\mu$ is sketched in a three-dimensional input space. The point with $\mu = 3$ is labeled (dashed arrow). The vector $\vec{e}_1$ ("principal component") points in the direction of the maximal variance of the data distribution. Formally, $\vec{e}_1$ is defined as the eigenvector of the covariance matrix C^0 with the largest eigenvalue.

Third, while the direction of the weight vector moves in the direction of the principal component, the *norm* of the weight vector grows without bounds. However, we can use additional terms in the expansion indicated in equation 6.2 so as to guarantee a normalization of the weight vector. Let us recall Oja's rule from equation 6.7. After the presentation of n patterns the weight vector has a weight $w_j(n)$ and the next update step is

$$\Delta w_j = \gamma \nu^{\text{post}} \nu_j - \gamma w_j(n)\,(\nu^{\text{post}})^2. \qquad (6.21)$$

In order to see that Oja's learning rule selects the first principal component, we show that the eigenvectors $\{\vec{e}_1, \ldots, \vec{e}_N\}$ of C are fixed points of the dynamics. For any fixed weight vector $\vec{w}$ we can calculate the expectation of the weight change in the next time step by averaging over the whole ensemble of input patterns $\{\vec{\xi}^1, \vec{\xi}^2, \ldots\}$. Repeating the steps of equation 6.15 we find from equation 6.21

$$\langle \Delta\vec{w}\rangle = \gamma C\vec{w} - \gamma\vec{w}[\vec{w}\cdot C\vec{w}]\,. \qquad (6.22)$$

We claim that any eigenvector $\vec{e}_i$ of the correlation matrix C is a fixed point of equation 6.22. Indeed, if we substitute $\vec{w} = \vec{e}_i$ in the above equation, we find that $\langle \Delta\vec{w}\rangle = 0$.

6.4 *Spike Timing Dependent Plasticity*

Traditional models of Hebbian learning have been formulated at the level of firing rates or abstract variables describing the activity of pre- and postsynaptic neurons. However, neuronal activity consists of spikes, short electrical pulses (action potentials). We show that the formulation of Hebbian learning rules at the level of spikes necessarily leads to the notion of STDP with a window of simultaneity that describes the co-occurrence of spikes.

Models of STDP can be formulated on several levels. We do not discuss the description of detailed biophysics-based models, mostly based on calcium, but focus directly on phenomenological or "minimal" models. STDP can be understood as a causal interaction between traces left at the synapses by spikes of presynaptic neurons (e.g., neurotransmitter) and action potentials propagating backward into the dendrite of the postsynaptic neuron. However, the voltage also plays an important role when there is no action potential.

6.4.1 Spikes versus Rates: The Learning Window

At the beginning of the chapter we emphasized two aspect of Hebbian learning: locality and correlation. When we formulated Hebbian learning on the level of rates, this led us to a family of Hebbian rules. Locality implied that the rule can depend only on pre- and postsynaptic activity; correlation implied that post- and presynaptic firing rates are important *together*.

What changes if we turn to spikes? In a spiking neuron model, there are many more local variables than just the rate (see chapter 2). There is the postsynaptic voltage; there are postsynaptic spikes; there may also be explicit variables for postsynaptic calcium. The local voltage at the synapse might be different from that at the soma, and the same is true for calcium, and all of these aspects may play a role. However, let us keep things simple and focus on a point neuron and neglect the spatial extension of the neuron. If the neuron emits a spike, the peak in the time course $u(t)$ is immediately felt at the location of the synapse. (In fact, even in neurons with spatial extension there is an action potential traveling back up the dendrite that could inform a synapse about a spike at the soma; Magee and Johnston, 1997; Larkum et al., 1999.) Spikes are events that can be defined, for example, as the onset of an action potential in real neurons or at the moment of threshold crossing in an integrate-and-fire neuron. As far as locality is concerned, a Hebbian learning rule could then depend on the spike timings t_j^f and t_i^f of a presynaptic neuron j and a postsynaptic neuron i.

Let us now turn to the second aspect, the notion of "correlation" that we substituted for Hebb's wording "takes part in firing it." A first, naive, approach would be to say that a synaptic change happens if pre- and postsynaptic firing occurs "at the same time." However, since firing times are threshold-crossing events, they are infinitely short, so that it is extremely rare that two firings occur at the exact same time: mathematically this probability tends to zero, unless special artificial conditions are imposed. To solve this issue, we need to set, somewhat arbitrarily, a time window or temporal resolution for our definition of simultaneity: if pre- and postsynaptic spikes occur within less than γ ms, then we call the events simultaneous and a synaptic change occurs.

Such a definition can be visualized as a rectangular time window of width 2γ centered symmetrically around the postsynaptic spike. However, why should the window be rectangular? And why should it be symmetrical? In fact, the Hebbian formulation that the presynaptic neuron "takes part in firing" the postsynaptic one suggests a causal relation and corresponds to a temporal order "pre-before-post." Hence, simultaneity in the Hebbian sense should be defined as a temporal window that is shifted slightly to the left as depicted in figure 6.9A, middle. Moreover, the sharp borders of such a window of coincidence look artificial. Why should a spike arriving at $(\gamma - \varepsilon)$ ms before the postsynaptic firing have a dramatically different effect than one arriving $(\gamma + \varepsilon)$ ms? Therefore, the coincidence window can be replaced by a smooth "learning window"—on purely theoretical grounds.

In the context of rate models of Hebbian learning, we have already seen that Hebb did not specify any conditions for depression of synapses. We are therefore free to complement the asymmetrical learning window by a negative part for post-before-pre timing (figure 6.9A, bottom), and this is indeed what has been postulated by theoreticians (Gerstner et al., 1996). Just as for the rate models of Hebbian learning discussed earlier, a rule for depression of synapses can induce competition between synapses so that some synapses grow at the expense of others, making Hebbian learning more selective. Applying this reasoning to spiking neurons led to the postulate of a biphasic learning window with millisecond resolution (Gerstner et al., 1996)—as found in experiments (Markram et al., 1997; Bi and Poo, 1998; Caporale and Dan, 2008). Hebbian plasticity at the level of spikes (figure 6.9B) was later termed spike timing dependent plasticity (Abbott and Nelson, 2000) and can be analyzed mathematically (Kempter et al., 1999) using methods analogous to those of section 6.3.3.

6.4.2 Standard STDP Model

STDP with an asymmetrical learning window can be described mathematically as a window function $W(s)$ that is positive for $s < 0$ and negative for $s > 0$ (Gerstner et al., 1996; Kempter

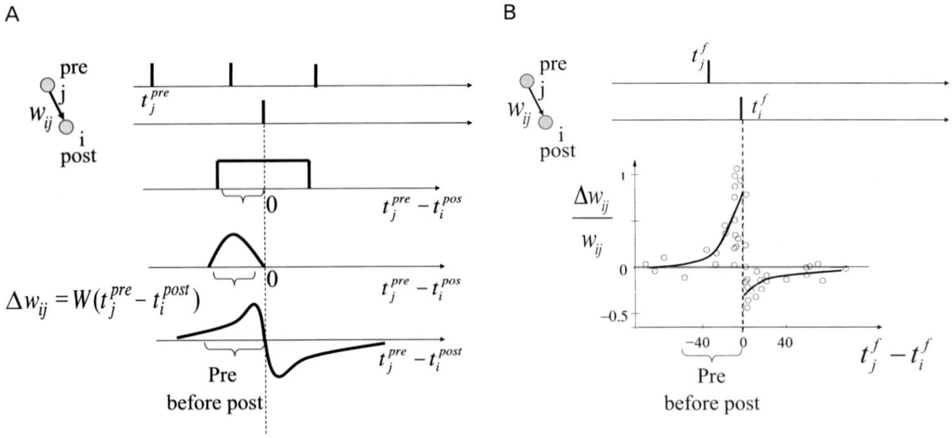

FIGURE 6.9 (A) Coincidence of a presynaptic spike with a postsynaptic one can be defined by a learning window which could be symmetrical and rectangular (top), asymmetrical (middle), or even contain two phases (bottom). The window is anchored at the postsynaptic spike at time t_i^{post}. If a presynaptic spike t_j^{pre} falls into this window, plasticity happens by an amount $\Delta w_{ij} = W(t_j^{pre} - t_i^{post})$. (B) An experimentally found spike timing dependent plasticity window shows two phases: if a presynaptic spike occurs repeatedly a few milliseconds before the postsynaptic one, the synaptic weight increases; for inverse timing, the weight decreases; figure adapted from Bi and Poo (1998).

et al., 1999; Song et al., 2000). The simplest choice is a learning window which decays exponentially (Song et al., 2000; Bi and Poo, 1998) with time constants τ_+ and τ_-, respectively

$$W(s) = \begin{cases} A_+ \exp(s/\tau_+), & \text{if } s < 0 \\ A_- \exp(-s/\tau_-), & \text{if } s > 0 \end{cases} \tag{6.23}$$

with $A_+ > 0$ and $A_- < 0$.

Apart from the learning window, synaptic plasticity on the level of spikes is implemented analogously to rate-based plasticity. Let us describe a presynaptic spike train $S_j(t) = \sum_f \delta(t - t_j^f)$ as the sequence of presynaptic firing times and analogously the postsynaptic spike train as $S_i(t) = \sum_f \delta(t - t_i^f)$. Synaptic weights change whenever presynaptic spikes arrive or when postsynaptic action potentials are triggered,

$$\begin{aligned} \frac{\mathrm{d}}{\mathrm{d}t} w_{ij}(t) &= a_0 + a_1^{\text{pre}} S_j(t) + a_1^{\text{post}} S_i(t) \\ &+ S_j(t) \int_0^\infty W(s) S_i(t-s)\, \mathrm{d}s \\ &+ S_i(t) \int_0^\infty W(-s) S_j(t-s)\, \mathrm{d}s. \end{aligned} \tag{6.24}$$

Analogously to the rate picture, presynaptic spikes alone could cause a change proportional to a_1^{pre} and postsynaptic spikes along proportional to a_1^{post}, and there could be a spontaneous decay or increase proportional to a_0. The essential feature of STDP is implemented by the two terms containing the learning window W. The first one accounts for post-before-pre timings, the second one for pre-before-post. Equation 6.24 is an online implementation of the standard model of STDP. Online means that the updates depend only on present and past spikes, not future activity. All coefficients in the above equations, that is, a_0, a_1^{pre}, a_1^{post}, as well as the window $W(s)$, may also depend on the weight w_{ij}.

6.4.3 STDP VERSUS RATE-BASED HEBBIAN LEARNING In order to show the relation between a spike-based and a rate-based formulation of plasticity rules, let us integrate equation 6.24 over a long period from time $t = 0$ to T. The result is

$$\begin{aligned} \Delta w_{ij} &= a_0 T + a_1^{\text{pre}} N_j(t) + a_1^{\text{post}} N_i(t) \\ &+ \int_0^T \int_{-\infty}^{\infty} W(s) S_i(t-s) S_j(t)\, \mathrm{d}s\mathrm{d}t \end{aligned} \tag{6.25}$$

where we have assumed that the duration T is much longer than the temporal extent of the learning window, so that border effects can be neglected. N_j and N_i are the number of pre- and postsynaptic spikes emitted over a time T.

Under the assumption that spiking occurs stochastically and independently with rates ν_i and ν_j that are constant over one time interval T (rates may change between one time interval and the next, though), then the update can be written (Kempter et al., 1999)

$$\frac{\Delta w_{ij}}{T} = a_0 + a_1^{\text{pre}} \nu_j(t) + a_1^{\text{post}} \nu_i(t) + \int_{-\infty}^{\infty} W(s) \mathrm{d}s \nu_i \nu_j, \tag{6.26}$$

which is equivalent to the rate model in equation 6.2. Note that the integral over the learning window, $\int W(s)\mathrm{d}s$, plays the role of the "Hebbian" coefficient c_2^{corr} in the rate formulation.

6.4.4 A Minimal Model of STDP The correlation condition in Hebb's postulate suggests that at least two components are involved in the induction of LTP. We do not wish to speculate on the nature of these components (which could be related to molecular or biochemical processes) but simply call them a and b. We assume that the first component is produced by a chemical reaction chain triggered by presynaptic spike arrival. In the absence of further input, the component a decays with a time constant τ_a back to its resting level $a = 0$. A simple way to describe this process is

$$\frac{\mathrm{d}}{\mathrm{d}t}a = -\frac{a}{\tau_a} + \alpha_+ \sum_f \delta(t - t_j^f), \tag{6.27}$$

where the sum runs over all *presynaptic* firing times t_j^f, $f = 1, 2, 3\ldots$ Equation 6.27 states that a is increased at each arrival of a presynaptic spike by an amount α_+. Loosely speaking, a presynaptic spike leaves a "trace" at the synapse. The trace could be the amount of glutamate bound to the postsynaptic receptors or something else.

Since it is a linear equation, it can be integrated (assuming an initial condition $a = 0$):

$$a(t) = \sum_f \alpha_+ \exp\left(-\frac{t - t_j^f}{\tau_a}\right) = \int_0^\infty \alpha_+ \exp\left(-\frac{s}{\tau_a}\right) S_j(t - s)\,\mathrm{d}s. \tag{6.28}$$

A high level of a sets the synapse in a state where it is susceptible to changes in its weight. The variable a by itself, however, does not yet trigger a weight change.

To generate the synaptic change, another substance b is needed. The production of b is controlled by a second process triggered by *postsynaptic* spikes,

$$\frac{\mathrm{d}}{\mathrm{d}t}b = -\frac{b}{\tau_b} + \frac{1}{\tau_b}\sum_f \delta(t - t_i^f), \tag{6.29}$$

where τ_b is another time constant. The sum runs over all postsynaptic spikes t_i^f. Note that the second variable b does not need to be a biochemical quantity; it could, for example, be the electrical potential caused by the postsynaptic spike itself. In the following, we assume that the time constant $\tau_b \ll \tau_a$ is so short that the process b can be considered as instantaneous. In the limit $\tau_b \to 0$, the process b can be approximated by a sequence of short pulses at the moment of the postsynaptic spikes, $b(t) = S_i(t) = \sum_f \delta(t - t_i^f)$.

Hebbian learning needs both "substances" to be present at the same time; thus,

$$\frac{\mathrm{d}}{\mathrm{d}t}w_{ij}^{\text{corr}} = \gamma a(t) b(t), \tag{6.30}$$

with some rate constant γ. The upper index "corr" is intended to remind us that we are dealing only with the correlation term on the right-hand side of equation 6.24.

If the process b is fast, then the process b simply "reads" out the value of $a(t)$ at the moment of each postsynaptic spike:

$$\frac{\mathrm{d}}{\mathrm{d}t}w_{ij}^{\text{corr}} = \gamma a(t) S_i(t). \tag{6.31}$$

Using our previous result for $a(t)$, we have

$$\frac{\mathrm{d}}{\mathrm{d}t}w_{ij}^{\text{corr}} = S_i(t)\int_0^\infty \gamma\alpha_+ \exp\left(-\frac{s}{\tau_a}\right) S_j(t - s)\,\mathrm{d}s, \tag{6.32}$$

which corresponds to the pre-before-post term in equation 6.24. It does not need a lot of work to convince ourselves that we need another set of substances c and d so as to construct a process with post-before-pre timing (see Exercise 3).

We may summarize equations 6.27 and 6.31 by saying that each presynaptic spike leaves a "trace" which is then read out at the moment of postsynaptic firing. Such an interpretation enables not just a rapid implementation of STDP in simulations but also the generalization to more complex STDP models (Morrison et al., 2008; Pfister and Gerstner, 2006).

6.4.5 Voltage and Spike Timing The learning window shows the amount of plasticity as a function of the time difference between pre- and postsynaptic spikes. The typical STDP plot (Gerstner et al., 1996; Bi and Poo, 1998; Abbott and Nelson, 2000), however, overemphasizes the role of pairs of spikes.

There are several important aspects that need to be highlighted. First, in experiments on cortical neurons (Senn et al., 2001; Sjöström et al., 2001) (as opposed to hippocampal slices or other preparations; Bi and Poo, 1998), using a single pair consisting of one presynaptic spike followed by one postsynaptic spike has no measurable effect. Even 50 or 60 pairs of pre-before-post, if given at a repetition frequency of 1 Hz or less, cause no potentiation of the synapses (Senn et al., 2001; Sjöström et al., 2001). Hence, one should consider models where the basic element is not a pair of spikes but triplets of spikes such as post–pre–post or pre–post–post (Pfister and Gerstner, 2006). Such extensions of the standard STDP model can be formulated in the framework outlined in this chapter. The basic idea is that each presynaptic spike leaves a "trace" and each postsynaptic spike leaves two "traces," one on a short and another one on a longer timescale. Potentiation requires that (1) a presynaptic spike has left a trace; (2) the slow postsynaptic trace is nonzero; and (3) the postsynaptic neuron presently emits a spike. For depression, the normal pair model can be used: an update happens when a presynaptic spike occurs after the postsynaptic firing has left a "fast" trace. Interestingly, this model is the spiking equivalent (Pfister and Gerstner, 2006) of the classical BCM model (Bienenstock et al., 1982) reviewed above.

Second, potentiation of synapses is possible even in the absence of postsynaptic spikes if the postsynaptic neuron is sufficiently depolarized (Artola et al., 1990; Ngezahayo et al., 2000). Moreover, isolated pairs of spikes with pre–post timing do induce potentiation, if they are preceded by a weak depolarization of the postsynaptic membrane (Sjöström et al., 2001). Hence, the more fundamental quantity is probably not spike timing but postsynaptic voltage. Indeed, a model wherein the local variable at the postsynaptic site of the synapse is not spike timing but the (low-passed-filtered) voltage is capable of explaining a large body of experimental results (Clopath et al., 2010). Since action potentials correspond to short but very pronounced peaks of the postsynaptic voltage, the spike-timing dependence of Hebbian plasticity follows from these models. In the voltage-based Clopath model, STDP is just one of many possible manifestations of synaptic plasticity (Spruston and Cang, 2010; Sjöström et al., 2001).

While the exact signal processing chain that biophysically generates the change of the synapse is still a mystery, it is clear from many experiments (reviewed, e.g., in Lisman and Zhabotinsky, 2001) that the time course of the postsynaptic calcium concentration plays a role. This observation has led to several calcium-based models of synaptic plasticity (Shouval et al., 2002; Lisman and Zhabotinsky, 2001; Rubin et al., 2005; Graupner and Brunel, 2007).

6.4.6 Functional Consequences of STDP For stationary Poisson firing, STDP models can be mapped exactly to rate-based models. The main difference is that STDP models have a causal effect at the level of spikes: if the presynaptic neuron fires, an EPSP is triggered in the postsynaptic neuron which makes the postsynaptic neuron more likely to fire. This additional effect can be included in an equivalent rate model. For a mathematical treatment, see chapter 11 in Gerstner and Kistler (2002) or Kempter et al. (1999).

If neurons fire not in a stationary Poisson fashion but in a precise temporal order, the main effect of STDP can be easily understood. Consider a network of 10 neurons with all-to-all connectivity and weight values distributed randomly between 0.5 and 1.5. Suppose an external input makes the neurons fire precisely timed spikes in the order $1 \rightarrow 2 \rightarrow 3 \rightarrow \ldots 10 \rightarrow 1\ldots$ In this case lateral connections between these neurons develop a unidirectional ring structure. Thus, the temporal order defined by a coding scheme with millisecond precision is reflected in the directionality of the connections (Clopath et al., 2010). In a pure rate picture of Hebbian learning, where rates are defined as the number of spikes in a time window of 100 ms, the same stimulus would strengthen all the weights symmetrically.

The paper of Clopath et al. can be seen as an example of how computational neuroscience contributes to clarifying concepts such as Hebbian learning and helps to unify a wide spectrum of different experimental data. Such conceptual work may eventually influence the way experimentalists think about fundamental ideas in neuroscience, even if no specific experimental predictions are made from the modeling.

6.5 *Reward-Modulated Learning*

6.5.1 Learning Depends on Feedback At the beginning of this chapter we mentioned different types of tasks. Learning to ski, to skateboard, or to ride a bicycle all involve feedback. It hurts if you fall down using the skateboard, and you get the praise of your peers if you master a new trick.

Hebbian learning, as discussed in this chapter, is an unsupervised learning rule: there is no notion of good or bad, successful or unsuccessful, rewarding or painful. Hebbian learning is suitable to detect correlations in the input and can therefore be used to explain developmental processes such as the formation of receptive fields.

Most tasks where you learn a novel activity, however, have the notion of success or reward attached to them. Note that we do not suppose a teacher who instructs every single neuron with respect to what it should do. Rather, the feedback you receive through reward (chocolate, money) or the feeling of success (esteem of peers, achievement of a self-set target) is a coarse

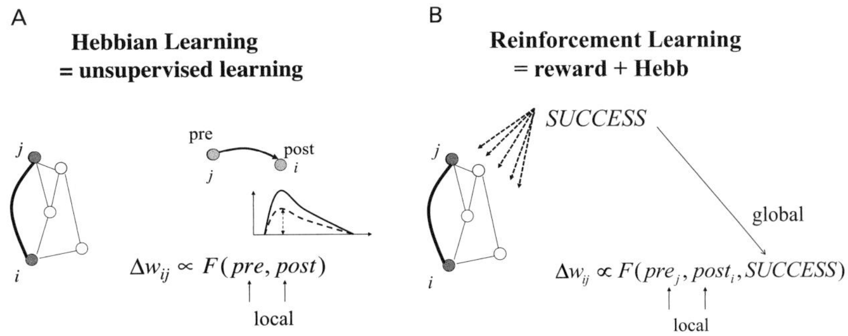

FIGURE 6.10 (A) Unsupervised Hebbian learning depends only on the activity (filled circles) of the neurons i and j that are connected. (B) In addition plasticity depends on neuromodulators that distribute information about the success (reward minus expected reward) of an action globally across the brain or a brain area.

signal, which we can describe as a single scalar variable. Let us discuss the traditional classification of learning paradigms.

6.5.2 UNSUPERVISED VERSUS REINFORCEMENT LEARNING In the field of machine learning, one often distinguishes three types of learning tasks: unsupervised learning, supervised learning, and reinforcement learning. Unsupervised learning detects correlations in the stream of input signals. The output of unsupervised learning can be to find the first principal component of the signal, an independent component, or the center of a cluster of data points, for example (Hertz et al., 1991). Hebbian learning falls into the class of unsupervised learning rules (figure 6.10A).

In supervised learning the neural network or artificial machine has to give, for every input signal, an output which is as close as possible to a target output specified by a "teacher" (and it should generalize to novel inputs, too).

Reinforcement learning does not specify a target output explicitly. Rather, it receives occasional but sparse feedback from the environment. If we take the task of riding a bicycle as an example, in supervised learning the teacher would demonstrate how to turn the handlebar at every moment in time whereas in reinforcement learning the only feedback from the environment might be negative reinforcement for failure (i.e., it hurts when you fall) or positive reinforcement for skill (i.e., you receive praise). Reinforcement learning is the topic of chapter 15.

6.5.3 REWARD-BASED HEBBIAN LEARNING Supervised learning methods, as studied in machine learning, suffer from the fact that it is difficult to see how they could be implemented biologically. Reinforcement learning methods, however, can be envisaged to be realized in the brain.

The basic idea is that we take a Hebbian learning rule which has two local factors (i.e., presynaptic and postsynaptic activity) and make the weight change proposed by the Hebbian learning rule subject to the presence of a neuromodulator signal S (figure 6.10B). The neuromodulator can signal "reward" or more generally "success S," defined as reward minus expected reward. Indeed, the neuromodulator dopamine, which is diffusively transmitted across broad areas of the brain, transmits a reward-related signal (Schultz et al., 1997). Experimental evidence indicates that Hebbian plasticity (be it STDP or other protocols) is influenced by neuromodulators (Reynolds and Wickens, 2002; Pawlak and Kerr, 2008; Seol et al., 2007; Zhang et al., 2009; Pawlak et al., 2010).

It has been shown in many modeling studies that Hebbian learning can be used for reward-based learning if the Hebbian learning rule, or STDP, is modulated by neuromodulatory signals (Sheynikhovich et al., 2009; Frémaux et al., 2010; Legenstein et al., 2008; Loewenstein and Seung, 2006; Loewenstein, 2008; Izhikevich, 2007). However, to achieve successful learning of nontrivial tasks, it is important that the success signal S defined as "reward minus expected reward" be perfectly balanced, that is, the brain needs an internal module capable of precisely estimating the reward expected for a given stimulus (Loewenstein, 2008; Frémaux et al., 2010).

6.5.4 LEARNING NOVEL ITEMS Finally, let us return to the question of how to form memories of facts or events and reconsider the example of the banana discussed at the beginning of the chapter. The Hebbian learning rules discussed in this chapter are capable of forming an internal memory trace of the banana, but they will also form memory traces of everything else we see or feel or think: our direction of gaze changes two or three times per second so that even during a single day more than 100,000 perceived images would have to be stored. It can be shown that such a continued bombardment with sensory items leads to rapid overwriting of memories, with the result that what we have memorized earlier is quickly washed out and forgotten (Fusi, 2002). In order to make memory storage and retrieval stable, one needs to postulate a gating mechanism that decides which perception is novel or interesting and worth storing. Similar to the situation discussed for reward-based Hebbian learning, neuromodulators could play the role of a global, brain-wide gating signal (Pawlak et al., 2010). Hebbian learning of a synapse from neuron j to i depends then not only on the activities of those two neurons but also on the presence of a more global neuromodulator that enables the initiation or stabilization of Hebbian weight changes (Hasselmo, 1999; Clopath et al., 2008; Frey and Morris, 1998). The topics sketched here in the last few paragraphs will be discussed in greater depth in the chapters on "Reinforcement Learning" and "Neuromodulation."

6.6 *Core Set of Readings*

Neuronal Dynamics: From Single Neurons to Networks and Models of Cognition, by W. Gerstner, W. M. Kistler, R. Naud, & L. Pansinski, chapters 10–12 (Cambridge University Press, 2014).

Theory of Cortical Plasticity, by L. N. Cooper, N. Intrator, B. S. Blais, & H. Z. Shouval (World Scientific, 2004).

Timing is not everything: Neuromodulation opens the STDP gate, by V. Pawlak, J. R. Wickens, A. Kirkwood, & J.N.D. Kerr (*Frontiers in Synaptic Neuroscience, 2,* 2010).

Spike-timing dependent plasticity, by J. Sjöström & W. Gerstner (Scholarpedia, 2010).

REFERENCES

Abbott, L. F., & Nelson, S. B. (2000). Synaptic plastictiy—Taming the beast. *Nature Neuroscience, 3,* 1178–1183.

Abbott, L. F., Varela, J. A., Sen, K., & Nelson, S. B. (1997). Synaptic depression and cortical gain control. *Science, 275,* 220–224.

Artola, A., Bröcher, S., & Singer, W. (1990). Different voltage dependent thresholds for inducing long-term depression and long-term potentiation in slices of rat visual cortex. *Nature, 347,* 69–72.

Bi, G., & Poo, M. (1998). Synaptic modifications in cultured hippocampal neurons: Dependence on spike timing, synaptic strength, and postsynaptic cell type. *Journal of Neuroscience, 18,* 10464–10472.

Bienenstock, E., Cooper, L., & Munroe, P. (1982). Theory of the development of neuron selectivity: Orientation specificity and binocular interaction in visual cortex. *Journal of Neuroscience, 2,* 32–48.

Bliss, T., & Lomo, T. (1973). Long-lasting potentiation of synaptic transmission in the dentate area of anaesthetized rabbit following stimulation of the perforant path. *Journal of Physiology, 232,* 351–356.

Bliss, T. V. P., & Collingridge, G. L. (1993). A synaptic model of memory: Long-term potentiation in the hippocampus. *Nature, 361,* 31–39.

Buonomano, D., & Merzenich, M. (1998). Cortical plasticity: From synapses to maps. *Annual Review of Neuroscience, 21,* 149–186.

Caporale, N., & Dan, Y. (2008). Spike timing-dependent plasticity: A Hebbian learning rule. *Annual Review of Neuroscience, 31,* 25–46.

Clopath, C., Busing, L., Vasilaki, E., & Gerstner, W. (2010). Connectivity reflects coding: A model of voltage-based spike-timing-dependent-plasticity with homeostasis. *Nature Neuroscience, 13,* 344–352.

Clopath, C., Ziegler, L., Vasilaki, E., Busing, L., & Gerstner, W. (2008). Tag-trigger-consolidation: A model of early and late long-term-potentiation and depression. *PLoS Computational Biology, 4,* e1000248.

Cooper, L., Intrator, N., Blais, B., & Shouval, H. Z. (2004). *Theory of Cortical Plasticity.* Singapore: World Scientific.

Frémaux, N., Sprekeler, H., & Gerstner, W. (2010). Functional requirements for reward-modulated spike-timing-dependent plasticity. *Journal of Neuroscience, 40,* 13326–13337.

Frey, U., & Morris, R. (1998). Synaptic tagging: Implications for late maintenance of hippocampal long-term potentiation. *Trends in Neurosciences, 21,* 181–188.

Fusi, S. (2002). Hebbian spike-driven synaptic plasticity for learning patterns of mean firing rates. *Biological Cybernetics, 87,* 459–470.

Gerstner, W., Kempter, R., van Hemmen, J., & Wagner, H. (1996). A neuronal learning rule for sub-millisecond temporal coding. *Nature, 383*(6595), 76–78.

Gerstner, W., & Kistler, W. K. (2002). *Spiking Neuron Models: Single Neurons, Populations, Plasticity.* Cambridge: Cambridge University Press.

Graupner, M., & Brunel, N. (2007). STDP in a bistable synapse model based on CaMKII and associated signaling pathways. *PLOS Computational Biology,* 3, e221.

Grossberg, S. (1969). On learning, information, lateral inhibition, and transmitters. *Mathematical Biosciences, 4,* 255–310.

Gütig, R., Aharonov, S., Rotter, S., & Sompolinsky, H. (2003). Learning input correlations through nonlinear temporally asymmetric Hebbian plasticity. *Journal of Neuroscience, 23*(9), 3697–3714.

Hasselmo, M. (1999). Neuromodulation: Acetylcholine and memory consolidation. *Trends in Cognitive Sciences, 3,* 351–359.

Hebb, D. O. (1949). *The Organization of Behavior.* New York: Wiley.

Hertz, J., Krogh, A., & Palmer, R. G. (1991). *Introduction to the Theory of Neural Computation.* Redwood City, CA: Addison-Wesley.

Hubel, D. H. (1988). *Eye, Brain, and Vision.* New York: Freeman.

Hubel, D. H., & Wiesel, T. (1962). Receptive fields, binocular interaction and functional architecture in the cat's visual cortex. *Journal of Physiology, 160*, 106–154.

Izhikevich, E. (2007). Solving the distal reward problem through linkage of STDP and dopamine signaling. *Cerebral Cortex, 17*, 2443–2452.

Kempter, R., Gerstner, W., & van Hemmen, J. L. (1999). Hebbian learning and spiking neurons. *Physical Review E: Statistical Physics, Plasmas, Fluids, and Related Interdisciplinary Topics, 59*, 4498–4514.

Kohonen, T. (1982). Self-organized formation of topologically correct feature maps. *Biological Cybernetics, 43*, 59–69.

Larkum, M. E., Zhu, J. J., & Sakmann, B. (1999). A new cellular mechanism for coupling inputs arriving at different cortical layers. *Nature, 398*(6725), 338–341.

Legenstein, R., Pecevski, D., & Maass, W. (2008). A learning theory for reward-modulated spike-timing-dependent plasticity with application to biofeedback. *PLoS Computational Biology, 4*, e1000180.

Levy, W. B., & Stewart, D. (1983). Temporal contiguity requirements for long-term associative potentiation/depression in hippocampus. *Neurosci, 8*, 791–797.

Linsker, R. (1986a). From basic network principles to neural architecture: Emergence of orientation columns. *Proceedings of the National Academy of Sciences of the United States of America, 83*, 8779–8783.

Linsker, R. (1986b). From basic network principles to neural architecture: Emergence of orientation selective cells. *Proceedings of the National Academy of Sciences of the United States of America, 83*, 8390–8394.

Lisman, J., & Zhabotinsky, A. (2001). A model of synaptic memory: A CaMKII/PP1 switch that potentiates transmission by organizing an AMPA receptor anchoring assembly. *Neuron, 31*, 191–201.

Loewenstein, Y. (2008). Robustness of learning that is based on covariance-driven synaptic plasticity. *PLoS Computational Biology, 4*, e1000007.

Loewenstein, Y., & Seung, H. (2006). Operant matching is a generic outcome of synaptic plasticity based on the covariance between reward and neural activity. *Proceedings of the National Academy of Sciences of the United States of America, 103*, 15224–15229.

MacKay, D. J. C., & Miller, K. D. (1990). Analysis of Linsker's application of Hebbian rules to linear networks. *Network, 1*, 257–297.

Magee, J. C., & Johnston, D. (1997). A synaptically controlled associative signal for Hebbian plastiticy in hippocampal neurons. *Science, 275*, 209–213.

Malenka, R. C., & Nicoll, R. A. (1999). Long-term potentiation—A decade of progress? *Science, 285*, 1870–1874.

Markram, H., Lübke, J., Frotscher, M., & Sakmann, B. (1997). Regulation of synaptic efficacy by coincidence of postysnaptic AP and EPSP. *Science, 275*, 213–215.

Markram, H., Sjostrom, J., & Gerstner, W. (2011). A history of spike-timing dependent plasticity. *Frontiers in Synaptic Neuroscience, 3*, 4.

Markram, H., Wu, Y., & Tosdyks, M. (1998). Differential signaling via the same axon of neocortical pyramidal neurons. *Proceedings of the National Academy of Sciences of the United States of America, 95*, 5323–5328.

Miller, K. D., & MacKay, D. J. C. (1994). The role of constraints in Hebbian learning. *Neural Computation, 6*, 100–126.

Morris, R. (2003). Long-term potentiation and memory. *Philosophical Transactions of the Royal Society of London. Series B, Biological Sciences, 358*, 643–647.

Morrison, A., Diesmann, M., & Gerstner, W. (2008). Phenomenological models of synaptic plasticity based on spike timing. *Biological Cybernetics, 98*, 459–478.

Ngezahayo, A., Schachner, M., & Artola, A. (2000). Synaptic activation modulates the induction of bidirectional synaptic changes in adult mouse hippocamus. *Journal of Neuroscience, 20*, 2451–2458.

Oja, E. (1982). A simplified neuron model as a principal component analyzer. *Journal of Mathematical Biology, 15*, 267–273.

Pawlak, V., & Kerr, J. (2008). Dopamine receptor activation is required for corticostriatal spike-timing-dependent plasticity. *Journal of Neuroscience, 28*, 2435–2446.

Pawlak, V., Wickens, J., Kirkwood, A., & Kerr, J. (2010). Timing is not everything: Neuromodulation opens the STDP gate. *Frontiers in Synaptic Neuroscience, 2*, 146.

Pfister, J.-P., & Gerstner, W. (2006). Triplets of spikes in a model of spike timing-dependent plasticity. *Journal of Neuroscience, 26*, 9673–9682.

Recanzone, G. H., Schreiner, C. E., & Merzenich, M. M. (1993). Plasticity in the frequency representation of primary auditory cortex following discrimination training in adult owl monkeys. *Journal of Neuroscience, 13*, 87–103.

Reynolds, J., & Wickens, J. (2002). Dopamine-dependent plasticity of corticostriatal synapses. *Neural Networks, 15*, 507–521.

Rochester, N., Holland, J., Haibt, L., & Duda, W. (1956). Tests on a cell assembly theory of the action of the brain, using a large digital computer. *I.R.E. Transactions on Information Theory, 2*, 80–93.

Rubin, J., Gerkin, R., Bi, G.-Q., & Chow, C. (2005). Calcium time course as a signal for spike-timing-dependent plasticity. *Journal of Neurophysiology, 93*, 2600–2613.

Schultz, W., Dayan, P., & Montague, R. (1997). A neural substrate for prediction and reward. *Science, 275*, 1593–1599.

Senn, W., Tsodyks, M., & Markram, H. (2001). An algorithm for modifying neurotransmitter release probability based on pre- and postsynaptic spike timing. *Neural Computation, 13*, 35–67.

Seol, G., Ziburkus, J., Huang, S., Song, L., Kim, I., Takamiya, K., et al. (2007). Neuromodulators control the polarity of spike-timing-dependent synaptic plasticity. *Neuron, 55*, 919–929.

Sheynikhovich, D., Chavarriaga, R., Strosslin, T., Arleo, A., & Gerstner, W. (2009). Is there a geometric module for spatial orientation? Insights from a rodent navigation model. *Psychological Review, 116*, 540–566.

Shouval, H. Z., Bear, M. F., & Cooper, L. N. (2002). A unified model of NMDA receptor dependent bidirectional synaptic plasticity. *Proceedings of the National Academy of Sciences of the United States of America, 99*, 10831–10836.

Sjöström, J., & Gerstner, W. (2010). Spike-timing dependent plasticity. *Scholarpedia, 5*, 1362.

Sjöström, P., Turrigiano, G., & Nelson, S. (2001). Rate, timing, and cooperativity jointly determine cortical synaptic plasticity. *Neuron, 32,* 1149–1164.

Song, S., Miller, K., & Abbott, L. (2000). Competitive Hebbian learning through spike-time-dependent synaptic plasticity. *Nature Neuroscience, 3,* 919–926.

Spruston, N., & Cang, J. (2010). Timing isn't everything. *Nature Neuroscience, 13,* 277–279.

Stent, G. (1973). A physiological mechanism for Hebb's postulate of learning. *Proceedings of the National Academy of Sciences of the United States of America, 70,* 997–1001.

von der Malsburg, C. (1973). Self-organization of orientation selective cells in the striate cortex. *Kybernetik, 14,* 85–100.

Willshaw, D. J., & von der Malsburg, C. (1976). How patterned neuronal connections can be set up by self-organization. *Proceedings of the Royal Society (London). Series B, 194,* 431–445.

Wimbauer, S., Gerstner, W., & van Hemmen, J. (1998). Analysis of a correlation-based model for the development of orientation-selective receptive fields in the visual cortex. *Network, 9,* 449–466.

Zhang, J., Lau, P., & Bi, G. (2009). Gain in sensitivity and loss in temporal contrast of STDP by dopaminergic modulation at hippocampal synapses. *Proceedings of the National Academy of Sciences of the United States of America, 106,* 13028–13033.

7 Neuromodulation

JEAN-MARC FELLOUS, CARMEN CANAVIER, AND MICHAEL E. HASSELMO

7.1 *Neuromodulation: What Is It?*

Our current conceptual understanding of computations in neural systems is dominated by the concepts of excitation and inhibition. Whether explaining the neural basis of perception, cognitive processing, or action, researchers express computations as intricate sequences of excitation and inhibition between and within brain areas. While often satisfactory at first, such a description of neural computation becomes quickly insufficient, as more data become available.

A key aspect of neural computation is the ability of a given anatomical configuration of cells to implement multiple computations. Classic examples can be found in the neuromodulator-controlled pattern-generation abilities of the stomatogastric ganglion (STG) among other systems (Dayan, 2012; Marder, 2012). In these systems endogenous chemical substances such as dopamine or norepinephrine change the conductances of key subsets of membrane currents, which, in turn, change the manner in which neurons integrate their inputs and produce their outputs. Most of the research in mammalian systems has focused on the role of synaptic plasticity in implementing this pluri-functionality. With a careful change of synaptic strengths a network can switch from one function to another, encode one item or another. However, synaptic plasticity requires an exquisite level of control, be it spatial (e.g., which synapses are activated) or temporal (e.g., which signal is pre- or postsynaptic within 40 ms). Some other phenomena are temporally slower and spatially more diffuse, such as the change in computation when attention enters into play or when under the influence of a drug of abuse or alcohol.

Decades of detailed studies of the mechanisms of synaptic transmission and of dendritic information processing have revealed a large number of biophysical processes that serve to modulate the size and time course of postsynaptic potentials, or the manner with which these potentials are summated. Some of these biophysical processes can be activated by endogenous modulators including acetylcholine (ACh), norepinephrine (NE), histamine, serotonin (5-HT), dopamine (DA), and a variety of neuropeptides. These modulatory effects often alter the response to the fast monopolar excitatory postsynaptic potentials (EPSPs) caused by glutamate and the inhibitory potentials caused by GABA. The contribution of these substances to behavior is rooted in evolution, and in most cases they are considered key to understanding specific functions (Kelley, 2005). For example ACh is often involved in learning and memory, DA in reward processing, NE in attention and 5-HT in aggression and depression. The biophysical properties of these neuromodulatory substances have been recently reviewed (Marder, 2012). Although these substances are known to act at different types of receptors, originate from different brain areas, and have different spatial distributions and time courses of action, they have at least one of the following three functional effects: modulation of intrinsic neural properties (such as resting membrane potential, threshold, or bursting vs. tonic modes), modulation of integrative properties (such as changing the manner in which synaptic signals travel within the dendritic tree), or modulation of synaptic properties (such as presynaptic modulation of vesicle release at the synapse). Neuromodulation has been studied at various levels of details, from single cells (Kaczmarek and Levitan, 1987) to invertebrate (Harris-Warrick and Marder, 1991) and mammalian (Hasselmo, 1995) networks, experimentally in vivo or in vitro, and through computer modeling techniques (Fellous and Linster, 1998; Dayan, 2012).

Neuromodulation can be characterized by its spatial (intrinsic or extrinsic to the neural circuit) and temporal (timescale) properties. The core computational difficulty in simulating neuromodulatory effects is to determine the proper level of modeling, for the particular neural computation under consideration. In the following, we give a broad description of the different spatial and temporal extents of neuromodulation. We then focus on two examples of neuromodulatory systems. The first, dopamine, will illustrate the importance of understanding the firing properties of neuromodulatory neurons at the single-cell level. The second, acetylcholine, will focus on a few network-level effects of neuromodulation.

7.2 *Extrinsic and Intrinsic Neuromodulations*

A first class of neuromodulatory signals may originate from an area *extrinsic* to the neural substrate whose computation is under study, so that lesioning the neuromodulatory center does not usually perturb the function itself but only modifies the manner with which it is achieved (e.g., speed, reliability; figure 7.1). In some cases, however, extrinsic neuromodulation can have a gating function (as in dopamine in the basal ganglia). The computational effects of such extrinsic neuromodulation have been empirically found to be somewhat global, in that they simultaneously influence many functionally different sites—though different classes of receptors will allow different neurons to respond to the signal in somewhat distinctive ways. A second class consists in *intrinsic* neuromodulations that either originate in the relevant substrate itself or in an anatomically distant site but are tightly controlled by the substrate. In such systems, neuromodulation is an integral part of the computation. Examples of such phenomena are (1) cotransmission (Brezina, Orekhova, et al., 1996; Tan and Bullock, 2008), the ability for single synaptic vesicles to contain multiple neurotransmitters; (2) presynaptic receptor activation (Starke, Gothert, et al., 1989; Marder, 1996), the ability for neurotransmitters to bind and regulate their own release; (3) glial modulation (Hansson and Ronnback, 1994), the ability for glial cells to sequester and influence neurotransmitters and ions in the extracellular space; or (4) volume transmission (Fuxe and Agnati, 1991; Ridet, Rajaofetra, et al., 1993), which allows for nonsynaptic interactions between cells. The functions of such *intrinsic* modulations are usually specific to the substrate under consideration (Katz and Frost, 1996; Katz, 1998).

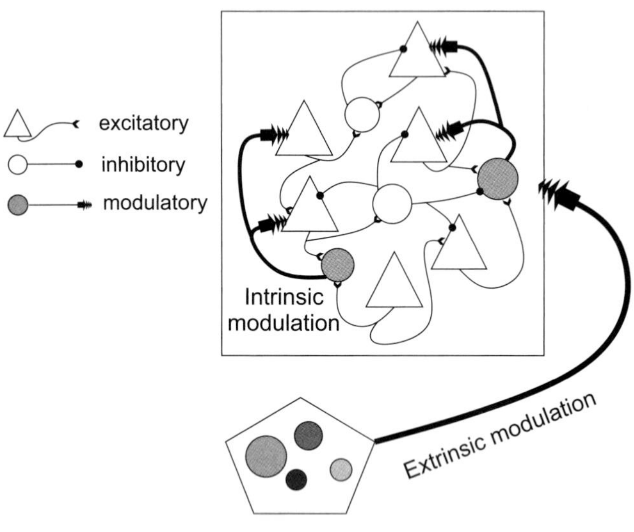

FIGURE 7.1 Intrinsic and extrinsic modulations. Neurons can be broadly classified as excitatory, inhibitory, and neuromodulatory. Intrinsic neuromodulation is characterized by a synaptically integrated population of neuromodulatory neurons, in a tight bidirectional loop with the computational substrate. Extrinsic neuromodulation typically involves a synaptically, and often anatomically, separate group of specialized neuromodulatory cells. Adapted from Marder (2012).

7.2.1 Intrinsic Neuromodulation

In some interesting instances, it is not possible to isolate the neuromodulatory phenomenon from the system it modulates. In such cases, neuromodulation is intrinsic to the network whose computation is under study. There have been a large number of insightful computational models of intrinsic neuromodulation (Marder and Thirumalai, 2002; Nadim, Brezina, et al., 2008; Brezina, 2010; Hudson and Prinz, 2010), but our concern in this section is to visit intrinsic neuromodulation only briefly before we turn to extrinsic neuromodulation, the focus of this chapter. Some experimental evidence for intrinsic neuromodulation is reviewed elsewhere (Katz and Frost, 1996); here we will briefly mention two examples.

In the STG of the lobster, an afferent axon (SNAX1) has been characterized as both a participant in the rhythmicity of the gastric mill network and as a conveyor of modulatory information (Nusbaum, Weimann, et al., 1992). SNAX1 receives inhibitory synaptic inputs from the STG and is capable of initiating action potentials (intrinsically, within the STG and not near the cell body, a few centimeters away) which generate EPSPs on the STG elements, participating therefore in the generation of the rhythm. However, because SNAX1 is also electrically coupled with key neurons of the central pattern generator (CPG), its level of depolarization (whether or not action potentials are present) influences the spike generation in other cells. This influence is carried by a shared voltage rather than a common neuromodulatory substance. This form of neural influence can functionally be as important as that of "classical" neuromodulators and can be part of a broad definition of a neuromodulatory phenomenon.

Similarly, in the CPG of tritonia, DSI (a serotonergic neuron) is known to enhance synaptic transmission presynaptically at synapses made by another key CPG neuron (Katz and Frost, 1995a; Katz and Frost, 1996). DSI elicits both a fast, neurotransmitter-like EPSP, and a slow neuromodulatory-like EPSP (Katz and Frost, 1995b), both pharmacologically separable. DSI, therefore, modulates the oscillatory pattern it is contributing to.

A large number of neuromodulatory phenomena in invertebrate organisms are intrinsic. It seems that as nervous systems became larger and as behavioral demands for flexibility and variety increased, computational circuits evolved from implementing computational flexibility with intrinsic neuromodulatory neurons to implementing flexibility through synaptic plasticity and/or population-wide dynamical computations. As theoretical work has repeatedly demonstrated, synaptic plasticity based computations or population-wide dynamical behaviors requires indeed a critical mass of neurons and synapses, typically available only in larger nervous systems (Chialvo, 2010; Massobrio, de Arcangelis, et al., 2015). It is likely then that intrinsic neuromodulatory neurons were no longer required to achieve a particular set of functions. Instead, they may have left the realm of elementary computations to system-wide computations, involving multiple neural circuits in physically distant brain areas. Neuromodulatory neurons formed specialized centers with diffuse efferent connections and complex afferent feedback. Neuromodulation became predominantly extrinsic while intrinsic modulation became essentially homeostatic.

7.2.2 Extrinsic Neuromodulation This form of neuromodulation depends on the parallel activity of functionally distinct systems. Such is the case of most neuromodulatory centers releasing specific neuroactive substances that modify the cellular and synaptic properties of their targets. We will review below two examples. The first will focus on the dopaminergic output of the ventral tegmental area and substantia nigra on the basal ganglia (section 7.4). The second will focus on the influence of the cholinergic output of the septum on hippocampal computations (section 7.5).

It is, of course, possible to envision dual extrinsic and intrinsic neuromodulations, whereby the former would express state or stimulus dependency and the latter would be activity dependent. An example of this dual modulation can be found in the motor pattern generator for biting in *Aplysia*, where intrinsic neuromodulation is mediated by an interneuronal peptide and extrinsic neuromodulation is mediated by an external serotonergic center (Morgan, Perrins, et al., 2000).

In the computational framework of modeling studies, extrinsic neuromodulations can be easily implemented by choosing appropriate sets of parameters (*tuning*) whereas intrinsic neuromodulations require that the neuromodulatory mechanisms be *regulated* by the specific computations under consideration.

7.3 Time Course of Neuromodulation

Depending on the function implemented, neural computation may follow different time courses, from a few milliseconds to several minutes or hours. Correspondingly, neuromodulation is also implemented at different timescales (Schultz, 2007). We will discuss in section 7.4 a model of a neuromodulatory neuron that can implement fast (phasic) and slow (tonic) forms of neuromodulations.

7.3.1 Fast Computations, Slow (Tonic) Modulation Most computational studies that feature neuromodulation implement it as a slow and diffuse process, tonically changing membrane or synaptic properties.

Many models use volume transmission (Umbriaco, Garcia, et al., 1995; Descarries, Gisiger, et al., 1997) in which a neuromodulator is released in a diffuse manner to cause relatively uniform activation of receptors throughout the tissue. Microdialysis samples of acetylcholine levels in the hippocampus taken over many minutes indicated slow changes, with high acetylcholine levels during active exploration or sleep (Marrosu, Portas, et al., 1995), sustained attention (Sarter, Hasselmo, et al., 2005), and movement-related theta oscillations (Monmaur, Collet, et al., 1997; Buzsaki, 2002). Acetylcholine therefore is involved in the modulations of different computations at different times (e.g., normalizing synaptic strength during sleep, Tononi and Cirelli, 2006; increasing thalamic sensory inputs during an attentive state; enabling oscillation-mediated place coding during spatial navigation). However, recent studies show fast, phasic changes in acetylcholine after stimulus events using the higher temporal resolution of amperometry (a technique involving carbon fibers able to detect specific ions in solution in vivo) (Parikh, Kozak, et al., 2007). We will discuss below the computational roles of acetylcholinergic modulation in enhancing encoding and attention for sensory input.

Dopamine modulation is also a good example in that dopamine cells are able to burst (fast timescale; see below) and fire tonically (slower timescale) depending on the behavioral or cognitive demands. The neuromodulatory effects of tonic levels of dopamine are thought to be important in both movement and cognition (Maia and Frank, 2011). In the basal ganglia, tonic levels are hypothesized to facilitate actions that are appropriate for the current state of the animal by increasing excitability in the direct pathway that facilitates action and by decreasing excitability in the indirect pathway that inhibits action, whereas, in the prefrontal cortex, they appear to stabilize representations that are important in attention, cognitive control, and working memory.

7.3.2 Slow Computations, Fast (Phasic) Modulation Computations may be influenced by fast modulations. In a classic series of experiments in monkeys, Schultz and his colleagues showed that during the learning of an association between a neutral stimulus and a reward, dopaminergic neurons of the ventral tegmental area (VTA) or substantia nigra pars compacta (SNC) exhibited transient increases in activity lasting less than 150 ms in response to behaviorally relevant signals associated with reward delivery (Schultz, Apicella, et al., 1993; Schultz, 2013). Monkeys were tested in several behavioral paradigms including a delayed response task in which they were taught to memorize one of two spatially distant targets for a variable amount of time (2.5–3.5 seconds) and to respond by arm movement at the onset of a trigger signal. Correct responses were rewarded.

During learning, VTA activity increased transiently after the presentation of the target signal and shortly after the delivery of the reward, regardless of the delay introduced. This firing behavior was in marked contrast to that of VTA targets such as the striatum or prefrontal cortex, in which neural activity may be tonically increased during the entire delay period (Goldman-Rakic, Lidow, et al., 1992). Moreover, the transient increase of VTA activity after reward appeared only during learning, not after the animal had acquired the task. These data suggest that the transient actions of dopamine after reward delivery may be specifically involved in learning. The precise duration of the postsynaptic effects of the release of dopamine in the prefrontal cortex during such a task is not known, but it might be as short as 100–200 ms (Jay, Glowinski, et al., 1995) and as long as 20 minutes (Brady and O'Donnell, 2004; Lapish, Kroener, et al., 2007). Other experimental and theoretical studies have related the response of the VTA and SNC neurons to positive and negative reward prediction errors (Schultz, 2007, 2010). Insofar as one considers that the performance of the delayed response task lasts up to 4 seconds, a 150-ms phasic involvement of the dopaminergic system appears as a fast modulatory process influencing a longer sequence of neural computations. The major theoretical framework used to explain the phasic behavior of dopamine neurons is that of temporal difference learning (Sutton and Barto, 1990) (see chapter 10, "Reinforcement Learning"). In its original form, the theory assumed that a stimulus was explicitly represented at each time step after onset (Montague, Dayan, et al., 1996). This strong assumption was later abandoned because it could not explain new data obtained when reward delivery was withheld (Ludvig, Sutton, et al., 2008; Hazy, Frank, et al., 2010). Nonetheless, there is a broad consensus that phasic dopamine release serves as a learning signal for classical and instrumental conditioning (Hazy, Frank, et al., 2010), likely via the well-documented neuromodulatory effects of dopamine on the late form of long-term potentiation (LTP) at corticostriatal and hippocampal synapses.

7.4 Dopaminergic Modulation

7.4.1 Anatomy of Dopamine Dopamine is a ubiquitous neurotransmitter widely found in invertebrate and vertebrate nervous systems. In the mammalian brain, there are 10 main groups of dopaminergic neurons. Groups A8 (retrorubral field) and A9 (SNC) project to the striatum and are involved in motor behaviors. Group A10 (VTA) projects to the nucleus accumbens, prefrontal and temporal cortices, and the basal forebrain and is mainly involved in the modulation of learning and emotional memory (figure 7.2). Groups A11 and A13 are in the hypothalamus and regulate sympathetic preganglionic neurons. The A12, A14, and A15 groups of dopaminergic neurons are part of the neuroendocrine system. Other dopaminergic neurons can be found in the olfactory bulb (A16) and retina (A17). We will focus here on groups SNC and VTA (Nair-Roberts, Chatelain-Badie, et al., 2008).

The behavioral roles of dopamine are numerous, from purely motor to highly cognitive. The loss of midbrain dopamine neurons results in the symptomatology of Parkinson's disease (Bernheimer, Birkmayer, et al., 1973), and abnormal dopaminergic signaling has been implicated in schizophrenia (Weinberger, 1987) and addiction (Koob, Le, et al., 1987). (See chapter 23, "Brain Diseases.") Dopamine neuromodulation is likely to be critically involved in the neural substrate of persistent firing and working memory (see chapter 11, "Short-term, Long-term, and Working Memory") and in the manner in which multiple systems interact at various phases of reinforcement learning (see chapter 10, "Reinforcement Learning").

7.4.2 Neuromodulatory Roles of Phasic and Tonic Firing in Dopaminergic Neurons Midbrain dopamine neurons are hypothesized to signal motivational value. In other words, an outcome that is better than predicted causes an increase in their firing rate and an outcome that is worse than expected causes a decrease in their firing rate (Daw, chapter 10). In a slice preparation that does not preserve the synaptic inputs

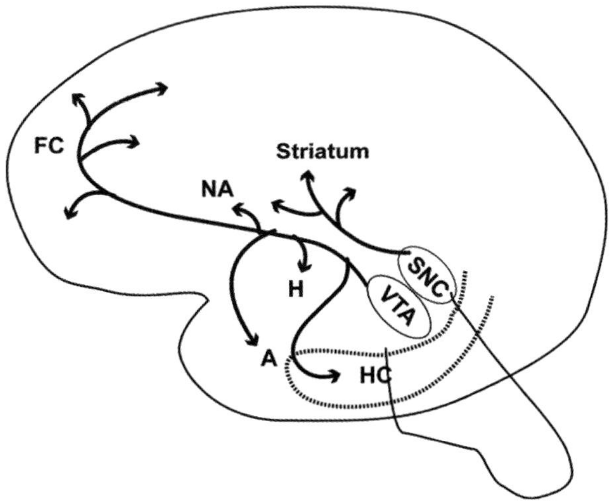

FIGURE 7.2 Targets of the substantia nigra and ventral tegmental area (VTA) dopaminergic neurons. The substantia nigra pars compacta (SNC) projects via the nigrostriatal pathway to the dorsal striatum. The VTA projects to the nucleus accumbens (NA), which is also called the ventral striatum, as well as to the amygdala (A), hypothalamus (H), and hippocampus (HC) via the mesolimbic pathway and to the frontal cortex (FC) via the mesocortical pathway.

normally received by these neurons, they nevertheless fire spontaneously in a regular fashion (Grace and Onn, 1989). Their intrinsic pace-making ability (Rubin and Rinzel, chapter 3, and Vierling-Claassen and Jones, chapter 4) allows these neurons to support a tonic signaling mode at 1–8 Hz (Grace and Bunney, 1984a,b) in the intact animal without requiring external excitation. During tonic firing, the extracellular concentration of dopamine in the target projection areas depends mostly upon the average firing rate rather than on the timing of individual spikes; thus, it varies slowly. The underlying cause of the slow variation is that the dopamine reuptake transporter removes dopamine in the extracellular space at a rate that allows the concentration to equilibrate in response to a tonic firing pattern (Gonon, 1988; Chergui, Nomikos, et al., 1996; Heien and Wightman, 2006).

In an intact animal, tonic signaling is not the only signaling mode. Instead, tonic firing is punctuated by bursts and pauses that cause transient firing rate increases and decreases, respectively (Grace and Bunney, 1984a, 1984b; Tepper, Martin, et al., 1995; Hyland, Reynolds, et al., 2002). Bursts are operationally defined as beginning with an interspike interval (ISI) < 80 ms and terminating with an ISI > 160 ms in order to differentiate this mode from tonic firing at 1–8 Hz. Dopamine neurons in freely moving rats fire about 20% of their spikes in bursts, but individual dopamine cells can differ greatly in their tendency to burst (Hyland, Reynolds, et al., 2002). Both bursts and pauses are considered to be phasic signals which cause extracellular dopamine concentrations in some target areas to vary rapidly with time. During phasic bursts, the dopamine uptake mechanism is overwhelmed by the release of dopamine (Heien and Wightman, 2006), which allows a burst firing pattern to produce a much greater transient increase in dopamine concentration than a tonic one (Gonon, 1988; Chergui, Nomikos, et al., 1996). High extracellular concentrations of dopamine in target areas are more likely if multiple dopamine neurons burst in unison, as they do in response to reward and reward-related cues (Schultz, 1998a,b). Under the motivational value hypothesis, phasic signals convey time-locked information regarding rewards. Although midbrain dopamine neurons were once thought to be relatively homogeneous in their responses to stimuli (Schultz, 1998a, 1998b), they are actually heterogeneous in their afferents, projections, electrophysiology, and bursting mechanisms (Roeper, 2013). In addition to the motivational value hypothesis, distinct subpopulations of dopamine neurons have been hypothesized to use bursts to code for motivational salience (including aversive stimuli), novelty, and start and stop signals for action selection (Roeper, 2013).

Figure 7.3A mimics experimental measurements of the extracellular dopamine concentration in the nucleus accumbens in response to electrical stimulation of the VTA and illustrates schematically how the reuptake mechanism can be overwhelmed. The rate of change of the concentration of dopamine in the extracellular cleft can be modeled (Wightman, May, et al., 1988) as

$$\frac{d[DA]}{dt} = \delta(t_n) - \frac{\mathrm{V}_{\mathrm{max}}[DA]}{\mathrm{K}_{\mathrm{m}} + [DA]} \tag{7.1}$$

where [DA] is the dopamine concentration, V_{max} is the maximum rate (5 μM/s) of dopamine uptake, K_m is the concentration (0.2 μM) at which this rate is half maximal, and $\delta(t_n)$ represents a fixed increase of 0.125 μM attributable to the nth stimulation. The gray trace illustrates an equilibrium for tonic stimulation, and the black trace shows an increase in concentration for a stimulation frequency that exceeds the ability of the reuptake to produce an equilibrium. Figure 7.3B illustrates the concept of tonic versus phasic signaling. The parameter $\delta(t_n)$ was reduced to represent the contribution of a single neuron. The baseline shows a population of neurons firing asynchronously at an average rate of 10 Hz, resulting in a relatively flat, tonic signal in the dopamine concentration. At the arrow, a simulated

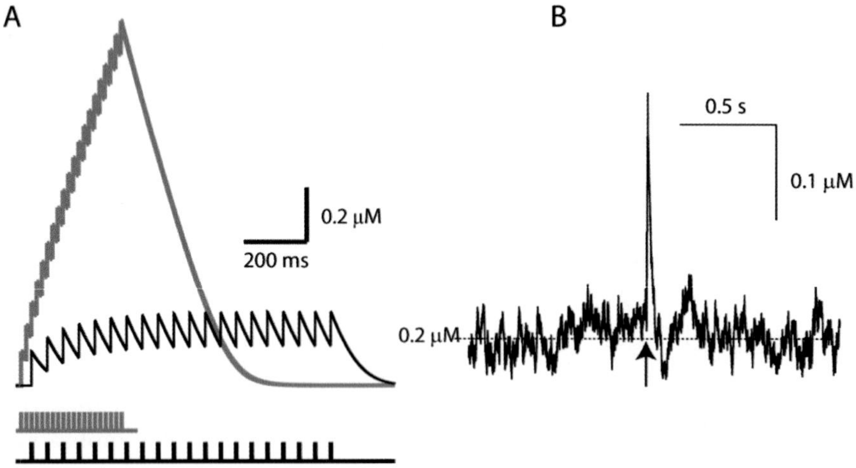

FIGURE 7.3 Tonic and phasic signals in extracellular dopamine concentration. (A) Simulation of extracellular dopamine concentration in the target area (nucleus accumbens, or ventral striatum) in response to 20-Hz (black pulses and traces) stimulation of the ventral tegmental area. (B) Simulation of the variation in extracellular dopamine concentration in response to asynchronous low-frequency tonic firing compared to a phasic response to a burst of five synchronous population spikes starting at the time indicated by the arrow.

burst of five synchronous population spikes at 50 Hz briefly increased the extracellular dopamine concentration, resulting in a phasic signal that is time-locked to the burst. Compared to the nucleus accumbens and the dorsal striatum, the prefrontal cortex shows low levels of dopamine reuptake transporter (lower V_{max}). Consequentially, dopamine levels remain elevated longer in the prefrontal cortex than in the striatum after a simulated burst, decreasing temporal resolution for phasic signals. Therefore, the contributions of tonic and phasic signaling vary between target regions (Lapish, Kroener, et al., 2007). In section 7.4.3, we describe how dopamine neurons can produce intrinsic tonic signaling (section 7.4.3.2), and how their intrinsic properties interact with their synaptic inputs during phasic signaling (section 7.4.3.3). The contributions of tonic and phasic dopaminergic signaling to reinforcement learning, working memory, and synaptic plasticity are described in section 7.4.4.

7.4.3 MODELING THE DEPENDENCE OF TONIC AND BURSTING FIRING MODES ON INTRINSIC AND SYNAPTIC PROPERTIES The intrinsic properties of dopamine neurons that allow pacemaking to support a basal level of tonic firing are necessary to enable signaling of worse than expected outcomes (such as omitted rewards) by decreases in firing rate from some nonzero average level. Moreover, given the importance of the balance between single-spike (tonic) and burst firing, it is important to understand some of the factors controlling burst generation. Although there is diversity between subpopulations of dopamine neurons, in order to gain some understanding of tonic and burst firing, we present a generic model of dopamine neurons based on studies of SNC neurons.

7.4.3.1 General Model Description In order to simulate the activity of dopamine neurons that provides tonic and phasic neuromodulatory signals to their targets, we build upon the conductance-based models first used by Hodgkin and Huxley (1952) (see Brunel, chapter 2, on neural models) to model the generation of action potentials. Dopamine neurons have more complex firing patterns than the original Hodgkin–Huxley model can accommodate, and their models require additional features, including additional currents, models of ionic species accumulation within the cytosol, and multiple cylindrical compartments representing distinct somatic and dendritic regions. Dopamine neurons have smooth, sparsely branched dendrites that have a similar set of currents as the soma (Hausser, Stuart, et al., 1995), so we model the dopamine neuron as a set of cylinders representing the soma, the proximal dendrites, and the distal dendrites (figure 7.2). In the simple model described in this chapter, we examine only a somatic compartment, proximal dendritic compartment, and a distal dendritic compartment. In the model of pacemaking described in section 7.4.3.2, this allows us to observe three distinct time courses for calcium dynamics (the smaller the diameter, the faster calcium can accumulate and be removed). In the model of bursting described in section 7.4.3.3, this allows us to honor the differential distribution of synapses, with inhibitory synapses primarily located peri-somatically and excitatory synapses located more distally. For simplicity, this model does not include an axon.

Like the Hodgkin–Huxley model (see Brunel, chapter 2, "Basic Neuron and Network Models"), the dynamics of the membrane potential V(t) in our dopamine neuron models are governed by a first-order differential equation

$$C_M \frac{dV_i}{dt} = -I_L(V_i) - I_{Na}(V_i) - I_{K,DR}(V_i) + \mathrm{I}_{\mathrm{ext}}(\mathrm{t}) - \mathrm{I}_{\mathrm{K,A}}(\mathrm{V_i}) - I_{K,SK}(\mathrm{V_i}, [\mathrm{Ca}^{2+}]_{\mathrm{i}}) - \mathrm{I}_{\mathrm{Ca,L}}(\mathrm{V_i}) - \mathrm{I}_{\mathrm{axial}}(\mathrm{V_i}, \mathrm{V_j}) \tag{7.2}$$

where C_M is the membrane capacitance per unit area. The subscripts *i* and *j* are needed because we use distinct compartments to represent different parts of the neuron. In table 7.1, we describe each current in equation 7.2 with the subscript x, where g_x is a conductance, I_x is a current, E_X is the reversal potential, and V is the membrane potential. The first four terms in equation 7.2 represent the same currents as in equation 2.15 of Brunel's chapter (chapter 2): an Ohmic leakage current $I_L(V)$, the fast sodium current $I_{Na}(V)$, the delayed rectifier potassium current $I_{K,DR}(V)$, and an external current Iext(t), which here describes a current injected through an electrode. As before, the sodium current has both an activation variable *m* and an inactivation variable *h*, whereas the delayed rectifier potassium current has only an activation variable *n*. Previously, the delayed rectifier current was simply denoted $I_K(V)$. However, in this chapter, we introduce two additional currents carried by potassium ions. The voltage-dependent A-type potassium current $I_{K,A}(V)$ is distinguished from the Hodgkin–Huxley potassium current, called the delayed rectifier, because unlike the delayed rectifier, the A-type potassium current is transient rather than sustained because it has an inactivation variable (q) in addition to its activation variable (p). The second additional potassium current does not have an explicit voltage dependence but instead is activated by an increase in intracellular calcium concentration $[Ca]_i$. We also introduce the L-type Ca^{2+} current, which is activated in the subthreshold regime in these neurons (Durante et al., 2004) and is an important source of both subthreshold depolarization and calcium influx in the model. Finally, I_{axial} is the current flowing into the compartment *i* from adjacent compartments *j*. The axial conductance parameter g_{axial} depends upon the axial resistivity, the distance between the midpoints of the compartments, and their respective diameters, as well as the branching configuration.

TABLE 7.1

Model intrinsic current descriptions

The Ohmic leak current I_L helps to set the rest potential and the passive properties.		
$I_L = g_L\ (V - E_K)$	no activation	no inactivation
The fast sodium current I_{Na} is responsible for the action potential upstroke.		
$I_{Na} = g_{Na}m^3h\ (V - E_{Na})$	$dm/dt = (m_\infty(V) - m)/\tau_m(V)$	$dh/dt = (h_\infty(V) - h)/\tau_h(V)$
The delayed rectifier potassium current is responsible for action potential repolarization.		
$I_{K,DR} = g_{K,DR}\ n^3\ (V - E_K)$	$dn/dt = (n_\infty(V) - n)/\tau_n(V)$	no inactivation
The transient outward A-type potassium current helps set the pacemaker rate in the model		
$I_{K,A} = g_{K,A}\ p^3q\ (V - E_K)$	$dp/dt = (p_\infty(V) - p)/\tau_p(V)$	$dq/dt = (q_\infty(V) - q)/50.0$
The SK potassium current produces the calcium-dependent AHP.		
$I_{K,SK} = g_{K,SK}\ (V - E_K)/(1+ K_{M,SK}/[Ca^{2+}])^4$	no voltage-dependent activation	no inactivation
In the model, the L-type calcium current brings the membrane up to the action potential threshold and provides Ca^{2+} entry to activate the SK channel.		
$I_{Ca,L} = g_{Ca,l}\ d\ (V - E_{Na})$	$dd/dt = (d_\infty(V) - d)/\tau_d(V)$	no inactivation
Axial current I_{axial} flows between any two adjacent cylindrical compartments *i* and *j* in the model.		
$I_{axial} = g_{axial}\ (V_i - V_j)$	no activation	no inactivation

Note. The activation (middle column) and inactivation (right column) variables have steady-state values $x_\infty(V)$ and time constants $\tau_x(V)$ that are in general functions of membrane potential. The activation variables turn the current on with increasing depolarization, whereas the inactivation variables, which are usually slower, turn the current off with increasing depolarization. A differential equation of the form $dx/dt = (x_\infty(V) - x)/\tau_x(V)$ describes the rate of change of each activation and inactivation variable.

In order to simulate a calcium-dependent component of the after-hyperpolarization (AHP) following the action potential, the model tracks the concentration of intracellular free calcium ions [Ca^{2+}] in the cytosol (figure 7.4). The rate of change of cytosolic free calcium concentration[1] in each compartment is given by

$$\frac{d[Ca]_i}{dt} = \frac{2 f_{Ca} \bullet (I_{Ca,L}(V_i) + I_{CaP}([Ca]_i) + I_{L,Ca}(V_i))}{DF} \tag{7.3}$$

where $I_{L,Ca}$ is the portion of the leak current carried by Ca^{2+} ions, F is Faraday's constant to convert charge to moles, 2 is the charge on a calcium ion, D is the diameter of the cylinder, and f_{Ca} is the fraction of free calcium (the remainder is assumed to be bound to buffers and unavailable). A nonelectrogenic calcium pump I_{CaP}, in which the efflux of positive charges on calcium ions is offset by the influx of other charged particles, is used to model the removal of calcium ions from the cell:

$(I_{CaP} = I_{CaP,max,}/(1 + K_{M,CaP}/[Ca^{2+}]))$

where $K_{M,CaP}$ is a dissociation constant.

7.4.3.2 Multicompartmental Spontaneous Pacemaker Model The Hodgkin–Huxley model required an applied stimulus current in order to produce an action potential whereas dopamine neurons fire spontaneously with no input. This is likely because in order to signal worse than expected outcomes, dopamine neurons must maintain a baseline level of firing so that phasic pauses are possible. Moreover, the mechanisms that support tonic firing interact with the mechanisms that support phasic firing, as we will show in section 7.4.3.3. Therefore we illustrate a putative mechanism for slow spontaneous pacemaking in dopamine neurons

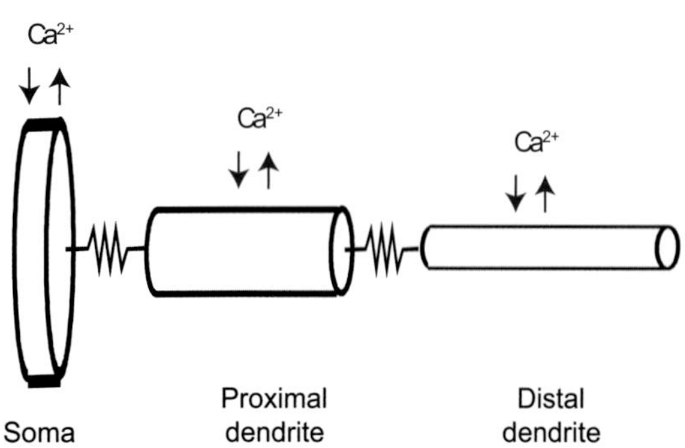

FIGURE 7.4 Model dopamine neuron. Schematic representation of the multicompartmental model and the calcium balance.

using a model dopamine neuron (Kuznetsova, Huertas, et al., 2010) http://senselab.med.yale.edu/ModelDB/ShowModel.cshtml?model=127507. The full model available for download has much higher spatial resolution (more spatial compartments) than the simpler model we present here but gives nearly identical results. In vitro, dopamine neurons exhibit spontaneous regular pacemaking. In the model (upper trace, figure 7.5A, 7.5B), the pacemaking cycle can be divided into three phases: the action potential itself, a calcium-dependent AHP (figure 7.5C), and a voltage-dependent depolarizing phase (figure 7.5D) leading to the next action potential. Figure 7.5C shows that $I_{K,DR}$ is the predominant outward current in the early phase of the AHP. The predominant outward current during most of the AHP is I_{axial} because of a contribution from $I_{K,SK}$ in the dendrites. The soma has a lower surface-to-volume ratio; therefore, calcium ions accumulate more slowly than in the smaller dendrites, so the SK current is more effectively recruited in the dendrites. The dendritic SK current flows axially to the soma and manifests itself as an outward component of this current. The hyperpolarization due to the outward currents causes I_{CaL} channels to close, resulting in de-activation of dendritic $I_{K,SK}$. Figure 7.5D shows that I_{axial} reverses direction. At this point, dendritic $I_{K,SK}$ no longer produces a somatic hyperpolarization, and the soma begins to depolarize. A positive feedback loop then drives the voltage-dependent phase of the ISI. This loop is because subthreshold depolarization opens $I_{Ca,L}$ channels, which in turn causes more depolarization. Concomitant increases in $I_{K,A}$ cause the net current to be small during the ISI, so the frequency is low.

Dopamine neurons exhibit other types of interesting oscillatory activity. For example, blocking the fast upstroke of the action potential by applying tetrodotoxin, a specific antagonist for the fast sodium channel, results in a subthreshold oscillatory potential (SOP) (Ping and Shepard, 1996). The model recapitulates this result as shown in figure 7.6A. The SOP is accompanied by an oscillation (Wilson and Callaway, 2000) in intracellular free Ca^{2+} (simulated by the dashed trace in figure 7.6B). As stated in chapter 3 (Rubin and Rinzel), oscillations can arise from the combination of a fast regenerative process to create a sudden rise or upstroke in activity and slower negative feedback to quench the activity. Similar to the case for pacemaking in figure 7.5D, the rising or depolarizing phase of the SOP in the model is also driven by the L-type calcium current. In the SOP, there is no action potential, so the Ca^{2+} entering during this depolarization eventually activates sufficient calcium-activated SK potassium channels (Amini, Clark, et al., 1999) to initiate the repolarizing

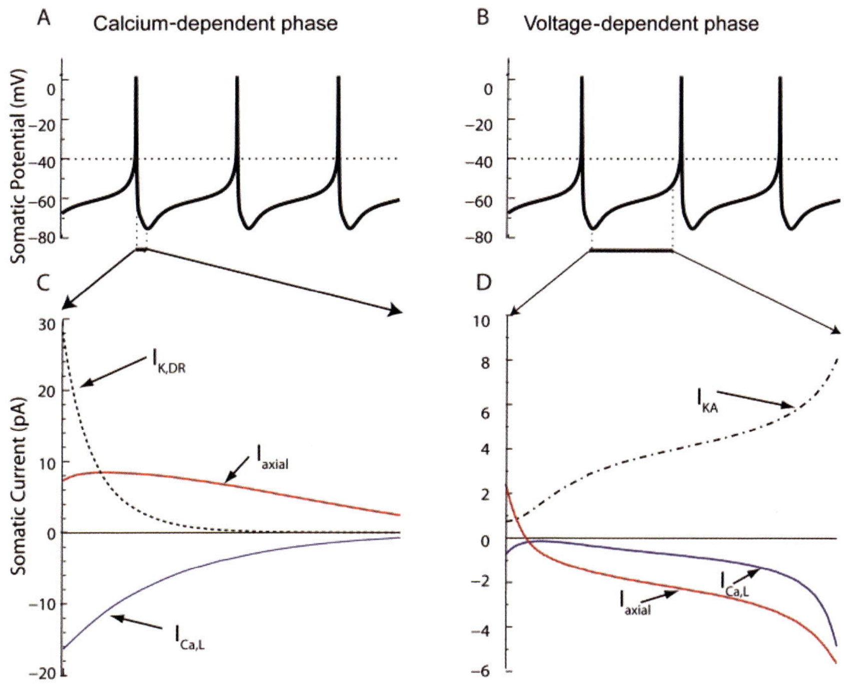

FIGURE 7.5 Model currents responsible for pacemaking. Somatic membrane potential, with the bar indicating the calcium-dependent phase of the interstimulus interval (A) or the voltage-dependent phase (B). (C) Somatic current during the calcium-dependent phase. (D) Somatic current during the voltage-dependent phase.

phase and "quench" the activity. Similar to the pacemaking case, during the repolarizing phase, the L-type channel turns off, and after the Ca^{2+} concentration decays sufficiently, the SK current decays sufficiently to allow the L-type calcium channel to reactivate, and the cycle repeats. The influx and efflux of calcium ions is proportional to the surface area of each compartment, but the rate of accumulation is slower for larger volumes. Compartments with smaller diameters have faster changes in free calcium ion concentration since they have larger surface-to-volume ratios. Therefore, during pacemaker firing, a single dopamine neuron constitutes a set of coupled oscillators (Wilson and Callaway, 2000) as in figure 7.4. In this model of the slow oscillatory potential (dashed traces in figure 7.6), the dendritic Ca^{2+} first reaches the level at which sufficient SK channel current is recruited to begin the repolarizing phase of the slow oscillation in Ca^{2+} (dashed curve in figure 7.6B).

During pacemaking, the slow depolarization preceding a spike looks very similar to the corresponding portion of the depolarization that does not precede a spike (i.e., during a SOP). When the spike threshold (dotted horizontal black line) is reached, an action potential is produced that is followed by an AHP (vertical arrow in figure 7.6A). This AHP is sufficient to turn off the L-type calcium current, which results in a net efflux of Ca^{2+} across the membrane, and $[Ca^{2+}]$ starts to decrease (vertical arrow in figure 7.6B). The underlying oscillation in Ca^{2+} (dashed curve in figure 7.6B) has effectively been reset from a point at which Ca^{2+} is increasing up to the same value of Ca^{2+} but when Ca^{2+} is decreasing (see horizontal arrow in figure 7.6B). The premature quenching, or truncation, of the underlying SOP by an action potential decorrelates the frequency of pacemaking from that of the SOP (Guzman, Sanchez-Padilla, et al., 2009; Kuznetsova, Huertas, et al., 2010; Drion, Massotte, et al., 2011), consistent with experimental data. Since the oscillation in calcium concentration does not drive the pacemaking frequency, what function does it serve? The answer follows from the observation that tonic signaling mediated by pacemaking tendencies in dopamine neurons is only part of the story: tonic signaling is maintained in vivo in the presence of synaptic inputs and is punctuated by phasic bursts. The next section addresses the roles of the currents that drive the oscillation in calcium concentration, namely, $I_{Ca,L}$ and $I_{K,SK}$, in producing phasic signals.

7.4.3.3 Multicompartmental Model of Phasic and Tonic Firing in Vivo In this section, we add synaptic inputs (see table 7.2) $I_{syn}(t,V)$ to a model of an isolated dopamine neuron in order to approximate the firing

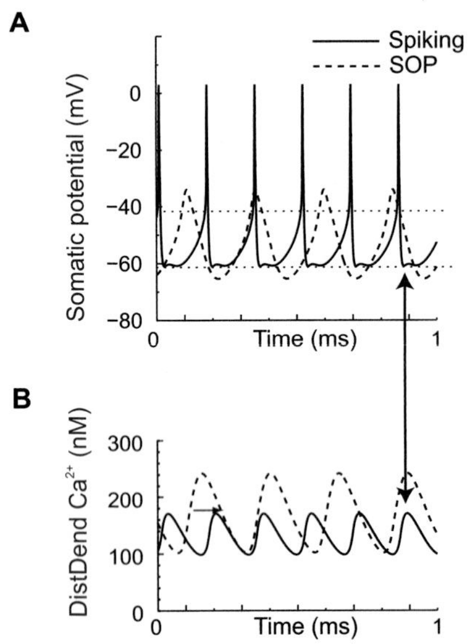

FIGURE 7.6 Comparison of subthreshold oscillatory potentials (SOPs) and pacemaker spiking. (A) Membrane potential. The oscillations in membrane potential are shown for pacemaker spiking under control conditions (black curve) versus the subthreshold oscillation (dashed curve) that is unmasked by setting the sodium conductance to zero to simulate the application of tetrodotoxin. In the presence of the fast sodium current, the slow, large amplitude subthreshold oscillations are truncated at the spike threshold, leading to faster pacing. (B) The corresponding excursions in free Ca^{2+} in the distal dendrites (DistDend). The after-hyperpolarization that follows a spike (vertical arrow in A) turn causes the Ca^{2+} oscillation to enter its decreasing phase (vertical arrow in B), truncating the amplitude of the oscillation. Parameters are as in Kusnetsova et al. (2010) except I_{stim} is 100 pA and the Ca^{2+} free fraction is 0.02.

patterns exhibited in vivo, which constitutes an additional term on the right-hand side of equation 7.1:

$$I_{syn}(t,V) = I_{NMDA}(t,V) + I_{AMPA}(t,V) + I_{GABA}(V).$$

The model presented in this section is again a three-compartment model, quite similar to the one in the previous section, available at https://senselab.med.yale.edu/ModelDB/ShowModel.cshtml?model=84612. Since dopamine neurons in vivo are subject to a barrage of inhibitory inputs, as a first approximation inhibition was modeled as a constant current. The inhibitory conductance g_{GABA} was much greater in the somatic compartment than in the dendritic compartments, consistent with experimental data and previous modeling results (Komendantov, Komendantova, et al., 2004). Time dependence was introduced in the excitatory inputs, which were located exclusively on the dendrites, in order to produce irregularity and variability as observed in vivo. The AMPA kinetics are conceptually

TABLE 7.2

Model synaptic current descriptions

The inhibitory current is modeled as a first approximation as tonic.

$I_{GABA} = g_{GABA}\ (V - E_{Cl})$

The excitatory current mediated by two subtypes of glutamate receptor has a fast, linear AMPA and a slow, nonlinear NMDA component.*

$I_{AMDA} = g_{AMPA} R_{AMDA} (V - E_{AMPA})$	$dR_{AMPA}/dt = [T]\alpha_A(1 - R_{AMPA}) - \beta_A R_{AMPA}$
$I_{NMDA} = g_{NMDA} R_{NMDA} (V - E_{NMDA}) / (1+\exp(-V/k)\ K_{M,Mg}/[Mg^{2+}])$	$dR_{NMDA}/dt = [T]\alpha_N(1 - R_{NMDPA}) - \beta_N R_{NMDA}$

Note. The GABA A receptor current is inhibitory because the reversal potential (E_{Cl}) of the chloride ions that carry this current is modeled as −70 mV. The reversal potential of the AMPA and NMDA receptors for glutamate is approximately 0 mV, so they are excitatory.

*The Komendantov et al. (2004) model actually utilized a more complicated Goldman–Hodgkin–Katz expression for the driving force for the NMDA receptor current in order to simulate the permeability to different ions more accurately, but for simplicity we present all driving forces for synaptic currents as Ohmic.

similar to those presented in Vierling-Claassen and Jones's chapter (chapter 4). The factor [T] represents the neurotransmitter glutamate in the synaptic cleft and is nonzero only for 1 ms after each synaptic input arrives. The rate constants for binding and unbinding of glutamate to the receptor are slower for NMDA receptors compared to AMPA receptors (α_A = 1,100 $s^{-1}mM^{-1}$ and β_A = 190 s^{-1}, compared to α_N = 72 $s^{-1}mM^{-1}$ and β_N = 6.6 s^{-1}). The intervals between successive synaptic inputs were exponentially distributed to simulate a random Poisson process (see Brunel, chapter 2). The NMDA receptor current is unique among the ligand-gated conductances because most ligand-gated conductances have only a linear, Ohmic dependence on membrane potential, similar to a leak conductance. In contrast, physiological concentrations of magnesium ($[Mg^{2+}]$) block the NMDA receptor conductance, but depolarization relieves the block. Therefore this current, like the $I_{Ca,L}$ current also included in the model and described in the previous section, has a regenerative dependence on membrane potential that can lead to a positive feedback loop and oscillations.

Certain signals may be specifically encoded by activation of NMDA receptors in dopamine neurons; in fact, disrupting burst signaling selectively by genetically inactivating NMDA receptors in dopamine neurons selectively impaired learning in cue-dependent tasks (Zweifel, Parker, et al., 2009). Experiments have shown that activation of NMDA receptors contribute to burst firing (Chergui, Nomikos, et al., 1996; Zweifel, Parker, et al., 2009) whereas activation of $GABA_A$ receptors reduces the tendency to burst (Paladini and Tepper, 1999). Moreover, pharmacological manipulations that enhance either the intrinsic L-type Ca^{2+} current or the synaptic NMDA receptor current can under some conditions induce repetitive burst firing in vitro (Johnson, Seutin, et al., 1992; Johnson and Wu, 2004). The regenerative voltage dependence (Rubin and Rinzel, chapter 3) of the intrinsic L-type Ca^{2+} current and the synaptic NMDA receptor current explains their likely role in producing phasic bursts. Both are inward currents that activate with depolarization and do not inactivate. Thus in the "on" or activated state, they can support depolarized plateau potentials in dopamine neurons in vitro, during which a sustained depolarization can be observed until these currents are switched off by membrane hyperpolarization. However, as shown in the previous section, the calcium-activated SK potassium channel normally limits spontaneous depolarizations to producing a single action potential per depolarization (Seutin, Johnson, et al., 1993; Canavier and Landry, 2006). Blocking this current weakens the AHP after a spike, which can prevent the L-type Ca^{2+} channels from closing before another spike is generated. This has the effect of extending the depolarization so that it can supports multiple action potentials, leading to repetitive burst firing (Ping and Shepard, 1996) in some cases. Depolarized plateau potentials are likely contributors to burst activity in vivo (Grace and Bunney, 1986).

In vivo, a balanced state between excitation and inhibition may stabilize single-spike firing. Dopamine neurons receive GABAergic inputs from the rostromedial tegmental nucleus (Hong and Hikosaka, 2011), the striatum, the substantia nigra pars reticulata, and the globus pallidus (Tepper and Lee, 2007). They receive glutamatergic inputs from the subthalamic nucleus (Kita and Kitai, 1987) and glutamatergic (and cholinergic) inputs (Floresco, West, et al., 2003) from the pedunculopontine nucleus. Figure 7.7 demonstrates the effect of balanced excitation and inhibition in the model. The AMPA/NMDA ratio was held constant as the average level of NMDA receptor activation was increased by decreasing the mean interevent interval between synaptic inputs. The tonic GABAergic conductance was also varied. Strong excitation relative to inhibition resulted in an increasing fraction of the spikes occurring in bursts (gray) whereas the converse led to silence (black). When the two were balanced as in the (red) diagonal band in figure 7.7, tonic single-spike firing was exhibited. From this balanced state, consistent with experimental data (Lobb, Wilson, et al., 2010, 2011), either a transient increase in excitation or a transient decrease in inhibition could evoke a burst. Thus, phasic burst signaling could be theoretically achieved via either mechanism. The balance between excitation and inhibition is very important since only dopamine neurons that are tonically active can emit a phasic burst in response to reward related cues (Lodge and Grace, 2006).

In addition to the balance between synaptic inhibition and excitation, modulation of intrinsic current also can affect the balance between tonic and phasic signaling. For example, downregulating the calcium-activated SK potassium conductance increases the tendency of dopamine neurons to emit bursts in vivo (Waroux,

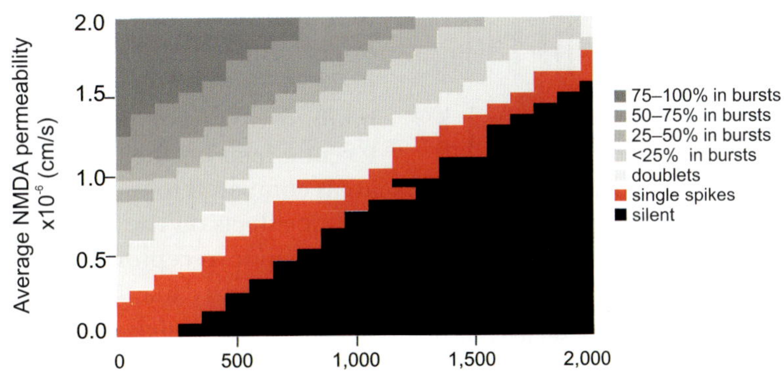

FIGURE 7.7 Balanced excitation and inhibition promotes single-spike firing. The activation of the linear $GABA_A$ synaptic receptor suppresses firing, leading to hyperpolarized silence (indicated by the black region), whereas the activation of nonlinear NMDA receptors promote plateau potentials that support bursting (gray shaded regions). In between (red region) a tonic but irregular single-spiking mode persists. Only active neurons in this region can be recruited for phasic signaling according to Lodge and Grace (2006). Adapted from Canavier and Landry (2006).

Massotte, et al., 2005; Ji and Shepard, 2006), whereas in vitro, SK blockade often tends to make the firing less regular (Shepard and Bunney, 1988). Figure 7.8 (Drion, Massotte, et al., 2011) shows that the same model is able to mimic and explain three experimentally observed firing patterns by varying the noise level and the level of the SK channel conductance. The low synaptic noise case produced with the SK channel conductance set to control values produced regular pacemaker-like firing intended to mimic the situation in a slice. The high synaptic noise case was intended to mimic the situation in vivo, and at control levels of SK channel conductance decreased the regularity of the tonic single-spike firing. At the same, high synaptic noise levels simulating down-regulation of the SK channel conductance resulted in phasic burst firing.

7.4.3.4 Summary and Additional Considerations The following observation suggests that NMDA receptors may play an additional role beyond contributing to plateau potentials underlying bursts: simple depolarization is not sufficient to induce a high-frequency burst in dopamine neurons in vitro, because they go into depolarization block (Richards, Shiroyama, et al., 1997) at frequencies above 10 Hz. At that frequency, the sodium channels cannot sufficiently recover from inactivation between action potentials, and spiking fails (Kuznetsova, Huertas, et al., 2010). Early models of dopamine neurons utilized burst mechanisms based on

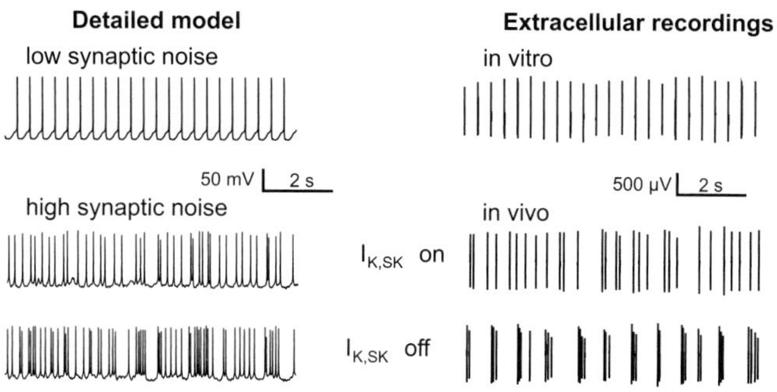

FIGURE 7.8 The model can mimic tonic and phasic firing. Top trace: With the SK channel conductance set to control values in the presence of low synaptic noise, the model exhibits the regular pacemaker-like firing observed in vitro with experimental data obtained from dopaminergic neurons. Middle trace: With the SK channel conductance set to control values in the presence of high synaptic noise, the model mimics the tonic single-spike firing observed in vivo with experimental data obtained from dopaminergic neurons. Bottom trace: With the SK channel conductance set to zero to mimic channel block and in the presence of high synaptic noise, the model mimic the phasic burst firing observed in vivo with experimental data obtained from dopaminergic neurons. Modified from Drion et al. (2011).

oscillatory plateaus or transient depolarizations (Canavier, 1999; Komendantov, Komendantova, et al., 2004; Canavier and Landry, 2006; Drion, Massotte, et al., 2011; Oster and Gutkin, 2011) and did not explain how the depolarization block observed in vitro is overcome in vivo to account for the higher frequencies (~50 Hz) observed during bursts in vivo (Hyland, Reynolds, et al., 2002). More recent models (Qian, Yu, et al., 2014) focus on how the regenerative nature of NMDA receptor currents may allow them to functionally augment the sodium current during the initiation and upstroke of action potentials, thereby enabling fast spiking at more depolarized levels.

Some salient hypotheses regarding tonic and phasic signaling in dopamine neurons are as follows:

1. In the absence of afferent input, dopamine neurons exhibit spontaneous regular pacemaker firing; this background level of tonic firing ensures that both positive and negative signals can be transmitted.
2. Dopamine neurons can only fire rapid bursts in the presence of NMDA synaptic activation, a key player in synaptic plasticity. Blocking the SK potassium channel responsible for the calcium-mediated AHP facilitates bursting.
3. In vivo, a balanced state between excitation and inhibition impinging upon a given dopamine neuron enables tonic single-spike firing, whereas an imbalance can result in a phasic signal in the form of one or more bursts.
4. The complex contributions of the intrinsic currents to the firing dynamics allow for rich modulation of the propensity of dopamine neurons to transmit tonic or phasic signals.

In summary, the complex intrinsic dynamics of the membrane currents of dopamine neurons provide for rich neuromodulatory possibilities and multiplexing opportunities for the signal-processing capabilities of these neurons with respect to both their excitatory and inhibitory inputs. Since phasic burst firing has very different effects on the dopamine concentration in target areas compared to tonic firing, which, in turn, will have drastic effects on the binding of dopamine on pre- and postsynaptic receptors (Dreyer, Herrik, et al., 2010), this complexity may have substantive implications for systems-level models of dopaminergic signaling. Any drug or phenomenon capable of modulating specific intrinsic conductances of dopaminergic neurons can change their activity from tonic to bursting and can, in turn, have drastic effects on the neural computations of dozens of brain areas to which the dopaminergic neurons project, likely affecting behavior, as reviewed below.

7.4.4 Postsynaptic Effects of Dopamine and Behavioral Functions Dopamine exerts its neuromodulatory effects via two receptor populations, those in the D1 and D2 families. The two receptor types play distinct roles in the two major areas to which SNC and VTA neurons project, the striatum and prefrontal cortex. Moreover, D2 autoreceptors on the dopamine neurons have a self-inhibitory effect. D2 receptors have a higher affinity for dopamine than D1 receptors. The fraction of occupied receptors in equilibrium with a particular concentration of dopamine is $[DA]/([DA]+EC_{50})$ where [DA] is the extracellular dopamine concentration, and EC_{50} indicates the concentration at which half the receptors are occupied, which is 1 μM for D1 receptors but only 0.01 μM for D2 (Dreyer, Herrik, et al., 2010).

7.4.4.1 Role of Dopamine Neuromodulation in Striatal Plasticity and Reinforcement Learning A rich body of computational work focuses on the role of dopamine in reinforcement learning (chapter 10). Dopamine is an almost direct biophysical correlate of reinforcement delivery, and the manners in which it influences synapses, single cells, and networks of different sizes give computational insights into the neural mechanisms underlying this specific form of learning (Samson, Frank, et al., 2010).

In the striatum, cortical glutamatergic input on medium spiny neurons can cause either LTP or long-term depression (LTD) to occur, depending upon the strength of the dopaminergic input (Reynolds and Wickens, 2002), which translates to the level of extracellular concentration. Striatal neurons in the direct pathway that facilitates movement express D1 receptors whereas those in the indirect pathway that inhibits movement express D2 receptors (Daw, chapter 10). A framework has been proposed for functionally integrating tonic and phasic signals (Dreyer, Herrik, et al., 2010) based on the observation that D2 receptors have a higher affinity for dopamine than D1 receptors. The theoretical framework for integrating tonic and phasic signaling makes the following assumptions. During tonic firing, extracellular dopamine levels are sufficient to ensure most D2 receptors are occupied, but most D1 receptors are not (Heien and Wightman, 2006). Bursts, therefore, preferentially increase the occupancy of D1 receptors. On the other hand, pauses preferentially decrease the occupancy of D2 receptors. Thus, time-locked single-population bursts and pauses alter D1 and D2 receptor signaling, respectively, and mediate synaptic plasticity that updates predictions of future reward (Daw, chapter 10). Interestingly, episodes of sustained phasic firing, which include both bursts and pauses, translate to higher average D1 occupancy and lower average D2occupancy compared to a tonic episode of the same duration and overall average firing rate. Episodes of phasic firing in the SN lead to novelty-induced exploration (Schiemann, Schlaudraff, et al., 2012), which might be explained by the resultant changes in the D1/D2 receptor balance in the downstream target, the striatum.

7.4.4.2 Role of Dopamine in Cortical Plasticity In prefrontal cortex, phasic dopamine release is required to trigger long-term plasticity in response to high-frequency stimulation in vitro; however, tonic levels of dopamine control the direction of plasticity, such that LTD occurs in the presence of low tonic dopamine and LTP occurs at high dopamine levels (see Sheynikhovich, et al., 2011). A recent model (Sheynikhovich, Otani, et al., 2011) addresses the role of phasic and tonic dopamine in long-term plasticity in cortex. The baseline dopamine concentration due to tonic dopamine levels is treated as a constant to mimic bath application of dopamine in the in vitro experiments and can take on different values in the simulations. The phasic dopamine signal is generated by a fixed increment in concentration each time a pulse in a high-frequency (50 Hz) pulse train is applied, and it decays back to baseline with a time constant on the order of 1 second.

The tonic value is sensed by a slowly activating kinase on a timescale of minutes. The level of kinase determined by tonic dopamine flips a switch between LTP and LTD. If the kinase level is below a certain threshold, then synapses receiving high-frequency stimulation are tagged for LTD, whereas if it is above, they are tagged for LTP. The tags are induced probabilistically by coincident presynaptic conductance activation and postsynaptic depolarization and decay stochastically. The synapses are bistable with a persistently depressed and a persistently potentiated state and are initialized such that both states are represented. Tagging the synapse induces a transient change in synaptic weight. However, maintenance, in the form of a sustained switch in the bistable configuration of the synapse, requires protein synthesis. Phasic dopamine is required for protein synthesis because, in the model, the product of the phasic signal in dopamine concentration and the level of kinase determines the rate of protein synthesis, and the protein has a constant rate of degradation. Thus, tonic dopamine determines the direction of the plasticity in the model and contributes to its amplitude, but phasic dopamine is required for permanent changes. The dependence of long-term plasticity on phasic dopamine is consistent with the hypothesized role of phasic dopamine in reinforcement learning (chapter 10).

7.4.4.3 Role of Dopamine in Working Memory The neuromodulatory effects of dopamine acting at D1 and D2 receptors have been intensively investigated in the medial prefrontal cortex. Computational models of working memory (chapter 11) suggest that activation of D1 receptors stabilizes representations in working memory, making them robust to distraction and noise (Lapish, Kroener, et al., 2007). These models suggest that D1-mediated effects enhance strong, persistent activity associated with ongoing representations of specific information in working memory that are supported by strong recurrent network activity. The D1-mediated enhancement of a persistent Na^+ current and NMDA receptors as well as downregulation of slowly inactivating K^+ currents on pyramidal cells contributes to the enhancement of recurrent activity. The suppression of weak activity is thought to result from D1-mediated increase in interneuron firing probability and decrease in glutamate release probability. On the other hand, D1-mediated effects diminish the effects of weaker or brief inputs, minimizing the effect of distractors. D2 receptor activation in general has the opposite effect on ionic currents and, therefore, has opposite effects on representations in working memory, destabilizing them and potentially allowing the working memory buffers to incorporate new information. The timescale of the postsynaptic dopamine-mediated neuromodulation can be on the order of minutes to hours, suggesting that the function of this neuromodulatory activity may be too slow to provide temporally precise information about the mismatch between predicted and actual rewards (Lapish, Kroener, et al., 2007). In the most common versions of the working memory task, animals have to hold a memory for a few seconds (delay period) before making a decision. The basic experimental finding is that the sustained firing of prefrontal neurons observed during this delay phase is modulated by dopamine in an inverted-U-shaped manner, whereby too little or too much dopamine modulation (D1 receptors) decreases the activity of prefrontal neurons and the performance of the animal (Goldman-Rakic, Muly, et al., 2000; Floresco, 2013). The way in which this "sweet spot" emerges and is maintained has been the central focus of most of these models. Some of these models have proposed that D1-receptor-mediated modulation of intrinsic currents (e.g., increase in persistent Na current) and synaptic currents (increase in NMDA and $GABA_A$, decrease in AMPA) of prefrontal pyramidal cells could explain much of the experimental data (Lisman, Fellous, et al., 1998; Durstewitz, Seamans, et al., 2000). Others have focused on dopaminergic modulation of the inhibitory neuron population to offer a complementary perspective, reproducing the inverted-U-shaped observed experimentally (Brunel and Wang, 2001; Lew and Tseng, 2014).

7.4.4.4 Integration of Dopaminergic Signaling into Larger Frameworks Interestingly, and going beyond reinforcement learning, dopamine has been suggested to act in concert with other neuromodulators to achieve higher level forms of learning (Daw, Kakade, et al., 2002; Doya, 2002; Seymour, Daw, et al., 2012). One suggestion is that dopamine controls the magnitude of the reward prediction error, serotonin controls the reward discounting amount (how much the reward value is decreased if it is expected late), noradrenaline controls the randomness of the policy (i.e., mapping from the state of the animal to the next action to take), and acetylcholine controls the overall learning rate (rate at which synapses are modified). Although no explicit computational model is proposed, this theory nicely demonstrate that different families of neuromodulators may work together to control a particular function, at the system level (Doya, 2002).

Others have used mean-field computational models of the VTA to show how nicotine may be motivationally reinforcing. In this model, nicotine acts through specific types of endogenous acetylcholine receptors to directly increase VTA activity, or to indirectly disinhibit it through GABAergic neuron deactivation (Graupner, Maex, et al., 2013).

Another theory proposed a more general role for these concerted neuromodulatory actions in generating the full range of emotional states in humans and animals (Fellous, 1999). Others have suggested that neuromodulators allow for the conversion (or readout) of the phylogenetically old motivational circuitry activity into what we commonly call emotions (Kelley, 2005). To date no modeling work has yet articulated these ideas.

Whether in the context of working memory or reinforcement learning, little if any attention is given to the tonic and phasic firing modes of VTA neurons and the manner in which they are controlled. At most, the account is phenomenological or unidirectional (effect of a specific dopaminergic firing mode on neural computation). Notably, a significant effort is underway at the connectionist level to place the dopaminergic system within the larger computational loops underlying behaviors, be it in the context of working memory or reinforcement learning (Frank and Claus, 2006; Hazy, Frank, et al., 2010; Dominey, 2013; Schroll, Vitay, et al., 2014). See also chapter 22, "Integrative Functions of the Corticostriatal System." A thorough understanding of the dynamics of the firing patterns that underlie tonic and phasic signaling is likely to give significant additional insights into the computations involved in reinforcement learning, working memory, and the other functions known experimentally to depend on dopamine, such as category learning (Moustafa and Gluck, 2011a,b), decision-making (Frank, 2006; Gupta, Balasubramani, et al., 2013; Guthrie, Leblois, et al., 2013), and various diseases including drug addiction, Parkinson's disease, and schizophrenia (Frank, Seeberger, et al., 2004; Spiga, Lintas, et al., 2010; Moustafa and Gluck, 2011a,b; Keramati and Gutkin, 2013; Lopez-Huerta, Carrillo-Reid, et al., 2013; Piray, Zeighami, et al., 2014; Schroll, Vitay, et al., 2014; see also chapter 23, "Brain Diseases").

Finally, while researchers are still trying to understand the computational role of single neuromodulators, it will soon become necessary to make explicit how different neuromodulatory systems interact with each other and with target substrates. It is, for example, clear that the dopaminergic system is in a dynamical computational loop with the hippocampus, an area traditionally studied for its involvement in cognitive processing such as spatial navigation or short-term memory (Lisman and Grace, 2005). How the hippocampus controls the VTA is still largely unknown. The same applies for the relationship between dopaminergic neurons and the prefrontal cortex, which are known to be in reciprocal interactions (Patton, Bizup, et al., 2013). It will be a challenge for future modeling work to explore ways in which neuromodulation is controlled; only then, we argue, will we truly understand its computational value.

7.5 *Cholinergic Modulation*

7.5.1 Anatomy of Acetylcholine Acetylcholine plays an essential role in cognition. Acetylcholine appears to be important for the encoding of new memories and for regulating attention to behaviorally relevant stimuli, based on the fact that drugs that block acetylcholine effects impair these cognitive functions of memory and attention.

Because this textbook offers a path from neuron to cognition, we will focus on how the anatomy and physiology of acetylcholine in the central nervous system relate to behavior and cognition. Cortical structures receive broadly distributed cholinergic innervation from populations of cholinergic neurons in the basal forebrain, including the nucleus basalis of Meynert, the diagonal band of Broca, and the medial septum (Descarries, Gisiger, et al., 1997) (figure 7.9). These basal forebrain structures are regulated by brainstem nuclei containing cholinergic neurons including the pedunculopontine nucleus and the laterodorsal nucleus. The cortex contains a tiny number of local cholinergic interneurons, but there are large populations of cholinergic interneurons in the basal ganglia that are not discussed further here.

7.5.2 Physiological Effects of Acetylcholine The striking cognitive effects of blocking acetylcholine receptors raise interesting questions about the role of acetylcholine in setting the physiological properties of neural circuits. In this section, we review these properties; in the next section, we will report on studies at the behavioral and cognitive level, and then establish bridges between these levels. As a broad overview, these physiological effects appear to enhance memory encoding and attention by enhancing the influence of feedforward afferent input to the cortex, enhancing the response of cortical circuits to features of the external world, while decreasing excitatory feedback activity (Hasselmo, 1995, 2006). These changes in circuit dynamics result from effects at different subtypes of acetylcholine receptors that include two major receptor categories named after individual examples of drugs that selectively activate these receptors. Nicotinic receptors respond to the drug nicotine (as well as other nicotinic drugs) but do not respond to the drug muscarine. Muscarinic receptors respond selectively to the drug

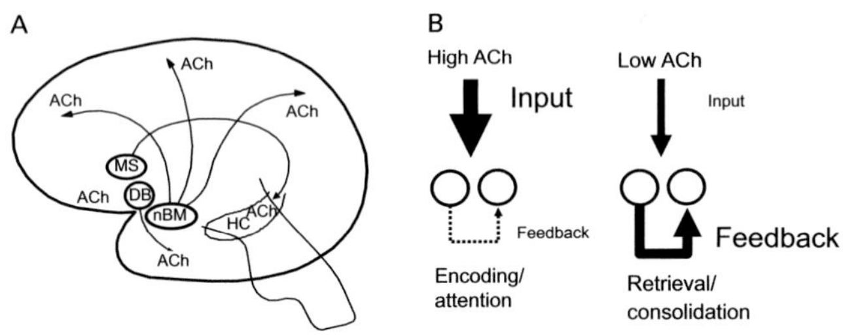

Figure 7.9 (A) Origin of fibers releasing acetylcholine (ACh) in cortical structures. Cholinergic innervation of the hippocampus (HC) arises from the medial septum (MS). Cholinergic innervation of parahippocampal structures (not shown) arises from the MS and the diagonal band of Broca (DB). Cholinergic innervation of the neocortex arises from the nucleus basalis of Meynert (nBM). (B) Summary of the overall effect of cholinergic modulation in cortical structures (Hasselmo, 2006). High levels of acetylcholine enhance memory encoding and attention by increasing the spiking response to afferent input and reducing excitatory feedback in the cortex. Lower levels of acetylcholine result in less influence of afferent input and stronger influence of excitatory feedback that could enhance consolidation of previously encoded memories (see also figure 7.14).

muscarine (and other muscarinic drugs) but not to the drug nicotine.

7.5.2.1 Enhancement of Spiking Response to Afferent Input Activation of nicotinic receptors increases the amplitude of excitatory synaptic potentials induced by thalamic input to neocortex but not at excitatory feedback (i.e., intracortical) synapses (Gil, Connors, et al., 1997). Nicotinic receptors also enhance the effect of synapses bringing afferent input to the hippocampus but not recurrent synaptic potentials in hippocampal region CA3 (Giocomo and Hasselmo, 2005).

Acetylcholine also activates muscarinic receptors to enhance the spiking response of cortical pyramidal cells to afferent input (reviewed in Hasselmo, 1995). This effect results from activation of postsynaptic muscarinic M1 receptors that cause depolarization of pyramidal cells and allow the neuron to generate more spikes during depolarization (Barkai and Hasselmo, 1994) (figure 7.10).

7.5.2.2 Enhancement of Synaptic Modification Because blockade of acetylcholine receptors is associated with impairments of encoding in long-term memory, a number of studies have tested the role of acetylcholine in regulation of synaptic modification, usually measured by effects on LTP in the hippocampus. The hippocampus plays an important role in encoding of new episodic memories, as shown by impairments of such encoding associated with damage to the hippocampus. The focus on LTP was motivated by previous research showing that the LTP in the hippocampus can last for weeks, consistent with the time course of long-term memory, and that drugs that block LTP in the hippocampus can impair the formation of new memories. Thus, modulators that regulate LTP could regulate the encoding of new memories. Consistent with a role of acetylcholine in regulating encoding of memory, physiological studies show that acetylcholine enhances LTP in areas including the hippocampus (Huerta and Lisman, 1995) and piriform cortex (Hasselmo and Barkai, 1995).

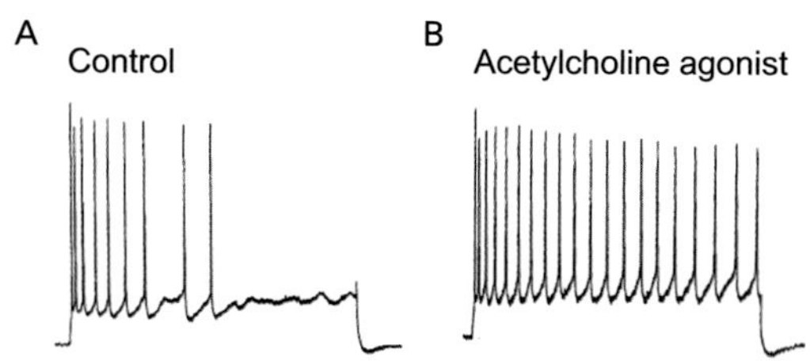

FIGURE 7.10 (A) In control conditions, a neuron responds to depolarizing current injection by initially generating repetitive spiking that slows and eventually stops as the current injection continues. (B) In contrast, when muscarinic receptors on the neuron are activated by a drug that resembles acetylcholine, the neuron responds with stronger, sustained spiking throughout the duration of the depolarizing current injection. From Barkai and Hasselmo (1994).

7.5.2.3 Muscarinic Presynaptic Inhibition Along with enhancing the effect of afferent input, acetylcholine also reduces the effects of excitatory synaptic feedback in the cortex (Hasselmo, 1995, 2006). As shown in figure 7.11, acetylcholine is released from axonal varicosities in the cortex and spreads via volume transmission to activate both presynaptic and postsynaptic receptors. The middle of figure 7.11 shows acetylcholine acting at muscarinic presynaptic receptors on excitatory feedback connections. Activation of these receptors by acetylcholine causes inhibition of the release of the transmitter glutamate from glutamatergic terminals, thereby reducing the size of glutamatergic excitatory synaptic potentials induced in the

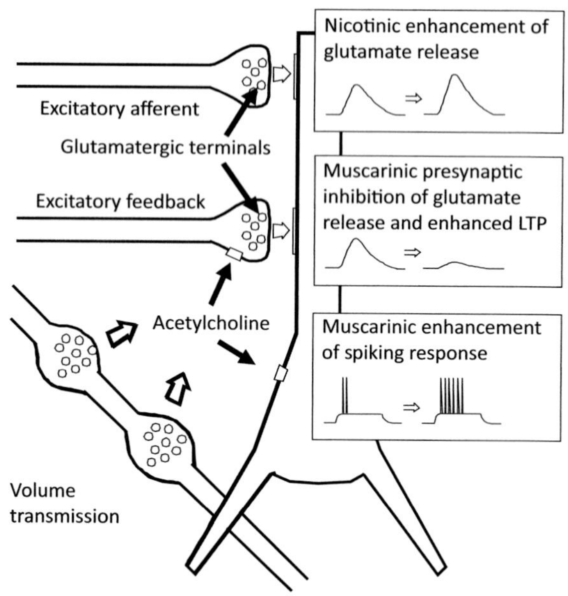

FIGURE 7.11 Acetylcholine released from axonal varicosities via volume transmission causes postsynaptic increases in the spiking response of neurons (bottom) to depolarizing input (Hasselmo, 1995). Acetylcholine also causes nicotinic enhancement of glutamate release at afferent synapses (top) and causes selective muscarinic presynaptic inhibition of glutamate release at feedback synapses, thereby reducing synaptic potentials evoked by stimulation of excitatory feedback synapses (middle) (Hasselmo and Bower, 1992; Hasselmo, Schnell, et al., 1995; Hasselmo, 2006). During the presynaptic inhibition of synaptic potentials, acetylcholine also enhances postsynaptic induction of long-term potential (LTP), so that, when measured at a later time, the excitatory feedback synapses have increased in strength more than when no acetylcholine was present.

postsynaptic neuron. Presynaptic inhibition of glutamatergic synaptic potentials by acetylcholine was first shown in the hippocampus (Yamamoto and Kawai, 1967). The selective presynaptic inhibition of excitatory feedback synapses by acetylcholine was first shown in the primary olfactory cortex (Hasselmo and Bower, 1992), where acetylcholine selectively reduces excitatory recurrent feedback potentials (figure 7.11, middle), while having a much weaker effect on synaptic potentials induced by excitatory input (figure 7.11, top). The selective effect on excitatory feedback but not afferent input has also been shown in both region CA3 and region CA1 of the hippocampus (Hasselmo and Schnell, 1994; Hasselmo, Schnell, et al., 1995). The initial description of selective cholinergic presynaptic inhibition inspired subsequent studies in the neocortex showing selective suppression of excitatory feedback synapses in the neocortex (Gil, Connors, et al., 1997) and nicotinic enhancement of glutamate release (figure 7.11, top).

7.5.2.4 Modulation of Inhibition and Theta Rhythm Oscillations Acetylcholine is associated with increases in amplitude of theta rhythm oscillations within the hippocampus. The theta rhythm is an oscillation with a frequency around 4–10 Hz of the local field potential that reflects the activity across a large number of synapses in the hippocampus and other cortical structures (Lee, Chrobak, et al., 1994; Buzsaki, 2002). (See also chapter 4, "Neural Rhythms.") Theta rhythm is most prominent when animals explore the environment or attend to behaviorally relevant stimuli and decreases when an animal performs behaviors such as eating or grooming (Buzsaki, 2002). The presence of theta rhythm correlates with better learning in conditioning tasks and encoding of memories, and the loss of theta rhythm correlates with impaired encoding (Givens, 1996). Theta rhythm is reduced by damage to the cholinergic (Lee, Chrobak, et al., 1994) and GABAergic input to the hippocampus from the medial septum. Cholinergic modulation of theta rhythm could involve hippocampal interneurons, which are directly depolarized by acetylcholine (McQuiston and Madison, 1999), though muscarinic receptors also cause presynaptic inhibition of GABA release (Pitler and Alger, 1992). The combination of depolarization and presynaptic inhibition appears paradoxical, but computational modeling demonstrates that these combined effects reduce background activity while heightening the response to suprathreshold sensory stimuli (Patil and Hasselmo, 1999). Such transient increases in signal to noise have also been proposed for dopamine (Kroener, Chandler, et al., 2009), and the two modulators can act together to enhance attention and memory encoding. The cholinergic regulation of theta rhythm could contribute to separation of encoding and retrieval as described below.

7.5.2.5 Enhancement of Persistent Spiking Acetylcholine also enhances the persistent spiking of individual cortical neurons, which could provide a possible cellular mechanism for working memory (see chapter 11, "Short-term, Long-term, and Working Memory") by allowing neurons activated by an input to maintain the spiking activity representing that input over a period of time. Persistent spiking has been shown in entorhinal cortex (Klink and Alonso, 1997a,b), as well as other regions. In standard control conditions, cortical neurons respond to an intracellular depolarizing current injection by generating spiking activity that stops after the end of current injection. However, in the presence of drugs that activate acetylcholine receptors, neurons persist in spiking activity for many minutes after current injection ends (Klink and Alonso, 1997a,b; Fransen, Alonso, et al., 2002). This effect appears in other cortical regions including perirhinal cortex (Navaroli, Zhao, et al., 2012) and prefrontal cortex (Haj-Dahmane and Andrade, 1998).

7.5.3 BEHAVIORAL FUNCTION OF ACETYLCHOLINE The physiological effects of acetylcholine underlie important roles of acetylcholine in cognitive function as demonstrated by studies on effects of drugs and lesions (reviewed in Hasselmo, 1995, 2006). For example, the drug scopolamine blocks muscarinic acetylcholine receptors. Behavioral studies show that injections of scopolamine in humans clearly impair the encoding of lists of words for subsequent free recall (Ghoneim and Mewaldt, 1977) or cued recall (Atri, Sherman, et al., 2004) but have no significant effect on the retrieval of words encoded before administration of scopolamine. These data indicate an important role for acetylcholine in the encoding of new information within neural circuits, but not for retrieval. The impaired encoding of new memories caused by scopolamine can be offset by drugs that increase acetylcholine levels by blocking the breakdown of acetylcholine by acetylcholinesterase (Ghoneim and Mewaldt, 1977), and such drugs are used as current treatments for Alzheimer's disease. Scopolamine selectively impairs episodic memories, the encoding of items stored at a specific time and place, but has little effect on performance in tasks testing semantic memory (memory for general facts) (Broks, Preston, et al., 1988) or procedural memory (memory for habitual actions) (Nissen, Knopman, et al., 1987). Some studies show that scopolamine also impairs working memory in delayed matching tasks in humans (Robbins, Semple, et al., 1997).

In rats, scopolamine impairs performance in the Morris water maze task in which a platform is hidden under the surface of opaque water and moved at the start of each testing day, requiring the animal to discover and then remember where the platform is on a given day. Scopolamine appears to impair encoding of the location of a new platform location each day (Whishaw, 1985; Buresova, Bolhuis, et al., 1986). The impairment on maze tasks is only observed when scopolamine is given prior to the encoding period when the rat first encounters the hidden platform but not when given during the delay between that initial encoding and the test of retrieval (Buresova, Bolhuis, et al., 1986). Localized infusions of scopolamine into perirhinal cortex impair encoding of stimuli for subsequent recognition in both monkeys (Tang, Mishkin, et al., 1997) and rats (Winters and Bussey, 2005). These effects might be due to impairments of mechanisms for the active maintenance of information during encoding into episodic memory. Support for the effect of scopolamine on active maintenance of information for encoding also comes from studies on trace conditioning, in which animals associate a tone with a shock presented several seconds after the tone ends. Infusion of drugs that block muscarinic receptors during the learning phase of trace conditioning reduces the effect of trace conditioning on rat freezing behavior (Bang and Brown, 2009). Interestingly, encoding in trace conditioning tasks is enhanced when stimuli are presented during periods of theta rhythmic oscillations in the hippocampal EEG (Griffin, Asaka, et al., 2004).

Local infusions of scopolamine into hippocampal region CA3 cause selective impairments of encoding but not retrieval of spatial trajectories in the Hebb–Williams maze with reconfigurable walls (Rogers and Kesner, 2003) and impair encoding of platform location on each day of the Morris water maze described in the previous paragraph (Blokland, Prickaerts, et al., 1992). Infusions of carbachol into the medial septum, which increase levels of hippocampal acetylcholine, also impair memory, possibly by interfering with consolidation of previously stored memories (Bunce, Sabolek, et al., 2004) as described below. Selective lesions of the cholinergic neurons projecting to entorhinal cortex in rats cause impairments in delayed non-match to sample for odors that were novel (not previously experienced) versus previously experienced familiar odor stimuli (McGaughy, Koene, et al., 2005), and cholinergic lesions of the perirhinal cortex in monkeys cause impairments in visual delayed match to sample performance (Turchi, Saunders, et al., 2005).

Acetylcholine also appears to play a role in attention. In humans, blockade of muscarinic receptors causes impairments in brief stimuli detection tasks requiring sustained attention (Wesnes and Warburton, 1983), and selective lesions of the cholinergic innervation of cortex impairs sustained attention in rats (McGaughy, Kaiser, et al., 1996). Acetylcholine also appears to be important in divided attention, as scopolamine impairs divided attention (McGaughy, Turchi, et al., 1994) and lesions of cortical cholinergic innervation impair performance in divided attention tasks (Robbins, Everitt, et al., 1989; Turchi and Sarter, 1997). These effects are also consistent with the physiological effects of acetylcholine for enhancing afferent input and causing presynaptic inhibition of excitatory feedback connections.

7.5.4 Cellular Models of Persistent Spiking

Computational modeling has addressed the effects of cholinergic modulation on cellular properties. Different models have addressed potential intrinsic mechanisms for persistent spiking (Durstewitz, 2003; Teramae and Fukai, 2005). Computational modeling of the entorhinal cortex (Fransen, Alonso, et al., 2002) demonstrates how cholinergic modulation of intrinsic persistent spiking could underlie spiking activity during the delay period of delayed matching tasks in rats (Young, Otto, et al., 1997) and monkeys (Suzuki, Miller, et al., 1997). In the models, cholinergic modulation causes activation of a nonspecific cation current, that is, a channel that increases the conductance to both sodium and potassium. The increase in conductances to both ions will push the membrane potential to a reversal potential of 0 mV (between the Nernst potential of these two ions), resulting in strong depolarization. This depolarization causes spiking that activates voltage-sensitive calcium channels that allow calcium influx. The influx of calcium further activates the same cation current which is calcium sensitive, thereby causing further spiking. These interactions form a regenerative process that maintains persistent spiking for extended periods as shown in models (Fransen, Alonso, et al., 2002). Modeling also demonstrates potential mechanisms for neurons in deep layers of entorhinal cortex to maintain graded persistent spiking frequencies that can be shifted up and down by periods of depolarizing or hyperpolarizing current injection (Fransen, Tahvildari, et al., 2006).

Most models of working memory focus on persistent spiking maintained by network mechanisms involving excitatory recurrent connections. However, novel stimuli would not match the existing pattern of previously strengthened synapses and, therefore, may not use this mechanism. Instead, working memory for novel stimuli might depend upon the intrinsic cellular mechanisms described here for persistent spiking that are enhanced by acetylcholine (Hasselmo and Stern, 2006).

This hypothesis is consistent with the behavioral data described above. Infusions of cholinergic antagonists impair conditioning of a "freezing" response due to an association between a tone and a shock stimulus when there is a gap between the tone and shock (trace conditioning) but not when the shock occurs during the tone (delay conditioning), suggesting that acetylcholine is required to maintain information about the tone through the *trace period* so it can be associated with a shock (Bang and Brown, 2009). The role of acetylcholine in maintaining neural activity during a delay period appears to be a general principle of cortical function present in studies on humans. In humans, scopolamine reduces parahippocampal functional magnetic resonance imaging (fMRI) activity observed during the delay period after presentation of novel sample stimuli in a delayed match to sample task (Schon, Atri, et al., 2005). Loss of this persistent activity during the delay period could underlie the impairment of delayed matching function in humans caused by scopolamine (Robbins, Semple, et al., 1997). Persistent spiking in perirhinal and entorhinal cortex may be important for encoding of information into long-term episodic memory, consistent with scopolamine causing a reduction of delay period activity correlated with subsequent memory in humans (Schon, Atri, et al., 2005) and with local infusions of scopolamine causing impairments of encoding for subsequent recognition in monkeys (Tang, Mishkin, et al., 1997). Acetylcholine levels appear to increase with a salient arousing stimulus, allowing gating of persistent spiking that actively maintains the stimulus for encoding or trace conditioning.

7.5.5 Network Models of Cholinergic Enhancement of Encoding Behavioral data show an important role for acetylcholine in the encoding of new information, but the link to specific features of the physiological data is not always clear. As shown in this section, computational modeling helps clarify how the physiological effects of acetylcholine may set appropriate dynamics for encoding new memories. These effects have been modeled in a number of papers with dynamics representing biological structures (Hasselmo and Bower, 1992; Barkai, Bergman, et al., 1994; Hasselmo, 1995; Hasselmo, Bodelon, et al., 2002). This section will focus on presenting a simple model showing the basic principles involved.

In models, the muscarinic depolarization of neurons and the nicotinic enhancement of afferent synaptic inputs allow a stronger influence of sensory inputs on cortical activity during encoding. Along with this enhancement of afferent inputs, muscarinic enhancement of Hebbian synaptic modification increases the formation of long-term associations. Modeling addresses the initially paradoxical question of why the same synapses that show muscarinic enhancement of Hebbian long-term synaptic modification (LTP—see chapter 6, "Hebbian Plasticity and Learning") (Huerta and Lisman, 1995) should also show strong muscarinic presynaptic inhibition of glutamate release (Hasselmo and Bower, 1992; Hasselmo and Schnell, 1994). Cholinergic presynaptic inhibition reduces the magnitude of synaptic potentials to less than half their initial size.

Why would one need to suppress transmission at the synapses undergoing modification during encoding? Computational modeling allows understanding of this combination of effects in terms of changes in cortical dynamics for encoding and retrieval. The simple example presented here has the structure of a standard associative memory model (Hasselmo and Bower, 1992; Hasselmo and Schnell, 1994). (For more on associative networks, and models addressing different aspects of hippocampal function, see chapter 18, "Hippocampus: From Spatial Processing to Episodic Memory.") This model forms associations between a pattern of presynaptic activity aj(t) and the pattern of postsynaptic activity $a_i(t + 1)$ by modifying the strength of synaptic connectivity weights W. The model uses the following learning rule:

$$W_{ij}(t+1) = W_{ij}(t) + a_i(t+1)a_j^{(k)}(t)$$

where W is the pattern of connections between the two regions, $a_i(t + 1)$ is the postsynaptic activation and $a_j^{(k)}$ is the presynaptic activation for input pattern k.

The application of this learning occurs after the activation rule has been applied for one step. In this example model, the spread of activity across the synapses is represented by the same activation rule during both encoding and retrieval. The activation rule is as follows:

$$a_i(t+1) = a_i^{(k)}(t) + (1 - c(t))\sum_j W_{ij}a_j^{(k)}(t)$$

where $a_i^{(k)}$ is the direct afferent input to the postsynaptic neurons, $a_j^{(k)}$ is the direct afferent input to the presynaptic neurons, and $c(t)$ represents the selective presynaptic inhibition by acetylcholine of the modifiable synapses W (but not afferent input). The cholinergic modulation takes the value of 1 when acetylcholine is present (causing presynaptic inhibition of glutamatergic transmission) or 0 when acetylcholine is not present (causing no presynaptic inhibition).

7.5.5.1 Problem: Interference during Encoding Cholinergic presynaptic inhibition of excitatory feedback can prevent the problem of interference during encoding. This problem appears in any model in which Hebbian synaptic modification of excitatory feedback connections encodes associations between different elements of the afferent input (Hopfield, 1982; Kohonen, 1984). An example is presented here using a paired associates task (or cued recall task). In a paired associates task, subjects are presented with a set of related word pairs, such as leather–holster. After encoding, the subject is cued with the first word (e.g., leather) and must recall the second word (e.g., holster).

Figure 7.12A–C shows how this function works in models (Hasselmo, 1995; Hasselmo and Wyble, 1997). Filled circles represent populations of neurons activated in region CA3 for the two words in context 1. Hebbian modification of synapses between spiking neurons causes strengthening of synapses, as represented by thicker black lines in the middle plot. After encoding, the network is cued with the word Leather in context 1, and the activity spreads across previously modified synapses to cause spiking in the population of neurons representing the word Holster.

Now the same example in the figure will be presented using the equations described above. Assume the synapses between the neurons start out at zero strength. The first association k = 1 is presented to the network at time t = 1, resulting in postsynaptic input $a_i^{k=1}(t=1) = \begin{bmatrix}1 & 0\end{bmatrix}$ and presynaptic input $a_j^{k=1}(t=1) = \begin{bmatrix}1 & 1 & 0\end{bmatrix}$. The vectors used in each discrete time step of the model represent the mean firing of neurons over 1 or 2 seconds. For this step we will assume the absence of cholinergic suppression (therefore, $c = 0$). The activations of the postsynaptic neurons are computed after one time step as follows:

$$a_i(t=2) = \begin{bmatrix}1\\0\end{bmatrix} + (1-0)\begin{bmatrix}0 & 0 & 0\\0 & 0 & 0\end{bmatrix}\begin{bmatrix}1\\1\\0\end{bmatrix} = \begin{bmatrix}1\\0\end{bmatrix}.$$

The learning rule is used to store the first association:

$$\begin{aligned}W_{ij}(2) &= W_{ij}(1) + a_i(2)a_j(1)\\ &= \begin{bmatrix}0 & 0 & 0\\0 & 0 & 0\end{bmatrix} + \begin{bmatrix}1\\0\end{bmatrix}\begin{bmatrix}1 & 1 & 0\end{bmatrix} = \begin{bmatrix}1 & 1 & 0\\0 & 0 & 0\end{bmatrix}.\end{aligned}$$

The activation rule can now retrieve this association by giving just one portion of the presynaptic activity

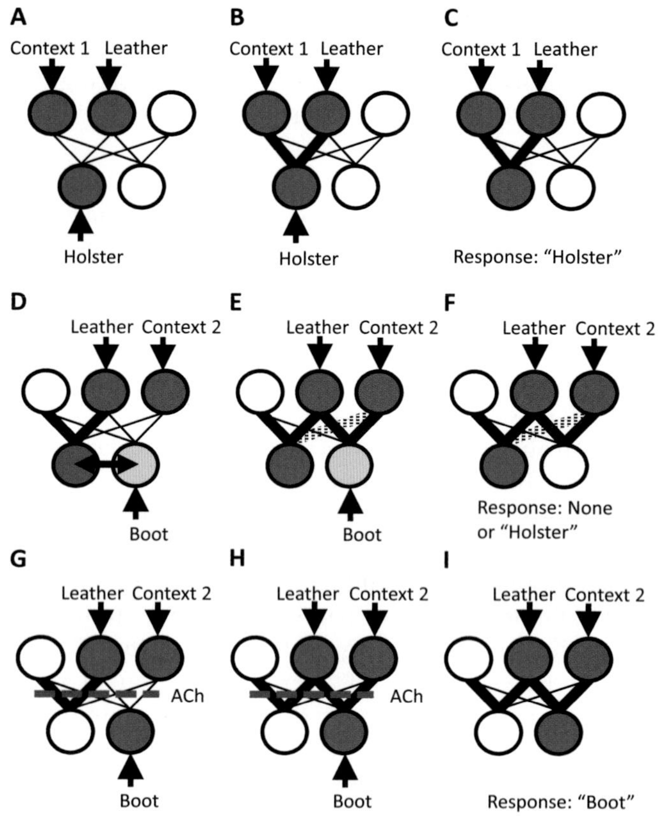

FIGURE 7.12 (A) During encoding, input to hippocampus (arrows) activates neurons coding the words leather and holster and the first context (context 1) (Hasselmo and Wyble, 1997). (B) Hebbian synaptic modification strengthens synapses between the active neurons (thick black lines). (C) During retrieval, the cue word activates the population of neurons coding the word Leather, and activity spreads across previously strengthened synapses (thick black lines) to activate the neurons coding the word Holster. (D) During encoding of the overlapping word pair, input to hippocampus activates neurons coding the words Leather and Boot and context 2. (E) Interference during encoding results from the spread of activity across previously strengthened synapses into the neurons coding the previous word Holster. This causes strengthening of undesired connections between context 2 and the first word Holster (thick dotted line). (F) During retrieval, the cue word Leather causes spread of activity across the undesired connection, resulting in additional retrieval of the incorrect response (Holster) or no response due to response competition. This reduces the likelihood of retrieval of the correct response Boot in context 2. (G) Encoding of the word pair leather–boot in the presence of acetylcholine (ACh). Cholinergic presynaptic inhibition reduces the activity spread across previously strengthened synapses (dashed line across middle) but allows response to the new afferent input. (H) Reduction of retrieval combined with enhancement of synaptic modification allows strengthening of only the connections from the word leather and context 2 with the new word Boot (thick black lines). (I) After encoding with acetylcholine, the cue word leather in the second context can retrieve the correct response (the word Boot).

pattern in the absence of postsynaptic input. This differs from figure 7.12C by giving only context 1 for clarity. If we assume that postsynaptic activity is a step function with maximum output of 1, then this retrieval results in the previously associated postsynaptic activity pattern:

$$a_i(t+1) = \begin{bmatrix} 0 \\ 0 \end{bmatrix} + (1-0)\begin{bmatrix} 1 & 1 & 0 \\ 0 & 0 & 0 \end{bmatrix}\begin{bmatrix} 1 \\ 0 \\ 0 \end{bmatrix} = \begin{bmatrix} 1 \\ 0 \end{bmatrix}.$$

Encoding a single word pair is not a problem for associative memory models. However, difficulties occur when overlapping word pairs are learned—for example, if the first word Leather is then associated with a different second word (e.g., Boot). Figure 7.12D–F illustrates the problem of interference during encoding. When the new word pair is heard, the word Leather activates the same neurons but with a new context representation (context 2). The new second word Boot activates a different population of neurons. However, the activity induced by the word Leather causes synaptic transmission across previously strengthened synapses that triggers spiking in the neurons that previously represented the word Holster. Hebbian synaptic modification then results in strengthening of an additional undesired connection (gray line) between the new context (context 2) and the word Holster from the first context. This is an incorrect association, as the word Holster did not appear in the new context.

To show this effect in the equations, we present the second association k = 2 to the network at time t = 3, resulting in postsynaptic input $a_i^{k=2}(t=3) = [0 \quad 1]$ and presynaptic input $A_j^{k=2}(t=3) = [0 \quad 1 \quad 1]$. Again, we will assume the absence of cholinergic suppression (therefore, $c = 0$). The activations of the postsynaptic neurons are computed after one time step at t = 4. Note that the postsynaptic activity now includes retrieval of the previous association:

$$a_i(t=4) = \begin{bmatrix} 0 \\ 1 \end{bmatrix} + (1-0)\begin{bmatrix} 1 & 1 & 0 \\ 0 & 0 & 0 \end{bmatrix}\begin{bmatrix} 0 \\ 1 \\ 1 \end{bmatrix} = \begin{bmatrix} 0 \\ 1 \end{bmatrix} + \begin{bmatrix} 1 \\ 0 \end{bmatrix} = \begin{bmatrix} 1 \\ 1 \end{bmatrix}.$$

The learning rule then modifies synapses based on presynaptic and postsynaptic activity:

$$\begin{aligned} W_{ij}(4) &= W_{ij}(3) + a_i(4)a_j(3) \\ &= \begin{bmatrix} 1 & 1 & 0 \\ 0 & 0 & 0 \end{bmatrix} + \begin{bmatrix} 1 \\ 1 \end{bmatrix}[0 \quad 1 \quad 1] = \begin{bmatrix} 1 & 2 & 1 \\ 0 & 1 & 1 \end{bmatrix}. \end{aligned}$$

Now we can use the activation rule to try to retrieve this association by giving just the nonoverlapping component of presynaptic activity pattern in the absence of postsynaptic input. This differs from figure 7.12F by giving only context 2 for clarity. However, due to interference, this retrieval results in retrieval of both of the associated postsynaptic activity patterns:

$$a_i(t+1) = \begin{bmatrix} 0 \\ 0 \end{bmatrix} + (1-0)\begin{bmatrix} 1 & 2 & 1 \\ 0 & 1 & 1 \end{bmatrix}\begin{bmatrix} 0 \\ 0 \\ 1 \end{bmatrix} = \begin{bmatrix} 1 \\ 1 \end{bmatrix}.$$

This could result in response competition in which feedback inhibition overcomes the correct retrieval and causes retrieval of the wrong word Holster or prevents sufficient activity to retrieve either word.

7.5.5.2 Solution: Suppress Retrieval during Encoding The solution to the problem of interference during encoding is to suppress the retrieval of previously encoded associations during the encoding of new associations. This can be achieved by selective presynaptic inhibition of all excitatory recurrent synapses to prevent them from having a strong influence on postsynaptic activity, while cholinergic modulation enhances Hebbian synaptic modification based on the remaining transmission. The selective suppression of the spread of activity (retrieval) during encoding of new patterns was a standard feature of associative memory models (Anderson, 1972; Hopfield, 1982; Kohonen, 1984), and cholinergic effects provide justification for this feature (Hasselmo and Schnell, 1994; Hasselmo, 1995). This model fits the physiological data, as the same synapses that show muscarinic enhancement of synaptic modification commonly show muscarinic presynaptic inhibition of glutamatergic synaptic transmission (Hasselmo and Schnell, 1994; Hasselmo, 1995).

Acetylcholine effects in the model prevent retrieval of previously stored associations from interfering with encoding, as shown in figure 7.12G–I (Hasselmo and Wyble, 1997). Spiking is still induced by the word Leather and the word Boot, but presynaptic inhibition by acetylcholine at all excitatory recurrent connections prevents the retrieval of prior associations (the word Holster) while new associations are being stored with the context 2 input. As described below, feedback regulation of acetylcholine levels can set appropriate dynamics for encoding novel stimuli while reducing acetylcholine in response to familiar patterns.

Even though acetylcholine causes presynaptic inhibition, it does not completely prevent glutamate release. Acetylcholine also enhances calcium flow through postsynaptic NMDA channels activated by the combination of glutamate release and postsynaptic depolarization. These NMDA currents have been shown to trigger synaptic modification to cause stronger subsequent potentials (LTP). This enhancement of NMDA currents allows new encoding at the time when presynaptic inhibition helps reduce interference by reducing synaptic transmission associated with retrieval of prior association during encoding of a new association. Computational modeling shows how cholinergic modulation in region CA3 and CA1 could allow effective separation of the dynamics of encoding and retrieval for different memory items represented by overlapping input patterns (Hasselmo and Schnell, 1994; Hasselmo, 1995). In these models, feedback regulation of cholinergic modulation based on the match between retrieval and afferent input allows regulation of acetylcholine levels. If there is no match, initial postsynaptic activity is low, acetylcholine levels stay high, and the information is encoded as new. If a pattern has been stored, the match between retrieval and input is strong, resulting in strong postsynaptic activity that suppresses the levels of acetylcholine and allows stronger synaptic transmission to mediate retrieval (Hasselmo and Schnell, 1994; Hasselmo, 1995, 2006). These mechanisms were modeled with feedback regulation of global acetylcholine levels in the hippocampus but might involve more local regulation of acetylcholine.

To show how cholinergic modulation prevents interference in the equations, we assume that acetylcholine levels are high when we present the second association k = 2 to the network at time t = 3, resulting in postsynaptic input $a_i^{k=2}(t=3)=[0 \quad 1]$ and presynaptic input $A_j^{k=2}(t=3)=[0 \quad 1 \quad 1]$. In contrast to the previous example, we now assume the presence of cholinergic suppression (therefore, c = 1). The activations of the postsynaptic neurons are computed after one time step at t = 4. Note that it no longer includes retrieval of the previous association:

$$a_i(t=4)=\begin{bmatrix}0\\1\end{bmatrix}+(1-1)\begin{bmatrix}1&1&0\\0&0&0\end{bmatrix}\begin{bmatrix}0\\1\\1\end{bmatrix}=\begin{bmatrix}0\\1\end{bmatrix}+\begin{bmatrix}0\\0\end{bmatrix}=\begin{bmatrix}0\\1\end{bmatrix}.$$

The learning rule then modifies synapses based on presynaptic and postsynaptic activity:

$$\begin{aligned}W_{ij}(4)&=W_{ij}(3)+a_i(4)a_j(3)\\&=\begin{bmatrix}1&1&0\\0&0&0\end{bmatrix}+\begin{bmatrix}0\\1\end{bmatrix}[0 \quad 1 \quad 1]=\begin{bmatrix}1&1&0\\0&1&1\end{bmatrix}.\end{aligned}$$

Now we can use the activation rule to try to retrieve this association by giving just the nonoverlapping component of the presynaptic activity pattern in the absence of postsynaptic input. This differs from figure 7.12I by giving only context 2 input for clarity. Note that we remove the cholinergic modulation. Due to cholinergic modulation that was present during encoding (but not during retrieval), the retrieval can now yield just the correct postsynaptic activity pattern for the second association:

$$a_i(t+1)=\begin{bmatrix}0\\0\end{bmatrix}+(1-0)\begin{bmatrix}1&1&0\\0&1&1\end{bmatrix}\begin{bmatrix}0\\0\\1\end{bmatrix}=\begin{bmatrix}0\\1\end{bmatrix}.$$

7.5.5.3 Experimental Test of Interference during Encoding Based on the example shown in the figures, scopolamine should cause greater interference during encoding of overlapping paired associates than during encoding of nonoverlapping paired associates. This prediction (Hasselmo and Wyble, 1997) motivated a study of the effects of scopolamine on cued recall that found that scopolamine caused an impairment of the encoding of new paired associates that was stronger for overlapping paired associates than for nonoverlapping pairs (Atri, Sherman, et al., 2004).

Other data are consistent with this model. Consistent with the models of cholinergic presynaptic inhibition in hippocampus, local infusion of cholinergic antagonists in hippocampus causes an increase in background spiking activity in place cells, which normally fire when a rat is in a single location but fire at a broader range of locations during acetylcholine blockade (Brazhnik, Muller, et al., 2003). In the primary visual cortex, acetylcholine reduces the extent of spatial integration in the visual field, shifting the spatial tuning of visual cells toward shorter visual bar stimuli (Roberts and Thiele, 2008). The enhancement of afferent input may also mediate attentional modulation of the responses of visual cortex neurons, which show enhancement by acetylcholine and are reduced by scopolamine (Herrero, Roberts, et al., 2008). This principle applies to human visual areas, where increases in acetylcholine levels induced by the acetylcholinesterase blocker donepezil (Aricept) cause a reduction in the spread of activity as measured by fMRI (Silver, Shenhav, et al., 2008).

All the models show that the effects of interference during encoding are cumulative, building up as more overlapping memories are stored. Figure 7.12G–I showed how an additional synapse is strengthened for each element of an overlapping pattern. For multiple overlapping patterns, this causes the phenomenon of runaway synaptic modification, in which the total number of strengthened synapses increases not with the number of stored patterns but with the square of the number of stored patterns (Hasselmo, 1994).

Differences in the dynamics of encoding versus retrieval are defining characteristics of associative memory function. If the modifiable synapse is not the dominant influence on postsynaptic activity during application of the learning rule, then the synapse will store an association between the presynaptic activity and the separately generated postsynaptic activity. In contrast, if the modifiable synapse is the dominant influence on postsynaptic activity during the application of the learning rule, then the synapse will contribute to the self-organization of the representation.

7.5.5.4 Encoding and Retrieval on Phases of Theta Rhythm The example above addresses relatively slow modulatory changes regulated by changes in acetylcholine levels. However, modulatory effects could also rapidly shift dynamics between encoding and retrieval (figure 7.13). Behavioral data indicate a role for theta rhythm oscillations in encoding of new information in the hippocampal formation (Givens and Olton, 1994; Buzsaki, 2002). The theta rhythm is an oscillation in the local field potential, usually around 8 Hz, that appears when animals are actively moving or attending to salient sensory stimuli. Theta rhythm is associated with differences in the magnitude of synaptic currents and cellular firing at specific phases (Buzsaki, 2002). Physiological data suggest that changes in the dynamics during each cycle of the theta rhythm oscillations in the hippocampus could be appropriate for encoding on one phase and retrieval on a separate phase (Hasselmo, Bodelon, et al., 2002). This hypothesis is based on the following characteristics during different phases:

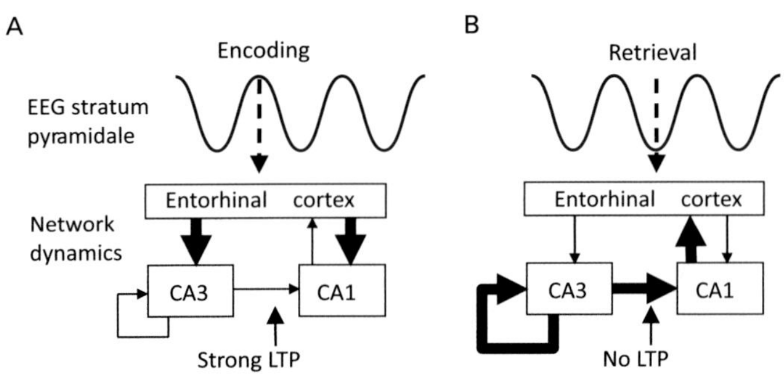

FIGURE 7.13 Schematic representation of the change in dynamics during hippocampal theta rhythm oscillations (Hasselmo et al., 2002). (A) Encoding. At the peak of the EEG in stratum pyramidale (s. pyr.) of CA1, synaptic transmission arising from entorhinal cortex is strong. Synaptic transmission arising from region CA3 is weak, but these same synapses show a strong capacity for long-term potentiation (LTP). This allows the afferent input from entorhinal cortex to set patterns to be encoded while preventing interference from previously encoded patterns on the excitatory synapses arising from region CA3. (B) Retrieval. At the trough of the EEG, synaptic transmission arising from entorhinal cortex is relatively weak (though strong enough to provide retrieval cues). In contrast, the synaptic transmission arising from region CA3 is strong, allowing effective retrieval of previously encoded sequences. During this phase, synapses in CA3 do not encode the retrieval activity because LTP at these synapses is weak; instead they show long-term depression or depotentiation.

Encoding phase: Current source density analysis of theta rhythm shows that the strongest input from entorhinal cortex (Brankack, Stewart, et al., 1993; Buzsaki, 2002) occurs during strong inhibition at the cell body (Kamondi, Acsady, et al., 1998). This could allow encoding of new information when spiking mediating retrieval is suppressed. Intracellular data show that dendritic spikes can induce LTP even when the soma is hyperpolarized to prevent spiking (Golding, Staff, et al., 2002). The existence of an encoding phase is supported by data showing that stimulation at different phases of theta has different effects on synaptic modification. LTP is induced when a tetanus is delivered on one phase of theta (Pavlides, Greenstein, et al., 1988) while stimulation on the opposite phase causes LTD (Huerta and Lisman, 1995; Holscher, 1997).

Retrieval phase: Current source density analysis shows that the strongest input from region CA3 is at a different phase of theta than the strongest EC input (Brankack, Stewart, et al., 1993). During this phase, CA1 pyramidal cells show the strongest firing due to depolarization of the postsynaptic membrane potential at this phase (Kamondi, Acsady, et al., 1998). This phase could allow synapses from region CA3 that were strengthened during encoding to drive retrieval when the cell body is depolarized enough to allow spiking output.

This model of the function of theta (Hasselmo, Bodelon, et al., 2002) accounts for the behavioral data on memory impairments after lesions of the medial septum or fornix that reduce theta rhythm oscillations in the hippocampus (Winson and Abzug, 1978; Givens and Olton, 1994). Consistent with this model, oscillations in human subjects show reset to different phases of the theta rhythm during encoding and retrieval (Rizzuto, Madsen, et al., 2006), spiking occurs at different phases of theta for match versus non-match stimuli (Manns, Zilli, et al., 2007), and higher frequency gamma oscillations of 30–100 Hz in hippocampal region CA1 show coherence with CA3 or with EC that involve bouts of gamma oscillations of different frequencies clustered on different phases of the theta rhythm (Colgin, Denninger, et al., 2009).

7.5.5.5 Neuromodulation and Consolidation During different stages of waking and sleep, cholinergic modulation may switch between high levels of acetylcholine appropriate for encoding and lower levels of acetylcholine appropriate for consolidation of previously encoded information (Marrosu, Portas, et al., 1995; Hasselmo, 1999) (figure 7.14). Here, consolidation refers to the formation of additional traces to make a memory more robust. Behavioral data demonstrate that recently encoded memories are more sensitive than older memories to the effects of partial lesions of the hippocampus (Zola-Morgan and Squire, 1990; Kim and Fanselow, 1992). This supports a two-stage theory of memory formation (Buzsaki, 1989) in which memories are initially encoded by Hebbian synaptic modification in the hippocampus during the first stage (active waking), and then memories are subsequently reactivated during a second stage (quiet waking or slow-wave sleep) to allow gradual strengthening of additional synaptic connections in the cortex (Buzsaki, 1989; McClelland, McNaughton, et al., 1995). High levels of acetylcholine during waking (Marrosu, Portas, et al., 1995) could prevent interference from previously stored memories whereas lower levels of acetylcholine during sleep would allow memories to reactivate in region CA3 and spread back to the cortex for consolidation (Buzsaki, 1989; Chrobak and Buzsaki, 1994; Hasselmo, 1999).

This theory predicts that consolidation of memory should be impaired by increases in acetylcholine levels, as supported by data showing impairments of consolidation caused by cholinergic infusions into medial septum after training in rats (Bunce, Sabolek, et al., 2004) and impairments of consolidation caused by injections of the acetylcholinesterase blocker physostigmine during slow wave sleep in humans (Gais and Born, 2004). Experimental data also support the prediction of the model that reducing acetylcholine levels should enhance consolidation (Rasch, Born, et al., 2006).

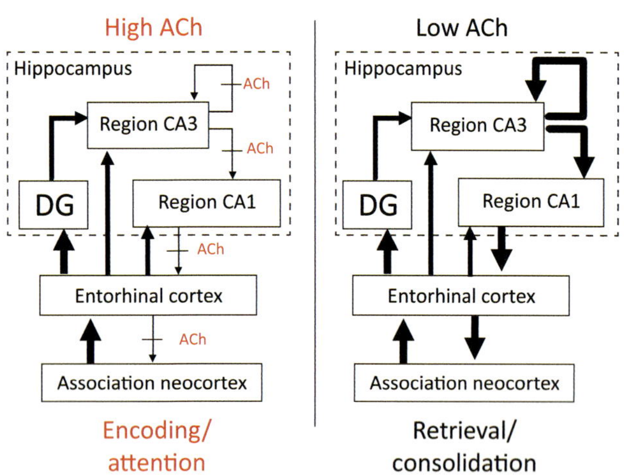

FIGURE 7.14 Role of acetylcholine in the memory formation process. Left: In the encoding (attentive) phase, levels of acetylcholine are elevated and selectively suppress the CA3 recurrent projections, the CA3 to CA1 projections, and the feedback projections from hippocampus to cortex. Right: In the retrieval and consolidation phases, when cholinergic levels are low, suppression of feedback is reduced and hippocampal activity can more strongly influence cortical activity for consolidation (Hasselmo, 1999). DG, dentate gyrus.

Modeling has also addressed why there might be two stages of memory formation, showing that the consolidation stage allows development of semantic memory representations that determine regularities in the information, independently from the associations in individual episodic memories (McClelland, McNaughton, et al., 1995). That model proposed that semantic representations in neocortical structures would be distorted if each new piece of information were inserted without reference to other existing relations. For example, the knowledge that birds can fly would be distorted if new episodic information about penguins that are birds that swim were learned in isolation. If the information about penguins were used to train the network in isolation, the association between bird and swim would be strengthened at the expense of the association between bird and fly, such that the network could generate the output that robins are birds that swim. In contrast, if the new information is initially encoded in hippocampus and then gradually presented to the cortex interleaved with information about other birds, the old semantic information would not be distorted by the new information. The new information could be incorporated with the old in an appropriate manner. In this framework, the modulatory effects of acetylcholine could switch the network between a state of acquiring new episodic information in the hippocampus during active waking and a state of updating the semantic representations in neocortex during the occurrence of sharp waves in quiet waking and slow-wave sleep.

7.6 Conclusions

This chapter reviewed some of the basic common computational issues pertaining to neuromodulation including its intrinsic and extrinsic nature, its timescale, and its spatial scale. All neuromodulators occupy a certain volume along those three dimensions, and the size and shape of this volume in turn constrains the computational functions of the substance.

Dopamine in the mammalian brain is extrinsic and spatially diffuse. Interestingly, its spatial scale spans a large area in that it can act at short timescales (bursts) and at long timescales (firing rate modulations). Section 7.4 showed how such apparently disparate timescales can coexist and can be controlled. We presented a detailed biophysical model of dopaminergic cells and discussed how synaptic and intrinsic mechanisms could switch the cell between bursting and tonic firing modes. How and why such changes occur remain fundamental questions and understanding their answers can give great insights into the normal and diseased brain.

Acetylcholine has both intrinsic (cortical ACh local interneurons) and extrinsic (e.g., from the septum) features. It can have subtle spatially local effects on selective subsets of synapses but not others on the same neuron. This property gives key insights into the function of this substance during memory tasks. As the modeling studies presented above clearly show, this substance may be crucial to the encoding of memory without interference from memory recall. As such it acts as a switch preventing the learning and recall neural mechanisms from interfering with each other. The timescale of ACh modulation can be fast (e.g., theta rhythm and learning) or sustained (e.g., attention and working memory).

While dopamine and acetylcholine have attracted a lot of attention from experimenters and modelers, there are, of course, many more neuromodulators. Substances such as serotonin or norepinephrine are known to have significant effects on cognitive and emotional processing, but there has not been extensive work to link these behavioral effects to the cellular effects of these neuromodulators on neural computations. The cellular effects of norepinephrine may mediate the enhancement of attention to behavioral stimuli by increasing the neural response to external afferent input. Norepinephrine may enhance the specificity of response to afferent input by reducing postsynaptic adaptation and causing presynaptic inhibition of glutamatergic transmission at excitatory feedback synapses. It will be a challenge for future work to build biophysical and network models of such systems to incrementally gain insights into their specific functions, and derive a general understanding of the role of neuromodulation in neural computations.

NOTE

1. The Kuznetsova et al. (2010) model also includes a balance on sodium ions, but the impact of the sodium balance is minimal and it is not described here.

REFERENCES

Amini, B., Clark, J. W., Jr., & Canavier, C. C. (1999). Calcium dynamics underlying pacemaker-like and burst firing oscillations in midbrain dopaminergic neurons: A computational study. *Journal of Neurophysiology, 82*(5), 2249–2261.

Anderson, J. (1972). A simple neural network generating an interactive memory. *Mathematical Biosciences, 14*, 197–200.

Atri, A., Sherman, S., Norman, K. A., Kirchhoff, B. A., Nicolas, M. M., Greicius, M. D., et al. (2004). Blockade of central cholinergic receptors impairs new learning and increases proactive interference in a word paired-associate memory task. *Behavioral Neuroscience, 118*(1), 223–236.

Bang, S. J., & Brown, T. H. (2009). Muscarinic receptors in perirhinal cortex control trace conditioning. *Journal of Neuroscience, 29*(14), 4346–4350.

Barkai, E., Bergman, R. E., Horwitz, G., & Hasselmo, M. E. (1994). Modulation of associative memory function in a biophysical simulation of rat piriform cortex. *Journal of Neurophysiology, 72*(2), 659–677.

Barkai, E., & Hasselmo, M. E. (1994). Modulation of the input/output function of rat piriform cortex pyramidal cells. *Journal of Neurophysiology, 72*(2), 644–658.

Bernheimer, H., Birkmayer, W., Hornykiewicz, O., Jellinger, K., & Seitelberger, F. (1973). Brain dopamine and the syndromes of Parkinson and Huntington: Clinical, morphological and neurochemical correlations. *Journal of the Neurological Sciences, 20*(4), 415–455.

Blokland, A., Prickaerts, J., & Raaijmakers, W. (1992). Reduced level of anxiety in adult Lewis rats after chronic ethanol consumption. *Physiology & Behavior, 51*(2), 245–248.

Brady, A. M., & O'Donnell, P. (2004). Dopaminergic modulation of prefrontal cortical input to nucleus accumbens neurons in vivo. *Journal of Neuroscience, 24*(5), 1040–1049.

Brankack, J., Stewart, M., & Fox, S. E. (1993). Current source density analysis of the hippocampal theta rhythm: Associated sustained potentials and candidate synaptic generators. *Brain Research, 615*(2), 310–327.

Brazhnik, E. S., Muller, R. U., & Fox, S. E. (2003). Muscarinic blockade slows and degrades the location-specific firing of hippocampal pyramidal cells. *Journal of Neuroscience, 23*(2), 611–621.

Brezina, V. (2010). Beyond the wiring diagram: Signalling through complex neuromodulator networks. *Philosophical Transactions of the Royal Society of London. Series B, Biological Sciences, 365*(1551), 2363–2374.

Brezina, V., Orekhova, I. V., & Weiss, K. R. (1996). Functional uncoupling of linked neurotransmitter effects by combinatorial convergence. *Science, 273*(5276), 806–810.

Broks, P., Preston, G. C., Traub, M., Poppleton, P., Ward, C., & Stahl, S. M. (1988). Modelling dementia: Effects of scopolamine on memory and attention. *Neuropsychologia, 26*(5), 685–700.

Brunel, N., & Wang, X. J. (2001). Effects of neuromodulation in a cortical network model of object working memory dominated by recurrent inhibition. *Journal of Computational Neuroscience, 11*, 63–85.

Bunce, J. G., Sabolek, H. R., & Chrobak, J. J. (2004). Intraseptal infusion of the cholinergic agonist carbachol impairs delayed-non-match-to-sample radial arm maze performance in the rat. *Hippocampus, 14*(4), 450–459.

Buresova, O., Bolhuis, J. J., & Bures, J. (1986). Differential effects of cholinergic blockade on performance of rats in the water tank navigation task and in a radial water maze. *Behavioral Neuroscience, 100*(4), 476–482.

Buzsaki, G. (1989). Two-stage model of memory trace formation: A role for "noisy" brain states. *Neuroscience, 31*(3), 551–570.

Buzsaki, G. (2002). Theta oscillations in the hippocampus. *Neuron, 33*(3), 325–340.

Canavier, C. C. (1999). Sodium dynamics underlying burst firing and putative mechanisms for the regulation of the firing pattern in midbrain dopamine neurons: A computational approach. *Journal of Computational Neuroscience, 6*(1), 49–69.

Canavier, C. C., & Landry, R. S. (2006). An increase in AMPA and a decrease in SK conductance increase burst firing by different mechanisms in a model of a dopamine neuron in vivo. *Journal of Neurophysiology, 96*(5), 2549–2563.

Chergui, K., Nomikos, G. G., Mathe, J. M., Gonon, F., & Svensson, T. H. (1996). Burst stimulation of the medial forebrain bundle selectively increase Fos-like immunoreactivity in the limbic forebrain of the rat. *Neuroscience, 72*(1), 141–156.

Chialvo, D. R. (2010). Emergent complex neural dynamics. *Nature Physics, 6*(10), 744–750.

Chrobak, J. J., & Buzsaki, G. (1994). Selective activation of deep layer (V–VI) retrohippocampal cortical neurons during hippocampal sharp waves in the behaving rat. *Journal of Neuroscience, 14*(10), 6160–6170.

Colgin, L. L., Denninger, T., Fyhn, M., Hafting, T., Bonnevie, T., Jensen, O., et al. (2009). Frequency of gamma oscillations routes flow of information in the hippocampus. *Nature, 462*(7271), 353–357.

Daw, N. D., Kakade, S., & Dayan, P. (2002). Opponent interactions between serotonin and dopamine. *Neural Networks, 15*(4–6), 603–616.

Dayan, P. (2012). Twenty-five lessons from computational neuromodulation. *Neuron, 76*(1), 240–256.

Descarries, L., Gisiger, V., & Steriade, M. (1997). Diffuse transmission by acetylcholine in the CNS. *Progress in Neurobiology, 53*(5), 603–625.

Dominey, P. F. (2013). Recurrent temporal networks and language acquisition—From corticostriatal neurophysiology to reservoir computing. *Frontiers in Psychology, 4*, 500.

Doya, K. (2002). Metalearning and neuromodulation. *Neural Networks, 15*(4–6), 495–506.

Dreyer, J. K., Herrik, K. F., Berg, R. W., & Hounsgaard, J. D. (2010). Influence of phasic and tonic dopamine release on receptor activation. *Journal of Neuroscience, 30*(42), 14273–14283.

Drion, G., Massotte, L., Sepulchre, R., & Seutin, V. (2011). How modeling can reconcile apparently discrepant experimental results: The case of pacemaking in dopaminergic neurons. *PLoS Computational Biology, 7*(5), e1002050.

Durstewitz, D. (2003). Self-organizing neural integrator predicts interval times through climbing activity. *Journal of Neuroscience, 23*(12), 5342–5353.

Durstewitz, D., Seamans, J. K., & Sejnowski, T. J. (2000). Dopamine-mediated stabilization of delay-period activity in a network model of prefrontal cortex. *Journal of Neurophysiology, 83*(3), 1733–1750.

Fellous, J.-M. (1999). The neuromodulatory basis of emotion. *Neuroscientist, 5*(5), 283–294.

Fellous, J.-M., & Linster, C. (1998). Computational models of neuromodulation. *Neural Computation, 10*, 771–805.

Floresco, S. B. (2013). Prefrontal dopamine and behavioral flexibility: Shifting from an "inverted-U" toward a family of functions. *Frontiers in Neuroscience, 7*, 62.

Floresco, S. B., West, A. R., Ash, B., Moore, H., & Grace, A. A. (2003). Afferent modulation of dopamine neuron firing differentially regulates tonic and phasic dopamine transmission. *Nature Neuroscience, 6*(9), 968–973.

Frank, M. J. (2006). Hold your horses: A dynamic computational role for the subthalamic nucleus in decision making. *Neural Networks, 19*(8), 1120–1136.

Frank, M. J., & Claus, E. D. (2006). Anatomy of a decision: Striato-orbitofrontal interactions in reinforcement learning, decision making, and reversal. *Psychological Review, 113*(2), 300–326.

Frank, M. J., Seeberger, L. C., & O'Reilly R. C. (2004). By carrot or by stick: Cognitive reinforcement learning in parkinsonism. *Science, 306*(5703), 1940–1943.

Fransen, E., Alonso, A. A., & Hasselmo, M. E. (2002). Simulations of the role of the muscarinic-activated calcium-sensitive nonspecific cation current INCM in entorhinal neuronal activity during delayed matching tasks. *Journal of Neuroscience, 22*(3), 1081–1097.

Fransen, E., Tahvildari, B., Egorov, A. V., Hasselmo, M. E., & Alonso, A. A. (2006). Mechanism of graded persistent cellular activity of entorhinal cortex layer V neurons. *Neuron, 49*(5), 735–746.

Fuxe, K., & Agnati, L. F. (1991). Volume transmission in the brain: Novel mechanism for neural transmission. *Advances in Neuroscience (Hindawi), 1*, 1–9.

Gais, S., & Born, J. (2004). Low acetylcholine during slow-wave sleep is critical for declarative memory consolidation. *Proceedings of the National Academy of Sciences of the United States of America, 101*(7), 2140–2144.

Ghoneim, M. M., & Mewaldt, S. P. (1977). Studies on human memory: The interactions of diazepam, scopolamine, and physostigmine. *Psychopharmacology, 52*(1), 1–6.

Gil, Z., Connors, B. W., & Amitai, Y. (1997). Differential regulation of neocortical synapses by neuromodulators and activity. *Neuron, 19*(3), 679–686.

Giocomo, L. M., & Hasselmo, M. E. (2005). Nicotinic modulation of glutamatergic synaptic transmission in region CA3 of the hippocampus. *European Journal of Neuroscience, 22*(6), 1349–1356.

Givens, B. (1996). Stimulus-evoked resetting of the dentate theta rhythm: Relation to working memory. *Neuroreport, 8*(1), 159–163.

Givens, B., & Olton, D. S. (1994). Local modulation of basal forebrain: Effects on working and reference memory. *Journal of Neuroscience, 14*(6), 3578–3587.

Golding, N. L., Staff, N. P., & Spruston, N. (2002). Dendritic spikes as a mechanism for cooperative long-term potentiation. *Nature, 418*(6895), 326–331.

Goldman-Rakic, P. S., Lidow, M. S., Smiley, J. F., & Williams, M. S. (1992). The anatomy of dopamine in monkey and human prefrontal cortex. *Journal of Neural Transmission. Supplementum, 36*, 163–177.

Goldman-Rakic, P. S., Muly, E. C., III, & Williams, G. V. (2000). D(1) receptors in prefrontal cells and circuits. *Brain Research. Brain Research Reviews, 31*(2–3), 295–301.

Gonon, F. G. (1988). Nonlinear relationship between impulse flow and dopamine released by rat midbrain dopaminergic neurons as studied by in vivo electrochemistry. *Neuroscience, 24*(1), 19–28.

Grace, A. A., & Bunney, B. S. (1984a). The control of firing pattern in nigral dopamine neurons: Burst firing. *Journal of Neuroscience, 4*(11), 2877–2890.

Grace, A. A., & Bunney, B. S. (1984b). The control of firing pattern in nigral dopamine neurons: Single spike firing. *Journal of Neuroscience, 4*(11), 2866–2876.

Grace, A. A., & Bunney, B. S. (1986). Induction of depolarization block in midbrain dopamine neurons by repeated administration of haloperidol: Analysis using in vivo intracellular recording. *Journal of Pharmacology and Experimental Therapeutics, 238*(3), 1092–1100.

Graupner, M., Maex, R., & Gutkin, B. (2013). Endogenous cholinergic inputs and local circuit mechanisms govern the phasic mesolimbic dopamine response to nicotine. *PLoS Computational Biology, 9*(8), e1003183.

Griffin, A. L., Asaka, Y., Darling, R. D., & Berry, S. D. (2004). Theta-contingent trial presentation accelerates learning rate and enhances hippocampal plasticity during trace eyeblink conditioning. *Behavioral Neuroscience, 118*(2), 403–411.

Gupta, A., Balasubramani, P. P., & Chakravarthy, V. S. (2013). Computational model of precision grip in Parkinson's disease: A utility based approach. *Frontiers in Computational Neuroscience, 7*, 172.

Guthrie, M., Leblois, A., Garenne, A., & Boraud, T. (2013). Interaction between cognitive and motor corticobasal ganglia loops during decision making: A computational study. *Journal of Neurophysiology, 109*(12), 3025–3040.

Guzman, J. N., Sanchez-Padilla, J., Chan, C. S., & Surmeier, D. J. (2009). Robust pacemaking in substantia nigra dopaminergic neurons. *Journal of Neuroscience, 29*(35), 11011–11019.

Haj-Dahmane, S., & Andrade, R. (1998). Ionic mechanism of the slow afterdepolarization induced by muscarinic receptor activation in rat prefrontal cortex. *Journal of Neurophysiology, 80*(3), 1197–1210.

Hansson, E., & Ronnback, L. (1994). Astroglial modulation of synaptic transmission. *Perspectives on Developmental Neurobiology, 2*(3), 217–223.

Harris-Warrick, R. M., & Marder, E. (1991). Modulation of neural networks for behavior. *Annual Review of Neuroscience, 14*, 39–57.

Hasselmo, M. E. (1994). Runaway synaptic modification in models of cortex: Implications for Alzheimer's disease. *Neural Networks, 7*, 13–40.

Hasselmo, M. E. (1995). Neuromodulation and cortical function: Modeling the physiological basis of behavior. *Behavioural Brain Research, 67*(1), 1–27.

Hasselmo, M. E. (1999). Neuromodulation: Acetylcholine and memory consolidation. *Trends in Cognitive Sciences, 3*(9), 351–359.

Hasselmo, M. E. (2006). The role of acetylcholine in learning and memory. *Current Opinion in Neurobiology, 16*(6), 710–715.

Hasselmo, M. E., & Barkai, E. (1995). Cholinergic modulation of activity-dependent synaptic plasticity in the piriform cortex and associative memory function in a network biophysical simulation. *Journal of Neuroscience, 15*(10), 6592–6604.

Hasselmo, M. E., Bodelon, C., & Wyble, B. P. (2002). A proposed function for hippocampal theta rhythm: Separate phases of encoding and retrieval enhance reversal of prior learning. *Neural Computation, 14*(4), 793–817.

Hasselmo, M. E., & Bower, J. M. (1992). Cholinergic suppression specific to intrinsic not afferent fiber synapses in rat piriform (olfactory) cortex. *Journal of Neurophysiology, 67*(5), 1222–1229.

Hasselmo, M. E., & Schnell, E. (1994). Laminar selectivity of the cholinergic suppression of synaptic transmission in rat hippocampal region CA1: Computational modeling and brain slice physiology. *Journal of Neuroscience, 14*(6), 3898–3914.

Hasselmo, M. E., Schnell, E., & Barkai, E. (1995). Dynamics of learning and recall at excitatory recurrent synapses and cholinergic modulation in rat hippocampal region CA3. *Journal of Neuroscience, 15*(7 Pt 2), 5249–5262.

Hasselmo, M. E., & Stern, C. E. (2006). Mechanisms underlying working memory for novel information. *Trends in Cognitive Sciences, 10*(11), 487–493.

Hasselmo, M. E., & Wyble, B. P. (1997). Free recall and recognition in a network model of the hippocampus: Simulating effects of scopolamine on human memory function. *Behavioural Brain Research, 89*(1–2), 1–34.

Hausser, M., Stuart, G., Racca, C., & Sakmann, B. (1995). Axonal initiation and active dendritic propagation of action potentials in substantia nigra neurons. *Neuron, 15*(3), 637–647.

Hazy, T. E., Frank, M. J., & O'Reilly, R. C. (2010). Neural mechanisms of acquired phasic dopamine responses in learning. *Neuroscience and Biobehavioral Reviews, 34*(5), 701–720.

Heien, M. L., & Wightman, R. M. (2006). Phasic dopamine signaling during behavior, reward, and disease states. *CNS & Neurological Disorders—Drug Targets, 5*(1), 99–108.

Herrero, J. L., Roberts, M. J., Delicato, L. S., Gieselmann, M. A., Dayan, P., & Thiele, A. (2008). Acetylcholine contributes through muscarinic receptors to attentional modulation in V1. *Nature, 454*(7208), 1110–1114.

Holscher, C. (1997). Nitric oxide, the enigmatic neuronal messenger: Its role in synaptic plasticity. *Trends in Neurosciences, 20*(7), 298–303.

Hong, S., & Hikosaka, O. (2011). Dopamine-mediated learning and switching in cortico-striatal circuit explain behavioral changes in reinforcement learning. *Frontiers in Behavioral Neuroscience, 5*, 15.

Hopfield, J. J. (1982). Neural networks and physical systems with emergent collective computational abilities. *Proceedings of the National Academy of Sciences of the United States of America, 79*, 2554–2558.

Hudson, A. E., & Prinz, A. A. (2010). Conductance ratios and cellular identity. *PLoS Computational Biology, 6*(7), e1000838.

Huerta, P. T., & Lisman, J. E. (1995). Bidirectional synaptic plasticity induced by a single burst during cholinergic theta oscillation in CA1 in vitro. *Neuron, 15*(5), 1053–1063.

Hyland, B. I., Reynolds, J. N., Hay, J., Perk, C. G., & Miller, R. (2002). Firing modes of midbrain dopamine cells in the freely moving rat. *Neuroscience, 114*(2), 475–492.

Jay, T. M., Glowinski, J., & Thierry, A. M. (1995). Inhibition of hippocampoprefrontal cortex excitatory responses by the mesocortical DA system. *Neuroreport, 6*(14), 1845–1848.

Ji, H., & Shepard, P. D. (2006). SK Ca2+-activated K+ channel ligands alter the firing pattern of dopamine-containing neurons in vivo. *Neuroscience, 140*(2), 623–633.

Johnson, S. W., Seutin, V., & North, R. A. (1992). Burst firing in dopamine neurons induced by N-methyl-D-aspartate: Role of electrogenic sodium pump. *Science, 258*(5082), 665–667.

Johnson, S. W., & Wu, Y. N. (2004). Multiple mechanisms underlie burst firing in rat midbrain dopamine neurons in vitro. *Brain Research, 1019*(1–2), 293–296.

Kaczmarek, L. K., & Levitan, I. B. (1987). *Neuromodulation: The Biochemical Control of Neuronal Excitability*. New York: Oxford University Press.

Kamondi, A., Acsady, L., Wang, X. J., & Buzsaki, G. (1998). Theta oscillations in somata and dendrites of hippocampal pyramidal cells in vivo: Activity-dependent phase-precession of action potentials. *Hippocampus, 8*(3), 244–261.

Katz, P. S. (1998). Comparison of extrinsic and intrinsic neuromodulation in two central pattern generator circuits in invertebrates. *Experimental Physiology, 83*(3), 281–292.

Katz, P. S., & Frost, W. N. (1995a). Intrinsic neuromodulation in the Tritonia swim CPG: Serotonin mediates both neuromodulation and neurotransmission by the dorsal swim interneurons. *Journal of Neurophysiology, 74*(6), 2281–2294.

Katz, P. S., & Frost, W. N. (1995b). Intrinsic neuromodulation in the Tritonia swim CPG: The serotonergic dorsal swim interneurons act presynaptically to enhance transmitter release from interneuron C2. *Journal of Neuroscience, 15*(9), 6035–6045.

Katz, P. S., & Frost, W. N. (1996). Intrinsic neuromodulation: Altering neuronal circuits from within. *Trends in Neurosciences, 19*(2), 54–61.

Kelley, A. (2005). Neurochemical networks encoding emotion and motivation. In J. A. Fellous & M. A. Arbib (Eds.), *Who Needs Emotions? The Brain Meets the Robot* (pp. 29–77). Oxford: Oxford University Press.

Keramati, M., & Gutkin, B. (2013). Imbalanced decision hierarchy in addicts emerging from drug-hijacked dopamine spiraling circuit. *PLoS One, 8*(4), e61489.

Kim, J. J., & Fanselow, M. S. (1992). Modality-specific retrograde amnesia of fear. *Science, 256*(5057), 675–677.

Kita, H., & Kitai, S. T. (1987). Efferent projections of the subthalamic nucleus in the rat: Light and electron microscopic analysis with the PHA-L method. *Journal of Comparative Neurology, 260*(3), 435–452.

Klink, R., & Alonso, A. (1997a). Ionic mechanisms of muscarinic depolarization in entorhinal cortex layer II neurons. *Journal of Neurophysiology, 77*(4), 1829–1843.

Klink, R., & Alonso, A. (1997b). Muscarinic modulation of the oscillatory and repetitive firing properties of entorhinal cortex layer II neurons. *Journal of Neurophysiology, 77*(4), 1813–1828.

Kohonen, T. (1984). *Self-Organization and Associative Memory*. Berlin: Springer-Verlag.

Komendantov, A. O., Komendantova, O. G., Johnson, S. W., & Canavier, C. C. (2004). A modeling study suggests complementary roles for $GABA_A$ and NMDA receptors and the SK channel in regulating the firing pattern in midbrain dopamine neurons. *Journal of Neurophysiology, 91*(1), 346–357.

Koob, G. F., Le, H. T., & Creese, I. (1987). The D1 dopamine receptor antagonist SCH 23390 increases cocaine self-administration in the rat. *Neuroscience Letters, 79*(3), 315–320.

Kroener, S., Chandler, L. J., Phillips, P. E., & Seamans, J. K. (2009). Dopamine modulates persistent synaptic activity and enhances the signal-to-noise ratio in the prefrontal cortex. *PLoS One, 4*(8), e6507.

Kuznetsova, A. Y., Huertas, M. A., Kuznetsov, A. S., Paladini, C. A., & Canavier, C. C. (2010). Regulation of firing frequency in a computational model of a midbrain dopaminergic neuron. *Journal of Computational Neuroscience, 28*(3), 389–403.

Lapish, C. C., Kroener, S., Durstewitz, D., Lavin, A., & Seamans, J. K. (2007). The ability of the mesocortical dopamine system to operate in distinct temporal modes. *Psychopharmacology, 191*(3), 609–625.

Lee, M. G., Chrobak, J. J., Sik, A., Wiley, R. G., & Buzsaki, G. (1994). Hippocampal theta activity following selective lesion of the septal cholinergic system. *Neuroscience, 62*(4), 1033–1047.

Lew, S. E., & Tseng, K. Y. (2014). Dopamine modulation of GABAergic function enables network stability and input selectivity for sustaining working memory in a computational model of the prefrontal cortex. *Neuropsychopharmacology, 39*, 3067–3076.

Lisman, J. E., Fellous, J.-M., & Wang, X.-J. (1998). A role for NMDA-receptor channels in working memory. *Nature Neuroscience, 1*(4), 273–275.

Lisman, J. E., & Grace, A. A. (2005). The hippocampal-VTA loop: Controlling the entry of information into long-term memory. *Neuron, 46*(5), 703–713.

Lobb, C. J., Wilson, C. J., & Paladini, C. A. (2010). A dynamic role for GABA receptors on the firing pattern of midbrain dopaminergic neurons. *Journal of Neurophysiology, 104*(1), 403–413.

Lobb, C. J., Wilson, C. J., & Paladini, C. A. (2011). High-frequency, short-latency disinhibition bursting of midbrain dopaminergic neurons. *Journal of Neurophysiology, 105*(5), 2501–2511.

Lodge, D. J., & Grace, A. A. (2006). The hippocampus modulates dopamine neuron responsivity by regulating the intensity of phasic neuron activation. *Neuropsychopharmacology, 31*(7), 1356–1361.

Lopez-Huerta, V. G., Carrillo-Reid, L., Galarraga, E., Tapia, D., Fiordelisio, T., Drucker-Colin, R., et al. (2013). The balance of striatal feedback transmission is disrupted in a model of parkinsonism. *Journal of Neuroscience, 33*(11), 4964–4975.

Ludvig, E. A., Sutton, R. S., & Kehoe, E. J. (2008). Stimulus representation and the timing of reward-prediction errors in models of the dopamine system. *Neural Computation, 20*(12), 3034–3054.

Maia, T. V., & Frank, M. J. (2011). From reinforcement learning models to psychiatric and neurological disorders. *Nature Neuroscience, 14*(2), 154–162.

Manns, J. R., Zilli, E. A., Ong, K. C., Hasselmo, M. E., & Eichenbaum, H. (2007). Hippocampal CA1 spiking during encoding and retrieval: Relation to theta phase. *Neurobiology of Learning and Memory, 87*(1), 9–20.

Marder, E. (1996). Neural modulation: Following your own rhythm. *Current Biology, 6*(2), 119–121.

Marder, E. (2012). Neuromodulation of neuronal circuits: Back to the future. *Neuron, 76*(1), 1–11.

Marder, E., & Thirumalai, V. (2002). Cellular, synaptic and network effects of neuromodulation. *Neural Networks, 15*(4–6), 479–493.

Marrosu, F., Portas, C., Mascia, M. S., Casu, M. A., Fa, M., Giagheddu, M., et al. (1995). Microdialysis measurement of cortical and hippocampal acetylcholine release during sleep-wake cycle in freely moving cats. *Brain Research, 671*(2), 329–332.

Massobrio, P., de Arcangelis, L., Pasquale, V., Jensen, H. J., & Plenz, D. (2015). Criticality as a signature of healthy neural systems. *Frontiers in Systems Neuroscience, 9*, 22.

McClelland, J. L., McNaughton, B. L., & O'Reilly, R. C. (1995). Why there are complementary learning systems in the hippocampus and neocortex: Insights from the successes and failures of connectionist models of learning and memory. *Psychological Review, 102*(3), 419–457.

McGaughy, J., Kaiser, T., & Sarter, M. (1996). Behavioral vigilance following infusions of 192 IgG-saporin into the basal forebrain: Selectivity of the behavioral impairment and relation to cortical AChE-positive fiber density. *Behavioral Neuroscience, 110*(2), 247–265.

McGaughy, J., Koene, R. A., Eichenbaum, H., & Hasselmo, M. E. (2005). Cholinergic deafferentation of the entorhinal cortex in rats impairs encoding of novel but not familiar stimuli in a delayed nonmatch-to-sample task. *Journal of Neuroscience, 25*(44), 10273–10281.

McGaughy, J., Turchi, J., & Sarter, M. (1994). Crossmodal divided attention in rats: Effects of chlordiazepoxide and scopolamine. *Psychopharmacology, 115*(1–2), 213–220.

McQuiston, A. R., & Madison, D. V. (1999). Muscarinic receptor activity has multiple effects on the resting membrane potentials of CA1 hippocampal interneurons. *Journal of Neuroscience, 19*(14), 5693–5702.

Monmaur, P., Collet, A., Puma, C., Frankel-Kohn, L., & Sharif, A. (1997). Relations between acetylcholine release and electrophysiological characteristics of theta rhythm: A microdialysis study in the urethane-anesthetized rat hippocampus. *Brain Research Bulletin, 42*(2), 141–146.

Montague, P. R., Dayan, P., & Sejnowski, T. J. (1996). A framework for mesencephalic dopamine systems based on predictive Hebbian learning. *Journal of Neuroscience, 16*(5), 1936–1947.

Morgan, P. T., Perrins, R., Lloyd, P. E., & Weiss, K. R. (2000). Intrinsic and extrinsic modulation of a single central pattern generating circuit. *Journal of Neurophysiology, 84*(3), 1186–1193.

Moustafa, A. A., & Gluck, M. A. (2011a). Computational cognitive models of prefrontal-striatal-hippocampal interactions in Parkinson's disease and schizophrenia. *Neural Networks, 24*(6), 575–591.

Moustafa, A. A., & Gluck, M. A. (2011b). A neurocomputational model of dopamine and prefrontal-striatal interactions during multicue category learning by Parkinson patients. *Journal of Cognitive Neuroscience, 23*(1), 151–167.

Nadim, F., Brezina, V., Destexhe, A., & Linster, C. (2008). State dependence of network output: Modeling and experiments. *Journal of Neuroscience, 28*(46), 11806–11813.

Nair-Roberts, R. G., Chatelain-Badie, S. D., Benson, E., White-Cooper, H., Bolam, J. P., & Ungless, M. A. (2008). Stereological estimates of dopaminergic, GABAergic and glutamatergic neurons in the ventral tegmental area, substantia nigra and retrorubral field in the rat. *Neuroscience, 152*(4), 1024–1031.

Navaroli, V. L., Zhao, Y., Boguszewski, P., & Brown, T. H. (2012). Muscarinic receptor activation enables persistent firing in pyramidal neurons from superficial layers of dorsal perirhinal cortex. *Hippocampus, 22*(6), 1392–1404.

Nissen, M. J., Knopman, D. S., & Schacter, D. L. (1987). Neurochemical dissociation of memory systems. *Neurology, 37*(5), 789–794.

Nusbaum, M. P., Weimann, J. M., Golowasch, J., & Marder, E. (1992). Presynaptic control of modulatory fibers by their neural network targets. *Journal of Neuroscience, 12*(7), 2706–2714.

Oster, A. M., & Gutkin, B. S. (2011). A reduced model of DA neuronal dynamics that displays quiescence, tonic firing and bursting. *Journal of Physiology, Paris, 105*(1–3), 53–58.

Paladini, C. A., & Tepper, J. M. (1999). GABA(A) and GABA(B) antagonists differentially affect the firing pattern of substantia nigra dopaminergic neurons in vivo. *Synapse (New York, N.Y.), 32*(3), 165–176.

Parikh, V., Kozak, R., Martinez, V., & Sarter, M. (2007). Prefrontal acetylcholine release controls cue detection on multiple timescales. *Neuron, 56*(1), 141–154.

Patil, M. M., & Hasselmo, M. E. (1999). Modulation of inhibitory synaptic potentials in the piriform cortex. *Journal of Neurophysiology, 81*(5), 2103–2118.

Patton, M. H., Bizup, B. T., & Grace, A. A. (2013). The infralimbic cortex bidirectionally modulates mesolimbic dopamine neuron activity via distinct neural pathways. *Journal of Neuroscience, 33*(43), 16865–16873.

Pavlides, C., Greenstein, Y. J., Grudman, M., & Winson, J. (1988). Long-term potentiation in the dentate gyrus is induced preferentially on the positive phase of theta-rhythm. *Brain Research, 439*(1–2), 383–387.

Ping, H. X., & Shepard, P. D. (1996). Apamin-sensitive Ca(2+)-activated K+ channels regulate pacemaker activity in nigral dopamine neurons. *Neuroreport, 7*(3), 809–814.

Piray, P., Zeighami, Y., Bahrami, F., Eissa, A. M., Hewedi, D. H., & Moustafa, A. A. (2014). Impulse control disorders in Parkinson's disease are associated with dysfunction in stimulus valuation but not action valuation. *Journal of Neuroscience, 34*(23), 7814–7824.

Pitler, T. A., & Alger, B. E. (1992). Cholinergic excitation of GABAergic interneurons in the rat hippocampal slice. *Journal of Physiology, 450*, 127–142.

Qian, K., Yu, N., Tucker, K. R., Levitan, E. S., & Canavier, C. C. (2014). Mathematical analysis of depolarization block mediated by slow inactivation of fast sodium channels in midbrain dopamine neurons. *Journal of Neurophysiology, 112*(11), 2779–2790.

Rasch, B. H., Born, J., & Gais, S. (2006). Combined blockade of cholinergic receptors shifts the brain from stimulus encoding to memory consolidation. *Journal of Cognitive Neuroscience, 18*(5), 793–802.

Reynolds, J. N., & Wickens, J. R. (2002). Dopamine-dependent plasticity of corticostriatal synapses. *Neural Networks, 15*(4–6), 507–521.

Richards, C. D., Shiroyama, T., & Kitai, S. T. (1997). Electrophysiological and immunocytochemical characterization of GABA and dopamine neurons in the substantia nigra of the rat. *Neuroscience, 80*(2), 545–557.

Ridet, J. L., Rajaofetra, N., Teilhac, J. R., Geffard, M., & Privat, A. (1993). Evidence for nonsynaptic serotonergic and noradrenergic innervation of the rat dorsal horn and possible involvement of neuron-glia interactions. *Neuroscience, 52*(1), 143–157.

Rizzuto, D. S., Madsen, J. R., Bromfield, E. B., Schulze-Bonhage, A., & Kahana, M. J. (2006). Human neocortical oscillations exhibit theta phase differences between encoding and retrieval. *NeuroImage, 31*(3), 1352–1358.

Robbins, T. W., Everitt, B. J., Marston, H. M., Wilkinson, J., Jones, G. H., & Page, K. J. (1989). Comparative effects of ibotenic acid- and quisqualic acid-induced lesions of the substantia innominata on attentional function in the rat: Further implications for the role of the cholinergic neurons of the nucleus basalis in cognitive processes. *Behavioural Brain Research, 35*(3), 221–240.

Robbins, T. W., Semple, J., Kumar, R., Truman, M. I., Shorter, J., Ferraro, A., et al. (1997). Effects of scopolamine on delayed-matching-to-sample and paired associates tests of visual memory and learning in human subjects: Comparison with diazepam and implications for dementia. *Psychopharmacology, 134*(1), 95–106.

Roberts, M. J., & Thiele, A. (2008). Spatial integration and its moderation by attention and acetylcholine. *Frontiers in Bioscience, 13,* 3742–3759.

Roeper, J. (2013). Dissecting the diversity of midbrain dopamine neurons. *Trends in Neurosciences, 36*(6), 336–342.

Rogers, J. L., & Kesner, R. P. (2003). Cholinergic modulation of the hippocampus during encoding and retrieval. *Neurobiology of Learning and Memory, 80*(3), 332–342.

Samson, R. D., Frank, M. J., & Fellous, J. M. (2010). Computational models of reinforcement learning: The role of dopamine as a reward signal. *Cognitive Neurodynamics, 4*(2), 91–105.

Sarter, M., Hasselmo, M. E., Bruno, J. P., & Givens, B. (2005). Unraveling the attentional functions of cortical cholinergic inputs: Interactions between signal-driven and cognitive modulation of signal detection. *Brain Research. Brain Research Reviews, 48*(1), 98–111.

Schiemann, J., Schlaudraff, F., Klose, V., Bingmer, M., Seino, S., Magill, P. J., et al. (2012). K-ATP channels in dopamine substantia nigra neurons control bursting and novelty-induced exploration. *Nature Neuroscience, 15*(9), 1272–1280.

Schon, K., Atri, A., Hasselmo, M. E., Tricarico, M. D., LoPresti, M. L., & Stern, C. E. (2005). Scopolamine reduces persistent activity related to long-term encoding in the parahippocampal gyrus during delayed matching in humans. *Journal of Neuroscience, 25*(40), 9112–9123.

Schroll, H., Vitay, J., & Hamker, F. H. (2014). Dysfunctional and compensatory synaptic plasticity in Parkinson's disease. *European Journal of Neuroscience, 39*(4), 688–702.

Schultz, W. (1998a). The phasic reward signal of primate dopamine neurons. *Advances in Pharmacology (San Diego, Calif.), 42,* 686–690.

Schultz, W. (1998b). Predictive reward signal of dopamine neurons. *Journal of Neurophysiology, 80*(1), 1–27.

Schultz, W. (2007). Multiple dopamine functions at different time courses. *Annual Review of Neuroscience, 30,* 259–288.

Schultz, W. (2010). Dopamine signals for reward value and risk: Basic and recent data. *Behavioral and Brain Functions, 6,* 24.

Schultz, W. (2013). Updating dopamine reward signals. *Current Opinion in Neurobiology, 23*(2), 229–238.

Schultz, W., Apicella, P., & Ljungberg, T. (1993). Responses of monkey dopamine neurons to reward and conditioned stimuli during successive steps of learning a delayed response task. *Journal of Neuroscience, 13*(3), 900–913.

Seutin, V., Johnson, S. W., & North, R. A. (1993). Apamin increases NMDA-induced burst-firing of rat mesencephalic dopamine neurons. *Brain Research, 630*(1–2), 341–344.

Seymour, B., Daw, N. D., Roiser, J. P., Dayan, P., & Dolan, R. (2012). Serotonin selectively modulates reward value in human decision-making. *Journal of Neuroscience, 32*(17), 5833–5842.

Shepard, P. D., & Bunney, B. S. (1988). Effects of apamin on the discharge properties of putative dopamine-containing neurons in vitro. *Brain Research, 463*(2), 380–384.

Sheynikhovich, D., Otani, S., & Arleo, A. (2011). The role of tonic and phasic dopamine for long-term synaptic plasticity in the prefrontal cortex: A computational model. *Journal of Physiology, Paris, 105*(1–3), 45–52.

Silver, M. A., Shenhav, A., & D'Esposito, M. (2008). Cholinergic enhancement reduces spatial spread of visual responses in human early visual cortex. *Neuron, 60*(5), 904–914.

Spiga, S., Lintas, A., Migliore, M., & Diana, M. (2010). Altered architecture and functional consequences of the mesolimbic dopamine system in cannabis dependence. *Addiction Biology, 15*(3), 266–276.

Starke, K., Gothert, M., & Kilbinger, H. (1989). Modulation of neurotransmitter release by presynaptic autoreceptors. *Physiological Reviews, 69*(3), 864–989.

Sutton, R. S., & Barto, A. G. (1990). Time-derivative models of Pavlovian reinforcement. In M. Gabriel & J. Moore (Eds.), *Learning and Computational Neuroscience.* Cambridge, MA: MIT Press.

Suzuki, W. A., Miller, E. K., & Desimone, R. (1997). Object and place memory in the macaque entorhinal cortex. *Journal of Neurophysiology, 78*(2), 1062–1081.

Tan, C. O., & Bullock, D. (2008). Neuropeptide co-release with GABA may explain functional non-monotonic uncertainty responses in dopamine neurons. *Neuroscience Letters, 430*(3), 218–223.

Tang, Y., Mishkin, M., & Aigner, T. G. (1997). Effects of muscarinic blockade in perirhinal cortex during visual recognition. *Proceedings of the National Academy of Sciences of the United States of America, 94*(23), 12667–12669.

Tepper, J. M., & Lee, C. R. (2007). GABAergic control of substantia nigra dopaminergic neurons. *Progress in Brain Research, 160,* 189–208.

Tepper, J. M., Martin, L. P., & Anderson, D. R. (1995). $GABA_A$ receptor-mediated inhibition of rat substantia nigra dopaminergic neurons by pars reticulata projection neurons. *Journal of Neuroscience, 15*(4), 3092–3103.

Teramae, J. N., & Fukai, T. (2005). A cellular mechanism for graded persistent activity in a model neuron and its implications in working memory. *Journal of Computational Neuroscience, 18*(1), 105–121.

Tononi, G., & Cirelli, C. (2006). Sleep function and synaptic homeostasis. *Sleep Medicine Reviews, 10*(1), 49–62.

Turchi, J., & Sarter, M. (1997). Cortical acetylcholine and processing capacity: Effects of cortical cholinergic deafferentation on crossmodal divided attention in rats. *Brain Research. Cognitive Brain Research, 6*(2), 147–158.

Turchi, J., Saunders, R. C., & Mishkin, M. (2005). Effects of cholinergic deafferentation of the rhinal cortex on visual recognition memory in monkeys. *Proceedings of the National Academy of Sciences of the United States of America, 102*(6), 2158–2161.

Umbriaco, D., Garcia, S., Beaulieu, C., & Descarries, L. (1995). Relational features of acetylcholine, noradrenaline, serotonin and GABA axon terminals in the stratum radiatum of adult rat hippocampus (CA1). *Hippocampus, 5*(6), 605–620.

Waroux, O., Massotte, L., Alleva, L., Graulich, A., Thomas, E., Liegeois, J. F., et al. (2005). SK channels control the firing pattern of midbrain dopaminergic neurons in vivo. *European Journal of Neuroscience, 22*(12), 3111–3121.

Weinberger, D. R. (1987). Implications of normal brain development for the pathogenesis of schizophrenia. *Archives of General Psychiatry, 44*(7), 660–669.

Wesnes, K., & Warburton, D. M. (1983). Effects of scopolamine on stimulus sensitivity and response bias in a visual vigilance task. *Neuropsychobiology, 9*(2–3), 154–157.

Whishaw, I. Q. (1985). Formation of a place learning-set by the rat: A new paradigm for neurobehavioral studies. *Physiology & Behavior, 35*(1), 139–143.

Wightman, R. M., May, L. J., & Michael, A. C. (1988). Detection of dopamine dynamics in the brain. *Analytical Chemistry, 60*(13), 769A–779A.

Wilson, C. J., & Callaway, J. C. (2000). Coupled oscillator model of the dopaminergic neuron of the substantia nigra. *Journal of Neurophysiology, 83*(5), 3084–3100.

Winson, J., & Abzug, C. (1978). Dependence upon behavior of neuronal transmission from perforant pathway through entorhinal cortex. *Brain Research, 147*(2), 422–427.

Winters, B. D., & Bussey, T. J. (2005). Removal of cholinergic input to perirhinal cortex disrupts object recognition but not spatial working memory in the rat. *European Journal of Neuroscience, 21*(8), 2263–2270.

Yamamoto, C., & Kawai, N. (1967). Presynaptic action of acetylcholine in thin sections from the guinea pig dentate gyrus in vitro. *Experimental Neurology, 19*(2), 176–187.

Young, B. J., Otto, T., Fox, G. D., & Eichenbaum, H. (1997). Memory representation within the parahippocampal region. *Journal of Neuroscience, 17*(13), 5183–5195.

Zola-Morgan, S. M., & Squire, L. R. (1990). The primate hippocampal formation: Evidence for a time-limited role in memory storage. *Science, 250*(4978), 288–290.

Zweifel, L. S., Parker, J. G., Lobb, C. J., Rainwater, A., Wall, V. Z., Fadok, J. P., et al. (2009). Disruption of NMDAR-dependent burst firing by dopamine neurons provides selective assessment of phasic dopamine-dependent behavior. *Proceedings of the National Academy of Sciences of the United States of America, 106*(18), 7281–7288.

8 Motor Pattern Generation

AUKE J. IJSPEERT, ANDREJ BICANSKI, JEREMIE KNUESEL, AND JEAN-MARIE CABELGUEN

8.1 *Pattern Generation and Motor Control*

8.1.1 Introducing Pattern Generators Animals are capable of amazing motor skills, from the efficient and agile swimming of tuna fish, to the skillful walking and climbing of goats in rocky terrains, to the multiple and diverse activities such as grasping and manipulating in octopus, to playing a music instrument or practicing sports in humans. All these skills require tight coordination between the nervous system, the musculoskeletal system, and the environment. The control loops underlying these skills are very complex because of the high number of degrees of freedom involved (large numbers of joints, muscles, and sensory channels), the highly nonlinear and complex dynamics of the musculoskeletal system, as well as uncertainties about the state of the environment.

Interestingly, as will be reviewed in this chapter, it has been shown that lower parts of the central nervous system (CNS) of vertebrates, namely, the spinal cord and the brainstem, play an essential role in producing the coordinated movements necessary for rich motor skills. For instance, completely isolated spinal cords can produce "fictive locomotion," that is, patterns of coordinated motor neuron outputs that are very similar to those of intact locomotion as shown in a large number of species (reviewed in Grillner, 1985; Orlovsky et al., 1999). Decerebrated animals (i.e., with a complete transection above the brainstem) can display stereotyped locomotor movements, spontaneously or in response to electrical or chemical stimulation of several brainstem locomotor areas, such as the mesencephalic locomotor region (MLR) (Armstrong, 1986; Grillner et al., 1997; Orlovsky et al., 1999; LeRay et al., 2011). For instance, in acute mesencephalic cats (precollicular–postmammillary section) placed on a moving treadmill belt, repetitive electrical stimulation of the MLR can suddenly trigger walking, and increasing the strength of the stimulus can evoke a trot or gallop (Shik et al., 1966) (figure 8.1). Importantly, MLR stimulation can induce fictive locomotion in paralyzed decerebrated cats, that is, in the absence of afferent feedback (Jordan et al., 1979; Arshavsky et al., 1986; Amemiya and Yamaguchi, 1984).

Isolated spinal cords are also able to produce fictive scratching (Deliagina et al., 1975; Robertson et al., 1985; Cuellar et al., 2009), fictive paw shaking (Yakovleff et al., 1995), fictive struggling (Green and Soffe, 1996), and even fictive ejaculation (Carro-Juárez et al., 2003). Also, neuronal networks located in the brainstem can generate several fictive motor patterns other than locomotion, such as those underlying respiratory behaviors (e.g., breathing, sighing, gasping, coughing, and sneezing), as well as swallowing, vomiting, sucking and chewing (reviewed in Ramirez and Richter, 1996; Jean, 2001; Feldman and Del Negro, 2006; Barlow and Estep, 2006; Bianchi and Gestreau, 2009; Morquette et al., 2012). Evidence also suggests that there are some shared central circuits for the multiple forms of breathing and oromotor movements (Grélot et al., 1993; Oku et al., 1994; Gestreau et al., 2000, 2005; Lieske et al., 2000; Baekey et al., 2001; Shiba et al., 2007).

Several studies suggest that the various rhythmic motor patterns exhibited by the limbs are produced by a combination of shared and specialized spinal circuits (see Berkowitz et al., 2010), as previously evidenced for rhythmic motor patterns in invertebrates (Marder and Calabrese, 1996; Marder et al., 2005).

The spinal cord is also involved in the generation of *discrete*, that is, point-to-point, and *targeted* movements. A spinalized frog can remove an irritant placed on the skin with precise movements of the leg and foot ("wiping reflex"), and the forms of wiping and movement trajectories vary with the location of the irritant (Fukson et al., 1980). Furthermore, if the irritant is placed on the same spot of the forelimb (elbow) and the forelimb is retracted or protracted, the hind limb of the spinal frog can accurately wipe the irritant regardless of the position of the forelimb, suggesting that the spinal cord can integrate complex proprioceptive inputs to plan wiping accurately. In other words, the spinal cord "knows" the scheme (or schema) of its body (Fukson et al., 1980; Poppele and Bosco, 2003). Interestingly, the turtle spinal cord can generate different forms of real or

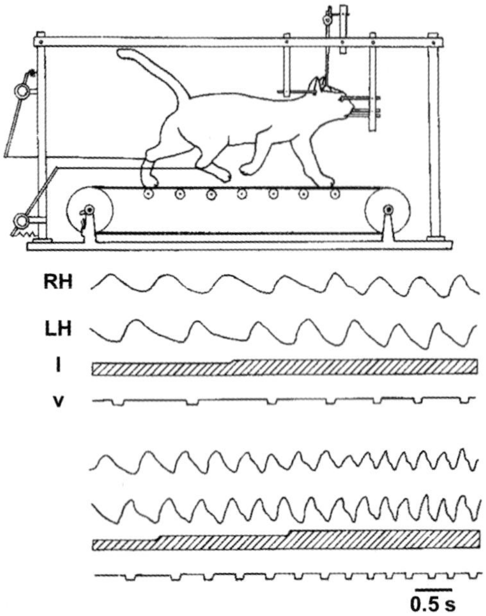

FIGURE 8.1 Locomotion in the mesencephalic cat evoked by electrical stimulation of the mesencephalic locomotor region (MLR) (adapted from Shik et al., 1966). The experimental arrangement is shown on the top. The mesencephalic cat is fixed in a stereotaxic device, with its four limbs walking on a treadmill belt which is free-wheeling. Movement recordings of the right (RH) and left (LH) hind limbs are made as strength of stimulation (*I*) is increased. The speed of the treadmill belt (*v*) is measured by a contact transducer. Note the switch from alternating to nonalternating gait.

fictive scratching depending on the location of the irritant on the shell (Mortin et al., 1985; Robertson et al., 1985).

In cats and rabbits, the system generating the muscle synergies that are relevant to postural reactions resides in the spinal cord ("spinal postural network") and consists of two relatively independent subsystems, one stabilizing the shoulder girdle and the other the pelvic girdle (Deliagina et al., 2008). Each subsystem includes two controllers (one for the left limb and one for the right limb), and each controller contains a local network driven both by mechanosensory inputs from its own limb and by tonic/phasic descending drives involved in postural adaptations to environment (Deliagina et al., 2012). It is likely that a modular organization of the spinal postural network increases its dynamics by increasing its number of degrees of freedom, thus allowing a better adaptation to complicated environmental conditions.

Altogether, the aforementioned studies show that the spinal cord is not just a relay that *executes* movements that are planned in other parts of the CNS, but it is an essential element of the control circuits, that is, it plays an important role in *generating* the movements. The underlying spinal circuits are called *central pattern generators* (CPGs), which are defined as circuits that can produce coordinated motoneuron activities without sensory feedback (hence the term *central* as opposed to *peripheral*). Note that, historically, the term CPG has mainly referred to the generation of *periodic* motor patterns. Sten Grillner proposed extending the notion of CPGs to cover both rhythmic and discrete pattern generation (Grillner, 2006), and we will do the same in this chapter (see also the discussion below). We will use the term *rhythmic CPG* when specifically referring to CPGs producing periodic motor patterns.

Complex neural mechanisms are used by vertebrates to stabilize or to shift gaze. Interestingly, studies in swimming tadpoles have provided compelling evidence that signals from the spinal locomotor CPG are used to produce eye movements appropriate to stabilize gaze (i.e., visual field) during locomotion (Stehouwer, 1987; Combes et al., 2008; Lambert et al., 2012). On the other hand, a central coupling between the locomotor and the respiratory CPGs has been evidenced in cats, birds, and lampreys (Funk et al., 1992; Kawahara et al., 1989, 1994; Gariépy et al., 2012). Thus, the central locomotor program includes not only the activation of the spinal locomotor CPG but also that of brainstem neurons and networks, which play an important role during ongoing locomotion. This underscores that, although it is easiest to consider the working of individual CPGs separately, it is important to remember that the motor behavior of an animal as a whole often requires that multiple CPGs and the motor outputs they generate be coordinated. Several cellular mechanisms of coordination between functionally related CPGs have been elucidated in invertebrates (see, e.g., Dickinson, 1995; Calabrese, 1998).

The CPGs are located in specific spinal segments (e.g., the limb locomotor CPGs), are located in brainstem subparts (e.g., the breathing CPG), or are distributed as discrete oscillators within the entire spinal cord (e.g., the body locomotor CPGs). They can be *activated* and *modulated* by a set of descending pathways from higher control centers (e.g., the motor cortex) so as to produce rich motor skills (Armstrong, 1986; Mori, 1987; Grillner et al., 1997; Orlovsky et al., 1999; Rossignol, 1996; Drew, 1991; Sinnamon, 1993; Jordan, 1998; Drew et al., 2008; Lemon, 2008). However, one should keep in mind that the control of the CPGs by descending pathways is not strictly hierarchical; the pattern-generating circuits have some control over the higher centers that input on them ("central feedback") (Arshavsky et al., 1984; Orsal et al., 1988).

Furthermore, the CPGs are subjected to a multiplicity of sensory signals. The sensory signals include those produced by the movements themselves ("reafference") and the unpredictable signals induced by external perturbations during ongoing movements (see, e.g., Rossignol et al., 1988; Pearson, 2000). In this chapter, because of the continuous dynamic interactions between CPGs, descending signals, and afferent inputs, we will distinguish the terms *central pattern generator* and *motor pattern generator* (MPG), with the CPG being the part of the motor circuit that can generate "open-loop" patterns, and the MPG being the complete motor circuits that include CPGs, sensory feedback loops, and modulation from descending pathways. However, note that there can be motor patterns without the involvement of CPGs—for example, through direct cortical connections onto motoneurons, as for finger movements in primates (Lemon, 2008).

Thanks to rhythmic and discrete pattern generators, the brainstem and spinal cord provide a (finite) vocabulary of movements available for learning and controlling rich motor skills. While this chapter will focus on rhythmic CPGs and locomotor movements, this view fits with the idea that the nervous system is modular and highly distributed and, as will be reviewed later in this chapter, with concepts such as muscle synergies and motor primitives.

This chapter is divided into three sections. In this first section, we now present the concept of MPG in more detail. In the second section, we will present a short overview of different types of CPG models that have been developed, from detailed biophysical models to abstract models that employ coupled oscillators. That section will also present in more detail models of the lamprey and salamander locomotor CPGs. These animals have been chosen because they are relatively simple, have been extensively investigated and modeled (in particular the lamprey), and provide the opportunity to investigate the changes in the spinal pattern generator circuits that accompanied the transition of vertebrates from aquatic habitats to land. Finally, a third section will present how robots can be used to test models of CPGs and how the concept of CPG can become a useful control technology in robotics.

8.1.2 Motor Pattern Generators for Locomotion We here review some of the main findings about MPGs with a particular focus on locomotion in lampreys and salamanders. These findings can be summarized in seven observations:

Observation 1. Isolated rhythmic CPGs are capable of producing coordinated rhythmic patterns without phasic (i.e., rhythmic) inputs from descending pathways or from sensory feedback. This observation, which is at the core of the definition of what a rhythmic CPG is, has been shown in many different animal preparations (for references, see above). In the lamprey, isolated spinal cords produce coordinated rhythmic motor activity when activated by bath application of glutamatergic agonists (Poon, 1980; Cohen and Wallén, 1980). The induced motor activity can be recorded from ventral roots, that is, from the axons of motoneuron pools (cf. figure 8.2). The motor pattern produced by a given spinal segment alternates between left and right sides, and the motor rhythm is the same for every spinal segment. Furthermore, there is a constant phase lag between the activations of successive spinal segments which leads to a traveling wave of bursts from head to tail, that is, the bursting activity propagates along the spinal cord (as opposed to a standing wave where all bursts would happen at the same time).

The similarity of the pattern of ventral root recordings with electromyography (EMG) and kinematics waves during intact swimming has led to the term "fictive swimming" (Wallén and Williams, 1984). Similarly, in vitro isolated spinal cords of salamander superfused with excitatory chemicals can generate fictive swimming (Delvolvé et al., 1999; Ryczko et al., 2010) (figure 8.3). From a control engineering point of view, this suggests that CPGs offer some kind of feedforward control (i.e., a controller that knows how to produce the right muscle activity patterns to get a given speed or mode of locomotion). (See chapter 9, "Computational Models in Motor Control," for more on feedback and feedforward.)

Observation 2. Rhythmic CPGs are distributed networks of multiple oscillatory pools coupled together, with typically one or two oscillatory pools per degree of freedom in the body. Through different studies, it has been shown that a rhythmic CPG is a network of multiple oscillatory pools connected together rather than a single large circuit. These oscillatory pools are groups of neurons that can produce oscillatory behavior in isolation from others. The pools are composed of interneurons and motoneurons (in the order of several thousand neurons per pool in the lamprey) and receive inputs from sensory neurons through the dorsal roots of the spinal cord and from descending pathways. In the lamprey and salamander, for instance, isolated segments and even hemisegments can produce oscillatory activity, such that the axial CPG (i.e., the CPG that controls the axial musculature for lateral undulations) in these animals is organized as a double chain of oscillatory pools connected together (Cangiano and Grillner, 2005; Ryczko et al., 2010; Cangiano et al., 2012). We will see later in this chapter how computational models use systems of

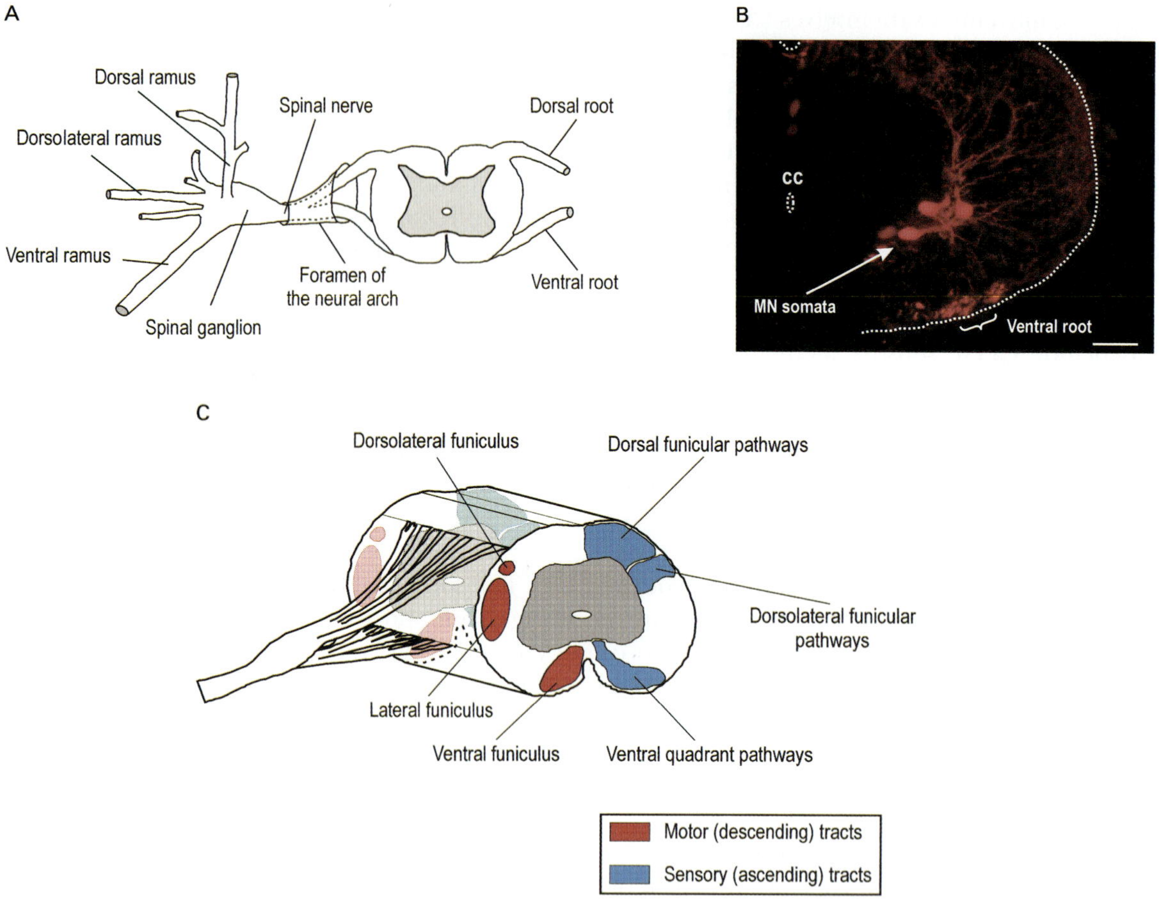

FIGURE 8.2 Organization of the spinal segments in salamanders. (A) Schematic drawing of a cross-section through the tenth spinal segment of *Salamandra salamandra* (adapted from Francis, 1934). (B) Photomicrograph illustrating the location of hind-limb motoneurons (MNs) in the lumbar spinal cord of *Notophthalmus viridescens* (Chevallier, 2004). MNs were labeled with a retrograde dye (Texas Red dextran amine) injected into the ipsilateral hind-limb muscles. Spinal outline is shown in dashed lines. cc, central canal. Scale bar: 50 μm. (C) Schematic drawing of a spinal segment showing the course of the main descending and ascending axons (based on Ten Donkelaar, 1982; Munoz et al., 1997; Nieuwenhuys et al., 1998; Sánchez-Camacho et al., 2001). For clarity, motor pathways have been drawn on the left half of the figure, and sensory pathways on the right half of the figure. Note that the supraspinal projections arise mainly from the reticular formation and descend via the ventral and the lateral funiculi. The red nucleus (absent in lampreys) projects to the spinal cord via the dorsolateral funiculus.

coupled oscillators or neural networks with similar topologies. On the other hand, Cheng et al. (1998) have discovered independent oscillatory centers for forelimb extensor and flexor motoneuron pools in salamanders. The two pools can produce rhythmic bursts independently despite the fact that there are reciprocal inhibitory interconnections between them.

Altogether, the aforementioned observations are in agreement with the concept that the burst-generating component of the locomotor CPGs are organized as coupled oscillatory pools, referred as "unit burst generators," with at least one unit per joint (i.e., per degree of freedom) (Grillner, 1981).

The fact that the multiple oscillatory pools produce coordinated patterns (as opposed to disorganized patterns) is evidence that well-tuned couplings exist between these pools to ensure coordination (i.e., to keep oscillations at the same frequency, with a fixed phase difference between pools). These couplings are provided by projections between interneurons in different pools. For instance, in lamprey, interneurons are known to project to neurons in other segments both rostrally (i.e., toward the head) and caudally (toward the tail) (Buchanan, 2001).

While in vertebrates the locomotor CPGs are distributed along the spinal cord, a (nonspinal) distributed organization is observed in invertebrates, for example, in insects where different oscillatory pools are located in different ganglia (Delcomyn, 1980; Pearson, 1993).

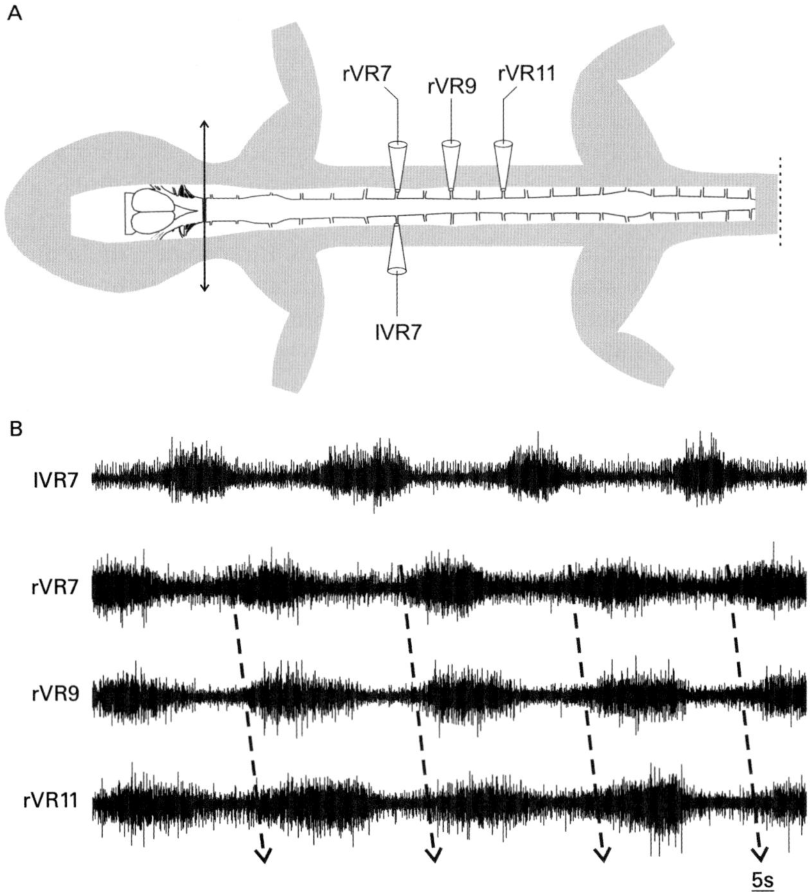

FIGURE 8.3 Fictive swimming in the salamander (adapted from Ryczko et al., 2010). (A) Schematic representation of the preparation. The spinal cord was disconnected from the brainstem by a transection at the level of the obex (vertical arrowed line). Extracellular recordings were made from three successive ventral roots on the right side (*rVR7*, *rVR9*, and *rVR11*) and from one ventral root on the left side (*lVR7*). (B) Example of rhythmic motor pattern induced by bath application of a mixture of *N*-methyl-D-aspartate (NMDA; 20 μM) and D-serine (10 μM). The pattern consists of left–right alternation (*lVR7* and *rVR7*) and waves of ventral root activity travelling caudally along the midtrunk (*rVR7 to rVR13*).

Observation 3. Rhythmic CPGs are shaped and entrained by sensory feedback. Although CPGs can produce sophisticated motor patterns in the absence of movement-related sensory feedback, in intact animals they receive modulation from sensory and descending signals and thus function as part of the larger circuitry of MPGs.

Several observations clearly indicate that some features of actual locomotion are not preserved under the conditions of fictive locomotion, thus stressing the importance of sensory signals (Rossignol et al., 1988; Boyd and McClellan, 2002, in lampreys). Another example is the locomotor circuit of the stick insect, where sensory feedback plays an essential role in interlimb coordination (Cruse et al., 1995).

Including sensory feedback in the overall motor pattern generation is clearly very important from a control point of view in order to handle uncertainties (e.g., from a complex terrain, changing body properties, etc.), up to situations where sensory feedback (and descending modulation) is so dominant that the contribution of the CPG might be negligible, for instance, when stepping from rock to rock along the shore.

In the lamprey, mechanically moving the body will change the *frequency* of the CPG to match the frequency of the mechanical stimulus (McClellan and Jang, 1993; Viana Di Prisco et al., 1990; Williams et al., 1990; see also Grillner and Wallén, 1977, in the dogfish). Such a phenomenon is called *entrainment*, and the range of frequencies at which the body CPG can be entrained above and below its nominal frequencies provides an estimation of the strengths of the feedback influence (the stronger the influence, the larger the entrainment frequency range). In the lamprey that range is within 25% to 50% of the rest frequency depending on the length of the spinal cord tested (Williams et al., 1990). As we will see next, computational models have been very useful in characterizing such dynamical behaviors.

In limbed vertebrates, proprioceptive inputs from the limbs (e.g., hip or shoulder position, extensor load) play an essential role in entraining or resetting the ongoing rhythm generated by the locomotor CPG (for mammals, see the in-depth review of Rossignol et al., 2006; for the salamander, see Cheng et al., 1998). Moreover, some sensory signals from the limb have critical effects on the pattern of particular muscle activity or of the whole limb during a part of the step cycle. One interesting example of sensory signals shaping the motor pattern generated by the locomotor CPG is the modification of the bursting activity recorded from some limb muscle nerves during fictive locomotion by tonic cutaneous stimulation (Perret and Cabelguen, 1980). These muscle nerves can display a flexor- or extensor-related activity, or some combination of the two, during the fictive step cycle. The two nerve bursts are controlled independently by cutaneous inputs from the ipsilateral limb: one of the two bursts may be deleted while the other is enhanced. Similarly, the EMGs of the corresponding limb muscles ("bifunctional muscles") can be swing related or stance related (or both) during real stepping, depending on the afferent feedback and form of locomotion (e.g., forward vs. backward; upslope vs. down slope) (Stein and Smith, 1997).

It is worth noting that most limb muscles of salamanders display a complex bursting activity during overground locomotion (Delvolvé et al., 1997; Ashley-Ross and Lauder, 1997), and it is likely that cutaneous inputs from the limbs play a crucial role in the shaping of their pattern of activity.

Taken together these results have led to the conclusion that the limb locomotor CPG can be subdivided into two functionally separate, but hierarchically connected, subcomponents. The first one could be a network which sets the locomotor rhythm and acts on a second component generating the individual pattern of activity within the cycle (Perret et al., 1988; McCrea and Rybak, 2008). A similar organization has been proposed and modeled for the brainstem CPG for breathing (Feldman and Smith, 1989; Smith et al., 2007).

Several studies provide further evidence that some of the interneurons of the segmental reflex pathways are part of either the rhythm generator network (e.g., the tendon organ afferent pathway from limb extensors) or the pattern generator network (e.g., the flexor reflex afferent pathway) (see Rossignol et al., 2006, for an in-depth review). The tight interactions between the limb/fin CPG and the afferent pathways induce an adjustment of reflex responses to the ongoing movement: that is, the reflexes have different effects, depending on their timing within the locomotor cycle ("phase-dependent modulation") (cat: Forssberg et al., 1977; Drew and Rossignol, 1987; dogfish: Grillner et al., 1977). Thus, the transmission through reflex pathways, which may modify the basic locomotor patterns, is phasically controlled by the CPG itself. Numerous gating mechanisms, at the cellular and network levels in the spinal cord, contribute to the phase modulation of reflexes. Importantly, the sensory modulation can occur early in the reflex pathways, at the level of the primary afferent terminals themselves ("presynaptic modulation") (cat: Dubuc et al., 1988; Gossard et al., 1989; 1990; lamprey: El Manira et al., 1996).

Interestingly, during rhythmic movements some proprioceptive inputs may shape the pattern of muscle activities during a cycle while others may affect the timing of the phases of the cycle (Rossignol et al., 2006). However, the adaptive control of fast movements requires some means of anticipating changing conditions rather than reacting to changes in the environment through error-detecting feedback loops. Thus, predictive central feedforward signals have been suggested to play an important role during ballistic reaching (Wolpert et al., 1995; Gritsenko et al., 2009) and the very fast eye movements (saccades) produced by a brainstem CPG (Scudder et al., 2002; Sparks, 2002). It is likely that predictive signals are integrated with feedback signals to produce optimal motor commands (Hoff and Arbib, 1991). (See also chapter 9, "Computational Models in Motor Control.")

Observation 4. Neuromodulators adapt the rhythmic CPGs to changing conditions. While rhythmic CPGs generate specific coordinated motor patterns, they need flexibility to produce motor patterns adapted to the changing environmental demands. In addition to afferent inputs, another way to achieve this flexibility is that the CPGs are subjected to neuromodulators, which can modify the intrinsic membrane properties and the synaptic connections of the constituent neurons. It is convenient to distinguish neuromodulators that are intrinsic to the CPG being considered from those that are delivered from an extrinsic source (Katz, 1995, 1998). (For a complementary perspective, see chapter 7, "Neuromodulation.")

A large number of studies in invertebrates (mainly using the crustacean stomatogastric system) have shown that several intrinsic and extrinsic neuromodulators contribute to the dynamics of the CPG networks by changing the frequency and/or the amplitude of the CPG output, or even by reconfiguring the CPG into a new functional circuit (recently reviewed in Marder and Thirumalai, 2002; Stein, 2009; Harris-Warrick, 2011; Marder, 2012; Taghert and Nitabach, 2012). For example, neuromodulation can determine whether or not a particular cell is a pacemaker.

In vertebrates, a variety of neuromodulators have been evidenced for the rodent respiratory CPG (Doi and Ramirez, 2008; Garcia et al., 2011) and the locomotor CPG in lampreys (LeBeau et al., 2005; El Manira et al., 2010), tadpoles (Sillar et al., 2002), salamanders (Jovanovic et al., 1996, 1999; Branchereau et al., 2000; Fok and Stein, 2002) and mammals (see Miles and Sillar, 2011, for a recent review). To date, very little is known about the targets and the cellular mechanisms by which neuromodulators modify the locomotor CPG activity in vertebrates (Harris-Warrick, 2011; Miles and Sillar, 2011). Conceptually, the neuromodulatory effects on the bursting frequency have been interpreted as occurring on the rhythm-generating part of the CPG network, and those on the amplitude of motor bursts as predominantly occurring at the level of the motoneurons or last-order interneurons of the CPG network (Miles and Sillar, 2012). Neuromodulation is not specific to the neurons of the CPG and motoneurons. Indeed, neuromodulatory substances (e.g., hormones) can alter the excitability of primary afferent terminals in the spinal cord, thus changing the encoding of sensory information (Marder and Thirumalai, 2002).

Altogether, the studies on neuromodulation strongly suggest that, although this is outside the scope of the present chapter, computational models of locomotor CPGs should now include a modeling of their neuromodulatory controls in order to increase the adaptability of their dynamics (see, e.g., Baxter et al., 1999). Furthermore, tonic versus phasic activation of neuromodulatory inputs should be distinguished.

Observation 5. Simple tonic electrical stimulation in the MLR can initiate locomotion and control the locomotor speed and gait transition. Many regions of the brain can be stimulated to elicit locomotion (reviewed in Jackson et al., 2007) (figure 8.4). In lampreys, these include the diencephalic locomotor region (El Manira et al., 1997), the MLR (Sirota et al., 2000), and a region located in the rostrolateral rhombencephalon (Paggett et al., 2004).

The MLR was found in several species (for reviews, see Grillner et al., 1997; LeRay et al., 2011), and, as mentioned in the introduction, electrical stimulation of the MLR elicited graded locomotor output and even gait transition. In lampreys and fish, the amplitude and frequency of swimming movements increase as the intensity or frequency of the stimulating current is increased (see, e.g., Kashin et al., 1974; Fetcho and Svoboda, 1993; Uematsu and Todo, 1997; Sirota et al., 2000). In salamanders, the MLR controls the two distinct locomotor modes (i.e., stepping and swimming) exhibited by these animals (Cabelguen et al., 2003). At low current strength stepping can be induced, and as the stimulation strength is subsequently increased, the stepping frequency first increases and then the locomotor mode switches to swimming. A further increase of the stimulation strength induces an increase of the frequency and amplitude of the swimming movements.

In mammals and lampreys, the MLR turns on the locomotor CPG through a monosynaptic activation of reticulospinal (RS) cells in the pons and medulla oblongata (Steeves and Jordan, 1984; for review, see McClellan, 1986; Jordan, 1998; Grillner et al., 1997; LeRay et al., 2011; Ryczko and Dubuc, 2013). A recent study in lampreys has further shown that the MLR provides bilaterally symmetrical inputs to RS cells on both sides (Brocard et al., 2010). This symmetrical organization of the MLR command at the brainstem level likely contributes to generating symmetrical bilateral locomotion with unilateral stimulation in the MLR (Sirota et al.,

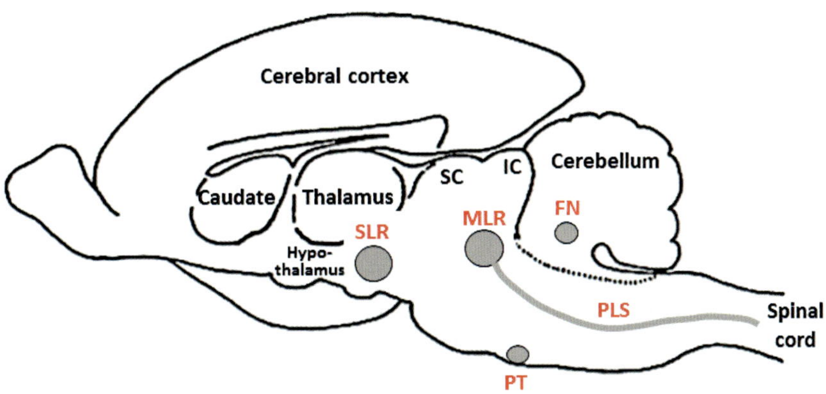

FIGURE 8.4 Regions in the brain of the cat initiating locomotion. Parasagittal presentation of the effective regions (gray). FN, fastigial nucleus; IC, inferior colliculus; MLR, mesencephalic locomotor region; PLS, pontomedullary locomotor strip; PT, pyramidal tract; SC, superior colliculus; SLR, subthalamic locomotor region.

2000). Taken together, these studies demonstrate that very simple signals control speed and gait transition, and this interesting property will be replicated in a coupled oscillator model presented below.

Two mechanisms by which the MLR controls the speed of locomotion have been found in lampreys. First, a sequential recruitment of two distinct groups of RS neurons as the MLR activation increases has been evidenced (Brocard and Dubuc, 2003). Second, MLR projects to some brainstem neurons that are included in a local feedforward loop which boosts the excitation of RS cells (Smetana et al., 2010). Interestingly, in the in vitro neonatal mouse lumbar spinal cord, fictive locomotor frequency is differentially modulated by the activation of diverse ionotropic glutamate receptors (i.e., NMDA and non-NMDA receptors) within the spinal cord (Talpalar and Kiehn, 2010). Non-NMDA receptors intrinsic to the locomotor CPG are indispensable for generating the highest locomotor speeds, with little or no contribution from NMDA receptors. Similarly, the fictive swimming frequency in the lamprey depends on the type of ionotropic glutamate receptors activated: NMDA and non-NMDA (e.g., AMPA) receptors control nonoverlapping frequency ranges (Brodin et al., 1985; Alford and Grillner, 1990). NMDA receptors mediate long-lasting plateau potentials (their excitatory postsynaptic potentials decay on timescales on the order of 100 ms). It is conceivable that a higher proportion is recruited in low-frequency regimes whereas more fast-acting AMPA synapses are needed for rapid control. This interpretation is also in line with the results obtained in mice (see above). By contrast, the mechanism by which the MLR controls gait transition is still unknown. Below we will present models that offer some hypotheses concerning gait transition in the salamander.

Observation 6. MPGs coupled to the musculoskeletal system can trigger gait transitions through sensory entrainment. Vertebrates commonly switch gait as they change their speed of locomotion (Hildebrand, 1976; Biewener, 2003). For example, salamanders can increase their locomotor speed on ground by switching from walking to trotting, and in water by switching from trotting to swimming (Delvolvé et al., 1997; Ashley-Ross et al., 2009; Deban and Schilling, 2009).

Spontaneous switches between fictive gaits have been observed in thalamic and high-spinal-paralyzed cats, thus revealing that central (especially spinal) circuits govern gait changes (Koehler et al., 1984; Orsal et al., 1990). However, several observations in invertebrates (see, e.g., Clarac and Barnes, 1985; Ritzmann and Büschges, 2007; Borgmann et al., 2009) and vertebrates (horses: Farley and Taylor, 1991; cats: see, e.g., Halbertsma et al., 1976; Perret, 1976; Forssberg et al., 1980; Bouyer and Rossignol, 2003; Pearson, 2004) suggest that the movement-related proprioceptive feedback from the limbs plays a crucial role in interlimb coordination during locomotion and, thus, in gait transitions. To date, there are no neurobiological data concerning the underlying mechanisms. Recently, we have investigated those mechanisms using a three-dimensional neuromusculomechanical model of salamander (Harischandra et al., 2011).

Observation 7. Steering is obtained by differentially activating left–right descending pathways. In swimming lampreys, the pattern of muscle activity during lateral turning is characterized by an increase in EMG burst amplitude and duration on the side toward which the animal is turning, as well as by an increase in cycle duration (Fagerstedt and Ullen, 2001). Similar changes in the locomotor parameters can be seen in the in vitro preparation of the isolated brainstem and spinal cord, in which fictive lateral turns can be evoked by stimulation of the skin of the head (Fagerstedt et al., 2001; McClellan and Hagevik, 1997). The basic components of the turn motor pattern are thus centrally generated.

The RS system plays an important role in the control of lateral turning in lampreys. Indeed, lateral turns are associated with an asymmetry in the bilateral activity of the RS nuclei: the activity on the ipsilateral (to turn) side being higher than that on the opposite side (Deliagina et al., 2000; Karayannidou et al., 2007). Directly related, in semi-intact preparations from larval lampreys, unilateral pharmacological microstimulation in certain parts of reticular nuclei was reported to elicit asymmetrical undulatory movements, while a bilateral stimulation in these areas can induce a sustained symmetrical swimming (Jackson et al., 2007). Similarly, a differential activation of higher-order locomotor areas (e.g., the MLR: Sirota et al., 2000; the ventromedial diencephalon, Paggett et al., 2004) making monosynaptic connections with RS neurons induces turning (Jackson et al., 2007; Kobayashi et al., 2009).

As will be reviewed below, some models of the lamprey CPG have replicated this behavior. Similarly, in salamanders the lateral turns during swimming are primarily produced by bending the body, and it is plausible that the neural mechanism underlying this kind of steering is similar to that evidenced in lampreys.

In limbed animals, turning movements during stepping further require appropriate adjustments between the movements of the different limbs in order to preserve an alternating gait when the limbs have to cover different amounts of ground. In cats, it has been shown that these adjustments include prolongation of stance and shortening of swing for limbs on one side while reciprocal changes occur in limbs of the other side (Forssberg et al., 1980; Halbertsma, 1983; Cruse and Warnecke, 1992). However, other adaptations of limb movements can occur. For example, salamanders actively use their forelimbs to sidestep, that is, to pull with the inner front limb and push with their outer front limb in a curve. By comparing the path followed when steering with sidestepping and pure body bending, using a musculomechanical salamander model (Harischandra et al., 2010), it is clear that sidestepping greatly improves maneuverability. It is still an open question how the nervous system generates appropriate sidesteps in salamanders. In decerebrate cats the contribution of asymmetrical signals from the labyrinths and the neck mechanoreceptors during steering have been suggested (Armstrong, 1986), and it is possible that a similar mechanism occurs during steering on the ground in salamanders.

8.1.3 Are CPGs Used in Human Locomotion? There is still a debate whether and how CPGs are used in human locomotion (Hultborn and Nielsen, 2007; MacKay-Lyons, 2002). However, the evidence increasingly leans toward CPG-based locomotion as in other vertebrates. For instance, different studies on patients with complete spinal lesion provided evidence suggesting that the spinal cord with intact sensory afferents can generate rhythmic locomotor-like output (Dimitrijevic et al., 1998), including recent studies using electrical stimulation of the spinal cord (Miniassan et al., 2007; Harkema et al., 2011). It was also shown that treadmill exercises for patients with spinal cord injuries improved their walking pattern (Barbeau and Rossignol, 1994; Dietz and Harkema, 2004; Edgerton et al., 2004; Rossignol et al., 2007; Wolpaw and Tennissen, 2001), which may be accounted for by the fact that CPGs can be trained to function independently of descending signals (Stein, 2008). Interestingly, Dietz et al. (2002) showed that in a setting with 100% body unloading (thus limiting the role of stretch reflexes), patterned leg movements could be elicited in patients with complete para-/tetraplegia. Moreover, studies of disabled patients have shown that in the absence of sensory information, gross movement control is preserved, even if peripheral information is necessary for precise movement organization and control (Jeannerod, 1988; Gandevia and Burke, 1992).

8.1.4 Rhythmic versus Discrete Movements So far, we have mainly reviewed the generation of rhythmic patterns while the generation of rich motor skills clearly needs other types of motor patterns, in particular patterns for discrete, that is, point-to-point, movements.

Navigation through challenging environments requires the incorporation of discrete limb movements into the gait cycle in order to reach specific points on ground. Several experiments in intact cats have shown that neurons in the motor cortex, including pyramidal tract neurons, play a crucial role in the execution of visually triggered gait modifications during ongoing locomotion (reviewed in Drew et al., 2008).

From these studies, a conceptual model based on groups of multiarticular synergistic muscles has been proposed to explain how the descending signals from the motor cortex may be integrated into the spinal locomotor circuits (Drew et al., 2008). Interestingly, sets of muscle synergies emerging from the interacting constraints and features of the nervous and musculoskeletal systems have been proposed for the production of a rich repertoire of postural responses to perturbations in humans (Ting and McKay, 2007), as well as during human locomotion (Ivanenko et al., 2004, 2005; Clark et al., 2010).

The concept of "muscle synergies" is quite similar to that of "spinal primitives" which has been developed by Bizzi and his collaborators (reviewed in Bizzi et al., 2008). By recording the forces produced by the frog's limb from different postures, they showed that stimulating a specific point in the spinal cord would activate a *force field* toward a specific rest posture. Stimulating another point would activate another force field toward another posture. Interestingly, stimulating the two points simultaneously would lead to a linear superposition of the two force fields, which is very useful from a control point of view as it implies that a whole range of movements can be generated by changing activation levels of a few force fields. Kargo and Giszter (2000) have demonstrated how a finite set of force fields could account for the natural wiping reflex in the frog, showing that the CNS could use these discrete MPGs as motor primitives to produce natural behaviors. Further, similar locomotor primitives have been evidenced in several animal species, suggesting that despite substantial phylogenetic distances and morphological differences, locomotion in several animal species is built starting from common primitives, perhaps related to a common ancestral locomotor network (Dominici et al., 2011). A deeper discussion of the mechanisms underlying the superimposition of discrete and rhythmic movements can be found in Degallier and Ijspeert (2010). A different approach to discrete movements, based on cortical networks rather than modulation of spinal CPGs, is provided in chapter 20, "Reach and Grasp: Control, Development, and Recognition."

8.2 *Modeling Pattern Generators*

Different levels of abstraction can be adopted in computational models of MPGs and CPGs. Depending on the scope of the model, it can often be appropriate to model entire populations of neurons (e.g., a spinal cord segment or hemisegment) as a single oscillatory element. Alternatively, each neuron in a population may be modeled individually. Independently of this choice, neuron models can be spiking (either explicitly or formally; see below) or nonspiking. At the lowest level of abstraction we find detailed biophysical multi-compartmental Hodgkin–Huxley (H–H) models (see chapter 2 for a brief introduction). Moving to higher levels of abstraction, we find 2-D models, e.g., the FitzHugh–Nagumo and Morris–Lecar models (FitzHugh, 1955, Nagumo et al., 1962, Morris and, Lecar 1981) as approximations of the H–H model, and formal spiking leaky integrate-and-fire neurons (Koch and Segev, 1999, Gerstner and Kistler, 2002), where formal spiking implies that the spike shape is not explicitly calculated (chapter 2). Rather a spike is said to have been fired when the membrane potential of the neuron reaches a given threshold. Foregoing the subthreshold membrane potential dynamics, firing rate models only consider the firing rate of individual neurons. Models of low complexity (i.e., a high level of abstraction) are often loosely referred to as connectionist models, to emphasize that any complex behavior of the modeled system is an emergent property due largely to the network structure (connections between populations which may be subject to learning) rather than the biophysical properties of neurons (their ionic channels, types of synapses, etc.). Note that although connectionist models usually employ simplified neurons, the distinction can be made between connectionist models, which do not necessarily conform to structural or performance criteria based on neurobiological constraints (e.g., in the domain of machine learning), and more biologically oriented simplified models. Finally, though analog neurons exist in biology, nonspiking leaky-integrator neurons and abstract nonlinear oscillators are typically used to represent the activity of entire populations of neurons.

This section gives a brief overview of the use of various types of models and then discusses three examples in more detail. Each example shows how such models are applied to vertebrate locomotor CPGs but focuses on different aspects: first, an example of a biophysically detailed model which focuses of the rhythm generation in spinal segments; second, a simplified neural model based on leaky integrator neuron models, which focuses on patterns of intersegmental coordination; and third, an abstract model based on coupled nonlinear oscillators, which deals with the transition between different locomotor modes (i.e., different patterns of body coordination) and investigates the case for the use of robots to advance our understanding of the vertebrate locomotor system.

8.2.1 Models of CPGs at Different Levels of Abstraction

Biophysical Models The most detailed models of CPGs are based on the H-H type of neuron models. These models may feature multiple compartments for a single neuron, for example, for the cell body and the dendrites. The activity of ionic channels is explicitly modeled and determines the membrane potential and the generation of action potentials. Such models have been instrumental in investigating *rhythmogenesis*, that is, how rhythmic patterns are generated in segmental circuits or even in single neurons (Grillner et al., 1988; Ekeberg et al., 1991; Wallén et al., 1992). They also have allowed the exploration of the neuromodulation of CPGs, that is, how neurotransmitters can change the functionality of CPGs by acting, for instance, on ionic channels in individual neurons, as, for example, in the stomatogastric system of the lobster (Goldman et al., 2001; see also chapter 8). Thanks to increasing computational power, biophysical models have also been used to investigate large populations of neurons. Recently Kozlov et al. (2009) have made increasingly complex models of the whole lamprey swimming CPG and explored the generation and modulation of traveling waves in a complete network of 100 segments. This network spanned the entire spinal cord and consisted of 10,000 neurons with 86 compartments each. An example of an H–H model for rhythm generation is presented below.

Simplified Neural Models CPGs have also been modeled with simpler neuron models than H–H neurons, for instance, using leaky-integrator neurons or integrate-and-fire neurons (Buchanan, 1992; Ekeberg, 1993). These models do not investigate the cellular properties underlying rhythm generation but rather focus on how network properties and interneuron connections affect rhythm generation and the coordination between multiple oscillatory centers, for example, for intra-axis and interlimb coordination. Such a model will be presented in more detail below.

Coupled-Oscillator Models At the most abstract level, CPGs have been investigated using systems of coupled oscillators. The idea is to represent the dynamics of a whole unit pattern generator circuit by a single mathematical oscillator model and to study the dynamics of the whole CPG circuit using several of these oscillators coupled together (Cohen et al., 1982; Collins and Richmond, 1994; Ijspeert et al., 2007; Kopell et al., 1991; Matsuoka, 1987; Schoner et al., 1990). These models allow one to study how the synchronization and phase lags within a population of oscillatory centers depend on interoscillator couplings as well as properties of the oscillators (e.g., their intrinsic frequency and the shape of the waveform they produce).

Although it might appear that such models are too remotely related to the underlying neural circuits, they are justified by a fact that is well-known in dynamical systems theory, namely, that the dynamics of populations of oscillatory centers depend mainly on the type and topology of couplings rather than on the local mechanisms of rhythm generation (Golubitsky and Stewart, 2002; Kuramoto, 2003). Furthermore, they can be based on existing neuroanatomical observations about the organization of CPGs, for example, by modeling a lamprey CPG as a double chain of oscillators with ranges of couplings that have been properly identified. In some cases, closed-form solutions or specific regimes (e.g., phase-locked regimes) can be analytically derived (Cohen et al., 1982; Kopell, 1995), but most systems are solved using numerical integration. A model of the salamander CPG based on coupled oscillators will be presented below.

Chapter 4, "Neural Rhythms," applies both H–H and coupled oscillator models to the very different types of neural rhythms seen in cerebral cortex of mammals, such as the alpha and gamma rhythms.

8.2.2 A Biophysically Detailed Model of Rhythm Generation A relatively simple yet instructive example of a biophysically detailed model can be found in the works of Ekeberg and Wallén (Ekeberg et al., 1991; Wallén et al., 1992). A five-compartment H–H neuron model was used in building a simplified segmental network of the spinal CPG in the lamprey. The neurons interact through two types of excitatory glutamatergic synapses (targeting AMPA and NMDA receptors, respectively) and inhibitory (glycinergic) synapses. Furthermore, the neurons can be activated by generic AMPA or NMDA conductances. These represent the combined effect of all nonspecific AMPA and NMDA receptors on a given neuron, which are not necessarily part of the circuit under study, but which may nevertheless be activated in experiments, when the spinal tissue is bathed in the glutamatergic agonists AMPA or NMDA. The generic AMPA and NMDA "bath conductances" thus mimic the effect of the addition of receptor-specific pharmacological agents to the solution that bathes the spinal tissue in physiological experiments. This mode of activation of the network corresponds to the network-wide activation of a specific type of receptor. The bath conductances are modeled as generic transmembrane conductances, multiplied by a dimensionless quantity representing the steady-state fraction of docked agonist.

The Neuron The model neuron is depicted in figure 8.5. It features five ionic channels and two generic conductances representing the NMDA or AMPA bath. The five membrane channels are the spike-generating sodium (Na) and potassium (K) channels, a calcium channel (Ca), and two calcium-dependent potassium channels (K_{Ca}, $K_{Ca\text{-}nmda}$). The calcium channel contributes to the depolarization of the neuronal membrane (together with the sodium channel). The potassium channels repolarize the membrane with a slight temporal delay, thus concluding the generation of the action potential. At the same time the calcium channel

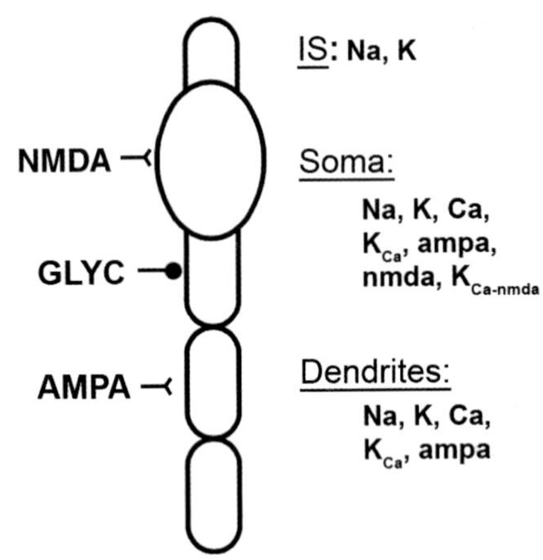

Figure 8.5 A schematic of the multicompartment Hodgkin–Huxley model used in Wallén et al. (1992). Synaptic inputs are shown on the left and ionic channels on the right: The initial segment (IS) of the axon is endowed exclusively with a high density of action potential–generating sodium (Na) and potassium (K) channels. In the soma, sodium and potassium channels are accompanied by NMDA and AMPA bath conductances, a membrane calcium channel (Ca) and two calcium-dependent potassium channels (K_{Ca}, $K_{Ca\text{-}nmda}$). The calcium channel is the source for the first intracellular calcium pool (driving the first calcium-dependent potassium channel). The NMDA-bath conductance and the NMDA synapse are the sources of the second calcium pool (driving the NMDA-calcium-dependent potassium channel; see text for further details). Dendrites lack NMDA-bath inputs and are not targeted by NMDA-type synapses. The dendrites are the targets of inhibitory glycinergic (GLYC) and excitatory (AMPA) synapses. NDMA synapses furthermore exhibit a voltage-dependent magnesium block. All excitatory connections are comprised of both AMPA and NMDA synapses. Adapted from Wallén et al. (1992).

increases the intracellular calcium contentration (see below), which is the driver of the first K_{Ca} channel. This channel is hyperpolarizing, and the effect of an increase in calcium concentration lags behind the action potential. The conductance of the second K_{Ca} channel ($K_{Ca\text{-}nmda}$) is modulated by a separate intracellular calcium pool (with slower inflow and decay parameters). Its concentration is increased with each synaptic event at an NMDA synapse and in the presence of the NMDA bath. The calcium concentration of each pool decays over time. The membrane potential and the membrane currents are determined by the following equation (for the full set of H–H equations and all parameters, the reader is referred to Wallén et al., 1992, and the accompanying MatLab model):

$$C\frac{dU}{dt} = \sum_i (U_i - U)g_{core} + \sum_j I_j + I_{leak} \qquad (8.1)$$

Here U is membrane potential of a given compartment. The first sum is taken over the adjacent compartments. The index j indicates summation over non-leak currents for the compartment. I_{leak}, C, and g_{core} represent the leak current, the capacitance, and the electrical coupling to the adjacent compartments, respectively. Standard H–H equations determine the compartmental currents,

$$I_j = g_j p^a q^b (U_i - E_{rev}), \qquad (8.2)$$

where g_j denotes the conductance of the ionic channel, and p^a and q^b the activation and inactivation variables with their respective exponents. E_{rev} is the reversal potential of the charge carriers. The reader is referred to Wallén et al. (1992) for the equations governing the activation and inactivation variables (p and q). Finally the calcium concentration in the cell is given by the equation

$$\frac{d[Ca]}{dt} = AI - B[Ca] \qquad (8.3)$$

where I is the calcium current, A the inflow rate, and B the decay rate.

The properties of this limited number of ionic channels already allow for various behaviors. Depending on the timescales of the calcium inflow and decay rates, the hyperpolarizing K_{Ca} channels can have diverse effects. For instance, before the calcium concentration stabilizes at a steady-state value for a given spiking frequency it slowly accumulates. Since the effect of the K_{Ca} channel is a hyperpolarization of the membrane, adding calcium delays the subsequent spike. Hence the instantaneous spike frequency (calculated from the current interspike interval) will decrease until the inflow and decay of calcium have reached a steady state. This constitutes a mechanistic explanation (a key feature of H–H models) for spike-frequency adaptation and a crucial mechanism for the rhythm generation in the simple segmental CPG network presented in this section. One way to visualize adaptation is to stimulate a neuron by injecting current for a short interval (long in relation to the adaptation time constant) and then to calculate the instantaneous frequency from the first, second, and last interspike interval. Plotting these frequencies in the same axes yields the the lower panel in figure 8.6, indicating the change in instantaneous frequency in the early phases of a spike train compared to the asymptotic value.

By changing the parametes A and B in equation 8.3 for the $K_{Ca\text{-}nmda}$ channel that is driven by the calcium entering through the NMDA channel, the neuron can be made to spike for a short period before the calcium concentration reaches a value that is sufficiently high to terminate spiking temporarily. Crucially, during this silent recovery period the neuron is still being stimulated while the calcium concentration decays (cf. equation 8.3). The result is a bursting neuron (figure 8.7). In the present model the synaptic and bath sources of calcium operate in this parameter regime.

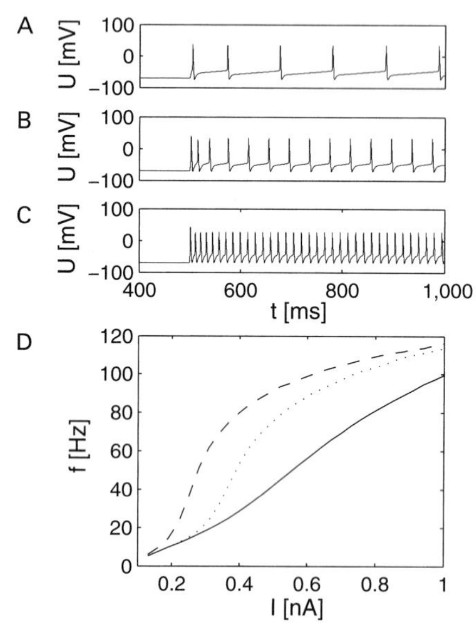

FIGURE 8.6 Spike-frequency adaptation. (A–C) Three spike trains due to injected current generated by the H–H model, illustrating spike frequency adaptation. Note the change in the interspike interval between subsequent spikes. At stronger injected currents, that is, at higher frequencies, the change in interspike interval becomes less pronounced. (D) Injected current versus the instantaneous frequency calculated from the first (dashed line), second (dotted line), and steady-state interspike interval. U refers to the voltage, t denotes time.

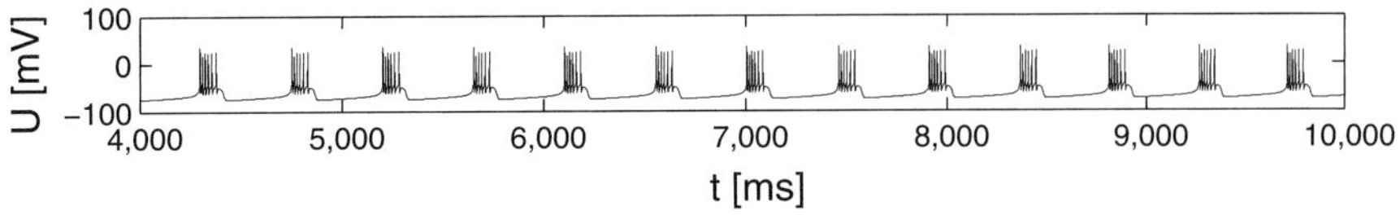

FIGURE 8.7 NMDA-bath-induced bursting in an isolated neuron. NMDA-bath stimulation is simulated by a tonic activation of the NMDA-bath conductance, representing the steady-state fraction of docked pharmacological agonists at NMDA receptors. Note that these receptors may belong to synapses other than the ones uses for communication in the network under study (i.e., they are not part of the model). Real neurons will be targeted by a multitude of synapses, and the bath conductances represent the mean effect the continuous docking and release of NMDA or AMPA, which must occur in experiments when a pharmacological bath is applied. Calcium entering through the NMDA-bath conductance accumulates during the active phase of the burst and modulates the calcium-dependent potassium conductance. The latter current terminates the burst. During the silent phase the intracellular calcium concentration decays. The stimulation (the NMDA bath) is constant throughout the cycle. *U* refers to the voltage, *t* denotes time.

Note that such a neuron will spike regularly when current is injected, or when it is activated by the generic AMPA bath conductance. It will only burst when activated by the NMDA bath or when mutually coupled to another neuron via an NMDA synapse, because these conductances are the sources of calcium, which is the driver of the burst-terminating slow K_{Ca} current (here we have implicitly assumed, for simplicity, that the calcium pool responsible for adaptation and the pool responsible for the NMDA bursting do not interfere with each other).

The Network The network built out of these neurons consists of two identical half-centers (Brown, 1914). The basic idea of the half-center oscillator is that the two subnetworks, which make up the oscillator, alternate their active and silent phases during an oscillation by mutually suppressing each other. Without this mutual suppression there need not be an oscillation. In the case of the lamprey network modeled here, identical subnetworks (one for each hemisphere of the spinal cord segment) are coupled via inhibitory synapses. Removing the inhibitory coupling, leading to isolated half-centers, will abolish the oscillation in the present model if the neurons are driven by an AMPA bath or by current injection. However, above we have seen that isolated neurons can exhibit pacemaker properties (e.g., they are intrinsically oscillating) when driven by the NMDA bath. Hence, the nature of the neural oscillator (a classical half-center vs. intrinsically generated oscillations) may depend on the biophysical properties of the neurons and how they react to different types of input. It should also be noted that, depending on the neuronal and synaptic parameters, isolated sets of neurons can exhibit oscillations when recurrent excitatory connections are included. In fact, isolated spinal cords of lampreys, when transected along the midline and activated by an AMPA bath, continue to exhibit oscillations (Cangiano and Grillner, 2003, 2005).

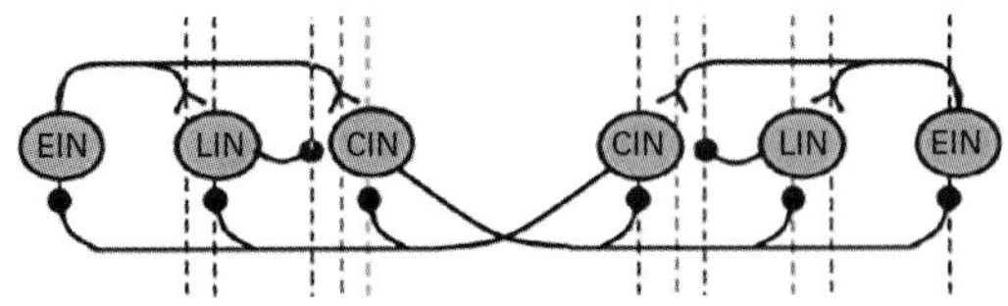

FIGURE 8.8 The segmental network module of the lamprey spinal central pattern generators. We show the interaction between two half-centers. Excitatory interneurons (EINs) drive lateral interneurons (LINs) and crossed inhibitory interneurons (CINs), which inhibit the contralateral side of the network. Note that motoneurons have been omitted in this model. Motoneurons are assumed to be driven by the output of the EINs.

In the present model each half-center contains one excitatory interneuron (EIN), one crossed inhibitory interneuron (CIN), and one lateral inhibitory interneuron (LIN). The slowly activating LIN inhibits the local half-center activity and thus helps to terminate the ipsilateral activity after some time. The CIN inhibits the contralateral half-center. (See figure 8.8.)

Network Activity and the Role of Ionic Currents Simulating the two different baths (AMPA and NMDA), that is, driving the present network with the two different types of generic conductances, yields the alternating segmental rhythms in the frequency range of in vivo lamprey swimming (figure 8.9). The mutual inhibition between the half-centers ensures the alternation between left and right active phases. To understand how the active burst on one side can be terminated we need to consider two mechanisms. First is the effect of spike-frequency adaptation (see figure 8.6). With an increasing duration of the activity in one half-center the interval between successive spikes will increase, thus leaving the contralateral side disinhibited for progressively longer times, eventually allowing the contralateral side to fire a first spike and start to inhibit the previously active side of the network. The second

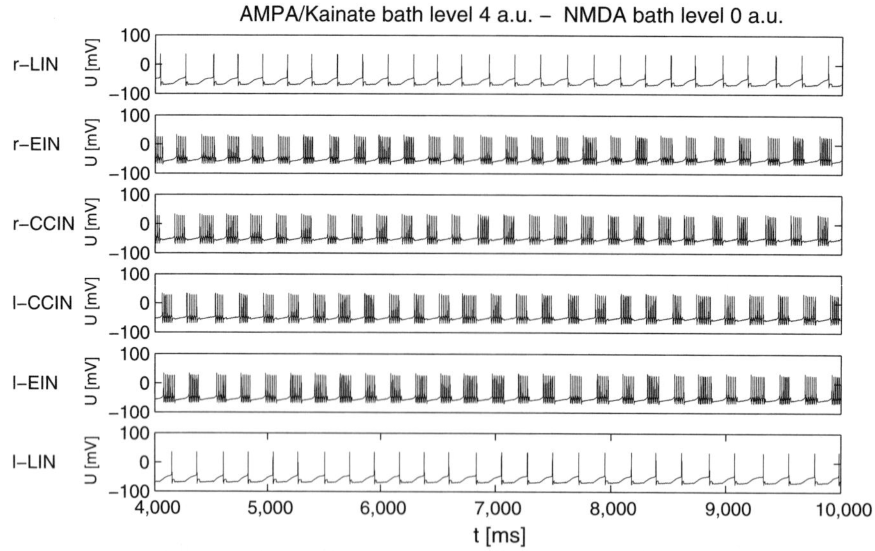

FIGURE 8.9 AMPA-bath-activated segmental network. The main burst-terminating mechanism is the accumulated effect of adaptation (accumulated after-hyperpolarization). At stronger bath levels lateral interneurons (LIN) spike and contribute to the termination of ipsilateral activity in crossed inhibitory interneurons (CCIN), to allow the contralateral side to become active (prefix *l* indicates the left-hand side of the network, prefix *r* indicates the right-hand side). EIN, excitatory interneurons.

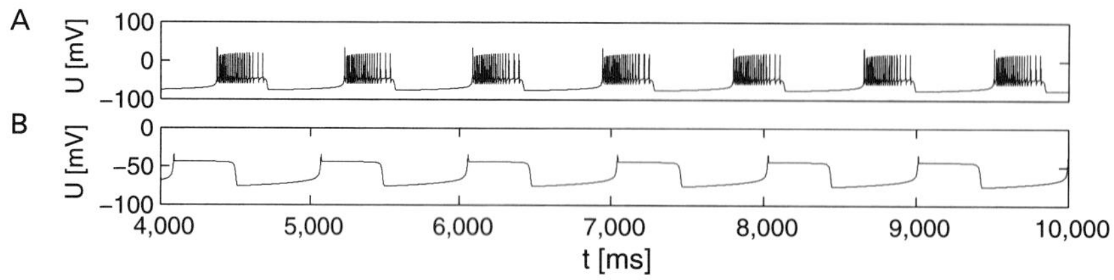

FIGURE 8.10 Tetrodotoxin-resistant membrane oscillations. (A) An NMDA-bath-induced bursting in an isolated neuron. (B) Blocking the action potential–generating sodium channel with the simulated addition of tetrodotoxin yields subthreshold membrane potential oscillations similar to physiological experiments.

mechansim, aiding the effect of spike-frequency adaptation, is the slowly activating LIN neuron, which helps terminate the burst of activity on its own side (compare figures 8.8 and 8.9).

However, such rhythms could also be generated in networks comprised of simpler neurons. The use of an H–H model is warranted when the added detail is useful to the study of the system at hand and if we are interested in mechanistic explanations. The present model allows us to investigate, for instance, what happens when the K_{Ca} conductance responsible for adaptation is scaled down. The channel that mediates adaptation is crucial for the rhythm generation in this model of the lamprey segmental network. In the animal this scaling down may occur through the release of neuromodulators like serotonin (Matsushima and Grillner, 1992).

H–H models lend themselves especially well to the reproduction of experimental findings where specific membrane channels have been (partially) blocked. Applying the drug tetrodotoxin (TTX) blocks the sodium channels that generate the action potentials. If a TTX affected neuron is stimulated with NMDA, long periodic membrane potential oscillations without spiking emerge (figure 8.10). The present model lends itself very well to the reproduction of such findings. The absence of the action potential–generating currents reveals the effect of the previously mentioned Ca-nmda/$K_{Ca\text{-}nmda}$ subsystem at work in the cell. In the presence of NMDA the membrane potential is raised to a plateau and calcium accumulates until the calcium-modulated potassium channel is strong enough to terminate the plateau. The calcium has to be removed through intracellular processes before another plateau can commence. This mechanistic explanation accounts well for observation under TTX. However, the close correspondence between the constituents of the system under study and the model elements (Ca-nmda/$K_{Ca\text{-}nmda}$) at this low level of abstraction does not warrant a categorical identification of the two. The calcium dynamics are the best model of TTX-resistant membrane potential oscillations, nothing more and nothing less.

8.2.3 A Simplified Neural Model of Rhythm Generation and Intersegmental Coordination As an example of a simplified neural model we present the neuromechanical model of lamprey swimming developed by Ekeberg (1993). This early work is notable for combining the CPG model with a muscle model and a mechanical simulation of the lamprey body in water, showing that the muscle activation signals generated by the CPG could indeed lead to forward swimming.

Because of the large wealth of studies concerning its neurophysiology and neuroanatomy, the lamprey is probably the vertebrate animal whose locomotor circuit has been modeled the most. The lamprey uses an anguilliform (eel-like) swimming mode in which a traveling wave of lateral body undulation is propagated from head to tail. The wavelength of the traveling wave corresponds roughly to one body length and is maintained approximately constant at different speeds of swimming (to which correspond different bursting frequencies).

The lamprey swimming CPG is distributed in the approximately 100 segments that constitute the spinal cord. It is organized as a double chain of oscillatory pools, with two oscillatory pools per segment that are coupled through contralateral projections of inhibitory interneurons (figure 8.11). As mentioned in the previous section ("A Biophysically Detailed Model of Rhythm Generation"), several types of interneurons contributing to the rhythm generation have been identified (Buchanan, 1982), namely, CIN (crossed inhibitory neurons), LIN (lateral inhibitory neurons) and EIN (excitatory interneurons). These neurons project to the motoneurons whose axons project to the muscle fibers through the ventral roots. In addition to different types of sensory input coming in through the dorsal roots, the lamprey possesses stretch-sensitive cells, called edge cells, within the spinal cord itself. The

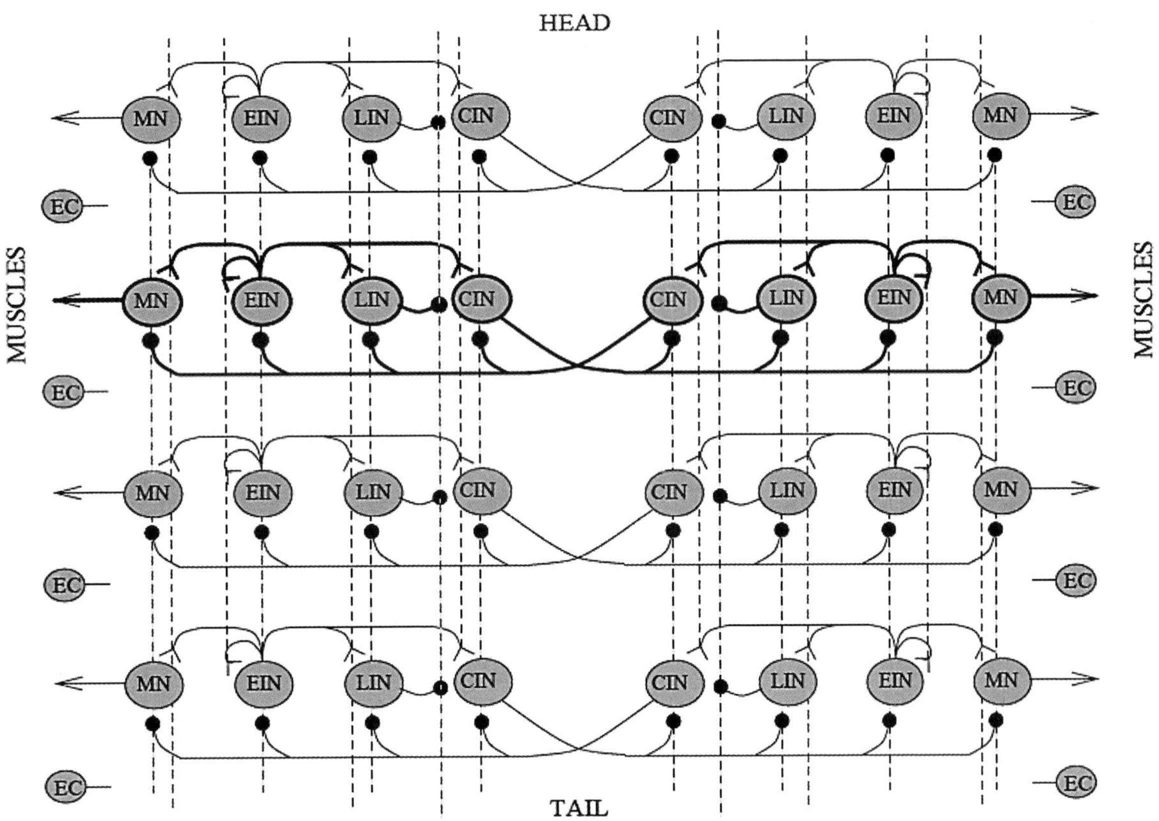

Figure 8.11 Schematic view of the lamprey network as modeled by Ekeberg (1993). Excitatory interneurons (EINs) drive lateral interneurons (LINs), crossed inhibitory interneurons (CINs), which inhibit the contralateral side of the network, and motoneurons (MNs), which provide activation for the axial muscles of the lamprey. Edge-cells (ECs) provide sensory feedback, with excitatory projections to ipsilateral interneurons and inhibitory projections to contralateral interneurons (projections not shown). Inhibitory synapses are indicated with a black disk, and excitatory synapses with an open triangle. Interneurons also project to neurons in neighboring segments (dashed lines), effectively implementing intersegmental coupling between segmental oscillators (see table 8.2).

segmental interneuron network is the same as in the previous biophysically detailed model (figure 8.8), but here motoneurons and sensory feedback through edge cells have been added.

Similarly to the animal, the model of Ekeberg (1993) employs a CPG made of coupled segmental networks (neural oscillators), with each segmental network consisting of 10 neuron models (figure 8.11, table 8.1). In the model, a neuron is a leaky-integrator neuron that represents the average firing rate of hundreds of real neurons (e.g., a CIN neuron model represents the average firing rate of all CIN neurons on one side of the segmental network). The complete controller is formed by connecting 100 copies of this segmental network. An intersegmental connection consists of extending the connection from one neuron to another in one segment to the corresponding neuron in neighboring segments (table 8.2). As the projections between segments in the lamprey were not well-known, Ekeberg chose a simplified coupling in which each segmental connection extends symmetrically in the caudal and rostral direction, except the connections from CIN neurons which have longer projections in the caudal direction. In order to limit the input of each neuron and to compensate neurons located in segments close to the extremities of the spinal cord, synaptic weights are rescaled and the weights of the connections to a neuron are divided by the number of segments it receives input from.

The output u of the neuron model corresponds to the mean firing frequency of the population it represents and is calculated as follows (leaky integrator with a saturating transfer function):

$$\dot{\xi}_+ = \frac{1}{\tau_D}\left(\sum_{i\in\Psi_+} u_i w_i - \xi_+\right), \tag{8.4}$$

$$\dot{\xi}_- = \frac{1}{\tau_D}\left(\sum_{i\in\Psi_-} u_i w_i - \xi_-\right), \tag{8.5}$$

$$\dot{\vartheta} = \frac{1}{\tau_A}(u - \vartheta), \tag{8.6}$$

$$u\begin{cases} 1-\exp\{(\Theta-\xi_+)\Gamma\}-\xi_- - \mu\vartheta & (u>0) \\ 0 & (u\le 0) \end{cases}, \tag{8.7}$$

where w_i are the synaptic weights, Ψ_- and Ψ_+ represent the groups of presynaptic excitatory and inhibitory neurons, respectively, ξ_- and ξ_+ are the delayed "reactions" to excitatory and inhibitory input, and ϑ represents neuronal frequency adaptation. Θ represents a threshold for activation while Γ is a gain constant. The parameters of each type of neuron are given in tables 8.1 and 8.2. Note that all neurons also receive tonic excitation from virtual brainstem neurons (these neurons are part of Ψ_+ and named BS in table 8.2, which provides the synaptic weights of the network).

The neuron parameters as well as the connection weights of tables 8.1 and 8.2 have been defined so

TABLE 8.1

Parameters of the neuron models used in Ekeberg (1993)

Neuron type	Θ	Γ	τ_D	μ	τ_A
EIN	–0.2	1.8	30 ms	0.3	400 ms
CIN	0.5	1.0	20 ms	0.3	200 ms
LIN	8.0	0.5	50 ms	0.0	—
MN	0.1	0.3	20 ms	0.0	—

Note. EIN, excitatory interneuron; CIN, crossed inhibitory interneuron; LIN, lateral inhibitory interneuron; MN, motoneuron.

TABLE 8.2

Parameters of the projections used in Ekeberg (1993)

To:	**From:**						
	EINl	**CINl**	**LINl**	**EINr**	**CINr**	**LINr**	**BS**
EINl	0.4 [2, 2]	—	—	—	–2.0 [1,10]	—	2.0
CINl	3.0 [2, 2]	—	–1.0 [5,5]	—	–2.0 [1,10]	—	7.0
LINl	13.0 [5, 5]	—	—	—	–1.0 [1,10]	—	5.0
MNl	1.0 [5, 5]	—	—	—	–2.0 [5, 5]	—	5.0
EINr	—	–2.0 [1, 10]	—	0.4 [2, 2]	—	–1.0 [5, 5]	2.0
CINr	—	–2.0 [1, 10]	—	3.0 [2, 2]	—	—	7.0
LINr	—	–1.0 [1, 10]	—	13.0 [5, 5]	—	—	5.0
MNr	—	–2.0 [5, 5]	—	1.0 [5, 5]	—	—	5.0

Note. The first number provides the synaptic weights. The numbers in brackets indicate to how many segments the neuron projects rostrally and caudally, respectively. EIN, excitatory interneuron; CIN, crossed inhibitory interneuron; LIN, lateral inhibitory interneuron; MN, motoneuron; l, left, r, right; BS, brainstem.

that the simulation of the model fits physiological observations (Ekeberg, 1993). The activity of a segmental network, or of the complete controller consisting of several segments, is calculated by integrating the system of coupled differential equations given by equations 8.4 to 8.7. When a segmental network with asymmetrical initial conditions receives adequate excitation from the brainstem (i.e., the activity of the brainstem is set to some value which leads to an excitation of the neurons through the brainstem–neurons connections), it oscillates regularly with the left and right neurons out of phase (figure 8.12). Burst termination is provided by the LIN neurons, which become active late in the cycle because of their high threshold and their high time constant. Once active, the LIN neurons inhibit the ipsilateral CIN neurons; this allows activity on the contralateral side to start. Increasing the external excitation has the effect of increasing the frequency of oscillation and the amplitude of the motoneurons. The frequency varies nearly linearly with the excitation and ranges from 1.7 to 5.6 Hz (figure 8.13).

When external excitation is applied to the complete network through the connections from the brainstem (with all segments receiving the same incoming connections from the brainstem; see table 8.2), all segments start to oscillate in synchrony, with the same frequency. Similarly to the single segment, the frequency of oscillation increases monotonically with the excitation. Applying extra excitation to the most rostral segments leads to small phase lags between segments and, therefore, to a wave of neural activity that travels from head to tail. The lag between segments relative to the period of oscillation is then constant over the whole spinal cord, except for the first ten and last ten segments at the extremities where it decreases toward zero. The value of the lag can be varied with the amount of extra excitation, and the higher the extra excitation, the larger the lag or the shorter the wavelength. This mechanism of phase lag generation can be understood intuitively. Providing extra excitation to the most rostral segments will lead to a slight increase of the bursting frequency in the whole network, as the faster rostral segments entrain their caudal neighbors (i.e., they induce a change of frequency due to intersegmental coupling), which will, in turn, entrain more caudal segments, until the whole chain is bursting at the same frequency. However each segment is entrained by its rostral neighbor with a delay, as it takes time for the early rostral burst to initiate and then terminate activity in the caudal segment. This delay introduces the observed phase lag. Interestingly, the frequency of oscillation and the wavelength of the undulation can be changed nearly independently.

The model can therefore reproduce the capacity of the real lamprey to cover a whole range of different frequencies of oscillations while keeping the wavelength of the undulation constant. The simulations show that frequencies between 1.6 and 5.5 Hz and

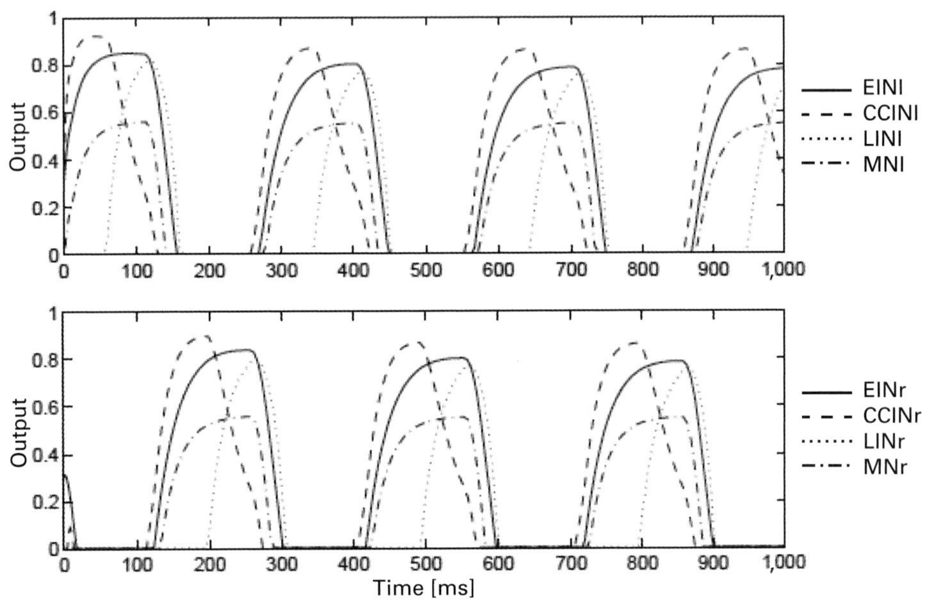

FIGURE 8.12 Oscillating activity of a lamprey segmental network in Ekeberg's (1993) model as re-implemented by the authors. The two panels show the activity of the left (top) and right (bottom) neurons in a segmental network (as re-implemented by the authors). The neuron labels are the same as in table 8.2. Note the anti-phase relationship between the two hemisegments.

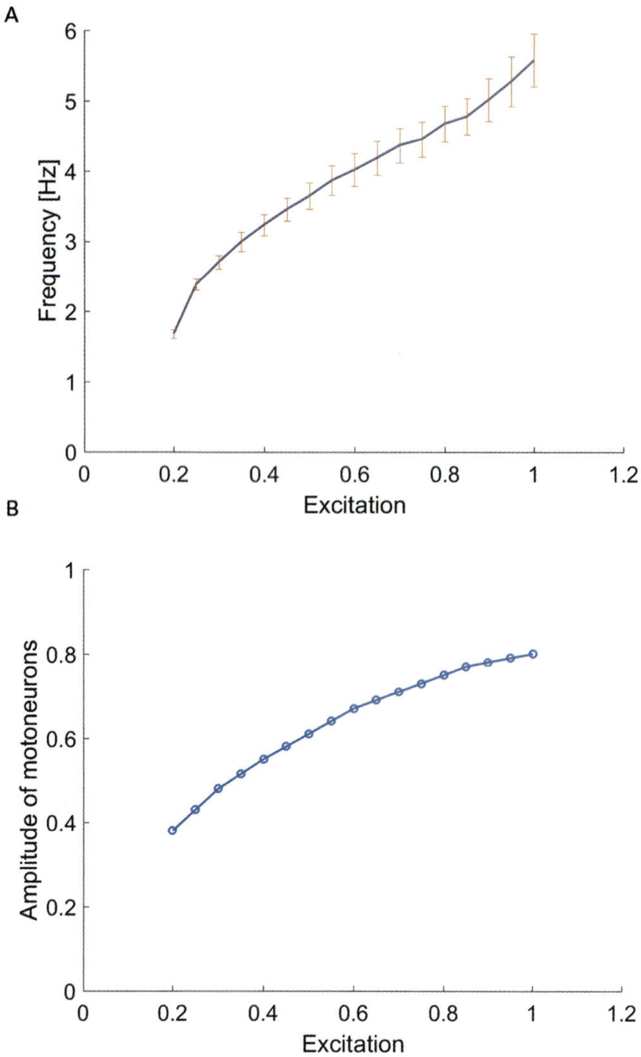

FIGURE 8.13 Change of frequency and motoneuron amplitude in a segmental network when excitation is increased.

phase lags per segment between 0.0% (no traveling wave) and 2.4% of the period can be obtained (as the wavelength corresponds to the inverse of the total phase lag between the first and the last segment, the wavelength is then smaller than 50% of the 100-segment body). Ekeberg also developed a simple 2-D mechanical model of the musculoskeletal system of the lamprey to illustrate how traveling waves of neural activity could lead to forward swimming (figure 8.14). The output of the CPG model was used to drive a simple muscle model, which calculated muscle forces generated at the joints of the simulated lamprey body. Hydrodynamic forces were calculated and applied to the lamprey body in response to the body motion. These forces led to the successful generation of forward swimming, a main result of the study. While in Ekeberg (1993) the effects of the edge cells on the CPG activity are negligible because of very low synaptic weights (e.g., little influence on frequency, phase lags, and amplitudes), in Ekeberg et al. (1995) it is shown that feedback can help the lamprey to cross a speed barrier (produced by water with local speed opposite to the direction of swimming) by coordinating the neural activity with the actual movements of the body and preventing a change of direction.

In summary, the connectionist model developed by Ekeberg demonstrated how coordinated rhythmic patterns, in particular traveling waves, could be generated by a circuit composed of interconnected segmental networks and lead to forward swimming in a simulated lamprey. By modulating the intensity of a few descending drives the wavelength and frequency (and hence the speed of swimming) of the traveling waves could be adjusted.

8.2.4 FROM LAMPREY TO SALAMANDER: OSCILLATOR MODELS AND ROBOTS The salamander, an amphibian capable of swimming and walking, offers an interesting link between research on the lamprey and research on tetrapods. Salamanders are capable of two basic locomotor modes, swimming and overground stepping. Similarly to the lamprey, forward swimming involves a traveling wave, exhibiting fast axial undulations that travel from the head to tail. The limbs are folded to the body. Out of the water, salamanders switch to a stepping gait (however note that underwater stepping has also been observed). In this so-called diagonal sequence walk the body makes an S-shaped standing wave with nodes at the girdles. The body movement is coordinated with the limb movement (diagonally opposed limbs move together) to optimize stride length (Frolich and Biewener, 1992; Ashley-Ross and Bechtel, 2004; Delvolvé et al., 1997).

Similarly to other organisms, a double chain of oscillatory pools of neurons is distributed along the spinal cord (for the lamprey, see Grillner et al., 1995; for the Xenopus embryo, see Combes et al., 2004). Activating an isolated preparation of this *axial CPG* with an excitatory (NMDA) bath generates traveling waves of fictive swimming (Delvolvé et al., 1999). The cervical spinal cord segments and the thoracolumbar segments contain the networks for limb movements—forming the *limb CPG* (Cheng et al., 1998). In intact (or semi-intact) salamanders locomotion (walking or swimming) can be induced by simple electrical stimulation of the MLR, located in the midbrain (Cabelguen et al., 2003). At low levels of stimulation the slow walking gait is induced, while at higher stimulation a rapid switch to the faster swimming mode occurs. The locomotor frequency is proportional to the strength of the stimu-

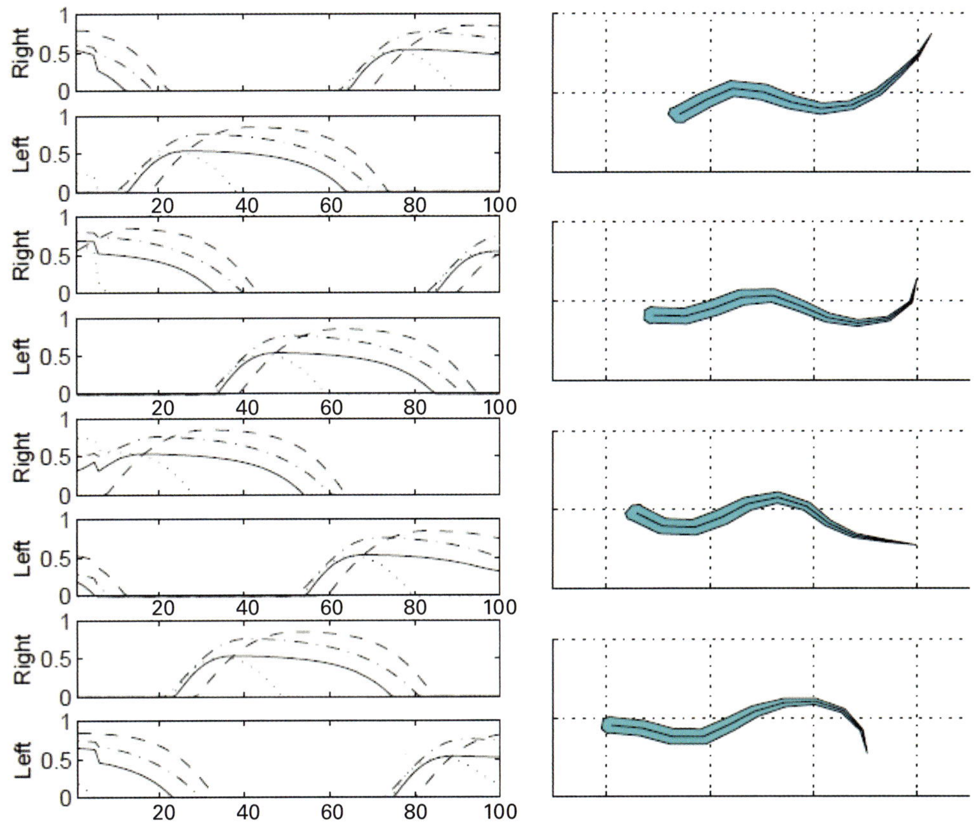

FIGURE 8.14 Neural activity and body snapshots. Left: Four snapshots of the neural activity of motoneurons, excitatory interneurons, crossed inhibitory interneurons, and lateral interneurons along the 100 segments of the lamprey model (same line type and legends as in figure 8.12). Right: corresponding movements in a 2-D mechanical model of the musculoskeletal system, as re-implemented by the authors.

lation in both modes, allowing the organism to modulate its speed.

Various models of the salamander locomotor infrastructure have been developed (Bem et al., 2003; Ermentrout and Kopell, 1994; Ijspeert, 2001; Ijspeert et al., 2005, 2007; Seo and Slotine, 2007). In our latest model (Ijspeert et al., 2007) we assume, similarly to previous models, that a lamprey-like swimming circuit can be extended by addition of a simplified limb CPG. It can generate swimming and walking patterns of intersegmental and limb coordination, and it provides a mechanistic explanation of the automatic transition from walking to swimming observed when the electrical stimulation of the MLR is increased.

The model builds on four hypotheses:

Hypothesis 1: The axial CPG is similar to the axial lamprey-CPG, capable of spontaneously generating traveling waves when tonically activated. The active limb CPG imposes the walking pattern on the entire CPG network (i.e., the axial + limb CPG), as previously proposed in Cohen (1988).

Hypothesis 2: The coupling strength of limb to axis connections among oscillators is stronger than the couplings among axial oscillators and from axial to limb oscillators. This permits the limb CPG to impose (slower) standing waves on the axis against the natural tendency of the axial CPG to produce traveling waves.

Hypothesis 3: Limb oscillators saturate at high frequencies and cease to oscillate at high levels of drive. This is the basis for automatically switching between walking and swimming when the drive is modulated (Cabelguen et al., 2003), with the assumption that the drive is adjusted through various sensory-feedback loops depending on whether the salamander is on land or in water. This hypothesis could furthermore explain why walking frequencies are systematically lower than swimming frequencies in freely moving salamanders.

Hypothesis 4: At the same drive, axial oscillators have higher intrinsic frequencies than the limb oscillators, offering an explanation for the rapid increase in frequency when the switch from walking to swimming occurs as well as explaining the gap between swimming and walking frequency ranges (Frolich and Biewener, 1992; Delvolvé et al., 1997).

Note that a salamander can exhibit swimming-like movements for brief periods of time (e.g., on slippery surfaces), and stepping-like movements underwater (underwater stepping), indicating that any influence of sensory information on the CPG (e.g., from cutaneous receptors when immersed in water) can be overridden by central commands. However this does not affect the frequency gap between swimming and stepping. Stepping at high (swimming) frequencies has so far not been observed in salamanders.

The whole CPG (axial CPG and limb CPG) is modeled as a system of coupled nonlinear oscillators (figure 8.15), similar to early lamprey models (Williams et al., 1990). Individual oscillators, whose outputs represent motoneuron activity in the organism, are modeled as phase oscillators with controlled amplitude:

$$\dot{\theta}_i = 2\pi v_i + \sum_j r_j w_{ij} \sin(\theta_j - \theta_i - \phi_{ij}), \tag{8.8}$$

$$\dot{r}_i = a_i\left(\frac{a_i}{4}(R_i - r_i) - \dot{r}_i\right), \tag{8.9}$$

$$x_i = r_i(1 + \cos(\theta_i)). \tag{8.10}$$

Here the state variables θ_i and r_i represent the phase and the amplitude of oscillator i, and v_i and R_i its intrinsic frequency and amplitude. a_i is a positive constant. The weights w_{ij} and phase biases ϕ_{ij} determine the couplings between oscillators. Finally, the positive oscillatory signal x_i represents the motoneuron activity produced by the center, which will produce periodic bursts (i.e., oscillate between low and high firing rates), similarly to the bursts produced by the biophysically detailed and simplified neural models presented above.

In the lamprey and the salamander, the strength of the brainstem stimulation determines the amplitude and frequency of bursts (Cabelguen et al., 2003; Grillner et al., 1998). Three domains can be distinguished in the motoneuron signals when increasing the drive: (1) a subthreshold domain without bursts, (2) the oscillatory domain in which frequency and amplitude (i.e., number of motoneurons spikes) of a burst increase with drive, and (3) a saturation domain where oscillations cease. In the present model this effect is reproduced by making the intrinsic frequency v_i, and the amplitude R_i, functions of a *drive* signal d_i. The functions are

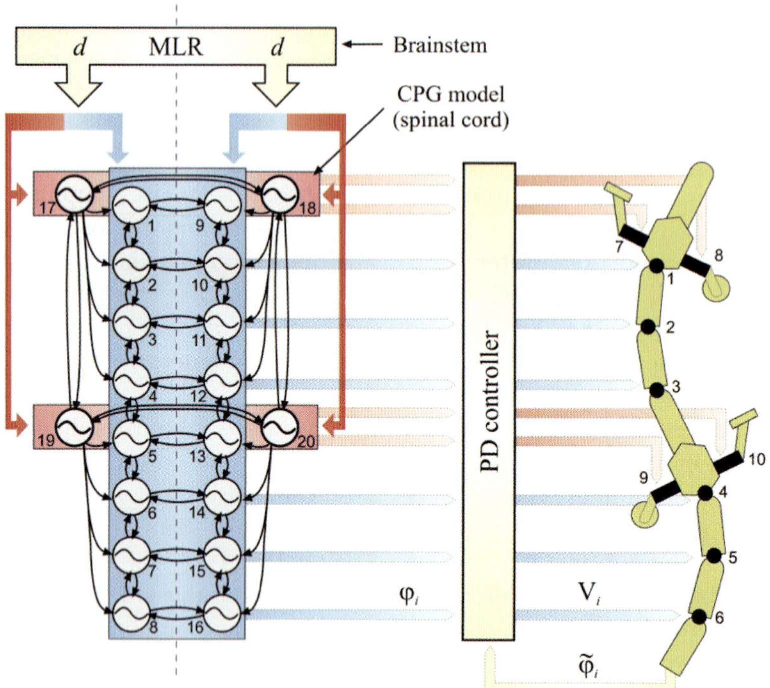

FIGURE 8.15 Salamander central pattern generator (CPG) model and robot schematic. Left: Circuit diagram of the CPG model used in Ijspeert et al. (2007). Four oscillators constitute the limb CPG (red) and are globally coupled to the axial CPG (blue). Reciprocal axial connections (between the left and right sides) and interlimb connections impose anti-phase oscillations between sender and receiver. Ipsilateral intra-axis connections maintain a uniform phase lag, and limb to axis connections enforce axial oscillations in phase with the limb oscillators. The oscillators receive an excitatory drive signal (d) from the mesencephalic locomotor region (MLR). Right: The CPG is used to drive a salamander-like robot through a proportional-derivative (PD) controller.

piece-wise linear *saturation functions* defined as follows (see also figure 8.15, left):

$$v = g_v(d) = \begin{cases} c_{v,1}d + c_{v,0} & \text{if} \quad d_{low} \le d \le d_{high} \\ v_{sat} & \text{otherwise} \end{cases}$$

$$R = g_R(d) = \begin{cases} c_{R,1}d + c_{R,0} & \text{if} \quad d_{low} \le d \le d_{high} \\ R_{sat} & \text{otherwise} \end{cases}.$$

The intrinsic frequency and amplitude are modulated by the saturation functions according to the drive signal between a lower oscillation threshold d_{low} and an upper one d_{high}. Limb and axial oscillators are subject to different saturation functions. For the same drive, limb oscillators oscillate at systematically lower frequencies than axial oscillators (hypothesis 4) and saturate at a lower threshold d_{high} (hypothesis 3; see figure 8.16. All oscillators receive the same drive $d_i = d$ except during turning, which is achieved with asymmetrical drive to the left and right parts of the CPG network.

Appropriate coupling parameters w_{ij} and ϕ_{ij} guarantee that the body CPG produces a traveling wave (hypothesis 1), while the limb CPG imposes a salamander stepping pattern when not saturated (i.e., at low drives). The limb CPG enforces the standing wave via unidirectional couplings to body oscillators whose

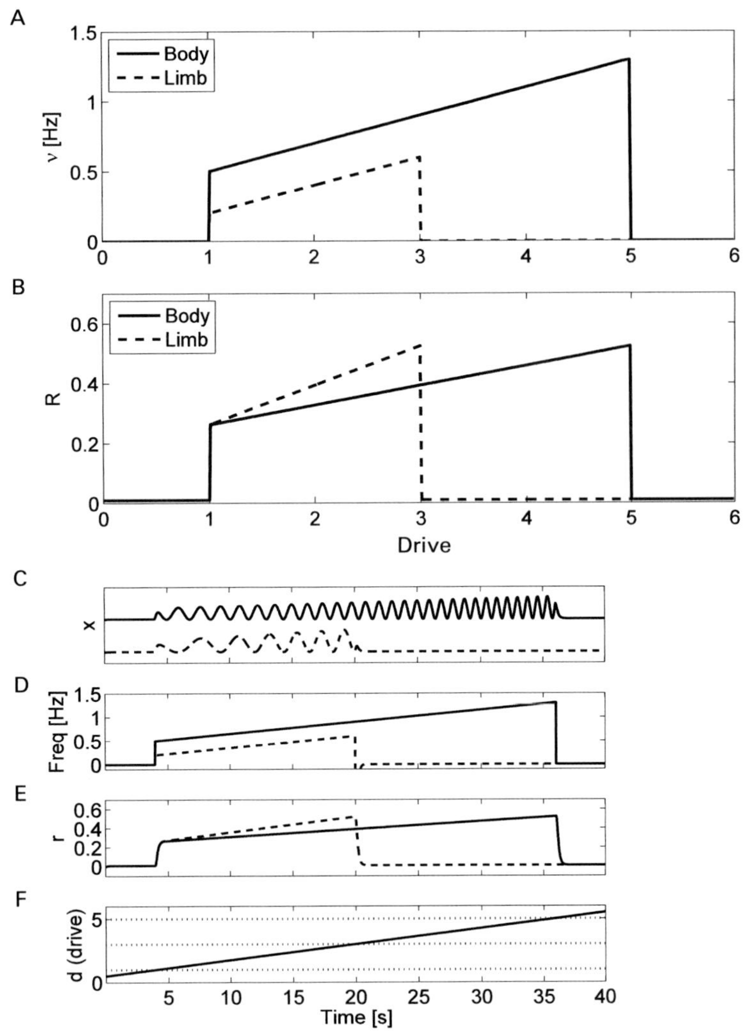

FIGURE 8.16 Saturation function and oscillations in isolated axial and limb oscillators. Saturation functions providing the intrinsic frequency ν (A) and amplitude R (B) for axial (body) and limb oscillators. As the drive increases above the lower thresholds (F), the axial and limb oscillators start to oscillate (C, solid and dashed lines, respectively) with increasing instantaneous frequencies, measured as $\dot{\theta}_i / 2\pi$ in cycles/s (Freq) (D) and amplitudes (E). When the drive passes the higher threshold for the limb oscillator, the oscillation stops and the instantaneous frequency and amplitude drop to zero. The same occurs in the axial oscillator after the drive increases beyond the axial higher threshold. Note the different oscillation frequencies (D) due to the absence of couplings between the oscillators.

coupling strengths are larger than those within the body CPG (hypothesis 2). The resulting stepping gait corresponds to a walking trot gait where front and hind limbs are in anti-phase, and where the trunk is bent at the right time such as to optimize stride length.

The swimming and walking patterns produced by the model (figure 8.17) were consistent with the CPG activity of real salamanders extracted from EMG recordings (Delvolvé et al., 1997). Stepping (i.e., at low drive), is characterized by a low frequency with an S-shaped standing wave enforced by the strong couplings from the limb to axial oscillators. The amplitude and frequency of oscillations increase proportionally with the drive. When the limb oscillators saturate, a rapid gait transition to the higher frequency swimming mode occurs through the release of a traveling wave for swimming, again reminiscent of the EMG recordings in the literature (Delvolvé et al., 1997). Amplitude and frequency again increase with the drive, until the body oscillators reach their upper oscillation threshold and stop oscillating. Finally, similarly to MLR stimulation experiments (Cabelguen et al., 2003), the model was able to produce an abrupt transition between gaits via a simple change in drive (figure 8.17).

These numerical results agree with detailed kinematic analyses of the gait transitions which found that traveling waves in the body axis are not observed simultaneously with limb movements (Ashley-Ross and Bechtel, 2004). Another important similarity with the MLR stimulation experiments and with recordings in tadpoles (Combes et al., 2004) is the step increase of frequencies during the transition from walking to swimming. In the model, the limb oscillators slow down the rhythms during walking, and once silent, rapidly release faster swimming rhythms due to the higher intrinsic frequencies of the body oscillators. This can also explain why salamander walking and swimming frequencies do not overlap (Frolich and Biewener, 1992; Delvolvé et al., 1997) but have distinct ranges with a gap between them (e.g., walking from 0.6 to 1.2 Hz, swimming from 1.6 to 2.9 Hz in the *Pleurodeles waltl*).

This spinal cord model was used to drive the motors of a salamander-like robot. This allowed us to investigate and observe aspects which require a "body" for validation and cannot be studied at the level of the nervous system. In particular, we demonstrated (1) that the CPG model could generate forward motion with variable speed and heading, (2) that the gaits generated

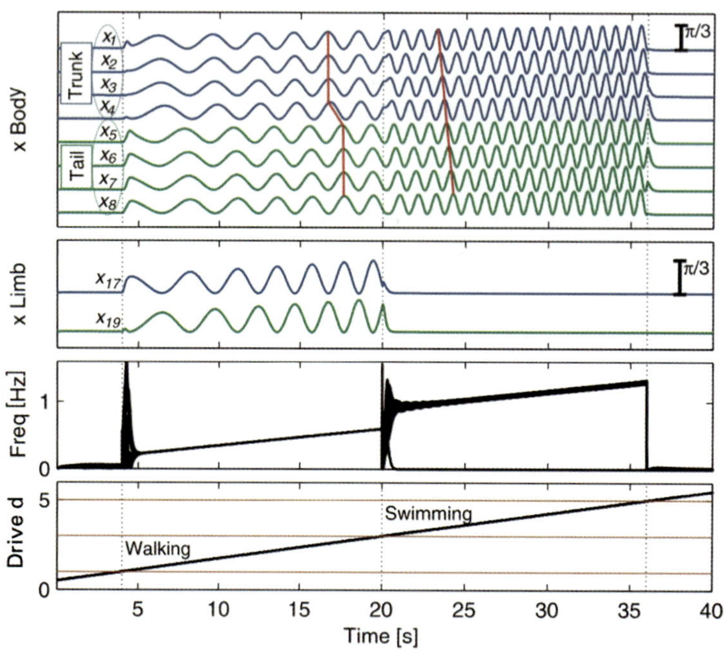

FIGURE 8.17 The gait transition from walking to swimming. Transition between walking and swimming patterns when the descending drive is increased (from Ijspeert et al., 2007). The top panel shows the outputs of the body central pattern generator oscillators on the left side of the double chain. The second panel from the top shows the outputs of the fore- and hind-limb oscillators of the left side. Notice the saturation (stop of oscillations) at t = 20 seconds when the drive reaches the limb oscillators' saturation thresholds. The third panel shows the instantaneous frequencies (Freq) of body and limb oscillators. The bottom panel shows the descending drive signal that is applied to all oscillators and that is linearly increased.

could be compared qualitatively to those of the real salamander, and (3) the concept of CPGs as robust locomotion controllers for robots with multiple articulated joints. In our model, walking and swimming frequencies range, respectively, from 0.2 to 0.6 Hz and from 0.9 to 1.3 Hz. The intrinsic frequencies of the model have been reduced compared to the real salamander to fit within the torque limits of the robot motors. While figure 8.17 shows an example with a simple linear increase of the drive, the model can readily deal with abruptly and continuously varied drives (as likely occur in a freely behaving animal) and modulate the velocity and type of gait accordingly. Experiments conducted with the robot driven by the CPG model showed that the model could successfully generate both walking and swimming gaits (figure 8.18) as well as transitions between them.

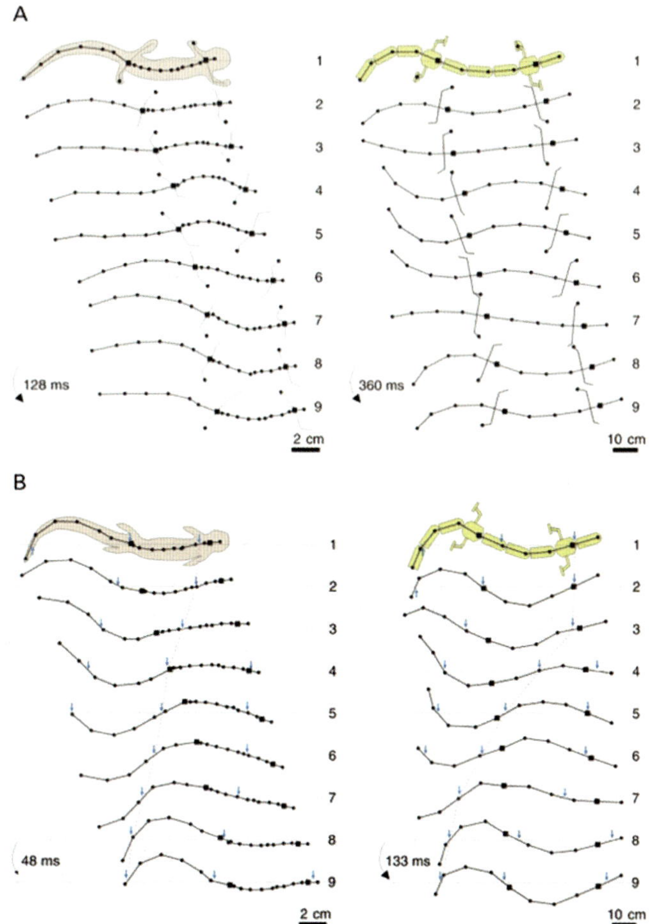

FIGURE 8.18 Qualitative comparison between walking and swimming gaits in the salamander (A) and the salamander-like robot (B). The figure shows the position of markers on the salamander and robot bodies for different parts of a locomotion cycle, as recorded from video cameras. Left: The standing wave in the trunk and tail during a stereotypical walking cycle. Note the nodes of the standing wave at the girdles (square markers), where deviations from the direction of locomotion are minimal. Right: The head-to-tail traveling wave during a stereotypical swimming cycle. Note the progression of the animal and the robot along the horizontal axis. From Ijspeert et al. (2007). The videos are provided in this book's website as supplementary material.

8.3 Pattern Generators and Robotics

CPG[1] models and reflex loops are now increasingly used to control various types of robots. This section briefly presents different examples of such projects. It shows how bio-inspiration can lead to new control technology in robotics, but also how robotics can serve as a useful tool for research in neuroscience—for instance, to understand the mechanisms of entrainment between a CPG and a mechanical system and to test hypotheses about gait modulation. For a more general perspective, see chapter 16, "Embodied Models and Neurorobotics."

8.3.1 CPGs FOR ROBOT LOCOMOTION As reviewed in Ijspeert (2008), the types of CPG models implemented in robots are typically more abstract than those used to develop biological models. These include connectionist models (Arena, 2000; Lu et al., 2005), vector maps (Okada et al., 2002), and systems of coupled oscillators (Crespi and Ijspeert, 2006; Ijspeert et al., 2007; Kimura et al., 1999; Williamson, 1998). Among the most biological-like models are the spiking neural network models developed by Lewis and colleagues (Lewis et al., 2005). While most models are numerically integrated on a microcontroller or a processor, some CPG models are directly realized in hardware, that is, on a chip (DeWeerth et al., 1997; Lewis et al., 2005; Nakada et al., 2003; Simoni and DeWeerth, 2007) or with analog electronics (Still and Tilden, 1998).

8.3.2 DIFFERENT TYPES OF ROBOTS AND LOCOMOTION Models of CPGs and reflex loops are increasingly used to control articulated and legged robots. Different types of robots and modes of locomotion have been explored, including hexapod and octopod robots inspired by insect and arachnid locomotion (Arena et al., 2004; Inagaki et al., 2006; Klaassen et al., 2002), anguilliform swimming of lamprey/eel robots (Arena, 2001; Crespi and Ijspeert, 2008; Stefanini et al., 2006; Wilbur et al., 2002; Conradt and Varshavskaya, 2003), quadruped robots (Kimura et al., 2007; Manoonpong et al., 2007; Tsujita et al., 2004), and biped robots (Aoi and Tsuchiya, 2005; Morimoto et al., 2006).

The work on anguilliform robots is closely related to the research on lamprey swimming CPGs presented in the previous sections. In particular, CPG models of coupled oscillators implemented on microcontrollers offer a useful way of controlling the swimming of eel-like robots (Crespi and Ijspeert, 2008). They are computationally light, and if the robot is constructed as a chain of modules, they can even be implemented in a distributed way with one or two oscillators per microcontroller in each robot module (Conradt and Varshavskaya, 2003).

As reviewed in Ijspeert (2008), CPG models present several interesting properties that make them useful for the control of locomotion in robots as an alternative to methods traditionally used in robotics. These properties include the following:

1. Robust limit cycle behavior. In dynamical systems theory, a stable limit cycle is a periodic regime that acts as an attractor for a dynamical system, that is, to which the system returns after a perturbation. This is an interesting property that provides robustness against perturbations during locomotion.
2. CPGs are well suited for distributed implementation, which, as mentioned above, might be interesting for modular robots; see the snake/eel robot (Conradt and Varshavskaya, 2003).
3. CPG models allow modulation of locomotion, for example, changes of speed, direction, or even type of gaits, by changing only a few control parameters. We have seen and discussed this in the previous section showing how modulation of a few descending drive signals from the MLR leads to changes in CPG activity. For instance, in the lamprey models, an increase of speed is obtained by increasing the descending tonic drive, which leads to an increase of the frequency and amplitude of bursts and, hence, of swimming speed in Ekeberg's (1993) neuromechanical model.
4. CPGs are well suited to integrate sensory feedback signals, which can be added as coupling terms in the differential equations. This is useful for the robot controller to adjust to natural frequencies of the robot mechanics (e.g., resonant frequencies of a compliant robot or natural frequencies of stepping movements). With proper couplings one can obtain *mutual entrainment,* that is, a common resulting frequency between the CPG and the mechanical body (Kimura et al., 1999; Taga, 1998; Verdaasdonk et al., 2006).
5. CPG models usually offer a good substrate for learning and optimization algorithms. This is related to the reduction of dimensionality offered by pattern generators; the learning algorithms do not need to explore high-dimensional actuator commands but only the low-dimensional descending control signals. See Lewis et al. (1993) and Ijspeert (2001) for examples of the use of evolutionary algorithms for setting CPG parameters.

Obviously CPG-based approaches in robotics also present several disadvantages:

1. It is still a kind of an art to design a CPG for a particular type of robot and a particular type of locomotion. Unlike some more theoretical, model-based, approaches to locomotion control in robotics, a sound design methodology for CPG-based controllers is yet missing. Here inspiration from the structure of biological CPGs, when available, can help the engineer to design controllers.
2. It is not yet clear how to obtain rich motor skills with CPGs. The dimensionality reduction offered by pattern generators (see point 5 above) also means that movements generated by them might be less varied and rich than those produced by control laws that directly act at the individual joint level.
3. Finally, a solid theoretical foundation for describing CPGs is still lacking—for instance, for proving the stability of the complete CPG-robot system from a dynamical systems point of view. This is a difficult problem because it is high-dimensional and nonlinear and because interactions with the environment are hard to model properly.

8.3.3 ROBOTS FOR NEUROBIOLOGICAL RESEARCH Similarly to computational models which are now essential tools in neuroscience (and in any scientific field), we believe that robots will increasingly be used as scientific tools to test hypotheses about animal motor control. Several projects specifically aim at doing so, such as the lobster robot project (Ayers and Crisman, 1993; Ayers and Witting, 2007), the lamprey robot projects (Stefanini et al., 2006; Wilbur et al., 2002), the stick insect robot (Schmitz et al., 2008), and our salamander robot (Ijspeert et al., 2007). See also Webb (2001, 2002) for a more general discussion of the use of robots in biological research.

We will here briefly analyze two examples of insights on pattern generators obtained using robots. The first example comes from the work of Hiroshi Kimura and his colleagues on cat-like locomotion (Kimura et al., 2001; Fukuoka et al., 2003; Kimura et al., 2007). Using a quadruped robot, they have explored different means of integrating sensory feedback in locomotion control (Kimura et al., 2001). They compared two cases: sensory feedback that acts independently from the CPG and sensory feedback that is fed through the CPG network. The CPG network was implemented as a Matsuoka oscillator (Mastuoka, 1987), and sensory feedback signals they looked at include body orientation, tendon force, and contact with the floor. They found that sensory feedback that is fed through the CPG network leads to the most stable locomotion in complex terrain compared to feedback that is independent of the CPG activity. Their conclusion is that the interplay between sensory feedback and the CPG is important because it allows proper adjustment of the cycle duration depending on the conditions (e.g., when going up or down a slope) and because it can make sensory feedback phase dependent (e.g., to make the reaction of a contact with the ground dependent on whether the limb is in swing or in stance) similarly to what is known in the cat (Pearson, 2004). In follow-up work, they added different types of reflexes and obtained some of the most robust outdoors locomotion to date (Kimura et al., 2007).

The second example comes from our own work on CPG models of salamander locomotion (Ijspeert et al., 2007). Even though the robot was only a first approximation of the salamander body (an undulating spine with four rotational limbs), it was very useful for validating various aspects of the mathematical model of the salamander CPG. With the robot we could show that our CPG model could produce effective gaits in water and on ground with variable speed and heading. Such a study requires a "body" for validation; it cannot be conducted at only a neuronal level. It also allowed us to compare the generated gaits to those of the real salamander and to demonstrate that the coordination between body undulation and limb movements during walking allows the salamander to increase its walking velocity compared to gaits that do not use body undulations (Crespi et al., 2013). We are currently using the robot to test a hypothesis proposed in Ijspeert et al. (2005) that sensory feedback plays a role in explaining the transition from traveling waves (in water) to standing waves (on ground) in body undulations due to differences in interaction forces with the environment.

More generally, robots can be useful in computational neuroscience because they allow models to be tested when coupled to a real body and embedded in a real environment (something that is difficult to properly model even in physics-based simulators). This means that models can be tested within complete sensing-to-acting loops. This is particularly important for studying animal motor control since some aspects of movements and locomotion might depend critically on the interaction with the environment. However, one should be aware that robots will always be only an approximation to real animals and that it is really difficult to properly replicate all the physical properties of musculoskeletal systems (e.g., the complex visco-elastic properties of muscles). Moreover, as explored in chapter 16, "Embodied Models and Neurorobotics," there is also the challenge of equipping the robot with realistic sensors. There is therefore a real risk that a robot will exhibit significantly different dynamics from the corresponding animal and hence draw attention away from true biological issues.

8.4 *Conclusion*

This chapter presented the concept of motor pattern generation and in particular of the CPGs underlying locomotion in vertebrates. It reviewed current knowledge about the underlying neural circuits and showed how different types of computational models have helped in understanding how they function.

The chapter demonstrated that pattern generators are important components of the CNS, representing core building blocks that can be activated and modulated by a set of descending pathways such as to produce the rich motor skills observed in animals. The chapter had a particular focus on the lamprey and salamander, that is, lower vertebrates which have been extensively studied by neurobiologists and whose locomotor circuits are among the best decoded (in particular the lamprey).

Computational models have played a key role in decoding how MPGs are organized, and we predict they will be increasingly needed in the future given the complex and dynamical nature of movement control. As reviewed in this chapter, models range from highly detailed neuronal models with H–H neurons to highly abstract models of coupled oscillators. They all share in common a key property of pattern generators, namely, that low-dimensional inputs are transformed into high-dimensional motoneuron output signals. Interestingly, pattern generators can therefore be seen as the inverse of *filters* (like the feature-detecting networks in the visual cortex), which take high-dimensional inputs and produce low-dimensional outputs. The chapter also discussed the usefulness of robots as scientific tools to test hypotheses concerning animal motor control. It presented some examples of how models of MPGs have been tested on board quadruped and salamander-like robots to explore questions related to sensory feedback integration and gait transitions. Even if the construction of realistic animal-like robots is a difficult task, we expect that such robots will increasingly be used in computational neuroscience in addition to neuromechanical simulations.

There remain many open questions that still need to be addressed. Future research should, for instance, investigate in more detail the activity, circuitry, and types of neurons involved in the MPGs, especially in mammals, including humans. In particular, it is not yet clear how circuits that can produce both discrete and rhythmic movements are implemented and interconnected. More research also needs to be done on decoding the interplay between different descending pathways, the CPG circuits, and multimodal sensory feedback. Finally, the role of pattern generators in rich motor skills (as opposed to relatively "stereotyped" movements) remains to be explored (compare chapter 20, "Reach and Grasp: Control, Development, and Recognition"). As shown in this chapter, pattern generators can produce a large variety of movements, richer than previously thought, but the exact mechanisms of how upper parts of the brain take advantage of pattern generators in complex movements, such as dogs running after a ball or humans playing a piano, remain to be elucidated. Computational models and robots building on those presented in this chapter will play a key role in addressing these challenges.

NOTE

1. Note that most CPG models in robotics include sensory feedback signals of some kind, usually similarly to reflex loops. In this section, the distinction between CPG and MPG is therefore less strict than in the previous sections, and we will mainly use the term CPG as it is now widely used in robotics, more than MPG.

REFERENCES

Alford, S., & Grillner, S. (1990). CNQX and DNQX block non-NMDA synaptic transmission but not NMDA-evoked locomotion in lamprey spinal cord. *Brain Research, 506*, 297–302.

Alstermark, B., & Isa, T. (2011). Circuits for skilled reaching and grasping. *Annual Review of Neuroscience, 35*, 559–578.

Alstermark, B., Isa, T., Pettersson, L. G., & Sasaki, S. (2007). The C3–C4 propriospinal system in the cat and monkey: A spinal pre-motoneuronal centre for voluntary motor control. *Acta Physiologica (Oxford, England), 189*(2), 123–140.

Alstermark, B., Lundberg, A., Norrsell, U., & Sybirska, E. (1981). Integration in descending motor pathways controlling the forelimb in the cat: IX. Differential behavioural defects after spinal cord lesions interrupting defined pathways from higher centres to motoneurones. *Experimental Brain Research, 42*(3–4), 299–318.

Alstermark, B., Pettersson, L. G., Nishimura, Y., Yoshino-Saito, K., Tsuboi, F., Takahashi, M., et al. (2011). Motor command for precision grip in the macaque monkey can be mediated by spinal interneurons. *Journal of Neurophysiology, 106*(1), 122–126.

Amemiya, M., & Yamaguchi, T. (1984). Fictive locomotion of the forelimb evoked by stimulation of the mesencephalic locomotor region in the decerebrate cat. *Neuroscience Letters, 50*, 91–96.

Aoi, S., & Tsuchiya, K. (2005). Locomotion control of a biped robot using nonlinear oscillators. *Autonomous Robots, 19*, 219–232.

Arena, P. (2000). The central pattern generator: A paradigm for artificial locomotion. *Soft Computing, 4*(4), 251–266.

Arena, P. (2001). A mechatronic lamprey controlled by analog circuits. In *Proceedings of the 9th IEEE Mediterranean Conference on Control and Automation.*

Arena, P., Fortuna, L., Frasca, M., & Sicurella, G. (2004). An adaptive, self-organizing dynamical system for hierarchical control of bio-inspired locomotion. *IEEE Transactions on Systems, Man, and Cybernetics. Part B, Cybernetics, 34*(4), 1823–1837.

Armstrong, D. M. (1986). Supraspinal contributions to the initiation and control of locomotion in the cat. *Progress in Neurobiology, 26*, 273–361.

Arshavsky, Y. I., Orlovsky, G. N., Pavlova, G. A., & Popova, L. B. (1986). Activity of C3–C4 propriospinal neurons during fictitious forelimb locomotion in the cat. *Brain Research, 363*(2), 354–357.

Arshavsky, Y. I., Orlovsky, G. N., & Popova, L. B. (1984). Activity of cerebellar Purkinje cells during fictitious scratch reflex in the cat. *Brain Research, 290*(1), 33–41.

Ashley-Ross, M. A., & Bechtel, B. F. (2004). Kinematics of the transition between aquatic and terrestrial locomotion in the newt *Taricha torosa. Journal of Experimental Biology, 207*, 461–474.

Ashley-Ross, M. A., & Lauder, G. V. (1997). Motor patterns and kinematics during backward walking in the Pacific giant salamander: Evidence for novel motor output. *Journal of Neurophysiology, 78*, 3047–3060.

Ashley-Ross, M. A., Lundin, R., & Johnson, K. L. (2009). Kinematics of level terrestrial and underwater walking in the California newt, *Taricha torosa. Journal of Experimental Zoology. Part A, Ecological Genetics and Physiology, 311*(4), 240–257.

Ayers, J., & Crisman, J. (1993). The lobster as a model for an omnidirectional robotic ambulation control architecture. In R. D. Beer, R. E. Ritzmann, & T. M. McKenna (Eds.), *Biological Neural Networks in Invertebrate Neuroethology and Robotics* (pp. 287–316). Boston: Academic Press.

Ayers, J., & Witting, J. (2007). Biomimetic approaches to the control of underwater walking machines. *Philosophical Transactions. Series A, Mathematical, Physical and Engineering Sciences, 365*(1850), 273–295.

Baekey, D. M., Morris, K. F., Gestreau, C., & Li, Z. (2001). Medullary respiratory neurones and control of laryngeal motoneurones during fictive eupnoea and cough in the cat. *Journal of Physiology, 534*(2), 565–581.

Barbeau, H., & Rossignol, S. (1994). Enhancement of locomotor recovery following spinal cord injury. *Current Opinion in Neurology,* 7(6), 517–524.

Barlow, S. M., & Estep, M. (2006). Central pattern generation and the motor infrastructure for suck, respiration, and speech. *Journal of Communication Disorders, 39*, 366–380.

Baxter, D. A., Canavier, C. C., Clark, J. W., & Byrne, J. H. (1999). Computational model of the serotonergic modulation of sensory neurons in Aplysia. *Journal of Neurophysiology, 82*(6), 2914–2935.

Bem, T., Cabelguen, J. M., Ekeberg, O., & Grillner, S. (2003). From swimming to walking: A single basic network for two different behaviors. *Biological Cybernetics, 88*, 79–90.

Berkowitz, A., Roberts, A., & Soffe, S. R. (2010). Roles for multifunctional and specialized spinal interneurons during motor pattern generation in tadpoles, zebrafish larvae, and turtles. *Frontiers in Behavioral Neuroscience, 4*, 36.

Bianchi, A. L., & Gestreau, C. (2009). The brainstem respiratory network: An overview of a half century of research. *Respiratory Physiology and Neurobiology, 168*(1–2), 4–12.

Biewener, A. A. (2003). *Animal Locomotion.* New York: Oxford University Press.

Bizzi, E., Cheung, V. C. K., d'Avella, A., Saltiel, P., & Tresch, M. (2008). Combining modules for movement. *Brain Research. Brain Research Reviews, 57*, 125–133.

Borgmann, A., Hooper, S. L., & Büschges, A. (2009). Sensory feedback induced by front-leg stepping entrains the activity of central pattern generators in caudal segments of the stick insect walking system. *Journal of Neuroscience, 29*(9), 2972–2983.

Bouyer, L. J. G., & Rossignol, S. (2003). Contribution of cutaneous inputs from the hindpaw to the control of locomotion: II. Spinal cats. *Journal of Neurophysiology, 90*(6), 3640–3653.

Boyd, M. R., & McClellan, A. D. (2002). Changes in locomotor activity parameters with variations in cycle time in larval lamprey. *Journal of Experimental Biology, 205*(23), 3707–3716.

Branchereau, P., Rodriguez, J. J., Delvolvé, I., Abrous, D. N., Le Moal, M., & Cabelguen, J.-M. (2000). Serotonergic systems in the spinal cord of the amphibian urodele *Pleurodeles waltl. Journal of Comparative Neurology, 419*, 49–60.

Brocard, F., & Dubuc, R. (2003). Differential contribution of reticulospinal cells to the control of locomotion induced by the mesencephalic locomotor region. *Journal of Neurophysiology, 90*, 1714–1727.

Brocard, F., Ryczko, D., Fénelon, K., Hatem, R., Gonzales, D., Auclair, F., et al. (2010). The transformation of a unilateral locomotor command into a symmetrical bilateral activation in the brainstem. *Journal of Neuroscience, 30*, 523–533.

Brodin, L., Grillner, S., & Rovainen, C. M. (1985). *N*-Methyl-D-aspartate (NMDA), kainate and quisqualate receptors and the generation of fictive locomotion in the lamprey spinal cord. *Brain Research, 325*, 302–306.

Brown, T. G. (1914). On the fundamental activity of the nervous centres: Together with an analysis of the conditioning of rhythmic activity in progression, and a theory of the evolution of function in the nervous system. *Journal of Physiology, 48*, 18–41.

Buchanan, J. T. (1982). Identification of interneurons with contralateral, caudal axons in the lamprey spinal cord: Synaptic interactions and morphology. *Journal of Neurophysiology, 47*(5), 961–975.

Buchanan, J. T. (1992). Neural network simulations of coupled locomotor oscillators in the lamprey spinal cord. *Biological Cybernetics, 66*, 367–374.

Buchanan, J. T. (2001). Contributions of identifiable neurons and neuron classes to lamprey vertebrate neurobiology. *Progress in Neurobiology, 63*, 441–466.

Cabelguen, J. M., Bourcier-Lucas, C., & Dubuc, R. (2003). Bimodal locomotion elicited by electrical stimulation of the midbrain in the salamander *Notophthalmus viridescens. Journal of Neuroscience, 23*(6), 2434–2439.

Calabrese, R. L. (1998). Cellular, synaptic, network, and modulatory mechanisms involved in rhythm generation. *Current Opinion in Neurobiology, 8*(6), 710–717.

Cangiano, L., & Grillner, S. (2005). Mechanisms of rhythm generation in a spinal locomotor network deprived of crossed connections: The lamprey hemicord. *Journal of Neuroscience, 25*, 923–935.

Cangiano, L., Hill, R. H., & Grillner, S. (2012). The hemisegmental locomotor network revisited. *Journal of Neuroscience, 210*, 33–37.

Carro-Juárez, M., & Rodríguez-Manzo, G. (2007). The spinal pattern generator for ejaculation. *Brain Research. Brain Research Reviews, 58*(1), 106–120.

Cheng, J., Stein, R. B., Jovanovic, K., Yoshida, K., Bennett, D. J., & Han, Y. (1998). Identification, localization, and modulation of neural networks for walking in the mudpuppy (*Necturus maculatus*) spinal cord. *Journal of Neuroscience, 18*(11), 4295–4304.

Chevallier, S. (2004). Plasticité Post-lésionnelle des Réseaux Médullaires Locomoteurs des Urodèles: Etude Electrophysiologique et Immunohistochimique. PhD Thesis n. 1178, Bordeaux University.

Clarac, F., & Barnes, W. J. P. (1985). Peripheral influences on the coordination of the legs during walking in decapod crustaceans. In B. M. H. Bush & F. Clarac (Eds), *Co-ordination of Motor Behaviour. Society for Experimental Biology, Seminar Series 24* (pp. 249–269). Cambridge: Cambridge University Press.

Clark, D. J., Ting, L. H., Zajac, F. E., Neptune, R. R., & Kautz, S. A. (2010). Merging of healthy motor modules predicts reduced locomotor performance and muscle coordination complexity post-stroke. *Journal of Neurophysiology, 103*(2), 844–857.

Cohen, A. H. (1988). Evolution of the vertebrate central pattern generator for locomotion. In A. H. Cohen, S. Rossignol, & S. Grillner (Eds.), *Neural Control of Rhythmic Movements in Vertebrates* (pp. 129–166). New York: Wiley.

Cohen, A. H., Holmes, P. J., & Rand, R. (1982). The nature of coupling between segmented oscillations and the lamprey spinal generator for locomotion: A mathematical model. *Journal of Mathematical Biology, 13*, 345–369.

Cohen, A. H., & Wallen, P. (1980). The neural correlate of locomotion in fish: "Fictive swimming" induced in a in vitro preparation of the lamprey spinal cord. *Experimental Brain Research, 41*, 11–18.

Collins, J., & Richmond, S. (1994). Hard-wired central pattern generators for quadrupedal locomotion. *Biological Cybernetics, 71*, 375–385.

Collins, S., Ruina, A., Tedrake, R., & Wisse, M. (2005). Efficient bipedal robots based on passive-dynamic walkers. *Science, 307*(5712), 1082–1085.

Combes, D., Le Ray, D., Lambert, F. M., Simmers, J., & Straka, H. (2008). An intrinsic feed-forward mechanism for vertebrate gaze stabilization. *Current Biology, 18*(6), R241–R243.

Combes, D., Le Ray, D., Lambert, F. M., Simmers, J., & Straka, H. (2008). An intrinsic feed-forward mechanism for vertebrate gaze stabilization. *Current Biology, 18*(6).

Combes, D., Merrywest, S. D., Simmers, J., & Sillar, K. T. (2004). Developmental segregation of spinal networks driving axial-and hindlimb-based locomotion in metamorphosing *Xenopus laevis. Journal of Physiology, 559*(1), 17–24.

Conradt, J., & Varshavskaya, P. (2003). Distributed central pattern generator control for a serpentine robot. In International Conference on Artificial Neural Networks.

Crespi, A., & Ijspeert, A. J. (2006). AmphiBot II: An amphibious snake robot that crawls and swims using a central pattern generator. In *Proceedings of the 9th International Conference on Climbing and Walking Robots* (pp. 19–27).

Crespi, A., & Ijspeert, A. (2008). Online optimization of swimming and crawling in an amphibious snake robot. *IEEE Transactions on Robotics, 24*(1), 75–87.

Crespi, A., Karakasiliotis, K., Guignard, A., & Ijspeert, A. J. (2013). Salamandra robotica II: An amphibious robot to study salamander-like swimming and walking gaits. *IEEE Transactions on Robotics, 99*, 1–13.

Cruse, H., Brunn, D. E., Bartling, C., Dean, J., Dreifert, M., Kindermann, T., et al. (1995). Walking: A complex behavior controlled by simple networks. *Adaptive Behavior, 3*(4), 385–418.

Cruse, H., & Warnecke, H. (1992). Coordination of the legs of a slow-walking cat. *Experimental Brain Research, 89*(1), 147–156.

Cuellar, C. A., Tapia, J. A., Juárez, V., Quevedo, J., Linares, P., Martínez, L., et al. (2009). Propagation of sinusoidal electrical waves along the spinal cord during a fictive motor task. *Journal of Neuroscience, 29*(3), 798–810.

Deban, S. M., & Schilling, N. (2009). Activity of trunk muscles during aquatic and terrestrial locomotion in *Ambystoma maculatum*. *Journal of Experimental Biology, 212*, 2949–2959.

Degallier, S., & Ijspeert, A. (2010). Modeling discrete and rhythmic movements through motor primitives: A review. *Biological Cybernetics, 103*(4), 319–338.

Delcomyn, F. (1980). Neural basis for rhythmic behaviour in animals. *Science, 210*, 492–498.

Deliagina, T. G., Beloozerova, I. N., Zelenin, P. V., & Orlovsky, G. N. (2008). Spinal and supraspinal postural networks. *Brain Research. Brain Research Reviews, 57*(1), 212–221.

Deliagina, T. G., Feldman, A. G., Gelfand, I. M., & Orlovsky, G. N. (1975). On the role of central program and afferent inflow in the control of scratching movements in the cat. *Brain Research, 100*(2), 297–313.

Deliagina, T. G., Zelenin, P. V., Fagerstedt, P., Grillner, S., & Orlovsky, G. N. (2000). Activity of the reticulospinal neurons during locomotion in freely behaving lamprey. *Journal of Neurophysiology, 83*, 853–863.

Deliagina, T. G., Zelenin, P. V., & Orlovsky, G. N. (2012). Physiological and circuit mechanisms of postural control. *Current Opinion in Neurobiology, 22*(4), 646–652.

Delvolvé, I., Bem, T., & Cabelguen, J. M. (1997). Epaxial and limb muscle activity during swimming and terrestrial stepping in the adult newt, *Pleurodeles waltl*. *Journal of Neurophysiology, 78*(2), 638–650.

Delvolvé, I., Branchereau, P., Dubuc, R., & Cabelguen, J. M. (1999). Fictive rhythmic motor patterns induced by NMDA in an in vitro brain stem-spinal cord preparation from an adult urodele. *Journal of Neurophysiology, 82*, 1074–1077.

DeWeerth, S. P., Patel, G. N., Simoni, M. F., Schimmel, D. E., & Calabrese, R. L. (1997). A VLSI architecture for modeling intersegmental coordination. In R. Brown & A. Ishii (Eds.), *Seventeenth Conference on Advanced Research in VLSI: Proceedings* (pp. 182–200). Los Alamitos, CA: IEEE Computer Society.

Dietz, V. (2003). Spinal cord pattern generators for locomotion. *Clinical Neurophysiology, 114*(8), 1379–1389.

Dietz, V., & Harkema, S. J. (2004). Locomotor activity in spinal cord-injured persons. *Journal of Applied Physiology, 96*(5), 1954–1960.

Dickinson, P. S. (1995). The contributions of motor neuronal and muscle modulation to behavioral flexibility in the stomatogastric system. *Integrative and Comparative Biology, 35*(6), 548–555.

Dimitrijevic, M. R., Gerasimenko, Y., & Pinter, M. M. (1998). Evidence for a spinal central pattern generator in humans. *Annals of the New York Academy of Sciences, 860*, 360–376.

Doi, A., & Ramirez, J.-M. (2008). Neuromodulation and the orchestration of the respiratory rhythm. *Respiratory Physiology & Neurobiology, 164*(1–2), 96–104.

Dominici, N., Ivanenko, Y., Cappellini, G., & d'Avella, A. (2011). Locomotor primitives in newborn babies and their development. *Science, 334*, 997–999.

Drew, T. (1991). Visuomotor coordination in locomotion. *Current Opinion in Neurobiology, 1*(4), 652–657.

Drew, T., Kalaska, J., & Krouchev, N. (2008). Muscle synergies during locomotion in the cat: A model for motor cortex control. *Journal of Physiology, 586*(5), 1239–1245.

Drew, T., Prentice, S., & Schepens, B. (2004). Cortical and brainstem control of locomotion. *Progress in Brain Research, 143*, 251–261.

Drew, T., & Rossignol, S. (1987). A kinematic and electromyographic study of cutaneous reflexes evoked from the forelimb of unrestrained walking cats. *Journal of Neurophysiology, 57*(4), 1160–1184.

Dubuc, R., Cabelguen, J. M., & Rossignol, S. (1988). Rhythmic fluctuations of dorsal root potentials and antidromic discharges of primary afferents during fictive locomotion in the cat. *Journal of Neurophysiology, 60*(6), 2014–2036.

Edgerton, V. R., Tillakaratne, N. J., Bigbee, A. J., de Leon, R. D., & Roy, R. R. (2004). Plasticity of the spinal neural circuitry after injury. *Annual Review of Neuroscience, 27*, 145–167.

Ekeberg, Ö. (1993). A combined neuronal and mechanical model of fish swimming. *Biological Cybernetics, 69*, 363–374.

Ekeberg, Ö., Lansner, A., & Grillner, S. (1995). The neural control of fish swimming studied through numerical simulations. *Adaptive Behavior, 3*, 363–384.

Ekeberg, Ö., Wallén, P., Lansner, A., Traven, H., Brodin, L., & Grillner, S. (1991). A computer-based model for realistic simulations of neural networks: I. The single neuron and synaptic interaction. *Biological Cybernetics, 65*, 81–90.

El Manira, A., Kyriakatos, A., & Nanou, E. (2010). Beyond connectivity of locomotor circuitry—Ionic and modulatory mechanisms. *Progress in Brain Research, 187*, 99–110.

El Manira, A., Tegnér, J., & Grillner, S. (1997). Locomotor-related presynaptic modulation of primary afferents in the lamprey. *European Journal of Neuroscience, 9*(4), 696–705.

Ermentrout, B., & Kopell, N. (1994). Inhibition-produced patterning in chains of coupled nonlinear oscillators. *SIAM Journal on Applied Mathematics, 54*(2), 478–507.

Fagerstedt, P., Orlovsky, G. N., Deliagina, T. G., Grillner, S., & Ullén, F. (2001). Lateral turns in the lamprey: II. Activity of reticulospinal neurons during the generation of fictive turns. *Journal of Neurophysiology, 86*(5), 2257–2265.

Fagerstedt, P., & Ullén, F. (2001). Lateral turns in the lamprey: I. Patterns of motoneuron activity. *Journal of Neurophysiology, 86*(5), 2246–2256.

Farley, C. T., & Taylor, C. R. (1991). A mechanical trigger for the trot-gallop transition in horses. *Science, 253*(5017), 306–308.

Feldman, J. L., & Del Negro, C. A. (2006). Looking for inspiration: New perspectives on respiratory rhythm. *Nature Reviews. Neuroscience, 7*(3), 232–242.

Feldman, J. L., & Smith, J. C. (1989). Cellular mechanisms underlying modulation of breathing pattern in mammals. *Annals of the New York Academy of Sciences, 563*, 114–130.

Fetcho, J. R., & Svoboda, K. R. (1993). Fictive swimming elicited by electrical stimulation of the midbrain in goldfish. *Journal of Neurophysiology, 70*(2), 765–780.

FitzHugh, R. (1955). Mathematical models of threshold phenomena in the nerve membrane. *Bulletin of Mathematical Biophysics, 17*(4), 257–278.

Fok, M., & Stein, R. B. (2002). Effects of cholinergic and noradrenergic agents on locomotion in the mudpuppy (*Necturus maculatus*). *Experimental Brain Research, 145*(4), 498–504.

Forssberg, H., Grillner, S., Halbertsma, J., & Rossignol, S. (1980). The locomotion of the low spinal cat: II. Interlimb coordination. *Acta Physiologica Scandinavica, 108*(3), 283–295.

Forssberg, H., Grillner, S., & Rossignol, S. (1977). Phasic gain control of reflexes from the dorsum of the paw during spinal locomotion. *Brain Research, 132*(1), 121–139.

Francis, E. T. B. (1934). *The Anatomy of the Salamander.* Oxford: Clarendon.

Frolich, L., & Biewener, A. (1992). Kinematic and electromyographic analysis of the functional role of the body axis during terrestrial and aquatic locomotion in the salamander *Ambystoma tigrinum. Journal of Experimental Biology, 162*, 107–130.

Fukson, O. I., Berkinblit, M. B., & Feldman, A. G. (1980). The spinal frog takes into account the scheme of its body during the wiping reflex. *Science, 209*, 1261–1263.

Fukuoka, Y., Kimura, H., & Cohen, A. H. (2003). Adaptive dynamic walking of a quadruped robot on irregular terrain based on biological concepts. *International Journal of Robotics Research, 3–4*, 187–202.

Funk, G. D., Steeves, J. D., & Milsom, W. K. (1992). Coordination of wingbeat and respiration in birds: II. "Fictive" flight. *Journal of Applied Physiology, 73*(3), 1025–1033.

Gandevia, S. C., & Burke, D. (1992). Does the nervous system depend on kinesthetic information to control natural limb movements? *Behavioral and Brain Sciences, 15*(04), 614–632.

Gao, K. Q., & Shubin, N. H. (2012). Late Jurassic salamandroid from western Liaoning, China. *Proceedings of the National Academy of Sciences of the United States of America, 109*, 5767–5772.

Garcia, A. J., III, Zanella, S., Koch, H., Doi, A., & Ramirez, J. M. (2011). Networks within networks: The neuronal control of breathing. *Progress in Brain Research, 188*, 31–50.

Gariépy, J. F., Missaghi, K., Chevallier, S., Chartré, S., Robert, M., Auclair, F., et al. (2012). Specific neural substrate linking respiration to locomotion. *Proceedings of the National Academy of Sciences of the United States of America, 109*(2), E84–92.

Gerstner, W., & Kistler, W. (2002). *Spiking Neuron Models. Single Neurons, Populations, Plasticity.* Cambridge: Cambridge University Press.

Gestreau, C., Dutschmann, M., Obled, S., & Bianchi, A. L. (2005). Activation of XII motoneurons and premotor neurons during various oropharyngeal behaviors. *Respiratory Physiology & Neurobiology, 147*(2–3), 159–176.

Gestreau, C., Grélot, L., & Bianchi, A. L. (2000). Activity of respiratory laryngeal motoneurons during fictive coughing and swallowing. *Experimental Brain Research, 130*(1), 27–34.

Goldman, M. S., Golowasch, J., Marder, E., & Abbott, L. F. (2001). Global structure, robustness, and modulation of neuronal models. *Journal of Neuroscience, 21*, 5229–5238.

Golubitsky, M., & Stewart, I. (2002). *The Symmetry Perspective: From Equilibrium to Chaos in Phase Space and Physical Space.* Basel: Birkhäuser.

Gossard, J. P., Cabelguen, J. M., & Rossignol, S. (1989). Intra-axonal recordings of cutaneous primary afferents during fictive locomotion in the cat. *Journal of Neurophysiology, 62*(5), 1177–1188.

Gossard, J. P., Cabelguen, J. M., & Rossignol, S. (1990). Phase-dependent modulation of primary afferent depolarization in single cutaneous primary afferents evoked by peripheral stimulation during fictive locomotion in the cat. *Brain Research, 537*(1–2), 14–23.

Green, C. S., & Soffe, S. R. (1996). Transitions between two different motor patterns in Xenopus embryos. *Journal of Comparative Physiology. A, Sensory, Neural, and Behavioral Physiology, 178*(2), 279–291.

Grélot, L., Milano, S., Portillo, F., & Miller, A. D. (1993). Respiratory interneurons of the lower cervical (C4–C5) cord: Membrane potential changes during fictive coughing, vomiting, and swallowing in the decerebrate cat. *Pflügers Archiv, 425*(3–4), 313–320.

Grillner, S. (1981). Control of locomotion in bipeds, tetrapods, and fish. In V. B. Brooks (Ed.), *Handbook of Physiology: The Nervous System: Motor Control* (Sect. I, Vol. 2, pp. 1179–1236). Bethesda, MD: American Physiological Society.

Grillner, S. (1985). Neural control of vertebrate locomotion—Central mechanisms and reflex interaction with special reference to the cat. In W. J. P. Barnes & M. H. Gladden (Eds.), *Feedback and Motor Control in Invertebrates and Vertebrates* (pp. 35–56). London: Croom Helm.

Grillner, S. (2003). The motor infrastructure: From ion channels to neuronal networks. *Nature Reviews. Neuroscience, 4*, 573–586.

Grillner, S. (2006). Biological pattern generation: The cellular and computational logic of networks in motion. *Neuron, 52*(5), 751–766.

Grillner, S., Buchanan, J. T., & Lansner, A. (1988). Simulation of the segmental burst generating network for locomotion in lamprey. *Neuroscience Letters, 89*, 31–35.

Grillner, S., Degliana, T., Ekeberg, Ö., El Marina, A., Lansner, A., Orlovsky, G. N., et al. (1995). Neural networks that co-ordinate locomotion and body orientation in lamprey. *Trends in Neurosciences, 18*(6), 270–279.

Grillner, S., Ekeberg, Ö., El Manira, A., Lansner, A., Parker, D., Tegnér, J., et al. (1998). Intrinsic function of a neuronal network—A vertebrate central pattern generator. *Brain Research. Brain Research Reviews, 26*(2), 184–197.

Grillner, S., Georgopoulos, A. P., & Jordan, L. M. (1997). Selection and initiation of motor behavior. In P. S. G. Stein,

S. Grillner, A. Selverston, & D. G. Stuart (Eds.), *Neurons, Networks and Motor Behavior* (pp. 3–19). Cambridge, MA: MIT Press.

Grillner, S., Rossignol, S., & Wallén, P. (1977). The adaptation of a reflex response to the ongoing phase of locomotion in fish. *Experimental Brain Research, 30*(1), 1–11.

Grillner, S., & Wallén, P. (1984). How does the lamprey central nervous system make the lamprey swims? *Journal of Experimental Biology, 112,* 337–357.

Grillner, S., Wallén, P., Dale, N., Brodin, L., Buchanan, J., & Hill, R. (1987). Transmitters, membrane properties and network circuitry in the control of locomotion in lamprey. *Trends in Neurosciences, 10,* 34–41.

Gritsenko, V., Yakovenko, S., & Kalaska, J. F. (2009). Integration of predictive feedforward and sensory feedback signals for online control of visually guided movement. *Journal of Neurophysiology, 102,* 914–930.

Halbertsma, J. M. (1983). The stride cycle of the cat: The modelling of locomotion by computerized analysis of automatic recordings. *Acta Physiologica Scandinavica. Supplementum, 521,* 1–75.

Halbertsma, J. M., Miller, S., Van der Meché, F. G. A., Herman, R. M., Stein, P. S. G., & Stuart, D. G. (1976). Basic programs for the phasing of flexion and extension movements of the limbs during locomotion. In R. M. Herman, S. Grillner, P. S. G. Stein & D. G. Stuart (Eds.), *Neural Control of Locomotion* (pp. 489–517). New York: Plenum Press.

Halbertsma, J. M. (1983). The stride cycle of the cat: the modelling of locomotion by computerized analysis of automatic recordings. *Acta Physiologica Scandinavica. Supplementum, 521,* 1.

Harischandra, N., Cabelguen, J.-M., & Ekeberg, Ö. (2010). A 3D musculo-mechanical model of the salamander for the study of different gaits and modes of locomotion. *Frontiers in Neurorobotics, 4,* 112.

Harischandra, N., Knuesel, J., Kozlov, A., Bicanski, A., Cabelguen, J.-M., Ijspeert, A. J., et al. (2011). Sensory feedback plays a significant role in generating walking gait and in gait transition in salamanders: A simulation study. *Frontiers in Neurorobotics, 5,* 3.

Harkema, S., Gerasimenko, Y., Hodes, J., Burdick, J., Angeli, C., Chen, Y., et al. (2011). Effect of epidural stimulation of the lumbosacral spinal cord on voluntary movement, standing, and assisted stepping after motor complete paraplegia: A case study. *Lancet, 377*(9781), 1938–1947.

Harris-Warrick, R. M. (2011). Neuromodulation and flexibility in Central Pattern Generator networks. *Current Opinion in Neurobiology, 21*(5), 685–692.

Hildebrand, M. (1976). Analysis of tetrapod gaits: General considerations and symmetrical gaits. In R. M. Herman, S. Grillner, P. S. G. Stein, & D. G. Stuart (Eds.), *Neural Control of Locomotion* (pp. 203–236). New York: Plenum Press.

Hoff, B., & Arbib, M. A. (1991). A model of the effects of speed, accuracy and perturbation on visually guided reaching. In R. Caminiti, P. B. Johnson, & Y. Burnod (Eds.), *Control of Arm Movement in Space: Neurophysiological and Computational Approaches* (pp. 285–306). Heidelberg: Springer-Verlag.

Hultborn, H., & Nielsen, J. B. (2007). Spinal control of locomotion—From cat to man. *Acta Physiologica, 189*(2), 111–121.

Ijspeert, A. J. (2001). A connectionist central pattern generator for the aquatic and terrestrial gaits of a simulated salamander. *Biological Cybernetics, 84*(5), 331–348.

Ijspeert, A. J., Crespi, A., & Cabelguen, J. M. (2005). Simulation and robotics studies of salamander locomotion: Applying neurobiological principles to the control of locomotion in robots. *Neuroinformatics, 3*(3), 171–196.

Ijspeert, A. J., Crespi, A., Ryczko, D., & Cabelguen, J. M. (2007). From swimming to walking with a salamander robot driven by a spinal cord model. *Science, 315*(5817), 1416–1420.

Ijspeert, A. J. (2008). 2008 Special Issue: Central pattern generators for locomotion control in animals and robots: A review. *Neural Networks, 21*(4), 642–653.

Inagaki, S., Yuasa, H., Suzuki, T., & Arai, T. (2006). Wave CPG model for autonomous decentralized multi-legged robot: Gait generation and walking speed control. *Robotics and Autonomous Systems, 54*(2), 118–126.

Ivanenko, Y. P., Cappellini, G., Dominici, N., Poppele, R. E., & Lacquaniti, F. (2005). Coordination of locomotion with voluntary movements in humans. *Journal of Neuroscience, 25*(31), 7238–7253.

Ivanenko, Y. P., Poppele, R. E., & Lacquaniti, F. (2004). Five basic muscle activation patterns account for muscle activity during human locomotion. *Journal of Physiology, 556*(1), 267–282.

Jackson, A. W., Pino, F. A., Wiebe, E. D., & McClellan, A. D. (2007). Movements and muscle activity initiated by brain locomotor areas in semi-intact preparations from larval lamprey. *Journal of Neurophysiology, 97*(5), 3229–3241.

Jean, A. (2001). Brain stem control of swallowing: Neuronal network and cellular mechanisms. *Physiological Reviews, 81*(2), 929–969.

Jeannerod, M. (1988). *The Neural and Behavioural Organization of Goal-Directed Movements.* Oxford: Clarendon Press.

Jordan, L. M. (1998). Initiation of locomotion in mammals. *Annals of the New York Academy of Sciences, 860,* 83–93.

Jordan, L. M., Pratt, C. A., & Menzies, J. E. (1979). Locomotion evoked by brain stem stimulation: Occurrence without phasic segmental afferent input. *Brain Research, 177*(1), 204–207.

Jovanovic, K., Petrov, T., Greer, J. J., & Stein, R. B. (1996). Serotonergic modulation of the mudpuppy (*Necturus maculatus*) locomotor pattern in vitro. *Experimental Brain Research, 111,* 57–67.

Jovanovic, K., Petrov, T., & Stein, R. B. (1999). Effects of inhibitory neurotransmitters on the mudpuppy (*Necturus maculatus*) locomotor pattern in vitro. *Experimental Brain Research, 129,* 172–184.

Karayannidou, A., Zelenin, P. V., Orlovsky, G. N., & Deliagina, T. G. (2007). Responses of reticulospinal neurons in the lamprey to lateral turns. *Journal of Neurophysiology, 97*(1), 512–521.

Kargo, W. J., & Giszter, S. F. (2000). Rapid correction of aimed movements by summation of force-field primitives. *Journal of Neuroscience, 20*(1), 409–426.

Kashin, S. M., Feldman, A. G., & Orlovsky, G. N. (1974). Locomotion of fish evoked by electrical stimulation of the brain. *Brain Research, 82*(1), 41–47.

Katz, P. S. (1995). Intrinsic and extrinsic neuromodulation of motor circuits. *Current Opinion in Neurobiology, 5*(6), 799–808.

Katz, P. S. (1998). Comparison of extrinsic and intrinsic neuromodulation in two central pattern generator circuits in invertebrates. *Experimental Physiology, 83*(3), 281–292.

Kawahara, K., Kumagai, S., Nakazono, Y., & Miyamoto, Y. (1989). Coupling between respiratory and stepping rhythms during locomotion in decerebrate cats. *Journal of Applied Physiology, 67*(1), 110–115.

Kawahara, K., Yamauchi, Y., Niizeki, K., & Yoshioka, T. (1994). Interactions between respiratory, cardiac and stepping rhythms in decerebrated cats: Functional hierarchical structures of biological oscillators. *Methods of Information in Medicine, 33*(1), 129–132.

Kimura, H., Fukuoka, Y., & Cohen, A. H. (2007). Adaptive dynamic walking of a quadruped robot on natural ground based on biological concepts. *International Journal of Robotics Research, 26*(5), 475–490.

Kimura, H., Fukuoka, Y., & Konaga, K. (2001). Adaptive dynamic walking of a quadruped robot using a neural system model. *Advanced Robotics, 15*(8), 859–878.

Klaassen, B., Linnemann, R., Spenneberg, D., & Kirchner, F. (2002). Biomimetic walking robot SCORPION: Control and modeling. *Robotics and Autonomous Systems, 41*(2), 69–76.

Kobayashi, N., Yoshida, M., Matsumoto, N., & Uematsu, K. (2009). Artificial control of swimming in goldfish by brain stimulation: Confirmation of the midbrain nuclei as the swimming center. *Neuroscience Letters, 452*(1), 42–46.

Koch, C., & Segev, I. (1999). *Methods in Neuronal Modeling : From Ions to Networks* (2nd ed.). Cambridge, MA: MIT Press.

Koehler, W. J., Schomburg, E. D., & Steffens, H. (1984). Phasic modulation of trunk muscle efferents during fictive spinal locomotion in cats. *Journal of Physiology, 353*, 187–197.

Kopell, N. (1995). Chains of coupled oscillators. In M. A. Arbib (Ed.), *The Handbook of Brain Theory and Neural Networks* (pp. 178–183). Cambridge, MA: MIT Press.

Kopell, N., Ermentrout, G. B., & Williams, T. L. (1991). On chains of oscillators forced at one end. *SIAM Journal on Applied Mathematics, 51*(5), 1397–1417.

Kozlov, A., Huss, M., Lansner, A., Kotaleski, J. H., & Grillner, S. (2009). Simple cellular and network control principles govern complex patterns of motor behavior. *Proceedings of the National Academy of Sciences of the United States of America, 106*, 20027–20032.

Kuramoto, Y. (2003). Collective behavior of coupled phase oscillators. In M. A. Arbib (Ed.), *The Handbook of Brain Theory and Neural Networks* (pp. 223–226). Cambridge, MA: MIT Press.

Lambert, F. M., Combes, D., Simmers, J., & Straka, H. (2012). Gaze stabilization by efference copy signaling without sensory feedback during vertebrate locomotion. *Current Biology, 22*(18), 1649–1658.

LeBeau, F. E., El Manira, A., & Griller, S. (2005). Tuning the network: Modulation of neuronal microcircuits in the spinal cord and hippocampus. *Trends in Neurosciences, 28*(10), 552–561.

Lemon, R. N. (2008). Descending pathways in motor control. *Annual Review of Neuroscience, 31*(1), 195–218.

Le Ray, D., Juvin, L., & Ryczko, D. (2011). Supraspinal control of locomotion: The mesencephalic locomotor region. *Progress in Brain Research, 188*, 51–70.

Lewis, M. A., Fagg, A. H., & Bekey, G. A. (1993). Genetic algorithms for gait synthesis in a hexapod robot. In Y. F. Zheng (Ed.), *Recent Trends in Mobile Robots* (pp. 317–331). World Scientific.

Lewis, M. A., Tenore, F., & Etienne-Cummings, R. (2005). CPG design using inhibitory networks. In *2005 IEEE International Conference on Robotics and Automation.* Barcelona, Spain. Piscataway, NJ: IEEE.

Lieske, S. P., Thoby-Brisson, M., Telgkamp, P., & Ramirez, J. M. (2000). Reconfiguration of the neural network controlling multiple breathing patterns: Eupnea, sighs and gasps. *Nature Neuroscience, 3*(6), 600–607.

Loeb, G. (2001). Learning from the spinal cord. *Journal of Physiology, 533*(1), 111–117.

Lu, Z., Ma, B., Li, S., & Wang, Y. (2005). Serpentine locomotion of a snake-like robot controlled by cyclic inhibitory CPG model. In *2005 IEEE–RSJ International Conference on Intelligent Robots and Systems* (pp. 96–101). Piscataway, NJ: IEEE.

MacKay-Lyons, M. (2002). Central pattern generation of locomotion: A review of the evidence. *Physical Therapy, 82*(1), 69–83.

Manoonpong, P., Pasemann, F., & Roth, H. (2007). Modular reactive neurocontrol for biologically inspired walking machines. *International Journal of Robotics Research, 26*(3), 301–331.

Marder, E. (2012). Neuromodulation of neuronal circuits: Back to the future. *Neuron, 76*(1), 1–11.

Marder, E., Bucher, D., Schulz, D. J., & Taylor, A. L. (2005). Invertebrate central pattern generation moves along. *Current Biology, 15*(17), R685–R699.

Marder, E., & Calabrese, R. L. (1996). Principles of rhythmic motor pattern generation. *Physiological Reviews, 76*, 687–717.

Marder, E., & Thirumalai, V. (2002). Cellular, synaptic and network effects of neuromodulation. *Neural Networks, 15*(4–6), 479–493.

Matsuoka, K. (1987). Mechanisms of frequency and pattern control in the neural rhythm generators. *Biological Cybernetics, 56*, 345–353.

Matsushima, T., & Grillner, S. (1992). Neural mechanisms of intersegmental coordination in lamprey: Local excitability changes modify the phase coupling along the spinal cord. *Journal of Neurophysiology, 67*(2), 373–388.

McClellan, A. D. (1986). Command systems for initiating locomotor responses in fish and amphibians—Parallels to initiation of locomotion in mammals. In S. Grillner, P. Stein, D. Stuart, H. Forssberg, & R. Herman (Eds.), *Neurobiology of Vertebrate Locomotion, Wenner-Gren Symposium Series* (Vol. 45, pp. 3–20). London: MacMillan.

McClellan, A. D., & Hagevik, A. (1997). Descending control of turning locomotor activity in larval lamprey: Neurophysiology and computer modeling. *Journal of Neurophysiology, 78*(1), 214–228.

McClellan, A. D., & Jang, W. (1993). Mechanosensory inputs to the central pattern generators for locomotion in the lamprey spinal cord: Resetting, entrainment, and computer modeling. *Journal of Neurophysiology, 70*(6), 161–166.

McCrea, D. A., & Rybak, I. A. (2008). Organization of mammalian locomotor rhythm and pattern generation. *Brain Research. Brain Research Reviews, 57*, 134–146.

Miles, G. B., & Sillar, K. T. (2011). Neuromodulation of vertebrate locomotor control networks. *Physiology (Bethesda, MD)*, *26*(6), 393–411.

Minassian, K., Persy, I., Rattay, F., Pinter, M. M., Kern, H., & Dimitrijevic, M. R. (2007). Human lumbar cord circuitries can be activated by extrinsic tonic input to generate locomotor-like activity. *Human Movement Science*, *26*(2), 275–295.

Mori, S. (1987). Integration of posture and locomotion in acute decerebrate cats and in awake, freely moving cats. *Progress in Neurobiology*, *28*(2), 161–195.

Morimoto, J., Endo, G., Nakanishi, J., Hyon, S., Cheng, G., Bentivegna, D., et al. (2006). Modulation of simple sinusoidal patterns by a coupled oscillator model for biped walking. In *Proceedings of the 2006 IEEE International Conference on Robotics and Automation* Orlando, Florida (pp. 1579–1584).

Morquette, P., Lavoie, R., Fhima, M. D., Lamoureux, X., Verdier, D., & Kolta, A. (2012). Generation of the masticatory central pattern and its modulation by sensory feedback. *Progress in Neurobiology*, *96*(3), 340–355.

Morris, C., & Lecar, H. (1981). Voltage oscillations in the barnacle giant muscle fiber. *Biophysical Journal*, *35*(1), 193.

Mortin, L. I., Keifer, J., & Stein, P. S. (1985). Three forms of the scratch reflex in the spinal turtle: Movement analyses. *Journal of Neurophysiology*, *53*(6), 1501–1516.

Munoz, A., Munoz, M., González, A., & ten Donkelaar, H. (1997). Spinal ascending pathways in amphibians: Cells of origin and main targets. *Journal of Comparative Neurology*, *378*(2), 205–228.

Nagumo, J., Arimoto, S., & Yoshizawa, S. (1962). An active pulse transmission line simulating nerve axon. *Proceedings of the IRE*, *50*(10), 2061–2070.

Nakada, K., Asai, T., & Amemiya, Y. (2003). An analog CMOS central pattern generator for interlimb coordination in quadruped locomotion. *IEEE Transactions on Neural Networks*, *14*(5), 1356–1365.

Nieuwenhuys, R., ten Donkelaar, H. J., Nicholson, C., & Smeets, W. J. A. J. (1998). *The Central Nervous System of Vertebrates (3 vols. with poster book)*. New York: Springer.

Okada, M., Tatani, K., & Nakamura, Y. (2002). Polynomial design of the nonlinear dynamics for the brain-like information processing of whole body motion. In *2002 IEEE International Conference on Robotics and Automation* (pp. 1410–1415). Washington DC, USA. Piscataway, NJ: IEEE.

Oku, Y., Tanaka, I., & Ezure, K. (1994). Activity of bulbar respiratory neurons during fictive coughing and swallowing in the decerebrate cat. *Journal of Physiology*, *480*(Pt 2), 309–324.

Orlovsky, G., Deliagina, T., & Grillner, S. (1999). *Neuronal Control of Locomotion: From Mollusc to Man*. New York: Oxford University Press.

Orsal, D., Cabelguen, J. M., & Perret, C. (1990). Interlimb coordination during fictive locomotion in the thalamic cat. *Experimental Brain Research*, *82*, 536–546.

Orsal, D., Perret, C., & Cabelguen, J. M. (1988). Comparison between ventral spinocerebellar and rubrospinal activities during locomotion in the cat. *Behavioural Brain Research*, *28*(1–2), 159–162.

Paggett, K. C., Jackson, A. W., & McClellan, A. D. (2004). Organization of higher-order brain areas that initiate locomotor activity in larval lamprey. *Neuroscience*, *125*, 25–33.

Pearson, K. G. (1993). Common principles of motor control in vertebrates and invertebrates. *Annual Review of Neuroscience*, *16*, 265–297.

Pearson, K. G. (2000). Plasticity of neuronal networks in the spinal cord: Modifications in response to altered sensory input. *Progress in Brain Research*, *128*, 61–70.

Pearson, K. G. (2004). Generating the walking gait: Role of sensory feedback. *Progress in Brain Research*, *143*, 123–129.

Perret, C. (1976). Neural control of locomotion in the decorticate cat. In R. M. Herman, S. Grillner, P. S. G. Stein, & D. G. Stuart (Eds.), *Neural Control of Locomotion* (pp. 587–615). New York: Plenum Press.

Perret, C., Cabelguen, J.-M., & Orsal, D. (1988). Analysis of the pattern of activity in "knee flexor" motoneurons during locomotion in the cat. In V. S. Gurfinkle, M. E. Ioffe, J. Massion, & J. P. Roll (Eds.), *Stance and Motion: Facts and Concepts* (pp. 133–141). New York: Plenum Press.

Poon, M. L. T. (1980). Induction of swimming in lamprey by L-DOPA and amino acids. *Journal of Comparative Physiology. A, Sensory, Neural, and Behavioral Physiology*, *136*(4), 337–344.

Poppele, R., & Bosco, G. (2003). Sophisticated spinal contributions to motor control. *Trends in Neurosciences*, *26*, 269–276.

Ramirez, J. M., & Richter, D. W. (1996). The neuronal mechanisms of respiratory rhythm generation. *Current Opinion in Neurobiology*, *6*(6), 817–825.

Ritzmann, R. E., & Büschges, A. (2007). Adaptive motor behavior in insects. *Current Opinion in Neurobiology*, *17*(6), 629–636.

Robertson, G. A., Mortin, L. I., Keifer, J., & Stein, P. S. (1985). Three forms of the scratch reflex in the spinal turtle: Central generation of motor patterns. *Journal of Neurophysiology*, *53*(6), 1517–1534.

Rossignol, S. (1996). Visuomotor regulation of locomotion. *Canadian Journal of Physiology and Pharmacology*, *74*(4), 418–425.

Rossignol, S., Dubuc, R., & Gossard, J. P. (2006). Dynamic sensorimotor interactions in locomotion. *Physiological Reviews*, *86*, 89–154.

Rossignol, S., Lund, J. P., & Drew, T. (1988). The role of sensory inputs in regulating patterns of rhythmical movements in higher vertebrates: A comparison among locomotion, respiration, and mastication. In A. H. Cohen, S. Rossignol, & S. Grillner (Eds.), *Neural Control of Rhythmic Movements in Vertebrates* (pp. 201–283). New York: Wiley.

Rossignol, S., Schwab, M., Schwartz, M., & Fehlings, M. G. (2007). Spinal cord injury: Time to move? *Journal of Neuroscience*, *27*(44), 11782–11792.

Ryczko, D., Charrier, V., Ijspeert, A., & Cabelguen, J.-M. (2010). Segmental oscillators in axial motor circuits of the salamander: Distribution and bursting mechanisms. *Journal of Neurophysiology*, *104*, 2677–2692.

Ryczko, D., & Dubuc, R. (2013). The multifunctional mesencephalic locomotor region. *Current Pharmaceutical Design*, *19*(24), 4448–4470.

Sánchez-Camacho, C., Marín, O., ten Donkelaar, H. J., & González, A. (2001). Descending supraspinal pathways in amphibians: I. A dextran amine tracing study of their cells of origin. *Journal of Comparative Neurology*, *434*(2), 186–208.

Schmitz, J., Schneider, A., Schilling, M., & Cruse, H. (2008). No need for a body model: Positive velocity feedback for the control of an 18-DOF robot walker. *Applied Bionics and Biomechanics, 5*(3), 135–147.

Schoner, G., Jiang, W. Y., & Kelso, J. A. S. (1990). A synergetic theory of quadrupedal gaits and gait transitions. *Journal of Theoretical Biology, 142,* 359–391.

Scudder, C. A., Kaneko, C. S., & Fuchs, A. F. (2002). The brainstem burst generator for saccadic eye movements: A modern synthesis. *Experimental Brain Research, 142*(4), 439–462.

Seo, K., & Slotine, J.-J. (2007). Models for global synchronization in CPG-based locomotion. In *Proceedings of the IEEE International Conference on Robotics and Automation* (pp. 281–286). Rome, Italy.

Shiba, K., Nakazawa, K., Ono, K., & Umezaki, T. (2007). Multifunctional laryngeal premotor neurons: Their activities during breathing, coughing, sneezing, and swallowing. *Journal of Neuroscience, 27*(19), 5156–5162.

Shik, M. L., Severin, F. V., & Orlovsky, G. N. (1966). Control of walking by means of electrical stimulation of the midbrain. *Biophysics, 11,* 756–765.

Sillar, K. T., McLean, D. L., Fischer, H., & Merrywest, S. D. (2002). Fast inhibitory synapses: Targets for neuromodulation and development of vertebrate motor behaviour. *Brain Research. Brain Research Reviews, 40*(1–3), 130–140.

Simoni, M., & DeWeerth, S. (2007). Sensory feedback in a half-center oscillator model. *IEEE Transactions on Biomedical Engineering, 54,* 193–204.

Sinnamon, H. M. (1993). Preoptic and hypothalamic neurons and the initiation of locomotion in the anesthetized rat. *Progress in Neurobiology, 41*(3), 323–344.

Sirota, M. G., Viana Di Prisco, G., & Dubuc, R. (2000). Stimulation of the mesencephalic locomotor region elicits controlled swimming in semi-intact lampreys. *European Journal of Neuroscience, 12,* 4081–4092.

Smetana, R., Juvin, L., Dubuc, R., & Alford, S. (2010). A parallel cholinergic brainstem pathway for enhancing locomotor drive. *Nature Publishing Group, 13*(6), 731–738.

Smith, J. C., Abdala, A. P. L., Koizumi, H., Rybak, I. A., & Paton, J. F. R. (2007). Spatial and functional architecture of the mammalian brain stem respiratory network: A hierarchy of three oscillatory mechanisms. *Journal of Neurophysiology, 98*(6), 3370–3387.

Sparks, D. L. (2002). The brainstem control of saccadic eye movements. *Nature Reviews. Neuroscience, 3*(12), 952–964.

Steeves, J. D., & Jordan, L. M. (1984). Autoradiographic demonstration of the projections from the mesencephalic locomotor region. *Brain Research, 307*(1–2), 263–276.

Stefanini, C., Orlandi, G., Menciassi, A., Ravier, Y., Spina, G. L., Grillner, S., et al. (2006). A mechanism for biomimetic actuation in lamprey-like robots. In *Proceedings of the First IEEE/ RAS-EMBS International Conference on Biomedical Robotics and Biomechatronics* (pp. 579–584) Rome, Italy.

Stehouwer, D. J. (1987). Compensatory eye movements produced during fictive swimming of a deafferented, reduced preparation in vitro. *Brain Research, 410*(2), 264–268.

Stein, P. S. (2008). Motor pattern deletions and modular organization of turtle spinal cord. *Brain Research. Brain Research Reviews, 57*(1), 118–124.

Stein, P. S. G., & Smith, J. L. (1987). Neural and biomechanical control strategies for different forms of vertebrate hindlimb motor tasks. In P. S. G. Stein, S. Grillner, A. I. Selverston, & D. G. Stuart (Eds.), *Neurons, Networks, and Motor Behavior* (pp. 61–73). Cambridge, MA: MIT Press.

Stein, W. (2009). Modulation of stomatogastric rhythms. *Journal of Comparative Physiology. A, Sensory, Neural, and Behavioral Physiology, 195*(11), 989–1009.

Still, S., & Tilden, M. W. (1998). Controller for a four legged walking machine. In L. S. Smith & A. Hamilton (Eds.), *Neuromorphic Systems: Engineering Silicon from* Neurobiology (pp. 138–148). Singapore: World Scientific.

Taga, G. (1998). A model of the neuro-musculo-skeletal system for anticipatory adjustment of human locomotion during obstacle avoidance. *Biological Cybernetics, 78*(1), 9–17.

Taghert, P. H., & Nitabach, M. N. (2012). Peptide neuromodulation in invertebrate model systems. *Neuron, 76*(1), 82–97.

Talpalar, A. E., & Kiehn, O. (2010). Glutamatergic mechanisms for speed control and network operation in the rodent locomotor CPG. *Frontiers in Neural Circuits, 4,* 19.

ten Donkelaar, H. J. (1982). Organization of descending pathways to the spinal cord in amphibians and reptiles. *Progress in Brain Research, 57,* 25–67.

Ting, L. H., & McKay, J. L. (2007). Neuromechanics of muscle synergies for posture and movement. *Current Opinion in Neurobiology, 17*(6), 622–628.

Tsujita, K., Toui, H., & Tsuchiya, K. (2004). Dynamic turning control of a quadruped robot using nonlinear oscillators. In *Proceedings of the IEEE/RSJ International Conference on Intelligent Robots and Systems* (pp. 969–974), Sendai Japan.

Uematsu, K., & Todo, T. (1997). Identification of the midbrain locomotor nuclei and their descending pathways in the teleost carp, *Cyprinus carpio. Brain Research, 773*(1–2), 1–7.

Verdaasdonk, B. W., Koopman, H. F. J. M., & van der Helm, F. C. T. (2006). Energy efficient and robust rhythmic limb movement by central pattern generators. *Neural Networks, 19*(4), 388–400.

Viana Di Prisco, G., Wallén, P., & Grillner, S. (1990). Synaptic effects of intraspinal stretch receptor neurons mediating movement-related feedback during locomotion. *Brain Research, 530,* 161–166.

Wallén, P., & Williams, T. L. (1984). Fictive locomotion in the lamprey spinal cord in vitro compared with swimming in the intact and spinal animal. *Journal of Physiology, 347,* 225–239.

Wallén, P., Ekeberg, Ö., Lansner, A., Brodin, L., Traven, H., & Grillner, S. (1992). A computer-based model for realistic simulations of neural networks: II. The segmental network generating locomotor rhythmicity in the lamprey. *Journal of Neurophysiology, 68,* 1939–1950.

Webb, B. (2001). Can robots make good models of biological behaviour? *Behavioral and Brain Sciences, 24*(6), 1033–1050.

Webb, B. (2002). Robots in invertebrate neuroscience. *Nature, 417,* 359–363.

Wilbur, C., Vorus, W., Cao, Y., & Currie, S. N. (2002). A lamprey-based undulatory vehicle. In J. Ayers, J. L. Davis, & A. Rudolph (Eds.), *Neurotechnology for Biomimetic Robots* (pp. 285–296). Cambridge, MA: MIT Press.

Williams, T. L., Sigvardt, K. A., Kopell, N., Ermentrout, G. B., & Remler, M. P. (1990). Forcing of coupled nonlinear oscillators: Studies of intersegmental coordination in the lamprey locomotor central pattern generator. *Journal of Neurophysiology, 64*, 862–871.

Williamson, M. M. (1998). Neural control of rhythmic arm movements. *Neural Networks, 11*(7–8), 1379–1394.

Wolpaw, J. R., & Tennissen, A. M. (2001). Activity-dependent spinal cord plasticity in health and disease. *Annual Review of Neuroscience, 24*(1), 807–843.

Wolpert, D. M., Ghahramani, Z., & Jordan, M. I. (1995). An internal model for sensorimotor integration. *Science, 269*(5232), 1880–1882.

Yakovleff, A., Cabelguen, J.-M., Orsal, D., Gimenez y Ribotta, M., Rajaofetra, N., Drian, M. J., et al. (1995). Fictive motor activities in adult chronic spinal rats transplanted with embryonic brainstem neurons. *Experimental Brain Research, 106*(1), 69–78.

9 Computational Models in Motor Control

NICOLAS SCHWEIGHOFER

Major advances in the field of human motor control have been made possible by insights gained in the engineering fields of control theory (including the subfields of linear control, optimal control, and state estimation), machine learning, and robotics. Thus, a discussion of engineering concepts is necessary for any up-to-date introduction to motor control. In this review, we provide a gentle overview of these concepts.

9.1 Introduction: The Motor Control Problem

As a starting point, consider the task of reaching to a spatial target. This action can be decomposed into five basic steps (Frey, Fogassi, et al., 2011): (1) determine the position of the target in the environment, often through vision; (2) determine the current position of the hand, which, for greater accuracy, can be determined by a combination of sensory inputs (visual, proprioceptive, or both) and prediction via an internal model (this concept will be developed in detail below); (3) compute the spatial relationship between the target and the hand, via transformation of sensory information into a common frame of reference; (4) form a plan that specifies the direction and distance from the current position of the hand to the target location; (5) finally, compute the motor commands that cause the hand to reach the target with sufficient precision within a reasonable period of time and energetic cost. (You can rehearse this sequence next time you want to start this really cool new application on your cell phone!)

Because we perform accurate and fast reaching movements numerous times during the day with no apparent effort, generating the motor commands for such movements may appear deceptively simple. However, four major characteristics of the human neuromuscular system, *nonlinearity*, *redundancy*, *delays*, and *noise*, which we will review in the next section (section 9.2), make the generation of reaching movement a very complex enterprise indeed (Franklin and Wolpert, 2011). Another complexity involves the constraints of the task for which reaching may be just a part. See, for example, chapter 20, "Reach and Grasp: Control, Development, and Recognition." And, of course, reaching is only one among many motor control problems: see also chapter 8, "Motor Pattern Generation," with its focus on locomotion, and chapter 19, "Saccades and Smooth Pursuit Eye Movements."

In section 9.3, we review how feedback control allows control of nonlinear systems that include noise, and how feedforward control allows control when feedback has long delays. In section 9.4, we discuss forward and inverse internal models, which are crucial for accurate and fast control. In section 9.5, we concentrate on the role of forward models for state estimation, which is necessary for feedback control with delays. In section 9.6, we discuss how internal models can be learned from experience, which is necessary because accurate internal models are important for accurate control and estimation. Finally, in section 9.7, we briefly review optimal control, which has recently been shown to account for many movement features and elegantly solves the redundancy problem, namely, how to choose one of many solutions to perform a movement in order to achieve the same goal.

9.2 The Complex Motor System

9.2.1 NONLINEARITY The mechanical properties of the limbs are highly complex and nonlinear. Unlike linear systems, the difficulty with nonlinear systems is that it is not possible to simply scale a motor input to obtain a proportionally scaled output. First, muscles exhibit highly nonlinear properties during movement, with respect to velocity and length. Second, moment arms are not constant and depend nonlinearly on the joint configuration; tendons have nonlinear responses as well. Finally, the dynamics of multiple-link systems like the human arm is highly complex. The mathematical computation of the dynamics of the human arm with its 7 degrees of freedom (DOFs) contains more than 3,000 mathematical terms; many of those are nonlinear and combining kinematic variables from different joints. Even a model of the dynamics of a simple two-dimensional arm contains enough complex nonlinear terms with cross coupling between joint variables (that account for inertial interaction, Coriolis, and centripetal torques) to scare off the typical beginning student. As noted above, we do not suggest that the brain performs such complex computations mathematically when generating a motor command; however, this does not remove the fact that the highly nonlinear dynamics is here to stay and cannot be ignored when generating accurate and fast arm movements.

Nonetheless, despite this high degree of complexity and nonlinearity, the nervous system can perform fast and accurate arm movement thanks to a sophisticated combination of feedback and feedforward controllers, as described below.

9.2.2 Redundancy Bernstein (1967) viewed redundancy as the central problem in motor control (Latash and Latash, 1994). To understand this, consider the many ways we can reach a target in front of us. First, either of the two arms, or even the trunk, can be used for the reach. Second, for each hand, an infinite number of trajectories and movement times are possible to reach the target. Such trajectories are due to the redundancy of each arm with respect to extrapersonal space: there are 7 DOFs in the arm to position and orient the hand in the six dimensions of external space, and many more DOFs if the trunk is used as well for the reach. Third, once the joints are selected, the problem is to select the right muscle combinations since many muscles span multiple joints. Fourth, for the selected muscles, there are infinite possible combinations of muscle coactivation pattern that can results in the same trajectories. Finally, there are vastly many more neurons in the central nervous system (CNS) than muscles. Despite the possibility of multiple solutions at all these levels, reaching patterns in (healthy) humans exhibit striking regularities in trajectories, joint angle patterns, and muscle contractions. As we will discuss below, there is increasing evidence showing that the nervous systems solves the redundancy problem by selecting the specific combinations of hand, joint, and muscle trajectories that minimize a "cost."

9.2.3 Delays Our nervous system must cope with transmission and conduction delays in the afferent sensory inputs, internal processing, and efferent motor outputs, as well as in the muscle and inertial responses. Such delays are long relatively to movement time, causing difficulties for control. It is thought that the shortest neural delays are in the monosynaptic stretch reflex (<40 ms), followed by delays in transcortical long-latency stretch reflex (40–80 ms; Marsden, Merton, et al., 1978), and by delays of feedback processing involving the visual system (around 100 to 150 ms). Thus, for arm reaching movements, total visual feedback delays can last as much as half the duration of the movements themselves. As a result, only when the arm is about midcourse toward the target can visual sensory feedback be used to modify the current motor command. In the case of saccadic eye movements, there is no possibility of correction based on sensory feedback because saccades are extremely fast (as fast as 20–30 ms), with the delays longer than the movements themselves. Such long delays limit the role of feedback control, that is, control based on actual sensory measurements, in fast biological movements. As we will discuss below, the nervous system can generate fast and stable movements despite such delays by relying on adaptive feedforward control.

9.2.4 Noise At all levels, the neural system is corrupted by noise—neural activity that is not carrying information about the task. Noise in the sensory system limits our ability to perceive accurately the position of our body, either visually or proprioceptively, and the position of objects in the world. Variability in synaptic transmission, neural firing, and recruitment of motor neurons are all thought to influence the actual movement. In addition, noise in the motor system is known to be, at least in part, signal-dependent: that is, the greater the mean motor signal, the greater the noise. In motor neurons for instance, the coefficient of variation (standard deviation divided by the mean firing) is between 10% and 25%. Signal-dependent noise explains why the faster one reaches, the greater the uncertainty in the final position of the hand (Harris and Wolpert, 1998). The converse of this is that one learns how much to slow down a movement when increasing accuracy is desired—contrast punching a ball and threading a needle. Note, however, that constant noise also contributes to movement variability (van Beers, Haggard, et al., 2004) and is thought to particularly affect slow movements (Todorov, 2005). Noise in the nervous system is thought to have both central (Churchland, Afshar, et al., 2006) and peripheral components (Jones, Hamilton, et al., 2002).

9.3 *Feedback and Feedforward Control*

What computational processes are used to generate the motor commands that bring the hand to the target in spite of high complexity and nonlinearity, redundancy, delays, and noise? Neuroscientists, robotics experts, and engineers distinguish between two general types of control—*feedback* and *feedforward*—each with its own unique advantages and drawbacks. As we will see in the following, the brain uses a combination of these two types of control, which, although they appear radically different at first, are blended seamlessly into a common form of combined feedforward and feedback control to achieve "best" performance. In this section, we will define "best" as the attainment of a desired target with minimal error (and in a reasonable amount of time). In the section on optimal control, we will refine our definition of "best" via the use of a performance cost that contains additional constraints besides end-point errors.

We first introduce the notions of state, dynamical systems model, and controller.

State: The state of a biomechanical dynamics system X is a set of physical quantities, the specification of which, together with external inputs, completely determines the trajectory of the system. For example, in a Newtonian system, the coordinates must be supplemented by the corresponding velocities to yield all or part of the state. It in some sense encapsulates the prior history of the system in compact form.

Dynamical system model: In a dynamical system model, the next state equation expresses how the state changes as a function of the current state and the input. There is also often an output, or measurement, equation, which relates the internal state to the measures. The general form of the (nonlinear) equation of a discrete-time, possibly nonlinear, system is given by the following:

$$\begin{aligned} X(t_{k+1}) &= f(X(t_k), u(t_k), t_k) \\ Z(t_k) &= g(X(t_k), t_k) \end{aligned}$$

where $X(t_k)$ is the state of the biomechanical systems at time instant t_k $(X \in R^n)$, $u(t_k)$ the motor command at the same instant, $Z(t_k)$ is the measurement at time t_k, and functions f and g are referred to as the next-state function and the output, or measurement function, respectively. Two observations: (1) in many cases, we may describe the system via differential equations in continuous time, rather than on a discrete timescale as shown here, and (2) in most cases, the f and g will *not* depend explicitly on time.

This formulation of a dynamical system is extremely general, as dynamical systems are descriptive models of the temporal evolution of an interrelated set of variables in a wide number of scientific fields (e.g., drugs kinetics, weather predictions, etc.). In motor control, let's first note that the "plant," that is, some body part together with the muscles that control it, is a dynamical system. In addition, the internal model (brain subsystem) that controls the plant is also a dynamical system. Thus, successful motor control involves the coupling of a neural dynamical system with a biomechanical dynamical system. For biomechanical systems, the next-state equation typically provides the dynamics of the system (say an arm), and the trajectory extracted from the values provided by the output equation the kinematics. The term "dynamics" relates forces or torques to movement variables (accelerations, velocity, and positions). For instance, for a single-joint arm rotating around one of its extremities, knowing the torque (the motor command) and the angular velocity and position (which together form the state) in the previous step determines the acceleration, position, and velocity on the next time step. The term "kinematics" refers to the transformation between coordinate systems (e.g., angular coordinates to Cartesian coordinates).

Analysis of nonlinear systems is often complicated. Therefore, one often locally linearizes the system around some nominal values of the state and motor command to simplify the analysis. Then, for instance, the next state of a linear, time invariant and discrete time system is given by the following:

$$\begin{aligned} X(t_{k+1}X(t_k)) &= AX(t_k) + Bu(t_k) \\ Z(t_k) &= HX(t_k) \end{aligned},$$

where $X(t_k)$ is the state of the system at time instant t_k with initial conditions $X(t_o)$, $(X \in R^n)$, $u(t_k)$the motor command at the same instant, $(u \in R^p)$, and $Z(t_k)$ is the measurement or output at time t_k. The matrices A, B are the state and control matrices, and H is the measurement matrix.

Controller: A controller is a device that affects the output variables of a given dynamical system, via the motor command , u which can be affected by adjusting certain input variables. In general, the controller will use measurements on the controlled system to estimate its internal state as the basis for determining how to choose u to get the system to behave in some desired way. In the next two subsections we will discuss the two kinds of controllers: feedback and feedforward.

9.3.1 Feedback Control In a feedback controller, feedback is used to control the states or the outputs of a dynamical system. The variable being controlled is measured and "fed back" into the controller to compute the motor command. Negative feedback is very often the most useful type of feedback, with the observed output value subtracted from the desired value (figure 9.1). Negative feedback control thus depends on error signals between the desired movement and the actual

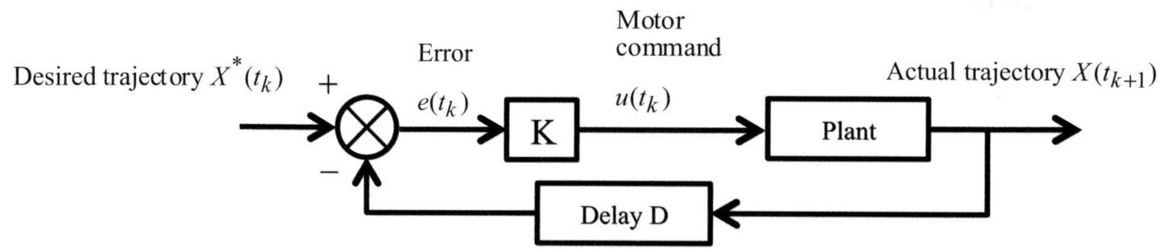

FIGURE 9.1 Negative feedback control. Example of a proportional controller. K is the gain, and D, the delays in the sensory feedback. The desired trajectory is compared with the delayed actual trajectory to create an error, which, after being multiplied by the feedback gain, generates the motor command. If the delays, D, is large, the stability of the system can only be ensured when the gains are relatively small (see text for details). Here the measurement process is such that measurement = state.

sensory consequences associated with its execution—the error can include, for instance, position error and velocity error. This error is used to compute the subsequent motor commands via the multiplication of the error signal by a set of (feedback) gains. Thus, if the error is large, then the control action is large. Mathematically, the so-called proportional feedback controller is given by the following:

$$u(t_k) = Ke(t_k) + u(0)$$

where $u(t_k)$represents the controller output, or feedback motor command, $e(t_k) = X^*(t_k) - X(t_k)$ represents the error, K represents the controller's gain, and $u(0)$ represents the steady-state control action (bias) necessary to maintain the variable at the steady state when there is no error. Note that other commonly used form is the PID controller: P depends on the present error, I on the accumulation of past errors, and D is a prediction of future errors. The I term eliminates the residual steady-state error that occurs with a pure proportional controller. The D term improves settling time and stability of the system.

Although negative feedback controllers such as the P or PID controller have only limited knowledge about the (possibly highly nonlinear) plant, they can, in some conditions, achieve very good performance in bringing the system quickly to the target. However, tuning the gains to obtain desirable regulation (staying at a given set point) or tracking (implementing set-point changes) can be a challenge. If the gains are too large, the control system can become instable, with or without oscillations, particularly in the presence of significant delay. In many industrial applications, such as robot control, delays are small and feedback control can be highly effective.

In contrast, as discussed above, delays are omnipresent in the nervous system. Feedback control of a system with long delays relatively to the movement times themselves can develop large oscillations as it attempts to self-correct. We have all experienced this phenomenon when trying to control the water temperature of an unfamiliar shower located at some distance from the water heater. You first turn the shower faucet, but after some delay, the temperature is hotter than you wish. You then attempt to adjust the temperature by increasing the flow of cold water but, shortly, the temperature becomes colder than you wish. With some luck, you may be successful on the third trial, but two or three more iterations may be needed to achieve the desired temperature. The difficulty here is that the change in water temperature (the consequence) is delayed relative to the time at which your movements occurred. The same phenomenon occurs in the nervous system: as a result of the time needed for signal conduction and transmission, the sensory consequences are delayed relative to our movements. As a result, discrepancies can develop between our motor plans and the actual sensory consequences of our actions. To alleviate the oscillations, it is possible to reduce the gains of the system. This reduces the amplitude of the motor commands, but movements then become slow and "sluggish." Feedforward control provides a potential solution to the delay problem.

9.3.2 FEEDFORWARD CONTROL Feedforward control can avoid the slowness of feedback control with low gains. In feedforward control, motor commands are generated directly from the goal of the action (e.g., reaching a target) (figure 9.2A). Although feedforward control does not suffer from the delay problem, a purely feedforward approach requires knowledge of the characteristics, such as the dynamics, of the controlled system. To achieve a high level of accuracy, biological feedforward controllers thus must be adaptive to learn about the controlled system from experience; just as you will (perhaps following several somewhat jolting experiences) eventually learn to rotate the faucet to the exact position necessary to produce the ideal water temperature.

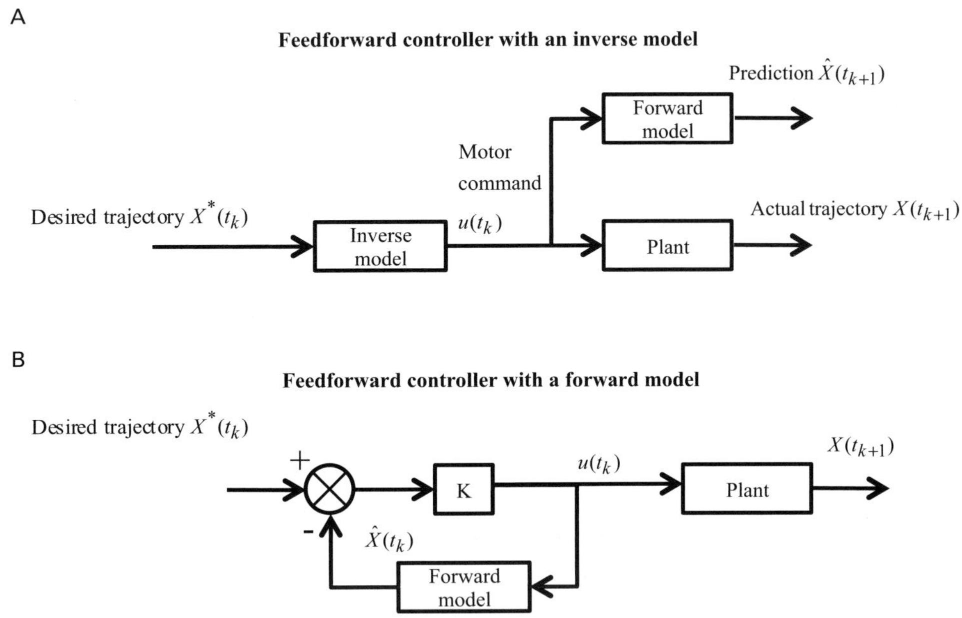

FIGURE 9.2 Feedforward control. (A) Feedforward controller based on an inverse model. This figure also illustrates how a forward model receives an efferent copy of the motor command to generate an estimate of the next state. Such an estimate could be used, for instance, in mental imagery or planning. Note that the actual trajectory does not influence the motor command. (B) Feedforward controller based on a forward model within an internal feedback loop.

Note that a feedforward system may use sensory information about the initial state of the plant in relation to the desired outcome of the action but will not use continual feedback-style monitoring of the system being directly controlled. (For instance, to compute the motor command to reach a target with the arm, the nervous system must know the initial position of the arm.) A good example of feedforward control in the nervous system is the vestibulo-ocular reflex, in which sensory information about head movements, detected via the vestibular canals, is used to control the eye movements to maintain fixed gaze in space as the head moves. Because no visual information is used to directly correct the eye position online, this is an (adaptive) feedforward system, not a feedback controller. Vestibular monitoring of the movement of the head provides the input to the controller, which commands compensatory movement of the eye. However, although this controller does not use visual feedback in generating control signals, it does employ "learning feedback"—it uses visual feedback on any errors in a movement to update itself to increase future accuracy.

As the feedforward controller does not use online feedback from sensors. As a result, the controller does not correct for (1) unanticipated disturbances from the environment or (2) noise in the motor command. But learning feedback can be employed to cumulatively and adaptively remove at least some inaccuracies in the knowledge of the dynamics of the controlled system.

Because of noise and disturbances in biological systems, movements resulting from a purely feedforward motor command will differ from what was planned; as a consequence, we may fail to reach the target. By combining both feedforward and feedback control, however, the nervous system can overcome these challenges and generate a motor command that brings the arm on target. Feedforward control produces fast movements in the face of long delays in neural transmission and in general brings the system close enough to its target that feedback control can rapidly correct movements when they deviate from the intended goal. Thus, we may see a fast feedforward phase followed by a slower feedback-dominated phase, as in the classic study of the reach to grasp (Jeannerod & Biguer, 1982; discussed in section 1.3.1).

9.4 *Internal Models: Forward and Inverse*

An important concept related to feedforward and feedback control is that of the internal model. Internal models can be classified into three conceptually distinct groups (Wolpert and Miall, 1996): forward models of body parts, forward models of external objects, and inverse models. In addition, although it is not completely clear whether the nervous system maintains such a distinction, it is useful to consider internal forward models of the kinematics (transformation between internal and external coordinate systems), forward models of the dynamics (transformation of motor commands to internal states), and forward models of internal states to sensory measurements, as well as all the respective inverse models. Note that this list is not exhaustive; for instance, internal models of muscles or muscle synergies can also be considered.

9.4.1 FORWARD MODELS OF BODY PARTS The first type of forward model mimics the causal flow of the physical process by predicting the consequences (e.g., position, velocity) of a motor command (Jordan and Rumelhart, 1992). Such dynamical forward models are neural representation of the future state of the physical system, such as the eye or the arm. Given the motor command, the forward model predicts the sensory consequence of this command, in effect mimicking the movements of the body in advance of actual movements. Thus, other parts of the brain can anticipate the effect of the movement and proceed accordingly; for example, one can proceed faster by initiating the next movement before the present movement is completed.

The equation of the forward model thus has the general form of the modeled dynamics. In its general discrete form,

$$\hat{X}(t_{k+1}) = \hat{f}(\hat{X}(t_k), u(t_k))$$

with the hat in $\hat{f}$ indicating that this is not the true dynamics of the system but an internal model within the nervous system, and the hat in $\hat{X}(t_k)$ indicating that the state is estimated (see section 9.5 below). Also note that there is no hat on the motor command $u(t_k)$; thus, the motor command is both sent to the actual biomechanical system and, in the form of an "efferent copy," sent to the input of the forward model.

For a linear time invariant system, we have the following:

$$\hat{X}(t_{k+1}) = \hat{A}\hat{X}(t_k) + \hat{B}u(t_k)$$

Because the internal model is necessarily learned (see below), the matrices $\hat{A}$ and $\hat{B}$ are not equal to the matrices of describing the actual dynamical system A and B. However, for simplicity and by convention, the matrices of the forward model are taken as equal to matrices describing the system. Thus, the equations governing the description of the actual dynamical system and that of the forward model are very similar looking, especially in computer simulations. However, the reader (and the beginning programmer of motor control systems) must be aware that these are very different entities. The forward models are implemented and learned within neural circuitry within the brain, presumably in the cerebellum and parietal cortex (Desmurget and Grafton, 2000). In contrast, the dynamical system represents an actual biomechanical system, such as an arm or an eye together with the attached muscles.

Forward models have many potential uses (Wolpert and Miall, 1996). First, in control, by replacing the actual delayed sensory input with predicted sensory inputs that are generated within the CNS, and hence with small conduction delays, a forward model can alleviate the delay problem faced by feedback control, as we will discuss below. Second, by predicting the sensory consequences of the movements in advance of the actual delayed sensory feedback that accompanies movement, forward models are used as part of the state estimation process, as we will discuss in more detail in section 9.5. Third, predictions from a forward model allow filtering of sensory information: expected sensory prediction can cancel out the sensory changes induced by self-motion, thereby enhancing more relevant sensory information. Fourth, forward models may be important in the sense of self, as they allow the actor to determine whether an external force was responsible for a movement. Fifth, forward models are also presumably used in mental imagery and motor planning by allowing comparison of the outcomes of alternative movements in advance of the actual movement.

9.4.2 FORWARD MODELS OF THE ENVIRONMENT The second group of internal models is conceptually similar to the forward models of body parts but instead represents the behavior of the external environment. For instance, the forward model could be that of the dynamics of common objects. The predictions from such models are used in motor planning and rapid movement corrections if the actual sensory measurements do not correspond to the expected sensory measurements. A vivid example of the use of such forward models in planning is the overshoot in lifting an object that we expect to be heavier than it actually is (e.g., an empty coffee cup that is believed to be full).

9.4.3 INVERSE MODELS An inverse model can be conceptualized as an inverted forward model (or quasi-inverse, as we see below). Given the desired sensory consequence (grasping the cup), and the current state of the body and environment, the inverse internal model computes a motor command to be sent to the body. Specifically, the inverse model generate a motor command $u(t_k)$ given the desired state, $X^*(t_k)$:

$$u(t_k) = h(X^*(t_k), X(t_k))$$

so that h and f have a *quasi*-inverse relationship

$$X(t_{k+1}) = f(X(t_k), h(X^*(t_k), X(t_k))).$$

Note that h is not necessarily unique and may not even exist for all $X^*(t_k)$, but this does not affect the model's formulation. Unlike forward models, inverse models are ill-posed when the dimension of the motor command is greater than that of the desired sensory consequence (which is typically the case in the human motor system).

In the case of moving a three-segment arm to position the hand in a 2-D space, for instance, the inverse kinematics model is ill-posed because the same hand position can be achieved by an infinite number of arm postures. Note how, in contrast, the forward kinematic model is well posed and causal, as there is a single hand position for a specific arm posture.

This simplest form of feedforward control (figure 9.2A) requires the inverse model to faithfully capture the characteristics of the actual physical systems (muscle properties, arm dynamics, etc.) it represents for movements to achieve the desired plan. However, as discussed above, environmental disturbances and internal noise will lead to unavoidable deviations from the desired movement.

In addition, feedforward control can also be achieved with forward models because an inverse model can be conceptualized as an inverted forward model. Specifically, a feedback control system with high gain is equivalent to an open-loop feedforward system that utilizes an explicit inverse model: if the gains are high and delays minimal, reaching the desired output can be very quick—see Jordan (1996) for a demonstration. (Remember however, that with delays, the system becomes instable.) The feedback loop is thus implicitly inverting the plant, which is intuitive because a high-gain feedback controller corrects errors as rapidly as possible. We can thus treat a feedback control system with a (high) finite gain and small delays as approximating the inverse model of the plant. It is also worth noting that the feedback control system with internal feedback via a forward model, as shown in figure 9.2B, can be considered in its own right as an implementation of a feedforward control system. If the forward model is an accurate model of the plant, then in the limit of high gain in this internal loop it is equivalent to an inverse model of the plant. Thus, the internal feedback loop is an alternative implementation of a feedforward controller.

9.4.4 Hybrid Feedback—Feedforward Controllers As discussed above, although feedforward control does not suffer from the delay problem (because there is no need to wait for sensory information), the internal models are never perfect, the environment is not static, and noise corrupts the motor command. The nervous system overcomes these challenges by combining both feedforward and feedback control: feedforward control (based either on inverse or forward models; see above) enables the fast movements whereas feedback control enables correction of these movements when they deviate from the intended goal. In its simplest implementation, the feedforward/feedback controller sums up the output of a feedback and of a feedforward controller (figure 9.3A).

As noted above, it is impractical to allow the gains to grow large in real feedback systems because of the delays in the feedback loop. It is however possible to increase the gain and achieve good performance with large delays if the delays themselves are modeled. The Smith control system uses the forward model simultaneously for control and state estimation (see figure 9.3B): the forward model in the internal feedback loop is used to approximate the inverse model for control (this loop can have high gains because of small delays in the internal loop). In addition, the forward model output, after passing through the delay model, is used to cancel out the predicted part of the feedback. Only the delayed feedback signal that deviates from that predicted by the forward model and delay are corrected for. This error now represents the true (delayed) disturbances that need to be canceled, without the effect of the delays between the sensory inputs and the (delayed) predictions. Note however that while modeling delays as long as 200 ms is quite trivial in computer simulations via "tapped" delay lines, in the nervous system modeling delay is a less trivial proposition although recurrent networks may be used.

This discussion suggests that the distinction between feedback and feedforward control in actual biological motor control may not be as crisp as it first appeared when we first introduced the two types of controllers. The balance between feedforward and feedback control depends on the degree to which the current state estimate is influenced by the internal prediction from the forward model or by sensory feedback. Early in the movement, before sensory feedback is available because of delays, the forward model within the internal feedback loop acts as a feedforward controller. Late in the

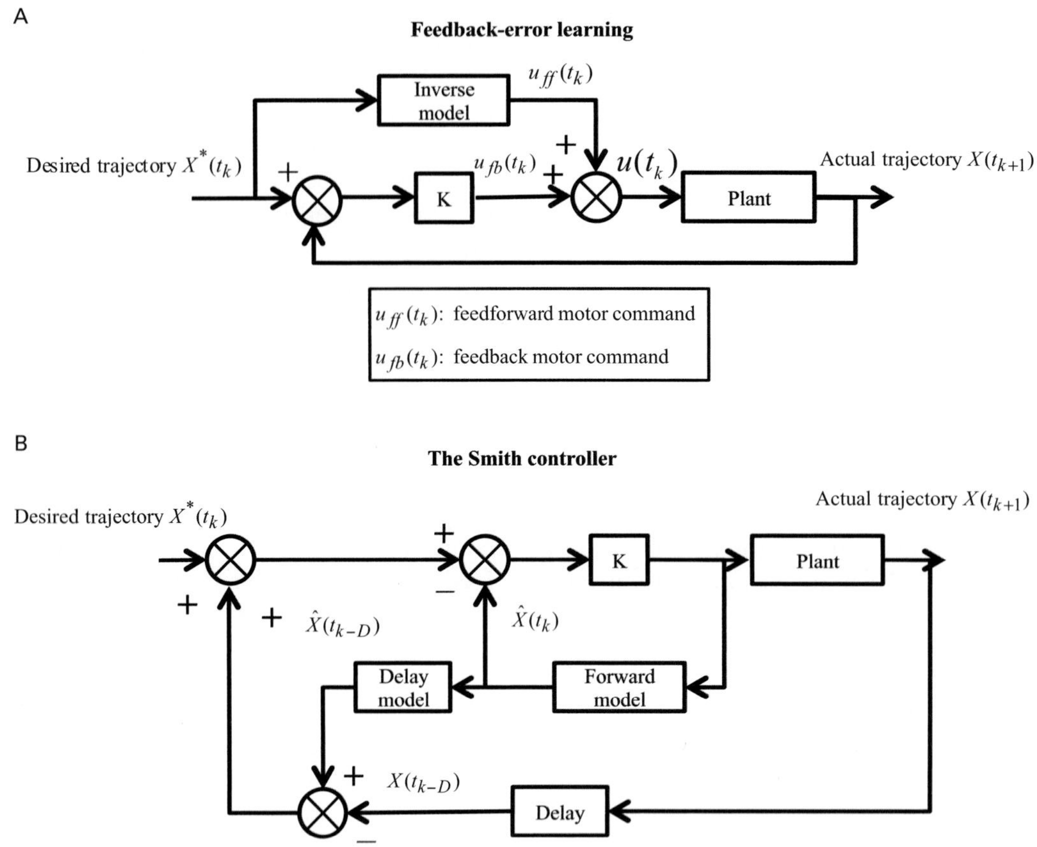

FIGURE 9.3 Hybrid feedback/feedforward control. (A) In the hybrid controller called "feedback error learning," the feedback error is used to learn the inverse model. (B) In the hybrid controller called the Smith predictor, the forward model is used both for control and sensory cancellation. See how the forward model and the delay model are used to cancel the delayed state, such that there is no temporal mismatch between the predicted and actual state.

movement the controller shifts more toward a feedback controller.

9.5 State Estimation

As discussed above, forward models are crucial in estimating the state before delayed sensory measurements are available. However, because forward model can be imprecise, or because of disturbances and the omnipresent noise in the neural and motor systems, it is important to incorporate actual sensory measurements to refine the state estimates. Thus, the role of the state estimator is to correct the predictions of the state derived from the forward model by computing the error between the actual sensory output and the predicted sensory output (this error is called the "innovation" or "measurement residual"). According to the optimal state estimation framework, noisy and delayed sensory feedback signals are combined with forward model output to provide state estimation (figure 9.4, dashed-line box), via a Kalman filter (Welch and Bishop, 2004). The Kalman filter has two modes of function: a "predict" and a "correct" mode. In the absence of any measurement of output variables, the filter is in "predict mode." The forward model is used to update the internal states and to generate predictions of the mean and the variance of the outputs. The variance of the internal states increases with time to reflect the growing uncertainties when the data are not available during the predict mode. Because prediction from such a model will undoubtedly deviate from actual measurements, the Kalman filter is designed to correct the output predictions with measured sensory data whenever available. The filter is then in "correct" mode: the measured data are optimally combined with prediction to minimize internal state estimate errors.

Using an experimental paradigm in which participants estimated the location of one of their hands at the end of movements made in the dark and under externally imposed forces, it was shown that humans estimate their hands' positions with a process that is

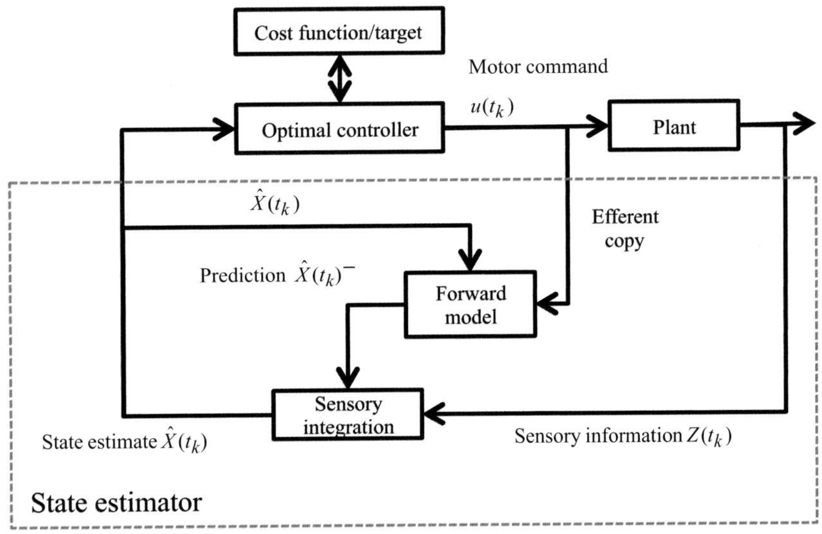

FIGURE 9.4 Optimal feedback controller with optimal state estimator (dashed-line box, lower part). Note how the goal of the movement is embedded within the cost function.

consistent with such an optimal state estimator (Wolpert, Ghahramani, et al., 1995).

9.6 *Optimal Control*

9.6.1 OPTIMALITY CRITERIA AND OPTIMAL CONTROL

Actual hand paths during point-to-point reaching movements are highly stereotypical within and across individuals, typically close to a straight line and with characteristic near smooth bell-shaped velocity profile. Why and how would the nervous system generate such stereotypical movements? More generally, how do humans choose one particular pattern of muscle coordination over the infinitely many other possible patterns that achieve the same task goal? Clearly, only specifying an end-point error is insufficient to solve this redundancy problem. Similarly, following a prespecified end-point trajectory solves only the trajectory redundancy problem and not the redundancy at the level of joints and muscles. The reader can conceive, however, that some movements would be more "economical" in some sense than other movements, much like a car following a straight route would consume less gasoline that a car following a windy route to arrive at the same location.

An *optimal controller* computes the motor commands that minimize an objective function (or "cost function"), which is a function of state and control variables (Todorov, 2005). Optimal control has a long history in engineering; for example, it has supported sending humans to the moon and back. Recently, these ideas have permeated the motor control field, and there is mounting evidence that the nervous system has devised solutions that are near optimal with respect to some cost. Much research is being undertaken to estimate what the optimality criteria in motor control may be. In most cases, the goal is to find the motor command u defined on the interval (t_0, t_N) which minimizes the expectation of a state-dependent cost (e.g., accuracy which is reduced the closer the final end point is to the desired target) and a command-dependent cost (e.g., effort), subject to the constraints of the stochastic dynamics. There are different methodological approaches for solving optimal control problems based on two optimality principles, namely, the Pontryagin maximization and the Bellman principle of optimality. In the following, we first briefly review open-loop optimal control models and then review optimal feedback optimal control models.

9.6.2 OPEN-LOOP OPTIMAL CONTROL MODELS

Optimal control schemes, often with different ad hoc cost functions, have been able to account for a wide range of experimental data such as kinematic regularities (i.e., near straight trajectory and bell-shaped velocity profiles), amplitude/duration scaling (Guigon, Baraduc, et al., 2007), and speed–accuracy trade-off via signal-dependent noise (Hoff, 1992). In a seminal study, Flash and Hogan (1985) proposed that the trajectories are generated by a controller that generates smooth movements by minimizing jerk (the derivative of acceleration). The resulting smooth trajectory is straight with bell-shaped velocity, as found experimentally. A feedback/feedforward controller can then ensure that the motor commands are appropriate for the arm to implement this trajectory. A problem with this design however was that the movement duration has to be specified in advance. Hoff (1994) extended this model by proposing that the nervous system would minimize a cost comprised of both the jerk and time.

What should the cost functions be in optimal control? The minimum jerk kinematics-only optimal control schemes have the important drawback that any change in dynamics will not affect the trajectory since only kinematics is considered in planning. To solve this issue, Uno, Kawato, et al. (1989) proposed a model of trajectory planning in which the dynamics was taken into account. The model, based on a cost that penalized torque change, could account for a number of reaching movement characteristics, such as slight deviations from straight lines. Starting from the consideration that it is unclear how the CNS should compute quantities such as jerk or torque derivatives, but instead should take the task goal into account when planning movement, Harris and Wolpert (1998) proposed that the only quantity that is taken into account during planning is the end-point variance of the reach. The key assumption was that the noise in the nervous system is signal dependent: the greater the motor command, the greater the noise. This "minimum variance" controller generates smooth optimal trajectories because abrupt changes in the trajectory of hand would require large driving signals, which would carry more signal-dependent noise and therefore be suboptimal.

9.6.3 Closed-Loop Optimal Control As discussed above, feedback control is necessary to accomplish tasks successfully in light of a changing and noisy neuromusculoskeletal system and environment. Thus, optimal control needs to be updated into optimal feedback control schemes (figure 9.4). Such schemes often include optimal state estimators to deal with sensory delays and noise.

In the above discussion on feedback control, we have assumed that some "desired output" was available at any time and the goal of control was to reduce the error between this trajectory and the actual trajectory. For instance, a desired trajectory of the hand would be sent to an inverse kinematic model and then be fed to a feedforward controller (implemented with, say, an inverse model) that would try to realize the desired trajectory. An alternative that does not require a fixed desired trajectory is an optimal feedback controller with an appropriate time sequence of the feedback gains. This "control policy" generates the motor command directly from estimated states and goals and yields the best possible performance when motor noise as well as sensory uncertainty and delays are taken into account (Todorov and Jordan, 2002; Todorov, 2004). Cost functions that are given by the sum of a final error component and a motor "effort" (an approximation of muscle metabolic cost) component can account well for the characteristics of reaching movements (Todorov and Jordan, 2002). The first component of this cost function encodes the external goal in terms of task-relevant states, such as hand position error. Because of redundancy in the motor system, this term alone does not define a unique solution, however. Therefore, a second component, which acts as a regularization term, is added that penalizes large motor commands, often taking the form of the weighted sum of the squared motor commands. Although the open-loop minimal variance model also accounts for Fitts's law (see above), optimal feedback control also naturally accounts for perturbations during the movement (Hoff, 1992), such as target shifts in midcourse, while still minimizing the optimality criterion. Finally, one of the strength of optimal control it is that it solves the redundancy problem (Todorov and Jordan, 2002). On average over repeated trials, the variability in internal coordinates often exceeds the variability in task space coordinates. For example, in point-to-point reaching movement with the unconstrained arm, the variability of joint angular trajectories is significantly larger than the variability of end-effector coordinates. Optimal feedback control provides "minimal intervention." From the viewpoint of task achievement, it is not necessary to suppress the variability for the unconstrained internal DOFs, such that they tend to exhibit higher variability, as seen biologically. And indeed, with a cost function such as the sum of the squares of the motor command over the movement, acting is expensive. Thus, a better solution, that is, one that minimizes the cost, is to do nothing about the deviation from the average behavior unless this affects task performance; that is, the optimal feedback controller does not act on the null space as it does not influence the task (Todorov and Jordan, 2002) (at a given configuration of a redundant arm, we have seen that there is an infinity of small joint displacements that do not affect the configuration of the hand—these displacement take place in the null space).

We conclude with two final notes of caution about optimal control models as models of human motor control. First, it is important to remember that any behavior can be considered "optimal" if the cost function is chosen without constraints. As a result, and to avoid circularity, it is important to set the cost function a priori before testing it with different experimental conditions. Second, the low-level motor commands (e.g., muscle activations) for complex motor acts (a tennis serve, for instance) cannot be learned from pure trial and error with learning methods such as reinforcement learning because of the hard credit assignment problem in high dimension (for instance, which muscle activation at what moment caused my serve to fail?). A hierarchical optimal control scheme may be able to solve this problem, with a higher level in low dimension that monitors progress and exploits the many different ways of achieving the goal, and a lower level in high dimension that provides automatisms and corrections—see Li, Todorov, and Pan (2004). As an alternative, there might be a "language" of more abstract building blocks in motor control, called motor primitives (or schemas, basis behaviors, macros, or options, depending on the field of study) (Bizzi and Cheung, 2013). Using primitives, discrete and rhythmic movement can be modeled with nonlinear differential equations that form stable points and limit cycle attractors, respectively. Such equations can be learned with optimization methods, such that even rather complex movements such as a tennis stroke can be represented by such differential equations. This creates a computational link between motor primitives and optimal control—see Schaal, Mohajerian, et al. (2007) for review and additional references.

9.7 *Motor Learning*

9.7.1 Self-Supervised Learning of Forward Models For both processes of state estimation and control, the forward models should accurately match the actual characteristics of the plant. Furthermore, because the properties of the sensorimotor system are not static but change throughout life, or with fatigue, the nervous system needs to learn how to control highly complex musculoskeletal systems that are constantly varying. Internal models must thus continuously be updated based on experience. In theory, the forward models can be updated via the body's own movements. In this process, known as self-supervised learning, the error signal generated by the difference between the actual state and the predicted state from the forward model is used to train the forward model (Jordan and Rumelhart, 1992). "Eligibility traces," temporary records in the occurrence of an event (potentially neurochemical traces in synapses—see, for instance, Schweighofer and Arbib (1998)), may solve the temporal credit assignment problem of matching the delayed sensory error with previous model predictions.

9.7.2 Learning of Inverse Models An inverse dynamic model transforms desired states to motor commands. However, errors resulting from a movement are typically available in extrapersonal space and do not tell the learner what muscle activation is needed to achieve the goal. Furthermore, the motor commands are often in much higher dimension than the sensory measurements. This is known as the distal credit assignment problem: because the errors in motor commands are not directly available for learning, the inverse model is a trickier proposition than learning the forward model. How then can we obtain an error in motor coordinates that can be used to train the inverse model? We now discuss two possible biological schemes for such training: feedback error learning and reinforcement learning.

Feedback Error Learning Kawato and colleagues provided an astute and simple solution to learning the inverse model that has received at least some biological support (Kawato, Furukawa, et al., 1987). In their formulation, called feedback error learning, a feedback controller (with low gains as to ensure stability) provides an error signal, in motor coordinates, that can be used to train the inverse model. The feedback error approximates the directions and magnitudes of the necessary modifications of the inverse model. As the inverse model becomes more accurate, the hybrid control system relies less and less on the feedback system; as a result the movements can become faster without compromising stability. Note that this method solves the ill-posedness problem because the feedback controller provides online control during training. Kawato and colleagues proposed computational models of cerebellar motor learning based on the feedback-error-learning scheme (Kawato and Gomi, 1992; Schweighofer, Arbib, et al., 1998). They proposed that the second major inputs to the Purkinje cells of the cerebellum, the climbing fiber, represent motor-command errors generated by feedback controllers at the spinal, brainstem, and cerebral levels. These error signals then train the Purkinje cells' output via known long-term depression such that different regions of the cerebellum learn to execute predictive and coordinative control of different types of movements.

Reinforcement Learning via Exploration In a situation in which no error is given, but only binary feedback such as "good/bad," humans can make use of reward information in motor learning. Here, we briefly outline an example of how controllers can be learned via reinforcement learning—see, for instance, Williams (1992) and Peters and Schaal (2008) for in-depth treatment of the topic. Because the motor errors are in general not directly available to the learner, one possibility is to estimate the motor error by perturbing the motor commands with noise and then comparing the outcome of the resulting movement to the averaged outcome of the movement over some previous time window. If the perturbation has improved the movement outcome, then the motor commands should be updated in the direction of the perturbation. If not, the motor commands should be updated in the other direction. Such "stochastic policy gradient" reinforcement learning methods can include modulation of the amplitudes of the perturbations as performance improves (Williams, 1992). Because no motor error is directly available, reinforcement learning systems are slow, especially in high dimensions. It has been proposed that hierarchical controllers, in which higher level control in lower dimensions is learned via reinforcement learning, may facilitate this spatial credit assignment problem (Peters and Schaal, 2008). Note that this "exploration noise" should ideally be tuned to find a balance between exploration (of novel motor commands) and exploitation (of already learned motor commands).

Current human motor control research is exploring how reinforcement learning can lead to (near) optimal control in movements such as reaching movement (Izawa and Shadmehr, 2011).

9.8 *Conclusions*

We conclude this review with some important remarks. As in most of the computational motor control literature, we used "model" to designate two very different creatures. Our first use of "model," as in the title, refers to a mathematical model, which is a description of a system using mathematical concepts and language. Mathematical models can take many forms, including, but not limited to, dynamical system models or statistical models. Often, the quality of a scientific field depends on how well the mathematical models agree with results of repeatable experiments. As we have seen in this review, the agreement between models of motor control and experimental results have been exceptional in the last two decades or so. Our second use of "model" refers to internal neural models. Like mathematical models of other complex systems, neural internal models capture important features of the systems that they mimic (e.g., the muscle properties, biomechanics, and dynamics of the arm and hand). However, neural internal models reside within the CNS, and, unlike mathematical models, they do not perform computations explicitly. For example, the prediction of the posture of an arm after a reaching movement via an internal model of the arm's dynamics can be performed via a neural computation that is learned through experience, without the nervous system's having the slightest (explicit) idea about Newton's mechanics. In fact, the nervous systems does not use single sensory and control variables, as in the mathematical formalism presented here. Instead these variables and computations are distributed over populations of neural activities. In other words, the n-dimensional vector of the mathematical representation are coded via the m >> n-dimensional vector of the neural encoding. For instance, in the motor cortex a population of neurons that are only broadly tuned to a particular direction in three-dimensional space can uniquely determine the direction of reaching movements in three-dimensional space (Georgopoulos, Schwartz, et al., 1986). This brings us back to the hardest problem of all, that of redundancy, with the following critical question presumably keeping researchers of motor control busy for decades to come. How does the CNS compute the activity of millions of motor neurons to control a few dozens of muscles that will move your finger to turn on this really neat application on your cell phone?

ACKNOWLEDGMENTS I thank Stefan Schaal, Emmanuel Guigon, Chunji Wang, Youngmin Oh, and Amar Bains for comments on a previous draft of the manuscript and Evangelos Theodorou for help with the mathematics. This work was in part supported by grants NSF BCS-1031899 and R01 HD065438–02.

REFERENCES

Bernstein, N. A. (1967). *The Coordination and Regulation of Movement (trans. from the Russian and German)*. Oxford: Pergamon.

Bizzi, E., & Cheung, V. C. (2013). The neural origin of muscle synergies. *Frontiers in Computational Neuroscience, 7*, 51.

Churchland, M. M., Afshar, A., & Shenoy, K. V. (2006). A central source of movement variability. *Neuron, 52*(6), 1085–1096.

Desmurget, M., & Grafton, S. (2000). Forward modeling allows feedback control for fast reaching movements. *Trends in Cognitive Sciences, 4*(11), 423–431.

Flash, T., & Hogan, N. (1985). The coordination of arm movements: An experimentally confirmed mathematical model. *Journal of Neuroscience, 5*(7), 1688–1703.

Franklin, D. W., & Wolpert, D. M. (2011). Computational mechanisms of sensorimotor control. *Neuron, 72*(3), 425–442.

Frey, S. H., Fogassi, L., Grafton, S., Picard, N., Rothwell, J. C., Schweighofer, N., et al. (2011). Neurological principles and rehabilitation of action disorders: Computation, anatomy, and physiology (CAP) model. *Neurorehabilitation and Neural Repair, 25*(5 Suppl), 6S–20S.

Georgopoulos, A. P., Schwartz, A. B., & Kettner, R. E., et al. (1986). Neuronal population coding of movement direction. *Science, 233*(4771), 1416–1419.

Guigon, E., Baraduc, P., & Desmurget, M. (2007). Computational motor control: Redundancy and invariance. *Journal of Neurophysiology, 97*(1), 331–347.

Harris, C. M., & Wolpert, D. M. (1998). Signal-dependent noise determines motor planning. *Nature, 394*(6695), 780–784.

Hoff, B. (1992). *A Computational Description of Organization of Human Reaching and Prehension* (doctoral dissertation). University of Southern California, Los Angeles.

Hoff, B. (1994). A model of duration in normal and perturbed reaching movement. *Biological Cybernetics, 71*(6), 481–488.

Izawa, J., & Shadmehr, R. (2011). Learning from sensory and reward prediction errors during motor adaptation. *PLoS Computational Biology, 7*(3), e1002012.

Jeannerod, M., & Biguer, B. (1982). Visuomotor mechanisms in reaching within extra-personal space. In D. J. Ingle, R. J. W. Mansfield, & M. A. Goodale (Eds.), *Advances in the Analysis of Visual Behavior* (pp. 387–409). Cambridge, MA: MIT Press.

Jones, K. E., Hamilton, A. F. D. C., & Wolpert, D. M. (2002). Sources of signal-dependent noise during isometric force production. *Journal of Neurophysiology, 88*(3), 1533–1544.

Jordan, M. I. (1996). Computational aspects of motor control and motor learning. In H. Heuer & S. Keele (Eds.), *Handbook of Perception and Action: Motor Skills.* New York: Academic Press.

Jordan, M. I., & Rumelhart, D. (1992). Forward models: Supervised learning with a distal teacher. *Cognitive Science, 16,* 307–354.

Kawato, M., Furukawa, K., & Suzuki, R. (1987). A hierarchical neural-network model for control and learning of voluntary movement. *Biological Cybernetics, 57*(3), 169–185.

Kawato, M., & Gomi, H. (1992). A computational model of four regions of the cerebellum based on feedback-error learning. *Biological Cybernetics, 68*(2), 95–103.

Latash, L. P., & Latash, M. L. (1994). A new book by N. A. Bernstein: "On dexterity and its development." *Journal of Motor Behavior, 26*(1), 56–62.

Li, W., Todorov, E., & Pan, X. (2004). Hierarchical optimal control of redundant biomechanical systems. In *Proceedings of the 26th Annual International Conference of the IEEE Engineering in Medicine and Biology Society* (pp. 4618–4621).

Marsden, C. D., Merton, P. A., Morton, H. B., Adam, J. E. R., & Hallett, M. (1978). Automatic and voluntary responses to muscle stretch in man. In J. E. Desmedt (Ed.), *Progress in Clinical Neurophysiology: Vol. 4. Cerebral Motor Control in Man: Long Loop Mechanisms* (pp. 167–177). Basel: Karger.

Peters, J., & Schaal, S. (2008). Reinforcement learning of motor skills with policy gradients. *Neural Networks, 21*(4), 682–697.

Schaal, S., Mohajerian, P., & Ijspeert, A. (2007). Dynamics systems vs. optimal control—A unifying view. *Progress in Brain Research, 165,* 425–445.

Schweighofer, N., & Arbib, M. A. (1998). A model of cerebellar metaplasticity. *Learning & Memory, 4*(5), 421–428.

Schweighofer, N., Arbib, M. A., & Kawato, M. (1998). Role of the cerebellum in reaching movements in humans: I. Distributed inverse dynamics control. *European Journal of Neuroscience, 10*(1), 86–94.

Todorov, E. (2004). Optimality principles in sensorimotor control. *Nature Neuroscience, 7*(9), 907–915.

Todorov, E. (2005). Stochastic optimal control and estimation methods adapted to the noise characteristics of the sensorimotor system. *Neural Computation, 17*(5), 1084–1108.

Todorov, E., & Jordan, M. I. (2002). Optimal feedback control as a theory of motor coordination. *Nature Neuroscience, 5*(11), 1226–1235.

Uno, Y., Kawato, M., & Suzuki, R. (1989). Formation and control of optimal trajectory in human multijoint arm movement: Minimum torque-change model. *Biological Cybernetics, 61*(2), 89–101.

van Beers, R. J., Haggard, P., & Wolpert, D. M. (2004). The role of execution noise in movement variability. *Journal of Neurophysiology, 91*(2), 1050–1063.

Welch, G., & Bishop, G. (2004). *An Introduction to the Kalman Filter.* Chapel Hill: University of North Carolina—Chapel Hill.

Williams, R. J. (1992). Simple statistical gradient-following algorithms for connectionist reinforcement learning. *Machine Learning, 8*(3–4), 229–256.

Wolpert, D. M., Ghahramani, Z., & Jordan, M. I. (1995). An internal model for sensorimotor integration. *Science, 269*(5232), 1880–1882.

Wolpert, D. M., & Miall, R. C. (1996). Forward models for physiological motor control. *Neural Networks, 9*(8), 1265–1279.

10 Reinforcement Learning

NATHANIEL D. DAW

10.1 Learning Behavior: Classical Conditioning and Prediction Errors

In this chapter we consider the problem of reinforcement learning (Sutton and Barto, 1998): trial-and-error learning that serves to increase the chance of gathering rewards and avoiding punishments—all in the absence of a teacher who specifies the "correct" strategy. This is the problem that faces an animal foraging for food, or a human practicing to ride a bicycle. In these tasks, and many others, the agent faces a number of options—different potential sources of food or different moves—and must learn by trial and error which one to choose.

As we will see, many computational models of this learning process assume that decisions between options are made based on predictions the agent makes about the value of each option (e.g., the amount of food or the chance of winning). On this view, control depends on prediction: that is, learning to predict biologically important events is an essential subcomponent of the full reinforcement learning problem of learning to make good decisions.

10.1.1 Classical Conditioning Interestingly, the prediction problem has been studied by psychologists in isolation from the decision problem for more than a century. Consider the famous classical conditioning experiments of Pavlov in which dogs could be trained, by repeated pairings in which a bell was sounded before food was presented, that a bell predicted food. That the animals were able to learn this predictive relationship was evidenced by the fact that they came, over training, to salivate to the bell, a stimulus (known as a "conditioned stimulus") with no particular previous biological significance. Indeed, they did so even prior to the presentation of the food or on trials when the food was not presented. Here, the salivation behavior (known as a "conditioned response") evidently serves as a direct behavioral readout of the learned association. That is, the animal is not salivating so as to obtain food; it is salivating because it expects food (the "unconditioned stimulus").

Thus, this general sort of experiment can be used to study prediction learning, separate from choice. In particular, having a behavioral window on animals' expectancy allows researchers to manipulate in many different ways the evidence they provide to the animal about a predictive association and to study how it affects the learned predictions. In modern times, these tasks have also provided researchers with a well-understood foundation for studying the neural substrates supporting these functions.

Many variations of this basic design have been studied, across species ranging from mollusks to humans, and the basic behavioral phenomena are well preserved. Expectation of different biologically important events—known as unconditioned stimuli—evokes different sorts of behaviors, often in an event- or species-specific manner. For instance, rats freeze in expectation of shock; many organisms blink in anticipation of noxious stimulation to the eye; and in humans, the galvanic skin response, related to sweating, is often used to measure expectancy of either appetitive or aversive events.

10.1.2 The Rescorla–Wagner Model We can formalize the prediction problem as follows. Consider a set of trials t on which an animal observes some stimulus s_t and receives some (potentially zero) reward r_t. (In this chapter, we speak in terms of rewards although the same models apply to aversive unconditioned stimuli.) We assume the animal maintains some prediction $V(s)$ (for "value") of the degree to which the reward is expected in the presence of stimulus s and that this controls the degree of conditioned responding to s. Thus, the modeling question is how to update the predictions V in light of each trial's experience.

An important proposal, from the model of Rescorla and Wagner (1972), is, when stimulus s_t is observed followed by reward r_t, to update $V(s_t)$ to a weighted average of its previous value and the observed reward:

$$V(s_t) \leftarrow (1-\alpha)V(s_t) + \alpha r_t.$$

The factor $0 < \alpha \leq 1$ is a free parameter known as the learning rate. This learning rule has the reasonable property of nudging the prediction incrementally toward the observation, with step size α. Asymptotically (as can be verified by recursively substituting the update expression inside itself), the prediction consists of the average of all rewards received previously in the presence of the stimulus, with the most recent reward weighted most heavily and the weight for prior rewards declining exponentially in their lag. In particular, the value is the sum over previous rewards with the d_{th} preceding reward weighted by $\alpha(1-\alpha)^d$. This exponentially (in $1-\alpha$) decaying dependence (figure 10.1A) is a key hallmark of this theory, which we will see verified in both behavioral and neural data later in this chapter. Learning in this way can capture basic learning curves for acquiring simple associations (figure 10.1B) or extinguishing previously acquired ones by nonreinforcement (figure 10.1C).

The learning rule can also be rearranged to reveal a particularly suggestive form:

$$V(s_t) \leftarrow V(s_t) + \alpha\delta_t. \tag{10.1}$$

This equation depends on the *prediction error*: the difference between received and predicted rewards:

$$\delta_t = r_t - V(s_t). \tag{10.2}$$

Viewed in this way, the update equation is seen to be adjusting V to reduce the discrepancy between the prediction and observation. Such a process is known as "error-driven learning."

10.1.3 BLOCKING AND PREDICTION-ERROR-DRIVEN LEARNING Evidence supporting the idea that learning in classical conditioning really is error driven arises from more involved experimental designs involving the interplay between multiple conditioned stimuli present on a trial. A key experiment of this sort entails blocking (Kamin, 1969). Blocking begins with a standard initial conditioning phase in which a stimulus like a bell is repeatedly paired with reward until the bell's prediction of food (as signaled by the animal's conditioned responding to it) has reached asymptote. This is followed by a second phase in which the bell is repeatedly presented paired with a second stimulus (e.g., a light) and this compound is also rewarded (but the light is not presented alone). The key question is what is learned about the second stimulus (the light) during the additional phase of training, as measured by conditioned responding to subsequent probes of the light alone.

The finding, now widely replicated across many experiments and species, is that learning about the light is attenuated ("blocked") by the previous learning about the bell, for example, as compared to a control condition in which the bell had not been pretrained (Kamin, 1969; Moore and Schmajuk, 2008). This result suggests strong constraints on the nature of the learning process; in particular it rules out learning by simply correlating each stimulus's occurrence with reward separately from the others (as by using the rule of equation 10.1 separately for each stimulus). Instead it shows that what is learned about the light depends on what was previously learned about the bell, as would be the case if the learning rule were, in effect, attempting to apportion responsibility for the reward between the stimuli. Rescorla and Wagner (1972) suggest that the interaction between the stimuli that produces blocking can be modeled as arising from the computation of a net value prediction when multiple conditioned stimuli are present: $V_{net} = \sum_i V(s_i)$, where the sum is over all the

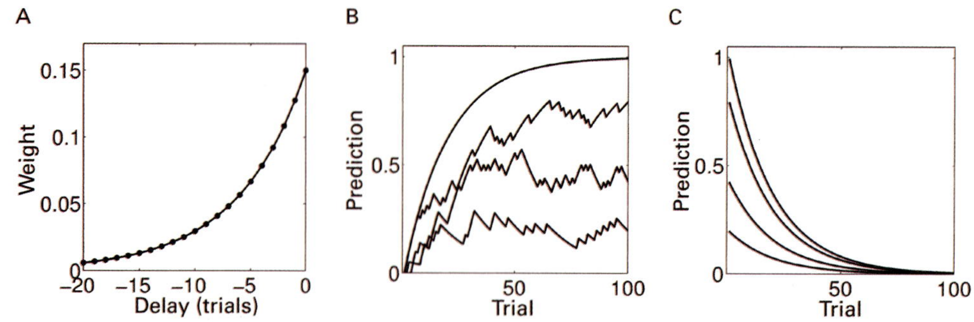

FIGURE 10.1 (A) The weights on rewards received at different past trials, according to the Rescorla–Wagner model. Weights decline exponentially into the past, with a steepness that depends on the learning rate parameter α. (B) Simulation of Rescorla–Wagner model learning about four different cues, which are reinforced (from top to bottom) 100%, 75%, 50%, and 25% of the time. Learning curves grow to asymptote; for the stochastically rewarded stimuli, the prediction is noisy (driven by random patterns of reward and nonreward) around the underlying average reward. (C) Simulation of Rescorla–Wagner model undergoing extinction of the four associations learned in panel (b) due to the stimuli being presented without reward.

stimuli present on the trial. This leads to a net prediction error $\delta_{net} = r_t - V_{net}$ which is then used to update each of the observed stimuli by equation 10.1. On this model, in blocking, there is no net prediction error (since the bell already predicts the reward) and therefore no additional learning about the light. This is the sense in which blocking is evidence supporting an error-driven learning rule.

The model of Rescorla and Wagner (1972) has many additional successes but also failures. One of its failures is particularly important for this chapter because it points the way toward a refined model for which there is also neural evidence. Imagine again that we follow training a bell to asymptote, with a second phase that in this case involves a few presentations of the bell together with a light, but this time unrewarded. According to the Rescorla–Wagner model, the prediction errors in the second phase will be strictly negative (since the reward predicted by the bell does not arrive), and therefore the effect of this training on the light will be to give it a *negative* prediction of reward. Such an effect, called conditioned inhibition, can indeed be observed (Rescorla, 1969).

However, in other experiments, *positive* responding to the light has been observed after this sort of training, as though it had inherited some of the bell's association with reward. This effect is known as *second-order conditioning* since it seems to reflect the bell's itself serving as a surrogate (second-order) reward. Second-order conditioning may arise after few unreinforced light–bell pairings whereas conditioned inhibition may emerge after more training (Yin et al., 1994). This effect is of some practical interest because it arguably represents a laboratory model of how outcomes that are not primary biological rewards (e.g., money) can nonetheless serve as surrogates for them. Second-order conditioning cannot be explained by the basic Rescorla–Wagner theory, a shortcoming that we will revisit and correct after we examine how prediction errors behave in the brain.

10.2 *Dopamine and Reinforcement Learning*

The idea that learning is error driven received powerful support from the report of an apparent neural correlate of a prediction-error signal. As we will see below, dopamine-containing neurons in the midbrain (see chapter 10), when recorded while animals perform classical conditioning tasks, carry a signal that appears similar to Rescorla and Wagner's prediction error. Moreover, additional differences between the observed signal and that predicted by the Rescorla–Wagner model suggest elaborated theories that explain second-order conditioning and connect these computations to a more sophisticated form of action choice.

10.2.1 Dopamine and the Basal Ganglia We begin with some background on the dopamine system. (See also chapter 8.) The neuromodulator dopamine is released primarily by neurons whose bodies cluster in two small midbrain nuclei, the substantia nigra pars compacta (SNc) and the ventral tegmental area (VTA) (figure 10.2). Though relatively few in number, these neurons deliver dopamine throughout a large area of the forebrain via extensive ascending projections. The diffuse nature of these projections makes them well suited for broadcasting a scalar signal like Rescorla–Wagner's net prediction error. For instance, an individual dopamine neuron's axon branches to form an extensive and dense "bush" innervating a rather large volume at target areas (Matsuda et al., 2009). Moreover, the response properties of different dopamine neurons are often reported to be relatively similar to one another

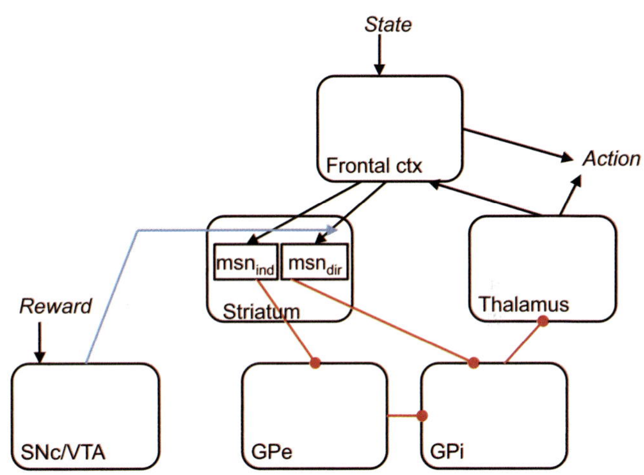

Figure 10.2 Hypothesized functional anatomy of dopaminergic influence on action learning, after Alexander and Crutcher (1990) and Frank et al. (2004). Medium spiny neurons in striatum project into either the "direct" pathway (msn_{dir}) to external segment of the globus pallidus (GPe), or the "indirect" pathway (msn_{ind}) to the internal segment of the globus pallidus (GPi), ultimately reciprocating their input from cortex (ctx) via thalamus. Because the indirect and direct pathways contain three and two inhibitory synapses (red), respectively, increased activity of medium spiny neurons in the indirect pathway has a net inhibitory effect on thalamus and cortex whereas activity into the direct pathway is disinhibitory. These effects are thought to inhibit and facilitate, respectively, thalamic and cortical movement commands. By influencing plasticity on either set of medium spiny neurons, dopamine may drive learning to produce or inhibit actions. SNc, substantia nigra pars compacta; VTA, ventral tegmental area.

(Cohen et al., 2012; Schultz, 1998; Ungless et al., 2004), even across the nuclei. Indeed, although there have recently been reports of systematic variation in dopaminergic neuron responses, these responses seem to cluster into a small number of classes (Brischoux et al., 2009; Matsumoto and Hikosaka, 2009). Overall, the sense is that these neurons collectively deliver a relatively homogenous signal to their targets.

And what are these targets? Dopaminergic neurons innervate cortex, particularly frontal cortex, and the basal ganglia. The brain's densest region of dopaminergic innervation is the basal ganglia's input nucleus, the striatum. What is the function of striatum, and of dopamine there? Hypotheses typically circle around two key areas: on the one hand, movement, and on the other hand, motivation and reward. On the motor side, damage to the basal ganglia produces a variety of movement impairments ranging from paralysis to tics (DeLong, 1990). Notably, Parkinson's disease results from the degeneration of the dopaminergic input to striatum; its symptoms primarily involve slowing of movements and difficulties in their initiation. At the same time, dopamine and its action in striatum have long been implicated in motivation and reward (Wise, 2008). Perhaps the most extreme example of this is dopamine's implication in drugs of abuse, more or less all of which appear to agonize dopamine, either directly (as with cocaine or amphetamines) or indirectly (as with nicotine).

Reconciling these two not obviously related functions—movement and reward—is a key puzzle posed by the basal ganglia and dopamine. One key concept (Mogenson et al., 1980) is that somehow these systems are the interface where reward influences action. For instance, if the basal ganglia are involved in the selection of actions (a widespread hypothesis), then rewards may influence which actions are chosen. Reward-related signals carried by dopamine may influence action selection in striatum—for instance, by affecting plasticity there (Reynolds and Wickens, 2002). To understand hypotheses about dopamine's role in learning to choose actions, we first return to its role in appetitive classical conditioning—that is, predicting motivationally relevant stimuli such as rewards.

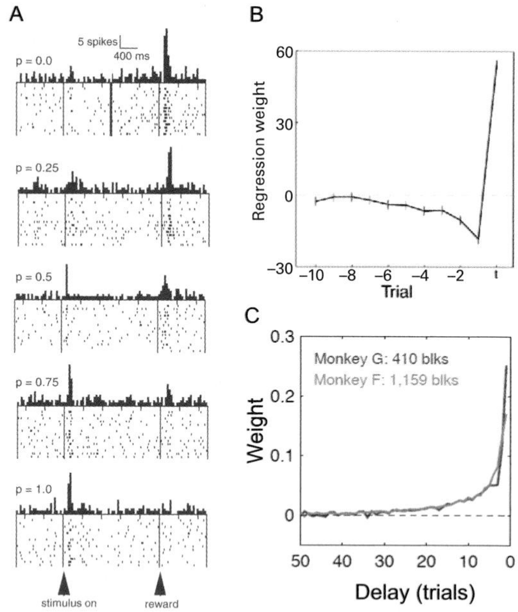

FIGURE 10.3 (A) Peri-stimulus time histograms from a primate dopaminergic neuron in a classical conditioning experiment, reproduced from Fiorillo et al. (2002) with permission. The five traces correspond to five cues trained with stochastic reinforcement at different probabilities. Only responses on rewarded trials are shown. The response at the time of a reward is larger the less expected the reward is; conversely, the better a cue predicts reward, the larger is the response to that cue. (B) Average regression weights, adapted from Bayer and Glimcher (2005) with permission, for a population of dopamine neurons. The weights were estimated so as best to capture the per-trial firing rate of the neurons as a weighted average of rewards received on the current and previous trials. The response is increasing in the current reward (time point 1, which is positive). However, it is decreasing in all previous rewards, with a decaying form. Overall, the response corresponds well to the difference between the current reward and an exponentially decaying average over previous trials' rewards, as expected for a reward-prediction error. (C) Functions relating choices and previous rewards in a decision task for two monkeys, from Corrado et al. (2004, with permission). These roughly follow an exponentially decaying form, consistent with an error-driven learning model. blks, blocks.

10.2.2 DOPAMINE AND CLASSICAL CONDITIONING Figure 10.3A illustrates the activity of a dopamine neuron recorded from the midbrain of a monkey performing a classical conditioning task (Fiorillo et al., 2003). The monkey had been extensively trained that different visual conditioned stimuli (colored fractal images) predicted delivery (vs. nondelivery) of liquid reward with different probabilities, ranging in steps from near-certain delivery to near-certain nondelivery. Following training to asymptote, the Rescorla–Wagner model predicts that the different stimuli should be associated with predictions V that scale linearly in the probability that they are rewarded (in particular, if we code r_t as binary, it is easy to verify that the asymptotic fixed point for a stimulus is just its reward probability, i.e., the average r_t observed in its presence). For the same reason, the prediction errors to reward delivery ($r_t = 1$

minus the predictions) should scale in the opposite direction: highest for the never rewarded stimulus to zero for the always rewarded stimulus.

Indeed, dopamine neurons (figure 10.3A) appear to show phasic excitation following reward receipt proportional to the prediction error it evokes (Houk et al., 1995; Montague et al., 1996; Schultz et al., 1997). Note that this is not a simple signal of reward since it is modulated by expectancy. Also, negative prediction errors, from trials where reward was not delivered, have a "negative" effect on dopamine neurons (not shown here): an inhibition of the background firing rate. In all, the phasic response of dopamine neurons at the end of a trial is well matched to the Rescorla–Wagner prediction error.

Similar findings have been reported in experiments ranging over different tasks, species, and measurement techniques. For instance, voltammetry can be used to detect transient changes in dopamine concentration in the striatum of rodents. These changes have similar prediction-error-like responses (Day et al., 2007). Also, human functional magnetic resonance imaging (fMRI) experiments have shown transient blood oxygenation level dependent (BOLD) responses in human striatum correlated with prediction errors in a manner resembling those seen in animal dopamine recordings (McClure et al., 2003a; O'Doherty et al., 2003). Of course, while there is some evidence that dopamine plays a role in these phenomena (Knutson and Gibbs, 2007), the BOLD signal is a nonspecific metabolic response.

A second feature of the data in figure 10.3A is that the dopamine neurons also show a response to the stimulus, which scales with reward probability in the opposite direction of the outcome prediction error (i.e., it follows the reward prediction itself). As we will see, these are not predicted by the Rescorla–Wagner model, but they can also be given a common interpretation as prediction errors, using a richer model that we will develop below.

First, let us consider the dopaminergic outcome response and its behavioral correlates in more quantitative detail. Figure 10.3B reproduces the results of an analysis in which Bayer and Glimcher (2005) expressed the firing rates of primate dopaminergic neurons on each trial as the weighted sum of the amount of juice the animal received on that trial and the immediately preceding ones. They used linear regression to estimate the weights that best explained the fluctuating dopaminergic response (figure 10.3B). Linear regression estimates the statistical relationship between a set of continuously valued measurements (y_i, here the firing rate on each trial i) and for each a corresponding set of explanatory variables x_{ij}. Each measurement is taken as the weighted sum of the explanatory variables, plus additive noise: $y_i = \sum_j \beta_j x_{ij} + \epsilon_i$. Linear regression estimates the weights β_j that best explain the data.

Bayer and Glimcher used linear regression to estimate the weights best accounting for neural firing y_i in terms of the set of recent rewards x_{ij}, indexed j over trials into the past. Recall that the Rescorla–Wagner model (figure 10.1A) implies that predictions V are derived from the weighted average over previous rewards, with the weights exponentially declining in delay as $\alpha(1-\alpha)^d$. The prediction error $r-V$, in turn, is the sum of the current reward r, weighted positively, and the negative expected reward (i.e., V, which is itself the sum over previous rewards weighted exponentially in $1-\alpha$, but all weighted negatively since the term is subtracted). The weights estimated by Bayer and Glimcher to explain dopaminergic responses closely follow these properties of a Rescorla–Wagner prediction error: they are positive for the current reward, and negative for the preceding rewards, decreasing in delay with a roughly exponential shape.

10.2.3 Prediction-Error-Driven Learning, Dopamine, and Choice So far we have talked only about prediction learning by itself. Are choices consistent with being driven by predictions made this way? And if dopamine influences predictions, does it also influence choice?

A standard laboratory test of learned decision-making is a so-called "bandit" task, in which subjects choose repeatedly between a set of options (say, two buttons, right and left) and receive reward probabilistically depending on their choice. The same sorts of reward-prediction models discussed above can perform such a task by learning to predict the reward expected following choice of each option $c \in (R, L)$, where R is the right-button choice and L is the left-button choice. The value of a choice is typically denoted $Q(c)$ to distinguish it from the value $V(s)$ of a stimulus. Whenever an option is chosen, and a reward received, the option's value can be updated by prediction-error-driven learning, exactly analogous to equation 10.1: $Q(c_t) \leftarrow Q(c_t) + \alpha\delta_t$, here with $\delta_t = r_t - Q(c_t)$.

Given predictions of the action values $Q(R)$ and $Q(L)$, it then seems reasonably straightforward to choose by comparing them. The main complication is a problem known as the "explore–exploit dilemma." Since the values are only predictions informed by limited experience, it can be risky to strictly choose the one you believe to be the best. It might turn out that the other one was even better, and if you maximize too strictly ("exploit") you may miss out on discovering this better outcome. Thus, it is important to "explore" seemingly suboptimal options as well, so as better to learn their true values.

At least in some classes of bandit problems, the additional value of exploration (in terms of the chance it will teach you something that will allow you to make better choices, and earn more rewards, later) can be quantified exactly, augmenting the predictions Q and giving rise to a deterministic, optimal choice rule (Gittins and Jones, 1974). A simpler and more widely applicable heuristic is to choose on the basis of Q alone but to inject some "randomness" into the choices so as to ensure adequate exploration (Hellman and Cover, 1970). A standard assumption is to assume that subjects choose stochastically with a bias toward the option predicted to be larger, for example, by the logistic "softmax" distribution (Daw et al., 2006b):

$$P(c=R)=\frac{\exp(\beta Q(R))}{\sum_c \exp(\beta Q(c))}. \tag{10.3}$$

Here the normalizing sum in the denominator is over all possible options (in this case, only two). The "inverse temperature" (so named for a physical analogy with the movement of particles in a gas) β is a free parameter that "softens" the selection of the value-maximizing option. It controls the extent that choices are focused exclusively on the best option. For $\beta = 0$, choices are entirely random; with increasing β, choices are more focused on the apparently better option, with strict maximization as $\beta \to \infty$.

This choice model, in which Q values give rise stochastically to choices via the softmax function, turns out to coincide with a standard model from statistics (Lau and Glimcher, 2005). In particular, if some continuously valued measurement is derived from the weighted sum of some explanatory variables, plus noise, then as in the Bayer and Glimcher study of dopaminergic responses as a function of previous rewards (figure 10.3B above), linear regression can be used to determine the weights that best explain the measurements. If the measurements are instead binary valued, as with the choices in a two-option bandit task, then linear regression is inapplicable (since the noise is not additive), but a generalization called logistic regression can be used to determine the weights on explanatory variables that best explain these binary measurements. In this case, the *log odds ratio* between the two possible outcomes is taken as a weighted sum over the explanatory variables: $\log(P(c=R)/P(c=L)) = \sum_j \beta_j x_{ij}$. Again, logistic regression allows estimating the coefficients β_j that best explain the data.

Taking a single explanatory variable $x_{i1} = Q(R) - Q(L)$ on trial i and solving for $P(c=R)$, the logistic regression equation is equivalent to the softmax of equation 10.3, with the single weight β_1 equal to the inverse temperature β. Now, if the values Q are themselves learned by a Rescorla–Wagner rule, then they should further arise from a weighted average over recently received rewards, with the weights following an exponential function (figure 10.1). We can use a logistic regression analogous to Bayer and Glimcher's study of dopaminergic signaling (figure 10.3B) to estimate what weighted average over previous rewards best explains choices (Lau and Glimcher, 2005). Figure 10.3C, from choices made by two monkeys in Sugrue et al.'s (2004) study, confirms the prediction that choices are driven by an exponentially weighted average over previous rewards. Similar results have been reported for experiments with humans and rodents.

These results support the idea that action choices are guided by reward-prediction errors, themselves signaled by dopamine. This idea goes some way toward resolving the central puzzle of dopamine with which this section began—its dual roles in motivation and action—and provides a theoretical counterpart to the suggestion of Mogensen et al. (1980) that the nucleus accumbens (the ventral part of the striatum, a dopaminergic target) is where motivational effects such as reward exert their influence over movement. (This is not to say that reward-prediction-error models provide a full account of the detailed motor symptoms of Parkinson's disease, which include tremor and slowed movement.)

10.2.4 Neural Models of Reward-Prediction Errors We have so far mostly discussed a model of behavior, and of one neural signal that appears to relate to a term in it. If dopamine neurons carry prediction errors, this raises two additional questions for neural modeling: where do they come from, and where do they go?

For the first question, there is a body of neurally detailed computational modeling work on how prediction errors are computed (Joel et al., 2002). Leveraging what is known about the dopaminergic neurons, researchers have also begun experimentally to investigate areas upstream of the dopaminergic neurons, which may be involved in producing the prediction error. For instance, a nucleus called the lateral habenula is upstream of the dopaminergic nuclei shown in figure 10.2: neurons there project to the VTA both directly and via an intermediate nucleus, called the rostromedial tegmental nucleus (RMTg; Jhou et al.,

2009). Neurons in the habenula respond in a fashion rather opposite that of dopamine neurons: they are excited by negative prediction errors (Matsumoto and Hikosaka, 2007). This excitation may be causally responsible for dopaminergic suppression by these same events. Lateral habenula neurons, when electrically stimulated, suppress dopamine neurons, probably via exciting inhibitory neurons in the RMTg (since the habenula neurons themselves are mostly excitatory) (Hong et al., 2011; Jhou et al., 2009). GABAergic neurons within the dopaminergic nuclei themselves also appear to participate in the generation of inhibitory responses at their dopaminergic neighbors (Cohen et al., 2012).

There is more computational modeling work on the second question: what happens downstream of the dopaminergic signal? In particular, if a dopaminergic prediction error trains action values, where might this plasticity occur? One prominent candidate is at the synapses where cortical neurons project onto medium spiny neurons in striatum, a synapse where dopamine is released and is known to affect plasticity (Reynolds and Wickens, 2002). But what do these neurons have to do with action? A classic conceptual model of the basal ganglia's role in movement (Alexander and Crutcher, 1990), illustrated in figure 10.2, holds that the medium spiny neurons, in turn, reciprocate the connection from their cortical inputs via either of two distinct pathways through the remaining nuclei of the basal ganglia: the direct and indirect pathways, with a different population of striatal neurons projecting into each. Due to different numbers of inhibitory synapses along the path, the direct and indirect pathways are believed to have a net facilitatory or inhibitory impact, respectively, on their cortical targets, in turn facilitating or inhibiting movements that motor neurons in cortex ultimately control. In a classic view of motor disorders (DeLong, 1990), imbalance between these two pathways can cause either poverty of movement—as in Parkinson's disease—or excess, as in tics or other undesirable movements seen in other disorders involving the basal ganglia. Before going on, we should note that this simple push–pull model of the basal ganglia is incomplete and oversimplified, both with respect to the anatomical connections and their functions (Nambu, 2008). For instance, it neglects the important role of neural oscillations and related physical tremor in Parkinson's disease (Rivlin-Etzion et al., 2006). Nevertheless, it is a useful and simple starting point.

Layering the concept of dopaminergically mediated reinforcement learning on the direct- and indirect-pathway architecture of figure 10.2, Frank et al. (2004) pointed out that dopaminergically mediated plasticity at these medium spiny neurons seems perfectly positioned to drive learning about action propensities. Imagine that via their projections into either the direct or indirect pathways, medium spiny neurons promote or oppose particular actions, respectively, and that their cortical inputs represent something about the current sensory state or situation. In this case, Frank et al. proposed that if an action in some situation was followed by positive prediction error, then dopamine could train action choice in either of two ways. First, by *strengthening* synapses onto the action-related medium spiny neurons in the direct (facilitating) pathway, it could increase the chance of choosing the action again. Conversely, by *weakening* synapses onto the indirect (inhibiting) pathway, it could *decrease* the chance of avoiding the action in the future. Negative prediction errors could have the opposite pattern of effects.

Putting this idea in terms of our previous models (and after a version of the model by Frank et al., 2007), we might consider the net action value to arise from the sum of the opposing contributions to action of both the direct and indirect pathways $Q^{net}(c) = Q^{+}(c) - Q^{-}(c)$, where the direct pathway's values Q^{+} are increased by positive prediction errors (e.g., successes), $Q^{+}(c) \leftarrow Q^{+}(c) + \alpha^{+}\delta_t^{+}$ and decreased by negative prediction errors (failures) $Q^{+}(c) \leftarrow Q^{+}(c) + \alpha^{-}\delta_t^{-}$. Here δ^{+} and δ^{-} are, respectively, positively and negatively rectified prediction errors, and α^{+} and α^{-} represent separate learning rates for learning from these positive or negative errors. Learning in the No-go pathway is similar but reversed: positive prediction errors should decrease avoiding while negative errors should increase it, with $Q^{-}(c) \leftarrow Q^{-}(c) - \alpha^{+}\delta_t^{+}$ and $Q^{-}(c) \leftarrow Q^{-}(c) - \alpha^{-}\delta_t^{-}$. Note that α^{+} is related to the effect of dopaminergic burst responses, and α^{-} to dopaminergic pauses, that is, the absence of dopamine; for this reason Frank et al. (2004) reasoned that disease or treatments that increase or decrease dopaminergic signaling would affect the balance between these two sorts of learning.

In an ingenious test of this model—and a clean piece of causal evidence for dopaminergic involvement in reinforcement learning in humans—Frank et al. (2004) studied learning to choose or avoid actions in human patients with Parkinson's disease playing a reinforcement learning game similar to the monkey task of figure 10.3C. Subjects were first trained to develop preferences between several different pairs of options—for instance in repeated choices between options *A* and *B*, *A* would pay off 80% of the time whereas *B* would pay off 20% of the time. Subjects also learned weaker preferences between other pairs *C* / *D* and *E* / *F*, which all paid off an intermediate amount. According to the model, the learned preference between A and B might occur due to either learning to choose A (high $Q^+(A)$) or to avoid B (high $Q^-(B)$), and this would depend on the relative strength of the parameters α^+ and α^-. To test this, subjects were asked to choose between novel pairings, involving either A or B, pitted for the first time against C, D, E, or F. The relative performance at choosing A in these novel pairings versus avoiding B reveals to what extent subjects learned the original preference via choosing or avoiding (in the model, via the direct or indirect pathway).

The experimenters tested Parkinson's patients on this task. They reasoned that the degeneration of the dopaminergic neurons in Parkinson's disease should attenuate positive fluctuations in dopamine, reducing learning from positive prediction errors (mostly from successful choices of A) while preserving downward fluctuations in remaining dopamine levels that report negative prediction errors (mostly from unsuccessful choices of B). This would bias patients to learn the tasks, if at all, primarily via avoidance. In the model as written above, this can be understood as the hypothesis that decreases in dopamine in Parkinson's would lead to decreased α^+, together perhaps with preserved or increased sensitivity to negative errors, α^-, due to reduced background levels of dopamine. By restoring positive signaling while also perhaps masking over "dips" in dopamine levels due to firing pauses, dopamine-restoring medication would be expected to invert these effects.

Accordingly, patients with Parkinson's disease, when tested off their dopamine-restoring medication, tilted toward learning to avoid the inferior action, but when tested on dopamine replacement therapy, this pattern reversed and medicated patients tended toward learning to choose the better action. These results are consistent with the idea that dopamine causally drives trial-and-error learning about choices, and with the more detailed implementation of this idea, suggested by the architecture of the basal ganglia, that this learning is layered onto mutually opposing direct and indirect pathways.

10.3 Sequential Predictions and Temporal-Difference Learning

So far, we have reviewed evidence and modeling suggesting a role for dopamine in signaling prediction errors to drive choices. However, the structure of these experiments was relatively simple: repeated, mutually isolated trials in which subjects chose an action (or observed a stimulus) and immediately received the associated reward, after which the next trial followed independently. Many real-world decisions have a more elaborate, interdependent structure, in which the consequences of any individual choice for reward may be deferred and also depend on subsequent choices and events.

Consider deciding what university to attend, what turn to make in a spatial maze, or what play to make in a game like American football. In all these tasks, each decision is followed by many others, and rewards (like points) may be accrued in an ongoing way on the basis of all of them together. Unlike in simple tasks such as that of figure 10.3C, in these tasks the decision maker must take into account not just the reward that is expected immediately for a choice but how that choice might set the stage for rewards or punishment that may not arrive until further down the line.

For instance, consider analyzing the game of football at a play-by-play level (e.g., the decision whether to pass, run, or kick at each down, not the movements within a play of each individual player) (Romer, 2006). In football, teams win points by moving the ball into the end zone. However, most plays do not score any points; instead, they move the ball up or down the field, increasing or decreasing the chance that the team will reach the end zone and score on subsequent plays, or that the opponents will. The plays thus mainly produce changes in what is called the "state" of the game—variables like the field position and who has possession of the ball at the beginning of each play—which has deferred consequences for scoring.

In such a task, predicting only the immediate reward earned by an action (the points scored on that play), and choosing on that basis, is clearly inappropriate. Instead, one would like to choose the action that maximizes the reward accumulated over the long run. Predicting this long-term quantity, taking into account the contingencies of subsequent states and actions, requires models with more technical sophistication. With these in place, we will see that the ability to assess the long-run reward consequences of states and actions—in essence, to project future points on the basis of field position—is linked to the aspects of behavioral and neural data that the Rescorla–Wagner model we have been using so far failed to explain. In particular, both second-order conditioning and the dopaminergic response to reward-predictive stimuli relate to the ability to assess future reward (like points) on the basis of signals (like field position) that bear only a predictive relationship to rewards.

10.3.1 The Markov Decision Process

To model predictions in tasks with sequential structure, we first need a formal characterization for such tasks. A common framework is the Markov decision process (MDP), a mathematical abstraction which captures many key features of tasks (Puterman, 1994; Sutton and Barto, 1998).

Imagine that the world consists of discrete time steps, indexed t, and a set of states S and of actions A. At each time step the world takes on some state s_t from S and delivers some scalar reward r_t based on that state. Finally, the agent chooses an action, a_t, from A. Finally, assume that at any t, the future of the process may be stochastic, but it depends only on the current state and action. (This is the "Markov property" for which the process is named.) That is, the new state is chosen according to a transition distribution $P(s_{t+1} \mid s_t, a_t)$. Reward can also be delivered probabilistically, conditional on the current state with some distribution $P(r_t \mid s_t)$. We will also define $r(s)$ as the expected reward in state s.

The goal of choice at a state, in general, is to maximize the reward expected in summation over a whole trajectory. As in the simple choice problems discussed earlier in the chapter, predicting this outcome will be useful for guiding choices. We can loosely write the objective, the expected sum of rewards following some state, as:

$$V(s) = E\left[r_t + \gamma r_{t+1} + \gamma^2 r_{t+2} + \ldots \mid s_t = s\right]$$

where the sum extends out to infinity (or until the process terminates—e.g., at the end of the maze). Here the time discount parameter γ allows for the possibility that delayed rewards might be less valuable than proximal ones. If the task does not terminate after a finite number of steps, then $\gamma < 1$ is required to ensure the value function is finite.

However, the expectation in this expression (over actions and future states) hides the dependence of value in a state both on the action you take there and in subsequent states. We could describe this more carefully by defining a "policy" $\pi(s)$ as a mapping from each state to an action and then spelling out the expectation over future states explicitly, assuming we follow policy π everywhere:

$$V^{\pi}(s) = r(s) + \gamma \sum_{s' \in S} P(s' | s, \pi(s)) \left[r(s') + \gamma \sum_{s'' \in S} P(s'' | s', \pi(s')) \left[r(s'') + \gamma \sum_{s''' \in S} \ldots \right]\right]. \qquad (10.4)$$

Here the nested sums reflect the branching structure for the expectation over the state transitions at each step, contingent on the previous one. The value V is now defined dependent on a particular policy π since the cumulative reward expected following state s depends on the action taken there and also the actions taken subsequently. This is reflected in the transition probabilities in the equation, which are conditioned on the series of action choices $\pi(s)$, $\pi(s')$,...

The nested structure of this expression allows it to be rewritten in a more manageable recursive form, which is also the key insight enabling the learning models we will consider:

$$V^{\pi}(s) = r(s) + \gamma \sum_{s' \in S} P(s' | s, \pi(s)) V^{\pi}(s'). \qquad (10.5)$$

This form of the function, called a Bellman equation (Bellman, 1957), defines the values of the states in terms of each other. It arises from observing that the long sum of rewards can be broken down as the first reward plus the rest, and the rest (the bracketed quantity in equation 10.4) is itself just the expected future value of the subsequent state, s'.

10.3.2 Temporal-Difference Learning With all of these quantities defined, we are in a position to define models for learning V from experience. Recall that in the Rescorla–Wagner model, in order to train the predictions V such that they equaled the immediate rewards r on average, $V(s_t) = E[r_t]$, we used an error-driven learning rule with a prediction error (equation 10.2) defined by the difference between the two sides of that equality: $r_t - V(s_t)$. Such an error measures the extent to which the equality fails to hold; it is zero (on average) when the predictions are correct.

To learn the long-run reward V^{π}, we can define an analogous error signal using the Bellman equation (equation 10.5) to define the desired equality. For this section, we will use notation that allows us to explicitly distinguish the agent's estimate of the value function, $\hat{V}^{\pi}$, from its true value as defined in equations 10.4 and 10.5 above. Imagine at some time step t the agent encounters state s_t, receives reward r_t, and transitions to state s_{t+1}. Expressing the Bellman equation in terms of these quantities, this requires that if the function is well learned, $\hat{V}^{\pi}(s_t) = E\left[r_t + \gamma \hat{V}^{\pi}(s_{t+1})\right]$, that is, the predicted values and obtained rewards should match up, on average (where the expectation is with respect to randomness in the state transitions and rewards). We can measure to what extent this relationship holds or fails to hold in any particular instance by substituting the obtained reward and state in the expectation, and subtracting both sides of this equation to produce an error signal:

$$\delta_t = r_t + \gamma\, \hat{V}^{\pi}(s_{t+1}) - \hat{V}^{\pi}(s_t) \qquad (10.6)$$

This is called the temporal-difference prediction error. By using it together with equation 10.1 to update the prediction $\hat{V}^{\pi}(s_t)$, we obtain a model called temporal-difference (TD) learning (Sutton, 1988).

Let's unpack this model a bit. First, the difference between the new model and Rescorla–Wagner is just the addition of the term $\gamma\hat{V}^{\pi}(s_{t+1})$ to the prediction error. This reflects the fact that we want to learn not merely the immediate reward, r_t, as in Rescorla–Wagner, but the sum of subsequent rewards as well. By the recursive Bellman decomposition, the subsequent state's value is a stand-in for the rest of the terms in the sum.

How does this work? The agent is trying to learn a long-run sum of rewards, $\hat{V}^{\pi}(s_t)$. One way to do this is to wait until the end of the episode, keeping track of all rewards received following s_t, then finally updating $\hat{V}^{\pi}(s_t)$ in light of their sum. But there is an alternative. In learning, the agent is updating a set of predictions V^{π}, one for each state. $\hat{V}^{\pi}(s_{t+1})$ is the agent's *own estimate* of the value of the new state s_{t+1}. That is, having observed that state s_t was followed by state s_{t+1}, the agent can use the estimated value of the new state as a proxy for the rewards remaining in the rest of the trajectory. This recursion allows the value be updated immediately toward an estimate of the desired value $r_t + \gamma\hat{V}^{\pi}(s_{t+1})$, rather than waiting to observe all the remaining rewards. Of course, this trick (known, evocatively, as bootstrapping) only works so far as the predictions $\hat{V}^{\pi}(s_{t+1})$ contain reliable information. During learning, when rewards are received, they first train states nearby them, and this information gradually spreads to states further from where reinforcement was actually received. This is a stochastic version of classic "dynamic programming" algorithms from computer science (Bellman, 1957), which progress recursively from the goals (e.g., endgame positions in chess) to compute values at the states that precede them.

Finally, the term $\gamma\hat{V}^{\pi}(s_{t+1}) - \hat{V}^{\pi}(s_t)$ is the temporal difference from which the model takes its name. This term is the whole prediction error whenever rewards are not immediately received, that is, $r_t = 0$. In the absence of rewards and if state transitions are predictable, the value function should change only smoothly, according to the discount factor γ. Fluctuations in $\hat{V}^{\pi}$ in the absence of actual reward occur when state transitions produce changes in reward expectation, which should drive learning to update the previous expectations.

Thus, if the temporal difference is positive, then the current reward expectation is higher than the previous prediction had anticipated. In football, this might happen if a particularly successful play led to a field position unexpectedly close to scoring. In this case, the previous value was too pessimistic and should be increased. Conversely, if the temporal difference is negative, future reward expectancy has dropped unexpectedly, and the previous prediction was too optimistic. In football, this might happen if an unexpected interception changes the possession of the ball, increasing the chance that the opposing team will score in upcoming plays.

10.3.3 Dopamine and TD Learning

This brings us back to the responses of dopamine neurons in classical conditioning (figure 10.3A). Initially, we viewed the neurons' responses at the time of outcomes as reward-prediction errors in the Rescorla–Wagner sense but had no explanation for their response to reward-predicting conditioned stimuli. Applied to the same experiment, the TD learning model preserves the account of the outcome responses but provides a unifying account of the stimulus responses as prediction errors as well.

In order to analyze the responses of the neurons over the course of a trial, we must break each trial down into notional "states." We will simply envision these as small intervals of time. Consider (for simplicity) that $\gamma = 1$ and the expected future value for time steps between trials is zero. Then at the end of the trial, when the reward arrives, the future value $\hat{V}^{\pi}(s_{t+1})$ is zero and the TD prediction error, from equation 10.6, is just $r_t - \hat{V}^{\pi}(s_t)$. This is the same as Rescorla–Wagner since the expected future reward $\hat{V}^{\pi}(s_t)$ in the interval following the stimulus is just the probability that the stimulus will be rewarded. Thus, the TD model offers the same explanation as before for the reward-time responses in figure 10.3A.

Meanwhile, at the start of the trial, when a stimulus first appears, there is no primary reward ($r_t = 0$). Also, in these experiments, the stimuli were themselves highly temporally unpredictable (which we will caricature by assuming, for simplicity, that the value expectation just prior to the stimulus is zero, $\hat{V}^{\pi}(s_t) = 0$). Thus, the arrival of the stimulus informs the monkey that reward is now expected according to the stimulus-specific probability: the value expectation $\hat{V}^{\pi}(s_{t+1})$ is now the reward probability. For this reason, TD predicts positive prediction error to the stimuli, related to the positive temporal difference between sitting in the dark between trials expecting nothing and seeing an unpredictable stimulus inducing an expectation that reward will soon occur with some probability. This is analogous to the case of a football play that ends in unexpectedly good field position. The TD error to the stimulus is just its reward probability, consistent with the stimulus responses in figure 10.3A.

Thus, the TD model explains both the cue- and outcome-time responses of the dopamine neurons as instances of a prediction error for not just immediate but also future reward (Houk et al., 1995; Montague et al., 1996). The model predicts, and experiments verify, that neural prediction errors can similarly, recursively transfer to stimuli separated from their rewards by multiple steps (Schultz et al., 1993; Seymour et al., 2004).

This model not only solves the Rescorla–Wagner model's problems explaining the anticipatory aspects of the dopamine response, but, for the same reason, it also solves a behavioral problem with the Rescorla–Wagner model as well: its failure to explain second-order conditioning. Since cues that predict future reward elicit reward-prediction error and activate dopamine, they can train positive reward predictions in preceding states, even if reward is ultimately omitted. This allows the TD algorithm to explain second-order conditioning, as a direct reflection of its recursive, bootstrapping strategy of training reward predictions on the basis of subsequent predictions.

10.3.4 ALTERNATIVE VIEWS OF THE ROLE OF DOPAMINE Although the models so far presented are reasonably prominent, there are a number of other models advancing different ideas about the function of dopamine. A first point is that dopamine has broad projections to many different parts of the brain, with some systematic variation in responses (Brischoux et al., 2009; Matsumoto and Hikosaka, 2009) and multiple types of receptors and local circuitry at targets such that even to the extent that a uniform message is broadcast, it will not be received the same way everywhere. It is thus reasonably clear, and not surprising, that dopamine serves different functions in different places. For instance, it is not inconsistent with TD models of dopamine's action in parts of striatum that dopamine projections to prefrontal cortex are also clearly involved in working memory, a function which is itself the subject of much computational modeling work.

Other research challenges the hypothesis of dopaminergic involvement in reinforcement learning more directly. Many of these turn on a number of additional findings about the responses of dopamine neurons that seem anomalous under the hypothesis that it carries just a reward-prediction error. In particular, dopamine neurons respond to a variety of surprising events that are not, strictly speaking, reward related, including salient novel stimuli, unrewarded stimuli that are similar to rewarded ones, and even (according to some reports) punishments or punishment-predicting stimuli. A number of authors have therefore suggested that dopamine has some more general role, perhaps in directing attention to surprising events. One version of this critique (Redgrave and Gurney, 2006; Redgrave et al., 2008) suggests that dopamine might have a role in discovering actions that have a causal effect on the sensory world, rather than reward per se.

Finally, some scientists, notably Berridge (2007), are sympathetic to the idea that dopaminergic signals in striatum are specifically involved in appetitive behavior but note that the causal evidence that they do so via driving learning is not as detailed or extensive as the merely correlational evidence from unit recordings. Models from this group (Zhang et al., 2009) envision that the causal effects of the dopaminergic signal are not via driving plasticity but instead are directing appetitive incentive motivation. These and other authors have also pointed out that among dopamine's most obvious causal effects are in motivating or invigorating ongoing behavior, as with the example of slow movements in parkinsonism. Although these observations are not easily explained in the reinforcement learning framework (and conversely, evidence like Frank's 2004 study of learning to choose and avoid in Parkinson's patients are hard to understand from a purely motivational standpoint), there have been efforts to build unified models of both sets of phenomena and to show how they may be related (McClure et al., 2003b; Niv et al., 2007).

10.4 *Dopamine and Sequential Choice*

Having extended reward prediction to temporally extended sequences, it remains to connect this sort of learning back to choices in sequential decision tasks. We have spoken so far in terms of learning value predictions under a fixed policy, $\hat{V}^{\pi}$, but not how a policy can be obtained, or how, conversely, the learned long-run value predictions can be used to guide choice. Here we describe such models and then discuss how they connect to what is known about animal behavior.

The learning algorithms discussed below all originate in machine learning and have theoretical guarantees to converge (e.g., to find the optimal policy) asymptotically, assuming various technical conditions and often in the limit of infinite experience (Bertsekas and Tsitsiklis, 1996). The requirement of infinite experience is of course absurdly unrealistic, but results of this sort at least ensure that the algorithms are doing something sound. Many other procedures are proven not to converge at all, or to converge to incorrect answers! That said, these algorithms are not necessarily practical when scaled up to large, real-world problems with many states and actions—for instance, because of the explosion in the difficulty of such problems due to the "curse of dimensionality." Understanding how the brain adapts these basic concepts to real-world control is an ongoing research question (e.g., Botvinick et al., 2009).

In any case, two ideas for applying learning to choice in sequential problems predominate in neuroscience. One way, starting with some initial policy π, is to define state *and action* specific value predictions using the Bellman equation $Q^{\pi}(s,a) = r(s) + \sum_{s'} P(s'|s,a) Q^{\pi}(s',\pi(s'))$. $Q^{\pi}(s,a)$ measures the value of taking action a in state s and then following policy π thereafter. These values can be learned directly using TD methods analogous to those derived previously for V^{π}; this algorithm is called SARSA (acronym for the sequence of state-action-reward-state-action that triggers a single update; Rummery and Niranjan, 1994). Then choosing an action in state s simply requires comparing the values $\hat{Q}^{\pi}(s,a)$, for example, by the softmax of equation 10.3. It is possible to prove that the policy implied by choosing according to Q^{π} is equal to or better than the policy that defined it, π, and that by iteratively improving a policy this way, one can arrive at the optimal policy Q^{*}.

An arguably cleverer variant, known as Q-learning, is based directly on a Bellman equation for the *optimal* policy, $Q^{*}(s,a) = r(s) + \sum_{s'} P(s'|s,a) \operatorname{argmax}_{a' \in A} Q^{*}(s',a')$, that is, the value of choosing action a in state s and then choosing the *best* action in the state s' where you arrive next. Again, this leads to a TD algorithm for learning the optimal action values and, implicitly, the optimal choice policy (Watkins and Dayan, 1992). Note that the argmax in this equation—the choice of the strictly best action—just controls which action's value at the subsequent state is used as the prediction for *learning* the previous action's value. The agent's choices themselves can still involve a mixture of exploitation and exploration, as by softmax, but regardless of the subject's actual exploration policy, because of the argmax, the algorithm will learn what all the actions are worth under the *optimal* policy.

Another variant, called the actor/critic, learns a state-value function V^{π} (the "adaptive critic" by the methods of the previous section, that is, with the prediction error from equation 10.6 above

$$\delta_t = r_t + \gamma V^{\pi}(s_{t+1}) - V^{\pi}(s_t)$$

and the update rule $V^{\pi}(s_t) \leftarrow V^{\pi}(s_t) + \alpha\delta_t$. These state values can then be used to help learn a policy. In particular, this model also maintains a separate set of state-action policy parameters $w(s,a)$ (the "actor") that are used to define a probabilistic choice policy, typically by a softmax operation over the probabilities at a state (i.e., $P(a_t = a) \propto \exp(\beta w(s,a))$. The learning rule for the actor, if action a_t is taken in state s_t resulting in reward-prediction error δ_t, is

$$w(s_t,a_t) \leftarrow w(s_t,a_t) + \eta\delta_t$$

where δ_t is the same prediction error as for the state values, from above and equation 10.6, and η is a second learning rate. This rule increases the probability of an action if it results in positive reward-prediction error, or decreases it for negative error. In this way, the values learned by the adaptive critic train the actor's policies via generating the reward-prediction error, thereby directing choices toward those expected to have high long-run value. Formally, rules of this sort can be understood as performing stochastic gradient ascent on the policy parameters with respect to the cumulative discounted reward obtained by the policy (Dayan and Abbott, 2000).

The actor/critic has been a particularly influential model in neuroscience because its functional separation between Pavlovian (state) value and instrumental (state-action) policy learning seems to mirror a functional breakdown between different subregions of striatum. As mentioned already, striatum receives connections from cortex and reciprocates them via loops; these connections are topographic such that different areas of striatum are associated with different areas of cortex, suggesting functional specialization. In particular, both the anatomy of the corticostriatal connections (Joel and Weiner, 2000; Voorn et al., 2004) and rodent lesion studies (Parkinson et al., 2002) suggest that the ventral part of striatum is associated with Pavlovian learning, reward, and motivation. Meanwhile, dorsal striatum (notably, the dorsolateral part) is connected with motor cortex, and lesions and recordings there suggest that it is more involved with movement per se and with stimulus–response learning (Jog et al., 1999; Yin et al., 2004). This suggests a mapping of critic and actor functions onto the ventromedial–dorsolateral axis of the striatum (figure 10.4). Also, it is noteworthy that even though these striatal subregions have apparently very different functions, the dopaminergic neurons innervating each of them apparently behave rather similarly to one another (Schultz, 1998), consistent with the idea that actor and critic learn from the same error signal.

10.4.1 The Actor/Critic and the Law of Effect

Another way to understand the actor/critic model is to recognize it as a formalization of a classic abstract model in psychology. Thorndike (1911) sought to identify laws of behavior akin to the laws of physics. This program, known as "behaviorism," was a

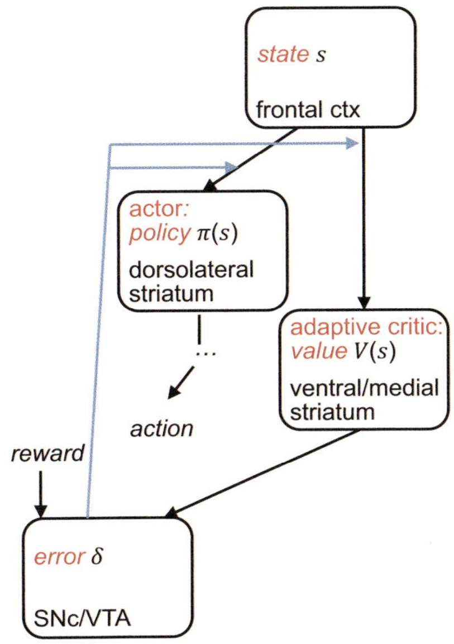

FIGURE 10.4 Hypothesized functional anatomy of the actor/critic. The ventromedial striatum ("adaptive critic") learns a value function, driving the prediction error in the dopaminergic nuclei that trains both the values and the action-selection policy in the dorsolateral striatum ("actor"). Downstream, both parts of striatum are associated with parallel corticostriatal loops in the manner shown in figure 10.2 (not illustrated here). ctx, cortex; SNc, substantia nigra pars compacta; VTA, ventral tegmental area.

counterreaction to earlier, introspective, trends in psychology and reflected an attempt to ground psychological theorizing in concrete, empirically observable events (e.g., links between stimuli and responses) without reference to subjective internal representations such as plans or beliefs.

Among Thorndike's (1911) laws of behavior was the "law of effect," which envisioned reinforcement learning occurring due to a process of strengthening links between states (or "situations") and responses:

> Of several responses made to the same situation, those which are accompanied or closely followed by satisfaction to the animal will, other things being equal, be more firmly connected with the situation, so that when it recurs, they will be more likely to recur.

If we identify the reinforcement learning abstractions "state" and "action" with Thorndike's "situation" (by which he had in mind sensory events or stimuli) and "response," then the actor/critic's policy parameters $w(s,a)$ are essentially equivalent to Thorndike's stimulus–response links: they represent learned connections between the two.

However, Thorndike's rule of strengthening these links based on temporally proximal "satisfaction" (which we might identify with immediate reward r_t) is clearly inappropriate for sequential tasks like MDPs, where rewards may be deferred. Instead, the actor's learning rule for strengthening these weights represents a version of the law of effect in which "satisfaction" is replaced by the TD error $r_t + \gamma \hat{V}^{\pi}(s_{t+1}) - \hat{V}^{\pi}(s_t)$. The last term of this update is a relatively inconsequential normalizing adjustment since it is the same whatever action is chosen. The remaining terms in the learning rule strengthen actions according to $r_t + \gamma \hat{V}^{\pi}(s_{t+1})$: not just immediate reward but long-term reward expectancy. This predictive notion of "satisfaction" extends the law of effect to sequential tasks but maintains the essential structure of what is learned: a mapping from states to responses. (Although we focus on the actor/critic, for all the points made here and below, models based on learning state-action values $\hat{Q}(s,a)$ are essentially equivalent.)

Thus, the TD models we have discussed so far essentially extend Thorndike's law of effect with a computational theory and a neural mechanism. However, the history of twentieth-century psychology largely represented the refutation of these views as a complete explanation of behavior. The "cognitive" movement that largely supplanted behaviorism leveled a number of criticisms at the behaviorist view. Some are relatively easy to address in TD models of the sort discussed thus far. For instance, not all behaviors are stimulus bound. Many features of sequential behaviors, for instance, indicate that they arise autonomously from a progression of internal states rather than some sort of cascade of stimulus–response reflexes whereby each action sets up the sensory stimulus that triggers the next (Lashley, 1951). Thus, in TD terms, the "states" that give rise to behaviors must incorporate subjective features such as memories or context rather than being strictly stimulus bound. This is a standard and straightforward assumption, also needed, for instance, to explain how dopamine neurons know when to "pause" for omitted rewards (Daw et al., 2006a; Schultz et al., 1997).

Even so, however, the focus on the end point of learning being the policy limits the scope and flexibility of the models. Tolman (1948) mocked Thorndike's stimulus–response model as a telephone switchboard view of cognition and showed that it leads to counterintuitive (and empirically wrong) predictions in a number of experimental settings. Consider a hungry animal trained in the maze of figure 10.5A. The reward values shown there—higher for states with food, lower for water—would lead the animal to learn high policy parameters (or Q values, or stimulus–response links) for the left branch at the start state.

However, what happens when the animal is made thirsty but not hungry and is returned to the maze for

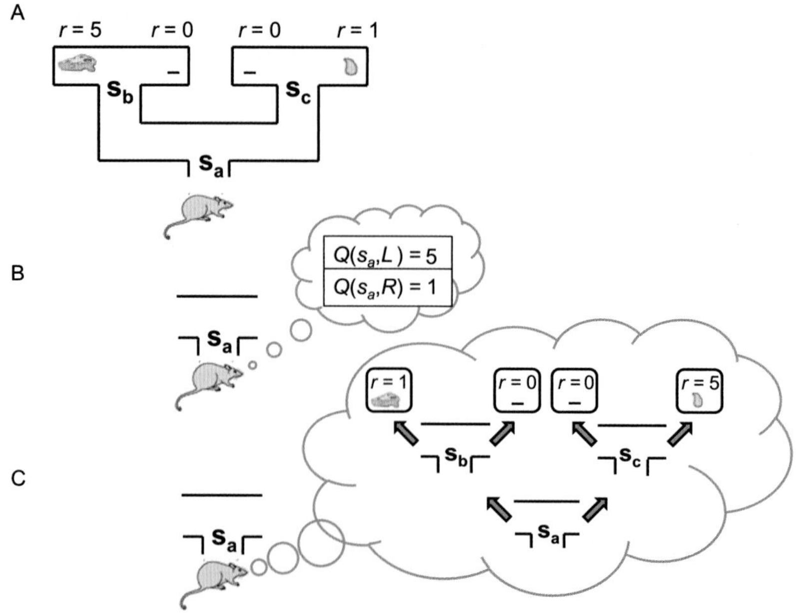

FIGURE 10.5 (A) An example of a Markov decision process that models a spatial maze with three states, two actions (left or right) at each state, and four terminal rewards. (B) A temporal-difference learner retrieves a stored policy or value function to choose actions and is thus insensitive to changes in outcome value in a devaluation task. (C) A model-based reinforcement learner can recompute action values if outcome values change.

the first time in this state? We can capture the new motivational state by assigning different rewards to the states in figure 10.5A: now higher for water (say, 5), and lower for food (1). We might expect a rational animal would now turn right, at the maze start, to get the water. However, the actor/critic (and other TD models) makes a clear prediction that, at least on the first trial, the animal will turn left: toward the food it does not want. This is because the policy learned by the actor contains no information about the particular reward identities that had favored one action over another (figure 10.5B). Thus, it does not have any way to update the values in light of the new motivational state.

Of course, the actor can *learn* to update its policy (from negative prediction error on encountering the worse-than-expected food), but this requires new experience in the maze. It cannot dream up the new policy just by reasoning about the new circumstances without additional learning. Note that the issue here is not whether or not the "state" includes the animal's drive or motivations. Clearly, a more careful accounting would define each state in figure 10.5A as also including the animal's internal drive state of hunger or thirst. (We omit this elaboration for notational simplicity below.) This would in effect make the maze into two different mazes and allow the model to learn separate behavioral policies by trial and error, appropriate to either motivational state. But what this strategy misses is the ability to train in the maze while hungry and then choose appropriately *the first time* the maze is encountered when thirsty.

For the same reasons, a simple TD agent cannot plan novel routes from scratch when a previously favored path is blocked or a new shortcut opens in a spatial maze, although rats can do these things (Tolman, 1948). These shortcomings are fundamental to the representational strategy of law-of-effect type theories. If an animal learns only which actions have been advantageous and then separately learns that the reward function or transition function has changed (e.g., because of a motivational shift or a reorganization of the maze), then it has no way to use this information to update the stored policy.

The above prediction has been tested in logically similar experiments involving shifts in motivation for rewards (Dickinson and Balleine, 2002). The answer is that animals (and humans) really do behave like law-of-effect theories predict, but only sometimes. In particular, rats trained to lever press for food while hungry will continue, when full, to work for food they demonstrably would not eat, but this is true only sometimes: notably, for rats who were *overtrained* to lever press in the first instance. Rats exposed to more moderate amounts of lever-pressing training will immediately shift their lever pressing when the food is unwanted, without reexposure to the food itself. Although the first result confirms the odd prediction of the law of effect, the second result is not compatible with it (or the actor/critic) and demonstrates that the decision to lever press, in these circumstances, is somehow mediated by a representation of the particular goal of the action (the food). For this reason, the devaluation-sensitive sort of behavior is known in psychology as "goal-directed." The devaluation-*insensitive* behavior is known as "habitual," a name that calls to mind the familiar phenomenon from our own lives that actions repeated enough times can become seemingly automatic and detached from their goals.

Lesions in rodents suggest that these two modes of behavior are also neurally dissociable, existing in parallel neural circuits that center on adjacent regions of striatum: dorsolateral striatum (and its dopaminergic input) for habits, and dorsomedial striatum for goal-directed actions (Yin et al., 2004; Yin et al., 2005). Altogether, these results are consistent with the actor/critic model but suggest that it is an incomplete account of the brain's systems for reinforcement learning. In particular, the much-modeled TD mechanism can explain only habitual, but not goal-directed, behavior. Can the latter also be understood formally?

10.4.2 Model-Based Reinforcement Learning and Goal-Directed Behavior

One candidate for a computational account of goal-directed behavior is a family of algorithms called *model-based reinforcement learning* (Daw et al., 2005). In computer science, such algorithms date back to early work in artificial intelligence on computer algorithms for games like checkers. Instead of learning a value function or policy directly, model-based algorithms are so named because they work by learning an *internal model* of the MDP: that is, a representation of the immediate rewards associated with each state $r(s)$, and the one-step state transition probabilities, $P(s_{t+1} \mid s_t, a_t)$. These are easy to learn: the former by Rescorla–Wagner, the latter by counting state transitions to fill in a table of conditional probabilities. Please note the unfortunate terminological collision: model-based reinforcement learning is a model (i.e., a quantitative theory) of learning, which relies on an internal model (i.e., an explicit representation of the task structure). TD models are known as *model-free* methods because they learn policies or value functions directly, without representing the reward or transition functions.

The reward and transition functions together constitute a complete description of the task: if they are learned, the value function (and hence the policy it implies) can directly be computed by the evaluation of the Bellman equation via iterative expansion of the recursion into a tree-structured set of nested sums (e.g.,. unrolling it back into the form of equation 10.4). For instance, if we initialize a matrix of values $Q^*(s,a)$ to zero and then repeatedly update it according to recursive applications of the Bellman equation,

$$Q^*(s,a) \leftarrow r(s) + \gamma \sum_{s' \in S} P(s'|s,a) \operatorname*{argmax}_{a' \in A} Q^*(s',a'), \qquad (10.7)$$

we arrive at the optimal values by recursively computing the forward sum. This algorithm is known as dynamic programming or value iteration (Sutton and Barto, 1998).

This process essentially amounts to exploring a "tree" of different possible state trajectories, accumulating rewards along them, and weighting them according to the relevant conditional probabilities. In this respect, it is reminiscent of mental simulation or planning: like the process of deliberating over a chess move.

Importantly, because model-based reinforcement learning computes its policy from a full representation of the task (figure 10.5C), it can update the policy appropriately whenever circumstances change. For instance, consider the experiment in which an animal learns the maze of figure 10.5A when hungry (thereby learning state transitions $P(s_{t+1} \mid s_t, a_t)$ and the rewards $r(s)$ appropriate to the state of hunger), if the animal then switches to a state of thirst (modeled as inducing a different reward function $r(s)$), then equation 10.7 will provide the correct action values and an animal choosing on that basis would be predicted correctly to turn right. Thus, model-based reinforcement learning captures the defining characteristic of goal-directed behavior. Note that the agent needs to *know* the new reward function, that is, that when the animal is thirsty, $r(s)$ is high for states involving the consumption of water and low for food. In the case of hunger and thirst, a typical animal has plenty of experience with this, though for more novel motivational situations, it can be shown that animals learn the reward value $r(s)$ of some outcome in some new motivational state from experience with it (a process called incentive learning; Balleine and Dickinson, 1991). And again, this stands in contrast to TD, which has no way of incorporating new knowledge about a changed reward function $r(s)$ into a previously learned policy.

Similarly, equation 10.7 can also correctly plan or replan actions on the basis of new information about the state transition contingencies (the distribution $P(s_{t+1} \mid s_t, a_t)$), which in the case of a spatial maze amounts to a map relating the various locations and actions. Consider an experiment championed by Tolman (1948), called latent learning. Here animals explore a maze (giving them the opportunity to learn $P(s_{t+1} \mid s_t, a_t)$) before any rewards are introduced and then are subsequently taught that a particular location is the rewarded goal. Model-based planning using equation 10.7 implies that animals should be able to use their previously learned knowledge of the maze layout to immediately plan a route to the new goal location. TD, on the other hand, predicts that animals will have to learn the navigational policy by trial and error, and in particular implies that this learning process will not be facilitated by unrewarded preexposure to the maze layout, which teaches TD nothing about the subsequently relevant policy. In fact, as model-based reinforcement learning predicts but TD does not, animals (and humans) who have experienced a maze's spatial layout are quicker to learn new routes when rewards are introduced (Gläscher et al., 2010; Tolman, 1948). In the same way, model-based planning can plan novel routes if previously favored routes are blockaded or new shortcuts are open—all behaviors that Tolman (1948) pointed out could be produced by rats but not by the law of effect—so long as the agent is able to incorporate information about these maze changes in its representation of the transition function $P(s_{t+1} \mid s_t, a_t)$. In such cases, inferring the available transition graph from space must be accomplished by analyzing visual input (to spot openings or blockades and work out where they lead) rather than learning the function experientially by trying out various actions in different states and memorizing their consequences (Guazzelli et al., 1998).

Although the correspondence between model-based reinforcement learning and the psychological category of goal-directed behavior is straightforward, the neural implementation of these computations is relatively unexplored. One insight is that the Bellman equation (equation 10.5) defines a dynamical system in V whose unique fixed point is the value function; thus it can be solved, in principle, by an attractor network (Suri, 2001; Sutton and Pinette, 1985). That is, if we assign one neuron per state and associate its firing rate with the state's value, then the value iteration equation ($V^\pi(s) \leftarrow r(s) + \gamma \sum_{s' \in S} P(s'|s,\pi(s)) V^\pi(s')$: the version of equation 10.7 for V^π) can equally be viewed as neural update rules. In this case, if we connect our value neurons to one another with weights determined by the transition matrix $P(s'|s,\pi(s))$ and then inject additional input to each neuron corresponding to its states reward $r(s)$, then the value iteration equation describes the total input received by each neuron at each time step. Iterating this repeatedly, the system carries out value iteration by relaxing to its unique attractor, the true values.

Finally, a different approach to solving some of these problems is to replace the net value function with a vector of separate value functions associated with different outcomes (Arbib and Bonaiuto, 2012; Dayan, 1993). TD methods can be used in parallel to learn the expected future food associated with each state and action, and, separately, the expected future water, and so on. Neurally, such a mechanism could leverage TD learning methods whose neural implementation is well understood (though note that this requires a *vector-valued* error signal). Then choices can be made appropriate to the current motivational state, without relearning policies when motivations change, simply by weighting these different functions according to the agent's current drives. In this way, such a model can capture goal-directed behavior in experiments involving revaluation and motivational shifts, as with the maze of figure 10.5C. However, because this scheme uses TD to learn aggregate long-run expectancies for each outcome, rather than deriving these at decision time from iterating a current representation of the one-step transition graph (equation 10.7), it cannot solve many problems involving changes in the transition function. For instance, it cannot plan novel routes when a shortcut is introduced.

10.5 Conclusion and Arbitration

We have arrived mostly at the end of our description of some of the brain's hypothesized mechanisms for reinforcement learning. In particular, we have described behavioral and neural evidence for an error-driven learning system that uses dopaminergic prediction errors to learn long-run reward predictions and action-selection policies that maximize long-term reward. Moreover, we have suggested that this system corresponds well to a subset of animal learning behavior—so-called habitual behaviors—and sits alongside a separate system for more deliberative "goal-directed" decision-making, which we have tentatively identified with model-based reinforcement learning. Together, these two systems appear to describe a reasonably plausible picture of reinforcement learning in the brain.

Or do they? One last, critical issue with dual-system solutions to the decision-making problem is that if two systems disagree, then they, in some sense, compound the decision problem with a new one: how to reconcile the recommendations, or which system to trust. Finally, then, a few models have tackled the "meta-decision" problem of how the brain might arbitrate between model-free and model-based controllers. The basic insight of these models is that the two reinforcement learning strategies represent different trade-offs between computational and statistical efficiency. Model-based reinforcement learning makes near-optimal use of experience to compute action values—it is able, for instance, to recompute new plans when circumstances change, but only at the expense of laborious evaluation of the Bellman equation by recursive expansion. In many circumstances—such as when actions have repeatedly and consistently been successful—such computation and recomputation is an unnecessary waste of time and calories. TD learning involves a computationally much easier choice rule: simply compare the stored values for the different actions. However, learning these values by local TD updates at particular states does not always make the best use of limited experience, as in the case when circumstances change but the stored values are not properly updated.

Models by Daw et al. (2005) and Keramati et al. (2011) propose ways that the brain might quantify these trade-offs, by using Bayesian variants of the learning algorithms (see also chapter 13) to characterize the uncertainty (or expected error) in their value estimates in different circumstances, and thereby measure whether additional computation in a model-based system is likely to produce a worthwhile gain in accuracy. In Keramati et al.'s version, the model-free value $Q(s,a)$ of an action at a state is learned via a Bayesian version of Q learning, so that it is represented as a posterior distribution over possible state action values, quantifying both the agent's expectations (e.g., the mean of $Q(s,a)$) and uncertainty (e.g., the variance of the posterior distribution over Q), measuring ignorance about its exact value. With this information, it is possible to quantify the value of computing a better estimate of Q using a more laborious model-based computation, and balance this against the opportunity cost of the time spent doing so. In brief, the value of having a better estimate of Q is that it might allow you to make a different, better decision than you would otherwise have made, earning more rewards. This can be quantified, at least approximately, by taking an expectation over the result of the choice with respect to uncertainty in Q. If Q is very uncertain, recomputing its value more accurately can cause a big shift in choices and rewards, whereas if it is already well-known, then knowing its value even better is unlikely to have much effect. These theories intuitively explain the progression from goal-directed to habitual behavior with overtraining, since when there is much experience about an action's value, the model-free values will be relatively certain and there is little additional value to model-based computation to weigh against the time spent doing so. The theories offer similar explanations for many other experimental manipulations that have been demonstrated to affect the trade-off.

Theories of this sort are beginning to bring the brain's relatively simple mechanisms for reinforcement learning into contact with higher level functions such as cognitive control—itself, in large part concerned with how to pursue goals in the face of conflict with prepotent responses (Botvinick et al., 2001). Similarly, another crucial function for control is the management of working memory, which some interesting modeling research has attempted to recast as itself a reinforcement learning problem (O'Reilly and Frank, 2006). The idea here is that strategically "gating" pieces of information into and out of working memory can be viewed as a set of actions, not unlike the actions in any other reinforcement learning problem except that they operate internally instead of externally. Then the same reinforcement learning machinery can be used to learn a gating policy that adaptively manages the brain's working memory resources, maintaining information that is relevant for future reward seeking and purging information that is not. Finally, the same reinforcement learning mechanisms used to learn simple problems can also be applied to control problems across a hierarchy comprising multiple levels of abstraction (Botvinick et al., 2009; Ribas-Fernandes et al., 2011). In all, the simple mechanisms described above may be broadly applicable as subsystems for a surprisingly rich set of functions.

REFERENCES

Alexander, G. E., & Crutcher, M. D. (1990). Functional architecture of basal ganglia circuits: Neural substrates of parallel processing. *Trends in Neurosciences, 13*, 266–271.

Arbib, M. A., & Bonaiuto, J. J. (2012). Multiple levels of spatial organization: World graphs and spatial difference learning. *Adaptive Behavior, 20*, 287–303.

Balleine, B., & Dickinson, A. (1991). Instrumental performance following reinforcer devaluation depends upon incentive learning. *Quarterly Journal of Experimental Psychology, 43*, 279–296.

Bayer, H. M., & Glimcher, P. W. (2005). Midbrain dopamine neurons encode a quantitative reward prediction error signal. *Neuron, 47*, 129–141.

Bellman, R. (1957). *Dynamic Programming*. Princeton, NJ: Princeton University Press.

Berridge, K. C. (2007). The debate over dopamine's role in reward: The case for incentive salience. *Psychopharmacology, 191*, 391–431.

Bertsekas, D. P., & Tsitsiklis, J. N. (1996). *Neuro-dynamic Programming*. Belmont, MA: Athena Scientific.

Botvinick, M. M., Braver, T. S., Barch, D. M., Carter, C. S., & Cohen, J. D. (2001). Conflict monitoring and cognitive control. *Psychological Review, 108*, 624–652.

Botvinick, M. M., Niv, Y., & Barto, A. C. (2009). Hierarchically organized behavior and its neural foundations: A reinforcement learning perspective. *Cognition, 113*, 262–280.

Brischoux, F., Chakraborty, S., Brierley, D. I., & Ungless, M. A. (2009). Phasic excitation of dopamine neurons in ventral VTA by noxious stimuli. *Proceedings of the National Academy of Sciences of the United States of America, 106,* 4894–4899.

Cohen, J. Y., Haesler, S., Vong, L., Lowell, B. B., & Uchida, N. (2012). Neuron-type-specific signals for reward and punishment in the ventral tegmental area. *Nature, 482,* 85–88.

Daw, N. D., Courville, A. C., & Touretzky, D. S. (2006a). Representation and timing in theories of the dopamine system. *Neural Computation, 18,* 1637–1677.

Daw, N. D., Niv, Y., & Dayan, P. (2005). Uncertainty-based competition between prefrontal and dorsolateral striatal systems for behavioral control. *Nature Neuroscience, 8,* 1704–1711.

Daw, N. D., O'Doherty, J. P., Dayan, P., Seymour, B., & Dolan, R. J. (2006b). Cortical substrates for exploratory decisions in humans. *Nature, 441,* 876–879.

Day, J. J., Roitman, M. F., Wightman, R. M., & Carelli, R. M. (2007). Associative learning mediates dynamic shifts in dopamine signaling in the nucleus accumbens. *Nature Neuroscience, 10,* 1020–1028.

Dayan, P. (1993). Improving generalization for temporal difference learning: The successor representation. *Neural Computation, 5,* 613–624.

Dayan, P., & Abbott, L. (2000). *Theoretical Neuroscience: An Introduction.* Cambridge, MA: MIT Press.

DeLong, M. R. (1990). Primate models of movement disorders of basal ganglia origin. *Trends in Neurosciences, 13,* 281–285.

Dickinson, A., & Balleine, B. (2002). The role of learning in the operation of motivational systems. In C. R. Gallistel (Ed.), *The Stevens' Handbook of Experimental Psychology* (pp. 497–534). Chichester: Wiley.

Fiorillo, C. D., Tobler, P. N., & Schultz, W. (2003). Discrete coding of reward probability and uncertainty by dopamine neurons. *Science, 299,* 1898–1902.

Frank, M., Moustafa, A., Haughey, H., Curran, T., & Hutchison, K. (2007). Genetic triple dissociation reveals multiple roles for dopamine in reinforcement learning. *Proceedings of the National Academy of Sciences of the United States of America, 104,* 16311–16316.

Frank, M. J., Seeberger, L. C., & O'Reilly, R. C. (2004). By carrot or by stick: Cognitive reinforcement learning in parkinsonism. *Science, 306,* 1940–1943.

Gittins, J. C., & Jones, D. M. (1974). A dynamic allocation index for the sequential design of experiments. In J. Gani (Ed.), *Statistics* (pp. 241–266). Amsterdam: North-Holland.

Gläscher, J., Daw, N., Dayan, P., & O'Doherty, J. (2010). States versus rewards: Dissociable neural prediction error signals underlying model-based and model-free reinforcement learning. *Neuron, 66,* 585–595.

Guazzelli, A., Bota, M., Corbacho, F. J., & Arbib, M. A. (1998). Affordances, motivations, and the world graph theory. *Adaptive Behavior, 6,* 435–471.

Hellman, M. E., & Cover, T. M. (1970). Learning with finite memory. *Annals of Mathematical Statistics, 41,* 765–782.

Hong, S., Jhou, T. C., Smith, M., Saleem, K. S., & Hikosaka, O. (2011). Negative reward signals from the lateral habenula to dopamine neurons are mediated by rostromedial tegmental nucleus in primates. *Journal of Neuroscience, 31,* 11457–11471.

Houk, J. C., Adams, J. L., & Barto, A. G. (1995). A model of how the basal ganglia generate and use neural signals that predict reinforcement. In J. C. Houk, J. L. Davis, & D. G. Beiser (Eds.), *Models of Information Processing in the Basal Ganglia* (pp. 249–270). Cambridge, MA: MIT Press.

Jhou, T. C., Fields, H. L., Baxter, M. G., Saper, C. B., & Holland, P. C. (2009). The rostromedial tegmental nucleus (RMTg), a GABAergic afferent to midbrain dopamine neurons, encodes aversive stimuli and inhibits motor responses. *Neuron, 61,* 786–800.

Joel, D., Niv, Y., & Ruppin, E. (2002). Actor–critic models of the basal ganglia: New anatomical and computational perspectives. *Neural Networks, 15,* 535–547.

Joel, D., & Weiner, I. (2000). The connections of the dopaminergic system with the striatum in rats and primates: An analysis with respect to the functional and compartmental organization of the striatum. *Neuroscience, 96,* 451–474.

Jog, M. S., Kubota, Y., Connolly, C. I., Hillegaart, V., & Graybiel, A. M. (1999). Building neural representations of habits. *Science, 286,* 1745–1749.

Kamin, L. J. (1969). Predictability, surprise, attention, and conditioning. In B. A. Campbell & M. R. Church (Eds.), *Punishment and Aversive Behavior* (pp. 279–296). New York: Appleton-Century-Crofts.

Keramati, M., Dezfouli, A., & Piray, P. (2011). Speed/accuracy trade-off between the habitual and the goal-directed processes. *PLoS Computational Biology, 7,* e1002055.

Knutson, B., & Gibbs, S. E. (2007). Linking nucleus accumbens dopamine and blood oxygenation. *Psychopharmacology, 191,* 813–822.

Lashley, K.S. (1951). The problem of serial order in behavior. In L. A. Jeffress (Ed.), *Cerebral Mechanisms in Behavior* (pp 112–146). New York: Wiley.

Lau, B., & Glimcher, P. W. (2005). Dynamic response-by-response models of matching behavior in rhesus monkeys. *Journal of the Experimental Analysis of Behavior, 84,* 555–579.

Matsuda, W., Furuta, T., Nakamura, K. C., Hioki, H., Fujiyama, F., Arai, R., et al. (2009). Single nigrostriatal dopaminergic neurons form widely spread and highly dense axonal arborizations in the neostriatum. *Journal of Neuroscience, 29,* 444–453.

Matsumoto, M., & Hikosaka, O. (2007). Lateral habenula as a source of negative reward signals in dopamine neurons. *Nature, 447,* 1111–1115.

Matsumoto, M., & Hikosaka, O. (2009). Two types of dopamine neuron distinctly convey positive and negative motivational signals. *Nature, 459,* 837–841.

McClure, S. M., Berns, G. S., & Montague, P. R. (2003a). Temporal prediction errors in a passive learning task activate human striatum. *Neuron, 38,* 339–346.

McClure, S. M., Daw, N. D., & Montague, P. R. (2003b). A computational substrate for incentive salience. *Trends in Neurosciences, 26,* 423–428.

Mogenson, G. J., Jones, D. L., & Yim, C. Y. (1980). From motivation to action: Functional interface between the limbic system and the motor system. *Progress in Neurobiology, 14,* 69–97.

Montague, P. R., Dayan, P., & Sejnowski, T. J. (1996). A framework for mesencephalic dopamine systems based on predictive Hebbian learning. *Journal of Neuroscience, 16,* 1936–1947.

Moore, J. W., & Schmajuk, N. A. (2008). Kamin blocking. *Scholarpedia, 3,* 3542.

NIV, Y., DAW, N. D., JOEL, D., & DAYAN, P. (2007). Tonic dopamine: Opportunity costs and the control of response vigor. *Psychopharmacology, 191*, 507–520.

O'DOHERTY, J. P., DAYAN, P., FRISTON, K., CRITCHLEY, H., & DOLAN, R. J. (2003). Temporal difference models and reward-related learning in the human brain. *Neuron, 38*, 329–337.

O'REILLY, R. C., & FRANK, M. J. (2006). Making working memory work: A computational model of learning in the prefrontal cortex and basal ganglia. *Neural Computation, 18*, 283–328.

PARKINSON, J. A., DALLEY, J. W., CARDINAL, R. N., BAMFORD, A., FEHNERT, B., LACHENAL, G., et al. (2002). Nucleus accumbens dopamine depletion impairs both acquisition and performance of appetitive Pavlovian approach behaviour: Implications for mesoaccumbens dopamine function. *Behavioural Brain Research, 137*, 149–163.

PUTERMAN, M. L. (1994). *Markov Decision Processes: Discrete Stochastic Dynamic Programming.* New York: Wiley.

REDGRAVE, P., & GURNEY, K. (2006). The short-latency dopamine signal: A role in discovering novel actions? *Nature Reviews. Neuroscience, 7*, 967–975.

REDGRAVE, P., GURNEY, K., & REYNOLDS, J. (2008). What is reinforced by phasic dopamine signals? *Brain Research. Brain Research Reviews, 58*, 322–339.

RESCORLA, R., & WAGNER, A. (1972). A theory of Pavlovian conditioning: Variations in the effectiveness of reinforcement and nonreinforcement. In A. H. BLACK & W. F. PROKASY (Eds.), *Classical Conditioning: Vol. 2. Current Research and Theory* (pp. 64–99). New York: Appleton-Century-Crofts.

RESCORLA, R. A. (1969). Pavlovian conditioned inhibition. *Psychological Bulletin, 72*, 77–94.

REYNOLDS, J. N., & WICKENS, J. R. (2002). Dopamine-dependent plasticity of corticostriatal synapses. *Neural Networks, 15*, 507–521.

RIBAS-FERNANDES, J. J. F., SOLWAY, A., DIUK, C., MCGUIRE, J. T., BARTO, A. G., NIV, Y., et al. (2011). A neural signature of hierarchical reinforcement learning. *Neuron, 71*, 370–379.

ROMER, D. (2006). Do firms maximize? Evidence from professional football. *Journal of Political Economy, 114*, 340–365.

RUMMERY, G., & NIRANJAN, M. (1994). On-line Q-learning using connectionist systems. Technical Report CUED/F-INFENG/TR166, Cambridge University.

SCHULTZ, W. (1998). Predictive reward signal of dopamine neurons. *Journal of Neurophysiology, 80*, 1–27.

SCHULTZ, W., DAYAN, P., & MONTAGUE, P. R. (1997). A neural substrate of prediction and reward. *Science, 275*, 1593–1599.

SUGRUE, L. P., CORRADO, G. S., & NEWSOME, W. T. (2004). Matching behavior and the representation of value in the parietal cortex. *Science, 304*, 1782–1787.

SURI, R. E. (2001). Anticipatory responses of dopamine neurons and cortical neurons reproduced by internal model. *Experimental Brain Research, 140*, 234–240.

SUTTON, R. S. (1988). Learning to predict by the methods of temporal differences. *Machine Learning, 3*, 9–44.

SUTTON, R. S., & BARTO, A. G. (1998). *Reinforcement Learning: An Introduction.* Cambridge, MA: MIT Press.

SUTTON, R. S., & PINETTE, B. (1985). The learning of world models by connectionist networks. *Proceedings of the Seventh Annual Conference of the Cognitive Science Society*, 55–64.

THORNDIKE, E. L. (1911). *Animal Intelligence: Experimental Studies.* New York: Macmillan.

TOLMAN, E. (1948). Cognitive maps in rats and men. *Psychological Review, 55*, 189–208.

UNGLESS, M. A., MAGILL, P. J., & BOLAM, J. P. (2004). Uniform inhibition of dopamine neurons in the ventral tegmental area by aversive stimuli. *Science, 303*, 2040–2042.

VOORN, P., VANDERSCHUREN, L., GROENEWEGEN, H., ROBBINS, T., & PENNARTZ, C. (2004). Putting a spin on the dorsal-ventral divide of the striatum. *Trends in Neurosciences, 27*, 468–474.

WATKINS, C. J. C. H., & DAYAN, P. (1992). Q-learning. *Machine Learning, 8*, 279–292.

WISE, S. P. (2008). Forward frontal fields: Phylogeny and fundamental function. *Trends in Neurosciences, 31*, 599–608.

YIN, H. H., KNOWLTON, B. J., & BALLEINE, B. W. (2004). Lesions of dorsolateral striatum preserve outcome expectancy but disrupt habit formation in instrumental learning. *European Journal of Neuroscience, 19*, 181–189.

YIN, H. H., OSTLUND, S. B., KNOWLTON, B. J., & BALLEINE, B. W. (2005). The role of the dorsomedial striatum in instrumental conditioning. *European Journal of Neuroscience, 22*, 513–523.

ZHANG, J., BERRIDGE, K. C., TINDELL, A. J., SMITH, K. S., & ALDRIDGE, J. W. (2009). A neural computational model of incentive salience. *PLoS Computational Biology, 5*, e1000437.

11 Short-Term, Long-Term, and Working Memory

STEFANO FUSI AND XIAO-JING WANG

11.1 General Definitions and Background

Memory is often defined as the faculty of bringing to mind information encountered at an earlier time. In neuroscience, memory has a broader meaning, and it does not necessarily imply some form of conscious recollection: past events are remembered whenever they can influence our behavior at the present time. Even more broadly, they are remembered when they can affect the behavior of individual neurons or, more generally, of any dynamical variable such as a key molecule in a subcellular signaling pathway within the nerve cell. Memories are encoded (the information is acquired), stored (the information is retained), and then retrieved at a later time (the stored information is accessed).

11.1.1 Short-Term, Long-Term, and Working Memory Memory is not a unitary phenomenon, and different types of memories have been identified and studied systematically for more than a century. William James, a pioneering American psychologist at the end of the nineteenth century, was among the first who attempted to define different types of memory. He believed that there are two qualitatively different types: primary and secondary memories. Primary memories are basically an extension of the present moment, and they last as long as they can affect our current state of mind through an uninterrupted series of connected conscious states ("conscious stream"). These memories typically last seconds to minutes, depending on the cognitive function that is considered. Secondary memories can last weeks, months, years, even a lifetime, and they are defined as "the knowledge of a former state of mind after it has already once dropped from consciousness; or rather it is the knowledge of an event, or fact, of which meantime we have not been thinking, with the additional consciousness that we have thought or experienced it before" (James, 1890).

More recently, at the end of the 1960s, Atkinson and Shiffrin introduced a three-stage model of memory inspired by studies on the architecture of the first computers (Atkinson and Shiffrin, 1968). In their model, selected components of sensory information are transferred to a short-term store with limited capacity. Typically the memories are retained in short-term memory for seconds although often the capacity is expressed in terms of the number of items that can be remembered. Only a further selection reaches a long-term memory store. The process of transfer of information from the short-term repository to the long-term memory store may require rehearsal of the information in the short-term store. This is a simple model of memory consolidation in which memories are first in a labile state and only later do they become long-lasting.

Finally, there is a third type of memory, called *working memory*, that complements short- and long-term memories. Short-term memory is assumed to be the result of a relatively passive process. In contrast, working memory, although it involves similar timescales, is more "active" as there are mental operations performed on the information currently held in memory. Therefore, working memory involves manipulation as well as storage of information for a short period of time (seconds).

11.1.2 Episodic, Semantic, and Procedural Memories Long-term memories are divided into different classes, depending on their content and on the way they are affected by memory disorders (i.e., some memory disorders affect specifically one type of memories and not the others). Consciously accessible memories are usually named *declarative* or *explicit* memories. They are further divided into *episodic* and *semantic* memories.

Episodic memories contain information about specific events and the time and the situation in which they occurred (e.g., autobiographical memories such as the memory of what I had for dinner two days ago). The medial temporal lobes, including the hippocampus and parahippocampal cortices (entorhinal and perirhinal cortex), are critical for storing episodic memories. Other cortical areas like prefrontal cortex seem also to be involved in episodic memories (see also the section about memory consolidation at the systems level).

Semantic memories contain information about general knowledge, typically about specific facts and concepts (e.g., what's the capital of the United States?). In contrast to episodic memories, the circumstances under which they are acquired are not part of the memory. We often combine episodic and semantic memories: for example, we often use our knowledge of the world to "compress" the information contained in our episodic memories. Moreover, we may remember the circumstances in which we learned about the capital of the United States if it happened recently or in otherwise memorable circumstances. Semantic memories seem to be stored in the same brain areas as episodic memories although there are studies that show that some semantic memories remain intact when hippocampus is damaged (see, e.g., Vargha-Khadem et al., 1997), similarly to what happens for consolidated episodic memories.

In contrast to explicit memories, *implicit memories* do not involve conscious recollection. They manifest themselves as any kind of influence of prior experiences on our current behavior. One typical example is priming, in which the previous exposure to a sensory stimulus affects our reaction times to a later stimulation. In priming experiments the subjects do not consciously remember the previous exposures, but their behavior is nevertheless influenced by them. Another example of implicit memories are the *procedural memories*, which are memories of typical sequences of cognitive or physical operations (e.g., rules to solve tasks and puzzles or motor skills like riding a bike, tying shoes, or playing a piano). Procedural memories are learned through repetition of a complex activity, and they are automatically retrieved and utilized. Brain regions involved in procedural memories include the basal ganglia and the cerebellum.

11.1.3 Studies on Memory Forgetting Forgetting is the gradual loss of information stored in memory, and it has been extensively studied in a large number of experimental works. The first experiments were conducted by Hermann Ebbinghaus at the end of the nineteenth century (Ebbinghaus, 1885). He taught himself to remember long lists of nonsense syllables (consonant–vowel–consonant trigrams). He then evaluated the strength of the memories after an interval ranging from tens of minutes to weeks by estimating the relearning time (i.e., better remembered lists could be quickly relearned). He discovered that (1) different memories may have different memory life spans, (2) repetition makes memories last longer, and (3) the rate of forgetting decreases with the passage of time. These findings have been confirmed by subsequent experimental studies. Today, it is widely believed that the empirical forgetting functions are most accurately described by a power function; namely, a memory trace decays with time t as $1/t^{\alpha}$ where α is a real number (see, e.g., Wixted and Ebbesen, 1997). Such power-law decay does not have a characteristic lifetime.

The pioneering work of Ebbinghaus focused on nonsensical memories as he was trying to reduce the effects due to previously stored memories. However, in most everyday experiences we do not memorize meaningless information like nonsense syllables. For this reason, Frederic Bartlett (Bartlett, 1932) started a series of studies aimed at understanding how we incorporate new memories into our existing body of knowledge. He asked subjects to read themselves stories and then, later, to recall them. Recalled stories became consistently more coherent than the original stories because subjects were using their general knowledge and cultural expectations ("schemas") to reconstruct the forgotten links. (See chapter 14, "Schema Theory and Neuropsychology.")

11.2 Neural and Synaptic Mechanisms Underlying Memory

In order to understand the storage mechanisms, it is fundamental to identify and characterize the physical substrate of memories. Almost any biochemical process has "memory" in the sense that the present state depends on what has happened in the past to the system. This is a general property of most physical systems. However, many of these memories are preserved for times that are relatively short (from milliseconds to hours); memories that last for decades probably require more sophisticated mechanisms. Moreover, only some of these biochemical processes can be effectively "read out" by a neural circuit (i.e., they can affect its dynamics) and be utilized to generate the proper behavior in response to a memory cue (memory retrieval). Biological systems are very complex and diverse, and for all of them memory is not a unitary phenomenon because it involves several distinct mechanisms that operate at different spatial and temporal levels. One general property of some of these mechanisms is that there seems to be a flow of information from fast to slow memory systems. In other words, memories are initially labile and subject to rapid degradation because they are stored in fast dynamical variables. With time, such memories are consolidated because they are transferred to slow dynamical variables and they become stable and highly resistant to memory interference (i.e., to the storage of correlated information that can disrupt similar previously stored memories). Consolidation occurs both in the form of molecular processes at individual synapses and in the form of system consolidation, which is a reorganization of the memory trace within different brain systems. These mechanisms protect memories both from the degradation due to the passage of time and from the overwriting caused by the storage of other memories. In the next section we will focus on the molecular mechanisms that make the synaptic state stable in the absence of events that generate new memories. As we will show, these mechanisms require some form of multistability, which then greatly affects how rapidly memories are overwritten by other memories (see the "Models of Short- and Long-Term Memory" section).

11.2.1 Synaptic Mechanisms of Memory Consolidation and Molecular Memories The consolidation and the maintenance of synaptic modifications require a complex molecular machinery that typically involves cascades of biochemical processes that alter the state of different molecules and eventually trigger protein synthesis. The specific mechanisms vary depending on the type of synapse that is considered, but they all seem to share similar dynamical properties. Typically the process of induction of long-term synaptic modifications starts with an alteration of some of the molecules that are locally present at the synapse. These modifications are induced by the activity of the presynaptic neuron, and they depend on the state of the postsynaptic neuron, as in the case of Hebbian plasticity (see chapter 6, "Hebbian Plasticity and Learning," for more details). This is how the synapses encode information about the memory that has to be stored.

For example, when the presynaptic excitatory neuron emits a spike, a neurotransmitter (glutamate) is released. The neurotransmitter diffuses across the synaptic cleft, binding to glutamate receptors on the postsynaptic membrane, causing the opening of ionic channels and the entry of positive ions which depolarize the postsynaptic cell. The amount of neurotransmitter that is released and the number of receptors on the postsynaptic side determine the strength of the synaptic connection.

The two most studied types of receptors that mediate excitatory connections are AMPA and NMDA receptors. Specific patterns of pre- and postsynaptic activity induce an alteration of the state of relatively complex molecules that are locally present. These alterations might directly modulate the strength of the synaptic connections or they can mediate the expression of long-term synaptic modifications. One of the molecules that is known to be involved in the expression of long-term potentiation (LTP) is the calcium/calmodulin-dependent protein kinase II (CaMKII). When glutamate is released by the presynaptic neuron and the postsynaptic neuron is sufficiently depolarized, the local calcium concentration changes because of the influx of calcium through NMDA channels. The entry of calcium then induces an alteration of the state of CaMKII. Each molecule of CaMKII is composed of two rings, each with six subunits. These subunits are phosphorylated (activated) in the presence of calcium, and their state affects the strength and the number of AMPA receptors and, hence, the strength of the synaptic connection. Each CaMKII subunit can then be considered as a memory switch which can be active or inactive, and the entire (small) population of the subunits of all CaMKII molecules that are present and immobilized in a dendritic spine (5–30 molecules) are an example of a molecular memory. CaMKII not only is necessary for LTP induction but can also, by itself, enhance the efficacy of AMPA synaptic transmission.

CaMKII is only one example of molecular memory, and there are several other molecules with similar properties that are involved in synaptic plasticity (e.g., protein kinase C [PKC], cAMP-dependent protein kinase [PKA], cGMP-dependent protein kinase [PKG]). One of the problems of these memories is molecular turnover (see, e.g., Crick, 1984): for example, CaMKII proteins are gradually destroyed and replaced by newly synthesized unphosphorylated proteins. As a consequence, the memory lifetime is limited by the lifetime of the protein, which is of the order of one day. Other molecules can last longer, but none of them can survive a lifetime. Thus, there must be an additional mechanism to maintain the molecular memories over timescales of decades.

Francis Crick hypothesized that there must be some form of interaction that allows the proteins that encode memories to copy their state to the incoming newly synthesized proteins (Crick, 1984). In the case of CaMKII proteins, Lisman and Zhabotinsky (Lisman, 1985; Lisman and Zhabotinsky, 2001; Miller et al., 2005) showed that the interplay of two mechanisms can solve the problem of protein turnover by making a population of CaMKII bistable (see figure 11.1). The first mechanism is mediated by an enzyme (a kinase) that activates, that is, phosphorylates, CaMKII subunits. The rate at which it activates them increases with the number of subunits that are already activated. In other words, if some CaMKII subunits are active, then others will be activated. The second mechanism is mediated by another enzyme (a phosphatase, named PP1) that operates in the opposite direction by inactivating CaMKII

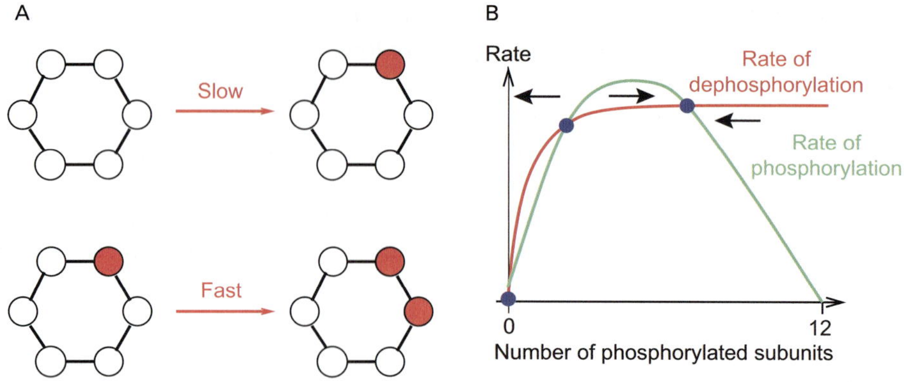

FIGURE 11.1 Bistability in synaptic plasticity. (A) The hexagons represent one of the two rings of CaMKII molecules. Each subunit (circle) can be in two possible states (phosphorylated, in red, or not phosphorylated, in white). Phosphorylation of the first subunit is slow at resting calcium concentration (top). Subsequent phosphorylation is faster because it is catalyzed by other active subunits (bottom). (B) Schematic figure indicating how bistability arises from the dependence of phosphorylation (green) and dephosphorylation (red) rates on the number of subunits phosphorylated. When the green curve is above the red curve, the number of phosphorylated subunits increases. It decreases in the other regions, and it remains constant at the intersections (blue circles). Stable states are at the left and right intersections of the two curves. The middle crossing is unstable.

subunits. The rate at which molecules are inactivated increases rapidly with the number of activated CaMKII subunits and then it saturates. The velocity of the reactions induced by the kinase and the phosphatase is such that each CaMKII molecule has two stable points, one in which none of the subunits is active, and another one in which a large proportion is active. As all CaMKII molecules behave similarly, the synaptic strength is also bistable.

The effects of protein turnover are greatly reduced by the interplay of these two mechanisms because when a new inactivated protein comes in, it remains unaltered if the majority of the existing CaMKII molecules are inactivated, and it is activated if they are in the active state. Indeed, when CaMKII molecules are activated, the phosphatase spends a significant amount of time bound to the existing CaMKII molecules and it is less available to inactivate the new molecules, which phosphorylate even at the basal calcium level (e.g., when both presynaptic and postsynaptic neurons are silent). In other words, the new molecules are likely to acquire the state of the existing ones, preserving the state of the molecular memory that is stored at the level of the population of CaMKII molecules.

CaMKII is only one example of molecular memory and is known to operate in the early phases of the induction of long-term synaptic potentiation (E-LTP). These molecules can then regulate gene transcription and protein synthesis which then lead to permanent changes in the morphology of the synaptic connections or to persistent molecular mechanisms that are known to underlie late LTP (L-LTP) maintenance. The pioneering and foundational work on these cascades of biochemical processes of Eric Kandel and colleagues is summarized in Squire and Kandel (1999). For a recent model of early and late LTP induction, see Clopath et al. (2008).

11.2.2 MEMORY CONSOLIDATION AT THE SYSTEM LEVEL Much of our knowledge about multiple memory systems and the principles of memory consolidation at the system level comes from studies on the famous patient HM (see, e.g., Squire, 2009). He suffered from a severe form of epilepsy which was treated by removing a large portion of the medial temporal lobe. The hippocampus, the amygdala, and the parahippocampal gyrus were completely or almost completely ablated bilaterally. As a consequence of such a lesion, HM became unable to form new memories (anterograde amnesia), as reported a few years later in Scoville and Milner (1957). Surprisingly, his intellectual and perceptual functions, his remote memories (i.e., memories of events that occurred before the surgery), and his working memory were intact. Moreover, only declarative memories were affected by the lesion of the medial temporal lobe as HM was still able to acquire new motor skills (i.e., he could form new procedural memories). These observations indicate that (1) there are multiple memory systems located in different brain areas, and (2) declarative memories require the medial temporal lobe in order to be stabilized, but once they are consolidated, the medial temporal lobe becomes unnecessary. A possible explanation is that the medial temporal lobe is a temporary repository for new memories. Then they are transferred to a permanent neocortical store. Many investigators believe that the transfer is mediated by the so-called *replay activity*, which is some form of spontaneous neural activity observed during rest and sleep that is highly correlated to the neural activity recorded during the creation of the memory (see, e.g., Kudrimoti et al., 1999; Diba and Buzsaki, 2007; Lee and Wilson, 2002).

Some theoretical studies (see, e.g., McClelland et al., 1995; O'Reilly and Frank, 2006; Káli and Dayan, 2004; Battaglia and Pennartz, 2011) on memory consolidation showed the importance of partitioning memory systems into multiple stages, typically two, that operate on different timescales. In particular, it is important to have a temporary repository when new semantic memories are inserted into an existing body of knowledge. Two-stage memory models were shown to be fundamentally important to avoid catastrophic forgetting. These studies focused mostly on "memory reorganization," as they presuppose that the memories are highly organized and that their correlations can rapidly destroy previously stored information. More recent studies (Roxin and Fusi, 2013) showed that memory consolidation at the system level can also solve a different class of problems which are related to the storage of uncorrelated memories in the case of synaptic variables that vary in a limited range. These problems will be discussed in detail in the rest of the chapter.

11.2.3 WORKING MEMORY Working memory, a concept introduced by Miller, Galanter, and Pribram in 1960 and developed by Baddeley in the 1970s in cognitive psychology (Baddeley, 2012), refers to the brain's ability to hold information of current relevance in the absence of external input. Working memory function has been hypothesized to depend on self-sustained persistent neural activity for online maintenance and manipulation of information (Hopfield, 1982; Amit, 1989). Different persistent activity patterns represent distinct items of information, and there is a need to be able to rapidly switch on and off between any of these memory states and a resting (background) state. Such stimulus-selective persistent activity has been observed widely in single neurons of the prefrontal cortex (Fuster and Alexander, 1971; Kubota and Niki, 1971; Funahashi et al., 1989; Miller et al., 1996; Romo et al., 1999), posterior parietal cortex (Gnadt and Andersen, 1988; Chafee and Goldman-Rakic, 1998; Powell and Goldberg, 2000; Constantinidis and Wang, 2004), and other brain areas, recorded from behaving monkeys performing tasks in which a behavioral response depends on working memory of a stimulus across a time delay of a few seconds. Human functional brain imaging studies also consistently show that the parietoprefrontal circuit plays a critical role in working memory (Courtney et al., 1997; Curtis and D'Esposito, 2003).

11.3 Models of Short- and Long-Term Memory

As explained in the previous section, there are three important aspects of memory: encoding, storage and retrieval. In this chapter we focus on storage and retrieval. Encoding is covered in chapter 6, "Hebbian Plasticity and Learning." In particular, we will first consider the problem of memory storage in long-term memory models. Then we will discuss short-term memory, and finally, in the next section, we will cover working memory models.

11.3.1 FORMALIZING THE MEMORY STORAGE PROBLEM Storing a memory involves several mechanisms that start with a reorganization of the information to be stored and end with long-term synaptic modifications. Most of the new memories are likely to bear some "resemblance" to previously stored ones. This is the typical situation in the case of semantic memories, which contain information about categorical and functional relationships between familiar objects. For this type of memory we can utilize our previous knowledge about the objects so that we can store only the information about the relationships between them (see, e.g., McClelland et al., 1995). In other words, we can clearly take advantage of the correlations between the new memory and the previously stored ones that encode the relevant objects. An efficient way of storing these memories is to exploit all possible correlations of this type and then store only the incompressible part of the information. Memories are probably actively and passively reorganized to separate the correlated and the incompressible part of the storable information. Modeling this process of reorganization is of fundamental importance, and it has been the subject of several theoretical studies (McClelland et al., 1995; O'Reilly and Frank, 2006; Káli and Dayan, 2004; Battaglia and Pennartz, 2011). However, in this chapter we will focus on the theories developed for the incompressible part, and for this reason, we assume that the memories to be stored, which are basically patterns of synaptic modifications (or modifications of any other variable that can preserve memories), are random and uncorrelated, and each memory is stored in a single exposure (a second exposure would be correlated with a previously stored memory). In other words, we assume that every time a new episode is stored, there are mechanisms to extract the novel information, or the information that is not already stored (i.e., the part that is not correlated). In the ideal case, this would be the only part of the information that is stored. The real memory capacity can only be worse in the nonideal case in which some of the correlated information is also memorized, and hence, assuming that only the uncorrelated part will be stored, we will derive an upper bound on the memory performance.

Under this assumption, the memory capacity problem is to determine how many random and uncorrelated memories can be stored in a neural circuit of a given size. The important questions we will try to answer are these: How does the memory capacity scale with the total number of synapses or the total number of modifiable variables? Can we build a synaptic model that can take advantage of the huge number of synapses that are available in biological brains? Interestingly, there are ways of answering these questions that do not depend on the details of the type of learning that is considered. Some of them will be reported in the next sections.

11.3.2 Memory Storage Capacity in Unbounded Synaptic Models At the beginning of the 1990s many of the theoretical neuroscientists investigating memory believed that the problem of memory storage was basically solved, and it was better to invest time on the study of the mechanisms underlying memory encoding and retrieval. Indeed, the memory storage capacity of simple synaptic models like the perceptron (Rosenblatt, 1958) or the Hopfield model (Hopfield, 1982) (see chapter 2 and below) is huge as it scales very favorably with the total number of synapses. Unfortunately this scaling breaks down when the synapses are restricted to vary in a limited range, as required for any realistic synapse.

In this section we derive a simple estimate of the memory capacity in the case of unbounded synapses and we show that it scales linearly with the total number of synapses. In the next section we will show what happens in more realistic situations and how most of the memory models are plagued by the problem of catastrophic forgetting as soon as we impose boundaries on the synaptic weights.

In order to characterize the memory performance of a neural system, we consider the problem of storing p random uncorrelated memories. Assume that there are p different events, each generating a memory. For example, a relevant event could be the presentation of a visual stimulus that has to be memorized. When an event is perceived, a specific pattern of neural activity is established (e.g., the neurons receive a strong input from the sensory systems, and their firing rate encodes the identity and the structure of the stimulus). This neuronal activity then produces a pattern of synaptic modifications. The sign and the extent of the modification depend on the statistics of the neural activity and on the details of the synaptic dynamics (see chapter 6 for more details). As previously discussed, we assume that these synaptic modifications are random and uncorrelated while the synapses remain unchanged between events.

More formally, we assume that a generic synapse w_{ij}, connecting neuron j to neuron i, is modified according to

$$w_{ij} \rightarrow w_{ij} + \Delta w_{ij}^{\mu}$$

following the storage of the μ-th memory. For example, in the case of the Hopfield model (Hopfield, 1982), the prescription for modifying a synapse follows the Hebb's rule (see again chapter 6) and the synaptic modification can be written as

$$\Delta w_{ij}^{\mu} = \xi_j^{\mu} \xi_i^{\mu},$$

where ξ_i^{μ} is the activity imposed on neuron i when memory μ is stored. More specifically, $\xi_i^{\mu} = \pm 1$ depending on whether the neuron is active or inactive.

In what follows, however, we will not make any assumption about the specific prescription for updating the synapses, but we will assume for simplicity that Δw_{ij} are random uncorrelated stochastic variables, which are ± 1 with equal probability (potentiations and depressions are balanced).

We now choose a generic memory, and we estimate its strength immediately after it is stored and then during the process of storage of other memories. The chosen memory is not special; it is just the memory that we decide arbitrarily to track. As a consequence, the other memories would all behave in the same way.

We take the point of view of an ideal observer who can access the value of all the synaptic strengths. This is a privileged point of view, and it can be very different from the point of view that individual neurons can read a limited number of synapses and only in the presence of presynaptic activity. However, the ideal observer approach allows us to derive bounds on the memory capacity that do not depend on the architecture of the network or the details of the neural dynamics. These bounds are also very informative as they often predict the scaling properties of actual neural networks.

We define the strength of the memory trace M^{μ} as the correlation between the synaptic matrix w_{ij} and the μ-th memory:

$$M^{\mu} = \frac{1}{N_{syn}} \sum_{i,j} w_{ij} \Delta w_{ij}^{\mu} - \left(\frac{1}{N_{syn}} \sum_{i,j} w_{ij} \right) \left(\frac{1}{N_{syn}} \sum_{i,j} \Delta w_{ij}^{\mu} \right),$$

where N_{syn} is the total number of synapses, and the sum extends over all pre- and postsynaptic neurons. The first term is the overlap between the synaptic matrix and the pattern of synaptic modifications. The subtracted term is the product of the average synaptic weight multiplied by the average synaptic modification, and it basically expresses the similarity (by chance) between the w_{ij} vector and the Δw_{ij}^{μ} when none of the memories that we are considering are stored in w_{ij}. The number of terms in each sum is N_{syn}. The memory trace is strong when M^{μ} is large, and weak when $M^{\mu} \sim 0$. In the case that we are considering, the average value of Δw_{ij}^{μ} is zero, and hence we can eliminate the subtracted term and rewrite M^{μ} as follows:

$$M^{\mu} = \frac{1}{N_{syn}} \sum_{i,j} w_{ij} \Delta w_{ij}^{\mu}.$$

The memory strength as defined here is directly related to the ability of a recurrent neural network to retrieve a memory (see Amit and Fusi, 1994, and below). As we will show later, it also represents the energy function of a Hopfield network.

Assume that we now track the first memory that is stored ($\mu = 1$). We start from a tabula-rasa synaptic matrix ($w_{ij} = 0$) and then we store memory 1: $w_{ij} = \Delta w_{ij}^{1}$. The memory strength immediately after its storage is given by the following:

$$M^{1} = \frac{1}{N_{syn}} \sum_{i,j} (\Delta w_{ij}^{1})^{2}.$$

As $\Delta w_{ij} = \pm 1$, then $M^{1} = 1$. We now store the other uncorrelated memories. As they are random, the synapses basically perform a random walk, and in particular they move up and down with the same probability as we assumed that potentiations and depressions are balanced. The value of the synaptic strength after the storage of p memories is

$$w_{ij} = \sum_{\mu=1}^{p} \Delta w_{ij}^{\mu}.$$

The memory strength of each memory is the sum of stochastic variables representing the synaptic modifications, and hence it is also a stochastic variable, which has an expected value, the signal $\mathcal{S}$, which, for memory 1 is equal to

$$\langle M^{1} \rangle = \frac{1}{N_{syn}} \left\langle \sum_{i,j} \sum_{\mu=1}^{p} \Delta w_{ij}^{\mu} \Delta w_{ij}^{1} \right\rangle,$$

where the angular brackets $\langle \ldots \rangle$ denote the operation of averaging over all realizations of random memories (i.e., the average over all possible sequences of synaptic modifications, each weighted by its probability of occurrence). As $\Delta w_{ij} = \pm 1$ with equal probability, it is easy to compute this quantity:

$$\mathcal{S} = \langle M^{1} \rangle = \frac{1}{N_{syn}} \sum_{i,j} \langle \Delta w_{ij}^{1} \Delta w_{ij}^{1} \rangle + \frac{1}{N_{syn}} \sum_{i,j} \sum_{\mu=2}^{p} \langle \Delta w_{ij}^{1} \Delta w_{ij}^{\mu} \rangle.$$

The second term contains terms of the following form, $\langle \Delta w_{ij}^{1} \Delta w_{ij}^{\mu} \rangle$, which are all zero because the memories are uncorrelated. The first term is simply equal to 1, as in the case of the initial memory strength. Hence the memory signal remains constant ($\mathcal{S} = 1$) when new memories are stored.

However, the noise due to the interference with other memories constantly increases as new memories are stored. Indeed, the variance of M^{1} is given by

$$\langle (M^{1} - \langle M^{1} \rangle)^{2} \rangle = \frac{1}{N_{syn}^{2}} \left\langle \left(\sum_{i,j} \sum_{\mu=2}^{p} \Delta w_{ij}^{1} \Delta w_{ij}^{\mu} \right)^{2} \right\rangle.$$

This is equal to the variance of the sum of $N_{syn}(p-1) \sim N_{syn}p$ independent $\pm$ 1 random variables, which is proportional to $N_{syn}p$. Hence, the *memory noise* $\mathcal{N}$, which is the square root of the variance of the memory trace is as follows:

$$\mathcal{N} = \sqrt{\langle (M^{1} - \langle M^{1} \rangle)^{2} \rangle} = \sqrt{\frac{N_{syn} p}{4N_{syn}^{2}}} = \frac{1}{2} \sqrt{\frac{p}{N_{syn}}}.$$

The memory signal and the memory noise are plotted in figure 11.2 (left). The exact number of memories that can be retrieved will in general depend on the details of the architecture of the network and on the neural dynamics. However, the scaling properties do not usually depend on these details and can be determined by choosing an arbitrary threshold for the $\mathcal{S}/\mathcal{N}$ ratio. For example, if we decide that memories are retrievable as long as $\mathcal{S}/\mathcal{N} > 1$ (any other value would produce the same scaling), then the memory capacity is as follows:

$$p \sim N_{syn}.$$

This is a huge capacity if we consider that in the human brain we have $\sim 10^{15}$ synapses. Notice that in this case the memory signal is constant, and it is the same for all memories. The only reason why the memories cannot be retrieved when the critical capacity is surpassed is not that the memories are forgotten but only that they are obscured by other memories.

This simple estimate is usually an upper bound for memory capacity, as it is based on an ideal observer

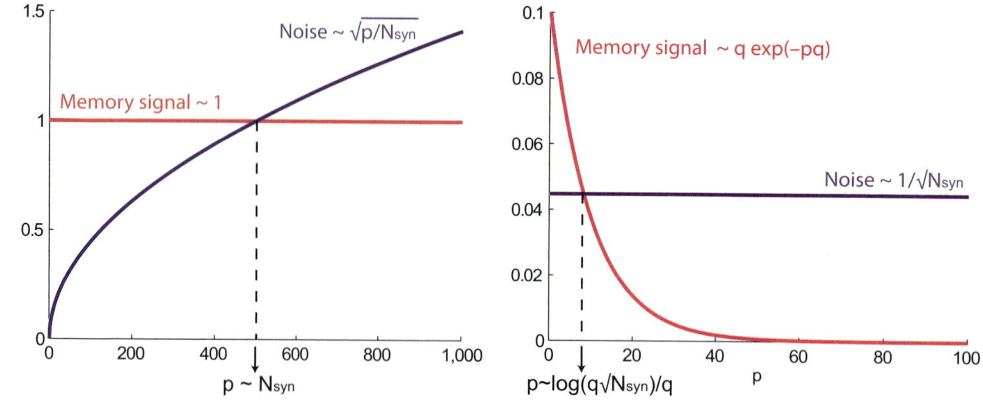

FIGURE 11.2 Memory signal (or memory strength) $\mathcal{S}$ and memory noise $\mathcal{N}$ as a function of the number of stored memories p. In the case of unbounded synapses (left), the memory signal is constant and the noise increases with p. They cross at $p \sim N_{syn}$. In the case of bounded synapses (right), the situation is drastically different because the memory noise is now approximately constant, and the memory signal decays exponentially with p when new memories are stored. They cross at $p \sim (\log qN_{syn})/q$.

approach. However, it is compatible with the results of more complex calculations of the storage capacity in specific models in which the process of retrieval is explicitly modeled. For example, the simple perceptron (Rosenblatt, 1958) (see chapter 2) is basically only one output neuron that has to remember all the patterns that it was trained to classify. It can store and retrieve (i.e., respond in the same way to the inputs as during training) a number of random uncorrelated memories which is on average $p = 2C$, where C is the number of afferent synaptic connections, and in this case, it coincides with the total number of adjustable synapses N_{syn}. This means that the number of storable and retrievable memories scales linearly with the total number of synapses, as in the simple estimate just derived.

In the case of the Hopfield network (see chapter 2 and below), memories are retrievable when their neural representations are attractors of the neural dynamics. The calculations of Amit, Gutfreund, and Sompolinsky (Amit et al., 1985; Amit, 1989) show that the critical capacity is $p = \alpha C$, where, again, C is the number of synapses per neuron, and $\alpha \sim 0.14$. Notice that in the case of fully connected attractor recurrent networks, the total number of synapses N_{syn} scales like C^2, as there are C neurons, each with C connections. The ideal observer approach would predict that the upper bound on the memory capacity scales linearly with the total number of synapses C^2, which is much larger than αC. However, the number of independent synaptic changes is again C because the patterns of neural activity seen by different neurons in the recurrent network are almost identical. Indeed each memory contains C independent components, which correspond to the different levels of activity of the C neurons. These components are stored in C^2 synapses. This means that the C^2 synaptic modifications Δw_{ij} must be correlated (the third order correlations are indeed different from zero). The ideal observer approach furnishes an estimate of the capacity only under the assumption that the synaptic modifications are uncorrelated. If they are not, as in the case of the Hopfield network, then one should consider only the actual number N_{syn} of uncorrelated synaptic modifications, which in this case scales like C. So also the scaling properties of the critical capacity of the Hopfield model are actually the same as those predicted by the ideal observer approach.

Interestingly, in the case of the Hopfield network, when the critical capacity is surpassed, the memories are not forgotten, but they become irretrievable. More specifically, there is an abrupt transition to a regime in which none of the memories, not even the most recently stored, can be retrieved. This is called the blackout catastrophe (see, e.g., Amit, 1989) and is clearly in contrast with what happens in biological brains in which old memories are gradually forgotten to make room for the new ones.

The storage capacity depends also on the representations of the memories. For example, if memories are stored as sparse patterns of activity (i.e., patterns of activity in which only a small fraction f of neurons are active), then the amount of information to be stored per memory decreases, and the memory capacity increases. In particular, Tsodyks and Feigel'man (1989) showed that it is possible to store and retrieve a number of patterns that scales like N^2 when $f \sim \log N/N$, where N is the number of neurons in the network. Neural representations in the real brain are probably sparse in the hippocampus ($f \sim 0.01$), and possibly in some cortical areas. However, the estimate of f is difficult as it depends on the sensory stimuli, on the way the neuronal activity is recorded (extracellular or intracellular), and on the animal that is considered. For a recent discussion of sparseness in the brain in relation to memory capacity, see Ben Dayan Rubin and Fusi (2007).

For all these models the synapses can vary unboundedly. For example, in the Hopfield model, the number of synaptic states (i.e., the number of different values of w_{ij}) increases with the number of stored memories. This is clearly unrealistic, because all variables characterizing the dynamics of biological synapses must vary in a limited range. It actually seems that biological synapses have a small number of stable states, in some cases as few as two to three (Petersen et al., 1998; O'Connor et al., 2005). This is probably a consequence of the disruptive effects of molecular turnover, which are discussed at the beginning of the chapter in the section on "Synaptic Mechanisms of Memory Consolidation and Molecular Memories." A possible solution to the problem of molecular turnover is to introduce active mechanisms that implement bistability or multistability, which probably greatly limit the number of different stable states.

What happens if boundaries are imposed on the synaptic weights? The answer depends on whether one considers online or offline learning algorithms and their implementations in terms of synaptic dynamics. In the case of offline learning, the synaptic weights are computed offline (i.e., not in the neural system) according to some prescription. For example the Hopfield learning rule prescribes

$$w_{ij} = \sum_{\mu=1}^{p} \xi_i^\mu \xi_j^\mu .$$

Once all the memories are stored, and the final synaptic weights are computed, they are transferred to the neural circuit that we intend to study. If in this case one reduces the number of stable states even to the minimal two states, the degradation in memory performance is rather limited. For example, say that at the end of the computation of the synapses, one performs the following operation:

$$w_{ij} = \text{sign}\left(\sum_{\mu=1}^{p} \xi_i^\mu \xi_j^\mu\right).$$

In practice only the sign of the final weight is kept. Sompolinsky (1986) proved that these synapses with only two states perform almost as well as the original Hopfield synapses when one evaluates the ability to retrieve memories. The sign operation is equivalent to introducing random and uncorrelated noise, and in the end, the storage capacity still scales linearly with C, the number of synapses per neuron. The estimate is $p = \alpha' C$, where $\alpha' \sim 0.1$. The memory capacity is actually reduced (for the "intact" Hopfield model $p \sim 0.14C$), but the scaling properties remain the same. It is actually easy to show that the scaling of the memory capacity with N_{syn} is preserved also in the case of the ideal observer approach. Indeed, the signal is

$$\mathcal{S} = \langle M^1 \rangle = \frac{1}{N_{syn}} \sum_{i,j} \left\langle \text{sign}\left(\sum_{\mu=1}^{p} \Delta w_{ij}^\mu\right) \Delta w_{ij}^1 \right\rangle.$$

As $\Delta w_{ij}^1 = \pm 1$, we can bring it inside the sign operation,

$$\begin{aligned}\mathcal{S} = \langle M^1 \rangle &= \frac{1}{N_{syn}} \sum_{i,j} \left\langle \text{sign}\left(\sum_{\mu=1}^{p} \Delta w_{ij}^\mu \Delta w_{ij}^1\right)\right\rangle \\ &= \frac{1}{N_{syn}} \sum_{i,j} \langle \text{sign}(1 + \zeta_{ij}) \rangle\end{aligned},$$

where the first term inside the sign function comes from $\Delta w_{ij}^1 \Delta w_{ij}^1 = 1$, and $\zeta_{ij} = \sum_{\mu=2}^{p} \Delta w_{ij}^\mu \Delta w_{ij}^1$ is approximately a Gaussian variable with zero mean and variance that scales linearly with p. The probability that the sign is positive is approximately $1/2 + 1/\sqrt{2\pi p}$. Indeed, the probability that ζ_{ij} is larger than -1 is $1/2$ plus the area under the Gaussian between the maximum, at 0, and -1. This area can be estimated as the height of the maximum $1/\sqrt{2\pi p}$ multiplied by 1. Hence the signal scales like

$$\mathcal{S} \sim \frac{1}{\sqrt{p}}.$$

Unlike the case of the Hopfield prescription, the memory signal decreases with p. However, the noise is smaller because the synapses are only $\pm$ 1 and they do not grow with p. Indeed the noise scales like $1/\sqrt{N_{syn}}$. As a consequence, the signal-to-noise ratio scales again like $\sqrt{N_{syn}/p}$, as in the case in which the synapses follow the fully linear Hopfield prescription.

Unfortunately this approach is not implementable in a biological system. It is indeed not possible to first compute the synaptic weights offline, using some other device with unbounded variables representing the synapses, and only at the end reduce the number of stable states. Biological systems can only perform online learning: whenever we store a new memory, we need to modify the synapses, and we cannot store the information in some temporary repository that is used at a later time to determine the synaptic weights. In the case of online learning and bounded synapses, the situation is drastically different, and it will be discussed in the next section.

11.3.3 Memory Storage Capacity in Online Learning Models with Bounded Synapses Let us now consider more realistic synapses whose dynamical variables are restricted to vary in a limited range. In contrast to what we described in the previous section, the memory signal decays exponentially fast with the number of stored memories. To illustrate the problem, we consider the case of synapses with two stable states, which is the minimal number of states to have memory. Basically we assume that the upper and the lower bounds, which are always present in any realistic model, coincide with the only two stable states. Although this may sound like a pathological case, it is actually representative of a wide class of realistic synaptic models (Fusi and Abbott, 2007). We will discuss a more general case at the end of this section.

As each synapse has only two states, it basically forgets all the previously stored information every time it is modified. Indeed, if the synapse is potentiated, it ends up in the state representing a strong synapse, regardless of its initial condition. In other words, the current state of the synapse is entirely determined by the very last modification, and previous ones do not affect the current synaptic weight. If all synapses are modified to store a new memory, then they can remember only that memory, and all previously stored information is erased. For this reason we are going to consider a model synapse that is modified only with probability $q < 1$, as suggested in Tsodyks (1990) and Amit and Fusi (1992, 1994). More specifically, we consider again the case of random and uncorrelated memories, and we assume that each synapse is either potentiated or depressed with equal probability ($\Delta w_{ij} = \pm 1$). Notice that now Δw_{ij}, with a slight abuse of notation, denotes the desired synaptic modification and not the actual difference between the synaptic weight w_{ij} before the update and w_{ij} after the update. For nonlinear saturating synapses Δw_{ij} is a specific function of the pre- and postsynaptic activity that indicates whether the synapse should be changed in one direction (potentiated or depressed) or left unchanged ($\Delta w_{ij} = 0$). The actual updating rule relating the synaptic weights to Δw_{ij} is as follows: when a new memory μ is stored, we consider for each synapse the desired modification Δw_{ij}^{μ}. If $\Delta w_{ij}^{\mu} = +1$, then we potentiate the synapse with probability q; otherwise, we depress it with the same probability,

$$w_{ij} \to w_{ij}(1-\zeta_{ij}^{\mu}) + \zeta_{ij}^{\mu}\Delta w_{ij}^{\mu},$$

where ζ_{ij}^{μ} are stochastic binary variables that are independent for all combinations of the indexes. In particular, $\zeta_{ij}^{\mu} = 1$ with probability q and $\zeta_{ij}^{\mu} = 0$ with probability $1 - q$. Now there are two types of stochastic variables, those coming from the random memories to be stored, and those that come from the intrinsically stochastic updating rule of the synapse.

In the case of binary synapses that we are considering, the memory signal $\mathcal{S}$ can be rewritten as

$$\mathcal{S} = \frac{1}{N_{syn}}\left\langle \sum_{i,j} w_{ij}\Delta w_{ij}^{1} \right\rangle = P_{w=\Delta} - P_{w\neq\Delta},$$

where $P_{w=\Delta} = P(w_{ij} = \Delta w_{ij}^{1})$ and, analogously, $P_{w\neq\Delta} = P(w_{ij} \neq \Delta w_{ij}^{1})$. Notice that in our case $P_{w\neq\Delta} = 1 - P_{w=\Delta}$, so we need to compute only $P_{w=\Delta}$. We now compute this probability immediately after the storage of memory $\mu = 1$, and then following the storage of the other memories:

$$P_{w=\Delta} = \frac{1}{2} + \frac{1}{2}q.$$

Indeed, the probability that Δw_{ij}^{1} is equal to random w_{ij} by chance is 1/2. These synapses will remain unchanged regardless the outcome of the stochastic synaptic updates. A fraction q of the remaining synapses are aligned to Δw_{ij}. This means that the initial signal is simply

$$\mathcal{S} = 2P_{w=\Delta} - 1 = q.$$

Intuitively, it is clear that the larger q, the stronger the initial memory signal and the larger the amount of information stored per memory. Following the presentation of any of the uncorrelated memories stored after the tracked memory, we need to update the probability $P_{w=\Delta}$ as follows:

$$P_{w=\Delta}(t+1) = P_{w=\Delta}(t)\left(1 - \frac{1}{2}q\right) + (1 - P_{w=\Delta}(t))\frac{1}{2}q.$$

Indeed, for every new memory, $\Delta w_{ij}^{t} = \Delta w_{ij}^{1}$ with probability 1/2, and only a fraction q of the synapses will be updated. If we rearrange the terms of the formula,

$$\Delta P_{w=\Delta} = P_{w=\Delta}(t+1) - P_{w=\Delta}(t) = -q\left(P_{w=\Delta}(t) - \frac{1}{2}\right),$$

it becomes clear that $P_{w=\Delta}$ converges exponentially fast to an equilibrium value of 1/2, which corresponds to zero signal. More specifically,

$$\mathcal{S}(t) \simeq q\exp(-qt).$$

This means that memories are forgotten exponentially fast with a time constant that depends on q. As for the noise, we can neglect the correlations between Δw_{ij}^{1}

and w_{ij} (see Amit and Fusi, 1994, for an explanation), and hence

$$\mathcal{N} \simeq \frac{1}{N_{syn}}\sqrt{N_{syn}\frac{1}{2}\left(1-\frac{1}{2}\right)} = \frac{1}{2\sqrt{N_{syn}}}.$$

So now the noise does not increase as in the case of unbounded synapses. The behavior of bounded synapses is summarized in figure 11.2 (right), where we plotted both the signal and the noise as a function of p. The point at which the two curves cross determines the maximum number of memories that are retrievable. In this case,

$$p < \frac{\log\left(q\sqrt{N_{syn}}\right)}{q}.$$

This is the catastrophic forgetting we were referring to at the beginning of this section. Memories are overwritten so rapidly that the number of retrievable items grows only logarithmically with the number of synapses N_{syn}. This is an extremely inefficient memory system, and the number of storable and retrievable memories can be orders of magnitude smaller than in the case of unrealistic unbounded synapses.

The memory capacity can be increased if the learning process is slowed down. Indeed, when $q \to 0$, the factor $1/q$ can greatly extend the number of storable memories. Unfortunately, when one reduces q, the amount of information stored per pattern also decreases. This is illustrated in figure 11.3, where we plotted the signal-to-noise ratio for a few values of q (see the inset). Fast synapses ($q \sim 1$) are good at storing new memories, but bad at retaining old ones. Indeed, the initial signal-to-noise ratio is large, and it increases with the number of synapses, but then it decays rapidly. When $q \sim 1$, then the initial signal-to-noise ratio scales like $q\sqrt{N_{syn}}$, but the memory lifetime is not an extensive quantity (i.e., it does not grow with N_{syn}, or it grows very slowly).

Slow synapses ($q \to 0$) are good at retaining memories over long timescales but bad at storing new ones. The longest memory lifetimes can be achieved by setting q to the minimal value that still allows the initial signal-to-noise ratio to remain finite (i.e., not become infinitesimally small). This condition is implemented when $q \sim 1/\sqrt{N_{syn}}$, for which the memory lifetimes scale like $\sqrt{N_{syn}}$. However, the price to be paid for these memory lifetimes is that the initial signal-to-noise ratio becomes independent from N_{syn} and is not an extensive quantity any longer. This trade-off is shared by most realistic synaptic models (Fusi, 2002), and it is similar to what is known as the stability–plasticity dilemma (Carpenter and Grossberg, 1991).

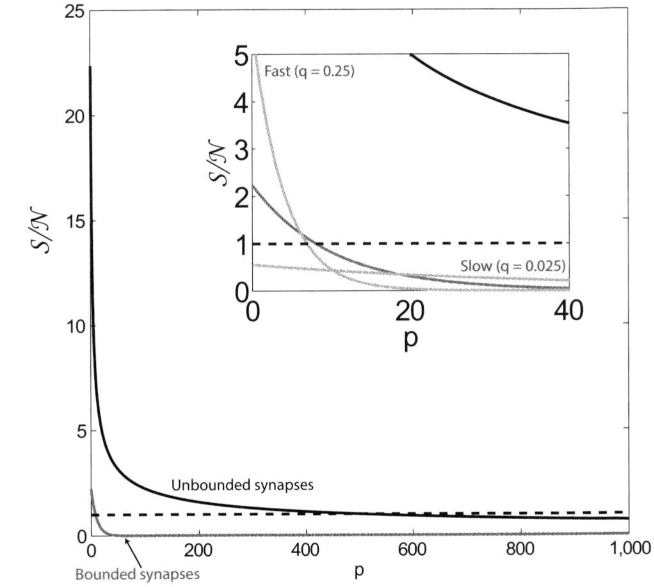

FIGURE 11.3 Memory signal-to-noise ratio as a function of the number of stored memories p for unbounded (black) and bounded (gray) synapses. For unbounded synapses the signal-to-noise ratio decreases as $1/\sqrt{p}$, significantly more slowly than for bounded synapses, for which the decay is exponential $\exp(-pq)$, where q is the learning rate. In the inset we show an enlargement of the small p range. The different gray lines represent the decay of the memory signal-to-noise ratio for different learning rates. Fast synapses (large q values) have a large initial signal-to-noise ratio, but then the memory trace decays very rapidly. Slow synapses retain a significantly large number of memories ($\sim 1/q$), but the amount of information stored per memory is greatly reduced (the initial signal-to-noise ratio is proportional to q).

11.3.4 LONG-TERM MEMORY MODELS WITH MULTIPLE TIMESCALES One possible solution to the stability–plasticity dilemma described in the previous section is to combine fast and slow synapses. Consider a heterogeneous population of synapses in which the learning rates are different for different synapses. If we divide the N_{syn} synapses into m different groups, each characterized by a different learning rate q_k ($k = 1, \ldots, m$), then the total signal is simply given by the weighted sum of the signals of the subpopulations (Roxin and Fusi, 2013):

$$\mathcal{S}(t) = \sum_{k=1}^{m} F(k) S_k(t),$$

where $F(k)$ is the fraction of synapses in subpopulation k, and $S_k(t) = q_k e^{-q_k t}$. Consider now the case in which all the subpopulations have the same size $F(k) = 1/m$, and the learning rates are equally spaced on a logarithmic axis, $q_k = q_s^{(k-1)/(m-1)}$, where q_s is the smallest learning rate. This choice of learning rates allows the synapses to cover learning rates that vary over multiple timescales, ranging from 1 to $1/q_s$. We can then rewrite the total signal as

$$\mathcal{S}(t) = \frac{1}{m}\sum_{k=1}^{m} q_k e^{-q_k t}.$$

If m is large enough (simulations show that $m \sim 4-5$ is sufficient), then the sum can be approximated by an integral over an auxiliary variable x,

$$\mathcal{S}(t) \simeq \int_0^1 q(x) e^{-q(x)t} dx,$$

where $q(x) = q_s^x$. In order to solve the integral, we change the variable from x to $q = q_s^x$. Given that $x = \log q/\log q_s$, and hence $dx = 1/(q \log q_s)$, we can then rewrite the signal as

$$\mathcal{S}(t) \simeq \frac{1}{\log q_s}\int_1^{q_s} e^{-qt} dq = \frac{1}{\log(1/q_s)}\frac{e^{-q_s t} - e^{-t}}{t}.$$

The noise is the same as in the homogeneous case, so the signal-to-noise ratio for sufficiently large t is

$$\frac{\mathcal{S}}{\mathcal{N}}(t) \simeq \frac{\sqrt{N_{syn}}}{\log(1/q_s)}\frac{e^{-q_s t}}{t}.$$

The initial signal-to-noise ratio can be computed from the nonapproximated formula, and it is given by

$$\frac{\mathcal{S}}{\mathcal{N}}(0) \simeq \frac{\sqrt{N_{syn}}}{\log(1/q_s)}.$$

These formulas show that the heterogeneous model is already a good compromise: when $q_s \sim 1/\sqrt{N_{syn}}$, then both the initial signal-to-noise ratio and the memory lifetime scale with N_{syn}. More specifically they scale as $\sqrt{N_{syn}}/\log\sqrt{N_{syn}}$ and $\sqrt{N_{syn}}$, respectively. Moreover, the memory trace decays as a power law $(1/t)$ over an extended range, up to $t \sim 1/q_s$, where the exponential starts to dominate (see figure 11.4).

Heterogeneity is the first important ingredient for solving the stability–plasticity dilemma. The simple example that we examined was introduced in the Discussion of Fusi et al. (2005) to show the importance of the diversity of timescales. It is possible to improve the memory performance even further by introducing the second important ingredient of memory consolidation models, which is some form of interaction between the processes operating on different timescales. In particular, when the information about memories is transferred from the fast synapses to the slow synapses, then it is possible to extend the memory lifetimes of the heterogeneous model with no interactions without

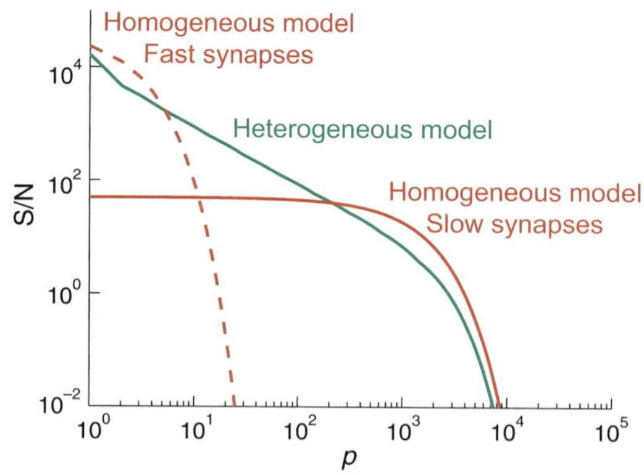

FIGURE 11.4 Memory trace (signal-to-noise ratio; $\mathcal{S}/\mathcal{N}$) as a function of time (new memories are stored at a fixed rate). Both the signal-to-noise ratio and time are on logarithmic axes. The memory trace of the heterogeneous model (green) decays with a power law over a wide range, and then, around $p = 10^3$, goes to zero exponentially. The memory traces for two homogeneous models (red) are also plotted for comparison. The homogeneous model with fast synapses (dashed line) has a very strong initial signal-to-noise ratio, but then it decays very rapidly. The memory trace of the homogeneous model with slow synapses is flat (on a log scale) over a long timescale, and then it decays rapidly. The heterogeneous model is a good compromise as it exhibits memory lifetimes that are comparable to those of the homogeneous model with slow synapses, and, at the same time, the initial signal-to-noise ratio is not much smaller than for the homogeneous model with fast synapses.

sacrificing the initial memory trace (Roxin and Fusi, 2013), or to increase the memory strength over the entire range of memory ages over which the memory decay is well described by a power law (Fusi et al., 2005). The information transfer can occur within a synapse (e.g., in the form of metaplasticity, reflecting the cascade of biochemical processes which underlies the expression of long-term modifications; see Fusi et al., 2005) or across synapses in different areas (memory consolidation at the system level; see Roxin and Fusi, 2013), as in the case in which memories are initially stored in a temporary repository where the synapses are fast (e.g., in the hippocampus) and then they are consolidated when transferred to brain areas (e.g., cortex) where the synapses are slow.

11.4 Models of Working Memory

11.4.1 ATTRACTOR MODELS The persistence time of sustained firing activity during working memory is orders of magnitude longer than the biophysical time constants (tens of milliseconds) of fast electrical signals in neurons and synapses. For this reason, persistent activity is believed to be generated by feedback dynamics, or reverberation, in a local circuit (Amit, 1995;

Goldman-Rakic, 1995; Wang, 2001, 2013). The idea of reverberation is made precise in theoretical work where persistent activity is described as "dynamical attractors" (Amari, 1977; Amari and Arbib, 1977; Amit, 1995; Wang, 2001; Brunel, 2005). The mathematical term "attractor" simply means any self-sustained and stable state of a dynamical system, such as a neural network. According to this picture, in a working memory system, the spontaneous state and stimulus-selective memory states are assumed to represent multiple attractors, such that a memory state can be switched on and off by transient inputs.

The Hopfield Model Figure 11.5A schematically illustrates a discrete attractor landscape, with multiple memory states in coexistence with a background state. The Hopfield model (Hopfield, 1982) represents a simple attractor network model that illustrated many of the important ideas of working memory. Consider N neurons that are either active ($s_i = +1$) or inactive ($s_i = -1$) ($i = 1, 2, \ldots, N$). Their dynamics is described by the following equation:

$$s_i(t+1) = \text{sign}\left(\sum_{j=1}^{N} w_{ij} s_j(t) + H_i\right),$$

where H_i is the external input, and w_{ij} is the synaptic strength of recurrent connections. However, unlike a classic McCulloch–Pitts network, the dynamics is asynchronous: only one neuron, selected at random, is updated at each time step. Suppose that p memory items ($\mu = 1, 2, \ldots, p$) are to be stored as attractor states; each one is a random pattern $\xi_i^\mu = \pm 1$. Different memories are uncorrelated, as in the examples discussed above. Using a Hebbian prescription,

$$w_{ij} = \frac{1}{N} \sum_{\mu} \xi_i^\mu \xi_j^\mu$$

for $i \neq j$. We set $w_{ij} = 0$ for $i = j$. These are examples of long-term memories as they are permanently stored in the synaptic connections (see previous sections). Each of these memories can be reactivated by stimulating the network. For example, when a partial cue, correlated to one of the previously stored patterns, is presented to the network, the network naturally relaxes into the state that represents the stored memory (memory retrieval). This state is a stable point of the dynamics, and it remains active until a new stimulus arrives. This is an example of working memory: one of the long-term memories has been selected by a stimulus, it is retrieved, and it remains active thanks to the neural interactions.

We now focus on the attractor dynamics with fixed synapses, that is, once the long-term memories have been stored and the synaptic couplings have been determined. What makes the Hopfield model appealing is the fact that, because w_{ij} is symmetrical and there is no self-coupling, there is a simple way of describing the collective dynamics of the network. Indeed, there exists a function

$$E = -\frac{1}{2} \sum_{i \neq j} w_{ij} s_i s_j$$

that, in the absence of external inputs ($H_i = 0$), decreases monotonically in time, when the network is updated asynchronously, one neuron at a time. This function is called energy, in analogy to what happens in physical systems. As this function always decreases, one can determine the stable states of the network by finding the minima of the energy, in analogy to what happens in physical systems. The dynamics is then a simple relaxation toward the minima. We show now that the energy function that we introduced always decreases in time. If neuron k is updated from t to $t + 1$, since

$$E(t) = -s_k(t) \sum_j w_{kj} s_j(t) - \frac{1}{2} \sum_{i \neq j \neq k} w_{ij} s_i s_j,$$

we have

$$E(t+1) - E(t) = -(s_k(t+1) - s_k(t)) \sum_j w_{kj} s_j(t).$$

If neuron k does not change its state, then the energy function remains the same. If it flips from −1 to +1, $\sum_j w_{kj} s_j(t)$ must be positive; if it flips in the opposite direction, $\sum_j w_{kj} s_j(t)$ must be negative. In either case, $E(t + 1) - E(t) < 0$. Therefore, the network evolves over time until it reaches a minimum of the energy function. Each energy minimum is an attractor state, in the sense that the network perturbed slightly away will converge back to it, just like a ball rolling down the valley of a mountainous landscape (figure 11.5A). The attractor network suggests a mathematically rigorous formulation of content-addressable associative memory, which can be retrieved by partial cues that lead to full memory recollection: a cue brings the memory network within the basin of attraction for a memory item, and the intrinsic network dynamics evolves to the attractor state, thereby completing memory recall (figure 11.5A).

Importantly, the existence of an energy function made it possible to analyze whether the attractor states indeed correspond to the stored items $\{\xi_i^\mu\}$. The answer depends on the total number of memory items p, as expected intuitively: if p is too large, different memory

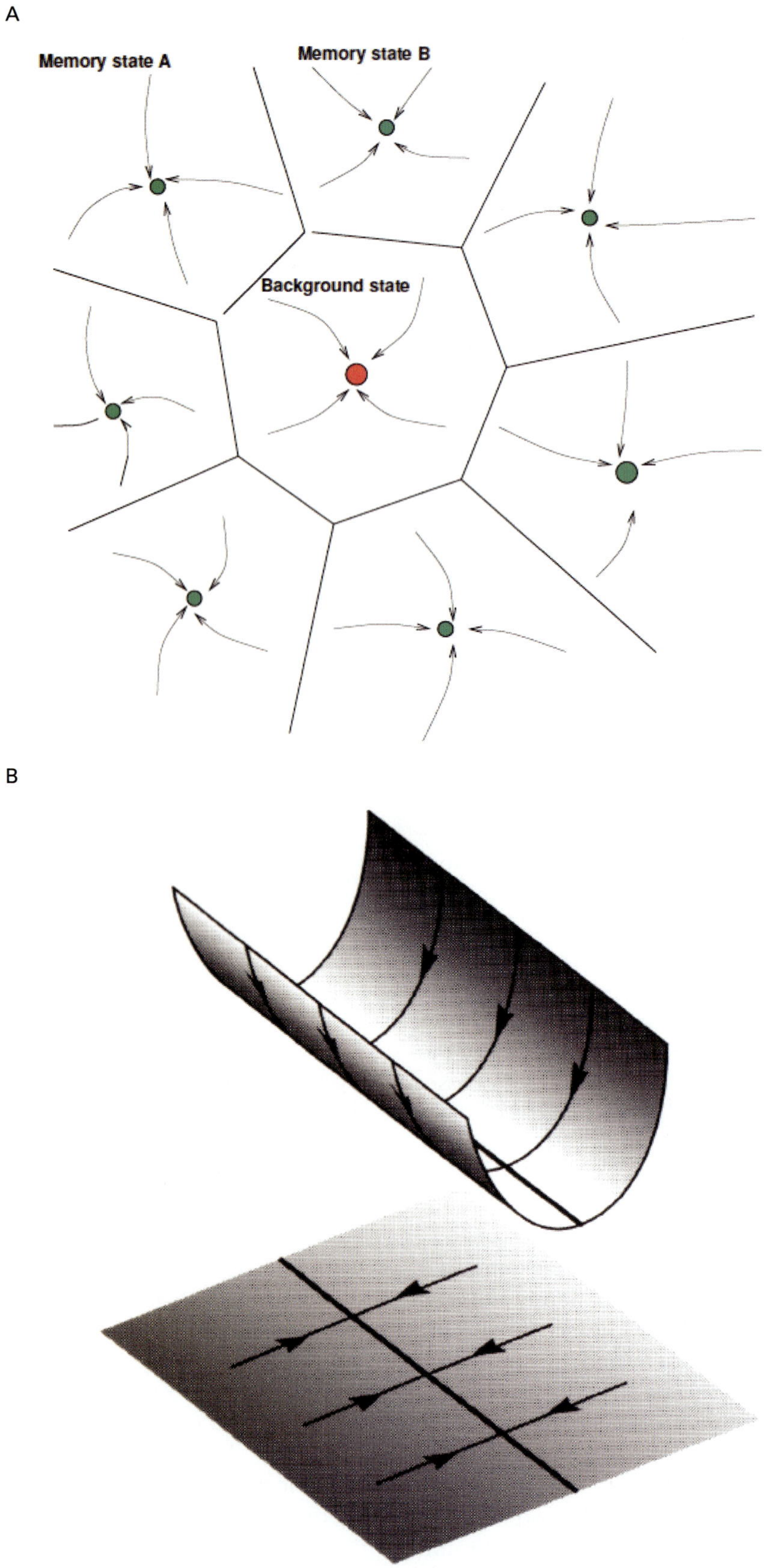

FIGURE 11.5 Schematic attractor model of working memory. (A) Discrete attractor network. The state space, depicted here on a plane of two neural firing rates, consists of a number of memory states (green) and a background state (red), The synaptic connections and other properties of the circuit, as well as external inputs, determine the contours (the basins of attraction, within which the system follows a trajectory toward the corresponding attractor state). In the Hopfield network, attractor states correspond to the minima of an energy function. (B) Line attractor network. All trajectories on the energy landscape flow down the walls of a trough toward a line of minimum energy at the bottom (a continuum, thick line), so that an analog quantity can be stored parametrically.

items will interfere with each other, and noise will induce confusion during memory retrieval. The calculations are the same as for the above analysis of long-term memory capacity. Notice that the memory signal $\mathcal{S}(\mu)$ of the ideal observer approach is minus the energy of the attractor states (i.e., the minimal energy):

$$\mathcal{S}(\mu) = -E(\{s\} = \{\xi^{\mu}\}) .$$

The stronger the memory signal, the more stable are the attractors.

As already discussed above, the activation of a memory item is instantiated by a persistent activity pattern (an attractor state). Thus, the working memory encoded in the persistent activity of the attractor could be viewed as activated long-term memory. This is why neurophysiological findings of elevated delay activity during working memory inspired computational neuroscientists to develop attractor network models of working memory of spiking neurons (Amit, 1995; Amit and Brunel, 1997; Wang, 1999; Durstewitz et al., 2000; Wang, 2001). Furthermore, it has been recognized that the brain often needs to hold information that is parametrically tuned, such as a spatial location (Funahashi et al., 1989), a frequency (Romo et al., 1999), or a gaze direction (Seung, 1996). An attractor network for such working memory should be endowed with a line upon which any point is a network state, a minimum of an energy function. This is called a line attractor (see figure 11.5B and the section "Bump and Line Attractors") (Ben-Yishai et al., 1995; Seung, 1996). (See also chapter 3, "Neurons and Neural Networks as Dynamical Systems.")

Biological Mechanism of Working Memory Circuits Peter Milner (a former colleague of Hebb) expressed doubts about the attractor theory, on the ground that a reverberatory neural assembly is "*liable to fire out of control*" (Milner, 1996). That is, positive feedback could drive neurons to fire at higher and higher rates until saturation is reached. This would be inconsistent with the observation that mnemonic persistent activity in the cortex occurs at moderate rates (typically 10 to 50 Hz), significantly above the spontaneous firing rate (a few Hertz) but well below the maximum firing capability of cortical neurons. When inhibition is incorporated to compensate for feedback excitation, "*inhibitory damping might prevent reverberation altogether*" (Milner, 1999).

These considerations can be translated into more quantitative statements by performing a simple mathematical analysis of the dynamics of a simple neural network. Consider first a population of excitatory neurons. Their dynamics can be approximately described by the following equation (see, e.g., Wilson and Cowan, 1972; Knight, 1972):

$$\tau_e \frac{d\nu_e}{dt} = -\nu_e + \Phi(w_{ee}\nu_e + H_e) ,$$

where $\nu_e(t)$ is the instantaneous firing rate of the excitatory neurons (i.e., the probability of emitting a spike in the interval t, $t + dt$ divided by dt, with dt small, typically of the order of 1 ms). τ_e is the time constant of the population dynamics, which is typically very short (a few milliseconds) in the absence of synaptic integration; otherwise, it is dominated by the synaptic dynamics (see, e.g., Renart et al., 2003). Φ is the f-I function which transforms the input current into an average firing rate. It has been computed for various types of integrate-and-fire neurons (see chapter 1), and it has been measured for real neurons (see, e.g., Rauch et al., 2003). w_{ee} is the effective recurrent coupling between excitatory neurons, and it depends on the average synaptic strength and the number of connected neurons. Finally H_e is an external input.

For almost any neuronal model, Φ is a monotonic function which is near to zero for negative or weak currents. It then grows supralinearly, it is approximately linear over a wide range of input currents, and then it saturates when the interspike interval becomes comparable to the refractory period (see chapter 2). The stability analysis illustrated in figure 11.6 (left) shows that, if w_{ee} is large enough, there are three fixed points: two are stable and one is unstable. The stable point at low frequencies represents spontaneous background activity, and it is typically at a few Hertz. The stable point at high frequencies is close to saturation, of the order of hundreds of Hertz.

We can now introduce a population of inhibitory neurons. The set of equations describing the neural system is now

$$\tau_e \frac{d\nu_e}{dt} = -\nu_e + \Phi_e(w_{ee}\nu_e - w_{ei}\nu_i + H_e) ,$$

$$\tau_i \frac{d\nu_i}{dt} = -\nu_i + \Phi_i(w_{ie}\nu_e - w_{ii}\nu_i + H_i) .$$

Let us assume for simplicity that $w_{ii} = 0$. If inhibition is fast enough, and the currents to the inhibitory neurons are tuned so that Φ_i can be approximated by a threshold-linear function (i.e., far from saturation), then $\nu_i \sim \phi(w_{ie}\nu_e + H_i)$ (ϕ is a constant that depends on the slope of the f-I curve) if the afferent current to the inhibitory neurons is strong enough (i.e., ν_e is large enough) and $\nu_i \sim 0$ if the total input current is weak

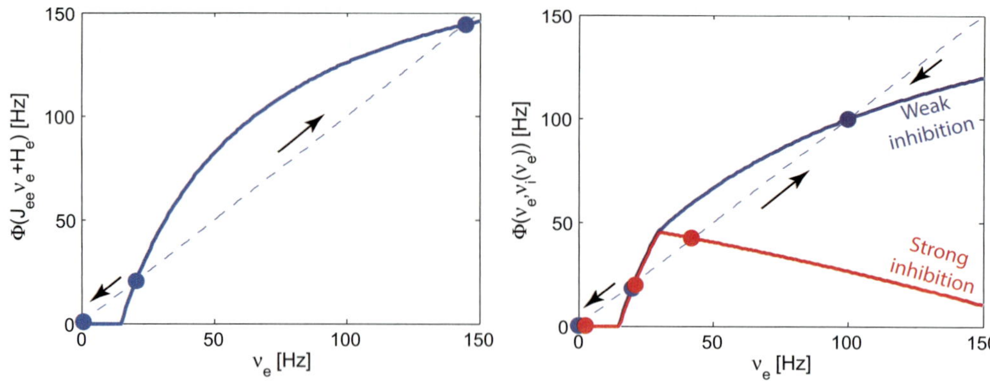

FIGURE 11.6 Stability analysis of a recurrently connected excitatory population of neurons (left) and of a network of excitatory and inhibitory neurons (right). For an initial firing rate ν_e (on the x-axis), we plotted the firing rate toward which the population converges (Φ in the differential equations described in the text). If this curve is above the diagonal (dashed line), the firing rate increases at the next time step. Similarly, when it is below the diagonal, it indicates that the firing rate decreases (see arrows). The crossing points (circles) are the fixed points of the dynamics. Left: for an excitatory population, Φ is a threshold-linear function with saturation due to the refractory period. In the absence of any external input ($H_e = 0$), there are three fixed points: two are stable (at $\nu_e = 0$ and at ν_e at high frequencies), and the third one is unstable (the fixed point in the middle). Right: Φ is plotted for two values of the inhibitory couplings (weak = blue and strong = red). For strong inhibition the Φ curve bends more rapidly and the high-frequency stable point moves to lower frequencies. Φ becomes nonmonotonic because inhibition dominates at high frequencies.

(when ν_e is below a critical threshold $\nu_0 = -H_i / w_{i.e.,}$). For simplicity we can assume that $\phi = 1$. We can then reduce the two dynamical equations to one:

$$\tau_e \frac{d\nu_e}{dt} = -\nu_e + \Phi_e((w_{ee} - w_{ie} w_{ei})\nu_e - H_i + H_e),$$

which is valid for $\nu_e > \nu_0$. For $\nu_e < \nu_0$, then inhibition is silent and we have the old equation for excitatory neurons:

$$\tau_e \frac{d\nu_e}{dt} = -\nu_e + \Phi_e(w_{ee}\nu_e + H_e).$$

The stability analysis is shown in figure 11.6 (right). When inhibition is strong enough ($w_{ei}w_{ie} > w_{ee}$), then the high-frequency fixed point is at much lower frequencies, thanks to the effects of a dominant inhibition. The fixed point at low frequencies remains essentially unaffected.

More realistic models with integrate-and-fire neurons and synaptic dynamics revealed that the problem of instability can pose a serious challenge. Fortunately, it was realized that cortical circuits with multiple attractor states are significantly more stable, if the recurrent excitation is slow compared to inhibition, for example, partially mediated by the NMDA receptors (Wang, 1999). The dynamics of NMDA receptors is qualitatively and quantitatively different from the dynamics of inhibitory (GABA) and fast excitatory (AMPA) conductances. The conductance of an individual synapse can be modeled with a simple first-order differential equation:

$$\tau_{AMPA} \frac{ds_{AMPA}}{dt} = -s_{AMPA} + \sum_k \delta(t - t_k),$$

where s_{AMPA} is the synaptic conductance, τ_{AMPA} is the integration time constant ($\tau_{AMPA} \sim 2$ ms), and the sum on the right-hand side extends over all presynaptic spikes, which arrive at times t_k. Analogously for GABA:

$$\tau_{GABA} \frac{ds_{GABA}}{dt} = -s_{GABA} + \sum_k \delta(t - t_k),$$

where τ_{GABA} is again relatively short (5 ms). For NMDA the equation for the conductance can be written in a simplified from as

$$\tau_{NMDA} \frac{ds_{NMDA}}{dt} = -s_{NMDA} + (1 - s_{NMDA})\left(\sum_k \delta(t - t_k)\right),$$

where τ_{NMDA} is significantly longer (100 ms). This long time constant helps to bridge the temporal gap between two consecutive presynaptic spikes, and it makes the fluctuations in the total input current significantly smaller, stabilizing the attractors. Moreover, the term $1 - s_{NMDA}$ indicates that the NMDA conductance saturates rapidly, also contributing to the bistability of the network.

Figure 11.7 illustrates the biophysics of an attractor network. In an object working memory model, subpopulations of neurons are selective to different object stimuli. When the strength of excitatory connections between neurons within each subpopulation is increased beyond a critical threshold, persistent activity appears as an all-or-none phenomenon (figure 11.7A, B). Below the critical threshold, only the spontaneous state exists. Above the threshold, the spontaneous activity state is still dynamically stable to small perturbations because at low firing rates excitation is effectively counteracted by feedback inhibition (figure 11.7C). However, if a stimulus generates a transient high activity in a neural subpopulation, now recurrent reverberation is sufficiently powerful to drive this group of cells to "escape" from the spontaneous state. A higher firing activity leads to an even larger recurrent synaptic excitation, which becomes sufficient to sustain a persistent active state after the stimulus is withdrawn. The firing rate is eventually stabilized by feedback inhibition and/or saturation of synaptic excitation; therefore, a stable attractor of persistent activity with an elevated firing rate is realized, which coexists with the stable spontaneous state (figure 11.7C).

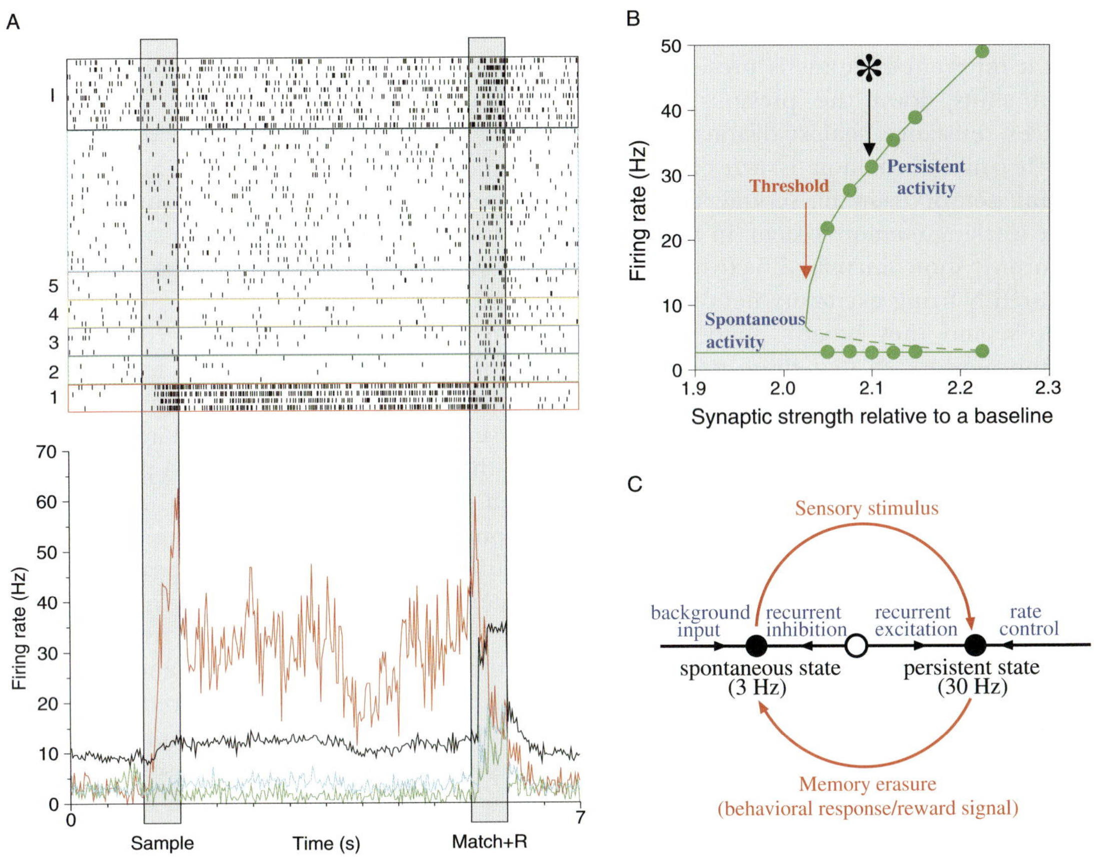

FIGURE 11.7 Attractor paradigm for persistent activity. (A) Simulation of an object working memory model. The network generates stimulus selective persistent activity, reproducing the observations in cortical recordings in delayed match-to-sample experiments in which the animal has to remember the sample stimulus across a delay of a few seconds and then compare it to a test stimulus. Neural subpopulations (labeled 1 to 5) are selective to different stimuli. Average activity from each neural group is plotted below the rastergram. Inhibitory (I) cell population is shown in black. In the simulation, the shown stimulus triggers persistent activity in a pyramidal cell group (red) at about 30 Hz, but not in other pyramidal groups. Delay period activity is switched off by a transient excitatory input generating a brief surge of activity in all neurons. (B) Dependence of network activity on the strength of recurrent synaptic connectivity. Solid lines: spontaneous (Spont.) state and persistent memory state; dashed line: unstable states. There is a critical threshold of synaptic strength, above which persistent activity appears abruptly as an all-or-none phenomenon. The example in (A) corresponds to the parameter indicated here by the star sign. (C) Schematic illustration of the biophysics underlying an attractor dynamics. An attractor is a neural firing state that is stable to perturbations: when a small input perturbs the network to a lower or higher activity level, there is a "restoring force" to bring the network back to the attractor state. In this case, the spontaneous state is stabilized from below by background inputs, and from above by feedback synaptic inhibition. The persistent activity state is stabilized from below by excitatory reverberation, and from above by various negative feedback "rate control" mechanisms. A transient sensory stimulus can switch the network from the spontaneous state to the memory state, and a behavioral response or reward signal can turn the network off and erase the memory. Adapted from Wang (2001) with permission.

In a working memory network presumably endowed with powerful recurrent excitation, feedback inhibition needs to be proportionally strong (Brunel and Wang, 2001; Lim and Goldman, 2013). In fact, in this model, feedback inhibition overweighs recurrent excitation (inhibition dominance) in the resting state and roughly balances with overall excitation (recurrent plus external background excitation) in both the resting state and memory states (Brunel and Wang, 2001). Inhibition underlies several important functions. First, it stabilizes the network dynamics. Second, it gives rise to stimulus selectivity so that different persistent activity patterns are tuned to different preferred stimuli. When a different stimulus is presented, it is inhibition that shuts down the previous attractor. Third, it induces resistance against distractors: while behaviorally relevant information is actively held in mind, irrelevant sensory stimuli should be denied entrance to the working memory system. In the model, inhibitory neurons show increased activity due to inputs of persistently firing excitatory neurons during working memory, and they broadly send enhanced inhibition to the network so that those neural pools selective for other stimuli are less excitable; thereby, the network is not sensitive to distractor stimuli. Fourth, a strongly recurrent working memory circuit may be prone to oscillations typically in the gamma (40-Hz) frequency range (Compte et al., 2000; Tegnér et al., 2002; Pesaran et al., 2002), and synaptic excitation plays an important role in inducing coherent network oscillations associated with cognition (Wang, 2013). Notice that network oscillations are not necessarily detectable at the level of individual neurons, which may fire at very different rates. Coherent oscillations of the network are due to the temporal modulation of the firing rate averaged over multiple cells. (See chapter 4, "Neural Rhythms.")

All these networks show examples of models that are significantly closer to biology. In the Hopfield model excitation and inhibition were not separated, active and inactive neurons were perfectly symmetrical, and there was no stable state corresponding to inactivity. The details of the more realistic models are extremely important as they helped us to understand the role of the biological components like NMDA integration. However, it is important to notice that all these models took inspiration from more abstract models like the Hopfield model, and they all share similar mechanisms based on the attractor dynamics proposed in the 1980s.

Bump and Line Attractors A well-known working memory paradigm is the delayed oculomotor response task, in which a subject is required to remember a visual cue (such as a directional angle or a visual target location) across a delay period in order to perform a memory-guided saccade (Funahashi et al., 1989; Chafee and Goldman-Rakic, 1998; Constantinidis and Wang, 2004). A network model for this spatial working memory experiment has the same circuit properties as that for object working memory but for a continuous network in which neurons are selective for stimulus direction as an analog feature (figure 11.8) (Compte et al., 2000). Synaptic excitation and inhibition collectively give rise to a "Mexican-hat" architecture: localized recurrent excitation between pyramidal cells with similar preference to spatial cues, and broader inhibition mediated by interneurons.

Figure 11.8B shows a model simulation of the delayed oculomotor task. Initially, the network is in a resting state in which all cells fire spontaneously at low rates. A transient input (in this case at 180 degrees) drives a subpopulation of cells to fire at high rates. As a result they send recruited excitation to each other via horizontal connections. This internal excitation is large enough to sustain elevated activity, so that the firing pattern persists after the stimulus is withdrawn. Synaptic inhibition ensures that the activity does not spread to the rest of the network, and persistent activity has a bell shape ("bump attractor"). At the end of a mnemonic delay period the cue information can be retrieved by reading out the peak location of the persistent activity pattern, and the network is reset back to the resting state. In different trials, a cue can be presented at different locations, and the firing activity of a single cell can be compared with the single-unit recording data from monkey's prefrontal cortex (Funahashi et al., 1989). At the network level, each cue triggers a persistent firing pattern of the same bell shape but peaked at a different location. A spatial working memory network thus requires a continuous family of "bump attractors," each encoding a potential location (Ben-Yishai et al., 1995; Compte et al., 2000). The instantiation of such a "line attractor" assumes that the system is effectively homogeneous, so that any part of the network is the same as any other. How can this be realized by a biological circuit where cellular or synaptic properties vary from neuron to neuron? Our modeling work showed that a continuous family of bump attractors can be rendered robust by regulatory homeostatic mechanisms in a biophysically realistic cortical network in spite of cellular heterogeneities (Renart et al., 2003b).

11.4.2 Slow Reverberation and Long Transients

A system with fast positive and slow negative feedbacks,

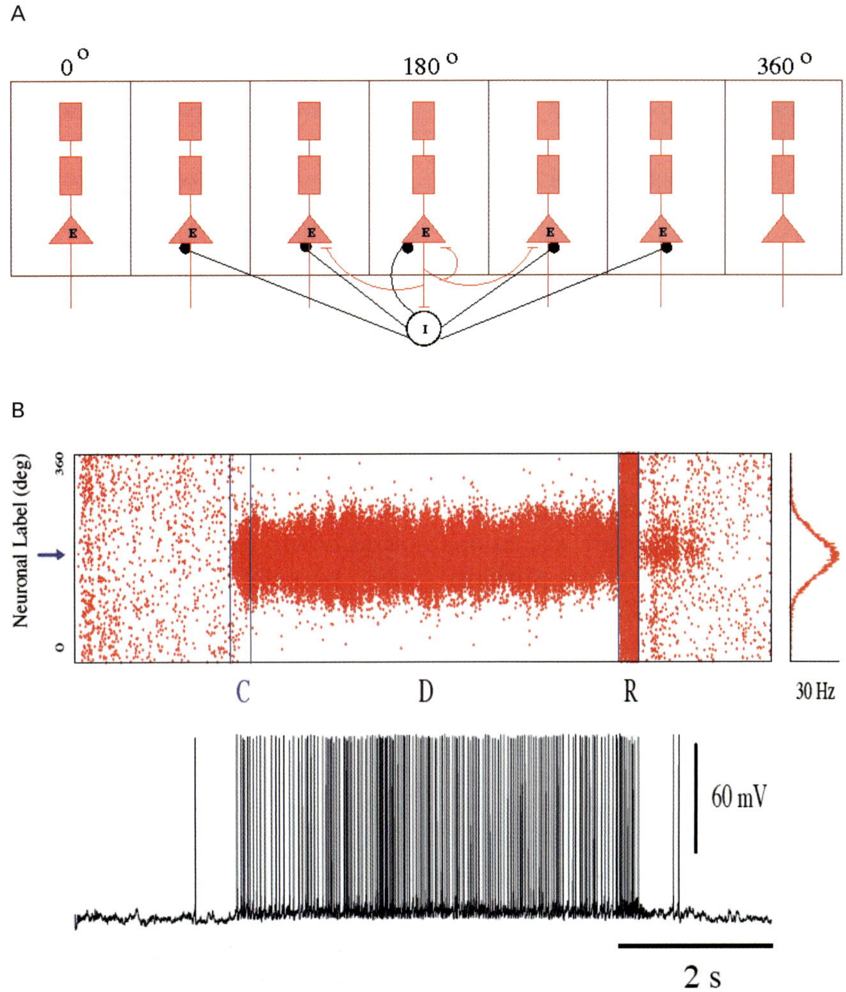

FIGURE 11.8 Working memory maintained by a tuned network activity pattern (a "bump attractor"). (A) Model architecture. Excitatory pyramidal cells are labeled by their preferred cues (0° to 360°). Pyramidal cells of similar preferred cues are connected through local E-to-E connections. Interneurons receive inputs from excitatory cells and send feedback inhibition by broad projections. (B) Spatiotemporal network activity (top) and membrane potential of a single cell (bottom) in a simulation of the delayed oculomotor experiment. Middle: a network simulation of delayed oculomotor response experiment. C, cue period; D, delay period; R, response period. Pyramidal neurons are labeled along the *y*-axis according to their preferred cues. The *x*-axis represents time. A dot in the rastergram indicates a spike of a neuron whose preferred location is at *y*, at time *x*. An elevated and localized neural activity is triggered by a transient cue stimulus and persists during the delay period.

both powerful, is prone to dynamical instability (Douglas et al., 1995; Wang, 1999). In biologically realistic models, if inhibition is significantly slower than excitation, persistent activity is often disrupted in the middle of a delay period; thereby, the memory is lost (Wang, 1999; Compte et al., 2000; Tegnér et al., 2002; Renart et al., 2003a). The same destabilization problem is present if negative feedback is instantiated by spike-frequency adaptation (McCormick et al., 1985) or short-term synaptic depression (Markram et al., 1998; Abbott and Regehr, 2004). Such instability does not occur, if the excitation is sufficiently slow, when compared to negative feedback, that is, when recurrent synapses are primarily mediated by NMDA receptors (time constant 50–100 ms) (Wang, 1999). Thus, the model yields a specific testable prediction that NMDA receptors at local (recurrent) excitatory synapses play a critical role in maintaining persistent activity in working memory. The model prediction of a critical role of NMDA receptors in a working memory system such as the prefrontal cortex has gained empirical support from both in vitro electrophysiological studies of the excitatory–excitatory local connections (Wang et al., 2008) and iontophoresis experiments with behaving monkeys that blocked NMDA receptors in the recorded neurons, leading to the abolishment of stimulus-selective persistent activity (Wang et al., 2013). In addition to the NMDA-receptor-mediated recurrent excitation, other cellular and synaptic processes may contribute to the generation of persistent activity. Three such processes are well-known and have been incorporated in models of working memory. First, single neurons may be endowed with intrinsic positive feedback leading to persistent activity (Camperi and Wang, 1998; Koulakov et al., 2002). For instance, accumulation of intracellular calcium during spiking activity activates a slow inward current, which, in turn, further increases the neuronal excitability, providing a positive feedback loop that can lead to spike firing outlasting a transient stimulus (Tegnér et al., 2002; Egorov et al., 2002; Loewenstein and Sompolinsky, 2003). Second, slow short-term facilitation between excitatory neurons dynamically enhances recurrent excitation and therefore can contribute to maintaining working memory (Hempel et al., 2000; Mongillo et al., 2008; Szatmary and Izhikevich, 2010). Third, depolarization-induced suppression of inhibition (DSI) refers to the phenomenon that depolarization of pyramidal neurons induces a reduced inhibition of the same neurons by a subclass of GABAergic cells through a process mediated by endogenous cannabinoids. As a result, the more active are pyramidal cells, the more are they disinhibited, leading to enhanced reverberation. A modeling study that incorporates DSI further revealed the balancing acts of positive and negative feedbacks in the complex dynamics of a working memory system (Carter and Wang, 2007). Taken together, these results led to the *slow reverberation hypothesis* about strongly positive feedback mechanisms required for the generation of persistent activity in a working memory circuit. In addition to ensuring circuit stability, slow reverberating neural activity also provides a time integration mechanism of critical importance for decision-making computations (reviewed in Wang, 2008).

It has been increasingly recognized that firing activity during working memory is widely heterogeneous across neurons and often displays complex temporal patterns during a delay period. Such high variability has inspired models based on the idea that memories may be encoded in the temporal trajectories of a neural network's dynamics. For this purpose the key is to have a mechanism to give rise to long memory traces. For instance, memory storage could involve propagation of activity pattern where each neural pool is active only transiently and the memory trace moves over time from one neural pool to another, either because there is a chain-like network architecture (Ganguli et al., 2008; Goldman, 2009) or as a result of the existence of heteroclinic/homoclinic orbits (i.e., paths in phase space which join two different/same equilibrium points) linking quasi-stable steady states of the system (Rabinovich et al., 2008).

A similar approach takes advantage of randomly connected neurons to generate complex patterns of nonstationary activity (Jaeger and Haas, 2004; Maass et al., 2002; Sussillo and Abbott, 2009; Buonomano and Maass, 2009). In this case, a to-be-remembered cue briefly perturbs the system. Analogously to the surface of a liquid (from which the name "liquid state machine" came), the network dynamics evolves with a rich, nearly chaotic dynamics, following trajectories that reflect the initial perturbation. The memory (i.e., the identity of the perturbative cue) is retrieved at a later time by training a downstream readout system to decode the complex trajectories. The random connectivity mixes the current state of the network with the external input in a way that partially preserves the information about the stimuli at previous time steps. Most importantly, it expands the dimensionality of the neural representations at each time step, making it possible for a linear readout to generate any function of the recent stimuli. In this sense, the liquid state machine is much more than a mere passive memory as the output can be a rather complex function of the entire temporal stream of inputs. Although random connections have been proposed since the time of David Marr, it was only recently that their role in dimensionality expansion was understood, taking inspiration from the theory of support vector machines.

These networks are extremely powerful at generating very complex waveforms that depend on the input or that are autonomously generated. Moreover, the readout does not need to be complex, and typically a linear readout can be trained to perform a huge number of tasks (see also the next section). However, it remains poorly understood what determines the timescale over which a memory trace can be reliably retrieved in such a system (Ganguli et al., 2008).

It is worth noting that, because reverberating excitation must be slow, as discussed above, the modern version of the attractor network model does not conform to the stereotypical image of a dynamical system, the behavior of which is dominated by steady states and rapid switches between them. Instead, attractor networks are compatible with very slow transients. Moreover, attractor states do not need to be fixed points but can display complex temporal dynamics (Laje and Buonomano, 2013). Alternatively, a working memory system may not be endowed with multiple attractor states at all, and memory traces are expressed as slowly decaying transients (Barak et al., 2013a). However, difficulties with this scenario are twofold. This form of neural representation would not be invariant with the duration of a mnemonic delay period, which is functionally undesirable. At the mechanistic level, very slow transients would require biophysical processes operating on the timescale of seconds. In this case, it would be difficult to switch on to, and off from, a memory state using brief (~ 200 ms) external inputs, which is needed for normal working memory function.

Rule-Based Behavior We have so far discussed persistent activity underlying working memory of sensory stimuli. However, internal representation is not limited to sensory information but can be more abstract. In general it can represent a mental state, which is basically any type of disposition to behavior. These mental states are actively maintained until task-relevant sensory stimuli induce transitions to other mental states. These transitions depend on both the previous state and the identity of the current sensory stimulus. A machine that can perform this type of task is more generally called an automaton, and it is one of the fundamental computational units in neural network theory, already proposed in the early studies of McCulloch and Pitts (1943), and Arbib (1961), and more recently in O'Reilly, Munakata, and McClelland (2000). Can the attractor network models of the type presented above be generalized to represent mental states and implement automata? Amit already proposed in the 1980s a general theoretical framework for encoding simple rules (Amit, 1988, 1989). Recently, it was found that simple, low-dimensional neuronal representations, which are typical of many attractor models, are inadequate for describing rule-based behavior (Rigotti et al., 2010, 2013). The difficulty arises from the fact that flexible context or rule-based behaviors involve neural computations akin to logic operations like exclusive OR (XOR, which is one when the two binary operands are different, and zero otherwise). For example, consider a simplified version of the Wisconsin Card Sort Task (WCST). Given a sensory cue (a colored shape, e.g., red circle), a subject selects one of two test stimuli that matches the cue either in color or shape, depending on the task rule (color or shape) (Mansouri et al., 2006). Presumably, the rule currently in play, say, color, is represented internally by persistent activity of "color-rule cells," which must be maintained across trials but switched off when the rule has changed, for example, from color to shape, signaled by negative feedback.

As illustrated in figure 11.9, the required state transitions of a color-rule cell as a function of its recurrent and feedback input amounts to a XOR operation. This makes it impossible to find the synaptic weights that implement the desired neural dynamics unless an intermediate layer of hidden neurons is introduced. This problem is a well-known case of nonlinear separability (i.e., the inputs cannot be separated by a linear classifier), and it plagues most neural networks implementing context-dependent tasks and, more generally, any network that has to integrate multiple sources of information (Barak et al., 2013b).

One possible solution (Rigotti et al., 2010) is similar to what has been proposed in the liquid state machines (see the previous section), and it is similarly based on the dimensional expansion that is traditionally performed by support vector machines (Cortes and Vapnik, 1995). Specifically, the dimensional expansion that solves the nonlinear separability is performed by a population of randomly connected neurons (figure 11.9B). The basic idea is that randomly connected neurons are naturally activated by a combination of synaptic inputs from external stimuli as well as rule-coding neurons (e.g., the color rule is currently in play *and* the network receives a negative feedback signal), and such mixed selectivity is exactly what is needed to solve the task. Surprisingly, in many realistic situations, randomly connected neurons perform as well as neurons which are carefully chosen for their response properties and that in a neural system would require some form of learning. This model provides a general framework for describing context- or rule-dependent tasks (Rigotti et al., 2010). Figure 11.9C and 11.9D show an implementation of such a network for the simplified WCST. It is important to notice that this class of models can in principle perform arbitrarily complex tasks provided that a sufficient number of neurons are available, but they do not explain how the connectivity patterns are learned. Understanding the human ability to rapidly and efficiently learn new tasks of similar complexity but different structure is still an open problem. Regardless, these models can reproduce many interesting features of the recorded activity, like the high degree of variability of firing activity, across cells and for a single neuron across task epochs. Heterogeneity and mixed selectivity are salient yet puzzling characteristics of frontal cortical neurons recorded from behaving animals (Asaad et al., 1998; Miller and Cohen, 2001; Sigala et al., 2008; Lapish et al., 2008). We recently proved that it is actually a signature of the high dimensionality of the neural representations (Rigotti et al., 2013). The model and this recent analysis of experiments suggest that mixed selectivity is computationally desirable as it allows for the representation of facts, memories, events, and importantly their combinations; the latter are critically important for enabling the prefrontal cortex to subserve context- and rule-dependent flexible behavior.

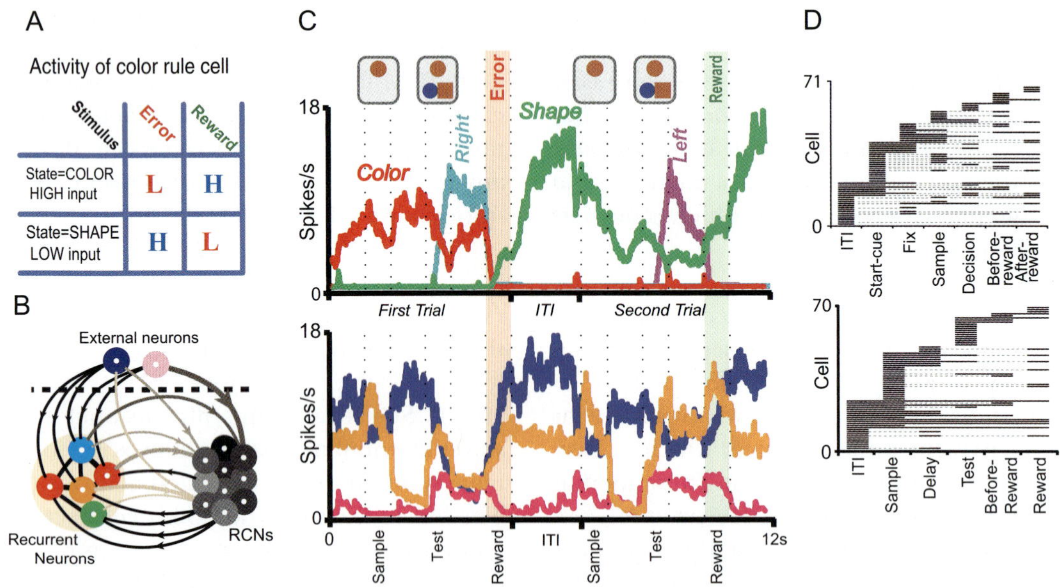

FIGURE 11.9 A decision-making network with the addition of randomly connected neurons (RCNs) for context- or rule-based behavior. (A) Exclusive OR (XOR) computation by a cell that encodes the rule "color" in a simple variant of the Wisconsin Card Sorting Task (see the text for more details). To illustrate the problem, assume that the neural activity (high or low, H or L) is determined by two types of inputs: recurrent drive which is high or low depending on whether the neuron is active or not (i.e., the internally maintained rule is color or shape), and a feedback signal which is positive (in which case the activity should stay) or negative (in which case the activity should switch). The required input–output mapping amounts to exclusive OR operation. (B) Neural network architecture: RCNs are connected both to a standard decision network of recurrent neurons and external neurons encoding sensory and feedback inputs, by fixed random weights (brown). Each RCN projects back to the recurrent network by means of plastic synapses (black). Not all connections are shown. (C) Simulation of a Wisconsin Card Sorting type of task after a rule shift. Simulated activity as a function of time of two sample neurons of the recurrent network that are rule selective (top panel) and three RCNs (bottom panel). The first neuron (red) is selective to "color" rule and the second (green) to "shape" rule. The events and the mental states for some of the epochs of the two trials are reported above the traces. (D) Rule selectivity pattern is heterogeneous over time and across neurons. Left panel: Rule selectivity for 70 simulated cells in the model: for every trial epoch (x-axis) we plotted a black bar when the neuron had a significantly different activity in shape- and in color-rule blocks. The neurons are sorted according to the first trial epoch in which they show rule selectivity. Right panel: Rule selectivity for spiking activity of single units recorded in prefrontal cortex of monkeys performing an analog of the WCST (Mansouri et al., 2006). Adapted from Rigotti et al. (2010) with permission.

11.5 Conclusions

Memory is a complex phenomenon that involves highly diverse processes that interact in a complicated way. These processes operate on a multitude of different temporal and spatial scales. The timescales range from milliseconds (e.g., the width of a spike) to years (e.g., long-term memory). The spatial scales range from the molecular level (e.g., the processes occurring in synaptic vesicles) to the entire brain. Currently, there are only a few models that explain how these processes are integrated to allow the nervous system to take full advantage of the diversity of its components.

Some of these models, which we tried to briefly review in this chapter, illustrate the general principles underlying memory consolidation, and they intentionally transcend the details of the biological implementation. An alternative approach to biological complexity is to attempt to simulate detailed models of all known biochemical processes in the brain and then observe and understand how computation may emerge from their interaction. We believe that it is probably premature to adopt this approach as we know only a tiny fraction of the biological processes that should be integrated in the model. Even if we knew all of them, simulating complex circuits would not necessarily provide us with a better understanding of the simulated processes. Our strategy is radically different as we believe it is important to start from fundamental computational problems in neuroscience, such as finding efficient synaptic models that have an efficient, scalable memory. We should then determine an efficient way of solving them using as simple a model as possible, identify the general and important principles behind the specific mechanisms of the model, and then generate a class of models whose constituents can be directly mapped onto known neuronal mechanisms and, more generally, onto biochemical elements. It is certainly extremely important to return to biology and to validate theoretical models both with experiments and with simulations studies that prove the robustness and generality of the model. However, a theoretical model should always reveal general principles; otherwise it would lose much of its explanatory and predictive power.

Understanding diversity and complexity in the nervous system is a grand challenge. Engineers typically make a huge effort to try to minimize component diversity when they build machines with a large number of simple components (e.g., they try to reduce as much as they can the mismatch between transistors in analog electronic chips). The brain seems to adopt a radically different strategy, as diversity and complexity dominate every biological component and every neural process. Understanding the computational advantages of the highly heterogeneous and complex machinery of the brain will require the development of a new theoretical and conceptual framework for designing new models and new experiments.

ACKNOWLEDGMENTS SF was supported by the Gatsby, Simons, Swartz and Kavli Foundations. XJW by R01MH062349, ONR grant N00014–13-1-0297, and STCSM grants 14JC1404900 and 15JC1400104. We are grateful to Omri Barak, Claudia Clopath, and Dani Martí for carefully reading the chapter and for many useful comments.

REFERENCES

Abbott, L. F., & Regehr, W. G. (2004). Synaptic computation. *Nature, 431,* 796–803.

Amari, S. (1977). Dynamics of pattern formation in lateral-inhibition type neural fields. *Biological Cybernetics, 27,* 77–87.

Amari, S., & Arbib, M. (1977). Competition and cooperation in neural nets. *Systems Neuroscience, 1,* 119–165.

Amit, D. (1989). *Modeling Brain Function.* New York: Cambridge University Press.

Amit, D. J. (1988). Neural networks counting chimes. *Proceedings of the National Academy of Sciences of the United States of America, 85,* 2141–2145.

Amit, D. J. (1995). The Hebbian paradigm reintegrated: Local reverberations as internal representations. *Behavioral and Brain Sciences, 18,* 617–626.

Amit, D. J., & Brunel, N. (1997). Model of global spontaneous activity and local structured activity during delay periods in the cerebral cortex. *Cerebral Cortex, 7,* 237–252.

Amit, D. J., & Fusi, S. (1992). Constraints on learning in dynamic synapses. *Network, 3,* 443.

Amit, D. J., & Fusi, S. (1994). Learning in neural networks with material synapses. *Neural Computation, 6*(5), 957–982.

Amit, D. J., Gutfreund, H., & Sompolinsky, H. (1985). Storing infinite numbers of patterns in a spin-glass model of neural networks. *Physical Review Letters, 55,* 1530–1531.

Arbib, M.A. (1961). Turing machines, finite automata, and neural nets. *Journal of the Association for Computing Machinery, 8,* 467–475.

Asaad, W. F., Rainer, G., & Miller, E. K. (1998). Neural activity in the primate prefrontal cortex during associative learning. *Neuron, 21,* 1399–1407.

Atkinson, R. C., & Shiffrin, R. M. (1968). Human memory: A proposed system and its control processes. In K. W. Spence & J. T. Spence (Eds.), *The Psychology of Learning and Motivation* (Vol. 2, pp. 89–195). New York: Academic Press.

Baddeley, A. (2012). Working memory: Theories, models, and controversies. *Annual Review of Psychology, 63,* 1–29.

Barak, O., Rigotti, M., & Fusi, S. (2013b). The sparseness of mixed selectivity neurons controls the generalization-discrimination trade-off. *Journal of Neuroscience, 33*(9), 3844–3856.

Barak, O., Sussillo, D., Romo, R., Tsodyks, M., & Abbott, L. F. (2013a). From fixed points to chaos: Three models of delayed discrimination. *Progress in Neurobiology, 103,* 214–222.

Bartlett, F. (1932). *Remembering: A Study in Experimental and Social Psychology.* Cambridge: Cambridge University Press.

Battaglia, F. P., & Pennartz, C. M. A. (2011). The construction of semantic memory: Grammar-based representations learned from relational episodic information. *Frontiers in Computational Neuroscience, 5,* 36.

Ben Dayan Rubin, D. D., & Fusi, S. (2007). Long memory lifetimes require complex synapses and limited sparseness. *Frontiers in Computational Neuroscience, 1,* 7.

Ben-Yishai, R., Lev Bar-Or, R., & Sompolinsky, H. (1995). Theory of orientation tuning in visual cortex. *Proceedings of the National Academy of Sciences of the United States of America, 92,* 3844–3848.

Brunel, N. (2005). Network models of memory. In C. C. Chow, B. Gutkin, D. Hansel, C. Meunier, & J. Dalibard (Eds.), *Methods and Models in Neurophysics* (pp. 407–476). Amsterdam: Elsevier.

Brunel, N., & Wang, X.-J. (2001). Effects of neuromodulation in a cortical network model of object working memory dominated by recurrent inhibition. *Journal of Computational Neuroscience, 11,* 63–85.

Buonomano, D. V., & Maass, W. (2009). State-dependent computations: Spatiotemporal processing in cortical networks. *Nature Reviews. Neuroscience, 10*(2), 113–125.

Camperi, M., & Wang, X.-J. (1998). A model of visuospatial short-term memory in prefrontal cortex: Recurrent network and cellular bistability. *Journal of Computational Neuroscience, 5,* 383–405.

Carpenter, G., & Grossberg, S. (1991). *Pattern Recognition by Self-Organizing Neural Networks.* Cambridge, MA: MIT Press.

Carter, E., & Wang, X.-J. (2007). Cannabinoid-mediated disinhibition and working memory: Dynamical interplay of multiple feedback mechanisms in a continuous attractor model of prefrontal cortex. *Cerebral Cortex, 17*(Suppl 1), 16–26.

Chafee, M. V., & Goldman-Rakic, P. S. (1998). Neuronal activity in macaque prefrontal area 8a and posterior parietal area 7ip related to memory guided saccades. *Journal of Neurophysiology, 79,* 2919–2940.

Clopath, C., Ziegler, L., Vasilaki, E., Busing, L., & Gerstner, W. (2008). Tag-trigger-consolidation: A model of early and late long-term-potentiation and depression. *PLoS Computational Biology, 4,* e1000248.

Compte, A., Brunel, N., Goldman-Rakic, P. S., & Wang, X.-J. (2000). Synaptic mechanisms and network dynamics underlying spatial working memory in a cortical network model. *Cerebral Cortex, 10,* 910–923.

Constantinidis, C., & Wang, X.-J. (2004). A neural circuit basis for spatial working memory. *Neuroscientist, 10,* 553–565.

Cortes, C., & Vapnik, V. (1995). Support-vector networks. *Machine Learning, 20*(3), 273–297.

Courtney, S., Ungerleider, L. G., Keil, K., & Haxby, J. V. (1997). Transient and sustained activity in a distributed neural system for human working memory. *Nature, 386,* 608–611.

Crick, F. (1984). Memory and molecular turnover. *Nature, 312,* 101.

Curtis, C. E., & D'Esposito, M. (2003). Persistent activity in the prefrontal cortex during working memory. *Trends in Cognitive Sciences, 7,* 415–423.

Diba, K., & Buzsaki, G. (2007). Forward and reverse hippocampal place-cell sequences during ripples. *Nature Neuroscience, 10,* 1241–1242.

Douglas, R. J., Koch, C., Mahowald, M., Martin, K. M., & Suarez, H. H. (1995). Recurrent excitation in neocortical circuits. *Science, 269,* 981–985.

Durstewitz, D., Seamans, J. K., & Sejnowski, T. J. (2000). Neurocomputational models of working memory. *Nature Neuroscience, 3,* 1184–1191.

Ebbinghaus, H. (1885). *Uber das Gedachtnis.* Leipzig: Duncker & Humblot.

Egorov, A. V., Hamam, B. N., Fransen, E., Hasselmo, M. E., & Alonso, A. A. (2002). Graded persistent activity in entorhinal cortex neurons. *Nature, 420,* 173–178.

Funahashi, S., Bruce, C. J., & Goldman-Rakic, P. S. (1989). Mnemonic coding of visual space in the monkey's dorsolateral prefrontal cortex. *Journal of Neurophysiology, 61,* 331–349.

Fusi, S. (2002). Hebbian spike-driven synaptic plasticity for learning patterns of mean firing rates. *Biological Cybernetics, 87*(5–6), 459–470.

Fusi, S., & Abbott, L. F. (2007). Limits on the memory storage capacity of bounded synapses. *Nature Neuroscience, 10,* 485–493.

Fusi, S., Drew, P., & Abbott, L. F. (2005). Cascade models of synaptically stored memories. *Neuron, 45*(4), 599–611.

Fuster, J. M., & Alexander, G. (1971). Neuron activity related to short-term memory. *Science, 173,* 652–654.

Ganguli, S., Huh, D., & Sompolinsky, H. (2008). Memory traces in dynamical systems. *Proceedings of the National Academy of Sciences of the United States of America, 105,* 18970–18975.

Gnadt, J. W., & Andersen, R. A. (1988). Memory related motor planning activity in posterior parietal cortex of macaque. *Experimental Brain Research, 70,* 216–220.

Goldman, M. S. (2009). Memory without feedback in a neural network. *Neuron, 61,* 621–634.

Goldman-Rakic, P. S. (1995). Cellular basis of working memory. *Neuron, 14,* 477–485.

Hempel, C. M., Hartman, K. H., Wang, X.-J., Turrigiano, G., & Nelson, S. B. (2000). Multiple forms of short-term plasticity at excitatory synapses in rat medial prefrontal cortex. *Journal of Neurophysiology, 83,* 3031–3041.

Hopfield, J. J. (1982). Neural networks and physical systems with emergent collective computational abilities. *Proceedings of the National Academy of Sciences of the United States of America, 79,* 2554–2558.

Jaeger, H., & Haas, H. (2004). Harnessing nonlinearity: Predicting chaotic systems and saving energy in wireless communication. *Science, 304,* 78–80.

James, W. (1890). *The Principles of Psychology.* New York: Henry Holt.

Káli, S., & Dayan, P. (2004). Off-line replay maintains declarative memories in a model of hippocampal-neocortical interactions. *Nature Neuroscience, 7*(3), 286–294.

Knight, B. W. (1972). Dynamics of encoding in a population of neurons. *Journal of General Physiology, 59,* 734–766.

Koulakov, A. A., Raghavachari, S., Kepecs, A., & Lisman, J. E. (2002). Model for a robust neural integrator. *Nature Neuroscience, 5,* 775–782.

Kubota, K., & Niki, H. (1971). Prefrontal cortical unit activity and delayed alternation performance in monkeys. *Journal of Neurophysiology, 34,* 337–347.

Kudrimoti, H. S., Barnes, C. A., & McNaughton, B. L. (1999). Reactivation of hippocampal cell assemblies: Effects of behavioral state, experience, and EEG dynamics. *Journal of Neuroscience, 19,* 4090–4101.

Laje, R., & Buonomano, D. V. (2013). Robust timing and motor patterns by taming chaos in recurrent neural networks. *Nature Neuroscience, 16*(7), 925–933.

Lapish, C. C., Durstewitz, D., Chandler, L. J., & Seamans, J. K. (2008). Successful choice behavior is associated with distinct and coherent network states in anterior cingulate cortex. *Proceedings of the National Academy of Sciences of the United States of America, 105,* 11963–11968.

Lee, A. K., & Wilson, M. A. (2002). Memory of sequential experience in the hippocampus during slow wave sleep. *Neuron, 36,* 1183–1194.

Lim, S., & Goldman, M. S. (2013). Balanced cortical microcircuitry for maintaining information in working memory. *Nature Neuroscience, 16,* 1306–1314.

Lisman, J. E. (1985). A mechanism for memory storage insensitive to molecular turnover: A bistable autophosphorylating kinase. *Proceedings of the National Academy of Sciences of the United States of America, 82*(9), 3055–3057.

Lisman, J. E., & Zhabotinsky, A. M. (2001). A model of synaptic memory: A CaMKII/PP1 switch that potentiates transmission by organizing an AMPA receptor anchoring assembly. *Neuron, 31,* 191–201.

Loewenstein, Y., & Sompolinsky, H. (2003). Temporal integration by calcium dynamics in a model neuron. *Nature Neuroscience, 6,* 961–967.

Maass, W., Natschlager, T., & Markram, H. (2002). Real-time computing without stable states: A new framework for neural computation based on perturbations. *Neural Computation, 14,* 2531–2560.

Mansouri, F. A., Matsumoto, K., & Tanaka, K. (2006). Prefrontal cell activities related to monkeys' success and failure in adapting to rule changes in a Wisconsin Card Sorting Test analog. *Journal of Neuroscience, 26,* 2745–2756.

Markram, H., Wang, Y., & Tsodyks, M. (1998). Differential signaling via the same axon of neocortical pyramidal neurons. *Proceedings of the National Academy of Sciences of the United States of America, 95,* 5323–5328.

McClelland, J. L., McNaughton, B. L., & O'Reilly, R. C. (1995). Why there are complementary learning systems in the hippocampus and neocortex: Insights from the successes and failures of connectionist models of learning and memory. *Psychological Review, 102,* 419–457.

McCormick, D., Connors, B., Lighthall, J., & Prince, D. (1985). Comparative electrophysiology of pyramidal and sparsely spiny stellate neurons in the neocortex. *Journal of Neurophysiology, 54,* 782–806.

McCulloch, W.S., and Pitts, W.H. (1943). A logical calculus immanent in nervous activity. *Bulletin of Mathematical Biophysics, 5,* 115–133

Miller, E. K., & Cohen, J. D. (2001). An integrative theory of prefrontal cortex function. *Annual Review of Neuroscience, 24,* 167–202.

Miller, E. K., Erickson, C. A., & Desimone, R. (1996). Neural mechanisms of visual working memory in prefrontal cortex of the macaque. *Journal of Neuroscience, 16,* 5154–5167.

Miller, P., Zhabotinsky, A. M., Lisman, J. E., & Wang, X. J. (2005). The stability of a stochastic CaMKII switch: Dependence on the number of enzyme molecules and protein turnover. *PLoS Biology, 3,* e107.

Milner, P. M. (1996). Neural representations: Some old problems revisited. *Journal of Cognitive Neuroscience, 8,* 69–77.

Milner, P. M. (1999). *The Autonomous Brain.* Mahwah, NJ: Erlbaum.

Mongillo, G., Barak, O., & Tsodyks, M. (2008). Synaptic theory of working memory. *Science, 319,* 1543–1546.

O'Connor, D. H., Wittenberg, G. M., & Wang, S. S.-H. (2005). Graded bidirectional synaptic plasticity is composed of switch-like unitary events. *Proceedings of the National Academy of Sciences of the United States of America, 102*(27), 9679–9684.

O'Reilly, R. C., & Frank, M. J. (2006). Making working memory work: A computational model of learning in the prefrontal cortex and basal ganglia. *Neural Computation, 18*(2), 283–328.

O'Reilly, R. C., Munakata, Y., and McClelland, J. (2000). *Computational Explorations in Cognitive Neuroscience: Understanding the Mind by Simulating the Brain.* Cambridge, MA: MIT Press.

Pesaran, B., Pezaris, J. S., Sahani, M., Mitra, P. P., & Andersen, R. A. (2002). Temporal structure in neuronal activity during working memory in macaque parietal cortex. *Nature Neuroscience, 5,* 805–811.

Petersen, C. C. H., Malenka, R. C., Nicoll, R. A., & Hopfield, J. J. (1998). All-or-none potentiation at CA3–CA1 synapses. *Proceedings of the National Academy of Sciences of the United States of America, 95,* 4732–4737.

Powell, K. D., & Goldberg, M. E. (2000). Response of neurons in the lateral intraparietal area to a distractor flashed during the delay period of a memory-guided saccade. *Journal of Neurophysiology, 84,* 301–310.

Rabinovich, M., Huerta, R., & Laurent, G. (2008). Neuroscience: Transient dynamics for neural processing. *Science, 321,* 48–50.

Rauch, A., La Camera, G., Luscher, H., Senn, W., & Fusi, S. (2003). Neocortical pyramidal cells respond as integrate-and-fire neurons to in vivo-like input currents. *Journal of Neurophysiology, 90,* 1598–1612.

Renart, A., Brunel, N., & Wang, X.-J. (2003a). Mean-field theory of recurrent cortical networks: Working memory circuits with irregularly spiking neurons. In J. Feng (Ed.), *Computational Neuroscience: A Comprehensive Approach* (pp. 432–490). Boca Raton, FL: CRC Press.

Renart, A., Song, P., & Wang, X.-J. (2003b). Robust spatial working memory through homeostatic synaptic scaling in heterogeneous cortical networks. *Neuron, 38,* 473–485.

Rigotti, M., Barak, O., Warden, M. R., Wang, X.-J., Daw, N. D., Miller, E. K., et al. (2013). The importance of mixed selectivity in complex cognitive tasks. *Nature, 497*(7451), 585–590.

Rigotti, M., Rubin, D. B., Wang, X.-J., & Fusi, S. (2010). Internal representation of task rules by recurrent dynamics: the importance of the diversity of neural responses. *Frontiers in Computational Neuroscience, 4,* 24.

Romo, R., Brody, C. D., Hernández, A., & Lemus, L. (1999). Neuronal correlates of parametric working memory in the prefrontal cortex. *Nature, 399,* 470–474.

Rosenblatt, F. (1958). The perceptron: A probabilistic model for information storage and organization in the brain. *Psychological Review, 65,* 386–408. Reprinted in J. A. Anderson & E. Rosenfeld (Eds.), *Neurocomputing: Foundations of Research.* Cambridge, MA: MIT Press.

Roxin, A., & Fusi, S. (2013). Efficient partitioning of memory systems and its importance for memory consolidation. *PLoS Computational Biology, 9*(7), e1003146.

Scoville, W. B., & Milner, B. B. (1957). Loss of recent memory after bilateral hippocampal lesions. *Journal of Neurology, Neurosurgery, and Psychiatry, 20,* 11–21.

Seung, H. S. (1996). How the brain keeps the eyes still. *Proceedings of the National Academy of Sciences of the United States of America, 93,* 13339–13344.

Sigala, N., Kusunoki, M., Nimmo-Smith, I., Gaffan, D., & Duncan, J. (2008). Hierarchical coding for sequential task events in the monkey prefrontal cortex. *Proceedings of the National Academy of Sciences of the United States of America, 105,* 11969–11974.

Sompolinsky, H. (1986). Neural networks with non-linear synapses and static noise. *Physical Review. A., 34,* 2571.

Squire, L., & Kandel, E. (1999). *Memory: From Mind to Molecules.* New York: Scientific American Library.

Squire, L. R. (2009). The legacy of patient H.M. for neuroscience. *Neuron, 61,* 6–9.

Sussillo, D., & Abbott, L. F. (2009). Generating coherent patterns of activity from chaotic neural networks. *Neuron, 63,* 544–557.

Szatmary, B., & Izhikevich, E. M. (2010). Spike-timing theory of working memory. *PLoS Computational Biology, 6,* e1000879.

Tegnér, J., Compte, A., & Wang, X.-J. (2002). The dynamical stability of reverberatory neural circuits. *Biological Cybernetics, 87,* 471–481.

Tsodyks, M. (1990). Associative memory in neural networks with binary synapses. *Modern Physics Letters B, 4,* 713–716.

Tsodyks, M., & Feigel'man, M. V. (1989). The enhanced storage capacity in neural networks with low activity level. *Europhysics Letters, 46,* 101.

Vargha-Khadem, F., Gadian, D. G., Watkins, K. E., Connelly, A., Van Paesschen, W., & Mishkin, M. (1997). Differential effects of early hippocampal pathology on episodic and semantic memory. *Science, 277*(5324), 376–380.

Wallis, J., Anderson, K., & Miller, E. (2001). Single neurons in prefrontal cortex encode abstract rules. *Nature, 411,* 953–956.

Wang, H., Stradtman, G. G., Wang, X. J., & Gao, W. J. (2008). A specialized NMDA receptor function in layer 5 recurrent microcircuitry of the adult rat prefrontal cortex. *Proceedings of the National Academy of Sciences of the United States of America, 105*, 16791–16796.

Wang, M., Yang, Y., Wang, C. J., Gamo, N. J., Jin, L. E., Mazer, J. A., et al. (2013). NMDA receptors subserve persistent neuronal firing during working memory in dorsolateral prefrontal cortex. *Neuron, 77*, 736–749.

Wang, X.-J. (1999). Synaptic basis of cortical persistent activity: The importance of NMDA receptors to working memory. *Journal of Neuroscience, 19*, 9587–9603.

Wang, X.-J. (2001). Synaptic reverberation underlying mnemonic persistent activity. *Trends in Neurosciences, 24*, 455–463.

Wang, X.-J. (2008). Decision making in recurrent neuronal circuits. *Neuron, 60*, 215–234.

Wang, X.-J. (2013). The prefrontal cortex as a quintessential "cognitive-type" neural circuit: Working memory and decision making. In D. T. Stuss & R. T. Knight (Eds.), *Principles of Frontal Lobe Function* (2nd ed., pp. 226–248). New York: Cambridge University Press.

Wilson, H. R., & Cowan, J. D. (1972). Excitatory and inhibitory interactions in localized populations of model neurons. *Biophysical Journal, 12*, 1–24.

Wixted, J. T., & Ebbesen, E. B. (1997). Genuine power curves in forgetting: A quantitative analysis of individual subject forgetting functions. *Memory & Cognition, 25*, 731–739.

12 Early Vision

ALAN L. YUILLE AND DANIEL KERSTEN

This chapter has six sections. Section 12.1 gives an overview of early vision and introduces the basic concepts. In section 12.2 we describe linear filter models of early vision and the types of computations which they can perform. Section 12.3 introduces probability models for neurons and for computational tasks. In section 12.4 we describe neural network models that take into account spatial context and can perform complex tasks. Section 12.5 discusses how different visual cues can be combined taking into account their dependencies. Finally, we summarize the chapter in section 12.6 and briefly sketch the relations to high-level vision. The first five sections are linked to demonstrations which we introduce at the start of each section.

12.1 Basic Concepts and Overview

We start by introducing the study of vision in section 12.1.1. Next we discuss visual tasks and modules in section 12.1.2. Then we discuss the visual system and the brain in section 12.1.3.

For this section, we suggest that the reader explore the fascinating demonstrations of visual phenomena on the website http://michaelbach.de/ot/. This website was created and maintained by Prof. Michael Bach, who accompanies the demos with descriptions and explanations of the phenomena. We particularly draw attention to (1) Hidden Figures, (2) Rotating Face Masks, (3) Ames Window, (4) Neon Color Spreading, (5) Dress Color Enigma, (6) Adelson's Checker-Shadow Illusion, and (7) Biological Motion. In addition, we encourage the reader to familiarize themselves with IPython Notebook in preparation for the interactive demos in later sections by going to the website http://www.nature.com/news/interactive-notebooks-sharing-the-code-1.16261.

12.1.1 What Is Vision? Why Is It Hard? How Is It Studied?

What Is Vision? Vision is the process of extracting information from retinal input, and we will here focus on images of visual scenes (whether natural or artificial). These images may capture a moment, or extend over time, as in a video. A human with normal vision can rapidly glance at an image and detect the objects in it and also decide what the object is made of, its distance and orientation relative to other objects, how it is moving (especially but not only if the image is dynamic), if it might collide with other objects, and whether to catch it or avoid it. Figure 12.1A illustrates how much information we can get from a single image, despite the fact that images are locally highly ambiguous, as shown in figure 12.1B. In short, humans can estimate a rough approximation to the three-dimensional scene that has generated the image. However, this is only one aspect of vision. In addition, humans have the ability to rapidly attend to different regions of the scene and ignore the rest. Vision is also used to enable actions, such as grasping objects or determining where to put your feet while hiking. In summary, vision performs a range of *visual tasks* which extract information from the scene in order to achieve goals.

Vision is extremely difficult. This is perhaps surprising because humans find it very easy. You simply open your eyes and you can understand the scene without any apparent effort. But this is only possible because a very large part of your brain is involved in doing vision. It is estimated that roughly 40% of neurons in the cortex are involved in visual processing. Despite decades of work, current machine vision systems perform significantly worse than humans, except for a few highly specialized tasks. However, probably most other animals, except monkeys and our other close relatives, get far less information from vision, judging by the much smaller numbers of neurons they devote to vision (e.g., the visual cortex of a mouse contains fewer neurons than a human or a monkey, by many orders of magnitude).

Why Is Vision Hard? The Complexity Problem and Natural/Ecological Constraints To get some appreciation of the difficulty of vision, consider how the image of the fox,

A B

FIGURE 12.1 (A) Humans can extract a lot of information from a single image. For example, "There is a young fox emerging from behind the base of a tree not far from the view point; it is heading right, stepping through short grass, and moving quickly. Its body fur is fluffy, reddish-brown, light in color, but with some variation. It has darker colored front legs and a dark patch above the mouth. Most of the body hairs flow from front to back." (B) Images are locally ambiguous. These two patches correspond to small parts of the fox's back and the side of the tree—see red circles in (A)—but are highly ambiguous without context.

A B

FIGURE 12.2 Why is vision hard? The raw input to the fox image (A) is the intensity values plotted as a function of spatial position (B). These intensity patterns vary depending on the pose of the fox, the lighting conditions, and other factors. The human visual system must decode this raw input, which is extremely difficult.

shown in figure 12.1, appears to the retina. Figure 12.2 shows the magnitude of the intensity of the image as a function of spatial position. These intensity magnitudes are proportional to the number of photons, or magnitude of light rays, that are imaged at different positions in the retina. The human visual system must somehow decode these intensity patterns and determine that they are caused by a fox emerging from behind a tree. But based on these intensity patterns, it is hard to perform *visual tasks* such as *segmenting the image* into regions corresponding to different objects, performing *object recognition* to determine that a region of the image corresponds to a fox and another region to a tree, or performing *depth estimation* to determine the positions of the objects in the visual scene. These tasks are particularly complex because the intensity patterns will change significantly if we make small changes to the visual scene. The patterns will vary greatly if we alter the pose of the fox, the lighting conditions, the viewpoint of the observer, and how much the fox is occluded by the tree.

Hence the main challenge of vision is due to the enormous complexity of natural images, and their local ambiguities shown in figure 12.1B. The number of possible images, or intensity patterns, that can be described by a small image array with 100×100 positions, or *image pixels*, is $(256)^{10,000}$ which is astronomically large (Kersten, 1987). These images are caused by the very large number of possible objects (estimated from 20,000 to 200,000) which can be arranged in a scene in an extremely large number of ways (in terms of the pose of the objects, their positions relative to each other) and be illuminated in an enormous number of different ways. Computer graphics studies how images can be generated if the scene is known. However, vision must be capable of performing the much harder *inverse inference* task of determining the scene from the image. From this perspective, it is almost miraculous that humans can simply open their eyes and recognize objects and visual scenes within a few hundred milliseconds (which is roughly the time to blink your eyes).

In order to perform these visual tasks, the visual system must be able to detect and exploit regularities in image patterns. These regularities are due to the structure of the world that we live in which causes the images which we observe. They include the assumption that surfaces are generally spatially smooth, that objects tend to move rigidly, and that most scenes contain a ground plane and objects touch the ground plane at contact points. These assumptions have been called ecological, or natural, constraints (Gibson 1986; Marr, 1982). Many of these constraints are "generic," in the sense that they are independent of the specific objects and object configurations in the scene and so can be applied to perform some visual tasks such as grasping an object without needing to recognize what the object is. It is speculated that humans have learned to exploit the structure of natural images and the world through evolution (Northcutt and Kaas, 1995), early development (Kellman and Arterberry, 2000), or learning later

FIGURE 12.3 This image gives the illusion of a flying carpet where the woman on the towel is perceived to be floating above the beach. This illusion shows that humans use constraints—about ground planes, shadows, and contact points—to interpret images. However, in this case, the constraints are violated because we incorrectly think that the shadow is being caused by the towel. Instead, the shadow is cast by a flag which is outside the image.

in life (Green and Bavelier, 2008). Vision science researchers can learn these image regularities by applying machine learning methods to image data sets. Figure 12.3 gives an example which illustrates both the ability of humans to exploit constraints about the three-dimensional world to perform inverse inference and the mistakes which can arise when the constraints are violated.

How Is Vision Studied? Broadly speaking, vision can be studied in three related ways: (1) at the "behavioral" level by studying how well humans, and animals, can perform visual tasks, (2) at the "neural" level to understand the neural mechanisms (by electrode recording or by noninvasive methods like fMRI), and (3) at the "computational" level by designing mathematical models and computer vision algorithms that can perform visual tasks. We distinguish between those mathematical models which attempt to describe how humans or animals see by accounting for behavioral or neural data, and those, called *computer vision*, whose goals are to perform visual tasks without attempting to model how humans, or animals, perform them. There is a complicated symbiosis between these two approaches to mathematical modeling of vision (which are done by different research communities). What they have in common is the need to address similar visual tasks and to deal with the complexity of image patterns. We note that computer vision researchers, even those who have no interest in biology, can nevertheless yield insights into human vision by developing algorithms which can perform the same visual tasks as humans in similar environments. This leads to a strategy of understanding the brain by reverse engineering.

The visual system is so complex that vision scientists must make simplifications to break it down into manageable pieces. These include (1) studying visual tasks in isolation instead of addressing the complete visual system, (2) simplifying the visual stimuli, (3) simplifying the models of neurons and neural circuits, and (4) simplifying the overall structure of the visual areas of the brain and how they interact with each other. Although these simplifications are necessary, they raise concerns which we will keep returning to in this chapter.

Firstly, vision researchers break vision down into different visual tasks which can be studied separately. These tasks include image segmentation, depth estimation, and object recognition. They are performed by *modules* which output *representations*. Modules, however, may not be localized to distinct parts of the brain. These simplifications assume that modules are semi-independent, which can be questioned (see the next section). They also raise the question of how the modules interact with each other. Marr's influential framework for vision (Marr, 1982) (see figure 12.4A) gives one way to address these issues. Marr proposed that the human visual system uses modules which compute a sequence of representations of the image which starts with a *primal sketch* of the image, proceeds to a 2 1/2–*D sketch* which represents the three-dimensional structure of the scene, and concludes with a *3-D* representation of objects. Hence the modules interact by outputting representations which are used as inputs to other modules. Marr's framework was never fully developed, but it does capture some important aspects of the visual system. It also illustrates the important classification of visual tasks into (1) *low-level* vision, which processes the image (e.g., produces the primal sketch), (2) *mid-level* vision, which estimates the structure and properties of geometric surfaces (e.g., produces the 2 1/2–D sketch), and (3) *high-level* vision, which recognizes objects and analyzes scenes. This classification is shown in figure 12.4B.

Secondly, the set of all possible stimuli is astronomically large. Hence it is impractical to study visual systems behaviorally by their response to all stimuli. In addition, when studying specific visual tasks, it is sensible to use stimuli which only contain information, or *visual cues*, which are specific for these tasks. Also, good experimental design requires, if possible, controlled stimuli so that the difficulty of performing a specific task can be quantified in terms of varying a small number of variables (e.g., by making the images darker or brighter).

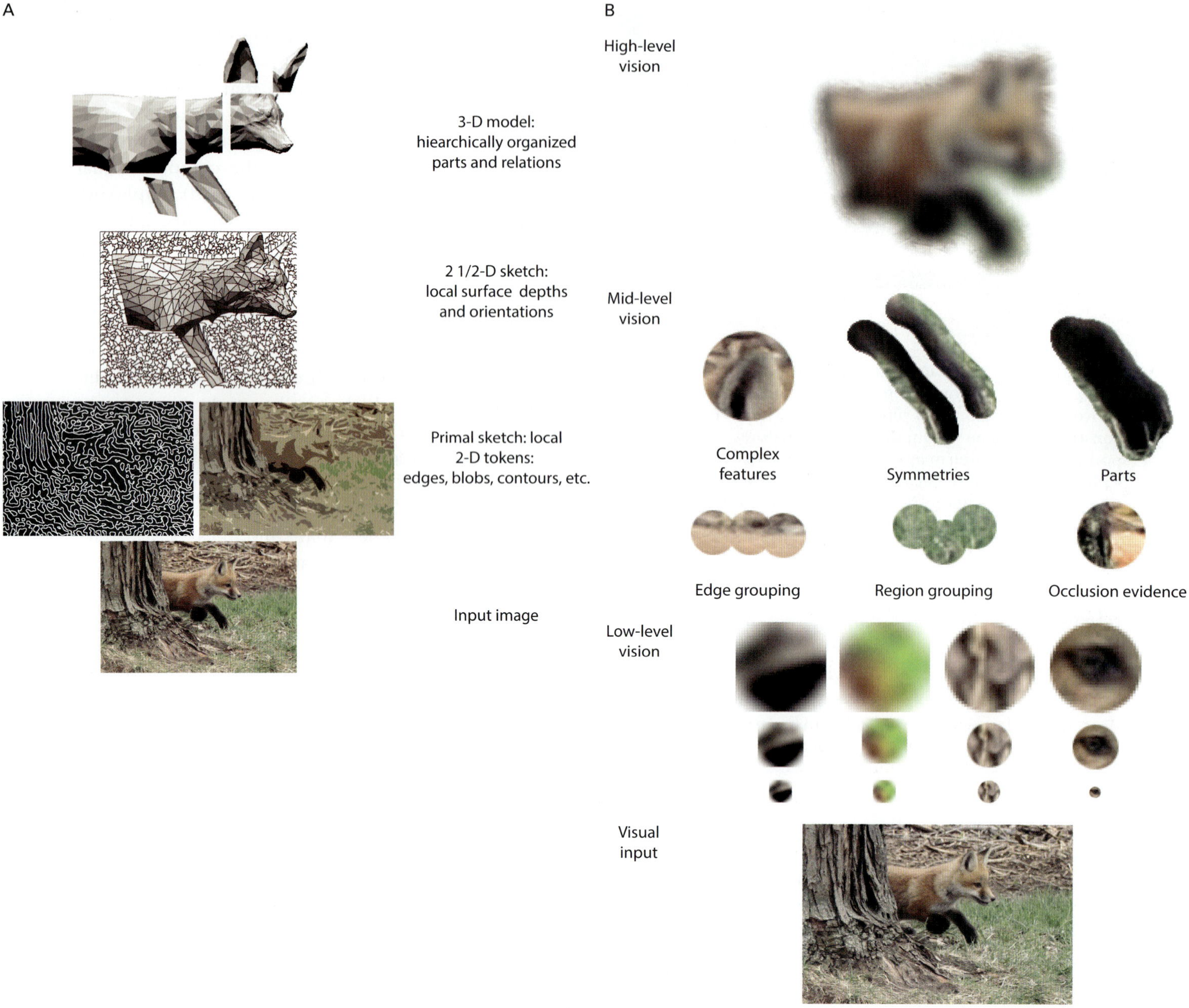

FIGURE 12.4 Marr's framework for vision (A) consists a series of representations. Visual tasks can be classified into low-, mid-, and high-level tasks (B). This classification roughly relates to Marr's framework.

For these reasons the study of vision is often simplified by using synthetic stimuli.

For example, the ability of humans to perceive depth from Julesz's random dot stereograms (Julesz, 1971) demonstrates that humans can perceive depth when objects are not present; see figure 12.5. Too much reliance on synthetic stimuli, however, can be misleading, and there is concern that experimental findings on synthetic stimuli may not generalize to human, or mammalian, abilities in more natural situations (Carandini et al., 2005; Yuille and Kersten, 2006). After all, human and animal visual systems have evolved to perform complex tasks on complex stimuli, and so understanding them requires probing them with stimuli which captures this complexity.

Thirdly, simplifications must be made when modeling neurons and neural circuits. The detailed biophysical properties of single neurons are very complicated, and so simpler models are used, particularly when modeling visual tasks. The most commonly used model is

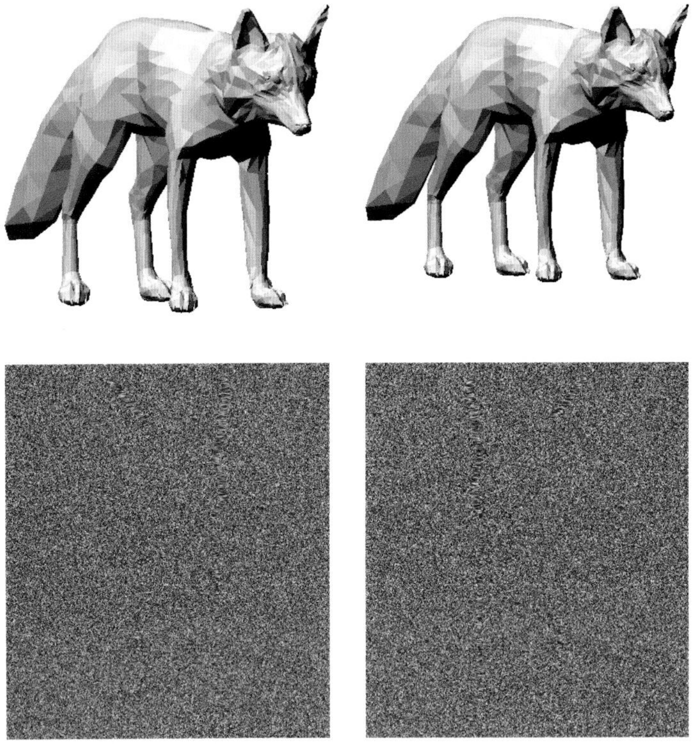

FIGURE 12.5 Binocular stereo of the fox for real images and for Julesz random dots stereograms. The top two images are a stereo pair (the left and right images) of a fox so that, when fused (e.g., by a stereo viewer presenting the left and right images to the right and left eyes, respectively), they yield the three-dimensional shape of the fox. The bottom two images are stereo pairs of random dot images of a fox. When fused, they also give the three-dimensional shape of the fox.

integrate and fire, when the neuron fires an action potential if a weighted linear sum of its inputs is above a threshold. We will use this model in this chapter, but it should be acknowledged that real neurons are more complicated. Their output may be a highly nonlinear function of the input, and they may signal information by a sophisticated "neural code" involving the precise timing of action potentials. Moreover, recent studies also show that there are large varieties of neurons which differ in their anatomy and function. There is also evidence that neural circuits can behave differently in different situations.

Fourthly, the numbers of neurons involved in visual perception is extremely large. This means that simplifications need to be done when studying the overall structure of visual areas of the brain and how they relate to each other. Although the broad structure of some visual areas are known, we are a long way from having wiring diagrams describing the connections between neurons within each visual area (there are, however, research programs to obtain such wiring diagrams for the mouse, whose visual cortex is many orders of magnitude smaller than that of a human or monkey).

12.1.2 VISUAL TASKS AND MODULES IN EARLY AND HIGH-LEVEL VISION This section gives a brief introduction to visual tasks, modules, and the distinctions between low-, mid-, and high-level vision. We discuss these tasks only at the behavioral level and postpone discussion of the brain until the next section. This chapter is about early vision, which we define to be low- and mid-level vision. But we also discuss how these early levels relate to high-level vision, which is described in chapter 17 by Lewis and Poggio.

Low-level visual tasks involve estimating local properties of the image. They include finding the boundary of an object (without deciding what the object is) and also include estimating the motion flow. Mid-level visual tasks estimate properties of geometrical surfaces, the shape and position of surfaces in the visual scene, and their depth ordering. High-level visual tasks estimate properties of objects, scene structures, the relationships between objects, and the actions of the objects. In addition, each level provides information which is passed on to the next level, as illustrated by Marr's theory. Another way to think of this organization is in terms of the knowledge available at different levels. For example, low-level vision can be performed by a system which knows only about regularities of image patterns (e.g., that images typically consist of regions where the intensity changes slowly which are separated by *edges* where the intensity changes rapidly). Mid-level vision processing knows about properties of geometric surfaces (e.g., that they tend to be spatially smooth) and that they can partially occlude each other. High-level vision knows about objects, the relationships between them, and actions. Hence the flow of information from low- to high-level vision is from generic to specific.

Low-Level Vision Low-level vision can be roughly defined to be the visual processing which can be done without explicit knowledge of objects and the three-dimensional structure of the world (although this additional knowledge, if available, will improve performance). Typical low-level vision tasks include determining differences within the image, such as detecting edges and performing segmentation; see figure 12.6. Low-level vision also involves extracting representations of image patterns which can be used for higher-level processing. As discussed later in the chapter, low-level vision can exploit statistical properties of images which are true for most images (e.g., that images tend to be piecewise smooth).

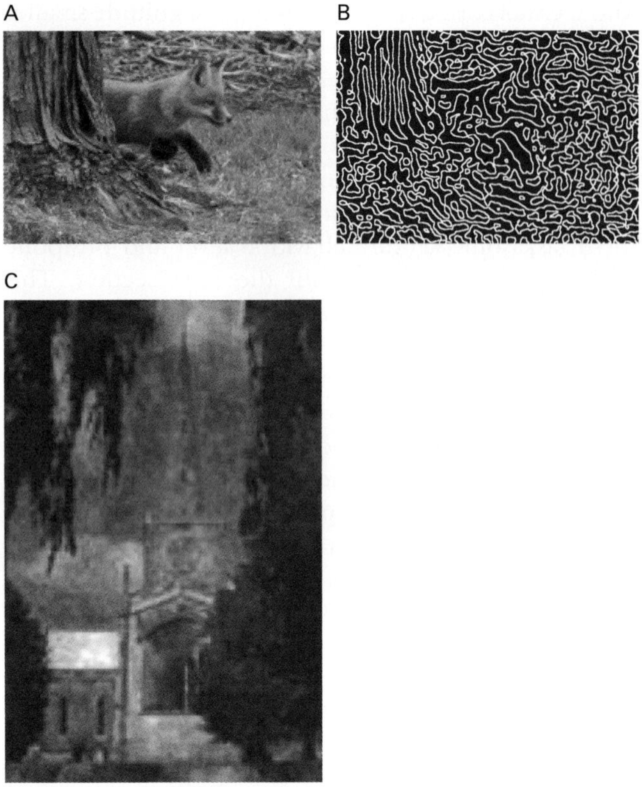

FIGURE 12.6 The edges of the fox image (A) detected by low-level processing (B). Some edges lie on the boundary of objects, like the fox and the tree, while others are due to properties of the textures (e.g., the grass or the bark of the tree). It is difficult to distinguish between these different types of edges. The church steeple (C), and the position of its edges, is obvious if you view the whole image. However, it is almost impossible to see locally because there is no strong local evidence for the edges of the steeple. This shows that sometimes edge detection is impossible except when done in conjunction with object detection. One possibility is that low-level vision proposes many possible edges which are validated, or rejected, by object models, as discussed in section 12.4.

Low-level vision also includes estimating the local motion of images. This involves finding the correspondence between points in images taken at different times. This is done by matching regions which have similar intensity properties. However, this has problems at places where there are many, equally good, matches. Figure 12.7 illustrates the *aperture problem* which is solved by making assumptions that the motion is usually slow and smooth.

Although these low-level visual tasks can be studied in isolation, there is much evidence that they interact with other higher level tasks. For example, it is usually impossible to detect the edges of the objects in an image without making mistakes using low-level processing alone. Instead, a better strategy is for low-level processing to *propose* a set of possible positions for edges, which can be validated by object models; see figure 12.6C. But there are many visual phenomena indicating the complexity of the interactions between low-level and high-level vision, such as the dalmatian dog illusion where the extreme ambiguity of low-level cues for edges make it very hard to detect the dalmatian; see http://michaelbach.de/ot/cog_dalmatian/.

Mid-Level Vision Mid-level vision builds on low-level vision. It corresponds, roughly speaking, to vision processing that knows about geometry, materials, and lighting. However, it does not know about specific objects or scene structures. For example, mid-level vision knows about surfaces of red metal but does not know about red cars. Mid-level vision includes many visual tasks, often formulated in terms of modules, and this section introduces some of them.

Mid-level vision includes inferences about depth ordering of surfaces and, in particular, knowing that surfaces can partially occlude each other. The Kanizsa figure 12.8 is perceived as three black disks which are partially occluded by a triangular surface. But close inspection shows that there is no direct evidence for the existence of the triangle, that is, there are no intensity edges at the boundary of the triangle. The Kanizsa triangle is an example of a *Gestalt* grouping phenomenon, many of which can be explained in terms of a human tendency to interpret images as simple geometric structures (Kanizsa, 1979).

Another example of how the human visual system uses geometry was shown earlier in figure 12.3. The human visual system assumes that most images contain a ground plane with objects standing on it (e.g., a man standing on a lawn). By the laws of *perspective projection*, if there are parallel lines in the image (such as the tracks of a railway line) then the projection of these lines in the image will converge at a *vanishing point*. Humans can use vanishing points to estimate the orientation of the ground plane. The contact points of the objects with the plane specify the positions of the object in the scene. More information about the scene can be extracted if the image contains several vanishing points corresponding to surfaces which are orthogonal in space.

Binocular stereo is another vision module which estimates the depth and orientation of surfaces. Humans have the ability to get depth from two eyes—hence the popularity of so-called 3-D movies. This is illustrated in figure 12.5. It requires solving a *correspondence problem* between features in the two eyes which are caused/imaged by the same point in space. If correspondence can be performed, then the depth can be estimated by

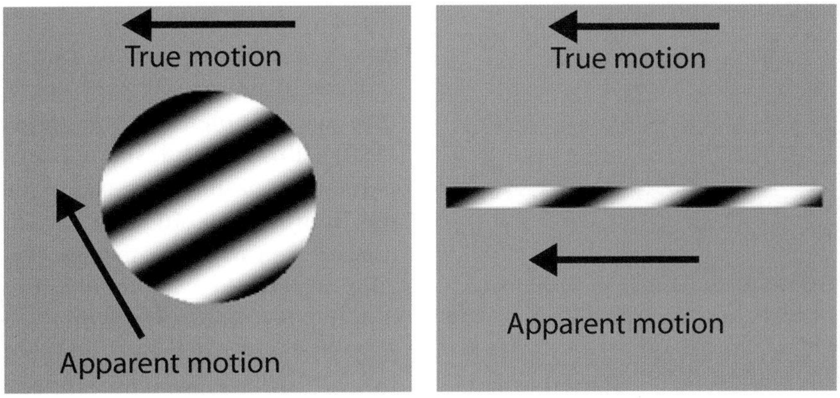

FIGURE 12.7 These images show black and white bars whose *true motion* is leftward viewed through two apertures (circular and rectangular). However the motion is locally ambiguous because we can only directly observe the motion component normal to the bars (we cannot detect if there is any motion tangential to the bars), and so the observed stimuli are consistent with many possible motions. The human visual system uses constraints to resolve these ambiguities. For these stimuli (left panel), humans assume that the motion is as slow as possible and hence is perpendicular to the bars (i.e., assuming that the unobservable tangential component is zero), as indicated by the *apparent motion*. More generally, as discussed in section 12.4.4, humans tend to assume that motion is slow and smooth. See demo 4e.

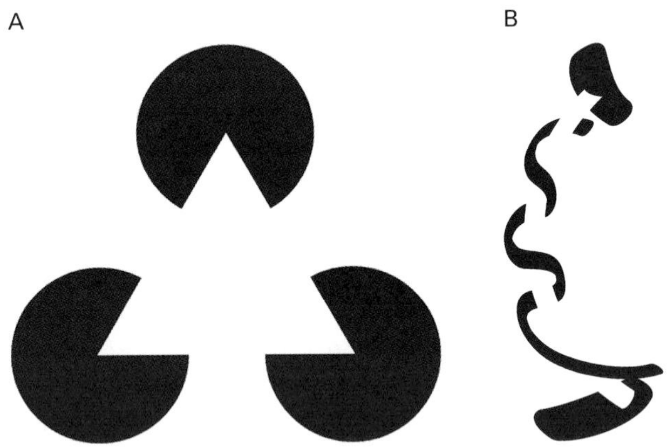

FIGURE 12.8 The Kanizsa triangle (A) is perceived as a white triangular surface which partially occludes three black disks; it shows the tendency of humans to interpret images in terms of geometric structures. Another example of this phenomenon (B) shows that human tend to "explain away" gaps in the black band by positing a white band which lies sometimes above and sometimes below the black band. For an interesting variant, see http://www.michaelbach.de/ot/cog-kanizsa/, which shows how the effect can disappear if other cues are present.

trigonometry. The correspondence problem is difficult (see section 12.4.3) and seems to assume that the geometric surfaces being viewed are spatially smooth. The correspondence problem is made easier because of the *epipolar line constraint* (see figure 12.9), which means that corresponding points require only searching in a one-dimensional direction. But knowledge of the epipolar lines requires knowing the direction of gaze of the cameras (maybe done by feedback from muscles controlling the eyes, or by *calibration*). Note that partial occlusion can happen, where part of the scene is only visible to one eye. Da Vinci was the first to point out that this was a useful visual cue.

Humans can also get three-dimensional shape information of surfaces from shading, texture, and even contours. These are often studied as separate modules. Typically models of shape from shading assume that surfaces can be modeled as matte (i.e., dull and un-shiny) and hence their reflectance properties (i.e., their tendency to reflect light) given by Lambert's law (Basri, 2003). This enables us to predict the image intensity of an object in terms of its geometry and the lighting conditions. In some conditions this can be inverted to estimate the shape of the object from the intensity patterns, which is called *shape from shading* (Basri, 2003). Shape from texture arises if a surface has a regular pattern of texture. This pattern will be distorted because of the shape of the object, which enables the shape of the surface to be estimated from the intensity patterns by *shape from texture* (Knill, 2001). Finally, certain contours naturally suggest shapes, which is called *shape from contour* (Knill, 2001).

In addition to detecting the shape of surfaces, humans are also able to estimate their material properties; see Buckingham (2009). This enables us to tell whether a teapot is metallic (e.g., by being shiny) or matte. This ability has many uses. For example, if a surface is metallic, then it could be part of an airplane, but it cannot be part of an animal. If a patch on a road is very shiny, then there is a possibility that it is ice, at least in winter,

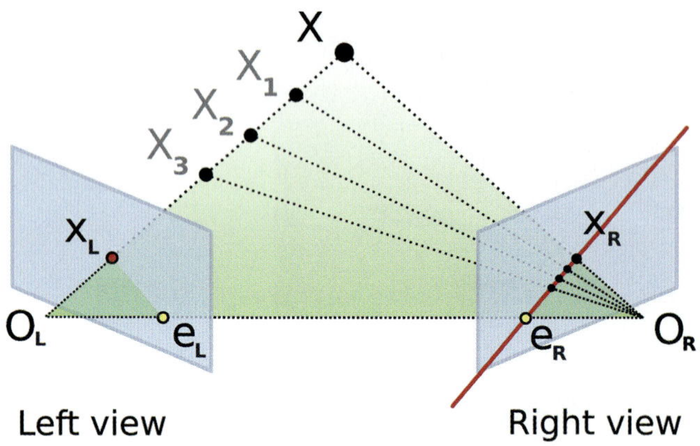

FIGURE 12.9 Stereopsis and epipolar lines. A point $\boldsymbol{x}$ in three-dimensional space gets projected onto positions x_L and x_R in the left and right eyes. This uses a pinhole camera model of each eye where the eye is specified as a plane (in gray) and O_L and O_R represent the centers of projection. Observe that all points on the plane defined by O_L, O_R and $\boldsymbol{x}$ get projected onto straight lines $\boldsymbol{e}_L$ and $\boldsymbol{e}_R$, the corresponding *epipolar lines*, in the two eyes. To illustrate this, we show the projections of a few points x_1, x_2, x_3 onto the right eye. If we alter the position of the point $\boldsymbol{x}$ in space, then we will get a family of corresponding epipolar lines. The *epipolar line constraint* states that points on an epipolar line in one eye can only be matched to a point in the other eye on the corresponding epipolar line. In this picture the eyes are fixating at a point. But if the eyes are parallel to each other, the fixation point will be at infinity and the epipolar lines will be parallel to each other in the two eyes (or images). Reprinted from https://en.wikipedia.org/wiki/Epipolar_geometry#/media/File:Epipolar_geometry.svg, Arne Nordmann, CC BY-SA 3.0.

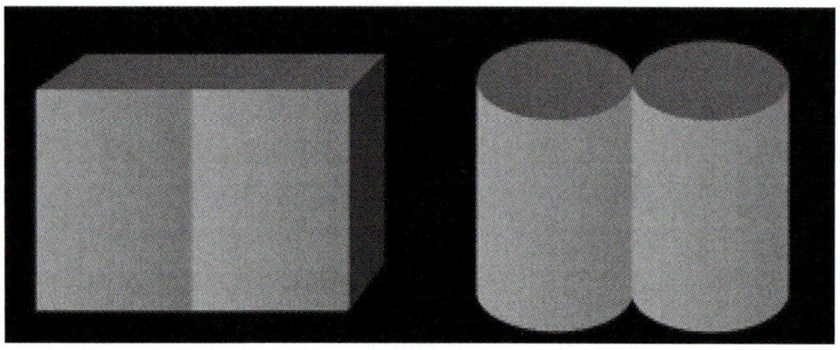

FIGURE 12.10 The shapes of the contours affect the perception of the shading patterns. The intensity patterns are the same in both stimuli, but the shape of the boundaries makes the perception of shading different in the two cases (Knill and Kersten, 1991).

and hence it is better not to step on it. In general, the ability to detect material properties and textures is very useful when performing actions—for example, is this object slippery? can I walk on it? will it make a sound if I do?—and as a precursor for object recognition and scene understanding.

These vision modules depend on specific properties of images which are called *cues*—for instance, the shadow on the sand in figure 12.3. These modules, and the visual cues they rely on, can be modeled and studied separately. However, many lines of evidence suggest that they are coupled. For example, Knill and Kersten (1991) illustrate that shape from contour can alter the perception of shading patterns and material properties; see figure 12.10. In addition, careful studies of how images are formed in terms of the surface of the scene (the objects, their geometry, how they reflect light, the lighting conditions, and the viewpoint) make it clear that the *cues* are often coupled; see section 12.5.1. Indeed, under certain lighting conditions it is extremely difficult to tell the color of an object being viewed; see http://michaelbach.de/ot/col-dress/. In addition, there is interaction between these modules and high-level vision. Humans perceive an inverted (i.e., concave) face mask to be a normal convex face even when binocular cues are present; see http://michaelbach.de/ot/fcs_hollow-face/. In general, object recognition trumps mid-level vision although Kanizsa (1979) shows some exceptions; see also http://michaelbach.de/ot/sze-AmesBallerina/.

12.1.3 THE VISUAL SYSTEM AND THE BRAIN This section briefly overviews what we know about how the brain does vision. It reviews the areas of the brain which perform visual processing, the relationships between them, the structures of these areas, and the visual tasks they perform. Because of the complexity of the visual system, our knowledge of these issues, though considerable, is limited. It is based on a combination of anatomical studies, electrophysiological recordings, and noninvasive imaging methods such as functional magnetic resonance imaging (fMRI).

Retina and Lateral Geniculate Nucleus Visual neural processing begins with the retina which transmits information to the visual cortex via the lateral geniculate nucleus (LGN). The anatomy and electrophysiology of the retina and LGN have been studied in detail (Merigan and Maunsell, 1993; Gollisch and Meister, 2010; Briggs and Usrey, 2011). Although we will not discuss it further in this chapter, another pathway from the retina leads to the superior colliculus in the midbrain. This plays an important role in, for example, eye movements (see chapter 19, "Saccades and Smooth Pursuit Eye Movements," and chapter 22, "Integrative Functions of the Corticostriatal System").

The retina converts intensity patterns—the light rays which reach the retina—into patterns of neural activity. This starts with *photoreceptors* which are directly activated by light. The photoreceptors have been studied in detail, and it can be shown that they are extremely efficient at "capturing" photons (Rieke et al., 1999). The remaining set of neurons in the retina, in particular *ganglion cells*, process the photoreceptors' output and encode it for transmission via the *optic nerve* to the rest of the brain (note that it is believed that this flow of visual information only goes one way although the brain does influence the muscles which control the eyes).

The retina is generally agreed to function as a sophisticated camera which captures the information in the incoming images and encodes it so that it can be transmitted to the visual cortex (although the control of eye movements means that the retina is not a passive device like a typical camera and instead is actively searching for information). To do this, the retina must face two challenges: (1) the enormous variability of intensity in natural images and (2) how to encode the images so they can be transmitted efficiently and robustly. Neural models of the retina are largely motivated by these two issues.

Firstly, the intensity of natural light varies enormously from faint starlight to bright sunlight dealing with intensity magnitudes ranging from 1 to 10^9. Moreover, the changes of intensity within specific images can also vary hugely (Demb, 2002). However, neurons have limited ranges of response, and hence they cannot encode these huge ranges of intensity. Therefore, many theories of the retina (see section 12.2) propose that the ganglion cells perform *gain control* and filter the images so that they capture only the *local contrast*—the differences of intensity between nearby parts of the retina—and hence reduce the need to represent the entire intensity range. Observe that digital cameras must perform a similar function since they convert incoming light patterns into digital representations where the intensity only takes 256 values from 0 to 255 (in each color channel).

Secondly, the retina must encode the image information so that it can be transmitted through the optic nerve to the rest of the brain for further processing. The image information is transmitted through a relatively smaller number of fibers in the optic nerve (compared with the number of photoreceptor cells). Information theory offers guidelines for how information can be encoded efficiently based on statistical knowledge of the stimuli. Researchers have applied this theory to predict retinal properties with some success, but this work is out of the scope of this chapter, and we refer the reader to the detailed description in Zhaoping (2014).

In addition, the human and animal visual system does not passively receive images and process them. Instead, it is active, and by using eye and head movements, and attentional processes, it seeks out information useful for performing tasks. The retina only has high spatial precision at the center in the *fovea*, and so high-resolution images can only be gathered there. Neurons in the periphery are more sensitive to motion and presumably alert the visual system to places to foveate to. This work is also out of the scope of this chapter and is addressed in Zhaoping (2014).

The studies of the retina illustrate the "simplification issues" which will reoccur throughout our chapter. At the computational level, the theories for describing how the retina deals with intensity are simpler than the engineering methods used by computer vision and image processing researchers who deal with the same challenges. At the experimental level, many of the findings about retinal neurons are based on simplified models of neurons obtained from studying their responses to synthetic stimuli. Moreover, although there is considerable knowledge of the anatomy, there has only recently been highly detailed studies of the wiring diagrams and characterization of the 50 or more anatomical types of neurons (Masland and Martin, 2007). It is also unclear why so many neurons are required to perform the two functions using current theories. Indeed it has been argued that the retina is considerably "smarter" than current theories suggest (Meister and Berry, 1999; Gollisch and Meister, 2010) and may require detecting motion, expansion, and extrapolation and more generally adapting to the complexity of image patterns.

The output from the retina is transmitted to the LGN and then to the visual cortex where it arrives in *visual area V1*. LGN is generally believed to have the limited function of serving as a way station on the route to the visual cortex. Hence current models of LGN neurons are fairly simple, as discussed in section 12.2. However, there is reason to believe that LGN is more complex. For example, there is substantial feedback from V1 to LGN (Briggs and Usrey, 2011), and connections between LGN and areas other than V1 (Sherman and Guillery, 2002; Nassi, Lyon, and Callaway, 2006).

Cortical Visual Areas and the Relationships between Them

The visual cortex can be decomposed into a number of visual areas based on anatomical and electrophysiological measurements (van Essen, Anderson and Felleman, 1992). Like all areas of the cortex, these regions have a standard six-layer structure. These visual areas—V1,V2,V4, medial temporal (MT), medial superior temporal (MST), and inferior temporal cortex (IT)—are illustrated in figure 12.11. There are numerous interactions between these visual areas (see figure 12.11), but it is common to concentrate on two *hierarchical streams*.

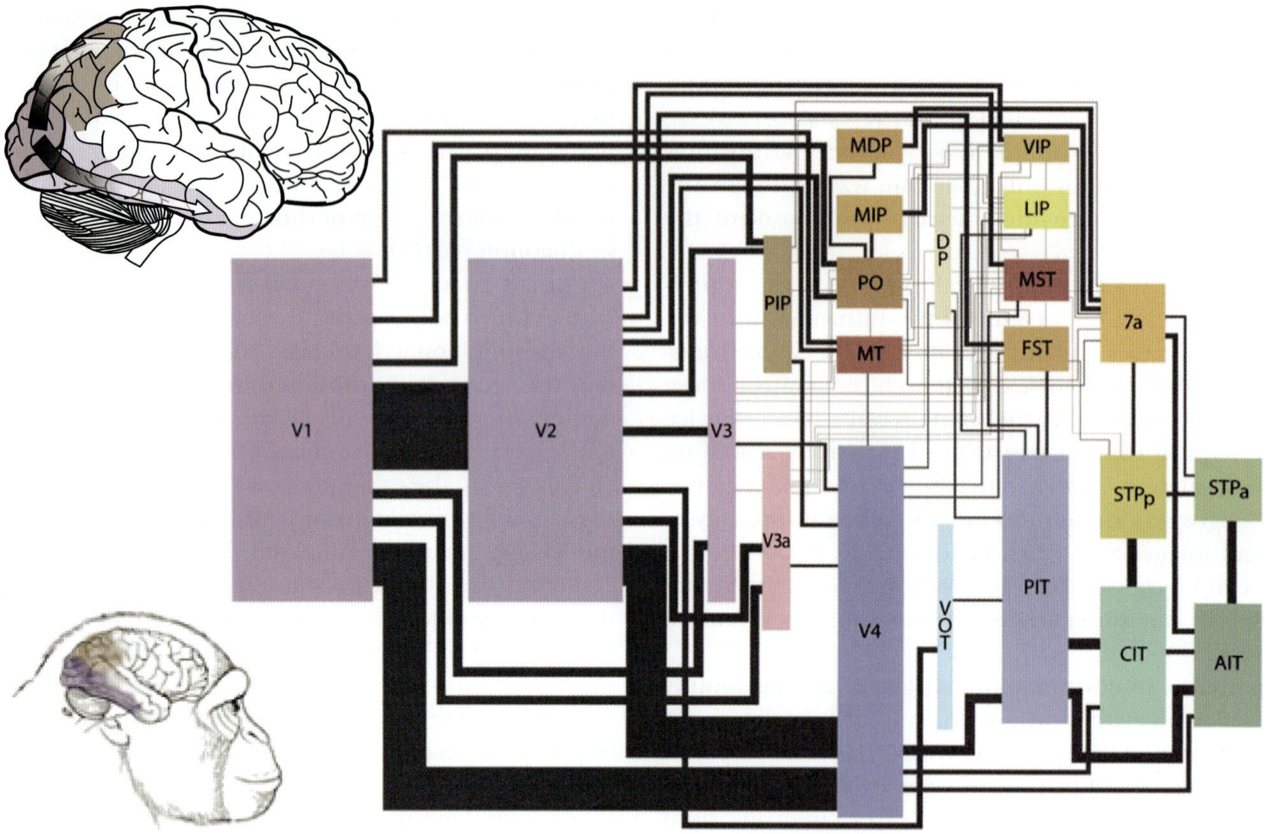

FIGURE 12.11 The colored regions of the left top and bottom panels illustrate human and monkey visual cortical areas, respectively. The right panel is a schematic of connections between visual cortical areas in the macaque monkey brain. The colored rectangles represent visual areas (see Felleman and van Essen, 1991). The black lines show the connections between areas, with the thickness proportional to estimates of the number of feedforward fibers. Areas in cool and warm tones belong to the ventral and dorsal streams, respectively. Figure reprinted with permission from P. Wallisch and J. A. Movshon. Structure and Function Come Unglued in the Visual Cortex. *Neuron, 60*(2), 194–197, October 2008. (See also Lennie, 1998.)

The first of these is the *ventral stream*, which consists of V1, V2, V4 (the functional organization of V3 has been under some debate) and then the IT areas of extrastriate cortex. This pathway is believed to perform object detection and scene understanding. The second stream is the *dorsal stream*, which goes from V1, MT, to the parietal cortex. It is believed that this is used for analysis of the movements and positions of objects as they relate to navigation and actions (Milner and Goodale, 2006). Although the distinction between ventral and dorsal pathways is well established (Lennie, 1998), it is also probable that the true situation is more complicated (Schenk and McIntosh, 2010).

The size of the visual areas varies greatly. The first two areas, V1 and V2, are enormous and together account for roughly 70% of the number of neurons in the visual cortex (hence 30% of the neurons in the entire cortex). The size of V1 is much bigger, by a factor of at least 200, than the number of fibers that leave the eye. Indeed it has been estimated that this is more by a factor of several hundred than the amount needed to represent the information conveyed by the LGN (Lennie, 1998), consistent with the idea that the purpose of V1 is to start interpreting the image instead of simply encoding it. Another major feature of the hierarchy of visual areas is that the sizes get progressively smaller as one rises in the hierarchy. For example, V4 is much smaller than V2, and visual areas within IT are considerably smaller than V4.

The early visual areas have regular structural organizations. These are studied by using electrophysiology to classify the receptive fields of neurons by probing their responses to synthetic stimuli with different *perceptual dimensions* such as position, orientation, color, texture, shape, sensitivity to input from both eyes, and motion. Individual neurons often show preferences for—that is, respond more strongly to—specific input stimuli and hence are said to be *tuned* to these stimuli. In particular, the positions of stimuli on the retina which cause the neuron to fire is called its *receptive field*. Neighboring neurons in early visual areas usually respond to similar regions of the image. Hence these areas are roughly *retinotopic* in the sense that their spatial organization is similar to that of the image at the retina, but with a nonlinear (log-polar) spatial transformation (Schwartz, 1980). This retinotopic structure is strongest in V1 and V2 and gets weaker at high visual areas (see following paragraphs). But although neurons often have small receptive fields, at least in V1 and V2, their tuning to other "perceptual dimensions" is often broader, and neurons can responds to a range of stimuli. It is helpful to broadly classify neurons by which specific "dimension," if any, those neurons responds to, but this can become problematic, particularly in higher areas (Roe, Chen, and Lu, 2009; Roe et al., 2012). Even in V1 most neurons respond to several dimensions (Lennie, 1998). Mapping has also been done using optical techniques (Lu and Roe, 2007; Kinoshita, Gilbert, and Das, 2009) which also show that most early visual areas are organized retinotopically although this is strongest in V1 and V2. Finally, we note that until the last few years it has only been possible to record from a neuron for a limited amount of time (e.g., from an hour to at most a couple of days). This meant it was only possible to test the response of a neuron to a limited number of stimuli and stimuli dimensions. Hence some of these findings may be revised when it becomes possible to test the response of neurons to a larger and more representative set of stimuli; see the discussion in section 12.2.3.

Other salient structures of V1 include *hypercolumns* (~1–2 mm) consisting of (1) a regular array of orientation columns, perpendicular to the cortical surface, in which orientation selectivity of neurons is approximately the same; the orientation tuning varies slowly parallel to the cortical surface; (2) ocular dominance columns (where the proportion of input from both eyes is constant within each column but varies smoothly between columns), and (3) a lattice of cytochrome oxidase blobs—sensitive to color (Hubel, 1982; Livingstone and Hubel, 1984). Sections 12.2, 12.3, and 12.4 describe computational models which exploit these structural properties. From a more abstract perspective, the organizational structures of hypercolumns can be partly explained by the need to map stimulus dimensions (e.g., retinal position, orientation, etc.) onto two-dimensional cortical surface while attempting to make the map as smooth as possible (this is not possible, on topological grounds, so discontinuities occur) (Durbin and Mitchison, 1990). Cells in V2 have larger receptive fields than in V1 but have other properties. For example, many cells in V2 seem capable of performing binocular integration to estimate depth. In addition, they appear to include neurons which respond to features which are linked spatially (von der Heydt and Peterhans, 1989; Peterhans and von der Heydt, 1989), which are sensitive to illusory contours (von der Heydt, Peterhans, and Baumgartner, 1984; Lee and Nguyen, 2001), and which relate to detecting occlusion and establishing border ownership (Zhou, Friedman, and von der Heydt, 2000; Fang, Boyaci, and Kersten, 2009). These issues are discussed in section 12.4.

A notable property of these visual areas is their hierarchical organization, which relates to the low-, mid-, and high levels discussed in section 12.1.2. Broadly speaking, V1 and MT seem to be involved in low-level processing, V2,V4, and MST in mid-level vision, and IT in high-level vision. Hence early vision is believed to be mostly performed in V1,V2,V4, MT, and MST. There is also a strong tendency for receptive fields to be larger as we ascend the visual hierarchy. Compared to those in V1, the receptive fields are 2–3 times larger in V2, 4–5 times larger in V3, and 7–10 times larger in MT. But, conversely, the receptive fields become increasingly specific to stimuli, and stimuli of greater complexity, as we move up the ventral stream. In summary, the receptive fields become more invariant to position and more specific to structure as we proceed up the ventral stream from V1 to V2 to IT (Rust and DiCarlo, 2010; Logothetis and Sheinberg, 1996).

Experimental Methods Many of the findings discussed above are based on electrophysiological studies of monkeys and noninvasive studies of monkeys and humans. These visual systems are fairly similar based on studies relating fMRI responses in humans and those in monkeys (where electrophysiology is performed). Relationships have been found in early visual areas V1, V2, V3 (Wandell, Dumoulin, and Brewer, 2007) but are not yet fully established at higher areas (Wandell and Winawer, 2011). However, it is noteworthy that a *face area* discovered in humans by fMRI studies (Kanwisher, 2010) corresponds to an analogous face area in monkeys which can be studied by fMRI and electrophysiology (Tsao, Moeller, and Friewald, 2008).

The interpretations of experimental findings are limited because of the simplifications discussed in earlier sections. Noninvasive studies like fMRI suffer from limited spatial and temporal resolution and currently can only observe coarse properties of the visual system. Electrophysiology is restricted to recording from a small number of neurons in response to a limited range of stimuli. See Carandini et al. (2005) for a detailed discussion of the problems of interpreting electrophysiological results in the early visual cortex. In general, even in V1 it is not easy to predict the response of neurons to natural stimuli. This may, however, be partially because neurons in the visual cortex interact with each other and may be better predicted by the activity of other neurons than by the image itself. All the visual areas contain lateral (i.e., sideways) structure (in addition to hierarchy), and neural connections within V1 extend as much as 8 mm. As will be discussed later in this chapter, it can be shown that these lateral connections may implement *nonclassical receptive fields* which relate to visual tasks such as linking, or grouping, features (Gilbert and Wiesel, 1990). There is considerable interest, and progress, in developing experimental methods which can probe the properties of neural circuits in much greater detail, such as optogenetics, which may revolutionize our understanding the early visual system.

12.2 Linear and Complex Filters

This section covers five topics. Firstly, we introduce simple linear models of neurons, describe how they are used to model the receptive fields of neurons in the retina, LGN, and *simple cells* in V1, and discuss the visual tasks they can be used for. We also discuss extensions of these models to deal with *complex cells* in V1. Secondly, we discuss an alternative perspective which thinks of these cells as representing images and introduces overcomplete bases and sparse encoding. Thirdly, we discuss how these receptive fields can be learned from images by unsupervised algorithms or estimated from neural data by regression. Fourthly, we describe how these receptive field models can be used for binocular stereo and for motion estimation. The section also introduces background material on linear filter theory and Fourier analysis.

The book's website includes three interactive demos: 2a, Linear Filters and Convolution; 2b, Gabor Filters; and 2c, Oja's Rule and Principal Component Analysis (PCA).

12.2.1 Linear Models of Neurons

This section introduces linear models of neurons. These models are used to describe the receptive field properties in the retina, the LGN, and simple cells in visual area V1. We introduce background material on linear filtering and Fourier analysis which is helpful in understanding the properties of these neural models.

Linear Models of Simplified Cells This section introduces a model of a simplified cell; see figure 12.12. The cell receives inputs $\mathbf{I} = (I_1, I_2, \dots, I_N)$ from *dendrites* which are weighted by *synaptic strengths* $\mathbf{w} = (w_1, w_2, \dots, w_N)$; these are summed at the *soma* (cell body) to obtain $\mathbf{w} \cdot \mathbf{I} = \sum_{i=1}^{N} w_i I_i$. The cells output a response $f(\mathbf{w} \cdot \mathbf{I})$ along its *axon*, indicated by the firing rate of the neuron. $f(.)$ is monotonic nonlinear function, which takes value 0 if the input is small, then increases linearly in the *linear regime* until it saturates at a maximum value. A typical choice of $f(.)$ is the sigmoid function $f(\mathbf{w} \cdot \mathbf{I}) = \sigma(\mathbf{w} \cdot \mathbf{I} - T)$, where T is a threshold and $\sigma(.)$ is a soft threshold as shown in figure 12.12, which we will see in later sections. Interactive demo 2a illustrates linear filters and convolution.

In this section, we ignore $f(.)$ (except when explicitly stated) and study the behavior of the model in the

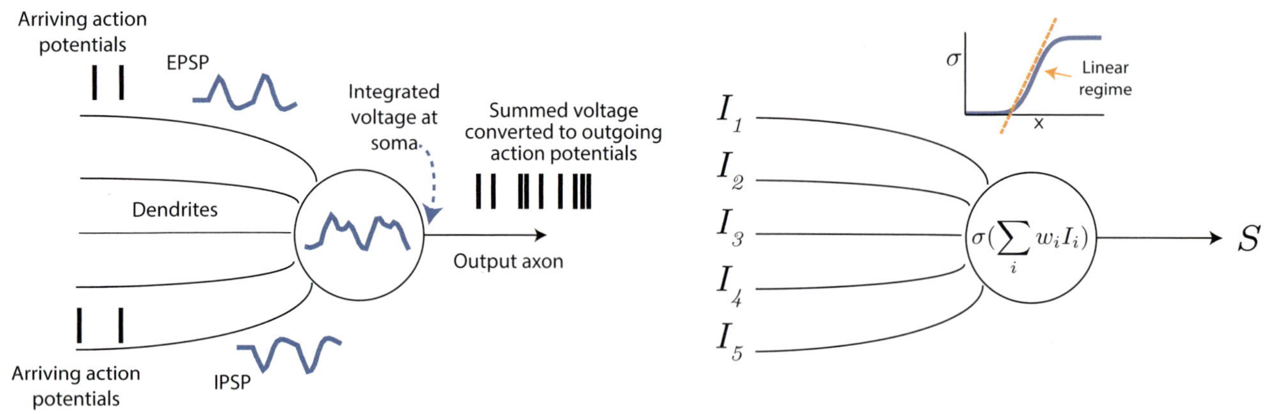

Figure 12.12 Left: A neuron receives input—action potentials from other neurons—at its dendrites which generate excitatory and inhibitory postsynaptic potentials (EPSPs and IPSPs, respectively) whose voltages are integrated at the soma and converted to outgoing action potentials. Right: A simplified model of a neuron. There are inputs $(I_1, \dots, I_5)$ at the dendrites, with synaptic strengths $w_1, \dots, w_5$, these are summed at the soma, $\Sigma_i w_i I_i$, and the output S is given by a sigmoid function $\sigma(\Sigma_i w_i I_i)$. The sigmoid function $\sigma()$ (top right) has a linear regime (brown line) and low and high thresholds.

linear regime. Cells in the retina and LGN are often modeled without the nonlinear function f (.), but adding instead a constant C to the output, to account for spontaneous firing of the cell, and yielding an output $\mathbf{w} \cdot \mathbf{I} + C$; see Zhaoping (2014).

Hence in this section, we model a simplified cell by

$$S = \mathbf{w} \cdot \mathbf{I} = \sum_{i=1}^{N} w_i I_i.$$

This model is linear in two respects. Firstly, it is linear in the input $\boldsymbol{I}$ so that if we double the input $\mathbf{I} \rightarrow 2\mathbf{I}$, then the output doubles also $S \rightarrow 2S$. Secondly, it is linear in the weights w. Most importantly, it obeys the *principle of superposition* so that if S^1, S^2 are the outputs to input $\mathbf{I}^1$, $\mathbf{I}^2$, respectively, then the output to input $\lambda_1 \mathbf{I}^1 + \lambda_2 \mathbf{I}^2$ is $\lambda_1 S_1 + \lambda_2 S_2$. This result is important for characterizing the response of simple neural cells since it implies that we can determine the output of the cell to any stimulus by observing its response to a limited set of input stimuli $\mathbf{I}$, and we will return to this issue later in this section. Note that this property still remains if we reintroduce the nonlinear function f(.), provided the function is known.

The retinotopic organization of the early visual system has two implications for these cells. *Firstly*, the weights of the cell depend on its retinotopic position $\boldsymbol{x} = (x_1, x_2)$ and the positions $\boldsymbol{y} = (y_1, y_2)$ of its dendrites. We replace the input I_i by $\boldsymbol{I}(\boldsymbol{y})$ and the weights w_i by $w(\boldsymbol{x} - \boldsymbol{y})$. The *receptive field* $w(\boldsymbol{x} - \boldsymbol{y})$ will typically be zero unless $|\boldsymbol{x} - \boldsymbol{y}|$ is small (this can be mapped by determining if image properties at position $\boldsymbol{y}$ in the image cause the cell to fire). Hence the neuron can be modeled by

$$S(\boldsymbol{x}) = \sum_{y} w(\boldsymbol{x} - \boldsymbol{y}) I(\boldsymbol{y}) = \mathbf{w} * \mathbf{I}.$$

Secondly, retinotopy implies that there are cells with similar properties (e.g., the same weights $\boldsymbol{w}$) arranged roughly evenly in spatial position (apart from the log-polar transformations [Schwartz, 1980]). This can be thought of as having "copies" of the same cell at all positions in space. In terms of linear filter theory (see later this section), these sets of cells are *convolving* the image $\boldsymbol{I}$ by a filter $\boldsymbol{w}$. Convolution is illustrated in figure 12.13.

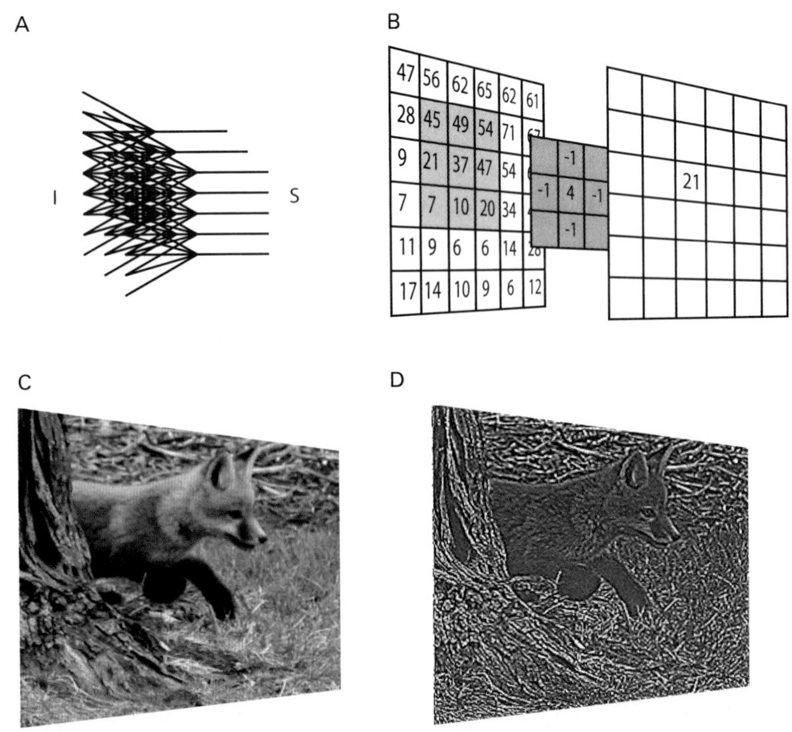

FIGURE 12.13 This figure shows the input–output of a center–surround cell (e.g., Laplacian of a Gaussian) in three different ways. The first way is in terms of the inputs and outputs of neurons (A). The second way is in terms of the digitized input image, the filter, and the digitized output (B). The output at each pixel is given by the product of the filter to the appropriate intensity values in the input image, for example, $4 \times 37 - 1 \times 49 - 1 \times 47 - 1 \times 10 - 1 \times 20 = 21$. The third way is in terms of the input and output images (C, D).

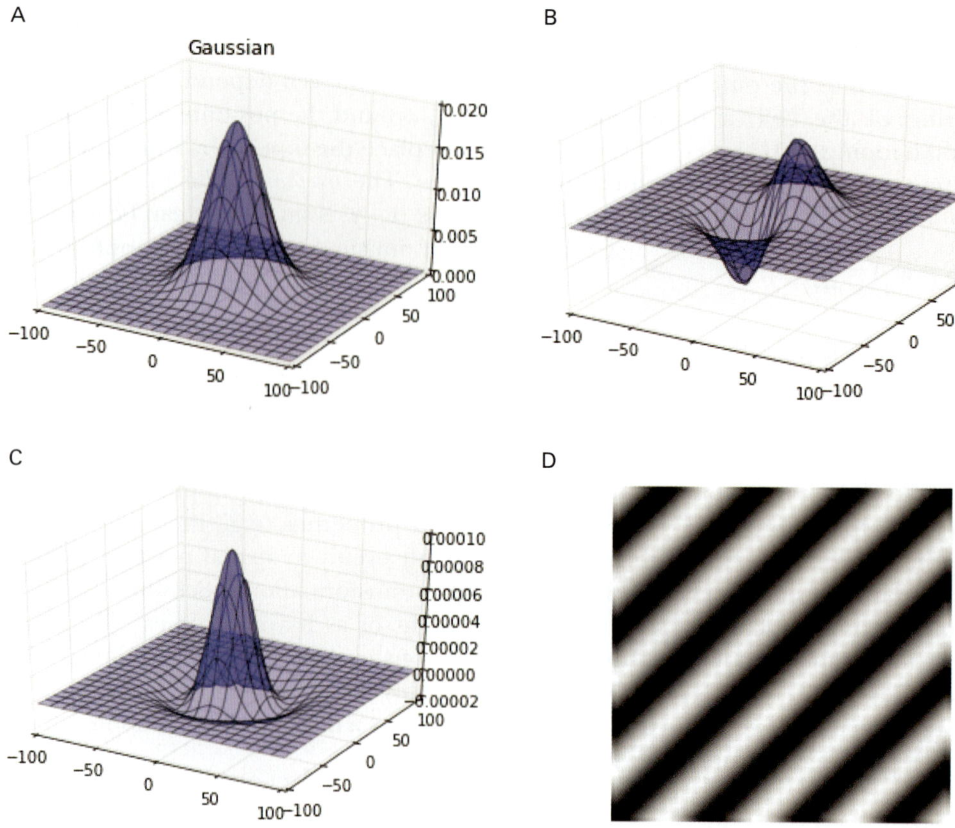

FIGURE 12.14 A Gaussian filter (A). The first derivative of a Gaussian (B). The Laplacian of a Gaussian or Mexican hat (C). A sinusoid (D).

Center–Surround Cells: Retina and LGN The receptive fields of the ganglion cells in the retina and cells in the LGN can be determined by measuring the firing rate of the neurons in terms of its response to different input stimuli $\boldsymbol{I}$ and estimating a model for the response, as discussed in the next section (we refer to Rieke et al. (1997) for a description of the photoreceptors). The experimental findings are that many simple cells have a characteristic receptive field called *center–surround.* However, these findings are done using synthetic stimuli, as we will briefly describe, and their response may be more complex if they are studied using natural stimuli; see section 12.2.3.

There are two different types: on-center and off-center. The receptive field weights $w(\mathbf{x}-\mathbf{y})$ are radially symmetric and take the form of a "Mexican hat" or inverted Mexican hat, for on-center and off-center cells, respectively (Marr, 1982); see figure 12.14C. These cell responses are usually thresholded, for example, by the sigmoid function, so that they usually only give positive responses. The weights $w(\mathbf{x}-\mathbf{y})$ can be approximated by the *Laplacian of a Gaussian* (LOG) or by its negative:

$$w_{LOG}(\mathbf{x}) = -\left\{\frac{\partial^2}{\partial x_1^2}+\frac{\partial^2}{\partial x_2^2}\right\}G(\mathbf{x}:\mathbf{0},\sigma^2) \qquad (12.1)$$

where $G(\mathbf{x}:\mathbf{0},\sigma^2)=\frac{1}{2\pi\sigma^2}\exp\{-(x_1^2+x_2^2)/(2\sigma^2)\}$; see figure 12.14A.

These cells have two important properties: Firstly, they are radially symmetric in the sense that $w_{LOG}(.)$ is invariant to rotation. For example, suppose we express position $\mathbf{x}$ in terms of radial components: $x_1 = r\cos\theta$, $x_2 = r\sin\theta$; then $w_{LOG}(r\cos\theta, r\sin\theta)$ is independent of θ. Secondly, the receptive field weights $w(.)$ sum up to zero. More precisely,

$$\sum_{\mathbf{x}} w_{LOG}(\mathbf{x}) = 0.$$

Note that center–surround cells are often modeled as the *differences of two Gaussians* $w_{DOG}(\mathbf{x}) = A_1G(\mathbf{x}:\mathbf{0},\sigma_1^2) - A_2G(\mathbf{x}:\mathbf{0},\sigma_2^2)$, where σ_1,σ_2 take different values (Zhaoping, 2014). This gives a similar model if $|\sigma_1-\sigma_2|$ and $|A_1-A_2|$ are small.

The purpose of these center–surround cells is believed to be to help deal with the large dynamic range of images. Suppose we can express the image locally as $I(\boldsymbol{x}) = C(\boldsymbol{x}) + B$ where $C(\boldsymbol{x})$ is the *contrast*, which describes the local details of the image, and B is the *background*. Then filtering an image by a center–surround cell, whose receptive field sums to 0, removes the background term and preserves part of the contrast. More precisely, using equation 12.1,

$$S(\boldsymbol{x}) = \sum_{\boldsymbol{y}} w_{LOG}(\boldsymbol{x}-\boldsymbol{y})I(\boldsymbol{y}) = \sum_{\boldsymbol{y}} w_{LOG}(\boldsymbol{x}-\boldsymbol{y})\,(C(\boldsymbol{y})+B)$$
$$= \sum_{\boldsymbol{y}} w_{LOG}(\boldsymbol{x}-\boldsymbol{y})C(\boldsymbol{y}).$$

Receptive fields of this type can also help efficiently encode the information at the retina in order to transmit it efficiently to the visual cortex. This can be studied using information theory and the statistics of natural images to predict properties of receptive fields and how they change in different environments (Atick and Redlich, 1992). This theory is beyond the scope of our chapter, and we refer the reader to the detailed exposition in Zhaoping (2014).

These models of cells in both the retina and the LGN are well studied. Although many of their properties were estimated using synthetic input data, it has been shown that in some cases the input image can be estimated from the response of cells in either the retina or the LGN using these types of models (Warland, Reinagel, and Meister, 1997; Carandini, 2005; Dan, Atick, and Reid, 1996). But other authors (Gollisch and Meister, 2010) argue that the retina is more complex and that, in particular, the neurons may act more as *feature detectors* (see section 12.2.2) instead of as spatial–temporal filters as described in this section. In particular, Gollisch and Meister (2010) describe many findings suggesting that the retina is more complex than the linear filtering model described above. It is known, for example, that if the light levels go down, then the receptive field size becomes larger (Zhaoping, 2014).

Note that we are also ignoring the temporal behavior of the cells and a more realistic model models the output as $S(\boldsymbol{x},t) = \sum_{\boldsymbol{y},\tau} w(\boldsymbol{x}-\boldsymbol{y},t-\tau)I(\boldsymbol{y},\tau)$, where $w(\boldsymbol{x}-\boldsymbol{y}, t-\tau)$ is a space–time filter. Broadly speaking there are two types of cells with very different temporal characteristics. These are (1) the M-cells, whose receptive fields are spatially large but temporally small (faster) and which project to the dorsal stream, and (2) the P-cells, whose receptive fields are spatially smaller but temporally larger (slower) and which project to the ventral stream. We also do not model the dependence of the cells on the wavelength of the input light. This would require a model $S(\boldsymbol{x}) = \int d\lambda w(\boldsymbol{x}-\boldsymbol{y})w_c(\lambda)I(\boldsymbol{x},\lambda)$, where λ denotes the wavelength and $w_c(\lambda)$ specifies the sensitivity of the cell to color (i.e., there are three types of sensitivity to color $w_{red}(.)$, $w_{green}(.)$, $w_{blue}(.)$). These types of models are described by Zhaoping (2014).

Studying Tuning by Response to Sinusoid Stimuli How do we know the receptive field of a neuron? One way is to study its response to a class of stimuli while varying the stimulus parameters (like the perceptual dimensions mentioned in the first section). In particular, we can find how well the neuron is *tuned* to particular stimulus parameters (typically the neuron will prefer a specific value of the parameter, and its response will decrease as the parameter changes). This is a classic way to study receptive field properties (Hubel, 1982). In this section, we analyze tuning when the stimuli are sinusoid gratings.

We probe the receptive field of a neuron by stimulating it by a sinusoid grating with intensity $I(\boldsymbol{x}) = A\cos(\boldsymbol{w}\cdot\boldsymbol{x}+\rho) + I_0$, where A is the *amplitude*, ρ is the *phase*, $\boldsymbol{w}$ is the *frequency*, and I_0 is the mean light level. The frequency is vector valued and specifies the orientation of the stimulus, by the unit vector $\hat{\boldsymbol{w}} = \boldsymbol{w}/|\boldsymbol{w}|$, and the period of the oscillation $|\boldsymbol{w}|$. The phase ρ shifts the center of the sinusoid; hence, if $\rho = 0$, the center occurs at $\boldsymbol{x} = 0$. We can reexpress $A\cos(\boldsymbol{w}\cdot\boldsymbol{x}+\rho) = A\cos(\boldsymbol{w}\cdot(\boldsymbol{x}-\boldsymbol{x}_0))$, where $\boldsymbol{x}_0 = -\rho\boldsymbol{w}/|\boldsymbol{w}|^2$ is the shift in position.

Then we can analyze the response of the neuron by assuming a functional form for the receptive field. For example, suppose we guess that the neuron is a center–surround cell and its receptive field is a Laplacian-of-a-Gaussian $w_{LOG}(\boldsymbol{x})$, given by equation 12.1. Then we can probe the neuron with sinusoid stimuli to determine if it is center–surround and, if so, to determine the parameter σ^2. In order to do this, we calculate the response of Laplacian-of-a-Gaussian to a sinusoid function. The Laplacian-of-a-Gaussian is a symmetric filter, so its Fourier transform is symmetric, and its response to the sinusoid is $A(\cos\rho)\hat{w}_{LOG}(\boldsymbol{w})$, where the Fourier transform of $w_{LOG}(.)$ is $\hat{w}_{LOG}(\boldsymbol{w}) = (\boldsymbol{w}\cdot\boldsymbol{w})\exp\{-(\sigma^2\boldsymbol{w}\cdot\boldsymbol{w})/2\}$. Hence the predicted response is

$$\int d\boldsymbol{x} w_{LOG}(\boldsymbol{x})A\cos(\boldsymbol{w}\cdot\boldsymbol{x}+\rho)$$
$$= A(\cos\rho)\,(\boldsymbol{w}\cdot\boldsymbol{w})\exp\{-(\sigma^2\boldsymbol{w}\cdot\boldsymbol{w})/2\}.$$

From this we can deduce three properties: (1) the response is biggest if the center of the sinusoid is aligned to the center of the cell, that is, $\rho = 0$, falling to zero at $\rho = \pi/2$, and (2) the cell responds best to frequencies with $|\boldsymbol{w}\cdot\boldsymbol{w}| = 2\sigma^2$ (this is obtained by maximizing the response with respect to $|\boldsymbol{w}|$), (3) the cell is insensitive to the orientation of the stimuli. If the real neuron obeys these properties, which we could find by stimulating it with sinusoids and measuring its firing rate, then we deduce that it is center–surround and estimate its parameter σ^2. If not—for example, if the neuron is sensitive to the orientation of the sinusoid—then we know the cell is not center–surround and needs a different model (see the next section). In short, important properties of the receptive field can be obtained by stimulating the cell with sinusoid gratings. This determines the sensitivity of the cell to the orientation θ, the frequency w, and the phase ρ. As we will show in the next section, simple cells in V1 are *tuned* to the orientation θ and frequency w of the local image patch.

Studying the tuning to stimuli parameters gives a way to characterize the receptive field properties of a cell. However, it is not sufficient to determine the receptive field uniquely. To do this requires studying the response of the neuron to many types of inputs (an advantage of studying tuning is that it can be done with comparatively few stimuli). Later, in section 12.2.3, we discuss more advanced ways of estimating receptive fields by regression which require making fewer assumptions about the receptive field.

Visual Cortex Area V1: Oriented Receptive Fields Now we turn to the receptive field properties of *simple cells* in area V1. These cells were first systematically studied by Hubel and Wiesel (1962, 1968), who showed that they were *tuned* to the orientation of edges and size of bars of light. (Later studies used gratings to show tuning to orientation and spatial frequency, which roughly relates to bar size.) Hubel and Wiesel also showed that these cells were spatially organized with hypercolumns and retinotopic organization. Further electrophysiological studies by Pollen and Ronner (1981) and Jones and Palmer (1987) showed that the receptive field properties of these cells could be approximately modeled by *Gabor filters* (Daugman, 1985), which are the product of Gaussians and sinusoids and can be thought of as performing local Fourier transforms; see figure 12.15. It was also reported that the receptive fields occur in quadrature pairs (Pollen and Ronner, 1981) so that neighboring cells are 90 degrees out of phase (e.g., a cosine Gabor is paired with a sine Gabor). These are illustrated in figure 12.16. Good fits, however, to the receptive fields can also be obtained using the derivatives of Gaussian filters (Young, Lesperance, and Meyer, 2001); see figure 12.14B. Interactive demo 2b illustrates Gabor functions and their properties.

Gabor functions can be expressed as the product of a Gaussian

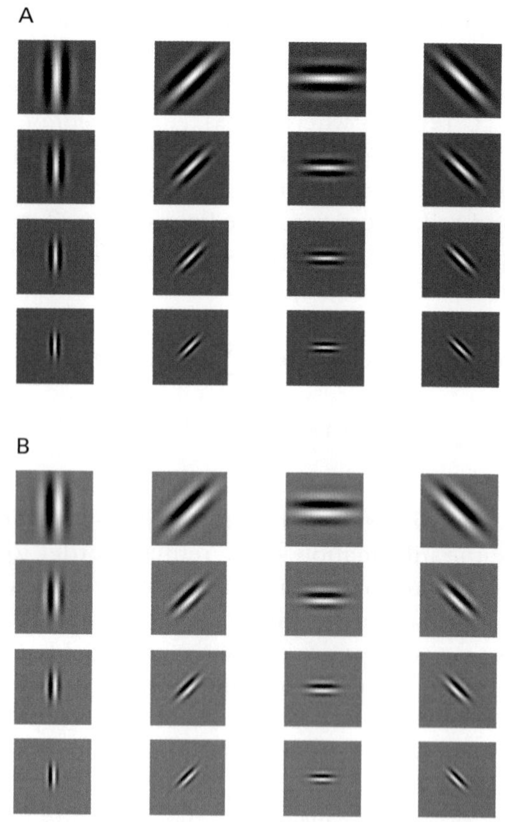

FIGURE 12.15 A family of Gabor receptive fields. The panels show cosine-Gabors (A) and sine-Gabors (B) at different orientations (rows) and different scales (columns). Observe that the cosine-Gabors have biggest responses at their centers (because cos0 = 1) while the sine-Gabors have small responses there (because sin0 = 0).

$$G(\boldsymbol{x};0,\Sigma) = \frac{1}{2\pi|\Sigma|^{1/2}} \exp\{-(1/2)\boldsymbol{x}^T \Sigma^{-1} \boldsymbol{x}\}$$

(Σ is the covariance of the Gaussian) by a sinusoid $\exp\{i\boldsymbol{w}\cdot\boldsymbol{x}\} = \cos \boldsymbol{w}\cdot\boldsymbol{x} + i \sin \boldsymbol{w}\cdot\boldsymbol{x}$. This gives two basic types of Gabors: (1) cosine-Gabors $G_{cos}(\boldsymbol{x}) = G(\boldsymbol{x}; \boldsymbol{0}, \Sigma) \cos \boldsymbol{w}\cdot\boldsymbol{x}$ and (2) sine-Gabors $G_{sin}(\boldsymbol{x}) = G(\boldsymbol{x}; \boldsymbol{0}, \Sigma) \sin \boldsymbol{w}\cdot\boldsymbol{x}$. These form a *quadrature pair*, because sin(.) and cos(.) are 90 degrees out of phase, which will be discussed and used later in this chapter (also phases can vary). The intuition for Gabor filters is they give a good trade-off between *localization* in position and in frequency. The Gaussian has good localization in position, in the sense that its response is very small if $|\boldsymbol{x}| > 2\sigma$. The sinusoid has perfect localization in frequency (due to the orthogonality of sinusoids) but is unable to localize in position (because a sinusoid does not tend to zero for large $\boldsymbol{x}$). More rigorously, Gabor derived the Gabor function by optimizing a criterion that balanced optimality in frequency with optimality in position.

FIGURE 12.16 A Gabor function aligned to the vertical axis (A). The image of a zebra (B). The response of the vertical Gabor filter on the zebra image (C).

A systematic study in the late 1990s concluded that many simple cells in V1 could be modeled by a family of Gabor filters with specific relationships between the parameters of the Gaussian and the sinusoid, Σ and $\boldsymbol{w}$ (Lee, 1996). The experimental data at that time suggested that only a restricted class of Gabor filters were implemented. To understand this restriction, express the frequency of the sinusoid as $\boldsymbol{w} = w(\cos\theta, \sin\theta)$, where θ specifies the orientation of the sinusoid (and its propagating direction) and ω the magnitude of its frequency. Then the Σ of the Gaussian is proportional to $(1/4)\,(\cos\theta, \sin\theta)\,(\cos\theta, \sin\theta)^T + (-\sin\theta, \cos\theta)\,(-\sin\theta, \cos\theta)^T$ (T denotes vector transform). So the aspect ratio of the Gaussian (the ratio between its major and minor axes) is fixed at 4. The sinusoid $\exp(i\boldsymbol{x}\cdot\boldsymbol{w})$ has its "propagating direction" along the shorter axis of the Gaussian, so the Gaussian smooths more in the direction perpendicular to the propagating direction, by a factor of $1/2 = \sqrt{1/4}$. This family of Gabor filters is illustrated in figure 12.15.

This family is specified as follows:

$$\psi(\boldsymbol{x};\omega,\theta,K) = \frac{\omega^2}{4\pi K^2}\exp\{-(\omega^2/8K^2)\{4(\boldsymbol{x}\cdot(\cos\theta,\sin\theta))^2 + (x\cdot(-\sin\theta,\cos\theta))^2\} \times \exp\{i\omega x\cdot(\cos\theta,\sin\theta)\}\exp\{(K^2/2)\}\} \tag{12.2}$$

In this family the orientation is specified by θ, the frequency by ω, and the variance is proportional to K^2. This is normalized so that $\int d\boldsymbol{x}\{\psi(\boldsymbol{x}; w, \theta, K)\}^2 = 1$. $K \approx \pi$ for a frequency bandwidth of one octave and $K \approx 2.5$ for a frequency bandwidth of 1.5 octaves ("octaves" are the log ratio of the frequency—see Zhaoping [2014]). This family can also be scaled to give a form

$$\psi_a(\boldsymbol{x};\omega,\theta,K) = \frac{1}{a}\psi_a(\boldsymbol{x}/a;\omega,\theta,K).$$

Gabor functions have many interesting properties. They respond well to periodic structure—that is, certain types of texture—provided the period of the Gabor filter is similar to the period of the texture structure; see figure 12.16. In addition, oriented sinusoid Gabors act like derivative filters, with their biggest responses on edges with the same orientation, and hence may function as components of edge detectors (see the next section). More generally, *filter banks of Gabor filters* offer local representations of image properties which can be used for many tasks including texture modeling, motion estimation, and binocular stereo (see later this section).

We can also study the tuning of Gabor cells by stimulating them with a family of stimuli of form $A\cos(\boldsymbol{w}\cdot\boldsymbol{x} + \rho)$ and varying $\boldsymbol{w}$ and ρ (similar to what we did for center–surround cells in the previous section). We define $w_x = \boldsymbol{w}\cdot(\cos\theta, \sin\theta)$ and $w_y = \boldsymbol{w}\cdot(\sin\theta, \cos\theta)$ to be the projections of the input sinusoid in the favored direction of the cell (i.e., $\boldsymbol{w}$) and in the orthogonal direction (i.e., $\omega_y = 0$ if the input sinusoid aligns perfectly with the orientation of the cell). The responses of the cosine-Gabor G_{cos} and the sine-Gabor G_{sin} are given by

$$\frac{A}{2}\cos\rho\exp\{-2K^2\omega_y^2/\omega^2\}\times\{\exp\{-(K^2/2\omega^2)(\omega+\omega_x)^2\}$$
$$+\exp\{-(K^2/2\omega^2)(\omega-\omega_x)^2\}\}\exp\{K^2/2\}, \tag{12.3}$$

$$\frac{A}{2}\sin\rho\exp\{-2K^2\omega_y^2/\omega^2\}\times\{\exp\{-(K^2/2\omega^2)(\omega+\omega_x)^2\}$$
$$-\exp\{-(K^2/2\omega^2)(\omega-\omega_x)^2\}\}\exp\{K^2/2\}. \tag{12.4}$$

These equations show that the cosine-Gabor cell is tuned to $\rho = 0$ and the tuning falls off as $\cos\rho$. The cell also favors sinusoid stimuli which are aligned to it (i.e., $\omega_y = 0$), and whose frequency $\omega_x = \pm\omega$. By contrast, the sine-Gabor prefers stimuli with $\rho = \pi/2$ (naturally, since this converts the stimulus into a sine function which aligns to the sine-Gabor) and has similar tuning to the frequency with $\omega_y = 0$ and $\omega_x = \pm\omega$.

Complex Cells Complex cells are sensitive to orientation, but they are less sensitive than simple cells to the spatial position of the stimuli. A standard theory of the ventral stream (see chapter 17 by Lewis and Poggio) suggests that visual processing proceeds up this stream using receptive fields similar to simple and complex cells, which are increasingly tuned to more complex structures and are less sensitive to the precise positions of the stimuli. Complex cells are the second stage after simple cells, forming a simple–complex cell module which gets repeated up the hierarchy.

We describe here the *energy model* where the complex cell receives input from two simple cells which are 90 degrees out of phase (i.e., cosine-Gabors and sine-Gabors). This is partly motivated by quadrature cells (Jones and Palmer, 1987) and because (see the following paragraphs) these cells are less sensitive than simple cells to the specific position of the stimuli. (See figure 12.17.) Note the word "energy" is based on analogy to physical systems.

More precisely, the energy model of a complex cell gives this response:

$$S(\boldsymbol{x}) = \{\psi_{\sin} * I(\boldsymbol{x})\}^2 + \{\psi_{\cos} * I(\boldsymbol{x})\}^2, \tag{12.5}$$

where $*$ indicates convolution (see section 12.2.1).

We can study the tuning of complex cells by measuring their response to stimuli $A\cos(\boldsymbol{w}\cdot\boldsymbol{x}+\rho)$ and varying $\boldsymbol{w}$ and ρ (as before). The findings show that these cells are, like simple cells, also tuned to orientation,

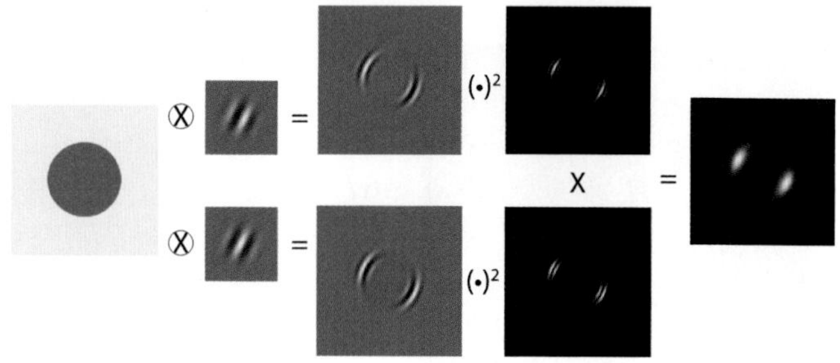

FIGURE 12.17 A complex cell can be modelled as a quadrature pair of Gabor filters. The stimulus is a gray circle on a much lighter background (far left). A quadrature pair of Gabor filters is applied to the stimulus giving the largest responses when the orientation of the Gabors matches the orientation of the edge of the circle. The responses of the Gabors are squared and then summed to yield the final output (far right).

frequency, and phase. However, their tuning, particularly to phase, is less precise. Hence complex cells are less sensitive to the precise position of the stimuli. The response is given by

$$\frac{A^2}{4}\exp\{K^2\}\exp\{-4K^2\omega_y^2/\omega^2\}\{\exp\{-(K^2/\omega^2)(\omega+\omega_x)^2\}$$
$$+\exp\{-(K^2/\omega^2)(\omega-\omega_x)^2\}$$
$$+2\cos2\rho\exp\{-(K^2/\omega^2)(\omega+\omega_x)^2\}$$
$$\exp\{-(K^2/\omega^2)(\omega-\omega_x)^2\}\}. \tag{12.6}$$

Observe that the dependence on the phase ρ is very small (the dominant term in the second line is independent of ρ).

Complex cells have several possible functions including stereo and motion, as we will show in later sections. They will also respond to oriented edges and can be used for edge detection. In addition, they can represent the local "energy" of textures.

This complex cell model has nice theoretical properties and has been extensively studied and used to construct models (see later in the lecture). However, there are other models where complex cells are built from simple cells in alternative ways (see chapter 17 by Lewis and Poggio) but where the complex cells retain their basic property of being tuned to orientation and frequency but being less sensitive to the position of the stimuli. Some researchers question whether complex cells receive input from single cells, arguing that the computations could be done by nonlinear neurons which exploit the complexity of the dendridic tree (Mel, Ruderman, and Archie, 1998). Other researchers

(Mechler and Ringach, 2002) argue that there is no sharp dichotomy between simple and complex cells but instead there is a continuum of cells with variable sensitivity to position. Observe that the change between simple and complex cells illustrates a general property of receptive fields in the ventral stream—namely, that receptive fields become more tuned to complex stimuli and less sensitive to the position of the stimulus (e.g., a cell tuned to faces is fairly insensitive to where the face lies in the visual field).

Linear Filtering, Basis Functions, and Fourier Analysis We now take a step backward and put these models into the context of the extensive mathematical literature on linear filtering and Fourier analysis. This is an advanced section which gives greater understanding but is not required for a basic introduction. Note that in this chapter we will often switch between treating $\boldsymbol{x}$ as a discrete variable and using summation, Σ_x, or treating it as a continuous variable and using integration, $\int dx$. Similar notation is used by Zhaoping (2014).

As discussed in the previous section, simple cell models apply *linear filters* to images and cells at different spatial locations and perform *convolution* $*$ by applying the same filter $\boldsymbol{w}$ across the image:

$$S(\boldsymbol{x}) = \boldsymbol{w} * \mathbf{I}(\boldsymbol{x}) = \sum_y w(\boldsymbol{x}-\boldsymbol{y})I(\boldsymbol{y}).$$

It is also convenient to approximate this (take the continuum limit) and express this as an integral:

$$S(\boldsymbol{x}) = \int_y w(\boldsymbol{x}-\boldsymbol{y})I(\boldsymbol{y})d\boldsymbol{y}.$$

This continuum limit is a good approximation if the summation Σ_y is over a dense set of positions $\boldsymbol{y}$, and enables certain types of analysis (e.g., showing that a center–surround cell model sums, approximately, to zero). Throughout this chapter we will often switch from continuous to discrete filters whenever it simplifies the analysis.

Convolving an image by a linear filter produces an output image $S(\boldsymbol{x})$ whose form depends on the type of filter $\boldsymbol{w}$. For example, if $w(\boldsymbol{x})$ is a Gaussian function

$$G(\boldsymbol{x};\sigma) = \frac{1}{2\pi\sigma}\exp\{-(x_1^2+x_2^2)/(2\sigma^2)\},$$

then convolution effectively just smooths the image by taking a linear weighted average. If $\boldsymbol{w}$ is a derivative of the Gaussian in the x_1 direction,

$$w(\boldsymbol{x}) = \frac{d}{dx_1}G(\boldsymbol{x};\sigma),$$

then this filter gives a large response to *edges*, positions $\boldsymbol{y}$ where the intensity $I(\boldsymbol{y})$ changes abruptly, and has small responses in places where the image intensity changes slowly.

We can better understand images, and linear filtering, by using *functional analysis*. This states that an image, or any signal, can be expressed uniquely as a weighted sum of *basis functions*:

$$I(\boldsymbol{x}) = \sum_i \alpha_i b_i(\boldsymbol{x}), \tag{12.7}$$

where the $b_i(\boldsymbol{x})$ are *basis functions* and the $\{\alpha_i\}$ are *coefficients*. These basis functions are usually chosen to be *orthonormal*, so that $\Sigma_{\mathbf{x}} b_i(\boldsymbol{x})\, b_j(\boldsymbol{x}) = \delta ij$ (= 1 if $i = j$ and = 0 if $i \neq j$). If the basis functions are orthogonal, then the coefficients α can be obtained by:

$$\alpha_i = \sum_x I(\boldsymbol{x})b_i(\boldsymbol{x}). \tag{12.8}$$

The *principle of superposition* states that we can determine the output S as a weighted combination of the outputs of the basis functions:

$$S(\boldsymbol{x}) = \sum_i \alpha_i S_i(\boldsymbol{x}), \text{ where } S_i(\boldsymbol{x}) = \sum_y w(\boldsymbol{x}-\boldsymbol{y})b_i(\boldsymbol{y}). \tag{12.9}$$

This implies that if we know the response $S_i(.)$ to each basis function $b_i(.)$, then we can predict the response to any input. This is an attractive property which, if it holds, enables us to measure the receptive field of a linear neuron, or a thresholded linear neuron, from a limited set of stimuli.

Fourier analysis deals with a special class of basis functions. These are sinusoids, that is, of form $\sin wx$, $\cos wx$. Then the α's are the *Fourier transform* of the image. If we restrict ourselves to an image defined on a lattice (i.e., so that x_1, x_2 each take a finite number of values, as on a digital camera), then this is the *discrete Fourier transform*. But if we allow x_1, x_2 to take continuous values, then we get the *Fourier transform*:

$$I(\boldsymbol{x}) = \frac{1}{2\pi}\int \hat{I}(\omega)\exp\{-i\omega\cdot\boldsymbol{x}\}d\omega, \tag{12.10}$$

$$\hat{I}(\omega) = \frac{1}{2\pi}\int I(\boldsymbol{x})\exp\{i\omega\cdot\boldsymbol{x}\}d\boldsymbol{x}. \tag{12.11}$$

Here $\exp\{i\boldsymbol{w}\cdot\boldsymbol{x}\} = \cos(\boldsymbol{w}\cdot\boldsymbol{x}) + i\sin(\boldsymbol{w}\cdot\boldsymbol{x})$. Note that if $I(.)$ is symmetric, $I(\boldsymbol{x}) = I(-\boldsymbol{x})$, then $\hat{I}(\boldsymbol{w})$ is also symmetric, $\hat{I}(-\boldsymbol{w}) = \hat{I}(\boldsymbol{w})$. Observe that equations 12.10 and 12.11 correspond to equations 12.7 and 12.8 for special choices of the basis functions (and changing from discrete to continuous $\boldsymbol{x}$).

Fourier analysis is particularly important because it gives us a way to represent nonlocal structure of images in terms of *frequencies* ω. The high frequencies (large $|\boldsymbol{w}|$) represent image patterns which change rapidly while the lower frequencies (small $|\boldsymbol{w}|$) represent slowly changing patterns. In particular, if an image pattern is *periodic*, like the stripes on a zebra, then it can be expressed in this form:

$$I(\boldsymbol{x}) = \sum_n A_n \cos(2\pi n\omega_0 \cdot \boldsymbol{x}),$$

where $\boldsymbol{w}_0$ is the basic frequency and n denote integers. Then the Fourier transform is only nonzero at integer multiples of the basic frequency $\boldsymbol{w} = \boldsymbol{w}_0$. Hence periodic image patterns, such as *textures*, have very simple descriptions in Fourier space.

If we blur the image, by convolving with a Gaussian $G(\boldsymbol{x}; \sigma)$, to obtain $G * I(\boldsymbol{x})$, then the high frequencies of the image $\boldsymbol{I}$ will be smoothed out. This follows from the *convolution theorem* which states that Fourier transform of $G * I(\boldsymbol{x})$ equals the product of the Fourier transforms of G and $\boldsymbol{I}$, and the fact that the Fourier transform of a Gaussian is also a Gaussian $\exp\{-|\boldsymbol{w}|^2 (\sigma^2/2)\}$. Hence we can express the convolved image as a weighted combination of sinusoids where the high-frequency weights are decreased by $\exp\{-|\boldsymbol{w}|^2 (\sigma^2/2)\}$:

$$\boldsymbol{I}(\boldsymbol{x}) = \frac{1}{2\pi}\int \hat{I}(\omega)\exp\{-i\omega\cdot\boldsymbol{x}\}\exp\{-|\omega|^2(\sigma^2/2)\}d\omega.$$

Hence, if we increase the amount of blurring by increasing the variance σ^2 of the Gaussian, we will make the coefficients of the high frequencies increasingly small. Blurring the image can be obtained by defocusing your eyes so that the image is seen out of focus. The receptive fields of cells occurs at a range of different scales, corresponding to convolving with Gaussians of different variances.

The superposition principle, combined with the use of basis functions, shows that we can determine the receptive fields of linear neurons by stimulating them with sinusoids. Sinusoids can be used as basis functions, and superposition can be used to predict the response to stimuli which have not been seen yet (i.e., as superpositions of those stimuli to which the response is known). This, however, is rarely done. We will return to studying ways to determine receptive fields from experiments in section 12.2.3.

12.2.2 SPARSITY, MATCHED FILTERS, AND NATURAL IMAGES This section considers receptive field models from different perspectives. This includes the use of *sparsity* to suggest receptive field properties based on the statistics of natural images and also the idea of *matched filters* which revert to an older idea of receptive fields as feature detectors (Lettvin et al., 1959). Sparsity was proposed by Barlow (1961) as a general principle for modeling the brain based on the observation that typically only a small number of neurons are active. It was developed as a way to predict receptive field properties by Olshausen and Field (1996). It is natural to ask whether the receptive fields of cells encode basis functions which somehow capture the typical structure of images and represent it in a form which is suitable for later processing.

Our starting point is the idea that images, and particularly local regions of images, can be represented as a linear combination of basis functions $I(\boldsymbol{x}) = \Sigma_i\alpha_i b_i(\boldsymbol{x})$; see equation 12.7.

Overcomplete Bases and Sparsity This section introduces the idea of *overcomplete* basis functions and *sparsity*. To motivate this idea, consider an image which consists partly of regions where the intensity varies spatially smoothly and others where the intensity is more jagged and consists of a number of bright spots, or *impulses*. The smoothly varying regions of the image can be represented by Fourier analysis efficiently, in the sense that we can approximate the intensity by only a small number of weighted sinusoids (in other words, the Fourier transform of the image is peaked at a limited number of frequencies). By contrast, the impulses are not well described by Fourier analysis because the Fourier transform is not zero for all frequencies (the Fourier transform of an impulse at $\boldsymbol{x}_0$ is $\exp\{i\boldsymbol{w}\cdot\boldsymbol{x}_0\}$, so the amplitude spectrum is constant at all frequencies). Instead, it would be better to represent the spikes in terms of a basis of impulse functions, but this representation would be very inefficient for the smoothly varying parts of the image. In short, different types of basis functions are suitable for different regions of the image. This suggests a strategy where we seek a representation in terms of an overcomplete set of basis functions, in this case sinusoids and impulse functions, and a criterion which selects an efficient representation so that only a small number of basis functions are activated for each image. This requirement is called *sparsity*.

More formally, we represent an image, or local image region, by

$$I(\boldsymbol{x}) = \sum_{i=1}^{N}\alpha_i b_i(\boldsymbol{x}),$$

where the $\{b_i\}$ are the basis functions (which are the same for all images and could include sinusoids and impulse functions) and the $\{\alpha_i\}$ are the coefficients of the bases (which depend on the image). The number N of bases is much bigger than the dimension of the image, and hence the bases are overcomplete. This differs from Fourier analysis where the data (e.g., an image) are expressed in terms of a set of basis functions which are mutually orthogonal, which enables the coefficients α for each image to be estimated by $\alpha_i = \sum_{\boldsymbol{x}} b_i(\boldsymbol{x}) \cdot \mathrm{I}(\boldsymbol{x})$. Overcompleteness implies that there are many ways to represent the image in terms of these basis functions (by different choices of the α's), and we need an additional criterion to select the α's. The *sparsity* criterion proposes that we favor representations which make $\sum_{i=1}^{N} |\alpha_i|$ small, which penalize the weights of the basis functions and encourages most coefficients to be 0.

More precisely, we represent an image $\boldsymbol{I}$ by the approximation $\sum_{i=1}^{N} \hat{\alpha}_i \boldsymbol{b}_i$, where the $\{\hat{\alpha}_i\}$ are chosen to minimize the function

$$E(\alpha) = \sum_{\boldsymbol{x}} \left(I(\boldsymbol{x}) - \sum_{i=1}^{N} \alpha_i b_i(\boldsymbol{x}) \right)^2 + \lambda \sum_{i=1}^{N} |\alpha_i|. \quad (12.12)$$

The first term penalizes the error of the approximation, and the second term, whose strength is weighted by a parameter λ, penalizes the coefficients $\{\alpha_i\}$. The solution $\hat{\alpha} = \arg\min_{\alpha} E(\alpha)$ cannot be specified in closed form (unlike the case for orthogonal basis functions), but $E(\alpha)$ is a *convex* function of α, and efficient algorithms exist for minimizing it to estimate $\hat{\alpha}$. The results of these algorithms can, for example, decompose an image into a sum of sinusoids and a sum of impulse functions.

These ideas give an alternative way to think about the receptive fields of cells in V1. Firstly, observe that V1 has far more cells than the retina or the LGN, and so it has enough neural machinery to implement overcomplete bases. Secondly, overcomplete bases can be designed for specific image structures of interest (e.g., impulse functions or edges), which enables us to start interpreting the image instead of simply representing it. Thirdly, it relates to the observation that cells in V1 fire *sparsely*, which suggests (Barlow, 1961) that they are tuned to specific stimuli and may relate to metabolic processes (firing a neuron takes energy which needs to be replenished). Hence the idea that the visual cortex seeks to obtain sparse, and hence presumably more easily interpretable, representations has intuitive appeal.

How does this discussion of overcompleteness and sparsity relate to our previous description of V1 cells in terms of Gabor filters? Gabor filters have some of the properties that this approach requires. Families of Gabor filters are built by taking a basic function and performing transformations on it which give an overcomplete basis. Hence they do not specify a unique representation of an image (i.e., any image can be represented many different ways in terms of Gabor functions). These issues, and the relations of Gabors to wavelets, are discussed in more detail in Lee (1996).

Sparsity and Natural Images Sparsity can also be used to derive the properties of receptive fields of cells in V1 if we assume that these cells are designed to be able to represent properties of natural images (Olshausen, 1996); see figure 12.18A. Hence instead of hypothesizing models of receptive fields (e.g., Gabor filters) we can try to predict these receptive fields from studying images. These predictions do give some justification for Gabor functions, but they also suggest other receptive field models which have also been experimentally observed.

This requires learning the basis functions $\{\boldsymbol{b}_i\}$ from a set of natural images $\{\boldsymbol{I}^{\mu}: \mu \in \Lambda\}$. This can be found by extending equation 12.12 to obtain a criterion $E(b,\alpha)$ for fitting basis functions b and coefficients α to the set of images:

$$E(b,\alpha) = \sum_{\mu \in \Lambda} \left(I^{\mu}(\boldsymbol{x}) - \sum_{i=1}^{N} \alpha_i^{\mu} b_i(\boldsymbol{x}) \right)^2 + \lambda \sum_{\mu \in \Lambda} \sum_{i=1}^{N} |\alpha_i|.$$

We estimate the basis functions $\hat{b}$ and the coefficients $\hat{\alpha}$ by minimizing $E(b,\alpha)$ to obtain

$$(\hat{b},\hat{\alpha}) = \arg\min_{(b,\alpha)} E(b,\alpha).$$

Note that the basis functions are the same for all images, but the coefficients vary for each image (hence they are indexed by the image μ as well as the basis coefficient i). This minimization is nonconvex, but there are efficient algorithms to perform it.

This criterion has been applied to natural images (where the $\boldsymbol{I}$ represent small image regions), and the resulting basis functions (see figure 12.18A) include filters which look like Gabor functions, but they also include other types of filters which are also observed in experiments (Olshausen, 1996).

We note that there are other methods for predicting receptive field properties from natural images using a similar image model, $I(\boldsymbol{x}) = \sum_{i=1}^{N} \alpha_i b_i(\boldsymbol{x})$, but imposing different assumptions on the form of the bases. In particular, independent component analysis (ICA) gives similar receptive field models (van Hateren and Ruderman, 1998). Hyvarinen (2010) explains this by showing that both types of models—L1 sparsity and ICA—both

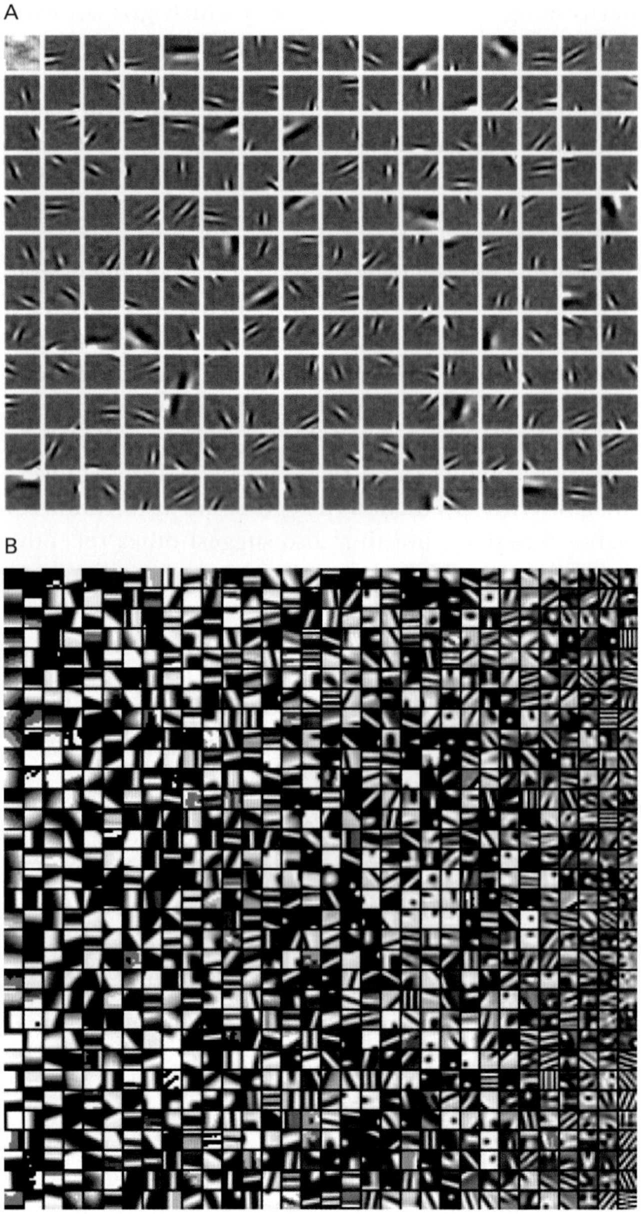

FIGURE 12.18 (A) The receptive fields learned using sparsity (Olshausen, 1996). Reprinted by permission from Macmillan Publishers Ltd: B. A. Olshausen and D. J. Field. Emergence of Simple-Cell Receptive Field Properties by Learning a Sparse Code for Natural Images. *Nature, 381*(6583), 607–609, Copyright 1996. (B) Receptive fields learned by matched filters, for example, see Papandreou, Chen, and Yuille (2014).

encourage the α_i to be strongly peaked at 0 but can occasionally have large nonzero values.

What happens if we remove the sparsity requirement and instead find the basis functions that minimize $\sum_{\mu\in\Lambda}\left(I^\mu(\boldsymbol{x})-\sum_{i=1}^{N}\alpha_i^\mu b_i(\boldsymbol{x})\right)^2$? The basis functions will be the eigenvectors of the correlation matrix of the images and can be found by PCA. Code for performing PCA is supplied in interactive demo 2c. It can be shown that the principal components of images will typically be sinusoids (provided the images are sufficiently representative of natural images). We return to this issue in section 12.2.3 when we describe unsupervised ways to learn receptive fields of neurons.

Matched Filter Interpretation This section gives an alternative, and earlier, model for receptive fields. The idea is the cells are feature detectors (Lettvin et al., 1959) which can be modeled by a set of matched filters, so that the cells indicate the type of image patch which is present. This can be thought of as an extreme form of sparsity because any image patch can be represented by a single matched filter (instead of a linear combination of them). Examples of matched filters are shown in figure 12.18B. We now describe the details of this approach.

Suppose we have a filter $\boldsymbol{W}$ and an input image patch $\boldsymbol{I}_p$. We want to find the best fit of the filter to the image by allowing us to transform the filter by $\boldsymbol{W}\to\alpha\boldsymbol{W}+b\boldsymbol{e}$, where $\boldsymbol{e}=\left(1/\sqrt{N}\right)(1,\ldots,1)$. This corresponds to scaling the filter by a and adding a constant vector b. If $\boldsymbol{W}$ is a derivative filter, then, by definition, $\boldsymbol{W}\cdot\boldsymbol{e}=0$. We normalize $\boldsymbol{W}$ and $\boldsymbol{e}$ so that $\boldsymbol{W}\cdot\boldsymbol{W}=\boldsymbol{e}\cdot\boldsymbol{e}=1$.

The goal is to find the best scaling/contrast a and background b to minimize the match:

$$E(a,b)=\left|\boldsymbol{I}_p-a\boldsymbol{W}-b\boldsymbol{e}\right|^2.$$

The solution $\hat{a},\hat{b}$ are given by (take derivatives of E with respect to a and b, recalling that $\boldsymbol{W}$ and $\boldsymbol{e}$ are normalized):

$$\hat{a}=\vec{W}\cdot I_p,\;\; \hat{b}=\boldsymbol{e}\cdot\boldsymbol{I}_p.$$

In this interpretation, the filter response is just the best estimate of the contrast a. The estimate of the background b is just the mean value of the image. Finally, the energy $E\left(\hat{a},\hat{b}\right)$ is a measure of how well the filter "matches" the input image. Receptive fields learned by matched filters are shown in figure 12.18B.

The idea of a matched filter leads naturally to the idea of having a "dictionary" of filters $\{\boldsymbol{W}^\mu:\mu\in\Lambda\}$, where different filters $\boldsymbol{W}^\mu$ are tuned to different types of image patches. In other words, the input image patch is encoded by the filter that best matches it. The dictionary of matched filters could be implemented by a set of cells (e.g., orientation columns). In this interpretation, the magnitude of the dot product $\boldsymbol{W}\cdot\boldsymbol{I}$ is less important than deciding which filter best matches the input $\boldsymbol{I}_p$. Matched filters can be thought of as an extreme case of sparsity. In the previous sections, an image was represented by a linear combination of basis functions whose weights were penalized by the *L1-norm*, $\Sigma_i|\alpha_i|$. By comparison, matched filters represent an image by a single basis function. This gives an ever sparser representation of the image, but at the possible cost of a much larger image dictionary. Matched filters can be thought of as *feature detectors* because they respond only to very specific inputs.

12.2.3 LEARNING We cover two topics in this section: (1) How can these receptive fields be learned using biologically plausible mechanisms? (2) How do we estimate receptive fields from experimental data?

Unsupervised Learning of the Receptive Fields Where do the receptive fields of simple cells come from? The most plausible explanation is that they are learned in an unsupervised self-organizing manner, but it is debatable whether this happens before birth, as a result of spontaneous activity in the brain, or in response to natural stimuli observed after birth. The methods in this section could work either way since the main property they rely on is that images are shift invariant. This section is based on computational studies performed in the 1980s (Linsker, 1986a,b; Yuille, Kammen, and Cohen, 1989); see Zhaoping (2014) for other references. These studies are based on modifications of the Hebb learning rule which has some experimental support. Interactive demo 2c illustrates PCA and Oja's rule (Oja, 1982).

The basic findings are that center–surround, orientation-selective, quadrature pairs, and disparity-sensitive cells (precursors to cells which can estimate depth from binocular stereo) could all be obtained by variants of the same learning rule.

We first describe a simple unsupervised learning model of this type for a single cells (Oja, 1982). The output $S(t)$ of the cell is a function of time t and is a weighted sum of the inputs $I_i(t)$, where the weights $\omega_i(t)$ are functions of time which are updated by Oja's rule (Oja, 1982):

$$S(t) = \sum_j w_j(t) I_j(t),$$
$$\frac{dw_i(t)}{dt} = S(t)\{I_i(t) - S(t)w_i(t)\}. \tag{12.13}$$

The first term of this update is the classic Hebb's term which increases the strength of a weight w_i if its input $I_i(t)$ is positively correlated with the output $S(t)$ (i.e., $< S(t) I_i(t) >> 0$), while the second term decreases the value of all weights by an amount proportional to their strength.

This can be expressed as a single update equation (substituting for $S(t)$):

$$\frac{dw_i(t)}{dt} = \sum_j w_j I_i(t) I_j(t) - \sum_{jk} w_i w_j w_k I_j(t) I_k(t). \tag{12.14}$$

Next we make the key assumption that the weights w_i change at a slower rate than the input images. This enables us to replace the terms $I_i(t) I_j(t)$ in equation 12.14 by their expectation $K_{ij} = < I_i(t) I_j(t) >$, which is the correlation function of the input. This gives an update rule:

$$\frac{dw_i(t)}{dt} = \sum_j w_j K_{ij} - \sum_{jk} w_i w_j w_k K_{jk}. \tag{12.15}$$

The fixed points of this equation, the values of w such that $(dw_i(t)) / dt = 0$, can be shown to be eigenvectors of the correlation function K_{ij}. A slight modification of this update rule (Yuille, Kammen, and Cohen, 1989) gives a rule that is guaranteed to converge to the global minimum of the cost function:

$$E(\boldsymbol{w}) = -(1/2)\sum_{i,j} K_{ij} w_i w_j + (k/4)\left(\sum_i w_i^2\right)^2.$$

It can be shown that this global minimum corresponds to the biggest eigenvalue of K_{ij}. It follows that if the correlation function K_{ij} decreases spatially, then the biggest eigenvalue is at frequency 0, so the cell is not tuned to any frequency. However, if the correlation function has the shape of a Mexican hat, then the biggest eigenvalue has a nonzero frequency, which implies that the cell is orientated. In fact, there are a set of eigenvalues with the same frequency modulus, each pointing in a different direction. Small changes to the input, or noise in the simulation, will determine which orientation is selected, which is like symmetry breaking in physics (Yuille, Kammen, and Cohen, 1989). The correlation function of natural images does decrease spatially, but Linsker showed that correlation functions similar to the Mexican hat arise if this learning procedure is applied to a sequence of layers (Linsker, 1986a,b).

This analysis yields receptive fields which are sinusoids and hence have no spatial fall-off, which is unrealistic. But receptive fields of neurons are limited by the geometrical positions of the dendrites. If these constraints are included, then the algorithms converge to receptive fields which are similar to Gabor functions. These methods can be expanded to other other properties of cells in the visual system. For example, quadrature pairs can be learned by introducing inhibition between neighboring output neurons (Yuille, Kammen, and Cohen, 1989).

How to Empirically Estimate Receptive Field Models by Regression This section describes how to estimate the receptive field properties of cells from electrical recordings of neurons by estimating the best model using *regression*. This is a standard method in statistics and machine learning. It requires making very few assumptions about the form of the receptive field (i.e., no need to assume Laplacian-of-Gaussian, or Gabor, or sum of squared Gabors).

We first recall, as discussed earlier, that the receptive field properties of neurons are traditionally found in simpler ways—for example, by choosing a set of basic stimuli, such as oriented bars, and then observing how the response of the cell changes as we move or rotate the bars. This approach can be extended by probing the receptive field response to different *perceptual dimensions*, such as position, orientation, color, contrast, motion, and whether it is sensitive to stimuli from one eye or both. This gives a classification of the type of the receptive field but does not specify its receptive field weights $\boldsymbol{w}$ unless strong assumptions are made (e.g., that the receptive field is a Gabor function).

The regression method described in this section makes fewer assumptions about the forms of the receptive field, but it does require more data. More formally, regression assumes that we have a stimulus data set of $S = \{(S^{\mu}, \boldsymbol{I}^{\mu}): \mu = 1, \ldots, N\}$ of inputs $\boldsymbol{I}^{\mu}$ and outputs S^{μ} (e.g., the firing rates). We assume a model $S = g(\boldsymbol{I}\text{: } \boldsymbol{w})$ where $\boldsymbol{w}$ specifies the parameters of the model and $g(.)$ is a nonlinear function. For example, we could choose the nonlinear function to be a sigmoid $g(\boldsymbol{I}\text{: } \boldsymbol{w}) = \sigma(\boldsymbol{w} \cdot \boldsymbol{I})$, where $\sigma(.)$ is a sigmoid function. Alternatively $g(.)$ could specify a Gabor filter where $\boldsymbol{w}$ specifies the parameters of the Gabor (e.g., frequency and scale).

Regression requires minimizing a cost function such as

$$F(\boldsymbol{w}) = \frac{1}{|S|}\sum_{\mu \in S}^{N} E(S^{\mu} - g(I^{\mu};\boldsymbol{w}))$$

where $E(.)$ is a penalty function such as $(S^{\mu} - g(I^{\mu}; T))^2$.

This minimization can be done by standard computer packages. It outputs an estimate of the model parameters $\hat{\boldsymbol{w}}$ and an error measure

$$F(\omega) = \frac{1}{|S|}\sum\nolimits_{\mu \in S} E(S^{\mu} - g(I^{\mu};\omega)).$$

Hence it not only outputs an estimate of the receptive field but it also produces an error measure, which can yield the variance of the output.

In practice, there are several complications. It is unrealistic to show the neuron all possible stimuli because there are so many possible image stimuli. Hence researchers have to choose a restricted set of stimuli. If neurons are linear, or a nonlinear function of a linear filter, then this should not matter because we can exploit the superposition principle and estimate the receptive field from a limited number of stimuli. But in reality, linearity is only an approximation, and in practice the choice of stimuli can matter considerably. One concern is that the stimulus set does not contain the types of stimuli that the neuron is most sensitive to, in which case regression will output unreliable estimates. Also, if the linear assumption is only partially correct, then there is no guarantee that the receptive field learned on one set of stimuli will predict the behavior well on another set of stimuli.

The complexities are illustrated by recent work by Talebi and Baker (2012). They describe in detail how receptive fields of neurons are estimated by regression techniques and also address the issue of how the results of regression differ depending on the stimulus set. Their findings (for neurons in the cortex of anesthetized cats) show that estimates of the receptive fields of neurons can depend heavily on the set of stimuli. They estimate receptive fields using three different stimulus sets: (1) white noise (WN), (2) oriented bars (B), and (3) natural images (NIs). This gives three estimates for the receptive fields $\boldsymbol{w}_{WN}$, $\boldsymbol{w}_B$, $\boldsymbol{w}_{NI}$ obtained using the three stimulus sets S_{WN}, S_B, S_{NI}. For each data set, they can compute the prediction errors F_{WN}, F_B, F_{NI}, which are the errors for that data set, for example, $F_{WN}(\omega_{\hat{W}N}) = (1/|S_{WN}|)\sum_{\mu \in S_{WN}} E(S^{\mu} - g(I^{\mu};\omega_{\hat{W}N}))$. These quantities show how well the models can fit each stimulus set. However, more interestingly, they can study how well the estimated receptive field from one stimulus set can predict the other data sets. This involves computing quantities such as FWN $F_{WN}(\hat{\omega}_B)$, $F_{WN}(\hat{\omega}_{NI})$, $F_B(\omega_{\hat{W}N})$, $F_{WN}(\omega_{\hat{N}I})$, $F_{NI}(\omega_{\hat{W}N})$, $F_{WN}(\hat{\omega}_B)$. Their results show, perhaps not surprisingly, that the receptive fields estimated on the natural image stimulus set were much better at predicting the responses on the other two stimulus sets.

12.2.4 Local Models for Binocular Stereo and Motion This section shows how these linear filter models of receptive fields can be used to perform local estimates of binocular stereo and motion. These involve having filter banks, or populations of filters, which are tuned to different properties of the stimuli so that estimates of depth and motion can be extracted from the population (Zhaoping, 2014).

Stereo Disparity Models Recall that we introduced binocular stereo in section 12.1.2. Depth is estimated by triangulation, provided we can solve the *correspondence problem* by finding which points in the left and right eyes correspond to the same point in three-dimensional space. This reduces to estimating the displacement, or *disparity*, between the images in the left and right eyes. This section introduces the disparity energy model, which estimates disparity based on local properties of the image. In section 12.4 we will discuss how nonlocal context can be used to improve disparity estimation.

The disparity energy model is formulated using Gabor filters and has some claim to biological plausibility. We follow the presentation in Qian (1994), which is based on experimental findings on the receptive field property of cells by Ohzawa et al. (1990). The model assumes that we have a large set of cells, receiving input from both images, which are tuned to different image frequencies and spatial phases. The disparity of the image can be computed from the response of these filters.

We give the presentation in one dimension for simplicity. This is allowed because of the epipolar line constraint; see figure 12.9. It assumes that the cell receives input from both left and right eyes with receptive fields $f_l\,(x) = \exp\{-x^2/(2\sigma^2)\}\cos(wx + p_l)$ and $f_r\,(x) = \exp\{-x^2/(2\sigma^2)\}\cos(wx + p_r)$. In other words, they are Gabors where the Gaussian has variance σ^2, tuned to frequency ω and with phases ρ_l, ρ_r. The linear response is

$$r = \int dx\{f_l(x)I_l(x) + f_r(x)I_r(x)\}. \qquad (12.16)$$

This filter is tuned to spatial frequency ω. Recall, from earlier sections, that we can express the image by a Fourier expansion. The filter is most sensitive to the image component at this frequency. Hence we can represent the image (approximately) by $\boldsymbol{I}(\boldsymbol{x}) = \rho\cos(wx + \theta)$.

Now suppose that the right image is a displaced version of the left image $I_r(x) = I_l(x + D(x))$, where $D(x)$ is the disparity. We assume that the disparity varies slowly so that we can approximate it locally as a constant D (over the size of the Gaussian, 2σ); see figure 12.19A. Then we make another approximation by ignoring the Gaussian to calculate r (this is similar to ignoring the spatial fall-off when we studied unsupervised learning algorithms).

This yields a response:

$$r_1 = \rho\{\cos(\theta - \rho_l) + \cos(\theta - \rho_r - \omega D)\}. \qquad (12.17)$$

which can be reexpressed (using trigonometric identities) by

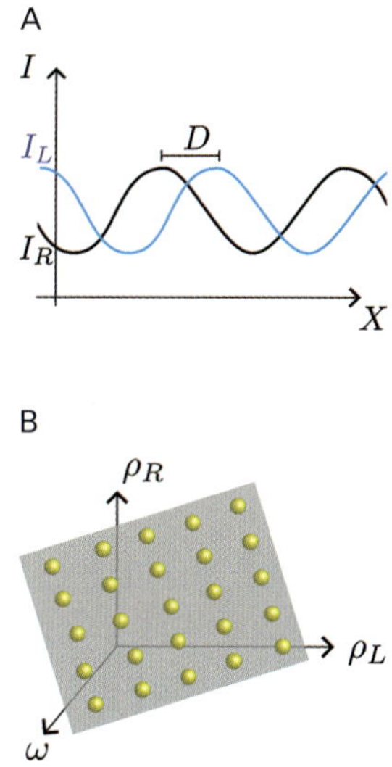

FIGURE 12.19 (A) The disparity D between the images in the two eyes corresponds to a change of phase if we approximate the intensities by sinusoids; see text. (B) The local disparity D is encoded by the feature response of cells tuned to frequencies which obey $\rho_l - \rho_r = \omega D$.

$$r_1 = 2\rho\cos\left(\theta - \frac{\rho_l + \rho_r}{2} - \frac{\omega D}{2}\right)\cos\left(\frac{\rho_l - \rho_r}{2} - \omega\frac{D}{2}\right). \qquad (12.18)$$

Hence the response of the cell depends on the disparity. In particular, when $\rho_l - \rho_r = \omega D$, then the second cosine takes its biggest value of 1. But the cell's response also depends on image properties, that is, the image phase θ which is an argument of the first cosine. This means it is unable to detect disparity by itself. However, disparity can be estimated from a population of quadrature cells of this type tuned to different frequencies.

To see this, suppose that we consider quadrature pairs of the two cells tuned to the same ω. Where one cell has phases ρ_l, ρ_r and the other has phases ρ_l', ρ_r', where $(\rho_l - \rho_r) = (\rho_l' - \rho_r')$ and $\rho_l' + \rho_r' = \rho_l + \rho_r + (\pi/2)$. Then the second cell has response

$$\begin{aligned} r_2 &= 2\rho\cos\left(\theta - \frac{\rho_l + \rho_r}{2} - \frac{\omega D}{2}\right)\cos\left(\frac{\rho_l - \rho_r}{2} - \omega\frac{D}{2}\right) \\ &= 2\rho\sin\left(\theta - \frac{\rho_l + \rho_r}{2} - \frac{\omega D}{2}\right)\cos\left(\frac{\rho_l - \rho_r}{2}\right). \end{aligned}$$

Hence if we square and add the responses of the two cells, we obtain

$$r_1^2 + r_2^2 = \cos^2\left(\frac{\rho_l - \rho_r}{2} - \omega\frac{D}{2}\right). \qquad (12.19)$$

This response depends only on the disparity D and the image frequency ω. It takes largest values when $\rho_l - \rho_r = \omega D$. Hence we can estimate D from a population of quadrature cells tuned to different phases ρ_l, ρ_r and frequencies ω; see figure 12.19.

A neural network for estimating D consists of two steps (there are many variants of this approach). In step 1 we define a set of disparity cells tuned to disparities $\{D_i: i = 1, \ldots, N\}$. The disparity cell tuned to disparity D_i receives input

$$\cos^2\left(\frac{\rho_l - \rho_r}{2} - \omega\frac{D_i}{2}\right)$$

from each quadrature pair (ρ_l, ρ_r, w) and sums these inputs together to compute a vote $v(D_i)$:

$$v(D_i) = \sum_{\rho_l, \rho_r, \omega} \cos^2\left(\frac{\rho_l - \rho_r}{2} - \omega\frac{D_i}{2}\right). \tag{12.20}$$

Step 2 uses a winner-take-all network to compute the disparity with the biggest vote by solving $\hat{D} = \operatorname{argmax}_{i=1,\ldots,N} v(D_i)$, so that $v(\hat{D}) \geq v(D_i)$ for $i = 1, \ldots, N$. There are many varieties of winner-take-all networks; see Maass (2000).

This theory of binocular stereo gives an example where the information of interest, in this case the disparity, is represented by the activity of a population of neurons. We will give another example for motion in the next section. There is plenty of evidence that the brain represents information by neural populations (Georgopoulos et al., 1983; McIlwain, 1991). There has also been much theoretical study of how populations of neurons could encode knowledge and perform computations (Pouget, Dayan, and Zemel, 2003; Ma et al., 2006).

Motion Measurement: SpatioTemporal Filters We now discuss how related models can be applied to estimate motion for sequences of images. Spatiotemporal filters are biologically plausible ways to measure motion which agree with properties of cells in the visual cortex. The standard model suggests two classes of cells where the first are spatiotemporal filters which are sensitive to the directions of motion while the second combine outputs of these filters to estimate the motion itself (Adelson and Bergen, 1985; Grzywacz and Yuille, 1990; Schrater, Knill, and Simoncelli, 2000).

Measuring the motion velocity assumes that locally the intensity can be modeled as a linear translating pattern; see figure 12.20A:

$$I(\mathbf{x}, t) = F(\mathbf{x} - \mathbf{v}t). \tag{12.21}$$

A

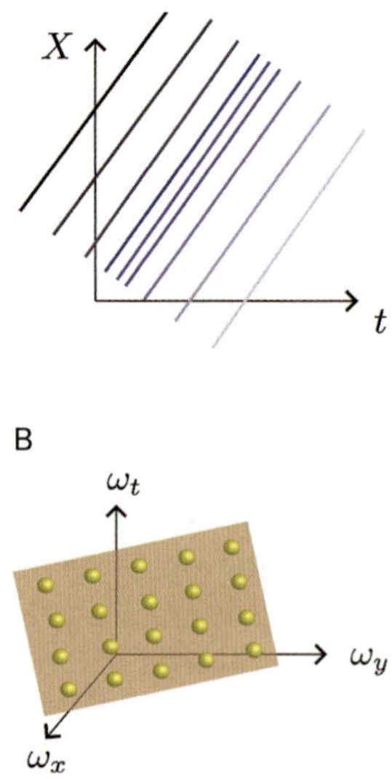

FIGURE 12.20 (A) This figure shows the space-time illustration of a signal traveling with constant velocity $I(X, t) = F(X - tv)$. This means that the intensity $I(X, t)$ is constant on the lines $X - tv = constant$. (B) A stimuli moving with velocity $\boldsymbol{v}$ will activate spatial–temporal filters $\boldsymbol{\omega}, \boldsymbol{\omega}t$ which lie on the plane $\boldsymbol{v} \cdot \boldsymbol{\omega} + \omega t = 0$. Hence the velocity can be estimated from the population of activity of the filters.

Differentiating with respect to $\mathbf{x}$ and t (using $\nabla I = \nabla F$ and $\partial I / \partial t = -\boldsymbol{v} \cdot \nabla F$), gives the *optical flow equation*:

$$\boldsymbol{v} \cdot \nabla I + \frac{\partial I}{\partial t} = 0. \tag{12.22}$$

This enables us to estimate one component of the motion $\boldsymbol{v}$ but suffers from the aperture problem discussed in section 12.1. One way to resolve this ambiguity is by applying a population of filters $\{G^{\mu}(\mathbf{x}, t): \mu = 1, \ldots, M\}$ indexed by μ (e.g., the $G^{\mu}(.)$ could be Gaussian filters). These filters introduce local context:

$$G^{\mu} * I(\mathbf{x}, t) = \int G^{\mu}(\mathbf{x} - \mathbf{y}, t - s) I(\mathbf{y}, s)\, ds\, d\mathbf{y}. \tag{12.23}$$

Hence each filter gives a constraint on the velocity,

$$\boldsymbol{v} \cdot \nabla G^{\mu} * I + \frac{\partial G^{\mu} * I}{\partial t} = 0. \tag{12.24}$$

We can then get an estimate of the velocity $\boldsymbol{v}$ by minimizing the cost function:

$$E(\boldsymbol{v}) = \sum_{\mu=1}^{M} \left(\boldsymbol{v} \cdot \nabla G^{\mu} * I + \frac{\partial G^{\mu} * I}{\partial t}\right)^2.$$

This minimization can be done using a similar neural network to that used for estimating disparity for stereo in the previous section. We have a set of cells tuned to different velocities $\{v_i: i = 1, \ldots, N\}$. The cell tuned to velocity v_i receives input $(v \cdot \nabla G^{\mu} * I + (\partial G^{\mu} * I / \partial t))^2$ from each filter μ and sums the responses to obtain $E(v_i)$. Then we use a variant of winner-take-all to compute $\hat{v} = \arg\min_{i=1,\ldots,N} E(v_i)$.

This approach assumes that there is enough local information to resolve the motion ambiguity, which may not be the case. For example, for the stimuli in figure 12.7 we can only locally estimate one component of the motion because of the aperture problem. To resolve this ambiguity we need to use more spatial or temporal context as described in section 12.4.4.

An alternative way to analyze this problem is by applying Fourier analysis to equation 12.21:

$$\hat{I}(\boldsymbol{\omega}, \omega_t) = \frac{1}{2\pi} \iiint \exp\{i(\boldsymbol{\omega} \cdot \boldsymbol{x} + \omega_t t)\} I(\boldsymbol{x}, t) d\boldsymbol{x} dt$$

$$\hat{I}(\boldsymbol{\omega}, \omega_t) = \frac{1}{2\pi} \iiint \exp\{i(\boldsymbol{v} \cdot \boldsymbol{x} + \omega_t)t\} \exp\{i\boldsymbol{\omega} \cdot (\boldsymbol{x} - \boldsymbol{v}t)\} F(\boldsymbol{x} - \boldsymbol{v}t) d\boldsymbol{x} dt$$

$$\hat{I}(\boldsymbol{\omega}, \omega_t) = \frac{1}{2\pi} \int \exp\{i(\boldsymbol{v} \cdot \boldsymbol{\omega} + \omega_t)t\} dt \iint \exp\{i\boldsymbol{\omega} \cdot \overline{\boldsymbol{x}}\} F(\overline{\boldsymbol{x}}) d\overline{\boldsymbol{x}}$$

$$\hat{I}(\boldsymbol{\omega}, \omega_t) = \delta(\boldsymbol{v} \cdot \boldsymbol{\omega} + \omega_t) \hat{F}(\boldsymbol{\omega})$$

where $\overline{\boldsymbol{x}} = \boldsymbol{x} - \boldsymbol{v}t$ is a change of variables in the integral.

This shows that if we have filters $\exp\{i(\boldsymbol{x}\boldsymbol{w} + w_t t\}$ tuned to spatial–temporal frequencies $\boldsymbol{w}$, wt, then the only filters which respond are those whose frequencies obey the equation $\boldsymbol{v} \cdot \boldsymbol{w} + wt = 0$ and hence lie on a plane in frequency space. Thus we can determine $\boldsymbol{v}$ from a population of filters by observing which filters are activated and finding the best fit plane; see figure 12.20B.

In practice, we cannot use filters tuned to frequency because these are not bounded in space and time. But instead we can impose bounds in space and time—for example, by converting the filters to spatiotemporal Gabors by multiplying sinusoids by Gaussians. The analysis for this case is more complicated because the Fourier transforms of the filters are no longer sharply localized in Fourier space. However, it can be shown (Grzywacz and Yuille, 1990) that if the filters are spatiotemporal Gabors, then the most active filters are those whose spatial–temporal tuning is centered on the plane $\boldsymbol{v} \cdot \boldsymbol{w} + wt = 0$. Hence the plane in frequency space can be estimated from a population of spatiotemporal filters and the velocity locally estimated.

This gives rise to a two-stage model of motion estimation where the first population of neurons are sensitive to the spatiotemporal frequency of the input image but not directly to the motion. The population, however, implicitly encodes the motion as described above. The second population of neurons extracts the motion information from the first population, and hence these neurons are tuned directly to motion. This is consistent with the experimental findings (Adelson and Bergen, 1985; Grzywacz and Yuille, 1990; Schrater, Knill, and Simoncelli, 2000). Similar models arise in related work on the fly and beetle visual systems (Hassenstein and Reichardt, 1956; Borst and Euler, 2011).

12.3 *Probabilities and Decision Theory*

The previous section has given examples about how we can combine the response of many features/filters to perform tasks like stereo or motion estimation. The filter responses were combined by specific mechanisms, for example, looking for the set of filters which have maximal response. This section describes a more principled approach based on probabilities and decision theory. The section also illustrates the importance of knowing whether filter responses, hence visual cues for the task, are dependent or independent.

We introduce the probabilities of filter responses by describing a classical experimental finding about natural image statistics. Intuitively, the intensities of neighboring pixels tend to be similar. This intuition can be captured by taking derivative filters of the image, that is, dI / dx or $(d^2 I) / (dx^2)$, and plotting their probability distribution, or histogram. Surprisingly, these probability distributions are very similar from image to image (Simoncelli and Olshausen, 2001). This can be verified from interactive demo 3a on natural image statistics. Interactive demo 3b explores the statistics of edge detection and illustrates decision theory.

12.3.1 Edge Detectors/Texture Detectors and Decisions

Consider the tasks of deciding whether an *image patch* at position x contains an *edge*, by which we mean the boundary of an object or a strong texture boundary (e.g., like the writing on a T-shirt). The previous section showed that some Gabor filters are tuned (i.e., respond strongly) to edges at specific orientations. However, such filters will also respond to other stimuli such as texture patterns, so how can we decide if their response is due to an edge? The simplest way is to *threshold* the response so that an edge, at a specific orientation, is signaled if the filter response is larger than a certain threshold value. But what should that threshold be? How do we do a trade-off to balance *false negative* errors, where we fail to detect a true edge in the image, with *false positive* errors, where we incorrectly label a pixel as an edge? Also, each filter in a filter bank contains some evidence about the presence of an edge, so how can we combine their evidence in an optimal manner? How can we formulate the intuition that some filters give *independent* evidence while others do not?

Decision theory gives a way to address these issues. The theory was developed as a way to make decisions in the presence of uncertainty. In this section we develop the key ideas of decision theory by addressing the specific task of edge detection. In the next section we give a more general treatment. We will only treat the case when we are detecting edges based on local evidence in the image. Later in this chapter we will extend to situations when we can use nonlocal, or contextual, information. Interactive demo 3b on decision theory and edge detection illustrates most of the ideas in these two sections.

To start with, we consider the evidence for the presence of an edge using a single filter $f(.)$ only. We assume we have a benchmarked data set so that at each pixel we have intensity $I(x)$ and a variable $y(x) \in \{\pm 1\}$ (where $y = 1$ indicates an edge, and $y = -1$ does the opposite). We apply the filter to the image to get a set of filter responses $f(I(x))$. If the filter is tuned to edges, then the response $f(I(x))$ is likely to be higher if an edge is present than if not. This requires selecting a filter $f(x)$, such as the modulus of the gradient of the intensity

$$|\nabla I(x)| = \sqrt{\frac{dI}{dx}^2 + \frac{dI}{dy}^2}$$

(since $|\nabla I(x)|$ is likely to be large on edges and small off edges).

To quantify this, we use the benchmarked data set to learn *conditional probability distributions* for the filter response $f(I)$ conditioned on whether there is an edge or not:

$$P(f(I) \mid y = 1), P(f(I) \mid y = -1).$$

Each distribution is estimated by computing the *histogram* of the filter response by counting the numbers of times the response occurs within one of N equally spaced bins and normalizing by dividing by the total number of responses. The histograms for $P(f(I)|y = 1)$ and $P(f(I)|y = -1)$ are computed from the filter responses on the points labeled as edges $\{f(I(x)): y(x) = 1\}$ and not-edges $\{f(I(x)): y(x) = 1\}$, respectively. Typical conditional distributions are shown in figure 12.21.

We can now perform edge detection on an image; see figure 12.22. At each pixel x we compute $f(I(x))$ and calculate the conditional distributions $P(f(I(x))|y = 1)$ and $P(f(I(x))|y = -1)$. These distributions give local evidence for the presence of edges at each pixel. Note, however, that local evidence for edges is often highly ambiguous; see figure 12.23. Spatial context can supply additional information to help improve edge detection, as discussed in a later section, and so can high-level knowledge (e.g., by recognizing the objects in the image).

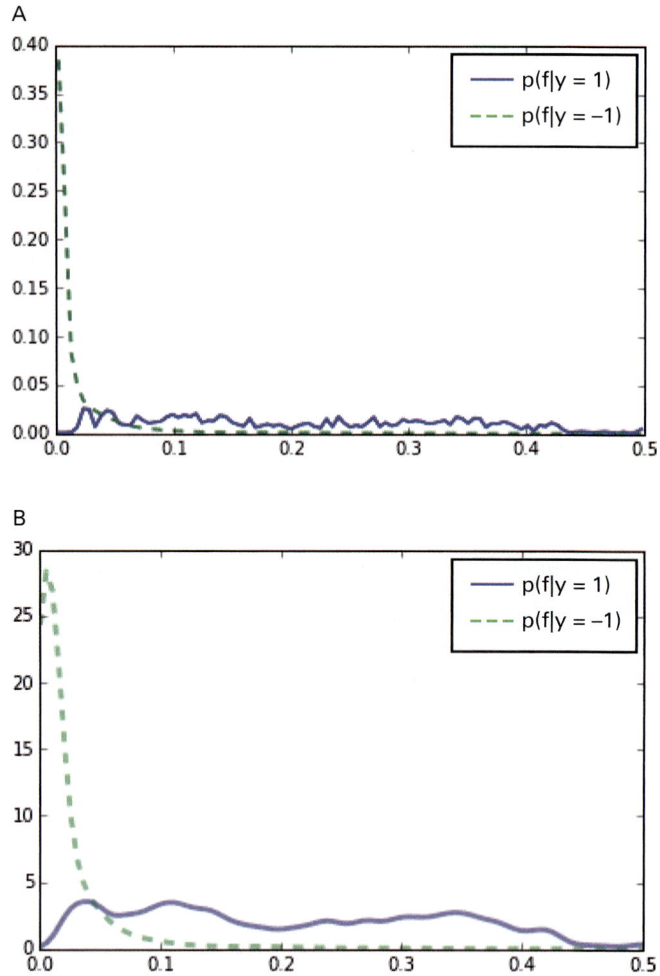

FIGURE 12.21 The probability of filter responses conditioned on whether the filter is *on* or *off* an edge – $P(f|y = 1)$, $P(f|y = -1)$, where $f(x) = |\nabla I(x)|$. (A) The probability distributions learned from a data set of images. (B) The smoothed distributions after fitting the data to a parametric model.

The log-likelihood ratio $\log[P(f(I(x)) \mid y = 1)] / [P(f(I(x)) \mid y = -1)]$ gives evidence for the presence of an edge in image I at position x. This ratio takes large positive values if $P(f(I(x))|y = 1) > P(f(I(x))|y = -1)$ (i.e., if the probability of the filter response is higher given an edge is present) and large negative values if $P(f(I(x))|y = -1) > P(f(I(x))|y = 1)$. So a natural decision criterion is decide that an edge is present if the log-likelihood ratio is greater than zero and that otherwise

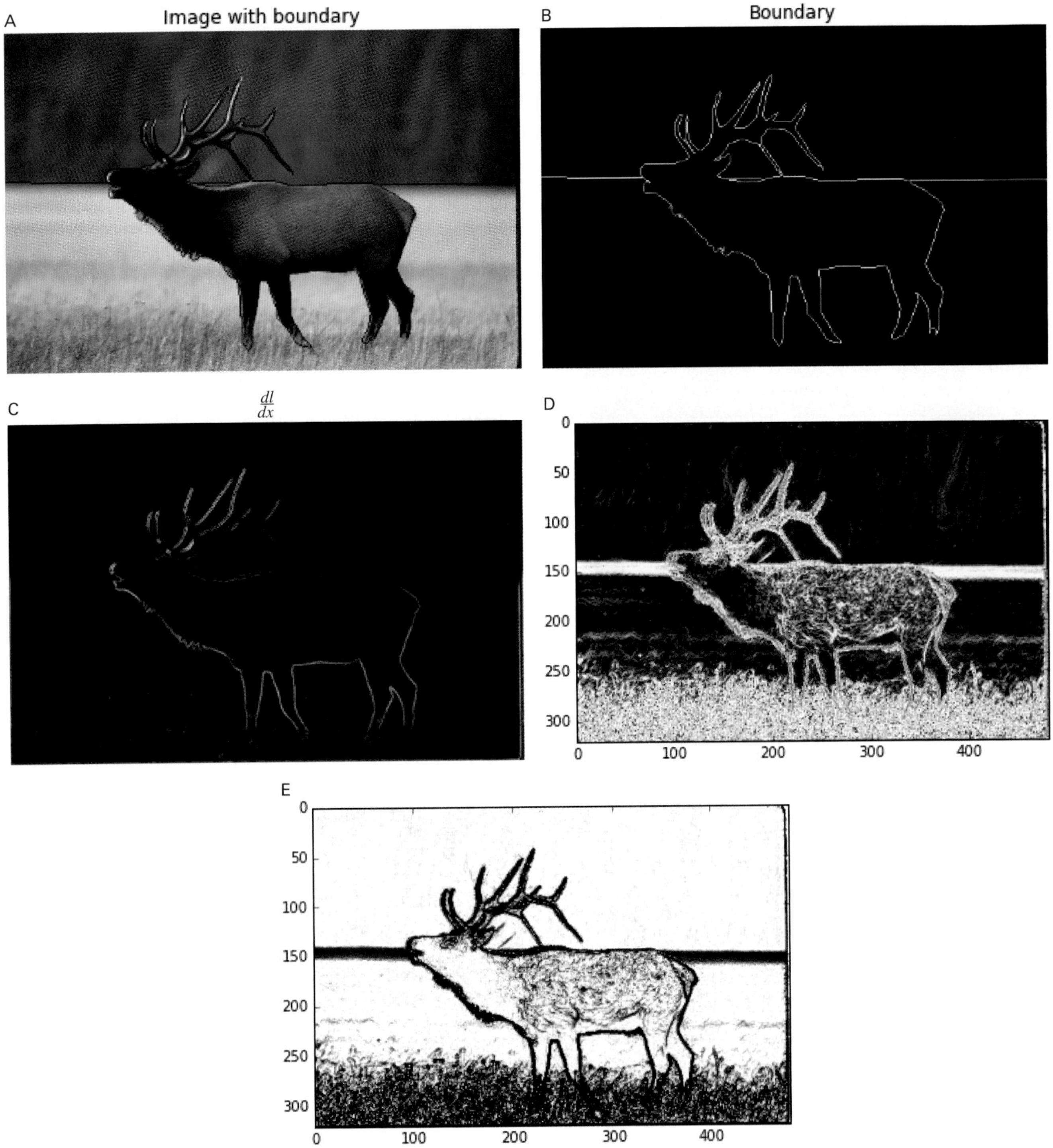

FIGURE 12.22 The input image and its ground truth edges (A and B). The derivative dI/dx of the image in the x direction (C). The probabilities of the local filter responses $P(f(I(x))|y=1)$ (D) and $P(f(I(x))|y=-1)$ (E) have their biggest responses on the boundaries and off the boundaries, respectively, hence the log-likelihood ratio $\log[P(f(I(x))\,|\,y=1)]/[P(f(I(x))\,|\,y=-1)]$ gives evidence for the presence of edges.

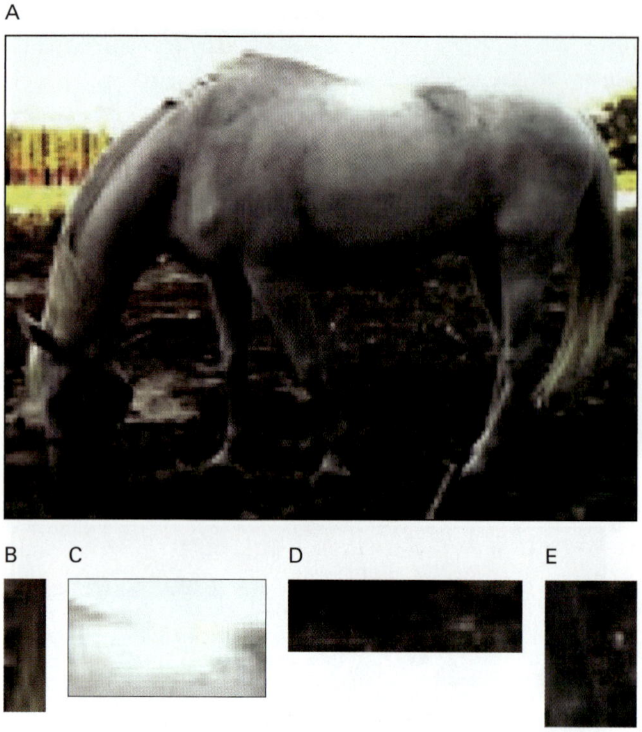

FIGURE 12.23 The local ambiguity of edges. An observer has no difficulty in detecting the entire boundary of the horse if the full image is available (A). However, it is much more difficult to detect edges locally (B–E).

there is no edge. This can be formulated as a *decision rule* $\alpha(x)$:

$$\alpha(x) = 1, \text{ if } \log \frac{P(f(I(x)) \mid y=1)}{P(f(I(x)) \mid y=-1)} > 0,$$
$$\alpha(x) = -1, \text{ if } \log \frac{P(f(I(x)) \mid y=1)}{P(f(I(x)) \mid y=-1)} < 0.$$

This can expressed, more compactly, as

$$\alpha(x) = \arg\max_{y \in \{\pm 1\}} y \log \frac{P(f(I(x)) \mid y=1)}{P(f(I(x)) \mid y=-1)}.$$

Note that this rule gives perfect results (i.e., is 100% correct) if the two distributions do not overlap, that is, if $P(f(I(x)) | y = 1) P(f(I(x)) | y = -1) = 0$ for all I. In this case it is impossible to confuse the filter responses to the different types of stimuli. But this situation is very unlikely to happen. Now consider a more general *log-likelihood ratio test* which depends on a threshold T; this gives a rule

$$\alpha_T(x) = \arg\max_{y \in \{\pm 1\}} y \left\{ \log \frac{P(f(I(x)) \mid y=1)}{P(f(I(x)) \mid y=-1)} - T \right\}.$$

By varying T, we get different types of mistakes. We can distinguish between the *false positives*, which are nonedge stimuli which the decision rule mistakenly decides to be an edge, and *false negatives*, which are edge stimuli which are mistakenly classified as not being edges. Increasing the threshold T reduces the number of false positives but at the cost of increasing the number of false negatives, while decreasing T has the opposite effect.

Making a decision requires a trade-off between these two types of errors. Bayes's decision theory says this trade-off should depend on two issues. The first issue is the *prior* probability that the image patch is an edge. Statistically, most image patches do not contain edges, so we would get a small number of total errors (false positives and false negatives) by simply deciding that every image patch is nonedge. This would encourage us to increase the threshold T (to $+\infty$ so that every image patch would be classified as nonedge). Secondly, we need to consider the *loss* if we make a mistake. If our goal is to detect edges, then we may be willing to tolerate many false positives provided we keep the number of false negatives small. This means we choose a decision rule, by reducing the threshold T, so that we detect all the real edges but also output "false edges," which we hope to remove later by using contextual cues (see the next section). Later we show how this approach can be justified using the framework of decision theory. (In the next section, we see that the log-likelihood ratio is justified as local *evidence* for the presence of an edge even if we are making the decision using nonlocal context.)

Now we consider combining several different filters $\{f_i(.) | i = 1, \ldots, M\}$ to detect an edge. Generalizing the analysis above, we must learn probability distributions for the *joint* response of all the filters $P(f_1, f_2, \ldots | y) = P(\{f_i(I(x))\} | y)$ *conditioned* on whether the image patch I at x is an edge $y = 1$ or not an edge $y = -1$. This leads to a decision rule:

$$\alpha_T(I(x)) = \arg\max_{y \in \{\pm 1\}} y \left\{ \log \frac{P(\{f_i(I(x))\} \mid y=1)}{P(\{f_i(I(x))\} \mid y=-1)} - T \right\}.$$

Methods of this type have been applied to edge detection and give good results (Konishi et al., 2003), but they have two related drawbacks. Firstly, the joint distributions require a large amount of data to learn, particularly if we represent the distributions by histograms. Secondly, the joint distributions are "black boxes" and give no insight into how the decision is made. Intuitively, filters tuned to edges at the same orientations will all respond strongly if there is an edge. However, this type of intuition is not obvious just from studying the joint distributions. So it is better to try to get a deeper understanding of how the different filters contribute to making this decision, by studying whether they are *statistically independent.*

The response of the filters is statistically independent if

$$P(\{f_i(I(x))\} \mid y) = \prod_i P(f_i(I(x)) \mid y) \text{ for each } y.$$

This implies that the distributions $P(f_i(I(x)) \mid y)$ can be learned separately (which decreases the amount of data) and also implies that the log-likelihood test can be expressed in the following form:

$$\alpha_T(x) = \arg\max_{y \in \{\pm 1\}} y \left\{ \sum_i \log \frac{P(f_i(I(x)) \mid y=1)}{P(f_i(I(x)) \mid y=-1)} - T \right\}.$$

Hence the decision rule corresponds to summing the evidence (the log-likelihood ratio) for all of the filters to determine whether it is above or below the threshold T. This means that each filter gives a "vote," which can be positive or negative, and the decision is based on the sum of these votes. This process is very simple, so it is easy to see which filters are responsible for the decision.

Unfortunately, very few filters are statistically independent. For example, the response of each filter will depend on the total brightness of the image patch, and so all of them will respond more to a "strong" edge than to a "weak" edge. This prevents them from being independent, but it suggests a weaker independence condition known as *conditional independence.* Suppose we add an additional filter $f_0(I(x))$ which, for example, measures the overall brightness. Then it is possible that the other filters are statistically independent conditioned on the value of $f_0(I(x))$:

$$P(\{f_i(I(x))\}, f_0(I(x)) \mid y) = P(f_0(I(x)) \mid y) \prod_i P(f_i(I(x)) \mid f_0(I(x)), y)$$

This requires only representing (learning) the distributions $P(f_i(I(x)) | f_0(I(x)), y)$ and $P(f_0(I(x)) | y)$. It also leads to a simple decision rule:

$$\alpha_T(x) = \arg\max_{y \in \{\pm 1\}} y \left\{ \log \frac{P(f_0(I(x)) \mid y=1)}{P(f_0(I(x)) \mid y=-1)} + \sum_i \log \frac{P(f_i(I(x)) \mid f_0(I(x)), y=1)}{P(f_i(I(x)) \mid f_0(I(x)), y=-1)} - T \right\}.$$

It has been argued (Ramachandra and Mel, 2013) that methods of this type can be implemented by neurons and may be responsible for edge detection. Note that the arguments here are general and do not depend on the type of filters $f_i(.)$ or whether they are linear or nonlinear. It has, for example, been suggested that edge detection is performed using the energy model of complex cells (Morrone and Burr, 1988).

The same approach can be applied to other visual tasks. For example, consider using local filter responses to classify whether the local image patch at x is "sky," "vegetation," "water," "road," or "other"; see figure 12.24. We denote these by a variable $y \in \mathcal{Y}$ (e.g., where $\mathcal{Y} = \{\text{"}sky\text{"}, \text{"}vegetation\text{"}, \text{"}water\text{"}, \text{"}road\text{"}, or \text{ "}other\text{"}\}$). We choose a set of filters $\{f_i(I(x))\}$ which are sensitive to texture and color properties of image patches. Then, as before, we learn distributions $P(\{f_i(I(x))\}|y)$ for $y \in \mathcal{Y}$. We select a decision rule of the form

$$\alpha(I(x)) = \arg\max_{y \in \mathcal{Y}} P(\{f_i(I(x))\} \mid y) T_y,$$

where T_y is a set of thresholds (which can be derived from decision theory).

Experiments on images show that this method can locally estimate the local image class with reasonable error rates for these types of classes (Konishi and Yuille, 2000), and computer vision researchers have improved these types of results using more sophisticated filters. It is unknown, however, whether human or monkey visual systems do make these types of classification.

We stress that the theories described in this section model edge detection *without context.* There are two types of context we will consider in this chapter. The first, discussed in section 12.4, uses spatial context and is low- or mid-level since it depends only on *generic* properties of images and surfaces. It exploits the idea that edges in natural images are often geometrically regular and colinear. The second type of context is high-level and is object specific. For example, if we detect a face in an image, then our knowledge about faces enables us to detect the boundaries of a face better than if we rely only on local edge cues. This second type of context is out of scope of this chapter but is briefly discussed in section 12.6.

12.3.2 Bayes's Decision Theory and Ideal Observers We now describe Bayesian analysis in more detail. This is advanced material which can be skipped on a first reading. Bayes's decision theory is a framework for making optimal decisions in the presence of uncertainty. We represent the input by $x \in \mathcal{X}$ and the output by $y \in \mathcal{Y}$ (e.g., for *edge detection* x is the filter response $f(I)$, and $y \in \{\pm 1\}$ indicates if an edge is present or not).

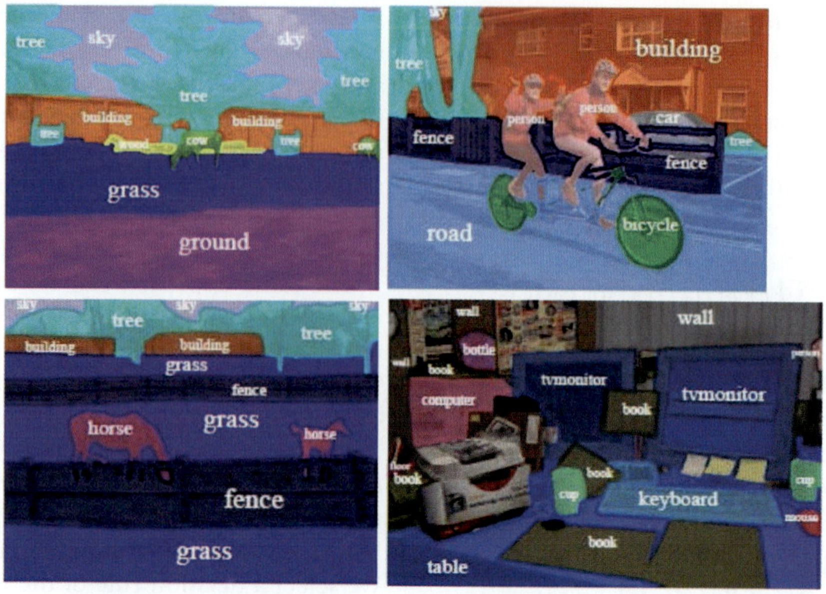

FIGURE 12.24 Classifying local image patches. The image shows the ground truth; see Mottaghi et al. (2014). Certain classes—sky, grass, water—can be classified approximately from small image patches.

We assume that there is a probability distribution $P(x, y)$ which generates the input and output. This can be expressed in terms of a *prior* $P(y)$ and a *likelihood* $P(x|y)$ by the identity $P(x, y) = P(x|y)P(y)$. A decision rule is expressed as $\hat{y} = \alpha(x)$. We specify a *loss function* $L(\alpha(x); y)$ which is the cost of making decision $\alpha(x)$ if the real decision should be y.

The *risk* is specified by

$$R(\alpha) = \sum_{x,y} P(x,y)L(\alpha(x),y).$$

The *Bayes rule* is $\hat{\alpha} = \text{argmin}_\alpha R(\alpha)$. The *Bayes risk* is $\min_\alpha R(\alpha) = R(\hat{\alpha})$ (except for a few highly unusual special cases).

The Bayes rule is the best decision rule you can make (subject to this criterion), and the Bayes risk is the best performance. Hence Bayes's decision theory can specify the optimal way to estimate y from input x. There are several important special cases. If the loss function penalizes all errors by the same amount, that is, $L(\alpha(x), y) = K_1$ if $\alpha(x) \neq y$ and $L(\alpha(x), y) = K_2$ if $\alpha(x) = y$ (with $K_1 > K_2$), then the Bayes rule corresponds to the *maximum a posteriori* estimator $\alpha(x) = \arg\max P(y|x)$, where $P(y \mid x) = [P(x \mid y)P(y) / P(x)]$ is the *posterior* distribution of y conditioned on y. If, in addition, the prior is a uniform distribution, that is, $P(y) = constant$, then Bayes rule reduces to the *maximum likelihood* estimate $\alpha(x) = \arg\max P(x|y)$.

For binary decision problems $y \in \{\pm 1\}$, the loss function is usually chosen to pay no penalty if the correct decision is made—that is, $\alpha(x) = y$—but has a penalty F_p for *false positives*, where $y = -1$ but $\alpha(x) = 1$, and F_n for *false negatives* where $y = 1$ but $\alpha(x) = -1$ (it is assumed here that the *target* is $y = 1$ and the *distracter* is $y = -1$, so a false positive occurs if we decide that a distracter is a target, and a false negative, if we decide that a target is a distracter). It follows that we can express the Bayes rule in terms of a log-likelihood ratio test $\log[P(x \mid y=1)] / [P(x \mid y=-1)] > T$, where T depends on the prior $p(y)$ and the loss function $L(\alpha(x), y)$.

More specifically, the Bayes risk is $R(\alpha) = \sum_x p(x) \sum_y L(\alpha(x), y) p(Y \mid x)$. Then we divide the data (x, y) into four sets: (1) the *true positives* $\{(x, y)$: s.t. $\alpha(x) = y = 1\}$, (2) the *true negatives* $\{(x, y)$: s.t. $\alpha(x) = y = -1\}$, (3) the *false positives* $\{(x, y)$: s.t. $\alpha(x) = 1, y = -1\}$, and (4) the *false negatives* $\{(x, y)$: s.t. $\alpha(x) = -1, y = 1\}$. These four cases correspond to loss function values $L(\alpha(x) = 1, y = 1) = T_p$, $L(\alpha(x) = -1, y = -1) = T_n$, $L(\alpha(x) = 1, y = -1) = F_p$, $L(\alpha(x) = -1, y = 1) = F_n$, respectively. Then the decision rule $\alpha_T(.)$ reduces to

$$\log \frac{P(x \mid y=1)}{P(x \mid y=-1)} > \log \frac{T_n - F_p}{T_p - F_n} + \log \frac{P(y=-1)}{P(y=1)}.$$

The intuition is that the evidence in the log-likelihood must be bigger than our prior biases while taking into account the penalties paid for different types of mistakes.

The results in the previous section on edge detection and texture classification can be derived from decision theory. The priors $P(y)$ specify the probability that an image patch contains an edge (empirically $P(y = 1) \approx 0.05$ and $P(y = -1) \approx 0.95$). The loss function should be chosen to specify the cost of making different types of mistakes. For texture classification, the variable y takes values in a set $\mathcal{Y}$, which is called multiclass decision. The same theory applies to tasks for which we need to make a set of related but nonlocal decisions, which we will address in the next section.

We now show that an important special case of *signal detection theory* (Green and Swets, 1966)—often used as a framework to model how humans make decisions when performing visual, auditory, and other tasks—can be obtained as a special case of Bayes's decision theory. We consider the two-class case, where $y \in \{\pm 1\}$, and suppose that the likelihood functions are specified by Gaussian distributions,

$$P(x \mid y) = \frac{1}{\sqrt{2\pi}\sigma_y} \exp\{-(x-\mu_y)^2 / (2\sigma_y^2)\},$$

which differ by their means (μ_1, μ_{-1}) and their variances $(\sigma_1^2, \sigma_{-1}^2)$. The Bayes rule can be expressed in terms of the log-likelihood ratio test:

$$\hat{\alpha}(x) = \arg\max_y y\{-(x-\mu_1)^2 / (2\sigma_1^2) - \log\sigma_1 + (x-\mu_{-1})^2 / (2\sigma_{-1}^2) + \log\sigma_2 - T\}.$$

This requires determining whether the data x is above or below a quadratic polynomial curve in x. In the special case when the standard deviations are identical $\sigma_1^2 = \sigma_2^2$ (so we drop the subscripts $_{1,-1}$), this reduces to making a decision based on whether the data point x satisfies

$$2x(\mu_1 - \mu_{-1}) + (\mu_1^2 - \mu_2^2) < 2T\sigma^2.$$

This special case, with $\sigma_1^2 = \sigma_{-1}^2$, is much studied in signal detection theory (Green and Swets, 1966). It means that the decision is based on a single function $d' = (\mu_1 - \mu_{-1})/\sigma$. This quantity is used to quantify human performance for psychophysical tasks.

This motivates the idea of an *ideal observer*. An observer like this has optimal performance, which requires exploiting the statistical properties of the distribution $P(x, y)$ of the data. A classic example of ideal observer theory shows that under certain conditions photoreceptors in the retina are almost *optimal* at detecting the photons which reach them (Barlow, 1962; Pelli, 1990). This takes into account the probability of the photoreceptors *firing* x if it receives a photon $P(x|y = 1)$, and the probability that the photoreceptor fires spontaneously $P(x|y = -1)$.

Ideal observers can also be defined for other vision tasks (Tjan et al., 1995; Gold, Mundy, and Tjan, 2012; Trenti, Barraza, and Eckstein, 2010; Geisler, 2011). The difficulty, however, is judging whether humans are adapted to doing the task. It is possible to define ideal observers where human performance is much worse than the ideal observers (Watson, Barlow, and Robson, 1983). Why can this happen? The task may provide information for which humans are not adapted (e.g., visual inspection of circuit boards to find deficits). Also, the ideal observers know the distributions $p(x, y)$ which, for synthetic stimuli, are those chosen by the scientist performing the experiment and may have little similarity to the natural statistics of stimuli of the world, which human vision has probably adapted to.

Another important concept is the receiver operating characteristic (ROC) curve. This allows us to study decisions when we do not want to restrict ourselves to specific priors and loss functions. Instead, we plot the *true positive rate* as a function of the *false positive rate* by allowing the decision threshold T to vary. For each value T of the threshold, we have a decision rule $\alpha_T(.)$ which results in a fraction of *true positives* $\sum_{x:\alpha_T(x)=1} P(x \mid y=1)$ and *false positives* $\sum_{x:\alpha_T(x)=1} P(x \mid y=-1)$. This gives a single point on the ROC curve. We plot the curve by allowing T to vary. Observe that for very large T (as $T \to \infty$), the true positive and false positive rates will tend to 0, while as T gets very small ($T \to -\infty$), both rates will tend to 1. Hence the ROC illustrates the trade-off between the two rates.

Note that this section presented Bayes's decision theory for binary classification. But the same framework can be extended in a straightforward manner if the output y takes multiple values. In particular, it applies to cases where we have a set of decision variables defined on each lattice site of an image (see later sections).

12.3.3 Divisive Normalization An important example is the use of probabilistic models (Wainwright and Simoncelli, 2000) to account for divisive normalization. Divisive normalization is a mechanism whereby cells mutually inhibit one another, effectively normalizing their responses with respect to stimulus inputs. Originally developed to explain nonlinear responses to contrast in V1 (Heeger, 1992), divisive normalization has been proposed as a basic cortical computation that underlies various effects of context (see the next section), as well as higher-level processes such as attention (Carandini and Heeger, 2011).

The probabilistic approach gives a theoretical justification for divisive normalization in V1. The main idea is that filters with similar preferences for orientation representing nearby spatial locations in a scene have striking statistical dependencies, which can be removed by divisive normalization. Specifically, if we plot the statistics of two linear filters f_c, f_s (center and surround) then the magnitudes of f_c, f_s are coordinated in a straightforward way, which has a characteristic shape of a bow tie.

This can be modeled by assuming there are hidden variables v which affect both responses and hence induce correlation between the responses (as discussed earlier for edge detection). For example, v could represent the local average image intensity, which could affect the response of both filters but after the filter response could be made independent by conditioning on the average intensity. Suppose v has a prior distribution $P(v) = v\exp\{-v^2/2\}$ for $v \geq 0$. We have a pair of filters $\{l_i: i = 1, 2\}$ which are related to Gaussian models $\{g_i: i = 1, 2\}$ (the analysis can be generalized to an arbitrary number of filters [Schwartz, Sejnowski, and Dayan, 2009]). The claim is that we can model the activation of the set of filter responses:

$$P(l_1, l_2) = \int dv P(v) \prod_{i=1}^{2} P(l_i \mid v, g_i) P(g_i), \quad (12.25)$$

where $P(l_i|v, g_i) = \delta(l_i - vg_i)$. In this model the filter responses are generated by independent processes, g_1, g_2, but then are multiplied by the common factor v. This is illustrated in figure 12.25.

In particular, for each filter we can compute $P(g_i|l_1, l_2)$. After some algebra, this is computed to be

$$P(g_1 \mid l_1, l_2) = \frac{g_1^{-1} \exp\left\{-\frac{g_1^2 l^2}{2\sigma^2 l_1^2} - \frac{l_1^2}{2g_1^2}\right\}}{B(0, l/\sigma)}, \quad (12.26)$$

where $l = \sqrt{l_1^2 + l_2^2}$, and $B(.,.)$ is a Bessel function (a well-known class of mathematical functions that arise in solving differential equations). To get some intuition, note that $g_1 = l_1/v$ and $g_1 = l_1/v$. So if v is small, then $|l_1|$ and $|l_2|$ are likely to be small together, while if v is large, then $|l_1|$ and $|l_2|$ are both likely to be large.

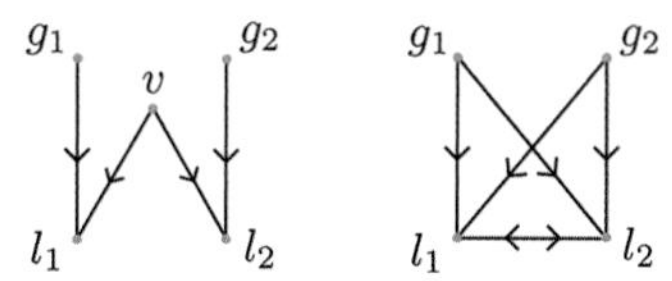

FIGURE 12.25 Left: The graphical structure of the divisive normalization model. The filter responses l_1, l_2 are generated from stimuli g_1, g_2, respectively, and by the common factor v. The distributions of l_1, l_2 are factorized if we condition on v. Right: But if we integrate out v, then almost all the variables become dependent as reflected by the complexity of the graph structure.

Assume that the goal of a model unit is to estimate the g_i from the observed filter responses $\{l_i: i = 1, 2\}$, which gives the nonlinear response of the cell. It follows, from the analysis above, that

$$E(g_1 \mid l_1, l_2) \propto \text{sign}\{l_1\}\sqrt{|l_1|}\sqrt{\frac{|l_1|}{\sqrt{l_1^2 + l_2^2 + k}}}. \quad (12.27)$$

The $\sqrt{l_1^2 + l_2^2 + k}$ term sets the gain and performs the divisive normalization.

The model has also been applied to explain the classic tilt illusion in perception (Schwartz, Sejnowski, and Dayan, 2009; Qiu, Kersten, and Olman, 2013). In the "simultaneous" tilt illusion, a set of vertically oriented lines appears to tilt right when surrounded by an annulus of lines tilted left—an effect called "repulsion." However, for large differences between the center orientation and surround (tilted left), the center vertical lines can appear to tilt left—an effect called "attraction." In their model, the population of neurons responding to the surround tilted lines contributes to divisive normalizing of the neurons responding to the center stimulus. This results in a change of their neural tuning curves which, together with the degree of coupling between center and surrounds, accounts for repulsion and attraction.

The suppressive effect of surround contrast on a central region is an example of spatial context. There are many phenomena like this whose effects vary in type and extent of their spatial interactions. The next section introduces a broader range of contextual phenomena and a set of computational techniques to model them.

12.4 *Context and Spatial Interactions between Neurons*

There is considerable evidence that low-level vision involves long-range spatial interactions so that human perception of local regions of an image can be strongly influenced by their spatial context. Psychophysicists have discovered many perceptual phenomena demonstrating spatial interactions. For example, local image regions which differ from their neighbors tend to "pop-out" and attract attention while, conversely, similar image features which form spatially smooth structures tend to get "grouped" together to form a coherent percept; see figure 12.26A. Image properties such as color tend to spread out, or fill-in regions, until they hit a boundary (Grossberg and Mingolla, 1985a; Sasaki, Watanabe, and Purves, 2004) as shown in figure 12.26B. In general, there is a tendency for low-level vision to group together similar image features and make breaks at places where the features change significantly. These

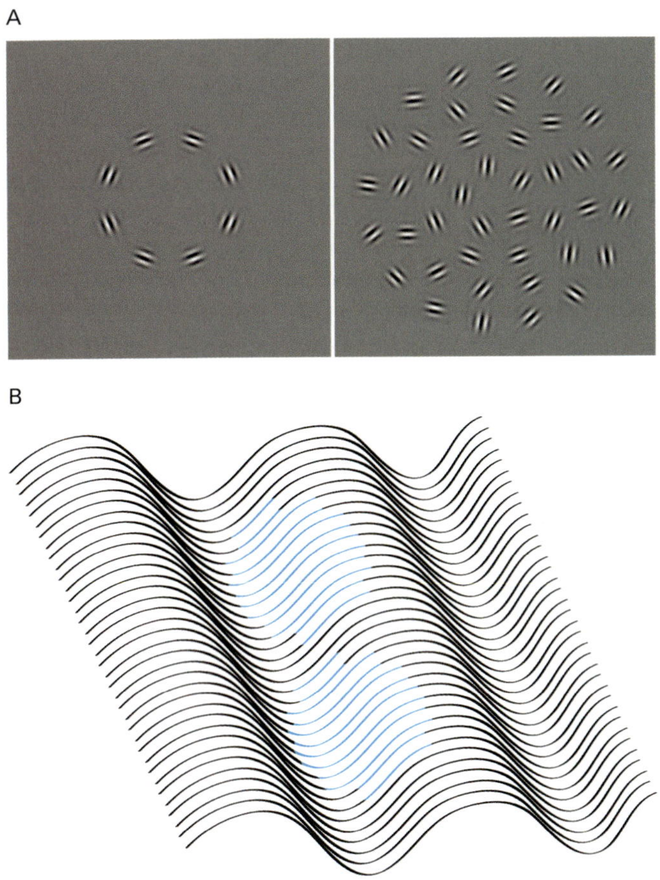

FIGURE 12.26 (A) Association fields. The circular alignment of Gabor patches (left) make it easier to see the circular form in the presence of clutter (right). (B) The neon color illusion. A bluish color appears to fill in the white regions between the blue lines creating the appearance of blue transparent disks.

perceptual phenomena are not surprising from a theoretical perspective since they correspond to low-level visual tasks such as segmentation and the detection of salient features. Segmenting an image into different regions is one of the first stages of object recognition (in the ventral stream) and a precursor to estimating the three-dimensional structure of objects, or surfaces, in order to grasp them or avoid them (dorsal stream). Detection of salient features has many uses including bottom-up attention (Itti and Koch, 2001). It has been suggested that many of these processes are performed in V1 (Zhaoping, 2014) although possibly this involves feedback and interactions between V1 and V2 (Shushruth et al., 2013). Note that, as discussed in the first section, low-level vision interacts with high-level vision, and it is very unlikely that tasks like segmentation are performed entirely by low-level processes.

These psychophysical and theoretical studies are supported by single-electrode studies (Lamme, 1995; Lee and Yuille, 2006) which show that the activities of neurons on monkey area V1 appear to involve spatial interactions with other neurons. When monkeys are shown stimuli consisting of a textured square surrounded by a background with a different texture, their responses over the first 60 ms are similar to those predicted by classic models (e.g., previous sections) but their later activity spreads in from the texture boundaries, roughly similar to the predictions of computational models (Lee and Yuille, 2006). There is also considerable literature on the related topic of nonclassical receptive fields (Kapadia, Westheimer, and Gilbert, 2000).

This section discusses neural networks models which address these phenomena. Although the models capture the essence of the phenomena, they are simplifications in three respects. Firstly, they use simple models of neurons, and it is currently not possible to compare them directly to real neural circuits. Secondly, these models are formulated in terms of lateral, or horizontal, connections. Thirdly, the performance of these models on natural images is significantly worse than humans'—although there are more advanced computer vision models, built on similar principles, whose performance starts to approach human vision (unless high-level cues are present, which humans can exploit).

We formulate these models in terms of probability distributions defined over graphs, where the nodes of the graph represent neurons. This differs from some of the standard "neural network" models for these types of phenomena; see Grossberg and Mingolla (1985b). However, our approach has several advantages. Firstly, this enables us to use a coherent framework which unifies the models in this section with those we will present in later sections. Secondly, it puts the models in a form where they can be directly related to a class of computer vision models. Thirdly, this probabilistic formulation is of increasing use in models of artificial intelligence and cognitive science and in the machine learning and statistical techniques used to analyze experimental neuroscience data. Fourthly, it is possible to derive many of these neural network models as approximations to the probability models.

We start by first introducing probabilistic models of neurons and showing how our previous linear filter models can be derived as approximations. Next we introduce neural network models and show their relationship to probability models. Then the following section uses this material to derive some specific models for a range of visual tasks. This section has three interactive demos: 4a, Gibbs Sampling; 4b, Mean Field Theory and Neural Models; and 4c Hopfield Networks and Stereopsis.

Single Neurons: Probabilistic Model and Integrate and Fire In the previous section, we described neurons as linear filters and briefly mentioned thresholds and nonlinearities. In this section we provide a more realistic model of a *stochastic neuron* where the neuron has a probability of firing an action potential. We will show how linear filters, thresholds, and nonlinearities can be obtained as approximations to this stochastic model. This stochastic model is, in turn, an approximation, and we refer the reader to the literature for more realistic models such as assuming that the probability of firing is specified by a Poisson process (Rieke et al., 1997). For simplicity, we restrict ourselves to the simpler stochastic *integrate-and-fire* model which is easier to analyze and to relate to computational models.

In the integrate-and-fire model a neuron i receives input I_j at each dendrite j. These inputs are weighted by the synaptic strengths w_{ij} and sent along the dendrites to the soma. At the soma, these weighted inputs are summed linearly to yield $\sum_j w_{ij} I_j$. The probability of firing, $s_i = 1$, or not firing, $s_i = 0$, is given by

$$P(s_i \mid \boldsymbol{I}) = \frac{\exp\left\{s_i\left(\sum_j w_{ij} I_j - T_i\right)\right\}}{1+\exp\left\{\sum_j w_{ij} I_j - T_i\right)\}}, \qquad (12.28)$$

where T_i is a threshold.

To relate this stochastic model to our earlier linear models, we calculate the probability that the neuron fires. In particular, the probability of firing ($s_i = 1$) is given by a sigmoid function:

$$\sum_{s_i=0}^{1} s_i P(s_i \mid \boldsymbol{I}) = \frac{1}{1+\exp\left\{\sum_j w_{ij} I_j - T_i\right\}} = \sigma\left(\sum_j w_{ij} I_j - T_i\right). \qquad (12.29)$$

Observe that this is also the *expected firing rate* $\sum_{s_i=0,1} s_i P(s_i \mid \boldsymbol{I})$ because

$$\sum_{s_i=0,1} s_i P(s_i \mid \boldsymbol{I}) = P(s_i = 1 \mid \boldsymbol{I}) = \sigma\left(\sum_j w_{ij} I_j - T_i\right). \qquad (12.30)$$

Hence, by computing the expected firing rate, we can obtain a deterministic approximation to a stochastic neuron. This is a sigmoid function of a linear weighted sum of the input (minus a threshold). The sigmoid function is approximately linear for small inputs, saturates at value 1 for large positive inputs, and suppresses large negative inputs to 0. Hence there is a linear regime where the probability of firing is $\sum_j w_{ij} I_j - T_i$. This enables us to recover the linear models used in the previous section as an approximation.

Next we modify the model so that it deals with nonlinear image features. This allows us to relate it to the types of computational models described in the previous section and will enable us to construct richer models of this type that can deal with spatial context.

To make this idea concrete, we consider detecting whether there is an edge or not at pixel x or, alternatively, classifying whether a pixel is foreground or background. In either task, we formulate the problem as Bayes estimation where we have conditional distributions $P(f(I(x))|s)$ and priors $P(s)$ for $s \in \{0, 1\}$. The posterior distribution $P(s|f(I(x)))$ can be expressed in the form

$$P(s \mid f(I(x))) = \frac{1}{Z}\exp\left\{s\left(\log\frac{P(f(I(x)) \mid s=1)}{P(f(I(x)) \mid s=0)} + \log\frac{P(s=1)}{P(s=0)}\right)\right\},$$

where Z is a normalization constant (chosen so that $\sum_{s=0}^{1} P(s \mid f(I(x))) = 1$). This shows that the posterior distribution for the presence of an edge, or a foreground object, can be expressed in the same form. The only difference is that the input is a nonlinear function of the image instead of the image itself. This claim can be justified by expressing $P(f(I(x))|s) = \{P(f(I(x))|s = 1)\}^s\{P(f(I(x))|s = 0)\}^{1-s}$, $P(s) = \{P(s = 1)\}^s\{P(s = 0)\}^{1-s}$, substituting these into the posterior $P(s|f(I(x))) = P(f(I(x))|s)P(s)/P(f(I(x)))$.

Probability Models with Context Now we consider generalizing the model for foreground–background classification so that it can include spatial context. Intuitively, neighboring pixels in the image are likely to belong to the same class, that is, are likely to be either all background or all foreground. This is a form of prior knowledge, or natural statistic, which can be learned by analyzing natural images.

We now specify neurons by spatial position $\boldsymbol{x}$ instead of index i. As above, we have distributions $P(f(I(\boldsymbol{x}))|s)$ for the features $f(I(\boldsymbol{x}))$ at position $\boldsymbol{x}$ conditioned on whether this is part of the foreground object $s(\boldsymbol{x}) = 1$, or not $s(\boldsymbol{x}) = 0$. We use the notation $\boldsymbol{S}$ to be the set of the states of all neurons $\{s(\boldsymbol{x})\}$. We also specify a prior distribution:

$$P(\boldsymbol{S}) = \frac{1}{Z}\exp\left\{-\gamma\sum_{\boldsymbol{x}}\sum_{\boldsymbol{y}\in N(\boldsymbol{x})}\{s(\boldsymbol{x}) - s(\boldsymbol{y})\}^2\right\},$$

where γ is a constant. This prior uses a neighborhood $N(\boldsymbol{x})$ which specifies those spatial positions which directly interact with $\boldsymbol{x}$ in the model. In graphical terms,

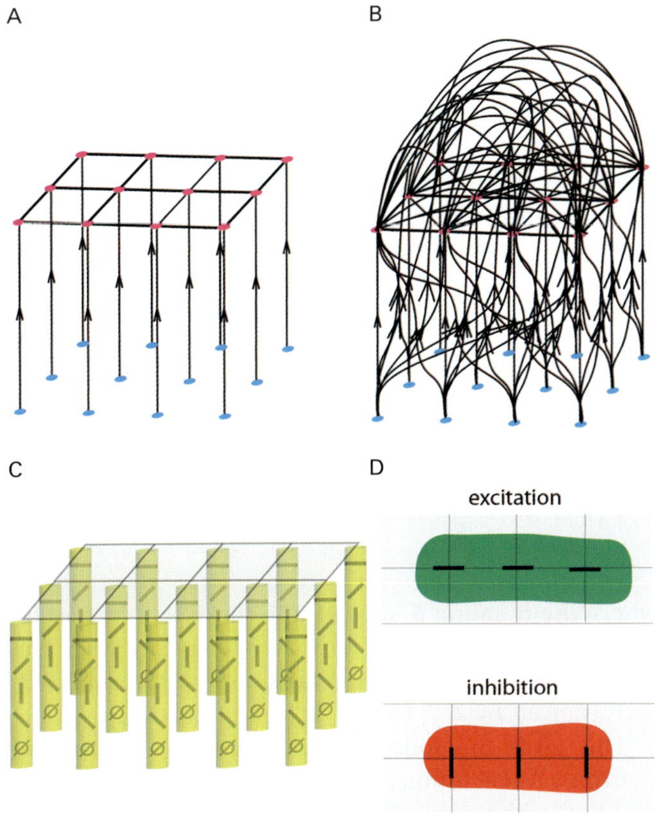

FIGURE 12.27 (A) The graphical structure of the Markov model with nearest neighbor connections. (B) A fully connected graphical model. (C) A hypercolumn structure where neurons within each column are tuned to different orientations and inhibit each other. (D) Edges have excitation (green) along the direction of the edge and inhibition (red) perpendicular to the edge.

the positions $\boldsymbol{x}$ are the nodes $\mathcal{V}$ of a graph $\mathcal{G}$ and the edges $\mathcal{E}$ specify which nodes are connected. This is illustrated in figure 12.27A.

Formally, the edges of the graph define the *Markov structure* of the probability distribution $P(\boldsymbol{S})$. It can be shown that the conditional distribution of the state $s(\boldsymbol{x})$ at one position depends *only* on the states of positions in its neighborhood $N(\boldsymbol{x})$. This is the *Markov condition*:

$$P(s(\boldsymbol{x})|\boldsymbol{S}/s(\boldsymbol{x})) = P(s(\boldsymbol{x})|\{s(\boldsymbol{y}): \boldsymbol{y} \in N(\boldsymbol{x})\}),$$

where $\boldsymbol{S}/s(\boldsymbol{x})$ denotes all states in $\boldsymbol{S}$ except $s(\boldsymbol{x})$. In real vision applications this type of prior, including the size of the neighborhoods, can be estimated from the statistics of natural images.

Next, we define a probability model for the observed image features at positions $\boldsymbol{x}$ in the image. We use the same models as before, at each position $\boldsymbol{x}$:

$$P(f(I(\boldsymbol{x}))|s) = \{P(f(I(\boldsymbol{x}))|s=1)\}^{s}\{P(f(I(\boldsymbol{x}))|s=0)\}^{1-s}.$$

We combine these, using independence assumptions, to get a distribution:

$$\begin{aligned} P(f(\boldsymbol{I})\,|\,\boldsymbol{S}) &= \prod_{\boldsymbol{x}} P(f(I(\boldsymbol{x}))\,|\,s) \\ &= \frac{1}{Z_l}\exp\left\{\sum_{\boldsymbol{x}} s(\boldsymbol{x})\left(\log\frac{P(f(I(\boldsymbol{x}))\,|\,s=1)}{P(f(I(\boldsymbol{x}))\,|\,s=0)}\right\},\right. \end{aligned}$$

where Z_l is a normalization term (which can be calculated directly).

These distributions $P(f(\boldsymbol{I})|\boldsymbol{S})$ and $P(\boldsymbol{S})$ can be combined to get the posterior distribution $P(\boldsymbol{S}|f(\boldsymbol{I}))$ which is of form:

$$P(\boldsymbol{S}\,|\,f(\boldsymbol{I})) = \frac{1}{Z_p}\exp\{-E(\boldsymbol{S})\},$$

where

$$\begin{aligned} E(\boldsymbol{S}) = &-\sum_{\boldsymbol{x}} s(\boldsymbol{x})\log\frac{P(f(I(\boldsymbol{x}))\,|\,s=1)}{P(f(I(\boldsymbol{x}))\,|\,s=0)} \\ &+\sum_{\boldsymbol{x}}\sum_{\boldsymbol{y}\in N(\boldsymbol{x})} \gamma\{s(\boldsymbol{x})-s(\boldsymbol{y})\}^2. \end{aligned}$$

The first term of $E(\boldsymbol{S})$ gives the local cues for foreground or background (the log-likelihood ratios of the features) while the second term adds the local context. This context encourages neighboring positions to be either all foreground or all background. Note that this method of specifying a distribution $P(\boldsymbol{S})$ in terms of a function $E(\boldsymbol{S})$ will keep reoccurring throughout this section.

This model specifies the posterior distribution for foreground–background classification using spatial context, and, as we will show, similar methods can be applied to other visual tasks. But there remains the issue of how to estimate the most probable states, that is, computing the Bayes estimator:

$$\hat{\boldsymbol{S}} = \arg\max P(\boldsymbol{S}|f(\boldsymbol{I})).$$

In the next two sections we will discuss neurally plausible algorithms which can do this. There are two types: (1) stochastic models which are natural extensions of the probabilistic neural models discussed earlier and which in the statistics literature are called *Gibbs* samplers (Liu, 2008) and (2) neural network models which are based on simplified biophysics of neurons but which can also, in certain cases, be related to *mean field approximations* to the stochastic models.

Probabilistic Models of Groups of Neurons In this section we introduce a more general probability distribution. It is also specified by a model defined over a graph where the nodes correspond to neurons and the edges to connections between them. However, we will not make any Markov restrictions on the edges, and so this model can be fully connected; see figure 12.27B.

More specifically, we have set of M neurons with states $\boldsymbol{S} = (s_1, \ldots, s_M)$ and with input $\boldsymbol{I} = (I_1, \ldots, I_N)$. We specify a *Gibbs probability distribution* over the set of activity of all neurons $\boldsymbol{S} = (s_1, \ldots, s_n)$ as follows. First we define an energy function:

$$E(\boldsymbol{S},\boldsymbol{I}:\boldsymbol{W},\boldsymbol{\theta}) = -\sum_{ij} W_{ij} s_i I_j + (1/2)\sum_{kl} \theta_{kl} s_k s_l.$$

This energy contains two types of terms: (1) those of form $s_i I_j$ which give the interactions between the states of the neurons $\boldsymbol{S}$ and the input $\boldsymbol{I}$ and (2) those which specify interactions between the neurons. This energy is used to specify a *Gibbs distribution*:

$$P(\boldsymbol{S},\boldsymbol{I}) = \frac{1}{Z}\exp\{-E(\boldsymbol{S},\boldsymbol{I}:\boldsymbol{W},\boldsymbol{\theta})\}. \qquad (12.31)$$

Here Z is a normalization constant chosen to ensure that $\sum_S P(\boldsymbol{S} \mid \boldsymbol{I}) = 1$. We note that Gibbs distribution originally arose in statistical physics where they specify the probability distribution of a physical system in thermal equilibrium. Here the physical energy of the system is E and the distribution can be derived using the maximum entropy principle.

The weights $\{w_{ij}\},\{\theta_{kl}\}$ specify the strength of the interactions between the neuron and the inputs, and between the neurons and each other. In particular, the *interaction term* $\sum_{kl} \theta_{kl} s_k s_l$ specifies the interactions between the neurons. If this term was not present, then the distribution simplifies and it can be expressed as a product of independent distributions:

$$P(\boldsymbol{S} \mid \boldsymbol{I}) = \frac{1}{Z}\exp\left\{\sum_{ij} w_{ij} s_i I_j\right\} = \prod_{i=1}^{n} P(s_i \mid \boldsymbol{I}). \qquad (12.32)$$

Hence in this special case the neurons act independently and are driven purely by the input (i.e., there is no context). As a technical point, in this case the normalization factor can be computed directly as $Z = \prod_i Z_i$, where $Z_i = \sum_{s_i=0}^{1} \exp\{\sum_j w_{ij} s_i I_j\}$.

Observe that we can recover the foreground–background model in the previous section by specifying values of the weights. This requires that we identify the nodes i with positions $\boldsymbol{x}$.

Now we specify stochastic dynamics on this model. These dynamics have two purposes: firstly, to describe the activities of sets of neurons interacting with each other; secondly, to give algorithms for estimating properties such as the most probable configurations of the states $\boldsymbol{S}$, which can be used for visual tasks and for making decisions.

To specify stochastic dynamics, we generalize the stochastic neural model (see equation 12.28) to deal with a set of neurons. A neuron receives input $\boldsymbol{S}$ from other neurons in addition to direct input from the stimulus $\boldsymbol{I}$. Consider only the activity of this neuron, fixing the states of all the others. Then the neurons will have total input of $\sum_j w_{ij} I_j$ plus input $\sum_k \theta_{ik} s_k$ from the other neurons. Then, extending equation 12.28, the probability that the cell i fires is

$$P(s_i \mid \boldsymbol{I},\boldsymbol{S}_{/i}) = \frac{1}{Z_i}\exp\left\{s_i\left(\sum_j w_{ij} I_j + \sum_{k\neq i} \theta_{ik} s_k\right)\right\}, \qquad (12.33)$$

where the notation $\boldsymbol{S}/i$ means the states $\{s_j : j \neq i\}$ of all the neurons except the neuron we are considering. The term Z_i is defined so that the distribution is normalized, so it is given by $Z_i = 1 + \exp\{\sum_j w_{ij} I_j + \sum_{k\neq i} \theta_{ik} s_k\}$.

This gives the following dynamics for a group of neurons. At each time, a neuron is selected at random and fires with a probability specified by equation 12.33. This model assumes that no neurons ever fire at the same time and ignores the time for a spike fired from one neuron to reach other neurons. This is illustrated in interactive demo 4a.

How does this stochastic dynamics relate to the Gibbs distribution specified above? From the statistical perspective, this is an example of *Markov Chain Monte Carlo* (MCMC) sampling (Liu, 2008). MCMC refers to a class of algorithms which explore the state space of $\boldsymbol{S}$ stochastically so that it will gradually move to configurations which have high probability $P(\boldsymbol{S}|\boldsymbol{I})$. More precisely, MCMC algorithms are guaranteed to give samples, $\boldsymbol{S}_1, \ldots, \boldsymbol{S}_M$, from the Gibbs distribution, $P(\boldsymbol{S}|\boldsymbol{I})$. The stochastic update rule in equation 12.33 is a special type of MCMC algorithm which is known as a *Gibbs sampler* because it samples from the conditional distribution $P(s_i|\boldsymbol{I}, \boldsymbol{S}_{/i})$. These samples enable us to estimate the most probable state of the system $\hat{\boldsymbol{S}} = \arg\max P(\boldsymbol{S}|\boldsymbol{I})$; hence they can estimate the *maximum a posteriori* (MAP) estimator of $\boldsymbol{S}$ and make optimal decisions for visual tasks.

To apply these models to visual tasks, we need to specify the weights. One strategy is purely data driven and consists of learning the weights from training examples; this is the *Boltzmann machine* (Ackley, Hinton, and Sejnowski, 1985), which is out of scope for this chapter. Another strategy is to specify distributions for specific visual tasks, and we will give examples in the next few sections.

Dynamical System Models of Neurons There is an alternative way to model sets of neurons using *dynamical systems* based on simplified models of their biophysics (Rieke et al., 1997; Dayan and Abbott, 2001). Pioneering work on this topic was done by Wilson and Cowan (1972), Grossberg and Mingolla (Grossberg, 1968; Grossberg and Mingolla, 1985b) Hopfield and Tank (1986), Abbott and Kepler (1990), and others. There is no space to cover the richness of these models and, in any case, our chapter concentrates on the probabilistic formulation. However, we will discuss an important subclass of dynamical models (Hopfield and Tank, 1986) which, as we will show, have very close relations to the probabilistic approach.

Following Hopfield and Tank, these dynamical systems are described as follows. A neuron is described by two (related) variables: (1) a continuous valued variable $u_i \in \{-\infty, \infty\}$, and (2) a continuous variable $q_i \in \{0,1\}$. Roughly speaking, u_i represents the input to the cell body (soma) due to the direct input and the input from other neurons, and q_i describe the probability that the cell will fire an action potential. These variables are related by the equations $u_i = \log(q_i / (1 - q_i))$ or, equivalently, by $q_i = \sigma(u_i)$ (where $\sigma(.)$ is the sigmoid function).

The dynamics of the neuron is given by:

$$\frac{du_i}{dt} = -u_i + \sum_j w_{ij} I_j + \sum_k \theta_{ik} q_k. \qquad (12.34)$$

Here, as before, $\sum_j w_{ij} I_j + \sum_k \theta_{ik} q_k$ represent the direct input and the input from the other neurons.

It can be shown (see the next section) that this dynamic system continually decreases a function $F(\boldsymbol{q})$, so that $(dF)/dt \leq 0$. The function F acts as a *Lyapunov function* for the system in the sense that it decreases monotonically as time t increases and is bounded below. The existence of a Lyapunov function for the dynamics guarantees that the system converges to a state which minimizes $F(\boldsymbol{q})$ (note that $F(\boldsymbol{q})$ will typically have many minima, and the system may converge to any one of them). This dynamical system is illustrated in interactive demo 4b.

Relations between Probabilistic Models and Dynamical System Models Perhaps surprisingly, there is a very close relationship between the dynamic systems in equation 12.34 and the stochastic update in equation 12.28. More specifically, the dynamic system is a *mean field approximation* to the stochastic dynamics. *Mean field theory* (MFT) was developed by physicists as a way to approximate stochastic systems.

To explain this relationship we first define *the mean field free energy* $F(\boldsymbol{q})$:

$$F(\boldsymbol{q}) = -\sum_{ij} W_{ij} I_j q_i - (1/2) \sum_{ij} \theta_{ij} q_i q_j + \sum_i \{q_i \log q_i + (1 - q_i) \log(1 - q_i)\}. \qquad (12.35)$$

Next we specify dynamics by performing steepest descent on the free energy (multiplied by a positive factor):

$$\frac{dq_i}{dt} = -q_i(1 - q_i) \frac{\partial F(\boldsymbol{q})}{\partial q_i}. \qquad (12.36)$$

Interestingly these are identical to the dynamical system in equation 12.34. This can be seen by introducing a new variable $u_i = \log q_i / (1 - q_i)$, which implies that $q_i = \sigma(u_i)$. Note that $\partial F / \partial q_i = -\sum_j W_{ij} I_j - \sum_j \theta_{ij} q_j + \log q_i / (1 - q_i)$, $u_i = \log q_i / (1 - q_i)$, and $dq_i / q_i(1 - q_i) = du_i$.

Equation 12.36 implies that the dynamical system decreases the free energy $F(\boldsymbol{q})$ monotonically with time t. This is because $dF / dt = -\sum_i (\partial F / \partial q_i)(\partial q_i / \partial t) = -\sum_i q_i (1 - q_i)(\partial F / \partial q_i)^2$. Hence $F(\boldsymbol{q})$ is a Lyapunov function for equations 12.34 and 12.36, and so the dynamics converges to a fixed point.

This shows that there is a close connection between the neural dynamical system and minimizing the mean field free energy. In turn, the mean field free energy is related to deterministic approximations to stochastic update methods like Gibbs sampling (Amit, 1992; Hertz, 1991). This connection is technically advanced and is not needed to understand the rest of this chapter. Briefly, the mean field free energy $F(\boldsymbol{q})$ is the *Kullback–Leibler divergence* $F(Q) = \sum_S Q(\boldsymbol{S}) \log[Q(\boldsymbol{S})] / [P(\boldsymbol{S} \mid \boldsymbol{I})]$ between the distribution $P(\boldsymbol{S}|\boldsymbol{I})$ and a factorized distribution $Q(\boldsymbol{S}) = \prod_i q_i^{S_i} (1 - q_i)^{1 - S_i}$ (plus an additive constant, this can be verified by substitution). Hence the dynamical system seeks to find the factorized distribution $\hat{Q}(\boldsymbol{S})$ which best approximates $P(\boldsymbol{S}|\boldsymbol{I})$ by minimizing the Kullback–Leibler divergence. In this approximation the response q_i is an approximation to the expected response $\sum_{S_1} S_i P(\boldsymbol{S} \mid \boldsymbol{I})$. The connections between MFT and neural models were described in Yuille (1987). For technical discussions about MFT and Gibbs sampling, see Yuille (2011).

12.4.1 The Line Process Model Our first example is the classic *line process* model (Geman and Geman, 1984; Blake and Ziserman, 2003; Mumford and Shah, 1989), which was developed as a way to segment images. It has explicit *line process* variables which "break" images into regions where the intensity is piecewise smooth. Our presentation will follow the work of Koch, Marroquin, and Yuille (1986), who translated it into neural circuits.

The model takes intensity values $\boldsymbol{I}$ as input and outputs smoothed intensity values. However, this smoothness is broken at places where the intensity changes are too high; see figure 12.28. The model has continuous variables $\boldsymbol{J}$ representing the intensity and binary-valued variables $\boldsymbol{l}$ for the line processes (or edges). The model is formulated as performing MAP estimation. The algorithm for estimating MAP is a neural network model which can be derived from the original Markov model (Geman and Geman, 1984) by MFT (Geiger and Yuille, 1991). Note that in this model the variables do not have to represent intensity. Instead, they can represent texture, depth, or any other property which is spatially smooth except at sharp discontinuities.

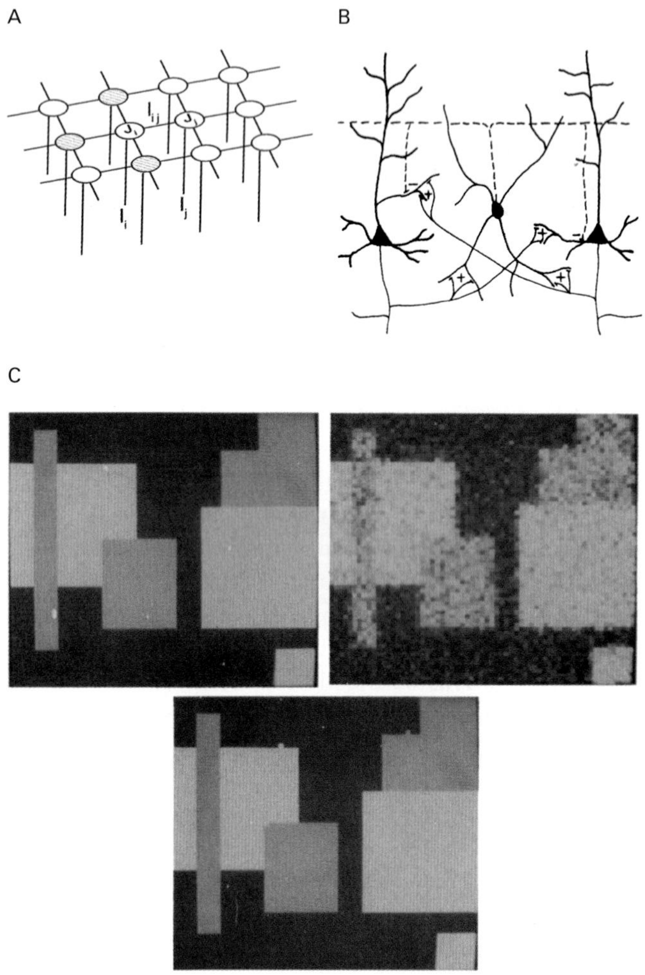

FIGURE 12.28 A representation of the line-process model (A) compared to a real neural network (B). On the right we show the original image (C, top left), the image corrupted with noise (C, top right) and the image estimated using the line-process model (C, bottom). (B) reprinted with permission from C. Koch, J. Marroquin, and A. Yuille. Analog "Neuronal" Networks in Early Vision. *Proceedings of the National Academy of Sciences of the United States of America, 83*(12), 4263–4267, June 1986; (C) reprinted with permission from S. Geman and D. Geman. Stochastic Relaxation, Gibbs Distributions, and the Bayesian Restoration of Images. *IEEE Transactions on Pattern Analysis and Machine Intelligence, 6*(6), 721–741, 1984.

For simplicity we present the weak membrane model in one dimension. The input is $\boldsymbol{I} = \{I(x)\colon x \in D\}$, the estimated, or smoothed, image is $\boldsymbol{J} = \{J(x)\colon x \in D\}$, and the line processes are denoted by $\boldsymbol{l} = \{l(x)\colon x \in D\}$, where $l(x) \in \{0, 1\}$.

The model is specified by a posterior probability distribution:

$$P(\boldsymbol{J},\boldsymbol{l} \mid \boldsymbol{I}) = \frac{1}{Z}\exp\{-E[\boldsymbol{J},\boldsymbol{l} : \boldsymbol{I}]/T\},$$

where

$$E[\boldsymbol{J},\boldsymbol{l} : \boldsymbol{I}] = \sum_x (I(x) - J(x))^2 + A\sum_x (J(x+1) - J(x))^2(1 - l(x)) + B\sum_x l(x).$$

The first term ensures that the estimated intensity $J(x)$ is close to the input intensity $I(x)$. The second encourages the estimated intensity $J(x)$ to be spatially smooth (e.g., $J(x) \approx J(x + 1)$) unless a line process is activated by setting $l(x) = 1$. The third pays a penalty for activating a line process. The result encourages the estimated intensity to be piecewise smooth unless the input $I(x)$ changes significantly, in which case a line process is switched on and the smoothness is broken. The parameter T is the variance of the probability distribution and has a default value $T = 1$.

This model can be implemented by a neural circuit (Koch, Marroquuin, and Yuille, 1986). The connections between these neurons is shown in figure 12.28. To implement this model, Koch, Marroquuin, and Yuille (1986) proposed a neural net model which is equivalent to doing MFT on the weak membrane MRF (as discussed earlier) by replacing the binary-valued line process variables $l(x)$ by continuous variables $q(x) \in [0, 1]$ (corresponding roughly to the probability that the line process is switched on).

This gives an algorithm which updates the regional variables $\boldsymbol{J}$ and the line variables $\boldsymbol{q}$ in a coupled manner. It is helpful, as before, to introduce a new variable $\boldsymbol{u}$ which relates to q by

$$q(x) = \frac{1}{1 + \exp\{-u(x)/T\}}$$

and $u(x) = T \log[q(x)]/[1 - q(x)]$:

$$\frac{dJ(x)}{dt} = -2(J(x)) - I(x)),$$

$$= -2A\{(1 - q(x))(J(x) - J(x+1)) + (1 - q(x-1))(J(x) - J(x-1))\}, \tag{12.37}$$

$$\frac{dq(x)}{dt} = \frac{1}{T} q(x)(1 - q(x)) \left\{A(J(x+1) - J(x))^2 - B - T \log \frac{q(x)}{1 - q(x)}\right\}, \tag{12.38}$$

$$\frac{du(x)}{dt} = -u(x) + A(J(x+1) - J(x))^2 - B. \tag{12.39}$$

The update rule for the estimated intensity $\boldsymbol{J}$ behaves like nonlinear diffusion which smooths the intensity while keeping it similar to input $\boldsymbol{I}$. The diffusion is modulated by the strength of the edges $\boldsymbol{q}$. The update for the lines $\boldsymbol{q}$ is driven by the differences between the estimated intensity; if this is small, then the lines are not activated.

This algorithm has a Lyapunov function $L(\boldsymbol{J}, \boldsymbol{q})$ (derived using MFT methods) and so will converge to a fixed point, with

$$L(\boldsymbol{J},\boldsymbol{q}) = \sum_x (I(x) - J(x))^2 + A\sum_x (J(x+1) - J(x))^2(1 - q(x)) + B\sum_x q(x) + T\sum_x \{q(x)\log q(x) + (1 - q(x))\log(1 - q(x))\}. \tag{12.40}$$

There is some evidence that a generalization of this model roughly matches the electrophysiological findings for those types of stimuli shown in figure 12.29. The generalization is performed by replacing the intensity variables $I(x)$ and $J(x)$ by a filter bank of Gabor filters so that the weak membrane model enforces edges at places where the texture properties change (Lee, Mumford, and Yuille, 1992). The experiments, and their relation to the weak membrane models, are reviewed in Lee and Yuille (2006). The initial responses of the neurons, for the first 80 ms, are consistent with the linear filter models described in section 12.2. However, after 80 ms the activity of the neurons changes and appears to take spatial context into account. The findings of the electrophysiological experiments are summarized as follows: (1) there are two sets of neurons where one set encodes regional properties (such as average brightness) and the other set codes boundary location (in agreement with the J and the l variable in the model, respectively); (2) the processes for computing the region and the boundary representations are tightly coupled, with both processes interacting with and constraining each other (as in the dynamical equations above); (3) during the iterative process, the regional properties diffuse within each region and tend to become constant, but these regional properties do not cross the region (in agreement with the model); and (4) the interruption of the spreading of regional information by boundaries results in sharp discontinuities in the responses across two different regions (in agreement with the model). The development of abrupt changes in regional responses also results in a gradual sharpening of the boundary response, reflecting increased confidence in the precise location of the boundary. These findings are roughly consistent with neural network implementations of the weak membrane model. However, other explanations are possible. For example, the weak membrane model requires lateral (sideways) interaction, and it is possible that the computations are done hierarchically using feedback from V2 to V1.

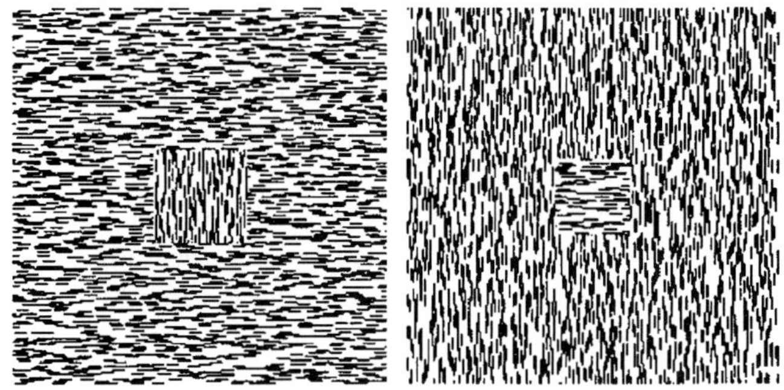

FIGURE 12.29 The stimuli for the experiments by Lamme (1995) and discussed by Lee and Yuille (2006).

While the weak membrane model is broadly consistent with the perceptual phenomena of segmentation and "filling in," the types of filling in, their dynamics, and the neural representations of contours and surface is complicated (von der Heydt, 2003; Komatsu, 2006). Exactly how contour and surface information is represented and processed in cortex is an active topic of research (Grossberg and Hong, 2006; Roe et al., 2012).

12.4.2 EDGE DETECTION WITH SPATIAL CONTEXT Our second example is to develop a model for detecting edges using spatial context. This relates to the phenomena known as association fields; see figure 12.26B, where Gabor filters which are spatially aligned (in orientation and direction) get grouped into a coherent form.

For this model, we have a set of neurons at every spatial position x, each tuned to a different angle θ_i: $i = 1, \ldots, 8$, and a default cell at angle θ_0. The first cells are designed to detect edges at each orientation—that is, they can be driven by the log-likelihood ratio of an edge detector at orientation θ_i at this position. The default cell is a dummy that is intended to fire if there is no edge present at this position. This organization forms a population of cells arrayed according to orientation (similar to a hypercolumn in V1). See figure 12.27C.

We define a Gibbs distribution for the activity s_{x,θ_i} of the cells. The energy function $E(\boldsymbol{s})$ contains four types of terms. First, it contains a term $\sum_x \sum_{i=0}^{8} s_{x,i}\varphi(f_1,\ldots,f_M)$ which represents the local evidence for an edge at each point and for its orientation. It is essentially the same term for local edge detection as discussed in the previous section where, if we ignore orientation so $s_x \in \{\pm 1\}$, then

$$\phi(f_1,\ldots,f_M) = \log \frac{P(f(I(x) \mid s=1)}{P(f(I(x) \mid s=0)}.$$

Second, there is a term $\sum_x (\sum_{i=0}^{8} s_{x,i} - 1)^2$ that is intended to ensure that only one cell is active at any spatial position. This corresponds to an inhibitory interaction between cells in the same hypercolumn. The cells in the hypercolumn give alternative, and inconsistent, interpretations of the input—hence only one of them can be correct. Third, there is a term that encourages edges to be continuous and for their directions to change smoothly. To define this term, we let $\boldsymbol{\theta}_i = (\cos\theta_i, \sin\theta_i)$ and $\boldsymbol{\theta}_i^T = (-\sin\theta_i, \cos\theta_i)$ denote the tangent to the edge and the normal. This term encourages there to be edges in the tangent direction, while the next term discourages them in the normal direction; see figure 12.27D. This term is motivated by the intuition that curves are spatially smooth and can be justified by the statistics of natural images (Geisler and Perry, 2009; Elder and Goldberg, 2002). We write it as $\sum_{x,y} \sum_{i,j=1}^{8} W^N_{(x,\theta_i),(y,\theta_j)} s_{x,i} s_{y,j}$, where

$$W^T_{(x,\theta_i),(y,\theta_j)} = -\exp\{-|\boldsymbol{\theta}_i - \boldsymbol{\theta}_j|/K_1\}\exp\{-|x-y|/K_2\} \exp\{-|\hat{xy} - \boldsymbol{\theta}_i|/K_3\} \tag{12.41}$$

and $\hat{xy}$ is the unit vector in direction $x - y$. This term encourages edges which are in similar directions (first term), nearby in position (second term), and where the edge orientation is similar to the difference $x - y$ between the two points. This term is excitatory. Fourth, the final term is inhibitory and discourages edges to be parallel to each other (if they are nearby). It is written as $\sum_{x,y} \sum_{i,j=1}^{8} W^N_{(x,\theta_i),(y,\theta_j)} s_{x,i} s_{y,j}$. Here

$$W^N_{(x,\theta_i),(y,\theta_j)} = \exp\{-|x-y|/K_4\}\exp\{-|\hat{xy} - \boldsymbol{\theta}_i^T|\}. \tag{12.42}$$

The first term says this interaction decreases with distance. The second term discourages edges which are parallel to each other.

This gives an overall energy:

$$\begin{aligned} E(s) = &\sum_x \sum_{i=0}^{8} s_{x,i}\phi(f_1,\ldots,f_M) + \hat{K}_0 \sum_x \left(\sum_{i=0}^{8} s_{x,i} - 1 \right)^2 \\ &+ \hat{K}_1 \sum_{x,y} \sum_{i,j=1}^{8} W^T_{(x,\theta_i),(y,\theta_j)} s_{x,i} s_{y,j} + \hat{K}_2 \\ &+ \sum_{x,y} \sum_{i,j=1}^{8} W^N_{(x,\theta_i),(y,\theta_j)} s_{x,i} s_{y,j}. \end{aligned} \tag{12.43}$$

This yields a probability:

$$P(s \mid f) = \frac{1}{Z}\exp\{-E(s)\}.$$

This model can be implemented in neural networks by defining either stochastic or deterministic neural dynamics (i.e., either Gibbs sampling or MFT). The resulting update equations are more complex than those defined for our earlier examples but have the same basic ingredients. Models of this type can qualitatively account for associative field phenomena.

12.4.3 STEREO MODELS This section introduces computational models for estimating depth by binocular stereo. The key problem is to solve the *correspondence problem* between the inputs in the two eyes to determine the *disparity*. Then the depth of the points in space can be estimated by trigonometry. (This presupposes that the eyes are *calibrated*, meaning that the distance between the eyes and the direction of gaze are known, which is beyond the scope of this chapter.) As discussed in section 12.1, Julesz (1971) showed that humans could perceive depth from stereo if the images consisted of random dot stereograms which minimize the effect of feature similarity cues, suggesting that human vision can solve this task by relying mainly on geometric regularities (assumed about the structure of the world). Other researchers (Bülthoff and Mallot, 1988) have studied human estimation of surface shape quantitatively and showed, among other things, bias toward frontoparallel surfaces.

Most stereo algorithm address the correspondence problem by assuming that (1) image features in the two eyes are more likely to correspond if they have similar appearance and (2) the surface being viewed obeys prior knowledge such as being piecewise smooth (e.g., like the weak membrane model). The first assumption depends on local properties of the images while the second assumption uses nonlocal context. In an earlier section we discussed how a population of Gabor filters could be used to match local image features. In this section we describe how context can be used to impose prior knowledge about the geometry of the scene. We will study classic models which assume that the surface is piecewise smooth. This leads to a Markov field model which includes excitatory connections, imposing the geometric constraints, with inhibitory connections which prevent points from one eye having more than one match in the second eye. This yields an algorithm which involves cooperation, to implement the excitatory constraints, and competition, to deal with the inhibitory constraints. This is consistent with findings from recent electrophysiological experiments (Samonds, Potetz, and Lee, 2009, 2012). These complement earlier experiments (Ohzawa, Deangelis, and Freeman, 1990) which tested the local stereo models described in section 12.2.

A Cooperative Stereo Model We now specify a computational model for stereo which, for simplicity, we formulate in one dimension. There is a long history of this type of model starting with the cooperative stereo algorithm (Dev, 1975; Marr and Poggio, 1976), and current computer vision stereo algorithms are mostly designed on similar principles.

We specify the left and right images by $\boldsymbol{I}_L$, $\boldsymbol{I}_R$ and denote features extracted from them by $f(\boldsymbol{I}_L) = \{f(x_L): x_L \in D_L\}$, $f(\boldsymbol{I}_R) = \{f(x_R): x_R \in D_R\}$. We define a discrete-valued correspondence variable $V(x_L, x_R)$, so that $V(x_L, x_R) = 1$ means that the features at x_L, x_R in the two images correspond, and hence the disparity is $x_L - x_R$. If the features do not match, then we set $V(x_L, x_R) = 0$. We encourage all data points to match one, but allow some data points to be unmatched and others to match more than once (by paying a penalty).

We specify a distribution

$$P(\boldsymbol{V} \mid f(\boldsymbol{I}_L), f(\boldsymbol{I}_R)) = \frac{1}{Z}\exp\{-E(\boldsymbol{V}; f(\boldsymbol{I}_L), f(\boldsymbol{I}_R))/T\},$$

where the energy $E(\boldsymbol{V}; f(\boldsymbol{I}_L), f(\boldsymbol{I}_R))$ is given by:

$$\begin{aligned} E(\boldsymbol{V}; f(\boldsymbol{I}_L), f(\boldsymbol{I}_R)) = {} & \sum_{x_L, x_R} V(x_L, x_R) M(f(x_L), f(x_R)) \\ & + A\sum_{x_L}\left(\sum_{x_R} V(x_L, x_R) - 1\right)^2 \\ & + A\sum_{x_R}\left(\sum_{x_L} V(x_L, x_R) - 1\right)^2 \\ & + C\sum_{x_L, x_R}\sum_{y_L \in N(x_L)}\sum_{y_R \in N(x_R)} V(x_L, x_R) \\ & V(y_L, y_R)\{(x_R - x_L) - (y_R - y_L)\}^2. \end{aligned} \tag{12.44}$$

The first term imposes that there are matches between image points with similar features, here $M(.,.)$ is a measure which takes small values if $f(x_L)$, $f(x_R)$ are similar and large values if they are different. We will discuss at the end of this section how $M(f(x_L), f(x_R))$ relates to the model for local stereo discussed earlier. The second two terms penalize image points which are either unmatched or are matched more than once. The third term encourages the disparities, $x_L - x_R$, to be similar for neighboring points (here $N(.)$ defines a spatial neighborhood as before). These models can be applied to two-dimensional images by solving the correspondence problem for each epipolar line separately (by maximizing $P(\boldsymbol{V}|f(\boldsymbol{I}_L), f(\boldsymbol{I}_R))$). This is shown in figure 12.30B. The parameter T is the variance of the model, as for the line process model, and has default value $T = 1$.

As for previous models, we can obtain a neural circuit model by performing MFT on $P(\boldsymbol{V}|f(\boldsymbol{I}_L), f(\boldsymbol{I}_R))$. (We restrict each point to have one or zero matches for this algorithm.) This replaces $V(x_L, x_R) \in \{0, 1\}$ by continuous-valued $q(x_L, x_R) \in [0, 1]$ and an associated variable

$$u(x_L, x_R) = T\log\frac{q(x_L, x_R)}{1 - q(x_L, x_R)} \quad \text{with}$$

$$q(x_l, x_R) = \frac{1}{1 + \exp\{-u(x_L, x_R)\}}.$$

The update equation is

$$\begin{aligned} \frac{du(x_L, x_R)}{dt} = {} & -u(x_L, x_R) - M(f(x_L), f(x_R)) \\ & - 2A\left(\sum_{y_R \neq x_R} q(x_L, x_R) - 1\right) - 2A\left(\sum_{y_L \neq x_L} q(x_L, x_R) - 1\right), \\ & - 2C\sum_{y_L \in N(x_L)}\sum_{y_R \in N(x_R)} q(x_L, x_R)\{(x_L - x_R) - (y_R - y_L)\}^2. \end{aligned} \tag{12.45}$$

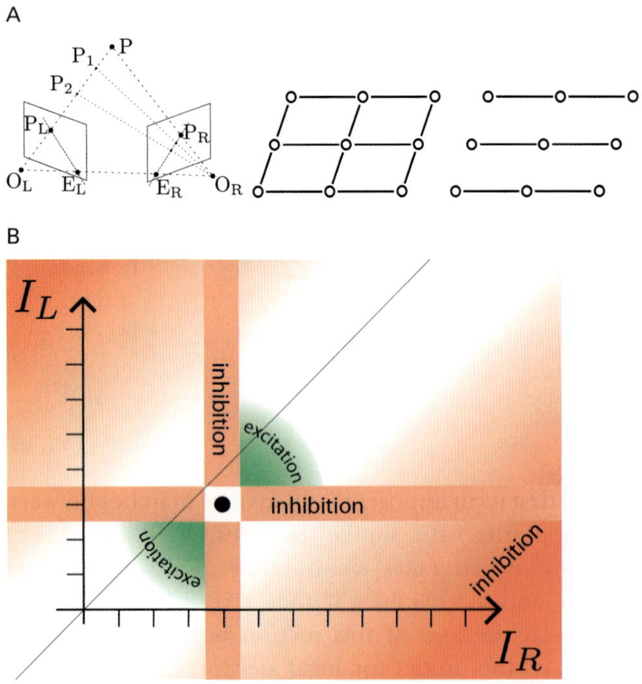

FIGURE 12.30 Stereo. (A) The geometry of stereo. A point P in 3-D space is projected onto points PL and PR in the left and right images. The projection is specified by the focal points OL, OR and the directions of gaze of the cameras (the camera geometry). The geometry of stereo enforces that points in the plane specified by P, OR, OL must be projected onto corresponding lines EL and ER in the two images (the epipolar line constraint). If we can find the correspondence between the points on epipolar lines, then we can use trigonometry to estimate their depth, which is (roughly) inversely proportional to the disparity, which is the relative displacement of the two images. (B) Binocular stereo requires solving the correspondence problem which involves excitation (to encourage matches with similar depths/disparities) and inhibition (to prevent points from having multiple matches).

This update includes the standard integration term (first term), and the second term encourages matches where the features agree. There is also inhibition between competing matches (the third and fourth term), and excitation for matches which are consistent with a smooth surface (last term).

There is a variant of this algorithm which is used in interactive demo 4c. This algorithm is a discrete Hopfield network which attempts to minimize the energy $E(\boldsymbol{V}; f(\boldsymbol{I}_L), f(\boldsymbol{I}_R))$ in equation 12.44. The algorithm starts by assigning initial values, 0 or 1, to each state variable $V(x_L, x_R)$. The algorithm proceeds by selecting a state variable, changing its value (e.g., changing $V(x_L, x_R) = 1$ to $V(x_L, x_R) = 0$), calculating if this change reduces the energy $E(\boldsymbol{V}; f(\boldsymbol{I}_L), f(\boldsymbol{I}_R))$, and keeping the change if it does. This process repeats until the algorithm converges (i.e., all possible changes raise the value of the energy).

How does the cooperative stereo algorithm relate to our earlier algorithm for computing stereo disparity locally? Recall that the algorithm estimated the disparity at a single point by having a set of neurons which were tuned to different disparities $\{D_i: i = 1, \ldots, N\}$, summing the votes $v(D_i)$ for each disparity by equation 12.20, and selecting the disparity with most votes. Using the cyclopean coordinate system (Julesz, 1971), we express the disparity by $D(x) = (1/2)(x_R - x_L)$ where $x = (1/2)(x_R + x_L)$. At each point x we specify a population of neurons which encode the votes $v(D(x))$ for the different disparities. Then, instead of using winner-take-all to make a local decision, we feed the responses $v(D(x))$ back into cooperative stereo algorithm by defining $M(f(x_L), f(x_R)) = \exp\{-v(1/2(x_R - x_L))\}$ (the negative exponential $exp\{-\}$ is required so the $M(f(x_L), f(x_R))$ is small if the vote for disparity $D(x) = 1/2(x_R - x_L)$ is large).

Analysis of electrophysiological studies (Samonds, Potetz, and Lee, 2009, 2012) were in general agreement with the predictions of this type of stereo algorithm. In particular, studies showed that neural populations' responses included excitation between cells tuned to similar disparities at neighboring spatial positions and inhibition between cells tuned to different disparities at the same position; see figure 12.31. In addition, Samonds et al. (2013) implemented a variant of the stereo algorithm described above and showed that it could account for additional phenomena such as sharper tuning to the disparity for larger stimuli, narrower tuning with time, and performance on anticorrelated stimuli (where the left and right images have opposite contrast polarity).

12.4.4 MOTION Similar models have been applied to a range of motion phenomena. Early computational studies (Ullman, 1979) showed that several perceptual phenomena of long-range motion could be described by a "minimal mapping" theory that uses a slowness prior. Subsequent work showed that smoothness priors accounted for findings on short-range motion (Hildreth, 1984), including the surprising fact that an ellipse rotating in the image plane is perceived to move nonrigidly. Yuille and Grzywacz (1988) qualitatively showed that a slow-and-smooth prior could account for a large range of motion perceptual phenomena—including motion capture and motion cooperation—both for short- and long-range motion. Weiss and his collaborators showed that slow (Weiss and Adelson, 1998) and slow-and-smooth priors (Weiss, Simoncelli, and Adelson, 2002) could explain other short-range

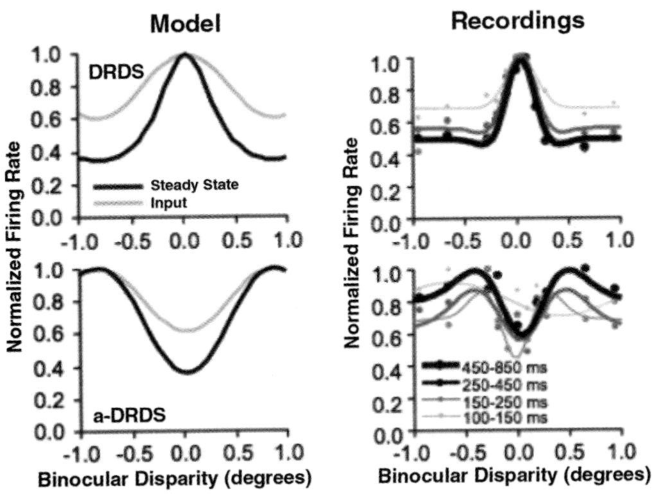

FIGURE 12.31 Model and actual neural firing rates as a function of binocular disparity in response to dynamic random dot stereograms. Tuning curves for a neuron in the model with recurrent connections and from monkey recordings become narrower over time (from gray to black) when presented with a standard correlated dynamic random dot stereograms (DRDS). Tuning curves in response to anti-correlated dynamic random dot stereograms (a-DRDS) do not sharpen in the same consistent way. Reprinted from Samonds, J. M., Potetz, B. R., Tyler, C. W., and Lee, T. S. (2013). Recurrent connectivity can account for the dynamics of disparity processing in v1. *Journal of Neuroscience, 33*(7), 2934–2946.

motion phenomena, such as how percepts can change dramatically as we alter the balance between the likelihood and prior terms (i.e., for some stimuli the prior dominates the likelihood and vice versa). All these models combine local estimates of the motion, such as those described in the previous section, with contextual cues implementing slow-and-smooth priors. They can be formulated using the same mathematical techniques. Look at http://www.michaelbach.de/ot/mot-motionBinding/ to see how spatial context can be affected by other cues such as occlusion. It is also possible to perceive three-dimensional structure by observing a motion sequence (somewhat similar to binocular stereo) as can be seen in http://michaelbach.de/ot/mot-ske/.

The perception of motion can be strongly influenced by its past history and not merely by the change of image from frame to frame. For example, Anstis and Ramachandran (1987) demonstrated perceptual phenomena where motion perception seems to require a temporal coherence prior in addition to the slow and smoothness priors described earlier in this section. Similarly, Watamaniuk et al. (1995) demonstrated that humans could detect a coherently moving dot despite the presence of many incoherently moving dots. These classes of phenomena can be addressed by models which make prior assumptions about how motion changes over time. These can be performed (Yuille, Burgi, and Grzywacz, 1998) by adapting the Bayes–Kalman filter (Kalman, 1960; Ho and Lee, 1964), which gives an optimal way to combine information over time.

The task of the Bayes–Kalman filter is to estimate the state x_t of a system at time t dependent on a set of observations $y_t, \ldots, y_1$ (e.g., x_t could be the position of an airplane and y_t a noisy measurement of the airplane's position at time t). The model assumes a probability distribution $P(x_{t+1}|x_y)$ for how the state changes over time and a likelihood function $P(y_t|x_t)$ for the observation.

This can be formulated by a Markov model; see figure 12.32A where the observations $y_t, \ldots, y_1$ and states $x_t, \ldots, x_1$ are represented by the blue and red dots, respectively (the lower and upper dots if viewed in black and white).

The purpose of Bayes–Kalman is to estimate the distribution $P(x_t|Y_t)$ of the state x_t conditioned on the measurements $Y_t = \{y_t, \ldots, y_1\}$ up to time t. It performs this by repeatedly performing the following two steps, which are called prediction and correction. The prediction uses the prior $P(x_{t+1}|x_t)$ to predict distribution $P(x_{t+1}|Y_t)$ of the state at $t+1$:

$$P(x_{t+1} \mid Y_t) = \int dx_t P(x_{t+1} \mid x_t) P(x_t \mid Y_t). \qquad (12.46)$$

The correction step integrates the new observation y_{t+1} to estimate $P(x_{t+1} \mid Y_{t+1})$ by:

$$P(x_{t+1} \mid Y_{t+1}) = \frac{P(y_{t+1} \mid x_{t+1}) P(x_{t+1} \mid Y_t)}{P(y_{t+1} \mid Y_t)}. \qquad (12.47)$$

Bayes–Kalman is initialized by setting $P(x_1|y_1) = P(y_1|x_1)P(x_1)/P(y_1)$ where $P(x_1)$ is the prior for the original position of the object at the start of the sequence. Then equations 12.46 and 12.47 are run repeatedly. The effect of prediction is to introduce uncertainty about the state x_t, while correction reduces uncertainty by providing a new measurement; see figure 12.32B.

12.4.5 SUMMARY OF MODELS WITH CONTEXT This section illustrated how neural networks and Markov models could be used to apply context to visual tasks. We concentrated on edge detection, segmentation, and binocular stereo. We stressed how context can include excitatory and inhibitory interactions. And how inference can be performed using stochastic neurons (e.g., Gibbs sampling) or dynamic neural networks (e.g., mean field approximations). These models have some relations to psychophysics and electrophysiology. However, we stress that detailed biological evidence in

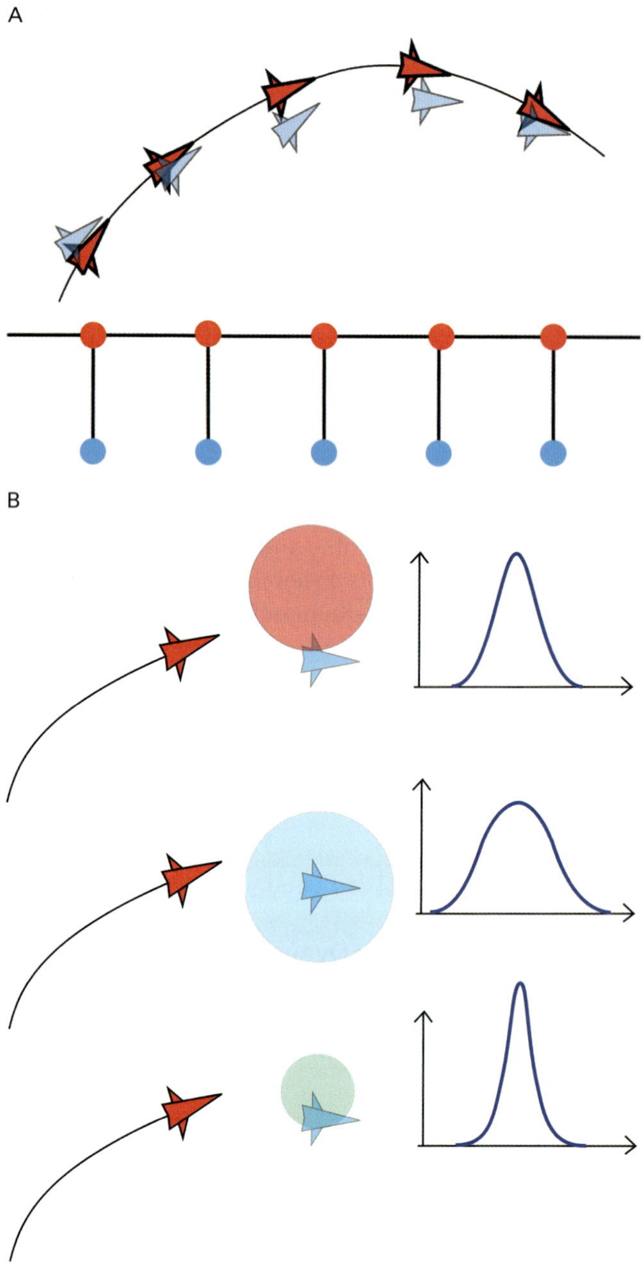

FIGURE 12.32 (A) Graph illustrating the unobserved states (red) and the observed states (blue) as a function of time. The airplane's true positions are shown in red and their observations (biased) are shown in blue. The Bayes–Kalman filter integrates observations to make estimate the true state using prior probabilities. (B) Bayes–Kalman updates a probability distribution for the estimated position of the target. The variance of the distribution is illustrated by the one-dimensional figure (on the right) and the size of the circle (red, blue, or green). In the prediction stage (middle panel) the variance becomes large and after the measurement the variance becomes smaller.

favor of these models remains preliminary because of the current limitations of experimental techniques. We note that current computer vision algorithms which address similar visual tasks are more complex although based on similar principles (Blake, Kohli, and Rother, 2011).

12.5 *Cue Coupling*

The ideas in this section are logical extensions of the ideas in the earlier sections. However, we are now addressing more complex aspects of vision, and so the techniques and the tools become more complex and more abstract as we begin to reason about surfaces, objects, and their relations. This section contains an interactive demo (demo 5) for combining cues by weighted averages.

12.5.1 VISION MODULES AND CUE COMBINATION At the behavioral level, psychophysicists have studied how humans combine different visual cues—such as shading, texture, binocular stereo, and structure from motion—to estimate depth, surface geometry, and other surface properties. For example, quantifiable psychophysics experiments are broadly consistent with the predictions of the types of models discussed in the previous two sections—see Bülthoff and Mallot (1988) and Cumming, Johnston, and Parker (1993)—but with some exceptions (Todd et al., 2001).

We now address how these cues can be combined. The most straightforward manner is to use a separate module for each cue to compute different estimates of the properties of interest, for example, the surface geometry, and then merge these estimates into a single representation. This was proposed by Marr (1982), who justified this strategy by invoking the principle of modular design. In particular, Marr proposed that surfaces should be represented by a *2 1/2–D sketch* which specifies the shape of a surface by the distance of the surface points from the viewer. A related representation, *intrinsic images*, also represents surface shape together with the material properties of the surface. This strategy of making separate estimates for different cues and then combining them has also been followed by computer vision researchers.

How should we combine cues computed independently? The natural way to combine these cues, from a probabilistic perspective, requires taking into account the uncertainty of the cues. This will be discussed in section 12.5.2 where we show that this can reduce to combining cues by taking weighted linear combinations, where the weights depend on the uncertainty. This strategy is able to account for some examples of cue combination (Landy et al., 1995; Jacobs, 1999; Ernst and Banks, 2002).

Next we consider the problem of cue combination from a deeper probabilistic analysis; see section 12.5.3 (Clark and Yuille, 1990). Theoretical analysis suggests that we need to distinguish between situations where the cues are statistically independent of each other and the cases where they are not. We need also need to determine whether cues are using similar, and hence redundant, prior information. These considerations lead to a distinction between *weak* and *strong* coupling, where weak coupling corresponds to the traditional view of modules while strong coupling considers more complex interactions. To understand strong coupling, it is helpful to consider the *causal factors* which generate the image. In addition, it involves the idea of *model selection*, called "competitive priors" in Yuille and Bülthoff (1996), which is a selection process that arises in two types of situations. Firstly, some image cues only apply to parts of the image and we need a process to select those parts (e.g., shape from texture cues are only valid within certain parts of an image). Secondly, for some images there may be several alternative ways to generate them, yielding several possible interpretations, and we need to select the most probable. Psychophysics experiments (e.g., Blake and Bülthoff, 1990; Sundareswara and Schrater, 2008) investigate this by setting up situations where small changes in the image are sufficient to switch from one interpretation to another.

To describe causal models, and strong coupling, in more detail requires studying how images are formed in terms of the observer's viewpoint, the structure and material properties of objects in the world, and the lighting conditions. Then we proceed to two examples by Körding et al. and by Knill in section 12.5.4.

There is strong evidence that high-level recognition affects the estimation of three-dimensional shape (e.g., a rigidly rotating inverted face mask is perceived as nonrigidly deforming face, while most rigidly rotating objects are perceived to be rigid). However, most visual cues are able to function even when images contain no objects. We briefly discuss this in section 12.6.

12.5.2 Combining Cues with Uncertainty

We first consider simple models which assume that the cues compute representations independently and then combine their outputs by taking linear weighted combinations. In some specific cases, these weights can be derived as measures of the uncertainly of the cues.

Combining Cues by Linear Weighted Sums Suppose there are two cues for depth, or some other scene property, which separately give estimates $\boldsymbol{S}_1^*, \boldsymbol{S}_2^*$. One strategy to combine these cues is by linear weighted combination yielding a combined estimate $\boldsymbol{S}^*$:

$$\boldsymbol{S}^* = \omega_1 \boldsymbol{S}_1^* + \omega_2 \boldsymbol{S}_2^*, \tag{12.48}$$

where ω_1, ω_2 are positive weights such that $\omega_1 + \omega_2 = 1$.

Landy, Maloney, and colleagues (1995) reviewed many early studies on cue combination and argued that they could be qualitatively explained by this type of model. They also discussed situations where the individual cues did not combine and "gating mechanisms" which require one cue to be switched off.

Special Case Where Weights Are Derived from Uncertainties An important special case of this model is when the weights are measures of the uncertainty, or reliability, of the two cues. In this case, the linear weighted combination rule is optimal for the restricted case of Bayes estimation when there are no priors (we will discuss priors in the next section). It also yields detailed experimental predictions which have been successfully tested for some types of cue coupling (Jacobs, 1999; Ernst and Banks, 2002) although there are some interesting exceptions (Cheng et al., 2007; Gori et al., 2008).

This special case associates each cue by an uncertainty σ_1^2, σ_2^2 and sets the weights to be $w_1 = \sigma_2^2 / (\sigma_1^2 + \sigma_2^2)$ and $w_2 = \sigma_1^2 / (\sigma_1^2 + \sigma_2^2)$. It follows that $w_1 / w_2 = \sigma_2^2 / \sigma_1^2$, so the cue with lowest uncertainty will have a higher weight. For example, if $\sigma_2^2 >> \sigma_1$ then $w_1 >> w_2$. This special case gives the linear combination rule:

$$S^* = \frac{\sigma_2^2}{\sigma_1^2 + \sigma_2^2} S_1^* + \frac{\sigma_1^2}{\sigma_1^2 + \sigma_2^2} S_2^*. \tag{12.49}$$

This combination rule can be shown to be optimal for the following conditions. Suppose the two cues are modeled by inputs $\{\boldsymbol{C}_i: i = 1, 2\}$ and outputs $\boldsymbol{S}$ which are related by conditional distributions $\{P(\boldsymbol{C}_i|\boldsymbol{S}): i = 1, 2\}$. We assume that these cues are *conditionally independent* so that $P(\boldsymbol{C}_1, \boldsymbol{C}_2|S) = P(\boldsymbol{C}_1|\boldsymbol{S})P(\boldsymbol{C}_2|\boldsymbol{S})$ and both distributions are Gaussians:

$$\begin{aligned} P(\boldsymbol{C}_1 \mid \boldsymbol{S}) &= \frac{1}{Z_1} \exp\left\{-\frac{|\boldsymbol{C}_1 - \boldsymbol{S}|^2}{2\sigma_1^2}\right\}, P(\boldsymbol{C}_2 \mid \boldsymbol{S}) \\ &= \frac{1}{Z_2} \exp\left\{-\frac{|\boldsymbol{C}_2 - \boldsymbol{S}|^2}{2\sigma_2^2}\right\}. \end{aligned} \tag{12.50}$$

In this case, the optimal estimates of the output $\boldsymbol{S}$, for each cue independently, are given by the maximum likelihood estimates:

$$\boldsymbol{S}_1^* = \arg\max_S P(\boldsymbol{C}_1 \mid \boldsymbol{S}) = \boldsymbol{C}_1, \boldsymbol{S}_2^* = \arg\max_S P(\boldsymbol{C}_2 \mid \boldsymbol{S}) = \boldsymbol{C}_2. \tag{12.51}$$

But if both cues are available, then the optimal estimate is given by

$$S_1^* = \arg\max_S P(C_1, C_2 \mid S) = \arg\max_S P(C_1 \mid S)P(C_2 \mid S) \quad (12.52)$$

$$= \frac{\sigma_2^2}{\sigma_1^2+\sigma_2^2} C_1 + \frac{\sigma_1^2}{\sigma_1^2+\sigma_2^2} C_2, \quad (12.53)$$

which reduces to the rule above using $S_1^* = C_1$ and $S_2^* = C_2$.

This shows that the optimal combination is a weighted linear sum of the two cues where the weights can be obtained from the uncertainties σ_1^2, σ_2^2 of the individual cues. This is the maximum likelihood estimate which, as discussed earlier, is the optimal Bayesian way to combine cues if there is no prior $P(S)$ (and the loss function is the default). Interactive demo 5 illustrates how the cues coupling result depends on the means and variances of each cue.

Extensions of Combinations Weighted by Uncertainties This section covers advanced material showing how the results in the previous section can be extended, as approximations, to a much richer class of models. These extensions require that conditional independence holds, $P(C_1, C_2|S) = P(C_1|S)P(C_2|S)$, and there is no prior. The best estimates for each cue individually are given by $S_1^* = \arg\max_S \log P(C_1 \mid S)$ and $S_2^* = \arg\max_S \log P(C_2 \mid S)$. We can locally approximate

$$P(C_1 \mid S) = \log P(C_2 \mid S^*) + \frac{1}{2}(S - S_1^*)^T \frac{\partial^2 \log P(C_1 \mid S)}{\partial S \partial S} S$$
$$= S_1^*(S - S_1^*)$$

(we use notation $S = S_1^*$ to indicate where we are evaluating the derivative). This is a Taylor series approximation of log $P(C_1|S)$ about its maximum values S_1^*. The first term in the expansion, $(\partial \log P(C_1 \mid S)S = S_1^*)/\partial S$ is zero because S_1^* is a maximum. The second term is the second order derivative of log $P(C_1|S)$ with respect to S_1^*. This reduces to $-(1/\sigma_1^2)$ for the Gaussian case discussed above; hence the second order derivative evaluated at the maxima S_1^* gives an estimate of the "local variance" for the likelihood function. If we make this Taylor series approximation for both likelihood functions, then we can estimate an approximation solutions S_{app}^* when both cues are present:

$$S^* = \left(\sum\nolimits_1^{-1} + \sum\nolimits_2^{-1}\right)^{-1}\left\{\sum\nolimits_1^{-1} S_1 + \sum\nolimits_2^{-1} S_2\right\}, \quad (12.54)$$

where $\sum_1^{-1}$ is the matrix for $\frac{\partial^2 \log P(C_1 \mid S)}{\partial S \partial S} S = S_1^*$

(and similarly for $\sum_2^{-1}$).

This Taylor series approximation is only valid if the results S^* is "close" to S_1^* and S_2^*. How close depends on the form of the functions log $P(C_1|S)$, log $P(C_2|S)$ and, more specifically, how well they can be approximated near S_1^*, S_2^* by quadratic functions. If these distributions are both Gaussians then their log-likelihoods are quadratic everywhere and so no approximation is required. In other situations, the nature of the cues determines how good an approximation we can obtain by this Taylor series expansion.

12.5.3 Bayesian Analysis: Weak and Strong Coupling This section discusses more complex models for coupling cues. It approaches the problem from a Bayesian perspective (Clark and Yuille, 1990; Yuille and Bülthoff, 1996). This approach emphasizes that the uncertainties of the cues are taken into account and the statistical dependencies between the cues are made explicit. Simple examples of cue coupling, where the cues and independent, are called "weak coupling" in this framework. In some situations, if priors are not used, weak coupling reduces to the types of models studied in the previous section. By contrast, "strong coupling" is required if the cues are dependent on each other. Determining the dependencies between different cues requires sophisticated modeling to determine the causal factors underlying the cues. This also includes *model selection* where the cues allow several alternative interpretations of the image and small changes of the cues can dramatically change the percept.

The Priors: Avoiding Double Counting As discussed in earlier sections, models of cues typically include prior probabilities about S. For example, most cues for estimating shape or depth assume that the viewed scene is piecewise smooth; see the section on binocular stereo. More generally, vision is ambiguous in the sense that there are many different ways to generate the identical image and a visual system must determine which is the most probable way. Ambiguity is greatly reduced if the input is a sequence of images, but humans have little difficulty interpreting single images correctly, except when there are unusual conditions (or when only small parts of the image are shown). But the bottom line, for cue integration, is that we typically need to consider the prior probabilities of S instead of neglecting them as we did in the previous section.

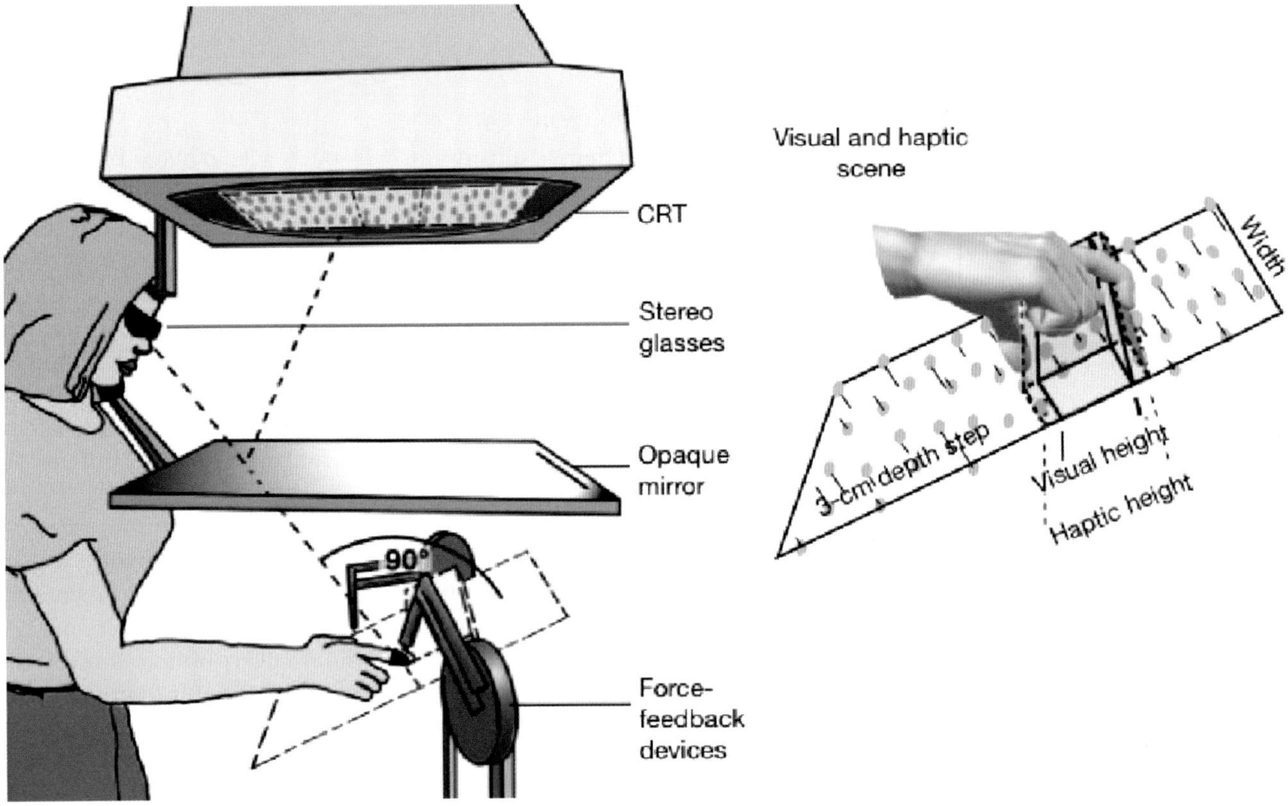

FIGURE 12.33 The work of Ernst and Banks shows that cues are sometimes combined by weighted least squares where the weights depend on the variance of the cues. Reprinted by permission from Macmillan Publishers Ltd: M. O. Ernst and M. S. Banks. Humans Integrate Visual and Haptic Information in a Statistically Optimal Fashion. *Nature, 415*(6870), 429–433, Copyright 2002.

We now revisit the linear weighted sum rule from the perspective that cues usually require priors. To get an intuitive feel for this, suppose we have two cues for estimating the shape of a surface, and suppose that both cues use the prior that the surface is spatially smooth. Simply taking a linear weighted sum of the cues will not be optimal because the prior would be used twice. All priors introduce a bias to perception, so we want to avoid doubling this bias. Experiments by Bülthoff and Mallot (1988) suggest that priors are shared in this manner (i.e., no double counting) and so cues are not combined by linear weighted averaging. In these experiments, subjects were asked to estimate the orientation of surfaces using texture and shading cues. If only one cue was available, then the subjects tended to underestimate the surface orientation, which is consistent with a prior bias toward smooth surfaces (Yuille and Bülthoff, 1996). But if both cues were present, then subjects estimated the surface orientation more accurately, which is inconsistent with combining the cues by a linear weighted sum.

We first model the two cues separately by likelihoods $P(\boldsymbol{C}_1|\boldsymbol{S})$, $P(\boldsymbol{C}_2|\boldsymbol{S})$ and a prior $P(\boldsymbol{S})$. (See figure 12.34.) For simplicity we assume that the priors are the same for each cue, which is reasonable because they specify assumptions about the world independent of any observations. This gives posterior distributions for each visual cue:

$$P(\boldsymbol{S}|\boldsymbol{C}_1) = \frac{P(\boldsymbol{C}_1|\boldsymbol{S})P(\boldsymbol{S})}{P(\boldsymbol{C}_1)}, \; P(\boldsymbol{S}|\boldsymbol{C}_2) = \frac{P(\boldsymbol{C}_2|\boldsymbol{S})P(\boldsymbol{S})}{P(\boldsymbol{C}_2)}. \quad (12.55)$$

This yields estimates of surface shape to be $\boldsymbol{S}_1^* = \arg\max_{S1} P(\boldsymbol{S}|\boldsymbol{C}_1)$ and $\boldsymbol{S}_2^* = \arg\max_{S2} P(\boldsymbol{S}|\boldsymbol{C}_2)$. The optimal way to combine the cues is to estimate $\boldsymbol{S}$ from the posterior probability $P(\boldsymbol{S}|\boldsymbol{C}_1, \boldsymbol{C}_2)$:

$$P(\boldsymbol{S}|\boldsymbol{C}_1, \boldsymbol{C}_2) = \frac{P(\boldsymbol{C}_1, \boldsymbol{C}_2|\boldsymbol{S})P(\boldsymbol{S})}{P(\boldsymbol{C}_1, \boldsymbol{C}_2)}. \quad (12.56)$$

If the cues are *conditionally independent*, $P(\boldsymbol{C}|\boldsymbol{S}) = P(\mathbf{C}_1|\boldsymbol{S})P(\boldsymbol{C}_2)|\boldsymbol{S})$, then this simplifies to

$$P(\boldsymbol{S}|\boldsymbol{C}_1, \boldsymbol{C}_2) = \frac{P(\boldsymbol{C}_1|\boldsymbol{S})P(\boldsymbol{C}_2|\boldsymbol{S})P(\boldsymbol{S})}{P(\boldsymbol{C}_1, \boldsymbol{C}_2)}. \quad (12.57)$$

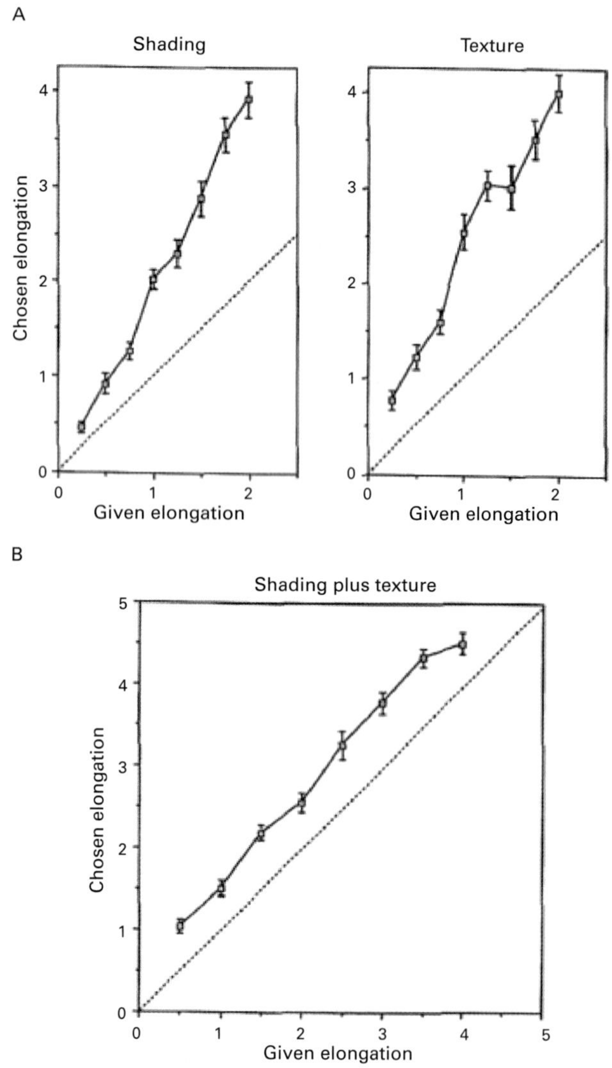

FIGURE 12.34 Cue coupling results which are inconsistent with linear weighted average (Bülthoff et al., 1990). (A, left) If depth is estimated using shading cues only, then humans underestimate the perceived orientation (i.e., they see a flatter surface). (A, right) Humans also underestimate the orientation if only texture cues are present. (B) But if both shading and texture cues are available, then humans perceive the orientation correctly. This is inconsistent with taking the linear weighted average of the results for each cue separately. Figure reprinted with permission from H. Bulthoff and H. A. Mallot (1990). Integration of stereo, shading and texture. In *AI and the Eye*, 11th European Conference on Visual Perception (pp. 119–146). Chichester: Wiley.

Coupling the cues in equation 12.57 cannot correspond to a linear weighted sum (as in the previous section). This is because a linear weighted sum would essentially be using the prior twice, once for each cue. In general, additional cues provide extra information and interact nonlinearly. To understand this, suppose the prior is

$$P(\boldsymbol{S}) = \frac{1}{Z_p}\exp\left\{-\frac{|S - S_p|^2}{2\sigma_p^2}\right\}.$$

Then, setting $t_1 = 1/\sigma_1^2, t_2 = 1/\sigma_2^2, t_p = 1/\sigma_p^2$, the optimal combination is $\boldsymbol{S}^* = (t_1C_1 + t_2C_2 + t_pC_p)/(t_1 + t_2\ t_p)$, hence the best estimate is a linear weighted combination of the two cues $\boldsymbol{C}_1$, $\boldsymbol{C}_2$ and the mean $\boldsymbol{S}_p$ of the prior. By contrast, the estimates using each cue individually are given by $\boldsymbol{S}_1^* = (t_1C_1 + t_pS_p)/(t_1 + t_2 + t_p)$ and $\boldsymbol{S}_2^* = (t_2C_2 + t_pC_p)/(t_1 + t_2 + t_p)$. Thus, any way to linearly combine the individual estimates $\boldsymbol{S}_1^*, \boldsymbol{S}_2^*$ will result in double counting the bias toward the mean $\boldsymbol{S}_p$ of the prior. (Note $t = 1/\sigma^2$ to make the connection to the previous notation.)

To summarize, visual perception for some cues can be predicted by assuming that the cues involve a prior and this biases human perception. But if more cues are available, then perception will generally become less biased. Theories of optimal cue coupling should take this into account and prevent "double counting" the priors.

Cue Dependence and Causal Structure The previous models in this chapter assumed that cues were conditionally independent; see equation 12.57. However, in practice cues are rarely independent. For many situations, such as the flying carpet example, the perception of depth is due to perspective, segmentation, and shadows cues interacting in a complex way. The perspective and segmentation cues must determine that the beach is a flat ground plane. Segmentation cues must isolate the person, the towel, and the shadow. Then the visual system must decide that the shadow is cast by the towel and hence presumably must lie above the ground plane. These complex interactions are impossible to model using the simple conditional independence model described above.

The conditional independence model is also problematic even for the simple stimuli used by Bülthoff and Mallot (1988) for studying shape from shading and texture. The model presupposes that it is possible to extract cues $\boldsymbol{C}_1$, $\boldsymbol{C}_2$ directly from the image $\mathbf{I}$ by a preprocessing step which computes $\boldsymbol{C}_1(\mathbf{I})$ and $\boldsymbol{C}_2(\mathbf{I})$. This requires decomposing the image $\mathbf{I}$ into texture and shading components. This may be possible (see figure 12.35) by exploiting the property that shading tends to be spatially smooth and texture is more jagged. In general, however, determining these cues from the image is not straightforward, and detailed modeling of it lies beyond the scope of this chapter.

The following set of experiments (Kersten, Mamassian, and Knill, 1997) suggest that visual perception

FIGURE 12.35 The input image (left) is decomposed into the base, or shading component (center), and the detail (right).

does seek to find causal relations underlying the visual cues. In Kersten's "ball-in-a-box" experiments an observer perceives the ball to rise off the floor of the box only if this is consistent with a cast shadow; see figure 12.36. To solve this task, the visual system must detect the surface and the orientation of the floor of the box (and decide it is flat), detect the ball, and estimate the light source direction and the motion of the shadow. It seems plausible that in this case, the visual system is unconsciously doing inverse graphics to determine the most likely three-dimensional scene that generated the image sequence. There seems to be no way to explain how these geometric (perspective projection) and shading cues combine to give this percept using only the linear weighted sum rule. Another example, but for color perception, is given by Bloj, Kersten, and Hurlbert (1999).

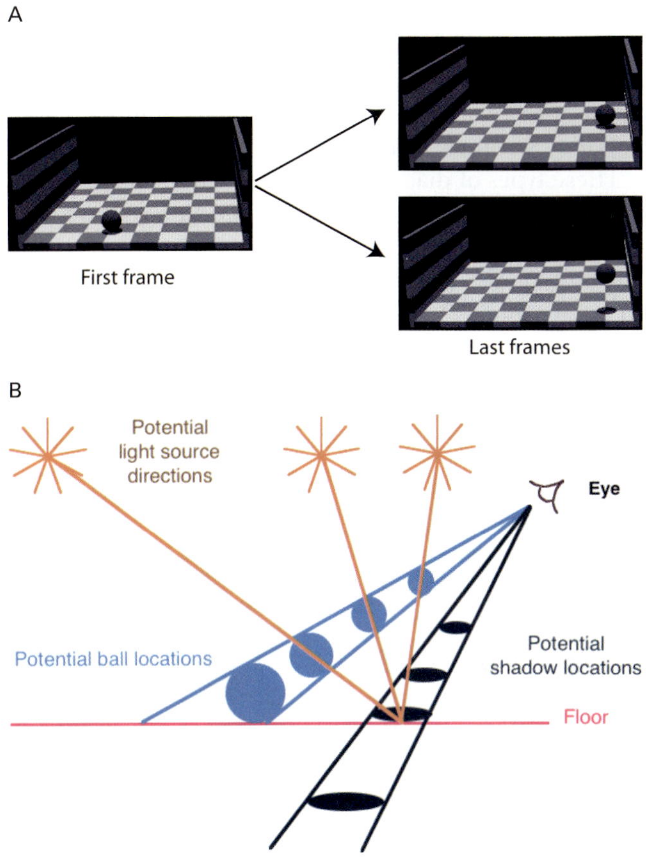

FIGURE 12.36 In the "ball-in-a-box" experiments the motion of the shadow affects the perceived motion of the ball. The ball is perceived to rise from the ground if the shadow follows a horizontal trajectory in the image but is perceived to move toward the back of the box if the shadow follows a diagonal trajectory. See http://youtu.be/hdFCJepvJXU. (A) The first frame and the last frames for the two movies. (B) The explanation is that the observer resolves the ambiguities in the projection of a three-dimensional scene to perceive the 3-D trajectory of the ball (Kersten, Mamassian, and Knill, 1997).

Directed Graphical Models Directed, or causal, graphical models (Pearl, 1988) offer a mathematical language to describe these phenomena. These are similar to the "undirected" graphical models in section 12.4, because the graphical structure makes the conditional dependencies between variables explicit, but differ because the edges between nodes are directed. See Griffiths and Yuille (2008) for an introduction to undirected and directed graphical models from a cognitive science perspective.

Formally *directed graphical models* are specified as follows. The random variables X_μ are defined at the nodes $\mu \in \mathcal{V}$ of a graph. The edges $\mathcal{E}$ specify which variables directly influence each other. For any node $\mu \in \mathcal{V}$, the set of parent nodes $pa(\mu)$ are the set of all nodes $\nu \in \mathcal{V}$ such that $(\mu,\nu) \in \mathcal{E}$, where (μ,ν) means that there is an edge between nodes μ and ν pointing to node μ. We denote the state of the parent node by $\boldsymbol{X}_{pa(\mu)}$. This gives a local *Markov property*—the conditional distribution $P(X_\mu|\boldsymbol{X}_{/\mu}) = P(X_\mu|\boldsymbol{X}_{pa(\mu)})$, so the state of X_μ is only directly influenced by the state of its parents (note $X_{/\mu}$ denotes the states of all nodes except for node μ).

Then the full distribution for all the variables can be expressed as follows:

$$P(\{X_\mu : \mu \in \mathcal{V}\}) = \prod_{\mu \in \mathcal{V}} P(X_\mu \mid \mathbf{X}_{pa(\mu)}). \qquad (12.58)$$

We have already seen two examples of directed graphical models in this chapter. The first example was when we studied divisive normalization and used them in figure 12.25 to represent the dependencies between the stimuli, the filter responses, and the common factor. The second example was when describing the Bayes–Kalman filter where the hidden state x_t at time t "causes" the hidden state x_{t+1} at time t and the observation y_t. Note that in some situations, the directions of the edges indicate physical causation between variables, but in others the arrows merely represent statistical dependence. The relationship between graphical models and causality is complex and is clarified in Pearl (2000).

These types of diagrams are helpful because they can be used to give a taxonomy between different ways that visual cues can be combined. For example, figure 12.37 shows two cases: (1) where two factors combine to cause an image and (2) where there is *common cause* of two cues.

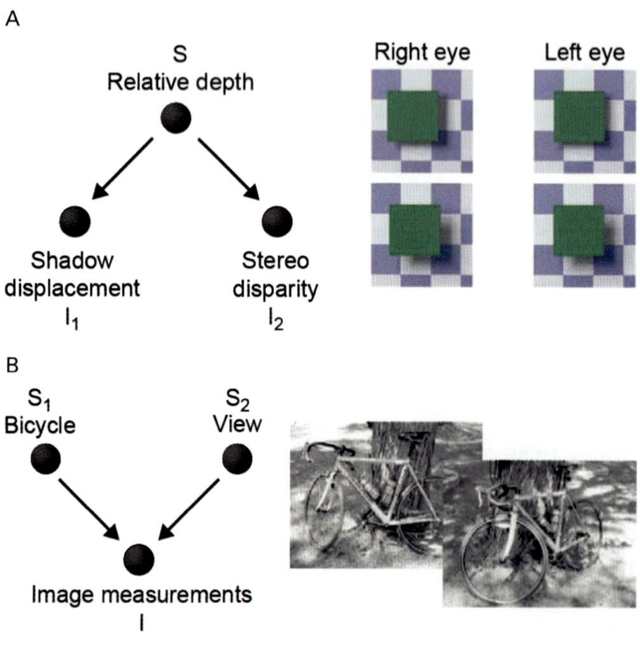

FIGURE 12.37 (A) An example of common cause. The shadow and binocular stereo cues are caused by the same event—two surfaces with one partially occluding the other. (B) The image of the bicycle is caused by the pose of the bicycle, the viewpoint of the camera, and the lighting conditions. Reprinted from D. Kersten and A. Yuille. Bayesian models of object perception. *Current Opinion in Neurobiology, 13*(2), 150–158, copyright 2003.

Graphical models were used by Pearl (1998) to illustrate the phenomenon of *explaining away*. This describes how our interpretations of events can change suddenly as new information becomes available. For example, suppose a house alarm A can be activated by either a burglary B or by an earthquake E. This can be modeled by $P(A|B,E)$ and priors $P(B)$, $P(E)$ for a burglary and an earthquake. In general, the prior probability of a burglary is much higher than the prior probability of an earthquake. So, if an alarm goes off, then it is much more probable to have been caused by a burglary, formally $P(B|A) >> P(E|A)$. But suppose, after the alarm has sounded, you are worried about your house and check the Internet only to discover that there has been an earthquake. In this case, this new information "explains away" the alarm, and so you stop worrying about a burglary.

Variants of this phenomenon arise in vision. In one version the "new information" is a more detailed analysis of the image. Suppose you see the "partly occluded *T*" where a large part of the letter *T* is missing. In this case there is no obviously reason why part of the *T* is missing, so the perception may be only of two isolated segments. On the other hand, if there is a gray smudge over the missing part of the *T* then most observers perceive the *T* directly. The presence of the smudge "explains away" why part of the *T* is missing. This is a common phenomenon which we will return to in the next section from a different perspective.

Directed Graphical Models and Visual Tasks Recall also that the human visual system performs a range of visual tasks and that the way cues are combined can depend on the tasks which are being performed. The language of directed graphical models is also useful for addressing this issue. For example, consider shape from shading, whose goal is to determine the shape of a surface from its shading. However, the shading depends both on the shape of the surface and on the light source direction. So an alternative task is to determine the light source direction. This can be formulated by a model $P(I|S, L)P(S)$, $P(L)$ where I is the observed image, S is the surface shape, and L is the light source direction. $P(I|S,L)$ is the probability of generating an image I from shape S with lighting L, and $P(S)$, $P(L)$ are prior probabilities on the surface shape and the lighting.

Suppose we only want to estimate the surface shape S. Then we do not care about the lighting L, and the optimal procedure is to integrate it out to obtain a likelihood $P(I \mid S) = \int dL P(I \mid S, L) P(L)$ which is combined with a prior $P(S)$ to estimate S. (If the loss function depends only on S, then Bayesian decision theory says it should be integrated out, if it is a continuous variable, or summed out if it is discrete.) Conversely, if we only want to estimate the lighting, then we should integrate out the surface shape to obtain a likelihood $P(I \mid L) = \int dS P(I \mid S, L) P(S)$ and combine it with a prior $P(L)$. On the other hand, if we want to estimate both the surface shape and the lighting, then we should estimate them using the full model $P(I|S,L)$ with priors $P(S)$ and $P(L)$.

This issue of "integrating out" nuisance, or generic, variables relates to the *generic viewpoint assumption* (Freeman, 1994). This assumption states that the estimation of one variable, such as the surface shape, should be insensitive to small changes in another variable (e.g., the lighting). The intuition is that interpretations of the image that depend on something which is very unlikely, such as observing it from a special viewpoint, should be discouraged. It can be shown (Freeman, 1994) that this assumption can be formulated in terms of integrating out nuisance variables, but this analysis is beyond the scope of this chapter.

Model Selection Certain types of cue coupling require *model selection.* While some cues, such as binocular stereo and motion, are usually valid in most places in the image, other cues only apply to some images and often only for subparts of each image. For example, shape from shading is a well-studied cue which relates the intensity values to the changing geometry of a surface. But the lighting conditions in real-world images are extremely complex, and it is extremely difficult, for example, to distinguish the image shading on a planar wall which is due to a nearby light source from the shading pattern of a curved object illuminated from a light source that is further away. Another example is shape from texture, which assumes that there is a homogeneous, or isotropic, texture painted on a three-dimensional surface so that the projection of the texture onto the image plane gives a cue for the shape of the surface. But the texture on most surfaces rarely has this type of regular pattern, and hence shape from texture can only be applied to a limited class of surfaces; see figure 12.38. Similarly perspective cues give very strong perceptions of shape, but they rely on the presence of straight lines in the image (as with the rectangular buildings of Manhattan) so they are valid for images of Manhattan but will not work in the jungle. These considerations show that cue combination often requires *model selection,* in order to determine in what parts of the image, if any, is the cue applicable.

Model selection also arises in situations where there are several alternative ways in which the image could have been generated. In other words, there are several possible interpretations. This requires doing model selection to determine the model which best describes the data. This was called "competitive priors" by Yuille and Bülthoff (1996). By using careful experimental design, it is possible to adjust the image so that small changes shift the balance between one interpretation and another. Examples include the experiments where there are two rotating planes, which can be arranged to have two competing explanations (Kersten et al., 1992). By making slight variations to the transparency cues, there are two surfaces which can be seen to either move rigidly together or to move independently; see figure 12.39B and http://youtu.be/gSrUBpovQdU.

The work by Blake and Bülthoff (1990) is a classic example where a sphere has a Lambertian (diffuse) reflection function and is viewed binocularly; see figure 12.39A. A specular component is adjusted so that it can lie in front of, between the center and the sphere, or

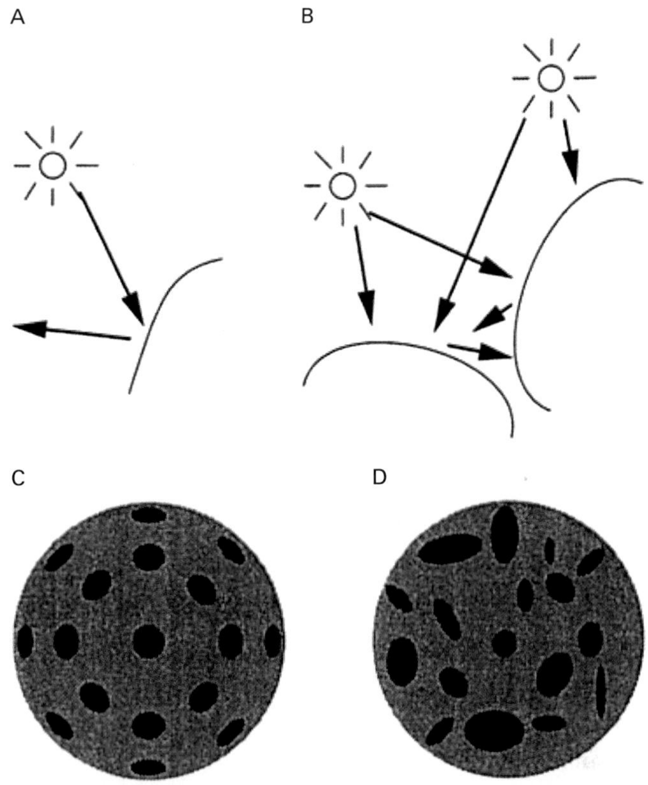

FIGURE 12.38 Model selection may need to be applied in order to decide if a cue can be used. Shape from shading cues will work for case (A) because the shading pattern is simply due to a smooth convex surface illuminated by a single source. However, for case (B) the shading pattern is complex—due to mutual reflection between the two surfaces—and so shape from shading cues will be almost impossible to use. Similarly, shape from texture is possible for case (C) because the surface contains a regular texture pattern but is much harder for case (D) because the texture is irregular. Figure reprinted with permission from Yuille and Bülthoff (1996).

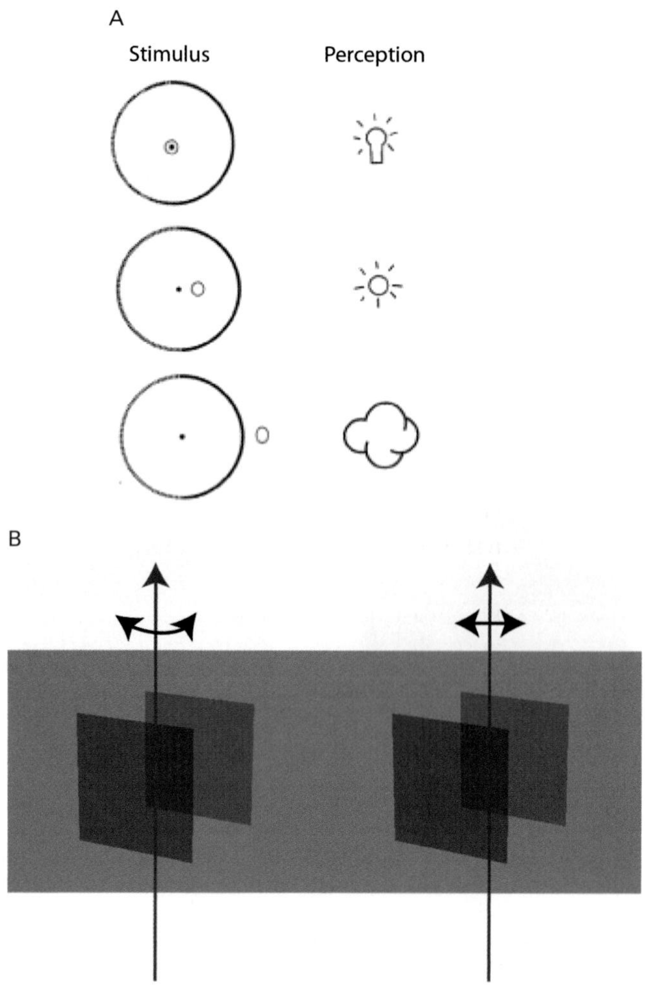

FIGURE 12.39 Examples of strong coupling with competitive priors. A sphere is viewed binocularly (A), and small changes in the position of the specularity lead to very different percepts (Blake and Bülthoff 1990). Similarly, altering the transparency of the moving surfaces (B) can make the two surfaces appear to rotate either rigidly together or independently.

at the center of the sphere. If the specularity lies at the center, then it is perceived as a light bulb and the sphere is perceived to be transparent. If the specularity is placed at the right position between the center and the sphere, then the sphere is perceived to be glossy. If the specularity lies in front of the sphere, then it is seen as a cloud floating in front of a matte (diffuse, Lambertian sphere). Yuille and Bülthoff (1996) interpreted both these phenomena as strong coupling with competitive priors.

We can also account for the "explaining away" examples in the previous section. We can consider two alternative models for the partial *T*. The first model is of two individual segments plus a smudge region. The second is a *T* which is partially hidden by a smudge. In this case the second model is more plausible since it would be very unlikely, an accidental viewpoint (or alignment), that the smudge happened to cover the missing part of the *T*, unless it really did occlude it. We can extend this argument to the Kanisza triangle; see figure 12.8. Humans perceive an illusory contour linking the three pac-men. But close inspection shows that there is no local edge evidence for the existence of the contour (except within the pac-men themselves). There are two possible interpretations of this figure. The first is that there are three partial circles which are arranged in a very special manner so that their edges are aligned (i.e., a straight line from one edge extends to another edge). The second is that there are three circles which are partially hidden by a triangular surface. In this second interpretation, the missing parts of the circles are *explained away* by the hypothesis that they are hidden by a triangle. Thus we have two possible models, and we can assign probabilities to each; then the phenomenon can be explained provided the explanation in terms of the triangular surface is more probable.

Finally, we can revisit the flying carpet illusion shown in figure 12.3. Like Kersten's ball-in-a-box experiments, it requires estimating the depth and orientation of the ground plane (i.e., the beach), segmenting and recognizing the woman and the towel she is standing on, detecting the shadow, and then using the shadow cues, which requires making some assumption about the lighting, to estimate that the towel is hovering above the ground. This is a very complex way to combine all the cues in this image. Observe that it relies on the generic viewpoint assumption, in the sense that it is unlikely for there to be a shadow of that shape in that particular part of the image unless it was cast by some object. The real object that cast the shadow (the flag) is outside the image, and so the visual system "attaches" the shadow to the towel, which then implies that the towel must hover off the ground.

12.5.4 EXAMPLES OF STRONG COUPLING This section gives two examples of strong coupling. The first example deals with coupling different modalities while the second example concerns the perception of texture.

Multisensory Cue Coupling The first example involves multisensory integration with structural uncertainty. Human observers are sensitive to both visual and auditory cues. Sometimes these cues have a common cause—for example, you see a dog moving and hear it barking. In other situations the auditory and visual cues are due to different causes—for example, a cat moves and a nearby dog barks (we ignore the possibility that the dog's barking is caused by the cat moving, or vice versa). Ventriloquists are able to fake these interactions by making the audience think that a puppet is speaking by associating the sound (produced by the ventriloquist) with the movement of the puppet. The ventriloquism effect occurs when visual and auditory cues have different causes—and so are in conflict—but the audience perceives them as having the same cause.

Körding and his collaborators (2007) developed an ideal observer model which determines whether two cues have a common cause or not. They formulated this using a meta-variable C; see figure 12.40. The common cause condition $C = 1$ means that the positions of the cues x_A, x_V are generated by the same process S (see figure 12.40, left) by a distribution $P(x_A, x_V|S) = P(x_A|S)P(x_V|S)$. Here $P(x_A|S)$ and $P(x_V|S)$ are normal distributions $N(x_A \mid S, \sigma_A^2)$, $N(x_V \mid S, \sigma_V^2)$—with the same mean S and variances σ_A^2, σ_V^2. It is assumed that the visual cues are more precise than the auditory cues so that $\sigma_A^2 > \sigma_V^2$. The true position S is drawn from a probability distribution $P(S)$, which is assumed to be a normal distribution $N(0, \sigma_p^2)$. By contrast, $C = 2$ means that the cues are generated by two different processes S_A and S_B, in which case we have $P(x_A|S_A)$ and $P(x_V|S_V)$ which are both Gaussian $N(S_A, \sigma_A^2)$ and $N(S_V, \sigma_V^2)$(see figure 12.40, right). We assume that S_A and S_V are independent samples from the normal distribution $N(0, \sigma_p^2)$. Note that this model involves model selection, between $C = 1$ and $C = 2$, and so in vision terminology is a form of strong coupling and competitive priors (Yuille and Bülthoff, 1996).

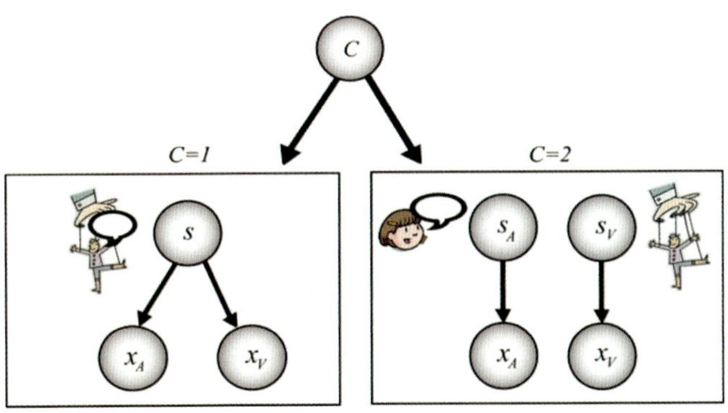

FIGURE 12.40 The subject is asked to estimate the position of the cues and to judge whether the cues are from a common cause—that is, at the same location—or not. In Bayesian terms the task of judging whether the cause is common can be formulated as model selection—are the auditory and visual cues more likely to be generated by a single cause (left) or by two independent causes (right). Figure reprinted with permission from K. P. Körding, U. Beierholm, W. J. Ma, S. Quartz, J. B. Tenenbaum, and L. Shams. Causal Inference in Multisensory Perception. *PLoS ONE*, *2*(9):e943, September 2007.

This model was compared to experiments where brief auditory and visual stimuli were presented simultaneously with varying amount of spatial disparity. Subjects were asked to identify the spatial location of the cue and/or whether they perceive a common cause (Wallace et al., 2004). The closer the visual stimulus was to the audio stimulus, the more likely subjects perceived a common cause. In this case subjects' estimate of its position is strongly biased by the visual stimulus (because it is considered more precise with $\sigma_V^2 < \sigma_A^2$). However, if subjects perceive distinct causes, then their estimate is pushed away from the visual stimulus and exhibits *negative bias*. Körding et al. (2007) argue that this bias is a selection bias stemming from restricting to trials in which causes are perceived as being distinct. For example, if the auditory stimulus is at the center and the visual stimulus at 5 degrees to right of center—then sometimes the (very noisy) auditory cue will be close to the visual cue and hence judged to have a common cause while in other cases the auditory cause will be further away (more than 5 degrees). Hence the auditory cue will have a truncated Gaussian (if judged to be distinct) and will yield negative bias. (See figure 12.41.)

More formally, the beliefs $P(C|x_A, x_V)$ in these two hypotheses $C = 1,2$ are obtained by summing out the estimated positions s_A, s_B of the two cues as follows:

$$\begin{aligned} P(C \mid x_A, x_V) &= \frac{P(x_A, x_V \mid C)P(C)}{P(x_A, x_V)} \\ &= \frac{\int dS P(x_A \mid S)P(x_V \mid S)P(S)}{P(x_A, x_V)}, \text{ if } C = 1, \\ &= \frac{\iint dS_A dS_V P(x_A \mid S_A)P(x_V \mid S_V)P(S_A)P(S_V)}{P(x_A, x_V)}, \text{ if } C = 2. \end{aligned} \tag{12.59}$$

There are two ways to combine the cues. The first is model selection. This estimates the most probable model $C^* = \arg\max P(C|x_V, x_A)$ from the input x_A, x_V and then uses this model to estimate the most likely positions s_A, s_V of the cues from the posterior distribution:

$$\begin{aligned} P(s_V, s_A) &\approx P(s_V, s_A \mid x_V, x_A, C^*) \\ &= \frac{P(x_V, x_A \mid s_V, s_A, C^*)P(s_V, s_A \mid C^*)}{P(x_V, x_A \mid C^*)}. \end{aligned} \tag{12.60}$$

The second way to combine the cues is by *model averaging*. This does not commit itself to choosing C^* but instead averages over both models:

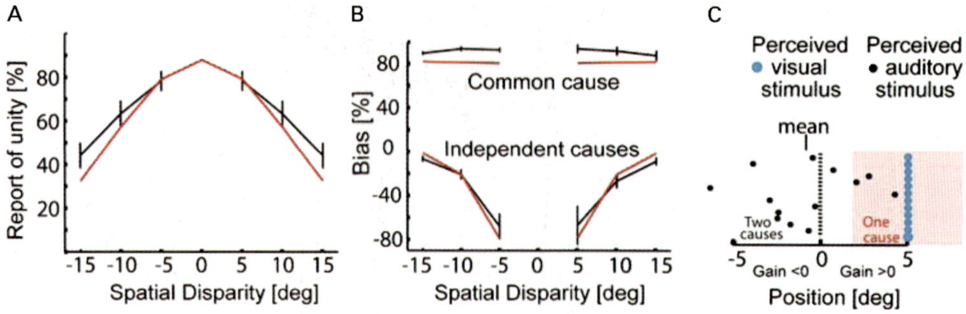

FIGURE 12.41 Reports of causal inference. (A) The relative frequency of subjects reporting one cause (black) is shown (reprinted with permission from Körding et al., 2007) with the prediction of the causal inference model (red). (B) The bias, that is, the influence of vision on the perceived auditory position, is shown (gray and black). The predictions of the model are shown in red. (C) A schematic illustration explaining the finding of negative biases. Blue and black dots represent the perceived visual and auditory stimuli, respectively. In the pink area people perceive a common cause. Reprinted from K. P. Körding, U. Beierholm, W. J. Ma, S. Quartz, J. B. Tenenbaum, and L. Shams. Causal Inference in Multisensory Perception. *PLoS ONE*, 2(9):e943, September 2007.

$$
\begin{aligned}
&P(s_V, s_A \mid x_V, x_A) \\
&= \sum_C P(s_V, s_A \mid x_V, x_A, C) P(C \mid x_V, x_A) \\
&= \sum_C \frac{P(x_V, x_A \mid s_V, s_A, C) P(s_V, s_A \mid C) P(C \mid x_V, x_A)}{P(x_V, x_A \mid C)},
\end{aligned}
\tag{12.61}
$$

where $P(C = 1|x_V, x_A) = \pi_C$ (the posterior mixing proportion).

Natarajan et al. (2008) investigated these issues further. In particular, they showed that human performance on these types of experiments could be better modeled by replacing the Gaussian distributions by alternative distributions which are less sensitive to rare events. As people who modeled the stock market learned to their cost in 2008, Gaussian distributions are nonrobust because the tails of their distributions fall off rapidly, which gives very low probability to rare events. Hence in many real-world applications distributions with longer tails are preferred. Following this reasoning, Natarajan et al. assumed that the observations x_A, Xx_V were generated by distributions with longer tails. More precisely, they assumed that the data are distributed by a mixture of a Gaussian distribution (as in the models above) and a uniform distribution which yields longer tails. More formally, they assume $x_A \pi N(x_A : s_A, \sigma_A^2) + [(1-\pi)/r_l]$ and $x_V \pi N(x_V : s_V, \sigma_V^2) + [(1-\pi)/r_l]$ where π is a mixing proportion and $U(x) = 1/r_1$ is a uniform distribution defined over the range r_1.

Homogeneous and Isotropic Texture The second example is by Knill (2003) and concerns the estimating of orientation in depth (slant) from texture cues. This relates to competitive priors because there are several alternative models for generating the image and the human observer must infer which is most likely; see figure 12.42. More formally, the data are generated by a mixture of models which enables nonlinear cooperative interaction between cues. In this example the data could be generated by isotropic homogeneous texture or by homogeneous texture only. Knill's finding is that human vision is biased to interpret image texture as isotropic, but if enough data are available, the system turns off the isotropy assumption and interprets texture using the homogeneity assumption only.

The posterior probability distribution for S is given by

$$
P(S \mid I) = \frac{P(I \mid S) P(S)}{P(I)}, \quad P(I \mid S) = \sum_{i=1}^{n} \phi_i P_i(I \mid S), \tag{12.62}
$$

where ϕ_i is the prior probability of model i, and $p_i(I|S)$ is the corresponding likelihood function.

More specifically, texture features T can be generated by either an isotropic surface or a homogeneous surface. The surface is parameterized by tilt and slant σ, τ. Homogeneous texture is described by two parameters α, θ, and isotropic texture is a special case where $\alpha = 1$. This gives two likelihood models for generating the data:

$$
P_h(T|(\sigma, \tau), \alpha, \theta), \; P_i(T|(\sigma, \tau), \theta). \tag{12.63}
$$

Here $P_i(T|(\sigma, \tau), \theta) = P_h(T|(\sigma, \tau), \alpha = 1, \theta)$.

Isotropic textures are a special case of homogeneous textures (also rigid motion is a special class of nonrigid motion). The homogeneous model has more free

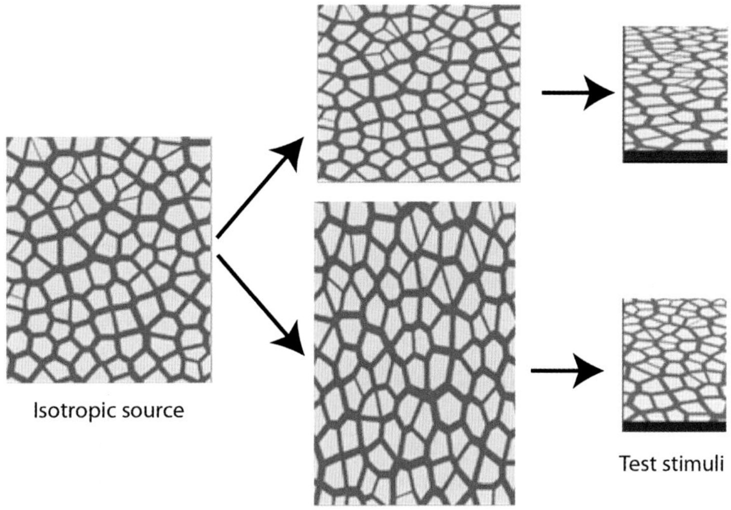

FIGURE 12.42 Generating textures that violate isotropy (Knill, 2003). An isotropic source image is either stretched (top middle) or compressed (bottom middle) producing texture maps that get applied to slanted surfaces shown on the right. A person that assumes surface textures are isotropic would overestimate the slant of the top stimulus and underestimate the slant of the bottom one. Figure adapted with permission from (Knill, 2003).

parameters and hence has more flexibility to fit the data, which suggests that human observers should always prefer it. However, the Occam factor (MacKay, 2003) means that this advantage will disappear if we put priors $P(\alpha)P(\theta)$ on the model parameters and integrate them out. This gives

$$P_h(T \mid (\sigma,\tau)) = \iint d\alpha d\theta P_h(T \mid (\sigma,\tau),\alpha,\theta),\ P_i(T \mid (\sigma,\tau)) \\ = \int d\theta P_i(T \mid (\sigma,\tau),\theta) \quad (12.64)$$

Integrating over the model priors smooths out the models. The more flexible model, P_h, has only a fixed amount of probability to cover a large range of data (e.g., all homogeneous textures) and hence has lower probability for any specific data (e.g., isotropic textures).

Knill describes how to combine these models using model averaging. The combined likelihood function is obtained by taking a weighted average:

$$P(T \mid (\sigma,\tau)) = p_h P_h(T \mid (\sigma,\tau)) + p_i P_i(T \mid (\sigma,\tau)). \quad (12.65)$$

where (p_h, p_i) are prior probabilities that the texture is homogeneous or isotropic. We use a prior $P(\sigma,\tau)$ on the surface and finally achieve a posterior:

$$P(\sigma,\tau \mid I) = \frac{P(I \mid (\sigma,\tau))P(\sigma,\tau)}{P(I)}. \quad (12.66)$$

This model has a rich interpretation. If the data are consistent with an isotropic texture, then this model dominates the likelihood and strongly influences the perception. Alternatively, if the data are consistent only with homogeneous texture, then this model dominates. This gives a good fit to human performance (Knill, 2003).

12.6 *Summary and the Relations of Early and High-Level Vision*

This chapter has given a rapid tour of early vision. In particular, we have provided a modern perspective and conceptualization of early vision in terms of probabilistic graphical models. In this final section we briefly mention how early vision relates to high-level vision.

Marr's (1982) theory of vision proposed that vision is done in a feedforward manner broken down into early vision, performed largely in visual areas V1 and V2, and high-level vision, performed in the IT lobe. A second feedforward class of theories model the ventral stream (visual areas V1, V2, IT) by a hierarchical neural network where, as we ascend the hierarchy, the receptive fields of neurons are tuned to increasingly complex visual structures but are increasingly less sensitive to the precise positions of the input features. Models of this type, such as Hmax, have been developed in detail by Poggio and his collaborators (cf. Riesenhuber and Poggio, 1999) and have been shown to correspond to many of the known aspects of the neuroscience and also have been shown to work well on some computer vision data sets. They concentrate on object detection and recognition and do not address tasks such as depth estimation or image parsing. They have limited representations, and so it is unclear how they can address a large range of visual tasks (Geman, 2006).

By contrast, Mumford (1992) argued for the importance of top-down processing, citing the large number of backprojections in the cortex and pointing out that this matched the "analysis by synthesis" pattern theoretic approach proposed by Grenander (1976, 1978). This class of theories has been developed by Lee and Mumford (2003). Related ideas are also discussed by Rao and Ballard (1999), who suggest that top-down processing can be used to implement predictive processing somewhat similar to the Bayes–Kalman models briefly discussed in this chapter. Ullman and others (Ullman, 1995; Epshtein, Lifshitz, and Ullman, 2008) have also proposed theories of vision which include bottom-up and top-down processing.

How do the models in this chapter fit into these three frameworks? There is clearly no problem in incorporating them into both classes of feedforward models. They can be used to compute the representations required by Marr's theory. They could also be used as the first stages of a feedforward network model like Hmax. The situation is more complicated for the third type of framework which combines bottom-up and top-down processing. However, this can also be formulated by extending the graphical model theories in this chapter so that they are hierarchical. In these models the low-level nodes represent elementary features, such as edges, and the intermediate-level nodes represent compositions of the lower-level features, such as the grouping of edges to form longer segments, or the grouping of parallel line segments. These intermediate-level structures are combined together to form larger structures such as objects and object parts. These theories are sometimes called compositional (Geman, Potter, and Chi, 2002; Zhu, Chen, and Yuille, 2011) because they build objects by composition and they are closely related to stochastic grammars (Zhu and Mumford, 2006; Mumford and Desolneux, 2010). For these classes of theories, the early and high levels of vision are strongly coupled (similar to strong coupling of cues in this chapter). Inference can be performed either bottom-up, where it is driven directly by the input image and low-level hypotheses are combined together to make hypotheses for more complex structures, or top-down, where high-level hypotheses drive the computation.

ACKNOWLEDGMENT A number of demos for the chapter are available on the book's website. We are grateful to Weichao Qiu, who has written almost all of the demos, and to John Flynn for drawing several of the figures.

REFERENCES

Abbott, L., & Kepler, T. (1990). Model neurons: From Hodgkin-Huxley to Hopfield. In L. Garrido (Ed.), *Statistical Mechanics of Neural Networks* (pp. 5–18). Berlin: Springer-Verlag.

Ackley, D. H., Hinton, G. E., & Sejnowski, T. J. (1985). A learning algorithm for Boltzmann machines. *Cognitive Science, 9*(1), 147–169.

Adelson, E. H., & Bergen, J. R. (1985). Spatiotemporal energy models for the perception of motion. *JOSA A, 2*(2), 284–299.

Amit, D. J. (1992). *Modeling Brain Function: The World of Attractor Neural Networks.* Cambridge: Cambridge University Press.

Anstis, S., & Ramachandran, V. (1987). Visual inertia in apparent motion. *Vision Research, 27*(5), 755–764.

Atick, J., & Redlich, A. (1992). What does the retina know about natural scenes? *Neural Computation, 4*(2), 196–210.

Barlow, H. (1961). Possible principles underlying the transformation of sensory messages. In W. A. Rosenblith (Ed.), *Sensory Communication.* Cambridge, MA: MIT Press.

Barlow, H. (1962). Measurements of the quantum efficiency of discrimination in human scotopic vision. *Journal of Physiology, 160*(1), 169–188.

Basri, R. (2003). Lambertian reflectance and linear subspaces. *IEEE Transactions on Pattern Analysis and Machine Intelligence, 25,* 218–233.

Blake, A., & Bülthoff, H. B. (1990). Does the brain know the physics of specular reflection? *Nature, 343*(6254), 165–168.

Blake, A., & Bülthoff, H. (1990). Does the brain know the physics of specular reflection? *Nature, 343*(6254), 165–168.

Blake, A., Kohli, P., & Rother, C. (2011). *Markov Random Fields for Vision and Image Processing.* Cambridge, MA: MIT Press.

Blake A., & Zisserman, A. 2003. *Visual Reconstruction (Artificial Intelligence Series).* Cambridge, MA: MIT Press.

Bloj, M. G., Kersten, D., & Hurlbert, A. C. (1999). Perception of three-dimensional shape influences colour perception through mutual illumination. *Nature, 402*(6764), 877–879.

Borst, A., & Euler, T. (2011). Seeing things in motion: Models, circuits, and mechanisms. *Neuron, 71*(6), 974–994.

Briggs, F., & Usrey, W. (2011). Corticogeniculate feedback and visual processing in the primate. *Journal of Physiology, 589*(1), 33–40.

Buckingham, G. (2009). Living in a material world: How visual cues to material properties affect the way that we lift objects and perceive their weight. *Journal of Neurophysiology, 102*(6), 3111–3118.

Bulthoff, H. H., & Mallot, H. A. (1988). Integration of depth modules: Stereo and shading. *Journal of the Optical Society of America. A, Optics and Image Science, 5*(10), 1749–1758.

Bülthoff, H., & Mallot, H. A. (1990). Integration of stereo, shading and texture. In *AI and the Eye*, 11th European Conference on Visual Perception (pp. 119–146). Chichester, UK: Wiley.

Carandini, M., Demb, J. B., Mante, V., Tolhurst, D. J., Dan, Y., Olshausen, B. A., et al. (2005). Do we know what the early visual system does? *Journal of Neuroscience, 25*(46), 10577–10597.

Carandini, M., & Heeger, D. J. (2011). Normalization as a canonical neural computation. *Nature Reviews. Neuroscience, 13*, 51–62.

Cheng, K., Shettleworth, S. J., Huttenlocher, J., & Rieser, J. J. (2007). Bayesian integration of spatial information. *Psychological Bulletin, 133*(4), 625–637.

Clark, J. J., & Yuille, A. L. (1990). *Data Fusion for Sensory Information Processing Systems*. Springer.

Cumming, B., Johnston, E. B., & Parker, A. (1993). Effects of different texture cues on curved surfaces viewed stereoscopically. *Vision Research, 33*(5–6), 827–838.

Dan, Y., Atick, J., & Reid, R. (1996). Efficient coding of natural scenes in the lateral geniculate nucleus: Experimental test of a computational theory. *Journal of Neuroscience, 16*(10), 3351–3362.

Daugman, J. (1985). Uncertainty relation for resolution in space, spatial frequency, and orientation optimized by two-dimensional visual cortical filters. *Journal of the Optical Society of America. A, Optics and Image Science, 2*, 1160–1169.

Dayan, P., & Abbott, L. (2001). *Theoretical Neuroscience: Computational and Mathematical Modeling of Neural Systems*. Cambridge, MA: MIT Press.

Demb, J. B. (2002). Multiple mechanisms for contrast adaptation in the retina. *Neuron, 36*(5), 781–783.

Dev, P. (1975). Perception of depth surfaces in random-dot stereograms: A neural model. *International Journal of Man-Machine Studies, 7*(4), 511–528.

Durbin, R., & Mitchison, G. (1990). A dimension reduction framework for understanding cortical maps. *Nature, 343*(6259), 644–647.

Elder, J. H., & Goldberg, R. M. (2002). Ecological statistics of Gestalt laws for the perceptual organization of contours. *Journal of Vision, 2*(4), 324–353.

Epshtein, B., Lifshitz, I., & Ullman, S. (2008). Image interpretation by a single bottom-up top-down cycle. *Proceedings of the National Academy of Sciences of the United States of America, 105*(38), 14298–14303.

Ernst, M. O., & Banks, M. S. (2002). Humans integrate visual and haptic information in a statistically optimal fashion. *Nature, 415*(6870), 429–433.

Fang, F., Boyaci, H., & Kersten, D. (2009). Border ownership selectivity in human early visual cortex and its modulation by attention. *Journal of Neuroscience, 29*(2), 460–465.

Felleman, D. J., & van Essen, D. C. (1991). Distributed hierarchical processing in the primate cerebral cortex. *Cerebral Cortex, 1*(1), 1–47.

Freeman, W. (1994). The generic viewpoint assumption in a framework for visual perception. *Nature, 368*(6471), 542–545.

Geiger, D., & Yuille, A. (1991). A common framework for image segmentation. *International Journal of Computer Vision, 6*(3), 227–243.

Geisler, W. S. (2011). Contributions of ideal observer theory to vision research. *Vision Research, 51*(7), 771–781.

Geisler, W. S., & Perry, J. S. (2009). Contour statistics in natural images: Grouping across occlusions. *Visual Neuroscience, 26*(01), 109–121.

Geman, S. (2006). Invariance and selectivity in the ventral visual pathway. *Journal of Physiology, Paris, 100*, 212–224.

Geman, S., & Geman, D. (1984). Stochastic relaxation, Gibbs distributions, and the Bayesian restoration of images. *IEEE Transactions on Pattern Analysis and Machine Intelligence, 6*(6), 721–741.

Geman, S., Potter, D., & Chi, Z. (2002). Composition systems. *Quarterly of Applied Mathematics, 60*(4), 707–736.

Georgopoulos, A. P., Caminiti, R., Kalaska, J. F., & Massey, J. T. (1983). Spatial coding of movement: A hypothesis concerning the coding of movement direction by motor cortical populations. *Experimental Brain Research*, Suppl. 7(32), 327–336.

Gibson, J. (1986). *The Ecological Approach to Visual Perception*. Hillsdale, NJ: Lawrence Erlbaum Associates.

Gilbert, C. D., & Wiesel, T. (1990). The influence of contextual stimuli on the orientation selectivity of cells in primary visual cortex of the cat. *Vision Research, 30*(11), 1689–1701.

Gold, J. M., Mundy, P. J., & Tjan, B. S. (2012). The perception of a face is no more than the sum of its parts. *Psychological Science, 23*(4), 427–434.

Gollisch, T., & Meister, M. (2010). Eye smarter than scientists believed: neural computations in circuits of the retina. *Neuron, 65*(2), 150–164.

Gori, M., Del Viva, M., Sandini, G., & Burr, D. C. (2008). Young children do not integrate visual and haptic form information. *Current Biology, 18*(9), 694–698.

Green, C. S., & Bavelier, D. (2008). Exercising your brain: A review of human brain plasticity and training-induced learning. *Psychology and Aging, 23*(4), 692–701.

Green, D. M., & Swets, J. A. (1966). *Signal Detection Theory and Psychophysics*. New York: Wiley.

Grenander, U. (1976). *Applied Mathematical Sciences: Vol. 18. Pattern Synthesis*. New York: Springer.

Grenander, U. (1978). *Pattern Analysis: Lectures in Pattern Theory* (Vol. 2). New York: Springer.

Griffiths, T., & Yuille, A. (2008). A primer on probabilistic inference. In N. Chater & M. Oaksford (Eds.), *The Probabilistic Mind: Prospects for Bayesian Cognitive Science* (pp. 33–57). New York: Oxford University Press.

Grossberg, S. (1968). Some physiological and biochemical consequences of psychological postulates. *Proceedings of the National Academy of Sciences of the United States of America, 60*(3), 758–765.

Grossberg, S., & Hong, S. (2006). A neural model of surface perception: Lightness, anchoring, and filling-in. *Spatial Vision, 19*(2), 263–321.

Grossberg, S., & Mingolla, E. (1985a). Neural dynamics of form perception: Boundary completion, illusory figures, and neon color spreading. *Psychological Review, 92*(2), 173.

Grossberg, S., & Mingolla, E. (1985b). Neural dynamics of perceptual grouping: Textures, boundaries, and emergent segmentations. *Attention, Perception & Psychophysics, 38*(2), 141–171.

Grzywacz, N. M., & Yuille, A. (1990). A model for the estimate of local image velocity by cells in the visual cortex. *Proceedings of the Royal Society of London. Series B, Biological Sciences, 239*(1295), 129–161.

HASSENSTEIN, B., & REICHARDT, W. (1956). Systemtheoretische Analyse der Zeit-Reihenfolgen- und Vorzeichenauswertung bei der Bewegungsperzeption des Rüsselkäfers *Chlorophanus. Zeitschrift fur Naturforschung, 11b*, 513–524.

HEEGER, D. J. (1992). Normalization of cell responses in cat striate cortex. *Visual Neuroscience, 9*(02), 181–197.

HERTZ, J. (1991). *Introduction to the Theory of Neural Computation* (Vol. 1). Basic Books.

HILDRETH, E. C. (1984). The computation of the velocity field. *Proceedings of the Royal Society of London. Series B, Biological Sciences, 221*, 189–220.

HO, Y., & LEE, R. (1964). A Bayesian approach to problems in stochastic estimation and control. *Automatic Control. IEEE Transactions on, 9*(4), 333–339.

HOPFIELD, J., & TANK, D. (1986). Computing with neural circuits: A model. *Science, 233*(4764), 625–633.

HUBEL, D. (1982). Evolution of ideas on the primary visual cortex, 1955–1978: A biased historical account. *Bioscience Reports, 2*(7), 435–469.

HUBEL, D., & WIESEL, T. (1962). Receptive fields, binocular interaction and functional architecture in the cat's visual cortex. *Journal of Physiology, 160*(1), 106–154.

HUBEL, D., & WIESEL, T. (1968). Receptive fields and functional architecture of monkey striate cortex. *Journal of Physiology, 195*(1), 215–243.

HYVÄRINEN, A. (2010). Statistical models of natural images and cortical visual representation. *Topics in Cognitive Science, 2*(2), 251–264.

ITTI, L., & KOCH, C. (2001). Computational modelling of visual attention. *Nature Reviews. Neuroscience, 2*(3), 194–203.

JACOBS, R. (1999). Optimal integration of texture and motion cues to depth. *Vision Research, 39*(21), 3621–3629.

JONES, J., & PALMER, L. (1987). The two-dimensional spatial structure of simple receptive fields in cat striate cortex. *Journal of Neurophysiology, 58*(6), 1187–1211.

JULESZ, B. *Foundations of Cyclopean perception.* (Anaglyphoscope in pocket.) The University of Chicago Press, Chicago, London, 1971.

KALMAN, R. (1960). A new approach to linear filtering and prediction problems. *Journal of Basic Engineering, 82*(1), 35–45.

KANIZSA, G. (1979). *Organization in Vision.* New York: Praeger Press.

KANWISHER, N. (2010). Functional specificity in the human brain: A window into the functional architecture of the mind. *Proceedings of the National Academy of Sciences of the United States of America, 107*(25), 11163–11170.

KAPADIA, M., WESTHEIMER, G., & GILBERT, C. D. (2000). Spatial distribution of contextual interactions in primary visual cortex and in visual perception. *Journal of Neurophysiology, 84*(4), 2048–2062.

KELLMAN, P. J., & ARTERBERRY, M. E. (2000). *The Cradle of Knowledge: Development of Perception in Infancy.* Cambridge, MA: MIT Press.

KERSTEN, D. (1987). Predictability and redundancy of natural images. *JOSA A, 4*(12), 2395–2400.

KERSTEN, D., BÜLTHOFF, H. H., SCHWARTZ, B., & KURTZ, K. (1992). Interaction between transparency and structure from motion. *Neural Computation, 4*(4), 573–589.

KERSTEN, D., MAMASSIAN, P., & KNILL, D. C. (1997). Moving cast shadows induce apparent motion in depth. *Perception, 26*(2), 171–192.

KERSTEN, D., & YUILLE, A. (2003). Bayesian models of object perception. *Current Opinion in Neurobiology, 13*(2), 150–158.

KINOSHITA, M., GILBERT, C. D., & DAS, A. (2009). optical imaging of contextual interactions in V1 of the behaving monkey. *Journal of Neurophysiology, 102*(3), 1930–1944.

KNILL, D. C. (2001). Contour into texture: Information content of surface contours and texture flow. *Journal of the Optical Society of America. A, Optics, Image Science, and Vision, 18*(1), 12–35.

KNILL, D. (2003). Mixture models and the probabilistic structure of depth cues. *Vision Research, 43*(7), 831–854.

KNILL, D. C., & KERSTEN, D. J. (1991). Apparent surface curvature affects lightness perception. *Nature, 351*, 228–230.

KOCH, C., MARROQUIN, J., & YUILLE, A. (1986). Analog "neuronal" networks in early vision. *Proceedings of the National Academy of Sciences of the United States of America, 83*(12), 4263–4267.

KOMATSU, H. (2006). The neural mechanisms of perceptual filling-in. *Nature Reviews. Neuroscience, 7*(3), 220–231.

KONISHI, S., & YUILLE, A. L. (2000). Statistical cues for domain specific image segmentation with performance analysis. In *Proceedings: IEEE Conference on Computer Vision and Pattern Recognition, 2000* (Vol. 1, pp. 125–132). Los Alamitos, CA: IEEE Computer Society.

KONISHI, S., YUILLE, A., COUGHLAN, J., & ZHU, S. (2003). Statistical edge detection: Learning and evaluating edge cues. *Pattern Analysis and Machine Intelligence. IEEE Transactions on, 25*(1), 57–74.

KÖRDING, K. P., BEIERHOLM, U., MA, W. J., QUARTZ, S., TENENBAUM, J. B., & SHAMS, L. (2007). Causal inference in multisensory perception. *PLoS One, 2*(9), e943.

LAMME, V. A. (1995). The neurophysiology of figure-ground segregation in primary visual cortex. *Journal of Neuroscience, 15*(2), 1605–1615.

LANDY, M. S., MALONEY, L. T., JOHNSTON, E. B., & YOUNG, M. (1995). Measurement and modeling of depth cue combination: In defense of weak fusion. *Vision Research, 35*, 389–412.

LEE, T. S. (1996). Image representation using 2d gabor wavelets. *Pattern Analysis and Machine Intelligence. IEEE Transactions on, 18*(10), 959–971.

LEE, T., & MUMFORD, D. (2003). Hierarchical Bayesian inference in the visual cortex. *Journal of the Optical Society of America. A, Optics, Image Science, and Vision, 20*(7), 1434–1448.

LEE, T., & NGUYEN, M. (2001). Dynamics of subjective contour formation in the early visual cortex. *Proceedings of the National Academy of Sciences of the United States of America, 98*(4), 1907–1911.

LEE, T. S., MUMFORD, D., & YUILLE, A. (1992). Texture segmentation by minimizing vector-valued energy functionals: The coupled-membrane model. In In G. SANDINI (Ed.), *Computer Vision ECCV '92* (pp. 165–173). Berlin: Springer.

LEE, T. S., & YUILLE, A. L. (2006). Efficient coding of visual scenes by grouping and segmentation. In K. DOYA, S. ISHII, A. POUGET, & R. P. RAO (Eds.), *Bayesian Brain: Probabilistic Approaches to Neural Coding* (pp. 145–188). Cambridge, MA: MIT Press.

LENNIE, P. (1998). Single units and visual cortical organization. *Perception, 27*, 889–935.

Lettvin, J. Y., Maturana, H. R., McCulloch, W. S., & Pitts, W. H. (1959). What the frog's eye tells the frog's brain. *Proceedings of the IRE, 47*(11), 1940–1951.

Linsker, R. (1986a). From basic network principles to neural architecture: Emergence of spatial-opponent cells. *Proceedings of the National Academy of Sciences of the United States of America, 83*(19), 7508–7512.

Linsker, R. (1986b). From basic network principles to neural architecture: Emergence of orientation-selective cells. *Proceedings of the National Academy of Sciences of the United States of America, 83*(21), 8390–8394.

Liu, J. S. (2008). *Monte Carlo strategies in scientific computing.* New York: Springer Science & Business Media.

Livingstone, M., & Hubel, D. (1984). Anatomy and physiology of a color system in the primate visual cortex. *Journal of Neuroscience, 4*(1), 309–356.

Logothetis, N., & Sheinberg, D. L. (1996). Visual object recognition. *Annual Review of Neuroscience, 19*(1), 577–621.

Lu, H., & Roe, A. (2007). optical imaging of contrast response in macaque monkey V1 and V2. *Cerebral Cortex, 17*(11), 2675–2695.

Ma, W. J., Beck, J. M., Latham, P. E., & Pouget, A. (2006). Bayesian inference with probabilistic population codes. *Nature Neuroscience, 9*(11), 1432–1438.

Maass, W. (2000). On the computational power of winner-take-all. *Neural Computation, 12*(11), 2519–2535.

MacKay, D. J. C. (2003). *Information Theory, Inference and Learning Algorithms* (1st ed.). Cambridge: Cambridge University Press.

Marr, D. (1982). *Vision.* W.H. Freeman.

Marr, D., & Poggio, T. (1976). Cooperative computation of stereo disparity. *Science, 194*(4262), 283–287.

Masland, R. H., & Martin, P. R. (2007). The unsolved mystery of vision. *Current Biology, 17*(15), R577–R582.

McIlwain, J. T. (1991). Distributed spatial coding in the superior colliculus: A review. *Visual Neuroscience, 6*(01), 3–13.

Mechler, F., & Ringach, D. L. (2002). On the classification of simple and complex cells. *Vision Research, 42*(8), 1017–1033.

Meister, M., & Berry, M. (1999). The neural code of the retina. *Neuron, 22*, 435–450.

Mel, B. W., Ruderman, D. L., & Archie, K. A. (1998). Translation-invariant orientation tuning in visual complex cells could derive from intradendritic computations. *Journal of Neuroscience, 18*(11), 4325–4334.

Merigan, W., & Maunsell, J. (1993). How parallel are the primate visual pathways? *Annual Review of Neuroscience, 16*(1), 369–402.

Milner, D., & Goodale, M. (2006). *The Visual Brain in Action* (2nd ed.). New York: Oxford University Press.

Morrone, M. C., & Burr, D. (1988). Feature detection in human vision: A phase-dependent energy model. *Proceedings of the Royal Society of London. Series B: Biological Sciences, 235*, 221–245.

Mottaghi, R., Chen, X., Liu, X., Cho, N.-G., Lee, S.-W., Fidler, S., et al. (2014). The role of context for object detection and semantic segmentation in the wild. In *2014 IEEE Conference on Computer Vision and Pattern Recognition (CVPR)* (pp. 891–898). Piscataway, NJ: IEEE.

Mumford, D. (1992). On the computational architecture of the neocortex. *Biological Cybernetics, 66*(3), 241–251.

Mumford, D., & Desolneux, A. (2010). *Pattern Theory: The Stochastic Analysis of Real-World Signals.* Natick, MA: A K Peters

Mumford, D., & Shah, J. (1989). Optimal approximations by piecewise smooth functions and associated variational problems. *Communications on Pure and Applied Mathematics, 42*(5), 577–685.

Nassi, J. J., Lyon, D. C., & Callaway, E. M. (2006). The parvocellular LGN provides a robust disynaptic input to the visual motion area MT. *Neuron, 50*(2), 319–327.

Natarajan, R., Murray, I., Shams, L., & Zemel, R. (2008). Characterizing response behavior in multisensory perception with conflicting cues. In *Advances in Neural Information Processing Systems NIPS.*

Northcutt, R.G., & Kaas, J. H. (1995). The emergence and evolution of mammalian neocortex. *Trends in Neurosciences, 18*(9), 373–379.

Ohzawa, I., DeAngelis, G. C., & Freeman, R. D. (1990). Stereoscopic depth discrimination in the visual cortex: Neurons ideally suited as disparity detectors. *Science, 249*(4972), 1037–1041.

Oja, E. (1982). Simplified neuron model as a principal component analyzer. *Journal of Mathematical Biology, 15*(3), 267–273.

Olshausen, B. A., & Field, D. J. (1996). Emergence of simple-cell receptive field properties by learning a sparse code for natural images. *Nature, 381*(6583), 607–609.

Papandreou, G., Chen, L.-C., & Yuille, A. L. (2014). Modeling image patches with a generic dictionary of mini-epitomes. In *2014 IEEE Conference on Computer Vision and Pattern Recognition (CVPR)* (pp. 2059–2066). Piscataway, NJ: IEEE.

Pearl, J. (1988). *Probabilistic Reasoning in Intelligent Systems: Networks of Plausible Inference* (1st ed.). San Mateo, CA: Morgan Kaufmann..

Pearl, J. (2000). *Causality: Models, Reasoning and Inference* (Vol. 29). Cambridge: Cambridge University Press.

Pelli, D. (1990). The quantum efficiency of vision. In C. Blakemore (Ed.), *Vision: Coding and Efficiency* (pp. 3–24). Cambridge: Cambridge University Press.

Peterhans, E., & von der Heydt, R. (1989). Mechanisms of contour perception in monkey visual cortex. II. Contours bridging gaps. *Journal of Neuroscience, 9*(5), 1749–1763.

Pollen, D. A., & Ronner, S. F. (1981). Phase relationships between adjacent simple cells in the visual cortex. *Science, 212*(4501), 1409–1411.

Pouget, A., Dayan, P., & Zemel, R. S. (2003). Inference and computation with population codes. *Annual Review of Neuroscience, 26*(1), 381–410.

Qian, N. (1994). Computing stereo disparity and motion with known binocular cell properties. *Neural Computation, 6*(3), 390–404.

Qiu, C., Kersten, D., & Olman, C. A. (2013). Segmentation decreases the magnitude of the tilt illusion. *Journal of Vision, 13*(13), 19.

Ramachandra, C. A., & Mel, B. W. (2013). Computing local edge probability in natural scenes from a population of oriented simple cells. *Journal of Vision, 13*(14), 19.

Rao, R. P. N., & Ballard, D. (1999). Predictive coding in the visual cortex: A functional interpretation of some extra-classical receptive-field effects. *Nature Neuroscience, 2*, 79–87.

Rieke, F., Warland, D., de Ruytervan Steveninck, R. R., & Bialek, W. (1997). *Spikes: Exploring the Neural Code.* Cambridge, MA: MIT Press. Paperback edition 1999.

Riesenhuber, M., & Poggio, T. (1999). Hierarchical models of object recognition in cortex. *Nature Neuroscience, 2,* 1019–1025.

Roe, A. W., Chelazzi, L., Connor, C. E., Conway, B. R., Fujita, I., Gallant, A. L., et al. (2012). Toward a unified theory of visual area v4. *Neuron, 74*(1), 12–29.

Roe, A. W., Chen, G., & Lu, H. D. (2009). Functional architecture of area v2. *Encyclyopedia of Neuroscience, 10,* 331–349.

Rust, N. C., & DiCarlo, J. J. (2010). Selectivity and tolerance ("invariance") both increase as visual information propagates from cortical area V4 to IT. *Journal of Neuroscience, 30*(39), 12978–12995.

Samonds, J. M., Potetz, B. R., & Lee, T. (2009). Cooperative and competitive interactions facilitate stereo computations in macaque primary visual cortex. *Journal of Neuroscience, 29*(50), 15780–15795.

Samonds, J. M., Potetz, B., & Lee, T. S. (2012). Relative luminance and binocular disparity preferences are correlated in macaque v1, matching natural scene statistics. *Proceedings of the National Academy of Sciences of the United States of America, 109*(16), 6313–6318.

Samonds, J. M., Potetz, B. R., Tyler, C. W., & Lee, T. S. (2013). Recurrent connectivity can account for the dynamics of disparity processing in v1. *Journal of Neuroscience, 33*(7), 2934–2946.

Sasaki, Y., Watanabe, T., & Purves, D. (2004). The primary visual cortex fills in color. *Proceedings of the National Academy of Sciences of the United States of America, 101*(52), 18251–18256.

Schenk, T., & McIntosh, R. D. (2010). Do we have independent visual streams for perception and action? *Cognitive Neuroscience, 1*(1), 52–62.

Schrater, P. R., Knill, D. C., & Simoncelli, E. P. (2000). Mechanisms of visual motion detection. *Nature Neuroscience, 3*(1), 64–68.

Schwartz, E. L. (1980). Computational anatomy and functional architecture of striate cortex: A spatial mapping approach to perceptual coding. *Vision Research, 20*(8), 645–669.

Schwartz, O., Sejnowski, T. J., & Dayan, P. (2009). Perceptual organization in the tilt illusion. *Journal of Vision, 9*(4), 19.1–20.

Sherman, S. M., & Guillery, R. W. (2002). The role of the thalamus in the flow of information to the cortex. *Philosophical Transactions of the Royal Society of London. Series B, Biological Sciences, 357*(1428), 1695–1708.

Shushruth, S., Nurminen, L., Bijanzadeh, M., Ichida, J. M., Vanni, S., & Angelucci, A. (2013). Different orientation tuning of near- and far-surround suppression in macaque primary visual cortex mirrors their tuning in human perception. *Journal of Neuroscience, 33*(1), 106–119.

Simoncelli, E. P., & Olshausen, B. A. (2001). Natural image statistics and neural representation. *Annual Review of Neuroscience, 24*(1), 1193–1216.

Sundareswara, R., & Schrater, P. (2008). Perceptual multistability predicted by search model for Bayesian decisions. *Journal of Vision, 8*(5), 12.1–12.19.

Talebi, V., & Baker, C. L. (2012). Natural versus synthetic stimuli for estimating receptive field models: A comparison of predictive robustness. *Journal of Neuroscience, 32*(5), 1560–1576.

Tjan, B. S., Braje, W. L., Legge, G. E., & Kersten, D. (1995). Human efficiency for recognizing 3-D objects in luminance noise. *Vision Research, 35*(21), 3053–3069.

Trenti, E. J., Barraza, J. F., & Eckstein, M. P. (2010). Learning motion: Human vs. optimal Bayesian learner. *Vision Research, 50*(4), 460–472.

Todd, J. T., Oomes, A. H., Koenderink, J. J., & Kappers, A. M. (2001). On the affine structure of perceptual space. *Psychological Science, 12*(3), 191–196.

Tsao, D. Y., Moeller, S., & Freiwald, W. A. (2008). Comparing face patch systems in macaques and humans. *Proceedings of the National Academy of Sciences of the United States of America, 105*(49), 19514–19519.

Ullman, S. (1979). *The Interpretation of Visual Motion.* Cambridge, MA: MIT Press.

Ullman, S. (1995). Sequence seeking and counter streams: A computational model for bidirectional information flow in the visual cortex. *Cerebral Cortex, 5*(1), 1–11.

Van Essen, D., Anderson, C. H., & Felleman, D. (1992). Information processing in the primate visual system: An integrated systems perspective. *Science, 255*(5043), 419–423.

van Hateren, J. H., & Ruderman, D. (1998). Independent component analysis of natural image sequences yields spatio-temporal filters similar to simple cells in primary visual cortex. *Proceedings of the Royal Society of London. Series B, Biological Sciences, 265*(1412), 2315–2320.

von der Heydt, R. (2003). Image parsing mechanisms of the visual cortex. In J. S. Werner & L. M. Chalupa (Eds.), *The Visual Neurosciences* (pp. 1139–1150). Cambridge, MA: MIT Press.

von der Heydt, R., & Peterhans, E. (1989). Mechanisms of contour perception in monkey visual cortex. I. Lines of pattern discontinuity. *Journal of Neuroscience, 9*(5), 1731–1748.

von der Heydt, R., Peterhans, E., & Baumgartner, G. (1984). Illusory contours and cortical neuron responses. *Science, 224*(4654), 1260.

Wainwright, M. J., & Simoncelli, E. (2000). Scale mixtures of Gaussians and the statistics of natural images. *Advances in Neural Information Processing Systems, 12*(1), 855–861.

Wallace, M. T., Roberson, G. E., Hairston, W. D., Stein, B. E., Vaughan, J. W., & Schirillo, J. A. (2004). Unifying multisensory signals across time and space. *Experimental Brain Research, 158*(2), 252–258.

Wallisch, P., & Movshon, J. A. (2008). Structure and function come unglued in the visual cortex. *Neuron, 60*(2), 194–197.

Wandell, B. A., Dumoulin, S. O., & Brewer, A. A. (2007). Visual field maps in human cortex. *Neuron, 56*(2), 366–383.

Wandell, B. A., & Winawer, J. (2011). Imaging retinotopic maps in the human brain. *Vision Research, 51*(7), 718–737.

Warland, D. K., Reinagel, P., & Meister, M. (1997). Decoding visual information from a population of retinal ganglion cells. *Journal of Neurophysiology, 78*(5), 2336–2350.

Watamaniuk, S. N., McKee, S. P., & Grzywacz, N. M. (1995). Detecting a trajectory embedded in random-direction motion noise. *Vision Research, 35*(1), 65–77.

Watson, A. B., Barlow, H., & Robson, J. G. (1983). What does the eye see best? *Nature, 302*(5907), 419–422.

WEISS, Y., & ADELSON, E. H. Slow and smooth: A Bayesian theory for the combination of local motion signals in human vision. Technical Report 1624, Massachusetts Institute of Technology, 1998.

WEISS, Y., SIMONCELLI, E. P., & ADELSON, E. H. (2002). Motion illusions as optimal percepts. *Nature Neuroscience, 5,* 598–604.

WILSON, H. R., & COWAN, J. D. (1972). Excitatory and inhibitory interactions in localized populations of model neurons. *Biophysical Journal, 12*(1), 1–24.

YOUNG, R., LESPERANCE, R., & MEYER, W. (2001). The Gaussian derivative model for spatial-temporal vision: I. Cortical model. *Spatial Vision, 3*(4), 261–319.

YUILLE, A. (1987). Energy functions for early vision and analog networks. *Biological Cybernetics, 61,* 115–123.

YUILLE, A. (2011). Loopy belief propagation, mean-field, and Bethe approximations. In A. BLAKE, P. KOHLI, & C. ROTHER (Eds.), *Markov Random Fields for Vision and Image Processing* (pp. 77–93). Cambridge, MA: MIT Press.

YUILLE, A., & BÜLTHOFF, H. H. (1996). Bayesian decision theory and psychophysics. In D. KNILL & W. RICHARDS (Eds.), *Perception as Bayesian Inference* (p. 123–162). Cambridge: Cambridge University Press.

YUILLE, A.L., & GRZYWACZ, N. M. (1988). A computational theory for the perception of coherent visual motion. *Nature, 333*(6168), 71–74.

YUILLE, A., BURGI, P. Y., & GRZYWACZ, N. M. (1998).Visual motion estimation and prediction: A probabilistic network model for temporal coherence. *Sixth International Conference on Computer Vision, 1998* (pp. 973–978). New York: IEEE.

YUILLE, A., KAMMEN, D., & COHEN, D. (1989). Quadrature and the development of orientation selective cortical cells by Hebb rules. *Biological Cybernetics, 61*(3), 183–194.

YUILLE, A., & KERSTEN, D. (2006). Vision as Bayesian inference: Analysis by synthesis? *Trends in Cognitive Sciences, 10*(7), 301–308.

ZHAOPING, L. (2014). *Understanding vision: Theory, Models, and Data.* Oxford: Oxford University Press.

ZHU, L., CHEN, Y., & YUILLE, A. (2011). Recursive compositional models for vision: description and review of recent work. *Journal of Mathematical Imaging and Vision, 41*(1–2), 122–146.

ZHOU, H., FRIEDMAN, H., & VON DER HEYDT, R. (2000). Coding of border ownership in monkey visual cortex. *Journal of Neuroscience, 20*(17), 6594–6611.

ZHU, S.-C., & MUMFORD, D. (2006). A stochastic grammar of images. *Foundations and Trends® in Computer Graphics and Vision, 2*(4), 259–362.

13 Neural Maps: Their Function and Development

JAMES A. BEDNAR AND CHRISTOPHER K. I. WILLIAMS

"I invoke the first law of geography: everything is related to everything else, but near things are more related than distant things." (Tobler, 1970)

13.1 *Overview*

In this chapter, we first define a *neural map* as a sheet of neurons systematically related to another population of neurons, including the special but common case of a *topographic map* as a spatially organized mapping of this type. We further define a *feature map* as a topographic mapping of some underlying features or patterns in the input data, beyond the strict anatomical arrangement of the input region.

Next, we review the main biological findings about the properties of these maps in adult animals, which will be expanded in more detail when discussing specific models of development and function later. We use well-established examples from the visual system but highlight similarities and differences with maps in other sensory modalities and in motor regions.

The remaining sections present models and analyses addressing a series of key questions about neural maps: *Where do neural map patterns come from? How do feature maps arise from neural mechanisms? What is the information-processing goal of neurons in maps?*

Together, the models and data suggest that systematic maps first emerge in sensory areas through genetic specification of initial topography, based on molecular guidance cues. The maps are then shaped by activity-dependent and experience-dependent processes in neurons and their connections that lead a neural population to have good coverage of possible stimuli, with neurons responding to the full range of stimuli encountered during development. These processes also typically result in locally smooth maps, though the functional significance of this organization remains unclear.

13.2 *Biological Background*

Many regions of the nervous system of vertebrates have been described as *neural maps*. A neural map is a sheet of neurons whose properties are systematically related either to the external world (e.g., visual or auditory space) or to activities in another set of neurons. This systematicity often takes the form of a *topographic map*, with nearby positions in the input (or output) space mapping to nearby positions in the neural sheet. In this section, we will describe the main features of such maps as found in adult animals, and in later sections we will focus on models and experimental studies of how such maps develop and function.

Perhaps the clearest examples of topographic maps in the brain are those corresponding to a sensory receptor surface, such as retinotopic maps (regions whose organization matches that of the retina in the eye), somatotopic maps (organized as a "homunculus" matching the surface of the body), cochleotopic maps (corresponding to the cochlea in the ear), rhinotopic maps (of the olfactory epithelium), or whisker barrel maps (of the whisker pad in rodents). Motor cortex also contains several "homunculus" maps, but mapping to the muscles and joints of the body, or perhaps to motor actions, rather than from touch receptors in the skin.

For a sensory surface that is itself organized in terms of some external quantity, such as the retina (mapping the visual field) or the cochlea (mapping auditory frequency), a topographic map can also be defined in functional terms, such as a visuotopic map (an area mapping the visual field) or a tonotopic map (an area mapping auditory frequency). However, note that distances in topographic neural maps typically correspond to the density of receptors on the receptor surface, not to distances in the external world or even distances across the receptor surface. For instance, retinotopic maps in humans typically devote most of their area to the central region of the eye where most of the axons in the optic nerve originate, and somatotopic maps have a much greater area devoted to receptor-rich areas like the hands and face rather than reflecting the skin surface area.

The visuotopic maps found throughout the vertebrate visual system are particularly well-studied examples of topographic maps. In a visuotopic map, the visual *receptive field* (RF) of a neuron (the area of the visual field or the pattern to which it responds; see chapter 1 and chapter 8) tends to be similar to those of its neighbors. A large fraction of the macaque monkey cortex can be divided into separate visuotopic maps, each with a representation of the visual field (figure 13.1). Specificity for visual location and overall organization for visuotopy are strongest in the regions most directly connected to the retina, but often remain clear enough elsewhere to be used as a defining criterion for a visual area (Van Essen and Gallant, 1994).

Other topographic maps cannot be directly traced to the geometry of a sensory surface, such as topographic *feature maps* of an abstract feature of the sensory input (also called *computational maps*). Figure 13.2 shows the oft-studied *orientation map* in the primary visual cortex. Here the largest-scale mapping is for retinotopy/visuotopy, but there is also locally smooth (though patchy) organization of tuning for edge orientation at each retinotopic location. As will be shown below, numerous other patchy or stripy feature maps have been measured across the surface of V1 in various species, including motion direction, spatial frequency, color, disparity, and eye preference maps. These maps consist of neurons with RFs selective along some or all of these visual feature dimensions, with each map revealing only one type of selectivity among the many that could be measured with appropriate stimuli. For instance, a cell might have strong connections via the lateral geniculate nucleus (LGN) from an oriented pattern of retinal ganglion cells of a certain type in one eye, giving it a specific preference for orientation, pattern size, and eye of origin, and neighboring neurons might have similar but slightly different connections and preferences. Emergent feature maps have similarly

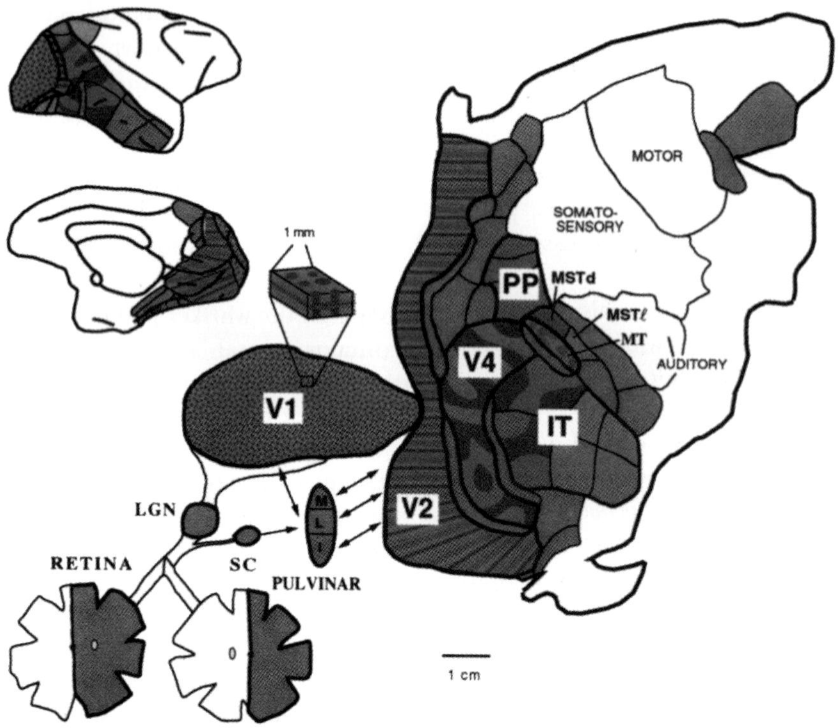

FIGURE 13.1 Visual maps in macaque. The surface of the macaque monkey cortex, viewed laterally (brain, top left), in a saggital cross-section (brain, middle left), or flattened to show the complete surface of the right hemisphere (middle and right) along with the subcortical paths from the left hemiretina of each eye (bottom left). When the eye is stimulated visually, there are well-organized responses arranged topographically in the lateral geniculate nucleus (LGN), superior colliculus (SC), pulvinar, and many cortical regions (including the V1, V2, V4, PP, MST, MT, and to some extent IT, regions labeled in this figure; see chapter 12, "Early Vision," for further details). Each of these areas has a map of all or a substantial part of the visual field, with nearby neurons in the map responding to nearby locations on the retina (and thus similar locations in the field of view). Areas farther from the retina (e.g., as measured by latency) have less clear visuotopy, with evidence for organization into more abstract categories like objects, faces, and places, but with some overall visuotopic organization and, moreover, local similarity in neural responses. Other sensory cortical areas are similarly mapped to their own receptor surfaces while primary motor areas are roughly mapped to the body's musculature. Organization of areas distant from the motor and sensory periphery is much less clear (e.g., "association" areas), whether because they have less organization, because they map more abstract quantities or categories, or because they are a combination of sensory inputs and potential motor outputs. Reprinted from Van Essen and Gallant (1994).

been reported for location in auditory space in auditory cortex, along with odotopic maps for odor similarity in the olfactory bulb, and gustotopic maps for taste similarity in gustatory cortex.

Historically, the first clear indications that cortex was organized into topographic maps were from electrode stimulation in epileptic humans by Wilder Penfield in the 1930s. The first evidence for feature maps in sensory cortex came from multiple recordings from an electrode moved along a straight line. These experiments revealed that neurons in somatosensory and visual cortex were locally similar in their responses, but that preferences for features like orientation varied across the surface of the cortex, with occasional jumps (Hubel and Wiesel, 1962; Mountcastle, 1957). In the 1980s, optical imaging techniques were developed that were able to show organization across a surface area of tens of square millimeters simultaneously, albeit at a low resolution that blurred responses from many adjacent neurons (see the example in figure 13.2).

Two-photon imaging of calcium signals developed in the 2000s now allows simultaneous imaging of activity in hundreds to thousands of individual cells across multiple layers (Ohki, Chung, Ch'ng, Kara, and Reid, 2005), albeit for a smaller surface area (typically less than a square millimeter). Studies using this technique have shown that feature maps are largely continuous (i.e., with similar RF) down to the single-neuron level in cat (figure 13.3B–E), ferret, monkey, and tree shrew V1, both across the surface of V1 and across the cortical laminae, thus validating the optical imaging results for these species. As first discovered electrophysiologically, the feature maps also contain local discontinuities (regions of quick change in preference) and are patchy overall, with the same feature preference repeated at multiple separate locations in the retinotopic map.

Interestingly, even though rodent V1 has many properties in common with nonrodent mammals, including orientation selectivity and retinotopic maps, there is no apparent continuous topographic or laminar organization for orientation or other emergent visual features in rodent V1 (figure 13.3F), suggesting that the continuity of the map may not be crucial for function. The *coverage* of the maps, that is, the extent to which they fully tile the multidimensional space of all combinations of features (e.g., orientation and retinotopic position), seems to be of more functional relevance. Evidence from cats suggests that the maps do provide good coverage (Swindale, Shoham, Grinvald, Bonhoeffer, and Hubener, 2000), though the methods for evaluating coverage are not necessarily straightforward to apply or interpret (Carreira-Perpiñán and Goodhill, 2002). The models in this chapter will focus on achieving good coverage via locally continuous, topographically organized feature maps, even though continuity does not apply to all maps in all species, because local continuity does provide important constraints on the possible map formation mechanisms. Once map formation has been established in these models, it is possible to account for results in rodents by relaxing these constraints (e.g., by reducing strengths or density of lateral interactions; Law, 2009), but precisely how to do so remains controversial.

Overall, the experimental data from adult animals indicate that neurons throughout much of the cortex and in many subcortical regions are systematically organized into large-scale neural maps, that their response properties together cover some multidimensional input space (and/or output space, in the case of motor neurons), that the maps are usually topographic at some scale, that feature maps are locally smooth and continuous in nonrodent mammals, and that the precise spatial organization varies depending on the modality, feature dimension, and species.

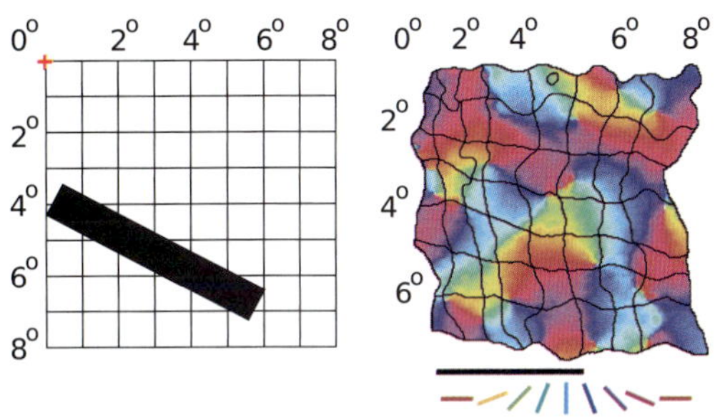

FIGURE 13.2 Visuotopic and orientation map in V1. Given a particular fixation point (marked with the cross at top left), the visual field seen by an animal can be divided into a regular grid, with each square representing a 1° × 1° area of visual space. In cortical area V1 of mammals, neurons are arranged into a visuotopic map, with nearby neurons responding to nearby areas of the visual field. As an example, the image on the right shows the visuotopic map on the surface of V1 of a tree shrew for an 8° × 7° area of visual space, measured using optical imaging of intrinsic signals (adapted from Bednar, 2009, and Bosking et al., 2002, with permission; scale bar is 1 mm). A stimulus presented in a particular location in visual space (such as the thick black bar shown at left) evokes a response centered around the corresponding grid square in V1 (3° right, 5.5° down). Which specific neurons respond within that general area, however, depends on the orientation of the stimulus. The V1 map is color coded with the preferred orientation of neurons in each location; for example, the black bar shown will primarily activate neurons colored in purple in the V1 grid squares around those covered by the stimulus. Similar maps could be plotted for V1 to reveal preference for other visual features, such as motion direction, spatial frequency, color, disparity, and eye preference, as will be shown later in figure 13.10.

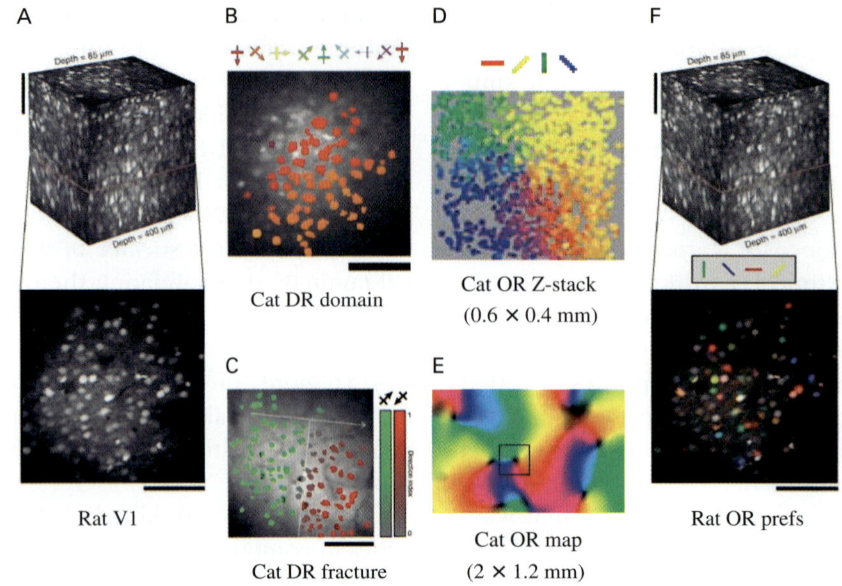

FIGURE 13.3 Single-cell organization of V1 feature maps. Two-photon calcium imaging provides single-cell-resolution measurements of neural responses within a volume containing thousands of cells. Plot (A) shows the change in calcium-dye fluorescence in response to visual stimuli, averaged over all stimuli, for rat V1 at each cortical depth. This process labels nearly every neuron in this region, as can be seen in the slice shown for depth 290 μm below the surface. In cat V1, the results show that neural maps are organized exquisitely smoothly with transitions between direction (DR; B, C) and orientation (OR; D) preferences precise down to the level of single cells, even when all cortical depths are combined together (D). These results validate previous low-resolution results from optical imaging (E); compare the square region in (E) with the individually identified neurons in (D). Interestingly, despite an overall retinotopic map and strong selectivity for orientation, orientation preference is randomly organized in rodents (F), which has not yet been explained adequately in models. Scale bars 0.1 mm. Reprinted from Ohki et al. (2005).

13.3 *Questions about Neural Maps*

Despite all of this evidence about the ubiquity and importance of neural maps, the following are still not completely clear:

- how and why the maps develop,
- why the maps are typically so highly organized,
- what determines the map pattern for different input features,
- what determines the range of values represented in a map,
- why the maps differ between species,
- why the maps differ between different neural regions processing the same modality,
- to what extent the maps are specified genetically versus developing based on experience,
- how multiple overlaid maps relate to each other,
- what precise computations are performed by the circuitry in the maps,
- whether the smooth, topographic organization is functionally significant, or
- to what extent any functional role may be specialized for a specific modality or range of sensory experience.

In each of these cases, there are some data available, and sometimes substantial data, but enough unknowns remaining that computational models can play an important role in formulating concrete hypotheses, evaluating consistency of possible explanations, making predictions to guide experiments, and articulating theoretical principles for interpreting empirical data. Many of these questions have yet to be answered, but the following sections consider a range of models of various types (see box 13.1) that address several of the most important ones:

- Where do neural map patterns come from?

  Examples: Mechanistic and geometrical developmental models mapping from an underlying multidimensional feature space into a two-dimensional sheet.
- How do feature maps arise from neural mechanisms?

  Examples: Mechanistic developmental models showing how maps, RFs, and lateral connectivity can arise from experience-driven local computations in neurons, and how these will behave during vision.
- What is the information-processing goal of neurons in maps?

> **Box 13.1 Descriptions of Models**
>
> There are many different types of neural map models with different goals and methods. Here we will discuss models of the following nonexclusive types:
>
> *Phenomenological:* Models designed to reproduce observed behavior, whether or not they work by the same mechanisms as the underlying biological system.
>
> *Geometrical:* Phenomenological models of the topology or geometric structure of neural maps, not necessarily addressing the behavior of neurons within the map.
>
> *Mechanistic:* Models with an explicit claim of isomorphism between model elements or mechanisms and the underlying neural hardware.
>
> *Developmental:* Models explaining how adult-like circuitry or mechanisms can emerge from a simpler starting point in a young animal.
>
> *Normative:* Models derived not from observed neural elements or circuits, but from an explicit specification that the model should achieve a functional criterion.
>
> *Generative:* Normative models built with the specific objective of being able to faithfully regenerate the likely sensory input from the map's activity pattern.

Examples: Normative generative models for the range and type of RF patterns observed in neural maps, based on an explicit objective about the function of this mapping.

In recent years, these three questions have been addressed mainly with very different types of models, developed in separate research communities with different goals, and so we consider each one separately. In each category, the models we consider in detail are those that are relatively simple, easily understood, and still in current use rather than either the most complex and realistic models currently available or models that were seminal yet are no longer used in current publications. Throughout and particularly in the concluding sections, we will suggest how these different models relate to each other and could potentially be combined to provide a comprehensive account of neural maps. The initial models we discuss focus on the most basic observations, of topography and of feature map organization, but in later sections we discuss more complex interdependencies between neurons in the maps that help reveal their role in sensory processing.

Figure 13.4 shows the basic architecture assumed by most of these models, along with a concrete example of a simple retinotopic map like those found in the LGN of cats or macaques. In the simplest case, a map model consists of two sets of neurons ("Input" and "Map" [output] in figure 13.4A), each arranged as a 2-D neural *sheet* (array of units) with topographic connections between them. More realistic models include multiple sheets and sets of connections (representing interconnected populations of cells in various brain regions) but with the same overall organization in terms of neurons in sheets receiving input from other sheets and from neighboring neurons in the same sheet. To make this type of large-scale model practical, in nearly all cases a neuron is modeled as a single-compartment (point) unit, with the response of a neuron represented by a single number standing for its firing rate. Each unit typically also represents many real units, often collapsing a vertical column of similar neurons into one computational unit for efficiency. Other abstractions are then made in specific cases below, while preserving the overall ability to represent the map computationally. We will first consider the simplest case of mapping a sensory surface and then consider emergent feature maps and the shape of RFs within them, followed by more detailed studies of how these neurons interact during sensory processing.

13.4 Where Do Neural Map Patterns Come From?

The most obvious question about neural maps is where their large-scale structure comes from—why do they have the patterns that they do? In this section we first focus on topographic maps, with their close relationship to a sensory surface, and then consider feature maps separately.

As a specific topographic-map starting point, how do the retinal ganglion cell axons from the eye form an ordered projection onto the optic tectum of a frog or the superior colliculus of a mammal (a *retinotectal map*)? This question was one of the first about maps to be investigated, and has involved a healthy exchange of ideas between modelers and experimenters ever since. Early experiments by Roger Sperry starting in the 1930s suggested that retinotectal maps develop by chemoaffinity, based on explicit matching of chemical signals between the retina and the target region. The main evidence was that axons growing from each part of the retina appear to target specific corresponding locations in the colliculus/tectum, even when the retinal tissue is moved around or transplanted. Initial ideas of absolute addressing eventually gave way to relative addressing, to account for experiments in which, for example, half of the retina was ablated and the remaining half innervated the entire tectum. However, theoretical considerations suggested that there was also a role for neural activity, and specifically correlations between neighboring retinal cells, as a cue for establishing local neighborhood relationships at the target. Other explanations

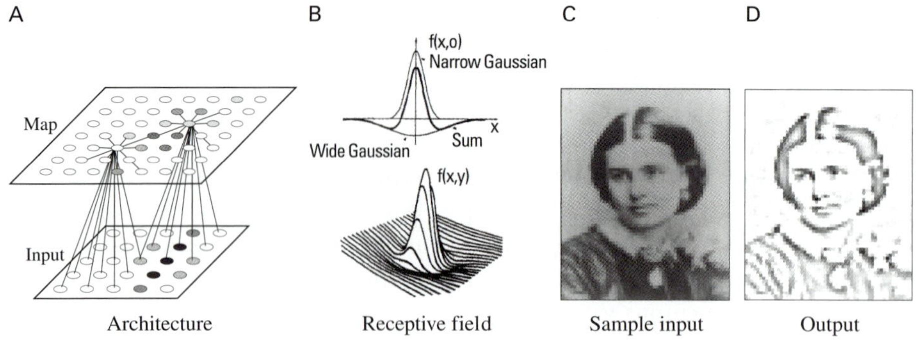

FIGURE 13.4 Basic map architecture and example. (A) Neurons in a topographic map get input from neurons in the corresponding region of an input sheet, along with input from their close (and possibly more distant) neighbors in the map. Connections to two such units in the map are shown. (B) For each neuron, the incoming connections each have different synaptic weights, which together form a receptive field (RF). For instance, the retinal RF for each lateral geniculate nucleus (LGN) neuron can be well approximated using a two-dimensional difference-of-Gaussians function like the one shown (reprinted from Rodieck, 1965; measured using reverse correlation), centered on the appropriate location in the photoreceptors. This ON neuron has a positive center and negative surround; the corresponding OFF neuron would have the same RF but with a negative center and positive surround. This function is an abstraction of a complex pattern of connectivity in the retina and LGN, but it captures much of the spatial preferences of a given class of LGN neurons (Rodieck, 1965). It is thus a good phenomenological model, even if highly simplified as a mechanistic model. For a simple network with no lateral (intramap) connections and topographically mapped difference-of-Gaussians input connections, the map performs a parallel convolution (D) of an input image (C), using an OFF RF in this example. That is, the activation of each output unit in the map is determined by multiplying the difference-of-Gaussians RF with the topographically corresponding input patch and then summing the results to give the value of one pixel in the plot. For this isotropic RF the overall result is simply edge enhancement, but other RF shapes will have different effects, leading to selectivity for specific visual features like orientation. Real neurons also have connections to other neurons across the map surface and feedback from higher areas, which lead to complex interdependencies such as the contextual modulation effects discussed later.

focused on competition between axons for some fixed synaptic resource at the target. Early models of all three types (e.g., Overton and Arbib, 1982; Prestige and Willshaw, 1975; Willshaw and von der Malsburg, 1976, 1979) helped drive subsequent experiments in this area, by clearly articulating possible alternative explanations.

As reviewed by Flanagan (2006), Hjorth, Sterratt, Cutts, Willshaw, and Eglen (2014), and Huberman, Feller, and Chapman (2008), these later experiments have now given detailed evidence for how these maps develop, and have shown clear roles for all three proposed mechanisms. As illustrated in figure 13.5, the chemical matching process is driven by orthogonal spatial gradients of families of molecules and receptors known as Ephs and ephrins found across the source and target regions, which establish a specific identity for each neuron relative to its neighbors. Once the axons arrive near their targets, neural activity appears to be important for local refinement of the projection because disruptions to spontaneous retinal activity change the specific patterns of connectivity at the topographically mapped location. This process also appears to require some form of competition for targets, such as axon–axon interactions (Triplett et al., 2011).

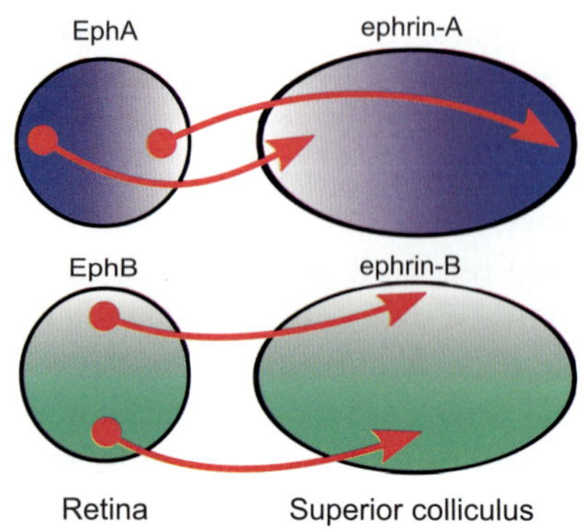

FIGURE 13.5 Retinotectal mapping via chemoaffinity. The retina and superior colliculus of mice establish chemical gradients of signaling molecules and receptors. Axons originating from neurons in different locations on the retina have different levels of EphA and EphB, forming a two-dimensional retinal coordinate system, and prefer corresponding levels of ephrins-A and -B on the targets. These molecules and receptors come in families with numerous genetically distinct variants in both source and target, with both attractive and repulsive effects, but only a pair of each is shown here for simplicity. Adapted from Wei et al. (2013).

As reviewed by Hjorth et al. (2014), there are now computational models of retinotectal development that can account for this process in detail, including matching the changes observed in the maps in a wide range of mutant mice that have disruptions to the gradient chemicals, retina structure, and/or spontaneous activity. These models include all three of the mechanisms listed above, that is, gradient matching, correlated retinal activity, and axonal competition (e.g., Triplett et al., 2011).

In the next section, we will look in detail at a model containing only the activity-dependent subset of these mechanisms, to highlight similarities and differences between explanations for topographic maps and feature maps. This approach is supported by findings that mechanisms based on activity and chemoaffinity are largely independent (Benjumeda et al., 2013). We will then consider how activity-dependent processes work alongside chemoaffinity and axonal competition, before investigating feature map development in later sections.

13.4.1 SOM Model of Topographic Map Development The many models that have been proposed for activity-dependent map formation are reviewed in Swindale (1996) and Goodhill (2007). Here we will primarily discuss models that use specific patterns of neural activity, whether spontaneous or visually evoked, rather than more abstract approaches without individual images or specific activity patterns (such as Carreira-Perpiñán, Lister, and Goodhill, 2005; Durbin and Mitchison, 1990; Linsker, 1986; Miller, 1994).

Many activity-dependent models can be traced back to von der Malsburg's original (1973) network, later extended to address retinotopy (Willshaw and von der Malsburg, 1976). This model and many others closely related to it (e.g., Grossberg, 1976; Kohonen, 1982; Obermayer, Ritter, and Schulten, 1990) use a set of 2-D input patterns presented to a network with initially random local connectivity, some form of lateral interactions in the target region that provide competition and cooperation between neurons, and a local Hebbian-based synaptic modification rule (see chapter 8).

The variant of von der Malsburg (1973) we will first consider here was proposed by Kohonen (1982) and is called the self-organizing-map (SOM) algorithm. Even though SOM is highly abstract, it has remained in common use (e.g., Obermayer et al., 1990; Yu, Farley, Jin, and Sur, 2005) because it is simple to implement, robust, computationally efficient, and faithfully reproduces the observed map patterns. SOM is thus a good phenomenological, geometrical model, if not a mechanistic model. In later sections we will present a more complex, mechanistic implementation that shares many of the same principles while being more easily related to the cortical circuitry.

The simplest Kohonen SOM model has an architecture like that in figure 13.4A, with two sheets of neural units: a map sheet (e.g., the retinal ganglion cells) and an input sheet (e.g., the optic tectum or superior colliculus), along with a set of connections from the full set of input units to each map unit. Starting with initially random weights (and thus no topography and no feature selectivity), the SOM is designed to develop map units that will respond to input patterns that are seen during development, achieving a balance between good coverage of the full range of input patterns and continuity between input pattern preferences of nearby neurons.

The SOM algorithm has two alternating phases: computing map activity (neural responses), and adjusting weight values. Given a 2-D array of input activity values s, for each map neuron (k, l) the SOM first computes the Euclidean distance v_{kl} between s and a 2-D weight vector (w_{kl}; initially random):

$$v_{kl} = \|s - w_{kl}\|. \tag{13.1}$$

In a process meant as an idealization of lateral interactions between neurons in the map, the initial activity of the map neurons is then constructed as a single isotropic "neighborhood kernel" function centered on the neuron (r, s) for which v_{rs} is smallest. Unit (r, s) is considered the "winner," as it has the weight vector most similar to the input pattern, and is considered to be responding most strongly. The final activity of each map unit (i, j) is then computed as a fixed function of its distance from (r, s), using a fixed kernel function (e.g., a 2-D Gaussian):

$$h_{rs,ij} = \exp\left(-\frac{(r-i)^2 + (s-j)^2}{\sigma_h^2}\right). \tag{13.2}$$

Each weight $w_{k,ij}$ from input unit k to map unit (i, j) is then updated by a variant of the Hebbian rule where the input is taken relative to the current weight:

$$w'_{k,ij} = w_{k,ij} + \alpha(s_k - w_{k,ij})h_{rs,ij}. \tag{13.3}$$

The result of this process is that the winning neuron and its distance-weighted neighbors now have weights that are more similar to this input pattern, a form of positive feedback that gradually drives neurons to become selective and drives nearby neurons to become more similar. As other neurons elsewhere in the map "win" for different input values, each patch of map neurons, in turn, becomes selective for some part of the input space. Implementations typically gradually reduce the cortical kernel width σ_h and learning rate α so that neurons can gradually differentiate, with different preferences in different locations in the map. The result provably converges to a smooth and complete mapping of the input space, in the limit of infinitely slow learning rate and radius reduction over the course of self-organization (Erwin, Obermayer, and Schulten, 1992). In practice, the mapping may actually be discontinuous or patchy, depending on the values of these parameters.

Note that if $h_{rs,ij}$ were set to a Dirac δ function (a neighborhood of just the winning neuron), this algorithm would reduce to k-means clustering (Hartigan, 1975). That is, the network would assign each input pattern to one of the k output neurons, updating its weight vector with an incremental estimate of the mean of all inputs activating that neuron (Bishop, 2006). Thus, one way to think of the SOM is as a form of clustering, plus a local smoothing value that achieves a continuous topographic map via lateral interactions.

For an example of this self-organization process, we can specify a set of input patterns with known properties and then see how the SOM responds. Consider a Kohonen SOM model with a 24 × 24 input sheet and a 40 × 40 map. For each input presentation, let us create random spatially localized training pattern images (e.g., small 2-D oriented Gaussians), such as might occur during retinal waves. This input is high dimensional (24 × 24 = 576) but actually varies in only two dimensions, the x and y of the pattern centers. That is, out of all of the possible image inputs, actual training patterns for this SOM lie only on a 2-D *manifold*, a lower-dimensional surface embedded in the higher-dimensional space. SOM (and many other manifold learning algorithms; see Bishop, 2006) can approximate this underlying ("latent") 2-D manifold using a discrete set of neurons arranged in 2-D, each responsible for part of the range of the input space over which the inputs varied during training (figure 13.6).

This example simulation shows that, in principle, neighborhood information provided by locally correlated retinal activity could be sufficient to drive topographic map formation in a target region (as suggested by Willshaw and von der Malsburg, 1976). We focused on this subset of the mechanisms because, as we will see in the following sections, similar mechanisms can also explain feature map formation. However, this example also serves to highlight the limitations of activity-only accounts of topographic map formation: (1) the model provides nothing to ensure the overall orientation of the map (such as whether up on the retina is to the right in the cortex), contrary to experimental observations of consistent orientation; (2) it assumes full initially random connectivity, which is likely only to be feasible for a very small network, and

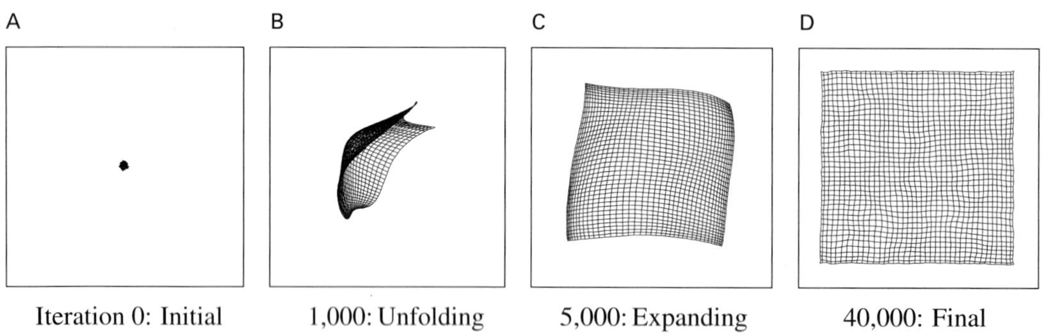

FIGURE 13.6 Unfolding to represent a 2-D space. For this SOM network with a 24 × 24 input, all input patterns have been chosen to be identical 2-D isotropic Gaussian spots, differing only in their (x, y) position. Each of these plots shows the full 24 × 24 input space, with a point for each map neuron located at the center of gravity of that neuron's weight vector. Each neuron is plotted with a line segment connecting it to its neighbors in the map array, yielding a grid-like plot if the map neurons have centers of gravity that smoothly tile the underlying 2-D space of pattern centers. (A) The initial weight vector for each neuron is random, and thus they are all mapped to the center of the input space (as the center of gravity of all of them is approximately equal). (B, C) Over time, the Hebbian learning process applied to winning neurons and their neighbors leads to neurons differentiating to cover the input space, while varying smoothly in their response properties due to the smooth neighborhood kernel. Weight values are adjusted to become more similar to the input patterns each time, and the centers of gravity gradually unfold to cover the 2-D manifold on which the inputs lie, with each map unit responding to some portion of the 2-D space. (D) The final result by iteration 40,000 is a smooth mapping covering the area of the input space that was encountered during training, with arbitrary final orientation and flipping. Here the mapping is flat, with equal cortical area for each location in the input space, but nonuniform distributions lead to mappings that reflect the probability density of each location in the input space (cf. Recanzone et al., 1992, for similar phenomena in somatosensory cortex). Reprinted from Miikkulainen et al. (2005).

moreover has not been observed experimentally; and (3) it does not provide explanations for how maps can develop under activity blockade, nor for the specific changes when genes related to the chemical gradients are knocked out.

As described above, addressing these limitations requires adding mechanisms explicitly incorporating chemical gradients and chemoaffinity processes, which establish the overall orientation of the map and the initial topography. Several such models are discussed by Hjorth et al. (2014). In the following sections, most of the models pick up where the gradient-based explanations leave off, modeling activity-dependent feature map formation once the overall topography has been established. As a first example of feature map development, in the next section we will consider a simple extension of this SOM model to address ocular dominance.

13.4.2 SOM MODELS OF FEATURE MAP DEVELOPMENT Given an overall topographic map, what explains the complex observed maps of neural feature preferences, such as orientation maps and other retinotopic visual feature maps? Some of these maps are already present at eye opening (including orientation and ocular dominance; reviewed in Huberman et al., 2008), suggesting that genetic mechanisms such as chemoaffinity could be involved in establishing feature maps as well. However, the development of feature maps appears to depend very strongly on neural activity, and these maps show widespread experience-driven plasticity during a postnatal *critical period* (see Kaas, 1991, for a review). Thus, most models of feature maps focus on the activity-dependent processes.

To explain the feature maps seen at birth, these models typically assume a prenatal phase of spontaneous activity, such as the spatiotemporally localized patterns of retinal waves. Indeed, disrupting spontaneous activity patterns does disrupt the formation of these maps though there is not yet consensus on the initial origin of the maps (reviewed in Huberman et al., 2008). In the rest of this chapter, we will focus primarily on models whose inputs could be either spontaneous or visually evoked activity, which lead to precise connectivity patterns through activity-dependent unsupervised learning rules (see chapter 8). As outlined below, the resulting feature map patterns are a close match to those observed in animals, suggesting that a similar process may be occurring in nature.

As a first example, we will again consider the SOM network from the previous section, this time focusing on the mapping betwen the LGN and V1. We saw previously how a SOM can recover an underlying manifold that is of the same dimensionality as the SOM map—a set of neurons arranged in 2-D forming a smooth map of an underlying 2-D space over which the pattern centers were chosen. We can now consider what happens when the underlying manifold is of a *higher* dimensionality than the SOM map and show how this process helps explain the patterns of emergent feature maps, not just the initial topography.

Let us now examine a SOM model where the inputs are chosen from a 3-D space of (x, y, e), such that the pattern is either in the left eye only $(e = 0)$, the right eye only $(e = 1)$, or in both eyes at different brightnesses $(0 < e < 1)$. Figure 13.7 illustrates the map pattern developed by such a SOM (cf. Miikkulainen et al., 2005; Ritter, Obermayer, Schulten, and Rubner, 1991), which resembles the stripy pattern of ocular dominance (eye preference) maps seen in V1 of macaque monkeys (LeVay, Hubel, and Wiesel, 1975).

SOMs are a type of dimension-reduction model, where the cortical surface is (typically) of a lower dimensionality than the input space (Durbin and Mitchison, 1990; Hastie and Stuetzle, 1989; Ritter et al., 1991). While folding to represent eye preference results in stripy or patchy patterns like ocular dominance maps, SOM folding to represent *cyclic* quantities like orientation generates realistic pinwheel-based orientation maps, in part due to geometric constraints on achieving map continuity for cyclic quantities in a 2-D plane (Obermayer et al., 1990). Interactions between multiple simultaneous features also appear to be realistic in SOM models, with feature values interacting such that areas with high selectivity and relatively low gradient (change in feature value across the map) for one feature are areas of low selectivity and high gradient for the others (Yu et al., 2005). Similar results have been found using other abstract geometric models using similar principles, such as the elastic net (Carreira-Perpiñán et al., 2005; Goodhill and Willshaw, 1990), and for SOM models of other modalities (e.g., the somatosensory homunculus; Ritter et al., 1991).

The results with SOM and related models have been very useful for understanding the geometrical process of map formation, in terms of input and output spaces. However, because of the abstract level at which these models are formulated, it is difficult to relate them to biological processes at any level of detail. For instance, the SOM process of picking a global winner requires unrealistic global knowledge of the network state, the process of neighborhood interaction is based on isotropic connectivity rather than the patchy connectivity seen in the visual cortex, the radius of lateral connectivity has not been found to decrease over time in animals, and the Euclidean distance response

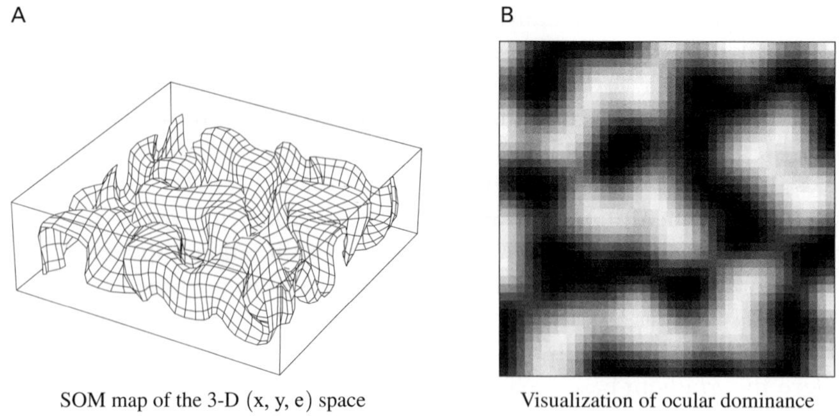

Figure 13.7 Mapping three dimensions onto two. This SOM model has developed a 2-D cortical map of a 3-D input feature parameter space (x, y, e). The first two dimensions can be interpreted as retinotopy, and the third dimension as ocular dominance (Ritter et al., 1991). (A) The full 3-D input space is indicated by the box outline. Weight vectors of the map units are plotted in this space as a grid, with adjacent units connected as in figure 13.6. The 2-D cortical space has developed a representation that effectively folds to fill up the 3-D input space, with each unit in the SOM responding best to some portion of the 3-D space, just as for the 2-D case. (B) The weight value for the height dimension e is visualized for each cortical neuron: Grayscale values from black to white represent continuously changing values from left to right eye preference. The resulting pattern resembles the ocular dominance stripes found in the visual cortex (compare to "OD [ocular dominance], Macaque" in figure 13.10), suggesting that this feature map could be the result of a self-organized 2-D mapping of a three-dimensional (or higher) parameter space. Reprinted from Miikkulainen et al. (2005).

function is difficult to relate to the mechanisms of input summation in real neurons. The following section will consider how some of the same dimension-reduction and folding principles could be implemented in models that can be related more directly to the circuitry of the cortex, in the hopes of revealing detailed explanations for cortical phenomena other than the map patterns themselves.

13.5 *How Do Feature Maps Arise from Neural Mechanisms?*

The type of model considered in the previous section focused on explaining feature map patterns in the abstract, glossing over the specific details of map formation and map function in animals. Including more of those details to build a functioning model sensory system leads to *mechanistic* models that can explain not just the map patterns but how the neurons in the map function when processing realistic inputs. For instance, models that can process natural images (e.g., Burger and Lang, 1999; Miikkulainen et al., 2005; Stevens, Law, Antolik, and Bednar, 2013) can be tested with any visual pattern used in an experiment, in order to determine how closely the model map matches the behavior of an animal's cortex. Models that include specific, modifiable lateral connections (Burger and Lang, 1999; Miikkulainen et al., 2005; Sirosh and Miikkulainen, 1994; Stevens et al., 2013) can be used to investigate phenomena like contextual modulation and surround supression that arise through interactions between large numbers of neurons in maps. Yet, as shown below, these models can still develop feature map patterns closely resembling the results of SOM and other more abstract models, unifying explanations of map formation and map function.

Several closely related mechanistic models of this type have been proposed, each using firing-rate single-compartment neurons, local Hebbian synaptic learning rules, and explicit patterns of connectivity rather than global picking of winners or optimization of abstract quantities (Barrow and Bray, 1992; Burger and Lang, 1999; Miikkulainen et al., 2005; Sirosh and Miikkulainen, 1994; Stevens et al., 2013). The idea behind each of these is to show how simple neuron models, plausible wiring patterns, and observed synaptic plasticity mechanisms could together lead to the observed properties of neural maps.

13.5.1 GCAL Model of Map Development and Function Here we will focus on the GCAL (gain-control/adaptive/lateral) model of cortical topographic feature maps, which has both a simple version suitable for analysis here (Stevens et al., 2013) and more complex implementations more closely tied to the neuroanatomy (Bednar, 2012). GCAL is a generalized variant of the LISSOM (laterally interconnected synergetically self-organizing map) family of cortical map models (Miikkulainen et al., 2005; Sirosh and Miikkulainen, 1994), replacing implausible mechanisms inherited from SOM with well-established local mechanisms such as homeostatic plasticity (described below). The resulting model shares general mechanisms with many other models and can be considered a more detailed elaboration of the original von der Malsburg (1973) model.

The simplest GCAL model of the pathway to V1 (Stevens et al., 2013) uses four 2-D sheets of neurons: Photoreceptors, LGN/RGC (retinal ganglion cells) ON, LGN/RGC OFF, and V1 (figure 13.8). Activity of the photoreceptors sheet is initialized using a bitmap image from an image database or from a synthetic distribution. Feature maps will only develop for features present in these input images; for example, if the input has only small, isotropic Gaussian patterns, no orientation map will develop. The ON and OFF sheets receive input from the photoreceptors and from neurons in the same sheet, as shown in figure 13.4A. These cells are meant to represent the complete pathway to V1's input layer phenomenologically, including processing in both the retina and in the LGN. The V1 sheet receives input from the ON and OFF sheets, as well as lateral connections from within the sheet. A full GCAL model of the known V1 feature maps is shown in figure 13.9, which includes numerous sheets representing different subcortical and cortical cell types and laminae (Bednar, 2012), but the simpler four-sheet version described here illustrates the key principles.

For GCAL, the subcortical pathways are treated as given, with each known type of ON or OFF RF implemented using a convolutional model from figure 13.4C–D with a different RF. This processing allows natural image input by reducing large areas of constant activation, as in figure 13.4D, and provides V1 with appropriate patterns of input activity that model outputs from various LGN cell types.

A GCAL V1 cell at location (k, l) receives afferent input from the ON or OFF cells near that topographic location, just as the ON or OFF cells receive input from the photoreceptors. That is, retinotopy is built in; GCAL simulates only the activity-dependent local reorganization in connectivity after the gradient-based processing described in the previous section has completed. In addition to the topographically mapped afferent connections, the V1 cells also receive lateral excitatory and lateral inhibitory input from their neighbors. Activity at

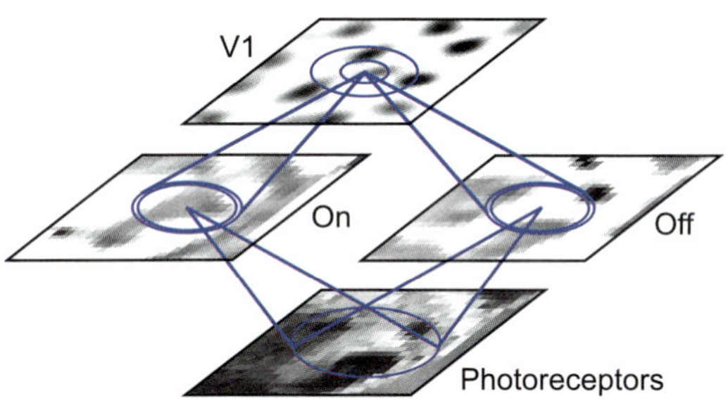

FIGURE 13.8 Basic GCAL (gain-control/adaptive/lateral) model architecture. The simplest GCAL model, for developing a retinotopic orientation map. The model consists of four neural sheets and eight separate projections between them. Each projection is illustrated with an oval showing the extent of the connection field in that projection. Afferent projections are shown with lines converging on the target of the projection, while lateral projections in the ON, OFF, and V1 layers connect within the same sheet. The single photoreceptor sheet represents input from one eye, processed by hardwired center–surround ON and OFF lateral geniculate nucleus/retinal ganglion cells, and driving a V1 sheet. The initial V1 response to an image is a blurred version of the contents of the ON and OFF sheets, because of the initially random weights, but V1's lateral connections settle the activity into discrete bubbles (multiple "winning" patches). Connections to the neurons remaining active strengthen via Hebbian learning, causing local patches of V1 to become selective for particular input patterns. Adapted from Stevens et al. (2013).

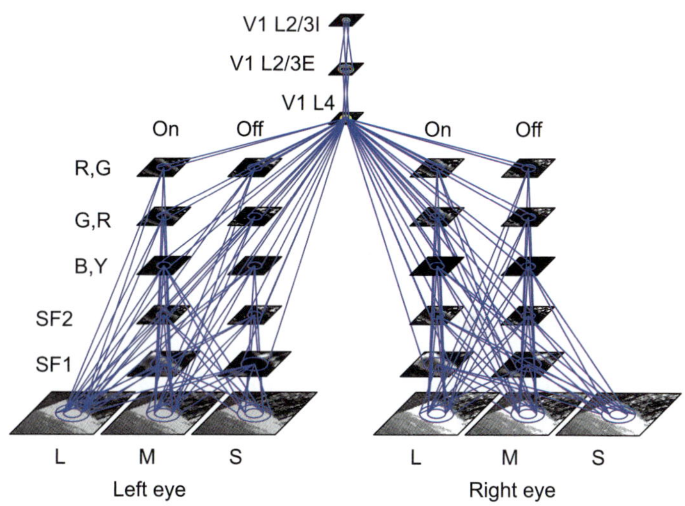

FIGURE 13.9 Comprehensive GCAL model architecture. GCAL model for simple and complex cells with surround modulation and maps for retinotopy, orientation, ocular dominance, disparity, motion direction, temporal frequency, spatial frequency, and color. The model consists of 29 neural sheets and 123 separate projections between them. The six L, M, and S sheets represent the three photoreceptor cone types in each eye, the 20 ON and OFF sheets are hardwired collections of different types of center–surround lateral geniculate nucleus/retinal ganglion cells, and the three V1 sheets represent cells of different V1 layers and/or cell types. Connections to V1 neurons adapt via Hebbian learning just as in the simpler GCAL model, allowing initially unselective V1 neurons to exhibit the full range of response types seen experimentally, by differentially weighting each of the subcortical and lateral inputs. Reprinted from Bednar (2012).

a given time is computed from each of these sources of input:

$$v_{kl} = \sigma\left(\sum_p \gamma_p x_{pkl} w_{pkl}\right). \tag{13.4}$$

Here $p \in \{A_{on}, A_{off}, E, I\}$ selects one of the four connection fields for this neuron in the simplest model (from the afferent ON, afferent OFF, lateral excitatory, and lateral inhibitory projections), γ_p is a fixed multiplicative factor scaling that projection, w_{pkl} is the weight vector in projection p to unit (k, l), and x_{pkl} is the topographically corresponding portion of the input vector. For a lateral projection, $x_{pkl} = v_{pkl}(t-1)$, that is, the activity of the topographically corresponding portion of this sheet in the previous time step. σ is a half-rectifying function to ensure that activities are positive, with a fixed gain of 1.0 but an adaptive threshold θ that is automatically adjusted to maintain a fixed target activity (Stevens et al., 2013):

$$\theta(t) = \theta(t-1) + \lambda(\overline{\eta_j}(t) - \mu). \tag{13.5}$$

where $\lambda = 0.01$ is the homeostatic adaptation learning rate, $\overline{\eta_j}$ is an average of this neuron's recent activity, and $\mu = 0.024$ is a target average firing rate.

In the simplest GCAL model (Stevens et al., 2013), as in nearly all other activity-dependent models of feature map development, the lateral connections have a Mexican-hat configuration (short-range excitation, longer-range inhibition). The result of such lateral interaction is to focus an intially diffuse pattern of activation around one or more hot spots that receive more excitation than inhibition, gradually resulting in spatially discrete "bubbles" of activity in response to a slowly changing input. Similar processes occur in the variants of the model that have more realistic connectivity patterns with long-range excitation, as long as disynaptic inhibition dominates at high contrasts (Bednar, 2012). These interactions are a locally implemented version of the global winner-picking and neighborhood kernel from the SOM, and just as for the SOM, they ensure that the maps achieve good coverage and continuity. Unlike the SOM, there can be many spatially distinct "winning" patches simultaneously (see the V1 sheet in figure 13.8), each learning properties from the retinotopically corresponding inputs.

As for a SOM, the weight values w_{pkl} in each projection (whether afferent or lateral) are initially isotropic or random, developing selective patterns only through Hebbian synaptic plasticity. Specifically, after the postsynaptic activity v_{kl} for unit (k, l) is computed using equation 13.4, a new weight vector w'_{pkl} is computed from the old weight vector w_{pkl}, the presynaptic activity vector x_{pkl}, the postsynaptic activity w_{pkl}, and the learning rate α:

$$w'_{pkl} = \frac{w_{pkl} + \alpha v_{kl} x_{pkl}}{\sum_k (w_{pkl} + \alpha v_{kl} x_{pkl})}, \tag{13.6}$$

where the denominator is the L1-norm of y, that is, the sum of all the weight magnitudes. This divisively normalized rule results in connections that reflect correlations between the presynaptic activity and the postsynaptic response, with competition enforced by the normalization so that weights are bounded while becoming selective.

Note that there is strong biological support for the various elements of the more elaborate version of this model (Bednar, 2012; Miikkulainen et al., 2005), but there is also a large number of other similar models that would also be compatible. For instance, there is little agreement about the particular form of divisive normalization used, but there is general agreement that some mechanism of this type is needed to keep values stable. Similarly, whether excitation and inhibition should combine additively, multiplicatively, or in some other way is not clear since the underlying operations of real neurons are a complex function of their detailed morphology and physiology. Rather than trying to incorporate all of these details, GCAL focuses on extracting a simple, plausible mechanism that is sufficient to replicate the observed behaviors and can be elaborated as needed when data are available.

When trained on natural images and with well-chosen values of parameters such as the distances and strength of lateral connectivity, this model develops map patterns that are a close match to those found in animals. Figure 13.10 shows GCAL and LISSOM maps from simulations of each of the spatial-feature dimensions for which maps have been found in V1, along with the corresponding experimental data. All of the maps show local continuity and smoothness, which in the model is due to local lateral connectivity that makes nearby neurons respond to similar patterns. The animal maps also each have unique properties that are reflected in the model results. For instance, orientation is represented by a smooth map interrupted by pinwheels and other discontinuities. Interestingly, the patterns of discontinuities in both animals and the model follow quantitative predictions from abstract principles of geometric pattern formation, rather than representing incomplete self-organization (Kaschube, Schnabel, Löwel, Coppola, White, and Wolf, 2010; Stevens et al., 2013).

Motion direction is mapped similarly while ocular dominance forms stripes alternating between left

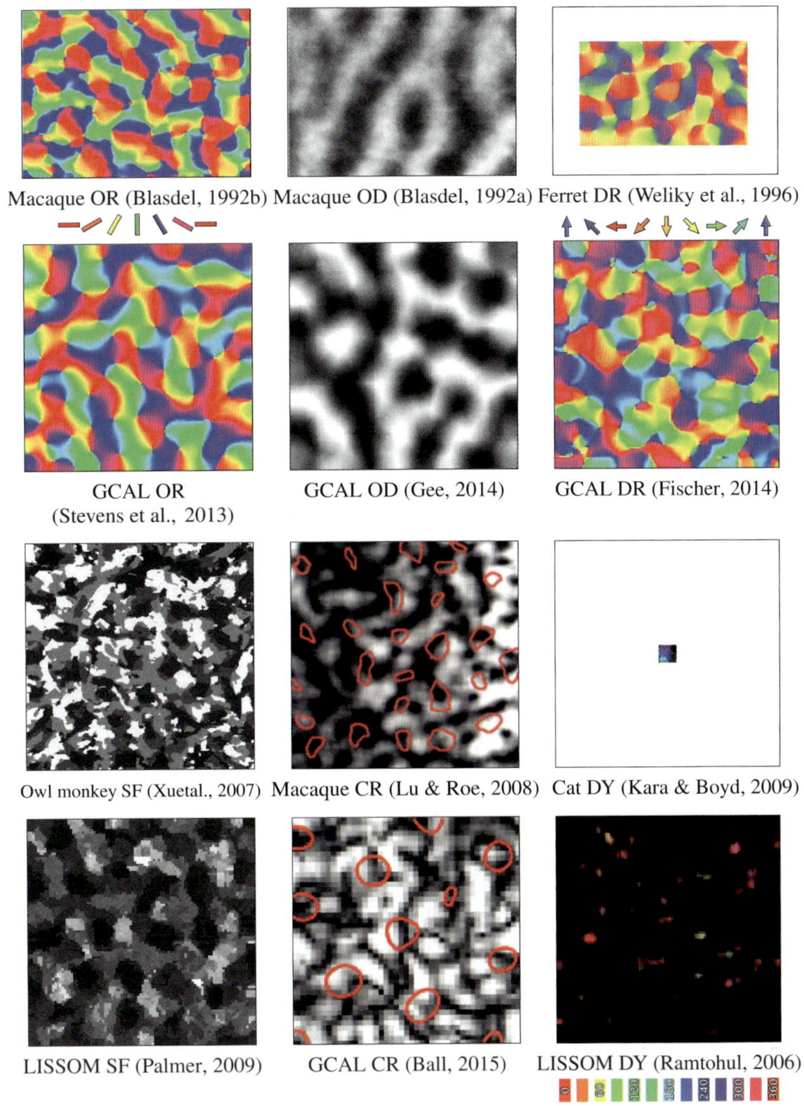

FIGURE 13.10 Animal and model V1 maps. Maps have been found in V1 in monkeys, ferrets, cats, and tree shrews for a large number of input features, including: orientation (OR), ocular dominance (OD), motion direction (DR), spatial frequency (SF), color (CR), and disparity (DY). Various maps for macaque, ferret, owl monkey, and cat are plotted above, each in a box representing a 4-mm-wide patch of V1, along with the species and citation for the corresponding publication, and a corresponding model plot (below each animal map). See the main text for further description.

eye (white) and right eye (black). Spatial frequency preferences occur in patches rather than stripes. Color responses are plotted as red outlines highlighting areas of strong response to a red–green isoluminant grating, compared to a monochromatic grating. Such color selectivity occurs in isolated patches, typically aligned with less selective regions of the orientation selectivity map (darker areas of grayscale background), indicating largely separate coding for color and orientation. Data for horizontal disparity is less clear, but disparity-selective cells tend to occur in small, isolated patches (of which only one is shown for the experimental data), due to horizontal disparity being detectable only for cells with a vertical orientation preference.

Note that each of these maps in principle can consist of the same set of physical neurons, just tested using different input patterns and analysis techniques. The overall result is that neurons are each selective for a small region in a large, multidimensional parameter space (e.g., for a pattern oriented in a specific direction, at a specific location on the retina, with a certain size, color, etc.), and together the neurons cover the space of possible patterns encountered during development (cf. Durbin and Mitchison, 1990). Although each neuron has some level of selectivity for every dimension, selectivities are typically orthogonal, such that a neuron highly selective in one dimension is less selective in others (cf. Yu et al., 2005).

Although not all of the dimensions in figure 13.10 have been tested using a SOM, it is likely that a SOM model could be built to yield similar map patterns, as for many other possible models of this general form. Unlike the SOM or other geometrical models, however, the GCAL model can be considered a functioning visual system and is thus able to be evaluated for visual illusions, aftereffects, transient responses, surround modulation, and in general tested with any visual pattern or experimental paradigm, as in the next section.

13.5.2 Contextual Modulation in Neural Maps The preceding sections have focused on how the feature preference and RF properties of neurons in neural maps can be generated. But once neurons have developed selectivity, treating them as a set of disconnected spatial filters (as in figure 13.4C–D) is at best a useful first step. Such neurons also exhibit very many "nonclassical" phenomena that involve interactions across large areas of the map. For example, retina and LGN neurons exhibit contrast gain normalization, such that responses are relative to the prevailing activation in that local area, not simply a function of the neuron's RF (Bonin, Mante, and Carandini, 2005). Cortical neurons also show such normalization and exhibit complex changes in RF and response properties when stimuli are present in locations surrounding the "classical" RF: perceptual completion or suppression effects, illusory contours, "pop-out" of certain stimuli from their background, and a variety of other contextual phenomena (for a review, see Graham, 2011).

These phenomena are generally considered to involve lateral connections from other neurons in the map and/or feedback from neurons in other regions, along with possible "read-out" mechanisms in downstream neurons receiving input from these areas (Angelucci and Bressloff, 2006). Many of the effects are specific to certain feature values, such as orientation. Importantly, long-range lateral connections have been shown to reflect the underlying feature maps, tending to connect neurons with similar orientation preferences in V1 (Bosking, Zhang, Schofield, and Fitzpatrick, 1997), which is consistent with a role in these effects.

Most models treat these lateral interactions as isotropic, primarily for simplicity and computational efficiency, which is a valid approximation only at very short ranges. The SOM analysis in terms of folding in previous sections suggests that seemingly complex patchy connectivity that respects feature maps could actually be compact, isotropic connectivity in an underlying multidimensional feature space, which then looks patchy only because the neural architecture is organized in only two dimensions. Interestingly, such connectivity emerges automatically from the mechanistic GCAL model in the previous section, driven by Hebbian learning (equation 13.6) that strengthens connections between neurons with correlated responses (Miikkulainen et al., 2005). Figure 13.11 shows examples of patchy lateral connectivity patterns like those developed by GCAL models, with neurons connected by similarity of responses to the multidimensional input space, not only by locality in the cortical space. Work with related but more complex models suggests that correlation-based connections can account for many of the nonclassical phenomena listed above (Bednar, 2012), including size tuning, surround modulation, and orientation illusions and aftereffects. Future work will be needed to investigate the role of feedback connections, which could have many of the same effects.

13.6 What Is the Information-Processing Goal of Neurons in Maps?

The previous sections outlined how map models could establish an initial topography, how feature map patterns could emerge through folding to represent a multidimensional latent input space, and how these processes could be related to neural mechanisms to provide detailed predictions and tests. The models also suggest explanations for the function of a feature map, such as that it provides good coverage of an underlying feature space (Swindale et al., 2000), that lateral connections help decorrelate the responses, and that contextual modulation and affereffects are the unavoidable result of these processes (Barlow, 1990; Miikkulainen et al., 2005).

However, in each case the functional aspects of the model are incorporated only indirectly, either feeding into the modeler's design choices or as a postfacto interpretation of the results. For a different class of models that can be considered *normative*, functional criteria are incorporated explicitly into the definition of the model. Doing so allows a hypothesis about function to be evaluated directly, independently of the details of specific neural mechanisms, which are often not known.

An influential series of normative models starts from an assumption that the computational goal of a sensory system is to recover information about the underlying structure of the world, presumably in order to guide actions relating to that structure. Given raw sensory

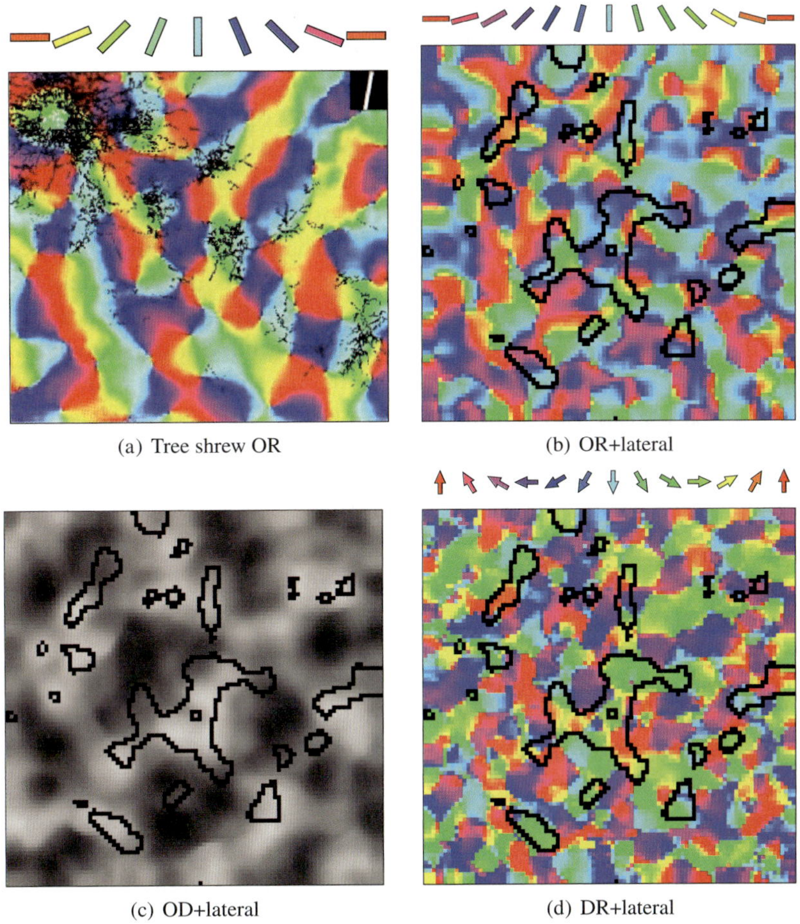

FIGURE 13.11 Orientation (OR), ocular dominance (OD), and motion direction (DR) lateral connections. (a) Reprinted from Bosking et al. (1997). Orientation map in 2.5 × 2-mm tree shrew V1, with black dots indicating areas labeled by a tracer injected in the white-dotted area in the upper-left corner. The patchy connections primarily connect to regions of the map with similar preferences. (b–d) Maps from a joint LISSOM model of orientation, ocular dominance, and direction, each with a contour threshold plot of lateral connection strength for the central neuron (denoted by a black square). The strongest connections to that neuron are those enclosed within each contour shape. This neuron is nearly monocular, and primarily connects to other right-eye-preferring neurons. The same pattern of connectivity can also be seen to respect the direction map, with long-range connections primarily to similar directions, and the orientation map, with long-range connections primarily to similar orientation and direction preferences. These connections reflect correlations between neurons during development, which in turn reflect each of the feature preference values for each neuron. The connections thus respect similarity in a multidimensional feature space, becoming patchy because that space has been projected down onto the 2-D cortical surface. Reprinted from Bednar and Miikkulainen (2006).

input, the objective of these models would be to *infer* a set of external causes that would explain the observed data (Bell and Sejnowski, 1997; Olshausen and Field, 1996).

The real causes of an image would comprise its constituent set of external 3-D objects, surfaces, and lighting sources. However, determining these is a highly complex and nonlinear inference process, and full reconstruction is not necessary for most of the tasks that an animal needs to do. Plus, many features of the world can only be discovered through interaction and thus require loops involving the motor system, placing them outside of the present account. A much more tractable goal, which we will focus on in the next two sections because it is locally realizable in V1, is to find a linear combination of *basis vectors* that can generate a small patch of an image.

13.6.1 Basis Vectors for Generating Images A *generative* model is one whose activity pattern can be used directly to generate an input patten. Starting with the idea of generating a small image patch is appropriate if the real causes of the scene can be estimated hierarchically, with V1 representations accounted for by causes at a higher level, as suggested by the multiple visuotopic maps in areas along the visual pathway. In this section we will first explain the general principles behind these models using simple examples, and then the following section will consider a specific normative generative model applied to explain neural maps and RFs in V1.

A basis set for an image patch x is a set of basis vectors v_i such that x can be approximated as a sum of the v_i weighted with scalar coefficients c_i:

$$x \approx \sum_i c_i v_i. \tag{13.7}$$

This model is represented graphically in figure 13.12. The coefficients c_i can be considered estimates of latent (or hidden) variables, specifying how much of each corresponding basis vector v_i is present in this image patch.

The generative approach allows explicit expression of an *objective function*, where a representation is iteratively optimized to minimize or maximize a mathematically specified criterion (which is what makes these models normative). For instance, the objective function might specify a low reconstruction error (to ensure that the representation is faithful), low levels of neural activity (to minimize metabolic cost), few neurons active (a *sparse* (mostly zero) representation that could simplify downstream processing), and so forth, or weighted combinations of any of these. Various automatic (though typically non-neural) techniques can then optimize the representation to achieve the specified objective, approximating the result of processes

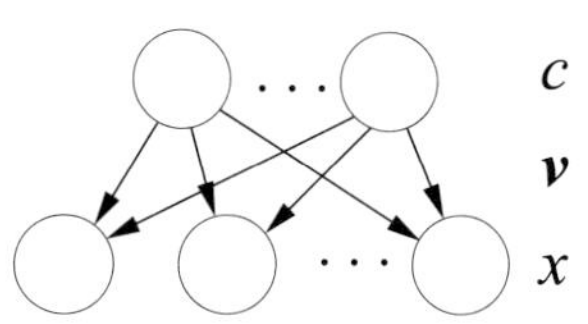

Figure 13.12 One-hidden-layer generative model. Here the inferred data vector x is obtained as a linear combination of the basis vectors v, with coefficients c. The arrowheads indicate that the basis vector acts in the "reverse" direction compared to the weights in the models previously discussed (e.g., figure 13.4), specifying the weights that could be used to reconstruct the input image, not the weights by which the coefficients c are calculated given an input image. That is, the values c have to be inferred approximately, when given an actual set of inputs, rather than calculated directly as in the mechanistic models from previous sections. Once the values c have been estimated, they provide a compact representation for the input image and can then be used directly for further processing. The actual input need not ever be reconstructed from c during sensory processing, as long as c provides a representation that has captured the important properties of the image patch well enough that it *could* be reconstructed with low error.

of evolution and development in animals. Note that specifying that reconstruction error should be low does not necessarily mean assuming that the visual system actually does reconstruct the input in practice—it simply specifies a criterion that the map's representation of the input should be rich enough to allow that in some form.

Perhaps the simplest generative model would be to set $c_j = 1$ for one particular index j, and set all of the other coefficients $c_i = 0$ (a winner-take-all approach achieving maximum sparsity in the cortical representation). If a suitable objective function were defined, such a model could do *template matching*, where the input is represented in terms of the closest-matching basis vector. For instance, each basis vector could be determined using the *k-means* clustering algorithm (Hartigan, 1975). Given a large set of image patches, *k-means* provides k vectors that are each averages of a cluster of image patches, with the clusters chosen iteratively to maximize within-cluster similarity. A novel input would then be mapped into the nearest cluster, as previously described as a limiting case of the SOM algorithm. Interestingly, such clustering does yield basis vectors similar to RFs found in V1 (Coates, Ng, and Lee, 2011). However, a single chosen template (akin to a single V1 neuron firing) is unlikely to be a good match to an arbitrary patch of a natural image, and thus this would be a poor generative model, with high reconstruction error.

A much richer representation is obtained by allowing multiple c_i coefficients to be nonzero at the same time, forming a *distributed representation* or *componential representation* of the input. That is, one can select multiple coefficients that *together* could reconstruct the input, echoing how GCAL has multiple "activity bubbles" in response to an image, instead of SOM's single winning neuron.

Figure 13.13 shows an example of how an image patch can be reconstructed using basis vectors developed by the Olshausen and Field (1996) "sparse coding" model of V1, to be discussed in more detail below. For now, we will just treat the basis set as given, and look at how a 14 × 14-pixel patch of an image can be reconstructed with relatively low error given a small number of basis vectors. Because natural images are only a limited subset of all possible patterns of pixels (forming a lower-dimensional manifold, as discussed in the context of dimension reduction above), only a small number of well-chosen basis vectors are needed to represent most natural image patches with high accuracy. For this example, the image patch is easily recognizable after only four basis vectors ($k = 1 \ldots 4$) have been summed, and barely distinguishable after 16, which means this specific input can be represented faithfully using 4 or 16 activation values c_i at the output layer rather than the 196 pixel values on the input layer.

In the context of image compression algorithms like JPEG and wavelet coding, achieving a good-quality reconstruction from a small number of coefficients (a *sparse* representation) is the final goal. In the visual system, of course, it is unlikely that images will actually ever need to be reconstructed, but requiring good reconstruction is a practical way of objectively ensuring that the algorithms have preserved the information in the image. In practice, the idea is that the sparse coefficients c_i themselves, rather than the reconstruction, will be an effective low-dimensional representation suitable for driving action or further computation.

13.6.2 Olshausen and Field Model of V1 RFs To form a full generative model, the basis-function approach can be combined with (1) an objective function that (in part) minimizes reconstruction error, (2) procedures for inferring the coefficients c_i in response to each image, and (3) procedures for developing the basis vectors v_i for a given image data set. For instance, the influential Olshausen and Field (1996) model defines an objective function as

$$\left| x - \sum_i c_i v_i \right|^2 + \lambda \sum_i g(c_i) \qquad (13.8)$$

with the first term expressing the reconstruction error (zero for perfect reconstruction), while in the second term $g(c_i)$ is a penalty function corresponding to the negative log sparse prior on c_i, and λ is a trade-off parameter determining the balance between reconstruction error and coefficient sparsity. Equation 13.8 is (up to constant terms) the negative log of the posterior probability $p(c|x)$ of the generative model. Here sparseness means that the prior distribution on the c_i favors values close to zero, reflecting a belief that "natural images may generally be described in terms of a small number of structural primitives" (Olshausen and Field, 1997). To allow a highly sparse representation, the model uses an *overcomplete basis set*, with more basis vectors than are required for a perfect linear reconstruction.

Given a set of basis vectors and an input image patch, the model must carry out *probabilistic inference* to determine the coefficients $\{\hat{c}_i\}$ that best explain the image rather than simply computing the map activity diretly as in the mechanistic models described in previous sections. The basis vectors associated with these coefficients can be interpreted as estimates of the *latent causes* or *latent variables* in the external world that led to the formation of that image. The basis vectors are learned by an iterative EM (expectation-maximization) style algorithm (Dayan and Abbott, 2001), where one alternately carries out inference to obtain $\hat{c}$, the set of c's that minimize equation 13.8, then adjusts the basis

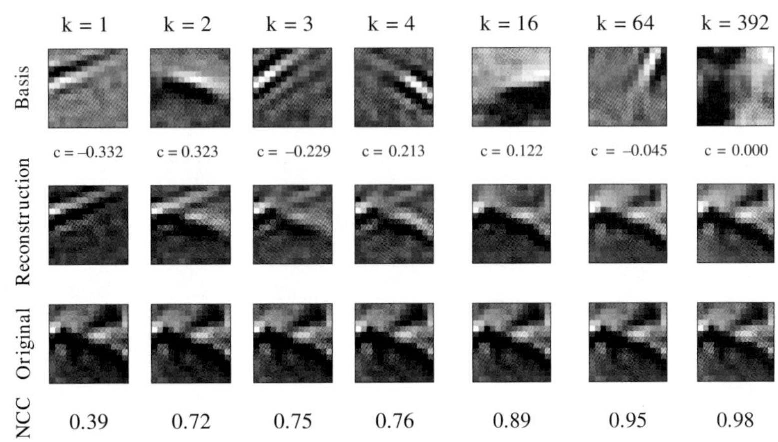

Figure 13.13 Reconstruction from basis vectors. The image patch that has been duplicated in each column of the row labeled "Original" can be reconstructed as a weighted linear combination of 392 basis vectors, some of which are shown in row "Basis" with their corresponding coefficient c. The basis vectors were developed by the Olshausen and Field (1996) model discussed below. The result is the reconstructed image patch in the column labeled k = 392 of the row "Reconstruction," which is nearly indistinguishable from the original. Using a single basis vector (column k = 1) amounts to template matching; even the best-matching single template is still a poor representation. However, most of the coefficients are small, and using, for example, the 16 basis vectors with the largest coefficients suffices for a high-quality reconstruction (high normalized cross correlation, row "NCC," and low visible difference, even if treating the other coefficients as being zero). The coefficients of those 16 vectors act as a compact and sparse but faithful representation of the original input patch.

vectors to minimize $|x - \sum_i \hat{c}_i v_i|^2$, eventually settling on a stable set of basis vectors for a given set of training images.

As an alternative to this general optimzation technique, Olshausen and Field (1997) also showed how one can design a locally implementable gradient descent scheme on the objective of equation 13.8 to achieve this computation. Interestingly, this physically realizable implementation shares important features in common with the mechanistic models like GCAL, suggesting that it may be possible to relate the two to show how a specific goal can be implemented in terms of neural mechanisms. For instance, the physical implementation involves *lateral interactions* between the c units, somewhat like the correlation-based connections in GCAL, where the strength of the interaction between units c_i and c_j depends on on the similarity of their weight vectors $v_i \cdot v_j$. This dependence is because two units with similar basis vectors that both could have contributed to the observed image structure need to compete if the sparsity term is to be minimized. Similarly, the inference step can be implemented using a feedforward "recognition" network analogous to the afferent weights in GCAL, predicting $\hat{c}$ from the input x (see, e.g., Zylberberg, Murphy, and DeWeese, 2011). Thus, mechanistic interpretations of these ideas do exist, though nearly all normative and generative models are implemented and presented in more abstract, nonmechanistic form.

In any case, once the model has developed such a set of basis vectors, they can be compared to data from animals, to see if this approach can explain anything about the structure and properties of V1 neurons. Because the basis vectors are used as a *projective field* from V1 back onto the LGN in the model, not a *receptive field* computing V1 activations from LGN activity, they are not directly comparable to V1 RFs. Even so, the model can be treated as a "black box" as in animal experiments, presenting patterns on the input and recording the responses, allowing an RF pattern to be estimated by reverse correlation for each input. The resulting patterns are similar to the projective fields, and so the projective fields (basis vectors) are typically compared as-is to V1 RFs (Olshausen and Field, 1997). In what follows, we will use the basis vector as an approximation to that neuron's RF because directly estimated RFs are not provided for most models of this type.

Figure 13.14 shows the basis vectors that resulted from iteratively maximizing the Olshausen and Field

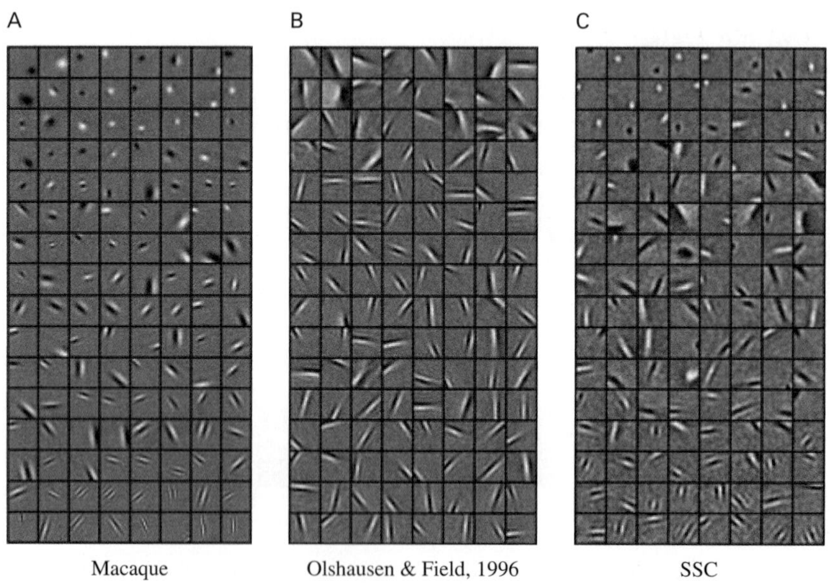

FIGURE 13.14 Macaque and model V1 receptive fields (RFs). (A) Macaque monkey RFs in V1 are typically shaped like Gabor functions (a sine grating masked by a circular Gaussian; Jones and Palmer, 1987), with various RF shapes amounting to different values of the Gabor's frequency, phase, and Gaussian-size parameters. These RFs cover a range of selectivities, orientations, spatial frequencies, and spatial locations. Gabor fits reprinted from Rehn and Sommer (2007); data from Ringach (2002), adjusted for eccentricity and sorted by RF shape. (The raw data are not available, but the authors found that the Gabor fits had low error.) (B) Basis vectors from the Olshausen and Field (1996) sparse coding model are also Gabors, although the range of sizes is more limited, with more multilobed patterns. Reprinted from Rehn and Sommer (2007); data from Olshausen and Field (1996) model. (C) A model with a slightly modified sparseness constraint (SSC) can reproduce the range of Gabor-shaped RFs found in macaque when the basis vector is taken as an approximation to the RF. Reprinted from Rehn and Sommer (2007).

(1996) objective function for a set of patches of whitened natural images, comparing them to results from monkey V1. The results resemble the RFs shaped like Gabor functions found in V1 in mammals, leading to a hypothesis that primary visual cortex neurons are wired to form a basis set that provides good coverage of the space of natural image patches using a small number of active units at any one time (sparse coding; Olshausen and Field, 1996).

Later work has shown that there are very many other ways to get Gabor-shaped weight matrices by training on natural image patches. For instance, models based on independent components analysis (ICA) also lead to Gabor-shaped weights (Bell and Sejnowski, 1997). ICA is nearly identical mathematically to the Olshausen and Field (1996) sparse coding model, but with a simplifying restriction that the number of input and output units are the same; a complete rather than overcomplete basis set. In this case, the basis vector weights are adjusted assuming that the underlying causes are statistically independent, hence the name. With this assumption, the inference step can be replaced with a simple feedforward operation, and the RFs can be calculated directly from the projective fields by matrix inversion. Even radically different models from these can also get similar results—simply clustering (i.e., grouping by similarity, then averaging) whitened image patches yields Gabor-shaped RFs (Barrow, 1987; Coates et al., 2011).

However, each of these algorithms results in a relatively narrow range of RF shapes, missing many of the RF types commonly observed in animals (compare figure 13.14B with figure 13.14A). More recent work has shown how to account for the full range of monkey RF shapes by using an alternative definition of sparseness (Rehn and Sommer, 2007; figure 13.14C) or by using very different approaches altogether (e.g., the undirected "energy-based" model from Osindero, Welling, and Hinton, 2006).

Note that nearly all of the work using normative models for neural maps has focused on a "bag of neurons" approach with no topography. Some such models have addressed topography as well, notably Topographic ICA (Hyvärinen and Hoyer, 2001; Hyvärinen, Hurri, and Hoyer, 2009), the energy-based model from Osindero et al. (2006), and a variant of ICA expressed in terms of mutual information (Kozloski, Cecchi, Peck, and Rao, 2007; Linsker, 1989). Each of these models adds some form of lateral interaction across a map surface, as in SOM and GCAL, causing neurons to become more similar to neighbors. However, so far the normative-model maps have only been simulated at a very coarse spatial scale, leaving it unclear whether the map shapes correspond to those seen in animal maps as has been shown for the geometrical and mechanistic models. Moreover, the required modifications to generate such maps are motivated primarily in terms of matching experimental data or in mechanistic terms, rather than showing how the topography satisfies a normative goal. Thus, normative approaches at present can account primarily for RF shapes and their diversity for neurons in neural maps rather than for topographic and feature maps, perhaps because there is no functional role for the topography (as discussed further below).

13.7 Nonvisual Maps

The preceding sections have covered a set of related topics from the early visual system in depth. The visual system is used as a model for map development for many reasons, such as having easily controllable stimuli, a receptor surface with a relatively simple and well-understood organization, a small number of synapses between the receptors and the first cortical map, and a cortical map that is easily accessible for imaging. The hope of much of this work is that similar principles will apply to other nonvisual modalities. For instance, there are numerous SOM-based models of other modalities (e.g., somatosensory cortex; Ritter et al., 1991), GCAL/LISSOM-based models of whisker-barrel maps (Wilson, Law, Mitchinson, Prescott, and Bednar, 2010), and sparse-coding or ICA-based models of a wide range of sensory neuron types. There is also corresponding data for maps and plasticity in the motor cortex (reviewed in Harrison and Murphy, 2014) as well as similar SOM modeling approaches for motor cortex (Morasso and Sanguineti, 1995).

As already shown for visual maps, different aspects of the stimuli will lead to very different map shapes and properties (such as ocular dominance stripes vs. isolated color blobs). Similarly, different organizations are expected for each modality. For instance, in auditory cortex there is a primary organization by tonotopy (matching the arrangement of receptors on the cochlea), but it is not yet well established which other dimensions of auditory stimuli might be represented. And although modeling has replicated the basic organization of somatosensory cortex, how the various touch, temperature, and pain receptors distributed across the skin surface affect the organizations in somatosensory cortex is not well understood. Similar work in auditory cortex is constrained by the huge number of auditory subcortical regions before cortical processing. Modeling of map development in these other sensory areas will require significant new experimental data on the subcortical properties and how they connect to the cortex, at which point the methods shown here can be applied and tested. Given the relatively little work done so far on applying such models to the motor system, it remains an open question how much of the same methods will apply, and again suitable experimental data are probably the constraining factor.

13.8 Concluding Discussion

The models covered in this chapter help explain how topographic mappings are established, how and why feature map patterns develop, how the context of neurons in feature maps affects their function, and what aspects of the information present in images are represented in the activity of neurons in maps. The models spanned a wide range of approaches, from abstract models focused on uncovering geometrical relationships, to mechanistic models expressed in terms of neural mechanisms, to normative models expressed in terms of their information-processing goals (refer to box 13.1).

Ideally, explanations of neural maps would address each of these levels at once, showing how the geometric patterns of maps and RFs emerge from biophysical mechanisms that satisfy clear information-processing goals. In practice, there are still significant gaps between the models of each type, and so no type has subsumed the others. For instance, there is a relatively small amount of data about the visual experience of developing animals, which makes it difficult to build fully mechanistic models of feature map development based on realistic stimuli, and so the more abstract geometrical models remain very relevant for these cases. However in principle there is no serious conflict between the geometrical and mechanistic map models, with the elastic net, SOM, and GCAL models all developing very similar types of feature map patterns through largely compatible mechanisms.

More interesting and problematic are the relationships between mechanistic and normative models, and between normative models and the neurobiological data. Normative models are primarily formulated at a level that is far removed from neural mechanisms, which achieves their purpose of revealing functional aspects but makes it more difficult to relate them to results from neurobiological experiments. As described above, some progress has been made on building mechanistic interpretations of normative models, with many of the required elements having a clear basis in known mechanisms. However, significant gaps still exist between all of the current normative models and what we know of neural architecture, which can be addressed in future modeling studies, as described below.

First, due in part to their computational complexity, none of the normative topographic map models have been run with a cortical area large enough to demonstrate a close match to the topographical properties of animal feature maps (other than local smoothness), such as regular periodicity, hierarchical relationships between multiple maps, and systematic topological defects (Blasdel, 1992). Making this topographical structure clear could help establish strong links between the normative models of uncovering latent structure and the geometrical models of dimensionality reduction, because both are driven by revealing underlying structure in the inputs. It should soon become feasible to run such large simulations, using a combination of efficient algorithms and general-purpose graphics-processing units such as those available on modern video cards. Meanwhile, it is not yet clear if the types of grouping implemented in these models correspond to the grouping processes presumably underlying animal maps.

Second, the existing normative topographic map algorithms (including at least Hyvärinen and Hoyer, 2001; Kozloski et al., 2007; Osindero et al., 2006) rely crucially on squaring or full-wave rectification of negative firing rates at some step of the calculation. This step is essential for these algorithms to avoid grouping by spatial phase so that they can develop complex cells by local pooling, but is difficult to reconcile with physical neural responses based on actual (positive) firing rates.

Third, the process of "development" in all of the normative topographic map models (and also in most of the nontopographic normative approaches) relies on non-neural optimization algorithms that require information distributed throughout the model network, rather than synaptically or neurally local learning rules that have a clear biological implementation. They typically are temporally nonlocal as well, with results that depend on batch learning to achieve smooth basis vector patterns (e.g., in Olshausen and Field, 1996), and on temporally separate inference and learning phases with no clear biological interpretation as a continuous system, even in the most mechanistic implementations of sparse coding built so far (Zylberberg et al., 2011). These conflicts between the normative and mechanistic approaches are worth investigating in future studies because they may reveal either previously unknown neural mechanisms or invalid assumptions about the goals of sensory processing. They could also represent implicit constraints provided by evolution, rather than actual neural mechanisms, which would also be important to establish. For now, all three types of explanation (geometrical, mechanistic, and normative models) need to be considered together to appreciate the range of phenomena that can be addressed by neural map models.

From a larger perspective, while the ubiquity of smooth, topographic maps suggests that they could be important, it has proven surprisingly difficult to demonstrate any functional advantage to having neurons organized spatially in maps. Conceptually, if one starts with a smooth, topographic feature map and randomly perturbs the position of each cortical neuron without changing the underlying connectivity pattern, then the pattern of connectivity, and hence much of the function, will remain unchanged. However, the lengths (and presumably latencies) of the connections *will* change, however, leading to a hypothesis that maps exist to optimize the total wiring length in the brain (which dates back at least to the seminal work of Cajal—see, for example, the second edition of his masterwork, Ramón y Cajal, 1911). Indeed, a smooth map appears to be at least a local optimum in wiring length, given the observed patterns of connectivity (Chklovskii and Koulakov, 2004).

A deeper issue is why the observed patterns of connectivity, that is, 2-D maps with primarily local connections but patchy long-range, feature-specific connections as well, occur in the first place. Although any answer to this question must be speculative, perhaps a 2-D cortex makes it easier to connect local neurons (using wiring in the third dimension), while not being much worse than the highest-dimensional physical arrangement possible (3-D), in comparison to the large, multidimensional space of variation in the external world. In turn, smoothness in the map presumably reflects the fact that properties that can be sensed (objects, sounds, odors, etc.) in the external world tend to be localized in space (as suggested by Tobler's first law of geography, quoted at the start of this chapter). With a smooth, topographic organization, sensory neurons with local RFs can then process meaningfully distinct underlying (latent) causes and interact with neurons somewhat more distant to improve discrimination and detection. In this context, smoothness could make the process of development more robust, allowing a large margin of error in axon guidance without affecting the results. As suggested for dimension-reduction models above, the observed patchiness in connectivity would then reflect a map that has extracted an underlying space with a dimensionality greater than 2-D.

Finally, model maps have been considered largely in isolation here, and primarily in terms of feedforward processing, but real maps are situated in the context of numerous other maps that are simultaneously involved in neural processing, including widespread feedforward and feedback connections between them (Van Essen and Gallant, 1994). Numerous models of feedback and attentional processes incorporate static (nondevelopmental) neural maps and can help us understand how the maps relate to systems-level phenomena and eventually to behavior. For instance, Rao and Ballard (1999) and Hyvärinen et al. (2009, chapter 14) have investigated the effect of feedback from higher regions on the activations of lower-level units. In their models higher-level units may identify longer-range structure, and feed back information about this to lower-level units, for example, to model contour integration. Understanding how lateral and feedback interactions affect representations in neural maps is an important topic for current research, because it can help bridge from the map level all the way to behavior.

13.9 Summary

- Neural maps are ubiquitous but diverse.
- Many maps are topographic.
- Topography is established under genetic control with activity-based refinement.
- Maps for other features appear to develop based on neural activity.
- Feature map patterns appear to be the result of a 2-D cortical space with preferences folding to cover an underlying multidimensional input space.
- Mechanistic implementation of such dimensionality reduction can explain contextual modulation, lateral connectivity, and aftereffects.
- Normative models suggest that neurons are uncovering latent causes in sensory inputs, forming a sparse representation suitable for further computation and for action.
- Determining the relationships between abstract geometric models, mechanistic models, and normative models of maps will be important.
- Lateral interactions across cortical space are crucial for all the classes of models.

REFERENCES

Anderson, J. A., & Rosenfeld, E. (Eds.). (1988). *Neurocomputing: Foundations of Research.* Cambridge, MA: MIT Press.

Angelucci, A., & Bressloff, P. C. (2006). Contribution of feedforward, lateral and feedback connections to the classical receptive field center and extra-classical receptive field surround of primate V1 neurons. *Progress in Brain Research, 154,* 93–120.

Ball, C. E. (2014). *Modeling the Emergence of Perceptual Color Space in the Primary Visual Cortex* (doctoral dissertation). School of Informatics, The University of Edinburgh, UK.

Barlow, H. B. (1990). A theory about the functional role and synaptic mechanism of visual after-effects. In C. Blakemore (Ed.), *Vision: Coding and Efficiency* (pp. 363–375). Cambridge: Cambridge University Press.

Barrow, H. G. (1987). Learning receptive fields. In *Proceedings of the IEEE First International Conference on Neural Networks* (San Diego, CA) (Vol. IV, pp. 115–121). Piscataway, NJ: IEEE.

Barrow, H. G., & Bray, A. J. (1992). An adaptive neural model of early visual processing. In I. Aleksander & J. G. Taylor (Eds.), *Artificial Neural Networks, 2: Proceedings of the 1992 International Conference on Artificial Neural Networks* (pp. 881–884). Amsterdam: North-Holland.

Bednar, J. A. (2009). Topographica: Building and analyzing map-level simulations from Python, C/C++, MATLAB, NEST, or NEURON components. *Frontiers in Neuroinformatics, 3*, 8.

Bednar, J. A. (2012). Building a mechanistic model of the development and function of the primary visual cortex. *Journal of Physiology, Paris, 106*, 194–211.

Bednar, J. A., & Miikkulainen, R. (2006). Joint maps for orientation, eye, and direction preference in a self-organizing model of V1. *Neurocomputing, 69*(10–12), 1272–1276.

Bell, A. J., & Sejnowski, T. J. (1997). The "independent components" of natural scenes are edge filters. *Vision Research, 37*, 3327.

Benjumeda, I., Escalante, A., Law, C., Morales, D., Chauvin, G., Muca, G., et al. (2013). Uncoupling of EphA/ephrinA signaling and spontaneous activity in neural circuit wiring. *Journal of Neuroscience, 33*, 18208–18218.

Bishop, C. M. (2006). *Pattern Recognition and Machine Learning*. Berlin: Springer.

Blasdel, G. G. (1992). Orientation selectivity, preference, and continuity in monkey striate cortex. *Journal of Neuroscience, 12*, 3139–3161.

Bonin, V., Mante, V., & Carandini, M. (2005). The suppressive field of neurons in lateral geniculate nucleus. *Journal of Neuroscience, 25*, 10844–10856.

Bosking, W. H., Crowley, J. C., & Fitzpatrick, D. (2002). Spatial coding of position and orientation in primary visual cortex. *Nature Neuroscience, 5*(9), 874–882.

Bosking, W. H., Zhang, Y., Schofield, B. R., & Fitzpatrick, D. (1997). Orientation selectivity and the arrangement of horizontal connections in tree shrew striate cortex. *Journal of Neuroscience, 17*(6), 2112–2127.

Burger, T., & Lang, E. W. (1999). An incremental Hebbian learning model of the primary visual cortex with lateral plasticity and real input patterns. *Zeitschrift für Naturforschung. C, Journal of Biosciences, 54*, 128–140.

Carreira-Perpiñán, M. Á., & Goodhill, G. J. (2002). Are visual cortex maps optimized for coverage? *Neural Computation, 14*(7), 1545–1560.

Carreira-Perpiñán, M. A., Lister, R. J., & Goodhill, G. J. (2005). A computational model for the development of multiple maps in primary visual cortex. *Cerebral Cortex, 15*(8), 1222–1233.

Chklovskii, D. B., & Koulakov, A. A. (2004). Maps in the brain: What can we learn from them? *Annual Review of Neuroscience, 27*, 369–392.

Coates, A., Ng, A., & Lee, H. (2011). An analysis of single-layer networks in unsupervised feature learning. *Journal of Machine Learning Research, Workshop and Conference Proceedings: AISTATS 2011, 15*, 215–223.

Dayan, P., & Abbott, L. F. (2001). *Theoretical Neuroscience: Computational and Mathematical Modeling of Neural Systems*. Cambridge, MA: MIT Press.

Durbin, R., & Mitchison, G. (1990). A dimension reduction framework for understanding cortical maps. *Nature, 343*, 644–647.

Erwin, E., Obermayer, K., & Schulten, K. J. (1992). Self-organizing maps: Ordering, convergence properties and energy functions. *Biological Cybernetics, 67*, 47–55.

Fischer, T. (2014). *Model of All Known Spatial Maps in Primary Visual Cortex* (master's thesis). The University of Edinburgh, UK.

Flanagan, J. G. (2006). Neural map specification by gradients. *Current Opinion in Neurobiology, 16*, 1–8.

Gee, A. (2014). *Neural Modelling of Congenitally Abnormal Visual Pathways* (master's thesis). The University of Edinburgh, UK.

Goodhill, G. J. (2007). Contributions of theoretical modeling to the understanding of neural map development. *Neuron, 56*(2), 301–311.

Goodhill, G. J., & Willshaw, D. J. (1990). Application of the elastic net algorithm to the formation of ocular dominance stripes. *Network (Bristol, England), 1*(1), 41–59.

Graham, N. V. (2011). Beyond multiple pattern analyzers modeled as linear filters (as classical V1 simple cells): Useful additions of the last 25 years. *Vision Research, 51*(13), 1397–1430.

Grossberg, S. (1976). On the development of feature detectors in the visual cortex with applications to learning and reaction-diffusion systems. *Biological Cybernetics, 21*, 145–159.

Harrison, T. C., & Murphy, T. H. (2014). Motor maps and the cortical control of movement. *Current Opinion in Neurobiology, 24*(1), 88–94.

Hartigan, J. A. (1975). *Clustering Algorithms*. Hoboken, NJ: Wiley.

Hastie, T., & Stuetzle, W. (1989). Principal curves. *Journal of the American Statistical Association, 84*, 502–516.

Hjorth, J. J. J., Sterratt, D. C., Cutts, C. S., Willshaw, D. J., & Eglen, S. J. (2014). Quantitative assessment of computational models for retinotopic map formation. *arXiv : 1408.6132*.

Hubel, D. H., & Wiesel, T. N. (1962). Receptive fields, binocular interaction and functional architecture in the cat's visual cortex. *Journal of Physiology, 160*, 106–154.

Huberman, A. D., Feller, M. B., & Chapman, B. (2008). Mechanisms underlying development of visual maps and receptive fields. *Annual Review of Neuroscience, 31*, 479–509.

Hyvärinen, A., & Hoyer, P. O. (2001). A two-layer sparse coding model learns simple and complex cell receptive fields and topography from natural images. *Vision Research, 41*(18), 2413–2423.

Hyvärinen, A., Hurri, J., & Hoyer, P. O. (2009). *Natural Image Statistics: A Probabilistic Approach to Early Computational Vision*. Berlin: Springer.

Jones, J. P., & Palmer, L. A. (1987). The two-dimensional spatial structure of simple receptive fields in cat striate cortex. *Journal of Neurophysiology, 58*(6), 1187–1211.

Kaas, J. H. (1991). Plasticity of sensory and motor maps in adult animals. *Annual Review of Neuroscience, 14*, 137–167.

Kara, P., & Boyd, J. D. (2009). A micro-architecture for binocular disparity and ocular dominance in visual cortex. *Nature, 458*(7238), 627–631.

Kaschube, M., Schnabel, M., Löwel, S., Coppola, D. M., White, L. E., & Wolf, F. (2010). Universality in the evolution of orientation columns in the visual cortex. *Science, 330*(6007), 1113–1116.

Kohonen, T. (1982). Self-organized formation of topologically correct feature maps. *Biological Cybernetics, 43*, 59–69.

Kozloski, J., Cecchi, G., Peck, C., & Rao, A. R. (2007). Topographic infomax in a neural multigrid. In D. Liu (Ed.), *Advances in Neural Networks—ISSN 2007* (Lecture Notes in Computer Science 4492, pp. 500–509). Berlin: Springer.

Law, J. S. (2009). *Modeling the Development of Organization for Orientation Preference in Primary Visual Cortex* (doctoral dissertation). School of Informatics, The University of Edinburgh, UK.

LeVay, S., Hubel, D. H., & Wiesel, T. N. (1975). The pattern of ocular dominance columns in macaque visual cortex revealed by a reduced silver stain. *Journal of Comparative Neurology, 159*(4), 559–576.

Linsker, R. (1986). From basic network principles to neural architecture: Emergence of orientation columns. *Proceedings of the National Academy of Sciences of the United States of America, 83*, 8779–8783.

Linsker, R. (1989). How to generate ordered maps by maximizing the mutual information between input and output signals. *Neural Computation, 1*(3), 402–411.

Lu, H. D., & Roe, A. W. (2008). Functional organization of color domains in V1 and V2 of macaque monkey revealed by optical imaging. *Cerebral Cortex, 18*(3), 516–533.

Miikkulainen, R., Bednar, J. A., Choe, Y., & Sirosh, J. (2005). *Computational Maps in the Visual Cortex.* Berlin: Springer.

Miller, K. D. (1994). A model for the development of simple cell receptive fields and the ordered arrangement of orientation columns through activity-dependent competition between ON- and OFF-center inputs. *Journal of Neuroscience, 14*, 409–441.

Morasso, P., & Sanguineti, V. (1995). Self-organizing body schema for motor planning. *Journal of Motor Behavior, 27*, 52–66.

Mountcastle, V. B. (1957). Modality and topographic properties of single neurons of cat's somatic sensory cortex. *Journal of Neurophysiology, 20*(4), 408–434.

Obermayer, K., Ritter, H., & Schulten, K. J. (1990). A principle for the formation of the spatial structure of cortical feature maps. *Proceedings of the National Academy of Sciences of the United States of America, 87*, 8345–8349.

Ohki, K., Chung, S., Ch'ng, Y. H., Kara, P., & Reid, R. C. (2005). Functional imaging with cellular resolution reveals precise micro-architecture in visual cortex. *Nature, 433*(7026), 597–603.

Olshausen, B. A., & Field, D. J. (1996). Emergence of simple-cell receptive field properties by learning a sparse code for natural images. *Nature, 381*, 607–609.

Olshausen, B. A., & Field, D. J. (1997). Sparse coding with an overcomplete basis set: A strategy employed by V1? *Vision Research, 37*, 3311–3325.

Osindero, S., Welling, M., & Hinton, G. E. (2006). Topographic product models applied to natural scene statistics. *Neural Computation, 18*(2), 381–414.

Overton, K. J., & Arbib, M. A. (1982). The extended branch-arrow model of the formation of retino-tectal connections. *Biological Cybernetics, 45*(3), 157–175.

Palmer, C. M. (2009). *Topographic and Laminar Models for the Development and Organisation of Spatial Frequency and Orientation in V1* (doctoral dissertation). School of Informatics, The University of Edinburgh, UK.

Prestige, M. C., & Willshaw, D. J. (1975). On a role for competition in the formation of patterned neural connexions. *Proceedings of the Royal Society of London. Series B, Biological Sciences, 190*(1098), 77–98.

Ramón y Cajal, S. (1911). *Histologie du systeme nerveux de l'homme et des vertebres.* Paris: A. Maloine (English translation by N. and L. Swanson, Oxford University Press, 1995).

Ramtohul, T. (2006). *A Self-Organizing Model of Disparity Maps in the Primary Visual Cortex* (master's thesis). The University of Edinburgh, UK.

Rao, R. P. N., & Ballard, D. H. (1999). Predictive coding in the visual cortex: A functional interpretation of some extra-classical receptive-field effects. *Nature Neuroscience, 2*, 79–87.

Recanzone, G. H., Merzenich, M. M., & Dinse, H. R. (1992). Expansion of cortical representation of a specific skin field in primary somatosensory cortex by intracortical microstimulation. *Cerebral Cortex, 2*, 181–196.

Rehn, M., & Sommer, F. T. (2007). A network that uses few active neurones to code visual input predicts the diverse shapes of cortical receptive fields. *Journal of Computational Neuroscience, 22*(2), 135–146.

Ringach, D. L. (2002). Spatial structure and symmetry of simple-cell receptive fields in macaque primary visual cortex. *Journal of Neurophysiology, 88*(1), 455–463.

Ritter, H., Obermayer, K., Schulten, K. J., & Rubner, J. (1991). Self-organizing maps and adaptive filters. In E. Domany, J. L. Hemmen, & K. Schulten (Eds.), *Models of Neural Networks* (pp. 281–306). Berlin: Springer.

Rodieck, R. W. (1965). Quantitative analysis of cat retinal ganglion cell response to visual stimuli. *Vision Research, 5*(11), 583–601.

Sirosh, J., & Miikkulainen, R. (1994). Cooperative self-organization of afferent and lateral connections in cortical maps. *Biological Cybernetics, 71*, 66–78.

Stevens, J.-L. R., Law, J. S., Antolik, J., & Bednar, J. A. (2013). Mechanisms for stable, robust, and adaptive development of orientation maps in the primary visual cortex. *Journal of Neuroscience, 33*, 15747–15766.

Swindale, N. V. (1996). The development of topography in the visual cortex: A review of models. *Network (Bristol, England), 7*, 161–247.

Swindale, N. V., Shoham, D., Grinvald, A., Bonhoeffer, T., & Hubener, M. (2000). Visual cortex maps are optimized for uniform coverage. *Nature Neuroscience, 3*(8), 822–826.

Tobler, W. R. (1970). A computer movie simulating urban growth in the Detroit region. *Economic Geography, 46*, 234–240.

Triplett, J. W., Pfeiffenberger, C., Yamada, J., Stafford, B. K., Sweeney, N. T., Litke, A. M., et al. (2011). Competition is a driving force in topographic mapping. *Proceedings of the National Academy of Sciences of the United States of America, 108*(47), 19060–19065.

Van Essen, D. C., & Gallant, J. L. (1994). Neural mechanisms of form and motion processing in the primate visual system. *Neuron, 13*(1), 1–10.

von der Malsburg, C. (1973). Self-organization of orientation-sensitive cells in the striate cortex. *Kybernetik, 15*, 85–100. Reprinted in Anderson & Rosenfeld (1988) (pp. 212–227).

Wei, Y., Tsigankov, D., & Koulakov, A. (2013). The molecular basis for the development of neural maps. *Annals of the New York Academy of Sciences, 1305*, 44–60.

Weliky, M., Bosking, W. H., & Fitzpatrick, D. (1996). A systematic map of direction preference in primary visual cortex. *Nature, 379*, 725–728.

Willshaw, D. J., & von der Malsburg, C. (1976). How patterned neural connections can be set up by self-organization. *Proceedings of the Royal Society of London. Series B, Biological Sciences, 194*, 431–445.

Willshaw, D. J., & von der Malsburg, C. (1979). A marker induction mechanism for the establishment of ordered neural mappings: Its application to the retinotectal problem. *Philosophical Transactions of the Royal Society of London. Series B, Biological Sciences, 287*, 203–243.

Wilson, S. P., Law, J. S., Mitchinson, B., Prescott, T. J., & Bednar, J. A. (2010). Modeling the emergence of whisker direction maps in rat barrel cortex. *PLoS One, 5*(1), e8778.

Xu, X., Anderson, T. J., & Casagrande, V. A. (2007). How do functional maps in primary visual cortex vary with eccentricity? *Journal of Comparative Neurology, 501*(5), 741–755.

Yu, H., Farley, B. J., Jin, D. Z., & Sur, M. (2005). The coordinated mapping of visual space and response features in visual cortex. *Neuron, 47*(2), 267–280.

Zylberberg, J., Murphy, J. T., & DeWeese, M. R. (2011). A sparse coding model with synaptically local plasticity and spiking neurons can account for the diverse shapes of V1 simple cell receptive fields. *PLoS Computational Biology, 7*(10), e1002250.

14 Schema Theory and Neuropsychology

RICHARD P. COOPER

14.1 Introduction

A great deal of the behavior of living organisms, whether they be slugs, frogs, monkeys, or humans, can be understood as being *schematic*, in the sense that at a certain level of abstraction the "same" stimulus generally leads to the "same" response. For example, for a frog, prey recognition leads to approaching the prey, regardless of the specific prey, while predator recognition leads to evading the predator, regardless of the specific predator. In schema theory this can be modeled by two kinds of schemas: *perceptual schemas* for prey and predator detection and *motor schemas* for approaching and evading behavior.

For humans, schemas are likely to be more complex, as when recognizing whether a stranger is a friend or a foe, or when a skilled driver takes evasive action to avoid an obstacle in his or her path. However, in both humans and frogs, the concept of a schema captures an abstraction over a set of similar situations, and invocation of a schema is a computationally efficient approach to either the recognition of what is common to those situations or the generation of a stereotyped response. Critically, while no two stimuli present identical retinal stimulation and no two responses require identical motor control, schemas abstract away the dissimilarities—the irrelevant perceptual information or the physics of environment in which the response is required—meaning that in familiar situations it is not necessary to analyze each feature of the perceptual input or to plan behavior from scratch. Instead, behavior may be generated by applying a known response schema in the context of the current situation. This is not to say there are no computational problems when behavior is viewed in schema-theoretic terms. For example, the location of a predator is highly relevant to evading it, so there is potentially the need for perceptual schemas to pass parameters to motor schemas.

Perception and motor control are not the only domains that may be analyzed in terms of schemas. The schema-based approach provides a general framework in which numerous higher-level functions such as perception and action, but also memory and language, may be described. This chapter begins with a discussion of schema theory in the context of amphibian perception and action before considering in more detail an example of a higher-level schema theory—the schema-based theory of routine sequential action proposed by Norman and Shallice (1980)—with computational implementations of the schema-based theory by Cooper and Shallice (2000) and of a rival theory by Botvinick and Plaut (2004). This chapter also illustrates a top-down approach to computational neuroscience, in which cognitive theory derived from empirical studies of behavior, rather than neural data, serves as the starting point.

Much of the evidence discussed in this chapter in support of the Norman and Shallice theory comes from neuropsychology—the study of effects of brain injury (through, e.g., stroke, brain tumor, or physical trauma to the head) on aspects of behavior. Neuropsychological evidence serves two functions. On the one hand, it provides additional constraints on theories of the processes underlying behaviors because such theories need to account not just for behavior of the normally operating system but also, when damaged in a way that is justified theoretically by the brain injury, for the behavior of the damaged system. On the other hand, neuropsychological evidence may provide insight into the neural basis of the affected behaviors. For example, a common finding is that (right-handed) patients who have suffered a stroke or traumatic head injury primarily affecting their left hemisphere tend to have difficulties with language. Whether those difficulties relate to language comprehension or language production (or both) depends largely on the precise location of the injury. Neuropsychology therefore provides one source of evidence for relating theories of the cognitive processes underlying behavior (e.g., theories of the interrelation between language comprehension and language production) to brain regions (e.g., regions of the left hemisphere). As touched on later in this chapter, functional brain imaging techniques provide another source.

The relation between cognitive theory, schema-theoretic accounts of cognitive faculties such as language, memory, and the like, and localization of theoretical constructs is a complex one. In the case of localization from neuropsychology and neuroimaging, both approaches make assumptions that may not be valid in any specific case. It is therefore when the neuropsychology and brain imaging techniques converge that localization evidence is at its strongest (see Shallice and Cooper, 2011). In some circumstances, hypotheses on localization may relate to specific (sets of) schemas in the brain; in other cases, it is not the schemas themselves that may be localized, but processes, including processes that may operate over schemas. In this latter case, specifying schemas may provide a crucial intermediary in the analysis of the underlying neural networks.

14.2 *Theoretical Background*

Head and Holmes (1911) first used the term "body schema" in reference to an internal model of body posture which was held to be used in the production of motor commands, but it is Bartlett (1932) who is generally credited with the first usage of the term "schema" in its modern sense, as a generic abstraction over a set of similar entities. In a study of human memory, Bartlett found that subjects "recalled" events that were typical of the general type of episode in question but that were not actually present in the specific episode being recalled. The suggestion was that human memory relied upon event schemas—abstractions over similar events—and that such schemas serve as shortcuts, eliminating the need to remember those details of a specific event that would normally be recoverable from the generic schema. Such schemas are clearly higher-level than those involved in perception and motor control, but they share the property of being an abstraction over a set of instances.

Bartlett's work also emphasized another property of schemas: to accurately encode a specific event or episode within a schema-based memory system, it is necessary to encode not only that the event is an instance of a specific schema but also the ways in which the specific event differs from the abstract or generic schema. One way to understand how the specifics of an instance of a schema might relate to the abstract schema is to conceive of schemas as knowledge structures with argument roles that, in the abstract case, are variables. For any instance of the schema these argument roles or variables will be assigned specific values. This notion that a schema has associated variables is general, applying not just to event schemas but also to perceptual schemas (e.g., specifying the distance to or size of prey) and motor schemas (e.g., specifying the speed with which a movement should be performed).

At about the same time as Bartlett invoked the concept of schema for memory, Piaget (1936/1952) developed the concept within the field that is now known as cognitive development. Piaget was particularly concerned with how children develop first perceptual schemas for categorizing objects within the perceptual world, then sensorimotor schemas for manipulating objects, and then more abstract schemas for thinking about objects, with each subsequent level grounded in the previous level. Two critical additional concepts introduced by Piaget were *assimilation,* whereby new instances (i.e., new perceptions, new sensorimotor interactions, and presumably new abstract thoughts) are understood as belonging to existing schemas, and *accommodation,* whereby existing schemas are modified, adapted, or refined to incorporate new information that is not consistent with the existing schemas.

In the examples so far, schemas have been presented as discrete atomic entities, but many schemas may be characterized in terms of a set of lower-level subschemas. This is clear in the case of events (e.g., the representation of a typical commute to work, which, if done by public transport, may include subevents such as traveling to the local train station, catching the train, and then walking from the destination station to one's office, each of which might itself be thought of as a schema), but it also applies to perceptual and motor schemas. Consider a schema for prey detection. At the lowest level this may draw upon feature detectors that encode orientation, color and texture, and so forth. These may feed into higher-level schemas (e.g., detecting the prey's body parts, such as its tail and its head), which in turn feed into higher-level schemas detecting specific types of prey (e.g., in the case of the cat, detecting birds vs. detecting mice). This characteristic of schemas raises two computational requirements of schema-based systems (see, e.g., Arbib, 1989). Firstly, schemas may operate cooperatively, such that multiple schemas support each other's operation. Secondly, schemas may also operate competitively: if an organism is to have two different responses to two types of prey, then it requires separate prey-detection schemas for the two types of prey, together with some mechanism to ensure that at any point in time behavior is under the control of one schema or the other, but not both.

The same requirements hold with respect to human perception. A variety of situations give rise to so-called categorical perception. Consider phoneme recognition. When an auditory stimulus is continuously varied from /ba/ to /pa/, humans perceive the stimulus as either /ba/ or /pa/, and not as something in between (Liberman et al., 1957). This may be accounted for if there is a /ba/ recognition schema and a /pa/ recognition schema which compete such that instances (i.e., auditory stimuli) are categorized as one or the other but not both. Similar results hold for categorization of colors (Bornstein, 1973), faces (Beale and Keil, 1995), and even facial expressions (Etcoff and Magee, 1992).

Cooperative and competitive interactions between schemas may be captured computationally by associating activation values with schemas. Activation values are real numbers between (say) zero and one that vary over time in response to excitatory or inhibitory influences. Cooperation may be operationalized through excitatory associations between schemas (so that if one schema is highly active, it will excite those schemas with which it cooperates) while competition may be operationalized through inhibitory associations (so that if one schema is highly active, it will inhibit those schemas with which it competes). The activation-based framework also requires a notion of "selection" whereby behavior is controlled by selected schemas. This may be achieved through the use of a selection threshold—a relatively high activation value (e.g., 0.8)—with schemas being selected when their activation exceeds the threshold. Note that this allows schemas to be active even if not selected.

The general utility of schema theory is illustrated by the fact that it also provides a natural framework in which to consider linguistic processing (e.g., Roy, 2005; see also chapter 24, "Language Processing")—both for comprehension and production. Thus, perceptual/linguistic schemas may detect various syntactic structures (e.g., for verb phrases comprising either an intransitive verb, or a transitive verb followed by a noun phrase, or any of a well-defined set of more complex verb phrases). Similarly, generation of linguistic output may be understood in terms of motor/linguistic schemas that produce sentences consisting of (e.g.) a noun phrase followed by a verb phrase, and noun phrase production schemas which consist of (e.g.) a determiner production schema followed by a noun production schema, and so forth. However, while there are some similarities in the abstract nature of perceptual, event, linguistic, and motor schemas, there are also domain-specific differences. Evans (1967) defined perceptual schemas for object categorization as "a set of rules which would serve as instructions for producing [...] a population prototype" (p. 87). In contrast, Schmidt (1975), in his theory of motor skill learning, argued that a motor response schema is a complex entity with four components, abstracting over a motor output's initial conditions, its response specifications, its sensory consequences, and its response outcome.

We have introduced schema theory without any overt reference to the neural localization of schemas. Presumably perceptual, event, linguistic, and motor schemas are encoded in different neural areas and may in fact be distributed across multiple regions, but the generic theory makes no commitment to neural localization. Indeed, Arbib (2003) presents schema theory as a general framework for cognitive neuroscience that provides a common language for presenting hypotheses about functional interactions (e.g., between perception and action) without *necessarily* committing to a specific implementation at either the neural or algorithmic levels. Schema-based accounts of various behaviors have also been implemented directly within simple mobile robots (e.g., Arkin et al., 2000), thus demonstrating the abstract nature of schema theory and allowing concrete evaluation of specific schema-theoretic accounts of the behaviors in question.

14.3 Perceptual and Motor Schemas in the Frog and Toad

14.3.1 A Simple Model of Predator/Prey Interactions Consider how the behavior of relatively simple animals—frogs and toads—might be described in schema-theoretic terms (cf. Arbib, 1989). On the perceptual side, such animals must detect prey and predators. They must also have other perceptual requirements in order to seek out suitable habitats, to mate, and so forth. On the motor side, they can orient to the left or right and they may hop forward or jump. They may also flick their tongue and swallow. All of these functions may be understood in terms of schema theory. Thus, when a small object is moved in front of a frog, it will snap at it. If the moving object is large, the frog will jump away, but if the object is stationary (regardless of size) then the frog does not respond. The way in which the stimulus moves, and its depth, is also important. Toads will orient toward a small rectangle moving in the direction of its major axis (in a manner similar to a worm) but not to one moving perpendicular to its major axis (see Ewert, 1972). When a toad is presented with a worm behind a partial barrier (e.g., a paling fence), it may move toward the prey or detour around the barrier. Finally, when hunting, the toad appears to execute a compound response sequence (fixate → snap → swallow) as a single uninterruptible schema. The toad then waits, with subsequent behavior requiring a new perceptual trigger.

Arbib (1989) interprets these and related findings as evidence for perceptual schemas for prey detection (which detect small objects moving in a biologically realistic fashion) and predator detection (which detect larger moving objects), and motor schemas for orienting, snapping, jumping, and detouring, and with hierarchical structuring of complex schemas (e.g., hunting) in terms of simpler subschemas. The neurophysiological evidence from toads is consistent with this approach. It moreover adds constraints on the localizations and interactions of schemas. For example, based on the above we might hypothesize perceptual schemas for the detection of small moving objects and large moving objects, with those schemas linked directly to snapping or evasion responses, as shown in figure 14.1A. However, toads with lesions to their pretectal area—a midbrain region that receives input via the optic nerve from retinal ganglion cells—will snap at all moving objects. This suggests the alternative configuration of perceptual and motor schemas, shown in figure 14.1B, where a perceptual schema that detects movement triggers the snap response, and a second perceptual schema that detects large objects inhibits the snap response and triggers an evasion response. There is a further possibility consistent with this data, where motor schemas compete, excitation of the evasion schema due to detection of a large moving object is greater than that of the snap schema due to detection of movement, and inhibition of the snap response arises as a by-product of activation of the evasion response (see figure 14.1C).

14.3.2 Extending the Model Frogs and toads also exhibit more complex behaviors. Collett (1982) describes the behavior of the toad when confronted with a barrier that partially occludes prey. When the barrier is not too wide, the toad will detour around the barrier. In the apparatus used by Collett, the prey was only visible from the toad's starting point. Therefore, once the toad began to detour around the barrier, the prey was occluded. Yet the position at which the toad stopped, and the direction in which it was pointing, depended on the distance between the barrier and the prey. Thus, the toad is able to integrate depth cues in "planning" its detour behavior. Arbib and colleagues (e.g., Lara et al., 1984; Arbib and House, 1987) have developed a series of schema-based computational models to account for this and related behaviors.

14.3.3 Internal Models and the Function of Perceptual and Motor Schemas It is useful here to introduce another idea, namely, that perceptual schemas serve to build or maintain an internal model of the world. For the amphibian these internal models may be minimal (e.g., there is a prey-like object behind a barrier), but for higher animals they may be more detailed and complex (e.g., on the desk to my right is a gray printer, beyond that is a white telephone that is on a telephone directory). Such models must be kept up to date if they are to play any useful role in behavior—and presumably this is their primary purpose (see chapter 9, "Computational Models in Motor Control"). Updating could be achieved in two ways. The internal

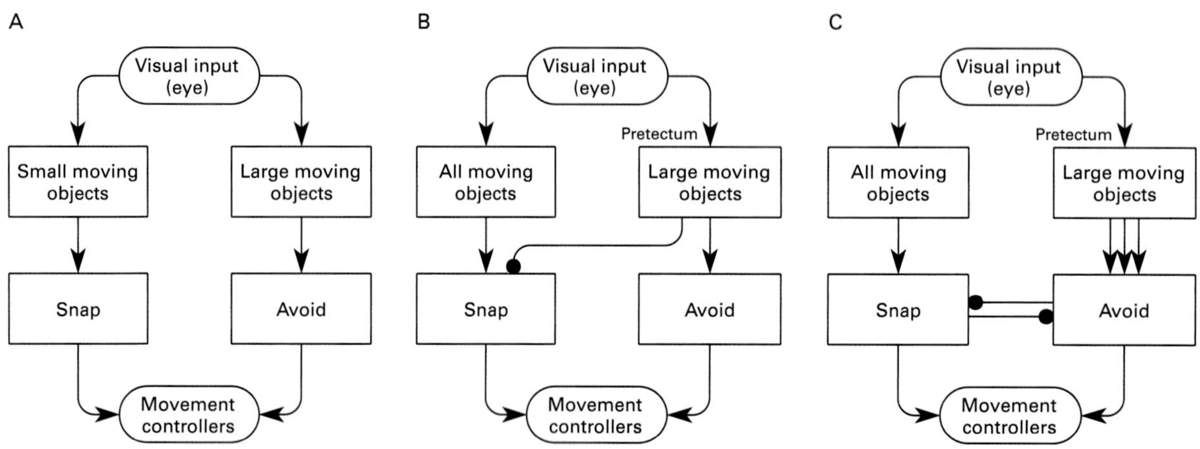

Figure 14.1 Possible relations between perceptual and motor schemas in the toad. Arrows with pointed heads indicate excitatory links. Arrows with circular heads indicate inhibitory links. (A) Two perceptual schemas for detecting prey and predators feed into two motor schemas for snapping or avoiding. Data from toads with lesions to the pretectum argue against this. (B) An alternative arrangement in which a perceptual schema that detects large moving objects inhibits the snapping response. This is consistent with data from toads with lesions to the pretectum. (C) A second alternative organization, where inhibition acts directly between the motor schemas rather than from a perceptual schema to the snap motor schema. Panels (A) and (B) adapted from Arbib (2003).

model may either (1) be continually updated on the basis of the current perceptual input, or (2) be updated based on the anticipated consequences of motor output (cf. Wolpert and Ghahramani, 2000), perhaps with additional reference to perceptual input. From a computational perspective, option 2 requires reentrant connections from the system in which motor response schemas are selected to the system that maintains the model of the world.

Conceivably for toads, maintenance of the internal model of the world may be achieved through option 1—via updates from perceptual schemas on completion of each motor schema. For higher animals, however, evidence suggests that internal models may be updated directly by motor schemas (Wolpert and Ghahramani, 2000). That is, the internal model of the current state of an animal's environment may be updated by the schemas that control the animal's actions without the animal's needing to explicitly perceive a change in the environment.

14.4 *Higher-Level Schemas for the Control of Thought and Action in Humans*

14.4.1 Contention Scheduling Theory Schema theory may also be applied to high-level behavior in humans. Norman and Shallice (1980) argue that there is a qualitative difference between the control of routine or automatic behavior and the control of nonroutine or deliberate (i.e., willed) behavior. Routine behavior, they argue, is the province of a schema-based system—contention scheduling—the operation of which is modulated in nonroutine situations by a second system—the supervisory system. We concentrate in this chapter on the contention scheduling system and its role in the control of action, but it is important to note that even in nonroutine situations or when behavior is under the control of the supervisory system, that control is held to operate on behavior indirectly by activating or deactivating (i.e., exciting or inhibiting) schemas within the contention scheduling system. Moreover, Norman and Shallice argue that contention scheduling is invoked not only in the control of action but also in the control of thought.

Contention scheduling theory posits that at any point in time behavior is under the control of an action (or thought) schema. Such schemas have all of the characteristics associated with schemas in the more general schema theory. Thus, schemas form a network with hierarchical structural relations between higher-level or "source" schemas and their component schemas, and at the lowest level "action" schemas have arguments (so an action such as *Pick Up* might be directed to a specific object, or carried out with a specific manner, such as hastily or carefully). Schemas are goal-directed in the sense that each schema achieves one or more goals, and goal/subgoal relations are echoed within the schema network by schema/subschema relations. A fragment of a schema network as might underlie the routine task of preparing instant coffee is shown in the upper part of figure 14.2. Here the *Prepare Instant Coffee* schema involves subschemas for boiling water and adding the various ingredients to a mug.

Within contention scheduling theory a schema's activation may be affected by several factors, including top-down excitation either from the supervisory system or from selected superordinate schemas, bottom-up excitation (or "triggering") from the internal model of the environment, and inhibitory influences from competing schemas. Full details of these influences are given in the following section, but they are shown in figure 14.2 as gray lines overlaid on the basic schema hierarchy and a fragment of the internal model of the world.

The triggering of schemas by representations of objects within the internal model of the world, as shown on the lower part of figure 14.2, together with the selection of objects as arguments for selected schemas, is a critical aspect of contention scheduling. Why for example, when the schema for *Pick Up* is selected, should this be applied to a specific coffee sachet rather than any other object in the immediate environment? While the details of how this might be achieved were not specified in the initial account of the contention scheduling theory, one possibility—and the possibility adopted in the implementation of Cooper and Shallice (2000) described below—is that processes of triggering and argument selection are also mediated by interactive activation. More precisely, object representations (within the internal model) are, like schemas, assumed to have associated activation values. These values reflect the salience of objects, which effectively compete for the argument roles of schemas. Object representations excite those schemas that can use them in proportion to their activation, and vice versa. This means that highly active schemas will tend to excite the representations of relevant objects while highly salient objects will tend to excite the schemas which might be applied to them. When a schema is selected, the most active object representations are then assigned, or "bound," to its argument roles.

14.4.2 The Supervisory System As noted above, in nonroutine situations—those situations for which an appropriate schema does not exist—the operation of contention scheduling is modulated by the supervisory

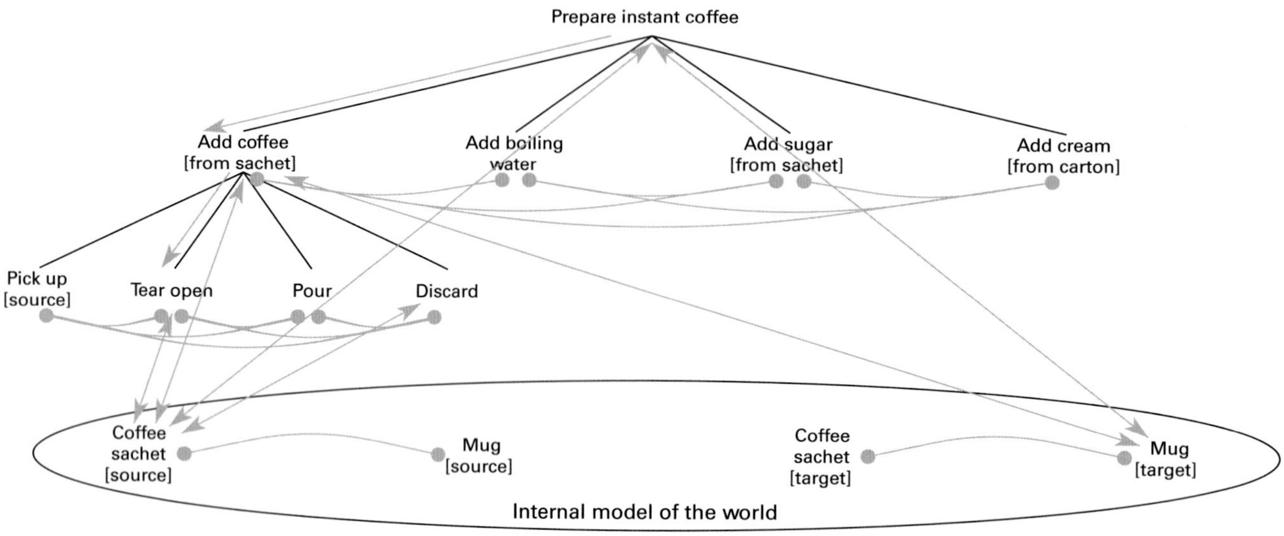

FIGURE 14.2 The partial hierarchical structure of schemas involved in preparing instant coffee, together with a partial representation of the internal model of the world and arrows showing influences (gray lines) on the activation of schemas and object representations at a point during preparation of instant coffee (when tearing open the coffee sachet). Within the schema network, each intermediate level schema consists of a (partially ordered) set of lower level schemas. Influences include: top-down influences of schemas on other schemas, bottom-up influences from the representations of objects in the internal model of the world (which are bidirectional), and lateral inhibitory influences between competing schemas. The coffee sachet (as a source) and the mug (as a target) are excited by the *Prepare Instant Coffee* and the *Add Coffee* schemas because these schemas require objects with those features in those roles. The coffee sachet (as a source) is additionally excited by the *Tear Open* schema because (at this point in task execution) it is held. It is excited by the *Discard* schema for the same reason. For clarity neither perceptual schemas nor bottom-up influences are shown.

system. This is held to be a complex of subsystems which work together to create a "temporary schema." This then feeds into the contention scheduling system to produce deliberate or planned behavior.

Figure 14.3 shows the various hypothesized subsystems of the supervisory system and their interaction. The system as a whole operates in a series of stages. Stage 1 constructs the temporary schema through processes such as problem solving or recall of a previously stored plan. The temporary schema consists of a (possibly novel) sequence of subschemas that are already known to contention scheduling, together with postconditions for at least some subschemas that allow progress of the temporary schema to be checked. In stage 2 the temporary schema is stored in a special purpose working memory while its subschemas are executed. Stage 3, which operates in parallel with stage 2, monitors the postconditions of the subschemas to ensure that the temporary schema is appropriate. If it is not, the temporary schema is rejected. This will lead to a further round of problem solving and generation of another temporary schema.

The supervisory system figure illustrates the box-and-arrow style of theory description that is common in cognitive psychology and related disciplines. The components of the system are hypothesized based on studies in standard experimental cognitive psychology and (once again) studies of the deficits of neurological patients (see Shallice and Burgess, 1996). A common criticism of the box-and-arrow style of description is its lack of precision in specifying individual components—a criticism that might be addressed by developing the components into functional computational models. Some work has been attempted along these lines for the supervisory system, but the operation of the supervisory system is not the focus of this chapter. Nevertheless, one final comment is in order. While the supervisory system's operation is not standardly conceived of in schema-theoretic terms, its components may be described in such a way. For example, "Problem Orientation" might be described in terms of a set of schemas that are sensitive to perceptual and working memory representations and that compete to set both the system's goal and its level of aspiration (i.e., a level of importance of the goal).

14.4.3 SUPPORT FOR THE CONTENTION SCHEDULING/SUPERVISORY SYSTEM THEORY Much of the evidence in support of the contention scheduling/supervisory system theory comes from two sources: the behavior of neurologically healthy individuals when they are distracted and the behavior of patients with specific

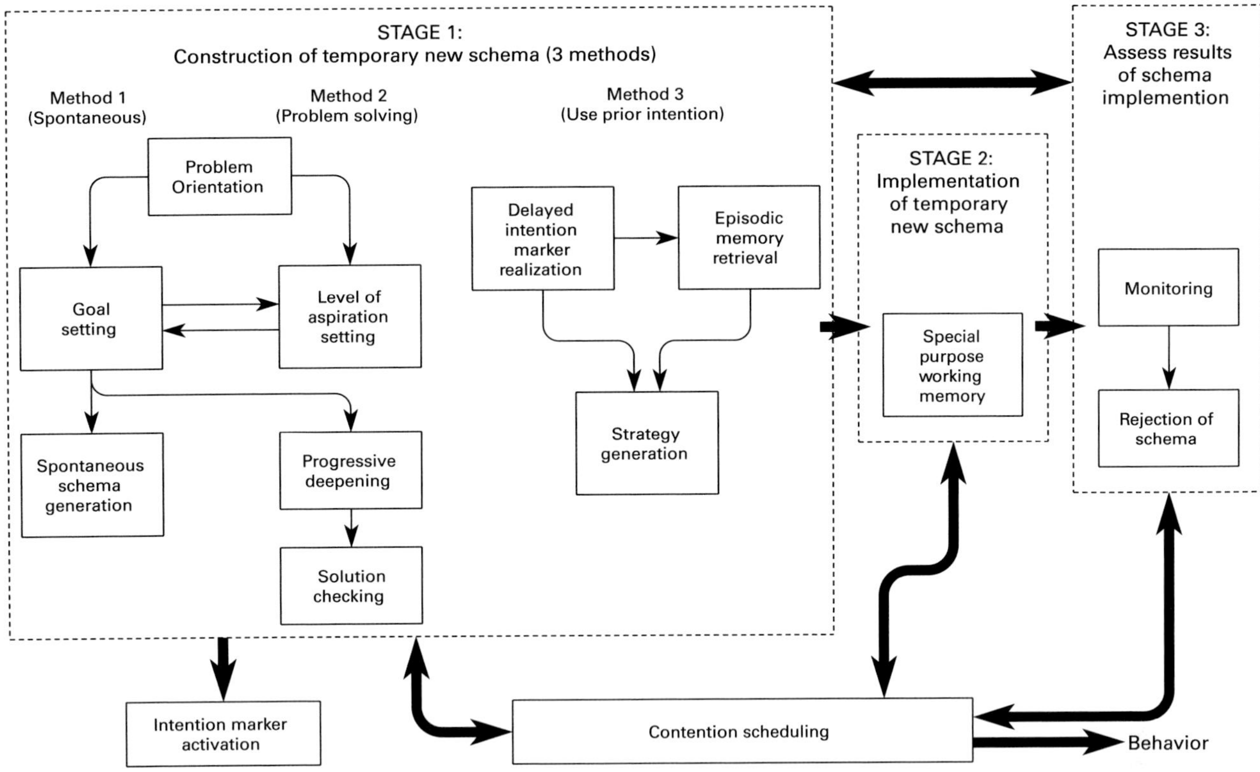

FIGURE 14.3 The functional components of the supervisory system model of Shallice and Burgess (1996). The primary interface between the supervisory system and contention scheduling is a special purpose working memory. This holds a representation of a temporary schema constructed by the supervisory system to guide behavior in a nonroutine situation (Stage 2). The temporary schema may be created in one of three ways (Stage 1), and execution of the schema is monitored in case it fails to yield the intended result (Stage 3). Adapted from Shallice and Burgess (1996).

deficits following brain injury. In the former case, diary studies, in which neurologically healthy individuals record their slips and lapses of action (e.g., Reason, 1979; Norman, 1981), show that a number of errors of action can be understood in terms of behavior being "captured" by an unintended behavioral routine (i.e., by a schema). Thus, Reason (1979) reports a subject who went to his garage intending to get the car out but, moments later, realized he had stopped as he passed through the porch and was instead putting on his gardening jacket and boots. Such errors suggest that common sequences of behavior (such as, for this subject, putting on his gardening jacket and boots) are encoded in schematic form within the brain, and that when attention is not focused on a given course of action, schemas consistent with the current state of the environment may capture behavior.

The neuropsychological evidence supports the above interpretation. For example, in *utilization behavior*, patients' behavior appears to be driven primarily by schemas triggered by the environment. Lhermitte (1983) described a patient who, when given a pair of spectacles, put the spectacles on his nose, even when he was already wearing one, or even two, pairs of spectacles. Shallice et al. (1989) described a similar patient who, while being questioned by clinicians, spontaneously picked up a packet of playing cards from the far end of the desk and dealt four cards to each person seated at the table. Such percept-driven activity is not only indicative of action being under the control of an automatized routine—an action schema—but also of the fact that much of ongoing behavior integrates perceptual input into an ongoing course of action. Utilization behavior is a relatively rare syndrome, but it has been argued to be associated with damage to prefrontal cortex, and particularly with bilateral damage to superior medial prefrontal areas (e.g., Boccardi et al., 2002; Bahrami Balani et al., 2012; see figure 14.4; but see also Iaccarino et al., 2014). Within the contention scheduling/supervisory system theory it is argued to reflect the operation of contention scheduling in the absence of appropriate supervisory control.

The Norman and Shallice theory garners additional support from a number of other behavioral disorders arising from neurological damage. For example, patients with lesions restricted to their frontal cortex

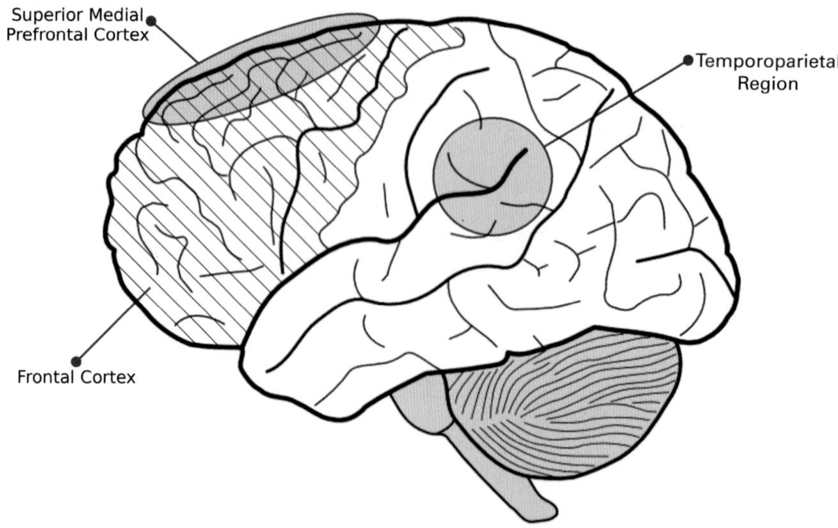

Syndrome or behavioral disturbance	Location of neural damage
Utilization behavior	Damage to superior medial prefrontal cortex
Action disorganization syndrome	Diffuse damage to the frontal lobe
Ideational apraxia	Lesion affecting the left temporoparietal region
Parkinson's disease	Dopamine depletion in the striatum

FIGURE 14.4 Regions and systems of the human brain which, when damaged, have been reported to lead to impairments of routine sequential action. The striatum (not shown) is one of a set of subcortical structures collectively known as the basal ganglia. These are located at the base of brain, deep inside the cortex.

(see figure 14.4), due, for example, to a circumscribed tumor or a gunshot wound, typically have difficulties in the production of organized goal-directed sequential behavior. They make errors reminiscent of the slips and lapses of neurologically healthy individuals, but the patient's errors are more frequent, more flagrant, and often claimed by the patient to be intentional. Thus, Luria (1966) described as typical of this class of patient behaviors such as lighting a candle (as instructed) but then attempting to smoke the candle as if it were a cigarette—what might appear to be an extreme form of capture error. Luria suggested that such a patient "could not retain the plan of the required action and was easily distracted by the more firmly established motor stereotype [i.e., schema]" (p. 237). Luria referred to the disorder as "frontal apraxia." More recently Schwartz et al. (1991) have used the phrase "action disorganization syndrome" to describe a range of patients with a similar behavioral disturbance. This term highlights the behavioral disorder while being neutral with regard to its neural localization.

A second relevant disorder is *ideational apraxia*.[1] Ideational apraxic patients are particularly prone to what have been called "conceptual errors" (see DeRenzi and Lucchelli, 1988; Rumiati et al., 2001). These may involve using tools inappropriately (e.g., attempting to cut an orange by pushing on a serrated knife, rather than by using a sawing motion) or producing a correct action but in a way that nevertheless fails to achieve the desired outcome (e.g., attempting to light a match by striking it on the wrong edge of a matchbox). On contention scheduling theory, the errors of ideational apraxics may be accounted for by selection of inappropriate lower-level schemas (e.g., sawing vs. slicing) or by the selection of inappropriate parameters (e.g., the wrong side of the match box) for those schemas. Cooper (2007) has shown how such errors may arise from impaired activation propagation between schema representations and object representations within the contention scheduling model.

A different form of action disorder is produced by patients with Parkinson's disease (PD). Here, the difficulty is not so much with errors of action, but with bradykinesia—slowed movement and particularly slowed initiation of sequences of actions. Physiologically, the condition is linked to a depletion of dopamine in the striatum (Robbins, 1991), part of the basal ganglia system (see chapter 22, "Integrative Functions of the Corticostriatal System"). A somewhat complementary condition of hyperactivity may arise with amphetamine use. Thus, Lyon and Robbins (1975) characterized the behavior of rats given high doses of amphetamine as being stereotyped with high response rates. Here the problem may be overstimulation of the action selection system, with stereotyped responses (rather than random behavior) being produced because of their strong associations with common features of the environment. At the neurophysiological level, amphetamine is known to increase dopamine concentration in the presynaptic cleft (Robbins, 1982), leading to heightened responses of postsynaptic neurons, but it is currently unclear whether the behavioral effects on action arise from dopamine overactivity or from interactions between dopaminergic and other neurotransmitter systems. (Compare the discussion of neuromodulation in chapter 7, "Neuromodulation.") These disorders of rate are potentially explicable within the contention scheduling theory if one assumes that the concentration of neurotransmitters such as dopamine affect the rate at which activation of schemas accumulates. We return to this issue in section 14.7.5.

Both neurological patients and neurologically healthy individuals produce a range of other errors in routine sequential action. They may omit actions within a sequence, perform actions out of sequence, or perseverate—repeating a subsequence unnecessarily and inappropriately. They may also make argument substitution errors—where an appropriate action is performed with the wrong argument (e.g., after making coffee, putting the coffee, instead of the cream, into the refrigerator). A significant challenge for computational models of routine sequential action selection is to account not just for capture errors and disorders of rate but also for the above types of error, which appear to arise from the breakdown of control processes (rather than the loss of specific schemas). The computational models described in the following two sections attempt to do just this.

14.5 The Interactive Activation and Competition Model of Contention Scheduling

14.5.1 The Beverage Preparation Task Cooper and Shallice (2000; see also Cooper et al., 1995) describe an implementation of contention scheduling theory in the context of the task of beverage preparation. The task is based on part of the daily routine of inpatients recovering from neurological damage. Schwartz et al. (1991; see also Schwartz et al., 1998) noted that some patients, particularly those who have sustained an injury affecting the frontal lobes of the brain, are prone to producing a range of errors of action in their daily routines. They described in detail the performance of one patient who had suffered a subarachnoid hemorrhage following a ruptured aneurism (i.e., bleeding on the brain following a burst blood vessel), resulting in extensive brain damage primarily affecting the patient's frontal lobe. The patient produced a range of errors when attempting to prepare instant coffee within the institutional setting, such as pouring cream into his orange juice, adding breakfast cereal to his coffee cup, and attempting to pour from a sealed container. The initial aim of the simulation work was to provide a qualitative account of these and related errors. In order to achieve this, a simplified version of the patient's task was developed. The goal of that task—known as the beverage preparation task—was to prepare a cup of instant coffee (with cream and sugar), given the following objects on the institutional breakfast tray: a mug containing hot water, a sugar sachet (sealed), a sugar bowl (open), a coffee sachet (sealed), a screw-top coffee jar (sealed), a cream carton (sealed), a spoon, and five other objects relevant for other subtasks of breakfasting.

The presence of the sugar sachet and the sugar bowl provides two different (and competing) means of adding sugar. Similarly, the presence of the coffee sachet and the coffee jar provides alternative means of adding dried coffee granules.

Cooper and Shallice assume that patients have available to them a range of action schemas at several levels of complexity, including the following:

- at the lowest level: *pick-up object, put-down object, tear-open sachet, unscrew jar, screw-closed jar, pour container, dip-spoon* (into jar), and *empty spoon* (into mug);
- at an intermediate level: *add sugar from sachet, add sugar from bowl, add cream from carton, add coffee from sachet*, and *add coffee from jar*, and
- at the highest level: *prepare instant coffee.*

For the purposes of implementation, schemas at the lowest level are assumed to correspond directly to basic or atomic actions such as "pick up" or "unscrew," but this is just a proxy for a model of the motor system. Thus, if the model were to interface with a model of the motor system, those atomic actions could well be implemented in terms of schemas with additional internal structure (e.g., extend forearm, rotate wrist, grasp, etc.: see chapter 20, "Reach and Grasp: Control, Development, and Recognition").

Action schemas are goal-directed. Thus, each schema has an associated goal (basically a state of the world) that will be achieved if the schema is performed successfully. Internally, schemas consist of a partially ordered set of subgoals. Thus, the *prepare instant coffee* schema, which achieves the high-level goal of producing a cup of coffee, comprises three mid-level subgoals: add coffee grounds, add sugar, and add cream. Cooper et al. (2005) assume that partial order is specified through preconditions associated with each subgoal. (Thus, a precondition of the add sugar subgoal is that the mug contains coffee grounds.) Postconditions are also associated with subgoals. Preconditions and postconditions provide a flexible way to allow for variation in the state of the world when a schema is selected (as, e.g., when the coffee jar has been left open and hence when it would be both unnecessary and inappropriate to attempt the *unscrew jar* schema).

14.5.2 Characteristics of Activation Propagation Each schema within the schema hierarchy is assumed to have an associated activation value which varies over time between zero and one in response to the net excitatory (positive) or inhibitory (negative) input to the schema. Moreover, it is assumed that sustained positive net input should drive activation toward one and that sustained negative net input should drive activation toward zero. This implies that with zero net input, activation should tend toward a resting value somewhere between zero and one.

In order to capture these requirements, the activation A of schema s at time $t + 1$ is calculated on the basis of its activation at time t and its net input $I_s(t)$ at that time according to equation 14.1:

$$A_s(t+1) = \bar{\sigma}\left(p \cdot \bar{\sigma}^{-1}(A_s(t)) + I_s(t)\right). \qquad (14.1)$$

Here, p (persistence) is a parameter that ranges from zero to one as described below, while the function $\bar{\sigma}(.)$ is based on the familiar logistic function with an offset, α:

$$\bar{\sigma}(x) = \frac{1}{1+e^{-(x-\alpha)}}. \qquad (14.2)$$

The offset is chosen somewhat arbitrarily so that with zero net input activation levels will converge to 0.1.

A little symbol manipulation shows that equation 14.1 is equivalent to equation 14.3:

$$A_s(t+1) = \bar{\sigma}\left(\sum_{i=0}^{t} p^i \cdot I_s(t-i)\right). \qquad (14.3)$$

Equation 14.3 demonstrates that the argument of the transformed logistic function is a decaying sum of the historic input to the schema (i.e., the input on the current time step, plus p times the input on the previous time step, plus p^2 times the input on the time step before that, etc.).

Equations 14.1 and 14.3 bear a passing resemblance to the equations that describe firing of a neuron in a McCulloch and Pitts model (see chapter 2, "Basic Neuron and Network Models"), with the argument of the transformed logistic function being equivalent to the presynaptic input to the neuron and the output of the logistic function being the postsynaptic neural potential. However, this resemblance is superficial. Equation 14.1 was designed purely to satisfy mathematical requirements of the model, and schemas within the interactive activation and competition (IAC) model are not intended to correspond directly to neurons. Moreover, as illustrated by equation 14.3, the input to the logistic function is not a weighted sum of the current inputs feeding into the schema as it would be in a McCulloch and Pitts model.

The parameter p (persistence) in equations 14.1 and 14.3 determines the extent to which the activations of schema nodes are more sensitive to their current input relative to their prior inputs (or prior activation, as in equation 14.1). When p is close to one, activations depend on the extended history of inputs and hence change slowly. If p is low (closer to zero), activation levels reflect more closely the input on the current time step (which, as described below, is determined by the activation of related schemas and object representations on the previous time step). Because the IAC model contains loops (with schema nodes exciting and inhibiting object representation nodes and vice versa), low values of p result in activations that change dramatically on each time step, leading to a system in which activations tend to oscillate. p can be understood as setting the granularity of time steps (i.e., the number of milliseconds between time t and time $t + 1$). When p is close to one, time steps are small (and hence very little decay occurs between steps) and equations 14.1 and 14.3 approximate a system that is continuous in time. When p is close to zero, time steps are large (and decay between steps can be large) and equations 14.1 and 14.3 alternate between overshooting and undershooting in the calculation of activation.

$I_s(t)$ in equations 14.1 and 14.3 is the net input to schema s at time t. It is the weighted sum of five influences:

$$I_s(t) = \tau \cdot T_s(t) + \beta \cdot B_s(t) + \sigma \cdot S_s(t) + \lambda \cdot L_s(t) + N(0,\eta^2). \qquad (14.4)$$

where τ, β, σ, and λ are parameters that weight the various contributions to the sum, and the remaining terms are as follows:

1. $T_s(t)$ (Top-down excitation): When a schema is intended, it receives strong top-down excitation from the supervisory system. Similarly, if (a) a schema corresponds to a subgoal of a selected schema, (b) that subgoal's preconditions are met, and (c) the subgoal's postconditions are not met, then the schema receives top-down excitation from the higher-level selected schema. Within the implementation of Cooper and Shallice (2000), this excitation is equal to the activation of the selected schema divided by the number of subgoals of the higher-level schema. Thus, the selected schema passes an equal proportion of its activation to subschemas that achieve each of its subgoals. This ensures that, when many subschemas are competing, top-down excitation is not so great as to override the influences (described below) of lateral inhibition and self-activation, which implement competition.
2. $B_s(t)$ (Bottom-up excitation): Schemas have associated triggering conditions. Schemas receive bottom-up excitation when their triggering conditions are compatible with the system's world model. The strength of bottom-up excitation is assumed to depend on the goodness of fit between the schema's triggering conditions and the system's world model.
3. $L_s(t)$ (Lateral inhibition): Schemas compete if they have overlapping resource requirements or correspond to alternative ways of achieving the same goal. Competition is implemented through lateral inhibition, such that a schema will inhibit each of its competitors with the degree of inhibition equal to the difference between the schema's activation and rest activation. Thus, if schemas S_1 and S_2 compete and S_1 is more active than S_2, then inhibition from S_1 on S_2 will be greater than that from S_2 on S_1. Consequently, and all other things being equal, the advantage of S_1 over S_2 will increase over time.
4. $S_s(t)$ (Self-activation): The effect of lateral inhibition (in the absence of top-down or bottom-up influences) would be to push the activations of all schemas toward zero. We therefore assume that schemas also experience self-excitation, with the strength of this influence being equal to the activation of the schema itself. Together with lateral inhibition, this implements a winner-take-all mechanism—in the absence of other influences, the activation of the most active schema from within a set of competitors will increase toward one while the activation of its competitors will decrease toward zero.
5. $N(0,\eta^2)$ (Noise): It is assumed that activations are also subject to noise. This is implemented by including normally distributed random noise (with mean zero and standard deviation η) in the weighted sum of influences.

Activated schemas therefore participate in an IAC network, where nodes excite or inhibit each other. As should be clear from the first influence in the above list, this network is hierarchical, with the top-down flow of activation within the hierarchy of schemas depending upon schema selection. A schema is selected when its activation exceeds a threshold (assumed to be 0.6 in the Cooper and Shallice, 2000, implementation). Selected schemas remain selected until (1) all postconditions of all of their subgoals are satisfied (if they have subgoals) or (2) the corresponding action is complete (for schemas at the lowest level, which by definition have no subgoals). On deselection, schemas are inhibited until their activation falls to rest.

The effect of the various influences on the activation of schemas, together with the modulation of activation flow by selection and deselection, leads to the kind of activation dynamics shown in figure 14.5. The hierarchical relationships between schemas are reflected in a kind of bracketing of the activation of lower-level schemas by that of higher-level schemas over time. First, the highest level becomes active (due to top-down excitation from the supervisory system). Once activation of this high-level schema exceeds the selection threshold, it is selected and activation flows to the schemas for its various subgoals. Assuming that these compete (which in the coffee-making task they do, because they all require use of a common resource—the hands), the activation of one mid-level schema will increase steadily until that schema is selected. Activation then passes to the schemas for the subgoals of the mid-level schema. This bottoms out with activation and selection of a schema at the lowest level, and subsequently with execution of the corresponding action (in this case, picking up the sachet of coffee). That low-level schema is then inhibited, allowing the next low-level schema to emerge from the competitive interaction of activation, and so on until all actions in the mid-level schema have been performed. That mid-level schema is then inhibited, allowing the activation of the next mid-level schema to rise, and so on. The resultant sequence of selection and behavior is shown in figure 14.6.

14.5.3 ARGUMENT SELECTION Why, in figure 14.6, is the first action "picking up coffee sachet with right hand," rather than "picking up teaspoon with left hand" (or any other variant of picking up)? As discussed

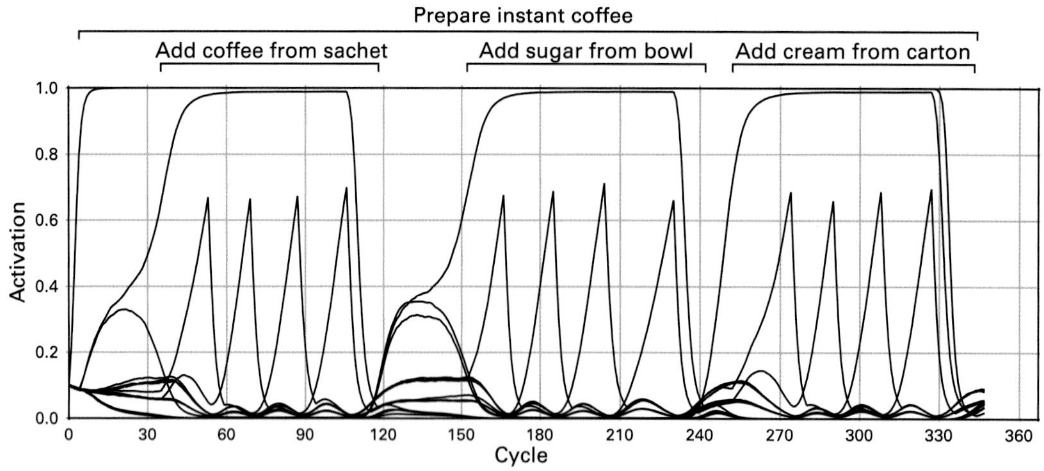

FIGURE 14.5 The profile of activations of schemas over time from the simulation of coffee making, as reported in Cooper and Shallice (2000). Activation is shown on the vertical axis while time is shown on the horizontal axis. The sharp peaks correspond to schemas for basic actions, with longer higher peaks corresponding to superordinate schemas. See figure 14.6 for a transcript of schema selections, schema deselections, and action selections corresponding to this plot.

earlier, within the model it is assumed that schemas have argument roles and that an internal model (established through perceptual schemas and possibly updated by motor schemas) makes object and effector representations available for binding to those argument roles. It is moreover assumed that there is a small set of such argument roles (paralleling the argument roles of agent, patient, theme, etc. from linguistic theory), and that object and effector representations compete for these roles through additional IAC networks. More precisely, each object in the coffee preparation domain was represented in the internal model with activation values for the argument roles of *source*, *target*, and *theme*. An activation value was also associated with each hand. All activation values were updated on each processing cycle according to equations 14.1 to 14.3, with the net input to object and effector (i.e., hand) representations calculated according to equation 14.4, with the following provisos:

- $T_{o,r}(t)$ and $T_e(t)$ (top-down excitation on object o in role r and top-down excitation on effector e) are zero, as the object and effector domains lack hierarchical structure.
- $B_{o,r}(t)$ and $B_e(t)$ correspond to the effect of schemas on object and effector representations. It is assumed that these effects are symmetrical, so if a schema excites/inhibits the representation of an object for a given role, then the object in that role will excite/inhibit the schema to the same degree.
- $L_{o,r}(t)$ and $L_e(t)$ operates across all items within each domain (i.e., across sources, across targets, across themes, and across effectors), so all object representations compete for each role and both hands compete with each other.
- $S_{o,r}(t)$ and $S_e(t)$ applies to nodes in the domains of object and effector representations just as it does in the schema domain.

The interactions between items within the schema network and the object and effector representation networks create feedback loops. Thus, in the coffee-making task the *prepare instant coffee* schema has an excitatory influence on the representations of the various coffee ingredients (coffee sachet, coffee jar, sugar sachet, sugar bowl, and cream carton) as sources, and on the representation of the coffee mug as a target. These, in turn, have excitatory influences on the *prepare instant coffee* schema (among others). A consequence of these feedback loops is that activating the *prepare instant coffee* schema will tend to activate its relevant arguments in their appropriate roles, which will tend to activate the original schema (and other schemas which use those objects in those roles), ensuring that both the appropriate schema and the relevant object representations will be highly active at the same time. It is this mechanism, propagated to all levels of the hierarchy of schemas and coupled with the assumption that when a schema is selected its argument roles are bound to the most active object/effector representations for those roles, that ensures the behavior shown in figure 14.6.

14.5.4 LEARNING WITHIN THE IAC MODEL The basic IAC model does not contain any learning mechanism. Action sequence knowledge is specified in terms of a fixed set of hardwired schemas which, for any schema,

4	*+PrepareInstantCoffee*	
35	*+AddCoffeeFromSachet*	
53	*+PickUp [Src]*	Picking up coffee sachet with right hand
59	*−PickUp [Src]*	
69	*+TearOpen*	Tearing open coffee sachet
76	*−TearOpen*	
87	*+Pour*	Pouring coffee from coffee sachet into coffee mug
94	*−Pour*	
106	*+Discard [Src]*	Putting down coffee sachet with right hand
115	*−Discard [Src]*	
118	*−AddCoffeeFromSachet*	
152	*+AddSugarFromBowl*	
166	*+PickUp [Imp]*	Picking up teaspoon with left hand
173	*−PickUp [Imp]*	
185	*+DipSpoon*	Scooping with teaspoon from sugar bowl
192	*−DipSpoon*	
204	*+EmptySpoon*	Emptying teaspoon into coffee mug
213	*−EmptySpoon*	
230	*+Discard [Imp]*	Putting down teaspoon with left hand
239	*−Discard [Imp]*	
242	*−AddSugarFromBowl*	
252	*+AddCreamFromCarton*	
274	*+PickUp [Src]*	Picking up cream carton with left hand
280	*−PickUp [Src]*	
290	*+TearOpen*	Tearing open cream carton
297	*−TearOpen*	
308	*+Pour*	Pouring cream from cream carton into coffee mug
316	*−Pour*	
327	*+Discard [Src]*	Putting down cream carton with left hand
336	*−Discard [Src]*	
343	*−AddCreamFromCarton*	
344	*−PrepareInstantCoffee*	

FIGURE 14.6 The sequential selection and deselection of schemas and actions (with arguments) from the same simulation of coffee making as shown in figure 14.5. "+" signs indicate schema selection. "–" signs indicate schema deselection. Numbers on the left indicate the cycle at which the selection or deselection occurred.

encodes (1) its goal, (2) its argument roles, (3) its subgoals, (4) any ordering constraints on those subgoals, and (5) the state of the environment that leads to the schema's excitation (i.e., its triggering conditions). While a complete computational account of how each of these parts of a schema might be acquired is lacking, Cooper and Glasspool (2001) demonstrate how standard reinforcement learning techniques (see chapter 10, "Reinforcement Learning") can address point 5 and some aspects of point 4, and Cooper and Shallice (2006) offer some speculations on how the other aspects of a schema might be learned.

Cooper and Glasspool (2001) augment the IAC model with a mechanism that learns associations between the states of objects (as represented in the internal model) and schemas by selecting a low-level schema at random and trying it. If it works (i.e., if it leads to a perceptible change in the state of the world), then the association (between the state and the corresponding schema) is strengthened. Otherwise it is weakened. This mechanism is sufficient to learn the class of order constraints that follow from the physics of the world (e.g., that an object can only be put down if it is held or that an object which is closed cannot be poured), but without additional mechanisms it results in a model that will perform one action over and over (rather like an infant practicing a behavior). An additional mechanism of novelty seeking, which must inhibit learned state–action associations and which would on the account of Shallice and Burgess (1996) be a function of the supervisory system, is required if the model is to explore a range of potential behaviors associated with its environment.

This simple associative mechanism could also learn goals (aspects 1–3 in the above list) of basic-level schemas by associating with the action the change in the world caused by attempting it, but in order to acquire higher-level schemas something else is required. Arguably this is provided by the supervisory system's building putative temporary schemas from existing simpler schemas in order to achieve simple goals and allowing these to be performed in a supervised manner by contention scheduling. With sufficient practice, contention scheduling should then internalize such temporary schemas. In this way a schema such as *add coffee from sachet* might be constructed from its parts. This internalization process is complex as it must also abstract over similar situations—not all instances of adding coffee from a sachet will feature the same actions in the same order. (Recall the Piagetian processes of assimilation and accommodation discussed in section 14.2.) However, once internalized, such an acquired schema can then form part of more complex temporary schemas, which may in turn be internalized, thus creating a hierarchy.

14.5.5 NEURAL LOCALIZATION OF SUBSYSTEMS RELATED TO THE IAC MODEL The IAC model has, so far, been presented in purely functional terms. The previous discussion has not attempted to relate subsystems of the model to regions or structures within the brain. This is a standard approach within cognitive neuropsychology (see Coltheart, 2006), in which theories are developed at a level that abstracts from their neural implementation. Nevertheless, neuropsychological and neuroimaging evidence may be used to map functional subsystems of the IAC model to regions and structures of the human brain. Consider first the evidence from neuropsychology, which suggests that three brain regions/structures are of particular interest within routine sequential action selection: frontal cortex, left temporoparietal cortex, and the basal ganglia (see figure 14.4). (For a somewhat related analysis, but grounded in macaque neurophysiology, see chapter 20, "Reach and Grasp: Control, Development, and Recognition.")

Consider first the role of frontal cortex. This large region of the brain has long been implicated in the control of goal-directed behavior (e.g., Luria, 1966; Duncan, 1986). Within the IAC model the region may be conceptualized as being involved in regulating the top-down flow of activation within the schema network, as this flow of activation is mediated by goal/subgoal relations. It has further been argued (see, e.g., Shallice et al., 2008; Shallice and Cooper, 2011) that one subregion of the frontal cortex, superior medial prefrontal cortex, is specifically involved in the "energization [...] necessary to activate operations not directly triggered in an overlearned fashion by perceptual and motivational inputs" (Shallice et al., 2008, p. 82). Within the contention scheduling/supervisory system framework this corresponds to the input from the supervisory system to contention scheduling.

A second function frequently associated with frontal cortex that may also be implicated in the control of sequential behavior is monitoring, which in general terms involves determining whether the cognitive system is functioning "adequately." The idea is that if the cognitive system is not functioning adequately, then some kind of reconfiguration is required. Monitoring has been invoked to understand how the cognitive system might respond to processing conflict (e.g., when two simultaneous stimuli demand different, incompatible, responses; see Botvinick et al., 2001) or more generally how it might compare expected and actual performance (Alexander and Brown, 2011). Intuitively, monitoring is involved in the IAC model in checking whether subgoal preconditions and postconditions are met. In a series of studies, Stuss and colleagues (e.g., Stuss et al., 2005; Picton et al., 2006; Shallice et al., 2008) have argued that the behavioral impairments of patients with damage to right lateral frontal cortex on a range of laboratory tasks may be understood in functional terms as an impairment of monitoring (or more broadly, checking). Note that if this is the case, then some errors of closed head injury (CHI) patients, who may have widespread damage to their frontal cortex, may also be associated with monitoring failure. This view answers a puzzle first raised by Schwartz et al. (1991). If contention scheduling can guide routine behavior in the absence of supervisory attention, then why is such behavior so prone to error when we are distracted? One possibility is that a great deal of what we have referred to as "routine behavior," particularly when performed within a changing and not fully predictable environment, may actually involve frequent monitoring together with top-down or supervisory intervention to assemble simpler routines and ensure that behavior remains task-appropriate and on track.

Turning to cortical areas beyond the frontal lobe, the left temporoparietal cortex has also been implicated in the control of routine action. De Renzi and Lucchelli (1998) attribute ideational apraxia to damage to the left temporoparietal junction (see also Rumiati et al., 2001). Cooper et al. (2005) speculate from the perspective of the IAC model that this region is involved in the triggering of schemas by object representations (or by what some authors describe as a "direct route to action"; Humphreys and Riddoch, 2003; Gonzalez Rothi et al., 1991). A similar localization of these processes is supported by brain imaging evidence. In one study Rumiati et al. (2004) contrasted regional cerebral blood flow in four experimental conditions: naming of objects, naming of actions, performing actions on presentation of objects, and imitating actions on presentation of actions. In comparison to all other conditions, including that in which actions were imitated, the condition in which actions were performed to objects led to significantly increased activation in the left inferior parietal lobule (slightly above the temporoparietal junction). In other words, the region that is typically damaged in patients with ideational apraxia shows, in neurologically healthy individuals, a strong response when subjects perform object-appropriate actions in response to those objects.

As noted earlier in this chapter, subcortical structures, and in particular the basal ganglia, are also implicated in the control of sequential behavior, with (e.g.) rate effects apparently related to dysfunction of dopamine regulation within the striatum (one of the subcortical structures that make up the basal ganglia). The IAC model is not committed to any specific function of the basal ganglia, but one possibility is that competition between schemas is resolved by the basal ganglia, which participate in corticosubcortical loops (see chapter 22, "Integrative Functions of the Corticostriatal System"). Redgrave et al. (1999) provide a comprehensive model of basal ganglia function which is compatible with this view.

14.5.6 Related Models The principles of interactive activation which are central to the IAC model—namely, the association of activation values with discrete entities and the interaction between those activation values through excitatory and inhibitory influences—are shared with many models across a range of domains, including models of sequential behaviors such as typing (Rumelhart and Norman, 1982), spelling (Houghton, 1990; Glasspoool and Houghton, 2005), and verbal short-term memory (Burgess and Hitch, 1999), as well as abilities such as reading (McClelland and Rumelhart, 1981) and processes involved in the regulation of cognitive processes (e.g., Cohen and Huston, 1994; Gilbert and Shallice, 2002). Many of these models differ in the equations used to calculate activation (i.e., equations 14.1 to 14.4), but the precise form of these equations is typically not critical at the abstract cognitive level at which these models are framed.

One feature of the IAC model that is critical to the sequential nature of the action domain is the inhibition of nodes corresponding to completed items. This mechanism also features in other interactive activation models of sequential behavior (e.g., Bonaiuto and Arbib, 2010; Bullock and Rhodes, 2003; Rumelhart and Norman, 1982; Houghton, 1990; Glasspoool and Houghton, 2005). It is necessary to prevent so-called perseverative behavior—where a behavior is inappropriately repeated immediately after it has been produced. One common difficulty with this form of inhibition concerns sequences that require a given action to be performed twice in succession. This is not an issue in the beverage preparation task, as immediate repetitions are not required. However, it is typically resolved by assuming that a special mechanism exists to support immediate repetition of a behavior (e.g., Houghton, 1990; Glasspoool and Houghton, 2005). Some authors (e.g., Rumelhart and Norman, 1982) conceptualize this mechanism as a "doubling schema," which excites (and consequently reactivates) the most recent response. While this has not been implemented within contention scheduling, it would seem to provide one possible resolution to the potential difficulty.

Perhaps more similar to the IAC model than the models of spelling and typing is the augmented competitive queuing (ACQ) model of Bonaiuto and Arbib (2010; see also Arbib and Bonaiuto, 2012; see also chapter 25, "Evolving the Language-Ready Brain"). In competitive queuing models (Houghton, 1990), a representation of a behavioral plan feeds into a bank of units that compete (typically through lateral inhibition and self-activation, as in the IAC model), and with each unit in the bank corresponding to a single action. Typically several units in that bank will be activated by the representation of the plan, but the bank acts as a competitive filter, such that only one of the competing units may be highly active at any moment in time, but once an action has been executed, it is removed from the next round of competition, and so on repeatedly, thus allowing a sequence to be activated. The ACQ model decomposes the input to the behavioral plan into two components, one encoding "executability" of current actions (in the context of the current state of the environment) and the other "desirability" with respect to the system's higher-level goal. Executability corresponds within the IAC model to the triggering of action schemas by object representations, while desirability corresponds to the top-down influence of selected schemas on those schemas that achieve the current subgoals. ACQ also includes a reinforcement learning mechanism (see chapter 10, "Reinforcement Learning") that exploits mirror neurons, that is, neurons that fire both when an animal performs an action and when the animal observes another performing the same action.

14.6 The Simple Recurrent Network Model of Routine Sequential Action Selection

A critical aspect of many IAC models, particularly those on the motor side of cognition, is that they produce a sequence of outputs, and they do this by maintaining a representation of some kind of plan that imposes differential excitation on its subcomponents. This approach can be traced to Lashley (1951). The behaviorist view, which held sway prior to Lashley's work, was that sequential behavior was generated by stimulus–response associations with one response serving as the stimulus for the next—so-called "chaining." Lashley argued that this approach cannot explain behaviors in which response R_1 is followed by response R_2 in one situation or task but by response R_3 in another. It also cannot explain fast motor behaviors such as playing a musical instrument, where there is insufficient time between the production of one response and the next for the first response to cue the second. Such behaviors suggest that a sequence of responses may be programmed in advance, and they have been taken as a key source of support for schema-based theories of the control of action (Cooper and Shallice, 2000; see Rosenbaum et al., 2007, for additional discussion).

Lashley's critique of associationist approaches to sequential behavior has been challenged by developments in recurrent connectionist networks. In particular, in a direct response to Cooper and Shallice's (2000) IAC model of contention scheduling, Botvinick and Plaut (2004) developed a simple recurrent network (SRN) model of the beverage preparation task that, the authors argue, operationalizes an associationist approach to sequential behavior. This section details that model and evaluates its performance.

14.6.1 The Architecture of Botvinick and Plaut's Model An SRN consists of three layers of neuron-like units (see figure 14.7A), where each unit has an activation value that varies as a function of its weighted input. The activation of units in the input layer is determined by a representation of the task input. These units feed, via weighted connections, to a hidden layer. The activation of units here is determined by the weighted sum of activations from input units, together with a weighted sum of recurrent activation from the hidden units on the previous time step. The hidden units may be understood as providing a kind of context or internal state within which the current input is to be interpreted. They feed, via further weighted connections, to the output layer. The activation of units here is determined by the weighted sum of activations of the hidden layer.

The Botvinick and Plaut model consists of 39 input units, 50 hidden units, and 19 output units (see figure 14.7B). Input to the model represents the current state of the environment. This consists of 19 units that represent what, if anything, is held by the model's virtual hand and 20 units that represent what the model is fixating on with its virtual eye. Two of these fixation units indicate the task instruction—whether the model is required to make coffee or to make tea. The encoding of held and fixated objects is feature based. That is, the input units correspond to various binary-valued features (whether an object is a sachet, whether it is closed, whether it contains sugar, etc.). Input units are set to one when these features are present and zero otherwise.

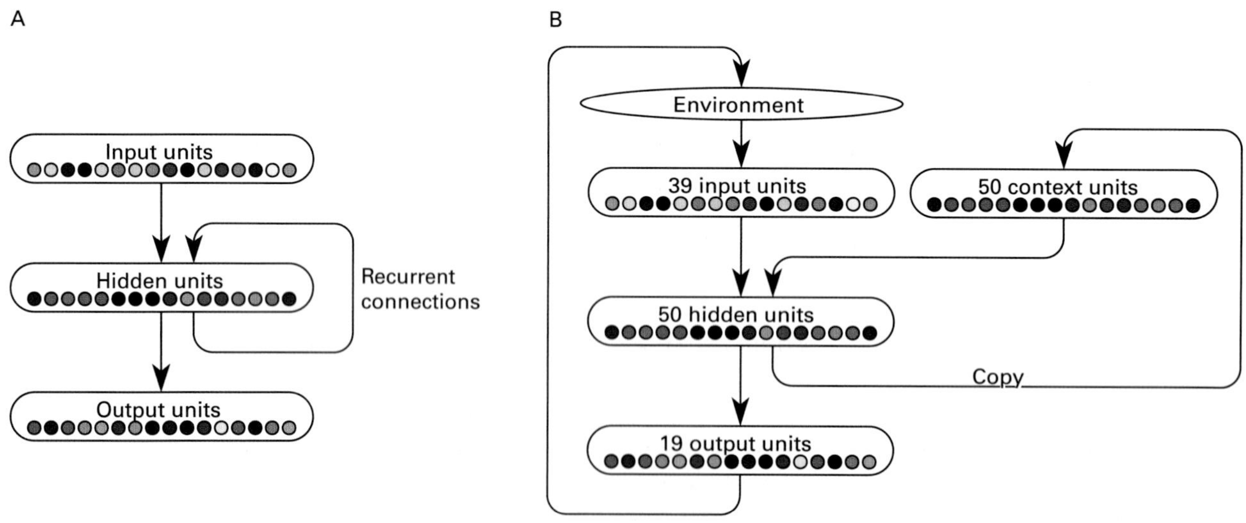

Figure 14.7 Simple recurrent network (SRN) models. Arrows represent full connectivity between layers. (A) A schematic SRN, showing recurrent connections from the hidden layer that feed the hidden layer activation pattern back into the hidden layer on the next processing step. (B) The specific implementation of Botvinick and Plaut (2004). Input units represent the currently fixated and held objects. Recurrent connections are implemented by a layer of context units that are used to feed the activation of the hidden layer on cycle $n - 1$ back to the hidden layer on cycle n. Output units represent actions. The most active action on cycle n affects the model's environment and hence the input to the model on the cycle $n + 1$.

Outputs units of the model represent the possible actions within a simulated world. Thus, one output node represents "fixate-mug," while another represents "pick-up." Given this approach to encoding action, if the model is to generate a sequence of actions to stir the contents of a mug, then it must perform a sequence of five actions: fixate on the spoon, pick it up, fixate on the mug, stir it, and then signal that stirring is complete. Note that input to the SRN is based on the model's representation of the world. When an action is performed, it may change the state of the world, and hence this representation, thereby changing the input on the next processing step.

The idea behind the SRN model is that inputs are first set for a specific task and configuration of the world. Activation is then propagated forward through the network with the activation of node i calculated as follows:

$$A_i = \sigma\left(\Sigma_j A_j \cdot w_{ji} + \beta_i\right) \quad (14.5)$$

where $\sigma(.)$ is the standard logistic function that maps $+\infty$ to 1.0, zero to 0.5, and $-\infty$ to 0.0, w_{ji} is the weight of the connection from node j to node i, and β_i is the bias of node i, as described below. There is some similarity between equation 14.5 and equations 14.1 and 14.3 of the IAC model, in that both equations involve an application of the logistic function to summed input, but the similarity is superficial. Unlike the case of the IAC, in the case of equation 14.5 the summed input can be viewed as a presynaptic potential that is transformed, via the logistic function, to a postsynaptic potential.

The bias term in equation 14.5 (β_i) effectively sets the responsiveness of the node to excitation. A node with a strong negative bias will need substantial positive input to activate it while a node with a strong positive bias will become activate with relatively little net input (and possibly even with net negative input). Biases can be "folded in" to the network by assuming that each non-input unit has an input from a bias unit whose value is always one. The bias for unit i then reduces to the weight of the connection between the bias unit and unit i.

A critical aspect of SRN models is their use of recurrent connections within the hidden layer. This means that input to the hidden layer on a given cycle of processing consists of excitation direct from the input units plus recirculated excitation from the units of the hidden layer on the previous time cycle—the so-called "context." This recurrence allows the network to generate sequential outputs where the output on each cycle is a function of both the current input and the state of the hidden layer on the previous cycle.

14.6.2 Learning by Backpropagation of Error

The trick within models of the type used by Botvinick and Plaut (2004), and indeed simpler feedforward networks, is to determine values for w_{ji} such that when the input units are set appropriately and activation is propagated through the network, the output units take on appropriate values. Normally this is done by specifying a training set—the desired pairs of input and corresponding output vectors—setting the weights to random values, and then using the training set to adjust the weights via the iterative application of a learning algorithm.

In the simple case of feedforward connectionist networks (i.e., networks with one or more hidden layers but no recurrent connections), this algorithm, backpropagation of error, comprises two phases. In the forward phase the output of the network is calculated for a given input. The difference between this actual output and the desired output is then determined, giving an error term for each output node. The aim of the training algorithm is to reduce this error, but because the change in activation of a node is not proportional to the change in the node's input, the error must be scaled by the gradient or slope of the activation function at its current value. Thus, a scaled error term—δ_k in the equations below—is calculated for each output node k. In the second phase this scaled error term is propagated back through the network as a correction to the weights, associating a proportion of the scaled error with each connection leading from the hidden units into each output node. This allows the calculation of a vector of target activations for the hidden units. Again, the difference between this target and the actual values can be propagated back to the connections leading into the hidden units, allowing the calculation of correction factors for all weights in the network. Following this approach, δ_k for output units may be calculated according to equations 14.6 and 14.7:[2]

$$\delta_k = (t_k - o_k).\sigma'\left(\Sigma_j A_j \cdot w_{jk}\right) \quad (14.6)$$

$$= (t_k - o_k) \cdot o_k \cdot (1 - o_k). \quad (14.7)$$

In these equations t_k is the target activation of unit k while o_k is that unit's actual activation as calculated by propagating activation forward in the network. $\sigma'(.)$ is the derivative of the $\sigma(.)$ function. For hidden units, the error term δ_j is calculated according to equations 14.8 and 14.9:

$$\delta_j = \left[\Sigma_k \delta_k \cdot w_{kj}\right] \cdot \sigma'\left(\Sigma_i A_i \cdot w_{ij}\right) \quad (14.8)$$

$$= \left[\Sigma_k \delta_k \cdot w_{kj}\right] \cdot o_j \cdot \left(1 - o_j\right). \quad (14.9)$$

Given equations 14.7 and 14.9, weight adjustments are then calculated according to equation 14.10:

$$w_{ji} \leftarrow w_{ji} + \eta \cdot \delta_i \cdot o_j \, . \tag{14.10}$$

Here, η is a parameter, the learning rate, which is typically small (e.g., 0.02). This is necessary because the learning algorithm will only converge if the changes in weights on each iteration of the learning rule are small.

The learning algorithm is applied for each input–output pair in the training set. This is known as one *epoch.* Weights may be adjusted by the application of equation 14.10 either after each training item or at the end of one epoch. In this latter case, weight changes must be accumulated over the full training set. Typically the algorithm must be repeated a large number of times (often many thousands of epochs) in order for the error term to fall to near zero and the weights to converge.

14.6.3 Learning by Backpropagation through Time Standard backpropagation is sufficient for training feedforward connectionist networks, but it must be adapted for recurrent networks of the type used by Botvinick and Plaut (2004) if they are to learn to produce a sequence of actions. The basic idea is to "unroll" the network (see Werbos, 1990). Figure 14.8 illustrates this for the stirring sequence of five actions described above. To learn a sequence of n actions, the network is copied $n - 1$ times to produce a network with n input layers, n output layers and $n + 1$ hidden layers, and where weights between copied units are constrained to be equal. (So the weight between input unit i and hidden unit j in the first row of figure 14.8 is the same as the weight between the corresponding units in the second row, and the third row, and the fourth row, and the fifth row.) The copied layers can be thought of as the layers of the original network at different steps in time in the action sequence. To train the network, inputs are set at each time step in the sequence (as shown in figure 14.8) and propagated forward. Target outputs are similarly set based on the required sequence. The difference between actual and target outputs can then be calculated and error propagated back as in standard backpropagation of error. Weights are then adjusted as in the standard backpropagation of error algorithm such that the change to each weight is the sum of changes over all steps of the sequence. (See chapter 1, "From Neuron to Cognition: An Opening Perspective," for more on backpropagation and backpropagation through time.)

This training procedure results in the hidden units' learning to represent a kind of context which, through recurrent connections, feeds together with the input representation on each step into the selection of an appropriate output. The hidden units on time step n are thus sometimes referred to as forming the context units for time step $n + 1$.

14.6.4 Neural Localization of Subsystems Related to the SRN Model The SRN model, as

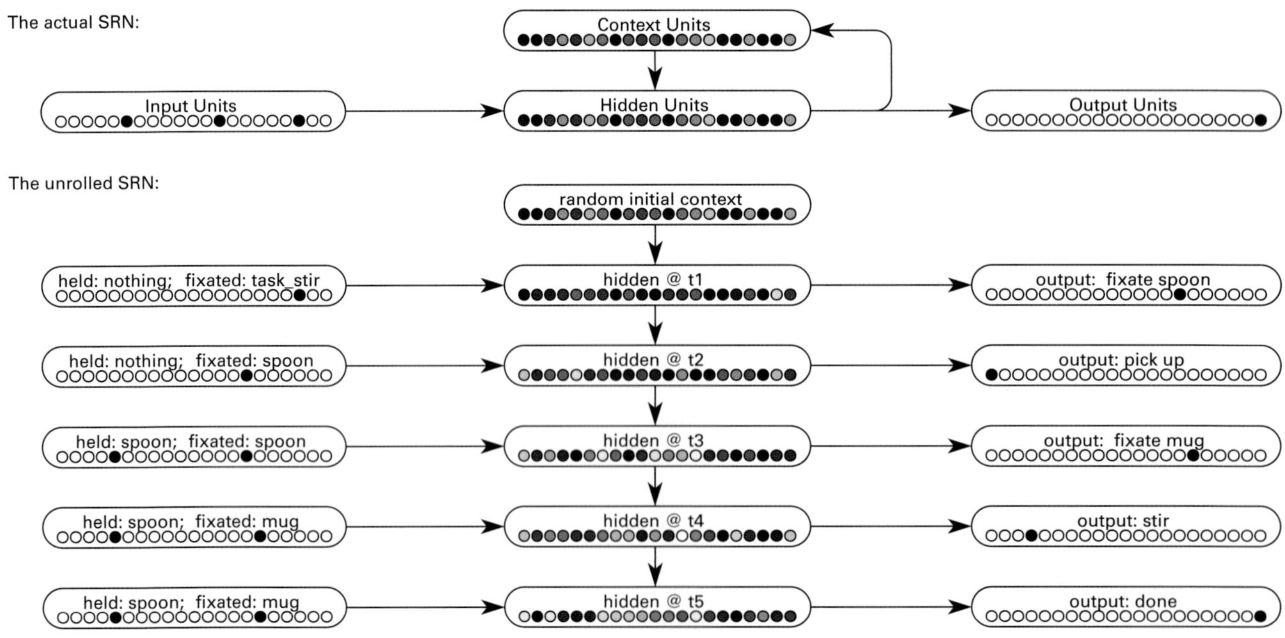

Figure 14.8 Schematic of the simple recurrent network (SRN) of Botvinick and Plaut (2004) and the "unrolled" network as required by backpropagation through time when learning a sequence of five actions—in this case the sequence for stirring.

originally presented by Botvinick and Plaut (2004), makes no firm commitment to the neural localization of its components, and no such commitment is required if the model is being used purely to understand normal and impaired behavior at the cognitive level. However, Botvinick (2007) discusses an elaboration of the SRN model in which the hidden/context layer is developed into a series of layers that learn increasingly abstract task representations. This is consistent with some current views of the prefrontal cortex, such as that of Badre and D'Esposito (2007) and Koechlin et al. (2003), in which regions closer to the front of the brain operate on more abstract representations than regions closer to the input (i.e., auditory and visual cortex) or the output (i.e., motor cortex). This work implies that the SRN model should be broadly conceived of as frontally localized, but beyond this the model makes no commitment to localization.

14.6.5 Performance of the SRN Model Botvinick and Plaut (2004) trained their network with 267 one-step sequences (corresponding to the single actions that could be performed in each possible input state), two 20-step sequences (corresponding to two ways of making tea), and four 37-step sequences (corresponding to four ways of making coffee). The network was trained with a learning rate of 0.001 for 20,000 epochs, by which time it was able to reproduce all multistep sequences in the training set. Thus, when initialized with a random activation vector for the context and an initial state corresponding to holding nothing but fixating on the mug and with the tea instruction unit set, the network produces a sequence of 20 steps appropriate for making tea. Similarly, when initialized with a random activation vector for the context and the same initial state but with the coffee instruction unit set, the network produces a sequence of 37 steps appropriate for making coffee.

Recall that the model was trained with two tea-making sequences and four coffee-making sequences. The difference between the two tea-making sequences concerns the way in which sugar is added to the beverage. Both use the fixate-sugar action, but this action is nondeterministic. It may leave the model fixating on a sugar sachet or on a sugar bowl. In the former case, the sachet is picked up and poured into the mug. In the latter case, the model removes the sugar bowl's lid and then uses the spoon to scoop sugar into coffee. Because the model is trained with these two alternatives, it is able to respond flexibly to the result of its fixate-sugar action, demonstrating that it can, in principle, be responsive to unpredictable or nondeterministic aspects of its environment.

The four coffee-making sequences on which the SRN model is trained involve the same two ways of adding sugar, crossed with different orders for adding sugar and cream (i.e., cream first or sugar first). The model is also therefore able to encode different orders of subsequences, demonstrating that associationist models with recurrent connections can generate behaviors beyond those possible with simple chaining accounts. A purely feedforward model of the task without recurrent hidden-to-hidden connections would not be able to handle the variability in sequencing required of the two possible orderings, that is, situations when, on some occasions, R_1 (e.g., stir in the coffee grinds) is followed by subsequence S_2 (add the sugar) and eventually S_3 (add the cream) and when, on others, R_1 is followed by S_3 first and then S_2. The hidden layer in the SRN model learns to provide a representation of context that, together with the input, allows the model to discriminate between these sequences.

Below we consider behavioral data that suggest that both the IAC and SRN models can account for basic facts relating to pathologies of action selection, but that the IAC model can account for a greater range of effects. The SRN model, though, has one great advantage over the IAC model—it acquires action sequence knowledge on the basis of experience with a training set. Notwithstanding the comments in section 14.5.4, in the IAC model, action sequence knowledge must be coded by hand in terms of schemas. One consequence of the SRN model's learning mechanism is that action schemas in that model are emergent behaviors, not explicit data structures that may be invoked in the control of behavior. While the originators of the SRN model do not claim that the learning mechanism is biologically or psychologically plausible—recall that 20,000 epochs of training was required to learn six sequences—they do argue against the explicit representation of action schemas. This difference reflects a fundamental theoretical difference between the approaches. Cooper and Shallice (2006) argue that higher-level processes must have access to some explicit, causally efficacious, representation of action schemas, for example, when recovering from error or when performing a deliberate modification of a routine sequence, as when preparing coffee for a guest who takes an extra measure of sugar. Botvinick and Plaut (2006) accept that higher-level processes such as planning make use of "abstract task representations" (p. 922) but argue against the involvement of such representations in action execution. This is problematic when actions must be assembled flexibly in response to a partially unpredictable environment. Thus, the SRN will fail to generate appropriate actions if confronted by a situation that has not been covered in its training (e.g., if someone has left the sugar bowl open, when the model has only been exposed to situations where the sugar bowl is initially closed). Moreover, it cannot sensibly assemble known action sequences so as to address novel goals. Thus, if trained to make coffee with one sugar, the SRN is not able to generalize from this to make coffee with two sugars—it must be trained as if from scratch on what (as far as the model is concerned) is an entirely novel task (but see Cooper et al., 2014, for an extended version of the SRN that includes an explicit representation of subgoals and hence that is able to generalize to this task). This is unfortunate, but in fact the situation is even worse than this. If this training focuses purely on the new task (of coffee with two sugars), the ability to perform the old task (coffee with one sugar) will be lost due to interference between the new training and the old training. Neither of these difficulties arise with the explicit representation of schemas assumed by the IAC model. Indeed, one of the virtues of schema theory is that one can rapidly learn a skill that involves combining known schemas in a novel way.

14.7 *Modeling Neurological Deficits*

The models described in the previous two sections may appear to be far removed from neuroscience. What is the relation between them and, for example, the neuropsychological data reviewed in section 14.4.3, and can neuroscience provide additional support for the models or help to discriminate between them?

14.7.1 Methodological Considerations The standard approach to relating high-level cognitive models to neuropsychological evidence is to consider performance of the model following an artificial "lesion." In patients, a brain lesion is an injury to brain structures or tissue, and such an injury may result in behaviors such as those described in section 14.4.3. Within a computational cognitive model a lesion may be simulated in a variety of ways. One simple approach in models that include numerical processes subject to noise is to increase the variance of that noise. This allows one to investigate the effects of generalized damage to the operation of the process under question. An alternative approach for models that include multiple interacting subsystems is to impair the level or degree of interaction between subsystems (or indeed to completely remove a subsystem). Both of these approaches may be related to the neural level if there is some putative mapping between model components and brain structures or regions, though as noted above such a mapping is not necessary if one is attempting to understand the cognitive system from a purely functional perspective (Coltheart, 2006).

Cooper and Shallice (2000) consider the effects of increasing their noise parameter and of altering the values of the parameters that govern the balance between the various influences on the activation of schemas (see also Cooper et al., 2005; Cooper, 2007). The former is argued to simulate generalized or nonspecific brain damage while the latter is justified by the claim that different parts of the IAC model are implemented by different brain regions, and lesions to these different regions (or connective fibers between the regions) effectively impair the efficiency of the corresponding functional component of the IAC model. Botvinick and Plaut (2004) adopt a somewhat related but less targeted approach, simulating a generalized impairment by adding normally distributed random noise with small variance to their hidden/context units on each processing cycle.

14.7.2 Utilization Behavior If the schema-based IAC model is lesioned by reducing the top-down influence within the schema network (i.e., reducing τ in equation 14.4), then the model exhibits utilization behavior. That is, hierarchically structured schema selection breaks down and behavior is driven more by the representation of the environment than by intentions. Such behavior also occurs if no high-level intention is set and the model is allowed to run in the absence of input from the highest level of the schema hierarchy. This is consistent with the assumption that the top-down influence within the IAC model is effected or mediated by the superior medial prefrontal cortex.

The SRN model is also able to show utilization behavior, but only under a restricted definition of utilization behavior. If the model is initialized without a specific instruction (i.e., with model fixated on the mug, both the "make-coffee" and "make-tea" instruction units set to zero, and with the hidden/context units randomized—equivalent to running the IAC model without a high-level intention) and then allowed to produce a sequence of actions, then the SRN will still perform a well-formed sequence of actions—resulting in successful preparation of either coffee or tea. In this situation, the actual sequence performed depends on the initial values of the context units (with some random contexts leading to tea-making and others to coffee-making). This is a very restricted form of utilization behavior—the model's behavior is determined not by any specific intention but by some action sequence that it has learned during its training history. What the SRN cannot do is show utilization behavior that is a function of the presence of objects in the environment. The actions of the SRN are driven by (1) the object on which it is fixated, (2) the object that it holds (if any), and (3) its internal representation of where it is up to in a task. If an unusual object *x* is present in the environment, that object can only influence the SRN model's behavior if the model happens to fixate on the object, but that can only happen if the model selects the corresponding "*fixate x*" action. For this to happen, the SRN would need to include exposure to *x* in its training, together with experience of how *x* might be manipulated. This is quite reasonable, but the real difficulty with the SRN is that even with such training there is no reason why *x* should capture the model's attention more than any objects that are relevant to the current task. The model has no concept of salience, and there is no way in which objects that are differentially salient might have differential effects on behavior.

While the SRN is largely agnostic about the neural localization of component processes, if one accepts the above restricted account of utilization behavior within the model, then it would appear that superior medial prefrontal cortex should be identified with intentional input to the action selection mechanism (i.e., the SRN model's instruction units).

14.7.3 Action Disorganization Syndrome Within the context of the IAC model, Cooper et al. (2005) suggest that action disorganization associated with CHI affecting the frontal lobes arises from nonspecific dysfunction of activation propagation between schema nodes within contention scheduling. It is further argued that this may be modeled though the addition of noise to the activations of nodes within the schema network. Simulations have shown that adding noise to these activations results in four characteristics also seen in the disordered action of CHI patients. First, these patients are particularly prone to omission errors—leaving out a subsequence within a longer sequence—but other errors are also frequent, including sequence ordering, action addition, and object substitution errors (Schwartz et al., 1998). Second, when severity of dysfunction is measured by the total number of errors across a battery of standardized tasks, severe CHI patients produce relatively more omission errors than mild CHI patients. Third, if one considers success at a task as the dependent measure, then successful patients produce fewer omission errors (as one would expect) but also fewer errors of commission (i.e., fewer errors of all other types). Finally, when task-irrelevant objects are present, CHI patients produce marginally more object substitution errors but significantly more omission errors. Thus, while task-irrelevant objects may distract the process of argument selection, their principal effect is to reduce the number of task-relevant actions performed by the patients.

Botvinick and Plaut (2004) simulate action disorganization syndrome through the addition of noise to the hidden/context layer within the SRN model. This results in the model's producing action sequences that contain numerous errors, and Botvinick and Plaut (2004) interpret the damaged model's behavior as being similar to that of Schwartz et al.'s (1998) patients. However, Cooper and Shallice (2006) argue that the errors produced by the SRN are all effectively capture errors. Thus, the SRN appears to produce an omission error when, for example, it performs the sequence for making coffee but without the subsequence that adds sugar. What appears to be happening here, however, is that the model begins to make coffee. Its behavior is then captured by another known sequence—tea-making or coffee-making—but at a point late in that sequence (where tea-making and coffee-making are identical), resulting in what appears to be an omission. Perseverative errors similarly result from capture, though in this case the error occurs late in the original sequence and the point at which behavior is captured is earlier in the capturing sequence, resulting in repetition of some elements of the sequence. Consistent with this interpretation, the SRN model produces very few object substitution errors.

14.7.4 Ideational Apraxia Rumiati et al. (2001) describe in detail the action selection errors of two patients with left parietal lesions characterized as ideational apraxic. Unlike CHI patients, the majority of errors produced by these patients were conceptual errors rather than sequential errors. Thus, rather than performing the correct actions in the wrong sequence (and hence, e.g., attempting to pour from a container before opening it), one patient tended to misuse tools (selecting the right tool for the task but using it in the wrong way) while the other tended to mislocate actions (e.g., attempting to light a candle at the wrong end). Cooper (2007) demonstrated that the misuse errors of the former patient could be simulated within the IAC model by partially lesioning the links from the internal model to the schema network (i.e., by reducing β in equation 14.4) while the mislocation errors of the latter patient could be simulated by partially lesioning the reverse links from the schema network to the internal model (i.e., by reducing the corresponding parameter involved in the calculation of the activation of object representations).

14.7.5 Disorders of Rate Deficits related to disorders of the rate of action production are beyond the scope of the SRN model. That model produces one output on each and every time step and so does not make contact with data related to differential action timing. However, this is not the case for the IAC model. In that model activation accumulates over time until a threshold is reached. A critical parameter affecting the rate of accumulation is the ratio of self-activation to lateral inhibition (i.e., the ratio of σ to λ in equation 14.4). If this ratio is low, activation accumulation (and consequently action selection) in the IAC model is slow—particularly at subtask boundaries where the model must select between multiple competing subtasks. The model's behavior is therefore reminiscent of the slowing in action selection associated with PD. If the ratio is high, activation accumulation is fast and the model exhibits the kind of hyperactivity associated with amphetamine use. (See Cooper and Shallice, 2000, for additional details.)

14.8 Concluding Discussion

This chapter has introduced schema theory, a general framework for describing both behavior and cognitive function. Schemas are abstractions over perceptual, linguistic, event, action, or motor situations. They may compete, or cooperate, in the production of behavior. Schemas may also be hierarchically structured, with complex schemas consisting of multiple simpler schemas. Schema theory is very general. It can be applied to low-level perceptual and motor behaviors (such as predator and prey detection, and snapping or evading behaviors), as well as high-level functions such as memory, language generation and understanding, and routine action control. Thus, while many behaviors of amphibians may be accounted for by schema theory, more complex behaviors, including routine sequential action selection in humans, may also be modeled within the theory.

The bulk of this chapter has considered this latter domain—the control of routine action—within the context of the contention scheduling theory of Norman and Shallice (1986). Two models, the IAC model of Cooper and Shallice (2000) and the SRN model of Botvinick and Plaut (2004), have received special attention. A key difference between the models concerns whether schemas are explicit control structures that might be directly accessed by other cognitive systems (as in the IAC model) or regularities over patterns of activity that emerge through learning (as in the SRN model). Both models can account for some deficits in action selection following neurological damage, though the latter model is more limited in the types of behavioral dysfunction that it can address. The chapter concluded by considering the possible neural location of elements of the contention scheduling theory and the way in which the two models account for behaviors resulting from neurological damage.

NOTES

1. While some researchers do not consider ideational apraxia to be behaviorally distinct from frontal apraxia (e.g., Buxbaum et al., 1998), the former is generally associated with damage to the left temporal and/or parietal cortices while the latter is associated with damage to frontal regions. Ideational apraxia is further to be distinguished from ideomotor apraxia, which is associated with clumsiness in movement and an inability to imitate gestures (particularly meaningless gestures). This is considered within the contention scheduling/supervisory system theory to be a disorder of motor control and is attributed to dysfunctional schemas and/or internal models within the motor system.
2. For a full mathematical derivation of the backpropagation learning algorithm, see, for example, chapter 8 of Rumelhart and McClelland (1986) or chapter 6 of Hertz, Krogh, and Palmer (1991).

REFERENCES

Alexander, W. H., & Brown, J. W. (2011). Medial prefrontal cortex as an action-outcome predictor. *Nature Neuroscience, 14*(10), 1338–1344.

Arbib, M. A. (1989). *The Metaphorical Brain 2: Neural Networks and Beyond.* New York: Wiley Interscience.

Arbib, M. A. (2003). *The Handbook of Brain Theory and Neural Networks* (2nd ed.). Cambridge, MA: MIT Press.

Arbib, M. A., & Bonaiuto, J. J. (2012). Multiple levels of spatial organization: World graphs and spatial difference learning. *Adaptive Behavior, 20*(4), 287–303.

Arbib, M. A., & House, D. H. (1987). Depth and detours: An essay on visually guided behavior. In M. A. Arbib and A. R. Hanson (Eds.), *Vision, Brain, and Cooperative Computation* (pp. 129–163). Cambridge, MA: MIT Press.

Arkin, R. C., Ali, K., Weitzenfeld, A., & Cervantes-Perez, F. (2000). Behavioral models of the praying mantis as a basis for robotic behavior. *Robotics and Autonomous Systems, 32*(1), 39–60.

Badre, D., & D'Esposito, M. (2007). Functional magnetic resonance imaging evidence for a hierarchical organization of the prefrontal cortex. *Journal of Cognitive Neuroscience, 19*(12), 2082–2099.

Bahrami Balani, A., Soto, D., & Humphreys, G. W. (2012). Separating top-down and bottom-up cueing of attention from response inhibition in utilization behaviour. *Neurocase, 18*(2), 98–111.

Bartlett, F. C. (1932). *Remembering.* Cambridge: Cambridge University Press.

Beale, J. M., & Keil, F. C. (1995). Categorical effects in the perception of faces. *Cognition, 57*(3), 217–239.

Boccardi, E., Della Sala, S., Motto, C., & Spinnler, H. (2002). Utilization behavior consequent to bilateral SMA softening. *Cortex, 38,* 289–308.

Bonaiuto, J. J., & Arbib, M. A. (2010). Extending the mirror neuron system model: II. What did I just do? A new role for mirror neurons. *Biological Cybernetics, 102*, 341–359.

Bornstein, M. H. (1973). Color vision and color naming: A psychophysiological hypothesis of cultural difference. *Psychological Bulletin, 80*(4), 257–285.

Botvinick, M. M. (2007). Multilevel structure in behavior and in the brain: A model of Fuster's hierarchy. *Philosophical Transactions of the Royal Society of London. Series B, Biological Sciences, 362*(1485), 1615–1626.

Botvinick, M. M., Braver, T. S., Barch, D. M., Carter, C. S., & Cohen, J. D. (2001). Conflict monitoring and cognitive control. *Psychological Review, 108*(3), 624–652.

Botvinick, M. M., & Plaut, D. C. (2004). Doing without schema hierarchies: A recurrent connectionist approach to normal and impaired routine sequential action. *Psychological Review, 111*, 395–429.

Botvinick, M. M., & Plaut, D. C. (2006). Such stuff as habits are made on: A reply to Cooper and Shallice (2006). *Psychological Review, 113*(4), 917–927.

Bullock, D., & Rhodes, B. J. (2003). Competitive queuing for planning and serial performance. In M. A. Arbib (Ed.), *The Handbook of Brain Theory and Neural Networks* (2nd ed., pp. 241–44). Cambridge, MA: MIT Press.

Burgess, N., & Hitch, G. J. (1999). Memory for serial order: A network model of the phonological loop and its timing. *Psychological Review, 106*(3), 551–581.

Buxbaum, L. J., Schwartz, M. F., & Montgomery, M. W. (1998). Ideational apraxia and naturalistic action. *Cognitive Neuropsychology, 15*, 617–643.

Chernova, S., & Arkin, R. C. (2007). From deliberative to routine behaviors: A cognitively inspired action-selection mechanism for routine behavior capture. *Adaptive Behavior, 15*(2), 199–216.

Cohen, J. D., & Huston, T. A. (1994). Progress in the use of interactive models for understanding attention and performance. *Attention and Performance XV: Conscious and Nonconscious Information Processing, 15*, 453–476.

Collett, T. S. (1982). Do toads plan routes? A study of the detour behavior of *Bufo viridis*. *Journal of Comparative Physiology. A, Neuroethology, Sensory, Neural, and Behavioral Physiology, 146*(2), 261–271.

Coltheart, M. (2006). What has functional neuroimaging told us about the mind (so far)? *Cortex, 42*, 323–331.

Cooper, R. P. (2007). Tool use and related errors in ideational apraxia: The quantitative simulation of patient error profiles. *Cortex, 43*, 319–337.

Cooper, R. P., & Glasspool, D. W. (2001). Learning action affordances and action schemas. In R. M. French & J. Sougné (Eds.), *Connectionist Models of Learning, Development and Evolution* (pp. 133–142). London: Springer-Verlag.

Cooper, R. P., Ruh, N., & Mareschal, D. (2014). The goal circuit model: A hierarchical multi-route model of the acquisition and control of routine sequential action in humans. *Cognitive Science, 38*(2), 244–274.

Cooper, R. P., & Shallice, T. (2000). Contention scheduling and the control of routine activities. *Cognitive Neuropsychology, 17*, 297–338.

Cooper, R. P., & Shallice, T. (2006). Structured representations in the control of behavior cannot be so easily dismissed: A reply to Botvinick and Plaut (2006). *Psychological Review, 113*, 929–931.

Cooper, R. P., Schwartz, M. F., Shallice, T., & Yule, P. (2005). The simulation of action disorganisation in complex activities of daily living. *Cognitive Neuropsychology, 22*, 959–1004.

Cooper, R. P., Shallice, T., & Farringdon, J. (1995). Symbolic and continuous processes in the automatic selection of actions. In J. Hallam (Ed.), *Hybrid Problems, Hybrid Solutions* (pp. 27–37). Amsterdam: IOS Press.

De Renzi, E., & Lucchelli, F. (1988). Ideational apraxia. *Brain, 111*, 1173–1185.

Duncan, J. (1986). Disorganisation of behavior after frontal lobe damage. *Cognitive Neuropsychology, 3*(3), 271–290.

Etcoff, N. L., & Magee, J. J. (1992). Categorical perception of facial expressions. *Cognition, 44*(3), 227–240.

Evans, S. H. (1967). A brief statement of schema theory. *Psychonomic Science, 8*(2), 87–88.

Ewert, J. P. (1972). Zentralnervose analyse und verarbeitung visueller sinnesreize. *Naturwissenschaftliche Rundschau, 25*, 1–11.

Gilbert, S. J., & Shallice, T. (2002). Task switching: A PDP model. *Cognitive Psychology, 44*(3), 297–337.

Glasspool, D. W., & Houghton, G. (2005). Serial order and consonant-vowel structure in a graphemic output buffer model. *Brain and Language, 94*(3), 304–330.

Gonzalez Rothi, L. J., Ochipa, C., & Heilman, K. M. (1991). A cognitive neuropsychological model of limb praxis. *Cognitive Neuropsychology, 8*(6), 443–458.

Head, H., & Holmes, G. (1911). Sensory disturbances from cerebral lesions. *Brain, 34*(2–3), 102–254.

Hertz, J., Krogh, A., & Palmer, R. G. (1991). *Introduction to the Theory of Neural Computation.* Reading, MA: Addison-Wesley.

Houghton, G. (1990). The problem of serial order: A neural network model of sequence learning and recall. In R. Dale, C. Mellish, & M. Zoch (Eds.), *Current Research in Natural Language Generation* (pp. 287–319). London: Academic Press.

Humphreys, G. W., & Riddoch, J. M. (2003). From vision to action and action to vision: A convergent route approach to vision, action, and attention. *Psychology of Learning and Motivation: Advances in Research and Theory : Cognitive Vision, 42*, 225–264.

Iaccarino, L., Chieffi, S., & Iavarone, A. (2014). Utilization behavior: What is known and what has to be known? *Behavioural Neurology, 2014*(297128), 1–9.

Koechlin, E., Ody, K., & Kouneiher, F. (2003). The architecture of cognitive control in the human prefrontal cortex. *Science, 302*, 1181–1185.

Lara, R., Carmona, M., Daza, F., & Cruz, A. (1984). A global model of the neural mechanisms responsible for visuomotor coordination in toads. *Journal of Theoretical Biology, 110*, 587–618.

Lashley, K. S. (1951). The problem of serial order in behavior. In L. A. Jeffress (Ed.), *Cerebral Mechanisms in Behavior* (pp. 112–136). New York: Wiley.

Lhermitte, F. (1983). Utilization behavior and its relation to lesions of the frontal lobes. *Brain, 106*, 237–255.

Liberman, A. M., Harris, K. S., Hoffman, H. S., & Griffith, B. C. (1957). The discrimination of speech sounds within and across phoneme boundaries. *Journal of Experimental Psychology, 54*(5), 358–368.

Luria, A. R. (1966). *Higher Cortical Functions in Man.* London: Tavistock.

Lyons, M., & Robbins, T. W. (1975). The action of central nervous system stimulant drugs: A general theory concerning amphetamine effects. In W. B. Essman & L. Valzelli (Eds.), *Current Developments in Psychopharmacology* (Vol. 2, pp. 80–163). New York: Spectrum.

McClelland, J. L., & Rumelhart, D. E. (1981). An interactive activation model of context effects in letter perception: I. An account of basic findings. *Psychological Review, 88*, 375–407.

Norman, D. A. (1981). Categorization of action slips. *Psychological Review, 88*, 1–15.

Norman, D. A., & Shallice, T. (1980). *Attention to Action: Willed and Automatic Control of Behavior. Chip Report 99*. San Diego: University of California.

Piaget, J. (1952). *The Origins of Intelligence in Children* (M. Cook, Trans.). New York: International Universities Press. Original work published 1936.

Picton, T. W., Stuss, D. T., Shallice, T., Alexander, M. P., & Gillingham, S. (2006). Keeping time: Effects of focal frontal lesions. *Neuropsychologia, 44*(7), 1195–1209.

Reason, J. T. (1979). Actions not as planned: The price of automatization. In G. Underwood & R. Stevens (Eds.), *Aspects of Consciousness* (pp. 67–89). London: Academic Press.

Redgrave, P., Prescott, T. J., & Gurney, K. (1999). The basal ganglia: A vertebrate solution to the selection problem? *Neuroscience, 89*, 1009–1023.

Robbins, T. W. (1982). Stereotypies: Addictions or fragmented actions? *Bulletin of the British Psychological Society, 35*, 297–300.

Robbins, T. W. (1991). Cognitive deficits in schizophrenia and Parkinson's disease: Neural basis and the role of dopamine. In P. Willner & J. Scheel-Krüger (Eds.), *The Mesolimbic Dopamine System: From Motivation to Action* (pp. 497–528). New York: Wiley.

Rosenbaum, D. A., Cohen, R. G., Jax, S. A., Weiss, D. J., & van der Wel, R. (2007). The problem of serial order in behavior: Lashley's legacy. *Human Movement Science, 26*, 525–554.

Roy, D. (2005). Semiotic schemas: A framework for grounding language in action and perception. *Artificial Intelligence, 167*, 170–205.

Rumelhart, D. E., & McClelland, J. L. (1986). *Parallel Distributed Processing. Explorations in the Microstructure of Cognition: Vol. 1. Foundations*. Cambridge, MA: MIT Press.

Rumelhart, D. E., & Norman, D. A. (1982). Simulating a skilled typist: A study of cognitive-motor performance. *Cognitive Science, 6*, 1–36.

Rumiati, R. I., Weiss, P. H., Shallice, T., Ottoboni, G., Noth, J., Zilles, K., et al. (2004). Neural basis of pantomiming the use of visually presented objects. *NeuroImage, 21*, 1224–1231.

Rumiati, R. I., Zanini, S., Vorano, L., & Shallice, T. (2001). A form of ideational apraxia as a selective deficit of contention scheduling. *Cognitive Neuropsychology, 18*, 617–642.

Schmidt, R. A. (1975). A schema theory of discrete motor skill learning. *Psychological Review, 82*(4), 225–260.

Schwartz, M. F., Montgomery, M. W., Buxbaum, L. J., Lee, S. S., Carew, T. G., Coslett, H. B., et al. (1998). Naturalistic action impairment in closed head injury. *Neuropsychology, 12*, 13–28.

Schwartz, M. F., Reed, E. S., Montgomery, M. W., Palmer, C., & Mayer, N. H. (1991). The quantitative description of action disorganisation after brain damage: A case study. *Cognitive Neuropsychology, 8*, 381–414.

Shallice, T., & Burgess, P. W. (1996). The domain of supervisory processes and temporal organization of behavior. *Philosophical Transactions of the Royal Society of London. Series B, Biological Sciences, 351*, 1405–1412.

Shallice, T., Burgess, P. W., Schon, F., & Baxter, D. M. (1989). The origins of utilization behavior. *Brain, 112*, 1587–1598.

Shallice, T., & Cooper, R. P. (2011). *The Organisation of Mind*. Oxford: Oxford University Press.

Shallice, T., Stuss, D. T., Picton, T. W., Alexander, M. P., & Gillingham, S. (2008). Mapping task switching in frontal cortex through neuropsychological group studies. *Frontiers in Neuroscience, 2*(1), 79–85.

Stuss, D. T., Alexander, M. P., Shallice, T., Picton, T. W., Binns, M. A., Macdonald, R., et al. (2005). Multiple frontal systems controlling response speed. *Neuropsychologia, 43*(3), 396–417.

Werbos, P. J. (1990). Backpropagation through time: What it does and how to do it. *Proceedings of the IEEE, 78*, 1550–1560.

Wolpert, D. M., & Ghahramani, Z. (2000). Computational principles of movement neuroscience. *Nature Neuroscience, 3*, 1212–1217.

15 Synthetic Brain Imaging

JACQUELINE A. GRIEGO, CARLOS R. CORTES, RANSOM WINDER, AND MALLE A. TAGAMETS

Our present knowledge about the neural basis of human behavior and cognition relies largely on the data from human neuroimaging studies and animal brain neurophysiology studies. The two different fields of study differ primarily with respect to spatial and temporal resolution, in which brain imaging is far cruder in both spatial (centimeters compared to micrometers) and temporal resolution (seconds and minutes compared to milliseconds). The challenge, then, is to pursue an explanation of neuroimaging results in terms of the possible function of a brain region and how this relates to the overall brain function in a specific task, that is, causal modeling at the "blobology" level. An even greater challenge is to attempt to see how groups of neurons influence each other in local circuits, nuclei, and larger complexes where data on the specifics of synaptic functioning are crucial to the imaging results.

Functional neuroimaging methods such as *positron emission tomography* (PET), *functional magnetic resonance imaging* (fMRI), and *electrical* and *magnetic encephalography* (EEG and MEG, respectively) are widely used to map specific cognitive functions to human brain function. In some cases, the tasks used to explore human cognitive function have also been studied in nonhuman primates and other animals using single- and multi-unit electrode recording methods. These electrophysiological studies provide direct measures of neuronal response to cognitive tasks. In contrast, human imaging methods are indirect, reflecting changes in blood flow or oxygenation in PET and fMRI, respectively, or field potentials for large ensembles of neurons in EEG and MEG. Relating the single- and multi-unit neuronal activity to the observed imaging data is the domain of synthetic brain imaging, using computational models based on neurophysiological data to simulate neuroimaging signals such as *regional cerebral blood flow* (rCBF; PET) and *blood oxygen level dependent* (BOLD; fMRI) imaging, which will be discussed in this chapter. Synthetic imaging techniques are also used on EEG and MEG data (e.g., Babajani-Feremi, Soltanian-Zadeh, and Moran, 2008; Barrès et al., 2013).

This chapter is divided into four sections. The first, "Introduction to Neuroimaging Methods and Issues Specific to Computational Modeling of Human Imaging Data," should provide a basic understanding of what PET and fMRI are measuring and how synthetic neuroimaging extends imaging methods. The example models introduced in this section demonstrate how synthetic neuroimaging enhances the interpretation of imaging data in terms of underlying neuronal activity. This basic knowledge and associated implementation examples are extended to a discussion of other models in the second section, "Models Linking PET/fMRI Data to Neurophysiological Data: Animal Data and Neuronal Dynamics." This section gives a detailed look into a large-scale model. The third section, "Models Linking PET/fMRI Data to Neuropsychological Data: Behavioral and Cognitive States," then explores additional modeling considerations involved in associating imaging results to cognitive processes. The final section, "Synthetic Imaging of Disease, Connectivity, and Data-Driven Learning," provides an example that demonstrates an approach to examining disease processes with synthetic neuroimaging. This section introduces the use of data-driven learning to match model connection weights to experimental data. It then examines how relations between local and global circuits affect connectivity, and demonstrates that reduced local connectivity can explain data obtained from people with schizophrenia. For a cellular-level model that examines mechanisms that underlie brain dynamics in schizophrenia, see chapter 24.

15.1 Introduction to Neuroimaging Methods and Issues Specific to Computational Modeling of Human Imaging Data

Brain function is characterized by both local regional specialization and global interaction. Synthetic PET (Arbib, Bischoff, Fagg, and Grafton, 1995) and synthetic fMRI (Tagamets and Horwitz, 1998) were introduced in the 1990s to answer questions regarding the interpretation of imaging data at both local and global levels. These computational techniques enabled questions previously unanswerable by imaging data to be addressed, such as the following:

1. What are the relative roles of local and interregional connectivity in generating both neuronal activity and functional imaging data?
2. How do local and global circuits support cognitive functions?
3. How and where do modulations of a task take place?

We will develop a model for *working memory*, a form of short-term memory that allows items to be held in memory and manipulated on a timescale of seconds to tens of seconds (see Baddeley, 2003, for a review and chapter 11 for further modeling). An example of the use of working memory is when numbers have to be kept in mind in order to add them.

15.1.1 Basic Neuroscience (Neuroanatomy and Neurophysiology) Methods Basic neuroscience methods describe and explain structural and functional brain characteristics. The primary excitatory neurons in the cerebral cortex are *pyramidal neurons*. These send excitatory connections to other neurons and are considered to be the main carriers of information in the cortex. Another major class of neurons are called *interneurons*. These neurons inhibit other neurons and play a major role in tuning the responses of the pyramidal neurons. Dissection and staining methods have illuminated structural characteristics such as *cytoarchitecture* (how neurons and their connections are arranged in the brain) and gross anatomy across many nonhuman and human animals. Circuitry may also be described using dissection, often paired with *anterograde* and *retrograde* tracers, which usually are injected into one location and travel along axons, thus marking the starting and ending locations of a neuronal pathway. Anterograde tracers are labeled substances that flow from the cell body outward, and thus are used to follow axons to identify the locations that they project to. Retrograde tracers flow in the opposite direction, and identify the sources of projections. Together, these methods help to elucidate how neurons are connected to form the communication networks that enable the brain to carry out computations. These networks are thought to be arranged in local clusters that are connected to each other with long-range connections. The local clusters are thought to carry out specialized functions, and their communication with each other is thought to be crucial to proper brain function. For instance, a local cluster that performs various aspects of visual object processing, for example, color and shape, is connected by long-range connections to frontal cortex, where the relevance of the stimulus and actions to be taken are processed in a cluster of neural populations dedicated to these operations.

In the *lesion method*, the specific function is inferred from damage to, or excision of, a given cluster or structure. In order to demonstrate the relationship between structure and function, damage to a specific part of the brain should produce a measurable functional deficit, and damage to other structures should not produce the same deficit. Lesions, or other structural deficits, can be naturally occurring through disease processes, engineered through genetic manipulation, or caused by chemical or surgical methods. An important caveat is that loss of a function on excision of a region R does not imply that R by itself supports the missing function (though this is true in some cases) but rather that other regions require R to jointly support it.

Functional properties are also investigated through electrophysiological recording methods that vary from a single unit to multiple units in extent. In cell recordings, an electrode may be placed through the cell membrane, pressure fit on the cell membrane, or placed in the extracellular space near the cell. Depending on the type of electrode and method, recordings can be made from one to many neurons. For a detailed look at basic neuroscience methods, see Kandel, Schwartz, and Jessel (2012).

15.1.2 Brain Imaging Methods Brain imaging methods are the primary methods for identifying function in the living human brain. There are several imaging methods that can be used in humans. They vary on characteristics such as what they measure, their spatial and temporal resolutions, settings where they are commonly used, and how invasive the method is. In EEG, electrodes are attached to the scalp to measure electrophysiological activity of primarily pyramidal neurons and synchronized brain currents, such as *gamma* and *alpha waves*, underneath the skull. Recent evidence suggests that synchronized oscillations of firing patterns among groups of neurons play an important role in cognition. Alpha and gamma waves are two of the key frequencies in this phenomenon (see chapter 5). EEG detects radial currents corresponding to both *gyral* and *sulcal* activations. Gyri are the ridges on the surface of the brain, and sulci are the fissures between the gyri. *MEG* also measures brain activity accompanied by magnetic fields using a *superconducting quantum interference device* (SQUID). The SQUID detects weak neuromagnetic fields tangential to the scalp. *Computed axial tomography* (CAT) uses x-rays projected around the head to build a 3-D structural brain image. *Magnetic resonance spectroscopy* (MRS) uses MRI technology to identify the spectra of peaks representing the structure and concentration of chemicals in the living brain. For example, one indicator of neuronal integrity, N-acetyl aspartate, is measured with MRS. *Diffusion tensor imaging* (DTI) uses MRI technology (diffusion weighted magnetic resonance scan) to assess both white matter connectivity (*tractography*) and integrity (*fractional anisotropy*, or FA) by measuring the direction and motion in water molecules.

In MRI, the head is placed in a strong magnetic field, where the rotations of hydrogen protons align with the field. Then a radio frequency pulse is sent through the field, which knocks the protons out of alignment. When the pulse stops, the protons relax and realign, or *precess*, to the original magnetic field. Signal-receiving coils in the MRI scanner measure the electrical signal at a particular point in time while the protons are returning to their original state. The rate at which they precess depends on their chemical composition, and particularly on the proportion of water in the tissue. Different structures have different compositions and can be distinguished from each other by measuring how far the precession has advanced at the time the measurement is taken. Structural features, for example, gray matter, white matter, and bone, are measured with MRI, and responses across time related to cognitive tasks are measured with fMRI, which is described in more detail in the next section.

15.1.3 PET AND FMRI Thus far, synthetic neuroimaging has primarily focused on simulating PET and fMRI methods. PET and fMRI share many characteristics. Both techniques are used to indirectly measure neural activity associated with cognitive states, and both have spatial and temporal scales that are considerably greater than single-neuron activity. By using appropriate experimental design, both methods can detect changes in neural activity with a temporal resolution of seconds to minutes. In both methods, the signal obtained is allocated into a two- or three-dimensional image that covers some area of the brain. The image is composed of three-dimensional "cubes," referred to as *voxels*, that reflect the signal in the corresponding cubes of brain. The voxels are on the order of several millimeters in each dimension.

Both PET and fMRI are thought to reflect synaptic activity more than neuron spiking activity (Jueptner and Weiller, 1995; Lauritzen, 2001; Lauritzen et al., 2012). However, this means that both excitatory and inhibitory synaptic activity may increase PET and BOLD activation, which can make the PET or fMRI signal difficult to interpret in terms of underlying neuronal activity. Computational models provide a way to more precisely describe how combinations of excitation and inhibition influence signal changes, as illustrated by the example at the end of this section.

PET is the most invasive of neuroimaging methods since radioisotopes have to be injected into the individual. PET measures the energy of gamma rays resulting from the collision of positrons and electrons. A positron emitting isotope is produced in a cyclotron and then attached to a compound that is normally present in the brain. The radioactive labeled tracer is injected and absorbed into the biologically active molecules. Depending on the tracer type, different types of metabolic activity can be measured, such as energy demands and neurotransmitter concentrations. For example, fluorodeoxyglucose ([18F]-labeled 2-deoxyglucose) is used to study *glucose* metabolism.

Before the development of fMRI, PET was commonly used to measure blood flow changes associated with cognitive tasks or states, but in recent years, PET usage primarily involves other types of measures, such as neurotransmitter concentration. However, one advantage that PET still holds over fMRI for cognition research is that measurements with PET are absolute, meaning that there is no need for comparing two different conditions. fMRI measurements typically are relative to a baseline, so that two different cognitive states (e.g., rest and an active task) need to be measured, and one must be compared against the other. However, PET studies typically do use comparative activity values to compute task-specific modulation of local activations (Arbib et al., 1995) or compare different groups of individuals (Fletcher et al., 1998). The spatial scale of PET is typically in centimeters, and the temporal scale is in minutes. Although more recent developments have improved both spatial and temporal resolution somewhat, PET still cannot attain the resolution that can be achieved in fMRI.

fMRI measures blood oxygen levels by detecting the energy emitted from precessing hydrogen atoms. Protons in different types of tissues and chemicals return at different rates, so a measurement at a specific time yields different intensities for different tissues or different states of fluids or tissues. The amount of *oxy-* versus *deoxyhemoglobin* in the blood affects the precession rate, and thus the signal intensity, and is the basis for the fMRI signal. The balance between oxy- and deoxyhemoglobin changes with changing neural activity, and thus deoxygenated hemoglobin acts as an *endogenous paramagnetic contrast agent*. In other words, the difference, or contrast, between signals from oxygenated and deoxygenated blood reflects differences that are due to changes in neural activity near the location where the signals are obtained. Inferences regarding activations in fMRI depend on relative measures between conditions, such a control and task comparison, and the effect size is commonly reported as the percent signal change, which is typically on the order of 1–3% of the total signal. The spatial scale is in millimeters; the temporal scale is typically about 2 seconds.

15.1.4 Relationship of Activations to Physiological Measures It is possible to relate neuronal activity to imaging data because the measures of blood flow and oxygenation in PET and fMRI are related to neuronal activity. Signals from PET are correlated with rCBF, which, in turn, is correlated with local synaptic activity. Synaptic activity within a location reflects the input from *afferent*, or incoming, connections plus the activity of synapses that are local to the area. The fMRI BOLD signal is correlated with *local field potentials* (LFPs; e.g., Logothetis et al., 2001), which are the summed synaptic currents in a local area of extracellular space. As in PET, this is the summation of afferent and local activity in the area. However, there is not an invariant relationship between the BOLD signal and LFPs. Relationships between spike rates, LFP, and BOLD may dissociate based on such factors as coding properties, anatomical layout of a brain region, dynamic properties of local circuits, or context (for reviews and discussion, see Ekstrom, 2010, and Lauritzen et al., 2012).

Both changes in blood flow and changes in metabolism contribute to the BOLD signal. Oxidative metabolism is thought to be needed to restore ionic concentrations following neural activity, with the greatest effects apparently occurring in the vicinity of the synapses (Attwell and Laughlin, 2001). Although neuronal spiking per se is thought to have little influence on the fMRI signal (for a review, see Ekstrom, 2010), its relationship to synaptic activity does allow interpretation of the signal in terms of locally active neurons, as is frequently done in fMRI studies.

Most studies of excitatory neuronal effects have suggested that there is a direct relationship between spiking activity and oxidative metabolism. In autoradiography studies, stimulation of specific pathways has been shown to cause increases in local glucose metabolism at sites to which the stimulated location projects, but not where the stimulated spiking activity occurs (Sokoloff, 1993). The increases were proportional to the spiking rate of the stimulated neurons. Studies suggest that much of the metabolic activity measured in imaging data can be attributed to *astrocytes*, which mediate metabolic demands of the transmitter reuptake processes at *glutamatergic* synapses (Magistretti and Pellerin, 1999). These are the synapses that emanate from the excitatory pyramidal neurons. It is thought that a similar mechanism may take place at *GABA*ergic synapses, which are the primary inhibitory synapses in the brain (Jueptner and Weiller, 1995; Magistretti and Pellerin, 1999). The consequence of the energy requirements of both types of synapses is that both excitatory and inhibitory synaptic activities have the potential to raise measures of rCBF or BOLD signal. The model described later in this section examines the conditions under which inhibition might be expected to cause increases in signal.

15.1.5 Imaging Experimental Designs and Analysis In PET experiments, the entire period of a scan usually involves only a single experimental condition although there may be several scans involving a different condition in each. These could be either a resting condition, during which no task demands are made of the participant, or an active task that requires participation, for example, viewing, hearing, or responding to stimuli. In fMRI, an image is acquired every few seconds, and a scan typically lasts for at least several minutes. The advantage over PET is that brain activity can be recorded continuously as task demands change. During this time, different task conditions are presented since fMRI requires comparisons of conditions within a scan. There are two basic design types for fMRI imaging experiments: block design and event-related designs. In a *block design* the same task type is repeated across a given time period, typically 15 to 45 seconds, and data are averaged within blocks. In an *event-related design*, multiple task or context manipulations are interleaved with just a few seconds between conditions. This allows individual events to be analyzed, usually against a *baseline*. A baseline is a period of time during a scan during which there are no task demands, and this can be either the time between individual task events or a longer period of time during which participants are asked to simply lie in the scanner and not do anything.

There are two common approaches to analyzing functional imaging data. The most frequently used method aims to identify brain areas that are activated in different experimental conditions. Both PET and fMRI data are commonly analyzed using *general linear models* (see Poline and Brett, 2012, for a review). These methods are extensions of regression techniques such as multiple regression, analysis of variance, *t* tests, and so forth and are most commonly used to contrast different conditions in order to localize brain regions that are more active in one condition than the other. As noted earlier, PET measures activation directly, without the need for control conditions, and any temporal information is collapsed across the session, which typically lasts for a few minutes. This data can also be analyzed using one-sample *t* tests. For example, Arbib et al. (1995) modeled a PET study that separately examined a simple saccade task and a saccade task requiring memory. In order to do the same comparison in fMRI, either both conditions would have to be in the same scan or a statistical contrast would have to be created between each of these two conditions and a baseline condition.

The second main approach to analysis is to examine how regions work together during different conditions. One such method is aimed at determining how the activations of two or more regions covary in the context of a particular task without regard to the direction of influence. This is known as *functional connectivity*. Functional connectivity assessments can be made with any technique for measuring covariance: simple correlations, semi-partial correlations, partial least squares and *seed voxel* analyses, among others (Büchel and Friston, 1998; Büchel and Friston, 2000; Bokde et al., 2001). Seed voxel analyses involve choosing a particular voxel, the "seed," and computing how this voxel covaries with all other voxels in the brain over time. This method identifies all regions that covary with the chosen seed voxel. Independent component analysis (Bartels and Zeki, 2005; Calhoun et al., 2001) can be used to extract a set of orthogonal components where each reflects a group of related signal sources across time. Although functional connectivity has been applied to PET data for around two decades, connectivity studies based on fMRI are more recent.

In contrast to functional connectivity, *effective connectivity* represents the functional strength of a particular connection from one region to another (Friston, 1994). Functional connectivity, as described above, does not necessarily arise from a direct connection between the regions. For example, correlations between two regions may arise from other influences, such as a common driving force. These are indirect influences in a statistical sense. Effective connectivity analysis methods identify strengths of specific connections while at the same time allowing for these influences. One disadvantage of effective connectivity methods as compared to functional connectivity is that specific *regions of interest* and their physical connections need to be specified before the analysis can take place. Once these are specified, these methods estimate connection strength for each of the physical connections.

The most commonly used effective connectivity methods are structural equation modeling (SEM; McIntosh and Gonzalez-Lima, 1994; McIntosh, 2012) and dynamic causal modeling (DCM, Friston, Harrison, and Penny, 2003). In SEM, one must define a set of regions and connections among them. The algorithm then attempts to find interregional connection weights for the model that best fit the covariance structure among the regions in the experimental data. Most implementations of SEM assume a linear relationship in the influence of regions on one another. DCM, on the other hand, does not assume a linear relationship, and in contrast to SEM, a DCM model can account for both interregional connection weights and modulation of these weights by external influences, such as changes in tasks. Both of these methods are widely used for estimating effective connectivity from PET and fMRI data. Unlike anatomical connectivity, both effective and functional connectivity are *task dependent*. For example inhibition from another region may block effective connectivity between two regions in one task yet be absent, thus allowing this effective connectivity, in another task. Moreover, in both SEM and DCM, the nodes in the models are black boxes with no internal structure. In essence, they represent averaged responses of specific brain regions.

Although the statistical analysis of imaging data yields quantified measures of activity, the results are in general descriptive. There is no information regarding how the underlying neuronal processes produced the activations. This makes imaging data interpretations problematic in several senses. As mentioned previously, there are differences in temporal and spatial scales between imaging and the physiological evidence that provides information regarding how neuronal processes produced the activations. The same regions can be activated in multiple tasks but may be performing different functions and be connected effectively/functionally to different networks in different tasks. It is also not possible to interpret the effects of disease states (e.g., too much dopamine [schizophrenia] or too little [Parkinson's disease]) on functional activations and connectivity. Furthermore, the significance of specific differences in activation or connectivity found between tasks or populations is difficult to interpret in a useful manner, such as with respect to its clinical utility. It is the goal of synthetic brain imaging to add models that incorporate detailed information of the dynamic interactions within each brain region. These models should include neuronal dynamics that support the observed task, and at the same time explain the PET or fMRI data. In some sense, then, it introduces "really causal" dynamic modeling to the study of brain imaging.

15.1.6 Basic Considerations for Modeling fMRI and PET at the Neuronal Level There are some basic issues to consider when modeling neuroimaging data involving the quantification of aspects that are not necessarily needed in other types of models. Both neuronal firing and synaptic activity need to be modeled, and a decision must be made how to computationally relate neuronal firing, synaptic activity, and imaging data. The neuronal firing models underlying aspects of neuronal activity, such as how stimuli are represented and how neurons implement cognitive operations such as maintaining and manipulating these representations. This is required in all models of neuronal activity, whether or not imaging data is to be modeled. In order to model imaging data, the synaptic activity must drive the neurons in these operations and at the same time generate values that can be related to the imaging signals. Proportions of different types of synaptic connections affect the total population synaptic activity, and this must be implemented in a manner that reflects the geometry and quantitatively matches the proportions of different types of synapses that are found in the brain. For a current discussion regarding various approaches to modeling synaptic activity, see Bonaiuto and Arbib (2014). Ultimately, the modeler has to decide the level of detail of the mechanisms underlying the neuronal electrical activity (Deco, Rolls, and Horwitz, 2004), its metabolic consumption mechanisms (Aubert and Costalat, 2002; Costalat et al., 2012; Sotero and Trujillo-Barreto, 2007), its coupling with the vascular response (Sotero and Trujillo-Barreto, 2007), or a combination of all of these (for a review, see Buxton, 2012).

For coupling to the vascular response in fMRI, the relationship of the *hemodynamic response* to neuronal activity has to be considered (see figure 15.1), as the vascular response to changes in neural activity is often delayed and dispersed, with a peak response at around 4–6 seconds (Friston, Jezzard and Turner, 1994). The hemodynamic response is the change in blood flow (in PET) or blood oxygenation (in fMRI) in response to local demands by neuronal activity. Both of these processes take considerably more time than the underlying neuronal activity that initiates the response. Conventional fMRI analyses account for this hemodynamic coupling by convolving a stimulus function (encoding evoked synaptic activity) with a hemodynamic response function (HRF), shown in figure 15.1. These linear convolution models implicitly account for the physiological blurring of neuronal responses that is encountered when looking at hemodynamic brain signals. Models of fMRI that simulate a time resolution on the order of seconds need to account for this by convolving the modeled neuronal output with an HRF. This is particularly important if correlations between regions are to be modeled. Models of longer duration signals, such as in a block fMRI design, do not necessarily require this, as the signal is averaged over time and the hemodynamic signal presumably reaches a steady state.

15.1.7 A Model of Factors Affecting PET and fMRI Signals: Single-Region Example

Tagamets and Horwitz (2001) used a single-region model to demonstrate how the nonlinear properties of local circuits and context may influence imaging data. The goal was to answer this question: Does neuronal inhibition, and the concomitant reduction in neuronal spiking activity, result in decreased PET and fMRI signals? As the metabolic demands of inhibition require energy, inhibitory synaptic input might also be expected to increase blood flow. Empirical studies show that stimulation of excitatory and inhibitory pathways in the rat cerebellum both resulted in increased CBF; however, there was decreased Purkinje cell spiking with inhibitory pathway stimulation (Mathiesen et al., 1998). Two human PET studies of the GABA agonist THIP have produced contradictory results. Roland and Friberg (1988) showed a reduction of metabolism in most cortical regions following the administration of THIP whereas Peyron et al. (1994) found an increase. Synthetic imaging can be used to examine factors that might underlie these results.

In the model of Tagamets and Horwitz (2001), a brain region is treated as a unit of functional imaging data with a resolution of about 1 cm^2. Each region is made up of a 9×9 array of basic units (see figure 15.1C and 15.2B), which represents a cortical patch ~0.5–1 cm in diameter. In the primate cerebral cortex there are ~100,000–150,000 neurons in 1 mm^2 (Carlo and Stevens, 2013) although estimates vary. Aguirre (2012) estimates 630,000 neurons in a 3-mm^3 typical functional voxel. Each of the basic units shown in figure 15.2A roughly represents a local circuit or cortical column, with a simple two-element (excitatory–inhibitory) unit with local connection strengths close to the proportions suggested by anatomical studies (Douglas et al., 1995). The excitatory unit represents the assembly of pyramidal neurons in the column while the inhibitory unit represents the population of inhibitory interneurons. This basic circuit quantitatively reproduces the proportions of excitatory and inhibitory synapses, derived from empirical data in nonhuman cortex (see Tagamets and Horwitz, 1998, and Douglas et al., 1995). Most connections in the cerebral cortex are excitatory, and most processing takes place in local circuits, as most of synapses into a neuron are from the surrounding area (Douglas et al., 1995). A consequence of these properties is that these local circuits have a high degree of excitatory recurrence. Afferents from other regions

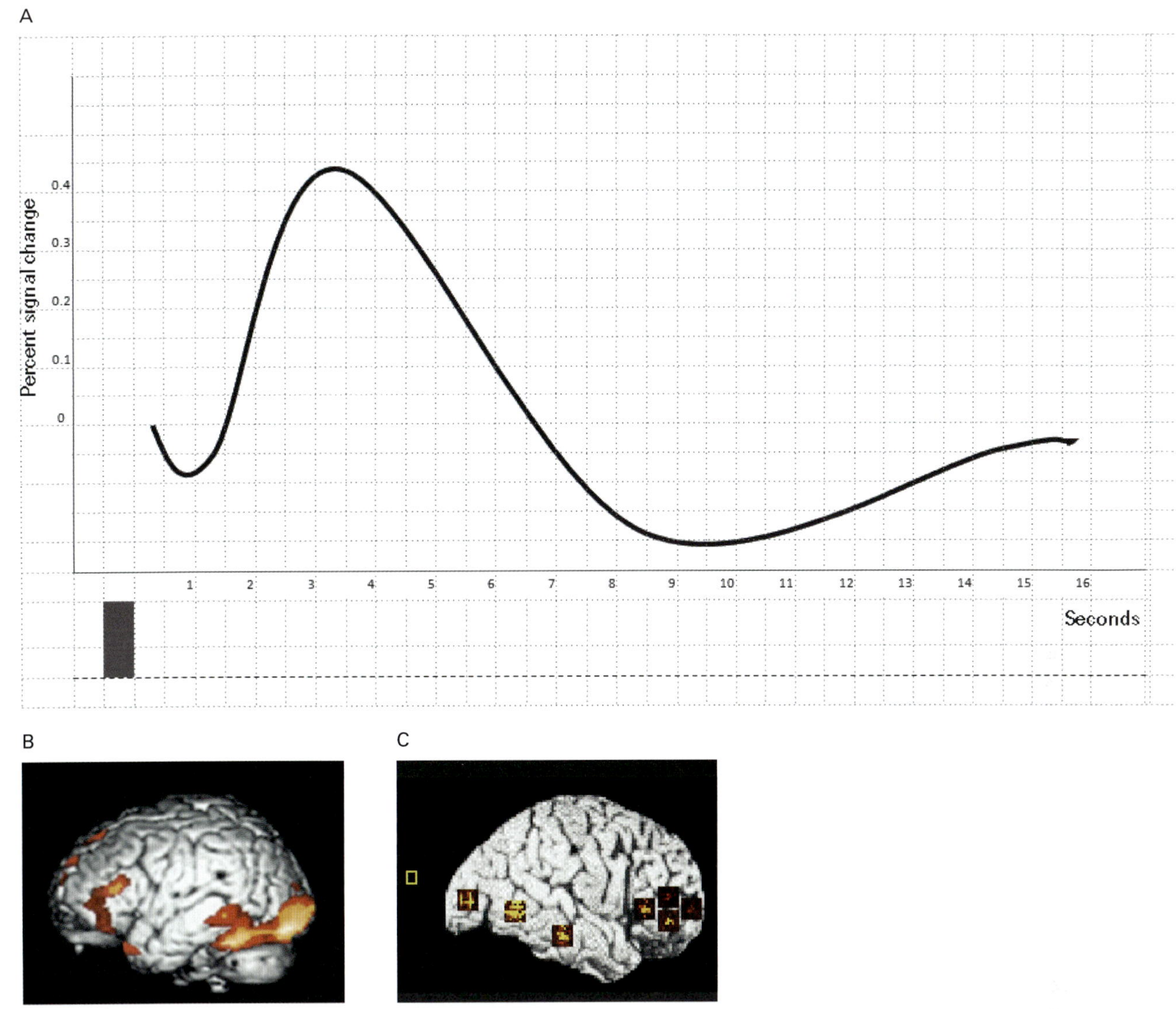

Figure 15.1 (A) Hemodynamic response function after a short neuronal event, shown as the black bar beneath the graph. The *x*-axis shows time in seconds; the *y*-axis shows the hemodynamic response as measured by functional magnetic resonance imaging (fMRI). This demonstrates that after a momentary neural event, it takes about 6 seconds for a peak response to occur in the fMRI signal. The signal shows an initial *undershoot,* the result of increased oxygen consumption, which changes the ratio of deoxygenated and oxygenated blood. The increased oxygen demand results in an inflow of oxygenated blood. After the response peak, the signal returns to prestimulus levels, often with a poststimulus *undershoot.* This undershoot reflects an increase in blood volume, resulting in an increased ratio of blood to oxygen, thus, blood is relatively deoxygenated. (B) An example of a whole-brain map of brain activation during a one-back matching task. Maps of whole-brain activations such as this are created by regressing all the voxels in the brain against a series of hemodynamic response functions that are aligned in time to the onsets of stimuli of interest. (C) Modeled brain activity using the large-scale model described in section 15.2 of this chapter. The model figure shows activity that corresponds to spiking activity of populations of neurons, the output of which is convolved with the hemodynamic response function to generate synthetic imaging data that matches the experimental data shown in B.

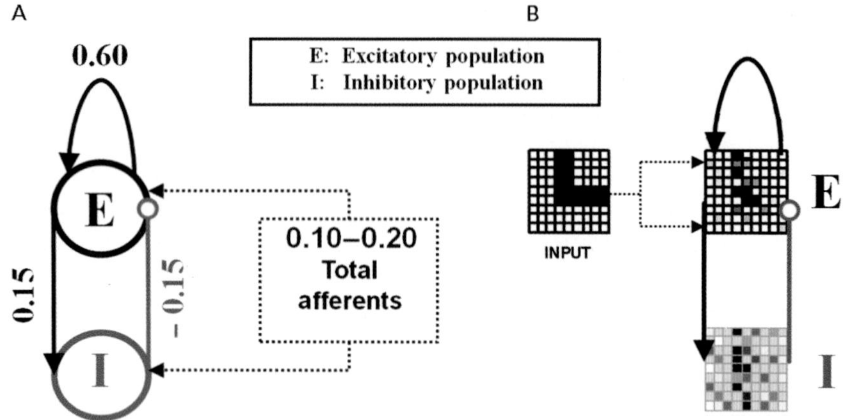

FIGURE 15.2 (A) The basic unit of the model. E represents the excitatory population and I the inhibitory population in a local assembly such as a cortical column. Local synaptic activity is dominated by the local excitation and inhibition, while afferents account for the smallest proportion, as indicated by the synaptic weights shown. (B) A cortical area is modeled by one or more 9×9 sets of basic units. The excitatory population is shown in bold lines above the inhibitory group, shown in lighter lines. Individual units in the excitatory and inhibitory populations within a group are connected as shown in A of this figure. From Tagamets and Horwitz (2000).

account for a relatively small proportion of the synapses. These characteristics mean a small afferent push into a local circuit is amplified within the circuit; in other words, there is a nonlinearly increasing response to linear input.

The model uses a sigmoid activation rule (equations 15.1 and 15.2) with different parameters (thresholds, τ, and steepness of the sigmoid, K) to model the electrical activity of the excitatory and inhibitory populations of neurons:

$$\frac{dE_i(t)}{dt} = \Delta\left[\frac{1}{1+e^{-K_E(w_{EE}E_i(t)-w_{IE}I_i(t)+in_{iE}(t)-\theta_E+N(t))}}\right] - \delta E_i(t), \tag{15.1}$$

$$\frac{dI_i(t)}{dt} = \Delta\left[\frac{1}{1+e^{-K_I(w_{EI}E_i(t)+in_{iI}(t)-\theta_I+N(t))}}\right] - \delta I_i(t). \tag{15.2}$$

$E_i(t)$ and $I_i(t)$ represent the electrical activations of the ith excitatory and inhibitory elements at time t, respectively. K_E and K_I are the gains, or steepness, of the sigmoid functions for excitatory and inhibitory units, respectively, θ_E and θ_I are the input thresholds of excitatory and inhibitory units, Δ is the rate of change, δ is the decay rate, and N (t) is an added noise term. The parameters Δ and δ determine the time resolution during the simulations.

As in any model, parameters must be chosen given the modeling goals. In this case, parameter values were chosen to achieve the desired dynamics with a single basic circuit (see figure 15.2 and equations 15.1 and 15.2). w_{EE}, w_{IE}, and w_{EI} are the weights within a unit: excitatory-to-excitatory (value = 0.6), inhibitory-to-excitatory (value = –0.15), and excitatory-to-inhibitory (value = 0.15), respectively. Values $in_{iE}(t)$ and $in_{iI}(t)$ are the total inputs coming from other areas into the excitatory and inhibitory units at time t:

$$\begin{aligned} in_{iE}(t) &= \Sigma_j w^E{}_{ji} E_j(t) + \Sigma_j w^I{}_{ji} I_j(t) \text{ and} \\ in_{iI}(t) &= \Sigma_j w^E{}_{ji} E_j(t) + \Sigma_j w^I{}_{ji} I_j(t), \end{aligned} \tag{15.3}$$

where w^E_{ji} and w^I_{ji} are weights coming from excitatory/inhibitory unit j in another area into the ith excitatory and inhibitory units, respectively. Electrical activations in the model range between 0 and 1 and can be interpreted as reflecting percentage of active units within a local population.

The excitatory (τ_E) and inhibitory (τ_I) thresholds are set such that if both excitatory (E) and inhibitory (I) elements are receiving half their possible inputs (0.5), the basic circuit will have an equilibrium state in which each element is 0.5, which is half the possible maximum value of 1. Manipulating the slope of the sigmoid activation functions K_E and K_I roughly corresponds to changing the responsiveness, or gain, of E and I elements in the circuit. These are chosen so that the equilibrium point is stable, but near the bifurcation point. In a system of differential equations, a bifurcation point is a point in the solution space where the qualitative behavior of the system suddenly changes. In a bistable system, the closer one is to a bifurcation point, the smaller the input needed to switch the system to one of the two attractors. As a concrete example, consider a book lying partly over the edge of a table. Two possible attractor states are that it is either on the table or on the floor, and it will remain in each of these states until some kind of force is applied to it. If the book is balanced so that exactly half is over the edge of the table, it is at a bifurcation point, and only an infinitesimally small push is needed to tip it onto the floor. If, on the other hand, most of the book is on the table, a larger push is required to tip it off the table. A smaller push will tip it toward the floor, but it will fall back onto the table. The system described by equations 15.1 and 15.2 is driven by the inputs in the argument of the exponential function, and the goal is to model a bistable system in which a small input can drive the system to either 0 or 1. The rationale for this choice is that neuronal populations that are not being driven by outside forces should be highly responsive to incoming inputs. In contrast, both high and low states should be most stable, reflecting an ongoing response to a stimulus that needs to be processed.

The electrical activity (i.e., spiking) in each E and I element is determined by a sigmoidal function of the summed synaptic inputs, which corresponds to the average spiking rates from single-cell recordings. The rCBF (for PET) or BOLD (for fMRI) value is computed within an area as the total sum of the absolute values of all synaptically weighted inputs to a given unit i, both from within the unit i itself and from all other sources at time t_i. This involves all the synaptic inputs from the E and I elements within a unit itself, and all the summed activities coming from all the units that connect to unit *i*. Ignoring the hemodynamic function in this initial model, the absolute synaptic activity is integrated over the time course for the study, for example, approximately 1 min for rCBF–PET and about 1 to 2 seconds for BOLD–fMRI. Although this sum can thus model either rCBF or BOLD, for simplicity we refer to it here as rCBF.

$$\mathrm{rCBF} = \sum_{t,i} \mathrm{IN}_i(t), \qquad (15.4)$$

where $\mathrm{IN}_i(\mathrm{t})$ is the total sum of absolute values of the inputs to both the E and I elements of unit i, both from within the unit and from all sources, at time t and is computed as follows:

$$IN_i(t) = \mathrm{w_{EE}E_i(t)} + \mathrm{w_{EI}E_i(t)} + |\mathrm{w_{IE}I_i(t)}| + \sum_{\mathrm{k,i}} \mathrm{w_{k,i}} E_k(t) + \sum_{m,i} |w_{m,i} I_m(t)| \qquad (15.5)$$

The first three terms on the right-hand side of equation 15.5 are the synaptic weights from within the unit i itself: $\mathrm{w_{EE}}$ is the weight of the recurrent excitatory connection, $\mathrm{w_{EI}}$ is the weight of the connection from E to I, and $\mathrm{w_{IE}}$ is the weight from I to E. The last two terms in equation 15.5 are the summed absolute values of activities coming from all other units which are connected to unit i. This is similar to the method used in Arbib et al. (1995) for computing simulated blood flow. For comparison to experimental PET data, both the simulated data and the PET data are normalized to a common reference area, for example, area V1, for one of the tasks. This allows us to examine both the relative amount of blood flow in different areas in various tasks in addition to computing the percent changes between tasks.

Another 9 × 9 array acts as an input layer. There are weighted connections from the input array to the cortical patch array. During the simulations, a subset of units in the input layer is clamped to a fixed value in order to simulate an input pattern. The simulations in this model evaluated three factors: (1) type of afferent, (2) context or task, and (3) local cortical connectivity. The first simulations examined the region's spiking frequency and simulated PET response to four different types of input, as shown in figure 15.3 and described below.

In any given context, afferent excitation of the E units can occur in one of two ways: directly, by synapses directly on the E units (figure 15.3, case A), or indirectly, by inhibition of the local inhibitory (I) units (case B). Likewise, inhibition can occur either directly, by inhibitory synapses on the E units (case C), or indirectly, by excitation of the I units (case D). These four cases are illustrated by the oval afferents to the right of the model regions in figure 15.3. Context is provided from the inputs shown on the left. An "active" context is simulated by presenting shaped patterns with high activity to the input area. A resting or passive context is simulated by setting all units in the input region to a uniformly low value. The latter reflects a case in which area A does not receive inputs from any other source. In order to examine afferent effects in these different contexts, the four types of afferents depicted in figure 15.3 were simulated during each of two task conditions: (1) in a low baseline (resting) condition and (2) with a high level of input coming from the input region. Simulations for each case are run for 200 iterations, with each iteration corresponding to approximately 5 ms of time.

Figure 15.4 shows the results of these simulations. Both simulated rCBF (solid lines) and the mean spiking levels in the region (dashed lines) are depicted in the

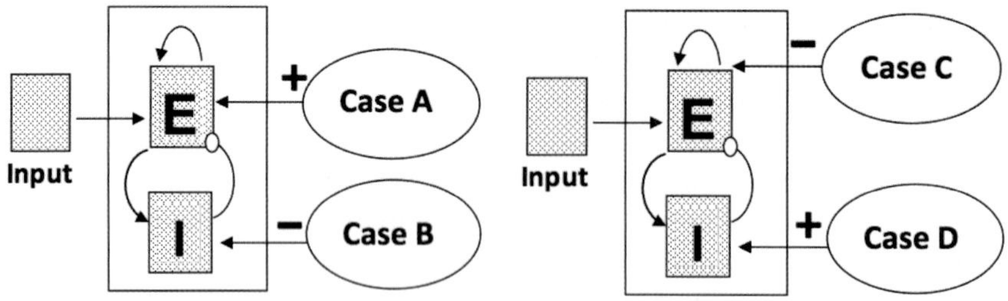

FIGURE 15.3 Four different types of afferents to a region. An input area (shown to the left of each local region) provides context, such as that experienced during sensory stimulation, for example. Connections from the input area on the left of each figure are topographically arranged and are all excitatory. Different types of excitatory (E) and inhibitory (I) afferents are shown on the right in ovals. These afferent connections are weak and synapse onto all units equally. (A) Two types of excitatory effects: Excitation by excitatory-to-excitatory connections (Case A) and indirect excitation of the excitatory units by inhibitory-to-inhibitory afferents (Case B). (B) Two types of inhibitory effects: Direct inhibition by inhibitory-to-excitatory connections (Case C) and indirect inhibition by excitatory-to-inhibitory afferents (Case D). From Tagamets and Horwitz (2001).

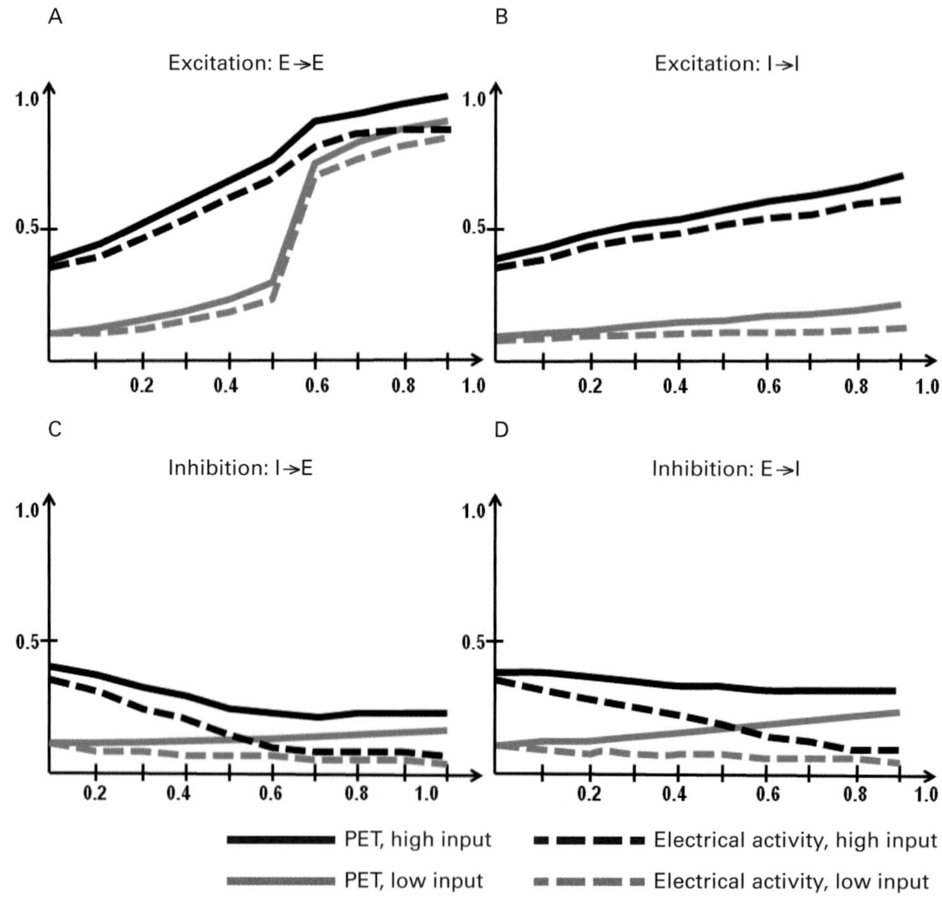

FIGURE 15.4 Electrical activity (dotted lines) and simulated cerebral blood flow (CBF) activity (solid lines) that result from the different afferent input types shown in figure 15.3. For each case, the amount of afferent excitation or inhibition (ovals to the right of the networks in figure 15.3) was systematically increased, indicated along the *x*-axis. Mean electrical activity and mean CBF activity was computed for the combined excitatory–inhibitory grouping that represents the local cortical region. (A, B) Result of excitation, corresponding to Case A and Case B, respectively, in figure 15.3A. (C, D) Inhibitory effects, corresponding to figure 15.3B, Case C and Case D, respectively. The low input condition, simulating rest, is shown in gray curves that begin at 0.1 on the vertical axis. The high-input condition, which simulates active inputs from other sources, is shown in the black curves that start at about 0.4 on the vertical axis. From Tagamets and Horwitz (2001).

graphs. The upper pairs of curves, beginning at $y = 0.4$ in each graph and shown in black, are the high-context (active) condition, and the lower two curves, starting at $y = 0.2$ and shown in gray, depict the baseline context, when there is no other activity entering the region. Along each curve, the context input is held constant, while the afferents from the ovals in figure 15.4 are systematically increased. The x-axis represents the amount of afferent input. Figure 15.4A and 15.4B show the two cases in which the net effect is excitation of the E units and correspond to the cases shown in figure 15.3, case A and case B, respectively. In both cases, the electrical activity and the rCBF measures both rise more or less concurrently with rising afferent modulation. This is consistent with the excitatory effects observed in animal studies. Figure 15.4C and 15.4D depict the two inhibitory cases. In the task context (black curves in figure 15.4C and 15.4D), when there is a high level of activity coming from other regions, both spiking and rCBF decline with increasing afferent inhibition, although there tends to be a relatively greater drop in spiking than in the rCBF as the inhibitory effect is increased. In the baseline conditions (gray curves in figure 15.4C and 15.4D), there is a dissociation between the modeled spiking activity, which drops, and the rCBF, which rises. This effect is particularly notable in the indirect inhibitory case (figure 15.4D), in which inhibition is achieved by excitatory afferent inputs to the I units. The latter mechanism corresponds to the indirect inhibitory pathway experimental condition in the study of Mathiesen et al. (1998), in which reduction of spiking in the Purkinje cells results from an indirect pathway through inhibitory neurons. The simulation results suggest that when a region does not receive extensive excitation from other sources, and given the local circuit properties shown in figure 15.2A, synaptic activity from afferent inhibition can exceed total reduction of local synaptic activity. The result is a rise in rCBF, even though spiking is reduced in local neurons.

The local circuit properties may also play a role in these effects. The modeling suggests that, given a moderate to high recurrent excitatory local connectivity, inhibition is more likely to result in increases of rCBF/metabolism in a context where the region does not receive excitation from other sources. The amount of local excitatory recurrence determines whether metabolism/rCBF reflects mainly local or mainly afferent synaptic activity. Anatomical studies suggest that recurrence in local cortical circuits is in the range in which our model suggests that inhibition may cause opposite effects between high- and low-activity contexts. Whereas the exact amount of recurrence in different cortical areas is not known, our model nevertheless suggests that under conditions that are similar to those found in cortical circuits, it is possible that the net effect of inhibition on imaging results can depend in subtle ways on the nature of the tasks and type of modulation.

The model described above involves incoming excitation and inhibition. This reflects release of either excitatory or inhibitory neurotransmitters, which act on the recipient units to either increase or decrease electrical activity, respectively. These effects are referred to as *postsynaptic* since they act on the recipient neurons after release has occurred at the synapse. A different form of inhibition can occur when the release of neurotransmitters is prevented before they reach the recipient neurons. This is referred to as *presynaptic* inhibition, and the net effect is reduction or withdrawal of incoming effects on the local population. In relation to effective connectivity, inhibitory brain mechanisms may contribute significantly to observed effective connectivity. Houk et al. (2007) illustrate an approach to explaining fMRI deactivations associated with presynaptic inhibition using a combination of a large-scale human network model and human imaging experiments. They modeled the computational features of basal ganglia processing loops as dynamic processing modules (see chapter 22 for further explanations on the anatomical and computational properties of the corticostriatial system). In the human task, subjects used a joystick to track sequences of visual cues. The execute contrast of interest was the difference between sensory-guided joystick movements and a rest period, and assessed the neural correlates of motor execution. The decode contrast was designed to assess the neural correlates of decoding by comparing execution in a memory-guided movement period versus the sensory-guided movement period. Simulations of the caudate BOLD under conditions of no inhibition, presynaptic inhibition, or postsynaptic inhibition demonstrated that presynaptic inhibition could account for the reduced caudate BOLD signal in the decode contrast whereas postsynaptic inhibition accounts for the increase in BOLD that was observed in putamen (Houk et al., 2007). Presynaptic inhibition presumably also requires energy, and thus increased blood flow and oxygenation, since a different type of neurotransmitter has to be released in order to prevent the release of the primary excitatory or inhibitory neurotransmitter. However, the work of Houk et al. suggests that these effects are small relative to those that occur at the recipient location once the main neurotransmitters have been released. This makes sense, given the high degree of recurrence in local circuits, as shown in the model of Tagamets and Horwitz that was described above.

15.1.8 Consideration of Modeling Goals Neuroimaging data can be modeled from a number of different perspectives with different goals. In the previous example, the goals were to examine how some basic properties of context and local circuits could influence imaging data. This was demonstrated by showing how synaptic inhibition, which is modeled using parameters derived from physiological data, could affect the hemodynamic response in PET and fMRI data. Another goal is to span anatomical and physiological levels across species. For example, Arbib et al. (1995) note that spanning levels from single-cell data depends on the assumption that a region in the animal brain has a homologue in the human brain. Alternatively, one may combine, for example, macaque neurophysiological insights with evolutionary hypotheses to derive neural network models of human brain function that can be tested against human brain imaging. We may also note the insights arising from comparative fMRI studies of macaque and human (e.g., Mantini et al., 2012, 2013; Peeters et al., 2009). Thus, homologies and inter species predictions can be generated by synthetic imaging data. In the next three sections, some of these approaches will be described in more detail.

15.2 Models Linking PET/fMRI Data to Neurophysiological Data: Animal Data and Neuronal Dynamics

This section introduces methods that attempt to relate animal data and human imaging. In a sense, these are "bottom-up" approaches that use known properties derived from basic neuroscience experiments in animals in an attempt to explain how these can modulate imaging data. The primary purpose of these models is to examine the relationship between the underlying physiological processes or anatomical features and imaging data.

Modeling is by nature the simplification of an abstraction from a more complex thing. A global consideration for imaging data, which typically assesses brain-wide activations, is how many and which regions will be included in the model. On one hand, a single region can be modeled in order to examine how underlying physiological processes affect imaging data, as was seen in the example in the previous section. On the other hand, cognitive tasks require multiple interacting brain regions, so adequate explanations need to include these global aspects. The motivation and empirical basis for the model can be data driven, informed by the imaging results, or driven by the results in the literature associated with the cognition or physiology being modeled. Often a combination is used.

The basic model building blocks must be chosen from a level appropriate for the simulation goals and neuronal activity being modeled. This can be from spiking neurons, from membrane potentials, from local populations, or on a regional level. Modeling imaging data requires quantification of the proportions of different types of synapses and anatomical connections between regions. At the regional level, there needs to be an architecture that defines how local populations connect to each other within a region and how different regions are connected. Next, a decision must be made about how to model the relationship between the physiological data and the neuroimaging output. For example, if the physiological level is spiking neurons, then some equation reflecting this must be tied to PET or fMRI data. The differences between spatial and temporal scales of the physiological data and imaging data must also be resolved in the implementation.

Tagamets and Horwitz (1998) modeled a *delayed match-to-sample* (DMS) task in a large-scale model that includes multiple regions of the brain. In a DMS task, a stimulus is first presented for a short time. The stimulus then disappears, and the participant must keep it in working memory for some period of time, usually for up to a few seconds to a few tens of seconds. After this delay period, a new stimulus appears, and the participant must indicate whether this matches the previous one. This type of experiment has been performed in both animals (using electrophysiology to record neuronal activity) and in human imaging studies. The goal of the model described here is to reproduce the electrical activity as recorded from animals, to generate outputs that indicate whether the stimuli match (i.e., reproduce appropriate behaviors), and to generate synthetic imaging data that is similar to that seen in the human experiments. The model includes four regions along the *ventral* visual pathway (Ungerleider and Mishkin, 1982; Ungerleider and Pasternak, 2004): primary visual cortex (V1), visual association cortex (V4), inferior temporal cortex (TEO/IT), and the prefrontal cortex (PFC). An overview of the model is shown in figure 15.5. This pathway, also referred to as the "what" pathway, was identified as being critical in object identification in nonhuman primates (Ungerleider and Mishkin, 1982). In the model, the PFC circuit performs the DMS task and was intended to capture the function of working memory, with neural responses that mimic the dynamics of single cells found in monkey prefrontal cortex during different task phases (Funahashi, Bruce, and Goldman-Rakic, 1990; Goldman-Rakic, 1995). The working memory circuit is composed of four different subclasses of neuronal populations found in primate studies of the DMS task. Four different components in

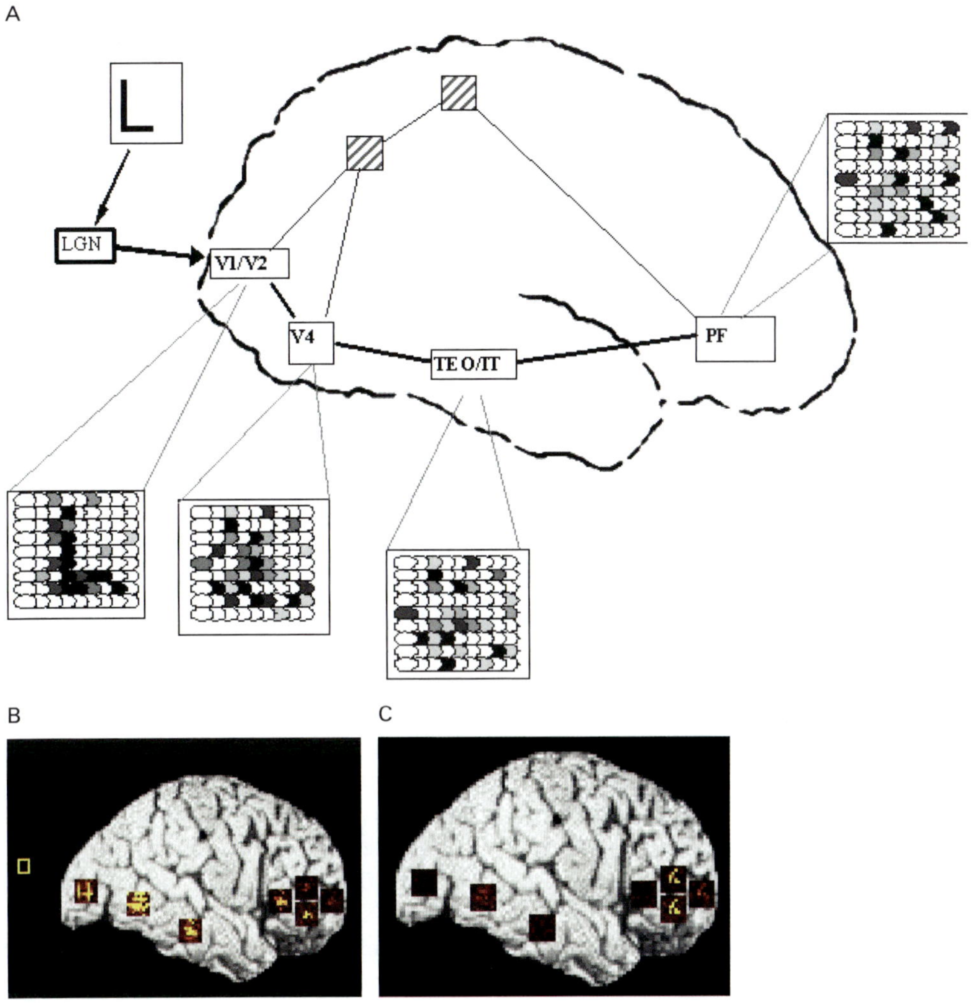

FIGURE 15.5 The model is made up of areas along the occipitotemporal pathway, which is thought to be the main pathway that is involved with object vision. (A) Different populations of units process incoming data in successively abstracted stages, indicated by the representations shown in response to an input of a letter L. LGN, lateral geniculate nucleus; V1/V2, early visual cortex; V4, secondary visual cortex; TEO/IT, inferior temporal cortex; PF, prefrontal cortex. From Tagamets and Horwitz (2000). (B) Modeled electrical activity in the regions while a stimulus is in view. (C) Modeled activity being maintained in prefrontal cortex after the stimulus has disappeared from view. In B and C, the different types of subpopulations of neurons in frontal cortex are depicted separately; the leftmost population consists of Cue units, the center two are the D1 (top) and D2 units, and the rightmost patch consists of Response units. These are described in detail in section 15.2 and in figure 15.6.

this model circuit reproduce the dynamics of these populations. Additionally, task performance is modulated by a parameter that models attention (see figure 15.6).

As in the previous example, the basic local circuit was used for building the model regions. The next consideration, then, is how to match physiological constraints to the construction of a large-scale network. This requires defining the architecture of two general types of connections: (1) connections between regions and (2) connections among units within each region. For most regions in our models, connections between regions are generated by an algorithm that depends on five parameters: the fan-out size, *anisotropy* (directional bias), the average weight value, the range of variability around this average, and the density, that is, percentage of the connections to be actually constructed. The specific parameter values used in the model are given in table A1 in Tagamets and Horwitz (1998). The total connectivity strengths between different areas were constrained to approximate the published data for known connections. Using a systematic rule-based method for determining connection topography between areas achieves the same type of receptive field properties that have been shown to exist in the ventral visual pathway and makes it possible to identify the effects of rules, such as increasing fan out (or divergence), on both electrical and CBF responses. In regards to temporal dynamics, the network must behave appropriately during a sequence of different phases such as those that are involved in each of the tasks. Responses need to be robust under conditions that vary from resting to extremely driven, and a memory circuit needs to implement the maintenance of a representation over the delay period, when no stimulus is present.

Connections from lateral geniculate nucleus (LGN) to V1 and from V1 to V4 are arranged in either a horizontal or a vertical direction in order to implement selectivity for either horizontal or vertical lines, respectively. (For simplicity, other types of feature detector are ignored.) In order for the model to distinguish among different stimuli and to achieve electrical response characteristics similar to those observed in monkey, with inferior temporal cortex (TEO/IT) tuning and shift invariance for specific shapes, an unsupervised competitive *Hebbian learning rule* is used to train feedforward connections from area V4 to area TEO/IT. As explained in chapter 6, Hebbian learning rules are based on the principle that neurons that fire "together" tend to strengthen connections between them (Hebb, 1949). The competition in the variant of the learning rule that is used in our model comes from the normalization step in equation 15.7, which causes more active connections to increase and less active connections to weaken.

$$\tilde{w}_{ji}(t+1) = w_{ji}(t) + d \cdot (s(a_j(t) - T_{pre}) s(a_i(t) - T_{post})) \quad (15.6)$$

Equation 15.6 establishes the amount by which the weight w_{ji} is to change prior to normalization, which is then given by

$$w_{ji}(t+1) = C \frac{\tilde{w}_{ji}(t+1)}{\sum_k \tilde{w}_{ki}(t+1)} \quad (15.7)$$

Equations 15.6 and 15.7 determine how the connection weights from a presynaptic region PRE to a postsynaptic region POST are trained. In equation 15.6, the new weight from presynaptic unit j to postsynaptic unit i at time t + 1 depends on the current value $w_{ji}(t)$ plus a change that is determined by the product of activations $a_j(t)$ and $a_i(t)$ of units j and i at time t. T_{pre} and T_{post} are the presynaptic and postsynaptic thresholds, respectively. s is the ramp function, such that $s(x) = 0$ if $x < 0$, and $s(x) = x$ when $x \geq 0$ .Thus, if either of the units has activity value below its threshold, the weight on this connection does not change at this time. However, the new weight is likely to decrease after normalization (equation 15.7) if other connections are active. The term d determines the learning rate. Equation 15.7 is the normalization step, where C is a constant that preserves the total weight between regions PRE and POST.

During training the stimuli are presented on the LGN input layer for 25 iterations, with a two-pixel shift of the entire stimulus in a random direction every 5 iterations. This procedure allowed the TEO/IT area to learn a shift invariant representation of the different stimuli. Hebbian learning works without a "teacher" (error signal) since the relationships between coactive pre- and postsynaptic-input-driven activations in specific units become represented by patterns of weights. If two units are both active simultaneously, the connection weight between them increases; if one or both have low activity, the weight decreases. After training, the area has a well-tuned distributed response to each of the training stimuli and weaker, less-tuned responses to other unlearned stimuli. The trained network allows the model to distinguish among different stimuli during the PET simulation of the match to sample task.

A memory circuit, shown in figure 15.6 (see also figure 15.7), allows the model to maintain representations of stimuli during the delay period and to generate a response if the current stimulus matches the previous one. The frontal cortex region of the model is composed of a 9 × 9 array of these memory circuits. Each element of a memory circuit is a basic unit such as that shown in figure 15.2A, and activity of these units in a model patch of brain is depicted in the frontal cortex area in figure 15.5B and 15.5C. Excitatory connections are all E to E, and inhibitory connections are E to I in the memory circuit.

To simulate the task, the model network is presented with a series of stimuli. An example stimulus (a square) is depicted outside the brain on the left in figure 15.5B. Activity travels along the ventral visual pathway (see figure 15.5A and 15.5B), from the LGN to area V1/V2, then to area V4, area TEO/IT, and finally to the PFC. The early visual areas V1/V2 have retinotopic representations of the stimulus, and, as is thought to occur in the brain, retinotopy is gradually lost as the signal travels forward through successive regions (see figure 15.5B). Area TEO/IT is thought to encode nonretinotopic abstract representations of objects, each of which consists of a distinct pattern of activated populations of neurons. Two subpopulations of the memory circuits in the frontal region mediate the matching task by holding a representation of the last seen stimulus in memory

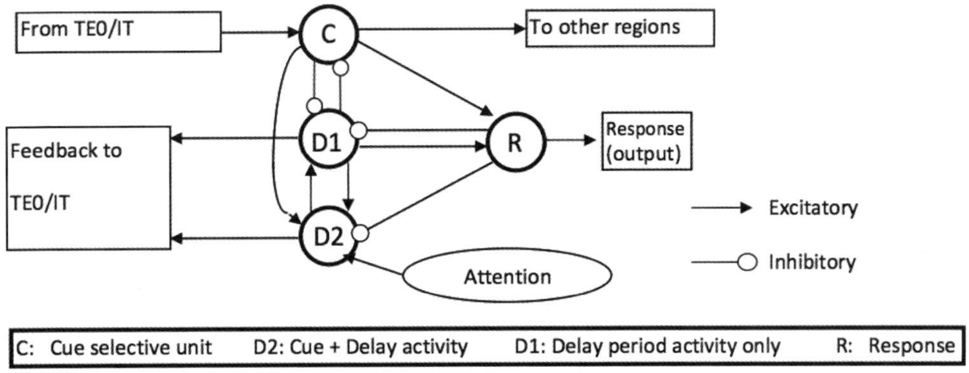

FIGURE 15.6 A memory is added to the prefrontal areas of the model as a local circuit composed of four different types of units, as identified in studies such as shown in figure 15.7. Each element of the circuit in the figure is a basic unit (shown in figure 15.2A). C, Cue unit. Input from visual pathways enters here, and it is only active as long as there is input. D1 and D2, Delay units. These comprise the memory in the circuit and can be active during a delay period, when there is no external stimulus coming into the circuit. D2 units are active both during the stimulus and during the delay period. D1 units become active only when the stimulus has disappeared from view. Recurrent connections between the D1 and D2 units maintain the memory of the previous stimulus if there is sufficient attention (see figure 15.5C). Modulation of attention is modeled as a diffuse, weak activity directed into the D2 units. R, Response units. These show a brief activation after presentation of the test stimulus if it matches the item currently in memory. This is the output from the circuit and indicates whether a new stimulus matches the previous one. Inhibitory connections are effected by excitatory connections onto inhibitory units. Both D1 and D2 units also are the source of feedback into earlier visual areas and can reactivate these regions based on the current memory. Activity of this working memory circuit during a simulated task trial is shown in figure 15.7. TEO/IT, inferior temporal cortex. From Tagamets and Horwitz (1998).

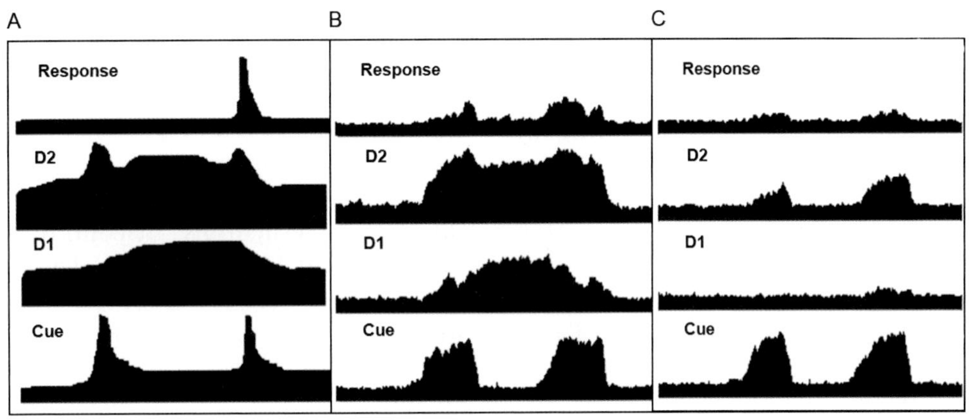

FIGURE 15.7 (A) The leftmost panel shows patterns of behavior of the different types of neurons that have been identified in prefrontal cortex of the macaque in spatial delayed-response experiments. Adapted from Funahashi et al. (1990); Goldman-Rakic (1995). (B, C) Behavior of averaged activity over five trials of the analogous units in the prefrontal area of the model, with attention (B), and without (C). In the model, the response units indicate whether a match has occurred. From Tagamets and Horwitz (1998).

during a delay period (D1 and D2, see figures 15.5C, 15.6, and 15.7B). When the signal reaches the PFC, it is first received by the Cue unit. The Cue unit passes the signal along to the D2 unit and at the same time inhibits the D1 unit. As a consequence, the D2 unit becomes active and the D1 unit remains inactive as long as the stimulus is present in the Cue unit. When the stimulus disappears, the Cue unit loses its activity, and this releases the inhibition on the D1 unit. At this point, the still active D2 unit is able to activate the D1 unit, and the representation of the stimulus is maintained by the recurrent excitatory connections between the D1 and D2 units. When a second stimulus is presented to the model, the signal travels along the same pathway as the first stimulus, activating the Cue unit again. At this point, if there is an existing representation of the same object in the frontal cortex memory circuit, the simultaneous activity of the Cue and D1 units activates the

Response unit, which then sends an inhibitory signal to the D1 and D2 units, shutting them off. At the same time, the activity of the Response unit signals that a match has occurred. Although not implemented in this model, this signal can then be used by other circuits that can make use of this information.

The electrical activity is determined by a sigmoidal function of the summed synaptic inputs that arrive at each E and I element of each unit (see equations 15.1 and 15.2 in the previous section). This corresponds to average spiking rates from single-cell recordings. In order to simulate an imaging experiment, two tasks are compared: the one-back matching task and passive viewing. The absolute values of all synaptic activities are summed, and the local blood flow or BOLD signal is computed by integrating these values over space and time using equation 15.3, as seen in the previous section. This provides the quantitative data for matching imaging studies. The simulations produced quantitatively similar PET activation to the human data across the different regions (see table 15.1).

We used the different stages of a DMS task and demonstrated a model of working memory that could mediate such a task and show appropriate behaviors during each stage. The model demonstrates how matching can be mediated by a single parameter that corresponds to attention and, in particular, suggests that the attentional component affects a particular type of neuron that responds both to a stimulus and during the delay, the D2 unit, which has been identified in delayed-response tasks in the monkey (Funahashi, Bruce, and Goldman-Rakic, 1990).

15.2.1 Different Approaches for Linking Physiological Data and Neuronal Dynamics to PET and fMRI

As implied earlier, building a model depends on the appropriate level for the simulation goals and neuronal activity being modeled. Arbib et al. (2000) review a related approach to that of Tagamets and Horwitz (1998) for modeling PET and fMRI in a large-scale network. Arbib et al. related the physiology underlying the metabolic demand in PET, and by extension fMRI, both to neural networks like those discussed above (developed more fully by Arbib et al., 2003) as well as to an approach known as *schema theory* (see chapter 14, "Schema Theory and Neuropsychology"). This approach combines ideas from the psychological roots of the theory with applications of schemas at the physiological level. In psychology, schemas are mental units that organize knowledge about a focused subject. A schema is an abstraction over instances (i.e., experiences). Action schemas maybe be invoked by a particular context, for example, getting into a car invokes a driving schema. Neural schemas are schemas whose activity can be neurally localized and related to nonhuman animal circuitry. The combination of schema networks with neural networks allows the model to span multiple granularities (Arbib et al., 2000).

Another example of a different approach is the exploration and proposal of mechanisms underlying the developmental relationship between working memory maturation and neuronal activity (Edin et al., 2007). The goal was to examine the relative contributions of synaptic strengthening, synaptic pruning, and myelination on brain activity and performance in a working memory task. The model was made up of two brain regions, parietal cortex and frontal cortex, each of which was composed of 128 pyramidal cells and 32 inhibitory interneurons. In contrast to the model of Tagamets and Horwitz (1998) described above, individual neurons were modeled at the biophysical level, where ion channels and receptor actions are modeled explicitly. The fMRI BOLD signal was modeled as total synaptic activity, as in the Arbib et al. and Tagamets and

Table 15.1

Comparison of simulated blood flow and experimental positron emission tomography data

Data	V1	V4	TEO/IT	Prefrontal
Simulated				
High attention	1	0.8958	0.8303	0.6964
Low attention	0.9696	0.8519	0.8103	0.6732
Percentage change	3.1	5.2	2.5	3.5
Experimental				
High attention	1	0.8924	0.805	0.7057
Low attention	0.9738	0.8258	0.7724	0.6776
Percentage change	2.7	8.1	4.2	4.1

Note. Data are normalized to the value of the high-attention V1 activations in each of the two cases (simulated and experimental). The experimental data are from those used by Haxby et al. (1995). TEO/IT, inferior temporal cortex. From Tagamets and Horwitz (1998).

Horwitz models. However, these synaptic activities were modeled as synaptic currents by summing the amounts of all the different receptors at the model synapses. Coding properties of the neurons were modeled by connectivity patterns, so that neurons that encode similar patterns are more highly connected than those with dissimilar feature responses. Differences due to developmental effects were then examined by creating several different models of a child brain and a single model of an adult brain and comparing brain activations between the child and adult models for each of the developmental effects of interest. Analogous experiments were run with groups of children and adults, and the modeled results were compared to the human data in order to determine which model best fit the experimental data.

The developmental changes due to synaptic strengthening, synaptic pruning, and myelination were each modeled by changing separate parameters in the child models. Changes in synaptic strength were modeled both within regions and between regions by changing parameters that modulate connection strengths, either within or between regions of the model. Synaptic pruning was modeled by redistributing synapses without changing total strength either within or between regions. In effect, loss of synaptic strength due to removal of some synapses was offset by increasing the strength of the remaining synapses. Changes in myelination were modeled by a parameter that controlled the speed of synaptic transmission between the two regions of the model. Modeled and experimental data comparisons were based on both activation and interregional connectivity. Additionally, the model was examined for the response properties of the neurons. For example, pruning could result in increased contrast, meaning that there is a bigger difference between the responses of neurons that are specific to a particular stimulus and neurons that do not code for that stimulus. Alternatively, pruning could result in increased specificity. In this case, the more similar the coding properties of two neurons are, the more similar their responses. In particular, increased contrast was taken to be more adult-like, in that the neurons are less susceptible to distracting stimuli. The results suggest that a combination of synaptic strength increases and synaptic pruning best distinguished the adult from the child brain activations. Neither of these effects alone was sufficient, nor did changes in myelination reproduce the experimental data.

15.2.2 SUMMARY This section focused on how to implement the transformation of single- and multiple-cell activity into synaptic activity and, in turn, use this synaptic activity to simulate imaging data that spans the activations in multiple brain regions. Tagamets and Horwitz (1998) implemented a large-scale model based on structural and functional properties of the ventral visual pathway. The cell types associated with working memory in the PFC in animal studies were implemented, as well as multiple regions in the pathway. Another important feature was how to represent the topography and geometry of connections. The model of Edin et al. (2007) demonstrates how a more detailed model at the biophysical level can be translated into quantitative measures that can be compared to human imaging data. The level of detail in this model was appropriate for the questions being addressed. In particular, the behavior of the individual neurons was one of the criteria that was used to distinguish the child from the adult brain in the model.

15.3 Models Linking PET/fMRI Data to Neuropsychological Data: Behavioral and Cognitive States

Whereas models in this section also use animal data to determine model design and parameters, the focus in this section is the relationship between imaging data and tasks, behaviors, cognitive states, and performance. This can include trying to explain empirical human imaging tasks from a cognitive perspective, or linking neurophysiological animal data to behavior, status, or performance in conditions that are comparable across species.

Many considerations from the previous sections also apply to cognitive modeling: what regions to select, what parameters to use, and how to relate physiological levels to imaging data. However, there are additional considerations when trying to relate behavior to imaging data, which introduce more design decisions. When choosing regions, there may not be any appropriate physiological data to guide selection, as, for example, in language tasks. The model must do the task and produce performance measures that can be related to human performance on the imaging task, which may be a challenging design problem. The model's task performance must be evaluated in terms of the experimental data, in addition to simulating the imaging results. A final consideration is how to compare the model and the imaging data, and relate them to the behavior or status of interest. The following two models illustrate approaches to these issues.

15.3.1 VISUAL ATTENTION MODULATION Corchs and Deco (2004) modeled how attention modulates regional responses as a function of specialization of object recognition in the ventral "what" pathway, and location and spatial relationships between objects in the dorsal "where" pathway (Ungerleider and Pasternak, 2004; see also the model description in the previous section, "Models Linking PET/fMRI Data to Neurophysiological Data: Animal Data and Neuronal Dynamics"). They simulated fMRI data from Saenz et al. (2002), who investigated whether feature based attention globally affects visual cortical responses to unattended stimuli that have an overlapping feature(s) with attended stimuli. The model instantiated the biased competition hypothesis (Desimone and Duncan, 1995) that attention modulates visual processing by enhancing features or locations of the attended stimulus and in concert reduces the suppressive interactions from neurons representing nearby distractors. See chapter 5 for a model of related data (Reynolds, Chelazzi, and Desimone, 1999). For a comprehensive evaluation of how inhibitory competition and top-down biasing influence the competitive and cooperative interactions of local and global processing, see Deco and Rolls (2004).

In the fMRI task (Saenz et al., 2002), subjects viewed overlapping fields of stationary red and green dots on one side of a central fixation point. A stationary field of single color red or green dots was on the other side of the fixation point. They attended to either the red or green dots in the overlapping field. There were six 40-second cycles in each 4 minutes of the fMRI scanning session. In each cycle, attention alternated between attending to a target field where the color was the *same* as the ignored field on the other side or attending to a target field which was a *different* color than the ignored field on the other side. Stimuli were presented for two sequential 1-second intervals with an interstimulus interval (ISI) of 100 ms. Only the fixation point was visible during the ISI.

In the fMRI study, response to the ignored stimulus was reported in areas V1, V2, V3, V3A, V4, and MT+. However, Corchs and Deco used a previously developed (Deco and Zihl, 2001; Deco and Rolls, 2004) computational framework for visual attention that included fewer visual areas than the fMRI study being modeled. This means that they decided to simplify the original to model to represent the visual cortex as V4, inferior temporal region IT (MT+ is in the posterior portion of the inferior temporal sulcus), and the posterior parietal (PP) cortex (see figure 15.8). They also added two regions in PFC, ventral and dorsal Brodmann area 46. These serve to direct attention to either the ventral (feature) or dorsal (location) pathways, respectively.

Regions were represented as modules with their own functional specialization, and global involvement in the ventral "what" and dorsal "where" streams. The original architecture (Deco and Zihl, 2001; Deco and Rolls, 2004) consisted of modules V4, IT, and v46 (ventral PFC) in the ventral stream and modules V4, PP, and d46 (dorsal PFC) in the dorsal stream. Many studies indicate that frontal and parietal regions modulate visual attention (see Buschman and Miller, 2007). The simulations in Corchs and Deco did not explicitly simulate regions v46 and d46, instead focusing on modulation from IT/TEO (object) into V4 and PP intoV4 (location).

The V4 module delivered both feature and spatial information to the dorsal and ventral streams. It consisted of pools of neurons representing hypercolumns organized topographically in a $P \times P$ matrix such that it matched a corresponding $N \times N$ pixel space in the input scene. In the ventral stream, the IT module was concerned with recognizing an object in the visual field. It consisted of C pools of neurons that learned to detect C objects. Each IT cell assembly (i.e., pool) is connected to those of V4, from which the representations enter the model. In the dorsal stream the V4 module is connected to the PP module, the latter implementing the spatial attention and object location mechanisms. The PP module has the same matrix configuration and mapping to the input image as module V4, but it encodes the location of an object in space rather than its identity. Inhibitory influences were modeled using a pool of inhibitory neurons interacting within elements of the hypercolumns in a nonspecific manner. These served to normalize the overall activity of the excitatory pools.

The model had two processing modes: a learning mode and a recognition mode. Learning consists of training synaptic connections between V4 and IT, again using Hebbian learning. The recognition mode simulates identifying an object by biasing the competition to the current object target with a top-down mechanism from v46 to IT. Intermodular attention via V4–IT connectivity enhances the activity of V4 pools associated with features of the specific object. The V4–PP interactions enable attention to bias activity in V4 pools toward the location of the object search target.

The model task was similar to the fMRI procedure. In the task simulation, input activity into V4 was presented as two matrices. The matrix elements represented the color presence or absence in a receptive field position. In each trial the stimulus was shown for 1 second, followed by 100 ms without input. Then the same stimulus was shown again followed by 1.2 seconds without input. There were blocks of six trials for each color. The relationship between simulation time and fMRI time was that each 100 iterations of simulation corresponded to 0.1 second. The neuronal activity of

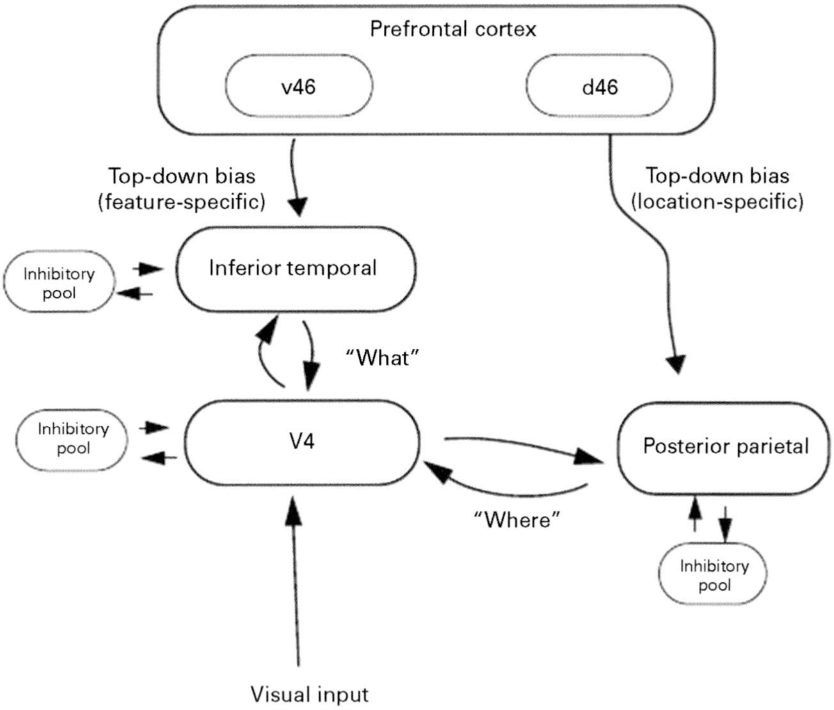

FIGURE 15.8 The Corchs and Deco model. From Corchs and Deco (2004), *NeuroImage*, *21*(1), 36–45. Key: v46, ventral Brodmann area (BA) 46; d46, dorsal BA 46. Brodmann area 46 is in the prefrontal cortex and is thought to send signals about goal-directed behavior to other areas of the brain. The experiment being modeled required subjects to pay attention to a specific object feature (red or green) and the location (left or right) of the stimulus array. In this model, v46 sends an attentional signal about the specific feature of the object to be attended to, for example, green. This enhances responsiveness of that feature's representation in the inferior temporal cortex. In a similar fashion, d46 sends a signal to posterior parietal cortex about the location to be attended to. Connections between the inferior temporal and posterior parietal regions serve to bind the two representation to each other and enhance the responsiveness of the appropriate combination of populations of neurons, such as the green dots on the left side.

each pool was computed by replacing the temporal average discharge rate of a cell with the population average (Brunel and Wang, 2001; Wilson and Cowan, 1972). The activity corresponding to the whole module V4 is computed by integrating neural activity in the model over time and space. A Poisson function representing the hemodynamic delay (see figure 15.1) was convolved with the simulated neural activity to produce the synthetic fMRI data. Of note, the hemodynamic response to sensory input in human V1 as a Poisson function has been validated (Boynton et al., 1996). In addition to simulating the BOLD response, Corchs and Deco also simulated single-cell response in V4 across the 220-second time course by using activity without global spatial averaging or convolution with an HRF. This procedure demonstrates how the same obtained signal may be teased apart to model different levels of the signal, regional BOLD, and single-cell spiking activity.

Comparisons between the simulated and actual fMRI BOLD response were qualitative. This means the rank order of the response strength by conditions was the same but the simulated BOLD and actual BOLD were in arbitrary units and not subject to quantitative evaluation. Simulated V4 response to the ignored dot field was stronger when the attended feature in the other dot field was the same as the ignored field. Thus, the simulated BOLD agreed with the fMRI finding that feature-based attention can influence the visual cortical response outside the attended location.

15.3.2 SUMMARY The model in this section illustrates how to make design choices and run simulations that mimic the tasks performed by human subjects in an fMRI scanner. Corchs and Deco (2004) were concerned with implementing characteristic structures and functioning of the dorsal and ventral visual streams in order to demonstrate top-down attentional modulation. The task initiated ventral stream deployment by object feature processing and dorsal steam deployment by spatial location processing. Model regions were chosen based on a simplified version of a previously validated visual attention model and did not have a one-to-one correspondence to regions from the fMRI study. Not only was the model used to generate synthetic fMRI data that could be compared to human fMRI data, it was also used to generate single-cell activity by using the matrix form of unit activity. The latter helps to explain potential mechanisms that mediate the tasks, and it can also be compared to single-cell or neural population activity in experimental animal studies. Thus, the model spans three different analysis levels of brain functioning. We saw how architecture, parameters, and processing protocols may be effectively used to reproduce human performance, BOLD signal, and single-cell data such as that recorded from animals.

15.4 Synthetic Imaging of Disease, Connectivity, and Data-Driven Learning

Considerable effort has been invested in imaging of brain-related disorders, such as schizophrenia, autism, bipolar disorder, and dyslexia, just to name a few (see chapter 23). These studies typically compare groups of patients with controls in order to identify brain regions or connections that differ between the groups. In particular, most PET and fMRI studies identify group differences during various cognitive tasks, such as those described in the previous sections, in order to isolate function-specific abnormalities. For many of these conditions, a number of abnormalities have been identified by these imaging methods, but the factors that underlie these differences are not known. Each disorder is associated with a particular set of symptoms, cognitive or psychological abnormalities, and patterns of brain activation or connectivity changes. However, in most cases, the underlying causes of these differences are not known. Modeling can be used to elucidate factors that effect specific patterns of changes and to help generate testable hypotheses about these. To date, most models that inform imaging studies of disease have been developed to generate hypotheses, and the imaging studies are used to test the hypotheses. However, the comparisons in general have been qualitative. It is only recently that synthetic brain imaging, in which imaging data are compared quantitatively to modeled data, has been applied to examining disease processes. We examine one such model in detail, in which factors that underlie connectivity changes in schizophrenia are examined. Data-driven training methods for matching model parameters to imaging data are introduced in this section. It is then shown how manipulating other model parameters can inform interpretations of imaging findings such as reduced interregional connectivity.

15.4.1 Examining Connectivity Changes in Schizophrenia Schizophrenia has perhaps the widest range of functional aberrations, as measured both by cognitive assessments and imaging studies. Cognitively, aberrations have been found in almost every domain, with the most commonly reported findings in language, working memory, attention, learning, cognitive control, social function, and emotion processing. Functional imaging studies have identified differences between patients with schizophrenia and controls in virtually every part of the brain, depending on the tasks being examined. Structural and chemical imaging studies also report abnormalities in various structures and chemical signatures. Despite this wealth of data, it has been difficult to pinpoint specific factors that underlie schizophrenia. One recent dominant theory is that the cognitive and symptomatic aspects of schizophrenia are best characterized as a disorder of connectivity, where the different parts of the brain do not communicate effectively with each other (Friston, 1998: Stephan et al., 2009). However, the factors that underlie this are not known.

As noted in earlier sections, diverse factors can cause changes in activations and connectivity. Changes in neurotransmitters or neuromodulators can affect both connectivity and activations. Likewise, structural changes in long-range connections can affect both, as can changes in local circuits. The nature of these changes can take multiple forms. For example, long-range connections are composed of bundles of axons that carry the signal from one part of the brain to another. The functional integrity of these bundles can be affected by the integrity of the axons themselves, by changes in the numbers of axons, by differences in myelination, or by several other factors. All of these can affect the transfer of electrical activity and subsequent activations in recipient regions. Likewise, reduced integrity can result in reduced covariance between the connected regions, resulting in reduced functional and effective connectivity. While long-range connections have been more widely studied in schizophrenia, changes in local circuit properties can have similar consequences. As noted in section 15.1, most functional imaging data reflects local synaptic activity, so changes in numbers of different types of synapses in a local area will affect both activation response and connectivity with other regions.

Winder et al. (2007) used a large-scale model to examine factors that might underlie effective connectivity reductions in schizophrenia. This work had two primary goals: (1) to examine how local circuit properties affect functional and effective connectivity and (2) to examine potential mechanisms that underlie the

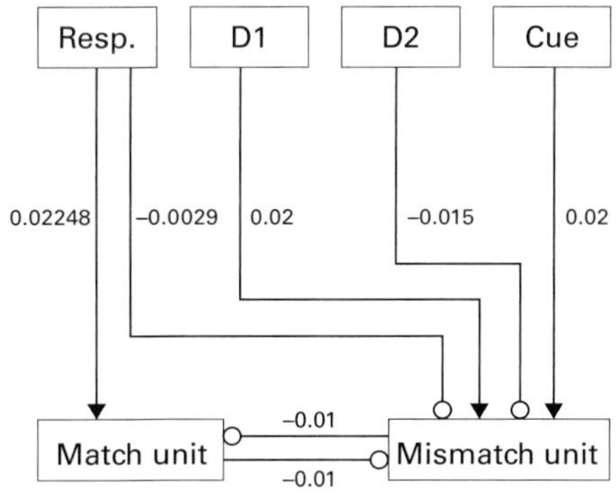

FIGURE 15.9 Configuration of the Match and Mismatch unit connections to the working memory circuit. The Match unit responds when the current stimulus matches that held in memory while the Mismatch unit fires when there is no match. Resp. (Response), D1, D2, and Cue are the units in the working memory circuit (see figure 15.6). The Response, Cue, D1, and D2 units are connected to each other as shown in figure 15.6, but these connections are omitted in this figure for clarity. From Winder et al. (2007).

reduced temporofrontal connectivity that is frequently seen in schizophrenia. The model from Tagamets and Horwitz (1998), described in section 15.2, was extended by adding a second hemisphere plus an additional region, the hippocampus, to each hemisphere. The working memory circuit was also extended to include both Match and Mismatch units, which respond when a stimulus either matches or does not match the one held in memory, respectively (see figure 15.9).

The task was adapted to match that used in an fMRI imaging study (Tagamets et al., 2000), which was a one-back matching task using words and other strings of letters. Toward this end, before using the model, connections from area V4 to area IT were trained on four-letter words using an unsupervised Hebbian learning rule, as described in more detail in section 15.2. The purpose of this was to use stimuli that are similar to those used in the human fMRI study, and which are "familiar" to the model, as words are to humans.

The Hebbian learning algorithm does not account for matching the modeled fMRI activation to the experimental data. Connection weights between units are adjusted according to whether or not units are simultaneously active. Therefore, it does not take externally imposed goals (e.g., matching regional activation values in the experiment) into account. However, if the goal is to explain specific differences between tasks or groups in an fMRI study, then one approach is to begin with a model of one condition that matches the experimental data for that condition and then examine model parameter changes that can explain the other condition.

In order to match the modeled data to the experimental data, a supervised learning method was introduced, referred to as *gains learning*. During gains learning, the interregional path weights are rescaled, without changing their topography, until the modeled fMRI signal in all regions matches data derived from an experimental study. Since gains learning modifies all individual connections between two regions by the same proportion, it preserves the topography and thus the coding properties among units in a region.

During the gains learning procedure, total modeled synaptic activity (v_A) is compared to a target experimental value u_A for each region A. The target u_A is defined as the mean within-task activity of the experimental fMRI in a region of interest. The u_A for all regions are normalized to the left V1 region by dividing all targets by the BOLD value of the left V1, and the modeled values v_A for all regions are likewise normalized to modeled v_A in the left V1 region of the model. This eliminates potential scaling differences between the modeled and experimental data and yields connection values relative to area V1. Before learning, the gains for all pathways that will learn are initialized to random values, though equal gain values are assigned for homologous left- and right-hemisphere connections in order to prevent artificial hemispheric bias. Consider a general case where the source region is labeled B and the destination region is labeled A. The gains learning algorithm updates the gain (g_{AB}) from region B to region A, based on differences between target values (u_A and u_B) and modeled values (v_A, and v_B). The modeled fMRI activity is the sum over N trials or time intervals x:

$$v_A = \frac{\sum_{x=1}^{N} v_{Ax}}{N}. \tag{15.8}$$

At each learning cycle, the gain from B to A, g_{AB}, changes as follows:

$$\begin{aligned} &\Delta g_{AB} = \varepsilon\alpha(u_A - v_A) - \lambda g_{AB}, \\ &\varepsilon = 0.5, \\ &\lambda = 0.05|1 - u_B / u_A|, \\ &\alpha = \begin{cases} u_B/v_B, & \text{if } u_A < v_A \\ v_B/u_B, & \text{otherwise.} \end{cases} \end{aligned} \tag{15.9}$$

Δg_{AB} is the change in gain, ε is the learning rate, λg_{AB} is a regularization term (described in more detail later),

Table 15.2

Connection weights learned by the model from functional magnetic resonance imaging data of healthy volunteers (HV) and volunteers with schizophrenia (SV)

		Left		Right	
Source	**Destination**	**HV**	**SV**	**HV**	**SV**
LGNs	V1/V2	0.141	0.141	0.129	0.124
V1/V2	V4	0.111	0.123	0.113	0.150
V4	TEO/IT	0.130	0.118	0.122	0.122
TEO/IT	PFC	0.058	0.015	0.033	0.066
TEO/IT	Hippocampus	0.079	0.087	0.080	0.076
Hippocampus	PFC	0.084	0.088	0.065	0.068

Note. In order to account for potential scaling differences, connection weights from left lateral geniculate nucleus (LGN) to V1 are fixed. TEO/IT, inferior temporal cortex; PFC, prefrontal cortex. From Winder et al. (2007).

and α is a momentum term that takes account of the relative errors between the source and destination activation levels. α computes the ratio between the target and modeled activation values, so that $\alpha = 1$ when they are equal, and greater than 1 otherwise.

The model is run over the N trials for one learning cycle, and the resulting error E_A for each region A is computed from the cumulative modeled fMRI activity as follows:

$$E_A = \sqrt{\left(\sum_x (u_A - v_{Ax}) \Big/ \mathrm{N}\right)^2 + \lambda \sum_B (g_{AB})^2} \, . \tag{15.10}$$

The gains learning algorithm moves the modeled fMRI activity v_A in the direction of the target value, u_A, in each region A, reducing the error toward zero over time. When all the regional errors are less than a specified error threshold, the algorithm terminates and records the learned gain values.

One issue that occurs in many learning methods for neural networks is that of uniqueness. Highly connected recurrent networks can have multiple solutions that satisfy the error minimization constraints. Regularization is one method that has been applied toward solving this uniqueness problem (Poggio, Torre, and Koch, 1985; Girosi, Jones, and Poggio, 1995). The regularization term enforces uniqueness by finding minimum weight values that satisfy the constraints. In our approach, the regularization term is modulated by the similarity of values of the source and the destination region. When activations of the two connected targets A and B are similar to each other, λg_{AB} is smaller and the emphasis on regularization is lower, allowing for a stronger connection. When the targets differ, λg_{AB} is larger and the effect of regularization is higher, and the connection between those regions is smaller. The rationale for this approach is that regions with a similar level of synaptic activity are more likely to have a strong influence on one another, while the more different they are, the more likely it is that their activities are modulated from other sources. The gains learning algorithm was tested by running it on a set of modeled networks with known connectivity values, and it was shown that for modeled data, estimation of connection weights was correct over 99% of the time.

In order to examine possible mechanisms for differences between healthy controls and persons with schizophrenia, gains learning was applied to data from each of the two groups. The results indicated a left-sided disconnection between TEO/IT and PFC, where the modeled connection strength in schizophrenia is less than 30% of that in the controls (see table 15.2). The other connections were largely similar to those of the controls. This is consistent with imaging studies that have found disconnections specific to frontal–temporal connections (Meyer-Lindenberg et al., 2001; Lawrie et al., 2002).

Since gains learning adjusts the overall strengths of connections between regions, these results could be interpreted as reflecting a dysfunction in these interregional connections. However, we wanted to examine what other factors could explain the data, hypothesizing that changes in the memory circuit within the frontal cortex could also explain the data. The rationale for this is threefold. Firstly, there is substantial evidence for local prefrontal circuit abnormalities in schizophrenia (Benes et al., 1991; Benes, 2000). Second, working memory impairment is one of the most robust findings in schizophrenia, and it is associated with PFC dysfunction (Perlstein et al., 2001; see also chapter 23). Furthermore, one study suggests that variations in the integrity of local circuits may play a substantial role in functional connectivity even when white matter tracts are intact (Steel et al., 2001). Together, these considerations suggest that the working memory circuit itself is likely to be compromised in schizophrenia. Based on these considerations, we hypothesized that changes in local connectivity in the working memory circuit could explain the data at least as well as changes in between-region connections. This hypothesis was examined by changing local recurrence in the working memory circuit of the model. Keeping the interregional connections at the values for the control subjects, recurrent excitatory connections from each of the D1 and D2 units in the working memory circuit were systematically changed in order to examine effects on activations and functional connectivity. It was found that increasing excitatory-to-excitatory recurrence within the D1 recurrent connections beyond 0.62 had the largest effect on correlations between PFC and area TEO/IT (figure 15.10B, dotted lines). Using this configuration, gains learning was again applied, and the learned interregional weights in the schizophrenia group were now closer to those of the healthy controls. The results suggest that reduced recurrent connectivity in local prefrontal circuits may underlie the frontotemporal disconnection and the working memory deficits frequently found in schizophrenia.

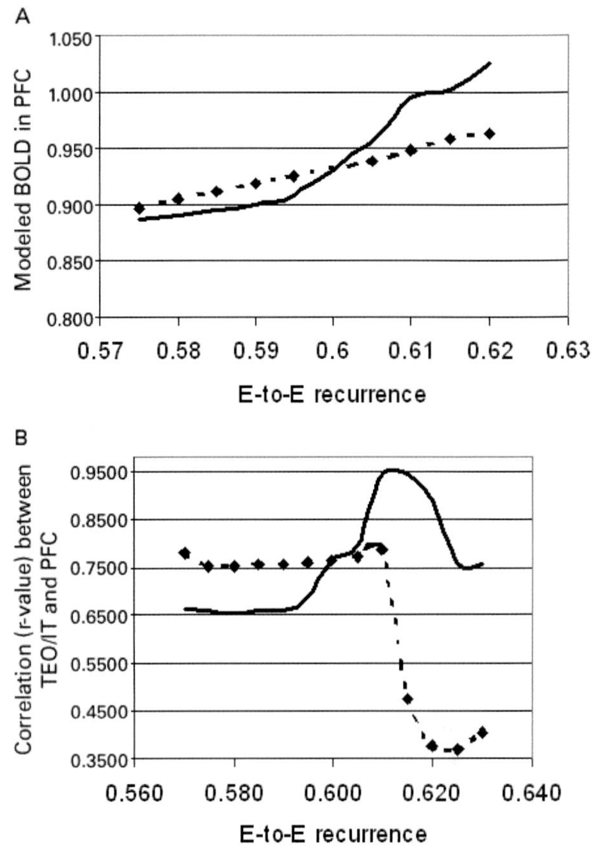

FIGURE 15.10 (A) Intensity of modeled blood oxygen level dependent (BOLD) signal in prefrontal cortex (PFC) versus the strength of recurrent connections in D1 and D2 units (see the circuit in figure 15.6). (B) Correlations between inferior temporal cortex (TEO/IT) and PFC BOLD signal versus the strength of recurrent connections in D1 or D2 units of area PFC. Dotted line: D1 self-recurrence strength; solid line: D2 self-recurrence strength. *x*-axis: strength of the excitatory recurrence from the unit to itself, *y*-axis: (A) modeled BOLD in PFC; (B) correlation coefficient (r-value) of the modeled activity between regions TEO/IT and PFC. From Winder et al. (2007).

15.4.2 SUMMARY This section introduced issues specific to using synthetic brain imaging for examining mechanisms of disease. A new learning method was introduced for quantifiably matching modeled data to imaging data from human subjects. This method was applied to an fMRI study of healthy control subjects and people with schizophrenia in a working memory task. The data-driven learning method found interregional connection values for each group, demonstrating a left-sided disconnection between frontal and temporal regions in schizophrenia, similar to that found previously in schizophrenia imaging studies. The model was then used to examine how local circuit properties affect both activations and interregional connectivity. It was shown that proportions of synapses in local circuits could explain both the activation and connectivity differences between healthy controls and people with schizophrenia, even when interregional connections are intact.

15.5 Chapter Summary

Recent advances in human imaging methods have provided a wealth of data about the function of the living human brain. In particular, PET and fMRI studies yield quantitative data about activity within and dynamic connectivity between brain regions, and these data can be related to cognitive operations or disease states. However, these data do not provide information about the underlying neuronal events that generate the data. The spatial scale at which imaging data is acquired is well above the resolution of single neurons, and the temporal scale is on the order of seconds. Conversely, most functional data in animal studies is acquired at the single-neuron level or below, providing significant information about neuronal coding properties and the functioning of synapses at the millisecond level. There are also differences in what is being measured. This chapter introduced some approaches for dealing with these issues.

One implication of the quantitative nature of imaging data is that, given the appropriate transformations, it should be possible to directly compare human imaging data with data from animal studies by using models that account for the appropriate transformations. The goals of the models in this chapter are to simulate the animal data at the neuronal level and at the same time produce quantifiable measures that can be directly compared to human imaging data. One of the main issues in this endeavor is that the primary measure used for determining neuron coding features in animal studies is spiking activity whereas imaging data does not reflect this directly. Rather, PET and fMRI are thought to be related to energy requirements of neuronal activity, which can include spiking as well as processes involved in transporting and recycling neurotransmitters at the synapses. Furthermore, these processes are not measured directly, but through changes in blood flow (in PET) or blood oxygenation (in fMRI). Most current models that simulate imaging data use some form of synaptic activity to model the imaging data, an idea that was first introduced by Arbib and colleagues (Arbib et al., 1995). Because different types of synapses can either increase or decrease spiking activity, it is possible that decreases in spiking can be reflected as increases in PET or fMRI. The first model introduced in this chapter demonstrates circumstances under which this can happen. A local circuit that represents a local population of neurons and that has proportions of excitatory and inhibitory connections that are similar to those found in animal studies was introduced. The effect of modifying these proportions under different activity levels was then examined. It was found that measures of imaging and single-neuron activity can change in opposite directions if there is low excitatory input and a high degree of recurrence in the local circuit. After introducing a larger model based on these local circuits, it was later shown in section 15.4 how changes in the local circuit could explain imaging data in people with schizophrenia: lower local recurrent connectivity in the frontal area of the model can explain both the reduced activation and reduced connectivity of frontal cortex with other brain regions, characteristics that are among the most robust findings in imaging studies of schizophrenia.

The models described above were mainly focused on examining how human imaging data relate to local circuit properties. Other approaches that have used modeling to explain imaging data are examined in sections 15.2 and 15.3. A model by Edin et al. (2007) examined whether changes in synaptic strength, synaptic pruning, or myelination of long-range connections could best explain differences between children and adults that have been found in imaging studies. Although this model also used synaptic activity to represent the imaging data, the model was based on biophysical properties of individual neurons. Thus, specific neurotransmitters can be modeled explicitly, a factor that could be of importance in examining some disease states. Their findings suggest that a combination of synaptic strength and pruning best explains the developmental data. Another model, by Corchs and Deco (2004), was aimed at examining the mechanisms behind selective attention. This model used a population-level measure of synaptic activity to model the imaging data. Although quantitative comparisons were not made between the model and experimental data, the qualitative results showed time series of activity that are similar to fMRI series, and the relative activations of the regions of the model were similar to the experimental data.

Most neuronal-level methods that are used in animals cannot be applied in humans, so there is a crucial need for methods that can allow interpretation of human data. By combining low-level animal data with human brain imaging data, synthetic brain imaging can shed light on underlying mechanisms that human imaging methods cannot reveal. This chapter has reviewed some issues that need to be accounted for in order to make quantitative comparisons across these modalities, and applications to normal function, brain development, and disease states were reviewed. Although some aspects of the relationship between neuronal events and PET and fMRI signals are known, as discussed here, a full accounting of this relationship is still not completely understood. As more knowledge is gained in this area, future models will hopefully be able to explain the human data in even more detail.

REFERENCES

Aguirre, G. (2012, September 12). Number of neurons in a voxel. https://cfn.upenn.edu/aguirre/wiki/public:neurons_in_a_voxel.

Arbib, M. A., Billard, A., Iacoboni, M., & Oztop, E. (2000). Synthetic brain imaging: Grasping, mirror neurons and imitation. *Neural Networks, 13*(8–9), 975–997.

Arbib, M. A., Bischoff, A., Fagg, A. H., & Grafton, S. T. (1995). Synthetic PET: Analyzing large-scale properties of neural networks. *Human Brain Mapping, 2*(4), 225–233.

Arbib, M. A., Fagg, A. H., & Grafton, S. T. (2003). Synthetic PET imaging for grasping: From primate neurophysiology to human behavior. In F. T. Sommer & A. Wichert (Eds.), *Exploratory Analysis and Data Modeling in Functional Neuroimaging* (pp. 232–250). Cambridge, MA: MIT Press.

Attwell, D., & Laughlin, S. B. (2001). An energy budget for signaling in the grey matter of the brain. *Journal of Cerebral Blood Flow and Metabolism, 21*, 1133–1145.

Aubert, A., & Costalat, R. (2002). A model of the coupling between brain electrical activity, metabolism, and hemodynamics: Application to the interpretation of functional neuroimaging. *NeuroImage, 17*(3), 1162–1181.

Babajani-Feremi, A., Soltanian-Zadeh, H., & Moran, J. (2008). Integrated MEG/fMRI model validated using real auditory data. *Brain Topography, 21,* 61–74.

Baddeley, A. (2003). Working memory: Looking back and looking forward. *Nature Reviews. Neuroscience, 4*(10), 829–839.

Barrès, V., Simons, A., & Arbib, M. A. (2013). Synthetic event-related potentials: A computational bridge between neurolinguistic models and experiments. *Neural Networks, 37,* 66–92.

Bartels, A., & Zeki, S. (2005). Brain dynamics during natural viewing conditions—A new guide for mapping connectivity in vivo. *NeuroImage, 24*(2), 339–349.

Benes, F. M. (2000). Emerging principles of altered neural circuitry in schizophrenia. *Brain Research. Brain Research Reviews, 31,* 251–269.

Benes, F. M., McSparren, J., Bird, E. D., SanGiovanni, J. P., & Vincent, S. L. (1991). Deficits in small interneurons in prefrontal and cingulate cortices of schizophrenic and schizoaffective patients. *Archives of General Psychiatry, 48,* 996–1001.

Bokde, A. L., Tagamets, M. A., Friedman, R. B., & Horwitz, B. (2001). Functional interactions of the inferior frontal cortex during the processing of words and word-like stimuli. *Neuron, 30*(2), 609–617.

Bonaiuto, J., & Arbib, M. (2014). Modeling the BOLD correlates of competitive neural dynamics. *Neural Networks, 49,* 1–10.

Boynton, G. M., Engel, S. A., Glover, G. H., & Heeger, D. J. (1996). Linear systems analysis of functional magnetic resonance imaging in human V1. *Journal of Neuroscience, 16*(13), 4207–4221.

Brunel, N., & Wang, X. (2001). Effects of neuromodulation in a cortical network model of object working memory dominated by current inhibition. *Journal of Computational Neuroscience, 11,* 63–85.

Büchel, C., & Friston, K. J. (1998). Dynamic changes in effective connectivity characterized by variable parameter regression and Kalman filtering. *Human Brain Mapping, 6*(5–6), 403–408.

Büchel, C., & Friston, K. (2000). Assessing interactions among neuronal systems using functional neuroimaging. *Neural Networks, 13,* 871–882.

Buschman, T. J., & Miller, E. K. (2007). Top-down versus bottom-up control of attention in the prefrontal and posterior parietal cortices. *Science, 315,* 1860–1862.

Buxton, R. B. (2012). Dynamic models of BOLD contrast. *NeuroImage, 62*(2), 953–961.

Calhoun, V. D., Adali, T., Pearlson, G. D., & Pekar, J. J. (2001). A method for making group inferences from functional MRI data using independent component analysis. *Human Brain Mapping, 14*(3), 140–151.

Carlo, C. N., & Stevens, C. F. (2013). Structural uniformity of neocortex, revisited. *Proceedings of the National Academy of Sciences of the United States of America, 110*(4), 1488–1493.

Corchs, S., & Deco, G. (2004). Feature-based attention in human visual cortex: Simulation of fMRI data. *NeuroImage, 21*(1), 36–45.

Costalat, R., Francoise, J. P., Menuel, C., Lahutte, M., Vallée, J. N., de Marco, G., et al. (2012). Mathematical modeling of metabolism and hemodynamics. *Acta Biotheoretica, 60,* 99–107.

Deco, G., & Rolls, E. T. (2004). A neurodynamical cortical model of visual attention and invariant object recognition. *Vision Research, 44*(6), 621–642.

Deco, G., Rolls, E. T., & Horwitz, B. (2004). "What" and "where" in visual working memory: A computational neurodynamical perspective for integrating fMRI and single-neuron data. *Journal of Cognitive Neuroscience, 16*(4), 683–701.

Deco, G., & Zihl, J. (2001). A neurodynamical model of visual attention: Feedback enhancement of spatial resolution in a hierarchical system. *Journal of Computational Neuroscience, 10*(3), 231–253.

Desimone, R., & Duncan, J. (1995). Neural mechanisms of selective visual attention. *Annual Review of Neuroscience, 18,* 193–222.

Douglas, R. J., Koch, C., Mahowald, M., Martin, K. A., & Suarez, H. H. (1995). Recurrent excitation in neocortical circuits. *Science, 269*(5226), 981–985.

Edin, F., Macoveanu, J., Olesen, P., Tegnér, J., & Klingberg, T. (2007). Stronger synaptic connectivity as a mechanism behind development of working memory-related brain activity during childhood. *Journal of Cognitive Neuroscience, 19*(5), 750–760.

Ekstrom, A. (2010). How and when the fMRI BOLD signal relates to underlying neural activity: The danger in dissociation. *Brain Research. Brain Research Reviews, 62*(2), 233–244.

Fletcher, P. C., McKenna, P. J., Frith, C. D., Grasby, P. M., Friston, K. J., & Dolan, R. J. (1998). Brain activations in schizophrenia during a graded memory task studied with functional neuroimaging. *Archives of General Psychiatry, 55*(11), 1001–1008.

Friston, K. J. (1994). Functional and effective connectivity in neuroimaging: A synthesis. *Human Brain Mapping, 2,* 56–78.

Friston, K. J. (1998). The disconnection hypothesis. *Schizophrenia Research, 30,* 115–125.

Friston, K. J., Harrison, L., & Penny, W. (2003). Dynamical causal modeling. *NeuroImage, 19,* 1273–1302.

Friston, K. J., Jezzard, P., & Turner, R. (1994). Analysis of functional MRI time-series. *Human Brain Mapping, 1*(2), 153–171.

Funahashi, S., Bruce, C. J., & Goldman-Rakic, P. S. (1990). Visuospatial coding in primate prefrontal neurons revealed by oculomotor paradigms. *Journal of Neurophysiology, 36,* 61–78.

Girosi, F., Jones, M., & Poggio, T. (1995). Regularization theory and neural networks architectures. *Neural Computation, 7,* 219–269.

Goldman-Rakic, P. S. (1995). Cellular basis of working memory. *Neuron, 14*(3), 477–485.

Haxby, J. V., Ungerleider, L. G., Horwitz, B., Rapoport, S. I., & Grady, C. L. (1995). Hemispheric differences in neural systems for face working memory: A PET-rCBF study. *Human Brain Mapping, 3,* 68–82.

Hebb, D. O. (1949). *The Organization of Behavior.* New York: Wiley.

Houk, J. C., Bastianen, C., Fansler, D., Fishbach, A., Fraser, D., Reber, P. J., et al. (2007). Action selection and refinement in subcortical loops through basal ganglia and cerebellum. *Philosophical Transactions of the Royal Society of London. Series B, Biological Sciences, 362*(1485), 1573–1583.

Jueptner, M., & Weiller, C. (1995). Review: Does measurement of regional cerebral blood flow reflect synaptic activity? Implications for PET and fMRI. *NeuroImage, 2*(2), 148–156.

Kandel, E. R., Schwartz, J. H., & Jessell, T. M. (2012). *Principles of Neural Science* (45th ed.). New York: McGraw-Hill.

Lauritzen, M. (2001). Relationship of spikes, synaptic activity, and local changes of cerebral blood flow. *Journal of Cerebral Blood Flow and Metabolism, 21*(12), 1367–1383.

Lauritzen, M., Mathiesen, C., Schaefer, K., & Thomsen, K. J. (2012). Neuronal inhibition and excitation, and the dichotomic control of brain hemodynamic and oxygen responses. *NeuroImage, 62*(2), 1040–1050.

Lawrie, S. M., Buechel, C., Whalley, H. C., Frith, C. D., Friston, K. J., & Johnstone, E. C. (2002). Reduced frontotemporal functional connectivity in schizophrenia associated with auditory hallucinations. *Biological Psychiatry, 51*(12), 1008–1011.

Logothetis, N. K., Pauls, J., Augath, M., Trinath, T., & Oeltermann, A. (2001). Neurophysiological investigation of the basis of the fMRI signal. *Nature, 412*(6843), 150–157.

Magistretti , P. J. & Pellerin, L. (1999). Astrocytes couple synaptic activity to glucose utilization in the brain. *News in Physiological Sciences*, 14, 177–182.

Mantini, D., Corbetta, M., Romani, G. L., Orban, G. A., & Vanduffel, W. (2013). Evolutionarily novel functional networks in the human brain? *Journal of Neuroscience, 33*(8), 3259–3275.

Mantini, D., Hasson, U., Betti, V., Perrucci, M. G., Romani, G. L., et al. (2012). Interspecies activity correlations reveal functional correspondence between monkey and human brain areas. *Nature Methods, 9*, 277–282.

Mathiesen, C., Caesar, K., Akgoren, N., & Lauritzen, M. (1998). Modification of activity-dependent increases of cerebral blood flow by excitatory synaptic activity and spikes in rat cerebellar cortex. *Journal of Physiology, 512*(2), 555–566.

McIntosh, A. R. (2012). Tracing the route to path analysis in neuroimaging. *NeuroImage, 62*(2), 887–890.

McIntosh, A. R., & Gonzalez-Lima, F. (1994). Structural equation modeling and its application to network analysis in functional brain imaging. *Human Brain Mapping, 2*(1–2), 2–22.

Meyer-Lindenberg, A., Poline, J. B., Kohn, P. D., Holt, J. L., Egan, M. F., Weinberger, D. R., et al. (2001). Evidence for abnormal cortical functional connectivity during working memory in schizophrenia. *American Journal of Psychiatry, 158*(11), 1809–1817.

Peeters, R., Simone, I., Nelissen, K., Fabbri-Destro, M., Vanduffel, W., Rizzolatti, G., et al. (2009). The representation of tool use in humans and monkeys: Common and uniquely human features. *Journal of Neuroscience, 29*(32), 11523–11539.

Perlstein, W. M., Carter, C. S., Noll, D. C., & Cohen, J. D. (2001). Relation of prefrontal cortex dysfunction to working memory and symptoms in schizophrenia. *American Journal of Psychiatry, 158*(7), 1105–1113.

Peyron, R., Le Bars, D., Cinotti, L., Garcia-Larrea, L., Galy, G., Landais, P., et al. (1994). Effects of $GABA_A$ receptors activation on brain glucose metabolism in normal subjects and temporal lobe epilepsy (TLE) patients: A positron emission tomography (PET) study: I. Brain glucose metabolism is increased after $GABA_A$ receptors activation. *Epilepsy Research, 19*(1), 45–54.

Poggio, T., Torre, V., & Koch, C. (1985). Computational vision and regularization theory. *Nature, 317*, 314–319.

Poline, J.-B., & Brett, M. (2012). The general linear model and fMRI: Does love last forever? *NeuroImage, 62*(2), 871–880.

Reynolds, J. H., Chelazzi, L., & Desimone, R. (1999). Competitive mechanisms subserve attention in macaque areas V2 and V4. The *Journal of Neuroscience, 19*(5), 1736–1753.

Roland, P. E., & Friberg, L. (1988). The effect of the GABA-A agonist THIP on regional cortical blood flow in humans: A new test of hemispheric dominance. *Journal of Cerebral Blood Flow and Metabolism, 8*(3), 314–323.

Saenz, M., Buracas, G. T., & Boynton, G. M. (2002). Global effects of feature-based attention in human visual cortex. *Nature Neuroscience, 5*(7), 631–632.

Sokoloff, L. (1993). Sites and mechanisms of function-related changes in energy metabolism in the nervous system. *Developmental Neuroscience, 15*(3–5), 194–206.

Sotero, R. C., & Trujillo-Barreto, N. J. (2007). Modelling the role of excitatory and inhibitory neuronal activity in the generation of the BOLD signal. *NeuroImage, 35*(1), 149–165.

Steel, R. M., Bastin, M. E., McConnell, S., Marshall, I., Cunningham-Owens, D. G., Lawrie, S. M., et al. (2001). Diffusion tensor imaging (DTI) and proton magnetic resonance spectroscopy (1H MRS) in schizophrenic subjects and normal controls. *Psychiatry Research, 106*, 161–170.

Stephan, K. E., Friston, K. J., & Frith, C. D. (2009). Dysconnection in schizophrenia: From abnormal synaptic plasticity to failures of self-monitoring. *Schizophrenia Bulletin, 35*(3), 509–527.

Tagamets, M. A., & Horwitz, B. (1998). Integrating electrophysiological and anatomical experimental data to create a large-scale model that simulates a delayed match-to-sample human brain imaging study. *Cerebral Cortex, 8*(4), 310–320.

Tagamets, M. A., & Horwitz, B. (2000). A model of working memory: Bridging the gap between electrophysiology and human brain imaging. *Neural Networks, 13*(8–9), 941–952.

Tagamets, M. A., & Horwitz, B. (2001). Interpreting PET and fMRI measures of functional neural activity: The effects of synaptic inhibition on cortical activation in human imaging studies. *Brain Research Bulletin, 54*(3), 267–273.

Tagamets, M. A., Novick, J. M., Chalmers, M. L., & Friedman, R. B. (2000). A parametric approach to orthographic processing in the brain: An fMRI study. *Journal of Cognitive Neuroscience, 12*(2), 281–297.

Ungerleider, L. G., & Mishkin, M. (1982). Two cortical visual systems. In J. Ingle, M. A. Goodale, & R. J. W. Mansfield (Eds.), *Analysis of Visual Behavior* (pp. 549–586). Cambridge, MA: MIT Press.

Ungerleider, L. G., & Pasternak, T. (2004). Ventral and dorsal cortical processing streams. In L. M. Chalupa & J. S. Werner (Eds.), *The Visual Neurosciences* (pp. 541–562). Cambridge, MA: MIT Press.

Wilson, H. R., & Cowan, J. D. (1972). Excitatory and inhibitory interactions in localized populations of model neurons. *Biophysical Journal, 12*, 1–24.

Winder, R., Cortes, C. R., Reggia, J. A., & Tagamets, M. A. (2007). Functional connectivity in fMRI: A modeling approach for estimation and for relating to local circuits. *NeuroImage, 34*(3), 1093–1107.

16 Embodied Models and Neurorobotics

TONY J. PRESCOTT, JOSEPH AYERS, FRANK W. GRASSO, AND PAUL F.M.J. VERSCHURE

Whereas most other chapters in this book study computer or mathematical models of various aspects of brain function, here we focus on the value of physically embodied models, in which simulated brain circuits are embedded within robotic bodies. The general goal in doing this is to better understand the interaction of the brain, the body, and the environment in generating the adaptive behavior we see in animals. We call these *embodied computational neuroscience* models (Prescott, Montes Gonzalez, et al., 2006), and the emerging domain of research in such systems *neurorobotics* (Ayers, Davis, and Rudolph, 2002; Dario, Carrozza, et al., 2005; Krichmar, 2008). Although the main emphasis of this chapter will on the value of such models to neuroscience, it is worth noting that the construction of these artifacts to address scientific aims can also lead to the availability of a new class of technologies that can address outstanding challenges in engineering (Prescott, Lepora, et al., 2014).

The structure of our chapter is as follows. First, we introduce the idea of developing embodied models of animals and their nervous systems, comparing and contrasting these with other types of models studied in neuroscience. Second, we discuss the relationship between brains and bodies, explaining why the notion of embodiment is critical if we are to fully understand the role of brains in determining or influencing behavior. Third, we explore, in some detail, several example robot models of invertebrate sensorimotor coordination, showing how useful models can be developed at different levels of description—for instance, in terms of information or in terms of circuits—and how the best models can combine multiple levels of description. Fourth, we investigate how the layered architecture of the mammalian brain can be approximated in robot models that incorporate key aspects of animal behavior and morphology. Finally, we consider how such models could be scaled up to address the challenge of understanding the whole mammalian brain.

16.1 Physical Models in Neurobiology

As explained in the introductory chapter, the goal of modeling is not to replicate as accurately as possible what we see in biology. Rather, it is to help us to understand the underlying principles that are exploited by biological systems to maintain life.

Today, modeling often takes the form of computer simulation; however, physical models have also played an important role in biological science. For instance, the discovery by Watson and Crick of the structure of our genetic material was based on the construction of a physical model comprised of little metal sheets. This three-dimensional structure allowed Watson and Crick to conceive of a shape—the double helix—that could account for the x-ray diffraction data of DNA gathered by Rosalind Franklin (Crick, 1990). The Watson and Crick model facilitated the scientific discovery process without looking to accurately mimic biological reality; rather, it employed a suitably defined abstraction.

In 1945, eight years before Watson and Crick's breakthrough, Rosenblueth and Wiener proposed that the criteria for a good model was not how accurately it captured all the details of the biological phenomenon of interest but how useful it was in casting new light on that phenomenon. The Watson and Crick model succinctly demonstrates this principle—to intuit the structure of the DNA molecule became much easier once they had a physical model they could manipulate.

A further strength of all good models is that they capture the major common features or properties of the system under study and summarize the individual variation. This is true not only of physical models, robots, and computer-based simulations but also of biological models. Neurobiologists, who can sometimes be dismissive of simulation as "only a model," should, if they were to be logically self-consistent, also dismiss the model systems they study themselves. For instance, someone might use one animal as a model of another (e.g., the rat as a model of a human), an animal in an altered condition (e.g., under anesthesia or awake but immobilized) as a model of the same animal but awake and behaving, or an isolated part (e.g., a brain slice or a muscle) as a model of that component within an intact, functioning system. Hence the neurobiologist, like the modeler, takes significant steps away from the actual system of interest—species difference, anesthesia/

immobility, dissection—for the practical benefits of having a system that is easier to study.

For similar reasons of accessibility and ease of study, Rosenblueth and Wiener argued that a good model should translate a question about the natural world into a domain that we understand better, and it should allow us to conduct experiments with relative ease. In this chapter we contend that—for exploring a range of relationships between brain, body, and behavior—robotics, which translates problems in biology into problems in computing and engineering, meets the requirements of being easier to study and to understand particularly well.

Let's begin from the standpoint of the neurobiologist looking at potential model systems for understanding the role of the brain in generating natural (ethologically relevant) behavior. Table 16.1 compares and contrasts neuroscientific models and robotic models in order to show some of their advantages and disadvantages. Our table illustrates that robotic models can nicely capture interesting properties of awake, behaving animals that are lost as we move to more accessible neurobiological models. In particular, robot models are more amenable to investigation of their internal processes, such as the current state of any calculations being performed, than any existing animal model. The main issue for robot models, though, is how to capture enough of the relevant properties of the neural and physical substrates of the target system to create a useful model.

Note that for each of the different neuroscientific model systems considered in table 16.1 we might also build a computer simulation to meet our requirement for accessibility and ease of experimentation. Indeed, in this age of high-power computer systems with real-time simulated physics engines, is it ever really worthwhile physically copying aspects of the animal? Our answer is "not always" or perhaps even "generally not." Depending on your question (and one issue, as we will see, is whether you are asking the best question), you may be better off staying in the world of simulation as many of the excellent computational models in this book show. In this chapter we will argue, nevertheless, that there are circumstances where a physical model is going to be a better model, in the sense of being easier to build, study, and understand, than a simulation.

As will become clear, tackling the problem of creating a physical model may sometimes also drive the researcher to ask a different set of questions that result in a new way of thinking about that system that is closer to biological reality. Building a physical model can also take us a step closer to creating a novel behaving system that may be of some real benefit in the world (and one of the goals of science must be to create this kind of useful artifact). In general, however, it is rarely a question of either/or. In developing robotic models, we will often use simulation as a tool for exploring the design space for biomimetic artifacts before we build them, and we will look to use our completed physical model to validate and extend the insights obtained through simulation.

16.2 *Embodiment Matters*

The existence of directed swimming in single-celled eukaryotes such as *Paramecia* (Jensen, 1959) reminds us that nervous conduction is not a prerequisite for movement or action. Indeed, the first nervous systems evolved some hundreds of millions of years after the appearance of the first active mobile animals (Bengston, 1994). What the nervous system certainly does, however, is to enable movement to be more rapid, more efficient, and better adapted to the environment.

TABLE 16.1

Advantages and disadvantages of different neuroscientific models compared to robotic (physical) models on an ordinal scale (more asterisks = better)

System	Properties				Amenable to Experimental Investigation of Internal Processes
	Similar Neural and Physical Substrates	Intact System	Embedded in the Environment	Expressing Ethologically Relevant Behavior	
Awake, free-moving organism (target)	*****	*****	*****	*****	*
Animal models					
Awake, restrained	*****	*****	*****	***	**
Under anesthetic	*****	*****	***	*	***
Brain slice or similar	*****	*	**	*	****
Robotic model	*	*****	*****	****	*****

Note. Adapted from Mitchinson, Pearson, et al. (2011).

The importance of the body in generating coordinated movement is beautifully illustrated in machines that exploit the natural dynamics of their parts for periodic motion. For example, McGeer (1990) showed that a bipedal walking machine, with no control system whatsoever, can generate a stable walking gait on a suitably sloped surface by relying on the passive dynamics of suitably configured mechanical parts. This synergy between form and function has also been called morphological computation (Pfeifer and Bongard, 2006). Animals provide control through their nervous systems in a manner that complements their natural body dynamics. In walking or running, for example, many legged animals exploit the pendulum-like natural motion of jointed limbs to help generate a suitable cyclic pattern. In a similar way, a parent pushing a child on a swing controls the back-and-forth motion by periodically interjecting energy into the natural dynamics of the system. By relinquishing some aspects of control to the body, animals also benefit from the energy recycling capability of elastic tissues. Muscles and tendons, for instance, convert kinetic energy to potential energy as the foot hits the ground, providing a store of energy to be released in the next step cycle. Designing controllers modeled on animal locomotion pattern generators that exploit these principles provides a very promising path for building efficient legged robots (see chapter 8, "Motor Pattern Generation"; Ijspeert, 2014; and section 16.3 below). Such controllers can be simpler—for instance, having fewer control parameters—than more traditional forms of continuous robot control and will entrain themselves to the dynamics of the body, making them highly adaptable. A broader lesson is that rather than thinking of the nervous system as the coordinator of the body, in the way that a conductor directs an orchestra, we might think of it more as "one of a group of players engaged in jazz improvisation" (Chiel and Beer, 1997, p. 555), influencing, but also being influenced by, the body and environment in which it is embedded.

Exploiting the physical properties of body parts leads to a simplification of the problem of locomotion for walking and running animals and machines. Similarly, exploiting the physical properties of the world allows animals and robots to generate apparently complex behavior through simple principles. This has been recognized many times, but some examples are worth noting.

The neurologist Grey Walter built one of the first brain-inspired robots in the 1950s when he created his robot "tortoises" Elmer and Elsie (Walter, 1953). Each tortoise (figure 16.1, left) had a single light-sensitive "eye," a shell that was sensitive to touch, and a pilot light that was configured to extinguish when a second light source was detected. Control was provided by simple analogue electrical circuits, termed relays, designed to mimic some of the properties of biological neural networks. As shown in figure 16.1 (center), these relays implemented a layered control hierarchy. The lowest level behavior provided forward movement (*explore*), if light was detected, this was suppressed and replaced by a movement toward the light (*phototaxis*), and if the shell touched against an obstacle, the robot would stop anything else it was doing and exhibit a simple escape maneuver (*wriggle*). By photographing his robot in darkness, with a lengthy shutter time, Grey Walter was able to record its trajectory when moving in environments such as that illustrated in figure 16.1 (right panel). Here, after interacting with various obstacles the robot eventually approaches a light source. In other photographs the robot can be seen homing in on its "hutch" in order to recharge its battery or traversing alongside a mirror with an oscillating locomotion movement pattern—alternately approaching and then moving away from its own reflection—that Grey Walter termed a "dance."

This idea of configuring a control system as a set of reflexes with a fixed priority scheme was adopted with enthusiasm in the 1980s by researchers interested in generating robust behavior for fast-acting mobile robots

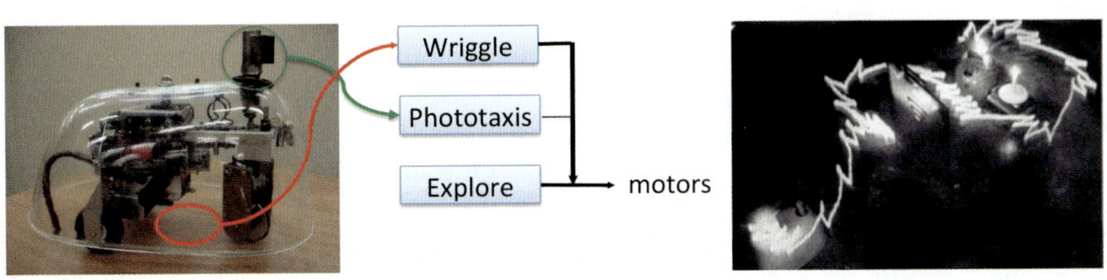

FIGURE 16.1 Grey Walter's tortoise "Elsie." The robot (left) generates apparently complex behavior as the result of a simple layered control system (center) interacting with an environment of obstacles and light sources. The photograph on the right illustrates the robot's trajectory recorded using a long shutter speed, with a light source placed on the robot's outer shell. From Walter (1953). The photograph on the left is of a reproduction tortoise built by Ian Horswill from Bristol Robotics Laboratory, the photograph on the right is reproduced courtesy of the Burden Institute.

(e.g., Brooks, 1986) that can engage in simple but reliable activities such as wall following (see also Mataric, 1997). Similar principles may also have underpinned the behavior of bilaterian creatures who left spiraling or meandering foraging trails that then became trace fossils in Precambrian rocks so providing our earliest evidence of the behavior of multicelled animals (see figure 16.2 and Prescott and Ibbotson, 1997). These fossil meanders belong to the class of *stigmergic* processes in which the behavior of the organism restructures its environment, which, in turn, affects the activity of that organism (Theraulaz and Bonabeau, 1995). Stigmergy can be thought of as a form of "information offloading" (Hutchins, 1995; Beer and Williams, 2015), whereby some aspects of the processing that are required to implement a behavior are externalized by altering restructuring the environment. In other words, we should think of the system formed by the *organism in its environment* as the substrate for the computation that results in the observed behavioral pattern, not just the organism and its nervous system.

A

B

FIGURE 16.2 Robot modeling of stigermic processes. (A) Fossilized animal from the Precambrian era likely to have been left by an early bilaterian animal, from Crimes and Anderson (1985), reproduced with permission from the Society for Sedimentary Geology. (B) A custom-built robot configured to leave a meandering paper trail using a fixed control hierarchy of reactive mechanisms. For details see Prescott and Ibbotson (1997), reproduced with permission from MIT Press.

Grey Walter's tortoises showed that the interaction of a reactive, but appropriately configured, control system with the environment, via suitable, but often primitive, sensors and actuators, can give rise to behavior that is surprisingly complex. Thirty years later, in a series of thought experiments, the neurobiologist Valentino Braitenberg (1986) imagined a series of robot "vehicles" of increasing sophistication, beginning with machines with only sensory and motor elements and finishing with variants that had simple artificial nervous systems capable of properties such as logic, memory, and abstraction. These internal processing elements hint at some likely functions of animal brains that go beyond reactive control. Indeed, in many animal nervous systems, theoretical and computational analyses point to the presence of "hybrid" control architectures that combine elements of reactive control with integrative mechanisms that operate both in space, coordinating different parts of the body, and in time, organizing behavior over multiple timescales (for discussion, see Verschure, Krose, et al., 1992; Arbib and Liaw, 1995; Prescott, Redgrave, et al., 1999; and sections 16.3 to 16.5 below).

Braitenberg relied on his imagination to conceive of how his vehicles might behave. The ability to do this degrades rapidly as internal or external complexity is added to the scenario envisaged; imagination is also susceptible to bias. Embodied computational neuroscience, on the other hand, via the medium of robotics, allows us to explore theories of brain architecture that reflect more of the true complexity of animal nervous systems. By embedding hypothesized control mechanisms into the dynamics of a real-world interaction, robotic models enable us to take seriously the role of body and environment in generating behavior. We can even do this in the same environment as that inhabited by the organism whose behavior we wish to understand. We present some specific examples of this possibility in section 16.2 below.

One of the first examples of neurorobotic modeling is the work by Edelman and coworkers on the NOMAD series of real-world artifacts (Edelman, Reeke, et al., 1992). These robot models, also known as *brain-based devices* (see also Verschure, Wray et al., 1995; Seth, McKinstry, et al., 2004; Krichmar, 2008; Krichmar and Wagatsuma, 2011), aimed at validating the theory of neuronal group selection, or "neural Darwinism," that

Edelman had proposed. In particular, they tested the putatively fundamental principle that differential selection of neuronal connections can lead to repertoires of neurons that are tuned to specific states of the world. Reentry or recurrence across these populations can allow the network to achieve higher-level function, and their behavior can be shaped through value-based learning such as that derived from reward and punishment.

Edelman's brain-based devices helped to demonstrate that the interaction with the environment is important for understanding not just how we control behavior but also how we *acquire* new forms of control. From birth, action shapes the animal's experience, which, in turn, has a profound influence on the coding systems used by the brain (Verschure, Voegtlin, et al., 2003; Lungarella, Pegors, et al., 2005). These self-organize so as to efficiently represent the kinds of signals that occur most frequently, and also, perhaps, those signals that will be most effective in predicting how the sensory world will unfold in the near future (see also Clark, 2013) or that are most behaviorally relevant to the activity in which the animal is currently engaged (Prescott, Diamond, et al., 2011). In the brain both of these forms of learning occur, and we will later describe *distributed adaptive control* (DAC) theory (Verschure, Krose, et al., 1992; Verschure, Voegtlin, et al., 2003; Maffei, Santos-Pata et al., 2015), which explores how different learning and memory systems, in the mammalian brain, work together to generate perception, cognition, and action. In chapter 13 ("Neural Maps: Their Function and Development"), Bednar and Williams also detail how self-organization shapes the function and development of neural maps that underlie the major mammalian perceptual systems.

16.2.1 The Role of Morphology in Computation

Over time, evolution selects physical morphologies that create and enhance behaviorally relevant signals prior to any processing by the nervous system.

For instance, the harbor seal has perfected the skill of tracking the hydrodynamic trails left by fish prey as they move through the water, allowing them to detect and follow an animal whose movement has disturbed the water up to several minutes previously (Dehnhardt, Mauck, et al., 2001). The sensory mechanism by which seals achieve this task is a specialization of the facial vibrissae, which in seals have richly innervated follicles that allow the detection of minute changes in water currents. The capacity to detect these changes is enhanced by an adaptation in the morphology of the whisker whereby the smoothly tapered vibrissal shaft of land mammals has been replaced by a nontapered and undulating shape (see figure 16.3A). Hydrodynamic tests of this whisker shape show that it has a remarkable capacity to eliminate the sensory noise due to the animal's own movement (Hanke, Witte, et al., 2010). Whereas a smooth and tapered whisker would bend as the animal swims through the water due to the effects of drag, the undulating shape of the seal whiskers creates microvortices that cancel and largely eliminate drag, giving a much improved signal-to-noise ratio for detecting disturbances in the water created by prey

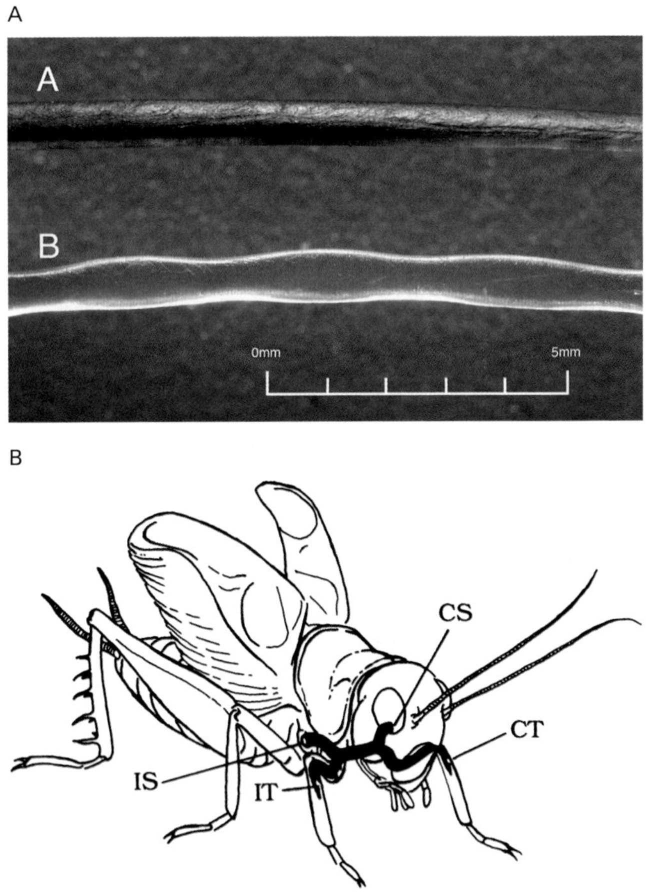

Figure 16.3 Examples of morphological computation in animal sensory systems. (A) The smooth whisker of a Californian sea lion (top) compared with the undulating profile of the whisker of a harbor seal (bottom), the presence of undulations can reduce turbulence caused by the whisker and thus improve sensitivity to hydrodynamic flow. Photograph from Murphy, Eberhardt et al. 2013). (B) Sound can reach the ear of the cricket at four locations: IT, the ipsilateral tympanum (the surface of the eardrum), IS, the ipsilateral spiracle (an opening in the leg connected to the ear via a tracheal tube), CT and CS—tubes connecting the contralateral tympanum and spiracle. The physical organization of this system provides enhanced directional-sensitivity to sounds at the wavelength produced by the male cricket. Reprinted with permission from Michelsen et al. (1994). Reprinted with permission from *J. Comp. Physiol. [A]*. Copyright 1994, Springer-Verlag.

animals. These changes to the sensory system of the seal are matched by adaptations of the body, such as streamlining for efficient swimming, that demonstrate how mechanisms for perception coevolve with those for action.

A second example of morphology simplifying the problem faced by the nervous system is provided by the mating behavior of the female cricket, which requires it to identify and track down a male of its species by attending to its distinctive chirping sound. The task of distinguishing and orienting toward a male "song" is facilitated by the physical structure of the cricket's tracheal tube, whose four openings (see Webb, 1998 and figure 16.3B) differentially amplify sounds arriving at the left and right ears depending on the sound-source direction (Michelsen, Popov, et al., 1994). The physics of this system is such that directional selectivity is greatest for auditory vibrations at the specific wavelength produced by the male animal, thus providing a nonneural mechanism that assists the female in orienting toward ethologically appropriate stimuli. Using robot models, Webb and her coworkers have explored the relative contributions of physical morphology and of neural circuits in generating an integrated phonotaxis behavior similar to that of crickets (see Webb, 2002, for a review). Further examples of the importance of morphology for simplifying perception and control in animals and robots are discussed by Chiel and Beer (1997) and Pfeifer and Bongard (2006).

16.2.2 When Is the Real World Better than Simulation? ENIAC, the first general purpose programmable electronic computer, was used by the U.S. military to calculate artillery firing tables based on measures such as the azimuth of the gun barrel, the weight and shape of the missile, and the strength and direction of the wind (Reed, 1952). Thus, since their origins, one of the important uses of computers has been to simulate the physics of the real world. Today, a range of simulated physics systems exist from the quick-and-dirty, but real-time, engines that drive gaming environments to computationally intensive programs that provide accurate, but domain-limited, simulation for specific problem spaces such as fluid dynamics, weather forecasting, or the dynamics of global warming. Simulators are available that can be configured to replicate the physical design and sensory and actuating systems of standard robot platforms (see, e.g., Michel, 1998; Koenig and Howard, 2004). Special purpose simulators have also been developed that capture aspects of the physical morphology of target animals such as the lamprey (Beauregard, Kennedy, et al., 2006), quadrupeds such as the salamander (see chapter 8 and Ijspeert, 2000), and, of course, humans (Fenner, Brook, et al., 2008; Tikhanoff, Cangelosi, et al., 2008). However, current simulations of real-world physics struggle to capture all of its mechanical and dynamical properties such as collision elasticity, surface friction, absorption, and so forth. Interactions between objects that are deformable, possess irregular surfaces, or are subject to environmental disturbances are a particular challenge (Ijspeert, 2014). Validation of these properties, which may be essential for trusting the scientific conclusions obtained with simulations, can be an immense undertaking.

The use of simulation for exploring nervous system, body, and environment interactions is therefore not always a "no-brainer." Simulators for games platforms are designed to give a believable impression of real-world physics but cut corners on the true physics in their pursuit of perceptual acceptability; special-purpose simulators are likewise generally not developed with the application of testing embodied brain models as their focus. Adapting any simulator to provide an embedded testing environment for a computational neuroscience model is therefore likely to require some work. Consider, for instance, the problem of how animals are able to track odor plumes in turbulent liquid flows. Experimental studies of animal behavior indicate that animals are sampling, and altering their behavior, at timescales (subsecond) and spatial scales (submillimeter) that are outside the limits of currently available simulators of real flows (see below). Below we explore how research in embodied robotics has allowed scientists to sidestep this problem by embedding suitably configured robots in real environments.

Although the world comes "for free" with a robot model, something that replicates key properties of the body has to be provided as the interface between model nervous system and the environment. Ideally, the properties of the physical robot model should match relevant biomechanical constraints, identified in the animal, at least as far as these are thought to be relevant for the behaviors targeted. One of the trade-offs in modeling is, then, to decide when the benefits of having an embodied model, in terms of access to real-world physics, justifies the cost of building and running a bespoke physical system. As we will see in the examples discussed below, valuable results can often be obtained by attending to a key set of morphological constraints rather than mimicking an entire creature. Indeed, holding to our view that a physical model should be useful rather than accurate, one usually abstracts away many of the physical details in order to get a good understanding of both the problem that a biological system is solving and the way in which this is accomplished. However, the challenge is also to be cogent and clear about these assumptions and how they affect the results and their interpretation (see Prescott, Montes Gonzalez, et al., 2006, for further discussion of this issue).

16.3 Embodied Models as Means to Understand Animal Sensorimotor and Information-Processing Capabilities

Often, when we look at the animal world, we see abilities that neither we, nor our technologies, possess. Hydrodynamic trail-tracking in seals is one example we have already mentioned; another is the memory abilities of Clark's nutcrackers, a bird who can recall the locations of up to 30,000 seeds stored in the summer the following winter over a 25-km area (Balda and Kamil, 1992). Such feats humble those of us who lose our car keys between the time we return home from work or school in the evening and the time we leave the next morning! The homing ability of salmon is another often cited, but underappreciated, example. These fish are able to swim thousands of miles across the (to us) featureless expanse of the Pacific Ocean to an exact location to which they had just a brief exposure years before, using learned magnetic, olfactory, and visual cues (Crossin, Hinch, et al., 2007). The secrets possessed by these birds and fish are not yet understood to a degree that permits their instantiation in designs for implementable technology. Therefore their study in streamlined simpler devices that are not encumbered by biological processes that are essential for the animal but not required for the task (waste elimination, reproduction, etc.) is an attractive approach to understanding how they do it.

16.3.1 Plume-Tracking in a Robot Lobster Nor are vertebrates the sole keepers of the secrets to information-processing abilities not realized by human technology. Invertebrates also possess abilities that we have yet to understand and that we would like to have in our technologies. American lobsters, as well as several other crustacean species, are able to locate the source of an attractive smell by walking to it, following the odor over distances of many meters (Grasso and Basil, 2002). This does not seem so startling until one realizes the complicated and unpredictable path that each whiff of aroma travels between the aromatic object and the lobster's "nose." The odor does not just make its leisurely slow way through molecular diffusion to the animal (this would take an inordinate amount of time) but it is carried by water flow. Thus, the lobster's behavior, also known as *rheotaxis*, is "go with (or against) the flow." In the real ocean, where the lobster makes its living, real water flows are also chaotic, or, in the language of fluid mechanics, they are *turbulent*.

Turbulence is not completely random but consists of eddies and vortices that appear and disappear on many scales and that interact with each other. The structure of turbulent flow is related to the Reynolds number, *Re*, where

$$Re = \frac{\rho U l}{\mu}$$

Here U is the mean velocity of the object relative to the fluid, l is a characteristic length used to specify spatial scale, is the dynamic viscosity of the fluid (its "stickiness"), and is the density of the fluid.

The value of *Re* gives an indication of the expected degree of turbulence. The ratio here is one of inertial forces (the numerator) to viscous forces (the denominator). If the value is less than one, then the viscous forces dominate and diffusion determines the rate and pattern of aroma transport from the source to the lobster's sensor. If it is greater than one, however, inertial forces dominate, and the flow patterns determine the rate and pattern of movement of aromas through space. The higher the Reynolds number, the more chaotic the flow. For each given range of numbers (say, 10–100, 100–1,000, etc.), however, the pattern of flow, and therefore the movement of aromas through space, is "dynamically similar" regardless of the medium. In other words, within each range, characteristic dispersal patterns emerge that are the same if the fluid is air, water, or liquid mercury or if the spatial scale is a few centimeters in a fish tank or the atmosphere of the planet Jupiter. A fuller discussion of turbulence and of experimental and simulation approaches to understanding turbulence is provided by Benzi and Frisch (2010).

It makes sense, then, that animal brains would evolve to make use of this universal patterning, and, indeed, they have (see Vogel, 1994, for a fascinating comprehensive tour of the relationship between fluid flow and the design of biological systems). In the case of the lobster, it gets *better* at tracking the longer it is engaged in following a given turbulent odor plume; specifically, it moves faster, as if more confident, and also reduces its steering error relative to the target. This suggests that memories of recent encounters with whiffs of odor inform its later decisions (Moore and Atema, 1991; Basil and Atema, 1994). Unfortunately, the principle of "dynamic similarity" does not allow us to model the exact time series of aromas encountered by the animal using a simulation; rather, it merely tells us that the patterns are "similar." Further, the more powerful Navier–Stokes equations, which describe all of turbulent flow, provide statistical averages of flows at points in space and time but not the *exact sequence* of inputs a single animal moving through a plume would receive. In the case of chemical-plume tracking in turbulent flows, then, standard modeling techniques cannot provide us with an accurate series of concentration values along the path an animal would take, only a series of independent estimates of the averages. A test of the hypothesis that the lobster remembers the aroma it encounters during tracking, if conducted in such a simulator, might miss out on the key information that the lobster uses and therefore be no test at all.

On the other hand, applying a suitable *robotic* model of lobster information processing should be an effective way to adequately test the hypothesis. Simply place the physical model into real plumes and let its behavior provide evidence for the effectiveness of the proposed model in explaining the information processing that occurs in the animal (figure 16.4). Indeed, when researchers did this, they found a role for memory in a robot lobster (Grasso and Atema, 2002) that might not have been obvious if the researchers had studied a simulated plume with a simulated lobster. Specifically, Grasso and Atema (2002) compared three different control schemes, exploiting (1) a single chemosensor, (2) a stereo sensor pair, and (3) a stereo pair with memory, as illustrated in figure 16.5. A measure of overall tracking performance compared with a lobster showed significant benefit of the stereo sensor pair, and of the additional memory component, although the best model fell some way short of lobster performance, suggesting that there may yet be better algorithms to be discovered (perhaps by looking in more detail at the neural substrates).

In this study the robot was a relatively simple device with two driven wheels capable of moving on the bottom of a fume or flow tank or a relatively smooth portion of the ocean floor as shown in figure 16.4, with sensors for flow detection that broadly replicated the sensitivity of lobster antennae rather than seeking to match specific chemosensory mechanisms. Key features such as the scale and positioning of the sensors relative to the flow were tuned to match the biological target. The results do not rule out alternative models of lobster rheotaxis but, at the same time, they do provide proof of principle of the memory-based method as a plausible strategy that can work with real-world flows. More generally, these results suggest that as the physics of the part of the world that is of interest becomes more difficult to model, an effective approach is to use the world as its own model and to test physical models of embodied sensorimotor processing against it.

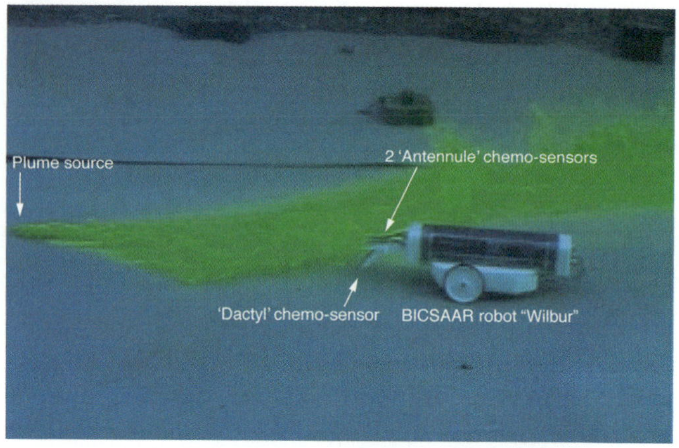

FIGURE 16.4 Robot "lobster" tracking a chemical plume. The problem of locating the plume source in this robot is decomposed into transforming a series of chemical concentration signals (visualized by dye in the photo) into a series of differential commands to the wheels of the robot which form a path in space. This robot demonstrated that a "memory" of recent concentration series led the robot to more efficient paths to the source, revealing the plausibility of the information-processing method and memory for the biological lobster. BICSAAR, biologically inspired chemical-sensing aquatic autonomous robot. Reprinted from Grasso and Basil (2002 with permission from Elsevier.

16.3.2 MODELING THE NEURAL CIRCUITS UNDERLYING BILATERIAN PATTERN GENERATION The example of lobster rheotaxis, described above, illustrates how robots can be used to evaluate theories of animal behavior couched at an algorithmic or informational level; however, embodied modeling can also be used to directly test the capacity of model neural circuits to generate adaptive behavior.

The nervous systems of all bilateral animals have been characterized as having an innate neuronal architecture based on *command* neurons, *coordinating* neurons, and *central pattern generators* (CPGs) modulated by phase and amplitude modulating reflexes (Kennedy and Davis, 1977; Stein 1978; Pearson, 1993; Stein, Grillner, et al., 1997; Jordan, Liu, et al., 2008). These network components appear to underlie innate behaviors such as the control of posture, repetitive movement, and intersegmental coordination (Ayers, 2002). In invertebrates, these components have been demonstrated and analyzed at the level of identifiable neurons and underlying molecular processes (Selverston and Ayers, 2006) and key underlying mechanisms have been shown to be conserved between invertebrates and vertebrates (Pearson, 1993; Grillner, Hellgren, et al., 2005a; Grillner, Markram, et al., 2005b). Biological

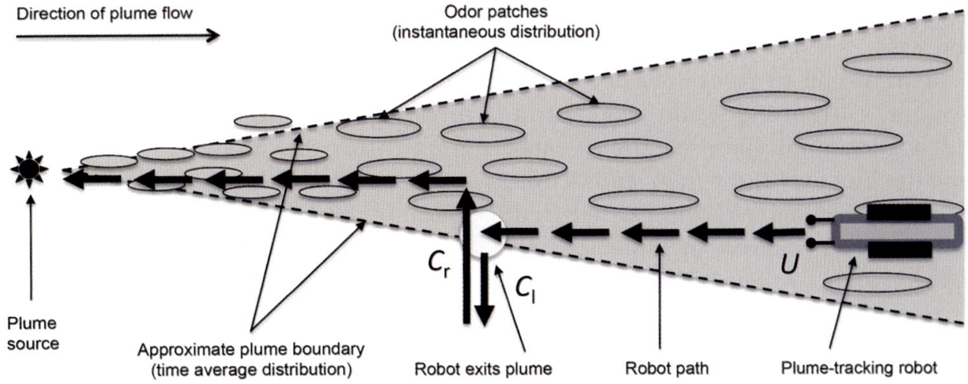

FIGURE 16.5 Illustration of the plume-tracking strategy used by the robot. Schematic illustration of the discrete *odor gated rheotaxis* strategy proposed by Grasso and Atema (2002). The sequence of motion is indicated by the progression of arrows. Chemical samples are collected at each successive location to inform the next move. Upstream motion (U), triggered by detection of the plume, is in a series of upstream steps. Failure to detect the plume at one of the sampling positions leads to an across-stream motion called a cast. The choice of casting left (C*l*) or right (C*r*) is one of the parameters that is under algorithmic control. A strategy using memory of past patterns of stimulation to control the direction of cast gave results closest to those found with living animals. Reprinted from Grasso and Atema (2002) with permission from Springer Science and Business Media.

cellular CPGs have been described in model animals where the relevant circuitry can be established by pairwise recordings between neurons (Selverston, Russell, et al., 1976; Buchanan and McPherson, 1995). An important aspect of these biological model systems is that they accommodate complex integrative phenomena such as neuromodulation (Hasselmo, 1995; Dickinson, 2006; Harris-Warrick, 2011) and that the details of their function can be characterized by anatomical (Weeks, 1981) and cellular dissection (Selverston, 1980). (Note that, in chapter 7, Fellous, Canavier, and Hasselmo describe a number of rather different functions of neuromodulation in mammals.)

Key to developing a robot model that can capture the functionality of these circuits is the capability to compute in real time and to achieve realistic mechanistic models of phenomena such as neuronal integration (Ayers, Rulkov, et al., 2010) and excitation/contraction coupling of artificial muscle (Witting, Ayers, et al., 2000) within a physical plant that adequately approximates the biomechanics of the animal. This can be achieved by using phenomenological models of neurons and their synapses, incorporated into a CPG control architecture organized around exteroceptive reflexes (i.e., those concerned with external stimuli), and embedded within a biomimetic robot body plan (Ayers and Rulkov, 2007; Ayers and Witting, 2007; Westphal, Rulkov, et al., 2011).

A critical component of bilaterian innate behavior is the CPG that drives central motor programs that mediate the behavioral contributions of single body segments (Selverston, 2010). Although some CPGs produce relatively fixed action patterns (Hume, Getting, et al., 1982), most are subject to extensive neuromodulation that allows the underlying networks to reconfigure to generate different behavioral acts (Heinzel, 1988). In others, gating synaptic input can switch between alternative patterns such as walking in different directions (Ayers and Davis, 1977; Ayers and Witting, 2007).

Using the lobster, lamprey, and honeybee as target animals, and focusing on the neural circuitry underlying locomotion, Ayers and Rulkov (2007) developed embodied models using electronic *discrete time map-based* (DTM) neurons whose membrane dynamics evolve according to the state-space plot illustrated in figure 16.6, which is intended to capture some of the

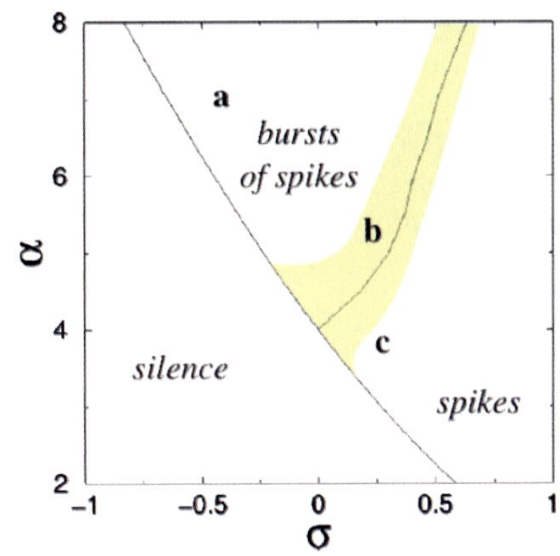

FIGURE 16.6 State space for discrete time map-based (DTM) neurons. The yellow region (b) indicates where chaotic firing occurs. (a) and (c) indicate regions of bursting and spiking discharge. Reprinted from Rulkov (2002) with permission from the American Physical Society.

dynamics of biological neurons. Here the membrane voltage in a given cycle $n + 1$ is specified relative to its value in cycle n according to two difference equations based on synaptic current input and two control parameters α and σ whose values are specific to each neuron type, shape the function of the map, and define the characteristics of individual neurons. A detailed specification of this model is given in box 16.1. As shown in the figure, when configured into different regions of α/σ space, neurons are either silent, tonically firing, bursting, or, in the highlighted saddle region between bursting and spiking, chaotically discharging (Shilnikov and Rulkov, 2003).

To create synthetic neural networks that could simulate locomotor CPGs, Ayers and Rulkov configured networks of electronic neurons in bursting mode (figure 16.6a) using predominantly inhibitory model synapses. The difference between postsynaptic voltage and the reversal potential of the spiking state of the presynaptic neuron was used to define the model synapses, with the topology of the network maintained in a look-up table of pre- and postsynaptic neurons. Because the neuron model is based on difference equations rather than differential equations, it is possible to control a large number of neurons and synapses in real time and tune their properties to behavioral context. This allows us to model phenomena such as neuromodulation (see below) using behaving robots (Ayers, Blustein, et al., 2012). Moreover, the option to vary chaos (figure 16.6b) allows us to model the adaptations underlying the wiggling and squirming behaviors observed in animals.

Box 16.1 Discrete Time Map-Based (DTM) Neurons

The DTM architecture is a map-based model that captures the dynamics of spiking and bursting neurons and their synaptic interactions (Rulkov, 2002). It allows the construction of complex sensorimotor networks that operate in real time on small digital signal processors (Westphal, Rulkov, et al., 2011; Ayers, Blustein, et al., 2012). It also allows dynamic tuning of synaptic strengths and time constants (Ayers, Rulkov, et al., 2010). Our goal with this model is to program the motor programs with neurons and synapses; hence, there are no algorithms in the controller. Each neuron and synapse is represented as an object in a look-up table. In a run-loop, the membrane voltage of a model cell in cycle n + 1 is determined as a function of its value in cycle n according to two difference equations based on synaptic current input and two control parameters σ (equation 16.1) and α (equation 16.2) and whose values specify the neuron type:

$$\begin{aligned} x_{n+1} &= f_\alpha(x_n, x_{n-1}, y_n + \beta_n), \\ y_{n+1} &= y_n - \mu(x_n + 1) + \mu\sigma + \mu\sigma_n, \end{aligned} \tag{16.1}$$

where x_n is the fast and y_n is the slow (due to $0 < \mu << 1$) dynamical variables The nonlinear function is written in the following form:

$$x_{n+1} = f_n(x_n, x_{n-1}, u) = \begin{cases} \alpha/(1 - x_n) + u, & x_n \le 0, \\ \alpha + u, & 0 < x_n < \alpha + u \ and \ x_{n-1} \le 0, \\ -1 & x_n \ge \alpha + u \ or \ x_{n-1} > 0, \end{cases} \tag{16.2}$$

where the third argument $u = y_n$ or a combination of input variables that depend on the model type.

The parameters α and σ (figure 16.6) shape the function of the map and define characteristics of individual neurons. Input variables β_n and σ_n incorporate the action of synaptic inputs I_{syn} and can be written as $\beta_n = \beta_E I_{syn}$, $\sigma_n = \sigma_E(I_{syn})$ where β_E and σ_E are constants that control how quickly neurons respond to the input and support dynamical mechanisms for spike frequency adaptation to DC pulses of current (Rulkov, 2002). If y is omitted, the 1-D model generates tonic spiking activity.

The simplest map-based model for a synaptic current can be written as:

$$I_{n+1}^{syn} = \gamma I_n^{syn} - \begin{cases} g_{syn}(x_n^{post} - x_{rp}), & \text{spike}_{pre}, \\ 0, & \text{otherwise}, \end{cases} \tag{16.3}$$

where g_{syn} is the strength of synaptic coupling, and indexes pre and post stand for the presynaptic and postsynaptic variables, respectively. Here, γ controls the relaxation rate of the synapse ($0 < \gamma < 1$), and x_{rp} defines the reversal potential and, therefore, the type of synapse: excitatory or inhibitory. The synaptic current from equation 16.3 is summed with equation 16.2, resulting in equation 16.1.

Synapse objects are used to connect neurons. The synapse equations evaluate the difference between postsynaptic voltage and reversal potential and the spiking state of the presynaptic neuron (equation 16.3). The topology of the network is defined in the look-up table of pre- and postsynaptic neurons for each synapse. Every neuron can project to a variety of neurons and can receive inputs from multiple neurons.

The model can be executed on a variety of processors including the Lego Mindstorms Brick (Westphal, Rulkov, et al., 2011; Blustein, Rosenthal, et al., 2013; Westphal, Blustein, et al., 2013) and can be implemented in LabView™ or procedural C. In procedural C, an initiation function sets the parameters for each neuron and synapse while a run-time loop updates all objects at a discrete time interval. The neurons and synapses otherwise proceed asynchronously.

Figure 16.7 illustrates CPGs configured in this way as either reciprocal half-center (figure 16.7A) or endogenous pacemaker inhibitory networks (figure 16.7B) to model lamprey-like swimming (Westphal, Rulkov, et al., 2011) and lobster-like walking, respectively (Ayers and Davis, 1977). In both types of networks, *synergies*—pools of neurons operating together—are formed by excitatory synapses between elements. Relative timing between synergies is then adjusted by differentially varying the control parameters α, σ, and the relative strength of the inhibitory synapses between synergies.

The half-center model of lamprey swimming, illustrated for a segment of the spinal cord in figure 16.7A, relies on reciprocal inhibition between the cross-caudal interneurons (CCs) and on postinhibitory rebound (i.e., increased neuronal excitability following the cessation of inhibition) in these cells (Perkel and Mulloney, 1974; Buchanan and McPherson, 1995; Buchanan, 1996). This mechanism is sufficient to generate proportional symmetrical alternating bursts driving bending of the tail to the left or right (Westphal, Rulkov, et al., 2011).

In the lobster walking model (figure 16.7B), an endogenous pacemaker-configured neuron, *elev* (elevator), establishes an oscillation with an antigravity synergy, *dep* (depressor), and a *stance* synergy that recovers later than *dep* from *elev* inhibition to create the late swing epoch (figure 16.7C). The *stance* synergy alternates with a *swing* synergy. *Elev* and *swing* generate the early phase while *swing* and *dep* generate the late swing phase (figure 16.7C). The *stance* phase of stepping is mediated by the *stance* and *dep* synergies. Command neurons gate off synapses between protractor and retractor bifunctional synergies to mediate forward and backward walking, and between *swing* and *stance* and extensor and flexor synergies during lateral walking (Ayers and Davis, 1977).

Ayers and coworkers have used DTM networks to control three biomimetic robots *RoboLobster, RoboLamprey*, and *RoboBee* (Ayers, Blustein, et al., 2012), as shown in figure 16.8. These robotic implementations have verified that the command neuron/coordinating neuron/CPG architecture, when instantiated with phenomenological models of neurons and synapses, can achieve adaptive behavior through swimming, walking, and flying (Ayers, Blustein, et al., 2012).

To achieve intersegmental coordination in a model system such as RoboLobster, interneurons are incorporated that pass information from one governing oscillator to another, causing the phase advances or delays that mediate gait (Ayers and Selverston, 1979). The system is also modulated parametrically by intersegmental commands that shift the elements into a bursting mode and control average frequency (Ayers and Witting, 2007). This architecture is illustrated in figure 16.9A.

Command neurons exist in a heterogeneous population that mediates a variety of behavioral acts (Bowerman and Larimer, 1974a, 1974b). In the lamprey some commands preferentially excite anterior segments to

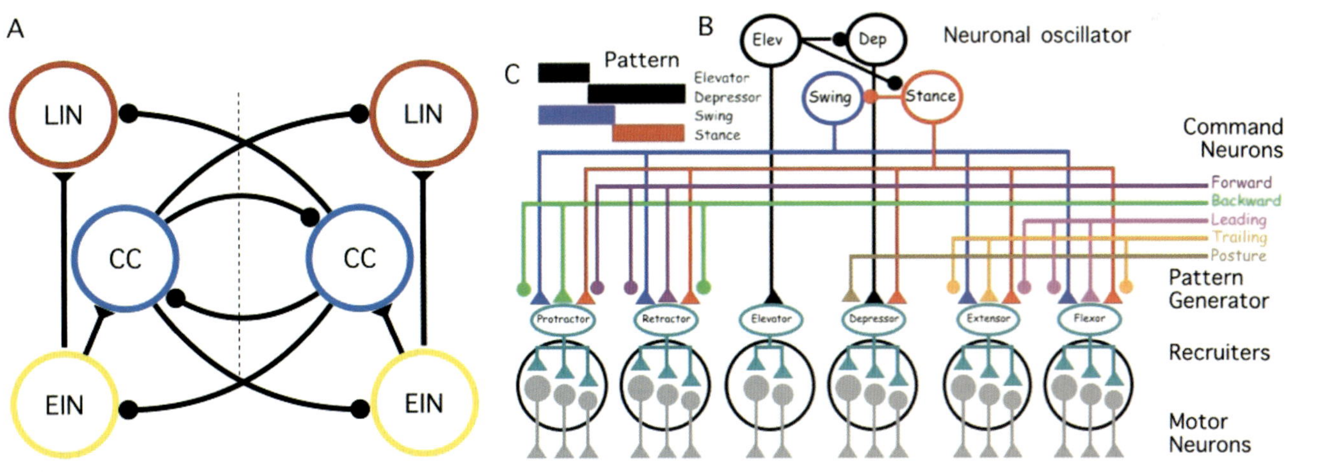

FIGURE 16.7 Central pattern generators for lamprey-like swimming (A) and lobster-like omnidirectional walking (B). In these diagrams the circles represent individual discrete time map-based neurons. (A) The contralateral synergies alternate because of intrinsic bursting mechanisms and reciprocal inhibition. Abbreviations: CC, cross-caudal interneurons; LIN, lateral inhibitory neurons; EIN, excitatory interneuron. Adapted from Westphal, Rulkov, et al. (2011). (B) Differential inhibitory synaptic strength between the elevator (Elev) synergy and depressor (Dep) or swing synergy creates a three-phase rhythm characteristic of walking in the four directions. Descending walking commands gate connectivity between swing and stance phase interneurons and bifunctional motor neurons, at a pattern-generating nexus, to mediate walking in different directions. (C) The three-phase pattern of walking is indicated.

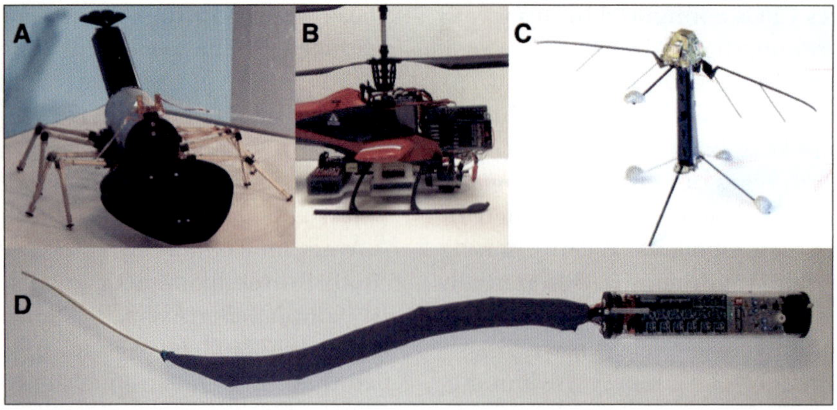

FIGURE 16.8 Biomimetic robots. (A) RoboLobster. (B) A helicopter proxy RoboBee. (C) RoboBee. (D) RoboLamprey. See Ayers, Blustein, et al. (2012) for further details. Photographs from Joseph Ayers.

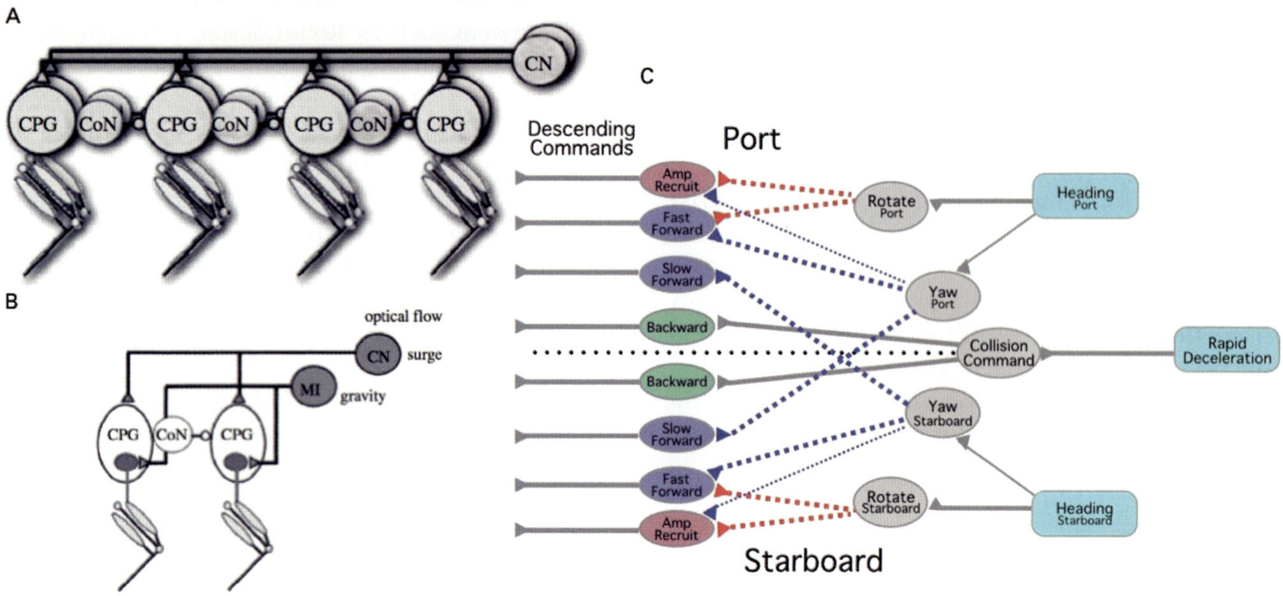

FIGURE 16.9 Intersegmental modulatory systems. (A) Command neuron, coordinating neuron central pattern generator (CPG) organization of RoboLobster. (B) Pathways for command (CN) and modulatory (MI) control of the RoboLobster CPGs by exteroceptive sensors. (C) Layered exteroceptive reflex architecture of RoboLamprey to mediate collision and heading control. A hierarchy of lamprey intersegmental and brain commands project the sensor inputs for heading deviation and rapid deceleration.

mediate forward swimming while others excite more posterior segments to mediate backward swimming (Matsushima and Grillner, 1992).

Exteroceptive reflexes link sensors to taxic and compensatory behavior (figure 16.8B and 16.8C) and can operate in parallel when triggered by environmental contingencies. For example, parallel exteroceptive reflex circuits for collision and heading control in RoboLamprey are illustrated in figure 16.8C. In robot implementations, small microprocessors are used to analyze analog sensor reports and generate "labeled-line" codes (Bullock, Orkand, and Grinnell, 1978). For example, to guide heading control, a heading estimate, provided by an analog compass, is compared with a desired heading; a current proportional to this difference is then used to activate a "heading deviation" neuron. Modulatory interneurons can target the neuronal oscillator or the motor neurons directly to modulate amplitude.

The exteroceptive reflex architecture works interchangeably between the three robot platforms, demonstrating its adaptability to different physical morphologies (Ayers, 2002). For example, a neuronal compass based on heading deviation neurons works equally well during walking in RoboLobster, swimming in RoboLamprey, or flying in RoboBee (Ayers, Blustein, et al., 2012; Westphal, Blustein, et al., 2013). Reflexes have also been implemented for optical flow (optical flow sensors), hydrodynamic flow (antennae), heading (compass), collision (accelerometer), gravity (inclinometer), odometry (optical flow sensors with neuronal integrators), and beacon tracking (short baseline sonar array), and these operate well in parallel (Westphal, Blustein, et al., 2013).

By approximating the biomechanics of the animal model, the dynamics of the robot vehicle behavior provides a good match to that of the living animal (Ayers, 2004). Furthermore, in underwater robots, the use of shape memory alloy actuators (Mohd Jani, Leary, et al., 2014) allows employment of heat to mediate excitation/contraction coupling much as living muscle uses intracellular calcium (Witting, Ayers, et al., 2000). Thus, the motor programs characteristic of the model neuromuscular system perform in an analogous fashion to that of the animal.

The processes that give rise to the motor rhythm (endogenous bursting, postinhibitory rebound, etc.) are also excellent models of the corresponding processes in the living networks (Selverston and Ayers, 2006). A key feature of these models is that because they seek to capture the nonlinear dynamical behavior of neurons, rather than being neuronal conductance models, they are simpler, can operate in real time, and are therefore suitable for robot control applications (Ayers and Rulkov, 2007). As shown in figure 16.10, which illustrates some example neural activity produced by the circuit illustrated in figure 16.7, the use of neuronal network models, instead of finite state systems, allows one to replicate in great detail the real behavior of the neurobiological system (a network) and, thanks to spiking nature of the models, provides a link between the electronic neurons and experimental measurements of neuronal activity from the animal.

Chemosensors based on principles of synthetic biology are currently under development that will allow tests of circuit-based models of odor-guided plume-following building on the more algorithmic approach described earlier (Grasso and Atema, 2002). By merging the information-processing hypotheses developed by Grasso and coworkers with the more mechanistic understanding developed by Ayers and his associates, it should be possible, in the future, to develop embodied models in which we have confidence that we are matching the target biological system at multiple levels of description. By matching multiple constraints in this way, we can achieve a form of *convergent validation* (Verschure, 1996). That is, we not only obtain a more complete understanding, we also increase the likelihood that our model captures how the living system actually works rather than merely how it "might" work.

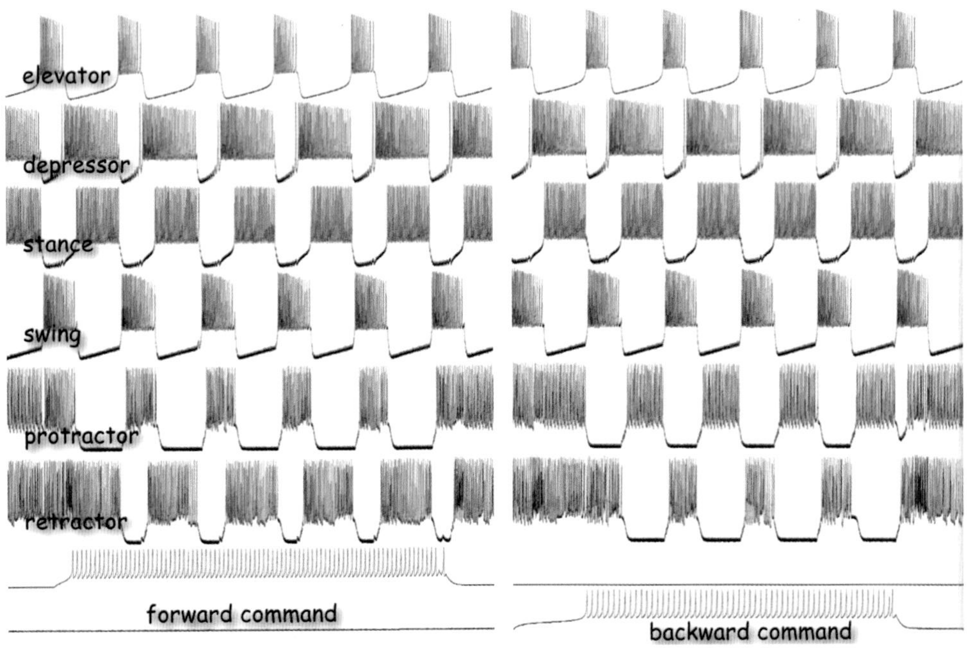

FIGURE 16.10 Motor programs for forward and backward walking selected from an electronic neuronal network shown in figure 16.7B. The forward command gates off synapses linking the stance phase and protractor neurons as well as those linking the swing phase and retractor neurons. See the text for a description of the different neural elements.

16.4 *Neurorobotics of the Mammalian Vibrissal System*

The above embodied models have focused largely on invertebrate target animals, but can neurorobotics also help us to gain insight into brains more similar to our own—that is, those of other mammals?

One of the most popular model systems in which to study processing in mammalian brains is the whisker, or *vibrissal*, sensorimotor system owing to its discrete organization from the sensory apparatus (the whisker shaft) all the way to the sensory cortex (see figure 16.11 and Diamond, von Heimendahl, et al., 2008), its relative ease of manipulation (for a living biological system), and, not least, its presence in laboratory rats and mice. Whisker signals are processed at multiple brain sites, crossing a minimum of two synapses, in the brainstem and in the thalamus, before reaching the sensory cortex. The whisker cortex, which is known as the *barrel cortex* because of the presence of cellular aggregates that have a one-to-one mapping with the facial vibrissae, is itself a target for huge research effort (see, e.g., Petersen, 2007). This is due to the ability to tweak a whisker at the periphery, in a known and quantifiable way, and then record in a precise area of cortex to determine the effect of that stimulus. Barrel cortex is therefore widely viewed as a preparation in which neuroscientific research could unlock the secrets of the "cortical microcircuit." That is, if we can understand the processing going on in the barrel field, then we may understand something about the generic processing capacities of mammalian six-layered cortex that is replicated, with some variation, across both of the cerebral hemispheres and in all mammalian species. It is no surprise, then, that new data on the barrel cortex are published on a weekly, if not daily, basis.

But what of the stimuli with which neuroscientists are probing the rat brain in order to understand what the barrel cortex is doing? Perhaps unsurprisingly, many of the studies of barrel cortical function have been performed in anaesthetized animals or in brain slices. Less frequent historically, but now increasing in number, are studies in immobilized animals (head-fixed) that are awake and able to move their whiskers. Because of the difficulty of stably recording from electrodes implanted in free-moving animals (where the recording device is usually connected via a springy umbilical cable), the number of studies that have looked at barrel cortex processing in more natural circumstances remains comparatively small. Moreover, the practical difficulties associated with recording in moving animals mean that such studies generally record extracellular activity—that is, they use electrodes that pick up activity in

A

B

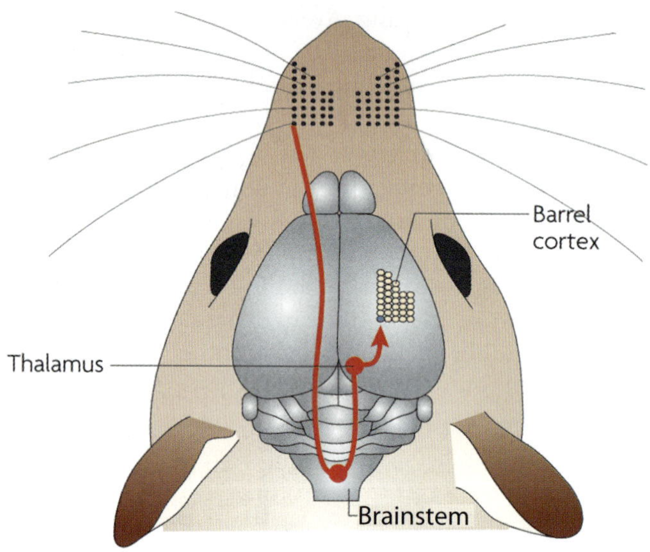

FIGURE 16.11 The rodent vibrissal system. The vibrissal system (A) of rats and mice is a major target of studies aimed at understanding mammalian brain architecture, partly due to the one-to-one correspondence (B) between single vibrissae on the snout and cellular aggregates in the vibrissal "barrel" cortex. (A) used with permission from the Active Touch Laboratory at Sheffield (ATL@S) (T. Prescott). (B) adapted from Diamond, von Heimendahl, et al. (2008) with permission from McMillan Publishers Ltd.

multiple nearby cells; the capacity to distinguish what is happening in single neurons is therefore limited. Overall, then, the picture is much as we have described it in table 16.1—as you move from the freely moving rat to more tractable experimental models—head-fixed, anaesthetized, or slice—the ability to interrogate the processing system improves but the capacity of those signals to tell you about the nature of processing in awake, naturally behaving animals falls away. This situation indicates that there could be scope for robotic models that can capture some of the physics of how whiskers interact with surfaces in exploring animals, and thus shed light on the nature of the processing occurring in this part of the mammalian sensorimotor system.

16.4.1 The Closed Loop of Active Vibrissal Touch There are two other strong motivations for investigating physical models of rodent vibrissal processing. First, as we have already encountered with lobster plume tracking, simulation of real sensory transduction can be difficult. In the case of whiskers, you have a tapered flexible shaft one end of which is embedded in a soft deformable body—the whisker follicle—packed with mechanoreceptors. Tactile sensing involves contact of the (usually) moving whisker with surfaces that often have complex microgeometry (e.g., texture), which may also be moving. Various simulation approaches have been developed including a recent dynamic model that includes effects of damping, inertia, and collisions (Quist, Seghete, et al., 2014); however, accurate simulation of the encounters of whiskers with interesting surface geometry remains an unsolved problem that may be more easily addressed by building a physical model.

The second motivation concerns the nature of vibrissal sensing itself. Like many natural sensing systems, vibrissae are not deployed passively to detect the consequences of objects brushing against them. Rather, many whiskered animals, and particularly rodents, actively move their whiskers against objects and surfaces of interest (see figure 16.12). Indeed, rats and mice move their whiskers back and forth many times per second (at a rate of about 8 Hz in rats), in a behavior known as "whisking," generating multiple touches of the whiskers against salient objects. The control of whiskers happens alongside orienting movements of the head and body (Grant, Mitchinson, et al., 2009; Grant, Sperber, et al., 2012); thus, we can think of whisking as one component of an active sensing system that uses the musculoskeletal system to help isolate and enhance the stimulus features of the environment that are of particular interest (Prescott, Diamond, et al., 2011). We can immediately begin to see a major compromise that studies of brain slices and anaesthetized animals have to make when they study barrel cortex. Electrical stimulation of a brain slice, or passive deflection of the whisker of a sleeping mouse or rat, are both poor substitutes for the kinds of signals that are ascending from the brainstem of an awake and active animal. Even in the head-restrained case, the animal is detecting with its whiskers what the researcher has chosen to expose it to rather than deciding for itself, through head and body movements, where to deploy its whiskers (Mitchinson and Prescott, 2013). Indeed, it is well-known that in head-fixed animals the normal whisking behavior is not generally expressed, and when it is evoked (e.g., by stimulation of the olfactory system), it may have somewhat different characteristics from the whisking of the free-moving animal.

How important is this for understanding the signals that are being processed in barrel cortex? This only becomes clear when we carefully study the nature of the whisker movements expressed by animals and the architecture of the neural circuits that process vibrissal signals. In the case of the latter, it is now well-known (e.g., Kleinfeld, Ahissar, et al., 2006) that the barrel cortex does not simply sit atop a feedforward circuit relaying whisker deflection systems up from the periphery as somewhat misleadingly implied in figure 16.11 above. Rather, as indicated in figure 16.12 (right), it is embedded within a complex architecture of nested sensorimotor loops at brainstem, midbrain, and cortical levels, each of which implements a relatively short-latency coupling between sensory input and motor output. Thus, both the behavior of the animal (movement of the whiskers) and the neural circuits themselves suggest that barrel cortical activity cannot be properly understood in isolation from the sensorimotor activity—exploratory whisking—that generates it. We return to consideration of this larger architecture after first homing in on the whiskers and their role in generating the sensory signals we find in the brain.

16.4.2 Toward a Robot Model—How Engineers Sometimes Ask the Right Questions What can we find out about the vibrissal system and processing in key brain areas such as the barrel cortex by building a robot model of the rat vibrissae? First off, we have to build such a model. How then should the robot move its whiskers? Back and forth, certainly, and at some speed, but when the whiskers touch an object, should they keep moving forward regardless or should they change their behavior? Surprisingly, when researchers began to develop robot vibrissae, the answer to this key question was not available in the scientific literature despite almost a century of effort to try to understand rat vibrissal sensing. Therefore, urged on by the robot engineers, Mitchinson, Martin, et al. (2007) carefully recorded free-moving rats interacting with an object using their whiskers and studied the trajectories and timing of whisker motion, before and after object contacts, in high-speed digital video recordings. These studies, and subsequent ones (Voigts, Sakmann, et al., 2008; Grant, Mitchinson, et al., 2009), have clearly demonstrated that rats modulate the movement of their

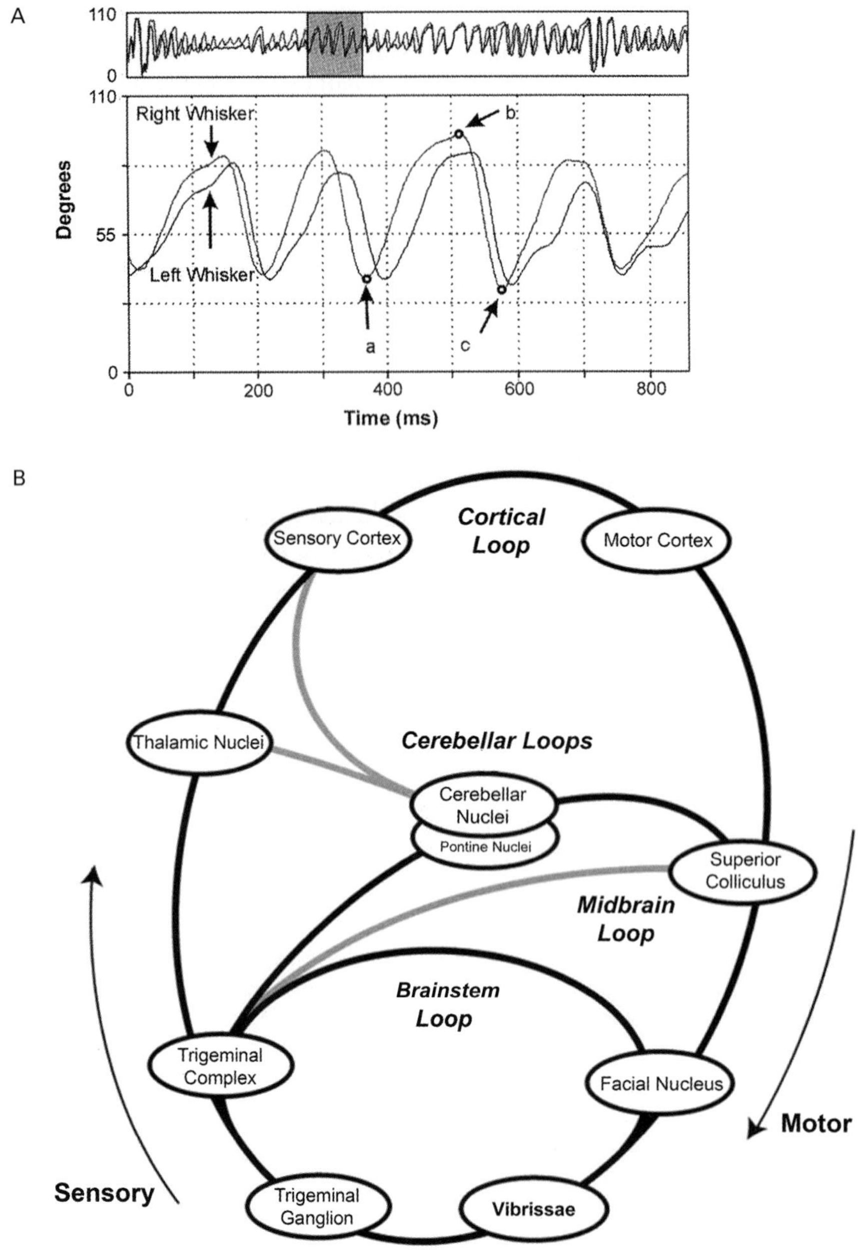

FIGURE 16.12 Active vibrissal touch is supported by a nested-loop brain architecture. (A) Many animals generate rhythmic whisking movements; shown here is a bout of rat whisking at two temporal scales as measured optoelectronically in a head-fixed animal. The upper trace shows the average left- and right-whisker movement over the course of an 8.5-second bout. The lower trace shows that the movement of the left and right whisker fields are usually strongly coupled. Reproduced from Gao, Bermejo, et al. (2001) with permission from the Society for Neuroscience. (B) Vibrissal signals are processed in multiple nested sensorimotor loops that show a tight coupling between sensing and actuation (whisker movement). Adapted from Kleinfeld, Ahissar, et al. (2006).

whiskers on a millisecond basis depending on the nature of the contacts they are making with objects and surfaces. That is, the neural circuits really are implementing some tight feedback control of the sensor apparatus. This stands as an example of a question that was important to ask, but for which an important impetus came from an attempt to build a physical model. A more general lesson, which was also recognized by Braitenberg (1986), is that when we approach a complex biological system with the aim of synthesizing an artifact that operates in a similar way, we are prompted to ask different questions than if we simply look at the target system and try to analyze its function.

16.4.3 Advancing Neurobiological Knowledge by Doing Experiments That Would Be Impossible in the Animal Knowing how the whiskers should move, the next step was to build an approximation to the whisker-processing apparatus found in the animal. We cannot *exactly* replicate the mechanics of rat whiskers, or of their embedding in the follicle, or the complex responses of the mechanoreceptors to whisker deflection, or the responses of the primary afferent cells in the brainstem that process the signals from these receptors (hundreds at a time). However, perhaps we could get close enough to learn something interesting. Mitchinson, Gurney, et al. (2004) implemented, in simulation, an electromechanical model of the whisker follicle (see figure 16.13) and fed it with inputs from an artificial whisker (a tapered, flexible plastic shaft) instrumented with strain gauge transducers. By suitable tuning of the parameters of this model they showed that they could reproduce, using deflections of the artificial whisker, patterns of firing in model primary afferent cells that were qualitatively similar to those recorded when the whiskers of anaesthetized animals are deflected (Mitchinson, Gurney, et al., 2004; Mitchinson, Arabzadeh, et al., 2008).

Now it becomes possible to perform an experiment that cannot be done in the animal (recall that a key criteria for a good model is that it should allow us to do interesting experiments with relative ease).

In order to understand the signals that are ascending from the vibrissae (as shown in figure 16.11) for processing in the barrel cortex of awake, behaving animals, it is clear that we need to know what effect the feedback control that animals exert on their whiskers is having on the ascending signals. The video studies showed that rats were modifying their whisker movements when they contacted objects. More precisely, the whiskers stop protracting (moving forward) and start retracting (moving backward) earlier when they contact surfaces than when whisking in free air (see figure 16.14). The consequence is that whiskers bend less against surfaces then they would otherwise—you could say that the rat explores surfaces using a "light touch" (Mitchinson, Martin, et al., 2007). The neural circuits that implement this feedback circuit in the animal are not well understood. Indeed, even if we knew where they were, experimentally interfering with them (e.g., attempting to block the feedback) would have unknown effects on the animal. Would an animal continue exploring objects as normal, or express anything like its natural behavior, if it were unable to regulate whisker movement in the usual way? However, the roboticists could perform the equivalent experiment in their model in a very simple way.

A whisker pattern generation model was implemented that generated a sinusoidal whisking motion similar to that seen in the animal, and modulation was

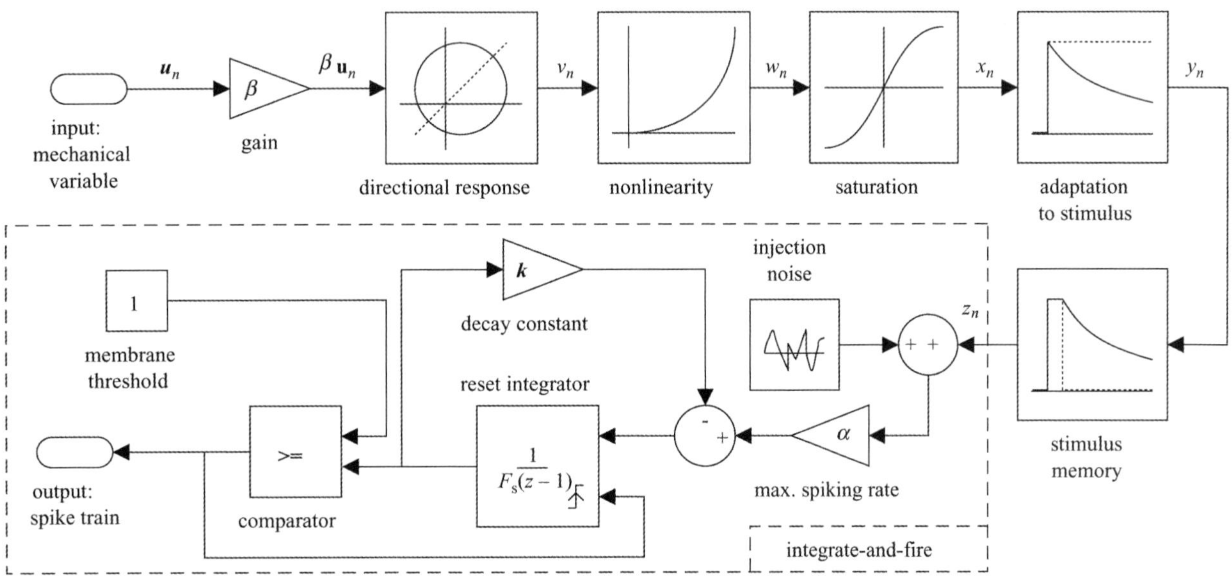

Figure 16.13 Using strain signals from an artificial whisker to model the response properties of vibrissal primary afferent neurons. The input (mechanical strain, u) is transformed, in turn, by processing units representing directional response, nonlinearity, saturation, adaptation to stimulus, and stimulus memory; the resulting signal (z) drives a conventional integrate-and-fire membrane model. See Mitchinson, Gurney, et al. (2004) for further details. max., maximum. Reprinted with permission from Royal Society Publishing.

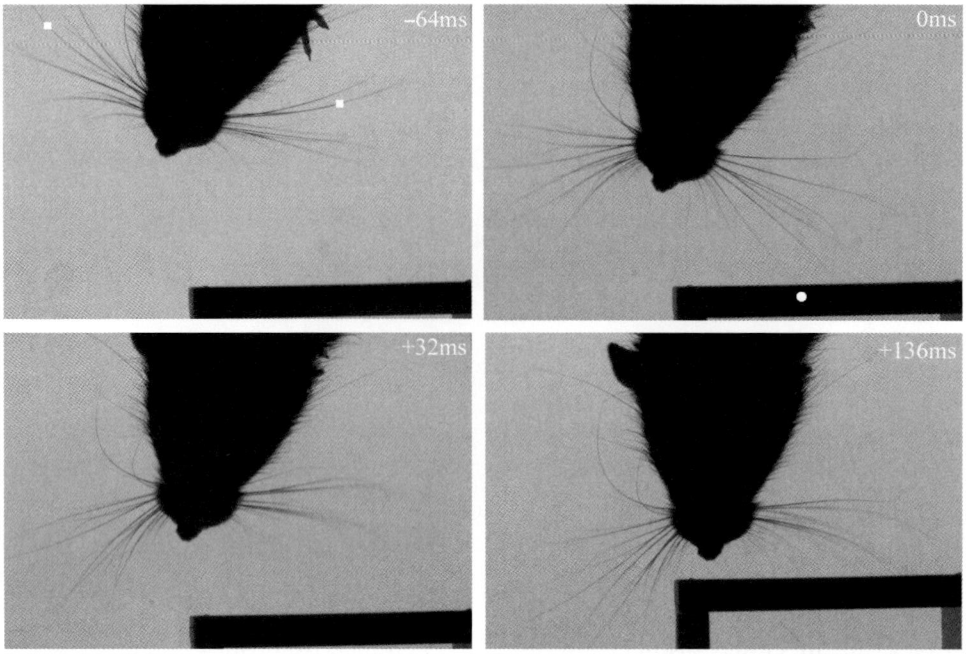

FIGURE 16.14 Effect of whisker–object contact on whisker motion. High-speed video recordings show that the protraction of the whisker often stops rapidly (12–14 ms) after contact with a surface. In this sequence of frames the right whisker field touches an object in frame 2 (this is indicated by the white dot on the object just below the contact point), frames 3 and 4 show that the whiskers on that side of the snout then begin to retract while those on the opposite side are still moving forward. The overall effect is that whiskers make relatively light contacts with surfaces. Figure adapted from Mitchinson, Martin, et al. (2007) (Royal Society Publishing), which should be consulted for more details of how this sequence was recorded).

imposed on this to suppress further protraction, and initiate earlier retraction, following contact (Pearson, Pipe, et al., 2007). Now this feedback mechanism could be easily turned on or off by toggling a variable in the control program. Figure 16.15 shows the effect of the presence or absence of feedback control on an important classes of model primary afferent neurons—so-called "rapidly adapting" (RA) cells, as simulated with the Mitchinson, Gurney, et al. (2004) electromechanical model. The figure shows that the effect of the feedback is to significantly alter the nature of the signals generated by these model cells (presented here as simulated spike rasters). Specifically, the responses of the cells were both briefer and less noisy when the feedback control was in place than when it was switched off. This outcome occurred, naturally enough, because the effect of the feedback was to produce a much smoother deflection pattern than when the feedback was deactivated. Of course, turning off protraction early should produce a smaller deflection, but what was interesting here was that despite having a smooth sinusoidal drive and touch against a smooth surface, in the no feedback condition, there was a highly variable deflection pattern for the duration of the contact, resulting in rather noisy model primary afferent responses.

Based on these results, and their experimental studies in animals, Pearson et al. (2011) have proposed, and further demonstrated experimentally, in a new whiskered robot *Shrewbot* (figure 16.15, bottom right), that the effect of feedback in this system could be to provide more uniform signals, within a constrained dynamic range, for relay up to cortical areas. This could be very useful information for experimentalists trying to design neuroethologically-valid deflection patterns or neural stimulus trains with which to investigate their animal models, and for theoreticians trying to understand the nature of vibrissal processing in the barrel cortex microcircuit.

Further work has developed models of other neural structures—superior colliculus, cerebellum, barrel cortex, hippocampus—and evaluated their performance in the context of sensory inputs derived from active artificial vibrissal sensors (Prescott, Pearson, et al., 2009; Anderson, Pearson, et al., 2010; Pearson, Mitchinson, et al., 2010; Mitchinson, Pearson, et al., 2011). A review of this broader work which is seeking to render an embodied model of the full sensorimotor loop architecture of the rodent vibrissal system shown in figure 16.13 is available in Prescott, Mitchinson, et al. (2015).

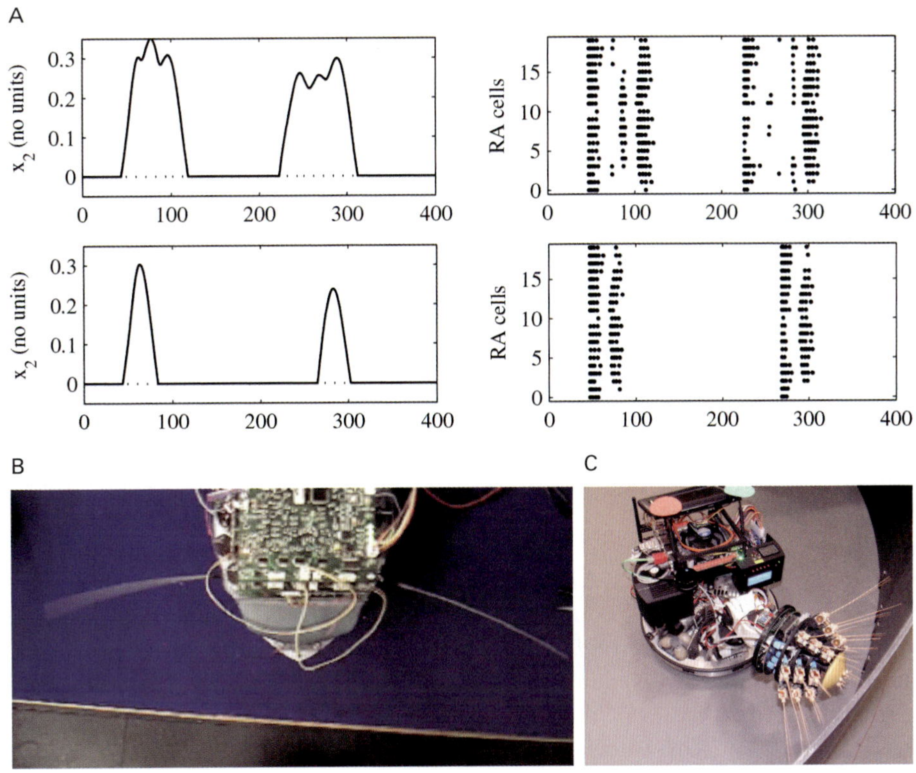

FIGURE 16.15 (A) Whisker deflection pattern (left) and activity in model rapidly adapting (RA) primary afferent neurons in the condition where feedback control of sensor movement is switched for the robot Whiskerbot (B). Reprinted from Mitchinson, Pearson et al. (2006) with permission from Springer Verlag. (C) The whiskered robot Shrewbot has recently been deployed to perform further neurorobotic investigations of vibrissal touch (Pearson, Mitchinson, et al., 2011) including tactile simultaneous localization and mapping (Pearson, Fox, et al., 2013) and modeling of whisker-based predator–prey pursuit (Mitchinson, Pearson, et al., 2014). Photographs from Martin Pearson of Bristol Robotics Laboratory, used with permission.

16.5 Scaling Up to the Whole Brain

We have argued throughout this chapter that an effective strategy for understanding the relationship between brain and behavior is through abstraction to a physical model of reduced but appropriate complexity. In this final section we argue that, at an appropriate level of abstraction, the strategy of neurorobotics can also scale up to understanding the whole brain.

An elaborate robot-based theory that seeks to extend embodied modeling to encapsulate the control architecture of the vertebrate brain is the DAC series of models proposed by Verschure and colleagues (see Verschure, Krose, et al., 1992; Verschure, 2012, for a review). DAC addresses the fundamental question that the theoretician and modeler faces: what level of abstraction is sufficient to allow a parsimonious description while in the meantime being sufficiently constrained by data to be meaningful? DAC solves this dilemma by advancing a dual agenda: an integrated systems-level architecture to identify the overall behavioral and functional constraints as described above, paired with anatomically and physiologically constrained models to test specific assumptions and predictions.

We earlier saw the notion of nested-loop architecture applied to the vibrissal system. The DAC theory aims at conceptualizing a wider organizational scheme for the whole brain from which we can extract general principles to be mapped both to behavior and to its underlying neuronal substrate. The DAC theory starts from the fundamental question of what the function of the brain could be and proposes that it can be seen as optimizing four fundamental objective functions to generate action in the physical world or the "How" of survival:

1. *Why:* the motivation for action in terms of needs, drives, and goals.
2. *What:* the objects in the world that actions pertain to.
3. *Where:* the location of objects in the world and the self.

4. *When:* the timing of action relative to the dynamics of the world.

We can call this the *H4W* problem for short, and each of the Ws designates a large set of subobjectives of varying complexity. Answering H4W requires the tight coordination of a number of processes ranging from control and sensing of the body through to the cognitive and integrative processes that underlie self-awareness. One can argue that a similar taxonomy could hold for any brain. This might be the case, and indeed the DAC theory has also been mapped to invertebrate systems such as ants and moths (Pyk, Bermudez, et al., 2006: Mathews, Lechón, et al., 2009; Mathews, Verschure, et al., 2010). However, vertebrate brains provide systems that explicitly process all aspects of H4W as opposed to invertebrates.

The DAC theory proposes that the brain solves H4W as an integrated control system comprising four layers called *Soma, Reactive, Adaptive,* and *Contextual.* Three functional columns intersect these layers delineating subsystems dealing with exosensing states of the *world,* endosensing states of the *self,* and the interfacing of world and self in *action* (figure 16.16). DAC proposes that we can think of the brain, and its organization along the neuroaxis, as a matrix. Further, DAC proposes that as we move from the reactive to the contextual layer, we advance from evolutionary ancient and predefined systems of the brainstem toward the memory-dependent decision making circuits of the frontal lobes of the neocortex.

An attentive reader might object that figure 16.16 is incomplete, or even incorrect, by raising concerns such as these: Why does one go through episodic memory to get to short-term memory? Why does working memory depend on long-term memory but episodic memory does not? Where is procedural memory? Why is value lower than goals? However, a two-dimensional diagram, that must fit on one page, clearly cannot express all of the relevant aspects of structural and functional brain organization. Rather, as the theory it reflects, it seeks to provide a suitable abstraction (Verschure, Pennartz, et al., 2014). Nevertheless, the DAC theory has generated some very precise predictions concerning the organization of the neuronal substrate underlying adaptive behavior that demonstrate its utility. We will describe a few of these after reviewing the basic organization of the theory. For a more detailed analysis of the mapping of the DAC theory to the brain structures underlying goal-oriented behavior, see Verschure, Pennartz, et al. (2014).

Within the DAC architecture, the *somatic layer* designates the body itself and defines three fundamental sources of information: *sensation* driven by external and internal sources of stimulation (or exo- and endo-sensing, respectively), *needs* defined by the essential variables that ensure survival (e.g., oxygen, glucose, carbohydrates, sleep, sex, etc.), and the skeletal-muscle system for *actuation* and the proprioception it provides. To keep the soma intact is the primary objective of the brain. This ability for self-maintenance was recognized by the nineteenth-century physiologists Claude Bernard and Ivan Pavlov under the notion of *homeostasis* (see Verschure, Pennartz et al., 2014). Later variations include the idea of *autopoiesis,* where the overall ability to maintain oneself is, in turn, a homeostatic variable (Varela, Maturana, et al., 1974). From a robotics perspective we can operationalize these essential variables as the need to optimize energy and minimize damage. However, we do need to keep in mind that the robots we consider are not literal animals but rather operational approximations that allow our models to generate behavior in the real world. The DAC theory is of interest because it suggests how self-regulation can be realized in the embodied brain considered as an integrated system.

The *reactive layer* (Verschure, Krose, et al., 1993) comprises fast predefined sensorimotor loops that support direct behaviors underlying basic functionality such as exploration, feeding, grooming, defense, orienting, mobility, sex, and so forth. Each of these reflexive behavior systems is triggered by low-complexity signals, sensing, and directly coupled to specific need dependent internal states of the agent, or drives. Hence, the primitive organizational elements of the reactive layer are sense–affect–act triads that are regulated following homeostatic principles (Sanchez-Fibla, Bernardet, et al., 2010). This layer can be seen as a more advanced version of the reflex-driven robot tortoise built by Grey Walter, the "vehicles" proposed by Valentino Braitenberg, or Brooks's subsumption architecture (see section 16.2 above). The key difference is that the DAC reactive layer is modeled after the primitive brainstem structures that govern these behavior systems in the vertebrate brain (Panksepp and Biven, 2012).

In the reactive layer every behavior system is conceived as a homeostatic control system that, in turn, is regulated by an integrative *allostatic* process. Whereas homeostasis considers a dynamic process as seeking stability, allostasis considers self-regulation, in general, as a continuous dynamic process—seeking stability through change (Sterling and Eyer, 1988). In our case this means that each individual behavioral homeostatic system, such as the need for fight or flight, is regulated in real time relative to the pertinent needs of the behaving system. The control model proposed here is

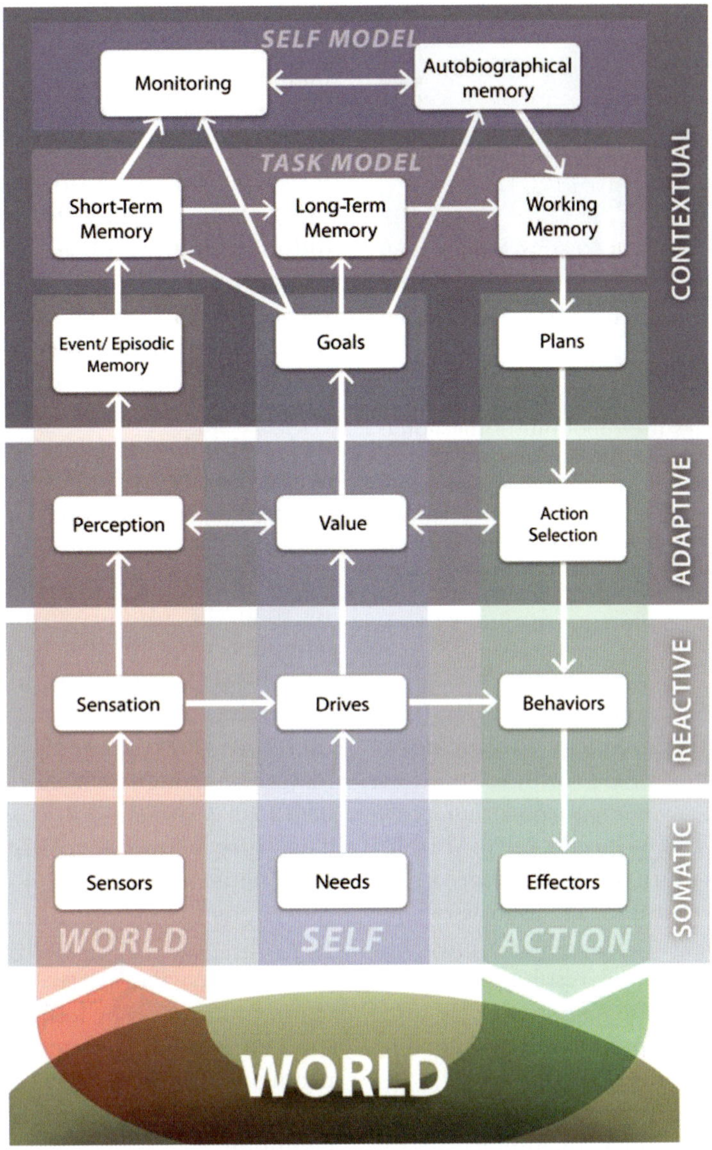

FIGURE 16.16 Conceptual scheme of the distributed adaptive control (DAC) architecture. DAC comprises four tightly coupled layers: Somatic, Reactive, Adaptive, and Contextual. Across these layers we distinguish three functional columns of organization: exosensing, defined as the sensation, perception, and retaining states of the *world* (left); endosensing, which is detecting, signaling, and retaining states of the physically instantiated *self* (center); and the interface to the world through *action* (right). The arrows show the primary flow of information mapping exo- and endosensing into action. See the text for further explanation. Figure from Verschure, Pennartz, et al. (2014) (Royal Society Publishing).

that behavior systems are individually tuned to specific properties of the environment, such as the home base for security or the open field for exploration, while being continuously tuned with respect to the overall behavioral needs of the agent. Robotic experiments show that these drive-reducing reactive control systems can give rise to complex patterns of environment exploration (Sanchez-Fibla, Bernardet, et al., 2010) similar to those measured in animals (see figure 16.17). Recent research has applied this model to provide allostatic control of a humanoid robot engaged in human–robot social interaction (Vouloutsi, Lallée, et al., 2013).

The *adaptive layer* of DAC is interfaced to the full sensorium of the agent, its internal needs and effector systems (Verschure and Coolen, 1991; Verschure, Krose, et al., 1992; Duff and Verschure, 2010), extending the sensorimotor loops of the reactive layer with acquired sensor and action states. Through learning, a state space of sensation and action helps the agent deal with the fundamental unpredictability of both the internal and the external environment. This is signified in

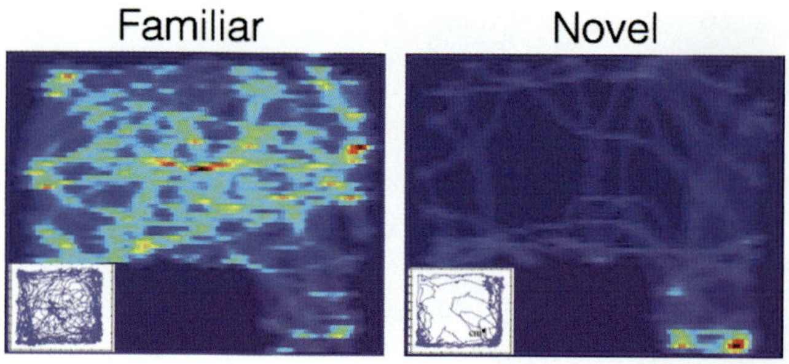

Figure 16.17 Behavior generated by the reactive layer of the distributed adaptive control model controlling a robot in the real world while *exploring* a familiar environment (left) or a *novel* environment (right). The familiar environment triggers exploration behavior that brings the robot out of the home base at the lower-right corner while the novel environment induces security seeking so that the robot moves primarily within the home-base region. Insets show example trajectories from rats under similar conditions. Figure adapted from Sanchez-Fibla, Bernardet, et al. (2010) with permission from WorldScientific publishing.

figure 16.16 by "Perception" and "Action Selection" which are both dependent on plasticity and memory. Moreover, the adaptive layer of DAC provides a solution to the fundamental *symbol-grounding problem* that real-world systems face (Searle, 1980; Harnad, 1990; Verschure, Krose, et al., 1992; Verschure, 1998). The core issue is that a system that needs to learn to act has to associate actions with states of the world. These states must be picked up with its sensors and the resulting internal representations grounded in the experience of the agent itself, as opposed to being predefined by an external entity such as the system programmer.

Learning in the adaptive layer occurs in a restricted temporal window of relatively immediate interaction. In order to go beyond "the now," other learning systems must be engaged that can expand this time window. The adaptive layer and the state-space it acquires thus provide a foundation for the more advanced learning and memory systems of the contextual layer. More specifically, the *contextual layer* of DAC develops behavioral plans or policies (Voegtlin and Verschure, 1998; Verschure and Voegtlin, 1999), the atomic elements of which are formed by the state space of exo- and endosensing built by the adaptive layer, and by its sensorimotor contingencies. The contextual layer forms sequential short- and long-term memory representations that reflect states of the environment and actions generated by the agent or its acquired sensorimotor contingencies. The acquisition and retention of these sequences is conditional on the goal achievement of the agent. The behavioral plans can be recalled through sensory matching and internal chaining among the elements of the retained memory sequences. The dynamic states that this process entails can be interpreted as DAC's working memory system.

The development of the DAC architecture has taken place through a close interaction between theoretical investigations of the neurobiology and psychology of living things and a wide variety of robotics models of different kinds, and at different levels of abstraction compared to the biological target. An example that shows the importance of embodied modeling, in the development of the DAC framework, is research that led to the understanding of the phenomenon of behavioral feedback in the coupling between perceptual and behavioral learning (Verschure, Voegtlin, et al., 2003). Behavioral feedback describes a process through which learning leads to the development of behavioral habits that, in turn, lead to the creation of an effective interaction space that is of lower complexity than the overall environment. This reduction in complexity means that the environment is subsampled and the perceptual learning systems further adjust to this reduced set of interactions. Behavioral feedback therefore directly couples behavioral and perceptual learning through behavior itself (Verschure, Voegtlin, et al., 2003) (see also discussion in section 16.2 above). This direct impact of situatedness on perception was first analyzed through embodied learning using a robotic model.

16.5.1 Answering the "Where" Question We conclude this section by looking at the practical impact of the DAC research strategy and by analyzing a specific problem—how to build a map of the world and localize oneself in it, or the problem of "where" of the H4W taxonomy.

Earlier we noted some of the remarkable feats of navigation and spatial knowledge shown by some species of birds and fish. For many animals, an essential aspect of survival is to build up appropriate knowledge of the environment and to know the position within it of the self and of other significant locations and objects. In robotics this is widely referred to as the *simultaneous localization and mapping* problem (SLAM; Leonard and Durrant-Whyte, 1991; Thrun, Burgard, et al., 2005). SLAM aims at simultaneously estimating locations of novel landmarks, and of the robot itself, while incrementally building a map of the environment. Solving the SLAM problem is seen as the technological breakthrough required for robots to perform autonomous tasks requiring mobility (Thrun, 2010). The most popular methods for SLAM are based on estimation-based approaches (Dissanayake, Newman, et al., 2001), grounded in adaptive filter techniques starting with the

Kalman filter (Kalman, 1960; the workhorse of many control engineering applications). However, solutions to SLAM face challenges due to various sources of uncertainty such as errors in path integration, sensor noise, and, above all, the dynamics of the world itself. Biological systems display robust solutions to SLAM; it is thus no surprise that attention has been paid to the ways in which animals solve spatial tasks (Milford and Wyeth, 2010; Sünderhauf and Protzel, 2010). Understanding how animals solve the SLAM problem could thus serve a dual purpose, both to provide good theories of animal spatial cognition and to provide a useful biomimetic advance in mobile robotics.

In mammals, spatial cognition is associated with the hippocampus and adjacent areas in the medial temporal lobe. Within this area of the brain a number of cell types have been described that relate to the position in space of the animal, such as place cells (O'Keefe and Dostrovsky, 1971; O'Keefe and Conway, 1978), head-direction cells (Taube, Muller, et al., 1990), and grid cells (Hafting, Fyhn, et al., 2005). Place cells show a response to a specific location in space, head-direction cells indicate the heading direction of the animal, while grid cells show a response field that includes a grid of precisely tuned location-specific responses. A common brain-based approach toward solving simple navigation tasks is to combine model place cells with reinforcement learning to associate actions to locations in space (e.g., Burgess, Donnet, et al., 1997; Foster, Morris, et al., 2000; Arleo and Gerstner, 2001). Recent studies have claimed that such hippocampus-inspired models are capable of competing with state-of-the-art SLAM algorithms in large-scale benchmarks (Milford and Wyeth, 2010). However, a limitation of these models is that they make the strong assumption that perception itself does not depend on behavior, for example, they often rely on a predefined discretized model of space and do not always take into account critical physiological and anatomical properties of the relevant brain structures.

We turn to a specific model developed within the DAC framework (Verschure, Krose, et al., 1992; Verschure, 2012) that illustrates how predictions formed at the level of the overall neuronal architecture can be mapped to the neuronal substrate. As described earlier, DAC postulates a contextual layer that operates on highly integrated, multimodal representations of sensorimotor couplets that comprise both sensing and action. This assumption constitutes a testable hypothesis, and the question thus becomes whether the brain operates on such integrated sensorimotor representations.

The hippocampus interfaces with the neocortex via the *entorhinal cortex* (EC), receiving sensory information via the lateral EC and spatial information from the grid cells of the medial EC. Given these inputs, Lisman (2005, 2007) has hypothesized that the *dentate gyrus* (DG), which is the first processing stage of the hippocampus, may form the sense–act couplets predicted by the DAC architecture (Lisman, 2007). Using an anatomically and physiologically constrained model, Rennó-Costa, Lisman, et al. (2010, 2014) have shown that representations in the lateral and medial EC may be multiplexed in the responses of the neurons in the DG (figure 16.18). The model proposes that the first step in the construction of sense–act couples predicted by DAC occurs in the fixed divergent mapping of the EC onto the DG. This model also explains the specific phenomenon of "rate remapping" of the population response in DG when visual cues are smoothly varying due to the "morphing" of the environment (e.g., when a square environment is gradually modified to become circular). Rate remapping describes the finding that the correlation of the population response of the DG in morphed environments smoothly degrades (remaps) as environments become more dissimilar (Leutgeb, Leutgeb, et al., 2007). The DAC model of the DG can account for this effect but, more importantly, also shows that the neurons in the DG provide an integrated encoding of action and sensory states, as is assumed for the DAC contextual layer, which drives place cell activity in the hippocampal CA3 region. Rennó-Costa, Luvizotto, et al. (2011) have also shown that this detailed neuronal model of mammalian spatial cognition can support robust SLAM in a mobile robot, providing additional evidence of the sufficiency of this account as a theory of animal navigation behavior.

16.6 Conclusion—Advancing Brain Theory through Neurorobotics

The approach of embodied modeling has a broad range of benefits and contributions to bring to brain science. Neurorobotics adds to the computational neuroscience approach the capacity to directly test models in the real world, avoiding the need to simulate in detail the physics of organism–environment interaction, and providing validation that our theoretical models can generate integrated and appropriate behavior in real time. Perhaps as importantly, the challenge to create neurorobotic models demands that we also close the loop between perception and action, and this serves to remind us that there is a tight coupling between brains, bodies, and the environment, and that behavior, and therefore any explanation of behavior, must take into account all three.

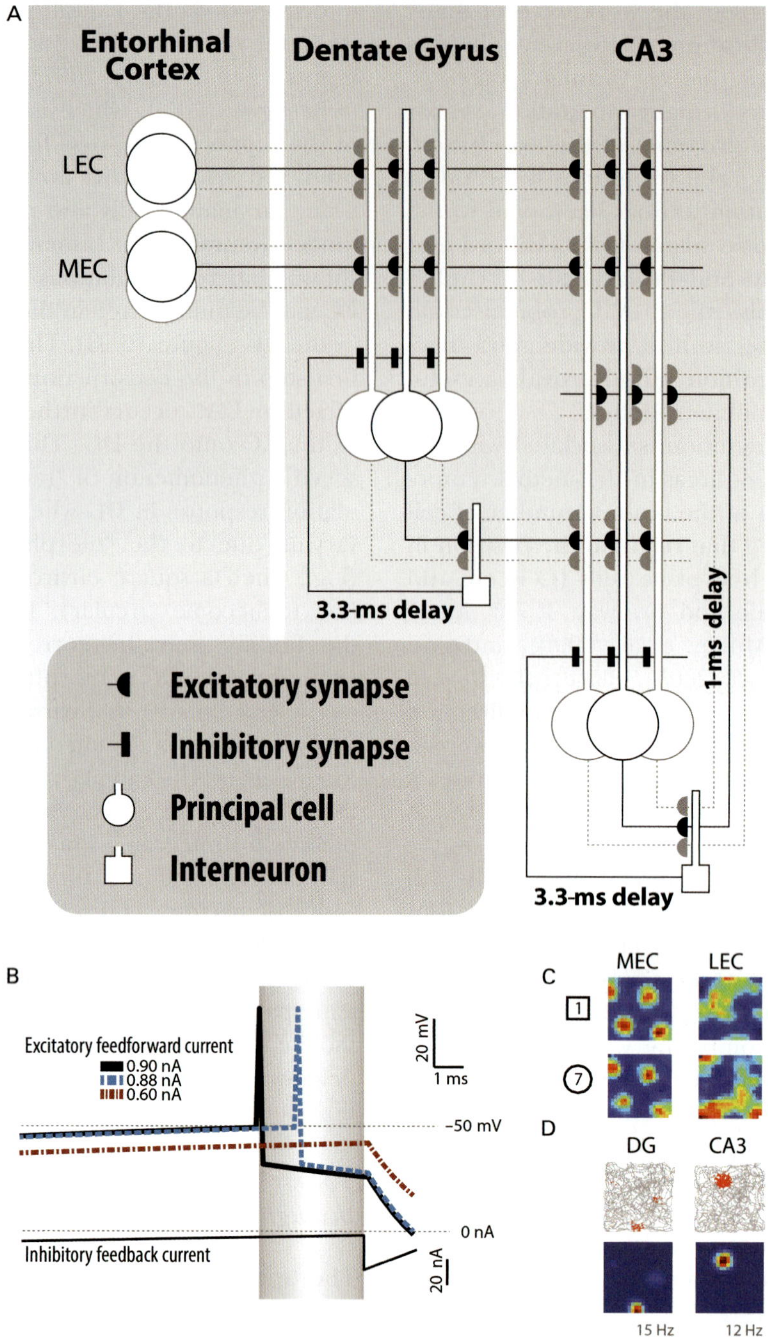

FIGURE 16.18 A computational model of the mammalian temporal lobe as a test of the sensorimotor multiplexing hypothesis of distributed adaptive control and its impact on spatial memory. (A) Excitatory granule cells of dentate gyrus (DG) receive convergent input from *entorhinal cortex* (EC) (convergence ratio of 2,700:1) with delayed inhibitory feedback (delay: 3.3 ± 0.4 ms) from local interneurons. Excitatory cells in hippocampal CA3 receive convergent input from both EC (ratio 2,900:1) and from *dentate gyrus* (DG) (ratio 50:1) together with delayed feedback inhibition from local interneurons (delay: 3.3 ± 0.4 ms) and recurrent excitatory input (delay: 1 ms). (B) Delayed feedback inhibition defines a winner-take-all competition that selects which cell fires in a given cycle. The diagram provides a trace of three sample cells with different strength of excitatory feedforward input. Time is represented by the horizontal axis; the gray area designates the window between the first spike and the onset of global inhibition. Cell voltage and input currents are shown on the ordinate. (C) Rate maps of sample EC grid cells exposed to two shapes of an environment. (D) Action potentials (red dots) with overlaid trajectory of a simulated agent (gray line) and equivalent rate maps of sample DG and CA3 cells showing a well-defined place cell response in CA3. MEC/LEC, medial/lateral entorhinal cortex. Figure from Rennó-Costa, Lisman et al. (2014) reprinted under Creative Commons license (CC BY 3.0).

Acknowledgments

The contributions of Tony Prescott and of Paul Verschure were partly supported by the EU Framework 7 Projects CSNII (601167), BIOTACT (215910), EFAA (270490), and WYSIWYD (612139). Joseph Ayers is supported by ONR MURI # N000141110725.

REFERENCES

Anderson, S. R., Pearson, M. J., Pipe, A., Prescott, T., Dean, P., & Porrill, J. (2010). Adaptive cancelation of self-generated sensory signals in a whisking robot. *Robotics, IEEE Transactions on, 26*(6), 1065–1076.

Arbib, M. A., & Liaw, J.-S. (1995). Sensorimotor transformations in the worlds of frogs and robots. *Artificial Intelligence, 72*(1–2), 53–79.

Arleo, A., & Gerstner, W. (2001). Spatial cognition and neuro-mimetic navigation: A model of hippocampal place cell activity. *Biological Cybernetics, 83,* 287–299.

Ayers, J. (2002). A conservative biomimetic control architecture for autonomous underwater robots. In J. Ayers, J. Davis, & A. Rudolph (Eds.), *Neurotechnology for Biomimetic Robots* (pp. 234–252). Cambridge, MA: MIT Press.

Ayers, J. (2004). Underwater walking. *Arthropod Structure & Development, 33*(3), 347–360.

Ayers, J., Blustein, D., & Westphal, A. (2012). A conserved biomimetic control architecture for walking, swimming and flying robots. *Biomimetic and Biohybrid Systems, 7375,* 1–12.

Ayers, J., & Davis, W. J. (1977). Neuronal control of locomotion in the lobster (*Homarus americanus*): I. Motor programs for forward and backward walking. *Journal of Comparative Physiology, 115,* 1–27.

Ayers, J., Davis, J., & Rudolph, A. (2002). *Neurotechnology for Biomimetic Robots.* Cambridge, MA: MIT Press.

Ayers, J., & Rulkov, N. (2007). Controlling biomimetic underwater robots with electronic nervous systems. In N. Kato & S. Kamimura (Eds.), *Bio-mechanisms of Animals in Swimming and Flying* (pp. 295–306). Tokyo: Springer-Verlag.

Ayers, J., Rulkov, N., Knudsen, D., Kim, Y.-B., Volkovskii, A., & Selverston, A. (2010). Controlling underwater robots with electronic nervous systems. *Appied Bionics and Biomimetics, 7,* 57–67.

Ayers, J., & Selverston, A. I. (1979). Monosynaptic entrainment of an endogenous pacemaker network: A cellular mechanism for von Holst's magnet effect. *Journal of Comparative Physiology. A, Neuroethology, Sensory, Neural, and Behavioral Physiology, 129,* 5–17.

Ayers, J., & Witting, J. (2007). Biomimetic approaches to the control of underwater walking machines. *Philosophical Transactions of the Royal Society of London. Series A, Mathematical and Physical Sciences, 365,* 273–295.

Balda, R. P., & Kamil, A. C. (1992). Long-term spatial memory in Clark's nutcracker, *Nucifraga columbiana. Animal Behaviour, 44*(4), 761–769.

Basil, J. A., & Atema, J. (1994). Lobster orientation in turbulent odor plumes: Simultaneous measurement of tracking behavior and temporal odor patterns. *Biological Bulletin, 187,* 272–283.

Beauregard, M., Kennedy, P. J., & Debenham, J. (2006). Fast simulation of animal locomotion: Lamprey swimming. *Professional Practice in Artificial Intelligence, 218,* 247–256.

Beer, R. D., & Williams, P. L. (2015). Information processing and dynamics in minimally cognitive agents. *Cognitive Science, 39,* 1–38.

Bengston, S. (Ed.). (1994). *Early Life on Earth.* New York: Columbia University Press.

Benzi, R., & Frisch, U. (2010). Turbulence. *Scholarpedia, 5*(3), 3439.

Blustein, D., Rosenthal N., & Ayers, J. (2013). Designing and implementing nervous system simulations on LEGO robots. *JoVE (Journal of Visualized Experiments) 75,* e50519–e50519.

Bowerman, R. F., & Larimer, J. L. (1974a). Command fibres in the circumoesophageal connectives of crayfish: I. Tonic fibres. *Journal of Experimental Biology, 60,* 95–117.

Bowerman, R. F., & Larimer, J. L. (1974b). Command fibres in the circumoesophageal connectives of crayfish: II. Phasic fibres. *Journal of Experimental Biology, 60,* 119–134.

Braitenberg, V. (1986). *Vehicles: Experiments in Synthetic Psychology.* Cambridge, MA: MIT Press.

Brooks, R. A. (1986). A robust layered control system for a mobile robot. *IEEE Journal on Robotics and Automation, RA-2,* 14–23.

Buchanan, J. T. (1996). Lamprey spinal interneurons and their roles in swimming activity. *Brain, Behavior and Evolution, 48*(5), 287–296.

Buchanan, J. T., & McPherson, D. R. (1995). The neuronal network for locomotion in the lamprey spinal cord: Evidence for the involvement of commissural interneurons. *Journal of Physiology, Paris, 89*(4–6), 221–233.

Bullock, T., Orkand, R., & Grinnell, A. (1978). *Introduction to Nervous Systems.* San Francisco, CA: Freeman.

Burgess, N., Donnet, J. G., Jeffery, K. J., & O'Keefe, J. (1997). Robotic and neuronal simulation of the hippocampus and rat naviagation. *Philosophical Transactions of the Royal Society of London. B, Biological Sciences, 352,* 1535–1543.

Chiel, H. J., & Beer, R. D. (1997). The brain has a body: Adaptive behavior emerges from interactions of nervous system, body and environment. *Trends in Neurosciences, 20,* 553–557.

Clark, A. (2013). Whatever next? Predictive brains, situated agents, and the future of cognitive science. *Behavioral and Brain Sciences, 36*(3), 181–204.

Crick, F. (1990). *What Mad Pursuit: A Personal View of Scientific Discovery.* New York: Basic Books.

Crimes, T. P., & Anderson, M. M. (1985). Trace fossils from late Precambrian–early Cambrian strata of southeastern Newfoundland (Canada): Temporal and environmental implications. *Journal of Paleontology, 59*(2), 310–343.

Crossin, G., Hinch, S., Cooke, S., Welch, D., Batten, S., Patterson, D., et al. (2007). Behaviour and physiology of sockeye salmon homing through coastal waters to a natal river. *Marine Biology, 152*(4), 905–918.

Dario, P., Carrozza, M. C., Guglielmelli, E., Laschi, C., Menciassi, A., Micera, S., et al. (2005). Robotics as a future and emerging technology: Biomimetics, cybernetics, and neuro-robotics in European projects. *Robotics & Automation Magazine, IEEE, 12*(2), 29–45.

Dehnhardt, G., Mauck, B., Hanke, W., & Bleckmann, H. (2001). Hydrodynamic trail-following in harbor seals (*Phoca vitulina*). *Science, 293*(5527), 102–104.

Diamond, M. E., von Heimendahl, M., Knutsen, P. M., Kleinfeld, D., & Ahissar, E. (2008). "Where" and "what" in the whisker sensorimotor system. *Nature Reviews. Neuroscience, 9*(8), 601–612.

Dickinson, P. S. (2006). Neuromodulation of central pattern generators in invertebrates and vertebrates. *Current Opinion in Neurobiology, 16*(6), 604–614.

Dissanayake, M. W. M. G., Newman, P., Clark, S., Durrant-Whyte, H. F., & Csorba, M. (2001). A solution to the simultaneous localization and map building (SLAM) problem. *Robotics and Automation, IEEE Transactions on, 17*(3), 229–241.

Duff, A., & Verschure, P. F. M. J. (2010). Unifying perceptual and behavioral learning with a correlative subspace learning rule. *Neurocomputing, 73,* 1818–1830.

Edelman, G. M., Reeke, G. N., Jr., Gall, W. E., Tononi, G., Williams, D., & Sporns, O. (1992). Synthetic neural modeling applied to a real-world artifact. *Proceedings of the National Academy of Sciences of the United States of America, 89*(15), 7267–7271.

Fenner, J. W., Brook, B., Clapworthy, G., Coveney, P. V., Feipel, V., Gregersen, H., et al. (2008). The EuroPhysiome, STEP and a roadmap for the virtual physiological human. *Philosophical Transactions. Series A, Mathematical, Physical and Engineering Sciences, 366,* 2979–2999.

Foster, D. J., Morris, R. G. M., & Dayan, P. (2000). A model of hippocampally dependent navigation, using the temporal difference learning rule. *Hippocampus, 10,* 1–16.

Gao, P., Bermejo, R., & Zeigler, H. P. (2001). Whisker deafferentation and rodent whisking patterns: Behavioral evidence for a central pattern generator. *Journal of Neuroscience, 21*(14), 5374–5380.

Grant, R. A., Mitchinson, B., Fox, C., & Prescott, T. J. (2009). Active touch sensing in the rat: Anticipatory and regulatory control of whisker movements during surface exploration. *Journal of Neurophysiology, 101*(2), 862–874.

Grant, R. A., Sperber, A. L., & Prescott, T. J. (2012). The role of orienting in vibrissal touch sensing. *Frontiers in Behavioral Neuroscience, 6,* 39.

Grasso, F. W., & Atema, J. (2002). Integration of flow and chemical sensing for guidance of autonomous marine robots in turbulent flows. *Environmental Fluid Mechanics, 2*(1), 95–114.

Grasso, F. W., & Basil, J. A. (2002). How lobsters, crayfishes, and crabs locate sources of odor: Current perspectives and future directions. *Current Opinion in Neurobiology, 12*(6), 721–727.

Grillner, S., Hellgren, J., Ménard, A., Saitoh, K., & Wikström, M. A. (2005a). Mechanisms for selection of basic motor programs—Roles for the striatum and pallidum. *Trends in Neurosciences, 28*(7), 364–370.

Grillner, S., Markram, H., De Schutter, E., Silberberg, G., & LeBeau, F. E. (2005b). Microcircuits in action—From CPGs to neocortex. *Trends in Neurosciences, 28*(10), 525–533.

Hafting, T., Fyhn, M., Molden, S., Moser, M. B., & Moser, E. I. (2005). Microstructure of a spatial map in the entorhinal cortex. *Nature, 436*(7052), 801–806.

Hanke, W., Witte, M., Miersch, L., Brede, M., Oeffner, J., Michael, M., et al. (2010). Harbor seal vibrissa morphology suppresses vortex-induced vibrations. *Journal of Experimental Biology, 213*(Pt 15), 2665–2672.

Harnad, S. (1990). The symbol grounding problem. *Physica, 42,* 335–346.

Harris-Warrick, R. M. (2011). Neuromodulation and flexibility in central pattern generator networks. *Current Opinion in Neurobiology, 21*(5), 685–692.

Hasselmo, M. E. (1995). Neuromodulation and cortical function: Modeling the physiological basis of behavior. *Behavioural Brain Research, 67*(1), 1–27.

Heinzel, H. G. (1988). Gastric mill activity in the lobster: II. Proctolin and octopamine initiate and modulate chewing. *Journal of Neurophysiology, 59*(2), 551–565.

Hume, R. I., Getting, P. A., & Del Beccaro, M. A. (1982). Motor organization of Tritonia swimming: I. Quantitative analysis of swim behavior and flexion neuron firing patterns. *Journal of Neurophysiology, 47*(1), 60–74.

Hutchins, E. (1995). *Cognition in the Wild.* Cambridge, MA: MIT Press.

Ijspeert, A. J. (2000). A 3-D biomechanical model of the salamander. *Virtual Worlds, 1834,* 225–234.

Ijspeert, A. J. (2014). Biorobotics: Using robots to emulate and investigate agile locomotion. *Science, 346*(6206), 196–203.

Jensen, D. D. (1959). A theory of the behaviour of *Paramecium aurelia* and the behavioral effects of feeding, fission and ultraviolet microbeam irradiation. *Behaviour, 15*(1), 82–122.

Jordan, L. M., Liu, J., Hedlund, P. B., Akay, T., & Pearson, K. G. (2008). Descending command systems for the initiation of locomotion in mammals. *Brain Research. Brain Research Reviews, 57*(1), 183–191.

Kalman, R. E. (1960). A new approach to linear filtering and prediction problems. *Transactions ASME Journal of Basic Engineering, 82,* 35–45.

Kennedy, D., & Davis, W. J. (1977). Organization of invertebrate motor systems. In S. R. Geiger, E. R. Kandel, J. M. Brookhart, V. B. Montcastle (Eds.), *Handbook of Physiology* (Vol. 1, part 2, pp. 1023–1087). Bethesda, MD: American Physiological Society.

Kleinfeld, D., Ahissar, E., & Diamond, M. E. (2006). Active sensation: Insights from the rodent vibrissa sensorimotor system. *Current Opinion in Neurobiology, 16*(4), 435–444.

Koenig, N., & Howard, A. (2004). Design and use paradigms for Gazebo, an open-source multi-robot simulator. *Proceedings of the IEEE/RSJ International Conference on Intelligent Robots and Systems (IROS 2004)* (Vol. 3, pp. 2149–2154).

Krichmar, J. (2008). Neurorobotics. *Scholarpedia, 3*(3), 1365.

Krichmar, J., & Wagatsuma, H. (Eds.). (2011). *Neuromorphic and Brain-Based Robots.* Cambridge: Cambridge University Press.

Leonard, J. J., & Durrant-Whyte, H. F. (1991). *Simultaneous map building and localization for an autonomous mobile robot.* In *IEEE/RSJ International Workshop on Intelligent Robots and Systems '91: Intelligence for Mechanical Systems.* New York: IEEE.

Leutgeb, J. K., Leutgeb, S., Moser, M.-B., & Moser, E. I. (2007). Pattern separation in the dentate gyrus and CA3 of the hippocampus. *Science, 315,* 961–966.

Lisman, J. E. (2005). The theta/gamma discrete phase code occuring during the hippocampal phase precession may be a more general brain coding scheme. *Hippocampus, 15,* 913–922.

Lisman, J. E. (2007). Role of the dual entorhinal inputs to hippocampus: A hypothesis based on cue/action (non-self/self) couplets. *Progress in Brain Research, 163,* 615–625.

Lungarella, M., Pegors, T., Bulwinkle, D., & Sporns, O. (2005). Methods for quantifying the informational structure of sensory and motor data. *Neuroinformatics, 3*(3), 243–262.

Maffei, G., Santos-Pata, D., Marcos, E., Sánchez-Fibla, M., & Verschure, P. F. M. J. (2015). An embodied biologically constrained model of foraging: From classical and operant conditioning to adaptive real-world behavior in DAC-X. *Neural Networks, 72*, 88–108.

Mataric, M. J. (1997). Behavior-based control: Examples from navigation, learning, and group behavior. *Experimental and Theoretical Artificial Intelligence, Special Issue on Software Architectures for Physical Agents, 9*(2–3), 323–336.

Mathews, Z., Lechón, M., Calvo, J. B., Dhir, A., Duff, A., Bermudez i Badia, S., & Verschure, P. F. M. J. (2009). Insect-like mapless navigation based on head direction cells and contextual learning using chemo-visual sensors. The 2009 IEEE/RSJ International Conference on Intelligent Robots and Systems (IROS), IEEE.

Mathews, Z., Verschure, P. F. M. J., & Bermudez i Badia, S. (2010). An insect-based method for learning landmark reliability using expectation reinforcement in dynamic environments. 2010 IEEE International Conference on Robotics and Automation (ICRA).

Matsushima, T., & Grillner, S. (1992). Neural mechanisms of intersegmental coordination in lamprey: Local excitability changes modify the phase coupling along the spinal cord. *Journal of Neurophysiology, 67*(2), 373–388.

McGeer, T. (1990). Passive dynamic walking. *International Journal of Robotics Research, 9*(2), 62–82.

Michel, O. (1998). Webots: Symbiosis between virtual and real mobile robots. In J.-C. Heudin (Ed.), *Virtual Worlds: First International Conference, VW'98 Paris, France, July 1–3, 1998 Proceedings* (pp. 254–263). Berlin: Springer.

Michelsen, A., Popov, A. V., & Lewis, B. (1994). Physics of directional hearing in the cricket *Gryllus bimaculatus*. *Journal of Comparative Physiology. A, Neuroethology, Sensory, Neural, and Behavioral Physiology, 175*(2), 153–164.

Milford, M., & Wyeth, G. (2010). Persistent navigation and mapping using a biologically inspired SLAM system. *International Journal of Robotics Research, 29*(9), 1131–1153.

Mitchinson, B., Arabzadeh, E., Diamond, M. E., & Prescott, T. J. (2008). Spike-timing in primary sensory neurons: A model of somatosensory transduction in the rat. *Biological Cybernetics, 98*(3), 185–194.

Mitchinson, B., Gurney, K. N., Redgrave, P., Melhuish, C., Pipe, A. G., Pearson, M., et al. (2004). Empirically inspired simulated electro-mechanical model of the rat mystacial follicle-sinus complex. *Proceedings. Biological Sciences, 271*(1556), 2509–2516.

Mitchinson, B., Martin, C. J., Grant, R. A., & Prescott, T. J. (2007). Feedback control in active sensing: Rat exploratory whisking is modulated by environmental contact. *Proceedings. Biological Sciences, 274*(1613), 1035–1041.

Mitchinson, B., Pearson, M., Melhuish, C., & Prescott, T. J. (2006). A model of sensorimotor coordination in the rat whisker system. In S. Nolfi, G. Baldassarre, R. Calabretta, et al. (Eds.), *From Animals to Animats 9* (pp. 77–88). Berlin: Springer.

Mitchinson, B., Pearson, M., Pipe, T., & Prescott, T. J. (2011). Biomimetic robots as scientific models: A view from the whisker tip. In J. Krichmar & H. Wagatsuma (Eds.), *Neuromorphic and Brain-Based Robots* (pp. 23–57). Cambridge: Cambridge University Press.

Mitchinson, B., Pearson, M. J., Pipe, A. G., & Prescott, T. J. (2014). Biomimetic tactile target acquisition, tracking and capture. *Robotics and Autonomous Systems, 62*(3), 366–375.

Mitchinson, B., & Prescott, T. J. (2013). Whisker movements reveal spatial attention: A unified computational model of active sensing control in the rat. *PLoS Computational Biology, 9*(9), e1003236.

Mohd Jani, J., Leary, M., Subic, A., & Gibson, M. A. (2014). A review of shape memory alloy research, applications and opportunities. *Materials & Design, 56*(0), 1078–1113.

Moore, P. A., & Atema, J. (1991). Spatial information in the three-dimensional fine structure of an aquatic odor plume. *Biological Bulletin, 181*(3), 408–418.

Murphy, C. T., Eberhardt, W. C., Calhoun, B. H., Mann, K. A., & Mann, D. A. (2013). Effect of angle on flow-induced vibrations of pinniped vibrissae. *PLoS One, 8*(7), e69872.

O'Keefe, J., & Conway, D. H. (1978). Hippocampal place units in the freely moving rat: Why they fire where they fire. *Experimental Brain Research, 31*(4), 573–590.

O'Keefe, J., & Dostrovsky, J. (1971). The hippocampus as a spatial map: Preliminary evidence from unit activity in the freely-moving rat. *Brain Research, 34*(1), 171–175.

Panksepp, J., & Biven, L. (2012). *The Archaeology of Mind: Neuroevolutionary Origins of Human Emotions.* New York: Norton.

Pearson, K. G. (1993). Common principles of motor control in vertebrates and invertebrates. *Annual Review of Neuroscience, 16*(1), 265–297.

Pearson, M. J., Fox, C., Sullivan, J. C., Prescott, T. J., Pipe, T., & Mitchinson, B. (2013). Simultaneous localisation and mapping on a multi-degree of freedom biomimetic whiskered robot. IEEE International Conference on Robotics and Automation (ICRA), Karlsruhe, Germany.

Pearson, M. J., Mitchinson, B., Sullivan, J. C., Pipe, A. G., & Prescott, T. J. (2011). Biomimetic vibrissal sensing for robots. *Philosophical Transactions of the Royal Society of London. Series B, Biological Sciences, 366*(1581), 3085–3096.

Pearson, M. J., Mitchinson, B., Welsby, J., Pipe, T., & Prescott, T. J. (2010). SCRATCHbot: Active tactile sensing in a whiskered mobile robot. In S. Doncieux, B. Girard, A. Guillot, J. Hallam, J.-A. Meyer, & J.-B. Mouret (Eds.), *From Animals to Animats 11:* 11th International Conference on Simulation of Adaptive Behavior, SAB 2010, Paris–Clos Lucé, France, August 25–28, 2010 (pp. 93–103). Berlin: Springer.

Pearson, M. J., Pipe, A. G., Melhuish, C., Mitchinson, B., & Prescott, T. J. (2007). Whiskerbot: A robotic active touch system modeled on the rat whisker sensory system. *Adaptive Behavior, 15*(3), 223–240.

Perkel, D. H., & Mulloney, B. (1974). Motor pattern production in reciprocally inhibitory neurons exhibiting post-inhibitory rebound. *Science, 185*, 181–183.

Petersen, C. C. (2007). The functional organization of the barrel cortex. *Neuron, 56*(2), 339–355.

Pfeifer, R., & Bongard, J. C. (2006). *How the Body Shapes the Way We Think: A New View of Intelligence.* Cambridge, MA: MIT Press.

Prescott, T. J., Diamond, M. E., & Wing, A. M. (2011). Active touch sensing. *Philosophical Transactions of the Royal Society of London. Series B, Biological Sciences, 366*(1581), 2989–2995.

Prescott, T. J., & Ibbotson, C. (1997). A robot trace-maker: Modeling the fossil evidence of early invertebrate behavior. *Artificial Life, 3*, 289–306.

Prescott, T. J., Lepora, N., & Vershure, P. F. M. J. (2014). A future of living machines? International trends and prospects in biomimetic and biohybrid systems. In *Proc. SPIE 9055, Bioinspiration, Biomimetics, and Bioreplication 2014.*

Prescott, T. J., Mitchinson, B., Lepora, N. F., Wilson, S. P., Anderson, S. R., Porrill, J., et al. (2015). The robot vibrissal system: Understanding mammalian sensorimotor co-ordination through biomimetics. In P. Krieger & A. Groh (Eds.), *Sensorimotor Integration in the Whisker System.* New York: Springer.

Prescott, T. J., Montes Gonzalez, F. M., Gurney, K., Humphries, M. D., & Redgrave, P. (2006). A robot model of the basal ganglia: Behavior and intrinsic processing. *Neural Networks, 19*(1), 31–61.

Prescott, T. J., Pearson, M. J., Mitchinson, B., Sullivan, J. C. W., & Pipe, A. G. (2009). Whisking with robots: From rat vibrissae to biomimetic technology for active touch. *IEEE Robotics & Automation Magazine, 16*(3), 42–50.

Prescott, T. J., Redgrave, P., & Gurney, K. N. (1999). Layered control architectures in robots and vertebrates. *Adaptive Behavior, 7*(1), 99–127.

Pyk, P., Bermudez, S., Bernardet, U., Knuesel, P., Carlsson, M., Gu, J., et al. (2006). An artificial moth: Chemical source localization using a robot based neuronal model of moth optomotor anemotactic search. *Autonomous Robots: Special Issue on Robot Olfaction, 20*(3), 197–213.

Quist, B. W., Seghete, V., Huet, L. A., Murphey, T. D., & Hartmann, M. J. Z. (2014). Modeling forces and moments at the base of a rat vibrissa during noncontact whisking and whisking against an object. *Journal of Neuroscience, 34*(30), 9828–9844.

Reed, H. L. (1952). Firing table computations on the Eniac. In *Proceedings of the 1952 ACM National Meeting (Pittsburgh)* (pp. 103–106). Pittsburgh, PA: ACM.

Rennó-Costa, C., Lisman, J. E., & Verschure, P. F. M. J. (2010). The mechanism of rate remapping in the dentate gyrus. *Neuron, 68*(6), 1051–1058.

Rennó-Costa, C., Lisman, J. E., & Verschure, P. F. M. J. (2014). A signature of attractor dynamics in the CA3 region of the hippocampus. *PLoS Computational Biology, 10*(5), e1003641.

Rennó-Costa, C., Luvizotto, A., Marcos, E., Duff, A., Sánchez-Fibla, M., & Verschure, P. F. M. J. (2011). Integrating neuroscience-based models towards an autonomous biomimetic synthetic. IEEE International Conference on RObotics and BIOmimetics (IEEE-ROBIO 2011). Phuket Island, Thailand, IEEE.

Rosenblueth, A., & Wiener, N. (1945). The role of models in science. *Philosophy of Science, 12*(4), 316–321.

Rulkov, N. (2002). Modeling of spiking-bursting neural behavior using two-dimensional map. *Physical Review E: Statistical, Nonlinear, and Soft Matter Physics, 65*, 041922.

Sanchez-Fibla, M., Bernardet, U., Wasserman, E., Pelc, T., Mintz, M., Jackson, J. C., et al. (2010). Allostatic control for robot behavior regulation: A comparative rodent-robot study. *Advances in Complex Systems, 13*(3), 377–403.

Searle, J. R. (1980). Minds, brains, and programs. *Behavioral and Brain Sciences, 3*, 417–457.

Selverston, A. I. (1980). Are central pattern generators understandable? *Behavioral and Brain Sciences, 3*, 535–571.

Selverston, A. I. (2010). Invertebrate central pattern generator circuits. *Philosophical Transactions of the Royal Society of London. Series B, Biological Sciences, 365*(1551), 2329–2345.

Selverston, A. I., & Ayers, J. (2006). Oscillations and oscillatory behavior in small neural circuits. *Biological Cybernetics, 95*(6), 537–554.

Selverston, A. I., Russell, D. F., King, D. G., & Miller, J. P. (1976). The stomatogastric nervous system: Structure and function of a small neural network. *Progress in Neurobiology, 7*(3), 215–290.

Seth, A. K., McKinstry, J. L., Edelman, G. M., & Krichmar, J. L. (2004). Active sensing of visual and tactile stimuli by brain-based devices. *International Journal of Robotics and Automation, 19*(4), 222–238.

Shilnikov, A., & Rulkov, N. (2003). Origin of chaos in a two-dimensional map modeling spiking-bursting neural activity. *International Journal of Bifurcation and Chaos in Applied Sciences and Engineering, 13*(11), 3325–3340.

Stein, P. S. G. (1978). Motor systems, with specific reference to the control of locomotion. *Annual Review of Neuroscience, 1*, 61–81.

Stein, P. S. G., Grillner, S., Selverston, A. I., & Stuart, D. G. (1997). *Neurons, Networks and Motor Behavior.* Cambridge, MA: MIT Press.

Sterling, P., & Eyer, J. (1988). Allostasis: A new paradigm to explain arousal pathology. In S. Fisher & J. T. Reason (Eds.), *Handbook of Life Stress, Cognition, and Health* (pp. 629-649). Chicester, UK: Wiley.

Sünderhauf, N., & Protzel, P. (2010). Learning from nature: Biologically inspired robot navigation and SLAM—A review. *KI-Künstliche Intelligenz, 24*(3), 215–221.

Taube, J. S., Muller, R. U., & Ranck, J. B., Jr. (1990). Head-direction cells recorded from the postsubiculum in freely moving rats: II. Effects of environmental manipulations. *Journal of Neuroscience, 10*(2), 436–447.

Theraulaz, G., & Bonabeau, E. (1995). Modeling the collective building of complex architectures in social insects with lattice swarms. *Journal of Theoretical Biology, 177*(4), 381–400.

Thrun, S. (2010). Toward robotic cars. *Communications of the ACM, 53*(4), 99–106.

Thrun, S., Burgard, W., & Fox, D. (2005). *Probabilistic Robotics.* Cambridge, MA: MIT Press.

Tikhanoff, V., Cangelosi, A., Fitzpatrick, P., Metta, G., Natale, L., & Nori, F. (2008). An open-source simulator for cognitive robotics research: The prototype of the iCub humanoid robot simulator. In *Proceedings of the 8th Workshop on Performance Metrics for Intelligent Systems*, pp. 57-61. New York: ACM.

Varela, F. G., Maturana, H. R., & Uribe, R. (1974). Autopoiesis: The organization of living systems, its characterization and a model. *Biosystems, 5*(4), 187–196.

Verschure, P. F. M. J., Wray, J., Sporns, O., Tononi, G., & Edelman, G. M. (1995). Multilevel analysis of classical conditioning in a behaving real world artifact. *Robotics and Autonomous Systems, 16*, 247–265.

Verschure, P. F. M. J. (1996). Connectionist explanation: Taking positions in the mind–brain dilemma. In G. Dorffner (Ed.), *Neural Networks and a New Artificial Intelligence.* (pp. 133–188). London: Thompson.

Verschure, P. F. M. J. (1998). Synthetic epistemology: The acquisition, retention, and expression of knowledge in natural and synthetic systems. In *Proceedings of the IEEE World Conference on Computational Intelligence, WCCI'98, Anchorage, Alaska* (pp. 147–153). Piscataway, NJ: IEEE.

Verschure, P. F. M. J. (2012). Distributed adaptive control: A theory of the mind, brain, body nexus. *Biologically Inspired Cognitive Architectures—BICA, 1*(1), 55–72.

Verschure, P. F. M. J., & Coolen, A. C. C. (1991). Adaptive fields: Distributed representations of classically conditioned associations. *Network, 2*, 189–206.

Verschure, P. F. M. J., Krose, B., & Pfeifer, R. (1992). Distributed adaptive control: The self-organization of structured behavior. *Robotics and Autonomous Systems, 9*, 181–196.

Verschure, P. F. M. J., Krose, B., & Pfeifer, R. (1993). Distributed adaptive control: The self-organization of structured behavior. *Robotics and Autonomous Systems, 9*, 181–196.

Verschure, P. F. M. J., Pennartz, C. M. A., & Pezzulo, G. (2014). The why, what, where, when and how of goal-directed choice: Neuronal and computational principles. *Philosophical Transactions of the Royal Society of London. Series B, Biological Sciences, 369*(1655). http://dx.doi.org/10.1098/rstb.2013.0483

Verschure, P. F. M. J., & Voegtlin, T. (1999). A bottom up approach towards the acquisition and expression of sequential representation applied to a behaving real-world device: Distributed adaptive control III. *Neural Networks, 11*, 1531–1549.

Verschure, P. F. M. J., Voegtlin, T., & Douglas, R. J. (2003). Environmentally mediated synergy between perception and behaviour in mobile robots. *Nature, 425*, 620–624.

Voegtlin, T., & Verschure, P. (1998). What can robots tell us about brains? A synthetic approach towards the study of learning and problem solving. *Reviews in the Neurosciences, 10*(3–4), 291–310.

Vogel, S. (1994). *Life in Moving Fluids: The Physical Biology of Flow.* Princeton, NJ: Princeton University Press.

Voigts, J., Sakmann, B., & Celikel, T. (2008). Unsupervised whisker tracking in unrestrained behaving animals. *Journal of Neurophysiology, 100*, 504–515.

Vouloutsi, V., Lallée, S., & Verschure, P. M. J. (2013). Modulating behaviors using allostatic control. In N. Lepora, A. Mura, H. Krapp, P. M. J. Verschure, & T. Prescott (Eds.), *Biomimetic and Biohybrid Systems. Lecture Notes in Artificial Intelligence 8064* (pp. 287–298). *Heidelberg: Springer.*

Walter, W. G. (1953). *The Living Brain.* London: Penguin Books.

Webb, B. (1998). Robots, crickets and ants: Models of neural control of chemotaxis and phonotaxis. *Neural Networks, 11*(7–8), 1479–1496.

Webb, B. (2002). Robots in invertebrate neuroscience. *Nature, 417*(6886), 359–363.

Weeks, J. C. (1981). Neuronal basis of leech swimming: Separation of swim initiation, pattern generation, and intersegmental coordination by selective lesions. *Journal of Neurophysiology, 45*(4), 698–723.

Westphal, A., Blustein, D., & Ayers, J. (2013). A biomimetic neuronal network-based controller for guided helicopter flight. In N. F. Lepora (Ed.), *Biomimetic and Biohybrid Systems* (pp. 299–310). Heidelberg: Springer.

Westphal, A., Rulkov, N., Ayers, J., Brady, D., & Hunt, M. (2011). Controlling a lamprey-based robot with an electronic nervous system. *Smart Structures and Systems, 8*(1), 37–54.

Witting, J., Ayers, J., & Safak, K. (2000). Development of a biomimetic underwater ambulatory robot: Advantages of matching biomimetic control architecture with biomimetic actuators. In G. McKee & P. Schenker (Eds.), Sensor Fusion and Decentralized Control in Robotic Systems III (Vol. 4196, pp. 54–61). Bellingham, WA: SPIE.

17 Object and Scene Perception

OWEN LEWIS AND TOMASO A. POGGIO

17.1 *Introduction*

The collection of questions humans can answer about the visual world is vast and diverse. Some can be answered at a glance: "Is this a picture of a car?" "Was this picture taken outdoors?" Others require more cognitive computation, more viewing time, and possibly links to nonvisual cognitive systems: "Is everyone is this picture looking at the camera?" "Are the people in this image friends?" "What will happen next in this scene?"

These two classes of questions are hypothesized to be handled by two different processing modes in the brain. As we will see, the brain's visual areas are organized into a hierarchy or chain, with each area passing information to its immediate neighbors. Fast visual processing is thought to occur in a single feedforward pass through this chain, with information moving exclusively from lower areas to higher ones. More extended processing, by contrast, brings feedback connections into play, allowing information to circulate among the visual areas in loops. After reviewing the anatomy of visual cortex in section 17.2, the organization of this chapter follows the feedforward/feedback distinction, with sections 17.3 and 17.4 focusing on feedforward processing, and section 17.5, on feedback.

The particular focus of our feedforward discussion is classification problems. In object classification, a model assigns to an image the category label of its principal object, for example "dog" or "car." In scene classification, the label identifies the picture's larger-scale gist: Is it an indoor or outdoor scene, a natural scene or an urban one? Classification is an important building block for complex visual tasks and is a rich and difficult problem in its own right: only very recently have models emerged that can begin to account for human recognition performance, and even today, important open problems remain.

Classification has also been a major focus of recent computer vision research, and the interaction between this field and computational neuroscience has been fruitful for both. Indeed, the currently dominant class of computer vision models, convolutional neural networks (CNNs), is directly inspired by the architecture of visual cortex. Partly due to this computer vision connection, many neural models of feedforward visual processing can handle difficult classification problems in large image data sets, an important engineering step that at the same time has the scientific advantage of facilitating direct comparison with humans' recognition performance.

The feedback models we present aim to go beyond classification, allowing all visual areas to interact recurrently, blending the different sorts of information extracted by each to form the kind of complete scene interpretation necessary to support a full range of visual tasks. This full scene interpretation is a more daunting problem than the classification, and our understanding of how it is solved in the brain is not yet as fully developed. While full scene interpretation in natural images remains beyond the grasp of the neural models we present, these models can tackle important subproblems, and they account for a range of physiological and imaging experimental results.

17.2 *Visual Cortex*

Humans' and animals' ability to process objects and scenes is supported by a large swathe of specialized neural machinery located in visual cortex, near the back of the brain (see figure 12.11 in chapter 12 of this volume). In this section, we lay out the functional and anatomical properties of visual cortex, giving a set of desiderata and structural constraints that will shape the models we develop in the rest of the chapter. This section's core message is that the visual cortex forms a hierarchy of interconnected areas, and that useful representations are built up sequentially, with low-level areas contributing the building blocks to the more high-level and semantically meaningful representations found further upstream.

Visual cortex is traditionally divided into two processing pathways: the dorsal, or "where," pathway, thought to represent the location and motion of objects and to support interaction with visual input via eye and body movements, and the ventral, or "what" stream, thought to support our core tasks of object and scene recognition. Our focus will be on the ventral stream, which encompasses four areas, V1, V2, V4, and inferotemporal cortex (IT).

Within each area, individual cells are responsive to stimuli presented in a small region of space, called the cell's receptive field. V1, V2, V4, and parts of IT show retinotopic organization, meaning that cells whose receptive fields are close to each other are themselves physically close together in the brain.

Across areas, the key organizational hypothesis is that the ventral stream forms a hierarchy; although all four areas are richly interconnected (Kravitz et al., 2013), they are commonly pictured as comprising a chain,

$$V1 \leftrightarrow V2 \leftrightarrow V4 \leftrightarrow IT\ ,$$

in which each area interacts primarily with its immediate neighbors. V1 takes input from the lateral geniculate nucleus (LGN), the structure that processes information immediately after it leaves the retina, and IT passes its output to nonvisual areas in prefrontal cortex. Within the chain, projections from earlier to later areas are called feedforward, and those from later to earlier areas are called feedback. For example, V2 sends feedforward projections to V4 and feedback projections to V1.

Justification of the hierarchical picture comes partly from anatomical studies tracing the projections that connect areas, and partly from functional considerations. For instance, the activation induced by a stimulus spreads sequentially up the hierarchy, emerging in V1 an average of 34–97 ms after stimulus onset, in V2 after 82 ms, and in V4 after 104 ± 23.4 ms in anaesthetized macaques (Schmolesky et al., 1998). A number of other quantities vary systematically along the ventral stream: receptive field sizes get larger from V1 to IT, and the number of neurons in successive areas becomes smaller. Most importantly, though, the hierarchical structure of the ventral stream is reflected in the representations it produces. As information passes through the ventral stream, each area computes a new representation, encoded in the activities of that area's cells. The higher the area, the better able its representation is to support tasks like classification.

A number of experimental techniques are available to study the brain's activity in general, and the representations computed along the ventral stream in particular. At the single-cell level, the tool of choice, electrophysiology, involves implanting electrodes in the brain of an animal—often a macaque monkey—and recording the electrical activity induced by presentation of different stimuli. Electrophysiological recordings have unparalleled temporal and spatial precision, but they are difficult and time-consuming to perform, meaning that most studies can examine at most a few hundred cells. Functional magnetic resonance imaging (fMRI) is a noninvasive neuroimaging technique that monitors the oxygen consumed by blocks of tissue a few millimeters on a side; oxygen intake is a correlate of neural activity. Compared to electrophysiology, fMRI sacrifices spatial and temporal resolution (fMRI scanners obtain a series of images every 2 seconds or so) but is quicker to perform and safe to use on awake human subjects as they perform behavioral tasks. Finally, magnetoencephalography (MEG) and electroencephalography (EEG) are other imaging techniques, also suitable for human subjects, that give millisecond temporal resolution. However, these methods rely on electrical or magnetic fields recorded at the scalp, making their spatial resolution poor.

Once one of these methods has been used to capture neural responses within an area, it remains to characterize the information that this area contains. For instance, we are often interested in determining whether an area's representation is sufficient to classify a stimulus based on its object or scene category. One relatively new analysis technique, neural decoding, underlies many of the studies reported in this chapter. The idea is that if a given representation contains category information, it should be possible to read this information out directly using a classifier. Concretely, suppose an experiment showed images $I_1, \ldots, I_n$, from categories $c_1, \ldots, c_n$, and that image I_i elicited neural response R_i. The classification problem is then to learn a mapping directly from neural response to category label. As usual in machine learning, the set of pairs (R_i, c_i) are divided into a training set used to train a classifier and a test set used to validate its performance. If good classification is possible, one may conclude that the representations R_i contain category information. Generally, decoding studies use linear classifiers, extracting category information in a way similar to how downstream neural populations may do so.

Several notes are in order. First, the fact that it is possible for a classifier to extract category information from an area's activity pattern does not mean that the brain itself extracts this information. Relatedly, the presence of category information does not rule out the possibility of other sorts of information being housed in a given region as well. Last, a negative decoding result, a failure to read out category from activity, does not necessarily imply an absence of information: using finer-grained information or different algorithms, the brain may access information our classifiers cannot.

17.2.1 Representation by Area The above collection of experimental and analytical techniques allows us to explore how information evolves as it passes along the ventral stream. Our particular focus in this section will be on the ability of each area's representation to support classification, though one may hypothesize that classification supporting representations also tends to carry semantic information important for fuller scene interpretation as well.

We begin with V1. The properties of the V1 representation are described in detail in chapter 12, but to briefly summarize, V1 cells are often modeled as Gabor filters, selective for bar-like stimuli at particular positions, orientation, and scales. The V1 representation is already more abstract than one using pixels, supporting edge detection and object segmentation, for instance, but it lacks the high-level semantic content and robustness needed for effective classification.

At the opposite end of the ventral stream, the IT representation contains a remarkable amount of category information. fMRI studies, for instance, have found subareas of IT specifically responsive to certain object classes, such as faces, places, and bodies. In a decoding study Hung et al. (2005) classified objects into one of eight categories using a representation consisting of electrophysiological recordings from approximately 250 IT neurons, achieving a high mean accuracy of 94% (with chance being 1/8 = 12.5%).

In a combined MEG and fMRI study, Cichy et al. (2014) explored the intermediate processing between V1 and IT. Taking advantage of MEG's millisecond temporal resolution, the authors were able to pose a large ensemble of decoding problems, one for each time point after stimulus onset. By examining the errors made in each of these problems, they conclude that representations at later times are better able to make fine classification judgments. For instance, decoding performance for the superordinate judgment of whether or not a stimulus was animate was highest at 157 ms, while the peak accuracy for the finer judgment of whether or not an animate stimulus contained a body occurred later, at 170 ms.

Exactly how these population-level decoding results arise from the response properties of individual cells in the various brain areas is not completely clear. A simplified picture has neurons in each successive area becoming responsive to more and more complex features, formed as combinations of the features represented at the layer below. V2, for instance, contains cells responsive to angles and junctions formed by combinations of the bar-like features preferred by V1 cells (Ito and Komatsu, 2004). Farther up, a number of studies have argued that V4 represents still more complex curvature features (Carlson et al., 2011). IT contains cells responsive specifically to certain high-level object classes, for instance faces. At each stage, though, most neurons are responsive to a variety of stimuli, arguing against a one-to-one correspondence between cells and high-level features. Overall, response properties of single cells are unlikely to tell the whole story: the full discriminative power shown by the higher-level ventral stream areas emerges from the conjunctive activity of the area's whole population.

17.2.2 Invariance To begin to get a more fine-grained picture of how neural representations come to support classification, we turn our attention to invariance to nuisance transformations such as translation and scaling. These transformations can radically change the pixel content of an image while preserving its class label, giving rise to difficult situations in which two images with very different pixel-level content should be assigned to the same category.

To isolate the effect of invariance on classification performance, Anselmi et al. (2013) looked at several classification problems in which all images were rectified to a canonical position, scale, and orientation before classification. In this regime, even the simplest pixel representations were able to achieve good classification performance. This suggests that, at least for the classes studied, much of the performance difference between lower and higher ventral stream representations may be due to the latter's invariance properties.

A particular benefit of invariant representations is that they reduce the sample complexity of the learning process, the number of labeled images a learner has to see before being able to categorize new examples. Intuitively, if a learner requires a separate example for each pose in which an object could appear, the total number of examples needed becomes extremely large. Anselmi et al. (2013) quantifies this intuition in the case of translation. Consider a collection of objects of $p \times p$ pixels, which may appear anywhere in an image of size $rp \times rp$. Given an invariant representation, it requires $O(p^2)$ examples to learn a linear classifier, as compared to $O(p^2r^2)$ in the noninvariant case.

Like category selectivity, invariance builds along the ventral stream. Decoding is again an important tool. In using decoding to test for invariance to position, say, the classifier is trained on representations elicited by images shown at positions $P_1, \ldots, P_{(n-1)}$ and tested on images shown at position P_n. Thus, the classifier is never shown examples of the position in which it is tested; good classification performance can only be attributed to information being preserved under positional shifts.

Using physiological data from IT, Hung et al. (2005) find decoding performance decreased by less than 10% in the presence of changes in translation and scale. In an MEG study Isik et al. (2014) show that invariance emerges in stages over the course of visual processing, with invariance to size arriving first, followed by invariance to position. In addition, invariance to smaller transformations precedes invariance to larger transformations.

Invariance also increases at the single-cell level, with responses in V1 being invariant to at most small transformations while some IT cells show invariance to fairly large degrees of rotation, position, and scale change (Ito et al., 1995; Hung et al., 2005). Additionally, Rust and DiCarlo (2010) show an increase in invariance from V4 to IT.

17.2.3 SUMMING UP: LESSONS FOR MODELS The experimental results summarized above pose a clear set of desiderata for the computational models we will study in the remainder of this chapter. At a minimum, the models should mimic cortex's hierarchical structure, arriving at the top level with a representation sufficient for classification. Experimental findings about invariance give an important clue about how these representations can be built. Sections 17.3 and 17.4 encode these insights in models of the visual system's initial feedforward processing stage, arriving at algorithms that mimic humans' ability to classify objects in scenes.

17.3 Feedforward Models

The past five years have seen dramatic progress in feedforward models of visual classification, yielding algorithms that can compete with humans on challenging tasks. Importantly, the class of models, convolutional neural networks (CNNs), responsible for this success, are inspired by the architecture of visual cortex: even in no-holds-barred computer vision, where performance, rather than biological fidelity, is the measure of success, neuroscientific insights have played an important developmental role.

Progress in CNNs, and the history of their rise to prominence, is encapsulated in the results of the ImageNet Large Scale Visual Recognition Challenge (ILSVRC), an annual contest in which models compete on a data set notable for its size, containing 1,000 image categories, each with between a few hundred and more than a thousand example images. In ILSVRC, models are judged on their ability to assign to an image a correct object-category label; example categories include "Afghan hound," "electric fan," "yurt." The competition uses natural images harvested from the Internet and not cropped to isolate the target item: to correct for the fact that an image may contain multiple object classes, models are allowed to produce five candidate labels and are considered to achieve a correct classification if any one of these five matches the target. All statistics reported here are from the survey paper (Russakovsky et al., 2014).

In 2010, the first year the contest was held, the winning model attained 28.2%, an impressive result given the difficulty of the task, but significantly short of the 5.1% error achieved by a human labeler. The following year saw a modest performance improvement, with the winning model's classification error dropping to 25.8%. Both this model and the 2010 winner are representative of the computer vision techniques popular at the time, representing an image in terms of the orientation statistics of its spatial gradients.

In the following year, 2012, CNNs entered the scene, cutting error rates dramatically, with Krizhevsky et al.'s (2012) AlexNet attaining 16.4% error. In subsequent years, CNNs have continued to dominate the

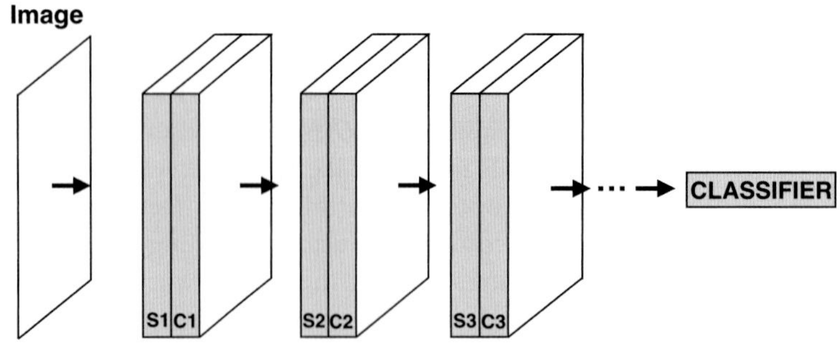

FIGURE 17.1 The convolutional neural network (CNN) architecture. Inspired by Hubel and Weisel's findings in V1, a CNN is composed of a series of layers, each with a simple (S) and complex (C) sublayer. Simple cells build selectivity for complicated features, and complex cells build translation invariance. As it passes through the network, an image undergoes a series of transformations, the end result of which is a representation that can be fed into a classifier. This applies more or less directly to HMAX and LeNet; more recent models include simple-only layers.

competition, becoming by far the most popular model class and achieving steadily decreasing error rates: 11.7% in 2013 and 6.4% in 2014. He et al. (2015) claimed 4.94% on the 2012 data set, beating reported human performance.

The success of CNNs is not limited to ILSVRC; it is now safe to say that they are the tool of choice for the field of object and scene recognition as a whole. They have also made an impact in related fields, for instance, being used to compute visual features for game playing (Mnih et al., 2015), and have attracted considerable investment from industry. As described below, CNNs also account well for humans' recognition behavior and provide unprecedented matches with electrophysiological data.

17.3.1 Convolutional Neural Networks CNNs have a long history prior to their recent climb to computer vision preeminence. The first CNN was Fukushima's 1980 model, the Neocognitron (Fukushima, 1980, 1988). In the 1990s, Yann LeCun and collaborators developed the LeNet family of CNNs (LeCun et al., 1998) and used them successfully for tasks like handwritten digit recognition. In 1999, Riesenhuber and Poggio proposed HMAX, a CNN specifically designed to model mammalian visual cortex (Riesenhuber and Poggio, 1999). HMAX continued to be developed in the 2000s (Serre et al., 2007b). While the different CNNs developed over the years are distinguished by various design choices, most share a core architecture: this will be the focus of this section. Section 17.3.3 presents HMAX, whose more explicitly biological orientation motivates a few important points of difference.

A series of classic experiments done by David Hubel and Thorsten Wiesel in the 1960s provide the principal neuroscientific inspiration for the CNN architecture. Hubel and Weisel recorded from V1 cells in anesthetized cats as the animals viewed patterns of light displayed on a screen. They found two main types of cells: simple cells responded to bar-like patterns at a particular orientation and position on the screen. Complex cells also responded to bars and also had a preferred orientation, but had a small degree of position invariance: their responses were preserved under small translations of the stimulus.

Mimicking the hierarchical organization of visual cortex, a CNN is composed of a succession of layers. Extrapolating Hubel and Wiesel's findings, each unit in the network is of either simple or complex type: simple cells build selectivity for complicated features whereas complex cells build translation invariance. Classically, the network is a composed of a series of layers, each with a simple and complex sublayer; complex cells take input from their own layer's simple cells, which, in turn, take input from the complex cells in the layer before. More recently (Krizhevsky et al., 2012; Szegedy et al., 2014), many models modify this schema by postfixing or interspersing some number of layers that contain only simple cells; we stick to the classical structure here.

Just as category information is exposed as an image passes through the areas of the ventral stream, a CNN's layers enact a similar series of transformations, culminating in a top-layer representation suitable for input to a classifier. This architecture is shown in figure 17.1. We now explore the complex and simple cell types in turn.

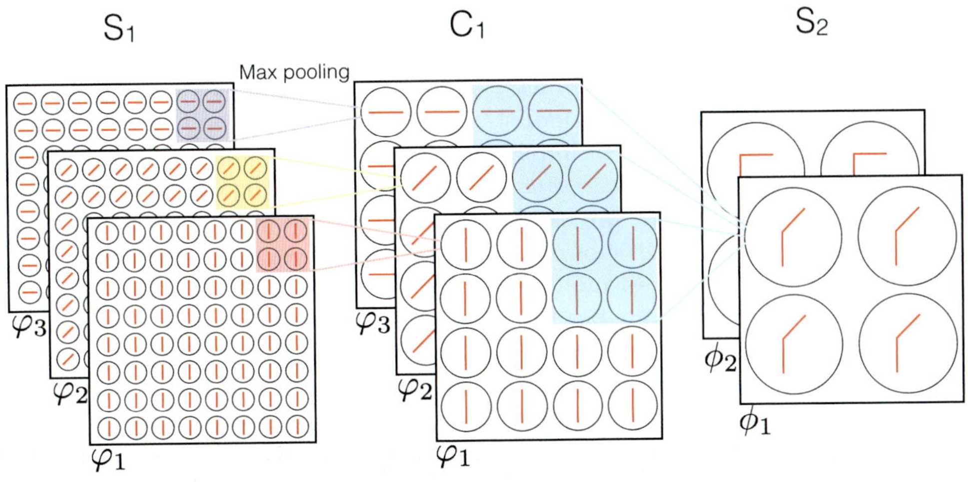

Figure 17.2 A segment from the early layers of a convolutional neural network. Each complex (C) and simple (S) sublayer consists of a collection of feature maps, shown in the figure as sheets, consisting of cells responsive to the same feature at different positions. Complex cells build position invariance by pooling over inputs from a single feature map in the previous sublayer. Simple cells build selectivity for complicated features by combing inputs from multiple feature maps. As shown in the figure, receptive field sizes increase farther up the network.

Consider first building a V1 complex cell like the ones Hubel and Wiesel observed; we will denote such a cell by $C_p^1(\theta_i)$, where the superscript "1" tells us that the cell lives in the first layer, and the subscript p indexes the cell's position within the layer. Because of the network's retinotopic organization, p also picks out a position in the input image where the cell's receptive field is centered. The argument θ_i is the orientation for which the cell is selective. To denote the activation of a cell, rather than its label, we use boldface: $\boldsymbol{C}_p^1(\boldsymbol{\theta}_i)$.

To make our task explicit, we need to wire up $C_p^1(\theta_i)$ in such a way that it activates to a bar at orientation θ_i appearing anywhere in its receptive field; we will denote the receptive field $R(p)$. Hubel and Wiesel proposed a simple way in which this could be done. They hypothesized that $C_p^1(\theta_i)$ receives input from a collection of simple cells, all also with orientation selectivity for θ_i, whose smaller receptive fields cover $R(p)$. Thus, if any one of these afferent cells becomes active, it indicates the presence of a θ_i-oriented bar somewhere in $R(p)$. All $C_p^1(\theta_i)$ has to do, then, is perform an OR-like operation over its input signals, becoming active if any one of them is active. In CNNs, the role of the OR operation is often played by the max function, yielding the so-called max-pooling activation function for V1 complex cells: $\boldsymbol{C}_p^1(\boldsymbol{\theta}_i) = Max_{q \in R(p)} \boldsymbol{S}_q^1(\boldsymbol{\theta}_i)$.

Complex cells in higher layers of the network work similarly. The complex cells in an arbitrary layer l perform max pooling over a localized collection of l's simple cells all tuned to the same feature ϕ_i. In higher layers, these features are more complicated than the orientations we have considered so far:

$$\boldsymbol{C}_p^l(\phi_i) = Max_{q \in R(p)} \boldsymbol{S}_q^l(\phi_i). \tag{17.1}$$

This expression requires that, for each position p, there does in fact exist a family of identically tuned simple cells at a range of positions in $R(p)$. This motivates a key architectural assumption of CNNs: simple cells are organized into "sheets," called *feature maps*, which tile the input plane with cells with the same tuning. In figure 17.2, feature maps are shown as stacked arrays. A single layer is composed of multiple feature maps. As also shown in figure 17.2, the fact that complex cells pool over multiple simple cells means that the complex cells' receptive fields are larger.

We now consider simple cells, again using an example of an early network layer, this time the second layer. Layer 2 simple cells are responsive to patterns such as corners and T-junctions, which can be constructed as combinations of the oriented bars selected for by their layer-1 complex cell afferents. Simple cells are similar to complex cells in that their afferents are spatially localized but differ in the fact that they take input from all of the previous layer's feature maps rather than just one.

In sum, a layer-2 simple cell's inputs come from complex cells detecting line segments at a full range of orientations and a small range of positions. To become selective to a feature like a corner, the cell has to activate strongly only if a certain combination of these segments, for instance a 90° one and 0° one, are present in the image. Whereas complex cells played an OR-like role, simple cells have to behave more like an AND over their inputs. To allow only particular inputs to activate our simple cell, we introduce connection weights modulating the influence these inputs have. Inputs corresponding to features included in our target pattern get high weights; other features get low ones. In real networks, as we will see, simple cell weights and the features they define are learned automatically rather than being set by hand.

Considering a simple cell $S_p^l(\phi_i)$ in an arbitrary layer l, suppose that the previous layer has n feature maps, corresponding to features $\varphi_1, \ldots, \varphi_n$. The afferents of $S_p^l(\phi_i)$ are then a subset of layer $l-1$'s complex cells, localized in space to $R(p)$, but spanning the full range of feature maps. Denote this set

$$A\left(S_p^l(\phi_i)\right) = \left\{C_q^{l-1}(\varphi_j) \mid q \in R(p),\ j = 1 \ldots n\right\}.$$

One of these afferent cells, $C_q^{l-1}(\varphi_j)$, is connected $S_p^l(\phi_i)$ via the weight $W_{(q,\varphi_j)\rightarrow(p,\phi_i)}$. With this (somewhat cumbersome) notation in place, we can present the simple cell activation function, the counterpart of equation 17.1:

$$\boldsymbol{S}_p^l(\phi_i) = \eta\left(\sum_{A\left(S_p^l(\phi_i)\right)} W_{(q,\varphi_j)\rightarrow(p,\phi_i)} \boldsymbol{C}_q^{l-1}(\boldsymbol{\varphi}_j) + b_{\phi_i}\right). \tag{17.2}$$

Here, η is a nonlinear function, such as a ReLU, sigmoid or hyperbolic tangent, and b_{ϕ_i} is a bias term. Overall, then, a simple cell sums the activations of its afferents, weighting each activation by the corresponding connection weight, adds a bias term, and applies a nonlinearity. As with complex cells, a simple cell's receptive field spans the receptive fields of its afferents and is therefore larger.

To take its place in the larger network architecture, $S_p^l(\phi_i)$ must itself be part of a feature map. In other words, there must exist simple cells $S_q^l(\phi_i)$ for all other positions q. Since ϕ_i-specificity is defined by a pattern of weights, a feature map is nothing other than an array of simple cells with the same pattern of input weights, placed at different positions in the layer; the weights are often said to be *tied*. Thus, computing the activations of all the cells in a feature map is mathematically equivalent to convolving the layer's input with the linear filter defined by its feature's weight pattern, adding the bias b to the result, and applying η. This is the origin of the name "convolutional neural network." Simple cell weights and biases are free parameters in a CNN and must be learned during a training period; the procedure is outlined below.

With simple and complex cell types in hand, a modeler can assemble an entire network by specifying a number of parameters: the number and sizes of its layers, the number of its feature maps, the sizes of its receptive fields, and so on. Classical CNNs such as HMAX, LeNet 5, and the Neocognitron make choices roughly similar to visual cortex, having around four layers and sticking to the alternating simple and complex sublayer scheme that we have used in this section. More recently, architectures have grown more exotic; GoogLeNet, winner of the 2014 ILSVRC, for instance, has 22 layers, and Simonyan and Zisserman (2014) present another high-performing model with 16–19. As described above, many modern models also include some number of simple-cell-only layers. The question of why such apparently nonbiological choices, particularly as regards layer numbers, yield state-of-the-art recognition performance is an interesting one that will require further investigation.

17.3.2 LEARNING IN CNNs Having specified the structure of a CNN, it remains to learn the simple cell connection weights. A number of learning techniques are possible. HMAX, for instance, uses a simple unsupervised learning scheme, described below. However, by far the most common approach in performance-optimized CNNs is to optimize a supervised learning objective function using gradient descent. These methods define an error function that measures the model's performance on a training data set of labeled images, compute the gradient of this error function with respect to each of the model's weights, and perform standard gradient descent.

The output layer of a CNN has one unit for each possible class label an input image could be assigned. The activation in the yth output unit in response to an image X represents the model's belief that y is X's correct label. We denote by $f(X;w)$ the output activations induced by an image X fed through a network, parameterized by weights w. Given an image-correct label pair (X_i, y_i) from the training set, a perfectly correct network would produce $f(X_i;w) = \delta_{y_i}$, a vector uniformly zero aside from a one in the y_ith position, indicating a complete concentration of belief on the correct label. A model accrues error to the extent to which it deviates from this desired output. Specifically, we define an error function $E(w)$ measuring the difference between $f(X_i;w)$ and δ_{y_i}. A simple squared l_2 distance illustrates the point, though more complex measures such as cross-entropy are generally used in practice.

$$E(w) = \sum_i \left\| f(X_i;w) - \delta_{y_i} \right\|_2^2 . \tag{17.3}$$

Given a choice of error function, optimization proceeds by gradient descent. A given weight w is updated by

$$w_j^{t+1} = w_j^{t+1} - \alpha \frac{\partial E}{\partial w_j} \tag{17.4}$$

where α is a learning rate. The actual computation of the gradient is usually done by an algorithm called backpropagation. While it works very well in practice, backpropagation is generally considered neurally implausible, and we do not present it here. Nevertheless, even backpropagation-trained networks can be neuroscientifically interesting, as they make it possible to test hypotheses about general network structure, and choice of learning objective, network properties that can be chosen independently of any specific learning process.

17.3.3 HMAX While the CNNs used for tasks like the ILSVRC are inspired by neuroscience, their main objective is classification performance. HMAX, by contrast, is explicitly designed as a cortical model, drawing design choices from the experimental literature wherever possible. HMAX differs from the CNN template presented so far in several ways. First, HMAX simple cells pool over scales as well positions, building invariance to both changes in size and translations. Second, while other models learn their simple cells' weights at every layer, HMAX layer-one simple cells are hand set to have the Gabor-like tuning found in V1. The first layer is composed of Gabor filters at 16 different scales, from 7 × 7 pixels to 37 × 37, and at four different orientations, 0°, 45°, 90°, and 135°.

In addition, HMAX's simple cell activation function is somewhat different from equation 17.2. To avoid the extra notation associated with different scales, we will drop some of our earlier indexing, considering an arbitrary simple cell S_i. As before, S_i receives inputs from a set $A(i)$ of complex cells from the previous layer. Its activation is given by

$$\boldsymbol{S}_i = \exp \frac{-1}{2\sigma^2} \sum_{j \in A(i)} (w_{ji} - \boldsymbol{C}_j)^2 . \tag{17.5}$$

The interpretation here is that the synaptic weights w_{ji} define a "template": S_i's excitation is when the activations in $A(i)$ match this template exactly, and it falls off in a Gaussian way as they deviate from it.

As measured by performance, probably the important distinguishing feature of HMAX is its learning rule. Proceeding with the activation function interpretation above, learning in HMAX consists of choosing an afferent template for each of its simple cells. HMAX accomplishes this with a simple, unsupervised sampling procedure. For each simple cell S_i, the model picks an (unlabeled) image from a data set provided, which, when presented to the network, induces a particular pattern of activation in the complex cells in S_i's afferent set, $A(i)$, with $\boldsymbol{C}_j$ being the induced activation in the jth simple cell. This activation pattern is stored as the template; $w_{ji}: = Cj$. In other words, each simple cell becomes tuned to the neural image of a particular image patch arising in the training set.

This learning scheme has two advantages over the supervised one presented above. First, it avoids the biological implausibility associated with backpropagation. Second, the fact that it is unsupervised means that it does not require a large database of natural images, likely better matching the conditions faced by a young human or animal during development. These advantages, though, come at the price of reduced performance. As we will see below, HMAX achieves respectable recognition performance, competing with humans on some tasks. However, it lags significantly behind performance-optimized CNNs in most classification problems. The difference between HMAX's unsupervised learning rule and the supervised backpropagation used by most other models is likely the main reason for this deficit.

17.3.4 Assessing CNNs A number of experimental methods are available to help us to study the performance and properties of CNNs and to assess the similarity of CNNs to the mammalian visual cortex.

A first question is behavioral: How well can CNNs account for human recognition performance? As we have already seen, CNN performance compares favorably to humans in the difficult ILSVRC recognition task. A more controlled experiment compared HMAX to humans on a simpler visual task, determining whether or not an image contains an animal. In difficult cases, humans can deploy a number of strategies to make this determination, for instance, moving their eyes for detailed examination of particular image regions or using prior knowledge about where in a natural scene an animal might hide. However, strategies like this require time and feedback processing and are outside the portion of visual processing that CNNs attempt to model, namely, the initial feedforward pass through the ventral stream. In an attempt to level the playing field by limiting humans to feedforward processing as well, Serre et al. (2007a) used a paradigm called *masking*, in which each target image is only shown for 20 ms and is followed by a mask, a noise image modified to share the target's low-level statistics. The motivation for this procedure is that lower visual areas will be occupied with feedforward processing of the mask images during the time period when feedback processing of the target could ordinarily occur, thereby preventing this feedback processing from taking place.

Under these conditions, human subjects achieved 80% accuracy on the animal/no-animal classification task. HMAX achieves a similar 82% accuracy. Moreover, the pattern of errors made by the model was qualitatively similar to that of humans: humans and the model tended to find the same images difficult.

To get a finer-grained picture of the properties of a trained CNN that supports recognition performance, Zeiler and Fergus (2014) use a clever algorithm to visualize the features preferred by units at different layers. Applying this algorithm to a CNN similar to the first ILSVRC winner (Krizhevsky et al., 2012), they find features qualitatively similar to those detected by human macaque visual cortex (figure 17.3). Unlike HMAX, in which the layer-1 Gabor filters are hand chosen, the weights at all layers of this CNN are learned during training. Even so, edge-like tuning still emerges in the network's first layer. The second layer selects for features involving multiple edge transitions, such as corners, similar to hypothesized V2 features. The highest layers of the network, 4 and 5, are selective for recognizable objects and object parts, such as animal faces in layer 4 and whole animals in layer 5, suggesting a similarity to mammalian IT. Roughly, then, the features a given CNN layer selects for are qualitatively similar to those selected for by a biological layer at a similar point in the hierarchy.

The feature-preference correspondence between biological neurons and artificial CNN units suggests a more direct comparison: How well can CNN activations predict actual physiological data? Yamins et al. (2014) recorded the responses of macaque IT and V4 units and compared the results to CNN activations induced by the same images. To make the comparison, Yamins et al. compare each biological neuron to the linear combination of artificial units that best matches its responses. Using this procedure, Yamins et al. find that the CNN's top layer explains 48.5 ± 1.3% of the explainable variance in the biological IT cells' firing patterns, and the model layer below explains 51.7 ± 2.3% of the variance of biological V4 firing rates. To put this number in context, it is comparable to the matches other encoding

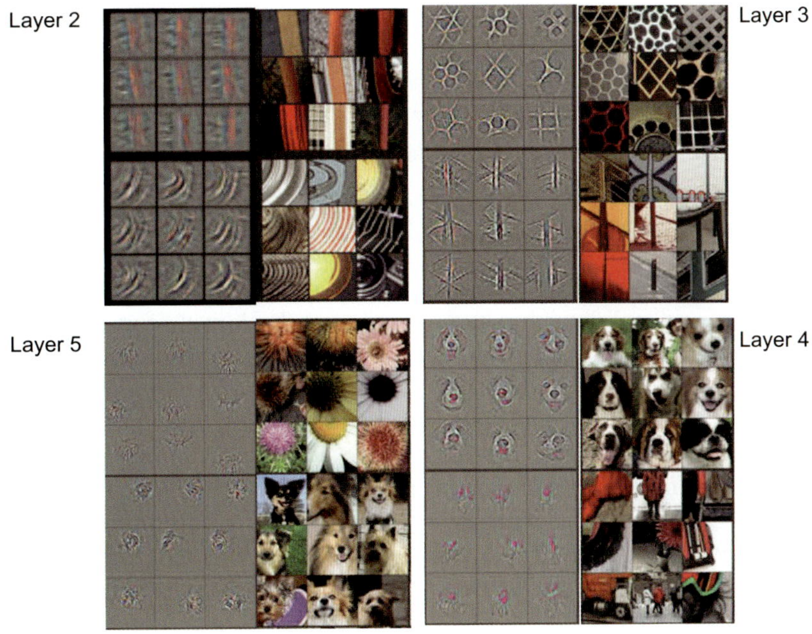

Figure 17.3 Zeiler and Fergus (2014) developed a deconvolution algorithm to visualize the features that best activate the units in a CNN. Here, the algorithm is applied to AlexNet, the best-performing CNN in ILSVRC 2012. Each 3 × 3 grid corresponds to one feature map. The images on the left of each panel show algorithmic reconstructions of the image features that cause high activations, and the color images on the right show patches in which these patterns occur. As in primate visual cortex, units in higher layers are responsive to complex stimuli. Reproduced with permission and modified from Zeiler and Fergus (2014).

models have achieved to the much better understood lower visual areas.

Another physiological question one can ask about CNNs concerns the biological reality of their basic operations. The simple cell activation function in equation 17.2 is a standard and widely accepted model of neural activation throughout the brain, but a CNN's complex cells' max-pooling operation is more exotic. Does it actually occur in cortex? Lampl et al. (2004) recorded from V1 complex cells and found that for 80% of them, a max-like activation function fit their responses better than a linear model. Knoblich et al. (2007) present a circuit-level model of how max pooling might occur.

Given these impressive results, it is natural to wonder if CNNs now offer a full account of feedforward object classification. While they undoubtedly make substantial progress toward this goal, a few caveats remain. First, while CNNs achieve human-like total error on tasks like the ILSRVC, the patterns of errors they make deviate somewhat from those made by humans (Russakovsky et al., 2014). Russakovsky compared human performance to GoogLeNet, the most recent ILSRVC winner. More than half of human errors are attributable to knowledge deficits as opposed to visual errors: humans may not realize that the correct label is an option and may struggle with fine-grained and obscure classes, specific species of dogs, for example. CNNs, by contrast, have more purely visual problems, finding it difficult to label objects that take up a small percentage of the image as a whole, and struggling with images distorted by Instagram-style filters. CNNs also struggle with abstract representations of objects such as drawings or statues. Taking these last two sources of error together, one might hypothesize that, compared to humans, a CNN is more likely to classify based on textural rather than structural features of an image. Relatedly, Szegedy et al. (2013) demonstrate that CNNs can be fooled by adversarial examples, images constructed by taking an image the model originally classified correctly, and adding a specially designed perturbation. While the resulting image is indistinguishable to human eyes from the original, the model nevertheless labels it incorrectly.

Second, achieving high classification accuracy with biological training mechanisms remains an open challenge. As mentioned, the popular backpropagation algorithm relies on non-neural mechanisms. More generally, commonly used supervised learning paradigms require very large amounts of labeled training data, arguably significantly more than human learners need. See Durbin and Rumelhart (1989), O'Reilly (1996), and Balduzzi et al. (2014) for proposed biologically plausible backpropagation alternatives and section 17.6.1 for further discussion of small-sample learning.

So feedforward classification is not without its loose ends. Nevertheless, CNN modeling has been a significant advance. Indeed, it is arguably the most successful effort to date in solving a difficult biological problem in an at least broadly biological way.

17.4 *Feedforward Scene Recognition*

Models like HMAX and other CNNs are generally used for object recognition, as in the animal-detection task described above, but in principle, nothing prevents them from categorizing scenes as well. Indeed, a natural conjecture is that scene recognition is nothing other than repeated object recognition—we might recognize a street scene, for instance, by realizing that it contains cars, pedestrians, and buildings in particular arrangements. But recent work gives compelling experimental evidence that this object-based view of scene perception is incorrect, or at least incomplete. Rather, the human object recognition apparatus is supplemented by a distinct set of processes housed in a distinct set of brain areas that quickly extract the global geometric properties of a scene, its "gist," that support fast recognition of a scene's semantic category and its functional properties. This section reviews this experimental work on gist extraction and then shows that its findings have been captured in computational models.

If the properties of a scene that support classification are not the identities of its constituent objects, what are they? The evidence suggests that scene classification relies on a collection of global geometric properties, such as a scene's openness, the extent to which the horizon is visible; its depth; and its navigability. In addition to their supporting role in classification, these properties have clear ecological relevance in their own right.

Greene and Oliva (2009) showed subjects masked images of natural scenes for variable amounts of time and examined which properties subjects were able to extract. They found that a scene's global properties became available earlier (mean = 34 ms) than did its basic-level semantic category ("mountain," "ocean," etc.; mean = 50 ms), consistent with a picture in which global properties underlie category judgments. To show the sufficiency of these properties for scene classification, Greene and Oliva (2006) had human subjects evaluate the prevalence of each of seven global properties in each database of natural scene images and trained a classifier to predict the category of a scene given only its properties, as coded by humans. They found a good correlation between human performance and that of the classifier, and a similar distribution of mistaken classifications.

Other timing studies examine when object information enters the mix, and they argue for the following approximate ordering: global properties, then scene category, then individual object categories. In a study by Fei-Fei et al. (2007), subjects generated free descriptions of scenes they had viewed for various durations. While the portions of their descriptions involving objects changed and became more detailed for longer viewing times, subjects' high-level scene classifications remained relatively constant. This is consistent with an interpretation in which the scene category is extracted quickly while object identities require longer viewing to emerge in full.

In addition to limiting the processing time available to viewers, other studies have restricted the information available in the frequency domain. Schyns and Oliva (1994) showed subjects images from which all frequencies above two cycles per degree had been eliminated. After this modification, objects generally appear only as unidentifiable blobs, but subjects' ability to identify the scene category was largely unimpaired. A similar low-pass-filtered image is shown in figure 17.4A.

Further evidence for scene processing as distinct from object processing comes from neuroimaging studies showing a distinct network of brain areas underlying scene perception. Imaging work has found at least three areas that preferentially respond to scenes or places over objects: the parahippocampal place area (PPA), retrosplenial cortex, and the occipital place area. In a particularly striking finding, Epstein and Kanwisher (1998) showed that PPA activation to an image of a room was more or less unchanged by removing all the moveable objects (furniture etc.), further evidence of a geometric rather than object-based encoding. Decoding studies with fMRI also support this view. For instance, Park et al. (2011) were able to decode scene information from all three of the scene-specific areas mentioned, as well as from the object-selective area lateral occipital cortex (LOC), but the pattern of errors the classifier made showed a sensitivity to spatial properties (open vs. closed) in the scene areas, as compared to content in LOC.

17.4.1 Feedforward Models of Scene Perception These experimental findings suggest some ways that a representation designed to support scene classification should differ from one for object classification. First, as indicated by humans' ability to classify scenes using only low spatial frequencies, a scene representation should need to be only weakly spatially localized. Second, and relatedly, the scene representation should be able to be relatively low dimensional, as shown by the good performance of the model in Greene

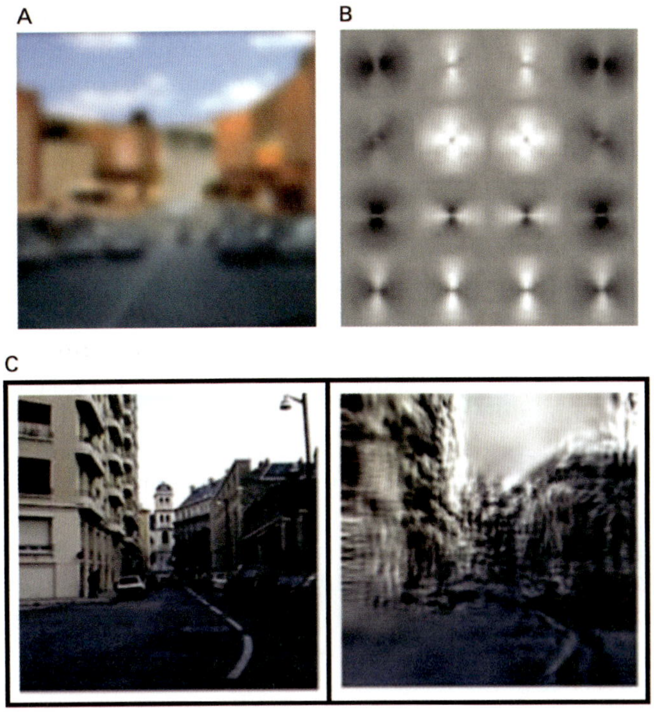

FIGURE 17.4 (A) Despite the fact that low-pass filtering has removed object information, it is still easy to recognize that the image shows a street scene. (B) A global feature template, one principal component of a multiscale filter-bank representation of a database of natural images. The image is divided into 16 subregions, and an output value is computed for each. The polar plot within subregions shows how the output is computed for different orientations and spatial frequencies: angle in the plot represents orientation, and distance along the radius represents spatial frequency, with low frequencies near the center. White regions correspond to positive outputs, black, to negative. (C) A noise image (right) coerced to share global features with a target image (left). The comparison shows that global features preserve the scene's large-scale geometry even while losing its finer details. Reproduced with permission and modified from Oliva and Torralba (2006).

and Oliva (2006) using only seven human-identified features.

Most models of scene processing stem from Oliva and Torralba (2001). Here we present a slightly more neural variant by the same authors, Oliva and Torralba (2006). The model begins with a V1-like representation given by the output of a bank of Gabor filters spanning a range of orientations and scales and positioned densely across the image. A first reduction step coarsens the spatial resolution of this representation, by dividing the image into $N \times N$ windows and averaging the output of each filter within each window. This reduces the dimensionality of the representation to $N \times N \times S \times R$, where S is the number of spatial scales in the filter bank, and R is the number of orientations.

The second reduction step computes this representation for each of a large number of images in a data set and compresses the resulting set of vectors with principal components analysis. Each principal component is a weighted combination of filter outputs at each of the $N \times N$ windows. These weighted combinations are then "fused" into what the authors call global feature templates (GFTs), which span the whole image; an example is shown in figure 17.4B. The first M of these GFTs can then be applied to a new scene image, by applying the original filters and weighting the results, and the output fed to a classifier to extract global properties like openness, depth, and so forth. To get a feel for the representation that the GFTs compute, figure 17.4C shows a noise image modified in such a way that it induces the same GFT responses as a target image of a natural scene. While the modified noise image loses the fine details present in the target, it maintains enough of its global gist to support easy classification.

Since Oliva and Torralba (2001), a number of authors have proposed different models of scene-level processing. See, for example, Renninger and Malik (2004) and Siagian and Itti (2007) and references therein.

Scene classification is a worthy end in itself, but scene information can also be useful in the context of object detection. An image's scene category contains a lot of information about which objects it is likely to contain and where they are likely to appear. A street scene, for instance, is likely to contain multiple cars, while a mountain or forest scene probably contains none. Furthermore, cars appear at predictable (vertical) locations in street scenes; a car is unlikely to appear in the sky, for instance. Torralba et al. (2010), building on Murphy et al. (2003), use scene-level information to adjust the outputs of local object detectors, ensuring that the final detections contain a plausible number of objects of each class and that these objects appear in plausible locations.

Similar intuitions underlie the model in Torralba et al. (2006), which uses scene features to predict human eye movements during visual search. Subjects were told to find an instance of a target object class, "car," for example, and their eye movements were tracked as they looked for it. If humans use their knowledge of the target image's class, together with quickly extracted scene gist features from the given image, then a model incorporating gist features should better predict their fixations than one without.

The tendency of a human to fixate a location X is modeled as

$$S(X) = L(X)^{-\gamma} P(X \mid G, O = 1) \qquad (17.6)$$

where $L(X)$ is the local saliency of the location X. Heuristically, L measures the extent to which X is unexpected given the rest of the image; salient regions stand out. This local contribution is traded off via the exponent γ with a global term, $P(X|G,O=1)$ denoting the probability that an instance of the target class is present at the location X, given the image's global scene-gist features G, and assuming that at least one target class instance is present in the image as a whole ($O = 1$). As expected, the combined global–local model significantly better predicted which image regions subjects were likely to fixate than a local-only one.

17.5 *Feedback Connections*

While the foundations of the feedforward models presented in section 17.3 were already in place by the end of the 1960s, an understanding of visual cortex's feedback connections has been more elusive. It is generally agreed that feedback connections mediate attentional processing, carrying signals about stimulus salience and task relevance from frontal and parietal areas (see chapter 19, "Saccades and Smooth Pursuit Eye Movements," and chapter 4, "Neural Rhythms"), but there has been less consensus about the role of feedback connections in this chapter's core themes of object and scene recognition. Here, we will focus on a class of models that has come to the fore over the last fifteen years or so, which views feedback connections as primarily generative in nature, enabling the brain to synthesize as well as consume images.

That the brain can accomplish this generation is intuitively clear from everyday experiences like dreaming and visual imagination and from clinical observations of visual hallucinations. Sufferers of Charles Bonnet syndrome, for instance, report extremely vivid hallucinations which they have trouble distinguishing from reality, and which imaging studies have shown to have very similar neural signatures to normal visual processing (Ffytche, 2005). But even granting the visual cortex's generative capacity, it still remains to explain how this capacity could be used to support recognition.

17.5.1 Generative Models for Recognition The spiritual father, arguably, of the generative approach to vision is Hermann von Helmholtz, who in 1867 advanced the theory of "perception as unconscious inference." On this view, the input the visual system receives is the result of a causal process, the imaging process mapping world states (particular configurations of objects, illumination, etc.) to two-dimensional images, or collections of LGN cell responses. The goal of visual cortex is then to invert this process, inferring from an image the world state that caused it. Note the ambiguity inherent in this problem: any image is, in general, consistent with many high-level causes. For instance, a percept of a given size might be caused by a large, distant object or by a small object close at hand.

The purely feedforward models we have studied so far approximate a solution to this inverse problem by learning a direct mapping from percept to causes, a purely bottom-up approach. Given a model of the imaging process, a purely top-down process is also possible: one could guess high-level causes and check them by running the imaging process and seeing if the result matched the observed percept. This approach has the advantage of being robust to low-level ambiguity, but in practice, of course, pure trial and error is hopelessly inefficient. Practical generative vision models use a mixed approach in which bottom-up and top-down information is progressively integrated.

Reflecting the ambiguity inherent in the inverse inference problem, generative models in vision are generally probabilistic. Sticking with our chain view of visual cortex, in which areas interact primarily with their immediate neighbors, allows us to factor the joint probability of the activities in all cortical areas, and of the input data in the LGN, as follows:

$$\begin{aligned}&P(LGN,\ V_1,\ V_2,\ V_4,\ IT)\\&\quad= P(LGN\ |V_1)P(V_1|V_2)P(V_2|V_4)P(V_4\ |\ IT)P(IT).\end{aligned} \tag{17.7}$$

The conditional probabilities $P(V_i|V_j)$ encode a model of the statistics of natural images given their content. For instance, $P(V_4\,|\,IT)$ indicates which V_4 responses the model expects given the presence of a particular configuration of objects, encoded as IT responses. These probabilities are acquired through experience; this is the learning problem. Once the relevant distributions are in place, we can ask Helmholtz's inversion question: which higher-level causes were likely to have generated a given LGN input? In the probabilistic context, this is called the inference problem, and it amounts to approximating or maximizing the following posterior distribution:

$$\begin{aligned}P(V_1,\ V_2,\ V_4,\ IT\ |\ LGN) &= \frac{P(V_1,\ V_2,\ V_4,\ IT,\ LGN)}{P(LGN)}\\&= \frac{1}{Z}P(V_1|V_2)P(V_2|V_4)P(V_4\ |\ IT)P(IT).\end{aligned} \tag{17.8}$$

Here, we have treated the probability of the LGN input as a normalizing constant and used equation 17.7 to perform the factorization in the last step.

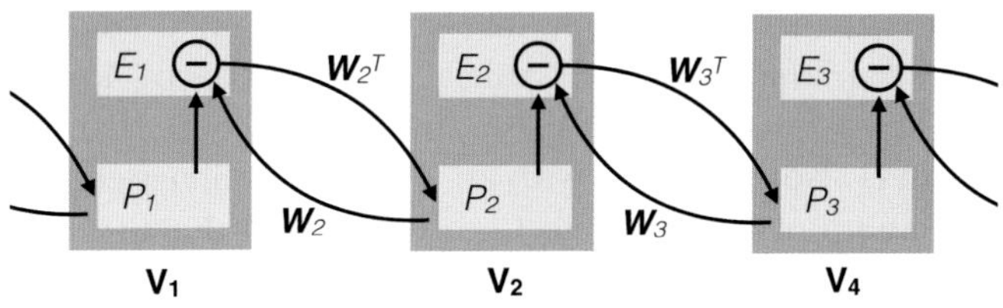

FIGURE 17.5 A possible neural architecture for predictive coding. Each area contains two populations of neurons: P_i predicts the activity of the downstream area, sending the predictions via the synaptic weights matrix W_i, and E_i computes the error made by the predictions coming from the next higher area.

In general, generative approaches to recognition share this basic structure and problem definition but differ in the way in which the optimization is performed and in how the distributions are parametrized.

17.5.2 PREDICTIVE CODING When you tell a friend about your day, you do not catalog routine minutiae. Rather, you only mention the few unusual events that distinguished this particular day from any other, relying on your friend's prior knowledge of your daily habits to fill in the details you left out. This is the intuition underlying predictive coding, a technique originally developed for speech processing in the 1960s: a sender need only transmit those parts of a signal that are unexpected given a predictive model possessed by the receiver.

Predictive coding underlies several models of visual cortex and other brain areas (Mumford, 1992; Friston, 2005). Our exposition most closely follows the model developed in Rao and Ballard (1999). In Rao and Ballard's model, higher visual areas play the role of the receiver in the sketch above, and lower areas, the sender. Higher areas use feedback connections to transmit their generative predictions of the next lower level's activity, and lower areas send up the errors in these predictions via feedforward connections.

In the simplest version of the model, the predictions made by the ith area is the linear function of the area's own activities:

$$\boldsymbol{W}_i V_i \tag{17.9}$$

where $\boldsymbol{W}_i$ is a matrix of synaptic weights. The $\boldsymbol{W}_i$ are structured so that each cell in area V_i sends connections to a spatially localized subset of cells in the layer below, the size of this subset increasing with i, corresponding to the fact that cells in higher areas have larger receptive fields.

With a Gaussian noise model, equation 17.9 gives the following conditional probabilities:

$$P(V_i|V_{i+1}) = \mathcal{N}(\boldsymbol{W}_{i+1}V_{i+1}, \operatorname{diag}(\sigma_i^2)) \tag{17.10}$$

where σ_i^2 is a layer-dependent noise variance. With this formulation, we can solve both the learning problem (finding $\boldsymbol{W}$) and the inference problem with gradient descent; we focus on the inference problem here. From equation 17.8, passing to log space and differentiating gives the following:

$$\frac{\partial}{\partial V_i}\log P(\{V\}_i|LGN) = \sum_k \frac{\partial \log P(V_k\,|\,V_{k+1})}{\partial V_i} \tag{17.11}$$

$$= \frac{\partial \log P(V_{i-1}\,|\,V_i)}{\partial V_i} + \frac{\partial \log P(V_i\,|\,V_{i+1})}{\partial V_i}.$$

After plugging in equation 17.10, we obtain the following update rule:

$$V_i \leftarrow V_i + \alpha\left(\frac{1}{\sigma_{i-1}^2}\boldsymbol{W}_i^T(V_{i-1} - \boldsymbol{W}_i V_i) + \frac{1}{\sigma_{i+1}^2}(\boldsymbol{W}_{i+1}V_{i+1} - V_i)\right). \tag{17.12}$$

This is a quantitative expression of the predictive coding intuition explained above. The second term in the sum is the error in the predictions coming from the higher area V_{i+1} and is available locally to V_i. The first term, the error in V_i's own predictions, has to be passed back up from V_{i-1} where it is computed.

These considerations suggest a neural implementation like the one shown in figure 17.5, adapted from Friston (2005). Each visual area is composed of two subpopulations: the prediction neurons P_i send predictions via feedback connections, and the error detecting neurons E_i compute the error in predictions coming down from the area above. For a more detailed attempt to map the predictive coding hardware onto neural machinery, see Bastos et al. (2012).

17.5.3 PREDICTIVE CODING: SIMULATIONS AND PREDICTIONS In simulations with a truncated version of the model above, encompassing LGN, V1, and V2, Rao and Ballard (1999) derive a number of properties of V1 cell responses, particularly the so-called extra-classical receptive field effects. As described in section 17.2, a cell's classical receptive field is the region of space in which a presented stimulus can evoke a response. Many V1 cells have in addition an extra-classical receptive field (ECRF), which is a larger area outside the classical receptive field where stimuli can modulate the cell's firing. Two examples are shown in figure 17.6. End-stopped cells are responsive to bar-like stimuli and show enhanced responses when a bar terminates in their ECRF. Another effect, surround suppression, occurs when a cell's firing decreases when the dominant orientation in its classical receptive field extends through its ECRF. In general, extra-classical effects occur at

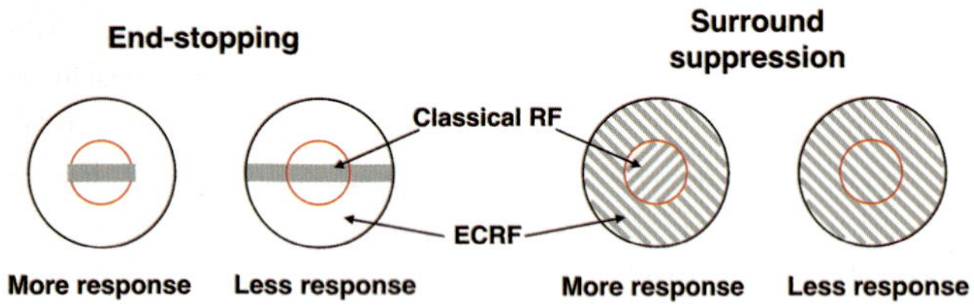

FIGURE 17.6 Extra-classical receptive fields (ECRFs). In both figures, the cell's classical receptive field (RF), the region in which a presented stimulus can directly induce firing, is shown in red, and its ECRF, a larger area in which stimuli can modulate firing, is shown in black. End-stopped cells show a decrease in firing when a bar extends beyond their classical RF. Surround suppression occurs when a cell fires less strongly when the orientation present in its classical RF continues in its ECRF.

around 80–100 ms after stimulus onset, as compared to 66 ms for classical V1 responses, suggesting a role for feedback, or at least lateral, processing.

After training their network on a database of natural images, Rao and Ballard found that a large proportion of their error-correcting neurons displayed end-stopping effects. An analysis of natural image statistics reveals why this should be so. Edges in natural images generally continue smoothly; short, isolated bar segments are rare. During training, the simulated V2 neurons adapt to this regularity and shape their predictions of V1 responses accordingly. Thus, short bars contained within a V1 cell's ECRF trigger a prediction error and cause increased activity in the V1 error-detecting cells. Recall that the receptive field of each simulated V2 neuron contains several V1 neurons; this group constitutes its central cell's ECRF.

The analysis for surround suppression is similar. By and large, the dominant orientation in a natural image changes slowly and smoothly over space, so this is what V2 cells come to predict. When this prediction is upheld, error-detecting neurons fire less, resulting in the observed suppression.

Subsequent studies have used fMRI to investigate predictive coding. Murray et al. (2004) used stimuli of a variety of types, each containing a collection of elements. In one class of stimuli, the elements could be interpreted as forming a coherent group of some kind, while, in the other class, they appeared essentially random. A collection of line segments forming a coherent two-dimensional shape, for instance, would be a member of the first class, while a collection of scattered segments would be a member of the second. In fMRI experiments, the groupable stimuli were found to induce increased response in higher visual areas, such as the object-responsive area LOC, and reduced responses in V1. This finding is consistent with predictive coding: groupable stimuli admit a higher-level explanation that is able to "predict away" lower-level responses. The authors point out, however, that the result is also consistent with an alternative "sharpening" explanation, in which lower-level responses that are consistent with higher-level predictions are enhanced, and others reduced, perhaps as a way to produce a sparser representation overall.

More directly related to our concern with object recognition, Egner et al. (2010) showed subjects images of either faces or houses. Before each image, subjects saw a colored frame that was stochastically predictive of the image category to follow: a face followed green frame with probability 0.25, a yellow frame with probability 0.5, and a blue frame with probability 0.75. Egner et al. recorded the activation in the FFA, a fusiform area selective for faces. Consistent with other studies, they found some FFA activity in response to house stimuli, but this activation was much higher, almost as much as for faces, when these stimuli were unexpected. The predictive coding explanation of this result is that the FFA activation is an error-correcting response induced by the violated face expectation. The authors rule out an alternative explanation that face expectation uniformly enhances FFA activation, perhaps as a result of anticipatory neural activity. On this explanation, activation on face-expected, face-present trials should be at least as high as on face-unexpected, face-present ones, the opposite of the actual finding that face-present activation increased the face expectation.

For a recent review of other predictive coding, including proposals about how it could function in cognitive domains other than vision, see Clark (2013).

17.5.4 OTHER APPROACHES TO INFERENCE Schemes like Rao and Ballard's are perhaps the best-developed approaches to inference in generative vision, but there is no shortage of other candidates. We focus in particular on one algorithm, known as belief propagation or message passing, in part because it has been proposed as a general algorithm for neural probabilistic inference, applicable to areas beyond visual cortex. As the name suggests, the message-passing algorithm is based on communication between probabilistically related variables, making it particularly amenable to neural interpretations. The specific approach we sketch here was developed in Lee and Mumford (2003). For another approach, see Rao (2007).

Lee and Mumford move to a slightly different formalism from the one presented above, replacing the conditional probabilities with potential functions $\phi(V_i, V_j)$, which measure the undirected compatibility of the representations in a pair of adjacent areas.

Lee and Mumford propose an inference mechanism that combines belief propagation with another algorithm popular in machine learning and computer vision called particle filtering. Particle filtering replaces the continuous distributions over V_i with a discrete approximation, taking the form of weighted samples: $\sum_{k=1}^{n} w_k \tilde{V}_i^k$ where the $\tilde{V}_i^k$, called particles, are a chosen set of possible activation values. Belief propagation then allows neighboring areas to update these distributions by passing messages. Specifically, each area passes up this message:

$$M_{V_i \to V_{i+1}}\left(\tilde{V}_{i+1}^k\right) = \sum_j M_{V_{i-1} \to V_i}\left(\tilde{V}_i^j\right)\phi(\tilde{V}_i^j, \tilde{V}_{i+1}^k). \qquad (17.13)$$

Similarly, the downward massages take the form

$$M_{V_i \to V_{i-1}}\left(\tilde{V}_{i-1}^k\right) = \sum_j M_{V_{i+1} \to V_i}\left(\tilde{V}_i^j\right)\phi(\tilde{V}_i^j, \tilde{V}_{i-1}^k). \qquad (17.14)$$

Focusing on the upward case, the kth component of the message $M_{V_i \to V_{i+1}}\left(\tilde{V}_{i+1}^k\right)$ can be understood as the expected compatibility of the particle $\tilde{V}_{i+1}^k$ with the activity in area V_i, where the expectation is taken with respect to incoming messages from the other direction. We have presented the standard version of the algorithm, called the sum product algorithm; Lee and Mumford use a variant called the max-sum algorithm, in which the sum is replaced with a max.

After one round of message passing, each particle $\tilde{V}_i^k$ has received "scores" from the messages coming from above and below. These are multiplied to give the particle's weight:

$$w_i^k = \frac{1}{Z} M_{V_{i+1} \to V_i}\left(\tilde{V}_i^k\right) M_{V_{i-1} \to V_i}\left(\tilde{V}_i^k\right). \qquad (17.15)$$

As the model's beliefs evolve, many of the original particles become overwhelmingly unlikely, as manifested in low weights. At each iteration, a new set of particles is formed by resampling with replacement particles according to their weights. This will tend to reinforce particles that are doing well while killing off ones that are doing poorly.

The basic form of the message-passing algorithm, propagation of locally computed information, makes it attractive from a neural point of view. One sticking point, though, is how the particles are represented. Since the algorithm requires each brain area to maintain multiple independent hypotheses (particles) simultaneously, the question arises how they can be kept internally coherent and mutually distinct. Lee and Mumford tentatively suggest a few options such as the synchronous firing, but at present no option constitutes a fully fleshed out proposal.

Comparing this picture with Rao and Ballard's model, presented above, we see that the top-down messages can still be interpreted as predictions, but the bottom-up messages are no longer directly interpretable as error signals. In addition, Lee and Mumford's model allows more complex interactions between areas than Rao and Ballard's. This makes the Lee and Mumford model more expressive, but at the cost of less tractable inference.

17.5.5 Feedback Processing and Priors In addition to providing an efficient representation, integrating generative top-down signals with bottom-up ones is an effective way of dealing with the ambiguities inherent in low-level cues. Yuille and Kersten (2006) argue that low-level determinations are difficult to make using only local low-level information but are relatively easy given a high-level analysis. McDermott (2004) asked human subjects to determine whether a junction of edges was present in a given image region. This is easy when the whole image is visible: an object's location determines the locations of its edges. However, when McDermott restricted his subjects' information to low-level cues by making them view the image through a small, 13 ×13 pixel window, their junction-recognition performance was poor.

A prediction, then, is that feedback processing should be especially prevalent when low-level cues are ambiguous or absent. The Kanizsa square (figure 17.7) is a well-known visual illusion, which admits two interpretations: a white square partially obscuring four black circles, or four mutilated circles. Lee and Mumford

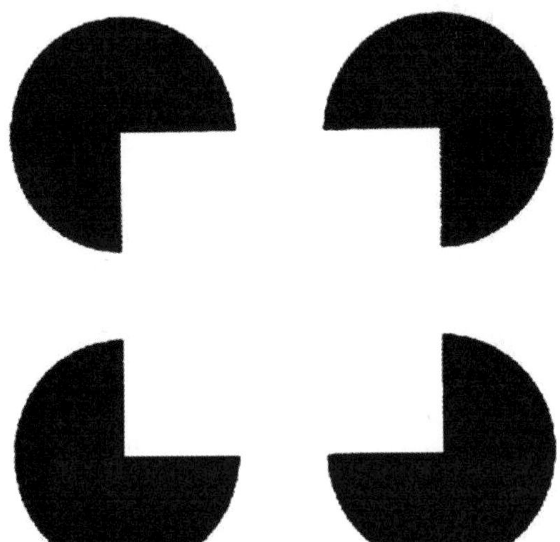

Figure 17.7 The Kanizsa square is an ambiguous stimulus, which can either be interpreted as a white square partially occluding four circles, or as four circles with missing pieces. Feedback processing induces V1 cells to respond to illusory contours defining the square.

recorded from V1 as monkeys viewed this stimulus, finding equally strong responses to the illusory contours between the circles as to the real ones within the circles, suggesting an interpretation in which higher-level processes settled on the "occluding square" interpretation and used feedback connections to fill in the "missing" contour information in V1. Importantly, V1 cells responded to illusory contours after V2 cells, consistent with the feedback story.

In a study using more natural stimuli, Tang et al. (2014) used subdural electrodes to record from the brains of human epilepsy patients undergoing surgery as they viewed two ordinary and heavily occluded natural images. Object recognition was possible in the occluded images—occlusion was calibrated so that subjects achieved approximately 80% accuracy in a behavioral identification task—but most (all but 18%, on average) low-level information was obliterated by the occluders. However, activity in category-selective IT cells was significantly delayed in the occluded condition, suggesting a reliance on preceding recurrent processing.

17.5.6 Feedforward and Feedback Processing: Summing Up Feedforward and feedback models present substantially different pictures of visual processing. How do the two fit together? The mainstream view is of a first, purely feedforward pass sufficient for relatively straightforward classification tasks followed by more sustained feedback processing supporting more difficult or ambiguous determinations.

A first line of evidence for this view comes from the timing of human recognition. In an EEG study, Thorpe et al. (1996) found category-selective information arising in the prefrontal cortex (after visual cortex) in only 150 ms. Historically, this timing has been a challenge for feedback models, given their more extensive processing demands. Lee and Mumford (2003) point out, however, that if recurrent processing occurs continuously in local feedback loops, rather than requiring multiple complete top-to-bottom passes, then recognition at the 150-ms timescale is not unrealistic. More recent time studies pose challenges for this argument, though. A recent behavioral study (Potter et al., 2014) used a rapid serial visual presentation (RSVP) paradigm in which a sequence of images was shown for only 13 ms each, after which subjects were asked whether an image matching a description (e.g., "bear catching fish") was present in the sequence. Even under these extremely demanding time constraints, subjects' performance was significantly better than chance. Short presentation times alone are consistent with feedback models, but the RSVP paradigm poses a challenge: in the time after presentation of the first image, early visual areas are occupied with analysis of subsequent images and are therefore unavailable for feedback processing of the first one.

Other evidence comes from physiology. In a decoding study, Hung et al. (2005), for instance, found that category information was present in the first IT spikes appearing in response to a stimulus, arguing against a feedback picture in which initial IT activity is fed back to and processed by earlier areas before a final interpretation is reached.

As argued above, though, fast, purely feedforward processing is insufficient for many tasks, particularly those with ambiguity, like occlusion. Precise characterization of the tasks supported by the two processing regimes is an important question that awaits future work.

17.6 *Conclusion*

As we have seen, several well-established frameworks exist for understanding the various processing stages in the ventral visual pathway. However, several important questions remain unanswered. Here we focus on two areas we think will be important for future research.

17.6.1 Digging Deep: Learning From Few Examples While the remarkable developments in computer vision over the last few years owe much to new algorithms and representation schemes, much credit must also go to the availability of huge databases of labeled images, and the computing power required to process them. This state of affairs contrasts markedly with human vision: we can often learn to recognize a new object (say an iPhone circa 2005) from only one example, not hundreds or thousands. Even as children, we hear our parents explicitly identify only a relatively modest number of objects in the world. It is an important challenge, therefore, both for the development of the next generation of computer vision systems and for genuine understanding and replication of the animal visual system, to come up with computational systems that can learn from similarly compact collections of data.

As we have seen, building in invariance to nuisance transformations is one route to data-efficient models. HMAX and other CNNs can build invariance to changes and scale, but we can hope for more gains by accounting for other transformation types. Anselmi et al. (2013) develops a more general theory of invariant representations.

Transformations like translation, scaling, and rotation are class-general; they apply to all object classes in the same way, a property that does not hold for transformations in general. Changes in expression, for instance, are a class of transformation that apply to faces alone. Less obviously, three-dimensional rotation in depth followed by projection onto the two-dimensional image plane is also class-specific: a long, thin object, for instance, rotates differently from an approximately spherical one. Anselmi et al. present a two-stage model to tackle class-specifc transformations. The first stage, corresponding to areas V1–V4, is an HMAX-style CNN. The second stage, corresponding to anterior IT, consists of a collection of modules each of which handles class-specific transformations for one important object class. The face patches in primate brains are hypothesized to be examples of such modules. The resulting architecture achieves competitive performance in face processing tasks and accounts for a number of experimental findings. It predicts, for instance, the existence of the mirror-symmetrical cells found in macaque face areas that respond equally to a face as to an axis-flipped version.

Another approach to improving data efficiency (Salakhutdinov et al., 2011) uses hierarchical probabilistic models. A nonvisual parable conveys the main idea. Suppose you are given a number of opaque bags of colored marbles. You empty out the first one, seeing that it contains only red marbles. Further exploration shows that all the marbles in second are blue, and all in the third are yellow. You then draw a single green marble from the fourth bag. Even without seeing any other bag-four marbles, you can be reasonably sure that they are green: after all, all of the other bags were monochromatic. Examples like this are formulated with hierarchical probability models: the individual bags are probability distributions, in this case distributions over marble colors, whose parameters are themselves drawn from a higher-level distribution. Thus, learning something about a few of the bags—for instance, that they are monochromatic—tells you something about the generative process for bags in general and allows you make hypotheses about new examples. Salakhutdinov et al. transfer this intuition to vision, replacing bags with object categories and marbles with image features. Thus, information about the feature distributions for a few classes is information about how features are distributed for classes in general. As with the single green marble in the example, this higher-level information, combined with a single image from a new class, constrains the new class's structure, making it easier recognize further examples.

17.6.2 Thinking Wide: Full Scene Interpretation in Natural Images We began this chapter with an overview of the range of questions humans can answer about the visual world but have focused the discussion so far on the areas best understood by neuroscience: feedforward object and scene classification and preliminary steps toward full scene interpretation made by feedback models. Here we discuss the prospects for modeling a wider range of visual tasks, ones requiring the analysis of scenes with multiple objects interacting in complex ways. We briefly survey promising computer vision approaches to these problems, approaches that may develop into fruitful sources of neuroscientific hypotheses.

A prerequisite, arguably, for human-level flexibility in visual scene analysis is the ability to recognize all the objects present in an image and to represent their spatial relations. One version of this task is called image parsing (Yao et al., 2010; Zhu and Mumford, 2007; Tu et al., 2005; Jin and Geman, 2006; Zhu et al., 2010; Socher et al., 2011). In language processing, one parses a sentence by recursively breaking it up into meaningful parts: a sentence is composed of phrases, which are themselves composed of smaller phrases, and so on. Images have similar hierarchical structure: an image depicts a scene, which is composed of objects, which are composed of parts, which are composed of primitive components like corners and edges. Given an input image, an image-parsing algorithm seeks to recover this hierarchal structure. Whereas the classification models we have seen so far assign an image a single label, image-parsing algorithms assign an image a whole parse tree.

Several models (Yao et al., 2010; Zhu and Mumford, 2007) define image grammars. Just as a string grammar is a generative model for the well-formed strings in a language, an image grammar is a rich, structured generative model for images. As in the feedback models surveyed above, image parsing is the task of inverting this generative model, going from an image to the sequence of choices that could have formed it. There are two main algorithmic challenges in this approach. In the learning problem, the model must infer the appropriate grammar to describe a collection of input images . In the inference problem, the model must find a parse to explain an image with respect to an already fixed grammar. Given the large number of possible grammars and parse trees, both of these problems are challenging.

Partly in response to these difficulties, other image-parsing approaches such as Zhu et al. (2010) and Jin and Geman (2006) eschew formal grammars, opting for more tractable recursive models amenable to efficient discriminative and dynamic programming algorithms. Socher et al. (2011) uses a neural network approach, though not one with direct correspondence to the brain. This model takes as input an image overseg-mented into regions, from each of which it extracts a feature vector. It then proceeds by merging regions pairwise, continuing until only one region remains in the image. The binary tree resulting from this process is the image's parse.

The heart of the model is a so-called recursive neural network (RNN) that takes as input two feature vectors and outputs a single merged feature vector of the same length. The RNN is used to evaluate candidate region merges. Given neighboring regions R_1 and R_2 with feature vectors F_1 and F_2, the model computes $F_3 = \mathrm{RNN}(F_1, F_2)$. A linear regression model then assigns a score to F_3, which measures the quality of merge. Given a current segmentation of the image into regions, the model considers each possible merge and chooses the one with the highest score, continuing in this way recursively. Each time a new merged region is created, it comes with its feature vector, F_3 above. In addition to being necessary for calculating further merges, this feature vector can be fed into a classifier assigning a category label to a region. Thus, each node in the parse tree can be assigned a category.

In the context of generative models like image grammars, one often hears the slogan "vision as inverse graphics," an expression of the Helmholtzian view of perception as unconscious inference. Some promising recent models (Mansinghka et al., 2013; Kulkarni et al., 2015) take this slogan literally, viewing images as arising not from a grammar but from a probabilistic graphics program, in which parameter choices such as object positions, camera angle, and lighting conditions are stochastic. Given a generative model of this kind, an inference algorithm can be used to find the collection of parameter choices that best explain a given input image. While this inference problem is difficult, advantages of the inverse graphics approach include the fact that it can model a scene's three-dimensional structure and can exploit the expressivity and flexibility of a full programming language, potentially obtaining richer scene interpretations than those possible with grammar-like formalisms.

Mapping structured scene interpretation models to neural computation is an important outstanding challenge. In addition to the purely visual aspects of this mapping, a full solution will have to address fundamental questions about how the brain represents structured information. In vision, these questions arise in even the simplest visual scenes. Consider, for instance, a ball inside a cup. We know how to represent the ball and the cup by feature vectors, and we may be able to extend the principle to find the vector for "inside" as well. The question, then, is how to combine all of these representations in a way that reflects the scene's structure, preserving, for instance, its distinctness from "cup inside ball." Considering neural representations of more complex objects, such as parse trees, only magnifies the problem. The question of neural representations of structured and relational data, sometimes called the "connectionist variable binding problem," has a long history, and a number of solutions have been proposed (Smolensky and Legendre, 2006; Shastri and Ajjanagadde, 1993; van der Velde and de Kamps, 2006). To date, though, none has emerged as a consensus solution. Open problems abound in this area, both for computational neuroscience in general and for fully structured vision in particular.

ACKNOWLEDGMENTS This work was supported by the Center for Brains, Minds and Machines (CBMM), funded by NSF STC award CCF-1231216.

REFERENCES

ANSELMI, F., LEIBO, J. Z., ROSASCO, L., MUTCH, J., TACCHETTI, A., & POGGIO, T. (2013). Unsupervised learning of invariant representations in hierarchical architectures. *arXiv preprint arXiv : 1311.4158.*

BALDUZZI, D., VANCHINATHAN, H., & BUHMANN, J. (2014). Kickback cuts Backprop's red-tape: Biologically plausible credit assignment in neural networks. *arXiv preprint arXiv : 1411.6191.*

BASTOS, A. M., USREY, W. M., ADAMS, R. A., MANGUN, G. R., FRIES, P., & FRISTON, K. J. (2012). Canonical microcircuits for predictive coding. *Neuron, 76*(4), 695–711.

CARLSON, E. T., RASQUINHA, R. J., ZHANG, K., & CONNOR, C. E. (2011). A sparse object coding scheme in area V4. *Current Biology, 21*(4), 288–293.

CICHY, R. M., PANTAZIS, D., & OLIVA, A. (2014). Resolving human object recognition in space and time. *Nature Neuroscience, 17*(3), 455–462.

CLARK, A. (2013). Whatever next? Predictive brains, situated agents, and the future of cognitive science. *Behavioral and Brain Sciences, 36*(3), 181–204.

DURBIN, R., & RUMELHART, D. E. (1989). Product units: A computationally powerful and biologically plausible extension to backpropagation networks. *Neural Computation, 1*(1), 133–142.

EGNER, T., MONTI, J. M., & SUMMERFIELD, C. (2010). Expectation and surprise determine neural population responses in the ventral visual stream. *Journal of Neuroscience, 30*(49), 16601–16608.

EPSTEIN, R., & KANWISHER, N. (1998). A cortical representation of the local visual environment. *Nature, 392*(6676), 598–601.

FEI-FEI, L., IYER, A., KOCH, C., & PERONA, P. (2007). What do we perceive in a glance of a real-world scene? *Journal of Vision (Charlottesville, Va.), 7*(1), 10.

FFYTCHE, D. H. (2005). Visual hallucinations and the Charles Bonnet syndrome. *Current Psychiatry Reports, 7*(3), 168–179.

Friston, K. (2005). A theory of cortical responses. *Philosophical Transactions of the Royal Society of London. Series B, Biological Sciences, 360*(1456), 815–836.

Fukushima, K. (1980). Neocognitron: A self-organizing neural network model for a mechanism of pattern recognition unaffected by shift in position. *Biological Cybernetics, 36*(4), 193–202.

Fukushima, K. (1988). Neocognitron: A hierarchical neural network capable of visual pattern recognition. *Neural Networks, 1*(2), 119–130.

Greene, M. R., & Oliva, A. (2006). Natural scene categorization from conjunctions of ecological global properties. In *Proceedings of the 28th Annual Conference of the Cognitive Science Society* (pp. 291–296). Mahwah, NJ: Erlbaum.

Greene, M. R., & Oliva, A. (2009). The briefest of glances: The time course of natural scene understanding. *Psychological Science, 20*(4), 464–472.

He, K., Zhang, X., Ren, S., & Sun, J. (2015). Delving deep into rectifiers: Surpassing human-level performance on imagenet classification. *arXiv preprint arXiv:1502.01852.*

Hung, C. P., Kreiman, G., Poggio, T., & DiCarlo, J. J. (2005). Fast readout of object identity from macaque inferior temporal cortex. *Science, 310*(5749), 863–866.

Isik, L., Meyers, E. M., Leibo, J. Z., & Poggio, T. (2014). The dynamics of invariant object recognition in the human visual system. *Journal of Neurophysiology, 111*(1), 91–102.

Ito, M., & Komatsu, H. (2004). Representation of angles embedded within contour stimuli in area V2 of macaque monkeys. *Journal of Neuroscience, 24*(13), 3313–3324.

Ito, M., Tamura, H., Fujita, I., & Tanaka, K. (1995). Size and position invariance of neuronal responses in monkey inferotemporal cortex. *Journal of Neurophysiology, 73*(1), 218–226.

Jin, Y., & Geman, S. (2006). Context and hierarchy in a probabilistic image model. In *2006 IEEE Computer Society Conference on Computer Vision and Pattern Recognition* (Vol. 2, pp. 2145–2152). Los Alamitos, CA: IEEE Computer Society.

Knoblich, U., Bouvrie, J., & Poggio, T. (2007). *Biophysical models of neural computation: Max and tuning circuits* (pp. 164–189). Berlin: Springer.

Kravitz, D. J., Peng, C. S., & Baker, C. I. (2011). Real-world scene representations in high-level visual cortex: It's the spaces more than the places. *Journal of Neuroscience, 31*(20), 7322–7333.

Kravitz, D. J., Saleem, K. S., Baker, C. I., Ungerleider, L. G., & Mishkin, M. (2013). The ventral visual pathway: An expanded neural framework for the processing of object quality. *Trends in Cognitive Sciences, 17*(1), 26–49.

Krizhevsky, A., Sutskever, I., & Hinton, G. E. (2012). *Imagenet classification with deep convolutional neural networks.* In B. Schölkope, J. Platt, & T. Hofmann (Eds.), Advances in Neural Information Processing Systems (pp. 1097–1105). San Mateo, CA: Morgan Kaufman; Cambridge, MA: MIT Press.

Kulkarni, T. D., Kohli, P., Tenenbaum, J. B., & Mansinghka, V. (2015). Picture: A probabilistic programming language for scene perception. 2015 IEEE Conference on Computer Vision and Pattern Recognition.

Lampl, I., Ferster, D., Poggio, T., & Riesenhuber, M. (2004). Intracellular measurements of spatial integration and the MAX operation in complex cells of the cat primary visual cortex. *Journal of Neurophysiology, 92*(5), 2704–2713.

LeCun, Y., Bottou, L., Bengio, Y., & Haffner, P. (1998). Gradient-based learning applied to document recognition. *Proceedings of the IEEE, 86*(11), 2278–2324.

Lee, T. S., & Mumford, D. (2003). Hierarchical Bayesian inference in the visual cortex. *JOSA A, 20*(7), 1434–1448.

Mansinghka, V., Kulkarni, T. D., Perov, Y. N., & Tenenbaum, J. (2013). *Approximate Bayesian image interpretation using generative probabilistic graphics programs.* In B. Schölkopf, J. Platt, & T. Hofmann (Eds.), Advances in Neural Information Processing Systems (pp. 1520–1528). San Mateo, CA: Morgan Kaufman; Cambridge, MA: MIT Press.

McDermott, J. (2004). Psychophysics with junctions in real images. *Perception, 33*(9), 1101–1128.

Mnih, V., Kavukcuoglu, K., Silver, D., Rusu, A. A., Veness, J., Bellemare, M. G., et al. (2015). Human-level control through deep reinforcement learning. *Nature, 518*(7540), 529–533.

Mumford, D. (1992). On the computational architecture of the neocortex. *Biological Cybernetics, 66*(3), 241–251.

Murphy, K., Torralba, A., & Freeman, W. T. (2003). Using the forest to see the trees: A graphical model relating features, objects, and scenes. In S. Thrun, L. K. Saul, & B. Schölkopf (Eds.), (pp. 1499–1506). Advances in Neural Information Processing Systems Cambridge, MA: MIT Press.

Murray, S. O., Schrater, P., & Kersten, D. (2004). Perceptual grouping and the interactions between visual cortical areas. *Neural Networks, 17*(5), 695–705.

Oliva, A., & Torralba, A. (2001). Modeling the shape of the scene: A holistic representation of the spatial envelope. *International Journal of Computer Vision, 42*(3), 145–175.

Oliva, A., & Torralba, A. (2006). Building the gist of a scene: The role of global image features in recognition. *Progress in Brain Research, 155*, 23–36.

O'Reilly, R. C. (1996). Biologically plausible error-driven learning using local activation differences: The generalized recirculation algorithm. *Neural Computation, 8*(5), 895–938.

Park, S., & Chun, M. M. (2009). Different roles of the parahippocampal place area (PPA) and retrosplenial cortex (RSC) in panoramic scene perception. *NeuroImage, 47,* 1747–1756.

Park, S., Brady, T. F., Greene, M. R., & Oliva, A. (2011). Disentangling scene content from its spatial boundary: Complementary roles for the PPA and LOC in representing real-world scenes. *Journal of Neuroscience, 31*(4), 1333–1340.

Potter, M. C., Wyble, B., Hagmann, C. E., & McCourt, E. S. (2014). Detecting meaning in RSVP at 13 ms per picture. *Attention, Perception & Psychophysics, 76*(2), 270–279.

Rao, R. P. (2007). Neural models of Bayesian belief propagation. In K. Doya, S. Ishii, A. Pouget, & R. P. N. Rao (Eds.), *Bayesian Brain: Probabilistic Approaches to Neural Coding* (p. 239). Cambridge, MA: MIT Press.

Rao, R. P., & Ballard, D. H. (1999). Predictive coding in the visual cortex: A functional interpretation of some extra-classical receptive-field effects. *Nature Neuroscience, 2*(1), 79–87.

Renninger, L. W., & Malik, J. (2004). When is scene identification just texture recognition? *Vision Research, 44*(19), 2301–2311.

Riesenhuber, M., & Poggio, T. (1999). Hierarchical models of object recognition in cortex. *Nature Neuroscience, 2*(11), 1019–1025.

Russakovsky, O., Deng, J., Su, H., Krause, J., Satheesh, S., Ma, S., et al. (2014). Imagenet large scale visual recognition challenge. *arXiv preprint arXiv : 1409.0575.*

Rust, N. C., & DiCarlo, J. J. (2010). Selectivity and tolerance ("invariance") both increase as visual information propagates from cortical area V4 to IT. *Journal of Neuroscience, 30*(39), 12978–12995.

Salakhutdinov, R., Torralba, A., & Tenenbaum, J. (2011). Learning to share visual appearance for multiclass object detection. In *2011 IEEE Conference on Computer Vision and Pattern Recognition (CVPR)* (pp. 1481–1488). IEEE.

Schmolesky, M. T., Wang, Y., Hanes, D. P., Thompson, K. G., Leutgeb, S., Schall, J. D., et al. (1998). Signal timing across the macaque visual system. *Journal of Neurophysiology, 79*(6), 3272–3278.

Schyns, P. G., & Oliva, A. (1994). From blobs to boundary edges: Evidence for time-and spatial-scale-dependent scene recognition. *Psychological Science, 5*(4), 195–200.

Serre, T., Oliva, A., & Poggio, T. (2007a). A feedforward architecture accounts for rapid categorization. *Proceedings of the National Academy of Sciences of the United States of America, 104*(15), 6424–6429.

Serre, T., Wolf, L., Bileschi, S., Riesenhuber, M., & Poggio, T. (2007b). Robust object recognition with cortex-like mechanisms. *IEEE Transactions on Pattern Analysis and Machine Intelligence, 29*(3), 411–426.

Shastri, L., & Ajjanagadde, V. (1993). From simple associations to systematic reasoning: A connectionist representation of rules, variables and dynamic bindings using temporal synchrony. *Behavioral and Brain Sciences, 16*(3), 417–451.

Siagian, C., & Itti, L. (2007). Rapid biologically-inspired scene classification using features shared with visual attention. *IEEE Transactions on Pattern Analysis and Machine Intelligence, 29*(2), 300–312.

Simonyan, K., & Zisserman, A. (2014). Very deep convolutional networks for large-scale image recognition. *arXiv preprint arXiv : 1409.1556.*

Smolensky, P., & Legendre, G. (2006). *The Harmonic Mind: From Neural Computation to Optimality-Theoretic Grammar: Vol. 1. Cognitive Architecture.* Cambridge, MA: MIT Press.

Socher, R., Lin, C. C., Manning, C., & Ng, A. Y. (2011). Parsing natural scenes and natural language with recursive neural networks. In *Proceedings of the 28th International Conference on Machine Learning (ICML-11)* (pp. 129–136).

Szegedy, C., Liu, W., Jia, Y., Sermanet, P., Reed, S., Anguelov, D., et al. (2014). Going deeper with convolutions. *arXiv preprint arXiv : 1409.4842.*

Szegedy, C., Zaremba, W., Sutskever, I., Bruna, J., Erhan, D., Goodfellow, I., et al. (2013). Intriguing properties of neural networks. *arXiv preprint arXiv : 1312.6199.*

Tang, H., Buia, C., Madhavan, R., Crone, N. E., Madsen, J. R., Anderson, W. S., et al. (2014). Spatiotemporal dynamics underlying object completion in human ventral visual cortex. *Neuron, 83*(3), 736–748.

Thorpe, S., Fize, D., & Marlot, C. (1996). Speed of processing in the human visual system. *Nature, 381*(6582), 520–522.

Torralba, A., Murphy, K. P., & Freeman, W. T. (2010). Using the forest to see the trees: Exploiting context for visual object detection and localization. *Communications of the ACM, 53*(3), 107–114.

Torralba, A., Oliva, A., Castelhano, M. S., & Henderson, J. M. (2006). Contextual guidance of eye movements and attention in real-world scenes: The role of global features in object search. *Psychological Review, 113*(4), 766–786.

Tu, Z., Chen, X., Yuille, A. L., & Zhu, S. C. (2005). Image parsing: Unifying segmentation, detection, and recognition. *International Journal of Computer Vision, 63*(2), 113–140.

Van der velde, F., & de Kamps, M. (2006). Neural blackboard architectures of combinatorial structures in cognition. *Behavioral and Brain Sciences, 29*(1), 37–70.

Yamins, D. L., Hong, H., Cadieu, C. F., Solomon, E. A., Seibert, D., & DiCarlo, J. J. (2014). Performance-optimized hierarchical models predict neural responses in higher visual cortex. *Proceedings of the National Academy of Sciences of the United States of America, 111*(23), 8619–8624.

Yao, B. Z., Yang, X., Lin, L., Lee, M. W., & Zhu, S. C. (2010). I2t: Image parsing to text description. *Proceedings of the IEEE, 98*(8), 1485–1508.

Yuille, A., & Kersten, D. (2006). Vision as Bayesian inference: Analysis by synthesis? *Trends in Cognitive Sciences, 10*(7), 301–308.

Zeiler, M. D., & Fergus, R. (2014). Visualizing and understanding convolutional networks. In *Computer Vision—ECCV 2014* (pp. 818–833). Springer.

Zhu, L., Chen, Y., Yuille, A., & Freeman, W. (2010). Latent hierarchical structural learning for object detection. In *2010 IEEE Conference on Computer Vision and Pattern Recognition* (pp. 1062–1069). New York: IEEE.

Zhu, S. C., & Mumford, D. (2007). *A Stochastic Grammar of Images.* Hanover, MA: Now.

18 Hippocampus: From Spatial Processing to Episodic Memory

EDMUND T. ROLLS

In this chapter a computational theory of how the hippocampus operates in memory is described (Kesner and Rolls, 2015; Rolls, 2008, 2010). The type of memory is episodic, referring to the memory of a particular event or linked group of events occurring typically at the same time and place. An example might be where dinner was yesterday, who was present, who sat where, what the menu was, and the discussion. This must be kept separate from, for example, what happened the day before that. Episodic memory almost always has a spatial component (Dere, Easton, Nadel, and Huston, 2008), and there is a spatial representation in the hippocampus that can be updated by self-motion to produce path integration (McNaughton et al., 1996; Rolls, 2008).

The theory is based on the remarkable neural architecture of the hippocampus, on the effects of damage to it, and on neuronal activity recorded in it. Once memories have been stored in the hippocampus, they may later need to be recalled to the neocortex, and a theory of the recall mechanism is part of the overall theory (Treves and Rolls, 1994) (section 18.3.5). Once recalled to the neocortex, the memories of particular events or episodes can be reported verbally, and hence this is a type of declarative memory (Squire and Wixted, 2011). The recalled information may also be combined with other information to be reorganized and stored semantically in the neocortex, that is, in a form that reflects meaning and structure, in contrast to the episodic memories captured as discrete memories by the hippocampus (McClelland, McNaughton, and O'Reilly, 1995). An example of a semantic representation might be a mental map that includes and describes the relations between the places to which one has made particular journeys. The interaction of theoretical and experimental approaches in the development of the theory is described. I start with a description of the underlying experimental neuroscience, use this to provide a firm foundation for the theory, and then show how the theory is being tested experimentally.

18.1 *Overview and Summary*

Some of the key points in the computational theory are as follows. The hippocampal CA3 system operates as a single attractor or autoassociation network to enable rapid, one-trial, associations between any spatial location (place, in rodents, or spatial view, in primates) and an object or reward, and to provide for completion of the whole memory during recall from any part. The theory is extended to associations between time and object or reward to implement temporal order memory, also important in episodic memory. The dentate gyrus performs pattern separation by competitive learning to produce sparse representations, producing, for example, neurons with place-like fields from entorhinal cortex grid cells. The dentate granule cells produce by the very small number of mossy fiber connections to CA3 a randomizing pattern separation effect important during learning but not recall that separates out the patterns represented by CA3 firing to be very different from each other, which is optimal for an unstructured episodic memory system in which each memory must be kept distinct from other memories. The direct perforant path input to CA3 projection is quantitatively appropriate to provide as a pattern association mechanism the cue for recall in CA3. The CA1 recodes information from CA3 to set up associatively learned backprojections to neocortex to allow subsequent retrieval of information to neocortex, providing a quantitative account of the large number of hippocampo–neocortical and neocortical–neocortical backprojections. Empirical tests of the theory, including hippocampal subregion analyses and selective hippocampal NMDA receptor knockouts, are described and support the theory.

18.2 *Structure and Function of the Hippocampal System*

18.2.1 The Effects of Damage to the Hippocampus In the patient HM, bilateral damage to the hippocampus performed to treat epilepsy produced an inability to remember "recent" events (those since the hippocampal and related damage) while leaving the memory of events that occurred prior to the hippocampal damage, and semantic and skill memory, relatively unimpaired (Corkin, 2002; Scoville and Milner, 1957). In tests to examine the exact brain regions that impair this memory for events, it has been shown that tasks that require objects to be associated with the place in which they are located are especially sensitive to hippocampal damage. Examples include memory for the location of an escape platform in a water bath in rats (Andersen, Morris, Amaral, Bliss, and O'Keefe, 2007; Morris and Frey, 1997) and for the location of an odor signifying the place where a food will be found in a cheese-board task (Kesner and Rolls, 2015). Temporal order memory for a sequence of places or objects is also impaired by hippocampal damage (Kesner and Rolls, 2015), and this functionality may be important in temporally linking a sequence of events within an episodic memory. In monkeys, analogous tasks involving object–place memory are impaired by hippocampal damage (Banta Lavenex and Lavenex, 2009), whereas damage to the overlying perirhinal cortex, which is connected to the inferior temporal cortex system involved in the computation of invariant object representations (Rolls, 2008, 2012c), impairs a different type of memory more closely related to perceptual functions, recognition memory for objects (Buckley, 2005).

These deficits produced by hippocampal damage underlie the importance of the hippocampus in event memory in which there is frequently an association of a place with an object, which provides a prototypical paradigm in which to analyze and conceptualize hippocampal computation.

18.2.2 Neurophysiology of the Hippocampus

The systems-level neurophysiology of the hippocampus shows what information could be stored or processed by the hippocampus. To understand how the hippocampus works, it is not sufficient to state just that it can store information—one needs to know what information.

In rodents, place cells, which respond when a rat is near a particular place, are found in the hippocampus (Hartley, Lever, Burgess, and O'Keefe, 2014; Jeffery, 2011; McNaughton, Barnes, and O'Keefe, 1983; O'Keefe and Dostrovsky, 1971). Place cells are found in regions CA3 and CA1 with smaller place fields in the dentate granule cells (Neunuebel and Knierim, 2012) (figure 18.1). The representation is allocentric (as contrasted with egocentric), in that the neurons fire whenever the rat is in the place field, typically independently of the head direction of the rat. In the medial entorhinal cortex, there are grid cells, which have regularly spaced peaks of firing in an environment, so that as a rat runs through an environment, a single neuron increases then decreases its firing a number of times as the rat traverses the environment (Fyhn, Molden, Witter, Moser, and Moser, 2004; Giocomo, Moser, and Moser, 2011; Hafting, Fyhn, Molden, Moser, and Moser, 2005; McNaughton, Battaglia, Jensen, Moser, and Moser, 2006). The grid-cell system appears to provide ring continuous attractors that would be useful not only for spatial path integration (computing position based on self-motion) (Giocomo et al., 2011; McNaughton et al., 2006), but also for the timing information useful in sequence encoding for nonspatial as well as spatial information (Kesner and Rolls, 2015), as described in section 18.3.7, Entorhinal Cortex Grid Cells.

In monkeys, used as a model to help understand human memory, there is a prominent representation of spatial view, the location at which the primate is looking (Rolls and Xiang, 2006). The representation of spatial view is allocentric, in that it is independent of the place where the monkey is located in the room or of eye position in the head (left gaze vs. right gaze) and of head direction, but depends on the location in space being viewed (Georges-François, Rolls, and Robertson, 1999). The spatial view cells are updated by self-motion (e.g., moving the eyes or running to a new part of the environment), indicating that path integration is implemented (Robertson, Rolls, and Georges-François, 1998).This type of representation is much more appropriate than that in rodents for a human memory system, for a human can remember where an object or person has been seen based just on looking at the place, without necessarily ever having been at that place (e.g., you may remember where you have seen a lecturer in a lecture theater, without ever having visited the place where the lecturer is standing). Moreover, some spatial view neurons respond to particular combinations of object and place, such as that object 1 is in place 1 on a screen (Rolls, Xiang, and Franco, 2005) or that reward 1 is at place 1 on a screen (Rolls and Xiang, 2005), and thus seem to encode what is necessary for an object–place memory representation system in the brain (Rolls and Xiang, 2006). Further, some of these neurons reflect the completion during recall of a whole memory from a part, for example, of the spatial location at which an object has previously been shown (Rolls and Xiang, 2006). Consistent with these findings, and with the computational theory, it has now been reported that human hippocampal neurons are activated during recall (Gelbard-Sagiv, Mukamel, Harel, Malach, and Fried, 2008). Although place cells might also be present in monkeys

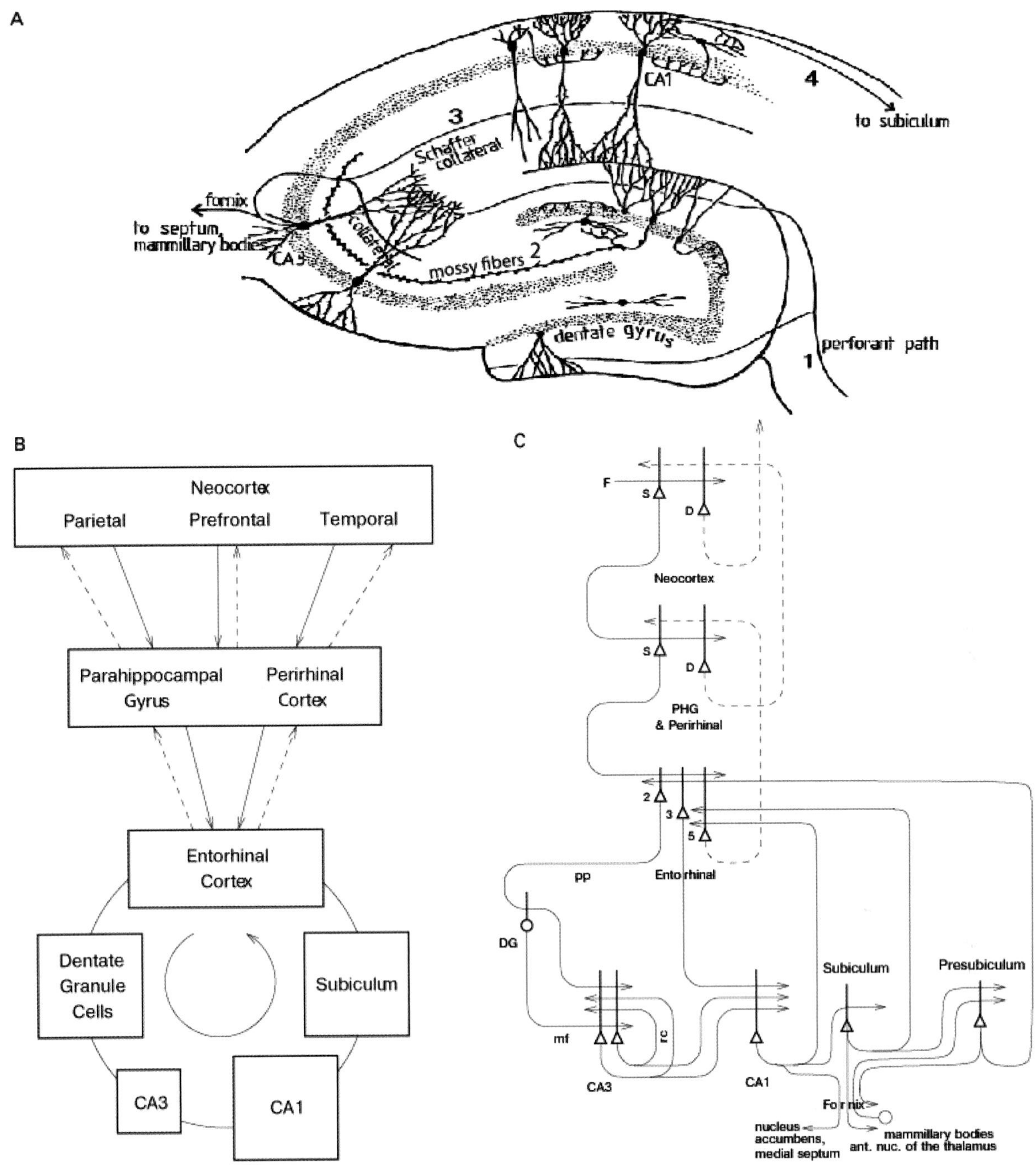

FIGURE 18.1 (A) Representation of connections within the hippocampus. Inputs reach the hippocampus through the perforant path (1) which makes synapses with the dendrites of the dentate granule cells and also with the apical dendrites of the CA3 pyramidal cells. The dentate granule cells project via the mossy fibers (2) to the CA3 pyramidal cells. The well-developed recurrent collateral system of the CA3 cells is indicated. The CA3 pyramidal cells project via the Schaffer collaterals (3) to the CA1 pyramidal cells, which, in turn, have connections (4) to the subiculum. (B, C) Forward connections (solid lines) from areas of cerebral association neocortex via the parahippocampal gyrus and perirhinal cortex, and entorhinal cortex, to the hippocampus; and backprojections (dashed lines) via the hippocampal CA1 pyramidal cells, subiculum, and parahippocampal gyrus to the neocortex. There is great convergence in the forward connections down to the single network implemented in the CA3 pyramidal cells; and great divergence again in the backprojections. (B) Block diagram. (C) More detailed representation of some of the principal excitatory neurons in the pathways. ant. nuc., anterior nucleus; D, deep pyramidal cells; DG, dentate granule cells; F, forward inputs to areas of the association cortex from preceding cortical areas in the hierarchy; mf, mossy fibers; PHG, parahippocampal gyrus and perirhinal cortex; pp, perforant path; rc, recurrent collateral of the CA3 hippocampal pyramidal cells; S, superficial pyramidal cells; 2, pyramidal cells in layer 2 of the entorhinal cortex; 3, pyramidal cells in layer 3 of the entorhinal cortex. The thick lines above the cell bodies represent the dendrites.

(Ludvig, Tang, Gohil, and Botero, 2004; Matsumura et al., 1999) (though we have not found any), care would be needed to establish by eye position recordings taken together with recordings of the place where the monkey is located during locomotion (Georges-François et al., 1999; Robertson et al., 1998; Rolls, 1999; Rolls, Robertson, and Georges-François, 1997a; Rolls, Treves, Robertson, Georges-François, and Panzeri, 1998) to establish that there are place cells in addition to spatial view cells in primates. In humans, places being viewed on a video monitor (i.e., spatial views being looked at), not places where the human is actually located, can activate hippocampal neurons (Ekstrom et al., 2003).

Results consistent with object–place neurons in primates (Rolls and Xiang, 2006; Rolls et al., 2005) are that in rats some hippocampal neurons respond on the basis of the conjunction of location and odor (Wood, Dudchenko, and Eichenbaum, 1999). Further, Diamond and colleagues have now shown, using the vibrissa somatosensory input for the "object" system, that rat hippocampal neurons respond to object–place combinations, objects, or places, and there is even a reward–place association system in rats similar to that in primates described below (Itskov, Vinnik, and Diamond, 2011). This brings the evidence from rats closely into line with the evidence from primates of hippocampal neurons useful for object–place episodic associative memory. There is also consistent evidence that rewards available in a spatial environment can influence the responsiveness of rodent place neurons (Hölscher, Jacob, and Mallot, 2003; Tabuchi, Mulder, and Wiener, 2003).

18.2.3 Systems-Level Anatomy To understand the functions of the primate hippocampus in event or episodic memory, it is necessary to understand which other parts of the brain it receives information from, and what it in turn connects to. The primate hippocampus receives inputs via the entorhinal cortex (Brodmann area 28) and the highly developed parahippocampal gyrus (areas TF and TH, two alphabetically labeled area of temporal cortex) as well as the perirhinal cortex from the ends of many processing streams of the cerebral association cortex, including the visual and auditory temporal lobe association cortical areas, the prefrontal cortex (PFC), and the parietal cortex (Lavenex and Amaral, 2000; Lavenex, Suzuki, and Amaral, 2004; Suzuki and Amaral, 1994; Van Hoesen, 1982) (see figure 18.1B). The hippocampus is thus by its connections potentially able to associate together object and spatial representations. In addition, the entorhinal cortex receives inputs from the amygdala, and the orbitofrontal cortex, which could provide reward/valence-related information to the hippocampus (Carmichael and Price, 1995; Pitkanen, Kelly, and Amaral, 2002).

The primary output from the hippocampus to neocortex originates in CA1 and projects to subiculum, entorhinal cortex, and parahippocampal structures (areas TF-TH) as well as PFC (Delatour and Witter, 2002; van Haeften, Baks-te-Bulte, Goede, Wouterlood, and Witter, 2003; Van Hoesen, 1982) (see figure 18.1), though there are other outputs (Kesner and Rolls, 2015). These are the pathways that are likely to be involved in the recall of information from the hippocampus back to the rest of the neocortex.

18.2.4 Internal Structure and Connectivity of the Hippocampus The internal hippocampal circuitry is shown in figures 18.1 and 18.2 (Amaral, Ishizuka, and Claiborne, 1990; Amaral and Witter, 1989; Andersen et al., 2007; Kondo, Lavenex, and Amaral, 2009). Projections from the entorhinal cortex layer 2 reach the granule cells (of which there are 10^6 in the rat) in the dentate gyrus, via the perforant path (Witter, 1993). The granule cells project to CA3 cells via the mossy fibers, which provide a *sparse* but possibly powerful connection to the 3.10^5 CA3 pyramidal cells in the rat. Each CA3 cell receives approximately 46 mossy fiber inputs, so that the sparseness of this connectivity is thus 0.005%. By contrast, there are many more—possibly weaker—direct perforant path inputs also from layer 2 of the entorhinal cortex onto each CA3 cell, in the rat of the order of 4.10^3. The largest number of synapses (about $1.2.10^4$ in the rat) on the dendrites of CA3 pyramidal cells is, however, provided by the (recurrent) axon collaterals of CA3 cells themselves (see figure 18.2). It is remarkable that the recurrent collaterals are distributed to other CA3 cells largely throughout the hippocampus (Amaral et al., 1990; Amaral and Witter, 1989; Amaral and Witter, 1995; Ishizuka, Weber, and Amaral, 1990; Witter, 2007), so that effectively the CA3 system provides a single network, with a connectivity of approximately 2% between the different CA3 neurons given that the connections are bilateral. The CA3–CA3 recurrent collateral system is even more extensive in macaques than in rats (Kondo et al., 2009). The neurons that comprise CA3, in turn, project to CA1 neurons via the Schaffer collaterals. In addition, projections that terminate in the CA1 region originate in layer 3 of the entorhinal cortex (see figure 18.1).

18.3 A Theory of the Operation of Hippocampal Circuitry as a Memory System

In this section, I consider how event or episodic memories might be stored in and retrieved by hippocampal

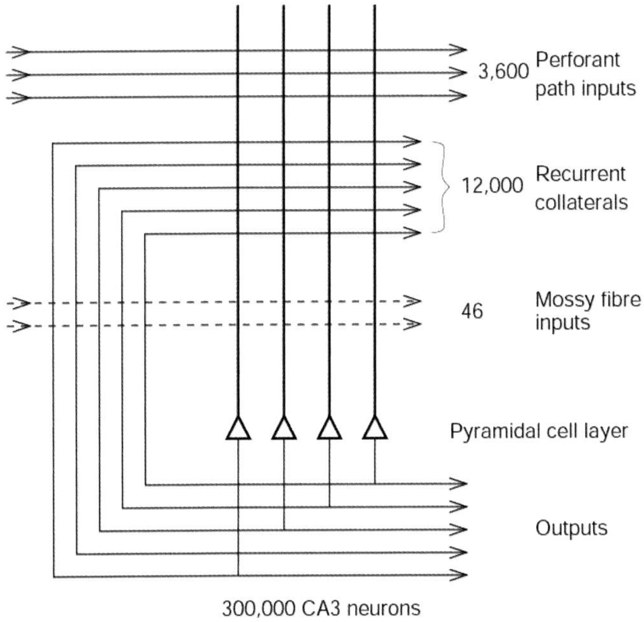

FIGURE 18.2 The numbers of connections from three different sources onto each CA3 in the rat. After Rolls and Treves (1998); Treves and Rolls (1992).

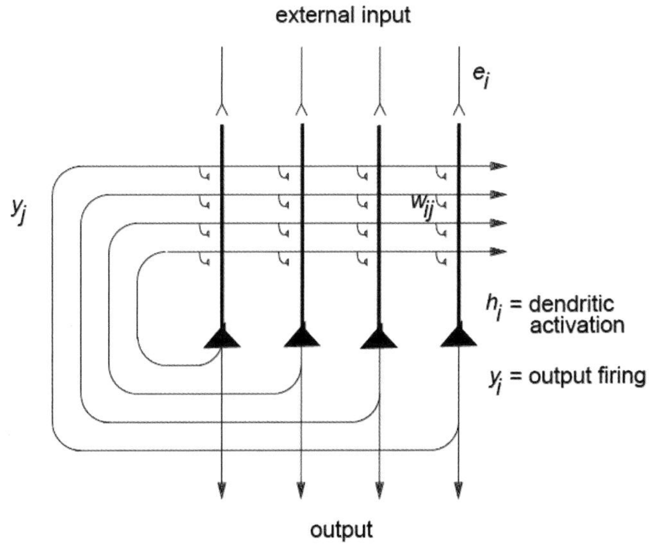

FIGURE 18.3 The architecture of an autoassociative neural network. The recurrent collateral synaptic weights are excitatory. Inhibitory neurons (not illustrated) must be present to control the firing rates. (See appendix B for further exposition.)

circuitry, and in addition retrieved back into the neocortex, where they may be incorporated into long-term semantic or autobiographical memory (Kesner and Rolls, 2015). The theory has been developed through many stages (Rolls, 1987, 1989b, 1995; Treves and Rolls, 1992, 1994), with fuller accounts and recent developments available (Kesner and Rolls, 2015; Rolls, 2008; Rolls and Deco, 2010). The theory illustrates the importance of taking into account the details of the circuitry involved in the development of theories of brain function, which has been emphasized by others, too (Abeles, 1991; Braitenberg and Schütz, 1991; Shepherd and Grillner, 2010). Some background is that many of the synapses in the hippocampus show associative modification as shown by long-term potentiation, and this synaptic modification appears to be involved in learning (Andersen et al., 2007; Takeuchi, Duszkiewicz, and Morris, 2014; Wang and Morris, 2010). Early seminal work by David Marr (1971) showed how associatively modified recurrent connectivity could support pattern completion, but he did not identify the CA3 network of the hippocampus as being the crucial network with an appropriate architecture for this to occur. This type of network became known as an autoassociation network (because a pattern is associated with itself using the recurrent collaterals) in further work by Kohonen (1977), and it also became known as an attractor network following the quantitative approach in which partial patterns could be attracted into a basin of attraction (Amit, 1989; Hopfield, 1982).

A summary description of the operation of autoassociation or attractor networks, pattern association networks, and competitive networks is provided in appendix B. A fuller description than is possible there is provided by Rolls (2008, 2016) in appendix B, which is available online: http://www.oxcns.org.

18.3.1 CA3 AS AN AUTOASSOCIATION OR ATTRACTOR MEMORY *Arbitrary Associations, and Pattern Completion in Recall* On the basis of the evidence summarized above on the connectivity of the hippocampus, on autoassociation networks, and on the role of synaptic modification in memory, Rolls (1987, 1989b) (at the Dahlem Conference in 1985 on The Neural and Molecular Bases of Learning; Rolls, 1987) and others (Levy, 1989; McNaughton, 1991; McNaughton and Morris, 1987) suggested that the CA3 stage acts as an autoassociation memory that enables episodic memories to be formed and stored in the CA3 network and that subsequently the extensive recurrent collateral connectivity allows for the retrieval of a whole representation to be initiated by the activation of some small part of the same representation (the cue). The crucial synaptic modification for this is in the CA3 recurrent collateral synapses (see figures 18.1,18.3).

The hypothesis is that because the CA3 operates effectively as a single network, it can allow arbitrary associations between inputs originating from very different parts of the cerebral cortex to be formed. These might involve associations between information originating in the temporal lobe visual cortex about the presence of an object (Rolls, 2012c) and information originating from scene-selective representations about where it is (Nasr et al., 2011). Each event might be stored in as little as 1 second (for autoassociation memories require only one-shot learning and long-term potentiation only brief inputs), and each event would consist of the vector of neurons firing at the inputs to the hippocampus. An episodic memory might be a single event or a sequence of events. Although there is some spatial gradient in the CA3 recurrent connections, so that the connectivity is not fully uniform in the rat (Ishizuka et al., 1990; Witter, 2007; though very interestingly there is less gradient in primates; Kondo et al., 2009), nevertheless the network will still have the properties of a single interconnected autoassociation network allowing associations between arbitrary neurons to be formed, given the presence of many long-range connections which overlap from different CA3 cells, and given the ability of attractor networks to operate with diluted connectivity shown in our computational studies prompted by this issue (Treves, 1990; Treves and Rolls, 1991). (The connectivity of the CA3 network is diluted, as contrasted with fully connected, in that there are only approximately 12,000 recurrent collateral synapses on each CA3 neuron, and 300,000 CA3 neurons, as shown in figure 18.2). Indeed, it has been suggested that diluted connectivity in autoassociation networks in the brain, for example, those in CA3 and those implemented by recurrent collaterals in the neocortex, may be advantageous by reducing the probability of multiple connections between any pair of neurons, which if present distort the basins of attraction and considerably reduce the number of memories that can be stored (Rolls, 2012a).

Crucial issues include how many memories could be stored in this system (to determine whether the autoassociation hypothesis leads to a realistic estimate of the number of memories that the hippocampus could store); whether the whole of a memory could be completed from any part; whether the autoassociation memory can act as a short-term memory, for which the architecture is inherently suited; whether the system could operate with spatial representations, which are essentially continuous because of the continuous nature of space; and when the system stores or recalls information. These and related issues are considered in the remainder of section 18.3.2 and in more detail elsewhere (Kesner and Rolls, 2015; Rolls, 2008, 2010, 2013a, 2013b).

Storage Capacity We have performed quantitative analyses of the storage and retrieval processes in the CA3 network (Treves and Rolls, 1991; Treves and Rolls, 1992). We have extended previous formal models of autoassociative memory (Amit, 1989; Hopfield, 1982) by analyzing a network with graded response units, so as to represent more realistically the continuously variable rates at which neurons fire, and with incomplete connectivity (Rolls, 2008; Rolls and Treves, 1998; Rolls, Treves, Foster, and Perez-Vicente, 1997b; Treves, 1990; Treves and Rolls, 1991). We have found that in general the maximum number $p_{\max}$ of firing patterns that can be (individually) retrieved is proportional to the number C^{RC} of (associatively) modifiable recurrent collateral synapses onto each neuron, by a factor that increases roughly with the inverse of the sparseness a of the neuronal representation (defined below). Each memory is precisely defined in the theory: it is a set of firing rates of the population of neurons (which represent a memory) that can be stored and later retrieved, with retrieval being possible from a fraction of the originally stored set of neuronal firing rates (Rolls, 2008). The neuronal population sparseness a of the representation can be measured by extending the binary notion of the proportion of neurons that are firing to any one stimulus or event as

$$a = \left(\sum_{i=1,N} r_i / N\right)^2 / \sum_{i=1,N} (r_i^2 / N) \qquad (18.4)$$

where r_i is the firing rate (e.g., spikes/s, typically in the range 0–100 spikes/s) of the ith neuron in the set of N neurons. The sparseness ranges from $1/N$, when only one of the neurons responds to a particular stimulus (a local or grandmother cell representation; Rolls and Treves, 2011), to a value of 1.0, attained when all the neurons are responding at the same rate to a given stimulus (Franco, Rolls, Aggelopoulos, and Jerez, 2007; Rolls and Treves, 2011; Treves and Rolls, 1991). The maximum number of patterns, $p_{\max}$, that can be stored and correctly retrieved is approximately

$$p_{\max} \cong \frac{C^{RC}}{a \ln(1/a)} k \qquad (18.5)$$

where C^{RC} is the number of recurrent collateral connections onto each neuron, and k is a scaling factor that depends weakly on the detailed structure of the rate distribution, on the connectivity pattern, and so forth, but is roughly in the order of 0.2–0.3 (Treves and Rolls, 1991). For example, for $C^{RC} = 12{,}000$ associatively modifiable recurrent collateral synapses onto each neuron, and $a = 0.02$, $p_{\max}$ is calculated to be approximately 36,000. This analysis emphasizes the utility of having a sparse representation in the hippocampus, for sparse distributed representations increase the number of different memories that can be stored (Treves and Rolls, 1991), essential for an episodic memory (Rolls, 2010).

In order for most associative networks to store information efficiently, heterosynaptic long-term depression, in which a synapse decreases in strength when the presynaptic term is low and the postsynaptic term is high (as well as long-term potentiation, an increase in synaptic strength when both the pre- and postsynaptic terms are high), is required (Collingridge, Peineau, Howland, and Wang, 2010; Rolls, 2008; Rolls and Deco, 2002; Rolls and Treves, 1990, 1998; Treves and Rolls, 1991). The long-term depression can effectively remove the effect of the positive-only firing rates in the brain by subtracting the mean firing rate. Simulations that are consistent with the analytic theory have been performed (Rolls et al., 1997b; Rolls and Webb, 2012; Simmen, Treves, and Rolls, 1996).

A number of points that arise, including measurement of the total amount of information (in bits per synapse) that can be retrieved from the network, the computational definition of a memory, the computational sense in which CA3 is an attractor network, and the possible computational utility of memory reconsolidation, are treated elsewhere (Rolls, 2008; Rolls and Kesner, 2006). Here I note that given that the memory capacity of the hippocampal CA3 system is limited, it is necessary to have some form of forgetting in this store, or other mechanism, to ensure that its capacity is not exceeded. (Exceeding the capacity can lead to a loss of much of the information retrievable from the network.) Heterosynaptic long-term depression could help this *forgetting*, by enabling new memories to overwrite old memories (Rolls, 2008). The limited capacity of the CA3 system does also provide one of the arguments that some transfer of information from the hippocampus to neocortical memory stores may be useful (see Treves and Rolls, 1994). Given its limited capacity, the hippocampus might be a useful store (e.g., for episodic memories) for only a limited period, which might be in the order of days, weeks, or months. This period may well depend on the acquisition rate of new episodic memories. If the animal were in a constant and limited environment, then as new information is not being added to the hippocampus, the representations in the hippocampus would remain stable and persistent. These hypotheses have clear experimental implications, both for recordings from single neurons and for the gradient of retrograde amnesia, both of which might be expected to depend on whether the environment is stable or frequently changing. They show that the conditions under which a gradient of retrograde amnesia might be demonstrable would be when large numbers of new memories are being acquired, not when only a few memories (few in the case of the hippocampus being less than a few hundred) are being learned (Rolls, 2008).

Recall and Completion A fundamental property of the autoassociation model of the CA3 recurrent collateral network is that the recall can be symmetrical, that is, the whole of the memory can be retrieved and completed from any part (Amit, 1989; Hopfield, 1982; Kesner and Rolls, 2015; Rolls, 2008, 2016; Rolls and Treves, 1998). For example, in an object–place autoassociation memory, a place could be recalled from an object retrieval cue (and potentially vice versa). In a test of this, Day, Langston, and Morris (2003) trained rats in a study phase to learn in one trial an association between two flavors of food and two spatial locations. During a recall test phase they were presented with a flavor which served as a cue for the selection of the correct location. They found that injections of an NMDA receptor blocker (AP5) or AMPA/kainate receptor blocker (CNQX) to the dorsal hippocampus prior to the study phase impaired encoding, but injections of AP5 prior to the test phase did not impair the place recall, whereas injections of CNQX did impair the place recall. The interpretation is that somewhere in the hippocampus NMDA receptors are necessary for forming one-trial odor–place associations, and that recall can be performed without further involvement of NMDA receptors. (The implication, consistent with investigations of long-term potentiation [Takeuchi et al., 2014], is that NMDA glutamate receptors are necessary for synaptic modification, but that recall may use the AMPA receptors modified by the learning. See further chapter 9.)

Evidence that the CA3 system is not necessarily required during recall in a reference memory (previously learned) spatial task, such as the water maze spatial navigation for a single spatial location task, is that CA3-lesioned rats are not impaired during recall of a previously learned water maze task (Brun et al., 2002; Florian and Roullet, 2004). However, if completion from an incomplete cue is needed (e.g., finding a place with only a few room cues), then CA3 NMDA receptors are necessary (presumably to ensure satisfactory CA3–CA3 learning) even in a reference memory task (Gold and Kesner, 2005; Kesner and Rolls, 2015; Nakazawa et al., 2002). Thus, the CA3 system appears to be especially needed in rapid, one-trial object–place recall, and when completion from an incomplete cue is required (see further section 18.5). (Note that an object–place task is a model of episodic memory, that episodic memory usually has a spatial component, and that place cells in rats and spatial view cells in primates provide the spatial representation that is needed; Rolls, 2008, 2016.)

Continuous, Spatial, Patterns and CA3 Representations The fact that spatial patterns, which imply continuous representations of space, are represented in the hippocampus has led to the application of continuous attractor models (see section 11.4.1 on Attractor Models in chapter 11; Rolls, 2016) to help understand hippocampal function. This has been necessary because space is inherently continuous, because the firing of place and spatial view cells is approximately Gaussian as a function of the distance away from the preferred spatial location, because these cells have spatially overlapping fields, and because the theory is that these cells in CA3 are connected by Hebb-modifiable synapses. This specification would inherently lead the system to operate as a continuous attractor network. Continuous attractor network models have been extensively studied (Amari, 1977; Battaglia and Treves, 1998; Rolls and Stringer, 2005; Samsonovich and McNaughton, 1997; Stringer and Rolls, 2002; Stringer, Rolls, and Trappenberg, 2004; Stringer, Rolls, Trappenberg, and Araujo, 2002a; Stringer, Trappenberg, Rolls, and Araujo, 2002b) and are described briefly next (see also Rolls, 2008).

A "Continuous Attractor" neural network (CANN) can maintain the firing of its neurons to represent any location along a continuous physical dimension such as spatial view, spatial position, head direction, and so forth. The network architecture is the same as that illustrated in figure 18.3, but each neuron has a peak of firing which gradually falls off the further away the current position is from the center of the spatial field, and each neuron has a spatial field that is offset from its neighbors' in the state space, as illustrated in figure 18.4. The CANN uses the excitatory recurrent collateral connections between the neurons which are set up by associative learning to reflect the distance between the neurons in the state space of the animal (e.g., place or spatial view or head direction). These networks can maintain the bubble or packet of neural activity constant for long periods wherever it is started to represent the current state (head direction, position, etc.) of the animal, and they are likely to be involved in many aspects of spatial processing and memory, including spatial vision and navigation. Global inhibition is used to keep the number of neurons in a bubble or packet of actively firing neurons relatively constant and to help to ensure that there is only one activity packet (see the example in figure 18.4).

Continuous attractor networks can be thought of as very similar to autoassociation or discrete attractor networks (Rolls, 2008) and have the same architecture, as illustrated in figure 18.3. The main difference is that the patterns stored in a CANN are continuous patterns, with each neuron having broadly tuned firing which decreases with, for example, a Gaussian function as the distance from the optimal firing location of the cell is varied, and with different neurons having tuning that overlaps throughout the space. Such tuning is illustrated in figure 18.4. The connections set up by associative learning between the neurons in the bubble of activity can later maintain that bubble of activity (Rolls, 2008). For comparison, autoassociation networks normally have discrete (separate) patterns (each pattern implemented by the firing of a particular subset of the neurons, which can overlap with other subsets), with no continuous distribution of the patterns throughout the space (see figure 18.4). A discrete sparse distributed representation with graded firing rates is used to encode and store representations of objects (Rolls, 2008; Rolls and Treves, 2011). A consequent difference from a discrete attractor network is that the CANN can maintain its firing at any location in the trained continuous space, whereas a discrete attractor or autoassociation network moves its population of active neurons toward one of the previously learned separate attractor states, and thus implements the recall of a particular previously learned pattern from an incomplete or noisy (distorted) version of one of the previously learned patterns.

Space is continuous, and object representations are discrete. If these representations are to be combined in, for example, an object–place memory (Leutgeb et al., 2005; Rolls and Xiang, 2005, 2006; Rolls et al., 2005), then we need to understand the operation of networks that combine these representations. Rolls, Stringer, and Trappenberg (2002) have shown that attractor networks can store both continuous patterns and discrete patterns (as illustrated in figure 18.4) and can thus be used to store, for example, the location in (continuous, physical) space (e.g., the place "out there" in a room represented by spatial view cells) where an object (a discrete item) is present. We showed this by storing associated continuous and discrete representations in the same single attractor network and then showing that the representation in the continuous space could be retrieved by the discrete object that was associated with that spatial position, and that the representation of the discrete object could be retrieved by providing the position in the continuous representation of space.

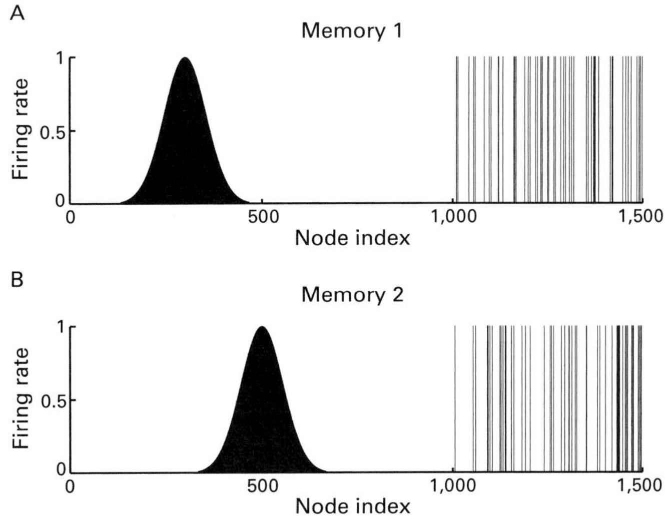

FIGURE 18.4 The types of firing patterns stored in continuous attractor networks are illustrated for the patterns present on neurons 1–1,000 for (A) Memory 1 (when the firing is that produced when the spatial state represented is that for location 300), and for (B) Memory 2 (when the firing is that produced when the spatial state represented is that for location 500). The continuous nature of the spatial representation results from the fact that each neuron has a Gaussian firing rate that peaks at its optimal location. This particular mixed network also contains discrete representations that consist of discrete subsets of active binary firing rate neurons in the range 1,001–1,500. The firing of these latter neurons can be thought of as representing the discrete events that occur at the location. Continuous attractor networks by definition contain only continuous representations, but this particular network can store mixed continuous and discrete representations and is illustrated to show the difference of the firing patterns normally stored in separate continuous attractor and discrete attractor networks. For this particular mixed network, during learning, Memory 1 is stored in the synaptic weights, then Memory 2, and so forth, and each memory contains part that is continuously distributed to represent physical space, and part that represents a discrete event or object. The spatial and object representations are bound together by being simultaneously present when the event is stored. From Rolls, Stringer, and Trappenberg (2002), where further details can be found.

If spatial representations are stored in the hippocampus, the important issue arises in terms of understanding memories that include a spatial component or context of how many such spatial representations could be stored in a continuous attractor network. The very interesting result is that because there are in general low correlations between the representations of places in different maps or charts (where each map or chart might be of one room or locale using, e.g., cues in the room), very many different maps can be simultaneously stored in a continuous attractor network (Battaglia and Treves, 1998).

We have considered how spatial representations could be stored in continuous attractor networks and how the activity can be maintained at any location in the state space in a form of short-term memory when the external (e.g., visual) input is removed. However, many networks with spatial representations in the brain can be updated by internal, self-motion (i.e., idiothetic), cues even when there is no external (e.g., visual) input. The ways in which path integration could be implemented in recurrent networks such as the CA3 system in the hippocampus or in related systems including the entorhinal cortex (see section 18.3.7) are described elsewhere (Giocomo et al., 2011; McNaughton et al., 2006; Samsonovich and McNaughton, 1997; Stringer et al., 2002a; Stringer et al., 2002b) and have been applied to primate spatial view cells by Rolls and colleagues (Rolls and Stringer, 2005; Stringer et al., 2004; Stringer, Rolls, and Trappenberg, 2005). "Cognitive maps" (O'Keefe and Nadel, 1978) can be understood by the operation of these attractor networks, and how they are updated by learning and by self-motion (Rolls, 2008). Those who have focused on spatial and navigation processing in the hippocampus rather than memory processing do though now envisage that attractor networks may be involved in hippocampal function (Hartley, Lever, Burgess, and O'Keefe, 2014).

Mossy Fiber Inputs to the CA3 Cells We hypothesize that the mossy fiber inputs force efficient information storage by virtue of their strong and sparse influence on the CA3 cell firing rates, to produce pattern separation, as described next. (The strong effects likely to be mediated by the mossy fibers were also emphasized by McNaughton and Morris, 1987, and McNaughton and Nadel, 1990.) We (Rolls, 1987, 1989c, 2008, 2013a; Rolls and Treves, 1998; Treves and Rolls, 1992) hypothesize that the mossy fiber input appears to be particularly appropriate in several ways. First, the fact that mossy fiber synapses are large and located very close to the soma makes them relatively powerful in activating the postsynaptic cell. Second, the firing activity of dentate granule cells appears to be very sparse (Jung and McNaughton, 1993; Leutgeb, Leutgeb, Moser, and Moser, 2007; Neunuebel and Knierim, 2012), and this, together with the small number of connections on each CA3 cell, produces a sparse signal, which can then be transformed into sparse firing activity in CA3 by a threshold effect. The hypothesis is that the mossy fiber sparse connectivity solution performs the appropriate function to enable learning to operate correctly in the CA3 to CA3 synaptic connections (Cerasti and Treves, 2010; Treves and Rolls, 1992). The perforant path input would, the quantitative analysis shows, not produce a pattern of firing in CA3 that contains sufficient information for learning (Treves and Rolls, 1992).

The particular property of the small number of mossy fiber connections onto a CA3 cell, approximately 46 (see figure 18.2), is that this has a *randomizing effect* on the representations set up in CA3, so that they are as different as possible from each other (Cerasti and Treves, 2010; Rolls, 1989b, 2008, 2013a; Rolls and Treves, 1998; Treves and Rolls, 1992). (This is a pattern separation effect. This means, e.g., that place cells in a given environment are well separated to cover the whole space, and that any new object–place associations formed are different from earlier episodic memories.) The result is that any one event or episode will set up a representation that is very different from other events or episodes because the set of CA3 neurons activated for each event is random. This is, then, the optimal situation for the CA3 recurrent collateral effect to operate, for it can then associate together the random set of neurons that are active for a particular event (e.g., an object in a particular place) and later recall the whole set from any part. It is because the representations in CA3 are unstructured, or random, in this way that large numbers of memories can be stored in the CA3 autoassociation system, and that interference between the different memories is kept as low as possible, in that they are maximally different from each other (Hopfield, 1982; Rolls and Treves, 1998; Treves and Rolls, 1991). (If some stored memory patterns were similar, they would tend to interfere with each other during recall. For an episodic memory, each stored memory pattern should be different from the others, so that each episode can be separately retrieved.)

The requirement for a small number of mossy fiber connections onto each CA3 neuron applies not only to discrete (Treves and Rolls, 1992) but also to spatial representations, and some learning in these connections, whether associative or not, can help to select out the small number of mossy fibers that may be active at any one time to select a set of random neurons in the CA3 (Cerasti and Treves, 2010). Any learning may help by reducing the accuracy required for a particular number of mossy fiber connections to be specified genetically onto each CA3 neuron. The optimal number of mossy fibers for the best information transfer from dentate granule cells to CA3 cells is in the order of 35–50 (Cerasti and Treves, 2010; Treves and Rolls, 1992). The mossy fibers also make connections useful for feedforward inhibition (Acsady, Kamondi, Sik, Freund, and Buzsaki, 1998), which may help to normalize the inputs and help stability (Rolls, 2008).

On the basis of these and other points, we predicted that the mossy fibers may be necessary for new learning in the hippocampus, but may not be necessary for the recall of existing memories from the hippocampus which can instead be implemented by the perforant path synapses that come directly from the entorhinal cortex, and which make many more connections onto each CA3 neuron and which are associatively modifiable (Rolls, 2008; Rolls and Treves, 1998; Treves and Rolls, 1992) (see the next section, "Perforant Path Inputs to CA3 Cells"). Experimental evidence consistent with this prediction about the role of the mossy fibers in learning has been found in rats with disruption of the dentate granule cells (Lassalle, Bataille, and Halley, 2000) (section 18.5).

We (Rolls and Kesner, 2006) have hypothesized that nonassociative plasticity of mossy fibers (i.e., synaptic potentiation that does not depend on the activity of the postsynaptic neuron; see Brown, Kairiss, and Keenan, 1990) might have a useful effect in enhancing the signal-to-noise ratio of the effects of the dentate input to CA3, in that a consistently firing mossy fiber would produce nonlinearly amplified currents in the postsynaptic cell, which would not happen with an occasionally firing fiber (Treves and Rolls, 1992). This plasticity, and also competitive learning in the dentate granule cells, would also have the effect that similar fragments of each episode (e.g., the same environmental location) recurring on subsequent occasions would be more likely to activate the same population of CA3 cells, which would have potential advantages in terms of economy of use of the CA3 cells in different memories, and in making some link between different episodic memories with a common feature, such as the same location in space. Consistent with this, dentate neurons that fire repeatedly are more effective in activating CA3 neurons (Henze, Wittner, and Buzsaki, 2002).

As acetylcholine turns down the efficacy of the recurrent collateral synapses between CA3 neurons (Giocomo and Hasselmo, 2007; Hasselmo, Schnell, and Barkai, 1995), then cholinergic activation also might help to allow external inputs rather than the internal recurrent collateral inputs to dominate the firing of the CA3 neurons during learning, as the current theory proposes (Rolls, 2013a; Rolls and Deco, 2015). If cholinergic activation at the same time facilitated long-term potentiation in the recurrent collaterals (as it appears to in the neocortex), then cholinergic activation could have a useful double role in facilitating new learning at times of behavioral activation and emotional arousal, when presumably it may be particularly relevant to allocate some of the limited memory capacity to new memories. Acetylcholine may also facilitate memory storage (vs. recall) by enhancing firing in dentate granule cells (see Kesner and Rolls, 2015).

Perforant Path Inputs to CA3 Cells By calculating the amount of information that would end up being carried by a CA3 firing pattern produced solely by the perforant path input and by the effect of the recurrent connections (i.e., without dentate input), we showed (Treves and Rolls, 1992) that an input of the perforant path type, alone, is unable to direct efficient information storage. Such an input is too weak, it turns out, to drive the firing of the cells, as the "dynamics" of the network is dominated by the randomizing effect of the recurrent collaterals. On the other hand, an autoassociative memory network needs afferent inputs to apply the retrieval cue to the network. We showed that the perforant path system is likely to be the one involved in relaying the cues that initiate retrieval in CA3. The concept is that to initiate retrieval, a numerically large input through associatively modified synapses is useful so that even a partial cue is sufficient, and that the retrieval cue need not be very strong, as the recurrent collaterals then take over in the retrieval process (Rolls, 2008; Treves and Rolls, 1992). In contrast, during storage, strong signals, in the order of millivolts for each synaptic connection, are provided by the mossy fiber inputs to dominate the recurrent collateral activations, so that the new pattern of CA3 cell firing can be stored in the CA3 recurrent collateral connections (Rolls, 2008; Treves and Rolls, 1992).

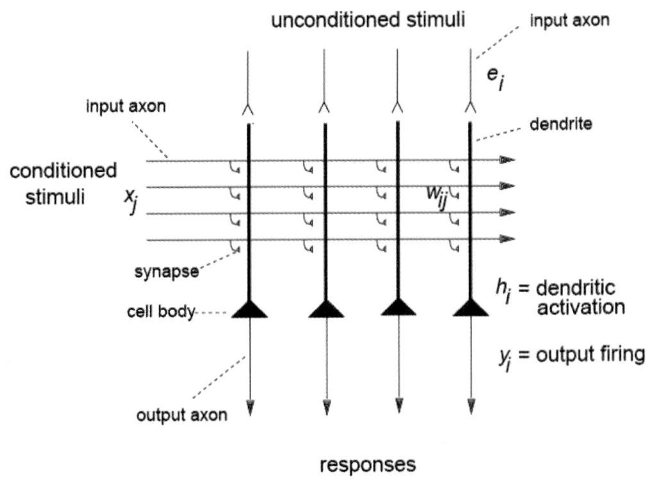

FIGURE 18.5 The architecture of a pattern association network. An unconditioned stimulus (US) has activity or firing rate e_i for the ith neuron and produces firing y_i of the ith neuron which is an unconditioned response (UR). The conditioned stimuli (CS) have activity or firing rate x_j for the jth axon. During learning, a CS is presented at the same time as a US, and the synaptic weights are modified by an associative synaptic learning rule $\delta w_{ij} = k \cdot y_i \cdot x_j$ where k is a learning rate constant, and δw_{ij} is the change in the synaptic weight from axon j to neuron i. During recall, only the CS is presented, and the activation h_i is calculated as a dot product between the input stimulus and the synaptic weight vector on a neuron $h_i = \Sigma_j x_j w_{ij}$ where the sum is over the input axons to each neuron, indexed by j. The output firing y_i is a function f of the activation $y_i = \text{f}(h_i)$. The activation function f should be nonlinear and may be, for example, binary threshold, linear threshold, or sigmoid. Inhibitory neurons not shown in the figure are part of the way in which the threshold of the activation function is set. The nonlinearity in the activation function enables interference from other pattern pairs stored in the network to be minimized. The pattern association network thus enables a CS to retrieve a response, the conditioned response (CR), that was present as a UR during the learning. An important property is that if a distorted CS is presented, generalization occurs to the effects produced by the closest CS during training. (See appendix B for exposition. Further details are provided in appendix B of Rolls (2008, 2016).)

The associatively modified synapses required in the perforant path to CA3 synapses makes this a pattern association network (figure 18.5). The architecture and properties of pattern association networks are described briefly in appendix B.2, and in more depth elsewhere (Rolls, 2008, 2016). These synapses need to be modified during the storage of an event memory, with the entorhinal input to CA3 becoming associated with whatever subset of neurons in CA3 is firing at that time (Treves and Rolls, 1992).

Noise in Memory Recall There is randomness (sometimes referred to as noise) in the spiking times of individual neurons. That is, for a given mean firing rate, the spike times often have a close to Poisson distribution. The noise arises from synaptic and neuronal processes in ion channels, the quantal release of transmitter, and so forth (Faisal, Selen, and Wolpert, 2008). A result is that in an autoassociation network, if one population of neurons for one attractor or memory state has by chance more spikes from its neurons than the other populations, then the memory with more spikes is more likely to be recalled, especially when the recall cue or cues for the different neurons are relatively similar in strength. The operation of such systems has been described in *The Noisy Brain: Stochastic Dynamics as a Principle of Brain Function* (Rolls and Deco, 2010) in the context of decision-making (Wang, 2008), but the approach applies equally to memory recall in an autoassociation memory, as the network architecture and operation for memory and for decision-making is the same (Rolls, 2008). It has been proposed that this noisy operation of the brain has many advantages, for example, in promoting the recall of different memories or different associations on different occasions even when the inputs are similar, and this, it is proposed, is an important contributor to original thought and creativity (Rolls, 2008; Rolls and Deco, 2010). Too much noise and therefore instability of memory and decision systems may promote the unstable attention and loose thought associations in schizophrenia (Loh, Rolls, and Deco, 2007; Rolls, 2012b; Rolls, Loh, Deco, and Winterer, 2008b). Too little noise and therefore too much stability may contribute to some of the symptoms of obsessive–compulsive disorder (Rolls, 2012b; Rolls, Loh, and Deco, 2008a). In both these cases, the combination of theoretical neuroscience approaches with experimental evidence on the transmitters present in these states is leading to interesting new approaches to understanding and perhaps better treating these disorders (Rolls, 2012b) and the cognitive effects in normal aging (Rolls and Deco, 2015).

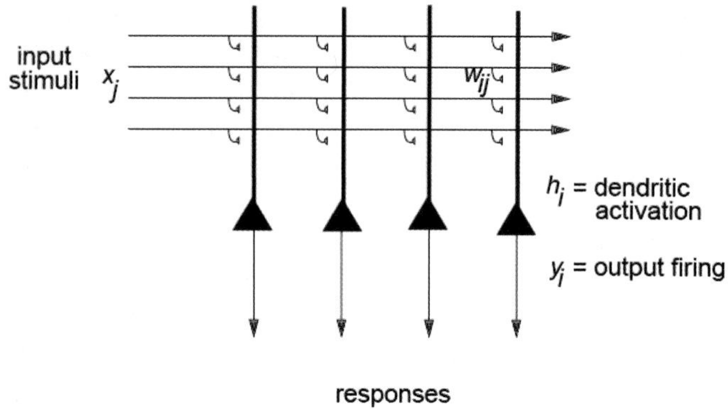

FIGURE 18.6 The architecture of a competitive network. During training, an input stimulus is presented to the synaptic matrix, which has random initial weights w_{ij} and produces activation h_i of the *i*th neuron calculated as a dot product between the input stimulus and the synaptic weight vector on a neuron $h_i = \Sigma_j x_j w_{ij}$ where the sum is over the input axons to each neuron, indexed by *j*. The output firing y_i is a function f of the activation $y_i = f(h_i)$. The activation function f should be nonlinear and may be, for example, binary threshold, linear threshold, or sigmoid. Inhibitory neurons not shown in the figure are part of the way in which the threshold of the activation function is set in which the threshold reflects the firing of all the output neurons. This or other competitive mechanisms result in a typically sparse set of output neurons having firing after the competition. (See appendix B for further exposition.)

18.3.2 Dentate Granule Cells

Pattern Separation We now turn to the hypothesis that the dentate granule cell stage of hippocampal processing which precedes the CA3 stage acts as a competitive network in a number of ways to produce during learning the sparse yet efficient (i.e., nonredundant) representation in CA3 neurons that is required for the autoassociation implemented by CA3 to perform well (Rolls, 1989b, 2008; Rolls, Stringer, and Elliot, 2006; Treves and Rolls, 1992). The properties of competitive networks are summarized in appendix B.3, and in more detail by Rolls (2008, 2016). An important property for episodic memory is that the dentate, by acting in this way, would perform pattern separation (or orthogonalization) (Rolls, 1989b, 2013a; Rolls et al., 2006; Treves and Rolls, 1992), enabling the hippocampus to store different memories of even similar events, and this prediction has been confirmed (Gilbert, Kesner, and Lee, 2001; Goodrich-Hunsaker, Hunsaker, and Kesner, 2008; Kesner and Rolls, 2015; Leutgeb and Leutgeb, 2007; McHugh et al., 2007; Rolls, 2008) (section 18.5). The term "pattern separation" refers to the property that the output patterns are less correlated with each other than the input patterns, that is, orthogonaliztion has been produced. This is one aspect of the categorization performed by competitive networks (figure 18.6), described in appendix B.3.

As just described, the dentate granule cells could be important in helping to build and prepare spatial representations for the CA3 network. The actual representation of space in the primate hippocampus includes a representation of spatial view (Rolls and Xiang, 2006), whereas in the rat hippocampus it is of the place where the rat is. The representation in the rat may be related to the fact that with a much less developed visual system than the primate's, the rat's representation of space may be defined more by the olfactory and tactile as well as distant visual cues present and may thus tend to reflect the place where the rat is. However, the spatial representations in the rat and primate could arise from essentially the same computational process as follows (de Araujo, Rolls, and Stringer, 2001; Rolls, 1999). The starting assumption is that in both the rat and the primate, the dentate granule cells (and the CA3 and CA1 pyramidal cells) respond to combinations of the inputs received. In the case of the primate, a combination of visual features in the environment will, because of the fovea's providing high spatial resolution over a typical viewing angle of perhaps 10–20 degrees, result in the formation of a spatial view cell, the effective trigger for which will thus be a combination of visual features within a relatively small part of space. In contrast, in the rat, given the very extensive visual field subtended by the rodent retina, which may extend over 180–270 degrees, a combination of visual features formed over such a wide visual angle would effectively define a position in space that is a place (de Araujo et al., 2001).

Entorhinal Cortex Grid Cells Transformed to Hippocampal Place and Spatial View Cells The entorhinal cortex contains grid cells, which have high firing in the rat in a two-dimensional spatial grid as a rat traverses an environment, with larger grid spacings in the ventral entorhinal cortex (Fyhn et al., 2004; Hafting et al., 2005). This may be a system optimized for path integration (McNaughton et al., 2006) which may self-organize during locomotion with longer time constants producing more widely spaced grids in the ventral entorhinal cortex (Kropff and Treves, 2008). How are the grid-cell representations, which would not be suitable for association of an object or reward with a place to form an episodic memory, transformed into a place representation that would be appropriate for this type of episodic memory? I have proposed that this could be implemented by a competitive network (Rolls, 2008) in the dentate gyrus which operates to form place cells, implemented by each dentate granule cell's learning to respond to particular combinations of entorhinal cortex cells firing, where each combination effectively specifies a place, and this has been shown to be feasible computationally (Rolls et al., 2006). The sparse representations in the dentate gyrus, implemented by the mutual inhibition through inhibitory interneurons and competitive learning, help to implement this "pattern separation" effect (Rolls, 1989b, 1989c, 2008; Rolls and Treves, 1998). Results of this competitive learning model are illustrated in figure 18.7. Figure 18.7A,B shows simulated (entorhinal cortex) grid cells with Gaussian firing rate response profiles. Figure 18.7C,D shows the place cells formed in the hippocampus by competitive learning (Rolls et al., 2006). Similar processes are involved in some later models of this transformation (Giocomo et al., 2011; Zilli, 2012).

In primates, there is now evidence that there is a grid-cell-like representation in the entorhinal cortex, with neurons having grid-like firing as the monkey moves the eyes across a spatial scene (Killian, Jutras, and Buffalo, 2012). Similar competitive learning processes may transform these entorhinal cortex "spatial view grid cells" into hippocampal spatial view cells and may help with the idiothetic (produced in this case by movements of the eyes) update of spatial view cells (Robertson et al., 1998).

In primates, spatial view cells represent a scene view allocentrically, as described in section 18.4. How could such spatial view representations be formed, in which the relative spatial position of features in a scene is encoded? I have proposed that this involves competitive learning analogous to that used to form place cells in rats, but in primates operating on the representations of objects that reach the hippocampus from the inferior temporal visual cortex (Rolls, Tromans, and Stringer, 2008c). We have shown that in complex natural scenes the receptive fields of inferior temporal cortex neurons become reduced in size and asymmetrical with respect to the fovea (Aggelopoulos and Rolls, 2005; Rolls, 2009), and we have demonstrated in a unifying computational approach that competitive network processes operating in areas such as the parahippocampal cortex, the entorhinal cortex, and/or the dentate granule cells could form unique views of scenes by forming a sparse representation of these object- or feature-tuned inferior temporal cortex ventral visual stream representations which have some spatial asymmetry which provides a foundation for building scene representations that incorporate the relative spatial positions of landmarks in a scene (Rolls et al., 2008c). In this theory, it is the spatial asymmetry with respect to the fovea of different neurons that solves the binding problem, for the neurons in fact respond to an object and to its location with respect to the fovea (Aggelopoulos and Rolls, 2005; Rolls, 2009; Rolls et al., 2008c). Another input to hippocampal spatial view cells may come from the parahippocampal place area (Nasr et al., 2011).

18.3.3 CA1 Cells

The CA3 cells connect to the CA1 cells by the Schaeffer collateral synapses. The associative modifiability in this connection helps the full information present in CA3 to be retrieved in the CA1 neurons (Rolls, 1995; Schultz and Rolls, 1999; Treves, 1995; Treves and Rolls, 1994). Part of the hypothesis is that the different subparts of an episodic memory, which must be represented separately in CA3 to allow for completion, can be combined

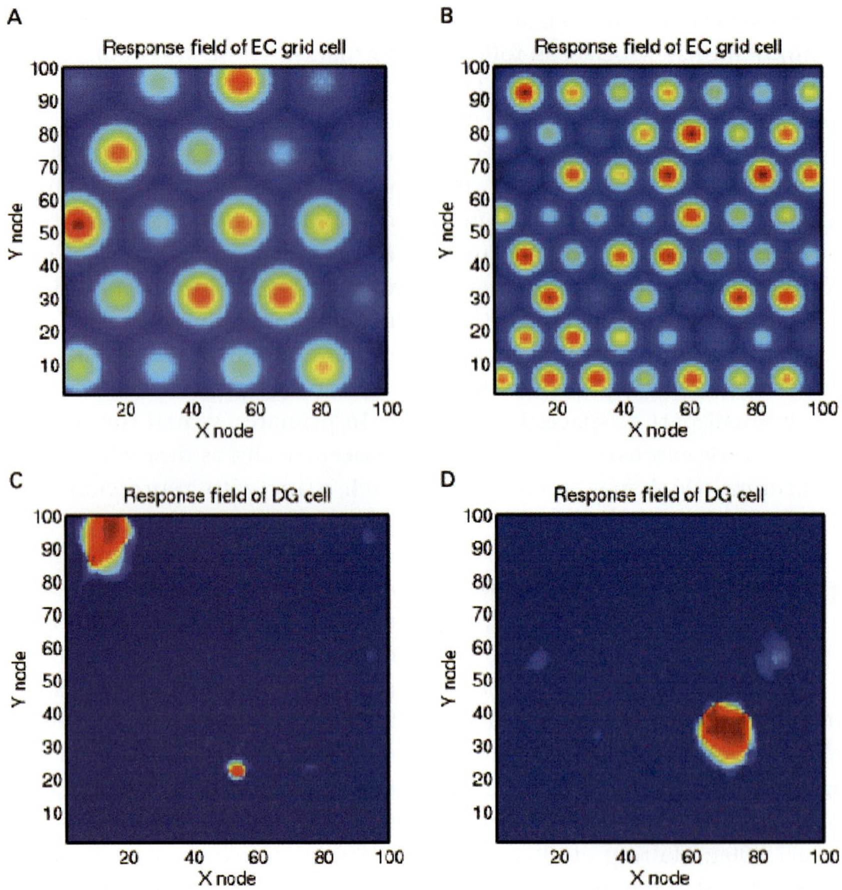

FIGURE 18.7 Simulation of competitive learning in the dentate gyrus to produce place cells from the entorhinal cortex grid-cell inputs. (A and B) Firing rate profiles of two entorhinal cortex (EC) grid cells with frequencies of 4 and 7 cycles. The colors show the firing rates with blue lowest and red highest in the testing environment (e.g., a room) in which the spatial coordinates are X and Y. (C and D) Firing rate profiles of two dentate gyrus (DG) cells after competitive learning. After Rolls, Stringer, and Elliot (2006).

by competitive learning in CA1 to produce an efficient retrieval representation for the recall via the backprojection pathways to the neocortex of memories stored in the neocortex (Rolls, 1989b, 2008; Treves and Rolls, 1994).

18.3.4 BACKPROJECTIONS TO THE NEOCORTEX, AND MEMORY RECALL The need for information to be retrieved from the hippocampus to affect other brain areas was noted in the introductory section. The way in which this could be implemented via backprojections to the neocortex is now considered.

It is suggested that the modifiable connections from the CA3 neurons to the CA1 neurons allow the whole episode in CA3 to be produced in CA1. The CA1 neurons would then activate, via their termination in the deep layers of the entorhinal cortex, at least the pyramidal cells in the deep layers of the entorhinal cortex (see figure 18.1). These entorhinal cortex layer 5 neurons would then, by virtue of their backprojections (Lavenex and Amaral, 2000; Witter et al., 2000) to the parts of cerebral cortex that originally provided the inputs to the hippocampus, terminate in the superficial layers (including layer 1) of those neocortical areas, where synapses would be made onto the distal parts of the dendrites of the (superficial and deep) cortical pyramidal cells (Rolls, 1989a, 1989b, 1989c, 2008) (see figure 18.1). The areas of neocortex in which this recall would be produced could include multimodal cortical areas (e.g., the cortex in the superior temporal sulcus which receives inputs from temporal, parietal, and occipital cortical areas, and from which it is thought that cortical areas such as Brodmann areas 39 and 40 related to language developed) and also areas of unimodal association cortex (e.g., inferior temporal visual cortex). The backprojections, by recalling previous episodic events, could provide information useful to the neocortex in the building of new representations in the multimodal and unimodal association cortical areas, which, by building new long-term and structured representations, can be considered as a form of memory consolidation (Rolls, 1989a, 1989b, 1989c, 1990a, 1990b, 2008) or useful in organizing actions.

The hypothesis of the architecture with which this multistage recall from the hippocampus to the neocortex would be achieved is shown in figure 18.1. The feedforward connections from association areas of the cerebral neocortex (solid lines in figure 18.1b) show major convergence as information is passed to CA3, with the CA3 autoassociation network having the smallest number of neurons at any stage of the processing. The backprojections allow for divergence back to neocortical areas. The way in which I suggest that the backprojection synapses are set up to have the appropriate strengths for recall is as follows (Rolls, 1989a, 1989b, 1989c). During the setting up of a new episodic memory, there would be strong feedforward activity progressing toward the hippocampus. During the episode, the CA3 synapses would be modified, and via the CA1 neurons and the subiculum, a pattern of activity would be produced on the backprojecting synapses to the entorhinal cortex. Here the backprojecting synapses from active backprojection axons onto pyramidal cells being activated by the forward inputs to entorhinal cortex would be associatively modified. A similar process would be implemented at preceding stages of neocortex, that is in the parahippocampal gyrus/perirhinal cortex stage, and in association cortical areas, as shown in figure 18.1b.

The concept is that during the learning of an episodic memory, cortical pyramidal cells in at least one of the stages would be driven by forward inputs from earlier cortical areas but would simultaneously be receiving backprojected activity (indirectly) from the hippocampus which would, by pattern association from the backprojecting synapses to the cortical pyramidal cells, become associated with whichever cortical cells were being made to fire by the forward inputs. Then later on, during recall, a recall cue from perhaps another part of neocortex might reach CA3, where the firing during the original episode would be completed. The resulting backprojecting activity would then, as a result of the pattern association learned previously in the hippocampocortical backprojections, bring back the firing in any cortical area that was present during the original episode. Thus, retrieval involves reinstating the neuronal activity that was present in different cortical areas during the learning of an episode. (The pattern association is also called heteroassociation, to contrast it with autoassociation. The pattern association operates at multiple stages in the backprojection pathway, as made evident in figure 18.1). If the recall cue was an object, this might result in recall of the neocortical firing that represented the place in which that object had been seen previously. As noted elsewhere in this chapter, and by McClelland et al. (1995), that recall might be useful to the neocortex to help it build new semantic memories, which might inherently be a slow process and is not part of the theory of recall from the hippocampus (see below).

Overall, this is thus a theory of how different events, linked together in CA3 during the formation of an episodic memory, could produce completion in CA3 if only one of those events is presented later in recall. This would then, in turn, via CA1 address by multistage pattern association the cortical areas in which activity was present during the original learning of the episodic memory, and reinstate the neocortical neuronal activity that was present when the episodic memory was formed. This theory is supported by a computational neuroscience model of the operation of the whole of this system (Rolls, 1995).

A plausible requirement for a successful hippocampo-directed recall operation is that the signal generated from the hippocampally retrieved pattern of activity, and carried backward toward neocortex, remain undegraded when compared to the noise due, at each stage, to the interference effects caused by the concurrent storage of other patterns of activity on the same backprojecting synaptic systems. That requirement is equivalent to that used in deriving the storage capacity of such a series of heteroassociative memories, and it was shown by Treves and Rolls (1991, 1994) that the maximum number of independently generated activity patterns that can be retrieved is given, essentially, by the same formula as in equation 18.5 above where, however, a is now the sparseness of the representation at any given stage, and C is the average number of (back-)projections each cell of that stage receives from cells of the previous one. (k' is a similar slowly varying factor to that introduced above.) If p is equal to the number of memories held in the hippocampal memory, it is limited by the retrieval capacity of the CA3 network, p_{max}. Putting together the formula for the latter with that shown here, one concludes that, roughly, the requirement implies that the number of afferents of (indirect) hippocampal origin to a given neocortical stage (C^{HBP}) must be $C^{HBP} = C^{RC} a_{nc} / a_{CA3}$, where C^{RC} is the number of recurrent collaterals to any given cell in CA3, the average sparseness of a representation is a_{nc}, and a_{CA3} is the sparseness of memory representations in CA3. (HBP refers to hippocampal backprojections. The implication of the argument is that there must be very many backprojection synapses onto each neocortical neuron; Treves and Rolls, 1994.)

The above requirement is very strong: even if representations were to remain as sparse as they are in CA3, which is unlikely, to avoid degrading the signal, C^{HBP} should be as large as C^{RC}, that is, 12,000 in the rat. If then C^{HBP} has to be of the same order as C^{RC}, one is led to a very definite conclusion: a mechanism of the type envisaged here could not possibly rely on a set of monosynaptic CA3-to-neocortex backprojections. This would imply that, to make a sufficient number of synapses on each of the vast number of neocortical cells, each cell in CA3 has to generate a disproportionate number of synapses (i.e., C^{HBP} times the ratio between the number of neocortical and that of CA3 cells). The required divergence can be kept within reasonable limits only by assuming that the backprojecting system is polysynaptic (i.e., involves several connected cortical stages), provided that the number of cells involved grows gradually at each stage, from CA3 back to neocortical association areas (Treves and Rolls, 1994) (cf. figure 18.1).

The theory of recall by the backprojections thus provides a quantitative account of why any neocortical area has as many backprojection as forward projection connections. Further aspects of the operation of the backprojecting systems are described elsewhere (Rolls, 2008).

The theory described by McClelland et al. (1995) is similar to the theory described above, except that it holds that the last set of synapses that are modified rapidly during the learning of each episode are those between the CA3 and the CA1 pyramidal cells (see figure 18.1). That theory also emphasizes the important point that the hippocampal and neocortical memory systems may be quite different, with the hippocampus specialized for the rapid ("one-shot") learning of unstructured single events or episodes, and the neocortex for the slower learning of semantic representations (structured representations in which the components are linked), which may necessarily benefit from the many exemplars needed to shape the semantic representation, a process that is helped by the recall of episodic memories from the hippocampus. The particular model on which they focus for the learning of semantic representations by interleaved learning is the connectionist model of Rumelhart (1990; Rumelhart and Todd, 1993) that is trained by error backpropagation (Rumelhart, Hinton, and Williams, 1986).

18.3.5 Temporal Order Memory in the Hippocampus, and Episodic Memory There has for some time been evidence that the hippocampus plays a role in temporal order memory, even when there is no spatial component (Kesner and Rolls, 2015). In humans, the hippocampus becomes activated when the temporal order of events is being processed (Lehn et al., 2009), and temporal context is important in episodic memory (Howard, Viskontas, Shankar, and Fried, 2012). One approach to how the hippocampus might be involved in temporal order memory is by encoding temporal order into each gamma cycle nested into a theta cycle (Lisman and Buzsaki, 2008; Lisman and Redish, 2009). A very different approach is to use firing-rate encoding in attractor networks (Rolls, 2010; Rolls and Deco, 2010) and is based on evidence that neurons in the rat hippocampus have firing rates that reflect which temporal part of the task is current (Macdonald, Lepage, Eden, and Eichenbaum, 2011). In particular, a sequence of different neurons is activated at successive times during a time delay period (see the example in figure 18.6). The tasks used included an object–odor paired associate nonspatial task with a 10-second delay period between the visual stimulus and the odor. The evidence also shows that a large proportion of hippocampal neurons fire in relation to individual events in a sequence being remembered (e.g., a visual object or odor), and some to combinations of the event and the time in the delay period (Macdonald et al., 2011).

These interesting neurophysiological findings indicate that rate encoding is being used to encode time, that is, the firing rates of different neurons are high at different times within a trial, delay period, and so forth (Macdonald et al., 2011). These findings suggest several possible computational processes (Kesner and Rolls, 2015; Rolls, 2010).

First, because some neurons fire at different times in a trial of a temporal order memory task or delay task, the time in a trial at which an object (e.g., a visual stimulus or odor) was presented could become encoded in the hippocampus by an association implemented in the CA3 recurrent collaterals between the neurons that represent the object (already known to be present in the hippocampus for tasks for which the hippocampus is required; Rolls and Xiang, 2006; Rolls et al., 2005) and the "time-encoding" neurons in the hippocampus (Macdonald et al., 2011). This would allow associations for the time at which the object was present to be formed. Given that time-encoding neurons are also found in the medial entorhinal cortex (Kraus et al., 2013), this could provide the source of the time information required by CA3. However, although lesions of CA3 impair temporal order–place representations, it is lesions of CA1 that impair temporal order–visual object and temporal order–odor representations (Kesner and Rolls, 2015). It is thus possible that temporal timing and object information is brought together by competitive learning in CA1 (Kesner and Rolls, 2015), which receives inputs not only from CA3, but also directly from the entorhinal cortex (see figure 18.1).

Second, these associations would provide the basis for the recall of the object from the time in a trial, or vice versa. The retrieval of object or temporal information from each other would occur in CA3 in a way that is analogous to that shown for recalling the object from the place, or the place from the object (Rolls et al., 2002), but substituting the details of the properties of the time-encoding neurons (Macdonald et al., 2011) for what was previously the spatial (place) component. Alternatively, if competitive learning in CA1 is the mechanism, generalization in the competitive learning (Rolls, 2008) from either the object or the temporal order cue would retrieve the whole representation. In addition, if the time-encoding neurons simply cycled through their normal sequence during recall, this would enable the sequence of objects or events associated with each subset of time-encoding neurons to be recalled correctly in the order in which they were presented.

Third, we need a theory of what the origin is of the temporal effect whereby different hippocampal (or potentially PFC) neurons fire in different parts of a trial or delay period. We can consider three hypotheses about how the firing of the time-encoding hippocampal neurons is produced. All utilize slow transitions between attractor states that can be a property of noisy attractor networks.

The first hypothesis is that an attractor network with realistic dynamics (modeled at the integrate-and-fire level with a dynamical implementation of the neuronal membrane and synaptic current dynamics, and with synaptic or neuronal adaptation) can implement a sequence memory, as shown by Deco and Rolls (2005). The hypothesis is that there are several different attractors, and that there are weak connections between the different attractors. In the model, adaptation produces effects whereby whatever sequence (order of stimuli) is presented on an individual trial, that order can be replayed in the same sequence because, as one attractor state dies as a result of the adaptation, the next attractor to emerge from the spontaneous firing because of the spiking-related noise is the one that has been active least recently, as it is the one that is least adapted (Deco and Rolls, 2005). The whole system operates at a rather slow timescale for the transitions between the attractors partly because of the time for the noise to drive the system from one attractor state to another, and the slow time course of the adaptation (Deco and Rolls, 2005; Rolls and Deco, 2010). This implements a type of order memory.

The second hypothesis is analogous and is also implemented in a recurrently connected system such as the hippocampal CA3 system or local recurrent circuits in the neocortex (Rolls and Deco, 2010). This second theory is that again there are several attractors, but that each attractor is connected by slightly stronger forward than reverse synaptic weights to the next. In previous work, we have shown that with an integrate-and-fire implementation with spiking noise, this allows slow transitions from one attractor state to the next (Deco, Ledberg, Almeida, and Fuster, 2005; Deco and Rolls, 2003). During learning of the synaptic weights in the network, adaptation might lead to each time-encoding population of neurons responding for only a limited period, helping to produce multiple sequentially activated populations of time-encoding neurons (Rolls and Deco, 2010). In this scenario, stronger forward than reverse weights between different attractors each consisting of a different population of time-encoding neurons would be the essence.

The third hypothesis is that the mechanism for the time-encoding neurons is in the entorhinal cortex where there are ring attractors, as described in section 18.3.7.

It is thus suggested that temporal order memory could be implemented in the hippocampus in this way and could make an important contribution to episodic memory in which several events linked in the correct order might form an episode. The theory shows how items in a particular temporal order could be separated from each other, a property we have referred to as the temporal pattern separation effect (Kesner and Rolls, 2015). The theory of episodic memory described here shows how events and sequences of events could be recalled from the hippocampus to the neocortex, and there a longer term more semantic representation of a recalled episode, such as what happened on one's fifth birthday, might be stored, and then accessed to describe the episode. For the order to be correctly implemented in the semantic neocortical store, a similar mechanism, involving, for example, stronger forward than reverse synaptic weights between long-term memory representations in attractors, could build an appropriate long-term order memory (Rolls and Deco, 2010).

18.3.6 Entorhinal Cortex Grid Cells The entorhinal cortex contains grid cells, which have high firing in the rat in a two-dimensional spatial grid as a rat traverses an environment (see figure 18.5), with larger grid spacings in the ventral entorhinal cortex (Fyhn et al., 2004; Hafting et al., 2005). Computational approaches to this system model it as a set of linked ring continuous attractors (Giocomo et al., 2011; Kropff and Treves, 2008). (These are the CANNs described above.) The concept is that as the rat locomotes, the peak of the firing in the continuous attractor moves, and after a certain distance has been moved, the place represented returns to the same set of neurons, completing the ring. The position of the peak in the ring continuous attractor is updated for example by self-motion, or possibly by time for at least some neurons. By having different ring attractors that cover large to small distances with one pass through the ring, the system provides, with its multiscale representation, information which when read out appropriately provides a coarse and fine representation of position. The phases of the different ring attractors must be locked for this to work. The use of ring attractors could in this way implement a representation of the position of the rat in a 2-D environment which would be self-generating and so would work in any environment, if it is updated by self-motion or time. Indeed, one theory of the underlying mechanism is that neuronal or synaptic adaptation could be used to make the continuous attractor move its peak of activity continuously round the ring as a function of time (Kropff and Treves, 2008). A fast adaptation mechanism would produce small rings for the grid, and a slow adaptation mechanism would produce large rings for the grid. Part of the interest in this suggestion is that grid cells formed using this adaptation process would effectively be time cells, different cells of which would fire at different times in a trial, which have now been described in the rat entorhinal cortex (Kraus et al., 2013) and also in the hippocampus itself (Kraus et al., 2013). A set of different modeling approaches for the grid cells are described by Giocomo et al. (2011). The system may be used therefore not only for spatial path integration (McNaughton et al., 2006) but also for the timing information useful in sequence encoding for nonspatial as well as spatial information (Kesner and Rolls, 2015).

18.4 Navigation and the Hippocampus

Spatial information is almost always part of an episodic memory, and spatial representations in the hippocampus may be useful for navigation. For example, episodic memories of particular journeys could help to build neocortical maps that would require many journeys to elaborate. Such maps may be found in the neocortex, given the evidence that lesions to the neocortex can produce topographical agnosia and inability to navigate (Kolb and Whishaw, 2009). Further, the right hippocampus in humans is activated during mental navigation in recently learned but not highly familiar environments (Hirshhorn, Grady, Rosenbaum, Winocur, and Moscovitch, 2012). Mental navigation in familiar environments activates cortical areas such as the lateral temporal cortex, posterior parahippocampal cortex, lingual gyrus, and precuneus (Hirshhorn et al., 2012). Given this, it is interesting to consider the role of the hippocampus in navigation.

First, any model of navigation based on place cells in rodents could not provide an adequate model of the role of the primate (including human) hippocampal cortex in navigation, in view of the presence of spatial view cells in primates, which by their firing provide a basis for representation of places other than where one is located, that is, for the representations of positions in scenes at which one is looking, even if the scene is being remembered, based, for example, on idiothetic (self-motion) update (Rolls and Xiang, 2006). Spatial view cells provide a basis for the representation of scenes, landmarks in scenes, and where objects and rewards are located in scenes (Rolls, 2008; Rolls and Xiang, 2006), and this type of representation is likely to be crucial in the computations involved in navigating to new places at which one has not been located previously in primates including humans. Moreover, spatial view neurons are found not only in CA3 and CA1 but also in the parahippocampal cortex (Rolls and Xiang, 2006).

Second, hippocampal place cells in rodents, and spatial view cells in primates, can be updated by idiothetic (self-motion) inputs, for example, by moving the eyes to a different location in a scene in the dark (Robertson et al., 1998), and the basis for this is the idiothetic update of attractor networks of grid cells on the entorhinal cortex (Giocomo et al., 2011). This may be useful not only in updating location for use in episodic memory but may also be used to update position for use in navigation (Burgess, 2008; Burgess and O'Keefe, 1996; Erdem and Hasselmo, 2012). These processes may utilize head direction cells found in the presubiculum of rodents and primates (Robertson, Rolls, Georges-François, and Panzeri, 1999; Taube, Goodridge, Golob, Dudchenko, and Stackman, 1996; Wiener and Taube, 2005) and cells that respond at the boundaries of an enclosure (Lever, Burton, Jeewajee, O'Keefe, and Burgess, 2009).

An example of a model of navigation is provided next. In the model, goal-directed navigation is based on forward linear look-ahead probe of trajectories in a network of head direction cells, grid cells, place cells, and PFC cells (Erdem and Hasselmo, 2012). The model allows selection of new goal-directed trajectories. In a novel environment, the virtual rat incrementally creates a map composed of place cells and PFC cells by random exploration. After exploration, the rat retrieves memory of the goal location, picks its next movement direction by forward linear look-ahead probe of trajectories in several candidate directions while stationary in one location, and finds the one activating PFC cells with the highest reward signal. Each probe direction involves activation of a static pattern of head direction cells to drive an interference model of grid cells to update their phases in a specific direction. The updating of grid-cell spiking drives place cells along the probed look-ahead trajectory similar to the forward replay during waking seen in place-cell recordings. Directions are probed until the look-ahead trajectory activates the reward signal, and the corresponding direction is used to guide goal-finding behavior (Erdem and Hasselmo, 2012). Links are being made from navigational theories of and approaches to hippocampal function to more memory-based approaches (Day et al., 2003; Redish and Touretzky, 1997).

18.5 *Tests of the Theory*

A useful theory should make predictions that can then be tested to substantiate the theory or show how it should be developed or modified. This section illustrates the important rich interplay between theory and experiment that is essential to understanding how the brain computes, and further developments have been described (Kesner and Rolls, 2015).

18.5.1 Dentate Granule Cells The theory predicts that the dentate granule cell mossy fiber system of inputs to the CA3 neurons is necessary to store spatial memories, but not to recall them (Rolls, 2008; Treves and Rolls, 1992, 1994). Lassalle et al. (2000) have obtained evidence consistent with this in rats with damage to the mossy fiber system, and there is further evidence consistent with this (Daumas, Ceccom, Halley, Frances, and Lassalle, 2009; Kesner and Rolls, 2015; Lee and Kesner, 2004).

The theory predicts that pattern separation is performed by the dentate granule cells. Evidence consistent with this has been found neurophysiologically in the small sparsely encoded place fields of dentate neurons (Jung and McNaughton, 1993; Leutgeb and Leutgeb, 2007) and their reflection in CA3 neurons (Leutgeb and Leutgeb, 2007). It has been shown that selective dentate lesions in rats (Gilbert and Kesner, 2003; Gilbert et al., 2001; Goodrich-Hunsaker et al., 2008; Kesner and Rolls, 2015; Rolls, 2008) or dentate NMDA receptor knockouts in mice (McHugh et al., 2007) impair spatial, object–place (or reward–place: remembering where to find a reward) association tasks especially when the places are close together and require pattern separation before storage in CA3.

If adult *neurogenesis* in the dentate gyrus (Clelland et al., 2009; Nakashiba et al., 2012) does prove to be functionally relevant, its computational role could be to facilitate pattern separation for new patterns by providing new dentate granule cells with new sets of random connections to CA3 neurons. Consistent with the dentate spatial pattern separation hypothesis (Rolls, 1989b, 1989c, 1996; Treves and Rolls, 1992, 1994), in mice with impaired dentate neurogenesis, spatial learning in a delayed non-matching-to-place task in the radial arm maze was impaired for arms that were presented with little separation, but no deficit was observed when the arms were presented farther apart (Clelland et al., 2009). Consistently, impaired neurogenesis in the dentate also produced a deficit for small spatial separations in an associative object-in-place task (Aimone and Gage, 2011; Clelland et al., 2009). Neurogenesis in this system may be useful because the role of the dentate granule cell/mossy fiber system is to produce pattern separation for the CA3 representations useful for making new episodic memories very different from previous episodic memories, and not to play a role by synaptic modification of the mossy fiber synapses in the retrieval of the information stored in CA3. In other cortical systems, the synapses involved in storage and recall are the same and are associatively modified (e.g., in neocortical pattern association, autoassociation, and competitive learning systems; Rolls, 2008), and neurogenesis is accordingly not present.

The theory predicts that the direct perforant path input from the entorhinal cortex to the CA3 cells (which bypasses the dentate granule cells) is involved in the recall of memory from the CA3 system, and Lee and Kesner (2004) have obtained evidence consistent with this in a Hebb–Williams maze recall task by showing that lesions of the perforant path impair retrieval (Lee and Kesner, 2004).

18.5.2 CA3 The theory predicts that the CA3 is especially important in object–place or reward–place tasks in which associations must be formed between any spatial location and any object (referred to as *arbitrary associations*). There is much evidence from subregion analyses involving disruption of CA3 that CA3 is necessary for arbitrary associations between places and objects or rewards (Gilbert and Kesner, 2003; Kesner and Rolls, 2015). Similar impairments were obtained following deletion of CA3 NMDA receptors in mice in the acquisition of an odor–context paired associate learning task (Rajji, Chapman, Eichenbaum, and Greene, 2006). If place or time is not a component, associative tasks such as odor–object association are not impaired (Kesner and Rolls, 2015), underlining the fact that the hippocampus is especially involved in episodic types of associative memory which typically involve place and/or time.

The theory predicts that the CA3 is especially important in object–place or reward–place *completion* tasks, in which associations must be completed from a part of the whole. As mentioned above, it has been shown that if completion from an incomplete cue is needed, then CA3 NMDA receptors are necessary (presumably to ensure satisfactory CA3–CA3 learning) even in a reference memory task (Gold and Kesner, 2005; Nakazawa et al., 2002).

The theory predicts that the CA3 system is especially needed in *rapid, one-trial object–place learning and recall.* It has been shown that hippocampal NMDA receptors (necessary for long-term potentiation to occur) are needed for one-trial flavor–place association learning, and that hippocampal AMPA/kainate receptors are sufficient for the recall, though the hippocampal subregion involved was not tested (Day et al., 2003). In subregion studies, Kesner and colleagues have shown that CA3 lesions produce chance performance on a one-trial object–place recall task (Kesner, Hunsaker, and Warthen, 2008) and other object–spatial tasks (Kesner and Rolls, 2001, 2015). For example, CA3 lesions produced chance performance on both a one-trial object–place recall and place–object recall task (Kesner et al., 2008). This is evidence that CA3 supports arbitrary associations as well as episodic memory based on one-trial learning. A control fixed visual conditional to place task with the same delay was not impaired, showing that it is recall after one-trial (or rapid, episodic) learning that is impaired (Kesner et al., 2008). CA3 NMDA receptors are as predicted by the theory necessary for rapid/one-trial spatial learning, as shown by a mouse knockout study by Nakazawa, Tonegawa and colleagues (Nakazawa, McHugh, Wilson, and Tonegawa, 2004; Nakazawa et al., 2003; Tonegawa, Nakazawa, and Wilson, 2003). As described in section 18.4, we have shown that primate hippocampal CA3 neurons reflect the computational processes necessary for one-trial object–place event memory, used as a model for episodic memory (Rolls and Xiang, 2006).

The theory predicts that if primates including humans can form an episodic memory in which objects or people are seen at particular locations even though the observer viewing the space has never been to those locations "out there" in space, there should be a neural system in CA3 that can support such associations between places "out there" in a scene and objects. Exactly this is provided by the spatial view neurons Rolls and colleagues have discovered that are present in CA3 (Georges-François et al., 1999; Robertson et al., 1998; Rolls, Robertson, 1997a; Rolls et al., 1998; Rolls and Xiang, 2005, 2006; Rolls et al., 2005). Place cells (Hartley et al., 2014; O'Keefe, 1984; O'Keefe and Dostrovsky, 1971) will not do for this type of episodic memory.

Another type of test of the autoassociation (or attractor) hypothesis for CA3 has been to train rats in different environments, for example, a square and a circular environment, and then test the prediction of the hypothesis that, when presented with an environment ambiguous between these, hippocampoal neurons will fall in an attractor state that represents one of the two previously learned environments, but not a mixture of the two environments. Evidence consistent with the hypothesis has been found (Wills, Lever, Cacucci, Burgess, and O'Keefe, 2005). In a particularly dramatic example, it has been found that within each theta cycle, hippocampal pyramidal neurons may, in an ambiguous environment, represent one or other of the learned environments (Jezek, Henriksen, Treves, Moser, and Moser, 2011). This is an indication, predicted by Rolls and Treves (1998), that autoassociative memory recall can take place sufficiently rapidly to be complete within one theta cycle (120 ms), and that theta cycles could provide a mechanism for a fresh retrieval process to occur after a reset caused by the inhibitory part of each theta cycle, so that the memory can be updated rapidly to reflect a continuously changing environment and not remain too long in an attractor state.

18.5.3 Recall via CA1 to Neocortex The theory shows quantitatively, analytically, how memories could be retrieved from the hippocampus to the neocortex (Treves and Rolls, 1994), and this has been shown by simulation of the multistage hippocampal system including the entorhinal cortex, dentate, CA3, CA1, and return to the entorhinal cortex to recall the memory to be quantitatively realistic (Rolls, 1995).

Many further tests of the theory are described elsewhere (Kesner, Morris, and Weeden, 2012; Kesner and Rolls, 2015; Rolls, 2008).

18.6 Conclusion

In conclusion, a theory of hippocampal function has been described, and this goes beyond a model by incorporating many analytic results, for example, on the importance of the number of synapses onto each CA3 neuron and the sparseness of the representation as providing an estimate of the memory capacity of the hippocampus, on the distinct roles of the mossy fiber and perforant path inputs to the CA3 neurons, and on how information of recently learned episodic information could be recalled to the neocortex from the hippocampus using the multistage corticocortical backprojection pathway (Rolls, 2008, 2010; Treves and Rolls, 1992, 1994). The approach underlines the importance of understanding spatial representations in the primate including hippocampus because they are different from those in rodents and are relevant to understanding episodic memory in humans, where memories can be formed of where objects and rewards are in an environment without having ever been present at the place. The difference may be related to the great importance and development of vision in primates, which has also had implications in understanding the cortical organization of many other processing systems, including those involved in taste processing and in emotion (Rolls, 2014). In primates, there has been great development of many neocortical areas, in part related to the great expansion of cortical visual computation (Rolls, 2012c), and those cortical developments are also important in understanding hippocampal function and its relation to episodic memory and to spatial processing. In addition, the approach has emphasized the importance of founding the theory on details of the anatomy, quantitative where possible, and of the neurophysiology (and I include functional neuroimaging) of the brain systems involved.

Acknowledgments

Different parts of the research described here were supported by Programme Grants from the Medical Research Council (MRC), by a Human Frontier Science program grant, by an EU BRAIN grant, by the MRC Oxford Interdisciplinary Research Centre in Cognitive Neuroscience, and by the Oxford McDonnell-Pew Centre in Cognitive Neuroscience. The author has performed the experimental and theoretical work which is incorporated in some of the ideas presented here on the hippocampus with many colleagues, including Alessandro Treves, Simon Stringer, Ray Kesner, Robert Robertson, Pierre Georges-François, and Shane O'Mara, and their contributions are sincerely acknowledged.

NOTE

A useful source for some of the papers referred to here is www.oxcns.org. Further citations to the rich research literature in this area are provided elsewhere (Kesner and Rolls, 2015).

REFERENCES

Abeles, M. (1991). *Corticonics—Neural Circuits of the Cerebral Cortex.* New York: Cambridge University Press.

Acsady, L., Kamondi, A., Sik, A., Freund, T., & Buzsaki, G. (1998). GABAergic cells are the major postsynaptic targets of mossy fibers in the rat hippocampus. *Journal of Neuroscience, 18*(9), 3386–3403.

Aggelopoulos, N. C., & Rolls, E. T. (2005). Natural scene perception: Inferior temporal cortex neurons encode the positions of different objects in the scene. *European Journal of Neuroscience, 22,* 2903–2916.

Aimone, J. B., & Gage, F. H. (2011). Modeling new neuron function: A history of using computational neuroscience to study adult neurogenesis. *European Journal of Neuroscience, 33*(6), 1160–1169. doi:10.1111/j.1460-9568.2011.07615.x.

Amaral, D. G., Ishizuka, N., & Claiborne, B. (1990). Neurons, numbers and the hippocampal network. *Progress in Brain Research, 83,* 1–11.

Amaral, D. G., & Witter, M. P. (1989). The three-dimensional organization of the hippocampal formation: A review of anatomical data. *Neuroscience, 31,* 571–591.

Amaral, D. G., & Witter, M. P. (1995). The hippocampal formation. In G. Paxinos (Ed.), *The Rat Nervous System* (pp. 443–493). San Diego, CA: Academic Press.

Amari, S. (1977). Dynamics of pattern formation in lateral-inhibition type neural fields. *Biological Cybernetics, 27,* 77–87.

Amit, D. J. (1989). *Modeling Brain Function.* Cambridge: Cambridge University Press.

Andersen, P., Morris, R. G. M., Amaral, D. G., Bliss, T. V. P., & O'Keefe, J. (2007). *The Hippocampus Book.* London: Oxford University Press.

Banta Lavenex, P., & Lavenex, P. (2009). Spatial memory and the monkey hippocampus: Not all space is created equal. *Hippocampus, 19*(1), 8–19.

Battaglia, F. P., & Treves, A. (1998). Attractor neural networks storing multiple space representations: A model for hippocampal place fields. *Physical Review E: Statistical Physics, Plasmas, Fluids, and Related Interdisciplinary Topics, 58,* 7738–7753.

Braitenberg, V., & Schütz, A. (1991). *Anatomy of the Cortex.* Berlin: Springer-Verlag.

Brown, T. H., Kairiss, E. W., & Keenan, C. L. (1990). Hebbian synapses: Biophysical mechanisms and algorithms. *Annual Review of Neuroscience, 13,* 475–511.

Brun, V. H., Otnass, M. K., Molden, S., Steffenach, H. A., Witter, M. P., Moser, M. B., et al. (2002). Place cells and place recognition maintained by direct entorhinal-hippocampal circuitry. *Science, 296,* 2243–2246.

Buckley, M. J. (2005). The role of the perirhinal cortex and hippocampus in learning, memory, and perception. *The Quarterly Journal of Experimental Psychology. B, Comparative and Physiological Psychology, 58*(3–4), 246–268. doi:P380Q67703124N71 [pii] 10.1080/02724990444000186

Burgess, N. (2008). Spatial cognition and the brain. *Annals of the New York Academy of Sciences, 1124,* 77–97.

Burgess, N., & O'Keefe, J. (1996). Neuronal computations underlying the firing of place cells and their role in navigation. *Hippocampus, 6,* 749–762.

Carmichael, S. T., & Price, J. L. (1995). Limbic connections of the orbital and medial prefrontal cortex in macaque monkeys. *Journal of Comparative Neurology, 346,* 403–434.

Cerasti, E., & Treves, A. (2010). How informative are spatial CA3 representations established by the dentate gyrus? *PLoS Computational Biology, 6*(4), e1000759.

Clelland, C. D., Choi, M., Romberg, C., Clemenson, G. D., Jr., Fragniere, A., Tyers, P., et al. (2009). A functional role for adult hippocampal neurogenesis in spatial pattern separation. *Science, 325*(5937), 210–213.

Collingridge, G. L., Peineau, S., Howland, J. G., & Wang, Y. T. (2010). Long-term depression in the CNS. *Nature Reviews. Neuroscience, 11*(7), 459–473. doi:nrn2867 [pii] 10.1038/nrn2867

Corkin, S. (2002). What's new with the amnesic patient H.M.? *Nature Reviews Neuroscience, 3*(2), 153–160. doi:10.1038/nrn726 nrn726 [pii]

Daumas, S., Ceccom, J., Halley, H., Frances, B., & Lassalle, J. M. (2009). Activation of metabotropic glutamate receptor type 2/3 supports the involvement of the hippocampal mossy fiber pathway on contextual fear memory consolidation. *Learning & Memory, 16*(8), 504–507.

Day, M., Langston, R., & Morris, R. G. (2003). Glutamate-receptor-mediated encoding and retrieval of paired-associate learning. *Nature, 424*, 205–209.

de Araujo, I. E. T., Rolls, E. T., & Stringer, S. M. (2001). A view model which accounts for the spatial fields of hippocampal primate spatial view cells and rat place cells. *Hippocampus, 11*, 699–706.

Deco, G., Ledberg, A., Almeida, R., & Fuster, J. (2005). Neural dynamics of cross-modal and cross-temporal associations. *Experimental Brain Research, 166*, 325–336.

Deco, G., & Rolls, E. T. (2003). Attention and working memory: A dynamical model of neuronal activity in the prefrontal cortex. *European Journal of Neuroscience, 18*, 2374–2390.

Deco, G., & Rolls, E. T. (2005). Sequential memory: A putative neural and synaptic dynamical mechanism. *Journal of Cognitive Neuroscience, 17*, 294–307.

Delatour, B., & Witter, M. P. (2002). Projections from the parahippocampal region to the prefrontal cortex in the rat: Evidence of multiple pathways. *European Journal of Neuroscience, 15*, 1400–1407.

Dere, E., Easton, A., Nadel, L., & Huston, J. P. (Eds.). (2008). *Handbook of Episodic Memory*. Amsterdam: Elsevier.

Ekstrom, A. D., Kahana, M. J., Caplan, J. B., Fields, T. A., Isham, E. A., Newman, E. L., et al. (2003). Cellular networks underlying human spatial navigation. *Nature, 425*, 184–188.

Erdem, U. M., & Hasselmo, M. (2012). A goal-directed spatial navigation model using forward trajectory planning based on grid cells. *European Journal of Neuroscience, 35*(6), 916–931. doi:10.1111/j.1460-9568.2012.08015.x.

Faisal, A. A., Selen, L. P., & Wolpert, D. M. (2008). Noise in the nervous system. *Nature Reviews. Neuroscience, 9*(4), 292–303.

Florian, C., & Roullet, P. (2004). Hippocampal CA3-region is crucial for acquisition and memory consolidation in Morris water maze task in mice. *Behavioural Brain Research, 154*, 365–374.

Franco, L., Rolls, E. T., Aggelopoulos, N. C., & Jerez, J. M. (2007). Neuronal selectivity, population sparseness, and ergodicity in the inferior temporal visual cortex. *Biological Cybernetics, 96*, 547–560.

Fyhn, M., Molden, S., Witter, M. P., Moser, E. I., & Moser, M. B. (2004). Spatial representation in the entorhinal cortex. *Science, 305*, 1258–1264.

Gelbard-Sagiv, H., Mukamel, R., Harel, M., Malach, R., & Fried, I. (2008). Internally generated reactivation of single neurons in human hippocampus during free recall. *Science, 322*(5898), 96–101.

Georges-François, P., Rolls, E. T., & Robertson, R. G. (1999). Spatial view cells in the primate hippocampus: Allocentric view not head direction or eye position or place. *Cerebral Cortex, 9*, 197–212.

Gilbert, P. E., & Kesner, R. P. (2003). Localization of function within the dorsal hippocampus: The role of the CA3 subregion in paired-associate learning. *Behavioral Neuroscience, 117*, 1385–1394.

Gilbert, P. E., Kesner, R. P., & Lee, I. (2001). Dissociating hippocampal subregions: Double dissociation between dentate gyrus and CA1. *Hippocampus, 11*, 626–636.

Giocomo, L. M., & Hasselmo, M. E. (2007). Neuromodulation by glutamate and acetylcholine can change circuit dynamics by regulating the relative influence of afferent input and excitatory feedback. *Molecular Neurobiology, 36*(2), 184–200.

Giocomo, L. M., Moser, M. B., & Moser, E. I. (2011). Computational models of grid cells. *Neuron, 71*(4), 589–603. doi:S0896-6273(11)00650-7 [pii] 10.1016/j.neuron.2011.07.023

Gold, A. E., & Kesner, R. P. (2005). The role of the CA3 subregion of the dorsal hippocampus in spatial pattern completion in the rat. *Hippocampus, 15*, 808–814.

Goodrich-Hunsaker, N. J., Hunsaker, M. R., & Kesner, R. P. (2008). The interactions and dissociations of the dorsal hippocampus subregions: How the dentate gyrus, CA3, and CA1 process spatial information. *Behavioral Neuroscience, 122*(1), 16–26.

Hafting, T., Fyhn, M., Molden, S., Moser, M. B., & Moser, E. I. (2005). Microstructure of a spatial map in the entorhinal cortex. *Nature, 436*, 801–806.

Hartley, T., Lever, C., Burgess, N., & O'Keefe, J. (2014). Space in the brain: How the hippocampal formation supports spatial cognition. *Philosophical Transactions of the Royal Society of London. Series B, Biological Sciences, 369*(1635), 20120510. doi:10.1098/rstb.2012.0510.

Hasselmo, M. E., Schnell, E., & Barkai, E. (1995). Dynamics of learning and recall at excitatory recurrent synapses and cholinergic modulation in rat hippocampal region CA3. *Journal of Neuroscience, 15*, 5249–5262.

Henze, D. A., Wittner, L., & Buzsaki, G. (2002). Single granule cells reliably discharge targets in the hippocampal CA3 network in vivo. *Nature Neuroscience, 5*(8), 790–795.

Hertz, J., Krogh, A., & Palmer, R. G. (1991). *Introduction to the Theory of Neural Computation*. Wokingham, UK: Addison-Wesley.

Hirshhorn, M., Grady, C., Rosenbaum, R. S., Winocur, G., & Moscovitch, M. (2012). The hippocampus is involved in mental navigation for a recently learned, but not a highly familiar environment: A longitudinal fMRI study. *Hippocampus, 22*(4), 842–852. doi:10.1002/hipo.20944.

Hölscher, C., Jacob, W., & Mallot, H. A. (2003). Reward modulates neuronal activity in the hippocampus of the rat. *Behavioural Brain Research, 142*, 181–191.

Hopfield, J. J. (1982). Neural networks and physical systems with emergent collective computational abilities. *Proceedings of the National Academy of Sciences of the United States of America, 79*, 2554–2558.

Howard, M. W., Viskontas, I. V., Shankar, K. H., & Fried, I. (2012). Ensembles of human MTL neurons "jump back in time" in response to a repeated stimulus. *Hippocampus, 22*(9), 1833–1847. doi:10.1002/hipo.22018.

Ishizuka, N., Weber, J., & Amaral, D. G. (1990). Organization of intrahippocampal projections originating from CA3 pyramidal cells in the rat. *Journal of Comparative Neurology, 295*, 580–623.

Itskov, P. M., Vinnik, E., & Diamond, M. E. (2011). Hippocampal representation of touch-guided behavior in rats: Persistent and independent traces of stimulus and reward location. *PLoS One, 6*(1), e16462. doi:10.1371/journal.pone.0016462.

Jeffery, K. J. (2011). Place cells, grid cells, attractors, and remapping. *Neural Plasticity, 182602.* doi:10.1155/2011/182602.

Jezek, K., Henriksen, E. J., Treves, A., Moser, E. I., & Moser, M.-B. (2011). Theta-paced flickering between place-cell maps in the hippocampus. *Nature, 278*, 246–249.

Jung, M. W., & McNaughton, B. L. (1993). Spatial selectivity of unit activity in the hippocampal granular layer. *Hippocampus, 3*, 165–182.

Kesner, R. P., Hunsaker, M. R., & Warthen, M. W. (2008). The CA3 subregion of the hippocampus is critical for episodic memory processing by means of relational encoding in rats. *Behavioral Neuroscience, 122*(6), 1217–1225.

Kesner, R. P., Morris, A. M., & Weeden, C. S. S. (2012). Spatial, temporal, and associative behavioral functions associated with different subregions of the hippocampus. In T. R. Zentall & E. A. Wasserman (Eds.), *Oxford Handbook of Comparative Cognition* (pp. 322–346). Oxford: Oxford University Press.

Kesner, R. P., & Rolls, E. T. (2001). Role of long term synaptic modification in short term memory. *Hippocampus, 11*, 240–250.

Kesner, R. P., & Rolls, E. T. (2015). A computational theory of hippocampal function, and tests of the theory: New developments. *Neuroscience and Biobehavioral Reviews, 48*, 92–147.

Killian, N. J., Jutras, M. J., & Buffalo, E. (2012). Spatial representation in the macaque entorhinal cortex. *Society for Neuroscience Abstracts,* 921.14.

Kohonen, T. (1977). *Associative Memory: A System Theoretical Approach.* New York: Springer.

Kolb, B., & Whishaw, I. Q. (2009). *Fundamentals of Human Neuropsychology* (6th ed.). New York: Worth.

Kondo, H., Lavenex, P., & Amaral, D. G. (2009). Intrinsic connections of the macaque monkey hippocampal formation: II. CA3 connections. *Journal of Comparative Neurology, 515*(3), 349–377.

Kraus, B. J., Brandon, M. P., Ronbinson, R. J., Connerney, M. A., Hasselmo, M. E., & Eichenbaum, H. (2013). Grid cells are time cells. *Society for Neuroscience Abstracts,* 769.719.

Kropff, E., & Treves, A. (2008). The emergence of grid cells: Intelligent design or just adaptation? *Hippocampus, 18*(12), 1256–1269.

Lassalle, J. M., Bataille, T., & Halley, H. (2000). Reversible inactivation of the hippocampal mossy fiber synapses in mice impairs spatial learning, but neither consolidation nor memory retrieval, in the Morris navigation task. *Neurobiology of Learning and Memory, 73*, 243–257.

Lavenex, P., & Amaral, D. G. (2000). Hippocampal-neocortical interaction: A hierarchy of associativity. *Hippocampus, 10*, 420–430.

Lavenex, P., Suzuki, W. A., & Amaral, D. G. (2004). Perirhinal and parahippocampal cortices of the macaque monkey: Intrinsic projections and interconnections. *Journal of Comparative Neurology, 472*, 371–394.

Lee, I., & Kesner, R. P. (2004). Encoding versus retrieval of spatial memory: Double dissociation between the dentate gyrus and the perforant path inputs into CA3 in the dorsal hippocampus. *Hippocampus, 14*, 66–76.

Lehn, H., Steffenach, H. A., van Strien, N. M., Veltman, D. J., Witter, M. P., & Haberg, A. K. (2009). A specific role of the human hippocampus in recall of temporal sequences. *Journal of Neuroscience, 29*(11), 3475–3484.

Leutgeb, J. K., Leutgeb, S., Moser, M. B., & Moser, E. I. (2007). Pattern separation in the dentate gyrus and CA3 of the hippocampus. *Science, 315*(5814), 961–966.

Leutgeb, S., & Leutgeb, J. K. (2007). Pattern separation, pattern completion, and new neuronal codes within a continuous CA3 map. *Learning & Memory, 14*(11), 745–757.

Leutgeb, S., Leutgeb, J. K., Barnes, C. A., Moser, E. I., McNaughton, B. L., & Moser, M. B. (2005). Independent codes for spatial and episodic memory in hippocampal neuronal ensembles. *Science, 309*, 619–623.

Lever, C., Burton, S., Jeewajee, A., O'Keefe, J., & Burgess, N. (2009). Boundary vector cells in the subiculum of the hippocampal formation. *Journal of Neuroscience, 29*(31), 9771–9777. doi:29/31/9771 [pii] 10.1523/jneurosci.1319-09.2009

Levy, W. B. (1989). A computational approach to hippocampal function. In R. D. Hawkins & G. H. Bower (Eds.), *Computational Models of Learning in Simple Neural Systems* (pp. 243–305). San Diego, CA: Academic Press.

Lisman, J., & Buzsaki, G. (2008). A neural coding scheme formed by the combined function of gamma and theta oscillations. *Schizophrenia Bulletin, 34*(5), 974–980.

Lisman, J., & Redish, A. D. (2009). Prediction, sequences and the hippocampus. *Philosophical Transactions of the Royal Society of London. Series B, Biological Sciences, 364*(1521), 1193–1201.

Loh, M., Rolls, E. T., & Deco, G. (2007). A dynamical systems hypothesis of schizophrenia. *PLoS Computational Biology, 3*(11), e228. doi:210.1371/journal.pcbi.0030228.

Ludvig, N., Tang, H. M., Gohil, B. C., & Botero, J. M. (2004). Detecting location-specific neuronal firing rate increases in the hippocampus of freely-moving monkeys. *Brain Research, 1014*(1–2), 97–109. doi:10.1016/j.brainres.2004.03.071S0006899304005657.

MacDonald, C. J., Lepage, K. Q., Eden, U. T., & Eichenbaum, H. (2011). Hippocampal "time cells" bridge the gap in memory for discontiguous events. *Neuron, 71*(4), 737–749. doi:S0896-6273(11)00609-X [pii] 10.1016/j.neuron.2011.07.012

Marr, D. (1971). Simple memory: A theory for archicortex. *Philosophical Transactions of the Royal Society of London. B, 262*, 23–81.

Matsumura, N., Nishijo, H., Tamura, R., Eifuku, S., Endo, S., & Ono, T. (1999). Spatial- and task-dependent neuronal responses during real and virtual translocation in the monkey hippocampal formation. *Journal of Neuroscience, 19*, 2318–2393.

McClelland, J. L., McNaughton, B. L., & O'Reilly, R. C. (1995). Why there are complementary learning systems in the hippocampus and neocortex: Insights from the successes and failures of connectionist models of learning and memory. *Psychological Review, 102*, 419–457.

McHugh, T. J., Jones, M. W., Quinn, J. J., Balthasar, N., Coppari, R., Elmquist, J. K., et al. (2007). Dentate gyrus

NMDA receptors mediate rapid pattern separation in the hippocampal network. *Science, 317*(5834), 94–99.

McNaughton, B. L. (1991). Associative pattern completion in hippocampal circuits: New evidence and new questions. *Brain Research. Brain Research Reviews, 16*, 193–220.

McNaughton, B. L., Barnes, C. A., Gerrard, J. L., Gothard, K., Jung, M. W., Knierim, J. J., et al. (1996). Deciphering the hippocampal polyglot: The hippocampus as a path integration system. *Journal of Experimental Biology, 199*, 173–185.

McNaughton, B. L., Barnes, C. A., & O'Keefe, J. (1983). The contributions of position, direction, and velocity to single unit activity in the hippocampus of freely-moving rats. *Experimental Brain Research, 52*, 41–49.

McNaughton, B. L., Battaglia, F. P., Jensen, O., Moser, E. I., & Moser, M.-B. (2006). Path integration and the neural basis of the "cognitive map." *Nature Reviews. Neuroscience, 7*, 663–678.

McNaughton, B. L., & Morris, R. G. M. (1987). Hippocampal synaptic enhancement and information storage within a distributed memory system. *Trends in Neurosciences, 10*(10), 408–415.

McNaughton, B. L., & Nadel, L. (1990). Hebb-Marr networks and the neurobiological representation of action in space. In M. A. Gluck & D. E. Rumelhart (Eds.), *Neuroscience and Connectionist Theory* (pp. 1–64). Hillsdale, NJ: Erlbaum.

Morris, R. G., & Frey, U. (1997). Hippocampal synaptic plasticity: Role in spatial learning or the automatic recording of attended experience? *Philosophical Transactions of the Royal Society of London. Series B, Biological Sciences, 352*(1360), 1489–1503.

Nakashiba, T., Cushman, J. D., Pelkey, K. A., Renaudineau, S., Buhl, D. L., McHugh, T. J., et al. (2012). Young dentate granule cells mediate pattern separation, whereas old granule cells facilitate pattern completion. *Cell, 149*(1), 188–201. doi:S0092–8674(12)00157–2 [pii] 10.1016/j.cell.2012.01.046

Nakazawa, K., McHugh, T. J., Wilson, M. A., & Tonegawa, S. (2004). NMDA receptors, place cells and hippocampal spatial memory. *Nature Reviews. Neuroscience, 5*, 361–372.

Nakazawa, K., Quirk, M. C., Chitwood, R. A., Watanabe, M., Yeckel, M. F., Sun, L. D., et al. (2002). Requirement for hippocampal CA3 NMDA receptors in associative memory recall. *Science, 297*, 211–218.

Nakazawa, K., Sun, L. D., Quirk, M. C., Rondi-Reig, L., Wilson, M. A., & Tonegawa, S. (2003). Hippocampal CA3 NMDA receptors are crucial for memory acquisition of one-time experience. *Neuron, 38*, 305–315.

Nasr, S., Liu, N., Devaney, K. J., Yue, X., Rajimehr, R., Ungerleider, L. G., & Tootell, R. B. (2011). Scene-selective cortical regions in human and nonhuman primates. *Journal of Neuroscience, 31*(39), 13771–13785. doi:10.1523/jneurosci.2792-11.2011 31/39/13771

Neunuebel, J. P., & Knierim, J. J. (2012). Spatial firing correlates of physiologically distinct cell types of the rat dentate gyrus. *Journal of Neuroscience, 32*(11), 3848–3858. doi:32/11/3848 [pii] 10.1523/jneurosci.6038-11.2012

O'Keefe, J. (1984). Spatial memory within and without the hippocampal system. In W. Seifert (Ed.), *Neurobiology of the Hippocampus* (pp. 375–403). London: Academic Press.

O'Keefe, J., & Dostrovsky, J. (1971). The hippocampus as a spatial map: Preliminary evidence from unit activity in the freely moving rat. *Brain Research. Brain Research Reviews, 34*, 171–175.

O'Keefe, J., & Nadel, L. (1978). *The Hippocampus as a Cognitive Map.* Oxford: Clarendon Press.

Pitkanen, A., Kelly, J. L., & Amaral, D. G. (2002). Projections from the lateral, basal, and accessory basal nuclei of the amygdala to the entorhinal cortex in the macaque monkey. *Hippocampus, 12*, 186–205.

Rajji, T., Chapman, D., Eichenbaum, H., & Greene, R. (2006). The role of CA3 hippocampal NMDA receptors in paired associate learning. *Journal of Neuroscience, 26*(3), 908–915.

Redish, A. D., & Touretzky, D. S. (1997). Cognitive maps beyond the hippocampus. *Hippocampus, 7*(1), 15–35. doi:10.1002/(sici)1098-1063(1997)7:1<15:aid-hipo3>3.0.co;2-6.

Robertson, R. G., Rolls, E. T., & Georges-François, P. (1998). Spatial view cells in the primate hippocampus: Effects of removal of view details. *Journal of Neurophysiology, 79*, 1145–1156.

Robertson, R. G., Rolls, E. T., Georges-François, P., & Panzeri, S. (1999). Head direction cells in the primate pre-subiculum. *Hippocampus, 9*, 206–219.

Rolls, E. T. (1987). Information representation, processing and storage in the brain: Analysis at the single neuron level. In J.-P. Changeux & M. Konishi (Eds.), *The Neural and Molecular Bases of Learning* (pp. 503–540). Chichester, UK: Wiley.

Rolls, E. T. (1989a). Functions of neuronal networks in the hippocampus and cerebral cortex in memory. In R. M. J. Cotterill (Ed.), *Models of Brain Function* (pp. 15–33). Cambridge: Cambridge University Press.

Rolls, E. T. (1989b). Functions of neuronal networks in the hippocampus and neocortex in memory. In J. H. Byrne & W. O. Berry (Eds.), *Neural Models of Plasticity: Experimental and Theoretical Approaches* (pp. 240–265). San Diego, CA: Academic Press.

Rolls, E. T. (1989c). The representation and storage of information in neuronal networks in the primate cerebral cortex and hippocampus. In R. Durbin, C. Miall, & G. Mitchison (Eds.), *The Computing Neuron* (pp. 125–159). Wokingham, UK: Addison-Wesley.

Rolls, E. T. (1990a). Functions of the primate hippocampus in spatial processing and memory. In D. S. Olton & R. P. Kesner (Eds.), *Neurobiology of Comparative Cognition* (pp. 339–362). Hillsdale, NJ: Erlbaum.

Rolls, E. T. (1990b). Theoretical and neurophysiological analysis of the functions of the primate hippocampus in memory. *Cold Spring Harbor Symposia on Quantitative Biology, 55*, 995–1006.

Rolls, E. T. (1995). A model of the operation of the hippocampus and entorhinal cortex in memory. *International Journal of Neural Systems, 6*, 51–70.

Rolls, E. T. (1996). A theory of hippocampal function in memory. *Hippocampus, 6*, 601–620.

Rolls, E. T. (1999). Spatial view cells and the representation of place in the primate hippocampus. *Hippocampus, 9*, 467–480.

Rolls, E. T. (2008). *Memory, Attention, and Decision-Making: A Unifying Computational Neuroscience Approach.* Oxford: Oxford University Press.

Rolls, E. T. (2009). The neurophysiology and computational mechanisms of object representation. In S. Dickinson, M.

Tarr, A. Leonardis, & B. Schiele (Eds.), *Object Categorization: Computer and Human Vision Perspectives* (pp. 257–287). Cambridge: Cambridge University Press.

Rolls, E. T. (2010). A computational theory of episodic memory formation in the hippocampus. *Behavioural Brain Research, 215*(2), 180–196. doi:10.1016/j.bbr.2010.03.027.

Rolls, E. T. (2012a). Advantages of dilution in the connectivity of attractor networks in the brain. *Biologically Inspired Cognitive Architectures, 1,* 44–54. doi:http://dx.doi.org/10.1016/j.bica.2012.03.003

Rolls, E. T. (2012b). Glutamate, obsessive-compulsive disorder, schizophrenia, and the stability of cortical attractor neuronal networks. *Pharmacology, Biochemistry, and Behavior, 100,* 736–751.

Rolls, E. T. (2012c). Invariant visual object and face recognition: Neural and computational bases, and a model, VisNet. *Frontiers in Computational Neuroscience, 6*(35), 1–70.

Rolls, E. T. (2013a). The mechanisms for pattern completion and pattern separation in the hippocampus. *Frontiers in Systems Neuroscience, 7,* 74.

Rolls, E. T. (2013b). A quantitative theory of the functions of the hippocampal CA3 network in memory. *Frontiers in Cellular Neuroscience, 7,* 98. doi:10.3389/fncel.2013.00098.

Rolls, E. T. (2014). *Emotion and Decision-Making Explained.* Oxford: Oxford University Press.

Rolls, E. T. (2016). *Cerebral Cortex: Principles of Operation.* Oxford: Oxford University Press.

Rolls, E. T., & Deco, G. (2002). *Computational Neuroscience of Vision.* Oxford: Oxford University Press.

Rolls, E. T., & Deco, G. (2010). *The Noisy Brain: Stochastic Dynamics as a Principle of Brain Function.* Oxford: Oxford University Press.

Rolls, E. T., & Deco, G. (2015). Stochastic cortical neurodynamics underlying the memory and cognitive changes in aging. *Neurobiology of Learning and Memory, 118,* 150–161.

Rolls, E. T., & Kesner, R. P. (2006). A computational theory of hippocampal function, and empirical tests of the theory. *Progress in Neurobiology, 79,* 1–48.

Rolls, E. T., Loh, M., & Deco, G. (2008a). An attractor hypothesis of obsessive-compulsive disorder. *European Journal of Neuroscience, 28,* 782–793.

Rolls, E. T., Loh, M., Deco, G., & Winterer, G. (2008b). Computational models of schizophrenia and dopamine modulation in the prefrontal cortex. *Nature Reviews. Neuroscience, 9,* 696–709.

Rolls, E. T., Robertson, R. G., & Georges-François, P. (1997a). Spatial view cells in the primate hippocampus. *European Journal of Neuroscience, 9,* 1789–1794.

Rolls, E. T., & Stringer, S. M. (2005). Spatial view cells in the hippocampus, and their idiothetic update based on place and head direction. *Neural Networks, 18,* 1229–1241.

Rolls, E. T., Stringer, S. M., & Elliot, T. (2006). Entorhinal cortex grid cells can map to hippocampal place cells by competitive learning. *Network, 17,* 447–465.

Rolls, E. T., Stringer, S. M., & Trappenberg, T. P. (2002). A unified model of spatial and episodic memory. *Proceedings. Biological Sciences, 269,* 1087–1093.

Rolls, E. T., & Treves, A. (1990). The relative advantages of sparse versus distributed encoding for associative neuronal networks in the brain. *Network, 1,* 407–421.

Rolls, E. T., & Treves, A. (1998). *Neural Networks and Brain Function.* Oxford: Oxford University Press.

Rolls, E. T., & Treves, A. (2011). The neuronal encoding of information in the brain. *Progress in Neurobiology, 95,* 448–490.

Rolls, E. T., Treves, A., Foster, D., & Perez-Vicente, C. (1997b). Simulation studies of the CA3 hippocampal subfield modelled as an attractor neural network. *Neural Networks, 10,* 1559–1569.

Rolls, E. T., Treves, A., Robertson, R. G., Georges-François, P., & Panzeri, S. (1998). Information about spatial view in an ensemble of primate hippocampal cells. *Journal of Neurophysiology, 79,* 1797–1813.

Rolls, E. T., Tromans, J., & Stringer, S. M. (2008c). Spatial scene representations formed by self-organizing learning in a hippocampal extension of the ventral visual system. *European Journal of Neuroscience, 28,* 2116–2127.

Rolls, E. T., & Webb, T. J. (2012). Cortical attractor network dynamics with diluted connectivity. *Brain Research. Brain Research Reviews, 1434,* 212–225.

Rolls, E. T., & Xiang, J.-Z. (2005). Reward-spatial view representations and learning in the hippocampus. *Journal of Neuroscience, 25,* 6167–6174.

Rolls, E. T., & Xiang, J.-Z. (2006). Spatial view cells in the primate hippocampus, and memory recall. *Reviews in the Neurosciences, 17,* 175–200.

Rolls, E. T., Xiang, J.-Z., & Franco, L. (2005). Object, space and object-space representations in the primate hippocampus. *Journal of Neurophysiology, 94,* 833–844.

Rumelhart, D. E. (Ed.). (1990). *Brain Style Computation: Learning and Generalization.* San Diego, CA: Academic Press.

Rumelhart, D. E., Hinton, G. E., & Williams, R. J. (1986). Learning internal representations by error propagation. In D. E. Rumelhart, J. L. McClelland, & T. P. R. Group (Eds.), *Parallel Distributed Processing: Explorations in the Microstructure of Cognition* (Vol. 1, pp. 318–362). Cambridge, MA: MIT Press.

Rumelhart, D. E., & Todd, P. M. (1993). *Learning and connectionist representations.* Paper presented at Attention and Performance XIV: Synergies in Experimental Psychology, Artificial Intelligence, and Cognitive Neuroscience.

Samsonovich, A., & McNaughton, B. L. (1997). Path integration and cognitive mapping in a continuous attractor neural network model. *Journal of Neuroscience, 17,* 5900–5920.

Schultz, S., & Rolls, E. T. (1999). Analysis of information transmission in the Schaffer collaterals. *Hippocampus, 9,* 582–598.

Scoville, W. B., & Milner, B. (1957). Loss of recent memory after bilateral hippocampal lesions. *Journal of Neurology, Neurosurgery, and Psychiatry, 20,* 11–21.

Shepherd, G., & Grillner, S. (Eds.). (2010). *Handbook of Brain Microcircuits.* New York: Oxford University Press.

Simmen, M. W., Treves, A., & Rolls, E. T. (1996). Pattern retrieval in threshold-linear associative nets. *Network, 7,* 109–122.

Squire, L. R., & Wixted, J. T. (2011). The cognitive neuroscience of human memory since H.M. *Annual Review of Neuroscience, 34,* 259–288. doi:10.1146/annurev-neuro-061010-113720.

Stringer, S. M., & Rolls, E. T. (2002). Invariant object recognition in the visual system with novel views of 3D objects. *Neural Computation, 14,* 2585–2596.

Stringer, S. M., Rolls, E. T., & Trappenberg, T. P. (2004). Self-organising continuous attractor networks with multiple activity packets, and the representation of space. *Neural Networks, 17,* 5–27.

Stringer, S. M., Rolls, E. T., & Trappenberg, T. P. (2005). Self-organizing continuous attractor network models of hippocampal spatial view cells. *Neurobiology of Learning and Memory, 83,* 79–92.

Stringer, S. M., Rolls, E. T., Trappenberg, T. P., & Araujo, I. E. T. (2002a). Self-organizing continuous attractor networks and path integration: Two-dimensional models of place cells. *Network, 13,* 429–446.

Stringer, S. M., Trappenberg, T. P., Rolls, E. T., & Araujo, I. E. T. (2002b). Self-organizing continuous attractor networks and path integration: One-dimensional models of head direction cells. *Network, 13,* 217–242.

Suzuki, W. A., & Amaral, D. G. (1994). Topographic organization of the reciprocal connections between the monkey entorhinal cortex and the perirhinal and parahippocampal cortices. *Journal of Neuroscience, 14,* 1856–1877.

Tabuchi, E., Mulder, A. B., & Wiener, S. I. (2003). Reward value invariant place responses and reward site associated activity in hippocampal neurons of behaving rats. *Hippocampus, 13,* 117–132.

Takeuchi, T., Duszkiewicz, A. J., & Morris, R. G. (2014). The synaptic plasticity and memory hypothesis: Encoding, storage and persistence. *Philosophical Transactions of the Royal Society of London. Series B, Biological Sciences, 369*(1633), 20130288. doi:10.1098/rstb.2013.0288.

Taube, J. S., Goodridge, J. P., Golob, E. J., Dudchenko, P. A., & Stackman, R. W. (1996). Processing the head direction signal: A review and commentary. *Brain Research Bulletin, 40,* 477–486.

Tonegawa, S., Nakazawa, K., & Wilson, M. A. (2003). Genetic neuroscience of mammalian learning and memory. *Philosophical Transactions of the Royal Society of London. Series B, Biological Sciences, 358,* 787–795.

Treves, A. (1990). Graded-response neurons and information encodings in autoassociative memories. *Physical Review A., 42,* 2418–2430.

Treves, A. (1995). Quantitative estimate of the information relayed by Schaffer collaterals. *Journal of Computational Neuroscience, 2,* 259–272.

Treves, A., & Rolls, E. T. (1991). What determines the capacity of autoassociative memories in the brain? *Network, 2,* 371–397.

Treves, A., & Rolls, E. T. (1992). Computational constraints suggest the need for two distinct input systems to the hippocampal CA3 network. *Hippocampus, 2,* 189–199.

Treves, A., & Rolls, E. T. (1994). A computational analysis of the role of the hippocampus in memory. *Hippocampus, 4,* 374–391.

van Haeften, T., Baks-te-Bulte, L., Goede, P. H., Wouterlood, F. G., & Witter, M. P. (2003). Morphological and numerical analysis of synaptic interactions between neurons in deep and superficial layers of the entorhinal cortex of the rat. *Hippocampus, 13,* 943–952.

Van Hoesen, G. W. (1982). The parahippocampal gyrus: New observations regarding its cortical connections in the monkey. *Trends in Neurosciences, 5,* 345–350.

Wang, S. H., & Morris, R. G. (2010). Hippocampal-neocortical interactions in memory formation, consolidation, and reconsolidation. *Annual Review of Psychology, 61,* 49–79, C41–44.

Wang, X. J. (2008). Decision making in recurrent neuronal circuits. *Neuron, 60*(2), 215–234.

Wiener, S. I., & Taube, J. S. (Eds.). (2005). *Head Direction Cells and the Neural Mechanisms of Spatial Orientation.* Cambridge, MA: MIT Press.

Wills, T. J., Lever, C., Cacucci, F., Burgess, N., & O'Keefe, J. (2005). Attractor dynamics in the hippocampal representation of the local environment. *Science, 308,* 873–876.

Witter, M. P. (1993). Organization of the entorhinal-hippocampal system: A review of current anatomical data. *Hippocampus, 3,* 33–44.

Witter, M. P. (2007). Intrinsic and extrinsic wiring of CA3: Indications for connectional heterogeneity. *Learning & Memory, 14*(11), 705–713.

Witter, M. P., Naber, P. A., van Haeften, T., Machielsen, W. C., Rombouts, S. A., Barkhof, F., et al. (2000). Cortico-hippocampal communication by way of parallel parahippocampal-subicular pathways. *Hippocampus, 10,* 398–410.

Wood, E. R., Dudchenko, P. A., & Eichenbaum, H. (1999). The global record of memory in hippocampal neuronal activity. *Nature, 397,* 613–616.

Zilli, E. A. (2012). Models of grid cell spatial firing published 2005–2011. *Frontiers in Neural Circuits, 6,* 16. doi:10.3389/fncir.2012.00016.

19 Saccades and Smooth Pursuit Eye Movements

ZIAD M. HAFED

19.1 Introduction

Vision, or the ability to use light impinging on the retina to interpret and interact with our environment, is a computationally challenging problem. In fact, the seemingly effortless ability to, say, infer the three-dimensional shape of an object from a two-dimensional retinal image is mathematically ill-posed (e.g., affording multiple rather than unique solutions) (Bertero, Poggio, and Torre, 1988). To solve such a problem, the human brain possesses several features that "regularize" vision and thus enable our successful reliance on this sensory modality. However, the ease with which we think we can see our environment masks even bigger challenges constantly faced by our visual system. One of these challenges has to do with the use of a foveated retina, and its associated brain machinery, in which a relatively tiny region of retinal image space (the fovea) receives preferential processing. While this specialization of processing allocates precious computational resources to images in the fovea, it also creates an incessant need to move the eyes, such that different regions of our environment are analyzed, at least sequentially, using powerful, high-resolution computational machinery. The purpose of this chapter is to understand how the brain generates and controls eye movements.

Humans employ two types of eye movements to voluntarily shift their gaze (figure 19.1). Saccades are rapid, jerk-like changes in eye position that quickly and accurately align the fovea with a new region in visual space. If an object is moving, smooth pursuit eye movements are utilized instead. These consist of continuous rotations of the eye such that images of moving objects are stabilized as much as possible, usually directly on the fovea. Both saccades and smooth pursuit may be thought of as closed-loop control systems in which an error signal is sensed and used to drive the eye muscles, through negative feedback, in order to minimize such error. Thus, models of saccades and smooth pursuit eye movements are not only important for understanding how the brain utilizes eye movements to perform visual processing, but, owing to the relative mechanical simplicity of the eye and its muscles, these models also form an ideal means to explore the principles of human motor control in general. In fact, the oculomotor system is one of the most well-studied motor systems in the brain, and discoveries in this system have often pioneered and informed subsequent results in skeletal motor systems.

In this chapter, we will learn about the principles governing saccade and smooth pursuit control. We will start with saccades, highlighting their kinematic properties and the underlying machinery that gives rise to such kinematics. Even though smooth pursuit eye movements appear fundamentally different from saccades, both types of eye movements share some control principles, and neuronal substrates, as we shall see. We will close this chapter by briefly discussing the control of a class of related eye movements, namely, the class of tiny eye movements that continue to occur even when we attempt to fix our gaze. These so-called fixational eye movements are less well-understood than saccades and smooth pursuit, but ongoing research suggests certain important similarities to their larger counterparts and shows that studying them can illuminate broader concepts related to oculomotor control in general.

It should be noted here that saccades and smooth pursuit constitute only a subset of all eye movements that we make. Other types of eye movements include visual or vestibular ocular reflexes, which act to stabilize gaze in the face of image motion (for visual ocular reflexes) or head motion (for vestibular ocular reflexes), or vergence eye movements, which allow us to shift gaze between objects at different distances from us. These eye movements are beyond the scope of this chapter, but they are described at length in other references (Carpenter, 1988). It should also be mentioned that the textbook of Goldberg and Wurtz (1989) includes a wealth of classic experimental data and models on the saccadic system, and it provides a valuable additional reference to this chapter.

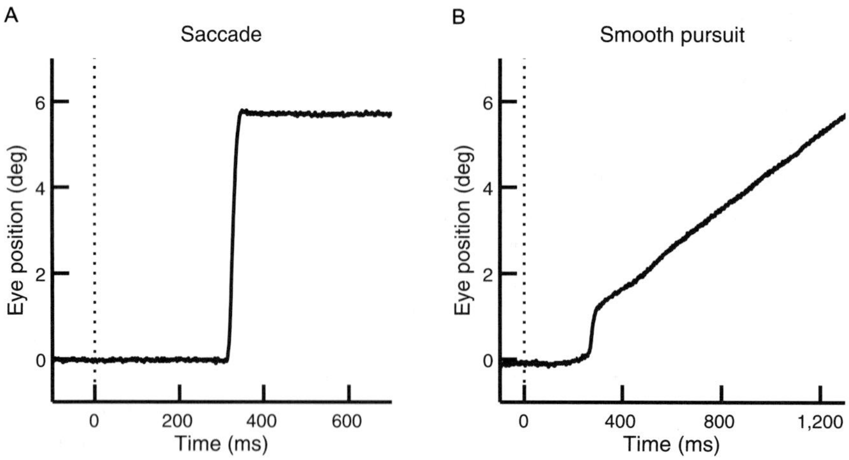

FIGURE 19.1 Example eye position traces showing a saccadic eye movement (A) and a smooth pursuit eye movement (B). Note how smooth pursuit initiation involves coordination with a small "catch-up" saccade to foveate the target.

19.2 Saccades

19.2.1 SACCADE KINEMATICS AND THE "EYE PLANT" Saccades are stereotypical eye movements. These eye movements are characterized by a rapid acceleration of the eye followed by a rapid deceleration to stop at the new intended orbital position (figure 19.1A). For a very brief moment in time during a saccade, the eye reaches a large peak velocity, up to ~800° per second (Bahill, Clark, and Stark, 1975). To appreciate the magnitude of this speed, a car traveling at ~110 km per hour across your visual field only 1 meter in front of you will span ~1,600–1,700° per second. Moreover, the relationships between saccade kinematic parameters, such as between movement amplitude and movement peak velocity, or between movement amplitude and movement duration, are lawful and are commonly referred to as the "saccade main sequence" (Bahill et al., 1975). Examples of the main sequence for peak eye velocity and for duration are illustrated in figure 19.2. As can be seen, both peak velocity and duration generally scale with saccade amplitude (for the shown amplitude ranges): the larger the movement being generated, the faster the eye moves and the longer it stays moving before reaching its target (but this relationship is not perfectly linear because it saturates for large movements).

The stereotypical nature of saccades suggests that these eye movements are controlled in a precise, repeatable manner, and for a large range of movement amplitudes. Using the language of control engineering, saccade control may specifically be thought of as an activation of eye muscles (the "actuator") to cause a change in eye position (the "plant"). The proper pattern of actuation involved in saccade control obviously depends on the mechanical response characteristics of the "plant." For example, if the eye acts as a sluggish overdamped system, then moving it will require very different patterns of actuator input than if it exhibits fast dynamics. Thus, the first step in understanding the control principles for saccades (as well as those for smooth pursuit) is to understand how the eye reacts to muscle innervation.

Because saccades appear ballistic, it may be thought that the eye plant has very rapid dynamics, and that a saccade (being a step-like change in eye position) is simply achieved by a step-like change in oculomotor neuron (abbreviated to MN in this chapter since we will not consider other types of motor neuron) activity driving the eye muscles. However, pioneering work by David A. Robinson (1964) has revealed that the passive components of the eye (e.g., the orbital fat, conjunctival tissue, noncontractile muscle components, etc.) in fact render the "oculomotor plant" severely overdamped (i.e., sluggish). In his work, Robinson measured the physical forces needed to move the eye, and based on these measurements, he modeled the eye plant mechanics as a fourth-order linear system. According to the frequency domain description of this system, it consisted of a single zero term and four pole terms (two real poles and one complex-pole pair). By assuming that the complex-pole pair contribution was small compared to other poles, the oculomotor plant could thus be modeled as follows (Optican and Miles, 1985; Robinson, 1964):

$$P(s) = \frac{(\tau_z s + 1)}{(\tau_1 s + 1)(\tau_2 s + 1)} \tag{19.1}$$

where the different τ parameters represent the time constants of the system (s is the Laplace transform variable; see box 19.1). Common values for τ_1 and τ_2 used

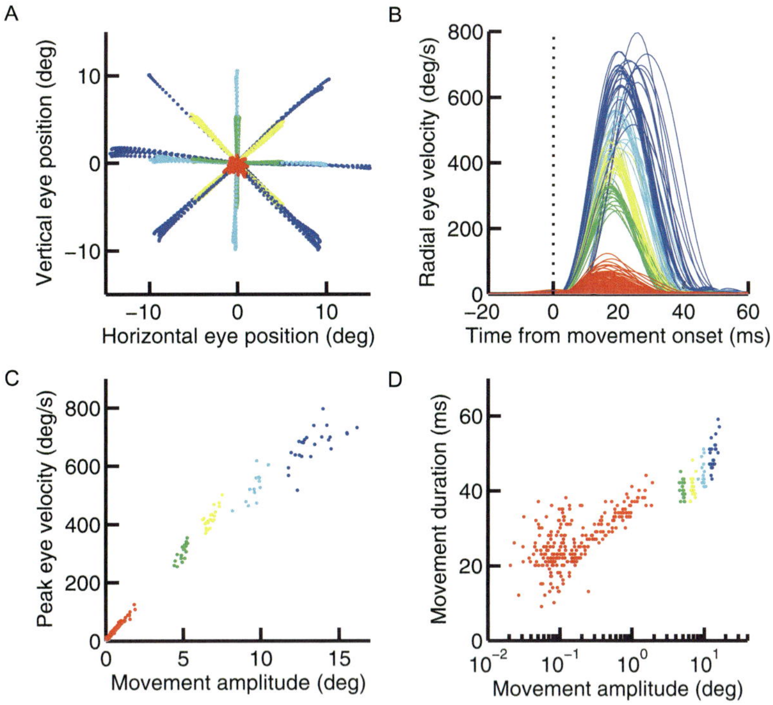

FIGURE 19.2 Saccades follow a "main sequence" relationship. (A) Eye position traces for a series of saccades (from the center out) of different directions and different amplitudes (different shades). (B) Eye velocity as a function of time from saccade onset for these same eye movements. As can be seen, the eye accelerates rapidly, reaches a large peak velocity, and then decelerates rapidly during a saccade. Larger eye movements reach larger peak velocities and last longer. (C) "Main sequence" relationship between saccade peak velocity and amplitude for the same saccades. Velocity increases with amplitudes but saturates for large saccades. (D) Similar relationship between duration and amplitude. The smallest movements (dots with amplitude < 1°) are a class of movements called "fixational saccades" or "microsaccades" because they occur during periods in which we attempt to prevent eye movements (section 19.4).

in saccade control models are ~12 ms and ~285 ms, respectively, which suggests that, left to its own devices, the eye would take up to several hundred milliseconds to reach a new stable position in response to a step-like input in MN activity. This is clearly much longer than normal saccade durations, and it is not validated by the known saccade kinematics (figure 19.2). Thus, achieving the ballistic nature of saccades with the sluggish nature of the eye plant requires a control input to the eye muscles that is more sophisticated than a simple step input. In fact, these observations led Robinson (1964) to predict that saccades are achieved not by a step change in MN activity but instead by a cascade of changes, in which an initial powerful "pulse" accelerates the eye and then a subsequent "slide" in activity occurs until it reaches the final "step" activity needed to maintain the eye at its new orbital position (figure 19.3). This control principle—of a strong initial "pulse" command before the subsequent final "step" command to keep the system at the new desired position—is often described by control engineers as "preemphasis," and it was validated through neurophysiological recordings from MNs (as we will see shortly).

Thus, saccades are stereotypical eye movements whose time course is much faster than the dynamics of the eye, suggesting that MNs innervate the eye muscles in a general "pulse–step" manner to overcome viscoelastic forces that render the eye plant sluggish. In what follows, we describe how the central nervous system arrives at such a pulse–step pattern of MN activity, first by describing the historical progress of the field and then the neuronal constraints on models provided by actual neurophysiological experiments.

19.2.2 SYSTEMS-LEVEL MODELS OF SACCADE GENERATION How are the different components of MN activity (e.g., "pulse" and "step" command) arrived at when generating a saccade? If one considers the "pulse" component as a signal derived from a "velocity" command for the eye plant, then the "step" component can simply be obtained through a "mathematical integration" of this velocity command. Thus, the oculomotor drive signal (MN activity) necessary for saccades may be achieved through a superposition of two related commands: a "velocity" command, and a second "position" command (an integral of velocity), which also serves

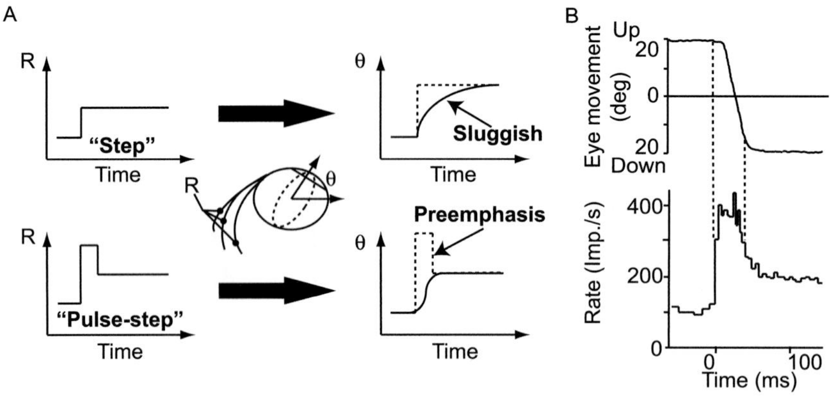

FIGURE 19.3 The eye acts as an overdamped mechanical system. (A) This figure shows the behavior of the "eye plant" for different patterns of hypothetical ocular muscle innervation (R). If a step input is provided to the ocular muscles (top), the eye will rotate (θ) to reflect the new state of contraction of the muscles, but it will do so in a sluggish manner due to its mechanical properties. Thus, to achieve rapid saccades (bottom), the muscle innervation consists of a brief, but strong, burst of muscle innervation to overcome the sluggish eye mechanics, followed by the pulse to maintain eye position. (B) Experimental recording from brainstem MNs validating the mechanism in A. The firing rate of the neuron exhibits a strong pulse and then a slide and final plateau to maintain eye position after saccade end. Panel B was adapted from (Robinson, 1970).

Box 19.1 Understanding Laplace Transform Notation

The Laplace transform is an integral transform that converts a function $f(t)$ to $F(s)$ using the following equation:

$$F(s) = \int_{0}^{\infty} f(t)e^{-st}dt \,.$$

The parameter s is a complex number ($\sigma + i\omega$) with σ and ω being real numbers.

The Laplace transform is similar to the Fourier transform in the sense that it decomposes a real function into a superposition of "moments." In the case of the Laplace transform, one speaks of converting f from being a function of time to a series of functions of complex angular frequency, s (in units of radians per second). As in the Fourier transform, the Laplace transform can be useful for simplifying the mathematical solution of linear differential equations because operations on f(t) can be rendered more simply when converted to "s-space." Specifically, linear differential equations can be converted to algebraic equations. For example, the derivative of f, df/dt, can be shown given the above equation to transform to $sF(s) - f(0)$ in the Laplace transform space. Thus, a differentiation is now converted to a multiplication. Similarly, convolution of two functions in time is equivalent to multiplying their two Laplace transforms.

In the text of this chapter, we often refer to the input–output relationship of a model using Laplace transforms. Consider, for example, a model whose output, $y(t)$, is defined by a differential equation like

$$dy/dt = -ay(t) + u(t)$$

where u(t) is the input to the model. One can take the Laplace transform of this equation to obtain

$$sY(s) - y(0) = -aY(s) + U(s)$$

The input–output relationship of the model would now be the equation given by Y(s)/U(s). Thus, when we use Laplace transforms to describe the input–output relationship of a model (as in, e.g., the eye plant model of figure 19.3 and equation 19.1), we are simply describing the differential equation that relates the input of this model to its output. If an input is now applied to the model, the model dynamics would follow the differential equation specified by the Laplace transform notation.

For more details, please see the textbook *Signals and Systems* by Oppenheim, Willsky, and Nawab (Prentice Hall, 1997).

the purpose of "gaze holding" in general. Note that variations of this scheme also exist in which a low-pass-filtered version of the velocity command is also input to the eye plant, and this serves as adding the "slide" component of the overall oculomotor command predicted by Robinson (1964) (Optican and Miles, 1985).

The amplitude and duration of the "velocity" command (the "pulse") determine the speed and duration of the eye movement, and the amplitude of the "position" command (the "step") determines the final position of the eye. This scheme of processing has served as the foundation for many systems-level models of saccade generation, with the different models differing in their predictions of how "control" of such velocity and position commands is achieved upstream of the MNs.

Early on, it was recognized that the trajectories of saccadic eye movements, and thus the velocity and position commands, cannot be controlled by continuous visual feedback (Young and Stark, 1963). This was so because visual correction of motor error is simply too slow to take effect during these extremely short-lived eye movements (sensory signals can take longer to traverse through the visual processing hierarchy in the brain than an actual saccade takes). This led to the belief that with regard to visual processing, saccade control is feedforward, that is, open loop (Young and Stark, 1963), and instead uses the position of a target in visual coordinates simply as an input specifying the desired saccade to a downstream control system that actually drives the eyes.

Having ruled out direct visual control of the eye velocity command and its integral, the position component, attention turned to how the so-called downstream saccade control system operates. Because of the ballistic nature of saccades, a class of early models hypothesized that the velocity and position commands may simply be fed forward to the MNs in a completely open-loop manner. Early examples of such open-loop models of saccade generation are presented in Robinson (1973) and Young and Stark (1963), and figure 19.4A shows their basic organization. In these models, a decision to initiate a saccade results in a sampling of the (visual) retinal error by a "pulse generator," which acts as the "controller" in this scheme. This controller programs the desired saccade, based on the input retinal error, and creates a "velocity" pulse that is fed to the eye plant both directly and through a so-called "neural integrator" (to provide the "position" command of figure 19.3A, bottom panels). Note that in this scheme, the "pulse generator" output is stereotyped, meaning that it is specified only once based on visual retinal error, and that it cannot be altered or modified once a saccade is initiated.

Later experimental observations revealed that saccade generation is, in fact, achieved in a closed-loop, feedback manner (i.e., that saccades are controlled precisely in flight). One important study of this kind was that presented in Jurgens, Becker, and Kornhuber (1981). In this work, the authors studied saccade accuracy as a function of variations (natural and drug induced) in eye movement velocity and duration, and they found that such accuracy was fairly independent of these variations. This meant that closed-loop control was the more plausible mechanism for saccade generation. For example, in the open-loop scheme of figure 19.4A, if the pulse generator is kept active for twice the duration that it normally is, the eye will continue moving for twice the duration and therefore overshoot its target. Other evidence supporting closed-loop theories of saccade control included the resumption of saccades in a fairly accurate manner after experimental interruption, say, by electrical stimulation (Keller, 1977).

The observations of accurate and/or corrected saccades despite variations in eye velocity/duration resulted in the idea of hybrid (feedforward + feedback) models that employed "local" feedback control (figure 19.4B) (the term "local" is used to differentiate this feedback from "global" visual feedback which would be needed to select the targets for subsequent saccades). As in the Hoff–Arbib model for reaching arm movements (compare chapter 20), one must note that "local" feedback here necessarily implies "estimated feedback" since only an estimate of eye position displacement is used (visual feedback is too slow to provide a real reference for how far the eye has moved). Thus, there is a delay in registering any inaccuracy in this estimate due to perturbations and so on (Hoff and Arbib, 1991). As a saccade proceeds, the local feedback pathway updates the estimate of eye displacement, resulting in a reduced "dynamic motor error," which then results in a smaller value of the "velocity" command and explains the deceleration of the eye toward saccade end.

Therefore, a hallmark of such models is that "desired eye displacement" is input to a negative-feedback control system whose function is to drive the eye plant by such a desired displacement. The feedforward path of this system is a high gain element ("pulse generator") that provides the eye "velocity" command. The ("local") feedback pathway may be described functionally as integrating this eye "velocity" command to provide an *estimate* of the "instantaneous displacement" of the eye during the saccade (although neurobiologically this feedback can be implemented differently; see the next section). The difference between the "desired displacement" and the "instantaneous displacement" is the "dynamic motor error," and it is this error that the control system aims to eliminate. Downstream of the local feedback loop, the rest of the model is similar to figure 19.4A: the "pulse–step" motor innervation needed for the saccade is compiled from the velocity command and input to the eye plant. Thus, for a given desired eye displacement, local feedback models use an estimate ("internal model" or "efference copy") of the instantaneous saccadic displacement to continuously adjust the eye velocity command during the saccade. Clearly, with local feedback, if the "pulse generator" output is

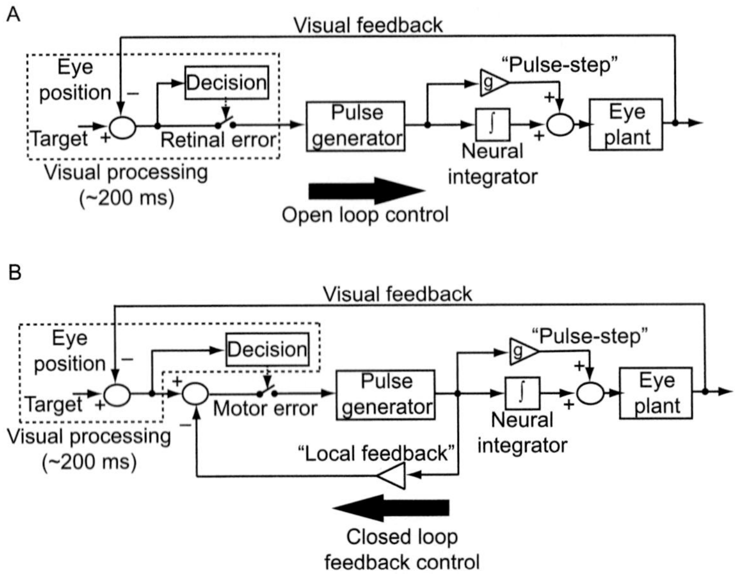

FIGURE 19.4 Systems-level models of saccade generation can replicate saccade trajectories. (A) An example of an "open-loop" model of saccade generation. When a "decision" is made to trigger a saccade, "retinal error" is sampled from the visual system. From there, the saccade proceeds in a completely open-loop manner: a high-gain "pulse generator" emits a strong pulse that is proportional to the size of the retinal error. This pulse is fed directly to the eye plant (through gain [g] element) as well as indirectly through a neural integrator, which will provide the signal to hold the eye in its new position. Thus, the output of the summing junction at the eye plant reflects the pulse–step component of MNs (figure 19.3). (B) An example of a closed-loop "local" feedback model of saccade generation. In this model, once a decision is made to trigger a saccade (the switch element in the figure), "retinal error" acts as a "desired eye displacement." The "pulse generator" still emits a "velocity" command to the MNs (both directly and through the neural integrator to give the "position command"). However, this signal is fed back through a second "local feedback" path, which functionally gives an estimate of "instantaneous eye displacement." Thus, the comparator at the input of the "pulse generator" estimates "dynamic motor error" and inputs this value to the "pulse generator." As the eye moves, the estimate of "current eye displacement" increases, thus decreasing the "dynamic motor error" and decreasing the velocity command.

artificially modified, saccades can still remain accurate despite variations in eye velocity and movement duration, so long as learning has appropriately calibrated the internal local feedback pathway. This is consistent with experimental evidence. Thus, systems-level models like the one in figure 19.4B capture the salient features of basic saccade control. We next turn to the constraints imposed on such models by the actual neurobiological implementation observed experimentally.

19.2.3 NEURAL IMPLEMENTATIONS OF SACCADE CONTROL The key concepts emerging from the systems-level models summarized above (figure 19.4B) are (1) the existence of local, negative feedback control; (2) the reliance on some form of "neural integration" (from a purely functional perspective) to convert velocity commands to position; and (3) the involvement of a decision or trigger signal to allow a saccade to begin or end (e.g., the "switch" in figure 19.4B, without which the oculomotor system would be unstable because it would instantaneously react to any nonzero error signal at any time). These concepts are sufficient to simulate realistic saccade trajectories. However, for the systems-level models to illuminate how the brain might actually control saccades, it is also necessary to ask whether the key concepts behind them possess a neural correlate in the activity of populations of neurons. For example, from a pure engineering perspective, controlling the saccade might intuitively involve reaching a desired eye position in the orbit. However, biologically, the largely retinotopic organization of the visual system suggests that the brain attempts to generate saccades by achieving a desired "eye displacement" from the current position. We thus now turn to the possible neural implementations of saccade control and how they put constraints on modeling such control.

The Brainstem Burst Generator Several groups of neurons distributed throughout the brainstem are believed to contribute to saccade control, and in a manner consistent with the general local, feedback scheme of figure 19.4B. These are often described aggregately as the "brainstem burst generator," and they comprise (1) the final MNs that innervate the eye muscles, as well as (2)

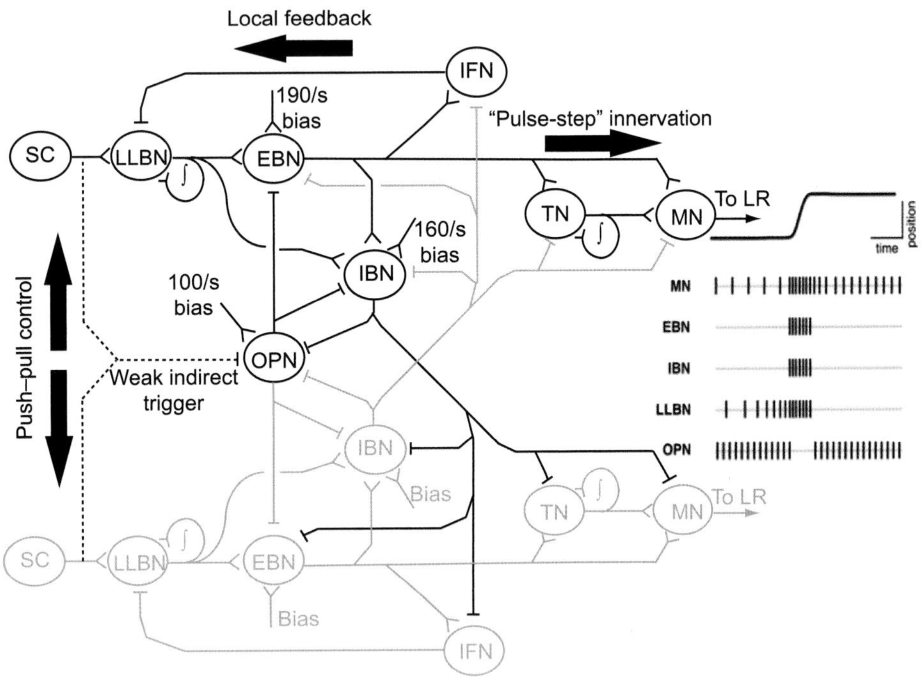

FIGURE 19.5 The brainstem burst generator implements several of the control principles of the model of figure 19.4B. This figure shows the burst generator model of Scudder (1988), which attempts to model the experimentally observed neural responses in the brainstem. Oculomotor neurons (MNs) show pulse–step discharge patterns as in the scheme of figure 19.3. Excitatory (EBN) and inhibitory (IBN) burst neurons implement the high-gain feedforward "velocity" command of the "pulse generator" in figure 19.4B. Long-lead burst neurons (LLBNs), whose discharge is similar to many superior colliculus (SC) neurons, may be part of the local feedback integrator as in Scudder (1988). Finally, omnipause neurons (OPNs) implement the switch or trigger mechanism and thus gate the occurrence of saccades. The scheme also highlights the bilateral push–pull nature of saccade control in the brainstem (other details as in the text). IFN, inhibitory feedback neuron; TN, tonic neuron; LR, lateral rectus muscle. The figure was adapted from Scudder (1988).

the premotor neurons that drive these motor neurons by compiling the pulse–step commands for saccades. Figure 19.5 summarizes the different classes of neurons in the brainstem burst generator and shows examples highlighting their discharge characteristics around the time of saccades. Some of these neurons appear directly correlated with specific control elements in the model of figure 19.4B, whereas the contribution of others to saccade control is less clear, as we learn below. In showing the neural discharges involved in saccade generation, figure 19.5 actually also shows the complete neural network model of Scudder (1988). This model is an example of how to simulate saccade generation and local feedback but in a manner consistent with the known neurobiological realization of it in the burst generator. In what follows, we describe the structure of this model in more detail.

At the bottom of the hierarchy, MNs innervating the eye muscles exhibit pulse–step (or, more generally, pulse–slide–step) discharges for saccades in their on-direction and pauses for saccades in their off-direction. For example, for a horizontal saccade in one direction, the motor neurons innervating the contracting muscle will burst to pull the eye in the desired direction, and the neurons innervating the antagonist (relaxing) muscle will pause to relax it. This suggests that oculomotor control in the brainstem occurs in a push–pull manner where "excitatory" drive to one set of muscles is complemented by "inhibitory" drive to the antagonist set. This observation is important because it has resulted in revised "push–pull" versions of the local, feedback models of figure 19.4B to more accurately reflect the brainstem circuitry (figure 19.5). After the saccade, MNs (for all muscles) maintain a tonic discharge rate, and this rate is different from that before the eye movement because such rate reflects the new eye position in the orbit. Thus, MNs implement the pulse–step command necessary to rapidly drive a saccade (figure 19.3), and existing models implement this command through a scheme like that in figure 19.5.

Medium-lead burst neurons (MLBNs), also referred to as excitatory burst neurons (EBNs) and inhibitory burst neurons (IBNs), are thought to provide the eye "velocity" command to the MNs, and they most directly correlate with the "pulse generator" in figure 19.4B.

These neurons exhibit a high-frequency burst of activity beginning immediately prior to saccade onset (by ~10 ms) and ending shortly before saccade end (Van Gisbergen, Robinson, and Gielen, 1981). For a given saccade, EBNs drive the MNs innervating the contracting muscle needed to pull the eye, and IBNs have cross projections that inhibit the (other) MNs of the antagonist eye muscle, which should relax. Thus, for a given MN, the (mathematical) difference between EBN excitation and IBN inhibition that it receives from the other side may be thought of as specifying the eye "velocity" command (Van Gisbergen et al., 1981), and this organization is again consistent with the idea of "push–pull" control evident at the MN level. As shown in figure 19.5 (Scudder, 1988), such a mathematical difference is also often the way that push–pull models of the brainstem burst generator are implemented. For example, instead of the pulse generator of figure 19.4B, a push–pull version might look like that included in the complete circuit of figure 19.5 (Scudder, 1988) or in the zoomed-in representation of figure 19.6 (based on Das, Gandhi, and Keller, 1995). In figure 19.6, the first stage is a low-pass filter to produce realistic saccades, u(t) is a rectification of the low-pass-filter output to 0 if omnipause neurons (OPNs) are active (see description of OPNs below), and $b_R(t)$ and $b_L(t)$ are the high-gain elements of the pulse generator, now implemented in push–pull fashion, given by the following:

$$b_R(t) = \begin{cases} b_m\left[1 - e^{\left(\frac{-(u(t)+e_0)}{b_k}\right)}\right] & u(t) > -e_0 \\ 0 & u(t) \leq -e_0 \end{cases}, \quad (19.2)$$

$$b_L(t) = \begin{cases} -b_m\left[1 - e^{\left(\frac{(u(t)-e_0)}{b_k}\right)}\right] & u(t) < e_0 \\ 0 & u(t) \geq e_0 \end{cases}. \quad (19.3)$$

In this formulation, the constants b_m, b_k, and e_0 influence the relationship between saccade amplitude (related to the "dynamic motor error" input to the burst neurons) and saccade peak velocity (related to the output of this pulse generator, the "velocity" command). Thus, according to this model, the nonlinear main sequence relationship between peak velocity and saccade amplitude is mainly caused by nonlinearities in the saccadic "pulse generator" (figure 19.2; but see section "The Superior Colliculus" below for a challenge to this idea). To summarize, EBNs and IBNs correspond to the "pulse generator" in figure 19.4B, with the specific implementation resembling that in figures 19.5 and 19.6 to reflect the push–pull nature of their influence on MNs.

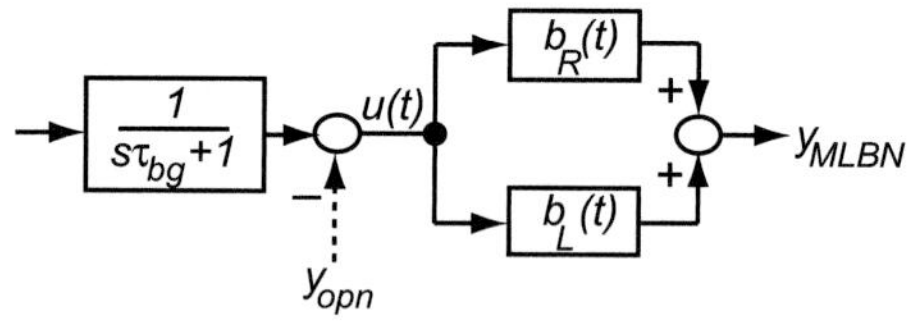

FIGURE 19.6 A revised model of the "pulse generator" component of figure 19.4B or figure 19.5, taking into account the push–pull nature of the "brainstem burst generator." In the depicted revised "pulse generator" scheme, "dynamic motor error" for horizontal saccades is input to both a right (R) and left (L) "burst generator." For a given saccade, equations 19.2, 19.3 indicate that one side "bursts" for contracting a muscle and the other "pauses" for relaxing the antagonist muscle. This scheme shows how the push–pull nature of oculomotor neurons and excitatory burst neurons/inhibitory burst neurons can be added to models of saccade control like those in figure 19.4B. MLBN, medium-lead burst neurons; opn, omnipause neurons; τ_{bg}, time constant of input filter.

The other major class of neurons in the brainstem burst generator is that of OPNs. These neurons are thought to correspond to the switch element present in the pulse generator of figures 19.4–19.6. OPNs fire at a constant tonic rate of ~100 spikes per second or more during fixation. Immediately before and during saccades of all sizes and directions, these neurons completely pause their activity. OPNs inhibit all burst neurons in the brainstem. Thus, these neurons' tonic activity acts to prevent or "gate" unwanted saccades during fixation. Without such a "gate," the saccade control system (e.g., figure 19.4B) would be unstable because it would instantaneously react to any nonzero motor error. When the OPNs pause, these neurons effectively trigger saccades by allowing the rest of the burst generator to drive MNs. OPNs are usually modeled as neuronal elements with a strong "bias" input, thought to reflect input from higher brain areas (including prefrontal cortex) inhibiting motor outputs (figure 19.5). These neurons also receive excitatory input from relatively large portions of the superior colliculus (SC), although the function of these projections has been the subject of debate, as we discuss later (see section 19.4).

The remaining brainstem neurons shown in figure 19.5 are the tonic neurons (TNs), the putative inhibitory feedback neurons (IFNs), and the long-lead burst neurons (LLBNs). The TNs exhibit tonic discharge that is dependent on eye position, and they are believed to receive the "velocity" command from EBNs and integrate it to provide the "position" command for the MNs (Scudder, Kaneko, and Fuchs, 2002). These neurons serve the purpose of "gaze holding" in general, without which we would have difficulty stabilizing our line of sight. However, the exact neural network implementation of how brainstem neurons (like TNs) might "integrate" in the mathematical sense is not very clear. In one proposal, Arnold and Robinson (1997) used physiologically plausible local, Hebbian-like learning algorithms to train an artificial neural network to perform integration. Other examples are provided in Nikitchenko and Koulakov (2008) and Scudder et al. (2002). In the goldfish oculomotor system, biophysical mechanisms of the neural integrator for eye position have been studied and modeled in great detail (Aksay, Gamkrelidze, Seung, Baker, and Tank, 2001; Aksay et al., 2007). However, regardless of implementation, in the particular scheme of figure 19.5, no explicit feedback "neural integrator" is included (or needed). Instead, the authors hypothesized the existence of a class of IFNs that innervate the LLBNs to implement the local feedback aspect of figure 19.4B. Neurophysiologically, these latter neurons exhibit early prelude activity long before saccade onset in addition to a saccade-related burst. These neurons may thus appear slightly more detached from direct saccade control because they are active long before movement onset. According to figure 19.5, these neurons help in the local feedback "integration" of figure 19.4B, as proposed by the model of Scudder (1988). These neurons may also be interneurons relaying signals from the SC to the brainstem burst generator (Scudder et al., 2002). The prelude activity of these neurons supports this view because many SC neurons also exhibit prelude activity. Thus, LLBNs are components of the brainstem burst generator, but assigning them to a specific control element in figure 19.4B is not entirely obvious.

In summary, the brainstem burst generator contains several classes of saccade-related neurons, and some of these classes appear directly correlated with the control concepts formalized by systems-level models (like that shown in figures 19.4B and 19.6). For example, EBNs and IBNs issue the velocity command, TNs are likely part of the neural integrator, and OPNs are part of the "switch" or trigger for saccades. That the function of some brainstem neurons (e.g., LLBNs) is less well-understood suggests that distributed neural implementations may deviate from proposals of "engineering control" while still supporting the desired function. For example, any given functional signal (like that of the feedback integrator of figure 19.4B or the "dynamic motor error" driving the pulse generator) may be an emergent property of a network of neurons (and their connections), few of which could explicitly exhibit such a signal in their own individual activity. In other words, the brain employs population codes to represent what, in a simple model, might be represented by activity in a single neuron, akin to the control variables of the original models. Some models involving the SC, described next, clearly demonstrate this idea.

The Superior Colliculus The SC contains a topographic representation of retinotopic space that is useful for identifying saccade endpoints (Lee, Rohrer, and Sparks, 1988; Robinson, 1972). Neurons located caudally in this structure represent peripheral eccentricities, and neurons located more rostrally represent foveal and parafoveal locations (figure 19.7A) (Robinson, 1972). For a typical saccade, neurons in the intermediate and deep layers of this structure exhibit firing rate increases prior to movement onset, a peak of discharge during the movement, and then firing rate decreases back to baseline after movement end (figure 19.7B) (Munoz and Wurtz, 1995; Wurtz and Optican, 1994). Not all neurons in the SC are activated for such a saccade, however, because individual neurons possess individual "movement fields" that are generally confined to a particular region of retinotopic space (figure 19.7C) (Goldberg and Wurtz, 1972; Lee et al., 1988; Wurtz and Goldberg, 1972; Wurtz and Optican, 1994): for a given saccade, a subset of SC neurons exhibits elevated activity, and there is a "hill" of activation in the SC map whose center of mass could be thought of as coding the saccade endpoint (Lee et al., 1988). Thus, the SC contains a spatial map of locations in which activity in one portion of this map rises for a saccade to the corresponding location in space, identifying the goal of the movement and contributing to its initiation and control. The SC is the dominant source of input to the brainstem burst generator (Scudder et al., 2002), but there is also cortical input (e.g., from the frontal eye fields; FEF).

The involvement of the SC in saccade generation raises several questions about the functional signals conveyed from this structure to the brainstem burst generator and how they are integrated for saccade control. First, signals in the SC are place-coded (e.g., a caudal neuron represents a larger eye movement) whereas the brainstem burst generator is rate-coded (e.g., a higher firing rate in a burst corresponds to a faster/larger eye movement). How is the spatial representation in the SC transformed into a temporal representation in the burst generator? Second, how does activity rise prior to saccade onset within the SC? Third, what is the functional significance of the drop in activity of SC neurons during saccades? Several models of saccade control by the SC have attempted to answer these and other questions, and these models demonstrate how distributed spatial representations can be combined computationally with downstream rate-coded representations in a biologically plausible manner.

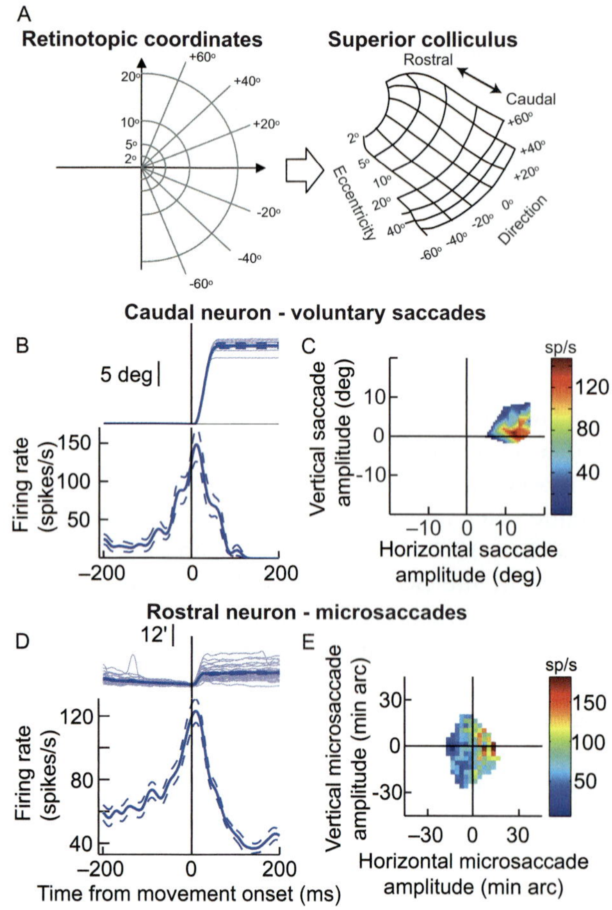

FIGURE 19.7 The superior colliculus (SC) contains a "motor map" for saccades of different amplitudes and directions. (A) Retinotopic representation of space in the SC. Neurons in the SC form a topographic representation with caudal neurons corresponding to peripheral eccentricities and rostral neurons corresponding to central locations. The figure shows iso-eccentricity and iso-direction lines to illustrate the anatomical mapping (right) of the visual coordinates on the left. Note that saccade amplitude is coded logarithmically on the SC map (as per equations 19.4, 19.5). (B) Activity of a sample SC neuron around saccades larger than approximately 10° directed to the neuron's preferred location in space. The top panel shows eye position aligned on saccade onset, and the bottom panel shows firing rate. The neuron exhibits movement-related activity patterns. (C) Activity of the same neuron in B during saccades to different locations. A subset of saccade endpoints is associated with higher activity than others, demonstrating the spatial specificity of the neuron, or its "movement field." (D) Activity of a sample rostral SC neuron for its preferred saccades, which happen to be very small ("microsaccades"). Note the great similarity to B except for the different eye movement amplitudes (12 min arc is equivalent to 0.2°). (E) Activity of the same neuron in D during microsaccades of different directions and amplitudes. Again, the neuron is similar to C except for the change in movement scale. That is, it possesses a "movement field" preference for a subset of microsaccades. This figure was taken with permission from (Hafed, 2011).

Van Gisbergen et al. (1987) proposed one of the earliest computational models of the SC and its role in saccade generation. In this model, the authors described an "ensemble coding" scheme of the SC, implementing the so-called spatial-to-temporal transformation. This scheme may be thought of as consisting of three stages: an "afferent mapping" stage, a "population activity profile" stage, and an "efferent mapping" stage. The afferent mapping function describes how locations in visual space map onto anatomical sites within the SC tissue, and it was modeled by Ottes, Van Gisbergen, and Eggermont (1986). Specifically, for a given location in visual space relative to the fovea (R,θ), the corresponding location in millimeters along the rostrocaudal (u) or mediolateral (v) extent of the SC tissue is as follows:

$$u(R,\theta) = B_u \cdot \ln\left(\frac{\sqrt{R^2 + 2AR\cos(\theta) + A^2}}{A}\right), \quad (19.4)$$

$$v(R,\theta) = B_v \cdot \arctan\left(\frac{R\sin(\theta)}{R\cos(\theta) + A}\right), \quad (19.5)$$

where B_u, B_v, and A are adjustable parameters. Using physiological data, such as the electrical stimulation results of Robinson (1972), these parameters are usually set to 1.4 mm, 1.8 mm/rad, and 3°, respectively. As can be seen from equations 19.4 and 19.5 and figure 19.7, the central 5° are overrepresented in the SC, much like foveal space is magnified in the primary visual cortex.

The next stage in describing how SC activity contributes to saccade control is to estimate the spatial extent of activity in this structure for a given saccade. As shown in figure 19.7, individual SC neurons have "movement fields" that can be broad. This means that for any given saccade, an extended population of neurons is active, and Ottes et al. (1986) found that using the afferent mapping function of equations 19.4 and 19.5, this population could simply be represented as a two-dimensional Gaussian. In other words, in terms of SC tissue space, the size of the active population for a given saccade is constant regardless of saccade size, meaning that the same number of neurons would be involved in generating any saccade.

According to Van Gisbergen et al. (1987) such "ensemble" or "population" code of saccade desired displacement is then projected to downstream premotor control circuitry through a final "efferent mapping" mechanism. This mechanism essentially specifies how each cell in the SC contributes to the total saccade, and it involves a linear vector summation of movement contributions from SC cells. These movement contributions vary across the SC map, with more caudal cells having larger contributions for saccades than more rostral counterparts. The model thus performs the spatial-to-temporal transformation in the strengths of projections from different spatial regions of the map to the brainstem burst generator. Note that going back to the scheme of figure 19.5, the model of Van Gisbergen et al. (1987) provides a role for the SC in driving the downstream local feedback circuitry for saccade control.

Later models of the SC's role in saccade generation can also be classified as closed-loop models employing some kind of local feedback mechanism. Some of these models suggest that the SC encodes the direction and amplitude of impending saccades and then projects this information to a feedback control circuit downstream from the SC as in figure 19.5 (Breznen and Gnadt, 1997; Jurgens, Becker, and Kornhuber, 1981; Moschovakis, 1994; Scudder, 1988; Walton, Sparks, and Gandhi, 2005). The SC is thus "outside" the local feedback loop in such models and simply projects the desired displacement to the brainstem burst generator.

Other models place the SC "inside" the local feedback loop. Being inside the loop means that the dynamic properties of saccades arise, at least in part, because of the SC. This, in turn, means that natural or drug-induced variations in SC activity can manifest themselves as variations in saccade dynamics. Such variations have been observed (Aizawa and Wurtz, 1998; Hikosaka and Wurtz, 1985, 1986; Lee et al., 1988). In addition, Waitzman et al. (1991) observed that some SC burst cells decrease their discharge during saccades in a manner that is consistent with the decreasing "dynamic motor error" (although such evidence is sometimes disputed based on electrical stimulation experiments that interrupt saccades). Thus, the SC may be part of the network performing the "comparator" function of figure 19.4B (i.e., the comparison between the "desired eye displacement" and instantaneous estimate of "current eye displacement"). We shall now learn about the specific details of a class of models in which spatial-to-temporal transformations, "comparator" functions, and SC burst generation can be an emergent property of distributed computational models of the SC and brainstem based on known anatomical connections. Even if more recent results dispute aspects such as computing dynamic motor error calculation within the SC, these details illuminate the main principles involved in computational models of saccade generation that include distributed spatial representations, such as those in the SC and FEF. After this description, we will close this section with a presentation of current models that directly challenge some of these classic observations.

An example model in which a distributed map can compute dynamic motor error is presented in figure 19.8 (Arai and Keller, 2005; Arai, Keller, and Edelman, 1994; Das, Keller, and Arai, 1996). In this model, two layers represent the SC: a superficial visual layer and a deeper motor layer. The visual layer has excitatory projections to the motor layer, and the latter has extensive lateral interconnections that are characterized by local excitation (including self-excitation) and more remote inhibition (which could be implemented externally from the SC; e.g., from the basal ganglia). The SC is proposed to directly control the brainstem burst generator, and it is believed to affect saccade triggering as well, through connections to model OPNs. Saccade dynamics are controlled by the dynamic properties of the model SC and burst-generator units, as well as by an external global inhibition (GI) signal. This signal inhibits each motor layer cell through shunting (divisive) inhibition, and it could represent inhibitory input from the basal ganglia. In such a collicular model, saccades are generated in the following manner. With GI being active, self-excitation of motor layer cells is very weak, making the motor layer act as a short-term memory. This means that this "memory" latches visual inputs from the visual layer for a finite amount of time (after which all activity dies because of lateral inhibition). When the GI signal eventually goes low, positive feedback due to the local excitation organization of the motor layer results in a "winner-take-all" (WTA) behavior in which the activity at the "remembered" location, when GI was high, increases dramatically and triggers a saccade. Thus, SC burst generation could arise through local excitation in a positive feedback manner (or WTA).

The SC motor layer in the model of figure 19.8 may be described by the following dynamic equations:

$$\tau_{srbn}\frac{dx_{srbni}}{dt} = -x_{srbni}(t) + \sum_j w_{ij}y_{srbnj} + y_{vi} - w_{fi}FB\,, \qquad (19.6)$$

$$y_{srbni}(t) = f_{srbn}(x_{srbni})\,, \qquad (19.7)$$

where x_{srbni} is the state of the ith *saccade-related* SC burst neuron (SRBN), τ_{srbn} is this neuron's time constant, y_{vi} is the output of the ith visual neuron, y_{srbnj} is the output of the jth SRBN, and *FB* is the dynamic feedback signal obtained from the brainstem burst generator (which could include both the velocity and position commands). w_{ij} and w_{fi} are connection strengths that are set to give rise to realistic motor layer activity, and f_{srbn} is a nonlinear output function given by

$$f_{srbn}(x) = \begin{cases} 0 & x < 0 \\ \dfrac{2}{1+e^{-0.8x}} - 1 & x \geq 0 \end{cases}. \qquad (19.8)$$

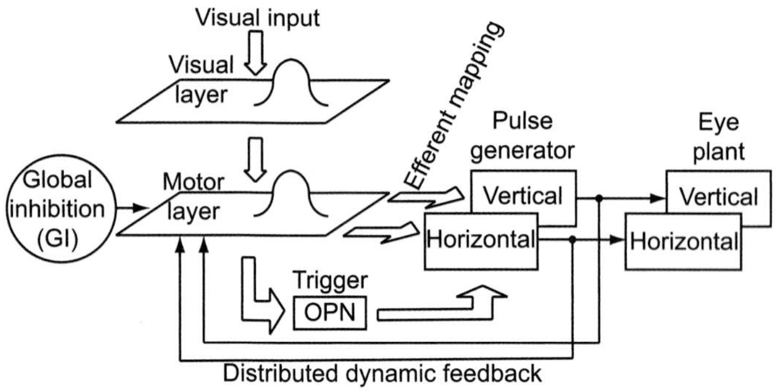

FIGURE 19.8 A neural network model demonstrates how the superior colliculus (SC) may participate in saccade control and implement functions in figure 19.4B using distributed neural elements (Arai and Keller, 2005; Arai et al., 1994; Das et al., 1996). In this model, the SC is represented by two layers, a visual layer and a motor layer. The motor layer exhibits saccade-related bursts that provide drive to the brainstem burst generator (downstream of the burst generator, the model is similar to figures 19.4B and 19.5 with a neural integrator driving the eye plant). In this scheme, the SC is inside the "local feedback" loop, receiving eye position or velocity feedback from the burst generator. Thus, comparison to figure 19.4B reveals that in this model, the SC provides "dynamic motor error" to the brainstem burst generator. Therefore, this model demonstrates how this quantity can be an emergent property of a distributed spatially coded network of neurons. The model SC neurons are modeled as in equations 19.6 and 19.7. OPN, omnipause neuron.

The global inhibitory input GI acts as shunting inhibition that reduces the self-excitation weight as follows:

$$w_{ii} = \frac{w0_{ii}}{1+GI}, \qquad (19.9)$$

where $w0_{ii}$ is the self-excitation weight when no inhibition is active and GI is an inhibition constant.

Thus, the model of figure 19.8 and equations 19.6–19.9 illustrates how a distributed spatial representation in the SC can be paired with the rate-coded brainstem burst generator to implement the spatial-to-temporal transformation. Moreover, the model provides an example of how a distributed network of model neurons (equations 19.6–19.9) can compute dynamic motor error during saccades using local feedback in an emergent manner. The model thus highlights some of the key features of neural implementations of saccade control in the brainstem.

This is not the only way the SC contributes to saccade control. In the most recent class of SC models (Goossens and van Opstal, 2006, 2012; van Opstal and Goossens, 2008a), a decidedly different, yet very fruitful, approach was taken. In particular, in all the models described so far, the burst generator downstream of the SC is assumed to be nonlinear (equations 19.2 and 19.3) and thus mediating the nonlinear main sequence relationship for saccades (figure 19.2). However, this assumption does not necessarily explain all aspects of saccade kinematics. For example, for oblique saccades having both a horizontal and a vertical eye displacement, it is known that the duration of the smaller eye displacement component is stretched in time to match the duration of the longer one (such "component stretching" ensures straight oblique saccades rather than curved ones). In the latest models of Goossens and van Opstal, the brainstem premotor circuitry downstream of the SC is presumed to be entirely linear, and the vectorial properties of saccade kinematics controlled by the SC. According to these models, every single SC spike during a saccade contributes a mini-saccade vector (according to its efferent mapping projection), and all such spikes are temporally integrated and fed to the downstream local feedback circuit. Thus, SC output during a saccade provides a dynamic vectorial velocity command. The difference from classic models, such as the one in Van Gisbergen et al. (1987), is that these newer models acknowledge the dynamic nature of SC movement fields and use this information as an instantaneous estimate of how far the saccade has progressed. A major strength of these models is that they directly link SC activity (and its dynamics) to saccade kinematic properties (figure 19.2) using a decidedly simple scheme, and completely linear premotor circuitry.

Other Areas Other cortical and subcortical areas also play a role in saccade generation. For example, the FEF, lateral intraparietal area (LIP), and supplementary eye fields (SEF) are all cortical areas that exhibit saccade-related modulations. Subcortically, influences of substantia nigra pars reticulata on SC activity for triggering saccades have also been described (Hikosaka and Wurtz, 1983a, 1983b, 1983c). The roles of these additional areas in saccade generation have been modeled computationally, as can be seen in more detail in chapter 22, "Integrative Functions of the Corticostriatal System." In general, these classes of models delve into higher-level aspects of saccade generation, such as the selection of the target for the saccade (Dominey and Arbib, 1992) (see box 19.2) or the detailed organization of the cortical "microcircuits" involved in saccade generation (Heinzle, Hepp, and Martin, 2007).

Cortical areas involved in saccades also often exhibit modulations that appear necessary for generating sequences of eye movements, as might happen when viewing a natural scene, and these have also been modeled (Dominey, Arbib, and Joseph, 1995; J. F. Mitchell and Zipser, 2003; J. Mitchell and Zipser, 2001).

Box 19.2 Target Selection, Decision-Making, and Attention for Eye Movements

Eye movement control is used to serve behavior. Thus, an important aspect of eye movement control is the selection of targets for eye movements. At any one moment in time, there are multitudes of potential targets in the scene to foveate—natural scenes in our surroundings are extremely rich in visual detail. Eye movements constitute a bottleneck for processing of such rich scenes because only one such target can be foveated by an eye movement at any one moment in time. Thus, functionally upstream of eye movement control, as described in the present chapter, the process of eye movement target selection plays an important role. The neural bases for target selection seem to be shared with the neural bases for eye movement control in many regards. For example, neural "maps" of visual space in lateral intraparietal area (LIP), frontal eye fields (FEF), and superior colliculus (SC) exhibit activity even without saccades or smooth pursuit. Such activity is thought to represent a real-time estimate of the "salience" and/or behavioral relevance of potential targets for eye movements. As some targets become more and more behaviorally relevant and salient, activity in these maps may rise enough to successfully select the targets and trigger eye movements. It is thus interesting that there is shared neural control for target selection and movement execution in this manner. This shared control has behavioral consequences. For example, studies of "visual attention" reveal that it is virtually impossible to attend to any scene location other than the location of the saccade target at the time of saccades. This suggests that activity in neural maps like FEF and SC rises in a winner-take-all manner at the time of saccade bursts, quenching the ability of any other scene location to attract our attention.

Such sequences of saccades create additional challenges for the visual system. Specifically, because visual areas are generally organized retinotopically, eye movements change the retinal images of objects in our environment even if the objects themselves remain stable. This creates a challenge for visual stability in the face of eye movements (see box 19.3). Experimental and computational studies related to this problem have revealed that visual areas might "predictively remap" their representations around the time of saccades (Duhamel, Colby, and Goldberg, 1992). In other words, a retinotopically organized neuron would momentarily alter the location of its "receptive field" in anticipation of the upcoming eye movement (a concept called "dynamic remapping"). Computationally, such malleability of visual receptive fields is an intriguing question, but it is one that is tractable neuronally, as was recently shown (Hamker, Zirnsak, Calow, and Lappe, 2008).

Box 19.3 The Influence of Eye Movements on Visual Perception

Because every eye movement moves images on the retina, the visual system must solve a form of "credit-assignment" problem whenever retinal images change: Is the change due to eye movement, or is it an external object that is moving in the environment? Thus, eye movements, despite being useful for allocating foveal resources of the brain to interesting scene locations, generate a serious challenge to vision. Without an ability to distinguish self-motion from motion of objects in the world, we would rarely have a moment of "perceptual stability" because our eyes (and bodies) are in constant motion. Computational models of the influences of eye movements on vision can help us understand how the brain solves this challenging problem. These models rely on the observation that there are distinct changes to vision around eye movements. For example, at the time of saccades, we are less sensitive to visual flashes than without saccades. In addition, we are poor at localizing brief flashes if the flashes appear peri-saccadically. Neural models of such changes in perception around the time of saccades suggest that there may be a corollary of saccade-related activity (say, in superior colliculus) that is broadcast to and modulates neural activity in visual "maps" of the brain. This modulation attempts to update the maps, since these maps are retinotopic (meaning that a stable object in world coordinates moves on these maps as the eye moves), in a predictive manner even before new visual information after saccades arrives at the maps. According to these models, such predictive dynamic remapping or gain modulation serves to bridge the transient period of eye movement to stabilize percepts. According to some of these models, brief transients around eye movements may be misperceived, as is observed in laboratory conditions with brief flashes. However, this is a decidedly small cost to perception compared to the benefit of stabilizing percepts across eye movements. After all, it is not too often in real life that very brief flashes happen in our natural environment right at the time of saccades.

19.2.4 Summary As we have seen, saccade generation models have spanned a range of approaches that attempt to explain how the brain controls these eye movements. The main hallmarks of these models are that brainstem nuclei play a prominent role in defining the dynamics of saccades. Models of saccade generation have evolved from ones exploring the constraints provided by the mechanical characteristics of the eye plant (e.g., figure 19.3) to ones studying the detailed neural network implementations of key control principles, such as local feedback or neural integration (e.g., figure 19.8).

Table 19.1 highlights the basic evolution of models of saccade control as a summary. The reader is also

Table 19.1
Saccade model evolution

Years	References	Concepts
1960s 1970s	Young and Stark (1963); Robinson (1973)	Open-loop saccade control
1975	Robinson (1975)	Local feedback (comparator computation: target position minus eye position), neural integrator
1981	Jurgens et al. (1981)	Resettable integrator in local feedback (comparator computation: desired minus current displacement instead of target position minus eye position)
1981	Van Gisbergen et al. (1981)	Bilateral, push–pull burst generator model
1988	Scudder (1988)	Desired displacement from the SC retinotopic map
1990s	Arai et al. (1994); Das et al. (1995); Das et al. (1996); Lefevre et al. (1998); Moschovakis (1994); Wurtz and Optican (1994)	SC involvement in feedback; spatial-to-temporal transformations; biomimetic distributed models; dynamic remapping (note the different kinds of models in this decade)
1990s	Lefevre et al. (1998)	Role of the cerebellum in local feedback
2000s	Walton et al. (2005)	Detailed role of the SC in saccade trajectory (e.g., saccadic curvature)
Current	Goossens and van Opstal (2006); van Opstal and Goossens (2008b)	"Dynamic vectorial velocity" command from the SC specifies saccade trajectory; brainstem burst generator completely linear

Note. SC, superior colliculus.

referred to a recent comprehensive review of the SC's motor functions (Gandhi and Katnani, 2011).

Looking forward, newer models of saccade generation need to address some of the remaining questions raised so far. For example, what is the role of other brain areas implicated in saccade generation? Of these areas, it has been suggested that the cerebellum plays a key role in the local feedback loop of saccade control (Lefevre, Quaia, and Optican, 1998). What are the details of such involvement, and how might it arise given that input from the brainstem burst generator to the cerebellum appears indirect at best (Scudder et al., 2002)? Future research on saccade generation at the neurophysiological, behavioral, and computational levels will answer these and other questions.

19.3 Smooth Pursuit

19.3.1 Smooth Pursuit Characteristics We now turn our attention to smooth pursuit eye movements, which constitute the second means of voluntary gaze shifting in humans. Smooth pursuit eye movements are relatively slow rotations of the eye to compensate for the motion of visual targets. These eye movements thus serve the purpose of stabilizing an image of a target on the retina, minimizing the detrimental effects of motion blur. Smooth pursuit eye movements are triggered by visual stimuli at latencies of ~100–200 ms, and they require a "percept" of motion to drive them. For example, if you attempt to move your eye smoothly from the left edge of this page to the right edge without explicitly experiencing motion, your eye movement will consist of a series of rightward saccades separated by periods of fixation rather than a smooth continuous trajectory. If, on the other hand, you move your index finger smoothly from left to right, your eye will track its position with a smooth movement. Thus, smooth pursuit eye movements are critically dependent on activity in areas of the brain related to motion perception. Moreover, because of the long neural delays that can occur when transducing visual motion signals to the oculomotor system, and because smooth pursuit deals with continuously moving objects, this class of eye movements also critically depends on extrapolation. If smooth pursuit waited for visual information, it would lag (predicted) changes in target trajectory by too long. Thus, mathematically, smooth pursuit deals with extrapolation of time series, as in classic engineering Wiener-type (Wiener, 1949) or Kalman-type (Kalman, 1960) filters.

Smooth pursuit eye movements are coordinated with saccades to track moving objects (e.g., figure 19.1B). Specifically, smooth pursuit eye movements can reach speeds of ~50°/s and are typically slower. So, if the "gain" of pursuit is not perfect (i.e., if eye velocity is only a fraction of target velocity), say, because the target is fast, then the eye will lag its target, resulting in both a "position" and "velocity" error. In this case, smooth pursuit alone may not be able to stabilize the target on the retina, and saccades (called "catch-up saccades") are triggered intermittently to foveate the moving target and eliminate the "position" error. Thus, smooth pursuit and saccades may be considered two parts of a single functional class of "tracking eye movements" in which coordination between the smooth and saccadic components is crucial.

19.3.2 Systems-Level Models of Smooth Pursuit

The mechanical characteristics of the eye plant are the same whether we generate saccades or smooth pursuit. Thus, the final motor pathway for smooth pursuit eye movements is that for saccades, and this is a feature of classic systems-level models of smooth pursuit eye movements. This suggests that just like saccades are driven by a pulse–step pattern of MN activity to overcome the sluggish dynamics of the eye (figure 19.3), smooth pursuit eye movements result from a pattern of MN activity consisting of a continuous change in firing rate (to reflect the continuous change in eye position during the movement) in addition to an increment to overcome the sluggish eye plant characteristics. Specifically, if the eye accelerates from fixation to a steady-state target velocity during smooth pursuit, the muscle force needed will involve a faster initial acceleration of force (than the actual eye movement), an observation made by Robinson (1965) when he repeated his classic saccade experiments we described above but for smooth pursuit. Specifically, figure 19.9 shows the muscle force needed to accelerate the eye to track a constant-speed target. As can be seen, the force accelerates faster than the eye. This is a second example of "preemphasis" that we saw exists for saccades as well (e.g., figure 19.3).

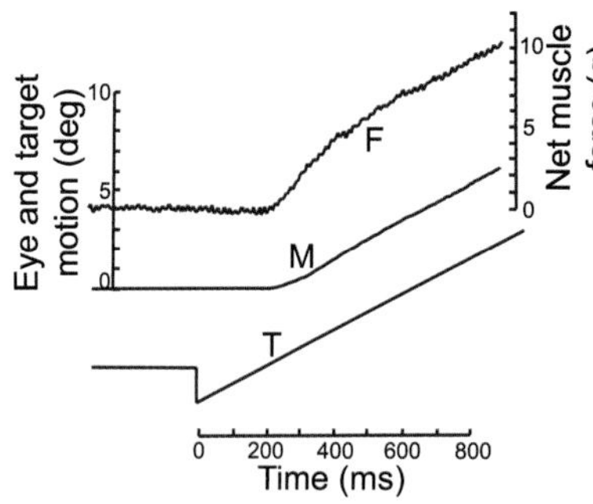

Figure 19.9 Muscle force for smooth pursuit eye movements also involves an additional component to compensate for the overdamped characteristics of the eye plant. In this figure, taken with permission from (Robinson, 1965), a target T steps at time zero and moves at constant speed in the opposite direction. The step is such that the eye movement (M) can track the target smoothly without saccades. The eye movement reaches a steady speed matched to the target. The measured net muscle force (F) on the horizontal muscles shows an initial acceleration of force that is faster than the actual acceleration of the eye, suggesting a form of "preemphasis" to overcome the mechanics of the eye, similar to the case of saccades. Note that this extra acceleration in the beginning is not necessarily to foveate the predicted target position, because under these experimental conditions, pursuit is initiated when the target hits the fovea. Rather, it seems to be a form of "preemphasis" to help move the sluggish eye plant.

Upstream of the final motor pathway, classic systems-level models of smooth pursuit involve a negative feedback loop to minimize the dynamic motor error during smooth pursuit eye movements, much like during saccades. However, in the case of smooth pursuit, these models emphasize the role of "velocity error" (the difference between "target velocity" and "eye velocity"), often referred to as "retinal slip velocity." In other words, because of the critical dependence of smooth pursuit on motion signals, classic models for generating these eye movements involve control over an error signal related to motion rather than to position (Krauzlis and Lisberger, 1989; Robinson, Gordon, and Gordon, 1986). An example of such models is shown in figure 19.10A (Krauzlis and Lisberger, 1989). In this model, "slip velocity," or the difference between target and eye velocity, is passed through a control element that generates an eye acceleration command (top pathway in the figure). This command is then integrated and input to a simplified version of the eye plant model (compare to equation 19.1 above) to drive the eye movement. In addition to "slip velocity," this model also differentiates this value to obtain "slip acceleration," or the rate of change of "slip velocity." Adding pursuit control signals based on "slip acceleration" helps make the pursuit response sensitive to either fast (middle pathway of figure 19.10A) or slow (bottom pathway of figure 19.10A) changes in target velocity.

However, models relying only on velocity suffer from a serious drawback: if the eye tracks a moving target perfectly, "retinal slip" will be zero (because eye velocity will match target velocity), thus nulling any "actuator" command to the eye plant. The eye will, by default, slow down. This is clearly not the case because smooth pursuit can continue even if eye velocity perfectly matches target velocity. Thus, subsequent systems-level models of smooth pursuit also involved a reconstruction of "target" velocity (based on an internal estimate of "eye velocity") added to "retinal slip" as the input to the control system, allowing this control system to continue driving the smooth eye movement even with perfect gain (i.e., even with zero "retinal slip") (Barnes, 2008; Barnes and Collins, 2008; Lisberger, Morris, and Tychsen, 1987; Nuding, Ono, Mustari, Buttner, and Glasauer, 2008; Robinson et al., 1986; Rothganger and Anastasio, 2009). Thus, just like saccades, the control of smooth pursuit involves the use of an "internal

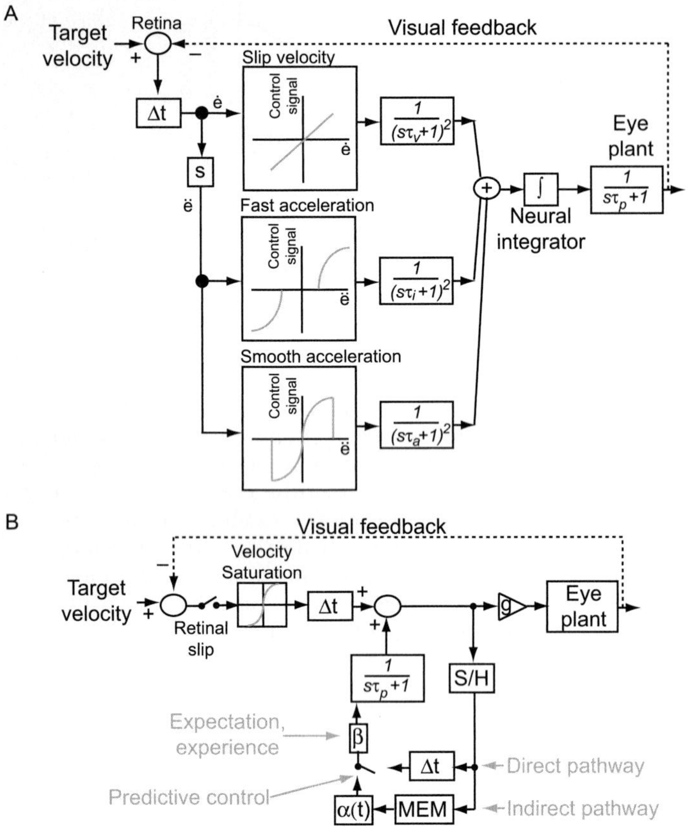

FIGURE 19.10 (A) Classic model of smooth pursuit relying on target velocity (Krauzlis and Lisberger, 1989). The top path shows that velocity error is passed through a control element and drives the eye with a velocity command. The middle and bottom paths use an estimate of acceleration error to allow the smooth pursuit system to be more responsive to changes in target velocity. (B) A more sophisticated model of smooth pursuit demonstrates how cognitive influences can manifest themselves in smooth pursuit. The model is from Barnes and Collins (2008) (figure replicated with permission) and it shows how "retinal slip" is added to an internal estimate of eye velocity to drive the smooth eye movement. The difference of this model from classic ones is that the "internal estimate" involves two alternate pathways. The "direct" pathway operates normally whereas the "indirect" pathway can explain why smooth pursuit can exhibit predictive capabilities.

model" of eye motion (or "efference copy") as an integral component of controlling the eye movement.

As mentioned earlier, because of neural delays in transducing visual motion signals to the oculomotor system, and because of the continuous nature of smooth pursuit, this class of eye movements also critically depends on predictive mechanisms. For example, if a target moves periodically in a predictable sinusoidal manner, smooth pursuit eventually will no longer lag the target and instead show a sinusoidal change in eye position that is almost perfectly synchronized with the target (i.e., with "zero phase lag"). Also, if a target of pursuit is transiently blanked (because of, e.g., occlusion), pursuit can proceed for some time and can show evidence of expectations and predictions depending on the behavioral state of the subject (Barnes, 2008; Barnes and Collins, 2008). How might this capability be achieved? The answer to this question again seems to lie in the use of the "internal" estimate of the ongoing eye movement. Specifically, according to smooth pursuit models, such use provides a means to implement "cognitive" predictions and expectations in the control of smooth pursuit at the level of the "internal" estimate, as shown explicitly in the model of figure 19.10B. In this model (Barnes and Collins, 2008), "retinal slip" is added (after a velocity saturation mechanism to reflect experimental observations) to an internal estimate of "target velocity" as mentioned above. However, in this particular model, the internal estimate is influenced by predictions and target expectations to explain experimental observations from manipulations such as transient target blanking during ongoing smooth pursuit (Barnes, 2008; Barnes and Collins, 2008). Thus, the model of figure 19.10B demonstrates how internal estimates of eye movements during smooth pursuit may be under the influence of higher cognitive processes related to expectations and predictions.

To summarize, we have seen how smooth pursuit commands need to rely not just on velocity error, but also on information about target/eye velocity, as well as mechanisms to extrapolate or predict target motion.

19.3.3 NEURAL IMPLEMENTATIONS OF SMOOTH PURSUIT CONTROL Neurophysiological studies of smooth pursuit control have identified brain circuits that presumably implement the different pathways highlighted above in systems-level models. A summary of these circuits is presented in figure 19.11. In panels A and B of this figure, adapted from Krauzlis (2004), the rhesus macaque brain is shown with the major areas involved in smooth pursuit control labeled. As can be seen, smooth pursuit recruits different areas to implement the different control pathways mentioned above and in figure 19.10. Panel C of figure 19.11, adapted from Tanaka and Lisberger (2001), demonstrates an example of how the roles of different areas in smooth pursuit came to be uncovered. The panel thus exemplifies the kind of neurophysiological experiment used to understand the neural control of smooth pursuit. In the particular experiment shown in the figure, electrical microstimulation was applied to the smooth-pursuit portion of FEF to mimic changes in eye movement gain during both fixation and smooth tracking. In what follows, we present more details about the neural

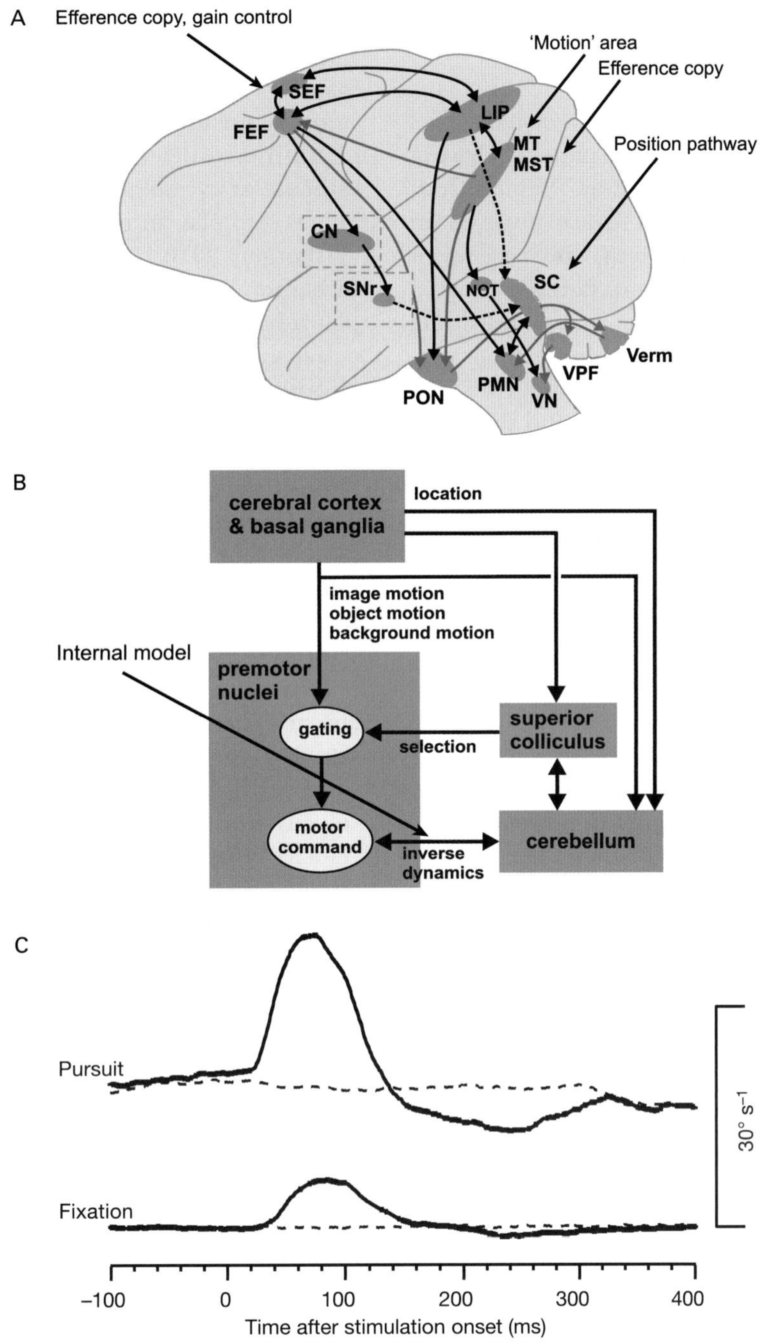

FIGURE 19.11 Neural implementation of smooth pursuit control. Panels A and B describe the mapping between systems-level and anatomical implementations of smooth pursuit control (using the rhesus macaque animal model). As can be seen, neural implementations cover motion pathways, position pathways, gain control, and extrapolation/prediction mechanisms described by systems-level models. The panels were adapted from Krauzlis (2004). Panel C demonstrates an example of the kind of neurophysiological experiments used to study neural control of smooth pursuit. In this example experiment, electrical microstimulation was applied to the frontal eye fields during fixation and smooth pursuit, to demonstrate the gain control aspect of the latter type of eye control. This panel was adapted from Tanaka and Lisberger (2001).

implementations of smooth pursuit control, and we relate them to systems-level models.

The "Motion Pathway" As suggested by systems-level models of smooth pursuit, the main pathway driving this eye movement is a "motion pathway." Signals from cortical middle temporal area (MT) are involved in estimating motion speed and direction, which are input to the smooth pursuit control system. Because an area like MT uses distributed retinotopic maps, individual neurons respond to motion signals within a small prescribed region in retinal space, and these neurons also exhibit specific tuning for motion speed and direction (Ilg and Thier, 2008; Lisberger, 2010; Lisberger et al., 1987; Thier and Ilg, 2005). This leads to the question of how the motion of a given pursuit target is decoded by the oculomotor system from the distributed representation of MT. This question is conceptually almost identical to the question of decoding distributed SC activity for saccades, which we alluded to above. In fact, a simple way to decode target motion from the population activity of MT is through "vector averaging" (Lisberger, 2010), similar to "population coding" in the SC. According to the vector averaging model of MT decoding, the motion of the smooth pursuit target may be described as the average motion of the active population of neurons in MT, weighted by the activity of each neuron. Thus, if the most active neuron prefers rightward motion, then this motion contributes a larger weight to the decoded signal than leftward motion, and so on.

Besides providing an estimate of "retinal slip" from MT, the cerebral cortex also contains several other areas involved in smooth pursuit. These include medial superior temporal area (MST), LIP, the smooth pursuit portion of FEF, and SEF (Fukushima, 2003; Ilg and Thier, 2008; Lisberger, 2010; Thier and Ilg, 2005). Each of these areas is believed to contain signals related to the classic systems-level models of smooth pursuit (e.g., figure 19.10B). For example, MST contains neurons that respond during pursuit of "imaginary" targets (Ilg, 2008). These neurons are thus thought to provide an extraretinal signal (or "internal estimate" of eye or target velocity) (as in figure 19.10B). Similarly, the FEF plays a crucial role in providing a "drive" signal for smooth pursuit, thus contributing to the "velocity" command that is input to the eye plant (Tanaka and Lisberger, 2001). In fact, the FEF is also thought to implement a "gain control" function during smooth pursuit (also highlighted in the model of figure 19.10B). Specifically, under normal circumstances, a change in visual motion signals can cause a much larger change in eye velocity if pursuit is ongoing as opposed to if the same change in visual motion happens during fixation. This suggests that the gain of a "visual-to-motor" transformation in the oculomotor system is enhanced during smooth pursuit, relative to fixation, probably because of the positive feedback associated with the internal estimate of eye velocity (e.g., figure 19.10), and FEF is believed to play an important role in such gain control (Lisberger, 2010; Lisberger et al., 1987; Tanaka and Lisberger, 2001) (figure 19.11C). Finally, SEF and LIP play a role in mediating the influence of cognitive processes such as attention, expectation, and prediction on smooth pursuit (Lisberger, 2010; Orban de Xivry and Lefevre, 2007).

Downstream of the cerebral cortex, neurons in the brainstem and cerebellum modulate descending motion-related signals to generate the required eye velocity command for smooth pursuit. Signals are relayed from the cerebral cortex to the cerebellum through brainstem pontine nuclei, which act as relay stations (Krauzlis, 2008; Lisberger, 2010). Areas MT and MST project to the ventral paraflocculus in the cerebellar cortex, and frontal areas send target and eye velocity signals to the dorsal vermis in the cerebellar cortex. The signals from the ventral paraflocculus are related to eye velocity and play an important role in providing the velocity command to the MNs (as well as the position command, through the neural integrator). On the other hand, the dorsal vermis may provide an acceleration as well as velocity signal to the brainstem. The need for an acceleration signal for smooth pursuit generation has, in fact, been proposed by some systems-level models of these movements, especially to account for the characteristics of smooth pursuit initiation in which the eye needs to accelerate from standstill (Krauzlis and Lisberger, 1989, 1991, 1994) (figure 19.10A).

Thus, the "motion pathway" for smooth pursuit involves a network of cortical–cerebellar areas encoding retinal slip, target velocity, eye velocity, and eye acceleration. There is a general similarity in the discharge characteristics of neurons in these areas, their activity being mainly correlated with "velocity" commands, suggesting that there are recurrent loops among neurons in these areas to implement smooth pursuit control in a multineuron, multinetwork fashion.

The "Position Pathway" and its Role in Coordination between Saccades and Smooth Pursuit Even though early descriptions of smooth pursuit eye movements have emphasized the role of the "motion pathway," computational models and experimental evidence have suggested that "position errors" (i.e., mismatches between the position of the eye and target) contribute to modifying smooth pursuit eye movements, in addition to their known role in affecting saccades. For example, the models of Lefevre and Galiana (1992) hypothesized that a position signal from the SC driving gaze saccades can under some circumstances cause a smooth change in eye position. This is an emergent property of these specific models, which fall in the general category of placing the SC inside the local feedback loop for saccades (see section 19.2 above). As for experimental evidence for an influence of "position error" on smooth pursuit eye movements, this includes the observation that stabilization of a target on a slightly eccentric position of the retina (resulting in a constant, but nonzero, "position error") consistently causes a smooth eye movement in the direction of the error. In addition, a brief flash during ongoing pursuit can cause deviations in the smooth eye movement that depend on the relative "position" between the retinal image of the flash and the fovea of the continuously moving eye. Thus, the "position pathway" commonly associated with saccades also influences smooth pursuit eye movements (Orban de Xivry and Lefevre, 2007).

Neural evidence for a role of the "position pathway" in smooth pursuit is also available (Krauzlis, 2004). For example, several of the brainstem areas involved in saccades exhibit modulations during smooth pursuit. At the level of the SC, neurons representing the retinotopic location of the pursuit goal (which is usually foveal and parafoveal) increase their activity when the goal enters their response fields. Thus, neurons in the SC exhibit modulations that may be viewed as a continuous, real-time representation of the retinotopic location of the pursuit goal (Hafed and Krauzlis, 2008; Krauzlis, Basso, and Wurtz, 1997). This representation may correlate directly with a "position error" signal for smooth pursuit. Moreover, manipulation of such SC activity using either electrical stimulation or reversible inactivation can influence smooth pursuit (Basso, Krauzlis, and Wurtz, 2000). Downstream of the SC, several premotor neurons in the brainstem burst generator exhibit modulations during smooth pursuit. For example, LLBNs show modulations in their tonic activity during smooth pursuit. Also, OPNs reduce their tonic activity during smooth tracking, and the reduction is proportional to the speed of the eye (Keller and Missal, 2003; Missal and Keller, 2002). In the limit, at saccadic speeds, OPNs pause completely. Thus, OPNs, which constitute the "gate" for allowing saccades to occur, may also be a critical component of regulating smooth pursuit.

All of these results suggest the existence of a "synergy" between saccades and smooth pursuit (Orban de Xivry and Lefevre, 2007). Although the functional significance of modulations in the saccade brainstem burst generator during smooth pursuit is still not clear, one possibility is that these modulations mediate the important ability to coordinate between both saccades and smooth pursuit during target "tracking." In other words, when we track a moving object, our visual and oculomotor systems attempt to optimize the stabilization of the target on the retina, and such an attempt requires the precise triggering of predictive "catch-up saccades" to overcome accumulated position and velocity errors. Thus, it could be that the target position and trajectory are read out by both the "motion" and "position" pathways for smooth pursuit and that brainstem and cerebellar neurons exhibiting shared modulations for both pursuit and saccades implement the decision to track the target either smoothly or saccadically (Krauzlis, 2008).

Is there explicit evidence for such a "synergy" between saccades and smooth pursuit? Lefevre and colleagues have shown this to be the case through several careful measurements of the coordination between the two types of eye movements. For example, if saccades and smooth pursuit are indeed coordinated precisely by the oculomotor system, then there should be a lawful relationship between when catch-up saccades are triggered during tracking and the instantaneous state of the eye "position" and "velocity" errors. In fact, Lefevre and colleagues (de Brouwer, Yuksel, Blohm, Missal, and Lefevre, 2002) found that a critical timing parameter can predict with high fidelity whether a catch-up saccade is triggered or not. This timing parameter is referred to as the "eye crossing time," or the time it would take the eye to intercept its target if the current smooth pursuit eye movement were to persist. This time may be estimated as the ratio between position error (PE) and retinal slip (RS) as follows:

$$T_{XE} = -\frac{PE}{RS}. \tag{19.10}$$

The logic of this eye crossing time parameter is summarized in figure 19.12. In this figure, target trajectory in space is given by the dashed line, and eye trajectory is given by the solid black line. The gray line is the extrapolation of eye trajectory if the current smooth pursuit eye movement were to persist. Panel A highlights the definition of the eye crossing time as the time it would take the eye to intercept the target with the current smooth movement. Panels B–D demonstrate how the value of this eye crossing time may influence whether a catch-up saccade (thick black) is triggered or not. If the eye crossing time is small enough (as in D), then the eye will intercept the target quickly and a catch-up saccade is not needed. If, on the other hand, the time is large or negative (indicating that the eye has

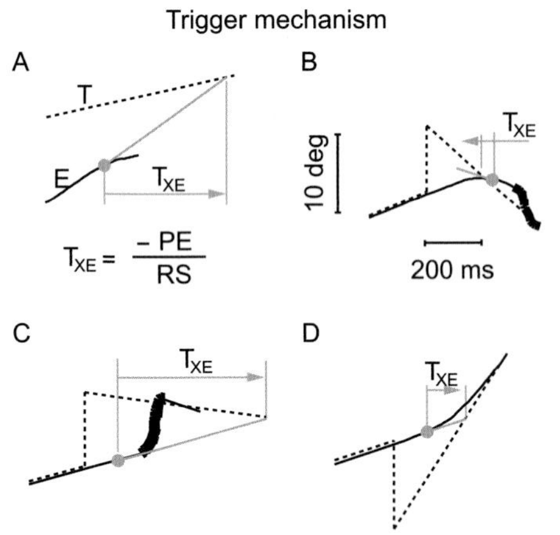

FIGURE 19.12 Triggering of catch-up saccades during ongoing pursuit may be related to how long it takes pursuit to intercept the target. (A) The definition of "eye crossing time" (T_{XE}) as the time it would take current ongoing pursuit to intercept the target. T is target trajectory, E is eye trajectory, and the gray line is the extrapolation of eye trajectory if current pursuit persists. The eye crossing time is the time it would take the gray trajectory to intercept the target. (B, C, and D) Different examples of eye crossing time. In B and C, a catch-up saccade (thick line) is triggered. However, in D, the eye crossing time is small enough such that no saccade is needed to intercept the target. This panel is taken with permission from Orban de Xivry and Lefevre (2007).

overshot the target), a catch-up saccade may be needed. With this simple parameter, these authors identified the range of eye crossing times that would be expected to trigger a catch-up saccade during target tracking and those that would not.

Thus, the shared neural pathways for saccades and smooth pursuit (Krauzlis, 2004), and the involvement of the "position pathway" in smooth pursuit, may play an important role in the necessary coordination between smooth pursuit and saccades in target tracking.

Other Slow Eye Movements As mentioned earlier, the eye can also move smoothly under other conditions, such as to stabilize gaze on a stationary object during head movements or to stabilize images of the entire scene when they move (as might happen if a person is sitting on a moving train and looking out the window). The mechanisms for some of these eye movements are related to the mechanisms of smooth pursuit. For example, rapid image stabilization reflexes rely on readout of activity from areas MT and MST describing global scene motion (Masson and Perrinet, 2012), and vestibular signals about head movement are ultimately relayed to the MNs to stabilize gaze. The neural coding and control principles related to these other eye movements are closely related to those described here, but they are beyond the scope of this chapter.

19.4 *Fixational Eye Movements*

Having discussed saccades and smooth pursuit, we shall now close this chapter by learning about a relatively less well-understood class of related eye movements and demonstrating how investigating them can help clarify some outstanding issues in oculomotor control, particularly about saccades.

19.4.1 CHARACTERISTICS OF FIXATIONAL EYE MOVEMENTS Fixational eye movements are small-amplitude displacements of the eye that paradoxically occur when we attempt to fix our gaze. These eye movements come in two main varieties: a slow ocular drift (sometimes referred to as "slow control") interspersed with brief step-like, saccadic displacements called "microsaccades." Drift has typical speeds of ~15 minutes of arc per second (0.25 °/s) and can last for several hundred milliseconds, only to be interrupted by microsaccades. Microsaccades are typically smaller than 15 minutes of arc in amplitude (0.25°), and they appear like scaled-down versions of larger saccades. Microsaccades, in fact, fall on the small-amplitude extension of the same main sequence relationships for saccades (see, e.g., figure 19.2). Microsaccades share other characteristics with larger saccades. For example, there is a correlation between microsaccades and visual attention (Hafed and Clark, 2002), or the selective processing of visual stimuli by the brain, consistent with a tight link between allocating attention and generating saccades in general. Moreover, microsaccades interact with slow drifts (Chen and Hafed, 2013; Cherici, Kuang, Poletti, and Rucci, 2012) and ocular following responses (Chen and Hafed, 2013), much like larger saccades do (Bahill, Hsu, and Stark, 1978; Kawano and Miles, 1986).

Less is known about the control of fixational eye movements than of larger ones. However, learning about such control can illuminate broader issues about both the oculomotor as well as visual systems. To illustrate this, we discuss below the specific case of microsaccade generation, because microsaccades are currently the component of fixational eye movements that is better understood in terms of both neural mechanisms and models of control.

19.4.2 MICROSACCADE GENERATION The available evidence supports the idea that the mechanisms for microsaccade generation may be fundamentally the same as those for larger saccades (Hafed, 2011), and this is consistent with the kinematic similarities of these two kinds of movements. For example, in one seminal paper describing the premotor and motor nuclei in the brainstem (described above for saccades), Van Gisbergen and colleagues (1981) recorded the activity of some neurons during some microsaccades. For their sample neurons, these authors found that MNs exhibit pulse–step patterns of discharge for microsaccades, just like for saccades. Moreover, these authors noted that premotor burst generator neurons (e.g., EBNs, IBNs, and OPNs) are also modulated during microsaccades, suggesting that the final pathway for saccade generation, consisting of the burst generator and the MN pulse–step command, is also responsible for microsaccade generation. These authors then simulated the generation of microsaccades using a brainstem burst generation model that is similar (in general terms) to that of figure 19.4B, demonstrating that microsaccades could in principle be generated by the brainstem burst generator. Similar evidence was also provided by a more recent model of microsaccade generation incorporating many of the same elements of saccade generation models described earlier in this chapter (Inagaki, Hirata, and Usui, 2011).

Other recent work is consistent with a similarity between microsaccade and saccade generation. At the level of the SC, it was found that the distributed spatial representation present in this structure (figure 19.7) may extend down to the smallest detectable microsaccades (Hafed, Goffart, and Krauzlis, 2009). Specifically, some neurons in the rostral portion of the SC, representing foveal regions of space, exhibit movement-related discharge patterns for microsaccades that are almost identical to those observed for larger saccades (figure 19.7D). That is, for each such neuron, there exists a set of microsaccade directions and amplitudes for which the neuron increases its firing rate in the period prior to and during microsaccades. Moreover, the neuron is specific in terms of which microsaccades cause such increases and, thus, possesses a "movement field" like caudal neurons do for larger saccades (figure 19.7E). So, prior to a given microsaccade, a hill of activity appears in the rostral (or foveal) SC that is centered on the neurons coding the movement amplitude and direction. Eventually, the activity builds up and a microsaccade is triggered. Thus, SC involvement in microsaccade generation is consistent with its involvement in saccade generation.

Hafed and colleagues (Hafed, 2011; Hafed, Goffart, and Krauzlis, 2008; Hafed et al., 2009; Hafed, Lovejoy, and Krauzlis, 2013) provided a possible model of the role of the SC in microsaccade generation. Their motivation was to specifically reconcile apparently contradictory findings about the role of the rostral (or foveal) portion of the SC map. Because rostral SC neurons exhibit tonic activity during fixation and pause for (some) large saccades (Munoz and Wurtz, 1993), the SC has often been described as issuing a "fixation" motor command from a putative "fixation zone" in its most rostral (or foveal) portion. According to this model, when the rostral zone is active, fixation is enforced; when it pauses, saccades are allowed. However, several pieces of experimental evidence, including the microsaccade results above, suggest that this role underestimates what the rostral SC does. For example, it is the same so-called "fixation" neurons that exhibit microsaccade-related increases in activity (Hafed et al., 2009). Clearly, these are all movements of the eye (i.e., breaks from fixation), but they are associated with increases in these neurons' activity rather than pauses. Second, the SC neurons that send excitatory projections to OPNs (see section 19.2 above) appear to be saccade related (i.e., they increase their activity for saccades when the OPNs actually pause) rather than fixation related (Gandhi and Keller, 1997). Third, the SC projections to OPNs arrive from a fairly large portion of the SC map (not just its most rostral portion) (Gandhi and Keller, 1997). Finally, rostral SC neurons are also modulated during smooth pursuit (see section 19.3 above).

How might these disparate observations be reconciled? One possibility is that low-frequency activity (i.e., not peri-saccadic bursts) in the rostral (as well as caudal) SC reflects the behavioral relevance of spatial locations. During any of of fixation, microsaccades, slightly larger saccades, or smooth pursuit eye movements, the most behaviorally relevant spatial locations are the foveal/parafoveal locations of the movement goal. Thus, low-frequency (nonburst) activity in the SC is expected to be predominantly located in the rostral portions of the map for each of these disparate gaze behaviors. Thus, according to this model, rostral SC activity contributes to microsaccade generation in particular, and eye movements (including fixation) in general, by defining the goal location for gaze position (figure 19.13), consistent with the putative role of the rest of the map. This would be consistent with the more general role of the SC as a potential salience map of behaviorally relevant locations, which can be used in natural viewing to select upcoming targets for foveation.

The hypothesis of figure 19.13 about the role of the rostral SC is worth investigating more in the future because it makes predictions about the functional role of the excitatory projections from the SC to OPNs that we alluded to earlier (figure 19.5). Because OPNs are tonically active during fixation and only turn off during saccades, evidence of excitatory projections from the

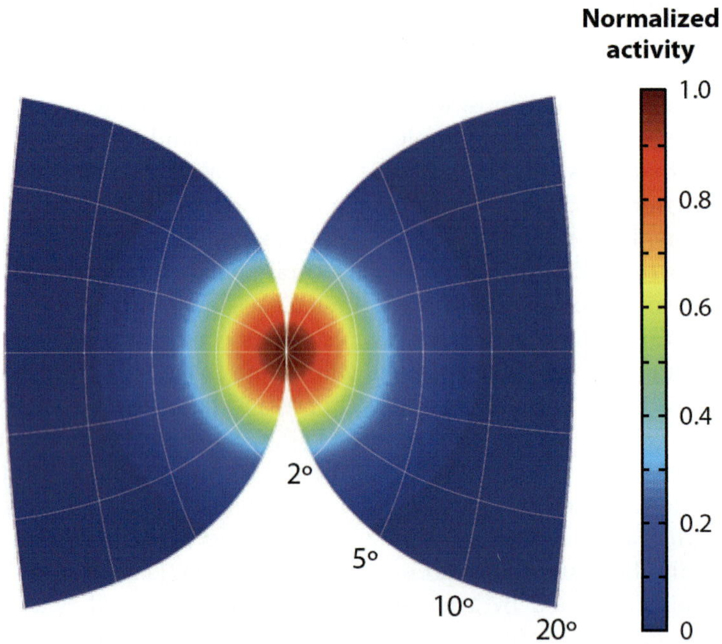

FIGURE 19.13 Investigating microsaccade generation by the superior colliculus (SC) can help clarify the role of the rostral (foveal) portion of this structure in gaze shifting and gaze fixation. The figure shows a bilateral model of the SC with the pseudocolor providing an estimate of ongoing activity during maintained fixation. According to the model of Hafed and colleagues (Hafed et al., 2008, 2009), such activity is balanced across the left and right visual fields. If activity starts to rise for a given microsaccade or small saccade, the profile shown in this figure will get skewed, resulting in one SC side's becoming more active than the other. This imbalance may be the trigger to pause omnipause neurons (OPNs) and allow eye movement. The advantage of this model is that it is consistent with experimental observations that the rostral SC remains active while OPNs pause for some eye movements.

SC to OPNs suggests that the SC in general (at least large parts of it) contributes to the maintenance of fixation, in addition to its putative contribution to the generation of saccades. How does this contribution take place? One possibility is that the contribution of the SC to fixation may not be in the form of an explicit motor command to fixate from its most rostral portion, but instead in the form of an indication of the current state of balance across the entire SC map (i.e., bilaterally). In other words, the projections to OPNs may allow these latter neurons to detect SC imbalances, instead of whether or not the rostral SC is active. If balance is achieved bilaterally, OPNs do not need to turn off and trigger a corrective movement. If, on the other hand, SC activity is biased to one side more than the other (as, e.g., when activity starts to build up for a given microsaccade), the center of mass of the activity specifying the "behavioral goal location" may deviate from zero ("straight ahead"), thus triggering an eye movement to refoveate this goal. This mechanism explains why fairly caudal neurons that would otherwise not be considered "fixation" neurons according to the "fixation zone" hypothesis often project to OPNs (Gandhi and Keller, 1997). A computation of balance across the SC that relies on more than just the most rostral portions of the map would allow stability of fixation in the face of neural noise, as well as flexibility of which neurons can be recruited to support a particular behavior. This mechanism also explains why peripheral stimulus onsets bias microsaccades (Hafed et al., 2013). In this case, OPNs might pause because peripheral activity on one side of the SC has shifted the population activity balance away from straight ahead. One set of interesting future experiments and models could be aimed at understanding the detailed computations performed by OPNs based on their inputs from the SC.

Thus, investigating the mechanisms for microsaccade generation has allowed for new testable hypotheses about how the oculomotor system maintains gaze fixation as well as how it potentially coordinates the switching between "gaze holding" (e.g., slow drift) and the triggering of small "fixational" microsaccades.

19.5 Conclusion

Because of the foveated nature of our retina, we rely on voluntary eye movements to sequentially shift our gaze from one scene location to the next. Saccades are the fast scanning eye movements that allow us to foveate new objects rapidly. These movements thus allow high-acuity inspection of stationary features or targets. When targets of interest are moving, smooth pursuit eye movements are used to match eye velocity to target velocity. These movements thus stabilize the image of the target on the retina, again allowing accurate perception without additional artifacts due to motion blur. Saccades and smooth pursuit are coordinated in target tracking in order to optimize perceptual performance in the face of tracking errors or unpredictable target trajectory deviations. Finally, during fixation, the eyes never remain still but continue to exhibit both slow deviations (similar to smooth pursuit in their smooth nature) and fast microsaccades (similar to saccades in their rapid kinematics). All of these types of eye movements share some neural mechanisms, suggesting synergistic interactions among the decisions to fixate, generate saccades, or smoothly control eye position that we need to make repeatedly in our everyday life as we interact with our visual environment.

While systems-level control models of eye movements provide a solid foundation for understanding eye movement control, the use of distributed neural network models has revealed interesting emergent properties of such networks. For example, models have informed us on how certain mathematical concepts, such as "integration," may be implemented by neurons. More importantly, distributed neural network models provide a basis on which to add higher-level inputs from the brain that are related to perception and cognition to the low-level control of eye movements. For example, using distributed models of the SC, it is possible to model how cognitive processes such as "attention" may enhance SC activity at particular attended locations versus others, and how such enhancement can influence the subsequent generation of saccades, smooth pursuit, or microsaccades. This can then allow teasing apart so-called bottom-up and top-down processes for guiding attentional resources in the brain. Thus, modeling the control of eye movements provides a basis for understanding more complex brain functions that involve processes such as plasticity, sensorimotor integration for goal-directed behavior, decision-making, attention, and perception.

REFERENCES

Aizawa, H., & Wurtz, R. H. (1998). Reversible inactivation of monkey superior colliculus: I. Curvature of saccadic trajectory. *Journal of Neurophysiology, 79*(4), 2082–2096.

Aksay, E., Gamkrelidze, G., Seung, H. S., Baker, R., & Tank, D. W. (2001). In vivo intracellular recording and perturbation of persistent activity in a neural integrator. *Nature Neuroscience, 4*(2), 184–193. doi:10.1038/84023.

Aksay, E., Olasagasti, I., Mensh, B. D., Baker, R., Goldman, M. S., & Tank, D. W. (2007). Functional dissection of circuitry in a neural integrator. *Nature Neuroscience, 10*(4), 494–504. doi:10.1038/nn1877.

Arai, K., & Keller, E. L. (2005). A model of the saccade-generating system that accounts for trajectory variations produced by competing visual stimuli. *Biological Cybernetics, 92*(1), 21–37. doi:10.1007/s00422-004-0526-y.

Arai, K., Keller, E. L., & Edelman, J. A. (1994). Two-dimensional neural network model of the primate saccadic system. *Neural Networks, 7*(6/7), 1115–1135.

Arnold, D. B., & Robinson, D. A. (1997). The oculomotor integrator: Testing of a neural network model. *Experimental Brain Research, 113*(1), 57–74.

Bahill, A. T., Clark, M. R., & Stark, L. (1975). The main sequence, a tool for studying human eye movements. *Mathematical Biosciences, 24*(3–4), 191–204.

Bahill, A. T., Hsu, F. K., & Stark, L. (1978). Glissadic overshoots are due to pulse width errors. *Archives of Neurology, 35*(3), 138–142.

Barnes, G. R. (2008). Cognitive processes involved in smooth pursuit eye movements. [Review.] *Brain and Cognition, 68*(3), 309–326. doi:10.1016/j.bandc.2008.08.020.

Barnes, G. R., & Collins, C. J. (2008). Evidence for a link between the extra-retinal component of random-onset pursuit and the anticipatory pursuit of predictable object motion. *Journal of Neurophysiology, 100*(2), 1135–1146. doi:10.1152/jn.00060.2008.

Basso, M. A., Krauzlis, R. J., & Wurtz, R. H. (2000). Activation and inactivation of rostral superior colliculus neurons during smooth-pursuit eye movements in monkeys. *Journal of Neurophysiology, 84*(2), 892–908.

Bertero, M., Poggio, T. A., & Torre, V. (1988). Ill-posed problems in early vision. *Proceedings of the IEEE, 76*(8), 869–889.

Breznen, B., & Gnadt, J. W. (1997). Analysis of the step response of the saccadic feedback: Computational models. *Experimental Brain Research, 117*(2), 181–191.

Carpenter, R. H. (1988). *Movement of the Eyes* (2nd ed.). London: Pion.

Chen, C. Y., & Hafed, Z. M. (2013). Postmicrosaccadic enhancement of slow eye movements. *Journal of Neuroscience, 33*(12), 5375–5386. doi:10.1523/jneurosci.3703-12.2013.

Cherici, C., Kuang, X., Poletti, M., & Rucci, M. (2012). Precision of sustained fixation in trained and untrained observers. *Journal of Vision, 12*(6). doi:10.1167/12.6.31.

Das, S., Gandhi, N. J., & Keller, E. L. (1995). Open-loop simulations of the primate saccadic system using burst cell discharge from the superior colliculus. *Biological Cybernetics, 73*(6), 509–518.

Das, S., Keller, E. L., & Arai, K. (1996). A distributed model of the saccadic system: The effects of internal noise. *Neurocomputing, 11*, 245–269.

de Brouwer, S., Yuksel, D., Blohm, G., Missal, M., & Lefevre, P. (2002). What triggers catch-up saccades during visual tracking? *Journal of Neurophysiology, 87*(3), 1646–1650.

Dominey, P. F., & Arbib, M. A. (1992). A cortico-subcortical model for generation of spatially accurate sequential saccades. *Cerebral Cortex, 2*(2), 153–175.

Dominey, P. F., Arbib, M. A., & Joseph, J.-P. (1995). A model of corticostriatal plasticity for learning oculomotor associations and sequences. *Journal of Cognitive Neuroscience, 7*(3), 311–336.

Duhamel, J. R., Colby, C. L., & Goldberg, M. E. (1992). The updating of the representation of visual space in parietal cortex by intended eye movements. *Science, 255*(5040), 90–92.

Fukushima, K. (2003). Frontal cortical control of smooth-pursuit. *Current Opinion in Neurobiology, 13*(6), 647–654.

Gandhi, N. J., & Katnani, H. A. (2011). Motor functions of the superior colliculus. *Annual Review of Neuroscience, 34*, 205–231. doi:10.1146/annurev-neuro-061010-113728.

Gandhi, N. J., & Keller, E. L. (1997). Spatial distribution and discharge characteristics of superior colliculus neurons antidromically activated from the omnipause region in monkey. *Journal of Neurophysiology, 78*(4), 2221–2225.

Goldberg, M. E., & Wurtz, R. H. (1972). Activity of superior colliculus in behaving monkey: I. Visual receptive fields of single neurons. *Journal of Neurophysiology, 35*(4), 542–559.

Goldberg, M. E., & Wurtz, R. H. (1989). *The Neurobiology of Saccadic Eye Movements.* Amsterdam: Elsevier North-Holland.

Goossens, H. H., & van Opstal, A. J. (2006). Dynamic ensemble coding of saccades in the monkey superior colliculus. *Journal of Neurophysiology, 95*(4), 2326–2341. doi:00889.2005 [pii] 10.1152/jn.00889.2005.

Goossens, H. H., & van Opstal, A. J. (2012). Optimal control of saccades by spatial-temporal activity patterns in the monkey superior colliculus. *PLoS Computational Biology, 8*(5), e1002508. doi:10.1371/journal.pcbi.1002508.

Hafed, Z. M. (2011). Mechanisms for generating and compensating for the smallest possible saccades. *European Journal of Neuroscience, 33*(11), 2101–2113. doi:10.1111/j.1460-9568.2011.07694.x.

Hafed, Z. M., & Clark, J. J. (2002). Microsaccades as an overt measure of covert attention shifts. *Vision Research, 42*(22), 2533–2545. doi:S0042698902002638 [pii].

Hafed, Z. M., Goffart, L., & Krauzlis, R. J. (2008). Superior colliculus inactivation causes stable offsets in eye position during tracking. *Journal of Neuroscience, 28*(32), 8124–8137. doi:28/32/8124 [pii] 10.1523/jneurosci.1317-08.2008.

Hafed, Z. M., Goffart, L., & Krauzlis, R. J. (2009). A neural mechanism for microsaccade generation in the primate superior colliculus. *Science, 323*(5916), 940–943. doi:323/5916/940 [pii] 10.1126/science.1166112.

Hafed, Z. M., & Krauzlis, R. J. (2008). Goal representations dominate superior colliculus activity during extrafoveal tracking. *Journal of Neuroscience, 28*(38), 9426–9439. doi:28/38/9426 [pii] 10.1523/jneurosci.1313-08.2008.

Hafed, Z. M., Lovejoy, L. P., & Krauzlis, R. J. (2013). Superior colliculus inactivation alters the relationship between covert visual attention and microsaccades. *European Journal of Neuroscience, 37*, 1169–1181. doi:10.1111/ejn.12127.

Hamker, F. H., Zirnsak, M., Calow, D., & Lappe, M. (2008). The peri-saccadic perception of objects and space. *PLoS Computational Biology, 4*(2), e31. doi:07-PLCB-RA-0514 [pii] 10.1371/journal.pcbi.0040031.

Heinzle, J., Hepp, K., & Martin, K. A. (2007). A microcircuit model of the frontal eye fields. *Journal of Neuroscience, 27*(35), 9341–9353. doi:10.1523/jneurosci.0974-07.2007.

Hikosaka, O., & Wurtz, R. H. (1983a). Visual and oculomotor functions of monkey substantia nigra pars reticulata: I. Relation of visual and auditory responses to saccades. *Journal of Neurophysiology, 49*(5), 1230–1253.

Hikosaka, O., & Wurtz, R. H. (1983b). Visual and oculomotor functions of monkey substantia nigra pars reticulata: II. Visual responses related to fixation of gaze. *Journal of Neurophysiology, 49*(5), 1254–1267.

Hikosaka, O., & Wurtz, R. H. (1983c). Visual and oculomotor functions of monkey substantia nigra pars reticulata: III. Memory-contingent visual and saccade responses. *Journal of Neurophysiology, 49*(5), 1268–1284.

Hikosaka, O., & Wurtz, R. H. (1985). Modification of saccadic eye movements by GABA-related substances: I. Effect of muscimol and bicuculline in monkey superior colliculus. *Journal of Neurophysiology, 53*(1), 266–291.

Hikosaka, O., & Wurtz, R. H. (1986). Saccadic eye movements following injection of lidocaine into the superior colliculus. *Experimental Brain Research, 61*(3), 531–539.

Hoff, B., & Arbib, M. A. (1991). A model of the effects of speed, accuracy and perturbation on visually guided reaching. In R. Caminiti, P. B. Johnson, & Y. Burnod (Eds.), *Control of Arm Movement in Space: Neurophysiological and Computational Approaches* (pp. 285–306). Heidelberg: Springer-Verlag.

Ilg, U. J. (2008). The role of areas MT and MST in coding of visual motion underlying the execution of smooth pursuit. *Vision Research, 48*(20), 2062–2069. doi:10.1016/j.visres.2008.04.015.

Ilg, U. J., & Thier, P. (2008). The neural basis of smooth pursuit eye movements in the rhesus monkey brain. [Review.] *Brain and Cognition, 68*(3), 229–240. doi:10.1016/j.bandc.2008.08.014.

Inagaki, K., Hirata, Y., & Usui, S. (2011). A model-based theory on the signal transformation for microsaccade generation. *Neural Networks, 24*(9), 990–997. doi:10.1016/j.neunet.2011.06.007.

Jurgens, R., Becker, W., & Kornhuber, H. H. (1981). Natural and drug-induced variations of velocity and duration of human saccadic eye movements: Evidence for a control of the neural pulse generator by local feedback. *Biological Cybernetics, 39*(2), 87–96.

Kalman, R. E. (1960). A new approach to linear filtering and prediction problems. *Journal of Fluids Engineering, 82*, 35–45.

Kawano, K., & Miles, F. A. (1986). Short-latency ocular following responses of monkey: II. Dependence on a prior saccadic eye movement. *Journal of Neurophysiology, 56*(5), 1355–1380.

Keller, E. L. (1977). Control of saccadic eye movements by midline brain stem neurons. In R. Baker & A. Berthoz (Eds.), *Control of Gaze by Brain Stem Neurons* (pp. 327–336). Amsterdam: Elsevier.

Keller, E. L., & Missal, M. (2003). Shared brainstem pathways for saccades and smooth-pursuit eye movements. *Annals of the New York Academy of Sciences, 1004*, 29–39.

Krauzlis, R. J. (2004). Recasting the smooth pursuit eye movement system. *Journal of Neurophysiology, 91*(2), 591–603. doi:10.1152/jn.00801.200391/2/591.

Krauzlis, R. J. (2008). Eye movements. In L. R. Squire (Ed.), *Fundamental Neuroscience* (3rd ed., pp. 775–792). Burlington, MA: Elsevier.

Krauzlis, R. J., Basso, M. A., & Wurtz, R. H. (1997). Shared motor error for multiple eye movements. *Science, 276*(5319), 1693–1695.

Krauzlis, R. J., & Lisberger, S. G. (1989). A control systems model of smooth pursuit eye movements with realistic emergent properties. *Neural Computation, 1*, 116–122.

Krauzlis, R. J., & Lisberger, S. G. (1991). Visual motion commands for pursuit eye movements in the cerebellum. *Science, 253*(5019), 568–571.

Krauzlis, R. J., & Lisberger, S. G. (1994). A model of visually-guided smooth pursuit eye movements based on behavioral observations. *Journal of Computational Neuroscience, 1*(4), 265–283.

Lee, C., Rohrer, W. H., & Sparks, D. L. (1988). Population coding of saccadic eye movements by neurons in the superior colliculus. *Nature, 332*(6162), 357–360. doi:10.1038/332357a0.

Lefevre, P., & Galiana, H. L. (1992). Dynamic feedback to the superior colliculus in a neural network model of the gaze control system. *Neural Networks, 5*, 871–890.

Lefevre, P., Quaia, C., & Optican, L. M. (1998). Distributed model of control of saccades by superior colliculus and cerebellum. *Neural Networks, 11*(7–8), 1175–1190. doi:S0893-6080(98)00071-9 [pii].

Lisberger, S. G. (2010). Visual guidance of smooth-pursuit eye movements: Sensation, action, and what happens in between. *Neuron, 66*(4), 477–491. doi:10.1016/j.neuron.2010.03.027.

Lisberger, S. G., Morris, E. J., & Tychsen, L. (1987). Visual motion processing and sensory-motor integration for smooth pursuit eye movements. *Annual Review of Neuroscience, 10*, 97–129. doi:10.1146/annurev.ne.10.030187.000525.

Masson, G. S., & Perrinet, L. U. (2012). The behavioral receptive field underlying motion integration for primate tracking eye movements. *Neuroscience and Biobehavioral Reviews, 36*(1), 1–25. doi:10.1016/j.neubiorev.2011.03.009.

Missal, M., & Keller, E. L. (2002). Common inhibitory mechanism for saccades and smooth-pursuit eye movements. *Journal of Neurophysiology, 88*(4), 1880–1892.

Mitchell, J., & Zipser, D. (2001). A model of visual-spatial memory across saccades. *Vision Research, 41*(12), 1575–1592.

Mitchell, J. F., & Zipser, D. (2003). Sequential memory-guided saccades and target selection: A neural model of the frontal eye fields. *Vision Research, 43*(25), 2669–2695.

Moschovakis, A. K. (1994). Neural network simulations of the primate oculomotor system: I. The vertical saccadic burst generator. *Biological Cybernetics, 70*(3), 291–302.

Munoz, D. P., & Wurtz, R. H. (1993). Fixation cells in monkey superior colliculus: I. Characteristics of cell discharge. *Journal of Neurophysiology, 70*(2), 559–575.

Munoz, D. P., & Wurtz, R. H. (1995). Saccade-related activity in monkey superior colliculus: I. Characteristics of burst and buildup cells. *Journal of Neurophysiology, 73*(6), 2313–2333.

Nikitchenko, M., & Koulakov, A. (2008). Neural integrator: A sandpile model. *Neural Computation, 20*(10), 2379–2417. doi:10.1162/neco.2008.12-06-416.

Nuding, U., Ono, S., Mustari, M. J., Buttner, U., & Glasauer, S. (2008). A theory of the dual pathways for smooth pursuit based on dynamic gain control. *Journal of Neurophysiology, 99*(6), 2798–2808. doi:10.1152/jn.90237.2008.

Oppenheim, A. V., Willsky, A. S., & Nawab, S. H. (1997). *Signals and Systems.* Upper Saddle River: Prentice Hall.

Optican, L. M., & Miles, F. A. (1985). Visually induced adaptive changes in primate saccadic oculomotor control signals. *Journal of Neurophysiology, 54*(4), 940–958.

Orban de Xivry, J. J., & Lefevre, P. (2007). Saccades and pursuit: Two outcomes of a single sensorimotor process. *Journal of Physiology, 584*(Pt 1), 11–23. doi:10.1113/jphysiol.2007.139881.

Ottes, F. P., Van Gisbergen, J. A., & Eggermont, J. J. (1986). Visuomotor fields of the superior colliculus: A quantitative model. *Vision Research, 26*(6), 857–873.

Robinson, D. A. (1964). The mechanics of human saccadic eye movement. *Journal of Physiology, 174*, 245–264.

Robinson, D. A. (1965). The mechanics of human smooth pursuit eye movement. *Journal of Physiology, 180*(3), 569–591.

Robinson, D. A. (1970). Oculomotor unit behavior in the monkey. *Journal of Neurophysiology, 33*(3), 393–403.

Robinson, D. A. (1972). Eye movements evoked by collicular stimulation in the alert monkey. *Vision Research, 12*(11), 1795–1808. doi:0042-6989(72)90070-3 [pii].

Robinson, D. A. (1973). Models of the saccadic eye movement control system. *Kybernetik, 14*, 71–83.

Robinson, D. A. (1975). Oculomotor control signals. In P. Bach-y-Rita & G. Lennerstrand (Eds.), *Basic Mechanisms of Ocular Motility and Their Clinical Implications* (pp. 337–374). Oxford: Pergamon.

Robinson, D. A., Gordon, J. L., & Gordon, S. E. (1986). A model of the smooth pursuit eye movement system. *Biological Cybernetics, 55*(1), 43–57.

Rothganger, F. H., & Anastasio, T. J. (2009). Using input minimization to train a cerebellar model to simulate regulation of smooth pursuit. *Biological Cybernetics, 101*(5–6), 339–359. doi:10.1007/s00422-009-0340-7.

Scudder, C. A. (1988). A new local feedback model of the saccadic burst generator. *Journal of Neurophysiology, 59*(5), 1455–1475.

Scudder, C. A., Kaneko, C. S., & Fuchs, A. F. (2002). The brainstem burst generator for saccadic eye movements: A modern synthesis. *Experimental Brain Research, 142*(4), 439–462. doi:10.1007/s00221-001-0912-9.

Tanaka, M., & Lisberger, S. G. (2001). Regulation of the gain of visually guided smooth-pursuit eye movements by frontal cortex. *Nature, 409*(6817), 191–194. doi:10.1038/35051582.

Thier, P., & Ilg, U. J. (2005). The neural basis of smooth-pursuit eye movements. [Review.] *Current Opinion in Neurobiology, 15*(6), 645–652. doi:10.1016/j.conb.2005.10.013.

Van Gisbergen, J. A., Robinson, D. A., & Gielen, S. (1981). A quantitative analysis of generation of saccadic eye movements by burst neurons. *Journal of Neurophysiology, 45*(3), 417–442.

Van Gisbergen, J. A., Van Opstal, A. J., & Tax, A. A. M. (1987). Collicular ensemble coding of saccades based on vector summation. *Neuroscience, 21*, 541–555.

van Opstal, A. J., & Goossens, H. H. (2008a). Linear ensemble-coding in midbrain superior colliculus specifies the saccade kinematics. *Biological Cybernetics, 98*(6), 561–577. doi:10.1007/s00422-008-0219-z.

van Opstal, A. J., & Goossens, H. H. (2008b). Linear ensemble-coding in midbrain superior colliculus specifies the saccade kinematics. *Biological Cybernetics, 98*(6), 561–577. doi:10.1007/s00422-008-0219-z.

Waitzman, D. M., Ma, T. P., Optican, L. M., & Wurtz, R. H. (1991). Superior colliculus neurons mediate the dynamic characteristics of saccades. *Journal of Neurophysiology, 66*(5), 1716–1737.

Walton, M. M., Sparks, D. L., & Gandhi, N. J. (2005). Simulations of saccade curvature by models that place superior colliculus upstream from the local feedback loop. *Journal of Neurophysiology, 93*(4), 2354–2358. doi:01199.2004 [pii] 10.1152/jn.01199.2004.

Wiener, N. (1949). *Extrapolation, Interpolation, and Smoothing of Stationary Time Series with Engineering Applications.* Cambridge, MA: MIT Press.

Wurtz, R. H., & Goldberg, M. E. (1972). Activity of superior colliculus in behaving monkey: III. Cells discharging before eye movements. *Journal of Neurophysiology, 35*(4), 575–586.

Wurtz, R. H., & Optican, L. M. (1994). Superior colliculus cell types and models of saccade generation. *Current Opinion in Neurobiology, 4*(6), 857–861.

Young, L. R., & Stark, L. (1963). Variable feedback experiments testing a sampled data model for eye tracking movements. *IEEE Transactions of the Professional Technical Group on Human Factors in Electronics, 4*, 38–51.

20 Reach and Grasp: Control, Development, and Recognition

JAMES J. BONAIUTO

This chapter describes computational models that address the control, development, and recognition of reaching and grasping movements. The first section introduces the various brain regions involved in reaching and grasping. The following section focuses on reaching and describes models of reach trajectory planning and development of reaching abilities in infancy. The next section focuses on grasping and splits the subject into grasp selection and planning, grasp development, and online control of grasp patterns. The next section describes several models that address how reaching and grasping movements are coordinated, and the section that follows it describes models of action recognition.

In order to manipulate objects, they must first be reached for and then grasped. This makes control of the arm and hand in reaching and grasping fundamental to understanding how primates interact with manipulable objects. In order to grasp an object the brain must convert visual information about the object's properties and location into a pattern of coordinated movements to transport the hand to the object, preshape the hand, and enclose the hand around the object.

Reaching movements require a representation of the object's location. Different brain regions encode the spatial location of objects in different reference frames (see chapter 1). A reference frame determines how spatial relations are to be encoded. For example, various parietal regions seem to encode object locations relative to the hand and must adjust their representations as the hand moves whereas retinotopic encoding encodes hand and object as they project to the retina.

Grasp selection and planning depend upon the perception and utilization of affordances. Affordances are defined by Gibson (1966) as opportunities for action that are directly perceivable without recourse to higher-level cognitive functions. For example, affordances for grasping a mug by the handle or around the cylinder can be perceived without recognizing that the object is a mug. In order to select and plan a grasp, the brain must extract affordances from an object's visual features and map these onto motor commands to shape the hand appropriately. This transformation is thought to be accomplished by a network of parietal and premotor brain regions centered around the anterior intraparietal area (AIP) and the ventral premotor area F5 (see below). Once contact with the object is made, haptic (touch) sensations are used to stabilize the grasp. For example, detecting the object slipping from the hand may trigger an adjustment of the hand shape or the application of more force. While these processes are important for the final stage of grasping and consequent manipulation of an object, the role of vision is emphasized in this chapter.

Action recognition is the result of mapping an observed action onto an internal motor or semantic representation. Area F5 contains two broad classes of neurons: canonical neurons and so-called "mirror" neurons. Canonical F5 neurons respond to the sight of a graspable object and fire during the performance of a grasp. F5 mirror neurons fire not only during performance of a grasp but also during the observation of someone else performing the same or similar grasps. Mirror neurons are therefore thought to be involved in action recognition in terms of mapping an observed action onto internal motor representations (but note that other brain regions are likely involved in mapping observed actions onto semantic representations).

The models discussed in this chapter fall into two main groups: mathematical or behavioral models, and neurobiologically inspired neural network models (see chapter 1). Mathematical models attempt to capture the computations that the brain must perform in order to generate the types of behavior observed in experiments that measure the kinematics or dynamics of movement. They typically do not attempt to localize any of these computations to any particular brain region, but instead provide a decomposition of the processes that underlie these computations. Most of the reaching models reviewed in this chapter fall into this category. In contrast, neurobiologically inspired neural network

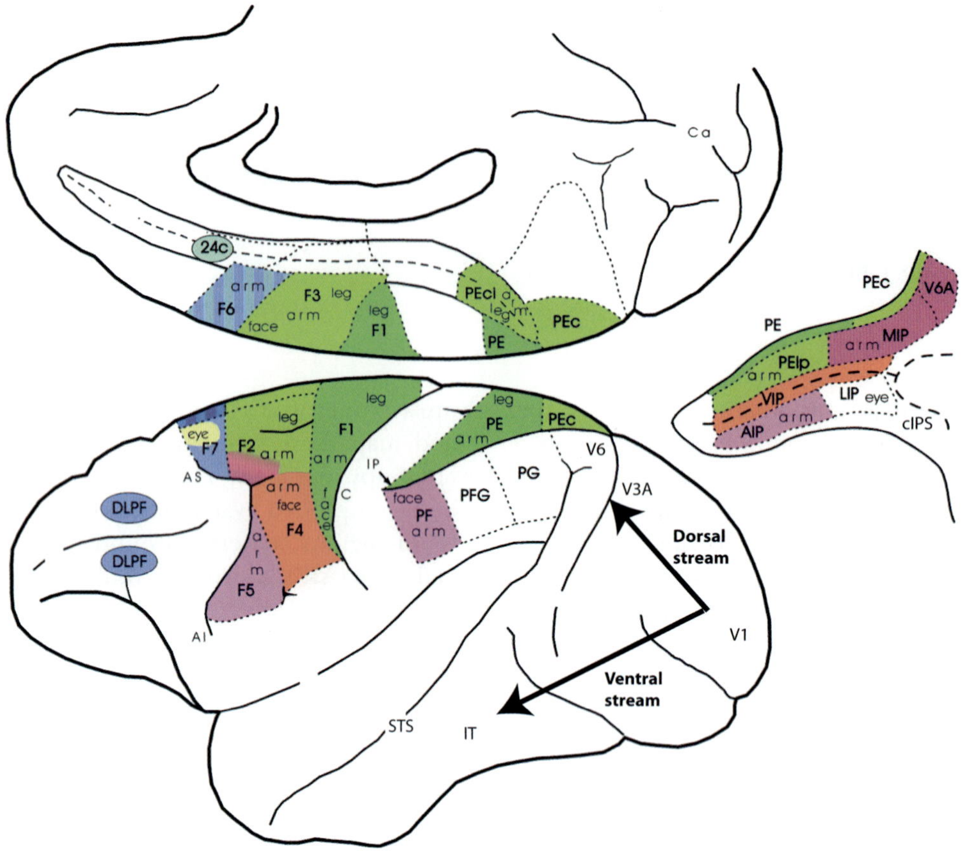

FIGURE 20.1 Mesial (top) and lateral (bottom) views of the macaque brain. The intraparietal sulcus has been opened so that regions inside of it are visible (right). AS, arcuate sulcus; AIP, anterior intraparietal area; C, central sulcus; Ca, calcarine fissure; Cg, cingulated sulcus; cIPS, caudal intraparietal sulcus; DLPF, dorsolateral prefrontal area; F1–F6, agranular frontal areas; IO, inferior occipital sulcus; IP, intraparietal sulcus, IT, inferior temporal cortex; L, lateral fissure; Lu, lunate sulcus; MIP, medial intraparietal area; P, principal sulcus; PF, PFG, PG=inferior parietal areas; PO, parieto-occipital sulcus; STS, superior temporal sulcus; V1, V3A, V6, V6A, visual areas. Courtesy of Giuseppe Luppino.

models attempt to model key brain regions thought to be involved in the behavior in question and try to show how neural circuits in the brain could perform the computations necessary to generate that behavior. Most of the models of grasping and action recognition described in this chapter are of this type.

20.1 Neural Control of Arm and Hand Movements

A dominant view holds that the hand transport (reach) and grasp components are controlled by separate but coordinated systems. This section gives an overview of the brain regions involved in controlling arm and hand movements. The regions involved are generally agreed upon; however, their functional roles and interactions are greatly debated. This chapter focuses on models of cortical control of the arm and hand (and some include the basal ganglia). See chapter 21 for models of the involvement of the cerebellum in motor control.

Visual information is first processed in cortex by primary visual area V1 and from there is processed by several other visual areas before separating into the dorsal and ventral visual streams (figure 20.1; see chapter 12). The ventral stream is thought to be primarily involved in processing visual features for the purposes of object recognition (it's sometimes called the "what" stream), while the dorsal stream is involved in processing visual information for controlling actions (also known as the "how" stream). The dorsal visual stream does not know "what" the object is—it can only see the object as a set of possible affordances—whereas the ventral stream is able to recognize what the object is.

Since the distinction was made between the dorsal and ventral visual streams (see chapter 1), the dorsal stream has been further subdivided into the dorsal–medial and dorsal–ventral streams (Rizzolatti and Matelli, 2003). It has been suggested that the dorsal–medial stream, involving superior parietal and intraparietal regions and the dorsal premotor cortex, controls reaching, while the dorsal–ventral stream, including inferior parietal and intraparietal regions and the ventral premotor cortex, controls grasping. The dorsal–medial stream receives its inputs from visual area V6, whose neurons encode object location, while the dorsal–ventral stream receives inputs from visual area V3a, whose neurons are sensitive to binocular disparity. The intraparietal sulcus and inferior parietal lobule are involved in the representation and transformation of sensory inputs in order to plan movements. Various premotor regions use these parietal representations of action affordances in order to plan behaviors such as reach and grasping.

Both premotor and parietal cortex of the macaque brain contain mirror neurons, each of which fires vigorously both when the monkey executes a certain limited set of actions and when the monkey observes some other perform a similar action. By contrast, canonical neurons in F5 fire vigorously when the monkey executes certain actions but not when it observes the actions of others.

20.1.1 Visual Areas The occipital cortex is responsible for processing visual information. Located at the back of the head, it is subdivided into the primary visual area V1 (also known as striate cortex) and several other extrastriate areas also specialized for vision. The primary visual area V1 receives input from the retina, relayed by the thalamus. Here we focus on the extrastriate cortices thought to provide inputs to the dorsal–ventral and dorsal–medial visual streams, as well as the superior temporal sulcus (STS) which provides visual inputs to the mirror system.

Area V3a Area V3a is an extrastriate visual area that receives weak direct projections from area V1 (Zeki, 1980) and stronger inputs from other extrastriate visual areas (Felleman, Burkhalter, et al., 1997; Gattass, Sousa, et al., 1997). V3a contains neurons that are sensitive to binocular disparity and have small, retinotopic receptive fields (Sakata, Tsutsui, et al., 2005). It projects to the caudal intraparietal sulcus (cIPS; see below) and is thought to provide the depth information needed by cIPS to compute three-dimensional object shape features.

Area V6a Area V6a is in the medial parieto-occipital cortex and contains mostly visual cells that are modulated by somatosensory stimulation (Galletti, Kutz, et al., 2003). So-called real-position cells are able to encode spatial location of objects in the visual scene in at least a head-centered reference frame, with visual receptive fields that remain anchored despite eye movements. Intermingled with real-position cells are retinotopic cells, whose visual receptive fields shift with gaze, suggesting that the region is involved in converting from retinotopic to head- or body-centered reference frames. Lesions of the region result in misreaching with the contralateral arm, and it has thus been suggested that the region is involved in encoding the direction of movement required to bring the arm to potential reach targets. Area V6a receives input from central and peripheral visual field representations in V6 and projects strongly to the dorsal premotor area F2, which we discuss below.

Superior Temporal Sulcus The STS is located in the temporal lobe and is thus part of the ventral visual stream. It contains cells which respond to the observation of biological actions such as walking, limb movements, and grasping. Many of these cells show great specificity for the type of action (Perrett, Harries, et al., 1989). A subset of cells in this region respond selectively to hand–object interactions and respond similarly to the sight of the monkeys' own movements as well as those of others (note that they are not mirror neurons as they do not respond during grasping in the dark—they have no motor function). STS is thought to provide the inferior parietal cortex with a visual analysis of the hand shape during grasp observation.

20.1.2 Parietal Areas There is converging evidence that affordances for grasping are extracted by posterior parietal cortex including the intraparietal sulcus, while the inferior parietal cortex has been found to be involved in action recognition (see below) in the macaque and is thought to provide visual input in terms of hand–object relations to mirror neurons area F5.

Caudal Intraparietal Sulcus The caudal intraparietal sulcus (cIPS) is a region located in the caudal part of the lateral bank and fundus of the intraparietal sulcus (it is also known as the posterior intraparietal area; PIP). The cIPS receives input mainly from V3a and projects primarily to the anterior intraparietal area (AIP; see below). Neurons in area cIPS have large receptive fields with no retinotopic organization (Tsutsui, Taira, et al., 2005). Two functional classes of neurons in area cIPS have been described: surface orientation selective neurons that are selective to the orientation of flat surfaces, and axis orientation selective neurons that respond best to an elongated object whose principal axis is oriented in a particular direction. Both types of cells respond best to binocular stimuli, are spatially intermingled, and include some neurons that are selective for the object's dimensions. It is thought that area cIPS extracts 3-D object shape features and provides this information to AIP for grasp affordance extraction.

Anterior Intraparietal Area It has been widely suggested that the anterior intraparietal area AIP is involved in the extraction of affordances relevant for grasping from visual object features. The area AIP is located on the lateral bank of the anterior intraparietal sulcus and contains visually responsive neurons selective for 3-D features of objects, motor dominant neurons that only respond during grasping, and visuomotor neurons that are activated by grasping and modulated by sight of the object (Sakata, Taira, et al., 1998). The region receives its main input from area cIPS, but also receives input from V6a and has strong reciprocal connections with the premotor region F5 (see below). In monkeys, neurons in AIP are responsive to 3-D features of objects relevant for manipulation such as shape, size, and orientation, regardless of their location (Murata, Gallese, et al., 2000).

Area 7 (PF, PG) Area 7 is located in the inferior parietal lobe. It has been subdivided into areas 7a and 7b, which roughly correspond to areas PG and PF in other nomenclatures. Both area 7a and 7b have reciprocal connections to the STS, and area 7b additionally receives input from 7a and AIP. This makes area 7 ideally situated to relate the visual features of an observed hand action to the object affordances it is directed toward.

20.1.3 FRONTAL AREAS In this chapter, frontal brain regions are referred to using the Parma nomenclature for the macaque brain, which gives each region a letter F (for frontal) and a number. Subregions are typically named by appending directional abbreviations to the region name, for example: F2vr is the ventrorostral subregion of F2.

The premotor cortex has been implicated in the control of both proximal (involving the arm) and distal (involving the hand or mouth) actions and in action recognition. An emerging view holds that premotor areas encode the goal of an intended movement which is transformed into the movements needed to achieve the goal by the primary motor cortex F1.

Primary Motor Cortex F1 The primary motor area F1 is thought to be the final cortical motor output region. The region corresponds to M1 in the human and can control muscle activity through direct projections to the spinal cord as well as to intermediate motor structures. This region receives massive inputs from the premotor regions F2 and F5 (see below) and is therefore uniquely situated to transform reach and grasp motor plans into patterns of muscle activation to appropriately control the arm and hand.

Dorsal Premotor Area F2 Within the premotor cortex, the dorsal region F2 is thought to encode reach movements. Visual inputs to area F2 come mainly from the superior parietal lobe including area V6A, while the main output of F2 projects to F1. The region contains two broad categories of cells: movement-related cells that discharge on movement onset and cells that show anticipatory activity prior to the start of movement (Cisek and Kalaska, 2005). Many of the cells in F2 are directionally tuned, and their population activity appears to encode a vector representing the direction of arm movement. In addition to reach target selection, the dorsal premotor cortex is implicated in wrist movements.

Ventral Premotor Area F5 The ventral premotor area F5 lies in the arcuate sulcus. Many of these neurons fire in association with specific types of manual action, such as precision grip, finger prehension, and whole hand prehension as well as tearing and holding (figure 20.2). Some neurons in F5 discharge only during the last part of grasping; others start to fire during the phase in which the hand opens and continue to discharge during the phase when the hand closes; finally, a few discharge prevalently in the phase in which the hand opens. Grasping appears, therefore, to be coded by the activity of populations of neurons, each controlling different phases of the motor act.

The two main types of cells in this region are referred to as "canonical" and "mirror" neurons. F5 canonical neurons begin to fire at the sight of a graspable object and continue to fire during grasping, even in the dark. Raos et al. (2006) found that grip posture rather than object shape determined F5 canonical neuron activity. For example, a small cube may be grasped with a precision pinch using the thumb and forefinger but could also be grasped using a power grasp. F5 canonical neurons selective for both grip type and wrist orientation maintained this selectivity when grasping in the dark. Simultaneous recording from F5 and F1 showed that F5 canonical neurons were selective for grasp type and phase while an F1 neuron might be active for different phases of different grasps (Umiltà, Brochier, et al., 2007). This suggests that F5 canonical neurons encode a high-level representation of the grasp motor schema while F1 neurons (or, at least, some of them) encode the component movements of a population

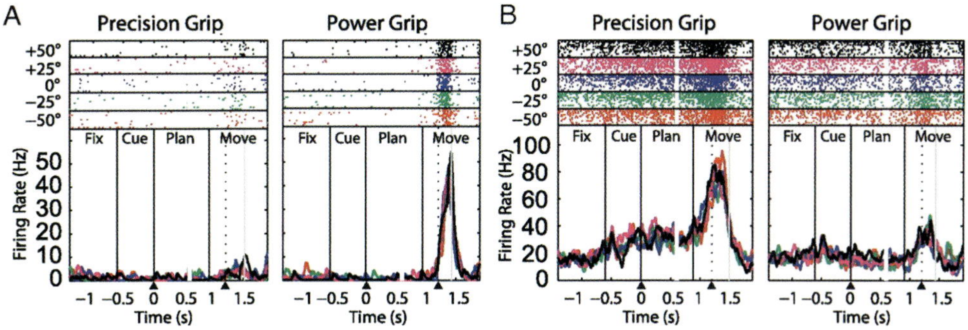

FIGURE 20.2 Two F5 neurons that fire during power (A) and precision (B) grasps regardless of the orientation of the object. The vertical lines in each plot show the time of the different phases of the task . Each orientation is shown with a different color. The cell in A responds during performance of power grasps, but only once the movement starts, while the cell in B only responds during performance of precision pinches, starting with the cue indicating the type of grasp to be performed. Modified from Fluet, Baumann, et al. (2010).

code for specific muscle activity irrespective of grasp phase.

In contrast to canonical neurons, F5 mirror neurons fire during the performance of specific grasp types (even in the dark) and also when observing another individual performing the same or similar grasp type. While F5 canonical neurons are thought to encode a high-level representation of a planned grasp, F5 mirror neurons are thought to recognize another individual's grasping actions by mapping the visual features of an observed grasp onto the same high-level grasp representation. In addition to projections from AIP, area F5 mainly receives visual input from the inferior parietal region 7b which is thought to relate visual features of an observed grasp with object affordance information.

Pre-supplementary Motor Area F6 Area F6, also known as the pre-supplementary motor area (pre-SMA), is located in the medial premotor cortex. It has no projections to the spinal cord, but does project to regions F2 and F5 (Luppino, Matelli, et al., 1993). Its function is not entirely clear, but it is activated during the production of sequences of motor actions and cells in this area respond during individual movements in a sequence but are also sensitive to their order in the sequence (Tanji and Shima, 1994).

20.2 Reaching

This section presents models that apply general techniques for motor control such as internal models of kinematics and dynamics to reaching (chapter 9). In order to reach for a target in space, the location of that target must be transformed into motor commands appropriate for moving the end effector (the hand or wrist) to that location. Although we will not pursue this point here, note that the target position of, say, the wrist will depend on the grasp used and where the fingers will touch the target object. (See the later section on "Grasp Development.") Two crucial computations in this transformation are the conversion of a desired hand position into the arm joint angles required to bring the hand to that position (inverse kinematics) and the computation of torques that must be applied to the joints to achieve these joint angles (inverse dynamics). A dominant view of this process is that of a hierarchy of systems which plan a smooth reach trajectory to the target (trajectory planning), convert each point along that trajectory to a set of target angles for the arm joints (inverse kinematics), compute the torques required to achieve those joint angles (inverse dynamics), and convert these into commands to motor neurons.

20.2.1 TRAJECTORY PLANNING

Given the initial position of the wrist and a target location, any number of trajectories could be used to reach the target. However, in the absence of obstacles most reaches follow a straight path to the target with a unimodal velocity profile. The vector integration to endpoint (VITE) model (Bullock and Grossberg, 1988) is an early model of trajectory generation that can reproduce many kinematic features of reaching movements. Given a target position vector, T, and the current position of the hand, $P(t)$, the VITE model computes a difference vector:

$$D(t) = T - P(t).$$

The difference vector represents a desired change in the hand position, which is multiplied by a gain signal, g, and then added to $P(t)$ to update the motor command:

$$P(t+1) = P(t) + D(t)g(t) .$$

The gain signal starts at zero during movement onset and monotonically increases, resulting in a gradual movement from the initial hand position to the target. This model was later extended to control multiple effectors and coordinate reach and grasp movements by using a common gain signal to ensure that the reach and grasp movements end at the same time (see the "Reach and Grasp Coordination" section).

Flash and Hogan (1985) found that the smoothness of hand trajectories could be described by a function that minimizes the derivative of acceleration, or *jerk*. In order to generate a minimum-jerk trajectory, you need to assign a cost function to each trajectory in the space of all possible trajectories and find one that minimizes that cost. Hogan suggested that this cost function should be the sum of the squared jerk over the trajectory:

$$C(x(t)) = \frac{1}{2}\int_{t=0}^{t} \dddot{x}^2 dt .$$

This model does indeed match most kinematic features of real straight and curved reach trajectories. Note that others have suggested various other cost functions (see Todorov, 2004, for examples).

But how does the brain generate minimum-jerk trajectories? It seems unlikely that it performs a search over all possible trajectories and then settles on an optimized trajectory entirely planned in advance. Moreover, this would be a case of pure *feedforward* whereas the control system should be open to feedback signals concerning departures from the *intended* trajectory. One approach uses what are called next-state planners to break up a reach movement into a series of smaller reaches. At each point in time, the next-state planner computes a target position for the wrist based on its current position and that of the goal (or, in the continuous time case, keeps updating the velocity of the hand), with a minimum-jerk trajectory being a an emergent feature of this process. This approach can more flexibly deal with object displacements and arm perturbations without requiring extensive replanning.

Hoff and Arbib (1993) showed how to derive a next-state planner that generates minimum-jerk trajectories. As input, this model receives the error vector (from the wrist to the reach target) and an estimate of the end-effector location, velocity, and acceleration. The model then produces the desired location, velocity, and acceleration of the end effector for the next time step in such a way that the minimum-jerk criterion is enforced. Note, however, that it this is a kinematic and not dynamic model. It does not compute the muscle forces required to achieve these targets.

To derive their model, Hoff and Arbib computed the solution to the minimum-jerk cost function for arbitrary initial and final end-effector states. This leads to the following expression for acceleration at time t:

$$\begin{aligned}\ddot{x}(t) = {} & \ddot{x}_0\left(1-9\varepsilon+18\varepsilon^2-10\varepsilon^3\right) \\ & + \dot{x}_0\left(-36\varepsilon+96\varepsilon^2-60\varepsilon^3\right)/D \\ & + (\hat{x}_t - x_0)\left(60\varepsilon+180\varepsilon^2+120\varepsilon^3\right)/D^2\end{aligned},$$

where x_0, $\dot{x}_0$, and $\ddot{x}_0$ are the initial position, velocity, and acceleration of the end effector, $\hat{x}_t$ is the position of the target, D is the duration of the movement (t_f–t_0 where t_f is the final time of the movement), and ε is the proportional elapsed time ($\varepsilon = (t-t_0)/D$). The jerk is then the derivative of this equation. Setting $\varepsilon = 0$ and $D = t_f - t$, and taking the state to be the vector x_{ee}, the equations for the change of state become

$$\begin{aligned}&\Delta x_d(t) = \dot{x}_d(t), \\ &\Delta \dot{x}_d(t) = \ddot{x}_d(t), \\ &\Delta \ddot{x}_d(t) = \left[-\frac{60}{D^3}(x_t - x_{ee}(t)) - \frac{36}{D^2}\dot{x}_d(t) - \frac{9}{D}\ddot{x}_d(t)\right].\end{aligned}$$

This set of equations describes a planner that can compute the change in the desired position, velocity, and acceleration of the end effector required to generate a straight, minimum-jerk movement.

In these equations the duration of the movement, D, is provided as input. Hoff and Arbib showed how this input could be generated by a function that predicts the optimum length of time required to perform the reach. For a reach starting from a stationary position, this movement time estimate is

$$D = \left(60\left(x_t - \hat{x}_{ee}\right)\right)^{1/3} r^{1/6},$$

where r is a parameter that manages the trade-off between smooth and fast movements.

While reach trajectories are generally straight, primates are capable of generating arbitrarily curved movements. Ijspeert et al. (2013) developed a model based on dynamic motor primitives, dynamical systems with parameters that can be set to generate any trajectory. This model uses a set of radial basis functions to modulate an otherwise straight trajectory. Radial basis functions are functions whose value depends on the distance of the encoded value from some center value, for example Gaussian functions. Sums of radial basis functions are used to approximate various functions and can be interpreted as a simple neural network where each neuron has a Gaussian response curve. Setting the time constants of this system will determine the speed of the movement, while setting the weights associated with each basis function alters the curvature of the movement. Planning a movement along a given trajectory then requires learning the values of these weights and time constants.

20.2.2 Reach Development Newborn infants aim their arm movements toward fixated objects (von Hofsten, 1982). These early arm movements have been related to the development of object-directed reaching (Bhat et al., 2005), leading to grasping, the development of which continues throughout childhood (Kuhtz-Buschbeck, Stolze, et al., 1998). The common thread to models of reach development is reinforcement-based learning of intrinsically motivated goal-directed actions based on exploratory movements, or motor babbling. In these models, movements are randomly generated in response to a target and the mechanisms generating the movements are modified via reinforcement. This process of variability and "opportunistic selection" is thought to introduce new behaviors into the motor repertoire.

Kuperstein (1988) developed an early model of reach development which learned to generate accurate reaches to visual stimuli with a multijoint arm. This model included a five-degree-of-freedom (DOF) arm and two controllable eyes. The model assumes that the circuit for directing the eyes to a target has already been trained but could also work with eye and arm circuits that are trained simultaneously. Initially, arm muscles are activated by randomly generated signals, and the eyes foveate on the resulting hand position. The eye muscle signals used to foveate the hand are represented as population codes in a gaze map. Visual input from the eyes is combined into a visual map, using binocular disparity to compute distance (see chapter 12). The gaze and visual maps are combined through modifiable weights to generate computed arm muscle signals. The learning rule adjusts each weight to minimize the difference between the computed and actual arm muscle signals. Once the weights have converged, the model is able to generate accurate reaches to targets in three-dimensional space when the random muscle activity generator is turned off. Given a target, the eyes foveate on it, and given the activity in gaze and visual maps, the model is able to generate arm muscle signals to reach to the target since the weights have been trained to represent the association between arm muscle signals and these maps.

Berthier's (1996) model of reach development hypothesizes that infants decompose reaching movements into a series of submovements in order to compensate for errors due to uncertainty in the effects of motor commands. In the jargon of control theory, this uncertainty could be due to inaccurate internal models (see chapter 9). Rather than develop associations between sensory and motor maps as the Kuperstein model does, this model focuses more on the structure of the movements themselves. In this highly simplified model, the state of the system is the planar coordinates of the hand (x,y) and the goal is to move the hand to a particular coordinate. The motor command output by the model is the desired displacement of the hand (x,y) and the desired average velocity, v. This output is applied to a model of the lower level motor controls and arm, with state transitions defined by

$$x(t+1) = x(t) + dx(t) + \eta\left(0, kv^2(t) + .2\right),$$
$$y(t+1) = y(t) + dy(t) + \eta\left(0, kv^2(t) + .2\right),$$

where $\eta\left(0,\sigma^2\right)$ is a Gaussian function with mean 0 and variance σ^2, and k is a parameter that controls the degree of uncertainty in control of the arm. The value of k is thought to decrease as the infant develops and learns more accurate internal models. Q-learning was used to learn which motor commands to execute in each state. Q-learning is a variant of reinforcement learning that learns the value of various actions in different states and thus operates on state–action pairs (Sutton and Barto, 1998).

Simulations with this model show that with a high value for k, representing an early stage of development, the optimal strategy is a series of high-speed submovements. As k decreases, simulating later stages of development, the optimal strategy changes to a single submovement to the target. This is in accord with data that show that as infants develop they reach with fewer, but longer, submovements (Von Hofsten, 1993). By investigating the transition in optimal strategies, Berthier found that before the model settled on a single high-speed movement, the strategy shifted from four or five high-speed submovements to one or two slow submovements. This predicts that a decrease in movement speed should accompany the developmental shift to fewer submovements. By fitting infant reaches to a series of minimum-jerk polynomials, Berthier was able to show that infant reaching movements consist of a sequence of minimum-jerk submovements and the model shows how a single minimum-jerk reach can develop from these submovements.

A later model extended this work to a dynamical arm model with a more sophisticated actor–critic reinforcement learning architecture (Berthier, Rosenstein, et al., 2005). Actor–critic architectures are composed of an actor which selects actions on the basis of their associated value and a critic which transforms externally received rewards into an internal reward signal (an estimate of expected reinforcement) which serves as the value used to train the actor (see chapter 10). This model used a simplified model of the arm with four DOFs and a simple linear spring muscle model. The muscle model generated joint torques required to move each joint to the configuration specified by the controller. Simulated movements with this arm model were shown to match recorded kinematics of infant reaches. The task of the controller was to learn a series of muscle rest lengths to produce joint torques that bring the hand to the target in the shortest possible time. In order to simplify the learning task, the authors assumed that infants take advantage of muscle synergies and limit the number of controlled DOFs. Only two DOFs were controlled by the model—flexion–extension and adduction-abduction of the shoulder (control of the elbow was not included). Berthier et al. showed that this model could learn optimal policies for reaching with varied initial and target locations and took advantage of uncertainty in the motor system to explore action space.

20.3 Grasping

This section is split into two subsections—grasp selection and planning, and grasp control. The selection and planning subsection discusses dorsal and ventral influences on grasp selection, grasp postures, and models of grasp selection and planning. The control subsection covers coordination of the reach and grasp and control of the preshape and enclosure phases of the grasp.

20.3.1 Grasp Selection and Planning Grasping in development seems to increasingly involve visual information in preprogramming the grasp. This involves mapping from a high-dimensional space of object features to another high-dimensional space of hand shapes for grasping. It has been suggesting that this mapping might utilize the concept of an opposition space, which is the space of possible contact surfaces on opposing sides of an object (Arbib, Iberall, et al., 1985). Two contact surfaces define an opposition axis through the object and a possible affordance for grasping it. The

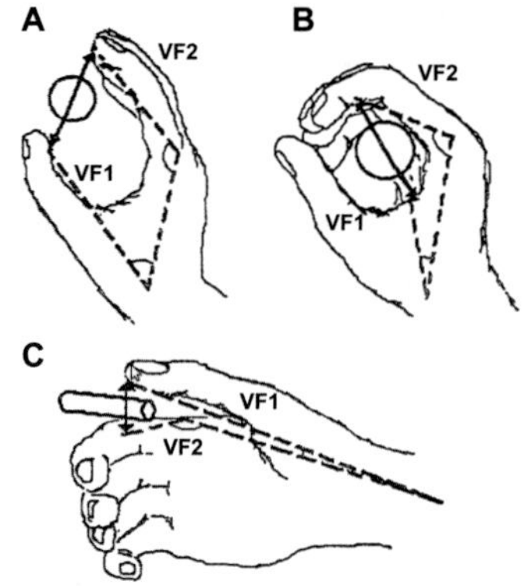

Figure 20.3 Examples of grasps involving two virtual fingers (VF1 and VF2). The dark lines indicate the opposition axis being grasped. The dotted lines show the virtual finger vectors. (A) Precision pinch. (B) Power grasp. (C) Side grasp. Modified from Iberall (1986).

virtual finger hypothesis states that grasping involves the assignment of real fingers to so-called virtual fingers or force applicators (Iberall, 1987), which define their own opposition axes. For example in a power grasp, one virtual finger might be the thumb and the other might be the palm (figure 20.3). The task of grasping is then to preshape the hand according to the selected virtual fingers and the size of the object and bring the opposition axis of the virtual fingers into alignment with the selected object surface opposition axis (grasp affordance). The real fingers composing a virtual finger act in unison to apply an opposing force on the object and against the other virtual fingers in a grasp. This concept replaces the analysis of the mechanical DOFs of individual fingers by the analysis of the functional roles of forces being applied in a grasp. The result is a projection from the high-dimensional space of possible configurations of real fingers to the lower-dimensional space of virtual finger opposition axes. Experimental evidence consistent with this hypothesis, also known as hierarchical control of prehension synergies, has been found (Smeets and Brenner, 2001; Zatsiorsky and Latash, 2004; Winges and Santello, 2005).

Computational models of grasp selection and planning typically use some variant of the virtual finger hypothesis and attempt to show how the brain can extract affordances from low-level object features and plan and control grasps appropriate for these affordances. The FARS model (named for Fagg, Arbib, Rizzolatti, and Sakata; Fagg and Arbib, 1998) is based on monkey neurophysiology and shows how grasp selection and control can occur via reciprocal connections between parietal area AIP, ventral premotor area F5, and supporting areas.

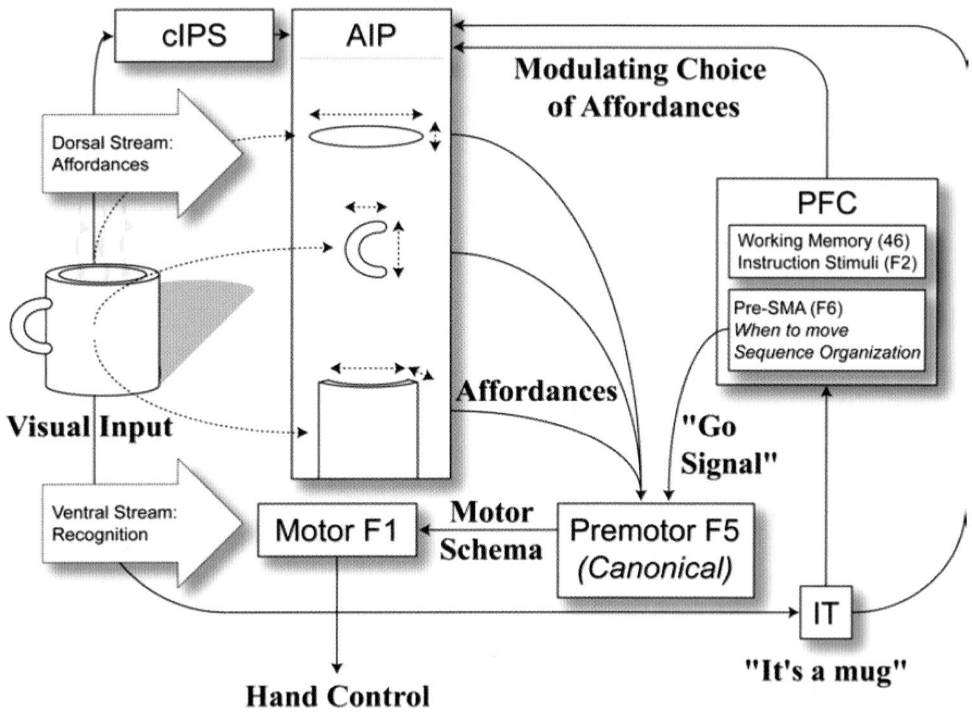

FIGURE 20.4 The original FARS diagram (from Fagg, 1996) is here modified to show prefrontal cortex (PFC) acting on anterior intraparietal area (AIP) rather than F5. The idea is that AIP does not "know" the identity of the object but can only extract affordances (opportunities for grasping for the object considered as an unidentified solid); PFC uses the inferotemporal cortex (IT) identification of the object, in concert with task analysis and working memory, to help AIP select the appropriate action from its "menu." Based on the output of AIP and sequence encoding output of the pre-supplementary motor area (pre-SMA), area F5 sends motor commands to the primary motor cortex (F1) for execution. From Arbib and Mundhenk (2005).

The FARS Model The basic planning and control system for grasping is modeled by the FARS model, which embeds F5 canonical neurons in a larger system (figure 20.4). In this model, the parietal area AIP processes visual information to implement perceptual schemas for extracting grasp parameters (affordances) relevant to the control of hand movements and is reciprocally connected with the canonical neurons of F5. Primary motor cortex (F1) formulates the neural instructions for lower motor areas and motor neurons. The dorsal visual stream (i.e., that which passes through AIP) does not know "what" the object is, it can only see the object as a set of possible affordances; however, the ventral stream (from primary visual cortex to inferotemporal cortex; IT) is able to recognize what the object is. This information is passed to prefrontal cortex (PFC), which can then, on the basis of the current goals of the organism and the recognition of the nature of the object, bias the affordance appropriate to the task at hand. AIP may represent several affordances initially, but only one of these is selected to influence F5. This affordance then activates the F5 neurons to command the appropriate grip once it receives a "go signal" from another region, F6 (pre-SMA), of PFC. As FARS was focused on the planning and sequencing of grasp phases and not on control of the hand, it used a highly simplified hand model which modulated the position of the thumb and index finger on the basis of F5 activity.

The FARS model was tested with an experimental protocol used to perform neurophysiology experiments on grasping in macaques (Taira, Mine, et al., 1990), referred to as the Sakata protocol. In this protocol, monkeys were trained to perform an object grasping task involving different types of objects and two cues indicated by differently colored lights. The monkeys were first shown the object without any cue (known as the Set phase). The first cue primed the monkey to start the movement to reach for and grasp the object (known as the Manipulation phase). The monkeys were trained to maintain the grip (known as the Hold phase) until the second cue appeared and then release it (known as the Release phase). This section focuses on the grasp selection and planning features of FARS that take place in the Set phase and initial part of the Manipulation phase while the "Online Control of Grasp Patterns" section discusses how the rest of the FARS model handles sequencing the different phases of the movement.

In FARS, the cIPS module is the first area in the dorsal visual stream. It contains populations of cells selective for different object shapes. Within each population, one subpopulation encodes the shape of the object and other subpopulations encode metric parameters of the shape in population codes. For example, the population of cells selective for cylindrical objects contains a subpopulation that is active whenever a cylindrical object is observed, one subpopulation that encodes the cylinder diameter in a population code, and one that encodes the cylinder length in a population code. Projections from cIPS to IT are hardwired for a simple set of objects (figure 20.5) so that IT can recognize the identity of a set of known object types based on their visual features.

The AIP module contains populations of neurons that are selective for affordances of each grasp type (precision pinch, power grasp, etc.). It receives input from cIPS encoding the object shape and from IT encoding the object identity. From these influences it represents the grasping affordances available and provides this information to F5. In FARS, the connections from cIPS to AIP are hardwired for a simple set of affordances. Clearly, this sets the stage for a learning model (see the grasp affordance extraction model described below). For example, cIPS neurons encoding cylinders with narrow diameters may excite AIP neurons selective for precision grasps while cIPS neurons that respond to large spheres might excite AIP power grasp neurons (figure 20.6). Similarly the projections from IT to AIP are hardwired such that object identity representations in IT activate grasp selective neurons in AIP when the identity of the object implies the type of grasp to use (e.g., lipstick case; figure 20.7).

Grasp selection can also be biased by task-dependent constraints rather than visual object features alone. In FARS the PFC, working in conjunction with F6, supplies task-dependent biases for grasp selection in F5 based on task requirements (such as which action will be performed after the grasp) or a working memory trace of

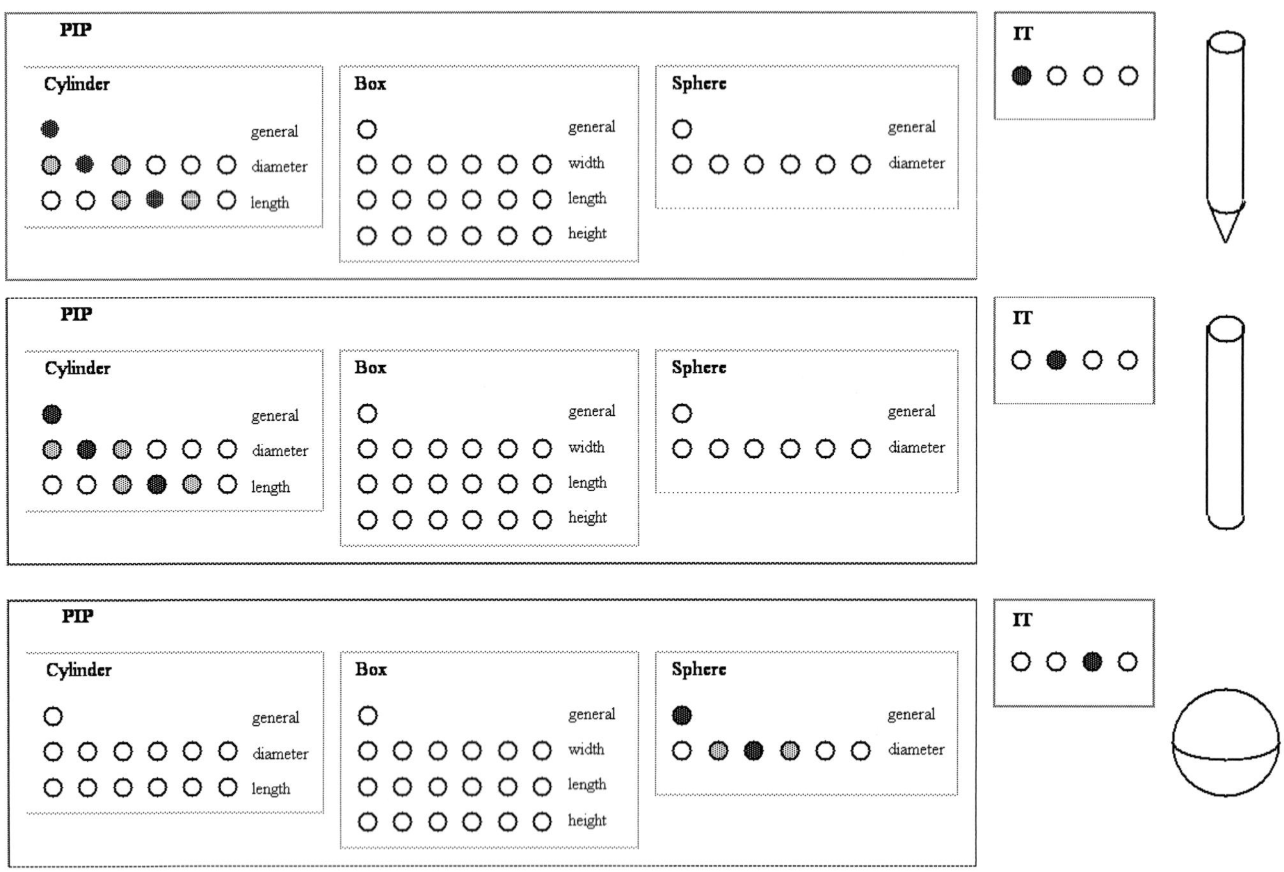

FIGURE 20.5 The caudal intraparietal sulcus (cIPS; also known as the posterior intraparietal area; PIP) module represents the shape and dimensions of an object. Projections to IT allow it to recognize a limited set of objects on the basis of this information. The cIPS representations of a pencil and cylinder (top and middle rows) are therefore similar, while their representations in IT are orthogonal. From Fagg (1996).

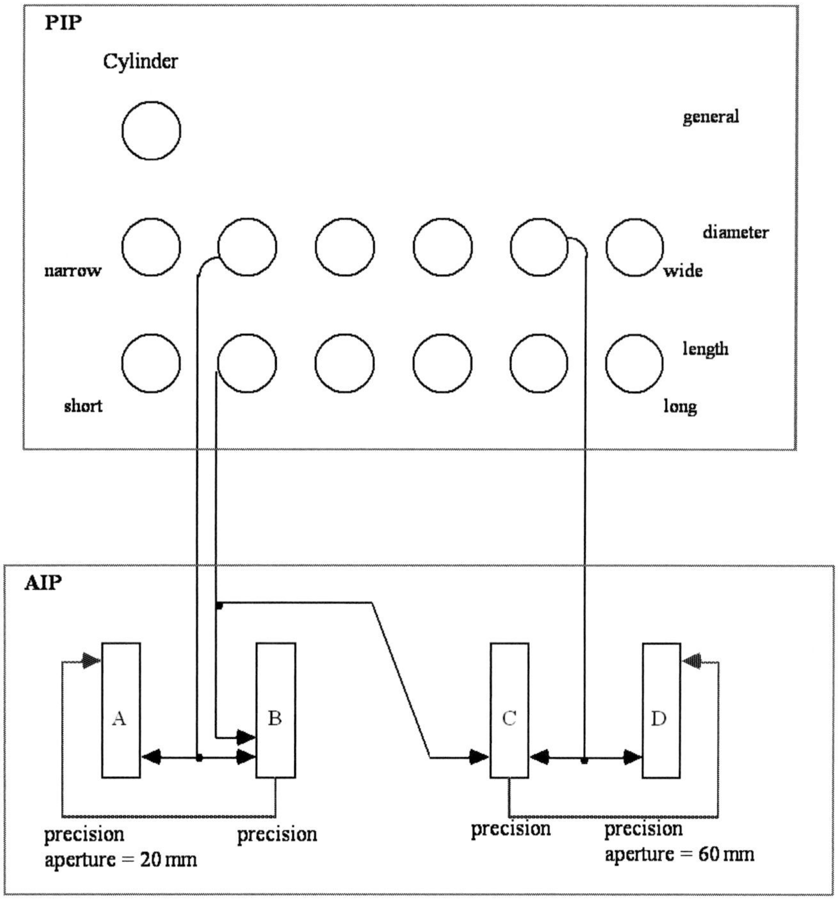

FIGURE 20.6 Projections from the caudal intraparietal sulcus (also known as the posterior intraparietal area; PIP) to anterior intraparietal area (AIP) are hardwired for a simple set of affordances. From Fagg (1996).

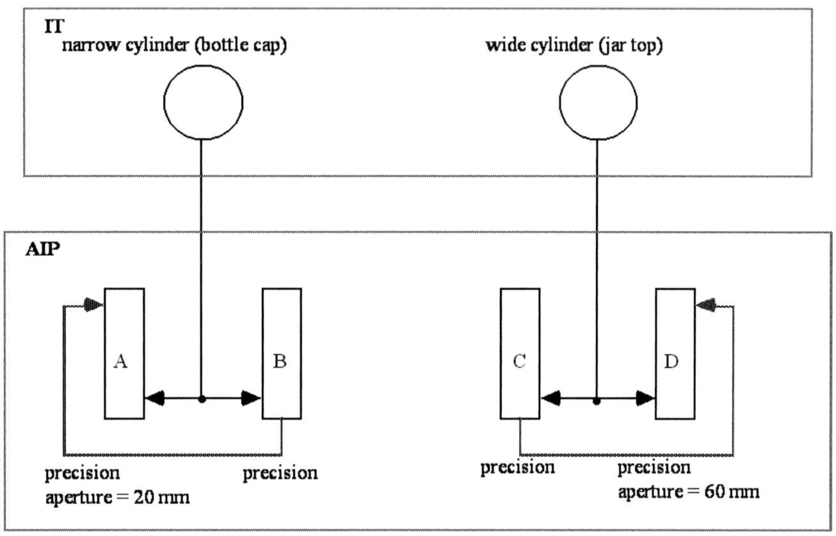

FIGURE 20.7 Projections from inferotemporal cortex (IT) to anterior intraparietal area (AIP) are hardwired for certain objects associated with a specific form of grasp required to use them. In this figure IT neurons selective for bottle caps activate precision grasp neurons in AIP selective for narrow aperture grasps, while IT jar top neurons active large aperture precision grasp neurons in AIP. From Fagg (1996).

a recently executed grasp. The biasing can be based on the class of grasp (e.g., power vs. precision) or include the parameters of the grasp (e.g., width of the aperture). The original FARS model suggested that the bias was applied by PFC to F5; however, subsequent neuroanatomical data (Rizzolatti and Luppino, 2001) suggest that PFC and IT may modulate action selection at the level of parietal cortex rather than premotor cortex. The "FARS Modificato" is a version of the FARS model updated to show this modified pathway.

In the FARS model, area AIP integrates visual (bottom-up) and contextual (top-down) information to select a set of affordances representing potential grasps while the final selection of the grasp occurs in F5. Each of these processes involves competition and cooperation, with the competition in the F5 circuit tuned to implement a winner-take-all process. In both AIP and F5, columns of neurons encoding the same or similar grasps cooperate via local excitatory projections (figure 20.8 and figure 20.9). Competition is implemented by loops through the basal ganglia that inhibit neural columns selective for different grasps. The result is a representation of a "menu" of possible grasp affordances in AIP that are appropriate for the object and the current task, and selection of a single grasp type in F5.

FARS also includes reciprocal projections from F5 back to AIP that target a subset of the neurons in that region. Projections between regions are generated probabilistically when the model is initialized, resulting a variety of connectivity profiles for neurons in AIP. These can range from neurons that solely receive input from cIPS and/or IT (visual), neurons that only receive feedback from F5 (motor), or some mixture of the two (visuomotor). This matches what is experimentally observed—about 40% of neurons in area AIP respond equally well when their preferred grasp is performed in the light or the dark (motor dominant neurons; Taira, Mine, et al., 1990). Other AIP neurons only respond during grasp performance or even just object fixation in the light (visual dominant neurons) while others respond during grasps in the dark but their activity is modulated by the sight of the object in the light condition (visual/motor neurons).

In monkeys about 50% of the neurons in AIP are grasp selective (Taira, Mine, et al., 1990), and many show object-specific activity. Other cells demonstrate specificity to other features of the object such as its size, position, and orientation. In order to investigate object and grasp coding in AIP and F5, the model was presented with two objects (a cylinder and a rectangular plate) that afford the same type of grasp (precision grip with 10-mm aperture, figure 20.10, left). Because the items are shaped differently, their representations in cIPS are nonoverlapping. The middle panel of figure 20.10 shows the responses of two visual-related AIP cells when presented with each object. Neuron A is active only for the condition in which the system is grasping the plate while neuron B is somewhat active while grasping the cylinder but is most active when grasping the plate. Two AIP cells with both visual- and (somewhat) motor-related components are shown in the right panel

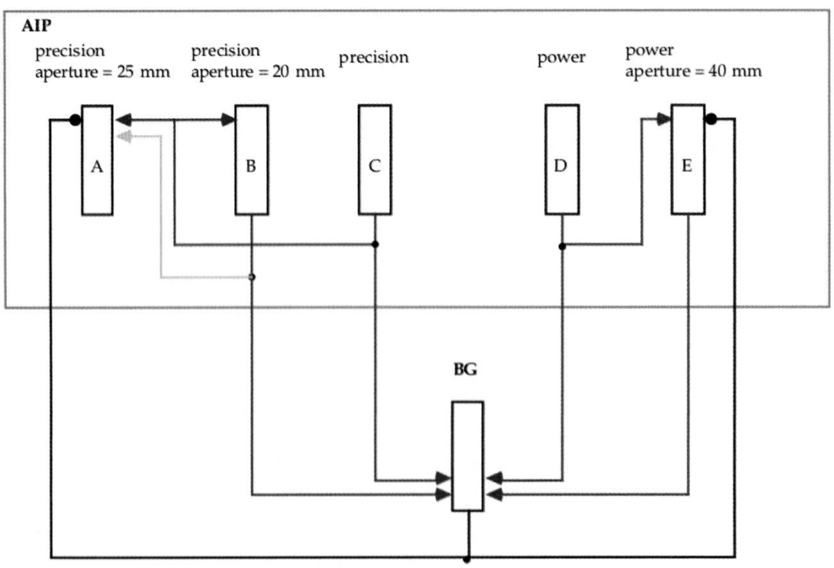

FIGURE 20.8 Grasp selection dynamics in anterior intraparietal area (AIP). Columns in AIP that encode the same or similar grasps exchange excitatory connections. Columns participating in different grasps inhibit each other via loops through the basal ganglia (BG). From Fagg (1996).

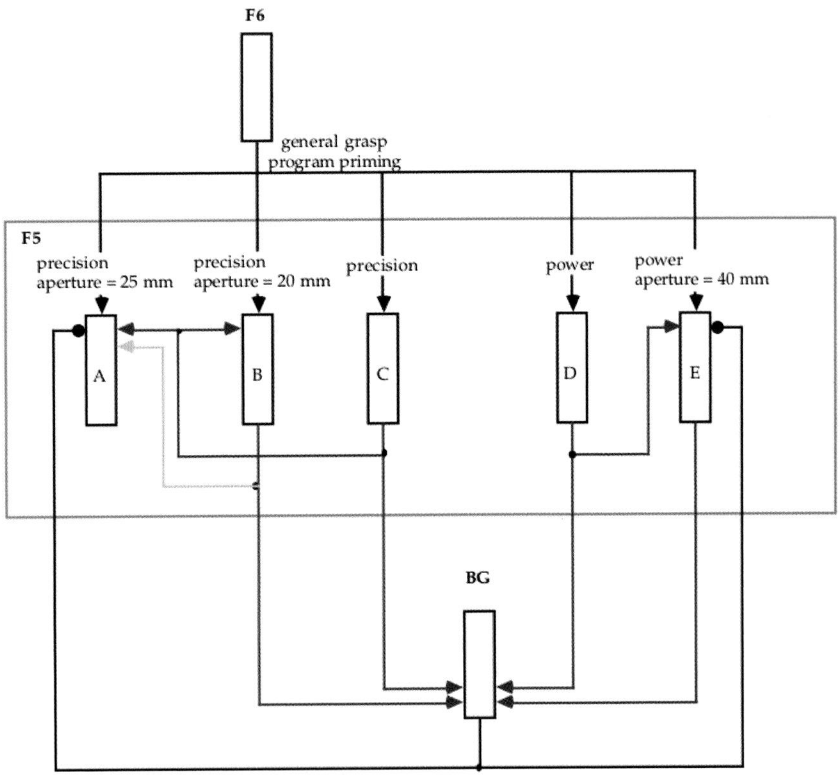

FIGURE 20.9 The final grasp selection process involves cooperation via local excitatory projections between F5 columns and competition via inhibitory loops through the basal ganglia (BG). From Fagg (1996).

of figure 20.10. Cell A responds more quickly to the cylinder when the object is first presented to the model while cell B responds more strongly when grasping the plate. The firing of both cells is modulated by the phase of the movement (see the "Online Control of Grasp Patterns" section).

The population responses in AIP and F5 while grasping the plate or the cylinder demonstrate the relationship between object and grasp specificity in these regions. When grasping in the light, AIP has a range of object responses, with some cells responding in the same way to both types of objects, and others showing a high level of object specificity. F5, however, shows mainly grasp selectivity, and each cell responds with almost the same level of activity to each object. In order to relate the object/grasp coding distinction with the visual/motor distinction in AIP, the model was tested during fixation only and grasping in the dark conditions. The fixation condition allows the visual response of the AIP population to be analyzed without the effects of the feedback projections from F5 (since no grasp is being performed). There are some subsets of visual and visuomotor AIP cells that receive input from the cIPS populations that encode both objects. However there are cells that respond only to the plate or the cylinder and therefore receive input exclusively from nonoverlapping cIPS populations. The grasping in the dark condition allows the effect of the feedback projections from F5 to be seen on the visuomotor and motor neurons in AIP. This analysis shows that none of the visuomotor or motor AIP cells show object specificity since their activity in this condition was determined by F5 activity which is grasp specific. This simulation makes the prediction that very few AIP cells classified as motor related should show object specificity in their visual responses.

As noted at the start of the section, the ventral path via IT and PFC can modify the task- or memory-dependent choice of affordance in AIP and/or grasp in F5. FARS was the first neural network model to address the link between AIP and F5 and their role in extracting grasp affordances and selecting appropriate grasps to exploit them. It explains many of the features of these two regions such as object and grasp selectivity in each region as well as the phase-specific activity of F5 canonical neurons and their possible role in sequencing the various phases of the movement (see the "Online Control of Grasp Patterns" section). The connectivity in FARS is all hardwired in order to generate the neural activity seen in macaque neurophysiology. This sets the

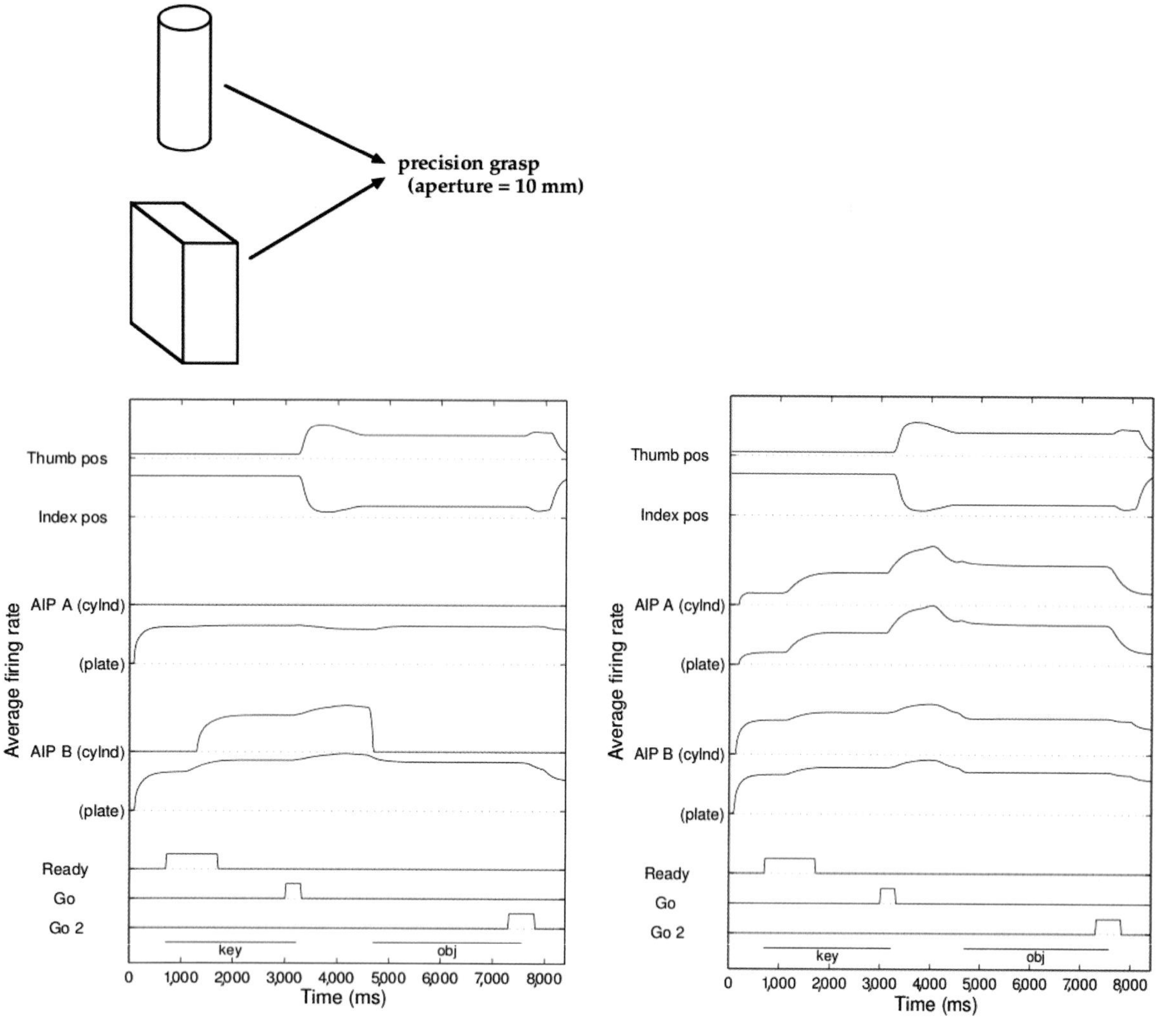

FIGURE 20.10 Top: Two different objects that afford the same type of grasp. Bottom left: Two visual anterior intraparietal area (AIP) cells that show significant object-specific modulation. The top two traces of each plot show the position (pos) of the thumb and index finger. The next four traces show the activity of two AIP neurons (A and B) in response to the cylinder (cylnd) or plate. Neuron A fires throughout the trial when presented with the plate and not at all for the cylinder. Neuron B fires during the early phases of grasping the cylinder but also fires more strongly and throughout the trial when presented with the plate. Bottom right: Two visual-related (with some motor) AIP cells that demonstrate only slight modulation as a function of the type of object that is presented. Each neuron is active throughout the whole trial with the plate and the cylinder and shows some modulation by the phase of the movement. Modified from Fagg (1996).

stage for a model which can modify connection weights on the basis of learning.

The TRoPICALS Model The FARS model was based heavily on data from the macaque brain, but it was hypothesized that it also describes how homologous regions in the human brain support the visually guided reach to grasp, and synthetic brain imaging (chapter 15) has been used to make predictions about positron emission tomography imaging of the brains of humans reaching to grasp different objects (Arbib, Fagg, et al., 2002). The two route, prefrontal instruction, competition of affordances, language simulation (TRoPICALS) model builds on FARS *considered as a model of the human brain* (since monkeys do not have language), not only to show how grasps appropriate to the affordances extracted by different objects can be selected but moreover ties this in with a simple model of language in order to study the effects of verbal cuing of grasps (chapter 24, "Language Processing," provides more details on the language aspects). As in the FARS model, the TRoPICALS model is based on the principles of dorsal–ventral neural pathways, top-down biasing, and neural dynamic competition but includes the priming influence of language on grasp selection (figure 20.11; Caligiore, Borghi, et al., 2010). When subjects are asked

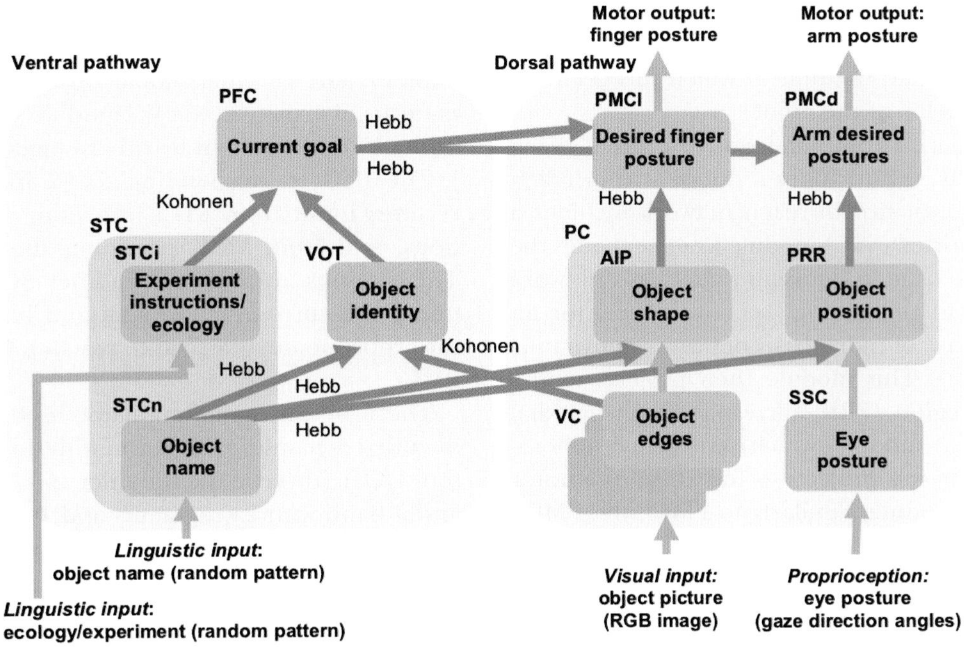

FIGURE 20.11 General architecture of the TRoPICALS model. Light arrows indicate connections which were hard-wired while dark arrows indicate connections which were updated by learning processes based on a Hebbian learning rule or a self-organizing map. Input to the model is provided by three RGB visual neural maps (VC) and a somatosensory map (SSC). The dorsal pathway implements sensorimotor transformations needed to perform an action on the basis of the current perceptual inputs, and the ventral pathway allows flexible control of behavior due to biasing effects exerted by the prefrontal cortex (PFC) on action selection. The PFC integrates linguistic input providing the experimental instructions with object identity input from the VOT which can be derived from visual input via VC and/or linguistic input from STCn. AIP, anterior intraparietal area; PC, parietal cortex; PMCd, dorsal premotor cortex; PMCl, lateral premotor cortex; PRR, parietal reach region; SSC, somatosensory cortex; STCi, superior temporal cortex for instructions; STCn, superior temporal cortex for nouns; VC, visual cortex; VOT, ventral occipitotemporal cortex. From Caligiore, Borghi, et al. (2010).

to observe an object and perform some action in response, their response time (RT) and error rate can be influenced by whether the action was appropriate for that object (congruent) or inappropriate for the object (incongruent). If the sight of the object automatically activates motor programs for interacting with it in a bottom-up manner, RTs will be shorter and error rates lower in congruent than in incongruent trials. This model provides a general framework to account for grasp compatibility effects between the visual and motor perception of objects.

Like FARS, the TRoPICALS model is split into dorsal and ventral pathways. The dorsal pathway is further subdivided into networks for controlling the reach and grasp, respectively. On the basis of visual input providing an image of the object and the gaze direction angle, the dorsal stream plans appropriate finger and arm target postures for reaching and grasping the object. The ventral stream includes a network for representing experimental instructions and object identity based on linguistic or visual input, and a PFC model that determines the current goal and can bias selection of the grasp or reach parameters in the dorsal stream.

In the dorsal stream grasp network, the visual cortex forms an edge representation from the raw visual input, area AIP uses this representation to extract object shape (affordance) information, and the lateral premotor cortex (PMCl, corresponding to area F5 in FARS) plans finger postures for grasps based on these affordances. The visual cortex module includes three maps that perform edge detection filtering on the visual input image in red, green, and blue. In order to represent object shape, the AIP module includes neurons that receive the average activation level of RGB neurons in visual cortex (VC) with similar receptive fields. The PMCl module computes the output to the simulated hand in terms of desired angles of the thumb and another virtual finger. Desired angles are selected based on the inputs to PMCl using a dynamic neural field (Erlhagen and Schoner, 2002). Dynamic neural fields use competition and cooperation between neurons that encode continuous values of some variable (see chapter 1). In TRoPICALS, the action is triggered when some cluster of neurons in the field reaches a threshold level of activation, and the time at which this occurs is then taken as the RT of the model.

While FARS lacks a computational account of reach planning, TRoPICALS includes a dorsal stream reach network. The dorsal stream reach network includes a somatosensory cortex (SSC) module which encodes the direction of gaze using a Gaussian radial basis network (see the "Trajectory Planning" section, above, for an example of the use of radial basis networks in dynamic motor primitives). This module then projects to the parietal reach region (PRR; corresponding to medial intraparietal area) module, which represents the location of the object and projects to the dorsal premotor cortex (PMCd) module. Similarly to PMCl, the PMCd module selects arm postures using a competition and cooperation process based on dynamic neural fields, and the RT is taken as the time in which some cluster within the field reaches an activation threshold. The model was tested on tasks involving only reaching or only grasping and therefore does not address the coordination of the reach and grasp movements (see the "Reach and Grasp Coordination" section).

The ventral visual stream determines the goal of the current action and on that basis, biases selection of the appropriate grasp and reach posture in the dorsal stream. The superior temporal cortex (STC) module handles linguistic input and contains two submodules: the superior temporal cortex for nouns (STCn) and the superior temporal cortex for instructions (STCi). Neurons in STCn encode linguistic input concerning object names and receive a different pattern of activity for each object. These neurons then project to the parietal cortex to prime object shape and position representations in AIP and PRR, respectively. Neurons in STCi encode linguistic input concerning the instructions of the task and were activated with one of two random patterns corresponding to those instructions.

The ventral occipitotemporal cortex (VOT; corresponding to IT in FARS) module receives input from STCn and VC and encodes object identity. In order to recognize (categorize) objects based on their visual features, the activity in the VC module is recoded into a self-organizing map (SOM). SOMs are neural networks trained using unsupervised learning to generate low-dimensional representations of some high-dimensional space while preserving its topography (see chapter 13, "Neural Maps: Their Function and Development," for an exposition). This means that representations that are close together in the high-dimensional input space are also close together in the low-dimensional output space of the network. In this model, the SOM in the VC module allows the module to encode object features as population codes which are learned in an unsupervised manner. The fact that similar objects will elicit similar patterns of activity in the VC module simplifies the task of the VOT module in learning object categorization.

The PFC (corresponding to F2 in FARS) module receives input from STCi, providing the task instructions, and from VOT, providing the object identity. These inputs are fed into another SOM which determines the current action goal and biases selection of the appropriate grasp and reach target postures in PMC.

The arm and hand simulator used by TRoPICALS included dynamics and an arm with three segments and four DOFs (three in the shoulder and one in the elbow) and a hand with 21 segments and 19 DOFs. Similar to the Berthier et al. (2005) model, a simple muscle model was used to apply torques to each joint given a desired joint angle. However, as the simulations involved the generation of grasps to single objects and reaches in a plane, only two DOFs of the hand and two DOFs of the arm were actually controlled by the model.

Whereas all of the neural projections in FARS were hardwired, TRoPICALS incorporates, as we have seen, several learning mechanisms to develop connectivity between its modules. The model was trained in three stages, roughly corresponding to stages of development.

In the first learning phase the model learned grasps appropriate for object shapes (anterior intraparietal area, AIP, → lateral premotor cortex, PMCl, weights), reaches appropriate for object locations (parietal reach region, PRR, → dorscal premotor cortex, PMCd, weights), and to recognize objects on the basis of their visual features (visual cortex, VC, → ventral occipitotemporal cortex, VOT, weights). The AIP → PMCl and PRR → PMCd connection weights were modified using a "motor babbling" phase, in which the model explores random movements, and a Hebb covariance learning rule (see chapter 6). This process resulted in mappings from the perceived shape of the object (AIP) to the hand posture required to successfully grasp it (PMCl), as well as mappings from the perceived location of the object (PRR) to the arm posture required to successfully reach it (PMCd). At the same time, projections from VC to VOT were modified using a SOM, resulting in clusters of neurons in VOT that respond to objects with similar visual features.

The second phase of learning simulated learning of object names by presenting an object to the model and activating the superior temporal cortex for nouns (STCn) module with the pattern of activity representing the object's name. This phase modified projections from STCn (encoding the linguistic input) to AIP (encoding the object shape) and VOT (encoding the object identity), using a Hebb covariance rule.

The last learning phase simulated learning to follow task instructions. In this phase, linguistic input was provided to superior temporal cortex for instructions (STCi), connections from STCi and VOT to PFC were modified using a SOM, and connections from PFC to PMC (dorsal and lateral) were modified using a Hebbian learning rule. Training the weights from STCi and VOT to PFC allowed the PFC module to represent the goal of the task given the instructions and object identity, while the PFC $\rightarrow$ PMC weights encoded the learned associations between goals and the actions required to achieve them.

The TRoPICALS model was tested in several simulation experiments involving congruent or incongruent linguistic and visual input. It was tested with various simulated objects including cylinders, spheres, and boxes of varying sizes by applying images of the target object to the VC module which encoded the RGB values of each pixel and performed edge detection, as described above. Here we review the results of the first simulation experiment that demonstrates the effects of top-down biasing by PFC on the grasp selection mechanism in PMC. This simulation was performed after the model was subjected to the first two phases of learning. After the first phase the model was able to execute grasps appropriate for objects based on their size and shape and could identify them on the basis of their visual features. The second learning phase simulated acquisition of the experimental task. When STCi was activated with the ecological task pattern, the model was required to perform a grasp appropriate to the object size. In other trials STCi was activated with the experimental task pattern, which required the model to perform a power grip for natural objects and a precision grip for human-made artifacts (or vice versa). On congruent trials the grip cued in the experimental condition matched that afforded by the object, and on incongruent trials the grip was inappropriate for the object. The model was able to reproduce the main result of Tucker and Ellis (2004), namely, that RTs were faster in congruent versus incongruent trials, suggesting that the appropriate grasp is automatically primed by the sight of the object (figure 20.12).

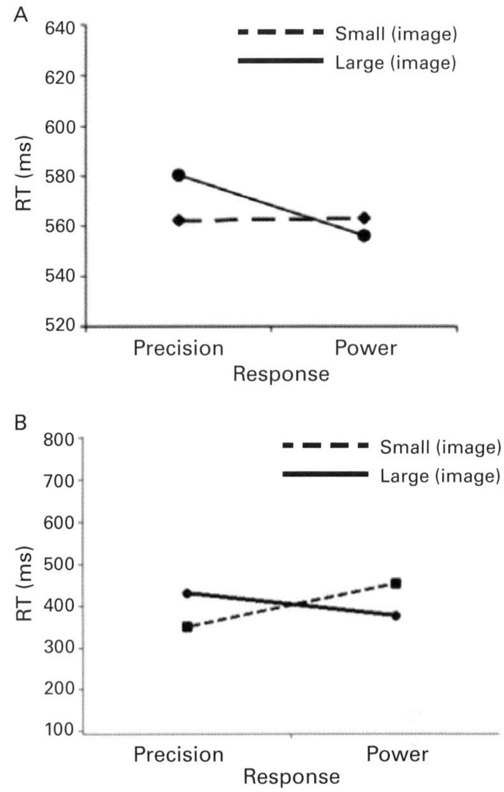

FIGURE 20.12 (A) Average reaction times (RTs) seen in human subjects (Tucker and Ellis, 2004). (B) Average reaction times generated by the TRoPICALS model. From Caligiore, Borghi, et al. (2010).

An examination of the activity of the model during congruent and incongruent trials provides an explanation of the mechanisms behind this effect. During incongruent trials, the ventral visual pathway activates a representation in PMC for an action which is different from the one activated by the dorsal pathway (figure 20.13A). The influence of PFC on PMC is stronger than that of AIP, and therefore the representation activated by the ventral stream wins the competition, but this increases the reaction time (RT). However during congruent trials, the ventral and dorsal streams activate the same action representations in PMC (figure 20.13B). This overlapping activation causes these neurons to reach the action-triggering threshold faster, reducing RT in the congruent condition.

The TRoPICALS model then goes beyond the published data to make predictions that could be tested with further experiments. One of these predictions concerns the effect of task overtraining on RTs. The basic hypothesis is that when a subject has performed a task enough times, the dorsal stream will begin to learn the task-specific knowledge encoded in the ventral stream. This should reduce compatibility effects on RTs in congruent versus incongruent conditions. To test this hypothesis, the model was trained for 16,000 trials on the same task. During incongruent trials, learning between AIP and PMC slowly influenced PMC to select actions consistent with the experimental task (and inappropriate for natural grasping of the object). The result was that interference between the dorsal and ventral streams in their influence on PMC gradually reduced and so did the compatibility effects seen in RTs.

20.3.2 Grasp Development The infant learning to grasp model (ILGM; Oztop, Bradley, et al., 2004) demonstrates the use of population codes and winner-take-all dynamics with reinforcement learning to learn the motor parameters needed to plan successful grasps.

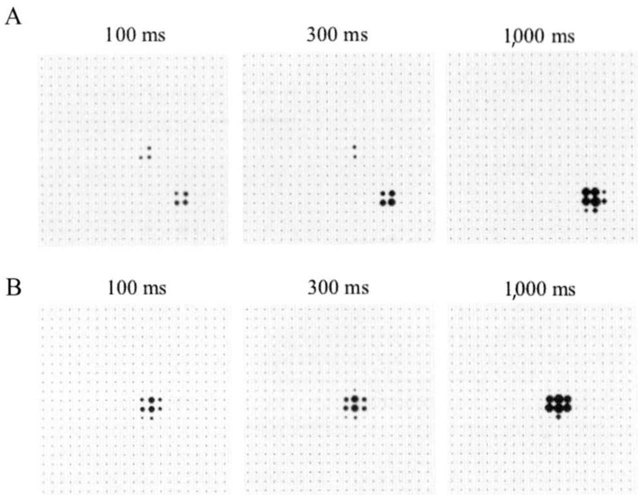

Figure 20.13 (A) Premotor cortex (PMC) activity during an incongruent trial. Conflicting inputs from prefrontal cortex (PFC) and parietal cortex (PC) cause two different action representations to compete until the one activated by the PFC suppresses the cluster activated in a bottom-up manner by the PC. (B) PMC during a congruent trial. Overlapping inputs from PFC and PC activate only one action representation. From Caligiore, Borghi, et al. (2010).

ILGM provides a learning mechanism that complements the FARS and TRoPICALS models. It addresses the earliest stages of such learning—the initial open-loop reaching strategies, subsequent feedback adjustments, and emergent feedforward strategies employed by infants in the first postnatal year (figure 20.14). Infants quickly progress from a crude reaching ability at birth to finer reaching and grasping abilities around 4 months of age, learning to overcome problems associated with reaching and grasping by interactive searching (see Berthier's developmental models of reaching, described above). At first, the poorly controlled arm, trunk, and postural movements make it very difficult for the young infant; nonetheless, experience eventually yields a well-established set of grasps, with preshaping to visual affordances by 12–18 months of age (von Hofsten and Fazel-Zandy, 1984). This model simulates reinforcement learning of motor parameter values for grasping through trial-and-error learning during a period of motor babbling.

ILGM is a systems-level model based on the broad organization of primate visuomotor control with visual features extracted by the parietal cortex and used by the premotor cortex to generate high-level motor signals that drive lower motor centers, thereby generating movement. It assumes that initial reaches are inherently variable and provide a basis for reinforcement learning. The reinforcement signal is related to the extent to which the current action yields a stable grasp of an object. The requirements of the model are that infants (1) will execute a reflex grasp when triggered by contact of the inside of the hand with an object encountered while reaching, (2) can sense the effects

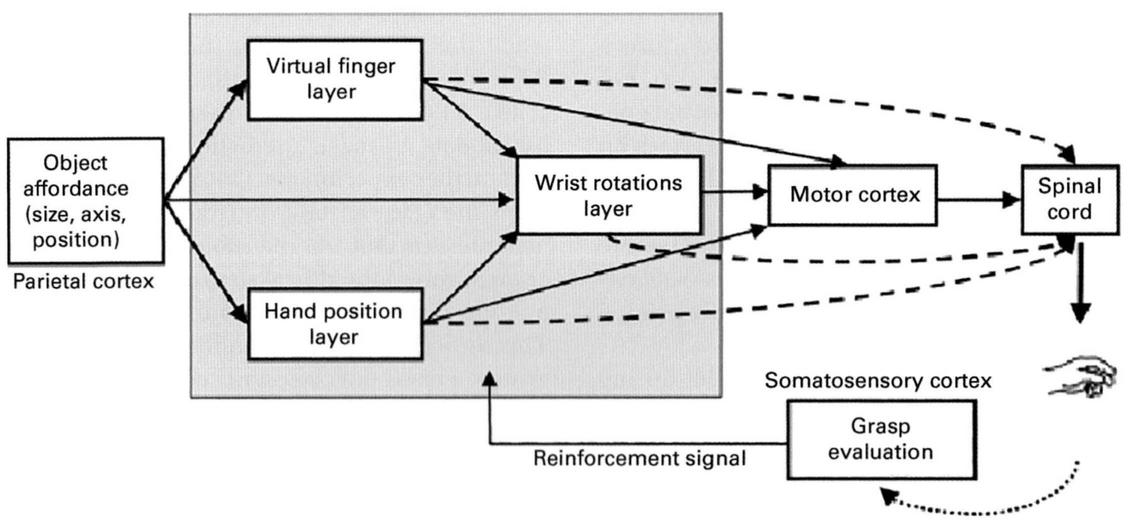

Figure 20.14 The structure of the infant learning to grasp model. The individual layers inside the gray box are trained based on a *Joy of Grasping* reinforcement signal arising from somatosensory feedback concerning the stability of the current grasp.

of their actions, and (3) will use reinforcement learning to adjust their movement planning parameters.

The reinforcement from object contact is used to modify grasp generation mechanisms within the premotor cortex. The model consists of four modules: the input module (Parietal Cortex), the grasp learning module (Premotor Cortex), the movement generation module (Motor Cortex and Spinal Cord), and the grasp evaluation module (Somatosensory Cortex). The grasp learning module formulates a grasp plan and instructs the movement generation module, located in both the spinal cord and motor cortex. The movement generation module completes task execution, and the sensory stimuli generated by the execution of the plan are then integrated by the movement evaluation module, located in the primary SSC. Output of the SSC, the "joy of grasping" reinforcement signal, is used to adapt the connection strength between Parietal–Premotor connections (figure 20.14). Implementation of the reinforcement signal is defined simply in terms of grasp stability: a grasp attempt that misses the target or yields an inappropriate object contact produces a negative reinforcement signal. Prior to learning, grasps are generated by a grasp reflex, which encloses the fingers once the hand makes contact with the object. As learning progresses, these reflex-triggered grasps are gradually replaced by learned grasps.

In contrast to TRoPICALS, ILGM used a kinematic simulator (no forces were modeled) with a 24 DOF arm and hand. The shoulder and wrist had three DOFs, and the elbow had one. Each finger contained two joints with one DOF, and the thumb was modeled with one joint with three DOFs. Reaching was implemented using inverse kinematics techniques from robotics. To move the arm, the joint angles were incrementally updated until they reached the target joint angles given by the inverse kinematics algorithm. Tactile sensation of the object was generated by object collision detection. The simulation environment was therefore detailed enough to interface with a kinematic model such as ILGM (as well as the mirror neuron system [MNS] and MNS2 models for which it was used as well, described below), but not as detailed as the environment used by TRoPICALS which included dynamics (although that model did not directly control the dynamics).

During early visually elicited reaching, while infants are reaching toward visual or auditory targets, they explore the space around the object and occasionally touch the object (Clifton, Muir, et al., 1993). ILGM models the process of grasp learning starting from this stage. The model represents infants' early reaches using an object-centered reference frame and includes random disturbances during initial reach trials so that the space around the object is explored (recall the role of exploration in Berthier, Rosenstein, et al., 2005). The *Hand Position* layer specifies the hand approach direction relative to the object, restricting the angles from which the object can be grasped and touched on the basis, for example, of the location of the object relative to the infant's shoulder; the *Wrist Rotation* layer learns the possible wrist orientations given the approach direction specified by the Hand Position layer; and the *Virtual Finger* layer indicates which fingers will move together as a unit.

In ILGM, the parietal and premotor cortices are modeled as layers of neural populations whose firing rates provide a probability code. In contrast to population codes which encode a single value in the activity of many neurons (see chapter 1), probability codes represent a probability distribution over multiple values. The neurons in each layer fire at time t with a probability proportional to the sum of their weighted synaptic input. The intralayer connectivity is center-surround, so a Gaussian distribution of activity forms around the first-firing neuron. The weighted synaptic input to the layer can be interpreted as a probability distribution on the parameter it encodes. This distribution can easily be biased by input from other regions. This architecture has the advantage that given a stimulus, it can represent a menu of possible values with varying probability of choosing each.

In order to test the workings of the model, it was run in a simulation experiment in which a sphere was placed in different locations in the work space. In this simulation the grasp plan consisted of only the object-centered wrist offset vector—the hand enclosure speed was fixed, and the palm was automatically oriented toward the object. The goal of this simulation is for the model to learn which offset vectors (which define the hand approach direction) will support a stable grasp. Figure 20.15A shows a grasp generated by the model during a trial in which the object was elevated relative to the shoulder in the work space. The simulated arm could not grasp the object from the top, and therefore the model learned to only approach from underneath and front/right. The Hand Position layer encodes a three-dimensional probability distribution over the azimuth, elevation, and radius parameters (α, β, r), which is projected onto the (α, β) plane for visualization purposes by summing over the dimension encoding the radius. The model learned to encode the approach directions $\beta < 0$ (below the object) and $-90 < \alpha < 90°$ (right and/or front of the object) with a higher probability, indicating that a grasp plan using these parameters is more likely to result in a stable

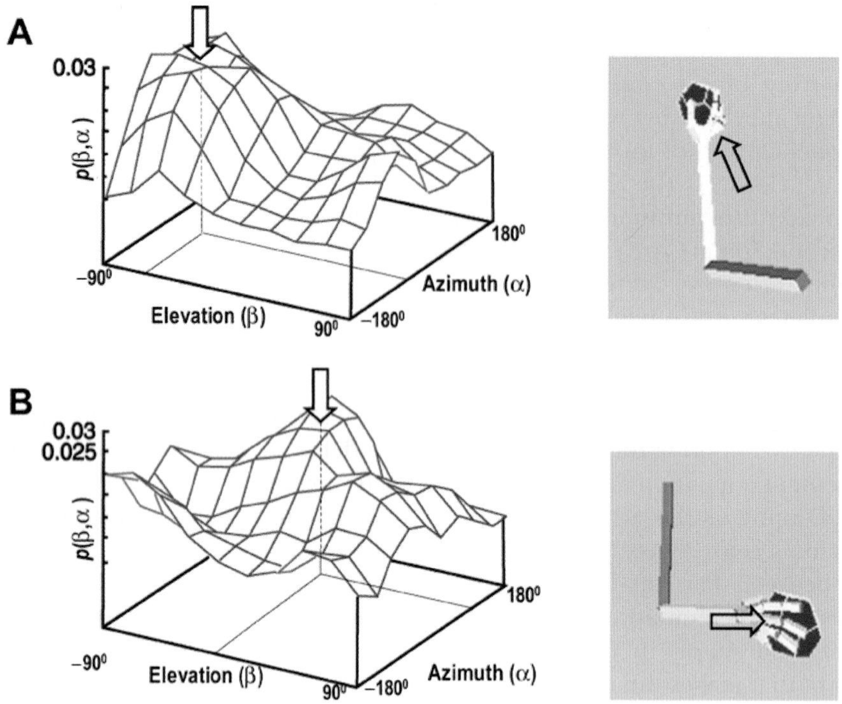

FIGURE 20.15 Learning to grasp with automatic palm orientating. Probability distribution encoded by the Hand Position layer (left) and a sample generated grasp (right) to an object located above the shoulder (A) and an object located to the left and below the shoulder (B). From Oztop, et al. 2004).

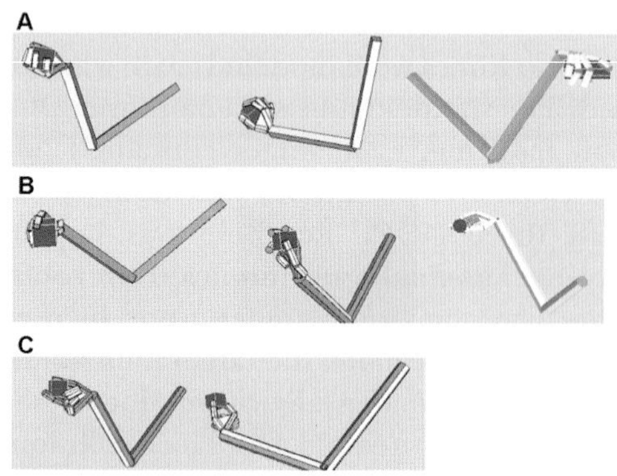

FIGURE 20.16 Learning to grasp without automatic palm orienting. (A) Examples of power grasps performed to different objects. (B) Precision grasps of a cube using the thumb and two or more fingers. (C) Two-finger precision grasps of a cube (from Oztop, et al., 2004).

grasp. In another trial the object was located to the left and below the shoulder. In these trials the model learned that a grasp directed to the right side ($\alpha > 0°$) of the object is likely to result in a stable grasp (figure 20.15B).

Without automatic palm-orienting, the model was still able to learn to generate power and precision grasps (power being most frequent). Figure 20.16A shows examples of power grasps for a cube, a sphere, and a cylinder (generated after separate learning sessions for each object). Examples of generated precision grasps using the thumb and two or more fingers are shown in figure 20.16B, and figure 20.16C shows examples of two-finger precision grasps. These simulation results demonstrate that precision grips can be generated from open-loop grasp learning, and that the palm orienting behavior can develop through learning rather than being innate.

Although infants younger than 7 months old can grasp an object once they make contact with it, they do not appear to use visual affordance information. In order to test the effect of affordance availability to the model, it was presented with a cylinder in horizontal, diagonal, or vertical orientations. The orientation encoding was disabled in order to simulate grasp learning in younger infants, and enabled in order to simulate older infants' learning. The Affordance layer was modeled as a population code encoding the orientation of the cylinder. In this simulation, different cylinder orientations afford different hand orientations, and therefore without affordance information, the model was not able to learn a perfect grasping strategy. With affordance-encoding enabled (simulating the 9-month-old infant), the model learned to perform grasps by associating the affordance (the orientation) of the cylinder with suitable grasp parameters.

After learning to grasp objects at varying positions, ILGM was able to learn to generalize over different locations, generating grasps to objects at locations not used during training. When the model was run again with exactly the same object locations, it generated a different set of grasping postures. This demonstrates that ILGM combines the ability to generalize over object affordance information as well as represent multiple grasp plans for a given object.

The ILGM model explained the development of neurons with properties similar to F5 canonical neurons. A related model, the grasp affordance extraction model (GAEM, Oztop, Imamizu, et al., 2006a) uses competitive supervised learning to extract affordances from low-level object features and map them to the motor parameters learned by ILGM. It builds on ILGM in order to train a model of AIP to extract object features relevant for grasping (affordances) from low-level visual features. This model simulates the development of affordance extraction through experience in grasping objects. In this model, AIP provides object affordance information to the canonical neurons of area F5 for the selection of grasp motor parameters. It is likely that the affordance extraction circuit modeled by GAEM and the grasp planning circuit modeling in ILGM are trained simultaneously. However, to simplify the implementation, GAEM uses a staged learning approach. In this implementation, GAEM learning is run using the successful set of grasps generated by ILGM—once ILGM learning is complete and its weights are fixed. More recent work offers an integrated model of the learning of grasps and affordances (Bonaiuto & Arbib, 2015).

20.3.3 Online Control of Grasp Patterns The models in the previous section address selecting and planning a grasp but for the most part do not simulate control of an ongoing pattern of grasping. However, FARS takes a step in this direction by including mechanisms to sequence different phases of the grasp such as extension of the fingers and flexion (enclosure), as well as the movements required for the hold and the release segments of the Sakata protocol. There are few neural models of such control—here we review elements of the FARS model relevant to the topic and then several efforts from the field of robotics.

Earlier, we saw how in the FARS model, ventral premotor region F5 contains populations of neurons that are selective for each grasp type (precision pinch, power grasp, etc.). In addition to cells that select which grasp to perform, area F5 also contains cells selective for the phase of each particular type of grasp. Area F6 and the basal ganglia collectively represent the high-level wait–reach/grasp–hold–release motor program and gate the flow of activity in F5 (see chapter 22, "Integrative Functions of the Corticostriatal System," for more on modeling the basal ganglia). The basal ganglia module contains neurons selective for each phase of all grasp types (set, extension, flexion, hold, and release). Based on inputs from F5 indicating the current grasp phase, it primes F5 cells selective for the next grasp phase and inhibits F5 cells selective for the last one. The F6 module detects the ready and go signals and biases the selection of grasp type in F5 according to the current task. Projections from F5 back to AIP continuously update AIP's representation of the grasp actually being performed (figure 20.17). The secondary somatosensory cortex (SII) detects when the hand aperture reaches the predicted maximal value for the planned grasp and triggers grasp enclosure by activating the appropriate Flexion subpopulation in F5.

The activation time course of several F5 neurons is shown in figure 20.18. When a particular object is presented to the model, it gives rise to activation patterns in PIP (cIPS) and IT which encode its visual features and activate visual-dominant cells in AIP. The AIP network implements a competitive process, selecting grasps which are appropriate for the object. The neural pools representing these grasps prime corresponding groups of neurons in F5. When F6 signals detection of the Ready cue (fifth trace from the bottom in figure 20.18A), F5 set-related neurons are activated and prime the F5 neurons selective for the extension phase. When the first go signal is received (fourth trace from the bottom in figure 20.18A), F6 activates the extension-related neurons in F5 to fire, resulting in a preshape of the hand for the chosen grasp. Simultaneously, set-related F5 cells are inhibited and flexion-related cells are primed via the basal ganglia loops described above. SII detects when the fingers reach an aperture that is appropriate for the size of the object (third trace from the bottom in figure 20.18A) and activates the flexion-related F5 neuron, which results in the enclosure of the hand. Again, at the same time extension-related cells (corresponding to the previous phase) are inhibited, hold-related cells (corresponding to the next phase) are primed. Once SII detects that the hand has contacted the object (second trace from the bottom in figure 20.18A), it activates hold-related F5 neurons, inhibiting flexion-related cells and priming release-related cells. When the second go signal is detected by F6 (bottom trace in figure 20.18A) the hold-related F5 neurons are inhibited and the release-related neurons are activated. During each phase, F5 neurons selective

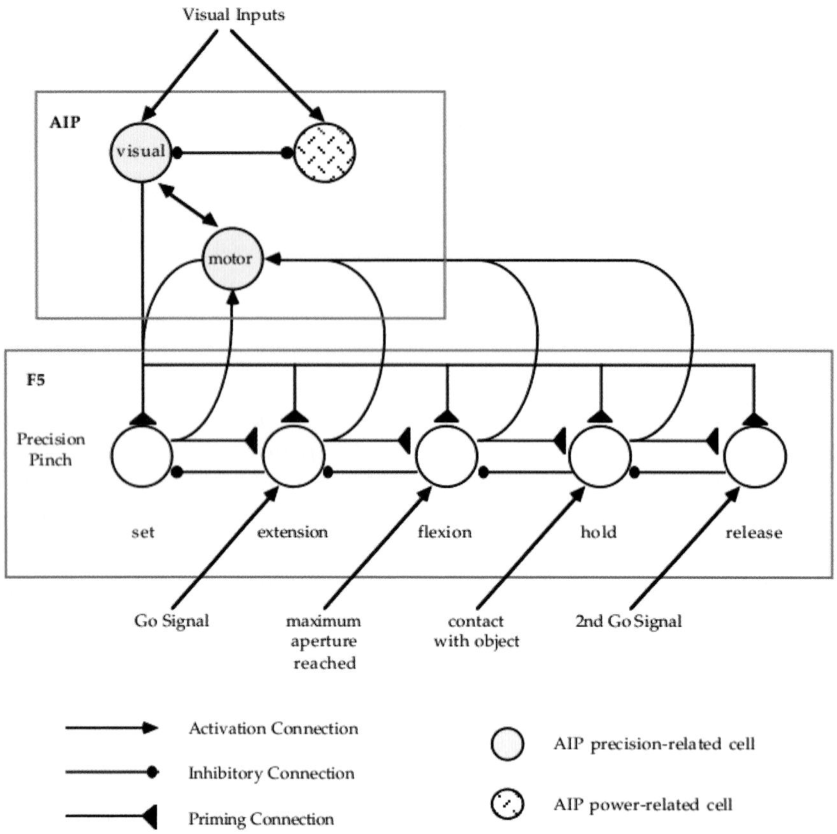

FIGURE 20.17 Interaction between anterior intraparietal area (AIP) and F5 populations (circles) during execution of the reach/grasp–hold–release sequence. Activity within F5 cascades from left to right as the program is executed. At each program phase, the state is reported back to the AIP motor-type population. However, the actual circuitry is more complex: the effective connections between program states are not coded within F5 but are managed by the combined action of pre-supplementary motor area (F6) and the basal ganglia. From Fagg and Arbib (1998).

for that phase send excitatory feedback to motor-related cells in AIP. These recurrent connections form a working memory trace of the active grasp, ensuring that during its execution only F5 neurons selective for phases of that grasp type are activated.

Most related grasping models focus on learning inverse kinematics for the hand, selecting contact points on the object's surface and developing feedback-based control of the hand. Most models plan the grasp in terms of kinematics, but at least one model stresses control of grasp forces. While many models use trial-and-error learning, they are not developmental models like ILGM in the sense that they begin learning at a phase corresponding to grasping development in infants 9 months and older. Some models are loosely based on neurophysiological data or relate the operation of their submodules to brain regions, but many are built purely using machine learning and robotics techniques.

Several models stress learning inverse kinematics transformations for the hand and plan grasps in terms of finger contact points on the object. Molina-Vilaplana et al. (2007) use heuristics to select contact points on the object surface for the fingers. The model first learns the inverse kinematics functions of the fingers and then learns to associate object properties with grasp postures (a function they relate to AIP/F5 functionality). The model first learns inverse kinematics for the thumb, index, and middle fingers so that it knows the relationships between finger motor commands and their sensory consequences (proprioceptive and visual) and then learns to associate object features with grasp postures with a local network called GRASP. The input to GRASP is a seven-dimensional vector encoding object shape (cube, sphere, or cylinder), object dimensions, and whether to grasp with two or three fingers. Similarly, Rezzoug and Gorce (2003) have a model with a modular architecture that first learns inverse kinematics for the fingers using backpropagation and then learns hand configuration for grasping using reinforcement. The model then learns finger postures given the desired position of the fingertip.

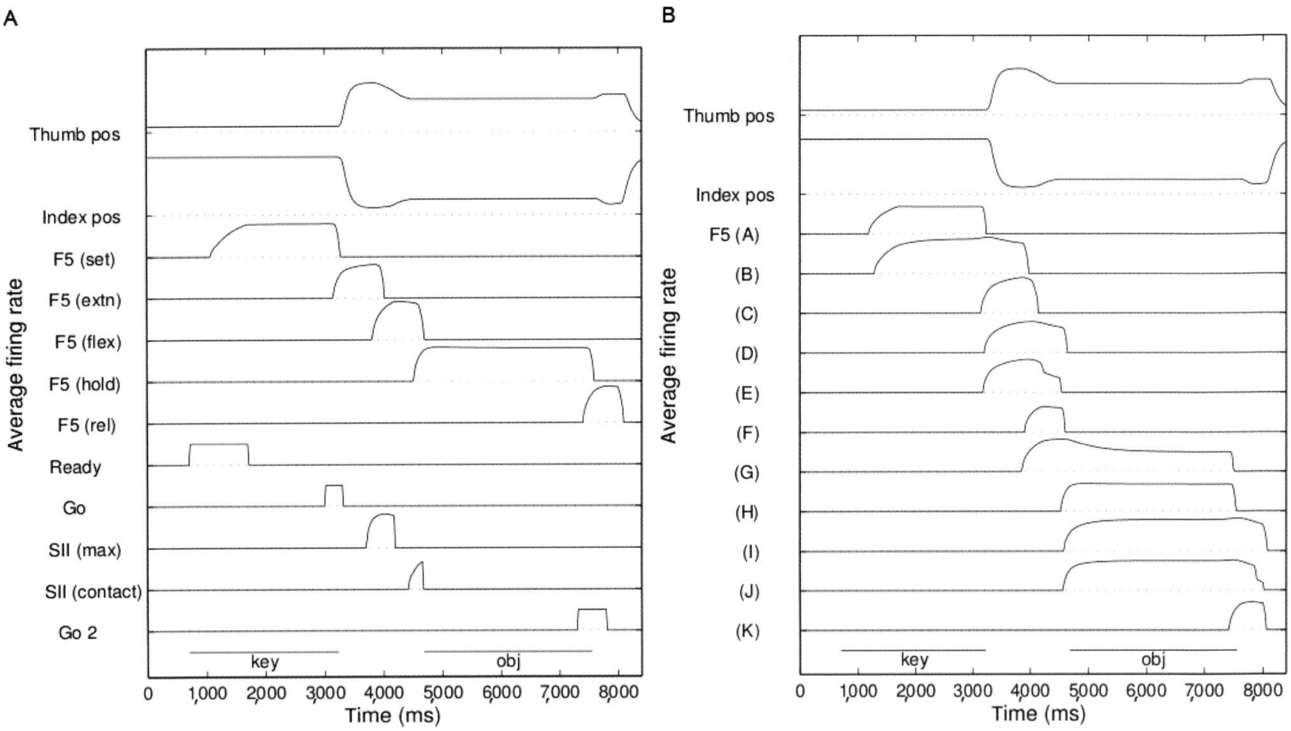

FIGURE 20.18 Model activity during performance of a precision grasp (A). The two traces at the top show the angle of the two proximal joints of the thumb and index finger (pos, position). The next five traces show the mean firing rate of F5 neurons tuned to different phases of the grasp: set, extension (extn), flexion (flex), hold, and release (rel). The five traces at the bottom show the various triggering signals: ready, go, secondary somatosensory cortex (SII) maximum aperture, SII contact, go 2. Model activity during the same grasp (B). F5 neurons A–K are active during more than one phase of the grasp. From Fagg and Arbib (1998).

Rezzoug and Gorce's (2003; Gorce and Rezzoug, 2004) model learns contact locations for each finger. Kamon, Flash, et al. (1996) split the problem of grasp learning into two problems: choosing grasping points and predicting the quality of the grasp. Each grasp type has a certain set of location and quality parameters used to select grasp locations and predict quality, which are supplied beforehand as task-specific knowledge. Grasp quality is predicted by the angles between the fingers and the surface normals at the contact points, and the distance between the opposition axis and the center of mass. They suggest that the grasp specification and evaluation modules run in an alternating manner until a suitable grasp is selected. Faldella, Fringuelli, et al. (1993) present an interesting model where a neural mechanism for matching object geometry to hand shape interacts with a symbolic rule-based expert system. The symbolic system performs some geometric analysis such as identifying curvature type, selects candidate hand contact positions, identifies symmetrical situations (to reduce neural module input), and ranks the selected grasp according to task constraints. Multilayer perceptrons trained using backpropagation were used to determine potential grasps based on geometric information.

For Grupen and Coelho (2000), grasping is primarily a force domain task, emphasizing *force* closure around an object over *form* (of the hand) closure. Their model used closed-loop control, utilizing tactile feedback to reposition contact forces based on models of interaction between the contacts and object surface. Q-learning is used to select a sequence of controllers to maximize object stability without knowing the object's identity, geometry, or pose. This model is complementary to those that focus on visually based kinematic grasping in that it learns to use haptic feedback in order to reposition contact forces to stabilize the object. It is one of the few models to utilize haptic feedback in grasp stabilization and thus sets forth future challenges for brain models.

20.4 Reach and Grasp Coordination

Natural reach and grasp movements involve smooth coordination of each phase of the grasp with the ongoing reach. An early suggestion was that the timescale of the reach temporally coordinates the phases of the grasp. Early models of reach and grasp coordination therefore focused on temporal coupling between the two movements.

The Hoff and Arbib (1993) model utilized temporal coupling prior to movement onset. The time needed for the reach (based on the distance to the object) and grasp (based on the size of the object) movements are compared, and the duration of the slower movement is used to scale the time course of execution for the other movement. Zaal et al. (1998) used a dynamical system to model this temporal coupling in which the hand aperture was continuously adjusted during the reach using the estimated time to object contact.

Haggard and Wing (1995) investigated how reach and grasp movements were coordinated after mechanical perturbations to the reach. They found that perturbation of the reach resulted in adjustment of the hand aperture which seemed to realign its posture to a stereotyped spatial relationship between hand aperture and reach progression. This spatial relationship consisted of a slow opening of the hand during most of the reach followed by a fast enclosure of the hand during the final centimeters of the reach. Based on these results, they present a model in which coupled feedback systems continuously enforce spatial, but not temporal, coupling between the two movements.

Ulloa and Bullock (2003) used the VITE model in order to enforce continuous spatial and temporal coupling of the reach and grasp. The model includes reach (T), grasp (A), and wrist orientation (O) circuits that work in parallel and receive a common gain signal (GO). The cells T_T, T_A, and T_O represent the target values for the reach position, hand aperture, and wrist orientation, respectively. Differences between these values and their present values (P_T, P_A, and P_O) are encoded in separate cells D_T, D_A, and D_O. The GO signal initiates the movement and activates the corresponding velocity cells V_T, V_A, and V_O, whose outputs are integrated by P_T, P_A, and P_O. Another set of cells, C_{TA} and C_{OA}, represent cross-coupling between the reach/wrist orientation movements and the grasp movement. P_A also includes self-inhibition in order to passively return the hand to a relaxed position even without voluntary movement.

20.5 Action Recognition and the Mirror System

Earlier we contrasted canonical and mirror neurons in F5—mirror neurons are active not only when the monkey performs a certain type of action but also when the monkey observes another monkey or human performing a more or less similar action (di Pellegrino et al., 1992; Gallese, Fadiga, et al., 1996). The discovery of the mirror neuron system in the 1990s was exciting in that it potentially provided a neural basis for action understanding, though it might better be said that mirror neurons provide a motoric basis for enriching such understanding as supported by the ventral stream. Visual activation of mirror neurons in observing another's action requires either vision of both hand and object, or recent vision of how the hand is moving toward the object, or (for a subclass, the audiovisual neurons) the sound of the action if that is sufficiently distinctive. In contrast to neurons in the STS which respond to the observation of different classes of actions (Perrett, Harries, et al., 1989), F5 mirror neurons are also active during the performance of actions even in the dark. Mirror neurons are therefore thought to match incoming visual (and auditory) information about another's actions to one's own motor representations, possibly allowing the observer to understand the other's action.

Neural models of the mirror system typically focus on the response of mirror neurons during action observation, with F5 canonical neurons providing a training signal which is used to shape their activity. Robotics systems including a model of mirror neurons tend to link them to internal models of actions such as forward or inverse models (see chapter 9) for action observation and execution (Demiris and Hayes, 2002).

The MNS family of models simulates the shaping of mirror neuron response properties through experience. The MNS model (Oztop and Arbib, 2002) of the monkey mirror system associated F5 canonical neuron activity encoding the type of grasp with visual input encoding the trajectory of observed hand–object relations. The model is organized around the idea that the AIP → $F5_{canonical}$ pathway emphasized in the grasp selection and planning models discussed above (such as FARS, TRoPICALS, and ILGM) is complemented by PF → $F5_{mirror}$ projections which provide F5 with a visual analysis of the hand's motion relative to the object. The upper diagonal in figure 20.19 corresponds to the basic pathway AIP → $F5_{canonical}$ → M1 (primary motor cortex) of the FARS model. The lower right diagonal controls the reach, transporting the hand to the object. The rest of figure 20.19 shows the core elements of the MNS model. Mirror neurons do not fire when the monkey sees the hand movement or the object in isolation—it is the sight of the hand moving appropriately to grasp an object that is required for the mirror neurons selective for the given action to fire. The model therefore includes schemas for the recognition of both the shape of the hand and analysis of its motion (performed in the model by STS) and for representation of the

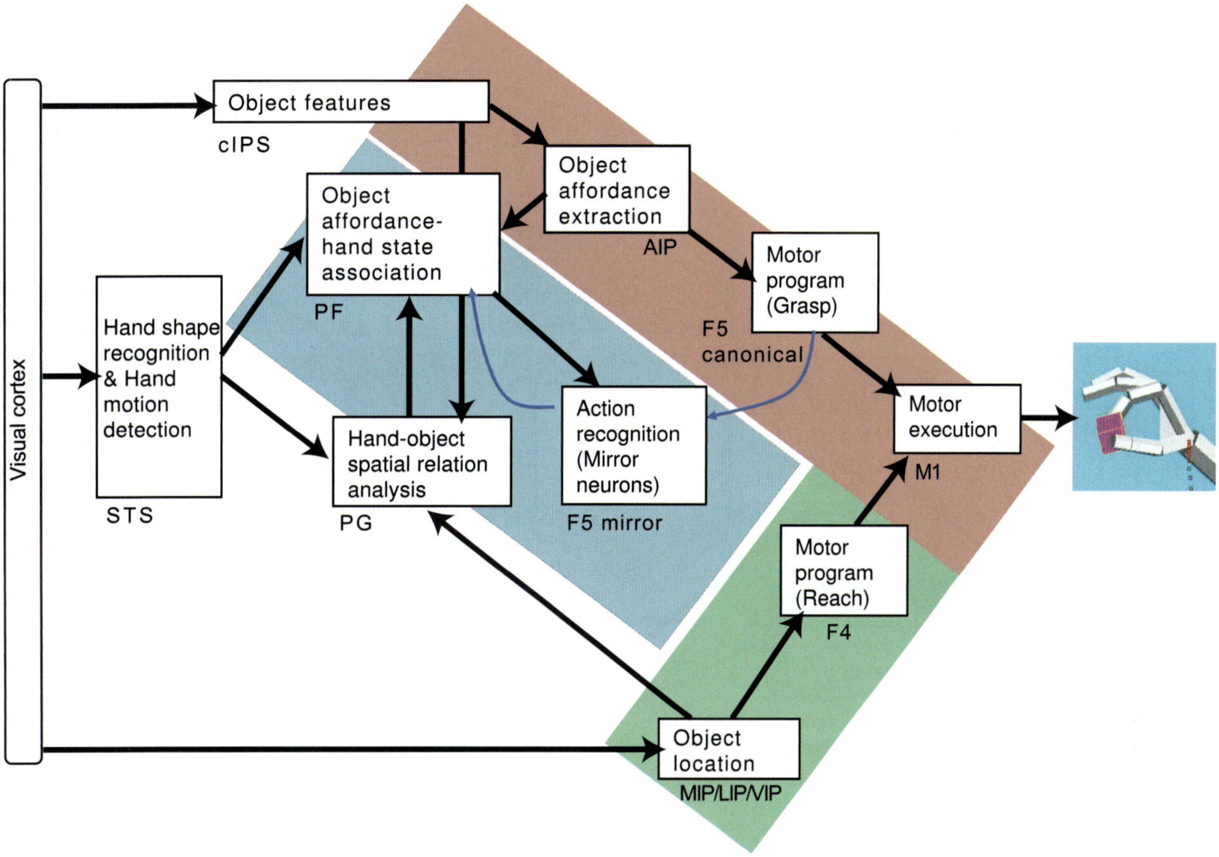

FIGURE 20.19 A schematic view of the mirror neuron system (MNS) model. The reach and grasp circuits (green and red rectangles) extract the object location and affordances and use this information to generate reach and grasp motor parameters. The core mirror circuit (blue rectangle) associates the hand shape and relation to the object affordances with the motor program represented in F5 canonical neurons during grasp performance, allowing it to recognize grasps performed by others. AIP, anterior intraparietal area; cIPS, caudal intraparietal sulcus; LIP, lateral intraparietal area; MIP, medial intraparietal area; STS, superior temporal sulcus; VIP, ventral intraparietal area. From Demiris et al. (2014).

relation of the hand to the location and affordance of the object.

In the MNS model, the *hand state* was defined as a vector whose components represented the movement of the wrist relative to the object's location, as well as the hand shape relative to the affordances of the object. As the hand state is composed of kinematic variables, and, indeed, only kinematic variables can be directly observed when watching another person perform an action, the model used the same kinematic arm and hand simulator used by ILGM. The arm simulator was used to generate grasps (power, precision, or side, depending on the type afforded by the object). The simulator provided visuospatial information about the arm and hand to the visual input schemas of the model. Oztop and Arbib (2002) showed that an artificial neural network could be trained to recognize the grasp type from the *hand state trajectory*, often well before the hand reached the object. During training, the output of the F5 canonical neurons, acting as a code for the grasp being executed by the monkey at that time, was used as the training signal for the output layer of the neural network, corresponding to F5 mirror neurons, to enable them to learn which hand–object trajectories corresponded to the encoded grasps. As a result of this training, the appropriate mirror neurons come to fire in response to viewing the appropriate trajectories even when the trajectory is not accompanied by F5 canonical firing. Importantly, this training allows the F5 mirror neurons to respond to trajectories of hand–object relations even when the hand is of the "other" rather than the "self" because the hand state is encoded in an object-centered reference frame, meaning that the trajectories are viewing angle invariant.

F5 canonical neurons were modeled as an array of units whose activity determined the type of grasp executed. The learning mechanism was backpropagation

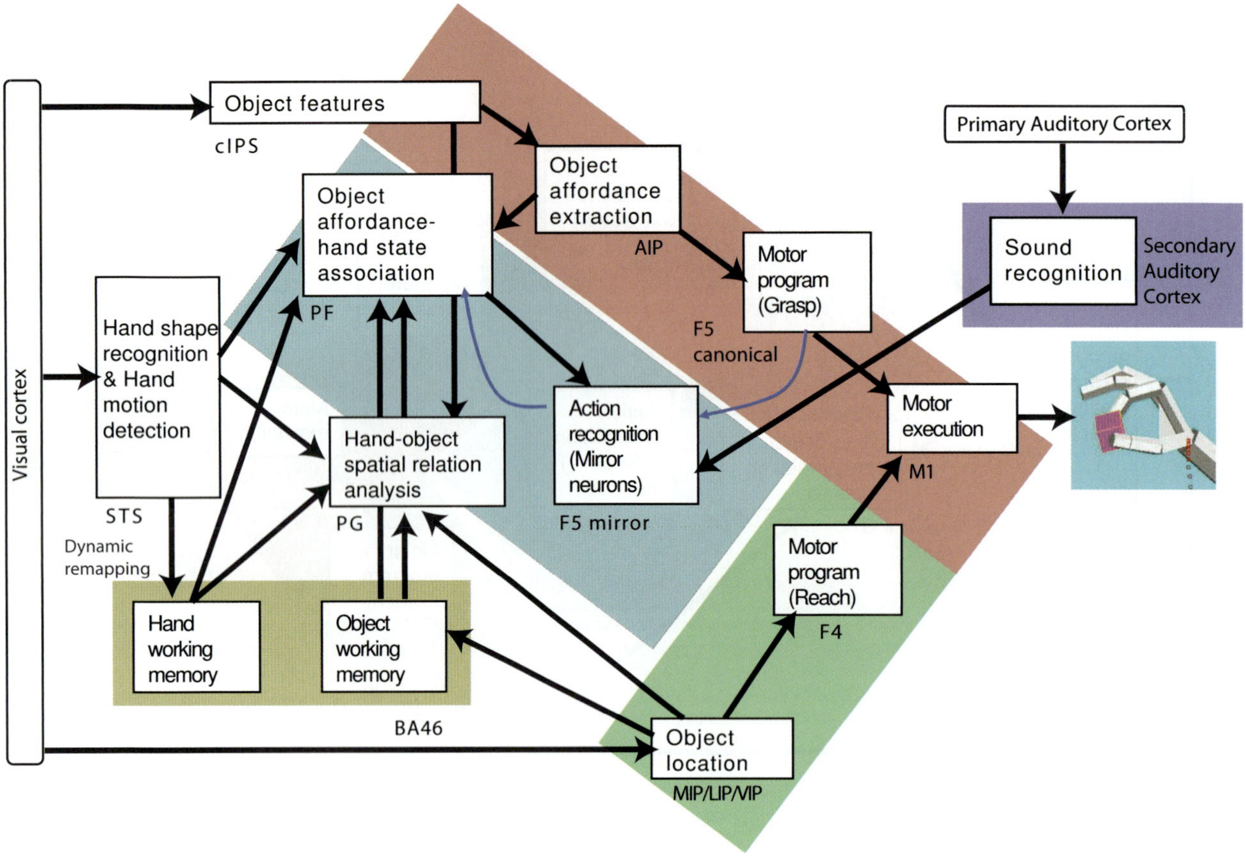

FIGURE 20.20 A schematic view of the mirror neuron system 2 (MNS2) model. Hand and object working memory with dynamic remapping allows the model to recognize partially obscured grasps. Sound recognition with Hebbian association allows mirror neurons to respond to the sound of unseen actions. AIP, anterior intraparietal area; cIPS, caudal intraparietal sulcus; LIP, lateral intraparietal area; MIP, medial intraparietal area; STS, superior temporal sulcus; VIP, ventral intraparietal area. From Demiris et al. (2014).

on a feedforward network with one hidden layer. Hand state information for a grasp—the relation of hand to object—was represented as a seven-dimensional trajectory encoding certain key relations. At each time point, the initial segment of the trajectory up to that time was fitted by a cubic spline and then sampled at 30 times spanning the segment to produce a 210-dimensional input vector to the network. In this way, the temporal trajectory of the hand state was preprocessed in order to be encoded in a spatial representation for input into the network. This system was shown to correctly classify observed grasps, often before the hand contacted the object. In addition, the network yielded neurophysiological predictions concerning the time course of mirror neuron activation and activity during the resolution of an ambiguous grasp.

The major drawback of the MNS model was its treatment of the trajectory. The MNS2 (Bonaiuto, Rosta, et al., 2007) model used a recurrent architecture that is biologically more plausible than that of the original MNS model and can handle raw hand–object affordance relations as input, rather than requiring preprocessing of the input as the MNS model did. It also included audio input which is associated with mirror neuron activity, and a spatial working memory with dynamic remapping of its representations (figure 20.20). These features allowed MNS2 to extend the capacity of the original model to address data on audiovisual mirror neurons, that is, mirror neurons that respond to the sight and sound of actions with characteristic sounds such as paper tearing and a nut cracking (Kohler, Keysers, et al., 2002), and on response of mirror neurons when the target object was recently visible but is currently hidden (Umiltà, Kohler, et al., 2001).

The main components of MNS2 are the audio and visual recurrent neural networks (RNNs), cochlear model, hand and object working memory, and the hand–object association schema. The hand–object association schema supplies input to the visual RNN in the form of hand–object relations such as axis disparities, joint angles, distance, and velocity. The hand and object working memory schemas relay the current visual hand and object signals to the hand–object association schema, but if either information is not available because the hand or object is invisible, a stored copy of the last observed values is passed. In the case of the hand working memory, this copy can be remapped using visual information about the movement of the arm.

Umiltà et al. (2001) have shown that mirror neurons in the macaque monkey can recognize a grasp if the monkey has seen the target object which was then hidden, but they cannot recognize the action lacking current or recent input on the affordances and location of the object. In their experiments, each monkey was shown an object that was then obscured by a screen. When the monkey observed the experimenter reaching behind the screen to grasp the object, the same mirror neurons were activated that responded to a visible grasp of the same object. The same neuron does not respond to a reach when no object is visible, or if the human reaches behind a screen that was not previously shown to conceal an object. MNS2 incorporated working memory and dynamic remapping components which allowed the model to recognize grasps even when the final stage of object contact is hidden and must be inferred (figure 20.21). Before being hidden, the object position and its affordance information were stored in working memory. Once the hand is no longer visible, the working memory of wrist position was updated using the still-visible forearm position. If the model observed an object which is then hidden by a screen and then observed a grasp that disappeared behind that screen, the wrist trajectory was extrapolated and the grasp was be classified accordingly.

Kohler et al. (2002) found that some of the mirror neurons in area F5 of the macaque premotor cortex that are responsive for the observation of actions associated with characteristic noises (such as breaking a peanut shell and ripping paper) are just as responsive for the sounds of these actions. MNS2 associated each sound with a distinct pattern of activity, and audio input units were fully connected to the output layer of the recurrent neural network, corresponding to a direct connection from auditory cortex to F5. These connection weights were modified using Hebbian learning. In this way, any sound that was consistently perceived during multiple occurrences of an executed action became associated with that action and incorporated into its representation (figure 20.22).

Two models focus on the role of F5 mirror neurons in feedback control of grasping. Metta, Sandini, et al. (2006) present a model where F5 canonical and mirror neurons form an internal model with mirror neurons as part of a feedback loop in action control. In this model, F5 is part of a larger circuit involving various parietal areas, STS, and other premotor and frontal areas. As in FARS, motor program representations in F5 are activated on the basis of object features represented in AIP and contextual signals. Inverse models are represented in projections from visual areas to the cerebellum, which compares them with a delayed output from F5 mirror neurons, which implement forward models. They note that these models and object affordances can be learned by trial and error. During action observation, visual information goes to the inverse models and the F5 mirror forward models in parallel, with the F5 canonical neurons filtering the F5 mirror responses. Maistros and Hayes (2004) present work inspired by FARS where mirror and canonical neurons are involved in grasping, with mirror neurons providing feedback for fine control. Their mirror system component consists of a SOM connected to a set of motor schemas. The system operates in learning mode where no output is produced and also in a recall phase. The motor system in this model is a simple inverse model. In imitation experiments, a noisy set of observed joint angles is used as input to the SOM of the mirror system. Motor schemas receive a series of joint angle postures through their linked SOM node.

Billard (2001) presents a model of human imitation incorporating a dynamic simulation of a 65 DOF avatar which approximates the motions of a humanoid. It takes as input the Cartesian coordinates of each joint of the demonstrator's limbs in an exocentric frame of reference and transforms these coordinates to a new set of coordinates relative to an egocentric frame of reference. The motor neurons in the simulation activate the avatar's muscles (each represented as a spring and a damper) using two muscles per degree of freedom per joint. Movement learning uses the dynamical recurrent associative memory architecture (DRAMA; Billard and Hayes, 1999) which allows learning of times series and of spatiotemporal invariants in multimodal inputs. Each connection of the network is associated with not only a weight (as usual) but also a time parameter. Both are updated during learning. The time parameters represent the decay of neural activity along the synapses while the weights keep a memory of the frequency of neural activation (weights are increased following classical Hebbian rules, i.e., whenever two neurons are simultaneously coactivated). The DRAMA network provides a general ability for learning complex time series by recording explicitly the temporal and spatial aspect of the patterns in its two different sets of connection

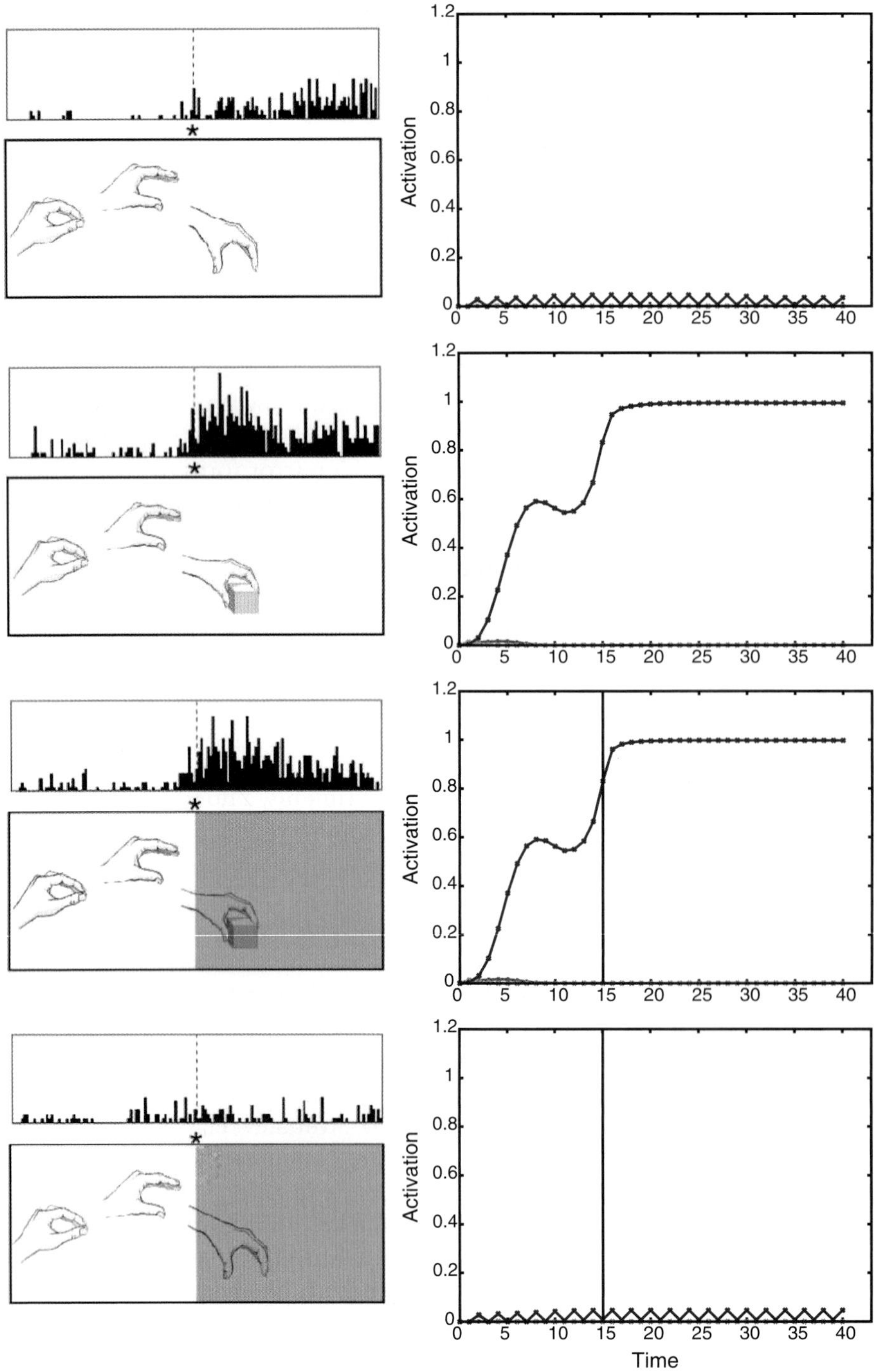

FIGURE 20.21 Left: Mirror neuron activation for (from top to bottom) visible pantomimed grasp, visible grasp, partially hidden grasp, and partially hidden pantomimed grasp conditions. From Umiltà, Kohler, et al. (2001). Right: Activation of the model's output units (mirror neurons) under the same conditions. The black vertical lines indicate the time step at which the hand was no longer visible to the network. The only output unit showing a significant level of activity in any plot is the one encoding power grasps. From Bonaiuto, Rosta, et al. (2007).

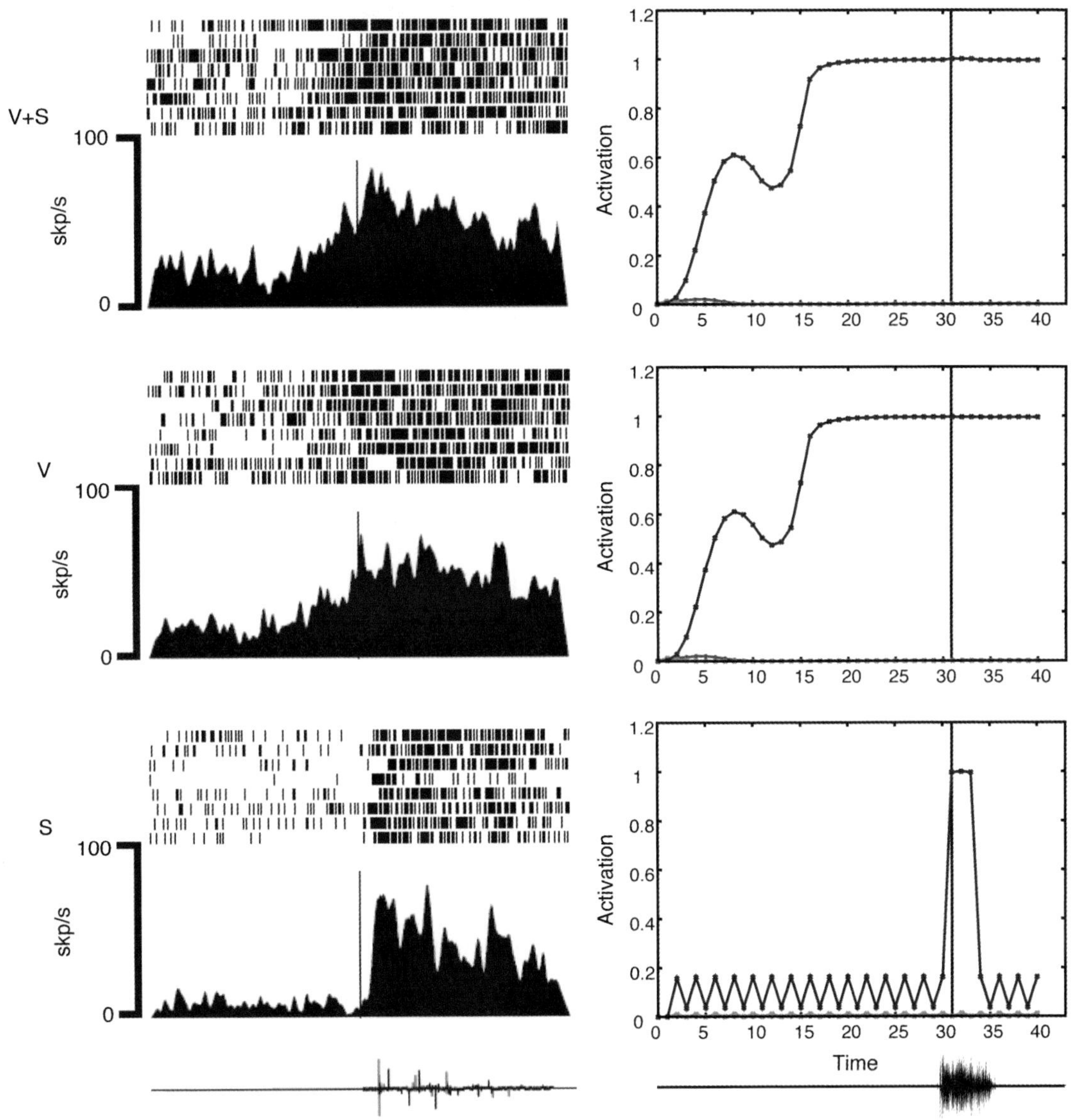

FIGURE 20.22 Left: Activation (in spikes per second [spk/s]) from Kohler et al. (2002) of an audiovisual mirror neuron responding to (from top to bottom) the visual and audio components, visual component alone, and audio component alone of a peanut-shell-breaking action. Right: Activation of the model's output layer when presented with a precision grasp sequence containing (from top to bottom) visual and audio (V + S), visual only (V), and audio only (S). The black vertical lines indicate the time step at which the hand made contact with the object. The unit encoding the precision grasp shows the greatest level of activation in all conditions. At the bottom is an oscillogram of the sound associated with the precision grasp (from Bonaiuto, Rosta, et al., 2007).

parameters. The Billard model implements a direct visuomotor mapping—it transforms the observed trajectory into a pattern of muscle activation which will yield it, and then learns the resultant time series.

Demiris and Hayes (2002) employ a battery of behaviors paired with forward models. Each "behavior" is a control system which takes the current state and goal and provides the motor output required to achieve that goal (it is thus what an "inverse model" models) while its forward model transforms the code for the motor output (and the goal) to generate what the next state would be if the corresponding behavior were executed. The task of the architecture is to observe a particular trajectory of states of a system and determine which behavior in its own repertoire best matches it. "Imitation" consists, then, of activating the chosen behavior and thus does not address how one may acquire a novel behavior through observation of its being performed by another. Specifically, the perceived states of observed behaviors are provided in parallel as input to the behavior modules which generate motor commands. The forward models use these motor commands to predict the next state of the behavior. These predictions are then compared with the next observed actual state. The prediction errors are then used to compute confidence values for each behavior (module), and determine which behavior best matches the observed behavior (the one with the highest confidence value). The best matching behavior is then selected for imitation. When an observed behavior is not in the observer's repertoire, the prediction error is high for each of the existing behaviors, and therefore a new behavior is added to the repertoire. The architecture can be related to mirror neurons by viewing the confidence values of the behavioral modules as mirror neuron activation (Oztop, Kawato, et al., 2006b).

Demiris, Aziz-Zadeh, and Bonaiuto (2014) demonstrate the use of the Brain Operation Database (BODB) to compare the MNS and MNS2 models with the Demiris and Hayes robotics model of the mirror system. This was accomplished by functionally decomposing each model into brain operating principles (BOPs; see chapter 5) and then comparing each model on the basis of their component mechanisms. Additionally, BODB requires that models be linked to summaries of the experimental data used to build or test them. Demiris et al. (2014) utilize BODB's model benchmarking functionality to further compare these models according the experimental data that they address and suggest directions for future models of the mirror system.

20.6 *Conclusion*

This chapter introduced several models of reaching, grasping, reach and grasp coordination, and action recognition. In many cases the models were based on macaque neurophysiology and used to explain human motor behavior. Models of reaching focused on the planning and generation of reach trajectories subject to some cost function, with developmental models demonstrating how single, smooth reach trajectories might develop from series of submovements in infancy. Grasping models such as FARS and TRoPICALS focused on the influences of the ventral and dorsal visual streams on the extraction and selection of grasp affordances, as well as planning grasps appropriate to utilize those affordances. Developmental grasping models such as ILGM and GAEM demonstrated how affordance extraction and grasp planning could be learned. Common to models of reach and grasp development are a process of motor babbling where movements are generated more or less randomly and successful actions are reinforced. Reaching and grasping are not controlled by entirely separate processes, and several models discussed in this chapter address how the two types of actions might be coordinated. Finally, models of the mirror system have shown how neural circuits for the generation and control of manual actions might be used to train systems for the recognition of observed actions.

Future challenges in modeling reaching and grasping involve the role of haptics as well as understanding the output of the ventral premotor cortex. As mentioned in the introduction, many neural models of grasping ignore the role of haptics in stabilizing a grasp once contact with an object has been made. Some progress has been made in developing robotic systems that utilize haptic information for grasping (Moisio, Leon, et al., 2013), but there is a scarcity of neural models of how the brain might accomplish this. Finally, it has been shown that the ventral premotor cortex projects not only to the primary motor cortex but also to the spinal cord (Shimazu, Maier, et al., 2004). It has been suggested that these projections may underlie the partial recovery that gradually takes place in many stroke patients after motor cortex lesions, but the role of these projections in the normal functioning of the brain is poorly understood. It has recently been found that many mirror neurons in the premotor cortex project directly to the spinal cord and that many of them are active during the performance of an action but inhibited during observation of the same action (Kraskov, Dancause, et al., 2009). These properties may allow them to inhibit automatic imitative responses, but this possibility as well as the possible role of mirror neurons in imitation has yet to be determined.

REFERENCES

ARBIB, M. A., FAGG, A. H., & GRAFTON, S. (2002). Synthetic PET imaging for grasping: From primate neurophysiology to human behavior. In F. T. SOMMER & A. WICHERT (Eds.), *Exploratory Analysis and Data Modeling in Functional Neuroimaging* (pp. 231–250). Cambridge, MA: MIT Press.

ARBIB, M. A., IBERALL, T., & LYONS, D. (1985). Coordinated control programs for movements of the hand. In A. W. GOODWIN & I. DARIAN-SMITH (Eds.), *Hand Function and the Neocortex* (pp. 111–129). Berlin: Springer Verlag.

ARBIB, M. A., & MUNDHENK, T. N. (2005). Schizophrenia and the mirror system: An essay. *Neuropsychologia, 43*(2), 268–280.

BERTHIER, N. E. (1996). Learning to reach: A mathematical model. *Developmental Psychology, 32*(5), 811–823.

BERTHIER, N. E., ROSENSTEIN, M. T., & Barto, A. G. (2005). Approximate optimal control as a model for motor learning. *Psychological Review, 112*(2), 329–346.

BHAT, A., HEATHCOCK, J., & GALLOWAY, J. C. (2005). Toy-oriented changes in hand and joint kinematics during the emergence of purposeful reaching. *Infant Behavior and Development, 28*, 445–465.

Billard, A. (2001). Learning motor skills by imitation: A biologically inspired robotic model. *Cybernetics and Systems, 32*(1–2), 155–193.

Billard, A., & Hayes, G. (1999). DRAMA, a connectionist architecture for control and learning in autonomous robots. *Adaptive Behavior, 7*(1), 35–63.

Bonaiuto, J. J., & Arbib, M. A. (2015). Learning to grasp and extract affordances: The Integrated Learning of Grasps and Affordances (ILGA) model. *Biological Cybernetics, 109*(6), 639–669.

Bonaiuto, J., Rosta, E., & Arbib, M. A. (2007). Extending the mirror neuron system model: I. Audible actions and invisible grasps. *Biological Cybernetics, 96*(1), 9–38.

Bullock, D., & Grossberg, S. (1988). Neural dynamics of planned arm movements: Emergent invariants and speed-accuracy properties during trajectory formation. *Psychological Review, 95*(1), 49–90.

Caligiore, D., Borghi, A. M., Parisi, D., & Baldassarre, G. (2010). TRoPICALS: A computational embodied neuroscience model of compatibility effects. *Psychological Review, 117*(4), 1188–1228.

Cisek, P., & Kalaska, J. (2005). Neural correlates of reaching decisions in dorsal premotor cortex: Specification of multiple direction choices and final selection of action. *Neuron, 45*(5), 801–814.

Clifton, R. K., Muir, D. W., Ashmead, D. H., & Clarkson, M. G. (1993). Is visually guided reaching in early infancy a myth? *Child Development, 64*(4), 1099–1110.

Demiris, Y., Aziz-Zadeh, L., & Bonaiuto, J. (2014). Information processing in the mirror neuron system in primates and machines. *Neuroinformatics, 12*(1), 63–91.

Demiris, Y., & Hayes, G. (2002). Imitation as a dual-route process featuring predictive and learning components: A biologically-plausible computational model. In K. Dautenhahn & C. Nehaniv (Eds.), *Imitation in Animals and Artifacts* (pp. 327–361). Cambridge, MA: MIT Press.

di Pellegrino, G., Fadiga, L., Fogassi, L., Gallese, V., & Rizzolatti, G. (1992). Understanding motor events: A neurophysiological study. *Experimental Brain Research., 91*(1), 176–180.

Erlhagen, W., & Schoner, G. (2002). Dynamic field theory of movement preparation. *Psychological Review, 109*(3), 545–572.

Fagg, A. H. (1996). *A Computational Model of the Cortical Mechanisms Involved in Primate Grasping* (doctoral disseration). University of Southern California, Los Angeles.

Fagg, A. H., & Arbib, M. A. (1998). Modeling parietal-premotor interactions in primate control of grasping. *Neural Networks, 11*(7–8), 1277–1303.

Faldella, E., Fringuelli, B., & Zanichelli, F. (1993). *A Hybrid System for Knowledge-Based Synthesis of Robot Grasps.* IEEE/RSJ International Conference on Intelligent Robots and Systems.

Felleman, D., Burkhalter, A., & Van Essen, D. C. (1997). Cortical connections of areas V3 and VP of macaque monkey extrastriate visual cortex. *Journal of Comparative Neurology, 379*, 21–47.

Flash, T., & Hogan, N. (1985). The coordination of arm movements: An experimentally confirmed mathematical model. *Journal of Neuroscience, 5*(7), 1688–1703.

Fluet, M., Baumann, M. A., & Scherberger, H. (2010). Context-specific grasp movement representation in macaque ventral premotor cortex. *Journal of Neuroscience, 30*(45), 15175–15184.

Gallese, V., Fadiga, L., Fogassi, L., & Rizzolatti, G. (1996). Action recognition in the premotor cortex. *Brain, 119*, 593–609.

Galletti, C., Kutz, D. F., Gamberini, M., Breveglieri, R., & Fattori, P. (2003). Role of the medial parieto-occipital cortex in the control of reaching and grasping movements. *Experimental Brain Research, 153*(2), 158–170.

Gattass, R., Sousa, A., Mishkin, M., & Ungerleider, L. G. (1997). Cortical projections of area V2 in the macaque. *Cerebral Cortex, 7*, 110–129.

Gibson, J. J. (1966). *The Senses Considered as Perceptual Systems.* Boston: Houghton Mifflin.

Gorce, P., & Rezzoug, N. (2004). A method to learn hand grasping posture from noisy sensing information. *Robotica, 22*, 309–318.

Grupen, R., & Coelho, J. J. (2000). *Structure and growth: A model of development for grasping with robot hands.* IEEE/RSJ International Conference on Intelligent Robots and Systems.

Haggard, P., & Wing, A. (1995). Coordinated responses following mechanical perturbation of the arm during prehension. *Experimental Brain Research, 102*(3), 483–494.

Hoff, B., & Arbib, M. (1993). Models of trajectory formation and temporal interaction of reach and grasp. *Journal of Motor Behavior, 25*, 175–192.

Iberall, T. (1986). *A Neural Model of Human Prehension* (doctoral dissertation). Department of Computer and Information Science, University of Massachusetts, Amherst.

Iberall, T. (1987). *The nature of human prehension: Three dextrous hands in one.* IEEE International Conference on Robotics and Automation.

Ijspeert, A., Nakanishi, J., & Schaal, S. (2013). Dynamical movement primitives: Learning attractor models for motor behaviors. *Neural Computation, 25*, 328–373.

Kamon, I., Flash, T., & Edelman, S. (1996). *Learning to grasp using visual information.* IEEE International Conference on Robotics and Automation, Minneapolis, MN.

Kohler, E., Keysers, C., Umiltà, M. A., Fogassi, L., Gallese, V., & Rizzolatti, G. (2002). Hearing sounds, understanding actions: Action representation in mirror neurons. *Science, 297*(5582), 846–848.

Kraskov, A., Dancause, N., Quallo, M. M., Shepherd, S., & Lemon, R. N. (2009). Corticospinal neurons in macaque ventral premotor cortex with mirror properties: A potential mechanism for action suppression? *Neuron, 64*(6), 922–930.

Kuhtz-Buschbeck, J. P., Stolze, H., Johnk, K., Boczek-Funcke, A., & Illert, M. (1998). Development of prehension movements in children: A kinematic study. *Experimental Brain Research, 122*(4), 424–432.

Kuperstein, M. (1988). Neural model of adaptive hand-eye coordination for single postures. *Science, 239*(4845), 1308–1311.

Luppino, G., Matelli, M., Camarda, R., & Rizzolatti, G. (1993). Corticocortical connections of area F3 (SMA-proper) and area F6 (pre-SMA) in the macaque monkey. *Journal of Comparative Neurology, 338*(1), 114–140.

Maistros, G., & Hayes, G. (2004). Towards an imitation system for learning robots. In *Methods and Applications of Artificial Intelligence: Proceedings* (Vol. 3025, pp. 246–255). Berlin: Springer-Verlag.

Metta, G., Sandini, G., Natale, L., Craighero, L., & Fadiga, L. (2006). Understanding mirror neurons—A bio-robotic approach. *Interaction Studies: Social Behaviour and Communication in Biological and Artificial Systems, 7*(2), 197–232.

Moisio, S., Leon, B., Korkealaakso, P., & Morales, A. (2013). Model of tactile sensors using soft contacts and its application in robot grasping simulation. *Robotics and Autonomous Systems, 61*(1), 1–12.

Molina-Vilaplana, J., Feliu-Batlle, J., & Lopez-Coronado, J. (2007). A modular neural network architecture for step-wise learning of grasping tasks. *Neural Networks, 20*(5), 631–645.

Murata, A., Gallese, V., Luppino, G., Kaseda, M., & Sakata, H. (2000). Selectivity for the shape, size, and orientation of objects for grasping in neurons of monkey parietal area AIP. *Journal of Neurophysiology, 83*(5), 2580–2601.

Oztop, E., & Arbib, M. A. (2002). Schema design and implementation of the grasp-related mirror neuron system. *Biological Cybernetics, 87*(2), 116–140.

Oztop, E., Bradley, N. S., & Arbib, M. A. (2004). Infant grasp learning: A computational model. *Experimental Brain Research, 158*(4), 480–503.

Oztop, E., Imamizu, H., Cheng, G., & Kawato, M. (2006a). A computational model of anterior intraparietal (AIP) neurons. *Neurocomputing, 69*, 1354–1361.

Oztop, E., Kawato, M., & Arbib, M. A. (2006b). Mirror neurons and imitation: A computationally guided review. *Neural Networks, 19*(3), 254–271.

Perrett, D. I., Harries, M. H., Bevan, R., Thomas, S., Benson, P. J., Mistlin, A. J., et al. (1989). Frameworks of analysis for the neural representation of animate objects and actions. *Journal of Experimental Biology, 146*, 87–113.

Raos, V., Umiltà, M. A., Murata, A., Fogassi, L., & Gallese, V. (2006). Functional properties of grasping-related neurons in the ventral premotor area F5 of the macaque monkey. *Journal of Neurophysiology, 95*(2), 709–729.

Rezzoug, N., & Gorce, P. (2003). A biocybernetic method to learn hand grasping posture. *Kybernetes, 32*(3–4), 478–490.

Rizzolatti, G., & Luppino, G. (2001). The cortical motor system. *Neuron, 31*(6), 889–901.

Rizzolatti, G., & Matelli, M. (2003). Two different streams form the dorsal visual system: Anatomy and functions. *Experimental Brain Research, 153*(2), 146–157.

Sakata, H., Taira, M., Kusunoki, M., Murata, A., Tanaka, Y., & Tsutsui, K. (1998). Neural coding of 3D features of objects for hand action in the parietal cortex of the monkey. *Philosophical Transactions of the Royal Society of London. Series B, Biological Sciences, 353*(1373), 1363–1373.

Sakata, H., Tsutsui, K., & Taira, M. (2005). Toward an understanding of the neural processing for 3D shape perception. *Neuropsychologia, 43*(2), 151–161.

Shimazu, H., Maier, M. A., Cerri, G., Kirkwood, P. A., & Lemon, R. N. (2004). Macaque ventral premotor cortex exerts powerful facilitation of motor cortex outputs to upper limb motoneurons. *Journal of Neuroscience, 24*(5), 1200–1211.

Smeets, J. B., & Brenner, E. (2001). Independent movements of the digits in grasping. *Experimental Brain Research, 139*(1), 92–100.

Sutton, R. S., & Barto, A. G. (1998). *Reinforcement Learning: An Introduction.* Cambridge, MA: MIT Press.

Taira, M., Mine, S., Georgopoulos, A. P., Murata, A., & Sakata, H. (1990). Parietal cortex neurons of the monkey related to the visual guidance of hand movement. *Experimental Brain Research, 83*(1), 29–36.

Tanji, J., & Shima, K. (1994). Role for supplementary motor area cells in planning several movements ahead. *Nature, 371*(6496), 413–416.

Todorov, E. (2004). Optimality principles in sensorimotor control (review). *Nature Neuroscience, 7*(9), 907–915.

Tsutsui, K., Taira, M., & Sakata, H. (2005). Neural mechanisms of three-dimensional vision. *Neuroscience Research, 51*(3), 221–229.

Tucker, M., & Ellis, R. (2004). Action priming by briefly presented objects. *Acta Psychologica, 116*(2), 185–203.

Ulloa, A., & Bullock, D. (2003). A neural network simulating human reach-grasp coordination by continuous updating of vector positioning commands. *Neural Networks, 16*, 1141–1160.

Umiltà, M. A., Brochier, T., Spinks, R. L., & Lemon, R. N. (2007). Simultaneous recording of macaque premotor and primary motor cortex neuronal populations reveals different functional contributions to visuomotor grasp. *Journal of Neurophysiology, 98*(1), 488–501.

Umiltà, M. A., Kohler, E., Gallese, V., Fogassi, L., Fadiga, L., Keysers, C., Rizzolatti, G. (2001). I know what you are doing: A neurophysiological study. *Neuron, 31*(1), 155–165.

von Hofsten, C. (1982). Eye-hand coordination in the newborn. *Developmental Psychology, 18*(3), 450–461.

von Hofsten, C. (1993). Prospective control: A basic aspect of action development. *Human Development, 36*(5), 253–273.

von Hofsten, C., & Fazel-Zandy, S. (1984). Development of visually guided hand orientation in reaching. *Journal of Experimental Child Psychology, 38*(2), 208–219.

Winges, S. A., & Santello, M. (2005). From single motor unit activity to multiple grip forces: Mini-review of multi-digit grasping. *Integrative and Comparative Biology, 45*(4), 679–682.

Zaal, F. T., Bootsma, R. J., & van Wieringen, P. C. (1998). Coordination in prehension: Information-based coupling of reaching and grasping. *Experimental Brain Research, 119*(4), 427–435.

Zatsiorsky, V. M., & Latash, M. L. (2004). Prehension synergies. *Exercise and Sport Sciences Reviews, 32*(2), 75–80.

Zeki, S. (1980). A direct projection from area V1 to area V3A of rhesus monkey visual cortex. *Proceedings of the Royal Society of London. Series B, Biological Sciences, 207*(1169), 499–506.

21 Cerebellar Adaptation and Supervised Learning in Motor Control

JOHN PORRILL AND PAUL DEAN

21.1 *Basic Concepts: The Purkinje Cell as a Signal Synthesis Device*

In humans the cerebellum is a relatively small structure tucked under the back of the brain (see figure 21.1A) that, although folded differently, looks somewhat like a smaller version of cerebral cortex (in fact, the word cerebellum means "little brain"). The functional status of the cerebellum is rather peculiar. Lesions of, say, motor areas in the cerebellum rarely lead to paralysis in the way that similar lesions in motor cortex would do. Instead they reduce the level of performance of the affected motor function. The advantage of having a cerebellum seems to be a little like the difference between being drunk and being sober: whereas a person who is drunk has to think through the individual movements of a task like putting a key in a lock, a sober person can perform this task automatically. Although the cerebellum is often thought of as primarily a motor structure, it is becoming evident that it underpins skilled behavior and automaticity in a very wide range of sensory, motor, and even cognitive tasks (Ito, 2008).

The cerebellum is an old structure in evolutionary terms. Although not present in invertebrates it appears in the simplest vertebrates and seems to have arisen from structures such as the lateral line organs and vestibular mechanisms of aquatic animals in order to maintain equilibrium in the turbulent marine environment. In some species with extraordinary sensory capabilities, such as electric fish, it occupies most of the volume of the brain. Even in humans, where it accounts for only 10% of the volume, around 80% of the neurons in the brain are in the cerebellum. Indeed in primates the evolution of cerebellar capacity seems to have tracked that of cortex, indicating that the cerebellum must supply some essential computational features that it has not proved possible to duplicate in cortex (Eccles, 1969).

A wide range of evidence strongly supports the idea that its role is to process a variety of types of information about the current sensorimotor context of the organism in order to increase the accuracy, reliability, automaticity, and so forth of behaviors actually initiated by other brain areas. An extension of this role was proposed by Brindley (1964, p. 80): "the message sent down by the fore-brain in initiating a voluntary movement is often insufficient ... it needs to be elaborated by the cerebellum in a manner that the cerebellum learns with practice.... The cerebellum is thus a principal agent in the learning of motor skills." This suggests that the cerebellum acts something like an operating system, allowing high-level control systems to issue simple task-related commands that initiate smooth, coordinated series of movements. Current computational theories in which the cerebellum implements high-level modules such as forward and inverse models (see chapter 9, "Computational Models in Motor Control") are extensions of this basic idea (Ito, 2008).

A feature that has always been of particular interest to computational neuroscientists is the relatively simple, ordered structure of the cerebellum. It consists of a repeating microcircuit arranged in a homogeneous structure that is so regular that is has been described as crystalline (see figure 21.1B). The basic anatomy and electrophysiology of this circuit was identified in the 1960s, when it was described as a "neuronal machine" (Eccles, Ito, and Szentágothai, 1967). The homogeneous, repeating nature of the circuit and the wide range of behaviors in which it is known to play a role strongly suggested that it performed some essential, widely applicable computation which could be accessed by other neural systems to improve their performance. Uncovering the algorithm implemented by the cerebellar microcircuit has remained a challenge ever since.

A possible interpretation of the main features of the microcircuit was proposed by Marr and Albus almost as soon as the circuit itself was identified (Albus, 1971; Marr, 1969). We will present a reinterpretation of this model due to Fujita, the adaptive-filter model, which emphasizes temporal processing (Fujita, 1982). It is only fair to say that the interpretation we present is still disputed, and that other accounts exist; these accounts

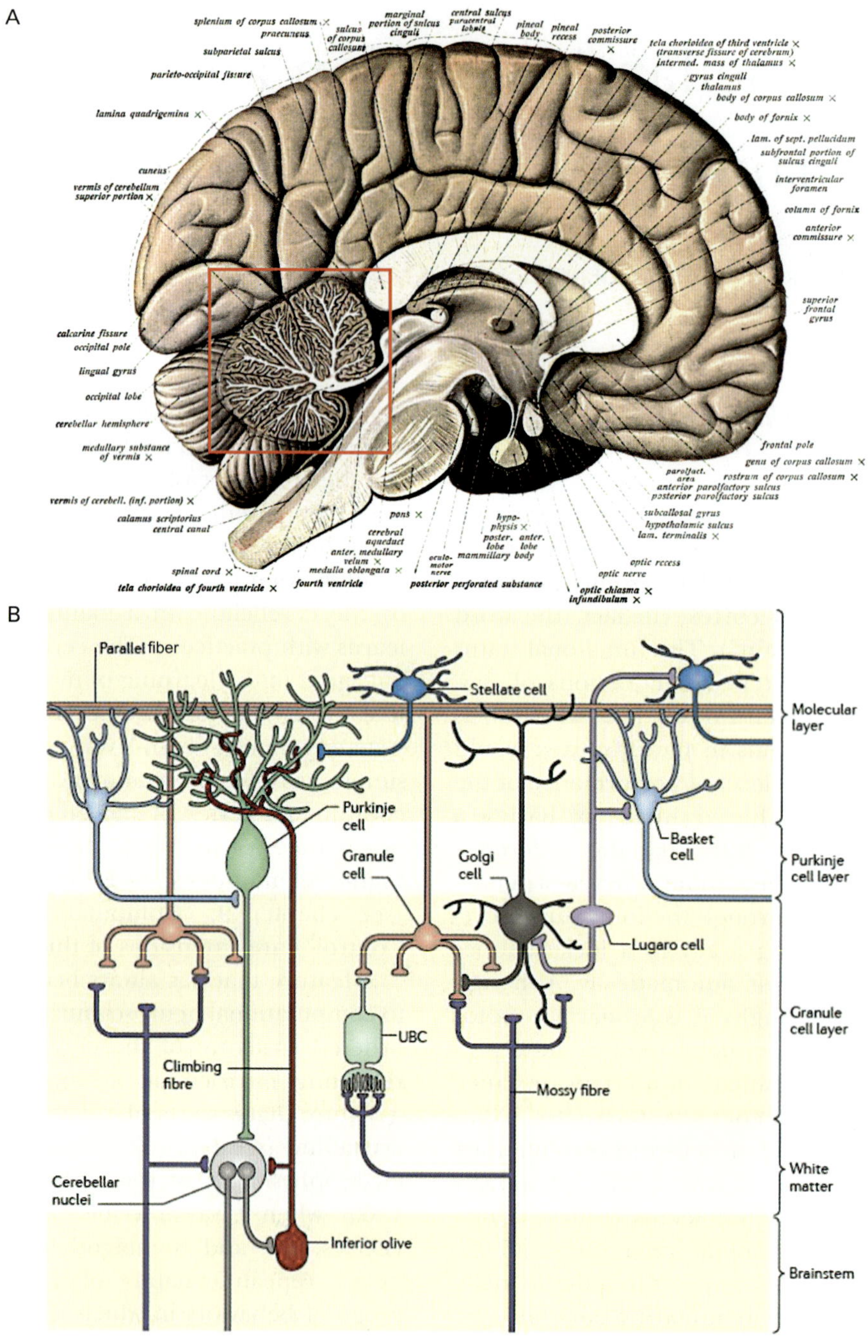

FIGURE 21.1 The structure of the cerebellum. (A) The cerebellum is the small fan-like structure behind the brainstem and under the rear part of the cerebral cortex highlighted by the red rectangle. This cross-section clearly shows the extensive foliation of the cortical surface of the cerebellum, which gives it a total surface area comparable to that of the volumetrically larger cerebral cortex. Adapted from "Sobo 1909 624" by Dr. Johannes Sobotta. Licensed under Public Domain via Wikimedia Commons. (B) A small segment of cerebellar cortex shown in cross-section. A basic microcircuit is repeated throughout the cortex. We will concentrate on the following aspects of this microcircuit: mossy fibers rise into the granular layer, where they make contact with the four to six dendrites of granule cells. Granule-cell axons then rise into the molecular layer where they bifurcate and run in parallel along the cortical surface, making excitatory contact with the fan-like dendritic trees of the large Purkinje cells. Parallel-fiber activity can also inhibit Purkinje cells via contacts with inhibitory interneurons, the stellate and basket cells. It can also affect granule-cell activity in a recurrent inhibitory loop via the Golgi cells in the granular layer. Climbing fibers are a second source of cerebellar input, rising into the Purkinje-cell layer and winding around the Purkinje-cell soma and dendritic trees making extensive synaptic contacts. Adapted from Cerminara et al. (2015), figure 1c.

often emphasize aspects of cerebellar phenomenology which do not play a central functional role in the Marr–Albus model. However we will focus here on the adaptive-filter version of the Marr–Albus framework because, to put it bluntly, no coherent account has yet been given of the exact nature of the computation these other phenomena could embody, and of which behavioral tasks they would be well adapted to performing. In sharp contrast we will show that the adaptive filter is "computationally adequate" for a very wide range of tasks.

There is another way of looking at this editorial decision. Before learning quantum mechanics, every physicist is presented with an account of the Bohr "toy model" of the hydrogen atom. This is not because the model is accurate—it is almost ridiculously simple-minded compared to the full quantum description—but because it gives an elegant and comprehensible account of several remarkable features of atomic behavior that point naturally to the more complete quantum theory: in fact, the Bohr model gives a clue to what a quantum theory of a system has to look like. We believe that the Marr–Albus adaptive-filter model of the cerebellum is also a good example, given our present level of understanding, of what a really good computational model of a brain region might look like.

21.1.1 Connectivity of the Cerebellum The cerebellum is composed of a layer of gray matter (cerebellar cortex) lying over internal white matter composed of myelinated axons in which are embedded three pairs of deep nuclei. Inputs and outputs to the cerebellum travel along a mass of fibers in three pairs of symmetrical tracts called the cerebellar peduncles.

The majority of cerebellar output is transmitted via the deep cerebellar nuclei (DCN) located in the cerebellum, but the phylogenetically oldest pathway travels via the vestibular nuclei located in the brainstem. Since much of our detailed knowledge of the cerebellum is obtained from vestibular experiments, this difference in output target may prove to be theoretically important.

Cerebellar cortex is divided by deep transverse fissures into three major lobes: anterior, posterior, and flocculonodular. In addition two longitudinal furrows separate the cerebellum into a central strip, the vermis, and into left and right lateral hemispheres. Further shallow fissures divide cerebellar cortex into ten named lobules in all. Cerebellar cortex (like cerebral cortex) is also corrugated by narrow transverse folia, with 85% of the surface area hidden in the sulci. These extensive corrugations mean that the surface area of cerebellar cortex is large for its size, about 75% of that of the volumetrically much larger cerebral cortex.

This macroscopic organization is further subdivided into smaller scale zonal stripes, which have recently been shown to have distinct molecular identity and specific afferent inputs, and then into microzones, which are small areas of cortex, with specific climbing-fiber connectivity and output targets, believed to be associated with specific functions. The exact nature and role of this level of topographical organization is not yet well understood.

21.1.2 The Basic Cerebellar Microcircuit Cerebellar cortex is three layered, and consists of the granular layer, the Purkinje-cell (PC) layer, and the molecular layer. The organization of the basic cerebellar microcircuit which traverses these layers is shown in figure 21.2A. There are two sources of inputs to the cerebellum, mossy-fiber inputs and climbing-fiber inputs. Mossy fibers arise from a variety of brain-stem nuclei and from neurons in the spinal cord. Climbing-fiber inputs come from a remarkable structure in the medulla called the inferior olive, and their function poses some of the most interesting questions about the behavior of the cerebellum, as we will see below. The sole output neurons of cerebellar cortex are the Purkinje cells. These are large cells, packed tightly together, with an extensive, flat, dendritic tree in the molecular layer. Their output is inhibitory on the cerebellar nuclei, with GABA as the neurotransmitter.

The mossy fibers excite small neurons in the granular layer called granule cells. These neurons take input from four to five mossy fibers. There is an enormous number of granule cells (more than 10^{11} in humans), and they make up the majority (in some estimates up to 80%) of the neurons in the brain. Axons from granule cells rise up through the Purkinje-cell layer, first making excitatory synaptic contact with nearby Purkinje cells and then bifurcating into parallel fibers (PFs) which extend laterally in the molecular layer up to a few millimeters along the folia, making further excitatory contacts with the dendrites of the Purkinje cells over which they pass. Each Purkinje cell makes approximately 150,000 synaptic contacts with passing parallel fibers, and each fiber will contact about 700 Purkinje cells. Since PCs are inhibitory, this provides an inhibitory pathway by which mossy-fiber activity can affect the activity of the deep cerebellar or vestibular nucleus neurons targeted by Purkinje cells.

Parallel fibers also synapse with inhibitory interneurons in the molecular layer called basket and stellate cells, and the axons of these cells then synapse on Purkinje cells. These interneurons provide an alternative,

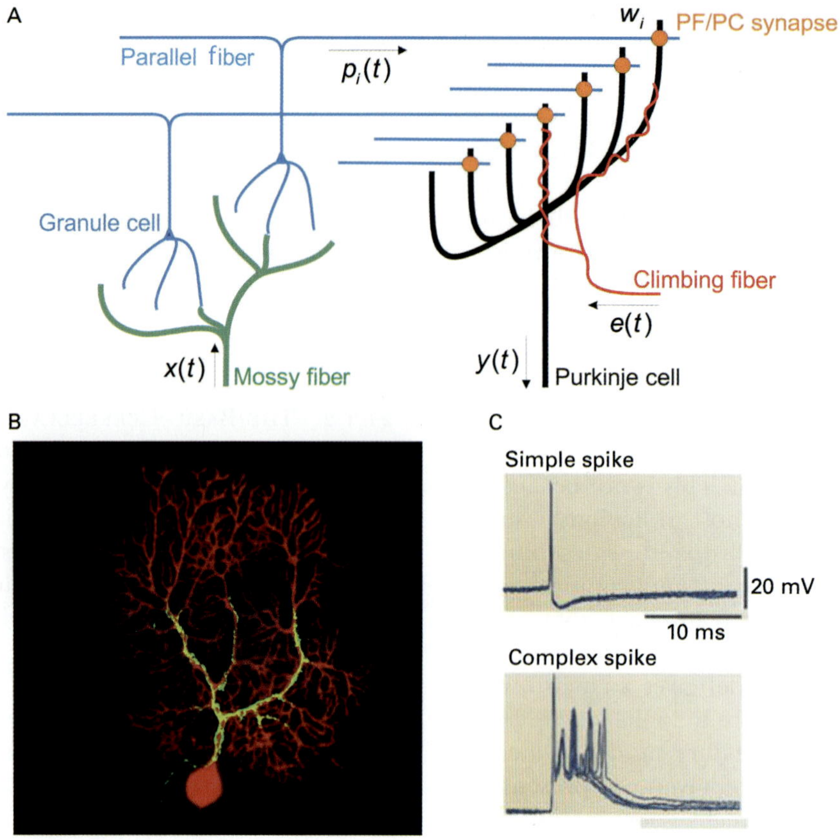

FIGURE 21.2 (A) A simplified schematic microcircuit showing mossy-fiber input being recoded by granule cells carried along the parallel fibers, and synthesized by the Purkinje cell (the inhibitory interneurons and the recurrent pathway via Golgi cells are not shown). Each climbing fiber innervates tens of Purkinje cells, but each individual Purkinje cell only takes input from a single climbing fiber. Adapted from Porrill et al. (2004), figure 1a. (B) Fluorescence imaging of a Purkinje cell showing the climbing fiber (yellow) winding around the soma and dendritic tree of the Purkinje cell (red). Adapted with permission from Kreitzer et al. (2000), figure 2E. (C) Top: Simple spikes are conventional action potentials with a single narrow peak generated intrinsically or via parallel-fiber excitation. Bottom: Complex spikes generated by climbing-fiber activity have a more complex spike shape with multiple peaks and are associated with a massive influx of calcium into the cell. Adapted from Dean et al. (2010), box 2A.

excitatory (more precisely: inhibition of inhibition) pathway from mossy fibers to the cerebellar nuclei. A further class of inhibitory interneurons, the Golgi cells, receive most of their input from parallel fibers, and their axons synapse onto granule cells, thus adding an inhibitory recurrent loop into the feedforward pathways above. It is becoming clear that all the synapses in these pathways are plastic and subject to activity-dependent modification.

The climbing fibers are the second source of excitatory input to Purkinje cells (figure 21.2A and B). This input is of a very peculiar kind. A climbing fiber winds around the soma and dendrites of a Purkinje cell, making hundreds of synaptic contacts. Each climbing fiber innervates tens of Purkinje cells, but each individual Purkinje cell only takes input from a single climbing fiber.

21.1.3 COMPLEX AND SIMPLE SPIKES Without synaptic inputs Purkinje cells fire conventional action potentials, called simple spikes (see figure 21.2C), at a tonic rate of about 40 spikes/s. The two main types of excitatory input, from parallel fibers and climbing fibers, have very different effects on this baseline firing.

Parallel fibers carry spikes generated by granule cells, with firing rates of that can be greater than 500 spikes/s. Although these individually have little effect on Purkinje-cell firing rate, spatial summation of around 150 such excitatory inputs over the dendritic tree input can increase the rate of firing of simple spikes, and PC firing rates of more than 200 spikes/s can be achieved.

The second source of input to Purkinje cells is very different in character. Climbing fibers arise from neurons in the olive which have rather low firing rates, about 1 spike/s. However because of the many strong synapses a climbing fiber makes with a target Purkinje cell, every climbing-fiber spike fires the Purkinje cell (called 1–1 firing). This input causes a dramatic neuronal depolarization which propagates through the soma and dendrites and is associated with a large calcium influx throughout the dendritic tree. This produces an action potential followed by a high-frequency burst of smaller action potentials, a grouping called a complex spike (see figure 21.2C, bottom graph). The result is that activity recorded from Purkinje cells consists of relatively high-frequency trains of simple spikes interleaved with occasional complex spikes.

21.1.4 Some Observations in Search of an Explanation In many ways the cerebellum is a conventional neural assembly, with input and output neurons, interneurons, synaptic plasticity, and so forth. However some of the features we have described are sufficiently surprising that they cry out for a theoretical explanation based on functional requirements peculiar to cerebellar processing. These features include the following:

- The dramatic morphology of the Purkinje cell
- The exceptionally large number of potential PF inputs
- A high proportion of silent synapses
- The unusual climbing-fiber input and 1–1 firing of complex spikes it causes
- The exceptionally low rate of climbing-fiber activity
- The high degree of homogeneity and the organization into microzones
- The large number of granule cells, and the recurrent dynamics of the granule-cell layer
- The differential information content of simple spike and complex spike outputs.

21.1.5 The Marr–Albus Interpretation Marr and Albus independently proposed a similar explanation for these features, usually referred to as the Marr–Albus model (Albus, 1971; Marr, 1969). The original framework was very perceptron like, with time being treated as a discrete entity, advancing in snapshots. This viewpoint is rather unnatural and limiting for time-dependent signal processing and motor control problems, so we will work with a more appropriate reformulation of the basic Marr–Albus framework proposed by Fujita (1982), which emphasizes time dependence and which we will call the adaptive-filter model. All the observations listed above find a very satisfying explanation in this model (summarized in figure 21.3) which has four basic hypotheses:

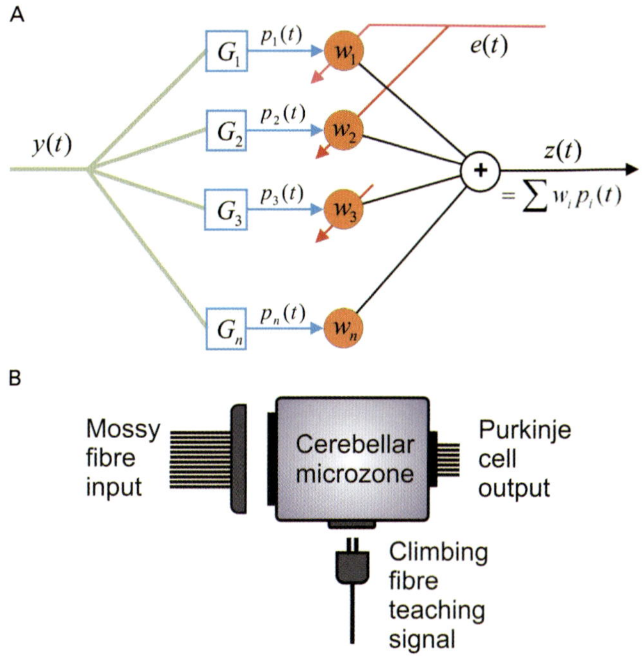

Figure 21.3 Cerebellar adaptive filter. (A) The adaptive-filter representation of the simplified cerebellar microcircuit shown in figure 21.2A. Input signal are analyzed by a bank of filters G_i, and the resulting signals are resynthesized by a linear element with adjustable weights w_i. A teaching signal $e(t)$ related to task performance is available to guide the weight adjustment. (B) The adaptive-filter interpretation of the basic cerebellar microcircuit is consistent with the "cerebellar chip" hypothesis (Ito, 1997) in which the chip embodies a universal algorithm which can be applied to a range of tasks; the particular task specifies the connectivity of its inputs and output. Mossy-fiber input is regarded as a relatively undifferentiated "bus" of context information, so that any individual input may or may not be relevant to the task in hand. The climbing-fiber input port is supplied with an appropriate teaching signal related to current task performance. Finally, the Purkinje-cell output is connected to an appropriate target which affects task performance. Each cerebellar chip can thus be identified with a specific area of cerebellar cortex with the same output target and essentially identical climbing-fiber input, called a microzone. Adapted from *Handbook of the Cerebellum and Cerebellar Disorders*, Manto et al., Eds. (2013), figure 58.9.

1. *The granule-cell layer provides an efficient recoding of the input signal:* Incoming mossy fibers carry information about the sensorimotor context of a task. The recurrent connectivity in the granule-cell layer analyzes this information into a convenient, massively expanded, representation which is then broadcast along the cerebellum along the parallel fibers.

2. *The Purkinje cell acts as a synthesis filter, and its informative output is carried by simple spike activity:* The Purkinje cell recombines the different components of the recoded sensorimotor context information carried

on the parallel-fiber inputs to form a task-related output signal. This is produced by spatial and temporal integration of the parallel-fiber inputs on the PC dendrites, each weighted by the synaptic efficacy of PF/PC synapses (so that learning to produce the correct output requires adaptation of these synaptic weights).

3. *The purpose of the climbing-fiber input is to supply a teaching signal:* The activity on the climbing fibers is a teaching signal related to task performance and is used to drive learning at PF/PC synapses. Complex spikes are an artifact of the cellular implementation of this learning mechanism and do not form part of the informative PC output. Both Marr and Albus also speculated about the nature of the learning rule at the PF/PC synapse. Marr hypothesized it was Hebbian, Albus that it was anti-Hebbian—Ito, Sakurai, and Tongroach (1982) supported the latter, establishing the widely accepted postulate that simultaneous PF and climbing-fiber activity should lead to long-term depression (LTD) of synapses.

4. *The resulting computational element implements a universal algorithm:* The cerebellar microcircuit behaves like a "cerebellar chip" (Ito, 1997). Performance on a broad range of tasks can be improved by plugging in the cerebellar chip with the correct macroscopic connectivity, that is, supplying the relevant inputs on the mossy fibers, an appropriate teaching signal on the climbing fibers, and connecting its output to an appropriate target nucleus where it can influence the given task (Porrill et al. 2013).

The Marr–Albus adaptive-filter hypothesis accords well with our previous discussion of the role of the cerebellum. An analysis–synthesis filter of this kind potentially has the power and flexibility required to combine the multiple sources of information required to contribute to the wide range of behaviors in which the cerebellum is implicated. The model also provides a coherent framework for supervised motor learning. To cope with environmental variations such as varying loads, and internal changes such as fatigue or damage, motor systems must be able to adapt. Adaptive learning of this kind is necessarily supervised, and the climbing-fiber teaching signal hypothesis provides the cerebellum with a biologically plausible signal to drive supervised learning. (Recall the general discussion of supervised learning in chapter 1.)

21.1.6 MODELING STUDY: USING THE PURKINJE CELL FOR ADAPTIVE SYNTHESIS Our first computational modeling study will concentrate on the central component of the microcircuit, the Purkinje cell. We will develop a simple mathematical model embodying hypotheses 2 and 3 above in concrete form. For the moment we postpone addressing the details of hypothesis 1 although the example codings we have chosen are closely related to the "spectral coding" model (Bartha, Thompson, and Gluck, 1991) and the "internal clock" model (Yamazaki and Tanaka, 2005).

We make the simplifying assumption that the parallel fibers carry a bank of signals which are responses to a trigger signal presented at time $t = 0$ carried on the mossy fibers. The task of the Purkinje cell is to learn to synthesize these inputs to produce a given output profile.

In our first example each parallel fiber carries a response of duration which is a sinusoid of fixed frequency and zero phase

$$p_k(t) = \sin 2\pi f_k t \quad 0 \le t \le T \tag{21.1}$$

where the frequencies $f_k = kf$ are harmonics of the fundamental frequency $f = 1/T$ (see figure 21.4A).

To allow parallel fibers to convey negative signal values, we assume that signals are rate coded by modulating a tonic firing rate, allowing both positive and negative signals to be transmitted

$$\mathrm{FR}_k(t) = \mathrm{FR}_{\mathrm{tonic}} + \mathrm{FR}_{\mathrm{sensitivity}}\, p_k(t) \tag{21.2}$$

This trivial linear recoding will usually be ignored in descriptions at the signal processing level although we will discuss some implications of spike coding in the next section.

These parallel-fiber inputs are assumed to be linearly combined with appropriate synaptic weights so that the Purkinje-cell output signal (also coded as a modulation of tonic PC firing rate) is

$$z(k) = \sum w_i p_i(t) \tag{21.3}$$

which is a waveform with fundamental frequency f.

Now suppose that our desired output signal $z_d(t)$ is the profile shown in figure 21.4B (blue plot). We first need to specify an appropriate teaching signal for this task (finding a suitable teaching signal will turn out to be one of the main difficulties in wiring the cerebellar chip). We assume here that we have direct access to output error, that is, the difference between actual and desired Purkinje-cell output,

$$e(t) = z(t) - z_d(t) \tag{21.4}$$

and we will use this as our teaching signal.

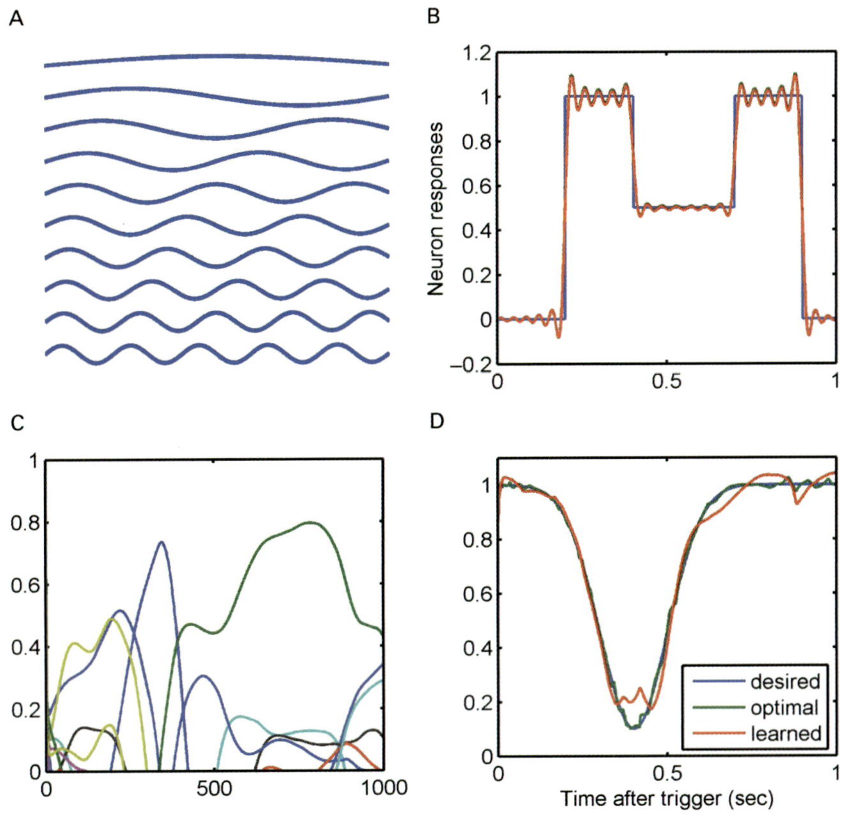

FIGURE 21.4 Modelling study: Adaptive filter learning to synthesize an output from a given input basis (simulation parameters can be found in the coding notes). (A) Sample from a basis of sinusoidal inputs $p_i(t)$ of increasing frequency presented on the parallel-fiber inputs to the Purkinje cell. (B) Castle shaped profile (blue) synthesized from 50 sinusoidal inputs showing the optimal reconstruction (green) almost exactly overlaid by the learned reconstruction (red). With more basis sinusoids the desired output profile would be more accurately reproduced. (C) Sample from a chaotic basis produced by recurrent inhibitory network neurons as a response to a trigger input. These basis functions alternate between off and on states and can be regarded as a population coding of elapsed time after a stimulus. (D) Timed Gaussian pause in Purkinje cell output (blue) synthesized from 100 chaotic inputs showing the optimal reconstruction (green) and learned reconstruction after a moderate learning time (red). With this highly correlated basis a very long learning time is required to obtain a near-optimal reconstruction.

The final component of the model is the learning rule. We use the covariance learning rule (Sejnowski, 1977)

$$\delta w_i = -\beta \langle e(t) p_i(t) \rangle \tag{21.5}$$

where the brackets <> indicate the expected value over a batch of data (generally speaking a batch should be chosen to be long enough to contain a representative statistical sample of the signal to be synthesized; however, there are circumstances, such as when Lyapounov convergence results apply, when this may not be necessary). Here β is a constant factor specifying the learning rate. This rule, which will be discussed in more detail later, can be justified heuristically on the assumption that "correlation implies causation": if the activity on a parallel-fiber input $p_i(t)$ is positively correlated with the error $e(t)$ over the batch duration, it is assumed to have caused that error; hence, its associated synaptic weight is decreased.

With this learning rule the filter quickly learns to synthesize a good approximation to the required profile (figure 21.4B, red plot). In principle it can synthesize any desired response to the trigger; in fact, you may recognize this as a poor man's Fourier analyzer: the learned weights w_i are the coefficients in the Fourier sine series of the desired response.

Although this example is rather artificial, it does demonstrate that the ability to combine a large number of inputs with adaptable weights conveys great computational power. This property lies at the heart of what we have called the "computational adequacy" of the adaptive-filter hypothesis. This example also makes a significant technical point. For example, it is clear that the completeness (in the sense of linear algebra) of the basis representation provided by the parallel-fiber inputs is essential. For instance, it is easy to check that accurate synthesis of the desired waveform is only possible if a large number of input frequencies are available.

Although mathematically convenient, the sinusoidal responses assumed above are not very plausible biologically. In our second example we use instead the outputs from a randomly connected recurrent inhibitory network or "reservoir" proposed as a granular-layer processing model by Yamazaki and Tanaka (2005). Figure 21.4C shows a sample of the chaotic responses of the network to a trigger input, and figure 21.4D shows how these can be synthesized by the Purkinje cell to produce an accurately timed Gaussian response to the trigger.

We also show the optimal reconstructions for both of these examples (figure 21.4B and 21.4D, green plots) where the best fit weights are calculated directly using the method of least squares (this is equivalent to presenting an infinite number of training examples). The uncorrelated sinusoidal basis functions can be seen to have learned the optimal reconstruction after a relatively small number of presentations whereas the more highly correlated chaotic basis learns much more slowly. This illustrates another fundamental theoretical point: whereas all complete bases are equivalent in their ability to represent complex outputs, highly correlated bases will yield much slower learning. It is likely that some of the plasticity in the granular layer is aimed at producing the optimal, uncorrelated, basis functions required for fast learning (e.g., see Coenen, Arnold, Sejnowski, and Jabri, 2001).

21.2 *The Learning Rule: Application to Noise Cancellation*

In the previous section we concentrated on the description of the Purkinje cell as a synthesis device, and the associated modeling study showed how this device could learn to produce a stereotyped output. This section begins by looking more closely at the nature of the learning rule required to allow supervised learning, and how it is related to plasticity of PF/PC synapses. The modeling study for this section then looks at a generic signal processing problem, noise cancellation, which highlights the ability of the adaptive filter to synthesize outputs from more general context signals than simple triggers.

21.2.1 Defining a Cost Function We begin by showing how the learning rule used in the last section can be derived by the method of gradient descent on a cost function which measures performance. If the desired neuron output is $z_d(t)$, we can define the neuron output error as the difference between actual and desired neuron output:

$$e(t) = z(t) = z_d(t) = \sum w_i p_i(t) - z_d(t) \tag{21.6}$$

Then, the mean square output error over a long sequence of data of length T,

$$\langle e^2 \rangle = \frac{1}{T}\int_{t_0}^{t_0+T} e(t)^2\, dt \tag{21.7}$$

provides a well-behaved measure of performance over that batch (we use angle brackets to express expected or mean values). Hence, we can define a cost function

$$E(w) = \frac{1}{2}\langle e^2 \rangle \tag{21.8}$$

which quantifies the performance of the neuron for any given value of the weight vector $w = (w_1, w_2, \cdots, w_n)$ (a factor of a half is conventionally included to simplify formulas involving the derivatives of E since it cancels when the squared error is differentiated—see below, and by analogy with physical quantities such as kinetic energy). The optimal weight estimate (in the sense of least squares, for that batch of data) minimizes this cost function. There are many direct techniques available to solve such minimization problems; however, we are looking for a biologically plausible algorithm that can be implemented as a synaptic learning rule.

21.2.2 Deriving a Learning Rule by Gradient Descent Given a well-behaved cost function, it is always possible to derive a learning law using the method of gradient descent. A small change Δw to the weight vector w changes the cost function by an amount which can be calculate approximately using the first order Taylor expansion

$$E(w + \Delta w) \approx \mathrm{E(w)} + \frac{\partial E}{\partial w}\Delta w = E(w) + \sum \frac{\partial E}{\partial w_i}\Delta w_i \tag{21.9}$$

Hence, for a learning rate parameter β which is chosen positive and small enough, the changes

$$\Delta w_i = -\beta \frac{\partial E}{\partial w_i} \tag{21.10}$$

are guaranteed to reduce the cost function since

$$\Delta E \approx \sum \frac{\partial E}{\partial w_i}\Delta w_i = -\beta \sum \left(\frac{\partial E}{\partial w_i}\right)^2 < 0 \tag{21.11}$$

unless the gradient is zero, in which case we are already at the global minimum (note that although gradient descent usually finds only a local minimum, the cost function we have chosen is quadratic in weights, so—for this particular synthesis problem—the endpoint will be a global minimum). The learning rule (equation 21.11) is called a gradient descent learning rule since weight changes are made in the opposite direction to the gradient of the cost function. To use this rule in practice, we need to calculate the gradient of the cost function

$$\frac{\partial E}{\partial w_i} = \frac{1}{2}\left\langle \frac{\partial e^2}{\partial w_i} \right\rangle = \left\langle e \frac{\partial e}{\partial w_i} \right\rangle \tag{21.12}$$

This depends on derivative of output error, easily calculated from (equation 22.6) as

$$\frac{\partial e}{\partial w_i} = \frac{\partial}{\partial w_i}\left(\sum w_i p_i - z_d\right) = p_i \tag{21.13}$$

Hence, the gradient descent learning rule can be written as

$$\Delta w_i = -\beta \langle e p_i \rangle \tag{21.14}$$

Since this change in weight is proportional to the expected product or covariance of $p_i(t)$ and $e(t)$, this rule is called a covariance learning rule; it is the learning rule we used in the previous modeling study. Clearly, learning stops when the covariance is zero, that is, when parallel-fiber activity has zero correlation with climbing-fiber activity, hence this process is also called decorrelation learning (Dean, Porrill, and Stone, 2002).

21.2.3 Comparison with Hebbian Learning The learning rule above looks a lot like a standard "neurons that fire together wire together" Hebbian learning rule, and the Hebb rule is often implemented in a stabilized correlational form as

$$\Delta w_i = \beta \langle z p_i \rangle \tag{21.15}$$

so that inputs p_i positively correlated with neuronal output z increase their weights.

There are two main differences between the rules in equation 21.14 and equation 21.15. The first is a change in overall sign which means that Purkinje neuron weights decrease rather than increase when the synaptic input $p_i(t)$ is positively correlated with the teaching signal $e(t)$; for this reason the new rule is said to be anti-Hebbian.

The second difference concerns locality. The Hebbian learning rule is strictly local because weight change at a given synapse depends only on the relevant synaptic input and the output of the neuron to which it belongs. The covariance rule (equation 21.14) is partially local, in that learning at the i-th PF synapse depends on the input to that synapse and not on inputs to other PF synapses. However it has a potentially fatal nonlocality in that learning depends on the external teaching signal $e(t)$ rather than just on neuronal output $z(t)$ as in the Hebb rule.

This is a generic problem in supervised learning: the associated neural learning rules require a mechanism by which the required teaching signal can be transmitted to the learning synapse. This biological constraint is a critical obstacle to implementing many artificial neural networks in biological hardware, ruling out, for example, backpropagation networks because of their need to transmit a backpropagated error signal to internal layers. The Marr–Albus climbing-fiber hypothesis solves the *physical* problem of carrying teaching signals from external sensors to the Purkinje cell at a stroke (we will discuss later how the cerebellum deals with the further problem of the nature of this signal, the so-called distal error problem). Since climbing-fiber spikes produce whole-cell complex spike activity via multiple synapses on the Purkinje cell, the teaching signal is made available locally at every parallel-fiber synapse. This mechanism is said to be heterosynaptic (rather than homosynaptic, as in the Hebb rule) since learning involves interactions between inputs at more than one synapse—parallel-fiber and climbing-fiber synapses in this case. To summarize, the covariance learning rule we have derived requires heterosynaptic, anti-Hebbian synaptic plasticity.

21.2.4 Driving a Continuous Time Learning Rule The covariance rule specified above is a batch learning rule; as such, it requires the storage of long sequences of data before the expected value can be calculated, making it unsuitable for biological learning systems. A more plausible continuous time learning rule can be obtained by dividing the long batch of data into small intervals δt and approximating the time-averaged integral $\langle e p_i \rangle$ as a sum

$$\Delta w_i = -\beta \langle e p_i \rangle \approx -\frac{\beta}{T}\sum e(t) p_i(t) \delta t \tag{21.16}$$

The batch increment Δw_i can therefore be expressed as a sum of individual increments

$$\delta w_i = -\frac{\beta}{T} e(t) p_i(t) \delta t \tag{21.17}$$

at each time step. Dividing through by δt and taking the limit gives a continuous update rule (which we will also call the covariance learning rule)

$$\frac{dw_i}{dt} = -\beta' e(t) p_i(y) \qquad (21.18)$$

with learning rate parameter $\beta' = \beta/T$. Note that this derivation assumes that learning is slow (in the sense that only small weight changes are made after each batch of data of length T) and that the signals are stationary (so that a sufficiently long batch is a representative sample of the whole signal).

21.2.5 An Underlying STDP Learning Rule? We can also sketch what a plausible spiking mechanism underlying the covariance rule might look like. When both PF and climbing-fiber signals have higher than normal values in a small time range, equation 21.18 requires that the weight decrease; this implies that near simultaneous climbing-fiber and PF spikes should produce LTD of the PF/PC synapse. This effect, predicted by Albus, was observed experimentally (Ito, Sakurai, and Tongroach, 1982) and was originally considered to be the main contributor to cerebellar learning. However it is clear that unaccompanied LTD would lead to eventual silencing of all synapses, so it must be balanced by some form of long-term potentiation (LTP). More recent experimental evidence suggests that LTP is induced by parallel-fiber spikes which are less closely correlated with complex spikes (Safo and Regehr, 2008).

This combination of LTD and LTP can be summarized in the spike-timing-dependent plasticity profile illustrated in figure 21.5, in which a strong LTD dip is produced for nearly coincident spikes surrounded by weak LTP side lobes for non-coincident spikes. It can be shown this STDP profile implements the covariance learning rule for suitable rate coding schemes. (Although there is a tendency to associate STDP with coding schemes based on precise spike timing, their role as a mechanism for the implementation of correlational learning rules for rate coded signals is probably just as important. (See Gerstner and Kistler, 2002, and chapter 6, "Hebbian Plasticity and Learning," for details.)

Parallel-fiber and climbing-fiber signals at their tonic rates carry no signal information, and hence should not produce learning. In this case the LTD and LTP processes must be exactly balanced. As a consequence climbing-fiber spikes which are "missing" from the tonic background and which are close in time to a PF spike fail to produce their expected LTD, and these spike "holes" effectively drive an equal and opposite LTP contribution. Hence, despite its seeming emphasis on LTD, the spiking learning rule is in practice symmetrical between LTP and LTD: positive signal correlations produce LTD while negative signal correlations produce equal and opposite LTP. This symmetry is very important—for example, it supplies a mechanism via which previously silent synapses can be recruited to perform a task.

In the derivations above we assumed that teaching signal $e(t)$ was instantaneously available for learning at the relevant synapses. In motor systems the teaching signal can be subject to large transmission delays (e.g., around 100 ms for the retinal slip signal we consider in the next section). This means that the parallel-fiber and climbing-fiber signals will not match exactly in time, and this mismatch can be shown to lead to unstable learning at high frequencies. To correct this, since we cannot advance the climbing-fiber signal, we must delay the effect of the parallel-fiber signal in the learning

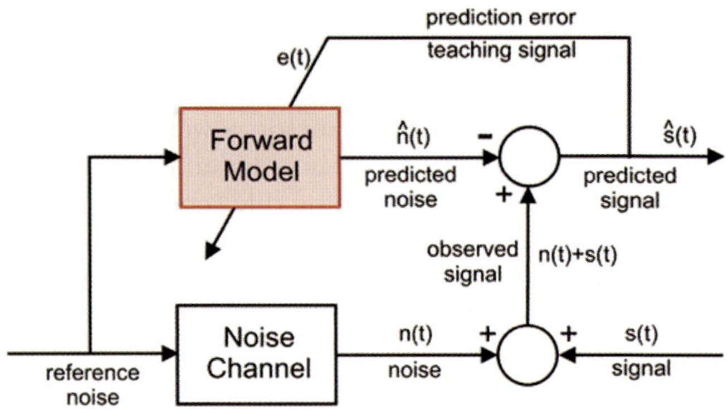

Figure 21.5 Spike-timing-dependent plasticity (STDP) implementation of covariance learning. This figure shows a qualitative STDP profile that is consistent with the covariance learning rule for rate coded signals. The profile is centered on a climbing-fiber spike (top row of sparse red spikes). Those parallel-fiber spikes (bottom row of blue spikes) which are closely aligned in time with the climbing-fiber spike produce strong long-term depression (LTD); those which are not closely correlated in time produce weaker long-term potentiation (LTP). The magnitude of the contribution on the interspike interval is given by the spike-timing-dependent profile, with a strong central LTD dip surrounded by broad, weak LTP lobes. If LTD and LTP contributions are balanced at tonic rates of firing, then over the long term this profile adjusted weights proportional to the negative of the correlation of the firing rate modulations, $\Delta w = -\langle ep \rangle$, over a frequency band whose lower limit is fixed by the width of the LTP lobes and upper limit by that of the LTD dip. The dip in the STDP profile can be advanced to provide an eligibility trace compensating for delay in the climbing-fiber signal.

rule. A filter which affects such a delay is called an eligibility trace, and it can be incorporated into the STDP profile by situating the LTD dip at an interspike time corresponding to the transmission delay, an effect for which there is some experimental evidence (Wang, Denk, and Häusser, 2000).

Finally, we mention a seeming limitation imposed by the low spike rate of climbing fibers. How can such a firing rate of around 1 Hz carry useful information about movement errors with frequencies up to 5 Hz and above? The answer is that the error signal is used not for instantaneous feedback but for learning, so, over the relevant training period, we only require that the expected value on the right-hand side of the learning rule is properly calculated. The theoretical constraint is that, over multiple trials, climbing-fiber activity should provide an unbiased estimate of the teaching signal at the relevant frequencies. It is clear that an "event-triggered average" of climbing-fiber activity over many trials can contain frequencies higher than 1 Hz, and in practice cerebellar learning rates are slow enough to provide substantial effective high-frequency content in the teaching signal.

21.2.6 Modeling Study: Noise Cancellation

Every organism capable of movement faces the problem of distinguishing the predictable sensory impressions caused by its own movements from the novel information produced by its environment. A homely illustration of our ability to cancel the predictable sensory consequences of our own motor activity is supplied by our inability to successfully tickle ourselves (Blakemore, Wolpert, and Frith, 1999). The removal of predictable noise from signals is a generic signal processing problem in biological and engineering systems. A practical example is given by noise-canceling headphones.

The noise-cancellation task may provide a clue to the evolutionary provenance of the cerebellum. Some species have cerebellar-like structures which are plausible evolutionary precursors of the cerebellum itself and which seem to perform exactly this function. For example, some neurons in the electrosensory lateral line lobe of the mormyrid electric fish are able to learn to form a negative image of the electric organ discharges produced during active electrosensing in order to cancel them from the output of its passive electrosensory receptors (Bell, Han, and Sawtell, 2008).

Our second modeling study will show how to use the adaptive filter to perform adaptive noise cancellation. The architecture we shall use is shown in figure 21.6. In its most basic form adaptive noise-cancellation allows us to recover a signal of interest $s(t)$ from a

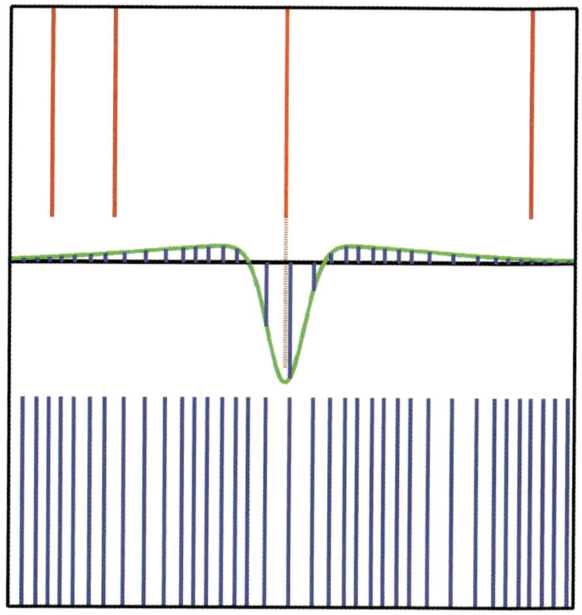

Figure 21.6 Adaptive filter used for acoustic noise cancellation. The reference noise signals $n_1, n_2, \ldots, n_k$ pass through a noise channel in which they are combined with unknown coefficients a_i to form the contaminating noise signal n. This combines with the signal of interests s to give the observed signal $s+n$. They are also passed through an adaptive filter (labeled Forward Model in the diagram) in which they are combined with weights w_i to provide a noise estimate $\hat{n}$ which is fed to a comparator to provide a signal estimate $\hat{s}$. The negative of the signal estimate is also the teaching signal e for the adaptive filter. After learning the adaptive-filter weights w_i approximate the nose channel coefficients a_i, and the adaptive filter provides a forward model of the noise channel (hence the label). Adapted from Porrill et al. (2013), figure 5A.

measurement $x(t)$ corrupted by additive noise sources $n_1(t), \cdots, n_k(t)$

$$x(t) = s(t) + \sum a_i n_i(t) = s(t) + n(t) \qquad (21.19)$$

when independent measurements of the noise signals are available but their relative contributions a_i are unknown. In the noise-cancellation architecture (figure 21.6) the adaptive filter takes as input the reference noise signals $n_i(t)$ and combines them to produce an output

$$\hat{n}(t) = \sum w_i n_i(t) \qquad (21.20)$$

which is an estimate of the noise $n(t)$. This is then subtracted from the contaminated signal to give an estimate of the signal of interests $s(t)$

$$\hat{s}(t) = x(t) - \hat{n}(t) = s(t) + (n(t) - \hat{n}(t)) = s(t) + e(t) \qquad (21.21)$$

where $e(t)$ is the output error

$$e(t) = \hat{n}(t) - n(t) = \sum (a_i - w_i) n_i(t) \qquad (21.22)$$

The aim of learning is for the synaptic weights w_i to reproduce the amplitudes a_i of the noise components; this can be regarded as learning a forward model of the noise channel (justifying the labeling in figure 21.6). If we use the mean square output error cost function (equation 21.8) and repeat the gradient descent calculation, the required teaching signal for this system is clearly this output error. Unfortunately it is unavailable to the biological system since it does not have access to the actual noise $n(t)$ to calculate $\hat{n}(t) - n(t)$ (and, if it did, it would use this signal directly to perform the cancellation rather than building a filter to learn it).

To get around this problem, we use an ingenious trick: we choose as cost function the mean square estimated signal

$$E = \frac{1}{2}\langle \hat{s}^2 \rangle \qquad (21.23)$$

This trick works because the basic assumption of noise cancellation is that the noise signals $n_i(t)$, and hence also $e(t)$, which is a combination of these noise signals by equation 21.22, are independent of the signal of interest $s(t)$. Hence, the correlation $\langle se \rangle$ is zero, and so by a simple calculation

$$E = \frac{1}{2}\langle (s+e)^2 \rangle = \frac{1}{2}\langle s^2 \rangle + \frac{1}{2}\langle e^2 \rangle \qquad (21.24)$$

This is (up to a constant shift) equal to the mean square output error suggested previously, so the optimal weights can equally well be found by minimizing this alternative cost function. Repeating the gradient descent calculation for this new cost function gives

$$\frac{\partial E}{\partial w_i} = \frac{1}{2}\left\langle \frac{\partial \hat{s}^2}{\partial w_i} \right\rangle = \left\langle \hat{s}\frac{\partial \hat{s}}{\partial w_i} \right\rangle = \langle -\hat{s}p_i \rangle \qquad (21.25)$$

(where we have used equation 21.22 in the last step) leading to a covariance learning rule

$$\Delta w_i = -\beta \langle \tilde{e}(t) p_i(t) \rangle \qquad (21.26)$$

with a new teaching signal $\tilde{e}(t) = -\hat{s}(t)$.

The external wiring of the "cerebellar chip" in figure 21.3B required for this noise-cancellation task is thus as follows: (1) parallel fibers carry the noise sources which are the context for this task, (2) Purkinje-cell output goes to a comparator to cancel the noise in the signal, and (3) a negative copy of the comparator output is used as teaching signal on the climbing fiber.

It is this peculiar character of the noise-cancellation teaching signal—that it is actually the signal of interest—that makes the simple cerebellar-like noise-canceling neurons possible. In these cells the context noise signals are supplied on parallel-fiber inputs to the apical dendrites, but the noisy signal is also supplied as an input on basal dendrites. The comparator stage can thus occur within the cell, with neuronal output as the teaching signal, eliminating the need for a climbing fiber. The required learning rule is homosynaptic and anti-Hebbian; for more discussion, see Roberts (2000).

This noise-cancellation calculation also makes an important theoretical point about the nature of the climbing-fiber signal. In textbook discussions of the Marr–Albus model the climbing-fiber signal is almost always described as an error signal, but this is not necessarily the case, and the term teaching signal is often more appropriate. Here the climbing-fiber signal is clearly not an error signal (in fact, it is a copy of the noise-canceled signal). An important consequence is that this climbing-fiber signal is not predicted to decay to zero after successful training as it would for an error signal. This possibility also applies to other adaptive learning tasks, such as optimal control tasks, where the training signal may be a straightforward error signal. Hence, the observation of a climbing-fiber signal whose modulation does not decay to zero over long periods of training (e.g., Catz, Dicke, and Thier, 2005) does not necessarily falsify the Marr–Albus interpretation of the climbing-fiber signal.

21.2.7 Noise-Cancellation Simulation Figure 21.7 shows the results of a simulation in which a sample of music is corrupted by a 440-Hz tone signal of unknown amplitude and phase. The noise-canceling filter takes as input a cos and sin pair of reference signals with frequency 440-Hz signals and learns the unknown amplitudes with which these contribute to the contaminating noise.

The signal is initially dominated by noise contribution (figure 21.7A), but the mean square error reduces systematically during training (figure 21.7B—in the figure the learning rates are deliberately chosen to be low so as to illustrate the progress of learning), and at the end of the 9-second sample the noise is accurately canceled.

This example shows that learning can take place in real time: data need not be presented in batches as they were for the triggered output in the previous modeling example. This allows the filter to follow slow changes in the amplitude and phase of the contaminating signal if the learning rate is well chosen. This characteristic of

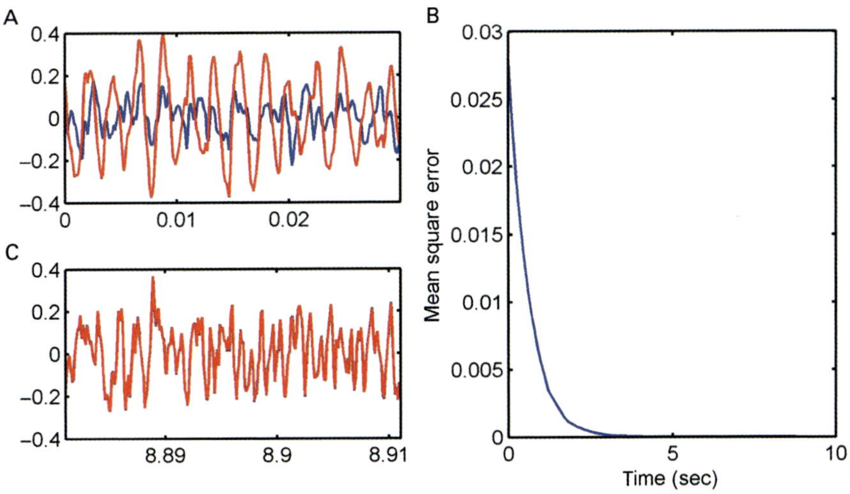

FIGURE 21.7 Noise cancellation applied to a sound signal. (A) Initial segment of a 9-secone sample from the Hallelujah chorus (blue curve) contaminated by a 440-Hz tone with unknown amplitude and phase; the noisy signal is shown in red. Before learning the hum clearly dominates the signal. (B) Plot of mean square error during learning; the learning rate is chosen to be slow enough to allow the approximately exponential convergence to zero to be clearly seen. (C) A segment take from the end of the learning process; the estimated signal (red) now accurately overlays the true signal (blue).

the adaptive filter lends it much of its computational power.

21.3 The Adaptive Filter: Application to Motor Control

We have shown thus far that the Purkinje cell has features which are well adapted to adaptive signal synthesis, and we applied the synthesis filter to a generic signal processing task, noise cancellation. In this section we will introduce a generic motor control task, adaptive plant inversion, and relate the requirements of this task to the structure of the granular layer.

21.3.1 Motor Systems: Adaptive Inverse Control

In very general terms all motor control problems reduce to the choice of control signals (sometimes called motor commands) to the muscles or actuators of a system in order to produce a desired movement (see chapter 9, "Computational Models in Motor Control"). To be specific, we will consider a motor system in which a control signal $u(t)$ drives the position $x(t)$ of an end effector (see figure 21.8). The processes relating control input to end-effector position will be lumped into a transformation P, called the motor plant,

$$P : u(t) \rightarrow x(t) \quad \text{or} \quad x = P[u] \qquad (21.27)$$

As an example control task, we will require the end effector to trace out a desired path $x_d(t)$. This can be achieved by designing a controller K that converts the desired motion $x_d(t)$ into appropriate control signals

$$K : x_d(t) \rightarrow u(t) \quad \text{or} \quad u = K[x_d] \qquad (21.28)$$

If control is perfect, then $x_d = PK[x_d]$ for all desired motions; hence, the required controller is called a (possibly nonunique) inverse plant

$$K = P^{-1} \qquad (21.29)$$

For engineering systems an inverse controller can sometimes be designed by identifying a sufficiently accurate mathematical model of the plant and then inverting it analytically. This kind of control design is unusual in practice since it results in a feedforward or "ballistic" controller: if anything disturbs the system output, there is no way to correct the resulting errors. Because of this drawback, some form of high-gain feedback control design is usually preferred. However in biological systems the relatively long latency of sensory information from the proprioceptive and visual systems limits the application of feedback control, since the sensory information arrives too late to control the motion, and the use of high gains can lead to control instability. Hence, biological systems often employ some element of ballistic control to achieve fast, high-frequency, motion. As shown in chapter 9, this may be coupled with feedback with prediction compensating in part for the delay.

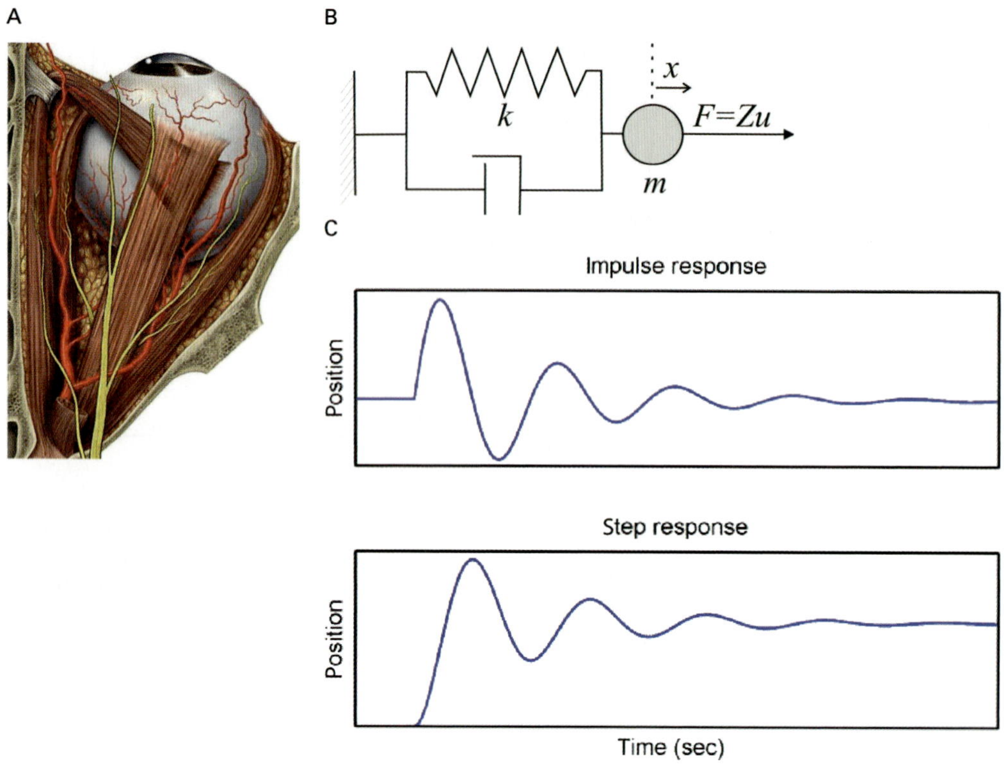

FIGURE 21.8 Motor plant description. The skeletomotor plant transforms control signals initiated in sites such as motor cortex into physical movements of the limbs. However, the complexity of the skeletomotor system makes this a difficult system in which to investigate basic control issues. (A) The oculomotor system presents many of the conceptual problems of skeletomotor control in a simpler (but as can be seen in this diagram, still rather complex) system. https://www.flickr.com/photos/patrlynch/450141893, reproduced under Creative Commons 2.5 license. (B) Many of the basic concepts of oculomotor control can be illustrated by a second order model in which the inertial, viscous, and elastic properties of the system are lumped into simple linear mass, spring, and damper elements and the actuator is represented as an ideal linear device producing force proportional to the control input (note that in the oculomotor system we should strictly refer to moments of inertia and torques etc., rather than masses and forces). (C)) The properties of linear systems such as the second order model in (B) are completely described by the impulse response or the step response. The plots show typical responses for an underdamped second order system.

There is a further complicating factor, in that biological motor systems (much more than manufactured devices) are subject to variations over time due to development, damage, fatigue, and so forth. These changes require a corresponding change in the properties of the controller; hence, biological inverse control systems are necessarily adaptive. In this section we will investigate how the cerebellum could implement an adaptive inverse controller, that is, a control module $K = P^{-1}$ able to adapt to changes in the motor plan P. This process is often referred to as learning an inverse model of a system and has many applications besides the simple trajectory control described here; for many examples, see Widrow and Walach (2008).

21.3.2 THE OCULOMOTOR SYSTEM For biological motor systems the term plant refers to the dynamics of two cascaded processes, the production of force by muscle tissue, and the transmission of that force through tendons and joints to the end effector. Both processes can be extremely complex, particularly when many degrees of freedom are involved. In order to concentrate on the underlying principles of adaptive control, it therefore makes sense to work with as simple a motor plant as possible. For neuroscientists the motor system of choice is often the oculomotor system.

The oculomotor system (illustrated in figure 21.8A) has many advantages over the skeletomotor system from a modeling point of view. Compared with control of, for example, limb movements, it has a restricted number of degrees of freedom (only three: horizontal, vertical, and torsional rotations of the globe). In addition, since the system producing horizontal eye movements is reasonably well separated both mechanically and in terms of neural circuitry, it is often possible to work with a system with just one degree of freedom controlled by just two muscles (the lateral and medial rectus), and a single control signal producing reciprocal innervation of these muscles. In addition there is a relatively small repertoire of eye movement types, which includes fast eye movements to targets called saccades and the vestibulo-ocular reflex (VOR), which is part of the automatic stabilization of the retinal image in response to head movements, and these different types of movement behavior can often be isolated in experiments (see also chapter 19, "Saccades and Smooth Pursuit Eye Movements").

21.3.3 Second Order Plant Model A one degree-of-freedom plant like this can often be adequately represented by lumping its mass, elasticity, and viscosity into single parameters m, k, and c (see figure 21.8B) and assuming that force generated is proportional to control input, $F = Zu$. By choosing units appropriately, we can set the strength $Z = 1$. With these assumptions we arrive at the second order plant model whose equation of motion of this system is the second order differential equation

$$m\ddot{x} + c\dot{x} + kx = u \qquad (21.30)$$

We can recover the required inverse model directly from this differential equation. Given a desired trajectory $x_d(t)$, the control signal required is

$$u_d(t) = m\ddot{x}_d + c\dot{x}_d + kx_d \qquad (21.31)$$

If we provide desired position, velocity, and acceleration as signals on the parallel fibers

$$p_1(t) = x_d(t), \quad p_2(t) = \dot{x}_d(t), \quad p_2(t) = \ddot{x}_d(t) \qquad (21.32)$$

the desired control signal can be calculated by a model Purkinje cell

$$u(t) = w_1 p_1(t) + w_2 p_2(t) + w_3 p_3(t) \qquad (21.33)$$

(see figure 21.9A) where the desired values of the weights are

$$w_1 = k \quad w_2 = c, \quad w_3 = m \qquad (21.34)$$

The analysis of the previous section tells us that the teaching signal required for gradient descent learning is the error in Purkinje-cell output

$$e_m(t) = u(t) - u_d(t) \qquad (21.35)$$

where the subscript m indicates that this is motor error (i.e., error in the control input).

This is an important result: in order to learn an inverse model directly, the climbing fiber should ideally carry a motor error signal. It is also a very inconvenient result, for three reasons. Firstly, motor error is not directly available to any sensor in the way that, for example, a visual sensor might directly code the error in end-effector position $e_s(t) = x(t) - x_d(t)$ (we will call this the sensory error signal). Secondly, since motor error is related to sensory error by the equation

$$m\ddot{e}_s + c\dot{e}_s + ke_s = e_m \qquad (21.36)$$

deriving motor error from sensory error requires an inverse model of the plant. This is a vicious circle, since in order to learn an inverse plant directly, we must already have an inverse model of the plant available. Thirdly, if motor error were available or derivable, why would we not simply add it to the current inaccurate motor command to obtain the desired motor command? This "motor error problem" is a particular example of the "distal error problem," a classic and generic problem in the theory of neural learning systems (Jordan, 1996) in which the teaching signal required for the internal layers of a network is proximal error, error in output at that level, rather than the more readily available distal error, that is, errors at the output layer.

Despite these difficulties, direct learning of inverse models using this architecture can be useful in certain situations. This is because there is a get out clause: although the gradient-descent error signal is motor error, other signals than motor error can produce stable learning if they satisfy a "strict positive realness" (SPR) condition (put simply, they are never more than 90° out of phase with the true motor error signal at any frequency; this condition guarantees that the correlation in the learning law always has the correct sign for all inputs). For simple systems, or limited classes of desired trajectories, this means that an approximate model of the inverse plant, called a reference model, can provide a suitable climbing-fiber signal (see figure 21.9B). This architecture is investigated in the modeling study below. For further discussion of this approach and an example application to the control of a robot arm, see Schweighofer, Arbib, and Kawato (1998a) and Schweighofer, Spoelstra, Arbib, and Kawato (1998b). The case for an alternative recurrent architecture for motor control in which the required teaching signals is sensory rather than motor error is described in Porrill and Dean (2004).

21.3.4 Modeling Study: Adaptive Plant Inversion The modeling study for this section uses the architecture described above (figure 21.9A and 21.9B) to learn the inverse controller for the second order plant (figure 21.8C).

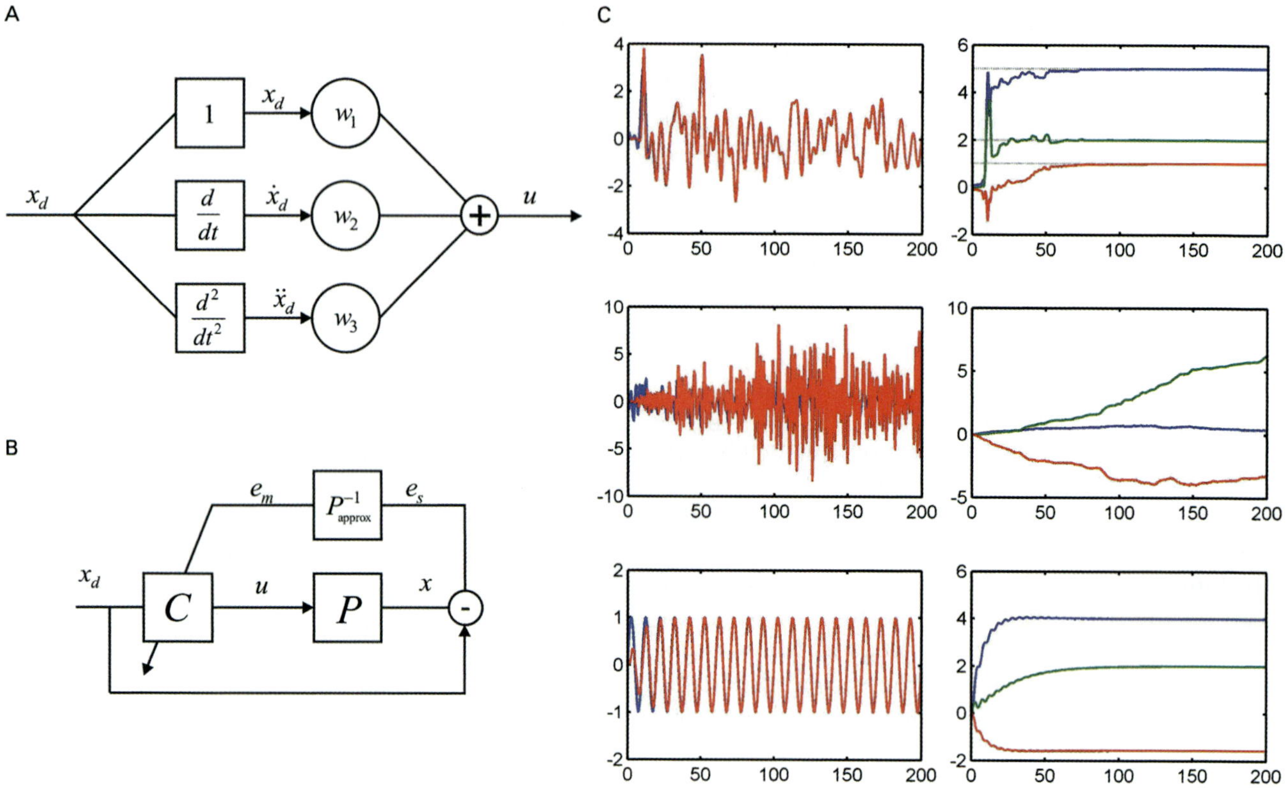

FIGURE 21.9 (A) Cerebellar element implementing the direct adaptive inverse controller for mass spring-damper system. The mossy-fiber input is desired position and position; velocity and acceleration signals are carried on the parallel fibers. (B) This panel shows the direct plant inverse architecture with an adaptive cerebellar element C as shown in panel b. If learning is successful, it will learn an inverse plant model $C = P^{-1}$. The ideal teaching signal carried on the climbing-fiber input is motor error e_m, a signal which is not directly unavailable. However if an approximation P^{-1}_{approx} to the inverse plant model is available, motor error can sometimes be successfully recovered from the sensory error e_s as shown in the diagram. (C) Three training runs are shown. Bottom row: Trained with a single sinusoid, it successfully learns the desired trajectory (left plot) but with incorrect parameter values (right plot). This learning model will not generalize correctly to other frequencies. Top: Trained with band-passed white noise containing frequencies for which the plant obeys the strict positive realness (SPR) condition, it successfully learns to generate the desired trajectory and, because the stimulus is persistently exciting, also learns the correct parameter values. Middle: Trained with band-passed white noise containing a wider range of frequencies for which SPR fails, the learning rule is divergent, as can be seen in the generated trajectory and the learned parameter values.

The first experiment specifies desired position x_d as a low-frequency sinusoid (figure 21.9C, top panels). It can be seen that the adaptive filter learns to track the motion accurately but does not learn the correct values for mass, elasticity, and viscosity, and it is easy to check that it fails to track sinusoidal inputs at other frequencies. A fundamental result in adaptive control theory emphasizes the need for *persistent excitation* during adaptation, that is, we must use inputs that explore the whole space of parameters of the model. It is common in both experimental and modeling studies to test only restricted classes of inputs such as sinusoids at a small number of frequencies. This is a dangerous practice since it can produce learning outcomes which are highly input dependent and ecologically unlikely.

The second experiment uses a much richer training signal consisting of band-passed white noise with a well-chosen high-frequency cutoff. This class of signals is much closer to the frequency content of natural movements, for example, head movements. It can be seen that the system again learns to track the desired motion, and that the model parameters now converge to the correct values.

Learning is successful in these examples despite the substitution of sensory error for the ideal motor error; this is because the input signals are in the frequency band where the phase shift of the second order plant is less than 90° so the teaching signal satisfied the SPR condition. To show the importance of this condition, the frequency cutoff for the band-passed noise is

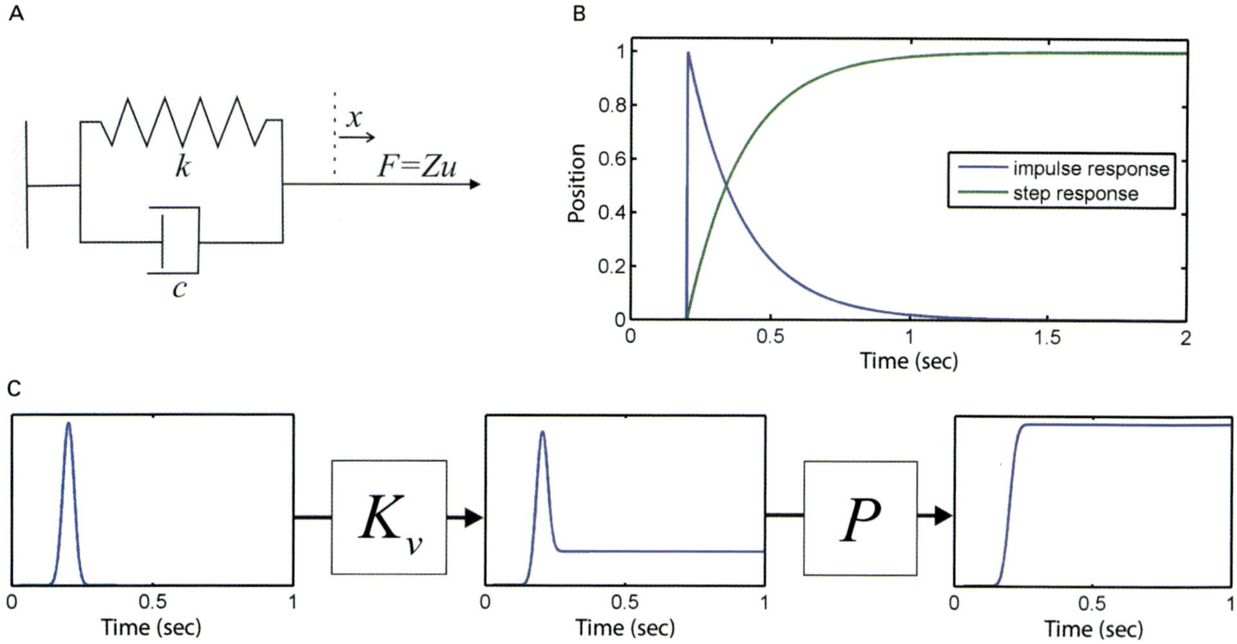

FIGURE 21.10 Pulse–step matching. (A) First order spring-damper approximation to the highly overdamped oculomotor plant. (B) The impulse and step responses of the first order model. The impulse response (blue) provides a fast movement to a target, but eye position subsequently decays to primary position. The step response (green) is slow to reach the target but is stable once the target is acquired. Neither provides a satisfactory profile for a saccadic eye movement. (C) Passing a delta function input (shown here as a smooth pulse to avoid scaling problems) through the (velocity) inverse controller produces a combined pulse–step control input which provides fast target acquisition, and then holds the eye on target. Only the position output of the plant is shown; the velocity output would exactly match the input.

increased in the third experiment beyond the SPR threshold. The contribution of some frequencies to the covariance term in the learning rule now has the wrong sign, and it can be seen that learning becomes unstable. With sufficient knowledge of the plant a reference model (see figure 21.9B) could be chosen which would extend the frequency limit for stable learning.

21.3.5 THE PULSE–STEP MATCHING PROBLEM The approach described above requires only three signals (position, velocity, and acceleration) to be available on the parallel fibers; these signals might be already available on the mossy fibers, or they might be obtained from a position signal on the mossy fiber by processing in the granular layer. However it is important to understand that this result only applies to a very limited class of inverse models.

In this section we will motivate a different and more flexible point of view by looking at a classic problem, namely, specifying the command to the eye muscles required to produce a fast movement (saccade) to a target (see chapter 19). The size of a saccade is specified by a fast burst of spikes (the "pulse") produced by neurons in premotor nuclei. It can be shown that the globe is so light that in most cases its inertia can be ignored (eye movements are viscosity dominated). The model becomes first order as shown in figure 21.10A with equation of motion

$$c\dot{x} + kx = u \tag{21.37}$$

which describes a leaky integrator. Its time constant $T = c/k \approx 200$ ms determines the time it takes for the eye to return to primary position when displaced (the displacement decays like $e^{-t/T}$).

The problems of controlling this plant to produce a fast movement can be illustrated by considering two naive strategies. First, the premotor burst ("pulse") alone could be used. Since this is effectively a delta function input to the system, the resulting eye movement is given by the impulse response of the plant, shown in figure 21.10b (blue curve). Although this command is effective in driving the eye quickly to a new position, the elasticity of the oculomotor plant causes the eye to relax to its original position over times of the order of 200 ms. Not only does the eye fail to stay fixated on the target, it is in motion as it returns to primary position, with a consequent drastic loss in acuity. An alternative strategy is to apply a step control input, producing the step response shown in figure

21.10B (green curve). The eye now successfully remains on target but takes rather longer than 200 ms get there. Since we make around three saccadic eye movements every second, and vision is seriously degraded while the eye is in motion, we would be effectively blind if either of these two strategies were used.

It looks as if some combination of the two strategies might work, but how are we to find the right combination? To answer this question, it is convenient to think of the burst as specifying desired eye velocity, and designing a controller for velocity $v(t) = \dot{x}(t)$, rather than position. From equation 21.37 we find that the required control signal is a combination of a scalar gain and an integrator

$$u = c\dot{x} + kx = cv + k\int v dt \tag{21.38}$$

Its impulse response is therefore a combination of a delta function pulse and a step (see figure 21.10C, middle plot). The scalar term produces the fast pulse of innervation to the muscles required to drive the eye quickly to a new position, and the integrator produces the steady command required to hold the eye in the new position. Clearly, an adaptive element is required to shape this pulse–step impulse response to match the physical characteristics of the plant.

21.3.6 Vestibulo-Ocular Reflex We now want to look in more detail at the implementation of the pulse–step controller using the known circuitry of the cerebellum. Unfortunately, saccadic control is not a straightforward inverse control problem because the eye's instantaneous velocity during saccades is not directly specified. Instead, saccades are required to be as fast and as accurate as possible given the limitations of the oculomotor plant. This is really an optimal control problem, and determining the role of the cerebellum in optimal control is a complex issue which is very much a topic of current research. Luckily, there is another eye movement type, the VOR, which corresponds almost exactly to the plant inversion problem we are discussing.

The function of the VOR is to maintain a stable retinal image by counterrotating the eyes in response to movements of the head signaled by the vestibular system. Since the eye movement produced by the reflex has no effect on the input vestibular signal, this reflex must operate in feedforward or open-loop mode. Long-term calibration is therefore required to ensure accuracy, and there is extensive evidence that implicates the floccular region of the cerebellum in this calibration process. There is also a clear candidate for the teaching signal in this case. An inadequate VOR will result in the eye's moving with the head relative to the world, and this produces movement of the image across the retina, called retinal slip (evidence that the climbing fiber carries retinal slip information will be discussed in the last section). The visual system is able to estimate the horizontal, vertical, and torsional components of retinal slip, and these three signals can be used as teaching signals for calibrating the VOR. For these reasons adaptation of the VOR has become a central experimental paradigm for investigating cerebellar function.

The VOR is explicitly a plant inversion problem: the head velocity signal from the vestibular system directly specifies the counterrotational eye velocity required, so the task of the controller is to convert desired velocity into a motor command signal to the oculomotor plant. A highly schematic diagram of the basic circuitry is shown in figure 21.11A. Two main pathways contribute to the reflex: the first is a three-neuron arc via the brainstem which causes vestibular signals to produce oppositely directed head movements; the second is a side path via the floccular region of the cerebellum. It is the cerebellar pathway that is usually regarded as adaptable (in fact, there is plasticity in both pathways, but to discuss why this might be so would take us beyond the aims of this chapter).

Cerebellar lesions reveal that a nontrivial contribution to the plant inversion process takes place in the direct brain-stem pathway. After such lesions the response of the VOR to fast head movements is slightly too small and gaze decays back to primary position, but more slowly than suggested by figure 21.10B. This suggests that the brain-stem pathway implements an undergained and leaky version $B \approx K_v$ of the ideal controller (equation 21.38). The task of the cerebellar side path is to compensate for the limitations in brainstem processing, suggesting the systems diagram shown in figure 21.11B. The controller element combines the outputs of the brainstem and (inhibitory) cerebellar pathways to give $K_v = B - C$; hence, the desired inverse controller $K_v = P_v^{-1}$ requires a cerebellar contribution

$$C_d = B - P_v^{-1} \tag{21.39}$$

Note that, when the brain-stem pathway is taken into account, the cerebellum is no longer predicted to learn a complete inverse model, but only a partial inverse model which supplements the processing already available in the brain-stem path. The estimated cerebellar impulse response is shown in figure 21.11B. Even for this relatively simple problem, the control problem is best described as one of "sculpting" this complex time-dependent desired impulse response.

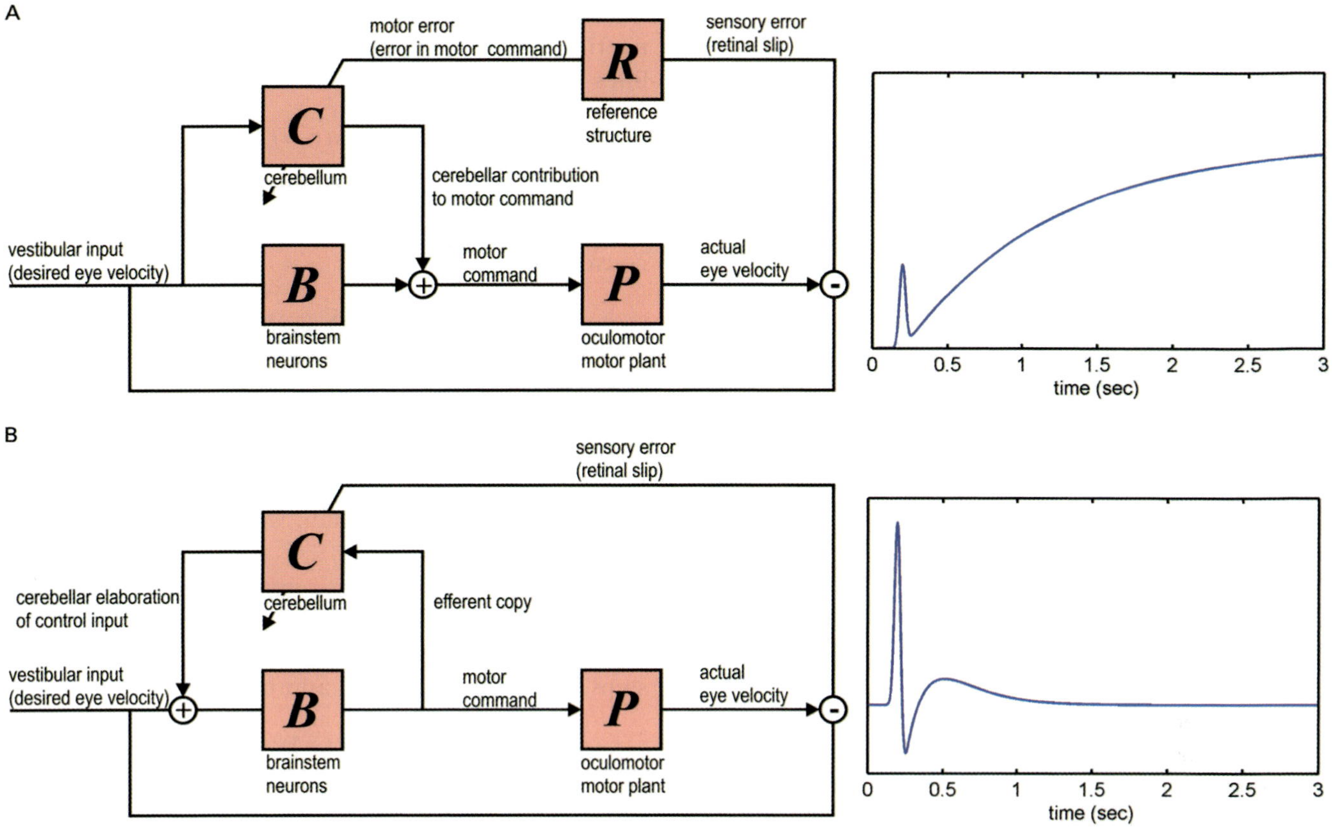

FIGURE 21.11 Vestibulo-ocular reflex (VOR) architectures. (A) Systems diagram of the conventional VOR circuitry showing the direct path though the brainstem and a forward side path via the cerebellum which receives vestibular information on the mossy fibers. Some processing is carried out in the direct path by a brain-stem component with a slightly undergained and leaky version of the ideal response shown in the middle panel of figure 21.10C. The cerebellar filter supplies a corrective contribution to the motor command. The impulse response of the desired cerebellar filter is shown in the right panel: it must combine a pulse correction to correct the gain and a slow rise to a step, to correct for the leaky output provided by the brainstem. (B) A systems model for recurrent connectivity, in which the cerebellum receives motor efferent copy on the mossy fibers, and provides a contribution to the vestibular input. The impulse response of the desired cerebellar filter is shown in the right panel. It is harder to interpret than in (A) because of the recurrent loop. One advantage of this architecture is that the desired cerebellar impulse response decays to zero and does not require very long time constant dynamics in the cerebellum as in (A).

There another feature of this system that is worth mentioning. Floccular inputs predominantly carry efferent copies of motor commands, suggesting that the recurrent connectivity shown in figure 21.11C may be a better description of this circuit. In this case the controller is $K_v^{-1} = B^{-1} - C$ (as can be seen by reading the diagram backward) requiring a cerebellar contribution

$$C_d = P_v - B^{-1} \tag{21.40}$$

so that with this architecture the cerebellum learns a partial forward plant model rather than an inverse model. The cerebellar impulse response for the recurrent case is shown in figure 21.11C. Since recurrent architecture is not subject to the motor error problem (the training signal required can be shown to be sensory error), it has been argued that it be more prevalent in motor systems than forward architecture (Porrill and Dean, 2004), and indeed, recurrent connectivity seems to be a very common feature in cerebellar circuits (Kelly and Strick, 2003). (See chapter 9 for a general discussion of the relation between inverse and forward models.)

At this point we draw some theoretical conclusions. Firstly, the commonly used language of forward and inverse models, although very suggestive, may be misleading in its details when discussing the cerebellum. Since it rarely acts alone, it is clear that in almost all situations it will only make an incremental contribution to a preexisting partial model. The nature of this contribution will also depend on the external connectivity:

for example, in the inverse control problem we have described the cerebellar contribution could either be a partial inverse model or a partial forward model (or, more likely, a combination of the two).

Secondly, neuroscientists often talk about control in terms of "gain and timing" tasks, sometimes assigning these two control components to different brain areas. The task "make a fast movement to a new position," might be thought to be of this type, where the system simply has to choose an appropriate gain/amplitude for a movement and time it accurately. However careful analysis of the control problem has shown that the controller must actually generate a complex, time-varying, impulse response. The simplified language of gain and timing does not seem to be helpful here.

Finally, it is important to note that once the inverse controller is available, the interface to the high-level control system is much simplified. For example, any sequence of saccades can now be specified by a series of pulse commands, and the controller takes care of the "elaboration" required to produce the combination of pulses and steps required by the oculomotor plant. The controller therefore acts like an operating system, hiding the complexities of the low-level system from the high-level user.

21.3.7 The Adaptive Filter The discussion above shows that to control even very simple motor systems, a controller needs to produce complex responses to simple inputs. To control a general system, we must be able to sculpt the impulse response of the controller to match that of the inverse plant, producing a time-varying signal of essentially arbitrary shape.

We have seen that a synthesis filter is able to produce any desired output as long as it has access to a rich basis of inputs. In the Marr–Albus adaptive-filter model the task of preprocessing the mossy-fiber input to produce this basis is assigned to the granule-cell layer. The schematic diagrams in figure 21.3A and B show a single mossy-fiber input carrying a signal $x(t)$ entering the granular layer. The adaptive-filter model of the cerebellum assumes that the effect of processing by the recurrent loops in the granule-cell layer is to recode this input into many parallel-fiber signals

$$p_i(t) = G_i[x(t)] \qquad (21.41)$$

where we will refer to the G_i as basis filters. The square-bracket operator notation $G_i[x(t)]$ is used to indicate that these are general causal filters, that is, that their outputs $p_i(t)$ may depend on the previous history of $x(t)$.

The most commonly used adaptive-filter basis in engineering applications consists of a set of tapped-delay lines

$$p_i(t) = G_i[x(t)] = x(t-i) \quad i = 0, \cdots, N \qquad (21.42)$$

where each operator G_i delays the signal by i time steps. In this case the output of the filter can be written as a convolution

$$z(t) = \sum w_i p_i(t) = \sum w_i x(t-i) = w * x(t) \qquad (21.43)$$

Since the time-varying impulse response of this filter is described by the series of coefficients $w = (w_0, w_1, \cdots, w_N)$, the tapped-delay line implementation allows an adaptive filter to represent any linear finite impulse response filter, whose length is smaller less the maximum delay.

Note that the derivatives required for direct inverse control (equation 21.31) can be specified using a tapped delay basis with just three lags, 0, 1, and 2 times steps, using the numerical approximations

$$\begin{aligned} \dot{x}(t) &\approx x(t) - x(t-1) \\ \dot{x}(t) &\approx x(t) - 2x(t-1) + x(t-2) \end{aligned} \qquad (21.44)$$

Tapped delay lines are not considered to be a biologically plausible mechanism in the cerebellum (Braitenberg, 1961, suggested that tapped delays might be implemented in cerebellar cortex by transmission delays along parallel fibers, but the delays involved are too small to be useful). However there are alternative, more plausible schemes that might be implemented by the granular layer. For example, it could implement approximate delays, in which the signal becomes increasingly delayed and broadened in time as it passes repeatedly through the circuit; this would produce a so-called spectral timing basis which is adequate for many tasks. These questions are discussed further in the final section.

Finally, we note that the adaptive-filter architecture is not restricted to linear problems. For example, if a nonlinear filter is required in engineering applications, a discrete Volterra series can be used. These extend the tapped delay line basis by including basis filters which are products of delayed signals, that is, with terms such as $p(t) = x(t-i)^2 x(t-j)$. In principle any well-behaved filter can be represented by a Volterra series in this way. A biologically plausible nonlinear basis, might well be produced by the granular layer. The split between a complex representation layer (granular layer) and a linear reconstruction layer (perhaps corresponding to the plastic synapses between granule cells and Purkinje cells or molecular-layer interneurons) is similar to the structure employed in reservoir computing (Lukoševičius and Jaeger, 2009). See chapter 22, "Integrative Functions of the Corticostriatal System," for further discussion.

21.3.8 DISCUSSION With the addition of an appropriate input-layer processing stage, the adaptive filter becomes a very powerful tool applicable to a wide range of adaptive signal processing and control problems as you can verify by flicking through Widrow and Stearns (1985). We have described this capability as *computational adequacy* and regard it as a necessary feature of a candidate cerebellar model: it must have, at least in principle, the capability to contribute to the wide range of motor and signal processing tasks known to involve the cerebellum. If the adaptive-filter model is correct, then details of external connectivity, choice of basis, and the precise nature of the teaching signal required will vary from application to application, but the basic elements of analysis stage, synthesis stage, and learning rule will be reproduced in every case. In a detailed survey of computational models of the cerebellum (Dean, Porrill, Ekerot, and Jörntell, 2010), among those models that successfully performed a nontrivial task, we found that the vast majority did implement an adaptive-filter algorithm combining these basic elements.

21.4 A Framework for Validating Cerebellar Models

This final section moves away from seeing how an adaptive filter *could* work in a variety of biological sensorimotor tasks to asking whether the cerebellum *does, in fact,* work in this way. In other words, we wish to see how the adaptive-filter hypothesis can be tested. The suggestions we make here apply not just to the adaptive-filter model but to functional cerebellar models in general. In particular they are concerned with how to bridge the gap between the (possibly beautiful) abstraction of a mathematically secure system-level model, and the (possibly messy) details of actual neuronal chemistry, firing, and connectivity.

21.4.1 FRAMEWORK There is an extensive body of experimental data which relates to the adequacy or otherwise of the adaptive-filter model, but organizing such a large amount of evidence turns out to be far from straightforward. We suggest that there should be two main approaches to evaluating a putative algorithm for the cerebellar microcircuit, a suggestion derived from the cerebellar chip metaphor (figure 21.3B). The first approach is to determine how far the model is consistent with experimental evidence concerning the cerebellar microcircuit itself. Since a great deal of microcircuit structure and function appears to be similar over the entire cerebellum (hence the chip metaphor), the evidence to be accounted for can come from studies almost anywhere in the cerebellum. The second approach is to assess whether the model is consistent with experimental evidence concerning how the external connectivity of the cerebellum underpins behavior. This connectivity varies from region to region of the cerebellum (see below) and so can only be addressed on a case-by-case basis where the relevant circuit details have been elucidated.

Each of these approaches to evaluation is now briefly considered in turn, and its significance for understanding of the adaptive-filter model related to the general problems that arise in assessing any model of the cerebellum. Further details and references can be found in Dean and Porrill (2011).

21.4.2 APPROACH 1: MICROCIRCUIT PROPERTIES One of the attractions of the adaptive-filter model is that it accounts in what seems to be a simple and intuitively satisfying way for a number of distinctive features of the cerebellar microcircuit. We consider first some fundamental features of the biological microcircuit that can be explained by the adaptive-filter model.

Microcircuit Properties Explained—Familiar Data *Why so many granule cells?* We have seen that the power of an analysis-synthesis filter depends crucially on the analysis stage providing an adequate range of basis signals which can be combined to produce the desired output. The need for a rich coding of all possibly relevant information (including the possibility that nonlinear combinations may be required) suggests a massive expansion–recoding at this stage. In fact, 10^{11} granule cells may hardly be enough to deal with the implied combinatorial explosion.

Why the peculiar PC morphology? Given the varying requirements of the environment, it may not be possible to hard wire in advance the particular inputs a microzone needs to perform its task. Instead a large sample of possibly relevant information must be presented, from which the PC chooses (by synaptic adaptation) the signals it needs. The flat, highly branched dendritic tree of a PC allows dense packing while allowing it to sample as much of the passing PF information as possible.

Why the climbing fiber? Explaining how information about task performance obtained at the periphery is relayed to internal neurons to supervise learning is a major problem in computational neuroscience.

Conveying this teaching signal on the climbing fiber is a dramatic biological trick to achieve this.

Why complex spikes? Suppose for concreteness that the teaching signal indicates a task error. When this signal arrives on the climbing fiber, PF/PC synapses must be adapted to prevent similar future errors, and all synapses are possible candidates for learning. Since synaptic weight changes are driven by changes in local properties, such as calcium concentration, the effect of the teaching signal input must be felt throughout the entire dendritic tree—hence, the winding fiber, the many synapses, and the massive burst of electrochemical activity produced by on a one-to-one basis by climbing-fiber input.

Why the low complex spike rate? Biology is not like engineering. It is not possible to train a module and then slot it in elsewhere, plug-and-play style. Part of the beauty of Hebb's unsupervised learning rule is that synaptic weight changes are driven by the natural output activity of the neuron; hence, a neuron can both learn and perform simultaneously. In contrast the climbing-fiber supervised teaching signal described above has a fundamental flaw. Every climbing fiber spike fires a complex spike, so, with a high rate of climbing-fiber input, the PC output would be dominated by complex spike activity. Thus, it would simply pass on the olivary input to target nuclei. To avoid this, the climbing fibers stochastically sample the teaching signal at a low spiking frequency of around 1 spike/s. Complex spikes are therefore rare enough (compared with a typical simple spike firing rate of 50 Hz) to have only a small effect on performance.

Why is the dynamics of the inferior olive so complex? The internal organization of the inferior olive is complex, with groups of neurons interconnected by electrical synapses exhibiting synchronous firing that can be modulated by external influences (for references, see, e.g., Van Der Giessen et al., 2008). These complexities may be related to the rate-lowering mechanism outlined above, which places extraordinary demands on the sampling algorithm. It must produce a statistically unbiased estimate of a teaching signal which might carry information in the 10-Hz frequency band using a very low spike rate of about 1 spike/s. Chaotic models of this sampling process have been developed which go some way toward explaining the complex dynamics of the olive (Schweighofer et al., 2004).

Microcircuit Properties Explained—New Data Although the congruence between the adaptive model and basic features of the cerebellar microcircuit is very encouraging, new data concerning microcircuit details continue to flood in. We now review the model's ability to explain this new experimental evidence and discuss whether the subsequent experiment–theory loop will refine our understanding of the adaptive-filter model or lead to a fundamental reappraisal of its status. In this section we consider two examples of new experimental evidence that appear to support the hypothesis: subsequent sections give examples of evidence that is apparently either irrelevant to the hypothesis or in flat contradiction to it. The first "favorable" example to be considered here concerns the properties of molecular-layer interneurons (MLIs). The second concerns the presence of a high proportion of silent granule-cell synapses, on both Purkinje neurons and MLIs, which will be the subject of the modeling study in this section.

Molecular-Layer Interneurons In the system-level version of the adaptive filter, "synaptic" weights can take either positive and negative values (figure 21.3). However, actual synapses are either excitatory or inhibitory and never switch between the two. It appears therefore that a biological implementation of the adaptive filter requires two pathways, one excitatory and one inhibitory, to convey the basis-function outputs from granule cells to the Purkinje cells. Moreover, the learning rule would seem to require a form of plasticity in the inhibitory pathway with a covariance learning rule of opposite sign to that given by equation 21.14 for the excitatory pathway.

There is a candidate inhibitory projection from granule cells to Purkinje cells via the MLIs (also known as basket and stellate cells—see figure 21.12). Synapses

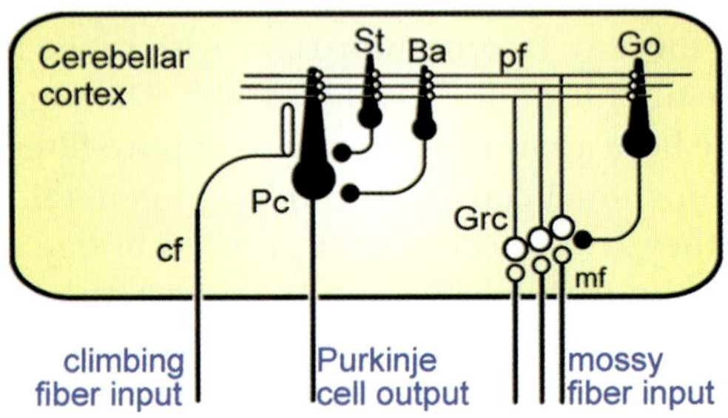

FIGURE 21.12 A more detailed schematic of microcircuit connectivity. In this diagram the molecular-layer interneurons—stellate (St) and basket (Ba) cells—are added to the basic microcircuit of figure 21.2A. The projections from these interneurons to Purkinje cells (Pc) provide a second inhibitory pathway which can implement the negative weights required by the adaptive-filter model. This diagram also shows the recurrent connectivity of granule cells (Grc) via the Golgi cells (Go) in the granular layer that may contribute to the processing power of the analysis stage of the adaptive filter. pf, parallel fiber; cf, climbing fiber; mf, mossy fiber. Adapted from Lepora et al. (2010), figure 1.

from granule cells to MLIs are excitatory, and those from MLIs to Purkinje cells are inhibitory, so the net effect of granule-cell output on Purkinje-cell firing is inhibitory, as required. However only recently has evidence concerning plasticity in this pathway become available. Very briefly, this evidence indicates that (1) the synapses of granule cells on MLIs are plastic, (2) this plasticity is controlled by climbing-fiber input, and (3) increased climbing-fiber activity increased the efficacy of these synapses (i.e., LTP) in contrast to the LTD produced by pairing climbing-fiber and parallel-fiber inputs at the Purkinje cells themselves (Dean et al., 2010; Jörntell and Ekerot, 2002). These are just the properties needed to implement the adaptive filter and its covariance learning rule.

21.4.3 Modeling Study: Silent Synapses It has been demonstrated experimentally that a very large proportion, perhaps as many as 90%, of the synapses between parallel fibers and Purkinje cells, have low or zero synaptic weights (Isope and Barbour, 2002). Why should so many of these synapses be functionally silent?

This phenomenon may be explained in terms of a fundamental optimality property of the covariance learning rule. Learned filter weights are optimal in the sense of least squares, enabling the filter to perform the task as accurately as possible. As a result, when signals carried on the parallel fibers are corrupted by noise, the filter chooses an overall combination of weights which is optimal for eliminating the disturbances causes by that noise.

21.4.4 Modeling Study This can be illustrated by considering a simple case where there is a single signal $s(t)$ of interest and a subset of PFs, each carrying a combination $p_i(t) = a_i s(t) + n_i(t)$ of the signal of interest $s(t)$ with amplitude to a_i and independent noise signals $n_i(t)$ with power σ_i^2. It is easy to show that the expected square output error is minimized when the corresponding weights have the ratios

$$w_i : w_j = \left({a_i}/{\sigma_i^2} \right) : \left({a_j}/{\sigma_j^2} \right) \tag{21.45}$$

Hence, the optimal weight for each channel is proportional to signal amplitude and inversely proportional to noise power.

The time course for learning these optimal weights is illustrated in the simulation shown in figure 21.13A. Weights changes are shown for three parallel fibers, two of which, PF_1 and PF_2, carry equal levels of the signal of interest, with the first having lower noise, while the third, PF_3, carries noise alone. In the early stage of learning the weights on PF_1 and PF_2 (initially zero) are driven to the same value because they carry equal amounts of signal. Later on in learning, to minimize the noise in the output, these weights are driven to the optimal values determined by the formula above, with higher weight being assigned to the lower-noise input PF_1. In addition learning drives the weight on the all-noise channel PF_3 (which was initially set to be nonzero) to its optimal value of zero (since it is all noise and no signal).

Experimentalists should note that the optimal values are only taken after lengthy training, and that at intermediate stages the weights take values for which performance is unbiased (has zero mean error) but is suboptimal (has raised error variance). It is likely that most time-limited cerebellar learning experiments leave synaptic weights in this intermediate state, rather than in the overtrained optimal state which is probably typical of the behavioral situation.

This optimality principle applies both to "genuine" noise, for example, intrinsic noise generated in the granular layer, and to situations in which the noise is an irrelevant sensory signal. For example, the weight histories shown in figure 21.13A apply to the idealized tactile learning task shown in figure 21.13B which depends on a spatially localized sensory input. Before training, the location relevant to the task (black cross) lies in the receptive field of PF_3 (red), so that this input has high synaptic weight. Weights on the irrelevant PF_1 and PF_2 (blue and green, respectively) are zero. If the task is modified so that the relevant input is now located at the black dot, then the input from PF_3 is irrelevant (effectively all noise). Relevant inputs now come from PF_2 (large green receptive field: always responds to a correct stimulus but often responds to irrelevant stimuli, hence also has a high noise content) and from PF_1 (small blue receptive field: also carries the relevant signal but rarely responds to irrelevant stimuli, hence has low noise content). Thus, the weight for PF_3 is driven to zero by long-term depression whereas PF_1 and PF_2 learn through LTP and stabilize at values that maximize the overall signal-to-noise ratio.

21.4.5 Silent Synapses as a Consequence of Optimality If parallel fibers carry a widespread array of information about the entire sensorimotor context in which motor activity takes place, then, since there are so many (~170,000) parallel-fiber inputs to a given Purkinje cell, it is likely that most parallel fibers will carry information which is only weakly related (low signal to noise) or is totally unrelated (all noise) to any given task. The analysis above shows that the long-term optimal synaptic weights for such synapses is necessarily

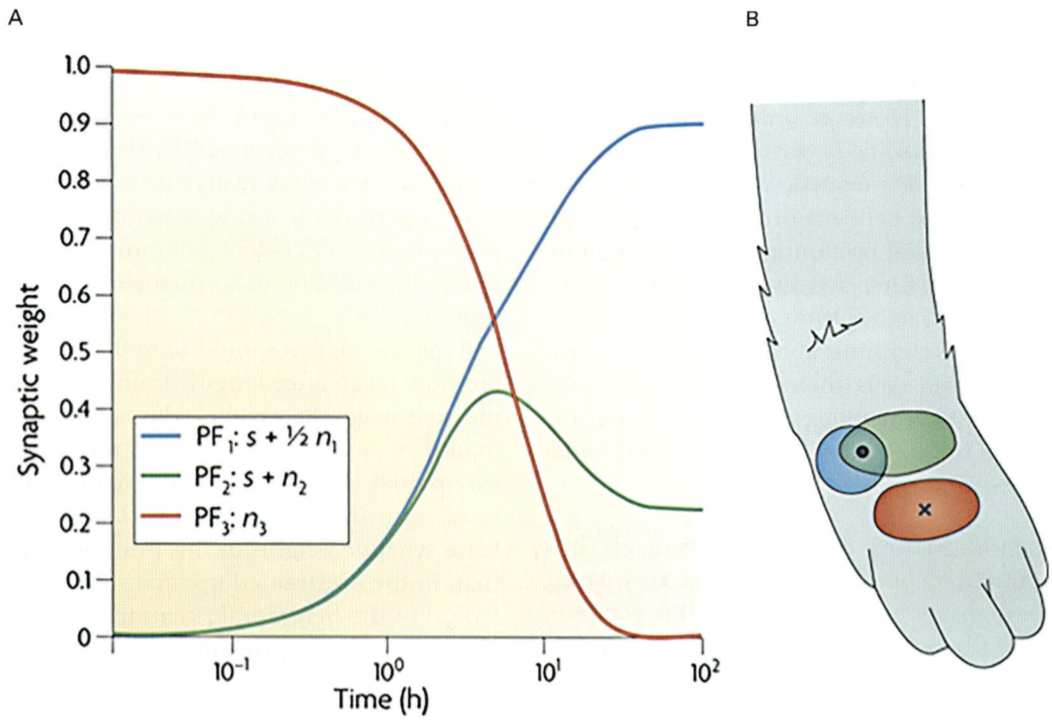

FIGURE 21.13 Time course for learning optimal synaptic weights. (A) A simulation in which the synaptic weights for parallel fibers (PFs) which carry different combinations of signal and noise converge to the optimal values. The weights for PF1 and PF2 which carry equal amounts of signal achieve similar values early in training, but eventually the PF with lower noise makes a larger contribution, reducing the noise level in the output. The initially nonzero synaptic weight on PF3 is driven to silence since it is irrelevant to the task. The learning has been chosen to approximate vestibulo-ocular reflex learning rates. Note the logarithmic scale: the time scale for achieving task optimality may be much longer than that for achieving a basic level of task performance. (B) The noise signals in a could be signals currently irrelevant to the current task rather than intrinsic noise signals. In the text a tactile learning situation is described in which the weight profiles plotted in panel (A) correspond to parallel fibers carrying information from the tactile receptive fields of the corresponding color. Adapted from Dean et al. (2010), figure 3.

small or zero. Hence, it is a consequence of the optimal performance of the system that most synapses between parallel fibers and Purkinje cells are expected to be silent, consistent with experimental evidence.

Although most synapses are silent, they may at some point become functionally relevant, as, for example, in classical conditioning in which a normally irrelevant tone stimulus can be made to drive a blink response by repeatedly pairing it with an aversive air puff or shock (unconditioned stimulus). Reviving a silent synapse in this way can only be achieved via LTP, either on PCs or on MLIs, emphasizing the importance of LTP in the learning process and providing evidence against the textbook dogma that the cerebellum learns via LTD alone.

21.4.6 MICROCIRCUIT PROPERTIES NOT EXPLAINED—BUT IRRELEVANT Besides these new microcircuit phenomena that are apparently computationally consistent with the adaptive-filter model, there are other phenomena which may simply be irrelevant to the model's evaluation. These might relate to properties of neurons that are concerned, for example, with the neuron's general survival rather than its particular computational task. Thus, it has been suggested that LTD observed at synapses between parallel fibers and Purkinje cells does not implement some form of learning rule but instead ensures that Purkinje cells remain in a viable operating range (e.g., do not fire too much and die). Synaptic plasticity with this kind of function is sometimes referred to "homeostatic" synaptic plasticity (Turrigiano, 1999). The key problem here is the difficulty of determining the computational significance of any particular form of plasticity. This is an example of an enormous topic facing computational neuroscientists—every day brings in new experimental data, whose computational relevance is very often simply unclear. It has been suggested that multiscale modeling may help with this problem—for example, developing system-level models like the adaptive-filter model alongside more detailed phenomena-based models.

21.4.7 Microcircuit Properties Not Explained—Possible Trouble We consider two examples here, concerning granular-layer processing and Purkinje-cell firing.

Granular-Layer Processing As indicated in figure 21.2, adaptive-filter models are generally thought to predict recoding of mossy-fiber input by the granular layer. The details of this recoding are unclear; those we have mentioned include sinusoids, tapped delay lines, and spectral timing bases. All these schemes agree that a wide variety of different granule-cell firing patterns needs to be generated. Technical difficulties have meant that recording of granule cell responses has only become possible relatively recently, and initial studies (Jörntell and Ekerot, 2006) have found evidence suggesting rather limited processing of mossy-fiber signals, with, for example, high-fidelity transmission of mossy-fiber bursts and little mixing of different mossy-fiber signals. In the somatosensory system, it has been suggested that "regardless of input type, the temporal patterns of granule-cell spike activity, both spontaneous and evoked, appeared to primarily follow the activity in the presynaptic mossy fibers." (see Jörntell and Ekerot, 2006, abstract)

If these findings prove to be correct, can they be reconciled with the adaptive-filter model? They may not pose a problem as far as the flocculus is concerned, because floccular mossy fibers themselves carry a wide variety of signals (see the next section). Moreover, in vivo recordings from floccular granule cells in anesthetized mice suggest that "MF-GC [mossy fiber–granule cell] synapses do not simply relay peripheral spike rates … but appear to present vestibular information to be integrated with other stimulus features" (Arenz et al., 2008, p. 979) There is also preliminary evidence for complex granular-layer transformation of mossy-fiber input in the flocculus of anesthetized rabbits (Van Dorp et al., 2009). These findings are broadly consistent with the evidence above that suggests the flocculus does operate like an adaptive filter.

However, the issue of generality persists. Current evidence does not rule out the possibility of differences in granular-layer processing between microzones. Input-segregation and high-fidelity transmission might be characteristic of some microzones outside the flocculus although this would appear to raise questions about the functional utility of granular-layer processing. Alternatively, it may be the case that recently described complex aspects of granular-layer processing in these microzones, such as "time window matching" and oscillations (D'Angelo et al., 2009; Dugué et al., 2009), provides recoding mechanisms more compatible with an adaptive filter. Thus, further characterization of granular-layer processing will prove crucial for assessing the generality of the adaptive-filter model.

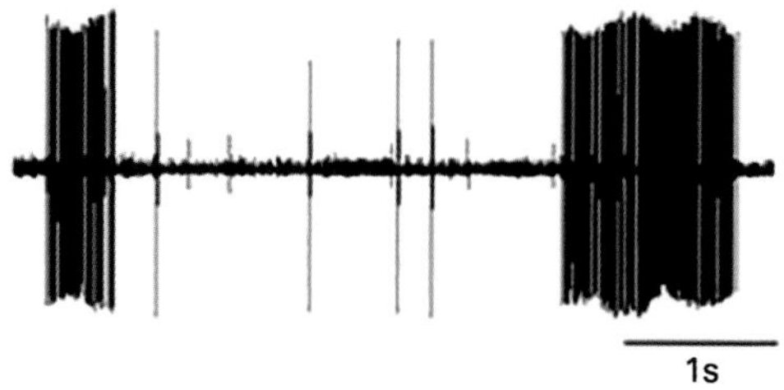

Figure 21.14 Purkinje-cell bistability. A segment of an extracellular recording from a Purkinje cell showing evidence for bistability. The recoding shows fast simple spike firing during the "on" state and an extended pause in firing during the "off" state. Adapted with permission from Loewenstein (2005), figure 3a.

Purkinje-Cell Bistability We now present an example of a microcircuit feature, Purkinje-cell bistability, which is apparently completely at odds with the adaptive filter. Bistability refers to a phenomenon in which Purkinje cells show both an "up" state with depolarized membrane potential and simple-spike firing and a "down" state with hyperpolarized membrane potential and no simple spikes (see figure 21.14). The presence of these down states would appear on the face of it to be incompatible with the adaptive-filter model.

If we are to assess what functional role this phenomenon might subserve, we must first ascertain in which areas it is found behaviorally. It seems that this feature may not play a major role in the flocculus, at least when the flocculus is actively engaged in image stabilization. For example, few long pauses are found in mouse floccular PCs during optokinetic response or VOR adaptation. This result is compatible with the large number of analyses relating primate PC firing rates in flocculus to eye movements since such analyses would be severely compromised by any major intrusion of down states.

It is possible that PC bistability is primarily a feature of anesthetized or in vitro preparations, and it has been argued that PCs in awake animals are almost always in the up state; however, an alternative and intriguing alternative possibility is that down-state frequency varies between microzones. For example, a recent study recording from the orofacial region of vermal lobule VI found long pauses (mean = 700 ms) in the simple-spike firing of 47% of Purkinje cells in awake cats (Yartsev, Givon-Mayo, Maller, and Donchin, 2009). As yet, however, the functional significance of bistability in these cerebellar regions is unclear, and the authors report no obvious behavioral correlates of these pauses.

It is possible that bistability becomes really important only in more lateral regions of cerebellum which receive input from principal olive (PO; the flocculus receives from dorsal cap of Kooy; DCK). PO neurons differ from DCK neurons in morphology and electrophysiological properties and show subthreshold oscillations and robust rhythmicity at 6–12 Hz, perhaps reflecting the different requirements of oculomotor and skeletomotor control. These differences in properties may be relevant to the bistability issue since there is evidence that a switch between Purkinje-cell states can be triggered by complex spikes (Lowenstein et al. 2005, Yartsev et al. 2009). Thus, the data on bistability might possibly be explained by supposing that DCK input to flocculus does not readily trigger switches between Purkinje-cell states whereas PO input to lateral hemisphere does.

In that case, it could be that the lateral hemispheres implement a different algorithm for motor control to cope with the extreme complexity either of the multijoint organization of the somatomotor system or of putative cognitive functions, requiring a new role for olivary input. It has been suggested that this involves direct driving of motor output by generating temporal patterns by switching between Purkinje-cell states (e.g., Jacobson et al., 2008). Such a proposal would be consistent with an evolution of cerebellar function from an early adaptive filter (vestibulocerebellum) to a new computational mode (lateral hemispheres). From the computational perspective, it is not yet clear how such an algorithm could, in fact, be used for complex tasks such as the control of a multijoint robot arm nor whether the new algorithm would replace the adaptive filter or in some sense complement it. Answering these questions is clearly vital for assessing the generality of the adaptive-filter model.

21.4.8 Approach 2: Connectivity The second type of test faced by the adaptive-filter model concerns the signal transformations carried out by individual functional regions of the cerebellum. As mentioned earlier in the context of the chip metaphor (figure 21.3B), different parts of the cerebellum have unique connectivity, in contrast to the generally uniform structure of the cerebellar microcircuit itself. A simplified diagram of cerebellar organization is shown in figure 21.15 (adapted from Apps and Hawkes, 2009).

Individual regions of the inferior olive send climbing fibers to strips of cortex running in the sagittal plane. The sagittal or longitudinal strips, referred to as zones, are labeled from A at the midline through to D2 at the most lateral. Each zone also projects to its own target region in the deep cerebellar or vestibular nuclei, each of which projects both to a specific set of target structures in the rest of the brain and spinal cord and to a specific region of the inferior olive. It has been shown that this olivary region is identical to the region that projects to the original cortical zone, so that a loop is formed suggesting a functional subunit. Zones can be further subdivided further into microzones on the basis of connectivity, giving rise to a (very crudely) estimated 5,000 functional subunits.

The microzone, together with its inputs and outputs, is the organization that appears to correspond to the idea of a cerebellar "chip" that always carries out the same algorithm, but with differing connectivities adapted to a range of functional tasks. Such an organization is entirely consistent with the adaptive-filter model since adaptive filters are used for many different signal processing tasks depending on how they are connected (Widrow and Stearns, 1985). Hence, for a given cerebellar task, we need to know the specific inputs and outputs of the microzone concerned to assess the model properly. We consider here two examples, first the flocculus (related to VOR adaptation as discussed in section 21.3), and secondly a region of the cerebellar hemispheres concerned with classical conditioning, which is the subject of this section's modeling exercise.

21.4.9 Flocculus The region of the cerebellum that has been most investigated in this regard is probably the flocculus (posterolateral cerebellum, figure 21.15). This is the region concerned with the retinal-image stabilization task described in section 21.3, and the evidence relating its function to that of an adaptive filter can be briefly summarized as follows:

1. *Function:* The involvement of the flocculus in image stabilization is well established, both by classical inactivation and lesion studies in a variety of species and by more recent studies in mutant mice.
2. *Mossy-fiber inputs:* One of the main types of mossy-fiber input to the flocculus conveys vestibular information exactly as required by the forward VOR architecture discussed in section 21.3 (figure 21.11A and B). The second main type of input, and, in fact, the predominant type, is efference-copy information about eye-movement commands, which is required for the recurrent architecture previously discussed (figure 21.11C).
3. *Climbing-fiber inputs:* The climbing-fiber input to flocculus signals retinal slip, as required.
4. *Output:* Floccular Purkinje cells project to ocular motoneurons, with the most direct projections having

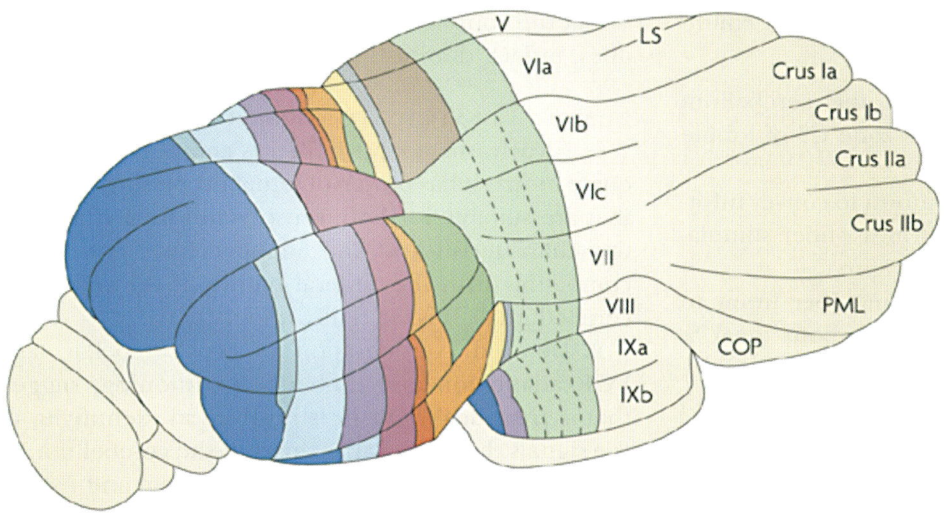

	Hemisphere			Paravermis				Vermis					
	D2	D0	D1	C3	C2	CX	C1	A2	B	X	AX	A	
	dlPO	dmPO	vlPO	vfDAO	rMAO	iMAO (med)	vfDAO	cMAO (subnuc b'/c)	dfDAO	iMAO (lat)	cMAO (subnuc a)	cMAO (subnuc b)	Inferior olive (input)
	NL (parvo)	DLH	NL (magno)	NIA	NIP	ICG; NIP	NIA	DLP	LVN	ICG; NIP	MedN (lat): ICG	MedN (lat); ICG	Cerebellar nuclei (output)

FIGURE 21.15 Organizational structure of the cerebellum. Regions of the inferior olive send their climbing fibers to sagittal strips of cerebellar cortex, termed zones. For example, the diagram of a dorsal view of rat cerebellum (top panel) shows a light green strip that corresponds to zone A of the vermis, which receives climbing fibers from the light green area (cMAO, caudal medial accessory olive) of the schematized inferior olive. Each of the 12 sagittal zones shown also projects to a specific target in a cerebellar or vestibular nucleus (medial nucleus in the case of zone A). A sagittal zone is thus defined by olivary inputs and corticonuclear outputs, and by the type of afferent system that activates the climbing fibers. A zone can in turn be divided into constituent microzones, each a narrower sagittal strip of cortex that can span several contiguous folia of the cerebellar cortex, and which receives climbing-fiber afferents corresponding to very specific peripheral inputs. The regions marked eyeb, V, and sacc (highlighted in red) are associated with eyeblink conditioning, vestibulo-ocular reflex, and saccadic adaptation, respectively. These are some of the very few regions for which there is a basic level of understanding of connectivity and function. Adapted from Apps and Hawkes (2009), figure 4.

either two or three synapses. Their simple-spike firing rates are related to eye movements.

5. *Plasticity:* When the requirements for image stabilization change, for example, in an apparatus that mimics a changed gain in the VOR (VOR adaptation), Purkinje-cell firing appears to change in the way predicted by the adaptive-filter algorithm. This is a complex issue, partly because VOR adaptation also induces plasticity in the brainstem (component *B* in figure 21.11B and C of section 21.3).

In summary, the evidence concerning the floccular inputs related to image stabilization and their transformation into floccular output appears to be broadly consistent with the flocculus acting as an adaptive filter. However, it should be noted that the mapping of engineering-style circuit diagrams onto actual neural circuits can become very complex at a detailed level and raises important conceptual and practical research questions (Porrill, Dean, and Anderson, 2013).

21.4.10 MODELING STUDY: CLASSICAL CONDITIONING Perhaps the second most studied behavioral task involving the cerebellum is classical conditioning of eyeblink. In this task, a conditioned stimulus (CS; often an auditory tone, 200–500 ms long) precedes an unconditioned stimulus (US; either an air puff to the cornea or a mild shock to the skin round the eye). The US elicits a brief eyeblink (the unconditioned response; UR), and after a number of CS–US pairings the CS also elicits a blink (the conditioned response; CR). In delay conditioning the CS and US terminate at the same time.

The involvement of the cerebellum in classical conditioning can be summarized as follows:

1. *Function:* The main region of the cerebellum involved in classical conditioning is zone C3 of lobule VI in the posterior cerebellum.

2. *Mossy-fiber inputs:* These are known to convey information about the CS, and indeed mossy-fiber stimulation can be used as a CS in its own right.

3. *Climbing-fiber inputs:* The climbing-fiber input to the eyeblink zone carries information about the US. Stimulation of this input can act as a US in its own right.

4. *Output:* Purkinje cells in the eyeblink region project to the anterior interpositus nucleus, which, in turn, projects, via the red nucleus, to the motoneurons that control the eyelids and nictitating membrane.

5. *Plasticity:* Conjunctive stimulation of mossy fibers and climbing fibers teaches eyeblink zone Purkinje cells to pause at an appropriate timing, relieving inhibition on the eyeblink mechanism (Hesslow, Jirenhed, Rasmussen, and Johansson, 2013).

An interesting feature of eyeblink conditioning is that application of the adaptive-filter model immediately encounters a problem. In the VOR and other motor tasks the adaptive filter receives a teaching signal that is related to movement error, and by reducing this signal, the filter learns accurate commands. But in classical conditioning of a movement there is apparently no signal related to movement inaccuracy because by definition the US is delivered independently of the subject's response. This arrangement leads to an immediate problem because it seems that a meaningful endpoint to learning can never be reached.

One popular solution to this problem is to move the comparison between motor response and actual outcome from the physical world outside the brain to an internal comparator implemented by the brain itself, as shown in figure 21.16. The basic idea is to send an inhibitory copy of the CR motor command to part of the brain which also receives excitatory information about the US, and use the difference between the two as the teaching signal. Before learning starts there is no CR motor command, so the US itself is the error signal. But as learning proceeds the CR command grows larger, until eventually it cancels out the US. At this point the error signal reaches zero, and learning stops. This mechanism has been proposed in a number of papers (see general references to eyeblink conditioning in Lepora, Porrill, Yeo, and Dean, 2010), with the comparator usually being identified with the inferior olive itself. The CR command copy arrives at the olive via the nucleo–olivary pathway, an inhibitory projection from the DCN. One reason for the mechanism's popularity is that the complex-spike signal to zone C3 in lobule VI does, in fact, diminish as learning proceeds.

21.4.11 Connectivity—Conclusions

1. Clearly, the adaptive filter is computationally adequate in that it has powerful signal processing capacities that are suitable for both sensory and motor tasks and the characteristics of the adaptive filter are broadly compatible with the general nature of the role of the cerebellum as it is currently understood.

2. Detailed consideration of two cerebellar tasks, VOR adaptation and eyeblink conditioning, suggests that a great deal of work is involved in identifying (1) the signals that arrive at any part of the cerebellum, (2) how these signals change during learning, and (3) how these signals are related to behavior. Even in these two cases where in broad terms the adaptive-filter model appears plausible there are still important questions left unanswered.

3. The number of regions where even this information is available appears to be surprisingly small. Part of the vermis is well-known to be involved in saccadic adaptation (figure 21.15), and the relevant external circuitry is in part identified. Unfortunately there are relatively few instances where the connectivity of a cerebellar microzone is understood well enough to allow us to assess the crucial feature of the input–output relationship: how a zone's output effects its climbing-fiber input. Indeed the relevant connectivity is understood for well under 1% of the ~5,000 cerebellar microzones (estimated in cat).

21.5 General Conclusions

We hope that we have convinced you that the cerebellum is a fascinating brain region, worthy of the effort that has been put into understanding it. The adaptive-filter model has been presented as an exemplar of what a successful systems-level model of a brain region might look like. We have tried to illustrate, using modeling examples, its computational adequacy, and we have explained in some detail the remarkable level of explanatory power it already incorporates. We have also attempted to stand back and present an honest appraisal of the future of this model in the light of the much broader range of evidence that is now flooding in from experimental studies. At the very least we hope that we have presented a case for these new findings to be assessed for compatibility with the adaptive-filter model before, or at least alongside, the development of new and exotic models inspired by the dramatic, but possibly irrelevant, features of new phenomenology.

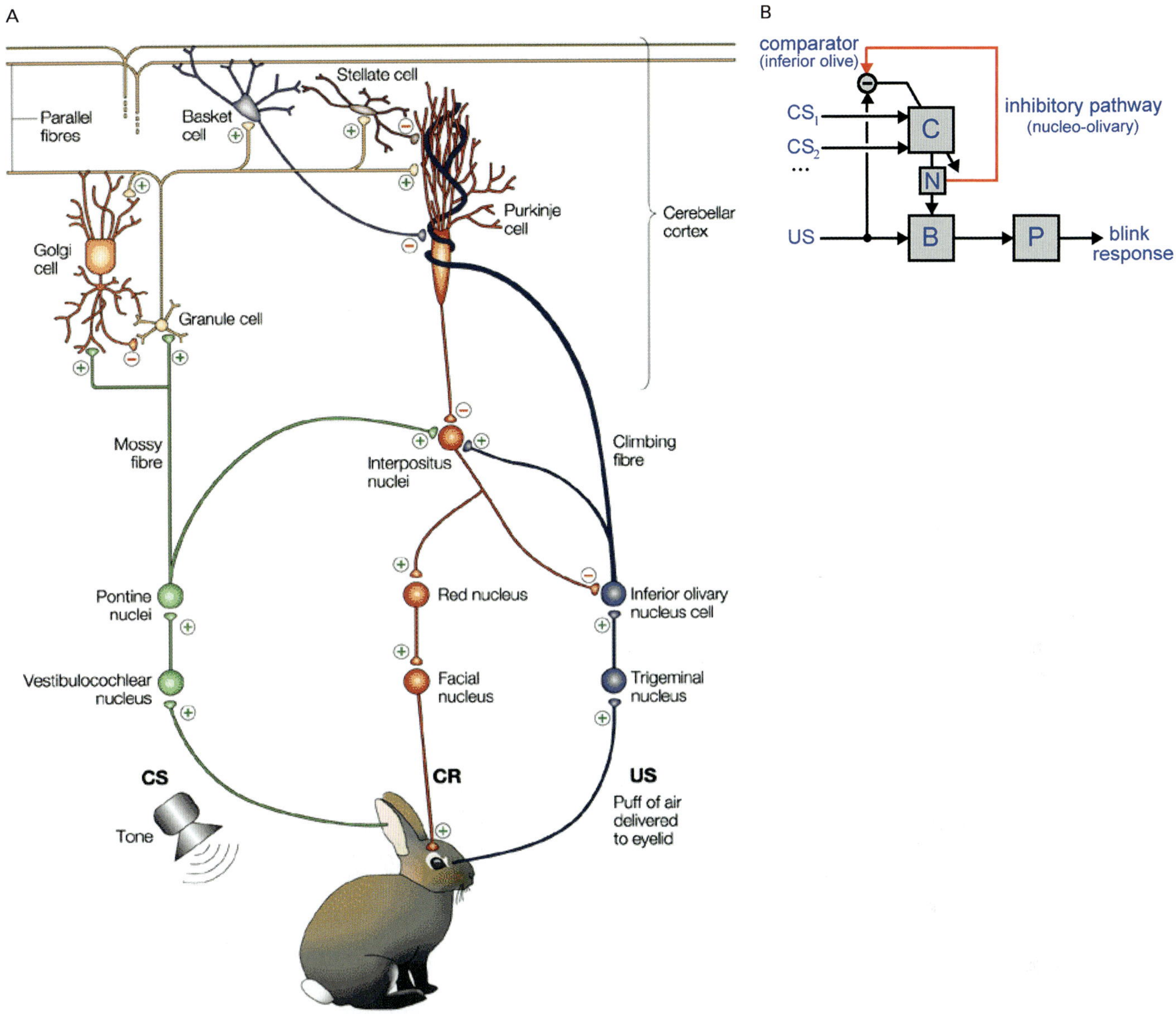

FIGURE 21.16 Cerebellar role in classical conditioning. (A) Schematic diagram of the connectivity of the circuitry underlying classical condition of eyeblink. The nuclear-olivary projection which allows Purkinje-cell output to influence firing in the olive has not appeared in previous diagram. Adapted from Medina et al. (2002), figure 1. (B) A systems diagram of a classical conditioning model based on the adaptive filter. This has two anatomical pathways driving reflexes: (1) the direct pathway, in which the unconditioned stimulus (US) drives premotor sites in the brainstem (B) that then drive the eyeblink motor plant (P) to produce the unconditioned response, and (2) the indirect pathway, with one or more conditioned stimuli (CSs) conveyed to the cerebellar cortex (C), the output of which is routed by the deep cerebellar nucleus (N) to converge with the direct pathway. Cerebellar learning is driven by a comparison in the inferior olive between the US-drive and cerebellar output. Adapted from Lepora et al. (2010), figure 1.

REFERENCES

Albus, J. S. (1971). A theory of cerebellar function. *Mathematical Biosciences, 10*, 25–61.

Apps, R., & Hawkes, R. (2009). Cerebellar cortical organization: A one-map hypothesis. *Nature Reviews. Neuroscience, 10*(9), 670–681.

Apps, R., & Hawkes, R. (2009). Cerebellar cortical organization: A one-map hypothesis. *Nature Reviews. Neuroscience, 10*(9), 670–681.

Arenz, A., Silver, R. A., Schaefer, A. T., & Margrie, T. W. (2008). The contribution of single synapses to sensory representation in vivo. *Science, 321*(5891), 977–980.

Bartha, G. T., Thompson, R. F., & Gluck, M. A. (1991). Sensorimotor learning and the cerebellum. In M. Arbib & J.-P. Ewert (Eds.), *Visual Structures and Integrated Functions* (Vol. 3, pp. 381–396). Berlin: Springer-Verlag.

Bell, C. C., Han, V., & Sawtell, N. B. (2008). Cerebellum-like structures and their implications for cerebellar function. *Annual Review of Neuroscience, 31*, 1–24.

Blakemore, S. J., Wolpert, D., & Frith, C. (1999). The cerebellum contributes to somatosensory cortical activity during self-produced tactile stimulation. *NeuroImage, 10*(4), 448–459.

Braitenberg, V. (1961). Functional interpretation of cerebellar histology. *Nature, 190*(4775), 539–540.

Brindley, G. S. (1964). The use made by the cerebellum of the information that it receives from sense organs. *International Brain Research Organization Bulletin, 3*, 80.

Catz, N., Dicke, P. W., & Thier, P. (2005). Cerebellar complex spike firing is suitable to induce as well as to stabilize motor learning. *Current Biology, 15*(24), 2179–2189.

Cerminara, N. L., Lang, E. J., Sillitoe, R. V., & Apps, R. (2015). Redefining the cerebellar cortex as an assembly of non-uniform Purkinje cell microcircuits. *Nature Reviews. Neuroscience, 16*(2), 79–93.

Coenen, O. M. D., Arnold, M., Sejnowski, T., & Jabri, M. (2001). Parallel fiber coding in the cerebellum for life-long learning. *Autonomous Robots, 11*(3), 291–297.

D'Angelo, E., Koekkoek, S. K. E., Lombardo, P., Solinas, S., Ros, E., Garrido, J., et al. (2009). Timing in the cerebellum: Oscillations and resonance in the granular layer. *Neuroscience, 162*(3), 805–815.

Dean, P., & Porrill, J. (2011). Evaluating the adaptive-filter model of the cerebellum. *Journal of Physiology, 589*(Pt 14), 3459–3470.

Dean, P., Porrill, J., Ekerot, C. F., & Jörntell, H. (2010). The cerebellar microcircuit as an adaptive filter: Experimental and computational evidence. *Nature Reviews. Neuroscience, 11*(1), 30–43.

Dean, P., Porrill, J., & Stone, J. V. (2002). Decorrelation control by the cerebellum achieves oculomotor plant compensation in simulated vestibulo-ocular reflex. *Proceedings of the Royal Society of London. Series B, Biological Sciences, 269*(1503), 1895–1904.

Dugué, G. P., Brunel, N., Hakim, V., Schwartz, E., Chat, M., Lévesque, M., et al. (2009). Electrical coupling mediates tunable low-frequency oscillations and resonance in the cerebellar Golgi cell network. *Neuron, 61*(1), 126–139.

Eccles, J. C. (1969). The development of the cerebellum of vertebrates in relation to the control of movement. *Naturwissenschaften, 56*(11), 525–534.

Eccles, J. C., Ito, M., & Szentágothai, J. (1967). *The Cerebellum as a Neuronal Machine.* Berlin: Springer-Verlag.

Fujita, M. (1982). Adaptive filter model of the cerebellum. *Biological Cybernetics, 45*, 195–206.

Gerstner, W., & Kistler, W. (2002). *Spiking Neuron Models: Single Neurons, Populations, Plasticity.* Cambridge: Cambridge University Press.

Hesslow, G., Jirenhed, D.-A., Rasmussen, A., & Johansson, F. (2013). Classical conditioning of motor responses: What is the learning mechanism? *Neural Networks, 47*, 81–87.

Isope, P., & Barbour, B. (2002). Properties of unitary granule cell → Purkinje cell synapses in adult rat cerebellar slices. *Journal of Neuroscience, 22*(22), 9668–9678.

Ito, M. (1997). Cerebellar microcomplexes. *International Review of Neurobiology, 41*, 475–487.

Ito, M. (2008). Control of mental activities by internal models in the cerebellum. *Nature Reviews. Neuroscience, 9*(4), 304–313.

Ito, M., Sakurai, M., & Tongroach, P. (1982). Climbing fibre induced depression of both mossy fibre responsiveness and glutamate sensitivity of cerebellar Purkinje cells. *Journal of Physiology, 324*, 113–134.

Jacobson, G. A., Rokni, D., & Yarom, Y. (2008). A model of the olivocerebellar system as a temporal pattern generator. *Trends in Neurosciences, 31*, 617–625.

Jordan, M. I. (1996). Computational aspects of motor control and motor learning. In H. Heuer & S. Keele (Eds.), *Handbook of Perception and Action: Vol. 2: Motor Skills* (pp. 71–120). London: Academic Press.

Jörntell, H., & Ekerot, C. F. (2002). Reciprocal bidirectional plasticity of parallel fiber receptive fields in cerebellar Purkinje cells and their afferent interneurons. *Neuron, 34*(5), 797–806.

Jörntell, H., & Ekerot, C. F. (2006). Properties of somatosensory synaptic integration in cerebellar granule cells in vivo. *Journal of Neuroscience, 26*(45), 11786–11797.

Kelly, R. M., & Strick, P. L. (2003). Cerebellar loops with motor cortex and prefrontal cortex of a nonhuman primate. *Journal of Neuroscience, 23*(23), 8432–8444.

Kreitzer, A. C., Gee, K. R., Archer, E. A., & Regehr, W. G. (2000). Monitoring presynaptic calcium dynamics in projection fibers by in vivo loading of a novel calcium indicator. *Neuron, 27*(1), 25–32.

Lepora, N., Porrill, J., Yeo, C. H., & Dean, P. (2010). Sensory prediction or motor control? Application of Marr–Albus models of cerebellar function to classical conditioning. *Frontiers in Computational Neuroscience, 4*(140), 1–16.

Loewenstein, Y., Mahon, S., Chadderton, P., Kitamura, K., Sompolinsky, H., Yarom, Y., et al. (2005). Bistability of cerebellar Purkinje cells modulated by sensory stimulation. *Nature Neuroscience, 8*, 202–211.

Lukoševičius, M., & Jaeger, H. (2009). Reservoir computing approaches to recurrent neural network training. *Computer Science Review, 3*(3), 127–149.

Manto, M., Gruol, D. L., Schmahmann, J., Koibuchi, N., & Rossi, F. (Eds.). (2013). *Handbook of the Cerebellum and Cerebellar Disorders.* Dordrecht: Springer Netherlands.

Marr, D. (1969). A theory of cerebellar cortex. *Journal of Physiology, 202*, 437–470.

Medina, J. F., Nores, W. L., & Mauk, M. D. (2002). Inhibition of climbing fibers is a signal for the extinction of conditioned eyelid responses. *Nature, 416*, 330–333.

Porrill, J., & Dean, P. (2004). Recurrent cerebellar loops simplify adaptive control of redundant and nonlinear motor systems. *Neural Computation, 19*(1), 170–193.

Porrill, J., Dean, P., & Anderson, S. (2013). Adaptive filters and internal models: Multilevel description of cerebellar function. *Neural Networks, 47*, 134–149.

Roberts, P. D. (2000). Modeling inhibitory plasticity in the electrosensory system of mormyrid electric fish. *Journal of Neurophysiology, 84*(4), 2035–2047.

Safo, P., & Regehr, W. G. (2008). Timing dependence of the induction of cerebellar LTD. *Neuropharmacology, 54*(1), 213–218.

Schweighofer, N., Arbib, M. A., & Kawato, M. (1998a). Role of the cerebellum in reaching movements in humans: I. Distributed inverse dynamics control. *European Journal of Neuroscience, 10*(1), 86–94.

Schweighofer, N., Doya, K., Fukai, H., Chiron, J. V., Furukawa, T., & Kawato, M. (2004). Chaos may enhance information transmission in the inferior olive. *Proceedings of the National Academy of Sciences of the United States of America, 101*(13), 4655–4660.

Schweighofer, N., Spoelstra, J., Arbib, M. A., & Kawato, M. (1998b). Role of the cerebellum in reaching movements in humans: II. A neural model of the intermediate cerebellum. *European Journal of Neuroscience, 10*(1), 95–105.

Sejnowski, T. J. (1977). Storing covariance with nonlinearly interacting neurons. *Journal of Mathematical Biology, 4*, 303–321.

Shadmehr, R., & Wise, S. P. (2005). *The Computational Neurobiology of Reaching and Pointing: A Foundation for Motor Learning*. Cambridge, MA: MIT Press.

Turrigiano, G. G. (1999). Homeostatic plasticity in neuronal networks: The more things change, the more they stay the same. *Trends in Neurosciences, 22*(5), 221–227.

Van Der Giessen, R. S., Koekkoek, S. K., van Dorp, S., De Gruijl, J. R., Cupido, A., Khosrovani, S., et al. (2008). Role of olivary electrical coupling in cerebellar motor learning. *Neuron, 58*(4), 599–612.

Van Dorp, S., Hensbroek, R. A., Van Beugen, B. J., Maruta, J., Ruigrok, T. J., & Simpson, J. I. (2009). *Signal processing by cerebellar granule cells. In 2009 Abstract Viewer/Itinerary Planner*. Washington, DC: Society for Neuroscience.

Wang, S. S.-H., Denk, W., & Häusser, M. (2000). Coincidence detection in single dendritic spines mediated by calcium release. *Nature Neuroscience, 3*(12), 1266–1273.

Widrow, B., & Stearns, S. D. (1985). *Adaptive Signal Processing*. Englewood Cliffs, NJ: Prentice-Hall.

Widrow, B., & Walach, E. (2008). *Adaptive Inverse Control: A Signal Processing Approach*. Hoboken, NJ: Wiley.

Yamazaki, T., & Tanaka, S. (2005). Neural modeling of an internal clock. *Neural Computation, 17*(5), 1032–1058.

Yartsev, M. M., Givon-Mayo, R., Maller, M., & Donchin, O. (2009). Pausing Purkinje cells in the cerebellum of the awake cat. [Original Research.] *Frontiers in Systems Neuroscience, 3*, article 2.

22 Integrative Functions of the Corticostriatal System

PETER FORD DOMINEY AND PIERRE ENEL

The corticostriatal system defines one of the principal components of the system-level functional neuroanatomy of the primate brain. Essentially all of the neocortex projects onto the striatum, which is made up of the caudate (CD) and putamen, defining the principal input nuclei of the basal ganglia (Alexander et al., 1986). The cortex is organized with a spatial layout or topography, and the projection from cortex to the striatum (the corticostriatal projection) roughly preserves this topography yet at the same time provides an intricate matrix for the integration and overlap of projections that allows for recombination of cortical input. The detailed functional significance of this intermixing remains a challenge for neuroscience to this day (Haber et al., 2000; Haber et al., 2006; Selemon and Goldman-Rakic, 1985).

A principal notion in the functioning of the corticostriatal system is that of "disinhibition" (Chevalier and Deniau, 1990; Chevalier et al., 1985; Deniau and Chevalier, 1985). In a sequential chain of connections that are, respectively, excitatory, inhibitory, and inhibitory, excitation of the first element in the (cortex) chain excites the second (striatum) which inhibits the third (SNr), thus removing its inhibition on the next element (thalamus), hence the notion of disinhibition (see figure 22.1B). Cortical inputs to the striatum are processed in the subsequent elements of the basal ganglia and are finally reintegrated in cortex via the thalamus. The output of the basal ganglia is therefore expressed as selective modulation of the thalamus (and thus cortex via the thalamus) and subcortical structures including the superior colliculus for generation of orienting gaze actions. Basal ganglia inputs to the thalamus are tonic and inhibitory, based on the inhibitory neurotransmitter GABA. Thus, the principal influence of the basal ganglia on the thalamocortical system is to inhibit the system and then selectively, under the right conditions as indicated by cortex, to release that inhibition (Chevalier and Deniau, 1990; Chevalier et al., 1985; Deniau and Chevalier, 1985). In the domain of motor control, this will allow action selection based on the behavioral context that is defined by cortical activity and its projection to the striatum and will also implement working memory via cortico-striato-thalamo-cortical reciprocal interaction.

The system is inherently adaptive so that through learning, different behavioral contexts can lead to the selection of appropriate action. The primary mechanism for plasticity in the corticostriatal system is at the level of the corticostriatal synapse, where the influence of dopamine can produce long-term synaptic changes (Centonze et al., 2003). The release of dopamine at the corticostriatal synapse is intricately linked to reward (Schultz et al., 1993; Schultz et al., 1997). (See section 22.3 below, chapter 10 by Daw on "Reinforcement Learning," and chapter 7 by Fellous, Canavier, and Hasselmo on "Neuromodulation.") Thus, corticostriatal plasticity, under the control of reward-related dopamine, should provide a major component in the adaptive behavioral repertoire of the primate. In this context, an idea that will permeate the chapter is the notion of convergence from multiple functional cortical areas within the striatum, and the possible integrative role for this convergence. Given this overview, section 22.1 will continue in more detail with the neuroanatomy and function of the cortex and basal ganglia, including the generic loop structure and a model example of this organization for the oculomotor system. Section 22.2 will then examine how cortex can code visual and sequential properties of perceptual inputs, and how plasticity in the corticostriatal system can be used to bind these representations to implement conditional behavior and sequence learning. This section will introduce the modeling tool kit developed by Pierre Enel, PyRates, for simulation of rate coding neurons that we use for implementing two models introduced in this section. Section 22.3 further examines systems-level models of corticostriatal system, and section 22.4 examines future challenges.

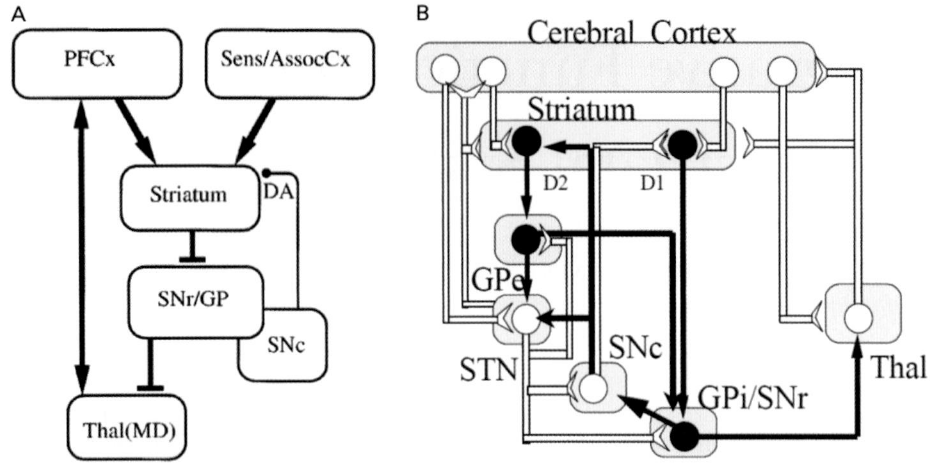

FIGURE 22.1 Simplified organization of human cortex and basal ganglia nuclei, including the striatum, made up of the caudate and putamen. Essentially all of cortex projects in an orderly manner onto the striatum. This so-called "corticostriatal" projection can be considered one of the major white matter systems of the human brain. (A) Simple schematic of "direct pathway": cortex excites striatum which inhibits SNr/GP which inhibits thalamus which excites cortex. (B) Cortex influences thalamus via the "direct pathway," cortex–striatum–GPi/SNr–thalamus, and via "indirect" pathways including cortex–striatum–GPe–STN–GP/SNr–thalamus. Additional connections link GPe, SNc. PFCx, prefrontal cortex; Sens/AssocCx, sensory and associative cortices; Thal(MD), medio-dorsal thalamus; D1, D2, dopamine receptor subtypes; GPe, globus pallidus externus; GPi/SNr, globus pallidus internus, substantia nigra pars reticulata; STN, subthalamic nucleus. Two "substantia nigra" components: SNc, substantia nigra pars compacta, with dopamine-producing cells project to striatum; SNr, receives inhibitory input from striatum, and itself inhibits the thalamus.

22.1 *Functional Overview of the Basal Ganglia Corticostriatal System Illustrated with the Oculomotor Saccade System*

The complexity of the corticostriatal system is vast, and an attempt to model such a system can be compromised if the space of exploration is not somehow constrained. In this context we have developed a "cognitive systems engineering approach" which provides a structured method for introducing the necessary constraints (Dominey and Arbib, 1992). The essential notion is to identify a specific behavior and then to study the anatomy and neurophysiology of the brain systems that have been studied in the context of that behavior. This then provides the material for the modeling component. The goal of the modeling is to construct a system that can explain the relevant behavior, in a manner that is consistent with the observed neurophysiology and neuroanatomy. Ideally, such a model can be used to synthesize a body of neurophysiological results, and also to predict new results for future experiments.

One of the most basic forms of behavior that has been studied in this context is the selection of orientation behaviors, that is, orientation of the eyes and head to specific targets in visual space. Different types of such orienting saccadic eye movements are illustrated in figure 22.2. Over the years saccadic eye movement has been a fruitful paradigm for at least two reasons. First, the timing of the experimental stimuli and behavior can be precisely controlled, and the orienting behavior can be easily characterized in terms of timing and accuracy. Second, because of the *relatively* direct input–output or sensorimotor mappings, the physiology of the motor system for these orienting movements has been well characterized. Hafed's chapter 19, "Saccades and Smooth Pursuit Eye Movements," explains the basic mechanisms in brainstem and superior colliculus involved in eye movements. Our concern here is to understand what cerebral cortex and basal ganglia add to the story. Previewing the link to disinhibition, attentional gaze should not be distracted, and so one would want to actively inhibit saccades and then selectively remove that inhibition to allow saccades to selected targets.

In a series of experiments in the rat and cat (Chevalier and Deniau, 1990; Chevalier et al., 1981; Chevalier et al., 1984; Chevalier et al., 1985) and monkey (Hikosaka and Wurtz 1983a, 1983b, 1983c, 1983d) it was demonstrated that orienting movements of the eyes and head controlled by the superior colliculus (SC) are under the "disinhibitory" influence of the corticostriatal system. That is, a tonic inhibitory input to the

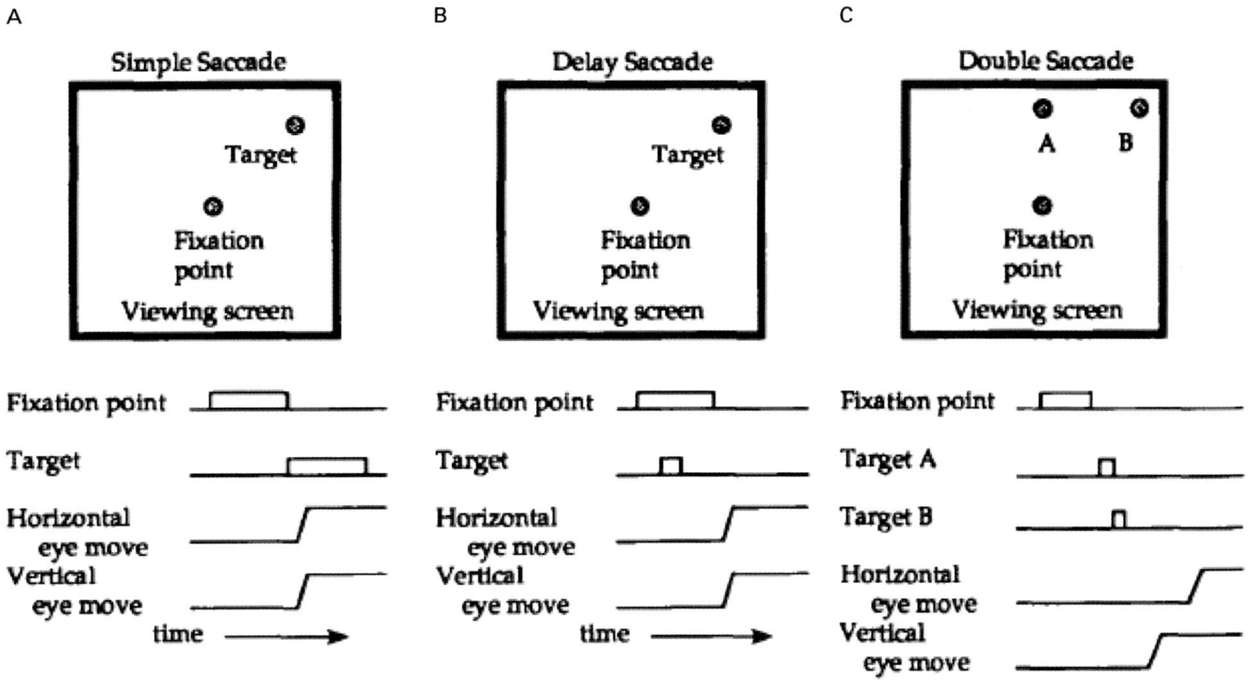

FIGURE 22.2 Oculomotor saccade tasks. Modified from Dominey and Arbib (1992). (A) Simple saccade: Peripheral target onset and central fixation offset trigger a saccade to the target. (B) Delay or memory saccade: Target presented during fixation must be remembered until offset of fixation point. (C) Double saccade: Two successive saccade targets are presented before the first saccade. The retinal location of target B must be "remapped" to account for the movement to A.

SC can be temporarily inhibited, thus allowing SC to become activated by inputs from cortex and directly from the retina. This temporary or "phasic" inhibition of the tonic inhibition is referred to as *disinhibition.* Thus, the substantia nigra pars reticulata (SNr) has a tonic inhibitory (GABA) influence on the SC which prevents movement execution. Cortical input to the striatum activates inhibitory GABAergic neurons in the striatum which, in turn, inhibit the inhibitory neurons in substantia nigra pars reticulata, globus pallidus internus (SNr/GPi). This disinhibition can be patterned, thus giving certain movements more scope than others to be executed, allowing a selective release of inhibition on the motor system for selection of the appropriate movement. This disinhibitory circuit for the oculomotor system is illustrated in figure 22.3: the path from frontal eye fields (FEF) to the CD thence to the SNr which tonically inhibits (and thence can disinhibit) SC.

22.1.1 OVERVIEW OF CORTICOSTRIATAL NEUROANATOMY Figure 22.3 illustrates a neural network model of the corticostriatal oculomotor system and its operation in a "simple" visually guided saccade. The goal in building this model was to account for extensive data that had been recorded in the monkey from the parietal cortex, FEF, CD, substantia nigra, thalamus, and SC in the generation of voluntary saccades. This neurophysiological data established functional links between firing rates of single units in these areas and different oculomotor behaviors in the context of the disinhibition described above. Thus, the simulations were performed using neurons whose activity was coded as firing rate. Topographic visual input traverses the retina and visual system and projects to the posterior parietal cortex (PP) and FEF. FEF projects with a topography that preserves saccade amplitude to the caudate nucleus of the striatum (CD). The CD projects with inhibitory (GABAergic) connections to the SNr, which itself projects with tonic inhibitory input to the SC and the thalamus. Each neural structure was modeled as a 9×9[1] array of leaky integrator neurons with sigmoid output functions. During execution of a visually guided saccade, visual input via PP drives spatially localized activation of FEF, which, in turn, activates the corresponding saccade-related neurons in CD. These, in turn, inhibit the saccade-related neurons in SNr, thus removing the tonic inhibition from SC allowing the appropriate saccade to be generated. In figure 22.3B we see data from the primate substantia nigra and SC during a visually guided saccade. Note that when the tonic activity in the SNr is interrupted, the SC phasic burst of activity takes place,

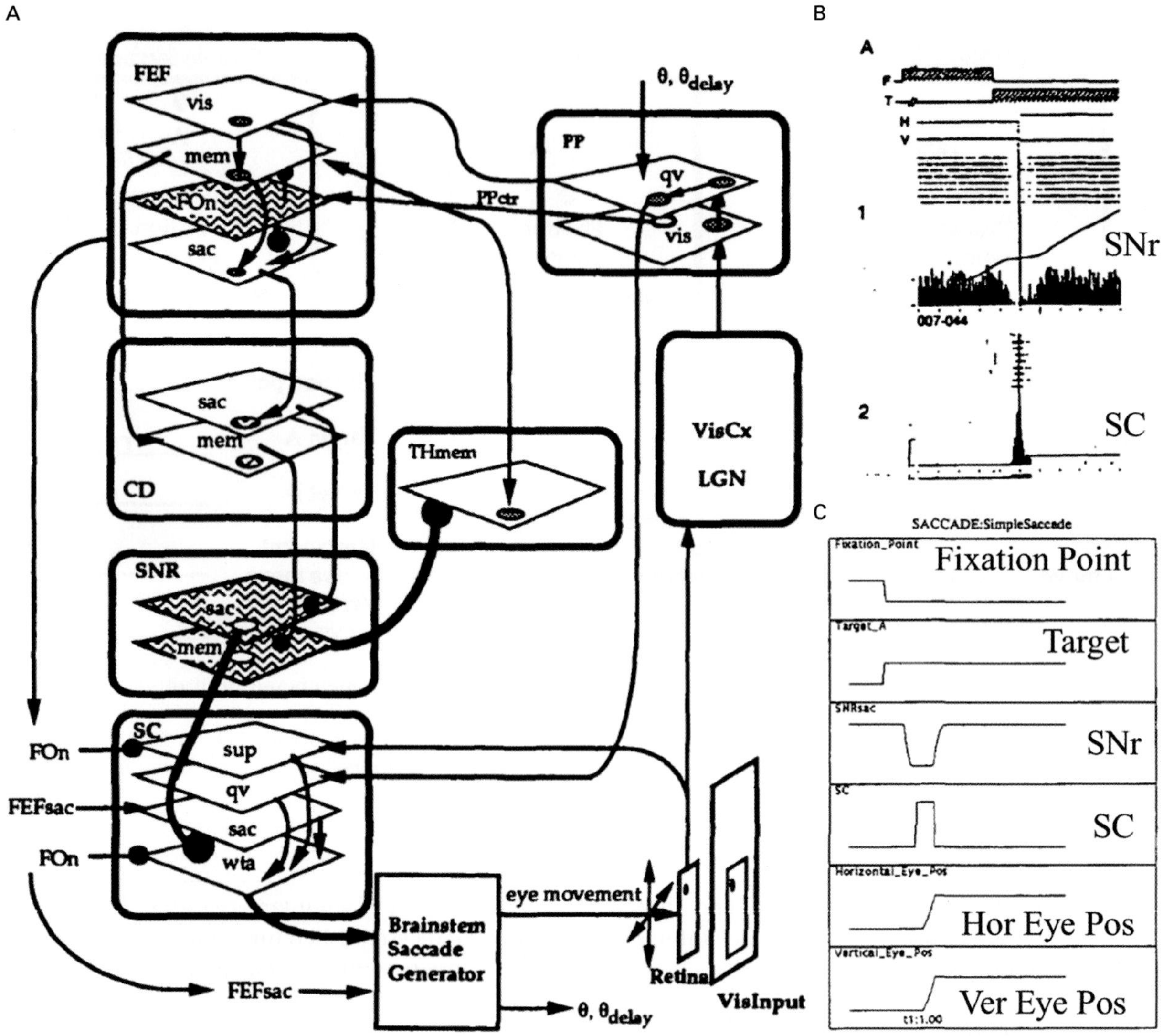

FIGURE 22.3 Corticostriatal organization of the oculomotor system. Modified from Dominey and Arbib (1992). (A) Model organization. From lower right: Eye movement displaces the retina over the visual input. Retinal signal passes via LGN and visual cortex (VixCx) and arrives in posterior parietal cortex (PP) and on to frontal eye fields (FEF) and superior colliculus (SC). FEF functional layers represent visual, memory, fixation on, and saccade responses that are communicated to the caudate (CD) and substantia nigra (SNr). These functional classes are observed in physiological recordings. Thalamus (THmem) participates in a memory loop with FEFmem. Disinhibition of SC in this circuit allows voluntary saccades. (B) Single-unit data from SNR and SC. (C) Time line of visual stimulus, neural activity, and motor behavior for a visually guided saccade. Fixation point onset is followed by target onset and fixation point offset. The retinotopically corresponding FEF saccade–related (FEFsac) neuron becomes active, yielding activation of the CDsac cell, which, in turn, inhibits the tonically active SNRsac cell allowing the FEFsac input to SC to activate the saccade generator and effect the saccade as revealed by the horizontal eye position trace. Sup, superficial layer; qv, quasi-visual; wta, winner take all. See text for further details.

and with it the saccade. Figure 22.3C illustrates the model simulation.

This circuitry can also be used to implement working memory, where instead of disinhibiting a motor function in SC, we disinhibit a memory loop between FEF and thalamus. This foreshadows corticostriatal processing in executive function (Hazy et al., 2007), addressed below. Figure 22.4 illustrates the model activation during a "memory-guided" saccade, in which the SNr influence on mediodorsal thalamus is disinhibited by FEF input, thus allowing the long loop from FEF–CD–SNr–thalamus–FEF, and the direct FEF–thalamus bidirectional connections, to generate a recurrent loop that stores the target of a future saccade in memory until the go signal (offset of the fixation point) is provided. The neural activation in the model allows a coherent

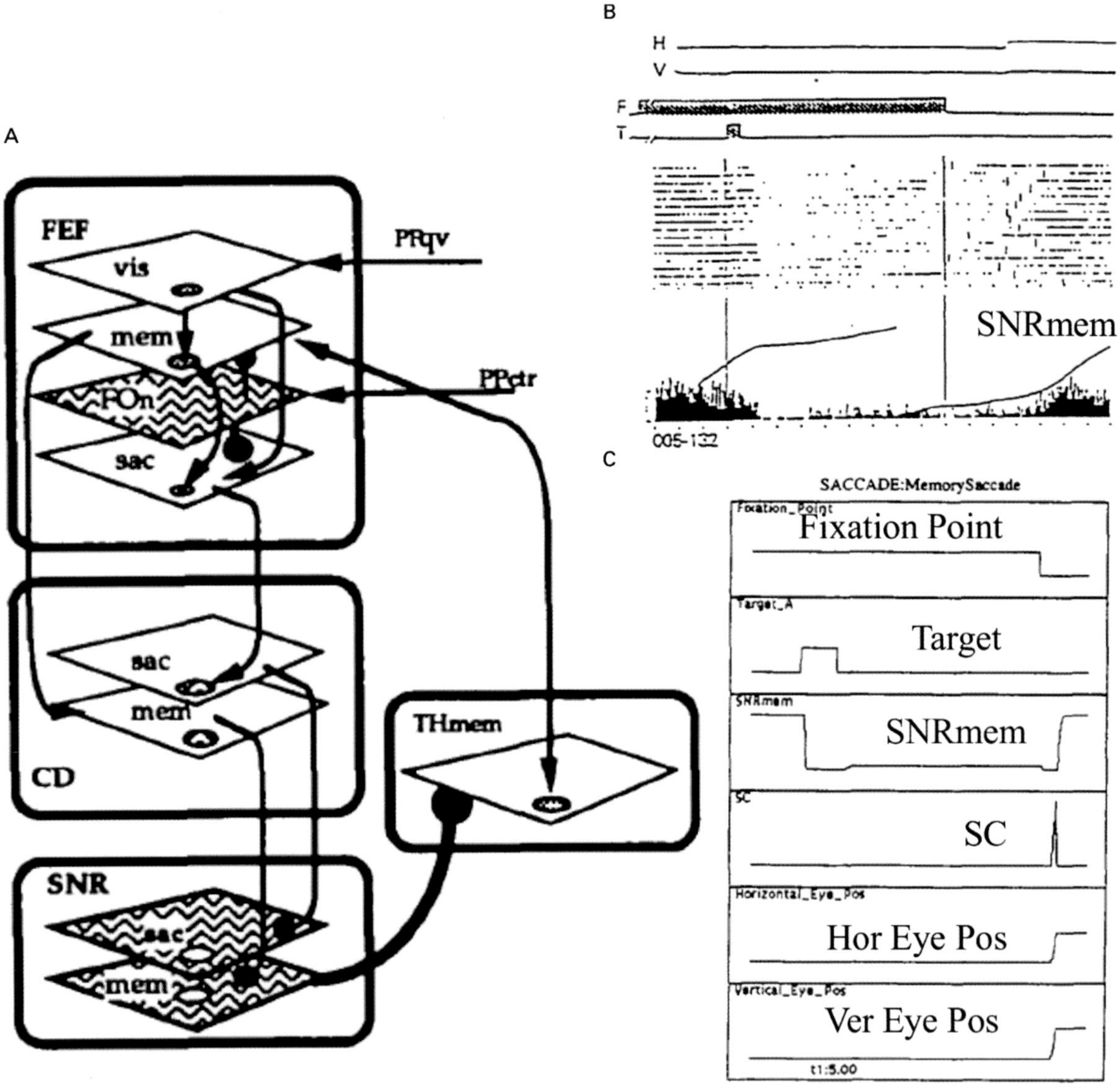

FIGURE 22.4 Memory-guided saccade. (A) Relevant model components. Visual input to frontal eye fields (FEF) activates FEFmem units, which activate caudate (CD) memory units (CDmem), inhibiting substantia nigra pars reticulata (SNr) memory units (SNRmem), disinhibiting thalamus memory units (THmem), thus playing back into FEFmem in a loop. (B) Data from a spatially selective SNr cell that displayed tonic inhibition of response during the delay period in a memory saccade. This corresponds to the SNRmem cells in A that will disinhibit thalamus (THmem) and allow the tonic activity in the FEFmem loop to persist, thus encoding the remembered saccade location. (C) Simulation results of a memory-guided saccade displaying time courses for fixation point, peripheral target, SNRmem activity, the superior colliculus (SC) burst, and horizontal and vertical eye position. The FEF activation disinhibits SNr and installs the cortico–striato–nigro–thalamo–cortical memory loop until the fixation point offset triggers the saccade.

explanation of sustained memory-related response in the FEF (Bruce and Goldberg, 1985), CD (Hikosaka et al., 1989a, 1989b), SNr (Hikosaka and Wurtz, 1983c), and thalamus (Fuster and Alexander, 1973). Thus we see an illustration of corticostriatal function for action selection and working memory. It is noteworthy that this notion of cortico–striato–thalamo–cortical loops is generalized to phylogenetically older subcortical brain-stem structures as well, including the SC in looped circuits with the basal ganglia (McHaffie et al., 2005). McHaffie et al. suggest that the common point in such loops is that of selection, where competing inputs can bid for resources in a central selection mechanism.

Figure 22.2C illustrates the double-step saccade which introduces new constraints. Targets A and B are presented prior to the first saccade to A. After the saccade to A, the initial retinal representation of B (referred to as "retinal error") no longer corresponds to the spatial location of B since the eye has moved. In order account for this movement, the "retinal error" must be updated, in a dynamic remapping process. In such a process, during the first saccade, activation on the map of the retinal or motor error is dynamically remapped or shifted to the location on the map that codes for the second saccade, in motor error coordinates (i.e., encoding the position where the second target would fall on the retina were it still visible after the first saccade). That is, the vector coding for the first movement is subtracted from the vector coding for the second movement. Referring again to figure 22.2C, the vector FP–A is subtracted from the vector FP–B, leaving the resulting vector A–B, that is, a saccade to the right. Neural activity reflecting this dynamic remapping as a form of "quasi-visual" (QV) activity has been observed in the SC of the cat (Munoz et al., 1991) and macaque (Mays and Sparks, 1980) and in the FEF (Goldberg and Bruce, 1990) and PP (Gnadt and Andersen, 1988) of the macaque. Dominey and Arbib (1992) reproduced this QV remapping in a two-dimensional layer of neurons, where lateral connections are modulated by eye velocity to favor the spatial flow of activation on the map in the sense opposite to that of the eye movement, thus compensating for that movement.

The oculomotor system provides a prototypical example of the general pattern of organization of the corticostriatal system. In this context we considered only the cortical projections from the FEF to the CD. More generally, essentially all of neocortex, including the primary auditory and visual cortices, projects to the striatum (Yeterian and Pandya, 1998) in a manner that globally preserves the topography of the distinct cortical areas. The corticostriatal projection neurons principally innervate medium spiny neurons (MSNs) which make up the large majority of striatal neurons. A second class of tonically active neurons (TANs) in the striatum display a low level of tonic activity and halt their activity in response to motivating stimuli (Apicella, 2007). The MSNs project to the output nuclei of the basal ganglia directly (in the so-called "direct pathway") or indirectly via the external segment of the globus pallidus (GPe) (in the so-called "indirect pathway"), illustrated in figure 22.1B. MSNs are inhibitory neurons that use the neurotransmitter GABA. Activity through the "direct pathway" from the striatum to the output basal ganglia nuclei (SNr and the internal globus pallidus; GPi) provides powerful disinhibitory control of these basal ganglia output neurons. In contrast, activity through the parallel "indirect pathway" leads to increased activity of excitatory glutamatergic neurons in the subthalamic nucleus (STN), which induces strong excitation of the SNr and GP (Di Filippo et al., 2009). In most of this chapter, we focus on the contributions of the direct pathway but do briefly discuss "Bringing in the Indirect Pathway" in section 22.3.2.

In 1986 Alexander, Delong, and Strick (Alexander et al., 1986) published an influential review article in which they summarized neuroanatomical and neurophysiological data that allowed them to characterize a generic cortico–striato–nigro–thalamo–cortical circuit, or loop, that is reduplicated throughout the corticostriatal system. This is illustrated in figure 22.5. Different cortical areas project to specific subdomains in the striatum and follow their segregated course through the GPi/SNr (internal segment of the globus pallidus and substantia nigra pars reticulatta) and thalamus back to cortex. Each of the five circuits contributes to its named function: thus, in the motor circuit, striatal function is correlated with movement parameters including direction, amplitude, and load (DeLong et al., 1986). The occulomotor circuit is involved in voluntary saccadic eye movements, which we describe in this chapter. The dorsolateral prefrontal circuit is associated with higher cognitive function, working memory, and cognitive control (Hazy et al., 2007). The lateral orbitofrontal circuit is likely involved in aspects of motivation, emotion, and decision-making (Bechara et al., 2000). The anterior cingulate circuit has been associated with aspects of evaluation and control in voluntary behavior (Shenhav et al., 2013).

Within and across these circuits, it is important to note that the notion of segregation was already known by these authors to be a simplification, and that, indeed, throughout the system there is the opportunity for information from different input sources to converge. In order to characterize this integrative convergence, Selemon and Goldman-Rakic (1985) undertook an extensive study of the corticostriatal projection in 18 monkeys. The animals were injected with anterograde tracers that followed the corticostriatal projection from one or two distinct areas per animal. Analysis of the tracer labeling in the striatum revealed two major and novel findings. First, rather than being organized in an anterior to posterior topography, the distinct cortical areas project to longitudinal bands that cover much of the anterior to posterior extent of the striatum. Second, while the territories of the corticostriatal projections were largely specific for each cortical area, there was significant overlap and interdigitation between these territories.

This is illustrated in figure 22.6, from Selemon and Goldman-Rakic (1985). This figure illustrates anterograde tracer (which goes from cell bodies in the

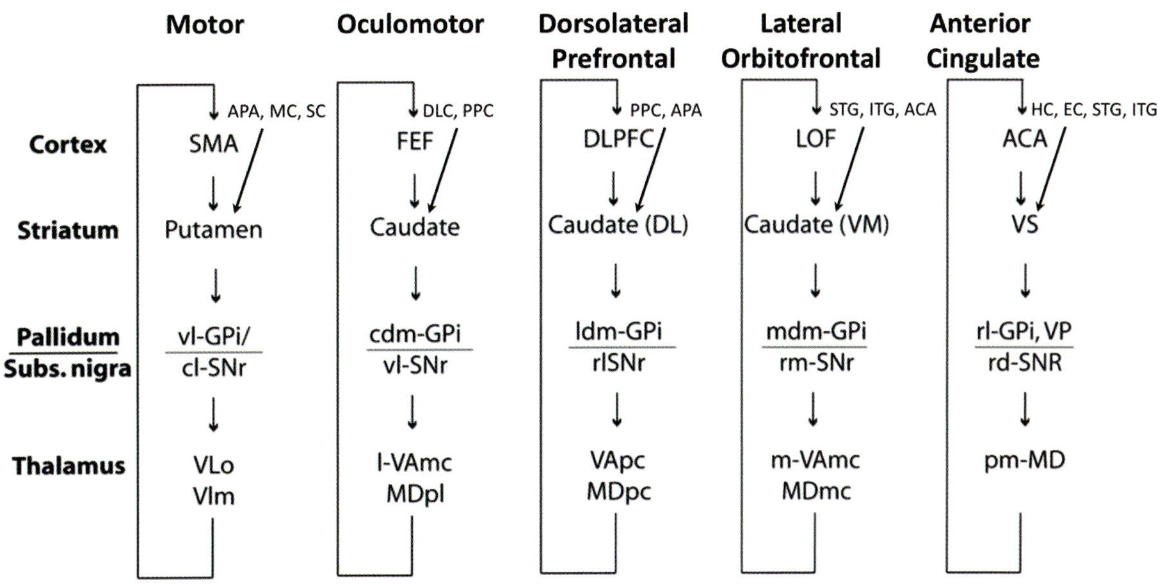

FIGURE 22.5 Parallel loop organization of the corticostriatal system. Note that the notion of convergence is already present here as in several of the "segregated" pathways there is input from some of the other pathways. SMA, supplementary motor area; DLPFC, dorsolateral prefrontal cortex, LOF, lateral orbitofrontal cortex; ACA, anterior cingulate area; MC, motor cortex; APA, arcuate premotor area; HC, hippocampal cortex; EC, entorhinal cortex; SC, somatosensory cortex; PPC, posterior parietal cortex; STG, superior temporal gyrus; ITG, inferotemporal gyrus. Modified from Alexander, DeLong, and Strick (1986).

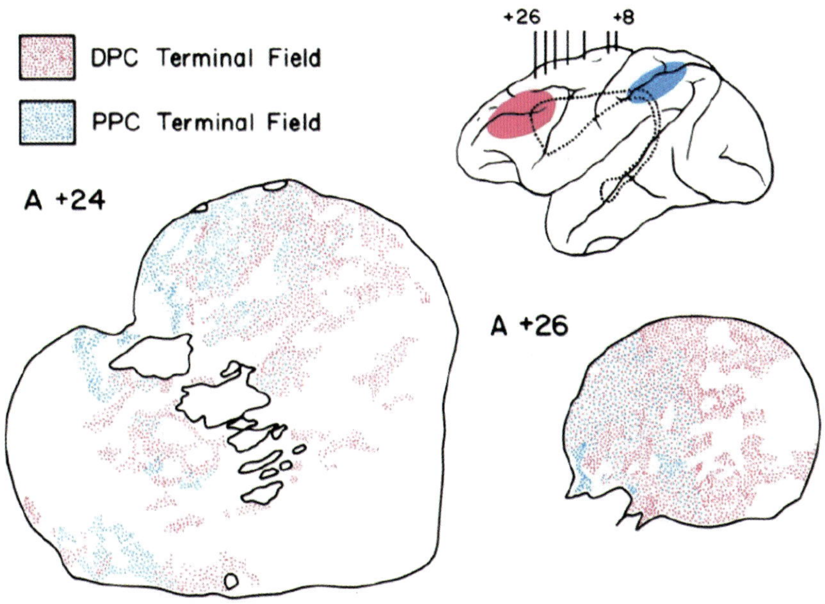

FIGURE 22.6 Corticostriatal projection labeling. Blue label was injected in posterior parietal cortex (PPC), red label in dorsolateral prefrontal cortex (DPC). Terminal zones of these cortical area projections are shown in the anterior striatum. A +24 and +26 correspond to the anterior–posterior location of the slices, indicated in the whole brain schema (upper right). Note that while there are segregated projection zones (areas with only a single colored label), there is also extensive overlap where projections from frontal and parietal areas converge in the striatum. From Selemon and Goldman-Rakic (1985).

injection site to the terminals of those cells, thus identifying the projection targets of cells in the injected area) injected in the frontal and parietal cortex of macaque monkey and transported to the projection terminals in the striatum. We see that while there are distinct components of these two projections, there is also substantial overlap. If overlapping projections terminate on common striatal neuron dendritic fields, then these striatal neurons can learn associations between these different cortical inputs.

This has significant impact in terms of how information in the corticostriatal signal can potentially be integrated in the striatum. Similar results have been demonstrated, using diffusion tensor imaging, in the human corticostriatal system (Lehericy et al., 2004), though with less precision. More recently, Haber et al. (2006) have demonstrated in the monkey overlapping projections from cognitive and reward-related cortical areas in the rostral and dorsal striatum which they interpret as providing the basis for incentive-based learning. By examining more diffuse projections than those in Selemon and Goldman-Rakic (1985), Haber et al. (2006) further advanced the argument for large-scale convergence and integration in the corticostriatal projections. A similar convergence of information is visible in the GABAergic striatonigral projection and the dopaminergic nigrostriatal projection (Haber et al., 2000). Thus, while the labeled-line, segregated pathway metaphor is useful, it is imperative to keep in mind that throughout the loop there is the possibility for interaction between different pathways.

22.2 Modeling Plasticity and Integrative Functions of the Corticostriatal System

What could be the advantage of allowing such a mixing of cortical inputs to the striatum? One possibility is that, based on the convergence of different cortical inputs in the striatum, an adaptive mechanism could bind the corresponding representations together in order to allow the formation of new behaviorally relevant associations. This could allow implementation of the notion of stimulus–response habit formation, where one can learn to arbitrarily associate a given stimulus with a given response. The neurophysiology supports this hypothesis. Reading et al. (1991) demonstrated that selective lesions of the CD lead to a specific impairment in visual stimulus–response learning, and Robbins et al. (1990) demonstrated that this learning required an intact nigrostriatal dopamine projection. The proposal that reward-related dopamine is involved in synaptic plasticity in the corticostriatal synapse that would allow such learning has now been clearly established (Calabresi et al., 2009; Di Filippo et al., 2009; Yin and Knowlton, 2006). Thus, from our "cognitive systems engineering" perspective, we have the target behavior of associative learning, we have the neuroanatomy (corticostriatal convergence) and neurophysiology (corticostriatal plasticity)—so we can now elaborate the mechanism for binding sensory and motor representations in the striatum.

22.2.1 Corticostriatal Convergence and Plasticity—Binding State Representation to Action for Sensorimotor Habit Formation As we saw above, the CD nucleus of the striatum participates in the preparation and execution of voluntary, context-dependent saccades (Hikosaka, 1989, 2007; Hikosaka et al., 2006; Hikosaka et al., 1989a, 1989b, 1989c; Nakamura and Hikosaka, 2006) and receives projections containing external context information from the occipitotemporal pathway—including inferotemporal cortex (IT) and spatial movement information from posterior PP and FEF (Selemon and Goldman-Rakic, 1985; Selemon and Goldman-Rakic, 1988; Stanton et al., 1988). The overlap of projections from IT and FEF onto CD neurons involved in the motor control (execution) of saccadic eye movements would allow the possibility of binding nonspatial visual representations in IT onto spatial codes from FEF in the oculomotor area of the CD for movement generation. In order to model this approach, we (Dominey et al., 1995) enhanced the corticosubcortical saccade model by adding the IT and its striatal projections, as illustrated in figure 22.7.

We first describe the model that was implemented in the Neural Simulation Language (NSL) (Weitzenfeld et al., 2002) and then present the updated version in the Python PyRates simulation environment which accompanies this chapter. In the model, retinal input codes aspects of the *what* and *where* visual pathways. In the *where* pathway, retinal topography that corresponds to motor error for spatial saccades is preserved throughout the visual input–posterior parietal cortex–FEF–caudate–SNr–thalamus–SC saccade circuit. In the *what* pathway, shape and color are represented as feature vectors in V4 and IT. Corticostriatal projections from IT to caudate are modifiable by reward-related dopamine from substantia nigra pars compacta (SNc).

A trial is initiated by the onset of the central fixation point. Once the model initiates fixation, the fixation point is replaced by a colored shape (cue). The "form vector" (color and shape) of the cue is projected by "V4" to IT, and the IT activity projects to CD via the modifiable IT–CD synapses. After a delay, the two peripheral targets appear, leading to the activation of PP, which, in turn, signals to FEF and then to CD. Each

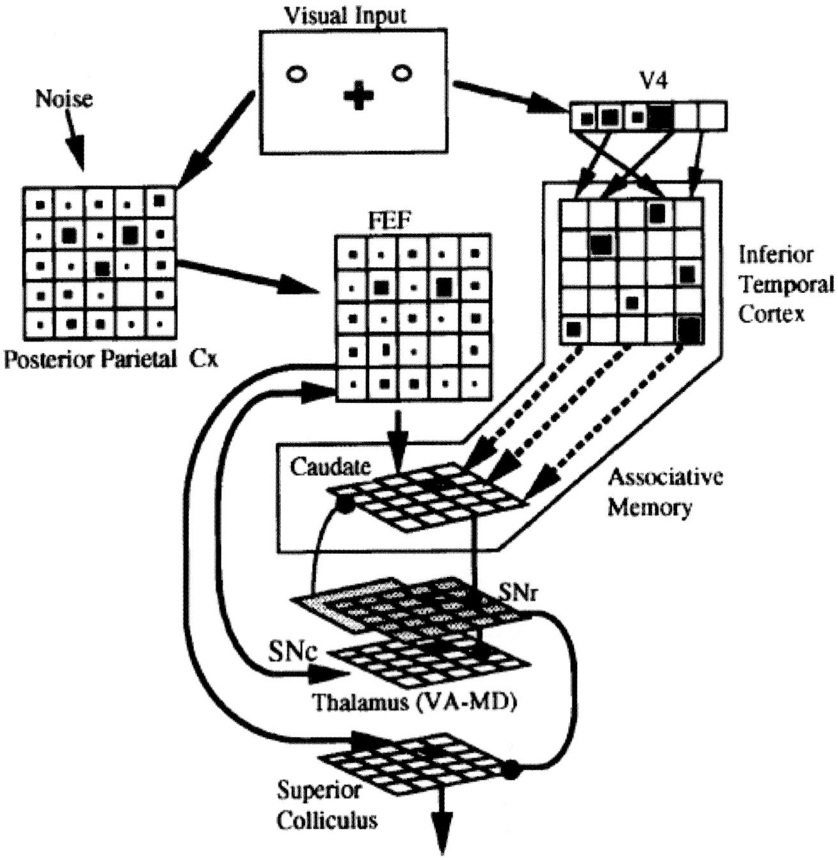

FIGURE 22.7 Corticostriatal model for learning conditional visual discrimination tasks in which the model should learn to associate Color/Shape central cues with saccades to either the left or right targets. A retinal topography that corresponds to motor error for spatial saccades is preserved throughout the visual input–posterior parietal cortex (Cx)–frontal eye fields (FEF)–caudate–substantia nigra pars reticulata (SNr)–ventral anterior and mediodorsal thalamus [Thalamus (VA-MD)]–superior colliculus saccade circuit. Shape and color are represented in V4 and inferotemporal cortex (IT). Corticostriatal projections from IT to caudate are modifiable by reward-related dopamine from substantia nigra pars compacta (SNc). When a correct (incorrect) saccade is made for a given cue, the reward (absence) causes synapses between IT cells coding that cue and caudate cells coding the saccade to be strengthened (weakened), increasing (decreasing) the probability that the correct (incorrect) saccade will be made in the future. From Dominey, Arbib, & Joseph (1995).

element of PP receives its topographic spatial input, to that is added a "noise pattern" that provides a "symmetry breaking" function, or form of guessing between the two targets during initial learning. During the period of cue and target overlap, the combined activity of FEF and IT produces "preparatory" activity in the CD, leading to its inhibition of SNr, which, in turn, disinhibits its thalamic nuclei that project back to FEF. Reciprocal excitation between disinhibited thalamic and FEF cells, combined with the lateral inhibition in CD, tends to amplify slight differences in corticostriatal inputs, favoring the stronger and reducing the weaker. This is similar to the reciprocal FEF–thalamic interactions that maintain spatial memory in Dominey and Arbib (1992), but in the present case, the reciprocal activity contributes to both selection and memory functions, as we will see below.

At the Go signal (removal of the cue), inhibitory fixation-related cells in FEF (FEFOn) are deactivated, allowing SC to respond to inputs from FEF and SNr. In the naive state, the cue-driven IT activity projects with random strength to CD, while the effects of noise in PP are seen in CD and downstream and drive the winner-take-all (WTA) mechanism in SC resulting in a saccadic "guess." After a correct "guess" the IT–CD synapses between cue and saccade-related cells are strengthened, increasing the probability that the cue, when repeated, will drive the correct CD cells. Incorrect trials result in a reduction of the active corticostriatal synapses. Here we anticipated the presence of long-term potentiation (LTP) and long-term depression (LTD) in the corticostriatal synapse (Calabresi et al., 2007). After significant learning, the IT influence on CD dominates that of noise, and the performance approaches 100%. A

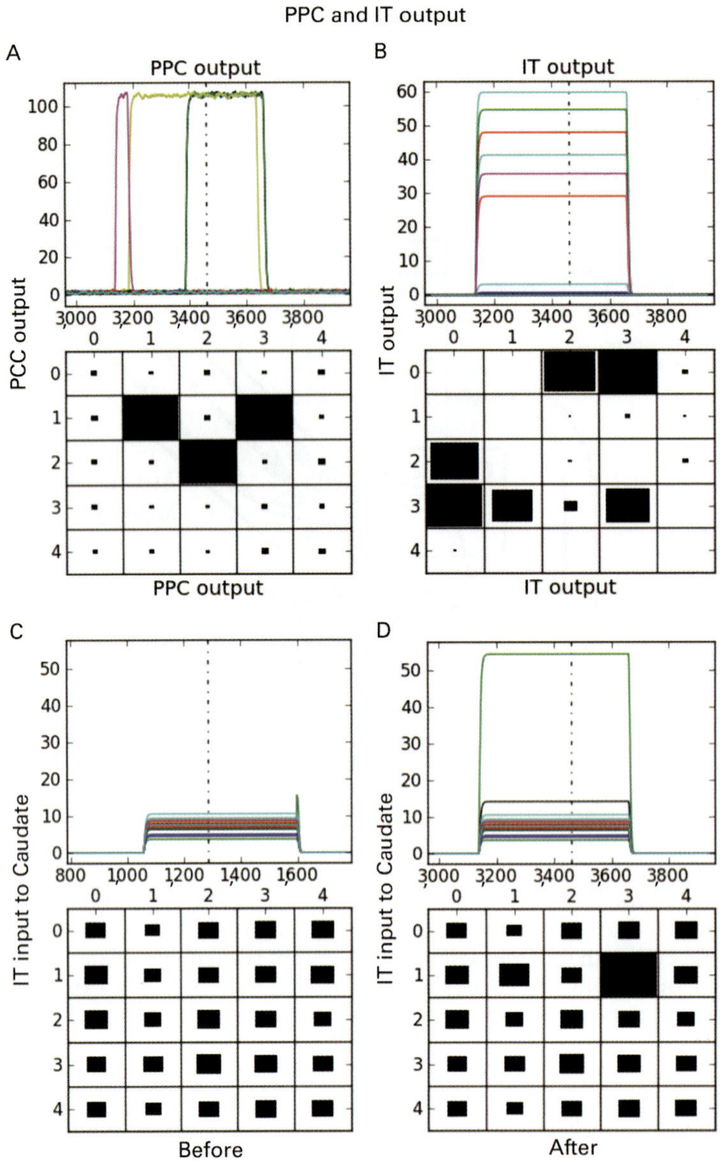

FIGURE 22.8 Conditional visual discrimination model simulation in PyRates. (A) Posterior parietal cortex (PPC) outputs display the temporal activation of all units during the trial, and the spatial activation when the fixation target is present, and two peripheral targets are visible, to the left and right, above the central fixation point. (B) Inferotemporal (IT) outputs display the temporal and spatial code of the IT units in response to a cue. (C) Before learning: IT input to caudate displays the activity in the caudate that is due to IT input, via the caudate_to_IT weight matrix (which will be modified by learning). Before—there is no spatial bias. (D) After learning: IT input to caudate displays how learning has associated IT activity for this cue with activation of the target that is paired with this cue (the right target).

comparison of cells in the model with neurons recorded in the primate striatum demonstrate similar cue and saccade-related activity in both (Dominey, Arbib, and Joseph, 1995).

Figure 22.8 illustrates neural activity in the model parietal cortex and inferotemporal cortex (PPC and IT, respectively) during presentation of one trial in the conditional visual discrimination task, and the influence of IT on CD before and after learning. We see that after learning, the cue activity in IT for this cue drives CD to choose the right target.

Modeling Details The model is implemented in the Python PyRates environment developed by Enel, based on the associative learning model in Dominey, Arbib, and Joseph (1995). As in the original model, the computational elements are leaky integrator neurons that model the membrane potential of the neurons as a capacity that leaks out its potential over time. This membrane potential is described by a differential equation where the derivative of the membrane potential is expressed as a function of the membrane potential itself. In the experiments described here, the differential equation is as follows:

$$\tau_m \, dm(t)/dt = -m(t) + S_m(t)$$

where m(t) is the membrane potential of the neuron, $S_m(t)$ the input that this neuron receives from other neurons, and τ_m the time constant of the leak, whose inverse $1/\tau_m$ is called the leak rate. In the PyRates simulator this differential equation is solved using the Euler method (discretization of time):

```
deltaState = (1. / tau) * (input + restingState-state)
state = state + deltaState
```

Here the state variable corresponds to the membrane potential. The difference in membrane potential between time step t and time step t + 1 is computed and saved in `deltaState`, which is added to the `state` variable to update the membrane potential.

The output of the modeled neurons here is their firing rate, computed as a function of the membrane potential. This function is commonly called the transfer, or activation, function of the neurons. It simulates the response of a neuron that varies according to its level of excitation, and in our case is a positive sigmoid function:

```
M = sigmoid(m, min_input, max_input, min_output, max_output)
```

where M is the firing rate and m the membrane potential. When m is bounded to the [min_input, max_input], the output firing rate is given by the following nonlinear sigmoid function:

```
M = (max_output-min_output) * [S² * (3-2S)] + min_output
```

where

```
S = (m-min_input)/(max_input-min_input)
```

Learning A striking feature of the dopaminergic neurons activity is that it is briefly modified at the time of reward (see chapter 10, "Reinforcement Learning," for further details). If the reward was unexpected, a short burst of activity in the dopaminergic activity is observed when the reward is delivered. Conversely, when a reward was expected but is not present, the baseline activity shows a short dip at the expected time of reward. As stated above we can postulate that dopamine release acts on the corticostriatal connections by modifying the synapses between the neurons that were active and thus participated in the choice that elicited or did not elicit the reward. The dopamine signal can be viewed in our model as a learning signal that will guide the modification of synapses. When the previous trial choice was correct, the agent is rewarded and dopamine activity is briefly enhanced, allowing the weights of the connection between the most active neurons that participated to the choice to be strengthened. On the contrary, if the previous choice was wrong, a dip in the DA (dopamine) activity will weaken the same connection weights. To avoid modifying synapses of all the neurons in the caudate that did not have an influence on the choice, only the neuron corresponding to the previous saccade has its weights modified. In the model, the modification of the synapses from all neurons of the cortical layer to the neuron n of the caudate is computed as follow:

```
cxCdProduct = dot(recGrpActivity, sendGrpActivity) * (weights !=
0) if reward:
DAactivity = posReinforcement
else:
DAactivity = negReinforcement
newWeights[n] = weights[n] + DAModulation * DAactivity * C1 *
cxCdProduct[n]
```

`cxCdProduct` corresponds to the product of the activity at the time of saccade of the neurons in the caudate layer which is the receiving neuron group of the connection (`recGrpActivity`) and the cortical layer which is the sending neuron group of the connection (`sendGrpActivity`). The Boolean matrix (`weights != 0`) is multiplied to the activity product matrix to remove the activity of neuron pairs that do not have connections. So the modification of the weights between each pair composed of a cortical neuron and the CD neuron n that triggered the saccade will depend on how much both of these neurons were active when the saccade was initiated. If both neurons were highly active, their weight will be strongly increased or decreased depending on the presence/absence of a reward—the reward contingency. The value of `DAactivity` will be positive or negative if there is a reward or not, corresponding to the burst and dip of activity in dopaminergic neurons observed at the expected time of reward. This allows the strengthening or weakening of the weights according to whether they contributed to a correct choice or not. The `DAModulation` variable described above is in this equation to influence the modification of weights and lower them if the input to caudate is too high. It acts as a second balance mechanism to avoid overexcitation of the CD neurons. C1 is a constant term that specifies how much the weight should be modified; in other words, it is the learning rate. It is set to $2.5 * 10^{-5}$ for both tasks. With successive modifications the weights could indefinitely increase and saturate. To avoid this, we limit the weight values to the interval [0, 0.2].

As is standard in many models of synaptic plasticity (see, e.g., chapter 6, "Hebbian Plasticity and Learning"), each cortical neuron has its weights normalized after each modification to mediate the competition between neurons for connections from input layers. It means that if a weight of one cortical neuron is strengthened, all the other weights of this neuron are weakened:

```
for n in range(sendingGroup.nbUnits):
newWeights[:,n] = newWeights[:,n] * sum(weights[:,n])
/sum(newWeights[:,n])
```

Here n represents the index of the neuron in the cortical layer. So `newWeights[:,n]` represent the vector of all weights from neuron n of the cortical layer to each neuron of the caudate layer. Thus, each vector of weights is normalized so the sum of all weights in this vector stays the same. Initial weights are randomly generated between 0 and 0.1 with a uniform distribution.

The results of this learning mechanism can be seen in the "Before" (before learning) and "After" (after learning) frames in the lower panels of figure 22.8. After learning, the IT activity associated with an arbitrary cue has now become bound to activation of the rightward element in the caudate, thus building a sensorimotor association. This learning provides a functional explanation for the utility of convergence of distinct cortical projections in striatum by allowing representations in one domain (in this case visual cues in IT) to become bound to another (in this case spatially coded saccade outputs).

Now we have the corticostriatal network architecture and a learning mechanism that will allow us to develop models that exploit the convergence of different cortical representations onto the striatum. We use this for tasks in which the goal is to discriminate between visual elements presented on a screen, whether they are cues with different shapes and color or sequence of cues presented at different locations on the screen. We first focused on a simpler task involving cue discrimination with the conditional visual discrimination and now proceed with sequence discrimination and learning.

22.2.2 Corticostriatal Convergence and Plasticity—Toward Organization of Sequential Behavior The model of IT–FEF convergence in CD, and reward-related dopamine-based learning, thus allows the system to acquire behavioral links between the visual features of objects and movement. This is of behavioral value, but it lacks an important component which is the temporal organization of sequential behavior. Extensive studies in Parkinson's disease indicate that intact basal ganglia function is required for many aspects of sequential behavior (Benecke et al., 1987; Dominey and Jeannerod, 1997; Dominey et al., 1997; Doyon, 2008; Shin and Ivry, 2003). This poses the following question: How can the corticostriatal system contribute to the sequential organization of behavior?

The response to this question has two components with two corresponding questions: How can sequential structure be represented in cortex, and how can this cortical representation be bound to action for sequential behavior? We have just seen that the response to the second question is provided in the corticostriatal convergence model framework. The cortical representation of sequence context will be bound to the appropriate motor (or cognitive) behavior via dopamine-modulated corticostriatal plasticity that we saw at work above.

This leaves the question concerning the cortical representation of sequential structure. Neurophysiological data responding to this question was provided by Barone and Joseph (1989) in the first study that recorded prefrontal cortical activity during the execution of sequential behavior in the monkey. The monkey was trained to observe the presentation of a three-element sequence on a spatial display during the "static" phase and then to reproduce the sequence by touching buttons on the display during the "dynamic" phase. Three targets (U, L, and R corresponding to up, left, and right) allow six possible three-element sequences (LRU, LUR, URL, ULR, RLU, RUL). Barone and Joseph observed that neurons in the prefrontal peri-arcuate sulcus region displayed classic spatial receptive fields but that these responses were modulated by the rank of targets within the sequence, with two predominate response types. Visual-tonic cells displayed a spatially selective response to the presentation of spatial targets during the static phase that was modulated by their rank in the sequence. Context cells displayed a spatially selective response to spatial targets during execution in the dynamic phase, modulated by the context of previous targets/saccades in the sequence. Together, these two types of cells provide the necessary ingredients for PFC to encode the sequence during the static phase, and its ongoing state of execution during the dynamic phase. The question remains, what neural coding or connectivity scheme could provide for these properties in prefrontal cortex?

To respond to this question, we look to the neuroanatomy. Goldman-Rakic (1987) notes that one of the characteristic features of prefrontal cortex is the abundance of local recurrent connectivity. That is, cortical neurons are connected to local neighbors in such a way that recurrent loops can be formed. Indeed, such local recurrent connectivity is a general property of cortex (Markov et al., 2011; Markov et al., 2013). Such recurrent systems have inherent sequence coding properties (Elman, 1991; Jaeger and Haas, 2004). Dominey, Arbib, and Joseph (1995) thus set out to model such sequence learning with two components—a prefrontal cortex as a network of recurrently connected leaky integrator neurons which would encode sequence state, and corticostriatal plasticity which would allow the binding of these states to the appropriate behavior components of the sequence. The model in figure 22.9 illustrates how these two components are assembled to form a model of sensorimotor sequence learning.

The prefrontal cortex is modeled as two layers of leaky integrator neurons (PFC and damped-PFC) that are reciprocally connected with randomized projections varying between –0.6 and 0.4, which provides the crucial recurrence that yields the dynamical system that will be seen to be sensitive to serial and temporal structure. PFC receives visuospatial input from PP and the reciprocal self-input (damped-PFC). The recurrent connections allow for complex recurrent memory loops, and also for the inhibitory interactions that will produce order selectivity. The negative bias (i.e., weights ranging between –0.6 and 0.4) prevents a runaway self-excitatory saturation of PFC. Post saccade input from SC via SC–PFC, and the PP–PFC are connection matrices with synaptic strengths that vary between –0.5 and 0.5, respectively, providing a balanced inhibitory and excitatory signal of SC and PP activity, respectively. Damped-PFC gets topographic input from PFC. Its 25 cells are split into groups of five that have five different time constants (0.1, 0.6, 1.1, 1.6, 2.1). This diversity of time constants was determined to accommodate different durations of target presentations and thus place different requirements on the temporal sensitivity of PFC. Details can be found in Dominey, Arbib, and Joseph (1995).

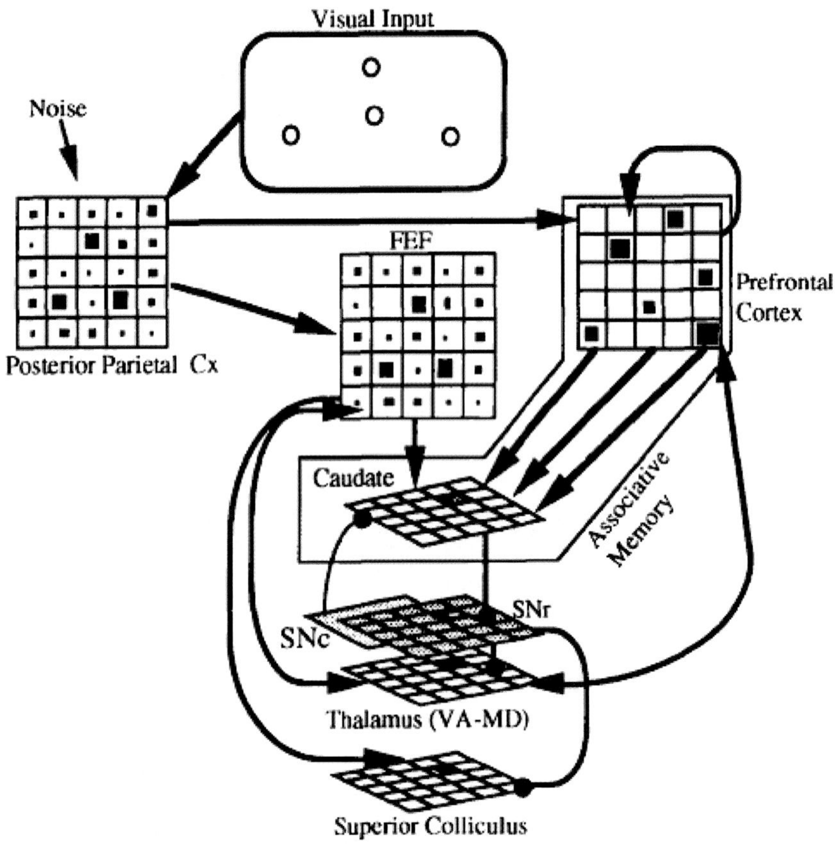

FIGURE 22.9 Corticostriatal model for sensorimotor sequence learning. Prefrontal cortex (PFC) is modeled as two reciprocally connected layers of neurons. This PFC network is a dynamical system that encodes the history of previous events. At any moment in sequence perception or execution, PFC encodes the current state. This state can be associated with appropriate motor responses via plasticity in the PFC–caudate synapses based on reward-related dopamine signals. Based on the sensitivity to temporal and serial structure provided by the recurrent network, we refer to the model as the temporal recurrent network. Cx, cortex; FEF, frontal eye fields; SNc, substantia nigra pars compacta; SNr, substantia nigra pars reticulate; VA-MD, ventral-anterior and mediodorsal.

Dynamic activity in the PFC network makes a temporal trajectory of neural activity. Points in this trajectory—activity patterns in PFC at a given time—become linked to motor responses for sequence learning. Via learning and normalization rules that apply reward or punishment after each saccade, the patterns of PFC activity that precede each saccade become associated with activation of CD saccade cells so that each pattern in PFC at the appropriate moment in sequence execution activates the correct saccade-related cells. This leads to inhibition of the SNr, and disinhibition of SC for production of the correct saccade. Inputs from PFC to CD are via the modifiable synapses in PFC–CD–synapses.

A crucial property of this PFC architecture is at each point in the execution of each of the sequences, there will be a unique pattern of activity in PFC. We can visualize this in figure 22.10. Consider the two sequences right–left–up (RLU) and right–up–left (RUL). Though the first saccade is the same in both sequences, the resulting patterns of cortical activity in PFC must be different if these patterns are to be reliably associated with correct second saccades appropriate for these two different sequences (L and U, respectively). Due to the order preservation provided by the mixed excitatory and inhibitory PFC inputs, and the recurrent connections in PFC, perception of these two sequences produces different patterns in PFC that remain different even after identical saccades to R, and identical visual input resulting from this saccade.

Figure 22.10 illustrates the PFC and CD activity after the saccades to R in these two sequences. Note the different patterns of PFC activity and the resulting preference in CD for the U and L targets, respectively, produced by these divergent patterns. Recall that the conditional visual discrimination learning model presented above (figure 22.7) has demonstrated that a unique pattern of activity (generated by a visual cue) can be associated with the correct one of two targets via corticostriatal trial-and-error learning. In the current simulation, the PFC sequence learning model (figure 22.9) learns to associate PFC activity patterns (generated by the context of the sequence) with the correct one of three targets, again via corticostriatal reinforcement learning.

The model training progresses gradually, based on the shaping used in primate training, moving progressively to sequences of one, two, and then three items. In a typical run of our simulation of the sequence learning task, the model attained over 95% correct performance after exposure to approximately 800 trials of the six three-item sequences. After the learning had taken place, we plotted the responses of the 25 PFC cells during all six sequences, analyzed them according to the criteria defined by Barone and Joseph (1989), and found a similar distribution of cells coding for visual target and sequence rank properties.

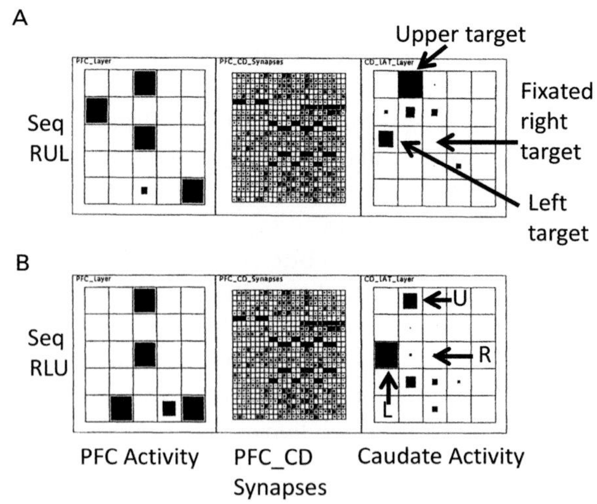

FIGURE 22.10 Snapshots of prefrontal cortex (PFC) activity, PFC–caudate (PFC–CD) synapses, and caudate activity after the first saccade to R in the sequences (Seq) RUL (A) and RLU (B). In both cases the model is now fixating R, so targets U and L are shifted 2 units left and 1 unit up on the updated retinal image, as indicated in caudate activity, labeled in right panel. In both RUL and RLU the model must select from these two targets. Though the first saccades and the subsequent retinal inputs were identical for RUL and RLU, the resulting patterns of cortical activity in PFC are different (encoding the different respective input sequences RUL and RLU). These patterns are transformed to caudate activity via the PFC–CD–synapses. In caudate, we see that for sequence RUL (A), the upper target is favored, while in RLU (B), the left target is favored, as required.

The model has been re-implemented in the PyRates environment, and demonstrated using a simpler sequence discrimination task. Here, instead of reproducing each sequence, the model should learn to associate each sequence with a motor response, similar to the cue discrimination task, where each cue was associated with a specific motor response. The six sequences are each associated with a single output response as indicated:

```
s1 = [['U', 'R', 'L'], 'U']
s2 = [['U', 'L', 'R'], 'U']
s3 = [['R', 'U', 'L'], 'R']
s4 = [['R', 'L', 'U'], 'R']
s5 = [['L', 'U', 'R'], 'L']
s6 = [['L', 'R', 'U'], 'L']
```

In the sequence discrimination task, after the presentation of a sequence, the three targets are presented simultaneously, and the model should respond by selecting the target above the central fixation point (U), below and to the left (L), or below and to the right (R). Figure 22.11 illustrates the activity in PFC, and the

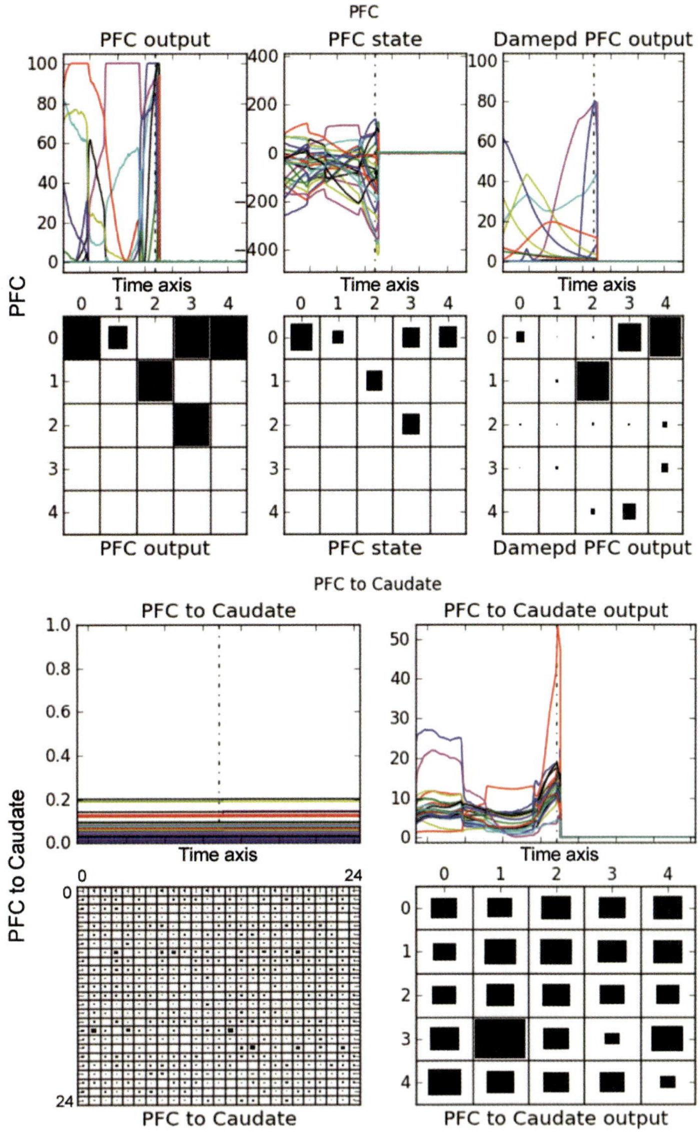

FIGURE 22.11 Activation of prefrontal cortex (PFC) layers, and the PFC to caudate output activity, which has been influenced by learning. PFC is modeled as two recurrently connected layers, PFC and damped PFC. Activity in the network after presentation of sequence "LUR," with model trained to produce a response to the left target L. PRC to caudate: Left, illustration of PFC to caudate weight matrix; Right, projecting PFC activity through PFC to caudate weights generates a larger activation in the topographic PFC to caudate array, at the target L, below and left of the central element, corresponding to the correct choice in the discrimination task.

input from PFC to caudate after the presentation of the sequence s5 after training. We see that the map element corresponding to the left choice (the element below and to the left of the central element in the map) is more activated than the elements for the upper and rightward movements. This activation contributes to the correct response to the leftward target in the discrimination task.

In the PyRates simulations, the same learning mechanism, based on modification of corticostriatal synapses, is used to bind cortical representations of visual features, or of sequential contexts, to motor responses that can be considered as the output in a discrimination task, or as the successive elements in a sequence reproduction task. This illustrates the flexibility of this corticostriatal plasticity mechanism, and how it can be extended from discrimination of static elements to discriminating representations' sequential context, when recurrent connections are included in the cortical architecture.

Abstract Sequence Processing and the "Language Circuit" Within the context of sequential organization of behavior, it is of interest to examine the capabilities of the corticostriatal system in processing more complex sensorimotor sequences. Dominey (1995) thus demonstrated that the model of Dominey, Arbib, and Joseph (1995) could learn complex sequences of the form ABCDABCEABCFABCGABCH. This is a complex sequence in the sense that the repeating subsequence ABC is followed by D, E, F, G, or H, depending on the context defined by the n-4th element. It was further demonstrated that the model is capable of learning multiple sequences, dependent on their length and complexity, for example, three simple sequences of length 20; six simple sequences of length 3; versus two complex sequences of length 12. Likewise, the timing delay between a motor response and the subsequent Go signal can be increased up to 200% with no perturbation for simple sequences while an increase of 50% impairs performance for complex sequences. This suggested a link between serial order and the temporal structure of the sequences.

To further investigate corticostriatal function in temporal sequencing, Dominey (1998b) tested the temporal recurrent network (TRN) sequence learning model from figure 22.9 on a modified version of the serial reaction time (SRT) task that manipulated temporal structure. In the standard SRT task, the model/subject should respond to visual targets as rapidly as possible. Unknown to the subject, the targets are presented in a repeating sequence. Sequence learning is revealed as reaction times (RTs) decrease for elements in the repeating sequence and then increase for element presented in a random order. Dominey (1998a) demonstrated that the sequence learning model displayed this behavior for the standard SRT task because, as corticostriatal learning proceeded, striatal neurons became activated more rapidly because of the increased learning-related inputs for elements in learned sequences. In the modified task, two versions of the same sequence B–C–B–D–C–A–D–A–C–D were used. The difference was that delays of 500 ms were systematically inserted after elements A and B in sequence 1 and after C and D in sequence 2. Thus the serial order or structure the sequences were identical, while their temporal structure, or rhythms, were different. In the simulation experiments, the model displayed the characteristic learning (with reduced RTs) for sequence 1. When sequence 2 was presented, the RTs increased, though to a lesser degree than for a random series. Validation in human experiments confirmed this prediction: changing the temporal structure changes the identity of the sequence and yields significantly increased RTs (Dominey, 1998b). The sensitivity of the model to temporal structure was also demonstrated by the system's ability to discriminate different prosodic attitudes in human speech (Blanc and Dominey, 2003).

In addition to serial and temporal structure, a third dimension of sequential structure is abstract structure. The sequences ABCBAC and DEFEDF have distinct serial structure, but they share the same internal repetitive structure (123213) that Dominey et al. (1998) refer to as abstract structure. Dominey et al. (1998) developed a SRT task in which the sequences contained both serial (surface) and abstract structure. The corticostriatal TRN model learned the serial structure but failed to learn the abstract structure. Only when the model was equipped with an additional capability to store the previous sequence elements in a working memory, which allowed comparison with the current element, could sequences then be recoded in terms of their internal repetitive—or abstract—structure, yielding the abstract recurrent network (ARN). This foreshadows more recent related models of working memory and executive function (Hazy et al., 2007). This requirement for an additional working memory system predicts that surface and abstract structure are processed by neurophysiologically distinct systems, as confirmed in human experiments (Dominey et al., 1998). Interestingly, Dominey and Georgieff (1997) observed that patients with schizophrenia, who are impaired in aspects of explicit processing, learned the serial but not abstract structure of these sequences. More interestingly, Lelekov et al. (2000) demonstrated that this impairment in abstract structure processing was highly correlated with an impairment in syntactic comprehension in language. Finally, Dominey et al. (2003) demonstrated that agrammatic aphasic patients also displayed this correlation between syntactic comprehension and abstract structure processing, and that they learned the serial but not abstract structure processing in the abstract SRT task.

Sensitivity to serial, temporal, and abstract structure of sound sequences has been demonstrated in human infants (Marcus et al., 1999; Nazzi et al., 1998; Saffran et al., 1996) and may provide part of the fundamental tools for language acquisition. Dominey and Ramus (2000) demonstrated that the combined TRN and ARN could capture these three forms of sensitivity. This led to the development of a model of corticostriatal function in language processing in which a dorsal cortical system including pars opercularis would contribute to abstract structure processing, and ventral system including pars triangularis would contribute to the integration of semantic content in language processing (Dominey and Hoen, 2006; Dominey et al., 2006; Dominey et al., 2009; Hoen et al., 2006). One of the outcomes of this research is the prediction of a corticostriatal "language loop." Language processing is further developed in chapter 24 by Cangelosi in this volume.

22.2.3 Related Models of Sequence Processing

The use of recurrent connections to encode sequence context has been exploited in several models of corticostriatal function in sequence processing. Berns and Sejnowski (Berns and Sejnowski, 1998) developed a similar model of sequence processing and generation inspired by basal ganglia anatomy and neurophysiology which exploits recurrence and reciprocal connections. The model exploited recurrent connections between the GPe and the STN to create a form of working memory that can also take into account sequential context. For output generation, tonic inhibition of the thalamus by GPi is selectively released as GPi selects an action. Within the recurrent loop, the connections between STN and GPe were modified via an associative learning rule to implement the sequence learning.

Houk et al. (2007) used recurrent networks made up of long loops linking PFC, CD, GPi, and thalamus, and short loops between PFC and thalamus, and demonstrated sequence-selective activity in PFC and CD. In the sequence learning, a WTA action selection "actor" generated responses, and the "critic"—a dopamine-based error prediction signal—was used to learn the appropriate responses. In this actor–critic architecture for reinforcement learning (see section 22.3 below and chapter 10, "Reinforcement Learning," for further exposition), Suri and Schultz (1998) demonstrated that a suitable critic could be modeled after the response characteristics of dopamine neurons, allowing the model to successfully learn a seven-element sequence. Doya (1999) similarly evokes a reinforcement learning function for the basal ganglia.

Nakahara et al. (2001) developed a dual basal gangliathalamocortical pathway approach to sensorimotor sequence learning, in order to account for data that indicates that sensorimotor sequences are learned as separate but integrated sensory and motor sequences. Thus, visual sensory sequences are learned in a loop that includes the dorsolateral prefrontal cortex and the anterior basal ganglia, and motor sequences are learned in a loop that includes the supplementary motor area and posterior basal ganglia. During a given sequence learning task, both systems will learn their appropriate dimensions (in this case visual and motor) in parallel. The model accounts for partial transfer when a sequence learning task acquired with one hand is then shifted to the other hand.

Analogies between the Corticostriatal System and Reservoir Computing A powerful sequence learning capability can thus be provided by simulation of PFC as a population of neurons linked by fixed recurrent connections, with corticostriatal connections modifiable by a reward-related dopamine signal. Sequence context is encoded in the recurrent network, and this context is read out in the striatum. It has recently come to light that there is a deep relation between this view of the corticostriatal system and a domain of machine learning referred to as *reservoir computing* (Lukosevicius and Jaeger, 2009).

Reservoir computing is characterized by two central notions. The first is that a network of computing units (rate coding or spiking neurons) that are connected with fixed recurrent connections can generate a form of dynamical system that encodes the history of its inputs. The second notion is that the contents of this representation can be "read out" by neurons connected to the reservoir units by modifiable synapses, in order to learn the appropriate transformation from the input signal to the output signal. In both learning and recall, the appropriate state of the reservoir is activated by the preceding stimuli. This is illustrated in figure 22.10 and figure 22.11. Lukosevicius and Jaeger (2009) review the history of echo state networks and liquid state machines under the encompassing heading of reservoir computing, in which a recurrent neural network (the reservoir) is generated randomly and only a readout is trained. They note that Dominey and Ramus (2000) first spelled out this reservoir principle (there in the context of PFC as the recurrent network and CD as the readout), which was also independently developed by Jaeger (Jaeger, 2001; Jaeger and Haas, 2004) and Maass (Maass et al., 2002). Based on its sensitivity to temporal as well as serial structure, the corticostriatal sequence learning model has, as we have seen, been dubbed the "temporal recurrent network," or TRN. It is of potential interest that the powerful mechanism of reservoir computing (Jaeger and Haas, 2004) has fundamental links to the corticostriatal system, with the cortex corresponding to the reservoir, and the striatum corresponding to the readout, with corticostriatal plasticity corresponding to the learning mechanism.

In their paper on liquid state machines Maass et al. (2002) demonstrated that the inherent dynamics of a high-dimensional dynamical system formed by a sufficiently large neural circuit could serve as universal analog fading memory that is sensitive to the sequential history of its inputs. They further showed that readout neurons can learn to extract information about current and past inputs from the current state of such recurrent neural circuits and can use this information for diverse tasks including sequence learning. Maass et al. (2002) note that the model of Dominey, Arbib, and Joseph is a special case within their framework. Later it was recognized that a class of systems that employ fixed recurrent connections and modifiable readout connections could be identified in the context of reservoir computing (Verstraeten et al., 2007).

Reservoir computing is making a significant impact in cognitive neuroscience. In particular, it is being recognized that cortical processing appears to have reservoir properties including nonlinear mixtures of task-related variables (or mixed selectivity) (Enel et al., 2016; Rigotti et al., 2010, 2013) and fading memory (Nikolic et al., 2009). Reservoir computing has also been demonstrated to be a viable method for implementing higher cognitive functions including learning the grammatical structure of language (Hinaut and Dominey, 2013; Hinaut et al., 2015; Tong et al., 2007).

22.3 Dopamine Function and System-Level Models of Corticostriatal System

22.3.1 NEUROMODULATION AND LEARNING Dopamine plays at least two crucial roles in the corticostriatal system. The first is related to modulation of excitability of striatal neurons from cortical inputs, and the second is related to corticostriatal plasticity. With the depletion of nigrostriatal dopamine in Parkinson's disease, both of these functions are impaired (Buhmann et al., 2004). Here we focus on plasticity in the intact condition.

It is clear that distinct components of the corticostriatal system can play fundamental roles in the encoding of behavioral context, and adaptation to changing environmental contingencies. The dopamine signal is a prominent component in this context with its relation to predicted reward. Schultz noted clearly that dopamine neurons display increased activity for unpredicted rewards, and decreased activity when predicted rewards did not occur (Ljungberg et al., 1991; Ljungberg et al., 1992; Schultz, 1986; Schultz et al., 1993). Interestingly, Schultz observed that when a novel cue was followed by a reward, the dopamine response shifted from the reward itself to the cue that predicted the reward. This signal is strikingly similar to the reward prediction error signal that is employed in the temporal-difference approach to reinforcement learning (Sutton, 1988; Sutton and Barto, 1998), as has been noted by Contreras-Vidal and Schultz (1999), Hollerman and Schultz (1998), Schultz and Dickinson (2000), and Suri and Schultz (1999, 2001), among others. Botvinick, Niv, and Barto (2009) thus note historically how the reinforcement learning paradigm (Barto, 1994) provided insight into this dopamine function in the context of the temporal-difference learning paradigm which formed a framework for interpreting temporal profiles of dopaminergic activity (Schultz et al., 1997). (See chapter 10 by Daw on "Reinforcement Learning" for more on temporal-difference learning.) In this context, the nigrostriatal dopamine system would act as the critic (i.e., the system that supplies reinforcement-related signals that evaluate the current action), and the striatum as the actor (in that it decides which action to disinhibit), based on cortical input. Botvinick et al. (2009) further note how actor–critic architectures for reinforcement learning help to provide new interpretations of functional dissociations in the corticostriatal system (Joel et al., 2002). Daw, Niv, and Dayan (2005) further advance the distinction that in the context of reinforcement learning, the dorsal striatum provides a form of model-free reinforcement learning, and the prefrontal cortex a form of model-based reinforcement learning. They propose a Bayesian system that performs a kind of trade-off analysis and assigns priority to one of the two systems according to the structure of uncertainty in the problem at hand.

Doya (2002) has extended the analysis of dopamine function to a multiple-system-level framework in which dopamine signals the error in reward prediction, serotonin controls the timescale of reward prediction, noradrenaline controls the randomness in action selection, and acetylcholine controls the speed of memory update. (See chapter 7 on "Neuromodulation" for further discussion.)

22.3.2 BRINGING IN THE INDIRECT PATHWAY Gurney et al. (2001a, 2001b) contest the traditional view of the direct (striatum–SNr–thalamus) and indirect (striatum–GPe–STN–SNr–thalamus) pathways in the basal ganglia, in which the direct pathway allows cortical inputs to yield a net excitatory effect on thalamus, and the indirect pathway a net inhibitory effect on thalamus (via SNr). They instead propose a new configuration with distinct pathways for action selection and control. Part of the novelty of this approach is the use of STN as part of the action selection system. A related model addressing the basic architecture of sensorimotor cortex, motor cortex, and loops through basal ganglia provides an extended investigation of thalamocortical and thalamic reticular neurons in the context of action selection (Humphries and Gurney, 2002).

In this context, Prescott et al. (2006) have recently used an extended version of this model in an embodied robotics context. (For diverse approaches to the linkage of computational neuroscience and robotics, see chapter 8 on "Motor Pattern Generation," chapter 16 on "Embodied Models and Neurorobotics," and chapter 24 on "Language Processing.") The action selection model successfully learned to perform behaviorally relevant sequences of actions and displayed behavior in the face of complexity that was comparable to that observed in related animal studies.

Frank et al. (2004) propose a model in which the direct "Go" pathway is responsible for action execution via disinhibition of the thalamus, and the indirect "No-go" pathway inhibits action via the inhibition of the thalamus due to the inclusion of an additional inhibitory component (the GPe). They exploit the respective excitatory and inhibitory effects of dopamine on D1 and D2 receptors in the direct and indirect pathway. In this way, reward (and its absence) allows specific learning-related changes in the Go and No-go pathways to more optimally adapt to the structure of the task. Interestingly, the model made a prediction that was subsequently confirmed in human studies, that in Parkinson's disease, where DA levels are impaired by the disease and artificially modified by medication, nonmedicated patients would make better use of negative feedback than either patients on medication or healthy seniors. Cohen and Frank (2009) reexamine this model in the context of new data on the role of the STN in a "hyperdirect" pathway, and interestingly make the comparison between neural network versus mathematical models of basal ganglia function. An important issue in all this is the origin of the dopamine response. Brown et al. (1999) propose a model in which an excitatory pathway generates CS–induced excitatory SNc dopamine bursts, and an inhibitory pathway prevents dopamine bursts in response to predictable reward-related signals. They note that the comparison of the dopamine response in SNc to the reward prediction error in the temporal-difference reinforcement learning context has provided significant insight into the nature of the dopamine response. However, they also note that this approach does not provide insight on the underlying biological mechanisms that actually compute the temporal-difference signal and how they do it.

O'Reilly and Frank (2006) address a complex sequence and working memory task that leads to a reconsideration of the origin of the dopaminergic neuron signal. In the 1–2-AX task, the inputs 1 and 2 set contexts where the target sequences are AX and BY, respectively. Each time the target sequence for the current context is presented, the subject should respond with R, and otherwise with L. This task thus places strong challenges on gated working memory: the system must learn what to store in working memory, and how to use the contents in the sequence recognition context. These constraints can be reformulated as a temporal credit assignment problem, that is, which prior events were critical for good (or bad) performance, and a structural credit assignment problem which involves what to store in working memory. Thus, temporal credit assignment depends on the contents of the working memory in PFC, but this depends, in turn, on the structural credit assignment problem of the choice of what to store in PFC. To address this circular learning problem and the associated problem of reward prediction, the DA training signal is generated by a Pavlovian associative learning mechanism based on primary and learned reward values and forms the critic, which trains the striatum which serves as the actor and gating mechanism. In this context, the model displays impressive performance on the 1–2-AX problem.

Humphries et al. (2006) developed a detailed model of basal ganglia function in action selection using spiking neurons. The use of spiking neurons has the corresponding advantage that they could address issues of oscillatory behavior and generate local field potentials. Indeed, the model was able to account for three bodies of experimental data corresponding to behavioral correlates of tonic rates, low-frequency oscillatory activity, and gamma-band oscillations. Thus, the use of spiking units allows this model to address phenomena including low-frequency and gamma-band oscillatory activity that cannot be directly addressed by firing rate models. (For more on "Neural Rhythms," see chapter 4 by Vierling-Claassen and Jones.)

22.3.3 Distinct and Common Factors across These Approaches Despite the variety of modeling approaches that have been taken toward understanding the functional neurophysiology of the corticostriatal system, it is clear that there is a baseline set of features that underlies many of these approaches. Perhaps the most obvious is the consensus that dopamine clearly plays a role in adaptive behavior, including its role in plasticity at the corticostriatal synapse, and its role as a neuromodulator. Another clear common element is the notion of disinhibition, with SNr/GPi holding a tonic inhibition over thalamus that is selectively released based on specific cortical input to the striatum. The notion of the role of the direct and indirect pathways is less coherent, with some models proposing specific and distinct roles (e.g., the Go and No-go systems), others holding that the indirect pathway modulates corticostriatal function, and others being more agnostic or concentrating only on the direct pathway. A number of models exploit the notion of using recurrence both for working memory and/or sequence context, either at the cortical level and/or at the long loop cortex–basal ganglia–thalamus–cortex level.

One of the important contributions of these models is the extension from simple action selection to higher cognitive function including learning to tasks that are considered to require executive function. As suggested by Hazy and colleauges (2007) in complex tasks, setting internal contexts for making future choices can be implemented by the same selection mechanisms used for action selection, thus extending corticostriatal function into cognitive control.

22.4 *Future Challenges*

Significant progress has been made in understanding and modeling the corticostriatal system, but significant challenges remain. Perhaps the most baffling of these open challenges concerns the functional significance of the anatomical convergence within the basal ganglia loops including that seen in the corticostriatal (e.g., Selemon and Goldman-Rakic, 1985; Haber et al., 2006) and nigrostriatonigral (e.g., Haber et al., 2000) projections. Selemon and Goldman-Rakic (1985) demonstrated that while it is indeed somewhat segregated, the corticostriatal projection system is characterized by substantial interdigitation of the terminal projections. Haber et al. (2006) further advanced this notion by examining more diffuse projections which demonstrated even more significant overlap in corticostriatal terminals. Dominey, Arbib, and Joseph (1995) made a first attempt at providing a functional exploitation for this convergence in the specific case of convergence of inputs from IT and PFC onto the oculomotor territory of CD, thus allowing the learning of oculomotor stimulus–response associations and sensorimotor sequences, respectively. In a related approach, Nakahara et al. (2001) consider the coordination of parallel sensory and motor loops. Still, a deep understanding of the functional significance of the patterns of overlap and interdigitation in the corticostriatal system remains a challenge.

An intriguing proposition in this context was made by Graybiel et al. (1994). They document how corticostriatal projections from a given node of sensorimotor cortex form a distributed set of projection modules in the striatum, and that a given module can receive somatotopically matched inputs from different sensory and motor areas. Projections from these striatal modules then reconverge in the SNr/GPi. Graybiel et al. (1994) suggest that under the control of TANs in the striatum, processing in these distributed modules can be coordinated. Grabiel et al. continue to investigate the role of TANs in adaptive behavior. The have recently demonstrated that TANs in the ventral striatum provide a bidirectional output signal during learning—increasing firing for rewarded trials and decreasing for unrewarded trials during acquisition and retaining only the negative responses after learning (Atallah et al., 2014).

The proposal of Graybiel et al. (1994) and their recent data (Atallah et al., 2014) suggests how somatotopically related modules, distributed in the striatum, could be coordinated by the global activity of TANs under the control of dopamine. Their framework also provides the basis for an alternative explanation for the functional significance of the overlapping structure of corticostriatal projections, within the context of a relative conservation of topography, as follows. When considered in the context of Haber's diffuse projections, the pattern of divergence of topographically related areas onto distributed striatal modules should provide a combinatorial lever, such that any arbitrary combination of convergence between two (and likely more) topographic areas should be represented in the striatum. But note that some associations will be strong (easy to exploit) whereas others will be weak and thus exploitable only through extensive training. Moreover, as the volume of the striatum is significantly less than that of the cortex, clearly the resolution of this representation will be reduced. Still, this combinatorial potential exists, and it is the challenge of the modeler to explain its potential. This raises the more general issue of the role of the detailed architecture of the MSNs and their dendritic arborization, and the function of the tonically activated TANs (as addressed in Graybiel et al., 1994; Gurney et al., 2001a; Gurney et al., 2001b).

Neural plasticity in the striatum is crucial for adaptation and can be achieved both via LTP and via LTD. These forms of learning can be selectively enhanced depending on the status of the spiny neurons' membrane potentials in an elevated up state or hyperpolarized down state (Calabresi et al., 2007). The effects of these states on learning has been addressed (Suri et al., 2001; Suri and Schultz, 2001), and this remains an avenue for future investigation.

Finally, as the problems that are addressed increase in complexity, more attention should be focused on the mapping of problems to neural architectures, and the identification of equivalence classes for types of problems that can be addressed by a given architecture. For example, what is the equivalence class of problems that can be addressed by the model of Cohen and Frank (2009)? Of the TRN, of the Reservoir system? It is likely that a new characterization of complexity—in which cognitive control is accounted for—is required. Typically, problem complexity is assessed within the context of the problem. In real life, part of the "problem" is determining what the *current problem* actually is, determining what dimensions of the perceptual array are relevant, determining whether one should shift or stay. Typically, models have been developed to address specific classes of behavior, and the notion of shifting between different problem sets, or learning how to solve multiple problems and how to determine what is the current problem has been less the focus. Aspects of this have been addressed. For example, in the 1–2-AX problem (O'Reilly and Frank, 2006), the system must learn to perform the proper gating of contents into working memory. This has been extended to a more general model of executive function (Hazy et al., 2007). Still, an extended form of this kind of flexibility and its management remains a challenge for future modeling of the corticostriatal system. In an example of such an approach (Bogacz and Gurney, 2007), it has been demonstrated that basal ganglia implement an asymptotically optimal statistical test for decision-making—the multiple sequential probability ratio test. Future research may extend this systems-level approach into the domain of cognitive robotics. Current neurophysiology continues to provide data concerning the cooperation of frontal cortical areas including dorsolateral prefrontal cortex and anterior cingulate cortex (Amiez et al., 2005; Amiez et al., 2006; Procyk and Goldman-Rakic, 2006; Quilodran et al., 2008; Sallet et al., 2007), which provide the basis for the development of systems-level implementation in robot control (Khamassi et al., 2011).

The research reviewed here suggests an integrative role for the corticostriatal system, with a form of combinatorial neural connectivity playing two distinct roles. First, the combinatorics of recurrent connections in cortical areas allow these areas to generate high-dimensional representations of their inputs over time. Second, through spatially complex patterns of interdigitation and overlap in the corticostriatal projections, the corticostriatal system provides a site where diverse cortical representations can be integrated and linked to behavior and further compositional representations in a selective manner via corticostriatal plasticity. This provides an open-ended system for adaptation to novel environmental contingencies.

ACKNOWLEDGMENTS The research reported in this chapter was supported by the European Union under the "Organic" and "WYSIWYD" grants, and by the French ANR under the "Comprendre" grant.

NOTE

1. A 9 × 9 visuomotor map was the minimum sufficient for allowing a realistic range of saccade targets in the single and double-step paradigms and was adopted to meet 1992 computing constraints.

REFERENCES

Alexander, G. E., DeLong, M. R., & Strick, P. L. (1986). Parallel organization of functionally segregated circuits linking basal ganglia and cortex. *Annual Review of Neuroscience, 9,* 357–381.

Amiez, C., Joseph, J. P., & Procyk, E. (2005). Anterior cingulate error-related activity is modulated by predicted reward. *European Journal of Neuroscience, 21,* 3447–3452.

Amiez, C., Joseph, J. P., & Procyk, E. (2006). Reward encoding in the monkey anterior cingulate cortex. *Cerebral Cortex, 16,* 1040–1055.

Apicella, P. (2007). Leading tonically active neurons of the striatum from reward detection to context recognition. *Trends in Neurosciences, 30,* 299–306.

Atallah, H. E., McCool, A. D., Howe, M. W., & Graybiel, A. M. (2014). Neurons in the ventral striatum exhibit cell-type-specific representations of outcome during learning. *Neuron, 82,* 1145–1156.

Barone, P., & Joseph, J. P. (1989). Prefrontal cortex and spatial sequencing in macaque monkey. *Experimental Brain Research, 78,* 447–464.

Barto, A. G. (1994). Reinforcement learning control. *Current Opinion in Neurobiology, 4,* 888–893.

Bechara, A., Damasio, H., & Damasio, A. R. (2000). Emotion, decision making and the orbitofrontal cortex. *Cerebral Cortex, 10,* 295–307.

Benecke, R., Rothwell, J. C., Dick, J. P., Day, B. L., & Marsden, C. D. (1987). Disturbance of sequential movements in patients with Parkinson's disease. *Brain, 110*(Pt 2), 361–379.

Berns, G. S., & Sejnowski, T. J. (1998). A computational model of how the basal ganglia produce sequences. *Journal of Cognitive Neuroscience, 10,* 108–121.

Blanc, J. M., & Dominey, P. F. (2003). Identification of prosodic attitudes by a temporal recurrent network. *Brain Research. Cognitive Brain Research, 17,* 693–699.

Bogacz, R., & Gurney, K. (2007). The basal ganglia and cortex implement optimal decision making between alternative actions. *Neural Computation, 19,* 442–477.

Botvinick, M. M., Niv, Y., & Barto, A. C. (2009). Hierarchically organized behavior and its neural foundations: A reinforcement learning perspective. *Cognition, 113,* 262–280.

Brown, J., Bullock, D., & Grossberg, S. (1999). How the basal ganglia use parallel excitatory and inhibitory learning pathways to selectively respond to unexpected rewarding cues. *Journal of Neuroscience, 19,* 10502–10511.

Bruce, C. J., & Goldberg, M. E. (1985). Primate frontal eye fields: I. Single neurons discharging before saccades. *Journal of Neurophysiology, 53,* 603–635.

Buhmann, C., Gorsler, A., Bäumer, T., Hidding, U., Demiralay, C., Hinkelmann K., et al. (2004). Abnormal excitability of premotor–motor connections in de novo Parkinson's disease. *Brain, 127,* 2732–2746.

Calabresi, P., Mercuri, N. B., & Di Filippo, M. (2009). Synaptic plasticity, dopamine and Parkinson's disease: One step ahead. *Brain, 132,* 285–287.

Calabresi, P., Picconi, B., Tozzi, A., & Di Filippo, M. (2007). Dopamine-mediated regulation of corticostriatal synaptic plasticity. *Trends in Neurosciences, 30,* 211–219.

Centonze, D., Gubellini, P., Pisani, A., Bernardi, G., & Calabresi, P. (2003). Dopamine, acetylcholine and nitric oxide systems interact to induce corticostriatal synaptic plasticity. *Reviews in the Neurosciences, 14,* 207–216.

Chevalier, G., & Deniau, J. M. (1990). Disinhibition as a basic process in the expression of striatal functions. *Trends in Neurosciences, 13,* 277–280.

Chevalier, G., Deniau, J. M., Thierry, A. M., & Feger, J. (1981). The nigro-tectal pathway. An electrophysiological reinvestigation in the rat. *Brain Research, 213,* 253–263.

Chevalier, G., Vacher, S., & Deniau, J. M. (1984). Inhibitory nigral influence on tectospinal neurons, a possible implication of basal ganglia in orienting behavior. *Experimental Brain Research, 53*, 320–326.

Chevalier, G., Vacher, S., Deniau, J. M., & Desban, M. (1985). Disinhibition as a basic process in the expression of striatal functions: I. The striato-nigral influence on tecto-spinal/tecto-diencephalic neurons. *Brain Research, 334*, 215–226.

Cohen, M. X., & Frank, M. J. (2009). Neurocomputational models of basal ganglia function in learning, memory and choice. *Behavioural Brain Research, 199*, 141–156.

Contreras-Vidal, J. L., & Schultz, W. (1999). A predictive reinforcement model of dopamine neurons for learning approach behavior. *Journal of Computational Neuroscience, 6*, 191–214.

Daw, N. D., Niv, Y., & Dayan, P. (2005). Uncertainty-based competition between prefrontal and dorsolateral striatal systems for behavioral control. *Nature Neuroscience, 8*, 1704–1711.

DeLong, M., Alexander, G., Mitchell, S., & Richardson, R. (1986). The contribution of basal ganglia to limb control. *Progress in Brain Research, 64*, 161–174.

Deniau, J. M., & Chevalier, G. (1985). Disinhibition as a basic process in the expression of striatal functions: II. The striato-nigral influence on thalamocortical cells of the ventromedial thalamic nucleus. *Brain Research, 334*, 227–233.

Di Filippo, M., Picconi, B., Tantucci, M., Ghiglieri, V., Bagetta, V., Sgobio, C., et al. (2009). Short-term and long-term plasticity at corticostriatal synapses: Implications for learning and memory. *Behavioural Brain Research, 199*, 108–118.

Dominey, P. F. (1995). Complex sensory-motor sequence learning based on recurrent state representation and reinforcement learning. *Biological Cybernetics, 73*, 265–274.

Dominey, P. F. (1998a). Influences of temporal organization on sequence learning and transfer: Comments on Stadler (1995) and Curran and Keele (1993). *Journal of Experimental Psychology: Learning, Memory, and Cognition, 24*, 234–248.

Dominey, P. F. (1998b). A shared system for learning serial and temporal structure of sensori-motor sequences? Evidence from simulation and human experiments. *Brain Research. Cognitive Brain Research, 6*, 163–172.

Dominey, P. F., & Arbib, M. A. (1992). A cortico-subcortical model for generation of spatially accurate sequential saccades. *Cerebral Cortex, 2*, 153–175.

Dominey, P. F., Arbib, M. A., & Joseph, J. P. (1995). A model of corticostriatal plasticity for learning oculomotor associations and sequences. *Journal of Cognitive Neuroscience, 7*, 311–336.

Dominey, P. F., & Georgieff, N. (1997). Schizophrenics learn surface but not abstract structure in a serial reaction time task. *Neuroreport, 8*, 2877–2882.

Dominey, P. F., & Hoen, M. (2006). Structure mapping and semantic integration in a construction-based neurolinguistic model of sentence processing. *Cortex, 42*, 476–479.

Dominey, P. F., Hoen, M., Blanc, J. M., & Lelekov-Boissard, T. (2003). Neurological basis of language and sequential cognition: Evidence from simulation, aphasia, and ERP studies. *Brain and Language, 86*, 207–225.

Dominey, P. F., Hoen, M., & Inui, T. (2006). A neurolinguistic model of grammatical construction processing. *Journal of Cognitive Neuroscience, 18*, 2088–2107.

Dominey, P. F., Inui, T., & Hoen, M. (2009). Neural network processing of natural language: II. Towards a unified model of corticostriatal function in learning sentence comprehension and non-linguistic sequencing. *Brain and Language, 109*, 80–92.

Dominey, P. F., & Jeannerod, M. (1997). Contribution of frontostriatal function to sequence learning in Parkinson's disease: Evidence for dissociable systems. *Neuroreport, 8*, iii–ix.

Dominey, P. F., Lelekov, T., Ventre-Dominey, J., & Jeannerod, M. (1998). Dissociable processes for learning the surface structure and abstract structure of sensorimotor sequences. *Journal of Cognitive Neuroscience, 10*, 734–751.

Dominey, P. F., & Ramus, F. (2000). Neural network processing of natural language: I. Sensitivity to serial, temporal and abstract structure of language in the infant. *Language and Cognitive Processes, 15*, 87–127.

Dominey, P. F., Ventre-Dominey, J., Broussolle, E., & Jeannerod, M. (1997). Analogical transfer is effective in a serial reaction time task in Parkinson's disease: Evidence for a dissociable form of sequence learning. *Neuropsychologia, 35*, 1–9.

Doya, K. (1999). What are the computations of the cerebellum, the basal ganglia and the cerebral cortex? *Neural Networks, 12*, 961–974.

Doya, K. (2002). Metalearning and neuromodulation. *Neural Networks, 15*, 495–506.

Doyon, J. (2008). Motor sequence learning and movement disorders. *Current Opinion in Neurology, 21*, 478–483.

Elman, J. (1991). Distributed representations, simple recurrent networks, and grammatical structure. *Machine Learning, 7*, 195–225.

Enel, P., Procyk, E., Quilodran, R., & Dominey, P. F. (2016). Reservoir Computing properties of neural dynamics in prefrontal cortex. *PLOS Computational Biology, 12*(6), e1004967.

Frank, M. J., Seeberger, L. C., & O'Reilly, R. C. (2004). By carrot or by stick: Cognitive reinforcement learning in parkinsonism. *Science, 306*, 1940–1943.

Fuster, J. M., & Alexander, G. E. (1973). Firing changes in cells of the nucleus medialis dorsalis associated with memory response behavior. *Brain Research, 61*, 79–91.

Gnadt, J. W., & Andersen, R. A. (1988). Memory related motor planning activity in posterior parietal cortex of macaque. *Experimental Brain Research, 70*, 216–220.

Goldberg, M. E., & Bruce, C. J. (1990). Primate frontal eye fields: III. Maintenance of a spatially accurate saccade signal. *Journal of Neurophysiology, 64*, 489–508.

Goldman-Rakic, P. S. (1987). Circuitry of primate prefrontal cortex and regulation of behavior by representational memory. In V. B. Mountcastle, F. Plum, & S. R. Geiger (Eds.), *Handbook of Neurophysiology. The Nervous System* (Sec. I, Vol. 5, pp. 373–417). Bethesda, MD: American Physiological Society.

Graybiel, A. M., Aosaki, T., Flaherty, A. W., & Kimura, M. (1994). The basal ganglia and adaptive motor control. *Science, 265*, 1826–1831.

Gurney, K., Prescott, T. J., & Redgrave, P. (2001a). A computational model of action selection in the basal ganglia: I. A new functional anatomy. *Biological Cybernetics, 84*, 401–410.

Gurney, K., Prescott, T. J., & Redgrave, P. (2001b). A computational model of action selection in the basal ganglia:

II. Analysis and simulation of behaviour. *Biological Cybernetics, 84,* 411–423.

Haber, S. N., Fudge, J. L., & McFarland, N. R. (2000). Striatonigrostriatal pathways in primates form an ascending spiral from the shell to the dorsolateral striatum. *Journal of Neuroscience, 20,* 2369–2382.

Haber, S. N., Kim, K. S., Mailly, P., & Calzavara, R. (2006). Reward-related cortical inputs define a large striatal region in primates that interface with associative cortical connections, providing a substrate for incentive-based learning. *Journal of Neuroscience, 26,* 8368–8376.

Hazy, T. E., Frank, M. J., & O'Reilly, R. C. (2007). Towards an executive without a homunculus: Computational models of the prefrontal cortex/basal ganglia system. *Philosophical Transactions of the Royal Society of London. Series B, Biological Sciences, 362,* 1601–1613.

Hikosaka, O. (1989). Role of basal ganglia in saccades. *Revue Neurologique, 145,* 580–586.

Hikosaka, O. (2007). Basal ganglia mechanisms of reward-oriented eye movement. *Annals of the New York Academy of Sciences, 1104,* 229–249.

Hikosaka, O., Nakamura, K., & Nakahara, H. (2006). Basal ganglia orient eyes to reward. *Journal of Neurophysiology, 95,* 567–584.

Hikosaka, O., Sakamoto, M., & Usui, S. (1989a). Functional properties of monkey caudate neurons: I. Activities related to saccadic eye movements. *Journal of Neurophysiology, 61,* 780–798.

Hikosaka, O., Sakamoto, M., & Usui, S. (1989b). Functional properties of monkey caudate neurons: II. Visual and auditory responses. *Journal of Neurophysiology, 61,* 799–813.

Hikosaka, O., Sakamoto, M., & Usui, S. (1989c). Functional properties of monkey caudate neurons: III. Activities related to expectation of target and reward. *Journal of Neurophysiology, 61,* 814–832.

Hikosaka, O., & Wurtz, R. H. (1983a). Visual and oculomotor functions of monkey substantia nigra pars reticulata: I. Relation of visual and auditory responses to saccades. *Journal of Neurophysiology, 49,* 1230–1253.

Hikosaka, O., & Wurtz, R. H. (1983b). Visual and oculomotor functions of monkey substantia nigra pars reticulata: II. Visual responses related to fixation of gaze. *Journal of Neurophysiology, 49,* 1254–1267.

Hikosaka, O., & Wurtz, R. H. (1983c). Visual and oculomotor functions of monkey substantia nigra pars reticulata: III. Memory-contingent visual and saccade responses. *Journal of Neurophysiology, 49,* 1268–1284.

Hikosaka, O., & Wurtz, R. H. (1983d). Visual and oculomotor functions of monkey substantia nigra pars reticulata: IV. Relation of substantia nigra to superior colliculus. *Journal of Neurophysiology, 49,* 1285–1301.

Hinaut, X., & Dominey, P. F. (2013). Real-time parallel processing of grammatical structure in the fronto-striatal system: A recurrent network simulation study using reservoir computing. *PLoS One, 8,* 1–18.

Hinaut, X., Lance, F., Droin, C., Petit, M., Pointeau, G., & Dominey, P. F. (2015). Corticostriatal response selection in sentence production: Insights from neural network simulation with reservoir computing. *Brain and Language, 150,* 54–68.

Hoen, M., Pachot-Clouard, M., Segebarth, C., & Dominey, P. F. (2006). When Broca experiences the Janus syndrome: An ER-fMRI study comparing sentence comprehension and cognitive sequence processing. *Cortex, 42,* 605–623.

Hollerman, J. R., & Schultz, W. (1998). Dopamine neurons report an error in the temporal prediction of reward during learning. *Nature Neuroscience, 1,* 304–309.

Houk, J. C., Bastianen, C., Fansler, D., Fishbach, A., Fraser, D., Reber, P.J., et al. (2007). Action selection and refinement in subcortical loops through basal ganglia and cerebellum. *Philosophical Transactions of the Royal Society of London. Series B, Biological Sciences, 362,* 1573–1583.

Humphries, M. D., & Gurney, K. N. (2002). The role of intra-thalamic and thalamocortical circuits in action selection. *Network, 13,* 131–156.

Humphries, M. D., Stewart, R. D., & Gurney, K. N. (2006). A physiologically plausible model of action selection and oscillatory activity in the basal ganglia. *Journal of Neuroscience, 26,* 12921–12942.

Jaeger, H. (2001). The "echo state" approach to analysing and training recurrent neural networks—With an erratum note (GMD Technical Report 148). Bonn: German National Research Center for Information Technology.

Jaeger, H., & Haas, H. (2004). Harnessing nonlinearity: Predicting chaotic systems and saving energy in wireless communication. *Science, 304,* 78–80.

Joel, D., Niv, Y., & Ruppin, E. (2002). Actor-critic models of the basal ganglia: New anatomical and computational perspectives. *Neural Networks, 15,* 535–547.

Khamassi, M., Lallee, S., Enel, P., Procyk, E., & Dominey, P. F. (2011). Robot cognitive control with a neurophysiologically inspired reinforcement learning model. *Frontiers in Neurorobotics, 5,* 1.

Lehericy, S., Ducros, M., Van de Moortele, P. F., Francois, C., Thivard, L., Poupon, C., et al. (2004). Diffusion tensor fiber tracking shows distinct corticostriatal circuits in humans. *Annals of Neurology, 55,* 522–529.

Lelekov, T., Franck, N., Dominey, P. F., & Georgieff, N. (2000). Cognitive sequence processing and syntactic comprehension in schizophrenia. *Neuroreport, 11,* 2145–2149.

Ljungberg, T., Apicella, P., & Schultz, W. (1991). Responses of monkey midbrain dopamine neurons during delayed alternation performance. *Brain Research, 567,* 337–341.

Ljungberg, T., Apicella, P., & Schultz, W. (1992). Responses of monkey dopamine neurons during learning of behavioral reactions. *Journal of Neurophysiology, 67,* 145–163.

Lukosevicius, M., & Jaeger, H. (2009). Reservoir computing approaches to recurrent neural network training. *Computer Science Review, 3,* 127–149.

Maass, W., Natschlager, T., & Markram, H. (2002). Real-time computing without stable states: A new framework for neural computation based on perturbations. *Neural Computation, 14,* 2531–2560.

Marcus, G. F., Vijayan, S., Bandi Rao, S., & Vishton, P. M. (1999). Rule learning by seven-month-old infants. *Science, 283,* 77–80.

Markov, N., Misery, P., Falchier, A., Lamy, C., Vezoli, J., Quilodran, R., et al. (2011). Weight consistency specifies regularities of macaque cortical networks. *Cerebral Cortex, 21,* 1254–1272.

Markov, N. T., Ercsey-Ravasz, M., Van Essen, D. C., Knoblauch, K., Toroczkai, Z., & Kennedy, H. (2013). Cortical high-density counterstream architectures. *Science, 342,* 1238406.

Mays, L. E., & Sparks, D. L. (1980). Dissociation of visual and saccade-related responses in superior colliculus neurons. *Journal of Neurophysiology, 43,* 207–232.

McHaffie, J. G., Stanford, T. R., Stein, B. E., Coizet, V., & Redgrave, P. (2005). Subcortical loops through the basal ganglia. *Trends in Neurosciences, 28*, 401–407.

Munoz, D. P., Pelisson, D., & Guitton, D. (1991). Movement of neural activity on the superior colliculus motor map during gaze shifts. *Science, 251*, 1358–1360.

Nakahara, H., Doya, K., & Hikosaka, O. (2001). Parallel cortico-basal ganglia mechanisms for acquisition and execution of visuomotor sequences—A computational approach. *Journal of Cognitive Neuroscience, 13*, 626–647.

Nakamura, K., & Hikosaka, O. (2006). Role of dopamine in the primate caudate nucleus in reward modulation of saccades. *Journal of Neuroscience, 26*, 5360–5369.

Nazzi, T., Bertoncini, J., & Mehler, J. (1998). Language discrimination by newborns: Toward an understanding of the role of rhythm. *Journal of Experimental Psychology: Human Perception and Performance, 24*, 756–766.

Nikolic, D., Hausler, S., Singer, W., & Maass, W. (2009). Distributed fading memory for stimulus properties in the primary visual cortex. *PLoS Biology, 7*, e1000260.

O'Reilly, R. C., & Frank, M. J. (2006). Making working memory work: A computational model of learning in the prefrontal cortex and basal ganglia. *Neural Computation, 18*, 283–328.

Prescott, T. J., Montes Gonzalez, F. M., Gurney, K., Humphries, M. D., & Redgrave, P. (2006). A robot model of the basal ganglia: Behavior and intrinsic processing. *Neural Networks, 19*, 31–61.

Procyk, E., & Goldman-Rakic, P. S. (2006). Modulation of dorsolateral prefrontal delay activity during self-organized behavior. *Journal of Neuroscience, 26*, 11313–11323.

Quilodran, R., Rothe, M., & Procyk, E. (2008). Behavioral shifts and action valuation in the anterior cingulate cortex. *Neuron, 57*, 314–325.

Reading, P. J., Dunnett, S. B., & Robbins, T. W. (1991). Dissociable roles of the ventral, medial and lateral striatum on the acquisition and performance of a complex visual stimulus-response habit. *Behavioural Brain Research, 45*, 147–161.

Rigotti, M., Barak, O., Warden, M. R., Wang, X.-J., Daw, N. D., Miller, E. K., & Fusi, S. (2013). The importance of mixed selectivity in complex cognitive tasks. *Nature, 497*, 585–590.

Rigotti, M., Rubin, D. B. D., Wang, X.-J., & Fusi, S. (2010). Internal representation of task rules by recurrent dynamics: The importance of the diversity of neural responses. *Frontiers in Computational Neuroscience, 4*, 24.

Robbins, T. W., Giardini, V., Jones, G. H., Reading, P., & Sahakian, B. J. (1990). Effects of dopamine depletion from the caudate-putamen and nucleus accumbens septi on the acquisition and performance of a conditional discrimination task. *Behavioural Brain Research, 38*, 243–261.

Saffran, J. R., Aslin, R. N., & Newport, E. L. (1996). Statistical learning by 8-month-old infants. *Science, 274*, 1926–1928.

Sallet, J., Quilodran, R., Rothe, M., Vezoli, J., Joseph, J. P., & Procyk, E. (2007). Expectations, gains, and losses in the anterior cingulate cortex. *Cognitive, Affective & Behavioral Neuroscience, 7*, 327–336.

Schultz, W. (1986). Responses of midbrain dopamine neurons to behavioral trigger stimuli in the monkey. *Journal of Neurophysiology, 56*, 1439–1461.

Schultz, W., Apicella, P., & Ljungberg, T. (1993). Responses of monkey dopamine neurons to reward and conditioned stimuli during successive steps of learning a delayed response task. *Journal of Neuroscience, 13*, 900–913.

Schultz, W., Dayan, P., & Montague, P. R. (1997). A neural substrate of prediction and reward. *Science, 275*, 1593–1599.

Schultz, W., & Dickinson, A. (2000). Neuronal coding of prediction errors. *Annual Review of Neuroscience, 23*, 473–500.

Selemon, L. D., & Goldman-Rakic, P. S. (1985). Longitudinal topography and interdigitation of corticostriatal projections in the rhesus monkey. *Journal of Neuroscience, 5*, 776–794.

Selemon, L. D., & Goldman-Rakic, P. S. (1988). Common cortical and subcortical targets of the dorsolateral prefrontal and posterior parietal cortices in the rhesus monkey: Evidence for a distributed neural network subserving spatially guided behavior. *Journal of Neuroscience, 8*, 4049–4068.

Shenhav, A., Botvinick, M. M., & Cohen, J. D. (2013). The expected value of control: An integrative theory of anterior cingulate cortex function. *Neuron, 79*, 217–240.

Shin, J. C., & Ivry, R. B. (2003). Spatial and temporal sequence learning in patients with Parkinson's disease or cerebellar lesions. *Journal of Cognitive Neuroscience, 15*, 1232–1243.

Stanton, G. B., Goldberg, M. E., & Bruce, C. J. (1988). Frontal eye field efferents in the macaque monkey: I. Subcortical pathways and topography of striatal and thalamic terminal fields. *Journal of Comparative Neurology, 271*, 473–492.

Suri, R. E., Bargas, J., & Arbib, M. A. (2001). Modeling functions of striatal dopamine modulation in learning and planning. *Neuroscience, 103*, 65–85.

Suri, R. E., & Schultz, W. (1998). Learning of sequential movements by neural network model with dopamine-like reinforcement signal. *Experimental Brain Research, 121*, 350–354.

Suri, R. E., & Schultz, W. (1999). A neural network model with dopamine-like reinforcement signal that learns a spatial delayed response task. *Neuroscience, 91*, 871–890.

Suri, R. E., & Schultz, W. (2001). Temporal difference model reproduces anticipatory neural activity. *Neural Computation, 13*, 841–862.

Sutton, R. S. (1988). Learning to predict by the methods of temporal differences. *Machine Learning, 3*, 9–44.

Sutton, R. S., & Barto, A. G. (1998). *Reinforcement Learning: An Introduction.* Cambridge, MA: MIT Press.

Tong, M. H., Bicket, A. D., Christiansen, E. M., & Cottrell, G. W. (2007). Learning grammatical structure with echo state networks. *Neural Networks, 20*, 9.

Verstraeten, D., Schrauwen, B., D'Haene, M., & Stroobandt, D. (2007). An experimental unification of reservoir computing methods. *Neural Networks, 20*, 391–403.

Weitzenfeld, A., Arbib, M. A., & Alexander, A. (2002). *The Neural Simulation Language: A System for Brain Modeling.* Cambridge, MA: MIT Press.

Yeterian, E., & Pandya, D. (1998). Corticostriatal connections of the superior temporal region in rhesus monkeys. *Journal of Comparative Neurology, 399*, 384–402.

Yin, H. H., & Knowlton, B. J. (2006). The role of the basal ganglia in habit formation. *Nature Reviews. Neuroscience, 7*, 464–476.

23 Brain Diseases

MOHAMED A. SHERIF AND WILLIAM W. LYTTON

One major approach to computational neuroscience utilizes multiscale modeling, covering molecular, cellular, and network phenomena as well as higher levels up to cognition and behavior. With growing appreciation of the enormous complexity of genome, proteome, and other -omes, multiscale modeling is becoming increasingly important in clinical medicine. These complexities are even more apparent when confronting brain diseases, dysfunctions of the most complex organ, an organ with manifestations that range up into the realms of cognition and behavior.

Different tools, used to explore different aspects of the brain, obtain data at different scales. Some of these are used only in experimental animals, such as visualization of intracellular calcium activation waves. Others can be used both experimentally and clinically in humans, such as electrophysiological recordings and functional or structural neuroimaging.

Additionally, different disease phenomena cover a range of temporal and spatial scales. This is true for both pathogenesis and pathophysiology (describing the mechanisms of how disease occurs) and treatment, whether it is pharmacological, behavioral, or surgical. Disease can start at a molecular or cellular level and then produce effects all the way up across different temporal and spatial scales, till reaching the systems level of the whole organism. Alternatively, disease may start at a high level (e.g., traumatic brain injury and stroke) and produce alterations at cellular and molecular levels. Similarly, bottom-up and top-down approaches to treatment may be used simultaneously. For example, drug therapy for psychiatric diseases is applied to molecular targets in concert with behavioral or cognitive therapy that is aimed at altering the function of brain maps and brain circuitry.

We will discuss four brain disorders. We start with epilepsy, arguably the most dynamically simple of brain diseases. We will discuss how computational neuroscience has aided in exploring the predictability of occurrence of seizures, and how it helped in investigating factors related to network excitability that may predispose someone to seizures. We then move to Parkinson's disease (PD), a neurodegenerative disease manifested by tremor and slowness of movement. Computational modeling of the basal ganglia circuits in the brain has helped to point out why many PD patients suffer from problems with learning new tasks, even after being treated. The next disease we explore is stroke, looking at models of the changes that occur following these major brain ablations. We look at how the death of neurons shifts the dynamics of cortical neurons, which helps reduce loss of function. Additionally, models of the effect of physiotherapy on patients with stroke made interesting predictions with implications for patterning physical therapy after stroke. Finally, we discuss computational modeling of schizophrenia. Further understanding of this devastating illness may help us shed light on the enigma of thought disorders and through this, better understand the processes of normal cognition.

For the reasons given above, this chapter will focus on multiscale modeling. However, we will say relatively little about high-level approaches that describe disease in terms of chaotic dynamics (Glass and Mackey, 1988). Although these and other top-down approaches have provided a number of important insights, we have chosen to focus on multiscale modeling because of its potential clinical applicability through new approaches to developing pharmacological treatments.

23.1 *Epilepsy*

23.1.1 PATHOPHYSIOLOGY OF SEIZURES Seizures are episodes of disturbed brain function due to abnormal prolonged neuronal firing. Seizures can cause simple changes in attention and behavior or a complete disruption of brain activity with loss of consciousness and abnormal movements of the limbs. The location of a localized (focal) seizure determines the brain function that is affected. For example, if the seizure location involves the motor cortex, then the disturbance of function will be in the form of involuntary jerking movements of the body part innervated by that area of the cortex. If the seizure location involves the sensory cortex, then it will present as abnormal sensations in the body region represented in that area. Generalized seizures involve the whole brain and result in loss of consciousness. When seizures occur repeatedly without provocation, the disease is called epilepsy. Usually, the abnormal firing during seizures will stop after a period of time, and so the symptoms will stop, too. However, if the symptoms persisted for a long period of time, it is called status epilepticus.

A common seizure presentation is the generalized tonic–clonic seizure. Tonic refers to sustained muscular contraction while clonic refers to alternating contraction and relaxation of muscles. During a generalized tonic–clonic seizure, the patient suddenly loses consciousness, followed by a generalized contraction of all body musculature for around 10–20 seconds. The patient's body becomes rigid, sometimes with arching of the back and neck (tonic phase). This is followed by alternating contraction and relaxation of body musculature (clonic phase), resulting in rhythmic recurrent movements of the limbs and body, known as convulsions. Convulsion frequency then decreases gradually until the seizure ends over a period of 1 to 2 minutes. If the seizure instead continues over many minutes, this is status epilepticus, and it is a life-threatening condition.

Multiple elements at different scales are involved in the pathophysiology of epilepsy. The occurrence of seizures is related not only to increased excitability of individual neurons but also to altered neuronal dynamics within the network. Multiscale modeling provides a method to meet the challenge of understanding how these many elements interact to produce the disorder, assimilating different areas of neuroscience, ranging from neurogenetics and ion channel dynamics to brain imaging, electroencephalography (EEG), and behavioral assessments. Because the presenting "behavior" (a convulsion) is extremely simple relative to motivated behaviors, the correlation between basic science and clinical phenomenology is closer in epilepsy than in other brain disorders. There has therefore been more success in linking molecular and cellular pathophysiology to clinical symptoms and signs in this disease. Modeling allows us to explore the effect of changes taking place along one or more of these space-time scales on the generation of network spontaneous repetitive activity.

The pathology responsible for the occurrence of seizures is rarely limited to one structure. The pathology can be in any structure ranging from the ionic channels within the neuronal membrane, going all the way up to the connectivity within a neuronal network or the connectivity between networks. For example, a group of diseases known as channelopathies (diseases of the channels) consists of mutations affecting the structure of the ion channels. This results in changes in their conductance, either increasing or decreasing it. The effect of such changes on neuronal excitability depends on the role of the particular channel in membrane depolarization/hyperpolarization. Another example is the effect of chronic alcohol (ethanol) use on $GABA_A$ receptors. Alcohol binds to the $GABA_A$ receptor complex, which is a chloride channel, augmenting its function and increasing the influx of chloride ions, resulting in inhibition of neuronal firing. Chronic alcohol exposure reduces $GABA_A$-receptor-mediated inhibition (Kang et al., 1996). Therefore, abrupt cessation of alcohol intake following chronic alcohol use can result in neuronal hyperexcitability with resultant seizures.

Abnormalities of network architecture have also been reported in epilepsy. Some cases of seizures can't be controlled with medications. In these cases, surgery is done to attempt to remove brain tissue that has been identified as the focus generating the seizure. Examination of the brain tissue removed reveals abnormalities of network architecture in humans. Similar abnormalities are also seen in animal models of epilepsy. These changes include sprouting of new fibers to make atypical connections, as well as the presence of cells in abnormal locations (Sutula et al., 1989; Parent et al., 1997).

23.1.2 Stochastic and Deterministic Modeling

Both stochastic and deterministic modeling have their role in investigating questions related to seizures and epilepsy. Underlying randomness may be real, as in the case of random opening and closing of individual ion channels. However, stochastic modeling is also often used when a system is too complex to consider modeling the underlying details, or when a system is subject to vagaries that cannot reasonably be modeled. This randomness is only apparent, due to incomplete information or due to sensitivity to initial conditions as in chaotic systems. Stochastic models are therefore helpful even in deterministic systems, capturing irreducible complexity that may determine specific aspects of epilepsy, such as the timing of seizures. For example, seizures are known to be more likely to occur after a night's sleep has been missed. Since we do not know exactly how this missed night's sleep is affecting the brain, it is not feasible to model all the details that could be affected. Such phenomena could be modeled using stochastic models.

The enormous dynamical complexity of the brain, like its structural complexity, can be considered as a set of subsystems. The existence of widespread distinct oscillatory frequencies in the EEG (see chapter 4, "Neural Rhythms") suggests that dynamical subsystems are detectable even at the highest scale. Hence, it has been proposed that the enormous dimensionality of the brain produces trajectories that lie in much lower-dimensional subspaces that can then be modeled by equivalent low-dimensional dynamical systems. Following this hypothesis, low-dimensional mean-field models (also called neuron population models or lumped models) have been widely utilized to simulate seizures.

These models look at the overall dynamics of a large ensemble of neurons—the lump. Depending on the model, this neural lump could be interpreted to be a cortical minicolumn, a column, a Brodmann area, a thalamic nucleus, and so forth. Alternatively, a lump could be a pool of similar neurons, typically an inhibitory pool coupled with an excitatory pool. These models have an advantage of simplicity compared to detailed neural-level modeling, making it is easier to simulate multiple brain regions.

Seizure spread is a clinically significant phenomenon, affected by multiple factors. Seizures can spread in a series of stages: a focal seizure with secondary generalization. For example, in temporal lobe epilepsy, a patient may experience "auras," unusual feelings such as déjà vu (a sense of familiarity to an unfamiliar situation), jamais vu (the opposite of déjà vu, a sense of unfamiliarity to familiar situations), or fear, at the beginning of abnormal neural firing in the temporal lobe. These are then followed by loss of consciousness as the wave of abnormal neuronal firing spreads broadly into the neocortex. Alternatively, a seizure can appear to arise suddenly in all areas, a primary generalized seizure. For example, in absence epilepsy, a classical generalized epilepsy usually seen in children, the child suddenly stops whatever he or she is doing, stares ahead, and becomes unresponsive for about a minute. Although seizure spread in generalized epilepsy occurs so quickly as to appear to be simultaneous across brain areas, there is evidence that generalized epilepsy instead involves very rapid spread.

23.1.3 Modeling Neuronal Populations

To investigate some of the mechanisms playing a role in the spread of a seizure, Larter et al. modeled the CA3 region of the hippocampus (Larter et al., 1999). The model consisted of connected modules, with each module consisting of a lumped excitatory population and a lumped inhibitory population. The excitatory population activation was described by differential equations which depended on calcium, potassium, and leak currents:

$$\frac{dV_i}{dt} = -g_{Ca}m_\infty(V_i - 1) - g_K W_i(V_i - V_i^K) - g_L(V_i - V^L) + 1 - \alpha_{inh} Z_i \quad (23.1)$$

where V is the voltage of the lumped pyramidal cell population and g_{Ca}, g_K, and g_L are the total conductances for calcium, potassium, and leak channels, respectively. W_i is the fraction of open potassium channels in the lumped pyramidal neurons. m_∞ is the fraction of open calcium channels and is voltage dependent. V_i^K is the Nernst equilibrium for potassium at node i while V^L is for leak potential. α_{inh} is the inhibitory synaptic strength on the lumped pyramidal neurons, while Z_i is the voltage of lumped inhibitory population at module i. The lumped inhibitory population was modeled similarly.

Activity spread in the model took place at different timescales based on different mechanisms. On a fast timescale, the activity of a module projected to other modules. At a slower timescale, activity results in increased extracellular potassium level, which then diffused to nearby neuronal groups. High extracellular potassium depolarizes the neurons, bringing them closer to threshold for firing, hence more excitable. The Nernst equation describes how different ion concentrations across the cell membrane creates the "batteries" that drive membrane electricity. Depolarization of the membrane with high potassium is then due to a shift of the potassium Nernst potential. To take that into account, V_i^K, in the second term of equation 23.1, was calculated to depend on the activity in the neighboring locations $\overline{V_j}$:

$$V_i^K = \left(\sum_{j=1}^{6} \frac{\overline{V_j}}{6}\right) - \frac{1}{2} \quad (23.2)$$

where

$$\overline{V_j} = \frac{1}{T}\int_{t_{int}/6}^{t_{int}} V_j(t)dt \quad (23.3)$$

The t_{int} parameter would determine the duration before the activity is transmitted from one group to another. A short t_{int} allowed the immediate synaptic effect to be more influential. A long t_{int} emphasized the effect of diffusion.

Exploring the range of values of t_{int} revealed two points of bifurcation (see chapter 3, "Neurons and Neuronal Networks as Dynamical Systems"). Below a certain value, all of the modules in the network would fire together, producing a seizure. Above that value, the network entered a periodic mode, with the modules firing out of phase. This mode was spatially dependent, such that closer modules fired at closer phase than those farther away. At the second point of bifurcation, the network entered a mode of random firing. The authors also explored the effects of inhibition on the firing synchronicity by a variable which governed the ratio of current going into excitatory and inhibitory modules. Lowering the input to inhibitory interneurons relative to excitatory interneurons produced less variability and so made the network more vulnerable to seizure-like activity.

Another factor which influences the spread of seizures is network connectivity. A graph theoretic approach can be used to describe how neurons in a network (or modules in a network of populations) are connected. In graph theory, random networks and small-world networks are the most heavily studied. In a random network, each node has the same probability of being connected to any other node. The notion of a small-world network comes from social networking and other applications where there are special nodes, called hubs, which have more connections to other nodes. In the case of social networks, these hubs are people with lots of friends. In the case of the airport system, these hubs are locations with flights to many other locations. In either case, the hubs provide short paths that connect any two nodes (two people or two airports) in the system.

Another arrangement is hierarchical connectivity. In this arrangement there are clusters of hubs: a cluster is a group of hubs which are heavily interconnected.

In the cerebral cortex, neurons self-organize via synaptic plasticity, producing network organizations that are far from random. However, they can be generally described by graph theory. Cortex is suborganized into subnetworks or modules, with more connections within the module (intramodular) than between the modules (intermodular), suggestive of hierarchical connectivity. The connections at various levels are described in terms of *connectomes* at various levels, within the column, between columns (within area), between areas (within system), and between systems (Sporns et al., 2005; Yoshimura et al., 2005).

Kaiser et al. (2007) compared seizure spread in three types of connectivity: random, small world, and hierarchical. The three types of networks were all of the same size: 1,000 vertices (nodes) and 12,000 edges (connections). Spread of activity was measured as the time to full activation of the whole network.

Spread across the random network topography was the most rapid. Network activity persisted for a longer duration in small-world and hierarchical networks. Compared to small-world networks, networks with hierarchical topography showed more limited activity spread with longer duration. In the case of hierarchical networks, the connections between clusters provided checkpoints of activation before activity could spread across the rest of the network. This suggested that the pace of activity spread in cortex might be limited by the delay in activation of individual columns or local areas before activity can spread further. In this model, there were no inhibitory connections, which shows that the topography of network can limit the spread of seizures even in the absence of inhibitory connections. The visual system has long been considered to be hierarchical, but the macaque data on which this interpretation has been based has been called into question lately by studies that identify a denser pattern of connectivity. Meanwhile human imaging data has suggested to some groups that there may be considerable small-world connectivity among some areas of the brain. Therefore, there remains considerable controversy as to the dominant pattern of connectivity of the connectome.

While the prior study emphasized seizure spread in cortex, other studies have looked at the dynamics of connections between cortex and subcortical structures such as basal ganglia and thalamus. Suffczynski et al. (2004) identified two attractors in a lumped study of interacting cortical and thalamic populations. Both the cortical and thalamic modules consisted of interconnected excitatory and inhibitory populations. Projections between cortex and thalamus were excitatory and projected onto both populations in the other area. The behavior of the model was assessed by evaluating mean activation of the pyramidal (excitatory) cell population of the cortical module.

The resulting system dynamics could maintain physiological activity corresponding to a low-amplitude attractor. Alternatively, a pathological pattern of ictal (seizure-like) activity was produced when dynamics switched over to a high-amplitude attractor. The physiological pattern demonstrated spindle-like oscillations of ~11 Hz frequency, which waxed and waned in amplitude. The pathological pattern showed high-amplitude oscillations of frequency ~9 Hz, which resembled seizure activity (figure 23.1). Transition from the normal nonseizure state to ictal activity occurred when the barrier between these two attractors was overcome. This occurred as the threshold for inhibitory synaptic activation was reached, augmenting oscillation amplitudes. Gaussian white noise introduced into the system could produce transitions from the physiological to the pathological attractor and back. In the absence of noise, the deterministic system remained in the physiological state.

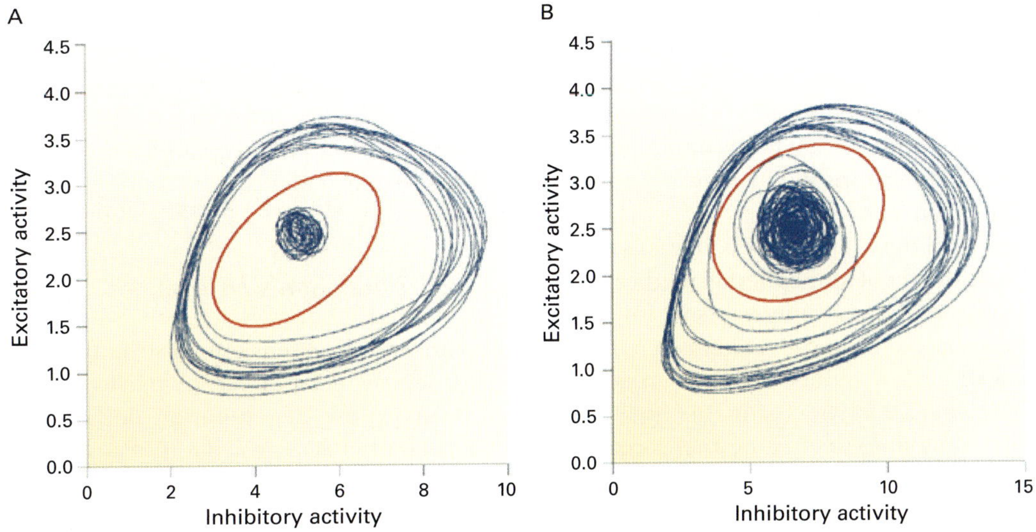

FIGURE 23.1 A lumped model of absence epilepsy. (A) Under deterministic conditions, trajectories (blue line) are stable within their own attractors: the inner attractor represents physiological activity while the outer attractor represents epileptic activity. The red line is the barrier (separatrix) between the attractors. (B) Trajectory jumps repeatedly between attractors as transitions from one attractor to the other are driven by noise. From Lopes Da Silva et al. (2003).

The importance of noise in switching between attractors in this type of model suggests why there might be limitations in the predictability of seizures. Two scenarios suggest themselves. If dynamical changes take place gradually—a physiological attractor evolving gradually into an ictal attractor—then seizures may be predictable. Such a gradual change is hypothesized to occur in temporal lobe epilepsy. By contrast, the dynamics of figure 23.1 shows a *jump* from one attractor to another. Here, the two attractors (ictal and normal) coexist, but are separated from each other by a barrier which is lower in patients suffering from epilepsy, making it easier for the brain state to jump from one attractor to the other. This would explain cases where the onset of seizures can happen very abruptly, as in the case of absence seizures. Additionally, since such jumps happen randomly, it becomes impossible to predict seizures far in advance. Note that in this case, the dynamics of the energy barrier will also play a role and could provide the possibility of prediction of periods of greater susceptibility.

23.1.4 NEURAL-LEVEL MODELING Regardless of the level of detail chosen, all models necessarily contain much simplification. Contrasting with the lumped models described above, more detailed neural-level models incorporate individual neurons but still must simplify in many different ways. For example, molecular interactions are usually not directly simulated. Although these can be helpful to provide direct application to neuropharmacology, simulation using simplified neurons can also be helpful in this respect.

Neural-level models may have both deterministic and stochastic aspects. Local neural activity is represented by differential equations, while stochastic driving is used to represent the variations in external activity patterns from other brain areas. Neural-level models have been used to compare the relative effects of intrinsic neuronal excitability and synaptic connectivity in the genesis of epileptiform activity. In these models, simplification occurs not only from the omission of molecular details but also from simplifications of connectivity due to the fact that there is little that is known about the wiring diagram (the microconnectome) among individual neurons.

As an example, we present here a simplified network consisting of excitatory, inhibitory, and driver neurons which was used to investigate the transition from normal activity to a clonic seizure (Lytton and Omurtag, 2007). The *driver neurons* drove the network through their spontaneous firing. Drivers were connected to both excitatory and inhibitory neurons via different types of excitatory and inhibitory synaptic connections. Individual neurons in this model are event-driven firing units, which fire at a point when the voltage state variable is driven above threshold. Intrinsic properties including the threshold, refractory period, and afterhyperpolarization potential were added to these units through a series of rules.

Two patterns of firing were noted in the network: low-amplitude activity, and prolonged continuous firing, suggestive of tonic epileptiform activity. It was observed that altering the firing timing of the driving neurons, without changing any of the other parameters, resulted in switching between these two patterns. Another finding of interest was that increasing the excitability of individual neurons did not necessarily result in increasing the excitability of the whole network. This was because the network excitability depended not only on the individual neurons but also on the dynamics between the neurons.

Because neural-level details were used, it was possible to grossly model the effect of anti-epileptic medications. One hypothesized molecular mechanism for anti-epileptic medications is blocking voltage-dependent sodium channels which will reduce the tendency of the neuron to spike (Xie et al., 1995). Using the model, this effect was simulated by simply reducing the excitability of the individual cells by reducing the number of spikes per burst for each unit. This resulted in reduction of the duration of each activation at the neural level but with no effect on the background activity at the population level. Note the similarity of pattern between the two field potentials (blue, untreated; red, treated) in figure 23.2. Note also that the tonic period of pathological activity (arrow) is eliminated in the treated model. Therefore, the tonic seizure was removed by treatment without affecting baseline activity.

When a clinical phenomenon is controlled by a large number of variables, as in the case of epilepsy,

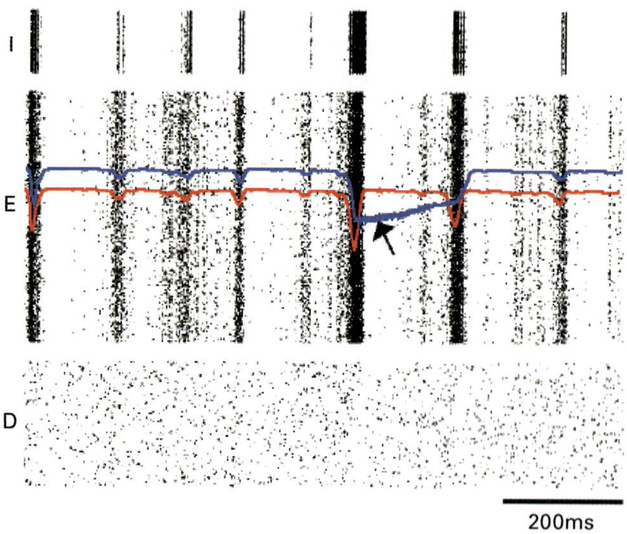

FIGURE 23.2 Effects of anti-epileptic medication. Lines show field potential generated by the activity of a set of expressor neurons (the excitatory neurons) (blue, without medication; red, with medication). Raster plot for the firing of different neuronal subpopulations (I, inhibitory; E, expressor; D, drive) with medication. Each point represents the firing of one cell in the network. The blue (untreated) trace dips at the time of the seizure (arrow); anti-epileptic medication prevented this from happening, without affecting baseline dynamics of the network. Modified from Lytton and Omurtag (2007).

computational modeling will aid in our understanding of the interacting and emergent effects of the different state variables at different scales. This would point toward influences that play a bigger role in the phenomenon, enabling research on therapies to focus on these particular state variables at these particular scales.

23.2 Parkinson's Disease

Parkinson's disease is one of several neurodegenerative disorders which affect a particular population of neurons whose degeneration deprives the rest of the brain of a needed projection or needed neuromodulator. In the case of PD, this population is the neurons of the substantia nigra pars compacta (SNc) in the brain stem. These neurons release the neurotransmitter dopamine and project strongly to the basal ganglia. Their degeneration results in reduction of the dopamine levels in the basal ganglia, producing a group of symptoms which are initially primarily motoric—movement problems. The disease then progresses to affect non-motoric brain functions as well. The major symptoms include difficulty initiating movements, slowed movement (bradykinesia) with concomitant slowed thought (bradyphrenia), increased muscle tone around the joints (rigidity), resting tremors, and difficulty coordinating bodily movements (postural instability).

The initial treatment of PD involves exogenous replacement of missing dopamine. This can be done by either providing a dopamine precursor (L-dopa) which gets converted into dopamine in the brain, or providing agonist drugs to activate dopamine receptors. Initially, this therapy produces improvement of several aspects of the disease, for example, reduced rigidity, faster spontaneous movements, and reduced tremor. Interestingly, there is sometimes an initial dramatic response to replacement, with the patient reporting a seemingly miraculous return to normal function. This is impressive since the original phasic release of the dopamine has been replaced by a continuous tonic dopamine "bath." Unfortunately, later on, the symptoms will start appearing again despite the treatment and then may be more severe than they otherwise would have been, perhaps because of downregulation of dopamine receptors (Hermanowicz, 2007). Such downregulation would be an example of homeostasis—maintenance of activity equilibrium in the face of change. In this case the abnormally high, continuous level of dopamine would reduce the sensitivity of the neurons to dopamine by reducing (downregulating) the number of dopamine receptors.

23.2.1 Anatomy and Function of Basal Ganglia

The basal ganglia consists of different nuclei and are divided anatomically into striatum, the major input portal to the basal ganglia, and globus pallidus, the output route. (See figure 23.3, and also chapter 22, "Integrative Functions of the Corticostriatal System.") The striatum receives inputs from multiple cortical areas and is further subdivided into caudate nucleus and putamen. In primates, the caudate nucleus receives cortical input from the prefrontal cortex while the putamen receives inputs from the somatosensory and motor cortices. The main cells in the striatum are GABAergic neurons—medium spiny projection neurons—that project to the globus pallidus interna (GPi) and substantia nigra pars reticulata (SNr) but not to SNc, the site of dopamine production. The GPi and SNr are main output pathways and also have GABAergic projection neurons. These areas convey signals from the striatum to the thalamus and thence back to cortex. This circuit therefore provides a cortico–striato–pallido–thalamo–cortical loop. Note that the primary projections pathways within and from striatum are GABAergic and therefore inhibitory; this differs from cortex and thalamus where the major projections are glutamatergic, excitatory projections. SNc and the neighboring ventral tegmental area (VTA) are the dopaminergic areas that provide projections to striatum (Gerfen and Bolam, 2010). An additional level of circuit complexity arises due to the existence of at least two major classes of striatal cells, D1 striatal cells and D2 striatal cells, which have two different types of dopamine receptors (D1 and D2) with very different cellular effects (Richfield et al., 1989).

The basal ganglia circuitry has been conceptualized as instantiating Go and No-go circuits.

GPi/SNr inhibits thalamus. Therefore, inhibition of GPi/SNr results in disinhibition of thalamus (Go signal). Stimulation of GPi/SNr results in inhibition of thalamus (No-go signal) (Chevalier and Deniau, 1990). So stimulation of Dl cells in striatum (which inhibits GPi/SNr; direct pathway) results in a Go signal for a *task* to be performed. Stimulation of D2 cells in the striatum (which disinhibits GPi/SNr through globus pallidus externa [GPe]; indirect pathway) results in a No-go signal: the *task* is less likely to be performed. Presumably, the No-go signal is not guaranteed to always override the Go signal generated. Since D2 receptors have higher affinity for dopamine than Dl receptors (Richfield et al., 1989), dopamine depletion in PD results in D2 receptors' being more active, which causes the indirect pathway (responsible for the No-go signal) to be more active.

The basal ganglia is also involved in reinforcement learning (RL). This is another place where modeling has been particularly helpful (see chapter 12 for an extended exposition) since learning theory is an important focus of computational neuroscience. RL takes place through variation in phasic release of dopamine. Phasic bursts of dopamine are related to receiving unexpected reward (positive feedback, which results in reinforcing the behavior). In contrast, phasic dips of dopamine are related to not receiving an expected reward (negative feedback, resulting in reduction of behavior). Bursts and dips follow positive and negative feedback based on prediction error. This is part of the reason why patients suffering from PD suffer from

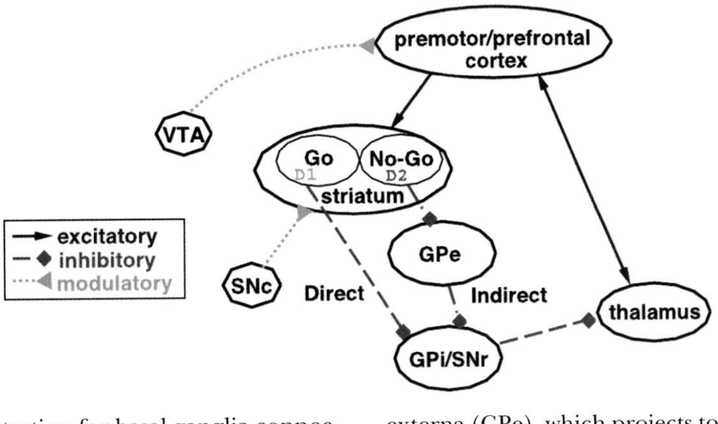

Figure 23.3 Schematic illustration for basal ganglia connections, showing the cortico–striato–pallido–thalamo–cortical loop. Premotor/prefrontal cortex projects to the striatum. D1 cells send inhibitory projections to globus pallidus interna (GPi)/substantia nigra pars reticulata (SNr) (direct pathway), while D2 cells send inhibitory projections to globus pallidus externa (GPe), which projects to GPi/SNr (indirect pathway). GPi/SNr then projects to the thalamus, which projects back to the cortex. Substantia nigra pars compacta (SNc) and ventral tegmental area (VTA) have modulatory dopaminergic projections to striatum; VTA also modulates cortex directly. From Frank (2005).

learning deficits and have impairment in tasks related to trial-by-trial error feedback.

Another way of viewing the basal ganglia is as a selection circuit and a control circuit, corresponding generally to the Go/No-go dichotomy mentioned above. GPi/SNr tonically inhibits thalamus. When a subset of GPi/SNr neurons themselves become inhibited by focused input from D1 striatal cells, activities mediated by that portion of the thalamus are allowed (figure 23.4). GPi/SNr neurons are under continuous stimulation by diffuse excitation from subthalamic nucleus (STN), and so those that are not inhibited from striatum continue to provide inhibition to these other portions of thalamus. This produces contrast between selected and non-selected neurons in the thalamus. In this way, focal activation of a group of D1 striatal cells by cortex selects certain circuits for ongoing activation. Meanwhile, inhibitory signals to GPi/SNr arise from the GPe, which receives inhibitory signals from D2 striatal cells (control circuit). This is, of course, a highly simplified model (Bevan et al., 1998; Nambu et al., 2002).

Given the complexity of basal ganglia circuitry, modeling approaches have been particularly valuable. As in other systems, we look for a relatively clear-cut task that requires the use of this system. In the case of basal ganglia, one such task is the initiation of Go and No-go signals. After initiation, these signals are sent to various brain circuits responsible for different brain functions: walk or not, switch from one task to another, think about one thing or another. This task also makes sense in terms of the pathology of PD, in that these patients

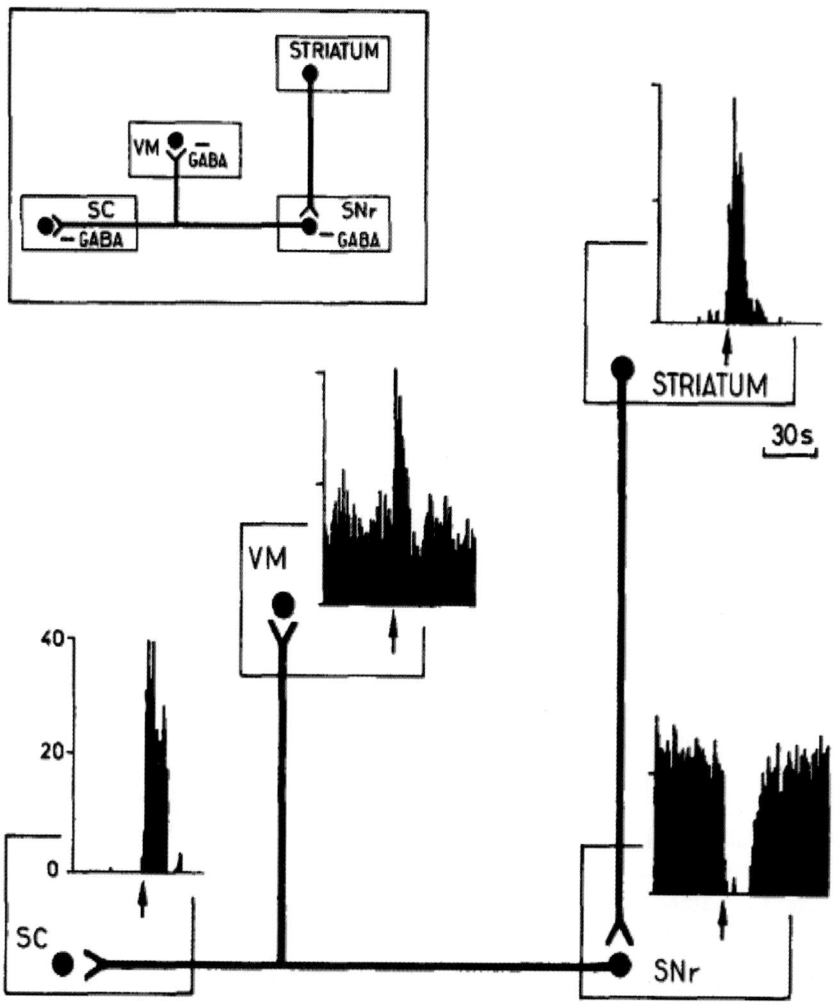

FIGURE 23.4 Disinhibition of thalamus following stimulation of striatum (time on *x*-axis, spikes per second on *y*-axis). In the upper right plot, the arrow shows applying glutamate to stimulate striatum. The arrows correspond to the same point in time in the four plots. The increase in firing in the striatum corresponds to reduction of firing in the substantia nigra pars reticulata (SNr), which results in increase in firing in the thalamus (VM, ventromedial nucleus of thalamus) and superior colliculus (SC) through disinhibition. This increase in firing would send an activation signal to the cortex. From Chevalier and Deniau (1990).

find it difficult to initiate movements or to abruptly stop (festination). It could also help in explaining the perseveration sometimes observed in these patients when they keep repeating certain phrases.

Similar to many clinical phenomena, PD crosses multiple spatial and temporal scales. As a disease process, PD starts at the cellular and molecular levels with degeneration of the neurons in the SNc, with the resultant reduction of dopamine levels at the basal ganglia. However, the symptoms appear at the behavioral level, in the form of problems in motoric functions and in learning. Treatment takes place at a molecular level, the level of dopamine receptors, while the improvement of symptoms takes place at a behavioral level. Computational neuroscience is needed to bridge these spatial and temporal scales.

23.2.2 Modeling the Role of Dopamine in Selection We give an example of an effort to explore the selection function of the basal ganglia circuitry and the role played by dopamine (Humphries et al., 2006). This study modeled five components of the basal ganglia circuitry: striatal D1 cells, striatal D2 cells, STN, globus pallidus (which represented GPe), and SNr (representing basal ganglia output). The leaky integrate-and-fire neurons within each of these structures were divided into three groups, corresponding to three different actions. Each group is connected to a corresponding group in each of the components, maintaining the topographic arrangement of fibers that is seen in the parallel loops that exist between striatum, GPe, GPi/SNr, thalamus, and cortex in the cortico–striato–pallido–thalamo–cortical circuit (figure 23.5)

Cortical input to the striatum (D1 and D2 cell populations) and STN was simulated using spike trains of different frequencies. This would correspond to saliency of a stimulus in real life, with the higher frequency representing a stimulus that has a higher salience. The stimulation protocol that was used to test action selection consisted of an input to group 1 at t = 1 second, followed by an input with a different frequency in group 2 at t = 2.5 seconds. The stimulation was terminated at 5 seconds. According to the action selection prediction, the basal ganglia model should be able to select a group at t = 1, the beginning of the first stimulus. The model should then only switch the selection to the second group, presented at t = 2.5, in cases when the second group is more salient, represented as higher frequency firing in that group. Selection of a group was considered to have taken place when the firing rate of an SNr group dropped below a threshold. This was taken to be the case because, physiologically, reduced firing in one SNr population will disinhibit the corresponding

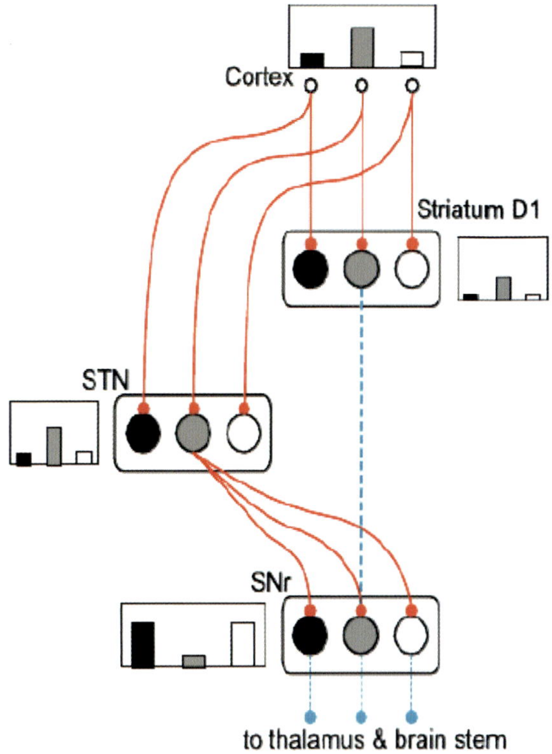

Figure 23.5 Groups of neurons in the selection circuit. The circuit responsible for selection of action in the model consists of the cortex, the D1 cells in striatum, subthalamic nucleus (STN), and substantia nigra pars reticulata (SNr). Each of those structures has three groups of neurons (black, gray, and white) corresponding to three different actions. Each group in each of the structures projects to the corresponding group (only connections of the gray population are shown in details), except for STN which projects diffusely (only the gray projections from STN are shown). Modified from Humphries et al. (2006).

population in thalamus, which will in turn stimulate the corresponding cortical population.

Different levels of simulated tonic dopamine on both D1 and D2 cells were used to simulate healthy basal ganglia circuitry, PD pathology, and PD with excess dopamine during excessive dopaminergic therapy. Dopamine's effect on synapses was modeled as follows:

$$I^S_{i\in\Omega S1} = (I^{ampa} + I^{nmda})(1+\lambda_{D1}), \qquad (23.4)$$

$$I^S_{i\in\Omega S2} = (I^{ampa} + I^{nmda})(1-\lambda_{D2}), \qquad (23.5)$$

where $I^S_{i\in\Omega S1}$ and $I^S_{i\in\Omega S2}$ are the currents in the soma of the i neuron that belongs to D1 ($\Omega S1$) or D2 ($\Omega S2$) cell populations in the striatum. I^{ampa} and I^{nmda} are the currents provided by excitatory synapses, mediated by AMPA and NMDA glutamatergic receptors, respectively. These receptors are involved in the classical driving of synapses, contrasted with neuromodulatory synapses such as those for dopamine, which depend on input from the cortex. λ_{D1} and λ_{D2} are the tonic levels of dopamine on D1 and D2 receptors, respectively, ranging from zero to 1. To simulate physiological amounts of dopamine, both λ_{D1} and λ_{D2} were set to 0.3. To simulate PD, values were set to zero. They were set to 1 for the case of excess dopamine under the effect of excess dopamine therapy. From equations 23.4 and 23.5, this variation will result in changes in the synaptic current. So, in comparison to the physiological state, in the case of PD, the activating synaptic currents from the cortex (carrying the signal), would reduce firing of D1 population and increase firing of D2 population. Both of these changes would disinhibit SNr cells (which receives GABAergic projections from Dl and GPe populations), increasing their firing and so the groups are considered not selected. In the opposite case, with excess dopamine (with λ_{D1} and λ_{D2} set to 1), this will result in increased excitatory synaptic current to the D1 cells and reduction in those currents to the D2 cells. Both these changes will result in reduction of SNr firing (which is considered as selection).

The physiological model was able to exhibit action (group) selection (blue circles in figure 23.6A): the SNr group was selected when the first input was given, and then switched to the other group when the more significant stimulus was introduced (the stimulus with the higher frequency). With low levels of dopamine (simulating PD), neither signal selection nor switching took place (figure 23.6B). The firing rate of the corresponding groups in SNr dropped slightly, but not sufficiently to indicate group selection. With excess dopamine (to simulate what happens with dopamine replacement therapy or dopamine agonists), switching becomes rare as competition between inputs results in selection of *both* first and second groups. With the introduction of the first stimulus, the group is selected. Then with introduction of the second stimulus through a second group, the second group is also selected, with no de-selection of the first one (green diamonds in figure 23.6C).

The above results points to the nonlinear mode of action of dopamine. Reduction of dopamine (as in PD) would result in dysfunction because of limited ability to select responses to stimuli. High levels of dopamine (like what happens with overtreatment using L-dopa or dopamine agonists) would also result in dysfunction because of inability to respond to a more salient stimulus presented after a less salient stimulus.

Here we have looked at some effects of dopamine on synaptic activation. Another related effect is the

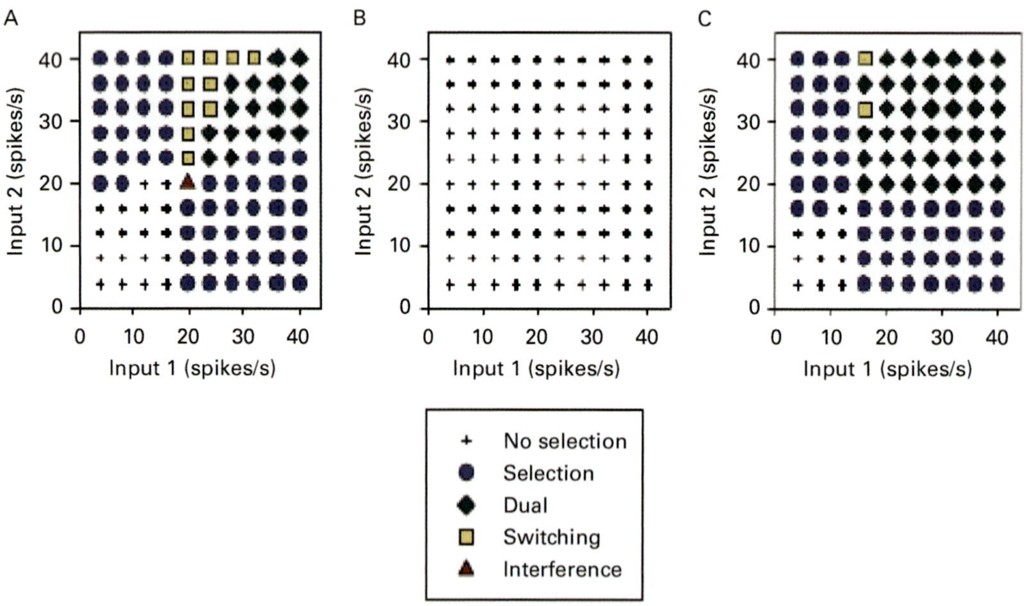

FIGURE 23.6 The grids A, B, and C represents the behaviors of the models under normal, low, and high dopamine levels, respectively. Each point in the grid represents a simulation. The *x*-axis and *y*-axis represents the spiking in channels 1 and 2, respectively. For each of the simulations, the model is presented with a stimulus along one channel, then with another more salient stimulus along the other channel. As seen in grid A (normal dopamine level), the stimulus with higher salience is selected (blue circles in the bottom-right and top-left corners of the grid). With low dopamine levels (as in PD, grid B), there is lack of selection of either group of neurons (crosses in the plot denoting "no selection"). With higher dopamine levels (grid C), there is competition, and switching takes place (yellow squares). With higher spiking rates, both channels are selected (dual selection, green diamonds). So, with high levels of dopamine, there is less switching and more dual representations of the channels. Modified from Humphries et al. (2006).

enhancement of synapses as part of RL, which we will explore in the next section.

23.2.3 Modeling Effects of Tonic and Phasic Dopamine Another computer model investigated the effects of dopamine depletion and repletion on both tonic and phasic dopamine inputs (Frank, 2005). The model consisted of elements representing cue input, striatum, GPe, GPi (as the output port), thalamus, premotor cortex (PMC, which projects to the output layer), and SNc (which supplied dopamine input to the striatum). These elements consist of units that behave like point neurons, with rate coding. The striatum had a Go and a No-go column for each of two possible responses, giving a total of four columns of activation "units." Each of the Go columns projected to GPi. Each of the No-go columns projected to GPe, which then projected to GPi. This way, based on competition in the level of activity of the units in each column in the striatum, GPi ended up with a level of activity which influenced the thalamic units in an inhibitory fashion. Thalamic units then projected to PMC units, which projected to the output element.

The model was used on two tasks. In the first, the network was presented with different cues in different combinations. Each cue had a probability to predict either "rain" or "sun." There were two learning phases. One phase was dependent on tonic dopamine stimulation from the SNc (the "minus" phase). During this phase, the network was presented with cues, and based on a given cue and the synaptic weights, a response was picked up. No feedback was given about whether the response was correct or not. During the second phase ("plus" phase), feedback was provided based on the response. A correct response resulted in an increase of dopamine stimulation, a dopamine "burst." This resulted in an enhancement of the Go signal and a reduction in the No-go signal, encouraging the network to select that response when presented with a similar set of cues next time. An incorrect response resulted in a decrease in dopamine stimulation ("dip"), reducing the Go signal and enhancing the No-go signal. The feedback modified the weights of the synapses from the input layer and from the PMC layer onto the striatum. Through this learning, the activation of the correct units of striatum was achieved.

The second task was a *probabilistic reversal task*. The model was presented with one of two cues and learned to associate one of two responses with each cue. Reversal then took place, and the network had to learn to associate the same cue with the opposite response (i.e., unlearn the previous learning).

In order to simulate PD pathology, three out of the four units representing SNc were silenced to form a "PD network." This resulted in a lower tonic level of dopamine, and also a smaller difference between the dopamine burst and dip.

Intact and PD networks were given both tasks. PD networks showed more errors in both tasks in comparison to the intact networks. To simulate the medicated state, a network with higher dopamine input during the tonic, burst, and dip phases was modeled. The effect of medication was compared on the network in the probabilistic reversal task. In comparison to the intact network, the medicated network was more prone to errors in the probabilistic reversal task during the part that required reversal of acquired learning. This happened because the high levels of dopamine suppressed the No-go units and so impaired the network's ability to learn No-go representations during negative feedback, important to allow for learning reversal. This model was able to replicate the phenomenon of patients suffering from PD being prone to different types of errors even when medicated.

As discussed, modeling can help in evaluating the effect that various levels of neurotransmitter can have on basal ganglia circuitry. Future modeling efforts will begin to look at the progression of PD by considering how alterations in dopamine levels over time will alter the expression of disease.

23.3 Stroke

23.3.1 Pathophysiology and Dynamics of Stroke The brain receives its blood supply through a network of blood vessels which supplies the brain in a pattern unrelated to the functions of the supplied areas. Therefore, strokes can produce peculiar patterns of deficit through effects on multiple brain areas related by proximity but not by functional interconnection. After the blood supply to a brain region stops, neurons suffer from reduction of oxygen and glucose supply. They become ischemic. If this continues for a prolonged period of time, the neurons start to die. A stroke develops when neurons die following the cessation or reduction of blood supply to a brain region. Note that such a region is determined by the blood supply, and has no relation to, for example, neuroanatomical regions in the sense of Brodmann areas.

The area surrounding the region where the blood supply is reduced, but where neurons have not yet died, is called the penumbra. The neurons within the penumbra show reduced metabolic and neural function and are sensitive to further insult either through further reduction in their metabolic supplies or through additional excessive excitation. This excessive excitation, termed excitotoxicity, can occur either through synaptic activation or through nonspecific stimulation of glutamate receptors by glutamate released into the extracellular space from other damaged cells.

Cortical representations of different body regions are dynamic. They change in response to alterations of received stimulus characteristics. For example, excessive stimulation of a body region will result in expansion of the sensory cortical area representing that body part (Buonomano and Merzenich, 1998). (See chapter 13, "Neural Maps: Their Function and Development.") In contrast, decreased stimulation of a body part—for example, by surgical removal of a finger—results in diminishing the size of the sensory cortical region representing the finger (Kaas et al., 1983). Following a stroke, changes occur in the other direction: the sensory organ is present, but the sensory cortical region representing that sensory organ is now missing. In order to understand changes following stroke, it is useful to think in terms of receptive fields (RFs) and projective fields. The RF of a neuron is the area from where the neuron is *receiving* its input (an area of a finger, in the above example, which will activate a particular neuron). A projective field is the set of neurons activated when a single location is stimulated, that is, the set of neurons onto which a particular location *projects.*

It is hypothesized that changes in the RFs of surviving cells may play a role in recovery following stroke: the function of the cells that are missing may be replaced in part by local cells with similar or overlapping RFs. The alternative, but not mutually exclusive, hypothesis is that other remote areas of the brain take over functions once handled by the ischemic area. The MT (middle temporal cortex, or V5) region which is involved in representing the motion of different targets in the visual field represents an interesting example. A lesion in the MT region in monkeys does not impair the monkey's ability to see stationary targets but does impair its ability to track moving objects when they are present in the area of the visual field represented by the lesioned part (Wurtz et al., 1990). Following a lesion to the MT cortex of a monkey, changes take place immediately within the RFs of remaining cells surrounding the ablated MT area. These surrounding cells would be comparable to the cells of the penumbra in a stroke. The changes occurring include both expansion and contraction as well as shifting of RFs. Expansion is often asymmetrical, with a tendency to expand primarily toward the lesion.

23.3.2 Modeling RFs Changes To investigate changes in RFs, a model was developed that used a one-layer recurrent model of MT which received convergent inputs from a receptive layer representing lower visual areas such as V1 (Lytton et al., 1999). The MT layer had lateral connections, in the form of near excitatory and far inhibitory connections (a "Mexican hat" distribution of connections, but with no self-excitation; figure 23.7).

Simulations on this network demonstrated that a process of unmasking could explain expansion and contraction of the RFs (figure 23.8). Cells in the network provided near excitatory, and far inhibitory, lateral inputs. After ablation, these influences were lost, resulting in loss of both the local excitatory and the distal inhibitory connections that had originated from the ablated region. The loss of nearby excitatory cells had a minimal effect. However, the loss of the more distant inhibitory connections resulted in substantial increased activation *(unmasking)* such that receptive inputs that were previously unable to activate a given MT unit could now do so, having less inhibition to counter. This resulted in expansion of the RF. A secondary consequence of this loss of inhibition was due to the disinhibition of inhibitory cells. This secondary effect resulted in contraction of the RFs for the principal cells that now received augmented inhibition. Inputs that previously would have stimulated these cells to threshold were no longer capable of doing so. Such expansion and contraction in the RFs represent an expanding wave of dynamical consequences resulting from the loss of projections from the lost neurons.

While the simulation was able to replicate both expansion and contraction of RFs, the contractions seen were much less than those found in the experimental animal model. This failure meant that there was a missing element, a missing piece in the model. This

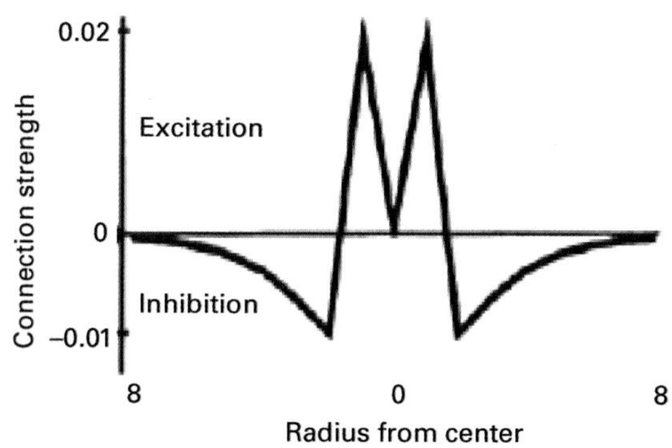

Figure 23.7 Lateral connections within the layer representing the middle temporal area. There are no self–self connections (connection strength = 0, at radius = 0). There are nearby excitatory connections and far inhibitory connections. From Lytton et al. (1999).

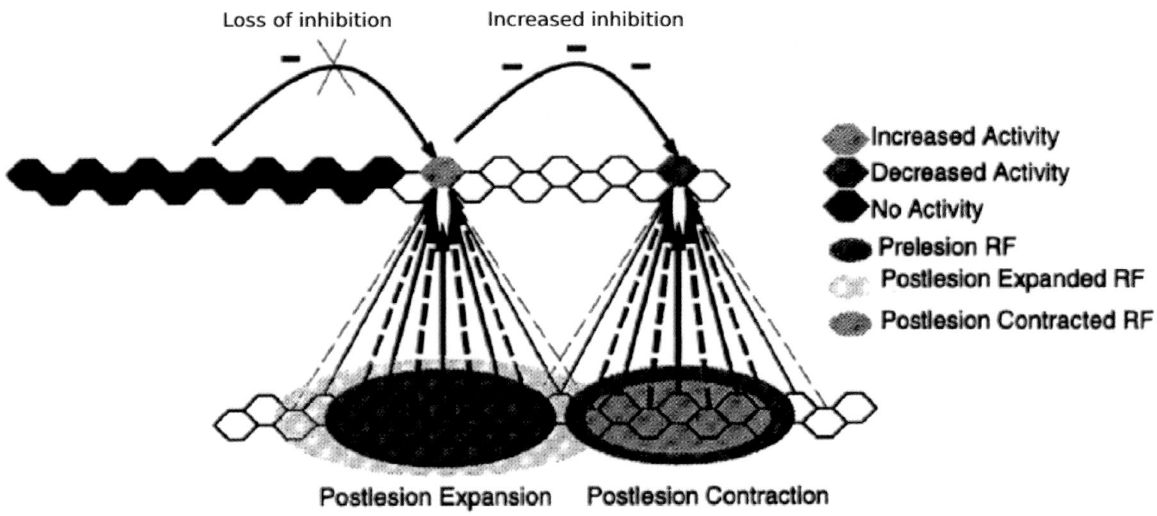

FIGURE 23.8 Unmasking can explain expansion and contraction of receptive fields (RFs) post-lesion. The top layer is middle temporal cortex (MT) while the lower one is V1. Inputs from V1 activate MT. In the MT layer, the left half is ablated. This results in loss of inhibition onto the region in the middle (light gray hexagon in top layer). This disinhibition reduces the activity in the right-most dark gray hexagon. Since the middle light gray hexagon is no longer inhibited, it can be activated by previously subthreshold input from wider regions of V1 (RF expansion in the bottom layer, on the left). Similarly, since the dark gray hexagon now receives more lateral inhibition, its threshold has increased and so will respond to narrower regions of V1 that can provide suprathreshold input (RF contraction in the bottom layer, on the right). From Lytton et al. (1999).

problem highlights the important role of failure (falsifiability) in modeling—the inability to explain some aspect of reality spurs the experimental search for additional critical facts about the system. This also highlights a contrast between simulation modeling and traditional closed-form physics modeling. Falsification in simulation generally requires that a model be augmented, whereas in closed-form modeling it may require that the model be entirely abandoned.

When the histology of the ischemic regions was looked at more carefully, it was found that there is reduced expression of $GABA_A$ receptors in the vicinity of stroke (Schiene et al., 1996), as well as increased excitability in these regions (Neumann-Haefelin et al., 1995). This suggests the presence of a halo of disinhibition around the ablated area. This is consistent with the inhibitory interneurons being more sensitive to ischemia, and so more of them are affected. The presence of this halo suggested that the penumbra region would show differential effects on the various kinds of cells found in the region.

When this halo was included in the model, it resulted in augmentation of both expansion and contraction of the RFs. Excitatory cells in the disinhibition halo were excitable by the relatively weak stimulation coming from distantly converging inputs resulting in expansion of their RFs. Also, the excessive activity in these cells produced increased inhibition in the inhibitory ring (the cells receiving the far inhibitory input), making it more difficult for units from the input layer to stimulate these units. This now resulted in more pronounced contractions, more similar to those seen experimentally. The mechanism of expansion and contraction of RF was the same in the simple model and the model with the disinhibited halo. However, the halo augmented the effects of each mechanism.

While the changes described in the model occurred immediately, cortical networks are dynamic on different timescales. This suggests that overall RF changes would likely take place in two phases. The first phase involves the immediate changes in neuronal dynamics described in this model which were the immediate consequence of cell death, with loss of the various excitatory and inhibitory connections on neighboring cells. During the second phase, synaptic plasticity changes would take place through changes in the synaptic weights among the surviving neurons. These changes will follow a pattern that would be determined by the new dynamics which resulted from the first phase. Naturally, synaptic plasticity changes will also result in changes in neuronal dynamics, closing the loop between these two interacting processes. Dynamic changes give rise to plasticity, and plasticity consists of structural changes that give rise to dynamic changes. Separating them into two phases is nonetheless justified by the vastly different time constants for each of them. While the neural dynamic changes take place over the course of milliseconds, the synaptic plasticity changes take place over the course of hours to days.

Perception depends on neural activity, and stroke results in alterations in encoding. If the decoding strategy leading to behavior were to remain unchanged, function would suffer. For this reason, full behavioral recovery will require changes in decoding strategy that match the alterations in encoding (see, e.g., Nudo, 2007).

To explore the effect of ablation on neural coding, the model was expanded so that input was represented as a stimulus with attributes of location, direction of motion, and speed. Also, the MT units (which respond to movements, with the attributes of direction and speed) were replaced with a set of units, each of which was assigned a preferred direction and speed. A winner-take-all algorithm was used to interpret neural activity. (Other algorithms that could have been used to determine the characteristics of the stimulus are vector sum or vector average.)

Animal stroke models and clinical experience suggest that a stroke in the MT visual area will produce a behavioral effect. In this case the effect would be a motion scotoma in the visual field: an area of the visual field with a perceptual deficit for moving target leading to underestimation of the speed of the target. This can be measured through measuring the lag of the eye movements, or the reduction in pursuit speed during following the target with the eyes. The model was able to reproduce this effect, producing underestimation of object speed within the region that corresponded to the ablated area.

Based on the model discussed, it is suggested that the expansion of the RF following ablation allows post-lesion stimuli to be misprocessed (as evidenced by inappropriate post-lesion behavior) rather than not processed at all. In the absence of these expansions, a portion of visual space would lose its cortical representation entirely because it would not fall in the RFs of any live neurons. Immediate RF expansion might produce a "Band-Aid effect" that is important both to immediate preservation of function and to the eventual shift and spread of RFs to fill the gap after cortical reorganization.

Hebbian synaptic plasticity suggests how physical therapies might improve functional recovery following stroke. According to this mechanism, synapses become strengthened when there is coincident activation of the presynaptic and postsynaptic neurons, and they get weakened when the activation of both neurons is not coincident. The model showed that stroke is likely to produce increased activation of a set of cortical neurons, which will then lead to increased synaptic strength with other neurons that are active. By purposefully activating specific pathways either by peripheral input (through physiotherapy) or cortical stimulation, one could produce desired synaptic connectivity to compensate for lost pathways. Presumably, specific patterns of exercises currently used in physical therapy determine synaptic reorganization in the brain.

Understanding the events underlying the immediate and the synaptic plasticity phases of changes following stroke points toward pharmacological and nonpharmacological interventions (e.g., physiotherapy) to minimize neuronal death after stroke (minimizing deficit) and promote synaptic plasticity (improving recovery). For example, it may be desirable to delay the onset of cortical reorganization to prevent formation of spurious connections caused by overexcited neurons of the ischemic penumbra. Or, an agent that can pharmacologically erase long-term potentiation, such as zeta inhibitory protein (Shema et al., 2007), might be used to allow cortical reorganization to occur during a period of intensive physical therapy which has been postponed till after medical therapy and recovery occur.

Computer models for the recovery process which takes place with physiotherapy provide clinical predictions. Subjects with a stroke affecting one hand will tend not to use it, and therapy may constrain the good hand to encourage the patient's use of the impaired hand. Han and colleagues (Han et al., 2008; Schweighofer et al., 2009) developed a model of the recovery process which predicted a functional threshold, beyond which improved use of the affected limb would lead to greater spontaneous use. This would then provide a virtuous circle which would provide continued improvement without the need for further explicit rehabilitation strategies.

They modeled motor cortex in both hemispheres and produced a stroke on one side. Neural reorganization in the motor cortex was modeled to include supervised, unsupervised, and reinforcement learning. The existence of the threshold could be explained by taking into account the complex interaction of learning dynamics across all three types of learning that were simulated. A subsequent study determined such a functional threshold clinically (Schweighofer et al., 2009). An important result of comparing model with clinical reality was that not all patients showed the same effects. Overall, the threshold could be shown in the group, with most of the patients improving above that threshold. Ideally, more detailed modeling could be used clinically to assess threshold for individual patients.

Modeling of the effect of ablation and stroke allowed the exploration of the dynamical changes that happen afterward, which play a role in recovery. We will now turn to investigating brain oscillations, their role in schizophrenia, and how modeling them can aid in pointing to targets for therapy.

23.4 Schizophrenia

23.4.1 Schizophrenia and Oscillations We have left schizophrenia for last as it remains one of the most puzzling of brain diseases and also one of the most challenging for modeling since it primarily affects thought processing. While the prior diseases are considered neurological, schizophrenia is considered psychiatric. This distinction reflects both historical and practical issues. Historically, schizophrenia was considered a disease of the mind rather than a disease of the brain. Hence, it was thought that it was a disease that could be cured by somehow replacing wrong thinking with right thinking, comparable to classical religious views of the origins of thought and behavior. Now it has become clear that schizophrenia is in fact a brain disease and that any treatment will have to address problems in the brain as well as trying to ameliorate learned behaviors. However, since psychiatry is considered to deal more with symptoms of the mind, schizophrenia is still treated by psychiatrists and in psychiatric settings. The dichotomy between neurology and psychiatry remains important logistically.

Patients with schizophrenia present with symptoms that belong to one of three domains: positive, cognitive, and negative symptoms. "Positive" symptoms include hallucinations (usually auditory, but they could also be visual), delusions (false fixed beliefs that the patient holds strongly, which are not shared by others of the same culture), and disorganized thought processes (where the forms of thought and speech become disrupted so that it is difficult to understand what the patient is trying to say or communicate). Cognitive symptoms include deficits of working memory (patients have difficulty holding information that is important to solve a task at hand, e.g., preparing a grocery list). Negative symptoms include lack of motivation and social isolation. Current available medications (antipsychotics) can control positive symptoms. However, the effects on improving cognitive and negative symptoms are still limited.

Different patients will manifest different combinations of symptoms. It is now accepted that both etiology (the cause, the why) and pathophysiology (the mechanism, the how) in schizophrenia are multifactorial. Etiologies include genetic susceptibility, intrauterine exposure to infections, and stressful life situations. No one of these factors is enough by itself to cause schizophrenia, but it is the cumulative effect of these factors that finally causes schizophrenia to occur. Progress in understanding and treating schizophrenia has been slow, in large part because of this multifactorial nature. No one set of studies holds the key to understanding schizophrenia, and no one researcher is likely to master all of the disparate threads that need to be pulled together. Computer modeling has been proposed as a technique that will eventually permit us to master this complexity.

Here, we focus on pathogenesis and look at one set of theories that has strong evidence from animal studies and neuropsychological testing: cognitive coordination (Phillips and Silverstein, 2003). According to cognitive coordination theory, different regions and networks of the brain are able to perform their function through coordinating their activities within themselves and between each other. Symptoms of schizophrenia arise because of failure of the coordination between brain networks (cognitive discoordination).

23.4.2 Schizophrenia and Oscillations Oscillations, ubiquitous in the brain, have been hypothesized to provide the neural coordination that underlies cognitive coordination (Bressler and Kelso, 2001). Oscillations within and between different brain networks may synchronize the firing of neuronal networks that are functionally related. Such synchronization has been hypothesized to underlie the "binding" of various attributes of an object to create the conscious perception of a unified object. In the context of schizophrenia, this would then suggest anomalies in binding that could lead to the cognitive and perceptual disorders associated with this disease.

Ketamine, one of the chemical relatives to PCP (phencyclidine, or angel dust) is an NMDA receptor (NMDAR) antagonist that is used as an anesthetic. As we mentioned above, NMDA is one of the glutamate receptors responsible for excitatory activation of the postsynaptic cell. When ketamine, in subanesthetic dose, is given to nonpsychotic human subjects (those who do not suffer from schizophrenia or related disorders), they start experiencing hallucinations, delusions, and thought process disturbances. When ketamine is given to patients with history of schizophrenia, it exacerbates and worsens their symptoms. Ketamine also produces changes in brain oscillations. In mice, these changes are an increase in the power of the gamma oscillations (30–100 Hz) and a decrease in the power of theta oscillations (3–12 Hz) (Lazarewicz et al., 2010). In humans, EEG changes have been recorded in healthy volunteers under subanesthetic doses of ketamine. The changes were increased power of gamma-band oscillations (40–85 Hz), and decrease in the power of delta-band oscillations (1–5 Hz) (Hong et al., 2009). Psychotic symptoms were found to be associated with similar changes in EEG power spectra (Uhlhaas and Singer, 2010). Based on this, treatment with subanesthetic doses of ketamine is used to produce an animal model for schizophrenia (or psychosis in general). The change in the power of different frequency bands could be one of the mechanisms responsible for generation of psychotic symptoms.

23.4.3 Modeling Changes in Oscillations Hippocampus and frontal cortex are two areas primarily affected in schizophrenia. Postmortem brain biopsies from patients who suffered from schizophrenia showed changes in hippocampal cytoarchitecture (Harrison, 2004). To investigate how ketamine produces the observed changes in the power spectrum, a neural-level model of hippocampal circuitry was developed, consisting of 800 pyramidal neurons, 200 inhibitory fast-spiking basket cells, and 200 inhibitory slow-spiking oriens-lacunosum moleculare (OLM) cells. Theta drive from medial septum (MS) was simulated by providing periodic inhibitory input to OLM and basket cells. The connectivity of the network is shown schematically in figure 23.9. (See chapter 4 for other models of "Neural Rhythms.")

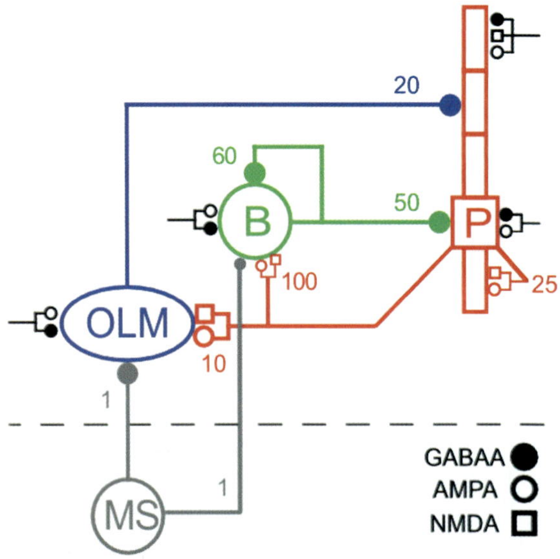

Figure 23.9 Schematic representation of the network representing the CA3 region of hippocampus. The network consists of 800 pyramidal neurons (P), 200 oriens-lacunosum moleculare inhibitory interneurons (OLM), and 200 basket cells (B). The numbers near the synapses represent convergence ratios, that is, number of inputs for an individual synapse: $GABA_A$ receptors (filled circles), AMPA receptors (open circles), NMDA receptors (open squares). External stimulation from other areas was modeled by synaptic bombardment (synapses with truncated lines). Externally generated theta oscillations from the medial septum (MS) were imposed on OLM and basket cells. From Neymotin et al. (2011).

Oscillations within the model were generated as pyramidal cells drove OLM and basket cells via both AMPA (short) and NMDA (longer) receptor activation. The OLM cells, in turn, inhibited the distal dendrites of pyramidal cells, while the basket cells inhibited the soma of pyramidal cells. Compared to the fast changes in the membrane potential of pyramidal somata caused by basket cells, dendritic filtering gave the OLM inputs longer time constants, allowing them to modulate pyramidal activity with a slower time course. Gamma generation has been grossly dichotomized as being due to interneuron network gamma (ING) or pyramidal interneuron network gamma (PING) (See chapter 4, "Neural Rhythms."). ING appeared to be the major gamma driver at the theta nadir (figure 23.10C), when pyramidal cell population activation was minimal (figure 23.10A). By contrast, during the theta upswing, gamma activity appeared to primarily emerge as a PING interplay, the pyramidal cells driving the basket cells which then coordinated population pyramidal cell activity through near-simultaneous basket cell inhibitory postsynaptic potentials on pyramidal cell somata.

Basket cells were entrained to theta at two levels: directly by the MS inputs and indirectly via the periodic firing of the OLM-disinhibited pyramidal cells. The basket cells then provided feedback onto pyramidal cells that augmented their theta response. Removal of the basket cell population greatly reduced theta strength, demonstrating that basket cells contributed strongly to theta, as well as to gamma.

Single-cell voltage traces reflected the dominant frequencies of basket cells and pyramidal cells. Individual OLM cells showed periodic firing on portions of the theta cycles after the recovery from MS inputs. The basket cell population fired at gamma frequency, but individual cells would only follow for 3–4 cycles at a time, and only at peak theta (figure 23.10D).

Ketamine blocks NMDA receptors. In the model, NMDA receptors are located on four locations: OLM soma, basket soma, pyramidal basal dendrite, and pyramidal apical dendrite. To simulate the blocking by ketamine at a particular site, conductance of NMDAR was set to zero at that site. Turning off all the NMDA synapses in the model reduced activity in all cell types, resulting in significant reduction in power for both theta and gamma. This did not match the experimental data (gamma up, theta down). Because different NMDA receptor subtypes that are expressed on different cell

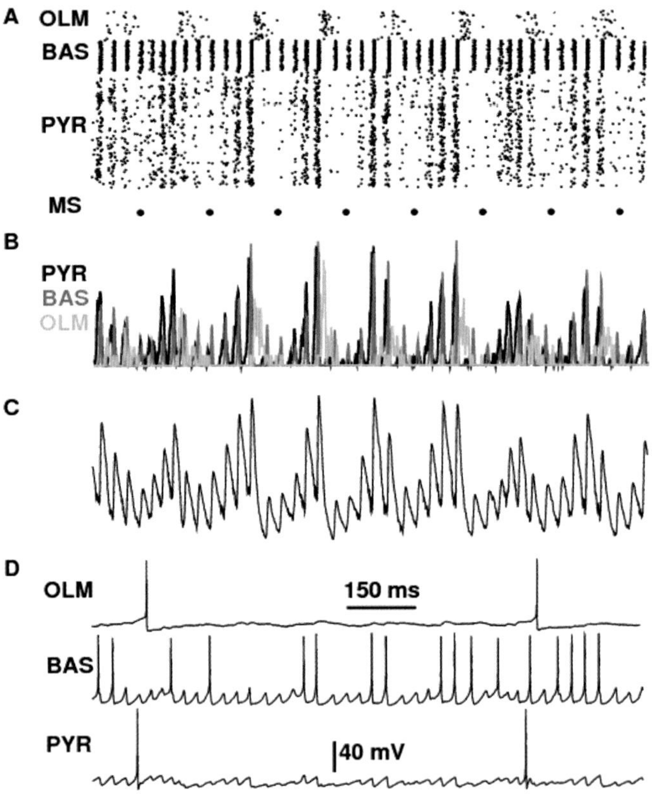

FIGURE 23.10 Network activity during baseline simulation. Network raster plot (A) shows the times of spikes in each cell organized into the three different cell populations—each dot on the raster is a single spike in a single cell. The periodic drive from medial septum (MS) is also shown. Spike densities (B) summarize the amount of spiking in each population over time by smoothing over the spikes in the raster plot (calculated using 1-ms bins and smoothing with 3-ms triangle filter). Local field potential (LFP; C) demonstrates the overall contribution of the pyramidal cells (PYR) as they would be seen by an electrode put into the brain. The single cell voltage traces (D) show the pattern of spikes and postsynaptic potentials (little bumps between spikes) for a representative cell for each of the three populations. Note that the fast frequency of activity seen in the LFP is an emergent population phenomenon based on many pyramidal cells firing together, despite each pyramidal cell firing infrequently. OLM, oriens-lacunosum moleculare cells. From Neymotin et al. (2011).

types have different sensitivity to NMDA-receptor antagonists (Bresink et al., 1995), it was hypothesized that the discrepancy might be due to different effects of ketamine on the different cell types. There were 16 combinations for blocking NMDAR on different sites. Changes in theta (3–12 Hz) and gamma (30–100 Hz) power were compared between the baseline and each of the combinations. Out of the 16 different combinations of NMDAR switched off, combinations that involved NMDAR being off at OLM cells replicated the experimental findings (Lazarewicz et al., 2010).

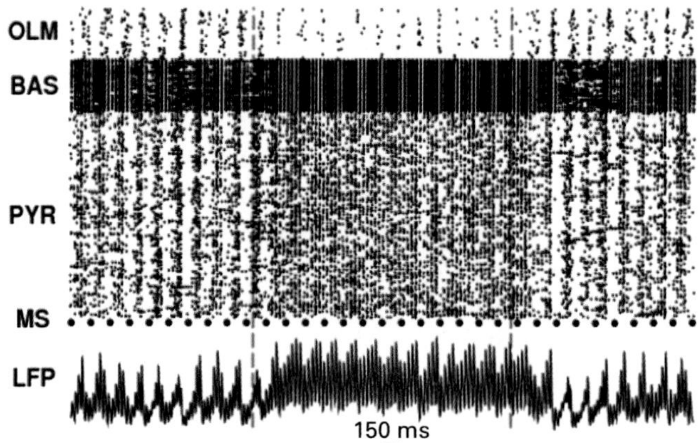

FIGURE 23.11 Raster plot (top) and local field potential (LFP) (bottom) showing baseline, ketamine wash-in, and wash-out from a single simulation. Vertical dotted lines mark times of ketamine wash-in and wash-out. During the period when ketamine is present, there is a great increase in the strength of fast (gamma) activity and a decrease in the strength of the slow (theta) activity. OLM, oriens-lacunosum moleculare cells; BAS, basket cells; PYR, pyramidal cells; MS, medial septum. From Neymotin et al. (2011).

Therefore, it could be predicted that the experimental effect of ketamine is based on primary blockage at NMDA receptors on the OLM cells (figure 23.11).

These simulations suggested a sequence of pathophysiological alterations that led from a reduction in activity at a particular receptor to spectral changes at the network level. In order to restore the physiological power spectra of oscillations, a selective current injection into the OLM cell population was used, which recovered the control oscillations by direct opposition at the primary pathological focus. Here a tonic activation was used to replace a missing periodic effect, suggestive of the situation in PD where tonic dopamine replaces the phasic release that has been lost. Such a prediction points to the role computational neuroscience can play in pointing out possible mechanisms of action for new treatments even in a complex illness such as schizophrenia.

23.5 *Summary and Conclusions*

In this chapter, we have discussed four major brain diseases. We have focused on these diseases since this is where most of the modeling has been done. These diseases also represent something of a cross-section of the types of problems that can affect the brain. Epilepsy, for example, is the prototypical dynamical disease. A variety of insults or abnormalities, at different levels of organization, can result in dynamical perturbations that result in common symptomatic forms. Hence the disease *is* the abnormal dynamics rather than being specific to one or the other of the possible causes. It is hypothesized that schizophrenia may represent a similar case, where multiple causes and levels of causation end up in a final common pathway of impaired neuronal dynamics, with the various manifestations being explained by the various circuits affected.

In epilepsy, there are multiple causes that converge on a relatively small set of manifestations. In PD, a single major disturbance diverges to produce several dynamical manifestations. For example, tremor and rigidity are not obviously related, an augmenting of movement contrasted with a decrementing. However, we are beginning to understand how this manifestation divergence might occur as expressions of underlying dynamical disturbances producing seemingly discrepant manifestations.

By contrast with the others, stroke is an ablative rather than a primarily dynamical disease and may have diverse loci, with varied effects accordingly. This ablation is imposed upon the brain, and the brain must react to it through alterations in activity that allow it to continue functioning at various levels. Although these alterations are also dynamical, it is reasonable to consider handling the pathology and responses to pathology as a series of snapshots at different points in time rather than necessarily following the dynamics in detail.

Because the brain is such a complex organ, with many underlying complex subsystems, its pathologies would also be expected to be extraordinarily complex and resistant to generalization. Thus, the presence of symptoms and signs that can be lumped together as a specific brain disease may be regarded as something of a surprise. However, the ways in which these symptoms and signs fit remains a major puzzle, a puzzle which we can hope to pull apart and then put back together using computational neuroscience. Nowhere is this puzzle more apparent than in schizophrenia, where bizarre ideas reflect thought disorders. Putting this shattered cognition back together again remains a far-off goal to which computational neuroscience can contribute, perhaps by yielding novel predictions that support the fitting of neural activity patterns back together through hitherto unexpected kinds of treatments.

REFERENCES

Bevan, M. D., Booth, P. A., Eaton, S. A., & Bolam, J. P. (1998). Selective innervation of neostriatal interneurons by a subclass of neuron in the globus pallidus of the rat. *Journal of Neuroscience, 18*(22), 9438–9452.

Bresink, I., Danysz, W., Parsons, C., & Mutschler, E. (1995). Different binding affinities of NMDA receptor channel blockers in various brain regions—Indication of NMDA receptor heterogeneity. *Neuropharmacology, 34*(5), 533–540.

Bressler, S. L., & Kelso, J. (2001). Cortical coordination dynamics and cognition. *Trends in Cognitive Sciences, 5*(1), 26–36.

Buonomano, D. V., & Merzenich, M. M. (1998). Cortical plasticity: From synapses to maps. *Annual Review of Neuroscience, 21*(1), 149–186.

Chevalier, G., & Deniau, J. (1990). Disinhibition as a basic process in the expression of striatal functions. *Trends in Neurosciences, 13*(7), 277–280.

Frank, M. (2005). Dynamic dopamine modulation in the basal ganglia: A neurocomputational account of cognitive deficits in medicated and nonmedicated parkinsonism. *Journal of Cognitive Neuroscience, 17*(1), 51–72.

Gerfen, C., & Bolam, J. (2010). The neuroanatomical organization of the basal ganglia. In H. Steiner & K. Tseng (Eds.), *Handbook of Basal Ganglia Structure and Function: A Decade of Progress* (Vol. 20). Access online via Elsevier.

Glass, L., & Mackey, M. (1988). *From Clocks to Chaos: The Rhythms of Life.* Princeton, NJ: Princeton University Press.

Han, C., Arbib, M., & Schweighofer, N. (2008). Stroke rehabilitation reaches a threshold. *PLoS Computational Biology, 4*(8), e1000133.

Harrison, P. (2004). The hippocampus in schizophrenia: A review of the neuropathological evidence and its pathophysiological implications. *Psychopharmacology, 174*(1), 151–162.

Hermanowicz, N. (2007). Drug therapy for Parkinson's disease. *Seminars in Neurology, 27,* 97–105.

Hong, L., Summerfelt, A., Buchanan, R., O'Donnell, P., Thaker, G., Weiler, M., et al. (2009). Gamma and delta neural oscillations and association with clinical symptoms under subanesthetic ketamine. *Neuropsychopharmacology, 35*(3), 632–640.

Humphries, M., Stewart, R., & Gurney, K. (2006). A physiologically plausible model of action selection and oscillatory activity in the basal ganglia. *Journal of Neuroscience, 26*(50), 12921–12942.

Kaas, J. H., Merzenich, M. M., & Killackey, H. P. (1983). The reorganization of somatosensory cortex following peripheral nerve damage in adult and developing mammals. *Annual Review of Neuroscience, 6*(1), 325–356.

Kaiser, M., Goerner, M., & Hilgetag, C. (2007). Criticality of spreading dynamics in hierarchical cluster networks without inhibition. *New Journal of Physics, 9*(5), 110.

Kang, M., Spigelman, I., Sapp, D., & Olsen, R. (1996). Persistent reduction of GABAA receptor-mediated inhibition in rat hippocampus after chronic intermittent ethanol treatment. *Brain Research, 709*(2), 221–228.

Larter, R., Speelman, B., & Worth, R. (1999). A coupled ordinary differential equation lattice model for the simulation of epileptic seizures. *Chaos: An Interdisciplinary Journal of Nonlinear Science, 9*(3), 795–804.

Lazarewicz, M., Ehrlichman, R., Maxwell, C., Gandal, M., Finkel, L., & Siegel, S. (2010). Ketamine modulates theta and gamma oscillations. *Journal of Cognitive Neuroscience, 22*(7), 1452–1464.

Lopes Da Silva, F., Blanes, W., Kalitzin, S., Parra, J., Suffczynski, P., & Velis, D. (2003). Dynamical diseases of brain systems: Different routes to epileptic seizures. *IEEE Transactions on Biomedical Engineering, 50*(5), 540–548.

Lytton, W., & Omurtag, A. (2007). Tonic-clonic transitions in computer simulation. *Journal of Clinical Neurophysiology, 24*(2), 175–181.

Lytton, W., Williams, S., & Sober, S. (1999). Unmasking unmasked: Neural dynamics following stroke. *Progress in Brain Research, 121,* 203–218.

Nambu, A., Tokuno, H., & Takada, M. (2002). Functional significance of the cortico–subthalamo–pallidal "hyperdirect" pathway. *Neuroscience Research, 43*(2), 111–117.

Neumann-Haefelin, T., Hagemann, G., & Witte, O. W. (1995). Cellular correlates of neuronal hyperexcitability in the vicinity of photochemically induced cortical infarcts in rats in vitro. *Neuroscience Letters, 193*(2), 101–104.

Neymotin, S., Lazarewicz, M., Sherif, M., Contreras, D., Finkel, L., & Lytton, W. (2011). Ketamine disrupts theta modulation of gamma in a computer model of hippocampus. *Journal of Neuroscience, 31,* 11733–11743.

Nudo, R. (2007). Postinfarct cortical plasticity and behavioral recovery. *Stroke, 38*(2), 840–845.

Parent, J. M., Timothy, W. Y., Leibowitz, R. T., Geschwind, D. H., Sloviter, R. S., & Lowenstein, D. H. (1997). Dentate granule cell neurogenesis is increased by seizures and contributes to aberrant network reorganization in the adult rat hippocampus. *Journal of Neuroscience, 17*(10), 3727–3738.

Phillips, W. A., & Silverstein, S. M. (2003). Convergence of biological and psychological perspectives on cognitive coordination in schizophrenia. *Behavioral and Brain Sciences, 26*(1), 65–82.

Richfield, E. K., Penney, J. B., & Young, A. B. (1989). Anatomical and affinity state comparisons between dopamine D_1 and D_2 receptors in the rat central nervous system. *Neuroscience, 30*(3), 767–777.

Schiene, K., Bruehl, C., Zilles, K., Q¨u, M., Hagemann, G., Kraemer, M., et al. (1996). Neuronal hyperexcitability and reduction of $GABA_A$-receptor expression in the surround of cerebral photothrombosis. *Journal of Cerebral Blood Flow and Metabolism, 16*(5), 906–914.

Schweighofer, N., Han, C., Wolf, S., Arbib, M., & Winstein, C. (2009). A functional threshold for long-term use of hand and arm function can be determined: Predictions from a computational model and supporting data from the extremity constraint-induced therapy evaluation (excite) trial. *Physical Therapy, 89*(12), 1327–1336.

Shema, R., Sacktor, T. C., & Dudai, Y. (2007). Rapid erasure of long-term memory associations in the cortex by an inhibitor of PKM zeta. *Science Signaling, 317*(5840), 951–953.

Sporns, O., Tononi, G., & Kotter, R. (2005). The human connectome: A structural description of the human brain. *PLoS Computational Biology, 1*(4), e42.

Suffczynski, P., Kalitzin, S., & Lopes Da Silva, F. (2004). Dynamics of non-convulsive epileptic phenomena modeled by a bistable neuronal network. *Neuroscience, 126*(2), 467–484.

Sutula, T., Cascino, G., Cavazos, J., Parada, I., & Ramirez, L. (1989). Mossy fiber synaptic reorganization in the epileptic human temporal lobe. *Annals of Neurology, 26*(3), 321–330.

Uhlhaas, P., & Singer, W. (2010). Abnormal neural oscillations and synchrony in schizophrenia. *Nature Reviews. Neuroscience, 11*(2), 100–113.

Wurtz, R., Yamasaki, D., Duffy, C., & Roy, J.-P. (1990). Functional specialization for visual motion processing in primate cerebral cortex. [Cold Spring Harbor Laboratory Press.] *Cold Spring Harbor Symposia on Quantitative Biology, 55,* 717–727.

Xie, X., Lancaster, B., Peakman, T., & Garthwaite, J. (1995). Interaction of the antiepileptic drug lamotrigine with recombinant rat brain type IIA Na+ channels and with native Na+ channels in rat hippocampal neurones. *Pflügers Archiv, 430*(3), 437–446.

Yoshimura, Y., Dantzker, J. L., & Callaway, E. M. (2005). Excitatory cortical neurons form fine-scale functional networks. *Nature, 433*(7028), 868–873.

24 Language Processing

ANGELO CANGELOSI

Language is an articulated cognitive capability relying on hierarchical structures, and based on a multitude of processes from sound and speech perception, to referential and lexical capabilities, complex recursive syntactic skills, up to social pragmatic phenomena. Moreover, language is not an isolated capability of the human cognitive system. In fact, it strictly integrates with sensorimotor processes and with social and cultural dynamics. As such, its computational investigation requires the simultaneous consideration of sensorimotor processes, abstract symbol handling skills, and social knowledge, each of which poses significant challenges for computational modeling.

Important progress in modeling ranges from the early contribution of connectionist neural networks on the simulation of (mostly) syntactic processing, to the more recent neuroanatomical models of language learning and to the embodied and neurorobotic models of action and language integration. In comparison with those fields of computational neuroscience where highly realistic models of the brain have been developed, most of the current neural computation models of language are little constrained by known brain physiology and anatomical findings. This is especially the case for connectionist neural networks, which are based on the functional neural properties of parallel and distributed processing. However, as will be shown in this chapter, the continuous extension and integration of such connectionist models into hybrid, neuroanatomically constrained modular architectures is making it possible to simulate neural processing data such as event-related potentials (ERPs) and functional magnetic resonance imaging (fMRI), thus leading toward the design of increasingly realistic computational neuroscience models of language.

This chapter provides a syncretic overview of such developments. We will start with a didactic introduction to linguistic paradigms and to the current understanding of language processing in the human brain (section 24.1). We will then look at connectionist approaches to language learning and to neuroanatomical models of brain involvement in linguistic functions (section 24.2). Finally, we present the most recent developments of neurorobotic models of the grounding of language on sensorimotor processes (section 24.3).

24.1 Language and the Brain

24.1.1 DESCRIBING LANGUAGE Human language is a multifaceted phenomenon involving a complex integration of numerous sensorimotor, cognitive, and sociocultural capabilities. Typically in the study of language we distinguish different (hierarchical) level of analyses. The starting point for spoken language, the *phonetic level*, concerns the investigation of the acoustic processes involved in speech perception, such as the identification of phonemes or syllables, the nontrivial problem of segmentation of a speech stream into individual words, and the prosodic analyses often related to the speaker's emotional states. Next is the *lexical level*, which refers to the identification of words and the meaning they refer to, and the organization of words into a lexicon. This level includes the understanding of human referential capabilities and lexical semantics, that is, the linking of a word with an internal meaning and the external referent in the world. The *syntactic level* focuses on the formal symbolic mechanisms involved in putting words together following grammatical rules. This is the most prototypical level of language study, given the uniquely human characteristic of syntax. This level involves word order, syntactic mechanisms, and morphological rules. The *semantic level* involves the (possibly ambiguous) identification of the meaning of a sentence or speech act, given its underlying lexical and syntactic analysis. On some accounts syntactic and semantic processing are integrated on a single level. Finally, the *pragmatic level* concerns the processes involved in the social context of communication, for example, the effects that a speaker intends to communicate to the hearer (perlocutionary effect) and the actual meaning deduced by the hearer (illocutionary effect).

Syntax distinguishes animal communication systems from human languages and has been the core of language analysis for most linguistics, psycholinguists, and language neuroscientists, resulting in a multitude of theories. Given the brevity of this introduction to language study, we distinguish here the two predominant approaches, or families of approaches, to syntax theory: Chomsky's grammar theories and cognitive constructivist linguistics. The first approach, initially proposed by the very influential and pioneer linguist Noam Chomsky, states that syntax consists of a series of abstract, uniquely human symbolic manipulation mechanisms. Chomsky's theory has changed significantly in the past 50 years starting from the initial theory of transformational generative grammar (Chomsky, 1957), then to the principle and parameters of government and binding theory (Chomsky, 1981), and lastly with the minimalist program (e.g., Chomsky, 1995). For example, the classical transformational generative grammar theory is based on a series of universal "rewriting rules" that people use to process language. A fundamental rule of English and many other human languages is "S → NP VP," which means that during language production a sentence S can be rewritten (substituted) as a combination of a noun phrase (NP) and a verb phrase (VP). The rules "NP → pronoun," "VP → verb," and recursive rules such as "VP → VP NP" can subsequently be used to rewrite further the sentence being built. This process stops when all nonterminals (S, VP, etc.) have been replaced with a word or words from the lexicon. For example to generate the sentence "You go," the last applied rules will be "Pronoun → You" and "Verb → go." Notwithstanding Chomsky's invaluable contribution to linguistics, his theoretical stance is controversial, especially for its focus on innate, universal linguistic knowledge. His theories hypothesize the existence of an innate "language acquisition device" in the brain, which, for example, might encode universal rewriting rules of generative grammar, or the "principles" in the subsequent government and binding theory ("parameters" are then tuned via cultural learning). Another controversial issue related to Chomsky's theory is whether concepts such as "rules" are simply a scientific formalism used to describe language or if, as Chomskians seem to imply, they actually refer to rules in some way genetically prefigured in the brain. This debate has significantly affected neural computation models of language, for example, with the contribution of connectionist models to provide an alternative "rule-less" explanation to language learning phenomena (see past tense debate in section 24.2.1).

As an exemplary alternative to Chomsky's nativist and formal linguistic theories we consider cognitive linguistics (Langacker, 1987) and its related theories of language development known as construction grammar and usage-based theories (Goldberg, 2006; Tomasello, 2003; MacWhinney, 1998). These emphasize the close relationship between semantics and grammar. They posit that syntactic categories and thematic roles (e.g., agent, patient, recipient, and more specific roles) emerge out of usage-based regularities in the semantic use of the language-speaking agent. Developmentally, this is based on the principle that the child does not require innate knowledge of universal syntactic principles. On the contrary, the child is seen as an active *constructor* of its own variant of the language system of the surrounding community through implicit observations and through the learning of statistical regularities and logical relationships between the meaning of words and the grammatical and semantic context in which the words are used. For example, in constructivist theory the general syntactic category of verbs emerges from more ad hoc categories of verb-like words during development from incremental and hierarchical refinement of similarities between verbs sharing common semantic features, as in Tomasello's (1992) verb island hypothesis (and the earlier work of Hill, 1983). In the early stages of grammatical development (around year 3), children are able to use a variety of verbs, but these appear to exist as independent syntactic groupings, called "verb islands," rather than a full-blown general grammatical category of verbs. For some verbs (e.g., "cut"), the child is only able to use very simple syntactic combinations of the same verb with a limited set of nouns as objects ("cut bread," "cut paper"). Other verb islands, instead, can have a more elaborate syntactic use, as with the verb "draw" that can be uttered in richer combinations such as "I draw," "Draw a picture," "I draw picture for you," and so forth. These differences in the level of complexity and maturity of different verb islands are due to usage-based experience. At this stage the child has not developed general syntactic and semantic categories like agent, patient, and instrument. Only later during development, at school age, does the child develop an adult-like general concept of verb, and its variations as transitive, intransitive, dative verbs. Moreover, in line with construction grammar theories, the general semantic categories such as agent, patient, instrument are too coarse for many verbs, and the syntactic subdivision of verbs in terms of applicable constructions can be quite detailed even at later developmental stages (Levin, 1993).

Cognitive constructivist theories of language have contributed to the development of computational models of language learning consistent with schema-based theories of cognition (e.g., Arbib, Conklin, and Hill, 1987), such as Arbib and Lee's (2008) neurolinguistic schema theory of language, Dominey's (2005) connectionist model of verb constructions, and Steels's (2012) fluid construction grammar framework.

24.1.2 Neuroanatomy of Language Processing The understanding of the human brain structures involved in language processing, and their rich network of integration with other sensorimotor and cognitive functions, has progressed significantly in recent decades as a result of the availability of numerous brain imaging methodologies, as well as the contribution of neuropsychological studies and brain cytoarchitectonic analyses. Among the numerous publications providing an overview of the brain's involvement in language processing are works such as Friederici (2009, 2011, 2012), Hagoort (2005, 2013), Pulvermüller (2003, 2013), and Vigneau et al. (2006). In this section we provide a brief summary of the main cortical areas responsible for the various aspects of language processing (in particular speech perception, semantics, sentence and syntactic processing, and mouth articulation), and the complex circuitry connecting these areas. For this overview we will use the Brodmann areas (BAs; Brodmann, 1909) classification of the cortical areas in summarizing the functional circuits proposed by Friederici (2009) and Vigneau et al. (2006). As language is characterized by a strong lateralization of functions in the dominant left brain, we will discuss only the left language cortical areas while noting the role of right hemisphere in prosody.

Figure 24.1 shows the main left-brain BAs involved in language processing, namely, frontal Broca's area and temporal Wernicke's area. These are also referred to as the perisylvian language areas, as they are around the Sylvian sulcus/fissure (or lateral fissure) separating the frontal and temporal areas. Broca's area, which is located in the inferior frontal gyrus, includes BAs 44 and 45. Broca's area is in general functionally correlated with language production tasks and syntax processing. Wernicke's area, which is located in the superior temporal gyrus (STG) and superior temporal sulcus (STS), includes the auditory areas BA 42 (belt, also known as AB) and BA 22 (parabelt, PB). Wernicke's area is functionally linked to language comprehension and semantic tasks. Another area of relevance to language and speech perception is BA 41, the primary auditory cortex (also known as A1).

However, rather than trying to localize linguistic functions in either of these two language areas, the intricate circuitry linking these two regions is studied to shed light on the contributions that each area brings to various language processing functions (Friederici, 2009). The proposed connectivity patterns between the language areas have been in part identified via the diffusion tensor imaging (DTI) method, which permits the identification and quantification in vivo of long-range connections between different brain areas. The main dorsal fibers directly connecting Broca's and Wernicke's areas are in the arcuate fasciculus. Rilling et al. (2008) offer an evolutionary perspective based on comparative DTI of the arcuate fasciculus in monkey, ape, and human. An additional indirect dorsal tract mostly connects Broca's BA 44 and Wernicke's lateral STG and middle temporal gyrus (MTG) via the superior longitudinal fasciculus (SLF). Two indirect ventral tracts mostly connect Broca's BA 45 with Wernicke's anterior STG via the uncinate fasciculus (UF) and the extreme capsule. The indirect dorsal route via SLF is known to be involved in the processing of syntactically complex and hierarchically structured sentences (e.g., nested sentences such as "The cat who chased the mouse died") (Bornkessel, Zysset, Friederici, von Cramon, and Schlesewsky, 2005). The indirect ventral pathway via UF is known to support phrase structure building (e.g., constructing a noun phrase with a determiner and a noun, such as "the cat"). Moreover, the connections between BA 45/47 and the mid part of STG/MTG are involved in semantic processing (Friederici, 2009).

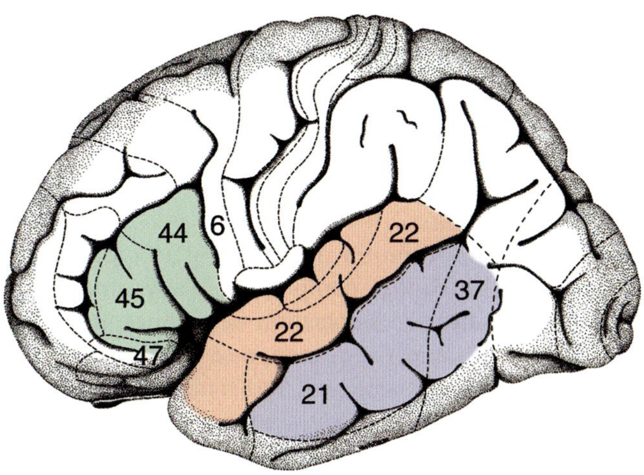

Figure 24.1 Brodmann areas (BAs) involved in language processing. This diagram, following Friederici's (2009) review of language brain areas, only shows the BAs directly related to linguistic functions referred to in the main text. The inferior frontal gyrus, also known as Broca's area, includes BAs 44 and 45. The superior temporal gyrus, also known as Wernicke's area, includes BAs 42 and 22. The middle temporal gyrus includes BAs 21 and 37. BA 47 is involved in semantic processing, together with BA 44. BA 6 corresponds to the premotor cortex. Figure courtesy of Angela Friederici. Reprinted with permission © *Trends in Cognitive Sciences.*

Vigneau et al.'s (2006) meta-analysis of the extensive literature on brain language areas has highlighted other functional roles of the left-brain areas and circuitry. Among numerous other findings, they report the existence of an elementary audiomotor loop for phonological processing, which supports a motor-sound-based representation for linguistic sounds. This loop involved Heschl's gyrus and the planum temporale in Wernicke's temporal lobe, for sound perception, and the frontal mouth motor area and inferior precentral cortex for the motor component. They also report the anteroposterior dissociation of phonological and semantic areas, and the presence of crossroad regions with overlap of phonological and semantic processing. These areas are modality dependent, with the auditory modality in the anterior temporal cortex and the visual modality in the posterior part of the inferior temporal gyrus. (But note that in the signed languages of the Deaf, phonology involves hand shapes, locations, and movements.) Vigneau et al. (2006) also demonstrate the existence of distinct working-memory networks, in frontoparietal reverberating loops, for each language component, though the data they summarize offer little insight into the interactions between these and other systems. The dorsal part of BA 44 has been found to be specialized for syntax processing. Finally, temporal areas also play a crucial role for sentence comprehension, with the processing of sentence meaning and syntactic constructions colocalized in the same region.

The overview of the specialization of left-brain areas and circuits for language-specific tasks does not mean that language processing is strictly modularized and acts separately from other nonlinguistic tasks. In fact, growing evidence from neuroscience and embodied cognition indicates a clear coactivation of brain areas involved in language and other cognitive functions, at least in certain tasks (e.g., Pulvermüller, 2003; Rizzolatti and Arbib, 1998; Pecher and Zwaan, 2005; Pezzulo et al., 2011). Hauk et al. (2004) studied the somatotopic representation of action words in the human motor and premotor cortex. They used event-related fMRI for passive reading tasks with words referring to face, arm, or leg actions (e.g., "lick," "pick," "kick"). These actions are controlled in spatially distinct areas in the prefrontal motor cortex, an area of the brain that has been unambiguously identified with motor control functions in a well-defined topological structure. fMRI results show that the purely linguistic task of reading a word differentially activated parts of the premotor area that were directly adjacent to, or overlapped with, the regions activated by actual movement of the tongue, the fingers, or the feet. Hauk et al. (2004) use this finding, and other evidence from related brain imaging studies (Pulvermüller, 2003), to claim that there is no need to look for a unified "meaning center" in the human brain or for the existence of a dedicated and separate neural system purely for language-specific functions. Instead, these findings support a dynamic view of language processing where words are processed by distributed neuronal assemblies, relying on cortical topographies that reflect word semantics. This dynamic view of language and sensorimotor knowledge processing is highly consistent with the embodied and neurorobotic models of language grounding described in section 24.3. However, it must be noted that, for example, somatotopy is not enough to distinguish, for instance, "punch" from "hit" or to link "hit" to the very different emotional valences of "He hit his daughter" and "He hit the ball." Again, it is unclear how premotor activity relates to a verb like "imagine," as distinct from imagining a specific action of the body. Thus, many challenges remain.

24.2 Neural Network Models of Language Learning

Progress in the neural modeling of language came initially from the field of artificial neural networks, in particular from the connectionist approach to cognitive modeling (Rumelhart et al., 1986; Christiansen and Chater, 2001). The connectionist approach focuses more on psycholinguistics rather than neurolinguistics, addressing the parallel and distributed subsymbolic processing of cognitive functions, rather than realistic neural network models of brain areas and their observed connectivity. However, this approach has contributed significantly to the understanding of the neural bases of language processing. More recently, through advances in large-scale neural simulation and neural modeling of other sensorimotor and cognitive functions, we have seen an increased development of computational neuroscience models. These tend to follow a "brain-constrained" approach to the neural modeling of language by constraining the network architecture to known language areas and connectivity circuits in the brain. In the following sections we will first look at the seminal contribution of connectionism to language learning modeling and, subsequently, explore the more recent computational neuroscience models.

24.2.1 Connectionist Modeling: From PDP to Brain-Inspired Neuropsycholinguistic Models

Connectionist models, also known as PDP (parallel distributed processing) models, consist of artificial neural network simulations that are loosely inspired by the brain mechanisms of parallel and distributed activation of nodes (artificial neurons), with connections (weights) between nodes trained using a variety of supervised (e.g., error backpropagation algorithm) and unsupervised (e.g., Hebbian rule, Kohonen competitive models) methods (Rumelhart et al., 1986; Levine, 2000). They have been extensively used for the modeling of language learning phenomena ranging from phonetics (e.g., reading) and lexicosemantic (e.g., object naming) simulation to syntax acquisition (grammar, morphology, recursion) studies. Here we will discuss some seminal examples of connectionist language simulations contributing to various aspects of linguistic functions. We will first consider the contribution of neural networks, through the English past tense debate, to providing alternative explanations of the use of rules in language (Rumelhart and McClelland, 1986). The role of a simple but elegant model of learning syntax with simple recurrent neural networks architecture will be examined (Elman, 1990, 1993). And the use of connectionist models for neuropsychology investigation of language deficits will also be discussed (Plaut and Shallice, 1993), as they have a more direct focus on the lesioning of brain regions in patients with impaired cognitive functions. For a more comprehensive review of the wide literature on connectionist models of language, see Christiansen and Chater (2001).

The case of the learning of the English past tense, with its typical classification into regular morphological forms (e.g., love → loved; book → booked) and irregular examples (e.g., go → went; do → did; wear → wore), has been used by psycholinguists and Chomskian linguists as evidence supporting the fact that people use rules in language processing. Pinker and Prince (1988, p. 73), in their analysis of the past tense debate, stated "researchers who could agree on little else [on language study] have all agreed on one thing: that linguistic knowledge is couched in the form of rules and principles." For instance, the rule "add suffix –ed" could be used to form the past tense of all regular verbs. Moreover, children follow a U-shaped learning trajectory in the acquisition of the English past tense. That is, initially children produce the correct past tense form for high-frequency irregular verbs, such as "went" and "came," and the correct –ed forms for the regular verbs. However, they subsequently go through a temporary stage of "overregularization" by producing incorrect forms such as "goed" and "comed." Eventually, they learn again to distinguish the multiple forms of irregular past tenses, in addition to the regular –ed form. This nonlinear developmental trajectory was used to support a dual-process theory of past tense: if stem is regular, use –ed rule; if stem is irregular (i.e., appears on a list of exceptions to the general rule), use the rote-memory-learned irregular form.

One of the very first connectionist models of language was developed by Rumelhart and McClelland (1986) specifically to simulate the learning of the English past tense by a simple feedforward neural network. In input the network encodes the phonological representation of the root form of each verb and in output the phonological representation of the past tense form. Rumelhart and McClelland (1986) used a phonological representation of words based on "Wickelphones" (Wickelgren, 1969), a context-sensitive method to represent linguistic sounds with a triplet of phonemes (preceding, actual, and successor phone), with the symbol # used to indicate the beginning or the end of a word. So, for instance, the verb "give," with its standard phonetic representation of, for example, using a *sequence* /g/ /i/ /v/, is encoded by a *set* of three Wickelphones: ${}_{\#}g_{i}$, ${}_{g}i_{v}$, ${}_{i}v_{\#}$. This encoding has the advantage that no two words share the same Wickelphone set. The network architecture (figure 24.2A) consists of a central two-layer simple perceptron to learn the association between the input Wickelphone verb roots and the output Wickelphone past tense forms. Each input–output layer has 460 units, which corresponds to the number of unique Wickelphones necessary to encode the set of verbs used in the experiment. Each of these two layers then has a paired linked layer for the encoding–decoding into standard phonological representations. Learning is implemented through the perceptron convergence procedure, which is a discrete version of the delta rule (Rosenblatt, 1962). After training with 506 regular and irregular past tense verbs, the network was able to learn both types of verbs and generalize the production of the correct Wickelphone form for 90% of the novel verb stems using the same learning mechanism and sharing the same connections and hidden layer units. The pattern of errors also resembles that of English speakers, for example, treating a novel test verb as irregular if it has many "neighbors" who are irregular. For example the past tense of the verb "sip" was produced as "sept" (instead of its correct form "sipped") presumably for its similarity to sleep–slept and keep–kept. Moreover, the networks' error curves and test performance during training followed the U-shaped patterns, with the production of overregularized forms for irregular verbs during the intermediate stages of learning. This model was used by Rumelhart and McClelland (1986) to reject the dual-process theory of language morphology learning, as the neural network is based on PDP mechanisms and does not implement any rule. Rumelhart and McClelland conclude: "We have, we believe, provided a distinct alternative to the view that children learn the rules of English past-tense formation in any explicit sense. We have shown that a reasonable account of the acquisition of past tense can be provided without recourse to the notion of a 'rule'

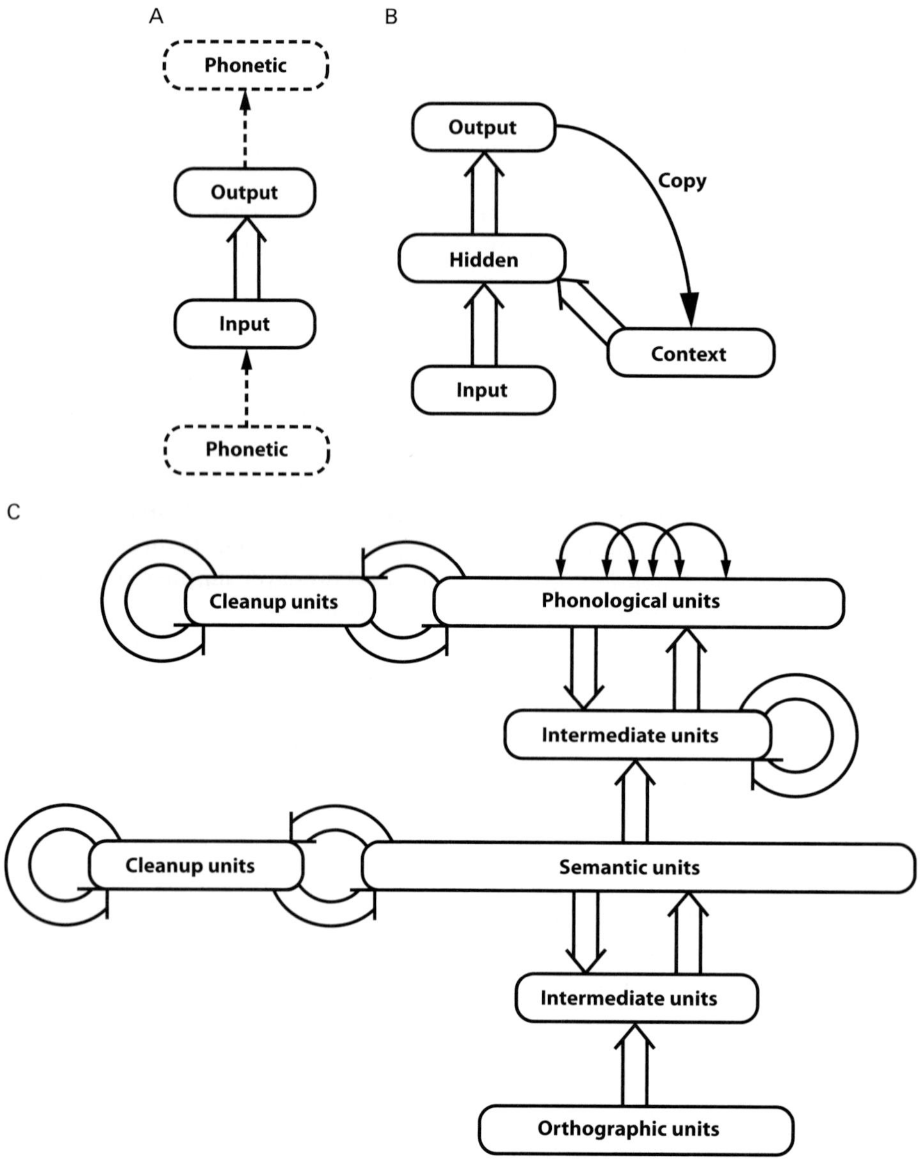

FIGURE 24.2 Examples of connectionist networks for language functions. (A) Architecture of the perceptron in Rumelhart and McClelland (1986). The two layers in continuous line constitute the main perceptron network whilst the two dotted line boxes only represent the phonetic encoding used by Rumelhart and McClelland. (B) Architecture of the simple recurrent network in Elman (1990). (C) Recurrent architecture of Plaut and Shallice's (1993) model of deep dyslexia for abstract and concrete words.

as anything more than a description of the language. [...] The child need not figure out what the rules are, nor even that there are rules. The child need not decide whether a verb is regular or irregular" (Rumelhart and McClelland, 1989, p. 267). This study initiated a long debate on rule versus connectionist models of language learning, as with the initial criticism by Pinker and Prince (1988) and with further, more elaborated connectionist models of the English past tense (e.g., Plunkett and Marchman, 1996). Overall, these connectionist simulations demonstrate that although the notion of "rule" might help the linguists to describe the behavior of children, and might be relevant for schoolchildren's and adults' metacognitive understanding of language and grammar and for second-language learning, it does not necessarily mean that rules play a role in generating most linguistic behavior, especially from a developmental perspective (Elman, 2001).

Another area where connectionist models have contributed to the neural modeling of language is that of word and syntax acquisition, in particular with the use of recurrent neural architectures. As language is highly characterized by time-based sequences, such as phoneme sequences to build a word, and word sequences to form a sentence, recurrent neural networks are suitable to integrate information over time and can learn to predict the next phoneme, or word, in a sentence. A seminal model that addressed such an issue with a simple but powerful recurrent architecture is that of Elman (1990). He developed the simple recurrent network (SRN), which is a standard three-layer feedforward perceptron with an additional set of units called a "memory" or "context" layer connected to form a loop (figure 24.2B). These memory nodes copy the state of activation of the hidden layer nodes at time step t and then send it back as input to the same hidden layer together with the new input stimuli at time $t + 1$. This is then used to predict the next sound/word at time step $t + 2$. The learning algorithm for the SRN is a standard error backpropagation algorithm (Rumelhart et al., 1986), as the memory layer connected to the hidden layer is considered just as an additional input layer in the calculations of the errors and deltas in the hidden nodes.

Elman (1990) used SRNs to simulate both the prediction of the next sound (letter) in words, as well as the prediction of the next word in a sentence. For the word learning example, 15 words based on the same four letters were selected and concatenated in 200 random sentences of four- to eight-word length. The input to the network encoded a single letter in a localist way (one unit used to represent each letter), and the task was to predict the next letter. The SRN architecture had five input units, 20 hidden units, five output units, and 20 context units. The model was able to learn the segmentation of the letter streams into words. While the prediction error of the first letter of each word remains high (as it is impossible to predict the initial letter of a random word), this decreases significantly with the next letters. This demonstrates that the network identifies the letters that form a word. The same methodology was used to predict words in a sequence and to discover lexical classes from word order. Twenty-nine words were used, each encoded localistically in the input and output layers (i.e., each word is associated to one node). These were combined to form 10,000 sentences using 15 templates of nouns with transitive and intransitive verbs. The SRN architecture was significantly more complex with respect to the single word prediction, with 31 input units, 150 hidden units, 31 output units, and 150 context units. After training the network to predict the next word in a sentence, the SRN was able to predict words correctly (except for high error at the first word). Moreover, an exploration of the hidden layer activations through hierarchical cluster analysis demonstrates that the SRN has extracted the lexical classes of the 29 words simply relying on word order information. The hidden layer representations are first separated for the classes of nouns and of verbs. Within the noun class, the words are separated for inanimate (food) and animate (humans and animals) entities and so forth. In a follow-up study, complex nested sentences were used, involving relative clauses and number agreement, as in "Boys who chase dogs see girls" (Elman 1993). The simulation shows that to be able to learn such complex sentences the SRN must go through maturational stages where its memory capability (number of previous cycles accumulated by the context layer before this is reset) grows over time.

This simple but powerful model of language learning has significantly impacted the field of neural network modeling of language—for example, in showing that PDP models can develop structured semantic–syntactic representations and demonstrating the principle of the "importance of starting small" in cognitive development (Elman 1993; Elman et al., 1996). Of course, this simplified model does not cover all aspects of language, such as the representation of meaning or the generation of sentences with much larger lexical items. Subsequent connectionist models have looked at the modeling of more properties of language, such as the one described below on semantic-related phenomena such as deep dyslexia.

The modeling of neuropsychological deficits is another area where connectionist networks have impacted the computational investigation of language, specifically in a field more relevant to computational neuroscience. Plaut and Shallice (1993) developed a connectionist model of deep dyslexia. This is an acquired reading disorder characterized by the production of semantic errors, as well as visual and orthographic ones. For example, when reading the word "NIGHT," the patient substitutes it with a semantically related word such as "sleep." Patients also demonstrate a better performance in concrete over abstract words. Building on a previous model of dyslexia by Hinton and Shallice (1991), Plaut and Shallice (1993) manipulate numerous properties of connectionist networks (e.g., network architecture, training protocol, encoding schemes for orthographic and semantic representations) to reproduce in greater detail the symptoms of deep dyslexic patients. In particular they use distributed orthographic and semantic representations for the word stimuli, implement gradient descent learning, include a richer semantic representation for concrete versus abstract words, and experiment with various alternative network topologies for the implementation of dynamic attractors for word meanings, instead of static hidden unit representations. For simplicity we only describe here one of the many network architectures used by Plaut and Shallice to model deep dyslexia for concrete and abstract words, to show the modeling details of a connectionist network of a language deficit. The network in figure 24.2C shows the most complex architecture used by Plaut and Shallice (1993) to model deep dyslexia. The 32 orthographic input units encode graphic features of the letters of a word, such as the presence of vertical, horizontal, diagonal, and curved strokes, of a closed part, and the horizontal and vertical symmetry. The intermediate layer of 98 semantic units encodes semantic features. These include 67 features adapted from Hinton and Shallice's (1991) semantic encoding for concrete words (e.g., main-shape-3d, found-in-woods, living-entities) and 31 features to represent abstract word meaning components (e.g., has-duration, relates-location, quality-difficulty). Through the recursive link with the cleanup layer, the activation of the semantic units facilitates the emergence of word meaning attractors. These attractors were proposed by Hinton and Shallice (1991) as a key component for modeling semantic-based language function and impairment. It is through the use of autoconnections, or the recursive link via the cleanup units, that the semantic hidden layer converges on more or less word-related patterns of semantic units, initially generated by the incomplete semantics supplied bottom-up. The output layer of the network uses 61 phonetic units. These include seven sets of position-specific and mutually exclusive phonetic units: three slots for phonemes from the initial consonant cluster, one for the vowel, and three for phonemes from the final consonant. For the training, 20 abstract and 20 concrete words were presented to the network repeatedly. Each pair of abstract–concrete word-only words differed by a single letter (e.g., "TART"–"TACT," "COAT"–"COST").

Following various types of simulated lesions to the networks (i.e., deletion of some units or connections, or adding of noise to some weights), they found, for example, a double dissociation in the reading of concrete versus abstract words. This is the result of assigning concrete words far more semantic features than abstract words, following the assumption that isolated concrete words evoke richer, more consistent semantic representations than do abstract words. The detailed simulation of a variety of symptoms of deep dyslexia patients demonstrates that PDP neural networks can explain the dissociation of neuropsychology syndromes (or symptom complexes) without having to rely on functional and modular box-and-arrow theories of normal and impaired cognitive capabilities (Plaut, 2003). In functional box-and-arrow models of cognitive functions, such as the well-known model of reading by Coltheart (1987), the researcher identifies separate functions contributing to a specific behavior. These are then connected in a serial and modular way, following serial human-information-processing strategies. Coltheart's (1987) model of reading is thus based on two separate, modular pathways for reading, namely, the orthographic–phonological module and the lexical access module. Damages to either of these modules can explain the double dissociation neuropsychology findings. In the distributed and parallel connectionist model by Plaut and Shallice (1993), as in the past tense model by Rumelhart and McClelland (1986), such separate modules do not exist, as, for example, the lexical access relies on the same units and connections used for the orthography–phoneme units. Although in an articulated connectionist architecture, as that of Plaut and Shallice, different layers specialize for orthographic, phonetic, and semantic representation, all these are involved in the processing of all words. Moreover, lesions of different intensity of the units and connections in the PDP network can explain symptoms of varying severity, while in the lesioning of the single arrow connecting functional boxes in functional modules, the symptom can only have a none-or-all effect.

The three main examples discussed above demonstrate the feasibility of connectionist networks to model some aspects of verb morphology, sentence processing, and deficits in reading. Connectionist models have covered a much wider spectrum of linguistic processes such as word recognition with the interactive activation and competition model (e.g., Rumelhart and McClelland, 1982), syntax and parsing (Chang et al., 2006; Vosse and Kempen, 2009), and language productivity and systematicity (Hadley, 1994). In addition, models have looked at aspects of semantics such as situated language comprehension (Mayberry et al., 2009), lexicon learning (Li et al., 2004; see also language acquisition connectionist and robotics models described in section 24.3.2), and also language processing beyond individual words and/or sentences, as in story and script processing models (Miikkulainen, 1993). However, a pure connectionist approach can only capture some general properties of the neural processing of language, specifically the use of distributed and parallel representations, or learning from experience with training stimuli and reproduction of nonlinear developmental phenomena, and the modeling of gradual loss of function due to unit or connection lesioning. Connectionist models cannot, on the other hand, model the specific neuroanatomical organization of brain areas and circuitry known to be involved in the various level of languages processing from speech perception, to lexical and meaning representation, to syntax and overall semantics. In the next section we will be how computational neuroscience methods can be employed to build neuroanatomical models of various language processing functions.

24.2.2 Neuroanatomical Models of Language Processing In neuroanatomical models, the architecture of the neural networks and their functional principles are constrained, as much as possible, by known brain anatomy and physiology. The development of such brain-constrained models is still in its early stages, as they build on the progress of connectionist models developed in the 1980s and 1990s, and they explicitly aim at validating recent neuroscience findings and hypotheses on language processing. To show how these models work, we will analyze in detail three key neuroanatomical models of various linguistic functions ranging from word learning (Garagnani, Wennekers, and Pulvermüller, 2008), to acoustic pattern recognition (Husain, Tagamets, Fromm, Braun, and Horwitz, 2004), and syntax and sequence learning (Dominey, Inui, and Hoen, 2009).

One neuroanatomical model that has attempted to simulate the various areas of the brain involved in language processing, and its link with perceptual (hearing the word) and motor (saying the word) functions, is that developed by Garagnani and collaborators (Garagnani et al., 2008; Garagnani and Pulvermüller, 2011). They implement a neuroanatomical neural network model of Broca's and Wernicke's perisylvian cortex areas to simulate the neural processes of early language acquisition and to model the EEG data of psycholinguistics experiments with familiar meaningful word and senseless pseudoword stimuli. The architecture of the neural network consists of a series of interconnected 2-D layers of units, each called a cell assembly, with Hebbian associative learning connections, simulating the connectivity between Broca's and Wernicke's areas, and auditory and motor areas, along the perisylvian cortex (figure 24.3). The model includes six different areas of the cortex, specifically the primary motor cortex (M1, approximately equivalent to BA 4), premotor cortex (PM, equivalent to BA 6), prefrontal cortex (PF, equivalent to Broca's areas BA 44 and 45), primary auditory cortex (A1, equivalent to BA 41), AB (equivalent to Wernicke's area BA 42), and PB (equivalent to Wernicke's area BA 22). Each area is implemented as a layer of 25×25 graded-response excitatory units and an underlying layer of 25×25 graded-response inhibitory units. These areas are linked with each other through long-distance corticocortical connections (see black arrows connecting cortical areas in figure 24.3 top) as well as having within-area lateral and recurrent connections.

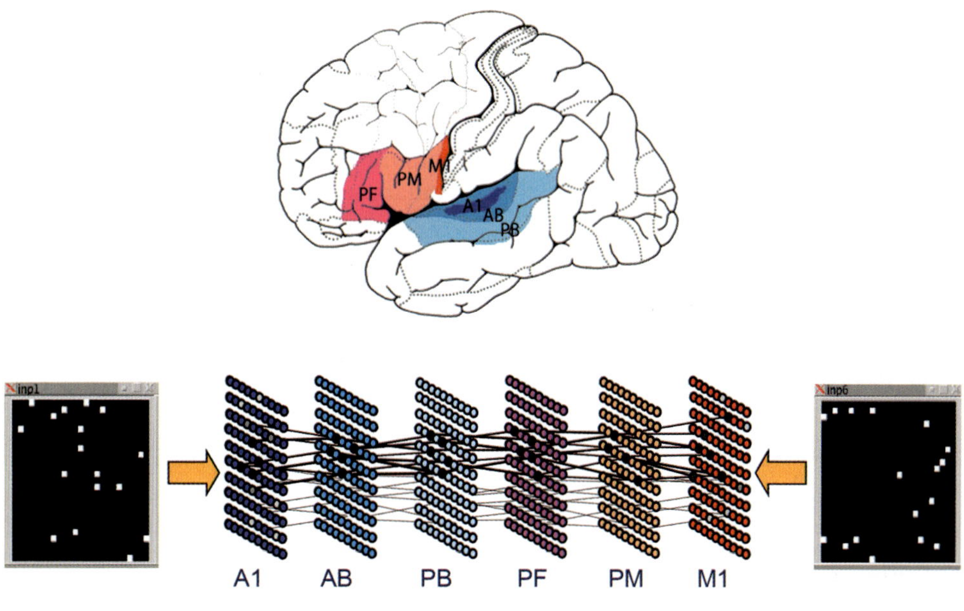

Figure 24.3 Simulated brain areas (top) and multilayer network architecture (bottom) in the Garagnani et al. (2008) model of word learning and mismatch-negativity detection. In the diagram of the six interconnected cortical layers, the small filled ovals represent excitatory neuronal pools, and gray ovals the coactivated cells. The areas are linked with each other through long-distance corticocortical connections as well as having within-area lateral and recurrent connections. The six cortical areas are: primary motor cortex M1 (approximately equivalent to BA 4), premotor cortex PM (BA 6), prefrontal cortex PF (Broca's areas BA 44 and 45), primary auditory cortex A1 (BA 41), belt (AB; Wernicke's area BA 42), and parabelt (PB; Wernicke's area BA 22). See text for explanation. Figure courtesy of Max Garagnani.

Within each area, a unit represents a cortical column, and its activity is based on mean field activation. Two main parameters are used for the activation of each mean field unit x at time t. The membrane potential $V(x,t)$ corresponds to the temporal low-pass properties of an assembly of neurons (equation 24.1). The output activation $O(x,t)$ (equation 24.2) transforms the membrane potential into the average firing rate through a piecewise linear approximation to a sigmoid function:

$$\tau \cdot \frac{dV(x,t)}{dt} = -V(x,t) + V_{In}(x,t), \qquad (24.1)$$

$$O(x,t) = \begin{cases} 0 & \text{if } V(x,t) \leq \theta \\ (V(x,t)-\theta) & \text{if } 0 < (V(x,t)-\theta) \leq 1 \\ 1 & \text{otherwise} \end{cases}. \qquad (24.2)$$

In equation 24.1 τ is the membrane time constant and $V_{In}(x,t)$ is the total input to unit x at time t, calculated as the sum of all excitatory and inhibitory postsynaptic potential connected to x. The low-pass dynamics of the units are integrated in equation 24.1 using the Euler scheme with step size Δt set at 0.5 arbitrary time units (where one Δt can be considered equivalent to about 20 ms of a real neuron activation).

For the output activation $O(x,t)$, representing the cumulative, graded output of the columnar neurons represented by the unit x, equation 24.2 produces an average firing rate between 1 and 0, with a linearized sigmoid shape using a slope between the lower and upper threshold, respectively, for the range θ to $\theta + 1$. The value of θ is initialized at 0 and varies according to the time-average of the unit's recent output, to adapt the unit's response to the input level. In particular, higher-than-average values of $O(x,t)$ lead to a gradual increase in θ, and lower-than-average values of $O(x,t)$ lead to its gradual decrease.

For learning of the connection between the excitatory units, a Hebbian plasticity mechanism is implemented, while the inhibitory units have fixed, randomly initialized connections. For the Hebbian learning, a simplified version of the Artola–Broecher–Singer (ABS) learning rule is used (equation 24.3), which models the neurobiological phenomena of long-term potentiation (LTP) and long-term depression (LTD) (Artola and Singer, 1993).

$$w_{t+1}(x,y) = \begin{cases} w_t(x,y) + \Delta w & \text{if } O(x,t) \geq \theta_{pre} \text{ and } V(y,t) \geq \theta_+ \\ w_t(x,y) - \Delta w & \text{if } O(x,t) \geq \theta_{pre} \text{ and } \theta_- \leq V(y,t) < \theta_+ \\ w_t(x,y) - \Delta w & \text{if } O(x,t) < \theta_{pre} \text{ and } V(y,t) \geq \theta_+ \\ w_t(x,y) & \text{otherwise} \end{cases} \qquad (24.3)$$

where $\Delta w = 0.0005$.

The new weight w of the connection between the two units x and y is obtained by adding–subtracting a constant Δw (equal to 0.0005, considering a very low learning rate) depending on the comparison of the output value and membrane potential with respect to fixed thresholds. Specifically, this implies the comparison of $O(x,t)$ with threshold θ_{pre} (equal to 0.05, as an arbitrarily minimum level of presynaptic activity required for LTP to occur) and $V_{In}(x,t)$ with θ_- and θ_+ (respectively, 0.15 and 0.25 as membrane potential minimum–maximum threshold). The three cases of changes of weight as in equation 24.3 implement the following Hebbian learning cases: (1) homosynaptic and associative LTP, (2) homosynaptic LTD, and (3) heterosynaptic LTD.

In the simulation, the model develops distributed functional circuits through the learning of perception–action mapping patterns, linking neural codes for the auditory and motor forms of words (the present model has no links to syntax or semantics). Hebbian associations are learned through the repeated, simultaneous activation of predetermined sets of cells in the primary auditory cortex (A1) and the primary motor cortex (M1), as a result of the simultaneous perception of auditory stimuli and its related articulatory pattern. The network's training consists of the cyclic presentation of four different pairs of patterns (randomly generated uncorrelated sparse binary configurations of 17 active units). As a model of bottom-up, stimulus-driven selective attention, a winner-take-all mechanism is used within a cell assembly. This is realized via a cortical activity regulation method based on the response of self-regulation, area-specific inhibitory loops between inhibitory and excitatory units. These area-specific self-regulation loops control and modulate the total activity within each cell assembly area. This is implemented with feedback inhibition (FI) connections of excitatory units to a single, area-specific inhibitory unit, and the corresponding connection from each of these inhibitory units back to all of the excitatory units of that area. This provides an amount of inhibition proportional to the total activity within that area to modulate attention. FI connections with small weights correspond to the provision of more attentional resources to various stimuli due to the lower inhibition of alternative cell assemblies dedicated to each stimulus. Larger FI weights, instead, correspond to the greater inhibition of competing cell assemblies and thus the availability of fewer attentional resources. The winner take-all-mechanism and the FI weights are used to implement a mechanism of attention-demanding task in language processing, which is one of the key features of the Garagnani et al. model.

This model, in addition to simulating the chain of cortical area activation during word learning, has been used for modeling specific brain imaging phenomena studied with ERPs. In particular this study focuses on the simulation of the ERP mismatch negativity for words and pseudowords. The mismatch negativity (latency 100–250 ms) is an early automatic brain response elicited under distraction, which is larger to meaningful stimuli (words) than senseless material (meaningless pseudowords) (Pulvermüller et al., 2001). A later response peaking at around 400 ms (N400) is an ERP that appears in attentional tasks and is larger for pseudowords than for matched meaningful words (Sinai and Pratt, 2002). Results of ERP signal simulations show that variation of the correlated of attentional mechanisms (i.e., area-specific inhibition) modulates the response to words and pseudowords. Specifically, weak FI connections, corresponding to greater attention, produces an N400-type response, that is, late activation differences with a stronger response to pseudowords. Stronger inhibitory connections for lower attention produce a mismatch-negativity response pattern, that is, early activation differences, with a stronger response to words than to pseudowords (figure 24.4). Through these simulation results the authors predict that the modulation effects of attention on the cortical responses to pseudowords can explain the reversed patterns of ERP activity observed experimentally under different conditions.

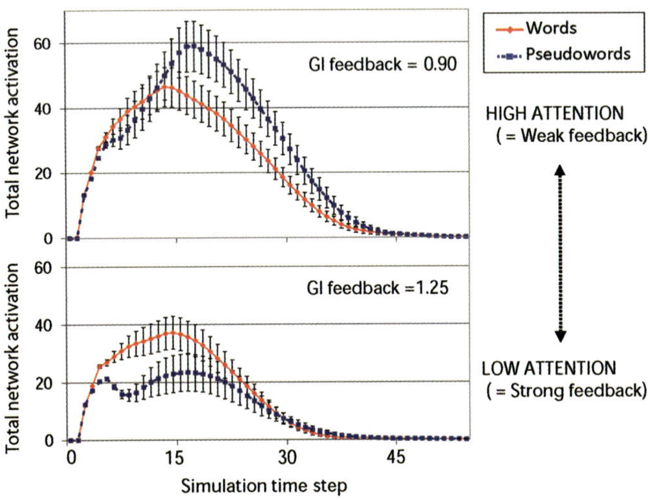

FIGURE 24.4 Simulated event-related potential (ERP) results in Garagnani et al. (2008) model. The top figure shows simulated ERP responses for weak global inhibition (GI) connections and high attention. This shows the presence of a late N400-like signal for pseudowords. The bottom figure shows simulated ERP responses for strong FI connections and low attention. This show a mistmatch-negativity response pattern for pseudowords, with the inversion of the signal in earlier stages. Figure courtesy of Max Garagnani.

Overall, Garagnani et al. (2008) and the follow-up simulations on auditory change detection and memory access (Garagnani and Pulvermüller, 2011) provide a computational neuroscience modeling framework for the investigation of cortical mechanisms of mismatch-negativity phenomena at the word level. This computational neuroscience framework has been shown to make valid predictions on the effects of attention on latency and magnitude of ERPs to words. This approach also supports the viability of using Hebbian associative learning, neuronal adaptation, and lateral inhibition in a multilayered neural network architecture for modeling language phenomena and ERP data on mismatch negativity, though in its current form it does not include the modeling of word meaning.

Other neuroanatomical models of language processing have been proposed for the simulation of functional neuroimaging data from fMRI and positron emission tomography (PET) (see also chapter 15 by Griego, Cortes, Winder, and Tagamets). Husain et al. (2004) developed a large-scale, neurobiologically realistic network model of auditory pattern recognition in the primary auditory cortex and prefrontal cortex. The main characteristic of this approach is the modeling of the electrical activities of the neuronal units, with the large-scale neural architecture constrained on data from the neurophysiological literature on the perception of auditory stimuli, specifically frequency modulated (FM) sweeps and tones. The primary auditory cortex acts as a set of feature detectors of sound patterns, which include the intensity, and rate of FM sweeps and tones, the duration of the sound, and the spatial location and direction of the sound source. In Husain's work, the focus is on the recognition of the direction of FM sounds (Kubovy and Van Valkenburg, 2001). Although this study does not directly model the learning of words of a lexicon but rather the response to acoustic stimuli and their processing in auditory cortical areas, it is relevant to neural models of language as it can be extended to the implementation and investigation of phonological aspects of language processing.

The architecture of Husain et al.'s neural model consists of groups of cortical areas from the auditory cortex and the prefrontal cortex (figure 24.5). The auditory signal is provided by the input module corresponding to the thalamic medial geniculate nucleus, with 81 basic units representing neuronal populations selective for a particular range of frequencies. This has oriented converging connections to the auditory areas for upward–downward sweeps. The auditory cortex consists of a complex network of feedforward and feedback connections for the primary auditory areas (Ai, broadly equivalent to BA 41, also known in the literature as A1),

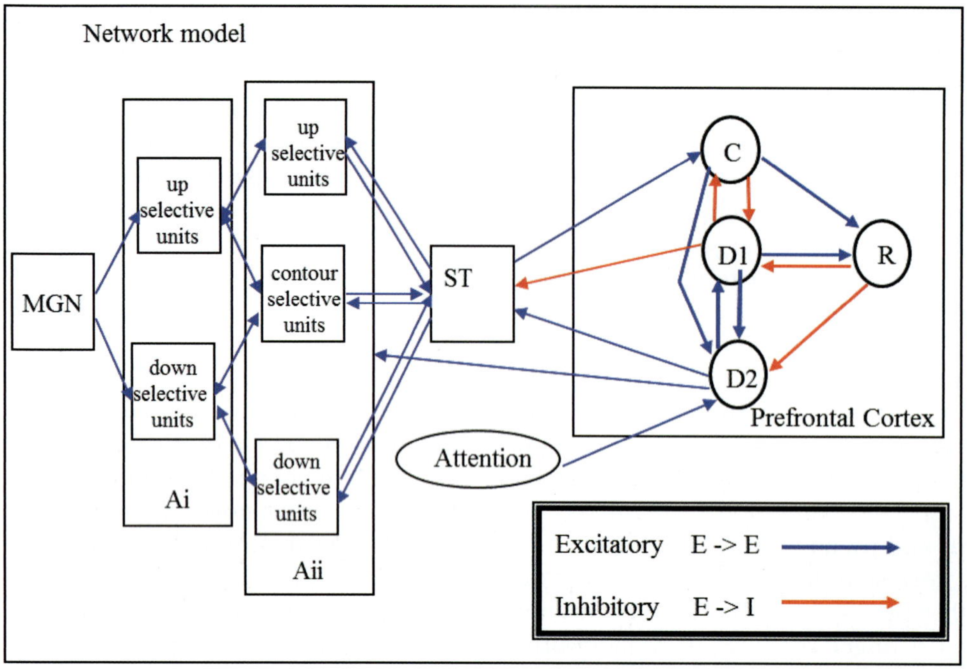

FIGURE 24.5 Architecture of Husain et al. (2004) model of auditory object processing. The auditory input comes from the medial geniculate nucleus (MGN), which is then encoded in the primary auditory areas A1 (Ai) and A2 (Aii) and passed to the superior temporal (ST) area. The prefrontal cortex area has four types of units: cue-sensitive units C to respond to external sound stimuli, delay units D1 and D2, and response units R to select the response. See text for full explanation. Figure courtesy of Fatima Husain. Reprinted with permission.

secondary auditory cortex (Aii, equivalent to BA 42 and 22, also known as A2), and areas of the STG and ST. The Ai area is simulated as units selective to upward-sweep and downward-sweep stimuli. Aii contains three types of units responsive for longer up–down sweep segments, as well as contour-selective units sensitive to changes in sweep direction. Ai and Aii units are tonotopically arranged, that is topographically arranged according to the frequency of the tone. The prefrontal cortex plays the function of short-term working memory. This consists of the area C of cue-sensitive units that respond to external sound stimuli, D1 and D2 for the delay units and R for the response units. The prefrontal cortex cell structure and its short-term memory system are inspired by the prefrontal cortex memory systems for visual stimuli (Funahashi et al., 1990). Each individual module in the network is composed of 81 basic units, each of which is composed of an excitatory and an inhibitory element, with connection weight values functionally equivalent to the proportion of synaptic connections. The network has both feedforward and feedback connections between different areas. As for intraregion connections, the units of the different subareas of the tonotopically arranged Ai and Aii regions have different types of lateral competition connections and within-area connections that lead to the upward or downward or contour-selective functionalities. A top-down (e.g., task instruction) attentional mechanism is implemented through the incoming activity to one of the delay modules D2. This corresponds to the level of attention with which the D2 units respond to a given stimulus, with higher attention corresponding to representations being maintained for longer.

The basic unit of all the modules of this architecture represents a simplified cortical column, containing both an excitatory element (corresponding to pyramidal cells) and an inhibitory element (for interneuron inhibitory cells). The excitatory component has stronger weights and contributes more to the overall activation of the unit. The activation function of both the excitatory and inhibitory elements of each basic unit is based on a sigmoidal rule that takes into consideration the input electrical activity, a decay component, and random spontaneous activity. The basic unit structure and activation function is based on previous models of visual cortex (see Tagamets and Horwitz, 1998).

The model is used to simulate the delayed match-to-sample (DMS) task of auditory pattern recognition. The DMS task consists in the presentation of an auditory stimulus, a delay, and the subsequent presentation of the same, or different, stimulus. The network must decide if the second stimulus is the same as the first, with the R units comparing the previous stimulus (stored in the D1 delay units) and the current stimulus (represented at the cue-selective C module) and decide whether the two stimuli match or not. The auditory stimuli used in the simulations were tonal contours composed of combinations of up- and down-frequency modulation sweeps and pure tones. The activations of areas in the model were compared with cortical area activation in an fMRI study with human participants on the same task. In the model, the spatially and temporally integrated electrical activities were used to simulate the BOLD signal to compare simulation data with the experimental observations by contrasting the response to tonal contours with that for pure tones. For example, results show a good match between cortical areas active in the model and those in the fMRI experiment. Only for the area Ai was the correspondence less consistent. Moreover, a better match was observed if the attention level during the tones task was set at a lower value than during the tonal contour task, consistent with the reaction time (RT) responses of human participants. The attention level (corresponding to top-down attentional requirements coming from the task instruction) is simulated as the modulation of the delay response units, with high attention corresponding to longer maintenance of the representation in the delay units. Overall, this computational neuroscience approach to the modeling of auditory processing provides a framework for linking physiological and hemodynamic–metabolic studies in both simulated and experimental investigation. The close correspondence between simulated and neuroimaging data permits the manipulation of the computational model to investigate specific hypothesis on the involvement of auditory areas in language tasks, by, for example, including other areas involved in language processing tasks, as in the Garagnani et al. (2008) study. Model predictions can then be further validation in new fMRI experiments, given the direct data comparison methodology.

Where the previous models have been at the auditory pattern level with no connection to syntax or semantics, Dominey and collaborators have proposed a neuroanatomical model of language learning that has been applied both to grammatical constructions (Dominey, 2005) and to sentence comprehension and nonlinguistic sequence learning (Dominey et al., 2009). Here we provide an overview of the sequential learning model used to investigate the role of brain language areas in both linguistic and nonlinguistic tasks. For the linguistic (syntactic comprehension) task, this model focuses on the assignment of thematic roles to noun phrases. In subject–verb–object (SVO) languages such as English, the canonical order in which thematic roles are assigned is agent–object–recipient. This applies to the canonical sentence "Fred introduced John to Bill," with "Fred" being the agent, "John" the object, and "Bill" the recipient. A noncanonical sentence representing the same situation, but with different order, would be "John was introduced to Bill by Fred." To perform thematic role assignment, a person must distinguish between content words (nouns, names) and function words ("by," "to"). The capability to assign semantic roles to the content words, guided by the function words, provides the basis for the assignment of thematic roles.

Dominey et al. (2009) use a modular neural architecture (figure 24.6) closely constrained on the model of the left-hemisphere brain areas involved in auditory sentence processing, as proposed by Friederici (2012). This architecture is based on dual processing streams for function words (closed class) and content words (open class), respectively corresponding to the lexical categorization of function words in STG and of content words in the medial temporal gyrus (MTG). BA 47 is modeled as a recurrent network encoding the temporal sequence of closed class words received in input from STG, while the network for the area BA 45 receives open class elements from MTG. The BA 47 network is based on fully recurrent connections with continuous update of the unit's activation (Dominey and Ramus, 2000; also see chapter 22 by Dominey and Enel in this volume) rather than an Elman-type simple recurrent network. BA 44/6 has the role of working memory to encode the sentence in terms of four semantic roles: predicate, agent, object, recipient. For example, in the input sentence "John was introduced to Bill by Fred," the verb "introduce" is represented in the BA 44/6 predicate units, "Bill" in the recipient population, "Fred" in the agent units, and "John" in the object population.

The input sentence is encoded through a grid of 5 × 5 units, with the first 13 elements reserved for the closed class words, and the next 12 for the open class words (localist encoding, one unit per word). In the modular architecture, each of the modules consists of a 5 × 5 population of neurons, through the use of leaky integrator neurons. This permits the analog representation of time processing, such as for the representation of the durations of the input sequence and the production of RTs, which are important in sequential learning tasks.

To investigate the neural bases of thematic role assignments, Caplan et al. (1985) use nine sentence types of varying syntactic complexity (five canonical and four noncanonical) to study thematic role assignment in aphasic patients. They reported a deficit in thematic role assignment that is highly selective for noncanonical sentences in patients with left-hemisphere lesions. In

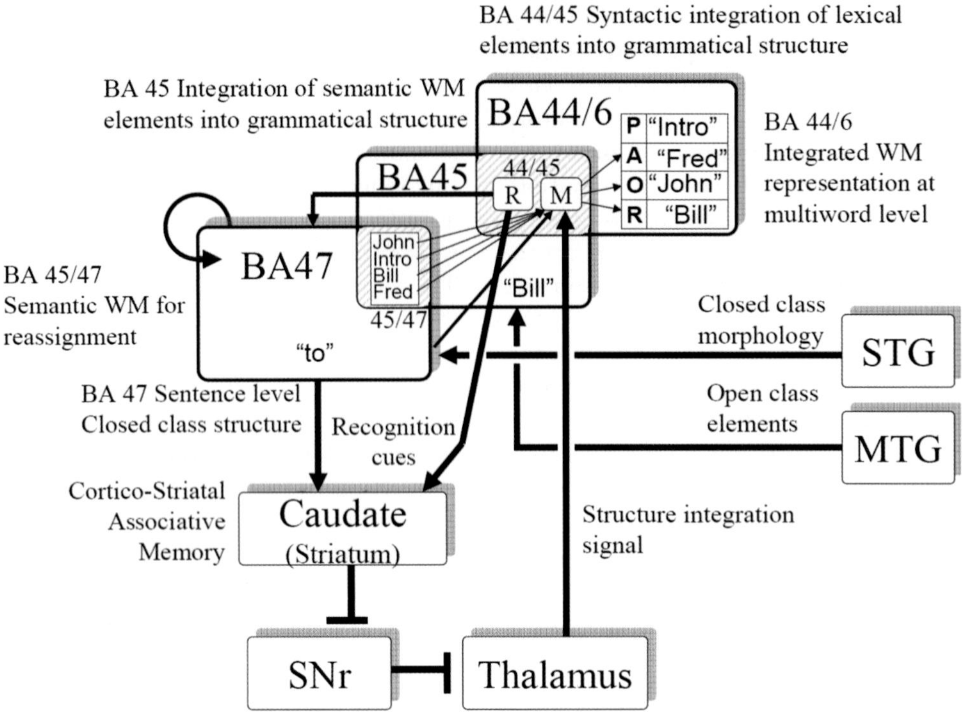

FIGURE 24.6 Neural architecture of Dominey et al.'s (2009) model of thematic role assignments and sequence learning. The diagram shows the correspondence between the modules of the neural network and the brain areas. Input is provided by the superior temporal gyrus (STG) module (closed class function words) directly to the BA 47 neural module, and by the middle temporal gyrus (MTG) module (open class content words) to BA 45. Working memory (WM) is implemented by the BA 44/6 module for the four meaning types: (P) predicate, (A) agent, (O) object, and (R) recipient. The function word activation in BA 47 is directly connected to the head of the caudate, and via the "canonical" cortico–striato–nigro–thalamo–cortico circuit, projects to BA 44/45. The structural unification of the sentence is represented in the transition along the interconnected BAs 47–45–44. Given the centrality of these areas, two functional subdivisions are considered. BA 44/45 M (modulation) modulates the binding of semantic content in BA 45 into the thematic frame in BA 44/6. BA 44/45R (recognition) recognizes the open class element in BA 45 and its role in the meaning coded in BA 44/6. SNr, substantia nigra pars reticulata. Figure courtesy of Peter Dominey. Reprinted with permission.

Dominey et al.'s (2009) simulation study, the neural network is trained on a supervised version of six of the Caplan task's sentence types.

The target of the network training is to learn the mapping between each sequence of open class elements in BA 45 and their correct thematic role in BA 44/6. Simulation results show that the model can perform the task correctly for the nine types of canonical and noncanonical sentences, including the generalization to three novel sentence types not used during training.

The same model was used to simulate a nonlinguistic task on sequence learning, using an artificial grammar learning task (Gomez and Schvaneveldt, 1994). In this methodology, participants are exposed to strings of letters generated by a small finite-state grammar. They are then able to classify new strings as belonging, or not, to the original artificial grammar class. This task shares with syntax comprehension the strategy that information contained within the sequence helps predicting the repetitive structure of the rest of the sequence. As a matter of fact, using the same architecture as in figure 24.3, the model was capable of replicating several aspects of the human participants' performance in the artificial grammar task.

Dominey et al. (2009) use this computational neuroscience model to support the hypothesis of the existence of a dual learning mechanism available for both syntactic and nonlinguistic sequence processing tasks. This assumes a direct corticocortical mechanism where syntactic structure information in BA 47 is used to modulate the semantic contents in BA 45. This results in the representation of the sentence level meaning in BA 44. These modeling results are consistent with brain imaging research, such as the work by Hoen, Pachot-Clouard, Segebarth, and Dominey (2006) on grammaticality judgment of natural language sentences and nonlinguistic sequences. This neuroimaging study observed the activation of the left frontal region adjacent to Broca's area (BA 6, 9, and 46) for both tasks.

Overall, these three sample studies show that it is possible to design complex neural network architectures that are based on the brain's language areas. Such models permit the validation of hypotheses on the role of the various cortical areas and connectivity patterns in the various aspects of language processing, including auditory stimulus recognition (Husain et al., 2004), word learning (Garagnani et al., 2008), and sentence–sequence learning (Dominey et al., 2009). One common aspect of these models is also the fact that all implement neural mechanisms dealing with top-down, or bottom-up, attention and for short-term memory systems. These models allow direct comparison of simulation and experimental brain imaging data (ERP, fMRI, PET) and behavioral responses.

24.3 *Language, Brain and Body*

Growing empirical and theoretical evidence exists that supports the view that at least some key aspects of language are intrinsically linked to embodiment and sensorimotor knowledge. This is normally referred to as the *grounding* of language in perception and action (Cangelosi, 2010; Barsalou, 2008; Glenberg and Kaschak, 2002; Pezzulo et al., 2011). This view is also consistent with embodied construction grammar theories linking grammar knowledge with the agents' experience of the world (Bergen et al., 2007; Feldman and Narayanan, 2004; Narayanan, 1997) and further extends this by stressing the influence on linguistic processing of more basic sensorimotor components.

A strong link between embodiment and language has been hypothesized to be crucial in the early stages of language acquisition (Smith, 2005). For example, Smith and Samuelson (2010) investigated the role of embodiment factors, such as the interaction with objects located in different parts of the child's peripersonal space, during the learning of first words. In experiments with 18-month-old children (typical age for the child's lexicon growth to take off) it was reported that infants can learn new names even in the absence of the referent objects when the new label is said while the child's posture is consistent with the left–right location where the object has appeared in precedence. In adult language speakers, various forms of action–language compatibility effects have been reported, supporting the grounding of language on action and perception. Glenberg and Kaschak (2002) demonstrate that pure linguistic tasks, such as yes–no judgments of the syntactic correctness of the sentences "Close the drawer" or "Liz told you the story," produce faster responses when the action required to press the yes–no button is consistent with the direction of the movement described by the sentence, both for concrete movements, as in closing a drawer, and more abstract movement as in telling ("giving out") a story to somebody. Borghi et al. (2004) reports similar action compatibility effects for up–down positions of the parts of objects, as in the association of the "head" or "feet" words after the sentence "There is a doll on the table in front of you." Moreover, the grounded view of the evolution of language (at least in its early stages) is consistent with the literature on mirror neuron systems, linking action perception and production, and the overlap between Broca area in the human's brain and the F5 mirror neuron area in the monkeys' brain (Rizzolatti and Arbib, 1998). For example, Rizzolatti and Arbib propose that pantomime might have been a bridge between actions and protosigns, later leading to more abstract speech and sign languages. This is also in line with the evidence that there is a linkage of certain action verbs with activation in motor cortex areas, as discussed with respect to the Hauk et al. (2004) brain imaging study (section 24.1.2). Overall, these numerous studies provide strong evidence on the grounding of language on sensorimotor knowledge, though the scientific debate on the extent and cognitive relevance of language and cognitive embodiment is still open.

Such developments in cognitive linguistics, developmental psychology, psycholinguistics, and neuroscience have been closely followed by an increased focus on the development of *grounded* computational models of language where agents (robots or embodied simulated agents) learn new names of objects and actions through embodied interactions with their world. This neurorobotic approach permits the modeling of embodiment factors both for higher order cognitive capabilities, such as language, as well as navigation and foraging tasks. In the following sections we will look at examples of the developmental acquisition of grounded language in humanoid robots and at embodied neurocomputational models of action–language grounding.

24.3.1 Embodied Models of Language Acquisition

Morse, Belpaeme, Cangelosi, and Smith (2010a) have directly addressed the modeling of embodiment factors in the developmental acquisition of language by replicating with humanoid robot experiments the child development study of Smith and Samuelson (2010). They use a developmental robotics approach (Cangelosi and Schlesinger, 2014) where a robot equipped with a neural control system gradually acquires linguistic and sensorimotor skills through interactions with the physical world and with a human tutor. Specifically, Morse et al. use the baby humanoid robot iCub and train it to learn the names of objects, to replicate the

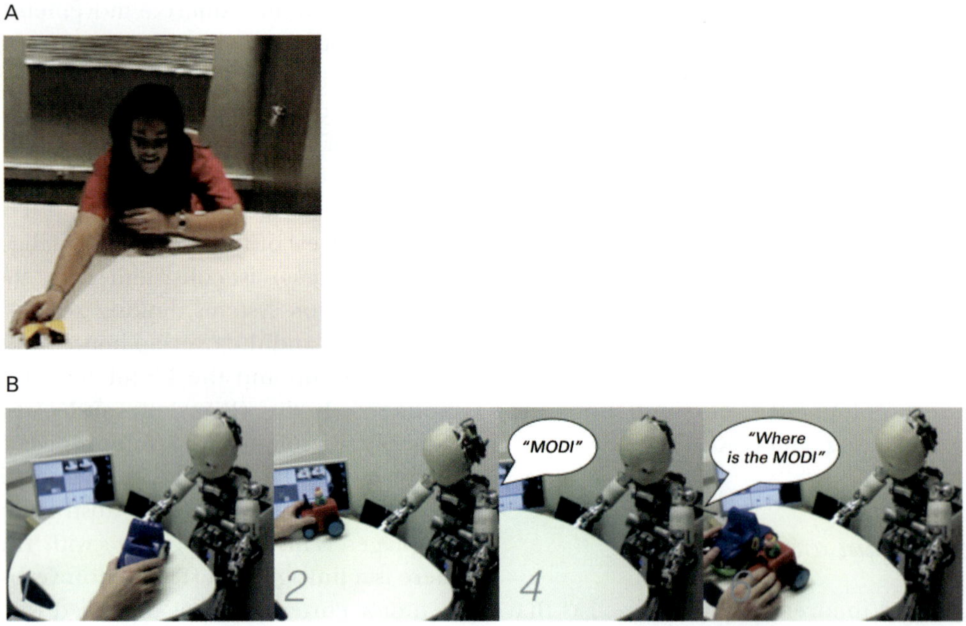

FIGURE 24.7 (A) Experimental setup of the Smith and Samuelson (2010) study with infants. (B) Experimental setup of the Morse et al. (2010a) study with robots replicating the child psychology study. Figure A courtesy of Linda Smith; figure B courtesy of Anthony Morse.

experimental setup used with infants (figure 24.7—see also the video in the book website). The iCub is an open source robotic platform recently developed as a benchmark platform for cognitive and developmental robotics experiments (Metta et al., 2010). It has 53 degrees of freedom, allowing experiments on visual, tactile, and proprioceptive perception, manipulation, and crawling. In this experiment, the degrees of freedom (joint angles) controlling the eyes, neck, and arms are used to allow the robot to gaze at objects and to point at them.

A tutor, in front of the robot, shows two novel objects, respectively, in the left and right location of a table. Initially the robot moves its posture to look at each object and learns to categorize it according to its visual features such as its color. Subsequently the tutor hides both objects, directs the robot's attention toward the right side where the first object was shown (the robot has an intrinsic motivation to gaze at the position of moving entities, such as when waving a hand), and says a new word: "Modi." In the test phase, both objects are presented simultaneously in the center of the table, the robot is asked "Find the Modi," and the robot must now look and point at the object that was presented in the right location. Four different experiments are carried out, as in Smith and Samuelson's child study. Two experiments differ for the consistency of the left–right location of the objects: the default condition, when each object always appears in the same location, and the switch condition, when the position of the two objects is varied, thus weakening the object–location spatial association. In the other two experimental conditions, the object is named while in sight, so as to compare the relative weighting of the embodiment spatial constraints and the time constraint.

The robot is equipped with a neural control system (figure 24.8) consisting of a series of pretrained Kohonen self-organizing maps (SOMs), connected through Hebbian learning weights that are trained online during the experiment. A SOM is an unsupervised neural network with a two-dimensional organization of the output units. After training based on a competitive learning and winner-take-all algorithm (see equations below), the 2-D output units self-organize into a topological representation of the input stimuli according to their category similarity (i.e., two stimuli belonging to similar categories activate nearby units while two stimuli belonging to very different categories activate two units in distant parts of the 2-D layout). SOMs were chosen in this model as they provide the robot with a (pre)categorical representation of the color of seen objects and of the robot's postures. The first SOM is a "color map" as it is used to categorize objects according to their color. This map is pretrained to process visual input from the iCub robot's cameras, taking the average RGB color of the fovea area and using this as an input to a 2-D SOM. The second SOM, the "auditory map," is used to represent the words

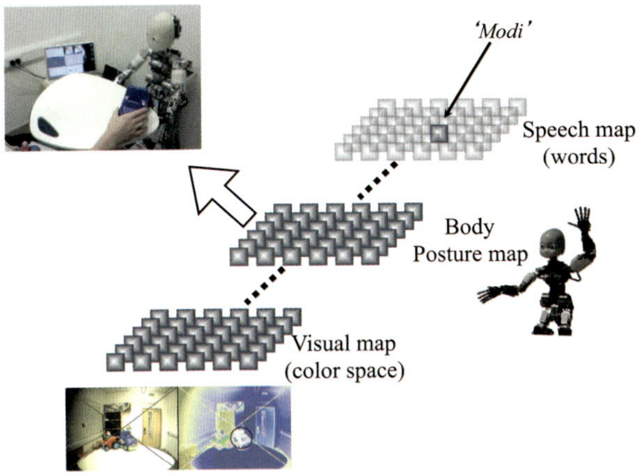

FIGURE 24.8 The robot's neural control architecture used in Morse et al. (2010a) study. The body posture self-organizing map, considered the hub of the system, encodes topological classification of the robot's joint angles states in various positions. The visual map classifies objects according to their color, with input from the robot eye's camera. The speech map uses a localist encoding of the words identified by a speech recognition system. Figure courtesy of Anthony Morse.

"heard" by the robot as the result of automatic speech recognition systems. The other SOM is the "body-hub map," and this is the key component of the robot's neural system that implements the role of embodiment. The body-hub SOM has four inputs, each being the angle of a single joint. Although the iCub robot has 53 degrees of freedom in total, for simplicity in the experiments detailed here only 2 degrees from the head (up–down and left–right motors), and 2 degrees from the eyes (up–down and left–right motors) are actually used. Embodiment is operationalized here as the posture of eye and head position when the robot has to look to the left and to the right of the scene. The "body-hub" SOM map for the body postures constitutes the key component to support the robot's learning of the object–label association.

The network activity across the interconnected SOMs follows a winner-take-all mechanism. For each SOM, the best matching unit (BMU) is identified as the unit that has the shortest Euclidean distance between its weights w_{ij} (initially random) and the input pattern a_j at that time (see equation 24.4). The BMU corresponds to the winner-take-all unit which will be active to represent the stimulus category within the topological 2-D layout. The weights of all units within the neighborhood of the BMU are then updated to move them closer to the input pattern, as with equation (24.5) calculating the Δw_{ij} weight update value. This depends on the distance *dist* of the current node from the BMU and on the square of the total number of nodes *a*, defined as network *size*:

$$BMU = Max_i\left(1-\sqrt{\sum(a_j - w_{ij})}\right), \tag{24.4}$$

$$\Delta w_{ij} = \alpha \exp\left(-\frac{dist^2}{2size}\right)(a_j w_{ij}). \tag{24.5}$$

The spreading of the activation within the network permits the activation *a* of the unit *i* to spread bidirectionally between the body posture SOM and the other maps, following standard interactive activation competition spreading activation (see equation 24.6). Δa_i is the variation of the activation of the unit, given the net input net_i calculate for the unit. The activation function uses the connections *w*, which are modified online with a Hebbian-like learning rule (see equation 24.7), and also uses the constant inhibitory connections within the same map. The Hebbian rule uses the learning rate λ. These Hebbian associative connections are only modified from the current active body posture node. Inhibitory competition between any simultaneously active nodes in the same map provides arbitration between multiple associated nodes. As the maps are linked together in real time based on the experience of the robot, strong connections between objects typically encountered in particular spatial locations, and hence in similar body postures, build up:

$$net_i = \sum w_{ij} a_j + \beta BMU_i. \tag{24.6}$$

$$\text{If } net_i > 0 \;\; \Delta a_i = (\max - a_i) net_i - decay(a_i - rest).$$

$$\text{Else } \Delta a_i = (a_i - \min) net_i - decay(a_i - rest).$$

$$\text{If } a_i a_j > 0 \;\; \Delta w_{ij} = \lambda a_i a_j (1 + w_{ij}). \tag{24.7}$$

$$\text{Else } \Delta w_{ij} = \lambda a_i a_j (1 - w_{ij}).$$

To replicate the four experimental conditions of Smith and Samuelson (2010), 20 different "robot participants" (i.e., the same robot, but with 20 different initialization of its SOM and Hebbian connections) were used in each condition. In each experimental trial, the results record which object, if any, was centered in the robot's view following the final step where the robot is asked to "Find the modi." Results show a very high match between the robots' data and the child experiment results, closely replicating the variations in the four conditions (see figure 24.9). For example, in the default condition, 83% of the trials resulted in the robots' selecting the spatially linked objects, while in the switch condition, where the space–object association was weakened, the robots' choices were practically due to chance at 55%. Smith and Samuelson (2010) reported that 71% of children selected the spatially

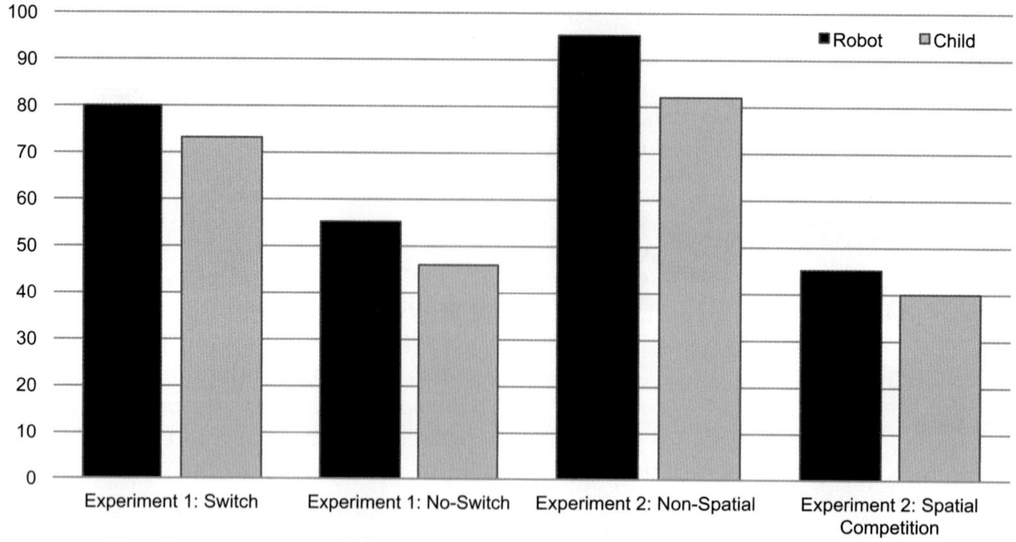

FIGURE 24.9 Results comparing the robot experiment data (dark gray bars) in Morse et al. (2010a) and the child experiment data (light gray bars) in Smith and Samuelson (2010). Figure courtesy of Anthony Morse.

linked object versus 45% in the switch condition. Moreover, the neurorobotic model has led to predictions on the role of variations of the learner's posture during the experiments, which have been further validated with new child psychology studies.

This model demonstrates that it is possible to build an embodied cognitive system that develops linguistic and sensorimotor capabilities through interactions with the world, closely resembling child development phenomena. Although the neural architecture in this study is not directly mapped into brain areas dedicated to vision, action, and language processing areas, the overall modeling framework, called Epigenetic Robotic Architecture (ERA; Morse, De Greeff, Belpaeme, and Cangelosi, 2010b) is suitable for the neural computation modeling of embodied language learning. The basic concept of a map connecting two or more other maps can constitute the building block of a hierarchical set of interconnected brain areas that can be adapted to investigate specific neural theories of language and cognition integration (see Morse et al., 2010b, for a detailed discussion on the extension of the ERA to model hierarchical brain structures). This cognitive architecture has been subsequently applied to other robot and infant language development models, as with the study of mutual exclusivity (Twomey, Horst, and Morse, 2013).

Other computational models with a direct focus on the developmental factors in the grounding of language on the agent's sensorimotor system have been recently proposed. These are mostly based on either a connectionist neural network approach or a developmental robotics framework. For example, Mayor and Plunkett (2010) use a neural architecture of Kohonen maps for object–word associations as a neurocomputational account of fast mapping in early word learning, that is, the learning of words after one or very few training examples. A review of developmental robotics models of language development is also available in Cangelosi and Schlesinger (2015).

Finally, another field of language study with a strong focus on embodiment is that of the multi-agent and robotic models of the evolution of language (Cangelosi and Parisi, 2002; Steels, 2011). These models look at the evolutionary learning of language, with a particular emphasis on cultural transmission. In particular, Steels (2012) has developed a model of the cultural evolution of syntax, based on fluid construction grammar (FCG), directly inspired by developmental constructivist theories of language. FCG has been tested in numerous experiments with the QRIO humanoid robot, such as, for example, for the acquisition of German spatial terms (Spranger, 2012). This models how the parameters of the FCG can be evolved as the result of cultural language game interaction between agents.

24.3.2 MODELS OF THE GROUNDING OF LANGUAGE

The use of neurorobotic and embodied models of language grounding permits the investigation of various phenomena on the integration between linguistic and sensorimotor capabilities. In this section we will analyze two examples of neural models of language grounding phenomena: one is an embodied neurocomputational model of action–language compatibility effects (Caligiore et al., 2010), and the other simulates the differences between noun and verb representations (Cangelosi and Parisi, 2004).

Caligiore and colleagues (2010) have recently proposed a general modeling framework for computational embodied neuroscience. This framework is based on four principles: (1) the two-route organization of the model to reflect the ventral and a dorsal neural pathways, (2) the role of top-down prefrontal signals (instructions) to guide action selection, (3) the competition between different active affordances led by prefrontal activations, and (4) the capability of language to trigger internal simulations of the words. Hence the acronym "TRoPICALS" reflecting the four principles of two route, prefrontal instruction, competition of affordances, and language simulation. These modeling design principles, especially the first three, are based on related computational neuroscience models (Fagg and Arbib, 1998).

Figure 24.10 provides an overview of the neural architecture used in Caligiore et al. (2010) to investigate a series of embodiment and language grounding phenomena. The architecture consists of a set of interconnected neural networks, each simulating brain areas known to be involved in these tasks. Some of these networks are Kohonen SOMs, used for classification tasks (as in the Morse et al., 2010b, ERA architecture), and others are population coding maps. The architecture is structurally and functionally grouped into two main neural routes, the ventral pathway (left of figure) for the "what" processing of object identity, and the dorsal pathway (right of figure) for the processing of "where" information such as object location (Milner and Goodale, 2008). The input to the model consists of three maps: an RGB visual neural map for the visual cortex (VC), a somatosensory cortex map (SSC), and linguistic inputs for object names and task instructions in the superior temporal cortex (STC). From the VC and the SSC input modules, the model divides into two main neural pathways. In the ventral pathway, this input leads to object identification in the ventral occipito-temporal (VOT) region and the prefrontal cortex for biasing effects on action selection. The dorsal pathway is formed by a pathway controlling grasping and reaching in the premotor cortex (PMC), which integrates top-down prefrontal cortex signals with bottom-up sensorimotor transformations of object shape in the anterior intraparietal sulcus (AIP) and object location in parietal reach region (PRR), respectively, for grasping and reaching. The networks are connected through Hebbian-learning weights or competitive Kohonen weights (see labels attached to light gray connection arrows) and fixed weights (dark gray arrows). The organization of the architecture into the ventral–dorsal (what–where) pathways and the selection of the various maps based on the brain areas and circuitry are chosen to constrain the model to the brain areas and circuitry

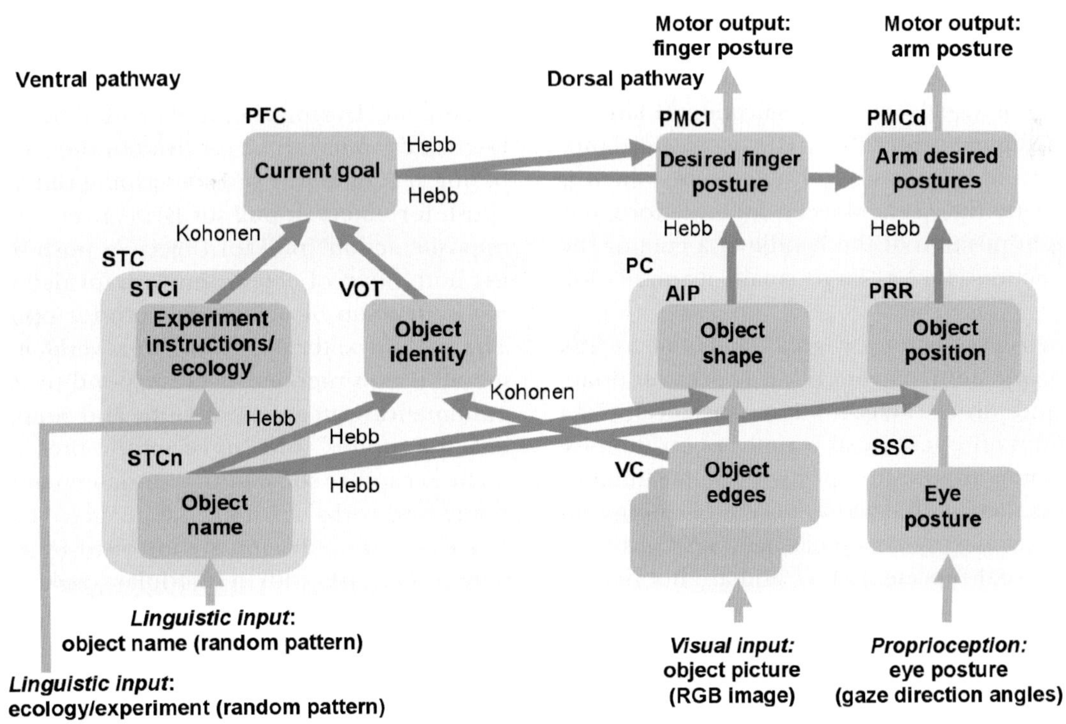

FIGURE 24.10 Two-route architecture of TRoPICALS model in Caligiore et al. (2010). See text for explanation. Figure courtesy of Daniele Caligiore. Reprinted with permission.

known to be involved in the sensorimotor and linguistic functions studied in Caligiore et al. (2010).

Such architecture has been tested on a variety of embodiment phenomena on object manipulation and language tasks such as Borghi et al.'s (2004) action–language compatibility effects and Tucker and Ellis's (2004) stimulus–response compatibility effects on microaffordance effects in visual categorization. A microaffordance is the hand–finger motor response spontaneously elicited by the graspable object. A simulation model of the iCub robot's arm is used for the experiments. For example, in the simulation of Borghi et al.'s experiments, 20 simulated robots are trained, through a "motor babbling" stage, to recognize the objects, and their names, using two large objects ("car," "doll"), four object parts ("roof," "wheels," "head," "feet"), and two unrelated objects ("plum," "mandarin"). The simulated robots first learn to produce a reaching movement to the up–central–down parts of the objects (e.g., central position to reach the "car" and the "plum," up for the "roof," and down for the "wheels"). Some robots are then trained to respond with a yes-is-up reaching movement, which is spatially compatible with upper object parts such as head and roof (the other half of the participants use a yes-is-down reaching response). After this training, the robot's RTs are recorded in the various experimental tests of the Borghi et al. study. Results show that robots produce action–language compatibility effects with significantly faster RTs with upward movements of upper parts than lower parts. The same neural architecture was used on a simulation of Tucker and Ellis's (2004) microaffordance effects for visual object classification (Caligiore et al., 2010; Caligiore et al., 2013). The microaffordance effects refer to the facilitation of a response which is compatible with the seen object's motor affordance (e.g., left–right position of the handle of a cup, or the grasp–grip action for a large–small manipulation object).

This model demonstrates a great potential to link sensorimotor phenomena with other processes in language use and visual categorization. Specifically, the model provides an operational hypothesis on the possibility that some important aspects of the meaning of words, such as the fact that whole–part associations and word priming result from the links between words and grounded internal representations and are not dependent solely on abstract symbolic word–word associations (Caligiore et al., 2010; Caligiore et al., 2013).

Computational models of language grounding have also been used to examine the role of embodiment in syntactic categories, such as in the case of nouns and verbs. In the neuroscience literature on the neural processing of verbs and nouns, Cappa and Perani (2003) reported a general agreement on the fact that the left temporal cortex plays a crucial role in lexical–semantic tasks related to the processing of nouns whereas the processing of words related to actions (verbs) involves additional regions of the left dorsolateral prefrontal cortex. Cangelosi and Parisi (2004) developed an embodied neural model to study the consequences of architectural and functional constraints in evolutionary neural networks that learn to process the different classes of nouns and verbs. The model uses synthetic brain imaging techniques to examine the internal organization of the neural networks and compares the results with data reported in the literature on language processing in the brain (Arbib et al., 2000).

The model consists of a population of 80 cognitive agents that have to evolve for their capability to manipulate objects and respond to linguistic instructions. The agent's motor system consists of a two-segment arm that has to perform two actions (push or pull) on two different objects (object A is a vertical bar, and object B a horizontal bar) appearing in turn in a 2-D grid. A genetic algorithm is used to model evolutionary phenomena (Goldberg, 1989). Agents are selected for reproduction according to their fitness, which measures their ability to manipulate the objects and to understand two-word utterances of nouns (objects' names) and verbs (push–pull actions). The selection procedure uses elitism, that is, 25% of the agents with highest fitness reproduce and the genotype of the offspring is subject to random mutation before the new generation is placed in the environment for their fitness assessment. During the agent's lifetime, five different linguistic inputs can be received by the agent: (1) noun of object A or B; (2) verb describing the default action (push for object A; pull for B); (3) verb describing the opposite action (pull for object A; push for object B); (4) both noun of object and verb for default action; or (5) both noun of object and verb for opposite action. The use of the terms "noun" and "verb" in the present model is not intended to correspond to the full-blown grammatical categories of verbs and nouns. Nevertheless, the "nouns" and "verbs" of the present simulation seem to capture some of the fundamental properties of nouns and verbs of human languages.

The agent's behavior is controlled by a feedforward neural network with a modular organization of its hidden layers. The object is perceived through a retina fovea, and the proprioceptive input encodes the state of the two pairs of muscles (extensor and contractor) of each of the arm's segments (figure 24.11A). The linguistic input is encoded localistically for the two nouns and two verbs. The network has two successive

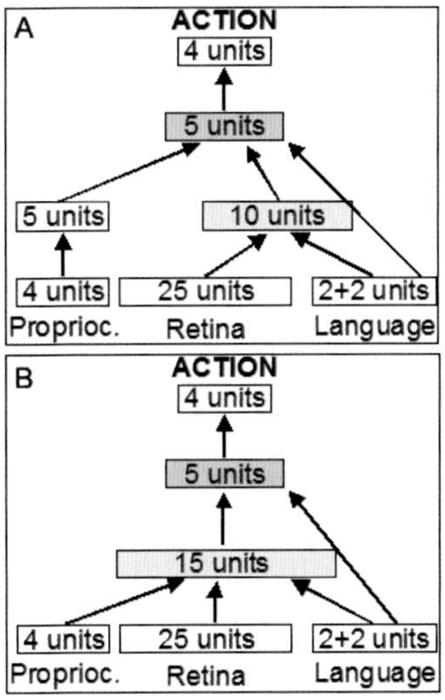

FIGURE 24.11 Neural network architecture in the evolutionary model of Cangelosi and Parisi (2004). The first hidden layer works as a sensory processing module separately processing motor (proprioceptive [Proprioc.]) and visual (retina) stimuli. The second layer works as a sensorimotor integration module combining visual and motor information. Figure reprinted with permission.

layers of hidden units. The first hidden layer is divided into two separate sensory-processing modules, one for processing the proprioceptive input and the other one for separately processing the visual input from the retina. Both modules are connected to a second hidden layer, which has the role of integrating sensorimotor knowledge, through the convergence of motor (proprioceptive) and visual information. The output layer contains four motor units controlling the extension–contraction of the four arm muscles. The connection weights of the agents' neural networks are evolved using the evolutionary algorithm.

After the agents are able to evolve good manipulation and language understanding skills, synthetic brain imaging techniques are used to examine the functional organization of the neural networks. One analysis considers the amount of synaptic input to a layer of neurons (integrated synaptic activity, or synthetic PET in Arbib et al., 2000). A second analysis is based on the neural activity of neurons (integrated neural activity, or synthetic fMRI in Arbib et al., 2000). Results comparing the various linguistic tasks show that nouns produce more integrated neural activity in the sensory processing hidden layer while verbs produce more integrated synaptic activity in the next hidden later layer where sensory information is integrated with proprioceptive input. Such findings are also qualitatively compared with human brain imaging data that indicate that nouns activate more the posterior areas of the brain related to sensory and associative processing while verbs activate more the anterior motor areas. The value of this model is that even simple modular neural networks can capture some of the functional and architectural principles of language processing when linguistic behavior is coupled with sensorimotor knowledge.

Other robotics models of language grounding have been developed to follow a similar methodology, as in the model of symbol grounding transfer for the acquisition of new concepts (Cangelosi and Riga, 2006) and the use of complex recurrent neural architectures for action and language compositionality (Yamashita and Tani, 2008; Peniak et al., 2011). As these models only simulate small lexicons, a key challenge in this field remains the stronger focus on complex syntactic processes.

24.4 *Conclusions*

This chapter has provided an overview of the significant contributions of neural network models to the computational investigation of language functions. The discussion of the various approaches and simulations shows a historical progression toward the development of increasingly brain-constrained and behaviorally rich models of language and cognitive processing. However, this overview of the literature shows that the majority of language models are mostly still based on connectionist neural network methodologies not heavily constrained by known brain structures and connectivity.

The field was initially characterized by the production of a large number of connectionist neural network models. These used simple neural architectures, only loosely inspired by the brain's functional mechanisms, which can simulate a variety of linguistic functions such as word learning, syntax processing, and reading disorders. The main contribution of connectionist models has been to challenge some important assumptions of classical psychological and linguistics theories of language. For example, the notion of a "rule" as an explicit mechanism used by people (and implemented in the brain) to build sentences and make grammatical decision has been seriously challenged. Rumelhart and McClelland's (1986) pioneering model of the English verbs' past tense shows that there is no need to use rules, or two distinct neural routes, to simulate the application of morphological changes and to replicate U-shaped learning phenomena. The neuropsychological models of language disorders also show that connectionist, distributed models can replicate a variety of neuropsychology symptoms.

Since the late 1990s, we have seen a shift toward two main directions. The first is on the use of modular, hybrid neural architectures that are constrained by the known cytoarchitectonic and connectivity structures observed in the brain. The second direction is toward the simultaneous simulation of language and sensorimotor grounding mechanisms. In brain-constrained computational neuroscience models of language learning, neural topologies are built to reflect the various areas involved in language processing such as Broca's and Wernicke's cortical regions. For example, Garagnani et al. (2008) propose a multilayer chain of six cortical areas linking auditory perception with motor representation, and passing through BA 44 and 45 (Broca's) and 42 and 22 (Wernicke). This permits the simulation of ERP responses to mismatch negativity in meaningful versus senseless word detection. Other methods, as with Husain et al. (2004), permit the simulation of functional brain imaging data in auditory tasks.

The second shift in neural language models concerns the focus on the grounding of language in sensorimotor knowledge. This permits the design of robotic models capable of implementing embodiment biases in early word learning (e.g., Morse et al., 2010a), and of brain-constrained models of action and language learning to simulate action–language compatibility effects (Caligiore et al., 2010). Moreover, the use of developmental robotics approaches, as in the iCub model of language acquisition, and its suitability to closely replicate child development phenomena shed important light on the developmental and maturational processes leading to the acquisition of complex linguistic skills. Such an embodied robotic approach to language learning can, for example, look at the way gestures provide a bridge to words in early language acquisition (Caselli et al., 2012) and at the sensorimotor bases of grammar (Marocco, Cangelosi, Fischer, and Belpaeme, 2010).

The existing variety of connectionist-type models and brain-constrained neuroanatomical models, and the more recent development of hybrid models combining connectionist components (e.g., SOMs) and computational neuroscience modules (e.g., population coding, or spiking neural networks), have the advantage of permitting the investigation of both high-level brain processing principles (distributed and parallel processing, competitive activation) and low-level neural mechanisms (spiking neurons). This is the case, for example, in the hybrid modular approach proposed by Caligiore et al. (2010) and Dominey et al. (2009). These hybrid models also permit the simulation of behavioral data, as in Caligiore's RT comparison, as well as ERP and fMRI data, as in Garagnani et al. (2008). Further work in the realization of more realistic computational neuroscience models of language can come from the integration of computational methodologies used in models of visual and auditory stimuli processing in animals, as in neural models of song production in birds (e.g., Fiete et al., 2004; Troyer and Doupe, 2000) and sensorimotor learning in monkeys and primates.

Overall, such a momentum toward the use of brain-constrained neural architectures and principles, the simulation of the various sensorimotor and cognitive processes interacting with language, and the exploration of developmental and learning dynamics in language acquisition provides a unique opportunity to improve our understanding of the intricate, diffuse, and multimodal mechanisms in the development of human cognition and its underlying neural mechanisms.

ACKNOWLEDGMENTS Cangelosi's work on the preparation of this chapter was supported by the European Union Framework Programme 7 (ICT Cognitive Systems and Robotics project "POETICON++") and the UK Engineering and Physical Sciences Research Council (EPSRC project "BABEL"). The author would also like to thank Max Garagnani for the preparation of figures 24.3 and 24.4 and for feedback on parts of the chapter.

REFERENCES

ARBIB, M. A., BILLARD, A., IACOBONI, M., & OZTOP, E. (2000). Synthetic brain imaging: Grasping, mirror neurons and imitation. *Neural Networks, 13*, 975–997.

ARBIB, M. A., CONKLIN, E. J., & HILL, J. C. (1987). *From Schema Theory to Language.* New York: Oxford University Press.

ARBIB, M. A., & LEE, J. (2008). Describing visual scenes: Towards a neurolinguistics based on construction grammar. *Brain Research, 1225*, 146–162.

Artola, A., & Singer, W. (1993). Long-term depression of excitatory synaptic transmission and its relationship to long-term potentiation. *Trends in Neurosciences, 16*, 480–487.

BARSALOU, L. W. (2008). Grounded cognition. *Annual Review of Psychology, 59*, 617–645.

BERGEN, B. K., LINDSAY, S., MATLOCK, T., & NARAYANAN, S. (2007). Spatial and linguistic aspects of visual imagery in sentence comprehension. *Cognitive Science, 31*, 733–764.

BORGHI, A. M., GLENBERG, A., & KASCHAK, M. (2004). Putting words in perspective. *Memory & Cognition, 32*(6), 863–873.

BORNKESSEL, I., ZYSSET, S., FRIEDERICI, A. D., VON CRAMON, D. Y., & SCHLESEWSKY, M. (2005). Who did what to whom? The neural basis of argument hierarchies during language comprehension. *NeuroImage, 26*, 221–233.

BRODMANN, K. (1909). *Vergleichende Lokalisationslehre der Großhirnrinde in ihren Prinzipien dargestellt auf Grund des Zellenbaues.* Leipzig: J. A. Barth.

CALIGIORE, D., BORGHI, A. M., ELLIS, R., CANGELOSI, A., & BALDASSARRE, G. (2013). How affordances associated with a distractor object can cause compatibility effects: A study with the computational model TRoPICALS. *Psychological Research, 77*(1), 7–19.

CALIGIORE, D., BORGHI, A. M., PARISI, D., & BALDASSARRE, G. (2010). TRoPICALS: A computational embodied neuroscience model of compatibility effects. *Psychological Review, 117*(4), 1188–1228.

Cangelosi, A. (2010). Grounding language in action and perception: From cognitive agents to humanoid robots. *Physics of Life Reviews, 7*(2), 139–151.

Cangelosi, A., & Parisi, D. (Eds.). (2002). *Simulating the Evolution of Language.* London: Springer.

Cangelosi, A., & Parisi, D. (2004). The processing of verbs and nouns in neural networks: Insights from synthetic brain imaging. *Brain and Language, 89*(2), 401–408.

Cangelosi, A., & Riga, T. (2006). An embodied model for sensorimotor grounding and grounding transfer: Experiments with epigenetic robots. *Cognitive Science, 30*(4), 673–689.

Cangelosi, A., & Schlesinger, M. (2015). *Developmental Robotics: From Babies to Robots.* Cambridge, MA: MIT Press.

Caplan, D., Baker, C., & Dehaut, F. (1985). Syntactic determinants of sentence comprehension in aphasia. *Cognition, 21,* 117–175.

Cappa, S. F., & Perani, D. (2003). The neural correlates of noun and verb processing. *Journal of Neurolinguistics, 16*(2–3), 183–189.

Caselli, M. C., Rinaldi, P., Stefanini, S., & Volterra, V. (2012). Early action and gesture "vocabulary" and its relation with word comprehension and production. *Child Development, 83,* 526–542.

Chang, F., Dell, G. S., & Bock, K. (2006). Becoming syntactic. *Psychological Review, 113*(2), 234–272.

Chomsky, N. (1957). *Syntactic Structures.* The Hague: Mouton.

Chomsky, N. (1981). *Lectures on Government and Binding: The Pisa Lectures.* Holland: Foris.

Chomsky, N. (1995). *The Minimalist Program.* Cambridge, MA: MIT Press.

Christiansen, M., & Chater, N. (Eds.). (2001). *Connectionist Psycholinguistics.* Westport, CT: Ablex.

Coltheart, M. (1987). Functional architecture of the language-processing system. In M. Coltherart, G. Sartori, & R. Job (Eds.), *Cognitive Neuropsychology of Language.* London: Erlbaum.

Dominey, P. F. (2005). From sensorimotor sequence to grammatical construction: Evidence from simulation and neurophysiology. *Adaptive Behavior, 13*(4), 347–361.

Dominey, P. F., Inui, T., & Hoen, M. (2009). Neural network processing of natural language: II. Towards a unified model of corticostriatal function in learning sentence comprehension and non-linguistic sequencing. *Brain and Language, 109*(2–3), 80–92.

Dominey, P. F., & Ramus, F. (2000). Neural network processing of natural language: I. Sensitivity to serial, temporal and abstract structure of language in the infant. *Language and Cognitive Processes, 15,* 87–127.

Elman, J. L. (1990). Finding structure in time. *Cognitive Science, 14,* 179–211.

Elman, J. L. (1993). Learning and development in neural networks: The importance of starting small. *Cognition, 48,* 71–99.

Elman, J. L. (2001). Connectionism and language acquisition. In M. Tomasello & E. Bates (Eds.), *Language Development: The Essential Readings* (pp. 295–306). Oxford: Blackwell

Elman, J. L., Bates, E. A., Johnson, M. H., Karmiloff-Smith, A., Parisi, D., & Plunkett, K. (1996). *Rethinking Innateness: A Connectionist Perspective on Development.* Cambridge, MA: MIT Press.

Fagg, A. H., & Arbib, M. A. (1998). Modeling parietal-premotor interactions in primate control of grasping. *Neural Networks, 11,* 1277–1303.

Feldman, J., & Narayanan, S. (2004). Embodied meaning in a neural theory of language. *Brain and Language, 89,* 385–392.

Fiete, I. R., Hahnloser, R. H. R., Fee, M. S., & Seung, H. S. (2004). Temporal sparseness of the premotor drive is important for rapid learning in a neural network model of birdsong. *Journal of Neurophysiology, 92,* 2274–2282.

Friederici, A. D. (2009). Pathways to language: Fiber tracts in the human brain. *Trends in Cognitive Sciences, 13*(4), 175–181.

Friederici, A. D. (2011). The brain basis of language processing: From structure to function. *Physiological Reviews, 91,* 1357–1392.

Friederici, A. D. (2012). The cortical language circuit: From auditory perception to sentence comprehension. *Trends in Cognitive Sciences, 16,* 262–268.

Funahashi, S., Bruce, C. J., & Goldman-Rakic, P. S. (1990). Visuospatial coding in primate prefrontal neurons revealed by oculomotor paradigms. *Journal of Neurophysiology, 63,* 814–831.

Garagnani, M., & Pulvermüller, F. (2011). From sounds to words: A neurocomputational model of adaptation, inhibition and memory processes in auditory change detection. *NeuroImage, 54*(1), 170–181.

Garagnani, M., Wennekers, T., & Pulvermüller, F. (2008). A neuroanatomically-grounded Hebbian learning model of attention-language interactions in the human brain. *European Journal of Neuroscience, 27*(2), 492–513.

Glenberg, A., & Kaschak, K. (2002). Grounding language in action. *Psychonomic Bulletin & Review, 9*(3), 558–565.

Goldberg, A. E. (2006). *Constructions at Work: The Nature of Generalization in Language.* Oxford: Oxford University Press.

Goldberg, D. E. (1989). *Genetic Algorithms in Search, Optimization and Machine Learning.* Redwood City, CA: Addison-Wesley.

Gomez, R. L., & Schvaneveldt, R. W. (1994). What is learned from artificial grammars? Transfer tests of simple association. *Journal of Experimental Psychology: Learning, Memory, and Cognition, 20,* 396–410.

Hadley, R. F. (1994). Systematicity in connectionist language learning. *Mind & Language, 9*(3), 247–272.

Hagoort, P. (2005). On Broca, brain, and binding: A new framework. *Trends in Cognitive Sciences, 9*(9), 416–423.

Hagoort, P. (2013). MUC (Memory, Unification, Control) and beyond. *Frontiers in Psychology, 4,* 416.

Hauk, O., Johnsrude, I., & Pulvermüller, F. (2004). Somatotopic representation of action words in human motor and premotor cortex. *Neuron, 41*(2), 301–330.

Hill, J. C. (1983). A computational model of language acquisition in the two-year-old. *Cognition and Brain Theory, 6,* 287–317.

Hinton, G. E., & Shallice, T. (1991). Lesioning an attractor network: Investigations of acquired dyslexia. *Psychological Review, 98*(1), 74–95.

Hoen, M., Pachot-Clouard, M., Segebarth, C., & Dominey, P. F. (2006). When Broca experiences the Janus syndrome: An ER-fMRI study comparing sentence comprehension and cognitive sequence processing. *Cortex, 42,* 605–623.

Husain, F. T., Tagamets, M. A., Fromm, S. J., Braun, A. R., & Horwitz, B. (2004). Relating neuronal dynamics for auditory object processing to neuroimaging activity: A computational modeling and an fMRI study. *NeuroImage, 21*(4), 1701–1720.

Kubovy, M., & Van Valkenburg, D. (2001). Auditory and visual objects. *Cognition, 80,* 97–126.

Langacker, R. W. (1987). Foundations of Cognitive Grammar: *Vol. 1 . Theoretical Prerequisites.* Stanford, CA: Stanford University Press.

Levin, B. (1993). *English Verb Classes and Alternations: A Preliminary Investigation.* Chicago, IL: University of Chicago Press.

Levine, D. (2000). *Introduction to Neural and Cognitive Modeling.* Mahwah, NJ: Erlbaum.

Li, P., Farkas, I., & MacWhinney, B. (2004). Early lexical development in a self-organizing neural network. *Neural Networks, 17,* 1345–1362.

MacWhinney, B. (1998). Models of the emergence of language. *Annual Review of Psychology, 49,* 199–227.

Marocco, D., Cangelosi, A., Fischer, K., & Belpaeme, T. (2010). Grounding action words in the sensorimotor interaction with the world: Experiments with a simulated iCub humanoid robot. *Frontiers in Neurorobotics, 4,* 7.

Mayberry, M., Crocker, M., & Knoeferle, P. (2009). Learning to attend: A connectionist model of situated language comprehension. *Cognitive Science, 33*(1), 449–496.

Mayor, J., & Plunkett, K. (2010). A neurocomputational account of taxonomic responding and fast mapping in early word learning. *Psychological Review, 117,* 1–31.

Metta, G., Natale, L., Nori, F., Sandini, G., Vernon, D., Fadiga, L., et al. (2010). The iCub humanoid robot: An open-systems platform for research in cognitive development. *Neural Networks, 23,* 1125–1134.

Miikkulainen, R. (1993). *Subsymbolic Natural Language Processing: An Integrated Model of Scripts, Lexicon, and Memory.* Cambridge, MA: MIT Press.

Milner, A. D., & Goodale, M. A. (2008). Two visual systems re-viewed. *Neuropsychologia, 46,* 774–785.

Morse, A. F., Belpaeme, T., Cangelosi, A., & Smith, L. B. (2010a). Thinking with your body: Modelling spatial biases in categorization using a real humanoid robot. In S. Ohlsson & R. Catrambone (Eds.), *Proceedings of the 32nd Annual Conference of the Cognitive Science Society, Portland* (pp. 1362–1368). Austin, TX: Cognitive Science Society.

Morse, A. F., de Greeff, J., Belpaeme, T., & Cangelosi, A. (2010b). Epigenetic robotics architecture (ERA). *IEEE Transactions on Autonomous Mental Development, 2*(4), 325–339.

Narayanan, S. S. (1997). *Knowledge-Based Action Representations for Metaphor and Aspect (KARMA).* Berkeley: University of California at Berkeley.

Pecher, D., & Zwaan, R. A. (2005). *Grounding Cognition: The Role of Perception and Action in Memory, Language, and Thinking.* Cambridge: Cambridge University Press.

Peniak, M., Morse, A., Larcombe, C., Ramirez-Contla, S., & Cangelosi, A. (2011). Aquila: An open-source GPU-accelerated toolkit for cognitive and neuro-robotics research. Paper presented at the 2011 International Joint Conference on Neural Networks, San Jose, CA.

Pezzulo, G., Barsalou, L. W., Cangelosi, A., Fischer, M. H., McRae, K., & Spivey, M. J. (2011). The mechanics of embodiment: A dialog on embodiment and computational modelling. *Frontiers in Psychology, 2*(5), 1–21.

Pinker, S., & Prince, A. (1988). On language and connectionism: Analysis of a parallel distributed processing model of language acquisition. *Cognition, 28,* 73–193.

Plaut, D. C. (2003). Interpreting double dissociations in connectionist networks. *Cortex, 39,* 138–141.

Plaut, D. C., & Shallice, T. (1993). Deep dyslexia: A case study of connectionist neuropsychology. *Cognitive Neuropsychology, 10,* 377–500.

Plunkett, K., & Marchman, V. (1996). Learning from a connectionist model of the English past tense. *Cognition, 61,* 299–308.

Pulvermüller, F. (2003). *The Neuroscience of Language.* Cambridge: Cambridge University Press.

Pulvermüller, F. (2013). How neurons make meaning: Brain mechanisms for embodied and abstract-symbolic semantics. *Trends in Cognitive Sciences, 17,* 458–470.

Pulvermüller, F., Kujala, T., Shtyrov, Y., Simola, J., Tiitinen, H., Alku, P., et al. (2001). Memory traces for words as revealed by the mismatch negativity. *NeuroImage, 14,* 607–616.

Rilling, J. K., Glasser, M. F., Preuss, T. M., Ma, X., Zhao, T., Hu, X., & Behrens, T. E. (2008). The evolution of the arcuate fasciculus revealed with comparative DTI. *Nature Neuroscience, 11,* 426–428.

Rizzolatti, G., & Arbib, M. (1998). Language within our grasp. *Trends in Neurosciences, 21,* 188–194.

Rosenblatt, F. (1962). *Principles of Neurodynamics.* Washington, DC: Spartan Books.

Rumelhart, D. E., & McClelland, J. L. (1982). An interactive activation model of context effects in letter perception: II. The context enhancement effect and some tests and extensions of the model. *Psychological Review, 89,* 60–94.

Rumelhart, D. E., & McClelland, J. L. (1986). On learning the past tenses of English verbs. In D. E. Rumelhart & J. L. McClelland (Eds.), *Parallel Distributed Processing: Explorations in the Microstructure of Cognition* (pp. 216–271). Cambridge, MA: MIT Press.

Rumelhart, D. E., McClelland, J. L., & the PDP Research Group. (1986). Parallel Distributed Processing: Explorations in the Microstructure of Cognition: Vol. 1. *Foundations.* Cambridge, MA: MIT Press.

Sinai, A., & Pratt, H. (2002). Electrophysiological evidence for priming in response to words and pseudowords in first and second language. *Brain and Language, 80,* 240–252.

Smith, L. B. (2005). Cognition as a dynamic system: Principles from embodiment. *Developmental Review, 25*(3–4), 278–298.

Smith, L. B., & Samuelson, L. (2010). Objects in space and mind: From reaching to words. In K. Mix, L. B. Smith, & M. Gasser (Eds.), *The Spatial Foundations of Language and Cognition* (pp. 188–207). Oxford: Oxford University Press.

Spranger, M. (2012). A basic emergent grammar for space. In L. Steels (Ed.), *Experiments in Cultural Language Evolution: Advances in Interaction Studies* (Vol. 3). Amsterdam: John Benjamins.

Steels, L. (2011). Modeling the cultural evolution of language. *Physics of Life Reviews, 8*(4), 339–356.

Steels, L. (2012). *Experiments in Cultural Language Evolution: Advances in Interaction Studies* (Vol. 3). Amsterdam: John Benjamins.

Tagamets, M. A., & Horwitz, B. (1998). Integrating electrophysiological and anatomical experimental data to create a large-scale model that simulates a delayed match-to-sample human brain imaging study. *Cerebral Cortex, 8,* 310–320.

Tomasello, M. (1992). *First Verbs: A Case Study of Early Grammatical Development.* Cambridge: Cambridge University Press.

Tomasello, M. (2003). *Constructing a Language.* Cambridge, MA: Harvard University Press.

Troyer, T. W., & Doupe, A. J. (2000). An associational model of birdsong sensorimotor learning: I. Efference copy and the learning of song syllables. *Journal of Neurophysiology, 84,* 1204–1223.

Tucker, M., & Ellis, R. (2004). Action priming by briefly presented objects. *Acta Psychologica, 116,* 185–203.

Twomey, K. E., Horst, J. S., & Morse, A. F. (2013). An embodied model of young children's categorization and word learning. In L. Gogate & G. Hollich (Eds.), *Theoretical and Computational Models of Word Learning: Trends in Psychology and Artificial Intelligence* (pp. 172–196). Hershey, PA: Information Science Reference.

Vigneau, M., Beaucousin, V., Hervé, P.Y., Duffau, H., Crivello, F., Houdé, O., Mazoyer, B., & Tzourio-Mazoyer, N. (2006). Meta-analyzing left hemisphere language areas: Phonology, semantics, and sentence processing. *NeuroImage, 30,* 1414–1432.

Vosse, T., & Kempen, G. (2009). The Unification Space implemented as a localist neural net: Predictions and error-tolerance in a constraint-based parser. *Cognitive Neurodynamics, 3*(4), 331–346.

Wickelgren, W. A. (1969). Context-sensitive coding, associative memory, and serial order in (speech) behavior. *Psychological Review, 76,* 1–15.

Yamashita, Y., & Tani, J. (2008). Emergence of functional hierarchy in a multiple timescale neural network model: A humanoid robot experiment. *PLoS Computational Biology, 4*(11), e1000220.

25 Evolving the Language-Ready Brain

MICHAEL A. ARBIB

25.1 Introduction

25.1.1 DRILL DEEPLY/THINK BROADLY Much success in science, and certainly in research funding, may come from specialization (Drill Deeply): finding a niche in which one can focus one's energies using well-defined methods to address questions which are well-defined in advance of their solution. A related form of specialization is the development of methods which require focused expertise, even if these methods may then be exploited by others to address varied problems. Nonetheless, what attracts many people to science are the Big Questions (Think Broadly): What is the origin of the universe? Is there such a thing as free will? How might we reduce global warming? How does the brain work? And what happens when it does not work "properly"? [see chapter 23, "Brain Diseases"]. An emergent paradigm to bridge these extremes (and one which is sometimes well funded) is to develop multidisciplinary teams which bring together specialists to answer questions which require varied expertise for their solution. The challenge for each specialist, then, is to know enough about the expertise of some other team members to be able to discuss their research at a level that can support collaboration and mutual progress.

The aim of this book has been to exhibit a range of specialties within computational neuroscience, in concert with samples of the experimental or clinical studies with which they engage. Readers planning a career in neuroscience can not only gain a sample of possible specialties but also develop an appreciation of the connections between the different topics.

This chapter focuses on one big question—How did the human brain evolve in such a way that, unlike any other species, humans have language?—and shows how at least some parts of an answer build upon several of the models and insights presented in earlier chapters. But it will also show the diversity of further disciplines that must be brought to bear, far more than any single multidisciplinary grant could support but nonetheless defining a community of interests that can spur conversation across many disciplines. Indeed, the question has informed a biennial conference on Evolution of Language (EvoLang, http://www.evolang.org) and many other multidisciplinary meetings. Unfortunately, many of the conversations pay little attention to neuroscience, but some sense of the range of issues discussed may be found in *The Oxford Handbook of Language Evolution* (Tallerman and Gibson, 2011) and *New Perspectives on the Origins of Language* (Lefebvre, Comrie, and Cohen, 2013).

25.1.2 COMPARATIVE NEUROPRIMATOLOGY AND THE STRUCTURE OF THE CHAPTER In the near-complete absence of single-cell recording in humans, the theory of language evolution emphasized in this chapter is grounded in the comparative study of monkeys, apes, and humans, informed by these questions:

- To what extent do language mechanisms in the human brain relate to those that support the integration of action and perception more generally?
- What can we learn about human language from the study of nonhuman primates?

Figure 25.1 provides an evolutionary tree for primates. One estimate suggests that the human evolutionary line diverged from that of macaques some 25 million years ago, from the gorillas 7 million years ago, and from chimpanzees, our closest extant nonhuman relatives, some 7 to 5 million years ago. *Comparative neuroprimatology* studies brain mechanisms and behaviors in, for example, macaques, chimpanzees, and humans. In evolutionary mode, we assess what we can infer from such data concerning the last common ancestor of human and macaque (LCA-m) and the last common ancestor of human and chimpanzee (LCA-c), and this, in turn, informs hypotheses about changes that might have occurred along the way from LCA-m via LCA-c to *Homo sapiens*.

Chapter 20 introduced a range of models of the control, development, and recognition of reach and grasping in both the human and the macaque. Of particular relevance for the present chapter is the neurophysiological discovery of *mirror neurons* in a region of

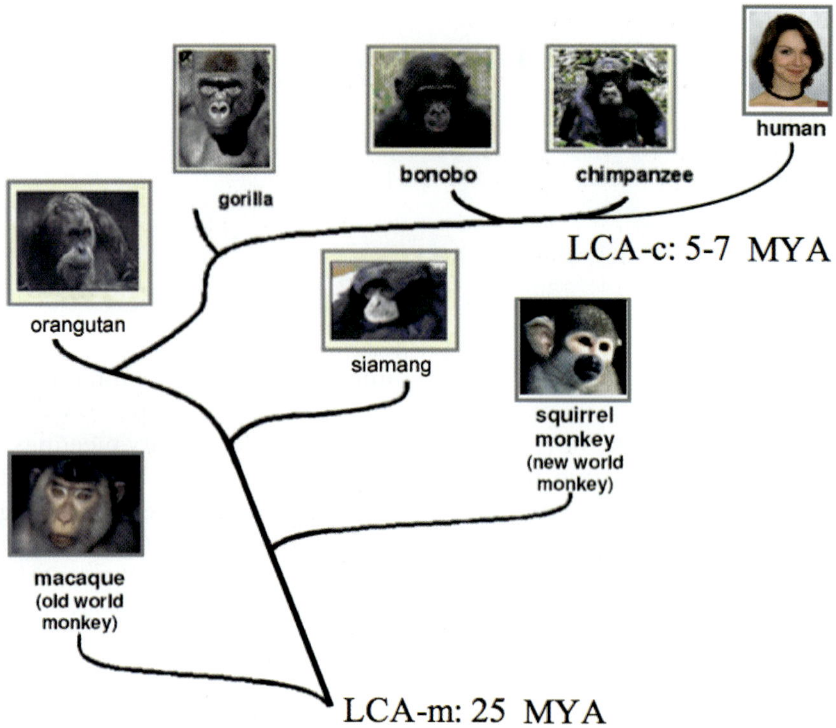

FIGURE 25.1 An evolutionary gallery of primates, including humans. This chapter links neurophysiological data on monkeys, primarily macaques, with data (mainly behavioral) on apes, primarily chimpanzees, to gain insights into the human brain mechanisms that support language. Adapted from Arbib, Liebal, and Pika (2008).

macaque premotor cortex called F5—each such neuron is active both when the monkey executes a specific set of manual actions (e.g., precision pinches but not power grasps) and when the monkey observes manual actions more or less similar to the actions in that set. This was followed by evidence from brain imaging (e.g., Grafton, Arbib, Fadiga, and Rizzolatti, 1996), rather than single-cell recording, that there is a human mirror system for grasping—that is, a brain region more activated for both grasping and observation of grasping an object as compared to just observing the object—in or near Broca's area, an area traditionally viewed as involved in speech production. However, the fact that aphasia of signed, and not just spoken, languages may result from lesions "embracing" Broca's area (Poizner, Klima, and Bellugi, 1987) supports the view that one should associate Broca's area and its environs with *multimodal* language production rather than with speech alone.

Such considerations led Arbib and Rizzolatti (1997; Rizzolatti and Arbib, 1998) to introduce the *mirror system hypothesis* (MSH), namely, that the basis for *language parity*—that to a great extent, the hearer "gets" the meaning intended by the speaker (and similarly for sign language)—evolved "atop" the mirror system for grasping, rooting speech in communication based on manual gesture. Section 25.2 provides an exposition of the core ideas of MSH as it has itself evolved since 1998. This summarizes the book-length exposition in *How the Brain Got Language* (Arbib, 2012). The reader wishing to probe further may wish to consult a special issue of *Language and Cognition* (volume 5, 2013, numbers 2 and 3) which contains twelve commentaries on the book along with my response (Arbib, 2013a). See also the earlier, less pedagogical version of the themes of the present chapter published in *Physics of Life Reviews* (Arbib, 2016a) along with a range of expert commentaries and an integrative response (Arbib, 2016b).

Very briefly, as we will see in section 25.2, MSH posits that LCA-m shared with modern macaques a mirror system for manual tasks and that LCA-c shared with modern chimpanzees the ability (lacking in monkeys) to learn a few novel gestures that ritualize interactions between a pair of animals. It then charts how biological evolution may have supported increasing skills in imitation on the path from LCA-c to modern humans, and how this in turn supported (and was supported by) the emergence of what we call *protolanguage*, with early gestural communication (protosign) scaffolding the emergence of protospeech. A further claim, then, is that all this was in place before *Homo sapiens* and that the first *Homo sapiens* had a "language-ready" brain but did not yet have the rich syntax and compositional semantics and accompanying conceptual structures that underlie modern human languages. Rather, the path from protolanguage to language was primarily one involving cultural rather than biological evolution.

Charting this last transition is outside the scope of this chapter (see Arbib, 2012, chapters 10–13; Arbib, 2016a). Instead we devote section 25.3 to recalling from chapter 20 the modeling of interaction between vision and action in macaque brain (suggesting what was available to LCA-m). Section 25.4 introduces dyadic brain modeling, focusing on the emergence of novel gestures (ontogenetic ritualization) from interactions between apes (suggesting what was available to LCA-c). Section 25.5 then offers an MSH-based view of brain mechanisms underlying language. Finally, section 25.6 outlines a schema-based model of the interactions between vision and language in the human brain.

With this, we offer a brief excursion into the methodology of brain modeling in general and computational comparative neuroprimatology in particular. We will then outline each of sections 25.3 through 25.6 in a little more detail before extending our review of MSH in section 25.2.

Key models of individual neurons were introduced in chapter 2 ("Basic Neuron and Network Models"), ranging from detailed models based on variations of the Hodgkin–Huxley equations through various simplified models based on spike generation, to models based on firing rates rather than spikes. We also encountered the logic-based model of Pitts and McCulloch. In this chapter, our neural network modeling will use the leaky integrator model of how firing rates are determined for each neuron in the circuitry. Our modeling of synapses will ignore fine details such as AMPA and NMDA receptors but employ *learning rules* (different forms of synaptic plasticity) that can cause synaptic "weights" to change with experience. These rules include backpropagation (appendix A) and, more biologically, Hebbian and reinforcement learning (chapters 6 and 10).

Above the neural level, one may proceed structurally or functionally. Structurally, one may aggregate neurons into circuits, columns, and/or layers to model a brain region; one can then combine several such models to assess how different brain regions work together in enabling the organism to carry out various tasks. Functionally, one may choose to delimit certain subfunctions employed in the tasks—these are what we call *schemas* (see chapter 1; chapter 14, "Schema Theory and Neuropsychology"; and sections 25.3.1 and 25.6.1 below)—and then assess how interactions within a network of these schemas may yield some overall pattern of behavior. In many studies of human cognition, including language, hypotheses about localization cannot be linked directly to data at the neuronal level. In some of these cases, it may be possible to define schemas in terms of interactions between subschemas for which hypotheses about their loci in the brain can be tested against neuropsychological data on patients with brain damage or brain imaging data, as exemplified in chapter 14. And in some of these cases, animal data may allow us to pursue hypotheses as to how various schemas are realized in neuronal circuits (see figure 25.2 and section 25.4.1).

The present chapter is an exercise in what we may call *computational* comparative neuroprimatology, seeking to learn more about the human brain by relating models of its function to models of the brains of other primates, all in the light of available data on brain and behavior. Chapter 15, "Synthetic Brain Imaging," includes examples of how to model human brain mechanisms for which one believes the relevant circuitry is similar to that revealed by animal (e.g., monkey) single-cell neurophysiology, and then process simulation results at the neural network level to infer predictions that can be tested against human functional magnetic resonance imaging or other noninvasive measures (Arbib, Billard, Iacoboni, and Oztop, 2000; Barrès, Simons, and Arbib, 2013; Bonaiuto and Arbib, 2014). One can also apply variations of this methodology to schema networks. Returning to biologically realistic neural networks, figure 25.2 illustrates an extension of that approach for mechanisms—such as those subserving language—for which nonhuman neural circuitry does not suffice: we use hypotheses about evolution of brain mechanisms to suggest how macaque circuitry (here the macaque brain is used as a "default model" of the brain of LCA-m, though we know that monkeys too have evolved greatly) is modified and expanded

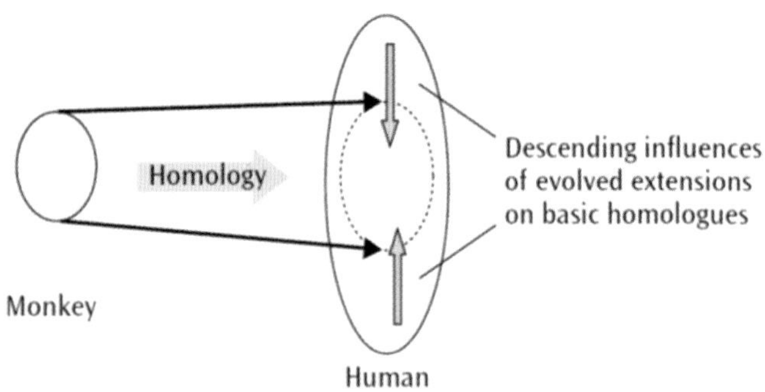

FIGURE 25.2 A strategy for modeling the human brain: use known circuitry in the macaque brain to suggest details of circuitry in homologous regions of the human brain; extend these models using hypotheses on brain evolution; then test the result against lesion data (cf. chapter 15) or human brain imaging studies by the use of synthetic brain imaging.

upon in the architecture of the human brain, and we then use the resultant model to make predictions for human brain imaging or for lesions.

We will turn to models of the macaque brain in section 25.3. In each model, the focus will be on the relation between vision and action—thus setting the stage for our later concern with the linkage between vision and language, especially in sections 25.5 and 25.6. We first recall, briefly, the FARS (Fagg and Arbib provided the model, based on the data of Rizzolatti and Sakata) and MNS (mirror neuron system) models introduced in chapter 20. The FARS model provides the neural elaboration of the schema model of the reach to grasp introduced in chapter 1, exploring how vision of an object can guide the grasping of the object. The MNS model looks at how the brain can learn to recognize manual actions on the basis of visual input (and auditory input for those actions, like tearing paper, that have distinctive sounds). Although primate neurophysiology offers a range of studies of mirror neurons, including those that were the basis for the MNS models, little attention has been paid to whether and how mirror neuron activity helps guide the monkey's own behavior. We thus introduce a new model, augmented competitive queuing (ACQ), which is an exception, showing how mirror neurons can play a key role in learning that serves the opportunistic scheduling of sequences of actions—primarily by assessing whether an executed action was successful and, if not, answering the question "What did I just do?"

To motivate our strategy in section 25.4, note that most chapters in this book have focused on some aspect of the function of a *single* brain (vision, memory, and action, for example), specific brain regions (hippocampus, basal ganglia, cerebellum, and various regions of cerebral cortex) as seen in the context of larger networks, focused studies of underlying mechanisms (dynamics of neural networks, Hebbian and reinforcement learning, etc.), or other general strategies of computational neuroscience (e.g., synthetic brain imaging). However, when we turn to communication in general and language more specifically, we are looking at *inherently social phenomena* which (whatever their importance to the cognition of a single individual) evolved to support interactions of two or more creatures. Recall our earlier statement that a key aim of MSH is to understand the phenomenon of *parity* in language—that to a great extent, the hearer "gets" the meaning intended by the speaker. The basic interaction of a speaker and a hearer (and the roles are reversed at each turn of the conversation) is well conveyed in figure 25.3, the semiotic cycle. The diagram shows two forms of interaction:

1. Interaction via sensorimotor activity whereby each actor can change the physical state of the world (which includes the bodies of both actors) and thus what each can attend to, which may involve joint attention to varying degrees at different stages of the interaction.

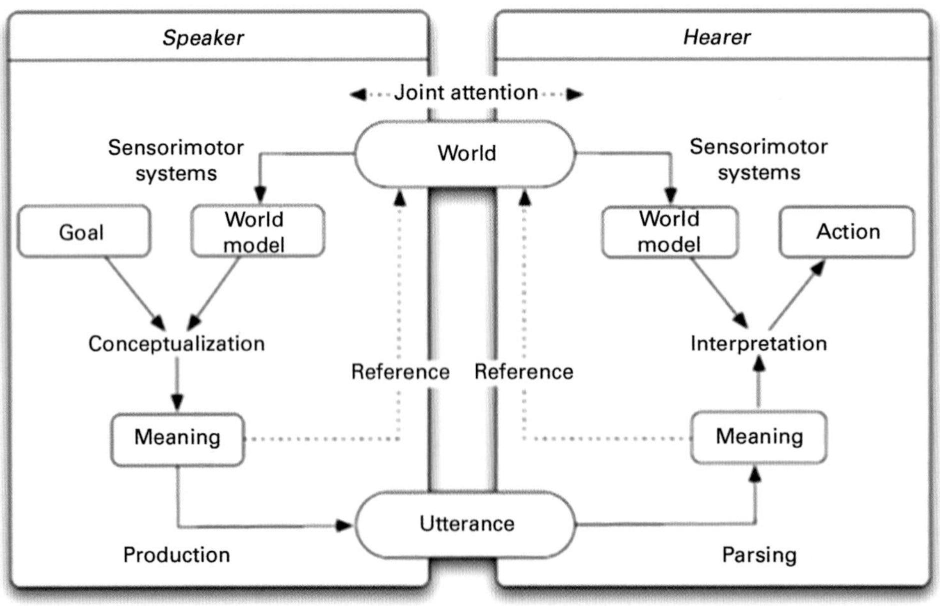

FIGURE 25.3 The semiotic cycle (Spranger, 2014) coupling shared attention and (not shown) interaction with the external world with the sharing of utterances that constitutes communication. Compare the action–perception cycle characterizing the single individual (chapter 1). (Courtesy of Michael Spranger and Luc Steels.)

2. Communicative interaction via "utterances," which may be spoken or signed and may include gestures of voice, face, hands, body posture, and more.

Before proceeding, note—crucially for the buildup to a discussion of evolution—that almost all of this applies to animals in general, not just to humans, with the important exception that nonhuman animals cannot master signed or spoken languages, and so we might want to replace the term "utterance" by "signal" in the general case. Admittedly, some apes have been taught to use a hundred or so signs based on American Sign Language (ASL), but there is no evidence that they can master grammar to support the ability to combine words in novel ways and have the novel meanings understood, the sine qua non of language in the sense in which we use it in this chapter.

Thus, where section 25.3 focuses on modeling the brain of a single monkey in the process of generating or recognizing an action or sequencing several actions, section 25.4 turns to a model of communication. Specifically, we want to understand how interaction between two apes can lead to the emergence of novel communicative gestures. We will address this concern in the realm of computational neuroscience by advocating *dyadic brain modeling* (figure 25.4), which focuses on what happens in the brains of two interacting agents, a dyad, where the actions of one influence the actions of the other, with both brains changing in the process. Luc Steels (e.g., Beuls and Steels, 2013) has used simulation of embodied agents in evolutionary games in a fashion relevant to studies of *cultural* language evolution (as distinct from the focus in MSH on brain evolution), implementing the semiotic cycle in computer programs (but not brain models) for interacting robots equipped with both visual and motor systems. Our innovation here is to provide the agents with "brains" based on prior work in brain modeling. The current implementation (section 25.4.3) is highly simplified but provides proof of concept for future work which can increasingly address data on the role of neural systems in primate social interaction and do so in an evolutionary context.

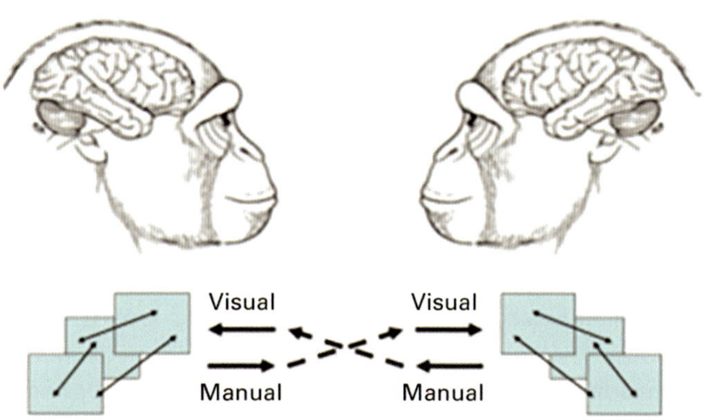

FIGURE 25.4 Schematic for dyadic brain modeling. Each brain has the same architecture, which it is our challenge to assess on the basis of empirical data and lessons learned from prior modeling. However, each agent's brain starts in a different state, and thus the actions and learning processes that unfold will differ between the two agents (Arbib, Ganesh and Gasser, 2014).

Section 25.5 offers a conceptual bridge between our concern with neural models of macaque and chimpanzee and our interest in the human language-ready brain, a bridge grounded on MSH. The focus is on how the mechanisms of FARS and MNS in the macaque may "lift" to a conceptual overview of the human brain—our discussion emphasizes the lifting of the vision–action connection to a vision–language connection and the lifting of mirror system properties to a partial explanation of language parity. The model here is conceptual—it thus provides a framework for future computational modeling (probably a mix of neural and schema networks), not an implemented model ready for detailed exploration through simulation and data analysis.

Indeed, not all neurocognitive models can be developed in terms of neural circuitry. For some of these a model at the level of interacting schemas may be more appropriate. In some cases, a schema-theoretic model may be refined to yield a neural network model, but this is not always the case. Unlike a neural network model, a schema model of diverse skills may readily redeploy prior schemas to perform novel tasks because these schemas are not "lost" in the hidden units of a neural network. But what is it about the neurobiology of our brains that makes different skills and schemas redeployable in such a fashion? The sad fact of the matter is that, despite decades of research (Arbib, 1988), there are areas of cognition where brain models that can be constructed satisfactorily at the schema network level do not transfer well to biologically realistic neural network models.

Thus, the models of high-level vision and language production and comprehension in section 25.6 are both schema-theoretic, and their linkage to more data from neuroscience—and thus their modification and/or elaboration in a continuing modeling–experimentation cycle (where "experimentation" includes clinical studies of humans and fieldwork to observe nonhuman behavior)—provides important challenges for both modeling and experimentation in neuroscience. Section 25.5 pursues the quest, introduced in section 25.2.5, for a suitable notion of grammar with which to answer the question "What evolved to make the brain language ready?"—but now with a particular concern for the connection between vision and language, building on our earlier attention to models of the processes linking vision and (manual) action. It should not surprise the reader that no consensus has emerged as to the best way to model such processes. Indeed, given the variety of neural models that are employed in computational neuroscience, it seems reasonable that different models may prove better suited to addressing different aspects of language readiness and language processing. Thus, in presenting template construction grammar (TCG) as a way to address the linkage between vision and language—both in the production and comprehension of utterances—the aim is to offer an encouraging example of research of this kind, not to foreclose alternatives.

But with this, the time has come to present MSH to provide a conceptual framework for all that follows.

25.2 *The Mirror System Hypothesis on the Evolution of the Language-Ready Brain*

Different brain regions (not individual neurons) may be implicated in the human brain as mirror systems for different classes of actions and even emotions, and many researchers have attributed high-level cognitive functions such as empathy, imitation, intention attribution, and language to human mirror regions (Rizzolatti and Sinigaglia, 2008). However, monkeys do not imitate or learn language. This suggests that any account of the role of human mirror systems in imitation and language must explain what the human brain offers that the monkey brain lacks. This section takes the position that such an explanation must include an account not only of the evolution of mirror systems but also of their interaction with more extended systems *beyond the mirror system* within the human brain.

Thus MSH argues that a language-ready brain resulted from the evolution of a progression of linkages of mirror systems with brain regions "beyond the mirror." It should be added that brain, body, and social structure coevolved. New social practices endowed certain changes in brain and body with a selective advantage that was not there before, but then certain social practices become newly adaptive (and this may include development and use of new tools). Though a full account must stress gene–culture coevolution, the key here is the phenotype which results from the development of the individual made possible by brain and body as well as the physical and cultural niche in which development occurs—thus the importance of considering *niche construction* (Iriki and Taoka, 2012; Odling-Smee, Laland, and Feldman, 2003).

MSH offers an evolutionary framework for the following data (see, e.g., Arbib et al., 2008, for a review of the relevant literature):

- Monkeys, apes, and humans all have great manual dexterity.
- Macaque mirror neurons respond only to observation of *transitive* manual actions in the sense of manual gestures that are directed to an actual object.
- Human imitation involves the mirror system ... but macaques don't imitate, while apes do though in a more limited way than humans.
- Monkeys and apes have genetically constrained call systems and no capacity for vocal learning.
- Apes can communicate using novel manual gestures which are *intransitive* (i.e., not directed at an object), though only to a limited extent.
- Humans can develop, with appropriate social interaction, equal facility for language in the manual and vocal domains—signed and spoken languages, respectively (and speakers gesture with their hands and faces as they talk).

25.2.1 Imitation We first need to engage in some *behavioral* comparative primatology contrasting the types of imitation exhibited by monkeys, apes, and humans. Despite the slogan "Monkey see, monkey do," monkeys have, at best, a very limited capacity for imitation (Visalberghi and Fragaszy, 2001), far overshadowed by what I call *simple imitation* as exhibited by apes. Byrne and Byrne (1993) found that gorillas learn complex feeding strategies but may take months to do so. Consider eating nettle leaves. Skilled gorillas grasp the stem firmly, strip off leaves, remove petioles bimanually, fold leaves over the thumb, pop the bundle into the mouth, and eat. The challenge for acquiring such skills is compounded because ape mothers seldom if ever correct and instruct their young (Tomasello, 1999) and because the sequence of "atomic actions" varies greatly from trial to trial. Byrne (2003) implicates *imitation by behavior parsing*, a protracted form of statistical learning whereby certain *subgoals* (e.g., nettle folded over the thumb) become evident from repeated observation as being common to most performances. Having recognized such subgoals, the young ape may acquire the skill over many months of trial and error to derive action strategies for achieving each subgoal.

By contrast with learning the overall structure of a specific feeding behavior over many, many observations, in many cases of *praxis* (i.e., skilled interaction with objects), humans need just a few trials to make sense of a relatively complex behavior if the constituent actions are familiar and the subgoals these actions must achieve are readily discernible, and they can use this perception to repeat the behavior under changing circumstances. We call this ability *complex imitation* (extending the definition of Arbib, 2002, to incorporate the goal-directed imitation of Wohlschläger, Gattis, and Bekkering, 2003).

Crucially, a mirror system does not provide imitation in itself. A monkey may have mirror neurons related to an action in its repertoire but cannot use observation of a novel action to add that action (movement + goal) to its repertoire. Thus, MSH hypothesizes that evolution extended a monkey-like mirror system in the last common ancestor of human and macaque (LCA-m) into more powerful systems in two stages: (1) a *simple imitation system for grasping*, shared with the last common ancestor of human and chimpanzees (LCA-c), and (2) a *complex imitation system for grasping* that developed in the hominid line since that ancestor. The complex imitation system involves an increased ability for both perceptual and motor chunking supporting the following:

- *complex action analysis*, the ability to recognize another's performance as resembling an assemblage of familiar actions,
- *the actual imitation*, grounded in complex action recognition, to repeat the assembled actions, and, more subtly,
- the recognition of another's performance as *resembling* an assemblage of familiar actions as a basis for the imitator to further attend to how novel actions differ from the ones they resemble, to facilitate becoming able to perform these variant actions.

The latter process may yield fast comprehension of (much of) the overall structure of the observed behavior yet with much practice required to yield truly skillful behavior.

The development of computational models of neural mechanisms underlying simple imitation and complex imitation lies outside the scope of this chapter, as is an assessment of the extent to which they build upon or diverge from the models of chapter 20 and section 25.3.1—the MNS models (which must be extended for assessing what the other is doing when the observed behavior is not in one's own repertoire) and the FARS model (for doing a new action oneself once one has acquired it). Indeed, relating the difference between simple and complex imitation to differences between chimpanzee and human brains is an open research issue. However, various publications (e.g., Custance, Whiten, and Bard, 1995; Oztop, Kawato, and Arbib, 2013; Sauser and Billard, 2006; Schaal, Ijspeert, and Billard, 2003) do address relevant issues.

Complex imitation was an evolutionary step of great advantage for sharing acquired praxic skills. It was thus adaptive for protohumans independent of any implications for communication. However, in modern humans it undergirds the child's ability to acquire language, while complex action analysis is essential for the adult's ability to rapidly comprehend novel compounds of "articulatory gestures" such as sentences.

We now turn briefly to ape communication, which combines an innate repertoire of vocalizations with a more flexible repertoire of manual gestures. Some authors (e.g., Hobaiter and Byrne, 2011) argue that ape gestures are either "species typical" or produced by only a single individual. However, a few group-specific gestures have been observed in ape populations, suggesting a role for social learning (Arbib et al., 2008). Tomasello and Call (1997) proposed *ontogenetic ritualization* as a means whereby (some) ape gestures could emerge:

- Individual A performs praxic behavior X and individual B consistently reacts by doing Y. (Here we use the term *praxic* behavior for practical interactions with the physical environment—which can include the body of another agent—as distinct from *communicative* actions.)
- Subsequently, B anticipates A's overall performance of X by starting to perform Y before A completes X.
- Eventually, A anticipates B's anticipation, producing a ritualized form X^R of X to elicit Y.

In section 25.4, we will introduce a model grounded in properties of the mirror system for the brain mechanisms that could support ontogenetic ritualization.

A word of warning, or encouragement: my current hypothesis is that it was the linkage of mirror systems to evolving systems beyond the mirror that underwrote the transition to increased ability of imitation manifested by apes and assumed to have been available to LCA-c. However, the remaining stages of MSH do not depend on whether or not mirror neurons play a key role in imitation. The disentangling of their contribution is a current challenge in comparative neuroanatomy of primates but is impeded by the fact that we cannot have single-cell recording data for apes or humans (with rare exceptions justified by human neurosurgery). In stressing the importance of imitation, MSH has much in common with Merlin Donald's (1991, 1998, 1999) theory of the evolution of culture and cognition which is based on *mimesis* and the exploitation of this capability to create mental representations with the critical property of voluntary retrievability.

25.2.2 Protolanguage versus Language Before going further, we need to clearly distinguish between protolanguage and true language.

A *language* is characterized by a large (and open-ended) stock of words (the *lexicon*) and a rich *grammar* which supports a *compositional semantics*—that is, knowing the meanings of words, we can combine them using the grammar not only to build phrases and sentences that express novel meanings but also to do so with a fair expectation that that these meanings will be understood.

A *protolanguage* here is a system of communication of the kind we hypothesize our distant ancestors had before true languages emerged—a protolanguage rests on "protowords" whose meaning can be established in a community by conventionalization (and in this sense is open-ended), but which has little or no grammar (Arbib and Bickerton, 2010).

We start with seven properties of "language readiness" that I see as supported by brain mechanisms that evolved during the emergence of protolanguage but prior to the emergence of language (Arbib, 2012, chapter 6):

- *Complex action recognition and complex imitation* (see above).
- *Intended communication.*
- *Symbolization.*
- *Parity:* What counts for the producer of one or more symbols must count, frequently, as approximately the same for the receiver of those symbols. The parity principle for communicative actions extends the role of the mirror neurons for grasping and other actions, "lifting" complex imitation from praxis to communication.
- *Linking hierarchical structuring to temporal ordering:* Animals can perceive the hierarchical structure of scenes to determine what actions to execute and when to execute them to achieve their goals.
- *Ability to go beyond the here and now 1:* The ability to recall past events or imagine future ones.
- *Paedomorphy and sociality:* A prolonged period of infant dependency, especially pronounced in humans, combines with the willingness of adults to act as caregivers and the consequent development of social structures to provide the conditions for complex social learning.

I claim that the mechanisms which underlie these seven properties were the result of biological evolution (though certainly in conjunction with "cultural evolution" constructing ever more complex niches for social interaction) and are supported by the genetic encoding of brain and body and the consequent space of possible social interactions. However, and perhaps more controversially, I claim that no changes in the genome were required for cultural evolution to yield the following four properties that distinguish language (though modern *Homo sapiens* may have these skills in larger measure than early *Homo* not only as a result of cultural evolution but also as a result of some measure of Baldwinian evolution [Baldwin, 1896; Waddington, 1953] that favored offspring with greater skill in exploiting these capacities):

- *Symbolization and compositionality:* The symbols become words in the modern sense, interchangeable and composable in the expression of meaning.
- *Syntax, semantics, and recursion:* The matching of syntactic to semantic structures grows in complexity, with the nesting of substructures making some form of recursion inevitable. (We will explore how the basic property "Linking hierarchical structuring to temporal ordering" operates at the level of language in section 25.5.)
- *Ability to go beyond the here and now 2:* Verb tenses or other language-specific tools express the ability to recall past events or imagine future ones.
- *Learnability:* To qualify as a human language, much of the syntax and semantics must be learnable by most human children.

In short, MSH claims that the "language-ready brain" of the first *Homo sapiens* supported basic forms of gestural and vocal communication (protosign and protospeech, as presented in section 25.2.3) but not the rich syntax and compositional semantics and accompanying conceptual structures that underlie modern human languages.

25.2.3 From Pantomime to Protosign and Protolanguage We now examine the transition our distant ancestors made to *protolanguage*, in the above sense of a communication system that is open to the addition of new "protowords" far more readily than the (almost) closed set of calls or gestures of a group of nonhuman primates—but still lacks the means, beyond mere juxtaposition of two or three protowords, to put protowords together to continually create novel utterances. MSH hypothesizes that the key transition here was the emergence of a brain and a social niche that could support the use of pantomime (as distinct from the occasional emergence of ape gestures that appear iconic). *Pantomime* is here the ability to use reduced forms of actions to intentionally convey aspects of other actions, objects, emotions, or feelings. Flapping one's arms could indicate an object (a bird), an action (flying) or both (a flying bird). Examples linked to feelings or emotions could include miming the brushing away of tears to feign sadness. The key point is that this form of pantomime provides a way to escape the strong semantic limitations imposed by the use of a small set of genetically determined signs. It allows communication based on an *open-ended semantics* system—any performance can be mimed to represent some associated object, action, or situation. Once the brain mechanisms *and social understanding* for pantomime are in place, it allows a group to consciously communicate information about novel requests and situations that could not otherwise have been indicated.

MSH suggests that brain mechanisms for imitation evolved further to first support pantomime of manual actions and, subsequently, pantomime of actions outside the mime's own behavioral repertoire. The latter would include the "flying bird" example. It would also involve a change of perspective from "something I do" to "something that an other does." Thus, an invitation to hunt for deer might involve pantomime of one's actions during the hunt (and thus might engage the whole body, not just the hands and face) or the movements typical of a deer. The former contains no symbol for "deer," the latter no symbol for "hunt"—both rely on context-based inference on the part of the observer (a process still important in language use today).

Not all concepts, situations, commands, or invitations are readily pantomimed. A pantomime may be costly to perform if ambiguity is to be reduced and even then may remain highly ambiguous. This creates the selective advantage for brain structures and social practices that support the invention and use of conventional gestures that simplify, disambiguate (e.g., to distinguish "bird" from "flying"), or extend pantomime. The result is *protosign*: a set of *conventionalized* gestures (signed protowords) used to formalize and disambiguate pantomime. To return to the "flying bird" example, starting from the same pantomime, it requires conventionalization for a community to converge on (differently simplified) separate gestures to represent "flying" and "bird" so that one may later use the new sign for "bird" as part of the expression for "dead bird."

A key point here is that MSH hypothesizes that *the path to speech is indirect.* There are far more actions, events, and objects that can be pantomimed than can be vocally imitated. Monkeys and apes have a limited repertoire of calls and cannot imitate human spoken words—and they have no capability for vocal learning. But monkeys and apes do have manual dexterity, and, as noted earlier, apes can learn to imitate some small fraction of manual signs from, for example, ASL but have no notion of grammar. Pantomime transcends the slow—and very limited—accretion of manual gestures by apes (as modeled in section 25.4), providing an "open semantics" for a large set of novel meanings (Stokoe, 2001). MSH hypothesizes that the capability for protosign—rather than elaborations intrinsic to the core vocalization systems—may then have provided the essential scaffolding for the vocal learning (lacking in other modern-day primates) that made extended protospeech possible as part of the evolution of the human language-ready brain. According to MSH, protosign and protospeech are involved in an expanding spiral. The mechanisms that evolved to support protosign (e.g., in Broca's area plus areas "beyond the mirror") extended collaterals to yield the control of the vocal apparatus that supported an increasingly precise control of vocalization needed to support speech—but that was only adaptive once protosign had built atop pantomime to develop the social practice of using conventionalized gestures to support an open-ended semantics.

The MSH claim that protosign played a necessary role in scaffolding protospeech remains controversial. The contrary "speech-only" view, if true, would make it harder to explain the signed languages of the Deaf or the use of gestures by blind speakers. But if we reject this view, we must still explain why and how speech arose and became dominant for most humans—a topic which will not be pursued further here (see Arbib, 2012, pp. 241–244, for discussion and references).

25.2.4 From Protolanguage to Language

We have now charted five stages posited by MSH in the evolution of the language-ready brain. The prehominid stages include the following:

A mirror system for grasping: This system was shared with the common ancestor of human and monkey (LCA-m).

A simple imitation system for grasping: This system was shared with the common ancestor of human and great apes (LCA-c).

The hominid evolution stages include the following:

A complex imitation system: This system combines the ability to recognize another's performance as a set of familiar movements with the ability to use this recognition to repeat the performance. More generally, it can recognize that another's performance involves actions similar to those already in the repertoire, with increasing practice in imitation yielding increasing skill.

Protosign: This was a manual-based communication system, breaking through the fixed repertoire of primate vocalizations to yield an open repertoire.

Protospeech and multimodal protolanguage: This advance rests on the "invasion" of the vocal apparatus by collaterals from the communication system based on F5/ Broca's area.

Note that the last two stages, at least, rest not only on new adaptations of the genetic basis of brain development but also on the development of new social practices (cultural evolution) that provide the context for further brain changes to have a selective advantage. By contrast, MSH hypothesizes that the brain of early *Homo sapiens* was "language ready," but that our ancestors of 100,000 or more years ago did not have languages (Noble and Davidson, 1996):

Language: The transition from protolanguage to languages supporting a grammar with a rich compositional semantics that supported the cultural coevolution of cognitive and linguistic complexity required little or no further change in the genes specifying the structure of brain and body.

The nature of the transition to language remains hotly debated, and the details are outside the scope of this chapter (but see Arbib, 2012, chapter 10, for details). Instead, in section 25.5 we will offer a schema-theoretic approach to how the modern human brain couples vision and language, complementing the connectionist approach of chapter 24, "Language Processing."

25.2.5 Introducing Construction Grammar

Much work in neurolinguistics is essentially atheoretical, being based on common-sense notions of syntax—"Does this word fill an appropriate grammatical relation?"—or semantics—"Does this word make sense where it sits in the sentence?" However, many researchers do use theories of grammar to guide data analysis, and a roadblock to integration of their insights is that the theories may be very different. The following are just three examples:

- Grodzinsky (2000) uses the grammar of government and binding (Chomsky, 1981) to explain Broca's aphasia by the trace-deletion hypothesis.
- Hagoort (2003) relates event-related potentials (ERPs) to a parser based on a unification grammar (Vosse and Kempen, 2000).
- Schlesewsky and Bornkessel (2004) structured their ERP findings by using representations from role and reference grammar (Van Valin, 2005). Such representations provided a convenient way to structure the form-to-meaning mapping as principally independent of constituent structures, but they are by no means the only way.

A crucial issue, then, is how to characterize grammar—and thus language—in a manner that supports modeling the mechanisms (whose evolution MSH and future such research seek to chart) that implement that grammar in both the production and comprehension of utterances. Idiomatic expressions like *kick the bucket* or *take the bull by the horns* are an integral part of the speaker's knowledge of his or her language, but their meaning cannot be inferred from the meaning of the individual words and highly general rules of grammar. This suggests that the meaning of each idiom must be stored in each speaker's mind. Fillmore, Kay, and O'Connor (1988) took a radical step: instead of adding idioms to a generative model which separates the lexicon from a separate syntax considered as a highly general and abstract set of rules for combining words, they argued that the tools they used in analyzing idioms could form the basis for *construction grammar* (Croft, 2001; Goldberg, 1995) as a new model of grammatical organization. Construction grammar comes in varied forms, but all blur the division between grammar and lexicon. What is common to these various efforts is that the grammar is based not on a small set of purely syntactic rules but instead on *constructions* which, like items in the lexicon, combine syntactic, semantic, and even in some cases phonological information.

In the framework of construction grammar, *He kicked the bucket* has two parsings. One yields an instance of the general formula *He X'd the Y* whose overall meaning varies with the meanings of *X* and *Y*. The other employs a construction whose meaning is *"He died."* As Hurford (2011, chapter 4) notes, there is evidence that on hearing an idiom, both the overall meaning (e.g., *die*) and the meanings of the individual parts (e.g., *kick* and *bucket*) can be primed. This fits with our general view of competition and cooperation of schemas (chapter 1) in that the initial activation of constructions for both parsings can have a priming effect even though just one eventually wins the competition determining our understanding of the sentence.

In the *He X'd the Y* construction, the role of *X* and *Y* can be filled by almost any transitive verb (for *X*) and noun phrase (for *Y*). Conversely, in the idiomatic *kick the bucket* construction, some syntactic variation is possible, but no substitutions may be made for *kick* or *bucket*. However, a core empirical foundation of construction grammar stems from closer analyses of constructions that reveal a continuum between rather syntactic and rather lexical construction, with many constructions being sui generis in terms of the constraints they impose on their parts. For example, the constraints on *Y* and *Z* in the goal-object construction

W X'd Y with Z as in "Sam filled the glass with water" and the theme-object construction *W X'd Z in Y* as in "Sam poured water in the glass" cannot be accounted for simply in terms of general syntactic classes. Indeed *fill/drench/soak/saturate/infuse* work with the goal-object and not with the theme-object construction and vice versa for *pour/drip/dribble/spill.* Such richness in constructional behavior is what led certain linguists to adopt constructions to represent grammatical knowledge, abandoning the duality between lexicon and grammar, as well as between syntactic and semantic components.

Clearly, the variation in such constructions is not genetically based but instead is acquired by individuals through their usage of the various constructions—and a usage-based emergentist account can also address how these constructions emerged with the language community in the first place (Bybee, 2013; MacWhinney and O'Grady, 2015). Here I simply want to note that construction grammar comes in many flavors and list four efforts under the general banner of construction grammar. Each provides ingredients that we need, but each is incomplete.

Dynamic construction grammar (Dominey and Inui, 2009—see the brief description in chapter 24, "Language Processing") was developed to assess how *brain models* can simulate the learning and use of the form–meaning mappings in constructions. It anchors modeling of the language system on previous modeling of the macaque oculomotor system which demonstrated the central role played by corticostriatal connections in sequence processing (see the background neural modeling in chapter 22, "Integrative Functions of the Corticostriatal System"). In order to link construction processing to sequential cognition, Dominey et al. (2006) used a simplified formulation of constructions, with each construction indexed by the sequence of function words it contains. For example, the passive construction used in *The ball is kicked by the boy* was characterized by the sequence *X Y by Z*, while the object-cleft construction used in *It is the ball that the boy kicked* was associated with the sequence *It is X that Y Z.* The authors showed (in principle) how an interlinked set of cortical and subcortical neural nets could learn to detect sequences of function words associated with different constructions and then use the construction so retrieved to map content words onto their proper thematic roles (*X* onto patient, *Y* onto the action, and *Z* onto agent in the passive construction). More recent work (Hinaut and Dominey, 2013) simulates prefrontal area BA47 as a recurrent network with plastic connections between cortex and striatum. The system is trained on sentence–meaning pairs, where meaning is coded as activation in the striatum corresponding to the roles that different nouns and verbs play in the sentences, and learns an extended set of grammatical constructions.

Embodied construction grammar (ECG) is rooted in a specific version of motor schemas called X-schemas that includes specification of what comes before and after the action and is within the "embodied cognition" tradition that views meaning of a sentence as extracted by mental simulation of the event which a sentence represents. An important feature of ECG is that the range of possible meanings is extended via metaphor (Feldman and Narayanan, 2004; Narayanan, 1997).

Template construction grammar (to be further developed in section 25.5) is anchored by perceptual schemas, assessing how the core of language processing may be captured by analysis of the production and comprehension of descriptions of visual scenes. As already noted, in presenting TCG as one way to address the linkage between vision and language, I in no way wish to exclude the search for alternative characterizations of grammar that may help us understand what, indeed, it is that evolved in the evolution of the language-ready brain.

Fluid construction grammar (FCG) has been developed in the context of interacting robots sharing visual access to a given environment as a basis for, for example, converging on a shared lexicon—and, indeed, this work defined the semiotic cycle of figure 25.3. It uses evolutionary games between embodied agents as the basis for studying the formation and dynamics of shared meaning (Beuls and Steels, 2013). This approach is now being advanced as a framework for the formal study of historical linguistics (cultural evolution).

Each of these frameworks has clear strengths, but each covers only a portion of the mechanisms whose evolution we need to understand if we are to develop an approach to language evolution that can make rich contact with studies of the interactions within and between brains. I see their integration as an important focus for future research. But whatever the appropriate grammatical framework for neurolinguistics, it is a tenet of MSH that languages emerged from protolanguages through a process of bricolage (tinkering), which added new constructions as well as (proto)words to prior protolanguages. Many, but not all, of these constructions became more or less regularized, with general "rules" emerging both consciously and unconsciously only as generalizations could be imposed on, or discerned in, an ad hoc accumulation. The result was a spiraling coevolution of communication and representation, extending the repertoire of achievable, recognizable, and describable actions, objects, and situations which could be talked about and thought about. We may speak of a protolanguage spectrum, with increasingly complex protolanguages shading into simple languages (Arbib, 2008).

As constructions emerged, so did the categories of words which could serve to fill their slots. As generalization integrates varied constructions, these categories

Increasing the Lexicon + More and More Powerful Constructions

Protolanguages: Simple → Complex//Simple → Complex: Languages

may themselves generalize to yield relatively abstract categories whose elements can fill the slots of diverse constructions. Thus, the categories may range from "highly semantic" (e.g., the set of want-able things) to "highly syntactic" (e.g., the category of nouns in a particular language which includes words-for-things as a particular subset).

25.3 Modeling the Macaque Brain

25.3.1 From Macaque Grasping to Mirror Neurons In chapter 1 we introduced schema theory with an example from the visual control of hand movements showing the integration of perceptual and motor schemas in a coordinated control program, reproduced here as figure 25.5. We now briefly recall from chapter 20 two models that carry the investigation from the realm of schema theory to the realm of computational neurobiology, focusing on the interactions between parietal and frontal cortex in the macaque monkey that are implicated in the visual control of manual behavior.

The first model, the FARS model (figure 25.6) of parietal–premotor interactions in primate control of grasping (Fagg and Arbib, 1998), links a subarea of parietal cortex named the anterior intraparietal cortex (AIP) to an area F5 of premotor cortex. AIP contains neurons responsive to visual properties of objects—but visual properties that indicated "affordances" for grasping, rather than the particular category of the object

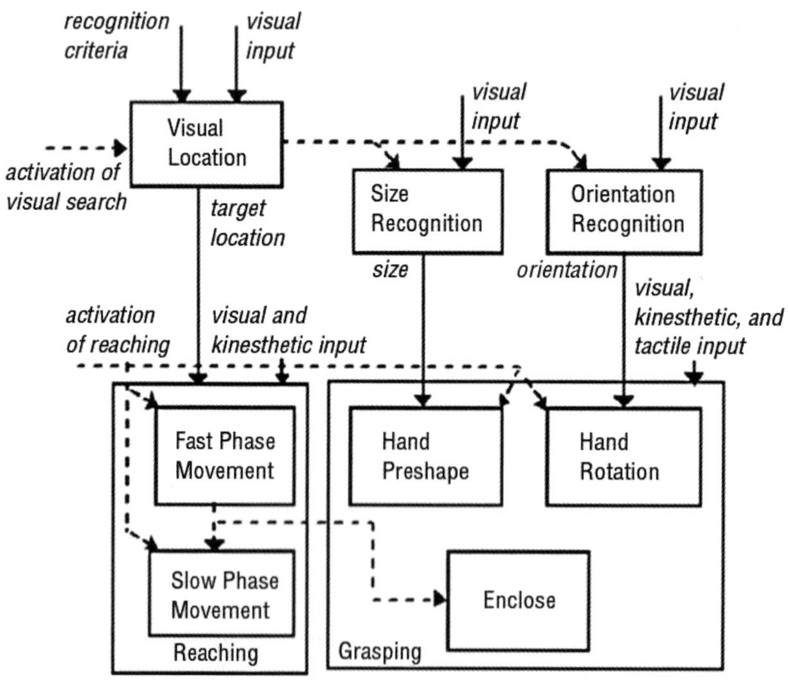

Figure 25.5 Hypothetical coordinated control program for reaching and grasping. Motor schemas control the arm (Reaching) and hand (Grasping = Preshape/Rotation Enclose). Different perceptual schemas (top half of figure) provide input for the motor schemas (bottom half of figure) for the control of "reaching" (arm transport ≈ reaching) and "grasping" (controlling the hand to conform to the object). Note, too, the timing relations posited here between subschemas within the "Reaching" motor schema and those within the motor schema for "Grasping." Dashed lines indicate activation signals; solid lines show transfer of data. Transport was hypothesized to involve a ballistic phase followed by a feedback phase, with the transition between the two also activating the Enclose motor schema. Adapted from Arbib (1981).

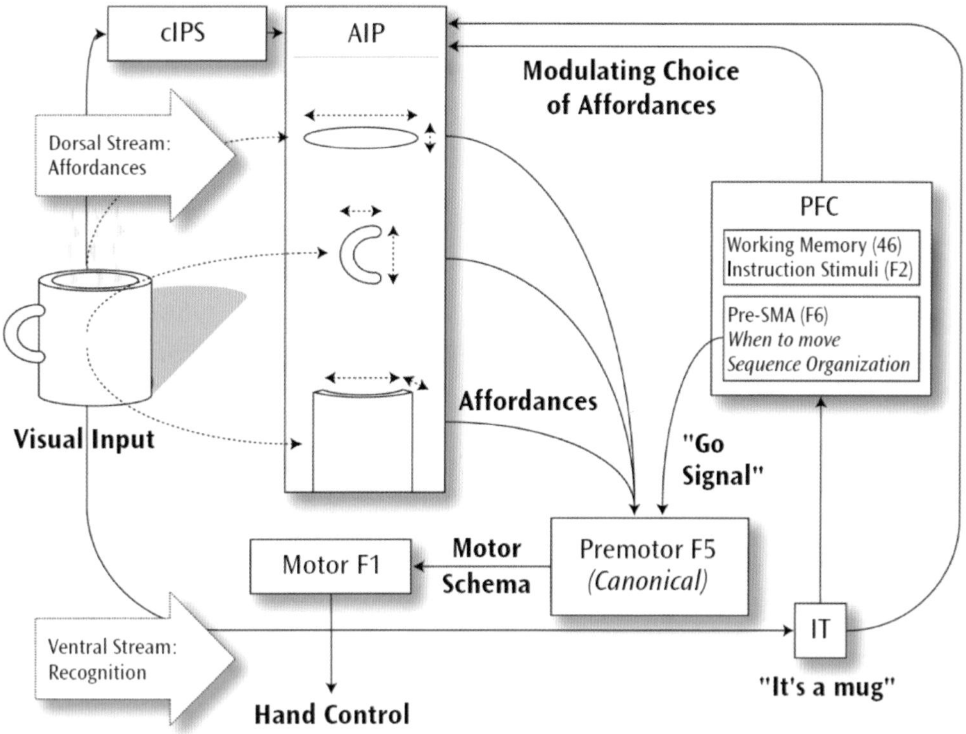

FIGURE 25.6 The dorsal stream of the FARS model takes input from the primary visual cortex and processes it via, especially, the anterior intraparietal area (AIP) of parietal cortex. The idea is that AIP does not "know" the identity of the object but can only extract affordances (opportunities for grasping the object considered as an unidentified solid), indicated here by portions of the object which are particularly graspable. The FARS model further integrates the dorsal stream with the ventral stream via inferotemporal cortex (IT). Prefrontal cortex (PFC) uses IT's identification of the object, in concert with task analysis and working memory, to help AIP select the appropriate affordance from its "menu." Given this affordance, F5 can select an appropriate motor schema—consider the difference between lifting a mug to drink from it and simply moving it out of the way—and instruct primary motor cortex to forward its execution. cIPS, caudal intraparietal sulcus.

(e.g., cube vs. sphere). Similarly, F5 contains neurons whose activity correlates with the macaque's execution of a particular type of grasp. As noted in chapter 1, one may contrast the dorsal "how" path from visual cortex through parietal cortex with the ventral "what" path through inferotemporal (IT) cortex, with IT characterizing the type of an object as a complement to AIP's characterization of its affordances. The dorsal stream of FARS, then, provides the "neuralization" of the schema model of figure 25.6.

Although we will not develop this theme here, it is worth noting that the FARS model was used as the basis for a "Synthetic Brain Imaging" study (see chapter 15) that developed predictions that were tested using positron emission tomography, with results that were fed back to suggest refinements to the original model (Arbib, Fagg, and Grafton, 2002; Grafton, Fagg, and Arbib, 1998).

We also recall from chapter 20 the MNS model (figure 25.7) of how mirror neurons may be formed by learning based on visual observation of one's own actions (Oztop and Arbib, 2002). Refinements in the MNS2 model (Bonaiuto, Rosta, and Arbib, 2007) explained data on audiovisual mirror neurons and on mirror neurons that could recognize a grasp even if the object had been recently obscured from view. Unlike FARS, the MNS models do not include processing in the ventral stream (another challenge for future modeling). FARS and MNS will be the basis for the chimpanzee model offered in section 25.4. In particular, we stress that, when presented with the view of an ongoing action, the MNS model as embedded in the ape model learns to strengthen recognition neurons earlier and earlier in the trajectory.

25.3.2 AUGMENTED COMPETITIVE QUEUING We here introduce a new model, the ACQ model (Bonaiuto and Arbib, 2010), which addresses the flexible scheduling of actions to achieve some overall goal. Crucially, and unlike most of the mirror neuron literature, *this model*

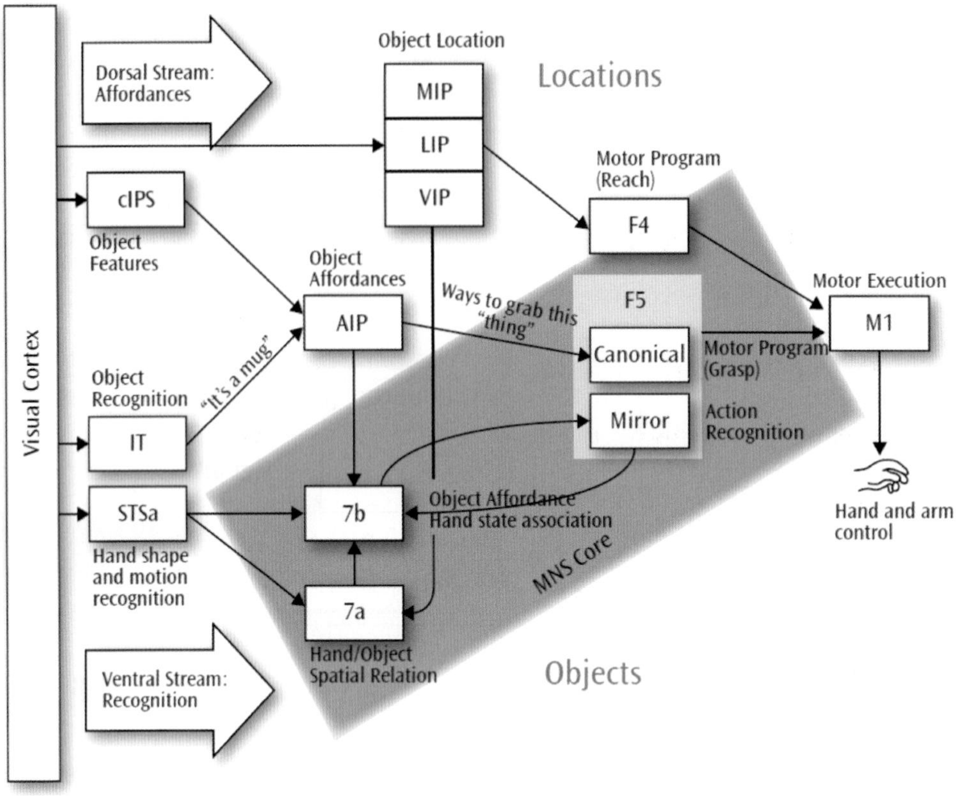

FIGURE 25.7 A schematic view of the mirror neuron system (MNS) model of the mirror system for grasping. Note that visual input is differentially processed in different streams to extract the affordances of an object and the shape and movement of a hand. The model describes mechanisms whereby, using the "grasp code" of the canonical neurons of F5 as training signal, particular mirror neurons come to recognize affordance-centered trajectories of the hand that correspond to specific types of grasp. Such a trajectory can thus equally well activate appropriate mirror neurons when the trajectory involves the hand of the self or the hand of another. Note, however, that the model does not address how attention comes to be directed to a particular hand and a particular object. From Arbib and Mundhenk (2005); adapted from the original system figure of Oztop and Arbib (2002).

emphasizes the utility of a mirror system in recognizing one's own action. But first we consider two key distinctions which compare and contrast ACQ with FARS and MNS:

Proximal versus distal goals: All the actions in a behavior may be steps toward a single shared goal, the *distal goal,* yet each may have a distinct *proximal goal* that shapes that particular action. In the FARS and MNS models, the focus is on a single goal: grasping an object based on one of its affordances, or recognizing an ongoing action. In ACQ, the emphasis shifts to scheduling a sequence of actions, with each linked to its own proximal goal which serves as a subgoal for the distal goal which is the overall goal of the current behavior.

Event-level versus trajectory-level modeling: In many psychological models, each event or trial or action is considered as an indivisible whole, and emphasis is placed on the stringing together of distinct behavioral events by decision processes. By contrast, *trajectory-level modeling* analyzes, for example, the trajectory of a hand movement and the temporal relation of neural firing to that trajectory. For example, the FARS and MNS models are trajectory level—assessing activity during the execution and recognition of an action—whereas ACQ is event level, scheduling complete actions under the assumption that other neural systems will manage their trajectories.

ACQ is defined in contrast to ordinary competitive queuing (CQ) (this contrast was already noted in chapter 14, "Schema Theory and Neuropsychology"). The idea of competitive queuing (Bullock and Rhodes, 2003; Houghton and Hartley, 1995) is that a neural array (the planning layer) holds a fixed pattern of activity, with each neuron coding the "priority" of a particular action. Applying a winner-take-all (WTA) network to its output selects the action with highest priority for execution. What distinguishes a CQ network from a WTA network is "inhibition of return": once an action is executed, the corresponding neuron is inhibited. Thus, reactivation of the network yields execution of the action with second highest priority. And so it goes: a spatial coding of priority is converted into a temporal sequence of actions. One of the most famous applications of this idea is in the scheduling of saccades where preprocessing of a visual image yields a map of "salience" of different regions of the image (Didday and Arbib, 1975; Itti and Koch, 2000; Koch and Ullman, 1985); CQ can then direct saccades to these regions in order of their salience. (See chapter 17, "Object and Scene Perception," and chapter 19, "Saccades and Smooth Pursuit Eye Movements." Chapter 22, "Integrative Functions of the Corticostriatal System," offers a different model of sequencing related to "reservoir computing.")

In ACQ, competition to determine the next action is again based on a priority map, but now this map is updated each time an action is executed. Specifically, separate subsystems in ACQ (figure 25.8) assess the *executability* of actions (e.g., the availability of suitable affordances) and their *desirability* (an estimate of the expectation that executing that action will lead on to attainment of the distal goal). The model is event level: the system chooses the action with the highest value of

$$\text{priority} = \text{executability} * \text{desirability}$$

and that action is then executed to completion. *Executability* and *desirability* are then updated, based in part on changes in the environment (distribution of affordances) following the preceding action, and the action with the currently highest priority is executed. This process continues until the distal goal is attained.

Moreover, ACQ contains a mirror system which, when monitoring *self*-actions, indicates whether an intended action proceeded as expected or whether the output resembled another action, the *apparent action*. In the current implementation of ACQ (Bonaiuto and Arbib, 2010) the mirror system is simplified, but to understand the issue here, consider the exposition of MNS (chapter 20). During training, canonical neurons controlling the monkey's grasp are posited to activate, via corollary discharge, a set of potential mirror neurons. This "canonical code" for a grasp serves as the training signal for the latter neurons to learn to recognize the corresponding hand-state trajectories (features coding the movement of the hand and its preshape relative to the selected object affordance). Eventually, the synapses formed under this training become powerful enough

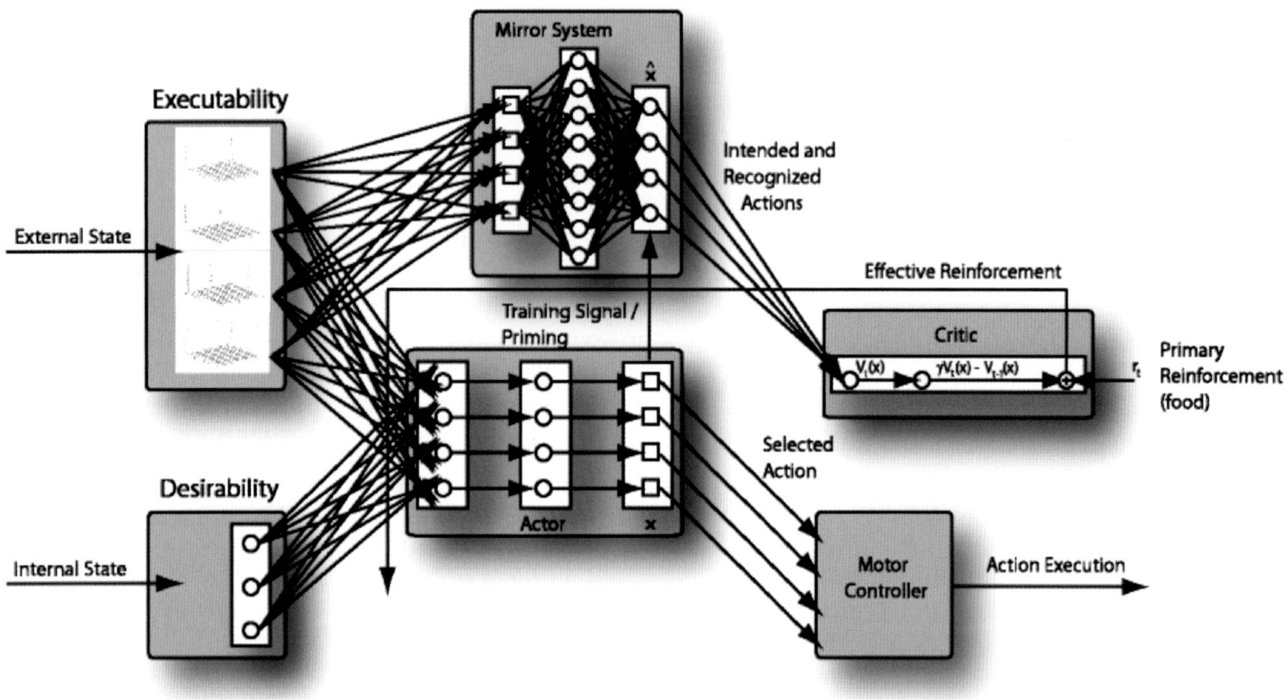

FIGURE 25.8 A view of the augmented competitive queuing (ACQ) model (Bonaiuto and Arbib, 2010). The agent works towards its distal goal by choosing the most desirable action that is currently executable. The mirror system monitors self-actions, indicating whether an intended action succeeded or whether the output instead resembled another action, the apparent action. The model includes learning mechanisms for updating estimates of executability of the intended action and desirability of the apparent action.

that these now-established mirror neurons can fire on observation of an appropriate hand-state trajectory even if it belongs to someone else—thus the mirror neuron property.

Now consider (recalling figure 25.5) what happens, according to the MNS model, for self-execution of an action after the mirror neurons have been trained. The mirror neurons for that action are then activated via two pathways: the canonical pathway brings an efferent copy of the canonical code for the movement while the hand-state trajectory is observed via the visual system (or via the sound of the action driving audiovisual mirror neurons for those actions with distinctive sounds). The new insight is this: in general, these two pathways will activate the same mirror neurons—*but only if the action is successful.* If the action fails, then two sets of mirror neurons may be activated: one set for the *intended* action and a distinct set for the *apparent* action if the unsuccessful action happens to resemble an action in the mirror system repertoire. This notion was motivated by analysis of cats that had lost their ability to grasp objects but were able to rapidly develop a new strategy to retrieve food from a glass tube (Alstermark, Lundberg, Norrsell, and Sybirska, 1981). Bonaiuto and Arbib (2010) demonstrate that the ACQ model with the features we next present could indeed explain this result: a (mirror) system may contribute to rapid reorganization of motor programs in the face of disruption when a known schema can be recognized as "filling the gap" for disrupted schemas. For this reason, Bonaiuto and Arbib titled their paper "What Did I Just Do? A New Role for Mirror Neurons."

The mirror system's evaluation of the end result of attempting the intended action is used to update the estimate of its executability, downgrading it if the apparent action departs too far from the intended action. Temporal-difference reinforcement learning (cf. the notion of *expected reinforcement* in chapter 10, "Reinforcement Learning") is used to update the desirability of the *apparently executed action*, whether it is the intended action or a different apparent action: the desirability of an action is greater the more likely it is to be on a path to attaining the distal goal and the closer it is along the path to the action that actually achieves it.

An extension of ACQ (Arbib and Bonaiuto, 2012) handles multiple possible distal goals, with the desirability of an action depending on the current selection of distal goal. This then requires modeling the processes setting the current distal goal. Achievement of a distal goal—for example, eat versus drink—may remove that goal, leading to promotion of another distal goal and consequent resetting of desirability, and so behavior.

25.4 *Grounding Ape Gestural Development in Macaque Brain Models*

We now focus on *ontogenetic ritualization* (defined in section 25.2.1), which was proposed as a means whereby some ape gestures could emerge through repeated interactions between pairs of individuals. Even though the role of this process remains controversial (Halina, Rossano, and Tomasello, 2013 vs. Hobaiter and Byrne, 2011), we argue that using computational neuroscience to hypothesize brain mechanisms that could support ontogenetic ritualization can generate new ideas for primatological as well as neurophysiological and neuroimaging studies, and may reset the debate in fresh ways.

The framework for both the conceptual model (section 25.4.1) and the reduced, implemented model (section 25.4.2) is provided by the dyadic brain modeling of figure 25.4.

25.4.1 The Conceptual Model For conspecifics and for now, we assume that the brain of each participant in the interacting dyad has the same overall architecture but that each agent/brain starts in a different state (in terms both of neural firing and synaptic encoding of skills and memories and goals), and thus the actions and learning processes that unfold will differ between the two agents. We first present the interactions of two apes which support the stages whereby, according to Gasser, Cartmill, and Arbib (2014), a gesture for *beckoning* arises through ontogenetic ritualization. (Of course, each "stage" merges into the next as ritualization proceeds.) We then introduce an extended *conceptual* model (Arbib, Ghanesh, and Gasser, 2014) to represent the brain of each member of the dyad. Section 25.4.2 presents a reduced, *computationally implemented* version of the model.

1. **Child (C) reaches out, grabs, and tugs on Mother (M), causing M to move toward C as a response.**

C's distal goal is "Bonding," and ACQ-C (the child's version of ACQ) schedules the actions

c1: Reach to grasp M's upper arm, and then

c2: Tug on M till she moves toward C. This combines a proximal goal of tugging with the distal goal of Bonding (or some such social goal).

Initially, M has no specific goal. However, once she recognizes C's distal goal, she creates her own distal goal of Bonding with proximal goal "Move close to C." and ACQ-M then schedules

m1: Move in the direction of C.
m2: Embrace C.

All four actions are assumed already in the repertoire of C or M as body-centered actions. MNS-M (Mother's action recognition system) recognizes C's distal goal on the basis of the proximal goal of c2, but this recognition is (in this example) at first based on the haptic input of C's tugging, not on visual recognition. This stage can be seen as the "praxic alternative" to a ritualized gesture since, if later a gestural communicative bout fails, C can always revert back to this praxic action sequence (Cartmill and Byrne, 2007).

2. **C reaches out and grabs M, and M moves toward C as soon as C begins to tug.**
This stage could proceed over many trials as MNS-M comes to recognize the distal goal represented by c2 more swiftly: with a visual stimulus, the MNS model learns to strengthen recognition neurons earlier and earlier in the trajectory. We now exploit this property during the *haptic* trajectory induced in M by c2. As M recognizes the proximal goal of c2, and thus C's distal goal, more quickly, she can respond earlier and earlier on the basis of visual input alone. Eventually, C no longer needs to complete c2. Here we exploit the notion that any action can be truncated when its proximal or distal goal is achieved.

3. **C reaches out and makes contact with M, and this is enough to get M to move toward C.**
Eventually C need only complete c1 for M to respond. ACQ-C continues to schedule c1, but now c2 is no longer executed because c1 alone achieves the distal goal. MNS-M then comes to associate c1, rather than c2, with C's distal goal. This stage, like stages 1 and 2, seems not to be a distinction between macaque and chimpanzee. As a further result of stage 2, the increasingly early termination of the intended c1, "reach to grasp firmly enough to tug," appears more and more similar to the action ĉ1, "reach to touch," and ACQ-C will increase the desirability of the latter action so that it comes to replace c2 in C's behavior. And MNS-M will recognize this modified action as also having Bonding as its distal goal *in this context.*

4. **C reaches out toward M, attempting to make contact, but M responds before contact is made.**
In this stage, contact with M is still C's proximal goal, and the action is still transitive. Whereas MNS-M linked tactile feedback to M's responses, we can now see robust visually guided adaptive behavior emerge. Seeing M's response, C can now truncate c1 (or ĉ1). However, C's action is still initiated as a movement toward M's arm and so may still be deemed transitive.

5. **C reaches out toward M, no longer attempting to make contact, yet M still responds by moving toward C.**
We now postulate that the brains of "apes" have far stronger pathways bringing proprioceptive input (as distinct from visual, haptic, and auditory input) than would be available in the macaque brain. This means that (in this scenario) as M responds earlier in recognition of C's goal, C comes to associate the distal (not the proximal) goal with the proprioceptive state (part way along the original transitive trajectory) of that response. It is not that C loses the ability to perform action c1 in appropriate circumstances but rather that a new action č emerges in C's repertoire: performing a reach that has achieving that proprioceptive state (rather than contact with M) as its proximal goal—but still with the distal goal (get M to come closer/bonding) that constitutes the meaning of what is now a novel gesture. In this way, there is a transition from a transitive action to an intransitive action.

6. **C "beckons" toward M.**
The ritualized gesture č is now part of the action repertoire of C so that C can "beckon" to elicit M's response. As in the above stage, M is capable of recognizing his gesture and responding appropriately. Note that apes tend to use visual-only gestures when they have the other's attention. Thus, C might retain the tug strategy when M is otherwise engaged but use č for times when M is attending visually. In other words, learning this gesture has not removed C's original behavior (c1 + c2) but has refined its context: use the gesture when M visually attends, but otherwise revert back to c1 + c2 for the same, albeit more effortful, result.

Having just postulated that the brains of "chimpanzees" have stronger pathways than are available in the macaque brain to bring proprioceptive input into control of reaching and hand shaping, we note that Hecht et al. (2013) used diffusion tensor imaging to conclude that chimpanzees have "stronger connections" between superior temporal sulcus and inferior parietal cortex (viewed as the posterior mirror system) than do monkeys and that this may allow more processing of the finer details of the spatial/kinematic structure of observed actions. How this relates to proprioceptive input requires further analysis. Additional challenges for related research include specifying what exactly is meant by such connectivity changes (and see Rilling et al., 2008, for changes relevant to language evolution), and how these data affect modeling projects: Do these results implicate new connections between modules, stronger existing connections between modules on a per-species basis, or actual processing differences between and/or within modules?

We now summarize changes required from the ACQ model (figure 25.8) to generate our new model of "ape" brain mechanisms shared by M and C (figure 25.9). However, the requirements below are *not* specific to beckoning and 1 through 4 in the following list represent abilities of general utility to macaques as well as chimpanzees:

1. There is a mechanism for social decision-making that can use recognition of the distal or proximal goal of another as a basis for possibly updating one's own goals.
2. Decision making occurs in two stages. A distal goal is set; this goal then sets up the desirability values used by ACQ to schedule a sequence of actions to achieve that goal.
3. The actions available to the new and extended version of ACQ are all trajectory level (unlike our present ACQ model). Depending on the stage of learning, a mirror system for an action may be sensitive to a goal and more or less sensitive to trajectory.
4. Each action in the repertoire of the agents of our dyad will be terminated when it is signaled that its proximal goal is reached. Moreover, ACQ can abort a whole behavior once its distal goal is reached.

We see the above additions as an extension of our models of what is common to macaque and chimpanzee. We now turn to a hypothesis on what is different about the chimpanzee:

5. Proprioceptive information becomes increasingly available to augment visual input in mechanisms underlying the learning of novel self-actions.
6. More finely tuned attention to visual and kinematic information about others' manual movements allows mirror system recognition of intransitive gestures.

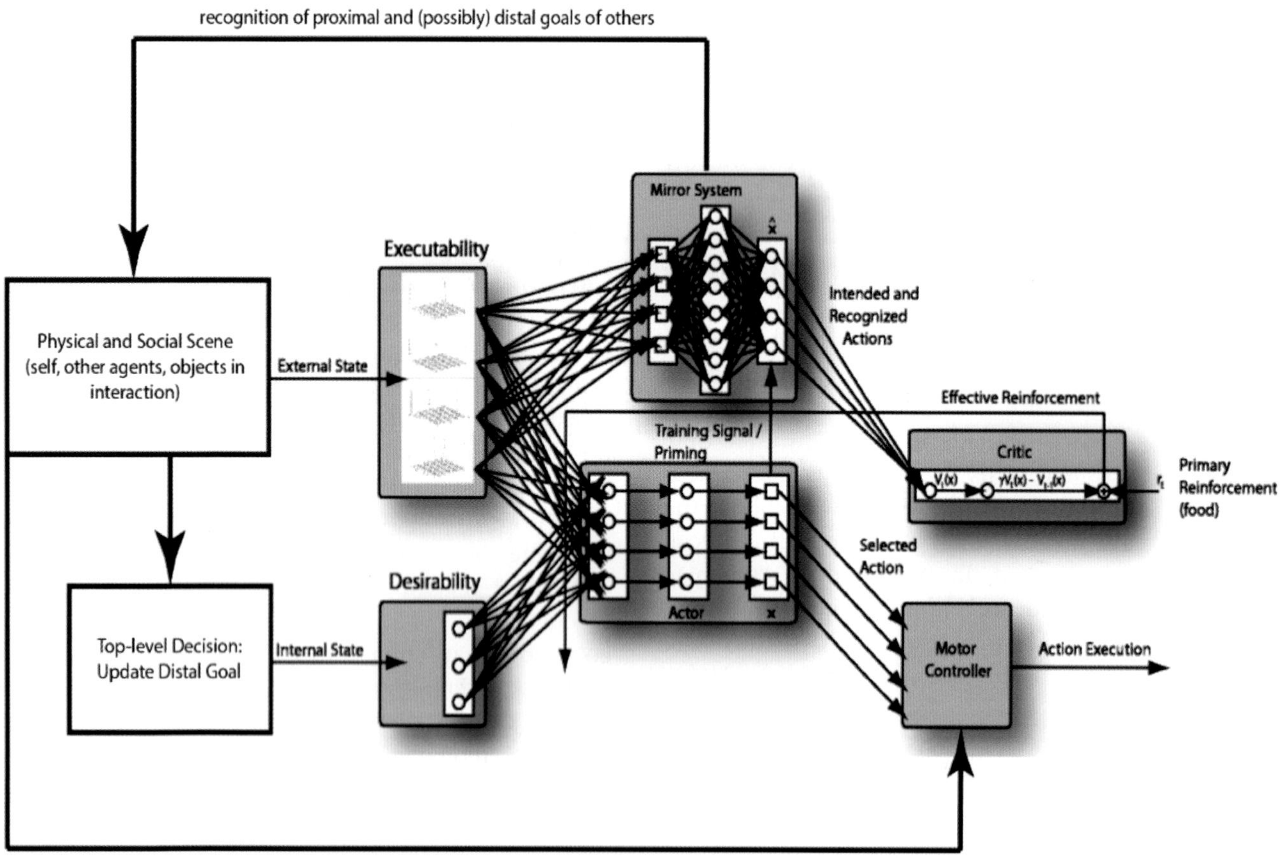

FIGURE 25.9 The overall model of the brain of each ape in a dyad. Updating the current distal goal is an explicit part of the model of each brain in the dyad, and this, in turn, controls desirability. The mirror system not only plays an augmented competitive queuing–like role in the analysis of self-actions but will also recognize the actions (movements and goals) of others. Perception of the physical and social scene sets up decisions on the current distal goal as well as providing input to diverse processes. Among the factors not shown here are that modeling is at the trajectory level, and that proprioception plays a key role in the generation and monitoring of one's own actions (Arbib, Ganesh, and Gasser, 2014).

25.4.2 The Reduced Model and Its Implementation We now describe a simulation framework for dyadic brain modeling and then present a simplified model we have implemented. The two model brains are implemented in separate Python scripts (figure 25.10), employing freely available Python packages, including numpy (numpy.org) for mathematical calculations and PyBrain (pybrain.org) for neural network algorithms. Each script is linked to the Maya animation software through a third Python script executed in the Maya Python shell. Each brain script can communicate to Maya, via this shell script, through file read–write cycles managed through their separate pipes. Each brain script implements functions from a "brain module" library—for action recognition, decision-making, and motor control—to maintain consistency in their implementation. Each script can read relevant joint and effector positions and information about the virtual world in their shared file, and write commands to it for the avatar to execute. After the avatars move in Maya, the new relevant positions are recorded in a file that can then be read by each separate brain script according to how its sensory systems are implemented. Each "brain" uses sensory data on the relation of its body parts to each other and the world, but Maya draws the bodies in a coordinate space appropriate to the view of an external observer.

We now turn to the simplifications of the model of figure 25.9 used in the present simulations (Arbib et al., 2014; readers are referred to this paper for the equations and parameters used in our simulations). The simulations are simplified by the fact that the Python script for each agent can read information about the current state of the other agent from the common file into which Maya records data. We thus avoid the need to simulate retinal input to each agent and consequent visual processing to extract state information about the other. In this sense, we finesse visual preprocessing while still taking it into account. However, simulation is still required to infer the action or goal of the other agent from the ongoing trajectory of observed state features plus haptic input. The overall architecture of each brain (figure 25.11A) was the same, but to reduce computation time, learning in the mirror system was implemented only for Mother, and motor learning for the reach-to-grasp only for Child. Decision-making processes were implemented similarly for both agents, and followed the ACQ paradigm (e.g., combining *executability* and *desirability* to yield *priority*). Our simulations consist of multiple episodes, each a single explicit case of interaction between the two agents. An episode may vary from (roughly) 50 to 100 simulation time steps (though the presumed interaction may last only a few seconds). Visual and haptic

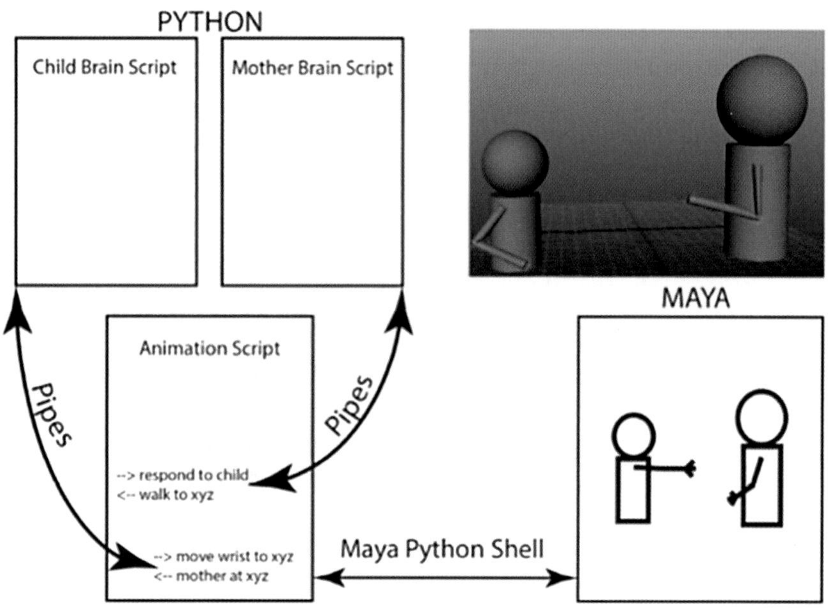

Figure 25.10 A general computational framework for dyadic brain simulation and visualization of the interaction of two agents. (Clearly, the methodology is not limited to simulating a Mother and Child, but could be adapted to model the neural underpinnings of social interactions of two or more agents.) (Top right) A view of the two avatars (small = Child; large = Mother in this example) as displayed by the animation software Maya. In the current simulations, all that changes in these displays is the position of each body in the ground plane, and the angles that define arm postures. (Rest of figure) Simulation overview: The two model brains are implemented as separate Python scripts that can only communicate through Maya's animation script (Arbib, Ganesh, and Gasser, 2014).

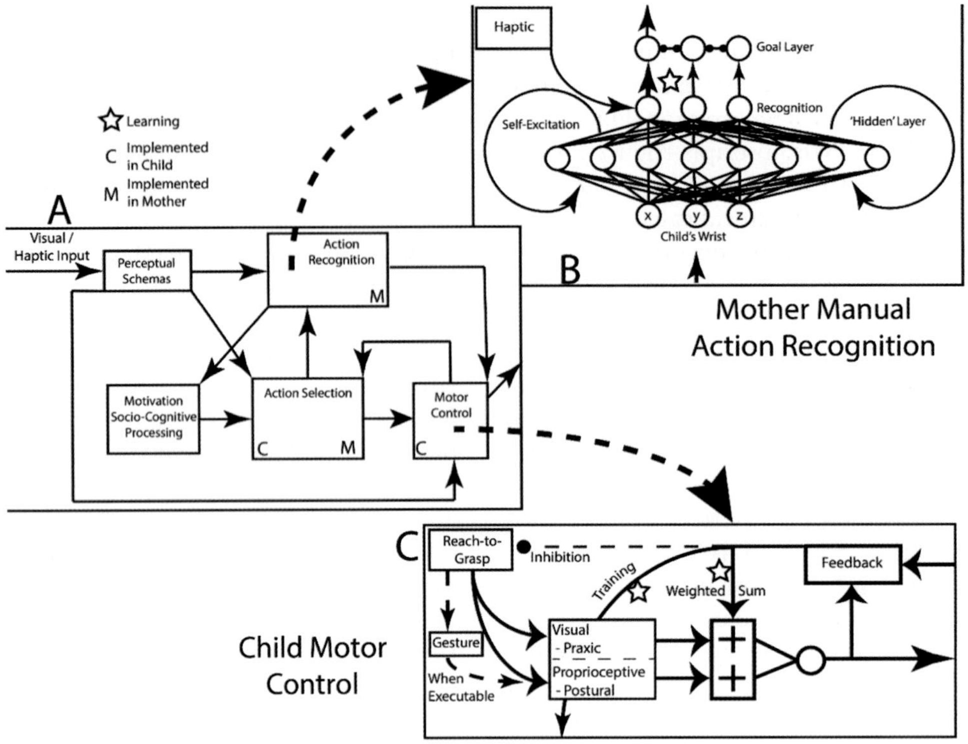

FIGURE 25.11 (A) Schematic of the reduced version of the brain architecture for both agents of the dyad. In the simple simulations reported here to demonstrate ontogenetic realization, only some parts are implemented as neural rather than algorithmic models, and differently for Mother and Child. "M" indicates modules implemented "neurally" for Mother, "C" for Child. Other modules are implemented algorithmically. (B) Action recognition for manual actions as implemented for Mother. (C) Motor control as implemented for Child (Arbib, Ganesh, and Gasser, 2014).

sensory inputs to each model represent the output of visual and somatosensory cortical areas, and these inputs may be the result of the others' ongoing motor output.

Action recognition modules, processing visual and haptic information to recognize self and others' actions, can affect social behavior by linking recognition of others' actions and goals with one's own. Such modules may also provide feedback to learn about the *trajectory* that achieves a goal. Motivational and socio-cognitive processing selects distal goals which yield new action policies (recall the extended ACQ model of figure 25.10). Each agent is initialized in different motivational states: the Child to "Bond" with the Mother (a social goal), and the Mother to be idle initially. Following the paradigm of ACQ, these different goal states lead to different behavioral policies for each of Mother and Child: Mother remains idle (we assume occupied in some other nonsocial behavior for the time being) while Child executes a sequence of actions to lead to Bonding. However, as a result of learning and recognition processes, these interactions adapt, and Mother's rapid updating of her goal and associative learning of mirror system output with this new goal lead to anticipatory responses that achieve the Bonding goal for both.

Figure 25.11B shows how manual action recognition is implemented for Mother. Action recognition is managed algorithmically for C in this implementation (meaning we do not simulate a neural network for this process but instead situate our code to achieve the desired behavior). C's wrist (x,y,z) information is input to a neural network with recurrent, self-excitatory hidden units to maintain temporal information about the inputs. These are converted from 3-D absolute coordinates in Maya to yield 3-D coordinates relative to C's posture: the wrist trajectory being relative to his starting posture. The recurrent network and its learning mechanism are based on MNS2, while we point out that at least some degree of input to the action recognition system here is independent of hand–object relations (i.e., is intransitive). The network's outputs activate to the degree to which they recognize the time-series data. These units, along with haptic units representing haptic feedback if present (i.e., that another has grasped one's arm, for example), then pass activation to goal-layer neurons, via modifiable weights, which compete between each other. If one passes some activation threshold, they switch the internal state of M, so that she shares the Bonding goal of C and will cooperate toward achieving that goal. More generally, recognition of another's goals may, but need not, lead to goal switching in the observer, and new goals may be quite different from the other's, leading not to cooperation, but to competition, for example (though we only consider cooperation in the present model).

M's motor control elements are managed algorithmically, as is walking in C. C's reach-to-grasp (figure 25.11C) is implemented as an inverse-kinematic controller managing transport of the arm to a target. We model learned target positions in shoulder-centered coordinates but achieve inverse-kinematic solutions through the animation software. Target data are based on a weighted-sum of visual and proprioceptive data, but learning processes may bias each modality's influence. An action will terminate when its proximal goal is reached, and an ongoing behavior will terminate when its distal goal is reached (which may involve recognition of the other's behavior).

With this, we turn to some further details of the figure 25.11 model. In each episode, C is initialized with Bonding with M as his distal goal, while M is at first passive. This leads to setting C's proximal goal and action. Since the avatars are initialized at some distance from each other, the Child must first walk toward his Mother. When *walk* is selected, a target position is computed for the agent to walk toward, namely, the threshold at which the Child can reach toward the Mother. When the Child reaches this threshold, the "reach-to-grasp" action becomes executable, and since it is more "desirable" in the sense that it is more closely associated with the goal, it is selected for execution. The controller for this action has two sources of sensory data, visual (which can suffice for object-directed actions) and postural (which would suffice for proprioceptively guided action). Each alone can determine a target for the reaching action.

Learning processes adapt the above to quickly generate new patterns of behavior. M's action recognition module (figure 25.11B) is centered around a recurrent neural network (RNN) fashioned after the MNS2 model (Bonaiuto et al., 2007). The RNN receives filtered visual input of C's wrist: the codes must be relative so recognition is possible in different locations. The network is trained through the backpropagation-through-time algorithm (Werbos, 1990), available through PyBrain, and is trained on a set of three different reaching actions of C—the two others being reaching to locations other than M's wrist—each corresponding to a single output unit of the RNN. During simulations, C's reach will activate a specific output unit to the degree to which its trajectory matches the training data (the first episode reach). In this implementation at least, it places a constraint then on the *visual form* of the gesture. As a result, it must (at least in the early stages of ritualization) sufficiently "overlap" in appearance with the original action that its form indicates from which action it is derived. (On a more extended timescale, similarity to recent instances of the gesture would allow changes whose accumulation could yield a gesture markedly different from a truncation of the original action.)

C maintains the potential for a full reach-to-grasp-to-pull but learns over time to (1) pull less and then (2) reach less effortfully. He learns at first that M responds to his pulling, and so pulls her for shorter distances—remaining sensitive to her responses—before learning to only intend to make contact with her wrist. Then, as C reaches, he learns postural targets he can move to with little or no activation of praxic control mechanisms. In each episode, he updates his *postural target* as the position at which he terminates his action. Simultaneously, he learns to value this postural target over his visual target. Because it is less effortful and so more desirable, the postural target comes to be selected as the "proximal" goal in service of the distal goal of Bonding. This leads to gesturing behavior that is not constrained by the peri-personal reach threshold: once mastered, the child can gesture from a greater distance removed from his mother, provided she attends to him visually.

Our simulations show how a "beckoning" gesture emerges over time, across a wide range of parameter settings. Figure 25.12 shows selected frame shots of the interaction of Mother and Child which demonstrate ontogenetic ritualization. Some parameters are artificially selected to yield more realistic patterns of behavior while others can be systematically varied to achieve faster (or slower) time courses of gestural acquisition, as well as longer or shorter gestural forms for C. Changes in certain parameters correspond to the speed of M's recognition and may correlate with sensitivity (or not) to gestural forms and understanding others' goals while others may correlate with C's ability to quickly learn postures and/or seek out less effortful options for action. Future work could more systematically explore this parameter space, replacing particular parameters with more nuanced control mechanisms, or else seek to relate these parameters, and their possible variation, with available data in primate communication and social cognition.

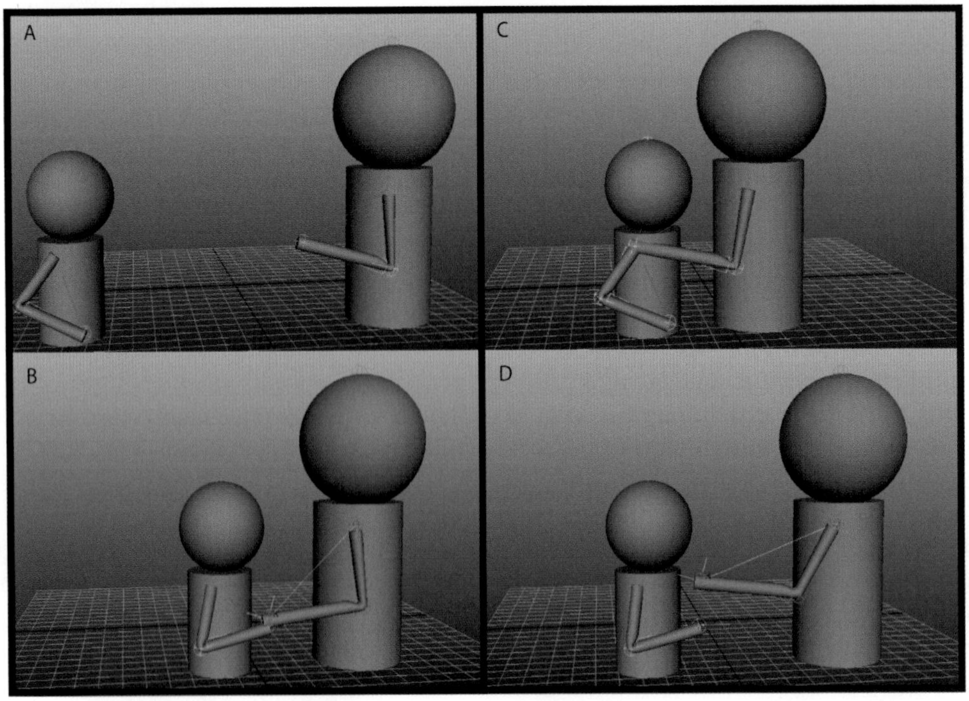

FIGURE 25.12 Simulation screen-grabs. The avatars are simple animated agents capable of "walking," moving limbs, and interacting with each other. (A) The agents initialized to start an episode. (B) The Child pulling the Mother toward him. (C) The Mother completing the Child's goal of Bonding. (D) The Mother responding to the Child's learned gesture (Arbib, Ganesh, and Gasser, 2014).

25.5 MSH and the Brain Mechanisms Underlying Language

How can we better assess what makes a brain language-ready as a target for better assessing how it evolved? Understanding the language system requires understanding not only the processing of grammatical constraints but also how such processes are integrated with evolutionarily more conserved systems that support our sensorimotor interactions with the world. In the so-called *visual world paradigm,* psycholinguists record the eye movements of a subject who is processing linguistic stimuli while simultaneously looking at a related visual scene. Altmann and Kamide (1999), among others, showed how subjects could incrementally combine linguistic, visual, and world knowledge information during a comprehension task. Faced with a visual display showing a cake, two toys, and a boy, for example, the subjects were faster to fixate the cake upon hearing the verb "eat" in the sentence "the boy will *eat*..." than upon hearing the more general verb "move" in the sentence "the boy will *move*...." The challenge is therefore to link language processes to the perception of "realistic" visual scenes while also accounting for the role of world knowledge.

25.5.1 MSH Revisited: A Modeling Perspective on the Steps toward Language

MSH has traced a complex path. Here, we clarify what is implied at each step along the way by building on the insights gained from the modeling of sections 25.3 and 25.4:

1. *A mirror system for manual action:* The ability to generate a range of manual actions and recognize when other's make actions already in one's own repertoire. Shared with LCA-m. Modeled by the FARS and MNS models, respectively.
2. *Scheduling of actions:* (a) Executing a learned schedule. Hidden within the key paper on the FARS model (Fagg and Arbib, 1998), there is an account of how the basal ganglia (not shown in figure 25.6) may schedule a sequence of actions of which the actual reach to grasp is but one element. Similarly, as explained in chapter 22 ("Integrative Functions of the Corticostriatal System"), Dominey, Arbib, and Joseph (1995) showed how interactions between prefrontal cortex (PFC) and basal ganglia could suffice to learn and control the execution of sequences of saccades. (b) Opportunistic scheduling. The ACQ model (section 25.3.2) showed how the basal ganglia, implicated in temporal-difference learning (chapter 10, "Reinforcement Learning"),

could update the desirability and executability of actions in relation to a particular distal goal, so that actions could be opportunistically scheduled by choosing the most desirable action that is currently executable.

3. *Simple imitation:* (a) The ability to recognize that another has achieved a desirable goal by using a short sequence of actions already in one's repertoire and being able to reproduce the sequence to achieve that goal oneself. (b) The ability to recognize that another has achieved a desirable (sub)goal using an action *not* already in one's repertoire and using trial and error to develop an action that can achieve the subgoal. Shared with LCA-c. We did not look at models of this, but the point to stress is that neither of these cases are achieved by having a mirror system alone without coupling to other subsystems.

4. *From action to gesture:* The dyadic brain modeling of section 25.4 suggested how augmenting the macaque-like mechanisms of FARS, MNS, and ACQ with the ability to use proprioceptive cues could support the generation and recognition of *intransitive* actions, and that this, in turn, could support the ontogenetic ritualization of actions of a dyad to yield novel communicative gestures.

5. *Complex imitation:* Two key differences between humans and apes, distinguishing complex imitation from simple imitation, are (a) overimitation, the focus on imitating a sequence even when evidence suggests that constituent actions are unrelated to the next subgoal, and (b) the ability to attend to the form (not just the proximal goal) of a novel action as a basis for short-cutting its acquisition, avoiding overly lengthy trial and error. Again, computational modeling awaits further research.

6. *Hierarchical behavior:* Still staying within the domain of praxic actions, before we turn to communication, it seems that humans are capable of mastering hierarchically organized action plans which involve a greater depth than those of apes, which may in general be limited to depth two.

6. *Pantomime:* Ontogenetic ritualization in apes may yield a very limited repertoire of gestures. Pantomime extends complex imitation into the gestural domain by developing a new social contract (a mixture of biological and cultural evolution): "If I perform some behavior, possibly in the absence of some of the objects or agents with which it would normally be performed, then you will exploit context to infer the related message I intend to convey." This then allows an open-ended communication about anything related to a group's behavioral repertoire.

7. *Protosign:* This could result from combining the "action to gesture" mechanisms of Step 4 with pantomime. Step 4 ritualized an action to achieve a distal goal more economically via a communicative, rather than praxic, action. Protosign emerges when a similar mechanism ritualizes a pantomime (one form of communicative action) to achieve a communicative gesture that is less effortful and less ambiguous.

8. *Protospeech:* The basic ability for communication with an open-ended semantics using conventionalized gestures then opens up an adaptive pressure for the evolution of the vocal learning abilities required to extend this capability from the visual–manual domain to the auditory–vocal domain.

9. *The emergence of grammar:* As argued elsewhere (Arbib, 2012, chapter 10), the capacity for language may have emerged by combining the capacity for protolanguage (protosign and/or protospeech) with the capacity for complex imitation to (a) fractionate performances into subparts that are correlated with agents, objects, or actions, and so forth, that are salient with respect to that which the performance indicates, and then (b) develop "constructions" for reassembling the pieces in such a way that novel combinations can be performed to convey meanings that have never been expressed before. Further cultural evolution then exploits the capacity for hierarchical behavior to support the production and comprehension of utterances to nested constructions of increased hierarchical depth.

With this, let's turn to a crucial distinction between words and what the words convey. MSH starts with manual actions that act upon the world but—of course—the language system that emerges is not limited to words that denote manual actions (i.e., a very limited set of "verbs")—just as the child's beckoning gesture denotes "Hug me," not the specific actions from which the gesture was ritualized, and a pantomime denotes "Come hunt deer with me," not the hunting actions that are being pantomimed. From the latter, fractionation and ritualization could yield protosigns for "Hunt," "Deer," and "Me." A great deal more cultural evolution would be required before there is a construction such as [AB with C], where A can be replaced by words for a variety of actions, B by a word for a variety of agents or objects, and C for a variety of agents.

Thus, with Saussure (1916) we must distinguish the *signifier* from the *signified* (or words from concepts) with the *sign* combining and linking them. My action-oriented view is that at least a range of basic concepts (the signifieds) are realized as the perceptual and motor schemas of a human acting in its world—but that the linkage is many to one, competitive, and contextual, so that appropriate words to express a schema may vary from occasion to occasion, both because of the assemblage in which the schema instance is currently embedded and in relation to the context in which the utterance occurs.

When we see or hear a word, though, we can regard it as a form, the *signifier*, devoid of meaning. We can simply write or pronounce it *as an action* and recognize a similar action made by others, as indeed we do for nonwords like *gosklarter*. However, in normal language use, the articulatory action is linked to the *signified*. Thus, a French speaker saying *cheval* and an English speaker saying *horse* have in mind the same kind of equine quadruped. In this section, our task is to build on our earlier exploration of the linkage between vision and action to address the relation between vision and language: relating a visual scene to a *structured signifier* (utterance) that describes it. But for this, we need a model of the *structured signified*, that is, of how visual perception supports recognition of agents, objects, actions, and their relations within a scene. We turn to this in section 25.5.2 before offering, in section 25.3, a conceptual view of the language-ready brain informed by sections 25.5.1 and 25.5.2.

25.5.2 Schema Theory Revisited: The VISIONS System Surprisingly, research in modeling high-level vision remains relatively unexplored. Indeed, such modeling is beyond the scope of chapter 12 ("Early Vision") and is still a step beyond the "Object and Scene Perception" of chapter 17 in which the scene perception refers more to getting the gist of a scene than to perceiving its details. Lacking a comprehensive model of brain mechanisms of high-level vision, then, we turn to a classic schema-theoretic model of scene understanding which establishes basic properties we need to ground our exploration of the vision–language interface.

How do we recognize a novel visual scene? The short answer here will be "By associating 'schema instances' with regions of the scene, with each instance linked to a region corresponding to the schema most activated for that region." We consider the VISIONS system (see Draper, Collins, Brolio, Hanson, and Riseman, 1989; Hanson and Riseman, 1978; and Arbib, 1989, section 5.3), a scene-understanding system that could move from a color photograph to a labeling of the various objects contained in the scene. It demonstrated how a schema network can be used to model some portion of world knowledge (how to recognize objects from low-level perceptual features extracted from an image, and how they relate spatially to their parts or frequently associated objects) and how this knowledge can be retrieved on the basis of activation values of already instantiated schemas. It was implemented on a serial computer, but the underlying computational architecture offered useful insights into the way I believe the brain operates, via *cooperative computation*, the competition and cooperation of a multitude of schema instances in coming together to compose a schema assemblage appropriate for the current scene: *competition and cooperation in the service of compositionality*. The reader may compare and contrast this with the role of perceptual schemas in parameterizing the readout of motor schemas in the preexisting coordinated control program of figure 25.5 (section 25.3.1) and the compositionality modeled by the ACQ model of section 25.3.2).

Before the schemas in VISIONS can operate, various processes of early vision (chapter 12) take an image of an outdoor visual scene and extract an *intermediate representation* including contours and surfaces tagged with features such as color, texture, shape, size, and location (see the middle panel of figure 25.13A). Perceptual schemas process different features of the intermediate representation to form confidence values for the presence of objects like houses, walls, and trees. (As already noted, chapter 17's account of "Object and Scene Perception" focused on the recognition of objects and the gist of a scene but had little to say about how the *particularities* of a scene come to be recognized.) The knowledge required for interpretation is stored in *long-term memory* (LTM) as a network of schemas while the state of interpretation of the particular scene unfolds in *working memory* as a network of schema *instances*. (Chapter 11 presents neural models of "Short-term, Long-term, and Working Memory.") Note that this visual working memory (WM) is not defined in terms of recency (as in very short-term memory) but rather in terms of continuing relevance. An example (increasingly rare in the age of smartphones) of working memory is the ability to remember a phone number long enough to enter it, but then forgetting it once it has been entered. However, the visual WM places its components in a spatial frame and includes relations between them, as distinct from the linear order of a phone number (though that linear order may be made easier to remember by chunking it—for a U.S. number—into an area code followed by three digits followed by four digits).

Interpretation of a novel scene starts with the data-driven instantiation of several schemas (e.g., a certain range of color and texture might cue an instance of the foliage schema for a certain region of the image). When a schema instance is activated, it is linked with an associated area of the image and an associated set of local variables (see figure 25.13B). Each schema

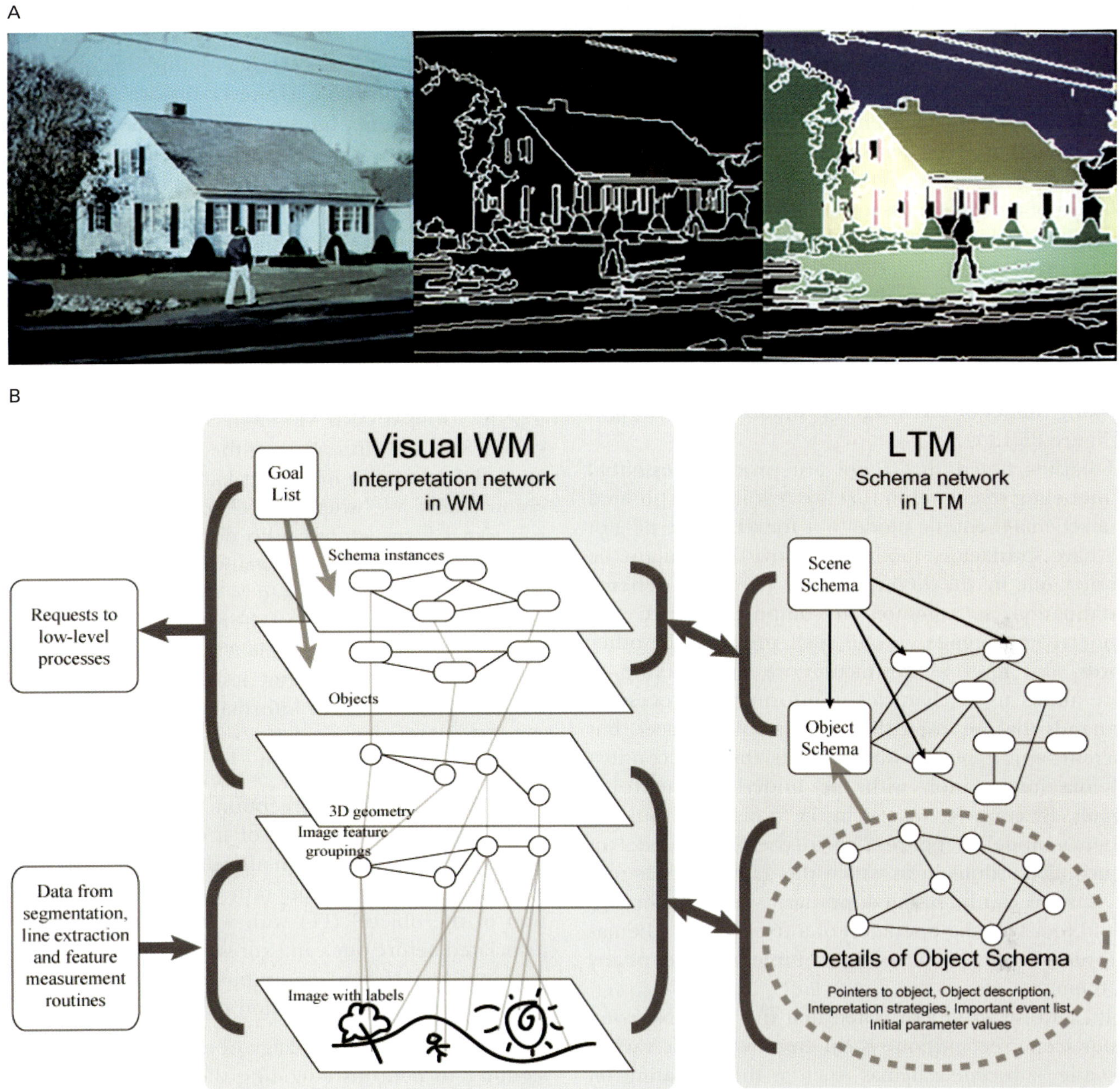

FIGURE 25.13 (A) Segmentation of a scene into candidate regions (middle) provides the bridge between the original image and the interpretation of a scene in VISIONS by associating regions of the image with schema instances. In this example, VISIONS classifies regions of the scene as sky, roof, wall, shutter, foliage, and grass but leaves other areas uninterpreted. (B) The visual working memory (WM) of VISIONS interprets the current scene by a network of parameterized instances of schemas from long-term memory (LTM). These schema instances are linked to the visual world via the intermediate database that offers an updatable analysis of the division of the world into regions that are candidates for interpretation as agents and objects, possibly in relation with each other (Arbib, 2012).

instance in working memory has an associated confidence level which changes on the basis of interactions with other instances in working memory. Once several schema instances are active, they may instantiate others in a "hypothesis-driven" way (e.g., recognizing what appears to be a roof will activate an instance of the house schema to seek confirming evidence such as the presence of walls in the region below that of the putative roof).

Ensuing computation is based on the competition and cooperation of concurrently active schema instances. Once a number of schema instances have been activated, the schema network is invoked to formulate hypotheses, set goals, and then iterate the process of adjusting the activity level of schemas linked to the image until a coherent scene interpretation of (part of) the scene is obtained. Cooperation (e.g., that between *roof* and *sky* if the spatial relations are appropriate) yields a pattern of "strengthened alliances" between mutually consistent schema instances that allows them to achieve high activity levels to constitute the overall solution of a problem. As a result of competition (e.g., the same region cannot be both "wall" and "sky"), instances which do not meet the evolving consensus lose activity and thus are not part of this solution (though their continuing subthreshold activity may well affect later behavior). The highly activated instances of perceptual schemas become part of the current working model of the image (as indicated to the right of figure 25.13A).

I earlier noted that there are processes essential to modeling of cognition that are readily incorporated in a schema-theoretic model but for which we do not yet have satisfactory models for how these might be carried out in the brain's neural networks. Schema instantiation, so common in computer science and artificial intelligence, is one such problem. In other words, the level of abstraction we are working at here allows us to characterize information-processing demands that we know the brain is able to meet, but current brain theory is limited in its ability to connect schema instantiation with the underlying neuronal details. By contrast, we saw in the transition from the schema model of figure 25.5 to the FARS model of figure 25.6 a domain in which the gap from schemas to neurons can be bridged, primarily because here we could focus on the operation of a fixed set of schemas without concern for handling multiple instances of any of them.

Itti and Arbib (2006) noted that the VISIONS computer scene recognition system employs a wide variety of basic "subscene schemas" such as those relating *sky* and *roof* (one way to recognize a roof is as a form of a certain geometry and size immediately below a region that has been recognized as sky) and *roof, house,* and *wall* (if you see a roof, check whether the region below it can be recognized as a wall to confirm that together they constitute views of parts of a house). This suggests that an approach to visual scene interpretation with a large but finite inventory of subscene schemas (*from the visual end*) may provide the linkage with constructions (*from the language end*) rich enough to encompass an exemplary set of sentences. This perspective informs the schema-theoretic approach to language offered here: as the perception of a scene unfolds incrementally, so does the verbal description of the scene. Conversely, the incremental comprehension of a sentence may be reflected in the eye movements a subject directs to various aspects of a related visual scene, as in the visual worlds paradigm. We take up this idea in developing TCG in section 25.6. However, first we need to build on both this section and section 25.5.1 in assessing the role of dorsal and ventral streams in the language system. Section 25.5.3 extends the account of mirror system activity for action and language as embedded in the integration of dorsal and ventral pathways (Arbib, 2010), which, in turn, extends an assessment (not offered here) of the relation between aphasia (disorders of language) and apraxia (disorders of praxic action) (Arbib, 2006).

25.5.3 Dorsal and Ventral Streams in the Language System

In reviewing the lesion studies dissociating the roles of human dorsal and ventral streams for vision ("how" vs. "where") in chapter 1, we saw that DF is a visual form agnosic who does not have access to conscious declarative information about form (which would normally be mediated by the temporal lobe) but can use form information (affordances, via the parietal lobe) to guide grasping, while AT can pantomime grasp but (because of a parietal lesion) cannot preshape based on current form information. The data suggest the following:

- Parietal "affordances" → preshape
- Inferotemporal "perception of object" → pantomime or verbal description of size; that is, one cannot pantomime or verbalize an affordance if it is only available via the dorsal path but rather one needs a "recognition of the object" (IT) with which attributes can be associated before one can express them. Note that these data put speech and pantomime on one side of the divide and object-based action on the other.

We now discuss the linkage of neurons encoding the signifier to neurons encoding the signified, though a less precise terminology will often be employed. The notion is that *mirror neurons for words encode recognition of the articulatory form (or written form, or gestural form in a signed language) but must be linked to other neural networks for the encoding of meaning.*

First recall the roles of the dorsal and ventral paths in the FARS model (figure 25.6):

The ventral path recognizes objects (IT) and provides inputs to PFC sufficient to relate the object to working memory and current tasks as the basis for choosing which affordance to exploit for the next action.

The dorsal path can then provide the exact parameters of the affordance to motor schemas in F5 which can then instruct F1 how to carry out the action.

Turning now to the MNS model (figure 25.7), we see that only a dorsal path has been implemented: given data about the trajectory of a hand relative to an affordance of an object, activate mirror neurons which represent the action. There are two shortcomings which need to be addressed in future modeling:

1. MNS currently models how to activate mirror neurons that best match the category of the observed action. However, in the spirit of FARS, one might expect the MNS to present a population code that provides a more detailed visuomotor representation of the action. Relevant to this, Aglioti, Cesari, Romani, and Urgesi (2008) find that expert basketball players not only have greater mirror system activation than other observers when viewing a video of a ball thrown at a basket but that they are uniquely skilled in being able to predict the success or failure of the throw from viewing the early frames of the video.

2. Unlike FARS, MNS has no explicit model of the dorsal path. Here relevant data comes from Buccino et al. (2004), who imaged humans watching three silent video clips: one of a human talking, one of a monkey making orofacial communicative gestures, and one of a dog barking. They saw significant mirror system activation for observing the human, just a little for observing the monkey, and none for observing the dog. My conclusion, then, is that a dorsal path activates a representation of any recognized action, but in the case of an action in one's own repertoire (or one that has been frequently observed), a more detailed representation can be evoked via the dorsal pathway.

Moreover, we must take account of the sequencing in both the FARS and ACQ models. As briefly noted, the full FARS model executes a sequence rather than a single action, but this is an overlearned sequence (i.e., the steps proceed in the same order on each trial, though the details of the executed actions will vary depending on which object is to be grasped), whereas in the ACQ model the sequencing of actions is opportunistic. In either case, though, interactions between PFC and the basal ganglia play a crucial role in determining which actions will next be executed.

The complementary challenge (and one vital to the modeling of imitation) has not been addressed: How does one extend MNS to recognize not only the individual actions but also the way in which they are sequenced to achieve proximal goals that lead to an overall, distal goal? Note that the recognition of an individual action here may be dorsal, ventral, or both.

The lower two boxes of figure 25.14 correspond to words and concepts, but we now make explicit that, following MSH, we postulate that a system for phonological expression (generating and recognizing the articulatory forms of words) evolved from the system for generating and recognizing praxic actions to serve communication integrating hand, face, and voice. We also postulate that the concepts—for diverse actions, objects, attributes, and abstractions—are now seen to be represented by a schema network stored in LTM, with our current "conceptual content" formed as an assemblage of schema instances in WM.

In section 25.5.2, we offered VISIONS as an example of how visual perception could represent a scene though activating instances of visual schemas in visual WM; complementary mechanisms engaging the basal ganglia

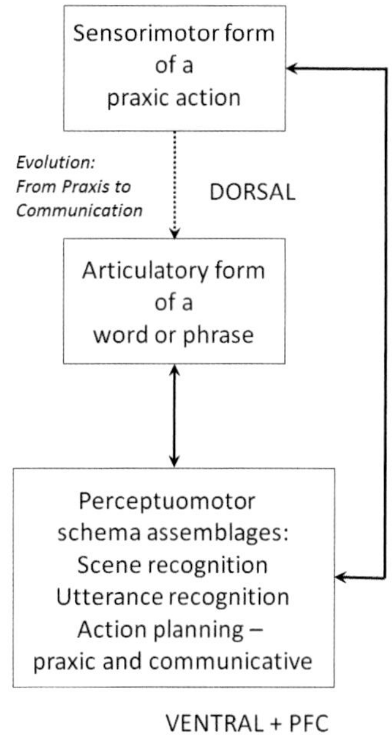

FIGURE 25.14 In this diagram, dorsal mechanisms mediate both the performance and recognition of the sensorimotor form of an action (top box). However, recognition of the setting in which an action occurs and the planning of an action require mechanisms in the ventral stream and prefrontal cortex (PFC; bottom box). The linkage is indicated by the bidirectional arrow linking the top and bottom boxes in the figure. According to the mirror system hypothesis, a system for the production and recognition of the *articulatory form* of words (middle box), whether signed or spoken, evolved from the system for the sensorimotor form of praxic actions. Thus, the arrow from the top box to the middle box expresses an evolutionary relationship, not an information flow in the brain. Instead, the word forms (signifiers) link to schemas related to their signifieds, not directly to the dorsal path for (praxic) actions. The bidirectional arrow linking the middle and bottom boxes shows that the dorsal form of words is in general linked to PFC schema assemblages. In the case in which a word is a verb describing an action within the person's repertoire, a direct link between dorsal representations of word and action may also exist—but most words are not of this form.

as well as PFC will then develop plans of action and schedule the actions involved, dynamically updating the plan as necessary (recall the action–perception cycle of chapter 1). Our task in section 25.6 will be to introduce a performance model of TCG, suggesting how such mechanisms may "lift" to generation and recognition of utterances, linking vision and language.

The dorsal–ventral division shown in figure 25.14 is reminiscent of that postulated by Hickok and Poeppel (2004) in their analysis of cortical stages of speech perception. They observe that early stages involve auditory fields in the superior temporal gyrus bilaterally (although asymmetrically) but offer evidence that this cortical processing system then diverges into two streams: A *dorsal stream* maps sound onto articulatory-based representations while a *ventral stream* maps sound onto meaning. The latter projects toward the posterior middle temporal gyrus, which serves as an interface between sound-based representations of speech and widely distributed conceptual representations.

The work on the ventral and dorsal streams in the visual system inspired related research on the macaque *auditory* system. Rauschecker and his colleagues (Rauschecker, 2005, 2011; Rauschecker and Scott, 2009; Rauschecker and Tian, 2000; Romanski et al., 1999), among others, offer evidence for a dorsal and a ventral stream which process auditory spatial information and auditory pattern/object information, respectively. Both streams eventually project to the frontal cortex, which integrates this information with visual and other modalities. This same anatomical substrate may support both the decoding of vocalizations in nonhuman primates and the decoding of human speech. In any case, modeling speech perception and the neural control of speech mechanisms is another fascinating topic in computational neuroscience that falls outside the scope of this book (Bohland, Bullock, and Guenther, 2009; Doupe and Kuhl, 1999; Fadiga and Craighero, 2003; Guenther, Ghosh, and Tourville, 2006; Hickok and Poeppel, 2004; Kotz and Schwartze, 2010; Moulin-Frier and Arbib, 2013).

Building on this and their own studies of neural correlates of sentence comprehension in humans adept in diverse languages, Bornkessel-Schlesewsky and Schlesewsky (2013) offer a conceptual model of sentence comprehension based on the role of the *auditory* dorsal and ventral streams in speech comprehension (but not production), rather than attending to the *visual* dorsal and ventral streams emphasized by MSH. Unfortunately, in attempting to pace this work in an evolutionary perspective, Bornkessel-Schlesewsky, Schlesewsky, Small, and Rauschecker (2015) not only choose to ignore the relevance of the dorsal and ventral visual streams to language (Arbib, 2006, 2010) but even go so far as to declare that

> … despite the obvious and extensive anatomical homologies between humans and nonhuman primates, neurobiological models of higher-order language have sought other candidate animal models (or no such models) to rely on, with songbird models as the most prominent. ... We have suggested that, to the contrary, the basic computational biological prerequisites for human language, including sentence and discourse processing, are already present in nonhuman primates. (Bornkessel-Schlesewsky, Schlesewsky, Small, and Rauschecker (2015, p. 148)

The arguments set forth in this chapter suggest that their paper is hardly innovative in stressing the relevance of the study of nonhuman primates and that the claim that "the basic computational biological prerequisites for … sentence and discourse processing, are already present in nonhuman primate" is simply wrong. Multiple transformations are required to transform those mechanisms shared with monkeys (and presumably LCA-m) into those necessary for the language readiness of the human brain.

Clearly, the present chapter falls short because it does not investigate auditory comprehension or vocal motor control. Where it is strong is that it does not simply look at aspects of auditory comprehension but looks at both comprehension *and production* (at a level which is neutral as to whether the signals are audio-vocal, visuo-manual, or multimodal) and—crucially—offers a preliminary account of the role vision plays in *providing things to talk about*, a linkage with the external world lacking in the account of Bornkessel-Schlesewsky, Schlesewsky, Small, and Rauschecker (2015).

25.6 Template Construction Grammar: Linking Language and Vision

In this section, we abandon evolutionary modeling and seek to fill in in some of the gaps in section 25.5 as we follow through on the discussion of construction grammar in section 25.2.5. We introduce Template Construction Grammar (TCG) to provide a plausible characterization of the linkage of vision and language, building a schema-theoretic implementation of construction grammar on the basis of the schema-based VISIONS model of visual scene understanding in section 25.5.2. However, as noted in section 25.2.5, neurolinguistics is still far away from a fully accurate model of grammar that links to brain mechanisms of human language processing, and so future research will require exploration of diverse grammatical theories in developing neurolinguistic models.

Section 25.6.1 introduces the notion of SemRep as an intermediary between a visual representation and a language representation of a scene, and it introduces TCG as the variant of construction grammar that builds on the SemRep representations of semantics. We then apply TCG to visual scene description. Here, a subject simultaneously gathers information from the image, fixating elements that become salient based on both bottom-up features and top-down hypotheses, and starts generating linguistic output based on relevant visual information. Under pressure to communicate she or he might start producing fragmented utterances while in more relaxed conditions the message might become composed of well-formed sentences packaging content collected through many fixations.

Section 25.6.2 then outlines the early stages of our attempt to extend TCG to support a model of scene comprehension. A key aim of ongoing research is to bring in brain mechanisms by looking at data from aphasia. Patients suffering from brain lesions impairing their capacity to use grammatical cues can compensate—in some cases but not others—by relying on their knowledge of the world to assign linguistic inputs to their correct semantic roles and correctly identify what picture matches the sentence they have just heard. This suggests a two-route theory of comprehension, as will be explained.

25.6.1 TCG and Scene Description We now introduce SemRep and TCG (Arbib and Lee, 2008; Barrès and Lee, 2014; Lee, 2012) to provide a visually grounded version of construction grammar. What is worth stressing, however, is that our approach is not an attempt to apply a prior construction grammar formalism to computational neurolinguistics—rather, it is based on the realization that early approaches to a schema-theoretic linguistics (e.g., Arbib, Conklin, and Hill, 1987) can indeed be understood by interpreting the type of schemas that we then called *templates* as being examples of *constructions* as broadly construed within variants of construction grammar.

Arbib and Lee (2008) introduced SemRep as a hierarchical graph-like "semantic representation" designed to link the semantic content of sentences to the representation of visual scenes. A SemRep is abstracted from schema assemblages such as those generated by VISIONS (recall section 25.5.1). Where the latter defines a "*cognitive structure*," a SemRep defines the associated "*semantic structure*." However, the following discussion postulates an extended visual system (more powerful than that modeled by VISIONS) in which actions and agents are recognized as well as objects and relations (clearly, this depends on the set of schemas embedded within the system for visual scene perception).

A SemRep abstracts from the current pattern of schema activation a set of nodes and relations which constitute one possible semantic structure for the current scene (figure 25.15, right). A given scene can be perceived in many different ways; it can have many SemReps. The properties of a recognized entity (an object or an action) are converted into a node linked to the perceptual schema instance in the schema assemblage, while the semantics of a relationship between entities (including semantic roles for an action) are converted into an edge. Each node or relation is associated with a *concept* which may later be translated into words by the language system. However, concepts are

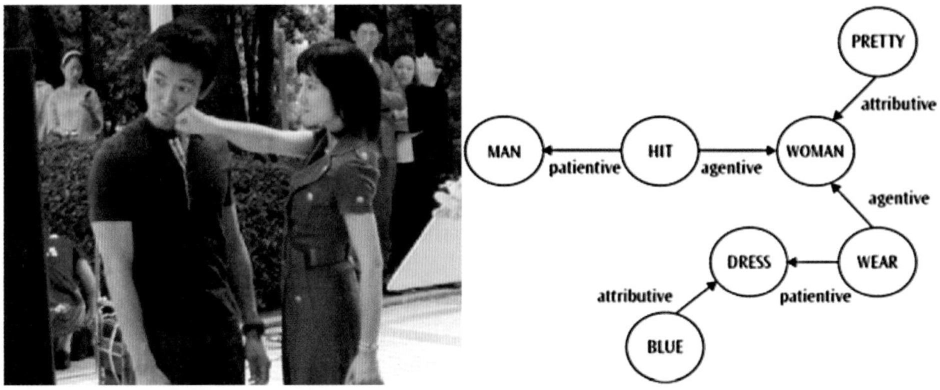

Figure 25.15 Left: A picture of a woman hitting a man (original image from "*Invisible Man Choi Jang Soo*," Korean Broadcasting System). Right: An example SemRep graph that could be generated from the scene. Note: the words on the nodes are labels of convenience for yet-to-be-verbalized concepts. These SemReps might yield such sentences as "*people are fighting*," "*a woman hits a man*," or "*a woman is wearing a blue frock*."

not word labels but more abstract descriptors, allowing the same graph to be expressed in multiple ways within a given language. Consider the SemRep shown on the right of figure 25.15. Here, the concept DRESS could yield a variety of phonological forms such as "dress" or "frock." A SemRep expresses semantic relations but with no commitment to word choice and can thus be the basis for description in any language once the appropriate grammar and lexicon are deployed.

A schema instance in the visual analysis may be associated with a number of parameters, some of which (such as size, shape, orientation, and location) may be relevant to possible interactions with what the schema represents and yet not be included in a verbal expression. We thus postulate that SemRep makes explicit very few parameters and can direct requests to visual WM when more information is required for explicit cognitive processing or verbal expression.

We now turn to the schema-based model of a grammar mapping the SemRep content onto verbal expression. Our version of construction grammar for language production, TCG, adopts two major policies common to all construction grammar frameworks: (1) each construction specifies a mapping between form and meaning, and (2) the systematic combination of constructions yields the whole grammatical structure. However, TCG also provides two key additions: (1) the semantic structure of an utterance is given as a SemRep, and (2) each construction is viewed as a schema and is therefore endowed with all the dynamic properties of cooperative computation.

The *lexical constructions* at the top of figure 25.16 exemplify the way in which the concept associated with a single node or edge of the SemRep can ground the selection of a word to express that concept. The constructions at the bottom of figure 25.16 move up the hierarchy to cover larger and larger parts of the SemRep. Thus, the IN_COLOR construction has the interesting property that it must recognize the node for CLOTHING but only uses it to license the construction which yields utterances like *pretty woman in blue*, where the nodes for WOMAN and PRETTY have already been covered by another construction to yield the phrase *pretty woman* to fill the first slot in IN_COLOR, which is linked to the node HUMAN.

During production, a SemRep may yield one or more utterances as TCG finds ways to "cover" the relevant portion of the given SemRep with a set of "small" subgraphs, where each is chosen such that there is a construction available in the language considered to express the content of that subgraph. Processing selects constructions that match the whole or subparts of the SemRep graph and can therefore participate in

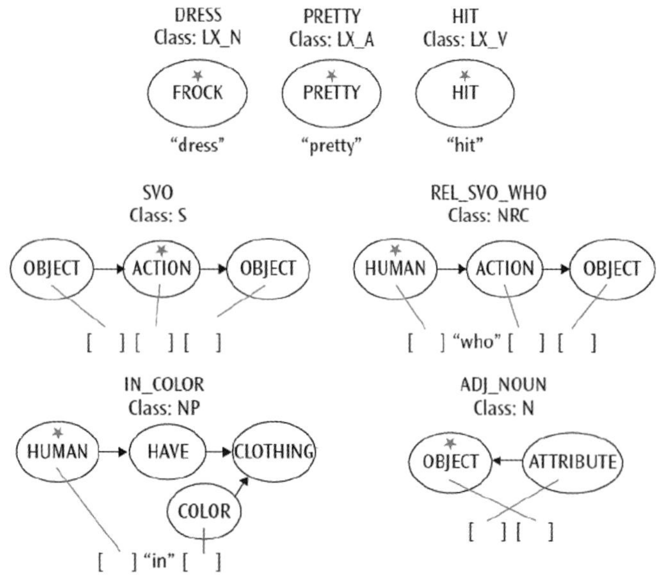

FIGURE 25.16 Examples of constructions. The name and class of each construction appears above it. Each construction includes a SemRep-like graph with either generic or specific labels on the edges and nodes, with each linked to a text or an empty slot. The node labels correspond to concepts, *not* words. Top: Constructions that correspond to elements in the lexicon, replacing a concept (small caps) with a word or, not shown here, phrase (lowercase). Bottom: Higher-level constructions used to encode grammatical information. For each slot there may be restrictions as to what can serve as slot fillers. The head of each construction is marked with a star (e.g., the node HUMAN in IN_COLOR). The word or phrase already associated with a node of the SemRep-like graph is then inserted into the corresponding slot of the lowercase string below the graph to yield the phrase that results from applying the construction.

expressing some of its content. Thus, constructions are applied recursively, starting with lexical constructions, which have no slots (figure 25.16, top), and then by applying higher-level constructions (figure 25.16, bottom) as based on the results of earlier application of constructions (see the figure 25.16 caption for details).

In this section, we treat lexical constructions the same way we treat other constructions. However, this does not mean that lexical and more schematic constructions are neurologically equivalent. Consider cases of double dissociation in language production (Goodglass, 1976) between anomia (patient has difficulties in naming objects or actions) and agrammatism (patient has difficulties producing syntactically correct utterances).

In characterizing the cooperative computation whereby TCG produces the description of a visual scene, the processes in VISIONS are "lifted" to a similar structure (figure 25.17) in which a linguistic WM (the working memory for the state of applying constructions

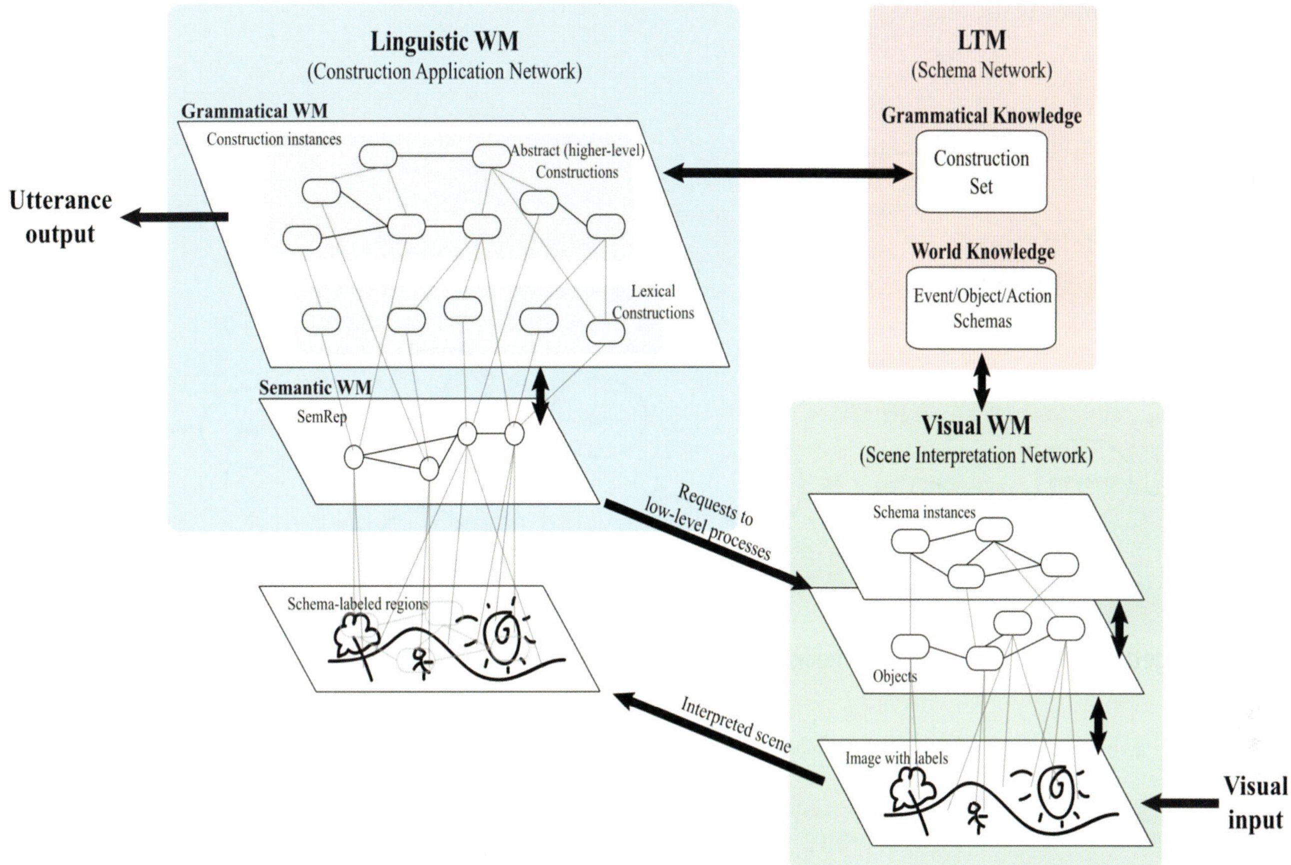

FIGURE 25.17 The structure of the SemRep/Template Construction Grammar model of scene description echoes the interaction of visual working memory (WM) and long-term memory (LTM) in VISIONS (see figure 25.3) but incorporates the language system by adding a linguistic WM and grammatical knowledge (a set of constructions). Based on visual attention and communicative goals, the system extracts a SemRep in semantic WM from the perceptual schemas active in visual WM to capture key entities and relations from the scene, thus discarding many details from visual perception less relevant for verbal description. The linguistic WM holds a hierarchical covering of the current SemRep by iterated applications of constructions from LTM. Lexical constructions are applied to associate nodes of SemRep with words, and higher-level constructions are applied to build either directly on nodes in SemRep or on an already partially completed construction assemblage. Linguistic WM not only provides a working memory for construction applications but also allows an utterance to be read off at any time. Just as VISIONS allows visual WM to request more data from low-level processes, our model allows the SemRep to be updated by requesting information from the vision system when completion of an utterance requires further attention to the visual scene. Adapted from Barrès and Lee, 2013.

to the current SemRep) provides the workspace for the operation of construction selection and attachment, thus providing a dynamic set of hierarchically organized construction structures with varying degrees of confidence. During production of a description, a number of constructions are activated simultaneously to build upon the unfolding SemRep. Constructions cooperate and compete with each other in order to produce a verbal description of a scene. The language system, which uses the linguistic working memory, applies constructions on SemRep hierarchically and reads off the well-formed sentence or sentence fragments that result.

As Spivey et al. (2005) note (in relation to the visual world paradigm generally, rather than to TCG), the visual environment can be treated as an external memory with eye movements being the typical access method. When necessary, the language system may generate requests for more details from the vision system. Constructions are applied based both on the "current" SemRep and on the state of the linguistic WM, where there are a number of partially (or fully) created construction assemblages. The system produces utterances as soon as some threshold is reached (see the discussion of figure 25.18). The sentence forming process is both (1) *incremental* since new constructions are constantly

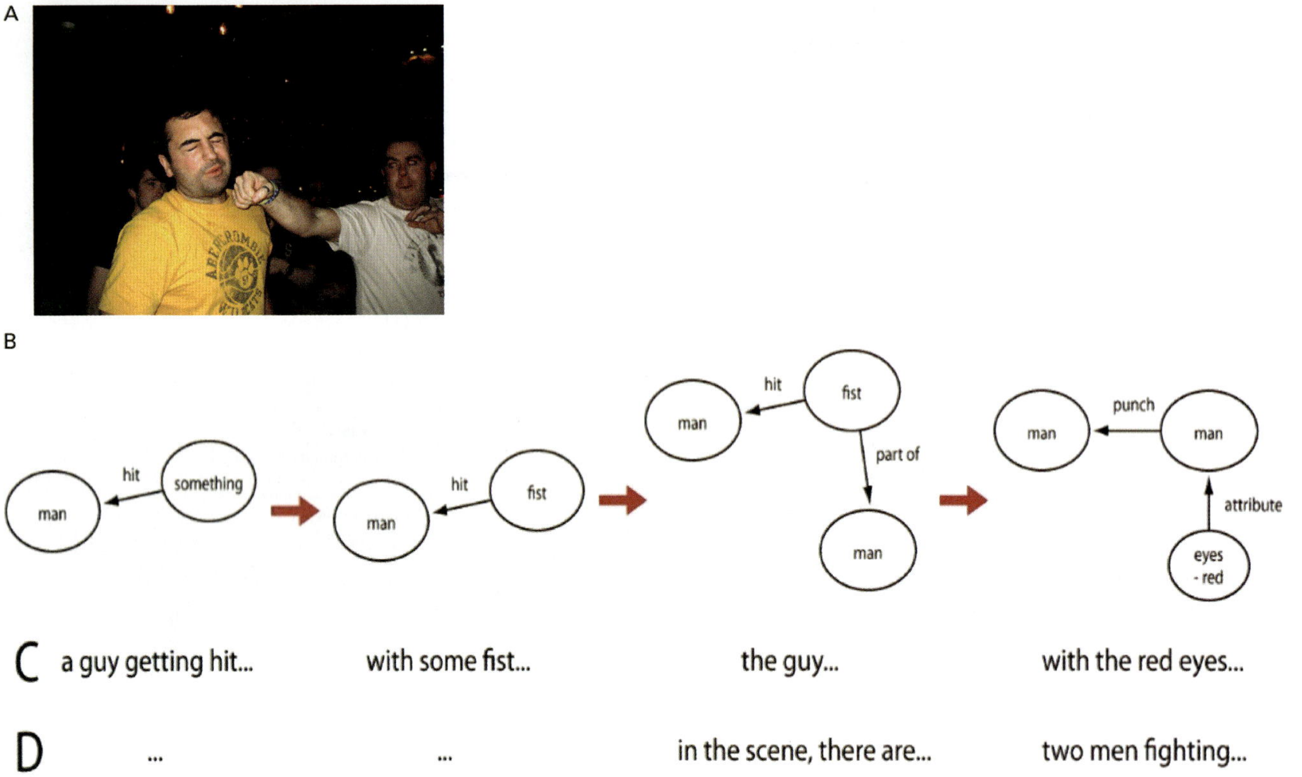

FIGURE 25.18 Descriptions of the indicated scene generated by two different speakers. The figure also shows the sequence of SemReps hypothesized to underlie the sequence of utterances produced by the low-threshold speaker, C (Lee, 2012).

being applied according to the current conceptual representation, SemRep, and (2) *hierarchical* since constructions may be applied atop other constructions to form a hierarchical structure anchored at its base by the SemRep abstracted from the visual processes. Note that this approach emphasizes the linkage of form and meaning—the "tower" of constructions built atop a (possibly dynamic) SemRep is thus qualitatively different from a syntactic tree built atop a sentence to parse it in terms of abstract syntactic categories. We see here the importance of incrementality and communicative goals.

An important feature of TCG is that it offers an account of language that explains how constructions may be combined to yield fragmentary utterances as well as complete sentences. As we see for speaker C in figure 25.18, a partial utterance may be read out at any time. Alternatively, D's "readout" was postponed until after more detail was discovered (a more comprehensive SemRep is formed) and further constructions have been applied. Lee (2012) explained this by saying that different speakers (and the same speaker under different time pressures) have a different *threshold* for producing utterances, such that the lower the threshold the sooner the constructions are read out and thus the more fragmented is the overt utterance.

25.6.2 TCG AS A MODEL OF COMPREHENSION Barrès and Lee (2014) proposed a conceptual extension of TCG as a model of language comprehension. The goal is twofold: (1) to incorporate the role that world knowledge plays in the incremental generation of semantic representations during comprehension, and (2) to do so in a way that accounts for neuropsychological data on comprehension performance in aphasic patients who display certain forms of agrammatism. They focus on the performances of agrammatic aphasics in *sentence–picture matching tasks* during which patients are asked to decide whether a sentence they hear matches a visual scene.

Agrammatic aphasics are patients suffering from brain lesions that result in the deterioration of their capacity to speak in a grammatically correct fashion. In contrast to a relatively unimpaired capacity to use the correct content words, agrammatic aphasics tend to omit function words, verbal inflections, and so forth. Caramazza and Zurif (1976) were among the first to show that agrammatic aphasics could also be impaired in their capacity to make use of syntactic cues during language *comprehension*. They were no different than normal subjects when asked to match canonical active sentences such as "The lion is chasing the fat tiger" with the appropriate picture from a set of related scenes (e.g., a picture of a lion chasing a tiger, a picture of a tiger chasing a lion, etc.) but were no better than chance for center-embedded object relatives such as "The tiger that the lion is chasing is fat." However, their performance was restored to the level of normal subjects when world knowledge cues were available to constrain the interpretation of embedded object relatives, as in "The apple that the boy is eating is red." This latter result led the authors to hypothesize a dissociation between a "heuristic" system based primarily on world knowledge information and an "algorithmic" system relying mainly on syntactic information. Later studies showed that the situation is more complex, but we ignore these complexities in this chapter—they remain a challenge for further research. Instead, we examine the schema-based model developed by Barrès and Lee (2014). Their focus of TCG/SemRep as the language-vision interface offers a platform well suited to simulate sentence–picture matching tasks. They adopt a two-route approach to comprehension, with a world knowledge route that may be left more or less intact while lesions are performed on a grammatical route. But first we must distinguish between the roles of two different types of semantic constraints.

The core tenet of construction grammar is that syntax and semantics are not dissociated into two different theoretical components (Croft and Cruse, 2005). However, the empirical results outlined above demand that we distinguish the world knowledge preserved in agrammatic aphasics from construction-related semantic constraints. We thus coin the terms *heavy semantics* and *light semantics* for world knowledge and construction-based semantics, respectively. World knowledge represents a source of information that plays a pervasive role in both visual scene and language comprehension (and in action more generally) and is *heavy* in terms of content since it spans motor and perceptual schemas as well as abstract knowledge that we can acquire through the very use of language. Such knowledge concerning specific agents, objects, actions, and more abstract entities contrasts with the *light* semantic content of constructions which develops through experiences of patterns of language about various categories of agents, objects, actions and more, determined by their differential roles as "slot fillers" in various constructions. Such roles may vary from the highly abstract (as in nouns vs. verbs providing a language-dependent syntactic elaboration of the semantic categories of objects vs. actions) to those strongly linked to sensory or motor experience as illustrated by the example of the IN-COLOR construction (see figure 25.15). For us, light semantics reflects this construction-related categorization, more or less abstracted from world knowledge in a usage-based language-laden way. It is "light" because only a few semantic features matter, and it cannot be refined and enriched by interacting with the world beyond the bounds set by a given language (although, of course, performances vary).

Based on this distinction, Barrès and Lee (2014) extended TCG to provide a conceptual (not yet implemented) model for the integration of light and heavy semantics underlying the dynamic interactions of world knowledge, linguistic, and visual information during language comprehension. In this case, a SemRep now serves as the output of the comprehension system and can thus serve as the basis for choosing which of several pictures corresponds to the given sentence. Crucially, the final SemRep can emerge through cooperative computation from both linguistic information and world knowledge. The general architecture of the TCG comprehension model is described in figure 25.19.

In the work on scene description, the SemRep was generated and dynamically updated by the visual system, with construction assemblages controlling the flexible generation of utterances corresponding to all or part of the SemRep. The TCG comprehension model complements this dynamics by allowing the SemRep to be built and updated not only by the vision system but also by two routes processing input utterances in parallel. These two routes, shown in figure 25.19, are the *heavy semantic route* (HS) that generates semantic representations from content words using world knowledge, and the *grammatical route* (G) that puts grammatical knowledge to work, using grammatical cues and constraints (including light semantics) to map the surface content of the utterance input to a SemRep graph representing its semantic content. Therefore, the SemRep, during comprehension, becomes the locus of interaction between three sources of information—vision, grammar, and world knowledge—that can enter into cooperative computation.

A key feature of the TCG comprehension model's architecture lies in the fact that each input word will have an effect on two different routes triggering two parallel processes to update the SemRep. Note that there is no claim that interpretation by one route has precedence over another. Rather, they may compete and cooperate. If one route reaches a high-confidence decision more quickly than another, then that route may produce the final interpretation, but in general both routes may serve to update the emerging SemRep

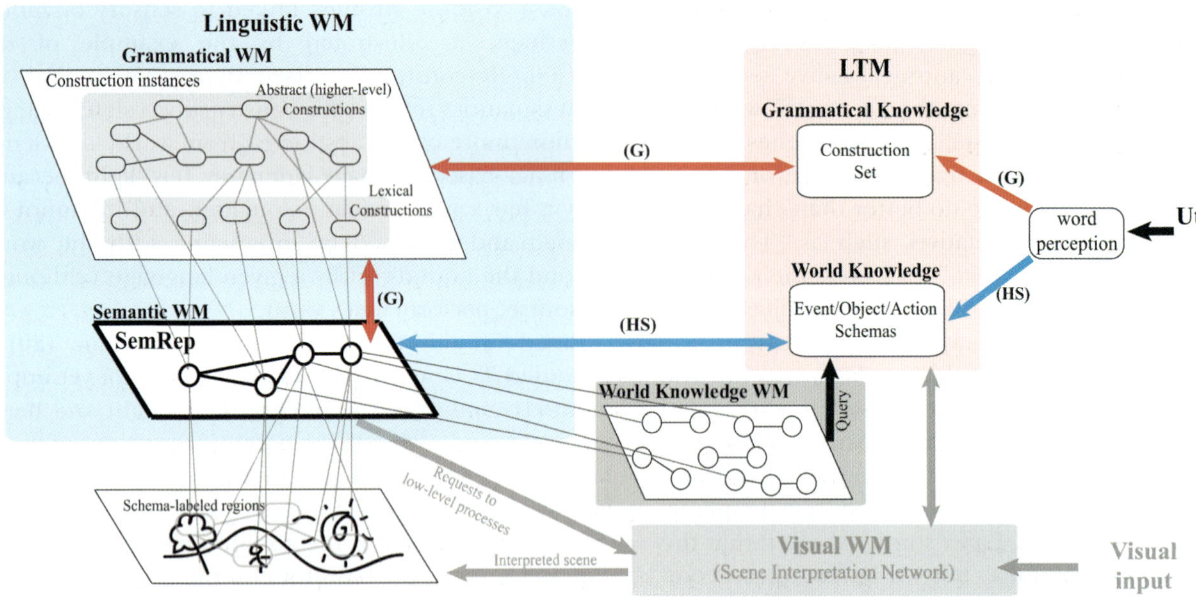

FIGURE 25.19 TCG as employed in a two-route model of language comprehension. The utterance input is fed in parallel to a grammatical route (G) that updates the SemRep indirectly through the creation of a construction schema assemblage in grammatical working memory (WM), and to a heavy semantic route (HS) that can generate SemRep nodes directly for content words but is not sensitive to grammatical cues. In addition, the heavy semantic route incorporates the possibility for the semantic WM to query the world knowledge WM to generate hypotheses about the plausible relations between SemRep nodes. The currently relevant hypotheses are kept active in the world knowledge WM, where they cooperate and compete to update the SemRep. Thus, the semantic WM becomes the locus of cooperative competition in which the construction assemblage, on the one hand, and the world knowledge hypotheses, on the other, compete and cooperate at each time to update the SemRep. The visual WM remains a source of input for the SemRep as in the production model. Adapted from Barrès and Lee, 2013.

as processing proceeds. We outline here how an input word W is processed by these two routes:

Heavy semantic route: If W is a content word, the heavy semantic route will allow W to create or verify an existing node in semantic WM by directly accessing its associated world knowledge content representation in LTM. If W is a function word, it is ignored by the heavy semantic route that is blind to grammatical cues. Complementary to this data-driven instantiation of world knowledge, the heavy semantic route allows active SemRep nodes to query the world knowledge through a world knowledge WM that builds plausible semantic relations between the nodes. For example, the word "eat" might not only instantiate a node for the concept EAT but also link it to a node for SOMETHING-EDIBLE. This world knowledge–based hypothesis can then verify, modify, or enrich the structure (nodes and edges) of the SemRep subgraphs currently active in Semantic WM or yield nodes which compete for later resolution. Working incrementally, upon receiving W the heavy semantics route updates the state of the Semantic WM, which is composed of a cloud of competing and cooperating SemRep subgraphs on which the next word W′ will be received, generating anticipations of what semantic content W′ will bring.

Grammatical route: Whether W is a content or a function word, it will result in the instantiation or verification of a construction, modifying the state of the construction assemblage in grammatical WM. In our previous production model, constructions are hierarchically applied in linguistic WM; this can be done with varying levels of confidence (competition and cooperation) with the final utterance read out from the construction assemblage that wins the competition. During comprehension, in contrast, the constructions are initially anchored on input words. At each time step, the unification of the SemRep-like subgraphs for the winning construction assemblage provides the (possibly still growing) SemRep yielding the current meaning associated with the word sequence received. Therefore, through the modification of the construction assemblage, the input W to the grammatical route will verify, update, or modify the SemRep graph structures in semantic working memory. By updating of the state of the semantic working memory upon receiving W, the grammatical route, like the heavy semantics route, modifies the semantic context on which the next word W′ will be received and generates anticipations. For the grammatical route, such anticipations and context modification also extend to the grammatical WM: the updated construction assemblage following the reception of W generates new grammatical expectations for W′.

Following the principle of cooperative computation, the SemReps graphs active in semantic working memory are also defined both in terms of their structure and in terms of the activation levels of their components (nodes and edges), activation levels that reflect the degree of confidence associated with a relation (edge) or with some semantic content (node). Any modification made by either the heavy semantics or the grammatical route on a SemRep subgraph is expressed in terms of a change in activation levels (that can result in piece of graph being discarded altogether if its activation level becomes too low). These can therefore register the competition and cooperation of both routes. As in VISIONS, all elements have updatable confidence levels; at some stage those nodes and edges with the highest confidence levels form the overall interpretation. Others are not part of that interpretation, but their subthreshold levels may nonetheless serve as priming for later processing.

The model of figure 25.19 can (conceptually) handle both normal and agrammatic comprehension performances for sentence–picture matching tasks used to test agrammatic comprehension (Sherman and Schweickert, 1989). The behavioral result of the sentence–picture matching trial rests on the complex interactions of three sources of information at the level of the SemRep. A bias toward one source of information or another can tip the cooperative computation in favor of one of the possible interpretations of the linguistic input. Discounting perceptual information while boosting the role of heavy semantics to compensate for the degradation of grammatical processing simulates the role world knowledge plays in agrammatic comprehension. However, the model puts at the forefront the fact that during a sentence–picture matching task, the impact of the perceptual content of such an image on the language comprehension system cannot be fully dissociated from that of the linguistic and heavy semantic content. Barrès and Lee (2014) develop this material in greater detail, though still only at a conceptual level, and also explore the challenges for neuroinformatics of linking the data and modeling required to advance our understanding of the language-ready brain.

25.7 *Challenges for Further Research*

This chapter was designed to round out the textbook by illustrating the interaction between drilling deeply and thinking broadly in approaching a specific Big Question, that of the evolution of the human language-ready brain—and to do so in a way that applied the findings of many previous chapters, in some cases placing them in a new perspective.

In particular, we introduced comparative neuroprimatology as an umbrella for this work, embracing multiple, distinct strands of empirical research including the following:

- studying the relation between neuroanatomy and neurophysiology in exploring the neural circuitry underlying the linkage of vision and manual action in the macaque;
- the use of fieldwork in studying the relation between behavior and gesture in apes, and the use of diffusion tensor imaging (not neurophysiology) to offer clues for comparison of chimpanzee brains with those of macaques and humans; and
- the use of eye-movement studies in psycholinguistic studies linking vision and language (the visual world paradigm), as well as both lesion studies of agrammatism and (although we had little to say about this) brain imaging studies in neurolinguistics.

Clearly, each subarea listed here can provide the scope for a whole career of digging deeply whether or not one is concerned with evolutionary questions, and clearly a concern with evolution brings other disciplines into play, including studies of genomics and niche construction.

At the same time, we saw the need for both neural network modeling and schema-theoretic approaches and suggested a methodology (figure 25.2) whereby insights into the details of macaque circuitry might be used in concert with (and in testing of) evolutionary hypotheses to derive computational models of the circuitry of the brains of apes and humans that cannot be tested against direct evidence on the cell-by-cell neurophysiology of those brains. Throughout, we tried not only to demonstrate models which offered steps forward in our understanding but to make clear that each success in turn opened up new challenges for further turns of the experiment–modeling cycle, as each model poses new challenges for empirical research, and each new set of interlinked data challenges us to question our existing models and, as appropriate, modify them or replace them.

In this spirit, having introduced the TCG model (and omitted many details), we stress that, as discussed in more detail elsewhere (Arbib, 2013b), neurolinguists have developed a number of competing *conceptual* models of language comprehension—for example, those of Bornkessel and Schlesewsky (2006), Friederici (2012), and Hagoort (2013)—which address a large body of neurolinguistic data going far beyond the aphasia data addressed so briefly here. Unfortunately, these models are not mutually consistent in their assumptions, and none is computationally implemented (though Hagoort does cite Vosse and Kempen, 2000, in advancing his views on the importance of unification). The development of an adequate model of the language-ready brain will remain a fascinating challenge for decades to come.

Note that section 25.6 is still in the "one brain at a time" mode, focusing separately on how one brain handles production or comprehension of utterances. Thus, the extension of our models to encompass dyadic brain modeling is a major research challenge. As already noted, fluid construction grammar has already been employed in the study of interacting agents, but with no attention to the brain, while dynamic construction grammar links constructions to the brain, but these constructions are etiolated. Embodied construction grammar pays much attention to motor schemas whereas TCG emphasizes vision schemas, and so it would be fruitful to link sentences describing courses of action to "motor constructions" (cf. the coordinated control program of figure 25.5) which may be related to sentence constructions as studied in the brain (Kemmerer, 2006). Thus, much is to be learned from each variant of construction grammar in building a more integrated and encompassing theory. And, of course, such a synthesis will also benefit from exploring the implications of nonconstruction grammars and of connectionist models such as those reviewed in chapter 24, "Language Processing" (and see also Chang, Dell, and Bock, 2006; and Crocker, Knoeferle, and Mayberry, 2010).

And so the quest to chart the path from neuron to cognition—and that particularly important dimension of cognition that incorporates language—continues, using computational comparative neuroprimatology in particular and computational neuroscience in general.

ACKNOWLEDGMENTS An earlier, less pedagogical version of the themes here was published in *Physics of Life Reviews* (Arbib, 2016a) along with a range of expert comments and an integrative response (Arbib, 2016b). This material is based in part on work supported by the National Science Foundation under Grant No. 0924674 and Grant No. BCS-1343544 "INSPIRE Track 1: Action, Vision and Language, and their Brain Mechanisms in Evolutionary Relationship" (Michael A. Arbib, Principal Investigator). My thanks to Brad Gasser and Victor Barrès for their research on topics presented here, and to them and to Pia Knoeferle for incisive comments on an earlier draft.

REFERENCES

Aglioti, S. M., Cesari, P., Romani, M., & Urgesi, C. (2008). Action anticipation and motor resonance in elite basketball players. *Nature Neuroscience, 11*(9), 1109–1116. doi:10.1038/nn.2182. Available at http://www.nature.com/neuro/journal/v11/n9/suppinfo/nn.2182_S1.html

Alstermark, B., Lundberg, A., Norrsell, U., & Sybirska, E. (1981). Integration in descending motor pathways controlling the forelimb in the cat: IX. Differential behavioural defects after spinal cord lesions interrupting defined pathways from higher centres to motoneurones. *Experimental Brain Research, 42,* 299–318.

Altmann, G. T. M., & Kamide, Y. (1999). Incremental interpretation at verbs: Restricting the domain of subsequent reference. *Cognition, 73*(3), 247–264. doi:10.1016/s0010-0277(99)00059-1

Arbib, M. A. (1981). Perceptual structures and distributed motor control. In V. B. Brooks (Ed.), *Handbook of Physiology—The Nervous System II. Motor Control* (pp. 1449–1480). Bethesda, MD: American Physiological Society.

Arbib, M. A. (1988). Schemas and neurons: Two levels of neural computing. In R. Eckmiller & C. v. d. Malsburg (Eds.), *Neural Computers* (pp. 331–320). New York: Springer-Verlag.

Arbib, M. A. (1989). *The Metaphorical Brain 2: Neural Networks and Beyond.* New York: Wiley Interscience.

Arbib, M. A. (2002). The mirror system, imitation, and the evolution of language. In K. Dautenhahn & C. L. Nehaniv (Eds.), *Imitation in Animals and Artifacts: Complex Adaptive Systems* (pp. 229–280). Cambridge, MA: MIT Press.

Arbib, M. A. (2006). Aphasia, apraxia and the evolution of the language-ready brain. *Aphasiology, 20,* 1–30.

Arbib, M. A. (2008). Holophrasis and the protolanguage spectrum. *Interaction Studies: Social Behaviour and Communication in Biological and Artificial Systems, 9*(1), 151–165.

Arbib, M. A. (2010). Mirror system activity for action and language is embedded in the integration of dorsal and ventral pathways. *Brain and Language, 112*(1), 12–24.

Arbib, M. A. (2012). *How the Brain Got Language: The Mirror System Hypothesis.* New York: Oxford University Press.

Arbib, M. A. (2013a). Complex imitation and the language-ready brain. *Language and Cognition, 5*(2–3), 273–312.

Arbib, M. A. (2013b). Mirror systems and the neurocognitive substrates of bodily communication and language. In C. Müller, A. Cienki, E. Fricke, S. Ladewig, D. McNeill, & S. Tessendorf (Eds.), *Body—Language—Communication* (pp. 445–460). Berlin: de Gruyter.

Arbib, M. A. (2016a). Towards a computational comparative neuroprimatology: Framing the language-ready brain. *Physics of Life Reviews.* Available at http://dx.doi.org/10.1016/j.plrev.2015.1009.1003.

Arbib, M. A. (2016b). Primates, computation, and the path to language. Reply to comments on "Towards a computational comparative neuroprimatology: Framing the language-ready brain." *Physics of Life Reviews.* Available at http://dx.doi.org/10.1016/j.plrev.2016.02.003.

Arbib, M. A., & Bickerton, D. (Eds.). (2010). *The Emergence of Protolanguage: Holophrasis vs Compositionality.* Philadelphia, PA: John Benjamins.

Arbib, M. A., Billard, A., Iacoboni, M., & Oztop, E. (2000). Synthetic brain imaging: Grasping, mirror neurons and imitation. *Neural Networks, 13*(8–9), 975–997.

Arbib, M. A., & Bonaiuto, J. J. (2012). Multiple levels of spatial organization: World graphs and spatial difference learning. *Adaptive Behavior, 20*(4), 287–303.

Arbib, M. A., Conklin, E. J., & Hill, J. C. (1987). *From Schema Theory to Language.* New York: Oxford University Press.

Arbib, M. A., Fagg, A. H., & Grafton, S. T. (2002). Synthetic PET imaging for grasping: From primate neurophysiology to human behavior. In F. T. Soomer & A. Wichert (Eds.), *Exploratory Analysis and Data Modeling in Functional Neuroimaging* (pp. 231–250). Cambridge, MA: MIT Press.

Arbib, M. A., Ghanesh, V., & Gasser, B. (2014). Dyadic brain modeling, ontogenetic ritualization of gesture in apes, and the contributions of primate mirror neuron systems. *Philosophical Transactions of the Royal Society. Series B, Biological Sciences, 369*(1644), 20130414.

Arbib, M. A., & Lee, J. Y. (2008). Describing visual scenes: Towards a neurolinguistics based on construction grammar. *Brain Research, 1225*, 146–162.

Arbib, M. A., Liebal, K., & Pika, S. (2008). Primate vocalization, gesture, and the evolution of human language. *Current Anthropology, 49*(6), 1053–1076. Available at http://www.jstor.org/stable/10.1086/593015.

Arbib, M. A., & Mundhenk, T. N. (2005). Schizophrenia and the mirror system: An essay. *Neuropsychologia, 43*(2), 268–280.

Arbib, M. A., & Rizzolatti, G. (1997). Neural expectations: A possible evolutionary path from manual skills to language. *Communication & Cognition, 29*, 393–424.

Baldwin, J. M. (1896). A new factor in evolution. *American Naturalist, 30*, 441–451, 536–533.

Barrès, V., & Lee, J. Y. (2014). Template construction grammar: From visual scene description to language comprehension and agrammatism. *Neuroinformatics, 12*(1), 181–208. doi:10.1007/s12021-013-9197-y

Barrès, V., Simons, A., & Arbib, M. A. (2013). Synthetic event-related potentials: A computational bridge between neurolinguistic models and experiments. *Neural Networks, 37*, 66–92. doi:10.1016/j.neunet.2012.09.021

Beuls, K., & Steels, L. (2013). Agent-based models of strategies for the emergence and evolution of grammatical agreement. *PLoS One, 8*(3), e58960. doi:10.1371/journal.pone.0058960

Bohland, J. W., Bullock, D., & Guenther, F. H. (2009). Neural representations and mechanisms for the performance of simple speech sequences. *Journal of Cognitive Neuroscience, 22*(7), 1504–1529. doi:10.1162/jocn.2009.21306

Bonaiuto, J. J., & Arbib, M. A. (2010). Extending the mirror neuron system model: II. What did I just do? A new role for mirror neurons. [Research Support, U.S. Gov't, Non-P.H.S.] *Biological Cybernetics, 102*(4), 341–359. doi:10.1007/s00422-010-0371-0

Bonaiuto, J. J., & Arbib, M. A. (2014). Modeling the BOLD correlates of competitive neural dynamics. *Neural Networks, 49*, 1–10.

Bonaiuto, J. J., Rosta, E., & Arbib, M. A. (2007). Extending the mirror neuron system model: I. Audible actions and invisible grasps. *Biological Cybernetics, 96*, 9–38.

Bornkessel, I., & Schlesewsky, M. (2006). The extended argument dependency model: A neurocognitive approach to sentence comprehension across languages. *Psychological Review, 113*(4), 787–821.

Bornkessel-Schlesewsky, I., & Schlesewsky, M. (2013). Reconciling time, space and function: A new dorsal–ventral stream model of sentence comprehension. *Brain and Language, 125*(1), 60–76. doi:10.1016/j.bandl.2013.01.010

Bornkessel-Schlesewsky, I., Schlesewsky, M., Small, S. L., & Rauschecker, J. P. (2015). Neurobiological roots of language in primate audition: Common computational properties. *Trends in Cognitive Sciences, 19*(3), 142–150. doi:10.1016/j.tics.2014.12.008

Buccino, G., Lui, F., Canessa, N., Patteri, I., Lagravinese, G., Benuzzi, F., et al. (2004). Neural circuits involved in the recognition of actions performed by nonconspecifics: An fMRI study. *Journal of Cognitive Neuroscience, 16*(1), 114–126.

Bullock, D., & Rhodes, B. J. (2003). Competitive queuing for planning and serial performance. In M. A. Arbib (Ed.), *The Handbook of Brain Theory and Neural Networks* (2nd ed., pp. 241–244). Cambridge, MA: MIT Press.

Bybee, J. (2013). Usage-based theory and exemplar representations of constructions. In T. Hoffmann, and G. Trousdale (Eds.), *The Oxford Handbook of Construction Grammar* (pp. 49–69). Oxford: Oxford University Press.

Byrne, R. W. (2003). Imitation as behavior parsing. *Philosophical Transactions of the Royal Society of London. Series B, Biological Sciences, 358*, 529–536.

Byrne, R. W., & Byrne, J. M. E. (1993). Complex leaf-gathering skills of mountain gorillas (*Gorilla g. beringei*): Variability and standardization. *American Journal of Primatology, 31*, 241–261.

Caramazza, A., & Zurif, E. B. (1976). Dissociation of algorithmic and heuristic processes in language comprehension: Evidence from aphasia. *Brain and Language, 3*(4), 572–582.

Cartmill, E., & Byrne, R. W. (2007). Orangutans modify their gestural signaling according to their audience's comprehension. *Current Biology, 17*, 1–14.

Chang, F., Dell, G. S., & Bock, K. (2006). Becoming syntactic. *Psychological Review, 113*(2), 234–272.

Chomsky, N. (1981). *Lectures on Government and Binding.* Dordrecht: Foris.

Crocker, M. W., Knoeferle, P., & Mayberry, M. R. (2010). Situated sentence processing: The coordinated interplay account and a neurobehavioral model. *Brain and Language, 112*(3), 189–201. doi:10.1016/j.bandl.2009.03.004

Croft, W. (2001). *Radical Construction Grammar: Syntactic Theory in Typological Perspective.* Oxford: Oxford University Press.

Croft, W., & Cruse, D. A. (2005). *Cognitive Linguistics.* Cambridge: Cambridge University Press.

Custance, D. M., Whiten, A., & Bard, K. A. (1995). Can young chimpanzees (*Pan troglodytes*) imitate arbitrary actions? Hayes & Hayes (1952) revisited. *Behaviour, 132*, 11–12.

Didday, R. L., & Arbib, M. A. (1975). Eye movements and visual perception: "Two visual systems" model. *International Journal of Man-Machine Studies, 7*, 547–569.

Dominey, P. F., Arbib, M. A., & Joseph, J.-P. (1995). A model of corticostriatal plasticity for learning oculomotor associations and sequences. *Journal of Cognitive Neuroscience, 7*(3), 311–336.

Dominey, P. F., Hoen, M., & Inui, T. (2006). A neurolinguistic model of grammatical construction processing. *Journal of Cognitive Neuroscience, 18*(12), 2088–2107.

Dominey, P. F., & Inui, T. (2009). Cortico-striatal function in sentence comprehension: Insights from neurophysiology and modeling. *Cortex, 45,* 1012–1018.

Donald, M. (1991). *Origins of the Modern Mind: Three Stages in the Evolution of Culture and Cognition.* Cambridge, MA: Harvard University Press.

Donald, M. (1998). Mimesis and the executive suite: Missing links in language evolution. In J. R. Hurford, M. Studdert-Kennedy, & C. Knight (Eds.), *Approaches to the Evolution of Language: Social and Cognitive Bases* (pp. 44–67). Cambridge: Cambridge University Press.

Donald, M. (1999). Preconditions for the evolution of protolanguages. In M. C. Corballis & E. G. Stephen (Eds.), *The Descent of Mind: Psychological Perspectives on Hominid Evolution* (pp. 138–154). Oxford: Oxford University Press.

Doupe, A. J., & Kuhl, P. K. (1999). Birdsong and human speech: Common themes and mechanisms. *Annual Review of Neuroscience, 22,* 567–631.

Draper, B. A., Collins, R. T., Brolio, J., Hanson, A. R., & Riseman, E. M. (1989). The schema system. *International Journal of Computer Vision, 2,* 209–250.

Fadiga, L., & Craighero, L. (2003). New insights on sensorimotor integration: From hand action to speech perception. *Brain and Cognition, 53*(3), 514–524.

Fagg, A. H., & Arbib, M. A. (1998. Modeling parietal-premotor interactions in primate control of grasping. *Neural Networks, 11*(7–8), 1277–1303.

Feldman, J., & Narayanan, S. (2004). Embodied meaning in a neural theory of language. *Brain and Language, 89*(2), 385–392.

Fillmore, C. J., Kay, P., & O'Connor, M. K. (1988). Regularity and idiomaticity in grammatical constructions: The case of let alone. *Language and Cognitive Processes, 64,* 501–538.

Friederici, A. D. (2012). The cortical language circuit: From auditory perception to sentence comprehension. *Trends in Cognitive Sciences, 16*(5), 262–268. doi:10.1016/j.tics.2012.04.001

Gasser, B., Cartmill, E., & Arbib, M. A. (2014). Ontogenetic ritualization of primate gesture as a case study in dyadic brain modeling. *Neuroinformatics, 12*(1), 93–109.

Goldberg, A. E. (1995). *Constructions: A Construction Grammar Approach to Argument Structure.* Chicago, IL: University of Chicago Press.

Goodglass, H. (1976). Agrammatism. *Studies in Neurolinguistics, 1,* 237–260.

Grafton, S. T., Arbib, M. A., Fadiga, L., & Rizzolatti, G. (1996). Localization of grasp representations in humans by PET: II. Observation compared with imagination. *Experimental Brain Research, 112,* 103–111.

Grafton, S. T., Fagg, A. H., & Arbib, M. A. (1998). Dorsal premotor cortex and conditional movement selection: A PET functional mapping study. *Journal of Neurophysiology, 79*(2), 1092–1097.

Grodzinsky, Y. (2000). The neurology of syntax: Language use without Broca's area. *Behavioral and Brain Sciences, 23,* 1–71.

Guenther, F. H., Ghosh, S. S., & Tourville, J. A. (2006). Neural modeling and imaging of the cortical interactions underlying syllable production. *Brain and Language, 96*(3), 280–301.

Hagoort, P. (2003). How the brain solves the binding problem for language: A neurocomputational model of syntactic processing. *NeuroImage, 20*(Suppl 1), S18–S29.

Hagoort, P. (2013). MUC (Memory, Unification, Control) and beyond. *Frontiers in Psychology, 4,* 416.

Halina, M., Rossano, F., & Tomasello, M. (2013). The ontogenetic ritualization of bonobo gestures. *Animal Cognition, 16*(4), 653–666.

Hanson, A. R., & Riseman, E. M. (1978). VISIONS: A computer system for interpreting scenes. In A. R. Hanson & E. M. Riseman (Eds.), *Computer Vision Systems* (pp. 129–163). New York: Academic Press.

Hecht, E. E., Gutman, D. A., Preuss, T. M., Sanchez, M. M., Parr, L. A., & Rilling, J. K. (2013). Process versus product in social learning: Comparative diffusion tensor imaging of neural systems for action execution—Observation matching in macaques, chimpanzees, and humans. *Cerebral Cortex, 23*(5), 1014–1024.

Hickok, G., & Poeppel, D. (2004). Dorsal and ventral streams: A framework for understanding aspects of the functional anatomy of language. *Cognition, 92*(1–2), 67–99.

Hinaut, X., & Dominey, P. F. (2013). Real-time parallel processing of grammatical structure in the fronto-striatal system: A recurrent network simulation study using reservoir computing. *PLoS One, 8*(2), e52946.

Hobaiter, C., & Byrne, R. W. (2011). The gestural repertoire of the wild chimpanzee. *Animal Cognition, 14,* 745–767. doi:10.1007/s10071-011-0409-2

Houghton, G., & Hartley, T. (1995). Parallel models of serial behavior: Lashley revisited. *Psyche, 2*(25), 1–25.

Hurford, J. R. (2011). *The Origins of Grammar II: Language in the Light of Evolution.* New York: Oxford University Press.

Iriki, A., & Taoka, M. (2012). Triadic (ecological, neural, cognitive) niche construction: A scenario of human brain evolution extrapolating tool use and language from the control of reaching actions. *Philosophical Transactions of the Royal Society of London. Series B, Biological Sciences, 367,* 10–23.

Itti, L., & Arbib, M. A. (2006). Attention and the minimal subscene. In M. A. Arbib (Ed.), *Action to Language via the Mirror Neuron System* (pp. 289–346). Cambridge: Cambridge University Press.

Itti, L., & Koch, C. (2000). A saliency-based search mechanism for overt and covert shifts of visual attention. *Vision Research, 40,* 1489–1506.

Kemmerer, D. (2006). Action verbs, argument structure constructions, and the mirror neuron system. In M. A. Arbib (Ed.), *Action to Language via the Mirror Neuron System* (pp. 347–373). Cambridge: Cambridge University Press.

Koch, C., & Ullman, S. (1985). Shifts in selective visual attention: Towards the underlying neural circuitry. *Human Neurobiology, 4,* 219–227.

Kotz, S. A., & Schwartze, M. (2010). Cortical speech processing unplugged: A timely subcortico-cortical framework. *Trends in Cognitive Sciences, 14*(9), 392–399.

Lee, J. Y. (2012). *Linking Eyes to Mouth: A Schema-Based Computational Model for Describing Visual Scenes* (doctoral dissertation). Computer Science, University of Southern California, Los Angeles.

Lefebvre, C., Comrie, B., & Cohen, H. (Eds.). (2013). *New Perspectives on the Origins of Language.* Amsterdam: John Benjamins.

MacWhinney, B., & O'Grady, W. (2015). *The Handbook of Language Emergence.* Chichester, UK: Wiley-Blackwell.

Moulin-Frier, C., & Arbib, M. A. (2013). *Recognizing Speech in a Novel Accent: The Motor Theory of Speech Perception Reframed.* Submitted for publication.

Narayanan, S. S. (1997). Knowledge-Based Action Representations for Metaphor and Aspect (KARMA) (Doctoral dissertation). University of California, Berkeley.

Noble, W., & Davidson, I. (1996). *Human Evolution, Language and Mind: A Psychological and Archaeological Inquiry.* Cambridge: Cambridge University Press.

Odling-Smee, F. J., Laland, K. N., & Feldman, M. W. (2003). *Niche Construction: The Neglected Process in Evolution.* Princeton, NJ: Princeton University Press.

Oztop, E., & Arbib, M. A. (2002). Schema design and implementation of the grasp-related mirror neuron system. *Biological Cybernetics, 87*(2), 116–140.

Oztop, E., Kawato, M., & Arbib, M. A. (2013). Mirror neurons: Functions, mechanisms and models. *Neuroscience Letters, 540,* 43–55. doi:10.1016/j.neulet.2012.10.1005

Poizner, H., Klima, E., & Bellugi, U. (1987). *What the Hands Reveal about the Brain.* Cambridge, MA: MIT Press.

Rauschecker, J. P. (2005). Vocal gestures and auditory objects. *Behavioral and Brain Sciences, 28*(02), 143–144. doi:10.1017/S0140525X05410031

Rauschecker, J. P. (2011). An expanded role for the dorsal auditory pathway in sensorimotor control and integration. *Hearing Research, 271*(1–2), 16–25. doi:10.1016/j.heares.2010.09.001.

Rauschecker, J. P., & Scott, S. K. (2009). Maps and streams in the auditory cortex: Nonhuman primates illuminate human speech processing. *Nature Neuroscience, 12*(6), 718–724. doi:10.1038/nn.2331.

Rauschecker, J. P., & Tian, B. (2000). Mechanisms and streams for processing of "what" and "where" in auditory cortex. *Proceedings of the National Academy of Sciences of the United States of America, 97*(22), 11800–11806. doi:10.1073/pnas.97.22.11800

Rilling, J. K., Glasser, M. F., Preuss, T. M., Ma, X., Zhao, T., Hu, X., et al. (2008). The evolution of the arcuate fasciculus revealed with comparative DTI. *Nature Neuroscience, 11*(4), 426–428.

Rizzolatti, G., & Arbib, M. A. (1998). Language within our grasp. *Trends in Neurosciences, 21*(5), 188–194.

Rizzolatti, G., & Sinigaglia, C. (2008). *Mirrors in the Brain: How Our Minds Share Actions, Emotions, and Experience* (F. Anderson, Trans.). Oxford: Oxford University Press.

Romanski, L. M., Tian, B., Fritz, J., Mishkin, M., Goldman-Rakic, P. S., & Rauschecker, J. P. (1999). Dual streams of auditory afferents target multiple domains in the primate prefrontal cortex. *Nature Neuroscience, 2*(12), 1131–1136.

Sauser, E. L., & Billard, A. G. (2006). Parallel and distributed neural models of the ideomotor principle: An investigation of imitative cortical pathways. *Neural Network, 19*(3), 285–298. doi:10.1016/j.neunet.2006.02.003.

Saussure, F. (1916). *Cours de linguistique générale* (C. Bally & A. Sechehaye, Eds., with the collaboration of A. Riedlinger). Lausanne and Paris: Payot (English translation by W. Baskin, Course in General Linguistics, Glasgow: Fontana/Collins, 1977).

Schaal, S., Ijspeert, A., & Billard, A. (2003). Computational approaches to motor learning by imitation. *Philosophical Transactions of the Royal Society of London. Series B, Biological Sciences, 358*(1431), 537–547.

Schlesewsky, M., & Bornkessel, I. (2004). On incremental interpretation: Degrees of meaning accessed during sentence comprehension. *Lingua, 114,* 1213–1234.

Sherman, J. C., & Schweickert, J. (1989). Syntactic and semantic contributions to sentence comprehension in agrammatism. *Brain and Language, 37*(3), 419–439. doi:10.1016/0093-934X(89)90029-1.

Spivey, M. J., Richardson, D. C., & Fitneva, S. A. (2005). Thinking outside the brain: Spatial indices to visual and linguistic information. In J. M. Henderson & F. Ferreira (Eds.), *The Interface of Language, Vision, and Action: Eye Movements and the Visual World* (pp. 161–190). New York: Psychology Press.

Spranger, M. (2014). *The Evolution of Grounded Spatial Language.* Berlin: Language Science Press.

Stokoe, W. C. (2001). *Language in Hand: Why Sign Came Before Speech.* Washington, DC: Gallaudet University Press.

Tallerman, M., & Gibson, K. R. (Eds.). (2011). *The Oxford Handbook of Language Evolution.* Oxford: Oxford University Press.

Tomasello, M. (1999). Emulation learning and cultural learning. *Behavioral and Brain Sciences, 21,* 703–704.

Tomasello, M., & Call, J. (1997). *Primate Cognition.* New York: Oxford University Press.

Van Valin, R. D., Jr. (2005). *Exploring the Syntax-Semantics Interface.* Cambridge: Cambridge University Press.

Visalberghi, E., & Fragaszy, D. (2001). Do monkeys ape? Ten years after. In K. Dautenhahn & C. Nehaniv (Eds.), *Imitation in Animals and Artifacts* (pp. 471–500). Cambridge, MA: MIT Press.

Vosse, T., & Kempen, G. (2000). Syntactic structure assembly in human parsing: A computational model based on competitive inhibition and a lexicalist grammar. *Cognition, 75*(2), 105–143.

Waddington, C. H. (1953). The "Baldwin effect," "genetic assimilation" and "homeostasis." *Evolution; International Journal of Organic Evolution, 7*(4), 386–387.

Werbos, P. J. (1990). Backpropagation through time: What it does and how to do it. *Proceedings of the IEEE, 78*(10), 1550–1560.

Wohlschläger, A., Gattis, M., & Bekkering, H. (2003). Action generation and action perception in imitation: An instance of the ideomotor principle. *Philosophical Transactions of the Royal Society of London. Series B, Biological Sciences, 358,* 501–515.

Appendix A: Basics of Connectionist Networks

MICHAEL A. ARBIB

In this appendix, we consider neurons operating on a discrete timescale t = {0, 1, 2, 3, ...}, with the output of each one falling in the range **M,** where **M** is {0,1} for binary neurons, and the set of non-negative real numbers (or possibly just the interval [0,1]) for a firing rate model. In each case, the neuron is memoryless, that is, we take the synaptically weighted sum $\Sigma w_i x_i(t)$ of its inputs at time t, and pass it through a threshold function in the binary case or a sigmoid function in the firing rate case to get the output y(t + 1) at time t. (See chapter 2 for further details.) We will briefly consider how a number of simple learning rules are employed in adapting the performance of a neural network.

A.1 *Connectionist Networks as Dynamical Systems*

As exemplified in figure A.1, we characterize a neural network by selecting *N* neurons and taking the axon of each neuron which may be split into several lines carrying identical output signals, and either connecting each line to a unique input of another neuron or feeding it outside the net to provide one of the K network output lines. Then every input to a given neuron must be connected either to an output of another neuron or to one of the (possibly split) L input lines of the network. Then the input set $X = \mathbf{M}^L$, the state set $Q = \mathbf{M}^N$, and the output set $Y = \mathbf{M}^K$. If the ith output line comes from the jth neuron, then the *output function* is determined by the fact that the ith component of the output at time t is the output of the jth neuron at time t.

In computer science, an *automaton* is a discrete-time system with discrete input, output, and state spaces. Formally, we describe an automaton by the sets X, Y, and Q of inputs, outputs, and states, respectively, together with the *next-state function* δ: $Q \times X \rightarrow Q$ and the *output function* β: $Q \rightarrow Y$. If the automaton is in state q and receives input x at time t, then its next state will be $\delta(q, x)$ and its next output will be $\beta(q)$. Clearly, a McCulloch–Pitts neural network (i.e., a network like that shown in figure A.1, but a discrete time network with binary elements; see section 2.3.4, "Binary and Analog Rate Neurons," for the definition of McCulloch-Pitts neurons) functions like a finite automaton, as each neuron changes state synchronously on each tick of the timescale t = 0, 1, 2, 3, ... Conversely, it can be shown (see Arbib, 1987, the result was essentially, though inscrutably, due to McCulloch and Pitts, 1943) that any finite automaton can be simulated by a suitable McCulloch–Pitts neural network.

With the previous discussion, we now have more than enough material to understand the crucial dynamic systems concept of *stability* and the related concepts of limit cycles and chaos. We want to know what happens to an "unperturbed" system, that is, one for which the input is held constant (possibly with some specific "null input," usually denoted by 0, the "zero" input in X). An *equilibrium* is a state q in which the system can stay at rest, that is, such that $\delta(q, 0) = q$ (discrete time) The study of stability is concerned with the issue of whether or not this rest point will be maintained in the face of slight disturbances. To see the variety of equilibria, we use the image of a sticky ball rolling on the "hillside" of figure A.2. We say that point *A* atop the "hillside" in this diagram is an *unstable equilibrium* because a slight displacement from A will tend to increase over time. Point B is in a region of *neutral equilibrium* because slight displacements will tend not to change further, while C is a point of *stable equilibrium,* since small displacements will tend to disappear over time. Note the word "small"—in a nonlinear system like that of figure A.2, a large displacement can move the ball from the *basin of attraction* of C (the set of states whose dynamics tends toward C) to another one. Clearly, the ball will not tend to return to C after a massive displacement which moves the ball to the far side of A's hilltop.

Nonlinear systems also have another interesting property: they may exhibit *limit cycles.* These are closed trajectories in the state space and, thus, may be thought of as "dynamic equilibria." If the state of a system follows a limit cycle, we may also say it oscillates or exhibits periodic behavior. A limit cycle is *stable* if a small displacement will be reduced as the trajectory of the

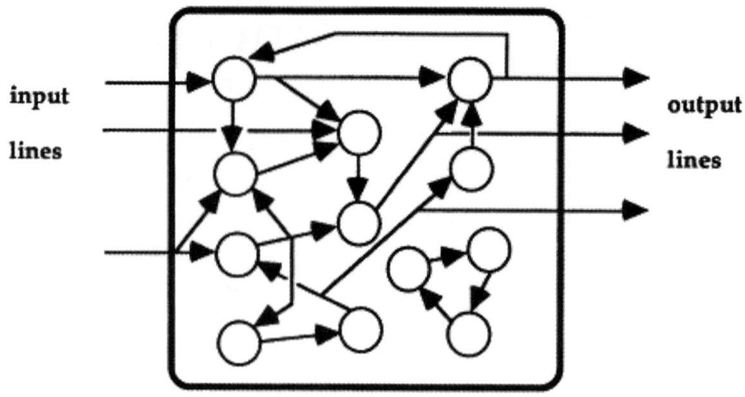

FIGURE A.1 A neural network viewed as a system. The input at time t is the pattern of firing on the input lines, the output is the pattern of firing on the output lines, and the internal state is the vector of firing rates (or other chosen neuron output variable) of all the neurons of the network.

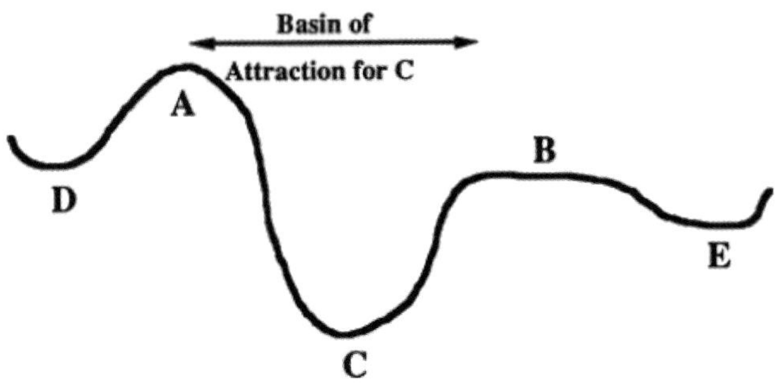

FIGURE A.2 An energy landscape: For a ball rolling on the "hillside," point A is an *unstable equilibrium*, point B lies in a region of *neutral equilibrium*, and point C is a point of *stable equilibrium*. Point C is called an attractor: the basin of attraction of C comprises all states whose dynamics tend toward C.

system comes closer and closer to the original limit cycle. By contrast, a limit cycle is *unstable* if such excursions do not die out. Research in nonlinear systems has also revealed what are called *strange attractors*. These are attractors which, unlike simple limit cycles, describe such complex paths through the state space that, although the system is deterministic, a path which approaches the strange attractor gives every appearance of being random because very small differences in initial state may be amplified with the passage of time so that differences that at first are not even noticeable will yield, in due course, states that are very different indeed. Such a trajectory has become the accepted mathematical model of *chaos*, and it is used to describe a number of physical phenomena, such as the onset of turbulence in a weather system, as well as a number of phenomena in biological systems.

A.2 *Hopfield Nets*

In a McCulloch–Pitts network, every neuron processes its inputs to determine a new output at each time step. By contrast, a *Hopfield net* is a net of such units with (a) *symmetrical* weights ($w_{ij} = w_{ji}$) and no self-connections ($w_{ii} = 0$), and (b) the nonbiological assumption of *asynchronous updating*. (See also section 2.5.2, "Networks of Binary Neurons.") For instance, let s_i denote the state (0 or 1) of the ith unit. At each time step, pick just one unit at random. If unit i is chosen, s_i takes the value 1 if and only if $\Sigma\, w_{ij}s_j \geq \theta_i$. Otherwise s_i is set to 0. Note that this is an *autonomous* (input-free) network: there are no inputs.

Hopfield (1982) defined a measure he called the *energy* for such a net:

$$E = -\Sigma_{ij}\, s_i s_j w_{ij} + \Sigma_i\, s_i\theta_i.$$

This is not the physical energy of the neural net but a mathematical quantity that, in some ways, does for neural dynamics what the potential energy does for Newtonian mechanics. In general, a mechanical system moves to a state of lower potential energy—just as, in figure A.2, the ball tends to move downhill. Hopfield showed that his symmetrical networks with asynchronous updating had a similar property.

For example, if we pick a unit and the foregoing firing rule does not change its s_i, it will not change E. However if s_i initially equals 0, and $\Sigma_j\, w_{ij}s_j \geq \theta_i$, then s_i goes from 0 to 1 with all other s_j constant, and the "energy gap," or change in E, is given by

$$\begin{aligned}\Delta E &= -\Sigma_j\,(w_{ij}s_j + w_{ji}\,s_j) + \theta_i \\ &= -\Sigma_j\, w_{ij}s_j\, s_j + \theta_i, \text{ by symmetry} \\ &\leq 0 \text{ since } \Sigma\, w_{ij}s_j \geq \theta_i.\end{aligned}$$

Similarly, if s_i initially equals 1, and $\Sigma\, w_{ij}s_j < \theta_i$ then s_i goes from 1 to 0 with all other s_j constant, and the energy gap is given by

$$\Delta E = \Sigma\, w_{ij}s_j - \theta_i < 0.$$

In other words, with every asynchronous updating, we have $\Delta E \leq 0$. Hence the dynamics of the net tends to move E toward a minimum. We stress that there may be different such states—they are *local* minima, just as, in figure A.2, both D and E are local minima (each of them is lower than any "nearby" state) but not global minima (since C is lower than either of them). Global minimization is not guaranteed.

The expression just presented for ΔE depends on the symmetry condition, $w_{ij} = w_{ji}$, for without this condition, the expression would instead be $\Delta E = -\Sigma_j (w_{ij}s_j + w_{ji} s_j) + \theta_i$, and in this case, Hopfield's updating rule need not yield a passage to energy minimum, but might instead yield a limit cycle, which could be useful in, for example, controlling rhythmic behavior (see, e.g., chapter 8, "Motor Pattern Generation"). In a control problem, a link w_{ij} might express the likelihood that the action represented by *i* would precede that represented by j, in which case $w_{ij} = w_{ji}$ is normally inappropriate.

The condition of *asynchronous* update is crucial, too. If we consider the simple "flip-flop" with $w_{12} = w_{21} = 1$ and $\theta_1 = \theta_2 = 0.5$, then the McCulloch–Pitts network will *oscillate* between the states (0, 1) and (1, 0) or will sit in the states (0, 0) or (1, 1); in other words, there is no guarantee that it will converge to an equilibrium. However, with $E = -\Sigma_{ij} s_i s_j w_{ij} + \Sigma_i s_i \theta_i$, we have $E(0,0) = 0$, $E(0, 1) = E(1, 0) = 0.5$, and $E(1, 1) = 0$, and the Hopfield network will *converge* to the global minimum at either (0, 0) or (1, 1).

Hopfield also aroused much interest because he showed how a number of optimization problems could be "solved" using *artificial* neural networks. (The quotes around "solved" acknowledge the fact that the state to which a neural network converges may represent a local, rather than a global, optimum of the corresponding optimization problem.) Such networks were similar to the "constraint satisfaction" networks that had already been studied in the computer vision community. (In most vision algorithms constraints can be formulated in terms of symmetrical weights, so that $w_{ij} = w_{ji}$ is appropriate.) The aim, given a "constraint satisfaction" problem, is to so choose weights for a neural network that the energy *E* for that network is a measure of the overall constraint violation. The techniques of simulated annealing and Boltzmann machines showed how noise may be added to "shake" a system out of a local minimum and let it settle into a global minimum. (Consider, e.g., shaking that is strong enough to shake the ball from D to A, and thus into the basin of attraction of C, in figure A.2, but not strong enough to shake the ball back from C toward D.)

A.3 *Learning in Connectionist Networks*

Most learning rules for connectionist networks take the form of schemes for adjusting the synaptic weights, the "w's." The two classic learning schemes for McCulloch–Pitts-type formal neurons are due to Hebb and Rosenblatt, and we now introduce these in turn.

A.3.1 Hebbian Plasticity and Network Self-Organization In Hebb's (1949) learning scheme (see chapter 6, "Hebbian Plasticity and Learning," for an extended treatment, addressing spiking as well as firing rate based neural models), the connection between two neurons is strengthened if both neurons fire at the same time. The simplest example of such a rule is to increase w_{ij} by the following amount:

$$\Delta w_{ij} = k\, y_i\, x_j$$

where synapse w_{ij} connects a presynaptic neuron with firing rate x_j to a postsynaptic neuron with firing rate y_i. The trouble with the original Hebb model is that every synapse will eventually get stronger and stronger until they all saturate, thus destroying any selectivity of association. One solution (von der Malsburg, 1973) is to normalize the synapses impinging on a given neuron. To accomplish this, one must first compute the Hebbian "update" $\Delta w_{ij} = kx_iy_j$ and then divide this by the total putative synaptic weights to get the final result which replaces w_i by

$$\frac{w_{ij} + \Delta w_{ij}}{\sum_k (w_{kj} + \Delta w_{kj})}$$

where the summation k extends over all inputs to the neuron. This new rule not only increases the strengths of those synapses with inputs strongly correlated with the cell's activity but also decreases the synaptic strengths of other connections in which such correlations did not arise.

von der Malsburg was motivated by the pattern recognition problem, and was concerned with how individual cells in his network might come to be tuned so as to respond to one particular input "feature" rather than another. This exposed another problem with Hebb's rule: a lot of nearby cells may, just by chance, all have initial random connectivity which makes them easily persuadable by the same stimulus; alternatively, the same pattern might occur many times before a new pattern is experienced by the network. In either case, many cells would become tuned to the same feature, with not enough cells left to learn important and distinctive features. To solve this, von der Malsburg introduced *lateral inhibition* into his model. In this connectivity, pattern, activity in any one cell is distributed laterally to reduce (partially inhibit) the activity of nearby cells. This ensures that if one cell—call it A—were most active, its connections to nearby cells would make them less active, and so make them less likely to learn, by Hebbian synaptic adjustment, those features that most excite A.

In summary, then, when the Hebbian rule is augmented by a normalization rule, it tends to "sharpen" a neuron's predisposition "without a teacher," getting its firing to become better and better correlated with a cluster of stimulus patterns. This performance is improved when there is some competition between neurons, so that if one neuron becomes adept at responding to a pattern, it inhibits other neurons from doing so. Thus, the final set of input weights to the neuron depends both on the initial setting of the weights and on the pattern of clustering of the set of stimuli to which it is exposed.

In the adaptive architecture just described, the inputs are initially randomly connected to the cells of the processing layer. As a result, none of these cells is particularly good at pattern recognition. However, by sheer statistical fluctuation of the synaptic connections, one will be slightly better at responding to a particular pattern than others are; it will thus slightly strengthen those synapses which allow it to fire for that pattern. Through lateral inhibition, this will make it harder for cells initially less well tuned for that pattern to become tuned to it. Thus, without any teacher, this network automatically organizes itself so that each cell becomes tuned for an important cluster of information in the sensory inflow. This is a basic example of the phenomenon of self-organization.

A.3.2 Perceptrons By contrast, perceptrons use an *error-correction rule* designed to change the weights of each response unit when it makes erroneous responses to stimuli that are presented to the network. We call the judge of what is correct the "teacher" although this may be another neural network, or some environmental input, rather than a signal supplied by a human teacher in the usual schoolroom sense. Consider the case in which a set R of input lines feeds a McCulloch–Pitts neural network whose neurons are called *associator units* and which, in turn, provide the input to a single McCulloch–Pitts neuron (called the *output unit* of the perceptron) with adjustable weights $(w_1, \ldots, w_d)$ and threshold θ. (In the case of visual pattern recognition, we think of R as a rectangular "retina" onto which patterns may be projected.) A *simple perceptron* is one in which the associator units are not interconnected, which means that it has no short-term memory. (If such connections are present, the perceptron is called *cross-coupled*. A cross-coupled perceptron may have multiple layers and loops back from an "earlier" to a "later" layer.) If the associator units feed the pattern $x = (x_1, \ldots, x_d)$ to the output unit, then the response of that unit will be to provide the pattern discrimination with discriminant function $f(x) = w_1x_1 + \ldots + w_dx_d - \theta$. The response unit "separates" patterns x for which $f(x) \geq 0$ from those for which $f(x) < 0$. In other words, the simple perceptron can only compute a *linearly separable* function of the pattern as coded by the associator units. The question asked by Rosenblatt (1958) was this: "Given a simple perceptron (i.e., only the synaptic weights of the output unit are adjustable), can we 'train' it to recognize a given linearly separable set of patterns by adjusting the 'weights' on various interconnections on the basis of feedback on whether or not the network classifies a pattern correctly?" The answer was "Yes: if the patterns are linearly separable, then there is a learning scheme which will eventually yield a satisfactory setting of the weights." The best known perceptron learning rule strengthens an active synapse if the efferent neuron fails to fire when it should have fired, and it weakens an active synapse if the neuron fires when it should not have done so:

$$\Delta w_{ij} = k\,(Y_i - y_i)\,x_j.$$

As before, synapse w_{ij} connects a presynaptic neuron with firing rate x_j to a postsynaptic neuron with firing rate y_i, but now Y_i is the "correct" output supplied by the "teacher." Notice that the rule does change the response to x_j "in the right direction." If the output is correct, $Y_i = y_i$ and there is no change, $\Delta w_j = 0$. If the output is too small, then $Y_i - y_i > 0$, and the change in w_j will add $\Delta w_j\, x_j = k\,(Y_i - y_i)\, x_j\, x_j > 0$ to the output unit's response to $(x_1, \ldots, x_d)$. Similarly, if the output is too large, then $Y_i - y_i < 0$, Δw_j will add $k\,(Y_i - y_i)\, x_j\, x_j < 0$ to the output unit's response. Thus, there is a sense in which the new setting $w' = w + \Delta w$ classifies the input pattern x "more nearly correctly" than w does. Unfortunately, in classifying x "more correctly," we run the risk of classifying another pattern "less correctly." However, the *perceptron convergence theorem* (see Nilsson, 1965) shows that Rosenblatt's procedure does not yield an endless seesaw but will eventually converge to a correct set of weights if one exists, albeit perhaps after many iterations through the set of trial patterns.

Gradient Descent and Credit Assignment The implication of the result on simple perceptrons is clear: if we limit the complexity of the units in a neural network, then in general we will need many adaptable layers, rather than a single layer, if the network is to have any chance of being trained to realize many "interesting" functions. This conclusion motivates the study of training rules for multilayer perceptrons of which the most widely used is *backpropagation.* Before describing this method, we first discuss two general notions of which it is an important exemplar: *gradient descent* and *credit assignment.*

In discussing Hopfield networks, we introduced the metaphor of an "energy landscape" (see figure A.2). The asynchronous updates move the state of the network (the vector of neural activity levels) "downhill," tending toward a local energy minimum. Our task now is to realize that the metaphor works again on a far more abstract level when we consider learning. In learning, the dynamic variable is not the network state but rather the vector of synaptic weights (or whatever other set of network parameters is adjusted by the learning rules). We now conduct *gradient descent in weight space.* At each step, the weights are adjusted in such a way as to improve the performance of the network. (As in the case of the simple perceptron, the improvement is a "local" one based on the current situation. It is, in this case, a matter for computer simulation to prove that the cumulative effect of these small changes is a network which solves the overall problem.)

But how do we recognize which "direction" in weight space is "downhill"? Suppose success is achieved by a complex mechanism after operating over a considerable period of time (e.g., when a chess-playing program wins a game). To what particular decisions made by what particular components should the success be attributed? And, if failure results, what decisions deserve blame? This is closely related to the problem known as the "mesa" or "plateau" problem (Minsky, 1961). The performance evaluation function available to a learning system may consist of large level regions in which gradient descent degenerates to exhaustive search, so that only a few of the situations obtainable by the learning system and its environment are known to be desirable, and these situations may occur rarely.

One aspect of this problem, then, is the *temporal* credit assignment problem. The utility of making a certain action may depend on the sequence of actions of which it is a part, and an indication of improved performance may not occur until the entire sequence has been completed. This problem was attacked successfully in Samuel's (1959) learning program for playing checkers. The idea is to interpret predictions of future reward as rewarding events themselves. In other words, neutral stimulus events can themselves become reinforcing if they regularly occur before events that are intrinsically reinforcing. Such *temporal-difference learning* (see chapter 10, "Reinforcement Learning") is like a process of erosion: the original uninformative mesa, where only a few sinkholes allow sudden descent to a local minimum, is slowly replaced by broader valleys in which gradient descent may successfully proceed from many different places on the landscape.

Another aspect of credit assignment concerns structural factors. In the simple perceptron, only the weights to the output units are to be adjusted. This architecture can only support maps which are linearly separable as based on the patterns presented by the preprocessors, and we have seen that many interesting problems require preprocessing units of undue complexity to achieve linear separability. We thus need multiple layers of preprocessors, and—since one may not know a priori the appropriate set of preprocessors for a given problem—these units should be trainable too. This raises the question "How does a neuron deeply embedded within a network 'know' what aspect of the outcome of an overall action was 'its fault' (or 'to its credit')?" This is the *structural* credit assignment problem. In the next section, we study the most widely used solution to this problem, called *backpropagation,* which propagates back to a hidden unit some measure of its responsibility.

Backpropagation Backpropagation is an "adaptive architecture": it is not just a local rule for synaptic adjustment; it also takes into account the position of a neuron in the network to indicate how the neuron's weights are to change. There is no evidence that backpropagation represents actual brain mechanisms. The task of backpropagation is to train a *multilayer* (feedforward) *perceptron* (or MLP), a loop-free network which has its units arranged in layers, with a unit providing input only to units in the next layer of the sequence. The first layer comprises fixed input units; there may then be several layers of trainable "hidden units" carrying an internal representation, and finally, there is the layer of output units, also trainable. (A simple perceptron then corresponds to the case in which we view the input units as fixed associator units, i.e., they deliver a preprocessed, rather than a "raw," pattern which connects directly to the output units without any hidden units in between.) For what follows, it is crucial that each unit *not* be binary: it has both input and output taking continuous values in some range, say, [0, 1]. The response is a sigmoidal function of the weighted sum. Thus, if a unit has inputs x_k with corresponding weights w_{ik}, the output x_i is given by $x_i = f_i(\Sigma w_{ik} x_k)$, where f_i is a sigmoidal function, say,

$$f_i(x) = \frac{1}{1+e^{-(x+\theta_i)}}$$

with θ_i being a bias or threshold for the unit.

The environment only evaluates the output units. We are given a training set of input patterns p and corresponding desired target patterns t^p for the output units. With o^p the actual output pattern elicited by input p, the aim is to adjust the weights in the network to minimize the error summed over all training patterns p and output neurons k. Rumelhart, Hinton, and Williams (1986) were among those who devised a formula for propagating back the gradient of this evaluation from a unit to its inputs. This process can continue by backpropagation through the entire net. At each trial, we fix the input pattern *p* and consider the corresponding "restricted error"

$$E_p = \Sigma_k (t_k - o_k)^2$$

where k ranges over designated "output units." The net has many units interconnected by weights w_{ij}. The learning rule is to change w_{ij} so as to reduce E by *gradient descent*:

$$\Delta w_{ij} = -\frac{\delta E}{\delta w_{ij}} = 2\sum_k (t_k - o_k)\frac{\delta o_k}{\delta w_{ij}}$$

Surprisingly, one can prove the following, by induction on how many layers back we must go to reach a unit:

Proposition: Consider a layered loop-free net with error $E = \Sigma_k (t_k - o_k)^2$, where k ranges over designated "output units," and let the weights w_{ij} be changed according to the gradient descent rule $\Delta w_{ij} = -\partial E / \partial w_{ij} = 2\, \Sigma_k (t_k - o_k)$.

Then the weights may be changed inductively, working back from the output units, by the rule Δw_{ij} is proportional to $\delta_i o_j$, where:

Basis Step: $\delta_i = (t_i - o_i) f_i'$ for an output unit.

Induction Step: If i is a hidden unit, and if δ_k is known for all units which receive unit i's output, then

$\delta_i = (\Sigma_k\, \delta_k w_{ki})\, f_i'$, where k runs over all units which receive unit i's output.

Thus, the "error signal" δ_i *propagates* back layer by layer from the output units. In $\Sigma_k \delta_k w_{ki}$, unit i receives error propagated back from a unit k to the extent to which i affects k. The theorem tells us how to compute Δw_{ij} for gradient descent. It does not guarantee that the above step size is appropriate to reach the minimum, nor does it guarantee that the minimum, if reached, is global. The backpropagation rule defined by this proposition is, thus, a heuristic rule, not one guaranteed to find a global minimum, but is still perhaps the most diversely used adaptive architecture for connectionist networks.

Simple Recurrent Networks[1] Standard backpropagation is sufficient for training feedforward connectionist networks, but it must be adapted for recurrent networks. The method is called *backpropagation through time.* The basic idea is to "unroll" the network (Werbos, 1990). To learn sequences of up to *n* actions, the network is copied $n - 1$ times to produce a network with *n* input layers, *n* output layers, and $n + 1$ hidden layers, and where weights between copied units are constrained to be equal. The copied layers can be thought of as the layers of the original network at different time steps in the action sequence. To train the network, inputs are set at each time step in the sequence and propagated forward. Target outputs are similarly set based on the required sequence. The difference between actual and target outputs can then be calculated and error propagated back as in standard backpropagation of error. Weights are then adjusted as in the standard backpropagation of error algorithm such that the change to each weight is the sum of changes over all steps of the sequence. This training procedure results in the hidden units' learning to represent a kind of context which, through recurrent connections, feeds together with the input representation on each step into the selection of an appropriate output. The hidden units on time step *n* are thus sometimes referred to as forming the context units for time step $n + 1$.

A.4 Attractor Networks

Appendix B, by Edmund T. Rolls, introduces an important class of connectionist networks, the attractor networks. Rolls applies them to brain modeling in chapter 18 ("Hippocampus: From Spatial Processing to Episodic Memory"), where different forms of such networks are used to explain properties of different regions of the hippocampus.

NOTE

1. This subsection is extracted from chapter 14, "Schema Theory and Neuropsychology."

REFERENCES

ARBIB, M. A. (1987). *Brains, Machines, and Mathematics.* New York: Springer-Verlag.

HOPFIELD, J. J. (1982). Neural networks and physical systems with emergent collective computational abilities. *Proceedings of the National Academy of Sciences of the United States of America, 79,* 2554–2558.

MCCULLOCH, W. S., & PITTS, W. H. (1943). A logical calculus of the ideas immanent in nervous activity. *Bulletin of Mathematical Biophysics, 5,* 115–133.

Minsky, M. L. 1961. Steps toward artificial intelligence. *Proceedings of the IRE, 49*, 8–30.

Nilsson, N. J. (1965). *Learning Machines.* New York: McGraw-Hill.

Rumelhart, D. E., Hinton, G. E., & Williams, R. J. (1986). Learning representations by back-propagation errors. *Nature, 323*, 533–536.

Samuel, A. L. (1959). Some studies in machine learning using the game of checkers. *IBM Journal of Research and Development, 3*, 210–229.

von der Malsburg, C. (1973). Self-organization of orientation sensitive cells in the striate cortex. *Kybernetik, 14*, 85–100.

Werbos, P. J. (1990). Backpropagation through time: What it does and how to do it. *Proceedings of the IEEE, 78*, 1550–1560.

Appendix B: Autoassociation, Pattern Association, and Competitive Networks

EDMUND T. ROLLS

For an extended exposition of the material summarized here, the reader is referred to appendix B in Rolls (2008, 2016). MATLAB programs to illustrate the operation of some of these networks are available at http://www.oxcns.org.

B.1 Summary of the Architecture and Operation of Autoassociation or Attractor Networks

The prototypical architecture of an autoassociation memory network is shown in figure B.1. The external input e_i is applied to each neuron i by unmodifiable synapses. This produces firing y_i of each neuron. Each neuron is connected by a recurrent collateral synaptic connection to the other neurons in the network, via associatively modifiable connection weights w_{ij}. This architecture effectively enables the vector of output firing rates to be associated during learning with itself. Later on, during recall, presentation of part of the external input will force some of the output neurons to fire, but through the recurrent collateral axons and the modified synapses, other neurons can be brought into activity. This process can be repeated a number of times, and recall of a complete pattern may be perfect. Effectively, a pattern can be recalled or recognized because of associations formed between its parts. This, of course, requires distributed representations.

Training for each desired pattern occurs in a single trial. The firing of every output neuron i is forced to a value y_i determined by the external input e_i. Then a Hebb-like associative local learning rule is applied to the recurrent synapses in the network:

$$\delta w_{ij} = k \cdot y_i \cdot y_j \tag{B.1}$$

where k is a learning rate constant, y_i is the activation of the dendrite (the postsynaptic term), y_j is the presynaptic firing rate, and δw_{ij} is the change in the synaptic weight from axon j to neuron i. This learning algorithm is fast, "one-shot," in that a single presentation of an input pattern is all that is needed to store that pattern.

It is notable that in a fully connected network, this will result in a symmetrical matrix of synaptic weights, that is, the strength of the connection from neuron 1 to neuron 2 will be the same as the strength of the connection from neuron 2 to neuron 1 (both implemented via recurrent collateral synapses).

During recall, an external input e is applied and produces output firing, operating through a nonlinear activation function. The firing is fed back by the recurrent collateral axons shown in figure B.1 to produce activation of each output neuron through the modified synapses on each output neuron. The activation h_i produced by the recurrent collateral effect on the ith neuron is, in the standard way, the sum of the activations produced in proportion to the firing rate of each axon operating through each modified synapses w_{ij}, that is,

$$h_i = \Sigma_j \, y_j \, w_{ij} \tag{B.2}$$

where the sum is over the input axons to each neuron, indexed by j.

The output firing y_i is a function of the activation produced by the recurrent collateral effect (internal recall) and by the external input e_i:

$$y_i = \mathrm{f}(h_i + e_i). \tag{B.3}$$

The activation function f should be nonlinear and may be, for example, binary threshold, linear threshold, sigmoid, and so forth. A purely linear system would not produce any categorization of the input patterns it receives and therefore would not be able to effect anything more than a trivial (i.e., linear) form of completion and generalization.

During recall, a part of one of the originally learned stimuli can be presented as an external input. The resulting firing is allowed to iterate repeatedly round the recurrent collateral system, gradually on each iteration recalling more and more of the originally learned pattern. Completion thus occurs. If a pattern is presented during recall that is similar to one of the previously learned patterns, then the network settles into a stable recall state in which the firing corresponds to that of the previously learned pattern. The network can thus generalize in its recall to the most similar previously learned pattern. Important results (cf. section "Recall

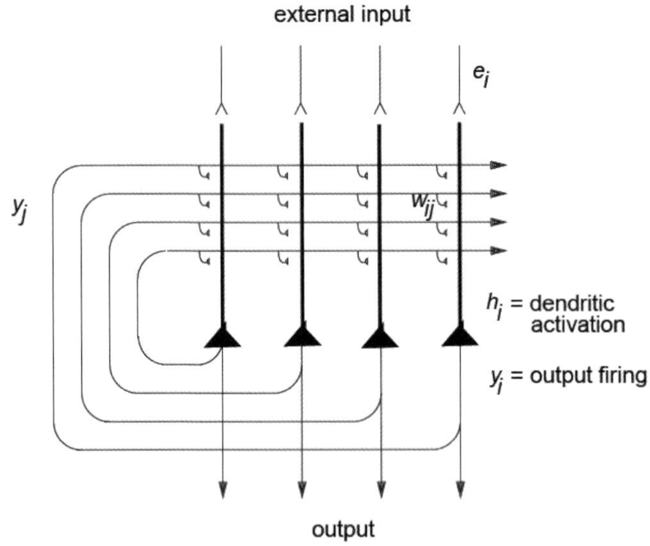

FIGURE B.1 The architecture of an autoassociative neural network. The recurrent collateral synaptic weights are excitatory. Inhibitory neurons (not illustrated) must be present to control the firing rates. From Rolls (2016).

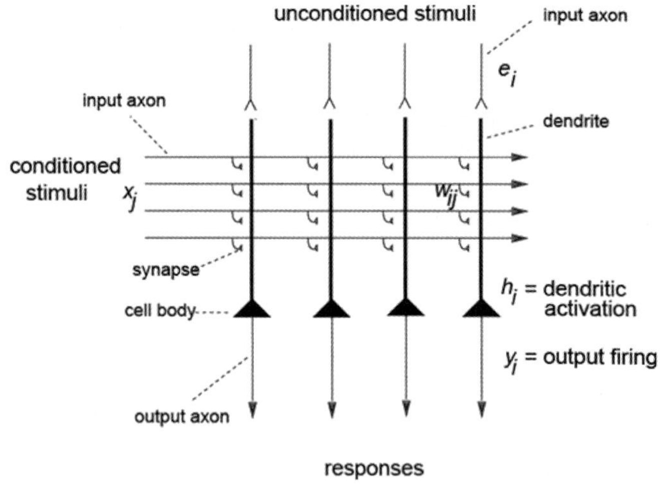

FIGURE B.2 The architecture of a pattern association network. From Rolls (2016).

and Completion" in chapter 18) characterize how many patterns can be stored in a network in this way without interference during recall.

B.2 Summary of the Architecture and Operation of Pattern Association Networks

Figure B.2 shows a pattern association network. We trace the emergence of a conditioned response (CR). An unconditioned stimulus (US) has activity or firing rate e_i for the ith neuron, and produces firing y_i of the ith neuron which is an unconditioned response (UR). The conditioned stimuli (CS) have activity or firing rate x_j for the jth axon. During learning, a CS is presented at the same time as a US, and the synaptic weights are modified by an associative synaptic learning rule $\delta w_{ij} = k \cdot y_i \cdot x_j$ where k is a learning rate constant, and δw_{ij} is the change in the synaptic weight from axon j to neuron i. During recall, only the CS is presented, and the activation h_i is calculated as a dot product between the input stimulus and the synaptic weight vector on a neuron $h_i = \Sigma_j x_j w_{ij}$ where the sum is over the input axons to each neuron, indexed by j. The output firing y_i is a function f of the activation $y_i = \mathrm{f}(h_i)$. The activation function f should be nonlinear and may be, for example, binary threshold, linear threshold, or sigmoid. Inhibitory neurons not shown in the figure are part of the way in which the threshold of the activation function is set. The nonlinearity in the activation function enables interference from other pattern pairs stored in the network to be minimized. The pattern association network thus enables a CS to retrieve a response, the CR, which was present as a UR during the learning. An important property is that if a distorted CS is presented, generalization occurs to the effects produced by the closest CS during training.

B.3 Summary of the Architecture and Operation of Competitive Networks

During training of a competitive network (figure B.3), an input stimulus is presented to the synaptic matrix which has random initial weights w_{ij} and produces activation h_i of the ith neuron calculated as a dot product between the input stimulus and the synaptic weight vector on a neuron $h_i = \Sigma_j x_j w_{ij}$ where the sum is over the input axons to each neuron, indexed by j. The output firing y_i is a function f of the activation $y_i = \mathrm{f}(h_i)$. The activation function f should be nonlinear and may be, for example, binary threshold, linear threshold, or sigmoid. Inhibitory neurons not shown in the figure are part of the way in which the threshold of the activation function is set in which the threshold reflects the firing of all the output neurons. This or other competitive mechanisms result in a typically sparse set of output neurons having firing after the competition.

Next, an associative synaptic modification rule is applied, while the presynaptic input and the postsynaptic output are both present, $\delta w_{ij} = k \cdot y_i \cdot x_j$ where k is a learning rate constant, and δw_{ij} is the change in the synaptic weight from axon j to neuron i.

Next, the synaptic weight vector on each neuron is normalized to ensure that no one neuron dominates the classification.

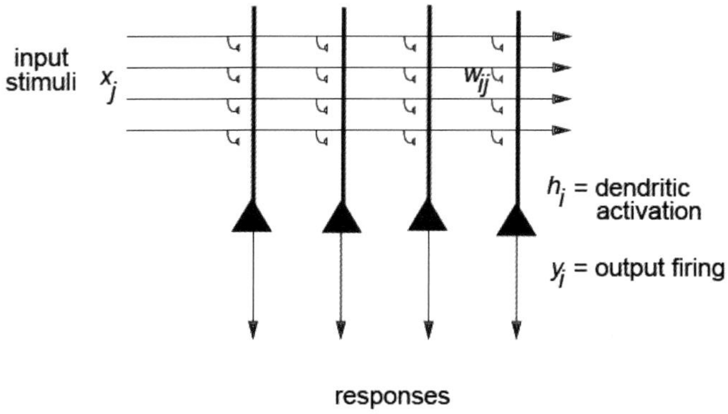

FIGURE B.3 The architecture of a competitive network. From Rolls (2016).

This process is repeated for every input pattern in random permuted sequence, and this process is repeated for a small number of training epochs.

After the training, each dendritic weight vector points toward a cluster of patterns in the input space. The competitive network has used self-organization with no teacher to categorize the inputs, with patterns close in the input space activating the same set of output neurons, and different clusters of inputs activating different sets of output neurons.

REFERENCES

Rolls, E. T. (2008). *Memory, Attention, and Decision-Making: A Unifying Computational Neuroscience Approach.* Oxford: Oxford University Press.

Rolls, E. T. (2016). *Cerebral Cortex: Principles of Operation.* Oxford: Oxford University Press.

Contributors

Michael A. Arbib, Computer Science and Neuroscience (Emeritus), USC; ABLE (Action, Brain, Language, Evolution) Project, La Jolla
Joseph Ayers, Marine Science Center, Northeastern University
James A. Bednar, School of Informatics, University of Edinburgh
Andrej Bicanski, Institute of Cognitive Neuroscience, University College London
James J. Bonaiuto, Sobell Department of Motor Neuroscience and Movement Disorders, University College London
Nicolas Brunel, Department of Statistics, University of Chicago
Jean-Marie Cabelguen, Biological Sciences, Université de Bordeaux
Carmen Canavier, Department of Cell Biology and Anatomy, Louisiana State University
Angelo Cangelosi, School of Computing and Mathematics, Plymouth University
Richard P. Cooper, Department of Psychological Sciences, Birkbeck, University of London
Carlos R. Cortes, School of Medicine, University of Maryland
Nathaniel D. Daw, Princeton Neuroscience Institute and Department of Psychology, Princeton University
Paul Dean, Department of Psychology, University of Sheffield
Peter Ford Dominey, Robot Cognition Laboratory, INSERM
Pierre Enel, Department of Neuroscience, Icahn School of Medicine at Mount Sinai
Jean-Marc Fellous, Department of Psychology, University of Arizona
Stefano Fusi, Center for Theoretical Neuroscience, Columbia University Medical Center
Wulfram Gerstner, Laboratory of Computational Neuroscience, École Polytechnique Fédérale de Lausanne
Frank W. Grasso, Department of Psychology, CUNY
Jacqueline A. Griego, Institute for Cognitive Science, University of Osnabrück
Ziad M. Hafed, Werner Reichardt Centre for Integrative Neuroscience, Tübingen University
Michael E. Hasselmo, Department of Psychology, Boston University
Auke J. Ijspeert, Biorobotics Laboratory, École Polytechnique Fédérale de Lausanne
Stephanie R. Jones, Department of Neuroscience, Brown University
Daniel Kersten, Psychology Department, University of Minnesota
Jeremie Knuesel, Biorobotics Laboratory, École Polytechnique Fédérale de Lausanne
Owen Lewis, Department of Brain and Cognitive Sciences, MIT
William W. Lytton, Physiology and Pharmacology Department, SUNY Downstate Medical Center
Tomaso A. Poggio, Department of Brain and Cognitive Sciences, MIT
John Porrill, Department of Psychology, University of Sheffield
Tony J. Prescott, Department of Psychology, University of Sheffield
John Rinzel, Center for Neural Science, New York University
Edmund T. Rolls, Department of Computer Science, University of Warwick
Jonathan E. Rubin, Department of Mathematics, University of Pittsburgh
Nicolas Schweighofer, Division of Biokinesiology and Physical Therapy, USC
Mohamed A. Sherif, Griffith School of Engineering, Griffith University
Malle A. Tagamets, School of Medicine, University of Maryland
Paul F.M.J. Verschure, Technology Department, Universitat Pompeu Fabra
Nathan Vierling-Claassen, Department of Neuroscience, Brown University
Xiao-Jing Wang, Center for Neural Science, New York University; NYU-ECNU Joint Institute of Brain and Cognitive Science, NYU-Shanghai, Shanghai, China
Christopher K. I. Williams, School of Informatics, University of Edinburgh
Ransom Winder, Center for Advanced Study of Language and Department of Computer Science, University of Maryland
Alan L. Yuille, Department of Statistics, UCLA

INDEX

Computational Neuroscience
Terrence J. Sejnowski and Tomaso A. Poggio, editors

Neural Nets in Electric Fish, Walter Heiligenberg, 1991

The Computational Brain, Patricia S. Churchland and Terrence J. Sejnowski, 1992

Dynamic Biological Networks: The Stomatogastic Nervous System, edited by Ronald M. Harris-Warrick, Eve Marder, Allen I. Selverston, and Maurice Moulins, 1992

The Neurobiology of Neural Networks, edited by Daniel Gardner, 1993

Large-Scale Neuronal Theories of the Brain, edited by Christof Koch and Joel L. Davis, 1994

The Theoretical Foundations of Dendritic Function: Selected Papers of Wilfrid Rall with Commentaries, edited by Idan Segev, John Rinzel, and Gordon M. Shepherd, 1995

Models of Information Processing in the Basal Ganglia, edited by James C. Houk, Joel L. Davis, and David G. Beiser, 1995

Spikes: Exploring the Neural Code, Fred Rieke, David Warland, Rob de Ruyter van Steveninck, and William Bialek, 1997

Neurons, Networks, and Motor Behavior, edited by Paul S. Stein, Sten Grillner, Allen I. Selverston, and Douglas G. Stuart, 1997

Methods in Neuronal Modeling: From Ions to Networks, second edition, edited by Christof Koch and Idan Segev, 1998

Fundamentals of Neural Network Modeling: Neuropsychology and Cognitive Neuroscience, edited by Randolph W. Parks, Daniel S. Levine, and Debra L. Long, 1998

Neural Codes and Distributed Representations: Foundations of Neural Computation, edited by Laurence Abbott and Terrence J. Sejnowski, 1999

Unsupervised Learning: Foundations of Neural Computation, edited by Geoffrey Hinton and Terrence J. Sejnowski, 1999

Fast Oscillations in Cortical Circuits, Roger D. Traub, John G. R. Jeffreys, and Miles A. Whittington, 1999

Computational Vision: Information Processing in Perception and Visual Behavior, Hanspeter A. Mallot, 2000

Graphical Models: Foundations of Neural Computation, edited by Michael I. Jordan and Terrence J. Sejnowski, 2001

Self-Organizing Map Formation: Foundations of Neural Computation, edited by Klaus Obermayer and Terrence J. Sejnowski, 2001

Neural Engineering: Computation, Representation, and Dynamics in Neurobiological Systems, Chris Eliasmith and Charles H. Anderson, 2003

The Computational Neurobiology of Reaching and Pointing, Reza Shadmehr and Steven P. Wise, 2005

Dynamical Systems in Neuroscience, Eugene M. Izhikevich, 2006

Bayesian Brain: Probabilistic Approaches to Neural Coding, edited by Kenji Doya, Shin Ishii, Alexandre Pouget, and Rajesh P. N. Rao, 2007

Computational Modeling Methods for Neuroscientists, edited by Erik De Schutter, 2009

Neural Control Engineering, Steven J. Schiff, 2011

Understanding Visual Population Codes: Toward a Common Multivariate Framework for Cell Recording and Functional Imaging, edited by Nikolaus Kriegeskorte and Gabriel Kreiman, 2011

Biological Learning and Control: How the Brain Builds Representations, Predicts Events, and Makes Decisions, Reza Shadmehr and Sandro Mussa-Ivaldi, 2012

Principles of Brain Dynamics: Global State Interactions, edited by Mikhail Rabinovich, Karl J. Friston, and Pablo Varona, 2012

Brain Computation as Hierarchical Abstraction, Dana H. Ballard, 2015

Visual Cortex and Deep Networks: Learning Invariant Representations, Tomaso A. Poggio and Fabio Anselmi, 2016

Case Studies in Neural Data Analysis: A Guide for the Practicing Neuroscientist, Mark A. Kramer and Uri T. Eden, 2016

From Neuron to Cognition via Computational Neuroscience, edited by Michael A. Arbib and James J. Bonaiuto, 2016